HANDBOOK OF
NANOPHYSICS

Handbook of Nanophysics

HANDBOOK OF NANOPHYSICS 2

Clusters and Fullerenes

Edited by

Klaus D. Sattler

CRC Press
Taylor & Francis Group
Boca Raton London New York

CRC Press is an imprint of the
Taylor & Francis Group, an **informa** business

CRC Press
Taylor & Francis Group
6000 Broken Sound Parkway NW, Suite 300
Boca Raton, FL 33487-2742

First issued in paperback 2020

© 2011 by Taylor and Francis Group, LLC
CRC Press is an imprint of Taylor & Francis Group, an Informa business

No claim to original U.S. Government works

ISBN 13: 978-1-138-11510-1 (pbk)
ISBN 13: 978-1-4200-7554-0 (hbk)

Library of Congress Cataloging-in-Publication Data

Handbook of nanophysics. Clusters and fullerenes / editor, Klaus D. Sattler.
    p. cm.
   "A CRC title."
   Includes bibliographical references and index.
   ISBN 978-1-4200-7554-0 (alk. paper)
   1. Microphysics--Handbooks, manuals, etc. 2. Nanoscience- Handbooks, manuals, etc. 3. Microclusters--Handbooks, manuals, etc. 4. Fullerenes--Handbooks, manuals, etc. I. Sattler, Klaus D.

QC173.4.M5H357 2009
530--dc22
                      2009047135

Visit the Taylor & Francis Web site at
http://www.taylorandfrancis.com

and the CRC Press Web site at
http://www.crcpress.com

Contents

PART I Free Clusters

v

PART II Clusters in Contact

PART III Production and Stability of Carbon Fullerenes

PART IV Structure and Properties of Carbon Fullerenes

PART V Carbon Fullerenes in Contact

PART VI Inorganic Fullerenes

Preface

The *Handbook of Nanophysics* is the first comprehensive reference to consider both fundamental and applied aspects of nanophysics. As a unique feature of this work, we requested contributions to be submitted in a tutorial style, which means that state-of-the-art scientific content is enriched with fundamental equations and illustrations in order to facilitate wider access to the material. In this way, the handbook should be of value to a broad readership, from scientifically interested general readers to students and professionals in materials science, solid-state physics, electrical engineering, mechanical engineering, computer science, chemistry, pharmaceutical science, biotechnology, molecular biology, biomedicine, metallurgy, and environmental engineering.

What Is Nanophysics?

Modern physical methods whose fundamentals are developed in physics laboratories have become critically important in nanoscience. Nanophysics brings together multiple disciplines, using theoretical and experimental methods to determine the physical properties of materials in the nanoscale size range (measured by millionths of a millimeter). Interesting properties include the structural, electronic, optical, and thermal behavior of nanomaterials; electrical and thermal conductivity; the forces between nanoscale objects; and the transition between classical and quantum behavior. Nanophysics has now become an independent branch of physics, simultaneously expanding into many new areas and playing a vital role in fields that were once the domain of engineering, chemical, or life sciences.

This handbook was initiated based on the idea that breakthroughs in nanotechnology require a firm grounding in the principles of nanophysics. It is intended to fulfill a dual purpose. On the one hand, it is designed to give an introduction to established fundamentals in the field of nanophysics. On the other hand, it leads the reader to the most significant recent developments in research. It provides a broad and in-depth coverage of the physics of nanoscale materials and applications. In each chapter, the aim is to offer a didactic treatment of the physics underlying the applications alongside detailed experimental results, rather than focusing on particular applications themselves.

The handbook also encourages communication across borders, aiming to connect scientists with disparate interests to begin interdisciplinary projects and incorporate the theory and methodology of other fields into their work. It is intended for readers from diverse backgrounds, from math and physics to chemistry, biology, and engineering.

The introduction to each chapter should be comprehensible to general readers. However, further reading may require familiarity with basic classical, atomic, and quantum physics. For students, there is no getting around the mathematical background necessary to learn nanophysics. You should know calculus, how to solve ordinary and partial differential equations, and have some exposure to matrices/linear algebra, complex variables, and vectors.

External Review

All chapters were extensively peer reviewed by senior scientists working in nanophysics and related areas of nanoscience. Specialists reviewed the scientific content and nonspecialists ensured that the contributions were at an appropriate technical level. For example, a physicist may have been asked to review a chapter on a biological application and a biochemist to review one on nanoelectronics.

Organization

The *Handbook of Nanophysics* consists of seven books. Chapters in the first four books (*Principles and Methods, Clusters and Fullerenes, Nanoparticles and Quantum Dots*, and *Nanotubes and Nanowires*) describe theory and methods as well as the fundamental physics of nanoscale materials and structures. Although some topics may appear somewhat specialized, they have been included given their potential to lead to better technologies. The last three books (*Functional Nanomaterials, Nanoelectronics and Nanophotonics*, and *Nanomedicine and Nanorobotics*) deal with the technological applications of nanophysics. The chapters are written by authors from various fields of nanoscience in order to encourage new ideas for future fundamental research.

After the first book, which covers the general principles of theory and measurements of nanoscale systems, the organization roughly follows the historical development of nanoscience. *Cluster* scientists pioneered the field in the 1980s, followed by extensive work on *fullerenes*, *nanoparticles*, and *quantum dots* in the 1990s. Research on *nanotubes* and *nanowires* intensified in subsequent years. After much basic research, the interest in applications such as the *functions of nanomaterials* has grown. Many bottom-up and top-down techniques for nanomaterial and nanostructure generation were developed and made possible the development of *nanoelectronics* and *nanophotonics*. In recent years, real applications for *nanomedicine* and *nanorobotics* have been discovered.

Acknowledgments

Many people have contributed to this book. I would like to thank the authors whose research results and ideas are presented here. I am indebted to them for many fruitful and stimulating discussions. I would also like to thank individuals and publishers who have allowed the reproduction of their figures. For their critical reading, suggestions, and constructive criticism, I thank the referees. Many people have shared their expertise and have commented on the manuscript at various stages. I consider myself very fortunate to have been supported by Luna Han, senior editor of the Taylor & Francis Group, in the setup and progress of this work. I am also grateful to Jessica Vakili, Jill Jurgensen, Joette Lynch, and Glenon Butler for their patience and skill with handling technical issues related to publication. Finally, I would like to thank the many unnamed editorial and production staff members of Taylor & Francis for their expert work.

Klaus D. Sattler
Honolulu, Hawaii

Editor

Klaus D. Sattler pursued his undergraduate and master's courses at the University of Karlsruhe in Germany. He received his PhD under the guidance of Professors G. Busch and H.C. Siegmann at the Swiss Federal Institute of Technology (ETH) in Zurich, where he was among the first to study spin-polarized photo-electron emission. In 1976, he began a group for atomic cluster research at the University of Konstanz in Germany, where he built the first source for atomic clusters and led his team to pioneering discoveries such as "magic numbers" and "Coulomb explosion." He was at the University of California, Berkeley, for three years as a Heisenberg Fellow, where he initiated the first studies of atomic clusters on surfaces with a scanning tunneling microscope.

Dr. Sattler accepted a position as professor of physics at the University of Hawaii, Honolulu, in 1988. There, he initiated a research group for nanophysics, which, using scanning probe microscopy, obtained the first atomic-scale images of carbon nanotubes directly confirming the graphene network. In 1994, his group produced the first carbon nanocones. He has also studied the formation of polycyclic aromatic hydrocarbons (PAHs) and nanoparticles in hydrocarbon flames in collaboration with ETH Zurich. Other research has involved the nanopatterning of nanoparticle films, charge density waves on rotated graphene sheets, band gap studies of quantum dots, and graphene foldings. His current work focuses on novel nanomaterials and solar photocatalysis with nanoparticles for the purification of water.

Among his many accomplishments, Dr. Sattler was awarded the prestigious Walter Schottky Prize from the German Physical Society in 1983. At the University of Hawaii, he teaches courses in general physics, solid-state physics, and quantum mechanics.

In his private time, he has worked as a musical director at an avant-garde theater in Zurich, composed music for theatrical plays, and conducted several critically acclaimed musicals. He has also studied the philosophy of Vedanta. He loves to play the piano (classical, rock, and jazz) and enjoys spending time at the ocean, and with his family.

Contributors

Victor V. Albert
Department of Physics
University of Florida
Gainesville, Florida

Manuel Alcamí
Departamento de Química
Universidad Autónoma
    de Madrid
Madrid, Spain

Julio A. Alonso
Departamento de Física Teórica,
    Atómica y Óptica
Facultad de Ciencias
Universidad de Valladolid
Valladolid, Spain

and

Departamento de Física
    de Materiales
Universidad del País Vasco

and

Donostia International
    Physics Center
Paseo M. de Lardizábal
San Sebastián, Spain

Ronaldo Junio Campos Batista
Departamento de Física
Instituto de Ciências
    Exatas e Biológicas
Universidade Federal de Ouro Preto
Preto, Brazil

A. John Berlinsky
Department of Physics and Astronomy
McMaster University
Hamilton, Ontario, Canada

Harry Bernas
Centre de Spectrométrie Nucléaire et de
    Spectrométrie de Masse
Centre national de la recherche
    scientifique
Université Paris-Sud 11
Orsay, France

Olle Björneholm
Department of Physics
Uppsala University
Uppsala, Sweden

Massimo Boninsegni
University of Alberta
Edmonton, Alberta, Canada

Jonathan A. Brant
Department of Civil and Architectural
    Engineering
University of Wyoming
Laramie, Wyoming

Yuriy V. Butenko
European Space Agency
European Space Research
    and Technology Centre
Noordwijk, the Netherlands

Florent Calvo
Laboratoire de Spectrométrie Ionique
    et Moléculaire
Centre national de la recherche
    scientifique
Université Lyon I
Villeurbanne, France

Giancarlo Cappellini
Dipartimento di Fisica
Università degli Studi di Cagliari

and

Sardinian Laboratory for Computational
    Materials Science
Istituto Nazionale di Fisica della Materia
Cagliari, Italy

Henrik Cederquist
Department of Physics
Stockholm University
Stockholm, Sweden

Hélio Chacham
Departamento de Física
Instituto de Ciências Exatas
Universidade Federal de Minas Gerais
Belo Horizonte, Brazil

Hassane Chadli
Laboratoire de Physique des Matériaux
    et Modélisation des Systèmes
Faculté des Sciences
Université MY Ismail
Meknes, Morocco

Ryan T. Chancey
Nelson Architectural Engineers, Inc.
Plano, Texas

Christian Chang
Zentrum für Astronomie und
    Astrophysik
Technische Universität Berlin
Berlin, Germany

Chia-Seng Chang
Institute of Physics
Academia Sinica
Taipei, Taiwan, Republic of China

Hai-Ping Cheng
Department of Physics and the Quantum
    Theory Project
University of Florida
Gainesville, Florida

Ya-Ping Chiu
Department of Physics
National Sun Yat-Sen University
Kaohsiung, Taiwan, Republic of China

Pieterjan Claes
Laboratory of Solid State Physics
    and Magnetism
Katholieke Universiteit Leuven

and

Institute for Nanoscale Physics
    and Chemistry
Katholieke Universiteit Leuven
Leuven, Belgium

Colm Connaughton
Centre for Complexity Science
University of Warwick
Coventry, United Kingdom

Walt A. de Heer
School of Physics
Georgia Institute of Technology
Atlanta, Georgia

Olof Echt
Department of Physics
University of New Hampshire
Durham, New Hampshire

and

Institut für Ionenphysik und
    Angewandte Physik
Leopold Franzens Universität
Innsbruck, Austria

Andrey N. Enyashin
Physikalische Chemie
Technische Universität Dresden
Dresden, Germany

and

Institute of Solid State Chemistry
Ural Branch of the Russian Academy
    of Sciences
Ekaterinburg, Russia

Roch Espiau de Lamaëstre
Electronics and Information Technology
    Laboratory
French Atomic Energy Commission
MINATEC
Grenoble, France

Yu-Qi Feng
Department of Chemistry
Wuhan University
Wuhan, People's Republic of China

Thomas Fennel
Institut für Physik
Universität Rostock
Rostock, Germany

André Fielicke
Fritz-Haber-Institut der Max-Planck-
    Gesellschaft
Berlin, Germany

René Fournier
Department of Chemistry
York University
Toronto, Ontario, Canada

Philipp Gruene
Fritz-Haber-Institut der Max-Planck-
    Gesellschaft
Berlin, Germany

Ricardo A. Guirado-López
Instituto de Física "Manuel Sandoval
    Vallarta"
Universidad Autónoma de San Luis Potosí
San Luis Potosí, México

Selçuk Gümüş
Department of Chemistry
Middle East Technical University
Ankara, Turkey

Nicole Haag
Department of Physics
Stockholm University
Stockholm, Sweden

Frank E. Harris
Department of Chemistry
University of Florida
Gainesville, Florida

and

Department of Physics
University of Utah
Salt Lake City, Utah

Chunnian He
School of Materials Science
    and Engineering
Tianjin University

and

Institute of Medical Equipment
The Academy of Military Medical Sciences
Tianjin, People's Republic of China

Thomas Heine
School of Engineering and Science
Jacobs University Bremen
Bremen, Germany

Shojun Hino
Department of Materials Science
    and Biotechnology
Graduate School of Science
    and Engineering
Ehime University
Matsuyama, Japan

Michael R. C. Hunt
Department of Physics
University of Durham
Durham, United Kingdom

Scott T. Huxtable
Department of Mechanical Engineering
Virginia Polytechnic Institute and State
    University
Blacksburg, Virginia

Atsushi Ikeda
Graduate School of Materials Science
Nara Institute of Science and Technology
Ikoma, Japan

Beatriz M. Illescas
Departamento de Química Orgánica
Facultad de Química
Universidad Complutense
Madrid, Spain

Bernd v. Issendorff
Fakultät für Physik
Universität Freiburg
Freiburg, Germany

Stanislav Jendrol'
Institute of Mathematics
Faculty of Science
Pavol Jozef Šafárik University
Košice, Slovakia

Catherine Kallin
Department of Physics and Astronomy
McMaster University
Hamilton, Ontario, Canada

František Kardoš
Institute of Mathematics
Faculty of Science
Pavol Jozef Šafárik University
Košice, Slovakia

Keun Su Kim
Department of Chemical Engineering
Université de Sherbrooke
Sherbrooke, Québec, Canada

Vitaly V. Kresin
Department of Physics and Astronomy
University of Southern California
Los Angeles, California

Peter Lievens
Laboratory of Solid State Physics
    and Magnetism
Katholieke Universiteit Leuven

and

Institute for Nanoscale Physics
    and Chemistry
Katholieke Universiteit Leuven
Leuven, Belgium

Fei Lin
Department of Physics
University of Illinois at Urbana-
    Champaign
Urbana, Illinois

Jonathan T. Lyon
Fritz-Haber-Institut der Max-Planck-
    Gesellschaft
Berlin, Germany

Giuliano Malloci
INAF-Osservatorio Astronomico
    di Cagliari and Laboratorio Scienza
Sestu, Italy

Tilmann D. Märk
Institut für Ionenphysik und
    Angewandte Physik
Leopold Franzens Universität
Innsbruck, Austria

and

Department of Experimental Physics
Comenius University
Bratislava, Slovak Republic

Fernando Martín
Departamento de Química
Universidad Autónoma
    de Madrid
Madrid, Spain

Nazario Martín
Departamento de Química Orgánica
Facultad de Química
Universidad Complutense
Madrid, Spain

José I. Martínez
Center for Atomic-scale Materials Design
Department of Physics
Technical University of Denmark
Lyngby, Denmark

Carlo Massobrio
Institut de Physique et Chimie des
    Matériaux de Strasbourg
Strasbourg, France

Masahiko Matsubara
Laboratoire des Colloïdes
Verres et Nanomatériaux
Université Montpellier II
Montpellier, France

Jochen Maul
Institut für Physik
Johannes Gutenberg-Universität
Mainz, Germany

Mário S. C. Mazzoni
Departamento de Fsica
Instituto de Ciências Exatas
Universidade Federal de Minas Gerais
Belo Horizonte, Brazil

Karl-Heinz Meiwes-Broer
Institut für Physik
Universität Rostock
Rostock, Germany

Masaaki Mitsui
Department of Chemistry
Faculty of Science
Shizuoka University
Shizuoka, Japan

Giacomo Mulas
INAF-Osservatorio Astronomico
    di Cagliari
Capoterra, Italy

Atsushi Nakajima
Department of Chemistry
Faculty of Science and Technology
Keio University
Yokohama, Japan

Masato Nakamura
Physics Laboratory
College of Science and Technology
Nihon University
Funabashi, Japan

Takashi Nakanishi
Organic Nanomaterials Center
National Institute for Materials Science
    (NIMS)
Tsukuba, Japan

and

MPI-NIMS International Joint
    Laboratory
Department of Interfaces
Max Planck Institute (MPI) of Colloids
    and Interfaces
Potsdam, Germany

and

Precursory Research for Embryonic
    Science and Technology
Japan Science and Technology
Chiyodaku, Tokyo

Vu Thi Ngan
Department of Chemistry
Katholieke Universiteit Leuven

and

Institute for Nanoscale Physics
    and Chemistry
Katholieke Universiteit Leuven
Leuven, Belgium

Minh Tho Nguyen
Department of Chemistry
Katholieke Universiteit Leuven

and

Institute for Nanoscale Physics
    and Chemistry
Katholieke Universiteit Leuven
Leuven, Belgium

Lene B. Oddershede
Niels Bohr Institute
Copenhagen University
Copenhagen, Denmark

Gunnar Öhrwall
MAX-lab
Lund University
Lund, Sweden

Jaisse Oviedo
Departamento de Química Física
Facultad de Química
University of Sevilla
Sevilla, Spain

Francesco Paesani
Department of Chemistry and
    Biochemistry
University of California
San Diego, California

Elke Pahl
Centre for Theoretical Chemistry
    and Physics
New Zealand Institute for Advanced
    Study
Massey University Albany
Auckland, New Zealand

Pascal Parneix
Institut des Sciences Molécularies d'Orsay
Centre national de la recherche
    scientifique
Université Paris Sud 11
Orsay, France

Beate Patzer
Zentrum für Astronomie und
    Astrophysik
Technische Universität Berlin
Berlin, Germany

Ronald A. Phaneuf
Department of Physics
University of Nevada
Reno, Nevada

Vladimir Popok
Department of Physics
University of Gothenburg
Gothenburg, Sweden

Abdelali Rahmani
Laboratoire de Physique des Matériaux
    et Modélisation des Systèmes
Faculté des Sciences
Université MY Ismail
Meknes, Morocco

R. Rajesh
Institute of Mathematical Sciences
Chennai, India

Jan-Michael Rost
Max-Planck-Institut für Physik
    komplexer Systeme
Dresden, Germany

and

Center for Free Electron Laser Science
Hamburg, Germany

Ulf Saalmann
Max-Planck-Institut für Physik
    komplexer Systeme
Dresden, Germany

and

Center for Free Electron Laser Science
Hamburg, Germany

John R. Sabin
Department of Physics
University of Florida
Gainesville, Florida

and

Institute of Physics and Chemistry
University of Southern Denmark
Odense, Denmark

Arta Sadrzadeh
Department of Mechanical Engineering
    and Materials Science
Rice University

and

Department of Chemistry
Rice University
Houston, Texas

Miguel A. San-Miguel
Departamento de Química Física
Facultad de Química
University of Sevilla
Sevilla, Spain

Javier F. Sanz
Departamento de Química Física
Facultad de Química
University of Sevilla
Sevilla, Spain

Yuta Sato
Nanotube Research Center
National Institute of Advanced Industrial
    Science and Technology

and

Core Research for Evolutional Science
    and Technology
Japan Science and Technology Agency
Tsukuba, Japan

Paul Scheier
Institut für Ionenphysik und
    Angewandte Physik
Leopold Franzens Universität
Innsbruck, Austria

Peter Schwerdtfeger
Centre for Theoretical Chemistry
    and Physics
New Zealand Institute for Advanced
    Study
Massey University Albany
Auckland, New Zealand

Gotthard Seifert
Physikalische Chemie
Technische Universität Dresden
Dresden, Germany

Yanfei Shen
MPI-NIMS International Joint
    Laboratory
Department of Interfaces
Max Planck Institute of Colloids
    and Interfaces
Potsdam, Germany

Nitin C. Shukla
Department of Mechanical Engineering
Virginia Polytechnic Institute
    and State University
Blacksburg, Virginia

Lidija Šiller
School of Chemical Engineering
    and Advanced Materials
Newcastle University
Newcastle upon Tyne, United Kingdom

Erik S. Sørensen
Department of Physics and Astronomy
McMaster University
Hamilton, Ontario, Canada

Gervais Soucy
Department of Chemical Engineering
Université de Sherbrooke
Sherbrooke, Québec, Canada

Wei-Bin Su
Institute of Physics
Academia Sinica
Taipei, Taiwan, Republic of China

Kazu Suenaga
Nanotube Research Center
National Institute of Advanced Industrial
    Science and Technology

and

Core Research for Evolutional Science
    and Technology
Japan Science and Technology Agency
Tsukuba, Japan

Detlev Sülzle
Zentrum für Astronomie und
    Astrophysik
Technische Universität Berlin
Berlin, Germany

Maxim Tchaplyguine
MAX-lab
Lund University
Lund, Sweden

Josef Tiggesbäumker
Institut für Physik
Universität Rostock
Rostock, Germany

Tien-Tzou Tsong
Institute of Physics
Academia Sinica
Taipei, Taiwan, Republic of China

Lemi Türker
Department of Chemistry
Middle East Technical University
Ankara, Turkey

Nele Veldeman
Laboratory of Solid State Physics
    and Magnetism
Katholieke Universiteit Leuven

and

Institute for Nanoscale Physics
    and Chemistry
Katholieke Universiteit Leuven
Leuven, Belgium

and

Vlaamse Instelling voor
    Wetenschappelijk Onderzoek
Mol, Belgium

Jiaobing Wang
MPI-NIMS International Joint
    Laboratory
Department of Interfaces
Max Planck Institute of Colloids
    and Interfaces
Potsdam, Germany

Yang Wang
Departamento de Química
Universidad Autónoma
    de Madrid
Madrid, Spain

Boris I. Yakobson
Department of Mechanical Engineering
    and Materials Science
Rice University
Houston, Texas

and

Department of Chemistry
Rice University
Houston, Texas

Qiong-Wei Yu
Department of Chemistry
Wuhan University
Wuhan, People's Republic of China

Oleg Zaboronski
Mathematics Institute
University of Warwick
Coventry, United Kingdom

Aristides D. Zdetsis
Division of Theoretical and
    Mathematical Physics, Astronomy
    and Astrophysics
Department of Physics
University of Patras
Patras, Greece

Henning Zettergren
Department of Physics and Astronomy
University of Aarhus
Aarhus, Denmark

Naiqin Zhao
School of Materials Science and
    Engineering
Tianjin University
Tianjin, People's Republic of China

I

Free Clusters

Nanocluster Nucleation, Growth, and Size Distributions

Harry Bernas
Université Paris-Sud 11

Roch Espiau de Lamaëstre
MINATEC

1.1 Introduction

A number of chapters in this handbook refer to, or rest on, the control of nanocluster (NC) size and size distribution, i.e., the first and second moments of the NC population. In fact, the entire, detailed shape of the NC size distribution is often very important. Its effect has actually been used for thousands of years. The colors of ancient or medieval glasses are generally due to the semiconducting or metallic NCs that they contain, and the varying shades of beautiful colors that we marvel at are most often due to the empirical control obtained by glassmakers, via complex heat treatments, over NC sizes and size distributions. Today, these parameters are known to bear critically on such important properties as the optical emission linewidth or the magnetic anisotropy distribution. The obtention of narrow size distributions is also crucial to the fabrication of three-dimensional NC arrays in solar cells, photodetectors, or other photonic devices. Control over these parameters requires that we understand and monitor the physical processes that determine them. Our purpose in this chapter is to summarize the present status in this regard.

In order to predict and control the NC average size and size distributions, we would ideally need (1) a full description of nucleation processes and of growth processes; (2) criteria to avoid overlap of nucleation and growth, since this obviously broadens the size distribution; and (3) a full description of coarsening (mass transport from small particles to their larger counterparts), i.e., of the particle size distribution's time evolution from an arbitrary initial distribution. This ideal situation is almost fully realized in techniques based on colloidal chemistry (e.g., Pileni 2001, 2003, Park et al. 2004, Weiss et al. 2008), which

have provided a broad range of metal and semiconductor NCs in solution, in a variety of shapes and sizes and with excellent control over the size distribution (at the 5% level or under). Studies of intrinsic NC properties, and a number of applications, benefit hugely from these techniques. Summarily, liquid-state chemistry is expected to provide more degrees of freedom in the NC synthesis process, and species diffusion does not limit interactions. However, most applications—notably involving inorganic solid matrices—require that the NCs be transferred from a solution to a solid host. This generally poses serious problems related to solubility, to possible reactions of the NC element(s) with host components, to clustering, or to coalescence. In solids, none of the requirements listed above are fully available for several reasons. The initiation of nucleation differs widely depending on the nature of the material in which the NCs are to be grown: e.g., in metals and many semiconductors, it usually depends on quasi-equilibrium thermodynamics and on diffusion, whereas in insulators and in some semiconductors, the primary aggregation process also depends on the charge state properties (i.e., the chemistry) of moving species, including electrons and holes. Growth processes are correspondingly also very different. As a result, except in the simplest cases, multiple mechanisms occur and interfere, a situation that generally leads, as we shall see, to difficulties in predicting how to control NC populations and narrow size distributions.

Our discussion starts with a short review of standard quasi-equilibrium thermodynamics nucleation and growth, which provides us with some terms of reference. Although it is by no means ideal for obtaining narrow size distributions, it has the advantage of being predictive in simple cases. The best example is that of the late-stage coarsening process, very well described by a scaling approximation. We then scrutinize nucleation conditions in various cases, since they determine most of an NC's "short and midterm" future. We summarize a generalization of the quasi-equilibrium thermodynamics approach, given by Binder, in terms of coupled rate equations. This provides a useful framework to evaluate conditions for the control of an NC population's evolution. It allows us to discern whether a study of the NC population moments' time and/or temperature evolution, in some of the more common NC nucleation and growth cases, provides information on the corresponding mechanisms. It also provides a framework to estimate the possibility of obtaining narrow size distributions from different classes of experimental synthesis techniques (near to, or far away from, equilibrium) as they are described elsewhere in this handbook.

Our topic involves many aspects of statistical physics, equilibrium, and nonequilibrium thermodynamics that cannot be covered in this chapter—detailed treatments are given in the references. The aim here is to provide a rather pragmatic guide to experimentalists interested in a critical view of the physics underlying attempts to control the synthesis of NCs.

1.2 Mechanisms of Nanocluster Nucleation and Growth

1.2.1 A Word on Precipitation

As is obvious from their very name, NCs are formed by progressively separating one or more constituents from a solid or a liquid solution or alloy, e.g., by cooling from the melt, modifying the system's pressure, or via the addition of a chemical reagent. The common term for this operation is "precipitation," a concept that covers a large variety of phenomena ranging from atmospheric rain droplet formation to solid-state nanoscale aggregation. Over more than a century, a rather general common description of all these occurrences has emerged, based on statistical physics and on the thermodynamics of phase transformation. Of course, there are limitations to the information derived from this description. From a theoretical viewpoint, precipitation is a highly nonlinear phenomenon, hence quite complex to formulate. Experimentally, material properties (e.g., rain versus second-phase nanoclustering) will obviously affect the behavior at some level of detail. In all instances, a major difficulty remains in discerning the initial nucleation stage, which is found to be crucial for control over precipitation. We will go into this later. First, we briefly review the more common physicochemical mechanisms of NC nucleation and growth, emphasizing those features that affect NC sizes and size distributions.

The essential driving force for precipitation of a new phase (e.g., ice from cooled water second-phase extraction from a liquid or a solid solution, etc.) is the lowering of the potential energy of a group of atoms (or molecules) when they bond together. Two rather different stages occur: nucleation and growth. Nucleation involves the formation of small clusters—it depends on an energy instability in the parent state, where lowering the temperature provides a driving force toward equilibrium, and this

driving force increases as the temperature is *reduced*. On the other hand, in order to cluster atoms must meet, so diffusion is essential, and diffusion increases at *increasing* temperatures. We recognize here that thermodynamics determines whether nucleation is possible (tending to minimize the system's free energy), whereas kinetics determine the nucleation rate. The two have opposite temperature dependences, so their multiplication leads to a maximum in the nuclei density temperature dependence. There is also, as we will see shortly, a condition for the formed clusters to be stable. If it is satisfied, growth of these clusters is diffusion controlled (i.e., increases with temperature). The total rate of NC formation is then the product of the nucleation rate and the growth rate. Let us now briefly quantify this picture.

1.2.2 Some Basic Thermodynamics

Quench the simplest binary system A–B (Figure 1.1, upper part): it will end up in a metastable state below a critical temperature, T_c (Porter and Easterling 1981, Binder and Fratzl 2001, Wagner et al. 2001). The locus of the T_c's for different compositions (coexistence line) defines a phase boundary. The evolution of a metastable system tends to minimize the Gibbs free energy, $\Delta G = \Delta H - T\Delta S$, where ΔH is the enthalpy, ΔS is the entropy, and T is the temperature. For an NC, the competition between

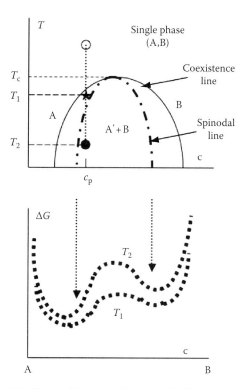

FIGURE 1.1 Upper: Schematic binary A–B alloy phase diagram, showing the coexistence line and the spinodal line. The region in which unmixing—hence homogeneous nucleation—occurs lies between these two lines. The region below the spinodal line is the unstable decomposition region, where spinodal decomposition (rather than nucleation) occurs. Lower: The free energy's composition dependence for the same system at two different temperatures (see text).

the formation enthalpy ΔH_f^0 (which is determined by the atomic interactions, hence size-dependent for small sizes) and the entropy term (also size-dependent) in the Gibbs free energy leads to a temperature (typically $T_c = \Delta H_f^0 / 2R$, where R is the perfect gas constant) below which B-rich domains may form. Quenching to lower temperatures, two possibilities occur depending on the composition- and temperature-dependent free energy of the system. The latter varies as shown for two temperatures in Figure 1.1 (lower part)—the locus of all the free energy curves' inflexion points is the so-called spinodal curve. Free energy fluctuations exist even at thermodynamic equilibrium. In both the metastable and unstable part of the phase diagram, some of these fluctuations can diverge and start growing. As seen by inspection of Figure 1.1 (lower), they have very different effects depending on the region of the phase diagram we consider. In the area under the spinodal (where the second derivative of the free energy is negative), any change in composition or temperature leads to an instability (there is *no energy barrier* and no phase transition at the spinodal line). If there is sufficient atomic mobility, the enhanced affinity of B for B, rather than for A, tends to form a solution with regions locally and randomly richer in B than in A. If kinetic blocking occurs (i.e., low temperatures freeze atomic movements in a region where typically $T < 0.1T_c$), diffusion constants are very low and separation is limited. Theoretical treatments describing how concentration fluctuations, particularly in the spinodal decomposition regime, can lead to nucleation and growth are summarized elsewhere (Binder and Fratzl 2001, Ratke and vorhees 2002). For an excellent introduction to relevant phase transition theories, see Binder (2001).

Outside the spinodal region, between the spinodal and coexistence lines, the second derivative of the free energy is positive, so that the system gains energy if a true second-phase nucleates (*with an energy barrier!*) and grows. This is the metastable region, in which unmixing occurs, producing locally organized precipitates, i.e., NCs. NCs generally grow in the low solute concentration/volume fraction part of the phase diagram, where unmixing occurs by metastable decomposition. Clusters form because of solute species mobility and chemical affinity (bond energies); monomers eventually bond to each other and form small aggregates that can grow or dissolve depending on their free energy. Once the NCs have reached a critical size (see below), the thermodynamic driving force induces further growth by capture of surrounding monomers, and later at the expense of surrounding precipitates (coarsening).

Microscopically, the very first stage of phase separation is described by (1) near-neighbor bond energies and (2) differing energy barriers for B and A atomic jumps as they diffuse and interact. These parameters determine the crucial macroscopic quantities—the diffusion coefficients and the solubility. As the corresponding cluster grows from 2, 3,… to n atoms, its energy changes relative to the surrounding solution, modifying its stability and internal and interface structure. The thermodynamics of binary systems provides a rather good initial approach (experimentally verified in many simple metallic systems) to the origin and main parameters determining the unmixing phase evolution that leads to NC formation in solids as well as liquids or on surfaces.

Whereas the phase diagram is obtained from studies of bulk solids, surface effects must be taken into account when NCs are involved. This leads to changes in the position of the boundaries in the unmixing diagram of Figure 1.1. Microscopically, this is due to the fact that the NC (or, more generally the second-phase domain) exchanges atoms with the primary phase through the interface, which—the NC being small—is curved and comprises a large fraction of NC atoms. The exchange rates depend on the curvature, so that the equilibrium concentration at any point in the interface vicinity also depends on the local curvature (Figure 1.2). Chemical and structural equilibrium conditions lead to the so-called Gibbs–Thomson relation, relating the concentration of B in the vicinity of a planar interface to that around a single NC of radius R:

$$c_{G-T}(R) = c_\infty \exp\left\{ \frac{R_c}{R} \right\}$$

where c_∞ is the impurity equilibrium concentration at a flat interface, the "capillary length" $R_c = 2\beta V \sigma$ depends on the atomic volume V and the surface tension σ, and $\beta = 1/kT$. The smaller the NC, the higher the equilibrium concentration at the NC surface. This will therefore shift the phase boundary upward in the unmixing diagram. These effects are important in the 1–10 nm size range. The Gibbs–Thomson relation has a major effect on the entire NC population: it shows that when in thermodynamical equilibrium with their embedding medium, NCs of different sizes will undergo correlated evolution. Due to the concentration gradient, the smaller ones tend to dissolve and the larger ones grow correspondingly. This process, known as Ostwald ripening, is rather widespread (Ratke and Vorhees 2002). It is, in fact, the only feature of the NC population evolution that has been modeled analytically to a reasonable approximation (based on scaling, see below). It is also the only one that is predictive as regards changes in the average NC size and

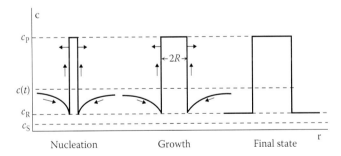

FIGURE 1.2 Schematic view of the solute concentration field in the host surrounding an NC with radius R and composition c_R, when in the unmixing regime. The NC is first shown in the initial, nucleation stage, then during growth, and lastly in its final, equilibrium state. The quantity c_s is the equilibrium concentration of the solute B in the host A. In the unstable spinodal regime, the progressively increasing concentration fluctuation amplitudes (with a characteristic wavelength) would ultimately—above a critical concentration—also lead to a precipitate structure. The long-term structures in both cases are essentially indistinguishable.

population, and which predicts a well-defined long-term size distribution.

A word on chemistry. As noted above, colloidal solution-based synthesis has met with huge success in those cases where the NCs produced may be manipulated and introduced into other media. Because the reacting system is liquid, it favors control over the way reagents are introduced as well as offers considerable versatility in exploring the influence of the environment. Chemical reagents are an excellent way to separate nucleation and growth (this approach is also used in glasses). Standard thermodynamics are at work here, and diffusion often plays a minor role as compared to chemical reaction rates. A popular analysis of the process is that of LaMer (LaMer and Dinegar 1950), often quoted by liquid solution chemists. It is a phenomenological description of the conditions required for obtaining narrow size distributions using a flash nucleation scheme—i.e., separating nucleation and growth. It has met with renewed interest because of the exciting properties of metallic and semiconducting NCs grown in solution. Transition probabilities are connected via the detailed balance condition, the free energies of the states involved in the transition are used, and the mass action law describes the chemical kinetics (implying proximity to thermodynamic equilibrium). This is essentially identical to the nucleation and growth model described below, with different notations.

1.2.3 Kinetics and Nucleation

NC nucleation is generally by no means an irreversible process smoothly evolving from a diatom to a several hundred- or thousand-fold configuration. Subtle quasi-equilibrium thermodynamics nucleation theories (Cahn and Hilliard 1958, 1959, Cook 1970, Binder and Fratzl 2001, Wagner et al. 2001) provide an adequate description of how concentration fluctuations lead to incipient nucleation—as long as atoms are free to move. Hence the importance of kinetics combined with thermodynamics (Philibert 1991). This feature cannot be overestimated: the very first steps of NC nucleation, their composition and structure at this early stage, are crucial to their evolution, since they determine the NC free energy (i.e., stability) and reactivity with their surroundings.

In order to obtain, after growth, as narrow a size distribution as possible, the initial NC population size distribution should itself be narrow—ideally, all nuclei should be formed simultaneously. This obviously depends on the nucleation speed, hence on an adequate combination of fast NC component diffusion, a large NC formation enthalpy, and free energy (the latter determining its stability). The ultimate NC density and average size, as well as the size distribution, all depend on how well this criterion is met. Now the metastable precipitation mechanism described above depends on an energy barrier, so it is relatively slow. As a result, in quasi-equilibrium thermodynamics conditions, a significant fraction of NCs are still undergoing formation as others grow—an effect that broadens the NC size distribution significantly. In some cases, such as growth of selected NCs in glasses (Borelli et al. 1987), this is circumvented by a two-stage anneal: first a low-temperature anneal to allow the slow nucleation process to develop, then—once the entire population of nuclei has been formed—a faster high-temperature anneal to induce growth. The efficiency of such procedures is limited by the phase diagram and diffusion properties. The effort to find tricks leading to a narrow, controlled initial NC population is one of the main reasons for developing techniques in which NCs are formed under far-from-thermodynamic equilibrium conditions.

What are the requirements for an NC to be stable? The primary one (notably in metals, many semiconductors, polymers, and liquids) is thermodynamical. In the binary system discussed above, for example, the volume term driving force for phase separation is the difference in B concentration between B-rich and B-poor domains: nucleation requires the existence of a steep concentration gradient. Forming these domains thus also leads to the formation of interfaces, hence to surface energetics terms. These constitute another crucial thermodynamic driving force responsible for the change of equilibrium concentration in an NC's vicinity (Gibbs–Thomson equation). This is schematized in Figure 1.3, which shows that the competition between the negative free energy volume term and the positive free energy surface (interface) term leads to a critical size below which the NC is unstable. There is an energy cost in constructing an interface; hence, above some free energy threshold, the system will tend to minimize the interfacial energy by increasing the NC size.

This greatly simplified presentation of the requirements for nucleation (Porter and Easterling 1981, Wagner et al. 2001) only considers the interplay between the surface and volume free energy terms in a binary system displaying homogeneous nucleation. Complications arise when aiming to synthesize compound NCs in the solid state. Clearly, working in a ternary or multielement phase diagram introduces new degrees of

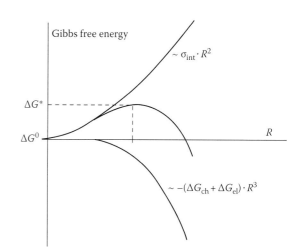

FIGURE 1.3 Competition between the negative free energy volume term and the positive free energy surface (interface) term for a coherent spherical precipitate in a host. The negative—energy-gaining—volume free energy is proportional to R^3. It includes a chemical driving force and an elastic coherence driving force. The surface term, proportional to the surface (or interfacial) energy and to R^2, is positive. Summing the two terms leads to a critical size R_c for growth.

freedom, such as the different diffusion coefficients of the individual species in the host or the competition between formation energies and stabilities of all possible compound NCs that may appear in the complex phase diagram, etc. The Gibbs–Thomson relation, for example, is substantially altered, involving diffusion of all alloy components and a competition between the chemical affinities of the components with different solubilities. There may also be strain- or defect-induced diffusion effects such as those observed, e.g., for transition metals in GaN or for chalcogenides in glass (e.g., see Espiau de Lamaestre et al. 2005). Similar or further complications arise when attempting to synthesize so-called core-shell NCs with differing compositions of the core volume and of near-surface layers. This has been very successful via colloidal chemistry or cluster beam deposition, but is far more difficult inside solid-state matrices. As discussed elsewhere in this volume, such structures are of considerable interest in order, e.g., to add a wide-gap semiconductor shell to a narrow-gap semiconductor core for exciton trapping (Hines and Guyot-Sionnest 1996, Peng et al. 1997), passivate semiconducting (Alivisatos 1998), or metallic (Skumryev et al. 2003, Morel et al. 2004) NCs by a surface oxide layer, e.g., to study magnetic exchange bias in NCs, or, by growing bimetallic NCs with different alloy compositions in the core and the shell, or to control the surface plasmon band emission wavelength (Mattei et al. 2009). But their nucleation and growth again require working in a multielement phase diagram that includes the host elements, with major difficulties in controlling NC evolution and size distribution. This is where nonequilibrium techniques become necessary.

Our discussion has largely remained so far within the realm of classical thermodynamics. As we have just indicated, however, other contributions to the energetics of an NC often come into play. These are quite diverse, as shown by two examples to which we return at the end of this chapter. (1) In a liquid or an insulator, the first stages of nucleation—typically involving just a few atoms—are largely a matter of chemistry. For a very small NC (say, well below a few hundred atoms), the free energy is a non-monotonic function of its size and depends on charge equilibrium in the liquid or solid host. The stability and mobility of reaction-induced ionic charge states are then important factors (possibly overriding, if there are only a few atoms). Also, in the same initial stage, redox effects usually dominate dynamics (reaction paths) and ultimate equilibrium among species. (2) If nucleation occurs in a host containing, say, interface dislocations or artificially introduced surface defects, these act as traps for heterogeneous nucleation. In such cases, surface energetics (including strain-induced diffusion) dominate the nucleation process, affecting the predictability and control of the growth process.

1.3 Growth and Coarsening

The first feature of growth according to thermodynamics is the concentration gradient around an NC nucleus (modified by the effect of NC surface curvature, Gibbs–Thomson relation). An isolated nucleus, surrounded by such a concentration gradient, progressively accretes atoms from its surroundings, i.e., growth is diffusion controlled. This should lead to a parabolic growth law ($R^2 \propto t$)—a prediction that is rarely verified experimentally, since dilution is never infinite and other processes intervene. These are notably the birth of new nuclei while others grow and precipitate interaction (coarsening). Another important possibility is that growth not be diffusion limited, but interface limited, i.e., determined by the free energy gained by transferring an atom from the host to the NC, leading to a linear growth rate.

The growth law may be deduced in a very general way from a conservation equation for the NC size density distribution. Let $f(R, t)$ be a population of NCs characterized by their size distribution. In size space, as long as there is no creation or destruction of NCs, the conservation of NC number around radius R (at time t) may be written as

$$\frac{\partial f}{\partial t} + \frac{\partial}{\partial \left(\frac{\mathrm{d}R}{\mathrm{d}t} \cdot f \right)} = 0$$

The term in parentheses is the flux density in size space. Were the rate $\mathrm{d}R/\mathrm{d}t$ of NC growth known, this continuity equation would predict the size distribution. Unfortunately, this is most often not the case: neither the number density nor the total NC mass are conserved, hence the difficulty in assessing how to control the size distribution.

If we now consider a population of grown NCs, it is clear that the system's total energy is enhanced by the existence of a large interfacial area. The largest relative contribution to the latter being provided by the smallest NCs, the total energy is reduced if the larger NCs grow at the expense of the smaller ones. This is the coarsening (Ostwald ripening) process involving competitive growth. An important, experimentally discovered feature of the process is the possibility of scaling: at sufficiently long times, the entire size distribution remains self-similar when normalized by an appropriate length scale such as the average NC size or the average interparticle distance. This has opened the way to the analytical treatment of Ostwald ripening (see below).

A similar treatment may be performed in the case of a related problem—that of coagulation. This concerns growth by aggregation, in the absence of mass transfer, of clusters whose sizes may be similar or different. This process often occurs in the case of NC diffusion on surfaces and dominates growth processes in polymers (e.g., those that encage metallic or semiconducting NCs).

Both coarsening and coagulation lead to an asymptotic size distribution; time and size scaling holds for both, so that an analytical treatment could be devised for the latter (Smoluchowski 1916) and for the former (Lifshitz and Slyozov 1961, Wagner 1961, commonly referred to as LSW). This was done in the framework of quasi-equilibrium thermodynamics, the evolution being due to a balance between thermodynamic and kinetic growth factors. We discuss these processes now, using the synthetic approach of Binder.

1.4 A General Description of Phase Separation

An extensive, generalized description of the phase separation in a binary mixture was proposed by Binder (Binder 2001, Binder and Fratzl 2001, Wagner et al. 2001). This is a microscopic aggregation model involving attachment and detachment of clusters or monomers, quite analogous to the chemical reaction theory. The main result is an equation describing the cluster size distribution's evolution in size and time space. Both the coagulation and condensation regimes are derived as a limiting behavior from this equation.

Coagulation involves reactions between clusters of any size, occurring, for example, in liquid systems or at relatively high concentrations, as opposed to condensation dealing with aggregation steps between a monomer and a cluster of any size.

In the case of the coagulation regime, the equation reads

$$\frac{\partial n_l(t)}{\partial t}\bigg|_{coag} = \frac{1}{2}\int_{lc}^{l} dl' W(l-l',l') \frac{n_{l'}(t)}{n_{l'}^{eq}} \frac{n_{l-l'}(t)}{n_{l-l'}^{eq}}$$

$$- \frac{n_l(t)}{n_l^{eq}}\int_{lc}^{\infty} dl' W(l,l') \frac{n_{l'}(t)}{n_{l'}^{eq}}$$

where

W are size-dependent kinetic factors
$n_l(t)$ is the population of *l*-size clusters at time *t*

This equation then reduces to that of Smoluchowski (Smoluchowski 1916), except that the latter used a discrete representation.

In the condensation regime, the equation reads

$$\frac{\partial n_l(t)}{\partial t}\bigg|_{cond} = \frac{\partial}{\partial l}\left[R_l \frac{\partial}{\partial l} n_l(t) - \frac{\partial}{\partial l}\left(\frac{\Delta F_l}{k_B T}\right) R_l n_l(t) \right]$$

where

R_l are the kinetic factors
ΔF_l is the free energy of a *l*-size cluster

This equation has the form of a conservation equation

$$\frac{dn_l(t)}{dt} + \frac{d}{dl} J_l = 0$$

where the cluster current in the size representation consists of two terms

1. A thermodynamically driven drift component that leads the system toward its minimum free energy. This is

$$J_{der} = -\frac{\partial}{\partial l}\left(\frac{\Delta F_l}{k_B T}\right) R_l n_l(t)$$

2. A diffusion component describing the contribution of fluctuations to the nucleation and growth process. This contribution is

$$J_{diff} = R_l \frac{\partial}{\partial l} n_l(t)$$

and always tends to *broaden* the size distribution.

The evolution of the size distribution is thus determined by a highly nonlinear set of coupled differential equations, among which various approximations allow us to identify the main contributions to either coagulation or condensation. For example, in a long-time approximation, the equation describing condensation provides the LSW expression for coarsening (late-stage growth):

$$f_{LSW}(R,t) = n_0 \left(\frac{R_{c0}}{R_c}\right)^3 3^4 2^{-5/3} e \left(\frac{R}{R_c}\right)^2 \left(\frac{R}{R_c} + 3\right)^{-7/3} \left(\frac{3}{2} - \frac{R}{R_c}\right)^{-11/3}$$

$$\times \exp\left(\frac{1}{\frac{2R}{3R_c} - 1}\right)$$

with

$$R_c(t) = \left(R_{c0}^3 + \frac{4}{9}\frac{2 V_m \sigma D}{\left(x_B^\beta\right)^2 k_B T} t \right)^{1/3}$$

the average radius (R_{c0} the initial mean radius, n_0 the initial cluster density, V_m the molar volume of the precipitates, σ the surface tension, D the diffusion constant, and x_B^β the fraction of B in the precipitate—usually close to 1). This expression is derived by assuming (see Section 1.4.2) an adequate scaling law for diffusion as well as mass conservation, the validity of the Gibbs–Thompson equation, and a very low cluster density, typically below 0.1 at%.

Binder's derivation is sufficiently general to account for a cluster growth mechanism that includes not only the number of constituent atoms but also, for example, the number of surface atoms or the NC ionic charge (if there is one); it may be extended to clusters including more than one chemical component, i.e., other contributions to their free energy. Unfortunately, due to the nonlinear nature of the problem, finding an analytical path to other than the long-term solution is difficult, hence the use of elaborate simulations (described in the references above). The fact that a general formulation of the nucleation, growth, and coarsening dynamics is obtained remains a significant advantage. In the latter two processes, the thermodynamical term dominates, whereas in the former, the diffusive term plays a crucial role in "igniting" the processes. We briefly review these two limiting cases.

1.4.1 Short-Term Behavior: Nucleation

The equation given above describes homogeneous nucleation, i.e., nucleation initiated only by intrinsic fluctuations of the system's free energy. Nucleation can also be initiated by external perturbations such as defects or impurities. We have seen that in order to obtain narrow size distributions, control over growth should attempt to carefully separate the nucleation and the growth stages: if all clusters start growing together, the resulting size distribution will be narrower than if new nuclei are continuously generated while older ones are growing.

Note that this rate equation approach is adequate for cluster precipitation in both liquid and solid-state matrices. In the former, it is often referred to as the method of LaMer (LaMer and Dinegar 1950, Park et al. 2007), in which reactants are rapidly introduced in the solvents to induce nucleation. We previously also mentioned the two-stage annealing technique (Borelli et al. 1987) well known to glass makers.

1.4.2 Long-Term Behavior: Scaling

The long-term behavior of a precipitate system is far easier to observe than the nucleation and growth processes. Can we deduce any information from it as concerns the inception and evolution of the system's essential features? As mentioned above, during long-term growth (the coarsening regime), the system's main parameters (e.g., density of clusters and size distribution) are self-similar. One can therefore clearly distinguish the system's time evolution, well described by the time dependence of a typical size, and other normalized topological observables such as the NC spatial and size distributions.

In the long-term LSW size distribution (Lifshitz and Slyozov 1961, Wagner 1961), the mean radius R_c is the scaling length. When the radius is normalized by R_c, the size distribution's shape is invariant in time:

$$\tilde{f}(z) = 3^4 2^{-5/3} e z^2 (z+3)^{-7/3} ((3/2) - z)^{-11/3} \exp\left(\frac{1}{(2z/3) - 1}\right)$$

where

$$f_{\mathrm{LSW}}(R,t) = n(t)\left[\frac{1}{R_c}\tilde{f}\left(\frac{R}{R_c}\right)\right]$$

This is referred to as the asymptotic form of the size distribution. It does contain some information about the system's precipitation physics. An example (Valentin et al. 2001) is shown in Figure 1.4. The entire nuclei population was first synthesized by ion irradiation (analogous to the first stage of the photographic process) and then all nuclei grew simultaneously under a thermal anneal. The conditions of this experiment were very close to LSW approximation conditions, and the resulting late-stage growth size distribution is in excellent agreement with the LSW prediction.

We have seen that the dynamics of R_c are determined by the diffusion constant, as expected, and also, more interestingly, by the surface tension that dominates coarsening in the LSW model. Note that $R_c^3 \propto t$ (this is also found in the experiment of

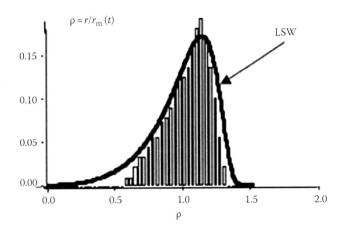

FIGURE 1.4 Comparison of the LSW distribution with an experiment in which all the nuclei of the very dilute NC population grew together after being simultaneously produced by an ion irradiation, whose effect is analogous to a photographic exposure, see text. (From Valentin, E. et al., *Phys. Rev. Lett.*, 86, 99, 2001.)

Figure 1.4). The influence of the microscopic mechanism is further illustrated by a comparison of the results above with those obtained when cluster condensation is no longer diffusion limited, but is limited by chemical reactions at the NC surface. This change in the microscopic conditions of aggregation leads to a broader size distribution (which still has a tail on the smaller-size side, again due to the surface tension mechanism) with a growth law exponent ½ instead of ⅓.

These examples reveal an influence of microscopic mechanisms on the final size distribution shape and growth exponents. However, although this agreement validates the assumptions regarding the system's long-term evolution, can one work backward and say anything at all about the earlier nucleation and growth mechanisms from the post-coarsening size distribution? Clearly, this is at best very difficult. For example, the long-term LSW solution is obtained whether or not there is an energy barrier for nucleation (spinodal or metastable decomposition). Other cases are worse, as shown by a scrutiny of the observable quantities.

1.4.2.1 Observable 1: Growth Time Exponent

Many experiments deal with the time-dependent growth law. However, its experimental determination is a very difficult task, requiring studies over several decades to obtain sufficient precision. When the experimental determination is based on a single decade, and since measurements are necessarily performed on small precipitate sizes, unequivocal results are unlikely. This difficulty is long known and needs to be carefully resolved in interpreting experiments. Worse still, agreement with a microscopic mechanism is sometimes claimed on the basis of a growth law exponent in contradiction to the observed size distribution shape. This illustrates the fact that a growth exponent is not unequivocally related to a growth mechanism. In fact, theoretical considerations show that there is a degree of universality in this exponent, i.e., it is representative of families of first order phase-separating systems (universality classes) rather than identifying a particular mechanism (Binder 1977).

1.4.2.2 Observable 2: Size Distribution Shape

We have seen, in the case of the LSW distribution, that changes in the shape of the size distribution were related to differences in the aggregation process. This is by no means a general result, as demonstrated by the example of lognormal size distributions. The lognormal distribution function reads

$$f(r) = \frac{1}{r \ln(\sigma)\sqrt{2\pi}} \exp\left(-\frac{\ln^2(r/\mu)}{2\ln^2(\sigma)}\right)$$

where

σ is the geometric standard deviation
μ is the geometric average

Both are dimensionless. Experimental results on NC size distributions after more or less complex, nonequilibrium synthesis techniques often display such shapes, which are usually noted by authors, but not discussed. Could they possibly provide any information on operative mechanisms?

Experiments on clusters provide evidence for lognormal distributions in very different nonequilibrium physical contexts (e.g., laser-, plasma-, or sputter-deposition, physical vapour deposition (PVD), ion implantation, etc.). For clusters synthesized by vacuum evaporation (Granqvist and Buhrman 1976), the observed lognormal shape originated in a single multiplicative stochastic process, i.e., the lognormal distribution of time spent by a nucleus in the region where growth occurred. Lognormal distributions were found in closed systems under coagulation of aerosols or colloids: Friedlander and Wang (1966) checked, by the approximate solution of the coagulation equation, that the asymptotic shape of the size distribution was close to the experimentally found lognormal shape. Note that, as in the case of the LSW distribution, this asymptotic size distribution shape is independent of the initial nucleation conditions. This was demonstrated theoretically (Hidy 1965). In fact, a perusal of the scientific literature shows that lognormal distributions are ubiquitous. For example, a lognormal distribution was even found when measuring the height distribution of British infantry soldiers in the late nineteenth century. It does not, apparently, tell us much about the birth and growth conditions of these young men.

1.4.3 Amount of Information Contained in the Size Distribution Shape

These results illustrate the difficulty in relating the shape of the size distribution to the existence of one or another growth mechanism. In order to estimate whether any information is obtainable at all, we may approach the problem in terms of information theory. The amount of information on a distribution is usually given by its entropy, defined by

$$S = \int f \ln f$$

For example, the entropy of a normalized Gaussian is $S_g = \ln\left(\sigma\sqrt{2\pi e}\right)$. The entropy increases with the size distribution width, confirming the intuition that the broader the distribution, the less controlled, or more disordered, the growth process.

This observation may be related to the example of LSW coarsening. We first note that the derivation of LSW is valid for the limit of a very low fraction of NCs. This assumption allows an analytic solution to the problem within a mean field approximation—clusters do not interact with one another, they are only subject to a mean solute concentration that fixes the rate of growth/dissolution with the help of the Gibbs–Thomson relation. If we now increase the volume fraction of NCs, they begin to interfere through dipole and higher order interactions. For a given NC, the rate of growth/dissolution differs depending on whether it is close to a large, or small, NC. The surroundings of a cluster are random because of the intrinsically random nature of nucleation, so the stochastic character of growth is enhanced. We then anticipate, from our initial comments to this section, a broadening of the size distribution. This is indeed the case, as confirmed by experiments and theory (Ardell 1972).

Information theory does help us to understand the significance of the lognormal shape. One of its principles (Jaynes 1957) is that entropy is maximized at equilibrium (the case here, since we consider long-term behavior) under a set of general constraints that can be written as

$$\int C_i(u) f(u) du = < C_i >$$

As concerns nucleation and growth, these constraints are, for example, the matter conservation equation

$$\int_0^\infty l n_l(t) dl = cste$$

and the size distribution's evolution equation given above, which is a conservation equation in size space:

$$\frac{dn_l(t)}{dt} + \frac{d}{dl} J_l = 0$$

These two equations fully determine the system's evolution. Each of them contains a different amount of information on the growth process, and the entropy maximization principle allows determination of the main one. Specifically, it may be shown (Rosen 1984) that the distribution function obtained by using the sole constraint of matter conservation is simply

$$\tilde{n} = e_l^{-v}$$

This equation is actually a very close approximation to the large-size tail of the lognormal size distribution in the case of, e.g., Brownian coagulation studied by Rosen, as well as in the case

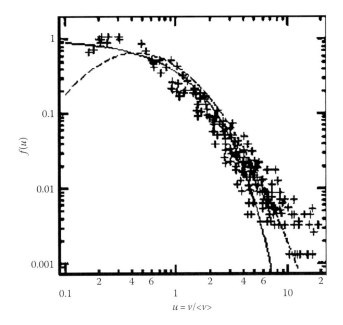

FIGURE 1.5 Plot of u, the reduced volume's probability density. Crosses are experimental data for semiconductor PbS, PbSe, CdSe, and PbTe nanocrystals grown after sequential implantation of the components into pure silica and long-term annealing in quite different conditions. Despite their differences in preparation, average sizes, and depth distributions, all these data fall on the universal curve corresponding to the maximum entropy distribution, e^{-u}, determined by the sole constraint of volume conservation. The dashed line is the best fit of experimental data ($u > 1$) to the reduced lognormal distribution $m = 1$: it is seen that the latter only deviates from the former when the NC sizes are extremely small. The existence of a lognormal distribution thus provides no information at all on the NC evolution. (From Espiau de Lamaestre and Bernas, H., *Phys. Rev.*, B73, 125317, 2006.)

(Figure 1.5) of NC syntheses in which multiple growth mechanisms combine and interfere (Espiau de Lamaestre and Bernas 2006). It was also shown (Gmachowski 2001) that even the standard deviation of the lognormal distribution is a fairly universal quantity, independent of the growth process.

Whereas the shape after LSW ripening reflected at least partially some aspects due to the initial nanocrystal population and its evolution, the existence of a long-term lognormal size distribution in an NC population reveals that any memory of its evolution mechanisms is lost in a maze of different (possibly interfering) nucleation and growth processes. The lognormal shape of the distribution is simply due to the existence of matter conservation. Its occurrence in a particular process signals that the nucleation and growth are too complex to control, other than by clever empiricism.

1.5 Perspectives and Conclusion: How to Narrow Size Distributions?

This analysis leads to several general remarks regarding control over NC synthesis. (1) Conceptually, the procedure that consists in separating nucleation and growth may be viewed as a means to avoid the interference of two distinct processes. (2) The versatility

and theoretical understanding provided by late-stage growth conditions are, unfortunately, less crucial to NC control than the nucleation stage conditions. Long-term size distributions are systematically broader and, as detailed above, are mostly controlled by general constraints such as the growth space dimensionality, matter conservation, and basic thermodynamical (including entropy) contributions. (3) The properties of the late-stage NC ensemble generally lose memory of the system's initial stage, i.e., the nucleation conditions. This emphasizes the radically different origin of the size distributions that are obtained in the two limiting stages, nucleation versus coarsening. For example, using recently developed techniques, one might wish to prepare an artificial nanostructured system with inhomogeneous (local) supersaturation and ordered nucleation centers. The discussion above shows that the long-term coarsening stage of such a system would behave exactly as one in which nucleation centers were initially random—all efforts made to control NC position and size would be lost. (4) Control of the nucleation stage offers the best conditions for size distribution control. As long as we stay in its vicinity, artificial structuring methods that localize (heterogeneous) nucleation or/and early growth (inhomogeneous supersaturation) tend to limit the disorder (entropy) increase during precipitation, leading to narrower size distributions. The price to pay for this is in the limitation of the NCs' average sizes, typically below or around 10 nm. For optical quantum dots this is within the range of the exciton Bohr radii, hence is not a drastic problem.

These considerations justify the broad interest in far-from-equilibrium techniques described elsewhere in this book. They are quite successful experimentally, but theoretical treatments for them are still rudimentary or absent. Here are a few general comments on NC population control in some of these cases.

First, what do we mean by nonequilibrium conditions? Consider our initial phase diagram: once thermodynamical conditions (formation enthalpy, free energies, surface energies, etc.) required for precipitation are fulfilled, the system can relax to an equilibrium state, provided that some atomic mobility is introduced. Mobility can of course be due to temperatures high enough to overcome activation energy barriers. But external sources may have an overriding influence: species mobility may also be due to a nonequilibrium concentration of interstitials and vacancies produced in a crystalline solid by irradiation, to 3D bulk diffusion in glasses, to diffusion on surfaces or in grain boundaries, and to convection in a liquid. Some of these processes involve random, homogeneous, nucleation; others (e.g., nucleation on lattice defects or grain boundaries) involve heterogeneous nucleation. The latter is viewed as disadvantageous when nucleation sites or the resulting precipitate size distribution are uncontrolled. However, techniques have been developed to produce ordered arrays of NCs for applications to magnetism or optics, as discussed elsewhere in this handbook. In such instances, nucleation of small NCs at defects such as surface defects produced by a focused ion beam impact (Bardotti et al. 2002), or by an ordered array of dislocations after surface (interface) strain relaxation (Romanov et al. 1999) or by ordered step formation on a specially chosen crystalline facet (e.g., Weiss et al. 2005) often allow control over the NC density. All these methods are based on trapping by external forces (e.g., the introduction of a dislocation array) rather than on the internal evolution of the system. But since the latter is required in order to nucleate and grow the NCs themselves, successful size and size distribution control (generally via temperature-controlled diffusion) has been, so far, essentially dependent on trial and error. More generally, due to the impact of molecular beam deposition (MBE) methods, the extension of quasi-equilibrium thermodynamics to clustering (mediated by atomic-scale diffusion from a supersaturated solution) on surfaces has been a major area of activity, for which we refer to the literature (e.g., Villain and Pimpinelli 1998). As evidenced by the Stranski–Krastanov or van der Merwe growth modes, the NC surface energetics include large strain or stress contributions due to differing lattice cell parameters. This has a major influence on the kinetics of NC formation when surface diffusion dominates, and requires simulations (e.g., Amar and Family 1995) to predict. As mentioned previously, the surface's detailed structural properties are also crucial to surface diffusion as well as to the trapping efficiency for different elements.

In glasses, NC synthesis is of importance not only for stained-glass window applications, but also for nanophotonics. This field has been repeatedly revisited, and it has become increasingly clear that—just as in liquids—redox chemistry plays a crucial role in both thermodynamical and kinetic effects. Specifically, moving charged species play a crucial role in the clustering process. Their relative stability and interactions (among themselves and with electrons or holes), as well as their chemical affinity for glass matrix components, all determine the nature and stability of the NC to be formed. In other words, NC formation is essentially analogous to the photographic process (Belloni and Mostafavi 1999, Espiau de Lamaestre et al. 2007). This means that, as in photography, it is possible to induce simultaneous nucleation of all incipient clusters (e.g., by irradiation with UV light, electrons, ions, etc.) and then—*separately!*—induce (and control) growth by a short thermal anneal.

A trick to obtain a rather dense, ordered NC array with a narrow size distribution is the nucleation and growth of NCs in nonreacting "cages" that may be inorganic (e.g., zeolites) or organic (polymers, biological systems, colloidal systems, see Sun et al. 2000). The size of the cage and the corresponding external forces may sometimes limit the amount of accretion to the growing NC. In some cases, this leads to a very narrow size distribution reflecting that of the cages (determined in turn by the chemical processing). However, in other instances such attempts have led to the lognormal-like size distributions described above, a strong indication that control over growth during the process was then mediocre.

Finally, a word on far-from-equilibrium techniques such as plasma deposition of cluster beams and UV, electron, or ion irradiations. In cryogenic plasma or thermal evaporation systems, as well as in techniques combining selective evaporation with incipient aggregation in a He jet, growth largely depends on (1) the mean time that a nucleus (typically a cluster of 2 to several atoms) spends in the region where it can grow and (2) the physical parameters governing the coagulation rate such as the monomer source temperature or plasma gas pressure. We then deal with an open system in which aggregates can be extracted from the growing zone and projected on a substrate on which they might be unstable. Growth on the substrate occurs essentially via coagulation (Smoluchowski, Binder), and if surface diffusion is involved, requires elaborate simulations (Politi et al. 2000). Experimentally, size sorting (via an electric or magnetic field) of charged aggregates allows deposition of NCs with a highly nonequilibrium, narrow size distribution. Depending on the electronic properties of the matrix, UV or electron and Ion beam irradiations may assist in controlling nucleation and growth in different ways. In metals, for instance, electron or, more usually, ion irradiation induces or accelerates species diffusion by producing a supersaturation of vacancies or interstitials. Processes that—at quasi-equilibrium—would occur at high temperatures then are active at temperatures that may be typically 200° lower. Playing on nucleation and growth with the combination of temperature and deposited energy density due to irradiation is a means of "driving" (and enhancing control of) the alloy system in a context where theoretical work relating driven alloys to thermodynamics has progressed (Martin and Bellon 1997, Averback and Bellon 2009). The "photographic process" in insulators mentioned above, or the use of defect creation and control to stabilize NC formation into well-defined arrays, are successfully implemented experimental techniques, but they remain to be systematically included in a predictive approach. If the reader comes away with the impression that each new NC synthesis technique enhances the need for a critical eye (and more theoretical work) on the conditions for size distribution control, we will have reached our purpose.

References

Alivisatos, A. 1998. Semiconductor quantum dots, *MRS Bull.* **23**(2), and refs. therein.

Amar, J.A. and F. Family. 1995. Critical cluster size: Island morphology and size distribution in submonolayer epitaxial growth, *Phys. Rev. Lett.* **74**: 2066.

Ardell, A.J. 1972. The effect of volume fraction on particle coarsening, *Acta Metall.* **20**: 61.

Averback, R.S. and P. Bellon. 2009. Fundamental concepts of ion beam processing, in *Materials Science with Ion Beams*, ed. H. Bernas, Chapter 1, Springer, Berlin, Germany.

Bardotti, L., B. Prevel, P. Jensen et al. 2002. Organizing nanoclusters on functionalized surfaces, *Appl. Surf. Sci.* **191**: 205.

Belloni, J. and M. Mostafavi. 1999. Radiation-induced metal clusters. Nucleation mechanisms and physical chemistry, in *Metal Clusters in Chemistry*, eds. P. Braunstein, L.A. Oro, and P.R. Raithby, p. 1213, Wiley, Weinheim, Germany, and refs. therein.

Binder, K. 1977. Theory for the dynamics of "clusters." II. Critical diffusion in binary systems and the kinetics of phase separation, *Phys. Rev.* **B15**: 4425.

Binder, K. 2001. Statistical theories of phase transitions, in *Phase Transformations in Materials*, ed. G. Kostorz, Chapter 4, Wiley-VCH, Weinheim, Germany.

Binder, K. and P. Fratzl. 2001. Spinodal decomposition, in *Phase Transformations in Materials*, ed. G. Kostorz, Chapter 6, Wiley-VCH, Weinheim, Germany.

Borelli, N.F., D. Hall, H. Holland, and D. Smith. 1987. Quantum confinement effects of. semiconducting microcrystals in glass, *J. Appl. Phys.* **61**: 5399.

Cahn, J.W. and J.E. Hilliard. 1958, 1959. Free energy of a nonuniform system, *J. Chem. Phys.* **28**: 258 and *J. Chem. Phys.* **31**: 688.

Cook, H.E. 1970. Brownian motion in spinodal decomposition, *Acta Metall.* **18**: 297.

Espiau de Lamaestre, R. and H. Bernas. 2006. Significance of log-normal size distributions, *Phys. Rev.* **B73**: 125317.

Espiau de Lamaestre, R., J. Majimel, F. Jomard, and H. Bernas. 2005. Synthesis of lead chalcogenide nanocrystals by sequential ion implantation in silica, *J. Phys. Chem.* **B109**: 19148.

Espiau de Lamaestre, R., H. Béa, H. Bernas, J. Belloni, and J.L. Marignier. 2007. Irradiation-induced Ag nanocluster nucleation in silicate glasses: Analogy with photography, *Phys. Rev.* **B76**: 205431 and refs. therein.

Friedlander, S.K. and C.S. Wang. 1966. The self-preserving particle size distribution for coagulation by brownian motion, *J. Colloid Interface Sci.* **22**: 126.

Gmachowski, L. 2001. A method of maximum entropy modeling the aggregation kinetics, *Colloid Surf. A: Physicochem. Eng. Aspects* **176**: 151.

Granqvist, C.G. and R.A. Buhrman. 1976. Ultrafine metal particles, *J. Appl. Phys.* **47**: 2200.

Hidy, G.M. 1965. On the theory of the coagulation of noninteracting particles in brownian motion, *J. Colloid. Sci.* **20**: 123.

Hines, M.A. and P. Guyot-Sionnest. 1996. Synthesis and characterization of strongly luminescing ZnS-capped CdSe nanocrystals, *J. Phys. Chem.* **100**: 468.

Jaynes, E.T. 1957. Information theory and statistical mechanics. II, *Phys. Rev.* **108**: 171.

LaMer, V.K. and R.H. Dinegar. 1950. Theory, production and mechanism of formation of monodispersed hydrosols, *J. Am. Chem. Soc.* **72**: 4847.

Lifshitz, I.M. and V.V. Slyozov. 1961. Kinetics of precipitation from supesaturated solid solutions, *J. Phys. Chem. Solids* **19**: 35.

Martin, G. and P. Bellon. 1997. Driven alloys, *Solid State Phys.* **50**: 189.

Mattei, G., P. Mazzoldi, and H. Bernas. 2009. Metal nanoclusters for optical properties, in *Materials Science with Ion Beams*, ed. H. Bernas, Chapter 10, Springer, Berlin, Germany.

Morel, R., A. Brenac, and R. Portement. 2004. Exchange bias and coercivity in oxygen-exposed cobalt clusters, *J. Appl. Phys.* **95**: 3757.

Park, J., K. An, Y. Hwang et al. 2004. Ultra-large-scale syntheses of monodisperse nanocrystals, *Nat. Mater.* **3**: 891.

Park, J., J. Joo, S.G. Kwon, Y. Jang, and T. Hyeon. 2007. Synthesis of monodisperse spherical nanocrystals, *Angew. Chem. Int. Ed.* **46**(25): 4630.

Peng, X., M.C. Schlamp, A.V. Kadanavich, and A.P. Alivisatos. 1997. Epitaxial growth of highly luminescent cdse/cds core/shell nanocrystals with photostability and electronic accessibility, *J. Am. Chem. Soc.* **119**: 7019.

Philibert, J. 1991. Atom movements: Diffusion and mass transport in solids, Ed. Physique, Les Ulis, France.

Pileni, M.P. 2001. Nanocrystal self-assemblies: Fabrication and collective properties, *J. Phys. Chem.* **B105**: 3358 and Pileni, M.P. 2003. The role of soft colloidal templates in controlling the size and shape of inorganic nanocrystals, *Nat. Mater.* **2**: 145.

Politi, P., G. Grenet, A. Marty, A. Ponchet, and J. Villain. 2000. Instabilities in beam epitaxy and similar growth techniques, *Phys. Rep.* **324**: 271.

Porter, D.E. and K.E. Easterling. 1981. *Phase Transformations in Metals and Alloys*, van Nostrand Reinhold, New York.

Ratke, L and P.W. Vorhees. 2002. *Growth and Coarsening*, Springer, Berlin, Germany.

Romanov, A.E., P.M. Petroff, and J.S. Speck. 1999. Lateral ordering of quantum dots by periodic subsurface stressors, *Appl. Phys. Lett.* **74**: 2280.

Rosen, J.M. 1984. A statistical description of coagulation, *J. Colloid Interface Sci.* **99**: 9.

Skumryev, V., S. Stoyanov, Y. Zhang, G. Hadjipanayis, D. Givord, and J. Nogués. 2003. Beating the superparamagnetic limit with exchange bias, *Nature* (*London*) **423**: 850.

Smoluchowski, M. 1916. Versuch einer mathematischen Theorie der Koagulationskinetik kolloider Lösungen, *Physik. Zeitschrift* **17**: 557.

Sun, S., C.B. Murray, D. Weller, L. Folks, and A. Moser. 2000. Monodisperse FePt nanoparticles and ferromagnetic FePt nanocrystal superlattices, *Science* **287**: 1989.

Valentin, E., H. Bernas, C. Ricolleau, and F. Creuzet. 2001. Ion beam "photography": Decoupling nucleation and growth of metal clusters in glass, *Phys. Rev. Lett.* **86**: 99.

Villain, J. and A. Pimpinelli. 1998. *Physics of Crystal Growth*, Cambridge University Press, Cambridge, U.K.

Wagner, C. 1961. Theorie de Alterung von Niederschlägen durch Umlösen (Ostwald-reifung), *Z. Elektrochemie* **65**: 581.

Wagner, R., R. Kampmann, and P.W. Vorhees. 2001. Homogeneous second-phase precipitation, in *Phase Transformations in Materials*, ed. G. Kostorz, Chapter 5, Wiley-VCH, Weinheim, Germany.

Weiss, N., T. Cren, M. Epple et al. 2005. Uniform magnetic properties for an ultrahigh-density lattice of noninteracting co nanostructures, *Phys. Rev. Lett.* **95**: 157204.

Weiss, E.A., R.C. Chiechi, S.M. Geyer et al. 2008. The use of size-selective excitation to study photocurrent through junctions containing single-size and multi-size arrays of colloidal cdse quantum dots, *J. Am. Chem. Soc.* **130**: 74 & 83.

2

Structure and Properties of Hydrogen Clusters

Julio A. Alonso
Universidad de Valladolid
Universidad del País Vasco
Donostia International
Physics Center

José I. Martínez
Technical University of Denmark

2.1 Introduction

Hydrogen is expected to play an important role in the future as an alternative to the present fuels for the massive production of energy. The first pillar, which still looks far away, will be the production of electricity in nuclear fusion reactors, once the problem of sustaining and controlling the reactions is solved. The second pillar, which looks closer, is the production of electricity by means of hydrogen fuel cells, and its widespread application in cars as an alternative to gasoline. For these reasons, the study of hydrogen becomes an important subject from the technological point of view. The basic aspects of the physics and chemistry of hydrogen are also interesting. Under normal conditions of pressure and temperature, hydrogen is a gas formed by H_2 molecules. The binding energy of the two H atoms in the molecule is strong, 4.8 eV. In this molecule, which is the simplest and more abundant molecule in the universe, the two electrons form a closed shell. The molecule exists in two isomeric forms differing in their nuclear spin configuration. In the para-hydrogen isomer, the nuclear spins of the two nuclei are in an antiparallel configuration, that is, they point in opposite directions, while in the ortho-hydrogen isomer, the two nuclear spins are parallel, that is, they point in the same direction. When the gas condenses, it forms a molecular liquid or a molecular solid in which the H_2 molecules interact weakly by van der Waals forces. The intensity of the H_2–H_2 interaction is intermediate between the He–He and Ar–Ar interactions. At low temperatures, the para-hydrogen isomer, which is the isomer with lower energy, is in its

ground rotational state ($J = 0$) and thus the molecule behaves as a boson of zero spin.

The condensation of the gas can also be forced by molecular beam techniques allowing the production of clusters formed by a finite number of H_2 molecules. In this chapter, clusters formed by N hydrogen molecules will be denoted as $(H_2)_N$. These clusters have attracted attention due to their peculiar properties, which arise from the coexistence of the strong intramolecular H–H bonding and weak H_2–H_2 intermolecular forces (Castleman et al. 1998, Alonso 2005). Part of the interest in para-hydrogen comes from the fact that it is considered to be the only natural species in addition to the He atom isotopes, which might exhibit superfluidity (Ginzburg and Sobyanin 1972).

Nuclear fusion reactions have been observed to occur by irradiating a dense molecular beam of large deuterium clusters (deuterium is an isotope of hydrogen with a nucleus formed by a proton and a neutron) with an ultra-fast high-intensity laser (Zweiback et al. 2000). The laser irradiation produces the ionization of the deuterium atoms and leads to a violent Coulombic explosion of the clusters; the nuclear fusion reactions occur in the collisions between the flying deuterium nuclei. Neutrons are produced in these nuclear reactions and tabletop neutron sources have been constructed based on this cluster beam technique. Many investigations of hydrogen clusters have focused on single-charged clusters of the family $(H_2)_N H_3^+$ with $N = 1, 2, \ldots$. The majority of the hydrogen clusters in the universe belongs to this family. On the other hand, these clusters are easily handled in the laboratory. In this chapter, a review is provided of the structure and properties of hydrogen clusters. Topics treated are the experimental production of hydrogen clusters in the laboratory, the structure of neutral and charged clusters, free and confined in cages, phase transitions, Coulombic explosions induced by laser irradiation, clusters on surfaces, and finally the manifestation of quantum effects and its possible relation to superfluidity.

The theoretical treatment of the clusters using state-of-the-art methods is given strong emphasis in this chapter because these methods provide important insights into the structure and energetics of the clusters, but connection to experiment is made in all possible cases. It is expected that the topics selected will give an idea of the wide reach of this field.

2.2 Production of Hydrogen Clusters in Cryogenic Jets

An efficient method to produce hydrogen clusters (Tejeda et al. 2004) consists in the expansion of extremely pure (99.9999%) H_2 gas, originally at a pressure P_0 of 1 bar, through a small hole (called nozzle) of diameter $D = 35$–$50\,\mu m$ into a second chamber at a lower pressure of 0.006 mbar. The first chamber is cooled by a helium refrigerator, which provides a source temperature T_0 of 24–60 K, regulated to within ±1 K. The vapor exiting the hole expands adiabatically into the vacuum, and the density

and temperature of the jet rapidly decrease as the distance z to the orifice increases. The expansion produces an extremely cold molecular jet and small para-H_2 clusters are formed by aggregation of the molecules in the jet. The analysis of the abundance of clusters of different sizes has been made applying Raman spectroscopy techniques using an Ar^+ laser. The spectrometer can be focused to different regions of the jet, that is, at different distances z from the expansion orifice. Figure 2.1a shows five Raman spectra measured at different reduced distances $\xi = z/D$ along the center line of the expanding jet. The reduced distances go from $\xi = 1$ to $\xi = 24$. The measuring time for each spectrum is between 4 and 15 min and it increases with z because the density of the jet is inversely proportional to ξ^2 and varies between 10^{20} and 10^{16} molecules cm^{-3}. The large peak at 4161.18 cm^{-1} is the Q(0) line of the para-H_2 molecule, characterizing the vibration of the molecule, and the small peak at 4155.25 cm^{-1}, marked by an asterisk, is due to the small amount (<1%) of ortho-H_2 in the jet. The interaction between the hydrogen molecules in the cluster shifts the Q(0) line of $(H_2)_N$ clusters to lower wave numbers, and each of the peaks at the left of the para-H_2 monomer line can be assigned to clusters of specific size N. Dimers, $(H_2)_2$, can already be identified at $\xi = 1$, but not larger clusters. Trimers, tetramers, and pentamers are visible at $\xi = 3$. Then, for $\xi = 5$, a broad peak appears at 4158 cm^{-1}, which was assigned to $N = 13$. The intensity and the width of this peak suggest the formation

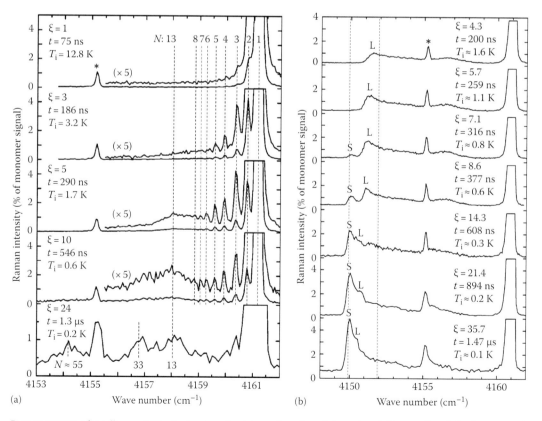

FIGURE 2.1 Raman spectra of small para-$(H_2)_N$ clusters in a free jet as a function of the reduced distance ξ and the flow time t from the orifice. T_i is the estimated ambient temperature at each distance ξ. Source conditions: (a) $T_0 = 46\,K$, $P_0 = 1$ bar, and $D = 50\,\mu m$; (b) $T_0 = 36.5\,K$, $P_0 = 2$ bar, and $D = 35\,\mu m$. L indicates liquid and S solid. (Reproduced from Tejeda, G. et al., *Phys. Rev. Lett.*, 92, 223401-1, 2004. With permission.)

of a first layer of H_2 molecules and also indicate that completion of that shell makes the cluster highly stable. Sizes corresponding to very stable clusters are usually called magic numbers (Alonso 2005). Measuring the spectrum at larger distances from the nozzle hole gives time for the formation of larger clusters. So, at $\xi = 24$, with the distribution approaching a steady state, additional peaks appear around $N = 33$ and $N = 55$. A good estimate of the abundance of the $(H_2)_N$ clusters relative to the monomers is obtained by dividing the area under each peak by N.

Using a lower source temperature and a higher pressure produces much larger clusters, as shown in Figure 2.1b. At $\xi = 4.3$, the maximum in the Raman spectrum is shifted to $4151.5\,cm^{-1}$, very close to the line for liquid para-H_2 at $18\,K$, which is the dashed line in the figure. In addition to this liquid peak (L), at $\xi = 7.1$, a new peak (S) appears at $4150\,cm^{-1}$, whose position agrees with the spectrum for the solid para-H_2 at $2\,K$. The intensity of the solid peak (S) grows at larger distances from the orifice, a feature that indicates that the cluster growth continues. The liquid peak is strongly shifted at large ξ toward the position of the solid peak. The coexistence of the two peaks suggests that the solid clusters have a significant liquid fraction, most likely located at the cluster surface.

2.3 Atomic Structure and Growth of Neutral Clusters

The depth of the interaction potential between two hydrogen molecules is very small, about 3 meV, but still several times larger than the interaction between two He atoms. Accurate ab initio quantum chemical calculations have been performed for the smallest clusters $(H_2)_2$, $(H_2)_3$, and $(H_2)_4$ using the Möller–Plesset (MP2) and Coupled Cluster (CC) methods (Diep and Johnson 2000, Carmichael et al. 2004). Those methods treat accurately the correlations between the electrons. The dimer, $(H_2)_2$, has a T-shaped structure, that is, the axes of the two molecules lie in the same plane and are perpendicular to each other, and the separation between the two mass centers is 6.55 a.u. (the atomic unit of length is equal to 0.529 Å). The T shape optimizes the interaction between the permanent electric quadrupoles of the two molecules. The binding energy E_b of a cluster $(H_2)_N$ can be defined as

$$E_b(N) = NE(H_2) - E((H_2)_N), \qquad (2.1)$$

where

$E((H_2)_N)$ is the total energy of the cluster
$E(H_2)$ is the energy of an isolated hydrogen molecule

The energy of the cluster is smaller than the sum of the energies of the N separated molecules, and the binding energy measures the stabilization gained by the system when the molecules are brought together to form the cluster. The calculated binding energy of the dimer is between 4 and 5 meV. The potential energy surface of $(H_2)_2$, that is, the interaction potential as a function of

the relative orientation of the molecules, shows some anisotropy, but interconversion to some other structural forms is easy. The trimer, $(H_2)_3$, is planar and cyclic (with C_{3h} symmetry), with each pair of molecules slightly offset from the ideal T-shape orientation (Carmichael et al. 2004). The binding energy of the trimer is nearly equal to three times the H_2–H_2 interaction; this means that three-body forces are very small. A structure of the tetramer, $(H_2)_4$, that preserves the T-shape arrangements for each pair of adjacent molecules can be formed as a square planar cluster, but this is not the structure of lowest energy. The ground state is a nonplanar, near tetrahedral structure in which the molecules still preserve to a large extent the T-shape arrangements among adjacent molecules (Carmichael et al. 2004). The stabilization gained by the nonplanarity is quite small. The binding energy of the tetramer is more than four times the H_2–H_2 interaction energy, and this reflects the contribution from quadrupole–quadrupole interactions between molecules on opposite corners of the structure.

For larger cluster sizes, the quantum chemical calculations become prohibitive, and only calculations employing less accurate methods have been performed. Carmichael et al. (2004) have used an empirical force field fitted to the ab initio results for the trimer and tetramer. The empirical total potential, $V = V_{elect} + V_{6-12}$, is the sum of a classical electrical interaction energy V_{elect} and a site–site term V_{6-12}. The site–site term is based on having a small number of selected sites distributed within the hydrogen molecule and a Lennard-Jones interaction

$$V_{ij} = \frac{C_{ij}}{r_{ij}^{12}} - \frac{D_{ij}}{r_{ij}^6} \qquad (2.2)$$

between every pair of sites i and j located in different molecules. r_{ij} represent the distances between sites i and j, and the site-specific parameters C_{ij} and D_{ij} specify the potential. The first term in Equation 2.2 is repulsive and the second term is attractive. This piece collectively represents nonelectrical effects, that is, van der Waals dispersion forces, quantum mechanical exchange (which arises from the antisymmetry of the many-electron wave function), and overlap between the electronic orbitals of different molecules. The electrical interaction energy V_{elect} includes the quadrupole–quadrupole permanent moment interaction, plus direct (non-mutual) polarization effects. The polarization effects account for the interaction of the permanent quadrupole on a molecule with the induced dipoles and induced quadrupoles in other molecules. Because of these induction effects, in a cluster of more than two molecules, V_{elect} implicitly includes three-body polarization effects. The calculations indicate that an increasing number of isomers with similar energies appear as the cluster size grows. The ground state of $(H_2)_5$ is a four-sided pyramid, nearly flat tetramer with a molecule overhead. $(H_2)_6$ is a bipyramid, that is, molecules are placed on both sides of the tetramer. Then, a pentagonal bipyramid is found for $(H_2)_7$. For larger clusters up to $(H_2)_{13}$, Carmichael et al. noticed the interesting trend that the energy of the ground state structure is close to the energies of many other local minima. The zero point energy reduces further the small binding energy of these clusters.

The observation of magic numbers (Tejeda et al. 2004) motivated a theoretical study of the structures of para-$(H_2)_N$ clusters (Martínez et al. 2007) using the density functional theory (DFT) with the local density approximation (LDA) for electronic exchange and correlation effects (Parr and Yang 1989). The calculations are valid at $T = 0$ K. An intramolecular H–H bond length of 1.38 a.u., close to the experimental value of 1.40 a.u. for the isolated molecule, is obtained independent of the cluster size. In the optimized structure of $(H_2)_2$ the two molecules are in a perpendicular configuration, a triangular structure is obtained for $N = 3$ and a bent rhombus for $N = 4$. Triangular, square, and pentagonal bipyramids are obtained for $N = 5$, $N = 6$, and $N = 7$, respectively. The optimized structures of these small clusters are similar, although not identical to the structures obtained with the ab initio MP2 and CC methods. Larger clusters grow following a pattern of icosahedral growth. The first icosahedron is completed for $N = 13$, where a central molecule is surrounded by other 12 molecules occupying the vertices of the icosahedron. This structure is characteristic of clusters formed by equal atoms (or molecules) with closed electronic shells, like the inert gases neon or argon (Alonso 2005). Due to this similarity, calculations for larger clusters were restricted to specific growth models. An icosahedron is formed by 20 triangular faces joined by 30 edges and 12 vertices. Molecules can then be added in two different ways to form a second shell. In a first type of decoration, molecules are added on top of edge (E) and vertex (V) positions. These provide a total of 42 sites (30 + 12), and a cluster with 55 molecules is obtained. This is called multilayer icosahedral (MIC) growth. Alternatively, the 12 V sites and the 20 sites at the center of the triangular faces (F sites) can be covered to obtain a cluster with 45 molecules. This second type of decoration is often called face-centered (FC) growth.

The calculated binding energies per molecule indicate that FC structures are more stable than MIC structures in $(H_2)_N$ clusters from $N = 14$ up to about $N = 27$. Between $N = 27$ and $N = 35$ (the largest cluster studied), the two structures are almost degenerate, and this trend suggests an imminent transition to MIC structures after $N = 35$. A similar transition occurs for inert gas clusters, although a bit earlier, for $N = 27–28$ (Alonso 2005). The quantity

$$\Delta e_b(N) = 2e_b(N) - \left[e_b(N+1) + e_b(N-1)\right] \qquad (2.3)$$

gives the relative stability of the $(H_2)_N$ cluster with respect to clusters with sizes $N + 1$ and $N - 1$, and has been plotted in Figure 2.2 for the FC and MIC families. In this equation, $e_b(N) = E_b(N)/N$ represents the binding energy per molecule of the cluster $(H_2)_N$. The peaks of $\Delta e_b(N)$ indicate the most stable clusters. For N larger than 13, the peaks reveal the progressive formation of the second layer. For instance, the FC peak at $(H_2)_{19}$ corresponds to the decoration of $(H_2)_{13}$ with a cap of six molecules forming a pentagonal pyramid. One molecule sits on a V site and the other

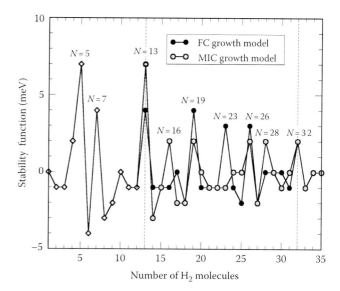

FIGURE 2.2 Stability function, $\Delta e_b(N)$, defined in Equation 2.3, from density functional calculations for MIC and FC growth models of $(H_2)_N$ clusters. Peaks denote clusters with enhanced stability. Vertical lines indicate features observed in a Raman spectroscopy experiment. (Reproduced from Martínez, J.I. et al., *Eur. Phys. J. D*, 43, 61, 2007. With permission.)

five molecules sit on the surrounding F sites. This structure is usually known as the double icosahedron. Filling adjacent caps leads to the structures of $(H_2)_{23}$, $(H_2)_{26}$, and so on. We notice that $(H_2)_{32}$ is a stability peak in both the FC and MIC structures. Some of the features of Figure 2.2 appear consistent with the Raman experiments. Large cluster abundance was observed for $N \approx 13$, 32, and 55, and as discussed above the DFT calculations provide an interpretation of the first two. The DFT calculations also suggest that a transition from FC to MIC structures occurs soon after $(H_2)_{35}$. The observed large abundance of $(H_2)_{55}$ is then trivially explained by the completion of the second icosahedral MIC shell.

The binding energy of $(H_2)_2$ obtained by the MP2 and CC methods is $E_b = 4–5$ meV and the LDA calculation gives 16.5 meV. The LDA overestimation of binding energies is well known, but considering the tiny binding energies of the $(H_2)_N$ clusters, the result can be considered acceptable and it is probably the most one can expect from a simple density functional. All clusters are subject to similar errors, and consequently trends in the binding energy as a function of N are more trustable than the absolute binding energies. Some recent implementations of van der Waals interactions in a DFT framework (Langreth et al. 2005) may improve matters. Zero point effects reduce substantially the binding energies. Accounting for anharmonicity in the zero point energy, a corrected value of $E_b = 2.75$ meV is obtained for $(H_2)_2$ in the LDA, and similar reductions affect other clusters. Due to those small binding energies, hydrogen is a molecular gas except at very low temperatures and high pressures. The small binding energies have other important consequences, as we discuss below.

In the quantum chemical calculations mentioned above, including DFT, the intermolecular interactions are evaluated explicitly by a quantum mechanical treatment of the electrons for each configuration of the molecules in the cluster, but the dynamics of the nuclei is classical. Other calculations of the structure of the clusters have been performed in which the bosonic nature of the para-H_2 molecules is explicitly considered (Cuervo and Roy 2006, Guardiola and Navarro 2006, Khairallah et al. 2007). These calculations make use of quantum Monte Carlo (MC) methods and include quantum effects by solving the Schrödinger equation to calculate the wave function for the system of N identical bosonic particles. However, the disadvantage is that the H_2–H_2 interactions are described by an effective two-body potential. In these quantum MC calculations, the H_2 molecules do not have precise positions in the cluster, and, as for any other system of quantum particles, one can only talk about probability distributions for finding the molecules in different regions of space. The quantum MC calculations of the structure at very low temperature of para-H_2 clusters having up to about 50 molecules employed the isotropic pairwise H_2–H_2 interactions modeled by Buck et al. (1983) and by Silvera and Goldman (1978). The chemical potential

$$\mu_N = E(N-1) - E(N) \qquad (2.4)$$

obtained from the ground state total energies of the clusters with $N-1$ and N molecules, is the energy required to remove a hydrogen molecule from the $(H_2)_N$ cluster, and gives a sensitive measure of its stability. The calculated μ_N shows a clear maximum for $N = 13$. This feature indicates the special stability of that cluster. The chemical potential shows a less pronounced maximum at $N = 19$, and Khairallah et al. (2007) also obtained features for clusters with $N = 23, 26, 29,$ and 32. The density of particles reveals that the molecules arrange in shells with a shape close to spherical. The radial thickness of those shells is about 2 Å. For clusters with sizes close to $N = 13$ and $N = 50$, the density of particles has a large peak near the center of mass of the cluster. This peak indicates that a molecule sits at the center of mass. More precisely, clusters with N between 7 and 16, and also with N between 43 and 50 have a molecule at the center, as indicated in Figure 2.3 (data for N larger than 50 was not reported). In contrast, for cluster sizes between $N = 17$ and $N = 42$, the molecules form two concentric spherical shells around an empty center. The radii of the shells and their population increase with increasing cluster size.

A comparison with the density functional results is not a simple task because of the intrinsic delocalization of the molecules in the quantum MC calculations. Nevertheless, similarities between the results of both methods can be established and are now discussed. $(H_2)_{13}$ is the best example. In the DFT calculations, this cluster has a central molecule surrounded by 12 other molecules at equal distances from the center, forming a shell, in agreement with the quantum MC result. DFT clusters with sizes near $N = 13$ also have a molecule at the center, like the quantum MC clusters (see Figure 2.3). Near $N = 19$, the DFT calculations

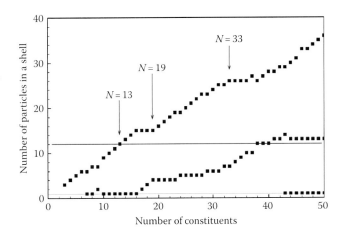

FIGURE 2.3 The number of molecules in each shell as a function of the size of the cluster obtained in quantum Monte Carlo calculations. Horizontal lines correspond to 1 and 12 molecules. (Reproduced from Guardiola, R. and Navarro, J., *Phys. Rev. A*, 74, 025201-1, 2006. With permission.)

predict structures with no central molecule, just like quantum MC. For instance, for $N = 19$, the DFT structure is the double icosahedron, a structure having an inner shell formed by two molecules. Near $N = 55$, the expected DFT structures are related to the icosahedron formed by a central molecule surrounded by two shells. The overall structural features in this size region are the same obtained in the quantum MC calculations, as shown in Figure 2.3. In summary, interesting similarities exist between the DFT and quantum MC structures. The similarities refer to the presence (of absence) of a molecule at the center of the cluster and to the number of shells surrounding that molecule (or empty center site), but these should not be overstated.

It is intriguing that a pattern of cluster growth similar to that of the quantum clusters in Figure 2.3 was found years ago for a simple cluster model, the SAPS (spherically averaged pseudopotential) model, applied to simple metallic clusters (Lammers et al. 1990). In the SAPS model, the total ionic pseudopotential acting on the electrons is spherically averaged around the cluster center. The model forces these clusters to be rather spherical, and the resulting structures calculated by minimizing the total cluster energy are formed by well-defined concentric shells of atoms, like shells in an onion. For instance, Cs_N clusters with N smaller than 7 do not have an atom at the center of the cluster. Clusters with N between 7 and 18 have a central atom surrounded by a single shell whose radius increases as N grows. The cluster center is again empty between $N = 19$ and $N = 39$ and the cluster is formed by two shells. Between $N = 40$ and $N = 62$, two shells surround an atom placed at the cluster center, and so on. The similarity with the quantum MC structural features of Figure 2.3 is striking, although the two systems are physically different. In the case of the SAPS model, the structural pattern arises from the interplay between electron–electron, electron-ionic pseudopotential, and ion–ion interactions in a constrained geometry imposed by the spherically averaged ionic pseudopotential. For the H_2 clusters, we propose a tentative explanation. A hollow spherical shell geometry with the center empty appears to be consistent with a wave function representing a system of N identical bosonic particles. Then, as the size N grows, the increase of the surface energy associated to the increasing radius of the shell promotes the formation of a new shell in the inner region of the cluster, and so on.

The influence of thermal effects on the stability and abundance of clusters at finite temperature has been studied by Guardiola and Navarro (2008). They calculated cluster excitations with angular momentum from $L = 0$ to $L = 13$ for sizes $N = 3, \ldots, 40$. For each value of L (L is the angular momentum of the wave function of the N-particle system), the Schrödinger equation was solved using the diffusion Monte Carlo (DMC) method, describing the H_2–H_2 interactions by the pair potential of Buck et al. (1983). Besides the excited states with angular momentum L, vibrational excitations characterized by a quantum number n were also considered. The excited states are then represented by the pair of quantum numbers (n, L), with energies $E_n^{(L)}$. After calculating the ground state energy $E_0^{(0)}(N)$ and the excited state energies $E_0^{(L)}(N)$ and $E_1^{(0)}(N)$, the excitation energies are defined as

$$\Delta E_L(N) = E_0^L(N) - E_0^{(0)}(N), \qquad (2.5)$$

$$\Delta E_0(N) = E_1^{(0)}(N) - E_0^{(0)}(N). \qquad (2.6)$$

Comparison of the excitation energies with the chemical potential $\mu(N) = E_0^{(0)}(N-1) - E_0^{(0)}(N)$ indicates which excited states are stable (those with energies smaller than μ; if the excitation energy is larger than μ, then the cluster will dissociate). Stable excitations exhibit size thresholds, as shown in Figure 2.4. All clusters with $N \geq 3$ exhibit stable excitations ($n = 0, L = 2$), ($n = 0, L = 3$), and ($n = 1, L = 0$). The excited level ($n = 0, L = 4$)

starts to be bound at $N = 4$, and the excited levels with $L = 5$ and $L = 1$ appear at $N = 6$. Then, the next L levels appear at regularly increasing size thresholds. In the size range studied, the highest stable excited state is $L = 13$, which occurs for $N \geq 31$. The quadrupolar, $L = 2$, is the lowest excitation for most sizes, except at $N = 26, 28, 30,$ and 37, for which the lowest excitation is the octupolar, $L = 3$. The vibrational excitation ($n = 1, L = 0$), not shown in the figure, displays a smooth behavior with N, and its energy lies in between the energies of the excited ($n = 0, L$) levels. Another useful observation from this figure is the local maximum of the chemical potential for $(H_2)_{13}$ and a less pronounced maximum for $(H_2)_{36}$.

The knowledge of the excitation spectra allows for the analysis of thermal effects using the partition function F_N of statistical mechanics. In particular, the temperature-dependent energy of the cluster, $E_N(T)$, becomes (Guardiola and Navarro 2008)

$$E_N(T) = E_0^{(0)}(N) + \frac{1 + \Sigma_L (2L+1)\, \Delta E_L(N)\, e^{-\Delta E_L(N)/kT}}{1 + \Sigma_L (2L+1)\, e^{-\Delta E_L(N)/kT}}, \qquad (2.7)$$

where k is Boltzmann's constant. Noticeable effects on the energy values begin to appear at temperatures close to the energy of the first excited state, $L = 2$. For instance, at $T = 1\,K$ the energy changes are minimal, with the exception of the clusters with $N \approx 20$, which is related to a minimum of ΔE_2 occurring at this size (see Figure 2.4).

The effect of the temperature on the mechanism of cluster formation in a free jet expansion is also interesting. We assume that cluster growth in the jet is dominated by the chemical equilibrium reaction

$$(H_2)_{N-1} + H_2 + X \Leftrightarrow (H_2)_N + X, \qquad (2.8)$$

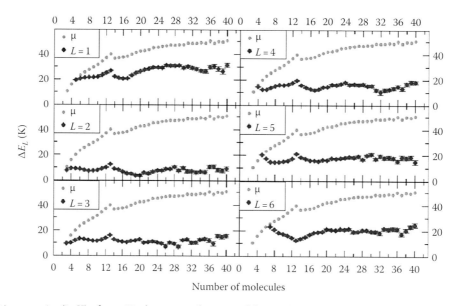

FIGURE 2.4 Excitation energies (in K) of para-H_2 clusters as a function of the number N of molecules, for states (0, L) with $L = 1$–6, obtained from DMC calculations. The chemical potential μ is also plotted to indicate the stability limit. (Reproduced from Guardiola, R. and Navarro, J., *J. Chem. Phys.*, 128, 144303-1, 2008. With permission.)

where X is a spectator particle needed for the conservation of energy and momentum. The equilibrium constant K_N for the reaction is obtained from the partition functions F as $K_N = F_N/(F_{N-1} F_1)$ and K_N is dominated by the ratio

$$\frac{F_N}{F_{N-1}} = \left(\frac{N}{N-1}\right)^{3/2} e^{\mu(N)/kT} \frac{1 + \sum_L g_L e^{-\Delta E_L(N)/kT}}{1 + \sum_L g_L e^{-\Delta E_L(N-1)/kT}}, \quad (2.9)$$

where the degeneracy factor for the excited state L is given by $g_L = 2L + 1$. Evidently, the ratio F_N/F_{N-1} reflects the non-smooth behavior of the chemical potential and the excitation energies as a function of N. A plot (Guardiola and Navarro 2008) of F_N/F_{N-1} at several temperatures, lower than the source temperature of the experiments of Tejeda et al. (2004), exhibits peaks at $N = 13$, 31, and 36, and a less pronounced peak at $N = 26$ in rough correlation with the experimental abundance maxima at $N = 13$ and 33. This points out the influence of the excited states on the abundance of hydrogen clusters produced in free jet expansions of pressurized gas.

2.4 Charged Clusters

Mass spectrometric experiments on positive cluster ions (Kirchner and Bowers 1987) have revealed that H_m^+ clusters with an odd number of atoms, $m = 3, 5, 7,\ldots$ are much more abundant than those with even-number of atoms, $m = 2, 4, 6, \ldots$ The ionization of a neutral $(H_2)_N$ cluster occurs in two steps. First, a H_2 molecule of the cluster is ionized, and the charged molecule reacts with a neighbor neutral molecule, producing a trimer carrying the positive charge and a neutral H atom

$$H_2^+ + H_2 \rightarrow H_3^+ + H. \quad (2.10)$$

This reaction is exothermic and the energy released, 1.7 eV, is enough to eject the H atom out of the cluster. The H_3^+ cation is stabilized by the surrounding H_2 molecules, which form solvation shells around the charged trimer (Bokes et al. 2001; Chermette and Ymmud 2001; Prosmiti et al. 2003). The composition of the cluster cations can be viewed as $H_3^+ (H_2)_N$. The fact of being charged makes these clusters easier to handle experimentally (Farizon et al. 1998). These clusters can grow easily (Clampitt and Towland 1969). The gas-phase clustering reaction

$$H_3^+(H_2)_N + H_2 \rightarrow H_3^+(H_2)_{N+1} \quad (2.11)$$

has been experimentally studied and the enthalpy ΔH of the reaction has been determined (Hiraoka and Mori 1989). The measurements, shown in Figure 2.5, indicate a stepwise decrease in the enthalpies at particular sizes. The gain in binding energy by adding a H_2 molecule drops substantially after $N = 3$ and $N = 6$, and this indicates that $H_3^+ (H_2)_3$ and $H_3^+ (H_2)_6$ are specially stable clusters. This stability has been interpreted as indicating shell formation.

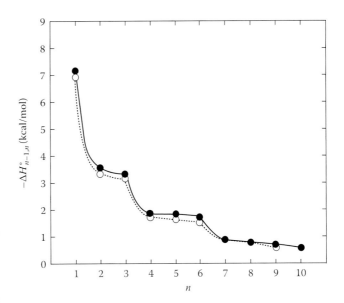

FIGURE 2.5 Measured enthalpies for the gas phase reaction $H_3^+(H_2)_{n-1} + H_2 \rightarrow H_3^+(H_2)_n$. (Reproduced from Hiraoka, K. and Mori, T., *Chem. Phys. Lett.*, 157, 467, 1989. With permission.)

Charged hydrogen clusters are important active species in the stratosphere and in interstellar clouds. The charged trimer, H_3^+, and its deuterated variant D_3^+ have attracted a lot of attention. The unusual nature of its bonding leads to an exceptional roto-vibrational spectrum (Kostin et al. 2003). Experiments and calculations (Tennyson and Miller 1994) have shed light on the electronic structure, the infrared photodissociation spectrum, and the classical and quantal behavior of the molecule at its dissociation limit. H_3^+ is present in any environment where molecular hydrogen gas is ionized: It has been detected in the atmospheres of Jupiter, Saturn, and Uranus. It was also identified in the supernova SN1987A (Miller et al. 1992) and in the interstellar medium (McCall et al. 1998). H_3^+ is, in fact, the main agent responsible for the formation of complex molecules in the reaction network of the interstellar medium.

The charge state has influence on the binding energy of the $H_3^+ (H_2)_N$ clusters. Taking $H_3^+ (H_2)_5$ as an example, the average binding energy of the H_2 molecules in the cluster, which can be defined as

$$e_{av}(\text{cation}) = \frac{E(H_3^+(H_2)_5) - E(H_3^+) - 5E(H_2)}{5} \quad (2.12)$$

is equal to 60 meV in an LDA calculation. For comparison, the average binding energy of five H_2 molecules attached to a sixth H_2 molecule to form the neutral $(H_2)_6$ cluster,

$$e_{av}(\text{neutral}) = \frac{E((H_2)_6) - 6E(H_2)}{5} \quad (2.13)$$

is 25 meV. So, the effect of the charge localized at the cluster center is to increase the binding energy substantially. Zero-point corrections lower those binding energies.

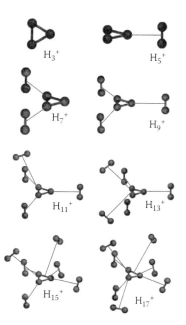

FIGURE 2.6 Structure of $H_3^+ (H_2)_N$ clusters calculated by ab initio quantum chemical methods. (Reproduced from Bokes, P. et al., *Int. J. Quantum Chem.*, 83, 86, 2001. With permission.)

Accurate theoretical studies of the ground-state configuration and binding energies of the $H_3^+ (H_2)_N$ clusters have been performed using different levels of theory. The results for $N = 1–7$ obtained using ab initio quantum chemical methods (Bokes et al. 2001) are shown in Figure 2.6. In $H_3^+ (H_2)$, $H_3^+ (H_2)_2$, and

$H_3^+ (H_2)_3$, each H_2 molecule is chemically attached to one of the H atoms of H_3^+. The center of mass of each of those H_2 molecules is in the plane of H_3^+, and the molecular axes are perpendicular to that plane. A first shell is completed at $H_3^+ (H_2)_3$ with the three H_2 molecules forming an external triangle with the same form as that of the internal H_3^+ core. The experiments show that $H_3^+ (H_2)_3$ is especially stable (notice the drop after $N = 3$ in Figure 2.5). Addition of more H_2 molecules builds up a second shell with the molecules at a longer distance from H_3^+. The binding of these molecules to H_3^+ is weaker compared to those in the first shell. Figure 2.6 shows the calculated structure for H_3^+ $(H_2)_4$, $H_3^+ (H_2)_5$, $H_3^+ (H_2)_6$, and $H_3^+ (H_2)_7$. There are small differences of detail among different calculations concerning the location of the added molecules (Bokes et al. 2001; Prosmiti et al. 2003; Seo et al. 2007), but the structures reported here are sufficiently representative. Those clusters show several isomers with similar energies (Seo et al. 2007). For instance, for $H_3^+ (H_2)_4$, Bokes et al. (2001) found six different isomers, three with the additional H_2 molecule above the H_3^+ moiety and three with the molecule below. A second shell appears to be completed for H_3^+ $(H_2)_6$, as indicated by the measured drop of the binding energy of an additional molecule (see Figure 2.5), also reproduced by the calculations. The three molecules of the second shell are on the same side of the H_3^+ plane in Figure 2.6, but in other calculations two molecules are on one side and one molecule on the other side (Seo et al. 2007). Although the enthalpy appears to indicate that a new shell is opened at $H_3^+ (H_2)_7$, the distance from the last added molecule to the H_3^+ core is similar to those for the

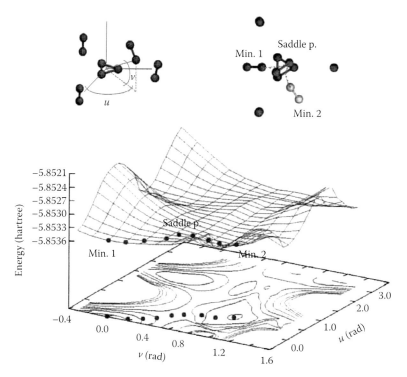

FIGURE 2.7 Potential energy surface of $H_3^+ (H_2)_4$ calculated by the Hartree-Fock method. The insets in the upper part show the coordinates of the potential energy surface and the molecular motion along the minimum energy path connecting two neighboring minima (from minimum 1 to minimum 2 through the saddle point). (Reproduced from Bokes, P. et al., *Int. J. Quantum Chem.*, 83, 86, 2001. With permission.)

molecules in the second shell. Further work should be welcome to clarify if a shell is completed or not at $H_3^+ (H_2)_6$.

Because of the weak bonds in these clusters, thermal and quantum fluctuations can cause fluxional behavior of the system. As an example, we summarize the study of Bokes et al. (2001) of the behavior of the most weakly bound H_2 molecule in $H_3^+ (H_2)_4$. That molecule is the first one in the second shell around H_3^+. For this purpose, Figure 2.7 shows the potential energy surface of the system obtained by scanning the two angle parameters u and v defined in the inset of this figure. In calculating the energy for each (u, v) point, the rest of the structural parameters of the cluster are free to relax. The calculations were performed with the Hartree–Fock method. There are three equivalent minima above the H_3^+ plane and three below. The transition path with the lowest energy barrier connecting two neighboring minima corresponds to the H_2 molecule moving, as indicated in the inset of the figure (from minimum 1 to minimum 2 through the saddle point). The barrier height is $\Delta E = 0.078$ kcal mol^{-1}, and a more accurate, coupled-cluster (CC) calculation gives a similar value, $\Delta E = 0.084$ kcal mol^{-1}. The lowest vibrational energy of the hydrogen molecule in the potential well is $(1/2)\hbar\omega_{min} = 0.066$ kcal mol^{-1}, and the thermal energy at temperature T is $kT = 0.0020T$ kcal mol^{-1}. Therefore, we can expect that zero-point motion fluctuations and temperature effects for $T = 10$ K and higher will enable the outermost H_2 molecule to pass the barriers from one minimum of the potential energy surface to another, resulting in fluctional behavior. H_2 molecules in the successive coordination shells will be even more mobile.

2.5 Liquid-to-Gas Phase Transition

A good identification of a first- or second-order phase transition in a cluster is provided by the specific shape of the caloric curve, that is, the thermodynamic temperature as a function of the total energy. The caloric curve of size-selected hydrogen clusters has been determined in high energy collision experiments (Gobet et al. 2001, 2002), and has been interpreted as indicating the transition from a bound cluster to the gas phase. In those experiments, the hydrogen clusters are first formed in a cryogenic cluster expansion source, then ionized using a high-performance electron ionizer and finally size-selected in an ion accelerator. The collisions between size-selected $H_3^+ (H_2)_N$ clusters with N smaller or equal to 14, accelerated to kinetic energies of 60 keV amu^{-1}, and a helium gas target were analyzed. Collisions lead to the fragmentation of the clusters

$$H_3^+(H_2)_N + He \rightarrow a\ H_3^+(H_2)_k + b\ H_3^+ + c\ H_2^+ + d\ H^+$$
$$+ e\ H_2 + f\ H, \qquad (2.14)$$

where $a - f = 0, 1, \ldots$, and for each collision event, a multidetector records simultaneously the number (multiplicity) of each mass-identified fragment ion resulting from the reaction (neutral species larger than H_2 are absent). The construction of the caloric curve requires the simultaneous determination of the energy

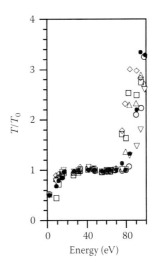

FIGURE 2.8 Caloric curves for hydrogen cluster fragmentation induced by collisions with helium. Reduced temperature T/T_0 (T_0 is the temperature in the plateau of the curve) is given as a function of the energy deposited on the clusters $H_3^+ (H_2)_N$, with $N = 6$ (open squares), $N = 8$ (open circles), $N = 9$ (triangles), $N = 11$ (diamonds), $N = 12$ (inverted triangles), and $N = 14$ (filled circles). (Reproduced from Gobet, F. et al., *Phys. Rev. Lett.*, 89, 183403, 2002. With permission.)

and the temperature of the system. The cluster energy, that is the energy deposited into the cluster by the collision with a He atom, is determined by the nature and multiplicity of the products in reaction (2.14). The temperature of the cluster prior to decay is obtained using a relationship (Fisher 1967), tested successfully in nuclear physics collisions, between the characteristic shape of the fragment mass distribution and the temperature of decaying nuclei (Belkacem et al. 1995).

The results for $H_3^+ (H_2)_N$ with $N = 6, 8, 9, 11, 12$, and 14 (Gobet et al. 2001, 2002) are shown in Figure 2.8. The caloric curves show three parts: after an initial rise, a plateau follows before the curve rises again. The curves show the typical features of a first-order phase transition. According to Gobet et al., the curves show back-bending, that is, a negative heat capacity, which has been predicted to be possible for small systems (Gross 1997). This feature becomes more clear by plotting a single curve with the geometric means for all the clusters. There is, however, some controversy on the method of constructing the caloric curves (Chabot and Wohrer 2004).

2.6 Laser Irradiation and Coulomb Explosion

The advances in laser technology permit the study of the interaction between matter and ultrafast lasers with intensities higher than 10^{14} W cm^{-2} and pulse duration below 100 fs. Available lasers achieve intensities exceeding the electric field created by an atomic nucleus, and the timescale of femtoseconds is also typical of the electron motion. This field of research allows for the exploration of the nonlinear response of atoms to intense laser pulses, leading to the observation of new processes. The behavior of molecules and clusters under similar laser conditions offers new challenges due to the existence of additional degrees of freedom, such as the nuclear motion or the presence of intramolecular and intermolecular forces. This gives rise to complex phenomena: above-threshold dissociation, bond softening, and enhanced ionization. Some of these phenomena are followed by a Coulomb explosion. That is, when molecules or clusters are multiply ionized by laser pulses of very short duration, the unbalanced positive charges are sufficiently close together to cause a repulsion-induced explosion of the nuclear skeleton (Poth et al. 2002, Heidenreich et al. 2007).

In a series of interesting experiments, Ditmire and coworkers irradiated a dense molecular beam of large deuterium clusters, $(D_2)_N$, with intense femtosecond lasers (Ditmire et al. 1999, Zweiback et al. 2000). The irradiation induces the multiple ionization of the clusters, which then explode due to the repulsive Coulomb forces between the bare nuclei of the ionized atoms of the cluster. Some of those flying nuclei collide with nuclei ejected from other clusters in the plasma, and when the kinetic energies of the colliding nuclei are higher than a few keV, nuclear D-D fusion processes $D + D \rightarrow {}^3He + n$ can occur with high probability. The kinetic energies of the bare deuterium nuclei depend only on the size of the original cluster, and for the cluster sizes in the experiments of Ditmire et al., the resulting kinetic energies are high enough to produce nuclear fusion reactions. Apart from the obvious interest for future thermonuclear devices, this technique has led to the development of tabletop neutron sources (Hartke et al. 2005): the D + D fusion reaction produces a neutron (n) with energy of 2.45 MeV, and those neutrons could potentially be used in neutron radiography and in materials research.

Motivated by those works, two groups (Ma et al. 2001, 2005; Isla and Alonso 2005, 2007) have studied the dynamical response of deuterium clusters irradiated by an intense femtosecond laser. This study simulates the first stages in the experiments of Ditmire and coworkers. The time-dependent density functional theory (TDDFT) (Gross et al. 1996) was the method used for the simulations. The TDDFT gives the response of the system to a time-dependent external perturbation $V_{ext}(\mathbf{r}, t)$, in the present case the laser field, by directly solving the fundamental equations of the formalism, the time-dependent Kohn–Sham equations. This gives directly the time-evolution of the electronic orbitals. One advantage of explicitly propagating the time-dependent Kohn–Sham equations is that it permits to couple the electronic system to the ionic background, which can be, for many purposes, treated classically. It is then possible to perform a combined dynamics of electrons and nuclei. The method allows to study both linear and nonlinear excitations. The simulation of the laser irradiation of $D_3^+ (D_2)_5$ and D_3^+ is now described.

The structure of the free $D_3^+ (D_2)_5$ cluster is shown in the leftmost panels of Figure 2.9 (the spheres represent the hydrogen atoms, and two mutually perpendicular views of the cluster are presented). The cluster then was perturbed by a laser pulse, whose time-dependent electric field $E(t)$ has a cosinoidal envelope,

$$E(t) = A_0 \cos\left(\frac{\pi}{2} \frac{t - 2\tau_0 - t_0}{\tau_0} \right) sen(\omega t)\hat{e}, \quad |t - t_0| < \tau_0, \quad (2.15)$$

where

ω is the frequency of the field
$\hat{e}$ is the polarization vector
$2\tau_0$ is the total duration of the pulse
A_0 its amplitude

The frequency of the laser is an important parameter in order to achieve an efficient coupling between the laser radiation and the cluster. The first peak in the calculated photoabsorption spectrum of the cluster occurs at a frequency $\omega = 0.352$ a.u. ($\hbar\omega = 9.58$ eV, where $\hbar$ is the reduced Planck constant). The results of a simulation in which a laser pulse of this resonant frequency is applied to the cluster are shown in Figure 2.9. The duration of the pulse is 9.6 fs and the amplitude of the field is

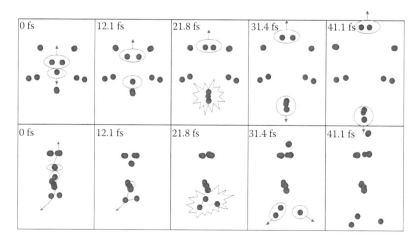

FIGURE 2.9 Snapshots of the structure of the cluster $D_3^+ (D_2)_5$ at different times after application of a short laser pulse of resonant frequency $\hbar\omega = 9.58$ eV, intensity of 1.4×10^{13} W cm^{-2}, and duration 9.6 fs. A slow fragmentation of the cluster occurs. Two mutually perpendicular views are presented for each snapshot.

0.02 a.u., giving a pulse intensity of 1.4×10^{13} W cm^{-2}. A few snapshots showing the evolution of the structure of the cluster in time are shown in that figure. The cluster maintains its original structure during the initial 10 fs of the simulation approximately, due to the inertia of the atoms and the time the cluster needs to absorb the necessary energy to break bonds. The absorbed energy causes the splitting of the inner D_3^+ trimer in two fragments, a D_2 molecule and a D atom, which are emitted in opposite directions. As the emitted molecule moves upward, it passes near two D_2 molecules, and these two molecules are set in motion and move apart. The intramolecular bond lengths of those two molecules oscillate but the bonds remain intact, that is, the molecules do not dissociate. On the other hand, the D atom moving downward collides with a molecule of the solvation shell, and an atom is exchanged in the collision. As the cluster dissociates, the two D_2 molecules originally most distant from the trimer remain little affected. This dissociation mode of the cluster can be characterized as a slow fragmentation. The laser frequency has influence on the results. For pulses of the same intensity as above, but half of the resonance frequency, the atoms oscillate around their equilibrium positions, but the cluster does not dissociate.

The behavior of the irradiated cluster changes drastically for a laser intensity five times larger, that is, 7×10^{13} W cm^{-2} (with the same resonant frequency, $\hbar\omega = 9.58$ eV, and pulse duration, 9.6 fs, as above). The evolution of the cluster structure is shown in Figure 2.10. The laser pulse produces a massive ionization of the deuterium atoms, and this occurs simultaneously in all regions of the cluster. As a consequence, the nuclei repel each other due to the Coulomb interaction and the cluster explodes. Coulomb explosion is a violent dissociation process that generally occurs in molecules and clusters when these are multiply ionized by femtosecond laser pulses. In small clusters, stripping just two electrons may be sufficient to induce Coulomb fragmentation. In fact, cluster size is an important parameter determining whether the cluster will follow this decay channel. This is the mechanism corresponding to the fast dissociation process shown in Figure 2.10. The irradiation of the cluster first produces a localized plasma of electrons and nuclei, and in a short interval, roughly corresponding to the duration of the laser pulse, the plasma loses five electrons which fly away. The loss of the remaining electrons continues afterward, at a slightly slower rate. Besides the size of the cluster, there is another requirement for a pure Coulomb explosion to take place: the cluster must be almost stripped of all their electrons in a timescale of

only a few femtoseconds. This is achieved with the very high power femtosecond lasers now available.

The kinetic energy of the nuclei resulting from the Coulomb explosion of D_3^+ $(D_2)_5$ is about 13 eV. In most nuclear fusion processes, from controlled fusion reactors to solar reactions, the reacting particles have kinetic energies of a few keV, enough to overcome the Coulomb barrier for fusion. This indicates that the Coulomb explosion of D_3^+ $(D_2)_5$ delivers kinetic energies that are very small compared to those required to produce nuclear D–D fusion. Of course, this is expected, since the cluster studied in these simulations is very small, whereas the clusters in the experiments of Ditmire and coworkers (Ditmire et al. 1999, Zweiback et al. 2000) are much larger: the average number of atoms is 1000 times larger. The maximum kinetic energy is proportional to R^2, where R is the radius of the cluster, and a beam of deuterium clusters with radii greater than 50 a.u. is necessary to produce ions with the multi-keV energies required for nuclear fusion. Last and Jortner (2001a,b) have proposed and confirmed by molecular dynamics simulations that very energetic deuterium or tritium nuclei (D^+ or T^+) can be produced by the Coulomb explosion of D_2O and T_2O clusters, similar to water clusters. These clusters will provide substantially higher fusion reaction yields than the homonuclear deuterium or tritium clusters of the same size.

The charged trimer D_3^+ forms the central core of the D_3^+ $(D_2)_N$ clusters. The ground state geometry of D_3^+ is a near equilateral triangle. Its calculated ionization potential, that is, the energy required to form D_3^{2+}, is 35.35 eV, and this high value arises from the unscreened attraction of the electrons by the nuclei. The photoabsorption spectrum of D_3^+ shows two dominant features in the UV range (Isla and Alonso 2007). The first one is a double-peak formed by two near degenerate excitations at 17.5 and 17.8 eV, and its average value is in very good agreement with the lowest excitation energy of H_3^+, 17.8 eV, found in the experiments of Wolff et al. (1992). The second feature is a peak at 20.1 eV with lower absorption strength. The results of simulations of the cluster excitation with laser pulses of 9.6 fs and different frequencies and intensities are now discussed. First, the laser frequency is tuned to the resonant absorption peak at 17.65 eV (to perform this experiment in practice, this high frequency would require multiphoton absorption). In response to a pulse of intensity 10^{12} W cm^{-2}, the atoms of the trimer oscillate in the plane of the cluster and the motion resembles a breathing mode. However, when a high-intensity pulse of 10^{15} W cm^{-2} is applied, the cluster undergoes a Coulomb explosion. This occurs in two steps. First, the system reaches

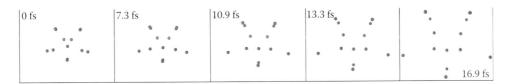

FIGURE 2.10 Snapshots of the structure of the cluster D_3^+ $(D_2)_5$ at different times after application of a laser pulse of resonant frequency $\hbar\omega = 9.58$ eV, intensity 7×10^{13} W cm^{-2}, and duration 9.6 fs. A Coulomb explosion occurs.

a transient nanoplasma-like state; this state is short lived, and the two electrons quickly escape. The repulsion between the positive nuclear charges then causes the Coulomb explosion of the cluster. When the laser frequency is tuned to match the other absorption peak at 20.1 eV, the dynamical response of the cluster to a pulse of intensity 10^{12} W cm^{-2} is again an oscillatory motion of the atoms in the plane of the trimer; however, the amplitude of the oscillations is smaller. For a pulse of 10^{15} W cm^{-2}, a Coulomb explosion occurs but some differences can be noticed compared to the case of frequency $\hbar\omega = 17.65$ eV. The transient plasma-like state has a longer lifetime and ionization is slower. Consequently, the velocities of the flying nuclei are 25% slower. A last example corresponds to a nonresonant laser frequency of 5 eV. Atomic vibrations occur for low and high pulse intensities; however, the absorption of energy is not large enough to break the bonds.

The electronic response in the linear domain, which is the case for the low-intensity laser field, can be analyzed by following the time evolution of the dipole moment of the cluster, as shown in Figure 2.11. For a laser frequency $\hbar\omega = 17.65$ eV, the dipole moment is greatly amplified by resonance with the external field, and strong dipole oscillations are observed long after the laser pulse is switched off (the pulse duration is 9.6 fs; see above). The amplification benefits from the fact that both, the electrons and the nuclei, oscillate in the plane of the nuclei. The behavior for a laser frequency $\hbar\omega = 20.1$ eV is different. A field of this frequency induces oscillations of the electronic cloud perpendicular to the plane of the nuclei. Then, the electrons follow closely the excitation field during the approximately 10 fs that the pulse is acting, although the electronic response is small (this can be noticed by comparing the scales of the left and center panels of Figure 2.11). The amplitude of the dipole oscillations is much less than in the previous case because the absorption strength of this peak is smaller and there is no enhancement due to the nuclear vibrations. The extreme situation is found in the nonresonant case at $\hbar\omega = 5$ eV (Figure 2.11c), where the dipole moment first follows closely the laser field, returning practically to its initial value when the field is turned off.

2.7 Endohedrally Confined Hydrogen Clusters

Solid hydrogen at low temperature and normal pressure is a molecular solid. Theoretical calculations have predicted a metallic state at very high pressures; see, for instance, the work by Johnson and Ashcroft (2000). However, the prediction is controversial. Shock-wave, pulsed-laser, and diamond anvil-cell experimental techniques have played an important role in studying hydrogen under high pressures (Loubeyre et al. 2002). The metallic state has not yet been achieved for pressures up to about 300 GPa, and recent DFT calculations (Pickard and Needs 2007) have confirmed that hydrogen remains insulating at 400 GPa. Loubeyre et al. (2002) estimated that pressures of at least 450 GPa will be required to achieve the closing of the electronic gap.

DFT calculations have also been performed to explore the effect of pressure on the structure and the properties of hydrogen clusters (Santamaria and Soullard 2005, Soullard et al. 2008). In these studies, the clusters were confined in model containers, rigid fullerene-like cages built of 60 hydrogen atoms, and compression was simulated by reducing the radius of those model containers. The initial radius was chosen large enough so as to exert only a small pressure on the confined hydrogen molecules and to avoid the formation of bonds between the confined molecules and the atoms forming the cage. At any of the simulated pressures, the encapsulated hydrogen molecules assemble into clusters, with equilibrium structures which depend strongly on the number of molecules and the pressure exerted by the cage. Two pressure ranges have been identified. In the first one, with pressures up to 100 GPa, formation of molecular $(H_2)_N$ clusters occurs. Figure 2.12 shows the preferred packing of the $(H_2)_N$ clusters with $N = 8$–13 and 15, at pressures below 100 GPa. The equilibrium geometries can be described as antiprisms and capped antiprisms, with an approximate rotational symmetry axis changing from C_4 (for $N = 8, 9, 10, 11$) to C_5 (for $N = 12, 13$) and C_6 (for $N = 15$). The growth pattern for $N = 8$ to $N = 11$ is based on the structure of $(H_2)_8$, which is a square antiprism. $(H_2)_9$ and $(H_2)_{10}$ are obtained by capping the two opposite basal

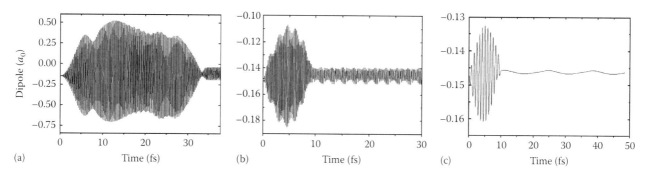

FIGURE 2.11 Evolution of the electric dipole moment of D_3^+ with time after laser irradiation in the linear domain (laser intensity = 10^{12} W cm^{-2}) with pulses of 9.6 fs. (a) Laser frequency corresponding to the excitation peak at 17.65 eV in the absorption spectrum. (b) Frequency corresponding to the excitation peak at 20.0 eV. (c) Nonresonant frequency at 5.0 eV. Notice the different scales in the dipole axis. (Reproduced from Isla, M. and Alonso, J.A., *J. Phys. Chem. C*, 111, 17765, 2007. With permission.)

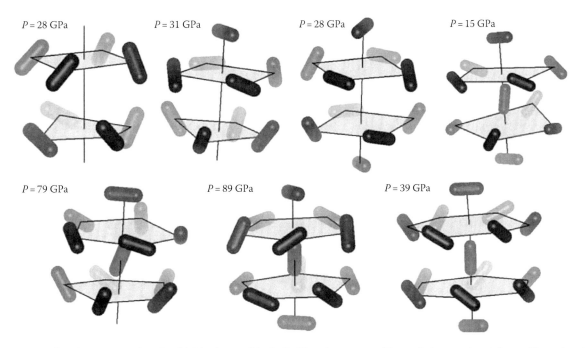

FIGURE 2.12 Preferred structures of confined $(H_2)_N$ clusters ($N = 8$–13, 15) under pressure. The confining cage is not shown. The main rotational symmetry elements (planes and axes) are indicated. (Reproduced from Soullard, J. et al., *J. Chem. Phys.*, 128, 064316, 2008. With permission.)

faces on the antiprism. Placing an extra molecule in the center of the antiprism leads to $(H_2)_{11}$. The structure of the antiprism changes from square to pentagonal in $(H_2)_{12}$, which can also be considered as an icosahedron with a molecule at the center and a missing molecule on the surface. $(H_2)_{13}$ is a perfect icosahedron with a molecule at the center. The configuration of $(H_2)_{15}$ is similar to that of $(H_2)_{13}$ but with the fivefold molecular rings replaced by sixfold molecular rings. The structures of encapsulated $(H_2)_{12}$ and $(H_2)_{13}$ coincide with those calculated for free $(H_2)_{12}$ and $(H_2)_{13}$ using DFT (see Section 2.3), but not for other clusters. This indicates that pressure affects the structure of the clusters. Only the magic cluster $(H_2)_{13}$ and its immediate neighbor $(H_2)_{12}$ appear to be insensitive to this effect. It is interesting to notice that the structure of $(H_2)_{15}$ is based on a hexagonal antiprism, while the structure of free $(H_2)_{15}$ in a similar calculation (see Section 2.3) is a capped icosahedron. This provides a nice visualization of the confinement effect. The capped icosahedron is a prolate (elongated) structure, and transformation to the more spherical structure of the hexagonal antiprism is favorable inside the spherical cage of decreasing diameter (this tendency toward spherical structures is further discussed below).

Application of pressures higher than 100 GPa results in a gradual dissociation of the H_2 molecules, as evidenced by the increasing interatomic distances, and in the formation of mixed clusters of atoms and molecules, with a molecular–atomic ratio dependent on the pressure. As pressure increases, the electronic HOMO–LUMO gap (the difference between the energy of the lowest unoccupied electronic state of the cluster and that of the highest occupied state) decreases. This shows a tendency toward metallization, although for the pressures applied in the simulations, the metallization of the clusters is not yet

achieved. Analysis of the pressure–volume curve for encapsulated $(H_2)_{13}$ and $(H_2)_{15}$ leads to a zero-temperature equation of state (Santamaria and Soullard 2005), which was found to be consistent with the experimental data for a macroscopic hydrogen crystal (Hemley et al. 1990).

Motivated by the urgent search for efficient containers of hydrogen for automotive applications, the encapsulation of hydrogen clusters in carbon cages like fullerenes and closed nanotubes has been studied. Komatsu et al. (2005) reported a chemical method to encapsulate hydrogen in the C_{60} fullerene by organic synthesis. Using a semiempirical electronic structure method, the MNDO (modified neglect of diatomic overlap), Barajas-Barraza and Guirado-López (2002) have studied the clustering of hydrogen molecules encapsulated in the fullerenes C_{60} and C_{82}. Although hydrogen encapsulation in fullerene molecules is not directly relevant for gas storage applications, its study may help us to get insight into the structure and properties of small molecules encapsulated in confined environments. On the other hand, porous carbon, a promising hydrogen storage material, contains interconnected pores of nanometric size, some of them with spheroidal shape. The number of encapsulated molecules studied by Barajas-Barraza et al. varied from 1 to 23 for encapsulation in C_{60}, and from 1 to 35 for C_{82}. The two upper limits define the maximum storage capacity for those two fullerenes. Beyond that, the cage breaks up. Inside C_{60}, the structure of $(H_2)_6$ is an octahedron, and $(H_2)_{13}$ is an icosahedron with an atom at its center, as shown in Figure 2.13. Encapsulated $(H_2)_{19}$ forms a spherical structure around a central molecule, different from the elongated (prolate) structure of free $(H_2)_{19}$. Actually, the structures of encapsulated $(H_2)_6$ and $(H_2)_{13}$ are also rather spherical. To understand this tendency for spherical structures,

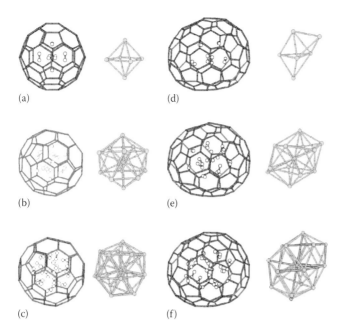

FIGURE 2.13 Calculated lowest energy structures of (a) $(H_2)_6@$ C_{60}, (b) $(H_2)_{13}@C_{60}$, (c) $(H_2)_{19}@C_{60}$, (d) $(H_2)_6@C_{82}$, (e) $(H_2)_{13}@C_{82}$, and (f) $(H_2)_{19}@C_{82}$ (first and third columns), together with the spatial distribution of the centers of mass of the encapsulated hydrogen molecules (second and fourth columns). (Reproduced from Barajas-Barraza, R.E. and Guirado-López, R.A., *Phys. Rev. B*, 66, 155426, 2002. With permission.)

one can first notice that the equilibrium distance between the two hydrogen molecules in $(H_2)_2$ obtained in different calculations is 3.4 Å, very close to the radius of the C_{60} fullerene, 3.5 Å. In addition, the hydrogen molecules maximize their attractive interaction with the carbon cage by occupying positions close to the inner wall of the fullerene. These two features explain the structure of the encapsulated clusters. The axes of hydrogen molecules in encapsulated $(H_2)_6$ and $(H_2)_{13}$ show orientational order, which disappears in encapsulated $(H_2)_{19}$. Evidently, as the number of molecules increases, the enclosed molecules feel an increasing pressure that would eventually lead to the breaking of the carbon cage.

These ideas are confirmed by the analysis of the structures of the clusters encapsulated in C_{82}. The cage of this fullerene shows some distortion from sphericity, and the structures of the encapsulated clusters become deformed, compared to their structures inside C_{60}, in order to profit from the interaction with the inner wall of the cage. This is clear in the panels d, e, and f of Figure 2.13. Again, the case of $(H_2)_{19}$ is interesting. The deformed cage allows $(H_2)_{19}$ to adopt a structure which is similar, although not identical, to that of free $(H_2)_{19}$. The structure of free $(H_2)_{19}$ is the double icosahedron (see Section 2.3), and it looks like a cylinder formed by three parallel rings (each ring having five molecules) plus one molecule at each end of the cylinder and two molecules inside the cylinder. The encapsulated $(H_2)_{19}$ is a similar cylinder, but with only a single internal molecule, and the middle ring is formed by six molecules. An additional set of calculations performed for $(H_2)_6$, $(H_2)_{13}$, and $(H_2)_{19}$ encapsulated in a

particular isomer of C_{82} with spherical symmetry (not shown in the figure) indicate that those hydrogen clusters adopt spherical structures, nearly the same as the ones found inside C_{60}. In summary, for hydrogen clusters encapsulated inside fullerenes, the interactions of the hydrogen molecules with the carbon walls of the cage dominate over the H_2–H_2 interactions and control the structure of the encapsulated clusters. This supports the expectation that porous carbons can act as efficient containers for hydrogen storage.

The same rules apply to the structures of hydrogen clusters encapsulated in closed nanotubes (Barajas-Barraza and Guriado-López 2002). A (5,5) nanotube closed at both ends can be constructed with 110 carbon atoms. When the number of encapsulated hydrogen molecules is very small, the cluster adopts a linear shape with well-defined orientational ordering of the various molecular axes. Due to the finite length of the nanotube, the structure of the cluster becomes two-dimensional for five H_2 molecules, and three-dimensional for seven molecules. For encapsulation of larger quantities, for example 18 molecules, these form a shell with tubular shape, characterized by a strongly correlated orientation of the molecular axes. The structure changes when the number of encapsulated molecules is about 30: the cylindrical hydrogen shell adsorbed on the nanotube internal wall accommodates a one-dimensional chain of H_2 molecules in its inner channel.

Experimental work to fill fullerenes with hydrogen is in the early stages, although encapsulation of a single molecule has been achieved (Komatsu et al. 2005). But the pressures produced inside by the encapsulated hydrogen and the maximum amount of encapsulated hydrogen before the carbon cage breaks have been studied theoretically. An interesting result obtained by Pupysheva et al. (2008) from DFT calculations is that the energy of formation of the encapsulated cluster

$$\Delta E = E(H_n@C_{60}) - E(C_{60}) - \frac{n}{2}E(H_2) \qquad (2.16)$$

is negative only for one or two molecules, and positive for four molecules and more. That is, all the structures with more than three encapsulated molecules are metastable, although once formed, dissociation of the system may have a substantial activation barrier. This critical number of three molecules may be underestimated, because the DFT calculations of Pupysheva et al. used the generalized gradient approximation (GGA) to electronic exchange and correlation, and there is some evidence that the GGA tends to underestimate the attractive interaction between H_2 and graphite-like carbon surfaces. Actually, Tada et al. (2001) obtained a purely repulsive interaction between H_2 and a graphene layer or a (6,6) nanotube using the GGA approximation. The last metastable structure of the $H_n@C_{60}$ family in the study of Pupysheva et al. contains $29\,H_2$ molecules. Above that size, the pressure becomes too high, the system becomes unstable, and the carbon cage spontaneously breaks up. For a number of encapsulated atoms n smaller than 20, all the hydrogen inside the fullerene exists only in molecular form. For *n* larger than 20, a part of the hydrogen remains in molecular form, but a few H_2 molecules dissociate, a few triangular H_3 molecules form, and some hydrogen atoms form covalent bonds with the carbon atoms of the cage. The fullerene cage deforms near the covalent C–H bonds in order to favor electronic sp^3 hybridization of those particular C atoms. At the same time, the hydrogenized carbon atoms no longer contribute to the conjugated π-electron system of C_{60}, and this breaks the fullerene aromaticity. Consequently, the cage stability is weakened, and breakage occurs for *n* = 58. The estimated hydrogen pressure in $H_{58}@C_{60}$ is 1.3 Mbar. This pressure is of the order of magnitude of the hydrogen pressure in the giant planets Jupiter and Saturn.

2.8 Supported Clusters

Under normal circumstances, the hydrogen clusters are formed by weakly interacting H_2 molecules, as stressed in the previous sections of this chapter. But, by directly depositing hydrogen atoms on a graphite surface, supported two-dimensional clusters have been obtained in which hydrogen is in the atomic state. Atomic deuterium has been deposited on graphite at 210 K in experiments performed under ultra-high-vacuum conditions, and scanning tunneling microscopy (STM) was used to analyze the results (Hornekaer et al. 2006). At very low deuterium coverage (0.03%), STM images show only adsorbed isolated atoms. Those D atoms are in a chemisorbed state, covalently bonded to a C atom of the graphite surface. When the coverage increases to 1%, the STM images show a dominance of deuterium dimers. However, these are not D_2 molecules. The two D atoms of a dimer are chemisorbed on C atoms, and the nearest-neighbor C–C distance is 1.41 Å, much larger than the bondlength, 0.79 Å, of the H_2 molecule. As shown in Figure 2.14, the most stable configuration of the adsorbed dimer is not the ortho configuration in which the two D atoms are adsorbed on first nearest neighbor C atoms, but the para configuration in which the two D atoms are bonded to C atoms on opposite vertices of a hexagon (third nearest neighbors). The para configuration is 0.2 eV more stable than the ortho, and another configuration where the two D atoms are bonded to C atoms in second neighbor sites is 1.2 eV less stable than the para configuration. The formation of dimers and larger clusters was not ascribed to thermal diffusion effects. The barrier for surface diffusion of a D atom, 1.14 eV, is larger than the activation barrier for desorption, 0.9 eV, so an isolated D atom would desorb, rather than diffuse, under heating, and this is corroborated by the experiments, which show that the atoms are immobile at 200 K and that heating to room temperature reduces the surface coverage. Instead, the mechanism of formation of dimers and larger deuterium clusters is preferential sticking. As shown by the interaction energy curves of Figure 2.14, constructed from DFT calculations, the chemisorption of the first D atom has a small barrier of 0.15 eV. Then, the barrier for chemisorption of the second D atom in the ortho configuration is only

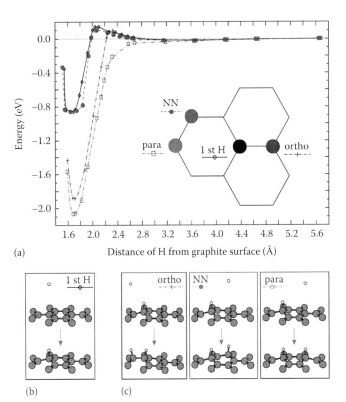

FIGURE 2.14 (a) Potential energy curves for the chemisorption of a single H atom (solid line), shown in (b), calculated using the density functional formalism, and for the three dimer configurations shown in (c), the ortho-dimer (dash-dotted line), the second neighbor site (dashed line), and the para-dimer (dash-double dotted line). (Reproduced from Hornekaer, L. et al., *Phys. Rev. Lett.*, 97, 186102, 2006. With permission.)

0.10 eV, and there is no barrier for chemisorption in the para configuration. This means that para-dimer configurations should dominate at low coverage because there is no barrier for sticking into this state, and the experiments show that this is the case.

At higher coverages, 3% or more, the majority of the adsorbed D atoms form larger clusters, and the structures are again dominated by preferential sticking. When a dimer is in the para configuration shown in Figure 2.14a, the calculations for the adsorption of a third atom show reduced barriers (of 0.10 eV) when the third atom is in any of the five positions on the right of the inset. One of those positions forms an ortho configuration with one of the atoms of the original dimer, other two positions form para configurations, and the final two are second neighbor positions. For the sticking of a fourth atom, sites with reduced or vanishing barriers for adsorption also exist. In summary, the structures of the two-dimensional hydrogen clusters formed by deposition of atomic hydrogen on graphite are very different from those of the usual molecular clusters. First of all, the basic unit is the chemisorbed H atom, covalently bonded to a surface C atom, and not the H_2 molecule; second, the clusters are not compact and its structure is controlled by preferential sticking.

2.9 Quantum Effects in Hydrogen Clusters

Superfluidity is a fascinating manifestation of quantum behavior at a macroscopic scale. In a superfluid liquid, the viscosity vanishes. Helium becomes superfluid at very low temperature, below its lambda (λ) point. This lambda point depends on pressure, and for He at a pressure of 0.05 atm, it occurs at 2.17 K. A relevant question is which condensed material, besides helium, can display superfluidity. Molecular para-hydrogen is a candidate because of the bosonic character and the light mass of the para-H_2 molecules. But, unlike helium, bulk para-H_2 solidifies at low temperature, because the interaction between two hydrogen molecules is substantially more attractive than the interaction between two He atoms. However, the lowering of the melting point compared to the bulk is a well known and rather general phenomenon in clusters (Alonso 2005), and this has motivated the interest in studying the possible superfluidity in para-H_2 clusters. Using infrared spectroscopy to study the rotational spectrum of a dopant molecule embedded in clusters of ^{4}He or para-H_2, the observation of the decoupling of the rotation of the dopant molecule from the surrounding medium gives evidence of the superfluidity of the cluster. Using this technique for a linear carbonyl sulfide (OCS) chromophore molecule surrounded by 14–16 para-H_2 molecules, all inside large He droplets, Grebenev et al. (2000) obtained confirmation of the occurrence of superfluidity in liquid para-H_2.

Mezzacapo and Boninsegni (2006, 2007) have presented a comprehensive theoretical study of (para-H_2)$_N$ clusters using path-integral quantum MC simulations and the pair potential of Silvera and Goldman (1978). Tests with other established potentials were also performed and gave similar results. The calculated chemical potential $\mu(N)$, or energy to remove one molecule from the cluster (H$_2$)$_N$ (see Equation 2.4), has local maxima for $N = 13$ and $N = 26$. The first magic number, $N = 13$, was ascribed to the completion of the first shell around a central atom (which is not inconsistent with liquid behavior; see below), and the peak at $N = 26$ was ascribed to solid-like behavior. The results for the superfluid character of these clusters are summarized in Figure 2.15, where the superfluid fraction ρ_S is plotted as a function of the cluster size at a temperature $T = 1$ K. The superfluid fraction is the fraction of molecules in the superfluid phase, that is, the fraction of the system that decouples from an externally induced rotation. Three regions can be identified in this figure. In the first region, formed by the clusters with less than 22 hydrogen molecules, the clusters are liquid-like and superfluid at low temperature. In these clusters, the superfluid fraction decreases monotonically as T increases. For instance, (H$_2$)$_{20}$ is entirely superfluid ($\rho_S = 1$) at temperatures below 1.25 K, and its ρ_S drops fast with increasing T, reaching a value of 0.2 at $T = 2.5$ K. In the transition region of Figure 2.15, formed by clusters with sizes between $N = 22$ and $N = 30$, the evolution from liquid-like character to solid-like character does not occur continuously. The superfluid properties depend sensitively on the

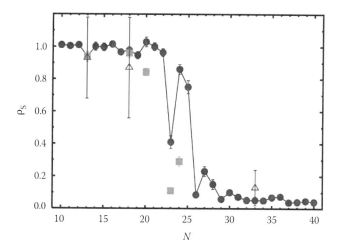

FIGURE 2.15 Solid symbols are the superfluid fraction of (H$_2$)$_N$ clusters as a function on N, at $T = 1$ K, obtained by quantum MC simulations. Other symbols are results of a previous calculation. (Reproduced from Mezzacapo, F. and Boninsegni, M., *Phys. Rev. A*, 75, 033201, 2007. With permission.)

cluster size, and strong differences occur for clusters differing by just one single molecule. This was interpreted as indicating the alternating liquid-like (superfluid) character or solid-like (insulating) character of the clusters. For instance, plots of the atomic structure of the clusters with $N = 25$ and $N = 26$ at $T = 1$ K indicate a clear solid-like structure of (H$_2$)$_{26}$; that is, the hydrogen molecules show a high degree of spatial localization. Quantum exchanges between molecules are suppressed and the superfluid response is weak. In contrast, the molecules in (H$_2$)$_{25}$ are more delocalized and their positions cannot be clearly identified. That delocalization promotes quantum exchanges responsible for the large superfluid response.

A peculiar behavior was observed for some clusters in this region. A good example is (H$_2$)$_{23}$, for which the superfluid fraction shows a local minimum in Figure 2.15. A detailed analysis of the computer simulations reveals the coexistence of two phases in this cluster, a liquid-like superfluid phase and a solid-like phase. This means that in its time evolution at a given temperature, the system visits ordered solid-like configurations and liquid-like superfluid configurations and the value of ρ_S in Figure 2.15 is the time average. A fascinating feature of this cluster is that, on lowering the temperature, the liquid-like phase becomes dominant as T approaches 0 K, that is, this phase is observed during a higher fraction of the simulation time: the cluster melts at a low temperature. Melting in this case is due to zero-point motion which induces quantum exchanges of the molecules. This behavior, which is driven by Bose statistics, was called quantum melting by Mezzacapo and Boninsegni (2006, 2007). On the other hand, for high T, the quantum exchanges are suppressed and the system solidifies. This behavior becomes reflected in the form of the radial density of particles $\rho(r)$, that is, the density of particles taking the cluster center as the origin of coordinates. At $T = 2$ K, $\rho(r)$ of (H$_2$)$_{23}$ displays a well-defined structure of two shells, a sharp peak at a distance $r \approx 2\,\text{Å}$ from the center of mass, and a broader peak centered at 5.5 Å; this is consistent with Figure 2.3. Lowering the temperature to $T = 0.75\,\text{K}$, the first peak experiences a substantial broadening and a lowering of its height. This indicates that the molecules are less localized and that quantum exchanges between shells as well as within shells increase. $(H_2)_{23}$ is not the only cluster in the transition region of Figure 2.15 showing quantum melting. Some other clusters in this region show the same features. The behavior just discussed is different from that of a cluster which is liquid but not 100% superfluid. In this case, which corresponds to the region $N < 22$, the clusters are liquid at all temperatures and the superfluid fraction grows at low T, but there is no coexistence between liquid and solid phases. Finally, Figure 2.15 shows that the superfluid response is significantly depressed above $N = 30$, consistent with the expectation that a crystalline phase has to appear at large N.

The nucleus of the deuterium atom has total angular momentum $J = 1$, and molecular ortho-deuterium (ortho-D_2) occurs in nature as a mixture of $S = 0$ and $S = 2$ spin states. The $S = 0$ state is a boson similar to para-H_2 but with a larger mass. The mass difference is responsible of some subtle differences in the properties of para-H_2 and ortho-D_2. Quantum effects are weakened in ortho-D_2, or stated in an alternative way, the behavior of these clusters is more classic (Mezzacapo and Boninsegni 2007). The binding energy $e_b(N)$ and the chemical potential $\mu(N)$ show peaks at $N = 13$ and $N = 19$, and those features are sharper than in para-H_2. The structure of $(D_2)_{13}$ and $(D_2)_{19}$ at $T = 0.5\,\text{K}$ is solidlike. Actually, the calculated structure of $(D_2)_{19}$ is quite similar to a classical double icosahedron.

Very recently, the distribution of the superfluid response across the clusters has been discussed with opposite views. Khairallah et al. (2007) have concluded that the superfluid response is largely confined at the surface of small clusters and arises from the exchange cycles involving surface molecules. On the other hand, Mezzacapo and Boninsegni (2008) have presented persuasive arguments that the small clusters are uniformly superfluid, that is, the superfluidity response is not localized at the cluster surface. In summary, the quantum properties of hydrogen clusters are a fascinating subject and intensive and interesting work on these nanoclusters is expected to continue in the near future.

2.10 Conclusions

Hydrogen clusters can be obtained by condensation of hydrogen gas in supersonic expansions. The clusters formed are molecular clusters with formula $(H_2)_N$. The experiments show that clusters of some particular sizes N are more abundant that other neighbor sizes, and for this reason those are often called magic clusters. It is difficult to obtain direct experimental information of the structure of the hydrogen clusters, that is, of the geometrical arrangement of the molecules, and theoretical calculations help a lot in this task. The calculations assign the first, and more clear magic number, $N = 13$, to the completion of a first shell

around a central molecule. Both experiment and theory suggest the existence of other higher magic numbers. If one of the molecules of a hydrogen cluster is ionized, the structure of the cluster suffers a drastic change: Instead of having a charged H_2^+ molecule immersed in the cluster, it is more favorable to eject a neutral H atom and form a charged trimer solvated in the cluster, $H_3^+ (H_2)_n$. Confining the hydrogen clusters in small cages, like fullerenes and nanotubes, affects the structure of the encaged clusters. The cluster roughly adapts its structure in order to maximize the interaction with the inner wall of the cage, and this effect may have consequences for the important technological problem of hydrogen storage. Clusters produced by direct deposition of atomic hydrogen on a substrate form two-dimensional structures where the relevant units are the chemisorbed atoms, not the molecules. Nuclear fusion between colliding deuterium nuclei has been achieved in experiments in which dense molecular beams of large deuterium clusters were irradiated with strong femtosecond lasers. The mechanism generating the flying energetic nuclei is the Coulombic explosion which follows the massive ionization of the deuterium clusters. The atomic part of the process, that is, the interaction between the cluster and the laser leading to ionization and Coulomb explosion is a fascinating area. At very low temperatures, hydrogen clusters display quantum effects, and a beautiful manifestation of these is superfluidity, already detected experimentally in some hydrogen clusters.

Acknowledgments

Work supported by MEC of Spain (Grant MAT2005-06544-C03-01 and MAT2008-06843-C03-01) and Junta de Castilla y León (Grants VA039A05, VA017A08 and GR23). The Center for Atomic-Scale Materials Design (CAMD) is sponsored by the Lundbeck Foundation. JAA acknowledges the hospitality of the University of the Basque Country and the Donostia International Physics Center, and an Ikerbasque Fellowship of the Basque Foundation of Science.

References

Alonso, J. A. 2005. *Structure and Properties of Atomic Clusters*. Imperial College Press, London, U.K.

Barajas-Barraza, R. E. and Guirado-López, R. A. 2002. Clustering of H_2 molecules encapsulated in fullerene structures. *Phys. Rev. B*: 66, 155426: 1–12.

Belkacem, M., Latora, V., and Bonasera, A. 1995. Critical evolution of a finite system. *Phys. Rev. C*: 52, 271–285.

Bokes, P., Štich, I., and Mitas, L. 2001. Electron correlation effects in ionic hydrogen clusters. *Int. J. Quantum Chem.*: 83, 86–95.

Buck, U., Huisken, F., Kohlhase, A., Otten, D., and Schaefer, J. 1983. State resolved rotational excitation in $D_2 + H_2$ collisions. *J. Chem. Phys.*: 78, 4439–4450.

Carmichael, M., Chenoweth, C., and Dykstra, C. E. 2004. Hydrogen molecule clusters. *J. Phys. Chem. A*: 108, 3143–3152.

Castleman, A. W. and Keese, R. G. 1998. Gas-phase clusters. Spanning the state of matter. *Science*: 241, 36–42.

Chabot, M. and Wohrer, K. 2004. Comment on "Direct experimental evidence for a negative heat capacity in the liquid-to-gas phase transition in hydrogen cluster ions: Backbending of the caloric curve." *Phys. Rev. Lett.*, 93, 039301-1.

Chermette, H. and Ymmud, I. V. 2001. Structure and stability of hydrogen clusters up to H_{21}^+. *Phys. Rev. B*: 63, 165427: 1–10.

Clampitt, R. and Towland, L. 1969. Clustering of cold hydrogen gas on protons. *Nature*: 223, 815–816.

Cuervo, J. E. and Roy, P. N. 2006. Path integral ground state study of finite-size systems: Application to small (parahydrogen)$_N$ (N = 2–20) clusters. *J. Chem. Phys.*: 125, 124314: 1–6.

Diep, P. and Johnson, J. K. 2000. An accurate H_2-H_2 interaction potential from first principles. *J. Chem. Phys.*: 112, 4465–4473.

Ditmire, T., Zweiback, J., Yanovsky, V. P., Cowan, T. E., Hays, G., and Wharton, K. B. 1999. Nuclear fusion from explosions of femtosecond laser-heated deuterium clusters. *Nature*: 398, 489–492.

Farizon, B., Farizon, M., Gaillard, M. J., Gobet, F., Carré, M., Buchet, J. P., Scheier, P., and Märk, T. D. 1998. Experimental evidence of critical behavior in cluster fragmentation using an event-by-event data analysis. *Phys. Rev. Lett.*: 81, 4108–4111.

Fisher, M. E. 1967. The theory of equilibrium critical phenomena. *Rep. Prog. Phys.*: 30, 615–730.

Ginzburg, V. L. and Sobyanin, A. A. 1972. Can liquid molecular hydrogen be superfluid? *JEPT Lett.*: 15, 242–244.

Gobet, F., Farizon, B., Farizon, M., Gaillard, M. J., Buchet, J. P., Carré, M., and Märk, T. D. 2001. Probing the liquid-to-gas phase transition in a cluster via a caloric curve. *Phys. Rev. Lett.*: 87, 203401:1–4.

Gobet, F., Farizon, B., Farizon, M., Gaillard, M. J., Buchet, J. P., Carré, M., Scheier, P., and Märk, T. D. 2002. Direct experimental evidence for a negative heat capacity in the liquid-to-gas phase transition in hydrogen cluster ions: Backbending of the caloric curve. *Phys. Rev. Lett.*: 89, 183403: 1–4.

Grebenev, S., Sartakov, B., Toennies, J. P., and Vilesov, A. F. 2000. Evidence for superfluidity in para-hydrogen clusters inside helium-4 droplets at 0.15 Kelvin. *Science*: 289, 1532–1535.

Gross, D. H. E. 1997. Microcanonical thermodynamics and statistical fragmentation of dissipative systems. The topological structure of the N-body phase space. *Phys. Rep.*: 279, 119–201.

Gross, E. K. U., Dobson, J. F., and Petersilka, M. 1996. Density functional theory of time-dependent phenomena. *Topics in Current Chemistry*, Vol. 81. Editor R. F. Nalewajski, Springer Verlag, Heidelberg, Germany, pp. 81–172.

Guardiola, R. and Navarro, J. 2006. Structure of small parahydrogen molecules. *Phys. Rev. A*: 74, 025201-1–025201-4.

Guardiola, R. and Navarro, J. 2008. Excitation levels and magic numbers of small parahydrogen clusters (N ≤ 40). *J. Chem. Phys.*: 128, 144303: 1–7.

Hartke, R., Symesa, D. R., Buersgens, F., Ruggles, L. E., Porter, J. L., and Ditmire, T. 2005. Fusion neutron detector calibration using a table-top laser generated plasma neutron source. *Nucl. Instrum. Methods A*: 540, 464–469.

Heidenreich, A., Jortner, J., and Last, I. 2007. Cluster dynamics transcending chemical dynamics toward nuclear fusion. *Proc. Natl. Acad. Sci. U.S.A.*: 103, 10589–10593.

Hemley, R. J., Mao, H. K., Finger, L. W., Jephcoat, A. P., Hazen, R. M., and Zha, C. S. 1990. Equation of state of solid hydrogen and deuterium from single-crystal x-ray diffraction to 26.5 GPa. *Phys. Rev. B*: 42, 6458–6470.

Hiraoka, K. and Mori, T. 1989. Thermochemical stabilities of D_3^+ $(D_3)_n$ with n = 1–10. *Chem. Phys. Lett.*: 157, 467–471.

Hornekaer, L., Rauls, E., Xu, W., Sljivancanin, Z., Otero, R., Stensgaard, I., Laegsgaard, E., Hammer, B., and Besenbacher, F. 2006. Clustering of chemisorbed H(D) atoms on the graphite (0001) surface due to preferential sticking. *Phys. Rev. Lett.*: 97, 186102: 1–4.

Isla, M. and Alonso, J. A. 2005. Fragmentation and Coulomb explosion of deuterium clusters by the interaction with intense laser pulses. *Phys. Rev. A*: 111, 17765: 1–8.

Isla, M. and Alonso, J. A. 2007. Interaction of the charged deuterium cluster D_3^+ with femtosecond laser pulses. *J. Phys. Chem. C*: 111, 17765–17772.

Johnson, K. and Ashcroft, N. W. 2000. Structure and band gap closure in dense hydrogen. *Nature*: 403, 632–635.

Khairallah, S. A., Sevryuk, M. B., Ceperley, D. M., and Toennies, J. P. 2007. Interplay between magic number stabilities and superfluidity of small parahydrogen clusters. *Phys. Rev. Lett.*: 98, 183401: 1–4.

Kirchner, N. J. and Bowers, M. T. 1987. An experimental study of the formation and reactivity of ionic hydrogen clusters: The first observation and characterization of the even clusters H_4^+, H_6^+, H_8^+ and H_{10}^+. *J. Chem. Phys.*: 86, 1301–1310.

Komatsu, K., Murata, M., and Murata, Y. 2005. Encapsulation of molecular hydrogen in fullerene C_{60} by organic synthesis. *Science*: 307, 238–640.

Kostin, M. A., Polyansky, O. L., Tennyson, J., and Mussa, H. Y. 2003. Rotation-vibration states of H_3^+ at dissociation. *J. Chem. Phys.*: 118, 3538–3542.

Lammers, U., Borstel, G., Mañanes, A., and Alonso, J. A. 1990. Electronic and atomic structure of Cs_N and Cs_NO clusters. *Z. Phys. D*: 17, 203–208.

Langreth, D. C., Dion, M., Rydberg, H., Schroder, E., Hilgaard, P., and Lundqvist, B. I. 2005. *Int. J. Quantum Chem.*: 101, 599–610.

Last, I. and Jortner, J. 2001a. Nuclear fusion induced by Coulomb explosion of heteronuclear clusters. *Phys. Rev. Lett.*: 87, 033401: 1–4.

Last, I. and Jortner, J. 2001b. Nuclear fusion driven by Coulomb explosion of homonuclear and heteronuclear deuterium- and tritium-containing clusters. *Phys. Rev. A*: 64, 063201: 1–11.

Loubeyre, P., Ocelli, F., and LeToullec, R. 2002. Optical studies of solid hydrogen to 320 GPa and evidence for black hydrogen. *Nature*: 416, 613–617.

Ma, L. M., Suraud, E., and Reinhard, P. G. 2001. Laser excitation and ionic motion in small clusters. *Eur. Phys. J. D*: 14, 217–224.

Ma, L. M., Reinhard, P. G., and Suraud, E. 2005. Dynamics of H_9^+ in intense laser pulses. *Eur. Phys. J. D*: 33, 49–58.

Martínez, J. I., Isla, M., and Alonso, J. A. 2007. Theoretical study of molecular hydrogen clusters. Growth models and magic numbers. *Eur. Phys. J. D*: 43, 61–64.

McCall, B. J., Geballe, T. R., Hinkle K. H., and Oka, T. 1998. Detection of H_3^+ in the diffuse interstellar medium toward cygnus OB2 No. 12. *Science*: 279, 1910–1913.

Mezzacapo, F. and Boninsegni, M. 2006. Superfluidity and quantum melting of p-H_2 clusters. *Phys. Rev. Lett.*: 97, 045301: 1–4.

Mezzacapo, F. and Boninsegni, M. 2007. Structure, superfluidity and quantum melting of hydrogen clusters. *Phys. Rev. A*: 75, 033201: 1–10.

Mezzacapo, F. and Boninsegni, M. 2008. Local superfluidity of parahydrogen clusters. *Phys. Rev. Lett.*: 100, 145301: 1–4.

Miller, S., Tennyson, J., Lepp, S., and Dalgarno, A. 1992. Identification of features due to H_3^+ in the infrared spectrum of supernova 1987A. *Nature*: 355, 420–422.

Parr, R. G. and Yang, W., 1989. *Density Functional Theory of Atoms and Molecules*. Oxford University Press, Oxford, U.K.

Pickard, C. J. and Needs, R. J. 2007. Structure of phase III of solid hydrogen. *Nat. Phys.*: 3, 473–476.

Poth, L., Wisniewski, E. S., and Castleman, A. W. 2002. Cluster dynamics: Fast reactions and Coulomb explosion. *Am. Sci.*: 90, 342–349.

Prosmiti, R., Villarreal, P., and Delgado-Barrio, G. 2003. Structures and energetics of H_n^+ clusters (n = 5–11). *J. Phys. Chem. A*: 107, 4768–4772.

Pupysheva, O. V., Farajian, A. A., and Yakobson, B. I. 2008. Fullerene nanocage capacity for hydrogen storage. *Nanoletters*: 8, 767–774.

Santamaria, R. and Soullard, J. 2005. The atomization process of endohedrally confined hydrogen molecules. *Chem. Phys. Lett.*: 414, 483–488.

Seo, H.-I., Sun, J.-Y., Shin, C.-H., and Kim, S.-J. 2007. Structure and dissociation energy of weakly bound H_{2n+1}^+ (n = 5–8) complexes. *Int. J. Quantum Chem.*: 197, 988–997.

Silvera, I. F. and Goldman, V. V. 1978. The isotropic intermolecular potential for H_2 and D_2 in the solid and gas phases. *J. Chem. Phys.*: 69, 4209–4213.

Soullard, J., Santamaria, R., and Jellinek, J. 2008. Pressure and size effects in endohedrally confined hydrogen clusters. *J. Chem. Phys.*: 128, 064316: 1–7.

Tada, K., Furuya, S., and Watanabe, K. 2001. Ab initio study of hydrogen adsorption to single-walled carbon nanotubes. *Phys. Rev. B*: 63, 155405: 1–4.

Tejeda, G., Fernández, J. M., Montero, S., Blume, D., and Toennies, J. P. 2004. Raman spectroscopy of small para-H_2 clusters formed in cryogenic free jets. *Phys. Rev. Lett.*: 92, 223401-1–223401-4.

Tennyson, J. and Miller, S. 1994. H_3: From first principles to Jupiter. *Contemp. Phys.*: 35, 105–116.

Wolff, W., Coelho, L. F. S., and Wolf, H. E. 1992. Collisional destruction of fast H_3^+ ions in noble gases. *Phys. Rev. A*: 45, 2978–2982.

Zweiback, J., Smith, R. A., Cowan, T. E., Hays, G., Wharton, K. B., Yanovsky, V. P., and Ditmire, T. 2000. Nuclear fusion driven by Coulomb explosions of large deuterium clusters. *Phys. Rev. Lett.*: 84, 2634–2637.

3

Mercury: From Atoms to Solids

Elke Pahl
Massey University Albany

Peter Schwerdtfeger
Massey University Albany

3.1 Introduction

Elemental mercury is the only metal that is liquid at room temperature (melting temperature $T_M = -38.83°C$); the only other low melting metal is gallium ($T_M = 29.76°C$) [1]. This has been known since ancient times and is reflected in the Greek name *Hydrargyrum* meaning "watery silver" and the Latin *Argentum vivum* meaning "quick silver." It is speculated that its low melting point is due to relativistic effects [2–4], which energetically lowers the 6s band substantially [5], thus making mercury a very hard atom with a static dipole polarizability of only $\alpha_D = 5.025(50)$ Å^3 [6,7]. It is clear that such strong and pronounced relativistic effects for the 6s shell [5], which are close to the relativistic 6s maximum of gold [4,8], makes mercury rather unique in chemical and physical properties among the other Group 12 metals zinc and cadmium. Some of the physical and chemical properties of the Group 12 elements are compared in Table 3.1, and in many of these properties we see anomalies that, most likely, are caused by relativistic effects. For the next Group 12 element (below Hg) with nuclear charge $Z = 112$ (Copernicium, Cn), even larger relativistic effects are predicted [9–11], and the dipole polarizability of this element is now the lowest of all the Group 12 elements ($\alpha_D(112) = 3.8$ Å^3) [12].

Mercury occurs in many natural materials including coal, gas, and oil in small quantities; it is estimated that by burning fossil fuels about 2400 tons per year of mercury are released into the atmosphere [13]. Thus, the contamination of the atmosphere and the bioaccumulation of it, especially in fish, present large-scale problems. Since oxidized mercury is easier removed than mercury in its elemental form, the search for efficient and regenerable oxidation catalysts like noble metals is important [14,15]. For humans, mercury is toxic and leads to damage to the nervous tissue, kidneys, and liver. It is therefore important that mercury be handled with care, which is the main reason why mercury is not the metal of choice in the study and design of new nanomaterials. In fact, most of the nanoscience research here concentrates on the development of sensors for the detection of small amounts of mercury [16].

Mercury is, however, an interesting element to be studied by theoreticians, as mercury clusters range from van der Waals bonded to more covalent systems, before finally reaching the metallic state at larger cluster size [17]. The dissociation energies (D_e) therefore vary widely from about 0.05 eV for the dimer [18] to 0.67 eV (cohesive energy E_{coh}) for the bulk metal. For an ideal Lennard-Jones system, we have the simple relation $E_{coh} = 8.61D_e$ [19], which is obeyed for either the face-centered cubic fcc or the hexagonal close-packed hcp solid. Instead, $E_{coh} = 13.4D_e$, i.e., mercury does not behave like an ideal Lennard-Jones system. It is therefore a challenge to simulate mercury in the gas, liquid, or the solid phase.

The first appearance of the 6s–6p gap closure to the metallic state at a specific cluster size is still a matter of intense debate [20,21]. Early measurements by Rademann gave an estimated band gap closure for Hg$_n$ at $n \approx 70$ [22,23], Singh obtained $n \approx 80$ [24,25], Pastor et al. $n \approx 135$ [26–28], and a recent photoelectron study on negatively charged mercury clusters by Busani et al. gave $n = 400 \pm 30$ [29–31]. In liquid mercury, the single 6s–6p gap opens at a density of $\rho = 8.8$ g cm^{-3} (the density of liquid mercury under standard conditions is 13.59 g cm^{-3} [32]). This agrees with experiment where a gradual transition from metallic to semiconducting and insulating properties has been observed at elevated temperatures and pressures [33]. An interesting comparison is to superheavy element 112; due to a very strong relativistic 7s stabilization element 112 never becomes metallic as recent solid-state calculations show [10].

There is also a very large contraction of the Hg–Hg equilibrium bond length r_e when going from the van der Waals bonded dimer ($r_e = 3.69$ Å) [18] to the solid state ($r_e = 3.01$ Å) [1], which is not found in the rare gas elements. Such large differences indicate

TABLE 3.1 A Comparison of Chemical and
Physical Properties of the Group 12 Elements

Property	Zn	Cd	Hg
I_1 (eV)	9.394	8.994	10.438
I_2 (eV)	17.96	16.908	18.757
I_w (eV)	3.63	4.08	4.48
α_D (Å³)	5.75	7.36	5.02
T_M (°C)	419.53	321.07	−38.83
T_B (°C)	907	767	356.62
T_c (°C)	0.85	0.517	3.95
λ (W m⁻¹ K⁻¹)	116	96.8	8.34
ρ (10⁻⁸ Ω m)	5.9	7.6	95.78
M	0.38	0.38	1.0
B (GPa)	70	42	25
Structure	hcp	hcp	rhomb

Notes: I_1 and *I_2* are the first and second ionization
potentials of the atom; *I_w* is the work function of the
bulk; α_D the static electric dipole polarizability; *T_M*
and *T_B* are the melting and boiling points; *T_c* the
superconducting transition temperature; λ the thermal conductivity; ρ the specific resistance; *M* the
electron–phonon coupling constant; *B* the bulk modulus, and for the bulk structures; hcp refers to hexagonal closed packing; rhomb to rhombohedral. Results
are taken from Ref. [1].

the importance of many-body forces beyond the two-body force
such as the two-body Lennard-Jones potential [34]. This clearly
distinguishes mercury from rare gas interactions, which are reasonably well described by two-body forces only [19]. Further,
there is no bonding interaction at the Hartree–Fock level starting
from the Hg dimer and up to the solid state, that is the interaction between Hg atoms without taking electron correlation into
account is purely repulsive [35,36]. This situation now resembles
the bonding behavior in the rare gases. And finally, relativistic
effects shorten the bond distance in Hg₂ considerably (by 0.2 Å)
[37], and solid Hg changes the crystal structure to hcp when
neglecting relativistic effects, the bond distance in the crystal
also being somewhat larger (3.06 Å at the nonrelativistic level of
theory) than in the rhombohedral relativistic structure (3.00 Å)
[38]. The correct description of both relativistic and electron
correlation effects is thus very important for the simulation of
mercury in clusters, the liquid or the solid state.

  Several groups have studied neutral and charged Hg$_n$ clusters in the gas phase, in particular the transitions in chemical
bonding from van der Waals to metallic [17]. For small van der
Waals clusters ($n < 13$) the band gap is large and little $6s$–$6p$
hybridization is observed. After a transition region, the bonding
becomes covalent ($30 < n < 100$); $6s$–$6p$ hybridization leads to an
increase of the binding between the atoms beyond the normal
two-body interaction. At much larger cluster size we enter the
metallic phase. In this process both the bond distances and the
ionization potential decrease toward the bulk value. This transition for a finite system has been described as being somewhat
similar to a Mott transition for the bulk [17]. It is clear that mercury is a complex system to be studied by both theoretical and

experimental methods. Here, we want to outline the progress
made in the last two decades in simulating Hg clusters, the liquid and the bulk system, and how well the results compare with
experiment.

3.2 The Mercury Atom

We briefly mention both, theoretical and experimental work
on the Hg atom, mainly to discuss the importance of relativistic effects necessary to understand the physical properties of
the Hg clusters. The most accurate calculations for Hg come
from Kaldor's group using Fock-space coupled-cluster theory
starting from a Dirac–Coulomb Hamiltonian including Breit
interactions in the low-frequency limit [39,40]. They reproduce experimental ionization potentials and electronic excitations [1,41] within a few 100 cm⁻¹. In fact, such calculations
are now so precise that quantum electrodynamic effects have
to be taken into account to produce results of experimental
accuracy [42]. Figure 3.1 shows ionization potentials and
static dipole polarizabilities for all group 12 metals. It is clear
that relativistic effects in such properties are large and cannot
be neglected anymore. The relativistic increase in the ionization potential and the subsequent decrease in the polarizability has led to the conclusion, that both Hg and element 112
are chemically inert [9]. The (indirect) relativistic expansion
of the core $5d$ orbitals also leads to a larger $5d/6s$ mixing [5].
Finally, spin–orbit coupling splits both the $5d$ and $6p$ levels
in Hg. The splitting between the $^3P_0/^3P_2$ states of neutral Hg
is 0.79 eV, while the $^2D_{3/2}/^2D_{5/2}$ splitting in Hg⁺ is 1.86 eV [41].
The $6p_{1/2}$ spin–orbit stabilization will reduce the $6s/6p$ band
gap and therefore shift the onset of metallicity to smaller Hg
clusters. Further, in a mercury vapor lamp the UV light produced at 253.7 nm comes from the spin-forbidden $^3P_1 \rightarrow {}^1S_0$
transition, which becomes allowed in the spin–orbit coupled
case [45].

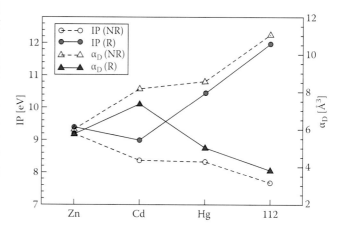

FIGURE 3.1 A comparison between experimental (or relativistic
coupled cluster) ionization potentials and static dipole polarizabilities
for the group 12 metals with (R) and without (NR) relativistic effects.
Experimental values for Zn, Cd, and Hg are taken from Refs. [1,6,43,44],
relativistic coupled-cluster calculations come from Refs. [12,39].

3.3 Mercury Clusters

It is obvious that any theoretical method able to accurately describe Hg clusters from small to large sizes has to reproduce experimental values reasonably well for the smallest cluster, the dimer Hg_2, as well as for the solid metal. For Hg_2, accurate experimental results are scarce. Table 3.2 shows a comparison between different theoretical results with experiment. As the interaction between two Hg atoms in Hg_2 is of van der Waals (dispersive) type with possibly a rather small dissociation energy of $400\,cm^{-1}$ [46], it is currently not easy to obtain accurate values from both theory or experiment. Currently, the most precise calculations come from Peterson [47] using a small-core relativistic pseudopotential for Hg, including spin–orbit corrections and extrapolating to the basis set limit using correlation consistent basis sets at the coupled-cluster level of theory. Relativistic effects are important, as they decrease the bond distance by about $0.2\,Å$, and slightly lower the dissociation energy [18,37,48]. Table 3.2 clearly shows the huge stabilization effect on removing an electron out of the antibonding orbital in Hg_2, i.e., the dissociation energy increases by a factor of 29 when going from Hg_2 to Hg_2^+. This rather large effect will diminish with increasing cluster size toward the solid state. Excited states of Hg_2 and Hg_2^+ were also studied in the past by both experimental [51–53,55,58–60] and theoretical methods [61–64]. Also, some excited states of Hg_2 are found to be more stable due to the excitation of an electron out of the antibonding orbital; the dissociation energy increases up to a factor of 25 and the bond length is strongly reduced to about $3\,Å$, the distance found in solid Hg (see Ref. [63] and references therein). We note here that the importance of relativistic effects in optical transitions in Hg-containing alloys was already discussed in 1972 by Kisiel and Lee [65].

The study of large mercury clusters to, for example, probe the convergence toward the bulk is a nontrivial task for both theoreticians and experimentalists. As calculations show, for the simple mercury dimer as well as for the solid state, relativistic effects have to be included [24,37,48]. For larger clusters, the computer time becomes prohibitively large for all electron methods at the Hartree–Fock or density functional level of theory, so naturally the relativistic pseudopotential approximation (also called effective core potential method) is used, which has been proven to be very accurate compared to all electron calculations if care is taken for the proper choice of the core and in the adjustment procedure [66]. While the inclusion of scalar relativistic effects is more or less straightforward, spin–orbit effects, electron correlation, and the basis set superposition error present a much larger problem [18,47,67]. Theoretical studies of clusters have therefore been limited to rather small cluster sizes, where the extrapolation to the bulk limit is questionable. Nevertheless, a number of interesting computational studies on mercury clusters have appeared in the past decade.

Dolg and coworkers have studied small neutral and charged mercury clusters [49,68]. An ELF (electron localization function) analysis showed predominantly van der Waals type bonding for Hg_n ($n \leq 4$) [68]. They also investigated the size dependence of ionization potentials, electron affinities, and binding energies with increasing cluster size up to Hg_{15} and found significant covalent bonding character for these clusters. For medium-sized clusters, Dolg and coworkers suggested a hybrid model consisting of a pairwise additive dispersion potential proportional to R^{-6} together with a two-valence electron relativistic pseudopotential. The latter is a so-called large-core pseudopotential which includes core-polarization effects describing dynamic correlation from the $5d$ core of Hg, and finally core–core repulsion effects between the Hg atoms [69]. This hybrid model is a good approximation for Hg_2 as comparison with more accurate coupled-cluster calculations using a small-core pseudopotential, which does not include the $5d$ electrons as core electrons, for Hg show [69]. A simulated annealing procedure confirmed the Lennard-Jones behavior of these structures with icosahedral structures for Hg_{13} and Hg_{55} [69]. However, in a later paper they used a genetic algorithm procedure when searching for the global minima and found very unusual structures from Hg_7 to Hg_{13}, which deviate substantially from the compact Lennard-Jones like shapes [70,71], which for small- to medium-sized closed atom-shell clusters are Mackay icosahedra [72]. As two-body interactions favor compact structures due to maximizing the number of interacting pairs, this result would imply that the many-body expansion of the interaction potential does not converge smoothly. This was indeed found in a recent paper by

TABLE 3.2 Spectroscopic Properties for the $Hg_2\ ^1\Sigma_g^+$ and $Hg_2^+\ ^2\Sigma_g^+$ Ground States from Experimental and Theoretical Work

Molecule	Method	r_e	D_0	ω_e	$\omega_e x_e$	I.P.	Refs.
Hg_2	Theo.	3.687	0.048	19.8	0.23	$8.85^a/9.60^b$	[47,49]
	Exp.	3.69 ± 0.01	0.047 ± 0.002	19.6 ± 0.3	0.26 ± 0.05	9.0–9.5	[23,50–55]
Hg_2^+	Theo.	2.74	1.40	112	—	—	[49,56]
	Exp.	—	1.6 ± 0.2	—	—	$14.5^{a,c}$	[57]

Notes: The equilibrium bond distance r_e is given in Å, the dissociation energy D_0 in eV (zero-point vibrational energy correction included), the harmonic vibrational frequency ω_e and the anharmonicity correction $\omega_e x_e$ in cm^{-1}, and the ionization potential I.P. in eV.

[a] Adiabatic I.P.

[b] Vertical I.P.

[c] Own unpublished results at the MP2 level of theory. The local minimum of metastable Hg_2^{2+} was taken as a reference. The vertical I.P. is only slightly different to the adiabatic value.

Moyano et al. [35]. However, more recent accurate coupled-cluster calculations show that, beside the importance of such many-body effects, the mercury clusters do follow the usual growth pattern of compact cluster structures [73], which contradicts the original results [70,71].

Moyano et al. used two-body and effective three-body interactions between the mercury atoms in a simulated annealing approach to obtain global minimum structures for larger mercury clusters up to Hg_{40} [35]. Again these clusters show rather compact structures similar to the Lennard-Jones structures with magic cluster numbers of 6, 13, 19, 23, 26, and 29 atoms, in agreement with diatomic-in-molecules (DIM) calculations by Kitamura [74]. These values also coincide with the mass distribution of mercury–cesium cluster ions observed by Ito et al. [75]. The calculations of Moyano et al. also reveal a fast convergence of the polarizability toward the bulk limit in contrast to the singlet–triplet gap or the ionization potential. However, the cluster sizes were far too small to accurately predict the onset of metallicity.

It is now evident that it remains a challenge to accurately describe electronic properties for global minimum structures for larger mercury clusters, and, what is required at finite temperatures, to perform molecular dynamic simulations to obtain properties, which can be compared to the experiment. Nevertheless, a number of theoretical studies appeared in the past dealing with neutral or charged mercury clusters in ground and excited electronic states [56,76–78]. Hg_3 is the smallest cluster which can bind an extra electron (electron affinity of 0.13 eV at the coupled-cluster, CCSD(T), level of theory) [77]. Gaston et al. studied the photoabsorption spectra of cationic mercury clusters [56]. The experimental photoabsorption spectra of singly charged cationic mercury clusters Hg_n^+ carried out by Haberland and coworkers show a sharp change in behavior at cluster size $n = 6$ [60,79]. It has been interpreted as the onset of a plasmon-like resonance in the $6s$–$6p$ transition. Both, relativistic density functional theory DFT and wavefunction-based methods revealed that the onset of a plasmon-like resonance corresponds to a structural change from linear to three-dimensional cluster isomers, i.e., a change from single electron–hole excitations in small linear clusters to plasmon-like collective transitions for the larger three-dimensional clusters [56].

Medium- to large-sized mercury clusters have been studied extensively in the past by experimental methods [20–23,26–31, 80–84], mainly to answer the question at what cluster size the transition from van der Waals to metallic bonding occurs [26]. An excellent review on cluster size effects and the extrapolation to the bulk for metallic clusters is given by Johnston [21]. Here we mention the most recent work of Busani et al. [29]. They measured photoelectron spectra of the mass-resolved negatively charged mercury clusters up to Hg_{250}^-. Upon photoexcitation, the $6p$ electron can be detached leaving the resulting neutral cluster in its electronic ground state. Alternatively, electrons can be detached from the $6s$ band of the mercury cluster leaving the resulting neutral cluster in an "electron–hole pair" excited state. The difference between both results is a direct measure for the

HOMO–LUMO ($6s$–$6p$) gap in the photoelectron spectrum of the negatively charged cluster, which approximately provides the excitation band gap for the corresponding neutral cluster. Large changes in the structure of mercury clusters due to the excess electron are not expected at larger cluster size as the charge will be smeared out. Extrapolation to higher cluster sizes indicates a band gap closure at the size range of $n = 400 \pm 30$, a considerably larger value than previously reported (see Section 3.1) [29].

Bescós et al. studied time-resolved ultrafast multiphoton ionization and fragmentation dynamics of mercury clusters Hg_n ($n \le 110$) with femtosecond pulses [81]. At laser intensities of 10^{11} W cm^{-2}, they observed singly, doubly, and triply charged mercury clusters. Ionization potentials and electron excitation energies for mercury clusters up to Hg_{109} were determined by photoelectron and UV/vis-photoabsorption spectroscopy by Rademann et al. [22,23,84]. Their estimate for the onset of the metallic phase is at much smaller cluster size compared to Busani et al. [29]. Blanc et al. looked at the stability of triply charged mercury clusters, Hg_n^{3+} [82]. They found Hg_{60}^{3+} to be stable with respect to fragmentation into Hg_m^{2+} and Hg_n^+ ($n + m = 60$), but not Hg_{50}^{3+}.

3.4 Liquid Mercury and the Mercury Surface

Along with the lowest melting point $T_m = -38.83°C = 234.32$ K [1] of all metals, assumed but not yet proven to be due to large relativistic effects lowering the $6s$ band substantially [2–5], mercury also shows the lowest critical temperature of all metals. Accepted experimental values for the liquid–gas critical point are a critical temperature $T_c = 1751$ K at a critical density $\rho_c = 5.8$ g cm^{-3} and a critical pressure $p_c = 1673$ bar [85,86]. Since these values are experimentally accessible, mercury presents an ideal system to investigate liquid properties close to the critical region. Therefore, over the last decades, extensive research on the experimental as well as on the theoretical side of the liquid–vapor coexistence curve up to the critical region has been taken place. Most of the studies focus on the interesting region of the metal to nonmetal (M–NM) transition which occurs when fluid mercury is expanded close to the critical values. At ambient conditions, the density of liquid Hg is 13.6 g cm^{-3}; the M–NM transition takes place at densities around 9 g cm^{-3} (see, e.g., the review on experimental evidence in Ref. [87] and theoretical calculations in Refs. [32,88]). In recent years, development in experimental techniques, especially the use of x-ray scattering methods based on third-generation synchrotron sources, as well as on the theoretical side (particularly in computational simulations) allowed for an improved understanding of the underlying mechanism of this M–NM transition (see, e.g., Refs. [89–97]) as well as for a better understanding of liquid metal surfaces where the existence of surface-induced atomic layering in liquid mercury was established [98,99]. New developments include the use of liquid Hg as a novel substrate for the deposition of Langmuir monolayers of organic substances [100–103] and the investigation of geometrically confined liquid Hg in nanopores [104–108] or carbon nanotubes [109,110]. Of interest are also the study of the liquid mercury–water interface [111–113] due to its importance in electrochemical cells and the adsorption of Hg on metal surfaces [114] in connection with the search for Hg oxidation catalysts.

Despite the fundamental, unsolved question, why mercury is liquid at ambient conditions, to our knowledge, there has only been one molecular dynamics study by Sumi et al. [115] trying to determine the melting temperature. They found a melting point of 232 K in accidentally good agreement with experiment, regarding the facts that their study relied on an *ab initio* potential curve of Hg_2 which considerably exceeds the bond length of the Hg dimer by 0.6 Å and that many-body effects were totally neglected. In later work, they used a scaled version of their potential in order to match the bond length better allowing them to reproduce experimental data at the M–NM transition region, but they did not calculate the melting point anew [116]. A promising ansatz to get information about the melting temperature is the cluster approach: Here, the melting temperature of nanoclusters, with several complete shells of atoms around a central atom therefore showing an enhanced stability, are extracted out of MC simulations and are extrapolated to the bulk value. This has recently been shown to be successful for neon and argon [117] where bulk melting points could be determined with high accuracy in a pure *ab initio* treatment. Of course, for Hg, great care of the treatment of the interparticle potential has to be taken, since the many-body expansion is known to fail (see below).

All computer simulations like molecular dynamics or Monte Carlo methods for liquids need, of course, information about the electron distribution of the system. The electronic distribution can either be modeled by a form of density functional theory (for liquid Hg, see, e.g., Refs. [32,118]) or by modeling or approximating directly the potential energy [119]. In most computational simulations of liquid Hg, the latter ansatz is chosen and the interparticle potential V_{int} is divided into pairwise $V^{(2)}$ and higher terms:

$$V_{int} = \sum_n V^{(n)} = \sum_{i<j} V^{(2)}(r_{ij}) + \sum_{i<j<k} V^{(3)}(r_{ij}, r_{ik}, r_{jk}) + \cdots \quad (3.1)$$

The sums run over all particles in the simulation cell or up to a chosen cutoff distance and r_{ij} is the interparticle distance. Unfortunately, already the three-body terms are becoming quickly too expensive and, for Hg, the expansion is known to be not converging [35,73]. As a result, recent molecular dynamics simulations on liquid metallic mercury using a Lennard-Jones potential cannot reproduce the observed structure factor [120].

Therefore, at the present stage, theoretical studies rely on the construction of effective two-body potentials [121–126] or, most recently, effective three-body potentials [96,97,127–129] based on the accurate *ab initio* Hg_2 potential curve by Schwerdtfeger et al. [18] (see also a recent parameterization of Tang and Toennies [130]). The effective potentials include the many-body effects in an approximate way and are dependent on the temperature. Effective pair potentials have been constructed in a number of different ways and applied with good success to a variety of the structure

and properties of liquid mercury in wide ranges of temperatures and pressures. These works used the link between pair potentials and the pair distribution function, which itself is related by Fourier transform to the experimentally determined static structure factor. Either one starts with guesses for the potential that is refined using experimental data [124,125] or a guessed initial configuration is refined using the experimental structure data like in the reverse Monte Carlo method [121,122,126,131]. The effective two-body potentials are characterized by a steeply rising repulsive branch at short interatomic distances and a relatively weak oscillating branch at larger distances. Effective three-body potentials were constructed by introducing a semiempirical $C_9(T)/r^9$ term fitted to experimental data [127,128] or through quantum-chemical calculations of cohesive energies for selected geometries of clusters and bulk crystals yielding effective many-body potentials that depend on the coordination number and the nearest-neighbor distance [96,97].

In Figure 3.2, the gas–liquid coexistence curve close to the critical region is depicted, showing the onset of the M–NM transition in the liquid region at densities around 9 g cm^{-3} corresponding to a temperature of about 1670 K at the gas–liquid coexistence line. First indications for the M–NM transition were found by Franck and Hensel [132]. Different mechanisms for the transition have been proposed, which can be roughly divided into a homogeneous expansion mechanism (dating back to the pseudogap model proposed by Mott [133–135]), where the nearest-neighbor distance is gradually increasing while the coordination number is constant, and a heterogeneous expansion mechanism [88,136] making a decrease in the average coordination of the Hg atoms in the liquid with an (almost) constant next-nearest-neighbor distance responsible for the transition. After extensive experimental and theoretical investigations, the heterogeneous mechanism first proposed by Mattheis and Warren [136] and in the Franz model [88] is now the accepted

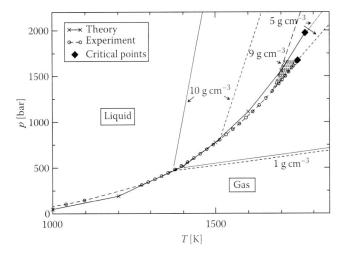

FIGURE 3.2 Gas–liquid coexistence curve and selected isochores at densities of 10, 9, 5, and 1 g cm^{-3}. The solid lines are theoretical data from Ref. [98]; the dashed lines refer to experimental data (Refs. [87,138]). The filled diamonds show the experimental and theoretical critical points. The shaded area presents the nonmetallic liquid region.

one and was refined in the last years due to experimental (see work of Hensel [86,87,137] and references therein as well as newer experimental work in Refs. [91,92,138]) and theoretical [32,96,97,121–123,126,129] progress. Exact measurements of the static structure factor and calculations of its Fourier transform, the pair distribution function $g(r)$ (which gives the relative probability of an atom having a neighbor at distance r) proved a clear decrease in the average coordination number while the first peak of $g(r)$ and thus the nearest-neighbor distance remains almost constant. In more detail: Below the M–NM transition, the local structure of liquid Hg is assumed to be similar to the crystal structure of α-Hg (going back to the early experimental work by Kaplow et al. [139]): There are six hexagonally arranged atoms at 3.5 Å (close to the van der Waals dimer distance) around a central atom and six atoms at 3.0 Å (corresponding to the metallic distance in the solid) in an upper as well as lower plane three by three (see Figure 3.3). Around the M-NM transition, atoms at the shorter distances were shown to be selectively taken away causing the transition to the nonmetallic phase [91,96,97,126,129].

New experiments of the dynamic structure factor that allow the study of dynamical properties of expanded fluid Hg by using inelastic x-ray scattering have shed new light on the M–NM transition [93,94], and suggest that the transition is not a gradual one but a first-order transition at which a new style of fluctuations between metallic and nonmetallic domains appears [94]. Thermodynamically, and thus in all experiments measuring static properties, the discontinuous density jumps in the first-order transition are obscured by strong thermal agitation and structural disorder and feign a gradual transition. Earlier indications for a first-order transition was given in a theoretical investigation of the volume dependence of the free energy by Kitamura [95], who suggested an irregular mixing of high-density metallic domains and low-density nonmetallic domains in the M–NM transition range. Local transformation between the metallic and nonmetallic domains on a slow timescale is caused by thermal fluctuations, supporting experimental evidence of a slow structural relaxation process [89] and an anomalous sound adsorption [90] at the M–NM transition.

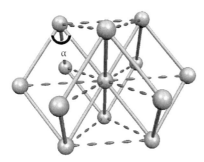

FIGURE 3.3 Rhombohedral lattice of Hg showing the 12 next-nearest neighbors: 6 mercury atoms are arranged in a hexagon around the central atom with a distance of 3.5 Å while the other 6 atoms are located closer to the central atom with a distance of 3.0 Å above and below the plane three by three.

High-resolution inelastic neutron scattering measurements of the dynamic structure factor were also performed at room temperature in order to investigate the microscopic dynamics of liquid mercury [140–142]. It was shown that the so-called cage diffusion plays a dominant role in the collective dynamics of liquid mercury: While moving diffusively, particles find themselves locked up in a cave formed by their nearest neighbors [142]. This behavior was confirmed by recent molecular dynamics simulation by González et al. [143].

Experimental progress also allowed to study the crystal structure and properties of liquid Hg under geometrical confinement by embedding Hg into nanoporous matrices like porous glass, synthetic opals, or zeolites [104–108]. Even a 1D confinement can be realized by filling carbon nanotubes with Hg, which is achieved as a result of electrowetting [109]. Large interest in these nanostructured composites stems from the fact that fundamental questions of condensed matter physics related to finite-size effects can be addressed as well as applications in nanoengineering and nano-fluidics are to be expected. Experiments on Hg in porous glasses and other nanoporous materials have shown that the crystal structure of bulk Hg (α-Hg) is identical to that of confined Hg [104], whereas the melting/freezing temperatures are shifted toward lower temperatures and the transitions are broadened in comparison to bulk Hg [105,106]. These trends are enhanced with decreasing pore size [107]. These effects caused by the finite size are in analogy to the observed behavior in clusters where one also finds decreasing melting temperatures and line broadening with decreasing cluster size (see, e.g., Ref. [117]). Also the Knight shift and thus the electronic susceptibility was shown to decrease with decreasing pore size [108]. For the carbon nanotubes similar effects are predicted by a molecular dynamics simulations [110]. In this study, an ordering of the Hg atoms near the walls of the nanotubes and density oscillations, which extend several atomic diameters into the bulk, are suggested, even resulting in a close-packed cylindrical shell structure for the smallest nanotubes.

Metallic liquids exhibit a complex surface structure in which the atoms are stratified parallel to the liquid–vapor interface persisting into the bulk for a few atomic diameters and leading to density oscillations along the surface normal. This phenomenon of surface layering has been predicted theoretically [144], but could not be proven unambiguously in experiments [145] until experimental improvement due to access to third-generation x-ray sources finally allowed to resolve these oscillations, first in Hg [98,146] and Ga [147]. In Hg, the layering was found to have a spacing of 2.72 Å and a decay length of less than 5 Å. Surface layering seems to be a more universal property of liquid surfaces in general [148–152] but the reasons behind are not yet clearly identified; possibilities include surface tension, geometrical confinement effects due to the necessarily rapid decay of the density at the liquid–vapor interface [152] and a low ratio of melting and critical temperature T_m/T_c [148–151].

Another interface of great interest due to its importance in electrochemical cells is the one between liquid Hg and water. This interface was studied by molecular dynamics simulations with pure water [111] as well as in the presence of alkali cations (see Refs. [112,113] and references therein). In analogy with the just discussed surface layering at the liquid mercury–vapor interface, one finds comparable far-ranging density oscillations in the mercury phase as well as a change in the water density profile up to distances of about 10 Å. The mobility of the adsorbed water is equally restricted parallel and perpendicular to the interface. The bulk densities are, however, almost identical to the pure liquids [111]. When alkali cations are present, it was shown, that the ions are adsorbed within the first layer of mercury losing part of their hydration shell and becoming less mobile. Otherwise, the structure of the hydration shell as well as of the mercury surface are not much changed in its main features [112,113].

Yet another example for the importance of liquid mercury interfaces is the use of liquid Hg as a novel substrate for the deposition of Langmuir monolayers of organic molecules like alkylthiols, fatty acids, and alkanes [100,103]. Langmuir monolayers are studied as models of 2D matter and as a route to nanoengineering and molecular electronics [153]. In contrast to the use of solid metals or of water as substrates, liquid mercury shows no steps or structural defects at the surface and the surface is atomically smooth due to the high surface tension and lack a long-range order of its own, but strong chemical bonds to the functional groups exist. Thus, new structures for the organic monolayers on liquid Hg have been found in synchrotron x-ray scattering experiments, especially until then unknown layers of surface-parallel molecules [101–103]. When the coverage is increased, one finds phase transitions to ordered phases of molecules with surface-normal orientation [101,103].

As already mentioned in the introduction, huge amounts of Hg are released into the atmosphere by combustion of fossil fuels. The removal of Hg in power plants is therefore an important issue. Since oxidized Hg can be removed quite efficiently, the search for oxidation catalysts is of great interest with the noble metals being regarded as hopeful candidates. Thus, a fundamental understanding of the adsorption of Hg to these metal surfaces is important. It was shown in experimental studies [154,155] as well as in theoretical studies [114,156] that Hg atoms adsorb quite strongly to the metal surfaces. In a recent DFT study [114], Steckel calculated binding energies for adsorption of mercury at Ag, Au, Cu, Ni, Pt, and Pd surfaces (on the 001 and the 111 face) for different amounts of coverage finding binding energies of up to about 1 eV per atom for Pt and Pd. The calculated values should provide a lower estimate to the actual values due to the used (GGA) density functionals, which is also confirmed by comparison with available experimental data (see Refs. [154,155] and references therein). The adsorbed Hg atoms are found in the fourfold or threefold follow positions of the 001 or 111 faces, respectively. This is in contrast to recent findings of Sarpe-Tudoran et al. [156] who found the bridge position as the most stable one.

3.5 Solid Mercury

Mercury freezes at T_m = −38.83°C adopting a rhombohedral crystal structure with lattice constants of a = 3.005 Å and an angle of α = 70.53° [1]. The lattice has three equal crystallographic axes inclined to each other at the angle α (an angle of α = 60° corresponds to a face-centered cubic [fcc] lattice). Thus, Hg deviates from the hexagonal close-packed (hcp) structure found for the lighter group IIB metals zinc and cadmium. The unusual structure is another anomaly of Hg caused by relativistic effects. This was recently shown unambiguously by Gaston et al. [38]. Until recently, not only the lattice constants, but also the cohesive energy E_{coh} and the bulk modulus could be calculated with increasing accuracy of up to about 1.5% of the experimental data [36,38,73,157,158]. On exerting pressure, one finds a multitude of high-pressure phases for Hg. Besides the rhombohedral α-Hg, a tetragonal β-phase, an orthorhombic γ-phase, and finally a hcp lattice structure for δ-Hg have been discovered [159,160].

Already in the 1960s the general features of the density of states (DOS) and the band structure of Hg were known by experiment giving detailed information on the Fermi surface by measuring the de Haas–van Alphen effect [161] as well as by calculations [162,163]: At the Fermi energy E_F, one finds a broad sp band (width about 10 eV), which mixes at its low-energy part with the narrow d band. This mixing is possible due to indirect relativistic effects. The strong relativistic contraction of the s bands [24,136,162,164] causes the interatomic Hg–Hg spacing in the solid [38], which in turn leads to an energy increase of the d bands. Just above E_F, a structure-induced minimum in the DOS separates the sp band from empty states of predominantly p character. The band structure was also calculated by Deng et al., emphasizing the existence of flat and steep bands at the Fermi level, what they used as a possible explanation for the superconductivity of Hg observed below 4 K [165]: The pairing of electrons occurs in the flat bands, whose position relative to E_F is periodically changed by lattice distortions (phonons). When the top of the band lies above E_F, the electron pairs are scattered into the steeper bands, whereas when the band maximum lies beyond E_F, the bands are filled up out of the electron reservoir of the steep bands. At the critical temperature, the pairs become stable and a superconducting state results [165].

The rhombohedral lattice structure of α-Hg in contrast to the hcp structures found for zinc and cadmium was studied intensely over the last decades. It was shown early on that the fcc structure is unstable with respect to a rhombohedral distortion and metastable with respect to a tetragonal distortion of the body-centered cubic (bcc) lattice [163], which was confirmed in later studies by Singh within a DFT framework including the energetically relatively high-lying d-electrons explicitly as valence electrons [24,164] and by Kresse and Hafner, who calculated the total energy of Hg as function of a rhombohedral or tetragonal distortion of an fcc or bcc lattice, respectively [32].

Nevertheless, theoretical predictions of the Hg structure that give accurate values for the lattice parameters or the cohesion energy remain a challenge. Whereas a pure mean-field (Hartree–Fock) treatment yields no binding at all [35,36], density functional approaches fail badly when attempting to optimize the lattice structure of Hg or getting sensible values for the cohesive energy. Depending on the underlying functionals, one gets anything from strong overbinding to severe underbinding

[35,38,73]. An alternative approach was developed in the last years in the form of a wavefunction-based correlation treatment [36,38,157,158] using the incremental scheme dating back to Stoll [166]. In this approach, the cohesive energy is split into the mean-field (Hartree–Fock) part, calculated for the infinite solid system, and the correlation energy E_{corr} computed with the following many-body expansion using the wavefunction-based coupled-cluster approach:

$$E_{corr} = \sum_i \varepsilon_i + \sum_{i<j} \varepsilon_{ij} + \sum_{i<j<k} \varepsilon_{ijk} + \cdots \qquad (3.2)$$

E_{corr} is expanded in one-body increments ε_i, two-body increments ε_{ij}, and so on and the sums extend over groups of localized orbitals. The energy increments are calculated for finite embedded clusters, which mimic the environment found in the solid. Details of this embedding are crucial and have been discussed in detail in Ref. [158]. While Hartree–Fock and the one-body increments of the correlation energy lead to no binding at all, the main contributions to the binding were shown to come from the two-body increments, half of that originating from core–valence correlation of the d-shell. Three-body increments are required and account for about 10% of the correlation energy [36,38]. Very high accuracy could be achieved by that treatment yielding lattice constants of $a = 2.96$ Å and $\alpha = 69.5°$, a cohesive energy of $E_{coh} = -0.649$ eV, and a bulk modulus $B = 0.360$ Mbar [38], compared to the experimental values $a = 3.005$ Å, $\alpha = 70.53°$, $E_{coh} = -0.67$ eV [1], and $B = 0.382$ Mbar [164].

Solid mercury exists in at least four different phases if subjected to pressure. At ambient pressure, Hg is liquid at room temperature, showing a local structure of α-Hg [139] with six hexagonally arranged atoms at 3.5 Å around a central atom and six atoms at 3.0 Å in an upper as well as lower plane three by three (See Figure 3.3). At $T_m = 234$ K, Hg solidifies in this rhombohedral α-Hg structure and may exhibit a tetragonal phase at lower temperatures (below $T = 77$ K), but only if this phase is formed at high pressures. This transition to the β-phase was already discovered very early by Bridgman [167] and a high potential barrier for the transformation from α to β-Hg was made responsible for the fact that α-Hg remains metastable at temperatures below 77 K [168]. Interestingly, also the β-phase becomes a superconductor at slightly lower temperatures, at $T_c = 3.94$ K compared to 4.15 K for α-Hg [169]. At room temperature, mercury becomes solid at a pressure of $p = 12$ kbar crystallizing in the α-form. Further increase of the pressure leads to the body-centered tetragonal β-phase at $p = 37$ kbar, the orthorhombic γ-phase at $p = 120$ kbar [170], and finally, at 370 kbar, Hg transforms into a hcp structure named δ-Hg [159,160]. In contrast to this observed polymorphism, Zn and Cd stay in the hcp structure over the whole pressure range [159].

3.6 Conclusions

The electronic configuration of Hg is [Xe] $4f^{14}5d^{10}6s^2$ with closed $5d$ and $6s$ shells. Due to large relativistic $6s$ contraction, the energy of the $6s$ orbitals/bands is lowered significantly leading to large changes in almost all properties observed, from the Hg atom over the clusters to the liquid or solid bulk phase. This sets Hg apart from its lighter homologues zinc and cadmium, with mercury behaving sometimes more like a rare gas. This causes anomalies in almost all atomic properties, like the small electric dipole polarizability or the high ionization potential, in the weak van der Waals binding of the Hg dimer, over to the complicated binding behavior in the clusters, to finally the rhombohedral crystal structure of the solid and low melting and gas–liquid critical temperatures. The unusual high superconducting transition temperature, which led Kamerlingh Onnes to the discovery of this phenomenon, is, most likely, also a manifestation of relativistic effects. The theoretical description of the interaction between Hg atoms is complicated by the necessity of a proper relativistic treatment, and by the need to explicitly consider the $5d$ electrons as valence electrons, as the corresponding orbitals/bands are energetically elevated by indirect relativistic effects mixing into the $6s$ orbitals/bands. Furthermore, the many-body decomposition of the interaction potential between Hg atoms is not converging smoothly, i.e., a simple treatment in terms of two-body and perhaps three-body potential terms is not adequate in contrast to the rare gas elements. Nevertheless, progress was made in the last years, giving detailed explanations for the M–NM transition in liquid Hg, and allowing the calculation of lattice parameters and the cohesion energy of solid Hg with high accuracy. Still, a number of fundamental problems remain unsolved, the most prominent are certainly the questions, why Hg is liquid at room temperature, or what causes the high superconducting transition temperature of 4 K.

Beside these fundamental questions, experimental as well as theoretical progress has opened up new exciting fields of research with possible application in different areas of nanoscience. Examples are the use of liquid mercury as a novel substrate for Langmuir monolayers of organic molecules and the investigations of liquid/solid Hg embedded in porous nano-matrices or carbon nanotubes. These fields are still at their initial stages and need a detailed understanding of the fundamental properties of Hg from the dimer over the clusters to the bulk.

Acknowledgments

The authors are grateful to H. Kitamura for providing the data for the figure about gas–liquid coexistence curve, and to N. Gaston for the discussion.

References

1. D. R. Lide (ed.), *CRC Handbook of Chemistry and Physics* (CRC Press, Boca Raton, FL, 2008).
2. L. J. Norrby, Why is mercury liquid? Or, why do relativistic effects not get into chemistry textbooks? *J. Chem. Educ.*, 68, 110–113 (1991).
3. P. Pyykkö, Relativistic quantum chemistry, *Adv. Quantum Chem.*, 11, 353–409 (1978).
4. P. Pyykkö and J. P. Desclaux, Relativity and the periodic system of elements, *Acc. Chem. Res.*, 12, 276–281 (1979).

5. P. Pyykkö, Relativistic effects in structural chemistry, *Chem. Rev.*, 88, 563–594 (1988).

6. D. Goebel and U. Hohm, Dipole polarizability, Cauchy moments, and related properties of Hg, *J. Phys. Chem.*, 100, 7710–7712 (1996).

7. P. Schwerdtfeger, Atomic static dipole polarizabilities, in *Computational Aspects of Electric Polarizability Calculations: Atoms, Molecules and Clusters* (ed. G. Maroulis, IOS Press, Amsterdam, the Netherlands, 2006), pp. 1–32.

8. P. Schwerdtfeger and M. Lein, Theoretical chemistry of gold, in *Gold Chemistry. Current Trends and Future Directions* (ed. F. Mohr, Wiley, Weinheim, Germany, 2009).

9. K. S. Pitzer, Are elements 112, 114 and 118 relatively inert gases? *J. Chem. Phys.*, 63, 1032–1033 (1975).

10. N. Gaston, I. Opahle, H. W. Gäggeler, and P. Schwerdtfeger, Is Eka-mercury (element 112) a group 12 metal? *Angew. Chem. Int. Ed.*, 46, 1663–1666 (2007).

11. R. Eichler, N. V. Aksenov, A. V. Belozerov, G. A. Bozhikov, V. I. Chepigin, S. N. Dmitriev, R. Dressler et al., Thermochemical and physical properties of element 112, *Angew. Chem. Int. Ed.*, 47, 3262–3266 (2008).

12. M. Seth, P. Schwerdtfeger, and M. Dolg, The chemistry of the superheavy elements. I. Pseudopotentials for 111 and 112 and relativistic coupled cluster calculations for (112) H^+, $(112)F_2$ and $(112)F_2$, *J. Chem. Phys.*, 106, 3623–3632 (1997).

13. E. B. Swain, P. M. Jakus, G. Rice, F. Lupi, P. A. Maxson, J. M. Pacyna, A. Penn, S. J. Spiegel, and M. M. Veiga, Socioeconomic consequences of mercury use and pollution, *Ambio*, 36, 45–61 (2007).

14. A. A. Presto and E. J. Granite, Survey of catalysts for oxidation of mercury in flue gas, *Environ. Sci. Technol.*, 40, 5601–5609 (2006).

15. Y. Zhao, M. D. Mann, J. H. Pavlish, B. A. F. Mibeck, G. E. Dunham, and E. S. Olson, Application of gold catalyst for mercury oxidation by chlorine, *Environ. Sci. Technol.*, 40, 1603–1608 (2006).

16. D. P. Ruys, J. F. Andrade, and O. M. Guimarães, Mercury detection in air using a coated piezoelectric sensor, *Anal. Chim. Acta*, 404, 95–100 (2000).

17. H. Haberland, H. Kornmeier, H. Langosch, M. Oschwald, and G. Tanner, Experimental study of the transition from van der Waals, over covalent to metallic bonding in mercury clusters, *J. Chem. Soc. Faraday Trans.*, 13, 2473–2481 (1990).

18. P. Schwerdtfeger, R. Wesendrup, G. E. Moyano, A. J. Sadlej, J. Greif, and F. Hensel, The potential energy curve and dipole polarizability tensor of mercury dimer, *J. Chem. Phys.*, 115, 7401–7412 (2001); *Corrigendum*, 117, 6881, (2002).

19. P. Schwerdtfeger, N. Gaston, R. P. Krawczyk, R. Tonner, and G. E. Moyano, Extension of the Lennard-Jones potential: Theoretical investigations into rare-gas clusters and crystal lattices of He, Ne, Ar, and Kr using many-body interaction expansions, *Phys. Rev. B*, 73, 064112-1–064112-19 (2006).

20. B. von Issendorff and O. Cheshnovsky, Metal to insulator transitions in clusters, *Ann. Rev. Phys. Chem.*, 56, 549–580 (2005).

21. R. L. Johnston, The development of metallic behaviour in clusters, *Phil. Trans. R. Soc. Lond. A*, 356, 211–230 (1998).

22. K. Rademann, B. Kaiser, U. Even, and F. Hensel, Size dependence of the gradual transition to metallic properties in isolated mercury clusters, *Phys. Rev. Lett.*, 59, 2319–2321 (1987).

23. K. Rademann, Photoelectron spectroscopy and UV/Vis-photoabsorption spectroscopy of isolated clusters, *Z. Phys. D*, 19, 161–164 (1991).

24. P. P. Singh, Relativistic effects in mercury: Atom, clusters, and bulk, *Phys. Rev. B*, 49, 4954–4958 (1994).

25. P. P. Singh and K. S. Dy, Evolution of bulk-like properties in mercury microclusters, *Z. Phys. D: Atom. Mol. Clust.*, 17, 309–311 (1990).

26. G. M. Pastor, P. Stampfli, and K. H. Bennemann, On the transition from Van der Waals- to metallic bonding in Hg-clusters as a function of cluster size, *Phys. Scr.*, 38, 623–626 (1988).

27. G. M. Pastor, P. Stampfli, and K. H. Bennemann, Theory for the transition from van der Waals to metallic bonding in small (Mercury) clusters, *Phase Trans.*, 24–26, 371–373 (1990).

28. G. M. Pastor, Electronic properties of divalent-metal clusters, *Z. Phys. D*, 19, 165–167 (1991).

29. R. Busani, M. Folkers, and O. Cheshnovsky, Direct observation of band-gap closure in mercury clusters, *Phys. Rev. Lett.*, 81, 3836–3839 (1998).

30. R. Busani, M. Folkers, and O. Cheshnovsky, Three attempts and one success in addressing the bandgap closure in mercury clusters, *Philos. Mag. B*, 79, 1427–1436 (1999).

31. R. Busani, R. Giniger, T. Hippler, and O. Cheshnovsky, Auger recombination and charge-carrier thermalization in Hg_n^--cluster photoelectron studies, *Phys. Rev. Lett.*, 90, 083401-1–083401-4 (2003).

32. G. Kresse and J. Hafner, Ab initio simulation of the metal/nonmetal transition in expanded fluid mercury, *Phys. Rev. B*, 55, 7539–7548 (1997).

33. A. R. Miedema and J. W. F. Dorleijn, The exceptional properties of mercury and other divalent metals, *Philos. Mag. B*, 43, 251–272 (1981).

34. A. Hermann, R. P. Krawczyk, M. Lein, P. Schwerdtfeger, I. P. Hamilton, and J. J. P. Stewart, Convergence of the many-body expansion of interaction potentials: From van der Waals to covalent and metallic systems, *Phys. Rev. A*, 49, 4954–4958 (1994).

35. G. E. Moyano, R. Wesendrup, T. Söhnel, and P. Schwerdtfeger, Properties of small to medium-sized mercury clusters from a combined *ab initio*, density-functional, and simulated-annealing study, *Phys. Rev. Lett.*, 89, 103401-1–103401-4 (2002).

36. B. Paulus and K. Rosciszewski, Metallic bonding due to electronic correlations: A quantum chemical ab initio calculation of the cohesive energy of mercury, *Chem. Phys. Lett.*, 394, 96–100 (2004).

37. P. Schwerdtfeger, J. Li, and P. Pyykkö, The polarisability of Hg and the ground-state interaction potential of Hg_2, *Theor. Chim. Acta.*, 87, 313–320 (1994).

38. N. Gaston, B. Paulus, K. Rosciszewski, P. Schwerdtfeger, and H. Stoll, Lattice structure of mercury: Influence of electronic correlation, *Phys. Rev. B*, 74, 094102-1–094102-9 (2006).

39. E. Eliav, U. Kaldor, and Y. Ishikawa, Transition energies of mercury and ekamercury (element 112) by the relativistic coupled-cluster method, *Phys. Rev. A*, 52, 2765–2769 (1995).

40. N. S. Mosyagin, E. Eliav, A. V. Titov, and U. Kaldor, Comparison of relativistic effective core potential and all-electron Dirac-Coulomb calculations of mercury transition energies by the relativistic coupled-cluster method, *J. Phys. B*, 33, 667–676 (2000).

41. C. E. Moore, *Atomic Energy Levels, National Bureau of Standards*, (U.S.) Circ. No. 467, U.S. GPO, Washington, DC (1958).

42. L. Labzowsky, I. Goidenko, M. Tokman, and P. Pyykkö, Calculated self-energy contributions for an *ns* valence electron using the multiple-commutator method, *Phys. Rev. A*, 59, 2707–2711 (1999).

43. D. Goebel and U. Hohm, Dispersion of the refractive index of cadmium vapor and the dipole polarizability of the atomic cadmium $^1S_0$ state, *Phys. Rev. A*, 52, 3691–3694 (1995).

44. D. Goebel, U. Hohm, and G. Maroulis, Theoretical and experimental determination of the polarizabilities of the zinc $^1S_0$ state, *Phys. Rev. A*, 54, 1973–1978 (1996).

45. A. B. Callear and R. G. W. Norrish, The metastable triplet state of Mercury, Hg $6(^3P_0)$: A study of its reactions by kinetic spectroscopy, *Proc. R. Soc. Lond. Ser. A Math. Phys. Sci.*, 266, 299–311 (1962).

46. V. Lukeš, M. Ilčin, V. Laurinc, and S. Biskupič, On the structure and physical origin of van der Waals interaction in zinc, cadmium and mercury dimers, *Chem. Phys. Lett.*, 424, 199–203 (2006).

47. K. A. Peterson, Correlation consistent basis sets with relativistic effective core potentials. The transition metal elements Y and Hg, in *Recent Advances in Electron Correlation Methodology*, (eds. A. K. Wilson and K. A. Peterson, ACS, Washington, DC, 2007).

48. L. Bučinský, S. Biskupič, M. Ilčin, V. Lukeš, and V. Laurinc, On relativistic effects in ground state potential curves of Zn_2, Cd_2, and Hg_2 dimers. A CCSD(T) study, *J. Comp. Chem.*, 30, 65–74 (2009).

49. M. Dolg and H.-J. Flad, Size dependent properties of Hg_n clusters, *Mol. Phys.*, 91, 815–825 (1997).

50. K. P. Huber and G. Herzberg, *Molecular Spectra and Molecular Structure IV*, (Van Nostrand Reinhold, New York, 1979).

51. J. Koperski, J. B. Atkinson, and L. Krause, The $G0_u^+(6^3P_1) \leftarrow X0_g^+$ spectrum of Hg_2 excited in a superonic jet, *Chem. Phys. Lett.*, 219, 161–168 (1994).

52. J. Koperski, J. B. Atkinson, and L. Krause, Molecular beam-spectroscopy of the $1_u(6^3P_2)\text{-}X_g^+$ and $1_u(6^3P_1)\text{-}X0_g^+$ transitions in Hg_2, *Can. J. Phys.*, 72, 1070–1077 (1994).

53. J. Koperski, J. B. Atkinson, and L. Krause, The $G0_u^+(6^1P_1)\text{-}X0_g^+$ excitation and fluorescence spectra of Hg_2 excited in a superonic jet, *J. Mol. Spectrosc.*, 184, 300–308 (1997).

54. B. Cabaud, A. Hoareau, and P. Melinon, Time-of-flight spectroscopy of a supersonic beam of mercury. Intensities and appearance potentials of Hg_x aggregates, *J. Phys. D*, 13, 1831–1834 (1980).

55. R. D. van Zee, S. C. Blankespoor, and T. S. Zwier, Direct spectroscopic determination of the Hg_2 bond length and an analysis of the 2540 Å, *J. Chem. Phys.*, 88, 4650–4654 (1988).

56. N. Gaston, P. Schwerdtfeger, and B. van Issendorff, The photoabsorption spectra of cationic mercury clusters, *Phys. Rev. A*, 74, 094102-1–094102-9 (2006).

57. B. van Issendorff, PhD thesis (Albert-Ludwigs-Universität, Freiburg, Germany, 1994).

58. M. Stock, E. W. Smith, R. E. Drullinger, and M. H. Hessel, Relaxation of the first excited 1_u state of Hg_2, *J. Chem. Phys.*, 68, 4167–4175 (1978).

59. C. R. González, S. Fernández-Alberti, J. Echave, J. Helbing, and M. Chergui, Simulations of the absorption band of the D-state of Hg_2 in rare gas matrices, *Chem. Phys. Lett.*, 367, 651–656 (2003).

60. H. Haberland, B. von Issendorff, J. Yufeng, and T. Kolar, Transition to plasmonlike absorption in small Hg clusters, *Phys. Rev. Lett.*, 69, 3212–3215 (1992).

61. K. Balasubramanian, K. K. Das, and D. W. Liao, Spectroscopic constants and potential energy curves for Hg_2, *Chem. Phys. Lett.*, 195, 487–493 (1992).

62. E. Czuchaj, F. Rebentrost, H. Stoll, and H. Preuss, Calculation of ground- and excited-stat potential energy curves for the Hg_2 molecule, in a pseudopotential approach, *Chem. Phys.*, 214, 277–289 (1997).

63. D. Figgen, G. Rauhut, M. Dolg, and H. Stoll, Energy-consistent pseudopotentials for group 11 and 12 atoms: Adjustment to multi-configuration Dirac-Hartree-Fock data, *Chem. Phys.*, 311, 227–244 (2005).

64. H. Kitamura, Theoretical potential energy surfaces for excited mercury trimers, *Chem. Phys.*, 325, 207–219 (2006).

65. A. Kisiel and P. M. Lee, Relativistic effects on energies of optical transitions in $Cd_xHg_{1-x}Te$ alloys, *J. Phys. F: Met. Phys.*, 2, 395–402 (1972).

66. P. Schwerdtfeger, Relativistic pseudopotentials, in *Progress in Theoretical Chemistry and Physics—Theoretical Chemistry and Physics of Heavy and Superheavy Elements*, (eds. U. Kaldor and S. Wilson, Kluwer Academic, Dordrecht, the Netherlands, 2003), pp. 399–438.

67. N. Gaston, P. Schwerdtfeger, T. Saue, and J. Greif, The frequency-dependent dipole polarisability of the mercury dimer from four-component relativistic density functional theory, *J. Chem. Phys.*, 124, 044304-1–044304-7 (2006).

68. H.-J. Flad, F. Schautz, Y. Wang, M. Dolg, and A. Savin, On the bonding of small group 12 clusters, *Eur. Phys. J. D*, 6, 243–254 (1999).

69. Y. Wang, H.-J. Flad, and M. Dolg, Realistic hybrid model for correlation effects in mercury clusters, *Phys. Rev. B*, 61, 2362–2370 (2000).

70. B. Hartke, H.-J. Flad, and M. Dolg, Structures of mercury clusters in a quantum-empirical hybrid model, *Phys. Chem. Chem. Phys.*, 3, 5121–5129 (2001).

71. B. Hartke, Structural transitions in clusters, *Angew. Chem. Int. Ed.*, 41, 1468–1487 (2002).

72. D. J. Wales and J. P. K. Doye, Global optimization by basin-hopping and the lowest energy structures of Lennard-Jones clusters containing up to 11 atoms, *J. Phys. Chem. A*, 101, 5111–5116 (1997).

73. N. Gaston and P. Schwerdtfeger, From the van der Waals dimer to the solid state of mercury with relativistic *ab initio* and density functional theory, *Phys. Rev. B*, 74, 024105-1–024105-12 (2006).

74. H. Kitamura, Cohesive properties of mercury clusters in the ground and excited states, *Eur. Phys. J. D*, 43, 33–36 (2007).

75. H. Ito, T. Sakurai, T. Matsuo, T. Ichihara, and I. Katakuse, Detection of electronic-shell structure in divalent-metal clusters $(Hg)_n$, *Phys. Rev. B*, 48, 4741–4745 (1993).

76. T. Bastug, W.-D. Sepp, D. Kolb, E. J. Baerends, and G. Te Velde, All-electron Dirac-Fock-Slater SCF calculations for electronic and geometric structures of the Hg_2 and Hg_3 molecules, *J. Phys. B*, 28, 2325–2331 (1995).

77. Y. Wang, H.-J. Flad, and M. Dolg, Structural changes induced by an excess electron in small mercury clusters, *Int. J. Mass Spectrom.*, 201, 197–204 (2000).

78. H. Kitamura, Analysis of excited mercury clusters with diatomic potential energy curves, *Chem. Phys. Lett.*, 425, 205–209 (2006).

79. H. Haberland, B. von Issendorff, J. Yufeng, T. Kolar, and G. Thanner, Ground state and response properties of mercury clusters, *Z. Phys. D: Atom. Mol. Clust.*, 26, 8–12 (1993).

80. M. E. Garcia, D. Reichardt, and K. H. Bennemann, Theory for the ultrafast melting and fragmentation dynamics of small clusters after femtosecond ionization, *J. Chem. Phys.*, 109, 1101–1110 (1998).

81. B. Bescós, B. Lang, J. Weiner, V. Weiss, E. Wiedemann, and G. Gerber, Real-time observation of ultrafast ionization and fragmentation of mercury clusters, *Eur. Phys. J. D*, 9, 399–403 (1999).

82. J. Blanc, M. Broyer, Ph. Dugourd, P. Labastie, M. Sence, J. P. Wolf, and L. Wöste, Photoionization spectroscopy of small mercury clusters in the energy range from vacuum ultraviolet to soft x ray, *J. Chem. Phys.*, 102, 680–689 (1995).

83. G. B. Griffin, A. Kammrath, O. T. Ehrler, R. M. Young, O. Cheshnovsky, and D. M. Neumark, Auger recombination dynamics in Hg_{13}^{-} clusters, *Chem. Phys.*, 350, 69–74 (2008).

84. B. Kaiser and K. Rademann, Photoelectron spectroscopy of neutral mercury clusters Hg_x ($x \leq 109$) in a molecular beam, *Phys. Rev. Lett.*, 69, 3204–3207 (1992).

85. W. Götzlaff, U. Schönherr, and F. Hensel, Thermopower behavior and conductivity behavior near the gas–liquid critical-point of mercury, *Z. Phys. Chem. NF*, 156, 219–224 (1988).

86. W. Götzlaff, Zustandsgleichung und elektrischer Transport am kritischen Punkt des fluiden Quecksilbers, PhD thesis (Philipps-Universität Marburg, Marburg, Germany, 1988).

87. F. Hensel, The liquid–vapour phase transition in fluid mercury, *Adv. Phys.*, 44, 3–19 (1995).

88. J. Franz, Metal–nonmetal transition in expanded liquid mercury, *Phys. Rev. Lett.*, 57, 889–892 (1986).

89. H. Kohno and M. Yao, Anomalous sound attenuation in the metal–nonmetal transition of liquid mercury, *J. Phys.: Condens. Matter*, 11, 5399–5413 (1999).

90. H. Kohno and M. Yao, Slow-structural relaxation in the metal–nonmetal transition range of liquid mercury: I. Experimental evidence, *J. Phys.: Condens. Matter*, 13, 10293–10305 (2001).

91. M. Inui, X. Hong, and K. Tamura, Local structure of expanded fluid mercury using synchrotron radiation: From liquid to dense vapor, *Phys. Rev. B*, 68, 094108–1–9 (2003).

92. M. Inui and K. Tamura, Static and dynamic structures of expanded fluid mercury, *Z. Phys. Chem.*, 217, 1045–1063 (2003).

93. D. Ishikawa, M. Inui, K. Matsuda, K. Tamura, S. Tsutsui, and A. Q. R. Baron, Fast sound in expanded fluid Hg accompanying the metal–nonmetal transition, *Phys. Rev. Lett.*, 93, 097801-1–097801-4 (2004).

94. M. Inui, K. Matsuda, D. Ishikawa, K. Tamura, and Y. Ohishi, Medium-range fluctuations accompanying the metal-nonmetal transition in expanded fluid Hg, *Phys. Rev. Lett.*, 98, 185504-1–185504-4 (2007).

95. H. Kitamura, Mesoscopic simulation of slow structural relaxation in fluid mercury near the metal–nonmetal transition, *J. Phys.: Condens. Matter*, 15, 6427–6446 (2003).

96. H. Kitamura, The role of attractive many-body interaction in the gas–liquid transition of mercury, *J. Phys.: Condens. Matter*, 19, 072102-1–072102-7 (2007).

97. H. Kitamura, Equation of state for expanded fluid mercury: Variational theory with many-body interaction, *J. Chem. Phys.*, 126, 134509-1–134509-8 (2007).

98. O. M. Magnussen, B. M. Ocko, M. J. Regan, K. Penanen, P. S. Pershan, and M. Deutsch, X-ray reflectivity measurements of surface layering in liquid mercury, *Phys. Rev. Lett.*, 74, 4444–4447 (1995).

99. J. Penfold, The structure of the surface of pure liquids, *Rep. Prog. Phys.*, 64, 777–814 (2001).

100. O. M. Magnussen, B. M. Ocko, M. Deutsch, M. J. Regan, P. S. Pershan, D. Abernathy, G. Grubel, and J.-F. Legrand, Self-assembly of organic films on a liquid metal, *Nature*, 384, 250–252 (1996).

101. H. Kraak, B. M. Ocko, P. S. Pershan, E. Sloutskin, and M. Deutsch, Structure of a Langmuir film on a liquid metal surface, *Science*, 298, 1404–1407 (2002).

102. H. Kraak, B. M. Ocko, P. S. Pershan, L. Tamam, and M. Deutsch, Temperature dependence of the structure of Langmuir films of normal-alkanes on liquid mercury, *J. Chem. Phys.*, 121, 8003 (2004).

103. B. M. Ocko, H. Kraak, P. S. Pershan, E. Sloutskin, L. Tamam, and M. Deutsch, Crystalline phases of alkyl-thiol monolayers on liquid mercury, *Phys. Rev. Lett.*, 94, 017802-1–017802-4 (2005).

104. Yu. A. Kumzerov, A. A. Nabereznov, S. B. Vakhrushev, and B. N. Savenko, Freezing and melting of mercury in porous glass, *Phys. Rev. B*, 52, 4772–4774 (1995).

105. B. F. Borisov, E. V. Charnaya, P. G. Plotnikov, W.-D. Hoffmann, D. Michel, Yu. A. Kumzerov, C. Tien, and C.-S. Wur, Solidification and melting of mercury in a porous glass as studied by NMR and acoustic techniques, *Phys. Rev. B*, 58, 5329–5335 (1998).

106. E. V. Charnaya, P. G. Plotnikov, D. Michel, C. Tien, B. F. Borisov, I. G. Sorina, and E. I. Martynova, Acoustic studies of melting and freezing for mercury embedded into Vycor glass, *Physica B*, 299, 56–63 (2001).

107. B. F. Borisov, A. V. Gartvik, F. V. Nikulin, and E. V. Charnaya, Acoustic study of melting and freezing of mercury nanoparticles in porous glasses, *Acoust. Phys.*, 52, 138–143 (2006).

108. C. Tien, E. V. Charnaya, M. K. Lee, and Yu. A. Kumzerov, Influence of pore size on the Knight shift in liquid tin and mercury in a confined geometry, *J. Phys.: Condens. Matter*, 19, 106217-1–106217-8 (2007).

109. J. Y. Chen, A. Kutana, C. P. Collier, and K. P. Giapis, Electrowetting in carbon nanotubes, *Science*, 310, 1480–1483 (2005).

110. A. Kutana and K. P. Giapis, Contact angles, ordering, and solidification of liquid mercury in carbon nanotube cavities, *Phys. Rev. B*, 76, 195444-1–195444-5 (2007).

111. J. Böcker, Z. Girskii, and K. Heinziger, Structure and dynamics at the liquid mercury–water interface, *J. Phys. Chem.*, 100, 14969–14977 (1996).

112. G. Tóth, E. Spohr, and K. Heinzinger, SCF calculations of the interactions of alkali and halide ions with the mercury surface, *Chem. Phys.*, 200, 347–355 (1995).

113. G. Tóth, Molecular dynamics study of the alkali cations at the water–liquid mercury interface, *J. Electroanal. Chem.*, 443, 73–80 (1997).

114. J. A. Steckel, Density-functional theory study of mercury adsorption on metal surfaces, *Phys. Rev. B*, 77, 115412-1–115412-13 (2008).

115. T. Sumi, E. Miyoshi, Y. Sakai, and O. Matsuoka, Molecular-orbital and molecular-dynamics study of mercury, *Phys. Rev. B*, 57, 914–918 (1998).

116. T. Sumi, E. Miyoshi, and K. Tanaka, Molecular-dynamics study of liquid mercury in the density region between metal and nonmetal, *Phys. Rev. B*, 59, 6153–6158 (1999).

117. E. Pahl, F. Calvo, L. Koči, and P. Schwerdtfeger, Genaue Schmelztemperaturen für Neon und Argon aus Ab-initio-Monte-Carlo-Simulationen, *Angew. Chem.*, 120, 8329–8333 (2008); Accurate melting temperatures for neon and argon from Ab initio Monte Carlo simulations, *Angew. Chem. Int. Ed.*, 47, 8207–8210 (2008).

118. W. Jank and J. Hafner, Structural and electronic properties of the liquid polyvalent elements. II. The divalent elements, *Phys. Rev. B*, 42, 6926–6938 (1990).

119. M. P. Allen and D. J. Tildesley, *Computer Simulation of Liquids* (Clarendon Press, Oxford, U.K., 1987).

120. K. Hoshino, S. Tanaka, and F. Shimojo, Dynamical structure of fluid mercury: Molecular-dynamics simulations, *J. Non-Cryst. Solids*, 353, 3389–3393 (2007).

121. V. M. Nield and P. T. Verronen, The structure of expanded mercury, *J. Phys.: Condens. Matter*, 10, 8147–8153 (1998).

122. T. Arai and R. L. McGreevy, RMC modelling of the structure of expanded liquid mercury along the co-existence curve, *J. Phys.: Condens. Matter*, 10, 9221–9230 (1998).

123. S. K. Bose, Electronic structure of liquid mercury, *J. Phys.: Condens. Matter*, 11, 4597–4615 (1999).

124. G. Tóth, An iterative scheme to derive pair potentials from structure factors and its application to liquid mercury, *J. Chem. Phys.*, 118, 3949–3955 (2003).

125. D. K. Belashchenko, The simulation of liquid mercury by diffraction data and the inference of interparticle potential, *High Temp.*, 40, 212–221 (2002).

126. X. Hong, Structural variation of expanded fluid mercury during M–NM transition: A reverse Monte Carlo study, *J. Non-Cryst. Solids*, 353, 3399–3404 (2007).

127. G. Raabe and R. J. Sadus, Molecular simulation of the vapor–liquid coexistence of mercury, *J. Chem. Phys.*, 119, 6691–6697 (2003).

128. G. Raabe, B. D. Todd, and R. J. Sadus, Molecular simulation of the shear viscosity and the self-diffusion coefficient of mercury along the vapor–liquid coexistence curve, *J. Chem. Phys.*, 123, 034511-1–034511-6 (2005).

129. H. Kitamura, Predicting the gas–liquid transition of mercury from interatomic many-body interaction, *J. Phys.: Conf. Ser.*, 98, 052010-1–052010-4 (2008).

130. K. T. Tang and J. P. Toennies, The dynamical polarisability and van der Waals dimer potential of mercury, *Mol. Phys.*, 106, 1645–1653 (2008).

131. A. Yamane, F. Shimojo, and K. Hoshino, Effects of long-range interactions on the structure of supercritical fluid mercury: Large-scale molecular-dynamics simulations, *J. Phys. Soc. Jpn.*, 75, 124602-1–124602-8 (2006).

132. E. U. Franck and F. Hensel, Metallic conductance of supercritical mercury gas at high pressures, *Phys. Rev.*, 147, 109–110 (1966).

133. N. F. Mott, Electrical properties of liquid metals, *Phil. Mag.*, 13, 989 (1966).

134. N. F. Mott, Evidence for a pseudogap in liquid mercury, *Philos. Mag.*, 26, 505 (1972).

135. M. H. Cohen and J. Jortner, Inhomogeneous transport regime in disordered materials, *Phys. Rev. Lett.*, 30, 699–702 (1973).

136. L. F. Mattheis and W. W. Warren Jr., Band model for the electronic structure of expanded liquid mercury, *Phys. Rev. B*, 16, 624–638 (1977).

137. F. Hensel and W. W. Warren Jr., *Fluid Metals* (Princeton University Press, Princeton, NJ, 1999).

138. K. Tamura, M. Inui, I. Nakaso, Y. Ohisho, K. Funakoshi, and W. Utsumi, X-ray diffraction studies of expanded fluid mercury using synchrotron radiation, *J. Phys.: Condens. Matter*, 10, 11405–11417 (1998).

139. R. Kaplow, S. L. Strong, and B. L. Averbach, Radial density functions for liquid mercury and lead, *Phys. Rev.*, 138, A1336–A1345 (1965).

140. L. E. Bove, F. Sacchetti, C. Petrillo, B. Dorner, F. Formisano, and F. Barocchi, Neutron investigation of the ion dynamics in liquid mercury: Evidence for collective excitations, *Phys. Rev. Lett.*, 87, 215504-1–215504-4 (2001).

141. L. E. Bove, F. Sacchetti, C. Petrillo, B. Dorner, F. Formisano, M. Sampoli, and F. Barocchi, Dynamic structure factor of liquid mercury, *J. Non-Cryst. Solids*, 307, 842–847 (2002).

142. Y. S. Badyal, U. Bafile, K. Miyazaki, I. M. de Schepper, and W. Montfrooij, Cage diffusion in liquid mercury, *Phys. Rev. E*, 68, 061208-1–061208-7 (2003).

143. L. E. Gonzáles, D. J. Gonzáles, L. Calderín, and S. Şengül, On the behavior of single-particle dynamic properties of liquid Hg and other metals, *J. Chem. Phys.*, 129, 171103-1–171103-4 (2008).

144. M. P. D'Evelyn and S. A. Rice, Structure in the density profile at the liquid–metal–vapor interface, *Phys. Rev. Lett.*, 47, 1844–1847 (1981).

145. B. N. Thomas, S. W. Barton, F. Novak, and S. A. Rice, An experimental study of the in-plane distribution of atoms in the liquid–vapor interface of mercury, *J. Chem. Phys.*, 86, 1036–1047 (1986).

146. E. DiMasi, H. Tostmann, B. M. Ocko, P. S. Pershan, and M. Deutsch, X-ray reflectivity study of temperature-dependent surface layering in liquid Hg, *Phys. Rev. B*, 58, R13419–R13422 (1998).

147. M. J. Regan, E. H. Kawamoto, S. Lee, P. S. Pershan, N. Maskil, M. Deutsch, O. M. Magnussen, B. M. Ocko, and L. E. Berman, Surface layering in liquid gallium: An x-ray reflectivity study, *Phys. Rev. Lett.*, 75, 2498–2501 (1995).

148. E. Chacón, M. Reinaldo-Falagán, E. Velasco, and P. Tarazona, Layering at free liquid surfaces, *Phys. Rev. Lett.*, 87, 166101-1–166101-4 (2001).

149. E. Velasco, P. Tarazona, M. Reinaldo-Falagán, and E. Chacón, Low melting temperature and liquid surface for pair potential models, *J. Chem. Phys.*, 117, 10777–10788 (2002).

150. J.-M. Bomont and J.-L. Bretonnet, An effective pair potential for thermodynamics and structural properties of liquid mercury, *J. Chem. Phys.*, 124, 054504-1–054504-8 (2006).

151. J.-M. Bomont and J.-L. Bretonnet, A molecular dynamics study of density profiles at the free surface of liquid mercury, *J. Phys.: Conf. Ser.*, 98, 042018-1–042018-5 (2008).

152. B. G. Walker, N. Marzari, and C. Molteni, *Ab initio* studies of layering behavior of liquid sodium surfaces and interfaces, *J. Chem. Phys.*, 124, 174702-1–174702-10 (2006).

153. V. M. Kaganer, H. Möhwald, and P. Dutta, Structure and phase transitions in Langmuir monolayers, *Rev. Mod. Phys.*, 71, 779–819 (1999).

154. S. Soverna, R. Dressler, C. E. Dullmann, B. Eichler, R. Eichler, H. W. Gäggeler, F. Haenssler et al., Thermochromatographic studies of mercury and radon on transition metal surfaces, *Radiochim. Acta*, 93, 1–8 (2005).

155. P. R. Poulsen, I. Stensgaard, and F. Besenbacher, The adsorption position of Hg on Ni(100)—A transmission channeling study, *Surf. Sci. Lett.*, 310, L589–L594 (1994).

156. C. Sarpe-Tudoran, B. Fricke, and J. Anton, Adsorption of superheavy elements on metal surfaces, *J. Chem. Phys.*, 126, 174702-1–174702-5 (2007).

157. B. Paulus, K. Rosciszewski, N. Gaston, P. Schwerdtfeger, and H. Stoll, The convergence of the *ab-initio* many-body expansion for the cohesive energy of solid mercury, *Phys. Rev. B*, 70, 165106-1–165106-9 (2004).

158. E. Voloshina, N. Gaston, and B. Paulus, Embedding procedure for ab initio correlation calculations in group II metals, *J. Chem. Phys.*, 126, 134115-1–134115-10 (2007).

159. O. Schulte and W. B. Holzapfel, Effect of pressure on the atomic volume of Zn, Cd, and Hg up to 75 GPa, *Phys. Rev. B*, 53, 569–580 (1996).

160. F. Jona and P. M. Marcus, First-principles study of the high-pressure hexagonal-closed-packed phase of mercury, *J. Phys.: Condens. Matter*, 19, 036103-1–036103-5 (2007).

161. G. B. Brandt and J. A. Rayne, de Haas-van Alphen effect in mercury, *Phys. Rev.*, 148, 644–656 (1965).

162. S. C. Keeton and T. L. Loucks, Electronic structure of mercury, *Phys. Rev.*, 152, 548–555 (1966).

163. D. Weaire, The structure of the divalent simple metals, *J. Phys. C (Proc. Phys. Soc.)*, 1, 210–221 (1968).

164. P. P. Singh, From hexagonal close packed to rhombohedral structure: Relativistic effects in Zn, Cd, and Hg, *Phys. Rev. Lett.*, 72, 2446–2449 (1994).

165. S. Deng, A. Simon, and J. Köhler, Superconductivity and chemical bonding in mercury, *Angew. Chem. Int. Ed.*, 37, 640–643 (1998).

166. H. Stoll, Correlation energy of diamond, *Phys. Rev. B*, 46, 6700–6704 (1992).

167. P. W. Bridgman, Polymorphism, principally of the elements, up to 50,000 kg/cm², *Phys. Rev.*, 48, 893–906 (1935).

168. C. A. Swenson, Phase transition in solid mercury, *Phys. Rev.*, 111, 82–91 (1958).

169. J. E. Schirber and C. A. Swenson, Superconductivity of *beta* mercury, *Phys. Rev. Lett.*, 2, 296–297 (1959).

170. K. Takemura, H. Fujihisa, Y. Nakamoto, S. Nakano, and Y. Ohishi, Crystal structure of the high-pressure gamma phase of mercury: A novel monoclinic distortion of the close-packed structure, *J. Phys. Soc. Jpn.*, 76, 023601 (2007).

4

Bimetallic Clusters

René Fournier
York University

4.1 Introduction

In scientific literature, various terms are used to denote small metal particles. There are no universally accepted definitions, but the various terms usually imply particles with different numbers of atoms N: *clusters* ($N \leq 100$), *nanoclusters or nanoparticles* (diameter in the 1–100 nm range, $\log_{10} N \approx 2$–7), *small particles* ($\log_{10} N \approx 7$–11). This chapter concerns primarily clusters ($N \leq 100$), that is, particles at the lower size limit of nanophysics. A further classification can be made according to the nature of the surface of clusters: *free*, *supported*, or *passivated* clusters. A *free* bimetallic cluster $A_m B_n$ contains nothing else than atoms of the two metals. These clusters are inherently unstable because of their high surface energy, which makes them coalesce into larger particles or react with other species. *Supported* clusters are strongly bound to a solid normally chosen to be a high surface area material like alumina (Al_2O_3). In this kind of cluster, some of the metal atoms are bound to the support, usually through strong covalent bonds to oxygen atoms, and the other metal atoms are at the surface. These clusters are marginally stable: The surface metal atoms have a strong tendency to oxidize but can be kept in the metallic (zero oxidation) state by maintaining a high H_2 pressure or high temperature in a flow reactor. Supported metal clusters, whether elemental or bimetallic, are mainly used as heterogeneous catalysts in large-scale industrial reactors, for cracking and ammonia synthesis for instance. *Passivated* clusters have a central metallic core, which is often icosahedral or crystalline and roughly spherical in shape and entirely covered by ligands. A common small ligand for passivating metal clusters is carbon monoxide; $Ni_{24}Pt_{14}(CO)_{44}^{4-}$ is one of many examples of CO-passivated

bimetallic clusters [1]. Bigger ligands are more often used. Thiols, $CH_3–(CH_2)_n–SH$, make good passivating ligands for elemental or bimetallic clusters that have a group 11 metal (Cu, Ag, Au) at the surface because of their high binding affinity to these metals. The "metal end" of the ligand (the S atom for a thiol) strongly binds to surface metal atoms. The "solvent end" is normally designed to assure solubility, e.g., a simple hydrocarbon chain for solubility in nonpolar solvents. But all sorts of designs are possible at the solvent end of a ligand: It can be tailored for purposes like self-assembly, covalent binding to surfaces, molecular recognition of DNA or other biomolecules, or for certain chemical reactions by inclusion of functional groups. As a group, passivated clusters are stable and highly versatile—they make logical building blocks for cluster-assembled materials [2].

Practical applications of clusters depend on their stability, and this *increases* on going from *free* to *supported* and *passivated* clusters. On the other hand, the number of low-lying excited electronic states, which are responsible for the properties unique to metals (conductivity, magnetism, versatile chemistry and catalytic activity, etc.) *decreases* in the order *free, supported, passivated*. Therefore, applications that depend specifically on *metallic properties* of *small* clusters (e.g., bimetallic catalysts) present special challenges. In molecular orbital (MO) theory, metallic properties are associated with a zero (or very small) highest occupied MO–lowest unoccupied MO (HOMO–LUMO) gap in orbital energies. In sufficiently large passivated metal clusters, interior atoms do not "see" the passivating ligands, they have bulk-like electronic states with a nearly zero HOMO–LUMO gap, and have metallic properties. But the atoms at the metal–ligand interface have a local density of electronic states that resembles a semiconductor. In small passivated metal clusters, the HOMO–LUMO gap is large not only for surface atoms, but also interior atoms [3,4]. In that sense, small passivated metal clusters are not metallic.

This chapter gives a brief description of the three kinds of bimetallic clusters and the way they are made (Section 4.2), and their current and potential applications (Section 4.3). Here, the emphasis is on *free* clusters (Section 4.4, and for bimetallics, Sections 4.5 through 4.8) because this is where the metallic character is most pronounced, and because the other types of clusters can be made into cluster-assembled materials, which are discussed in other chapters.

4.2 Types of Bimetallic Clusters and How They Are Made

Free clusters contain no atom other than the two metals. They can be made in the gas phase by laser ablation of the metals (an alloy rod, or two rods, or a powder mixture), or by evaporating the two metals with an intense electrical discharge or otherwise. The vaporized atoms condense into growing clusters as they are carried away from the source by an inert gas. Supersonic expansion rapidly cools the clusters and produces a narrow distribution of cluster velocities. This effectively stops the growth and gives a cluster beam [5]. Free metal clusters are inherently unstable vis-à-vis coalescence and various chemical reactions: They survive only in the low-pressure environments of molecular beams, or get deposited on surfaces [5–7], or in a matrix [8,9], or in a cryogenic matrix by co-condensation with an inert gas (e.g., Ar) [10]. In ultra-high-vacuum (UHV) experiments, clusters are often formed with kinetic energies on the order of a few eV/cluster. But mass selection imparts much larger kinetic energies as the cluster ions get accelerated in an electrical field. In order to avoid fragmentation upon cluster deposition on a substrate, two complementary techniques can be used. First, one can apply a retardation potential in the region between mass selection and substrate and bring kinetic energies down to roughly 1 eV/atom. Second, one can condense multiple layers of a rare gas on the substrate prior to deposition. Then, upon deposition, the clusters' kinetic energy is gradually dissipated by heat transfer to desorbing rare gas atoms and to the substrate. This so-called soft-landing method can yield cluster fragmentation rates lower than 15%–20% [11].

Supported clusters are formed or deposited on a surface. One method is to take metal complexes with small and relatively weakly bound ligands, bind them to silica, and heat for a few hours at ~200°C in vacuum. One can get very small (~10–20 atoms) bare metal clusters bound to the silica [12]. Electrochemical synthesis has also been used [13]. The PdPt clusters of Ref. [13] are both supported and passivated, so these two categories are not mutually exclusive. *Supported bimetallic catalysts* often combine an oxophilic transition metal and a noble metal in a particle that is bound to a high surface area material such as alumina, silica, zeolite, or a carbon powder, in

order to achieve high *dispersion* (see Section 4.4.1) and maximize activity. The oxophilic metal binds strongly to the oxygen atoms of the support, which helps achieve both high dispersion and stability. If A is oxophilic and the free A_mB_n cluster is mixed (alloyed), interaction with the support will promote segregation of A toward the oxygen atoms of the support. The support itself often plays a nontrivial role for the chemistry. The exposed surface of the cluster is normally kept in the metallic state by maintaining a H_2 atmosphere and high temperature, which help reduce any metal oxide that may form. Keeping a cluster catalyst active (keeping the metallic surface intact) is always a concern. The *turnover number* is defined as the mean number of reaction events catalyzed per unit surface area of catalyst before it gets chemically modified and becomes inactive. Ideally, a catalyst would have an infinite turnover number, but in reality, catalysts become deactivated after some time. Some catalysts are simply discarded and replaced after they are deactivated. Others can be regenerated by chemical means: sometimes deactivation is so rapid that a reactor must alternate between two modes of operation—running the catalytic reaction and regenerating [14]. The methods of preparations of supported bimetallic clusters and their applications in catalysis have been described by Alexeev and Gates [15].

Passivated bimetallic clusters are normally made in solution, often at high temperature, by reduction or decomposition of organometallic precursors in the presence of surfactants. The size and composition of bimetallic clusters made this way can be controlled by adjusting the concentrations and molar ratios of the organometallic precursors. For example, passivated FePt colloids with tunable size in the 3–10 nm range, and with Fe molar fractions of 0.48, 0.52, and 0.72, have been made by simultaneous reduction of $Pt(acac)_2$ (acac = acetylacetonate) by a diol and thermal decomposition of $Fe(CO)_5$ in the presence of oleic acid and oleyl amine [16]. Co-reduction of a mixture of metal salts generally gives core-shell clusters $A_m@B_n$ where A, the metal with highest redox potential, makes a central core surrounded by a shell of B atoms. But some choices of surfactants can produce the reverse core-shell ($B_n@A_m$) structure [17]. Bimetallic clusters have been made by irradiating aqueous solutions of metal ions with γ rays (radiolysis). The γ rays ionize water molecules producing solvated electrons, which go on to reduce metal ions to neutral metal atoms. Atoms then coalesce and get capped by a polymer like polyvinyl alcohol. Doudna et al. used this method to make AgPt nanoparticles [18]. They got very high aspect ratio wirelike structures with lengths up to 3500 nm and diameters between 3 and 20 nm, but were also able to get spherical clusters with some choices of counterions, mole ratio of the two metals, and capping polymer. Intense ultrasounds also produce solvated electrons that can reduce metal ions in aqueous solutions. This is the basis for *sonochemical synthesis*, which has been used, for instance, in the preparation of core-shell Au@Pd clusters with an average diameter of 8 nm. Radiolysis and sonochemical synthesis generally produce clusters with different size and morphology [19]. In smaller clusters, the passivating layer, which imparts stability, opens up a gap in the density of electronic states [3,4] and this affects metallic properties. The control afforded by the choice of ligands and possible subsequent chemistry with these ligands is a big advantage for this type of clusters. Passivated clusters can sometimes be assembled into 3-dimensional materials (crystals of nanocrystals) [2,20]. This can be done, for example, by the "tri-layer technique of controlled oversaturation" [21]. In this technique, three immiscible liquids are put on top of each other, in decreasing order of nanoclusters solubility: high solubility (a solution of nanoclusters) at the bottom, slight solubility (buffer layer), and no solubility (nonsolvent) on top. As the nonsolvent and nanoclusters slowly diffuse into the buffer layer, single crystals of nanoclusters nucleate and grow.

4.3 Applications

Bimetallic clusters have current or potential applications in catalysis, biodiagnostics, electronics, and ultra-high-density magnetic recording media [22]. The properties of bulk alloys can be tuned by changing the composition. In bimetallic clusters, more variables can be controlled. As mentioned earlier, it is sometimes possible to control the size, shape, or morphology of the clusters, in addition to their overall composition. Sometimes, one can control the surface composition by making a reverse core-shell cluster $B_n@A_m$ and doing a partial transformation to the thermodynamically favored form ($A_m@B_n$, or mixed alloy) by a short, controlled annealing. For example, $Co_m@Pt_n$ core-shell clusters ($m/(m + n) = 0.45$, size 6.4 nm) have been synthesized by redox metallation [23]. Upon annealing for 12 h at 600°–700°, the diffusion of Co and Pt atoms led to a mixed alloy with very different magnetic properties from the initial Co_nPt_m clusters. Heating by laser irradiation offers better controlled annealing [24]. Pairs of elements that are not miscible in the bulk often mix and form alloy-like structures in very small clusters like RuCu.

Small bimetallic clusters are qualitatively different from mixtures of the clusters of the two pure metals. This opens up possibilities, especially for catalysis. A catalytic reaction normally involves a sequence of events at the molecular scale, or, steps. Different metal catalysts have different effects on the rate of the individual steps. So, one can in principle speed up *two or more* rate-limiting steps and achieve superior catalytic activity by using a catalyst having two (or more) metals in close proximity of each other [25]. Combining two metals can have a more profound effect: the electronic structure of each metal is altered by the presence of the other. For instance, Pt, a good catalyst in itself, can be modified and have its catalytic activity toward some reaction optimized by adding the right amount of a chosen metal partner [26]. In the case of highly structured bimetallic particles (e.g., alloyed or core-shell), completely different electronic or geometric structures could emerge. Supported bimetallic cluster catalysts have been studied by Sinfelt and coworkers since the 1960s. It has been observed that the activity of a catalytically active metal can be drastically reduced by the presence of a group 11 metal for some reactions, but not others, which then makes these bimetallic catalysts highly selective [27]. The MPt (M = Re, Sn, Ir) catalysts are very important and have replaced pure Pt catalysts for large-scale naphtha reforming [27]. Bimetallics like RhPt are used for conversion of automobile exhaust [15]. Very small (~10–20 atoms) supported RuX bimetallic clusters (X = Cu, Pd, Ag, Sn, Pt) exhibit high activity and high selectivity for hydrogenation reactions [12]. Supported bimetallic clusters are comparatively costly, but the possibility of varying composition, size, and support makes them interesting for applications where high selectivity is needed, for example, in the pharmaceutical industry or for fine chemicals. Several PtX clusters (X = Cr, Mn, Fe, Co, Ni, Mo, Ru, Sn) in the 1–10 nm size range supported on high surface area carbon powders have been used as catalysts in fuel cells [28].

Some bimetallic clusters have potential applications associated with their unusual magnetic properties. Vapor-deposited granular films of ferromagnetic–noble metal combinations, like Fe_nAg_m, display giant magnetoresistance effects, i.e., large changes in resistance upon application of a magnetic field [29]. It has been seen, in a variety of systems, that the orbital magnetic moment and magnetic anisotropy energy (MAE) increase for atoms with a low coordination [30], and such atoms make up a significant fraction of nanostructures. On the other hand, the addition of Pt to Co nanoparticles was shown to increase the volume contribution to the MAE [30]. Three-dimensional lattices of FePt clusters (3–10 nm) also display large MAE and other interesting properties [16]. It has been estimated that magnetic nanoclusters as small as 4 nm could maintain their magnetic moment orientation at room temperature, so they could form the basis for ultra-high-density recording with densities on the order of terabits per square inch [31]. For this reason, there are ongoing efforts to improve synthetic control over the size and structure of CoPt and FePt nanoparticles [32].

Passivated bimetallic clusters can be used as probes in biomedical applications. The ligands on the outer shell of the clusters can be tailored to optimize their binding affinities to specific biomolecular targets. In one example, it was shown that oligonucleotide-modified 13 nm gold clusters interact with a specific DNA sequence and aggregate under its influence. Upon aggregation of the clusters, the color of the solution changed from red to blue, giving a simple colorimetric detection test for this specific DNA [33]. But the small number and small size of the Au nanocluster colorimetric probes gives this method a low sensitivity. The sensitivity can be increased by several orders of magnitude by catalytic reduction of Ag onto the Au nanoclusters that are attached to the target DNA because the Au@Ag clusters are much easier to detect [34]. In principle, one can detect and analyze several different DNA strands simultaneously by a variation of this method: use bimetallic clusters of various size, shape, and composition, each having its own optical properties (its own "signature"), and each being attached to one specific type of oligonucleotide. The ability to make in a controlled way many different clusters, each with its own intense and unique signal (optical, mass spectrometric, …), is the key in these kinds of applications.

4.4 Free Metal Clusters

4.4.1 Dispersion, Surface Energy

We now turn our attention to free clusters. We will first look at ideas that apply to elemental as well as bimetallic clusters. The *dispersion D* of a particle is the fraction of its atoms that are at the surface. In metal clusters, as in metallic solids, atoms have a tendency to arrange in a compact way. If a metal particle is quasi-spherical with diameter $d = 2R$, the interatomic nearest-neighbor distance is $R_{NN} = 2r$, and the packing is compact and similar to fcc, then the number of atoms N and dispersion D can be estimated as follows:

$$N \approx f(R/r)^3; \quad D \approx 1 - (R_2/R_1)^3 \quad (4.1)$$

$$\text{where } R_1 = [(N/f)^{1/3} - \sqrt{2}/2]r \quad \text{and} \quad R_2 = R_1 - \sqrt{2}r \quad (4.2)$$

and $f \approx 0.7$ is an effective packing fraction. These formulas assume a spacing between concentric layers of atoms close to that in close-packed crystals ($\approx \sqrt{2}r$). Taking a nearest-neighbor distance of $2r = 2.7$ Å (a typical value for metals) gives diameters of roughly 0–1.4 nm for clusters ($N \leq 100$), 1.4–70 nm for nanoparticles, and 70–1400 nm for small particles ($N \geq 10^7$). A rough calculation of surface energy shows a strong thermodynamic force to destroy free clusters via coalescence, oxidation, or passivation with ligands. Empirically, the surface energy per metal atom is roughly 0.16 times the cohesive energy [14]. We will assume a surface dominated by a square lattice (e.g., fcc(100) facets) and typical transition metal values for the nearest-neighbor distance ($R_{NN} \approx 2.7$ Å) and cohesive energy ($E_c \approx 4.5$ eV). Then, the surface contribution to the total energy of a quasi-spherical N-atom metal particle is roughly ($0.16E_cDN$). Using Equations 4.1 and 4.2, one can calculate particle diameters $d = 2R_1$ in nm, the dispersion D, and the surface energy on a per atom basis in eV/atom, ($0.16E_cD$). This is shown below

$\log_{10} N$	1	2	3	4	5	6	7	8	9
d (nm)	0.7	1.4	3.0	6.5	14	30	65	140	300
D	0.99	0.68	0.35	0.17	0.08	0.037	0.017	0.008	0.004
($0.16E_cD$)	0.72	0.49	0.25	0.12	0.058	0.027	0.013	0.006	0.003

Typical surface energies are 1–3 J/m² for metals but only 0.02–0.08 J/m² for organic liquids, with 0.03 J/m² being a typical value for long alkanes [35]. This is because organic molecules interact through van der Waals interactions and these are much weaker than metal–metal bonds. It is clear that passivating a cluster brings about a large drop in surface energy in absolute terms (in eV). But the relative energy change (in eV/atom) is rather small (compared to k_BT) for $\log_{10} N > 7$. The energy decrease associated with coalescence of two N-atom clusters into one $2N$-atom cluster is also very large in absolute term, but it gets smaller, on a per atom basis, as N increases. This energy

is roughly $(2 - 2^{2/3}) = 0.41$ times the surface energy ($0.16E_cD$) in the above table. Other contributions to cluster energies scale in different ways. In particular, the energy associated with spin pairing, electronic shell closings, and details in geometry (e.g., small displacement of one or few atoms) are on the order of 1 eV and do not scale up with N. There are no clear demarcations, but roughly speaking, the energy terms that do not scale up with N (those caused by the molecular aspects of geometric and electronic structure) become negligible compared to the surface energy at around $N \approx 100$, and the surface energy itself becomes small (relative to k_BT) at around $N \approx 10^7$. These general considerations explain why at small N *every atom counts*, why small metal clusters can survive only under special conditions or stabilized by a support or ligands, and why only larger particles can be in the form of a pure metal (and without special care even these usually develop an oxide skin).

4.4.2 Size Evolution of Properties

The atomization energy (AE) of a cluster X_n is the energy of the reaction $X_n \rightarrow nX$, with the cluster X_n and atoms X in their ground states. Consider the AE in a family of metal clusters. The size dependence of AEs can be fitted rather well by the formula

$$AE(N) \approx AE_{fit}(N) = E_iN + E_sN^{2/3} + E_eN^{1/3} \quad (4.3)$$

In this formula, the coefficients E_i, E_s, and E_e represent energy contributions from interior (or "bulk") atoms, surface atoms, and edge atoms respectively. It is difficult to determine E_e accurately by fitting the AEs of very small clusters ($N \lesssim 30$), but the ($E_eN^{1/3}$) term is surely a lot smaller than the other two. Setting $E_e = 0$, one finds that the ratio of the coefficients E_i/E_s varies between −0.8 and −1.1 among 11 series of elemental metal clusters (Li, Be, Mg, Al, V, Fe, Co, Ni, Nb, Ag, Au) with data that includes a mix of experimental [36] and theoretical results from various sources. So, in a typical case, $E_i/E_s \approx -1$ and the cohesive energy of a small N-atom metal cluster is roughly $E_c(N) \approx (1 - N^{-1/3})E_{c,bulk}$. The difference between the actual and fitted AEs, or residual E_{res}, is of interest because it shows the effect on energy of many nontrivial things such as the precise geometrical structure of clusters, electron spin pairing, and the symmetry and filling of orbitals (electronic shell closings). For the 11 elemental clusters above, the range over which E_{res} varies is roughly 0.5 eV (i.e., ±0.25 eV) for Li clusters, 1 eV for Al, Ag, and Au, and 1.5 eV for transition metals and group 2 metals. Figure 4.1 shows E_c and E_{res} (calculated relative to a fit using Equation 4.3 with $E_e = 0$) as a function of N for two very different metals, Ni and Mg. The data used for Ni is from collision-induced dissociation experiments [37], and data for Mg comes from density functional theory (B3PW91 method) calculations [38].

It is apparent from these and other examples that the shape of the $E_{res}(N)$ curve depends a lot on the position of the element in the periodic table. For example, the $E_{res}(N)$ curve of Co resembles Ni; Ag and Au are similar and have much bigger even–odd

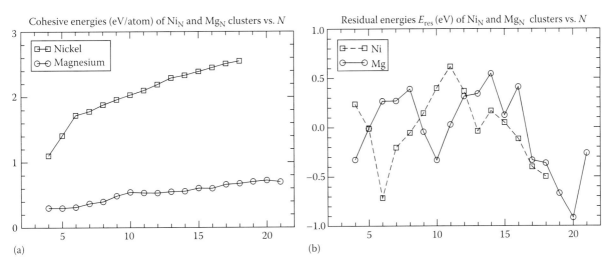

FIGURE 4.1 Cohesive energies (eV/atom) and difference between AE data and its fit (eV) for Ni and Mg clusters as a function of number of atoms N.

oscillations than other elements; alkali metals have local minima in $E_{res}(N)$ at $N = 8, 20$, instead of $N = 4, 10, 20$ for Mg and Be; etc. But in any case, the magnitude of E_{res} is rarely larger than 1 eV. So, as a fraction of the total, E_{res} is appreciable only for clusters with $N \leq 100$.

Variations on the order of 1 eV are common for ionization energies (IE) and lowest excitation energies (E_g) among isomers, and among clusters that differ only by one or two atoms. Those kinds of energy differences can have dramatic consequences for a number of other properties, most notably for chemical reactivity [39]. For this reason, it is often said that in small clusters *every atom counts*. Properties other than the AE can also be fitted with a formula like Equation 4.3, and one observes that different properties converge toward the bulk value in different ways. Calculations on clusters with up to 225 atoms confirmed the $N^{-1/3}$ scaling of cohesive energies but also showed a very different scaling for the mean nearest-neighbor interatomic distance, R_{NN}, as a function of N: R_{NN} varies *linearly with the mean coordination number* [40]. Assuming a coordination of roughly 6 or 7 for surface atoms and 12 for interior atoms, and using the formula given earlier for dispersion D, one can estimate the mean coordination number c as a function of N. For $N < 1000$, c increases by roughly 0.55 every time N doubles, so it is roughly 6.5 for $N = 10$, 7.4 for $N = 40$, 8.7 for $N = 160$, and 9.8 for $N = 640$. For any property P, there is a certain value of N beyond which P changes monotonically with N. This is called the *scalable regime* for that cluster and property. It has been shown that the key features of the density of electronic states reaches the scalable regime near $N = 80$ [40]; adsorption energies on Pd clusters also get in the scalable regime at around $N = 80$ [41]. All of the above also holds for bimetallic clusters $A_m B_n$, but in addition, qualitative changes in geometric structure can occur at relatively large values of $(m + n)$. For example, holding (m/n) constant, there could be a transition from alloy structures to core-shell $A_m@B_n$ segregated structures at some critical value of $N = (m + n)$. In such a case, one would normally require two separate curves and fit parameters to describe the size variations for each structure type.

4.4.3 Magic Clusters

In elemental clusters, some cluster sizes show special stability as evidenced by a higher abundance in mass spectra [42] or by cohesive energies that lie above polynomial fits, as shown in the previous section. These have been called "magic clusters." Much effort has gone into identifying and explaining magic numbers in elemental clusters. Two simple models that have been used a lot are *electronic shells* [42,43] and *atomic shells* [44,45]. These models have limited reliability and scope but remain useful conceptually because no simpler model can explain so much about metal clusters.

The *electronic shells* model is based on the idea that some metal clusters are liquid-like, their geometric configuration is not essential, and the number of valence electrons N_e is the critical variable. Electronic shells are usually predicted using the *ellipsoidal jellium model* (EJM) [43,46]. In the EJM, clusters consist of a hypothetical uniform positive charge background bounded by an ellipsoid (instead of point charge nuclei) and N_e electrons. Calculations show that electrons fill cluster orbitals of $ns, np, nd, nf, \ldots$ type where $n \geq 1$ irrespective of angular momentum. The usual order of filling these orbitals is 1s 1p 1d 2s 1f 2p 1g 2d 1h 3s …It is reproduced, except for inversions of (3s, 1h) and (4s, 1i), by drawing oblique lines from upper right to lower left across the orbital types arranged in this manner:

1s	1p	1d	1f	1g	1h	1i
	2s	2p	2d	2f	2g	…
		3s	3p	3d	3f	…
			4s	4p	4d	…
				…		…

This produces electronic shell closings at $N_e = 2,8,(18),20,(34), 40,58,(68),(90),92$, where minor shell closings are indicated by parentheses. Electronic shells and the EJM have been successful in accounting for unusually high atomization and ionization energies in clusters of group 1, group 2, and group 11, and a few other elements, and also in accounting, in an average sense, for the energetically preferred shapes of these clusters [47]. In bimetallic clusters, the electronic shell closings generally follow the same pattern if the two elements have similar valence electron densities, for example, the combinations (K, Na), (Na, Ba), or (Na, Li) [48]. But when A and B have very different electron densities, such as K and Mg, the basic assumption of the jellium model (uniform potential and density) breaks down and other electron counts must be used. If A and B separately satisfy the EJM assumptions of a free metal, one still has EJM orbitals but the order of their filling is different: 1s 1p 2s 1d ...giving the sequence of magic numbers (8), 10, 20, ...Also, in $A_{N-1}B$ clusters where A is well described in the jellium model (e.g., a group 1 or 11 metal) and B is a transition metal, the electron count corresponding to shell closing is not trivial. If B is rather well described by the jellium model (e.g., Pt) it can be seen as bringing a single *s* electron to the system and normal rules apply. But most transition metals can also be described as a +Z ion with Z valence electrons (e.g., Z = 4 for Zr). In that case, B often has a higher cohesive energy than A and takes a central location in the cluster. Then, the effective potential felt by the electrons is a lot deeper in the center of the cluster than elsewhere and, overall, the potential is intermediate between those of a jellium and of a hydrogenoid atom. Evidence for this effect comes from the observed stability of the 18-electron species $Au_{16}Y^+$ and $Au_{16}In^+$ [49], and $ScCu_{16}^+$ [50], and theoretical predictions [51]. We confirmed this effect with calculations on a series of icosahedra doped with multivalent central atoms: They show small HOMO–LUMO gaps for 20-electron species like $Fe@Ag_{12}$ ($N_e = 20$), and large HOMO–LUMO gap for 18-electron species like $Mo@Ag_{12}$ ($N_e = 18$). Another complicating factor is that only delocalized electrons should be counted. For instance, a Nb atom is probably best seen as a Nb^+ ion plus one loosely bound *s*-type

electron that can easily delocalize, not a 5+ ion with 5 electrons. Magic numbers in mass abundance spectra of Au_nX^+ clusters (X = Sc to Ni) suggest the following rules [52]: (1) Prominent shell closings occur at $N_e = 18, 34$, but not 20; (2) To calculate N_e, one considers that Sc and Ti contribute 3 and 4 electrons to the delocalized system of electrons, respectively; (3) heavier TM elements contribute either one or two *s*-type electrons, whichever produces an even total number of electrons; (4) as a result of (3) the magnitude of even–odd oscillations in cluster stabilities correlate with the magnitude of the dopant's $s − d$ excitation energy, e.g., large for Mn and small for Ni.

The *atomic shells* model supposes that, as clusters grow in size, the atoms pack according to a specific pattern that minimizes, or nearly minimizes, the exposed surface area. From that point of view, the optimal packing is triangle, tetrahedron, trigonal bipyramid (TBP), a face-capped TBP, and so on, producing polytetrahedral clusters. This packing can continue at $n = 7$ (pentagonal bipyramid, PBP) only by admitting small distortions in the tetrahedra, and it can only continue at $n = 13$ (the icosahedron, ICO) and beyond by having some strain, i.e., with interatomic distances that differ between intralayer and interlayer bonds. This growth sequence is illustrated in Figure 4.2 for $N = 4$–8, 12, 13, and 55. Note that the lowest energy structure at $N = 6$ for pair potentials is normally the octahedron, not the structure in Figure 4.2c. In the ICO, the interlayer distance is 5% smaller than the intralayer distance. The number of atoms of ICO or fcc-packed clusters with K layers is

$$N_K = \frac{10}{3}K^3 - 5K^2 + \frac{11}{3}K - 1,$$

so the first few magic numbers in that model are 1, 13, 55, 147, 309, 561, 923, 1415, ..., and $K \approx 0.7 N_K^{1/3}$ when N_K is large. The mass

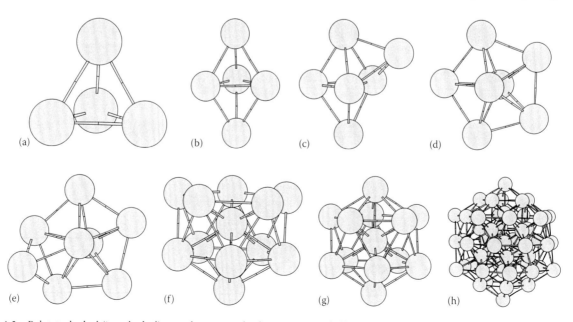

FIGURE 4.2 Polytetrahedral (icosahedral) growth sequence leading to atomic shells in clusters with *N* atoms, *N* = 4, 5, 6, 7, 8, 12, 13, and 55 (a–h).

spectra of van der Waals clusters, in particular, rare gas atoms do show several of these magic numbers [53].

Optimization of bimetallic clusters structures with respect to a semiempirical potential energy suggests that polyicosahedral structures are still favored [54]. However, the pattern in atomic shells and magic numbers generally depends on composition and that complicates matters. In particular, a large size mismatch between atoms of A and B favors segregation of the larger atoms to the surface. Extra stability may ensue when A has a higher cohesive energy, and a somewhat smaller size than B. In such a case, a perfect $A_m@B_n$ core-shell arrangement should correspond to enhanced stability as this structure removes the strain inherent in pure polyicosahedral clusters. For example, results of a computational study suggest that $Cu_{13}@Ag_{27}$ minimizes the excess energy (see Equation 4.10 below) of 40-atom mixed Ag–Cu clusters [55].

4.5 Geometric Structure of Bimetallic Clusters

In the bulk, two metals A and B will mix to form a homogeneous solid A_xB_{1-x}, or will not mix, depending on A, B, the mole fraction x, temperature T, and pressure P. When they do mix, different phases are possible, depending on A, B, x, T, P. Some phases have definite crystal structures while others are mixed in a somewhat random way. Alloy phase diagrams summarize this kind of information and have been compiled for many AB binary solids. But clusters are quite different from the bulk. Metals that do not mix in the bulk can form mixed clusters; it is often thermodynamically favorable for A_{2m} and B_{2m} clusters to mix and give two A_mB_m clusters for elements A and B that do not mix in the bulk [56]. Some combinations of elements have a strong tendency to form just one kind of mixed or segregated structure. This is driven by thermodynamics and mostly tied to the relative cohesive energy, surface energy, and electronegativity of the two elements (see Section 4.7). But chemical ordering also depends on the method of preparation. For instance, three kinds of AgAu bimetallic clusters (mixed, $Ag_n@Au_m$, $Au_n@Ag_m$) have been made by simply changing the sequence of Ag and Au deposition on an alumina substrate [57]. In fact, all free clusters

are unstable (or metastable) and must be made with some kind of kinetic control. So the composition and structure of possible A_mB_n clusters is limited primarily by the degree of control that can be achieved.

Theoretically, the number of possible isomers of a A_mB_n cluster is much greater than for an elemental cluster X_p. Take an arbitrary configuration of the p atoms of X_p. By substituting m of the atoms of X by atoms of A, and the rest by B, one gets one A_mB_n homotop (also called *permutational isomer*) of the parent structure. If there is no symmetry, there are $p!/(m!n!)$ ways to make these substitutions, i.e., $p!/(m!n!)$ distinct homotops [58]. Classification of A_mB_n cluster structures is broken in these two parts: (1) parent structure (sometimes called "isomer" in the literature) and (2) homotop. There is already a rich diversity of structures among elemental metal clusters (parent structures) [59], but we will discuss only the mixing aspect (homotops). A lot of experimental and theoretical work pertains to fairly large quasi-spherical clusters. Homotops that have been observed or computed usually resemble one of these ideal forms [22]: maximally mixed (MM), left–right segregated (LRS), core-shell (CS), and "onion layers structures" (OLS) [60,61]. The MM, LRS, and CS types are illustrated in Figure 4.3 using the 55-atom Mackay icosahedron as the parent structure.

Different variables have been used to quantify mixing in A_mB_n, such as the number of A–B bonds M, the weighted mean distance between pairs of unlike atoms δ^n, the distance between centers of mass D, and difference in mean distances of A and B atoms to the center of mass, S:

$$M = \sum_{i \in A} \sum_{j \in B} f(d_{ij}/d_0) \tag{4.4}$$

where $d_0 = r_A + r_B$ and $f(x) = 1$ if $x < 1.2$, $f(x) = 0$ otherwise

$$\delta^n = \frac{\sum_{i \in A} \sum_{j \in B} (d_{ij})^{1-n}}{\sum_{i \in A} \sum_{j \in B} (d_{ij})^{-n}} \tag{4.5}$$

$$D = \left| (1/m) \sum_{i \in A} \vec{R}_i - (1/n) \sum_{j \in B} \vec{R}_j \right| \tag{4.6}$$

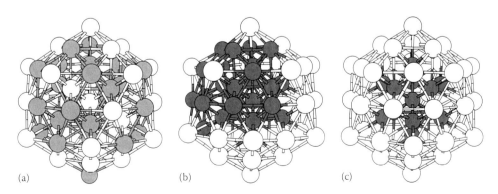

FIGURE 4.3 Maximally mixed (MM), left–right segregated (LRS), and core-shell segregated (CS) types of homotop for a 55-atom Mackay icosahedron.

$$S = (1/m) \sum_{i \in A} |\vec{R}_i - \vec{R}_{cm}| - (1/n) \sum_{j \in B} |\vec{R}_j - \vec{R}_{cm}| \qquad (4.7)$$

where

d_{ij} is the distance between atoms i and j

$i \in A$ means the sum is over atoms of element A

$\vec{R}_{cm}$ is the cluster center of mass obtained after assigning a fictitious unit mass to every atom

A convenient way to characterize a cluster structure is to first calculate M, δ^n, D, and S for that structure *and for all of its homotops*, and then normalize the variable in one of two possible ways. We illustrate this for M:

$$\overline{M} = \frac{M - M_{avg}}{\sigma(M)} \quad \text{or} \quad \overline{M} = \frac{M - M_{min}}{M_{max} - M_{min}} \qquad (4.8)$$

where M_{avg}, M_{min}, M_{max}, and $\sigma(M)$ are, respectively, the average, minimum, maximum, and standard deviation, of M evaluated over the set of all homotops of M.

4.6 Properties of Bimetallic Clusters

Properties of $A_m B_n$ vary in complex ways with A, B, m, n and structure. Few generalizations can be made. Some of the unique properties of bimetallic clusters were mentioned in Section 4.3. Here, we will make a few more comments about how properties of $A_m B_n$ may or may not relate to properties of the elemental clusters, and also about optical properties.

A central question in studies of elemental clusters X_p is this: *How do properties evolve with size p?* This is also relevant to bimetallic clusters, but two other questions come up. The first is *How do the properties of $A_m B_n$ (m + n = p) compare to those of A_p and B_p?* The second, slightly different, question is *For a given p = n + m, how do the properties of $A_m B_n$ vary as a function of (m/p)?* Answers vary a lot according to property and cluster, but there are standard ways to approach these questions. In particular, one is often interested in the *excess property* associated with a property P, which can be defined in at least two ways.

$$P_{excess}(A_m B_n) = \frac{P(A_m B_n) - (m/p)P(A_p) - (n/p)P(B_p)}{(m/p)P(A_p) + (n/p)P(B_p)} \times 100\% \qquad (4.9)$$

$$\Delta(A_m B_n) = \frac{P(A_m B_n) - (m/p)P(A_{bulk}) - (n/p)P(B_{bulk})}{p^{2/3}} \qquad (4.10)$$

A comparison of clusters with same size but different compositions can be made with P_{excess}, whereas Δ is useful for comparing clusters that differ in both size and composition [56]. Some properties depend much more on surface atoms than interior atoms, e.g., surface energy, chemical reactivity, collective resonance in photoabsorbtion spectra [62], and in many clusters one of the elements, say "A," segregates to the surface. In such a case, $P(A_m B_n)$

should be closer to $P(A_p)$ than to $P(B_p)$. This can be accounted for by modifying the above equations so that only surface atoms are counted in all of the species involved (clusters and bulk solid). In most cases P_{excess} is at most a few percent.

It seems natural to classify bimetallic cluster properties into three categories: (1) those where P_{excess} calculated with all atoms is a few percent or less; (2) those where P_{excess} calculated using only the surface atoms is a few percent or less; (3) others having properties that deviate a lot from the weighted average of the two pure metals, no matter what kind of atoms (surface, interior, both) are counted. It is dangerous to generalize, but it appears that energy, surface energy, bond lengths (or density), and polarizability normally fall in categories (1) or (2). Some other properties, for instance, chemical reactivity and magnetization, are more complex, more sensitive to composition, and often fall in category (3).

In theory, the energy can sometimes be decomposed into a sum of atomic contributions, in which case it is convenient to consider the *mixing energy* E_{mix} and mixing coefficient M_E for a $A_m B_n$ cluster with nuclear configuration **X** defined like this [58]:

$$E_{mix}(A_m B_n; \mathbf{X}) = E(A_m B_n; \mathbf{X}) - E(A_m \text{ in } A_m A_n; \mathbf{X})$$
$$- E(B_n \text{ in } B_m B_n; \mathbf{X}) \qquad (4.11)$$

$$M_E = E_{mix} / E(A_m B_n) \qquad (4.12)$$

The optical spectra of small clusters have several peaks, like molecules. But larger clusters are characterized by a single intense peak, called plasmon [63]. The plasmon can be understood as a collective excitation of the electrons: To a first approximation, the change in electric dipole is due to a rigid displacement of the electronic cloud (valence electrons) relative to the positions of the ions. The shape and frequency of the plasmon resonance depend strongly on the size and shape of the particle [63]. Studies of $Ag_n Au_m$ have shown that the plasmon resonance also varies a lot with composition and morphology [64]. Laser irradiation or annealing of core-shell $Au_n@Ag_m$ clusters leads to mixing of the two metals, which gives a way to tune optical properties [24]. The UV-visible absorption of small clusters can be modeled in detail by theory. A fairly standard technique for this is time-dependent density functional theory (TDDFT). A recent application of TDDFT showed that there are many absorption peaks in the spectra of $Ag_n Ni_{p-n}$ ($p \leq 8$) clusters, and many nontrivial differences between those spectra as a function of composition and structure [65].

4.7 Structure–Energy Principles for Bimetallic Clusters

It has been shown that for a cluster X_{13} where atoms interact via a Lennard-Jones (LJ) potential, $E = \sum_{j>i} 4\epsilon \left[(\sigma/d_{ij})^{12} - (\sigma/d_{ij})^6 \right]$, there are more than 1500 distinct geometric isomers (energy minima) [66]. If a pair potential can model a system adequately, then the number of minima depends on the range of the pair interactions, with a shorter range giving more minima [67]. There are probably *more* isomers for clusters described by accurate energy methods (*ab initio*, DFT), compared to a LJ potential, considering the different possible atomic hybridizations and cluster spin states. Neglecting symmetry, the number of homotops for a small clusters like A_7B_6 is $(7!6!/13!) = 1716$. This gives an estimated total number of distinct minima on the order of 10^6. Obviously, it is very difficult to find, without making *a priori* assumptions, the experimental or computed lowest energy structures of even moderate-size bimetallic clusters. Therefore, it is desirable to have *structure–energy principles* that would allow quick educated guesses of cluster structures. It must be said at the outset that no simple set of rules allows reliable detailed predictions of cluster structure. The structure–energy principles that we are about to show summarize empirical observations, or correlations, drawn from a variety of experimental and computational sources and they are good only for qualitative predictions.

One can relate some aspects of cluster structure to certain properties of the constituent atoms. On the atom side of the problem, we have the ionization energy (IE) and electron affinity (EA), or, electronegativity χ and hardness η:

$$\chi = \frac{(IE + EA)}{2}; \quad \eta = \frac{(IE - EA)}{2} \qquad (4.13)$$

The atomic radius R_A and cohesive energy $E_{c,A}$ of an element A are also critical, and so is the ratio $(E_{c,A}/R_A^2)$ (it correlates with the surface energy of that element). Another important thing is whether the atom's d orbitals are partially filled (sometimes the precise d electronic configuration matters). On the cluster side, it is useful to consider geometry descriptors such as the number of nearest-neighbor pairs or the surface area (these two things are strongly correlated); the cluster shape (roughly spherical, or oblate, or prolate), which may be quantified using the three moments of inertia; the distribution function $g(d)$ of interatomic distances; and the distribution function $h(\theta)$ of *ijk* angles, where atom j is a neighbor to both atoms i and k. Note that the first two peaks in $g(d)$, at d_1 and d_2, quickly characterizes familiar structures: the ratio (d_2/d_1) is 1.414 for fcc, 1.155 for bcc, and 1.050 for an ideal 13-atom icosahedron. The width of the first peak w_1 (and to a lesser extent, widths of other peaks) is also informative: a small w_1 indicates little or no strain, or, high "crystallinity." Relative peak intensities in $h(\theta)$ (especially at 60°, 90°, and 108°) also show very clear differences between the different structure types like fcc, bcc, ICO, etc. Finally, note that in a pair-potential model, the range of interaction is correlated to the IE of the atom because the long-range form of an atom's electronic density depends essentially on IE [68] (a small IE correlates with a long-range and "soft" pair potential).

There are some rules that seem to govern *parent structures* of elemental [59] (and also bimetallic) clusters. (1) There is a general tendency to minimize the exposed surface area, or, maximize the number of bonds. (2) Structures where the first peak in $g(d)$ is narrow (i.e., the strain is small) are favored, especially for

elements with a large IE. (3) Elements with a large atomic hardness η, or a large IE, tend to favor structures where all atoms have nearly the same number of neighbors: in extreme cases, this gives cage structures, as in several gold clusters, for example. (4) Atoms that have a ground state electronic configuration with a number of valence d electrons that differs from 0, 5, or 10 often exhibit 90° bond angles. (5) Clusters of elements of groups 1, 2, and 11, and other elements that are well described in the jellium model (e.g., Al) tend to adopt shapes that conform to the EJM [42,43]. (6) There is sometimes a competition between high-spin, high-symmetry structures, and low-spin, EJM-distorted structures, Li_{13} being one example of that.

There are some rules about *homotops*. Elements A and B have a tendency to (1) mix if they have very different electronegativities χ; (2) segregate if they have very different cohesive energies E_c, or very different ratio $(E_{c,X}/R_X^2)$, with the element with smallest E_c (or smallest $(E_{c,X}/R_X^2)$) going to the surface; (3) segregate into a core-shell structure with B outside if R_B is much larger than R_A [54] since that decreases the strain; (4) segregate if one element (A or B) has a stronger interaction to the environment (support, ligand, or ambient gas); (5) when the surface is composed of A and B atoms, the atom with the lowest E_c tends to occupy the sites with lowest coordination; and (6) when the difference $(E_{c,A} - E_{c,B})$ is large and positive, the structure of A_mB_n has a compact A_m core with a structure that maximizes the number of A–A bonds, whereas the total number of bonds is much less important.

4.8 Special Bimetallic Clusters

Some clusters should be relatively stable in theory because they satisfy two or more stability criteria simultaneously. We refer to them as *doubly magic clusters* (DMC) even if they satisfy more than two criteria. The interest in free DMCs is largely theoretical because, for metals, even the most stable-free DMCs of size $N \leq 100$ are probably not stable enough to be used for materials. But there is also a practical side to DMCs. The metal core part of stable passivated clusters are probably DMC cations in many cases. For example, the stable thiolate-protected cluster $Au_{25}(RS)_{18}$ has a metallic icosahedral core Au_{13}^{5+} with closed atomic and electronic shells [4] (the 12 other Au atoms have nonzero oxidation numbers). Conversely, passivation of neutral DMCs or of their charged isovalent analogs could yield truly stable clusters. Also, the polyanions of known stable Zintl phases, like Ge_9^{4-} ($N_e = 40$), may be viewed as magic clusters. This suggests the possibility of stable core-shell bimetallic clusters $A_m@B_n$ where A and B have a large difference of electronegativity, though maybe not as large as in the true Zintl phases [69,70]. What follows is a short and partial list of free clusters having unusual stability. Each of these has many isovalent (neutral or charged) bimetallic analogs, which may also be stable.

The unusual stability of MAl_4^- clusters (M = Li, Na, Cu) was explained by proposing the concept of *metal aromaticity* that would apply to the square planar Al_4^{2-} unit [71]. The Na_6Pb cluster is an octahedron with the Pb atom at the center and with the $N_e = 10$ shell closing characteristic of inverted EJM orbital order ($1s^2 2p^6 2s^2$) [72,73]. The Al_{13}^- icosahedral cluster satisfies the EJM closing at $N_e = 40$, whereas neutral Al_{13} can be seen as a kind of "halogen superatom" [74]. Relatively stable bimetallic species can be derived by reaching $N_e = 40$ in different ways, e.g., $Al_{13}Ag$ [75], $Sn@Al_{12}$, etc. Generally, many doped (bimetallic) icosahedra $A@B_{12}$ have been predicted to have large HOMO–LUMO gaps on the order of 2–3 eV: $Al@Pb_{12}^+$ [76], WAu_{12} [77], $MoCu_{12}$ [78], etc. The Au_{20} cluster is a pyramidal fragment of the fcc crystal with a large HOMO–LUMO gap (1.8 eV) that was identified by the combined use of photoelectron spectroscopy and computations [79]. Many tetrahedral clusters with $N_e = 18$ or 20 and special stability can be derived from it. Removal of the four apex atoms gives Au_{16} which is interesting in its own right because, in the anion, the four symmetry equivalent atoms move outward giving a cage structure [80]. Substitution of these four atoms by a divalent metal gives a series A_4B_{12} of clusters. In particular, Mg_4Ag_{12} has the tetrahedral structure as its global minimum and a HOMO–LUMO gap of approximately 2.5 eV [81]. In addition to its compact strain-free geometric structure and closed-shell electron count, the electrons and nuclei in Mg_4Ag_{12} make coincident electrically neutral shells (CENS): the Mg^{2+} and Ag^+ ions' spatial distribution has high symmetry (T_d) and yields cumulative ionic charges that match two successive EJM shell closings (8 and 12 + 8 = 20), which gives extra stability to the T_d homotop. The CENS property might be useful as an additional shell closing criterion. The basic T_d M_{16} structure can be either fcc-like (compact) or cage-like and, in the latter case, it can easily accommodate a central atom impurity. If the impurity brings the electron count to 18 (not 20), special stability ensues as in, for example, $Sc@Cu_{16}^+$ [50].

4.9 Summary and Conclusion

The properties of clusters are often very different from the bulk, which leads to various applications, particularly in catalysis, bioanalysis (as probes), and for magnetic materials. Compared to elemental clusters, bimetallic clusters obviously offer more possibilities for design and applications. Small clusters are characterized by a high surface area. Generally speaking, this makes them unstable. However, there exist several methods for synthesizing small bimetallic clusters either in the gas phase (sometimes followed by deposition in or on a substrate), or in solution, or by using precursor metal complexes on a support, or by solid-state chemistry. Synthetic methods, and the tools and instruments for characterization of clusters, are improving rapidly and give increasingly better control and understanding of the shape and composition of bimetallic clusters. It is possible, in some cases, to make core-shell structures $A_m@B_n$ or $B_m@A_n$, or mixed A_mB_n structures, for a given A–B combination by using different synthetic approaches. Some principles help rationalize the geometric structure and relative stability of bimetallic clusters, in particular: special stability is associated with electronic shell closings (e.g., electron counts of 8, 20, 40, ...in the jellium model); clusters tend to minimize surface energy, therefore the element with smaller surface tension tends to segregate to the surface; clusters tend to minimize strain, so bigger atoms tend to segregate to the surface; A and B tend to mix if they have very different electronegativities; and there is extra stability associated with the simultaneous closing of electronic and atomic shells. Improvements to synthetic methods, characterization tools, theory and modeling, and multidisciplinary approaches to design and prototyping for specific applications have been, and will continue to be, essential for progress in the field of bimetallic clusters.

Acknowledgments

I am grateful to the European Science Foundation for supporting the *Workshop on Computational Nanoalloys* where participants discussed many of the topics of this chapter, and to NSERC-Canada for research funds.

References

1. C Femoni, M C Iapalucci, G Longoni, and P H Svensson, *Chem. Commun.* (2001) 1776.

2. G A Ozin and A C Arsenault, *Nanochemistry, A Chemical Approach to Nanomaterials*, RSC Publishing, Cambridge, U.K. (2005).

3. H Grönbeck and J M Thomas, *Chem. Phys. Lett.* 443 (2007) 337.

4. M W Heaven, A Dass, P S White, K M Holt, and R W Murray, *J. Am. Chem. Soc.* 130 (2008) 3754–3755; J Akola, M Walter, R L Whetten, H Häkkinen, and H Grönbeck, *J. Am. Chem. Soc.* 130 (2008) 3756–3757.

5. C Binns, *Surf. Sci. Rep.* 44 (2001) 1.

6. R A Ristau, K Barmak, L H Lewis, K R Coffey, and J K Howard, *J. Appl. Phys.* 86 (1999) 4527.

7. M A Gracia-Pinilla, E Pérez-Tijerina, J A García, C Fernández-Navarro, A Tlahuice-Flores, S Mejía-Rosales, J M Montejano-Carrizales, and M José-Yacamán, *J. Phys. Chem. C* 112 (2008) 13492.

8. P Liu, Y Huang, Y Zhang, M J Bonder, G C Hadjipanayis, D Vlachos, and S R Deshmukh, *J. Appl. Phys.* 97 (2005) 10J303.

9. M Hillenkamp, G di Domenicantonio, and C Félix, *Rev. Sci. Instrum.* 77 (2006) 025104.

10. F Conus, J T Lau, V Rodrigues, and C Félix, *Rev. Sci. Instrum.* 77 (2006) 113103.

11. J T Lau, A Achleitner, H-U Ehrke, U Langenbuch, M Reif, and W Wurth, *Rev. Sci. Instrum.* 76 (2005) 063902.

12. J M Thomas, B F G Johnson, R Raja, G Sankar, and P A Midgley, *Acc. Chem. Res.* 36 (2003) 20.

13. U Kolb, S A Quaiser, M Winter, and M T Reetz, *Chem. Mater.* 8 (1996) 1889.

14. G A Somorjai, *Introduction to Surface Chemistry and Catalysis*, Wiley, New York (1994).

15. O S Alexeev and B C Gates, *Ind. Eng. Chem. Res.* 42 (2003) 1571.

16. S Sun, C B Murray, D Weller, L Folks, and A Moser, *Science* 287 (2000) 1989.

17. D V Goia and E Matijević, *New J. Chem.* 22 (1998) 1203.

18. C M Doudna, M F Bertino, F D Blum, A T Tokuhiro, D Lahiri-Dey, S Chattopadhyay, and J Terry, *J. Phys. Chem. B* 107 (2003) 2966.

19. Y Mizukoshi, T Fujimoto, Y Nagata, R Oshima, and Y Maeda, *J. Phys. Chem. B* 104 (2000) 6028.

20. E V Shevchenko, D Talapin, A L Rogach, A Kornowski, M Haase, and H Weller, *J. Am. Chem. Soc.* 124 (2002) 11480.

21. E Shevchenko, D Talapin, A Kornowski, F Wiekhorst, J Kötzler, M Haase, A Rogach, and H Weller, *Adv. Mater.* 14 (2002) 287.

22. R Ferrando, J Jellinek, and R L Johnston, *Chem. Rev.* 108 (2006) 845.

23. J I Park, M G Kim, Y W Jun, J S Lee, W R Lee, and J Cheon, *J. Am. Chem. Soc.* 126 (2004) 9072.

24. J H Hodak, A Henglein, M Giersig, and G V Hartland, *J. Phys. Chem. B* 104 (2000) 11708.

25. A P J Jansen and C G M Hermse, *Phys. Rev. Lett.* 83 (1999) 3673.

26. V Stamenkovic, B S Mun, K J J Mayrhofer, P N Ross, N M Markovic, J Rossmeisl, J Greeley, and J K Nørskov, *Angew. Chem. Int. Ed.* 45 (2006) 2897.

27. J H Sinfelt, *Catal. Today* 53 (1999) 305.

28. A Russell and A Rose, *Chem. Rev.* 104 (2004) 4613.

29. K Sumiyama, K Suzuki, S A Makhlouf, K Wakoh, T Kamiyama, S Yamamuro, T Konno, Y F Xu, M Sakurai, and T Hihara, *J. Non-Cryst. Solids* 192/193 (1995) 539.

30. S Rohart, C Raufast, L Favre, E Bernstein, E Bonet, and V Dupuis, *Phys. Rev. B* 74 (2006) 104408.

31. M L Plumer, J van Eck, and D Weller, *The Physics of Ultrahigh-Density Magnetic Recording*, Springer, Berlin, Germany (2001).

32. F Tournus, A Tamion, N Blanc, A Hannour, L Bardotti, B Prével, P Ohresser, E Bonet, T Épicier, and V Dupuis, *Phys. Rev. B* 77 (2008) 144411, and references therein.

33. C A Mirkin, R L Letsinger, R C Mucic, and J J Storhoff, *Nature* 382 (1996) 607.

34. N L Rosi and C A Mirkin, *Chem. Rev.* 105 (2005) 1547.

35. J Israelachvili, *Intermolecular and Surface Forces*, 2nd Edn., Academic Press, London, U.K. (1991).

36. S K Loh, D A Hales, L Lian, and P B Armentrout, *J. Chem. Phys.* 90 (1989) 5466; D A Hales, C X Su, L Lian, and P B Armentrout, *J. Chem. Phys.* 100 (1994) 1049.

37. L Lian, C-X Su, and P B Armentrout, *J. Chem. Phys.* 96 (1992) 7542.

38. A Lyalin, I A Solov'yov, A V Solov'yov, and W Greiner, *Phys. Rev. A* 67 (2003) 063203.

39. M B Knickelbein, *Annu. Rev. Phys. Chem.* 50 (1999) 79.

40. A Roldán, F Viñes, F Illas, J M Ricart, and K M Neyman, *Theor. Chem. Acc.* 120 (2008) 565.

41. I V Yudanov, R Sahnoun, K M Neyman, and N Rösch, *J. Chem. Phys.* 117 (2002) 9887.

42. W D Knight, K Clemenger, W A deHeer, W A Saunders, Y M Chou and M L Cohen, *Phys. Rev. Lett.* 52 (1984) 2141.

43. W A DeHeer, *Rev. Mod. Phys.* 65 (1993) 611; M Brack, *Rev. Mod. Phys.* 65 (1993) 677.

44. O Echt, K Sattler, and E Rechnagel, *Phys. Rev. Lett.* 47 (1981) 1121.

45. T P Martin, *Phys. Rep.* 273 (1996) 199.

46. K Clemenger, *Phys. Rev. B* 32 (1985) 1359.

47. R Fournier, *J. Chem. Phys.* 115 (2001) 2165.

48. J A Alonso, *Structure and Properties of Atomic Nanoclusters*, Imperial College Press, London, U.K. (2005).

49. W Bouwen, F Vanhoutte, F Despa, S Bouckaert, S Neukermans, L T Kuhn, H Weidele, P Lievens, and R E Silverans, *Chem. Phys. Lett.* 314 (1999) 227.

50. N Veldeman, T Höltzl, S Neukermans, T Veszprémi, M T Nguyen, and P Lievens, *Phys. Rev. A* 76 (2007) 011201.

51. Y Gao, S Bulusu, and X C Zeng, *J. Am. Chem. Soc.* 127 (2005) 156801.

52. S Neukermans, E Janssens, H Tanaka, R E Silverans, and P Lievens, *Phys. Rev. Lett.* 90 (2003) 033401.

53. O Echt, O Kandler, T Leisner, W Miehle, and E Recknagel, *J. Chem. Soc. Faraday Trans.* 86 (1990) 2411.

54. A Rapallo, G Rossi, R Ferrando, A Fortunelli, B C Curley, L D Loyd, G M Tarbuck, and R L Johnston, *J. Chem. Phys.* 122 (2005) 194308.

55. R Ferrando, A Fortunelli, and R L Johnston, *Phys. Chem. Chem. Phys.* 10 (2008) 640.

56. R Ferrando, A Fortunelli, and G Rossi, *Phys. Rev. B* 72 (2005) 085449.

57. W Benten, N Nilius, N Ernst, and H-J Freund, *Phys. Rev. B* 72 (2005) 045403.

58. J Jellinek and E B Krissinel, *Chem. Phys. Lett.* 258 (1996) 283.

59. Y Sun, M Zhang, and R Fournier, *Phys. Rev. B* 77 (2008) 075435.

60. D Ferrer, A Torres-Castro, X Gao, S Sepúlveda-Guzmán, U Ortiz-Méndez, and M José-Yacaman, *Nano Lett.* 7 (2007) 1701.

61. F Baletto, C Mottet, and R Ferrando, *Phys. Rev. Lett.* 90 (2003) 135504.

62. A Bol, G Martin, J L López, and J A Alonso, *Z. Phys. D* 28 (1993) 311.

63. K L Kelly, E Coronado, L L Zhao, and G C Schatz, *J. Phys. Chem. B* 107 (2003) 668.

64. J P Wilcoxon and P P Provencio, *J. Am. Chem. Soc.* 126 (2004) 6402.

65. M Harb, F Rabilloud, and D Simon, *Chem. Phys. Lett.* 449 (2007) 38.

66. S F Chekmarev, *Phys. Rev. E* 64 (2001) 036703.

67. P A Braier, R S Berry, and D J Wales, *J. Chem. Phys.* 93 (1990) 8745.

68. M M Morrell, R G Parr, and M Levy, *J. Chem. Phys.* 62 (1975) 549.

69. J D Corbett, *Angew. Chem. Int. Ed.* 39 (2000) 670.

70. W van der Lugt, *J. Phys. Condens. Matter* 8 (1996) 6115.

71. X Li, A E Kuznetsov, H-F Zhang, A I Boldyrev, and L-S Wang, *Science* 291 (2001) 859.

72. C Yeretzian, U Röthlisberger, and E Schumacher, *Chem. Phys. Lett.* 237 (1995) 334.

73. L C Balbás and J L Martins, *Phys. Rev. B* 54 (1996) 2937–2941.

74. D E Bergeron, P J Roach, A W Castleman, Jr, N O Jones, and S N Khanna, *Science* 307 (2005) 231.

75. V Kumar and Y Kawazoe, *Phys. Rev. B* 64 (2001) 115405.

76. S Neukermans, E Janssens, Z F Chen, R E Silverans, P V R Schleyer, and P Lievens, *Phys. Rev. Lett.* 92 (2004) 163401.

77. P Pyykkö and N Runeberg, *Angew. Chem. Int. Ed.* 41 (2002) 2174.

78. Q Sun, X G Gong, Q Q Zheng, D Y Sun, and G H Wang, *Phys. Rev. B* 54 (1996) 10896.

79. J Li, X Li, H-J Zai, and L-S Wang, *Science* 299 (2003) 864.

80. S Bulusu, L-S Wang, and X C Zeng, *Proc. Natl. Acad. Sci.* 103 (2006) 8326.

81. Y Sun and R Fournier, *Phys. Rev. A* 75 (2007) 063205.

5

Endohedrally Doped Silicon Clusters

Nele Veldeman
Katholieke Universiteit Leuven

and

Vlaamse Instelling voor
Wetenschappelijk Onderzoek

Philipp Gruene
Fritz-Haber-Institut der
Max-Planck-Gesellschaft

André Fielicke
Fritz-Haber-Institut der
Max-Planck-Gesellschaft

Pieterjan Claes
Katholieke Universiteit Leuven

Vu Thi Ngan
Katholieke Universiteit Leuven

Minh Tho Nguyen
Katholieke Universiteit Leuven

Peter Lievens
Katholieke Universiteit Leuven

5.1 Introduction

The ongoing trend toward miniaturization in microelectronics has triggered interest in small particles made of silicon, the most used element in the semiconductor industry. An important discovery of the unique nanoscale properties of silicon has been the observation of bright luminescence from porous silicon nanostructures (Canham 1990), whereas bulk silicon is a bad emitter of light due to its indirect band gap. This phenomenon has attracted much attention toward its quantum mechanical origin. By downsizing bulk silicon to the nanoscale, there is hope to induce a more efficient luminescence and to shift the emission wavelength from 1.1 eV into the visible region (Canham 1990, Iyer and Xie 1993, Ledoux et al. 2000). Possible visible-light emitters made out of silicon are of great relevance because of their potential for new functional silicon-based optoelectronic devices. Moreover, the tunability of the emission wavelength by changing only their particle size is still one of the open demands associated with a large variety of applications.

Elemental silicon clusters, however, are unsuitable as building blocks for future nanomaterials, since their dangling bonds make them chemically reactive (Röthlisberger et al. 1994). Contrary to carbon, the formation of silicon fullerenes and nanotubes is disfavored due to the tendency of silicon to prefer sp^3-rather than sp^2-like hybridization. A possible approach to stabilize a silicon cage is to locate a guest atom in the center of the cluster.

In 1987, Beck succeeded in preparing mixed metal-silicon cluster ions by laser vaporization and found strikingly high abundances of $Si_{15}M^+$ and $Si_{16}M^+$ for three different transition-metal dopants M, namely chromium, molybdenum, and tungsten, and low abundances for other sizes (Beck 1987, 1989). However, despite a valuable attempt based on a topological model (King 1991), the magic behavior of these structures remained unexplained for a long time, presumably due to the fact that understanding the structures and properties of bare silicon clusters was itself evolving as an active field of research and that the investigation of clusters doped with transition-metal atoms was a difficult and challenging experimental task.

Almost a decade past by before the first quantum mechanical study on doped silicon cages appeared. Inspired by small C_{28} fullerenes that were stabilized by a central guest atom, as for example in $C_{28}U$, Jackson and coworkers calculated a strikingly large binding energy of 15.2 eV for a Zr atom located inside an icosahedral fullerene-like Si_{20} cage, which is not a stable isomer for the bare silicon cluster (Jackson and Nellermoe 1996).

Another 5 years later, Hiura and coworkers showed experimentally that many types of transition-metal ions can be introduced into silicon cages and that the smallest possible cage size crucially depends on the dopant atom. In addition to the experiments, ab initio calculations were performed to identify the precise structure of $Si_{12}W$, which resulted in a regular hexagonal Si_{12} cage with a tungsten atom in the center (Hiura et al. 2001). The finding of this dopant-encapsulating cluster in combination with the progress made in understanding the structures and properties of bare silicon clusters attracted renewed interest in the very stable $Si_{15}M^+$ and $Si_{16}M^+$ (M = Cr, Mo, W) species, detected by Beck 15 years earlier. Kumar and Kawazoe (2002) computationally attributed the strong stability of these species to highly symmetric cages with exceptionally pronounced highest occupied–lowest unoccupied molecular orbital (HOMO-LUMO) gaps and large embedding energies of the dopant atom.

Today, the interest in doped silicon clusters is so high that theoretical studies have investigated Si_nM structures for dopants from almost every group of the periodic table. Recent investigations on metal-encapsulated silicon cages have proposed novel functional silicon devices (Hiura et al. 2001, Khanna et al. 2002, Kumar and Kawazoe 2003b, Kumar 2006). As compared to elemental silicon clusters, doped clusters are predicted to bear higher stabilities and more symmetric structures, making them attractive for cluster-assembled materials. The novel interest by theoreticians has been quickly followed by experimentalists' efforts. Today, a variety of methods, which go beyond simple mass spectrometry, are applied to study doped silicon clusters.

The focus of this chapter lies on the developments of the experimental approaches. Section 5.2 provides a brief summary of the different models that explain the enhanced stabilities of certain cluster sizes and stoichiometries, stressing that it is generally the interplay between geometric and electronic factors that results in "magic" clusters. Section 5.3 reviews the history of the investigation of doped silicon clusters by different means. In Section 5.3.1, mass spectrometric studies are covered starting with the pioneering work of Beck and arriving at more recent contributions that found some general trends for especially stable species. Several reactivity probes have been developed, especially for investigating the transition from exo- to endohedrally doped silicon clusters. These approaches are presented in Section 5.3.2. Due to these experiments, doped silicon clusters came into the focus of quantum mechanical studies, which are reviewed in Section 5.3.3. Exemplifying the successful interplay between theory and experiment, nowadays innovative instrumental techniques are used to probe the theoretical predictions. Section 5.3.4 shows how the electronic structure of doped silicon clusters

can be analyzed in detail by means of photoelectron spectroscopy (PES). For the investigation of the geometric structure, infrared (IR) spectroscopy is a powerful method. Section 5.3.5 demonstrates how the vibrational fingerprint of doped silicon clusters is obtained upon infrared multiple photon dissociation (IR-MPD) spectroscopy of their complexes with rare-gas atoms.

5.2 Stabilization of Clusters

In the mass spectrometric detection of clusters, their size distribution is not necessarily smooth. Often certain clusters are much more pronounced in intensity than their adjacent sizes. Such modulations often reflect the thermodynamic stability of the clusters under investigation. The origin of an enhanced stability can either be due to geometric or to electronic reasons. Both are briefly discussed in the following section.

5.2.1 Geometric Stabilization

When in 1981 Echt and coworkers recorded the mass spectrum of an effusive beam of xenon clusters, they observed enhanced intensities for clusters containing 13, 55, 147, etc., atoms (Echt et al. 1981). These "magic" sizes reflect the tendency of the cluster's atoms to maximize their number of nearest neighbors and their average coordination, while minimizing their surface area. The numbers can be rationalized by geometric shell closings of icosahedral clusters, known as Mackay icosahedra. Icosahedra are not closed-packed structures and, e.g., in a 13-atom icosahedron, the surface bonds are approximately 6% longer than the radial bonds from the center to the surface atoms. This frustration increases for larger clusters and eventually favors closed-packed geometries. A way to stabilize an icosahedron is the substitution of the central atom by a smaller dopant. One example is the magic $AlPb_{12}^+$ cage (Neukermans et al. 2004). It is remarkably pronounced in the mass spectrum of cationic aluminum-doped lead clusters (see Figure 5.1), and its closed-packed I_h geometry is calculated to be the most stable isomer by not less than 1.96 eV to the next stable isomer.

Another example of a particularly stable high symmetric geometry is neutral Au_{20} (Li et al. 2003, Gruene et al. 2008a). Here the cluster is a pyramid of perfect tetrahedral symmetry and represents a fragment of the bulk fcc structure with four 111 surfaces.

Clusters not necessarily need to be closed-packed in order to gain from a geometric stabilization. The Buckminster fullerene is a hollow cage but is drastically stabilized because it is the smallest cluster size in which all five-membered carbon rings are separated from each other by six-membered rings (Kroto et al. 1985).

5.2.2 Electronic Stabilization

Similar to the mass spectrum of xenon clusters, the spectrum of sodium clusters is highly structured with pronounced peaks for clusters containing 8, 20, 40, 58, and 92 atoms (Knight et al. 1984). These findings were immediately rationalized by assuming a one-electron shell model. In this phenomenological shell

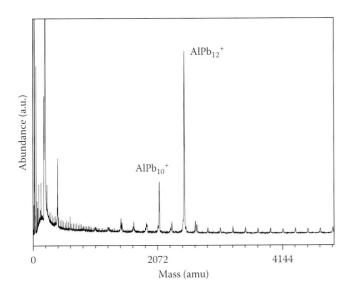

FIGURE 5.1 Mass abundance spectrum of mixed $AlPb_n^+$ clusters, showing the extremely enhanced stability and magic character of $AlPb_{10}^+$ and $AlPb_{12}^+$ (see Neukermans et al. 2004).

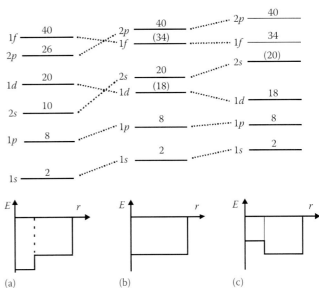

FIGURE 5.2 Schematic representation of the single-particle energy levels in the standard jellium model (b) and in the two-step jellium model with a depression (a) and an increase (c) in the central part of the background potential. Note the magic numbers 10 and 26, caused by the lowering of the 2s and the 2p levels in (b), and the enhanced magic number 18 by lifting the 2s level in (c) (see Baladron and Alonso 1988; Janssens et al. 2004).

model, which bears a lot of resemblance with the shell model as it is well known in nuclear physics, the 3s electrons of the sodium atoms are independently delocalized and move in a spherical potential. Obviously this model can describe only clusters with a spherical geometry, while for species with lower symmetries, the model must be modified accordingly (Clemenger 1985, de Heer 1993). Meanwhile, various implementations of shell models for electrons exist, including the so-called jellium model where the electron problem is solved quantum mechanically, assuming a smeared-out homogeneous positively charged ion distribution.

For binary clusters, a two-step spherical jellium model has been developed (Baladron and Alonso 1988, Janssens et al. 2004) to explain the occurrence of the magic electron number 10 in case of electronegative dopants and 18 for electropositive dopants. For an electronegative dopant in the center of the cluster, the effective potential is more attractive in the vicinity of the dopant. This stabilizes the orbitals that have a high electron density in the center of the cluster, like s and p orbitals, while d and f orbitals are destabilized. Accordingly, the 2s orbital is more favorable than the 1d orbital in case of electronegative dopants, resulting in a shell closing at 10 valence electrons. The opposite is true for electropositive dopants, giving rise to the magic number 18 (Figure 5.2). This model works nicely for spherical main group elements. A good example is the $CuAu_{16}^-$ anion with a compact structure and a very high adiabatic detachment energy of 4.12 eV (Wang et al. 2007).

For transition-metal dopants, it is not always clear how many d electrons are donated to the quasi-free electron gas. While radially extended d-orbitals bear a high propensity to hybridize strongly with s and d electrons from the host, this behavior decreases along a row. Transition-metal-doped clusters are often better described by concepts originating from organometallic chemistry. The 18-electron rule takes into account that a total of 18 electrons is needed to fill the s, p, and d orbitals of the valence

shell of a transition-metal atom. WAu_{12} obeys the 18-electron rule and was calculated to be a very stable cluster (Pyykkö and Runeberg 2002). Shortly afterward, the anion WAu_{12}^- was synthesized in the gas phase and studied by means of PES (Li et al. 2002). The large HOMO-LUMO gap of 1.68 eV measured with PES support the electronic stability of this species. Other examples are 16-atom gold cluster cations endrohedrally doped with a trivalent metal, that all exhibit extraordinary stability with respect to neighboring sizes, as demonstrated with photofragmentation mass spectrometry (Bouwen et al. 1999, Neukermans et al. 2003, Veldeman et al. 2008).

More recently, the concept of aromaticity has been introduced for explaining the stability of metal clusters (Li et al. 2001, Boldyrev and Wang 2005). Also for doped clusters, aromaticity has been used to account for certain especially stable stoichiometries, with Au_5Zn^+ being a prominent example (Tanaka et al. 2003). This planar cluster has 6 valence electrons ($4n + 2$ with $n = 1$, according to a Hückel-aromatic molecule) and a strong magnetic shielding inside and above the structure caused by a diamagnetic ring-current, an effect ascribed to its aromaticity. Aromaticity is not limited to planar species but can also account for enhanced stabilities of spherical clusters (Chen and King 2005).

5.2.3 Combined Effects

It has been realized early on that enhanced stability is achieved best when both geometric and electronic stabilization conditions are fulfilled (Khanna and Jena 1992, Chen et al. 2006). For example, Al_{13}^- is a compact icosahedral structure and its

40 valence electrons result in an electronic shell closing within the jellium model (Bergeron et al. 2004). Similarly, the stability of Au_{20} lies in its tetrahedral geometry, as well as 20 valence electrons satisfying the shell model. The WAu_{12} cluster is stabilized by three factors: the above-mentioned 18-electron rule, a relativistic contraction with strengthening of the W–Au and Au–Au bond, and aurophilic interactions in the gold shell. $AlPb_{12}^{+}$ is stable due to its compact geometric structure but also due to the spherical aromaticity of the Pb_{12}^{2-} cage, which surrounds the formal Al^{3+} dopant.

5.3 Investigation of Doped Silicon Cages

In this section, it is shown how the concepts from Section 5.2 have been used to explain the enhanced intensity of certain doped silicon clusters in mass spectra (Section 5.3.1). The first simple mass spectrometric investigations have been followed by reactivity studies and a wealth of quantum mechanical calculations (Sections 5.3.2 and 5.3.3). In the last years, innovative experimental methods have been applied to doped silicon clusters in order to probe those theoretical predictions (Sections 5.3.4 and 5.3.5).

5.3.1 Mass Spectrometry

Contrary to carbon fullerenes and nanotubes, the formation of a silicon hollow cage or tube is very unlikely. Since the dangling bonds in sp^3-hybridized silicon clusters make them chemically reactive and thus unstable, it is very important to quench these bonds for allowing the formation of cluster-assembled materials. One of the strategies for doing so is the doping of a suitable metal atom into a pure Si cluster, as was first experimentally revealed in 1987 (Beck 1987).

Beck first reported mass spectra for cationic transition-metal-doped silicon clusters Si_nM^+ (M = Cr, Mo, W). Silicon clusters were formed by laser ablation from a silicon wafer. When the carrier gas was seeded with the corresponding transition-metal carbonyl, doped clusters were formed with identical stoichiometries, irrespective of the dopant element. This is shown in Figure 5.3, where high-fluence (1 mJ/cm²) 193 nm ionization mass spectra of bare and doped silicon clusters are compared. Spectrum 3a was obtained using pure helium as carrier gas and shows bare silicon clusters. The spectrum is dominated by clusters containing 6–11 atoms. These cluster cations are particularly stable and result from fragmentation of larger silicon clusters upon absorption of multiple UV photons. Figure 5.3b through d presents the spectra obtained when the helium carrier gas was seeded with metal carbonyls. These latter spectra exhibit additional peaks, darkened in the figure, which were assigned to clusters containing one transition-metal atom and varying amounts of silicon atoms. It is striking how the most prominent new peaks are always due to $Si_{15}M^+$ and $Si_{16}M^+$ clusters (M = Cr, Mo, and W) and that the observed abundances of other sizes of metal-doped silicon clusters are very low. The spectra clearly reveal that incorporating a single metal atom in the Si_{15} and Si_{16} clusters stabilizes these sizes with respect to photofragmentation. In order to

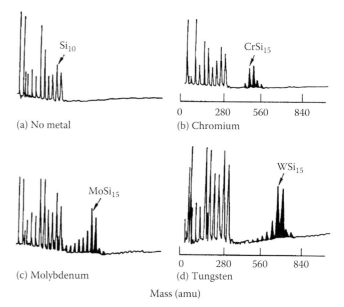

FIGURE 5.3 193 nm (6.4 eV) ArF laser photoionization time-of-flight mass spectra showing: (a) bare silicon clusters formed by laser vaporization of a silicon wafer followed by supersonic expansion, (b) reaction products formed in a supersonic nozzle between chromium atoms and silicon to form Si_nCr clusters, (c) Si_nMo, and (d) Si_nW clusters. The metal-atom-silicon cluster peaks in each spectrum are darkened for emphasis. Undarkened peaks represent bare silicon clusters. ArF laser fluence in each case was about 1 mJ/cm². At this fluence, very little intensity is seen for elemental silicon clusters with more than 11 atoms. (Reproduced from Beck, S.M., *J. Chem. Phys.*, 90, 6306, 1989. With permission.)

understand this behavior, Beck postulated a possible scenario where the metal atom acts as a seed and silicon atoms form a shell structure with a specific number of atoms around the metal dopants (Beck 1989). While the observed mass spectra for the three group 6 metals Cr, Mo, and W with silicon are very similar, doping with copper provided different results (Beck 1989). First, no copper-silicon reaction products containing less than six silicon atoms were observed. Further, the most intense product peak in the spectrum was due to $Si_{10}Cu^+$.

Since Beck described the remarkable formation of $Si_{15}M^+$ and $Si_{16}M^+$, more mass spectrometric investigations have been presented. The enhanced stability of $Si_{15,16}Cr^+$ and $Si_{10}Cu^+$ was confirmed by Neukermans and coworkers who presented a mass spectrometric stability investigation of Cr, Mn, Cu, and Zn-doped Si_n, Ge_n, Sn_n, and Pb_n clusters (Neukermans et al. 2006). Moreover, significantly enhanced abundances for $Si_{15}Mn^+$ and $Si_{16}Mn^+$ were found.

Ohara and coworkers studied mass spectrometrically anionic silicon clusters doped with the group 4 transition metals titanium and hafnium and the group 6 transition metals molybdenum and tungsten (Ohara et al. 2003). Typical time-of-flight mass spectra were dominated by pure silicon cluster anions with $n = 4$–11. In addition to the pure Si_n clusters, Si_n anions doped with a single transition-metal atom were observed in the mass spectra again with a predominant formation of clusters containing 15 and 16 silicon atoms. They thus revealed that the striking stability of certain cationic silicon clusters doped with group 6 metals, as observed by Beck (1987), also persists in the case of anionic clusters and in the case of dopants of other groups. This behavior rather pointed to a geometric stabilization and a special role of the Si_{16} cage, as the number of electrons seemed to be of minor importance.

However, in 2005 Koyasu and coworkers presented an experimental investigation of scandium-, titanium-, and vanadium-doped silicon clusters (Koyasu et al. 2005). This study revealed very clearly that the electronic structure also plays an important role. Neutral $Si_{16}Ti$ species could selectively be produced by fine-tuning the cluster source, as shown in Figure 5.4b. The mass spectra of neutral scandium- and vanadium-doped silicon clusters, however, hardly showed any preference for the size containing 16 silicon atoms. Only when isoelectronic species were formed, magic clusters could be detected; thus, a change of the charge state was required. The three magic sizes $ScSi_{16}^-$, $TiSi_{16}$, and VSi_{16}^+ (see Figure 5.4a through c) point toward a closed electron configuration. Koyasu and coworkers proposed that the transition-metal dopants each contribute their 4 valence

electrons in order to yield, together with the 16 $3p_z$ electrons of the sp^2 hybridized Si atoms, a closed electronic shell of 20 electrons according to the jellium model.

Koyasu and coworkers continued their investigations by systematically changing the dopant atoms among the group 3, 4, and 5 elements (Koyasu et al. 2007). Figure 5.4 shows the typical mass spectra of the transition-metal atom-doped clusters $Si_nM^{-/0/+}$ (M = Sc, Y, Lu, Ti, Zr, Hf, V, Nb, and Ta). The neutral clusters of Si_nTi, Si_nZr, and Si_nHf were photoionized by a pulsed F_2 laser (7.90 eV) for mass spectrometric detection, while the charged metal-silicon clusters were directly accelerated in a pulsed electric field. In all of the mass spectra of $Si_nM^{-/0/+}$, bimodal size distributions were observed. The two distributions are indicated in Figure 5.4 by solid squares (distribution I, $n = 6–11$) and open circles (distribution II, $n = 12–20$), respectively. As confirmed by a chemical-probe method (described in detail in Section 5.3.2), the bimodal distributions correspond to the structural transition from exohedral to endohedral geometries, i.e., a transition from structures where the dopant is located on the cluster's surface to

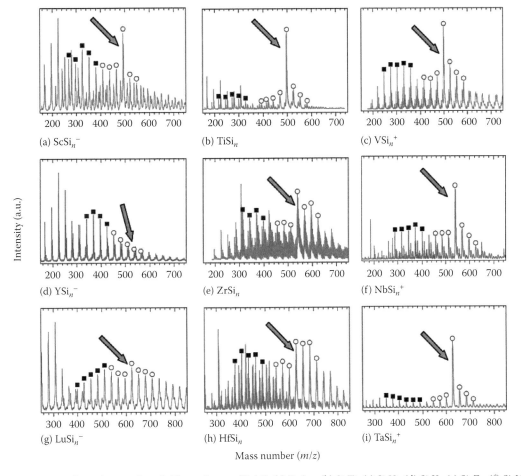

FIGURE 5.4 Mass spectra of metal-atom-doped silicon clusters (Si_nM): (a) Si_nSc^-, (b) Si_nTi, (c) Si_nV^+, (d) Si_nY^-, (e) Si_nZr, (f) Si_nNb^+, (g) Si_nLu^-, (h) Si_nHf, and (i) Si_nTa^+. Left three columns are anionic Si_nM^- (M = Sc, Y, and Lu), middle three columns are neutral Si_nM (M = Ti, Zr, and Hf), and right three columns are cationic Si_nM^+ (M = V, Nb, and Ta). Bimodal distributions were typically observed. The solid squares and open circles represent the two distributions (I and II). Distribution I was similar to that of pure Si_n in their charge state, while distribution II had a prominent peak at $Si_{16}M$, indicating that these species are particularly stable. Arrows indicate the mass peaks of the $Si_{16}M^{-/0/+}$ clusters. (Reproduced from Koyasu, K. et al., *J. Phys. Chem. A*, 111, 42, 2007. With permission.)

structures in which the dopant is fully surrounded by the silicon atoms (Koyasu et al. 2007). Indeed, the abundance of smaller Si_n clusters ($n = 6$–11) is relatively rich, while larger-sized Si_n clusters ($n \geq 12$) are formed to a much smaller extent; a finding very similar to the one in Beck's early work. The similarity of the size distribution for small ($n = 6$–11) bare and doped silicon clusters was presumed to be due to the stability of the bare Si clusters and indicated that clusters of distribution I exhibit an exohedral dopant. Distribution II, on the other hand, was completely different from that of pure Si clusters. It exhibited a prominent peak at $Si_{16}M^{-/0/+}$, whereas Si_{16} was hardly formed at all. In particular, doping with the transition-metal atoms Sc, Ti, V, Zr, Nb, Hf, and Ta resulted in the magic behavior of $Si_{16}M$, while the intensities of $Si_{16}Y^-$ and $Si_{16}Lu^-$ were hardly more intense than their neighboring sizes with $n = 15$ and 17.

5.3.2 Reactivity and Chemical-Probe Studies

The work of Hiura and coworkers constitutes another landmark in the investigation of doped silicon clusters (Hiura et al. 2001). They introduced metal ions into an ion trap by resistive heating of a metal wire and subsequent ionization by electron irradiation. As a silicon source, SiH_4 gas was introduced into the trap. The transition-metal cation reacted subsequently with silane, resulting in the formation of $Si_nMH_x^+$ species. It was observed that the sequential growth of $Si_nMH_x^+$ species almost stopped when n approached a specific number m. Moreover, a clear dependence of m on the metal element M was noted, e.g., $m = 14$ for Hf, 13 for Ta, 12 for W, 11 for Re, and 9 for Ir. Furthermore, for elements of the same group in the periodic table, the same termination number m was found. When $n = m$, the clusters tended to lose all of their H atoms, and dehydrogenated clusters, Si_mM^+, were highly abundant in the mass spectrum. This finding suggested that the clusters were not chainlike but that the metal atom was highly coordinated and saturated all dangling bonds of the silicon atoms. Strikingly, the sum of m and the atomic number of the transition-metal element was often found to keep a constant value of 86, which corresponds to the atomic number of the rare gas Rn. Thus, the maximum amount of silane molecules that would be added is to a large extent determined by the 18-electron rule (see Section 5.2.1). In $Si_{12}W$, each Si atom in the cluster donates a single electron to the central tungsten atom, which then possesses 18 electrons in total (12 electrons from the Si_{12} cage and 6 valence electrons by the tungsten atom itself), resulting in a closed electronic shell. Three valence electrons on each silicon atom are left and thus allow for the formation of a silicon polyhedron without appreciable dangling bonds. The same argument accounted for the termination numbers for M = Hf, Ta, Re, and Ir as well (Hiura et al. 2001). Hiura and coworkers based their discussion of the termination numbers mainly on electronic arguments. They mentioned that probably geometric factors also add to the enhanced stabilities; however, they did not make any claims whether the experimental findings corresponded to the smallest cage possible. Since most probably the transition-metal atom constitutes the reactive site toward silane, the reduced reactivity from a certain size onward could also mark the onset of endohedrally doped silicon cages.

This question was explicitly tackled by another method by Nakajima and coworkers (Ohara et al. 2002). This group produced doped silicon clusters by ablation from two target rods by means of two independent lasers. In the work from 2002, anionic terbium-doped silicon clusters were produced and allowed to react with H_2O vapor downstream of the source in a flow reactor. Upon exposure to H_2O, doped clusters that contained less than 10 silicon atoms were found to react away, while larger clusters remained unreactive. Presuming that an exohedral terbium atom was the reactive site for the adsorption reaction with water, the low reactivity of $TbSi_n^-$ for $n \geq 10$ indicated cage clusters, in which silicon completely surrounds the terbium dopant. Correspondingly, at least 10 silicon atoms were therefore needed to fully cover a terbium atom (problems with impurities were noted later, which, however, should not have affected the size range of the basket-to-cage transition [Ohara et al. 2007]). For Si_nTi^+ clusters, the decrease upon reaction toward H_2O was observed at $n = 7$–11, whereas clusters with $n = 13$–17 remained practically unaffected (Ohara et al. 2003).

After improvements in their source design that allowed for a better mixing of the two plasmas in the cluster formation process, the Nakajima group extended their reactivity studies to all stable group 3, 4, and 5 transition-metal doped silicon clusters in their cationic, neutral, and anionic form (Koyasu et al. 2007). The results are summarized in Table 5.1. An element and charge-state dependence can be noted. Obviously, within the same row of the periodic system and the same charge state, the threshold sizes for dopant encapsulation decreases with increasing atomic number. This can be understood in terms of atomic radii, which

TABLE 5.1 Smallest Doped Silicon Cages MSi_n for Cationic, Neutral, and Anionic Species for Various Transition-Metal Dopants

	Cation	Neutral	Anion	Cation	Neutral	Anion	Cation	Neutral	Anion	Cation	Cation	Cation
M		Sc			Ti			V		Cr	Co	Cu
N	17[a]	15[a]	15[a]	13[a,b]	13[a]	11[a]	12[a,b]	10[a]	9[a]	11[b]	8[b]	12[b]
M		Y			Zr			Nb				
N	21[a]	20[a]	20[a]	15[a]	14[a]	12[a]	13[a]	12[a]	11[a]			
M		Lu			Hf			Ta				
N	21[a]	16[a]	18[a]	14[a]	14[a]	12[a]	13[a]	10[a]	11[a]			

Sources: Koyasu, K. et al., *J. Phys. Chem. A*, 111, 42, 2007; Janssens, E. et al., *Phys. Rev. Lett.*, 99, 063401/1, 2007.

Note: The threshold sizes have been determined experimentally upon their reactivity with water vapor (a) or argon (b).

generally decrease with increasing atomic number in one row. A larger dopant obviously requires more silicon atoms to be fully surrounded than a smaller one. As for the charge-state dependence, the threshold sizes decrease going from cations to neutrals to anions, although the charge-state dependence of anionic Si_nLu^- and Si_nTa^- is exceptional. The threshold size dependence on the charge state was rationalized by metal encapsulation as well. According to mobility experiments on pure silicon clusters, the anions have systematically longer drift times, and hence smaller mobilities, than the cations (Hudgins et al. 1999). Consequently, the anions are effectively larger than the cations. This systematic shift was assigned not to a structural change, but to the extra charge causing the surface electron density of the anions to spill out further than for the cations. Here, however, it seems that the anionic silicon cages are actually larger as compared to the cationic ones and afford for transition-metal encapsulation already at smaller cluster sizes. The species that were richly abundant in the mass spectra (e.g., $Si_{16}Sc^-$, $Si_{16}Ti$, $Si_{16}V^+$, $Si_{16}Zr$, $Si_{16}Nb^+$, $Si_{16}Hf$, and $Si_{16}Ta^+$, as shown in Figure 5.4) all exhibit no adsorption reactivity, pointing toward especially stable cages.

To investigate the structure of doped Si clusters, not only molecules such as SiH_4 and H_2O can be used in chemical-probe techniques. This was recently shown by Janssens and coworkers who reported that the physisorption of argon atoms acts as a structural probe for transition-metal-doped silicon clusters (Janssens et al. 2007). Ar is an ideal probe as it has a negligible influence on the cluster and merely serves as a spectator atom. Further, it has no internal degrees of freedom and represents a good compromise in terms of polarizability. It can be attached to the cluster ion without much perturbation of the electronic structure. The study focused on cationic silicon clusters doped with 3d transition metals. Doped clusters were formed in a dual-target source (Bouwen et al. 2000) using a mixture of 1% of Ar in He as a carrier gas. All clusters passed through a thermalization channel that was cooled by a flow of liquid nitrogen to 80 K. In the mass spectra of mixed $Si_nM_m^+$ (M = Ti, V, Cr, Co, Cu) clusters, the highest ion signals were recorded for bare Si_n^+ and singly doped Si_nM^+ clusters, but also a fraction of $Si_nM_2^+$ clusters was formed. For both singly and doubly doped clusters,

$Si_nM_{1,2}^+ Ar_{1,2}$ complexes were formed, whereas for bare Si_n^+ no argon complexes were observed at all. Moreover, the abundance of the $Si_nM_{1,2}^+ Ar_{1,2}$ complexes was strongly size dependent and collapsed after a certain critical number of Si atoms in the cluster. These effects are represented best by plotting the total fraction of Ar complexes as a function of n, as is shown in Figure 5.5. Both for singly and doubly doped silicon clusters, critical sizes for argon attachment, which depend on the dopant element, are observed. Figure 5.5 shows how, in agreement with the observations of Koyasu et al., for singly doped silicon clusters the threshold size (defined as the smallest size that does not form complexes) changes along the 3d row, i.e., $Si_{13}Ti^+$, $Si_{12}V^+$, $Si_{11}Cr^+$, and Si_8Co^+. The threshold number for copper doping increases for $Si_{12}Cu^+$ in agreement with the increasing covalent radius, when going from cobalt to copper (Cordero et al. 2008). With the knowledge that Si_n^+ clusters do not form stable complexes with Ar at 80 K, it was assumed that Ar must bind to the dopant. As for SiH_4 and H_2O, binding to the transition-metal atom was only feasible if the dopant was located on the surface of the host cluster. If the dopant resided in the interior of a Si_n cage, the Ar atom could only interact with Si surface atoms; thus, no Ar complexes were formed. The disappearance of the Ar complexes marked the formation of endohedral clusters. Additionally, for doubly doped silicon clusters, the threshold size also decreases along the 3d row: $Si_{20}Ti_2^+$, $Si_{18}V_2^+$, $Si_{17}Cr_2^+$, and $Si_{14}Co_2^+$, as seen in Figure 5.5. The gradual transition of the Ar sticking probability for $Si_nV_2^+$ and $Si_nCo_2^+$ was attributed to the possible coexistence of endohedral and exohedral isomers.

5.3.3 Quantum Chemical Calculations

In 1996, Jackson and Nellermoe published their observation of a remarkably strongly bound endohedral zirconium atom inside a Si_{20} cage. Based on local-density approximation calculations, they found a binding energy of 11.2 eV with respect to an isolated Zr atom and the most stable Si_{20} isomer known at the time (Jackson and Nellermoe 1996). Interestingly, they were obviously unaware of the experimental study by Beck almost 10 years earlier, since they encourage experimentalists to start studying doped silicon clusters without citing Beck's work. For 5 years, the call for

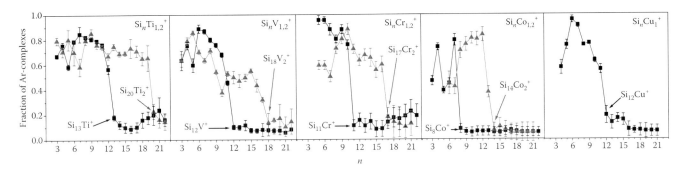

FIGURE 5.5 Fraction of argon complexes formed for Si_nM^+ (M = Ti, V, Cr, Co, Cu) as a function of cluster size. A critical size for which argon complexes no longer are formed is found for both singly and doubly doped species. The corresponding sizes are the smallest clusters, exhibiting endohedrally doped silicon cage structures (see Janssens et al. 2007).

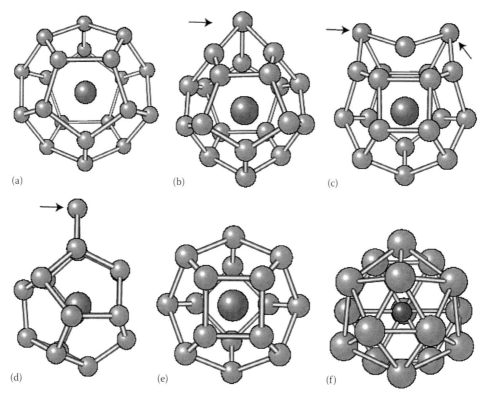

FIGURE 5.6 Shrinkage of the Si cage. (a) Dodecahedral Zr-encapsulated Zr@Si$_{20}$, as found by Jackson and Nellermoe (1996), (b)–(e) optimized structures of Zr@Si$_{20}$, Zr@Si$_{19}$, Zr@Si$_{17}$, and Zr@Si$_{16}$, respectively. The arrows indicate the atoms that were removed. (f) The Frank-Kasper polyhedral structure of M@Si$_{16}$ (M = Ti and Hf). (Reproduced from Kumar, V. and Kawazoe, Y., *Phys. Rev. Lett.*, 87, 045503/1, 2001. With permission.)

experimental verification remained unheard until Hiura et al. triggered again the interest in doped silicon clusters. A few months later, Kumar and Kawazoe used the Jackson and Nellermoe work as a basis for the reinvestigation of Zr-doped silicon clusters (Kumar and Kawazoe 2001). They found that the cage proposed by Jackson is not stable but that the cage shrinks with one silicon atom sticking out (Figure 5.6a and b). The removal of this atom and reoptimization lead to a ZrSi$_{19}$ cage with two silicon atoms sticking out (Figure 5.6c). The further removal of these two atoms and reoptimization lead to ZrSi$_{17}$, a cage from which again one atom sticks out (Figure 5.6d). Finally, the removal of this atom yielded a compact, fullerene-like ZrSi$_{16}$ cage (Figure 5.6e). The embedding energy of the Zr atom was calculated to be almost 14 eV. For titanium as the dopant element, a Frank-Kasper (FK) polyhedron (Figure 5.6f) was found to be the most stable isomer with an exceptionally large HOMO-LUMO gap of 2.36 eV (Kumar and Kawazoe 2001, 2003a).

Since the Hiura and Kumar work, many theoretical studies have addressed the structures of Si$_n$M clusters with dopants from almost every group of the periodic table. Additional quantum chemical investigations have focused on periodic trends when doping silicon clusters. A review of all theoretical work on doped silicon clusters would go far beyond the scope of this chapter. Instead, only some selected studies are chosen, which deal with systems that have been presented in the Sections 5.3.1 and 5.3.2.

While it seems that alkaline-doped clusters containing up to 20 silicon atoms are always more stable when the dopant

is situated on the surface of the cluster (Sporea and Rabilloud 2007), transition-metal doped silicon clusters form cage-like geometries from a certain size onward (Guo et al. 2008).

Beck had already assigned the enhanced stability of the group-6-doped cluster cations MSi$_{15}^+$ and MSi$_{16}^+$ to a highly coordinated central transition-metal atom. Their structures remained unknown until Kumar and Kawazoe confirmed that metal-encapsulated Si$_n$ cage structures account for their unusual stability, performing ab initio calculations on Si$_n$M clusters (M = Cr, Mo, and W; n = 14–17) (Kumar and Kawazoe 2002). For CrSi$_{15}$, different isomers have been proposed as ground-state structures over the years by the same group. While initially a structure was proposed that was constructed by placing one silicon atom on the face of a body-centered cubic structure (Kumar and Kawazoe 2002), in a later study three very similar structures based on a decahedral isomer were found to be almost degenerate in energy (Kawamura et al. 2004). In a more recent work, a FK-type polyhedron has been proposed to be the ground-state geometry (Kumar 2006). The calculated global minimum of CrSi$_{16}$ was constructed by capping a higher-energy fullerene-like isomer of CrSi$_{15}$. Kumar found competing growth models that gave rise to rather different structures of group-6-encapsulated 15- and 16-atom silicon clusters and account for their simultaneous strong abundances (Kumar and Kawazoe 2002).

Also, the findings of Koyasu and coworkers can be understood better with the help of theory. As mentioned above,

$TiSi_{16}$ has been found to adopt an FK polyhedral geometry with a remarkable HOMO-LUMO gap of 2.36 eV (Kumar and Kawazoe 2001, 2003a). This isomer was found to lie 0.18 eV lower in energy than the fullerene-like isomer (Figure 5.6e), while the zirconium-doped fullerene-like isomer has been calculated to be more than 1 eV more stable than the FK isomer (Kumar et al. 2003). Reveles and Khanna (2006) found that the compact FK isomer was also the ground-state structure for $ScSi_{16}^-$ and VSi_{16}^+. Whenever the number of valence electrons was different from 20, the structures distorted. However, the stability cannot be explained only by adding together the number of valence electrons of the dopant and the number of silicon atoms, since, e.g., $ScSi_{17}$ or $TiSi_{17}$ were not found to be more pronounced in the mass spectra. It was proposed that it was necessary to identify the silicon sites, which contribute an electron to the free-electron gas of the cluster. Such sites were identified upon the bond critical points in the charge density. Only in the compact structures, all silicon atoms were directly coordinated to the metal atom and could therefore contribute an electron. Thus, both the geometry and the electron counting needed to be considered to correctly predict stable clusters (Reveles and Khanna 2006). Recently, it was found that the perfectly symmetric FK polyhedron might not be the global minimum for the 16-atom cage but that slightly distorted geometries lie a bit lower in energy (Torres et al. 2007).

Reactivity studies, as described in Section 5.3.2, mainly addressed the question at what size the transition from exo- to endohedrally doped silicon clusters occurs. Hiura et al. (2001) explained their findings for the 5d transition-metal dopants mainly on the basis of an 18-electron rule. However, the reduced reactivity from a certain size onward could also be explained by endohedral cages, since the sizes for which the reactivity dropped are the same as the threshold sizes identified for cationic Hf and W-doped silicon clusters by their reactivity with water vapor by Koyasu and coworkers (Table 5.1) (Koyasu et al. 2007). In the case of Hf-doped clusters, theory does not reproduce the experimental transition size precisely but calculates basket-like structures up to $HfSi_{12}$ and cage-like structures for larger species (Kawamura et al. 2005). Studies by Guo and coworkers found the smallest cages for cationic and neutral tantalum-doped silicon clusters at $TaSi_{12}$ (Guo et al. 2004, 2006), and therefore did not reproduce the experimental findings neither for the cations nor for the neutrals (see Table 5.1). The reduced reactivity of WSi_{12}^+ is in agreement with a highly symmetric D_{6h} cage structure, as calculated for neutral WSi_{12} (Miyazaki et al. 2002, Lu and Nagase 2003).

Some better agreement has been achieved between theory and the experimental studies for 3d transition-metal dopants (Janssens et al. 2007, Koyasu et al. 2007). The proposed basket-shaped structures for Si_nTi ($n = 8-12$) and endohedral systems for larger sizes (Kawamura et al. 2005) agree nicely with the experiment, though Guo and coworkers predicted an endohedral structure already for $Si_{12}Ti$ (Guo et al. 2007). Also for $Si_{12}V$, Guo found a cage, which is in agreement with the critical size obtained for the cationic cluster but not with the threshold size measured for neutral vanadium-doped silicon clusters (Table 5.1) (Guo et al. 2008). It was predicted by Khanna and coworkers that Cr is small enough to occupy the center of Si_n ($n = 11-14$) cages (Khanna et al. 2002). Other groups, however, found caged structures for Si_nCr from $n \geq 12$ and open basket-like structures for $n < 12$ (Kawamura et al. 2004, Guo et al. 2008). The argon-physisorption experiments identified $Si_{11}Cr^+$ as the smallest endohedral Si_nCr^+ cluster (see Figure 5.5) (Janssens et al. 2007). In the case of similar structures for cations and neutral clusters, these findings would favor the findings of Khanna et al. The exohedral structure of neutral Si_9Co found by different groups (Lu and Nagase 2003, Guo et al. 2008) is different from the disappearance of the Ar complexes from Si_8Co^+ onward. Ma and coworkers are one of the very few groups that have applied a global-minimum search based on a genetic algorithm (Ma et al. 2005). The genetic algorithm used a tight-binding model and resulted mainly in structures in which the cobalt atom prefers to sit in the center of the silicon cluster. The identified low-lying isomers were further optimized with density functional theory (DFT) within the generalized gradient approximation. They identified Si_9Co as the smallest cage, while Si_8Co was a cluster in which cobalt is almost fully surrounded by silicon atoms (Ma et al. 2005). In that regard, the results are in quite good agreement with the critical size found for argon physisorption (Figure 5.5).

Hagelberg and coworkers have been interested in copper-doped silicon clusters since the late 1990s. They have shown that $Si_{12}Cu$ is most stable in a cage geometry, while in $Si_{10}Cu$ the Cu atom occupies a surface site (Xiao et al. 2002). Again, the argon-physisorption experiments support the computations.

In general, quantum chemical calculations have helped a lot in the understanding of the physical and chemical properties of doped silicon clusters. The enhanced stabilities of certain cluster sizes have been reproduced or, e.g., in the case of $TiSi_{16}$, even predicted. Stabilization has been shown to be mainly due to geometric shell closings and especially compact structures, like the FK polyhedron and the 20 electron system of $TiSi_{16}$. Also, the onset of cage formation is often predicted correctly. However, cages are only formed for rather large particles, and the potential energy surface of such species is characterized by an enormous number of local minima. The binary composition of doped silicon clusters enhances the complexity of the problem even further. In general, an effective sampling of the potential energy surface of binary clusters for identifying its global minimum requires global-optimization techniques (Johnston 2003, Ferrando et al. 2008). Probably due to the high computational costs, only very few studies on doped silicon clusters have employed global-minimum search schemes (Ona et al. 2004, Ma et al. 2005, Wu and Hagelberg 2006). Still, these studies have been limited by the computational method that has been used for the global optimization or they have been restricted to a rather small size range.

Even if a putative global minimum can be identified, the error associated with the theoretical method still makes definite assignments difficult. The situation is somewhat different once detailed experimental data that characterize the species produced under laboratory conditions are available. In such cases, the search can be stopped once an isomer is found whose calculated characteristics fit the experiment. Mass spectrometry and reactivity studies have added a great deal to the knowledge of doped silicon clusters. However, they are somewhat limited regarding the content of detailed information on the electronic and geometric structure of the particle. For such insights, two methods have been proven to be of major importance, namely photoelectron and vibrational spectroscopy.

5.3.4 Photoelectron Spectroscopy

The credit for the first photoelectron spectrum of an anionic transition-metal-doped silicon cluster goes again to Nakajima and coworkers. They published spectra for mixed terbium-silicon clusters containing between 6 and 16 silicon atoms (Ohara et al. 2002). It was found that the electron affinities changed remarkably from ~2.2 eV for $TbSi_9$ to 3.6 eV for $TbSi_{10}$. In agreement with their reactivity studies (see Section 5.3.2), they interpreted this result as support for the onset of cage formation. A drastic drop in the electron affinities for clusters of masses corresponding to species containing 12 and more silicon atoms could not be explained and was later shown to be due to Tb_3OSi_n clusters. With better mass resolution and cleaner spectra, the electron affinities of clusters $TbSi_n$ with $n \geq 12$ were determined to be around 3 eV (Ohara et al. 2007).

Ohara and coworkers extended their photoelectron studies to silicon cluster anions doped with Ti, Hf, Mo, and W (Ohara et al.

2003). Differences between the spectra of bare silicon clusters and the doped species immediately pointed to structural rearrangements upon doping. Local minima in the electron affinities for doped clusters containing 16 silicon atoms were noted. However, the authors did not comment on the already visible large HOMO-LUMO gap of these cluster sizes but supposed a geometric stabilization upon cage formation (Ohara et al. 2003).

A large HOMO-LUMO gap of 1.9 eV was reported explicitly by the Nakajima group in 2005 to account for the enhanced intensity of $TiSi_{16}$ in the mass spectrum (Koyasu et al. 2005). The electron affinity of $TiSi_{16}$ was found to be only 2.03 eV (Figure 5.7). The large gap confirmed the stability of a closed 20 electron shell. The experimental findings were in reasonable agreement with theoretical predictions of an electron affinity of 1.91 eV and a HOMO-LUMO gap of 2.35 eV (Kumar and Kawazoe 2001, Kumar et al. 2003). For the anions $ScSi_{16}^-$, VSi_{16}^-, and $TiSi_{16}F^-$ instead, the HOMO-LUMO gap was too small to be detected, further confirming the unique stability of $TiSi_{16}$ (Figure 5.7) (Koyasu et al. 2005).

Remarkably large HOMO-LUMO gaps were measured also for $ZrSi_{16}$ and $HfSi_{16}$ with 1.36 and 1.37 eV, respectively. Compared to the 1.9 eV in the case of titanium doping, the trend of these values reproduces the order of the atomic radii, which are 1.45 Å for titanium, 1.59 Å for zirconium, and 1.56 Å for hafnium (Koyasu et al. 2007). Extending the studied size range, local maxima in the electron affinity were found for group-3-doped silicon cluster anions containing 10 silicon atoms. This effect was rationalized by an especially stable electronic configuration of the assumed Si_{10}^{4-} frame (Koyasu et al. 2008).

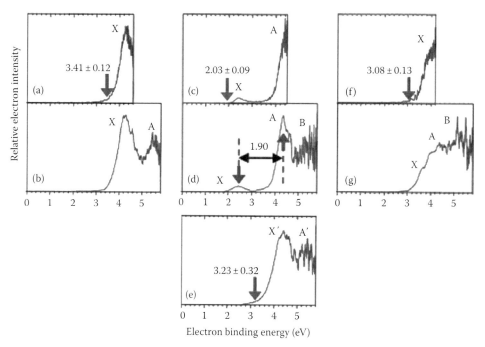

FIGURE 5.7 Photoelectron spectra of $ScSi_{16}^-$ (a and b), $TiSi_{16}^-$ (c and d), and VSi_{16}^- (f and g) at two different photon energies, at 266 nm (4.66 eV; top three spectra) and at 213 nm (5.82 eV; bottom four spectra). The HOMO-LUMO gap can be assigned by comparing the photoelectron spectra of $TiSi_{16}^-$ with that of $TiSi_{16}F^-$ (e). Photoelectrons were analyzed with a magnetic bottle-type photoelectron spectrometer calibrated using a known spectrum of Au^-. (Reproduced from Koyasu, K. et al., *J. Am. Chem. Soc.*, 127, 4998, 2005. With permission.)

A second group that has studied doped silicon clusters by means of PES is the one of Bowen and coworkers. They produced Cr-doped silicon cluster anions upon laser ablation from a chromium-coated silicon rod (Zheng et al. 2005). Their photoelectron spectrum of $CrSi_{12}^-$ started with a smooth onset and then exhibited a very sharp narrow peak. This peak has been argued to represent the transition from the ground state of the anion to the ground state of the neutral and confirmed that the geometry of the anions and the neutral species are similar. Also, their measured vertical detachment energy of 3.18 eV nicely reproduced the predicted value of 3.11 eV (Zheng et al. 2005). Unfortunately, they were not able to confirm the predicted HOMO-LUMO gap of ~1 eV.

Using a two-laser vaporization source, Bowen and coworkers recently managed to produce sufficient amounts of europium-doped silicon cluster anions to record their photoelectron spectra (Grubisic et al. 2008). They noted a large increase in the electron affinity from 1.9 to 2.8 eV, when going from $EuSi_{11}$ to $EuSi_{12}$. This finding was similar to the one by Ohara and coworkers for terbium-doped silicon clusters and has been consequently attributed to a structural rearrangement from exo- to endohedrally doped silicon clusters (Grubisic et al. 2008).

5.3.5 Infrared Spectroscopy on Doped Silicon Clusters

PES yields valuable information on the electronic structures of gas-phase clusters. However, it is not always clear in how far structural rearrangements are responsible for measured changes in the electronic structures. Vice versa, the similar photoelectron spectra of different clusters do not necessarily mean that the geometric structure is similar.

IR spectroscopy is very sensitive to the cluster's internal structure as molecular vibrations directly reflect the arrangements of atoms in the cluster and the forces acting between them. Unfortunately, achievable cluster densities in the gas phase are usually too low for the detection of direct absorption in the IR. One way to circumvent this problem is to deposit and accumulate clusters inside a cryogenic matrix. This technique has been used to assign the structures of small bare silicon clusters (Honea et al. 1993) but has not been applied to doped silicon clusters yet.

If one wants to study true gas-phase clusters, coupling spectroscopic techniques with mass spectrometry allows, in principle, for detection efficiency close to unity. The use of mass spectrometric detection requires a reaction of the cluster in response to the absorption of IR light, which can, e.g., be fragmentation. A similar approach has been used by Duncan and coworkers in the visible range, using a 532 and 355 nm laser to dissociate cationic silicon clusters doped with copper, silver, and chromium atoms (Jaeger et al. 2006). The clusters that contained seven silicon atoms tended to lose primarily the dopant atom, indicating exohedral structures. The same was true for $AgSi_{10}^+$ and $CuSi_{10}^+$. For $CrSi_{15}^+$ and $CrSi_{16}^+$, however, dissociation occurred mainly through the loss of silicon atoms, confirming endohedral cages, as predicted earlier (Jaeger et al. 2006).

Tunable table-top laser systems do not provide sufficient fluence in the IR to excite strongly-bound clusters to a point where fragmentation can occur. Only with the access to powerful free-electron lasers like the Free Electron Laser for Infrared eXperiments (FELIX, see [Oepts et al. 1995]), a variety of innovative means of gas-phase spectroscopy became possible (Asmis et al. 2007).

In IR-MPD spectroscopy, either the depletion of parent ions or the formation of the photofragments is monitored to probe the absorption process. Probing the fragments relies on an initial mass selection and gives rise to almost background-free spectra (Asmis and Sauer 2007). The measurement of depletion has the advantage that all complexes in the molecular beam are probed, and since the detection method is mass-selective, the simultaneous measurement of IR spectra for different cluster sizes is possible (Lyon et al. 2009). Further, also neutral molecules can be probed with FELIX radiation (Fielicke et al. 2005, Gruene et al. 2008a). A disadvantage of measuring the depletion is that the spectra are not background-free and can suffer from a limited signal-to-noise ratio.

The vibrational transitions of doped silicon clusters lie in the far-infrared (FIR), typically between 150 and 600 cm^{-1}, which corresponds to an energy per photon of only ~20–75 meV. On the other hand, the clusters are rather strongly bound with bond dissociation energies of around 4 eV for bare silicon clusters (Jarrold 1995). Even with the high laser power that is provided by FELIX, IR-MPD of such species has not been observed. This problem can be overcome by using the so-called messenger method, in which a loosely bound ligand that is supposed to have a minor to negligible influence on the structure and vibrational properties, is attached to the species that is to be analyzed (Fielicke et al. 2004).

As has been shown in Section 5.3.2, argon binds to exohedrally doped silicon clusters at 80 K. Therefore, the vibrational spectra of small exohedral vanadium- and copper-doped silicon cluster cations have been obtained upon IR-MPD of their complexes with one argon atom (Gruene et al. 2008b). Figure 5.8 shows the experimental and theoretical vibrational spectra of Si_nCu^+ and Si_nV^+ ($n = 6$–8). Only the theoretical spectrum of the particular isomer that reproduced the experimental spectrum the best is shown here. In general, the peak positions are in good agreement while the peak intensities deviate between theory and experiment. In particular, the low-energy absorptions around 300 cm^{-1} are less pronounced in the experiment, which could be due to the larger number of photons needed for photodissociation. Furthermore, it has to be kept in mind that the IR-MPD spectra do not correspond directly to linear absorption spectra (Oomens et al. 2006). In almost all cases, the experimental spectrum was reproduced best by the calculated lowest-energy structure. For Si_6V^+, theory found a Si-capped octahedron as the lowest-energy structure, while the experiment was reproduced much better by the spectrum of a triplet-state pentagonal bipyramid with vanadium in an equatorial position, which was calculated to be 0.03 eV higher in energy. Interestingly, the experiment did not show any features that would point to the coexistence

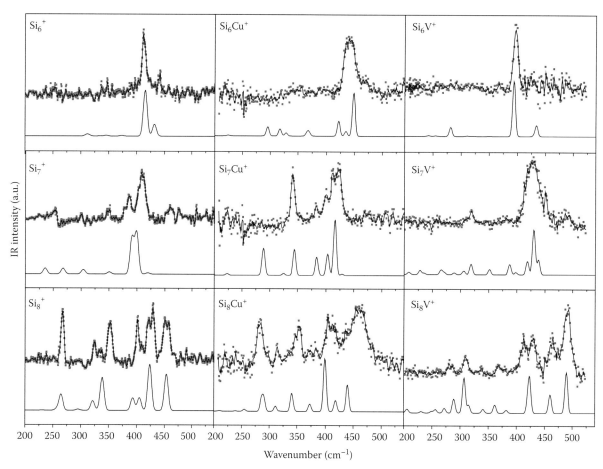

FIGURE 5.8 Vibrational spectra of Si_n^+ (left), Si_nCu^+ (center), and Si_nV^+ (right) ($n = 6–8$). The upper traces in each panel show the experimental IR-MPD spectra of the corresponding complex with one argon atom. The experimental data points are overlaid with a three-point running average to account for the bandwidth of FELIX. They are compared to the calculated vibrational spectra of the best fitting isomer. The calculated stick spectra are folded with a Gaussian linewidth function of $5\,cm^{-1}$ full-width at half-maximum for ease of comparison (see Gruene et al. 2008b and Lyon et al. 2009).

of a second isomer, although the isomers were calculated to be extremely close in energy. The assigned structures for the copper- and vanadium-doped silicon clusters are shown in Figure 5.9. They can be compared among themselves as well as with the predicted structures of bare cationic silicon clusters in order to elucidate the influence of the dopant. IR-MPD has led to definite assignments for the structures of small cationic silicon clusters (Lyon et al. 2009). For Si_6^+, an edge-capped trigonal bipyramid was found to be the lowest-energy structure. Si_7^+ is a distorted pentagonal bipyramid, while Si_8^+ is an edge-capped pentagonal bipyramid (Figure 5.9 left; experimental and computed IR spectra for the corresponding cationic bare Si clusters are included in Figure 5.8). In principle, three types of doped silicon structures are possible. The dopant can (1) add to, or (2) substitute a silicon atom in a bare silicon cluster structure, or it can (3) induce a complete geometric reconstruction. Si_7Cu^+ and Si_8Cu^+ are of the first type, simply adding to the Si_7^+ and Si_8^+, respectively. Si_6V^+ and Si_7Cu^+ belong to the second type and can be constructed by substituting atoms in the ground-state structures of Si_7^+ and Si_8^+. Si_6Cu^+, Si_7V^+, and Si_8V^+ bear entirely new structures that cannot be constructed directly from the bare clusters.

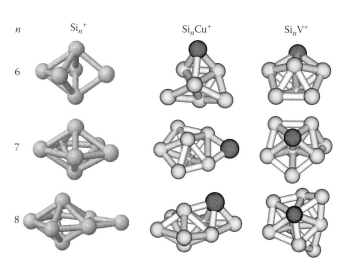

FIGURE 5.9 Structures of Si_n^+ (left), Si_nCu^+ (center), and Si_nV^+ (right) ($n = 6–8$) of which the calculated vibrational spectra fit best the experimental findings.

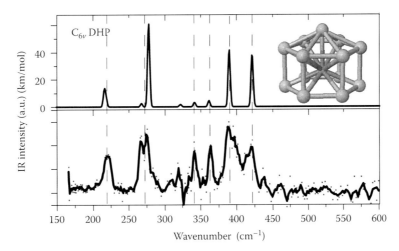

FIGURE 5.10 Vibrational spectrum of $Si_{13}V^+$. The lower trace shows the experimental IR-MPD spectrum of the complex with one xenon atom. The experimental data points are overlaid with a three-point smoothing to account for the bandwidth of FELIX. The experimental data are compared to the calculated vibrational spectrum of the best fitting isomer, shown in the upper trace. The main features fit well with those of the computed ground-state structure of C_{6v} symmetry, a hexagonal prism with one hexagonal face capped by the 13th silicon atom.

While argon does not bind to endohedrally doped silicon clusters at 80 K, xenon attaches also to the silicon atoms of the cluster and is, thus, much less sensitive to the position of the dopant. Therefore, it allows for the IR-MPD of bare silicon clusters (Lyon et al. 2009) and of endohedrally doped silicon clusters as well. Figure 5.10 shows the vibrational spectrum of $Si_{13}V^+$ obtained upon IR-MPD of its complex with isotopically enriched ^{129}Xe. Its main features are reproduced very well by a singlet isomer of C_{6v} symmetry that consists of a hexagonal prism whose hexagonal face is capped by an additional silicon atom. Another low-lying isomer (C_s-singlet) can be derived from the hexagonal prism motif and lays 0.37 eV higher than the C_{6v}-singlet. However, its IR spectrum fits the experiment much worse and can be excluded, which shows nicely how vibrational spectra can discriminate also between somewhat similar geometries.

5.4 Summary

The study of doped silicon clusters has been a remarkably active research field over the past years. This is in part due to the importance of silicon in many high-technological applications but also because it is a good example of the fruitful interplay between experimental and theoretical approaches. The interest in doped silicon clusters has been provoked by rather simple mass spectrometric investigations. These studies lead to a wealth of quantum mechanical investigations from which it followed that the stability of certain stoichiometries is due to both geometric and electronic stabilization. Reactivity studies have shown that every transition-metal dopant falls into the center of a silicon cage once the dopant-specific cluster size has been reached. The number of silicon atoms, which is needed for cage formation, mainly depends on the covalent radius of the transition-metal atom. Today, the electronic and geometric structure can be probed in detail by sophisticated experimental tools, like photoelectron and vibrational spectroscopy.

Such techniques confirm that compact geometries and closed electronic shells are responsible for the occurrence of magic clusters, and that the properties of silicon clusters can be influenced by carefully controlling the cluster's composition and size.

References

Asmis, K. R. and J. Sauer. 2007. Mass-selective vibrational spectroscopy of vanadium oxide cluster ions. *Mass Spectrom. Rev.* 26: 542–562.

Asmis, K. R., A. Fielicke, G. von Helden, and G. Meijer. 2007. Vibrational spectroscopy of gas-phase clusters and complexes: In *The Chemical Physics of Solid Surfaces*, Vol. 12: *Atomic Clusters: From Gas Phase to Deposited Atomic Clusters*, ed. D. P. Woodruff, pp. 327–375. Amsterdam, the Netherlands: Elsevier.

Baladron, C. and J. A. Alonso. 1988. Stability and magic numbers of hetero-atomic clusters of simple metals. *Phys. B: Condens. Matter* 154: 73–81.

Beck, S. M. 1987. Studies of silicon cluster-metal atom compound formation in a supersonic molecular beam. *J. Chem. Phys.* 87: 4233–4234.

Beck, S. M. 1989. Mixed metal-silicon clusters formed by chemical reaction in a supersonic molecular beam: Implications for reactions at the metal/silicon interface. *J. Chem. Phys.* 90: 6306–6312.

Bergeron, D. E., A. W. Castleman, Jr., T. Morisato, and S. N. Khanna. 2004. Formation of $Al_{13}I^-$: Evidence for the superhalogen character of Al_{13}. *Science* 304: 84–87.

Boldyrev, A. I. and L. S. Wang. 2005. All-metal aromaticity and antiaromaticity. *Chem. Rev.* 105: 3716–3757.

Bouwen W., F. Vanhoutte, F. Despa et al. 1999. Stability effects of $Au_nX_m^+$ (X = Cu, Al, Y, In) clusters. *Chem. Phys. Lett.* 314: 227–233.

Bouwen, W., P. Thoen, F. Vanhoutte et al. 2000. Production of bimetallic clusters by a dual-target dual-laser vaporization source. *Rev. Sci. Instrum.* 71: 54–58.

Canham, L. T. 1990. Silicon quantum wire array fabrication by electrochemical and chemical dissolution of wafers. *Appl. Phys. Lett.* 57: 1046–1048.

Chen, Z. and R. B. King. 2005. Spherical aromaticity: Recent work on fullerenes, polyhedral boranes, and related structures. *Chem. Rev.* 105: 3613–3642.

Chen, Z., S. Neukermans, X. Wang et al. 2006. To achieve stable spherical clusters: General principles and experimental confirmations. *J. Am. Chem. Soc.* 128: 12829–12834.

Clemenger, K. 1985. Ellipsoidal shell structure in free-electron metal clusters. *Phys. Rev. B* 32: 1359–1362.

Cordero, B., V. Gomez, A. E. Platero-Prats et al. 2008. Covalent radii revisited. *Dalton Trans.* 2832–2838.

de Heer, W. A. 1993. The physics of simple metal clusters: Experimental aspects and simple models. *Rev. Mod. Phys.* 65: 611–676.

Echt, O., K. Sattler, and E. Recknagel. 1981. Magic numbers for sphere packings: Experimental verification in free xenon clusters. *Phys. Rev. Lett.* 47: 1121–1124.

Ferrando, R., A. Fortunelli, and R. L. Johnston. 2008. Searching for the optimum structures of alloy nanoclusters. *Phys. Chem. Chem. Phys.* 10: 640–649.

Fielicke, A., A. Kirilyuk, C. Ratsch et al. 2004. Structure determination of isolated metal clusters via far-infrared spectroscopy. *Phys. Rev. Lett.* 93: 023401/1–4.

Fielicke, A., C. Ratsch, G. von Helden, and G. Meijer. 2005. Isomer selective infrared spectroscopy of neutral metal clusters. *J. Chem. Phys.* 122: 091105/1–4.

Grubisic, A., H. Wang, Y. J. Ko, and K. H. Bowen. 2008. Photoelectron spectroscopy of europium-silicon cluster anions, $EuSi_n^-$ ($3 \leq n \leq 17$). *J. Chem. Phys.* 129: 054302/1–5.

Gruene, P., D. M. Rayner, B. Redlich et al. 2008a. Structures of neutral Au_7, Au_{19}, and Au_{20} clusters in the gas phase. *Science* 321: 674–676.

Gruene, P., A. Fielicke, G. Meijer et al. 2008b. Tuning the geometric structure by doping silicon clusters. *ChemPhysChem* 9: 703–706.

Guo, P., Z.-Y. Ren, F. Wang, J. Bian, J.-G. Han, and G.-H. Wang. 2004. Structural and electronic properties of $TaSi_n$ ($n = 1$–13) clusters: A relativistic density functional investigation. *J. Chem. Phys.* 121: 12265–12275.

Guo, P., Z.-Y. Ren, A. P. Yang, J.-G. Han, J. Bian, and G.-H. Wang. 2006. Relativistic computational investigation: The geometries and electronic properties of $TaSi_n^+$ ($n = 1$–13, 16) Clusters. *J. Phys. Chem. A* 110: 7453–7460.

Guo, L.-J., X. Liu, G.-F. Zhao, and Y.-H. Luo. 2007. Computational investigation of $TiSi_n$ ($n = 2$–15) clusters by the density-functional theory. *J. Chem. Phys.* 126: 234704/1–7.

Guo, L.-J., G.-F. Zhao, Y.-Z. Gu, X. Liu, and Z. Zeng. 2008. Density-functional investigation of metal-silicon cage clusters MSi_n (M = Sc, Ti, V, Cr, Mn, Fe, Co, Ni, Cu, Zn; $n = 8$–16). *Phys. Rev. B* 77: 195417/1–8.

Hiura, H., T. Miyazaki, and T. Kanayama. 2001. Formation of metal-encapsulating Si cage clusters. *Phys. Rev. Lett.* 86: 1733–1736.

Honea, E. C., A. Ogura, C. A. Murray et al. 1993. Raman spectra of size-selected silicon clusters and comparison with calculated structures. *Nature* 366: 42–44.

Hudgins, R. R., M. Imai, M. F. Jarrold, and P. Dugourd. 1999. High-resolution ion mobility measurements for silicon cluster anions and cations. *J. Chem. Phys.* 111: 7865–7870.

Iyer, S. S. and Y. H. Xie. 1993. Light emission from silicon. *Science* 260: 40–46.

Jackson, K. and B. Nellermoe. 1996. $Zr@Si_{20}$: A strongly bound Si endohedral system. *Chem. Phys. Lett.* 254: 249–256.

Jaeger, J. B., T. D. Jaeger, and M. A. Duncan. 2006. Photodissociation of metal-silicon clusters: Encapsulated versus surface-bound metal. *J. Phys. Chem. A* 110: 9310–9314.

Janssens, E., S. Neukermans, and P. Lievens. 2004. Shells of electrons in metal doped simple metal clusters. *Curr. Opin. Solid State Mater. Sci.* 8: 185–193.

Janssens, E., P. Gruene, G. Meijer, L. Wöste, P. Lievens, and A. Fielicke. 2007. Argon physisorption as structural probe for endohedrally doped silicon clusters. *Phys. Rev. Lett.* 99: 063401/1–4.

Jarrold, M. F. 1995. Drift tube studies of atomic clusters. *J. Phys. Chem.* 99: 11–21.

Johnston, R. L. 2003. Evolving better nanoparticles: Genetic algorithms for optimising cluster geometries. *Dalton Trans.* 4193–4207.

Kawamura, H., V. Kumar, and Y. Kawazoe. 2004. Growth, magic behavior, and electronic and vibrational properties of Cr-doped Si clusters. *Phys. Rev. B* 70: 245433/1–10.

Kawamura, H., V. Kumar, and Y. Kawazoe. 2005. Growth behavior of metal-doped silicon clusters Si_nM (M = Ti, Zr, Hf; $n = 8$–16). *Phys. Rev. B* 71: 075423/1–12.

Khanna, S. N., and P. Jena. 1992. Assembling crystals from clusters. *Phys. Rev. Lett.* 69: 1664–1667.

Khanna, S. N., B. K. Rao, and P. Jena. 2002. Magic numbers in metallo-inorganic clusters: Chromium encapsulated in silicon cages. *Phys. Rev. Lett.* 89: 016803/1–4.

King, R. B. 1991. Topological models of the chemical bonding in transition metal-silicon clusters. *Z. Phys. D At. Mol. Clusters* 18: 189–191.

Knight, W. D., K. Clemenger, W. A. de Heer, W. A. Saunders, M. Y. Chou, and M. L. Cohen. 1984. Electronic shell structure and abundances of sodium clusters. *Phys. Rev. Lett.* 52: 2141–2143.

Koyasu, K., M. Akutsu, M. Mitsui, and A. Nakajima. 2005. Selective formation of MSi_{16} (M = Sc, Ti, and V). *J. Am. Chem. Soc.* 127: 4998–4999.

Koyasu, K., J. Atobe, M. Akutsu, M. Mitsui, and A. Nakajima. 2007. Electronic and geometric stabilities of clusters with transition metal encapsulated by silicon. *J. Phys. Chem. A* 111: 42–49.

Koyasu, K., J. Atobe, S. Furuse, and A. Nakajima. 2008. Anion photoelectron spectroscopy of transition metal- and lanthanide metal-silicon clusters: MSi_n^- ($n = 6$–20). *J. Chem. Phys.* 129: 214301/1–7.

Kroto, H. W., J. R. Heath, S. C. O'Brien, R. F. Curl, and R. E. Smalley. 1985. C_{60}: Buckminsterfullerene. *Nature* 318: 162–163.

Kumar, V. 2006. Alchemy at the nanoscale: Magic heteroatom clusters and assemblies. *Comput. Mater. Sci.* 36: 1–11.

Kumar, V. and Y. Kawazoe. 2001. Metal-encapsulated fullerene-like and cubic caged clusters of silicon. *Phys. Rev. Lett.* 87: 045503/1–4.

Kumar, V. and Y. Kawazoe. 2002. Magic behavior of $Si_{15}M$ and $Si_{16}M$ (M = Cr, Mo, and W) clusters. *Phys. Rev. B* 65: 073404/1–4.

Kumar, V. and Y. Kawazoe. 2003a. Erratum: Metal-encapsulated fullerenelike and cubic caged clusters of silicon [*Phys. Rev. Lett.* 87, 045503 (2001)]. *Phys. Rev. Lett.* 91: 199901/1–1.

Kumar, V. and Y. Kawazoe. 2003b. Hydrogenated silicon fullerenes: Effects of H on the stability of metal-encapsulated silicon clusters. *Phys. Rev. Lett.* 90: 055502/1–4.

Kumar, V., T. M. Briere, and Y. Kawazoe. 2003. Ab initio calculations of electronic structures, polarizabilities, Raman and infrared spectra, optical gaps, and absorption spectra of $M@Si_{16}$ (M = Ti and Zr) clusters. *Phys. Rev. B* 68: 155412/1–9.

Ledoux, G., O. Guillois, D. Porterat et al. 2000. Photoluminescence properties of silicon nanocrystals as a function of their size. *Phys. Rev. B* 62: 15942–15951.

Li, X., A. E. Kuznetsov, H.-F. Zhang, A. I. Boldyrev, and L.-S. Wang. 2001. Observation of all-metal aromatic molecules. *Science* 291: 859–861.

Li, X., B. Kiran, J. Li, H.-J. Zhai, and L.-S. Wang. 2002. Experimental observation and confirmation of icosahedral $W@Au_{12}$ and $Mo@Au_{12}$ molecules. *Angew. Chem. Int. Ed.* 41: 4786–4789.

Li, J., X. Li, H.-J. Zhai, and L.-S. Wang. 2003. Au_{20}: A tetrahedral cluster. *Science* 299: 864–867.

Lu, J. and S. Nagase. 2003. Structural and electronic properties of metal-encapsulated silicon clusters in a large size range. *Phys. Rev. Lett.* 90: 115506/1–4.

Lyon, J. T., P. Gruene, A. Fielicke et al. 2009. Structures of silicon cluster cations in the gas phase. *J. Am. Chem. Soc.* 131: 1115–1121.

Ma, L., J. Zhao, J. Wang, Q. Lu, L. Zhu, and G. Wang. 2005. Structure and electronic properties of cobalt atoms encapsulated in Si_n (n = 1–13) clusters. *Chem. Phys. Lett.* 411: 279–284.

Miyazaki, T., H. Hiura, and T. Kanayama. 2002. Topology and energetics of metal-encapsulating Si fullerenelike cage clusters. *Phys. Rev. B* 66: 121403/1–4.

Neukermans, S., E. Janssens, H. Tanaka, R. E. Silverans, and P. Lievens. 2003. Element- and size-dependent electron delocalization in Au_NX^+ clusters (X = Sc, Ti, V, Cr, Mn, Fe, Co, Ni). *Phys. Rev. Lett.* 90: 033401/1–4.

Neukermans, S., E. Janssens, Z. F. Chen, R. E. Silverans, P. v. R. Schleyer, and P. Lievens. 2004. Extremely stable metal-encapsulated $AlPb_{10}^+$ and $AlPb_{12}^+$ clusters: Mass-spectrometric discovery and density functional theory study. *Phys. Rev. Lett.* 92: 163401/1–4.

Neukermans, S., X. Wang, N. Veldeman, E. Janssens, R. E. Silverans, and P. Lievens. 2006. Mass spectrometric stability study of binary MS_n clusters (S = Si, Ge, Sn, Pb, and M = Cr, Mn, Cu, Zn). *Int. J. Mass Spectrom.* 252: 145–150.

Oepts, D., A. F. G. van der Meer, and P. W. van Amersfoort. 1995. The free-electron-laser user facility FELIX. *Infrared Phys. Technol.* 36: 297–308.

Ohara, M., K. Miyajima, A. Pramann, A. Nakajima, and K. Kaya. 2002. Geometric and electronic structures of terbium-silicon mixed clusters ($TbSi_n$; $6 \leq n \leq 16$). *J. Phys. Chem. A* 106: 3702–3705.

Ohara, M., K. Koyasu, A. Nakajima, and K. Kaya. 2003. Geometric and electronic structures of metal (M)-doped silicon clusters (M = Ti, Hf, Mo and W). *Chem. Phys. Lett.* 371: 490–497.

Ohara, M., K. Miyajima, A. Pramann, A. Nakajima, and K. Kaya. 2007. Geometric and electronic structures of terbium-silicon mixed clusters ($TbSi_n$; $6 \leq n \leq 16$). *J. Phys. Chem. A* 111: 10884–10884.

Ona, O., V. E. Bazterra, M. C. Caputo, M. B. Ferraro, P. Fuentealba, and J. C. Facelli. 2004. Modified genetic algorithms to model atomic cluster structures: CuSi clusters. *J. Mol. Struct. (Theochem)* 681: 149–155.

Oomens, J., B. G. Sartakov, G. Meijer, and G. von Helden. 2006. Gas-phase infrared multiple photon dissociation spectroscopy of mass-selected molecular ions. *Int. J. Mass Spectrom.* 254: 1–19.

Pyykkö, P. and N. Runeberg. 2002. Icosahedral WAu_{12}: A predicted closed-shell species, stabilized by aurophilic attraction and relativity and in accord with the 18-electron rule. *Angew. Chem. Int. Ed.* 41: 2174–2176.

Reveles, J. U., and S. N. Khanna. 2006. Electronic counting rules for the stability of metal-silicon clusters. *Phys. Rev. B* 74: 035435/1–6.

Röthlisberger, U., W. Andreoni, and M. Parrinello. 1994. Structure of nanoscale silicon clusters *Phys. Rev. Lett.* 72: 665–668.

Sporea, C., and F. Rabilloud. 2007. Stability of alkali-encapsulating silicon cage clusters. *J. Chem. Phys.* 127: 164306/1–7.

Tanaka, H., S. Neukermans, E. Janssens, R. E. Silverans, and P. Lievens. 2003. σ-Aromaticity of the bimetallic Au_5Zn^+ cluster. *J. Am. Chem. Soc.* 125: 2862–2863.

Torres, M. B., E. M. Fernández, and L. C. Balbás. 2007. Theoretical study of isoelectronic Si_nM clusters (M = Sc⁻,Ti,V⁺; n = 14–18). *Phys. Rev. B* 75: 205425/1–12.

Veldeman N., E. Janssens, K. Hansen, J. De Haeck, R. E. Silverans, and P. Lievens. 2008. Stability and dissociation pathways of doped Au_nX^+ clusters (X = Y, Er, Nb). *Faraday Disc.* 138: 147–162.

Wang, L.-M., S. Bulusu, H.-J. Zhai, X.-C. Zeng, and L.-S. Wang. 2007. Doping golden buckyballs: $Cu@Au_{16}^-$ and $Cu@Au_{17}^-$ cluster anions. *Angew. Chem. Int. Ed.* 46: 2915–2918.

Wu, J. and F. Hagelberg. 2006. Equilibrium geometries and associated energetic properties of mixed metal-silicon clusters from global optimization. *J. Phys. Chem. A.* 110: 5901–5908.

Xiao, C., F. Hagelberg, and W. A. Lester, Jr. 2002. Geometric, energetic, and bonding properties of neutral and charged copper-doped silicon clusters *Phys. Rev. B* 66: 075425/1–23.

Zheng, W., J. M. Nilles, D. Radisic, and K. H. Bowen, Jr. 2005. Photoelectron spectroscopy of chromium-doped silicon cluster anions. *J. Chem. Phys.* 122: 071101/1–4.

The Electronic Structure of Alkali and Noble Metal Clusters

Bernd v. Issendorff
Universität Freiburg

6.1 Introduction

A crude, but nevertheless very successful description of the electronic system of a metal is the free-electron model, which assumes that the valence electrons move through the infinite metal lattice like free particles. In this model, the electronic wavefunctions are just plane waves; any wavelength is allowed. This changes dramatically if the metal is shrunk down to nanoscopic dimensions. Now, the wavefunctions form standing waves between the surfaces of the particle, which is possible only for certain wavelengths. A direct consequence is the discretization of the electronic density of states: the continuous valence band breaks up into a finite number of states. This is the so-called quantum size effect, which can lead to significant changes of the metal particle properties. One can expect that such effects can be most clearly seen for a metal that comes close to ideal free-electron behavior. The alkali metal sodium is a very promising candidate: it exhibits a band structure with a dispersion very close to that of free electrons (Figure 6.1); its Fermi surface deviates from a sphere only by less than 0.1% (Ashcroft and Mermin, 1976). It is therefore not surprising that the first observation of a quantum size effect in free metal particles was made on sodium clusters.

Twenty-five years ago, Knight et al. (1984) published the mass spectrum shown in Figure 6.2. It shows a strongly size-dependent abundance of the clusters. The fact that some sizes have a higher intensity indicates that they must be more stable than the other sizes. The reason for the enhanced stability of these "magic" sizes was immediately clear: in spherical clusters, the electrons occupy orbitals, which, as in an atom, can be characterized by a radial and an angular momentum quantum number. With growing cluster size, these orbitals get gradually filled with electrons, just like the atomic orbitals in the periodic system with increasing order number. Enhanced stability occurs whenever a shell (a set of energetically degenerate orbitals) is completely full, which is the case for the "magic" cluster sizes visible in Figure 6.2. In this sense the "magic" sizes are equivalent to the rare gas atoms in the periodic system. This impressive demonstration of a simple quantum mechanical effect inspired a wealth of studies on a broad range of cluster systems. Similar quantum size effects have been found for many metals in measurements of cluster stabilities, ionization potentials, polarizabilities, absorption spectra, and photoelectron spectra (de Heer, 1993; Haberland, 1995; Ekardt, 1999). Especially the latter method is very well suited for characterizing the electronic systems of clusters, as it (at least in principle) yields a direct image of the electronic density of states. Therefore, this whole chapter will be devoted solely to photoelectron spectroscopy studies of simple metal clusters, namely, of sodium, copper, silver, and gold clusters. All measurements presented here were performed on negatively charged clusters, but only for a purely technical reason: only charged clusters can be size selected, and because of the lower ionization potentials of anions a larger part of the cluster density of states can be studied with available laser photon energies.

Photoelectron spectroscopy on size-selected metal cluster anions has a rather long history. The very first experiments were done by the group of Lineberger in 1987 in a study of small copper clusters (Leopold et al., 1987). Shortly thereafter

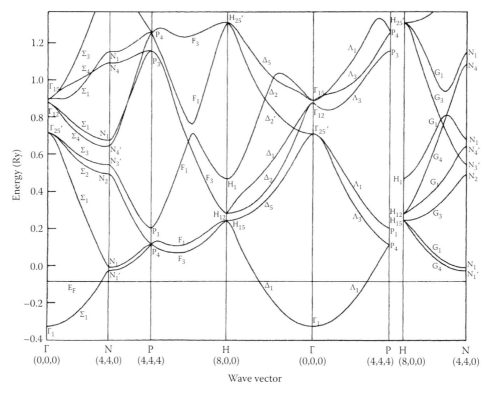

FIGURE 6.1 Valence electron band structure of sodium. The relation between state energies and wave vectors is close to that of free electrons, especially for the occupied states below the Fermi energy. (From Ching, W.Y. and Callaway, J., *Phys. Rev. B*, 11, 1324, 1975. With permission.)

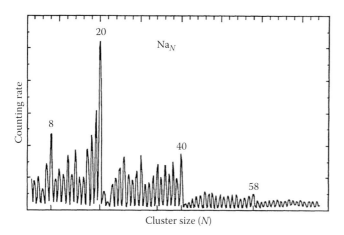

FIGURE 6.2 Mass spectrum of sodium clusters with 3–74 atoms. High intensities indicate "magic" sizes, that is, clusters with enhanced stability due to electronic shell closure. (From de Heer, W.A., *Rev. Mod. Phys.*, 65, 611, 1993. With permission.)

the group of Smalley started to study larger copper clusters (Pettiette et al., 1988), followed by silver and gold clusters (Taylor et al., 1992). At the same time, Ganteför et al. (1988) studied aluminum and silver clusters (Ganteför et al., 1990). Cha et al. (1993) were the first to demonstrate the successive occupation of single orbitals in small copper clusters. The group of Bowen pioneered photoelectron spectroscopy on alkali metal cluster anions, measuring spectra of small

sodium, potassium, rubidium, and cesium clusters with up to seven atoms (McHugh et al., 1989). Soon after this, they demonstrated the existence of an electron shell structure in potassium clusters with up to 19 atoms (Eaton et al., 1992). Since then many more results have been obtained for alkali and noble metal clusters; this chapter attempts to summarize the most important findings.

6.2 Theoretical Background

Here a brief introduction into the electronic structure of metal clusters and its interplay with their geometric structure will be given. Most of the time very simple models will be used, which can help to understand many of the cluster properties in a rather intuitive way. One should not forget, however, that the clusters are in fact complicated multiparticle systems, and that the simple models used are only approximations, which can fail to make correct predictions.

In the following, first, the origin of the simple one-electron models will be sketched, after which the main findings of these models with respect to the electronic-level structure and the cluster shape will be described. Then, the influence of the atomic structure onto the electronic structure will be discussed from a perturbation viewpoint. Finally, a very brief introduction into "real" quantum mechanics, namely, density functional theory (DFT)-calculations, will be included, as this is now the standard method to model metal clusters.

6.2.1 Derivation of an Effective Single-Particle Potential

Like for any molecular or bulk system, the total Hamiltonian for a cluster is given by

$$H = -\sum_i \frac{\hbar^2 \nabla_i^2}{2m_e} - \sum_\alpha \frac{\hbar^2 \nabla_\alpha^2}{2m_N} + \frac{e^2}{4\pi\varepsilon_0}$$

$$\times \left(\sum_{\alpha<\beta} \frac{Z^2}{|\mathbf{R}_\alpha - \mathbf{R}_\beta|} + \sum_{i<j} \frac{1}{|\mathbf{r}_i - \mathbf{r}_j|} - \sum_{i,\alpha} \frac{Z}{|\mathbf{R}_\alpha - \mathbf{r}_i|} \right) \quad (6.1)$$

where m_N, $\mathbf{R}_\alpha$, m_e, and $\mathbf{r}_i$ are the masses and coordinates of nuclei and electrons, respectively. The first term in the parenthesis describes the nuclei–nuclei interaction, the second one the electron–electron interaction, and the third the electron–nuclei interaction. The associated wavefunction is a multiparticle wavefunction describing both nuclei and electrons. Although this Hamiltonian already is an approximation as it neglects relativistic effects, the corresponding Schrödinger equation is utterly unsolvable. Several additional approximations have to be made so that one can handle it. The first one is the Born–Oppenheimer approximation, which separates electronic and nuclear wavefunctions (Born and Oppenheimer, 1927). In this approximation, the electrons move in the static electric fields produced by the immobile nuclei. The next approximation assumes that the total wavefunction of the electrons can be written as a product of single-particle wavefunctions. This allows one to speak of "orbitals" that can be "occupied" by single electrons. Furthermore, one can make a distinction between more strongly bound inner shell electrons and valence electrons; the inner electrons are then treated as rigid negative-charge distributions that partly screen the positive charges of the nuclei, leaving just the valence electrons to be described by the Schrödinger equation. The next far-reaching approximation assumes that one can neglect the correlated movements of the electrons and can treat the interaction of a given electron with all other electrons as an interaction with a static charge distribution. Using an iterative Hartree or Hartree–Fock approach (only in the latter the correct antisymmetry of the total wavefunction as requested by the Pauli principle is obeyed), leads to a so-called self-consistent field for the single-valence electrons. This field is the field of the ionic background, screened by the other valence electrons (here and in the following the term "nucleus" is used for the naked atom cores, while the term "background ion" is used for the nucleus plus the inner shell electrons). In metallic systems with their strongly delocalized valence electrons, the screening of the background ions is very effective; therefore, in the self-consistent field the Coulomb potential of the ions is strong only in their direct vicinity. This suggests that it might be possible to neglect the localized ion charge altogether and mimic the influence of the ions by using a homogenous positive-charge background (a strict mathematical justification of such a procedure is rather involved, though). This is the so-called jellium model (Brack, 1993). The next approximation used is straightforward; as already the internal structure of the cluster is neglected, one can go on and neglect details of its surface structure as well, assuming a spherical (or spheroidal) shape. In the jellium approximation, the potential produced by a spherical ionic background is then given by

$$V(r) = \begin{cases} \dfrac{Zen}{6\varepsilon_0}(r^2 - 3r_0^2) & r \leq r_0 \\[2ex] -\dfrac{Zen\,r_0^3}{3\varepsilon_0\,r} & r > r_0 \end{cases} \quad (6.2)$$

where

Ze is the ion charge

n is the ion density (usually that of the bulk)

$r_0 = \sqrt[3]{\dfrac{3}{4\pi n}N}$ is the radius of a cluster with N atoms

With this potential the Hamiltonian for the valence electrons reads:

$$H = \sum_{i=1}^{NZ} \left(-\frac{\hbar^2 \nabla_i^2}{2m_e} + V(|\mathbf{r}_i|) \right) + \sum_{i<j} \frac{e^2}{4\pi\varepsilon_0 \, |\mathbf{r}_i - \mathbf{r}_j|} \quad (6.3)$$

As mentioned above, the interaction between the electrons can be treated in an averaged (mean field) way; a single electron therefore interacts with a spherical (or spheroidal) smeared out ion background and a negative-charge distribution of the same symmetry. This negative charge compensates most of the positive charge of the background and also alters the radial dependence of the potential.

Instead of doing a self-consistent quantum calculation (which will be discussed below), one can now try to guess adequate potential functions, as this already allows to obtain some useful conclusions about the electronic system of a cluster. Assuming that the negative-charge distribution is uniform (similar to the charge of the ion background), a harmonic potential inside the cluster results. Assuming that the inside of the cluster is completely field-free (as it should be the case within a metal) leads to flat potential within the cluster, which should then increase more or less abruptly at the cluster border. Although the true potential of the cluster should have a finite depth, it is easier to extrapolate these potentials to infinity, that is to use an infinite harmonic potential or a spherical box potential with infinitely high walls. The Schrödinger equation for a single electron in one of these potentials (or in any other spherically symmetric potential) is given by

$$E\psi(r,\theta,\varphi) = \left(-\frac{\hbar^2}{2m_e}\nabla^2 + V(r) \right) \psi(r,\theta,\varphi) \quad (6.4)$$

This allows to separate the electron wavefunction into angular and radial components:

$$\psi(r,\theta,\varphi) = Y_{lm}(\theta,\varphi)R(r) \qquad (6.5)$$

where the Y_{lm} are the spherical harmonics, which means that the wavefunctions are angular momentum eigenfunctions with angular momentum quantum numbers l and m. The radial part of the wavefunctions is obtained by solving the radial Schrödinger equation

$$ER(r) = \left(-\frac{\hbar^2}{2m_e}\left(\frac{d^2}{dr^2} + \frac{2}{r}\frac{d}{dr} - \frac{l(l+1)}{r^2}\right) + V(r)\right)R(r) \qquad (6.6)$$

For the harmonic potential the solutions are based on Laguerre functions; the corresponding energies are

$$E = \hbar\omega\left(2k + l + \frac{3}{2}\right) = \hbar\omega\left(n + \frac{3}{2}\right) \qquad (6.7)$$

where

- k and l are the radial and angular momentum quantum numbers
- $n = 2k + l$ is the principal quantum number of the three-dimensional harmonic oscillator

The energetically equidistant solutions are highly degenerate: there are always $(n + 1)(n + 2)/2$ different states for a given value of n.

For the box potential with infinitely high walls, the radial solutions are the spherical Bessel functions:

$$R(r) = cj_l(kr) \qquad (6.8)$$

with an appropriate normalization constant c. Here the energies are given by

$$E_{ln} = \frac{\hbar^2 k^2}{2m_e} = \frac{\hbar^2 u_{li}^2}{2m_e r_0^2} \qquad (6.9)$$

where

- r_0 is the radius of the spherical box potential
- u_{li} is the ith zero of the spherical Bessel function $j_l(r)$

The degeneracy of the different states is just that of the angular momentum eigenstates, that is $2l + 1$.

A more realistic potential than the former two is the Wood–Saxon potential, a finite-depth box potential with rounded walls:

$$V(r) = -\frac{\varepsilon}{1 + \exp(a(r - r_0))} \qquad (6.10)$$

Here

- ε is the potential depth
- r_0 its radius
- a determines the steepness of the potential wall

For this potential no analytical solutions exist; the Schrödinger equation has to be solved numerically. The levels obtained for all three potentials are shown in Figure 6.3. Levels with the same angular momentum are connected. One can see that the details of the level sequence depend on the radial potential. This has an influence on the "magic" numbers, that is on the electron numbers for which clusters are especially stable. On the levels, the accumulated electron numbers are given: 2 electrons fill the 1s level, 8 electrons fill 1s and 1p, and so on. The most stable clusters are those with a completely filled level (called a closed shell) and a large gap between the highest occupied level and the lowest unoccupied level. So the electron numbers 34 and 58 are "magic" numbers of the square potential, but not for the Wood–Saxon potential. The numbers 20 and 40 are strongly "magic" for the harmonic oscillator and the Wood–Saxon potential, but less so for the box potential. So which cluster sizes are especially stable depends on the details of the radial potential, which can be different for different metals. The numbers observed for sodium clusters in Figure 6.2 (8, 20, 40, 58) seem to indicate that their effective single-particle potential is somewhere between a box and a Wood–Saxon potential. But most importantly the agreement of these numbers with the simple model predictions underlines that the free-electron model is able to describe the electronic system of sodium clusters.

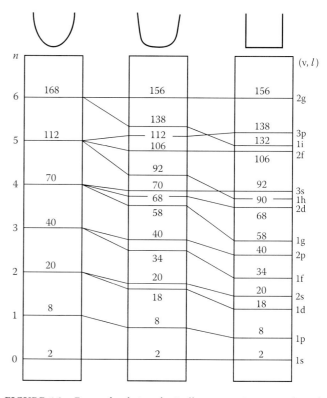

FIGURE 6.3 Energy levels in spherically symmetric potentials with different radial dependencies (harmonic, Wood–Saxon and box potential), with assignment of harmonic oscillator and angular momentum quantum numbers. The numbers on the levels indicate the accumulated electron count. (From de Heer, W.A., *Rev. Mod. Phys.*, 65, 611, 1993. With permission.)

6.2.2 Deformation

The assumption that clusters are spherical is not completely correct; even jellium clusters can often lower their total energy by undergoing a deformation. This is certainly true for open shell clusters, which are subject to the Jahn-Teller effect. But even some closed shell clusters have a deformed ground state. Good insight into this effects can be obtained from a simple model from nuclear physics, which has been adopted to metal clusters by Clemenger (1985). It treats spheroids with cylindrical symmetry, for which a deformation parameter is defined by

$$\delta = 2\frac{R_z - R_x}{R_z + R_x} \tag{6.11}$$

where

R_z is the radius along the rotational symmetry axis
$R_x (=R_y)$ are the radii perpendicular to it

These are then used in a three-dimensional harmonic oscillator model where the oscillator frequencies along the symmetry axis and perpendicular to it are proportional to the respective inverse radii. Additionally a small perturbation is included that lifts the degeneracy of states with the same harmonic oscillator quantum number, but different angular momentum quantum numbers.

The single-electron levels obtained in this model are shown in Figure 6.4. For spherical clusters ($\delta = 0$), a level sequence very similar to those shown in Figure 6.3 is obtained; upon oblate ($\delta < 0$) or prolate ($\delta > 0$) deformation, these states then split up. The total electronic energy of a cluster is just the sum of the energies of all occupied levels; for every electron number, a certain distortion exists for which this total energy has its minimum value, as indicated in Figure 6.4. One can see that most closed shell sizes (e.g., electron numbers 2, 8, 20, 40) tend to adopt a spherical shape, while the open shell sizes tend to be deformed, typically with a prolate deformation for a less-than-half-filled uppermost shell and an oblate one for an almost filled shell. Prominent exceptions are the electron numbers 18 and 34, which are closed shell sizes, but nevertheless adopt an oblate shape. The reason is that here the difference in angular momentum between the uppermost filled shell and the lowest empty shell (1d-2s for size 18, 1f-2p for size 34) is $\Delta l = 2$. A spheroidal deformation can be described as a quadrupolar perturbation potential, which couples states with $\Delta l = 2$. In the two clusters, the deformation therefore produces an avoided crossing between the highest occupied state and the lowest unoccupied state, due to which the total electronic energy of the distorted clusters is lower than that of the spherical ones.

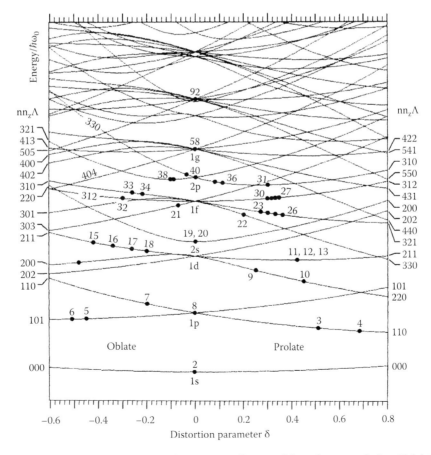

FIGURE 6.4 Energy levels in a modified three-dimensional harmonic oscillator model as a function of spheroidal deformation. The numbered dots indicate the energetically most favorable deformation for a given number of electrons in the system. (From Clemenger, K., *Phys. Rev. B*, 32, 1359, 1985. With permission.)

The same mechanism can also lead to higher-order deformations: a 40 electron cluster, e.g., which has $\Delta l = 3$ between the highest occupied state and the lowest unoccupied state, undergoes an octupole deformation (Rytkönen et al., 1998).

This last statement already shows that the restriction to spheroidal shapes cannot capture the full story. An obvious extension is the introduction of three independent axes, allowing ellipsoidal deformation (Yannouleas and Landman, 1995), or a combination of such triaxial and higher-order deformations (Reimann et al., 1995). These calculations were mostly done using empirical models for the electronic system; DFT calculations (which are more precise, but also numerically much more demanding) were used to study a combination of quadrupolar, octupolar, and hexadecapolar deformations (Montag et al., 1995). A jump toward infinite-order deformation was done in the so-called ultimate jellium model, where in a DFT calculation the ion background was allowed to deform without any restrictions (Koskinen et al., 1995). The results of this study will be compared to experimental results below.

6.2.3 Influence of the Atomic Structure

The treatment above assumes that the atomic structure of a cluster can be completely neglected. This is certainly not completely true; the atoms have to be arranged in some definite way, which restricts the choice of possible cluster shapes. More importantly, there must be some interaction of the electrons with the localized charge of the nuclei. As the charge is screened, this interaction is not very strong and in a simple approach can be treated as a perturbation of the spherically symmetric potential. The structured ion background of course does not have spherical symmetry, so this perturbation will lift the degeneracy of the angular momentum eigenstates. Strictly speaking, in the presence of ions, angular momentum is not a conserved quantity and therefore not a good quantum number anymore. The new eigenstates are irreducible representations of the symmetry group of the atomic structure of the cluster. This is demonstrated in Figure 6.5 for the case of a cluster with octahedral or icosahedral symmetry. In the middle column, the unperturbed eigenstates of a spherically symmetric potential are shown; on the left and the right are the new eigenstates in an octahedral or icosahedral structure. States of related character are connected by lines. Icosahedral symmetry allows up to fivefold degeneracy. s, p, and d states therefore do not split up at all; f states split up into a fourfold (g) and a threefold (t) degenerate state; g states into a fivefold (h) and a fourfold (g) degenerate state. So a cluster with an icosahedral atomic structure will obviously still have a highly discrete density of states (DOS) even in case of rather strong electron–ion interaction. The lower the symmetry of the atomic structure, the more complex the density of states will be. This makes it necessary to know the exact structure of a cluster if one wants to understand the details of its electronic structure. But it also means that a measured DOS of a cluster can actually be used as a fingerprint of its geometric structure, that is as a sensitive test for structure calculations.

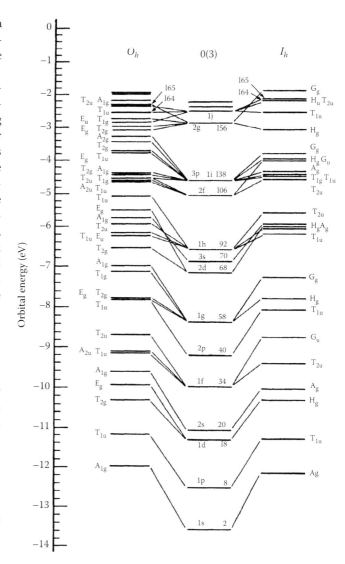

FIGURE 6.5 Correlation diagram for the levels in a spherically symmetric potential (center column) and in potentials of octahedral (left column) and icosahedral symmetry (right column), here calculated for the case of an aluminum cluster with 55 atoms. (From Cheng, H. et al., *Phys. Rev. B*, 43, 10647, 1991. With permission.)

6.2.4 DFT Calculations

The problem that in a multielectron system the movement of each electron is correlated with that of any of other electrons, which in principle makes it impossible to define single-particle states, can be circumvented by density functional theory (DFT). This theory is based on the Hohenberg–Kohn theorem, which states that it is sufficient to know the spatial distribution of the total electron density to calculate the system energy $E = E[\rho(r)]$ (Hohenberg and Kohn, 1964). The ground state of an electronic system can therefore be found by varying the electron density distribution until the energy minimum is found. In fact this would be the case if the functional to calculate the energy was known; unfortunately this is not the case. A broad selection of functionals are in use, and many give good results for a range of problems; but a single functional which is the optimum choice for any problem does not seem to have been found yet.

The minimization of the energy through variation of the electron density can be shown to be equivalent to solving the so-called Kohn–Sham equations (Kohn and Sham, 1965). Here the electron density is represented by a superposition of basis functions

$$\rho(\mathbf{r}) = \sum_{i=1}^{N_e} |\phi_i(\mathbf{r})|^2 \qquad (6.12)$$

and the equations are given by

$$\varepsilon_i \phi_i = \left(-\frac{\hbar^2}{2m}\nabla^2 + V(\mathbf{r}) + \int \frac{\rho(\mathbf{r}')}{|\mathbf{r}-\mathbf{r}'|}d^3\mathbf{r}' + V_{xc} \right)\phi_i \qquad (6.13)$$

The potential $V(r)$ can be a jellium potential, but also a set of (pseudo-) potentials centered at the position of the ions. Except for the "exchange correlation term" V_{xc} the Kohn–Sham equations look like Schrödinger equations in the Hartree approximation, and therefore are similarly easy to solve. The total energy of the system is then given by the sum over the ε_i:

$$E = \sum_{i=1}^{N_e} \varepsilon_i \qquad (6.14)$$

This seems to indicate that one can interpret the functions ϕ_i as single-particle wavefunctions, and the values ε_i as the corresponding single-particle energies. Although this is not true in a strict mathematical sense, it has become a very common assumption, and a very practical one, too (Stowasser and Hoffmann, 1999). As will be shown below, the DFT density of states (the distribution of the so-called Kohn–Sham energies ε_i) often is in very good agreement with measured photoelectron spectra.

The earliest DFT calculation of sodium clusters within the jellium model was done by Ekardt (1984). Figure 6.6 shows the results for a neutral Na_{20} cluster. In the upper panel, the homogenous ion density and the resulting electron density are shown; in the lower panel one sees the self-consistent single-particle potential, where the energy levels occupied by the 20 electrons are indicated. The increased electron density in the cluster center leads to a "wine bottle" shape of the potential, that is to a local maximum in the center. Apart from this hump, the potential is close to a smeared out finite box potential, similar to the Wood–Saxon potential discussed above. This underlines that the use of these simple potentials is justified.

If the ion positions are explicitly included in the DFT calculation, one can use it to determine the geometrical structure of the clusters (by trying to find the structure with the lowest total energy). The rapid increase of available computing power has quite an impact here: while up to a few years ago such geometry optimizations were only feasible when empirical (classical) potentials were used for the atomic interactions, now full DFT calculations are done for larger and larger systems.

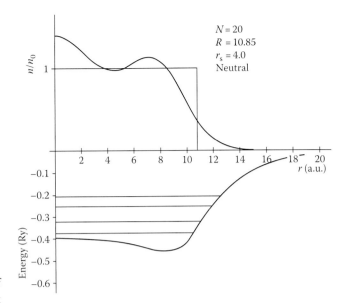

FIGURE 6.6 Results of a DFT calculation for Na_{20} in the spherical jellium model. The upper panel shows the homogenous density of the ion background and the resulting density of the electrons; the lower panel shows the corresponding self-consistent single-particle potential and the energies of the occupied levels (1s, 1p 1d, 2s). (From Ekardt, W., *Phys. Rev. B*, 29, 1558, 1984. With permission.)

6.3 Experiment

In this section, first, some principles of photoelectron spectroscopy will be discussed, after which the actual experimental setup will be described.

6.3.1 Principle of the Measurement

The fundamental process used in photoelectron spectroscopy is the photoelectric effect: absorption of a photon of sufficient energy by an atom or atom ensemble can lead to the emission of an electron. Depending on the initial charge state of the system, this process is called photoionization (neutral or positively charged systems) or photodetachment (negatively charged systems). For cluster anions it can be written as

$$M_n^- + h\nu \rightarrow M_n^* + e^- \qquad (6.15)$$

The star indicates that the neutralized cluster is not necessarily in its ground state. The kinetic energy of the electron is given by

$$E_{\mathrm{kin}} = h\nu - EA - E\left(M_n^*\right) \qquad (6.16)$$

where

   EA is the adiabatic electron affinity of the cluster (the energy difference between the lowest neutral state and the lowest anionic state)

   $E(M_n^*)$ is the excitation energy of the neutral cluster after the electron detachment

This excitation can be both vibrational and electronic. Electronic excitation in an effective single-particle picture results from the emission of an electron out of an orbital energetically below the highest occupied one, which is equivalent to detaching an electron from the uppermost orbital and exciting an electron from the lower orbital to this one. This directly leads to the statement of Koopman's theorem (Koopmans, 1934): the kinetic energy of a photoelectron emitted out of an orbital ϕ_n is given by

$$E_{kin} = h\nu - \varepsilon_n \qquad (6.17)$$

where ε_n is the binding energy of the electron in the respective orbital. This is a rather crude approximation, as it neglects the relaxation of the states of all other electrons caused by the missing charge of the emitted electron. Even more severe is the assumption that the excited state of the remaining system (here: the neutral cluster) can actually be described within a single-particle picture, which is never exactly the case. Nevertheless Koopman's theorem is used throughout molecular and solid-state photoelectron spectroscopy, and has proven to be such an useful approach that one sometimes forgets that it is only an approximation (Cederbaum et al., 1977). Measured photoelectron spectra are often very close to calculated DOS, which shows that unlike bound excited states, which especially in metallic systems can be highly correlated multielectron states (Bonačić-Koutecký et al., 1995), the excited states produced by electron emission in a decomposition into single-particle states have dominant contributions from just one excited state. The same pragmatic view can be employed for the comparison of photoelectron spectra and the DOS as obtained from DFT calculations. The latter consists of the energies of the Kohn–Sham wavefunctions, which have been introduced as mathematical basis functions to describe the electron density and therefore do not necessarily have a physical significance. One can, however, show that Kohn–Sham states can indeed be interpreted as single-particle states (Stowasser and Hoffmann, 1999; Chong et al., 2002). Most importantly there are now so many cases documented where there is a very good agreement between DFT DOS and measured photoelectron spectrum that this comparison is well justified.

When doing such a comparison, one nevertheless has to be aware that there are a couple of effects that lead to differences between the photoelectron spectrum and the DOS. One is the partial photodetachment cross-section. Every orbital has a slightly different detachment cross-section, due to which the intensity of a peak in a photoelectron spectrum does not reflect the number of electrons in the related orbitals. Fortunately the differences between states close in energy are usually not very strong. A general trend, which is always present, however, is that lower-lying states have smaller cross-sections and therefore are less visible in the spectra. Another reason for a deviation of the measured spectrum from the DOS has been mentioned already: photodetachment can also lead to vibrational excitation. Usually

the ground-state structure of the negatively charged cluster is not identical to that of the neutral one; the larger this difference is, the more the neutral cluster will be vibrationally excited after the electron detachment. As these vibrations cannot be resolved in the photoelectron spectrum (except for very small sizes), this leads to broadening of the peaks in the spectrum. It is more pronounced if the cluster is initially hot, as in this case many more initial and final vibrational states can contribute; a spectrum measured on cold clusters is therefore closer to the true electronic DOS. But even in cases where the cluster initially is in its vibrational ground state and the geometries of the anion and the neutral are identical, some broadening remains. The reason is that the excited state of the neutral cluster produced by the emission of an electron out of any but the highest orbital has a finite lifetime; especially in metal systems, this can be as short as a few femtoseconds. This leads to a rather strong lifetime broadening of the photoelectron peaks of lower-lying states, as the lifetime decreases with increasing excitation energy. Finally photoelectron spectra can be more complex than the DOS because there are electronically excited states produced by the photoemission that cannot be described as a single-particle excitation. For smaller clusters, this leads to additional peaks in the photoelectron spectrum. For larger clusters, this usually produces a certain background of very low kinetic energy photoelectrons (sometimes referred to as "inelastically scattered photoelectrons"). All this means that a comparison between measured photoelectron spectra and the calculated DOS is more meaningful close to the threshold, and should put more weight onto peak positions than onto peak heights.

6.3.2 Experimental Setup

A schematic of the apparatus used to measure the photoelectron spectra shown in this chapter is depicted in Figure 6.7. It consists of a so-called gas aggregation cluster source producing the charged clusters, a radio frequency (RF) trap for intensity enhancement and cluster thermalization, a time-of-flight mass spectrometer for size selection, and a photoelectron spectrometer. The cluster source is a liquid-nitrogen-cooled double-wall tube with an inner diameter of 100 mm and a length of 0.5 m, in which the cluster material is evaporated into a stream of helium with a pressure of about 0.5 mbar. Evaporation of the materials is done either thermally out of a crucible (used here in the case of sodium) or by magnetron sputtering (used for the noble metals). For the mass selection cluster ions are needed; therefore, in the case of sodium, a weak-pulsed gas discharge in the aggregation tube is used to charge the clusters. This is not necessary in the case of the magnetron, which itself produces a sufficient amount of charge carriers. The mean cluster size produced by the source can be tuned by varying the distance of the magnetron or crucible to the exit aperture, the buffer gas flux and composition, the amount of metal evaporated, and the size of the exit aperture. Clusters leaving the source are stored for about 1 s in a RF-trap attached to a cold head, which can be cooled to about 10 K. The

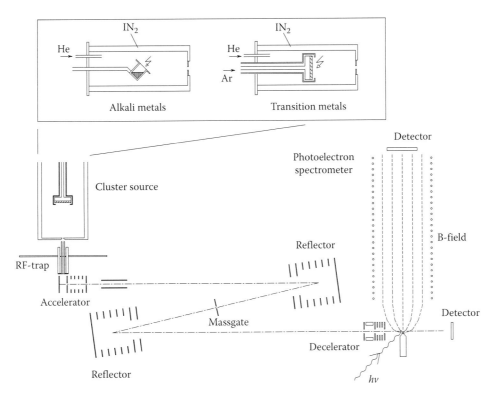

FIGURE 6.7 Schematic of a typical experimental setup for photoelectron spectroscopy on size-selected cluster anions of controlled temperature. It consists of a cluster source, a thermalization radio frequency multipole ion trap, a high-resolution tandem reflectron time-of-flight mass spectrometer, and a magnetic bottle-type time-of-flight photoelectron spectrometer. Depending on the cluster materials in the cluster source, either thermal evaporation or magnetron sputtering is applied to bring the material into the gas phase.

trap is a linear 12 pole trap, with a design very similar to the 22 pole trap developed by Gerlich (1995); the clusters stored in it thermalize by collisions with helium buffer gas having a density of about 10^{13} cm³. Bunches of cluster ions are extracted from the trap and inserted into a double-reflectron time-of-flight mass spectrometer, where a multiwire ion beam chopper (the "mass-gate") in between the two reflectors allows to select a certain cluster size with a resolution of $m/dm = 2000$. The second reflector refocuses the ion package into the interaction region of the photoelectron spectrometer. Directly before entering this region the ions are decelerated from a typical beam energy of 1100 eV down to about 20 eV by pulsed momentum deceleration (Markovich et al., 1994) in order to avoid doppler broadening of the photoelectron energies. The photoelectron spectrometer is a magnetic bottle-type time-of-flight spectrometer based on the design of Kruit and Read (1983). The electrons are detached from the cluster ions by a laser pulse in the strongly divergent magnetic field of a permanent magnet, which goes over into a weak guiding magnetic field in a drift tube of 1.6 m length. The electrons follow the magnetic field lines and are detected by a channeltron at the end of the drift tube. From the measured flight time, the electron kinetic energy is then calculated. The collection efficiency of the spectrometer is close to 100%; the energy resolution is about $E/dE = 50$, that is, at an electron kinetic energy of 2 eV the minimum peak width in a spectrum is about 40 meV. Typical electron

intensities are 10 electrons per laser shot; spectra are usually averaged over 20,000 shots at a repetition rate of 100 Hz.

6.4 Results

6.4.1 Small Sodium Clusters

Typical results obtained for warm (close to room temperature) small sodium cluster anions Na_n^- are shown in Figure 6.8. Here like in all spectra shown in this chapter, the measured electron intensity is plotted as a function of binding energy (photon energy minus kinetic energy, Equation 6.17). In the spectra different bands are visible, which one can relate to the angular momentum eigenstates discussed above. The respective radial and angular momentum quantum numbers are indicated in the graph. The size dependence of the DOS follows a simple pattern, which results from the gradual filling of these electron shells. The smallest cluster shown is Na_{17}^-, which accommodates 18 valence electrons. In the spherical potential model, therefore, the states 1s, 1p, and 1d should be completely filled. Indeed in the photoelectron spectrum, one can see a large peak at 1.75 eV and a smaller one at 2.6 eV binding energy, which can be identified as the 1d shell and the 1p shell. The 1s orbital is not visible anymore for such a size because of its small intensity and the strong broadening of lower-lying states; it can only be detected for sizes

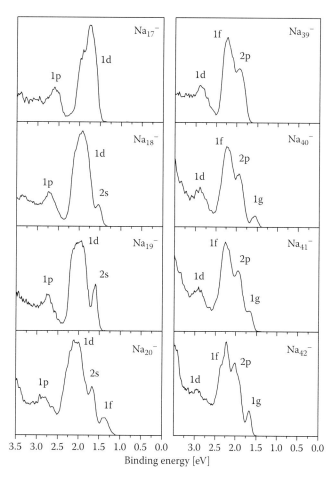

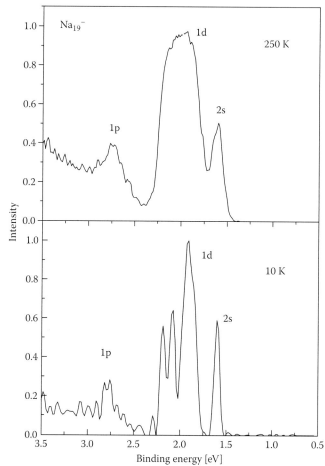

FIGURE 6.8 Photoelectron spectra of room temperature clusters Na_n^-, recorded at a photon energy of 4.02 eV. The visible bands are labeled by the quantum numbers of the corresponding states in the spherical free-electron model. Note the appearance of new shells for total valence electron numbers of 19, 21, and 41.

FIGURE 6.9 Temperature dependence of the photoelectron spectrum of Na_{19}^-, measured at a photon energy of 4.02 eV. At the low temperature, the resolution is much higher due to the vanishing initial vibrational excitation. This reveals that the broad band assigned as the 1d-shell consists of five distinguishable peaks, indicating that the degeneracy of the d-state is completely lifted.

up to about $n = 14$ (Figure 6.10). If now another electron is added to the 18 electron system, it has to occupy the next higher shell, namely, 2s. Indeed in the spectrum of Na_{18}^-, a new peak appears. In the next higher cluster, Na_{19}^-, this 2s orbital is occupied by two electrons; the peak in the spectrum therefore is about twice as high. An s-orbital can only accommodate two electrons; so the next electron has to occupy the next higher shell, 1f. Accordingly in the spectrum of Na_{20}^-, again a new peak appears. This 1f shell gets filled at size $n = 33$; at size 39 also the next higher shell 2p is complete. Indeed in the photoelectron spectrum of Na_{39}^- three peaks can be identified, indicating the 1d, the 1f, and the 2p shell. Starting from size 40 then the 1g shell is getting filled. This good agreement with the predictions of the spherical model seems to indicate that sodium clusters are indeed perfect free-electron systems. This is, however, only almost true. One can notice that in Figure 6.8 the peaks related to the different shells are rather broad bands. This is not due to the limited resolution of the spectrometer, and only partly to vibrational excitation. The main reason is the interaction with the lattice, which lifts the degeneracy of the angular momentum eigenstates. This can be clearly

seen in Figure 6.9. Here the photoelectron spectrum of Na_{19}^- is shown for two temperatures, for $T = 250$ K (at which the cluster is highly vibrationally excited and potentially even liquid), and at $T = 10$ K, where the cluster should have a well-defined geometry (and practically no vibrational excitation in the initial state). One can see that while the peak of the 2s orbital just gets narrower, the peak of the 1d shell breaks up into at least four separate peaks (a fit with Gaussians of identical widths actually reveals five different states in this band). This splitting is not due to a nonspherical shape of the cluster—Na_{19}^- is very close to spherical, as will be shown below. It is the ionic background of the cluster that perturbs the angular momentum eigenstates. As the energies of the new eigenstates of the system should strongly depend on the positions of the ions, this "fine structure" of the photoelectron spectra can serve as a fingerprint of the cluster geometry.

Indeed the comparison of measured and calculated photoelectron spectra has become one of the major tools for the determination of cluster structures. One example is shown in Figure 6.10. Here the measured spectra of small sodium cluster anions

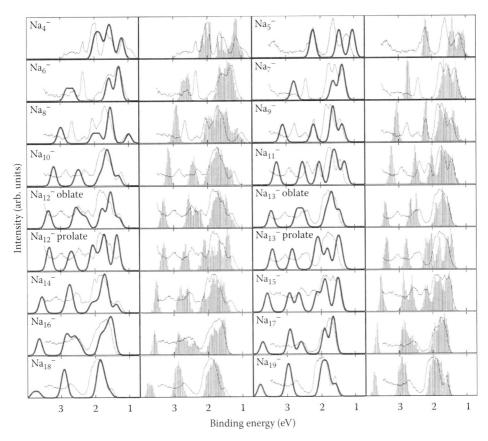

FIGURE 6.10 Measured photoelectron spectra of room temperature sodium cluster anions (thin lines) compared to the broadened DFT density of states of the probable ground-state structures shown in Figure 6.11 (thick lines) and to histograms of the electronic Kohn–Sham energies obtained by DFT during molecular dynamics simulation of the room temperature clusters. For Na_{12}^- and Na_{13}^- two quasi degenerate isomers exist (an oblate and a prolate one), which do not interconvert on the time scale of the simulation. (From Moseler, M. et al., *Phys. Rev. B*, 68, 165413, 2003.)

with 4–19 atoms are compared to the DOS calculated by DFT in two ways. In the first and third column, the DOS (the distribution of the Kohn–Sham energies) is shown as calculated for the ground-state structure of the cluster (more precisely the lowest energy structure found in an extensive search). Here the states have been broadened by Gaussians of 80 meV width to simulate the broadening effects and the finite resolution of the spectrometer. In the second and fourth column, histograms of Kohn–Sham energies are shown, which have been sampled during a finite-temperature (300 K) molecular dynamics simulations. In most cases, the resulting simulated DOS are rather similar, but for some sizes (e.g., $n = 7, 8, 9$) during the thermal movement obviously different structures are adopted, which changes the DOS significantly. This allows one to conclude that as long as the temperature is not high enough to allow isomerization, a simple broadening of the ground-state DOS is a rather good approximation. The rather good agreement of the measured photoelectron spectra and the calculated DOS in Figure 6.10 indicates that for most sizes the correct cluster structures have been found. These structures are depicted in Figure 6.11. One can see that from size 7 the clusters grow by attaching more and more atoms to one side of a quadratic bipyramid. At size 12 and 13 an alternative motif appears. In addition to this prolate structure, both cluster

sizes can adopt a rather oblate structure, which has practically the same energy. For larger clusters this oblate structure prevails, until the clusters finally get more compact again. Note that in this size range, icosahedral structures do not play a role.

The knowledge of the geometrical structures of the clusters allows one now to test one of the predictions of the simple free-electron model, the size-dependent shape deformation. In order to characterize the overall shape of a cluster, its moments of inertia have been calculated from the ion positions, which allows to represent the cluster by an ellipsoid with sodium bulk density and the same moment of inertia tensor. The extensions of this ellipsoid along its three principal axes are then taken as the cluster radii. The results are shown in Figure 6.12a, where the three radii are plotted as a function of the number of atoms in the cluster. One can immediately observe that the three values are close to identical only for $n = 7$ (8 electrons) and $n = 18/19$ (19 and 20 electrons). This is in perfect agreement with the free-electron model: at the electron numbers 8 and 20 closed shell configurations are present, where the electronic system prefers to be spherical (Figure 6.4). Even more impressive is the behavior in between these sizes. One can see that starting with $n = 8$ the clusters first get a prolate shape (two short radii, one long radius), become triaxially deformed (three different radii)

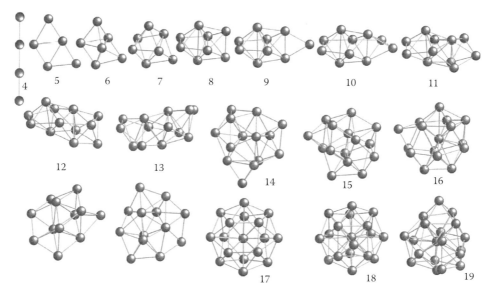

FIGURE 6.11 Lowest energy structures of sodium cluster anions found by DFT calculations. The resulting electronic density of states is shown in Figure 6.10. Two quasidegenerate isomers are given for the sizes 12 and 13. (From Moseler, M. et al., *Phys. Rev. B*, 68, 165413, 2003. With permission.)

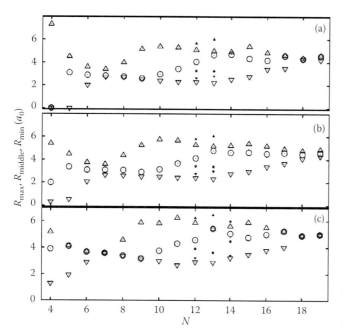

FIGURE 6.12 Comparison between the outer shapes of the sodium cluster anions and simple model predictions, as a function of atom number N. (a) The three principal radii of the ground-state cluster structures are shown in Figure 6.11 as obtained from their moments of inertia. Small dots indicate additional isomers. (b) Same as (a), but averaged over time during a molecular dynamics simulation of room temperature clusters. (c) Radii calculated from the reported Hill-Wheeler parameters of ultimate jellium calculations for electron numbers $N + 1$. (Data taken from Koskinen, M. et al., *Z. Phys. D*, 35, 285, 1995; From Moseler, M. et al., *Phys. Rev. B*, 68, 165413, 2003. With permission.)

around $n = 11$, get prolate (two long radii) at $n = 14$, and finally spherical again at $n = 17$. In Figure 6.12b, the results for the cluster shapes are shown as obtained from a room temperature

molecular dynamics simulation. Obviously there is almost no difference, which indicates that the shapes of the ground-state clusters are already close to those which are optimal for a certain electron number; the additional floppiness caused by the elevated temperature does not lead to strong changes. Indeed the cluster shapes are in perfect agreement with simple model predictions. In Figure 6.12c, the results of an ultimate jellium DFT calculation are shown, where the energy was optimized without any restrictions for the deformation of the jellium background (Koskinen et al., 1995). The experimental and the simple model results are close to identical; even the occurrence of two shape isomers for the sizes 12 and 13 is reproduced. So obviously the ion background manages to adopt structures with shapes close to the ideal free-electron ones. In this size range, the geometrical structure is therefore dominated by quantum effects and cannot be understood within models that treat the interaction between the atoms by classical potentials. Recently a similar analysis of the shapes of copper and silver clusters in the same size range has been done (Yang et al., 2006a,b), which came to the same conclusion. Going back to the comparison of the measured photoelectron spectra and the calculated DFT DOS in Figure 6.10, one can make a general remark about the validity of such an approach. Clearly for some sizes the agreement is good; the general patterns of the measured spectrum are very well reproduced by the calculation. Nevertheless, there is quite a disagreement with respect to the peak spacing: the DFT calculation strongly overestimates the total bandwidth of the cluster valence states. It has been checked carefully whether this disagreement is a consequence of the specific functional used (Mundt et al., 2006). It was found that although the total bandwidth of the DOS does depend on the details of the calculation, it was not possible to reproduce the measured spectrum correctly. A second observation that can be made is that the DFT calculation works less well for clusters with an odd number of electrons, where spin-dependent calculations are necessary. These calculations seem to overestimate the singlet-triplet splitting of some levels (like the splitting of the 1s orbital in Na_6^-, which is not seen in the experiment), and to overestimate the binding energy of the singly occupied orbital, like that of the 2s orbital in Na_{18}^-. This demonstrates the limits of the comparison between a photoelectron spectrum and the DFT DOS: as described above, the photoelectron spectrum indicates the distribution of excited states of the neutral cluster, which can well deviate from the energies of single-particle states in the ground state of the anion. Especially for small sizes a better agreement is indeed achieved if really the excited states are calculated (Bonačić-Koutecký et al., 1990). This is, however, numerically much more demanding than the DFT ground-state calculation, and may not be absolutely necessary as long as perfect quantitative agreement is not required. An optimum solution would be of course if an easy extension of the DFT calculation could solve

such problems—first attempts in this direction have already been made (Massobrio et al., 1995; Li et al., 2005).

6.4.2 Large Sodium Clusters

In Figure 6.13, the same type of comparison is done for larger clusters sizes ($n = 39$–309). Again the measured photoelectron spectra are compared to the DOS calculated for the lowest energy structures found. The column on the right shows the comparison for wrong isomers, that is, for structures which are higher in total energy. Most of the structures found are also ground-state structures in calculations using classical interaction potentials between the atoms (Noya et al., 2007). The letters indicate for which potential the respective structure is lowest in energy (G: Gupta, MM: Murrell-Mottram). The letter D indicates that the structure is the ground state only for the DFT calculation. Several observations can be made. First of all the agreement between the DOS and the

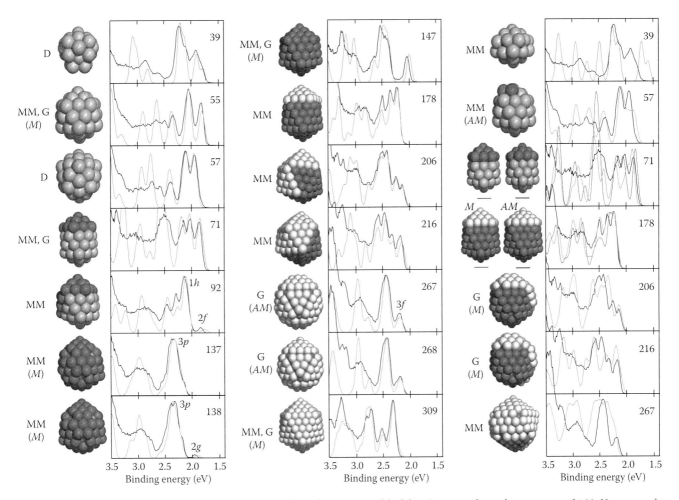

FIGURE 6.13 Photoelectron spectra of cold ($T \approx 100\,K$) sodium cluster anions (black lines) measured at a photon energy of 4.02 eV, compared to the calculated density of states (gray lines) of the lowest energy isomers found in a DFT-based structure search (Kostko et al., 2007). The structures used in the calculations are indicated (left and center column: assigned ground-state structures; right column: wrong structure candidates). The letters give the origin of the structures (MM/G: ground-state structure for Murrell-Mottram or Gupta potential (Noya et al., 2007), D: ground state only within DFT) and indicate the structural motif of the icosahedral structures (M/AM: Mackay/anti-Mackay overlayer [Northby, 1987]). (From Kostko, O. et al., *Phys. Rev. Lett.*, 98, 043401, 2007; Noya, E.G. et al., *Eur. Phys. J. D*, 43, 57, 2007; Northby, J.A., *J. Chem. Phys.*, 87, 6166, 1987.)

photoelectron spectra is very good in the two left columns, and very bad in the right one. This demonstrates that even for these rather large sizes the electron DOS is very sensitive for the details of the geometric structure. This gives strong evidence that in cases where the agreement is good the correct structure has been found. Secondly, one can see that the cluster in this size range now follows an icosahedral growth motif. Between the closed-packed perfect icosahedra ($n = 55, 147, 309$) the clusters grow by forming well-defined capping layers. This study revealed that a prediction of an empirical potential calculation (Noya et al., 2007) was correct: the capping layers tend to undergo a small twist rotation, which avoids the occurrence of less densely packed (100) facets (as are present in some of energetically unfavored structures in the right column). Obviously the clusters always try to maintain a well-packed surface. This predominance of well-packed structures, and even more so the fact that classical calculations correctly predict these structures, shows that in this larger size range quantum effects do not play a strong role anymore as far as the geometric structure is concerned.

This should of course not lead to the conclusion that these clusters are bulk-like pieces of matter. In Figure 6.14, the photoelectron spectrum of $Na_{309}{}^-$ is shown again, together with the DFT DOS. Here now the full valence DOS is depicted. One can see the perfect agreement of the uppermost part of the DOS with the measured spectrum, which gives strong evidence that the lower-lying part is also correct. Obviously, the electronic system of even such a big particle is still highly discretized; the envelope of the DOS follows $\sqrt{E}$ (with E being the kinetic energy of the electrons) as one expects for the valence band of a free-electron metal, but the DOS itself consists of a number of highly degenerate states. One can assign electron shell quantum numbers to these states, as done in Figure 6.14 (only the assignments for the series with radial quantum number 1 is shown in order to keep

the figure simple). But one should keep in mind that these are only approximate angular momentum eigenstates: the reason why the DOS looks like this is only partly due to the validity of the free-electron model; the main reason is that this cluster has a highly symmetric atomic structure.

6.4.3 Noble Metal Clusters

The noble metals copper, silver, and gold are in principle very similar to alkali metals: they are monovalent, with the single-valence electron occupying an atomic s-orbital. In the bulk this leads to the formation of a half-filled s-band similar to that in the alkalis. There is an important difference, though: in the noble metals, the s-band is intersected by a band formed from the fully occupied atomic d-orbitals. In copper and gold, the upper edge of this d-band lies about 2 eV below the Fermi energy; in silver at about 4 eV. For the case of copper, the resulting electronic structure is shown in Figure 6.15. The s-band, which reaches from almost −9 to 0 eV with respect to the Fermi energy, exhibits a free-electron-like parabolic dispersion very similar to that seen for the valence band of sodium (Figure 6.1). It gets intersected by the d-band between about −5.5 and −2 eV, the flat dispersion of which indicates rather strongly bound electrons. As the d-band is occupied by 10 electrons per atom, it dominates the projected DOS shown on the right. But more importantly it has a strong influence on the structural properties of the noble metals: through hybridization with the s/p-band the d-band states get a bonding character, which is the reason why the bonding in noble metal is much stronger and more directional than in the alkali metals. A second consequence of this hybridization is that the delocalized s-band states get some admixture of rather localized d-band states, which increases the effective electron–ion interaction.

For noble metal clusters one would therefore expect that the s-band exhibits a similar discretization into approximate angular momentum eigenstates as seen in sodium clusters, with maybe a stronger perturbation of these states. This is indeed observed. In Figure 6.16, photoelectron spectra of copper, silver, and gold cluster anions with 50–58 atoms are shown; for comparison the corresponding sodium cluster spectra are shown as well. Note that the binding energy scales are different: while the noble metal cluster spectra span about 3 eV, the sodium one spans only 1.5 eV. As this is about the ratio of the Fermi energies of the materials, the same fraction of the valence band is visible in each case. Several observations can be made. For copper at about 2 eV and for gold at about 1.5 eV below threshold the intensity increases abruptly; these are the edges of the d-bands discussed above. The region between the onset of the d-band and the threshold should consequently be the upper part of the s-band. Here practically identical patterns can be seen for copper and silver clusters for all sizes shown except $n = 50$. The gold clusters, however, exhibit completely different patterns. This directly indicates that the copper and silver clusters have identical geometrical structures (except for $n = 50$), while the gold clusters are different. For copper and silver size $n = 55$ is obviously special: here both

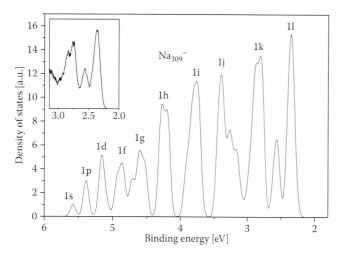

FIGURE 6.14 Calculated density of states (Kohn–Sham energies broadened by Gaussians with a width of 100 meV) of icosahedral $Na_{309}{}^-$. The assignment of the major shell series (shells with radial quantum number 1) is given. The inset shows the measured photoelectron spectrum of $Na_{309}{}^-$. (From Kostko, O. et al., *J. Phys.: Conf. Ser.*, 88, 012034, 2007. With permission.)

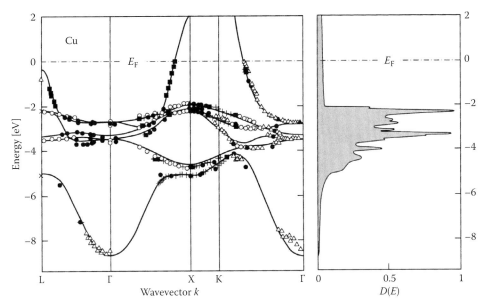

FIGURE 6.15 Left panel: valence electron band structure of copper. The symbols indicate experimental results. (Data taken from Courths, R. and Hüfner, S., *Phys. Rep.*, 112, 53, 1984.) while the lines are the results of calculations. (Data taken from Eckardt, H. et al., *J. Phys. F: Met. Phys.*, 14, 97, 1984.) Right panel: projected density of states. Note the high density of d-band states superimposed onto the low density of states of the broad s-band. (From Hunklinger, S. 2007. *Festkörperphysik*, Oldenbourg, München, Germany. With permission.)

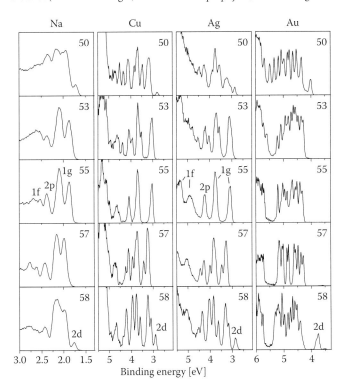

FIGURE 6.16 Photoelectron spectra of sodium, copper, silver, and gold cluster anions with 50–58 atoms, measured at photon energies of 4.02 eV (Na) and 6.4 eV (Cu, Ag, Au). (From Häkkinen, H. et al., *Phys. Rev. Lett.*, 93, 093401, 2004; Kostko, O. et al., *Phys. Rev. Lett.*, 98, 043401, 2007.) For the icosahedral clusters Na_{55}^-, Cu_{55}^-, and Ag_{55}^-, the visible peaks are assigned according to the spherical free-electron model; note the splitting of the 1f and 1g shell, which is in agreement with group theory predictions (Figure 6.5). At size 58 (electron number 59) for all systems a new shell (2d) appears.

exhibit the cleanest spectra, which consist only of a few highly degenerate states. The reason is the same as for sodium clusters: these 55 atom clusters are closed-packed icosahedra and therefore highly symmetric. Their DOS can directly be read off the diagram in Figure 6.5: in icosahedral symmetry the 2p shell stays degenerate, while 1f and 1g split up into two states each. This is exactly the same electronic structure as Na_{55}^- has; only for Cu_{55}^- and Ag_{55}^- it is much better resolved due to the higher rigidity of the noble metals (less vibrational excitation upon photodetachment) and their higher Fermi energy (larger peak spacing). The better resolution also allows to see the effect of a slight perturbation of the symmetry: when two atoms are added to, or taken away from, the perfect icosahedron, the degeneracy of the states is lifted; the corresponding peaks split up. Therefore, the spectra of the 53 and 57 atom clusters are still similar to those of the 55 atom clusters, but there are shoulders visible on all peaks. As the spectrum of Au_{55}^- does not exhibit any strong degeneracy at all, one can directly conclude that it must have a very low symmetry structure. In particular, it cannot have the cuboctahedral structure that has been proposed for neutral, passivated Au_{55} clusters (Häkkinen et al., 2004; Schmid, 2008). At size 58 (electron number 59) for all four metals, a new peak appears at threshold. This is in perfect agreement with the free-electron model: the 1f shell is filled at electron number 58, the additional electron has to occupy the next higher shell, 2d. Here gold is quite special again. It exhibits the largest gap between the new 2d shell and the rest of the DOS.

Similar findings have been made for other size ranges as well. Small copper and silver cluster anions up to size $n = 18$ have been shown to have very similar photoelectron spectra (Cha et al., 1993; Handschuh et al., 1995), indicating identical geometrical structures. The spectra are related, but not identical to those of sodium cluster anions in this size range. It is not clear whether the reason for this discrepancy are different geometrical structures or the influence of the d-bands in the case of the copper and silver clusters (which lead to a partial breakdown of Koopman's theorem [Massobrio et al., 1995]). Recent calculations indicate that there are indeed some structural differences between small ($n \leq 11$) copper and sodium clusters (Li et al., 2006). In any case in the size range $n = 20$–40 sodium and copper cluster anions exhibit very similar photoelectron spectra for most sizes, indicating identical structures. In all of this size range, silver cluster anions show different spectra. For most larger sizes up to at least $n = 100$ sodium, copper, and silver cluster anions have close to identical spectra. So the three materials sodium, copper, and silver exhibit all in all a rather similar behavior. Gold is different, however; here relativistic effects change the nature of the bonding, due to which gold clusters tend to adopt completely different geometrical structures than other metal systems (Häkkinen et al., 2002). Gold cluster anions, e.g., are planar up to size $n = 12$ (Furche et al., 2002). Au_{16}^- is a tetrahedral cage (Bulusu et al., 2006), while Au_{20}^- has the unusual form of a perfect tetrahedron, which can be seen as a piece cut out of the bulk fcc lattice (Li et al., 2003). Au_{34}^- is related to an cuboctahedron, but exhibits a chiral surface reconstruction (Lechtken et al., 2007). As mentioned above, Au_{55}^- does not have a regular, symmetric structure (Häkkinen et al., 2004). Only for significantly larger sizes gold clusters adopt the regular structures known from other metals (Koga et al., 2004; Li et al., 2008).

In summary, the electronic structure of copper and silver clusters is rather similar to that of sodium clusters, except of course for the region of the DOS where the d-band dominates. The better-resolved photoelectron spectra of noble metal clusters show the limits of the free-electron model more clearly: although an electron shell structure is visible, in most cases it is quite strongly perturbed by the interaction with the ion background. In gold clusters, this perturbation is so strong that for sizes with low symmetry structures it can lead to a complete vanishing of the electron shells; here the free-electron model simply does not work anymore.

6.4.4 Angle-Resolved PES: Sodium Clusters

The observation of the rather strong perturbation of the electronic states by the ion background raises the question how good the assignment of the observed shells as angular momentum eigenstates actually is. It is clear that any perturbing potential of nonspherical symmetry will lead to a mixing of states of different angular momenta. The big question is how strong this mixing is, that is, whether the new eigenstates still have dominant contributions from a single-angular momentum. One possibility to get information about the angular momentum of an electronic state is to use angle-resolved spectroscopy. This will be briefly explained now.

The angular distribution of electrons emitted in a single-photon process out of an nonoriented molecule or cluster can be described by the following expression (in dipole approximation):

$$f(\theta) = 1 + \beta \left(\frac{3}{2}\cos^2\theta - \frac{1}{2} \right) \qquad (6.18)$$

where

θ is the angle with respect to the polarization direction of the light

β is the anisotropy parameter, which can have values between -1 (emission perpendicular to the polarization) over 0 (isotropic emission) to 2 (emission parallel to the polarization)

As has been shown already by Bethe (1933) for spherical single-electron systems and by Cooper and Zare (1968) for many-electron systems, this parameter strongly depends on the angular momentum of the initial state of the electron. One could therefore expect that different electronic shells in a cluster exhibit different angular distributions. One example of an angle-resolved photoelectron spectrum is presented in Figure 6.17, which has been measured on Na_{58}^- at a photon energy of 2.48 eV. In the top part, the angle-integrated photoelectron spectrum is shown, which is very similar to the one in Figure 6.16. As explained already, one can recognize three electron shells: the largest peak (or band) was identified as the completely filled 1g shell, the small peak at higher binding energy is the 2p shell, and the small peak at threshold is the 2d shell. The image beneath shows the angle-resolved photoelectron spectrum, where the angle is

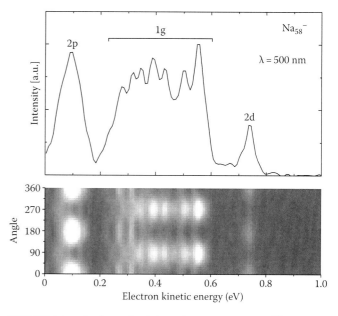

FIGURE 6.17 Angle-resolved photoelectron spectrum of Na_{58}^-, measured at a photon energy of 2.48 eV. In the upper panel, the angle-integrated spectrum is shown; in the lower panel, the spectra are plotted as a function of the emission angle with respect to the light polarization. Note that electrons from the 2p and the 2d shell are emitted preferentially into the direction of the light polarization (0° and 180°), while electrons out of the 1g shell are emitted perpendicularly to it. (From Bartels, C. et al., *Science*, 323, 1323, 2009.).

measured with respect to the polarization of the laser light. It turns out that the three shells exhibit distinct differences; while the electron emission out of both the 2p shell and the 2d shell is peaked parallel to the laser polarization, the emission out of the 1g shell is perpendicular. Even more surprising is the uniformity of this behavior. Although the shells are clearly not fully degenerate (the 1g shell is split into a broad band of almost 0.4 eV width), all sublevels of one band exhibit very similar angular distributions. This demonstrates that the electron–lattice interaction is strong enough for lifting the degeneracy of the angular momentum eigenstates, but not for causing strong hybridization of the different angular momentum states, which would blur the differences between the shells. This is a very promising result as it shows that the free-electron model for sodium clusters is really justified, in the sense that the wavefunctions of the electrons in the clusters largely retain the character of truly free-electron wavefunctions.

6.5 Summary and Outlook

We have shown that the electronic structure of sodium, copper, and silver clusters is close to what one would expect for a simple free-electron model, but that details can only be understood if the atomic structure is taken into account. Here, the symmetry of the structure is very important; the clearest electron shell structure is seen for the cluster sizes with perfect icosahedral symmetry. Gold is different: it adopts different geometries, and due to the stronger electron–ion interaction the electronic structure of the clusters is more complex than for the other systems discussed here. The further directions of research are straightforward: the other alkali metals have to be studied, which is interesting because they are less good free-electron metals than sodium (Ashcroft and Mermin, 1976). Angle-resolved photoelectron spectroscopy of these and noble metal clusters will show how much the electronic wavefunctions are perturbed; this is an especially interesting question in the case of gold. Another question is how these almost free-electron systems will interact with impurities, for example, by incorporation of a magnetic atom into the cluster (Janssens et al., 2005). Here a size and temperature-dependent photoelectron spectroscopy study might reveal some precursor of the Kondo effect (Neel et al., 2008). Finally the simple metal clusters, which are now well characterized, can serve as model systems for the study of dynamics, like finite-size phase transitions (Haberland et al., 2005) or ultrafast relaxation processes (Maier et al., 2006). The question arises whether the findings discussed here can also be used for applications. Sodium clusters are unfortunately very reactive, so despite their very interesting properties they will most probably play a role only in fundamental science. Noble metal clusters, on the other hand, have promising properties for many fields like, e.g., photonics (Quinten et al., 1998) or catalysis; in the latter field, the discovery of the reactivity of small gold clusters has caused quite a bit of excitement (Haruta, 1997). For a detailed understanding especially of such effects the knowledge of the geometric structure and the electronic DOS of the particles are of high importance (Yoon et al., 2007); here cluster spectroscopy can contribute significantly.

Spectroscopy on size-selected gas phase clusters is therefore a very lively field, and it seems a safe prediction that many exciting discoveries are still to be made.

References

Ashcroft, N. W. and Mermin, N. D. 1976. *Solid State Physics*. Saunders College, Philadelphia, PA.

Bartels, C., Hock, C., Huwer, J., Kuhnen, R., Schwöbel, J., and van Issendorff, B. 2009. Probing the angular momentum character of the valence orbitals of free sodium nanoclusters. *Science*, 323: 1323–1327.

Bethe, H. 1933. In Geiger, H. and Scheel, K., editors, *Handbuch der Physik*, pp. 475–484. Springer, Berlin, Germany.

Bonačić-Koutecký, V., Fantucci, P., and Koutecký, J. 1990. Theoretical interpretation of the photoelectron detachment spectra of Na_{2-5}^- and of the absorption spectra of Na_3, Na_4, and Na_8 clusters. *J. Chem. Phys.*, 93: 3802–3825.

Bonačić-Koutecký, V., Pittner, J., Fuchs, C., Fantucci, P., Guest, M. F., and Koutecký, J. 1995. Ab initio predictions of structural and optical response properties of Na_n^+ clusters: Interpretation of depletion spectra at low temperature. *J. Chem. Phys.*, 104: 1427–1440.

Born, M. and Oppenheimer, J. R. 1927. Zur Quantentheorie der Molekeln. *Ann. Phys.*, 389: 457–484.

Brack, M. 1993. The physics of simple metal clusters: Self-consistent jellium model and semiclassical approaches. *Rev. Mod. Phys.*, 65: 677–732.

Bulusu, S., Li, X., Wang, L.-S., and Zeng, X. C. 2006. Evidence of hollow golden cages. *Proc. Natl. Acad. Sci. U.S.A.*, 24: 8326–8330.

Cederbaum, L. S., Schirmer, J., Domcke, W., and von Niessen, W. 1977. Complete breakdown of the quasiparticle picture for inner valence electrons. *J. Phys. B*, 10: L549–L553.

Cha, C.-Y., Ganteför, G., and Eberhardt, W. 1993. Photoelectron spectroscopy of Cu_n^- clusters: Comparison with jellium model predictions. *J. Chem. Phys.*, 99: 6308–6312.

Cheng, H., Berry, R. S., and Whetten, R. L. 1991. Electronic structure and binding energies of aluminum clusters. *Phys. Rev. B*, 43: 10647–10653.

Ching, W. Y. and Callaway, J. 1975. Energy bands, optical conductivity, and Compton profile of sodium. *Phys. Rev. B*, 11: 1324–1329.

Chong, D. P., Gritsenko, O. V., and Baerends, E. J. 2002. Interpretation of the Kohn–Sham orbital energies as approximate vertical ionization potentials. *J. Chem. Phys.*, 116: 1760–1772.

Clemenger, K. 1985. Ellipsoidal shell structure in free-electron metal clusters. *Phys. Rev. B*, 32: 1359–1362.

Cooper, J. and Zare, R. N. 1968. Angular distribution of photoelectrons. *J. Chem. Phys.*, 48: 942–943.

Courths, R. and Hüfner, S. 1984. Photoemission experiments on copper. *Phys. Rep.*, 112: 53–171.

de Heer, W. A. 1993. The physics of simple metal clusters: Experimental aspects and simple models. *Rev. Mod. Phys.*, 65: 611–676.

Eaton, G., Kidder, L. H., Sarkas, H. W., McHugh, K. M., and Bowen, K. H. 1992. Photodetachment spectroscopy of alkali cluster anions. In P. Jena, S. N. Khanna, and Rao, B., editors, *Physics and Chemistry and Finite Systems: From Clusters to Crystals*, volume 374 of *Part I of NATO ASI Series C*, pp. 493–507. Kluwer Academic, Dordrecht, the Netherlands/ Boston, MA.

Eckardt, H., Fritsche, L., and Noffke, J. 1984. Self-consistent relativistic band structure of the noble metals. *J. Phys. F: Met. Phys.*, 14: 97–91.

Ekardt, W. 1984. Work function of small metal particles: Self-consistent spherical jellium-background model. *Phys. Rev. B*, 29: 1558–1564.

Ekardt, W., editor. 1999. *Metal Clusters*. Wiley & Sons Ltd, Chichester, U.K.

Furche, F., Ahlrichs, R., Weis, P., Jacob, C., Gilb, S., Bierweiler, T., and Kappes, M. M. 2002. The structures of small gold cluster anions as determined by a combination of ion mobility measurements and density functional calculations. *J. Chem. Phys.*, 117: 6982.

Ganteför, G., Gausa, M., Meiwes-Broer, K.-H., and Lutz, H. 1990. Photoelectron spectroscopy of silver and palladium cluster anions: Electron delocalization versus localization. *J. Chem. Soc., Faraday Trans.*, 86: 2483–2488.

Ganteför, G., Meiwes-Broer, K. H., and Lutz, H. O. 1988. Photodetachment spectroscopy of cold aluminum cluster anions. *Phys. Rev. A*, 37: 2716–2718.

Gerlich, D. 1995. Ion-neutral collisions in a 22-pole trap at very low energies. *Phys. Scr.*, T59: 256–263.

Haberland, H., editor. 1995. *Clusters of Atoms and Molecules I.* Springer, Berlin, Germany.

Haberland, H., Hippler, T., Donges, J., Kostko, O., Schmidt, M., and van Issendorff, B. 2005. Melting of sodium clusters: Where do the magic numbers come from? *Phys. Rev. Lett.*, 94: 035701.

Häkkinen, H., Moseler, M., and Landman, U. 2002. Bonding in Cu, Ag, and Au clusters: Relativistic effects, trends, and surprises. *Phys. Rev. Lett.*, 89: 033401.

Häkkinen, H., Moseler, M., Kostko, O., Morgner, N., Hoffmann, M. A., and van Issendorff, B. 2004. Symmetry and electronic structure of noble-metal nanoparticles and the role of relativity. *Phys. Rev. Lett.*, 93: 093401.

Handschuh, H., Cha, C.-Y., Bechthold, P. S., Ganteför, G., and Eberhardt, W. 1995. Electronic shells or molecular orbitals: Photoelectron spectra of Ag_n^- clusters. *J. Chem. Phys.*, 102: 6406–6422.

Haruta, M. 1997. Size- and support-dependency in the catalysis of gold. *Catal. Today*, 36: 153–166 (Copper, Silver and Gold in Catalysis).

Hohenberg, P. and Kohn, W. 1964. Inhomogeneous electron gas. *Phys. Rev.*, 136: B864–B871.

Hunklinger, S. 2007. *Festkörperphysik*, Oldenbourg, München, Germany.

Janssens, E., Neukermans, S., Nguyen, H. M. T., Nguyen, M. T., and Lievens, P. 2005. Quenching of the magnetic moment of a transition metal dopant in silver clusters. *Phys. Rev. Lett.*, 94: 113401.

Knight, W. D., Clemenger, K., de Heer, W. A., Saunders, W. A., Chou, M. Y., and Cohen, M. L. 1984. Electronic shell structure and abundances of sodium clusters. *Phys. Rev. Lett.*, 52: 2141–2143.

Koga, K., Ikeshoji, T., and Sugawara, K. 2004. Size- and temperature- dependent structural transitions in gold nanoparticles. *Phys. Rev. Lett.*, 92: 115507.

Kohn, W. and Sham, L. J. 1965. Self-consistent equations including exchange and correlation effects. *Phys. Rev.*, 140: A1133–A1138.

Koopmans, T. C. 1934. Über die Zuordnung von Wellenfunktionen und Eigenwerten zu den einzelnen Elektronen eines Atoms. *Physica*, 1: 104–113.

Koskinen, M., Lipas, P. O., and Manninen, M. 1995. Electron-gas clusters: The ultimate jellium model. *Z. Phys. D*, 35: 285–297.

Kostko, O., Huber, B., Moseler, M., and van Issendorff, B. 2007. Structure determination of medium-sized sodium clusters. *Phys. Rev. Lett.*, 98: 043401.

Kostko, O., Bartels, C., Schwöbel, J., Hock, C., and Issendorff, B. V. 2007. Photoelectron spectroscopy of the structure and dynamics of free size selected sodium clusters. *J. Phys.: Conf. Ser.*, 88, 012034.

Kruit, P. and Read, F. H. 1983. Magnetic field paralleliser for 2π; electron-spectrometer and electron-image magnifier. *J. Phys. E*, 16: 313–324.

Lechtken, A., Schooss, D., Stairs, J. R., Blom, M. N., Furche, F., Morgner, N., Kostko, O., van Issendorff, B., and Kappes, M. M. 2007. Au_{34}^-: A chiral gold cluster? *Angew. Chem. Int. Ed. Engl.*, 46: 2944–2948.

Leopold, D. G., Ho, J., and Lineberger, W. C. 1987. Photoelectron spectroscopy of mass-selected metal cluster anions. I. Cu_n^-, $n = 1$–10. *J. Chem. Phys.*, 86: 1715–1726.

Li, J., Li, X., Zhai, H.-J., and Wang, L.-S. 2003. Au_{20}: A tetrahedral cluster. *Science*, 299: 864–867.

Li, S., Alemany, M. M. G., and Chelikowsky, J. R. 2005. Ab initio calculations of the photoelectron spectra of transition metal clusters. *Phys. Rev. B*, 71: 165433.

Li, S., Alemany, M. M. G., and Chelikowsky, J. R. 2006. Real space pseudopotential calculations for copper clusters. *J. Chem. Phys.*, 125: 034311.

Li, Z. Y., Young, N. P., Vece, M. D., Palomba, S., Palmer, R. E., Bleloch, A. L., Curley, B. C., Johnston, R. L., Jiang, J., and Yuan, J. 2008. Three-dimensional atomic-scale structure of size-selected gold nanoclusters. *Nature*, 451: 46–48.

Maier, M., Wrigge, G., Hoffmann, M. A., Didier, P., and van Issendorff, B. 2006. Observation of electron gas cooling in free sodium clusters. *Phys. Rev. Lett.*, 96: 117405.

Markovich, G., Pollack, S., Giniger, R., and Cheshnovsky, O. 1994. Photoelectron spectroscopy of Cl⁻, Br⁻, and I⁻ solvated in water clusters. *J. Chem. Phys.*, 101: 9344–9353.

Massobrio, C., Pasquarello, A., and Car, R. 1995. First principles study of photoelectron spectra of Cu_n^- clusters. *Phys. Rev. Lett.*, 75: 2104–2107.

McHugh, K. M., Eaton, J. G., Lee, G. H., Sarkas, H. W., Kidder, L. H., Snodgrass, J. T., Manaa, M. R., and Bowen, K. H. 1989. Photoelectron spectra of the alkali metal cluster anions: $Na_{n=2-5}^-$, $K_{n=2-7}^-$, $Rb_{n=2-3}^-$, and $Cs_{n=2-3}^-$. *J. Chem. Phys.*, 91: 3792–3793.

Montag, B., Hirschmann, T., Meyer, J., Reinhard, P.-G., and Brack, M. 1995. Shape isomerism in sodium clusters with $10 \leq Z \leq 44$: Jellium model with quadrupole, octupole, and hexadecapole deformations. *Phys. Rev. B*, 52: 4775–4778.

Moseler, M., Huber, B., Häkkinen, H., Landman, U., Wrigge, G., Hoffmann, M. A., and van Issendorff, B. 2003. Thermal effects in the photoelectron spectra of Na_N^- clusters ($N = 4$–19). *Phys. Rev. B*, 68: 165413.

Mundt, M., Kümmel, S., Huber, B., and Moseler, M. 2006. Photoelectron spectra of sodium clusters: The problem of interpreting Kohn-Sham eigenvalues. *Phys. Rev. B*, 73: 205407.

Neel, N., Kroger, J., Berndt, R., Wehling, T. O., Lichtenstein, A. I., and Katsnelson, M. I. 2008. Controlling the Kondo effect in $CoCu_n$ clusters atom by atom. *Phys. Rev. Lett.*, 101: 266803.

Northby, J. A. 1987. Structure and binding of Lennard-Jones clusters: $13 \leq N \leq 147$. *J. Chem. Phys.*, 87: 6166–6177.

Noya, E. G., Doye, J. P. K., Wales, D. J., and Aguado, A. 2007. Geometric magic numbers of sodium clusters: Interpretation of the melting behaviour. *Eur. Phys. J. D*, 43: 57–60.

Pettiette, C. L., Yang, S. H., Craycraft, M. J., Conceicao, J., Laaksonen, R. T., Cheshnovsky, O., and Smalley, R. E. 1988. Ultraviolet photoelectron spectroscopy of copper clusters. *J. Chem. Phys.*, 88: 5377–5382.

Quinten, M., Leitner, A., Krenn, J. R., and Aussenegg, F. R. 1998. Electromagnetic energy transport via linear chains of silver nanoparticles. *Opt. Lett.*, 23: 1331–1333.

Reimann, S. M., Frauendorf, S., and Brack, M. 1995. Triaxial shapes of sodium clusters. *Z. Phys. D*, 34: 125–134.

Rytkönen, A., Häkkinen, H., and Manninen, M. 1998. Melting and octupole deformation of Na_{40}. *Phys. Rev. Lett.*, 80: 3940–3943.

Schmid, G. 2008. The relevance of shape and size of Au_{55} clusters. *Chem. Soc. Rev.*, 37: 1909–1930.

Stowasser, R. and Hoffmann, R. 1999. What do the Kohn-Sham orbitals and eigenvalues mean? *J. Am. Chem. Soc.*, 121: 3414–3420.

Taylor, K. J., Pettiette-Hall, C. L., Cheshnovsky, O., and Smalley, R. E. 1992. Ultraviolet photoelectron spectra of coinage metal clusters. *J. Chem. Phys.*, 96: 3319–3329.

Yang, M., Jackson, K. A., and Jellinek, J. 2006a. First-principles study of intermediate size silver clusters: Shape evolution and its impact on cluster properties. *J. Chem. Phys.*, 125: 144308.

Yang, M., Jackson, K. A., Koehler, C., Frauenheim, T., and Jellinek, J. 2006b. Structure and shape variations in intermediate-size copper clusters. *J. Chem. Phys.*, 124: 024308.

Yannouleas, C. and Landman, U. 1995. Electronic shell effects in triaxially deformed metal clusters: A systematic interpretation of experimental observations. *Phys. Rev. B*, 51: 1902–1917.

Yoon, B., Koskinen, P., Huber, B., Kostko, O., van Issendorff, B., Häkkinen, H., Moseler, M., and Landman, U. 2007. Size-dependent structural evolution and chemical reactivity of gold clusters. *Chem. Phys. Chem.*, 8: 157–161.

7

Photoelectron Spectroscopy of Free Clusters

Maxim Tchaplyguine
Lund University

Gunnar Öhrwall
Lund University

Olle Björneholm
Uppsala University

7.1 Introduction

7.1.1 What Are Clusters?

Clusters are aggregates of a *finite* number of atoms or molecules [1], bridging the gap between the isolated atom/molecule and the macroscopic solid state of matter, as schematically shown in Figure 7.1. For large clusters, the development of properties is typically a smooth change with the size, often connected to the surface/volume ratio. For small clusters, this smooth development is replaced by increasingly strong variations in the property. This is the so-called quantum size regime in which each atom makes a difference. In some sense, such small clusters can be regarded as molecules, with each size being a unique entity. If a cluster is larger than just a few atoms, it often adopts the geometry that minimizes the surface-atom fraction: the surface atoms generally have lower coordination (number of nearest neighbors) than the inner, bulk atoms, and the cluster thus reaches the lowest energy structure by maximizing the bulk-atom fraction. This has been shown to lead to the peculiarly stable geometric shell structures, first observed as "magic numbers" in mass spectra for rare gas clusters [2]. In the cluster formation process (discussed in more detail below), those aggregates are more likely to be created which have enclosed, onion-like, close-to-spherical layers, or shells. Such geometry gives certain well-defined numbers of atoms per cluster in a stable configuration—the "magic"

numbers. The phenomenon of magic numbers is already observed in the case of a cluster containing 13 atoms, for which the most stable geometry is often an icosahedron—a geometric structure with all the vertices on a spherical surface. The 13th atom is in the center of such a spherical cluster. The number of atoms N in a cluster with m enclosed completed icosahedral shells can be calculated exactly:

$$N = \frac{1}{3}(2m^m - 1)(5m^2 - 5m + 3) \qquad (7.1)$$

with the number of atoms N_m in each mth shell defined by the formula

$$N_m = 10m^2 - 20m + 12, \quad m > 1 \qquad (7.2)$$

Thus, the magic numbers for the icosahedral geometry are 13, 55, 147, 309, 561, 923, etc. It is the clusters with these amounts of atoms per unit that are more abundant in the mass spectra of inert gas clusters. It should be noted here that the same formulas are valid for the shells of cuboctahedral geometry in which case shells are not any more spherical.

With decreasing size, the relative importance of the surface atoms over the bulk ones increases. To illustrate this, the left-hand panel of Figure 7.2 shows the fraction of atoms in the surface

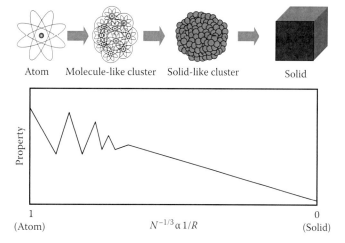

Atom   Molecule-like cluster   Solid-like cluster   Solid

FIGURE 7.1 Schematic presentation of dimensional transformation from atoms via small and large clusters to the solid state. In the small-cluster-size regime, each extra atom matters, so the properties change in a "zigzag" way with the size. When the clusters become large, their properties change smoothly.

versus the total number of atoms in the cluster for the closed-shell structures defined by Equations 7.1 and 7.2. Macroscopic objects typically have surface fractions of 10^{-8} or less, but as seen from Figure 7.2, this fraction is drastically increased for the smaller size, nanoscale objects. This means that those phenomena for which macroscopic solids are of importance only for the surface may largely determine the properties of the whole cluster,

and possibly also of the assemblies of clusters. This specificity of clusters may allow making novel cluster-based materials with tunable properties.

As mentioned above, the surface atoms have lower coordination than the bulk ones. These coordination differences also depend on the size, as illustrated in the right panel of Figure 7.2, where the relative fraction of different surface sites with different coordination numbers is depicted as a function of size for the closed-shell clusters. The example is for the cuboctahedral structure, which can be seen as a suitably truncated fcc structure, the latter being characteristic for many solids. It should be noted that this is by no means the only possible structure for small clusters, especially since not all clusters have closed shell geometries, but is meant to illustrate that smaller clusters generally have a higher fraction of less coordinated surface atoms. These low-coordinated surface atoms often have special properties, for instance, higher chemical reactivity or magnetic moments. It is therefore especially important to study small clusters with surface-sensitive methods.

For the clusters of metal elements, with freely mobile valence electrons, it is also the energy structure defined by the number of shared, delocalized electrons, which determines whether the cluster is stable or quickly falls apart, and which gives birth to another type of "magic numbers," first seen for Na clusters [3]. As mentioned above, the changes in the electronic and geometric structures are intimately connected to the size dependence of chemical and physical properties. Probably, the most spectacular example of these changes is the emergence of metallicity

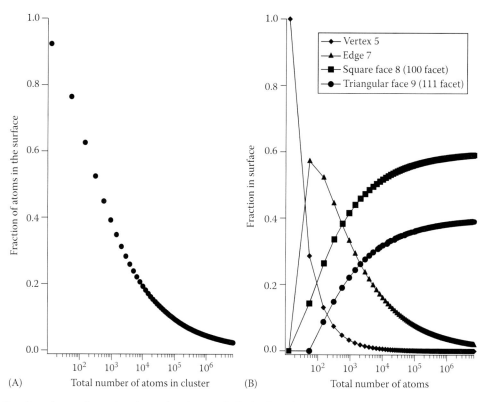

FIGURE 7.2 (A) Fraction of atoms located at the surface for icosahedral/cuboctahedral clusters with enclosed shells. (B) Fractions of surface atoms with chemically nonequivalent environments (different sites) in a cuboctahedral cluster depending on the size.

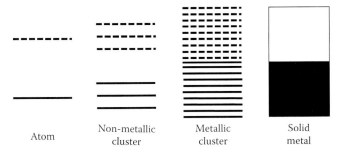

Atom   Non-metallic   Metallic   Solid
       cluster   cluster   metal

FIGURE 7.3 Schematic presentation of the energy structure transformation from metal atoms via small and large clusters to the solid metal. Below a certain cluster size, at which the empty and the populated bands merge, a cluster of metal atoms may have semiconductor-like energy structure.

in the clusters of metal atoms of a certain size, as schematically depicted in Figure 7.3. The highest occupied and the lowest unoccupied levels of an atom form bonding–antibonding combinations in small, molecular-like nonmetallic clusters. As the number of atoms increase, so will the number of levels, until the splitting between occupied and unoccupied levels becomes smaller than the thermal energy kT, at which point the cluster starts exhibiting metallic behavior. For the sizes larger than this critical size of transition, the metallic clusters will behave as small pieces of the corresponding macroscopic metal, but with a size-dependent high surface/bulk ratio, as discussed in connection to Figure 7.2 [4]. Below this critical size, the clusters of metal elements can have semiconductor-like properties.

To fully understand the development with the size, it is crucial to study *free* clusters when the influence of the massive substrate on the nanoscale objects is absent. For *supported* clusters—either deposited from a beam of atoms/molecules/clusters or grown by chemical methods on a substrate—both the geometry and the energy-level structures are distorted or obscured by the interaction with this substrate.

While some clusters (like rare gas clusters) are purely artificial creations, free molecular clusters exist in nature, especially in atmosphere where, for example, water clusters are intermediates in cloud formation and contribute significantly to the IR absorption in some wavelength regions [5,6]. Other molecules, such as NH_3, SO_x, and NO_x, and organic molecules form mixed clusters with water, which are considered to be important condensation centers in the upper atmosphere [7,8].

Free clusters can be created and studied in different ways, some of which will be described below. To create free clusters, one needs a dedicated apparatus that generates a *beam* of clusters propagating in vacuum. Most of such devices, often referred to as cluster sources, produce clusters not of a certain, unique number N of atoms/molecules per cluster (cluster size), but a wide distribution of sizes, which has to be established experimentally. This is why one of the first methods of cluster studies occurred to be mass spectroscopy. It has been experimentally shown that most of the cluster sources generate the so-called log-normal distribution of sizes in a beam

$$L(N) = \frac{1}{N\sigma\sqrt{2\pi}} \exp\left[-\frac{(\ln N - \mu)^2}{2\sigma^2}\right] \quad (7.3)$$

with the mean cluster size $\langle N \rangle = \exp[\mu]$ and σ as a standard deviation from the mean size.

In many experiments, in order to separate clusters of different sizes/masses present in the primary distribution, the clusters are ionized. However, the ionization process imposes significant changes on the cluster size distribution as well as on each cluster. The details and consequences of the cluster ionization will be discussed later in the text.

7.1.2 Basics of Electron Spectroscopy, Its Subfields, and Acquisition Techniques

The cluster ionization studies developed in a separate subfield, where such properties and phenomena as size-dependent ionization potentials, singly and multiply ionized cluster fragmentation, and lifetimes of excited states became the subjects of investigation. The ionization tools included widely used lasers and spectroscopic lamps of different wavelengths, beams of electrons and ions. In an ionization event, electrons are ejected by the system under investigation. These electrons contain detailed information on the energy structure of the sample, and the detection and analysis of the emitted electrons, the so-called electron spectroscopy, brought a lot of knowledge on atoms, molecules, and solid phase of matter. When it is the irradiation that causes the electron ejection, the method of studies is referred to as photoelectron spectroscopy (PES). The basic physical law defining the kinetic energy E_{kin} of the electron coming out from a sample irradiated by the photons of a certain energy $h\nu$ is Einstein's law of photoeffect:

$$h\nu = E_{kin} + E_{bin} \quad (7.4)$$

The E_{bin} is the binding energy of the electron in the sample, and it is as a rule this energy which is extracted in a photoelectron spectroscopy experiment, and which contains various information on the system. A primary result of a PES probing is an energy spectrum, reflecting all kinetic/binding energies of the electrons in the sample in the range under investigation. The spectral intensity, *I*, is defined by the ionization probability which can, in principle, be calculated using a quantum mechanics approach, if the wavefunctions Ψ_i and Ψ_f describing the initial and the final states of the object under investigation are known, as well as the interaction between the light and the system. The interaction causing the transition from the neutral to the ionized state in the PES case is often approximated by the so-called dipole operator μ:

$$I \propto \langle \Psi_i | \mu | \Psi_f \rangle \quad (7.5)$$

For a separate atom, the E_{bin} can be just one single energy value or, more correctly, a narrow interval of energies around a certain value. For molecules, a binding energy spectrum can contain discrete progressions of peaks reflecting vibrations in the ground, neutral state, and in the final, ionized states. For a solid sample, a spectrum can be of continuous intensity in a large interval of energies—a band, for which the onset, the width, and the shape can be determined in an experiment. Over the years, photoelectron spectroscopy has proved to be a valuable tool in the studies of electronic and geometric structure of isolated atoms, molecules, surfaces, and solids. As recognition of its importance, a Noble Prize was awarded to Professor Kay Siegbahn in 1981 for the development of the electron spectroscopy as a technique, and for his and his colleagues' achievements in the field. An important role in the electron spectroscopy progress belongs to the x-ray radiation made available at the synchrotron facilities. This radiation is an invaluable tool, since for many elements and substances already the first ionization potential lies beyond the reach of conventional laboratory light sources such as lasers. Some spectroscopic lamps do provide ultraviolet (UV) and the so-called vacuum-ultraviolet (VUV) radiation, whose photon energy ($\approx 3\,eV < h\nu < \approx 10\,eV$ for the UV, and $\geq 10\,eV$ for the VUV) is high enough to ionize the outermost, valence energy levels. However, already for the studies of the so-called inner-valence states—with electron binding energies $>20\,eV$, not talking about the so-called core levels with energies often of hundreds and more electron volt, there is nowadays practically no competitor for the x-ray radiation of the synchrotrons. It will be shown below how the properties of this radiation allow addressing cluster-specific problems in a unique in the accessed information way.

In this chapter, the development and current status of *free* cluster studies with photoelectron spectroscopy is described with the emphasis on the synchrotron-radiation-based research. The covering starts with simple model systems, such as rare gas clusters, and proceeds via the clusters out of molecular gases and liquids to the clusters out of solid materials like salts, semiconductors, and metals. In a most general sense, the electron spectroscopy technique includes valence, also called "ultraviolet" photoelectron spectroscopy (UPS), x-ray photoelectron spectroscopy (XPS), Auger electron spectroscopy (AES), resonant Auger electron spectroscopy (RAS), and x-ray absorption spectroscopy (XAS) [9–11]. Figure 7.4 shows a schematic overview of various PES methods: UPS, XPS, AES, RAS, and XAS.

In some cases of UPS, which probes the valence levels, UV and VUV lasers and lamps can be used for ionization of the sample. The light sources implemented in cluster UPS have been of both pulsed and continuous operation. The former allowed efficiently using cluster sources based on a pulsed adiabatic expansion and on laser ablation (see Sections 7.2 and 7.5). Also a time-structured acquisition method, time-of-flight (TOF) electron spectroscopy could be fruitfully used in this case. The function principle of the TOF electron spectrometers is based on the proportionality between the ejected electron kinetic energy and the arrival time at the detector after a flight in the so-called drift tube in which there is no electric field.

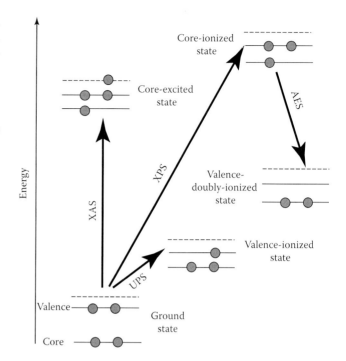

FIGURE 7.4 Schematic overview of the relations between some core-level spectroscopies. XAS (x-ray absorption spectroscopy) probes unoccupied levels by core-electron excitation, XPS (x-ray photoelectron spectroscopy) studies core-ionized states, UPS (ultraviolet photoelectron spectroscopy)-valence-ionized states, and AES (Auger electron spectroscopy) probes the final states after the core-hole recombination.

At the same time, electron kinetic energies with which TOF spectrometers can work are quite low: the electron arrival times on the detector become too short at higher kinetic energies to be resolved for the close in energy electrons. Slowing down of the electrons by an electrostatic lens can somewhat improve the situation but the accessible kinetic energy anyway remains limited to a few tens of electron volts. Another modification of such a TOF spectrometer is in applying magnetic field of a certain configuration to guide the electrons toward the detector. A coil around the drift tube of such a spectrometer creates a field along the tube axis that makes the electrons precess around the axis without changing their kinetic energy. This modification increases the collection efficiency of the TOF spectrometers. In the case of high electron kinetic energies in the range of tens and hundreds of electron volts, the so-called electrostatic energy analyzers are more appropriate. The basic principle of these spectrometers is in forcing the electrons to fly in an electrostatic field of such a configuration that leads to different landing positions on the detector for the electrons with different initial kinetic energies. A widely spread type of these spectrometers utilizes spherical electrostatic field. A detector of such a spectrometer is split into spatial channels, each channel being associated with certain, known kinetic energy. Thus, the impact of an electron on such a channel allows to state that the system ejects the electrons corresponding to this channel kinetic energy. Both types of electron spectrometers—TOF and electrostatic—have been used in PES experiments on clusters.

7.1.3 Relevant to Clusters Details and Consequences of Interaction between the Ionizing Radiation and Matter

7.1.3.1 The Case of the Valence-Level Ionization: UPS

As mentioned above, the electronic structure of a sample can be divided into valence levels and core levels. The valence electrons, responsible for the interatomic bonding, often form more or less delocalized molecular orbitals or energy bands. The valence electronic structure is intimately connected to the macroscopic properties of the system. In the UPS method, the electronic density of the valence states is derived via the measurements of the binding energy (BE) and the signal intensity of the electrons populating different valence levels, which in large clusters are developed into solid-like energy bands. Small clusters can be often treated as molecules, for which their geometry is the decisive factor forming the energy structure. As mentioned above, for metal elements, the metallicity emerges in clusters of a certain size, which is reflected in photoelectron spectra. A *metallic* cluster can be approximated by a microscopic metal sphere with the lowest electron BE at the Fermi level (BE_{Fermi}), which corresponds to the valence ionization potential or the work function. When an electron is removed from the Fermi level, the resulting lowest BE ionized final state will have a positive charge delocalized over the cluster surface, just as for a macroscopic metal sphere.

7.1.3.2 The Case of the Core-Level Ionization: XPS

The core levels are affected little by the interatomic bonding, so they remain atomic-like and localized in compounds. As a consequence, the core-level probing provides information about which elements the sample contains, and even about the number of chemically inequivalent bonding situations within the sample. Core-level spectroscopy can, for instance, separate surface and bulk spectral features, and even different types of surface sites (vertex, edge, face). This capability is of special importance for clusters, since many of peculiar phenomena associated with clusters are connected to the size-dependent surface-to-bulk ratio and to the accompanying changes in the electronic and geometric structure. The main observable in XPS is the binding energy of the core-level electrons. For clusters, it is often fruitful to study the core-level BE difference between the monomer and the cluster, ΔBE_{m-c}, also called the monomer-to-cluster BE shift. For an isolated monomer, its BE can be seen as an intrinsic property. When N monomers of the same type form a cluster, a difference in energy of a free and bound monomer appears for both the initial, neutral state and for the final, ionic state: ΔE_i, and ΔE_f, correspondingly. The binding energy difference between the monomer and the cluster, ΔBE_{m-c}, is then $\Delta BE_{m-c} = \Delta E_i - \Delta E_f$. ΔE_i, is the cohesive energy in a cluster of N monomers. For describing ΔE_f, a concept of screening, i.e., the response of the system to the core ionization, is used (see for details further down). Both the initial state contribution ΔE_i and the final state contribution ΔE_f are determined by different bonding mechanisms. For van der Waals-bonded systems, like rare gas clusters, the initial state contribution ΔE_i is practically negligible (~0.1 eV). It can be more important for clusters out of molecular gases and liquids or out of metal, semiconductor and dielectric solid materials. For example, mutual orientation/packing of polar molecules in a cluster increases ΔE_i. In van der Waals systems, as well as in other nonconducting materials, the screening appears due to the polarization of the surrounding when one atom is core-ionized. The polarization field reduces the final state energy relative to the free monomer situation. Since the surface atoms/molecules have fewer nearest neighbors than the bulk sites, the screening is less efficient at the surface than in the bulk. The first shell of neighbors around the ionized monomer in the nonconducting material contributes with roughly 2/3 of the maximum screening reached in the infinite bulk case. More distant parts of the sample contribute less to the screening, yielding for clusters the $1/R$ dependence of ΔBE_{m-c}. Besides, monomers of different kinds, i.e., different atoms or molecules, have different polarizabilities and contribute differently to the screening. All these considerations are valid also for the clusters out of metal elements in the size regime prior to the metallicity emergence. For *metallic* clusters, both ΔE_i and ΔE_f are considerable and are of a comparable magnitude. For ΔE_f, a major difference relative to the nonmetallic cases is the mobility of the valence electrons, resulting in *complete* screening. Since free atoms of metallic elements are open shell systems (the valence electronic s or/and p shells are not completely filled), their XPS spectra are rather complex: a multiplet of ionic states arises from the interaction between the disturbed core-electronic shell and the valence electrons. It is thus often more fruitful to compare clusters and the corresponding solid, as will be illustrated on several examples below.

7.1.3.3 The Case of the Core-Level Excitation: XAS

In XAS, also known as NEXAFS (near-edge x-ray absorption fine structure) the core electrons are not removed from the system, but excited into normally unfilled levels, which requires tunable photon energy provided by synchrotron radiation [11]. Thus XAS supplies the information about the unoccupied electronic levels/bands. What is collected in XAS is the electrons ejected as the result of the following core-excitation Auger-like decay. Experimentally all the electrons—without energy resolution—are collected while the photon energy is scanned in the vicinity of the threshold. Thus the collected signal indirectly reflects the absorption efficiency in the energy range under investigation. If the cluster is initially neutral, then the final state in XAS is also neutral, so the screening considerations discussed above for XPS do not generally apply. When comparing XAS for free monomers on one side, and clusters and solids on the other, it is often fruitful to consider the spatial distribution of the core-excited-state electron clouds in relation to the surroundings. If the core-excited-state electron cloud is significantly smaller than the first coordination shell radius, there is either no change in the state energy relative to the monomer, or there is an upward shift due to the electron cloud spatial confinement within the shell of the nearest neighbors (the first coordination shell) in the "lattice." If the core-excited-state electron density distribution spreads significantly further out into the cluster beyond the first coordination shell, the interaction between the ionic core and the excited electron is reduced by the screening from the first coordination shell, leading to a downward shift in energy. If instead the energy overlap of the core-excited state with the surrounding electronic structure is significant, the excited electron delocalizes into the surrounding electronic structure. Reality is often in between these simplified limiting cases.

7.1.3.4 The Case of the Core-Level Excitation Decay: Auger Electron Spectroscopy

In AES, the radiationless core-hole decay following the core-level ionization is studied. This decay results in two vacancies (holes) in the valence level. These two-hole states contain information about the valence electronic structure. When considering AES of non-metallic clusters, the same considerations concerning the charged states as in XPS apply. The polarization screening scales with the square of the charge, so the localized doubly charged two-hole final states in AES have four times larger screening contribution than the singly charged core-ionized state. As will be discussed below, some bonding mechanisms in nonconducting clusters allow charge delocalization, significantly reducing the final state energy.

7.1.3.5 The Case of the Core-Level Excitation Decay: Resonant Auger Spectroscopy

In a simplified way, resonant Auger spectroscopy (RAS) can be viewed as a two-step process [10,12]. The first step is a selective resonant core excitation as in XAS. The second step is an Auger-like decay from the core-excited state. Since this state is neutral, the decay results in a singly charged final state as in UPS. RAS can thus be used to probe the valence levels with elemental selectivity via the resonant core excitation, and is therefore sometimes denoted as resonant photoemission.

7.2 Cluster Sources for Electron Spectroscopy

There are significant practical difficulties in applying PES to the studies of free clusters. Clearly, it is desirable to study clusters out of solid materials like metals and semiconductors. A part from demonstrating the fundamental changes with the size, these clusters also promise practical applications. Probing them by PES is however more difficult than the clusters out of gaseous or liquid materials. For each aggregation state of macroscopic matter—gaseous, liquid, or solid—the cluster production demands a separate, dedicated source. Common for all three phases of matter is that primarily the substance should be *vaporized*, and then the vapor must be deeply cooled. The latter is valid if one plans to produce clusters larger than just dimers and trimers. As a rule, creating concentration of vapor out of solids comparable with that of gases or even liquids is out of the question. Sample handling is also

much more troublesome than in the case of gases and liquids. The need for a high cluster flux and often for their continuous production put severe demands on the cluster sources, so their development has been closely connected to the scientific progress in the field of cluster photoelectron spectroscopy. The substances gaseous at normal conditions are already available in the "vapor" phase, so they have often been the first choice of the cluster researchers. For substances that are normally liquids, there is a larger variety of compounds with some of them not even necessary to heat in order to create sufficiently high vapor concentration. Sometimes, additional cooling is also not needed: the work is done by the expansion of the gas/vapor into vacuum. Both "gas" and "liquid" cluster sources are based on the adiabatic expansion through a tiny hole of ~100 µm diameter (a nozzle) from a high (stagnation) pressure volume into vacuum. Moreover, as a rule it is not just a hole but a divergence toward vacuum long cone of a small opening angle. This conical geometry of the nozzle facilitates larger cluster production. But even with a flat nozzle, low stagnation pressure, and room temperature of the nozzle, there is often a certain percentage of small clusters formed in a *molecular* beam. This small-cluster presence in a gas jet was utilized in early days of the gas-phase electron spectroscopy when a He-lamp-based setup made it possible to record a valence spectrum of diatomic and triatomic molecules of inert gases. Adiabatic expansion cluster sources create a supersonic beam containing clusters of a wide size range with the size distribution centered at a certain value defined by the gas pressure P on the high-pressure side, nozzle temperature T, nozzle geometry (diameter d and divergence angle Θ), and the type of gas (a constant k calculated from the sublimation enthalpy K and density of the solid). For the inert gases and for some molecular gases, it became possible to obtain a numerical correlation between all these experimental parameters and the cluster size via the so-called scaling parameter Γ^* introduced in molecular beam gas dynamics. The following equation illustrates this connection:

$$\Gamma^* \sim \frac{kd^\alpha P}{T^\beta} \tag{7.6}$$

where
$\alpha = 0.85$
$\beta = 2.2875$

Decimal logarithm of the average cluster size is in a long range of values linearly proportional to the logarithm of this scaling parameter Γ^*. The next step in the complexity in the sense of production procedure—after the inert gas clusters—is the clusters out of liquids. Water is not the easiest object to handle in vacuum at the demands set by the electron spectroscopy experiment but it is so attractive that it was often the first to be attempted when a corresponding cluster source was assembled. In such an apparatus, a liquid-containing vessel with a ~100 µm conical nozzle is placed inside a vacuum chamber. The vessel and the nozzle should have independent heating circuits to avoid building up of the condensate in the opening of the nozzle, which happens due to the cooling of the vapor by its adiabatic expansion into vacuum. A useful option is to be able to seed the vapor with some inert gas, as a rule He. This seeding has shown to be an efficient tool in changing the ratio "monomer/cluster" in the beam in favor of clusters by creating high total pressures without too much of heating.

In many of the works published up to now, the studies of free clusters out of solids were performed using the so-called gas-aggregation cluster source. Such a source has been widely used in the ionization mass spectroscopy on cluster beams. The basic principle is relatively simple: a furnace converting solid to vapor is placed inside a cryostat kept at a cryogenic temperature. A flow of an inert gas (He, Ar) through the cryostat cools the vapor by transferring the heat from the vapor to the cold cryostat walls and takes clustering material through the exit hole of the cryostat—the nozzle—into vacuum. Expansion through the nozzle again takes place, but not at all as abrupt and strong as in the case of gaseous primary substances, since the pressure inside the cryostat (behind the nozzle) is of the order of ~1 mbar (compared to ~1000 mbar for the gases) and typical nozzle diameters are in the order of 1 mm.

A promising but also a more complex variation of the gas-aggregation cluster source utilizes the magnetron sputtering process for solid vaporization. The furnace is replaced by the magnetron head for which the so-called sputtering targets out of different solid materials can be used. Magnetron sputtering sources are known to create neutral as well as charged particles, both negatively and positively charged. Discharge plasma created in such a process also contains a corresponding amount of electrons. A flow of ions and electrons from the source builds additional concentration of charge in the ionization volume what influences the performance of the electron spectrometer. In spite of these and some other difficulties, it is the sputtering-based sources that are among the most promising—both in the fundamental and practical sense—providing the largest variety of samples. This is probably why during the latest years this type of cluster sources has been winning its place in many cluster labs.

7.3 Rare Gas Clusters: Model Systems for Pioneering Studies

7.3.1 Valence-Level Photoelectron Spectroscopy: From Diatomic Molecules to Excitons

The progress of experimental equipment and techniques was vitally important for the free clusters to become feasible to study by photoelectron spectroscopy. The small-cluster presence in an atomic beam experiment was utilized in early days of the gas-phase electron spectroscopy when, as mentioned above, a He-lamp-based setup made it possible to record a valence spectrum of diatomic molecules of inert gases [13,14]. These clusters manifest themselves as several features in the close vicinity (±0.3 eV) of the two atomic lines dominating the spectrum, the latter due to the $^2P_{1/2}$ and $^2P_{3/2}$ ionic states. The inert gas dimer spectra have been interpreted in terms of various possible molecular ionic states [15]. A publication of a Xe *trimer* valence photoelectron spectrum recorded with

an electron-ion coincidence setup came next [16,17]. To record a valence photoelectron spectrum of argon clusters containing 10 atoms required 10 years of further development and acquisition times of up to 15 h [18]. The upper size limit possible to reach with a pulsed-nozzle-based adiabatic expansion cluster source [19] in these latter experiments has been ~100 atoms. There, the spectra were explained by a trimer ionic core formed within a cluster as a stabilizing unit for the smallest sizes (<N> ≤10), and by Ar_{13}^+ for the larger ones. Two years later the same group published valence ionization spectra of Kr and Xe clusters interpreted in similar terms [20]. Some years before the first PES studies of rare gas dimers were performed, photoelectron spectroscopy experiments and related calculations [21] on the solid rare gases had been pioneered. Their results could be explained using the concepts of solid-state physics—a fully occupied valence band separated from the empty conduction band with a large gap of ~10 eV. The solid-state approach worked also well for the assignment of the features in the *cluster* spectra observed at ~12 eV above the 3p level for argon and ~17 eV above the 2p for neon [22]. Similar features were known from the studies of solid rare gases where they were explained as due to the excitonic states. In [22], the exciton states were populated as the result of inelastic losses experienced by the photoelectrons from the outer valence p-states on their way out of the clusters. The size of the clusters in [22] was 200–300 atoms. Around the same time, valence ionization spectra were recorded for heavy rare gas clusters containing thousands of atoms [23], as shown in Figure 7.5. These spectra resemble very much those of the solid rare gases [21] with the shape attributed to the spin-orbit

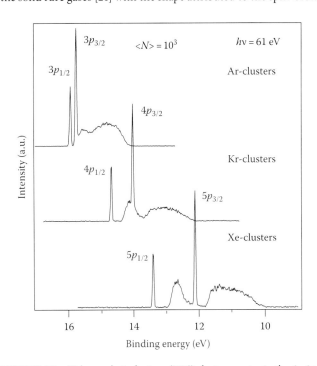

FIGURE 7.5 Valence photoelectron (UPS) cluster spectra in the Ar 3p, Kr 4p, and Xe 5p regions recorded at 61 eV photon energy. The mean cluster size in the beam is ~10^3 atoms/cluster. Sharp lines are due to the atomic ionized np states which are spin-orbit split into 3/2 and 1/2 components. (Based on Tchaplyguine, M. et al. *Eur. Phys. J. D*, 90: 343, 2004.)

splitting into $np_{3/2}$ and $np_{1/2}$ states. The broader band originating from the $np_{3/2}$ atomic states has been interpreted to be due to the removed degeneracy in clusters.

7.3.2 Up in Photon Energy: The First XAS and XPS Studies

With the advance of the synchrotron radiation sources, it became possible to collect ion yields of the cluster fragments (partial or total ion yield—PIY and TIY) while scanning the synchrotron x-ray radiation in the vicinity of the *core-level* ionization threshold. These ion yield spectra can be seen as the first XAS measurements for free clusters. The first total *electron* yield (TEY), recorded similar to the ion yield way, has been obtained for Ar clusters at Ar $2p$ ($\approx 250\,eV$) and Ar $1s$ ($\approx 3200\,eV$) thresholds [24,25]. As discussed in the introduction, these yield spectra are generated as the result of the electronic transitions from the *core* levels to the unoccupied valence states (Rydberg states in rare gases) with consequent emission of Auger electrons. In these publications, the average size of argon clusters in a continuous jet out of a cryogenically cooled nozzle reached up to 750 atoms, so the cluster yield spectra were close to those of the solid argon. The lowest energy cluster feature in the spectra was shifted $\approx 1\,eV$ up relative to the corresponding atomic line in both cases due to the $1s \rightarrow 4p$ and to the $2p \rightarrow 4s$ electronic transitions. The interatomic distance of 3.8 Å has been determined from the EXAFS-like oscillations in the $1s$ spectrum. At that time, the experimental resolution at synchrotron facilities was not yet enough to distinguish between the responses of the surface and the inner or "bulk" atoms of the clusters, neither in TEY, nor in the TIY/PIY.

The first *core-level photoelectron* spectrum of free clusters has been recorded for Ar clusters [26] with the largest average size of ≈ 4000 atoms. The ionization of the Ar $2p$ core-level has been investigated in detail, as shown in Figure 7.6. The cluster XPS features have been observed as two humps, each at equal separation from the parent atomic lines. (In a free-argon-atom spectrum, there are two peaks due to the spin-orbit splitting in the final $2p^{-1}$ ionic state.) In each hump, there have been two identifiable subcomponents—interpreted as due to the bulk and due to the surface atoms of the clusters. The bulk peaks were about 1 eV below the corresponding atomic lines (for the largest size) and the surface separation was reflecting a smaller coordination number of the surface atoms: 9 relative to 12 in the bulk. The number of neighbors come from the dominating geometry in rare gas clusters: icosahedral or cuboctahedral [27]. One should mention here that in the XPS studies of solid rare gases, the bulk-surface separation was not clearly experimentally demonstrated by that time, though it was always assumed to be present. In one of the few works on that topic [28] in the fit of the Xe adsorbate XPS spectra (Xe $4d$ core-level), two separate peaks have been used—due to the bulk and due to the surface responses—though the spectrum itself was lacking these features' resolution. The problem with the rare gas adsorbates is their dielectric nature that leads to charge accumulation in

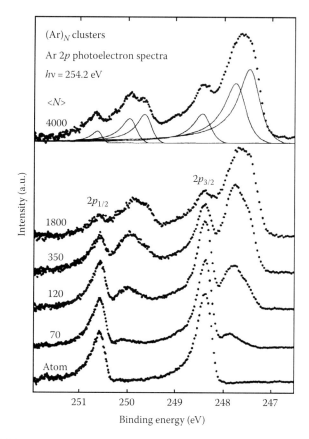

FIGURE 7.6 Ar $2p$ XPS spectra for Ar clusters of different sizes recorded at 254 eV photon energy. Atomic spectrum is shown at the bottom. Decomposition into bulk and surface substructure is shown at the top. (Reprinted from Björneholm, O. et al., *Phys. Rev. Lett.*, 74, 3017, 1995. With permission.)

the sample under the ionizing irradiation. In its turn, it causes the drift of the peaks' energy positions during the acquisition. In respect to this problem, one can say that free clusters in a beam—where the sample is continuously renewed—are advantageous for a high-resolution study. Moreover, precise absolute energy positions can be determined using the reference atomic lines present in the spectrum due to the ionization of the uncondensed atoms in the gas jet. The transformation of the spectra with the size has been also demonstrated in the work [26]: the bulk feature was moving further away from the parent atomic line down in binding energy, and the bulk intensity was growing relative to the surface response. A TOF electron spectrometer capable of working only at low kinetic energies has been used so that the photon energy could be just 2 eV above the higher $2p_{1/2}$ ionization threshold. As a result the bulk-surface resolution was to a large extent smeared out by the peak asymmetry caused by the so-called post-collision interaction (PCI) between the photoelectron and the Auger electron emitted as the result of the energy relaxation in the system. Such asymmetry is well known to appear in free atom spectra when an outgoing photoelectron is considerably slower than the Auger electron [29]. The article [26] did contain the first observation of the PCI phenomenon in free clusters.

Apart from XPS studies, the authors of [26] also carried out the Ar 2*p* partial ion yield (PIY) measurements on clusters monitoring the argon dimer signal (the main fragmentation channel after the core-excitation) while scanning the photon energy across the 2*p* threshold, and made it with the resolution that allowed to separate the bulk and surface in this XAS spectrum. As in the XPS series, the changes with size in this XA/PIY spectrum were showing the transformation toward the solid: the atom-bulk separation was increasing and the bulk relative intensity was growing. There [26] the explanation of the opposite in direction energy shifts in the cluster core-ionized (XPS) and core-excited (XA) spectra was given, the explanation made in analogy to the solid dielectrics. In the case of the core-excited states in rare gases, the lowest-in-energy excitation leaves the electron within the shell of the nearest neighbors whose electrons repel the excited core-electron. This repulsion, often referred to as spatial confinement, increases the potential energy of the system relative to the free atom case, so the level shifts up in binding energy. A study of Kr cluster core-excited levels [30] appeared a pair of years later after the work on argon clusters: the Kr 3*d* (≈95 eV) TEY spectra for different cluster sizes were presented, in which many features were well resolved and interpreted. First of all, close resemblance of the Kr XA spectrum due to the $3d_{5/2} \rightarrow 5p$ excitation to the Ar XA cluster spectrum due to the $2p_{3/2} \rightarrow 4s$ excitation [26] has been observed. In both elemental clusters, the response of the lowest energy $3d_{5/2}^{-1}5p$ and $2p_{3/2}^{-1}4s$ core-excited levels occurred to be free from the overlaps with the other states. In the Kr cluster spectrum, the $3d_{5/2}^{-1}5p$ core-excited bulk and surface states have been observed at higher binding energies (blueshifted) relative to the corresponding atomic line with the bulk being further away from it than the surface. The bulk response was increasing with the size relative to the surface. Using the calculations performed by Knop et al. [30], the changing abundance of different surface sites in the $3d_{5/2}^{-1}5p$ cluster state was discussed. For the assignment of the higher lying core-excited states, an heuristic hypothesis was made: the higher-in-energy features were due to the *redshifted* core-excited states. Though these states were neutral, the direction of the shift was toward the lower energy, as in the core-ionized states. The observation of the blueshifts for the first core-to-Rydberg excitations was explained as due to the spatial confinement. The opposite behavior for the higher core-to-Rydberg excitations, a redshift, was qualitatively explained as due to these Rydberg orbitals being larger than the first coordination shell, which then screens the interaction between the Rydberg electron and the ion core. These conclusions have been confirmed by theoretical calculations [31–33].

The summarizing overview of the core-level electronic structure–related studies on free clusters performed by the end of the twentieth century has been given in a special issue of the *Journal of Electron Spectroscopy* [34].

The first XPS studies of Kr and Xe spectra were published more than 5 years after the XPS work on Ar clusters [26] had been done. The Kr- and Xe-cluster study has been carried out at a third-generation synchrotron facility with an electrostatic electron energy analyzer. Also Ar 2*p* cluster spectra have been remeasured in a wide range of photon energies—up to 200 eV above the Ar 2*p* threshold [35,36]. Similar to the argon XPS spectra, Kr 3*d* and Xe 4*d* cluster core-ionization spectra contained well-resolved bulk and surface responses. Performed at about the same time, an extensive series of XPS photon-energy-dependence measurements showed that the bulk-to-surface intensity ratio in the Ar and Kr cluster spectra changed with the photoelectron kinetic energy, as could be expected from the knowledge of solid-state physics. Due to the varying photoelectron escape depth that depends on the electron kinetic energy, the bulk response is high close to the ionization threshold, then it goes down and reaches its minimum at several tens of electron volts above the threshold, and finally resumes growing again at higher energies. This behavior is often referred to as the "universal curve," since the electron escape depth dependence on the photon energy is similar for many materials. The minimum in the curve in nonconducting substances is connected with the kinetic energy reaching the exciton formation threshold, which lies more than 10 eV above the top of the filled valence band for the rare gas clusters [22]. Above the exciton threshold, the flux *F* of escaping from the cluster bulk photoelectrons monotonously grows. The attenuation what is usually expressed by an exponential attenuation law [28]

$$F \sim \exp\left(\frac{-x}{\lambda}\right) \qquad (7.7)$$

where

x is the electron path to the surface
λ is the characteristic escape depth, or the mean-free-path related constant

For a cluster of a fixed size, the surface response remains approximately constant with the photon energy, so the increasing bulk electron flux can be characterized by the dimensionless ratio of the bulk-to-surface intensity I_b/I_s. As just mentioned above, in the first photon-energy-dependence studies of argon and krypton clusters, the I_b/I_s curves were close to the expectations, while in Xe clusters, this ratio was growing fast with the photon energy in the region where one would normally see the minimum, as shown in Figure 7.7 [37]. These were the measurements when the electron spectrometer was placed at 90° to the horizontal polarization plane of the x-ray radiation. The understanding came after the experiments at the so-called magic angle (about 55°), at which the angular effects for the ionization cross section of free atoms are canceled out. The problem with Xe was in the rapid change of the anisotropy in the photoelectron cross-section angular distribution due to the existence of the so-called Cooper minimum in the 4*d* ionization cross section at around 105–110 eV above the 4*d* threshold (the latter is at ≈65 eV). The angular distribution of the electrons emitted from the 4*d* level is strongly anisotropic with the dominating direction along the electric vector of the radiation. Thus, the observation at 90° to this vector was efficiently discriminating the signal from the uncondensed Xe atoms in the beam and from the cluster surface, as expected, but this was not the case for the bulk electrons [37].

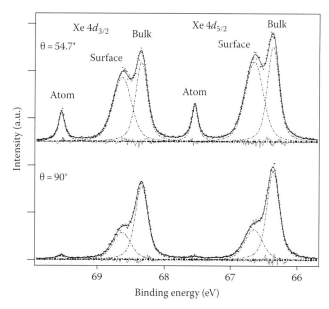

FIGURE 7.7 Xe 4*d* XPS spectrum for Xe clusters recorded at two different spectrometer orientations: at the magic angle (top) and perpendicular to the light polarization plane (bottom). Decomposition into atomic, surface, and bulk features is shown. (From Öhrwall, G. et al., *J. Phys. B: At. Mol. Opt. Phys.*, 36, 3937, 2003. With permission.)

With the spectrometer set at the magic angle, the bulk-to-surface intensity ratio in Xe clusters started behaving "normally" (Figure 7.8) similar to the Ar and Kr cases in the experiments in the "perpendicular" geometry. The stronger bulk response in the "90°-case" was due to the elastic scattering of the outgoing bulk photoelectrons on the inner cluster atoms, the effect which was making the bulk photoelectron angular distribution more uniform [37]. The effect was not noticed in the Ar and Kr cluster cases because there were no as drastic changes in the anisotropy in the Ar 2*p* and Kr 3*d* cross sections as in the Xe 4*d* one, so the overall photon energy dependence was not strikingly different in the 90°- and magic-angle geometries. Recently, the anisotropy of electron distribution from Xe clusters was studied in a wide range of angles, and the value of the anisotropy parameter was obtained [38]. The size and photon energy dependences of the XPS spectra of three heavy rare gas clusters were presented 2 years earlier in [39], where a model for determining the cluster size using the experimental bulk-to-surface intensity ratio was also suggested. Having applied this model, the authors estimated the photoelectron escape-depth values which occurred to be somewhat higher than those for solid rare gases scarcely found in literature [28]. In these XPS measurements, the lowest average cluster size <*N*> was not below ≈100 atoms per cluster. In another XPS study with an electrostatic electron spectrometer, it was shown that also small Kr clusters (<*N*> ≤30) could be informatively studied, as shown in Figure 7.9. The full-scale publication [40] also included theoretical calculations of the binding energies for different surface sites—vertices, corners, and faces. The instrumental resolution—below the inherent core-hole-lifetime-determined spectral widths—reached in the XPS spectra of rare gas clusters has allowed a low-kinetic-energy

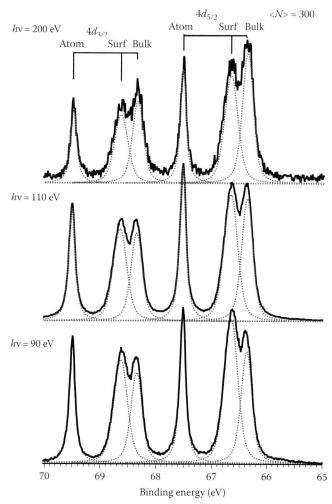

FIGURE 7.8 Xe 4*d* XPS spectrum for Xe clusters with <*N*> ≈300 recorded at three different photon energies: 90, 110, and 200 eV. Decomposition into atomic, surface, and bulk features is shown by dotted lines. (From Tchaplyguine, M. et al., *J. Chem. Phys.*, 120, 345, 2004. With permission.)

investigation in order to study the PCI phenomenon in clusters. It has resulted in deriving numerical PCI-related asymmetry parameters of cluster features, which occurred to be quite different for the bulk and for the surface peaks [41]. The surface asymmetry was practically the same as for the free-atom values, while the bulk one was considerably smaller. In a simplified picture of the core-ionization specific case, when the photoelectron is slow, and the Auger electron is fast, the latter takes over the slow photoelectron at a certain time, so the photoelectron starts seeing a doubly ionized atomic core. The resulting stronger electrostatic field slows down the photoelectron, and the energy lost by it is given over to the passing-by Auger electron which then sees only a singly ionized core. Photoelectron slowing manifests itself in a spectral tail toward lower kinetic energies in an XPS spectrum. Correspondingly, Auger-electron acceleration creates asymmetry toward higher kinetic energies in the Auger electron spectral features. For free atoms, it is enough to consider the Coulomb interaction of the charges involved in order

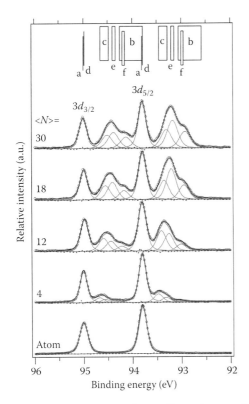

FIGURE 7.9 Kr $3d$ XPS spectra for small Kr clusters of different sizes recorded in Ref. [40]. Decomposition into different surface substructures is shown: d: dimer, a: atom, c: center, e: edge, f: face, b: base. (Reprinted from Hatsui, T. et al., *J. Chem. Phys.*, 123, 154304, 2005. With permission.)

to obtain the asymmetry parameter—a numerical measure of the phenomenon. For a free atom, the electrons are moving in vacuum, but it is not the case for clusters and solids. In a dielectric medium, e.g., a rare gas cluster, the Coulomb interaction is weakened by the medium polarization characterized by the dielectric permittivity. It has been shown in [41] that with such a correction to the electrostatic interaction, the calculated asymmetry becomes very close to that of the bulk electron experimental value. One can say here that the work [41] was not only the first adequate *experimental* treatment of the PCI effect in clusters, but also in dielectric solids. This is because of the difficulties involved in studying supported dielectrics with electron spectroscopy methods. As mentioned above, it was not easy to resolve bulk and surface features, leave alone recording different asymmetries in each of them.

Writing about the cluster XPS studies, one cannot but mention a puzzling observation of different core-level shifts in the $4d$ and $3d$ levels of free Xe clusters. The 15% larger bulk-feature separation from the corresponding atomic line in the $3d$ case remains so far unexplained [42]. This larger separation is transferred to the $3d$ Auger spectrum, where the bulk-to-atom shift should be three times the corresponding separation in the XPS spectrum (why it is so will be discussed later). It can be emphasized here that such a difference in the chemical shifts may not have been noticed in a study on a

supported dielectric solid, also because the absolute energy calibration in the free cluster case is more precise and stable due to the spectral response of the uncondensed atoms in the gas jet.

7.3.3 Inner Valence Studies and Interatomic Coulombic Decay

Around the fall of the new century, the attention of the cluster scientists was attracted by the so-called inner-valence levels in rare gas clusters: to the filled with 2 electrons s-subshell lying under the delocalized p-subshell [43]. A comparative photoemission study of the inner valence states in Ar, Kr, and Xe followed [23]. There, it has been shown that the Ar $3s$ inner valence cluster spectra were very much like the core-ones: the bulk and surface responses could be clearly resolved at lower than the $3s$ atomic line (29.2 eV) binding energy. This should be the case when the level is localized, atomic-like, and is not chemically coupled to the outer-valence delocalized electron band. The inner-valence spectrum in Xe clusters looked rather different from the Ar cluster case, namely, much more like the delocalized valence level spectra: below the $5s$ atomic line (23.4 eV) there was a single hump without any clear structure. The absence of a clear structure was interpreted as a consequence of s-electrons' delocalization, or, at least, of the delocalized $6p$ electrons sufficient influence on the inner-valence levels. Kr cluster $4s$ spectrum looked like a case intermediate between Ar and Xe: some hint for the bulk surface-like separation could be traced, but not at all as clear as in Ar clusters.

Two years earlier a theoretical investigation predicted the existence of a new (relative to the monomers) inner-valence relaxation channel in some weakly bound systems, with neon clusters among them [44,45]. In a free Ne atom, the $2s$ photo-ionization (48.4 eV) is followed by a radiative decay: the inner valence vacancy is filled with a $2p$ valence electron, which releases a quantum of energy taken away by a photon, since the energy is not enough to eject the second electron from the same atom: the double ionization potential is too high. In other words no Auger-like decay is possible. Photon emission is a slow (nanoseconds) process, so the Lorentzian width of the inner-valence line is undetectably small for conventional electron spectroscopy. If the ionized atom is surrounded by neighbors in a compound then the released energy quantum might be enough for an electron *from a neighbor* to be ejected. This extra-atomic relaxation mechanism was called interatomic coulombic decay (ICD). Soon the ICD in Ne clusters was experimentally observed [46]. It was not trivial to directly detect it since the ejected from the neighbor-atom electron had quite low kinetic energy—just a pair of eV. This is a range difficult to access with an electrostatic spectrometer: its detection efficiency is low and rapidly changes with energy here. Besides, the vacuum in an ionization chamber is full of slow electrons due to various scattering processes obscuring the ICD electrons. The effect manifested itself as a structureless feature on top of the so-called zero-kinetic edge tail in the spectrum, as seen in Figure 7.10. Later, with the

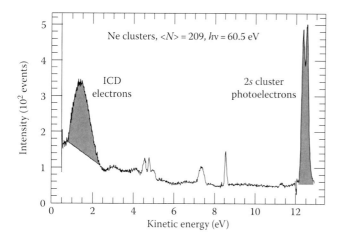

FIGURE 7.10 Electron intensity vs. kinetic energy of the Ne cluster contribution. A scaled and energetically shifted monomer spectrum was subtracted from the original spectrum. The ICD and the Ne $2s$ cluster photo-line signals are represented by the shaded areas. (Reprinted from Barth, S. et al., *Chem. Phys.*, 329, 246, 2006. With permission.)

improved acquisition procedure, it became possible to quantify the ICD in Ne clusters and to state that it was responsible for almost 100% of relaxation [47]. One of the consequences of switching from photons to electrons in the channel of relaxation is its three orders of magnitude shorter time. Theory predicted some tens of femtoseconds [45]. This change which should manifest itself in the broadening of the Lorentzian width of the cluster photoemission features, the width defined by the lifetime of the $2s$ vacancy, became possible to deduce from the Ne cluster $2s$ photoelectron spectrum, as shown in Figure 7.11 [48]. The ICD was later observed in the smallest-size clusters—in the dimers of Ne and Ar, and several new publications on it appeared [49–51]. Also, in ArKr dimers the phenomenon was detected [52]. In the

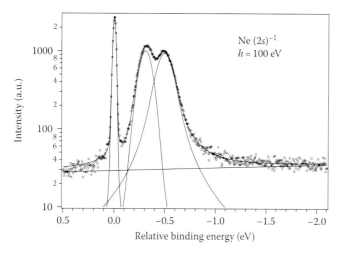

FIGURE 7.11 Ne $2s$ inner valence photoelectron spectrum for clusters with $\langle N \rangle \approx 900$. The spectrum is shown on a logarithm-of-intensity scale. A fit using a Voigt line profile is also presented. The full width at half maximum of the atomic line, at 0 eV relative binding energy, is ≈ 30 meV. (From Öhrwall, G. et al., *J. Phys. B: At. Mol. Opt. Phys.*, 36, 3937, 2003. With permission.)

case of Ar_2 and ArKr dimers, it was not the inner valence state which was primarily ionized, but the $2p$ core level. In one of the Auger channels after this $2p$ ionization, a vacancy is created in the $3s$ inner valence shell, whose decay was shown to proceed via the ICD. Similar sequential ICD was recorded in argon trimers. All these experiments on the smallest clusters were performed by multiple coincidence technique.

The use of bicomponent clusters in the experimental studies of ICD introduced an important practical advantage: by a proper choice of constituents, the ICD-electron kinetic energy could be increased so that it became easier to detect by an electrostatic spectrometer. Such an idea has been realized using NeAr clusters [53], and later the study was extended [54]. When Ne $2s$ inner valence level was ionized, an electron from a neighbor Ar atom was ejected in the de-excitation process. Since Ar valence ionization energy is lower than Ne by several electron volts, the same quantum of energy released in the Ne $2s$ relaxation process provides argon ICD electrons with a higher kinetic energy than the ones from Ne would get. The possibility of what was called "resonant ICD" in Ne clusters has been also demonstrated [55]. There the $2s$ level electrons were resonantly excited to the unoccupied Rydberg states just below the $2s$ threshold. The excitation energy put into the system was then also enough to ionize a neighbor atom in the cluster.

7.3.4 Normal Auger Spectroscopy

The ICD is an example of a delocalized electronic decay, which takes place when the interatomic bonding and the ionization energies are in a certain relation. Conventional electronic decay in inert gas clusters, the so-called normal Auger decay, has been also investigated [42,56,57]. As mentioned above, in the case of the decay localized on the same atom as the core ionization, the changes in the energies of the cluster final states relative to the corresponding atomic states can be predicted. It has been discussed in the introduction that the nature of these changes is in the same polarization screening as in the core-ionized states. If for the latter single-charge states the change in energy $-\Delta E_{\text{XPS}}$-is known (determined in the XPS studies), the shift in energy for the final doubly valence ionized states, ΔE_{2+}, of the same atom can be well approximated as four times ΔE_{XPS} [28,56], since the induced-polarization energy should scale with the square of the charge. Thus the kinetic energies of the cluster Auger electrons, determined as the difference of the final and initial state energies, will vary from the corresponding atomic ones by

$$\Delta E_{2+} - \Delta E_{\text{XPS}} = 3\,\Delta E_{\text{XPS}} \tag{7.8}$$

The first Auger study for free clusters proving the localized character of the Auger electronic decay in weakly bound systems has been performed using a beam of Ar clusters in which the $2p$ core level was ionized. The calibration could be again conveniently made using the atomic features in the spectrum, and the shifts between the atomic and cluster features has been accurately determined [56]. If the most common core states studied by XPS in rare gas clusters are again addressed, the intense Auger lines are generated when both electrons are leaving the outermost p sub-shell of the valence shell. In Ar, this is the $2p^53p^6 \rightarrow 2p^63p^4$ transition, in Kr it is the $3d^94p^6 \rightarrow 3d^{10}4p^4$ transition, and in Xe it is the $4d^95p^6 \rightarrow 4d^{10}\,5p^4$ transition. For these three rare gases, Auger spectra look very similar. Each of 10 atomic lines (two initial spin-orbit-split states and five final states) is accompanied by a corresponding cluster surface and bulk feature, as illustrated for Kr in Figure 7.12. The change of the cluster size in the Auger spectra is manifested in a way analogous to that seen in the XPS cluster spectra: the bulk-to-surface relative response increases with the size. However, it has been established [56] that the Auger method was more sensitive to the cluster surface layers than XPS: at comparable kinetic energies of Auger and photoelectrons (≈ 200 eV), the surface-to-bulk ratio was 15% higher in the Auger spectra in comparison to the XPS spectra. Various possible reasons have been discussed but no definite conclusion was made.

In [58] the shape of the cluster Auger spectra has been shown to depend on the photon energy used for the core-level ionization: with the photon energy approaching the ionization threshold the higher-kinetic-energy part of the spectrum was clearly growing in relative intensity. As one of the possible reasons for this behavior, the so-called recapture of a slow photoelectron into an unoccupied excited state was suggested. Such a phenomenon is known to influence atomic Auger spectra of rare gases [59] causing the line asymmetry toward higher kinetic energies:

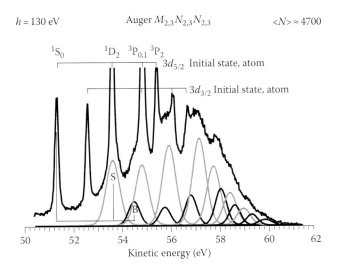

FIGURE 7.12 $M_{2,3}N_{2,3}N_{2,3}$ normal Auger spectrum for Kr clusters with the average size $\langle N \rangle \approx 4700$ at 130 eV photon energy. Sharp lines are due to atomic transitions from the core-ionized $3d_{5/2}$ and $3d_{3/2}$ states to the doubly valence-ionized $3d^{10}4p^4$ states in different configurations: $^1S_0$, $^1D_1$, and $^3P_{0,1,2}$. Decomposition into separate cluster peaks is tentatively suggested. Gray-colored features are due to the transitions in the surface atoms, and black peaks—due to those in the bulk. A bracket illustrates the separation between the parent $^1S_0$ atomic state and corresponding surface (S) and (B) states. (From Peredkov, S. et al., *Phys. Rev. A*, 72, 021201(R), 2005. With permission.)

The energy released due to the photoelectron recapture is given over to the Auger electrons.

7.3.5 Heterogeneous Cluster Composition Disclosed in XPS Studies

Various experimental techniques applied to different electronic levels provide complimentary information on the cluster electronic and geometric structure. One more dimension to the information field is added when clusters of mixed composition are created. A few examples have already been briefly discussed when the ICD studies were reviewed. A series of works on free, heterogeneous inert-gas clusters have been performed using also the XPS method. It came out that this approach was capable to shed light on the distribution of components in binary rare gas clusters created via a self-assembling mechanism. When two different gases are mixed in the co-expansion process, clusters of different geometric structures can, in principle, be formed. There could be uniform mixing of constituent atoms, or, for example, no clustering for one of the components. If the components are similar in properties, binary clusters with smooth gradients of concentrations are likely to be created. In the case of somewhat different properties of constituents, such phenomena as segregation and layering known from the epitaxial growth of solids could be expected. The structure to be formed is determined by the tendency to reach the lowest energy configuration for the system if the conditions allow. For binary rare gas clusters, similar to many other complex materials, the element with the lowest cohesive energy can be expected to be pushed out to the surface, so the other element with the larger cohesive energy would be responsible for the formation of the bulk where each atom has more bonds per atom than on the surface. A suitable method for studying segregated composition is XPS, since, applied to rare gas clusters, it gives well-separated responses from the bulk and surface atoms. In the heavy rare gas series (Ar, Kr, and Xe), it is the Ar–Xe pair in which the properties of constituents differ the most (cohesive energies per atom for solid Ar, Kr, and Xe are 0.080 eV, 0.116 eV, and 0.16 eV [60]). This pair has been the first choice in the XPS studies of the clusters formed in the adiabatic expansion of a premixed bicomponent gas [61]. In a detailed comparative analysis of the changes in the XPS spectra (Ar 2p versus Xe 4d responses), for various primary mixing compositions, shown in Figure 7.13, it has become possible to disclose the structure of the binary clusters formed: their core was dominated by Xe atoms, while Ar was covering this core with one or few layers. A clear response in the spectrum has been seen from the interface layer—the outermost Xe-core shell for which the energy shift from the parent atomic line was smaller than in the pure Xe bulk, but larger than in the pure Ar bulk. In the Ar–Kr co-expansion case the difference in the cohesive energies is lower, and the radial distribution of two components in Ar–Kr clusters, obtained from the spectral analysis, has been shown to be more gradual than in Ar–Xe case. Nevertheless, the core of the clusters was built out of Kr atoms, but no clear interface peak was observed [62]. Later, the calculations of the binding energies

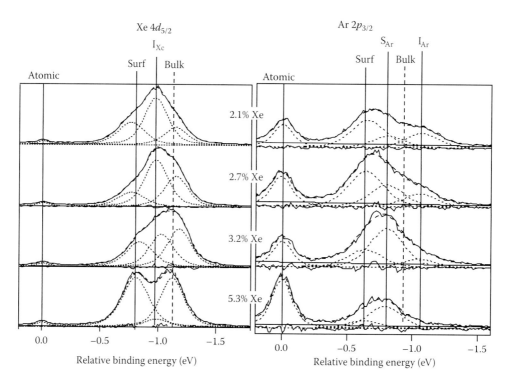

FIGURE 7.13 XPS spectra of Ar $2p_{3/2}$ and Xe $4d_{5/2}$ core levels for clusters produced from different primary gas mixing ratios. The peak at $\Delta E_B = 0$ corresponds to the atomic peak with the same total momentum as the cluster features. Decomposition into separate peaks is suggested. Apart from "pure" surface and bulk features also the "interface" peaks I_{Xe} and I_{Ar} are shown. S_{Ar} stays for the response of Ar atoms on the surface coordinated to Xe. (From Tchaplyguine, M. et al., *Phys. Rev. A*, 69, 031201(R), 2004.)

for various surface sites combined with the experimental studies [63] showed that when Kr atoms were present on the surface of such heterogeneous clusters, they occupied high-coordination sites, edges, and faces, as shown in Figure 7.14.

When Ne was added to the group of rare gases in the investigation of the self-assembling mechanism for binary clusters, the gap in cohesive energies became multifold: Solid neon cohesive energy is only 0.02 eV [60], which is four times lower than in argon. For binary mixture co-expansion, practical considerations forced choosing argon as a partner for neon, since clustering conditions for heavier than rare gases are so much different from those for neon [54]. As in the Ar–Xe case, the element with the higher cohesive energy—Ar in the Ar–Ne pair—was forming the core of the cluster with as low as monolayer coverage by Ne in the case of lower Ne concentrations in the primary mixture.

In all these experiments co-expansion of a binary mixture has been shown to lead to the minimal energy, close-to-equilibrium distribution of components in heterogeneous clusters. Another way of creating binary clusters is based on the so-called pick-up technique, when a preformed beam of monocomponent clusters is let through a cloud of a secondary gas. If the primary clusters are sufficiently cold for this secondary gas to condensate on them, binary clusters with the so-called core-shell geometry, inverted to the equilibrium composition, can be formed for Ar–Xe and Ar–Kr pairs [64]. To demonstrate the possibility of creating far-from-equilibrium structure, the primary clusters were made of argon,

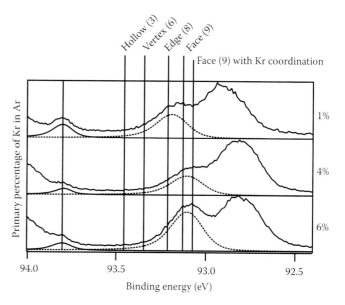

FIGURE 7.14 Kr $3d_{5/2}$ binding-energy region for the case of co-expanded Ar and Kr. The primary mixing ratio was varied from 1% to 6% Kr in Ar while the stagnation pressure was fixed at 2500 mbar. The surface feature fit is shown with dotted black lines. The calculated positions of surface peaks for differently coordinated atoms are denoted with vertical bars. (From Lundwall, M. et al., *Phys. Rev. A*, 74, 043206-1, 2006. With permission.)

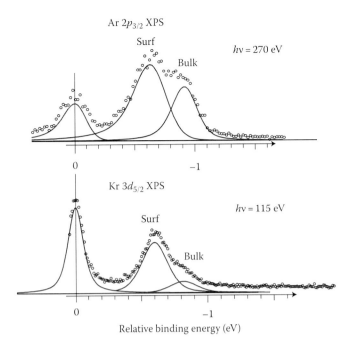

FIGURE 7.15 XPS spectra of Ar $2p_{3/2}$ and Kr $3d_{5/2}$ for the same clustering and doping conditions presented on the relative binding energy scale. For the zero position atomic line with the same momentum is taken. The surface and bulk feature fits are suggested. Kr cluster spectrum consists of practically only surface, what is interpreted as due to the argon core covered by Kr atoms. (Courtesy of A. Lindblad.)

which, in the co-expansion case is pushed to the surface. When the preformed large argon clusters were exposed to a flow of Kr or Xe, the atoms of these latter gases were getting stuck to the surface of the "host" clusters, as shown in Figure 7.15. In the Kr $3d$ cluster spectra, the specific structure was manifesting itself in the strong domination of the Kr surface response, which was appearing at noticeably (~30%) lower energies than in pure Kr clusters. This observation has been interpreted as due to the Kr covering the surface and coordinated to the argon bulk. Similar results have been obtained when Xe was blown over the preformed argon clusters. In the case of heterogeneous clusters created by doping the host clusters, atoms of the elements with the larger cohesive energy (Kr, Xe) occupied the sites with higher coordination on the surface [65] similar to the co-expansion case.

7.3.6 Resonant Auger Spectroscopy on Inert Gas Clusters

Already the early x-ray electron spectroscopy approaches to free clusters where the third-generation synchrotron was used included the attempts to apply the resonant Auger (RA) method [35]. It came out that this method was a unique tool for the assignment of multiple *core-excited* cluster states whose population reflects the x-ray absorption efficiency in the vicinity of the ionization thresholds. In the first works [35,36] a RAS study has allowed disentangling atom and cluster responses in the x-ray absorption spectrum of a jet containing both uncondensed argon

atoms and clusters. Excitation on top of a chosen absorption resonance leads to the population of only definite final states which are then singly ionized and—in rare gases—excited. The core-electron initially transferred to an unoccupied level stays there, and two p electrons from the outermost valence shell participate in the decay: one fills the core-hole, the other is ejected. The final state spectra, or resonant Auger (RA) spectra as they are often called, carry different, mutually complimentary bits of information when they are recorded either on the binding or kinetic energy scale. In spite of the resonant character of the process, it is not always clear from what core excitation comes this or the other feature in a RA spectrum (Figure 7.16). In order to clarify the situation, one can define then various, narrow energy windows in such a RA spectrum, the windows in which the unidentified features show up, and then record a partial electron yield spectrum collecting the RA electrons only in the chosen window [36]. Such a "feedback" approach increases the selectivity of the probing method as a whole to an even higher degree, and this is how the atomic and cluster features in only one excited, neutral $2p_{3/2}^{-1}4s$ state were disentangled in the $2p$ XA absorption spectrum of argon clusters, as shown in Figure 7.17. The identification of the RA spectra themselves can be attempted based on the fact that the cluster final state spectra are a replica of the atomic

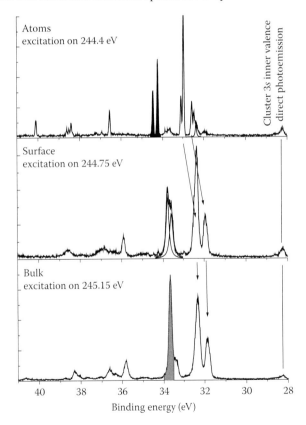

FIGURE 7.16 Resonant Auger spectra excited on top of the $2p^{-1}4s$ absorption resonances: for atoms at 244.4 eV, for cluster surface at 244.75 eV; for cluster bulk at 245.15 eV. Filled area on the "bulk" plot indicates the energy window in which electrons were collected in the $4s$ PEY spectrum in Figure 7.17. (From Tchaplyguine, M. et al., *Chem. Phys.*, 289, 3, 2003. With permission.)

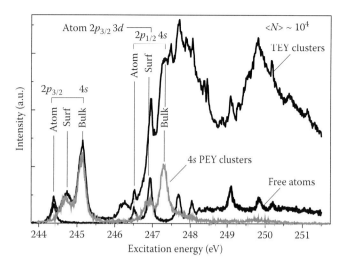

FIGURE 7.17 2p XAS (TEY) spectra for Ar clusters with <N> ~10⁴ and Ar atoms, and 4s PEY (gray) for clusters. For the TEY cluster and atom spectrum the electrons were collected in a 50 eV energy window between 8 and 58 eV binding energy. For the cluster PEY spectrum the electrons were collected between 33.5 and 34.0 eV, so only the *2p⁻¹4s* cluster response is observed. (From Tchaplyguine, M. et al., *Chem. Phys.*, 289, 3, 2003. With permission.)

ones (the relative spacing between the lines is the same as for the atoms), but just broadened and shifted down in binding energy [66]. There for argon clusters the shift for the lowest excited $3p^4 4s$ final state was shown to be ≈0.8 eV and ≈1.0 eV for the $3p^4 3d$ state. Several tens of millielectron volts have been reported to separate the bulk and surface features in the $3p^4 4s$ final state. The higher in the core-excitation energy one goes the more complex, overlapping becomes the pattern in a corresponding RA spectrum. To add the complexity, the second Rydberg series—due to the $2p_{1/2}^{-1} nl$ states—is partly above the $2p_{3/2}$ ionization threshold, so the normal Auger (NA) decay channel opens when in the XA-scan the photon energy exceeds the threshold. The kinetic energies of both RA and NA electrons become very close. Moreover, in the case of rare gas clusters, even at the excitations below the threshold the normal Auger decay has been shown to take place, with the doubly ionized final states populated [66]. These NA-caused states have been possible to identify by plotting the final state spectra on the kinetic energy scale. In this case the features due to the NA decay do not change their energy when the photon energy is detuned from the resonance. At the same time, the features due to the RA decay stay constant on the binding energy scale at detuning, which means they shift on the kinetic energy scale. The presence of the NA features with the excitation below the threshold has been interpreted [66] by the involvement of the cluster conduction band in both the excitation and decay processes. As it has been established at a rather early stage of solid-inert-gas studies, the localized core-excited states overlap in energy with the conduction band delocalized states. Thus photons of the same energy can transfer core electrons into both types of states. In rare gas solids, the excitation into the conduction band has been called "internal ionization" since the local decay proceeds in the absence of one electron

"lost" in the conduction band. Due to this, the features constant in kinetic energy have been present in the final state spectra [67].

One of the observations made yet in the first RA studies on free rare gas clusters has been the growth—with the main quantum number—of the atom-to-cluster energy shifts in the final excited states populated as the result of the core-excitation decay. The nature of these shifts is similar to the one discussed above for the core-ionized states and for the doubly-ionized final states reached after the normal Auger decay: It is in the polarization screening of the ion by the neighbor-atom valence electrons. The peculiarity of the energy structure for the singly ionized *excited* states is that the promoted (yet in the core-excitation process) to a Rydberg orbital electron is quite far away in space from the remaining ionic core of the atom. For free Kr *atoms*, the electron density distributions of various Rydberg states have been calculated [68]: already for the lowest $4p^4 5p$ orbital, its radial density maximum is between 5 and 6 Å, which is larger than the interatomic separation in clusters (≈3 Å). For the next Rydberg orbital with $n = 6$, the electron density maximum is 10 Å away, which is beyond at least the second shell of neighbors in a cluster. This tendency develops with n. Theoretical treatment has showed that in a similar case of excitations in an argon nanocrystal, the excited electron density is spread over a large volume of the crystal lattice [31]. Thus, this electron influence on the energy of the final state is weak. For krypton clusters, it has been shown experimentally [57] that the singly ionized state energy approaches more and more that of the doubly ionized state when the degree of excitation increases. Moreover, similar large spatial separation of the Rydberg electron exists also in core-excited neutral states in Kr. For the lowest $3d^{-1} 5p$ state the orbital is yet within the shell of the neighbors, so it is not the screening like in an ionic case, but the orbital spatial confinement which defines the shift for clusters. Thus the state energy becomes larger. Starting from the next core-excited state, the $3d^{-1} 6p$ state, the promoted electron is out of the nearest neighbor shell and is spread over the crystal; hence the main phenomenon defining the energy of this yet neutral state is screening, as in the ionic case. So the state energy is lower relative to the parent atomic state, and the cluster $3d^{-1} 6p$ features shift down in binding energy—in contrast to the shifted up $3d^{-1} 5p$ features, as shown in Figure 7.18. Further studies of the core-excited states in Kr clusters with the photon energies just above the ionization threshold have been suggesting the involvement of the cluster conduction band into the excitation decay [69]. Comparing the final state spectra on the binding and kinetic energy scales has allowed identifying the features with the probable origin due to the "internal" ionization–delocalization of the core-excited electron in the conduction band. The competition between the delocalized and localized decays in the time domain has been suggested to explain the relative intensities of the features due to these two de-excitation channels.

As shown in this section, rare gas clusters have been important systems for the initial explorations of free clusters using valence

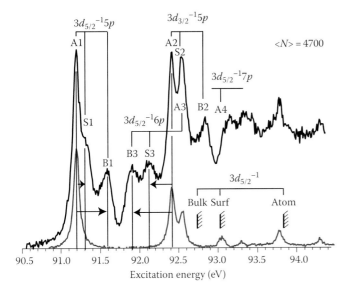

FIGURE 7.18 $3d$ XAS spectrum (black) for Kr clusters with $<N>$ ≈4700 atoms per cluster. Also the corresponding atomic spectrum is shown (gray). Core-excited neutral state response is marked with vertical bars, and denoted as A_n for free atoms, S_n for the surface, and B_n for the bulk. The arrows show opposite directions of the energy shifts in the cluster $3d^{-1}5p$ and $3d^{-1}6p$ states. Also the $3d_{5/2}$ ionization energies for free atoms, surface and bulk are marked on the graph. (From Peredkov, S. et al., *Phys. Rev. A*, 72, 021201(R), 2005. With permission.)

and core-level electron spectroscopy. The main reason for this has been the relative simplicity of producing and controlling the size of rare gas clusters. They can be regarded as relatively simple systems, for which the understanding of the relations between the experimental spectra and the cluster structure has been developed. In this perspective, the findings from rare gas clusters form a necessary basis for addressing more complex systems such as clusters of molecules and solids.

7.4 Clusters Out of Molecular Gases and Liquids: Increased Complexity and Relevance

Molecular systems are bound by strong covalent bonds within the molecules and weaker intermolecular bonds. The latter range from van der Waals bonding to significantly stronger hydrogen bonding. The relatively weak intermolecular bonds often make it possible to use adiabatic expansion cluster sources similar to those used for rare gases with the possible addition of an oven in which a molecular liquid is heated to produce sufficiently high vapor pressure [70]. Thus, clusters out of molecular gases and liquids represent a natural development from rare gas clusters. They are clearly more complex, due to, e.g., the presence of intramolecular vibrations or directionality of the intermolecular bonds. In this section, some examples of valence and core-level studies of these clusters, starting out with weakly bonded van der Waals systems, and progressing toward more strongly bonded systems such as hydrogen-bonded water clusters will be discussed.

As with rare gas clusters, the first photoelectron spectroscopy studies performed on molecular clusters investigated the valence levels. Dimers of several molecules such as $(NO_2)_2$ [71], $(H_2O)_2$ [72], $(HCOOH)_2$ [73], $(NO)_2$ [74], and $(NH_3)_2$ [75] were the first to be studied using resonance lamp radiation. The interpretations of the spectra were done using conventional *ab initio* calculations, which were tractable for dimer systems at that time.

The first van der Waals-bonded "molecular" clusters to be studied by valence photoelectron spectroscopy were nitrogen clusters [76]. The outer valence spectra showed three bands, corresponding to the $X^2\Sigma_g^+$, the $A^2\Pi_u$, and the $B^2\Sigma_u^+$ states in the monomer, shifted down 0.5–0.7 eV in vertical binding energy. The fact that cluster spectrum was essentially the same as for the dimer, except for a further shift in binding energy, which could be attributed to increased relaxation after ionization, was interpreted as an indication that the dimer ion served as the ionization chromophore, even in the clusters and the solid. This would imply that the dimer ion formation would have to occur on the same timescale as the photoionization event, 10–100 fs according to the authors [76]. The shifts observed for the cluster were interpreted using a model taking into account the electronic relaxation due to the polarization of a dielectric medium. The cluster was considered as a sphere consisting of a dielectric medium, and the ion was considered as a hole inside this sphere [76]. From this model and published values of the gas–solid binding energy shifts, the size of the studied clusters was estimated to be in the range of 10 molecules.

Also for oxygen clusters, the valence band spectra were interpreted in terms of a dimer ion being formed upon ionization due to the similarity between the spectra for dimers, larger clusters, and solid oxygen [77]. A modern study of large oxygen clusters has revealed fine structure in the outermost band for clusters corresponding to the $X^2\Pi_g$ state in the monomer [78]. This was not observed in the early experiment, and shows that the earlier interpretation needs to be amended.

For the van der Waals-bonded "molecular" clusters, the earliest core-level studies were the XAS ones. In the process of x-ray photon absorption, both the initial and the final states are neutral, which often leads to only moderately changed interaction between the molecules after the excitation. This resulted in only small differences in the cluster XA spectra relative to those of the monomers. For example, in N_2 clusters [79,80], the vibrationally resolved $1s\rightarrow1\pi_g$ core-to-valence excitation band of clusters shows a redshift of 6 ± 1 meV relative to the isolated molecule, and the vibrational structure and linewidths are essentially unchanged. This shift was assigned to dynamic stabilization of $1s^{-1}1\pi_g$ excited molecules in clusters, arising from the dynamic dipole moment generated by the core-hole localization in the low-symmetry cluster field. The lowest core-excited states below the $N\,1s$ ionization energy ($1s^{-1}3s$ and $1s^{-1}3p$) are blueshifted relative to the molecular Rydberg transitions, whereas the others ($1s^{-1}3d$ and $1s^{-1}4p$) show a redshift. Results from *ab initio* calculations on model clusters clearly indicate that the molecular orientation within a cluster is critical to the spectral shift, where bulk sites as well as inner- and outer-surface sites are characterized by different core-level absorption energies. Small, but distinct, spectral shifts, line broadening, and changes in intensity between monomers and clusters, are expected to occur generally in "molecular" clusters and in the corresponding condensed phase. This was explained in terms of variations in motion of the core-excited molecule within the molecular clusters [81], and has further been explored for molecules such as SF_6 [82].

Ultrafast dissociation is a process in which a molecule is core-excited to a dissociative state, in which it fragments to a significant degree during the core-hole lifetime, i.e., a few femtoseconds [83]. This is observed in Resonant Auger spectroscopy as contributions to the spectrum from the Auger decay in both dissociated fragments and the parent molecule. An O_2 molecule is known to undergo ultrafast dissociation after its excitation to the repulsive $1s^{-1}3\sigma^*$ state [84]. In the case of clusters the situation is more complex. The existence of neighboring molecules could, for example, suppress the ultrafast dissociation channel by caging the fragments inside the lattice or intermolecular hybridization could reduce the antibonding character of the $1s^{-1}3\sigma^*$ state. Resonant Auger spectra for van der Waals-bonded O_2 clusters display features that remain constant in kinetic energy as the photon energy is detuned [78]. The energy separation in the RA spectra between the known atomic fragment features and the cluster features is consistent with that observed for atoms and clusters in singly charged states in direct photoemission. These findings are a strong evidence for the existence of molecular ultrafast dissociation processes within the clusters or on their surface.

Another example of a van der Waals-bonded system is methane clusters. Methane was the first molecule for which vibrational structure was observed by XPS [85], and C $1s$ XPS spectra of methane clusters exhibit well-resolved surface and bulk features as well as vibrational fine structure [86], as shown in Figure 7.19. Methane clusters are characterized by strong covalent intramolecular bonds and weak intermolecular bonds. The vibrational structure in the cluster signal is well reproduced by a model that assumes independent contributions from inter- and intramolecular modes, and where the intramolecular contribution to the vibrational line shape is taken equal to that of the monomer in the gas phase, while the intermolecular part is simplified to induce only line broadening. Methane clusters represent a favorable case, with very weak van der Waals intermolecular bonds and large vibrational splitting. In general, clusters of more strongly bonded, polar molecules exhibit neither resolvable vibrational structures nor discernable surface-bulk splitting. Qualitatively this is due to large and inhomogeneous initial and final state contributions, the role of which has been studied for clusters of methanol and methyl chloride [87,88]. Their experimental C $1s$ XPS spectra have been interpreted by means of theoretical models based on molecular dynamics simulations for obtaining cluster geometries, and on a polarizable force-field for computing site-specific chemical shifts in the ionization energies and the linewidths. The data has been used to explore to what extent core-level photoelectron spectra may provide information on the bonding mechanism and the geometric structure of clusters of

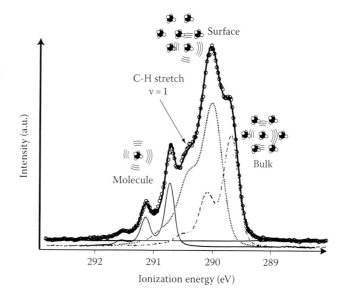

FIGURE 7.19 C $1s$ XPS spectrum of methane clusters. The different curves indicate different contributions; molecule (solid line), cluster surface (dotted line) and bulk (dashed line). (From Bergersen, H. et al., *Chem. Phys. Lett.*, 429, 109, 2006. With permission.)

polar molecules. The experimental cluster-to-monomer shifts in the C $1s$ XPS spectra of methanol and methyl chloride are quite similar, but the modeling allows a deep understanding of their nature. In the methyl chloride case, the shift is dominated by the polarization effects in the ionized state, despite the significant inherent dipole moment of the molecule. In methanol clusters, however, the ability to form strong and directionally specific hydrogen bonds changes this picture, and a significant contribution to the cluster-to-monomer shift originates from the permanent electrostatic terms in the initial state. The modeling has also shown that the larger width of the cluster feature in methanol clusters as compared to methyl chloride clusters is partly due to the structured surface of methanol clusters, with the –OH group sticking into the surface to form hydrogen bonds, and the –CH_3 group sticking out of the surface. Unlike water, which with its ability for two donor and two acceptor hydrogen bonds can form three-dimensional hydrogen-bonded networks, methanol with its single –OH group can only form one donor and one acceptor bond. With two hydrogen bonds per molecule, methanol has a tendency to form two-dimensional structures. Figure 7.20 shows C $1s$ XPS of methanol clusters belonging to two distinct size regimes, which are due to changes of the expansion conditions [89]. Conditions creating small clusters give rise to a shoulder (A) shifted less than 0.5 eV toward lower binding energy relative to the main monomer peak just below 293 eV. As conditions are changed to favor the creation of large clusters, structure A is gradually replaced by structure B with a significantly larger shift. Using the above described modeling procedure, the larger size regime can be well described by the line shapes calculated for clusters consisting of hundreds of molecules—three-dimensional nanodroplets. The smaller size regime has been found to correspond to small methanol clusters: chain-like and ring-like

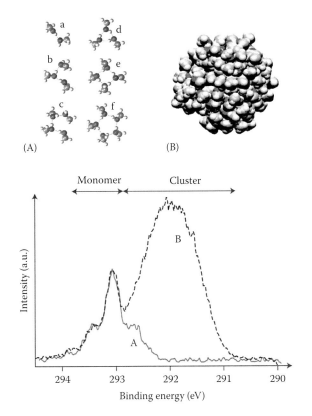

FIGURE 7.20 C 1*s* XPS spectra for expansion conditions creating small clusters (A), and large clusters (B), respectively. Selected structures for some small oligomers generated by simulations are also shown in A, a–c being linear ones, d–f ring-like ones. (From Bergersen, H. et al., *J. Chem. Phys.*, 125, 184303-1, 2006. With permission.)

pronounced than for the core level, but its presence shows that the polarization screening dominates the binding energy shift even in the valence band.

The above examples illustrate that XPS is sensitive to intermolecular orientations. An example provided in an XAS is a study of benzene clusters [93]. Compared to the above-mentioned van der Waals-bonded case of N_2, the intermolecular bonding is stronger for benzene, which is a liquid at normal conditions. Consequently, the corresponding XAS spectra exhibit somewhat larger, but still small changes from monomer to cluster than N_2 clusters. The results have been assigned using the *ab initio* calculations on model structures of dimers, trimers, and tetramers. In an isolated benzene molecule, the carbon atoms are equivalent. The calculations of the cluster geometry yield structures with many different relative orientations of the molecules in a cluster. These asymmetric surroundings remove the equivalence of the carbons atoms, and different carbon sites in the clustered molecules have been found to give rise to distinct spectral shifts.

Adsorption of molecules on clusters is an important phenomenon in atmospheric chemistry. In order to model such processes, the adsorption of polar molecules such as chloromethane and bromomethane on rare gas clusters has been studied as a model situation [94,95]. CH_3Cl and CH_3Br molecules were adsorbed on preformed rare gas clusters using a doping setup. By recording XPS spectra for all relevant core levels, and comparing to the spectra of the pure CH_3Cl and CH_3Br clusters [90,91], it has been seen that the low initial temperature of the rare gas clusters made the adsorbates remain on the surface, but that gradual cluster heating induced by further adsorption leads to some diffusion into the bulk.

Water clusters have been among the first molecular clusters to be studied. Core-level XAS and XPS spectra of small free water clusters presented in [96] show a weak but gradual change with the cluster size. Comparisons to the spectra of an isolated molecule and solid ice have indicated that water molecules had a lower average coordination in clusters than in the bulk solid. This work has been followed up by a later study of the electronic structure of relatively large, free water clusters consisting of ~10^3 molecules probed by XPS and AES [97]. Due to the continuous sample renewal, the spectra were free from charging effects, which are a problem for nonconducting samples such as condensed water. This case exemplifies an approach in which large clusters are used as an approximation of the infinite system. Figure 7.21 shows the AES spectrum of large water clusters. There are two contributions, which could be identified using total energy arguments possible due to the accurate energy calibration provided by the presence of the separate molecule response in the spectra. These arguments relied on the model, which has been used for van der Waals-bonded rare gas clusters in the assignment of their Auger spectra: the kinetic energy difference ΔE_{final} between the Auger electrons from the molecules and clusters is three times the difference in the core-level binding energy (ΔE_{XPS}) for the molecules and clusters, $\Delta E_{\text{final}} = 3 \, \Delta E_{\text{XPS}}$. One of the two contributions to the cluster AES spectrum could be well attributed to

oligomers. The rings have been found to be more stable due to the additional hydrogen bond relative to the linear structures of the same size. By comparing the model-generated C 1*s* spectra for the different oligomers and the experimental spectra, it has been concluded that the clusters produced were predominantly two-dimensional rings.

In the clusters of *polar* molecules, such as chloromethane and bromomethane, a distinct difference in the monomer-to-cluster shift between the C 1*s* and Cl 2*p*/Br 3*d* core levels has been observed. The shift in the C 1*s* is larger by 20% or more than those of the core levels in the halides, 1.05 eV versus 0.85 eV for chloromethane clusters [90] and 1.25 eV versus 0.92 eV for bromomethane clusters [91]. This has been interpreted as an evidence of the methyl group being surrounded by halide atoms, which have higher polarizability than the methyl group, and therefore should lower the binding energy of the core electron more. This is consistent with the antiparallel packing of the molecules in the clusters, found earlier for the dimer [92]. This specific mutual orientation of the molecules is also manifested in the valence band energy structure, where a high degree of localization is present for certain orbitals. The 2e orbital is a halide lone-pair orbital, and the 1e orbital is concentrated on the methyl group, and as for the core levels, the orbital localized at the methyl site exhibits a larger shift [90,91]. The effect is less

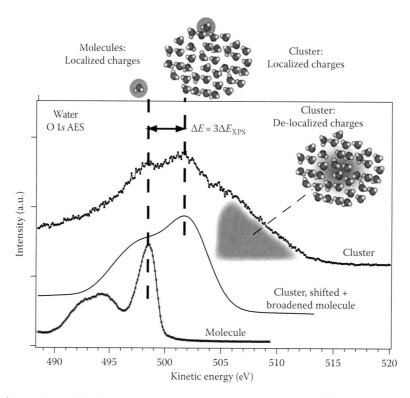

FIGURE 7.21 O 1s AES of water clusters. The bottom curve shows the molecular spectrum, and the upper curve shows the cluster spectrum. A broadened and shifted copy of the molecular spectrum, modeling the localized decays, is shown in the middle. (From Öhrwall, G. et al., *J. Chem. Phys.*, 123, 054310-1, 2005. With permission.)

the final states with both valence holes on the same molecule, i.e., to the *localized* final states similar to those of rare gas clusters discussed above. The second contribution is at higher kinetic energies than allowed for the localized final states. This second contribution to the AES spectrum has been possible to explain as due to the *delocalized* final states in which the two valence holes were located at different water molecules. These conclusions based on simple arguments have been corroborated by ab initio and density-functional calculations. These calculations show how the delocalized states obtain intensity using the established Auger theory, and how the holes are delocalized mainly to the surface molecules. This AES study of water clusters gives an interesting insight into the charge delocalization dynamics in a hydrogen-bonded system. The intermolecular coupling is sufficiently strong to make the delocalized states participating in the Auger decay on the core-hole-lifetime scale which is in the low femtosecond range. In contrast to this, similar delocalized states are not observed in the Auger spectra of van der Waals-bonded systems. The O 1s XPS spectrum of water clusters is typical for many molecular clusters in the sense that it is relatively broad, and lacks easily discernable subcomponents. Information about the clusters has however been extracted by theoretical modeling of the cluster geometries and resulting spectra [98] where a water cluster with 200 molecules has been simulated at 215 K. The model based on the dominating tetrahedral orientation of the molecules in the cluster bulk and less coordinated molecules on the surface led to a good agreement between the calculated

and the measured XPS spectra. The examples presented above illustrate the progress which has been achieved in the studies of "molecular" clusters with electron spectroscopy. Compared to the rare gas cases in Section 7.3, the spectra for molecular clusters are often less structured and more difficult to interpret. With advanced modeling, a considerable amount of information can however be obtained.

7.5 Clusters Out of Solid Materials

7.5.1 Clusters Out of Metallic Solids

7.5.1.1 Electron Spectroscopy Using Visual-Range, UV, and VUV Photons

The research on mercury clusters has played a noticeable role in the early days when photoelectron spectroscopy and related methods were first introduced into the free metal cluster field. This special role should probably be attributed to the easiness of Hg-cluster production in a wide range of sizes, reaching hundreds of atoms per cluster. Remarkably, one of the very first works here was a core-level study in which the transition from van der Waals bonding to metallic bonding with the Hg-cluster size has been probed [99,100]. Inner-shell autoionization, or PIY spectra of mass-selected clusters have been obtained for Hg_N, $N \leq 40$, while synchrotron radiation was scanned in the vicinity of the 5d ionization threshold (8.5–11.5 eV). Before the ionization the neural clusters had been mass-selected by a quadruple mass spectrometer. For small clusters with $N \leq 12$ the spectral behavior as a function of size has been interpreted in terms of an excitonic model. For larger cluster sizes, a gradual deviation from van der Waals bonding has been observed. The asymmetry of the line profiles reflected the progressive development of the *sp*-electron hybridization toward metallic band structure. This pioneering study relied on the high cross section of the resonant excitation, and, as briefly mentioned earlier, on the possibility to efficiently produce clusters by heating Hg metal and letting the vapor undergo adiabatic expansion, similarly to the rare gas and "molecular" clusters. Also, some of the first photoelectron spectra mapping the valence density of states (DOS) for a wide size range ($2 \leq N \leq 109$) of *neutral* metal clusters have been recorded for mercury [101], Figure 7.22. There, an adiabatic expansion cluster source has been used in a junction with a VUV discharge

lamp and an electron–ion coincidence setup comprising a time-of-flight mass spectrometer and a magnetic-bottle electron spectrometer. The DOS distribution has been shown to broaden with size, reaching ≈5 eV when mercury clusters contained ≈100 atoms. The discharge lamp with the highest photon energy of 10.6 eV has probably not allowed to see the whole populated valence band: In bulk mercury metal, the DOS is more than 9 eV wide. Another type of single-photon ionization light sources used in metal cluster photoelectron spectroscopy are lasers. They provide light pulses of considerably higher power than discharge lamps and synchrotrons, but with the photon energy not exceeding 7.9 eV (fluorine laser). Nevertheless, it has been these light sources which made it possible to monitor the evolution of the DOS toward the bulk-like with the size for several metals, first for copper [102] (Figure 7.23) and then for silver, gold, [103] and mercury clusters [104] (Figure 7.24). For the coinage metal cluster production, a laser ablation source has been used allowing probing these high-melting-point metals. In these works, the new horizons opened in the studies of *negatively charged* metal clusters by means of photoelectron spectroscopy have been demonstrated. Working with initially charged clusters allows

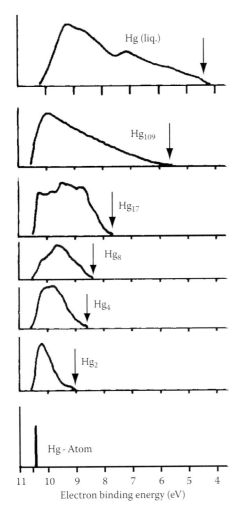

FIGURE 7.22 Photoelectron-photoion coincidence spectra of mercury clusters with up to 109 atoms recorded at 10.6 eV photon energy. The top graph is the spectrum of liquid bulk mercury recorded at 10.2 eV. The spectra display the photoelectron intensity as a function of electron binding energy relative to the vacuum level. The vertical arrows indicate the ionization potentials for clusters, and the work function for bulk mercury. (Reprinted from Kaiser, B. and Rademann, K., *Phys. Rev. Lett.*, 69, 3204, 1992. With permission.)

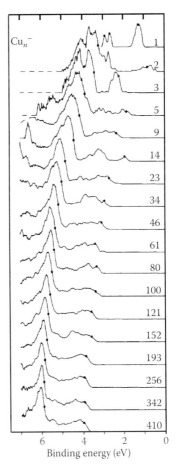

FIGURE 7.23 UPS of negatively charged copper clusters in the 1–410-atom size range mass-selected from a supersonic cluster beam, taken with F_2 excimer laser at 7.9 eV. (Reprinted from Cheshnovsky, O. et al., *Phys. Rev. Lett.*, 64, 1785, 1990. With permission.)

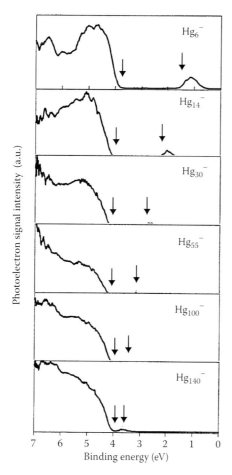

FIGURE 7.24 Selected photoelectron spectra of negatively charged Hg_n^- clusters taken with 7.9 eV laser excitation. The arrows mark the evaluated energies of the HOMO and LUMO binding energies. (Reprinted from Busani, R. et al., *Phys. Rev. Lett.*, 81, 3836, 1998. With permission.)

mass-selecting metal clusters, a beam of which originally contains species of a wide size distribution. In a photoelectron spectrum an excess charge—an electron—marks the position of the unoccupied energy band (LUMO, using molecular terms), which is a precursor for the metal conduction band. Since this excess electron energetically lands at the bottom of the unoccupied band, the difference between its binding energy, or vertical electron affinity, and that of the electron at the top of the occupied band (HOMO) gives an estimate of the energy gap between the valence and conduction bands for each cluster size. (The latter energy approximates the vertical ionization potential of a cluster.) Following the gap evolution with size, one can determine when clusters become metallic. In the above-mentioned copper cluster studies, the formation of the valence DOS out of $4s$ and $3d$ electrons has been traced up to 410 atoms per cluster—for a long time a record in the cluster photoelectron spectroscopy studies [102].

In mercury clusters, the gap between the valence and conduction bands has been extrapolated to disappear at the cluster size of ≈ 400 derived [104]. At the same time, for Mg cluster anions

the filled *s*-band and the empty p-band have been shown to merge at much smaller sizes—at $N = 18$ [105]. Photoelectron spectroscopy on mass-selected negatively charged metal clusters allowed also mapping the density of states in the clusters of sodium—an almost ideal free-electron metal—in a wide range of sizes, starting with a few atoms per cluster [106,107] and up to 500 atoms [108]. For small Na clusters, as well as for Cu, Ag, and Au with $N \leq 10$, the presence of the "quantum-size" regime has been experimentally confirmed, where such properties as electron affinity and ionization energy derived from the photoelectron spectra have been shown to change in a zigzag manner. The UPS spectra of these small clusters, as well as those of aluminum, gallium [109,110], and niobium [111] obtained using a laser ablation source, laser ionization, and TOF-electron spectroscopy have been found reflecting various isomer structures and the buildup of a continuous occupied valence band out of separated energy levels. UPS spectra of larger clusters with the size of tens of atoms, like those of Al [110], Cu, Ag, and Au [112], around a hundred like that of Zn [113], and even several hundreds of atoms, like Na [108] have been successfully assigned as due to the electronic shell development and closures, Figure 7.25. In the latter work, the jellium-model-based calculations have been shown to give a good description of the experimental spectra with N up to 100. Simpler model using a spherical box potential has been shown to describe the spectra with $N \approx 300$ well. The wide cluster-size range has been possible to create and study due to the introduction of the gas-aggregation source. As mentioned above, in such a source, solid metal is first vaporized either in an oven (for volatile materials) or by magnetron sputtering (for less volatile metals) [114–118]. This vaporization takes place inside a liquid-nitrogen-cooled cryostat, and the metal vapor is swept along by a flow of a cold rare gas, He, Ar, or a mixture of them. The gas flow passes through a narrow nozzle into vacuum. Due to the efficient clusterization provided by a magnetron-based source, Al clusters with the unthinkable otherwise size of $3 \cdot 10^4$ atoms have become possible to study by the valence photoelectron spectroscopy. The authors of this latter work have developed further their acquisition technique by introducing a momentum-imaging detector into the TOF spectrometer [119]. Such a modification has allowed obtaining information on the angular distribution of photoelectrons, and thus getting a deeper insight into the electronic structure of metal clusters. It has become possible to experimentally disentangle spectral features due to the electrons with different orbital momenta, which has given more detailed experimental confirmation of the jellium theory validity for clusters.

7.5.1.2 XAS, XPS, and Auger Spectroscopy on Metal Clusters

One of the first studies of metal clusters in the x-ray photon energy range has been carried out for free titanium clusters using XAS technique [120]. There, Ti clusters have been produced in an arc-discharge plasma source. Among nanostructured systems, the systems out of titanium and titanium oxide are of strategic importance for applications in several technological fields such

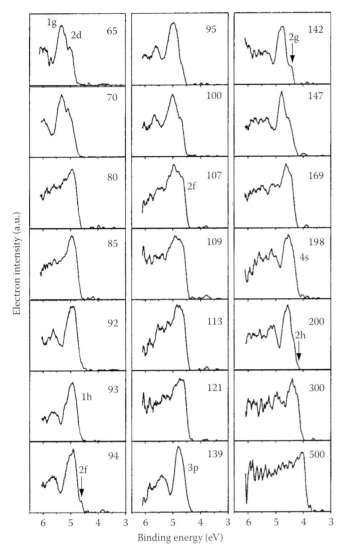

FIGURE 7.25 Photoelectron spectra of Na_n^+ ($n = 31$–60) at 6.42 eV photon energy. The peak labels give the quantum numbers of the corresponding electron shells. Shell closings with subsequent openings of additional shells can be seen for $n = 35$ (34 electrons), $n = 41$, and $n = 59$. The exceptional sharpness of the Na_{55}^+ spectrum is an indication for a spherical symmetry of this cluster. (Reprinted from Wrigge, G. et al., *Phys. Rev. A*, 65, 063201, 2002. With permission.)

as photocatalysis, solar energy conversion, energy storage, gas sensing, and as biocompatible coatings. With this in mind, free titanium, titanium oxide, and titanium hydride clusters have been studied by XAS [121] providing the first insights into the structure and reactivity of the isolated nanoparticles. In particular, free titanium clusters were found to exhibit noncompact structures opening possibilities for even higher surface-to-volume ratios and thus enhanced chemical activity. Being one of the XAS varieties, PIY measurements have been performed on free size–selected $3d$ transition metal clusters [122]. This method allows, in principle, determining the sample magnetic moment by comparing the $2p_{3/2}$ and $2p_{1/2}$ responses in the absorption spectrum. The relative intensity of these two spin-orbit components

has been found to be strongly size dependent for the clusters under investigation—in the size range up to 200 atoms.

The first successful XPS experiments have used a gas aggregation source [123] for metals that combined high production efficiency (high vapor pressure or sputtering rate) and shallow core levels with a high ionization cross section. Such a choice of materials, in combination with a high-performance, tunable x-ray radiation source at a third generation synchrotron facility, has allowed a series of XPS studies of the core levels of free Na clusters [123], Pb clusters [124], K clusters [125], and Bi and Sn clusters [126] clusters. For Cu and Ag clusters, a study on valence band photoemission at photon energies above 60 eV has been published [127]. The x-ray range of photon energies has made it possible to map also the d-bands, which are not accessible in the laser ionization experiments. In the first study carried out on free Na clusters, the $2p$ XPS and AES spectra for the large clusters of 10^4–10^6 atoms have been recorded, as shown in Figure 7.26. An atomic XPS spectrum, shown in Figure 7.26, has a multiplet pattern due to the coupling between the open $2p$ and $3s$ orbitals. The multiplet lines are spread in an approximately 1 eV wide interval around 38 eV. The spectrum of the dimer [128], Figure 7.26, differs substantially from the atomic case in both binding energy and shape. The dimer feature is approximately 2 eV lower in binding energy than that of the monomer, and consists of two components—due to the spin-orbit splitting of the core level—with a total width of approximately 0.5 eV. For large clusters, the XPS feature is shifted down by $\approx$4.5 eV relative to the atomic case, the total width is larger than in both the atomic and dimer cases. The cluster $2p$ binding energy is very close—within $\sim$0.1 eV—to that of the supported solid Na sample. The multiplet pattern of the atomic case is replaced by the spectral shape similar to that of supported massive Na. The spectrum consists of two spin-orbit doublets due to the ionization of the surface and bulk atoms. The line shape of the individual components also changes with the system size. In the atomic case, the inherent line shape is a very narrow Lorentzian due to the long lifetime of the $2p^{-1}$ state, which can only decay radiatively. The large clusters exhibit asymmetric Doniac-Sunjic line shapes characteristic of metals due to the excitations of the electron–hole pairs across the Fermi level. Also similar to supported metal samples, the core ionized state in metallic clusters can be seen as completely screened by the quasi-localization of one of the free valence electrons, with the resulting vacancy in the valence band delocalized over the surface of the cluster just as for a metal sphere with a radius R. This leads to a $1/R$ dependence of the core-level electron binding energy similar to the case of the valence ionization:

$$E_{cl}(Z_i, R) = E_\infty + (Z_i + \alpha) \cdot \frac{e^2}{R_{eff}} \tag{7.9}$$

where Z_i is the initial charge of the clusters, which in the case of Na cluster study [123] case has been equal to zero. The α coefficient has been a matter of debate: 1/2 and 3/8 have been used in literature [129–131], and later quantum calculations have provided similar

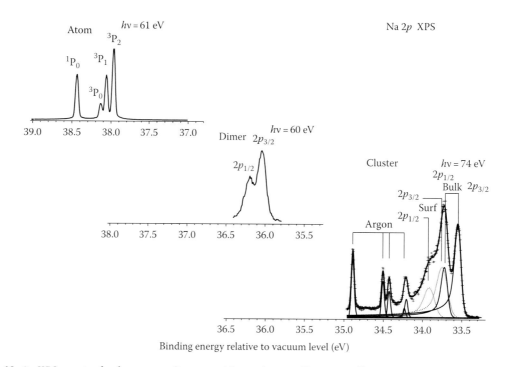

FIGURE 7.26 Na $2p$ XPS spectra for free atoms, dimers, and large clusters. (From Peredkov, S. et al., *Phys. Rev. B*, 75, 235407-1, 2007; Rander, T. et al., *Phys. Rev. A*, 75, 032510-1, 2007. With permission.)

in value material-dependent coefficients [132]. As in the case of valence ionization studies, the number of atoms per cluster can be estimated from Equation 7.2. For the core-level case, there is another approach (discussed above for the inert gas clusters) that uses the experimental ratio between the surface and bulk response

intensities in the XPS spectrum. These two methods have yielded reasonably consistent results for large Na clusters [123].

Alkali metals are peculiar in the sense that no Auger decay is possible in the case of a free atom, as only one electron, the $3s$ for Na, exists outside the core hole in the $2p$ shell for Na. For the dimer, the $3s$ electrons from two Na atoms combine to form a σ orbital enabling the Auger decay of the core-ionized $2p^{-1}$ state. For the nanoscale Na clusters, this Auger decay leads to a $\approx 12\,eV$ broad feature. This largely reflects the increased width of the $3s$ band in the solid and the involvement of other relaxation channels. Similar studies have also been performed for K. Additionally, in K cluster XPS spectra, the bulk and surface plasmon-caused satellites of the main lines have been observed, as shown in Figure 7.27 [125]. The values of the plasmon energies derived from the spectra occurred to be indistinguishably close to those of the supported bulk sample.

In the XPS experiments with a magnetron-based gas-aggregation source the responses of both initially neutral and initially charged clusters have been possible to record, as shown in Figure 7.28 [124]. In the $5d$ XPS spectra for Pb clusters, each of the spin-orbit components has shown to have three subcomponents, with the middle one being the most intense. The two middle peaks have been interpreted as due to the ionization of neutral clusters $-Pb_N$, and the minor peaks have been shown to arise from the ionization of initially charged clusters, i.e., Pb_N^{+1} and Pb_N^{-1}. The presence of an extra electron in the Pb_N^{-1} case leads to lower binding energies, whereas an extra positive charge for Pb_N^{+1} leads to higher binding energies (Equation 7.2). The binding energy splitting $\Delta E(R)$ between the clusters with $Z_i = 0$ and $Z_i = \pm 1$ initial charges should be equal to e^2/R. Using the

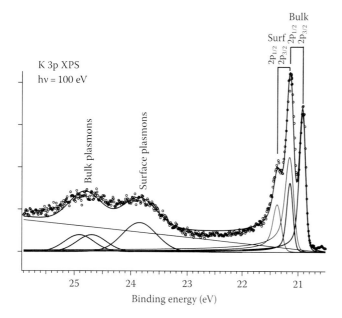

FIGURE 7.27 K $3p$ XPS spectrum of K clusters showing the decomposition into the main surface and bulk lines around 21 eV binding energy, and the bulk and surface plasmon-loss satellites in the 23–25 eV region. The cluster size is $\sim 10^3$ atoms/cluster. (From Rosso, A. et al., *Phys. Rev. A*, 2008. 77, 043202-1, 2008. With permission.)

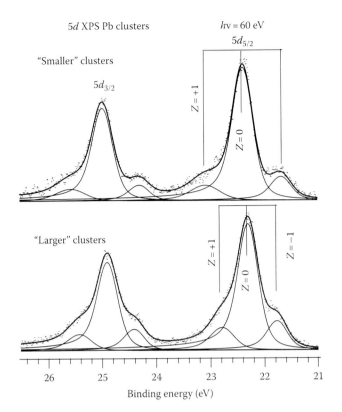

FIGURE 7.28 $5d$ XPS spectra of "smaller" and "larger" Pb clusters. The different initial charge states are indicated for the $5d_{5/2}$ component in the "smaller" cluster case. The separation between the neutral and the charged cluster response is marked for both cluster sizes. (From Peredkov, S. et al., *Phys. Rev. B*, 76, 081402-1, 2007. With permission.)

experimental splitting between features with different Z_i and the density of the solid, this approach has allowed estimating the cluster dimensions: a radius of $\approx$2–3 nm has been obtained corresponding to ~10^3 atoms/cluster.

The main observable in the XPS method is the core-level binding energies. For metallic elements, there exists a general treatment of the core-level binding-energy changes or the shifts relative to the free atom positions in the spectrum. This treatment assumes a fully screened final state in the metallic case and the so-called $(Z + 1)$ approximation for the screening of the core-ionized site by the valence charge redistribution [133]. The screened core-ionized atom is then treated as a neutral impurity in an otherwise perfect metal. The combination of the complete screening picture and the $(Z + 1)$ approximation makes it possible to introduce a Born–Haber cycle which connects the initial state with the final state of the core-ionization process. The XPS studies on free metallic clusters has allowed probing the validity of this model for the finite-size unsupported nanoparticles of several metallic elements [126]. For the cluster sizes investigated so far, this model has been able to account for the observed binding energy shifts. In the future, the model may be useful to analyze smaller and/or mixed cluster systems, and derive quantitative information for them similar to what has been obtained for macroscopic mixed/surface/interface systems. This could, for instance, mean size-dependent solvation and cohesive energies.

All XPS studies on metal clusters discussed so far have been performed in the nanoscale regime when the cluster properties resemble those of the infinite solid and do not significantly change anymore. XPS probing of the quantum size regime and with the size selection has been hindered by very low cluster flux in the case when preliminary size selection is made. Even at the best synchrotron radiation sources, the light intensity is not sufficient. The advent of the free electron lasers providing very high photon fluxes in the x-ray photon energy range may change this. The first XPS experiments have been carried out at the first free electron laser for the size-selected Pb clusters (with N well below 100) using their shallow $5d$ state [134].

7.5.2 Clusters Out of Covalent and Ionic Solids

The covalently bonded solids are characterized by strong directional bonds. This class includes important examples such as silicon and carbon. Small clusters can have geometric and electronic structures significantly different to those found in the infinite solid, as exemplified by fullerenes.

One of the first investigations of electronic structure for nonmetallic clusters was partial ion yield measurements for the size-selected antimony clusters with $4 \leq N \leq 36$ in the vicinity of the Sb $4d$ core level [135]. The initially neutral clusters have been produced in a gas-aggregation source. The x-ray radiation from a synchrotron radiation source has been scanned in the 25–120 eV photon energy range. The near-threshold structure of the unoccupied $5p$ core-excited states has been mapped as well as the cluster response in the region of the giant shape resonance above the threshold. The spectra have been shown to depend on the symmetry of the cluster geometry.

Different sulfur-cluster oligomers—chains and rings—in the size range up to $N = 9$ have been shown to define the photoelectron spectra in the valence energy region. Cluster anions have been produced using a pulsed arc source and have been ionized with a VUV laser [136].

The first studies of covalent clusters that utilized XAS to probe the differences in geometric structure between small clusters relative to the macroscopic solid have been performed for selenium and sulfur. For small Se clusters, simultaneous measurements of EXAFS and photoelectron–photoion coincidences have been utilized [137]. By modeling both the x-ray absorption and de-excitation processes, structural information could be obtained, showing the dimer to be contracted relative to somewhat larger oligomers.

For sulfur, an S $1s$ XAS investigation of variable-size sulfur clusters has been focused on S_2 and S_8, which are relatively abundant in vapor phase [138]. It has been shown that the electronic structure of S_2 was similar to that of O_2, but that there were also distinct differences, especially in the σ^*-state excitation regime as a result of different core–valence- and valence–valence-exchange interaction. The intermediate size species S_n with $3 < N < 7$ of lower abundance were concluded to have electronic structures not too different from that of S_2 or S_8.

With the discovery of fullerenes, much attention has been attracted to carbon clusters of different sizes and geometries.

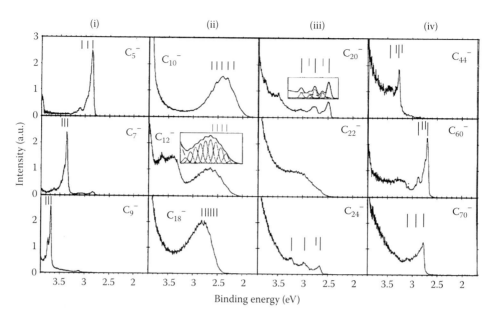

FIGURE 7.29 Examples of UPS spectra of mass-selected carbon cluster anions C_N^- ($h\nu \approx 4.0\,\text{eV}$). The insets show fits of the spectra at the positions of the bars. (Reprinted from Handschuh, H. et al., *Phys. Rev. Lett.*, 74, 1095, 1995. With permission.)

Various stable configurations of cluster anions produced by a laser-vaporization source have been studied by photoelectron spectroscopy for sizes up to 70 atoms. Different vibrational modes observed in C clusters have been connected to the adopted by the clusters specific geometric structures, like chains, rings, and fullerenes [139,140], Figure 7.29.

A specific type of carbon related to diamond is called diamondoids. Internally these have a three-dimensional diamond-like structure, but the dangling bonds of the surface are passivated by hydrogen. Small diamondoid clusters ranging in size from adamantane ($C_{10}H_{16}$) to cyclohexamantane ($C_{26}H_{30}$) have been probed using C 1s XAS [141]. It has been seen that in contrast to the other group IV semiconductors, the bulk-related absorption onset did not exhibit a size-dependent blue shift as expected within the quantum confinement model. Furthermore, additional spectral intensity has been observed below the bulk absorption onset and within the band gap of bulk diamond, which has been empirically correlated to the hydrogenated species of the surface termination. The experimental results show that these small diamondoids differ significantly from bulk diamonds, and therefore require a more molecular description in their ultimate size limit.

Introduction of a magnetron-based gas-aggregation cluster source into the photoelectron spectroscopy field allowed reaching sufficient abundance for the nanoscale-size clusters out of nonmetallic substances. A valence electron spectroscopy study on silicon clusters with $N \approx 10^3$ has shown—by mapping the density of states—that at this size the clusters were not yet semiconductors [118].

Alkali halides are prototypical ionic compounds, and small clusters of NaCl and CsI produced using pick-up of alkali halide vapor by a beam of pre-formed Ar clusters have been studied by XAS. Under some conditions, a part of the Ar cluster remains resulting in alkali halide clusters embedded in Ar. Effects like charge transfer from the Ar shell has been shown to influence the ionization and fragmentation of alkali halide clusters [142,143]. For NaCl, information about the bond length development with the cluster size has been obtained, showing that NaCl clusters have a distorted cubic structure with the bond length that is halfway between that of the molecule and the solid [144–146].

7.6 Summary and Conclusions

The research on free clusters using electron spectroscopy and related methods has progressed tremendously during the last decade due to a continuous development of the cluster sources, and to a wider implementation of the synchrotron radiation, allowing probing both the valence and core levels. In this review, some basic concepts useful for connecting the spectra to the cluster properties have been outlined. As exemplified above, PES makes it possible to determine important parameters such as cluster size, composition, geometry, and energy structure. The changes in the latter from the isolated monomer, over small clusters for which each size can be regarded as a separate molecule-like entity, to large clusters which essentially are small pieces of the infinite solid, are clearly reflected by the method. The field is now developing into a new phase, where the achievements of the pioneer period may be used to address specific problems. This will include further fundamental studies, but also studies of naturally occurring clusters, such as complex molecular clusters in the atmosphere, and technically important clusters such as functional nanoparticles.

References

1. Haberland, H. (Ed.). 1994–1995. *Clusters of Atoms and Molecules I & II*, Springer Series in Chemical Physics, Vol. 52, Springer, Berlin, Germany.

2. Echt, O., Sattler, K., and Recknagel, E. 1981. Magic numbers for sphere packings: Experimental verification in free xenon clusters. *Phys. Rev. Lett.* 47: 1121–1124.

3. Knight, W. D., Clemenger, K., de Heer, W. A., Saunders, W. A., Chou, M. Y., and Cohen, M. L. 1984. Electronic shell structure and abundances of sodium clusters. *Phys. Rev. Lett.* 52: 2141–2144.

4. von Issendorff, B. and Cheshnovsky, O. 2005. Metal to insulator transitions in clusters. *Annu. Rev. Phys. Chem.* 56: 549–580.

5. Chylek, P. and Geldart, D. J. W. 1997. Water vapor dimers and atmospheric absorption of electromagnetic radiation. *Geophys. Res. Lett.* 24: 2015–2018, and reference therein.

6. Vaida, V., Daniel, J. S., Kjaergaard, H. G. et al. 2001. Atmospheric absorption of near infrared and visible solar radiation by the hydrogen bonded water dimmer. *Q. J. R. Meteorol. Soc.* 127: 1627–1643, and reference therein.

7. Kulmala, M. Lehtinen, K. E. J., and Laaksonen, A. 2006. Cluster activation theory as an explanation of the linear dependence between formation rate of 3 nm particles and sulphuric acid concentration. *Atmos. Chem. Phys.* 6: 787–793, and reference therein.

8. Kulmala, M., Riipinen, I., Sipilä, M. et al. 2007. Toward direct measurement of atmospheric nucleation. *Science* 318: 89–92, and reference therein.

9. Hüfner, S. 2003. *Photoelectron Spectroscopy*, 3rd edition, Springer, Berlin, Germany.

10. Eberhardt, W. (Ed.). 1994. *Applications of Synchrotron Radiation: High-Resolution Studies of Molecules and Molecular Adsorbates on Surfaces*, Springer Series in Surface Sciences, Springer, Berlin, Germany.

11. Stöhr, J. 1992. *NEXAFS Spectroscopy*, Springer Series in Surface Sciences, Springer, Berlin, Germany.

12. Piancastelli, M. N. 2000. Auger resonant Raman studies of atoms and molecules. *J. Electron Spectrosc. Relat. Phenom.* 107: 1–26.

13. Dehmer, P. M. and Dehmer, E. L. 1977. Photoelectron spectrum of the Xe_2 van der Waals molecule. *J. Chem. Phys.* 67: 1774–1775.

14. Dehmer, P. M. and Dehmer, E. L. 1978. Photoelectron spectra of Ar_2 and Kr_2 and dissociation energies of the rare gas dimer ions. *J. Chem. Phys.* 69: 125–133.

15. Dehmer, P. M. and Dehmer, E. L. 1978. Photoelectron spectrum of Xe_2 and potential energy curves for Xe_2^+. *J. Chem. Phys.* 68: 3462–3470.

16. Poliakoff, E. D., Dehmer, P. M., Dehmer, E. L. et al. 1981. The photoelectron spectrum of Xe_3 by the photoelectron-photoion coincidence technique. *J. Chem. Phys.* 75: 1568–1569.

17. Poliakoff, E. D., Dehmer, P. M., Dehmer, E. L. et al. 1982. Photoelectron-photoion coincidence spectroscopy of gas-phase clusters. *J. Chem. Phys.* 76: 5214–5224.

18. Carnovale. F. 1989. Photoelectron spectroscopy of argon clusters: Evidence for an Ar_{13}^* ionization chromophore. *J. Chem. Phys.* 90: 1452–1459.

19. Milani, P. and Iannotta, S. 1999. *Cluster Beam Synthesis of Nanostructured Materials.* Springer, Berlin, Germany.

20. Carnovale, F., Peel, J. B., and Rothwell, R. G. 1991. Photoelectron spectroscopy of krypton and xenon clusters. *J. Chem. Phys.* 95: 1473–1478.

21. Schwentner, N., Himpsel, F.-J., Saile, V. et al. 1975. Photoemission from rare-gas solids: Electron energy distributions from the valence bands. *Phys. Rev. Lett.* 34: 528–531.

22. Hergenhahn, U., Kolmakov, A., Riedler, M. et al. 2002. Observation of excitonic satellites in the photoelectron spectra of Ne and Ar clusters. *Chem. Phys. Lett.* 351: 235–241.

23. Feifel, R., Tchaplyguine, M., Öhrwall, G. et al. 2004. From localized to delocalized electronic states in free Ar, Kr and Xe clusters. *Eur. Phys. J. D* 90: 343–351.

24. Rühl, E., Heinzel, C., Hitchcock, A. P. et al. 1993. Ar $2p$ spectroscopy of free argon clusters. *J. Chem. Phys.* 68: 2653.

25. Rühl, E., Heinzel, C., Hitchcock, A. P. et al. 1993. K-shell spectroscopy of Ar clusters. *J. Chem. Phys.* 68: 6820–6826.

26. Björneholm, O., Federmann, F., Fössing, F., and Möller. T. 1995. Core level photoelectron and x-ray absorption spectroscopy of free argon clusters. *Phys. Rev. Lett.* 74: 3017–3020.

27. Benfield, R. E. 1992. Mean coordination numbers and the non-metal-metal transition in clusters. *J. Chem. Soc. Faraday Trans.* 88: 1107–1110.

28. Chiang, T.-C., Kaindl, G., and Mandel, T. 1986. Layer-resolved shifts of photoemission and Auger spectra from physisorbed rare-gas multilayers. *Phys. Rev. B* 33: 695–711.

29. van der Straten, P., Morgenstern, R., and Niehaus, A. 1988. Angular dependent post-collision interaction in Auger processes. *Z. Phys. D* 8: 35–45.

30. Knop, A., Wassermann, B., and Rühl, E. 1998. Site-specific excitation in free krypton clusters. *Phys. Rev. Lett.* 80: 2302–2305.

31. Gauyacq, J. P. 2005. Core-excited $Ar^*(2p_{3/2}^{-1}\, nl)$ atoms inside an Ar crystal: Effect of Ar neighbors on the excited state. *Phys. Rev. B* 71: 115433-1–115433-9.

32. Gauyacq, J. P. and Kazansky, A. K. 2006. Core excited $Ne^*(1s^{-1}nl)$ atoms inside bulk Ne. *Chem. Phys.* 327: 300–310.

33. Gauyacq, J. P. 2007. Core-excited $Ne^*(1s^{-1}nl)$ states on a Ne surface. *Chem. Phys.* 333: 194–200.

34. Rühl, E. and Jortner, J. 2000. Foreword. *J. Electron Spectrosc. Relat. Phenom.* 106: 107.

35. Björneholm, O. 2002. Resonant core level studies of molecules and clusters. *Surf. Rev. Lett.* 9: 3–12.

36. Tchaplyguine, M., Feifel, R., Marinho, R. R. T. et al. 2003. Selective probing of the electronic structure of free clusters using resonant core-level spectroscopy. *Chem. Phys.* 289: 3–13.

37. Öhrwall, G., Tchaplyguine, M., Gisselbrecht, M. et al. 2003. Observation of elastic scattering effects on photoelectron angular distributions in free Xe clusters. *J. Phys. B: At. Mol. Opt. Phys.* 36: 3937–3949.

38. Rolles, D., Zhang, H., Pešić, Z. D. et al. 2007. Size effects in angle-resolved photoelectron spectroscopy of free rare-gas clusters. *Phys. Rev. A* 75: 031201(R)–031204(R).

39. Tchaplyguine, M., Marinho, R. R. T., Gisselbrecht, M. et al. 2004. The size of neutral free clusters as manifested in the relative bulk-to-surface intensity in core level photoelectron spectroscopy. *J. Chem. Phys.* 120: 345–356.

40. Hatsui, T., Setoyama, H., Kosugi, N. et al. 2005. Photoionization of small krypton clusters in the Kr 3d regime: Evidence for site-specific photoemission. *J. Chem. Phys.* 123: 154304-1–154304-6.

41. Lindblad, A., Fink, R. F., Bergersen, H. et al. 2005. Postcollision interaction in noble gas clusters: Observation of differences in surface and bulk line shapes. *J. Chem. Phys.* 123: 211101–211104.

42. Lundwall, M., Fink, R. F., Tchaplyguine, M. et al. 2006. Shell-dependent core-level chemical shifts observed in free xenon clusters. *J. Phys. B* 39: 5225–5235.

43. Hergenhahn, U., Kolmakov, A., Loefken, O. et al., 1999. Evidence for excitonic satellites and interatomic Coulomb decays in photoelectron spectra of Ne clusters. *HASYLAB Annual Report*.

44. Cederbaum, L. S., Zobeley, J., and Tarantelli, F. 1997. Giant intermolecular decay and fragmentation of clusters. *Phys. Rev. Lett.* 79: 4778–4781.

45. Santra, R., Zobeley, J., Cederbaum, L. S., and Moiseyev, N. 2000. Interatomic Coulombic decay in van der Waals clusters and impact of nuclear motion. *Phys. Rev. Lett.* 85: 4490.

46. Marburger, S., Kugeler, O., Hergenhahn, U., and Möller, T. 2003. Experimental evidence for interatomic Coulombic decay in Ne clusters. *Phys. Rev. Lett.* 90: 203401-1–203401-4.

47. Barth, S., Marburger, S., Kugeler, O. et al. 2006. The efficiency of Interatomic Coulombic Decay in Ne clusters. *Chem. Phys.* 329: 246–250.

48. Öhrwall, G., Tchaplyguine, M., Lundwall, M. et al. 2004. Femtosecond interatomic Coulombic decay in free neon clusters: Large lifetime differences between surface and bulk. *Phys. Rev. Lett.* 93: 173401–173404.

49. Jahnke, T., Czasch, A., Schöffler, M. S. et al. 2004. Experimental observation of interatomic Coulombic decay in neon dimers. *Phys. Rev. Lett.* 93: 163401-1–163401-4.

50. Morishita, Y., Liu, X.-J., Saito, N. et al. 2006. Experimental evidence of interatomic coulombic decay from the Auger final states in argon dimers. *Phys. Rev. Lett.* 96: 243402-1–243402-4.

51. Jahnke, T., Czasch, A., Schöffler, M. et al. 2007. Experimental separation of virtual photon exchange and electron transfer in interatomic Coulombic decay of neon dimers. *Phys. Rev. Lett.* 99: 153401-1–153401-4.

52. Morishita, Y., Saito, N., Suzuki, I. H. et al. 2008. Evidence of interatomic coulombic decay in ArKr after Ar 2p auger decay. *J. Phys. B* 41: 025101.

53. Barth, S., Marburger, S. P., Joshi, S., Ulrich, V., and Hergenhahn, U. 2006. Interface identification by non-local autoionization transitions. *Phys. Chem. Chem. Phys.* 8: 3218–3222.

54. Lundwall, M., Tchaplyguine, M., Öhrwall, G. et al. 2007. Self-assembled heterogeneous argon/neon core-shell clusters studied by photoelectron spectroscopy. *J. Chem. Phys.* 126: 214706.

55. Barth, S., Joshi, S., Marburger, S. et al. 2005. Observation of resonant Interatomic Coulombic decay in Ne clusters. *J. Chem. Phys.* 122: 241102–241104.

56. Lundwall, M., Tchaplyguine, M., Öhrwall, G. et al. 2005. Enhanced surface sensitivity in AES relative to XPS observed in free argon clusters. *Surf. Sci.* 594: 12–19.

57. Peredkov, S., Kivimäki, A., Sorensen, S. et al. 2005. Ioniclike energy structure of neutral core-excited states in free Kr clusters. *Phys. Rev. A* 72, 021201(R)–021204(R).

58. Lundwall, M., Lindblad, A., Bergersen, H. et al. 2006. Photon energy dependent intensity variations observed in Auger spectra of free argon clusters. *J. Phys. B* 39: 3321–3333.

59. Helenelund, K., Hedman, S., Asplund, L., Gelius, U., and Siegbahn, K. 1983. An improved model for post-collision interaction (PCI) and high resolution Ar LMM Auger spectra revealing new PCI effects. *Phys. Scr.* 27: 245–253.

60. Kittel, C. 2005 *Introduction to Solid State Physics*, 8th edition, University of California, Berkeley, CA.

61. Tchaplyguine, M., Lundwall, M., Gisselbrecht, M. et al. 2004. Variable surface composition and radial interface formation in self-assembled free, mixed Ar/Xe clusters. *Phys. Rev. A* 69: 031201(R)–031204(R).

62. Lundwall, M., Tchaplyguine, M., Öhrwall, G. et al. 2004. Radial surface segregation in free heterogeneous argon/krypton clusters. *Chem. Phys. Lett.* 392: 433–438.

63. Lundwall, M., Bergersen, H., Lindblad, A. et al. 2006. Preferential site occupancy observed in coexpanded argon-krypton clusters. *Phys. Rev. A* 74: 043206-1–043206-7.

64. Lindblad, A., Bergersen, H., Rander, T. et al. 2006. The far from equilibrium structure of argon clusters doped with krypton or xenon. *Phys. Chem. Chem. Phys.* 8: 1899–1905.

65. Lundwall, M., Lindblad, A., Bergersen, H. et al. 2006. Preferential site occupancy of krypton atoms on free argon-cluster surfaces. *J. Chem. Phys.* 125: 014305-1–014305-7.

66. Kivimäki, A., Sorensen, S., Tchaplyguine, M. et al. 2005. Resonant Auger spectroscopy of argon clusters at the 2p threshold. *Phys. Rev. A* 71: 033204-1–033204-7.

67. Wurth, W., Rocker, G., Feulner, P. et al. 1993. Core excitation and deexcitation in argon multilayers: Surface- and bulk-specific transitions and autoionization versus Auger decay. *Phys. Rev. B* 47: 6697–6704.

68. Aksela, H., Aksela, S., Pulkkinen, H. et al. 1989. Shake processes in the Auger decay of resonantly excited $3d^9 4s^2 4p^6 np$ states of Kr. *Phys. Rev. A* 40: 6175–6180.

69. Tchaplyguine, M., Kivimäki, A., Peredkov, S. et al. 2007. Localized versus delocalized excitations just above the 3d threshold in krypton clusters studied by Auger electron spectroscopy. *J. Chem. Phys.* 127: 1–8.

70. Brudermann, J., Lohbrandt, P., Buck, U., and Buch, V. 1998. Surface vibrations of large water clusters by He atom scattering, *Phys. Rev. Lett.* 80: 2821.

71. Yamazaki, T. and Kimura, K. 1976. He I photoelectron spectrum of dinitrogen tetraoxide (N_2O_4). *Chem. Phys. Lett.* 43: 502–505.

72. Tomoda, S., Achiba, Y., and Kimura, K. 1982. Photoelectron spectrum of the water dimer. *Chem. Phys. Lett.* 87: 197–200.

73. Carnovale, F., Livett, M. K., and Peel, J. B. 1979. A photoelectron spectroscopic study of the formic acid dimmer. *J. Chem. Phys.* 71: 255–258.

74. Carnovale, F., Peel, J. B., and Rothwell, G. 1986. Photoelectron spectroscopy of the NO dimmer. *J. Chem. Phys.* 84: 6526–6527.

75. Carnovale, F., Peel, J. B., and Rothwell, G. 1986. Ammonia dimer studied by photoelectron spectroscopy. *J. Chem. Phys.* 85: 6261.

76. Carnovale, F., Peel, J. B., and Rothwell, G. 1988. Photoelectron spectroscopy of the nitrogen dimer $(N_2)_2$ and clusters $(N_2)_n$: N_2 dimer revealed as the chromophore in photoionization of condensed nitrogen. *J. Chem. Phys.* 88: 642–650.

77. Carnovale, F., Peel, J. B., and Rothwell, G. 1991. Photoelectron spectroscopy of the oxygen dimer and clusters. *Org. Mass Spectrom.* 26: 201.

78. Rander, T., Lundwall, M., Lindblad, A. et al. 2007. Experimental evidence for molecular ultrafast dissociation in O_2 clusters. *Eur. Phys. J. D* 42: 252–257.

79. Flesch, R., Pavlychev, A. A., Neville, J. J. et al. 2001. Dynamic stabilization in $1\sigma_u \rightarrow 1\pi_g$ excited nitrogen clusters. *Phys. Rev. Lett.* 86: 3767–3770.

80. Flesch, R., Kosugi, N., Bradeanu, I. L. et al. 2004. Cluster size effects in core excitons of 1s-excited nitrogen. *J. Chem. Phys.* 121: 8343–8350.

81. Pavlychev, A. A., Flesch, R., and Rühl, E. 2004. Line shapes of 1s$\rightarrow\pi^*$ excited molecular clusters. *Phys. Rev. A* 70: 015201-1–015201-4.

82. Pavlychev, A. A., Brykalova, X. O., Flesch, R., and Rühl, E. 2006. Shape resonances in molecular clusters: the $2t_{2g}$ shape resonances in S 2p-excited sulfur hexafluoride clusters. *Phys. Chem. Chem. Phys.* 8: 1914.

83. Morin, P. and Nenner, I. 1986. Atomic autoionization following very fast dissociation of core-excited HBr. *Phys. Rev. Lett.* 56: 1913–1916.

84. Schaphorst, S. J., Caldwell, C. D., Krause, M. O., and Jiménez-Mier, J. 1993. Evidence for atomic features in the decay of resonantly excited molecular oxygen. *Chem. Phys. Lett.* 213: 315–320.

85. Asplund, L., Gelius, U., Hedman, S. et al. 1985. Vibrational structure and lifetime broadening in core-ionized methane. *J. Phys. B* 18: 1569–1579.

86. Bergersen, H., Abu-samha, M., Lindblad, A. et al. 2006. First observation of vibrations in core-level photoelectron spectra of free neutral molecular clusters. *Chem. Phys. Lett.* 429: 109–113.

87. Abu-samha, M., Børve, K. J., Sæthre, L. J. et al. 2006. Lineshapes in carbon 1s photoelectron spectra of methanol clusters. *Phys. Chem. Chem. Phys.* 21: 2473–2482.

88. Abu-samha, M., Børve, K. J., Harnes, J. et al. 2007. What can C1s photoelectron spectroscopy tell about structure and bonding in clusters of methanol and methyl chloride? *J. Phys. Chem. A* 111: 8903–8909.

89. Bergersen, H., Abu-samha, M., Lindblad, A. et al. 2006. Two size regimes of methanol clusters produced by adiabatic expansion. *J. Chem. Phys.* 125: 184303-1–184303-5.

90. Rosso, A., Lindblad, A., Lundwall, M. et al. 2007. Synchrotron radiation study of chloromethane clusters: Effects of polarizability and dipole moment on core level chemical shifts. *J. Chem. Phys.* 127, 024302–024306.

91. Rosso, A., Rander, T., Bergersen, H. et al. 2007. The role of molecular polarity in cluster local structure studied by photoelectron spectroscopy. *Chem. Phys. Lett.* 435, 79–83.

92. Futami, Y., Kudoh, S., Ito, F., Nakanaga, T., and Nakata, M. 2004. Structures of methyl halide dimers in supersonic jets by matrix-isolation infrared spectroscopy and quantum chemical calculations. *J. Mol. Struct.* 609: 9–16.

93. Bradeanu, I. L., Flesch, R., Kosugi, N. et al. 2006. C 1s$\rightarrow\pi^*$ excitation in variable size benzene clusters. *Phys. Chem. Chem. Phys.* 8: 1906–1913.

94. Rosso, A., Pokapanich, W., Öhrwall, G. et al. 2007. Adsorption of polar molecules on krypton clusters. *J. Chem. Phys.* 127: 084313-1–084313-5.

95. Rosso, A., Öhrwall, G., Tchaplyguine, M. et al. 2008. Adsorption of chloromethane molecules on free argon clusters. *J. Phys. B* 41: 085102–085107.

96. Björneholm, O., Federmann, F., Kakar, S., and Möller, T. 1999. Between vapor and ice: Free water clusters studied by core level spectroscopy. *J. Chem. Phys.* 111: 546–550.

97. Öhrwall, G., Fink, R. F., Tchaplyguine, M. et al. 2005. The electronic structure of free water clusters probed by Auger electron spectroscopy. *J. Chem. Phys.* 123: 054310-1–054310-10.

98. Abu-samha, M. and Børve, K. J. 2008. Surface relaxation in water clusters: Evidence from theoretical analysis of the oxygen 1s photoelectron spectrum. *J. Chem. Phys.* 128: 154710-1–154710-6.

99. Bréchignac, C., Broyer. M., Cahuzac, P. et al. 1985. Size dependence of linear-shell autoionization lines in mercury clusters. *Chem. Phys. Lett.* 120: 559.

100. Bréchignac, C., Broyer. M, Cahuzac, P. et al. 1988. Probing the transition from van der Waals to metallic mercury clusters. *Phys. Rev. Lett.* 60: 275–278.

101. Kaiser, B. and Rademann, K. 1992. Photoelectron spectroscopy of neutral mercury clusters Hg_x ($x \leq 109$) in a molecular beam. *Phys. Rev. Lett.* 69: 3204–3207.

102. Cheshnovsky, O., Taylor, K. J., and Smalley, R. E. 1990. Ultraviolet photoelectron spectra of mass-selected copper clusters: Evolution of the 3d band. *Phys. Rev. Lett.* 64: 1785–1788.

103. Taylor, K. J., Petiette-Hall, C. L., Cheshovsky, O., and Smalley, R. E. 1992. Ultraviolet photoelectron spectra of coinage metal clusters. *J. Chem. Phys.* 96: 3319–3329.

104. Busani, R., Folkers, M., and Cheshnovsky, O. 1998. Direct observation of band-gap closure in mercury clusters. *Phys. Rev. Lett.* 81: 3836–3839.

105. Thomas, O. C., Zheng, W., Xu, S. et al. 2002. Onset of metallic behavior in magnesium clusters. *Phys. Rev. Lett.* 89: 213403–213401.

106. McHugh, K. M., Eaton, J. G., Lee, G. H. et al. 1989. Photoelectron spectra of the alkali metal cluster anions: $Na_{n=2-5}^-$, $K_{n=2-7}^-$, $Rb_{n=2-3}^-$, and $Cs_{n=2-3}^-$. *J. Chem. Phys.* 91: 3792–3793.

107. Handschuh, H., Cha, C. Y., Möller, H. et al. 1994. Delocalized electronic states in small clusters. Comparison of Na_n, Cu_n, Ag_n, and Au_n clusters. *Chem. Phys. Lett.* 227: 496–502.

108. Wrigge, G., Astruc Hoffmann, M., and Issendorff, B. v. 2002. Photoelectron spectroscopy of sodium clusters: Direct observation of the electronic shell structure. *Phys. Rev. A* 65: 063201-1–063201-5.

109. Cha, C. Y., Ganteför, G., and Eberhardt, W. 1994. The development of the $3p$ and $4p$ valence band of small aluminum and gallium clusters. *J. Chem. Phys.* 100: 995–1010.

110. Li, X., Wu, H., Wang, X. B., and Wang, L. S. 1998. S-p hybridization and electron shell structures in aluminum clusters: A photoelectron spectroscopy study. *Phys. Rev. Lett.* 81: 1909–1912.

111. Kietzmann, H., Morenzin, J., Bechthold, P. S. et al. 1996. Photoelectron spectra and geometric structures of small niobium cluster anions. *Phys. Rev. Lett.* 77: 4528–4531.

112. Häkkinen, H., Moseler, M., Kostko, O. et al. 2004. Symmetry and electronic structure of noble-metal nanoparticles and the role of relativity. *Phys. Rev. Lett.* 93: 093401-1–093401-4.

113. Kostko, O., Wrigge, G., Cheshnovsky, O. et al. 2005. Transition from a Bloch-Wilson to a free-electron density of states in Zn_n^- clusters. *J. Chem. Phys.* 123: 221102-1–221102-4.

114. Haberland, H., Mall, M., Moseler, M. et al. 1994. Filling of micron-sized contact holes with copper by energetic cluster impact. *J. Vac. Sci. Technol. A* 12: 2925–2930.

115. Zimmermann, U., Malinowski, N., Näher, U. et al. 1994. Producing and detecting very large clusters. *Z. Phys. D* 31: 85–93.

116. Ellert, C., Schmidt, M., Schmitt, C., Reiners, T., and Haberland, H. 1995. Temperature dependence of the optical response of small, open shell sodium clusters. *Phys. Rev. Lett.* 75: 1731–1734.

117. Eastham, D. A., Hamilton, B., and Denby, P. M. 2002. Formation of ordered assemblies from deposited gold clusters. *Nanotechnology* 13: 51–54.

118. Astruc Hoffmann, M., Wrigge, G., von Issendorff, B. et al. 2001. Ultraviolet photoelectron spectroscopy of Si_4 to Si_{1000}. *Eur. Phys. J. D* 16: 9–11.

119. Kostko, O., Bartels, C., Schwöbel, J. et al. 2007. Photoelectron spectroscopy of the structure and dynamics of free size selected sodium clusters. *J. Phys. Conf. Ser.* 88: 012034–012041.

120. Piseri, P., Mazza, T., Bongiorno, G. et al. 2006. Core level spectroscopy of free titanium clusters in supersonic beams. *New J. Phys.* 8: 136-1–136-12.

121. Mazza, T., Piseri, P., Bongiorno, G. et al. 2008. Probing the chemical reactivity of free titanium clusters by x-ray absorption spectroscopy. *Appl. Phys. A* 92: 463–471.

122. Lau, J. T., Rittmann, J., Zamudio-Bayer, V. et al. 2008. Size dependence of $L_{2,3}$ branching ratio and 2p core-hole screening in x-ray absorption of metal clusters. *Phys. Rev. Lett.* 101: 153401–153403.

123. Peredkov, S., Schulz, J., Rosso, A. et al. 2007. Free nanoscale sodium clusters studied by core-level photoelectron spectroscopy. *Phys. Rev. B* 75: 235407-1–235407-8.

124. Peredkov, S., Rosso, A., Öhrwall, G. et al. 2007. Size determination of free metal clusters by core-level photoemission from different initial charge states. *Phys. Rev. B* 76: 081402-1–081402-4.

125. Rosso, A., Öhrwall, G., Bradeanu, I. et al. 2008. Photoelectron spectroscopy study of free potassium clusters: Core-level lines and plasmon satellites. *Phys. Rev. A* 77: 043202-1–043202-5.

126. Tchaplyguine, M., Peredkov, S., Rosso, A. et al. 2008. Absolute core-level binding energy shifts between atom and solid: The Born-Haber cycle revisited for free nanoscale metal clusters. *J. Electron Spectrosc. Relat. Phenom.* 166–167: 38–44.

127. Tchaplyguine, M., Peredkov, S., Rosso, A. et al. 2007. Direct observation of the non-supported metal nanoparticle electron density of states by x-ray photoelectron spectroscopy. *Eur. Phys. J. D* 45: 295–299.

128. Rander, T., Schulz, J., Huttula, M. et al. 2007. Core-level electron spectroscopy on the sodium dimer Na 2p level. *Phys. Rev. A* 75: 032510-1–032510-4.

129. Brechignac, C., Cauzac, Ph., Carlier, F., and Leygnier, J., 1989. Photoionization of mass-selected K_n^+ ions: A test for the ionization scaling law. *Phys. Rev. Lett.* 63: 1368–1371.

130. Macov, G., Nitzan, A., and Brus, L. E., 1988. On the ionization potential of small metal and dielectric particles. *J. Chem. Phys.* 88: 65076–65085.

131. De Heer, W. A. and Milani, P. 1990. Comment on "Photoionization of mass-selected K_n^+ ions: A test for the ionization scaling law". *Phys. Rev. Lett.* 65: 3356.

132. Seidl, M., Perdew, J. P., Brajczewska, M. et al. 1998. Ionization energy and electron affinity of a metal cluster in the stabilized jellium model: Size effect and charging limit. *J. Chem. Phys.* 108: 8182–8189.

133. Johansson, B. and Mårtensson, N. 1980. Core-level binding-energy shifts for the metallic elements. *Phys. Rev. B* 21: 4427–4452.

134. Senz, V., Fischer, T., Oelßner, P. et al. 2006. Core-level photoelectron spectroscopy on mass-selected metal clusters using VUV-FEL radiation. *Hasylab Annual Report*, 393–394.

135. Bréchignac, C., Broyer, M., Cahuzac, Ph. et al. 1991. Shape resonance in $4d$ inner-shell photoionization spectra of antimony clusters. *Phys. Rev. Lett.* 67: 1222–1225.

136. Hunsicker, S., Jones, R.O., and Ganteför, G. 1995. Rings and chains in sulfur cluster anions S^- to S_9^-: Theory (simulated annealing) and experiment (photoelectron detachment). *J. Chem. Phys.* 102: 5917–5936.

137. Nagaya, K., Yao, M., Hayakawa, T. et al. 2002. Size-selective extended x-ray absorption fine structure spectroscopy of free selenium clusters. *Phys. Rev. Lett.* 89: 243401-1–243401-4.

138. Rühl, E., Flesch, R., Tappe, W. et al. 2002. Sulfur 1s excitation of S_2 and S_8: Core-valence- and valence-valence-exchange interaction and geometry-specific transitions. *J. Chem. Phys.* 116: 3316–3322.

139. Handschuh, H., Ganteför, G., Kessler, B. et al. 1995. Stable configurations of carbon clusters: Chains, rings, and fullerenes. *Phys. Rev. Lett.* 74: 1095–1098.

140. Gunnarsson, O., Handschuh, H., Bechthold, P. S. et al. 1995. Photoemission spectra of C_{60}^-: Electron-phonon coupling, Jahn-Teller effect, and superconductivity in the fullerides. *Phys. Rev. Lett.* 74: 1875–1878.

141. Willey, T. M., Bostedt, C., van Buuren, T. et al. 2005. Molecular limits to the quantum confinement model in diamond clusters. *Phys. Rev. Lett.* 95: 113401-1–113401-4.

142. Kolmakov, A., Löfken, J. O., Nowak, C. et al. 1999. Observation of small metastable multiply charged CsI clusters embedded inside rare gas clusters. *Eur. Phys. J. D* 9: 273–276.

143. Kolmakov, A., Löfken, J. O., Nowak, C. et al. 2000. Argon coated alkali halide clusters: The effect of the coating on the ionization and fragmentation dynamics. *Chem. Phys. Lett.* 319, 465–471.

144. Nowak, C., Rienecker, C., Kolmakov, A. et al. 1999. Inner shell photoionization spectroscopy of NaCl clusters. *J. Electron Spectrosc. Relat. Phenom.* 101: 199–203.

145. Riedler, M., de Castro, A. R. B., Kolmakov, A. et al. 2001. Na 1s photoabsorption of free and deposited NaCl clusters: Development of bond length with cluster size. *Phys. Rev. B* 64: 245419-1–245419-9.

146. Yalovega, G., Soldatov, A. V., Riedler, M., Mark, R. P. et al. 2002. Geometric structure of $(NaCl)_4$ clusters studied with XANES at the chlorine L-edge and at the sodium K-edge. *Chem. Phys. Lett.* 356: 23–28.

8

Photoelectron Spectroscopy
of Organic Clusters

Masaaki Mitsui
Shizuoka University

Atsushi Nakajima
Keio University

8.1 Introduction

8.1.1 Molecular Clusters

A cluster produced in the gas phase is a finite nanoscale aggregate consisting of two to several thousands of atoms or molecules with a diameter of several nanometers or less. Because of the intermediate nature of the cluster between individual molecules and those of condensed matter, the cluster often exhibits unique size-specific characteristics that are rather different from those of their corresponding bulk materials. Also, gaseous clusters enable us to explore the gradual size evolution of the structural, electronic, optical, thermodynamic, and chemical properties from an isolated molecule to condensed matter systems under conditions that are completely free of chemical and physical impurities including environmental factors. The number of atoms or molecules making up the cluster (i.e., cluster size or n) can be apparently defined using mass spectrometer, allowing us to probe the transition of such properties in a stepwise fashion. Among the class of clusters, "molecular clusters," where constituents aggregate via weak van der Waals forces, have been regarded as an ideal model system to provide informative, molecular-level description of many physical and chemical processes occurring in bulk and interfacial environments. Most importantly, molecular clusters can provide a direct insight into intermolecular van der Waals interactions, which play an important role in determining the structures and properties of molecular assemblies in chemistry, biology, and soft-material science. Over the last two decades, therefore, an enormous amount of experimental and theoretical studies pertinent to molecular clusters have been performed, and many excellent review articles on the study of molecular clusters have already been published in the last 20 years.[1-23] Since this chapter focuses only on the recent topics of molecular clusters consisting of π-conjugated organic molecules, the reader is recommended to read those review articles for an in-depth understanding of various aspects of molecular clusters.

One of the ultimate goals in molecular cluster study is to realize a quantitative evaluation of intermolecular potential energy surfaces, hence small clusters such as the dimer and the trimer have been principal targets in the study. For this reason, the application of many laser-based spectroscopic methods developed so far have been limited to small-sized clusters with less than 10 constituent molecules, except for water and some small molecules. Another fundamental interest in the molecular cluster study is to trace how the structural, electronic, and thermodynamic properties of condensed phases begin to emerge as the constituent molecular number increases. To observe this transition in a stepwise fashion, size-selective investigations on a broad size range of molecular clusters are imperative. Some past studies have already addressed this subject. Representative examples are the spectroscopic works conducted around 1990: the electronic spectroscopy of $(benzene)_n$ $(n \leq 60)$[24,25] and $(H_2O)_n^-$ $(n = 6-50)$,[26] and the photoelectron spectroscopy of $(H_2O)_n^-$ $(n = 2-69)$,[27,28] $(NH_3)_n^-$ $(n = 41-1100)$,[28,29] $X^-(H_2O)_n$ $(n = 1-60)$,[30,31] and $X^-(CH_3CN)_n$ $(n = 1-55; X = Cl, Br, and I)$.[32] In recent years, many interesting experimental results have been reported for large hydrated clusters, such as $SO_4^{2-}\text{-}(H_2O)_n$ $(n = 4-40)$,[33] $H^+(H_2O)_n$ $(n \leq 27)$,[34-36] and $(H_2O)_n^-$ $(n \leq 200)$,[37-40] and large anionic methanol clusters ($n \sim 145$–535)[41] using the coupling of the state-of-the-art molecular beam sources and optical and imaging methods with mass spectrometry. It is also noted that, while no size-selection has been performed, infrared predissociation spectroscopy of giant water clusters (average sizes: $\langle n \rangle$ = 20–1960) has also been reported by Buck and coworkers.[42]

8.1.2 Clusters of π-Conjugated Organic Molecules

As mentioned above, the size-selective spectroscopic study establishing a link from an isolated molecule to the bulk made significant progress for clusters of small σ-type molecules, like water. In contrast, π-conjugated organic molecules, especially polycyclic aromatic hydrocarbons (PAHs), rather fall behind in such a cluster approach and remain almost completely undeveloped to date, though their bulk counterpart has been widely studied up to now.[43–47] Important microscopic aspects, such as excimer-formation dynamics and charge-resonance interactions, have been elucidated in the past experimental studies on π-conjugated organic clusters.[48–53] However, almost all those studies have been limited to the cluster size range of $n < 10$, which is too small to bridge the gap between a molecule and the bulk.

As will be described in Section 8.3.1, the efficient formation of large clusters requires a relatively high vapor pressure of the sample as well as a high stagnation pressure of the carrier gas in the supersonic jet expansion by continuous nozzles or pulsed valves. Unfortunately, most of π-conjugated organic molecules are solid at room temperature, have a relatively high melting temperature, and are nonvolatile. Moreover, the intense gas expansion with high pressure into a vacuum chamber needs a large vacuum-pumping system. Thus, it is generally very difficult to produce sufficient amounts of large clusters of π-conjugated organic molecules to implement size-selective spectroscopic investigations. Second, due to the significant improvements in *ab initio* and density functional theory (DFT) algorithms, and rapid progress in computational capability, acceptably accurate theoretical calculations have become available even for larger aggregation systems. However, a realization of this is still extremely difficult at present even for the relatively small clusters of π-conjugated organic molecules, because a large number of constituent carbon atoms in those clusters inevitably result in a large basis set number and the importance of dispersion interactions (i.e., electron correlation effects) between highly polarizable and non-dipolar molecules in the clusters requires very accurate and costly molecular orbital calculations.

The experimental obstacle described above was recently conquered by using a high-pressure, high-temperature, ultra-short-pulsed valve (Even–Lavie valve)[54] capable of providing very cold molecular beam conditions within a reasonable gas loading for a vacuum system. We succeeded in producing large cluster anions of many π-conjugated organic molecules, such as benzene,[55,56] oligoacenes,[57–61] oligothiophenes,[62] oligophenylenes,[63,64] and other PAHs,[58] by using this valve. Besides, their electronic structures were size-selectively examined with anion photoelectron

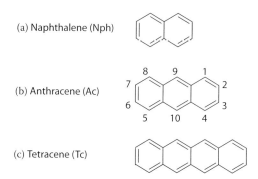

FIGURE 8.1 Structures of some of the π-conjugated organic molecules which have been studied by anion photoelectron spectroscopy: (a) naphthalene (Nph), (b) anthracene (Ac), and (c) tetracene (Tc). The number of the alkyl-substitution positions is shown in (b).

spectroscopy (PES) up to the size of over 100-mer and the comparison of the results of the large clusters with their bulk counterparts was conducted.[55–64] In this chapter, we describe the formation of large anionic clusters of π-conjugated organic molecules, especially linear oligoacenes—i.e., naphthalene (Nph), anthracene (Ac), and tetracene (Tc) (their chemical structures shown in Figure 8.1), and their anion PES from an isolated molecule to the 100-mer.

8.2 Polarization Effects in π-Conjugated Organic Aggregates

The electronic structure of organic crystals and amorphous organic solids has been the object of intensive research for almost half a century,[43–47] because it is directly connected with the charge transport property that is one of the most important functions in organic materials. In general, the electronic structure of crystalline and amorphous organic solids is governed by the distributions of site and state energies that occur due to the polarization of nanoscale local environments surrounding charge carriers. Therefore, it is especially important to understand how the magnitude of polarization energy correlates with the microscopic structure of molecular aggregations.

π-Conjugated organic molecules are typically highly polarizable and interact through weak intermolecular van der Waals forces. This situation allows charge carriers (electrons or holes) to become localized on individual sites or on a small number of molecules in the solids. As is well known, such a localized charge exists as a polaron-type quasi-particle—an electron (or hole) "dressed" in the electronic polarization cloud of polarized neighboring molecules.[43–47] In other words, charge-induced dipole interactions (i.e., electronic polarization) occur instantaneously upon charge carrier formation in organic solids. To our knowledge, although no direct experimental evidence has been presented by the condensed phase study, the polarization effects are generally believed to extend over several nanometers from the charge core.[46,65]

As shown in Figure 8.2, the polarization effects give rise to so-called gas-to-solid shift, and the total polarization energies W_+ for cations (holes) and W_- for anions (electrons) are the

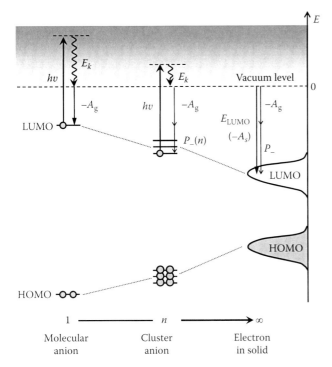

FIGURE 8.2 Schematic energy diagram of the electronic structure of an isolated molecular anion, finite cluster anion, and bulk solid. Probing of the excess-electron occupying LUMO of cluster anions allows an exploration of energetics of the LUMO level on going from an isolated molecule to the corresponding bulk solid in a stepwise manner.

quantities that determine the charge transport energy levels; the highest occupied molecular orbital (HOMO), and the lowest unoccupied molecular orbital (LUMO) in organic molecular aggregates. The total polarization energies ($W_\pm$) contain not only the electronic polarization contributions ($P_{id,\pm}$) but also charge–quadrupole interaction energies ($P_{q,\pm}$) and intra- and intermolecular relaxation energies of the nuclear coordinates, i.e., molecular polarization energies ($\lambda_{mol,\pm}$) and lattice relaxation energies ($\lambda_{lat,\pm}$), respectively. Thus, the total polarization energies are given by

$$W_\pm = P_{id,\pm} + P_{q,\pm} + \lambda_{mol,\pm} + \lambda_{lat,\pm} \qquad (8.1)$$

Lattice relaxation energies $\lambda_{lat,\pm}$ in organic crystals are generally negligibly small (~0.01 eV) because of the rigidity of the crystal lattice relative to the localized charge.[46] For instance, a lattice relaxation value of 15 meV was theoretically obtained for the localized charge in the anthracene crystal.[66] For organic crystals, therefore, the total polarization energies can be approximated by

$$W_\pm \approx P_\pm + \lambda_{mol,\pm} \qquad (8.2)$$

where the effective polarization energies $P_\pm$ represent the sum of $P_{id,\pm}$ and $P_{q,\pm}$, and are negative.

Thus far, the HOMO energy levels in organic aggregates have been investigated to the last detail using various methods, such as photoemission yield spectroscopy (PYS) and ultraviolet photoelectron spectroscopy (UPS). In particular, detailed UPS data are increasingly available from the sub-monolayer to ~10 nm organic films.[67–69] In contrast, little experimental information is available about the LUMO energy levels in π-conjugated organic aggregates, because their direct observation has been experimentally more difficult than that of the occupied levels. Recently, inverse photoemission spectroscopy (IPES) for thin films of various π-conjugated organic molecules was performed to directly observe their unoccupied electronic levels, and the energetics of the LUMO, such as solid-state electron affinity (A_s) and negative effective polarization energy (P), have been evaluated.[69,70] If we assume that the excess electron is localized on a single molecule in the solid, the solid-state electron affinity will be given by

$$A_s = A_g - P_- \qquad (8.3)$$

where A_g represents a gas-phase electron affinity and includes the contribution of intramolecular reorganization energy in the anion state (i.e., negative molecular polarization energy $\lambda_{mol,-}$). Here, the lattice relaxation energy $\lambda_{lat,-}$ is ignored.[46,66]

As mentioned above, the polarization energy associated with a given site in organic solids depends on its local environment, so that it is of particular importance to be able to account for the microscopic correlation between geometric and electronic structures of the medium surrounding a charge carrier. In this regard, ionic clusters of π-conjugated molecules are expected to serve as an ideal microscopic model for describing polarization and charge localization phenomena in crystalline and amorphous organic solids. As mentioned below, the PES of cluster anions enables us to trace the stepwise increase in the negative polarization energy P_- as a function of cluster size n (see Figure 8.2), since the mass selection of cluster anions can be readily performed during the PES measurement. Due to the higher energy resolution of anion PES (e.g., ~50 meV at 1 eV electron kinetic energy) compared to IPES (typical energy resolution: 400–500 meV),[69,70] the PES of cluster anions can provide direct information on the electron-vibrational coupling and, in some cases, charge delocalization (or charge resonance) effects at a specific molecular number and intermolecular geometry.[53,57,71]

8.3 Experimental Methodology

8.3.1 Production of Large Molecular Cluster Anions

Molecular clusters are usually produced by using supersonic jet expansion into a high vacuum through continuous nozzles or pulsed valves at ambient or elevated temperatures, which is now a very familiar method.[72,73] Therefore, we focus here only on the salient points pertaining to the production of "large" molecular clusters.

By means of a supersonic expansion of a mixture of molecules diluted in a carrier gas (typically, a noble gas), both efficient cooling and formation of molecular clusters were easily achieved in the gas phase. As is well known, there are some important factors in the production of large molecular clusters, e.g., the partial pressure of the seed molecules (P_{seed}), the total pressure of the supersonic expansion ($P_{total} = P_{carrier} + P_{seed}$, where $P_{carrier}$ represents the partial pressure of the carrier gas), the seed ratio ($R_{seed} = P_{seed}/P_{total}$), and the shape of the nozzle. To increase the total number of collisions between seed molecules during expansion, it is necessary to obtain a reasonable P_{seed} by heating both the sample reservoir and the valve. Concurrently, a higher $P_{carrier}$ is needed to sufficiently remove the condensation energy generated in the clustering process. Since neat expansions (i.e., $R_{seed} = 1$) tend to produce metastable clusters via evaporative cooling and produce a wide energy distribution, they are commonly not preferred to produce cold stable clusters. Hence, the supersonic expansion at higher P_{seed} and $P_{carrier}$ with an optimized seed ratio R_{seed} is needed to generate internally cold, large molecular clusters.

Even et al.[54] have developed a high-pressure, high-temperature, ultra-short-pulsed valve (Even–Lavie valve), and have successfully produced ultracold aromatic molecules with the rotational temperature below 1 K. In addition, they have produced He-solvated clusters using this valve.[54,74] A high operating pressure (up to 120 bar) and temperature (up to ~300°C), and ultra-short pulse (<10 μs) of the Even–Lavie valve make it possible to produce cold molecular beams under adiabatic expansion conditions with a high P_{seed} and P_{total} for a wide variety of molecular systems with small gas loading. Thus, it is very well suited to produce internally cold, large molecular clusters for a conventional vacuum chamber.

The determination of the size distribution of clusters is generally implemented by ionization combined with mass spectrometry. In the case of experiments on size-selected cluster ions, the ionization process is included in a part of production.[11,13] In the production of cluster anions, a method of electron attachment early in the supersonic expansion is widely used,[3,13] because cluster growth, reorganization, and cooling after electron attachment can be promoted by multiple collisions in the high-density zone of the expanding jet. As shown in Figure 8.3, we incorporated the

Even–Lavie valve with an electron gun in order to challenge the efficient production of large molecular cluster anions.[55] In the high-energy electron impact ionization method, the formation of cluster anions results from an attachment of slow secondary electrons (kinetic energy: <1 eV) generated by the ionization of the carrier gas. In the electron attachment process, a considerable amount of excess energy can also be imparted to the clusters, and this energy may be rapidly dissipated into the dense manifold of intermolecular, vibrationally excited states of a cluster, causing reorganization, evaporative cooling, and/or fragmentation of the cluster during the expansion. In this method, electrons were injected into the dense area of the supersonic expansion with the distance (z) between the ionization region and the nozzle orifice variable from 1 to 50 mm. The cluster anions, thus formed, were skimmed into an ion packet, extracted coaxially with a pulsed electric field and detected with an in-line Wiley–McLaren time-of-flight mass spectrometer (TOF-MS).

8.3.2 Mass Spectrometry of Oligoacene Cluster Anions

As an example of mass spectrometry, mass spectra of cluster anions measured for two linear oligoacenes, that is, Nph and Ac, that are solid at room temperature and have different melting points (Nph: 80°C and Ac: 218°C), are shown in Figure 8.4. These cluster anions were formed by expanding sample vapor seeded in 10–100 atm He-carrier gas into a vacuum chamber and subsequent electron-impact ionization. The efficient production of large cluster anions over $n = 100$ are identified in the mass spectra. In the case of Ac cluster anions, $(Ac)_n^-$, continuous formation of $(Ac)_n^-$ from $n = 1$ to ~140 was confirmed. In our TOF-MS measurement, the detectable mass was limited to ≤25,000 a.u. (i.e., $(Ac)_n^-$ with $n \leq 140$) due to the limited sensitivity of the ion detector's lower velocities of larger cluster anions. In the photoelectron measurement, however, a photoelectron signal was detected up to $n \approx 300$, suggesting the formation of $(Ac)_n^-$ clusters with a value of n of approximately 300 (ca. 55,000 a.u.).

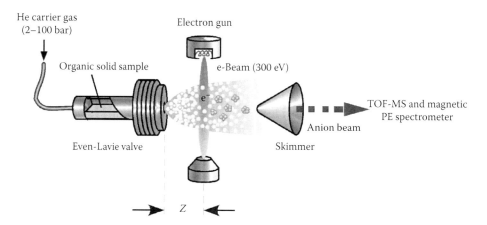

FIGURE 8.3 Schematic drawing of an anion production method: high-energy electron impact ionization where the formation of cluster anions results from an attachment of slow secondary electrons (<1 eV).

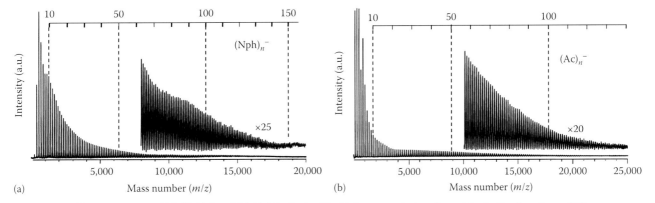

FIGURE 8.4   TOF mass spectra of (a) $(Nph)_n^-$ and (b) $(Ac)_n^-$, obtained by high-energy electron impact ionization (see Figure 8.3).

A characteristic mass distribution with intensity anomalies sometimes provides us a hint about the structural motifs of cluster ions. Anionic clusters of highly anisotropic "rod-like" π-conjugated molecules show this trend markedly.[60,64] One example is the mass spectrum observed for Tc cluster anions, $(Tc)_n^-$, displayed in Figure 8.5. This mass spectrum exhibits some striking features, involving the even–odd alternation ($3 < n < 14$); the strong magic numbers of $n = 5$, 14, and 19; and the periodical intensity oscillation of $\Delta n = +2$ or $+3$ (e.g., $n = 14$, 16, 19, 21, 24, 27, and 30). These local maxima are different from those predicted on the basis of isotropic Lennard-Jones interactions between hard spheres,[75] for example, $n = 13$, 19, 23, 26, 29, and 32. In general, the intermolecular interactions between molecules with anisotropic charge distributions induce a preferred orientation of the molecules. Considering a highly anisotropic aromatic framework of Tc, a plausible structural motif for its relatively small clusters is one with two-dimensional (2D) ordering, where all Tc molecules interact with their long molecular axes in parallel.

As also illustrated in Figure 8.5, it has been proposed that the $n = 14$ cluster adopts a geometrically closed 2D structure, corresponding to a local structure of the 2D herringbone layer found in the Tc crystal. In this structure, all the long molecular axes are parallel, and the central four molecules are surrounded by 10 other molecules; namely, this 14-mer unit constitutes the first 2D shell closing for the central tetramer unit. For the size range

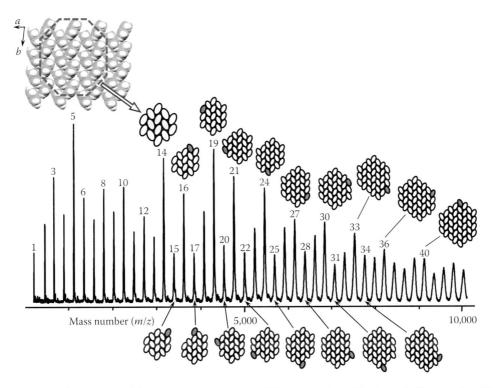

FIGURE 8.5   Mass spectrum of $(Tc)_n^-$, recorded at a stagnation pressure of 70 atm, together with schematic diagrams for a Tc crystal structure along the a–b plane and 2D herringbone ordering of $n = 14$–40, perpendicular cross section with respect to the long molecular axes of Tc. Proposed structures for local maxima and local minima of the anion distribution are displayed. See Section 8.3.2 for further details.

of $n = 15–30$, additional molecules, labeled 15–30, two-dimensionally enclose the 14-mer unit in a stepwise manner, filling the second 2D shell for the tetramer core, as shown in Figure 8.5.

In fact, this picture could perfectly explain the characteristic mass distribution of $(Tc)_n^-$. When the second 2D shell is partly filled and this yields a more or less geometrically closed structure, it is often called a subshell closing.[76] In Figure 8.5, such a condition is encountered at the numbers of $n = 16, 19, 21, 24, 27,$ and 30, where all the molecules of the second 2D shell interact with three or four neighboring molecules. All of these cluster numbers appear as local maxima in the ion distribution of $(Tc)_n^-$. On the other hand, the $n = 15, 17, 20, 22, 25,$ and 28 clusters, which appear as local minima in the mass spectrum, possess one constituent molecule that interacts with only the two nearest neighbors (molecule 15, 17, 20, 22, 25, or 28). This is an origin for the less abundant production of these clusters. For $n > 30$, the observed oscillations in mass intensity can also be interpreted in terms of subshell closings. However, their amplitudes reduce gradually with cluster size because of the population reduction of the 2D clusters at larger sizes, which has been confirmed by the anion PES.[60]

8.3.3 Anion Photoelectron Spectroscopy

8.3.3.1 Principle

Anion PES is a powerful method for a direct observation of the electronic states of molecules and their clusters.[13] It is conducted by crossing a mass-selected beam of anions with a fixed-wavelength laser light ($h\nu$) and measuring the kinetic energy (E_k) of the resultant photodetached electrons. The outstanding virtues of anion PES are its capability of allowing mass selection prior to photodetachment, which is extremely difficult in conventional PES via cation $\leftarrow$ neutral transitions, and of directly measuring the electron affinities of isolated molecules and their cluster anions.

Figure 8.6 illustrates the principle of anion PES, based on photodetachment from an anion ground state (M^-) to a neutral one (M). Like conventional PES, it is a direct approach to measure the electron-binding energy (eBE), since it relies on the energy conserving relationship

$$h\nu = eBE + E_k \tag{8.4}$$

In the case of an isolated molecule, the gas-phase electron affinity, A_g, is defined as the energy difference between the ground states of neutral and anion:

$$A_g = E(M, \nu = 0) - E(M^-, \nu' = 0) \tag{8.5}$$

Here, the energies (E) are taken as negative, so that the A_g is positive. The structures of the lowest-energy states of neutral and anion are so different that the corresponding vibrational wave functions $|\nu\rangle$ and $|\nu'\rangle$ do not have an overlap, i.e., $\langle \nu | \nu' \rangle = 0$. This overlap controls the strength of an optical transition; its square

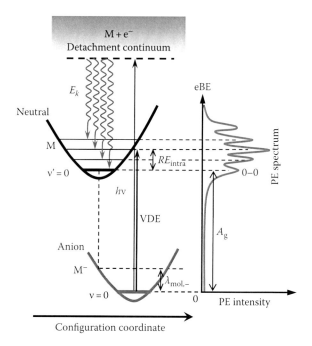

FIGURE 8.6 Scheme of the photodetachment transition in anion PES. On the right side, the corresponding photoelectron spectra are displayed. Notice that the 0–0 transition represents A_g. See text for further details.

is called the Franck–Condon factor. If the Franck–Condon factor is zero between the two states, the A_g cannot be optically measured. However, for higher vibrational states ($\nu > 0$), the Franck–Condon factor can become finite. The Franck–Condon factor increases, reaches a maximum and decreases afterwards. This is directly reflected in the E_k of the detached photoelectrons. The vertical detachment energy (VDE) corresponds to the most probable part of this transition. In other words, the VDE corresponds to the energy difference between the anion and the neutral at the equilibrium geometry of the anion, as shown in Figure 8.6. The energy difference given by VDE $- A_g$ is often called intramolecular nuclear reorganization energy in the neutral state (denoted as RE_{intra} below). As already noted above, the intramolecular reorganization energy in the anion state ($\lambda_{mol,-}$) is involved in A_g. For an isolated molecule, thus, the VDE is expressed by

$$VDE = A_g + RE_{intra} \tag{8.6}$$

On the other hand, the VDE at a given cluster size n [VDE(n)] contains additional contributions coming from the electronic polarization energy [$P_{id,-}(n)$], the charge–quadrupole interaction energy [$P_{q,-}(n)$], and the intermolecular reorganization energy [$RE_{inter}(n)$] in the neutral state. Thus, VDE(n) is defined as

$$VDE(n) = A_g - P_-(n) + RE(n) \tag{8.7}$$

where the cluster polarization energy $P_-(n)$ represents the sum of $P_{id,-}(n)$ and $P_{q,-}(n)$, and the cluster reorganization energy $RE(n)$ is the sum of RE_{intra} and $RE_{inter}(n)$. When the equilibrium geometry in the anion state is far from that of the neutral, the cluster reorganization energy $RE(n)$ can make a considerable contribution to the VDE.

8.3.3.2 Magnetic Bottle Photoelectron Spectrometer

A magnetic bottle PE spectrometer[77,78] is often used in the anion PES, because of an extremely high photoelectron (PE) collection efficiency (near 100%). Although the energy resolution of the magnetic bottle PE spectrometer depends on some instrumental parameters, e.g., the length of the drift region, the energy resolution is limited by a Doppler effect of the velocity of the anion beam that causes the energy of the detached electron to be spread,[77] because photoelectrons with the same speed in the center-of-mass frame of reference will have different speeds in the laboratory frame. In our anion PES measurements, the cluster anions with <4,000 u accelerated by a pulsed electric field of −1.5 keV were decelerated to minimize the Doppler effect. The resultant energy resolution was about 50 meV for 1 eV photodetached electrons. On the other hand, the larger cluster anions with ≥4,000 u have relatively lower velocities, irrespective of the high acceleration voltage (E_I) of ~4 keV, and the Doppler broadening[77] has been, for example, estimated to be ca. 90 meV at $E_k = 1$ eV for 4,000 u or ca. 40 meV at $E_k = 1$ eV for 25,000 u (at $E_I = 4$ keV). These values are similar to our instrumental resolution (ca. 50 meV at $E_k = 1$ eV) attained by ion deceleration. Indeed, the PE spectral feature for

the large mass anions (≥4,000 u) was not changed by varying the acceleration voltage.

8.3.4 Anion Beam Hole-Burning Technique

Even with size selection, there is often more than one species (i.e., isomers) contributing to the signal at a given size, requiring additional dissection of the observed PE spectra. In most cases, isomers have different electron-binding energies from one another, so that they are often observed as plural distinct bands in the PE spectrum. In this case, one can employ an anion beam hole-burning technique so as to verify whether the observed bands originate from different anionic isomers or not. The first application of this technique was reported by Tsukuda et al.[79] for $(CO_2)_6^-$, showing that the $(CO_2)_6^-$ cluster fluctuates between the two isomeric forms of $CO_2^- \text{-} (CO_2)_5$ and $C_2O_4^- \text{-} (CO_2)_4$. More recently, Akin and Jarrold[80] have applied this method for $Al_3O_3^-$, where two structural isomers: a kite-shaped isomer coexists with a bent rectangle isomer of a higher electron affinity.

The experimental scheme for anion beam hole-burning is schematically depicted in Figure 8.7. As shown in the inset of Figure 8.7, if eBE (or VDE) of isomer A is much lower than that of isomer B, and if no transformation between two isomers takes place on the timescale of the measurement, the lower VDE "isomer A" can be selectively photodetached by using a first laser (hereafter denoted as the HB laser, hv_1) and bleached from the

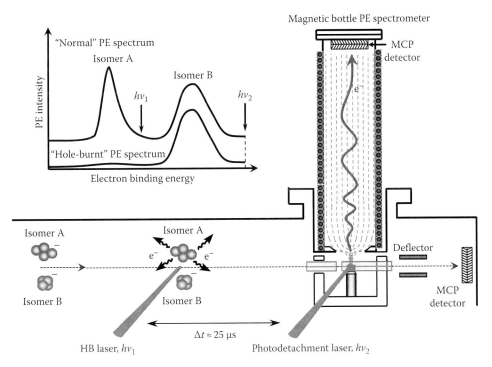

FIGURE 8.7 Experimental setup of anion beam hole-burning PES. As exemplified in the inset, if *eBE* of isomer A is much lower than that of isomer B and if no transformation between two isomers takes place on the timescale of the measurement, the lower eBE "isomer A" can be selectively photodetached by using the HB laser (hv_1). After the photodetachment by the HB laser, the remaining anions enter the magnetic bottle PE spectrometer, where the "Hole-burnt" photoelectron spectrum is measured by intersecting the photodetachment laser (hv_2). The two laser pulses were temporally separated from each other by ~25 μs.

ion beam just prior to the photoelectron spectroscopy interaction region. The photodetachment of a fraction of the isomer A in the ion beam by the HB laser pulse is ensured by monitoring the resultant fast-neutral clusters. After the photodetachment by the HB laser, the remaining anions and fast-neutral clusters enter the magnetic bottle PE spectrometer, where the PE spectrum is measured by intersecting the second photodetachment laser pulse ($h\nu_2$).

8.4 Anion Photoelectron Spectroscopy of Oligoacene Cluster Anions

Among the class of organic solids, linear oligoacenes (e.g., Nph, Ac, and Tc) have in particular experienced thorough experimental and theoretical investigations since the 1960s, and are a prototypical model system for the study of charge-localization and transport.[43–47] Recent recognition of their promising potential as organic semiconductor materials has further generated a renewed interest in their electronic structure relevant to charge-transport, i.e., the HOMO and LUMO levels. In this section, we would like to show that oligoacene cluster anions can be indeed useful models to help understand the electronic structure and polarization phenomena of related solid-state systems.

8.4.1 Coexistence of Isomers

Recent systematic anion PES studies[57–60] of oligoacene cluster anions ($n = 1$–100) revealed the coexistence of disordered and crystal-like anionic clusters. Figure 8.8 shows the PE spectra for the $(Nph)_n^-$, $(Ac)_n^-$, and $(Tc)_n^-$ clusters ($n = 1$–100). The two distinct bands, labeled "I" and "II-1" ($10 \leq n < 50$) or "I" and "II-2" ($n \geq 60$) in the PE spectra, are identified in a wide-size range.

For $(Nph)_n^-$, the anion beam HB experiments were performed to show that the bands labeled I and II-1/II-2 originate from different anionic isomers.[58,59] In the PE spectra for $(Nph)_n^-$, the VDEs of bands II-1 and II-2 are energetically separated so distinctly from band I that the HB laser light of 1064 nm (1.165 eV) could selectively photodetach electrons from just the isomer with the lower VDE (band II-1 or II-2) without detaching electrons from cluster anions yielding band I. After the photodetachment by the HB laser, the surviving cluster anions entered the PE spectrometer, where the PE spectrum was measured at 532 nm (2.331 eV). Figure 8.9 shows a typical PE spectra for the $(Nph)_{30}^-$ and $(Nph)_{100}^-$ clusters taken with the HB laser off (open circles; ●) and on (filled circles; ○). As shown in Figure 8.9, a clear depletion of bands II-1 and II-2 is obtained with the HB laser (~50 mJ/pulse) on confirming that bands I and II-1/II-2 originate from different anionic isomers. Also, these results suggest that

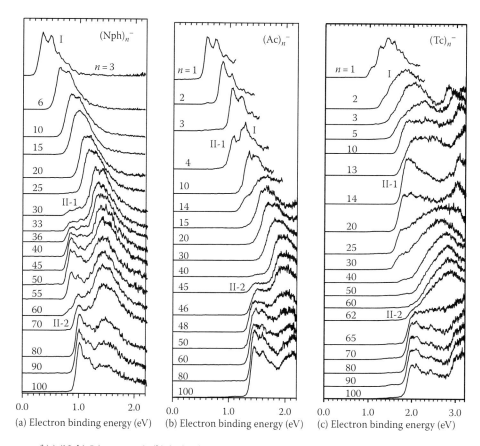

FIGURE 8.8 PE spectra of (a) $(Nph)_n^-$ ($n = 3$–100), (b) $(Ac)_n^-$ ($n = 1$–100), and (c) $(Tc)_n^-$ ($n = 1$–100) measured at 532 nm (2.331 eV). Note that all the photoelectron spectra were measured under a cold ion source condition which was optimized to maximize the intensity of band II-1/II-2.

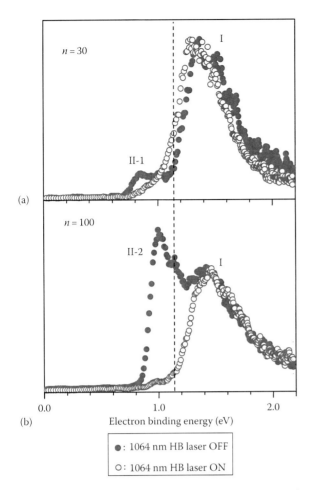

FIGURE 8.9 PE spectrum of $(Nph)_{30}^-$ and $(Nph)_{100}^-$ measured with (○) and without (●) 1064 nm (1.165 eV) photodetachment prior to the photoelectron measurements at 532 nm. The arrow indicates the maximum electron binding energy accessible with a 1064 nm photodetachment.

there is an activation energy barrier between isomers I and II-1/II-2 prohibiting isomerization on the timescale of the measurement (~25 μs).

8.4.2 Characteristics of Isomers I, II-1, and II-2

Since the natures of each isomer are not immediately apparent, we focus on this issue below. Noticeable characteristics of the isomers I, II-1, and II-2, which are common to the $(Nph)_n^-$, $(Ac)_n^-$, and $(Tc)_n^-$ clusters ($n = 1–100$),[57–60] are summarized as follows:

1. The VDEs of isomer I consistently increase over the entire size range examined.
2. The VDEs of isomers II-1 and II-2 remain constant and are 0.4–0.6 eV lower than those of isomer I.
3. The isomer II-2 is rather favored in a stronger supersonic expansion that can produce a colder anion beam, while warmer ion source conditions preferably form only isomer I. (Note that all the PE spectra presented in Figure 8.8 were measured under colder beam conditions.)

4. The PE spectra of isomer II-2 exhibit a relatively sharp profile showing a discernible intramolecular vibrational structure, while those of isomer I are broader and less structured.
5. Isomers II-1 and II-2 could be classified into a sub-group within the common isomer II, because the formation of isomers II-1 and II-2 is sensitively correlated to the geometrical perturbation of the molecular shape and the substituted position of the alkyl group in the aromatic ring. In contrast, isomer I is always formed in every organic cluster anion examined so far (over 20 PAHs).

As a clear illustration of (1) and (2), –VDE values for $(Ac)_n^-$ are plotted against $n^{-1/3}$ in Figure 8.10, and the VDE data of the clusters are compared with the corresponding bulk data.[57] Together with the VDE data, the energies of the LUMO level with respect to the vacuum level (E_{LUMO}) are also shown in the figure—this quantity was determined by IPES for a highly ordered Nph, Ac, and Tc thin film solid on an Ag(111) surface.[81]

The solid-state electron affinity A_s is given by Equation 8.3 so that

$$E_{LUMO} = -A_s = -A_g + P_- \qquad (8.8)$$

A linear variation of P_- against $n^{-1/3}$ in the Ac crystal has been theoretically predicted using the cluster approach in the limit

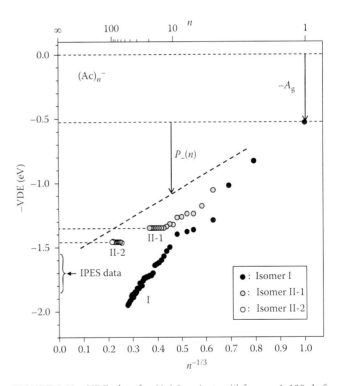

FIGURE 8.10 VDE plots for $(Ac)_n^-$ against $n^{-1/3}$ for $n = 1–100$. Left arrow shows the energy of the LUMO level with respect to the vacuum level (E_{LUMO}) of –1.7 ± 0.15 eV, determined by IPES for a highly ordered Ac thin film on Ag(111).[81] The dash line is obtained by adding the $-A_g$ value of an Ac molecule to the calculated polarization energy (P_-) for a monomer anion in the Ac single crystal reported in Ref. [81].

of no intermolecular overlap, where a spherical cluster is centered on the unit cell that contains an Ac molecular anion.[82] This dependence indicates that the charge-induced dipole interaction in the crystal is considerably long-ranged. The size-dependence of E_{LUMO} on $n^{-1/3}$ can be obtained by adding the $-A_g$ value of the Ac molecule to the size-dependent polarization energy (P_-),[83] and it is also shown in Figure 8.10 as a dash line.

Based on the characteristics of (1)–(5), the assignments of isomers I, II-1, and II-2 are conducted in the following subsections.

8.4.2.1 Isomer I: Disordered Cluster Anions

The $-VDE$ values of isomer I increase almost linearly with $n^{-1/3}$, which is similar to the linear dependence of E_{LUMO} on $n^{-1/3}$ in Figure 8.10. However, they are much lower than the predicted E_{LUMO} values (the corresponding dash-line in Figure 8.10). In addition, the slope of the VDE plot is steeper than that of the E_{LUMO} plot, resulting in larger energy differences between the VDE and E_{LUMO} with increasing cluster size. These discrepancies indicate that, in isomer I, the cluster reorganization energy $RE(n)$ makes a considerable contribution to the VDE; namely, isomer I is ascribed to the "disordered" cluster anion in which the surrounding neutral molecules are fully reorganized around the monomeric anion core.[57]

In the production of cluster anions, electrons were injected into the dense area of the supersonic expansion, and after the electron attachment, the growth, the reorganization, and the cooling of cluster anions are efficiently promoted through multiple collisions of molecules and the He carrier gas. For the formation of disordered cluster anions, therefore, two plausible processes can be considered as follows. One is the successive condensation of neutral molecules to a monomer anion, and the other is the electron attachment to pre-formed neutral clusters followed by structural reorganization, evaporative cooling, and/or fragmentation of the cluster. It is expected that the former process mainly occurs in smaller size clusters, whereas the latter process becomes dominant in the larger sizes. Consistently, for $n > 40$, disordered cluster anions are preferentially produced at lower stagnation pressures. In other words, electron attachment to warmer (or liquid-like) neutral clusters allows substantial intermolecular rearrangements of neutral molecules that encompass the monomer anion three-dimensionally. Since oligoacene molecules do not have a permanent dipole moment but do have a relatively large quadrupole moment,[84] the unexpectedly large $RE(n)$ probably originates from charge–quadrupole interactions.

8.4.2.2 Isomer II-1: 2D Crystal-Like Cluster Anions

As clearly seen in isomer I, polarization energy in molecular clusters increases correspondingly with the cluster size. For $(Ac)_n^-$ and $(Tc)_n^-$, however, the VDEs of isomer II-1 become constant independently for the cluster size above $n = 14$. The constant VDEs signify that the stepwise stabilization energy added per molecule in the anion is the same as that in the neutral. The size dependence is surprising because the ion-neutral electrostatic interactions are generally much stronger than the

neutral-neutral interactions. Since factors such as the degree of structural order and the spatial distribution[85] and location[86,87] of the excess charge within a cluster can influence the polarization energies, one can infer the following four origins for the unvarying nature of the VDE values:

1. "Efficient charge screening" in isomer II-1.
2. "Structural rigidity" of the clusters with $n \geq 14$, because the magnitude of VDE is dominated not only by the energetics, but also by structural factors, i.e., $RE(n)$ [see Equation 8.7], meaning that the amount of structural relaxation incurred in the change from the equilibrium geometry of the anion to that of the neutral no longer depends on the cluster size above $n = 14$.
3. "Charge delocalization" in isomer II-1, because a larger spatial distribution of an excess charge was shown to give a lower stabilization energy from polarization per neutral molecule.[84]
4. "Surface state," where an excess electron is localized on a molecule residing at the cluster surface rather than on an interior molecule in the cluster (i.e., internal state). In the surface state, the $RE(n)$ has been theoretically shown to be much smaller than in the internal state.[86]

In the case of $(Tc)_n^-$, since isomer II-1 is prominently observed under the experimental condition of Figure 8.5, a structural motif for isomer II-1 is the 2D crystal-like herringbone-type ordering, as described in Section 8.3.2, in detail. The establishment of the densely packed, geometrically shell-closed 2D structure at $n = 14$, where 10 molecules completely surround and efficiently screen an excess charge delocalized on a central tetramer (see Figure 8.5), can simultaneously explain the aforementioned origins of (1)–(3), while it cannot do the same for (4).

At first glance, the fourth origin of (4) also seemingly explains the observation that the VDEs of isomers II-1 are 0.4–0.6 eV lower than those of isomer I that must be an internal state, as mentioned above. However, the alkyl-substitution effects on the formation of isomer II-1 casts strong doubt upon this interpretation.[57] The alkyl-substitutions produce steric hindrance effects on the cluster geometry, while it seems evident that the electron affinity and the distribution of the excess electron are barely influenced by the substitutions. By the alkyl-substitutions for anthracene molecules it has been demonstrated that isomer II-1 could not be produced in cluster anions ($n = 2$–100) of 1-methylanthracene (1-MA), 9-methylanthracene (9-MA), 9,10-dimethylanthracene (DMA), and 2-*tert*-butylanthracene (2-TBA) under any experimental conditions, but could be produced only in 2-methylanthracene (2-MA) cluster anions.[57] The alkyl-substitution positions in Ac is shown in Figure 8.1b. These Ac methyl-derivatives possess similar positive A_g values confined to the range from 0.47 to 0.59 eV, so that if isomer II-1 is the surface state, it should be formed in all the anionic cluster systems. Thus, the alkyl-substitution effects on the formation of isomer II-1 strongly support the 2D structural motif of isomer II-1.

When the H atom in the C1 or C9 position of Ac (see Figure 8.1) is substituted by a methyl-group, it is conceivable that the effective π–π interactions are substantially hindered and reduced within the 2D herringbone layer, giving rise to the instability of the 2D structure. The similar behavior between $(2\text{-MA})_n^-$ and $(\text{Ac})_n^-$ is likely because the methyl-group in the C2 position can be located outside the 2D layer and does not greatly retard the π–π interactions within the 2D layer. In the case of 2-TBA, isomer II-1 did not appear because of the higher steric hindrance of a *tert*-butyl group, even at the C2 position, that would strongly disturb the 2D structural motif. Hence, isomer II-1 is assignable to a finite 2D herringbone-type aggregate of the type shown in Figure 8.5.

8.4.2.3 Isomer II-2: Multilayered Crystal-Like Cluster Anions

Like isomer II-1 ($n \geq \sim 14$), the −VDE value of isomer II-2 does not increase at all over the entire size range and remains constant at −0.99 eV for Nph,[59] −1.47 eV for Ac,[59] and −2.0 eV for Tc.[60] Furthermore, the relatively sharp PE profile of isomer II-2, exhibiting a clear intramolecular C–C stretching vibrational structure like that of the monomer anion; for Ac clusters at larger sizes of $n = 50$–100, the ν_6 vibrational progression due to a neutral Ac molecule(s) is clearly observed in band II, and no shift of band II is discernible. This behavior indicates that isomer II-2 can be confidently assigned to the anion formed from pre-formed neutral clusters without (or with much less) intermolecular geometrical relaxation (i.e., $\text{RE}(n) \approx 0$). Like isomer II-1, these facts signify that isomer II-2 should also have some kind of rigid stable structural motif with a common anion core throughout the entire size range.

Since isomer II-2 appears only under colder beam conditions (≥ 30 atm), it is presumed that the formation of isomer II-2 results from electron attachment to internally cold crystal-like neutral clusters. This idea is compatible with the fact that the lattice relaxation energy of the oligoacene crystals is quite small, only about ~0.01 eV.[46,66] Furthermore, we find that the −VDE values of isomer II-2 are comparable to the corresponding E_{LUMO} values of a highly ordered oligoacene thin-film solid on Ag(111),[81] as shown in Figure 8.10. These literature values of E_{LUMO} may be about −0.2 eV larger than the intrinsic bulk value of E_{LUMO} because of the effect of image charges in the metal substrate.[88] Therefore, it is reasonable to assert that the constant −VDE values of isomer II-2 have already converged to the bulk values of E_{LUMO} and isomer II-2 can be viewed as a cluster analogue of polaron-type quasi-particles in crystals. The excellent agreement between the −VDE values of isomer II-2 and the E_{LUMO} of a oligoacene thin-film solid is a great surprise because the electronic polarization in crystal has been theoretically shown to extend over more than 10^3 molecules around the charge core.[46,65] Hence, the agreement clearly demonstrates the definite localized nature of the electronic polarization in the crystal-like cluster (isomer II-2), where only about 40–60 molecules provide a charge-stabilization energy comparable to that of the bulk.[57–60]

The invariable nature of the VDE, as well as the correlated appearance of isomers II-1 and II-2, implies that these isomers have a similar local geometrical environment. Thus, isomer II-2 likely possesses an identical anion core to isomer II-1 for $n \geq 14$. However, the constant VDE of isomer II-2 is higher by about 0.1–0.2 eV than that of isomer II-1, showing the presence of some additional charge-stabilization effects in isomer II-2. Isomer II-2 appears only in much larger clusters than isomer II-1, which have a single 2D layer. It is inferred that the isomer II-2 clusters possess a finite multilayered herringbone packing structure that is analogous to the local structure of oligoacene crystal. The additional small stabilization would thus arise from the electronic polarization of neutral molecules located in the adjacent layer(s).

Notwithstanding the fact that isomer II-2 appears on the ~0.5 eV lower-binding energy side of isomer I, colder beam conditions prefer to produce isomer II-2. This behavior points out that the crystal-like anion (isomer II-2) is energetically (or adiabatically) more stable than the coexisting disordered one (isomer I), though the energy difference between the adiabatic levels of isomers I and II-2 is predicted to be very small. As is well-known, the magnitude of the polarization energy is intimately related to the morphology of molecular aggregation and hence to the molecular packing density (p).[89] In an ultraviolet photoelectron spectroscopy (UPS) study of pentacene solid film, for example, Sato et al.[89] have observed an increase in the value of the positive polarization energy (P_+) of 0.3 eV on going from the amorphous to the crystalline state, due to an increase in p. Hence, it is expected that the magnitude of P_- for the crystal-like state (isomer II-2) is somewhat larger than that for the disordered state (isomer I). Interestingly, isomer II-2 does not appear in $(1\text{MA})_n^-$, $(9\text{MA})_n^-$, $(\text{DMA})_n^-$, and $(2\text{TBA})_n^-$, though the overall VDE energetics of isomer I in these cluster anions are similar to that of isomer I in $(\text{Ac})_n^-$. These facts suggest that in the case of $(\text{Ac})_n^-$ and $(2\text{MA})_n^-$, there are some additional charge stabilization effects in the crystal-like state (isomer II-2), which are absent in the other alkyl-substituted cluster anions. As discussed above, when the formation of the multimer anion core occurs in isomer II-2 (also isomer II-1), the charge resonance stabilization could contribute to the lowering of the adiabatic energy level of isomer II-2 (as well as isomer II-1).

In contrast, the substitution of a methyl group in an Ac molecule should mainly change the effective intermolecular distances and orientations. The resultant decline of the effective intermolecular π-orbital overlaps should decrease polarization (P_-) and charge resonance stabilization in isomer II-2.[85,89] Consequently, the crystal-like anion appears only in the π-conjugated molecular systems that can naturally self-assemble into a crystal-like (herringbone-type) aggregation structure.

Finally, let us discuss the correlation of isomers II-1 and II-2 of the oligoacene group. The appearance threshold sizes of isomers II-1 and II-2 for oligoacene cluster anions are different from each other.[57–60] For benzene clusters, no isomers II-1 and II-2 have been observed up to $n \sim 120$. The appearance threshold size ($n \geq 28$) of isomer II-1 of Nph cluster anions is much larger than that ($n \geq 4$) of the Ac and the Tc ones. This difference n in the threshold size means that more Nph molecules are required to form the ordered crystal-like structure than Ac and Tc. Although the threshold sizes of isomer II-1 for Ac and Tc are almost the same, it is experimentally much easier to produce isomer II-1 for Tc compared to that for Ac. For Ac cluster anions, the intensity ratio between isomers I and II-1 is changed sensitively by the cluster source condition. For Tc cluster anions, in contrast, isomer II-1 is usually more preferentially formed in comparison to that of Ac cluster anions leading to the characteristic mass distribution with prominent magic numbers and periodical intensity oscillations, as already described in Section 8.3.2.[60] The trend mentioned above is consistent with the fact that the strength of anisotropic π–π interaction increases as the number of benzene rings in oligoacenes increases and the melting point of the crystal increases remarkably from benzene to tetracene. Overall, it appears that the formation of crystal-like anion states is governed by the strength of anisotropic π–π interaction between constituent molecules.

8.5 Summary and Future Perspective

In this chapter, we described the production of nanoscale cluster anions of π-conjugated organic molecules and systematic anion PES studies on the size effects for the correlation between their structural morphologies and their electronic structures. The two distinctive anionic isomers, i.e., disordered and crystal-like cluster anions, were shown to coexist over a broad size range of the cluster anions of oligoacenes (Nph, Ac, and Tc). Especially, the energetic of crystal-like isomers provided a direct evidence for a highly localized nature of polarization in a cluster analogue of π-conjugated organic crystals. Also, the propositions for the structural motifs, the spatial distribution and the location of the excess charge within a cluster were given for the coexisting isomers, while they are difficult parameters to be characterized experimentally. For verification of these propositions, a synergy between experiment and theory is necessary in the future.

The cluster temperature could not be quantitatively characterized; the varieties in the cluster source conditions (in particular, stagnation pressures) drastically influence the relative abundance of the disordered and crystal-like isomers. This behavior implies that liquid-like (nonrigid) and solid-like (rigid) neutral clusters temporarily coexist in a supersonic beam under nonequilibrium conditions. In larger clusters, a vast number of the internal quantum states, e.g., intermolecular vibrational modes, should be populated at a finite temperature, so that the concept of free energy becomes more important.[90] In the large size region, therefore, the vibrational entropy contribution to the free energy may dominate the relative abundance of isoenergetic isomers at a given size. Apparently, a key step to future progress necessitates realizing the reliable control and measurement of cluster temperature.

Acknowledgments

We are grateful to Professor Uzi Even and Nachum Lavie for a newly designed pulsed valve. We are also very indebted to people from the Nakajima group for conducting the experiments: Shinsuke Kokubo, Dr. Naoto Ando, and Yukino Matsumoto (Keio University). This work is supported partly by CREST (Core Research for Evolutional Science and Technology) of Japan Science and Technology Agency (JST), and partly by Grant-in-Aids from the Ministry of Education, Culture, Sports, Science, and Technology of Japan (MEXT) on Priority Area "Molecular Theory for Real Systems," No. 19029041 and No. 20038043, by those for Young Scientist (B), No. 14740332 and No. 17750016.

References

1. Bartell, L. S. 1986. Diffraction studies of clusters generated in supersonic flow. *Chem. Rev.* 86: 491–505.
2. Celii, F. G. and Janda, K. C. 1986. Vibrational spectroscopy, photochemistry, and photophysics of molecular clusters. *Chem. Rev.* 86: 507–520.
3. Castleman, A.W. Jr. and Keesee, R. G. 1986. Ionic clusters. *Chem. Rev.* 86: 589–618.
4. Nesbitt, D. J. 1988. High-resolution infrared spectroscopy of weakly bound molecular complexes. *Chem. Rev.* 88: 843–870.
5. Leutwyler, S. and Bösiger, J. 1990. Rare-gas solvent clusters: Spectra, structures, and order-disorder transitions. *Chem. Rev.* 90: 489–507.
6. Jortner, J. 1992. Cluster size effects. *Z. Phys. D: At. Mol. Clusters* 24: 247–275.
7. Hobza, P., Selzle, H. L., and Schlag, E. W. 1994. Structure and properties of benzene-containing molecular clusters: Nonempirical ab Initio calculations and experiments. *Chem. Rev.* 94: 1767–1785.
8. Felker, P. M., Maxton, P. M., and Schaeffer, M. W. 1994. Nonlinear Raman studies of weakly bound complexes and clusters in molecular beams. *Chem. Rev.* 94: 1787–1805.
9. Neusser, H. J. and Krause, H. 1994. Binding energy and structure of van der Waals complexes of benzene. *Chem. Rev.* 94: 1829–1843.
10. Dethlefs, K. M., Dopfer, O., and Wright, T. G. 1994. ZEKE spectroscopy of complexes and clusters. *Chem. Rev.* 94: 1845–1871.
11. Castleman, A. W. Jr. and Bowen, Jr. K. H. 1996. Clusters: Structure, energetics, and dynamics of intermediate states of matter. *J. Phys. Chem.* 100: 12911–12944.
12. Sun, S. and Bernstein, E. R. 1996. Aromatic van der Waals clusters: Structure and nonrigidity. *J. Phys. Chem.* 100: 13348–13366.
13. Boesl, U. and Knott, W. J. 1998. Negative ions, mass selection, and photoelectrons. *Mass Spectrom. Rev.* 17: 275–305.
14. Buck, U. and Huisken, F. 2000. Infrared spectroscopy of size-selected water and methanol clusters. *Chem. Rev.* 100: 3863–3890.
15. Brutschy, B. 2000. The structure of microsolvated benzene derivatives and the role of aromatic substituents. *Chem. Rev.* 100: 3891–3920.

16. Neusser, H. J. and Siglow, K. 2000. High-resolution ultraviolet spectroscopy of neutral and ionic clusters: Hydrogen bonding and the external heavy atom effect. *Chem. Rev.* 100: 3921–3942.

17. Desfrançois, C., Carles, S., and Schermann, J. P. 2000. Weakly bound clusters of biological interest. *Chem. Rev.* 100: 3943–3962.

18. Caroline, E. H. D. and Dethlefs, K. M. 2000. Hydrogen-bonding and van der Waals complexes studied by ZEKE and REMPI spectroscopy. *Chem. Rev.* 100: 3999–4022.

19. Lardeux, C. D., Grégoire, G., Jouvet, C., Martrenchard, S., and Solgadi, D. 2000. Charge separation in molecular clusters: Dissolution of a salt in a salt-(solvent)$_n$ cluster. *Chem. Rev.* 100: 4023–4037.

20. Kim, K. S., Tarakeshwar, P., and Lee, J. Y. 2000. Molecular clusters of π-systems: Theoretical studies of structures, spectra, and origin of interaction energies. *Chem. Rev.* 100: 4145–4186.

21. Zwier, T. S. 2001. Laser spectroscopy of jet-cooled biomolecules and their water-containing clusters: Water bridges and molecular conformation. *J. Phys. Chem. A* 105: 8827–8839.

22. Sinnokrot, M. O. and Sherrill, C. D. 2006. High-accuracy quantum mechanical studies of π−π interactions in benzene dimers. *J. Phys. Chem. A* 110: 10656–10668.

23. Verlet J. R. R. 2008. Femtosecond spectroscopy of cluster anions: Insights into condensed-phase phenomena from the gas-phase. *Chem. Soc. Rev.* 37: 505–517.

24. Hahn, M. Y., Paguia, A. J., and Whetten, R. L. 1987. The ultraviolet spectra of size-resolved (benzene)$_n$, n = 2–60. *J. Chem. Phys.* 87: 6764–6765.

25. Easter, D. C., El-Shall, M. S., Hahn, M. Y., and Whetten, R. L. 1989. Spectroscopic manifestations of structural shell filling in (benzene)$_n$ clusters, N = 1–20. *Chem. Phys. Lett.* 157: 277–282.

26. Ayotto, P. and Johnson, M. A. 1997. Electronic absorption spectra of size-selected hydrated electron clusters: $(H_2O)_n^-$, n = 6–50. *J. Chem. Phys.* 106: 811–814.

27. Coe, J. V., Lee, G. H., Eaton, J. G., Arnold, S. T., Sarkas, H. W., and Bowen, K. H. et al. 1990. Photoelectron spectroscopy of hydrated electron cluster anions, $(H_2O)_{n=2-69}^-$. *J. Chem. Phys.* 92: 3980–3982.

28. Lee, G. H., Arnold, S. T., Eaton, J. G., Sarkas, H. W., Bowen, K. H., and Ludewigt, C. et al. 1991. Negative-ion photo-electron-spectroscopy of solvated electron cluster anions, $(H_2O)_n^-$ and $(NH_3)_n^-$. *Z. Phys. D: At. Mol. Clusters* 20: 9–12.

29. Sarkas, H. W., Arnold, S. T., Eaton, J. G., Lee, G. H., and Bowen, K. H. 2002. Ammonia cluster anions and their relationship to ammoniated (solvated) electrons: The photoelectron spectra of $(NH_3)_{n=41-1100}^-$. *J. Chem. Phys.* 116: 5731–5737.

30. Markovich, G., Giniger, R., Levin, M., and Cheshnovsky, O. 1991. Photoelectron spectroscopy of iodine anion solvated in water clusters. *J. Chem. Phys.* 95: 9416–9419.

31. Markovich, G., Pollack, S., Giniger, R., and Cheshnovsky, O. 1994. Photoelectron spectroscopy of Cl⁻, Br⁻, and I⁻ solvated in water clusters. *J. Chem. Phys.* 101: 9344–9353.

32. Markovich, G., Perera, L., Berkowitz, M. L., and Cheshnovsky, O. 1996. The solvation of Cl⁻, Br⁻, and I⁻ in acetonitrile clusters: Photoelectron spectroscopy and molecular dynamics simulations. *J. Chem. Phys.* 105: 2675–2685.

33. Wang, X. B., Yang, X., Nicholas, J. B., and Wang, L. S. 2001. Bulk-like features in the photoemission spectra of hydrated doubly charged anion clusters. *Science* 294: 1322–1325.

34. Headrick, J. M., Diken, E. G., Walters, R. S., Hammer, N. I., Christie, R. A., Cui, J. et al. 2005. Spectral signatures of hydrated proton vibrations in water clusters. *Science* 308: 1765–1769.

35. Miyazaki, M., Fujii, A., Ebata, T., and Mikami, N. 2004. Infrared spectroscopic evidence for protonated water clusters forming nanoscale cages. *Science* 304: 1134–1137.

36. Shin, J.-W., Hammer, N. I., Diken, E. G., Johnson, M. A., Walters, R. S., Jaeger, T. D. et al. 2004. Infrared signature of structures associated with the $H^+(H_2O)_n$ (n = 6 to 27) clusters. *Science* 304: 1137–1140.

37. Bragg, A. E., Verlet, J. R. R., Kammrath, A., Cheshnovsky, O., and Neumark, D. M. 2004. Hydrated electron dynamics: From clusters to bulk. *Science* 306: 669–671.

38. Paik, D. H., Lee, I.-R., Yang, D.-S., Baskin, J. S., and Zewail, A. H. 2004. Electrons in finite-sized water cavities: Hydration dynamics observed in real time. *Science* 306: 672–675.

39. Verlet, J. R. R. Bragg, A. E. Kammrath, A. Cheshnovsky, O. Neumark, D. M. 2005. Observation of large water-cluster anions with surface-bound excess electrons. *Science* 307: 93–96.

40. Weber, J. M., Kim, J., Woronowicz, E. A., Weddle, G. H., Becker, I., Cheshnovsky, O. et al. 2001. Observation of resonant two-photon photodetachment of water cluster anions via femtosecond photoelectron spectroscopy. *Chem. Phys. Lett.* 339: 337–342.

41. Kammrath, A., Griffin, G. B., Verlet, J. R. R., Young, R. M., and Neumark, D. M. 2007. Time-resolved photoelectron imaging of large anionic methanol clusters: (Methanol)$_n^-$ (n ~ 145–535). *J. Chem. Phys.* 126: 244306-1–6.

42. Steinbach, C., Andersson, P., Kazimirski, J. K., Buck, U., Buch, V., and Beu, T. A. 2004. Infrared predissociation spectroscopy of large water clusters: A unique probe of cluster surfaces. *J. Phys. Chem. A* 108: 6165–6174.

43. Gutmann, F. and Lyons, L. E. 1967. *Organic Semiconductors*. New York: Wiley.

44. Pope, M. and Swenberg, C. E. 1982. *Electronic Processes in Organic Crystals*. London, U.K.: Oxford Univ. Press.

45. Silinsh, E. A. 1980. *Organic Molecular Crystals: Their Electronic States*. Berlin, Germany: Springer.

46. Silinsh, E. A. and Čápek, V. 1994. *Organic Molecular Crystals: Interaction, Localization, and Transport Phenomena*. New York: American Institute of Physics.

47. Wright, J. D. 1987. *Molecular Crystals*. New York: Cambridge University Press.

48. Beitz, T., Laudien, R., Löhmannsröben, H.-G., and Kallies, B. 2006. Ion mobility spectrometric investigation of aromatic cations in the gas phase. *J. Phys. Chem. A* 110: 3514–3520.

49. Benharash, P., Gleason, M. J., and Felker, P. M. 1999. Rotational coherence spectroscopy and structure of naphthalene trimer. *J. Phys. Chem. A* 103: 1442–1446.

50. Piuzzi, F., Dimicoli, I., Mons, M., Millie, P., Brenner, V., Zhao, Q. et al. 2002. Spectroscopy, dynamics and structures of jet formed anthracene clusters. *Chem. Phys.* 275: 123–147.

51. Saigusa, H. and Lim, E. C. 1995. Excited-state dynamics of aromatic clusters: Correlation between exciton interactions and excimer formation dynamics. *J. Phys. Chem.* 99: 15738–15747.

52. Song, J. K., Han, S. Y., Chu, I., Kim, J. H., Kim, S. K., Lyapustina, S. A. et al. 2002. Photoelectron spectroscopy of naphthalene cluster anions. *J. Chem. Phys.* 116: 4477–4481.

53. Song, J. K., Lee, N. K., and Kim, S. K. 2003. Multiple ion cores in anthracene anion clusters. *Angew. Chem. Int. Ed.* 42: 213–216.

54. Even, U., Jortner, J., Noy, D., Lavie, N., and Cossart-Magos, C. 2000. Cooling of large molecules below 1 K and He clusters formation. *J. Chem. Phys.* 112: 8068–8071.

55. Mitsui, M., Nakajima, A., Kaya, K., and Even, U. 2001. Mass spectra and photoelectron spectroscopy of negatively charged benzene clusters, (benzene)$_n^-$ (n = 53–124). *J. Chem. Phys.* 115: 5707–5710.

56. Mitsui, M., Nakajima, A., and Kaya, K. 2002. Negative ion photoelectron spectroscopy of (benzene)$_n^-$ (n = 53–124) and (toluene)$_n^-$ (n =33–139): Solvation energetics of an excess electron in size-selected aromatic hydrocarbon nanoclusters. *J. Chem. Phys.* 117: 9740–9749.

57. Ando, N., Mitsui, M., and Nakajima, A. 2007. Comprehensive photoelectron spectroscopic study of anionic clusters of anthracene and its alkyl derivatives: Electronic structures bridging molecules to bulk. *J. Chem. Phys.* 127: 234305-1–13.

58. Ando, N., Mitsui, M., and Nakajima, A. 2008. Photoelectron spectroscopy of cluster anions of naphthalene and related aromatic hydrocarbons. *J. Chem. Phys.* 128: 154318-1–8.

59. Mitsui, M., Kokubo, S., Ando, N., Matsumoto, Y., Nakajima, A., and Kaya, K. 2004. Coexistence of two different anion states in polyacene nanocluster anions. *J. Chem. Phys.* 121: 7553–7556.

60. Mitsui, M., Ando, N., and Nakajima, A. 2007. Mass spectrometry and photoelectron spectroscopy of tetracene cluster anions, (tetracene)$_n^-$ (n = 1–100): Evidence for the highly localized nature of polarization in a cluster analogue of oligoacene crystals. *J. Phys. Chem. A* 111: 9644–9648.

61. Mitsui, M. and Nakajima, A. 2007. Formation of large molecular cluster anions and elucidation of their electronic structures. *Bull. Chem. Soc. Jpn.* 80: 1058–1074.

62. Mitsui, M., Matsumoto, Y., Ando, N., and Nakajima, A. 2005. Negative ion photoelectron spectroscopy of 2,2′-bithiophene cluster anions, (2T)$_n^-$ (n = 1–100). *Eur. Phys. J. D* 34: 169–172.

63. Mitsui, M., Matsumoto, Y., Ando, N., and Nakajima, A. 2005. Formation and photoelectron spectroscopy of nanoscale cluster anions of biphenyl, (BP)$_n^-$ (n = 2–100). *Chem. Lett.* 34: 1244–1245.

64. Mitsui, M., Ando, N., and Nakajima, A. 2008. Mass spectrometry and photoelectron spectroscopy of o-, m-, and p-terphenyl cluster anions: The effect of molecular shape on molecular assembly and ion core character. *J. Phys. Chem. A* 112: 5628–5635.

65. Čápek, V. and Silinsh, E. A. 1995. Dynamics of electronic polarization in molecular crystals. *Chem. Phys.* 200: 309–318.

66. Brovchenko, I. V. 1997. Lattice relaxation in molecular crystals with localized charges. *Chem. Phys. Lett.* 278: 355–359.

67. Sato, N., Seki, K., and Inokuchi, H. 1981. Polarization energies of organic-solids determined by ultraviolet photoelectron-spectroscopy. *J. Chem. Soc., Faraday Trans. II* 77: 1621–1633.

68. Salaneck, W. R., Seki, K., Kahn, A., and Pireaux, J.-J. (Eds.) 2001. *Conjugated Polymer and Molecular Interfaces.* New York: Marcel Dekker.

69. Seki, K. and Kanai, K. 2006. Development of experimental methods for determining the electronic structure of organic materials. *Mol. Cryst. Liq. Cryst.* 455: 145–181.

70. Hill, I. G., Kahn, A., Soos, Z. G., and Pascal, R. A. Jr. 2000. Charge-separation energy in films of π-conjugated organic molecules. *Chem. Phys. Lett.* 327: 181–188.

71. Ando, N., Kokubo, S., Mitsui, M., and Nakajima, A. 2004. Photoelectron spectroscopy of pyrene cluster anions, (pyrene)$_n^-$ (n = 1–20). *Chem. Phys. Lett.* 389: 279–283.

72. Scoles, G. 1986. *Atomic and Molecular Beam Methods.* Oxford, U.K.: Oxford University Press.

73. Bernstein, E. R. 1990. *Atomic and Molecular Clusters, Studies in Physical and Theoretical Chemistry.* Amsterdam, Netherlands; New York: Elsevier.

74. Even, U., Al-Hroub, I., and Jortner, J. 2000. Small He clusters with aromatic molecules. *J. Chem. Phys.* 115: 2069–2073.

75. Farges, J., Deferaudy, M. F., Raoult, B., and Torchet, G. 1985. Cluster models made of double icosahedron units. *Surf. Sci.* 156: 370–378.

76. Näher, U., Zimmermann, U., and Martin, T. P. 1993. Geometrical shell structure of clusters. *J. Chem. Phys.* 99: 2256–2260.

77. Cheshnovsky, O., Yang, S. H., Pettiette, C. L., Craycraft, M. J., and Smalley, R. E. 1987. Magnetic time-of-flight photoelectron spectrometer for mass-selected negative cluster ions. *Rev. Sci. Instrum.* 58: 2131–2137.

78. Handschuh, H., Ganteför, G., and Eberhardt, W. 1995. Vibrational spectroscopy of clusters using a "magnetic bottle" electron spectrometer. *Rev. Sci. Instrum.* 66: 3838–3843.

79. Tsukuda, T., Johnson, M. A., and Nagata, T. 1997. Photoelectron spectroscopy of (CO$_2$)$_n^-$ revisited: Core switching in the $2 \leq n \leq 16$ range. *Chem. Phys. Lett.* 268: 429–433.

80. Akin, F. A. and Jarrold, C. C. 2003. Separating contributions from multiple structural isomers in anion photoelectron spectra: Al$_3$O$_3^-$ beam hole burning. *J. Chem. Phys.* 118: 1773–1778.

81. Frank, K. H., Yannoulis, P., Dudde, R., and Koch, E.-E. 1988. Unoccupied molecular orbitals of aromatic hydrocarbons adsorbed on Ag(111). *J. Chem. Phys.* 89: 7569–7576.

82. Tsiper, E. V. and Soos, Z. G. 2001. Charge redistribution and polarization energy of organic molecular crystals. *Phys. Rev. B* 64: 195124-1–12.

83. Schiedt, J. and Weinkauf, R. 1997. Photodetachment photoelectron spectroscopy of mass selected anions: Anthracene and the anthracene-H$_2$O cluster. *Chem. Phys. Lett.* 266: 201–205.

84. Bounds, P. J. and Munn, R. W. 1981. Polarization energy of a localized charge in a molecular crystal. II. Charge-quadrupole energy. *Chem. Phys.* 59: 41–45.

85. Bouvier, B., Brenner, V., Millie, P., and Soudan, J.-M. 2002. A model potential approach to charge resonance phenomena in aromatic cluster ions. *J. Phys. Chem. A* 106: 10326–10341.

86. Barnett, R. N., Landman, U., Cleveland, C. L., and Jortner, J. 1988. Electron localization in water clusters. II. Surface and internal states. *J. Chem. Phys.* 88: 4429–4447.

87. Makov, G. and Nitzan, A. 1994. Solvation and ionization near a dielectric surface. *J. Phys. Chem.* 98: 3459–3466.

88. Amy, F., Chan, C., and Kahn, A. 2005. Polarization at the gold/pentacene interface. *Org. Electron.* 6: 85–91.

89. Sato, N., Seki, K., Inokuchi, H., and Harada, Y. 1986. Photoemission from an amorphous pentacene film. *Chem. Phys.* 109: 157–162.

90. Kuo, J.-L. and Klein, M. L. 2005. Structure of protonated water clusters: Low-energy structures and finite temperature behavior. *J. Chem. Phys.* 122: 024516-1–9.

Vibrational Spectroscopy of Strongly Bound Clusters

Philipp Gruene
Fritz-Haber-Institut der
Max-Planck-Gesellschaft

Jonathan T. Lyon
Fritz-Haber-Institut der
Max-Planck-Gesellschaft

André Fielicke
Fritz-Haber-Institut der
Max-Planck-Gesellschaft

9.1 Introduction

The changes of material properties on the nanoscale and the appearance of size-dependent phenomena are just the observables of corresponding developments in the electronic and geometric structures. Consequently, the investigation of structural properties is one of the central topics in cluster research. Vibrational spectroscopy is a classical spectroscopic method for structural determination as it probes the arrangements of atoms and the forces acting between them. Through vibrational spectroscopy and, in particular, tunable infrared (IR) laser spectroscopy, structural information has been obtained in the past on many different types of transient species like free radicals, ions, van der Waals molecules, and clusters (Miller 1992, Bernath 2002, Curl and Tittel 2002).

This chapter gives an introduction into the methods for obtaining structural information via vibrational spectroscopy for isolated, strongly bound clusters containing only a few, typically less than 20 atoms. This includes, in particular, clusters of the transition metals, of carbon and silicon, and also compound clusters, for instance, oxides, or carbides. The focus is especially on "isolated" clusters in well-controlled environments, which generally refers to gas-phase species under high dilution. In the beginning, an overview on the experimental techniques is given, before highlighting in the subsequent section more recent developments in the application of free electron lasers (FELs) for the IR spectroscopy of clusters.

The primary technique for the measurement of vibrational spectra is *infrared absorption spectroscopy* wherein the frequency-dependent attenuation of IR light by the absorbing medium is determined. Generally speaking, in absorption spectroscopy one detects what the molecules do to the light. The ratio $\varphi(\nu)/\varphi_0(\nu)$ of the light intensity changed upon absorption at a certain frequency ν is described by Beer's law

$$\frac{\varphi(\nu)}{\varphi_0(\nu)} = e^{-\sigma(\nu)\cdot n\cdot l}$$

with
    $\sigma(\nu)$ the absorption cross section
    n the particle density
    l the thickness of the irradiated sample

In absorption spectroscopy, the detected signal $\varphi(\nu)/\varphi_0(\nu)$ is determined by the number of absorbing particles but is independent of the initial light intensity $\varphi_0(\nu)$ (as long as the intensity change can be quantified). If absolute particle densities are known, the absolute absorption cross section can be calculated or vice versa. Clusters are usually produced as mixtures with wide size distributions and can, in most cases, only be handled in the gas phase, either in molecular beams or in ion trap experiments. Thereby, the achievable densities are usually low (typically below 10^6 cm^{-3}) and do not allow for the application of standard techniques of absorption spectroscopy. Only very few examples of direct absorption measurements on gas-phase clusters in the IR have been reported, for instance, by applying cavity ring-down spectroscopy (CRDS) (Section 9.2.2.1.2), which might be seen as a multipass absorption experiment. Besides the sensitivity issue, the assignment of spectral features to a specific cluster is difficult when dealing with size distributions. Mass discrimination, or selection, requires the coupling of the spectroscopic experiment with the techniques of mass spectrometry. An elegant example of such a combination is achieved by the spectroscopy of mass-selected and accumulated clusters in inert matrices (Section 9.2.1).

In the case of gas-phase clusters, instead of measuring the attenuation of radiation, i.e., studying what the particles do to the light, one often detects what the light does to the particles: changes in the medium induced by the absorption of photons are used to obtain spectroscopic information. As this relies on an action of the medium, such a technique is usually called "action spectroscopy." The ratio of the population $n(\nu)/n_0$ of the initial state of the absorbing species with and without interaction with the light can be expressed as

$$\frac{n(\nu)}{n_0} \propto e^{-\sigma(\nu)\cdot\varphi(\nu)}$$

and is determined by the absorption cross section $\sigma(\nu)$ and the laser fluence $\varphi(\nu)$ (in photons/cm^2) only. Obviously, one can gain in the action spectroscopic signal by using light sources with the highest possible fluence. This is the origin of the sensitivity of action spectroscopic techniques that also allows to measure spectra if only very low particle densities can be obtained or even only a countable number of particles are under investigation. There are different kinds of action that can be used to follow the absorption process. Absorption of a photon leads to excitation into higher states, which may be followed by a detectable fluorescence; the depopulation of the (vibrational) ground state can be traced by state-specific photoionization schemes (ion-dip spectroscopy). Mass spectrometric detection is amenable to ionization or fragmentation processes. In a wider sense, even photoemission spectroscopy is an "action spectroscopy" technique. Here, the energy distribution of the electrons emitted upon photoexcitation of a species contains the information on the (ro-)vibronic states. Charged species have the advantage that they can be mass selected before photoexcitation. This forms the basis for the application of anion photoelectron spectroscopy in the investigation of cluster structures (Section 9.2.2.2.1).

In general, the energy that is carried by a single IR photon is by far too low to induce fragmentation or ionization of a cluster. For instance, the bond dissociation energies of typical transition metal clusters are on the order of 2–5 eV, their ionization potentials are in the 4–7 eV range, but the vibrational fundamentals are below 500 cm^{-1} (0.06 eV). How can then, e.g., a fragmentation process be used for detecting the absorption of IR photons by a cluster? There are two solutions to this problem: (1) the cluster itself is not fragmented but rather its complex with a very weakly bound messenger (Section 9.2.2.3), or (2) the absorption of multiple, often very many, IR photons helps to overcome the fragmentation barrier (Section 9.3.3.2). For strongly bound clusters with relatively low ionization energies, this multiple photon excitation may even lead to an ionization of the neutral cluster (Section 9.3.3.3).

In most cases, structural information only emerges upon comparison of the experimental vibrational spectra with data obtained by means of theory. For structural assignments, one aims for a quantitative prediction of the fundamental vibrational frequencies and the corresponding IR and Raman intensities. Today, this is often obtained from quantum chemical calculations that are used to investigate the configurational space for a certain cluster composition. Often, different structural models are considered and the assignments are based on a comparison of the experimental spectra with those predicted by theory. As the cluster size increases, the number of theoretically possible structural isomers increases dramatically, and sophisticated search schemes are needed. Among other methods, genetic algorithms have been found to be particularly useful to identify the structure that corresponds to the global energy minimum (Johnston 2003, Santambrogio et al. 2008). A more detailed discussion of the variety of methods for predicting vibrational spectra goes beyond the scope of this chapter but can be found in a recent review (Meier 2007).

9.2 Experimental Techniques

9.2.1 Clusters in Matrix Isolation

Embedding clusters in an inert host matrix makes it possible to accumulate these species without their further aggregation as the matrix environment isolates the clusters and suppresses diffusion and subsequent reactions. The matrix usually consists of cryogenically solidified rare gases or chemically inert molecules like N_2. Naturally, the matrix must be transparent in the spectral region of interest. Rare gases have a wide spectral window from the far IR up to the vacuum ultraviolet (VUV); therefore, the limits are usually set by the substrate material where the matrix is grown on.

The technique of matrix isolation was first proposed by Pimentel et al. in order to accumulate a significant quantity of a reactive species in an inert environment so that it could be studied with spectroscopy (Whittle et al. 1954). The technique has now been widely used to study multiple species with a number of different spectroscopic methods (Bondybey et al. 1996). In principle, any molecular species can be studied with this technique, provided that the species itself, or a photolytic precursor, can be produced initially in the gas phase. In matrix isolation, it is assumed that the interactions between the guest species and host matrix are sufficiently weak that the spectra differ only slightly from the corresponding gas-phase spectra.

In the past, small metal clusters consisting typically of up to three atoms have been grown directly in the matrix by co-deposition of atomic metal vapor with neon or argon, which forms the matrix. However, as this in-matrix cluster formation always produces a distribution of sizes, it is intrinsically not a straightforward approach for obtaining cluster size-specific spectra. A more recent development of mass-selective deposition of clusters into the matrix (Figure 9.1) allows for a more direct assignment of the size-specific vibrational spectrum. However, in depositing preformed clusters in matrices, there exists a possibility for the metal cluster to fragment upon landing, a topic that has been discussed in detail previously (Harbich 1999). A major challenge of cluster characterization via matrix isolation is to obtain high densities in the matrix. This requires mass-selected cluster beams of high intensities

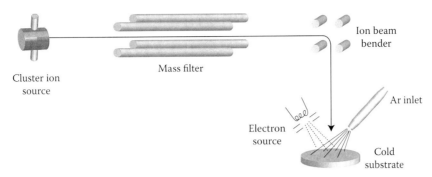

FIGURE 9.1 Typical experimental setup for matrix deposition of mass-selected clusters. Cluster cations emitted, e.g., by a laser ablation source, are mass-filtered to select a specific cluster size. An electrostatic beam bender separates the ionic from the neutral species and the cluster ions are co-deposited with the matrix (here Ar) onto the cold substrate that is mounted onto a cryostat. The sample is flooded with low-energy electrons to neutralize the cationic species and to prevent charging. Finally, the produced sample can be exposed to the beam path of a Raman or IR spectrometer.

and usually long (several hours) accumulation times. For the most part, matrix isolation has so far been a suitable technique to study the vibrational modes of mainly neutral rather than ionic clusters.

9.2.1.1 Infrared Absorption Spectroscopy

There are only a few examples for the characterization of clusters in a matrix environment by IR absorption spectroscopy, which is limited by the sensitivity issues described in Section 9.1. For example, small silicon clusters have been detected after deposition into a noble gas matrix using Fourier transform infrared (FTIR) spectroscopy (Li et al. 1995). Only very few trimeric clusters of transition metals deposited into an inert matrix have been studied in the IR region of the spectrum (Nour et al. 1987, Guo et al. 2002).

In addition to the aforementioned studies of clusters in inert matrices, there have been reports on the growth of metal clusters in zeolites, detected via the appearance of bands in the far-IR that are assigned to metal cluster modes (Baker et al. 1985).

9.2.1.2 Raman Spectroscopy

To date, Raman spectroscopy appears to be by far the preferred technique for the study of the vibrational properties of metal clusters isolated in matrices. At a first glance, this might be surprising as Raman scattering occurs with comparably low cross sections σ_R on the order of only 10^{-29} cm^2. However, Raman spectroscopy is a kind of "action spectroscopy" as one measures the inelastically scattered photons that reflect changes in the vibrational quantum state of the matrix-embedded species. The integrated Raman signal I_R, which is the amount of total Stokes scattered light, scales linearly with the intensity I of the excitation light: $I_R = \sigma_R \cdot n_A \cdot I$. Therefore, the use of intense excitation lasers makes the detection of the Raman-shifted photons possible, even if the achievable density n_A of clusters in the matrix is low. With a typical power of 100 mW for an Ar ion laser emitting at about 500 nm and a cluster density n_A of 10^{15} cm^{-2}, this yields a total Raman signal of 10^{-15} W, or about 2500 Stokes photons/s. Out of these, however, most of the time only a fraction can be collected and detected for technical reasons.

By tuning the excitation wavelength in resonance with an electronic transition, the intensity of vibrational transitions associated with the excited electronic state can be enhanced by a factor of up to $\sim10^5$ compared to normal Raman spectra. This resonance-enhanced Raman spectroscopy has been intensively used to study the vibrational modes and force constants of small metal and lanthanide clusters (Lombardi and Davis 2002). Surface-enhanced Raman spectroscopy (SERS) (Moskovits 1985) is even more sensitive and has been used to study carbon and silicon clusters in cryogenic matrices (Honea et al. 1999, Rechtsteiner et al. 2001).

Figure 9.2 depicts the structural assignments made for small silicon clusters based on SERS spectra. In SERS, the electric field surrounding a silver or gold substrate increases as the surface plasmons are excited by a laser. As Raman intensities are proportional to the electric field, this technique causes a large increase (up to 10^{11}) in the signal compared to conventional Raman spectroscopy.

9.2.2 Gas-Phase Clusters

9.2.2.1 Direct Absorption Measurements

As explained before, direct absorption measurements on cluster beams are difficult mainly due to (1) the low particle density and (2) the presence of a distribution of different cluster sizes that complicates the assignment of any spectral feature to a certain species. However, for a few cases, beams of sufficient cluster density have been produced to investigate them with highly sensitive methods that provide rotational resolution. In these cases, definite assignments to a specific species become possible through the determination of the rotational constants.

9.2.2.1.1 High-Resolution Diode Laser Spectroscopy

Infrared diode lasers have been intensively employed in the IR spectroscopy of carbon clusters (Giesen et al. 1994, Van Orden and Saykally 1998). In these experiments, the dense supersonic beams of carbon clusters are formed via laser ablation and clustering within a He carrier gas. The molecular beam is intersected by multiple (~20) passes of a focused IR beam produced by a lead

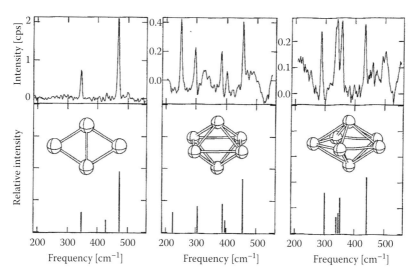

FIGURE 9.2 Raman spectra of matrix-isolated neutral silicon clusters and comparison to spectra predicted by ab initio quantum chemical calculations. (Reprinted from Honea, E.C. et al., *J. Chem. Phys.*, 110, 12161, 1999. With permission.)

salt (PbSe) diode. Sensitivity can be improved by frequency modulation and subsequent lock-in detection on a photoconductive IR detector to up to 10^{-7} in the absorption measurement (Bernath 2002). By this means, rotationally resolved IR spectra for linear carbon clusters up to C_{13} have been obtained. Assignments to a certain cluster size rely, for instance, on predictions of the rotational constants based on expected C–C distances within the chain.

9.2.2.1.2 Cavity Ring-Down Spectroscopy

CRDS is a highly sensitive variant of absorption spectroscopy (Berden et al. 2000). In pulsed CRDS, a short monochromatic light pulse with a typical duration of ~10 ns is coupled into an optical cavity formed by two highly reflective mirrors that enclose the sample. The light pulse is reflected back and forth within the cavity, thereby multiplicatively passing through the sample. The intensity I of the light within the cavity decays exponentially with time $I(t) = I_0 \cdot \exp(-t/\tau(\nu))$ by absorption in the sample and by the inherent losses of the cavity, i.e., due to the limited reflectivity of the mirrors. The ring-down times $\tau(\nu)$ are on the order of microseconds or even more, and $1/\tau(\nu)$ is proportional to the absorption coefficient $\sigma(\nu)$. The analysis of the time-dependent intensity decay via the small fraction of light leaking out of the cavity allows for a sensitive measurement of the absorption. Spectral information is obtained from the dependence of $\tau(\nu)$ on the frequency of the light coupled into the cavity. In the visible and near-IR range, where mirrors with up to 99.999% reflectivity are available, CRDS is now widely used for trace detection, for instance, in atmospheric gases. It has also yielded absorption spectra of small metal and carbon clusters in the UV and visible range. In the mid and far-IR, the reflectivity of available mirrors is significantly lower, which is, together with the lower absorption cross sections, one of the limiting factors in the application of CRDS to the IR spectroscopy of clusters. However, there has been a recent report

of high-resolution pulsed IR CRDS on the ν_6 band of the C_9 cluster in a supersonic molecular beam plasma. The minimum detectable particle density has been estimated to be 10^9 cm^{-3} (Casaes et al. 2002). Unfortunately, this is still a factor of about a 1000 higher than the densities achievable for mass-selected clusters in the gas phase.

9.2.2.2 Vibrational Resolution in Electronic Transitions

Electronic excitations occur usually in the visible to UV range of the spectrum. Using different single or multiple photon excitation schemes, the energies of electronic transitions can be sensitively probed with resolutions down to the rovibrational sublevels. Many different experimental approaches, for instance, resonant two-photon ionization (R2PI) or fluorescence spectroscopy, can provide optical excitation spectra with vibrational resolution. In the following, only two methods are discussed that are more frequently applied to the investigation of strongly bound clusters.

9.2.2.2.1 Anion Photoelectron Spectroscopy

Photoelectron spectroscopy was originally developed for the investigation of surfaces and gas-phase molecules (Siegbahn 1982), and shortly thereafter extended to also study anionic species. Figure 9.3 shows a scheme of the principle of anion photoelectron spectroscopy (Neumark 2008). An anion is excited from an initial state by means of an UV/vis laser with photon energy $h\nu$. If the photon energy exceeds the binding energy of the electron (electron affinity, EA), the electron is detached and any photon energy in excess of that needed for electron detachment is carried by the outgoing electron (if the recoil of the remaining neutral species is neglected) in the form of kinetic energy E_{kin} according to

$$E_{kin} = h\nu - EA - E_{exc}.$$

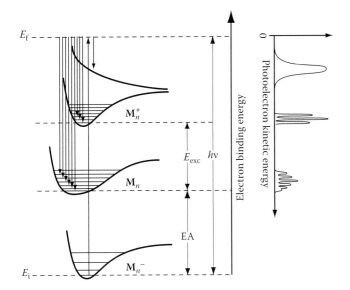

FIGURE 9.3 Schematic diagram depicting an electron detachment process. A cluster anion M_n^- is excited from the ground state upon absorption of a photon with energy $h\nu$. If $h\nu$ is larger than the electron affinity of the neutral cluster, an electron can be detached. The electron leaves a neutral cluster behind, which can remain in the electronic ground state M_n or, provided that the excitation energy is sufficient, in an excited state M_n^*. Within an electronic state, the neutral cluster can be vibronically excited. All excess energy is taken away by the outgoing electron in the form of kinetic energy. The probability of the transitions is given by the Franck–Condon factors of the involved states. When the kinetic energy of the photoelectron is measured (right), a photoelectron spectrum is obtained, which reflects the energies of the electronic and vibrational levels of the neutral cluster in the ground-state geometry of its anion. Characteristic values like the electron affinity (EA) of M_n or the excitation energy (E_{exc}) from the ground state to the excited states M_n^* can be derived directly from the spectrum.

The detachment process is fast in comparison to structural rearrangements of the nuclei upon removal of the electron. That is the reason why it is generally stated that photoelectron spectroscopy of anions probes the electronic structure of the neutral species in the ground-state geometry of its anion. The detachment may not only lead to the ground state of the neutral ($E_{exc} = 0$), but also to rovibrational or electronically excited states ($E_{exc} > 0$). The transition probabilities from the ground state of the anion to the various electronic states of the neutral (including their rovibrational sublevels), and therefore their spectral intensity, are determined to a large extent by the corresponding Franck–Condon factors. The kinetic energy of the detached electron depends on the rovibrational state of the anion and the final electronic and rovibrational state of the remaining neutral species.

In this version, the technique offers several advantages over spectroscopic techniques that probe the electronic and vibrational structure of neutral species directly. It is mass-selective as the anions can be mass-selected before electron detachment. Since it is a linear absorption process but goes into the continuum, it has less strict selection rules than, for instance, electric-dipole-allowed rovibrational transitions. In general, no tunable

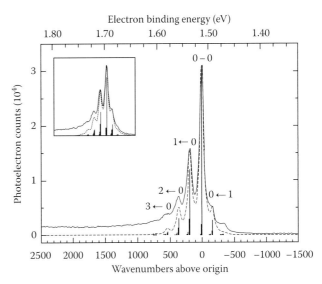

FIGURE 9.4 Parts of the photoelectron spectrum of Nb_8^- together with one-dimensional harmonic Franck–Condon fits (dashed lines and sticks). Spectra are measured at room temperature (inset) and with a thermalization channel cooled to 80 K (main). The analysis reveals a progression in a mode with a frequency of 180 cm^{-1} in Nb_8 and 165 cm^{-1} in Nb_8^-. Assignments ($\nu_{neutral} \leftarrow \nu_{anion}$) indicate the main contribution to each peak. (Reprinted from Marcy, T.P. and Leopold, D.G., *Int. J. Mass Spectrom.*, 196, 653, 2000.)

laser is required to obtain spectroscopic information over a broad frequency range. Given a sufficient resolution, detailed information about the electronic structure, possibly including a vibrational substructure, of the neutral species can be gained. Typical electron spectrometers provide resolutions of 5–10 meV. More recently, the introduction of the velocity map imaging (VMI) technique has led to a significant improvement of the energy resolution for slow electrons down to a few tenths of a micro-electronvolt (Neumark 2008). In addition, the VMI technique images the angular distribution of photoelectrons that is determined by the symmetry of the orbital from which the electron is removed and the final state produced by photodetachment.

Photoelectron spectroscopy has been used since the mid-1980s for the study of the electronic structure of clusters (Cheshnovsky et al. 1987). For small cluster anions like dimers and trimers, the vibrational structure could be resolved early on. Ever-improving instrumental resolution allowed for the observation of a vibrational substructure in the photoelectron spectra also for larger clusters like Si_{3-7}^- (Xu et al. 1998), $Au_{4,6}^-$ (Handschuh et al. 1995), and Nb_8^- (Figure 9.4) (Marcy and Leopold 2000).

9.2.2.2.2 Zero Electron Kinetic Energy Spectroscopy

In photoelectron spectroscopy, the detached electrons are usually produced with finite kinetic energies. However, if the energy of the detachment photon is tuned in resonance with an excitation transition, electrons with vanishing little kinetic energy are emitted. This is the basic idea behind *zero electron kinetic energy* (ZEKE) *spectroscopy*, or as it is also called *threshold photoelectron spectroscopy* (Müller-Dethlefs and Schlag 1998). As slow electrons are very sensitive to any electric and magnetic stray fields, they have to be avoided, which is experimentally challenging. Finally, one has to discriminate the slow ZEKE electrons from the ones having higher energies. Such an approach requires the application of tunable lasers, whereas in photoelectron spectroscopy, the electron kinetic energy distribution is measured at a fixed laser wavelength. In practice, the electron is often not detached directly, but initially a highly excited Rydberg state is prepared from which the electron is separated via pulsed-field ionization (PFI). ZEKE spectroscopy has a superior resolution compared to normal anion PES that is limited mainly by the bandwidth of the excitation laser, which is on the order of $0.1\,cm^{-1}$ (typical bandwidth of a pulsed dye laser).

Despite their relatively high resolution, the interpretation of the ZEKE spectra of clusters is not always straightforward as it requires a multidimensional Franck–Condon simulation to entangle the contributions of different vibrational modes. This becomes even more challenging with increasing cluster sizes. In addition, as the neutral clusters are formed in a distribution of different sizes, the assignment of the spectral signals to a particular cluster is difficult. A partial mass selection has been obtained by using the velocity slip between different sized neutral clusters in a molecular beam and has led to vibrationally resolved PFI-ZEKE spectra for a number of small transition metal clusters and their adducts, for instance, with O, N, and C atoms (Yang and Hackett 2000).

Originally, ZEKE spectroscopy was applied to neutral species, but also the electron detachment from anions at the threshold is possible. One advantage of such an approach is the possibility of an initial mass selection of the charged species (Neumark 2008). In a variant of neutral ZEKE spectroscopy called mass-analyzed threshold ionization (MATI), the resulting ZEKE cations, instead of electrons, are detected. This again has the benefit of a possible identification of the nature of the ionized species, i.e., the cluster size, via determination of its mass. Using this technique, vibrationally resolved excitation spectra have been recently reported for V_3^+ (Ford and Mackenzie 2005).

9.2.2.3 Photodissociation Spectroscopy and the Messenger Technique

Photodissociation techniques have been extensively used before in the spectroscopy of clusters in the UV and visible spectral range. For instance, direct photodissociation

$$\mathbf{M}_n \xrightarrow{\;h\nu\;} \mathbf{M}_{n-1} + \mathbf{M}$$

of bare metal clusters $\mathbf{M}_n$ has been used to probe their electronic absorption spectra. Vibrationally resolved optical spectra have been observed, e.g., for small copper clusters (Morse et al. 1983, Knickelbein 1994). Obviously, this method is limited to clusters for which the height of the barrier for dissociation is smaller than the photon energy; at least as long as only single photon absorption processes are concerned. In the case of more strongly bound systems, the introduction of a weakly bound spectator ligand $\mathbf{X}$ that is vaporized off the cluster upon excitation according to

$$\mathbf{M}_n \cdot \mathbf{X} \xrightarrow{\;h\nu\;} \mathbf{M}_n + \mathbf{X}$$

allows to study the absorption spectra and to extend the spectral range into the IR (Okumura et al. 1986). The sensitivity towards thermal excitation of the cluster is particularly high when using physisorbed messengers, e.g., N_2 molecules or rare gas atoms that have low binding energies to the cluster. The basic idea behind this approach is that the cluster forms the chromophore, which is only marginally disturbed by the presence of the messenger species. However, one has to be aware that this assumption is not always justified. A highly polarizable rare gas atom as a messenger can have significant effects on the observed absorption cross sections (Gruene et al. 2008). In highly fluxional systems, even the attachment of weakly interacting messengers might change the structure as compared to the bare cluster (Knickelbein 1994). Finally, it has been observed that the attachment of messengers can alter the relative energetic order of cluster isomers (Asmis and Sauer 2007). Typical binding energies for rare gas–cluster complexes are on the order of 0.1–$0.2\,eV$ for ionic species. By using He atoms as messengers, many of the disturbing effects can be practically avoided due to their low polarizability and their particularly low binding energy to the clusters (Asmis and Sauer 2007). A small binding energy, however, implies that such a complex can only be formed at rather low temperatures. This can be achieved, for instance, in cryogenic ion trap experiments that allow for a thermalization down to ~10 K for mass-selected charged clusters.

Figure 9.5 displays three different experimental schemes for the photodissociation spectroscopy of charged and neutral clusters. If a cluster distribution is irradiated by monochromatic light, a particular species may fragment upon excitation. The relative intensity of that species compared to reference experiments where the clusters are not irradiated is measured from the corresponding mass spectra. The wavelength dependence of the intensity change yields the depletion spectrum for this species. This works well if the depleting species cannot be formed by fragmentation of other species. Otherwise, it becomes difficult to unravel the contributions from the dissociation and the formation out of heavier species. Therefore, depletion spectroscopy is particularly suited for cluster complexes with very weakly bound messengers like rare gas atoms, where only loss of the messenger occurs but no fragmentation of the cluster. This has the advantage that if the full mass spectrum is recorded for each wavelength step, data for all cluster complexes present are measured at once within a single wavelength scan. The disadvantage is that as the depletion of the ion intensity is measured, it is inherently sensitive to fluctuations of the cluster intensity between the reference and the dissociation experiment. Therefore, either sources that produce very stable cluster distributions or a large amount of data averaging is required. If the dissociation is performed on mass-selected ions, the depletion of the selected ion, and also the appearance of the fragment, can be determined. The latter can be detected very sensitively against almost zero background. Depletion spectroscopy can also be applied to neutral species. However, an ionization step needs to be included to allow for mass spectrometric detection. The ionic intensities do not directly map the initial size distribution of

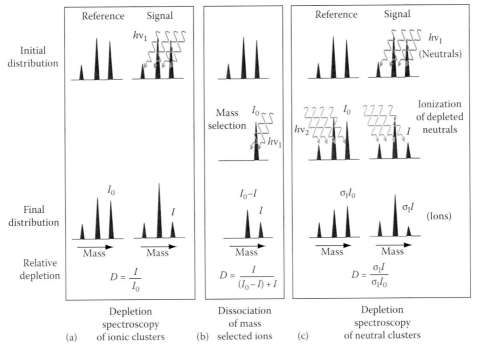

FIGURE 9.5 Schemes for three different variations of depletion spectroscopy on ionic and neutral species. The upper line displays the initial mass distributions. From the three different species, only the heaviest specie absorbs photons of the energy $h\nu_1$ and dissociates forming the species associated with the middle peak. The lowest line resembles the resulting final distributions. Relative intensity changes in a distribution of ionic clusters can be determined by comparison with a reference experiment (a). Mass selection (b) often allows for an internal reference by measuring the abundances of parent and fragment ions in a single experiment. Changes in the abundance of neutral species (c) cannot be directly analyzed using mass spectrometry. Therefore, they are transformed before the mass selective detection into cations by ionization with, e.g., UV photons of energy $h\nu_2$ with a species-specific ionization cross section σ_i. This cross section, however, cancels out in the relative depletion. (From Marcy, T.P. and Leopold, D.G., *Int. J. Mass Spectrom.*, 196, 653, 2000. With permission.)

the neutral clusters as the ionization cross sections depend on the species, e.g., cluster size. For instance, the probability to ionize the cluster–messenger complex and the bare cluster may be different, related to different ionization energies. However, this does not cause a problem as in depletion spectroscopy only changes of the relative intensities need to be determined.

Absorption spectra can be obtained by converting the measured depletions to one photon absorption cross sections $\sigma(\nu)$ using the analogue of the Lambert–Beer law

$$\sigma(\nu) = \frac{1}{\varphi(\nu)} \cdot \ln \frac{I_0}{I(\nu)}$$

where

$I(\nu)$ and I_0 are the intensities of a certain cluster–messenger complex with and without irradiation, respectively

$\varphi(\nu)$ is the photon fluence

This approach assumes a perfect overlap of the cluster beam (or the cloud in an ion trap) with the excitation laser, which is in practice difficult to realize. Therefore, often only relative cross sections are reported.

The UV, optical, and near-IR absorption spectra of neutral and charged transition metal clusters have been measured via the dissociation of weakly bound messenger complexes with conventional laser systems (for IR lasers see Section 9.3.1). However, the lack of easily accessible intense and widely tunable light sources in the mid- and far-IR has limited so far the use of this method for the direct measurement of cluster vibrational spectra. This has changed with the application of IR FELs for the spectroscopy of clusters. These light sources produce tunable IR radiation over a wide spectral range, including the mid- and far-IR. Figure 9.6 shows IR spectra of two aluminum oxide clusters obtained by photodissociation of their He complexes in the 550–1100 cm^{-1} range and their assigned structures (Santambrogio et al. 2008). Further examples of the IR spectroscopy of clusters using IR FELs will be given in Section 9.3.

9.2.2.4 Spectroscopy in He Droplets

The recently developed technique of embedding species in superfluid He droplets is a new and very promising merger of matrix isolation and dissociation spectroscopy (Toennies and Vilesov 2004). Clusters can be formed via subsequent pick-up of single metal atoms by the droplet from the gas phase. These species are instantaneously thermalized to the temperature of the droplet (0.37 K for ^{4}He) resulting in a stabilization of structures that are not necessarily the thermodynamically most stable ones. The superfluid He interacts only very weakly with the embedded species and any excitation is followed by

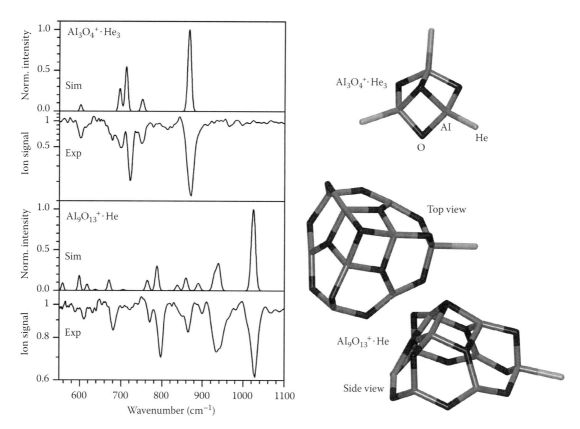

FIGURE 9.6 Comparison of experimental IR depletion spectra of the He complexes of $Al_3O_4^+$ and $Al_9O_{13}^+$ formed in a cryogenic ion trap with simulated linear absorption spectra for the structures shown (Santambrogio et al. 2008). Note the logarithmic scale for the depleted ion signal to compare with the predicted IR intensities. The identified oxide cluster structures correspond to the global minima found by probing the configurational space via a genetic algorithm.

the evaporation of He atoms from the droplet. The absorption of photons by embedded species can be sensitively monitored, e.g., via bolometric detection of changes in the particle flux. First examples for the vibrational spectroscopy of the complexes of metal clusters within He nanodroplets have been reported (Choi et al. 2006).

9.3 Free Electron Laser-Based Infrared Spectroscopy

9.3.1 Comparison of IR Laser Sources

Absorption spectroscopy in many cases relies on the availability of laser light sources. Lasers emitting in the IR range of the electromagnetic spectrum can be based on rather different physical principles, as, for instance, transitions between the rovibrational levels of a molecule in the gas phase, nonlinear optical frequency conversion, electron-hole recombination in a semiconductor, or light emission from relativistic electrons (Miller 1992, Bernath 2002, Curl and Tittel 2002). Table 9.1 gives an overview on the specifications of some typical IR lasers that are exploited in the spectroscopy of clusters. For spectroscopic applications, tunability over a certain frequency range plays an important role. Here, one can distinguish between lasers that are only line tunable, like the CO_2 laser or the far-IR gas lasers, and those which are

continuously tunable. Particularly wide tuning ranges are realized by frequency conversion using optical parametric oscillators and amplifiers (OPO/OPA) or difference frequency mixing (DFM) and by IR FELs.

Driving multiple photon absorption, which forms the basis for the action spectroscopies discussed later in this section, becomes only possible with high laser pulse energies. For instance, using line tunable CO_2 lasers, mass-selective IR multiple photon dissociation (IR-MPD) has been used in the past to obtain IR spectral information for some metal cluster complexes (Knickelbein 1999). However, the application of CO_2 lasers for the investigation of clusters is limited as they emit only in a rather restricted spectral range around $10\,\mu m$. FELs are by far more widely tunable and intense and have recently found applications in the vibrational spectroscopy of clusters in the mid- and far-IR.

9.3.2 FEL as IR Source

An FEL (Marshall 1985) is a unique light source, as the lasing medium are free electrons, whereas in conventional laser systems electrons are bound to atoms, molecules, or crystals. Obviously, the electrons are not completely free, but are under the influence of magnetic forces, which cause them to radiate. Figure 9.7 shows the generic layout of an FEL. A relativistic beam of electrons,

TABLE 9.1 Comparison of the Specifications of Selected Commercially Available IR Lasers (based in Part on (Miller 1992)) and an IR-FEL

Type	Pump Laser	Tuning Range (cm^{-1})	Bandwidth (cm^{-1})	Mode of Operation	Typical Pulse Energies (mJ)/Average Power (mW)
Diode laser (PbSe)	—	400–3400 <50 cm^{-1} range for single laser	<10^{-4}	cw	0.1–1 mW
Quantum cascade laser	—	800–1800 <10 cm^{-1} range for single laser	0.01–0.3 (single mode)	Pulsed (10–10^5 Hz) cw	<0.05 mJ 50 mW
CO laser	—	Line tunable 1670–1920	<3 × 10^{-5}	cw	1 W
CO$_2$ laser	—	Line tunable 920–1090	<3 × 10^{-5}	Pulsed (<100 Hz) cw	<50 mJ >10 W
Far-IR laser	CO$_2$ laser	Line tunable 10–250 (depending on molecule)	<10^{-5}	cw	<1–500 mW
Raman-shifted lasers	Dye laser	1200–vis	0.02	Pulsed cw	<10 mJ
OPO/OPA	Nd:YAG	2000–vis	<0.01–10	Pulsed/10 Hz	10 mJ
DFM (LiNbO$_3$)	Dye laser	1950–vis	0.05–1	Pulsed/10 Hz	3 mJ (3000 cm^{-1}) In combination with OPA ~5 mJ
DFM (AgGaSe$_2$)	OPA	900–2000	<1	Pulsed/10 Hz	<0.1 mJ
IR-FEL (FELIX)	—	40–2200	0.3%–5% of central frequency	Pulsed/10 Hz	<100 mJ (mid-IR) <50 mJ (far-IR)

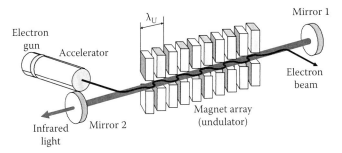

FIGURE 9.7 Schematic view of an FEL consisting of an electron gun, an accelerator, and an undulator (with period λ_U) placed in a resonator. Electrons oscillate between the undulator magnets and emit radiation, which is stored and amplified in the optical cavity. A fraction of the light is coupled out through a hole in one of the resonator mirrors.

produced by an accelerator, is injected into a resonator, consisting of two high-reflectivity mirrors at each end of the undulator. The magnetic field in the undulator is perpendicular to the direction of the electron beam and periodically changes polarity a (large) number of times along its length. This causes a periodic deflection, a "wiggling" motion, of the electrons while transversing the undulator. The transverse motion is quite analogous to the oscillatory motion of electrons in a stationary dipole antenna, and hence results in the emission of radiation with a frequency equal to the oscillation frequency. This oscillation frequency is given by the ratio of the velocity of the electrons to the path length traveled by the electrons per period of the undulator. This path length is larger than the period (λ_U) of the magnetic field by a factor $(1 + K^2)$ due to the transverse motion of the electrons induced by the magnetic field; the dimensionless factor K is a measure of the strength of the magnetic field. The overall motion of the electrons in the undulator resembles the motion of the electrons in a dipole antenna moving close to the speed of light. This high velocity results in a strong Doppler shift. Together with the relativistic

contraction, this leads to an upshift of the frequency of the radiation emitted in the forward direction, as measured in the laboratory frame by a factor $\approx 2\gamma^2$, where γ, the Lorentz factor, is a measure of the electron kinetic energy in units of its rest mass. The wavelength of the emitted radiation then becomes

$$\lambda_0 = \frac{\lambda_U}{2\gamma^2}\left(1 + K^2\right).$$

The radiation, referred to as spontaneous emission, is usually very weak though. This is a consequence of the fact that the electrons are typically spread out over an interval that is much larger than the radiation wavelength, and will therefore not emit coherently. But on subsequent round trips in the resonator, this weak radiation will be amplified by fresh electrons until saturation sets in at a power level that is typically 10^6–10^8 times that of the spontaneous emission.

The aforementioned principles are generally applicable to FELs emitting in wavelength ranges that permit operation of an optical resonator for amplification. At short wavelength, i.e., in the vacuum UV and x-ray range where reflectivity losses at the mirrors would become too high, FELs rely on single-pass amplification schemes like self-amplified stimulated emission (SASE). The following describes in more detail the characteristics of a specific IR FEL, the Free Electron Laser for Infrared eXperiments (FELIX) at the FOM Institute for Plasma Physics in Nieuwegein, the Netherlands (Oepts et al. 1995). In the past, this IR laser has been intensively used for the vibrational spectroscopy of strongly bound clusters (von Helden et al. 2003, Asmis et al. 2007). FELIX uses linear radio-frequency-driven (normal conducting) accelerators that produce macropulses of electron bunches with up to 10 μs length at a repetition rate of up to 10 Hz. These macropulses contain a train of electron bunches with a repetition rate of either 25 MHz or 1 GHz. With electron beam energies in the 15–45 MeV range, FELIX produces radiation between 4.5 and

250 μm when lasing on the fundamental. The amplification of the lasing on the third harmonics also allows covering the 2.7–4.5 μm range. The wavelength of the laser light in the FEL can be tuned by changing the electron-beam energy as well as the magnetic field periodicity and strength. In principle, all of these parameters can be modified. However, while changing the periodicity of the magnets requires major changes in the hardware of the FEL, modifications in the electron-beam energy require readjustments of large parts of the electron beam optics. Varying the strength of the magnetic field on the optical axis changes the K-value and is easily achieved at FELIX by changing the gap between the pairs of magnets by means of stepper motors.

The temporal structure of the emitted light follows the structure of the electron beam. Through the spectral bandwidth, the micropulse length can be adjusted between 300 fs and several picoseconds. The length of the micropulse determines the bandwidth of the light through Fourier limitation. At the above micropulse duration, the bandwidth usually ranges between 0.3% and 1% of the central wavelength. The typical output energies of FELIX can be up to 100 mJ/macropulse in the mid-IR and 15–50 mJ/macropulse in the far-IR.

9.3.3 Multiple Photon Excitation in the IR

9.3.3.1 Mechanism of Multiple Photon Excitation

Exploiting the high intensity of the radiation delivered by an IR-FEL, it becomes possible to induce multiple photon processes, i.e., infrared multiple photon excitation (IR-MPE). It has been realized early on that IR-MPE induced by an IR-FEL cannot be a purely coherent multiphoton process, given the large amount of photons that need to be absorbed to induce ionization or dissociation (von Helden et al. 1997). In view of the width of the IR radiation and typical anharmonicities in vibrational potentials, climbing the vibrational ladder of the excited mode ($v_i = 0 \rightarrow v_i = 1 \rightarrow v_i = 2 \rightarrow \cdots$) is unrealistic due to the anharmonicity bottleneck. The excitation mechanism is better described by subsequent absorption of single IR photons.

At low internal energies, i.e., in the initial phase of the multiple photon excitation, the cluster interacts with the IR radiation and climbs up the discrete energy levels of a vibrational ladder. This resonantly enhanced multiple photon absorption process is governed by the selection rules for direct photoabsorption. One must keep in mind that although the molecules are concentrated at the vibrational ground state, its quanta are distributed over a large number of rotational and vibrational-rotational sublevels. This results in the rotational compensation of anharmonic shifts and allows the cluster to climb up the vibrational ladder to a certain degree, despite the anharmonic bottleneck. Further shifts can be compensated by the anharmonic splitting of excited degenerate states. The anharmonicity can lead to a red-shift of the experimentally observed transition. The overall cross section for climbing the vibrational ladder in the resonant region up to the quasi-continuum depends on many parameters: the absorption efficiency of the specific mode, its anharmonicity parameters, the initial vibrational and rotational distribution

in the cluster, the quasi-continuum limit of the cluster, and the laser frequency, peak intensity, and bandwidth.

In case of small clusters, e.g., trimers, the anharmonic shifts are large, the number of sublevels is small, and the quasi-continuum limit lies very high. Therefore, it can be difficult to drive the IR-MPE with consecutive single-photon absorptions. In this case, very high laser powers might be necessary in order to excite via two- and three-photon vibrational excitations.

With every absorbed photon, the internal energy E_i of the cluster is increased. Since the density of vibrational states increases roughly with E_i^N, where N is the number of vibrational degrees of freedom ($3n - 6$ for a nonlinear n-atom cluster), even a cluster that contains only few atoms quickly reaches a density of many states per cm^{-1} at an internal energy of >1000 cm^{-1}. Very quickly, the densities of states are reached where their coupling due to vibrational anharmonicities results in fast internal vibrational redistribution (IVR), typically on the order of picoseconds, which is short in comparison to typical macropulse lengths of an FEL. IVR rapidly removes the population from the excited state into the bath of vibrational background states so that the molecule can escape the anharmonic bottleneck and is ready for the next photon absorption (Oomens et al. 2006).

While small clusters need to absorb several photons before reaching the quasi-continuum, for species containing tens of atoms it is reached upon absorption of the very first photon. At that point, the IR-MPD spectrum often resembles the linear absorption spectrum because the absorption of the first photon is the rate-determining step in the excitation mechanism (Oomens et al. 2003, 2006). The energy deposition in the quasi-continuum region is mainly through stepwise incoherent one-photon transitions. Although the name "quasi-continuum" of vibrational states suggests that photons of any wavelength can be absorbed, this region is characterized by semiresonant absorption in zones near the original fundamental transition. The absorptions in the quasi-continuum are generally red-shifted due to cross-anharmonicities, which are due to the anharmonic interaction between the vibrational mode absorbing with the populated background vibrations in a hot molecule.

9.3.3.2 IR Multiple Photon Dissociation

As a competing process to IR-MPE, the cluster tries to lower its internal energy by either the emission of photons, electrons, or by fragmentation. The photon emission can be approximated by the Stefan-Boltzmann law and scales with the fourth power of the internal energy. It is the dominant cooling channel at low internal energies. The rate constants for fragmentation and electron ejection grow exponentially and will thus dominate at high energies. The branching ratio of these three processes depends on the internal energy and the specific properties of the cluster, like its ionization potential (IP, for neutral molecules) and dissociation energy (DE). Mostly, emission of (neutral) fragments is the faster process, as the DEs are generally lower than their IPs (second IPs in case of cations, electron affinity in case of anions). Once the internal energy rises over the DE, the dissociative channels open up. If the cluster is excited to an unbound state, it will dissociate. If the cluster is excited to a bound state above the DE, it will dissociate as well, since the bound state is now metastable due to coupling to the unbound states. The cluster can in principle also be excited further but dissociation quickly becomes dominant if the internal energy is further increased.

In practice, IR-MPD spectra are recorded, as described in Section 9.2.2.3. The clusters are exposed to the intense IR light either in a molecular beam or after mass selection in an ion trap, and the IR-laser-induced changes in their abundances are measured as a function of the frequency of the IR light. IR-MPD spectroscopy has allowed to elucidate the structures, for instance, for a variety of cationic and anionic transition metal oxide clusters (Janssens et al. 2006, Asmis and Sauer 2007, Asmis et al. 2007). Typical fragmentation channels for these systems are, depending on the stoichiometry of the oxide cluster, the loss of small neutral oxide species or of O_2.

In a similar way, information on species adsorbed to the surface of a cluster can be probed using IR-MPD. Recently, for instance, the interaction of carbon monoxide (CO) with transition metal clusters has been intensively studied as it acts as a model system for many relevant catalytic applications. The C–O-stretching frequency, $\nu(CO)$, is highly sensitive to the nature of the binding site and its local electron density. Measuring $\nu(CO)$ by means of IR spectroscopy has long been used in order to study binding sites of CO on transition metal surfaces and on technical catalysts. The IR-MPD experiments on the cluster complexes in the gas phase lead to insights into the cluster size and charge state dependence of the CO binding to the metal in terms of, e.g., binding geometry and degree of activation of the C–O bond (Fielicke et al. 2009).

Finally, the far-IR radiation emitted by IR-FELs also allows for directly exciting internal vibrational modes of bare metal clusters. The investigation of the vibrational properties of metal clusters by action spectroscopy becomes difficult for mainly two reasons: (1) the generally low IR absorption cross sections of the clusters, which show rather low polarity and (2) the low energy of single far-IR quanta. For the latter reason, even photodissociation of rare gas complexes, which are bound by typically 0.1–0.2 eV, requires the absorption of multiple photons. Using far-IR-MPD of such rare gas complexes, the vibrational spectra of cationic and neutral clusters of several transition metals have been measured in the range of their vibrational fundamentals (Fielicke et al. 2004, Asmis et al. 2007). As an example, Figure 9.8 shows the far-IR spectra of neutral Au_7, Au_{19}, and Au_{20} clusters that have been obtained upon IR-MPD of their complexes with a Kr atom. For

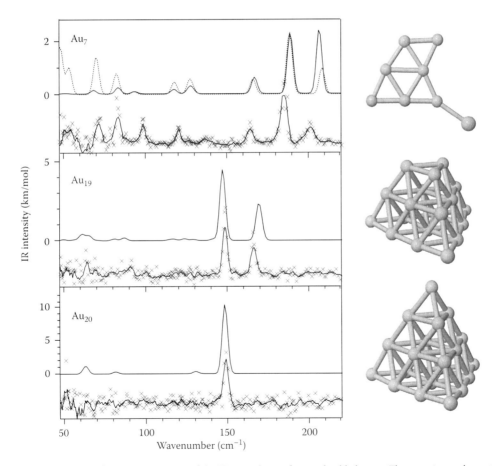

FIGURE 9.8 Far-IR multiple photon dissociation spectra of the Kr complexes of neutral gold clusters. The experimental spectra are compared to calculated spectra of the lowest lying isomers of the gold clusters. Note the change in the IR intensities in the calculated spectrum of the Au_7-Kr complex (dashed line), leading to better agreement with the experiment. The spectrum of the T_d symmetric Au_{20} cluster is dominated by an intense band at 150 cm^{-1}. In Au_{19}, this band splits as the symmetry is lowered to C_{3v} due to the missing corner atom.

mass spectrometric detection, the cluster complexes are ionized with 7.9 eV VUV photons. Comparison with spectra predicted by means of density functional theory allows for unambiguous assignments of the gas-phase structures (Gruene et al. 2008).

9.3.3.3 IR Multiple Photon Ionization

In some clusters that are strongly bound but have comparably low IPs with respect to the dissociation energy, the thermionic emission of electrons is favored over the fragmentation processes. In these cases, the vibrational spectra of the neutral clusters can be obtained via IR resonance-enhanced multiple photon ionization (IR-REMPI) (von Helden et al. 1997). To induce ionization, up to several hundreds of IR photons need to be absorbed by a single cluster. Recording the ion signal as a function of IR wavelength yields the IR-REMPI spectrum of the corresponding neutral cluster. Few systems are so exceptionally stable that thermionic ionization can be observed. Therefore, this technique has more limited applications. Examples for investigated systems include fullerenes (Figure 9.9), clusters of refractory metal oxides like MgO or ZrO_2, and clusters of transition metal carbides. For the latter, two different types of structures are formed by several metals: metallo carbohedrenes (met-cars) with the stoichiometry M_8C_{12} and cubic nanocrystals with close to 1:1 stoichiometry that resemble pieces of the bulk structure. This is why relatively small nanocrystals, e.g., $Ti_{14}C_{13}$ (a $3 \times 3 \times 3$ cube), show an IR-REMPI spectrum that strongly resembles the vibrational spectrum of the bulk carbide (von Helden et al. 2003).

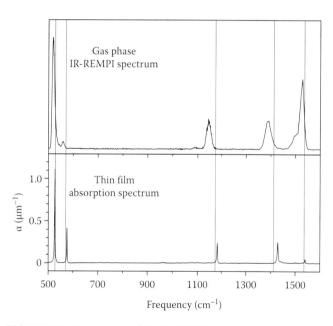

FIGURE 9.9 Comparison of the IR-REMPI spectrum of C_{60} with the linear thin-film IR absorption spectrum. The bands observed in the IR-REMPI spectrum are red-shifted in comparison to the linear absorption spectrum, which is a consequence of the anharmonic interaction between the vibrational mode absorbing with the populated background vibrations in hot C_{60}. A similar, albeit smaller shift is seen in the band positions of the IR emission spectrum of gas-phase C_{60} at 875 K (vertical lines). (Adapted from von Helden, G. et al., *J. Phys. Chem. A*, 107, 1671, 2003.)

9.4 Summary

Vibrational spectroscopy has a universal potential for obtaining detailed structural information on clusters. The various "action spectroscopy" techniques result in high sensitivities that allow for the characterization of strongly dilute media, like clusters in molecular beams. In particular, the introduction of the IR FEL for the spectroscopy of strongly bound clusters has recently led to a boost in the knowledge of cluster structures; the different detection schemes of (multiple) photon excitation allow to access the vibrational spectra of various types of clusters, in cationic, anionic, as well as neutral forms, and over a wide range of cluster sizes. Thereby, vibrational spectroscopy adds to the understanding of the unique properties of materials on the nanoscale, and allows unraveling how structures evolve as the particle size increases.

Acknowledgments

The authors wish to thank Gerard Meijer for his continuing support of our work and many useful suggestions and comments. We gratefully acknowledge the contributions from all coauthors of the original publications, the "Stichting voor Fundamental Onderzoek der Materie (FOM)" in providing beam time on FELIX and the skillful assistance of the FELIX staff, in particular, A. F. G. van der Meer and B. Redlich. K. R. Asmis and G. von Helden are acknowledged for providing the drawings of Figures 9.6 and 9.9, respectively. This work is supported by the Cluster of Excellence "Unifying Concepts in Catalysis" coordinated by the Technische Universität Berlin and funded by the Deutsche Forschungsgemeinschaft. J.T.L. thanks the Alexander von Humboldt Foundation for funding.

References

Asmis, K. R. and J. Sauer. 2007. Mass-selective vibrational spectroscopy of vanadium oxide cluster ions. *Mass Spectrom. Rev.* 26: 542–562.

Asmis, K. R., A. Fielicke, G. von Helden, and G. Meijer. 2007. Vibrational spectroscopy of gas-phase clusters and complexes, In *Atomic Clusters: From Gas Phase to Deposited*, ed. D. P. Woodruff, pp. 327–375. Amsterdam, the Netherlands: Elsevier.

Baker, M. D., J. Godber, and G. A. Ozin. 1985. Watching silver clusters grow in zeolites: Direct probe Fourier transform-far-infrared spectroscopy of the red form of fully silver ion exchanged zeolite A. *J. Phys. Chem.* 89: 2299–2304.

Berden, G., R. Peeters, and G. Meijer. 2000. Cavity ring-down spectroscopy: Experimental schemes and applications. *Int. Rev. Phys. Chem.* 19: 565–607.

Bernath, P. F. 2002. Infrared laser spectroscopy of transient species. In *An Introduction to Laser Spectroscopy*, eds., D. L. Andrews and A. A. Demikov, pp. 211–232. New York: Kluwer Academic.

Bondybey, V. E., A. M. Smith, and J. Agreiter. 1996. New developments in matrix isolation spectroscopy. *Chem. Rev.* 96: 2113–2134.

Casaes, R., R. Provencal, J. Paul, and R. J. Saykally. 2002. High resolution pulsed infrared cavity ringdown spectroscopy: Application to laser ablated carbon clusters. *J. Chem. Phys.* 116: 6640–6647.

Cheshnovsky, O., S. H. Yang, C. L. Pettiette, M. J. Craycraft, and R. E. Smalley. 1987. Magnetic time-of-flight photoelectron spectrometer for mass-selected negative cluster ions. *Rev. Sci. Instrum.* 58: 2131–2137.

Choi, M. Y., G. E. Douberly, T. M. Falconer, W. K. Lewis, C. M. Lindsay, J. M. Merritt, P. L. Stiles, and R. E. Miller. 2006. Infrared spectroscopy of helium nanodroplets: Novel methods for physics and chemistry. *Int. Rev. Phys. Chem.* 25: 15–75.

Curl, R. F. and F. K. Tittel. 2002. Tunable infrared laser spectroscopy. *Annu. Rep. Prog. Chem., Sect. C: Phys. Chem.* 98: 217–270.

Fielicke, A., A. Kirilyuk, C. Ratsch, J. Behler, M. Scheffler, G. von Helden, and G. Meijer. 2004. Structure determination of isolated metal clusters via far-infrared spectroscopy. *Phys. Rev. Lett.* 93: 023401.

Fielicke, A., P. Gruene, G. Meijer, and D. M. Rayner. 2009. The adsorption of CO on transition metal clusters: A case study of cluster surface chemistry. *Surf. Sci.* 603: 1427–1433.

Ford, M. S. and S. R. Mackenzie. 2005. Preparing transition-metal clusters in known structural forms: The mass-analyzed threshold ionization spectrum of V_3. *J. Chem. Phys.* 123: 084308.

Giesen, T. F., A. van Orden, H. J. Hwang, R. S. Fellers, R. A. Provençal, and R. J. Saykally. 1994. Infrared laser spectroscopy of the linear C_{13} carbon cluster. *Science* 265: 756–759.

Gruene, P., D. M. Rayner, B. Redlich, A. F. G. van der Meer, J. T. Lyon, G. Meijer, and A. Fielicke. 2008. Structures of neutral Au_7, Au_{19}, and Au_{20} clusters in the gas phase. *Science* 321: 674–676.

Guo, R., K. Balasubramanian, X. Wang, and L. Andrews. 2002. Infrared vibronic absorption spectrum and spin-orbit calculations of the upper spin-orbit component of the Au_3 ground state. *J. Chem. Phys.* 117: 1614–1620.

Handschuh, H., G. Ganteför, and W. Eberhardt. 1995. Vibrational spectroscopy of clusters using a "magnetic bottle" electron spectrometer. *Rev. Sci. Instrum.* 66: 3838–3843.

Harbich, W. 1999. "Soft landing" of size-selected clusters in chemically inert substrates. *Philos. Mag. B* 79: 1307–1320.

Honea, E. C., A. Ogura, D. R. Peale, C. Felix, C. A. Murray, K. Raghavachari, W. O. Sprenger, M. F. Jarrold, and W. L. Brown. 1999. Structures and coalescence behavior of size-selected silicon nanoclusters studied by surface-plasmon-polariton enhanced Raman spectroscopy. *J. Chem. Phys.* 110: 12161–12172.

Janssens, E., G. Santambrogio, M. Brümmer, L. Wöste, P. Lievens, J. Sauer, G. Meijer, and K. R. Asmis. 2006. Isomorphous substitution in bimetallic oxide clusters. *Phys. Rev. Lett.* 96: 233401.

Johnston, R. L. 2003. Evolving better nanoparticles: Genetic algorithms for optimising cluster geometries. *Dalton Trans.* 4193–4207.

Knickelbein, M. B. 1994. Photodissociation spectroscopy of Cu_3, Cu_3Ar, and Cu_3Kr. *J. Chem. Phys.* 100: 4729–4737.

Knickelbein, M. B. 1999. Reactions of transition metal clusters with small molecules. *Annu. Rev. Phys. Chem.* 50: 79–115.

Li, S., R. J. Van Zee, W. Weltner Jr., and K. Raghavachari. 1995. Si_3-Si_7. Experimental and theoretical infrared spectra. *Chem. Phys. Lett.* 243: 275–280.

Lombardi, J. R. and B. Davis. 2002. Periodic properties of force constants of small transition-metal and lanthanide clusters. *Chem. Rev.* 102: 2431–2460.

Marcy, T. P. and D. G. Leopold. 2000. A vibrationally resolved negative ion photoelectron spectrum of Nb_8. *Int. J. Mass Spectrom.* 196: 653–666.

Marshall, T. G. 1985. *Free-Electron Lasers,* New York: Macmillian Publishing Company.

Meier, R. J. 2007. Calculating the vibrational spectra of molecules: An introduction for experimentalists with contemporary examples. *Vib. Spectrosc.* 43: 26–37.

Miller, R. E. 1992. Infrared laser spectroscopy, In *Atomic and Molecular Beam Methods*, ed. G. Scoles, pp. 192–212. New York: Oxford University Press.

Morse, M. D., J. B. Hopkins, P. R. R. Langridge-Smith, and R. E. Smalley. 1983. Spectroscopic studies of the jet-cooled copper trimer. *J. Chem. Phys.* 79: 5316–5328.

Moskovits, M. 1985. Surface-enhanced spectroscopy. *Rev. Mod. Phys.* 57: 783–826.

Müller-Dethlefs, K. and E. W. Schlag. 1998. Chemical applications of zero kinetic energy (ZEKE) photoelectron spectroscopy. *Angew. Chem. Int. Ed.* 37: 1346–1374.

Neumark, D. M. 2008. Slow electron velocity-map imaging of negative ions: Applications to spectroscopy and dynamics. *J. Phys. Chem. A* 112: 13287–13301.

Nour, E. M., C. Alfaro-Franco, K. A. Gingerich, and J. Laane. 1987. Spectroscopic studies of nickel and iron clusters at 12 K. *J. Chem. Phys.* 87: 4779–4782.

Oepts, D., A. F. G. van der Meer, and P. W. van Amersfoort. 1995. The free-electron laser user facility FELIX. *Infrared Phys. Technol.* 36: 297–308.

Okumura, M., L. I. Yeh, J. D. Myers, and Y. T. Lee. 1986. Infrared spectra of the cluster ions $H_7O_3^+ \cdot H_2$ and $H_9O_4^+ \cdot H_2$. *J. Chem. Phys.* 85: 2328–2329.

Oomens, J., A. G. G. M. Tielens, B. Sartakov, G. von Helden, and G. Meijer. 2003. Laboratory infrared spectroscopy of cationic polycyclic aromatic hydrocarbon molecules. *Astrophys. J.* 591: 968–985.

Oomens, J., B. G. Sartakov, G. Meijer, and G. von Helden. 2006. Gas-phase infrared multiple photon dissociation spectroscopy of mass-selected molecular ions. *Int. J. Mass Spectrom.* 254: 1–19.

Rechtsteiner, G. A., C. Felix, A. K. Ott, O. Hampe, R. P. Van Duyne, M. F. Jarrold, and K. Raghavachari. 2001. Raman and fluorescence spectra of size-selected, matrix-isolated C_{14} and C_{18} neutral carbon clusters. *J. Phys. Chem. A* 105: 3029–3033.

Santambrogio, G., E. Janssens, S. Li, T. Siebert, G. Meijer, K. R. Asmis, M. Döbler, M. Sierka, and J. Sauer. 2008. Identification of conical structures in small aluminum oxide clusters: Infrared spectroscopy of $(Al_2O_3)_{1-4}(AlO)^+$. *J. Am. Chem. Soc.* 130: 15143–15149.

Siegbahn, K. 1982. Electron spectroscopy for atoms, molecules, and condensed matter. *Rev. Mod. Phys.* 54: 709–728.

Toennies, J. P. and A. F. Vilesov. 2004. Superfluid helium droplets: A uniquely cold nanomatrix for molecules and molecular complexes. *Angew. Chem. Int. Ed.* 43: 2622–2648.

Van Orden, A. and R. J. Saykally. 1998. Small carbon clusters: Spectroscopy, structure, and energetics. *Chem. Rev.* 98: 2313–2357.

von Helden, G., I. Holleman, G. M. H. Knippels, A. F. G. van der Meer, and G. Meijer. 1997. Infrared resonance enhanced multiphoton ionization of fullerenes. *Phys. Rev. Lett.* 79: 5234–5237.

von Helden, G., D. van Heijnsbergen, and G. Meijer. 2003. Resonant ionization using IR light: A new tool to study the spectroscopy and dynamics of gas-phase molecules and clusters. *J. Phys. Chem. A* 107: 1671–1688.

Whittle, E., D. A. Dows, and G. C. Pimentel. 1954. Matrix isolation method for the experimental study of unstable species. *J. Chem. Phys.* 22: 1943–1944.

Xu, C., T. R. Taylor, G. R. Burton, and D. M. Neumark. 1998. Vibrationally resolved photoelectron spectroscopy of silicon cluster anions Si_n^- ($n = 3$–7). *J. Chem. Phys.* 108: 1395–1406.

Yang, D.-S. and P. A. Hackett. 2000. ZEKE spectroscopy of free transition metal clusters. *J. Electron Spectrosc. Relat. Phenom.* 106: 153–169.

10

Electric and Magnetic Dipole Moments of Free Nanoclusters

Walt A. de Heer
Georgia Institute of Technology

Vitaly V. Kresin
University of Southern California

10.1 Introduction

This chapter discusses the behavior and detection of electric and magnetic dipole moments in atomic and molecular nanoclusters. We concentrate on studies of free particles (i.e., clusters produced, probed, mass selected, and detected in beams [Pauly 2000]), because in such work the focus is on exploring the inherent physical properties of these nanoparticles. In this way, clusters can be studied with their size and composition precisely known, and with their features unperturbed by substrate interactions.

A fundamental characteristic of a molecule or nanoparticle is the arrangement and distribution of electrical charges within it, as well as that of electronic orbital and spin angular momenta. Electric and magnetic dipole moments are observables, which directly reflect these distributions, and are, therefore, very useful experimental parameters for the direct assessment of models and theories of particle structure, bonding, and internal dynamics. In addition, of course, polar and magnetic nanoclusters carry significant practical promise as building blocks for materials with novel magnetic and ferroelectric properties.

Although the principles underlying electric and magnetic ordering in nanoclusters may be distinct, the experimental signatures of such ordering in cluster beam work have a lot in common, in particular, the patterns of beam deflection under the influence of externally applied fields. It is for this reason that they are discussed jointly in this chapter.

10.2 Definitions

As a starting point, it is appropriate to cite the textbook expression for the total electric dipole moment:

$$\vec{p} = \int \rho(\vec{r})\vec{r}\,\mathrm{d}^3 r. \tag{10.1}$$

Here $\rho(\vec{r})$ is the charge density (including both electrons and nuclei) at position $\vec{r}$ within the particle, and the integral is over the entire particle volume. It is important to keep in mind that the magnitude of the electric dipole moment is independent of the choice of the origin if, and only if, the particle on the whole is electrically neutral.

Electric dipole moments are usually quoted in Debye (D) units ($1\,\mathrm{D} = 3.34 \times 10^{-30}\,\mathrm{C \cdot m} = 0.21\,e \cdot \text{Å} = 0.39$ a.u.).

The corresponding definition of the magnetic dipole moment is

$$\vec{\mu} = \frac{1}{2c}\int \vec{r} \times \vec{J}(\vec{r})\,\mathrm{d}^3 r. \tag{10.2}$$

This is written in the Gaussian (cgs) system of units. $\vec{J}$ is the volume current density and c is the speed of light. Since current flow is associated with the motion, or velocity, of charges, the vector product is obviously related to the charges' angular momentum. The resulting quantum-mechanical expression for the magnetic moment of a system of electrons is

$$\hat{\vec{\mu}} = \frac{e\hbar}{2m_e c}\left(\hat{\vec{L}} + 2\hat{\vec{S}}\right). \tag{10.3}$$

The hats emphasize the fact that we are dealing with quantum-mechanical operators of the orbital and spin angular momentum. A similar equation would also describe nuclei; however, because the mass entering the denominator would be so much greater, the discussions and measurements of cluster magnetic dipole moments have so far focused on the dominant electronic contribution.

Magnetic dipole moments are usually quoted in Bohr magneton (μ_B) units ($1\,\mu_B = e\hbar/2m_e c = 9.27 \times 10^{-21}$ erg/G $= 9.27 \times 10^{-24}$ J/T).

As regards notation, one should be aware that the electric dipole moment is frequently denoted by $\vec{\mu}$, especially in the literature on molecular physics and physical chemistry, and also by $\vec{d}$; magnetic moments are sometimes denoted by $\vec{m}$.

10.3 Forces and Deflections Produced by External Fields

Many recent experimental measurements of the dipole moments of free clusters have been performed by the technique of beam deflection, as sketched in Figure 10.1. The main idea is that a narrow, accurately collimated beam of nanoclusters is directed through the region of a static inhomogeneous electric or magnetic field.

The potential energy of a permanent dipole in an external electric field is

$$U = -\vec{p}\cdot\vec{E}, \tag{10.4}$$

and by differentiating this, one finds (Griffiths 1999) the force experienced by a polar particle:

$$\vec{F} = (\vec{p}\cdot\vec{\nabla})\vec{E}. \tag{10.5}$$

(It is assumed, of course, that the dimensions of the dipole are much smaller than the scale of spatial variation of the field.)

The above equation reflects the fact that a homogeneous field will not exert a net force on the dipole (the forces on the negative and positive regions will balance), and therefore the apparatus field must be designed with a well-defined gradient. A similar equation can be written down for the magnetic force:

$$\vec{F} = (\vec{\mu}\cdot\vec{\nabla})\vec{B}. \tag{10.6}$$

Thus, as a cluster particle passes between the electrostatic plates or magnet poles, two things happen: the field applies a torque to the dipole ($\vec{\tau} = \vec{p}\times\vec{E}$ or $\vec{\mu}\times\vec{B}$), attempting to align it along the field lines; at the same time, the particle is given an impulse by the collinear component of the field gradient.

Different quantum states (rotational, vibrational, electronic) of a cluster or a molecule usually possess different effective dipole moments; therefore, one also needs to express the force in terms of the field-induced shift of state energy, i.e., in terms of the Stark (electric) or Zeeman (magnetic) shift. Thus for a state $|i\rangle$ with energy $\varepsilon_i(E)$ one has

$$\vec{F}_i = -\nabla\varepsilon_i(E) = \left(-\frac{\partial\varepsilon_i}{\partial E}\right)\left(\frac{\partial\vec{E}}{\partial z}\right) \tag{10.7}$$

(and analogously for magnetic fields and moments), where for conciseness we assumed that the field varies only along one coordinate, which is indeed the case for the most common experimental arrangements. In view of Equation 10.5, the first factor in Equation 10.7 can be viewed as the projection of the dipole moment in state $|i\rangle$ on the field direction, $p_{i,z}$ (or $\mu_{i,z}$). States with negative Stark or Zeeman shifts (negative derivatives) are called high-field-seeking states, and those with positive shifts are called low-field-seeking states.

One established field geometry comes from an equidistant arrangement of a set of electric or magnetic poles around a circle, in a cylindrical-type assembly (e.g., a "hexapole"). This creates a radial field near the cylinder axis (Parker and Bernstein 1989, Pauly 2000), which acts to defocus or refocus an axially collimated beam of molecules, depending on their quantum state, serving as a kind of thick lens for polar species.

For cluster beams, a much more common arrangement is a set of two parallel, curved plates or poles producing the so-called Rabi "two-wire" deflecting field visible in Figure 10.1, for which the field and its gradient are collinear and their product is approximately constant over a relatively large region around the

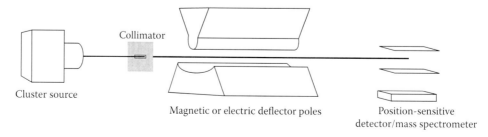

Cluster source

Collimator

Magnetic or electric deflector poles

Position-sensitive detector/mass spectrometer

FIGURE 10.1 Outline of a configuration for electric or magnetic deflection measurements for nanocluster beams.

beam axis (Ramsey 1956, Pauly 2000, Tikhonov et al. 2002). This type of field produces a sideways force, $F_z = p_z(\partial E/\partial z)$ or $F_z = \mu_z (\partial B/\partial z)$, and a corresponding sideways deviation of the particle:

$$d_z = a\langle p_z \rangle \frac{(\partial E/\partial z)}{m_c v^2}, \qquad (10.8)$$

where

$\langle p_z \rangle$ is the time-averaged projection of the cluster dipole moment in the direction of the field

a is a geometrical constant of the apparatus (encompassing such things as the length of the deflection plates and the flight distance from the deflection region to the detector; it is usually determined by calibrating the apparatus with a beam of well-characterized polarizable species)

m_c is the mass of the nanocluster

v is the beam velocity (one power of v for the time spent in the field region, and another for the flight time to the detector; it is evident that accurate measurements require a good knowledge of beam velocity)

An identical form can be written for the magnetic case.

In a real experimental situation, one needs to average the above expression over all populated internal states and initial orientations of the entering particles, and convolute with the finite width of the incoming beam. Then, one can finally extract information about the inherent electric or magnetic dipole moment of the nanocluster.

10.4 Beam Deflection and Broadening: "Rigid" and "Floppy" Polar Particles

10.4.1 Beam Deflection

It is evident from the foregoing that a critical bridge between the experimental signature and the intrinsic cluster moments p and μ lies in elucidating the average orientation of the former during passage between the electric plates or the magnetic poles. The two most commonly encountered parameters are the average deflection and the broadening of the beam.

The former corresponds to the average of $\langle p_z \rangle$ or $\langle \mu_z \rangle$ for the whole population of particular size clusters in the beam, or equivalently to the average orientation cosine, i.e., the expectation value of the angle between the molecular dipole axis and the field direction: $\langle \cos\theta \rangle = \langle p_z \rangle/p$. The averaging takes place over the flight time through the field and over the internal state distribution of the nanoclusters.

If the particles were in thermal equilibrium at some temperature T, the answer would be well known. For an ensemble of polar molecules in a gas in the presence of an external field, it was shown by Langevin (1905) that $\langle p_z \rangle = p[\coth x - 1/x] \equiv p\mathcal{L}(x)$, where $x \equiv pE/k_BT$ (or replace p, E by μ, B). $\mathcal{L}(x)$ is called the Langevin function. If $x \ll 1$, the polarization can be approximated by $\langle p_z \rangle = p^2E/3k_BT$, generally referred to as the Langevin–Debye

law. The textbook derivation is usually done for a point dipole, but in *tour de force* calculations Van Vleck (1932) and Niessen (1929) proved that the result is valid (under realistic conditions) both in classical and in quantum mechanics for a general system possessing a dipole moment, with the understanding that p^2 represents a statistical mean over the phase space of the system.

Unfortunately, the situation for a transiting beam of polar nanoclusters is qualitatively different. Here the particles' state distributions are acquired in an earlier zero-field region (e.g., the rotational and vibrational temperatures T_{rot}, T_{vib} produced at the beam source), and do not change, i.e., re-thermalize, upon the relatively slow, adiabatic entry into the field. (Nonadiabatic, a.k.a. "Majorana," transitions between close-lying quantum states coupled by the field may complicate the situation further but will be neglected here.) Although the fact that low-field response should still scale with the ratio pE/T remains expected on dimensional grounds (Schnell et al. 2003), the numerical correlation between the magnitude of the dipole and the average force on the particle ensemble (and hence its deflection) is no longer universal.

As a result, experimental work and analysis have so far focused on, or sometimes simply assumed, certain treatable cases: (1) Clusters whose dipole moment rigidly rotates with the cluster's reference frame and does not interact with vibrations ("rigid" or "locked" dipole, or "strong magnetic anisotropy"). (2) Clusters within which the orientation of the dipole can relatively easily readjust with respect to the particle axis ("floppy," "fluxional," or "superparamagnetic" cases). Situations with intermediate coupling strengths and anisotropy energies remain a challenge.

As mentioned above, in both cases the contribution of the permanent dipole will scale as pE/T or $\mu B/T$, but a qualitative difference immediately arises: for "rigid" clusters the relevant T will be the rotational temperature of the beam, and for "floppy" clusters it will be their vibrational ("internal") energy. This is critically important, because for clusters in molecular beams the two temperatures do not at all have to be (and commonly are not) equal.

For floppy clusters, one normally assumes that their internal degrees of freedom are able to serve as a canonical heat bath for the dipole moment. For such systems, one typically observes that the beam profile is deflected uniformly, as would a beam of particles possessing only an induced dipole moment: in Equation 10.8 for $\langle p_z \rangle$ ($\langle \mu_z \rangle$) one has $\langle p_z \rangle = \chi E$ ($\langle \mu_z \rangle = \chi B$) with the linear susceptibility

$$\chi = \frac{p^2}{3k_BT_{int}}, \qquad (10.9)$$

(or $\mu^2/3k_BT_{int}$) and the deflection is characteristically proportional to the square of the field intensity (since in deflection setups the field gradient and the field strength are proportional). For practical field strengths, the linear Langevin–Debye limit of the susceptibility is adequate. Examples of such behavior will be presented in subsequent sections. It is important to appreciate that the thermal fluctuations implicit in Equation 10.9 can disguise quite strong dipole moments as weak susceptibilities: for example, a $p = 1\,\mathrm{D}$ electric dipole moment in a practical $E = 100\,\mathrm{kV/cm}$ electric field at $T = 300\,\mathrm{K}$ is effectively suppressed by a factor of $pE/3\,k_B T \approx 3 \times 10^{-3}$.

As the vibrational temperature of the cluster decreases below the dipole fluctuation (or anisotropy) energy, the end result is the locked-moment regime, i.e., the problem of a polar rigid body rotating within an external field. The subject is recognizable from the classical and quantum-mechanical treatments of the rotational dynamics of rigid bodies, molecules, nuclei, etc.; the theoretical treatments and results are interestingly varied, and not all questions are yet resolved.

The most important variation arises based on the symmetry of the cluster in question. A rigid top can be characterized by its three principal moments of inertia $I_{a,b,c}$. Often their reciprocals are used: rotational constants A, B, C defined, e.g., as $B = \hbar^2/2I_b$, and by convention assigned as $A \geq B \geq C$. (The rotational constant B should not be confused with the magnetic field denoted by the same symbol.) If $A = B = C$, the top is called spherical; if $A = B$ it is symmetric; and if all constants are different this is an asymmetric top. As is well known, the free motion of a spherical or symmetric top has an exact solution, while the problem of the asymmetric top is non-integrable and can exhibit unstable dynamics.

For symmetric rotor clusters, in the second order of perturbation theory, one finds

$$\langle p_z \rangle = \frac{p^2 E}{k_B T_{\mathrm{rot}}} z(\kappa) \qquad (10.10)$$

(and the same for μ in place of p)

where
- $\kappa = (C - A)/C$ for a prolate top (with $B = C < A$ and therefore $\kappa < 0$)
- $\kappa = (A - C)/A$ for an oblate top (with $B = A > C$ and $0 < \kappa < 0.5$)
- $z(\kappa)$ is a somewhat bulky function written out in Bulthuis et al. (2008)

For small deviations from a spherical rotor, i.e., for small κ, one finds $z(\kappa) = \frac{2}{9}\left(1 - \frac{1}{5}\kappa + \cdots\right)$. For a spherical rotor cluster, $z(0) = 2/9$ (instead of the universal factor of $1/3$ for the Langevin–Debye thermal-equilibrium value). For deflecting fields stronger than those treatable by perturbation theory, numerical evaluation may become necessary (Dugourd et al. 2001, Ballentine et al. 2007). From this result, one already sees that for extracting reliable values of the dipole moment from a deflection profile, even for a rigid symmetric-top system, it is important to have good knowledge of the shape parameter κ of the cluster as well as of its rotational temperature. Otherwise the information obtained will at best qualify as "semiquantitative."

For asymmetric clusters, the situation is not clear-cut. No analytical solution for adiabatic-entry orientation for asymmetric tops is available, and even a second-order perturbation approach is intractable, hence in general one has to resort to numerical diagonalization of the rotation + field Hamiltonian

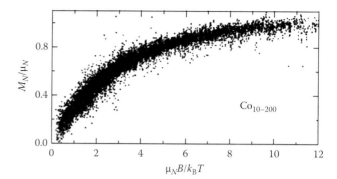

FIGURE 10.2 Normalized magnetization projections of Co_N clusters ranging in size from $N = 10$–200, plotted as a function of $x = \mu_N B/k_B T$, from Xu et al. (2008). The plot combines 63 separate data sets collected at nine temperatures from 25 to 100 K and seven fields from 0 to 2 T, representing about 10,000 data points in total, all collapsing into a universal curve. The trend is consistent with the Langevin function, although the latter approaches saturation slightly slower than the data points. The initial slope is consistent with Equation 10.15. The underlying mechanism in this case is not internal spin relaxation but rotational–precessional dynamics and the relevant T is the rotational temperature of the cluster beam.

(Abd El Rahim et al. 2005, Moro et al. 2007), or to classical molecular dynamics simulations, which are applicable for large angular momenta (Dugourd et al. 2006). However, there are indications that out of complexity there may arise simplicity. Qualitatively (to our knowledge, there are no rigorous proofs of these conjectures), the non-integrability of asymmetric top rotational motion implies a nascent chaotic behavior, which in turn hints at ergodicity and thermodynamics. As a result, the statistical Langevin–Debye behavior is restored (see Figure 10.2 and Section 10.6.2). This can take place when sufficiently high rotational levels come into play (Bulthuis et al. 2008), in particular when the Stark (or Zeeman) diagram of the rotational levels is replete with avoided crossings (Xu et al. 2008). The latter has been identified as one of the marks of quantum chaos, and can be enhanced by perturbation of the top by weak collisions with background gas molecules or by rotational–vibrational coupling (Abd El Rahim et al. 2005, Antoine et al. 2006). Full theoretical criteria for the onset of such simplifying behavior have not yet been established, however, so individual deflection patterns of asymmetric nanoclusters need to be assessed on a case-by-cases basis and/or treated numerically.

10.4.2 Beam Broadening

The average shift of the cluster beam discussed above reflects the overall field-induced reorientation of the particles' dipole moments, in other words, the net orientational polarization of the cluster ensemble induced by the deflecting field. Another important observable is the significant amount of (generally asymmetric) broadening displayed by beams of "locked-moment" clusters. Indeed, the very presence of such broadening is taken as a qualitative indicator that the nanoclusters possess a strong "frozen-in" permanent dipole moment (in contrast, as mentioned above, beams of "floppy" clusters tend to deflect uniformly without broadening, appearing as particles with an enhanced polarizability). The broadening originates from the simple fact that particles enter the deflection field region with randomly oriented dipole moments. Those clusters whose dipole has a component along the field will experience a force opposite to those whose moment points against the field, and as a result the beam spreads out and broadens. For weak fields that do not induce a strong reorientation of the cluster population, the broadening will be symmetric (equal numbers of moments pointing along and against the field); increasing field intensity is accompanied by a growing polarization of the cluster ensemble and an increasingly asymmetric broadening (as well as an increasing overall shift of the beam's "center of mass" discussed above), as shown in Figure 10.3b.

For weak fields, to first order the broadening of the beam can be taken to be proportional to the zero-field population (i.e., that which exists prior to entering the field region) of various p_z (or μ_z) projections of the cluster dipole moment (see Equation 10.8). For rigid spherically symmetric tops, the rotational energy is given by

$$\varepsilon = BJ(J+1) \tag{10.11}$$

where
  B is the rotational constant
  J is the angular momentum quantum number

The component of the dipole moment along the z axis is (Townes and Schawlow 1975)

$$p_z = p\frac{MK}{J(J+1)} \tag{10.12}$$

where M and K are projections of the angular momentum on the space- and body-fixed axes. For a fixed J, the population of each M, K sublevel is the same, $P_0(M,K) = 1/(2J+1)^2$, and so

the probability of p_z having a certain value is proportional to the number of ways that integers M and K, each ranging from $-J$ to $+J$, can be combined into a product equal to $J(J+1)(p_z/p)$. Under the assumption that mostly $J \gg 1$ states contribute to the beam population and therefore $J(J+1) \approx J^2$ and $(2J+1)^2 \approx 4J^2$, one finds the following zero-field probability distribution of the dipole moment projections:

$$P_0(p_z) = \frac{1}{2p}\ln\left(\frac{p}{|p_z|}\right) \tag{10.13}$$

and similarly for magnetic moments μ_z, as shown in Figure 10.3a. The singularity at $p_z = 0$ is just an artifact of the continuum approximation (Bertsch and Yabana 1994, Liang 2009). Since expression (10.13) is independent of J, the entire zero-field moment projection distribution for the whole cluster beam ensemble has the same form.

As stated above, to lowest order, the effect of the deflecting field is to displace each dipole according to its original orientation. The weak-field beam profile $b(z)$ is therefore a convolution of the profile of the entering beam $b_0(z)$ and the moment component distribution (10.13): $b(z) \approx b_0(z)*P_0(p_z)$. Since the variance of a convolution is the sum of two individual variances (Jaynes 2003), and the variance of the distribution (10.13) is $\sigma_p^2 = p^2/9$ (as can also be derived directly from Equation 10.12 (Schäfer, Assadollahzadeh et al. 2008)), it follows that the magnitude of the dipole moment can be deduced from the degree of beam broadening as

$$p = 3\sqrt{\sigma_b^2 - \sigma_{b_0}^2}. \tag{10.14}$$

It is important to keep in mind the relatively narrow region of applicability of this expression: spherically symmetric tops in the weak-field (uniform broadening) regime. For rigid symmetric rotors, the low-field σ_p becomes a function of the asymmetry parameter κ encountered in Equation 10.10 (Bulthuis and Kresin, to be published).

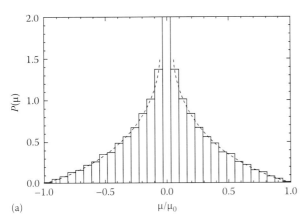

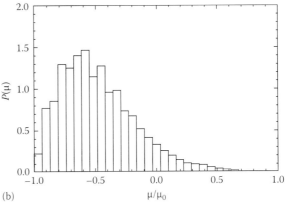

(a) μ/μ_0 (b) μ/μ_0

FIGURE 10.3 Distributions of magnetic moment projections for an ensemble of spherically symmetric rigid rotors with a magnetic moment μ_0. The histograms show numerical results for states with $J_{max} \le 40$ and rotational temperature $T_{rot} = 100B$, where B is the cluster rotational constant. (a) Zero magnetic field; the dashed line is the approximation Equation 10.13. (b) Adiabatic entry into a magnetic field of intensity equal to $4T_{rot}/\mu_0$. (From Bertsch, G.F. and Yabana, K., *Phys. Rev. A*, 49, 1930, 1994. With permission.)

For asymmetric tops, the response to external fields is dominated by the aforementioned avoided crossing phenomenon, and one finds that beam broadening should decrease with increasing rotational temperature (Xu et al. 2008): $\sigma_\mu^2 \approx \mu^2 B / \left[3k_B T \sqrt{(\gamma - 1)} \right]$ (written for magnetic dipoles for which the result was originally derived), where it is assumed that the density of states is proportional to some power of the excitation energy: $\rho(\varepsilon) \sim \varepsilon^\gamma$, which is often the case. The electric-field deflection of asymmetric molecules provides qualitative support to this prediction (Carrera et al. 2008).

10.4.3 Résumé

In summary, the deflection pattern of a beam of polar clusters provides two main variables that in principle can be correlated with the value of the dipole moment: the average displacement and (for locked-moment species) the amount of broadening. For certain families and in certain limits, these correlations can be derived analytically; for other situations, such as asymmetric-top particles and strong deflecting fields, one has to resort to numerical simulation of the beam profile. The analytical expressions play another important role: they point out the fact that unless one begins with a consistent physical picture of the cluster's shape, structure, and temperature (rotational or vibrational, as required by the situation), quantitative errors can result in extracting its electric and magnetic dipole moments from field deflection profiles.

10.5 Electric Dipole Moments in Selected Cluster Families

This section outlines experimental information on the electric dipole moments of several selected cluster families obtained via beam measurements. (An analogous overview of magnetic moments of free clusters is presented in Section 10.6.)

10.5.1 Metal-Fullerene Clusters

A beautiful symmetric-top polar system illustrating "textbook" response behavior ranging from the floppy to the rigid is a fullerene C_{60} with a metal atom (usually an alkali) riding on the outside of the cage (Figure 10.4). These beam-deflection experiments are reviewed in detail by Broyer et al. (2002, 2007). Because of charge transfer from the atom to the C_{60} cage, a strong dipole is formed.

At low temperatures, the alkali atom (and hence the dipole moment) is bound to the cage and the electric deviator induces significant beam broadening, as shown in Figure 10.5a. Analogous behavior is found for the strongly bound TiC_{60} (Dugourd et al. 2001). As the temperature increases, the alkali atom becomes able to move freely around the cage (the hopping barrier has been found to be 0.02 eV [Dugourd et al. 2000]), and the beam profile changes into a pure deflection pattern, as shown in Figure 10.5b. The dipole moments of MC_{60}, determined from

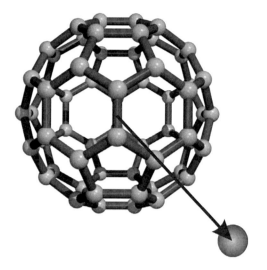

FIGURE 10.4 A polar metal-fullerene cluster. (Adapted from Dugourd, Ph. et al., *Phys. Rev. A*, 62, 011201R, 2000.)

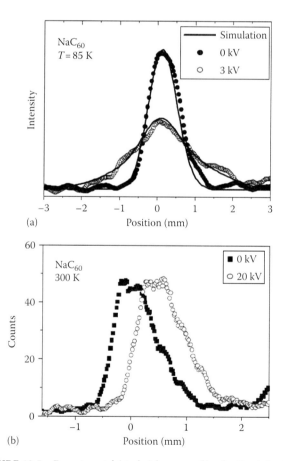

FIGURE 10.5 Experimental (circles) beam profiles for the deflection of NaC_{60}, (a) at 85 K and (b) at 300 K. On the left, the solid line is a calculation for a rigid locked-moment rotor with $p = 14.8$ D. On the right, the deflection pattern reflects a fluctuating dipole resulting from the mobility of the alkali atom on the fullerene cage. (From Broyer, M. et al., *Phys. Scr.*, 76, C135, 2007. With permission.)

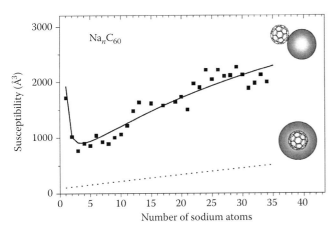

FIGURE 10.6 Susceptibility of Na_nC_{60} clusters. Dark squares: experimental values. Dashed line: values calculated assuming that the sodium atoms form a metal shell around the fullerene. Full line: values calculated assuming formation of a metal droplet on the surface. (From Broyer, M. et al., *Phys. Scr.*, 76, C135, 2007. With permission.)

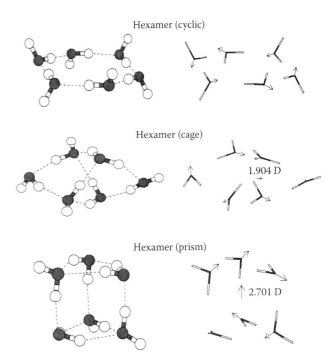

FIGURE 10.7 Three theoretical structures of the water hexamer and their dipole moments. (The cyclic isomer has no dipole moment.) (From Gregory, J.K. et al., *Science*, 275, 814, 1997. With permission.)

the beam profiles, increase with the size (and hence decreasing ionization potential) of the alkali atom M, ranging from 12.4 D for M = Li to a huge 21.5 for M = Cs.

The experiments were extended to fullerenes carrying not just one, but a whole cluster of alkali molecules on the surface (Na_nC_{60}). Interestingly, it appears that for n up to ≈ 8, the atoms are spread around on the surface of the cage ("wetting"), but larger sodium clusters tend to collect into a droplet (non-wetting) (see Figure 10.6). Note that in the case of a few Na atoms on the surface, in the lowest-energy state they may arrange themselves symmetrically around the cage and thereby produce only a low net dipole moment, but at finite temperatures they can slide around on the surface and give rise to a much higher dipole. Thus the time-averaged dipole moment that enters Equation 10.9 is itself temperature-dependent (Broyer et al. 2007), an effect that may be labeled "vibronic-induced polarization."

10.5.2 Water Clusters

The high dipole moment of the isolated water molecule, 1.855 D (Clough et al. 1973), which in a liquid is further enhanced to ~3 D (at T = 300 K) by intermolecular interactions (Gubskaya and Kusalik 2002), is responsible for the chemical and solvent properties of water and, as one corollary, for the form of life as we know it. Despite its relatively simple molecular structure, water in the condensed phase presents a major challenge to theory and simulations because of its very high rotational and hydrogen-exchange mobility and the propensity to form a network of crucially important but fleeting and fluctuating hydrogen bonds. Water clusters have been extensively studied as model systems for understanding such correlated networks, and the dipole moment of a nanodroplet or nanoicicle of a finite number of water molecules is an important characteristic of such an assembly.

High-resolution laser measurements of the Stark effect of so-called vibration–rotational tunneling spectra of very cold clusters in a beam were performed by Gregory et al. (1997). They obtained an electrical dipole moment of ≈ 1.9 D for the cage isomer of the $(H_2O)_6$ species, in agreement with calculations (see Figure 10.7).

Beam-deflection experiments were carried out by Moro et al. (2006). In this case, the clusters were estimated to be at a temperature of ~200 K, and the beam exhibited a uniform deflection proportional to the square of the electric field, a picture fully consistent with orientational polarization of a "floppy" dipole. Using Equation 10.9 and subtracting the electronic polarizability contribution, values of $p \approx 1.3$ D for $(H_2O)_{3-8}$ and $p \approx 1.6$ D for $(H_2O)_{9-18}$ were obtained. In this situation, the p value represents the time-averaged total dipole moment of a nano-assembly of rotationally labile polar water molecules. Repeating the measurement under colder expansion conditions yielded χ values shifted upward in agreement with Equation 10.9 implying that the average magnitude of p, and therefore the landscape of molecular fluctuations, remained unchanged down to ~120 K, and therefore suggesting that at this temperature there still had been no freezing transition to a state of molecular orientational ordering.

10.5.3 Alkali Halide Clusters

The alkali halide salts are prototypes of ionically bonded polar structures. In fact, the very first electrostatic beam-deflection experiment was performed in 1927 on the KI molecule (Wrede 1927,

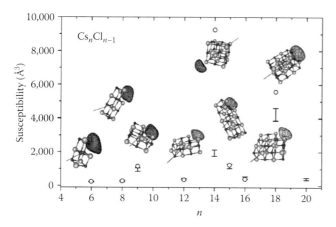

FIGURE 10.8 Calculated (circles) and experimental (bars) electrical susceptibilities of Cs_nCl_{n-1} clusters. For each size, the theoretical lowest energy structure is also shown, together with the direction of the electric dipole and a surface of constant charge density for the excess electron. (From Jraij, A. et al., *Chem. Phys.*, 322, 298, 2006. With permission.)

Ramsey 1956). The primary structural motif of alkali halide clusters is believed to be that of cubic nanocrystals analogous to the bulk lattice structure. Deflection experiments have been performed on a number of M_nX_{n-1} clusters (M = alkali atom, X = halogen atom) with n in the range from 2 to 20 (Rayane et al. 2002, 2003; Jraij et al. 2006). Many of these are considered to be cuboids with a vacancy, in particular a corner vacancy, which is occupied by the electron contributed by the excess metal atom (see Figure 10.8). This gives rise to strong dipole moments, $p \sim 3–10\,D$.

An interesting feature of the ionic nanocrystal data is that in the temperature range from 77 K up to 500 K, the cluster deflection profiles are uniformly displaced, unbroadened, and with an average shift proportional to the square of the electric field, corresponding to a susceptibility of the form given in Equation 10.9. This looks counterintuitive, because these clusters would be expected to be of the "rigid dipole" type. The response dynamics of these systems is not fully understood, although simulations suggest that it is strongly affected by the presence of vibrational excitations (Rayane et al. 2002), and it would also be interesting to explore the implications of the avoided crossing model (Xu et al. 2008).

10.5.4 Metal Cluster Dipole Moments and Metal Cluster Ferroelectricity

One of the most characteristic properties of metals is that they cannot support voltage differences. This implies the absence of electric fields in bulk metals. Noting that even a relatively small dipole moment of 1 D will produce local electric fields of the order of 10^7 V/cm in a 1 nm particle, one might expect that electric dipole moments are rare in metal clusters. This is indeed found to be the case. With very few exceptions, the electric dipole moments in metal clusters are immeasurably small ($\ll 1\,D$). For example, recent measurements of sodium clusters up to Na_{200} show that all of them have dipole moments below $5 \times 10^{-3}\,D$ per atom, with the only possible small exceptions of Na_3 and Na_6

(Liang 2009, Bowlan et al., to be published). Apparently even very small clusters behave like bulk metal particles, which cannot have electric dipole moments.

The general absence of dipole moments in metal clusters draws attention to those few species that violate this rule. For example, rhodium clusters in a beam, $Rh_{5–28}$, have been investigated by Beyer and Knickelbein (2007). Most exhibited small uniform deflections in the high-field direction, corresponding to purely polarizable behavior. However, Rh_7 and Rh_{10} manifested anomalously high susceptibility values, suggesting the presence of permanent dipole moments. The Rh_7 beam behaved as that of a rigid rotor, displaying relatively symmetric broadening and a centroid shift compatible with Equation 10.10. Using this equation with the assumption $\kappa \approx 0$ yielded an electric dipole moment of 0.24 D. Rh_{10}, on the other hand, differed in producing only a uniform shift toward the high field, and in being sensitive to variation of cluster vibrational temperature by the source expansion conditions. This behavior was interpreted as being characteristic of a spatially fluctuating (i.e., floppy) dipole. The suggestion of two closely related nanoparticles displaying two different types of dipole moment coupling is interesting and highlights the sensitivity of cluster polarization to cluster structure.

A somewhat analogous picture emerged in a study of lead clusters $Pb_{7–38}$ by Schäfer et al. (2008). In this case, $Pb_{12,14,18}$ exhibited anomalous deflections, which were interpreted as evidence for their possessing a polar structure (and in the case of Pb_{12} for there being two isomer populations in the beam, one polar and the other nonpolar).

Besides these few and rare cases, there is also an entire class of metal clusters that display unusual polarization. Clusters of vanadium, niobium, and tantalum, as shown by Moro et al. (2003), as well as several alloys of these metals, behave normally at room temperature, but exhibit very large dipole moments at cryogenic temperatures (typically below 25 K). The origin of this effect is very different from the electric dipole moments of nonmetallic clusters. The fact that they only have dipole moments at low temperatures and not at higher temperatures indicates that they undergo a phase transition from a normal state to one that has an electric dipole moment. Materials with this property are called ferroelectrics.

In the bulk, ferroelectric materials are always nonmetals. Hence, this form of metallic ferroelectricity is a cluster property. Moreover, not only are the dipole moments large, they are ubiquitous and in all cases show several distinctive properties (Xu et al. 2007). One is that the dipole moment has an odd–even effect. That is, the dipole moments of the even clusters are systematically larger than those of the adjacent odd clusters. Neither the dipole moment nor the even–odd effect show a tendency to diminish with increasing cluster size. Moreover, careful experiments in which "impurity" atoms were imbedded in the clusters demonstrated that the odd–even effect is directly related to the number of electrons in the cluster (Yin et al. 2008). These unusual effects indicate that the electric dipole moments are not directly related to the atomic structure of the clusters, but rather to the electronic structure of the materials. Further measurements showed that the dipole moments themselves are not rigidly attached to the cluster framework, which further supports this interpretation.

A full understanding of the nature of ferroelectricity in metallic clusters is still lacking, however it is intriguing to take note of a correlation between metals that exhibit the ferroelectric effect and those that are superconductors in the bulk.

10.5.5 Extremely Polar Chains Assembled in Helium Nanodroplets

Finally, we ought to mention a different kind of nanocluster assembly, namely, one that involves polar impurities in helium nanodroplets. The technique involves atoms and molecules picked up, entrapped, transported, and cooled to sub-Kelvin temperatures by a beam of superfluid $^4\mathrm{He}_n$ nanodroplets ($10^2 \lesssim n \lesssim 10^5$) generated by supersonic expansion of pure helium gas through a small cryogenically cold nozzle (Toennies and Vilesov 2004, Stienkemeier and Lehmann 2006).

A unique feature of this approach is that the picked up molecules are very rapidly quenched to the nanodroplet temperature, which is ~0.37 K. As a result, very interesting metastable arrangements with unusual properties can be produced and "frozen" into long-term existence. Nauta and Miller (1999) discovered that upon sequential pickup, linear polar HCN molecules ($p_{\mathrm{HCN}} = 3.0\,\mathrm{D}$) are steered by long-range electrostatic (dipole–dipole) forces to self-assemble into long chains, aligned head to tail inside the nanodroplet. An analogous observation was made for HCCCN, $p_{\mathrm{HCCCN}} = 3.7\,\mathrm{D}$ (Nauta et al. 1999). The length of the chain seems to be limited only by the droplet diameter. In normal environments such a system would be unstable as the orientation effects would be weak relative to the large thermal rotational energy of the molecules. But within the very cold and inert liquid helium droplet, this becomes a long-lived configuration.

The structure of the chains was verified by high-resolution vibrational spectroscopy (Figure 10.9) of doped droplets passing

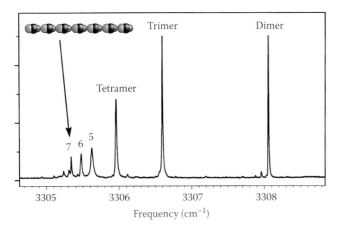

FIGURE 10.9 Spectrum of the free C–H stretching region of HCN chains in helium nanodroplets, showing peaks corresponding to clusters up to at least the heptamer ($n = 7$). A highly polar linear heptamer chain is shown in the inset. (From Nauta, K. and Miller, R.E., *Science*, 283, 1895, 1999. With permission.)

through a region of strong uniform electric field. The field enforces so-called brute force orientation of the polar impurities so that their rotations are replaced by pendulum-like oscillations about the field direction ("pendular state spectroscopy").

Very interesting aspects of these systems are the facts that these are electric dipoles of unprecedented magnitude in the molecular and cluster research fields (e.g., a chain of seven lined up HCN molecules will have a dipole moment of as much as $\approx 20\,\mathrm{D}$, and possibly even higher due to enhancement by nearest neighbor interactions), and that by virtue of being embedded in the superfluid nanodroplet, the chains are not only aligned, but also vibrationally and rotationally cold.

10.6 Cluster Ferromagnetism

10.6.1 Ferromagnetism

Ferromagnetism is the bulk property in which a material spontaneously acquires a magnetic moment (for an introduction see, e.g., Kittel 1953). Ferromagnets are typically metals and the magnetic moment is associated with a spontaneous alignment of the electronic spins. This is mediated by an interaction between the electrons, which favors mutual alignment of the spins. Physically, this tendency reflects a closely related atomic property where electrons in close proximity tend to align their spins in order to minimize the wave function overlap and thereby to minimize their mutual electrostatic repulsion. Hence, spin alignment tends to reduce the potential energy of the system. This somewhat subtle property of electrons is called the exchange interaction.

The tendency to reduce the potential energy comes at the cost of increasing the kinetic energy of the electrons. In a dilute electron gas, electrons favor a spin up–spin down arrangement resulting in no net magnetic moment. The reason is that this configuration minimizes the total kinetic energy: Electrons can occupy the same electronic orbits if their spins are opposite. In contrast to a dense electron gas, in a dilute electron gas the resulting large overlap of the electronic clouds does not significantly increase the potential energy.

In all metals itinerant valence electrons spend some time near the atomic cores where the electronic densities are relatively high and they interact with each other rather strongly causing the spins to align. They also spend time between the cores where the electronic density is low and here the low spin configuration is favored. In ferromagnets, the tendency to align the spins dominates and therefore the metal spontaneously polarizes.

The magnetic moment that arises from ferromagnetism has no unique preferred direction. In some cases, the magnetic moment direction is isotropic so that it can point in any direction. In other cases, it has a weak tendency toward a preferred axis; in this situation the magnetic moment may be either aligned or anti-aligned with that direction. Consequently, in contrast to those electric dipole moments that are rigidly attached to the geometrical structure, the magnetic dipole moment is relatively free to change its direction while keeping its magnitude constant.

At sufficiently high temperatures (the Curie temperature of the material), the magnetic moment vanishes. At this point, thermal vibrations become competitive with the energies associated with the mutual alignment of the spins, thereby reducing and ultimately extinguishing the magnetic moment.

10.6.2 Single-Sided Deflections of Ferromagnetic Clusters

Ferromagnetism also manifests itself in clusters of ferromagnetic metals, as first demonstrated by de Heer et al. (1990). The most striking property of free ferromagnetic clusters is that their magnetic moments appear to spontaneously align with an applied magnetic field. The effect reveals itself as single-sided deflection of a beam of ferromagnetic clusters in an inhomogeneous magnetic field (a Stern–Gerlach magnet), in contrast to the expected symmetric deflections of paramagnetic clusters (as seen, for example, in the classical Stern–Gerlach experiment on silver atoms). This property is shown in Figure 10.10 contrasting Stern–Gerlach deflections of paramagnetic atoms with those of paramagnetic clusters.

For clusters, it is experimentally found (see Figure 10.2) that the average magnetization $\langle M_z \rangle$ of the clusters in weak magnetic fields corresponds to

$$\langle M_z \rangle = \frac{\mu^2 B}{3k_B T}, \tag{10.15}$$

where

μ is the magnetic moment
T is the temperature of the cluster source
B is the magnetic field strength

reminiscent of the low-field limit of the Langevin equation. These deflections are observed for all clusters, even at cryogenic temperatures. However, as emphasized in Section 10.4.1, the clusters are not in contact with a heat bath since they are isolated when they pass through the magnet. Moreover, many are small and cold enough that an internal statistical temperature cannot be defined. For example, a cluster with five atoms emerging from a cluster source with $T = 20\,\text{K}$ contains on average much less than one phonon (the quantum of lattice vibrations).

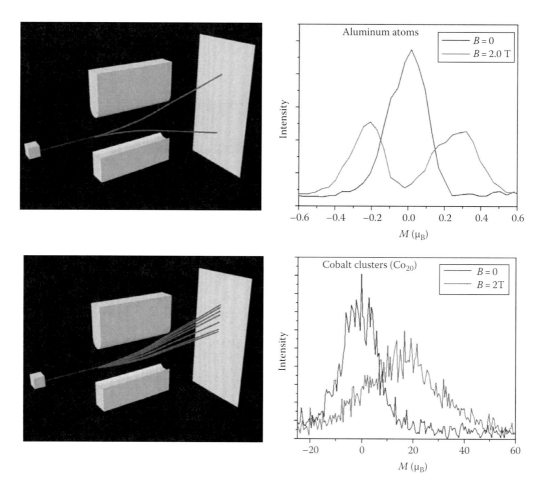

FIGURE 10.10 Stern–Gerlach deflections of atoms and clusters. Aluminum atom deflections exhibit two peaks corresponding to the possible spin orientation with respect to the magnetic field, as shown schematically at top left and experimentally at top right. Ferromagnetic clusters, on the other hand, exhibit single-sided deflections (bottom panels).

Thus this limit corresponds to the aforementioned case of asymmetric rigid dipolar species. As discussed by Xu et al. (2008), the single-sided deflections ultimately result from the alignment of the magnetic moment of each specific cluster in a magnetic field. When a cluster with a magnetic moment is placed in a magnetic field, it will precess much like a spinning top, and this will cause the magnetic moment on average to slightly align with the magnetic field. The degree of alignment depends on the rotational state of the cluster. Each individual cluster rotates at a different rate that follows a distribution given by the temperature in the source. In the case of a dense spectrum, it has been shown by Xu et al. (2008) that quite generally the average deflection of the entire cluster ensemble will be given by Equation 10.15 for a wide range of cluster sizes and temperatures.

10.6.3 Ferromagnetic Cluster Magnetic Moments

Cluster magnetic moments can be extracted from the beam deflection data using Equation 10.15. They have been found by Billas et al. (1994) to be significantly larger than the expected moments extrapolated from the bulk (Figure 10.11). The enhanced magnetic moments were explained in terms of an enhancement caused by the relatively large surface area in clusters, since it is known that the magnetic moments at ferromagnetic surfaces are enhanced. Specifically, for small clusters of iron, cobalt, and nickel the magnetic moments correspond approximately to 3, 2, and 1 μ_B per atom. These values are readily understood since they correspond to the total spin of the 3d electronic shells of the corresponding atoms. As the clusters increase in size, the magnetic moments decrease toward the corresponding bulk values.

Using the molecular beam deflection technique, the magnetic moments of a variety of ferromagnetic clusters have been measured, including alloys (see, e.g., Yin et al. 2007). Particularly interesting are metals that are nonmagnetic in the bulk but of which the clusters are ferromagnetic. Rhodium clusters (Cox et al. 1993) are an important example of this class of ferromagnetic clusters, which also exhibit unusual electric dipole polarizabilities, as mentioned above.

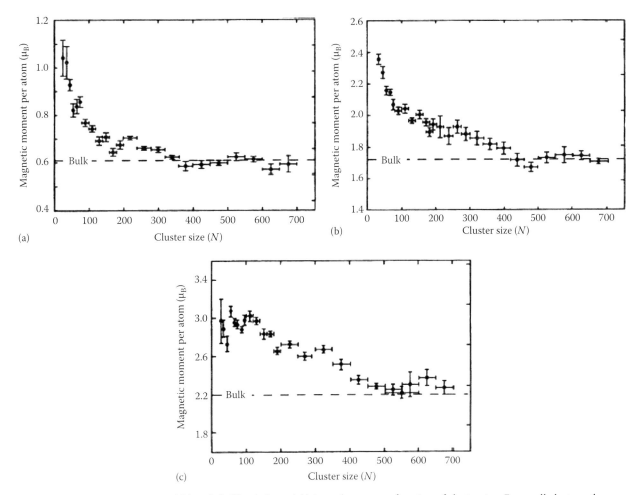

FIGURE 10.11 Magnetic moments of (a) nickel, (b) cobalt, and (c) iron clusters as a function of cluster size. For small clusters, the magnetic moments per atom are approximately 1, 2, and 3 μ_B, respectively. (From Billas, I.M.L. et al., *Science*, 265, 1682, 1994. With permission.)

10.7 Summary

Magnetic and electric dipole moments of cluster nanoparticles are observables displaying high sensitivity to the electronic and geometric structure of the system. Therefore, they are of considerable experimental and theoretical value in nanoscience. It is very appealing that both the intrinsic magnitudes of these quantities and their experimental manifestations (e.g., in beam deflection studies and in spectroscopy) involve a wide range of fundamental issues, for example, phase transitions, exchange interactions, long-range forces, chemical bonding, quantum and classical chaos, superconductivity, and superfluidity. At the same time, one has to be cognizant of the fact that the quantitative interpretation of experimentally measured patterns requires careful awareness of the underlying assumptions and mechanisms.

Acknowledgments

Useful discussions with J. Bowlan, J. Bulthuis, and A. Liang, and support from the U.S. National Science Foundation are sincerely appreciated.

References

Abd El Rahim, M., R. Antoine, M. Broyer, D. Rayane, and Ph. Dugourd. 2005. *J. Phys. Chem. A* 109: 8507–8514.

Antoine, R., M. Abd El Rahim, M. Broyer, D. Rayane, and Ph. Dugourd. 2006. *J. Phys. Chem. A* 110: 10006–10011.

Ballentine, G. E., G. F. Bertsch, N. Onishi, and K. Yabana. 2007. *Comput. Phys. Commun.* 178: 48–51.

Bertsch, G. F. and K. Yabana. 1994. *Phys. Rev. A* 49: 1930–1932.

Beyer, M. K. and M. B. Knickelbein. 2007. *J. Chem. Phys.* 126: 104301.

Billas, I. M. L., A. Châtelain, and W. A. de Heer. 1994. *Science* 265: 1682–1684.

Bowlan, J., A. Liang, and W. A. de Heer, to be published.

Broyer, M., R. Antoine, E. Benichou, I. Compagnon, Ph. Dugourd, and D. Rayane. 2002. *C. R. Physique* 3: 301–317.

Broyer, M., R. Antoine, I. Compagnon, D. Rayane, and P. Dugourd. 2007. *Phys. Scr.* 76: C135–C139.

Bulthuis, J. and V. V. Kresin, to be published.

Bulthuis, J., J. A. Becker, R. Moro, and V. V. Kresin. 2008. *J. Chem. Phys.* 129: 024101.

Carrera, Á., M. Mobbili, G. Moriena, and E. Marceca. 2008. *Chem. Phys. Lett.* 467: 14–17.

Clough, S. A., Y. Beers, G. P. Klein, and L. S. Rothman. 1973. *J. Chem. Phys.* 59: 2254–2259.

Cox, A. J., J. G. Louderback, and L. A. Bloomfield. 1993. *Phys. Rev. Lett.* 71: 923–926.

de Heer, W. A., P. Milani, and A. Châtelain. 1990. *Phys. Rev. Lett.* 65: 488–491.

Dugourd, Ph., R. Antoine, D. Rayane, E. Benichou, and M. Broyer. 2000. *Phys. Rev. A* 62: 011201.

Dugourd, Ph., I. Compagnon, F. Lepine, R. Antoine, D. Rayane, and M. Broyer. 2001. *Chem. Phys. Lett.* 336: 511–517.

Dugourd, Ph., R. Antoine, M. Abd El Rahim, D. Rayane, M. Broyer, and F. Calvo. 2006. *Chem. Phys. Lett.* 423: 13–16.

Gregory, J. K., D. C. Clary, K. Liu, M. G. Brown, and R. J. Saykally. 1997. *Science* 275: 814–817.

Griffiths, D. J. 1999. *Introduction to Electrodynamics,* 3rd edn. Upper Saddle River, NJ: Prentice-Hall.

Gubskaya, A. V. and P. G. Kusalik. 2002. *J. Chem. Phys.* 117: 5290–5302.

Jaynes, E. T. 2003. *Probability Theory: The Logic of Science.* New York: Cambridge University Press.

Jraij, A., A. R. Allouche, F. Rabilloud, et al. 2006. *Chem. Phys.* 322: 298–302.

Kittel, C. 1953. *Introduction to Solid State Physics.* New York: Wiley.

Langevin, P. 1905. *J. Phys. Theor. Appl.* 4: 678–693.

Liang, A. 2009. Ph.D. dissertation. Atlanta: Georgia Institute of Technology.

Moro, R., X. S. Xu, S. Y. Yin, and W. A. de Heer. 2003. *Science* 300: 1265–1269.

Moro, R., R. Rabinovitch, C. Xia, and V. V. Kresin. 2006. *Phys. Rev. Lett.* 97: 123401.

Moro, R., J. Bulthuis, J. Heinrich, and V. V. Kresin. 2007. *Phys. Rev. A* 75: 013415.

Nauta, K. and R. E. Miller. 1999. *Science* 283: 1895–1897.

Nauta, K., D. T. Moore, and R. E. Miller. 1999. *Faraday Discuss.* 113: 261–278.

Niessen, K. F. 1929. *Phys. Rev.* 34: 253–278.

Parker, D. H. and R. B. Bernstein. 1989. *Annu. Rev. Phys. Chem.* 40: 561–595.

Pauly, H. 2000. *Atom, Molecule, and Cluster Beams.* Berlin, Germany: Springer.

Ramsey, N. 1956. *Molecular Beams.* Oxford, U.K.: Oxford University Press.

Rayane, D., I. Compagnon, R. Antoine, et al. 2002. *J. Chem. Phys.* 116: 10730–10738.

Rayane, D., A. R. Allouche, I. Compagnon, et al. 2003. *Chem. Phys. Lett.* 367: 278–283.

Schäfer, S., B. Assadollahzadeh, M. Mehring, P. Schwerdtfeger, and R. Schäfer. 2008. *J. Phys. Chem.* A112: 12312–12319.

Schäfer, S., S. Heiles, J. A. Becker, and R. Schäfer. 2008. *J. Chem. Phys.* 129: 044304.

Schnell, M., C. Herwig, and J. A. Becker. 2003. *Z. Phys. Chem.* 217: 1003–1030.

Stienkemeier, F. and K. K. Lehmann. 2006. *J. Phys. B: At. Mol. Opt. Phys.* 39: R127–R166.

Tikhonov, G., K. Wong, V. Kasperovich, and V. V. Kresin. 2002. *Rev. Sci. Instrum.* 73: 1204–1211.

Toennies, J. P. and A. F. Vilesov. 2004. *Angew. Chem. Int. Ed.* 43: 2622–2648.

Townes, C. H. and A. L. Schawlow. 1975. *Microwave Spectroscopy.* New York: Dover.

Van Vleck, J. H. 1932. *The Theory of Electric and Magnetic Susceptibilities*. London, U.K.: Oxford University Press.

Wrede, E. 1927. *Z. Phys.* 44: 261–268.

Xu, X. S., S. Y. Yin, R. Moro, A. Liang, J. Bowlan, and W. A. de Heer. 2007. *Phys. Rev. B* 75: 085429.

Xu, X., S. Yin, R. Moro, and W. A. de Heer. 2008. *Phys. Rev. B.* 78: 054430.

Yin, S., X. Xu, A. Liang, J. Bowlan, R. Moro, and W. A. de Heer. 2008. *J. Supercond. Nov. Magn.* 21: 265–269.

Yin, S. Y., R. Moro, X. S. Xu, and W. A. de Heer. 2007. *Phys. Rev. Lett.* 98: 113401.

11

Quantum Melting of Hydrogen Clusters

Massimo Boninsegni
University of Alberta

Melting is a phase transition with which we all become familiar early on in our lives, and is also among the generally better understood ones, on basic grounds of statistical thermodynamics. At a sufficiently low temperature, the equilibrium phase of a generic condensed matter system is predicted by classical mechanics to be a crystalline solid, with atoms or molecules arranged in *orderly*, periodic arrays, typically referred to as *lattices*. This arrangement is that in which the potential energy of the system is minimum. In a crystal, particles (atoms or molecules) are bound to lattice positions, held in place by restoring forces that, for small displacements, can be regarded as harmonic, i.e., "spring-like"; particles only make small excursions away from their equilibrium positions, with a mean square amplitude proportional to the temperature T. As T is raised, thermal agitation increases, displacements away from equilibrium positions become significant, and, eventually, particles are no longer localized at lattice sites. A transition occurs to a disordered, isotropic phase, i.e., the liquid.

In this chapter, a different kind of melting will be illustrated, featuring similarities as well as marked differences with respect to conventional melting. The most important difference is that the melting considered here occurs in a finite system, whose elementary constituents are hydrogen molecules; their behavior at low temperatures is properly described by quantum mechanics. As will be shown, the competition between quantum effects and the tendency of the system to establish solid order locally gives rise to a much richer phase diagram than that observed in bulk hydrogen, including a superfluid phase. Arguably, the most surprising physical prediction described here is that at least some specific hydrogen *nano*-clusters, comprising a number of molecules as low as 23, display "solid-like" behavior in some range of low temperature, but then *melt* into a "liquid-like" phase as the temperature is lowered even further. This counterintuitive result is observed in first principles computer simulations and is a direct consequence of quantum mechanics.

11.1 Melting in Condensed Matter

As mentioned above, the main feature of the crystalline phase of matter is the presence of *long-range* order, which extends over distances much greater than the characteristic separation between neighboring particles. In a liquid, the type of order present in a crystal only exists over the short distance that separates a particle from its nearest neighbors, which is of the order of few tenths of a nanometer.* In the liquid phase, particles enjoy substantial mobility, and thus can diffuse throughout the system much more easily than in a crystal. Altogether, the liquid phase features a higher degree of symmetry than the crystalline one. Macroscopically, the most obvious differences between solid and liquid are the resistance of the former to shear and the ability of the latter to flow.

The melting of a crystal into a liquid at high temperatures is driven by *entropy*. Although crystalline particle arrangements (configurations) have lower potential energy and are therefore

---

* The distance between nearest neighboring particles is *generally* greater in the liquid than in the solid phase, but the difference rarely exceeds a few percent.

individually more probable than those that are prevalent in a liquid, the latter *vastly outnumber* the former, owing to their reduced symmetry. As a result, as the temperature is raised and particles become increasingly mobile, the system "samples" liquid configurations more and more frequently. Crystals of different substances melt at different temperatures, on account of the different interactions among their elementary constituents; however, melting occurs fundamentally in the same way in every crystal.

The above description is based on classical considerations, but quantum mechanics leaves it largely unaltered, at least qualitatively. The most important consequence of quantum mechanics is that particles execute another type of motion, known as *zero-point* (see, for instance, Ref. [1]), conceptually distinct from thermal vibrations. This is a purely quantum-mechanical effect, as a result of which particles cannot be considered "at rest" (in a classical sense) even at temperature $T = 0$. Rather, a particle confined in a spatial region of characteristic size a executes quantum oscillations, whose characteristic (kinetic) energy is of the order of $\hbar^2/ma^2$ where $\hbar$ is the quantum-mechanical Dirac constant and m is the mass of the particle. In a crystalline lattice, a is of the order of the lattice constant, i.e., the distance between adjacent lattice sites.

The presence of zero-point motion in condensed matter systems can be detected experimentally by measuring the mean kinetic energy per atom or molecule, for example, by means of neutron scattering experiments. The mean kinetic energy per particle exceeds the classical value* $3T/2$, the kinetic energy in excess arising from zero-point motion. The excess kinetic energy offers a measure of the importance of quantum effects in a solid or a liquid.

Zero-point motion can also be imaged by means of computer simulations.

Computational procedures through which figures such as Figure 11.1 can be produced are discussed later. Suffice to say for the moment that Figure 11.1 shows what can be regarded as an instantaneous† particle configuration in two solids at low temperatures, namely, neon (bottom) and hydrogen (top).

In both cases shown, particles (atoms in the case of neon, molecules for hydrogen) are mathematically modeled as point-like, but are *not* imaged as points. Rather, their positions, while centered at lattice sites corresponding to the specific crystallographic structure of these two solids, are "smeared" over a fuzzy region whose characteristic size represents the extent of quantum zero-point oscillations. As mentioned above, this effect is greater for less massive particles, in this case hydrogen molecules, whose mass is very nearly a tenth of that of a neon atom.

Because zero-point motion weakens the binding of particles to lattice sites, it causes a crystal to melt at a slightly lower

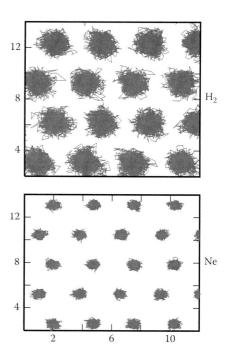

FIGURE 11.1 Computer-generated snapshots of particle configurations in solid neon (bottom) and molecular hydrogen (top), both at temperature $T = 1\,$K, in the same crystallographic planes. Both solids have the same crystalline structure (hexagonal close packed). Particles are represented by fuzzy "clouds," as their locations are smeared over a characteristic region, whose size represents the extent of quantum delocalization. Such an effect is substantially greater for a hydrogen molecule, whose mass is close to a tenth of that of a Ne atom.

temperature than that predicted based on classical calculations. However, this correction is fairly small in most cases, as the vast majority of substances relevant to daily life melt at temperatures at which the thermal energy dominates over the zero-point one. In general, zero-point energy is essentially unmeasurable at ordinary conditions of temperature and pressure, and even for those crystals that melt at relatively low temperatures (e.g., solids of rare gases, such as Ne or hydrogen), for which quantum effects are quantitatively large, melting largely still takes place qualitatively as understood classically.

The only known condensed matter system for which quantum effects profoundly alter the classical picture is *helium* (He), which does not solidify *at all* under the pressure of its own vapor, ostensibly not even at $T = 0$. The failure of He to form a crystal is a consequence of the large quantum zero-point motion of He atoms (due to their relatively light mass) coupled with the weakness of the interaction between them. The destabilization of the crystalline phase occurs because the potential energy gain arising from arranging atoms in a regular lattice would not offset the increase in zero-point energy due to particle localization.

Helium is the only known condensed matter system which escapes crystallization at low temperatures; one might think that a fluid of hydrogen molecules ought to endure the same fate, since the mass of a hydrogen molecule is half that of a helium atom, and therefore quantum zero-point motion should

---

* We adopt here a system of units wherein the Boltzmann constant K_B is set equal to 1. Therefore, all energies are expressed in Kelvin (K), and the value of the Dirac constant $\hbar$ is equal to 7.638×10^{-12} K s.

† Within the limitations that Quantum Mechanics imposes on this type of classical terminology.

be expected to be even more important than in He. Indeed, it was precisely this observation that led Ginzburg and Sobyanin [3], almost thirty years ago, to speculate that fluid hydrogen at low temperatures (approximately 6 K) would undergo a *superfluid transition*, i.e., just like helium (see, for instance, Ref. [2]), it would cease to behave as a *normal* liquid and turn into one capable of *persistent* flow. However, this is not what one observes experimentally, for liquid H_2 crystallizes at a low but finite temperature of 13.8 K due to the strength of the attraction between two molecules being significantly greater than that between two helium atoms.

If an external pressure of approximately 25 atm is applied, then helium too forms a crystal [2], whose temperature-driven melting transition does not differ significantly, at least qualitatively, from that of other substances. In summary, the melting transition in bulk solids, while displaying quantitative differences from one system to another, in some cases attributable to pronounced quantum behavior of the elementary constituents, for the most part can be described based on a classical picture.

11.2 Phase Transitions in Clusters

A cluster (sometimes also referred to as a *droplet*) is a *finite* system, i.e., one in which the contribution to the total energy associated to the *surface* (a contribution that is proportional to the power 2/3 of the number N of particles) is comparable to the *bulk* contribution, which is proportional to N. In the *thermodynamic limit* (i.e., $N \to \infty$), the surface contribution becomes negligible, and the physical properties of a cluster increasingly resemble those of bulk matter made of the same atoms or molecules; they are independent of the conditions at the surface. In a cluster, however, the competition between bulk and surface energy contributions can give rise to unusual and/or unexpected physical properties, often markedly different from that of bulk matter. The search for such novel behavior underlies the experimental and theoretical investigation of clusters, spanning now at least two decades. Its general aim is to understand how bulk properties arise and/or evolve from those of a cluster, as the size of the latter is increased.

An intriguing issue is whether or not one can meaningfully speak of "liquid" or "solid" phases, as well as of *phase transitions*, for a system that may comprise as few as, say, 20 atoms or molecules bound together. If the notion of melting makes sense for a system of such a small size, one may speculate that it should be qualitatively different than in bulk, mostly for two reasons: (a) the much lesser role played by entropy (b) the presence of the surface, which can frustrate the formation of an ordered structure, favoring instead a featureless, uniform arrangement of particles throughout the cluster.*

---

* Clearly, the notion of "surface," as opposed to the "interior" of the cluster, is well defined for clusters of sufficient size, say, 40 particles or more. For smaller clusters, any definition of surface inevitably *de facto* encompasses most of the particles in the system.

In principle, *no phase transition at all* can occur in a finite system, at least if one goes by a strict thermodynamic definition. Indeed, even the very notion of "phase" is not well defined for a cluster, as a thermodynamic phase is properly characterized by the presence (or lack) of some kind of *order* over distances comparable to the linear size of the system, assumed to be much greater than the mean distance between particles constituting it—a condition not holding for a cluster, especially one of the size considered here.

In spite of such difficulties, it might still be possible for a finite system to draw a clear distinction between configurations that are, for example, "solid-like," i.e., with particles executing only small displacements away from classical equilibrium positions corresponding to some well-defined geometrical structure (as shown in Figure 11.2, for example), and those that are "liquid-like," i.e., wherein particles are delocalized throughout the whole system, with no discernible geometrical order present. It would seem reasonable to seek a definition of phase of a finite system expressing more rigorously, from a formal standpoint, these simple intuitive considerations.

It must be recognized, however, that in a bulk thermodynamic system a free energy barrier ΔF of magnitude proportional to the number N of particles, separates two different phases. The observed stability of a thermodynamic equilibrium phase is formally expressed through its *lifetime* $\tau \sim \Delta F$, which for a macroscopic sample of condensed matter, comprising a number of particles $N \sim 10^{24}$, becomes very large, greatly exceeding the time over which a typical experiment is performed. In a finite system, on the other hand, all free energy barriers are necessarily finite, which means that any "phase" may only exist over a finite time interval. Thus, over a time interval comparable to the experimental time scale, a finite system will generally explore configurations belonging to all possible phases. This is something that a thermodynamic system clearly does not do, as only one phase is

FIGURE 11.2 Structure of a "solid-like" cluster, comprising 19 particles. The geometrical arrangement is clear; it consists of a central axis of symmetry along which four particles are placed, and three rings of five particles each, centered on the axis. This computer-generated image refers to a cluster of *ortho*-deuterium molecules, a heavier isotope of hydrogen, at a temperature $T = 1$ K. The smearing of the particles due to quantum effects is limited, owing to their relatively large mass.

in equilibrium at a given temperature.* This means that any talk of phase in the context of a finite system is necessarily approximate and semiquantitative, and is based on a comparison of the lifetime of the phase to the typical time scales that are experimentally relevant.

11.2.1 Cluster Melting and Freezing

Hereinafter, we restrict the introductory discussion of clusters to those comprising atoms of rare gases or hydrogen molecules. In these systems, constituents (atoms or molecules) can be regarded as elementary particles, owing to their very low chemical reactivity. The interactions among particles in the cluster can be accurately described by means of a pair-wise central potential. Over the past four decades, a number of theoretical works have addressed the issue of whether these clusters may melt and freeze in a way analogous to bulk systems [4–11]. An important role in the theoretical investigation has been played by computer simulations, based on Monte Carlo and Molecular Dynamics. These methods are rooted in general principles of classical statistical mechanics. They have been utilized to investigate equilibrium and dynamical properties as well as "melting" and "freezing" of clusters of rare gas atoms for which quantum-mechanical effects are negligible (e.g., argon) [12,13].

The main result of such an investigative effort may be summarized as follows: a cluster possesses two distinct melting (T_m) and freezing (T_f) temperatures, with $T_m > T_f$. At temperatures above T_m, the cluster displays essentially liquid-like behavior, whereas below T_f, it is solid-like. In the temperature range between T_f and T_m, the cluster fluctuates between the two regimes, as explained in Section 11.2, in what has come to be referred to as the Cluster Solid-Liquid Transition (CSLT). The CSLT is found to be especially prominent in some clusters, comprising specific numbers of particles referred to as "magic." Magic numbers are those for which the system features a pronounced potential energy minimum as a function of N, typically in correspondence of a particular geometrical structure that confers the cluster enhanced stability over clusters with one extra or less particle. Such structures have been investigated extensively in the context of classical clusters (see, for instance, Ref. [14]).

11.2.2 Quantum Clusters

Unlike the case of bulk, the cluster melting physical scenario is greatly enriched if quantum-mechanical effects are important, as in clusters of light atoms, such as helium, or molecules, such as hydrogen. Helium clusters, both pure as well as doped with impurities, have been the subject of vigorous experimental [15]

and theoretical [16] investigations aimed at gaining understanding of the phenomenon of superfluidity in systems of finite size. There is no CSLT transition for helium clusters, however, as they remain liquid-like at arbitrarily low temperatures,† i.e., in this sense they do not display significantly different physics than bulk He.

More interesting is the situation if clusters of other species are considered, for which solid- and liquid-like phases occur, for example hydrogen clusters. Because the physics of bulk H_2 is expected to be noticeably different than that of small clusters thereof (unlike the case of helium), the investigation of clusters of varying size allows one to study systematically the emergence of bulk physics.

Analogous to that observed in bulk solids, quantum delocalization effects cause a lowering of both cluster melting and freezing temperatures; this has been shown quantitatively for argon and neon clusters [17–21] as well as for clusters made of hydrogen molecules, both *para*-hydrogen (p-H_2) and the heavier *ortho*-deuterium (o-D_2) isotopes [22–24].

The rest of this chapter discusses how another quantum-mechanical effect can play a crucial role in shaping the physical properties of quantum clusters of hydrogen molecules. Indeed, for some of these clusters featuring specific numbers of particles, the phase diagram is completely altered with respect to what conventional wisdom would suggest, in that a *reentrant* low-temperature liquid-like phase is predicted to occur, underlain by *quantum-mechanical exchanges*. In other words, these clusters freeze at a characteristic temperature T_f, *but melt again and turn liquid-like at a lower temperature.*

11.2.2.1 Quantum Exchanges and Bose Condensation

The effect of exchange of indistinguishable particles is one of the most intriguing (and mind-boggling) manifestations of quantum mechanics. It takes place at a sufficiently low temperature in a system of identical particles. As mentioned above, as a result of zero-point motion, particles are delocalized over a finite region of space, whose size increases as the temperature is lowered. When the regions over which particles are delocalized begin to overlap significantly, particles lose their individual identities; if they obey *Bose statistics*,‡ (which is true of all cases of interest here) a finite fraction thereof ends up occupying the same, single-particle quantum-mechanical state. This phenomenon, known as *Bose-Einstein Condensation* (BEC), is presently

---

* It is true that coexistence of two (or more) phases is observed in a thermodynamic system, e.g., near a first-order phase transition. However, in this case, separate *macroscopic* regions of the system feature the qualities of only *one* of the various coexisting phases, e.g., water and ice, the relative proportions not changing with time. On the other hand, a finite system will display at any given time the physical behavior of *one* of the allowed phases, constantly morphing into a different one at a successive time.

† Given that condensed helium does not crystallize at low temperature, the same must *a fortiori* hold for finite systems, for which crystallization is further hampered by the presence of a surface.

‡ All particles known in Nature can be classified into two broad categories, namely *bosons* and *fermions*. Bosons feature integer values of their intrinsic angular momentum (spin), whereas fermions take on half-odd values. Bose Condensation is only possible for bosons. For fermions, it is prohibited by a fundamental restriction known as *Pauli exclusion principle*, which prevents any two identical fermions from occupying the same single-particle quantum-mechanical state. For fermions, superfluidity is believed to require *pairing* of particles, i.e., the formation of composite particles (pairs) comprising two fermions. The total spin of the pair is an integer, and therefore the pairs obey Bose statistics.

known to underlie superfluidity in the liquid phase of the most abundant isotope of helium (i.e., ^{4}He), and, in this context, it has been extensively investigated for decades, both experimentally as well as theoretically (see, for instance, Ref. [25]).

However, BEC can also take place in finite systems. Indeed, its investigation received novel impetus in 1995, with its observation in physical systems comprising a number (varying from several thousands to millions) of atoms of various alkali metals (e.g., Rb) in the gas phase, magnetically confined to a restricted region of space and cooled by means of evaporative and laser techniques (see, for instance, Ref. [26]). In these "artificial," weakly interacting many-body systems, the condensation of a macroscopic number of particles in the same quantum-mechanical single-particle state can be established unambiguously by direct imaging, owing to the high degree of control that experimenters enjoy.

11.2.3 BEC and Superfluidity of Clusters

The occurrence of BEC in clusters of helium was studied theoretically two decades ago by Lewart et al. [27], who estimated that, in the $T \to 0$ limit, approximately 27% of all helium atoms in a droplet of 70 should occupy the same single-particle state. However, the experimental methodology utilized for confined alkali gases, based on optical techniques, cannot be adopted to establish the presence of BEC in clusters of helium or hydrogen, which for the moment can only be inferred indirectly, through the observation of their superfluid properties.

Superfluidity is the property of a substance to flow without dissipation; once it is established, flow is experimentally observed to persist indefinitely, with no need of any external agent to sustain it (e.g., a pressure differential) [28]. The chief example of a superfluid is liquid ^{4}He, which transitions into a superfluid phase below the temperature $T_\lambda = 2.177$ K. However, superconductors, namely, materials capable of conducting persistent electric current without any voltage applied (below a certain transition temperature), can be regarded as essentially charged superfluids (see, for instance, Ref. [29]).

In many respects, superfluidity can be regarded as the most dazzling manifestation of Quantum Mechanics on a macroscopic scale, and it is indeed this macroscopic character that facilitates its experimental observation; spectacular displays of superfluid behavior visible to the naked eye can be obtained relatively straightforwardly in any modern low-temperature physics laboratory.

That a superfluid transition can take place in finite clusters of helium as well is known since the pioneering simulation work of Sindzingre, Klein, and Ceperley, who first predicted that a cluster of 64 ^{4}He atoms would turn superfluid at a temperature very close to T_λ (around 2 K) [16]. Observing experimentally superfluidity in a finite system, especially one of the size of a cluster of ~40 helium atoms, is a more involved proposition than for bulk liquid helium, but the goal can be accomplished by means of several remarkably clever procedures. For example, a recently introduced experimental technique, known as Helium NanoDroplet

Isolation spectroscopy (HENDI), allows one to investigate a single molecular impurity embedded in clusters comprising from a few to several thousand He atoms [15]. In these experiments, the superfluidity of the medium surrounding the molecule (i.e., the cluster) can be deduced by studying the rotational spectrum of the dopant molecule. If the cluster is in the so-called normal (i.e., non-superfluid) phase, spectral lines associated with the rotation of the molecule are broadened by dissipation due to drag with the medium. However, as soon as the temperature is brought below a characteristic "superfluid transition temperature" (which typically depends not only on the number of particles, but also on the type of dopant molecule), spectral lines become much sharper, and closely reproduce the expected theoretical result for a rigid, free rotator. In other words, it is as if the dopant molecule were rotating it *in vacuo*, with no effect of drag from the surrounding medium. The observed decoupling of the rotation of the molecule from that of the cluster is interpreted as a signature of the onset of superfluidity in the cluster.

The first evidence of superfluidity obtained in this way, was produced for a helium cluster surrounding a linear carbonyl sulfide (OCS) impurity, at a temperature $T < 1$ K [30]. Successively, analogous experiments in which the OCS impurity was replaced by an N_2O molecule showed similar superfluid behavior for clusters comprising as few as seven helium atoms [31].

11.2.3.1 Superfluidity of H_2 Clusters

The question of whether one may be able to observe a superfluid transition in clusters of H_2 molecules as well naturally arises. As mentioned in Section 11.1, condensed H_2 is believed to be a potential "second superfluid," owing to the light mass of its elementary constituents (approximately one half of that of a helium atom). However, the depth of the attractive well of the potential describing the interaction between two such molecules is such that condensed hydrogen solidifies at a temperature much higher than that at which BEC, and possibly superfluidity, ought to take place, estimated at roughly 6 K [3]. Therefore, the observation of a superfluid phase of H_2 hinges on stabilizing a (metastable) liquid phase at temperatures around ~10 K below that of crystallization.* Experimental and theoretical attempts to stabilize a low temperature metastable liquid phase of bulk H_2, e.g., making use of confinement and/or reduced dimensionality, have so far not met with success; moreover, there are good reasons to be skeptical about the long term prospects of success of this line of research (see, for instance, Ref. [32]).

On the other hand, as explained above, sufficiently small H_2 clusters may remain liquid-like at significantly lower temperatures, therefore leaving open the door to a possible superfluid transition, much as in ^{4}He clusters. Indeed, theoretical calculations carried out some 17 years ago predicted that clusters of p-H_2 with 13 and 18 molecules should be liquid-like, turning superfluid at temperatures of the order of 2 K [23]. However, it was also found that a cluster with 33 molecules is non-superfluid

---

* It is widely believed that BEC is incompatible with the presence of crystalline order.

at the same temperature (actually, down to $T = 1$ K), suggesting that perhaps superfluidity may be observable only in very small clusters (i.e., with less than ~20 particles).

The same type of experiment described above, based on the HENDI technique, did yield some (perhaps not fully conclusive) evidence of superfluid behavior in clusters of $N = 14$–16 H_2 molecules surrounding the same impurity used in early experiments with helium clusters, namely, OCS. In these experiments, H_2 clusters were enclosed in a larger helium droplet [33].

At the time of writing this chapter, no other experiments have reported the evidence of H_2 superfluidity, either in doped or pure H_2 clusters. However, this is an area of great current interest, and experiments are being both planned and performed, at the time of this writing [34]. Moreover, novel techniques based on Raman spectroscopy have been proposed, which may soon succeed at isolating pristine clusters (i.e., free of impurities) and characterizing their superfluid properties [35].

The next sections summarize the results of theoretical studies based on first principle computer simulations, predicting different degrees of superfluidity for H_2 clusters of size up to 36 molecules (the largest for which superfluidity has been clearly observed). These clusters have a characteristic size around 1 nm, which warrants the denomination of *nano*-clusters. A fundamental observation is that superfluid behavior of clusters strongly depends on their numbers of particles, with dramatic changes occurring even on adding or removing a single molecule. The characteristic superfluid transition temperature is generally around 2 K for clusters with fewer than 25 molecules, dropping almost an order of magnitude for bigger clusters.

The theoretical study of superfluidity of hydrogen clusters can yield important insight into their structural properties, as solid order is generally believed to be incompatible with superfluidity. In other words, clusters that are fractionally more superfluid are generally presumed to be in a liquid-like phase, although this concept has to be stated more precisely, as some *caveats* exist.

11.2.4 Mixed Hydrogen Clusters

At room temperature, molecular hydrogen gas in equilibrium is composed of a mixture of two species, namely, *para*-hydrogen (p-H_2, 25.1%), which has a total spin equal to zero, and *ortho*-hydrogen (o-H_2, 74.9%), which has a spin equal to one. At low temperatures, the higher-energy o-H_2 state is disfavored, and below $T \sim 20$ K hydrogen gas consists essentially entirely of p-H_2 (see, for instance, Ref. [36]). There exists also a heavier isotope, known as *Deuterium*, which has very nearly the same mass of a helium atom (i.e., twice that of H_2); unlike in the case of H_2, the species of spin zero is called *ortho*-deuterium (o-D_2).

To an excellent approximation, pair-wise interactions between molecules of the different isotopes are very nearly the same (see, for instance, Ref. [37]); thus, a mixed cluster of these components provides a simple experimental realization of a mixture of different bosons, either with the same mass (p-H_2 and o-H_2), or different masses (e.g., p-H_2 and o-D_2). A number of fundamental questions can therefore be addressed by studying these systems,

e.g., how does the presence of one component affect the superfluid properties of the other(s) or the structure of the cluster? In this sense, hydrogen is a much richer playground than helium, for which only two isotopes exist, and they have different quantum statistics (^{3}He atoms have spin 1/2 and therefore obey Fermi statistics). As will be shown, the study of hydrogen clusters in the presence of a single molecule of a different type of hydrogen (e.g., a single o-D_2 impurity in a cluster of p-H_2 molecules) provides interesting insight into the quantum melting of pure clusters.

11.3 Computer Simulation of Quantum Clusters

11.3.1 Model

The starting point of any theoretical investigation is a mathematical model for the system of interest. The one utilized in this work aims at providing as realistic a *microscopic* description as possible of a cluster of hydrogen molecules, based on first principles. Let us consider for definiteness a cluster of p-H_2 molecules; these are essentially spherical objects of size ~0.1 nm. A strong repulsion at short distances, arising from the overlap of their electronic clouds, keeps any two of them further apart than ~0.3 nm. Electronic exchange among molecules is virtually nonexistent, and therefore molecules can be regarded as elementary particles of spin zero, i.e., obeying Bose statistics. At the temperatures of interest here, of the order of 1 K, quantum-mechanical effects cannot be neglected.

The interaction among p-H_2 molecules can be, to a good approximation, described with a pair-wise central potential.* All of this information can be summarized in the following quantum-mechanical many-body Hamiltonian operator:

$$H = -\frac{\hbar^2}{2m} \sum_{i}^{N} \nabla_i^2 + V(\mathbf{R}) \tag{11.1}$$

where m is the mass of a H_2 molecule (very close to that of a helium atom, i.e., 4 amu). Henceforth, the short-hand notation $\mathbf{R}$ will be utilized to represent the collection of coordinates $\mathbf{r}_1, \mathbf{r}_2, \cdots, \mathbf{r}_N$ referring to all N molecules in the system, and $V(\mathbf{R})$ the total potential energy of the configuration $\mathbf{R}$.

As mentioned above, the potential energy V is expressed as a sum of pair-wise contributions, each represented by a spherically symmetric interaction:

$$V(\mathbf{R}) = \sum_{i<j} v\left(\left|\mathbf{r}_i - \mathbf{r}_j\right|\right) \tag{11.2}$$

All of the results presented here were obtained using, for $v(r)$, the Silvera-Goldman (SG) pair potential [38], which is the most commonly employed potential in previous investigations.

---

* Three-body interactions can be quantitatively important, especially in the presence of a strong dopant, but they are not expected to alter significantly the physical picture arising from calculations based on pair potentials only.

For comparison purposes, results were also obtained with different intermolecular potentials, e.g., that of Buck [39–41]; while there are sometimes significant quantitative differences between the results obtained with different potentials, the basic physics and general trend are not strongly dependent on the specific model interaction adopted. The basic features of any such potential are (a) a "hard core" repulsion for intermolecular distances less than ~0.2 nm (b) an attractive potential well, of depth ~35 K, centered at approximately 0.3 nm (c) an attractive tail, arising from interaction between mutually induced molecular electric dipoles, decaying as $1/r^6$ at long distances.

11.3.2 Methodology

11.3.2.1 Path Integrals

Feynman's path integral approach to quantum statistical mechanics provides general formalism for evaluating thermodynamic properties of quantum mechanical systems described by a many-particle Hamiltonian such as (11.1) (see, for instance, Ref. [42]). A thorough illustration of this methodology is beyond the scope of this chapter; only the basic ideas will be described here, by means of a simple comparison of classical and quantum systems.

If quantum effects were negligible, i.e., if a description based on classical mechanics were sufficiently accurate, then one could evaluate thermal averages of physical quantities of interest for a cluster of particles, in equilibrium at a temperature T, by computing expressions such as

$$\langle \mathcal{O} \rangle = \frac{\int d\mathbf{R}\, \mathcal{O}(\mathbf{R}) \exp\left[-V(\mathbf{R})/T\right]}{\int d\mathbf{R} \exp\left[-V(\mathbf{R})/T\right]} \qquad (11.3)$$

where the integrals are over all possible *configurations* $\mathbf{R}$ of the system (i.e., positions of the N particles), $\mathcal{O}(\mathbf{R})$ is the value of the physical quantity of interest (e.g., pressure, potential energy, some structural correlation function, etc.) corresponding to the configuration $\mathbf{R}$, and $V(\mathbf{R})$ is the total potential energy of the system for the same configuration. Integrals such as those at the numerator and denominator of (11.3) are called *configurational*. Every possible configuration "weighs in" with a weight proportional to the Boltzmann's factor $\exp[-V(\mathbf{R})/T]$; at low temperatures, configurations of lower potential energy become favored; for example, configurations in which any two particles are at a distance closer to the "hard core" of the potential $v(r)$ are strongly suppressed. In the $T \to 0$ limit, the system will be in its ground state, i.e., the state of lowest potential energy.

The passage from a classical to a quantum-mechanical description is formally accomplished by replacing the notion of "configuration" as a collection of particle positions, with that of particle *paths*. Specifically, to each particle one associates a path $\mathbf{r}(u)$. The variable u ranges in the $[0, \hbar/T]$ interval; it has formally the units of time, and is often referred to as such ("imaginary time") for reasons that need not concern

us here; however, no reference to any physical time is implied, nor should be understood. Thermal expectation values are now obtained as follows:

$$\langle \mathcal{O} \rangle = \frac{\int \mathcal{D}\mathbf{R}(u)\, \mathcal{O}\big(\mathbf{R}(0)\big) \exp\left[-S\{\mathbf{R}(u)\}\right]}{\int \mathcal{D}\mathbf{R}(u) \exp\left[-S\big(\mathbf{R}(u)\big)\right]} \qquad (11.4)$$

where the integrals are now taken over all possible paths[*] of all particles, i.e., $\mathbf{R}(u) \equiv (\mathbf{r}_1(u)\mathbf{r}_2(u)\ldots\mathbf{r}_N(u))$. Here, the quantity $S\{\mathbf{R}(u)\}$ takes the place of the potential energy in (11.3) and is known as the *Euclidean action* associated with the many-particle path $\mathbf{R}(u)$:

$$S\{\mathbf{R}(u)\} = \frac{1}{\hbar} \int_0^{\hbar/T} du \left[\sum_{i=1}^N \frac{m}{2} \left(\frac{d\mathbf{r}_i}{du}\right)^2 + V\big(\mathbf{R}(u)\big) \right] \qquad (11.5)$$

where $V(\mathbf{R}(u))$ is the potential energy of the system associated with the configuration $\mathbf{r}_1(u), \mathbf{r}_2(u) \ldots \mathbf{r}_N(u)$. Thus, the second term of (11.5) is simply the potential energy of the system averaged over the paths of all particles (physical interactions among different particles only exist at the same imaginary time).

Unlike the case for classical systems, configurations (i.e., many-particle paths) that are favored at low temperatures are not necessarily those of lower potential energy; as the temperature is lowered, paths extend over a larger spatial region, and there is a competition between the potential energy of the system averaged along the path and the first term of the action (11.5), which has the dimensions of kinetic energy, and is greater for paths of higher curvature. This term expresses zero-point motion.

Paths are periodic in the variable u, i.e., $\mathbf{R}(u + n\hbar/T) = \mathbf{R}(u)$, but this requirement does *not* extend to paths of individual particles, i.e., $\mathbf{r}_i(u + n\hbar/T)$ need *not* be equal to $\mathbf{r}_i(u)$; instead, a weaker condition is imposed, namely, $\mathbf{r}_i(u) = \mathbf{r}_{P(i)}(u + n\hbar/T)$, where $P(i)$ is a permutation of the particle indices $i = 1, 2,\ldots, N$. These ideas are schematically shown in Figure 11.3.

That particle positions should be replaced by extended objects like paths, ongoing from a classical to a quantum-mechanical formulation, is a consequence of the formalism of quantum mechanics, and working out the details would take us too far afield. Intuitively, however, this can be understood in terms of the spatial delocalization of particles associated with quantum zero-point motion. The fact that paths can become "entangled" as explained above simply expresses the indistinguishability of identical particles. Any quantum-mechanical computational scheme must incorporate in full exchanges among particles, for these underlie physical phenomena of interest such as Bose-Einstein condensation and superfluidity.

---

[*] The notion of "integration over paths" is one that requires some nontrivial mathematics, in order to be formulated precisely. However, all of the formal details can be worked out.

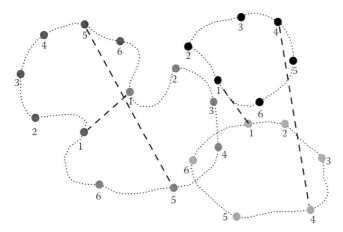

FIGURE 11.3 Representative path for a system of four identical particles. "Beads" of different shades of gray represent the positions in space of the four particles at six successive "imaginary times." Paths are shown by dotted lines. Dashed lines drawn between beads corresponding to the same times represent interaction between the two particles (only a few of them are shown for clarity). The particular path shown here includes the exchange of two particles, as the paths where the "dark" and "light" beads are placed merge into one another.

11.3.2.2 Monte Carlo Simulation

A straightforward analytical evaluation of Path Integral expressions such as (11.4) is out of question for essentially any nontrivial interacting quantum many-particle system. However, an effective numerical procedure, based on the Monte Carlo method, allows one to accomplish such a goal.*

The Monte Carlo method [43] is a well-established, general numerical procedure to compute integrals in many dimensions. In particular, Monte Carlo simulations have been for a long time the tool of choice to compute thermodynamic properties of many-particle systems based on classical statistical mechanics (i.e., ratios of multidimensional integrals as in Equation 11.3). The idea is to transform the computation of ratios, such as in (11.3), into a *statistical average* of the quantity of interest over a large set of random configurations {$\mathbf{R}_i$}, each generated with a probability proportional to $\exp[-V(\mathbf{R}_i)/T]$. This goal can be accomplished on a computer, by means of a random walk through the space of many-particle configurations, using the famous Metropolis Algorithm. The results obtained in this way are essentially *exact*, the only errors being statistical in nature, and therefore at least in principle reducible arbitrarily, provided that enough computer time be available.

As it turns out, the same basic procedure can be utilized to construct a Monte Carlo algorithm applicable to quantum many-particle systems, i.e., suitable to evaluate Path Integral expressions such as (11.4). There are only two additional complications with respect to a Monte Carlo procedure for classical systems.

The first is the fact that paths must be discretized in order to evaluate the Euclidean action (11.5); this introduces a discretization error, which, however, can be dealt with relatively easily by performing numerical extrapolations of results obtained with increasingly finer imaginary time grids.

The second challenge consists of devising a sampling methodology for the Metropolis random walk for many-particle paths, capable of generating with the proper frequency paths in which particles exchange. As mentioned above, this is a crucial ingredient of the quantum algorithm, as quantum exchanges underlie, for example, the superfluid properties of the clusters.

A recently proposed scheme, known as *Worm Algorithm*, has been shown to allow for efficient Monte Carlo evaluation of thermodynamic averages given by expressions such as (11.4). While the full power of this methodology is harnessed when performing large-scale calculations of bulk properties of matter (for which a number of particles of the order of 10^4 is well within reach [see, for instance, Ref. [44]]), calculations for finite clusters are also carried out with greater accuracy than that achievable with any other existing numerical methodology. The reader is referred to the original references [45,46] for a thorough illustration of the algorithm. Suffice to say here that quantum simulations of clusters of hydrogen molecules at low temperatures have been performed using such methodology, based on the model (11.1). These calculations are *numerically exact*, i.e., they contain no *ad hoc* physical assumption other than the (first principles) Hamiltonian (11.1). Results are only affected by a statistical error, which has been reduced to a point where it is essentially negligible. Accurate estimates have been obtained for cogent quantities, such as structural correlation functions, as well as the superfluid fraction. Based on the results, which will be illustrated in the remainder of this chapter, and on the insight that has been gained, some definite predictions regarding the intriguing physical behavior of these objects are made.

Before moving on to the discussion of the results, a remark is in order. There are several obvious differences between the results of an analytical calculation and those of a computer simulation. It is fully accepted that the power, intuitiveness, versatility, and elegance of closed-form mathematical expressions are unmatched by anything that a computer can produce and that one should resort to a simulation only if the problem is truly analytically intractable. Unfortunately, this is almost invariably the case for nontrivial physical problems, especially those involving several (as in, more than two) interacting particles, for which no general, reliable analytical methods exist. For this reason, computer simulations are becoming increasingly ubiquitous in modern theoretical physics. It should be noted, however, that besides raw numbers a computer simulation can also yield a different type of information, a more visual one, of the type similar to that which experiments often provide. It does not have the rigor and conclusiveness of an analytical calculation, but can often allow for no less valuable understanding. Examples of this will be shown in the next sections.

---

* It should be mentioned that this is only strictly true for many-particle systems obeying Bose statistics. No *exact* Quantum Monte Carlo method applicable to fermions is presently known.

11.4 Results for H₂ Clusters

In this section, results are described of Monte Carlo simulations at low temperatures of molecular hydrogen clusters of size up to $N = 40$. Before delving into the details of the numerical data, it is worth reminding the reader that the main objective of this work is to provide insight into the phase of clusters of hydrogen at low temperatures. Obviously, this requires that one be able to identify, for instance, solid- or liquid-like behavior. The question of how one could actually make such a distinction for a cluster immediately arises. After all, as explained in Section 11.2, the hallmark of the crystalline phase, namely, the presence of *long-range* order, is by definition foreign to a cluster of finite size.

As mentioned in Section 11.2.3, circumstantial evidence for liquid-like behavior can be obtained from the study of the superfluid properties. Intuitively, a large superfluid fraction is associated with greater particle mobility, typical of the liquid phase. It is important to keep in mind, however, that this criterion is not fool-proof.

Firstly, at sufficiently low temperatures in principle *any* finite system ought to display superfluid behavior. The physical mechanism that gives rise to a finite superfluid response for a cluster that features otherwise solid-like properties does not involve any structural transformation. Much more mundanely, the cluster ultimately behaves as a rigid rotator, and as soon as the temperature is lowered below the characteristic separation between the first two energy levels of a spherical rotator with the same classical moment of inertia I of the cluster, the system will no longer be able to convert thermal energy in rotational excitations, and will therefore rotate freely, i.e., be a "superfluid" according to the definition. The characteristic "transition temperature" in this case is of the order of $T_R = \hbar^2/I$; it is the temperature below which a cluster that displays solid-like behavior down to that temperature will also turn superfluid. Therefore, whenever the computed superfluid transition temperature for a cluster is close to, or of the order of T_R, the association of superfluidity with liquid-like

behavior is no longer justified, as one may simply be looking at the onset of free rotation. In a computer simulation, presumably the statistics of cycles of particle exchanges might give a clue to whether this is the regime that one has entered, as exchanges among particles is different, which are depressed in a rigid cluster, shells ought not to increase significantly below T_R.

Secondly, solid order is not necessarily incompatible with superfluidity; indeed, speculations about the possible existence of a *supersolid* phase, namely, one simultaneously featuring solid order and superfluidity, date back to the late 1960s [52–54] and have gained momentum in recent years after the provocative claim of its observation in solid ^{4}He by Kim and Chan [55]. In a finite system, given that the demarcation between solid- and liquid-like phases is blurred, the occurrence of a solid-like phase displaying some finite superfluid response is possible. Nonetheless, we shall tentatively make the assumption, at the outset, that clusters with a larger superfluid fraction should be liquid-like; conversely, clusters whose superfluid response is depressed are more likely to feature enhanced molecular localization, i.e., be solid-like. The results presented in the following sections are published in Refs. [47–51].

11.4.1 Superfluidity

The superfluid fraction ρ_s of a cluster is defined as the fraction of the system that decouples from an externally applied rotation; it saturates to a value of 1 (100%) for a cluster that is entirely superfluid and is zero (or close to zero) for a non-superfluid cluster. It can be computed by Quantum Monte Carlo simulations using a suitable estimator derived within the Path integral formalism, known as the "area" estimator, originally introduced in Ref. [16]. Results for ρ_s are shown in Figure 11.4 both for clusters of p-H₂ (a) as well as the heavier o-D₂ isotope (b) (from Ref. [48]).

The first observation is that clusters of p-H₂ (o-D₂) below a given size (approximately 23 for p-H₂ and 8 for o-D₂) are entirely superfluid (i.e., their superfluid fraction is essentially 100%) at

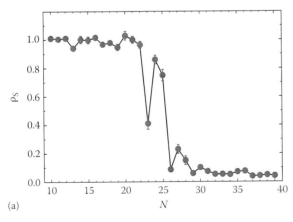

(a)

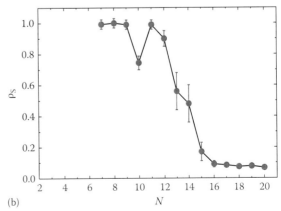

(b)

FIGURE 11.4 Superfluid fraction of p-H₂ (a) and o-D₂ (b) clusters with varying numbers of molecules, at $T = 1\,\mathrm{K}$ ($T = 0.5\,\mathrm{K}$ for o-D₂). When not shown explicitly, statistical error bars are of the order of the symbol size. (Reprinted from Mezzacapo, F. and Boninsegni, M., *J. Phys.: Condens. Matter*, 21, 164205, 2009. With permission.)

1 K (0.5 K); the superfluid fraction then decreases with *N*, but in a non-monotonic fashion, markedly so for *p*-H₂. In particular, in a size range between 23 and 30, the superfluid fraction oscillates sharply, even on adding a single *p*-H₂ molecule. It seems reasonable to surmise that such large oscillations ought to reflect structural changes that take place as the number of molecules is changed.* The superfluid response of *o*-D₂ clusters is considerably suppressed with respect to the corresponding clusters of *p*-H₂, due to the larger molecular mass. There are reasons to expect, however, that the fundamental physical picture may be very similar, only "shifted" to lower temperature. We shall henceforth restrict the discussion to clusters of *p*-H₂ only; *o*-D₂ will only be discussed again as an isotopic dopant for these clusters.

11.4.2 Structure

11.4.2.1 Clusters with *N* < 23

Clusters with fewer than 23 *p*-H₂ molecules show a behavior quite similar to that of helium clusters [16]. In order to gain quantitative understanding of the structure of these clusters, we compute the radial density profile, i.e., the spherically averaged distribution of mass around the center of mass of the cluster. Figure 11.5 shows this measure for a cluster with 20 *p*-H₂ molecules at *T* = 1 K. A floppy structure is observed, featuring an inner and an outer shell with no sharp separation between them.

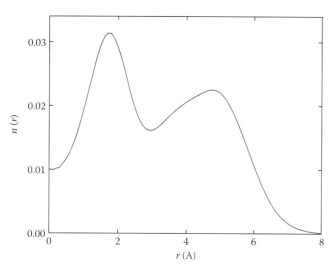

FIGURE 11.5 Radial density profile for a (*p*-H₂)₂₀ droplet, computed with respect to the center of mass of the system, at *T* = 1 K. Density is expressed in Å⁻³. (Reprinted from Mezzacapo, F. and Boninsegni, M., *J. Phys.: Condens. Matter*, 21, 164205, 2009. With permission.)

---

* The distinction between clusters of size less than 23 molecules and those of size between 23 and 40 pertains to results obtained with the Silvera-Goldman interaction potential. On using a weaker potential, such as the Lennard-Jones, clusters of size greater than 23 are liquid-like at *T* = 1 K. Conversely, if a stronger binding model interaction is adopted (e.g., the Buck potential), superfluidity is depressed. Qualitatively, however, the overall physical picture is unaltered.

Visual inspection of many-particle configurations generated by the simulation reveals no preferred molecular arrangement, consistently with liquid-like character.

At *T* = 1 K, these clusters are essentially entirely superfluid within the statistical errors of our calculation [47,48]. Their structural properties are found to depend relatively weakly on the temperature.

11.4.2.2 Clusters with *N* > 23

Figure 11.6 shows radial density profiles for clusters with *N* = 25, 26, and 27 molecules at *T* = 0.5 K. They are generally sharper than that of (*p*-H₂)₂₀, pointing to increased molecular localization and greater rigidity of the cluster inner shell, as evidenced by the absence of particles in the middle of the cluster.†

However, some relative differences are clear, even within these three systems. For instance, the radial density profile of (*p*-H₂)₂₆ displays a noticeably sharper peak than that of both other clusters, as well as a lower inter-shell minimum (albeit in no case are the two shells sharply separated), and an additional structure further away from the center. A sharper main peak generally implies greater rigidity, i.e., molecules in the inner shell are more localized. Thus, the addition of a single molecule to the (*p*-H₂)₂₅ has the effect of enhancing localization and the rigidity of the inner shell, whereas the addition of a second molecule (yielding the (*p*-H₂)₂₇ cluster) has the opposite effect, namely, that of "softening" the inner shell. The superfluid fraction of the (*p*-H₂)₂₅ cluster at this temperature is close to 85% and that of (*p*-H₂)₂₇ is around 25%, whereas that of the (*p*-H₂)₂₆ is less than 10%. In other words, the cluster with a more "rigid" inner shell is less superfluid, which is

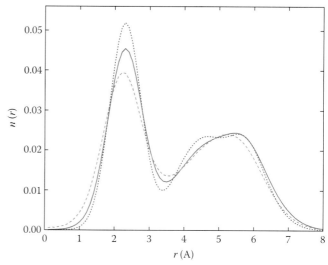

FIGURE 11.6 Radial density profile for a (*p*-H₂)₂₅ (lowest main peak), (*p*-H₂)₂₆ (highest) and (*p*-H₂)₂₇ (intermediate) droplets, computed with respect to the center of mass of the system, at *T* = 0.5 K. Density is expressed in Å⁻³. (Reprinted from Mezzacapo, F. and Boninsegni, M., *J. Phys.: Condens. Matter*, 21, 164205, 2009. With permission.)

---

† Henceforth, we shall adopt the short-hand notation (*y*)ₙ to refer to a cluster with *N* molecules of species *y*.

consistent with the general idea that superfluidity should go hand in hand with molecular mobility, i.e., more liquid-like behavior. There is more than that, however. So far we have referred to the superfluid fraction as a global quantity, characterizing the whole cluster, but one might wonder whether, for those clusters that are only partially superfluid, the superfluid response ought to be concentrated in some regions of the cluster, e.g., near the surface, where interactions among molecules are presumably less important. As it turns out, the superfluid response of a finite system can be spatially resolved, using an estimator recently proposed by Kwon et al. [56]. It is consistently observed [50] that these clusters are essentially *uniformly* superfluid, i.e., with no concentration of the response in particular regions. Because superfluidity is ultimately underlain by exchanges among identical particles, the results obtained for the three clusters discussed above point to the importance of exchanges involving particles in different shells. We shall come back to this point again later on.

In order to gain more visual understanding of the structure of the different clusters, we use a graphical representation of their density, obtained as proposed in Ref. [57]. Figure 11.7 shows p-H_2 clusters consisting of 25 and 26 molecules at $T = 1$ K. The cluster (p-H_2)$_{26}$ displays remarkably solid-like properties. Its structure consists of three rings of five molecules, with four other molecules linearly arranged along the axes of the rings, while the remaining seven molecules form an outer shell. Although their position is smeared by zero-point fluctuations, the various molecules in the cluster can be clearly identified, indicating that they enjoy a fairly high degree of spatial localization; consequently, exchanges among different molecules are highly suppressed (though not completely absent), and the superfluid response is weak. Conversely, the molecules in the cluster (p-H_2)$_{25}$ cannot be clearly identified, and the entire system appears amorphous. Because of their pronounced delocalization, molecules have a strong propensity to be involved in quantum exchanges, hence the larger superfluid response observed. On the other hand, in the $T \rightarrow 0$ limit *all* of these clusters, including (p-H_2)$_{26}$, turn superfluid. Indeed, the superfluid fraction of the (p-H_2)$_{26}$ cluster reaches essentially 100% below ~0.1 K.

For clusters with N in the range between 22 and 30, the transition temperature T_R associated with free rotation of a rigid rotator takes on a value of ~0.05 K. On the other hand, our results show onset of superfluid behavior at temperatures an order of magnitude higher than that for (p-H_2)$_{23}$ and for other clusters [e.g., (p-H_2)$_{27}$] that display quantum melting. This points to a structural transformation [41] taking place at low T from a classical, solid-like phase to a liquid "phase" of some of these clusters, rather than the onset of free rotation of a rigid cluster.

11.4.3 Magic Numbers

As discussed in Section 11.2.1, a cluster solid-to-liquid transition is typically found in classical clusters corresponding to specific numbers of particles, referred to as "magic." Magic clusters are energetically more stable than the others, in general, owing to a particular geometrical structure (e.g., several filled concentric shells). In order to explore the same issue in the context of hydrogen clusters, for which quantum effects are important, we introduce the *chemical potential* $\mu(N)$ associated with a cluster of N molecules is defined as

$$\mu(N) = E(N-1) - E(N) \qquad (11.6)$$

where $E(N)$ is the total energy of the cluster in the $T \rightarrow 0$ limit. A peak in the chemical potential for a particular value of N signals enhanced stability for that cluster, much like in the classical case.

Figure 11.8 shows computed values of $\mu(N)$ for clusters of p-H_2 and o-D_2 based on energy estimates obtained at finite T. In both cases, well-defined peaks can be seen at $N = 13$, and for o-D_2 clusters at $N = 19$ as well, using energy values computed at $T = 1$ K for p-H_2, 0.5 K for o-D_2 (filled circles in Figure 11.8). For o-D_2 clusters, energy estimates do not appreciably change on lowering the temperature (at least down to 0.25 K). For p-H_2 clusters, on the other hand, the additional structure seen for $25 < N < 34$ at $T = 1$ K is "washed out" at lower T. Indeed, our extrapolated ground state results are consistent with those of other authors, whose calculations only show the $N = 13$ peak in the 1–30 N range, at $T = 0$ [58,59]. The picture becomes more complex for larger clusters, as calculations at lower and lower temperatures are needed in order to ascertain whether some peaks that are present at finite T, persist in the $T \rightarrow 0$ limit. At the present time, our calculations for $N > 30$ show that $N = 34$ is a magic number, in agreement with Ref. [59] but in disagreement with Ref. [58].

In any case, this behavior suggests that, at temperatures of the order of 1 K, clusters of p-H_2 in the size range between ~23 and 30 behave as more classical objects than in the $T \rightarrow 0$ limit, wherein, as we shall see in some cases, distinct liquid-like behavior is observed.

11.4.4 Quantum Melting

This section exemplifies the type of information that a computer simulation can offer about the physics of the system of interest, information not readily accessible to analytical studies.

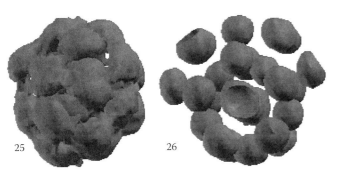

25   26

FIGURE 11.7 Three-dimensional representation of the molecular density of the two clusters (p-H_2)$_{25}$ and (p-H_2)$_{26}$. (Reprinted from Mezzacapo, F. and Boninsegni, M., *Phys. Rev. A*, 75, 033201, 2007.)

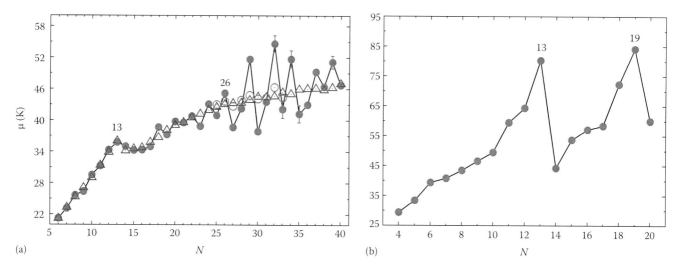

FIGURE 11.8 Chemical potential as a function of the number N of molecules in clusters of p-H$_2$ (a) and o-D$_2$ (b). Filled circles are estimates obtained at $T = 1$ K for p-H$_2$, 0.5 K for o-D$_2$. The latter remains unchanged at lower T. Open circles show estimates at $T = 0.25$ K, triangles refer to results of Ref. [58]. (Reprinted from Mezzacapo, F. and Boninsegni, M., *J. Phys.: Condens. Matter*, 21, 164205, 2009. With permission.)

Previously, when discussing the superfluid fraction at low T of clusters with $23 \leq N \leq 30$, we made a conceptual distinction between those that are superfluid, to which we attribute liquid-like properties, and those that are non-superfluid, which we refer to as *insulating*, i.e., solid-like. It might be objected that the fact that the superfluid fraction is low does not necessarily imply a solid-like arrangement of molecules and that the greater apparent rigidity of, say, $(p$-H$_2)_{26}$ compared to $(p$-H$_2)_{25}$ and $(p$-H$_2)_{27}$ does not, by itself, justify the conclusion that this cluster is solid-like. Indeed, clusters with a low value of ρ_S at the temperatures shown in Figure 11.4 may be no less liquid-like than those with a high value of ρ_S and simply have a lower superfluid transition temperature.

This is obviously a reasonable objection; the most convincing evidence in favor of the solid-like character of non-superfluid

clusters with $N > 23$, as opposed to liquid-like clusters that are above their superfluid transition temperature, is provided by the raw output of the simulation itself (the type of information that one gathers while "staring at the screen").

Figure 11.9 shows the values of the superfluid fraction and of the potential energy per molecule recorded during a typical Monte Carlo run for clusters of 18 and 23 p-H$_2$ molecules, the former at $T = 2$ K and the latter at both $T = 1.0$ and 1.4 and 1 K. In particular, we show consecutive block averages of ρ_S and V, each block consisting of 500 "sweeps" (see Ref. [48] for details).

For a $(p$-H$_2)_{18}$ cluster, both quantities simply fluctuate about their average values. On the other hand, for $(p$-H$_2)_{23}$, despite the large fluctuations, two different regimes can be easily identified, namely, one in which the superfluid fraction is high (with an average value close to 1) and the other characterized

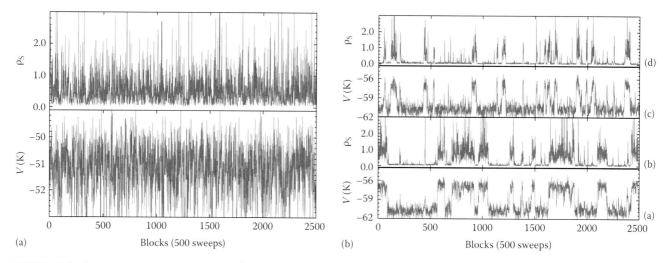

FIGURE 11.9 Potential energy per molecule and superfluid fraction observed during a typical Monte Carlo run (see text) for a cluster of $N = 18$ p-H$_2$ molecules at $T = 2$ K (a) and for one of 23 p-H$_2$ molecules (b) at $T = 1$ K (panels a and b) and $T = 1.4$ K (panels c and d). (Reprinted from Mezzacapo, F. and Boninsegni, M., *Phys. Rev. A*, 75, 033201, 2007. With permission.)

by low values of ρ_S (average value close to zero). The potential energy, correspondingly, takes on high (low) values when ρ_S is large (small).

Clearly, the behavior of the two clusters shown in Figure 11.9 is markedly different. The cluster with $N = 18$ does not undergo any structural change as the temperature is lowered; it is a liquid-like cluster which becomes more superfluid at low T, according to the basic understanding of the superfluid transition as it occurs in liquid ^{4}He. Conversely, the switching of the values of the potential energy between two different regimes, separated by some ~6 K, and the simultaneous switching of the superfluid fraction can be interpreted in terms of the system visiting relatively ordered, solid-like, insulating configurations (characterized by low potential energy) and disordered, liquid-like, superfluid ones. Of course, this type of "coexistence" between two distinct (disordered and ordered) phases is nothing new, *per se*. It is completely analogous to the observation made in Section 11.2.1 in the context of melting of classical clusters (see, for instance, Ref. [8]).

What is unique and remarkable here is that the *melting* of the $(p\text{-}H_2)_{23}$ cluster takes place as the temperature is *lowered*, due to quantum effects. On comparing the results for the cluster $(p\text{-}H_2)_{23}$ at the two different temperatures, we can see that the liquid-like superfluid phase becomes dominant at lower T. Correspondingly, a sharp increase is seen in the frequency of occurrence of permutation cycles involving *all* of the particles in the cluster. This process is ostensibly induced by zero-point motion and the ensuing exchanges of molecules, whose importance increases as T is lowered. In this sense, one could state that melting is driven by Bose statistics, i.e., it is associated with the energy contribution due to quantum exchanges. On the other hand, when T is sufficiently high, quantum exchanges are suppressed and the system "freezes" in a solid-like structure.

Additional evidence of this transformation can be obtained by looking at the evolution of the density profile of a cluster as a function of temperature (Figure 11.10). In this case, three different temperatures are considered, namely, $T = 4$ K, $T = 2$ K, and $T = 0.1$ K. At the highest temperature, there is a fairly broad main peak, and the system features two floppy shells, this structure essentially duplicating that shown in Figure 11.5 for a cluster with 20 molecules. Superfluidity is absent here. As the temperature is lowered, the main peak goes up by some 50% and becomes narrower; as well, the shell structure becomes better defined, with the inter-shell minimum becoming shallower and the density at the center of the cluster dropping to essentially zero. The system is still very much non-superfluid, and the tendency toward crystallization is clear. However, as the temperature is lowered even further, the density profile in many respects takes on similar features as at the higher temperature. Indeed, at $T = 0.1$ K, the main peak has lost almost all of the strength that it had acquired on lowering the temperature from 4 to 2 K and is almost as broad as at 4 K, while the density in the center of the cluster is again finite. This is consistent with higher delocalization of molecules in the inner shell, in turn physically interpreted as enhanced liquid-like behavior.

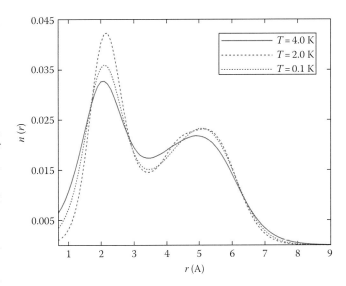

FIGURE 11.10 Radial density profile for a $(p\text{-}H_2)_{23}$ cluster, computed with respect to the center of mass of the system, at $T = 4$ K (solid line), $T = 2$ K (dashed line), and $T = 0.1$ K (dotted line). Density is expressed in Å^{-3}. Note how the main peak is lower and slightly broadened *both* at the highest as well as at the lowest temperature. The same effect is observed for $(p\text{-}H_2)_{27}$.

One might speculate that the same physical behavior may be present in other clusters in the range explored here, possibly even above $N = 30$. Indeed, although Figure 11.4 shows a depressed superfluid response at T down to 0.5 K for clusters with more than 27 $p\text{-}H_2$ molecules, simulations at lower temperatures (in this study, the lowest temperature considered is $T = 0.0625$ K) show marked superfluid behavior for a few clusters in the range $27 \leq N \leq 40$; the largest cluster for which superfluidity was observed in this work is $(p\text{-}H_2)_{36}$, for which $\rho_S > 0.5$ for $T < 0.125$ K (still a few times higher than T_R). However, only for the two clusters $(p\text{-}H_2)_{23}$ and $(p\text{-}H_2)_{27}$ has evidence of the type shown in Figure 11.9b been obtained so far.

More generally, *quantum melting* (QM) ought to take place at lower and lower temperature as the size of the cluster is increased (as the cluster approaches the crystal bulk limit). As shown by the results in Figure 11.4, however, the trend is non-monotonic. In our simulations, we have observed clear evidence of QM for at least two clusters (namely, $N = 23$ and 27) in the temperature range between 0.25 and 1.25 K. Using a different interaction potential does not eliminate the effect, it simply shifts it to a higher (lower) temperature, depending on which potential is utilized.

An examination of the frequency of occurrence of permutation cycles of a given length shows that the superfluid response of clusters with $N > 22$ crucially hinges on the onset of permutations involving all of the molecules in the clusters, *not* just those on the surface [50] or on any other specific region of the cluster. This is illustrated in Figure 11.11. Moreover, results obtained in the presence of an isotopic impurity (see below), showing that a single substitutional $o\text{-}D_2$ molecule can turn insulating and solid-like a $p\text{-}H_2$ cluster that is otherwise entirely superfluid

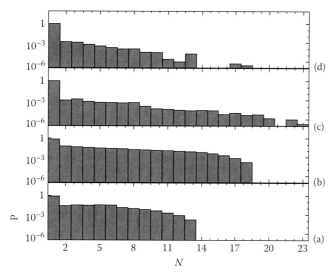

FIGURE 11.11 Frequency of occurrence of exchange cycles involving N particles observed in a Monte Carlo simulation of the following clusters (at $T = 1$ K): $(p\text{-}H_2)_{13}$ (a), $(p\text{-}H_2)_{18}$ (b), $(p\text{-}H_2)_{26}$ (c), and $(p\text{-}H_2)_{33}$ (d). The superfluid fraction of the two smaller clusters is close to 100%, that of the two larger ones is less than 10%. (Reprinted from Mezzacapo, F. and Boninsegni, M., *Phys. Rev. A*, 75, 033201, 2007. With permission.)

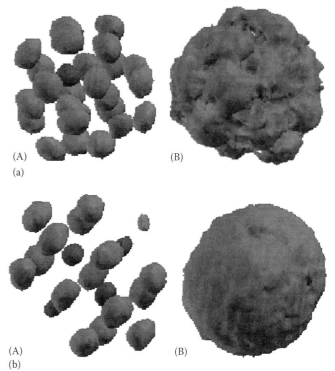

FIGURE 11.12 Three-dimensional representations of $p\text{-}H_2$ clusters doped with one (a) and four (b) $o\text{-}D_2$ molecules. (a) $(p\text{-}H_2)_{24}\text{-}(o\text{-}D_2)_1$ (A) and of $(p\text{-}H_2)_{25}$ (B). (b) $(p\text{-}H_2)_{16}\text{-}(o\text{-}D_2)_4$ (A) and of $(p\text{-}H_2)_{20}$ (B). Darker color is used for impurity molecules. (Reprinted from Mezzacapo, F. and Boninsegni, M., *Phys. Rev. A*, 76, 021201, 2007. With permission.)

when undoped, rules out a possible scenario of superfluidity arising from loosely bound surface molecules on top of an otherwise rigid cluster; for, the impurity sits in the center of the cluster, and therefore its effect on exchanges involving only surface molecules is negligible.

11.4.4.1 Doped Clusters and Exchange Cycles

In order to gain additional insight into the mechanism underlying superfluidity in these clusters and the importance of quantum exchanges, we have performed additional simulations of $p\text{-}H_2$ clusters, in which one or more $p\text{-}H_2$ molecules were replaced by molecules of $o\text{-}D_2$. As stated above, the only difference between $o\text{-}D_2$ and $p\text{-}H_2$ in our microscopic model is the mass of the substitutional particles.

In Figure 11.12, we make use of the same schematic three-dimensional representations used above to illustrate the structural changes that occur upon doping two different $p\text{-}H_2$ clusters, namely, $(p\text{-}H_2)_{25}$ and $(p\text{-}H_2)_{20}$ with substitutional $o\text{-}D_2$ impurities. The main effect in both cases is the same, i.e., the introduction of heavier impurities, whose energetically favored position is at (or, near) the geometrical center of the cluster, has the effect of rendering a liquid-like, superfluid pristine cluster *insulating and solid-like.*

Consider the case of a single $o\text{-}D_2$ impurity substituted into a $(p\text{-}H_2)_{25}$ cluster first (Figure 11.12a). The lone $o\text{-}D_2$ molecule sits in the central part of the cluster. The cluster turns from liquid (A) to solid-like (B); correspondingly, superfluidity is altogether depressed [49]. By contrast, if the dopant molecule is taken to have *the same mass* as all other molecules but *distinguishable* from them, and thus it is not allowed to take part in exchanges of $p\text{-}H_2$ molecules (this is actually a reasonable model for an $o\text{-}H_2$ dopant molecule), then the structure of the cluster and

its superfluid properties are much less affected by the substitution. Indeed, a doped cluster of 25 molecules retains most of its superfluidity, in this case. The physical difference arises from the fact that a lone $o\text{-}D_2$ molecule will sit in the central part of the cluster, where it has a strong inhibiting effect on long exchanges of $p\text{-}H_2$ molecules; the lighter $o\text{-}H_2$ dopant, conversely, is considerably more delocalized and indeed is found prevalently near the surface [49]. As a result, long exchanges of $p\text{-}H_2$ molecules, also involving those in the inner shell of the cluster, remain possible.

The effect of cluster "crystallization," induced in the $(p\text{-}H_2)_{25}$ clusters by the substitution of one or two $o\text{-}D_2$ impurities (and the ensuing suppression of superfluidity) can also be observed in smaller systems, but a greater number of substitutions is needed in order to produce the same dramatic change. Figure 11.12b shows the structures of the two clusters $(p\text{-}H_2)_{16}\text{-}(o\text{-}D_2)_4$ (part A of the figure) and $(p\text{-}H_2)_{20}$ (part B); the pristine cluster is featureless and liquid-like, entirely superfluid at $T \leq 1$ K. On the other hand, the superfluid component is small in the doped cluster, whose solid-like structure is evident, with a central axis surrounded by rings of molecules. Two of the four $o\text{-}D_2$ molecules are placed on the axis and the other two on the central ring. All of this shows that it is not only zero-point motion that underlies a liquid-like behavior of these systems, for quantum exchanges play actually a crucial role.

11.5 Conclusions

A systematic theoretical study of the physical properties of clusters of hydrogen molecules at low temperatures has been carried out, with the aim of understanding the interplay between superfluid and structural properties. We have considered both pristine clusters of p-H_2 and o-D_2, as well as clusters of p-H_2 doped with few substitutional o-D_2. While the more quantitative details depend on the specific choice of interaction potential, there are some interesting general aspects that emerge in numerical simulations based on any reasonable pair-wise, spherically symmetric interaction. We do not expect the inclusion of nonspherical corrections to alter the picture significantly.

Our results show that clusters with a number of molecules less than ~22 are essentially liquid-like at temperatures lower than ~4 K, and turn superfluid in the low T limit, much like clusters of helium atoms. To the extent that such a terminology can be accepted, theirs appears to be a conventional "normal-to-superfluid" transition. On the other hand, clusters of size between 22 and ~30 molecules display a richer behavior in many respects. Firstly, they behave classically in a range of low temperatures, at which they display a relatively ordered structure, with molecules arranged in some preferred geometrical fashion. This is more pronounced for some clusters (e.g., $N = 26$) than others. At sufficiently low temperature, however, all of these clusters appear to display superfluid properties. Although we cannot completely rule out a possible "supersolid" phase (i.e., one characterized by SF as well as orderly arrangement of molecules), we have seen so far little evidence of that, in *any* of the clusters studied. On the contrary, numerical evidence accumulated by others suggests that the ground state of all these clusters is liquid-like. However, further study is underway, to clarify this intriguing issue.

We have observed at least two cases in which a distinct change of physical character occurs as the temperature is lowered, going through a stage in which a solid-like non-superfluid "phase" of these droplets coexist with a liquid-like superfluid one, the latter becoming predominant at lower T. This conclusion is mostly based on the observation of large, concomitant oscillations, recorded during the computer simulation of the superfluid density and the potential energy per particle. This intriguing phenomenon, to which we have come to refer as *quantum melting*, might conceivably be experimentally observable in p-H_2 droplets, as well as in other physical systems, e.g., dipolar cold molecules confined in external harmonic potentials.

Examination of the statistics of permutation cycles, as well as the observation that the replacement of a single p-H_2 molecule with an o-D_2 can turn solid-like and insulating a cluster that is superfluid and liquid-like when undoped, directly point to the importance of long exchanges, especially involving particles in the inner and outer shells of the cluster, in stabilizing a liquid-like, superfluid phase of these objects. Information coming from structural quantities is also consistent with the melting of the clusters at low temperature, but the signal is smaller and sometimes ambiguous (especially at the lowest temperatures).

Acknowledgments

This work was supported in part by the Natural Science and Engineering Research Council of Canada under research grant 121210893, and by the Alberta Informatics Circle of Research Excellence.

References

1. R. W. Robinett, *Quantum Mechanics* (Oxford University Press, New York, 1997), p. 17.
2. J. Wilks, *The Properties of Liquid and Solid Helium* (Clarendon Press, Oxford, U.K., 1967).
3. V. L. Ginzburg and A. A. Sobyanin, *JETP Lett.* **15**, 242 (1972).
4. P. Jena, B. K. Rao, and S. N. Khanna, *The Physics and Chemistry of Small Clusters* (Plenum, New York, 1987).
5. H. L. Davis, J. Jellinek and R. S. Berry, *J. Chem. Phys.* **86**, 6456 (1987).
6. P. Labastie and R. L. Whetten, *Phys. Rev. Lett.* **65**, 1567 (1990).
7. S. Sugano, *Microcluster Physics* (Springer, Berlin, Germany, 1991).
8. R. E. Kunz and R. S. Berry, *Phys. Rev. Lett.* **71**, 3987 (1993).
9. R. S. Berry, *J. Chem. Phys.* **98**, 6910 (1994).
10. D. J. Wales and R. S. Berry, *Phys. Rev. Lett.* **73**, 2875 (1994).
11. S. Goyal, D. L. Schutt, and G. Scoles, *J. Chem. Phys.* **102**, 2302 (1995).
12. J. K. Lee, J. A. Barker, and F. F. Abraham, *J. Chem. Phys.* **58**, 3166 (1973).
13. W. D. Kristensen, E. J. Jensen, and R. M. J. Cotterill, *J. Chem. Phys.* **60**, 4161 (1974).
14. A. L. Mackay, *Acta Cryst.* **15**, 916 (1962).
15. S. Goyal, D. L. Scoles, and G. Scoles, *Phys. Rev. Lett.* **69**, 933 (1992).
16. P. Sindzingre, M. L. Klein, and D. M. Ceperley, *Phys. Rev. Lett.* **63**, 15 (1989).
17. C. Chakravarty, *J. Chem. Phys.* **120**, 956 (1995).
18. F. Calvo, J. P. K. Doye, and D. J. Wales, *J. Chem. Phys.* **114**, 7312 (2001).
19. C. Predescu, D. Sabo, and J. D. Doll, *J. Chem. Phys.* **119**, 12119 (2001).
20. C. Predescu, P. A. Frantsuzov, and V. A. Mandelshtam, *J. Chem. Phys.* **122**, 154305 (2005).
21. P. A. Frantsuzov, D. Meluzzi, and V. A. Mandelshtam, *Phys. Rev. Lett.* **96**, 113401 (2006).
22. T. L. Beck, J. D. Doll, and D. L. Freeman, *J. Chem. Phys.* **90**, 5651 (1989).
23. P. Sindzingre, D. M. Ceperley, and M. L. Klein, *Phys. Rev. Lett.* **67**, 1871 (1991).
24. D. Scharf, G. J. Martyna, and M. L. Klein, *Chem. Phys. Lett.* **197**, 231 (1992).
25. D. Pines and P. Nozières, *The Theory of Quantum Liquids*, Vol. II (Addison-Wesley, New York, 1992).

26. C. J. Pethick and H. Smith, *Bose-Einstein Condensation in Dilute Gases* (Cambridge University Press, Cambridge, U.K., 2002).

27. D. S. Lewart, V. R. Pandharipande, and S. C. Pieper, *Phys. Rev. B* **37**, 4950 (1988).

28. J. D. Reppy and D. Depatie, *Phys. Rev. Lett.* **12**, 187 (1964).

29. D. R. Tilley, *Superfluidity and Superconductivity*, (IOP Publishing, New York, 1990).

30. S. Grebenev, J. P. Toennies, and A. F. Vilesov, *Science* **279**, 2083 (1998).

31. Y. Xu, W. Jäger, J. Tang, and A. R. W. McKellar, *Phys. Rev. Lett.* **91**, 163401 (2003).

32. J. Turnbull and M. Boninsegni, *Phys. Rev. B* **78**, 144509 (2008).

33. S. Grebenev, B. Sartakov, J. P. Toennies, and A. F. Vilesov, *Science* **289**, 1532 (2000).

34. K. Kuyanov-Prozument and A. F. Vilesov, *Phys. Rev. Lett.* 205301 (2008).

35. G. Tejeda, J. M. Fernandez, S. Montero, D. Blume, and J. P. Toennies, *Phys. Rev. Lett.* **92**, 223401 (2004).

36. H. Wiechert. *Adsorption of Molecular Hydrogen Isotopes on Graphite and Boron Nitride*, (Springer-Verlag, Berlin, Germany, 2003).

37. M. Zoppi, *J. Phys. CM* **15**, 1047 (2003).

38. I. F. Silvera and V. V. Goldman, *J. Chem. Phys.* **69**, 4209 (1978).

39. U. Buck, F. Huisken, A. Kohlhase, D. Otten, and J. Schaeffer *J. Chem. Phys.* **78**, 4439 (1983).

40. M. J. Norman, R. O. Watts, and U. Buck, *J. Chem. Phys.* **81**, 3500 (1984).

41. E. Cheng and K. B. Whaley, *J. Chem. Phys.* **104**, 3155 (1996).

42. R. P. Feynman, *Statistical Mechanics: A Set of Lectures* (Addison-Wesley, New York, 1972).

43. N. Metropolis, A. Rosenbluth, M. Rosenbluth, A. Teller, and E. Teller, *J. Chem. Phys.* **21**, 1087 (1953).

44. L. Pollet, M. Boninsegni, A. Kuklov, N. Prokof'ev, B. Svistunov, and M. Troyer, *Phys. Rev. Lett.* **98**, 135301 (2007).

45. M. Boninsegni, N. Prokof'ev, and B. Svistunov, *Phys. Rev. Lett.* **96**, 070601 (2006).

46. M. Boninsegni, N. Prokof'ev, and B. Svistunov, *Phys. Rev. E* **74**, 036701 (2006).

47. F. Mezzacapo and M. Boninsegni, *Phys. Rev. Lett.* **97**, 045301 (2006).

48. F. Mezzacapo and M. Boninsegni, *Phys. Rev. A* **75**, 033201 (2007).

49. F. Mezzacapo and M. Boninsegni, *Phys. Rev. A* **76**, 021201 (2007).

50. F. Mezzacapo and M. Boninsegni, *Phys. Rev. Lett.* **100**, 145301 (2008).

51. F. Mezzacapo and M. Boninsegni, *J. Phys.: Condens. Matter* **21**, 164205 (2009).

52. A. F. Andreev and I. M. Lifshitz, *Sov. Phys. JETP* **29**, 1107 (1967).

53. G. Chester, *Phys. Rev. A* **2**, 256 (1970).

54. A. J. Leggett, *Phys. Rev. Lett.* **25**, 1543 (1970).

55. E. Kim and M. H. W. Chan, *Science* **305**, 1941 (2004).

56. Y. Kwon, F. Paesani, and K. B. Whaley, *Phys. Rev. B* **74**, 174522 (2006).

57. S. Baroni and S. Moroni, *Chem. Phys. Chem.* **6**, 1884 (2005).

58. R. Guardiola and J. Navarro, *Cent. Eur. J. Phys.* **6**, 33 (2008).

59. J. E. Cuervo and P.-N. Roy, *J. Chem. Phys.* **128**, 224509 (2008).

Superfluidity of Clusters

Francesco Paesani
University of California

12.1 Introduction

Superfluidity, which was discovered in 1938 simultaneously by Kapitsa [2], and by Allen and Misener [3], is a phase of matter characterized by the complete absence of viscosity [1]. Many studies have been devoted to the investigation of the peculiar properties associated with the superfluid phase. Among these are an unusually high thermal conductivity and a zero entropy as well as many other fascinating effects, such as the "fountain effect," film flow and creep, and quantized vortices [1]. As a result of vibrant research in this area, several Nobel Prizes have been awarded for investigations of superfluidity: [Landau (1962); Kapitsa (1978); Lee-Osherhoff-Richardson (1996); and Abrikosov–Ginzburg–Leggett (2003)].

On a macroscopic scale, superfluidity has only been observed so far for helium (He). Although eight isotopes of helium exist, only helium-4 (^{4}He), with an atomic mass $m = 4.0026$ amu, and helium-3 (^{3}He), with an atomic mass $m = 3.0160$ amu, are stable. Unlike any other element, under normal pressures helium remains liquid below the boiling point apparently down to absolute zero ($T = 0$ K, where T is the temperature). This behavior results from the combination of two different factors. The first is the extremely weak He–He interaction that, as shown in Figure 12.1, displays a single minimum with energy $\varepsilon \sim 7.5$ cm^{-1} at an interatomic distance of $R \sim 3.0$ Å [4]. The second factor is represented by the large zero-point energy arising from the small mass of the helium atoms. Importantly, at low temperature, the thermal wavelength associated to a helium atom ($\Lambda = \sqrt{\hbar^2/2\pi mkT}$, where $\hbar$ is the Planck constant divided by 2π, and k is the Boltzmann constant [5]), is comparable to the He–He interatomic distance, which implies that liquid helium manifests a quantum rather than a classical behavior. The ^{4}He isotope is a boson (i.e., it obeys the Bose–Einstein statistics [5]) with a nuclear spin $I = 0$, which undergoes a phase transition to a superfluid state at $T_\lambda = 2.17$ K (the so-called λ-transition) [1]. The phase diagram of helium-4 is shown in Figure 12.2. By contrast, the ^{3}He isotope is a fermion

(i.e., it obeys the Fermi–Dirac statistics [5]) with a nuclear spin $I = 1/2$, which also displays superfluid behavior at temperatures $T \leq 3 \times 10^{-3}$ K after formation of bosonic pairs [1].

The excitation spectrum of ^{4}He in the superfluid phase (Figure 12.3) is described by a particular dispersion curve that shows an initial linear increase from the origin followed by a maximum and then a minimum in energy as the momentum increases [6]. Excitations with momenta in the linear region are called phonons, while those with momenta close to the minimum and maximum are called rotons and maxons, respectively.

Recently, the existence of a helium-4 supersolid phase has also been suggested from the observation of an unexpected drop of the torsional oscillator period of solid ^{4}He at low temperatures [7,8]. Since the period is proportional to the square root of the moment of inertia, the origin of its decrease has been attributed to the decoupling of the ^{4}He solid mass from the rotating walls, resulting in the so-called nonclassical inertia [9]. Although this phenomenon has now been observed in several other experiments [10–14], the underlying physics still remains unclear, both theoretically and experimentally [15].

Compared to the large number of studies devoted to the analysis of macroscopic superfluid phenomena, early experimental investigations aimed at characterizing the corresponding properties at an atomic-scale level proved to be difficult. This was mainly due to the lack of appropriate microscopic probes. Liquid helium has indeed an intrinsic ability to cleanse itself of impurities, which either aggregate in the interior of the bulk phase or condense on its surface. Techniques based on laser ablation of materials inside liquid and solid helium were initially developed and applied to the study of metal atoms or ions [16]. However, because of the strong interactions that these species have with helium atoms, the measured electronic spectra display large line-shifts and very broad features, which prevent a detailed analysis in terms of microscopic superfluid properties.

Later, it was demonstrated that large liquid clusters (nanodroplets) containing up to $N = 10^3$–10^4 helium atoms (corresponding

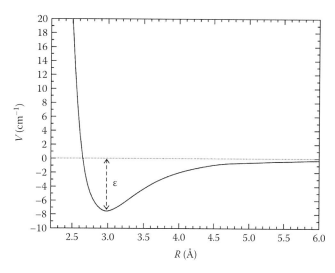

FIGURE 12.1 He–He potential energy curve. (Adapted from Jeziorska, M. et al., *J. Chem. Phys.*, 127, 124303, 2007.)

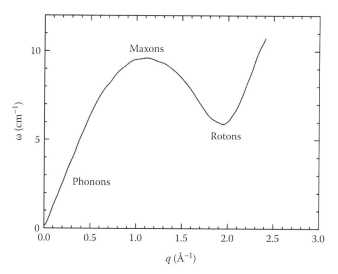

FIGURE 12.3 Excitation spectrum for superfluid helium-4. (Adapted from Henshaw, D.G. and Woods, A.D., *Phys. Rev.*, 121, 1266, 1961.)

to diameters in the range of 5–10 nm) can be formed by free jet expansions and can readily be doped with other atomic and molecular species [17–19]. Molecular beam techniques were then combined with spectroscopic measurements of the dopant species which, acting as microscopic probes, provided insights into the local properties of the nanodroplets [20,21]. Using this approach, atomic and molecular spectra have now been recorded for a large variety of dopant species attached to or embedded in ⁴He nanodroplets [22–25]. Some measurements have also been reported for the corresponding doped ³He nanodroplets [26] and for mixed nanodroplets containing small amounts of ⁴He solvated by a larger number of ³He atoms [27]. The measured electronic, infrared, and microwave spectra have provided a rapidly growing wealth of experimental data on electronic and vibrational shifts, rotational fine structure, phonon wing (PW) structure, and line shapes.

The nanodroplet temperatures are estimated to be $T \sim 0.38\,\text{K}$ in the case of helium-4, and $T \sim 0.15\,\text{K}$ for helium-3 [28], in good agreement with earlier theoretical predictions [29]. For the fermionic ³He isotope, the droplet temperature is higher than the corresponding bulk superfluid transition temperature, but still lies in the regime where the liquid manifests quantum rather than classical behavior. By contrast, since the temperature of ⁴He nanodroplets is well below the corresponding bulk superfluid transition temperature (T_λ), spectroscopic measurements on doped ⁴He nanodroplets offer the unique opportunity to investigate the superfluid properties of helium-4 at a microscopic level. In particular, the appearance of specific features in the electronic spectra, as well as of sharp lines in the infrared spectra of dopant molecules embedded in ⁴He nanodroplets provided the first experimental evidence of the onset of ⁴He superfluidity in finite systems [27,30].

More recent developments in the experimental techniques have also made possible the measurements of high-resolution infrared and microwave spectra of dopant species inside small clusters (⁴He$_N$ with $N < 80$) as ⁴He atoms are added one by one [31–38]. The analysis of the observed spectral features as a function of the cluster size has provided new insights into the evolution of the superfluid behavior of helium-4 from small clusters to nanodroplets.

From a theoretical point of view, the first quantum Monte Carlo calculations that were carried out in the 1980s indicated that the clusters of both helium isotopes are liquid-like [39]. A subsequent study, which employed the Path-integral Monte Carlo (PIMC) method, showed that small clusters with only few tens of ⁴He atoms display a significant superfluid component [40], while explicit quantum calculations of the excited states found evidence of the collective excitations characteristic of the superfluid state of bulk helium-4 shown in Figure 12.3 [41]. The first theoretical study of the structures and energetics of doped-helium clusters was reported in 1992, when quantum Monte Carlo simulations were performed for a single hydrogen

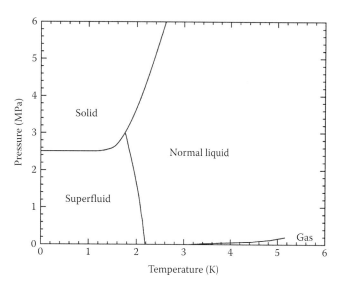

FIGURE 12.2 Phase diagram of helium-4.

molecule in small ^4He_N clusters with $N \leq 20$ [42]. This study indicated the existence of a quantum solvation structure in the local helium environment of the hydrogen molecule resembling that found around atomic impurities in bulk liquid helium-4 [43]. Subsequent simulations performed for ^4He_N clusters doped with a single sulfur hexafluoride (SF_6) molecule demonstrated that, although these density modulations can be quite large with peak heights exceeding the bulk solid density, the solvation shell structure nevertheless remains liquid-like [44]. A PIMC analysis of the structural properties of small clusters with $N \leq 64$ helium-4 atoms doped with single molecular impurities then showed the existence of a significant local superfluid density around the dopant species [45–47].

The rapid growth in recent spectroscopic experiments of atomic and molecular impurities in small helium clusters has also been accompanied by a large number of theoretical studies aimed at characterizing their structures and energetics [48–72]. Quantum calculations of the shifts on the dopant molecular vibrations as well as of the ^4He density distributions provided atomic-level insights into the preferential location of helium-4 atoms around the dopant [48,51,73]. On the other hand, a theoretical analysis of the rotational excitations of molecular impurities in small ^4He_N clusters as a function of N identified a transition from a molecular complex to a quantum solvated molecule [54]. This behavior was explained in terms of nonclassical rotational inertia and helium superfluidity, providing theoretical evidence for the onset of superfluid behavior in small ^4He_N clusters doped with single molecular impurities [56]. Quantum calculations for relatively larger doped ^4He_N clusters have also shown a coupling between the rotational motion of the molecular impurity and the elementary excitations of the surrounding superfluid helium-4 [57–60,74].

This chapter is primarily concerned with the recent progress made toward formulating a microscopic description of superfluid behavior in helium-4 clusters. Specific attention is devoted to the analysis of the structures, energetics, and dynamics of pure and doped ^4He_N clusters with a particular focus on the connection between spectroscopic measurements and theoretical calculations. The chapter is organized as follows: Section 12.2 provides an overview of the experimental studies on helium clusters. The theoretical progress to date is reviewed in Section 12.3, while future challenges are briefly discussed in Section 12.4.

12.2 Overview of Experimental Studies

The formation of helium clusters was first observed during the initial attempts at liquefying the gas [75]. Later, clusters were produced in cryogenic free jet-gas expansions [76], and scattering experiments demonstrated that helium nanodroplets (large ^4He_N clusters with $N = 10^4$–10^8) can be deflected by both atomic and molecular beams [77]. The drag coefficients obtained from such experiments were lower than those expected in the case of a complete momentum transfer, which led to the conclusion

that the impinging species could actually penetrate the clusters [78,79], although it was initially unclear whether they would be absorbed or would reemerge [80]. Size distributions of the ionized clusters were also determined using mass spectrometry [81,82].

In the late 1980s, the excitation spectra of metastable clusters showed that impurities such as molecular nitrogen (N_2) and oxygen (O_2) could be absorbed by helium clusters [83]. This discovery was followed by a series of experiments in which mass spectrometry was used to demonstrate conclusively that free helium clusters can pick up a variety of different atomic and molecular species [17–19,84]. The size distribution of neutral clusters was then obtained from small angle single-collision deflection measurements [85]. From these data an accurate relationship was established between cluster size and both nozzle temperature and stagnation pressure, which is still employed to determine average sizes in current experiments with doped ^4He_N clusters. Surprisingly, recent diffraction measurements from a 100 nm period transmission grating have revealed the existence of more probable cluster sizes (also known as "magic numbers") [86].

The demonstration that spectroscopic methods could be used to probe the dopant species represented an important breakthrough in the experimental investigations of the quantum behavior of helium clusters at a microscopic level. Combining molecular-beam techniques with spectroscopic interrogation by laser-induced evaporation [87], the first infrared absorption spectrum for a molecular impurity (SF_6) in helium-4 clusters was measured in 1992 [20].

A schematic representation of a typical apparatus used for performing spectroscopic measurements on dopant species embedded in or attached to small helium nanodroplets is shown in Figure 12.4. The helium nanodroplets are formed by adiabatic gas expansion at a temperature of 10–20 K and at a typical pressure of 20 bar (2 MPa), through a small orifice into the vacuum. Individual dopant species are then captured by the nanodroplets as they pass through a pickup scattering chamber. The energy transferred to such nanodroplets by the dopant species leads to the evaporation of several hundred helium atoms, after which each nanodroplet returns to its ambient temperature ($\sim$0.38 K for ^4He and $\sim$0.15 K for ^3He). The doped nanodroplets then interact with a laser beam, and the absorption is detected by monitoring the ion signal from a mass spectrometer. Different variations to the apparatus shown in Figure 12.4 have also been employed [23–25].

The first experimental evidence of superfluidity in ^4He nanodroplets was reported in 1996, based on the analysis of electronic excitation spectra measured for an embedded glyoxal ($C_2H_2O_2$) molecule [30]. These spectra show two distinct features: a zero-phonon line (ZPL), which is split into several lines corresponding to partially resolved rotational transitions in the glyoxal molecule, and a PW at higher frequencies arising from simultaneous excitation of the helium collective modes. Although the appearance of PWs is common in spectra measured for species trapped in low-temperature matrices, the PW observed in ^4He nanodroplets is unusual. It displays a sharp

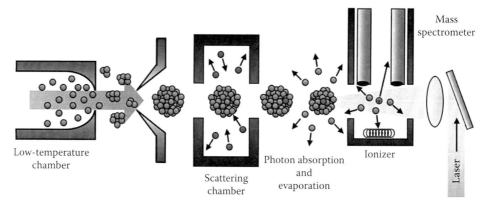

Mass spectrometer

Low-temperature chamber

Scattering chamber

Photon absorption and evaporation

Ionizer

Laser

FIGURE 12.4 Schematic diagram of a typical molecular beam apparatus for performing spectroscopic measurements on molecules embedded in or attached to helium nanodroplets. (Adapted from Toennies, J.P. and Vilesov, A.F., *Angew. Chem. Int. Ed.*, 43, 2622, 2004.)

rise and is separated from the ZPL by a distinct gap of about 6 cm^{-1}. These spectral features can be accurately reproduced considering the coupling between the molecular electronic excitations and the phonon–roton dispersion curve of the superfluid ^{4}He shown in Figure 12.3 [30]. The observed rise of the PW also indicates that the width of the roton energy levels in ^{4}He nanodroplets is much smaller than the roton gap. Since a well-defined roton gap is characteristic of the superfluid phase of helium-4 (Figure 12.3), the appearance of a similar gap in the electronic excitation spectra of $C_2H_2O_2$ provided the first evidence of superfluidity in large ^{4}He clusters [30]. This conclusion was corroborated by a subsequent analysis of electronic spectra measured at $T \sim 0.15$ K for a glyoxal molecule embedded in ^{3}He nanodroplets. As a consequence of the absence of superfluidity in helium-3 at this temperature, no gap between the ZPL and the PW was observed [26]. Subsequently, ZPLs and PWs have been observed for several other organic molecules in superfluid helium-4 nanodroplets [88,89]. The presence of sharp ZPL transitions in the electronic spectra of molecules in superfluid helium-4 has suggested a potential application of ^{4}He nanodroplets in the study of different possible structural conformers of floppy biological molecules such as tryptophan and tyrosine [90].

Additional insight into the superfluid behavior of helium clusters is obtained from the examination of the rotational motion of the embedded molecules [27,91]. High-resolution infrared spectra measured for a single SF_6 molecule in helium-4 nanodroplets reveal the presence of sharp absorption lines with evidence of a rotational fine structure [21]. By contrast, spectral lines corresponding to rotational excitations of molecular species in classical liquids can rarely be resolved because random collisions interrupt the development of rotational coherence [92]. The spectra measured in ^{4}He thus demonstrated that the sulfur hexafluoride molecule can freely rotate within the nanodroplet [21]. Even more importantly, the infrared spectrum measured in ^{4}He nanodroplets resembles the corresponding gas-phase spectrum with the only exception that the rotational lines are about a factor of three closer together. The latter can be fitted with extremely great accuracy assuming

the same Hamiltonian as that of the free molecule with energy levels given by

$$E_{rot} = \frac{\hbar^2}{2I_{eff}} j(j+1) = B_{eff} j(j+1), \quad (12.1)$$

where

I_{eff} and B_{eff} are the effective moment of inertia and the rotational constant of the SF_6 molecule in the nanodroplet, respectively,

j is its rotational quantum number.

The narrower rotational spacing observed in the spectrum thus indicates a considerably larger effective moment of inertia of the SF_6 molecule in ^{4}He nanodroplets than in the gas phase.

In principle, the appearance of sharp rotational lines in the SF_6 infrared spectra measured in helium-4 might simply be a consequence of the weak SF_6–He van-der Waals interaction. However, this possibility was ruled out in a series of elegant experiments in which the infrared spectra of a single carbonyl sulfide (OCS) molecule were measured in pure ^{4}He and ^{3}He, and mixed ^{4}He/^{3}He nanodroplets (Figure 12.5) [27]. Since the temperature of the helium-3 nanodroplets ($T = 0.15$ K) is above the superfluid transition temperature for the fermionic helium isotope, these nanodroplets involve the same weak molecular interactions as the corresponding ^{4}He nanodroplets but, contrary to the latter, cannot exhibit superfluid behavior. The spectrum in helium-4 (Figure 12.5—panel a) displays a sequence of narrow rotational lines with a full width at half maximum as small as 5×10^{-3} cm^{-1} [27]. As in the case of SF_6, this is indicative of the rotational coherence in the motion of the OCS molecule. By contrast, in pure ^{3}He nanodroplets, the rotational lines collapse into a broad single-band (Figure 12.5—panel b), which is very similar to those observed in infrared spectra of molecules immersed in classical liquids. Considering that the zero-point energy associated with ^{3}He atoms is even larger than that of ^{4}He atoms because of the smaller mass of the fermionic isotope, if the weakness of the molecular interactions was the reason for the observed

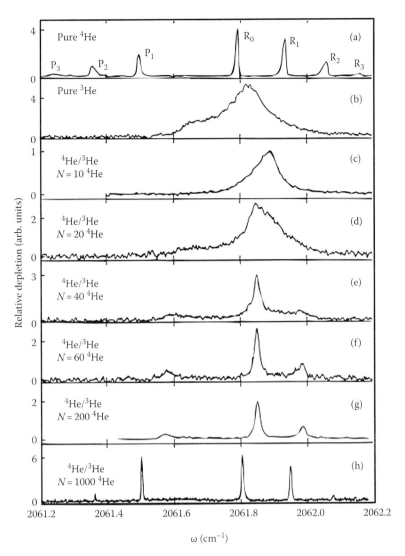

FIGURE 12.5 Infrared spectra measured for a single OCS molecule in pure ^4He nanodroplets (panel a), in pure ^3He nanodroplets (panel b), and in mixed $^4He/^3He$ nanodroplets with a variable number of ^4He atoms (panel cthrough h). (Courtesy of Prof. J. Peter Toennies.)

sharp rotational lines, the infrared spectrum measured in ^3He nanodroplets should display even sharper rotational lines. The infrared spectra measured in helium-4 and in helium-3 thus conclusively demonstrated that the ability of molecular impurities to rotate coherently in ^4He nanodroplets is directly connected to the superfluid properties of the nanodroplets themselves [27].

Furthermore, these spectroscopic measurements also shed light on the size dependence of superfluidity in helium-4 clusters. Using a second pickup chamber containing helium-4 gas at different pressures, mixed $^4He/^3He$ nanodroplets with a variable number of ^4He atoms were formed, and subsequently, doped with a single OCS molecule. Because of the difference in the zero-point energy, helium-4 interacts more strongly than helium-3 with the OCS molecule, and therefore, ^4He atoms diffuse within the nanodroplet replacing ^3He atoms from the solvation layers close to the OCS molecule. From the analysis of the spectral features measured for the mixed $^4He/^3He$ nanodroplets (Figure 12.5—panels c through h), it is possible to see that about

35 helium-4 atoms are necessary for the appearance of weak rotational lines, which become well resolved when about 60 helium-4 atoms are added to the ^3He nanodroplet [27].

Rotationally resolved infrared spectra have been subsequently measured for several molecular species embedded in ^4He nanodroplets [23–25]. In all cases, the rotational lines can be fitted with great accuracy using the gas-phase Hamiltonian of the dopant molecule. However, the effective rotational constants obtained from these fits display a large range of values depending on the physicochemical properties of each specific molecular impurity. For heavy molecules such as SF_6 [21,28], OCS [93], N_2O [94], and CO_2 [94], which are also slow rotors, the effective rotational constant in ^4He nanodroplets is about ~3–5 times smaller than the corresponding gas-phase value. In contrast, for light molecules (fast rotors) such as HF [95], HCCN [96], and HCN [97,98], B_{eff} shows almost no decrease. The effective rotational constants experimentally measured for several small molecules in helium-4 nanodroplets are compared in Table 12.1 with the

TABLE 12.1 Comparison of the Rotational Constants Measured in the Gas Phase and in ⁴He Nanodroplets

Molecule	Rotational Constant			
	Gas Phase (cm^{-1})	⁴He (cm^{-1})	Ratio Gas Phase/⁴He	References
HF	19.787	19.47	1.016	[95]
HCN	1.478	1.204	1.23	[97,98]
HCCH	1.1766	1.042	1.13	[96]
CO_2	0.39	0.154	2.53	[94]
CH_3CCH	0.2851	0.0741	3.85	[162]
OCS	0.2029	0.0732	2.78	[93]
CF_3CCH	0.0959	0.0355	2.7	[162]
SF_6	0.0911	0.034	2.68	[21]

corresponding gas-phase values. Interestingly, the ratio between the magnitude of the rotational constants measured in helium-4 nanodroplets and in the gas phase appears to be inversely proportional to the gas-phase value. As discussed in more detail in Section 12.3, different theoretical models have been developed to explain the observed phenomenon of the free rotation of molecules inside superfluid helium-4 nanodroplets. The observed larger increase of the effective moment of inertia for heavy molecules is now attributed to a strong coupling of the molecular rotation to the helium environment [46,54,56,99]. On the other hand, the absence of a significant reduction of B_{eff} for light molecules is explained in terms of their fast rotational motion and small anisotropy of the underlying molecular interactions, both of which enable a decoupling from the hindrance imposed by the surrounding helium-4 atoms [59,60,74,99,100].

Recent spectroscopic studies have focused on probing the underlying interactions between the dopant species and the surrounding helium atoms as well as the dynamics that takes place within the superfluid nanodroplet. Investigations on the C–H stretching vibration of the ethylene (C_2H_4) molecule in both its fundamental and first overtone regions indicate a large variation in broadening between different vibrational bands suggesting a different coupling to the surrounding superfluid environment [101,102]. The infrared spectrum measured for a single ammonia (NH_3) molecule in helium-4 nanodroplets shows a very small reduction (5%) of the rotational constants compared to the corresponding gas-phase values [103]. The tunneling splitting for the NH_3 umbrella motion is also found to be within 6% of the gas-phase value, which implies that only a small change in the helium-4 solvation structure around the ammonia molecule takes place upon inversion [103]. Infrared spectra measured for nitric oxide (NO) indicate that the magnitude of the effective rotational constant in ⁴He nanodroplets is about 76% of the gas-phase value placing NO in an intermediate regime among the molecules listed in Table 12.1 [104]. The experimental resolution also allowed resolving both the Λ doubling and the hyperfine splittings of NO. While the hyperfine interaction constant is unchanged in helium-4, the Λ doubling constant is 55% larger than the corresponding gas-phase value.

This behavior can be explained in terms of the confinement of the electronically excited state by the surrounding helium-4 environment as well as by considering the coupling between the molecular rotation and the elementary excitations in the superfluid nanodroplet [104]. A similar coupling of the molecular rotations and ⁴He phonons and rotons has also been proposed for interpreting the infrared spectra of a single carbon monoxide (CO) molecule in helium-4 nanodroplets [59]. In this context, the appearance of a PW in the infrared spectrum of carbon dioxide (CO_2) in helium-4 nanodroplets has been attributed to a strong coupling between the rotational motion of the molecule and that of the helium atoms located in the first solvation shell [105].

The unique properties of superfluid helium-4 nanodroplets also offer the opportunity to investigate the structural properties of molecular complexes in higher-energy conformations that are separated from the global minimum structure by even very small barriers. The self-assembly of long chains of polar molecules such as HCN and HCCN represents one of the most striking applications in this area [106,107]. Furthermore, the superfluid helium-4 environment has also allowed the observation of a cyclic structure of the water hexamer even though it is at a higher energy than the gas-phase cage conformation [108]. Lastly, the helium-4 superfluid medium has made possible the identification of specific structural features of small biological molecules [109].

Recent developments in the experimental techniques have enabled measurement of high-resolution spectra of molecules embedded in small clusters as helium-4 atoms are added one by one [32–34,36,37,110–115]. Rotationally resolved infrared and microwave spectra of OCS(⁴He)$_N$ clusters with $N = 2$–8 were initially recorded using an infrared diode laser spectrometer and a molecular beam Fourier-transform microwave spectrometer [32]. The clusters were generated using pulsed supersonic jet expansions of trace amounts (<0.1%) of OCS in helium-4 at backing pressures ranging from 10 to 30 atm. These measurements provide an unambiguous assignment of vibrational shifts and rotational constants of the OCS(⁴He)$_N$ clusters as a function of their size. For $N = 1$–5, the band origin displays a blue shift with an almost constant increment suggesting that the first five ⁴He atoms are located in equivalent positions around the OCS molecule. After reaching a maximum at $N = 5$, the frequency shift shows an essentially linear decrease with the addition of other three ⁴He atoms up to $N = 8$, which leads toward its ultimate limiting-value measured in ⁴He nanodroplets [27]. This nonmonotonic behavior clearly indicates that some structural change may be occurring between $N = 5$ and $N = 6$, which is discussed in more detail in Section 12.3. The combined infrared and microwave data can also be used to determine the rotational parameters of the OCS(⁴He)$_N$ clusters as a function of N. For $N = 2$, the spectrum corresponds to that of an asymmetric top molecule, while for $N = 3$–8 the rotational lines can be analyzed as $K = 0$ transitions using the Hamiltonian of a symmetric top-molecule [32]. In this model, the energy levels are expressed as a function of the rotational quantum number

j in terms of an effective rotational constant B_{eff}, and distortion parameters D and H,

$$E_{rot} = B_{eff} j(j+1) - Dj^2(j+1)^2 + Hj^3(j+1)^3. \quad (12.2)$$

However, it was found that for clusters with $N = 6-8$ it was necessary to vary as many parameters as observed transitions in order to achieve a good fit, which was interpreted as an indication of a very nonrigid behavior or slightly asymmetric top structures [32]. The effective rotational constant B_{eff} decreases from $N = 3$ to $N = 8$, at which point its value ($B_{eff} = 0.0483\,cm^{-1}$) is smaller than that measured in ^4He nanodroplets, $B_{eff} = 0.0732\,cm^{-1}$, suggesting the existence of a turnaround in the value of the effective rotational constant at a larger N. Experimentally, the turnaround in B_{eff} was observed for the first time in high-resolution infrared and microwave spectra for a single nitrous oxide (N_2O) molecule in ^4He_N clusters in the range $N = 3-12$ [33]. The numerical fit of the rotational lines using Equation 12.2 provided a large value for the centrifugal distortion parameter, D, that demonstrate the extreme floppiness of these clusters. On the other hand, the effective rotational constant was found to decrease up to $N = 6$, while an oscillatory behavior of B_{eff} was instead obtained from $N = 7$ to $N = 12$. These measurements thus provided a direct experimental evidence for the decoupling of the helium motion from the N_2O rotations as predicted by earlier theoretical studies [54,61,116]. Similar results have also been reported for $CO_2(^4He)_N$ clusters with N up to 17, which corresponds to the completion of the first solvation shell around CO_2 [34]. In this case, the value of B_{eff} derived from the analysis of the rotational transitions displays a pronounced turnaround at $N = 5$, and becomes approximately constant for $N > 13$ with a value about 17% larger than that measured in ^4He nanodroplets [94], which implies a non-negligible contribution from helium-4 atoms located in more distant solvation shells. More recently, high-resolution spectra for $OCS(^4He)_N$ clusters with N up to 39 in the microwave and N up to 72 in the infrared regions have been measured with line widths of about $15\,kHz$ and $0.001\,cm^{-1}$, respectively [35]. The analysis of the effective rotational constant as a function of the number of helium atoms confirms the existence of a turnaround in B_{eff}, which occurs at $N = 9$. For larger cluster sizes, broad oscillations are found in the value of B_{eff} with maxima at $N = 24$ and 47, and minima at $N = 36$ and 62. This behavior can be interpreted in terms of the *aufbau* of a nonclassical helium solvation structure, suggesting a complex evolution of superfluidity around a molecular impurity as ^4He atoms are added one by one. Similar behavior of the effective rotational constant has also been reported for larger $N_2O(^4He)_N$ [36] and $CO_2(^4He)_N$ [37] clusters with $N = 80$ and $N = 60$, respectively.

By contrast, qualitatively different infrared spectra are obtained for the CO molecule, a light and a fast rotor, embedded in clusters containing up to $N = 20$ ^4He atoms [111,112]. In particular, two series of transitions are observed, which correlate with the known *a*-type and *b*-type series evolving from the end-over-end and free molecule rotations of the binary He–CO complex. The intensity of the *b*-type series begins quite strong at $N = 2$ and then diminishes to disappearance at $N = 8$, while the *a*-type series shows the opposite behavior, gaining intensity as a function of N. Two critical regions are identified in the cluster-size evolution around $N = 7$ and $N = 15$, which appears to correlate with the calculated maximum and minimum, respectively, in the binding energy per helium atom [117].

While outside the formal scope of this chapter, it is important to note that superfluid helium-4 nanodroplets also offer a unique opportunity to serve as host for other quantum-fluid clusters, which might also display superfluid behavior. In this context, beyond the two helium isotopes, *para*-hydrogen (*p*-H₂ or *para*-H₂) is the only other known species that might exhibit superfluidity, as was discussed for the first time by Ginzburg and Sobaynin in 1972 [118]. *Para*-hydrogen molecules have zero total nuclear spin ($I = 0$), and because of their large rotational constant, at low temperature they are all in their rotational ground state ($j = 0$). Therefore, *p*-H₂ molecules are spinless indistinguishable bosons, and like ^4He atoms, might show superfluid behavior at low temperature. However, the interaction energy between *para*-hydrogen molecules ($\varepsilon \sim 24\,cm^{-1}$) is about three times stronger than the He–He interaction, which results in a triple point for bulk *p*-H₂ at $13.96\,K$. Since the superfluid transition for *para*-H₂ has recently been estimated to occur at $T \sim 2\,K$ [119], bulk *para*-hydrogen solidifies before cooling to a low enough temperature for the manifestation of superfluidity. In a series of recent experiments, the multiple pickup technique was employed to generate small *para*-hydrogen and *ortho*-deuterium (*o*-D₂ or *ortho*-D₂) clusters doped with a single OCS molecule embedded in larger ^4He and mixed $^4He/^3He$ nanodroplets [120]. It is important to note that *o*-D₂ molecules are also bosons but, because of their intrinsic large nuclear spin multiplicity ($I = 0$ and 2), are not indistinguishable. Consequently, *ortho*-deuterium is expected to exhibit superfluid behavior at a much lower temperature. The resolution of the infrared spectra measured for the $OCS(p-H_2)_N$ and $OCS(o-D_2)_N$ clusters in helium allows the resolution of the rotational bands as well as the determination of the vibrational shifts [120]. In particular, the spectra measured in pure ^4He droplets ($T = 0.37\,K$) for clusters containing $N = 14-17$ *para*-hydrogen or *ortho*-deuterium molecules, show the presence of sharp Q-branches indicating that the energy levels corresponding to rotations around the OCS symmetry axis of the resulting symmetric-top clusters are excited. At the lower temperature ($T = 0.15\,K$) of the mixed $^4He/^3He$ nanodroplets, the Q-branches are still observed in the *o*-D₂ clusters, but surprisingly disappear in the corresponding *p*-H₂ clusters. This evidence can be explained in terms of a large decrease of the effective moment of inertia of the $OCS(p-H_2)_N$ clusters resulting in rotational energy constants so large that rotations around the OCS axis cannot be excited at the low temperature of the $^4He/^3He$ droplets. Since a large decrease in the moment of inertia is considered to be a manifestation of superfluid behavior [121], the disappearance of Q-branches in the spectra of the $OCS(p-H_2)_N$ clusters in mixed $^4He/^3He$ droplets were interpreted as the first experimental evidence of the onset of superfluidity in a liquid other than helium [120].

12.3 Theoretical Studies

In this section, theoretical investigations of the properties of helium-4 clusters are reviewed with particular focus on the connection between the manifestations of superfluid behavior and spectroscopic measurements.

Since helium-4 clusters are highly quantum systems, calculations of their microscopic properties require the solution of the corresponding many-body time-independent Schrödinger equation, $\hat{H}\Psi = E\Psi$, where $\hat{H}$ and Ψ are the associated Hamiltonian and wavefuction, respectively [122]. The Hamiltonian corresponds to the total energy of the cluster and contains a potential model that describes the underlying molecular interactions. The latter is usually represented as a sum of all possible two-body interactions within the cluster which are obtained from quantum chemistry calculations [123]. The only theoretical approaches currently available for studying quantum systems with many degrees of freedom are quantum Monte Carlo methods, which employ stochastic techniques to solve the corresponding time-independent Schrödinger equation [124–126]. These methods can be distinguished into two categories: zero-temperature ($T = 0$ K) methods that include variational Monte Carlo (VMC) [127,128], diffusion Monte Carlo (DMC) [126,129], reptation quantum Monte Carlo (RQMC) [130] and the projection operator imaginary time spectral evolution (POITSE) method [131], and finite-temperature ($T > 0$ K) methods such as PIMC [132].

VMC is the simplest stochastic approach that applies the variational principle to approximate the ground state of the cluster allowing the calculation of its energetics and structural properties. Numerically exact results can instead be obtained using the DMC and RQMC methods. DMC exploits the isomorphism between the many-body time-independent Schrödinger equation and a multidimensional reaction-diffusion equation, which is then solved using the appropriate Green's function [126,129]. RQMC is a recent zero-temperature method for calculating the static and the dynamic properties of quantum systems, which overcomes some of the computational limitations of DMC (e.g., mixed-estimate and population-control biases). Calculations of excited-state energies of helium-4 clusters can also be computed using the POITSE method through numerical inversion of imaginary-time correlation functions obtained from a combination of VMC and DMC methods. Structural and thermodynamic properties of pure and doped helium-4 clusters at finite temperature ($T > 0$) can be computed accurately using the PIMC method. This approach is based on Feynman's original idea of mapping a system of N quantum particles into that of N interacting classical ring-polymers [133,134]. Within PIMC, it is also possible to quantify the superfluidity of a bosonic system in terms of exchange-coupled paths of the corresponding ring-polymers, which are comparable to the system size [132]. Within the linear response theory, the ^{4}He superfluid fraction for a homogeneous system is computed in terms of the response of the total helium angular momentum to infinitesimal changes in

the rotation rate imposed by an external field [132]. Recently, path-integral estimators of superfluidity for inhomogeneous systems have also been derived [135,136].

The present analysis of the properties of helium-4 clusters starts with the ^{4}He$_2$ dimer, which is also the most weakly bound of all naturally occurring molecular complexes with only one bound state at energy $E \sim -0.9$ cm^{-1} [137–140]. This particular feature is responsible for the remarkable behavior of the helium trimer (^{4}He$_3$) whose bound states become unbound as the strength of the two-body He–He potential is artificially increased (Efimov states) [141,142]. Contrary to ^{4}He$_2$ whose average internuclear distance is extremely large ($R \sim 50$ Å) [143], the ground state of the helium-4 trimer is considerably more compact with an average radius $R \sim 6$–7 Å and a noticeable contribution from nearly linear geometries [144,145]. The dimensions of larger ^{4}He$_N$ clusters with $N = 3$–7 have been found to decrease, which correlates with the corresponding increase on the relative binding energy per ^{4}He atom [145]. For $N > 7$, the average radius of the helium clusters increases again, which is consistent with a limit on compactness dictated by the hard-core nature of the He–He potential [145,146]. Theoretical calculations demonstrated that all clusters are liquid-like, with a diffuse surface region extending for about 6–7 Å, and an interior density that reaches the bulk value at $N \sim 70$ [147]. A detailed analysis of three-body correlation functions also showed the presence of significantly nonspherical structures in the interior of small clusters with $N < 10$ [145]. At the one-particle density level, however, it was pointed out that pure ^{4}He clusters are to be regarded as liquid-like complexes where the delocalization in the ground state or at low temperatures is purely quantum mechanical, deriving both from the low mass and from the Bose exchange symmetry [148]. As mentioned in Section 12.2, recent experimental measurements of the size distributions of pure ^{4}He$_N$ clusters have revealed "magic numbers" at $N = 10$, 11, 14, 22, 26, 27, and 44 [86]. Although the appearance of "magic numbers" is common in several types of molecular clusters indicating enhanced stabilities [149], this is not expected for quantum fluid clusters such as the ^{4}He$_N$ clusters [150]. A detailed analysis of the ground-state energies as well as of the radial and pair distribution functions for ^{4}He$_N$ clusters with $N = 3$–50 computed using the DMC method has shown that the appearance of such "magic numbers" in helium-4 is not caused by the presence of structures with enhanced stabilities. Instead, it can be explained in terms of stability thresholds related to those cluster sizes at which excited levels cross the chemical potential from above and become stabilized. These calculations also indicate that the compressibility coefficient of small ^{4}He$_N$ clusters with $N \leq 50$ is one order of magnitude smaller than the bulk compressibility [146].

The superfluid properties of helium clusters were first investigated using the VMC method which demonstrated the existence of significant condensate fractions in ^{4}He$_N$ clusters with $N = 20$–240 [39]. Finite temperature PIMC simulations were then carried out for clusters with $N = 64$ and 128 ^{4}He atoms for $T \leq$ 2.5 K. Manifestations of superfluid behavior were shown to exist already in clusters with 64 ⁴He atoms, while a remnant of the lambda transition of bulk helium-4 was found in clusters with 128 ⁴He atoms [40]. Importantly, the temperature dependence of the superfluid fraction in such clusters is very similar to that observed in the bulk phase. The excitation spectra of the compressional modes in ⁴He$_N$ clusters, with N = 20, 70, and 240, were calculated at a temperature of 0 K assuming that these clusters can be treated as quantum liquid drops [41]. The spectrum of the cluster with N = 240 was found to be very similar to that of superfluid helium-4 (Figure 12.3) with a well-defined roton structure. Although no roton minimum was obtained for N = 20, a weak feature was found to emerge in the spectrum of the ⁴He$_{70}$ cluster.

The experimental investigations discussed in the previous section have also been accompanied by a large number of theoretical studies aimed at characterizing the structural, dynamical, and superfluid properties of doped ⁴He clusters. Initial studies focused on the structural modifications of pure ⁴He$_N$ clusters induced by the presence of a foreign species that were shown to be particularly sensitive to the location of the latter. This primarily depends on two different factors. The first and most significant is the strength of the interaction between the dopant species and the helium atoms relative to the He–He interaction. Based only on this physical consideration, species that have weaker interactions than the He–He interaction are expected to be located at the surface of the cluster, while species that bind relatively strongly to He atoms are expected to be located in the interior where they can maximize the attractive interactions with the surrounding helium atoms. A second factor can also affect the location of foreign species in ⁴He$_N$ clusters. This is the change in exchange energy of helium-4 atoms, which derives from the loss of Bose exchange symmetry due to the presence of the foreign species. This loss of exchange energy drives the foreign species to the surface in order to minimize the number of neighboring helium atoms. Accurate quantum calculations including both factors show that in the cluster ground state, weakly-bound species such as Na and Li are located at the surface [151], while more strongly bound atoms such as Ag [152], and molecules, such as HF [48], SF$_6$ [44], OCS [51,54,61,73], N$_2$O [63], CO$_2$ [34], and CO [62] are located at the center.

Quantum simulations carried out for molecular impurities embedded in ⁴He$_N$ clusters have played a central role in identifying structural perturbations in the local helium-4 environment and their relationship to superfluid behavior. Variational and DMC calculations carried out for SF$_6$(⁴He)$_N$ clusters with N in the range 1–499 indicate that the SF$_6$ molecule induces a large degree of localization and a shell-like structure in the surrounding helium-4 cluster [44]. Subsequent quantum simulations reported on the size dependence of energetics, structures, and vibrational shifts of a single HF molecule embedded in clusters with up N = 198 helium-4 atoms [48]. The calculated shift of the HF vibrational frequency displays a strong initial size dependence and saturates at a value of about −2.7 ± 0.1 cm⁻¹ for clusters with over 100 ⁴He atoms, which is in remarkably good agreement with the experimental value of −2.65 ± 0.15 cm⁻¹ obtained

for clusters with over 1000 atoms. The helium density distribution around the HF molecule also shows significant structural changes with respect to pure helium-4 clusters. In particular, the density in the nearest shell exceeds the bulk helium density by a factor of two. By contrast, the second solvation shell has a density maximum very close to the bulk helium density. Despite the anisotropy of the underlying HF–He interaction, all clusters have nearly perfect spherical-helium density profiles. This evidence is interpreted as an indication that the behavior of the HF molecule inside helium-4 clusters is very close to that of a free rotor [48]. A similar weak anisotropy in the helium density distribution is also found in helium-4 clusters doped with HCN [52] and CO [50,62,117] molecules. In both cases, the dependence of the ground-state energy on the size of the cluster reflects the extremely smooth evolution of the He density. The energy per helium atom remains constant at a value of about −10 cm⁻¹ up to N ~ 15 for the HCN(⁴He)$_N$ clusters and of about −6.5 cm⁻¹ up to N ~ 12 for the CO(⁴He)$_N$ clusters, after which it starts to increase indicating completion of the first solvation shell. The radial density profiles shown in Figure 12.6, for the CO(⁴He)$_N$ clusters clearly indicate the presence of a quantum layering around the CO molecule and the completion of the first solvation shell

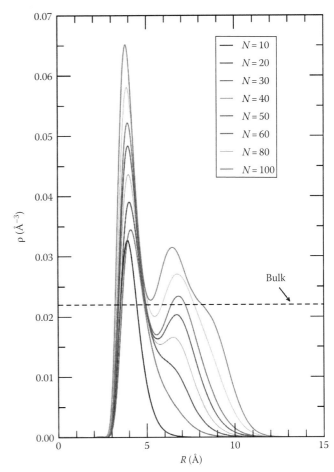

FIGURE 12.6 (See color insert following page 25-14.) ⁴He radial density distributions for selected CO(⁴He)$_N$ clusters. (Adapted from Paesani, F. and Gianturco, F.A., *J. Chem. Phys.*, 116, 10170, 2002.)

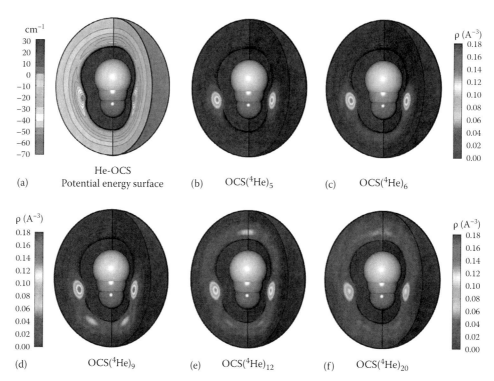

FIGURE 12.7   **(See color insert following page 25-14.)** (a) Potential-energy surface for the OCS–He complex [73]. (b–f) ⁴He density distributions for selected OCS(⁴He)$_N$ clusters. (Adapted from Paesani, F. and Whaley, K.B., *J. Chem. Phys.*, 121, 4180, 2004.)

between $N = 20$ and $N = 30$. Quantum calculations of the CO frequency shift carried out as a function of the cluster size using a highly accurate He–CO potential energy surface [153] predicted a red shift [117] that is in remarkably good agreement with more recent experimental measurements [111].

Highly anisotropic helium-4 densities are instead found for the OCS(⁴He)$_N$ clusters [51] as a result of the strong anisotropy of the underlying He–OCS potential energy surface [73,154–156]. The latter, shown in Figure 12.7, exhibits a global minimum of about −50 cm⁻¹ for a nearly T-shaped geometry. Two other local minima for collinear geometries are also found. The first, at the sulfur side of the OCS molecule has a well depth of about −26 cm⁻¹ and is separated from the global minimum by a saddle point at energy of about −15 cm⁻¹. The second local minimum at the oxygen side is very shallow with a well depth of about −24 cm⁻¹. A detailed analysis of the energetics and helium-4 density distributions demonstrates that the first five ⁴He atoms are located in the global minimum region forming a ring around the OCS molecular axis as shown in Figure 12.7. The additional three atoms in the OCS(⁴He)$_8$ cluster occupy the second local minimum at the oxygen end of the OCS molecule (Figure 12.7). Starting at $N = 9$, the OCS molecule begins to be completely solvated by helium-4 atoms, with completion of the first solvation shell occurring at $N \sim 20$. These structural features are strongly correlated to the frequency shift of the OCS ν_3 vibration [73]. From $N = 1$ to $N = 5$, each ⁴He atom added to the cluster contributes by an approximately constant value of +0.1 cm⁻¹. After reaching a maximum at $N = 5$, the computed

frequency-shift decreases as a function of N, becoming red at $N \sim 12$, and approaches the experimental value of −0.557 cm⁻¹ as measured in ⁴He nanodroplets. A comparison with recent experimental results for small OCS(⁴He)$_N$ clusters with $N =$ 1–70 [35] shows that the theoretical predictions slightly overestimates the red shift for $N \geq 10$, which may indicate small inaccuracies in the underlying He–OCS potential energy surface. Similar anisotropic distributions of helium-4 atoms around a molecular impurity are also found in the analogous N$_2$O(⁴He)$_N$ [55,63] and CO$_2$(⁴He)$_N$ [34,56,70] clusters. In the latter case, very recent PIMC simulations have provided an excellent agreement with the experimental measurements of the frequency shift of the asymmetric CO$_2$ vibration in clusters with $N \leq 17$ [70].

Without any doubt, one of the most striking features emerging from the spectroscopic studies on doped helium-4 clusters and nanodroplets is the observation of apparent free rotations of the embedded molecules. As discussed in Section 12.2, this phenomenon is accompanied by a large increase of the effective moment of inertia in the case of slow rotors such as SF$_6$, OCS, N$_2$O, and CO$_2$, while a very small increase (if any) is observed for fast rotors such as HF, HCN, and CO (see Table 12.1). This behavior is qualitatively correlated to the helium-4 solvation structure that is almost spherical around fast rotors but quite anisotropic for slow rotors. Several interpretations have been advanced for explaining the phenomenon of free rotations of the embedded molecules and the associated change in their effective moment of inertia, which include models based on the rigid attachment of helium atoms to the rotating molecule [21,93], phenomenological approaches based on the hydrodynamics of ideal fluids [157,158] or the microscopic two-fluid picture [99,159], and the quantum many-body theory [160]. Currently, the most complete interpretation of such phenomenon is based on accurate quantum mechanical calculations of the rotational excitations of the embedded molecules as well as on a direct analysis of the superfluid behavior of the surrounding helium-4 density [54,56,57,99,116,159].

The first quantum calculations of the rotational energy levels were carried out using the fixed-node DMC method for a single SF_6 molecule embedded in small clusters with $N = 1$–20 helium-4 atoms [116]. A monotonic increase of the effective moment of inertia from the gas-phase value is found up to $N \sim 8$, the size at which it approaches the experimental value measured in ^4He nanodroplets [21]. Since the first solvation for the $SF_6(^4He)_N$ clusters is only complete at $N \sim 22$–23 [44], these results suggest that only a fraction of the first solvation shell can follow the molecular rotation [116]. PIMC simulations performed for the $SF_6(^4He)_{39}$ cluster at low temperature show that the helium-4 density becomes gradually superfluid at a temperature between $T = 0.625$ K and $T = 1.25$ K [45]. Subsequent PIMC calculations were then carried out for different cluster sizes ($N = 4$–64) in the temperature range $T = 0.3$–0.75 K [46]. An extensive analysis of the superfluid density defined in terms of exchange paths involving ≥ 6 ^4He atoms shows the existence of a finite nonsuperfluid component in the first solvation shell around the SF_6 molecule, which persists even at the lowest temperature ($T = 0.3$ K) of the simulations. The nonsuperfluid fraction beyond the first solvation shell is instead found to decrease rapidly as a function of N, suggesting that for larger clusters, the helium-4 density outside the first solvation shell is entirely superfluid. More accurate calculations of the rotational excitations were performed using the POITSE, RQMC, and PIMC methods for the $OCS(^4He)_N$ clusters with $N \leq 20$ [54,61,67,73]. Depending on the underlying He–OCS potential energy surface employed in the calculations, the effective rotational constant is found to monotonically decrease up to $N \sim 6$–8, which is then followed by a turnaround and saturation to a value slightly larger than that measured in helium-4 nanodroplets [27]. A comparison between the theoretical predictions and recent experimental results [32,35,115] shows good agreement for $N \leq 12$, while some discrepancies exist for larger cluster sizes indicating possible inaccuracies in the underlying molecular interactions [123]. Importantly, the turnaround in the effective rotational constant occurs in the size range where the OCS molecule becomes completely solvated by helium atoms (Figure 12.7). A comparison of the POITSE results with the rotational energy levels computed assuming a quasi-rigid attachment of ^4He atoms to the OCS molecules indicates the occurrence of a transition between a molecular complex to a quantum solvated molecule, where the molecular complex is characterized as a near rigidly coupled van der Waals complex and quantum solvation results from the onset of permutation exchange paths that extend along the entire OCS axis [54].

A direct analysis of the dependence of the effective rotational constant on N in terms of helium-4 superfluidity was carried out for the $N_2O(^4He)_N$ clusters with $N \leq 16$ [55]. Similarly to the $OCS(^4He)_N$ clusters, the first five helium atoms are found to form an equatorial ring around the molecular axis, while the sixth helium atom is located in the local minimum of the N_2O–He potential-energy surface at the oxygen side of the molecule. $N = 9$ is the critical size at which there is an onset of helium solvation all along the molecular axis. The rotational spectroscopic constants computed with the POITSE method are in excellent agreement with the experimental data [33] as well as with previous calculations performed with a different He–N_2O potential energy surface [63]. Path-integral simulations show a direct relationship between the N-dependence of the effective rotational constant and the superfluid behavior of the surrounding helium-4 atoms. In particular, an essentially complete superfluid response to rotations about the molecular axis is found for $N \geq 5$ that explains the absence of a Q-branch in the corresponding infrared spectra. A significant perpendicular superfluid response to rotations around an axis perpendicular to the N_2O axis is also found for $N \geq 8$, which directly correlates with the oscillatory behavior of the effective rotational constant obtained for these cluster sizes.

A definitive theoretical evidence for the onset of superfluidity in doped helium-4 clusters at less than one solvation shell was obtained from a quantitative analysis of the spectroscopic constants for a CO_2 molecule in ^4He_N clusters that were computed in terms of nonclassical rotational inertia and helium superfluidity [56]. The PIMC method was used to calculate the anisotropic superfluid and the non-superfluid densities in $CO_2(^4He)_N$ clusters with $N \leq 16$, which are compared in Figure 12.8 with the corresponding total densities for selected cluster sizes. With a rigid attachment of helium as CO_2 rotates, the density close to the molecular axis should be more effective in reducing the effective rotational constant B_{eff} below its gas-phase value. For $N \geq 6$, Figure 12.8 clearly shows that the increasing helium density located in these regions is predominantly superfluid to rotation around an axis perpendicular to the CO_2 axis and, therefore, results in a large decrease in the nonclassical rotational inertia. In contrast, a complete rigid coupling is the situation for $N \leq 5$, for which the helium density is entirely nonsuperfluid to rotation around an axis perpendicular to the CO_2 axis. Figure 12.8 also shows that for $N \geq 5$, there is an essentially complete superfluid response to rotations around the CO_2 axis. This anisotropic superfluid response accounts for the nonmonotonic behavior of the effective rotational constant measured for small $CO_2(^4He)_N$ clusters (Figure 12.9), causing a turnaround in B_{eff} when the rotational inertia becomes nonclassical due to helium superfluidity. Calculations of the effective moment of inertia in terms of an anisotropic superfluid response were also carried out for the $(HCN)_n$ chains [135], and for the two linear complexes HCN•HCCCN and HCCCN•HCN [161] in large helium-4 clusters. The results, which are in qualitative agreement with the available experimental data, provide a theoretical explanation

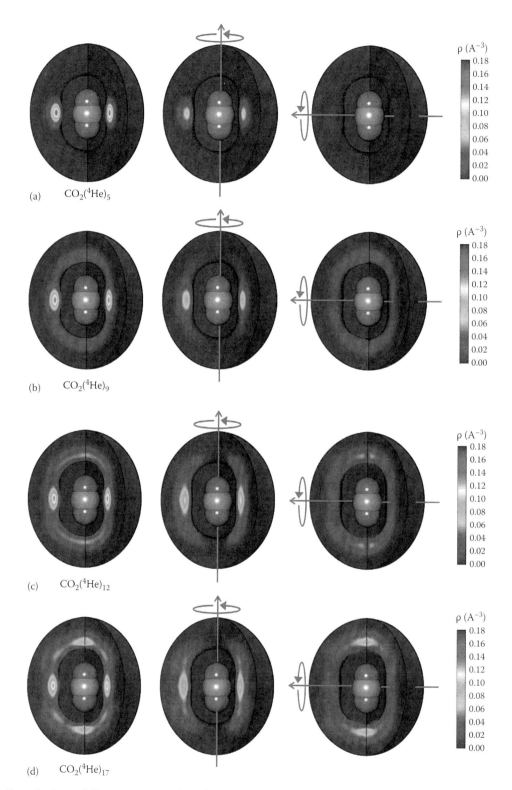

FIGURE 12.8 **(See color insert following page 25-14.)** ^{4}He density distributions for selected $CO_2(^4He)_N$ clusters. (Adapted from Paesani, F. et al., *Phys. Rev. Lett.*, 94, 2005.) Left column: total ^{4}He density; central column: superfluid density to rotations along the CO_2 axis; right column: superfluid density to rotations along an axis perpendicular to the CO_2 axis. (Adapted from Paesani, F., et al., *Phys. Rev. Lett.*, 94, 153401, 2005.)

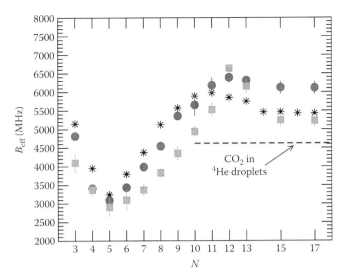

FIGURE 12.9 Effective rotational constant, B_{eff}, for the $CO_2(^4He)_N$ clusters. Stars: experimental values [34]; red circles: POITSE results [56]; green squares: estimates from nonclassical rotational inertia [56]; dashed blue line: experimental droplet value. (Adapted from Nauta, K. and Miller, R.E., *J. Chem. Phys.*, 115, 10254, 2001.)

for the presence of anomalous Q-branches in the measured infrared spectra [107,161].

Contrary to the behavior described above for heavy molecules, light molecules such as HF, HCN, HCCH, and CO show only a small reduction of rotational constant in helium clusters. Theoretical calculations have demonstrated that the local helium density cannot follow the motion of such relatively fast rotating molecules that results in a smaller reduction of their rotational constants [52,100]. Calculations of the rotational energy levels for the HCN [64] and CO [58,62] molecules in clusters containing up to $N = 50$ ^4He atoms indicate the presence of two series of excitations corresponding to the a-type and b-type transitions discussed in Section 12.2. While no experimental measurement has been reported yet for small $HCN(^4He)_N$ clusters, the theoretical results for $CO(^4He)_N$ clusters are in remarkably good agreement with the available experimental data [111,112]. The presence of two spectral series is related to the propensity of helium-4 atoms to be located on the same side of the molecular axis due to a stronger interaction with the oxygen atom of the CO molecule [62]. As more ^4He atoms are added to the first solvation shell, this propensity weakens resulting in a more isotropic helium-4 density distribution and the disappearance of the b-type line. Recent theoretical studies have established that for fast rotors, the mechanism of coupling to the surrounding helium-4 atoms is quite different from that observed for slow rotors [74]. In particular, a combination of the correlated basis set theory and the DMC allowed a microscopic analysis of the infrared spectrum of a single acetylene (HCCH) molecule in helium-4 clusters [74]. Calculations of the rotational excitations using the path-integral method provide a distortion constant, D, in excellent agreement with the corresponding experimental

value [96], which is anomalously large relative to values measured for slow rotors. This anomaly is now attributed to a strong direct coupling of the excited rotational energy levels of the HCCH molecule to the roton and maxon excitations of the surrounding superfluid helium-4 environment [74]. Similar results were also reported for the HCN and DCN molecules [57]. More recently, a combined experimental and theoretical study of the infrared spectrum of CO in helium-4 nanodroplets has shown that both the reduction of the effective rotational constant and the observed line broadening can be attributed to a coupling of the rotational motion of the CO molecule to the phonon excitations of helium-4 [59,60].

12.4 Summary

This chapter has reviewed the recent progress made toward a microscopic characterization of superfluid behavior in helium-4 clusters. The current understanding of superfluid properties at an atomic-level scale derives from a synergistic contribution of spectroscopic measurements and theoretical calculations. High-resolution spectra have now been measured for a large variety of molecules embedded in ^4He clusters at low temperature providing a huge body of experimental data on electronic and vibrational shifts, rotational fine structure, PW structure, and line shapes. Without any doubt, one of the most striking features that emerge from these measurements is the observation of free rotations of the embedded molecules that is accompanied by an increase of the effective moment of inertia. Accurate quantum calculations show that this phenomenon is a direct consequence of the superfluid character of the helium-4 clusters. The observed increase of the effective moment of inertia is now well understood in terms of the coupling between the rotational motion of the molecule and the excitations of the ^4He environment that may be local or collective, depending on the nature of the molecule and the anisotropy of the underlying interactions. Rotationally resolved infrared and microwave spectra accompanied with explicit calculations of molecular energy levels are also provided new insights into the evolution of superfluid behavior as ^4He atoms are added to the cluster one by one.

While excellent agreement between experimental observations and theoretical predictions is obtained for doped helium-4 clusters containing up to one ^4He solvation shell, some discrepancies still remain for larger cluster sizes. Experimental results indicate broad oscillations in the values of the effective moment of inertia after completion of the first ^4He solvation shell. This behavior suggests a complex evolution of the superfluid character from small clusters to nanodroplets, which, however, is not well reproduced by current quantum calculations. Therefore, an accurate description of both dynamical and superfluid properties of helium-4 clusters in this intermediate size regime remains an important challenge for theory, which may require further improvements in the computational methodologies as well as in the representation of the underlying interaction potentials.

References

1. D. R. Tilley and J. Tilley, *Superfluidity and Superconductivity* (Adam Hilger, Bristol, U.K., 1990).
2. P. Kapitsa, *Nature* **141**, 74 (1938).
3. J. F. Allen and A. D. Misener, *Nature* **141**, 75 (1938).
4. M. Jeziorska et al., *J. Chem. Phys.* **127**, 124303 (2007).
5. K. Huang, *Statistical Mechanics* (John Wiley & Sons, New York, 1987).
6. D. G. Henshaw and A. D. Woods, *Phys. Rev.* **121**, 1266 (1961).
7. E. Kim and M. H. W. Chan, *Nature* **427**, 225 (2004).
8. E. Kim and M. H. W. Chan, *Science* **305**, 1941 (2004).
9. A. J. Leggett, *Phys. Rev. Lett.* **25**, 1543 (1970).
10. A. S. C. Rittner and J. D. Reppy, *Phys. Rev. Lett.* **97**, 165301 (2006).
11. A. S. C. Rittner and J. D. Reppy, *Phys. Rev. Lett.* **98**, 175302 (2007).
12. Y. Aoki, J. C. Graves, and H. Kojima, *Phys. Rev. Lett.* **99**, 015301 (2007).
13. M. Kondo et al., *J. Low Temp. Phys.* **148**, 695 (2007).
14. A. Penzev, Y. Yasuta, and M. Kubota, *J. Low Temp. Phys.* **148**, 677 (2007).
15. M. H. W. Chan, *Science* **319**, 1207 (2008).
16. B. Tabbert, H. Günther, and G. Putlitz, *J. Low Temp. Phys.* **109**, 653 (1997).
17. A. Scheidemann, J. P. Toennies, and J. A. Northby, *Physica B* **165**, 135 (1990).
18. A. Scheidemann, J. P. Toennies, and J. A. Northby, *Phys. Rev. Lett.* **64**, 1899 (1990).
19. M. Lewerenz, B. Schilling, and J. P. Toennies, *J. Chem. Phys.* **102**, 8191 (1995).
20. S. Goyal, D. L. Schutt, and G. Scoles, *Phys. Rev. Lett.* **69**, 933 (1992).
21. M. Hartmann et al., *Phys. Rev. Lett.* **75**, 1566 (1995).
22. J. P. Toennies and A. F. Vilesov, *Annu. Rev. Phys. Chem.* **49**, 1 (1998).
23. J. P. Toennies and A. F. Vilesov, *Angew. Chem. Int. Ed.* **43**, 2622 (2004).
24. M. Y. Choi et al., *Int. Rev. Phys. Chem.* **25**, 15 (2006).
25. F. Stienkemeier and K. K. Lehmann, *J. Phys. B: At. Mol. Opt. Phys.* **39**, R127 (2006).
26. S. Grebenev et al., *Phys. B: Condens. Matter* **280**, 65 (2000).
27. S. Grebenev, J. P. Toennies, and A. F. Vilesov, *Science* **279**, 2083 (1998).
28. J. Harms et al., *J. Mol. Spectrosc.* **185**, 204 (1997).
29. D. M. Brink and S. Stringari, *Z. Phys. D At. Mol. Clusters* **15**, 257 (1990).
30. M. Hartmann et al., *Phys. Rev. Lett.* **76**, 4560 (1996).
31. W. Topic et al., *J. Chem. Phys.* **125**, 144310 (2006).
32. J. Tang et al., *Science* **297**, 2030 (2002).
33. Y. J. Xu et al., *Phys. Rev. Lett.* **91**, 163401 (2003).
34. J. Tang et al., *Phys. Rev. Lett.* **92**, 145503 (2004).
35. A. R. W. McKellar, Y. J. Xu, and W. Jager, *Phys. Rev. Lett.* **97**, 183401 (2006).
36. A. R. W. McKellar, *J. Chem. Phys.* **127**, 044315 (2007).
37. A. R. W. McKellar, *J. Chem. Phys.* **128**, 044308 (2008).
38. Y. J. Xu and W. Jager, *J. Chem. Phys.* **119**, 5457 (2003).
39. V. R. Pandharipande et al., *Phys. Rev. Lett.* **50**, 1676 (1983).
40. P. Sindzingre, M. L. Klein, and D. M. Ceperley, *Phys. Rev. Lett.* **63**, 1601 (1989).
41. M. V. Rama Krishna and K. B. Whaley, *Phys. Rev. Lett.* **64**, 1126 (1990).
42. R. N. Barnett and K. B. Whaley, *J. Chem. Phys.* **96**, 2953 (1992).
43. K. E. Kürten and M. L. Ristig, *Phys. Rev. B* **27**, 5479 (1983).
44. R. N. Barnett and K. B. Whaley, *J. Chem. Phys.* **99**, 9730 (1993).
45. Y. Kwon, D. M. Ceperley, and K. B. Whaley, *J. Chem. Phys.* **104**, 2341 (1996).
46. Y. Kwon and K. Birgitta Whaley, *Phys. Rev. Lett.* **83**, 4108 (1999).
47. Y. Kwon and K. B. Whaley, *J. Chem. Phys.* **115**, 10146 (2001).
48. D. Blume et al., *J. Chem. Phys.* **105**, 8666 (1996).
49. F. Paesani et al., *J. Chem. Phys.* **111**, 6897 (1999).
50. F. A. Gianturco et al., *J. Chem. Phys.* **112**, 2239 (2000).
51. F. Paesani, F. A. Gianturco, and K. B. Whaley, *J. Chem. Phys.* **115**, 10225 (2001).
52. A. Viel and K. B. Whaley, *J. Chem. Phys.* **115**, 10186 (2001).
53. A. Viel, K. B. Whaley, and R. J. Wheatley, *J. Chem. Phys.* **127**, 194303 (2007).
54. F. Paesani et al., *Phys. Rev. Lett.* **90**, 073401 (2003).
55. F. Paesani and K. B. Whaley, *J. Chem. Phys.* **121**, 5293 (2004).
56. F. Paesani, Y. Kwon, and K. B. Whaley, *Phys. Rev. Lett.* **94**, 153401 (2005).
57. R. E. Zillich and K. B. Whaley, *Phys. Rev. B* **69**, 104517 (2004).
58. R. E. Zillich et al., *J. Chem. Phys.* **123**, 114301 (2005).
59. K. von Haeften et al., *Phys. Rev. B* **73**, 054502 (2006).
60. R. E. Zillich, K. B. Whaley, and K. von Haeften, *J. Chem. Phys.* **128**, 094303 (2008).
61. S. Moroni et al., *Phys. Rev. Lett.* **90**, 143401 (2003).
62. P. Cazzato et al., *J. Chem. Phys.* **120**, 9071 (2004).
63. S. Moroni, N. Blinov, and P. N. Roy, *J. Chem. Phys.* **121**, 3577 (2004).
64. S. Paolini et al., *J. Chem. Phys.* **123**, 114306 (2005).
65. T. Skrbic, S. Moroni, and S. Baroni, *J. Phys. Chem. A* **111**, 12749 (2007).
66. T. Skrbic, S. Moroni, and S. Baroni, *J. Phys. Chem. A* **111**, 7640 (2007).
67. N. Blinov, X. G. Song, and P. N. Roy, *J. Chem. Phys.* **120**, 5916 (2004).
68. X. G. Song et al., *J. Chem. Phys.* **121**, 12308 (2004).
69. N. Blinov and P. N. Roy, *J. Low Temp. Phys.* **140**, 253 (2005).
70. Z. Li et al., *J. Chem. Phys.* **128**, 224513 (2008).
71. A. Sarsa et al., *Phys. Rev. Lett.* **88**, 123401 (2002).
72. H. Jiang et al., *J. Chem. Phys.* **123**, 224313 (2005).

73. F. Paesani and K. B. Whaley, *J. Chem. Phys.* **121**, 4180 (2004).

74. R. E. Zillich, Y. Kwon, and K. B. Whaley, *Phys. Rev. Lett.* **93**, 250401 (2004).

75. H. Kamerlingh Onnes, *Commun. Phys. Lab. Univ. Leiden* **105**, 744 (1908).

76. E. W. Becker, R. Klingelhöfer, and P. Lohse, *Z. Naturforsch. A* **16**, 1259 (1961).

77. J. Gspann and H. Vollmar, *J. Phys.* **39**, 330 (1978).

78. J. Gspann, in *Physics of Electronic and Atomic Collisions*, edited by S. Datz (North-Holland, Amsterdam, the Netherlands, 1982), p. 79.

79. J. Gspann, in *Rarefied Gas Dynamics*, edited by O. M. Beloserkovskii et al. (Plenum, New York, 1982), p. 1187.

80. J. Gspann, *Physica B* **108**, 1309 (1981).

81. A. P. J. van Deursen and J. Reuss, *J. Chem. Phys.* **63**, 4559 (1975).

82. P. W. Stephens and J. G. King, *Phys. Rev. Lett.* **51**, 1538 (1983).

83. J. P. Toennies, in *International School of Physics "Enrico Fermi", Course CVII,* Varenna, Italy, 1988 (North Holland, Amsterdam, the Netherlands, 1990), p. 597.

84. H. Buchenau et al., *J. Chem. Phys.* **92**, 6875 (1990).

85. M. Lewerenz, B. Schilling, and J. P. Toennies, *Chem. Phys. Lett.* **206**, 381 (1993).

86. R. Brühl et al., *Phys. Rev. Lett.* **92**, 185301 (2004).

87. T. E. Gough et al., *J. Chem. Phys.* **83**, 4958 (1985).

88. M. Hartmann et al., *PCCP* **4**, 4839 (2002).

89. A. Lindinger et al., *Z. Phys. Chem.-Int. J. Res. Phys. Chem. Chem. Phys.* **215**, 401 (2001).

90. A. Lindinger, J. P. Toennies, and A. F. Vilesov, *J. Chem. Phys.* **110**, 1429 (1999).

91. M. Hartmann et al., *Czech. J. Phys.* **46**, 2951 (1996).

92. J. Jang and R. M. Stratt, *J. Chem. Phys.* **113**, 5901 (2000).

93. S. Grebenev et al., *J. Chem. Phys.* **112**, 4485 (2000).

94. K. Nauta and R. E. Miller, *J. Chem. Phys.* **115**, 10254 (2001).

95. K. Nauta and R. E. Miller, *J. Chem. Phys.* **113**, 9466 (2000).

96. K. Nauta and R. E. Miller, *J. Chem. Phys.* **115**, 8384 (2001).

97. K. Nauta and R. E. Miller, *Phys. Rev. Lett.* **82**, 4480 (1999).

98. A. Conjusteau et al., *J. Chem. Phys.* **113**, 4840 (2000).

99. Y. Kwon et al., *J. Chem. Phys.* **113**, 6469 (2000).

100. M. V. Patel et al., *J. Chem. Phys.* **118**, 5011 (2003).

101. C. M. Lindsay and R. E. Miller, *J. Chem. Phys.* **122**, 104306 (2005).

102. I. Scheele et al., *J. Chem. Phys.* **122**, 104307 (2005).

103. M. N. Slipchenko and A. F. Vilesov, *Chem. Phys. Lett.* **412**, 176 (2005).

104. K. von Haeften et al., *Phys. Rev. Lett.* **95**, 215301 (2005).

105. H. Hoshina et al., *Phys. Rev. Lett.* **94**, 195301 (2005).

106. K. Nauta and R. E. Miller, *Science* **283**, 1895 (1999).

107. K. Nauta, D. T. Moore, and R. E. Miller, *Faraday Discuss.* **113**, 261 (1999).

108. K. Nauta and R. E. Miller, *Science* **287**, 293 (2000).

109. F. Dong and R. E. Miller, *Science* **298**, 1227 (2002).

110. J. Tang and A. R. W. McKellar, *J. Chem. Phys.* **119**, 5467 (2003).

111. J. Tang and A. R. W. McKellar, *J. Chem. Phys.* **119**, 754 (2003).

112. A. R. W. McKellar, *J. Chem. Phys.* **121**, 6868 (2004).

113. J. Tang and A. R. W. McKellar, *J. Chem. Phys.* **121**, 181 (2004).

114. A. R. W. McKellar, *J. Chem. Phys.* **125**, 164328 (2006).

115. A. R. W. McKellar, Y. J. Xu, and W. Jager, *J. Phys. Chem. A* **111**, 7329 (2007).

116. E. Lee, D. Farrelly, and K. B. Whaley, *Phys. Rev. Lett.* **83**, 3812 (1999).

117. F. Paesani and F. A. Gianturco, *J. Chem. Phys.* **116**, 10170 (2002).

118. V. L. Ginzburg and A. A. Sobyanin, *JETP Lett.* **15**, 242 (1972).

119. S. M. Apenko, *Phys. Rev. B* **60**, 3052 (1999).

120. S. Grebenev et al., *Science* **289**, 1532 (2000).

121. A. B. Migdal, *JETP Lett.* **37**, 249 (1959).

122. C. Cohen-Tannoudji, B. Diu, and F. Laloe, *Quantum Mechanics* (Wiley-Interscience, New York, 2006).

123. K. Szalewicz, *Int. Rev. Phys. Chem.* **27**, 273 (2008).

124. M. H. Kalos and P. A. Whitlock, *Monte Carlo Methods* (John Wiley & Sons, New York, 1986).

125. M. P. Nightingale and C. J. Umrigar, *Quantum Monte Carlo Methods in Physics and Chemistry*, Vol. 525 (Kluwer Academic, Dordrecht, the Netherlands, 1999).

126. B. L. Hammond, W. E. Lester Jr., and P. J. Reynolds, *Monte Carlo Methods in Ab Initio Quantum Chemistry* (World Scientific, Singapore, 1994).

127. W. L. McMillan, *Phys. Rev.* **138**, A442 (1965).

128. D. Ceperley, G. V. Chester, and M. H. Kalos, *Phys. Rev. B* **16**, 3081 (1977).

129. J. B. Anderson, *J. Chem. Phys.* **63**, 1499 (1975).

130. S. Baroni and S. Moroni, *Phys. Rev. Lett.* **82**, 4745 (1999).

131. D. Blume et al., *Phys. Rev. E* **55**, 3664 (1997).

132. D. M. Ceperley, *Rev. Mod. Phys.* **67**, 279 (1995).

133. R. P. Feynman and A. R. Hibbs, *Quantum Mechanics and Path Integrals* (McGraw-Hill, New York, 1965).

134. R. P. Feynman, *Statistical Mechanics* (Benjamin, New York, 1972).

135. E. W. Draeger and D. M. Ceperley, *Phys. Rev. Lett.* **90**, 065301 (2003).

136. Y. Kwon, F. Paesani, and K. B. Whaley, *Phys. Rev. B* **74**, 174522 (2006).

137. F. Luo et al., *J. Chem. Phys.* **98**, 3564 (1993).

138. W. Schollkopf and J. P. Toennies, *J. Chem. Phys.* **104**, 1155 (1996).

139. J. B. Anderson, C. A. Traynor, and B. M. Boghosian, *J. Chem. Phys.* **99**, 345 (1993).

140. A. R. Janzen and R. A. Aziz, *J. Chem. Phys.* **103**, 9626 (1995).

141. V. Efimov, *Phys. Lett. B* **33**, 563 (1970).

142. B. D. Esry, C. D. Lin, and C. H. Greene, *Phys. Rev. A* **54**, 394 (1996).

143. R. E. Grisenti et al., *Phys. Rev. Lett.* **85**, 2284 (2000).

144. M. V. R. Krishna and K. B. Whaley, *J. Chem. Phys.* **93**, 6738 (1990).
145. M. Lewerenz, *J. Chem. Phys.* **106**, 4596 (1997).
146. R. Guardiola et al., *J. Chem. Phys.* **124**, 084307 (2006).
147. S. Stringari and J. Treiner, *J. Chem. Phys.* **87**, 5021 (1987).
148. K. B. Whaley, in *Advances in Molecular Vibrations and Collision Dynamics*, edited by J. M. Bowman and Z. Bačić (JAI, Stamford, CT, 1998), p. 397.
149. J. P. K. Doye and F. Calvo, *J. Chem. Phys.* **116**, 8307 (2002).
150. R. Melzer and J. G. Zabolitzky, *J. Phys. A* **17**, L565 (1984).
151. A. Nakayama and K. Yamashita, *J. Chem. Phys.* **114**, 780 (2001).
152. M. Mella, M. C. Colombo, and G. Morosi, *J. Chem. Phys.* **117**, 9695 (2002).
153. T. G. A. Heijmen et al., *J. Chem. Phys.* **107**, 9921 (1997).
154. F. A. Gianturco and F. Paesani, *J. Chem. Phys.* **113**, 3011 (2000).
155. K. Higgins and W. Klemperer, *J. Chem. Phys.* **110**, 1383 (1999).
156. J. M. M. Howson and J. M. Hutson, *J. Chem. Phys.* **115**, 5059 (2001).
157. C. Callegari et al., *Phys. Rev. Lett.* **83**, 5058 (1999).
158. C. Callegari et al., *J. Chem. Phys.* **115**, 10090 (2001).
159. Y. K. Kwon and K. B. Whaley, *Phys. Rev. Lett.* **83**, 4108 (1999).
160. V. S. Babichenko and Y. Kagan, *Phys. Rev. Lett.* **83**, 3458 (1999).
161. F. Paesani et al., *J. Phys. Chem. A* **111**, 7516 (2007).
162. C. Callegari et al., *J. Chem. Phys.* **113**, 10535 (2000).

13

Intense Laser–Cluster Interactions

Karl-Heinz Meiwes-Broer
Universität Rostock

Josef Tiggesbäumker
Universität Rostock

Thomas Fennel
Universität Rostock

13.1 Introduction

The science of strong-field laser–matter interactions has experienced enormous developments due to the advent of the femtosecond (fs) optical laser technology (Brabec and Krausz, 2000). With the possibility to adjust the intensity, field envelope, carrier phase, and polarization of the pulses almost at will, methods are at hand to drive, explore, and even control complex nonlinear processes on time scales down to the laser oscillation period. The investigation of small particles and clusters exposed to such tailored pulses has become a hot topic of present-day research. In fact, subjecting free atomic and molecular clusters to intense radiation has revealed fascinating phenomena like the generation of highly charged atomic ions (Ditmire et al., 1997a,b; Köller et al., 1999; Lezius et al., 1998; Snyder et al., 1996), energetic electrons (Döppner et al., 2006; Kumarappan et al., 2003a; Springate et al., 2003), and x-rays (McPherson et al., 1994; Parra et al., 2000; Prigent et al., 2008). Due to an extreme charging, the systems decay via Coulomb explosion (Köller et al., 1999; Saalmann and Rost, 2003; Suraud and Reinhard, 2000) with ion recoil energies up to 10^6 eV (Ditmire et al., 1997b; Lezius et al., 1998). Even nuclear fusion has been observed in the Coulomb explosion of deuterium clusters (Ditmire et al., 1999).

The central issue of this chapter is to illuminate why the strong-field optical response of clusters and nanoparticles is quite different to atoms and bulk materials. It turns out that the finite system size introduces new aspects to the absorption and ionization dynamics. A key mechanism for the energy transfer from laser light into sub-wavelength-sized particles involves the driven coherent oscillation of many electrons. According to the well-known Mie formula (Mie, 1908), the corresponding dipole resonance or *surface-plasmon* has an angular frequency of

$$\omega_{\mathrm{Mie}} = \sqrt{\frac{e\rho_i}{3\epsilon_0 m_e}}, \qquad (13.1)$$

where

   e is the elementary charge
   ρ_i is the charge density of the ionic background
   ϵ_0 is the free space permittivity
   m_e is the electron mass

This classical picture of the surface-plasmon corresponds to the collective harmonic oscillation of the conduction electrons against the ionic background in a perfect metallic particle of spherical shape. If the photon energy of the exciting laser matches the plasmon energy, $\hbar\omega_{\mathrm{Mie}}$, the photoexcitation cross sections will significantly rise when compared to the nonresonant situation. With fs lasers, the excitation time can be even shorter than the ionic response time, e.g., the oscillation period of phonons. Thus, before the atomic cluster constituents have separated into independent atoms/ions, the nanosystems may gather tremendously high energy from the radiation field and turn into an extremely heated nanometer-sized plasma, as schematically depicted in Figure 13.1.

Depending on the intensity and envelope of an optical laser pulse, quite different properties and response mechanisms of these exotic systems can be probed. Exemplarily, Figure 13.1 illustrates typical decay channels that may be analyzed.

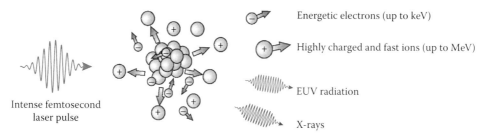

FIGURE 13.1 Upon strong-field laser excitation, clusters are transformed into highly nonstationary nanoplasmas. The variety of decay channels comprises electrons, ions, EUV radiation, and x-ray radiation.

For example, strong laser excitation of clusters leads to the emission of energetic electrons with kinetic energies up to the keV range. As a result of strong ionization, highly charged atomic ions induce a violent Coulomb explosion of the system, which can result in ions with up to MeV kinetic energy. Interestingly, the temporal structure of the laser pulse has strong impact on the ionization dynamics. For example, the energy of emitted electrons as well as the charge states and yields of ions can be tuned by the time delay between two intense fs-laser pulses. We will demonstrate that these effects are directly related to an interplay of collective plasma heating and ultrafast relaxation of the ionic system. Another consequence of the violent ultrashort excitation is the creation of inner-shell atomic vacancies. Upon recombination, radiation in a wide spectral domain, ranging from the ultraviolet to the hard x-ray region, can be emitted. There are prospects that with a fine tuning of laser parameters, one might be able to, e.g., control the yield of pulsed x-ray radiation in a selected wavelength region. This may lead to future technical applications, e.g., in the fields of optical lithography, new materials, or radiation therapy.

A fundamental issue concerns the wavelength dependence in the strong-field cluster excitations. Whereas optical fs lasers address the collective absorption regime, VUV-free electron lasers like FLASH with $\hbar\omega$ up to 200 eV allow coherent multiphoton inner-shell excitations of the cluster atoms (Wabnitz et al., 2002). By this, a rare-gas cluster can undergo *metallization* through simultaneously lifting many core electrons into the valence region, a scenario that might be realized in the near future. However, here we focus on the cluster response to optical laser fields with a typical wavelength of $\lambda = 800$ nm and restrict ourselves to a few side remarks on short wavelength lasers.

The chapter is organized as follows: General features of strong-field laser–cluster interactions are discussed in Section 13.2. Section 13.3 concentrates on the numerical methods to describe and analyze the many-particle dynamics in the clusters, namely semiclassical and molecular dynamics (MD) approaches. In Section 13.4, the basics of ultrashort laser pulse generation as well as the experimental techniques to generate intense cluster beams and to study their response to optical laser excitation are reviewed. In presenting selected results on strong-field dynamics in Section 13.5, we order the material in parts describing signatures from absorption and ionization, possibilities to interrogate the many-particle dynamics, as well as opportunities to control

the excitation by structured laser fields. Finally, Section 13.6 provides a brief outlook of the topic. There we address in particular aspects arising from VUV-/XUV-pulses and possibilities of the projected x-ray-free electron lasers.

13.2 Basics of Laser–Cluster Interactions

For the discussion of ionization and heating processes in clusters it is convenient to depart from processes on the atomic level, where two fundamentally different reaction channels for photoionization may be distinguished. The first is vertical excitation of a bound electron into the continuum as a result of single- or multiple photon absorption in a rapidly oscillating laser field, i.e., multiphoton ionization (MPI). In its most simple form, the reaction rate for an nth order process can be written as $\Gamma^{(n)} = \sigma^n I^n$, where I is the laser intensity and σ^n is the corresponding cross section. The second mechanism is optical field ionization (OFI), for which the laser can be treated as a quasistationary electric field. If the latter is sufficiently strong, a bound electron can tunnel through the barrier emerging from the combined potential of the residual q-fold charged ion and the laser field, i.e., $V(x) \propto -a/|z| - z$, with $a = (qe^2)/(4\pi\epsilon_0\epsilon_0)$, where ϵ_0 is the amplitude of the laser field. The probability for atomic tunneling ionization can be described, for example, by the well-known ADK rates found by Ammosov, Delone, and Krainov (Ammosov et al., 1986).

A useful measure for the significance of MPI over OFI is the Keldysh adiabaticity parameter (Keldysh, 1965; Radziemski and Cremers, 1989)

$$\gamma = \sqrt{\frac{E_{\mathrm{IP}}}{2U_{\mathrm{p}}}}, \tag{13.2}$$

which compares the ionization potential, E_{IP}, with the peak kinetic energy of a freely quivering electron. The latter is twice the ponderomotive potential U_{p}, which reads

$$U_{\mathrm{p}} = \frac{e^2 \epsilon_0^2}{4m_{\mathrm{e}}\omega_{\mathrm{las}}^2} \tag{13.3}$$

and specifies the cycle-averaged kinetic energy of a free electron in the laser field (no drift velocity). For convenience, the peak ponderomotive potential of a pulse can be expressed by $U_p = 9.33 \times 10^{-14}$ eV $\times I_0$ [W/cm²](λ [μm])², where I_0 is the peak intensity.

In terms of the Keldysh parameter, single- or multiphoton ionization dominates for $\gamma \gg 1$, where the energy an electron can capture per laser cycle is small compared to the ionization potential. For $\gamma \lesssim 1$, the binding energy can be overcome within a single laser cycle and OFI is promoted. An equivalent expression for the Keldysh parameter is $\gamma = \omega_{las}\tau_{tunnel}$, which gives roughly the ratio of the tunneling time $\tau_{tunnel} = \sqrt{2E_{IP}m_e/e^2\epsilon_0^2}$ and the optical period. Optical field ionization dominates when the tunneling time is comparable to or smaller than the optical period; MPI is the leading process otherwise.

In the tunneling regime ($\gamma \lesssim 1$), the ionization probability within one optical cycle approaches unity when the potential barrier is fully suppressed. For an atomic system, the onset of this barrier-suppression ionization (BSI) can be characterized by the so-called BSI intensity

$$I_{BSI} = \frac{\pi^2 c \epsilon_0^3}{2e^6} \frac{E_{IP}^4}{q^2} \approx 4 \times 10^9 \frac{(E_{IP} \text{ [eV]})^4}{q^2} \text{ W/cm}^2, \quad (13.4)$$

which provides a reasonable approximation for the appearance threshold intensity of highly charged ions in atomic gases, see e.g., Augst et al. (1989).

The above considerations apply to isolated atoms where mostly the laser parameters determine the dynamics. For extended systems, i.e., from the molecular level on, the structural details become increasingly important. A very convenient concept for the description of charging dynamics in clusters is the picture of inner and outer ionization (Last and Jortner, 1999). As indicated in Figure 13.2, electrons in the cluster may be classified into tightly bound, quasifree, and continuum electrons. Within this picture, *inner ionization* describes the excitation of tightly bound electrons to the conduction band, i.e., electrons are removed from their host ion but reside within the cluster. Correspondingly, the final excitation into the continuum is termed *outer ionization*,

which contributes to the net ionization of the system. At moderate laser intensities, systems with initially delocalized electrons, like metallic particles, may undergo outer ionization only. In any case, however, the energy span between the thresholds for inner and outer ionization grows with cluster charge, underlining the importance of quasifree or delocalized electrons for strong excitations. Irrespective of the particular system, ionization barriers are influenced by the fields from the neighboring ions, which, for example, gives rise to *charge-resonance-enhanced ionization* (CREI) well known from strong-field excitation of diatomic molecules (Seideman et al., 1995; Zuo and Bandrauk, 1995). Within this process, an appropriate internuclear separation results in a simultaneous lowering or suppression of inner and outer potential barriers and enhances ionization. For larger or smaller separations either the inner or the outer barriers increase and the ionization probability declines. As a truly cooperative effect, CREI has been considered also for very small clusters. Besides MPI and field ionization processes, inner and outer ionization can be driven by collisional ionization from impinging quasifree electrons and may be further supported by dynamic polarization fields and the cluster space-charge field. The onset and self-amplification of these additional ionization processes as a result of previously created quasifree electrons and cluster outer ionization has been described as *ionization ignition* (Rose-Petruck et al., 1997).

Once a *nanoplasma* is established, i.e., the cluster contains quasifree electrons and (multi-)charged atomic ions, substantial energy can be captured from a laser pulse. The term *nanoplasma* indicates that the system does no longer show much of the atomic or molecular behavior of its constituents, but responds to the laser field more like a localized microscopic plasma. As a result of electron scattering in the Coulomb field of the ions, electrons acquire energy from the laser field through Inverse Bremsstrahlung (IBS). IBS is a basic plasma volume-heating effect and relies on the conversion of purely laser-driven electron motion into thermal energy because of directional momentum redistribution within elastic collisions (Krainov, 2000; Krainov and Smirnov, 2002). When depolarization fields in the system can be neglected, the corresponding cycle-averaged energy-capture rate per electron follows as

$$\left\langle \frac{dE}{dt} \right\rangle_{IBS} = 2U_p \frac{\tau_{coll}\omega_{las}^2}{\tau_{coll}^2\omega_{las}^2 + 1}, \quad (13.5)$$

with the dephasing time τ_{coll} (inverse collision frequency), the laser frequency ω_{las}, and the ponderomotive potential U_p. This rate corresponds to the physical picture of laser-driven electron motion in the presence of collision-induced friction. For clusters, pure IBS heating dominates for laser frequencies far above the Mie plasmon. For laser frequencies comparable to or smaller than ω_{Mie}, the collective response of quasifree electrons in the cluster, i.e., restoring forces introduced by cluster polarization, must be taken into account. In the most simple case of a metallic sphere, the absorption rate per electron can be described by a Lorentz formula

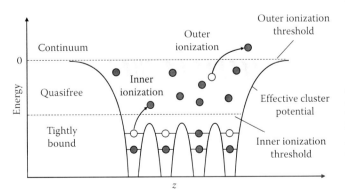

FIGURE 13.2 Schematic view of inner and outer ionization of a cluster based on the effective cluster potential. The degree of localization gives rise to the classification of electrons into tightly bound, quasifree, and continuum ones.

$$\left\langle \frac{dE}{dt} \right\rangle_{\mathrm{RES}} = 2U_{\mathrm{p}} \frac{\tau_{\mathrm{coll}} \omega_{\mathrm{las}}^4}{\tau_{\mathrm{coll}}^2 \left(\omega_{\mathrm{Mie}}^2 - \omega_{\mathrm{las}}^2\right)^2 + \omega_{\mathrm{las}}^2}, \qquad (13.6)$$

where ω_{Mie} reflects the characteristic eigenfrequency of the dipole mode of the system, cf. Equation 13.1. It should be noted that the expression in Equation 13.6 is equivalent to the heating rate in Dirmire's well-known nanoplasma model (Ditmire et al., 1996) and contains the simple IBS result from Equation 13.5 as the high-frequency limit. Further, excitation with $\omega_{\mathrm{las}} \approx \omega_{\mathrm{Mie}}$ results in a resonance-enhanced energy absorption and can lead to strong-field amplification in the system, see e.g. (Fennel et al., 2007a; Reinhard and Suraud, 1998). For $\omega_{\mathrm{las}} \ll \omega_{\mathrm{Mie}}$, on the other hand, heating is strongly suppressed, since the polarization fields produced from cluster electrons efficiently screen the external field. Although the above expression may suggest fully linear response behavior, resonance-enhanced heating remains crucial up to laser intensities, where the system response becomes strongly nonlinear. More specifically, for strong laser excitation the time and/or intensity dependencies of the dephasing time and the plasmon energy introduce additional nonlinear terms in Equation 13.6. Nonetheless, the efficient heating of quasifree electrons at resonance strongly supports the outer ionization leading to *plasmon-enhanced ionization*. In most cases, however, the cluster must expand significantly in order to maintain the required frequency matching (delayed resonance), as will be worked out in more detail below. With increasing laser intensity, plasmon-enhancement becomes important irrespective of the cluster material, i.e., for metal as well as rare-gas systems.

13.3 Theoretical Modeling of the Dynamics

The interaction of intense laser pulses with clusters leads to very high excitation and ionization of the system. A theoretical description thus requires dedicated methods that are able to resolve the coupled and highly non-adiabatic electron and ion dynamics for a many-particle system far out of equilibrium. Since a full quantum mechanical description is prohibitive under these circumstances, simplified methods, ranging from MD up to quantum approaches based on the time-dependent local-density approximation (Reinhard and Suraud, 2003; Saalmann et al., 2006), have to be used. A particularly challenging aspect for metal clusters is the presence of delocalized electrons and the fermionic nature of the dynamics. Starting from a fully degenerate state with complete Pauli blocking, the electronic system experiences highly nonequilibrium conditions with significant Coulomb coupling and partial degeneracy within the interaction process. As a result, electron–electron collisions (EEC), which require a quantum statistical treatment under these conditions, can also transiently become very important. Further, the highly nonlinear collective motion of electrons has to be incorporated self-consistently.

Along this line, Vlasov and Vlasov-Uehling-Uhlenbeck (VUU) methods are very a appealing for the treatment of metal clusters in strong fields. These semiclassical kinetic treatments, also termed $\hbar \to 0$ methods, represent a counterpart to orbital-based approaches and permit an approximate description of the quantum dynamics without directly referring to wavefunctions (Fennel and Köhn, 2008). In the following, a concise description of the basic methodic aspects and corresponding examples are presented for the energy absorption and the ionization process. The presented calculations for dual-pulse excitations of Na_N yield insight into the dynamics and serve to explain the experimental data, cf. Section 13.5.

For larger systems and very high laser intensity, where time-dependent density-functional methods are no longer feasible, MD approaches have proven to be versatile. They permit the treatment of much higher particle numbers, and deeply bound electrons can be taken into account for the cluster ionization and the nanoplasma dynamics. We will present the results from MD simulations on rare-gas clusters (Xe_N) for the discussion of ion charge-state distributions and the strong-field absorption dynamics.

13.3.1 Semiclassical Vlasov and VUU Methods

The starting point is the semiclassical approximation of the electron dynamics in the absence of EEC. On the mean-field level, the quantal time evolution of the electronic system can be described by the von Neumann equation

$$-i\hbar \frac{\partial}{\partial t}\rho = \left[\frac{-\hbar^2}{2m}\left(\nabla_{\mathbf{r}'}^2 - \nabla_{\mathbf{r}}^2\right) + V_{\mathrm{eff}}(\mathbf{r}') - V_{\mathrm{eff}}(\mathbf{r}) \right]\rho \qquad (13.7)$$

where

$\rho = \rho(\mathbf{r}, \mathbf{r}')$ is the one-body density matrix
$V_{\mathrm{eff}}(\mathbf{r})$ is an effective self-consistent potential containing interactions and external fields (Fennel and Köhn, 2008)

Using the Wigner transform

$$f_{\mathrm{w}}(\mathbf{r},\mathbf{p}) = \frac{1}{(2\pi\hbar)^3}\int d\mathbf{q}^3 e^{i\mathbf{p}\cdot\mathbf{q}/\hbar}\rho\left(\mathbf{r}+\frac{\mathbf{q}}{2}, \mathbf{r}-\frac{\mathbf{q}}{2}\right) \qquad (13.8)$$

Equation 13.7 can be rewritten as a transport equation

$$\frac{\partial}{\partial t}f_{\mathrm{w}} + \frac{\mathbf{p}}{m}\nabla_{\mathbf{r}}f_{\mathrm{w}} - \frac{2}{\hbar}f_{\mathrm{w}}\sin\left(\frac{\hbar}{2}\overleftarrow{\nabla}_{\mathbf{p}}\cdot\overrightarrow{\nabla}_{\mathbf{r}}\right)V_{\mathrm{eff}}(\mathbf{r}) = 0, \qquad (13.9)$$

where $f_{\mathrm{w}}(\mathbf{r}, \mathbf{p})$ has the meaning of a one-body phase-space distribution, but for the fact that it can be negative in some regions (Bertsch and Das Gupta, 1988). The semiclassical limit ($\hbar \to 0$) of Equation 13.9 follows from expanding the sine in the lowest order and introducing a smoothed nonnegative distribution function, $f(\mathbf{r}, \mathbf{p})$. This yields the Vlasov equation

$$\frac{\partial}{\partial t}f + \frac{\mathbf{p}}{m}\cdot\nabla_{\mathbf{r}}f - \nabla_{\mathbf{p}}f\cdot\nabla_{\mathbf{r}}V_{\mathrm{eff}}(\mathbf{r},t) = 0. \qquad (13.10)$$

Quantum effects are now solely contained in the effective potential and in the initial conditions for the distribution function. The latter can be determined from the self-consistent Thomas–Fermi ground state according to

$$f^0(\mathbf{r},\mathbf{p}) = \frac{2}{(2\pi h)^3}\Theta(p_{\mathrm{F}}(\mathbf{r}) - p), \qquad (13.11)$$

where

$p_{\mathrm{F}}(\mathbf{r}) = \sqrt{2m[\mu - V_{\mathrm{eff}}(\mathbf{r})]}$ is the local Fermi momentum
μ is the chemical potential

Based on that, Equation 13.10 describes the collisionless dynamics of the electron system.

To incorporate binary EEC, Equation 13.10 can be complemented by an Uehling-Uhlenbeck collision term (Uehling and Uhlenbeck, 1933), which results in the Vlasov-Uehling-Uhlenbeck equation

$$\frac{\partial}{\partial t}f + \frac{\mathbf{p}}{m}\cdot\nabla_{\mathbf{r}}f - \nabla_{\mathbf{p}}f\cdot\nabla_{\mathbf{r}}V_{\mathrm{eff}}(\mathbf{r},t) = I_{\mathrm{UU}}. \qquad (13.12)$$

The UU-collision integral reads

$$I_{\mathrm{UU}}(\mathbf{r},\mathbf{p}) = \int d\Omega d\mathbf{p}_1 \frac{|\mathbf{p} - \mathbf{p}_1|}{m}\frac{d\sigma\big(\theta,|\mathbf{p} - \mathbf{p}_1|\big)}{d\Omega}$$
$$\times \left[f_{\mathbf{p}'}f_{\mathbf{p}_1'}\big(1 - \tilde{f}_{\mathbf{p}}\big)\big(1 - \tilde{f}_{\mathbf{p}_1}\big) - f_{\mathbf{p}}f_{\mathbf{p}_1}\big(1 - \tilde{f}_{\mathbf{p}'}\big)\big(1 - \tilde{f}_{\mathbf{p}_1'}\big)\right] \qquad (13.13)$$

and represents a local gain–loss balance for elastic electron–electron scattering $(\mathbf{p},\mathbf{p}_1) \leftrightarrow (\mathbf{p}',\mathbf{p}_1')$ with the momentum-dependent differential cross section $d\sigma(\theta,|\mathbf{p}_{\mathrm{rel}}|)/d\Omega$, where the scattering angle, θ, defines the deflection of the relative momentum vector, the local phase-space densities $f_{\mathbf{p}} = f(\mathbf{r},\mathbf{p})$, and the Pauli blocking factors in parenthesis as functions of the relative phase-space occupation for paired spins $\tilde{f}_{\mathbf{p}} = f_{\mathbf{p}}(2\pi h)^3/2$. Equations 13.12 and 13.13 describe the electron dynamics including EEC. Note, because of the blocking factors the collision integral vanishes in the ground state, where the Vlasov description is recovered as the limit for weak excitations. The evaluation of the UU-collision integral requires the determination of scattering cross sections that can be obtained from the quantum scattering theory. Assuming a screened electron–electron interaction potential

$$V_{\mathrm{sc}}(r) = \frac{e}{4\pi\epsilon_0}\frac{e^{-r/r_{\mathrm{TF}}}}{r}, \qquad (13.14)$$

where $r_{\mathrm{TF}} = (\pi/3n_{\mathrm{e}})^{1/6}\sqrt{a_0}/2$ is Thomas–Fermi screening length for a fully degenerate Fermi gas and n_{e} is the electron density,

the cross sections can be computed by partial wave analysis. For details see Köhn et al. (2008).

Within the Vlasov or VUU dynamics a self-consistent mean-field potential of the form

$$V_{\mathrm{eff}}(\mathbf{r}) = \sum_i V_{\mathrm{ion}}(\mathbf{r} - \mathbf{R}_i(t)) + V_{\mathrm{Har}} + V_{\mathrm{xc}} + e\,\varepsilon(t)\cdot\mathbf{r} \qquad (13.15)$$

is considered, containing the sum over the ion potentials for the present configuration $\mathbf{R}_i(t)$, the electron Hartree potential V_{Har}, the LDA exchange-correlation potential V_{xc}, e.g., from Gunnarsson and Lundquist (1976), and the laser field in dipole approximation via the last term $\varepsilon(t)\cdot\mathbf{r}$. The Hartree term and the exchange-correlation potential are calculated from the actual total electron density, $n_{\mathrm{e}}(\mathbf{r},t) = \int d^3p f(\mathbf{r},\mathbf{p},t)$. To avoid the numerically expensive propagation of strongly localized states only valence electrons are treated explicitly in the model, while the interaction with nuclei and core electrons is described by a local pseudopotential for the sodium ions (Fennel et al., 2004). Classical motion is assumed for the ions.

In this form, the semiclassical approximation is valid for high electronic excitations $E_{\mathrm{ex}} \gg \Delta$, and/or dense electronic energy levels $\epsilon_{\mathrm{F}} \gg \Delta$, where E_{ex} is the excitation energy, ϵ_{F} the Fermi energy, and Δ the single particle level spacing. In particular for sodium clusters, the semiclassical method is well tested and predicts ground state geometries, optical spectra, and dynamics close to results from quantal density-functional calculations for cluster sizes $N > 10$ (Legrand et al., 2006; Plagne et al., 2000; Reinhard and Suraud, 2003).

The Vlasov/VUU equation can be efficiently solved using the test particle method. This method is, like many other concepts in cluster physics, inherited from nuclear physics (Bertsch and Das Gupta, 1988). The key idea of the test particle method is to sample the continuous distribution function with a swarm of fractional particles and to map the dynamics into classical equations of motion for the discrete samples. A straightforward way of representation is

$$f(\mathbf{r},\mathbf{p},t) = \frac{1}{N_s}\sum_i^{N_{pp}} g_r(\mathbf{r} - \mathbf{r}_i(t))\, g_p(\mathbf{p} - \mathbf{p}_i(t)), \qquad (13.16)$$

with the positions $\mathbf{r}_i$ and the momenta $\mathbf{p}_i$ of the test particles and the smooth weighting functions g_r and g_p in coordinate and momentum space, respectively. The parameter N_s sets the number of test particles per physical particle and defines the total number of test particles $N_{pp} = NN_s$. One possible choice for the weighting are normalized Gaussians

$$g(\mathbf{x}) = \frac{1}{\pi^{3/2}d^3}e^{-x^2/d^2}, \qquad (13.17)$$

where d is a numerical smoothing parameter. Using the test particle ansatz, the mean-field part of the electron propagation reduces to classical motion for the test particles according to

$$\dot{\mathbf{r}}_i = \frac{\mathbf{p}_i}{m} \quad \text{and} \quad \dot{\mathbf{p}}_i = \underbrace{-\int V_{\text{eff}}(\mathbf{r})\Delta_{\mathbf{r}_i} g_r(\mathbf{r}-\mathbf{r}_i)d^3\mathbf{r}}_{\mathbf{f}_i}. \qquad (13.18)$$

For the semiclassical treatment a smooth phase-space distribution is essential to suppress the tendency of classical thermalization, which requires a finite width of the test particles in practice (Jarzynski and Bertsch, 1996). However, this must not be a general shortcoming since the width parameter can be used as a parameter to express a semiclassical version of uncertainty.

13.3.2 Resonance-Enhanced Ionization with Dual Pulses

As an example, the impact of resonant charging on the cluster ionization is investigated for Na_{55} subject to dual laser pulse irradiation using the Vlasov propagation, i.e., without electron–electron collisions (Döppner et al., 2005). The cluster response is simulated for excitations with 50 fs linearly polarized Gaussian pulses at $\hbar\omega_{\text{las}} = 1.54$ eV having a peak intensity of $I_0 = 4 \times 10^{12}$ W/cm^2 for various pulse delays, Δt. In the ground state the semiclassical plasmon energy of the system is $\hbar\omega_0 = 2.84$ eV, which is well above the laser photon energy. At these laser parameters the highest cluster ionization occurs for a pulse delay of $\Delta t \approx 250$ fs. Selected observables from the corresponding simulation at optimal delay are displayed in Figure 13.3.

The leading pulse causes only a weak cluster ionization (c), as the excitation is far off-resonant. This is also reflected in a small amplitude (a) of the induced dipole moment as well as its small phase shift (b) with respect to the laser field. In response to the first pulse the excited cluster starts to expand, as can be seen from the increasing rms-cluster radius (d). According to the model of plasmon-enhanced charging high ionization is

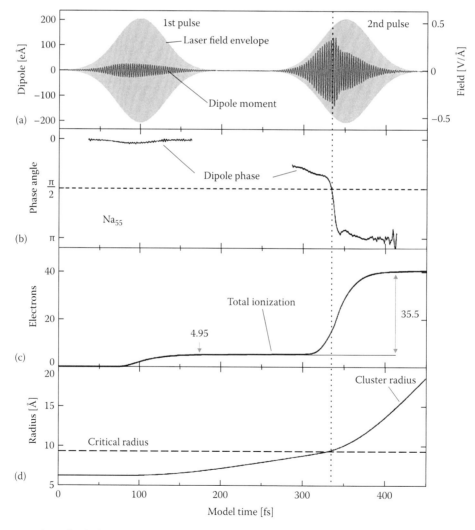

FIGURE 13.3 Response of Na_{55} for dual-pulse laser excitation with $I = 4 \times 10^{12}$ W/cm^2 at 800 nm and 250 fs optical delay. Shown are the laser field envelope (gray) and (a) the electron dipole amplitude, (b) the phase lag between laser field and dipole signal, (c) the total cluster ionization, (d) and the root-mean-square radius of the ion distribution. Note that the dipole phase angle passes $\pi/2$ as the rms-radius is close to the critical value, R_{crit}. (After Döppner, T. et al., *Phys. Rev. Lett.*, 94, 013401, 2005.)

expected if the second pulse excites the collective mode of the cluster resonantly, i.e., at its critical cluster radius R_{crit}. A rough estimate based on the classical Mie formula (cf. Equation 13.1) yields $R_{crit} = R_0(\omega_0/\omega_{las})^{2/3}$ with R_0 the initial cluster radius. For the given example this results in a value of about 1.5 R_0, indicated as the dashed horizontal line in Figure 13.3d. After reaching its critical radius the cluster is exposed to the second pulse, leading to 7-times increased ionization when compared to the effect of the first pulse (35 vs. 5 electrons). Strong resonant energy absorption from the second pulse is also indicated by the high dipole amplitude and the $\pi/2$ transition of the phase shift, c.f. (a) and (b).

A systematic theoretical analysis of the pulse delay effect on the cluster ionization is shown in Figure 13.4. Excitations of sodium clusters with sizes $N = 13, 55$, and 147 by 25 fs dual pulses

at $I = 8 \times 10^{12}$ W/cm^2 are considered and analyzed for a delay range $\Delta t = 50$–800 fs (Köhn et al., 2008). The figure displays the total cluster ionization as extracted from both, Vlasov and VUU calculations, as a function of Δt. For all cases, resonant collective excitations at optimal pulse delays induce a pronounced enhancement in energy deposition as well as in cluster ionization. As the values of the total ionization at the corresponding optimal delays are practically the same for Vlasov and VUU it can be concluded that the effect of EEC on the coupling at resonance is small. However, the time scales of the dynamics can be strongly affected. Two trends are obvious from the comparison of the optimal pulse delays as a function of cluster size: First, the time required to establish resonance increases with system size, irrespective of the consideration of EEC. This behavior reflects the buildup of a stronger space-charge potential in larger clusters that leads to reduced relative ionization and excitation by the off-resonant first pulse. Second, as indicated by the increasing ratio of the optimal delays from the Vlasov and VUU simulations, the importance of EEC for the dynamics grows with clusters size. This trend can be traced back to the competition between field ionization and laser heating of the cluster electrons. Optical field ionization is practically unaffected from EEC and the dominant nonresonant excitation mechanism for smaller clusters in this scenario. Their initial excitation is only weakly affected by the collision term. For larger systems, the heating of cluster electrons becomes increasingly important for the excitation energy and the expansion speed. The optimal delays are thus more affected from enhanced heating through electron–electron collisions.

From such simulations the significance of resonance-enhanced charging can be worked out and we find qualitative agreement also with the experimental trends, see Section 13.5. However, a description of high atomic charge states within such simulations remains distant, as the treatment of deeply bound electronic levels is numerically demanding and presently out of reach with this technique. Nevertheless, for model studies on a simplified system the semiclassical method is a powerful tool to obtain detailed insight into the microscopic dynamics of the strong-field excitation. For further discussion of the mechanisms behind resonant electron emission and the possibility of directed electron acceleration see Section 13.5.4.

13.3.3 Exploring High Cluster Charging with MD Simulations

Many experiments show that clusters from medium and high atomic number elements emit ions with high charge states up to 20–40 already for moderate laser intensities of 10^{14}–10^{15} W/cm^2. The mechanisms underlying the generation of these highly charged ions, however, are not well understood yet. Since their investigation requires the inclusion of core electrons, time-dependent density-functional methods can no longer be used. An alternative are MD methods, where core electrons are taken into account via effective ionization rates. MD methods further permit the treatment of larger clusters and become increasingly valid at high laser intensity. Still, the explanation of the very high

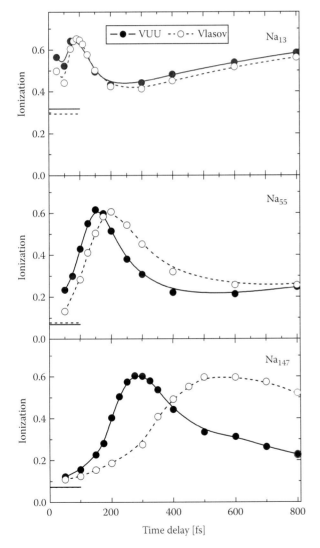

FIGURE 13.4 Total ionization per atom for Na$_N$ exposed to dual pulses as a function of pulse delay for different cluster sizes (Na$_{13}$, Na$_{55}$, Na$_{147}$). Both pulses are identical ($I = 8 \times 10^{12}$ W/cm^2, $\tau = 25$ fs FWHM). Lines at the left side correspond to excitation by the first pulse alone. (Adapted from Köhn, J. et al., *Phys. Rev. A*, 77, 033202, 2008.)

charge states remains challenging and is highly debated (Fennel et al., 2007b; Heidenreich et al., 2007).

For cluster excitation with near-infrared pulses, the dominant mechanisms for inner ionization are optical field ionization, i.e., tunnel- or barrier-suppression ionization, and electron-impact ionization. In order to describe these processes properly, the local fields and screening effects within the nanoplasma have to be incorporated. This is more or less straightforward for tunnel ionization, as the effective microfield resulting from space charges, cluster polarization, and the laser is accessible numerically. The effective field can then be used, for example, for the determination of tunneling ionization probabilities (Ammosov et al., 1986). More involved is the treatment of electron-impact ionization that is often described by the empirical cross sections from Lotz (1967). This requires atomic ionization potentials that are, however, much affected by the many-particle effects in the cluster, such as electronic screening and fields from neighboring ions (Bornath et al., 2007a; Fennel et al., 2007b; Gets and Krainov, 2006). In addition electron-ion recombination processes must also be considered for the comparison of results

from simulations with the experimental spectra. Usually it is assumed that only continuum electrons produced during the laser pulse contribute to the final ionization spectra and cluster-bound electrons fully recombine with cluster ions. Under experimental conditions, this might not be true. The extraction fields for time-of-flight (TOF) analysis release weakly bound electrons and thus hinder full recombination.

The effects of both, enhancement of electron-impact ionization trough IP lowering and frustrated recombination due to static external fields, have been investigated for Xenon cluster (Fennel et al., 2007b). The results show that the combined action of these effects strongly modify the ion charge-state distributions, see Figure 13.5, and can enhance the maximum ion charge states significantly. More specifically, the left column shows ion charge distributions calculated with bare atomic electron-impact ionization cross sections and total recombination. Including the local-field enhanced cross sections and the frustrated recombination due to ion extraction fields the spectra are strongly shifted to higher charge states, see the right-hand side column. The highest charge state for the Xe_{5083} simulation

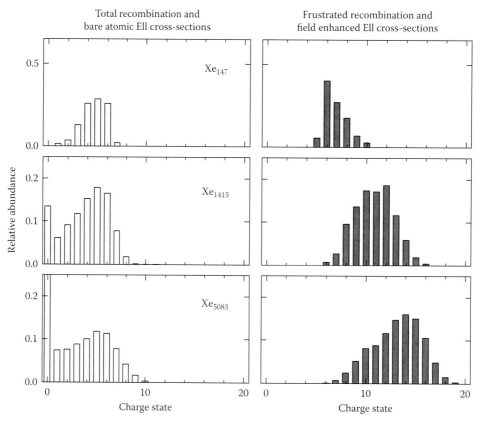

FIGURE 13.5   Ion charge-state distributions from Xe_N (N = 147–5083, as indicated) exposed to 250 fs laser pulses with peak intensity, 4×10^{14} W/cm^2, calculated from MD simulations. The results correspond to different treatments of electron-impact ionization (EII) and electron-ion recombination. Conventional and enhanced EII correspond to atomic and local-field corrected ionization thresholds, respectively. Total recombination assumes that cluster-bound electrons recombine with the closest ion after the laser pulse, while the long-term dynamics of quasifree electrons in the presence of an ion extraction field in the kV/m-range is taken into account for the frustrated recombination. (Adapted from Fennel, T. et al., *Phys. Rev. Lett.*, 99, 233401, 2007b.)

is $q = 19$, in reasonable agreement with experiments (Snyder et al., 1996), given the uncertainties in cluster size distribution, laser pulse profile, and peak intensity. Further contributions like excitation-autoionization processes or ionization via intermediate states, the importance of which is known for atomic electron-impact ionization (Griffin et al., 1984; Loch et al., 2008), have not been studied in detail yet.

13.4 Experimental Methods

With the modern molecular beam machines, the variety of radiation sources from the infrared to the x-ray regime, and the multiply parallel detection and data processing possibilities, challenging and highly sophisticated experiments can be performed. It is possible to prepare cluster targets with narrowed size distribution or even completely size-selected, partially at low or ultralow temperature. Vast literature exists on cluster production, e.g., Echt and Recknagel (1991), Haberland (1994), Milani and Ianotta (1999), Pauly (2000), and Whaley and Miller (2001). Optical single or collective electron excitation, in some cases also being followed by a probing ultrashort light pulse, has led to far-reaching insight into the fundamental processes of the light–matter interaction. In this section, rather than covering the vast multitude of experimental methods, we will review selected current techniques used for analyzing laser-induced dynamics on free clusters.

13.4.1 Femtosecond Lasers

Compared to linear response experiments on clusters, where large laser beam diameters (several mm) can normally be accepted, a tight focusing to some tens of micrometers is necessary to attain intensities of 10^{13}–10^{16} W/cm^2 in the interaction region. The corresponding field strengths are 10^8–10^{11} V/m. Note, that these values are on the order of the electric field strength in the hydrogen atom, thus in a regime where the binding conditions are strongly modified by the presence of the radiation. Such high intensities can most conveniently be supplied by fs lasers. As an advantage, the ultrashort pulses, in particular when using dual pulses, enable the investigation of the light-induced dynamics in real time.

Technically such radiation with power densities of up to about 10^{16} W/cm^2 is generated with *table-top* systems relying on amplification in Ti:Sa crystals, see Table 13.1 for typical realizations. A standard device for pulses in the sub-50 fs regime consists of a Ti:sapphire oscillator and, e.g., a multi-pass amplifier. The oscillator delivers pulses with durations of about 30 fs with a rate of 90 MHz, an energy of several nJ, and a spectral width up to 80 nm (FWHM). For amplification most systems use the technique of chirped pulse amplification (Maine et al., 1988) where the energy is increased up to some mJ with a reduced rate of 1 kHz. The initial width is nearly maintained and bandwidth-limited pulses of below 50 fs are then available for the experiment. The temporal shape of the pulse can be modified by detuning the distance of the gratings in the optical compressor. As a result stretched pulses with up to several tens of picoseconds can easily be generated. However, this modification necessarily goes along with a chirp, i.e., a time-dependent carrier frequency.

A straightforward way to generate pairs of pulses is offered by a Mach–Zehnder interferometer setup. With a beam attenuator introduced in one of the interferometer arms, pairs of pulses of adjustable optical delay (Δt) and intensity ratio can be produced. For strong-field laser–cluster interactions the subsequent pulse often not only probes the dynamics but dominates the excitation. We thus prefer the term *dual-pulse* technique instead of the commonly used notation *pump-probe*.

In the control studies, methods are established to shape the laser pulse in amplitude and phase. The concept of coherent control (Brumer and Shapiro, 1995; Tannor et al., 1986) can be used for self-optimization, i.e., to find laser parameters that lead, e.g., to a preselected ion charge-state distribution. Feedback algorithms are available that in connection with pulse shapers (liquid crystal arrays, acousto-optical modulators, and programmable dispersive filters or deformable mirrors, see e.g., Weiner (2000) for more details) can be used to modify laser pulses and control the interaction. Prominent examples are found in molecular physics, where adaptive control is used to selectively break chemical bonds (Assion et al., 1998). First such studies on strong-field excitation of clusters are on the way (Truong et al., 2010; Zamith et al., 2004).

TABLE 13.1 Typical Optical and Extended Ultraviolet-Free Electron Laser Systems Used for Spectroscopy on Clusters in the High Intensity Regime

Laser	Wavelength	Photon Energy	Repetition Rate	Pulse Duration	Pulse Energy	Intensity in the Focus
Table top Ti:Sa	800 nm	1.55 eV	1 kHz	30 fs	2 mJ	5×10^{16} W/cm^2
Ti:Sa	800 nm	1.55 eV	10 Hz	100 fs	500 mJ	1×10^{19} W/cm^2
Ti:Sa + SHG	400 nm	3.10 eV	1 kHz	50 fs	1 mJ	5×10^{15} W/cm^2
Ti:Sa + ArF	248 nm	4.65 eV	0.4 Hz	230 fs	400 mJ	1×10^{20} W/cm^2
FLASH	6–120 nm	10.0–200 eV	5 Hz	10–30 fs	10–100 µJ	5×10^{14} W/cm^2

Note: Ti:Sa, titanium sapphire laser; ArF, argon fluoride excimer laser amplifier; SHG, second harmonic generation; FLASH, free electron laser Hamburg.

13.4.2 Cluster Formation

The method of choice to form intense beams of clusters depends on the desired material and composition. With rare gas and molecular clusters a high rate of particles can be delivered by a strong supersonic expansion. Clusters are formed in the adiabatic cooling process when streaming from a high-pressure region before the nozzle exit into the low-pressure vacuum chamber. Supersonic expansion sources deliver particle densities of up to $10^{15}/cm^3$. However, this value is achieved only close to the nozzle exit where the gas load is too high to allow for any operation of a particle detector. In addition, the mean free path might be too small to suppress collisions with the buffer gas completely, which is a strong requirement in order to extract reliable values like particle energy and emission angle from the experiment. Cluster–cluster interactions might also contribute (cluster–matter effects). Therefore, in some configurations additional differential pumping stages are integrated downstream the molecular beam axis. By this, well-collimated beams of noninteracting clusters can be investigated, free of any background gas load (Table 13.2).

Solid materials have to be brought into the vapor phase first and then condensed in an appropriate environment to form clusters. Several methods are at hand, including heating the material in a crucible (Schulze et al., 1987), applying sputtering techniques (Haberland et al., 1994), laser vaporization (Dietz et al., 1981), and pulsed (Milani and Ianotta, 1999; Siekmann et al., 1991) or continuous (Kleibert et al., 2007; Methling et al., 2001) arcs. In all cases the vapor undergoes cooling and expansion in a stream of rare (or seeding) gas. This can be pulsed, allowing for a hard expansion of the seeded clusters into vacuum, or it is continuously streaming at lower pressure. In the first case, usually He pulses serve as seeding gas with an admixture of Ne or Ar at backing pressures of 2–20 bar. Particle growth proceeds via multiple collisions of the atoms with the high density buffer gas atoms. At low pressure, longer and cooled aggregation tubes support the condensation into clusters.

For experiments at ultralow temperatures, helium droplet pickup sources prove to be very versatile. He droplets are produced by the supersonic expansion of precooled helium gas with a stagnation pressure of 20 bar through a 5 μm diameter nozzle (Bartelt et al., 1996; Goyal et al., 1992; Tiggesbäumker and Stienkemeier, 2007). By choosing the temperature at the orifice (8–14 K), the log-normal droplet size distributions can be adjusted in the range of $\langle N \rangle = 10^3 – 10^7$ atoms. After passing differential pumping stages the beam enters the pickup chamber containing a gas target or a heated oven, where atoms are collected and aggregated to clusters inside the He droplets. With this setup it is possible to record clusters with up to 150 silver atoms (Radcliffe et al., 2004) or 2500 magnesium atoms (Diederich et al., 2005), respectively. Downstream another differential pumping stage laser light or an electron beam ionizes the doped droplets. The benefits of pickup sources rely on the feasibility to embed the cluster into a well-controlled environment. In the case of He, the embedding medium is superfluid, weakly interacting, and ultracold with a temperature of about 0.4 K (Hartmann et al., 1995), thus being an ideal matrix for spectroscopic studies (Whaley and Miller, 2001). Similarly, droplets or particles of other elements might serve as pickup medium, e.g., Ar, Kr, or Xe. Subsequent atom agglomeration also can lead to the formation of electronically excited species (Ievlev et al., 2000). One very important advantage of pickup

TABLE 13.2 Techniques for Cluster Generation

Source	Materials	Particle Size	Timing	Typical Applications
Knudsen cell	Low boiling-point materials; cluster formation in the thermodynamic equilibrium	Up to 1.2 nm	cw	Spectroscopy and deposition of small clusters, e.g., Sb_4 or fullerenes
Supersonic nozzle	Atomic and molecular gases	Up to 5 nm	cw or pulsed	Intense beams for spectroscopy (rare and molecular gases)
Seeded jet	All vaporized or gaseous/molecular, in addition a seeding gas (He, Ne, etc.)	Up to 2 nm	Usually cw	Intense beams for spectroscopy and deposition
Pickup sources	All vaporized or gaseous/molecular materials	Up to 1.5 nm	Usually cw	Growth and spectroscopy in a controlled environment, pickup in a host particle, e.g., metal vapor in He droplets
Gas aggregation	Like the seeded jet, but at low pressure and with cryogenic aggregation tube	Up to 10 nm	cw	Intense beams for spectroscopy and deposition
Sputtering (a) high ion energy, low pressure (b) high pressure	Solid materials, preferentially conducting	(a) Up to 1.5 nm (b) up to 20 nm	Usually cw	(a) Ultraclean environment, low intensity (b) intense beams for spectroscopy and deposition, universal tool
Laser vaporization source (LVS)	Most solid materials, liquid metals are possible	Up to 2 nm	Pulsed	Spectroscopy with pulsed lasers, alloy formation with two lasers and two targets
Pulsed arc cluster ion source (PACIS)	Conducting solid materials	Up to 2 nm	Pulsed	Spectroscopy with pulsed lasers, deposition
Arc cluster ion source (ACIS)	Conducting solid materials	Up to 30 nm	cw	Intense beams for deposition, good for low-cost materials as large targets are needed

Note: The given particle sizes are rough estimates and can strongly vary for particular setups and different modes of operation.

sources is their extraordinary temporal stability. In strong-field experiments with dual or shaped pulses, particle density and the size distribution have to be kept as constant as possible. The reason is the need to record the system response while tuning the laser parameters; some results below have been obtained with clusters embedded in helium droplets.

One example of a complete experimental setup is sketched in Figure 13.6. A Haberland-type magnetron gas aggregation source delivers a constant flux of particles, with variable cluster sizes from a few atoms up to particles with 15 nm diameter ($\langle N \rangle \approx 10^5$). With aerodynamic focusing the particle intensity at the interaction point can be enhanced whereby achieving a narrowing of the size distribution in the large nanometer regime (Passig et al., 2006).

After strong-field excitation the clusters completely disintegrate. The nascent ion signals comprise cluster fragments and atoms of various charge states. The detection of the fragment mass and the ion charge state is accomplished by mass spectrometry. We use a reflectron TOF mass spectrometer (Mamyrin, 1994) downstream the cluster beam to identify the products. In the reflectron, ions with low-recoil energies are detected with the highest yield and resolution. Highly charged ions are partially discriminated, such that the measured spectra do not fully reflect the real charge-state distributions. Ion recoil energies are monitored by the linear vertical TOF. Without any fields the flight times toward the detector are measured and converted into kinetic energy spectra. With the same instrument electrons can also be analyzed. Now a magnet under the interaction point and a magnetic guiding field complete a magnetic bottle-type

photoelectron spectrometer (Kruit and Read, 1983). Field-free electron detection is possible as well at the expense of the collection angle. With this configuration, angular-dependent photoemission can be analyzed by rotating the laser polarization with respect to the electron-detection axis.

For a thorough comparison with theory it is helpful to simultaneously analyze the ion charge state and energy. This is not possible with TOF measurements alone. A further parameter is needed, e.g., a magnetic deflection in order to resolve the ion momentum. One useful method goes back to Thomson (1907). Figure 13.6 sketches a Thomson analyzer used for the simultaneous measurement of the energy and the charge of ions expelled from exploding clusters. It consists of parallel electric and magnetic fields, followed by a field-free drift zone in connection with a position-sensitive ion detector. The experimentally obtained raw Thomson deflection data give the momentum and energy per charge. Thus they have to be transformed into energy *vs.* charge-state spectra, which can be achieved in a unique way. For Ag clusters, for example, it turns out that the energy distribution of the ions as a function of the charge state is rather narrow. Moreover, the maximum recoil energy rises almost linearly with the charge state (Döppner et al., 2003).

With these examples of experimental methods we now turn to selected results on strong-field dynamics. There are more decay channels like the emission of VUV and x-ray radiation. Also, the details of several optical methods applied nowadays go beyond the scope of this chapter. Among those are the briefly mentioned excitation with adaptively controlled laser pulses, the excitation with VUV-free electron lasers, or the combined and controlled action of pulsed infrared and VUV fs radiation.

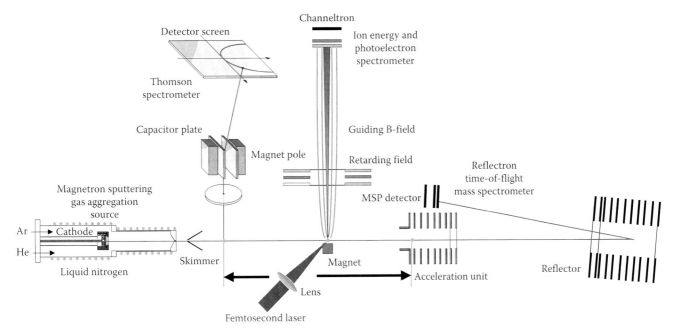

FIGURE 13.6 Experimental setup for the investigation of the strong laser pulse interaction with metal clusters. Magnetron sputtering serves to vaporize the target material. The magnetron is located within a LN_2-cooled tube and clusters form by gas aggregation in a 1 mbar mixture of helium and argon buffer gas. The beam expands into vacuum through a nozzle (0.5–3.0 mm). Ion and electron emission resulting from laser excitation at selected interaction points can be analyzed by different TOF methods. In addition, a Thomson spectrograph enables the simultaneous analysis of charge state and recoil energy of the ionic products.

13.5 Results on Strong-Field Dynamics

In the following results of experiments on laser–cluster dynamics in the high-intensity regime are reviewed. We first concentrate on basic findings on absorption and ionization processes. After that, selected results on the dynamics including angular-resolved electron emission in a dual-pulse setup will be described.

13.5.1 Absorption of Radiation

A remarkable property of clusters in intense laser fields is their capability to efficiently absorb radiation. One way to directly probe the energy capture of a cluster target is laser excitation close to the exit of the source at dense target conditions. The

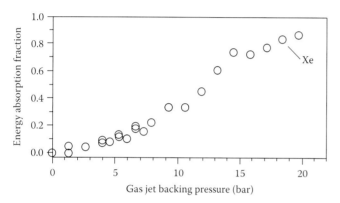

FIGURE 13.7 Energy absorption of Xe_N exposed to intense laser pulses (7×10^{16} W/cm²; 2 ps; 527 nm) as a function of backing pressure. Corresponding Xe_N sizes can be calculated from the Hagena formula (Hagena, 1987). The maximum cluster diameter is approximately 80 Å. A considerable fraction of the incoming laser energy (up to 90%) is absorbed by the cluster beam. (Adapted from Ditmire, T. et al., *Phys. Rev. Lett.*, 78, 3121, 1997a. With permission.)

energy absorption fraction $\mathcal{A}_L$, defined as the depletion ratio of the pulse, may be considered as a measure. The cluster size, density, material, and laser parameters influence the efficiency of the absorption. As an example, Figure 13.7 shows $\mathcal{A}_L$ of Xe_N as function of backing pressure, which is a measure of the cluster size.

Obviously the cluster target becomes increasingly opaque with backing pressure, whereas an uncondensed beam with the same average atomic density would remain practically transparent. This was also demonstrated for Ar_N and $(H_2)_N$ (Ditmire et al., 1997a). Further investigations show equivalent trends, i.e., high absorption with deuterium and krypton clusters (Ditmire et al., 1999; Miura et al., 2000). The impact of light scattering, which also acts as an extinction channel besides absorption, was reported to be insignificant under typical experimental conditions (Ditmire et al., 1997c). In all cases, similar to the behavior of Xe_N in Figure 13.7, the absorption scales almost linearly with the stagnation pressure and then saturates, see also Jha et al. (2006).

The value of $\mathcal{A}_L$ shows a strong dependence on the laser parameters like pulse duration and timing as well as on the cluster size. Early experimental evidence for the importance of the laser pulse timing for energy absorption was found in Zweiback et al. (1999) (see Figure 13.8) where the laser energy absorption in a Xe cluster beam is measured as a function of the laser pulse length and the time delay between the double pulses.

In panel (a) the laser pulse width is varied, leading to a clear peak in the absorption at an optimal length. This optimum is believed to reflect the most efficient conditions for resonant plasmon heating of the systems, see Section 13.2. The optimal duration increases with cluster size, showing that larger clusters require more time to maintain frequency matching. The panel (b) shows the result of a dual-pulse experiment. The leading pulse is assumed to result in moderate ionization and

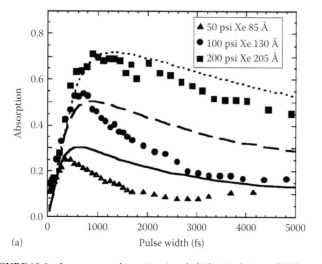

(a)

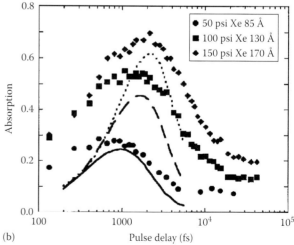

(b)

FIGURE 13.8 Laser power absorption (symbols) by Xe clusters of different size (as indicated) for (a) single-pulse and (b) dual-pulse excitation at λ = 810 nm. The single-pulse experiment was performed with a constant pulse energy of 6.5 mJ, resulting in a peak intensity of 2.3×10^{17} W/cm² at 50 fs pulse duration. The dual-pulse data were obtained with 50 fs pulses of peak intensities $I_{pump} = 1.6 \times 10^{16}$ W/cm² and $I_{probe} = 1.6 \times 10^{17}$ W/cm². The curves represent results using the hydrodynamic model from Ditmire. (Adapted from Zweiback, J. et al., *Phys. Rev. A*, 59, R3166, 1999. With permission.)

heating, which, in turn, leads to a steady cluster expansion due to the Coulomb forces and thermal pressure. At a certain time, the radius reaches its critical value, i.e., a value where the laser frequency matches the plasmon, leading to a strong absorption peak. And again, the optimum time increases with system size. The trends seen in both experiments are qualitatively reproduced by calculations using a hydrodynamic model, see curves in Figure 13.8.

The treatment of the optical cluster response in terms of a single and time-dependent collective mode is, however, strongly oversimplified. A more realistic picture of the absorption dynamics is attained in simulations that integrate the fully coupled electron and ion motion under the influence of the laser field. For high laser intensity this can be accomplished with MD simulations.

As an illustrative example, Figure 13.9 displays results from an MD simulation of Xe_{5083} exposed to a 250 fs pulse of 10^{15} W/cm², for methodic details see Fennel et al. (2007b). The upper panel shows the energy absorption (dashed line) and the effective absorption cross section derived therefrom (solid line). Similar to the results of Saalmann (2006), the total energy absorption is high, rising up to more than 50 keV per atom. The cross section shows an early feature at $t \approx -240$ fs that is related to the buildup of the nanoplasma in the initial phase due to an avalanche-like inner ionization, triggered by optical field ionization and further supported by electron-impact ionization. The quickly growing

number of quasifree electrons results in a fast increase of the critical radius. This is accompanied by a short period of resonant coupling when R_{crit} crosses the actual system radius (see panel b), leading to the first feature in the absorption cross section in the rising edge of the pulse. Some outer ionization and thermal excitation induce steady cluster expansion. The second broad maximum in the absorption then emerges when the mean cluster radius, $\tilde{R}$, approaches again the critical radius, leading to a maximum cross section of 0.75 Å² per atom. The strong thermal excitation of the nanoplasma during irradiation is evident from the high average kinetic energy of cluster electrons (electrons within the cluster radius), see dashed curve in Figure 13.9c. Values of more than 1 keV are attained shortly after the strong resonance. The maximum kinetic energies of up to 6 keV further indicate the transient development of a deep space-charge potential as a result of ionization and the thermally excited cluster electrons. We will come back to this point later on in connection with x-ray emission.

The main absorption feature observed in the MD simulations is relatively broad (FWHM ≈ 120 fs). This large width can to some extent be linked to an inhomogeneous expansion of the cluster, see e.g. (Milchberg et al., 2001), and a thus time-delayed resonant absorption in radial shells of critical density. A comparison of the radial position of the outermost ion R_{oi} with $\tilde{R}$ in Figure 13.9b displays that outer ions indeed expand more quickly. This is due to a less effective screening of ions near the cluster surface

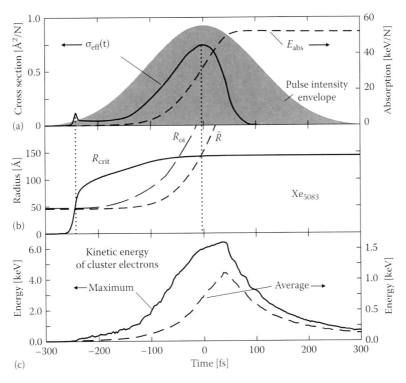

FIGURE 13.9 Simulated dynamics of Xe_{5083} exposed to a laser pulse (10^{15} W/cm²; 250 fs; 800 nm) using the MD code from Fennel et al. (2007b). (a) Pulse intensity profile, cycle-averaged energy capture, E_{abs}, and associated absorption cross section, $\sigma_{eff} = \dot{E}_{abs}/I$. (b) Mean cluster radius $\tilde{R} = \sqrt{5/3}\, R_{rms}$, where R_{rms} is the root-mean-square radius, and critical radius R_{crit} for resonant coupling. The vertical dotted lines mark the match between R_{rms} and R_{crit} giving a large energy absorption. The difference between $\tilde{R}$ and R_{oi} (radial position of the outermost ion) indicates inhomogeneous cluster expansion. (c) Average and maximum kinetic energy of electrons within the cluster radius $\tilde{R}$.

(Peano et al., 2006). Further, the driving of electrons beyond the cluster surface introduces broadening due to nonlinear damping (Jungreuthmayer et al., 2004; Megi et al., 2003). It should be noted, that for very high laser intensity a nonlinear resonance has also been discussed, see e.g., Kundu and Bauer (2006) and Mulser et al. (2005), which applies even for overcritical cluster density.

The quick drop in the cross section in Figure 13.9a after the pulse peak reflects partial depletion of the nanoplasma as well as the substantial decrease in the coupling efficiency at *undercritical* density (see panel b). Collective enhancement dies out and pure IBS heating of residual electrons is weak due to rare electron–ion collisions. Further, the expansion of the ionic background leads to an efficient electron cooling, see Figure 13.9c.

The above analysis illustrates that in general the rather complex absorption dynamics is dominated by collective energy absorption near the critical ion charge density. The expansion of the cluster to the radius R_{crit} therefore sets a crucial time scale for strong-field laser–cluster interactions in the IR regime, as will become even more apparent in the following sections. Only at very high intensities, where the laser field exceeds the restoring force from the background potential, such resonance effects can be disregarded (Heidenreich et al., 2007; Krainov and Smirnov, 2002).

13.5.2 Ionization Dynamics

The strong optical absorption leads to high ionization and usually complete disintegration of the clusters. Mostly ions with high ionization stages, q, are finally detected (see Figure 13.10 for an ion spectrum from the excitation of Pb_N).

The observed $q^{max} = 30$ is much higher than the charge states from atomic Pb ($q = 3–5$) under similar conditions. It should be noted that the shape of the distribution in Figure 13.10 may be affected by the TOF method and the intensity profile of the focused laser beam. Closer to the focus, the intensity increases on the expense of the irradiated volume. The measured distribution thus contains ion signals from regions of different

pulse intensities. Only recently, it has been demonstrated that various intensity contributions can be deconvoluted and be traced back to charge-state-resolved threshold intensities (Döppner et al., 2007b). Even higher ionized species with $q^{max} = 40$ from Xe_N were reported by (Ditmire et al., 1997a) for 65 Å particles. The cases of various metal clusters were furthermore explored in several papers (Köller et al., 1999; Radcliffe et al., 2005; Schumacher et al., 1999), leading to values for q^{max} between 15 and 28 for Ag_N, Au_N, Pt_N, and Pb_N at intensities below 10^{16} W/cm^2.

Dedicated studies on embedded atomic clusters have been performed to investigate the effect of the environment on ionization. Strong-field excited metal clusters in helium nanodroplets show evidence for charge transfer processes (Döppner et al., 2007a). Here the environment reduces the final ion charge state of the dopant by electron capture from helium. On the other hand, the environment may also contribute actively to the energy capture, i.e., due to strong heating through a collective resonance within the matrix (Mikaberidze et al., 2008).

Whereas most of the results have been obtained with optical lasers, first experiments are at hand making use of the VUV-free electron laser (VUV-FEL) (Laarmann et al., 2004; Wabnitz et al., 2002). Power densities of up to 3×10^{13} W/cm^2 at, e.g., 98 nm (=12.65 eV) were used in the studies on rare-gas clusters. Note that because of the low ponderomotive potential, IBS heating is strongly suppressed, suggesting multiphoton ionization conditions. Moreover, collective heating can be ruled out because of the high laser frequency. Ions with charge states up to Xe^{8+} and Ar^{6+} have been detected. The reason for the high charging, and the therefore required energy absorption, is still under discussion. Various concepts have been proposed, ranging from more realistic scattering potentials for IBS heating (Santra and Greene, 2003), over local-field enhancement of inner ionization by neighboring ions, to many-body heating effects (Bauer, 2004; Jungreuthmayer et al., 2005).

Due to the strong charging on the fs timescale substantial Coulomb energy is accumulated, leading to high recoil energies in the subsequent explosion of the system. This has opened the route to table-top experiments on cluster-based fusion (Ditmire

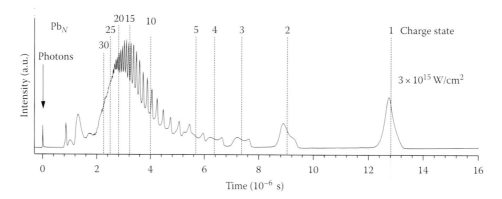

FIGURE 13.10 Mass spectrum of highly charged ions after exposure of Pb_N to intense laser radiation of 3×10^{15} W/cm^2. In this experiment, the charge states reach up to more than $q = 30$. (After Lebeault, M.A. et al., *Eur. Phys. J. D*, 20, 233, 2002. With permission.)

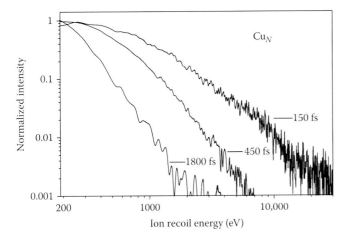

FIGURE 13.11 Normalized ion energy spectra of copper clusters ($\langle N \rangle \sim 1000$) exposed to laser pulses of different duration at constant fluence. The most energetic ions are obtained at short pulse conditions, i.e., 1.8×10^{16} W/cm^2 at 150 fs. (Adapted from Radcliffe, P. et al., *Contrib. Plas. Phys.*, 45, 424, 2005.)

et al., 1999). In one of the first experiments, atomic ions from Xe$_N$ with kinetic energies beyond 1 MeV were observed (Ditmire et al., 1997b). Figure 13.11 displays more recent recoil energy spectra obtained from Cu$_N$ for different laser pulse durations (Radcliffe et al., 2005).

The distributions are rather broad and most of the ions have energies in the range below 1 keV. The yield rapidly decreases toward higher energies, whereas the maximum energy increases with cluster size (not shown here). Several effects contribute to the shape of the spectra, i.e., the temporal and spatial laser intensity profiles, the degree of cluster ionization, and the cluster size distribution. These effects have been incorporated into a simple model (Islam et al., 2006) based on Coulomb explosion.

Considering a uniformly charged monoatomic cluster, the final kinetic energy E$_I$ of an atomic ion is determined by its initial potential energy (Last et al., 1997; Nishihara et al., 2001; Zweiback et al., 2000)

$$E_I(r) = \frac{4\pi}{3} n_I r^2 q^2 \times 14.4 \text{ eV Å} \qquad (13.19)$$

where

r is the initial radial position of the ion
q the charge state of the ions
n_I the number density of ions in the cluster

Therefore, ions located at the cluster surface acquire the highest recoil energies E$_I^{max}$, whereas this maximum value should also increase with cluster size. This trend was experimentally observed by several groups, e.g., for Xe$_N$ and Ar$_N$ by Ditmire et al. (1997a), Lezius et al. (1998), and Li et al. (2003). In Krishnamurthy et al. (2004), a monotonous rise of E$_I^{max}$ from below 500 eV to about 10 keV was found for nitrogen clusters when increasing the size from $N = 50$ to $N = 2300$. In Pb$_N$, an increase of the maximum

kinetic energy from 70 eV at $\langle N \rangle = 100$ to 180 keV at $\langle N \rangle = 500$ has been observed (Teuber et al., 2001). Interestingly, for rare-gas clusters a sharp rise in E$_I^{max}$ occurs as a function of laser intensity. For Xe$_N$, a threshold of about 5×10^{14} W/cm^2 was reported (Tisch et al., 2003), which is roughly comparable with the threshold intensity for BSI in atoms (see Equation 13.4) and thus supports the picture of an ignition-like ionization.

An enhancement in the recoil energies as well as the ion yield at a given energy was observed in clusters containing spurious amounts of appropriate dopants (Jha and Krishnamurthy, 2008; Jha et al., 2006; Purnell et al., 1994). For instance, Ar clusters ($\langle N \rangle = 2000$) containing about 60 H$_2$O molecules were considered in Jha et al. (2006). Under exposure to pulses with 1×10^{16} W/cm^2, the intensity in the recoil energy spectrum scales up by a factor of three at 100 keV, as compared to the dopant-free case. The authors attribute their findings to the reduced ionization thresholds of H$_2$O as compared to Ar.

13.5.3 Time-Resolved Particle and Photon Emission

Besides the mere occurrence of the highly charged ions, as shown above, their ionization stage distribution can significantly be controlled by the radiation field. Several groups reported on enhanced ionization yields as well as higher maximum charge states for certain pulse durations (Döppner et al., 2000, 2007b; Fukuda et al., 2003; Köller et al., 1999; Lebeault et al., 2002; Schumacher et al., 1999). As a typical result, Figure 13.12 displays the highest

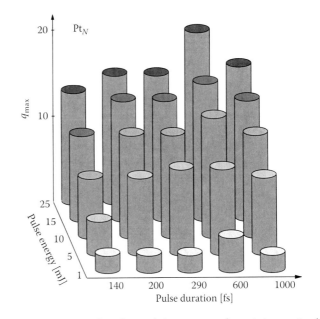

FIGURE 13.12 Highest detected charge states of atomic ions emitted from platinum clusters after exposure to laser pulses of variable pulse duration and given pulse energies (1–25 mJ). At the highest pulse energy, the laser peak intensity is 2.6×10^{16} W/cm^2 for the shortest pulse (140 fs), and 10^{15} W/cm^2 for the optimal duration (600 fs). The highest charge state of $q_{max} = 20$ is obtained with longer pulses, thus at reduced intensity. (Adapted from Köller, L. et al., *Phys. Rev. Lett.*, 82, 3783, 1999.)

ionization stages from small Pt_N s a function of pulse width while the *fluence* was kept constant in each trace (Köller et al., 1999). The shortest and most intense pulse (140 fs) yields atomic ions up to $q = 13$. With increasing pulse duration, the maximum charge state grows toward a maximum for an optimal pulse width of 600 fs, where atomic ions up to $q = 20$ can be identified. When further increasing the pulse duration, the maximum charge states (as well as the overall signal) degrade. Obviously, similar to laser energy absorption described in the preceding section, the occurrence of high charge states as well as the integral ion yields can be controlled by the laser pulse timing.

The efficient charging for a certain pulse duration is in accordance with the above sketched model of delayed resonant heating. Another mechanism for the enhanced ionization was proposed by Siedschlag and Rost (2003) and relies on the concept of CREI know from diatomic molecules (see Section 13.2), which was also considered for multiple ionization of clusters (Last and Jortner, 1998). Within this model, the increased ionization probability occurs for an optimal interatomic distance, where the tunneling barrier between neighboring ions and the outer cluster Coulomb barrier are reduced at the same time. The optimal pulse duration can thus be related to the instant at which the expanding cluster reaches the optimal interatomic distance. However, because of the large outer Coulomb barriers in highly charged clusters, enhanced ionization due to CREI is considered to be relevant primarily for small systems. Plasmon-enhanced ionization, on the other hand, applies to clusters of a broad size range (Döppner et al., 2005; Reinhard and Suraud, 2001; Saalmann, 2006; Saalmann and Rost, 2003; Suraud and Reinhard, 2000).

There are several other experiments showing the strong influence of pulse duration on the ionization yields and recoil energies. For example, in experiments on Xe_N it was found that the mean ion energy grows rapidly with pulse duration up to an optimal value of about 500 fs (Fukuda et al., 2003), in accordance with the picture of a delayed resonance. Moreover, also the temporal phase of the pulse plays an important role. A significant influence of the sign of the chirp on the ion energies has been found. A negative chirp, i.e., a decreasing laser frequency with time, enhances the mean ion energy by 60% over the result for positive chirp. Thus, the negative chirp induces higher total ionization or more efficient thermal excitation. This can qualitatively be explained by a longer time span for resonant collective absorption because of the joint gradual frequency red-shift of both laser pulse and resonance, the latter induced by the expansion of the system.

Not only the ionization but also the electron emission (Döppner et al., 2006; Kumarappan et al., 2002, 2003b; Springate et al., 2003) and the x-ray emission significantly depend on the laser pulse length (Chen et al., 2002; Issac et al., 2004; Jha and Krishnamurthy, 2008; Lamour et al., 2005; Parra et al., 2000; Zweiback et al., 1999). Since the x-ray spectrum shows line emission from transitions between certain atomic levels, they contain information on the transient charge-state distributions and the core levels being attacked during the interaction process. As an example, Figure 13.13 displays the Ar K-shell x-ray yield after excitation of Ar clusters (clean and doped with water).

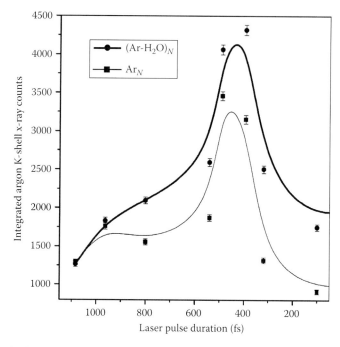

FIGURE 13.13 Ar K-shell x-ray yield from pure and water-doped Ar_N irradiated with stretched laser pulses of fixed energy at 800 nm wavelength. A pulse width of $\tau = 500$ fs corresponds to a peak intensity of 10^{15} W/cm². Lower symbols: clean Ar_{9000}, upper symbols: water-doped clusters. (After Jha, J. and Krishnamurthy, M., *Appl. Phys. Lett.*, 92, 191108, 2008. With permission.)

The figure shows a steep rise in the x-ray yield for pulse durations up to 400 fs, indicating a strong increase of the number of multi-keV electrons in the cluster at this certain pulse width. The same trend is found with the doped clusters, where the higher overall yield can be explained by the reduced ionization threshold of the dopant. In any case, fast cluster electrons are required to create the core-level vacancies by electron impact. Their particularly efficient creation at the instant of resonant heating was observed in the simulation results shown in Figure 13.9c for a comparable pulse length. Coming back to Figure 13.13, the signal substantially decreases behind the maximum, i.e., for even longer pulses. The thus arising optimal duration is a strong indication for the most efficient resonant collective electron heating (see also Parra et al. (2000) and Zweiback et al. (1999)). Similar trends are observed for EUV emission (Chen et al., 2002). The enhanced generation of cluster-bound multi-keV electrons required for x-ray emission under resonant coupling conditions is also theoretically supported by Saalmann and Rost (2005). An alternative electron heating mechanism, namely multiple large-angle electron-ion backscattering in phase with the laser field, was proposed in Deiss et al. (2006) in order to explain x-ray production with pulses that are too short for reaching resonant conditions.

However, there still are many pending questions. For example, details of ionization processes over intermediate multi-electron excited states, or the impact of multi-electron collisions on the production of core vacancies have not been resolved satisfactorily so far. Therefore, the physics behind x-ray emission from clusters remains a fascinating subject for further studies and possible technical applications. Nevertheless, in spite of the lack of a full theoretical understanding, laser-excited clusters already have proven to be powerful table-top soft x-ray sources. Their technical application was demonstrated, e.g., with CO_2 clusters that delivered photon fluxes in the range of 10^{10} photons/sr per pulse at 653.7 eV (Fukuda et al., 2008).

13.5.4 Dual-Pulse Excitation and Angular-Resolved Emission

The emission spectra discussed so far correspond to experiments with one single pulse of variable duration. According to the previous simulations, scenarios are conceivable with separated pulses using a pump-probe type of experimental setup. A nonresonant leading pulse moderately heats and inner ionizes the system, initiating an expansion of the cluster. By varying the optical delay of the subsequent pulse the time-dependent response is then mapped into the emission spectra. Results from a dual-pulse experiment on Ag clusters are given in Figure 13.14 showing the yield of Ag^{10+} and the maximum energy of the emitted electrons as a function of pulse delay. Both signals show pronounced maxima for delays of about 12 ps. Analogously to the findings from the semiclassical simulations in Section 13.3.1, the measurements show that cluster activation and enhanced ionization can be disentangled (see also the theoretical work of Bornath et al. (2007b), Martchenko et al. (2005), and Siedschlag and Rost (2005)). The coincidental appearance of high ion yield and energetic electrons in Figure 13.14 underlines the importance of resonant heating for both of the decay channels. It should further be noted, that the observed electron energies exceed

the ponderomotive potential (here $U_p = 4.8$ eV) by almost two orders of magnitude.

Besides yields and energies, also the angular distributions of emitted species provide valuable insight into the coupling dynamics. Ion energy spectra exhibit a clear directional asymmetry and higher kinetic energies for an emission along the laser polarization axis. This was reported for Xe_N (Kumarappan et al., 2002; Springate et al., 2000a) and for Ar_N (Hirokane et al., 2004; Kumarappan et al., 2001). Polarization enhancements in the ion energy between 15% and 40%, depending on pulse duration, are typical for rare gas as well as molecular systems (Krishnamurthy et al., 2004; Kumarappan et al., 2001; Li et al., 2005; Mathur and Krishnamurthy, 2006; Springate et al., 2000b). Three fundamentally different models were proposed to explain the asymmetry in ion emission. In Ishikawa and Blenski (2000), a mechanism is proposed where the additional acceleration of ions is a direct result of the laser field. Since the net effect of the laser averages out over a laser period for ions with constant q, one has to postulate rapid charge-state oscillations of ions at the surface such that higher charges appear during laser half-cycles with outward field components. Hence, maximum repulsion from the cluster space-charge field plus the laser occurs along the laser polarization axis. However, this mechanism is unlikely to fully explain the experimentally observed asymmetry since the rates for electron–ion recombinations are low at the typically high electron temperatures (Bethe and Salpeter, 1977). The second mechanism is an asymmetric Coulomb explosion due to inhomogeneous charging of ions in the cluster. This effect has initially been considered to explain the asymmetric ion emission from C_{60} (Kou et al., 2000). At the cluster poles (surface normal parallel to the laser field), the superposition of fields from laser, polarization, and space-charge can produce high total electric field gradients, which leads to increased ionization. Thus, ions located in this region experience stronger Coulomb repulsion. This view of enhanced

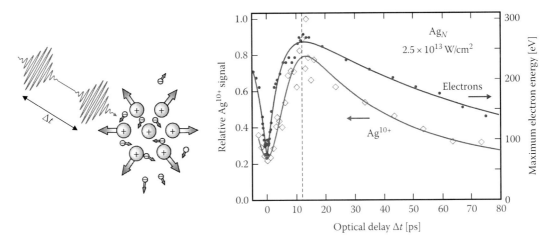

FIGURE 13.14 Comparison of the Ag^{10+} yield (diamonds, left axis) with the maximum kinetic energy of the emitted electrons (dots, right axis) following laser excitation of Ag clusters ($\langle N \rangle \approx 2 \times 10^4$) with dual 100 fs laser pulses. The curves are guides to the eye fits. Both the electron and the ion signals exhibit a strong dynamics and their maxima occur nearly simultaneously (see dashed vertical line). (After Döppner, T. et al., *Phys. Rev. A*, 73, 031202(R), 2006.)

ion acceleration along the polarization axis is supported by numerical simulations (Fennel et al., 2004; Jungreuthmayer et al., 2004) as well as the observed asymmetric high-q emission (Hirokane et al., 2004). Finally, additional forces from the cluster polarization field enhance asymmetric ion acceleration (Breizman et al., 2005; Fennel et al., 2004; Kumarappan et al., 2002). These are particularly large for resonant collective excitation and can thus explain the measured pulse length dependence. The driven oscillation of the electron cloud results in a nonvanishing asymmetric contribution to the radial component of the electric field at the cluster surface. In consequence, enhanced repulsion is supported for surface ions near the cluster poles (Breizman et al., 2005).

Compared with the ions, the asymmetry is much more pronounced with electrons. Strong alignment along the laser polarization axis has been observed by several authors, e.g., Kumarappan et al. (2002), Shao et al. (1996), and Springate et al. (2003). This preferential ejection is a direct marker for laser-assisted and nonthermal emission and turns out to be strongly dependent also on the pulse duration. On Xe_N (Kumarappan et al., 2003a) find a ratio of parallel to perpendicular electron yields of three for optimal pulse durations of about 1 ps. Almost isotropic emission and less energetic electrons are observed for the shortest and most intense pulses. The authors related this effect to resonant collective enhancement of the polarization field. The increase in asymmetry for optimal pulse conditions is also supported by simulations (Martchenko et al., 2005).

A particularly strong alignment effect has been observed in a dual-pulse experiment on Ag clusters (Fennel et al., 2007a). Two pulses with optimal separation yield simultaneously higher electron energies and stronger asymmetry as compared to single-pulse excitation, see Figure 13.15a. The comparison of parallel and perpendicular electron yields for different energy windows as a function of pulse delay (Figure 13.15b) shows that the asymmetry increases with electron energy. The strongest anisotropy of about 6.5:1 (parallel over perpendicular) is found for the most energetic electrons, see uppermost panel in Figure 13.15b. For all chosen energy windows a maximum yield is observed for similar delays, again supporting the presence of plasmon-enhanced ionization. VUU calculations on the model system Na_{147}, presented in Figure 13.16, show the same qualitative behavior (Fennel et al., 2007a). An off-resonance excitation induces low-energy electron emission and only a smaller asymmetry (Figure 13.16b), while a resonant dual-pulse excitation results in energetic electrons and a clear preference along the polarization axis (Figure 13.16a).

From an analysis of the trajectories it was found that the laser-assisted rescattering of certain electrons by the dynamic cluster potential is crucial for the high-energy part of the spectrum. The electrons gain their energy from the combined field of the laser and the dynamic cluster polarization within a final passage through the cluster. Since the process is linked to the phase of the laser field, the electron emission is strongly modulated and leads to energetic and short electron bursts. This process of surface-plasmon-assisted rescattering in clusters (SPARC) supports preferential ejection along the laser polarization axis and provides a principle explanation of high acceleration of electrons to kinetic energies far beyond the ponderomotive potential. Further it can explain the increased energies at resonance due to the higher polarization fields. In the resonant case in Figure 13.16a the peak of the effective field gradient from polarization and the laser field gradient reaches an impressive value of 35 GV/m. Note that this corresponds to an effective intensity 25 times larger than that of the laser field alone in this particular example.

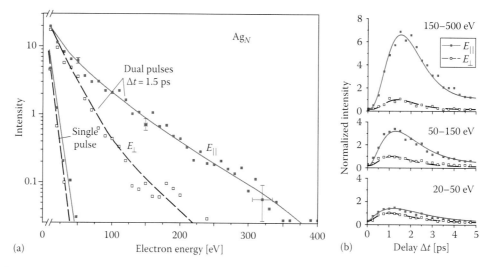

FIGURE 13.15 Photoemission spectra from silver clusters ($\langle N \rangle \approx 10^3$) exposed to 100 fs laser pulses with peak intensity 8×10^{13} W/cm² at 800 nm wavelength: (a) Energy-resolved emission parallel ($E_\parallel$) and perpendicular ($E_\perp$) to the laser polarization axis for excitation with a single pulse and dual pulses with an optimal temporal delay of $\Delta t = 1.5$ ps. (b) Integrated signals for three electron energy intervals (as indicated) and normalized to the maximum obtained for ($E_\perp$) as a function of pulse delay. (After Fennel, T. et al., *Phys. Rev. Lett.*, 98, 143401, 2007.)

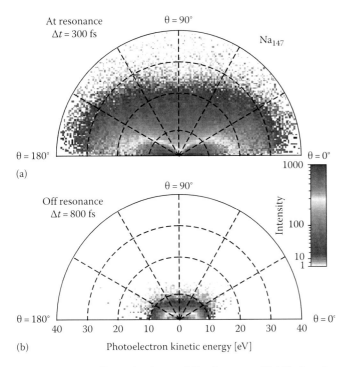

(a) At resonance $\Delta t = 300$ fs Na$_{147}$

(b) Off resonance $\Delta t = 800$ fs

Photoelectron kinetic energy [eV]

FIGURE 13.16 (See color insert following page 25-14.) Angular resolved electron emission spectra from Na$_{147}$ exposed to 25 fs laser pulses (800 nm) with peak intensity 8×10^{12} W/cm^2, as calculated from semiclassical VUU-MD simulations, cf. Section 13.3.1. The data correspond to dual-pulse excitations at (a) optimal and (b) a longer nonresonant delay. The emission angle, θ, is given with respect to the laser polarization axis. The plot is based on the data from Fennel et al. (2007a) but shown with a modified intensity scaling.

13.6 Conclusion and Outlook

Subjecting matter to intense electromagnetic radiation induces many interesting phenomena. Among those are collective electronic excitations, plasma formation, and nuclear reactions. Here we concentrated on the intensity range of 10^{13}–10^{16} W/cm^2, typically at 800 nm wavelength. These intensities, which can be supplied by table-top fs-laser systems, are particularly interesting as they allow the detailed study of electron and ion dynamics. The investigated systems of choice comprise free clusters of atoms that have been prepared in molecular beams. Equipped with appropriate detectors the experiments uncover details of the light–matter interaction as well as the nascent motion of atoms and electrons. Corresponding high-level numerical simulations go along with the measurements.

This chapter has shown that the scientific area of strong-field cluster excitations is in an actively developing state and many more interesting results can be expected. Several exciting prospects of *nanoplasma research* are fuelled by novel technical developments. Two promising directions emerge from recent progress in laser technology: First, pulses shaped in amplitude and phase offer new routes for the targeted control of the laser–matter interaction. Second, intense radiation

with high photon energy delivered by free electron lasers gives access to novel excitation regimes with very different heating and ionization mechanisms.

Shaping the pulse in amplitude and phase can be a fascinating tool to selectively steer the dynamics of charging, particle, or photon emission. Basic findings along this line are the control of the Coulomb explosion by varying the laser pulse length as well as the time delay in the dual-pulse experiments as outlined above. For example, Figure 13.14 has shown the dramatic effect of the adjustment of two ultrashort laser pulses on the charging efficiency and the energy of emitted electrons. In a first application of adaptive fs, control on clusters (Zamith et al., 2004) could optimize the Coulomb explosion of Xe$_N$. Here the signal of highly charged Xe^{q+} has been controlled by shaping an initially Fourier-transform-limited 100 fs pulse with the help of a genetic algorithm. The procedure converged to a field envelope containing two 120 fs pulses of similar amplitude separated by about 500 fs. It should be emphasized that this pulse structure, which is similar to that from dual-pulse experiments (Döppner et al., 2005), has been worked out by the algorithm starting from an 80-parameter unbiased configuration. Corresponding simulations within a semiclassical MD approach predicted that for selected combinations of cluster size, laser intensity, and laser wavelength, the ionization may be optimized by a three-pulse sequence (Martchenko et al., 2005). In another closed-loop optimal control experiment on rare-gas clusters, pulsed x-ray emission could significantly be enhanced (Moore et al., 2005). Only recently, similar studies could be extended to silver clusters embedded in helium droplets (Truong et al., 2010). Here the light fields shaped in amplitude and phase lead to a substantial increase of the Ag^{q+} yield for $q > 10$ when compared to bandwidth-limited and optimally stretched pulses. A remarkably simple double-pulse structure, compsed of a low-intensity prepulse and a stronger main pulse, turns ot to produce the highest atomic charge states up to Ag^{20+}. The phase and the corresponding instantaneous wavelength develop such that the photon energy appears to follow the temporal change of th plasmon. In other words, the control experiment finds the best solution for the time-dependent amplitude and phase structure in order to drive the system into the highest possible charging. Moreover, in the meantime it became possible to tailor the charge state distributions nearly at will. However, now the resulting laser field turn out to be extremely complex. Further experimental and theoretical effort is essential to bring progress in this exciting but challenging science.

The nature of the laser–cluster coupling fundamentally changes when going from the IR regime toward excitation with VUV, XUV, or even x-ray pulses. As indicated by the increasing Keldysh parameter (Equation 13.2), vertical photoionization of cluster constituents becomes dominant over field ionization processes. For the 100 nm wavelength pulses used in the first VUV experiments on rare-gas clusters (Wabnitz et al., 2002), collective effects can be disregarded as the required critical density cannot be reached. Consequently, mostly pure IBS heating prevails. Nevertheless, the observation of surprisingly high ion charge states in these experiments has sparked remarkable efforts in theoretically understanding the underlying heating and ionization processes (Bauer, 2004; Georgescu et al., 2007b; Jungreuthmayer et al., 2005; Ramunno et al., 2006; Saalmann et al., 2006; Santra and Greene, 2003; Siedschlag and Rost, 2004; Ziaja et al., 2007). When further increasing the laser frequency, IBS heating becomes more and more suppressed (cf. Equation 13.5) so that the photoexcitation of tightly bound electrons even becomes the leading energy-capture process. Signatures from this transition have recently been observed on Ar_N in intense fs XUV FEL pulses at $\hbar\omega_{las} = 38\,eV$ (Bostedt et al., 2008). The experimental photoelectron spectra could be well reproduced with Monte-Carlo simulations, in which IBS heating is explicitly switched off.

Another interesting issue concerns the possible time-resolved monitoring of the cluster excitation and the subsequent Coulomb explosion by combining different types of pulses. For example, the ionization of rare-gas clusters may be driven by VUV radiation, whereas a subsequent IR pulse could probe the collective electron response of the inner-ionized system (Siedschlag and Rost, 2005). A combination of VUV and XUV pulses was proposed to monitor the time-dependent ionization stages in small clusters (Georgescu et al., 2007a). Another not yet realized scheme suggests to apply x-ray radiation for time-resolved Thomson scattering on exploding clusters or droplets, which have been initially excited by strong IR pulses (Höll et al., 2007). By this, a fundamental understanding can be gained on highly nonstationary strongly coupled nanoplasmas. Thus, the advent of the x-ray free electron lasers will open direct access to the temporal development of such complex systems.

Acknowledgments

The authors gratefully acknowledge financial support by the DFG Collaborative Research Center 652 at the University of Rostock. Additional computer resources were provided by the High Performance Computing Center for North Germany (HLRN) and the High-Performance Computational Virtual Laboratory (HPCVL).

References

Ammosov, M., N. Delone, and V. Krainov, 1986, *Sov. Phys. JETP* **64**, 1191.

Assion, A., T. Baumert, M. Bergt, T. Brixner, B. Kiefer, V. Seyfried, M. Strehle, and G. Gerber, 1998, *Science* **282**, 919.

Augst, S., D. Strickland, D. Meyerhofer, S. Chin, and J. Eberly, 1989, *Phys. Rev. Lett.* **63**, 2212.

Bartelt, A., J. Close, F. Federmann, N. Quaas, and J. Toennies, 1996, *Phys. Rev. Lett.* **77**, 3525.

Bauer, D., 2004, *Appl. Phys. B* **78**, 801.

Bertsch, G. F. and S. Das Gupta, 1988, *Phys. Rep.* **160**, 191.

Bethe, H. and E. Salpeter, 1977, *Quantum Mechanics of One- and Two-Electron Atoms* (Plenum Press, New York).

Bornath, T., P. Hilse, and M. Schlanges, 2007a, *Contrib. Plasma Phys.* **17**, 591.

Bornath, T., P. Hilse, and M. Schlanges, 2007b, *Contrib. Plasma Phys.* **47**, 402.

Bostedt, C., H. Thomas, M. Hoener, E. Eremina, T. Fennel, K.-H. Meiwes-Broer, H. Wabnitz et al., 2008, *Phys. Rev. Lett.* **100**, 133401.

Brabec, T. and F. Krausz, 2000, *Rev. Mod. Phys.* **72**, 545.

Breizman, B., A. Arefiev, and M. Fomytskyi, 2005, *Phys. Plas.* **12**, 056706.

Brumer, P. and M. Shapiro, 1995, *Sci. Am.* **3**, 34.

Chen, L. M., J. J. Park, K. H. Hong, I. W. Choi, J. L. Kim, J. Zhang, and C. H. Nam, 2002, *Phys. Plas.* **9**, 3595.

Deiss, C., N. Rohringer, J. Burgdörfer, E. Lamour, C. Prigent, J. Rozet, and D. Vernhet, 2006, *Phys. Rev. Lett.* **96**, 013203.

Diederich, T., T. Döppner, T. Fennel, J. Tiggesbäumker, and K.-H. Meiwes-Broer, 2005, *Phys. Rev. A* **72**, 11.

Dietz, T., M. Duncan, D. Powers, and R. Smalley, 1981, *J. Chem. Phys.* **74**, 6511.

Ditmire, T., T. Donnelly, A. M. Rubenchik, R. W. Falcone, and M. D. Perry, 1996, *Phys. Rev. A* **53**, 3379.

Ditmire, T., R. A. Smith, J. W. G. Tisch, and M. H. R. Hutchinson, 1997a, *Phys. Rev. Lett.* **78**, 3121.

Ditmire, T., J. Tisch, E. Springate, M. Mason, N. Hay, R. Smith, I. Marangos, and M. Hutchinson, 1997b, *Nature* **386**, 54.

Ditmire, T., J. W. G. Tisch, E. Springate, M. B. Mason, N. Hay, J. P. Marangos, and M. H. R. Hutchinson, 1997c, *Phys. Rev. Lett.* **78**, 2732.

Ditmire, T., J. Zweiback, V. Yanovsky, T. Cowan, G. Hays, and K. Wharton, 1999, *Nature* **398**, 489.

Döppner, T., T. Diederich, A. Przystawik, N. X. Truong, T. Fennel, J. Tiggesbäumker, and K.-H. Meiwes-Broer, 2007a, *Phys. Chem. Chem. Phys.* **9**, 4639.

Döppner, T., T. Fennel, T. Diederich, J. Tiggesbäumker, and K.-H. Meiwes-Broer, 2005, *Phys. Rev. Lett.* **94**, 013401.

Döppner, T., T. Fennel, P. Radcliffe, J. Tiggesäumker, and K.-H. Meiwes-Broer, 2006, *Phys. Rev. A* **73**, 031202(R).

Döppner, T., J. Müller, A. Przystawik, J. Tiggesbäumker, and K.-H. Meiwes-Broer, 2007b, *Eur. Phys. J. D* **43**, 261.

Döppner, J., S. Teuber, T. Diederich, T. Fennel, P. Radcliffe, J. Tiggesbäumker, and K.-H. Meiwes-Broer, 2003, *Eur. Phys. J. D* **24**, 157.

Döppner, T., S. Teuber, M. Schumacher, J. Tiggesbäumker, and K.-H. Meiwes-Broer, 2000, *Appl. Phys. B* **71**, 357.

Echt, O. and E. Recknagel (eds.), 1991, *Small Particles and Inorganic Clusters* (Springer, Berlin, Germany).

Fennel, T., G. F. Bertsch, and K.-H. Meiwes-Broer, 2004, *Eur. Phys. J. D* **29**, 367.

Fennel, T., T. Döppner, J. Passig, C. Schaal, J. Tiggesbäumker, and K.-H. Meiwes-Broer, 2007a, *Phys. Rev. Lett.* **98**, 143401.

Fennel, T. and J. Köhn, 2008, in *Lecture Notes in Physics: Computational Many-Particle Physics*, H. Fehske, R. Schneider, and A. Weisse (eds.), (Springer, Berlin, Heidelberg), pp. 255–273.

Fennel, T., L. Ramunno, and T. Brabec, 2007b, *Phys. Rev. Lett.* **99**, 233401.

Fukuda, Y., A. Y. Faenov, T. Pikuz, M. Kando, H. Kotaki, I. Daito, J. Ma et al., 2008, *Appl. Phys. Lett.* **92**, 121110.

Fukuda, Y., K. Yamakawa, Y. Akahane, M. Aoyama, N. Inoue, H. Ueda, and Y. Kishimoto, 2003, *Phys. Rev. A* **67**, 061201(R).

Georgescu, I., U. Saalmann, and J. M. Rost, 2007a, *Phys. Rev. Lett.* **99**, 183002.

Georgescu, I., U. Saalmann, and J. M. Rost, 2007b, *Phys. Rev. A* **76**, 043203.

Gets, A. and V. Krainov, 2006, *J. Phys. B* **39**, 1787.

Goyal, S., D. Schutt, and G. Scoles, 1992, *Phys. Rev. Lett.* **69**, 933.

Griffin, D., C. Bottcher, M. Pindzola, S. Younger, D. Gregory, and D. Crandall, 1984, *Phys. Rev. A* **29**, 1729.

Gunnarsson, O. and B. Lundquist, 1976, *Phys. Rev. B* **13**, 4274.

Haberland, H. (ed.), 1994, *Clusters of Atoms and Molecules 1 and 2*, vols. 52 and 56 (Springer Series in Chemical Physics, Berlin, Germany).

Haberland, H., M. Mall, M. Moseler, Y. Qiang, T. Reiners, and Y. Thurner, 1994, *J. Vac. Sci. Technol. A* **12**(5), 2925.

Hagena, O., 1987, *Z. Phys. D* **4**, 291.

Hartmann, M., R. Miller, J. Toennies, and A. Vilesov, 1995, *Phys. Rev. Lett.* **75**, 1566.

Heidenreich, A., I. Last, and J. Jortner, 2007, *J. Chem. Phys.* **127**, 074305.

Hirokane, M., S. Shimizu, M. Hashida, S. Okada, S. Okihara, F. Sato, T. Iida, and S. Sakabe, 2004, *Phys. Rev. Lett.* **69**, 063201.

Höll, A., T. Bornath, L. Cao, T. Döppner, S. Düsterer, E. Förster, C. Fortmanna et al., 2007, *High Energy Density Phys.* **3**, 120.

Ievlev, D., I. Rabin, W. Schulze, and G. Ertl, 2000, *Chem. Phys. Lett.* **328**, 142.

Ishikawa, K. and T. Blenski, 2000, *Phys. Rev. A* **62**, 63204.

Islam, M., U. Saalmann, and J. M. Rost, 2006, *Phys. Rev. A* **73**, 041201.

Issac, R., G. Vieux, B. Ersfeld, E. Brunetti, S. Jamison, J. Gallacher, D. Clark, and D. Jaroszynski, 2004, *Phys. Plas.* **11**, 3491.

Jarzynski, C. and G. Bertsch, 1996, *Phys. Rev. C* **53**, 1028.

Jha, J. and M. Krishnamurthy, 2008, *Appl. Phys. Lett.* **92**, 191108.

Jha, J., D. Mathur, and M. Krishnamurthy, 2006, *Appl. Phys. Lett.* **88**, 041107.

Jungreuthmayer, C., M. Geissler, J. Zanghellini, and T. Brabec, 2004, *Phys. Rev. Lett.* **92**, 133401.

Jungreuthmayer, C., L. Ramunno, J. Zanghellini, and T. Brabec, 2005, *J. Phys. B* **38**, 3029.

Keldysh, L. V., 1965, *Sov. Phys. JETP* **20**, 1307.

Kleibert, A., J. Passig, K.-H. Meiwes-Broer, and J. Bansmann, 2007, *J. Appl. Phys.* **101**, 114318.

Köhn, J., R. Redmer, K.-H. Meiwes-Broer, and T. Fennel, 2008, *Phys. Rev. A* **77**, 033202.

Köller, L., M. Schumacher, J. Köhn, S. Teuber, J. Tiggesbäumker, and K.-H. Meiwes-Broer, 1999, *Phys. Rev. Lett.* **82**, 3783.

Kou, J., V. Zhakhovskiia, S. Sakabe, K. Nishihara, S. Shimizu, S. Kawato, M. Hashida et al., 2000, *J. Chem. Phys.* **112**, 5012.

Krainov, V., 2000, *J. Phys. B* **33**, 1585.

Krainov, V. and M. Smirnov, 2002, *Phys. Rep.* **370**, 237.

Krishnamurthy, M., D. Mathur, and V. Kumarappan, 2004, *Phys. Rev. A* **69**, 033202.

Kruit, P. and F. Read, 1983, *J. Phys. E* **16**, 313.

Kumarappan, V., M. Krishnamurthy, and D. Mathur, 2001, *Phys. Rev. Lett.* **87**, 085005.

Kumarappan, V., M. Krishnamurthy, and D. Mathur, 2002, *Phys. Rev. A* **66**, 033203.

Kumarappan, V., M. Krishnamurthy, and D. Mathur, 2003a, *Phys. Rev. A* **67**, 043204.

Kumarappan, V., M. Krishnamurthy, and D. Mathur, 2003b, *Phys. Rev. A* **67**, 063207.

Kundu, M. and D. Bauer, 2006, *Phys. Rev. Lett.* **96**, 123401.

Laarmann, T., A. de Castro, P. Gürtler, W. Laasch, J. Schulz, H. Wabnitz, and T. Möller, 2004, *Phys. Rev. Lett.* **92**, 143401.

Lamour, E., C. Prigent, J. Rozet, and D. Vernhet, 2005, *Nucl. Instrum. Methods Phys. Res. B* **235**, 408.

Last, I. and J. Jortner, 1998, *Phys. Rev. A* **58**, 3826.

Last, I. and J. Jortner, 1999, *Phys. Rev. A* **60**, 2215.

Last, I., I. Schek, and J. Jortner, 1997, *J. Chem. Phys.* **107**, 6685.

Lebeault, M. A., J. Vialln, J. Chevaleyre, C. Ellert, D. Normand, M. Schmidt, O. Sublemontier, C. Guet, and B. Huber, 2002, *Eur. Phys. J. D* **20**, 233.

Legrand, C., E. Suraud, and P.-G. Reinhard, 2006, *J. Phys. B* **39**, 2481.

Lezius, M., S. Dobosz, D. Normand, and M. Schmidt, 1998, *Phys. Rev. Lett.* **80**, 261.

Li, S., C. Wang, J. Liu, P. Zhu, X. Wang, G. Ni, R. Li, and Z. Xu, 2005, *Plas. Sci. Technol.* **7**, 2684.

Li, S., C. Wang, P. Zhu, X. Wang, R. Li, G. Ni, and Z. Xu, 2003, *Chin. Phys. Lett.* **20**, 1247.

Loch, S., M. Pindzola, and D. Griffin, 2008, *Int. J. Mass Spectrosc.* **271**, 68.

Lotz, W., 1967, *Z. Phys.* **216**, 205.

Maine, P., D. Strickland, P. Bado, M. Pessot, and G. Mourou, 1988, *J. Quantum Electron* **24**, 398.

Mamyrin, B. A., 1994, *Int. J. Mass Spectrom. Ion Proc.* **131**, 1.

Martchenko, T., C. Siedschlag, S. Zamith, H. G. Muller, and M. J. J. Vrakking, 2005, *Phys. Rev. A* **72**, 053202.

Mathur, D. and M. Krishnamurthy, 2006, *Laser Phys.* **16**, 581.

McPherson, A., B. Thompson, A. Borisov, K. Boyer, and C. Rhodes, 1994, *Nature* **370**, 631.

Megi, F., M. Belkacem, M. Bouchene, E. Suraud, and G. Zwicknagel, 2003, *J. Phys. B* **36**, 273.

Methling, R.-P., V. Senz, E.-D. Klinkenberg, T. Diederich, J. Tiggesbäumker, G. Holzhüter, J. Bansmann, and K.-H. Meiwes-Broer, 2001, *Eur. Phys. J. D* **16**, 173.

Mie, G., 1908, *Ann. Phys. (Leipzig)* **25**, 377.

Mikaberidze, A., U. Saalmann, and J. M. Rost, 2008, *Phys. Rev. A* **77**, 041201.

Milani, P. and S. Ianotta, 1999, *Cluster Beam Synthesis of Nanostructured Materials* (Springer Series in Cluster Physics, Berlin, Germany).

Milchberg, H. M., S. J. McNaught, and E. Parra, 2001, *Phys. Rev. E* **64**, 056402.

Miura, E., H. Honda, K. Katsura, E. Takahashi, and K. Kondo, 2000, *Appl. Phys. B* **70**, 783.

Moore, A., K. Mendham, D. Symes, J. Robinson, E. Springate, M. Mason, R. Smith, J. Tisch, and J. Marangos, 2005, *Appl. Phys. B* **80**, 101.

Mulser, P., M. Kanapathipillai, and D. Hoffmann, 2005, *Eur. Phys. J. D* **95**, 103401.

Nishihara, K., H. Amitani, M. Murakami, S. Bulanov, and T. Esirkepov, 2001, *Nucl. Instrum. Methods A* **464**, 98.

Parra, E., I. Alexeev, J. Fan, K. Kim, S. McNaught, and H. Milchberg, 2000, *Phys. Rev. E* **62**, 5931(R).

Passig, J., K.-H. Meiwes-Broer, and J. Tiggesbäumker, 2006, *Rev. Sci. Instrum.* **77**, 093302.

Pauly, H., 2000, *Atom, Molecule and Cluster Beams 2* (Springer, Berlin, Germany).

Peano, F., F. Peinetti, R. Mulas, G. Coppa, and L. Silva, 2006, *Phys. Rev. Lett.* **96**, 175002.

Plagne, L., J. Daligault, K. Yabana, T. Tazawa, Y. Abe, and C. Guet, 2000, *Phys. Rev. A* **61**, 33201.

Prigent, C., C. Deiss, E. Lamour, J. P. Rozet, D. Vernhet, and J. Burgdörfer, 2008, *Phys. Rev. A* **78**, 053201.

Purnell, J., E. M. Snyder, S. Wei, and A. W. Castleman Jr., 1994, *Chem. Phys. Lett.* **229**, 333.

Radcliffe, P., T. Döppner, M. Schumacher, S. Teuber, J. Tiggesbäumker, and K.-H. Meiwes-Broer, 2005, *Contrib. Plasma Phys.* **45**, 424.

Radcliffe, P., A. Przystawik, T. Diederich, T. Döppner, J. Tiggesbäumker, and K.-H. Meiwes-Broer, 2004, *Phys. Rev. Lett.* **92**, 173403.

Radziemski, L. and D. Cremers (eds.), 1989, *Laser-Induced Plasmas and Applications* (Dekkers, New York).

Ramunno, L., C. Jungreuthmayer, H. Reinholz, and T. Brabec, 2006, *J. Phys. B* **39**, 4923.

Reinhard, P.-G. and E. Suraud, 1998, *Eur. Phys. J. D* **3**, 175.

Reinhard, P.-G. and E. Suraud, 2001, *Appl. Phys. B* **73**, 401.

Reinhard, P. and E. Suraud, 2003, *Introduction to Cluster Dynamics* (Wiley, New York).

Rose-Petruck, C., K. J. Schafer, K. R. Wilson, and C. P. J. Barty, 1997, *Phys. Rev. A* **55**, 1182.

Saalmann, U., 2006, *J. Mod. Opt.* **53**, 173.

Saalmann, U. and J. M. Rost, 2003, *Phys. Rev. Lett.* **91**, 223401.

Saalmann, U. and J. M. Rost, 2005, *Eur. Phys. J. D* **36**, 159.

Saalmann, U., C. Siedschlag, and J. M. Rost, 2006, *J. Phys. B* **39**, R39.

Santra, R. and C. Greene, 2003, *Phys. Rev. Lett.* **91**, 233401.

Schulze, W., B. Winter, and I. Goldenfeld, 1987, *J. Chem. Phys.* **87**, 2402.

Schumacher, M., S. Teuber, L. Köller, J. Köhn, J. Tiggesbäumker, and K.-H. Meiwes-Broer, 1999, *Eur. Phys. J. D* **9**, 411.

Seideman, T., M. Y. Ivanov, and P. B. Corkum, 1995, *Phys. Rev. Lett.* **75**, 2819.

Shao, Y. L., T. Ditmire, J. W. G. Tisch, E. Springate, J. P. Marangos, and M. H. R. Hutchinson, 1996, *Phys. Rev. Lett.* **77**, 3343.

Siedschlag, C. and J. M. Rost, 2003, *Phys. Rev. A* **67**, 013404.

Siedschlag, C. and J. M. Rost, 2004, *Phys. Rev. Lett.* **93**, 043402.

Siedschlag, C. and J. M. Rost, 2005, *Phys. Rev. A* **71**, 031401(R).

Siekmann, H. R., C. Lüder, J. Faehrmann, H. O. Lutz, and K.-H. Meiwes-Broer, 1991, *Z. Phys. D* **20**, 417.

Snyder, E. M., S. A. Buzza, and A. W. Castleman Jr., 1996, *Phys. Rev. Lett.* **77**, 3347.

Springate, E., S. Aseyev, S. Zamith, and M. Vrakking, 2003, *Phys. Rev. A* **68**, 053201.

Springate, E., N. Hay, J. Tisch, M. Mason, M. H. T. Ditmire, and J. Marangos, 2000a, *Phys. Rev. A* **61**, 063201.

Springate, E., N. Hay, J. W. G. Tisch, M. B. Mason, T. Ditmire, J. P. Marangos, and M. H. R. Hutchinson, 2000b, *Phys. Rev. A* **61**, 44101.

Suraud, E. and P.-G. Reinhard, 2000, *Phys. Rev. Lett.* **85**, 2296.

Tannor, D., R. Kosloff, and S. Rice, 1986, *J. Chem. Phys.* **84**, 5805.

Teuber, S., T. Döppner, T. Fennel, J. Tiggesbäumker, and K.-H. Meiwes-Broer, 2001, *Eur. Phys. J. D* **16**, 59.

Thomson, J. J., 1907, *Philos. Mag. S. VI* **13**, 561.

Tiggesbäumker, J. and F. Stienkemeier, 2007, *Phys. Chem. Chem. Phys.* **9**, 4748.

Tisch, J., N. Hay, K. Mendham, E. Springate, D. Symes, A. Comley, M. Mason et al., 2003, *Nucl. Instrum. Methods B* **205**, 310.

Truong, N.X., P. Hilse, S. Göde, A. Przystawik, T. Döppner, Th. Fennel, Th. Bornath, J. Tiggesbäumker, M. Schlanges, G. Gerber, and K.-H. Meiwes-Broer, 2010, *Phys. Rev. A,* **81**, 013201.

Uehling, E. and G. Uhlenbeck, 1933, *Phys. Rep.* **43**, 552.

Wabnitz, H., L. Bittner, A. de Castro, R. Döhrmann, P. Gürtler, T. Laarmann, W. Laasch et al., 2002, *Nature* **420**, 482.

Weiner, A. M., 2000, *Rev. Sci. Instrum.* **71**, 1929.

Whaley, K. B. and R. E. Miller (eds.), 2001, Special issue on helium nanodroplets: A novel medium for chemistry and physics, *J. Chem. Phys.* **115**, 22.

Zamith, S., T. Martchenko, Y. Ni, S. Aseyev, H. Muller, and M. Vrakking, 2004, *Phys. Rev. A* **70**, 011201(R).

Ziaja, B., E. Weckert, and T. Möller, 2007, *Las. Part. Beams* **25**, 407.

Zuo, T. and A. D. Bandrauk, 1995, *Phys. Rev. A* **52**, R2511.

Zweiback, J., T. E. Cowan, R. A. Smith, J. H. Hartley, R. Howell, C. A. Steinke, G. Hays, K. B. Wharton, J. K. Crane, and T. Ditmire, 2000, *Phys. Rev. Lett.* **85**, 3640.

Zweiback, J., T. Ditmire, and M. D. Perry, 1999, *Phys. Rev. A* **59**, R3166.

Atomic Clusters in Intense Laser Fields

Ulf Saalmann

*Max-Planck-Institut für
Physik komplexer Systeme*

and

*Center for Free Electron
Laser Science*

Jan-Michael Rost

*Max-Planck-Institut für
Physik komplexer Systeme*

and

*Center for Free Electron
Laser Science*

14.1 Introduction

Clusters bridge the gap between atomic (microscopic) and condensed (macroscopic) objects (Haberland 1994). They allow for almost continuous interpolation from single atoms to nanodroplets by changing the number of atoms in the cluster. Clusters in intense laser fields form a fundamental yet quite advanced research topic, regarding the target that is exposed to light and regarding the light that differs quite a bit in its effect on the target from the photoeffect that one may consider as the fundamental interaction of light with matter. With this single photon dynamics, we contrast dynamics induced by intense laser fields that need to be incorporated in the dynamical description of matter non-perturbatively (Brabec and Kapteyn 2004).

In Section 14.2, we discuss the interaction of an electron with an intense field and focus on the question of energy transfer between the electron motion and the field. We briefly touch upon the fundamental atomic process when an electron bound to an atom is exposed to intense laser pulses. Its energy gain from the laser field, resulting, e.g., in *high harmonic generation* can be understood in very simple terms of classical mechanics (Lewenstein et al. 1994). In the next step, we describe, analytically, how for a cluster this picture gets replaced by a similarly simple one. It is akin to the mechanism by which spacecrafts gain velocity through gravitational slingshots (Prado 1996).

Section 14.3 describes briefly how clusters are produced in the experiment. Their evolution under a strong laser pulse can be divided into four phases, which are governed by characteristic dynamical processes. Finally, we discuss how this evolution can be probed in an experiment.

Due to the high local atomic density in clusters, multiple ionization by intense laser pulses results in the formation of nanoplasmas inside the irradiated clusters. These locally confined plasmas are transient in nature, fed through electrons ionized into the plasma from atomic bound states, and depleted through those electrons that leave the plasma and the cluster. A unique feature of these nanoplasmas is an enormously effective resonant energy absorption, which is discussed from a microscopic and a macroscopic perspective in Section 14.4. The resulting Coulomb explosion is discussed in terms of kinetic energy spectra measured for the fragment ions in Section 14.5. To compare them with ideal theoretical single-cluster spectra, one has to take into account the laser focus, cluster size distribution, and saturation. For all three effects, we give an analytical description, which allows one to understand easily the experimental spectra.

Finally, we use composite clusters in Section 14.6 to highlight an important dynamical feature which is universal for all clusters, namely, electron migration within the cluster. It can be identified in experiments with radiation of shorter wavelength than 800 nm, available at new free-electron laser (FEL) light sources (Feldhaus et al. 2007). Hence, this section contains already a preview of two possible directions of future cluster research; a more general perspective is provided in the outlook (Section 14.7).

14.2 Non-Perturbative Light–Matter Interaction

Linear coupling of light to matter, realized by single photon processes, is characterized by photoabsorption rates being independent of the intensity of the incident light. This is the regime of the photoeffect, where photoionization of an electron with ionization potential E_I, requires a photon energy $\hbar\omega$ such that $\hbar\omega - E_I > 0$. At moderate laser intensities ($I \sim 10^{12}$ W/cm^2), the response becomes nonlinear and atoms can be ionized at photon energies $\hbar\omega < E_I$ by absorbing more than one photon. While these processes can still be understood in terms of (high-order) perturbation theory, such a perturbative treatment of multiphoton absorption fails at intensities $I > 10^{14}$ W/cm^2. Here, typically the ponderomotive energy U_p becomes comparable (or larger) than the photon energy $\hbar\omega$. The ponderomotive energy

$$U_p = \frac{1}{T}\int_0^T dt\,\frac{\dot{\mathbf{x}}^2(t)}{2} \equiv \frac{F_0^2}{4\omega^2} \tag{14.1}$$

is the kinetic energy $\dot{\mathbf{x}}(t)^2/2$ of an electron oscillating in the laser field $\mathbf{F}(t) = \mathbf{F}_0 \cos\omega t$ with the velocity $\dot{\mathbf{x}}(t) = -(\mathbf{F}_0/\omega)\sin(\omega t)$ averaged over one laser cycle $T = 2\pi/\omega$. The velocity follows through classical mechanics from the Hamiltonian

$$H_F = \frac{\mathbf{p}^2}{2} + \mathbf{x}\mathbf{F}(t) \tag{14.2}$$

since H_F contains only linear and quadratic operators (Ehrenfest theorem). The Hamiltonian H_F is based on two approximations: Firstly, the light is coupled in dipole approximation (linear in $\mathbf{x}$), which is justified as long as the wavelength of the light is much larger than the electron orbitals it couples to. Since we deal with 800 nm light, this is certainly fulfilled. Secondly, we treat the light field classically as $\mathbf{F}(t)$, which is justified since the intensity is high enough to be in the limit of large photon numbers. Hence, the light field is essentially a source of photons, which does not change if it exchanges some of them with the atom.

In the presence of a potential, $H = H_F + V$ no longer fulfills the Ehrenfest theorem and the generally quite complicated strong field dynamics must be solved for quantum mechanically. However, another simplification arises: On the atomic timescale (introduced by the potential V), the light-induced dynamics is adiabatically slow with a photon energy of 1.2 eV (800 nm wavelength) corresponding to $\omega = 0.057$ atomic units. Atoms in this case are simply field ionized, either through tunneling or above-the-barrier ionization at suitable phases of the electric field of the laser (see Figure 14.1). The so-called ADK rate provides an analytical expression for the ionization rate in this regime (Ammosov et al. 1986).

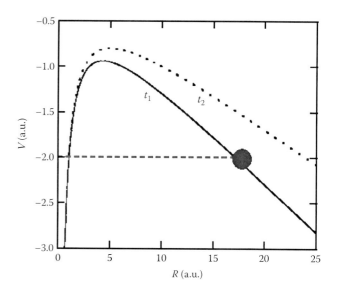

FIGURE 14.1 The principle of tunneling ionization. Snapshot of the potential for a one-electron atom (He$^+$, $E_I = 2$ a.u.) $H = H_F - 2|\mathbf{x}|^{-1}$ along the (linear) laser polarization at maximum field (t_1) and slightly later (dashed, t_2) and with the electron after tunneling indicated by a black bullet.

14.2.1 Atoms: The Principle of Energy Absorption from a Strong Laser Pulse

A free electron continuously exchanges energy with a light field without net gain. This can be easily seen by calculating the energy difference $\Delta E(t_f, t_i) = \int_{t_i}^{t_f} \dot{H}_F dt$, where we assume that the electron was initially at rest, i.e., $\dot{\mathbf{x}}(t) = -(\mathbf{F}_0/\omega)[\sin(\omega t) - \sin(\omega t_i)]$. Then

$$\Delta E(t_f, t_i) = -\int_{t_i}^{t_f} dt\,\dot{\mathbf{x}}\mathbf{F}(t) = 4U_p\int_{\phi_i}^{\phi_f} d\phi[\sin\phi - \sin\phi_i]\cos\phi$$

$$= 2U_p(\sin\phi_f - \sin\phi_i)^2 \equiv \beta U_p \tag{14.3}$$

with $\phi = \omega t$ and corresponding definitions for ϕ_i and ϕ_f. Clearly, for n full laser periods $t_f - t_i = 2\pi n/\omega$, we get $\Delta E = 0$. Hence, for any kind of net energy absorption, the coupling to the light must be limited to incomplete laser cycles. The energy gain has a maximum of $8U_p$ for $\phi_i = (k - 1/2)\pi$ and $\phi_f = (k + 1/2)\pi$ with some integer k.

One of the most important atomic processes in strong fields, namely, high harmonic generation, can be understood in this way (Corkum 1993). In this case, many photons of low frequency are absorbed by an atomic electron through tunneling ionization (see above). However, the field is so strong that it pushes the electron back to the ion with which it recombines releasing its kinetic energy and (ionization) potential energy by emitting a high-energy photon with a maximum energy of E_{HHG}. One can almost guess E_{HHG} from Equation 14.3: The electron should come back with maximum velocity, i.e., $\sin\phi_f = \pm 1$

and it should be initially released for optimum tunneling ionization close to maximum field strength, i.e., $\sin \phi_i \approx 0$ leading to $E_{HHG} \approx 2U_p + E_I$. Numerical calculation with Equations 14.2 and 14.3 leads to $E_{HHG} = 3.17\hbar\omega + E_I$, realized by a trajectory that starts slightly after the maximum of the laser field from the nucleus and returns to it at the next but one zero crossing of the electric field (Corkum 1993).

To summarize, the essence of energy gain from a periodic field is to interrupt the motion of the charged particle through some perturbation in order to effectively limit the exposure to the light to an incomplete field cycle. This may either be realized by a sudden release of the electron in the field (ionization) or by sudden stop of the free motion (collision with another particle). Only in such cases, there can be a net transfer of energy between the particle and the field. Over full cycles, the same amount of energy is absorbed and released by the particle leading to zero net energy exchange with the field, as shown in Equation 14.3. This simple picture is surprisingly accurate and universal and has been coined the "simple man's approach" (Lewenstein et al. 1994). The (Coulomb) potential does not play a crucial role. It is also interesting to note that the field property (ponderomotive energy U_p) and the atomic property (ionization potential E_I) enter additively in the energy gain of Equation 14.3. This differs qualitatively from the situation for clusters to be discussed next.

14.2.2 Clusters: Optical Slingshots

While the same quasi-classical principles of light absorption can be carried over to clusters, the simple rescattering from a point-like potential (ion) no longer happens. Instead, in an idealized way, an electron with velocity v_i traveling through an extended cluster potential can be thought of passing two potential steps: first, when entering the cluster, and second when leaving it again. Both these potential steps (and if the potential is constant otherwise *only* these steps) give rise to possible energy absorption since they interrupt the cycle of equal absorption and release of energy in the field. The classical dynamics of an electron in such a potential under a laser field is stepwise free motion and can be solved analytically in one dimension (Saalmann and Rost 2008). In the limit of a deep potential, $v \equiv \sqrt{2V} \gg v_i$ and $v \gg A$, the final momentum of the exiting electron v_f defines also the energy gain through $\Delta E = v_f^2/2$,

$$\Delta E(\phi) = 4Ap \sin\left[\frac{L\omega}{2v}\right] \cos\left[\frac{\phi}{2}\right], \quad (14.4)$$

where

    L is the extension of the potential (see Figure 14.2)
    $\phi = \phi_f + \phi_i$ is the sum of the phases of the field when the electron enters and exits the potential

As one can easily see, Equation 14.4 gives rise to a "double resonance," i.e., a maximum energy gain if both trigonometric

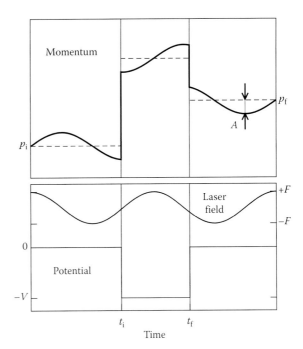

FIGURE 14.2 Schematic picture of the scattering process. The upper part shows the momentum of the electron, the lower part the potential V and the electric field $F(t)$ (assumed to be homogeneous in space) as a function of time. (After Saalmann, U. and Rost, J.M., *Phys. Rev. Lett.*, 100, 133006, 2008. With permission.)

functions are unity. One condition can be met with all potentials since it is a condition on the dynamics of the electron: energy gain is maximized for symmetric phases $\phi = 0$ with the result, rewritten in terms of U_p and V,

$$\Delta E_{max} = 4\sqrt{U_p}\sqrt{2V}\sin\left(\frac{L\omega}{\sqrt{8V}}\right). \quad (14.5)$$

In other words, absorption is maximal if the electron passes the potential around a field maximum.

If additionally $L\omega/\sqrt{8V} = \pi/2$, the absorption is resonant in the sense that from entrance to exit of the potential, exactly half a laser cycle is completed (as sketched in Figure 14.2), which results quite generally in optimum absorption, as shown in Equation 14.3. Compared to atomic systems, one immediately sees the qualitative difference: Now the field (U_p) and system (V) properties enter *multiplicatively* rendering clusters very efficient absorbers of intense light. Figure 14.3 illustrates the validity of this simple model by comparing experimental data and microscopic calculations with the outcome from the model, Equation 14.5.

In general, as apparent from Equation 14.3, a particle absorbs most efficiently energy from an external field if it has high velocity. Hence, when passing the center of a realistic cluster where the potential is deepest, the electron should feel the largest field. The phase $\phi = 0$ (Equation 14.5) ensures exactly this condition making the model quite robust.

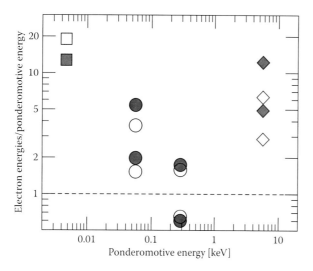

value in relation to the depth of the potential and the light, as can be seen from the argument of the sine function in Equation 14.5: Absorption is maximized for $L\omega/\sqrt{8V} = \pi/2$, i.e., if the potential of extension L is transversed by the electron with momentum $p = \sqrt{2V}$ in the time $(\pi\omega)^{-1}$, half a period of the light. This is reminiscent of the resonant heating in a large but finite plasmas (Taguchi et al. 2004).

14.3 Cluster Dynamics in Intense Fields

Before discussing the mechanism of light absorption by collective electron motion in atomic clusters, we briefly comment on their preparation and the typical scenario for clusters subjected to a strong laser pulse.

14.3.1 Cluster Preparation

Rare gas clusters can be easily produced experimentally. A gas is pressed through a nozzle and expands into vacuum adiabatically. As a result, its temperature drops and saturated vapor pressure becomes smaller than the gas pressure, which leads to the formation of clusters. Since the size of the cluster can be easily controlled by the pressure used for pushing the gas through the nozzle, one has experimental access to all cluster sizes from small ones ($10–10^3$ atoms) over medium-sized clusters ($10^3–10^5$ atoms) up to large clusters or droplets (10^5 and more atoms). Thus, one may study the intense field response from few atoms to bulk systems.

Rare gas clusters are bound by van der Waals forces with typical binding energies of a few meV only (Tang and Toennies 2003). The cluster geometries for small cluster, which can be found in a database (Wales et al. 2007), are obtained by minimizing clusters that interact pairwise by a Lennard-Jones interaction. The dominating motif is an icosahedral structure, which becomes most prominent at the magic numbers, where atoms are arranged in geometrically closed shells in the so-called Mackay icosahedra (Hoare 1979) shown in Figure 14.4.

FIGURE 14.3 Electron energies as a function of the ponderomotive energy U_p from various clusters. The filled symbols show E_{kin} for experiments and microscopic calculations (circles): Ag_{1000} (square, [Fennel et al. 2007]), Ar_{1700} and Ar_{33000} (diamonds, [Chen et al. 2002]), Xe_{1151} and Xe_{9093} (circles). The corresponding estimates ΔE from the rescattering model (Equation 14.5) are shown by open symbols. The dashed line indicates the atomic limit. (From Saalmann, U. and Rost, J.M., *Phys. Rev. Lett.*, 100, 133006, 2008. With permission.)

A very similar mechanism called "gravitational slingshot" is known in space science: The precious thrust for acceleration of a spacecraft is systematically applied when flying close by a planet that carries a large gravitational field, where the spacecraft has a large velocity (similar to the effect of a deep cluster potential on the electron). The thrust (in analogy to the acceleration caused by the laser field) yields optimum increase of speed if the velocity is already high (Prado 1996). Acknowledging the analogy, we call the optimum energy absorption from light fields "optical slingshots."

One can also view the resonant absorption in clusters from a more general perspective. In comparison to point-like atomic systems, extended systems have a length scale, the extension L which can give rise to *resonant* absorption when having the right

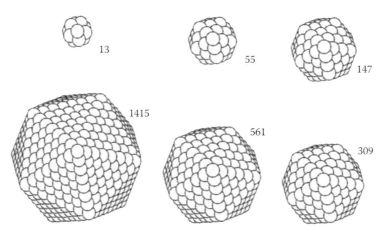

FIGURE 14.4 Closed shell rare gas clusters assume the geometry of Mackay icosahedra, see text.

14.3.2 Evolution

Common to all clusters subjected to strong laser pulses is the scenario sketched in Figure 14.5. In a first step (a), the light couples to the atoms as if they were isolated. Soon, neighboring ions become important since they facilitate the ionization of remaining neutral atoms (b) in the phase of *cooperative dynamics*. This effect was first discovered in diatomic molecules (Seideman et al. 1995, Zuo and Bandrauk 1995) and applies for small clusters in full analogy (Siedschlag and Rost 2002). In these small systems, electrons that are liberated from their mother atom can leave the molecule or cluster directly. There is an optimal interatomic distance where enhanced ionization occurs.

In larger clusters, this effect induces an avalanche of electrons, which was termed ionization ignition (Rose-Petruck et al. 1997). In contrast to small systems, however, the promotion of all these electrons to the continuum is prevented by the space charge, which is built up in the cluster (c). Electrons are trapped in the cluster volume and only additionally absorbed energy could drive these quasi-free or plasma electrons out of this trapping potential. *Collective dynamics* can give rise to very efficient resonant absorption in this phase, which we will discuss in Section 14.4. The two ionization stages—removal of electrons from the individual ions and from the entire cluster—are called *inner* and *outer* ionization, respectively (Last and Jortner 1999). Finally, during the last phase (d), energy is redistributed within the cluster, e.g., through recombination. The cluster completely disintegrates and the asymptotic (measurable) distribution of ions and electrons forms.

Concerning the theoretical description, we will restrict ourselves to the approach (Saalmann et al. 2006) we have followed, namely, classical molecular dynamics simulations of the cluster explosion where the coupling of the bound electrons to the laser light and/or existing electric fields is described by quantum rates.

14.3.3 Pump-Probe Techniques

Recent experiments have provided a time-resolved mapping of the absorption process by using a pump-probe setup. A first short laser pulse triggers the expansion followed after variable delay by a second pulse. Thereby, the absorption efficiency of the expanding cluster can be probed (Zweiback et al. 1999, Springate et al. 2000, Kim et al. 2003, Döppner et al. 2005) since the explosion of the cluster ions and hence the change of the cluster radius takes place on a femtosecond timescale.

14.4 Resonant Light Absorption by Clusters

Energy gain from the light field was discussed in Section 14.2 for a single electron in a point-like (atomic) potential and for the case of optical slingshots applying to extended (cluster) potentials. While optical slingshots can make single electrons very fast, any mechanism based on individual electrons will be dominated by a collective energy absorption if operational, as we will discuss next with an analytical model.

14.4.1 The Cluster as a Driven Damped Oscillator

As described in the previous section, during the initial phase of the laser pulse, electrons are outer ionized and (eventually multiply charged) ions form. When the electrons do not have sufficient kinetic energy to escape the considerable positive charge of the cluster ions, they remain as quasi-free electrons in the cluster, driven back and forth over the positive charge by the external laser field. This scenario resembles collective, driven periodic motion. Under the assumption of a spherical, uniformly charged cluster with total ionic charge Q and radius R, the potential inside the cluster is harmonic with an eigenfrequency

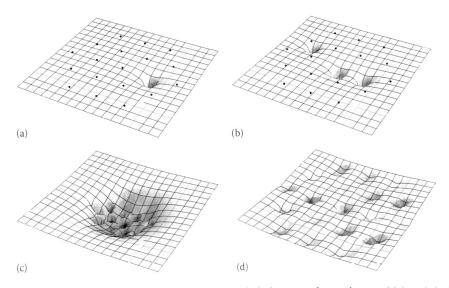

(a) (b)

(c) (d)

FIGURE 14.5 Sketch of the cluster potential as a function of time. Dots mark the location of neutral atoms. (a) Atomic ionization, (b) cooperative dynamics, (c) collective dynamics, and (d) explosion and relaxation.

$$\Omega_t = \sqrt{\frac{Q_t}{R_t^3}}. \tag{14.6}$$

Hence, the center of mass $X(t)$ coordinate of all these electrons along the (linear) laser polarization may be described by the dynamics of a harmonic oscillator, driven by the laser field $F(t)$ and in addition damped with rate Γ_t due to loss (outer ionization) and gain (inner ionization) processes:

$$\ddot{X}(t) + 2\Gamma_t \dot{X}(t) + \Omega_t^2 X(t) = F(t). \tag{14.7}$$

The subscript t indicates that due to ionization and expansion of the cluster, Ω_t and Γ_t depend parametrically on time with a much slower variation than the electronic coordinate $X(t)$.

For periodic driving $F(t) = F_0 \cos(\omega t)$, the dynamics is given by $X(t) = A_t \cos(\omega t - \phi_t)$ with (Landau and Lifschitz 1994)

$$A_t = \sqrt{\frac{U_p}{\left(\Omega_t^2/\omega - \omega\right)^2/4 + \Gamma_t^2}} \equiv a_t \sqrt{U_p}, \tag{14.8}$$

$$\phi_t = \arctan\left(\frac{2\Gamma_t \omega}{(\Omega_t^2 - \omega^2)}\right). \tag{14.9}$$

The energy balance of the dynamics (14.7) is characterized, on one hand, by energy loss due to the damping (i.e., induced inner ionization) and, on the other hand, by energy gain from the external laser field. The absorbed energy per cycle reads

$$\Delta E = 2\pi \omega a_t U_p \sin \phi_t. \tag{14.10}$$

If there is no phase lag, $\phi_t \approx 0$ of the electron motion relative to the laser field (which occurs for steep potentials, i.e., large Ω_t) or if the phase lag is $\phi_t \approx \pi$ (which occurs for weak potentials or free motion, i.e., small Ω_t), then no energy can be gained for the same reason, as explained in Section 14.2. In contrast to the discussion there, Equation 14.10 applies to the whole plasma inside the cluster. The total energy absorption is thus much larger.

On the other hand, maximum energy gain is achieved for $\phi_t \approx \pi/2$ which is the resonant case, well known from a driven oscillator. In fact, this is the only reliable indicator for resonant motion in the presence of damping, which may heavily suppress the raise of the amplitude most widely known as the signature of a resonance. Moreover, as evident from Equation 14.9, without damping there would be no phase lag and, therefore, no energy gain. The damping can also be thought of an expression of electron collisions transferring energy to and from the ions—this notion will be elaborated on when we discuss the macroscopic picture of energy absorption in the cluster.

Figure 14.6c shows the CM velocity $\dot{X}(t)$ during a flat laser pulse (160 fs plateau and 20 fs rise and fall time, respectively). No significant increase can be seen around the resonance ($t \approx 20$ fs) (see Figure 14.6d) indicating strong damping. However, a large increase of the absorption rate around the time of resonance is visible in the charging rate $\dot{Q}$ of the cluster (gray shaded area

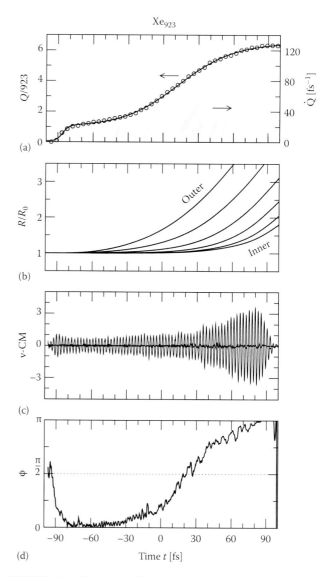

FIGURE 14.6 Dynamics of Xe$_{923}$ in a strong laser pulse ($\lambda = 780$ nm, $I = 9 \times 10^{14}$ W/cm, rise and fall time 20 fs, plateau for $t = -80$ to $+80$ fs). All quantities are shown as a function of time t. (a) Average charge per atom (*circles, left axis*) and corresponding rate (*gray filled line, right axis*). (b) Radii R of all cluster shells in units of their initial radii R_0 demonstrating that the shells explode sequentially. (c) Center-of-mass velocity v_{cm} of the electronic cloud inside the cluster volume. Note that the oscillations are spatially along the linear polarization of the laser whereas the electron velocity perpendicular to the laser polarization is very small and hardly to see in the figure. (d) Phase shift ϕ_t of the collective oscillation in laser direction with respect to the driving laser. (From Saalmann, U. and Rost, J.M., *Phys. Rev. Lett.*, 91, 223401, 2003. With permission.)

in Figure 14.6a). A clear effect of the resonance is also seen in Figure 14.7: The damping rate decreases sharply after the resonance since a lot of electrons are outer ionized during resonant energy absorption. A close inspection of Figure 14.7 reveals that the eigenfrequency Ω_t passes the laser frequency ω twice: However, on the way up the number of quasi-free electrons, N_t is still too small to contribute substantially to energy absorption.

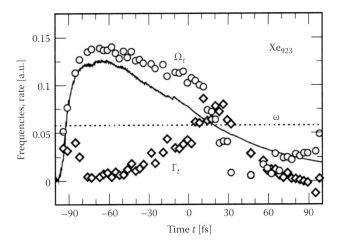

FIGURE 14.7 Parameters of the harmonic oscillator model (Equation 14.7) as calculated from the Xe_{923} dynamics in Figure 14.6. *Solid line:* eigenfrequency for a spherical, uniformly charged cluster $\Omega_t = \sqrt{Q_{ion}(t)/R(t)^3}$. *Circles:* derived eigenfrequency Ω_t. *Diamonds:* derived damping rate Γ_t. *Dotted line:* laser frequency ω. (From Saalmann, U. and Rost, J.M., *Phys. Rev. Lett.*, 91, 223401, 2003. With permission.)

Remarkably, the simple analytical model of the damped driven harmonic oscillator describes the microscopic dynamics very well: The eigenfrequency determined from the microscopic values for A_t and ϕ_t by inverting Equations 14.8 and 14.9 is in very good agreement with the eigenfrequency (solid line in Figure 14.7) determined from the cluster volume and the ion charges according to Equation 14.6.

The collective resonant absorption is by far the dominating mechanism, simply, due to its nature: if active, by definition many electrons participate, while other mechanisms always involve only a small subgroup of electrons. This resonant absorption mechanism does not contain quantum features and can, therefore, also be described macroscopically with the so-called nanoplasma model introduced by Ditmire et al. (1996).

14.4.2 Nanoplasma Model

A homogeneous plasma requires clusters larger than the Debye screening length $\lambda_D = \sqrt{kT_{el}/4\pi e^2 \rho}$. For a plasma of solid density $\rho = 10^{23}$ cm^{-3} at a temperature of $kT_{el} = 1000\,eV$, one gets $\lambda_D \sim 5\,\text{Å}$. Hence, only clusters with $R > 100\,\text{Å}$ can be described with the nanoplasma model (Ditmire et al. 1996). The macroscopic mechanism of energy absorption is quite similar to the microscopic one. However, the electric field itself contains the energy and its time derivative (polarization) plays the role of the velocity in Equation 14.3 leading to the energy absorption rate per unit volume of

$$\dot{E} = \frac{1}{4\pi}\mathcal{E}\frac{\partial(\varepsilon\mathcal{E})}{\partial t}. \tag{14.11}$$

Upon averaging over one laser cycle, the heating rate becomes

$$\langle\dot{E}\rangle = \frac{\omega}{8\pi}\Im(\varepsilon)\mathcal{E}^2, \tag{14.12}$$

while the field inside a spherical cluster is given by (Jackson 1998)

$$\mathcal{E} = \frac{3}{|2+\varepsilon(\omega)|}F(t) \tag{14.13}$$

with the vacuum electric field $F(t)$ and the dielectric constant ε (Ashcroft and Mermin 1976)

$$\varepsilon(\omega) = 1 - \frac{4\pi e^2\rho/m}{\omega(\omega+i\nu)}, \tag{14.14}$$

containing the ion–electron collision frequency ν. With the help of (14.13) and (14.14), the rate can be expressed in terms of ρ_t, ω, ν, and $F(t)$ (Ditmire et al. 1996). The time evolution of the density is calculated self-consistently: The electric field $\mathcal{E}$ inside the cluster depends via the dielectric function $\varepsilon(\omega)$ on the density ρ_t, and changes of the density ρ_t are caused by inner and outer ionizations due to $\mathcal{E}$. Given the electric field (14.13) in the cluster, dynamical processes such as ionization, energy absorption, or cluster expansion can be included by rate equations.

Energy absorption (14.12) becomes particularly large for large electric fields $\mathcal{E}$ in the cluster. They occur for $\varepsilon(\omega) \sim -2$, cf. (14.13), which corresponds to the resonance condition in the Drude model (14.14)

$$\omega^2 = \frac{4}{3}\pi e^2\frac{\rho}{m}. \tag{14.15}$$

Note that the collision frequency ν has taken the role of the damping in the damped driven oscillator model Equation 14.10 or the interruption of unperturbed electron motion in a laser field Equation 14.3 to induce energy absorption from the laser field: It represents the imaginary part of the dielectric constant ε, which is necessary for absorption, as shown in Equation 14.12.

The assumption of a homogeneous density $\rho(t)$ over the cluster volume (Ditmire et al. 1996) can be relaxed (Milchberg et al. 2001) allowing for a radial dependence of the density $\rho(r,t)$ and averaging the electric field over the solid angle $\mathcal{E}(r) = \langle\mathcal{E}(\mathbf{r})\cdot\mathcal{E}(\mathbf{r})\rangle^{1/2}$. Then, the dominant absorption mechanism is resonant absorption at the cluster *surface* where the density is at the critical value in Equation 14.15. This resonance moves inward and is maintained for a long time (typically for the whole pulse of a few hundred femtoseconds) until the maximum of the density at the inner part of the cluster falls below this value (Milchberg et al. 2001). Clearly, this mechanism requires a large cluster.

14.5 Coulomb Explosion

Once the cluster has been strongly charged, it will undergo Coulomb explosion, which can be studied by measuring the kinetic energy of the fragment ions. Data from these measurements are always averages from samples at different positions in the pulse, i.e., from different intensities due to the spatially inhomogeneous laser beam. An additional complication arises for clusters from the fact that under experimental conditions clusters have not a unique size with N atoms, but rather a distribution $g(N)$ where the mean N_0 depends on the way they have been generated, as shown in Figure 14.8.

The kinetic energy distribution of the ions (KEDI), i.e., the abundance of fragment ions as a function of their final kinetic energies, is a perfect observable to illustrate this effect (Islam et al. 2006). The KEDI differs substantially for a single cluster under a spatially homogenous laser beam (the theoretical standard) from the experimentally observed KEDI. If we start for simplicity with a homogenous charge distribution of ions in the cluster then three simple steps are sufficient to convert the single-cluster KEDI (Figure 14.9a) to the realistic one, namely, averaging over the spatial laser profile (Figure 14.9b), averaging over the cluster size distribution (Figure 14.9c), and taking into account saturation in the ionization (Figure 14.9e).

14.5.1 Single-Cluster Explosion

The basic mechanism underlying the KEDI in clusters is their Coulomb explosion. It converts the potential energy $E_{coul}(r)$ of a (partially) ionized cluster atom at a distance r from the cluster center into kinetic energy E. A homogeneously charged sphere expands self-similarly, i.e., the radius R increases, the particle (or charge) density decreases, but remains constant as a function of the distance from the cluster center. This can be seen from a trajectory of an infinitesimal shell of the charged sphere

(with a total charge Q) due to Coulomb repulsion (Hau-Riege et al. 2004). The equation of motion for a shell with the initial radius r_0 reads

$$\frac{dr}{dt} = \sqrt{2(E - U(r))}, \quad U(r) = \frac{Q(r_0/R)^3}{r}, \quad E = U(r_0) = \frac{Qr_0^2}{R^3}.$$
(14.16)

Integrating the above equation,

$$t = \frac{1}{\sqrt{2Q/R^3}} \frac{1}{r_0} \int_{r_0}^{r} \frac{dr'}{\sqrt{1 - r_0/r'}} = \frac{1}{\sqrt{2Q/R^3}} \int_{1}^{x} \frac{dx'}{\sqrt{1 - 1/x'}},$$
(14.17)

whereby the last equation was obtained by substituting $x = r/r_0$. This integral can be solved analytically (cf. caption of Figure 14.10). What is more important here is that the last expression of Equation 14.17 does not depend on r_0. In other words, each shell has expanded at a certain time t by the same factor x showing that the expansion indeed is self-similar and justifies a posteriori the equation of motion (14.16). However, this does not hold for arbitrary particle densities; if the density does not drop abruptly to zero at $r = R$ but decays slowly, inner shells may take over outer shells resulting in radial shock waves (Kaplan et al. 2003).

Since the Coulomb explosion converts the entire potential energy E_{coul} into kinetic energy E, with the energy depending on the initial position in the cluster, one can immediately write the probability distribution for the KEDI of a single cluster (Zweiback et al. 2002, Sakabe et al. 2004)

$$\frac{dP}{d\epsilon} = \frac{3}{2}\sqrt{\epsilon}\Theta(1 - \epsilon),$$
(14.18)

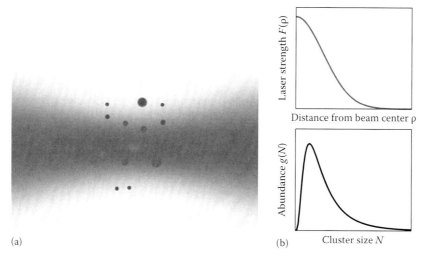

(a)   (b)

FIGURE 14.8 Sketch of a typical cluster experiment. (a) Spectra result from a distribution of various cluster sizes (bubbles) that are subjected to different laser intensities in the focused beam (shaded). (b) The functional dependence of the size distribution and the laser focus is shown on the right.

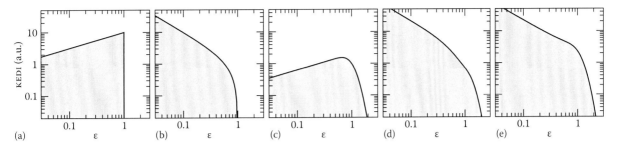

FIGURE 14.9 Kinetic energy distributions of ions (KEDI) for Coulomb exploding clusters as a function of scaled energy ε, see text (a) Single clusters, see Equation 14.18, (b) convoluted with a Gaussian laser profile, see Equation 14.19, (c) convoluted with a log-normal cluster size distribution, see Equation 14.21, (d) effect of laser profile and cluster size distribution combined, see Equation 14.22, and (e) including saturation of ionization in addition, see Equation 14.24. (From Mikaberidze, A. et al., *Phys. Rev. A*, 77, 041201, 2008. With permission.)

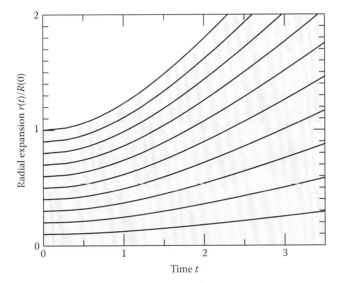

FIGURE 14.10 Self-similar expansion of a homogeneously charged sphere. The trajectories are given by the implicit Equation 14.17 with the integral $\int_1^x dx'/\sqrt{1-1/x'} = \log\left(\sqrt{(x-1)/x}+1\right) + \log(x)/2 + \sqrt{(x-1)x}$.

which is shown in Figure 14.9a. We have used the dimensionless quantity $\epsilon = E/E_R$ with the energy scale $E_R := E_{\mathrm{coul}}(R,q,N) = q^2 N/R$, given by the maximum kinetic energy acquired by ions at the cluster surface $r_0 = R$.

14.5.2 Pulse-Profile and Size-Distribution Effects

The spatial profile of the laser pulse is usually well described by a Gaussian function with field amplitude $F(\rho) = F_0 \exp(-\rho^2/2\xi^2)$, where ρ is the distance (radius) from the center of the laser beam in the plane perpendicular to the beam (see Figure 14.8). Along the laser beam, we assume a constant intensity since the experiments discussed later (Krishnamurthy et al. 2004) are performed with a narrow cluster beam, i.e., a radius smaller than the Rayleigh length (Siegman 1986) of the laser beam. This does not hold for the experiments where the cluster beam is irradiated near the output of the gas-jet nozzle (Hirokane et al. 2004).

The charging of the cluster is assumed to be proportional to the field strength, $q \propto F$, which applies for resonant charging of

the cluster (Saalmann and Rost 2003, Saalmann 2006). Hence, we obtain the spatial distribution of charge $q(\rho)$ by replacing $F(\rho)$ with $q(\rho)$ and F_0 with q_0, the maximum charge per ion reached in the laser focus $\rho = 0$. After integration over ρ, the laser profile–averaged KEDI reads in terms of the scaled energy $\epsilon = E/E_{\mathrm{coul}}(R, q_0, N)$

$$\frac{dP_{\mathrm{las}}}{d\epsilon} = \frac{\pi \xi^2 N}{2} \frac{1-\epsilon^{3/2}}{\epsilon} \Theta(1-\epsilon), \qquad (14.19)$$

as shown in Figure 14.9b. What has changed compared to the original KEDI from Equation 14.18 is the qualitatively different behavior with ϵ^{-1} instead of $\epsilon^{1/2}$ for small ϵ. The formal divergence of Equation 14.19 for $\epsilon \to 0$ can be cured at the expense of a more complicated expression by taking into account that beyond a maximum radius ρ_{max} the laser intensity is too weak to ionize.

The enhancement of small kinetic energies after averaging over the laser profile is easily understandable from the higher weight of laser intensities less than the peak intensity, which leads to less charging and, consequently, to more ions with smaller kinetic energy.

In most experiments, the laser beam interacts with clusters of different size N, which are log-normally distributed (Gspann 1982, Lewerenz et al. 1993) according to

$$g(N) = \frac{1}{\sqrt{2\pi}\nu N} \exp\left(-\frac{\ln^2(N/N_0)}{2\nu^2}\right) \qquad (14.20)$$

shown in Figure 14.8. Convoluting the single-cluster KEDI from Equation 14.18 with $g(N)$ yields in scaled units $\epsilon = E/E_{\mathrm{coul}}(R, q, N_0)$

$$\frac{dP_{\mathrm{size}}}{d\epsilon} = \frac{3}{4} N_0 \sqrt{\epsilon}\, \mathrm{erfc}\left(\frac{\frac{3}{2}\ln\epsilon}{\sqrt{2}\nu}\right). \qquad (14.21)$$

This size-averaged KEDI, shown in Figure 14.9c, reaches with its tail beyond the energy $\epsilon = 1$, as more as larger the width parameter ν of the cluster size distribution (14.20) is. The fastest fragments are those from the large clusters in the long tail of this

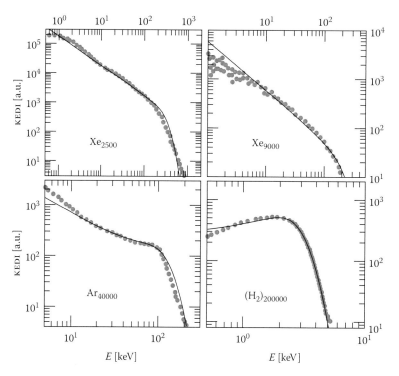

FIGURE 14.11 Ion energy spectra for Xe_{2500} (Ditmire et al. 1997), Xe_{9000} (Springate et al. 2000), Ar_{40000} (Kumarappan et al. 2001), and $(H_2)_{200000}$ (Sakabe et al. 2004) clusters from experiments (*circles*) and fits by our model, Equation 14.24 (*solid line*); (From Mikaberidze, A. et al., *Phys. Rev. A*, 77, 041201, 2008. With permission.)

distribution. Note that we have assumed the average charge q per fragment to be independent of the cluster size N.

Of course, for a realistic experimental KEDI, one has to take into account both the spatial profile of the laser beam and the cluster size distribution. This yields in a similar manner as for the other distributions,

$$\frac{dP_{both}}{d\epsilon} = \frac{\xi^2 \pi}{4} \frac{N_0}{\epsilon}$$
$$\times \left[\exp\left(\frac{v^2}{2}\right) \left(1 + \mathrm{erf}\left(\frac{2v^2 - 3\ln\epsilon}{2\sqrt{2}v}\right)\right) + \epsilon^{3/2} \mathrm{erfc}\left(\frac{3\ln\epsilon}{2\sqrt{2}v}\right) \right].$$
$$(14.22)$$

Here, we have used $\epsilon = E/E_0$, with the reference energy $E_0 = E_{coul}$ (R, q_0, N_0) defined as the maximum Coulomb energy of ions from clusters with the median size $N_0^{\star}$ at the laser focus (charge q_0). The corresponding distribution is shown in Figure 14.9d. Since the spatial laser profile modifies the low-energy part and the cluster size distribution the high-energy part of the ion distribution, it is possible to gain information from a measured KEDI on both effects separately.

The final phenomenon that must be taken into account to understand an experimental KEDI is saturation, i.e., the fact, that

independent of the laser intensity the charging cannot be higher than a certain maximum value q_{sat}, either because the next atomic shell has a much higher ionization potential or because the atoms are completely ionized. We can model the situation by changing our spatial charging function $q(\rho)$ to

$$q(\rho) = \begin{cases} q_{sat} & \text{for } \rho \le \rho_{sat}, \\ q_0 \exp(-\rho^2/2\xi^2) & \text{for } \rho > \rho_{sat}, \end{cases} \quad (14.23)$$

with the maximum charge q_{sat}, which is realized for clusters close to the center of the laser focus with $\rho < \rho_{sat}$. The saturation can be characterized by the dimensionless quantity $\eta := q_{sat}/q_0$ $\in [0,1]$. The radius of saturation in Equation 14.23 is given by $\rho_{sat} = \xi\sqrt{-2\ln\eta}$. The charging function Equation 14.23 amounts to averaging over the spatial profile only for $\rho > \rho_{sat}$ and suggests to define the energy scale as $\epsilon = E/E_{sat}$ with the saturation energy $E_{sat} = E_{coul}(R, q_{sat}, N_0)$. The result is the KEDI

$$\frac{dP_{sat}(\eta)}{d\epsilon} = \frac{dP_{both}}{d\epsilon} - \ln\eta \frac{dP_{size}}{d\epsilon}, \quad (14.24)$$

which develops a characteristic hump before $\epsilon = 1$, as can be seen in Figure 14.9e.

To illustrate how the presented expressions for KEDI apply, we fit them in Figure 14.11 to experimental data of very different situations. Whereas xenon clusters do not show any noticeable saturation effect ($\eta = 0.8$, upper two graphs in Figure 14.11), the large gap between the 1st and the 2nd shell of argon

---

* The median size N_0 separates the higher half of the distribution from the lower half. In a log-normal distribution it differs from the average size which is $\exp(v^2/2)N_0$.

is responsible for the hump seen in the KEDI ($\eta = 0.35$, lower left graph in Figure 14.11). Finally, hydrogen clusters are extreme cases, since only one electron per atom is available ($\eta = 0$, lower right graph in Figure 14.11). Of course, the KEDI derived can not only be used to interpret experimental spectra regarding mean size and saturation of the cluster but also serve as a general tool to enable the comparison of single-cluster KEDI obtained theoretically via the corresponding convolution to experimental results.

14.6 Composite Clusters and the Role of Charge Migration

14.6.1 Charge Migration in a Cluster

We have seen that electrons, trapped in the cluster potential, referred to as "quasi-free" electrons, play a key role in the dynamics of clusters under intense laser pulses. One may ask if these electrons migrate to a preferred location in the cluster. This is indeed the case, if the precondition for their localizability is given, i.e., if the cluster radius R is much larger than the quiver amplitude $x_F \equiv F_0/\omega^2$ (the typical distance electrons travel in a given external laser field), $R/x_F \gg 1$. In this case, the fairly general electron migration sequence holds, as sketched in Figure 14.12: (1) Electrons are (single) photoionized from the atoms by the laser and leave the cluster, cf. Figure 14.5a. (2) The substantial ionic charge developed during (1) generates a large electric field at the cluster surface, leading to field ionization of surface atoms. The ionized electrons move

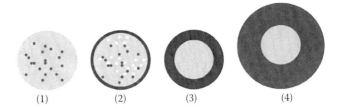

FIGURE 14.12 Sketch of different phases of electron migration in a cluster, see text. Negative charges are white, positive charges and charge density dark, neutral is light gray.

toward the cluster center. (3) As photoionization and the loss of electrons continues, the ionic charge becomes so large that photoionized electrons do not have sufficient energy to leave the cluster potential. This marks the onset of plasma formation, cf. Figure 14.5c. The electron plasma shields the core of the cluster and (4) the (highly charged) ions at the surface explode.

We illustrate the electron migration during the laser pulse with a calculation (Siedschlag and Rost 2004) for the first experiment at FLASH (Wabnitz et al. 2002), the FEL at DESY in Hamburg, for a small (80 atoms) Xenon cluster, exposed to laser short pulses of 12.7 eV photon energy. The radial density of ions and electrons at different times in Figure 14.13 shows how the net surplus of positive charge due to photoionization (step 1, Figure 14.13a) leads to the migration of the quasi-free electrons toward the center (step 2, Figure 14.13b and c), which is shielded while the outer unscreened ions explode (step 4, Figure 14.13c and d).

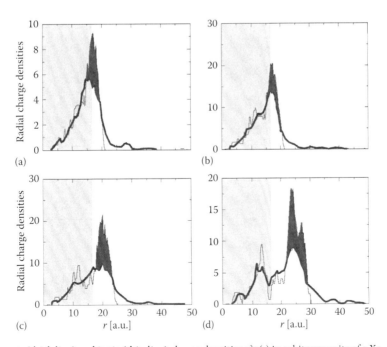

FIGURE 14.13 Radial electronic (thick line) and ionic (thin line) charge densities $r^2\rho(r)$ in arbitrary units of a Xe$_{80}$ cluster at different times during the interaction with the laser pulse at 1.2 fs (a), 3.6 fs (b), 7.3 fs (c), and 9.7 fs (d). The ionic surplus is marked with a dark shaded area, the gray shaded rectangles indicate the initial cluster size. The pulse parameters were $I = 7 \times 10^{13}$ W/cm^2, $T = 100$ fs, $\omega = 12.7$ eV. (After Siedschlag, C. and Rost, J.M., *Phys. Rev. Lett.*, 93, 043402, 2004.)

14.6.2 Two-Component Clusters

Experimentally, charge migration can be made visible by using two different kinds of atoms in the center and for the surface, so-called core–shell systems. In a recent experiment on strong laser pulses from FLASH impacting two-component cluster containing a xenon core and an argon surface, it was experimentally shown that one can completely suppress the occurrence of charged xenon ions (see Figure 14.14) if the parameters in the experiment are chosen such that the xenon core remains fully screened by quasi-free electrons (Hoener et al. 2008).

Composite clusters, containing at least two sorts of atoms or molecules give rise to quite spectacular effects when exposed to intense laser pulses. For instance, a strong enhancement of x-rays from laser-irradiated argon clusters doped by a few percent of water molecules was observed (Jha et al. 2005). In another setup, deuterons, fast enough to induce nuclear fusion (Ditmire et al. 1999) can be generated in heteronuclear clusters (Last and Jortner 2001, Hohenberger et al. 2005).

A similar electron migration was found in a xenon cluster surrounded by a helium droplet (Mikaberidze et al. 2008).

This is the most common two-component system, since helium embedding is an alternative to produce rare gas clusters compared to supersonic expansion. Moreover, embedding a cluster or another object of interest in a so-called tamper or "sacrificing layer" is an important step toward single molecule imaging, which is one goal with intense x-ray radiation at FELs (Hau-Riege et al. 2007).

14.7 Outlook

We have briefly touched upon the most relevant mechanisms for and consequences of the effective energy absorption by atomic clusters from short, intense laser pulses focusing on a wavelength of 800 nm.

Whereas pump-probe experiments have revealed atomic motion (cf. Section 14.3.3), electronic motion needs shorter pulses—in the range of attoseconds—to be seen. Due to a rapid progress in the last decade, laser technologies have reached this range (Corkum and Krausz 2007). An application to clusters would not only allow to study the transient state of the nanoplasma (Georgescu et al. 2007a) but, even more fascinating, its formation (Saalmann et al. 2008) which occurs on sub-femtosecond timescale.

A completely new research field of laser–cluster interaction is opened by the construction of intense *short-wavelength* light sources, namely, FELs operating at ultraviolet (Feldhaus et al. 2007) and x-ray (Feldhaus et al. 2005) wavelengths. The basic process of energy absorption differs qualitatively form what was discussed above. First FEL experiments with clusters showed surprisingly very high charge states (Wabnitz et al. 2002) although the interaction for these high frequencies was clearly of perturbative nature. The dynamics of the created nanoplasma under these conditions is a topic of current investigations (Georgescu et al. 2007b).

Acknowledgments

We would like to thank our collaborators Christian Siedschlag, Ranaul Islam, Ionut Georgescu, Alexey Mikaberidze, and Christian Gnodtke, as well as Vitali Averbukh for their input and discussions.

References

Ammosov M V, Delone N B, and Krainov V P (1986). *Sov. Phys. JETP* **64**, 1191.

Ashcroft N W and Mermin N D (1976). *Solid State Physics.* Saunders College, Philadelphia, PA.

Brabec T and Kapteyn H (2004). *Strong Field Laser Physics.* Springer Verlag, New York.

Chen L M, Park J J, Hong K H, Kim J L, Zhang J, and Nam C H (2002). *Phys. Rev. E* **66**, 025402 (R).

Corkum P B (1993). *Phys. Rev. Lett.* **71**, 1994.

Corkum P B and Krausz F (2007). *Nat. Phys.* **3**, 381.

Ditmire T, Donnelly T, Rubenchik A M, Falcone R W, and Perry M D (1996). *Phys. Rev. A* **53**, 3379.

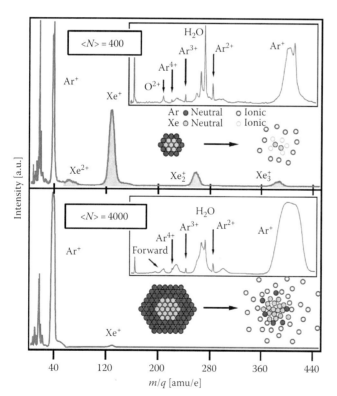

FIGURE 14.14 Mass over charge (m/q) spectrum of Xe core-shell clusters. For small clusters ($N \approx 400$, top) predominantly Xe and Ar ions are detected. For large clusters ($N \approx 4000$, bottom), more highly charged Ar ions are detected while almost no Xe was measured. Arrows in the insets mark the peak positions of high charge states without any initial kinetic energies, i.e., atomic contributions. Also shown are sketches of Ar (dark) and Xe (light gray) atoms (filled) and ions (open) to illustrate the processes. (From Hoener, M. et al., *J. Phys. B*, 41(18), 181001, 2008. With permission.)

Ditmire T, Tisch J W G, Springate E, Mason M B, Hay N, Smith R A, Marangos J, and Hutchinson M H R (1997). *Nature* **386**, 54.

Ditmire T, Zweiback J, Yanovsky V P, Cowan T E, Hays G, and Wharton K B (1999). *Nature* **398**, 489.

Döppner T, Fennel T, Diederich T, Tiggesbäumker J, and Meiwes-Broer K H (2005). *Phys. Rev. Lett.* **94**, 013401.

Feldhaus J, Arthur J, and Hastings J B (2005). *J. Phys. B* **38**, S 799.

Feldhaus J, Choi J Y, and Rah S (2007). *AIP Conf. Proc.* **879**, 220.

Fennel T, Döppner T, Passig J, Schaal C, Tiggesbäumker J, and Meiwes-Broer K H (2007). *Phys. Rev. Lett.* **98**, 143401.

Georgescu I, Saalmann U, and Rost J M (2007a). *Phys. Rev. Lett.* **99**, 183002.

Georgescu I, Saalmann U, and Rost J M (2007b). *Phys. Rev. A* **76**, 043203.

Gspann J (1982). In S Datz, ed., *Physics of Electronic and Atomic Collisions*. North-Holland Publishing Company, Amsterdam, the Netherlands, pp. 79–92.

Haberland H, ed. (1994). *Springer Series in Chemical Physics*, Vol. 52. Springer, Berlin, Germany.

Hau-Riege S P, London R A, and Szöke A (2004). *Phys. Rev. E* **69**, 051906.

Hau-Riege S P, London R A, Chapman, H N, Szoke A, and Timneanu N (2007). *Phys. Rev. Lett.* **98**, 198302.

Hirokane M, Shimizu S, Hashida M, Okada S, Okihara S, Sato F, Iida T, and Sakabe S (2004). *Phys. Rev. A* **69**, 063201.

Hoare M R (1979). *Adv. Chem. Phys.* **XL**, 49.

Hoener M, Bostedt C, Thomas H, Landt L, Eremina E, Wabnitz H, Laarmann T, Treusch R, de Castro A R B, and Möller T (2008). *J. Phys. B* **41**(18), 181001.

Hohenberger M, Symes D R, Madison K W, Sumeruk A, Dyer G, Edens A, Grigsby W, Hays G, Teichmann M, and Ditmire T (2005). *Phys. Rev. Lett.* **95**, 195003.

Islam R Md, Saalmann U, and Rost J M (2006). *Phys. Rev. A* **73**, 041201.

Jackson J D (1998). *Classical Electrodynamics*. Wiley Text Books, New York.

Jha J, Mathur D, and Krishnamurthy M (2005). *J. Phys. B* **38**, L 291.

Kaplan A E, Dubetsky B Y, and Shkolnikov P L (2003). *Phys. Rev. Lett.* **91**, 143401.

Kim K Y, Alexeev I, Parra E, and Milchberg H M (2003). *Phys. Rev. Lett.* **90**, 023401.

Krishnamurthy M, Mathur D, and Kumarappan V (2004). *Phys. Rev. A* **69**, 033202.

Kumarappan V, Krishnamurthy M, and Mathur D (2001). *Phys. Rev. Lett.* **87**, 085005.

Landau L D and Lifschitz E M (1994). *Mechanics*. Pergamon Press, Oxford, U.K.

Last I and Jortner J (1999). *Phys. Rev. A* **60**, 2215.

Last I and Jortner J (2001). *Phys. Rev. Lett.* **87**, 033401.

Lewenstein M, Balcou P, Ivanov M Y, L'Huillier A, and Corkum P B (1994). *Phys. Rev. A* **49**, 2117.

Lewerenz M, Schilling B, and Toennies J P (1993). *Chem. Phys. Lett.* **206**, 381.

Mikaberidze A, Saalmann U, and Rost J M (2008). *Phys. Rev. A* **77**, 041201.

Milchberg H M, McNaught S J, and Parra E (2001). *Phys. Rev. E* **64**, 056402.

Prado A F B A (1996). *J. Guid. Control Dyn.* **19**, 1142.

Rose-Petruck C, Schafer K J, Wilson K R, and Barty C P J (1997). *Phys. Rev. A* **55**, 1182.

Saalmann U (2006). *J. Mod. Opt.* **53**, 173.

Saalmann U and Rost J M (2003). *Phys. Rev. Lett.* **91**, 223401.

Saalmann U and Rost J M (2008). *Phys. Rev. Lett.* **100**, 133006.

Saalmann U, Siedschlag C, and Rost J M (2006). *J. Phys. B* **39**, R 39.

Saalmann U, Georgescu I, and Rost J M (2008). *New J. Phys.* **10**, 025014.

Sakabe S, Shimizu S, Hashida M, Sato F, Tsuyukushi T, Nishihara K, Okihara S et al. (2004). *Phys. Rev. A* **69**, 023203.

Santra R (2006). *Chem. Phys.* **329**, 357.

Seideman T, Ivanov M Y, and Corkum P B (1995). *Phys. Rev. Lett.* **75**, 2819.

Siedschlag C and Rost J M (2002). *Phys. Rev. Lett.* **89**, 173401.

Siedschlag C and Rost J M (2004). *Phys. Rev. Lett.* **93**, 043402.

Siegman A E (1986). *Lasers*. University Science Books, Sausalito, CA.

Springate E, Hay N, Tisch J W G, Mason M B, Ditmire T, Marangos J P, and Hutchinson M H R (2000). *Phys. Rev. A* **61**, 044101.

Taguchi T, Antonsen T M Jr., and Milchberg H M (2004). *Phys. Rev. Lett.* **92**, 205003.

Tang K T and Toennies J P (2003). *J. Chem. Phys.* **118**, 4976.

Wabnitz H, Bittner L, de Castro A R B, Döhrmann R, Gürtler P, Laarmann T, Laasch W et al. (2002). *Nature* **420**, 482.

Wales D J, Doye J P K, Dullweber A, Hodges M P, Naumkin F Y, Calvo F, Hernández-Rojas J, and Middleton T F (2007). The Cambridge Cluster Database. http://www-wales.ch.cam.ac.uk/CCD.html.

Zuo T and Bandrauk A D (1995). *Phys. Rev. A* **52**, R 2511.

Zweiback J, Ditmire T, and Perry M D (1999). *Phys. Rev. A* **59**, R 3166.

Zweiback J, Cowan T E, Hartley J H, Howell R, Wharton K B, Crane J K, Yanovsky V P, Hays G, Smith R A, and Ditmire T (2002). *Phys. Plasmas* **9**, 3108.

15

Cluster Fragmentation

Florent Calvo
Université Lyon I

Pascal Parneix
Université Paris Sud 11

15.1 Introduction

Atomic and molecular clusters in the gas phase have played a historical role in the emergence of nanotechnology, as the focus of intense fundamental research on the influence of size on the physical and chemical properties. Even today, nanoscience heavily relies on materials built from clusters in the gas phase that are subsequently deposited on a substrate or are self-assembled together into a bulk phase.

Any finite system in vacuum, without specific confinement, is metastable and due to decay after some time, as soon as its energy exceeds any dissociation threshold, or simply if it has a finite temperature. Thermodynamical metastability results from the lower free energy of the vapor phase with respect to the bound phase, when the allowed volume diverges. Fortunately, metastable clusters can be long lived and prone to efficient formation and detection in various apparatus. In particular, mass spectrometry is an invaluable tool to achieve size selection and investigate cluster properties at the mass resolution of single atoms. The necessity of ionizing the clusters for mass detection is one example of perturbation, which can induce fragmentation. More generally, the dissociation of a cluster is intimately dictated by several physical and chemical factors. The ability to keep the stored energy depends on the chemical nature and

bonding within the cluster that determine not only the ground state stability but also the complex nonadiabatic transitions between excited electronic states. Inert gas clusters will react very differently from, for example, ionic clusters for a same given excitation. The fragmentation of a cluster also depends highly on the character of its excitation and its magnitude. Collisions can be soft and transfer only a moderate amount of momentum (Dietrich et al., 1996; Jarrold et al., 1987; Su and Armentrout, 1993), or they can involve highly energetic projectiles that alter the electronic structure at both the valence and core levels (Gobet et al., 2001; Martinet et al., 2004). Laser irradiation may also be moderate with the purpose of heating the cluster (Fielicke et al., 2006), or range up to x-ray lasers (Gisselbrecht et al., 2008), ultra intense pulses (Ditmire et al., 1997; Lezius et al., 1998), or synchrotron radiation (Gisselbrecht et al., 2005). The timescales are a third crucial element, as fragmentation is an out-of-equilibrium process that may not have the time to take place within the experimentally available detection time. Alternatively, it may also be too fast and followed by other spurious decay processes that prevent the interesting species from being detected. Finally, although the focus here is on gas-phase systems, the environment of a cluster plays a role on the fragmentation process. For instance, clusters can be prepared in a heat bath or a supersonic expansion, can be surrounded by

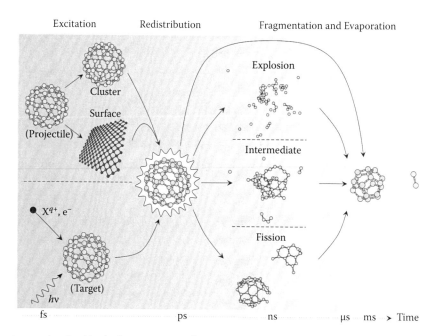

FIGURE 15.1 Important stages involved in the fragmentation of an atomic or molecular cluster, pictured here as buckminsterfullerene. From left to right, the cluster is first excited by collision of an incoming projectile, by a laser, or by collision onto another cluster or a surface. This excitation process is usually very fast (from femtoseconds to picoseconds). The energy is partially or entirely redistributed among the vibrational degrees of freedom within a typical few tens of picoseconds. Fragmentation during the next nanoseconds follows under various forms that differ in the number and sizes of the fragments. Eventually, the remaining clusters can evaporate atoms or small molecules over macroscopically long timescales.

other clusters, or can be transferred from vacuum into a fluid chamber where they may act as nucleation seeds. The interplay between these factors creates a rich variety of fragmentation behaviors.

The main stages of a cluster fragmentation event are depicted in Figure 15.1. Typical excitation processes can be of two types, the cluster being either a target or a projectile. Excitation by a laser (photoabsorption) or by impact of a small but energetic particle offers a rather precise estimation of the energy deposited in the cluster, as well as a way of ionizing for subsequent mass spectrometry analysis. Alternatively, clusters can be excited by throwing them against other clusters (Campbell and Rohmund, 2000; Farizon et al., 1997; Knospe et al., 1996) or onto a substrate. The latter case has been specifically investigated for potential applications in materials science (Moseler et al., 2000), or with the purpose of inducing chemical reactions at the cluster–substrate contact (Châtelet et al., 1996).

The excitation of a cluster by collision or photoabsorption is usually fast, and lies in the femtosecond to picosecond range, or even shorter. In most situations, the excitation energy is deposited electronically, and the conversion and redistribution of this electronic energy into the nuclear degrees of freedom constitutes the second major stage, before fragmentation itself takes place in the picosecond to microsecond timescales. Important decay modes range from fission into two main fragments, as can be found in the case of Coulomb dissociation at low fissility, to the explosive multifragmentation into many small clusters. In addition, and besides all possible intermediate situations, other decay

modes can sometimes contribute significantly to the release of the excess energy in the cluster. These most common decay modes, namely, electron emission, radiative cooling by photon emission (blackbody radiation), Auger processes and gamma emission, are not covered in this chapter, but are mentioned here for sake of completeness.

Several forms of fragmentation are also present for a given system at different timescales. Explosive fragmentations leave some clusters with relatively low internal energies that can still emit several atoms or molecules by evaporative cooling over the longer experimental timescales, extending beyond the millisecond.

The goal of this chapter is to provide an overview of the main classes of fragmentation behaviors in atomic and molecular clusters. We have chosen to focus on the theoretical methods that are useful to analyze fragmentation observables and on their relation with experimental measurements. In particular, fragmentation is a key process that has been shown to provide quantitative information about the dissociation energies and, more recently, thermodynamical properties. The chapter is organized into six main sections that cover the short and long timescales, the various statistical approaches to the unimolecular dissociation and multifragmentation phenomena, or the more specific features of Coulomb fragmentation and nucleation theories. Most of these methods are illustrated on experimental and simulation data taken from various groups. In the concluding section, some final comments are given on the possible combination of several such methods within a single, integrated approach capable of addressing the fragmentation problem from a multiscale perspective.

15.2 Short-Time Fragmentation Dynamics

The duration of an excitation that leads to fragmentation can vary over many orders of magnitude, for collisional excitations (10^{-15} to 10^{-12} s range), but especially for laser irradiation (10^{-16} to 10^{-9} s range). This section discusses some of the general features occurring under short timescales, which determine the possible future fragmentation of the cluster.

15.2.1 Impulsive and Electronic Excitations

Fragmentation in clusters is usually caused by a fast perturbation, the time needed to excite the system being much shorter than the typical time needed to actually dissociate. A possible exception to this rule is the exposure of a cluster to a picosecond or nanosecond laser, in which the pulse is long enough for the excitation to compete with dissociation. However, in most experiments, the fragmentation mechanisms strongly depend on the kind of excitation and on the nature and chemical bonding of the cluster. The three main mechanisms at play during the first stages after excitation are depicted in Figure 15.2. The sensitivity of fragmentation towards the details of the potential energy surface is particularly true in the case where the excitation has the function of ionizing the cluster for subsequent mass spectrometry analysis. Ionization itself can represent a dramatic perturbation for a weakly bound cluster made up of rare-gas atoms or closed shell molecules. In addition to stronger polarization forces, such cationic systems usually exhibit some degree of covalent bonding (Gadéa and Amarouche, 1990; Kuntz and Valldorf, 1988), whereas anionic systems are prone to forming dipole-bound structures (Jordan and Wang, 2003). In contrast, adding or removing an electron to a large covalent or metallic cluster will not alter its relaxation so significantly because these systems have bunches of closeby electronic states.

If the excitation is stored as nuclear momentum on the initial electronic ground state, as would be the case in a low-energy collision, *impulsive* fragmentation on adiabatic surfaces may occur (see Figure 15.2a). At the other end, excitation by photoabsorption or impact by an energetic particle will initially alter the electronic cloud. These *electronic* processes can lead to rapid dissociation if the excited surface is repulsive (see Figure 15.2b). However, if the cluster has a sufficient number of atoms or if many electronic surfaces are close to each other, a partial or even complete internal conversion of electronic energy into nuclear momenta can occur through multiple nonadiabatic transfers across conical intersections (see Figure 15.2c).

Generally speaking, both impulsive and electronic processes are present in the earliest stages following excitation. Collisions of inert atoms on metal clusters, for instance, will be more likely to transfer moderate nuclear kinetic energies at high impact parameters, but may also change the electronic populations for frontal collisions (Fayeton et al., 1998). Laser excitations, on the other hand, will first alter the electronic state, and the subsequent dynamics is strongly imprinted by the chemical bonding within the cluster.

Rare gas clusters provide a good example where the dominating mechanism varies depending on cluster size, electronic effects being especially sensitive in the smaller clusters. Kuntz and Hogreve (1991) were the first to recognize the importance of electronic processes involving nonadiabatic effects in the fragmentation of the cationic argon trimer. Besides them, fragmentation in rare gas clusters has been computationally studied by the groups of Gadéa (Amarouche et al., 1989; Calvo et al., 2003) in the

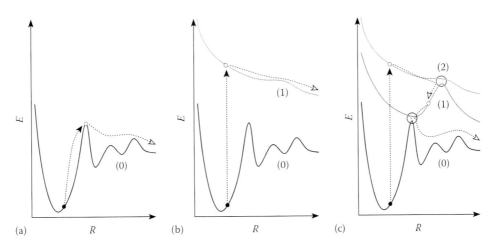

FIGURE 15.2 Schematic potential energy surfaces of clusters subject to an excitation, as a function of a typical dissociation coordinate R. In (a), the cluster is collisionally heated, remains on its electronic ground state surface (0), but gains in nuclear momenta. In (b), the cluster is excited on state (1), but this surface is repulsive, and the cluster dissociates on the excited state. This situation is typical in small rare-gas clusters undergoing photoabsorption at specific wavelengths. In (c), the cluster is excited on state (2), but an additional intermediate excited state (1) is coupled with both the ground state and the second excited state through conical intersections highlighted by empty circles. Through two nonadiabatic transitions, the cluster is brought back toward the ground state before fragmenting. This latter situation occurs for larger rare-gas clusters, or clusters bound by covalent or metallic forces.

case of photodissociation and more recently by Halberstadt and coworkers (Bonhommeau et al., 2007) who focused on fragmentation by electron impact. Experiments (Albertoni et al., 1987) on the photodissociation of Ar_3^+ in visible wavelengths have shown that the fragmentation exclusively takes place in the excited state (Ar^+ and two neutral Ar atoms) rather than on the ground state surface ($Ar_2^+ + Ar$). These experiments were interpreted as the manifestation of the purely repulsive surface reached upon excitation at these wavelengths, leading to an explosive dynamics. In contrast, internal conversion of electronic energy into kinetic nuclear energy plays an increasing role in larger clusters, as seen from the more numerous Ar_2^+ fragments (Nagata et al., 1991) found in experiments.

If the fragmentation is induced by a projectile, the region undergoing excitation can differ quite significantly from the Franck-Condon zone of photoexcitation. In addition, each collision can produce different excitations due to the varying impact parameter. This is, for instance, reflected on the relative proportions of Ar^+ and Ar_2^+ fragments found by Buck and Meyer (1986) for excitations by electron impact, and whose nontrivial variations with parent size have not been fully interpreted yet. One case where the excess energy brought by the collision can be accurately determined has recently been proposed by Chen et al. (2007). These authors used protons as projectiles to excite neutral fullerenes and measured the kinetic energy of the H^- ions produced on double-electron capture. Since these ions only exist in the ground electronic state, the energy deposited in the doubly cationic fullerene can be directly obtained from the measured kinetic energies.

15.2.2 Theoretical Modeling of Nonadiabatic Dynamics

One of the key issues in determining the outcome of the excitation is the importance of intramolecular vibrational relaxation. At the theoretical level, simulations must be able to account for the simultaneous presence of multiple electronic state surfaces and their possible crossings. Unfortunately, obtaining these surfaces and the couplings is a difficult task of computational chemistry, for which time-dependent density-functional theory (TDDFT) (Marques and Gross, 2003) stands as one of the very few but general methods for this purpose. TDDFT has been used by Suraud and coworkers (Calvayrac et al., 1998) in the local density approximation to study energy deposition in sodium clusters by intense laser pulses and their subsequent fragmentation.

In addition to the computationally costly density-functional methods, useful approximations have been developed for specific cluster systems, sometimes with a good accuracy compared with high-level ab initio calculations. The diatomic-in-molecules (DIM) approach (Kuntz and Valldorf, 1988), for instance, proceeds somewhat similarly to a valence-bond approach and performs really well not only for the rare gases but also for dedicated systems such as metals. The tight-binding method, with a comparable complexity as the DIM approach, also provides information about the excited state surfaces and their couplings.

Tight-binding models have been particularly successful for studying the interaction between lasers and covalent systems such as fullerenes (Jeschke et al., 2002).

The dynamics on multiple electronic state surfaces can be addressed using either a mean-field, Ehrenfest approximation for propagating the nuclear and electronic degrees of freedom at the same time (Heller, 1978), or trajectory surface hopping (TSH) techniques (Tully, 1990), in which a single electronic surface is populated at any time. The mean-field method has been used extensively by Gadéa and coworkers to study photofragmentation in argon clusters (Amarouche et al., 1989; Calvo et al., 2003). Briefly, the idea of the method is to solve the time-dependent Schrödinger equation

$$| \dot{\Psi}(t) \rangle = -i\hbar \hat{H}_{el} | \Psi(t) \rangle, \tag{15.1}$$

together with the classical nuclear equations of motion (written for atom i),

$$\dot{\mathbf{R}}_i = \frac{\mathbf{P}_i}{m_i}; \quad \dot{\mathbf{P}}_i = -\left\langle \Psi(t) \left| \frac{\partial \hat{H}_{el}}{\partial \mathbf{R}_i} \right| \Psi(t) \right\rangle. \tag{15.2}$$

In these equations, the electronic wavefunction Ψ of the electronic Hamiltonian $\hat{H}_{el}$ is expressed as a linear combination of the static basis set $\{\Psi_k\}$ with time-dependent coefficients (Amarouche et al., 1989; Heller, 1978). The method is efficient, even for larger systems; however, it suffers from the fact that the electronic wavefunction is always mixed between various states. As shown by Calvo et al. (2003), this leads to a significant underestimation of the extent of intramolecular vibrational relaxation in the photofragmentation of argon clusters containing nine or less atoms.

The TSH method, also known as molecular dynamics with quantum transitions, expands the electronic wave function in a similar way, but assumes that the system lies on a single electronic state at each time. Hops between surfaces near conical intersections are allowed with some probability described and discussed by Tully (1990). At regular time intervals δt, the switching probabilities $\{p_{jk}\}$ between the current state j and the other states k are computed, and a new state may be randomly drawn based on these probabilities. The probability p_{jk} involves the nonadiabatic coupling σ_{jk} that depends on the nuclear velocities $\dot{\mathbf{R}}$ and the wavefunctions and their gradient as (Tully, 1990)

$$\sigma_{jk} = \dot{\mathbf{R}} \cdot \left\langle \Psi_j \left| \frac{\partial \Psi_k}{\partial \mathbf{R}} \right\rangle. \tag{15.3}$$

Finally, a_{jj} denoting the electronic population in the initial state, the probability p_{jk} is given by (Tully, 1990)

$$p_{jk} = \frac{\delta t \times \sigma_{jk}}{a_{jj}}. \tag{15.4}$$

During these time intervals, the classical equations of motion are propagated for the nuclei on the current potential energy surface. The electronic amplitudes are propagated simultaneously, following the time-dependent Schrödinger equation. The time interval δt is chosen so as to yield a significant number of transitions during the simulation time. The TSH method has been used by Halberstadt and coworkers (Bonhommeau et al., 2007) for simulating rare gas clusters excited by electron impact and by Sizun and coworkers (Sizun et al., 2005) for small metal clusters undergoing collisions by a helium projectile. It was also employed by Fiedler et al. (2008), together with the Ehrenfest approach, for the photoexcitation dynamics in Xe_3^+. Although it does not have the mixing problem of the mean-field method, the TSH method has some stochastic character, which introduces some noise in the dynamics. However, as demonstrated by its very recent combination with TDDFT (Tapavicza et al., 2007) for studying photochemical processes in biomolecules, this method seems generally attractive for the fragmentation problem, especially when used with swarms of trajectories.

15.3 Mass Spectrometry and Cluster Stabilities

The stability of an atomic or molecular cluster depends on both its electronic and geometrical structures. Specially stable clusters are often determined as prominent peaks in mass spectra (the so-called magic number sizes), which indicate a greater resistance to dissociation (Knight et al., 1984; Mcelvany and Ross, 1992; Pedersen et al., 1991). A lot of work has been devoted to developing methods for estimating binding energies from mass spectrometry, and two significant examples of such methods are discussed in this section. More recently, it has become possible to extend the capabilities of these analyses, by getting insight into the temperature-dependent energetic properties of size-selected clusters in order to achieve nanocalorimetry measurements.

15.3.1 Evaporative Ensemble

In a typical mass spectrometry experiment, a cluster is first prepared in the gas phase using various possible methods. Depending on the material, it may be more convenient to expose a solid to a high intensity laser, to heat it into an oven or, if their vapor pressure is sufficiently high, to grow the cluster from individual atoms or molecules in a supersonic beam. All these methods generally produce neutral clusters with a broad distribution of sizes. In addition, they do not allow a stringent control of the internal energy when the clusters are formed from solid samples. Clusters are then ionized after interacting with a laser or an electron beam, before being accelerated in an electric field zone, entering a field-free zone and being eventually detected. Cluster ions with different charge/mass ratios arrive at the detector after different times-of-flight, allowing their separation.

As they enter the field zone, the clusters possess some internal energy that may be most likely dissipated by emission of a neutral atom or molecule, sometimes of larger clusters. The time needed for evaporating an atom X from cluster X_{n+1}^+ mainly depends on the internal energy of this parent cluster and the dissociating

energy D_n defined as the difference in binding energy between X_{n+1}^+ and X_n^+. If the dissociation takes place after the cluster has left the electrostatic field zone, the product cluster should be detected at about the same time as its parent because the kinetic energy released (KER) in the dissociation process is usually very small compared with the ion kinetic energy. This problem can be solved by isolating the clusters X_{n+1}^+ and X_n^+ by an additional electrostatic gate for accelerating or decelerating the ions in order to separate them. This method has been used in the Bréchignac group to determine branching ratios between the monomer and dimer evaporations in alkali clusters (Bréchignac et al., 1989, 1990, 1994a).

Suitable analyses of the mass spectra intensities I_n can then be used to estimate the dissociation energies. The pioneering method introduced by Klots (1987), known as evaporative ensemble, is followed here. The main idea underlying this method is that the experimental detection of a given cluster ion X_n^+ imposes constraints on the internal energy (or the temperature) of its parent at the time when it entered the electric field zone. Had the parent cluster been colder, it would not have evaporated in the time-of-flight. Conversely, had it been warmer, the product cluster itself would have evaporated.

Denoting k_{n+1} and k_n as the evaporation rates of the clusters X_{n+1}^+ and X_n^+, respectively, and E_n the internal energy of the product cluster, the detection of this cluster enforces that the time τ needed for the cluster to travel across the electric field zone is such that $k_n(E_n) \times \tau < 1$. Reciprocally, the knowledge of the relation $k_n(E)$ and the duration τ provides an estimate of the maximum internal energy $E_n^{\max}$ that can be carried by the cluster X_n^+ without evaporating before detection. A minimum internal energy $E_{n+1}^{\min} = E_n^{\min} + D_{n+1}$ of the parent cluster can be likewise estimated, by assuming that X_n^+ was formed from X_{n+1}^+ during the time τ, which leads to the inequality $k_{n+1}(E_{n+1}) \times \tau > 1$. The two extremal energies $E_n^{\min}$ and $E_n^{\max}$ turn out to depend on τ; however, their difference is close to the dissociation energy D_n being looked for.

To proceed further, an expression is needed for the dissociation rate k_n, and a convenient choice is provided by a simple Arrhenius-like expression (Hansen and Näher, 1999):

$$k_n(T_n) = A_n \exp\left(-\frac{D_n}{k_B T_n}\right), \tag{15.5}$$

in which T_n is the temperature of cluster X_n^+ and the preexponential factor A_n a constant. This Arrhenius form will be justified below in Section 15.4 from the more rigorous microcanonical ensemble. At the maximum energies $E_{n+1}^{\max}$ and $E_n^{\max}$, the parent and product clusters have the corresponding temperatures $T_{n+1}^{\max}$ and $T_n^{\max}$, respectively. Assuming for simplicity that the clusters behave harmonically, $E_n^{\max} = C_n T_n$, with $C_n = (3n-6)k_B$ the classical heat capacity, and a similar expression between $E_{n+1}^{\max}$ and T_{n+1}. (Note that $6k_B$ have been removed from C_n due to the conservation of both linear and angular momenta.) It should be mentioned here that the clusters are supposed to be large enough for the relation $C_{n+1} \approx C_n$ to hold. For a same time-of-flight τ, these temperatures satisfy the relation

$$k_{n+1}(T_{n+1}^{\max}) \times \tau \approx k_n(T_n^{\max}) \times \tau \approx 1. \tag{15.6}$$

Assuming further that the preexponential factors A_n do not depend on n, the dimensionless Gspann parameter G is introduced such that $A_n\tau = \exp(G)$ (Gspann, 1982). Together with the Arrhenius form of the dissociation rate constant and the caloric relation between temperature and energies, the difference between maximum energies reads

$$E_{n+1}^{\max} - E_n^{\max} = \frac{C_n}{k_B G}(D_{n+1} - D_n). \tag{15.7}$$

The ion signal $\mathcal{I}_n$ linearly depends on the range of internal energies allowing detection of the corresponding cluster ion, that is $E_n^{\max} - E_n^{\min}$. Since $E_n^{\min} = E_{n+1}^{\max} - D_n$, the relation between the dissociation energies and the intensities is finally obtained as

$$\mathcal{I}_n \propto \left\{ D_n - \frac{C_n}{k_B G}(D_{n+1} - D_n) \right\}. \tag{15.8}$$

It should be mentioned that, despite a long-time use in experiments, the Gspann parameter is not very well known. Its most recommended value for atomic clusters is 23.5 for time-of-flights close to 10 μs, as also found by Klots (1991) for fullerene ions. However, more recent measurements on such clusters have shown quite significant deviations, with G values in the 30–40 range (Foltin et al., 1998; Hansen and Campbell, 1996; Laskin et al., 1998; Tomita et al., 2001).

A limitation of the above procedure is that it relies to a large extent on an explicit formula for the dissociation rate constant. As is shown in Section 15.4, accurate unimolecular dissociation theories are not as simple as the Arrhenius model employed above. Recently, Vogel et al. (2001b) have proposed an alternative method in which the cluster ions are stored in a Penning trap and excited in a controlled way using photoabsorption. Trapping allows the dissociation process over much longer timescales with respect to standard time-of-flight apparatus. By looking separately at the single and double evaporation processes from a given parent, these authors were able to estimate the first dissociation energy without using any specific expression for the rate constants.

The principle of the method is to eject the ions from the trap at various times after exposing the sample to the laser, and to record the evolution of the mass spectra over a long timescale. Considering the evaporation of one and two monomers from the cluster X_{n+1}^+, the peak intensities $\mathcal{I}_{n+1}(t)$, $\mathcal{I}_n(t)$ and $\mathcal{I}_{n-1}(t)$ are measured for a given excitation energy E_{n+1}^* and increasing t. The same procedure is repeated, but for the parent X_n^+ undergoing a single monomer evaporation, for which the excitation energy E_n^* is set so as to yield the same peak intensities $\mathcal{I}_n$ and $\mathcal{I}_{n-1}$ as in the previous experiment. The time variations of the peak intensities straightforwardly provide the rate constants of the two evaporation processes, and the dissociation energy D_n is related to the excitation energies through the relation

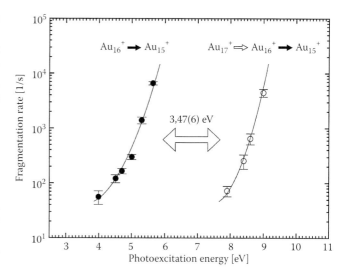

FIGURE 15.3 Dissociation rates for $Au_{16}^+ \rightarrow Au_{15}^+$ following either a direct photoexcitation of Au_{16}^+ or a higher photoexcitation of Au_{17}^+ in which two atoms are successively emitted. (Reprinted from Vogel, M. et al., *Phys. Rev. Lett.*, 87, 013401, 2001. With permission.)

$$D_n = E_{n+1}^* - E_n^* + (E_{n+1}^{\text{th}} - E_n^{\text{th}}) - \langle\varepsilon\rangle, \tag{15.9}$$

where

$E_{n+1}^{\text{th}} \approx C_{n+1}T_{n+1}$ is the (small) initial thermal energy prior to excitation

$\langle\varepsilon\rangle$ is the negligible part of the total energy released as kinetic energy in the first dissociation process

Reasonably accurate estimates of the thermal energies are provided by harmonic or Debye approximations (Vogel et al., 2001b). The method has been first applied to cationic gold clusters in a size range where only monomer evaporation is the dominating channel (Vogel et al., 2001b). On adjusting E_n^*, the evaporation rates obtained for $Au_{16}^+ \rightarrow Au_{15}^+$ display a nice shift of about 3.47 eV, as represented in Figure 15.3. The method has since been extended to measure the branching ratios between monomer and dimer emissions in gold clusters (Vogel et al., 2001a, 2002), as well as the sequential evaporation over timescales reaching the second (Schweikhard et al., 2005). It has also provided accurate dissociation energies for vanadium clusters (Hansen et al., 2005).

Trapping the clusters can also be achieved using storage rings, and this technique has been used notably by Andersen and coworkers to estimate the dissociation energies of fullerenes (Tomita et al., 2001, 2003) from the decay rates, assuming an Arrhenius form with a given prefactor. However, under these very long timescales (of the order of the milliseconds), it is important to account for the effects of radiative cooling when estimating the dissociation rates.

15.3.2 Kinetic Energy Release Spectra

In addition to raw intensities, the shape profile of mass spectra can be exploited to get insight into the KER of metastable clusters, which in turn is related to the dissociation energy. A pioneer of such approaches, Stace has determined the KER distribution in the evaporation of cationic carbon dioxide (Stace and Shukla, 1982) and argon (Stace, 1986) clusters in the field-free zone of the time-of-flight mass spectrometer. The measurement of accurate kinetic energies of size-selected clusters has been perfected by the groups of Lifshitz and Märk in the so-called mass-analyzed ion kinetic energy spectra (Cao et al., 2001; Gluch et al., 2004). Appropriate procedures transform the raw mass-analyzed ion kinetic energy spectra into KER distributions, $p(\varepsilon)$. These distributions usually have a bell shape that can be conveniently fitted by an exponential form

$$p(\varepsilon) \propto \varepsilon^{\gamma} \exp\left(-\frac{\varepsilon}{k_B T_n}\right). \tag{15.10}$$

As will be shown in Section 15.4, the two parameters γ and T_n are related to the interaction between the dissociating fragments and the microcanonical temperature of the product cluster, respectively. Knowledge of γ provides a linear relation between the product temperature T_n and the average KER $\langle \varepsilon \rangle = \gamma k_B T_n$. An estimate of the dissociation rate is used, together with the Arrhenius expression of Equation 15.5, to determine the dissociation energy D_n. Such analyses have been carried out with a very high accuracy for the sequential emission of carbon dimers from fullerene ions (Cao et al., 2001; Gluch et al., 2004).

15.3.3 Cluster Calorimetry from Mass Spectra

After fitting onto an Arrhenius-type expression, the shape of the KER distribution provides an estimate of the product temperature. If the internal energy can also be estimated, the caloric curve of the cluster can be reconstructed. This idea has been pursued by the Bréchignac group (Bréchignac et al., 2001, 2002), who measured the KER in the evaporation of sodium (Bréchignac et al., 2001) and strontium (Bréchignac et al., 2002) cationic clusters. In the latter work, the variations of the microcanonical temperature with internal energy exhibit a plateau, which the authors interpreted as the manifestation of the liquid–gas phase transition rounded by size effects. These caloric curves are shown in Figure 15.4 for Sr_{10}^+ and Sr_{11}^+.

The experimental determination of cluster caloric curves has also been achieved with an impressive accuracy by the Haberland group, who suggested to determine the internal energy of a cluster based on its fragmentation pattern (Schmidt et al., 1997). The method consists in producing size-selected clusters with a temperature controlled through equilibrium with a heat bath (Ellert et al., 1995), and to expose them to a laser beam. The clusters are heated either initially in the heat bath or by absorbing n photons of energy $h\nu$, and mass spectra are monitored as a function of both the number n and the initial temperature T (see Figure 15.5). For the metal clusters considered in this experiment, these optical excitations are efficiently converted into thermal energy. At a given temperature, the mass spectrum can be analyzed to determine the average number of photons absorbed, corresponding to an energy $\Delta E = nh\nu$. Once this number of photons

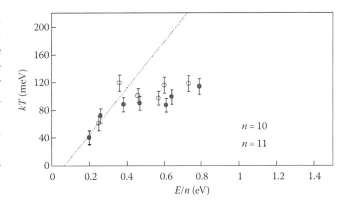

FIGURE 15.4 Microcanonical temperature of the Sr_{10}^+ and Sr_{11}^+ clusters versus internal energy, as inferred from kinetic energy release distributions in the evaporation of the parent clusters. (Reprinted from Bréchignac, C. et al., *Phys. Rev. Lett.*, 89, 203401, 2002. With permission.)

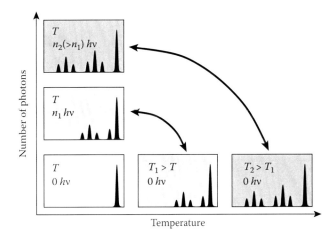

FIGURE 15.5 Schematic representation of the cluster calorimetry experiment of the Haberland group. Mass spectra of initially size-selected clusters are recorded as a double function of temperature (horizontal axis) and number of absorbed photons (vertical axis).

is known, the equivalent temperature increase ΔT needed to produce the same mass spectrum, but without laser exposure, is estimated by a trial-and-error procedure. The heat capacity $C_v = \Delta E / \Delta T$ is then estimated, which on integration on temperature provides the internal energy $U(T)$. This method, initially developed for sodium clusters (Schmidt et al., 1997), revealed unexpected variations in the melting point and latent heat of melting of clusters containing up to about 300 atoms (Schmidt et al., 1998). These complex size effects have since been shown to result mainly from the geometry, rather than from the electronic structure (Aguado and López, 2005; Haberland et al., 2005; Noya et al., 2007). By extending their method to the microcanonical ensemble, Schmidt et al. (2001a,b) were also able to find evidence for a possible negative heat capacity in the magic number cluster Na_{147}^+, and signatures of the liquid–gas transition at higher temperatures.

The above method is convenient for controlling the amount of energy deposited in the cluster, but becomes of limited use if the clusters under study are not prone to laser heating. Collision-induced dissociations (Dietrich et al., 1996; Jarrold et al., 1987; Su and Armentrout, 1993), on the other hand, do not generally provide such a clean control on energy but are more widely applicable to arbitrary systems. The use of collisions in cluster calorimetry as the way of heating the clusters has been first put forward in the Jarrold group (Breaux et al., 2003). The basic idea behind the method consists in varying the collision energy in a gas chamber, through which the clusters travel before being detected. The fragmentation patterns are again recorded as a double function of initial temperature and collision energy, leading to an estimate of the cluster heat capacity. The method has been used for gallium (Breaux et al., 2003, 2004) and aluminum (Breaux et al., 2005) clusters. A noteworthy improvement on this method has been recently proposed by Chirot et al. (2008), in which the collision occurs at low energy and is limited to the sticking of a single atom.

15.4 Unimolecular Dissociation Theories

As shown in Section 15.3.1, cluster dissociation energies can be estimated from mass spectrometry measurements; however, this determination often requires some assumptions about the dissociation rates or the KER distributions. The main theories of unimolecular dissociation, from which these quantities derive, are now presented. Contrary to Section 15.2 where only the first stages of the cluster dynamics immediately following excitation were considered, the focus here is on the long-time evaporation kinetics. In particular, it will be henceforth assumed that all excitation energy has been completely transferred and redistributed among all nuclear degrees of freedom. This statistical hypothesis is justified by the very long timescales involved in evaporation phenomena. Unimolecular dissociation through evaporation can be a very slow process, and the rate constant is expected to depend drastically on size. For instance, a diatomic molecule will dissociate as soon as its internal energy exceeds its binding energy, after a single vibrational period. In a bulk system, even if a material has a huge amount of energy stored in its vibrational modes (phonons) with respect to the binding energy of a single atom, dissociation of this atom is extremely unlikely due to the statistical repartition, or very slow with respect to the typical timescale of vibrational motion. At low energies, evaporation and thermal dissociation are thus fundamentally rare events in the time evolution of a cluster.

15.4.1 The Rice–Ramsperger–Kassel–Marcus Theory

A simplified approach to the dissociation of an atomic cluster X_{n+1} into $X_n + X$ is to treat them as sets of harmonic oscillators, following the seminal ideas of Rice and Ramsperger (1927) and Kassel (1928a). Denoting $g = 3n - 3$ the number of independent degrees of freedom of the parent cluster, the probability for the available energy E to be localized in one dissociative mode first requires this energy to exceed the dissociation value D_n. The number of ways to distribute the energy E over g oscillators is $E^{g-1}/(g - 1)!$, which yields the Rice–Ramsperger–Kassel (RRK) dissociation rate as

$$k_n(E) = v_0 \left(\frac{E - D_n}{E} \right)^{g-1}, \quad (15.11)$$

the prefactor v_0 being usually adjusted to reproduce experimental data. The above expression holds for classical systems, but Kassel has proposed a quantum version, more satisfactory in small systems or at low energy (Kassel, 1928b). Assuming for simplicity that all oscillators have the same frequency v, the number of energy quanta stored in the total and dissociation energies are $p = E/hv$ and $q = D_n/hv$, respectively, with h the Planck constant. The number of ways to distribute j quanta among g harmonic oscillators is given by $(j + q - 1)!/j!/(g - 1)!$; hence, the quantum dissociation rate is now expressed as

$$k_n(E) = v_0 \frac{p!(p - q + g - 1)!}{(p + g - 1)!(p - q)!}. \quad (15.12)$$

As expected, the classical expression of Equation 15.11 is straightforwardly recovered from this quantum expression by taking the limit $h \to 0$.

If the cluster is assumed to be in thermal equilibrium at temperature T, a temperature-dependent rate constant $k_n(T)$ can be estimated by convolution of the microcanonical rate $k_n(E)$ with the thermal distribution $P_{th}(E,T) \propto E^{g-1} \exp(- E/k_B T)$. The result is precisely the Arrhenius expression already met in Equation 15.5. Therefore, a justification of the Arrhenius form for the rate constant can be found in the harmonic assumption.

Unfortunately, statistical theories cannot be directly verified by experimental measurements because in most cases, the latter use the former for interpreting the results. Molecular simulation, on the other hand, provides a much better testing ground for these theories. The RRK model has been compared with the results of molecular dynamics trajectories in rare-gas (Weerasinghe and Amar, 1993) and metallic (López and Jellinek, 1994) clusters. Without using any input from these trajectories, and estimating the prefactors from the known vibrational frequencies of the parent and product clusters, the RRK theory turned out to strongly underestimate the evaporation rates in both cases (López and Jellinek, 1994; Weerasinghe and Amar, 1993).

The main criticism against the RRK model is the harmonic assumption, which is obviously wrong for a realistic dissociating system since atoms would not be able to escape from such a quadratic confining potential. These problems lead Marcus to introduce the concept of transition state (Marcus, 1952). The Rice–Ramsperger–Kassel–Marcus (RRKM) or transition state theory (TST) assumes that the dissociation of X_{n+1} into $X_n + X$ occurs through a hypersurface in configurational space that separates the parent from the products. A first postulate of TST is that all accessible states on this hypersurface are equiprobable. In addition, all energetic properties involving the products are calculated at the transition state, and no recrossing or energy transfer back to the parent is possible.

Introducing $E^\dagger$, the potential energy at the transition state, Marcus showed that the rate constant at the internal energy E can be written as (Marcus, 1952)

$$k_n(E) = \frac{W_n(E - E^\dagger)}{h\Omega_{n+1}(E)}, \quad (15.13)$$

where

$\Omega_{n+1}(E)$ is the density of vibrational states available in the parent cluster at energy E

$W_n(E - E^\dagger)$ is the total number of vibrational states available at the transition state energy

W_n can be cast as the integral of a density of states $\Omega_n^\dagger$

$$W_n(E - E^\dagger) = \int_0^{E-E^\dagger} \Omega_n^\dagger(E - E^\dagger - \varepsilon_t) d\varepsilon_t, \quad (15.14)$$

where ε_t represents the (translational) KER in the dissociation. The probability p of finding a dissociation event at total energy E with a KER of ε_t is given by the differential rate $\mathcal{R}(\varepsilon_t; E) = \Omega_n^\dagger(E - E^\dagger - \varepsilon_t)/h\Omega_{n+1}(E)$, up to a normalization factor which is nothing else but the rate constant $k_n(E)$:

$$p(\varepsilon_t; E) = \mathcal{R}(\varepsilon_t; E)/k_n(E). \quad (15.15)$$

If the transition state is assumed to be at the dissociation products, then $E^\dagger = D_n$. If the problem is furthermore simplified by using harmonic densities of states for both Ω_{n+1} and $\Omega_n^\dagger = \Omega_n$, explicit forms are found for the differential rate and the KER distribution, and the rate constant is then given by Equation 15.11. The RRK theory is thus a special case of TST.

A difficulty in applying TST lies in determining a precise location of the transition state. In the variational extension of TST (Miller, 1974; Wigner, 1937), the dividing surface is defined from the thermodynamical properties of the parent cluster. Variational TST has been used for investigating the evaporation of water clusters (see Section 15.7 below). Transition states are important in the case of multiply charged clusters, for which the repulsion between the charges is balanced with the cohesive forces, yielding a so-called Coulomb barrier (see Section 15.5). However, for neutral or singly-charged systems, statistical dissociation usually occurs with some orbital angular momentum that induces a centrifugal barrier. Angular momentum in the parent cluster can be accounted for by replacing the densities and numbers of vibrational states with the corresponding densities and numbers of rovibrational states (Hase, 1998). This procedure can also be applied to the simple RRK theory, at least by including the centrifugal contribution in the total available energy. Miller and Wales have followed this strategy in their computational study of evaporation in rotating clusters (Miller and Wales, 1996). However, these ad hoc corrections still neglect the barrier that originates from the orbital momentum.

15.4.2 Theories Based on Microreversibility

The RRK(M) theories consider dissociation from the point of view of the parent clusters. However, the alternative point of view of the products is possible as well. Theories based on the microscopic reversibility principle (also referred to as detailed balance) consist in determining the statistical observables by considering the phase space equilibrium between dissociation and sticking events. In other terms, and besides the dissociation from X_{n+1} into $X_n + X$ with rate k_n, the reverse process of nucleation of $X + X_n$ onto X_{n+1} is assumed to take place as well with the reaction rate k'. These ideas have been first developed by Weisskopf (1937) in relation with nuclear decay, and extended later in the field of chemical reaction kinetics under the name of phase space theory (PST). The latter approach, which has been highly detailed by Light and coworkers (Light, 1967; Pechukas and Light, 1965), Nikitin (1965a,b), Klots (1971, 1972), and Chesnavich and Bowers (1976, 1977a,b), among others, is briefly described in the following. In PST, the potential energy barrier is supposed to be at the products, and conservation of the angular momentum during dissociation is fully taken into account. An orbiting transition state arises from the finite value of orbital angular momentum and its balance with the interaction between the products.

The dissociation of a cluster with internal energy E and angular momentum J is considered from the above perspective. The microreversibility principle equates the forward and backward fluxes (in phase space) $\Phi(E, J)$ and $\Phi'(E, J)$ corresponding to the dissociation and reverse nucleation reaction. The forward flux Φ is the product of the dissociation rate k_n and the vibrational density of states Ω_{n+1} of the parent. In PST, an additional factor S_{rot} is included to account for the rotational degeneracy of the parent, which essentially depends on its symmetry properties:

$$\Phi(E, J) = k_n(E, J)S_{\text{rot}}\Omega_{n+1}(E - E_{\text{rot}}). \quad (15.16)$$

In this equation, E_{rot} denotes the rotational energy of the parent cluster. In the simplest cases of linear or spherical parents, $E_{\text{rot}} = BJ^2$ with B the rotational constant. For the inverse nucleation reaction, the backward flux Φ' is related to the probability of forming the parent X_{n+1} from the products X and X_n, which depends on both the translational and rotational energies of each collision event. An explicit form for Φ' is given by (Chesnavich and Bowers, 1977b)

$$\Phi'(E - D_n, J) =$$

$$\rho S'_{\text{rot}} \iint k'(\varepsilon_r, \varepsilon_t; J)\rho_t(\varepsilon_t) d\varepsilon_t \times \Omega_n(E - D_n - \varepsilon_t - \varepsilon_r) d\varepsilon_r.$$

$$(15.17)$$

where

ρ accounts for the symmetry factors of the parent and products

S'_{rot} is the rotational degeneracy factor of the products

$k'(\varepsilon_r, \varepsilon_t; J)$ is the differential rate for the collision to form the parent cluster with angular momentum J, at translational energy ε_t and rotational energy ε_r

The density of translational states $\rho_t(\varepsilon_t)$ can be exactly eliminated from the above equation. Equating the fluxes Φ and Φ' leads to a formal expression for the differential rate of dissociation as a function of the total KER $\varepsilon_{tr} = \varepsilon_t + \varepsilon_r$:

$$\mathcal{R}(\varepsilon_{tr}; E, J) = \frac{\rho S'_{rot}}{S_{rot}} \frac{\Omega_n(E - D_n - \varepsilon_{tr})\Gamma_{rot}(\varepsilon_{tr}, J)}{\Omega_{n+1}(E - E_{rot})}. \quad (15.18)$$

In this equation, the rotational density of states Γ_{rot} counts the number of available rotational states at fixed values of the total angular momentum and the kinetic energies released. In practice, this quantity is calculated by assuming specific shapes for the two products in order to use simple relations between the individual angular momenta and the rotational energy. The conservation of total energy and total linear and angular momenta during dissociation imposes constraints on the possible values that the orbital and products angular momenta can take. The interaction between the two products also affects the rate via the rotational density of states, because together with the orbital angular momentum, it determines the position and height of the centrifugal barrier (as in a Langevin picture). The densities of states involved in the differential rate do not depend explicitly on the angular momenta, except through the rotational energy E_{rot}. Strictly speaking, Ω_{n+1} should explicitly depend on J, however, and since the angular momentum of the product does not have a fixed value, the situation is less clear for Ω_n. To some extent, the dissociation energies should also have some dependence on angular momenta due to the correcting centrifugal energies (Calvo and Labastie, 1998).

The Weisskopf theory is essentially similar to the PST approach, except that it ignores constraints related to angular

momentum conservation (Weisskopf, 1937). The rotational density of states of the PST differential rate is then replaced by a simple function of the translational energy only, which represents a sticking cross-section $\sigma(\varepsilon_t)$:

$$\mathcal{R}(\varepsilon_t; E) \propto \varepsilon_t \frac{\sigma(\varepsilon_t)\Omega_n(E - D_n - \varepsilon_t)}{\Omega_{n+1}(E)}. \quad (15.19)$$

This relation can be made further explicit if the interaction between the products is known to have a simple radial form of the type $-C/r^p$, because then $\sigma(\varepsilon_t)$ behaves as $\varepsilon_t^{-2/p}$. Typical values for p are 4 and 6 for ion–neutral and neutral–neutral dissociations, respectively. Although the Weisskopf approach is generally more accurate than RRK(M) theories, the ability to predict the angular momentum of the products and its distribution makes PST an even more general formalism. Its performances have been checked with respect to the results of molecular dynamics simulations, which in this context can be considered as to provide numerically exact reference data. In addition to the simplest Lennard-Jones clusters (Calvo and Parneix, 2003; Parneix and Calvo, 2003, 2004; Weerasinghe and Amar, 1993), PST has been validated on metal clusters (Peslherbe and Hase, 1994, 1996, 2000) and molecular clusters (Calvo and Parneix, 2004). As an example of application, Figure 15.6a shows the distributions of total KER during the unimolecular evaporation of the cluster Ar_{14} at fixed total energy for two values of angular momentum (Calvo and Parneix, 2003). Similarly, in Figure 15.6b, the distributions of angular momentum of the product cluster are shown in the case of the evaporation of a methane cluster, again for two values of the initial angular momentum (Calvo and Parneix, 2004). For these two systems, the vibrational densities

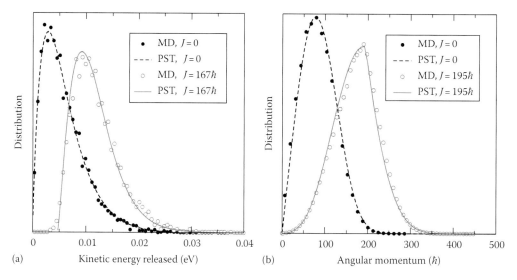

FIGURE 15.6 Comparison between the predictions of phase space theory and molecular dynamics trajectories for observables related to the unimolecular evaporation of van der Waals clusters. (a) Normalized distributions of kinetic energy released after evaporation of an atom from the Lennard-Jones cluster Ar_{14}, either nonrotating ($J = 0$) or with an angular momentum of $J = 167\hbar$, and for a total energy of 0.2 eV units above the ground state energy. (Adapted from Calvo, F. and Parneix, P., *J. Chem. Phys.*, 119, 256, 2003.) (b) Normalized distributions of angular momentum of the product cluster after evaporation of a methane molecule from the molecular cluster $(CH_4)_{14}$, either nonrotating ($J = 0$), or with an angular momentum of $195\hbar$, at 70 kcal/mol internal energy above the ground state. (Adapted from Calvo, F. and Parneix, P., *J. Chem. Phys.*, 120, 2780, 2004.)

of states were obtained from independent Monte Carlo simulations to include full anharmonic effects, and the rotational densities were calculated by treating the clusters as spherical tops, the products interacting with each other via a $-C/r^6$ dispersion potential (Calvo and Parneix, 2003, 2004). As can be seen from Figure 15.6, the agreement between molecular dynamics trajectories and the PST method is quantitative for the two properties considered.

Besides energy and angular momentum distributions, PST predicts rate constants in much better agreement than the RRK approach (Weerasinghe and Amar, 1993). It can also be used to calculate branching ratios, by comparing their absolute differential rates. This has been achieved for the mixed atomic cluster $KrXe_{13}$, which shows a low-temperature structural transition (Parneix et al., 2003). It should be emphasized here that predicting equilibrium distributions or average quantities is a much easier task for these theories, due to the difficulty in calculating the absolute densities of states in general and to the poor knowledge of the prefactors entering Equations 15.18 and 15.19.

The failure of the RRK model reported in many simulation studies (Calvo and Parneix, 2003, 2004; Parneix and Calvo, 2003; Weerasinghe and Amar, 1993) can be due either to its harmonic approximation or to its complete neglect of mechanical constraints. This situation can be clarified thanks to the model developed by Engelking (1986, 1987) for interpreting mass spectra experiments on cationic carbon dioxide clusters. The Engelking approach is closely related to the historical Weisskopf model, but assumes that the sticking cross-section σ does not depend on the kinetic energy, in a geometric approximation. Engelking also assumes that vibrational densities of states can be taken as harmonic.

In Figure 15.7, the variations of the average KER $\langle \varepsilon_{tr} \rangle$ during the unimolecular evaporation from a nitrogen cluster, as obtained

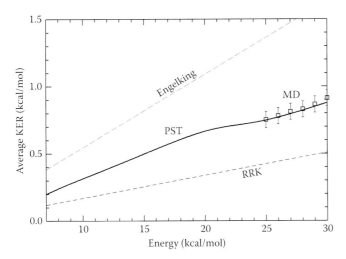

FIGURE 15.7 Average kinetic energy released during the evaporation of a nitrogen molecule from the $(N_2)_{14}$ cluster, as a function of its internal energy. The molecular dynamics results are compared with the predictions of phase space theory, the RRK model, and Engelking model. (Adapted from Calvo, F. and Parneix, P., *J. Chem. Phys.*, 120, 2780, 2004.)

from molecular dynamics trajectories and from the predictions of various statistical theories, are represented. Because $\langle \varepsilon_{tr} \rangle$ is a normalized quantity, it does not involve the density of states of the parent only but that of the product also. Among the three statistical theories, only PST accounts for anharmonic effects. Figure 15.7 shows that both the Engelking and RRK models are qualitatively wrong, while the PST calculation gives a quantitative account of the averaged KER at moderate and high energies. Similar results have been obtained previously for rare-gas clusters (Weerasinghe and Amar, 1993). The success of PST is a strong evidence that anharmonicities are important for reproducing dissociation-related properties.

The close relation between the KER distribution and the density of states of the products encourages these dissociation observables to be used for getting insight into the thermodynamical properties of the fragmented cluster, along similar lines as previously discussed in Section 15.3.3. For instance, as shown in Figure 15.7, PST predicts an inflection in the variations of $\langle \varepsilon_{tr} \rangle$ for the $(N_2)_{14}$ cluster at internal energies near 22 kcal/mol. This inflection signals the solid–liquid phase transition in the product cluster, rounded by size effects (Calvo and Parneix, 2004). The relation between average KER and thermodynamic properties can be further explicited by considering the microcanonical temperature T_n, which derives from Ω_n through $(k_B T_n)^{-1} = \partial \ln \Omega_n(E)/\partial E$. A Taylor expansion of the differential rate $R(\varepsilon_t; E)$ leads to an expression for the translational KER distribution with an explicit Arrhenius form of the type of Equation 15.10. In the Weisskopf theory, γ is related to the interaction parameter p between the products as $\gamma = 1 - 2/p$ (Hansen and Näher, 1999; Weisskopf, 1937). In the PST approach, a similar result is found (Klots, 1992) if appropriate approximations on the maximum rotational energy are made.

In small systems, the Arrhenius expression of Equation 15.10 may not be accurate (Calvo et al., 2006), and possible improvements have been discussed (Andersen et al., 2001; Calvo et al., 2006). The expansion of the density of states may be corrected at second order, introducing the microcanonical heat capacity C_{n-1} in the expression of the KER distribution. Finite-size corrections to the rotational energy have also been considered (Andersen et al., 2001; Calvo et al., 2006), and phenomenological expressions for the KER distribution, featuring the two aforementioned corrections, have been proposed by Calvo et al. (2006). Noteworthy, a Taylor expansion of the density of states has also been used by Andersen et al. (2001) who corrected for anharmonicities in a Weisskopf approach by introducing the microcanonical temperature.

15.5 Coulomb Fragmentation

A way to induce instability in a finite system is to ionize it sufficiently. Because a cluster has a limited spatial extension, each additional charge will tend to move away from the other charges, and the balance with the attractive binding energy will eventually be broken. This phenomenon known as Coulomb decay occurs at different length scales, in aerosols and mesoscopic droplets (Duft et al., 2003; Widmann et al., 1997), in optical molasses (Pruvost et al., 2000), and in nuclei (Bohr and Wheeler, 1939) alike. The spontaneous fragmentation due to Coulomb repulsion is also essential for the electrospray device, which aims at producing charged molecules in the gas phase for mass spectrometry. Experimental production of multiply charged clusters is usually achieved by relatively intense irradiation (Näher et al., 1994) or by collision of a highly charged projectile (Chandezon et al., 1995).

The appearance size $n_c(q)$ of a multiply charged cluster X_n^{q+} is its first characteristic property. It is defined for a given apparatus as the smallest size, below which the cluster cannot be observed under standard experimental conditions. The appearance size strongly depends on the system; however, the liquid drop model accounts reasonably well for most observations (Echt et al., 1988) (except perhaps neon [Mähr et al., 2007]).

15.5.1 The Fissility Parameter

The first systematic investigation of Coulomb instability in finite systems dates back to the seminal work by Rayleigh (1882), who considered the equilibrium of a liquid droplet in equilibrium between its cohesive surface tension and the Coulomb repulsion. Rayleigh (1882) introduced a fissility parameter χ defined as the ratio between the Coulomb repulsion energy and twice the cohesion energy:

$$\chi = \frac{E_{\text{Coulomb}}}{2E_{\text{surface}}}. \tag{15.20}$$

The fissility is related to the energy barrier that must be crossed for dissociating the cluster, as depicted in Figure 15.8a. In particular, for $\chi = 1$, the barrier vanishes and the cluster spontaneously dissociates. For fissilities below 1, the barrier turns Coulomb dissociation into a thermally activated process, which competes with other decay channels (see Figure 15.8b). Fissilities larger than 1 correspond to even less stable clusters that are prone to Coulomb explosion.

15.5.2 Low Fissilities and Fission Channels

The competition between the channels of neutral evaporation, symmetric fission, and asymmetric fission is a central issue of Coulomb fragmentation in clusters, as it was (and remains) in nuclear physics (Näher et al., 1997). This competition involves both thermodynamic and kinetic factors. Especially in very small systems, quantum effects and shell corrections can be very significant in stabilizing some products. In nuclear physics, studies based on the liquid drop model have shown that dissociation into two main fragments with comparable sizes and charges is favored when χ is close to the so-called Businaro-Gallone point $\chi = 0.39$ (Cohen and Swiatecki, 1963). Recent measurements (Duft et al., 2003) on mesoscopic glycol droplets have shown that the onset of instability occurs close to the critical value $\chi = 1$, and that the droplet shape resembles a lemon. These results have recently been verified theoretically (Giglio et al., 2008). In this case, however, the fission is highly asymmetric and the charged particles are emitted as jets along the symmetry axis.

Fission of multiply charged atomic clusters has been experimentally observed in small silver (Sweikhard et al., 1996), lithium (Bréchignac et al., 1994b), potassium (Bréchignac et al., 1994c), and strontium (Bréchignac et al., 1998) clusters. The role of electronic shell effects was demonstrated on gold by Saunders (1990, 1992) who found that even-electron clusters tend to exhibit fission decay, whereas odd-electron clusters evaporate neutral atoms. Single- or double-magic fission events, in which one or two products are magic number clusters, respectively, have been experimentally observed in alkalis (Yannouleas et al., 2002).

Several groups have addressed the fission problem from a theoretical point of view. A popular approach inspired by similar efforts in nuclear physics has been to focus on the dissociation pathway between two specific fragments, trying to minimize the energy along the path to determine the fission barrier. Several such calculations have been based on continuum-like descriptions or jellium models at the semiclassical (extended-Thomas-Fermi) level (Garcias et al., 1995) or by including explicit electronic shell effects in a Kohn-Sham approach

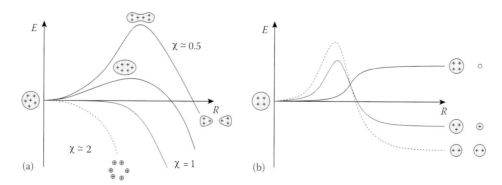

FIGURE 15.8   (a) Coulomb dissociation of a multiply charged cluster, for different values of the fissility parameter χ. The energy of the cluster is represented as a function of the elongation coordinate R. At $\chi = 1$, the dissociation is barrierless. (b) Competition between the three main decay channels in Coulomb fragmentation below the Rayleigh limit. The evaporation of a neutral atom or molecule is essentially barrierless (increasing curve). In the case depicted here, symmetric fission (dashed red line) is thermodynamically favored but kinetically disfavored over the asymmetric fission channel (dashed line).

(Garcias et al., 1994). Deformation of the cluster along the dissociation pathway has been accounted for either by considering two interpenetrating spheres (Engel et al., 1993; Garcias et al., 1994; Saito and Cohen, 1988) or by using more complex parametrized uniaxial shapes (Garcias et al., 1995; Vieira and Fiolhais, 1998; Yannouleas and Landman, 1995). A concern with these approaches is that the fission barrier (and to some extent, the products themselves) depends on the detailed cluster shape (Lyalin et al., 2000).

An explicit account of atomic structure is recovered in the cylindrically averaged pseudopotential scheme of Montag and Reinhard (1995), or in detailed first-principles calculations (Barnett et al., 1991; Blaise et al., 2001; Jena et al., 1992). These studies have in particular confirmed the favored emission of magic number clusters, by showing that the fission barrier is largely driven by the energetic stability of the products.

Further approximations are necessary to address the fragmentation dynamics in larger clusters or over longer timescales. An interesting approach is that of Fröbrich and Ecker (1998), who modeled the dissociation kinetics using a Kramers equation. Simplifying the interactions also provides useful insights. Last et al. (2002, 2005) theoretically investigated the mechanisms of Coulomb fragmentation in highly charged atomic clusters $(X^+)_n$ by varying the range of the Morse pairwise potential. The range of the interaction is a critical parameter that already affects the stability of the neutral system, clusters with short-range interactions exhibiting sublimation rather than melting (Calvo, 2001). As shown by Last et al. (2002, 2005), long-ranged potentials favor dissociation into a limited number of large fragments, especially near the Rayleigh threshold. In contrast, clusters bound by short-ranged potentials tend to emit many small particles, even at low fissilities (Last et al., 2002, 2005). Snapshots from typical molecular dynamics trajectories are illustrated in Figure 15.9.

The Morse potential, while generally satisfactory to model diatomic molecules, is probably not appropriate for modeling metals because of the delocalized character of metallic bonding. Using an empirical many-body potential, Li et al. (1998) found that the cluster Cs_{100}^{4+} symmetrically dissociates into C_{46}^{2+} and Cs_{54}^{2+}. These authors assumed that the charge was homogeneously distributed in the cluster as partial atomic charges, an approximation known to fail in conducting systems such as metals. Similar to the range of the potential, the charge distribution has a strong role in the Coulomb fragmentation process. This has been known for some time, and largely explored in studies based on the jellium model (Näher et al., 1997). The comparison between surface-charged metal clusters and homogeneously charged nuclear matter, in particular, has consistently showed why symmetric fission is favored in the latter case. The surface charge distribution in metal clusters can be mimicked using dedicated models (Calvo, 1999) or fluctuating charges (Mortier et al., 1986; Rappé and Goddard, 1991) that minimize a quadratic, geometry-dependent function. Together with a many-body potential similar to the one employed by Li et al., it is possible to study the systematic effects of charge distribution on the fragmentation mechanisms in multiply charged clusters

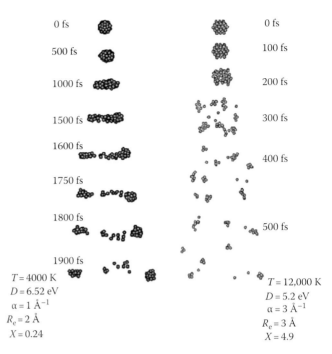

$T = 4000$ K
$D = 6.52$ eV
$\alpha = 1$ Å^{-1}
$R_e = 2$ Å
$X = 0.24$

$T = 12{,}000$ K
$D = 5.2$ eV
$\alpha = 3$ Å^{-1}
$R_e = 3$ Å
$X = 4.9$

FIGURE 15.9 Dissociation dynamics of $(X^+)_{55}$ clusters bound by a Morse potential with range α, equilibrium distance R_e, and dissociation energy D, as obtained from molecular dynamics simulations (Last et al., 2005). The excess temperature T and fissility X are reported. (Reprinted from Last, I. et al., *J. Chem. Phys.*, 123, 154301, 2005. With permission.)

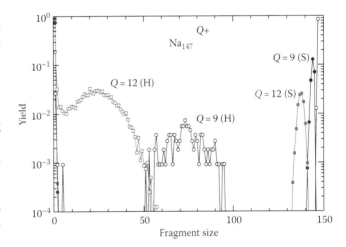

FIGURE 15.10 Fragment distributions obtained from molecular dynamics simulations of the cluster Na_{147}, initially thermalized at 300 K and suddenly ionized at charges $Q = +9$ and $+12$, assuming that the excess charge is distributed on the surface (S) or homogeneously (H). The distributions are recorded 30 ps after ionization and result from binning over 1000 independent trajectories.

(Calvo, 2006b). Figure 15.10 illustrates the fragment mass distributions recorded 30 ps after suddenly ionizing the cluster Na_{147} at charges $Q = +9$ and 12, assuming either homogeneous (H) or surface (S) charge distributions. The corresponding fissilities are 1.0 and 1.7 (homogeneous case) and 0.75 and 1.2 (surface case), respectively. The mass distributions highly contrast with each other. The charge state $Q = 9$ mostly leads to symmetric fission events if the cluster is assumed to be uniformly charged, but only emits a few small particles with one main remaining cluster if the charge is on the surface. By increasing the charge state to $Q = 12$ for uniformly charged clusters, the fission process becomes more asymmetric, and more smaller fragments are emitted. The geometric features of the fragmentation process have also been investigated numerically with this model (Calvo, 2006b). While the fission process into two main fragments is extremely anisotropic, the multiple emission of small particles was found to be essentially isotropic (Calvo, 2006b). Recent experimental measurements on charged rare-gas clusters near the Rayleigh limit (Hoener et al., 2008) have confirmed this transition from anisotropic to isotropic emission as the fissility decreases.

15.5.3 Large Fissilities and the Explosion Regime

High excitations and fissilities above the Rayleigh limit $\chi = 1$ can be reached by submitting a cluster to a collision with an energetic highly charged ion or by irradiation with an intense laser pulse. Fast ionization is achieved if the excitation is very short, within the tens of femtoseconds timescale. Sufficiently low excitations only affect the valence electrons, whose response depends on the material. In the case of metallic or covalent clusters, plasmon oscillations can take place especially near specific wavelengths, which then couple with the ionic motion after tens or hundreds of femtoseconds, possibly inducing evaporation or fragmentation. If the excitation is higher, ionization may also occur.

Very high intensities, with peaks above $10^{14} - 10^{18}$ W/cm^2, can nowadays be reached with ultraintense x-ray lasers (Ditmire et al., 1997; Lezius et al., 1998). A cluster placed into such laser fields loses a large part of its valence and core electrons, each atom reaching a high charge state (Mijoule et al., 2006; Ranaul Islam et al., 2006). Because the outer and inner electron ionization processes are very fast with respect to nuclear motion, the Coulomb repulsion energy between the remaining ions can be extremely high and even reach values near the mega electron volts, usually met in nuclear physics (Last and Jortner, 2000). The resulting explosion proceeds isotropically (Last and Jortner, 2005). However, the presence of several clusters close to each other before ionization, as they would be in a molecular beam, leads to multiple collisions between the ions of different clusters. These collisions are sufficiently energetic to trigger nuclear fusion, as was first experimentally shown by Zweiback et al. (2000) in the example of large $(D_2)_n$ clusters.

The rapid expansion following multiple ionization is particularly interesting for clusters containing light and reactive elements such as hydrogen, deuterium, or tritium, which can be stripped of their unique valence electron. However, as was initially suggested by Last and Jortner (2001), it could be advantageous to work with heterogeneous clusters such as $(DI)_n$, because the very high ionization level of the iodine atom will further enhance the kinetic energy of the deuterium ions and their radial expansion. Heavy water (Tes-Avetisyan et al., 2005) and

deuterated methane (Grillon et al., 2002; Madison et al., 2004) have been indeed found to provide other candidates for nuclear fusion experiments involving clusters.

The fusion reactions triggered by the multiple ionization of atoms in molecular clusters can serve as a basis for chain nucleosynthesis reactions that could be relevant in the astrophysical context (Adelberger et al., 1998). In their computational work, Last and Jortner have found spontaneous reactions between protons and ^{12}C^{6+}, ^{14}N^{7+}, and ^{16}O^{8+} nuclei upon Coulomb explosion of methane, ammonia, and water clusters, respectively (Last and Jortner, 2006, 2007). Such reactions could even be possible within a single large cluster (Last and Jortner, 2008).

15.6 Multifragmentation

High-energy excitations of clusters can lead to strong rearrangements and significant mass losses. For sufficiently large clusters, the short time of these violent processes, as compared with the detection timescale, often refers them as to multifragmentation events. Multifragmentation typically takes place in collision experiments on atomic nuclei, where it is sometimes possible to detect all dissociation products thanks to a 4π apparatus. These sophisticated detectors have also been used by several groups who exposed small clusters to highly energetic projectiles produced in particle accelerators (Gobet et al., 2001; Martinet et al., 2004). The analysis of such experiments cannot usually afford the detailed description allowed in molecular physics through mass spectrometry. The large number of fragments suggests another probe, which was precisely pioneered by the nuclear physics community (Richert and Wagner, 2001), namely, the fragment distribution. These distributions can be analyzed in various ways, but the general context of most efforts in the field is the relation with critical phenomena and the liquid–gas transition (Belkacem et al., 1995; Dorso et al., 1999; Elliott et al., 2000; Raduta et al., 2002). This section reviews the main approaches to the statistical study of multifragmentation that have been used more specifically for atomic and molecular clusters.

15.6.1 The Fisher Model

The global fragment multiplicity (or mass distribution), $\mathcal{P}(n,T)$, depends on the intensity of the excitation assumed to be described by a single temperature parameter T. The shape of $\mathcal{P}(n)$ at fixed T is generally found to be well represented by a simple expression function of three parameters only and known as the Fisher model. Historically, the Fisher model considered the equilibrium of a liquid droplet surrounded by its vapor (Fisher, 1967), and is closely related to classical nucleation theories (CNTs), discussed in the following section. The mass distribution $\mathcal{P}(n,T)$ derives from an expansion of the Gibbs free energy in terms of volume, surface, and curvature contributions. At equilibrium, the probability of formation of a droplet containing n particles is proportional to $\exp(-\Delta G/k_B T)$, where ΔG is the difference in Gibbs free energy corresponding to the formation of the liquid droplet with n particles from the vapor (Belkacem et al., 1995; Fisher, 1967):

$$\Delta G(n,T) = n(\mu_l - \mu_v) + 4\pi R^2 \gamma + T\tau \ln n. \quad (15.21)$$

In the above equation,

μ_v and μ_l stand for the chemical potential of the vapor and liquid phases, respectively

γ is the surface tension

τ is a parameter

$R \propto n^{1/3}$ is the droplet radius, and the last term accounts for the closure of the droplet surface onto itself (Fisher, 1967). It should be noted that this $\ln n$ term is phenomenological: a liquid drop expansion would yield a $n^{1/3}$ correction, but as noted by (Belkacem et al., 1995) $\ln n \sim n^{1/3}$ when n remains below a few hundreds.

The previous equation is straightforwardly rewritten to give the mass distribution

$$\mathcal{P}(n,T) \propto n^{-\tau} A^{n^{2/3}} B^n, \quad (15.22)$$

in which the coefficients A and B are functions of temperature. Different fragmentation regimes can be qualitatively described by the Fisher model. Above a critical temperature T_c, the surface tension and the associated term $A^{n^{2/3}}$ vanish. At the critical temperature itself, the chemical potentials are equal, $B = 0$ and the mass distribution is a power law. In practice, undercritical multifragmentation is characterized by a U-shaped distribution corresponding to the loss of many small fragments and a residual large droplet, while overcritical multifragmentation shows rapidly decreasing distributions with increasing fragment size (Belkacem et al., 1995; Fisher, 1967).

The different multifragmentation regimes have been numerically explored by Belkacem et al. (1995) who simulated high-energy fragmentation of model nuclei containing a few hundred of particles. In particular, these authors were able to locate the critical temperature T_c based on their molecular dynamics simulations, as the value at which the mass distribution decays as a power law with fragment size.

The Fisher model has been used to analyze the fragment distributions obtained in molecular dynamics simulations by Kondratyev and Lutz (1997) for atomic clusters, and by Beu and coworkers (Beu, 2003; Beu et al., 2003) for molecular clusters. The latter authors investigated the fragmentation of water and ammonia cluster containing 1000 molecules on heating above 1000 K, and found mass distributions that could be well approximated by power laws. These typical distributions are represented in Figure 15.11. From the simulated distributions, Beu et al. (2003) have also estimated the maximum fragment size and its variations with increasing parent size, which were found in reasonable agreement with experimental measurements using electron impact. The Fisher model has also been used by Gobet et al. (2002) to fit the mass distributions obtained in collision-induced fragmentation of hydrogen clusters. Using specific relations between the parameters A, B, and the temperature, these authors were also able to relate the collision energy to

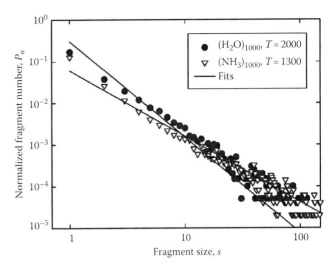

FIGURE 15.11 Normalized fragment size distributions obtained from molecular dynamics simulations of the clusters and $(H_2O)_{1000}$ and $(NH_3)_{1000}$, heated at 2000 and 1300 K, respectively. The straight lines are simple power law fits. (Reprinted from Beu, T.A. et al., *Eur. Phys. J. D*, 27, 223, 2003. With permission.)

a temperature and deduce a microcanonical caloric curve $T(E)$ (Gobet et al., 2002). The presence of a small backbending in this caloric curve was interpreted as the manifestation of the liquid–gas transition in the cluster (Gobet et al., 2002). Note, however, that the accuracy of the data from which this conclusion was reached has been debated (Chabot and Wohrer, 2004; Gobet et al., 2004).

Noteworthy, the Fisher model has some close relations with percolation theories (Debierre, 1997). The important parameter is the probability p to break a bond between two nearest neighbors on a two- or three-dimensional lattice. The cluster size distribution, or multiplicity, is found to depend on n as (Stauffer and Aharong, 1994)

$$\mathcal{P}(n,p) \propto n^{-\tau} \exp(-n|p - p_c|^{1/\Lambda}), \quad (15.23)$$

where

Λ is a parameter

p_c is the critical probability above which the infinite cluster (for an infinite lattice) vanishes from the system

A power law with exponent τ, which depends on the dimensionality of the lattice, is found for the size distribution at the critical probability. This provides a way of estimating p_c in experiments. Comparison between experimental mass distributions and percolation models has been attempted by Lebrun et al. (1994) in high-energy collisions of highly charged xenon ions with C_{60} targets. These authors found a power law in the mass distribution with exponent 1.3. In a theoretical study of collisions between fullerenes, Schulte (1995) also reported power law variations for the fragment size distributions, with exponent 1.5. However, as discussed by Bauer (1988, 2001), this observed exponent cannot be straightforwardly related to the quantity τ of Equation 15.23

because the experiments are inclusive and result from integrating over all possible values of the impact parameter.

15.6.2 Campi Scatter Plots

Another statistical way of interpreting mass distributions in multifragmentation relies on correlations between moments of various orders in event-by-event data analysis, following the approach pioneered by Campi (1986, 1988) in nuclear physics. The Campi method looks differently at the size distribution from the point of view of its moments, and consists in representing conditional moments of the fragment distribution against the largest fragment size (Campi, 1986, 1988). The moment $M_\alpha^{(i)}$ of event (trajectory) i and order $\alpha = 0, 1, 2, \ldots$ is defined by

$$M_\alpha^{(i)} = \sum_n n^\alpha \mathcal{P}_n^{(i)}, \tag{15.24}$$

where $\mathcal{P}_n^{(i)}$ is the multiplicity of fragment of size n for event i, the summation excluding the largest fragment (Campi, 1986, 1988). For the two first moments, the scaled second moment $S_2^{(i)} = M_2^{(i)}/M_1^{(i)}$ is determined for all events, and the Campi scatter plot consists in representing the largest fragment size $N_{\max}$ against S_2, in double logarithmic scale. Depending on temperature, the fragmentation events are located in different regions of the $(S_2, N_{\max})$ diagram. Above the critical temperature T_c, the Campi plot shows one main lower branch with increasing values of $N_{\max}$ and S_2. Below T_c, another upper branch is expected with higher values of $N_{\max}$, but decreasing with S_2. Events at the critical temperature are then determined by the locus where the two branches meet.

In collision-induced multifragmentation, a broad range of excitation energies are sampled, and Campi scatter plots may include events below, at, and above the critical point. Dorso and López (2001) suggested a further selection of events based on the so-called critical multiplicity $\mathcal{P}^{(i)}$, which should maximize the reduced variance $\gamma_2^{(i)} = M_2^{(i)} M_0^{(i)}/[M_1^{(i)}]^2$ (Campi and Krivine, 1992). The Campi method has been applied by Rentenier et al. (2005) to analyze the size distributions in collision-induced fragmentation of buckminsterfullene with high-energy protons. From the moments of the distributions, these authors were able to find clear evidence for undercritical and overcritical events, as shown in Figure 15.12.

Other statistical approaches have been used to address multifragmentation in atomic clusters. In the method developed independently by Levine and coworkers (Raz et al., 1995; Silberstein and Levine, 1981) in molecular physics and by Randrup and Koonin (1981) in nuclear physics, all possible fragments are listed and given statistical weights $W_{j,q}$ that depend on their size j and their charge q. The weights $W_{j,q}$ are determined by maximizing entropy, assuming known heats of formation and degeneracies, and by taking account of conservation laws on mass and charge. This method was applied to the problem of inverse collisions of C_{60}^+ on Ar targets by Campbell et al.

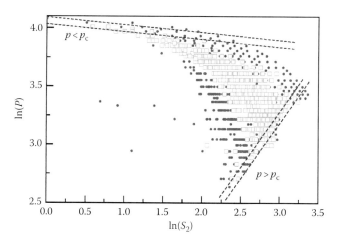

FIGURE 15.12 Campi scatter plot showing the correlation between the size of the largest fragment P as a function of the normalized moment S_2, in proton-C_{60} collisions at 40 keV, as obtained by Rentenier et al. (2005). The solid dots are the results over all events, while the open squares are restricted to events with multiplicity 4. The dashed lines emphasize the undercritical and overcritical branches for the two samples. (Reprinted from Rentenier, A. et al., *J. Phys. B: At. Mol. Opt. Phys.*, 38, 789, 2005. With permission.)

(1996), who showed that the total mass distributions and appearance energies were in good agreement with mass spectrometry experiments at low collision energies.

Along similar lines, more sophisticated models were developed by Gross and coworkers (Gross, 1997; Gross and Hervieux, 1995), who used a Monte Carlo scheme to sample the most probable partitionings between fragments. Briefly, the idea here is to write the total energy of a system of fragments as the sum over internal (binding) energies of individual species, plus their kinetic contributions (translational and rotational); interaction energies being calculated assuming simple physical rules (Gross, 1997; Gross and Hervieux, 1995). Monte Carlo moves consist of usual displacements, as well as annihilation and creation steps, with microcanonical acceptance criteria that fulfill the conservation of mass, charge, and energy. Again, such models have also been used in nuclear physics (Bondorf et al., 1995; Gross, 1990).

15.7 Nucleation Theories

Until now, only systems that are preformed and undergo fragmentation in vacuum, under strong out-of-equilibrium conditions, were considered. It is also possible to look at clusters in a restricted volume, for which the presence of surrounding atoms creates an equilibrium situation with a vapor. This point of view has been adopted in early investigations (Lee et al., 1973), followed by many others (Laasonen et al., 2000; Schaaf et al., 2001; Soto and Cordero, 1999; Zhukhovitskii, 1995), for its relevance in many areas and primarily in atmospheric science. Homogeneous nucleation, in which the clusters are formed from the vapor, has to be distinguished from heterogeneous nucleation where the cluster grows on a seed. However, although they are of greater practical importance

(Kulmala, 2003), the latter processes involve much more complex mechanisms.

At equilibrium, the homogeneous nucleation equilibrium can be summarized by the reactions given below:

$$X_n \xleftrightarrow[k_{cond}]{k_{evap}} X_{n-1} + X \qquad (15.25)$$

for different cluster sizes n. The kinetics of this set of reactions can be modeled using a master equation, whose solution requires knowledge of the condensation and evaporation rates. At equilibrium, the condensation and evaporation fluxes are equal and the rates are related to each other through the equilibrium concentrations $\mathcal{P}^{eq}(n, T)$

$$\frac{k_{cond}(n-1)}{k_{evap}(n)} = \frac{\mathcal{P}^{eq}(n, T)}{\mathcal{P}^{eq}(n-1, T)}. \qquad (15.26)$$

Nucleation theories aim at characterizing the equilibrium state of a cluster that can exchange atoms with the surrounding vapor, depending on the physical parameters (temperature, density, etc.) as well as the related out-of-equilibrium quantities, namely, the rates.

15.7.1 Classical Nucleation Theory and Variants

Similar to the Fisher model of multifragmentation, CNT (Becker and Döring, 1935; Feder et al., 1966; Volmer and Weber, 1926) relies on macroscopic thermodynamical concepts and considers the equilibrium between a spherical liquid droplet having n atoms and a supersaturated vapor of the same element. CNT assumes that the interface between the liquid and the vapor is sharp (capillarity approximation), and that the droplet shares the properties of the bulk liquid. The Gibbs free energy $\Delta G(n)$ corresponding to the formation of the droplet from the vapor has the expression already given in Equation 15.21, without the last correcting term in $\ln n$. The critical radius R^* is determined at the nucleation barrier. Introducing the liquid density ρ, the size is given by $n = 4\pi\rho R^3/3$ and R^* is such that $\partial\Delta G/\partial R = 0$ with $R^* > 0$, or

$$R^* = \frac{2\gamma}{\rho(\mu_v - \mu_l)}. \qquad (15.27)$$

In CNT, the critical radius is accessed from macroscopic properties, rather than from molecular data. The saturation vapor corresponding to a plane interface, p_{sat}, is corrected at the spherical interface of the droplet according to the Kelvin relation. Denoting $S = p/p_{sat}$ the ratio between the actual saturation vapor and its ideal bulk value, R^* is such that

$$\ln S = \frac{2\gamma}{\rho k_B T R^*}. \qquad (15.28)$$

The nucleation rate k_{cond} is determined by an exponential (Arrhenius) relation involving the nucleation barrier, that is the value of ΔG at $R = R^*$. Since $\mu_v - \mu_l$ is the variation in chemical potential when moving one atom from the liquid in equilibrium with the saturating vapor toward the vapor itself at pressure p, it is also equal to $k_B T \ln S$, hence

$$k_{cond} \propto \exp\left[-\frac{16\pi\gamma^3}{3\rho^2(k_B T)^3(\ln S)^2}\right], \qquad (15.29)$$

with a prefactor that depends on the surface area of the critical nuclei and on the second derivative of ΔG at $R = R^*$ (the so-called Zel'dovich factor) (Feder et al., 1966).

Unfortunately, CNT in its basic version is only acceptable for very simple fluids, but is usually far from quantitative measurements when compared with experimental measurements (Ford, 2004). This has motivated several improvements and many numerical studies aimed at providing reference data on model systems. Computer simulations, in particular, can be fruitfully used to determine the critical size of a cluster in a supersaturated vapor (Horsch et al., 2008; Schaaf et al., 2001; Zhukhovitskii, 1995), once a suitable criterion for defining a cluster is adopted (Harris and Ford, 2003; Pugnaloni and Vericat, 2002; Stillinger, 1963). Extensions and modifications of CNT have been described in the review by Laaksonen et al. (1995).

A first correction concerns the role of free translations that contribute to the global Gibbs free energy as prefactors in the partition functions (Blander and Katz, 1972; Reiss et al., 1998). The conservation of the center of mass of the droplet, also referred to as law of mass action, reduces the effective number of degrees of freedom in the cluster by 3, which amounts to reducing n by $n - 1$ in the first term of the right-hand side of Equation 15.21. Another phenomenological extension of CNT, called internal consistency, consists in replacing $n^{2/3}$ in the surface energy of the droplet by $n^{2/3} - 1$, as an attempt to correct for the behavior at $n \to 1$ (Girshick and Chiu, 1990).

One major criticism to CNT applied for small systems is that it is a macroscopic approach (Merikanto et al., 2007; Schaaf et al., 2001). In this respect, the assumptions that the cluster shape is perfectly spherical, that its energy only depends on the volume and the surface, and that the surface tension does not depend on size can be questioned. In the spirit of the liquid drop model, the Gibbs free energy can be corrected using an expansion in powers of $n^{1/3}$ (Dillmann and Meier, 1989). The dependence of surface tension on size occurs through a curvature effect as predicted by Tolman (1949)

$$\gamma(R) = \frac{\gamma(\infty)}{1 + 2\delta/R}, \qquad (15.30)$$

where the parameter δ is called the Tolman length. However, as noted by Laaksonen et al. (1994), this corrective term could also be accounted for in the liquid drop expansion. Surface corrections have been shown to improve significantly the predictions of CNT with respect to numerical calculations, even though no quantitative agreement with experiment could still be reached in the case of argon (Horsch et al., 2008).

15.7.2 The Sticking Cross-Section Issue

Another implicit assumption in CNT lies within the prefactor of the nucleation rate, which according to standard kinetic theory depends linearly on both the molecular flux and the sticking cross-section σ (Ford, 1997). In the vast majority of nucleation studies so far, σ is taken as the surface area, corresponding to a hard-sphere droplet. However, in the context of unimolecular dissociation theories based on microreversibility, the sticking cross section enters the evaporation rates as well (see Section 15.4 above). The difference in the latter approaches is that they account for the long-range attraction in $-C/r^p$, but neglect the spatial extension of the cluster, assumed to be small enough.

Size effects on the sticking cross section have been found in molecular dynamics simulations by Venkatesh et al. (1995), and they have been theoretically studied by Vigué et al. (2000) in the case of large clusters bound by dispersive interactions. In this work, the effective interaction between the cluster and the dissociating atom was taken as $-C_n/(r^2 - R_n^2)^3$ with $C_n \propto n$ and $R_n \propto n^{1/3}$, and the calculation was carried out assuming a Langevin model. At small sizes, the long-range interaction dominates and the cross section scales as $\sigma(n) \propto n^{1/3}$. Conversely, in large clusters, the centrifugal barrier is very close to the cluster radius and $\sigma(n) \propto n^{2/3}$, as would be expected from a pure geometrical basis. The failure of the geometric approximation has recently been confirmed experimentally (Chirot et al., 2007). Using mass spectrometry selection, Chirot et al. have been able to measure the sticking cross section of cationic sodium clusters crossing a gas chamber of neutral sodium atoms. The variations of σ with size found at different collision energies, shown in Figure 15.13, exhibit strong deviations with respect to the hard sphere model, whereas a Langevin model accounting for the smooth interaction between the cluster and the evaporating atoms performs significantly better (Chirot et al., 2007).

15.7.3 Other Approaches

CNTs are built from the perspective of a mesoscopic fluid droplet in contact with a supersaturated vapor. In the dynamical nucleation theory (DNT) developed by Schenter and coworkers (Kathmann et al., 1999; Schenter et al., 1999a,b), the gas-phase point of view of the dissociating cluster is adopted, in a way that is quite similar to the microreversible theories discussed previously. In particular, DNT requires many ingredients to be calculated from molecular simulations at equilibrium in order to be quantitatively accurate. The main originality of DNT is the use of variational TST (Miller, 1974; Wigner, 1937) to compute the evaporation rate $k_{evap}(n)$ for an n-particle cluster as (Kathmann et al., 1999; Schenter et al., 1999a,b)

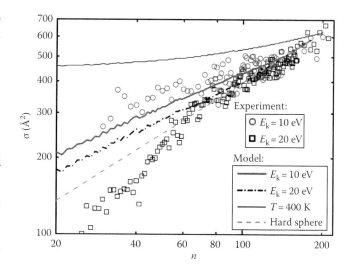

FIGURE 15.13 Variations of the sticking cross section of neutral sodium atoms on size-selected Na_n^+ clusters obtained experimentally and compared with the predictions of simple Langevin and hard-sphere models. (Reproduced from Chirot, F. et al., *Phys. Rev. Lett.*, 99, 193401, 2007. With permission.)

$$k_{evap}^{TST}(n) \propto -\frac{dF_n(\mathcal{V},T)}{dR_{cut}}, \qquad (15.31)$$

in which $F_n(\mathcal{V},T)$ is the Helmholtz free energy of an n-particle cluster kept within volume $\mathcal{V} = 4\pi R_{cut}^3/3$ and at canonical temperature T. The free energy is calculated by restricting configurations to remain within the sphere of radius R_{cut}. According to TST, the dividing surface should be chosen at the value where the reactive flux is minimized (Miller, 1974; Wigner, 1937). This dynamical definition of a cluster volume is at variance with the purely thermodynamical definition of CNTs; however, it should be stressed that its calculation is based on equilibrium thermodynamics, no dynamical simulation of the evaporation process being actually performed at any stage.

Helmholtz free energies play here the same role as the Gibbs free energies in CNT, and they can be used to provide mass distributions at equilibrium (Lee et al., 1973) that in turn allow an estimation of the condensation rates by application of Equation 15.26. In DNT, the free energies and mass distributions are obtained from statistical mechanics and Monte Carlo simulations (Kathmann et al., 1999; Schenter et al., 1999a,b). When applied to realistic molecular systems, this immediately raises the problem of the accuracy of the intermolecular interaction. This problem is especially acute in the crucially important case of water, for which no potential energy function widely stands out. This is manifested on the different values for the evaporation and condensation rates of small water clusters, when they are described using either a simple nonpolarizable (TIP4P) potential or the more accurate Dang-Chang model (Kathmann et al., 2002).

For simple fluids with pair interactions that can be written as the sum of a mainly repulsive part $V_0(r)$ plus a perturbative attractive part $V(r)$, the free energy can also be calculated using classical density-functional theory (Evans, 1979). This approach is particularly useful to determine equilibrium shapes of liquid droplets and to check whether the rigid interface approximation of CNT holds. Density-functional theory writes the Helmholtz free energy F relative to its value for the reference system V_0 by linear integration along a path joining the two fluids:

$$F[\rho] = \frac{1}{2}\int_0^1 d\alpha \iint d\mathbf{r}d\mathbf{r}'\rho_\alpha(\mathbf{r},\mathbf{r}')V(|\mathbf{r}-\mathbf{r}'|). \quad (15.32)$$

In this equation, and at a given point α along the path, the particles interact via the pair potential $V_\alpha(r) = \alpha V(r) + V_0(\mathbf{r})$, and $\rho_\alpha(\mathbf{r},\mathbf{r}')$ is the pair distribution function corresponding to the intermediate potential V_α. There are several difficulties in applying DFT for the nucleation problem (Zeng and Oxtoby, 1991). The pair distribution function can be decoupled using a random phase approximation, $\rho_\alpha(\mathbf{r},\mathbf{r}') \simeq \rho_\alpha(\mathbf{r}) \times \rho_\alpha(\mathbf{r}')$. For hard-sphere potentials, the reference free energy is satisfactory approximated using the local-density approximation.

The equilibrium density profile is determined from a variational condition, $\delta\Omega/\delta\rho(\mathbf{r}) = 0$, where Ω is the grand potential of the system. The solution of this equation, which is obtained iteratively, shows that the liquid–vapor interface is not sharp but increasingly smooth as temperature increases (Zeng and Oxtoby, 1991). The way to define the droplet radius under such conditions has been discussed by Talanquer and Oxtoby (1994). As shown by Zeng and Oxtoby (1991), the free energies obtained from density-functional theory lead to comparable variations in the nucleation rates with respect to CNT; however, the temperature dependence varies more significantly.

The results obtained by classical DFT have inspired several attempts at improving CNTs by revisiting the capillarity approximation. The diffuse interface method of Gránásy (1993), in particular, parametrizes the cross-interfacial free energies obtained for smooth profiles. Of interest is also the recent extended modified liquid drop model of Reguera and Reiss (2004) that combines the features of a smooth droplet profile and the volume definition of DNT, all together within a thermodynamical CNT framework. The extended modified liquid drop model lacks any atomistic input that makes it suitable for most substances for which the interaction potentials are not well characterized. It was found to predict accurate critical sizes and even mass distributions without additional fitting (Reguera and Reiss, 2004).

15.8 Conclusions and Outlook

The various aspects of fragmentation in atomic and molecular clusters previously discussed have illustrated many points of view on this topic. Several of them may seem at first poorly connected to each other, either because they cover different timescales of the fragmentation events or because they are relevant for specific classes of systems. However, beyond the particularities of each cluster in terms of experimental production, excitation, and detection, the important features of cluster fragmentation can generally be described by a limited number of parameters. For instance, a perturbation of given magnitude will be experienced much more strongly by a small cluster, possibly leading to nonstatistical dissociation, whereas a larger cluster will have more time to redistribute the excitation energy among the vibrational degrees of freedom. Also, short excitations (below hundreds of femtoseconds) will tend to alter directly the electronic structure, whereas low-energy collisions will mainly heat the cluster vibrations.

The occurrence of nonstatistical fragmentation is ruled by an uncomplete redistribution of the excitation energy into pure vibrations shared among all modes. However, even a good redistribution may not produce the energetically most stable dissociation products. This was illustrated in the case of multiply charged clusters, for which the Coulomb barrier to dissociation may kinetically favor other products.

Some ending remarks about the limitations of current theoretical approaches will now end this chapter. A first difficulty lies in the many timescales spanned by fragmentation, that all require different strategies. Following the schematic picture of Figure 15.1, the fragmentation of clusters is expected to produce smaller clusters unless the excitation energy is sufficient for a complete atomization. For a brief and intense excitation, the first stages of the dynamics are highly nonstatistical, but some redistribution can still take place after a few picoseconds or more. At this stage, a statistical description for the remaining lifetime of the cluster can be envisaged, by modeling the sequential evaporation of the cluster by kinetic Monte Carlo methods, as are frequently employed to model the disassembly of finite systems (Barbagallo et al., 1986; Calvo, 2006a; Calvo and Parneix, 2007; Casero and Soler, 1991; Richert and Wagner, 1990; Wagner et al., 1999), particularly the fullerenes (Chen et al., 2007; Tomita et al., 2003). Bridging the timescales from the short-time excitation to the long-time evaporative cooling, regime has been achieved only recently for realistic systems (Calvo et al., 2007). In summary, the method consists in treating the early excitation and redistribution stages at a fully atomistic level using nonadiabatic molecular dynamics trajectories, switching to adiabatic trajectories once the electronic ground state is reached, and eventually switching a second time to a purely kinetic description based on PST after a fixed period (Calvo et al., 2007). This approach has been tested on the electron impact ionization and excitation of argon clusters, modeled using a DIM model. The average size of argon clusters containing initially 20 and 30 atoms, ionized at $t = 0$, is represented in Figure 15.14. The variations in cluster sizes are nicely smooth at the second switching time, provided that the densities of vibrational states fully account for anharmonicities. Assuming that evaporative cooling takes place as soon as internal conversion is complete (i.e., before a single picosecond) leads to fragment distributions at variance with the results of the multiscale approach (Calvo et al., 2007), in agreement with the nonstatistical character of the earliest stages of the fragmentation dynamics.

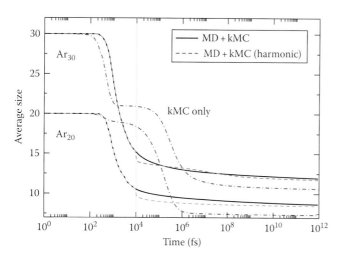

FIGURE 15.14 Average size of two argon clusters ionized and excited at time $t = 0$ by electron impact, obtained from the multiscale simulation method of Calvo et al. (2007). The molecular dynamics (MD) trajectories are performed until $t = 10\,\mathrm{ps}$, and switched to a statistical description based on kinetic Monte Carlo (kMC) at this time, assuming either harmonic densities of states (dashed lines) or anharmonic densities (solid lines). Performing an anharmonic statistical treatment immediately after internal conversion to the ground electronic state leads to the dashed-dotted lines. (Adapted from Calvo, F. et al., *Phys. Rev. Lett.*, 99, 083401, 2007.)

In the above example, the statistical theory was applied in a rather simple case where monomer evaporation dominates over all channels, which is reasonable for rare-gas clusters (Napari and Vehkamäki, 2004). In the general case, competing decay channels can also be dealt with in the kinetic Monte Carlo approach, but they require all corresponding differential rates to be characterized. This way of solving the master kinetic equations of multi-sequential decay has been developed by Richert and coworkers (Barbagallo et al., 1986; Richert and Wagner, 1990; Wagner et al., 1999) in the nuclear physics context.

Despite their many successes, statistical approaches to cluster fragmentation suffer from several practical limitations. The calculation of fully anharmonic densities of states is usually a significant numerical task, only achievable for rather small clusters (below about 1000 atoms) described with simple, explicit interactions. Possibly more severe is the classical assumption for the nuclear degrees of freedom used in the vast majority of the theoretical work reported so far. Highly quantum systems such as helium droplets can be treated specifically assuming continuous media approximations (Brink and Stringari, 1990; Lehmann and Dokter, 2004). However, for arbitrary clusters, the most common approach is to include quantum effects perturbatively, for instance, by correcting for zero-point energies, or through semiclassical approximations for the density of states (Peslherbe and Hase, 1994, 1996, 2000). Quantum densities are most efficiently calculated for separable oscillators, although recent progresses in Monte Carlo methods have allowed anharmonic couplings to be taken into account with accuracy (Basire et al., 2008). Another influence of quantum delocalization could be the possible tunneling through the transition state or centrifugal

barriers. Semiclassical theory (Child, 1991) again provides some ways of accounting for these effects. Yet the major issue concerns the validation of these various corrections because unlike in chemical kinetics, the experiments on clusters are not detailed enough to serve as a benchmark for testing statistical theories. Further improvements in quantum dynamical wavepacket methods (Buch, 2002) for large systems appear as mandatory in order to assess their relevance.

List of Variables

k	rate constant
E	internal energy
D	dissociation energy
T	temperature
G	Gspann constant
ε	kinetic energy released
g	number of degrees of freedom
Ω	density of vibrational states
W	sum of vibrational states
Γ	rotational density of states
σ	collision cross section
χ	fissility
ΔG	free-energy difference
R	cluster radius or coordinate
J	angular momentum
I	intensity
$\mathcal{P}, P$	probability distribution
$\mathcal{R}$	differential rate
ϕ	flux in phase space

Acknowledgments

The authors would like to thank T. Pino, B. Concina, and M.-A. Lebeault for useful remarks on the manuscript.

References

Adelberger, E. G., Austin, S. M., and Bahcall, J. N. (1998) Solar fusion cross sections, *Rev. Mod. Phys.*, 70, 1265–1291.

Aguado, A. and López, J. M. (2005) Anomalous size dependence in the melting temperatures of free sodium clusters: An explanation for the calorimetry experiments, *Phys. Rev. Lett.*, 94, 233401–233404.

Albertoni, C. R., Kuhn, R., Sarkas, H. W., and Castleman, A. W., Jr., (1987) Photodissociation of rare gas cluster ions: Ar_3^+, *J. Chem. Phys.*, 87, 5043–5044.

Amarouche, M., Gadéa, F. X., and Durup, J. (1989) A proposal for the theoretical treatment of multi-electronic-state molecular-dynamics-hemiquantal dynamics with the whole DIM basis (HWD)-a test on the evolution of excited Ar_3^+ cluster ions, *Chem. Phys.*, 130, 145–157.

Andersen, J. U., Bonderup, E., and Hansen, K. (2001) On the concept of temperature for a small isolated system, *J. Chem. Phys.*, 114, 6518–6525.

Barbagallo, C., Richert, J., and Wagner, P. (1986) Disassembly of an excited nuclear aggregate through sequential decay, *Z. Phys. A: Hadrons Nuclei*, 324, 97–106.

Barnett, R. N., Landman, U., and Rajagopal, G. (1991) Patterns and barriers for fission of charged small metal clusters, *Phys. Rev. Lett.*, 67, 3058–3061.

Basire, M., Parneix, P., and Calvo, F. (2008) Quantum anharmonic densities of states using the Wang-Landau method, *J. Chem. Phys.*, 129, 081101.

Bauer, W. (1988) Extraction of signals of a phase transition from nuclear multifragmentation, *Phys. Rev. C*, 38, 1297–1303.

Bauer, W. (2001) Common aspects of phase transitions of molecules, nuclei, and hadronic matter, *Nucl. Phys. A*, 681, 441c–450c.

Becker, R. and Döring, W. (1935) The kinetic treatment of nuclear formation in supersaturated vapors, *Ann. Phys. (Leipzig)*, 416, 719–752.

Belkacem, M., Latora, V., and Bonasera, A. (1995) Critical evolution of a finite system, *Phys. Rev. C*, 52, 271–285.

Beu, T. A. (2003) Fragmentation statistics of large H_2O and NH_3 clusters from molecular-dynamics simulations, *Phys. Rev. A*, 67, 045201.

Beu, T. A., Steinbach, C., and Buck, U. (2003) Model analysis of the fragmentation of large H_2O and NH_3 clusters based on MD simulations, *Eur. Phys. J. D*, 27, 223–229.

Blaise, P., Blundell, S. A., Guet, C., and Zope, R. R. (2001) Charge-induced fragmentation of sodium clusters, *Phys. Rev. Lett.*, 87, 063401.

Blander, M. and Katz, J. L. (1972) The thermodynamics of cluster formation in nucleation theory, *J. Stat. Phys.*, 4, 55–59.

Bohr, N. and Wheeler, J. A. (1939) The mechanism of nuclear fission, *Phys. Rev.*, 56, 426–450.

Bondorf, J. P., Botvina, A. S., Iljinov, A. S., Mishustin, I. N., and Sneppen, K. (1995) Statistical multifragmentation of nuclei, *Phys. Rep.*, 257, 133–221.

Bonhommeau, D., Halberstadt, N., and Buck, U. (2007) Fragmentation of rare-gas clusters ionized by electron impact: New theoretical developments and comparison with experiments, *Int. Rev. Phys. Chem.*, 26, 353–390.

Breaux, G. A., Benirschke, R. C., Sugai, T., Kinnear, B. S., and Jarrold, M. F. (2003) Hot and solid gallium clusters: Too small to melt, *Phys. Rev. Lett.*, 91, 215508–215511.

Breaux, G. A., Hillman, D. A., Neal, C. M., Benirschke, R. C., and Jarrold, M. F. (2004) Gallium clusters "magic melters", *J. Am. Chem. Soc.*, 126, 8628–8629.

Breaux, G. A., Neal, C. M., Cao, B., and Jarrold, M. F. (2005) Melting, premelting, and structural transitions in size-selected aluminum clusters with around 55 atoms, *Phys. Rev. Lett.*, 94, 173401–173404.

Bréchignac, C., Cahuzac, P., Leygnier, J., and Weiner, J. (1989) Dynamics of unimolecular dissociation of sodium cluster ions, *J. Chem. Phys.*, 90, 1492–1498.

Bréchignac, C., Cahuzac, P., Carlier, F., de Frutos, M., and Leygnier, J. (1990) Cohesive energies of K_n^+ $5 < n < 200$ from photoevaporation experiments, *J. Chem. Phys.*, 93, 7449–7456.

Bréchignac, C., Bush, H., Cahuzac, P., and Leygnier, J. (1994a) Dissociation pathways and binding energies of lithium clusters from evaporation experiments, *J. Chem. Phys.*, 101, 6992–7002.

Bréchignac, C., Cahuzac, P., Carlier, F., and de Frutos, M. (1994b) Shell effects in fission of small doubly charged lithium clusters, *Phys. Rev. B*, 49, 2825–2831.

Bréchignac, C., Cahuzac, P., Carlier, F., de Frutos, M., Barnett, R. N., and Landman, U. (1994c) Dynamics and energy release in fission of small doubly charged clusters, *Phys. Rev. Lett.*, 72, 1636–1639.

Bréchignac, C., Cahuzac, P., Kébaili, N., and Leygnier, J. (1998) Temperature effects in the Coulombic fission of strontium clusters, *Phys. Rev. Lett.*, 81, 4612–4615.

Bréchignac, C., Cahuzac, P., Concina, B. et al. (2001) Experimental evidence for anharmonic and rotational effects in the evaporation of sodium clusters, *Chem. Phys. Lett.*, 335, 34–42.

Bréchignac, C., Cahuzac, P., Concina, B., and Leygnier, J. (2002) Caloric curves of small fragmenting clusters, *Phys. Rev. Lett.*, 89, 203401–203404.

Brink, D. M. and Stringari, S. (1990) Density of states and evaporation rate of helium clusters, *Z. Phys. D: At. Mol. Clusters*, 15, 257–263.

Buch, V. (2002) Exploration of multidimensional variational Gaussian wave packets as a simulation tool, *J. Chem. Phys.*, 117, 4738–4750.

Buck, U. and Meyer, H. (1986) Electron bombardment fragmentation of Ar van der Waals clusters by scattering analysis, *J. Chem. Phys.*, 84, 4854–4861.

Calvayrac, F., Reinhard, P. G., and Suraud, E. (1998) Coulomb explosion of an Na_{12} cluster in a diabatic electron-ion dynamical picture, *J. Phys. B: At. Mol. Opt. Phys.*, 31, 5023–5030.

Calvo, F. (1999) Phenomenological model for the statistical study of charged metal clusters, *Phys. Rev. B*, 60, 15601–15604.

Calvo, F. (2001) Thermal stability of the solidlike and liquidlike phases of $(C_{60})_n$ clusters, *J. Phys. Chem. B*, 105, 2183–2190.

Calvo, F. (2006a) Kinetic approach to the cluster liquid-gas transition, *Phys. Rev. A*, 71, 041205.

Calvo, F. (2006b) Role of charge localization on the Coulomb fragmentation of large metal clusters: A model study, *Phys. Rev. A*, 74, 043202.

Calvo, F. and Labastie, P. (1998) Monte-Carlo simulations of rotating clusters, *Eur. Phys. J. D*, 3, 229–236.

Calvo, F. and Parneix, P. (2003) Statistical evaporation of rotating clusters. I. Kinetic energy released, *J. Chem. Phys.*, 119, 256–264.

Calvo, F. and Parneix, P. (2004) Statistical evaporation of rotating clusters. III. Molecular clusters, *J. Chem. Phys.*, 120, 2780–2787.

Calvo, F. and Parneix, P. (2007) Accurate modeling of sequential decay in clusters over long time scales: Insights from phase space theory, *J. Chem. Phys.*, 126, 34309.

Calvo, F., Galindez, J., and Gadéa, F. X. (2003) Internal conversion in the photofragmentation of Ar_n^+ clusters (n = 3–8), *Phys. Chem. Chem. Phys.*, 5, 321–328.

Calvo, F., Parneix, P., and Gadéa, F.-X. (2006) Temperature measurement from the translational kinetic energy release distribution in cluster dissociation: A theoretical investigation, *J. Phys. Chem. A*, 110, 1561–1568.

Calvo, F., Bonhommeau, D., and Parneix, P. (2007) Multiscale dynamics of cluster fragmentation, *Phys. Rev. Lett.*, 99, 083401.

Campbell, E. E. B. and Rohmund, F. (2000) Fullerene reactions, *Rep. Prog. Phys.*, 63, 1061–1109.

Campbell, E. E. B., Raz, T., and Levine, R. D. (1996) Internal energy dependence of the fragmentation patterns of C_{60} and C_{60}^+, *Chem. Phys. Lett.*, 253, 261–267.

Campi, X. (1986) Multifragmentation—Nuclei break up like percolation clusters, *J. Phys. A: Math. Gen.*, 19, L917–L921.

Campi, X. (1988) Signals of a phase-transition in nuclear multifragmentation, *Phys. Lett. B*, 208, 351–354.

Campi, X. and Krivine, H. (1992) Critical-behavior, fluctuations and finite size scaling in nuclear multifragmentation, *Nucl. Phys. A*, 545, C161–C172.

Cao, B. P., Peres, T., Cross, R. J., Saunders, M., and Lifshitz, C. (2001) Do nitrogen-atom-containing endohedral fullerenes undergo the shrink-wrap mechanism? *J. Phys. Chem. A*, 105, 2142–2146.

Casero, R. and Soler, J. M. (1991) Onset and evolution of "magic numbers" in mass spectra of molecular clusters, *J. Chem. Phys.*, 95, 2927–2935.

Chabot, M. and Wohrer, K. (2004) Comment on direct experimental evidence for a negative heat capacity in the liquid-to-gas phase transition in hydrogen cluster ions: Backbending of the caloric curve, *Phys. Rev. Lett.*, 93, 039301.

Chandezon, F., Guet, C., Huber, B. A. et al. (1995) Critical sizes against Coulomb dissociation of highly charged sodium clusters obtained by ion impact, *Phys. Rev. Lett.*, 74, 3784–3787.

Châtelet, M., Benslimane, M., DeMartino, A., Pradère, F., and Vach, H. (1996) Energy transfer during van der Waals cluster-surface collisions, *Surf. Sci.*, 352, 50–54.

Chen, L., Martin, S., Bernard, J., and Brédy, R. (2007) Direct measurement of internal energy of fragmented C_{60}, *Phys. Rev. Lett.*, 98, 193401–193404.

Chesnavich, W. J. and Bowers, M. T. (1976) Statistical phase space theory of polyatomic systems. Application to the cross section and product kinetic energy release distribution of the reaction $C_2H_4^+ + C_2H_4 \rightarrow C_3H_5^+ + CH_3$, *J. Am. Chem. Soc.*, 98, 8301–8309.

Chesnavich, W. J. and Bowers, M. T. (1977a) Statistical phase space theory of polyatomic systems. Application to the unimolecular reactions $C_6H_5CN^+ \rightarrow C_6H_4^+ + HCN$ and $C_4H_6^+ \rightarrow C_3H_3^+ + CH_3$, *J. Am. Chem. Soc.*, 99, 1705–1711.

Chesnavich, W. J. and Bowers, M. T. (1977b) Statistical phase space theory of polyatomic systems: Rigorous energy and angular momentum conservation in reactions involving symmetric polyatomic species, *J. Chem. Phys.*, 66, 2306–2315.

Child, M. S. (1991) *Semiclassical Mechanics with Molecular Applications*, Clarendon Press, Oxford, U.K.

Chirot, F., Labastie, P., Zamith, S., and L'Hermite, J.-M. (2007) Experimental determination of nucleation scaling law for small charged particles, *Phys. Rev. Lett.*, 99, 193401.

Chirot, F., Feiden, P., Zamith, S., Labastie, P., and L'Hermite, J.-M. (2008) A novel experimental method for the measurement of the caloric curves of clusters, *J. Chem. Phys.*, 129, 164514–164522.

Cohen, S. and Swiatecki, W. J. (1963) The deformation energy of a charged drop: Part V: Results of electronic computer studies, *Ann. Phys. (N. Y.)*, 22, 406–437.

Debierre, J.-M. (1997) Exact scaling law for the fragmentation of percolation clusters: Numerical evidence, *Phys. Rev. Lett.*, 78, 3145–3148.

Dietrich, G., Dasgupta, K., L.ützenkircha, K., Schweikhard, L., and Ziegler, J. (1996) Chemisorption of hydrogen on a V_5^+ cluster, *Chem. Phys. Lett.*, 252, 141–146.

Dillmann, A. and Meier, G. E. A. (1989) Homogeneous nucleation of supersaturated vapors, *Chem. Phys. Lett.*, 160, 71–74.

Ditmire, T., Tisch, J. W. G., Springate, E. et al. (1997) High-energy ions produced in explosions of superheated atomic clusters, *Nature (London)*, 386, 54–56.

Dorso, C. O. and López, J. A. (2001) Selection of critical events in nuclear fragmentation, *Phys. Rev. C*, 64, 027602.

Dorso, C. O., Latora, V. C., and Bonasera, A. (1999) Signals of critical behavior in fragmenting finite systems, *Phys. Rev. C*, 60, 034606.

Duft, D., Achtzehn, T., Müller, R., Huber, B. A., and Leisner, T. (2003) Coulomb fission—Rayleigh jets from levitated microdroplets, *Nature (London)*, 421, 128–128.

Echt, O., Kreizle, D., Recknagel, E., Saenz, J. J., Casero, R., and Soler, J. M. (1988) Dissociation channels of multiply charged van der Waals clusters, *Phys. Rev. A*, 38, 3236–3248.

Ellert, C., Schmidt, M., Schmidt, C., Reiners, T., and Haberland, H. (1995) Temperature dependence of the optical response of small, open shell sodium clusters, *Phys. Rev. Lett.*, 75, 1731–1734.

Elliott, J. B., Albergo, S., Bieser, F. et al. (2000) Statistical signatures of critical behavior in small systems, *Phys. Rev. C*, 62, 064603.

Engel, E., Schmitt, U. R., Lüdde, H.-J. et al. (1993) Accurate numerical study of the stability of Na_{19}-cluster dimers, *Phys. Rev. B*, 48, 1862.

Engelking, P. C. (1986) Photoinduced evaporation of charged clusters, *J. Chem. Phys.*, 85, 3103–3110.

Engelking, P. C. (1987) Determination of cluster binding energy from evaporative lifetime and average kinetic energy release: Application to $(CO_2)^+$ and Ar n^+ clusters, *J. Chem. Phys.*, 87, 936–940.

Evans, R. (1979) Nature of the liquid-vapor interface and other topics in the statistical-mechanics of nonuniform, classical fluids, *Adv. Phys.*, 28, 143–200.

Farizon, B., Farizon, M., Gaillard, M. J. et al. (1997) Mass distribution and multiple fragmentation events in high energy cluster-cluster collisions: Evidence for a predicted phase transition, *Int. J. Mass Spectrom. Ion Process.*, 164, 225–230.

Fayeton, J. A., Barat, M., Brenot, J. C., Picard, Y. J., Saalmann, U., and Schmidt, R. (1998) Detailed experimental and theoretical study of collision-induced dissociation of Na$_2^+$ ions on He and H$_2$ targets at keV energies, *Phys. Rev. A*, 57, 1058–1068.

Feder, J., Russell, K. C., Lothe, J., and Pound, G. M. (1966) Homogeneous nucleation and growth of droplets in vapours, *Adv. Phys.*, 15, 111–178.

Fiedler, S. L., Kunttu, H. M., and Eloranta, J. (2008) Application of mean-field and surface-hopping approaches for interrogation of the Xe$_3^+$ molecular ion photoexcitation dynamics, *J. Chem. Phys.*, 128, 164309–164319.

Fielicke, A., Rabin, I., and Meijer, G. (2006) Far-infrared spectroscopy of small neutral silver clusters, *J. Phys. Chem. A*, 110, 8060–8063.

Fisher, M. E. (1967) The theory of equilibrium critical phenomena, *Rep. Prog. Phys.*, 30, 615–730.

Foltin, V., Foltin, M., Matt, S. et al. (1998) Electron impact ionization of C$_{60}$ revisited: Corrected absolute cross section functions, *Chem. Phys. Lett.*, 289, 181–188.

Ford, I. J. (1997) Nucleation theorems, the statistical mechanics of molecular clusters, and a revision of classical nucleation theory, *Phys. Rev. E*, 56, 5615–5629.

Ford, I. J. (2004) Statistical mechanics of nucleation: A review, *Proc. Inst. Mech. Eng., Part C: J. Mech. Eng. Sci.*, 218, 883–899.

Fröbrich, P. and Ecker, A. (1998) Langevin description of fission of hot metallic clusters, *Eur. Phys. J. D*, 3, 237–243.

Gadéa, F. X. and Amarouche, M. (1990) Theoretical study of the visible photodissociation spectrum of Ar$_3^+$, *Chem. Phys.*, 140, 385–397.

Garcias, F., Alonso, J. A., Barranco, M., López, J. M., Mañanes, A., and Nemeth, J. (1994) Fission barriers for Na$_n^{2+}$ cluster dissociation, *Z. Phys. D: At. Mol. Clusters*, 31, 275–277.

Garcias, F., Mañanes, A., López, J. M., Alonso, J. A., and Barranco, M. (1995) Deformed-jellium model for the fission of multiply charged simple metal clusters, *Phys. Rev. B*, 51, 1897–1901.

Giglio, E., Gervais, B., Rangama, J. et al. (2008) Shape deformations of surface-charged microdroplets, *Phys. Rev. E*, 77, 036319.

Girshick, S. L. and Chiu, C.-P. (1990) Kinetic nucleation theory—A new expression for the rate of homogeneous nucleation from an ideal supersaturated vapor, *J. Chem. Phys.*, 93, 1273–1277.

Gisselbrecht, M., Lindgren, A., Burmeister, F. et al. (2008) Size dependent fragmentation of argon clusters in the soft x-ray ionization regime, *J. Chem. Phys.*, 128, 044317–044323.

Gisselbrecht, M., Lindgren, A., Tchaplyguine, M. et al. (2005) Photon energy dependence of fragmentation of small argon clusters, *J. Chem. Phys.*, 123, 194301–194307.

Gluch, K., Matt-Leubner, S., Echt, O., Concina, B., Scheier, P., and Märk, T. D. (2004) High-resolution kinetic energy release distributions and dissociation energies for fullerene ions C$_n^+$, 42 < = n < =90, *J. Chem. Phys.*, 121, 2137–2143.

Gobet, F., Farizon, B., Farizon, M. et al. (2001) Event-by-event analysis of collision-induced cluster-ion fragmentation: Sequential monomer evaporation versus fission reactions, *Phys. Rev. Lett.*, 86, 4263–4266.

Gobet, F., Farizon, B., Farizon, M. et al. (2002) Direct experimental evidence for a negative heat capacity in the liquid-to-gas phase transition in hydrogen cluster ions: Backbending of the caloric curve, *Phys. Rev. Lett.*, 89, 183403.

Gobet, F., Farizon, B., Farizon, M. et al. (2004) Gobet et al. Reply, *Phys. Rev. Lett.*, 93, 039302.

Gránásy, L. (1993) Diffuse interface approach to vapor condensation, *Europhys. Lett.*, 24, 121–126.

Grillon, G., Balcou, P., Chambaret, J.-P. et al. (2002) Deuterium-deuterium fusion dynamics in low-density molecular-cluster jets irradiated by intense ultrafast laser pulses, *Phys. Rev. Lett.*, 89, 065005.

Gross, D. H. E. (1990) Statistical decay of very hot nuclei-the production of large clusters, *Rep. Prog. Phys.*, 53, 605–658.

Gross, D. H. E. (1997) Microcanonical thermodynamics and statistical fragmentation of dissipative systems. The topological structure of the N-body phase space, *Phys. Rep.*, 279, 119–201.

Gross, D. H. E. and Hervieux, P. A. (1995) Statistical fragmentation of hot atomic metal-clusters, *ZPD*, 35, 27–42.

Gspann, J. (1982) Electronic and atomic impacts on large clusters, in *Physics of Electronic and Atomic Collisions*, S. Datz, (ed.), North-Holland, Amsterdam, the Netherlands, pp. 79–96.

Haberland, H., Hippler, T., Donges, J., Kostko, O., Schmidt, M., and von Issendorff, B. (2005) Melting of sodium clusters: Where do the magic numbers come from? *Phys. Rev. Lett.*, 94, 035701–035704.

Hansen, K. and Campbell, E. E. B. (1996) Radiative cooling of fullerenes, *J. Chem. Phys.*, 104, 5012–5018.

Hansen, K. and Näher, U. (1999) Evaporation and cluster abundance spectra, *Phys. Rev. A*, 60, 1240–1250.

Hansen, K., Herlert, A., Schweikhard, L., Vogel, M., and Walther, C. (2005) The dissociation energy of V$_{13}^+$ and the consequences for radiative cooling, *Eur. Phys. J. D*, 34, 67–71.

Harris, S. A. and Ford, I. J. (2003) A dynamical definition of quasi-bound molecular clusters, *J. Chem. Phys.*, 118, 9216–9223.

Hase, W. L. (1998) Some recent advances and remaining questions regarding unimolecular rate theory, *Acc. Chem. Res.*, 31, 659–665.

Heller, E. J. (1978) Quantum corrections to classical photodissociation models, *J. Chem. Phys.*, 68, 2066–2075.

Hoener, M., Bostedt, C., Schorb, S. et al. (2008) From fission to explosion: Momentum-resolved survey over the Rayleigh instability barrier, *Phys. Rev. A*, 78, 021201(R).

Horsch, M., Vrabec, J., and Hasse, H. (2008) Modification of the classical nucleation theory based on molecular simulation data for surface tension, critical nucleus size, and nucleation rate, *Phys. Rev. E*, 78, 011603.

Jarrold, M. F., Bower, J. E., and Kraus, J. S. (1987) Collision induced dissociation of metal cluster ions: Bare aluminum clusters, Al$_n^+$ (n = 3–26, *J. Chem. Phys.*, 86, 3876–3885.

Jena, P., Khanna, S. N., and Yannouleas, C. (1992) Comment on "patterns and barriers for fission of charged small metal clusters", *Phys. Rev. Lett.*, 69, 1471.

Jeschke, H. O., Garcia, M. E., and Alonso, J. A. (2002) Nonthermal fragmentation of C_{60}, *Chem. Phys. Lett.*, 352, 154–162.

Jordan, K. D. and Wang, F. (2003) Theory of dipole-bound anions, *Annu. Rev. Phys. Chem.*, 54, 367–396.

Kassel, L. S. (1928a) Studies in homogeneous gas reactions. I, *J. Phys. Chem.*, 32, 225–242.

Kassel, L. S. (1928b) Studies in homogeneous gas reactions. II. Introduction of quantum theory, *J. Phys. Chem.*, 32, 1065–1079.

Kathmann, S. M., Schenter, G. K., and Garrett, B. C. (1999) Dynamical nucleation theory: Calculation of condensation rate constants for small water clusters, *J. Chem. Phys.*, 111, 468–46978.

Kathmann, S. M., Schenter, G. K., and Garrett, B. C. (2002) Understanding the sensitivity of nucleation kinetics: A case study on water, *J. Chem. Phys.*, 116, 5046–5057.

Klots, C. E. (1971) Reformulation of the quasiequilibrium theory of ionic fragmentation, *J. Phys. Chem.*, 75, 1526–1532.

Klots, C. E. (1972) Quasi-equilibrium theory of ionic fragmentation: Further considerations, *Z. Naturforsch. A*, 27a, 553–561.

Klots, C. E. (1987) The evaporative ensemble, *Z. Phys. D: At. Mol., Clusters*, 5, 83–89.

Klots, C. E. (1991) Systematics of evaporation, *Z. Phys. D: At. Mol. Clusters*, 20, 105–109.

Klots, C. E. (1992) Arrhenius parameters for ionic fragmentations, *J. Phys. Chem.*, 96, 1733–1737.

Knight, W. D., Clemenger, K., de Heer, W. A., Saunders, W. A., Chou, M. Y., and Cohen, M. L. (1984) Electronic shell structure and abundances of sodium clusters, *Phys. Rev. Lett.*, 52, 2141–2143.

Knospe, O., Glotov, A. V., Seifert, G., and Schmidt, R. (1996) Theoretical studies of atomic cluster—Cluster collisions, *J. Phys. B: At. Mol. Opt. Phys.*, 29, 5163–5174.

Kondratyev, V. N. and Lutz, H. O. (1997) Signatures of critical phenomena in rare atom clusters, *Z. Phys. D: At. Mol. Clusters*, 40, 210–214.

Kulmala, M. (2003) How particles nucleate and grow, *Science*, 302, 1000–1001.

Kuntz, P. J. and Hogreve, J. J. (1991) Classical path surface-hopping dynamics. II. Application to Ar_3^+, *J. Chem. Phys.*, 95, 156–165.

Kuntz, P. J. and Valldorf, J. (1988) A DIM model for homogeneous noble-gas ionic clusters, *Z. Phys. D: At. Mol. Clusters*, 8, 195–208.

Laaksonen, A., Ford, I. J., and Kulmala, M. (1994) Revised parametrization of the Dillmann-Meier theory of homogeneous nucleation, *Phys. Rev. E*, 49, 5517–5524.

Laaksonen, A., Talanquer, V., and Oxtoby, D. W. (1995) Nucleation—Measurements, theory, and atmospheric applications, *Annu. Rev. Phys. Chem.*, 46, 489–524.

Laasonen, K., Wonczak, S., Strey, R., and Laaksonen, A. (2000) Molecular dynamics simulations of gas-liquid nucleation of Lennard-Jones fluid, *J. Chem. Phys.*, 113, 9741–9747.

Laskin, J., Hadas, B., Märk, T. D., and Lifshitz, C. (1998) New evidence in favor of a high (10 eV) C_2 binding energy in C_{60}, *Int. J. Mass Spectrom. Ion Process.*, 177, L9–L13.

Last, I. and Jortner, J. (2000) Dynamics of the Coulomb explosion of large clusters in a strong laser field, *Phys. Rev. A*, 62, 013201.

Last, I. and Jortner, J. (2001) Nuclear fusion induced by Coulomb explosion of heteronuclear clusters, *Phys. Rev. Lett.*, 87, 033401.

Last, I. and Jortner, J. (2005) Regular multicharged transient soft matter in Coulomb explosion of heteroclusters, *Proc. Natl. Acad. Sci. U.S.A.*, 102, 1291–1295.

Last, I. and Jortner, J. (2006) Tabletop nucleosynthesis driven by cluster Coulomb explosion, *Phys. Rev. Lett.*, 97, 173401.

Last, I. and Jortner, J. (2007) Nucleosynthesis driven by Coulomb explosion of nanodroplets, *Phys. Plasmas*, 14, 123102.

Last, I. and Jortner, J. (2008) Nucleosynthesis driven by Coulomb explosion within a single nanodroplet, *Phys. Rev. A*, 77, 033201.

Last, I., Levy, Y., and Jortner, J. (2002) Beyond the Rayleigh instability limit for multicharged finite systems: From fission to Coulomb explosion, *Proc. Natl. Acad. Sci. U.S.A.*, 99, 9107–9112.

Last, I., Levy, Y., and Jortner, J. (2005) Fragmentation channels of large multicharged clusters, *J. Chem. Phys.*, 123, 154301.

Lebrun, T., Berry, H. G., Cheng, S. et al. (1994) Ionization and multifragmentation of C_{60} by high-energy, highly charged Xe ions, *Phys. Rev. Lett.*, 72, 3965–3968.

Lee, J. K., Barker, J. A., and Abraham, F. F. (1973) Theory and Monte Carlo simulation of physical clusters in the imperfect vapor, *J. Chem. Phys.*, 58, 3166–3180.

Lehmann, K. K. and Dokter, A. M. (2004) Evaporative cooling of Helium nanodroplets with angular momentum conservation, *Phys. Rev. Lett.*, 92, 173401.

Lezius, M., Dobosh, S., Normand, D., and Schmidt, M. (1998) Explosion dynamics of rare gas clusters in strong laser fields, *Phys. Rev. Lett.*, 80, 261–264.

Li, Y., Blaisten-Barojas, E., and Papaconstantopoulos, D. A. (1998) Structure and dynamics of alkali-metal clusters and fission of highly charged clusters, *Phys. Rev. B*, 57, 15519–15532.

Light, J. C. (1967) Statistical theory of bimolecular exchange reactions, *Discuss. Faraday Soc.*, 44, 14–29.

López, M. J. and Jellinek, J. (1994) Fragmentation of atomic clusters: A theoretical study, *Phys. Rev. A*, 50, 1445–1458.

Lyalin, A., Solov'yov, A., Greiner, W., and Semenov, S. K. (2000) Hartree-Fock approach for metal-cluster fission, *Phys. Rev. A*, 65, 023201.

Madison, K. W., Patel, P. K., Price, D. et al. (2004) Fusion neutron and ion emission from deuterium and deuterated methane cluster plasmas, *Phys. Plasmas*, 11, 270–277.

Mähr, I., Zappa, F., Denifl, S. et al. (2007) Multiply charged neon clusters: Failure of the liquid drop model? *Phys. Rev. Lett.*, 98, 023401–023404.

Marcus, R. A. (1952) Unimolecular dissociations and free radical recombination reactions, *J. Chem. Phys.*, 20, 359–364.

Marques, M. A. L. and Gross, E. K. U. (2003) Time-dependent density functional theory, *Lect. Notes Phys.*, 620, 144–184.

Martinet, G., Díaz-Tendero, S., Chabot, M. et al. (2004) Fragmentation of highly excited small neutral carbon clusters, *Phys. Rev. Lett.*, 93, 063401–063404.

Mcelvany, S. W. and Ross, M. M. (1992) Mass-spectrometry and fullerenes, *J. Am. Soc. Mass Spectrom.*, 3, 268–280.

Merikanto, J., Zapadinsky, E., Vehkamäki, H., and Lauri, A. (2007) Origin of the failure of classical nucleation theory: Incorrect description of the smallest clusters, *Phys. Rev. Lett.*, 98, 145702.

Mijoule, V., Lewis, L. J., and Meunier, M. (2006) Coulomb explosion induced by intense ultrashort laser pulses in two-dimensional clusters, *Phys. Rev. A*, 73, 033203.

Miller, W. H. (1974) Quantum mechanical transition state theory and a new semiclassical model for reaction rate constants, *J. Chem. Phys.*, 61, 1823–1834.

Miller, M. A. and Wales, D. J. (1996) Structure, rearrangements and evaporation of rotating atomic clusters, *Mol. Phys.*, 89, 533–554.

Montag, B. and Reinhard, P.-G. (1995) Symmetric and asymmetric fission of metal clusters, *Phys. Rev. B*, 52, 16365–16368.

Mortier, W. J., Ghosh, S. K., and Shankar, S. (1986) Electronegativity equalization method for the calculation of atomic charges in molecules, *J. Am. Chem. Soc.*, 108, 4315.

Moseler, M., Rattunde, O., Nordiek, J., and Haberland, H. (2000) On the origin of surface smoothing by energetic cluster impact: Molecular dynamics simulation and mesoscopic modeling, *Nucl. Inst. Met. Phys. Res. B*, 164, 522–536.

Nagata, T., Horikawa, J., and Kondow, T. (1991) Photodissociation of Ar_n^+ cluster ions, *Chem. Phys. Lett.*, 176, 526–530.

Näher, U., Blørnholm, S., Frauendorf, S., Garcias, F., and Guet, C. (1997) Fission of metal clusters, *Phys. Rep.*, 285, 245–320.

Näher, U., Frank, S., Malinowski, N., Zimmermann, U., and Martin, T. P. (1994) Fission of highly-charged alkali-metal clusters, *Z. Phys. D: At. Mol. Clusters*, 31, 191–197.

Napari, I. and Vehkamäki, H. (2004) The role of dimers in evaporation of small argon clusters, *J. Chem. Phys.*, 121, 819–822.

Nikitin, E. (1965a) Statistical theory of endothermic reactions. I. Bimolecular reactions, *Theor. Exp. Chem.*, 1, 83–89.

Nikitin, E. (1965b) Statistical theory of endothermic reactions. II. Unimolecular reactions, *Theor. Exp. Chem.*, 1, 90–94.

Noya, E. G., Doye, J. P. K., Wales, D. J., and Aguado, A. (2007) Geometric magic numbers of sodium clusters: Interpretation of the melting behaviour, *Eur. Phys. J. D*, 43, 57–60.

Parneix, P., Bréchignac, P., and Calvo, F. (2003) Evaporation of a mixed cluster: $KrXe_{13}$, *Chem. Phys. Lett.*, 381, 471–478.

Parneix, P. and Calvo, F. (2003) Statistical evaporation of rotating clusters. II. Angular momentum distribution, *J. Chem. Phys.*, 119, 9469–9475.

Parneix, P. and Calvo, F. (2004) Statistical evaporation of rotating clusters. IV. Alignment effects in the dissociation of non-spherical clusters, *J. Chem. Phys.*, 121, 11088–11097.

Pechukas, P. and Light, J. C. (1965) On detailed balancing and statistical theories of chemical kinetics, *J. Chem. Phys.*, 42, 3281–3291.

Pedersen, J., Bjørnholm, S., Borggreen, J., Hansen, K., and Martin, T. P. (1991) Observation of quantum supershells in clusters of sodium atoms, *Nature (London)*, 353, 733–735.

Peslherbe, G. H. and Hase, W. L. (1994) A comparison of classical trajectory and statistical unimolecular rate theory calculations of Al_3 decomposition, *J. Chem. Phys.*, 101, 8535–8553.

Peslherbe, G. H. and Hase, W. L. (1996) Statistical anharmonic unimolecular rate constants for the dissociation of fluxional molecules: Application to aluminum clusters, *J. Chem. Phys.*, 105, 7432–7447.

Peslherbe, G. H. and Hase, W. L. (2000) Product energy and angular momentum partitioning in the unimolecular dissociation of aluminum clusters, *J. Phys. Chem. A*, 104, 10556–10564.

Pruvost, L., Serre, I., Duong, H. T., and Jortner, J. (2000) Expansion and cooling of a bright rubidium three-dimensional optical molasses, *Phys. Rev. A*, 61, 053408–053416.

Pugnaloni, L. A. and Vericat, F. (2002) New criteria for cluster identification in continuum systems, *J. Chem. Phys.*, 116, 1097–1108.

Raduta, A. H., Raduta, A. R., Chomaz, P., and Gulminelli, F. (2002) Critical behavior in a microcanonical multifragmentation model, *Phys. Rev. C*, 65, 034606.

Ranaul Islam, M., Saalmann, U., and Rost, J. M. (2006) Kinetic energy of ions after Coulomb explosion of clusters induced by an intense laser pulse, *Phys. Rev. A*, 73, 041201(R).

Randrup, J. and Koonin, S. E. (1981) The disassembly of nuclear matter, *Nucl. Phys. A*, 356, 223–234.

Rappé, A. K. and Goddard, W. A. (1991) Charge equilibration for molecular-dynamics simulations, *J. Phys. Chem.*, 95, 3358–3363.

Rayleigh, L. L. (1882) On the equilibrium of liquid conducting masses charged with electricity, *Philos. Mag.*, 14, 184–186.

Raz, T., Even, U., and Levine, R. D. (1995) Fragment size distribution in cluster impact: Shattering versus evaporation by a statistical approach, *J. Chem. Phys.*, 103, 5394–5409.

Reguera, D. and Reiss, H. (2004) Fusion of the extended modified liquid drop model for nucleation and dynamical nucleation theory, *Phys. Rev. Lett.*, 93, 165701.

Reiss, H., Kegel, W. K., and Katz, J. L. (1998) Role of the model dependent translational volume scale in the classical theory of nucleation, *J. Phys. Chem.*, 102, 8548–8555.

Rentenier, A., Moretto-Capelle, P., Bordenave-Montesquieu, D., and Bordenave-Montesquieu, A. (2005) Analysis of fragment size distributions in collisions of monocharged ions with the C_{60} molecule, *J. Phys. B: At. Mol. Opt. Phys.*, 38, 789–806.

Rice, O. K. and Ramsperger, H. C. (1927) Theories of unimolecular gas reactions at low pressures, *J. Am. Chem. Soc.*, 49, 1617–1629.

Richert, J. and Wagner, P. (1990) Sequential decay of excited nuclei and characteristic features of nuclear fragmentation, *Nucl. Phys. A*, 517, 399–412.

Richert, J. and Wagner, P. (2001) Microscopic model approaches to fragmentation of nuclei and phase transitions in nuclear matter, *Phys. Rep.*, 350, 1–92.

Saito, S. and Cohen, M. L. (1988) Jellium-model calculation for dimer decays of potassium clusters, *Phys. Rev. B*, 38, 1123–1130.

Saunders, W. A. (1990) Fission and liquid-drop behavior of charged gold clusters, *Phys. Rev. Lett.*, 64, 3046–3049.

Saunders, W. A. (1992) Metal-cluster fission and the liquid-drop model, *Phys. Rev. A*, 46, 7028–7041.

Schaaf, P., Senger, B., Voegel, J. C., Bowles, R. K., and Reiss, H. (2001) Simulative determination of kinetic coefficients for nucleation rates, *J. Chem. Phys.*, 114, 8091–8104.

Schenter, G. K., Kathmann, S. M., and Garrett, B. C. (1999a) Dynamical nucleation theory: A new molecular approach to vapor-liquid nucleation, *Phys. Rev. Lett.*, 82, 3484–3487.

Schenter, G. K., Kathmann, S. M., and Garrett, B. C. (1999b) Variational transition state theory of vapor phase nucleation, *J. Chem. Phys.*, 110, 7951–7959.

Schmidt, M., Kusche, R., Kronmüller, W., von Issendorff, B., and Haberland, H. (1997) Experimental determination of the melting point and heat capacity for a free cluster of 139 sodium atoms, *Phys. Rev. Lett.*, 79, 99–102.

Schmidt, M., Kusche, R., von Issendorff, B., and Haberland, H. (1998) Irregular variations in the melting of size-selected atomic clusters, *Nature (London)*, 393, 238–240.

Schmidt, M., Hippler, T., Donges, J. et al. (2001a) Caloric curve across the liquid-to-gas change for sodium clusters, *Phys. Rev. Lett.*, 87, 203402–203405.

Schmidt, M., Kusche, R., Hippler, T. et al. (2001b) Negative heat capacity for a cluster of 147 sodium atoms, *Phys. Rev. Lett.*, 86, 1191–1194.

Schulte, J. (1995) Time-resolved differential fragmentation cross sections of $C_{60}^+ + C_{60}$ collisions, *Phys. Rev. B*, 51, 3331–3343.

Schweikhard, L., Hansen, K., Herlert, A., Herráiz Lablanca, M. D., and Vogel, M. (2005) Photodissociation of stored metal clusters—Fragment-abundance spectra after multisequential decay of Au_{30}^+, *Eur. Phys. J. D*, 36, 179–185.

Silberstein, J. and Levine, R. D. (1981) Statistical fragmentation patterns in multiphoton ionization: A comparison with experiment, *J. Chem. Phys.*, 75, 5735–5743.

Sizun, M., Aguillon, F., and Sidis, V. (2005) Theoretical investigation of nonadiabatic and internal temperature effects on the collision-induced multifragmentation dynamics of Na_5^+ cluster ions, *J. Chem. Phys.*, 123, 074331–074341.

Soto, R. and Cordero, P. (1999) Cluster birth-death processes in a vapor at equilibrium, *J. Chem. Phys.*, 110, 7316–7325.

Stace, A. J. (1986) A measurement of the average kinetic energy released during the unimolecular decomposition of argon ion clusters, *J. Chem. Phys.*, 85, 5774–5778.

Stace, A. J. and Shukla, A. K. (1982) A measurement of the average kinetic energy released during the unimolecular decomposition of argon ion clusters, *Chem. Phys. Lett.*, 85, 157–160.

Stauffer, D. and Aharong, A. (1994) *Introduction to Percolation Theory*, 2nd edn., Taylor and Francis, London, U.K.

Stillinger, F. H. (1963) Rigorous basis of the Frenkel-Band theory of association equilibrium, *J. Chem. Phys.*, 38, 1486–1494.

Su, C.-X. and Armentrout, P. B. (1993) Collision-induced dissociation of Cr_n^+ ($n = 2 - 21$) with Xe: Bond energies, dissociation pathways, and structures, *J. Chem. Phys.*, 99, 6506–6516.

Sweikhard, L., Beiersdorfer, P., Bell, W. et al. (1996) Production and investigation of multiply charged metal clusters in a Penning trap, *Hyperf. Interact.*, 99, 97–104.

Talanquer, V. and Oxtoby, D. W. (1994) Dynamical density functional theory of gas-liquid nucleation, *J. Chem. Phys.*, 100, 5190–5200.

Tapavicza, E., Tavernelli, I., and Rothlisberger, U. (2007) Trajectory surface hopping within linear response time-dependent density-functional theory, *Phys. Rev. Lett.*, 98, 023001–023004.

Ter-Avetisyan, S., Schnürer, M., Hilscher, D. et al. (2005) Fusion neutron yield from a laser-irradiated heavy-water spray, *Phys. Plasmas*, 12, 12702.

Tolman, R. C. (1949) The effect of droplet size on surface tension, *J. Chem. Phys.*, 17, 333–337.

Tomita, S., Andersen, J. U., Gottrup, C., Hvelplund, P., and Pedersen, U. V. (2001) Dissociation energy for C_2 loss from fullerene cations in a storage ring, *Phys. Rev. Lett.*, 87, 073401–073404.

Tomita, S., Andersen, J. U., and Hvelplund, P. (2003) Stability of buckminsterfullerene, C_{60}, *Chem. Phys. Lett.*, 382, 120–125.

Tully, J. C. (1990) Molecular dynamics with electronic transitions, *J. Chem. Phys.*, 93, 1061–1071.

Venkatesh, R., Lucchese, R. R., Marlow, W. H., and Schulte, J. (1995) Thermal collision rate constants for small nickel clusters of size 2–14 atoms, *J. Chem. Phys.*, 102, 7683–7699.

Vieira, A. and Fiolhais, C. (1998) Shell effects on fission barriers of metallic clusters: A systematic description, *Phys. Rev. B*, 57, 7352–7359.

Vigué, J., Labastie, P., and Calvo, F. (2000) Evidence for $N^{-1/3}$ dependence of the sticking cross-section of atoms on small and medium-size van der Waals clusters, *Eur. Phys. J. D*, 8, 265–272.

Vogel, M., Hansen, K., Herlert, A., and Schweikhard, L. (2001a) Decay pathways of small gold clusters—The competition between monomer and dimer evaporation, *Eur. Phys. J. D*, 16, 73–76.

Vogel, M., Hansen, K., Herlert, A., and Schweikhard, L. (2001b) Model-free determination of dissociation energies of polyatomic systems, *Phys. Rev. Lett.*, 87, 013401–013404.

Vogel, M., Hansen, K., Herlert, A., and Schweikhard, L. (2002) Dimer dissociation energies of small odd-size clusters Au_n^+, *Eur. Phys. J. D*, 21, 163–166.

Volmer, M. and Weber, A. (1926) Nucleus formation in supersaturated systems, *Z. Phys. Chem. (Leipzig)*, 119, 277–301.

Wagner, P., Richert, J., Karnaukhov, V. A., and Oeschler, J. (1999) Is binary sequential decay compatible with the fragmentation of nuclei at high energy? *Phys. Lett. B*, 460, 31–35.

Weerasinghe, S. and Amar, F. G. (1993) Absolute classical densities of states for very anharmonic systems and applications to the evaporation of rare gas clusters, *J. Chem. Phys.*, 98, 4967–4983.

Weisskopf, V. (1937) Statistics and nuclear reactions, *Phys. Rev.*, 52, 295–303.

Widmann, J. F., Aardahl, C. L., and Davis, E. J. (1997) Observations of non-Rayleigh limit explosions of electrodynamically levitated microdroplets, *Aerosol. Sci. Technol.*, 27, 636–648.

Wigner, E. (1937) Calculation of the rate of elementary association reactions, *J. Chem. Phys.*, 5, 720–725.

Yannouleas, C. and Landman, U. (1995) Barriers and deformation in fission of charged metal clusters, *J. Phys. Chem.*, 99, 14577–14581.

Yannouleas, C., Landman, U., Bréchignac, C., Cahuzac, P., Concina, B., and Leygnier, J. (2002) Thermal quenching of electronic shells and channel competition in cluster fission, *Phys. Rev. Lett.*, 89, 173403.

Zeng, X. C. and Oxtoby, D. W. (1991) Gas-liquid nucleation in Lennard-Jones fluids, *J. Chem. Phys.*, 94, 4472–4478.

Zhukhovitskii, D. I. (1995) Molecular dynamics study of cluster evolution in supersaturated vapor, *J. Chem. Phys.*, 103, 9401–9407.

Zweiback, J., Smith, R. A., Cowan, T. E. et al. (2000) Nuclear fusion driven by Coulomb explosions of large deuterium clusters, *Phys. Rev. Lett.*, 84, 2634–2637.

II

Clusters in Contact

16

Kinetics of Cluster–Cluster Aggregation

Colm Connaughton
University of Warwick

R. Rajesh
Institute of Mathematical Sciences

Oleg Zaboronski
University of Warwick

16.1 Introduction

Consider a collection of particles (monomers) undergoing disordered motion as a result of diffusion, for example. This motion sometimes brings particles sufficiently close to each other to produce interaction. Suppose that the result of interaction is that two sufficiently close particles have a finite probability to stick together irreversibly to form a cluster which then continues to diffuse and may interact again with other particles. If we had started with a collection of particles of equal mass, the net result of many such interactions is to convert it into a collection of clusters with a distribution of different masses. The process whereby an initially monodisperse collection of particles is converted into a distribution of clusters of different sizes by localized aggregation of clusters is what we mean by cluster–cluster aggregation (CCA).

We should distinguish, at the outset, between CCA and diffusion-limited growth. The latter is a related problem that studies the (usually fractal) structure of a single aggregate, which grows by the slow accretion of single independently diffusing particles (Sander 2000). In diffusion-limited growth, aggregation occurs between the large cluster whose geometric structure is being investigated and many small particles. In CCA, we are not interested in the detailed structure of the individual clusters

but rather in the interactions of many clusters having a broad distribution of sizes.

CCA is relevant to many applications in physics and chemistry. These range from the growth of clusters on a substrate during epitaxial thin film growth (see Krapivsky et al. (1999) and the references therein) to the origin of the size distribution of galaxies (Silk and White 1978). At intermediate scales, CCA has been investigated in relation to the formation of raindrops (Falkovich et al. 2002) and the modeling of chemical reactors (Smit et al. 1994). From the perspective of nanophysics, CCA may be an undesirable process which is to be inhibited as, for example, in the case of a colloidal suspension that may become unstable to precipitation of the dispersed material resulting from CCA of the colloidal particles mediated by van der Waals forces (Fernández-Toledano et al. 2007). On the other hand, it may be useful as a tool in applications involving self-organization and self-assembly of nanoparticles where it is required to produce clusters of elementary units of a given size (Murri and Pinto 2002). Since aggregation events are irreversible, CCA is an intrinsically non-equilibrium phenomenon. In some applications, it might make sense to allow fragmentation so that clusters may break up into smaller components as well as coalesce to form larger ones. Such models may reach an equilibrium in which there is detailed balance between aggregation and fragmentation. We shall not consider such models here. The interested reader may refer to Ernst and van Dongen (1987).

The quantity that we want to understand or control is the cluster size distribution, which is denoted by $N(\vec{x}, m, t)$. $N(\vec{x}, m, t)$ tells the average density of clusters of mass m at a particular point in space, $\vec{x}$, at a given time. In these notes, we shall mostly consider spatially homogeneous systems, in which case we drop the $\vec{x}$ argument. We are not interested in individual clusters per se. Rather, we are interested in the cluster size distribution in situations where there are large numbers of clusters. In such contexts, it makes sense to adopt a statistical, or kinetic theory, point of view to describe the kinetics of the aggregation process. In most presentations of kinetic theory, the standard approach is to begin from a mean-field description which assumes that particles are uncorrelated and then discuss how to modify the mean-field theory to account for correlations. We shall take a different approach here. We shall begin, in Section 16.2, by discussing, without technical details, the *stochastic rate equation* describing the statistics of a microscopic lattice model of CCA. The stochastic rate equation encodes all information about the probability distribution of clusters. The price to be paid for encoding so much information in a single equation is that this equation is stochastic. The advantage is that it allows us to treat the mean-field and fluctuation-dominated regimes in a unified way. The mean-field kinetic equation—in this case, the well-known Smoluchowski equation—can be obtained in a transparent way from the stochastic rate equation in the limit where the noise is small. Meanwhile, the stochastic rate equation makes the conditions of applicability of this approximation clear and provides a starting point for the analysis when it fails. Having introduced the stochastic rate equation and derived from it the mean-field Smoluchowski kinetic equation in Section 16.2, we then proceed in a more traditional way and devote considerable time to discuss the properties of CCA at the mean-field level in Sections 16.3 through 16.5 before returning to deal with the question of fluctuations in Section 16.6.

Given that CCA does not reach an equilibrium state, characterized by detailed balance, two rather different scenarios are possible depending on whether or not new monomers are being constantly supplied. In the absence of an external supply of monomers, the system evolves for all time, producing larger and larger clusters. The typical cluster size increases monotonically in time while the total density of clusters decays as smaller clusters are continually converted into larger ones. We refer to such a scenario as a decay problem. No nontrivial stationary distribution is possible for such decay problems owing to the irreversibility of the aggregation process. We shall discuss such decay problems in more detail in Section 16.3. In the second scenario, often relevant in applications, mass is externally supplied to the system from a source of monomers. Small clusters lost by conversion to larger ones can then be replaced. In this situation, which we refer to as CCA with source, the typical cluster size again grows monotonically in time but the cluster density does not decay. In problems with source, a stationary state may be reached in which the loss of clusters of a given size due to

aggregation to produce larger ones is balanced by the production of clusters of that size coming from the aggregation of smaller ones. This balance is only possible in the presence of a source since we need to constantly replenish the small clusters. To be accurate, such a stationary state should be called quasi-stationary since time evolution of the cluster size distribution continues at the scale of the largest mass. Stationary state is achieved for cluster sizes much less than the largest mass. The stationary state attained in this way is not an equilibrium state. It lacks detailed balance and actually carries a flux of mass through the space of cluster sizes from small masses to large masses. In this sense, it is analogous to the kind of far-from-equilibrium constant flux stationary states commonly used to describe turbulence. We shall discuss problems with source, which have received relatively little attention in the literature, in more detail in Section 16.4.

Aggregating systems provide simple examples of several types of nonequilibrium transition. Two such transitions shall be considered here. The first, the so-called gelation transition, manifests itself at the mean-field level. We discuss this in Section 16.5. Gelation is a dynamic phase transition associated with the apparent loss of mass conservation in the system of aggregating clusters despite the explicit conservative nature of the elementary interactions. It is often interpreted as corresponding to the formation of a "gel" (an infinite sized cluster) if the process of cluster aggregation is sufficiently rapid. The second interesting transition that occurs in aggregating systems is the transition between a weakly and strongly fluctuating regime as the spatial dimension is decreased. This necessarily requires that we go beyond the mean-field description and is addressed in Section 16.6. There, we shall see that CCA has a critical dimension of 2. In three dimensions, diffusive fluctuations are relatively weak and the statistics are well described by mean-field arguments. In one dimension, however, diffusive fluctuations are always dominant for sufficiently heavy clusters and one cannot neglect spatial correlations between clusters. As a result, scaling exponents obtained by mean-field arguments are incorrect in one dimension. The two-dimensional case is marginal and mean-field predictions acquire logarithmic corrections due to fluctuations.

This is, of course, a vast subject. We cannot possibly hope to cover all possible topics of interest in these notes. What we have endeavored to do is to provide a readable introduction to kinetic theories of CCA from a theoretical physics perspective. Many details have been omitted and many simplifications have been adopted in favor of clarity of exposition. Where appropriate, we have provided references to the literature where the interested reader can find the details that we have chosen to skip over. We should briefly mention some of the most important simplifications and omissions made here. We have already mentioned that we are assuming there are no fragmentation events. We have also mentioned that we are not particularly interested in the spatial structure of the clusters. This latter point is rather important. It is not to say that we ignore entirely the spatial structure. It is true that we think of the clusters as points. We do, however, partially take into account their spatial structure by allowing the aggregation probability to be a function of the cluster masses. Thus, larger clusters typically have a larger aggregation rate, which one can interpret as being due to their greater spatial extent enhancing the probability that they will collide with other clusters. There are, of course, limitations to this indirect accounting for spatial structure. While it is likely to be a reasonable model when the cluster mixture is dilute, it is unlikely to make much sense if the cluster size becomes comparable to the diffusion length. This is a particularly potent issue when one attempts to interpret results from point-like models too literally in the context of gelation.

16.2 Modeling Cluster–Cluster Aggregation

16.2.1 A Lattice Model

For the purpose of both numerical simulations and theoretical analysis, it is convenient to replace d-dimensional space in which aggregating particles are transported by a d-dimensional lattice $\mathbf{Z}^d$.

Consider a system of massive particles on $\mathbf{Z}^d$. Each particle is characterized by a positive label m—its mass. Denote the hopping rate as $D(m)$. We allow multiple particles to reside at a given lattice point at any moment of time. Particles of masses m_1 and m_2 residing at the same lattice site can coagulate irreversibly with exponential rate $gK(m_1, m_2)$ to create a particle of mass $m_1 + m_2$. The constant g is added to help us keep track of dimensions. In addition, we can inject new particles randomly into the system. For example, particles of mass m_0 can be injected independently at each site at an exponential rate J/m_0. As for the initial distribution of particles, we will assume that particles of a given mass are distributed independently at each site according to a Poisson distribution with intensity $N_0(\vec{x}, m, t)$. Due to the irreversible nature of aggregation, the asymptotic properties of CCA do not depend on the particular choice of the initial conditions.

The model of CCA we have just described was originally introduced in Kang and Redner (1984). The purpose of their work was to study the kinetics of CCA while disregarding the effects of cluster geometry.

Formally speaking, this model is an example of an infinite-dimensional Markov chain. Its state at time t is specified by listing the number of particles of a given mass at each lattice site. All information about kinetics of the model is encoded in the master equation, which governs the evolution of the probability distribution over the state space:

$$\frac{dP(\alpha, t)}{dt} = -\sum_\beta R_{\alpha \to \beta} P(\alpha, t) + \sum_\beta R_{\beta \to \alpha} P(\beta, t), \quad (16.1)$$

where

α, β denote microscopic states
$R_{\alpha \to \beta}$ denotes the rate of going from α to β
$P(\alpha, t)$ is the probability of finding the system in state α at time t

The rate of an event is defined as the probability per unit time of the event occurring or equivalently, the time between two successive events is drawn from an exponential distribution with the mean of the distribution being equal to the rate.

16.2.2 The Stochastic Smoluchowski Equation

The master equation is first order and linear. Using this similarity between the master equation and the Schroedinger equation of many-body quantum mechanics, Doi and Zeldovich-Ovchinnikov (DZO) built a formal solution to a general master equation in the form of a path integral for some effective statistical field theory. The power of applying this approach to the theory of reaction diffusion systems was demonstrated by Lee and Cardy in the context of one- and two-species annihilation systems, see Lee and Cardy (1996), Cardy (1998).

For CCA, the final result of DZO transformation is the following: the average mass distribution at the lattice site $\vec{x}$ is obtained as the average of a random variable $P(\vec{x}, m, t)$:

$$N(\vec{x}, m, t) = \langle P(\vec{x}, m, t) \rangle_\xi \quad (16.2)$$

$P(\vec{x}, m, t)$ solves a special stochastic equation (to be described below), which is driven by a noise, ξ. The average, $\langle \cdot \rangle_\xi$, in Equation 16.2 is with respect to the distribution of this noise. The equation satisfied by $P(\vec{x}, m, t)$ is called the stochastic Smoluchowski equation. Written out in full, it is

$$\partial t P(\vec{x}, m, t) = N_0(\vec{x}, m)\delta(t) + \frac{J}{m_0}\delta(m - m_0) + D(m)\Delta P(\vec{x}, m, t)$$

$$+ \frac{g}{2}\int_0^m dm_1 K(m - m_1, m_1)P(\vec{x}, m - m_1, t)P(\vec{x}, m_1, t)$$

$$- g\int_0^\infty dm_1 K(m, m_1)P(\vec{x}, m, t)P(\vec{x}, m_1, t)$$

$$+ \text{Noise}(K, P). \quad (16.3)$$

Here Δ is the discrete Laplacian,

$$\Delta P(\vec{x}) = \sum_{\vec{y} \sim \vec{x}} \left(P(\vec{x}) - P(\vec{y}) \right), \quad (16.4)$$

with the notation $\vec{y} \sim \vec{x}$ denoting summation over the nearest neighbors of $\vec{x}$. For details of the derivation of Equation 16.3, the reader is referred to Zaboronski (2001) and Connaughton et al. (2006). Here, we simply discuss the *meaning* of each term on the right-hand side of the stochastic Smoluchowski equation, Equation 16.3, rather than their derivation. We shall proceed from left to right. The first term is the initial condition. It is often taken to be a Poisson distribution. The second term encodes the process of injection of particles of mass m_0 onto the lattice. The third linear term is just a discrete diffusion term, which describes the evolution of noninteracting random walkers on the lattice. The fourth (nonlinear) integral term describes the rate of increase of the number of particles of mass m at time t due to coagulations of lighter particles. The fifth (nonlinear) integral term describes the rate of decrease of the number of particles of mass m due to collisions of particles of mass m with particles of any other mass. The sixth term, the noise term requires more explanation. Without the noise term, Equation 16.3 would be the Smoluchowski kinetic equation, which we shall discuss in considerable detail in the following sections. It describes CCA in the mean-field limit. However, DZO transformation dictates, that there is an additional stochastic term in Equation 16.3, describing the evolution of mass distribution in CCA. The noise term encodes all fluctuation effects in cluster–cluster aggregation. Informally, it describes the process of a pair of particles coming to the same lattice site and then jumping away without interaction. Such events lead to emergence of spatial, temporal, and mass fluctuations in CCA.

Noise(K, P) is an *imaginary* multiplicative noise term, which contains all information about the correlations between diffusing–coagulating particles:

$$\text{Noise}\,(K, P) = iP(\vec{x}, m, t)\xi(\vec{x}, m, t), \tag{16.5}$$

where $\xi(\vec{x}, m, t)$, the noise defining the average in Equation 16.2, is mean zero white Gaussian noise in space and time, but correlated in mass space:

$$\langle \xi(\vec{x}_1, m_1, t_1)\xi(\vec{x}_2, m_2, t_2) \rangle = gK(m_1, m_2)\delta_{\vec{x}_1, \vec{x}_2}\delta(t_1 - t_2). \tag{16.6}$$

In the particular case, when the coagulation kernel is multiplicative,

$$K(m_1, m_2) = \beta(m_1)\beta(m_2),$$

the stochastic noise can be expressed in terms of mass-independent Gaussian noise:

$$\text{Noise}(K, P) = i\beta(m)\,P(\vec{x}, m, t)\xi(\vec{x}, t), \tag{16.7}$$

where

$$\langle \xi(\vec{x}_1, t_1)\xi(\vec{x}_2, t_2) \rangle = \delta_{\vec{x}_1, \vec{x}_2}\delta(t_1 - t_2). \tag{16.8}$$

The presence of imaginary i in the noise term leads to the two-point function of noise being negative, which leads to *anti-correlations* of particles in CCA, which makes physical sense: the presence of a particle at a given lattice site *reduces* the probability of finding another particle nearby.

If the transport mechanism is recurrent (as is the random walk in one dimension), the anti-correlations between particles become very important and solutions to Equation 16.3 become very different from solutions to the mean-field Smoluchowski equation. As a result, fluctuation effects in CCA in low dimensions are very important, whereas mean-field approximation often provides a good description of CCA in high dimensions. We discuss this point in considerable detail in Section 16.6. In principle, Equation 16.3 contains all information about the statistics of the lattice

model. In practice, Equation 16.3 is difficult to solve. We shall proceed, therefore, by looking at two complementary approaches, which have been very fruitful: direct Monte Carlo simulation of the lattice model and mean-field approximation of Equation 16.3.

16.2.3 Monte Carlo Simulation of Cluster–Cluster Aggregation

The master equation, Equation 16.1 can be simulated on a computer using Monte Carlo methods. These algorithms are often referred to in the literature as the Gillespie algorithms (Gillespie 1976). Suppose the system is in a state α. First, enumerate all the states β into which the system can evolve. The total rate of these transitions are $R(t) = \sum_{\beta} R_{\alpha \to \beta}$. One reaction out of these $R_{\alpha \to \beta}$'s is chosen with the state β being chosen with probability $R_{\alpha \to \beta}/R$. Time is then incremented by Δt where Δt is drawn from the probability distribution $R\exp(-R\Delta t)$.

Consider the lattice model introduced in Section 16.2.1. Each site can have any number of particles. Only particles on the same site are allowed to aggregate with each other. Given a certain configuration, the rate of an event occurring is given by

$$R(t) = \sum_{i=1}^{L} r_i, \tag{16.9}$$

where
r_i is the rate of reaction at site i
L denotes the total number of sites

Let there be n_i particles at site i. Let the masses on site i be denoted by m_{ij} with $j = 1, 2, \ldots, n_i$. Then the rate at site i is given by

$$r_i = \frac{J}{m_0} + Dn_i + g\sum_{j=1}^{n_i - 1}\sum_{k=j+1}^{n_i} K(m_{i,j}, m_{i,k}). \tag{16.10}$$

Then, as described in the previous paragraph, a certain reaction is chosen with the corresponding probability and the time increment is chosen from an exponential distribution.

For simulating CCA with input, the initial condition is chosen to be one where there are no particles. The system is evolved in time using the above Monte Carlo algorithm. Since mass is continually pumped into the system and there is no dissipation mechanism, the system reaches only a quasi-stationary steady state. Still, if one is interested in measuring the mass distribution up to a largest mass M, then the system is allowed to evolve until the distribution up to M does not change with time. This can be numerically achieved by tracking the sum of all masses in the system below M and waiting for it to reach a stationary value.

The data for mass distribution, and other higher order correlation functions are averaged over in the steady state. As we shall see, these are expected to show scaling behavior. The next step is to extract exponents. A unbiased method of extracting exponents is to use the "maximum likelihood estimation" method. A very readable account is given in Goldstein et al. (2004).

16.2.4 Mean-Field Theory: The Smoluchowski Equation

Averaging Equation 16.3 with respect to the noise, ξ, leads to the Hopf equation:

$$\partial_t N(\vec{x}, m, t) = D(m)\Delta N(\vec{x}, m, t)$$

$$+ \frac{g}{2} \int_0^m dm_1 K(m - m_1, m_1) C_2(\vec{x}, m_1, m - m_1, t)$$

$$- g \int_0^\infty dm_1 K(m, m_1) C_2(\vec{x}, m, m_1, t) + N_0(m)\delta(t)$$

$$+ \frac{J}{m_0} \delta(m - m_0). \quad (16.11)$$

where the second-order correlation function,

$$C_2(\vec{x}, m_1, m_2, t) = \langle P(\vec{x}, m_1, t) P(\vec{x}, m_2, t) \rangle, \quad (16.12)$$

measures the probability that two particles having mass m_1 and m_2 meet at the point $\vec{x}$ at time t. The mean-field approximation means that we assume that all clusters are independent of each other, or equivalently, correlations are negligible. If this is the case, then the second-order correlation function, Equation 16.12, factorizes as a product of the densities:

$$C_2(\vec{x}, m_1, m_2, t) \approx N(\vec{x}, m_1, t) N(\vec{x}, m_2, t) \quad (16.13)$$

so that Equation 16.11 becomes a closed kinetic equation for the cluster size distribution, known as the Smoluchowski equation:

$$\partial_t N(\vec{x}, m, t) = D(m)\Delta N(\vec{x}, m, t) + \frac{J}{m_0}\delta(m - m_0)$$

$$+ \frac{g}{2} \int_0^m K(m_1, m - m_1) N(\vec{x}, m_1, t) N(\vec{x}, m - m_1, t) dm_1$$

$$- g N(\vec{x}, m, t) \int_0^\infty K(m, m_1) N(\vec{x}, m_1, t) dm_1. \quad (16.14)$$

For the most part, we shall be interested in spatially homogeneous situations, in which case, averaged quantities have no explicit dependence on $\vec{x}$. From now on, unless we explicitly need to consider spatial structure, we shall drop all $\vec{x}$ labels. The Smoluchoswki equation then becomes:

$$\partial_t N(m, t) = \frac{g}{2} \int_0^m K(m_1, m - m_1) N(m_1, t) N(m - m_1, t) dm_1$$

$$- g N(\vec{x}, m, t) \int_0^\infty K(m, m_1) N(\vec{x}, m_1, t) dm_1 + \frac{J}{m_0}\delta(m - m_0). \quad (16.15)$$

We shall restrict ourselves to kernels, which are homogeneous functions, denoting the degree of homogeneity by λ:

$$K(am_1, am_2) = a^\lambda K(m_1, m_2). \quad (16.16)$$

g is a constant having dimension $[g] = M^{-\lambda}T^{-1}$. Many kernels which arise in practice are indeed homogeneous. The dimensions of the other quantities in Equation 16.15 are also worth stating explicitly since they will be used in what follows: $[N] = M^{-1}$, $[K(m_1, m_2)] = M^\lambda$ and $[J] = MT^{-1}$. We now note some families of kernels that provide archetypal models for various different behaviors observable in solutions of the Smoluchowski equation:

$$K(m_1, m_2) = (m_1 m_2)^{\frac{\lambda}{2}} \quad \text{Generalised product kernel}, \quad (16.17)$$

$$K(m_1, m_2) = m_1^\lambda + m_2^\lambda. \quad \text{Generalised sum kernel} \quad (16.18)$$

The constant kernel, having $K(m_1, m_2) = 1$ is a special case of Equation 16.17 with $\lambda = 0$.

16.3 Smoluchowski Equation and Scaling

16.3.1 The Scaling Hypothesis

One of the principal achievements of the theory of CCA is the existence of a practically complete classification of the possible scaling behaviors of the solutions of the Smoluchowski equation as the kernel is varied. This is a large subject in its own right which we shall only discuss in overview here. The interested reader is referred to the excellent review article of Leyvraz (2003) for a much more in-depth discussion of scaling and exact solutions of the Smoluchowski equation. The basic idea is that there is a typical cluster size, $s(t)$, which increases in time and the time evolution of the cluster size distribution is self-similar with respect to $s(t)$. That is to say, $N(m, t) = s(t)^\alpha \Phi(m/s(t))$ for some exponent, α. For now, we assume that there is no external injection of monomers so that $J = 0$. If we then assume[*] that the total mass in the system is conserved, we require

$$\int_0^\infty m N(m, t) dm = \int_0^\infty m s(t)^\alpha \Phi\left(\frac{m}{s(t)}\right) dm = s(t)^{\alpha - 2} \int_0^\infty \xi \Phi(\xi) d\xi$$

to be independent of time. This requires that we choose $\alpha = -2$. Thus, following Leyvraz (2003) we say that a solution of Equation 16.15 is of scaling form if it can be written as

---

[*] The conservation of mass in the absence of any external source of monomers might seem, at first sight, to be self-evident. We shall see in Section 16.5 that this is not, in fact, the case.

$$N(m,t) = \frac{C}{s(t)^2} \Phi\left(\frac{m}{s(t)}\right) \qquad (16.19)$$

where

   $s(t)$ is an increasing function of time
   C is a constant which keeps dimensions correct
   $\Phi(x)$ is a function known as the scaling function

The time evolution of a scaling solution simply consists of a continuous dilation and rescaling of the cluster size distribution described by the scaling function. A loose statement of the scaling hypothesis is that all sufficiently narrow (in the space of cluster sizes) initial conditions tend toward a scaling solution for large times.

Substituting Equation 16.19 into Equation 16.15 and assuming a homogeneous kernel, as defined by Equation 16.16, leads to the following condition on $s(t)$:

$$\frac{ds}{dt} = Cgs^\lambda \qquad (16.20)$$

while the scaling function, $\Phi(x)$, must satisfy the integro-differential equation

$$-x\frac{d\Phi(x)}{dx} - 2\Phi(x) = \frac{1}{2}\int_0^x K(y, x-y)\Phi(y)\Phi(x-y)dy$$

$$-\Phi(x)\int_0^\infty K(x,y)\Phi(y)dy. \qquad (16.21)$$

From Equation 16.20 we expect that the typical cluster size, $s(t)$, should exhibit three qualitatively different types of behavior depending on the value of λ (we set $Cg = 1$ for convenience and set $s(0) = 1$):

Case A $s(t) = \left(1 + (1-\lambda)t\right)^{\frac{1}{1-\lambda}}$ $\lambda < 1$

Case B $s(t) = e^t$ $\lambda = 1$

Case C $s(t) = \dfrac{1}{\left(1 - (\lambda-1)t\right)^{\frac{1}{\lambda-1}}}$ $\lambda > 1$.

In case A, the regular case, the typical cluster size grows as $t^{1/1-\lambda}$ for large times. In case B, the marginal case, the typical cluster size increases exponentially in time. In case C, the singular case, the typical cluster size diverges within a finite time. This is indicative of a singularity in the solution of Equation 16.15 for kernels having a degree of homogeneity, λ, greater than 1. This singularity is associated with a loss of mass conservation at the singular time and is often associated with a phenomenon called gelation. We shall return to this topic in detail in Section 16.5. For the remainder of this section, we shall restrict ourselves to the regular case.

If scaling is observed, then we know how the typical cluster size grows with time, $s(t) \sim t^{1/1-\lambda}$. To more fully understand the

dynamics, we would like to further know how the cluster size distribution behaves at small m for fixed t and how it decays at fixed m for large times. To this end, we introduce two new exponents, τ and w:

$$N(m, t) \sim m^{-\tau} \quad \text{for } m \ll s(t) \qquad (16.22)$$

$$N(m, t) \sim t^{-w} \quad \text{for } t \to \infty. \qquad (16.23)$$

The exponents τ and w for systems satisfying the scaling hypothesis is controlled by the asymptotic scaling properties of the kernel. Therefore, a lot of research has focused on the following kernel characterized by two exponents which, adopting the notation of Van Dongen (1987), Ernst and van Dongen (1988), we shall call μ and ν:

$$K(m_1, m_2) = m_1^\mu m_2^\nu + m_1^\nu m_2^\mu. \qquad (16.24)$$

Clearly $\mu + \nu = \lambda$. We shall further assume, at least until Section 16.5, that $\nu < 1$. For the reader interested in the physical relevance of these various exponents, a list of the homogeneity exponents of various kernels coming out of physical applications can be found in Ernst (1986). Given the restrictions on the kernel listed above, it can be shown that the *large x* behavior of the scaling function, $\Phi(x)$ is exponential. The *small x* behavior of $\Phi(x)$ turns out, unsurprisingly, to determine the exponent, τ. That is

$$\Phi(x) \sim x^{-\tau} \quad \text{as } x \to 0. \qquad (16.25)$$

Thus knowing τ requires knowledge of the scaling function. It can be shown relatively easily (see Leyvraz (2003)) that the exponents τ, w, and λ must satisfy a scaling relation:

$$\frac{2-\tau}{1-\lambda} = w. \qquad (16.26)$$

The principal aim of scaling theory is therefore to determine the small x behavior of $\Phi(x)$ as a function of μ and ν. From this, the exponent w follows from Equation 16.26. Given the complexity of Equation 16.21, this is not a trivial task. The interested reader is referred, again, to Leyvraz (2003) for a complete account.

16.3.2 Exact Solution for the Constant Kernel

For certain special cases, exact solutions of the Smoluchowski equation may be constructed. These are usually for the case of monodisperse initial conditions, $N(m, 0) = \delta(m - 1)$. For such initial conditions, only integer cluster sizes are ever produced so that the solution takes the form

$$N(m,t) = \sum_{i=1}^\infty c_i(t)\delta(m-i) \qquad (16.27)$$

While inevitably special and restricted to particularly simple kernels, exact solutions have played a key role in understanding the behavior of the solutions of the Smoluchowski equation. The scaling theory described above is based on a hypothesis that cannot be mathematically proven in general. At the same time, experimental realizations of cluster–cluster aggregation (see, for example, Broide and Cohen (1990)) tend to be too noisy to test the more subtle predictions, which come out of scaling theory. Exact solutions therefore provide a testing ground for scaling theory.

In these notes, we outline in some detail the exact solution for the case of constant kernel using the generating function technique. This is a commonly used approach, which illustrates several key points which will come in useful later. We provide references to some of the more complex, and indeed more interesting solutions which are known for more complicated kernels.

We define the generating function, $G(\mu, t)$, by taking the Laplace transform of the cluster size distribution:

$$G_\mu(t) = \int_0^\infty N(m,t)e^{\mu m}dm. \tag{16.28}$$

Note that $G_0(t)$ is the total density of clusters, $N(t) = \int_0^\infty N(m,t)dm$. Substituting Equation 16.28 into Equation 16.15 with $K(m_1, m_2) = 1$, absorbing g into t and performing some manipulations leads to

$$\frac{\partial G_\mu}{\partial t} = \frac{1}{2}G_\mu^2 - G_0 G_\mu \tag{16.29}$$

Note that setting $\mu = 0$, we can solve G_0 explicitly. Assuming that the initial density is 1,

$$\frac{dG_0}{dt} = -\frac{1}{2}G_0^2 \tag{16.30}$$

$$\Rightarrow \quad G_0(t) = \frac{2}{2+t}. \tag{16.31}$$

Note that Equation 16.30 says that the total density of clusters for constant kernel aggregation follows the rate equation for the single species reaction $A + A \rightarrow A$. We shall exploit the connections between these two systems, which are connected by the Laplace transform, later in Section 16.6. If we define,

$$\tilde{G}_\mu(t) = G_\mu(t) - G_0(t)$$

we obtain the simple differential equation

$$\frac{\partial \tilde{G}_\mu}{\partial t} = \frac{1}{2}\tilde{G}_\mu^2 \tag{16.32}$$

with the initial condition

$$\tilde{G}_\mu(0) = \phi_\mu = \int_0^\infty N(m,0)(e^{\mu m} - 1)\,dm. \tag{16.33}$$

This is easily solved to provide the generating function in the implicit form

$$G_\mu(t) = \frac{2\phi_\mu}{2 - t\phi_\mu} + \frac{2}{2+t} \tag{16.34}$$

For the case of monodisperse initial conditions, $\phi_\mu = e^\mu - 1$, and, comparing with Equation 16.27,

$$G_\mu(t) = \sum_{j=1}^\infty c_j(t)e^{j\mu}. \tag{16.35}$$

This allows us to convert Equation 16.34 into an explicit formula:

$$G_\mu(t) = \frac{2(e^\mu - 1)}{2 + t(1 - e^\mu)} + \frac{2}{2+t}$$

$$= \frac{4e^\mu}{(2+t)^2}\left[1 - \left(\frac{te^\mu}{2+t}\right)\right]^{-1}$$

$$= \frac{4e^\mu}{(2+t)^2}\sum_{j=0}^\infty \left(\frac{te^\mu}{2+t}\right)^j$$

$$= \sum_{j=1}^\infty \frac{4}{(t+2)^2}\left(\frac{t}{t+2}\right)^{j-1}e^{j\mu}.$$

Comparing with Equation 16.35 we finally obtain

$$c_i(t) = \frac{4}{(t+2)^2}\left(\frac{t}{t+2}\right)^{i-1} \tag{16.36}$$

By some manipulations, we can write this in the form

$$c_i(t) = \frac{4}{t^2}\left(1 + \frac{2x}{i}\right)\left[\left(1 + \frac{2x}{i}\right)^i\right]^{-1}, \tag{16.37}$$

where $x = i/t$. It is thus evident that the typical cluster size grows as $s(t) \sim t$. Let us now take the *scaling limit*, meaning that we take $i \rightarrow \infty$ keeping x fixed. The result is

$$c_i(t) \sim \frac{1}{s(t)^2}4e^{-2x}. \tag{16.38}$$

Thus, the scaling function for the constant kernel case is $\Phi(x) = 4e^{-2x}$. From these results, we can read off the exponents discussed in the previous section: $\tau = 0$ and $w = 2$ which clearly satisfy the scaling relation, Equation 16.26. Thus the scaling theory checks out in this case at least. Several other exact solutions and their relation to scaling can be found in Leyvraz (2003).

16.4 Cluster–Cluster Aggregation with Source

16.4.1 Differences between Decay and Source Problems

In this section, we consider the solutions of Equation 16.15 when J is not zero. This case, although relevant to applications, has attracted considerably less attention in the literature. The physics is rather different in this case. The scaling theory discussed in the previous section had as one of its main aims the determination of the large time and mass behavior of the cluster size distribution. As mentioned in the introductory discussion, a quasi-stationary state is reached for large times in the presence of a source of monomers. The properties of this stationary state then become of primary interest while scaling theory may be of use to describe how this state is approached in time. The stationary state is a nonequilibrium stationary state in which the loss of clusters of a given size, m, due to aggregation with other clusters is balanced by the generation of clusters of size m by the aggregation of smaller clusters having sizes $m - m_1$ and m_1. The source makes such a balance possible by ensuring that the smaller clusters are constantly replenished.

In the case of a constant kernel, the generating function approach with more work can be adapted to solve Equation 16.15 with the source term included. The result is

$$c_i(t) = \frac{\pi^2}{Jt^3} \sum_{j=-\infty}^{\infty} (2j+1)^2 \left[1 + \frac{\pi^2(2j+1)^2}{2Jt^2} \right]^{-(i+1)} \quad (16.39)$$

To the best of our knowledge, this is the only explicitly solvable example, which includes a source term. Some algebra yields the stationary state:

$$\lim_{t \to \infty} c_i = \sqrt{\frac{J}{2\pi}} \frac{\Gamma\left(i - \frac{1}{2}\right)}{\Gamma(i+1)}. \quad (16.40)$$

For large cluster sizes, $i \gg 1$, this has the asymptotic expansion (see, for example, Erdélyi and Tricomi (1951))

$$c_i \sim \sqrt{\frac{J}{2\pi}} i^{-\frac{3}{2}} \left(1 + O\left(\frac{1}{i}\right) \right). \quad (16.41)$$

For more general kernels, we may learn something from dimensional analysis. Given that the only parameters explicitly in the Smoluchowski equation are J, g, and m, the only dimensionally correct way to construct something having the dimensions of a mass density is

$$N(m) = \sqrt{\frac{J}{g}} m^{-\frac{\lambda+3}{2}}. \quad (16.42)$$

Is this the correct form for the stationary state? In this section, we show how the stationary state can be obtained analytically for an arbitrary homogeneous kernel.

16.4.2 The Stationary State for Arbitrary Homogeneous Kernel

Let us rewrite Equation 16.15 in the symmetric form:

$$\frac{\partial N(m,t)}{\partial t} = \frac{g}{2} \int K(m_1, m_2) N(m_1, t) N(m_2, t) \delta(m - m_1 - m_2) dm_1 dm_2$$

$$- \frac{g}{2} \int K(m, m_1) N(m, t) N(m_1, t) \delta(m_2 - m - m_1) dm_1 dm_2$$

$$- \frac{g}{2} \int K(m, m_2) N(m, t) N(m_2, t) \delta(m_1 - m_2 - m) dm_1 dm_2$$

$$+ \frac{J}{m_0} \delta(m - m_0). \quad (16.43)$$

Let us now look for a stationary solution,

$$N(m) = cm^{-x}, \quad (16.44)$$

where x and c are to be determined. This gives for $m > m_0$

$$0 = -\frac{\partial J(m,x)}{\partial m} = mc^2 \frac{g}{2} \int K(m_1, m_2)(m_1, m_2)^{-x} \delta(m - m_1 - m_2) dm_1 dm_2$$

$$- mc^2 \frac{g}{2} \int K(m, m_1)(mm_1)^{-x} \delta(m_2 - m - m_1) dm_1 dm_2$$

$$- mc^2 \frac{g}{2} \int K(m, m_2)(mm_2)^{-x} \delta(m_1 - m_2 - m) dm_2 dm_2. \quad (16.45)$$

We now apply the Zakharov transformations (further details in Connaughton et al. (2004)):

$$(m_1, m_2) \rightarrow \left(\frac{mm_1}{m_2}, \frac{m^2}{m_2} \right) \quad (16.46)$$

$$(m_1, m_2) \rightarrow \left(\frac{m^2}{m_1}, \frac{mm_2}{m_1} \right). \quad (16.47)$$

to the second and third integrals in Equation 16.45. The Jacobians of these transformations are m^3/m_2^3 and m^3/m_1^3 respectively. Using the homogeneity of $K(m_1, m_2)$, we arrive at the following

$$0 = mc^2 \frac{g}{2} \int K(m_1, m_2)(m_1, m_2)^{-x} (m^y - m_1^y - m_2^y)$$

$$\delta(m - m_1 - m_2) dm_1 dm_2, \quad (16.48)$$

where $y = \lambda - 2x - 2$. The integrand clearly vanishes only for $y = 1$, which gives the exponent of the stationary solution:

$$x = x_0 = \frac{\lambda + 3}{2}. \quad (16.49)$$

Scaling out the mass dependence by introducing new variables $m_1 = mz$ and $m_2 = mz'$ and integrating out z' we get

$$-\frac{\partial J(m,x)}{\partial m} = c^2 g m^{\lambda - 2x + 2} I(x) \tag{16.50}$$

where

$$I(x) = \frac{1}{2} \int_0^1 dz\, K(z, 1-z)(z(1-z))^{-x} \left[1 - z^{2x - \lambda - 2} - (1-z)^{2x - \lambda - 2} \right] \tag{16.51}$$

On the stationary distribution, $J(m,x)$ must be constant and equal to the input mass rate, J. Thus we have

$$\lim_{x \to x_0} \frac{c^2 g m^{\lambda - 2x + 3}}{\lambda - 2x + 3} I(x) = J. \tag{16.52}$$

The limit on the left-hand side must be taken using L'Hopital's rule which gives

$$\frac{c^2 g}{2} \left. \frac{dI}{dx} \right|_{x = x_0} = J.$$

Therefore, we get the final answer for the amplitude of the steady state:

$$c = \sqrt{\frac{2J}{g \left. \dfrac{dI}{dx} \right|_{x = x_0}}}. \tag{16.53}$$

We can sometimes compute the integral $\left. \dfrac{dI}{dx} \right|_{x = x_0}$, thus allowing us to evaluate the amplitude of the stationary state exactly. Consider, for example, the following two-parameter family of kernels often studied as models:

$$K(m_1, m_2) = m_1^\mu m_2^\nu + m_1^\nu m_2^\mu. \tag{16.54}$$

For this kernel, one can show that $\left. \dfrac{dI}{dx} \right|_{x = x_0}$ is a function of $|\mu - \nu|$ and is finite and real only for $|\mu - \nu| < 1$:

$$c = \sqrt{\frac{J \left(1 - (\mu - \nu)^2\right) \cos\left(\frac{\pi}{2}(\mu - \nu)\right)}{4\pi g}}. \tag{16.55}$$

Note that the stationary state vanishes when $\mu - \nu = 1$. It is instructive to consider the case of the generalized sum kernel, Equation 16.18 for which $\mu = 0$ and $\nu = \lambda$. It is clear that something dramatic happens when λ passes through 1. The vanishing of the amplitude of the stationary state is associated with the presence of an instantaneous gelation transition in the model for $\lambda > 1$. We now turn to a discussion of the gelation transition.

16.5 The Gelation Transition

16.5.1 Finite Time Gelation

Let us consider the total mass contained in a clustering system in the absence of an external source:

$$M = \int_0^\infty m N(m,t)\, dm. \tag{16.56}$$

Given that the elementary interactions between clusters explicitly conserve mass, it seems intuitively obvious that M should be conserved. Indeed, if we use Equation 16.43 with $J = 0$ to derive an equation for dM/dt, then some elementary relabeling of integration variables on the right-hand side demonstrate that this is formally so. Such formal manipulations do require, however, that the integrals in question do not diverge. That this should be the case for an arbitrary kernel is not at all obvious, and indeed it has been understood that when $\lambda > 1$ there do not exist solutions of the Smoluchowski equation, Equation 16.15, which conserve mass for all times. The qualitative difference between kernels having $\lambda < 1$ and those having $\lambda > 1$ is well illustrated by the scaling theory discussed in Section 16.3.1. There, it was suggested that the typical cluster size diverges in finite time for kernels having $\lambda > 1$. This divergence is indicative of a singularity in the solution of the Smoluchowski equation. We shall denote this singular time by $t^\star$. The loss of global mass conservation goes together with the divergence of the typical cluster size. The singularity at $t = t^\star$ then came to be interpreted as the loss of mass from the normal cluster distribution (referred to as the "sol" component of the system) to a ubiquitous infinite mass cluster (referred to as the "gel" component), which is formed at $t = t^\star$ by the runaway nature of the aggregation process. The process whereby such infinite clusters are generated at $t = t^\star$ is called "gelation." Typically, the sol mass is conserved until gelation occurs at $t = t^\star$ and then starts to decrease after $t^\star$. There is considerable scope for confusion in this interpretation since the gel component is formed with zero concentration and sits at the infinite mass end of the cluster size distribution, a rather singular configuration.

Gelation was first understood clearly in the case of the simple product kernel (Equation 16.17 with $\lambda = 2$) where an exact solution (Ziff and Stell 1980, Leyvraz and Tschudi 1981) illustrated that the solution exists and makes sense before and after the gel point. The loss of mass from the system for $t > t^\star$ corresponds to a flux of mass at $m = \infty$. This flux to infinity can then be interpreted as feeding the growth of the gel component. One of the key simplifications of the interpretation of the gelation process described above is that the gel component does not interact with the sol component. This is a rather serious simplification from the physical point of view. It also means that, from the perspective of the sol particles, the infinite clusters are simply removed from the system. Thus, if one does not like thinking of infinite clusters floating around in the system, one can equally well think of the gelation process as corresponding to the onset of intrinsic dissipation whereby the large clusters are simply removed from the system.

16.5.2 Instantaneous Gelation

It turns out that among kernels having $\lambda > 1$, there is a further difference between those having $\nu < 1$ and those having $\nu > 1$. The former behave as described above, exhibiting a gelation transition at some time, $t^* > 0$. The latter on the other hand seem to be even more singular (see van Dongen (1987)) in that the gelation time is widely believed to be zero for such kernels and the very existence of a consistent solution to the Smoluchowski equation for such kernels remains in question (see Lee (2000, 2001)). Despite the fact that kernels having $\nu > 1$ seem to exhibit rather pathological behavior, examples have been proposed as models in the literature, in particular in the context of aggregation due to differential sedimentation (Horvai et al. 2008) or gravitational attraction (Kontorovich 2001). Hence, understanding the nature of the instantaneously gelling regime is not an entirely academic issue despite the fact that such kernels are probably not relevant for colloidal systems because geometric constraints on the reactivity of a cluster limit the exponent ν.

16.5.3 Regularizing the Gelation Kinetics

Part of the reason that gelation phenomena, and especially instantaneous gelation, are so difficult to understand mathematically is that one is required to study the solution of a singular integro-differential equation without knowing, a priori, the form of the singularity. One way around this difficulty is to work instead with a regularized kinetic equation whose solutions clearly exist and conserve mass and study what happens as the regularization is removed. The other possibility is to work directly with a stochastic particle system and study how it behaves in the thermodynamic limit. This latter approach has been successfully carried through to completion in Lushnikov (2005) for the simple product kernel. We adopt the former approach here.

There are many possible ways to introduce a regularization to the Smoluchowski equation. We choose to introduce a mass cutoff, M_{max}. Given that we allow M_{max} to become large, we consider clusters having masses less than M_{max} to constitute the sol component. Mass that is transferred to clusters having mass greater than M_{max} is considered to be part of the gel component. We write Equation 16.15 as follows:

$$\frac{\partial N(m,t)}{\partial t} = \frac{g}{2} \int_0^m K(m_1, m-m_1) N(m_1,t) N(m-m_1,t) dm_1$$

$$-gN(m,t) \int_0^{M_{max}-m} K(m,m_1) N(m_1,t) dm_1$$

$$-gN(m,t) \int_{M_{max}-m}^{M_{max}} K(m,m_1) N(m_1,t) dm_1. \qquad (16.57)$$

The sol mass is

$$M_{sol} = \int_0^M m N(m,t) dm. \qquad (16.58)$$

It is conserved by Equation 16.57 in the absence of the third term. The third term causes M_{sol} to decrease. It describes the transfer of mass from the sol component to the gel. The mass of the gel, M_{gel} thus grows according to the equation

$$\frac{dM_{gel}}{dt} = g \int_0^M dm \, m N(m,t) \int_{M-m}^M K(m,m_1) N(m_1,t) dm_1. \qquad (16.59)$$

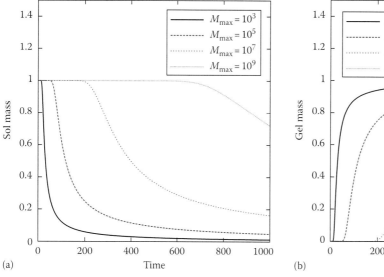

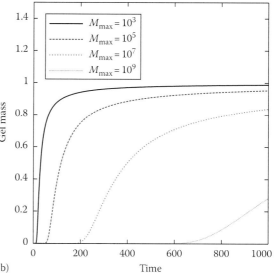

FIGURE 16.1 Sol mass (a) and gel mass (b) as functions of time obtained by solving Equation 16.57 with various values of M_{max} with the kernel Equation 16.17 having $\lambda = 3/4$.

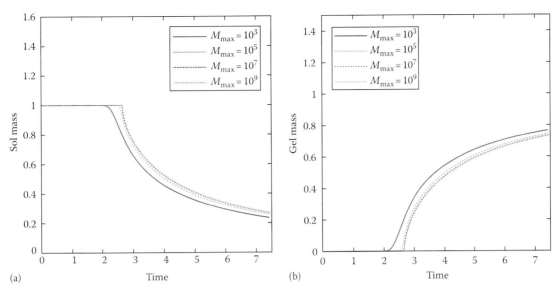

(a) Time (b) Time

FIGURE 16.2 Sol mass (a) and gel mass (b) as functions of time obtained by solving Equation 16.57 with various values of M_{max} with the kernel Equation 16.17 having $\lambda = 3/2$.

The total mass, $M_{sol} + M_{gel}$, is conserved by Equations 16.57 and 16.59. They have perfectly regular and unique solutions. The gelation transition can be understood by considering how the solutions of these equations behave as $M_{max} \to \infty$.

Figure 16.1 shows how M_{sol} and M_{gel} behave as M_{max} is increased for the generalized product kernel, Equation 16.17, with $\lambda = \frac{3}{4}$ and monodisperse initial conditions. Observe that M_{sol} is conserved over a longer time interval as M_{max} grows. Extrapolating this behavior would lead one to conclude that no gel component will be generated in finite time as $M_{max} \to \infty$, which is what one expects from the scaling theory.

Figure 16.2 shows the corresponding plots, again for the generalized product kernel, Equation 16.17, with monodisperse initial conditions, but this time having $\lambda = \frac{3}{2}$. The behavior in this case is completely different. Observe that the period over which M_{sol} is conserved becomes independent of M_{max} as M_{max} grows. The graphs illustrate that the gel component will be generated in finite time, $t^* \approx 2.5$, as $M_{max} \to \infty$. This corresponds to regular gelation.

Perhaps most interestingly of the three, Figure 16.3 shows the corresponding plots, for the generalized sum kernel, Equation 16.18, with monodisperse initial conditions and having $\lambda = \frac{3}{2}$. This case would be expected to undergo instantaneous gelation in the absence of regularization according to the classification of Van Dongen and Ernst. For the regularized system, there is, of course, no instantaneous generation of a gel component. M_{sol} is conserved over a finite time interval. However, the length

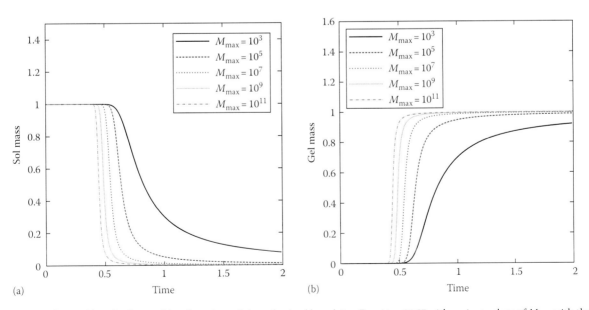

(a) Time (b) Time

FIGURE 16.3 Sol mass (a) and gel mass (b) as functions of time obtained by solving Equation 16.57 with various values of M_{max} with the kernel Equation 16.18 having $\lambda = 3/2$.

of this interval *decreases* as M_{max} increases. Extrapolating this behavior would lead one to expect that the gelation time tends to zero as the cutoff is removed, a behavior which may reasonably be interpreted as signifying instantaneous gelation. One would further expect that, in this case, the gelation will also be complete in the sense that it appears that once the gelation time is reached practically all of the sol mass is converted into gel. It is worth remarking that the decrease in the gelation time as M_{max} is increased is extremely slow. Based on the numerical results, it seems plausible that the decrease is logarithmically slow. This is consistent with recent work (Ben-Naim and Krapivsky 2003) on a simpler aggregation model (exchange driven growth), which also exhibits singular behavior analogous to instantaneous gelation. In this case, it was found that in a finite system, the gelation time decreases as the logarithm of the system size. This extremely slow dependence of the gelation time on the regularization explains why it has been so difficult to study instantaneous gelation numerically. It also illustrates that the use of instantaneously gelling kernels as models of physical systems may not cause such catastrophic difficulties as one might initially expect. This is because, in practical terms, there is always a limit to the applicability of the Smoluchowski equation which, at the simplest level may be modeled as a cutoff. In the presence of this cutoff, the regularized equations always behave reasonably, even if the corresponding unregularized system exhibits instantaneous gelation.

16.6 Strong Fluctuations in the Diffusion-Limited Regime

In our discussion of the strong fluctuation effects in CCA, we will concentrate on the simplest case of constant kernel aggregation. As we will see, this case is rich enough already to explain the essence of the effect fluctuations have on the statistics of mass distribution for spatially extended system of aggregating particles. Only at the very end, we present the only general non-mean-field result concerning CCA with nonconstant kernels known to us.

The simplicity of constant kernel CCA stems from its relation to the simplest reaction–diffusion model—$A + A \rightarrow A$. As we saw in Section 16.3.2, at the mean-field level, the total number of clusters in constant kernel aggregation satisfies the rate equation for $A + A \rightarrow A$. This remains true if fluctuations are included as can be seen by integrating the stochastic Smoluchowski equation, (Equation 16.3) with respect to mass. The result coincides with the stochastic rate equation of $A + A \rightarrow A$ derived in Cardy (1998). We therefore start our study of fluctuations in CCA with the statistics of the particle density.

16.6.1 Multiscaling of the Total Particle Density in Decaying CCA

The mean-field equation for the average particle concentration, $N(t) = \int_m dm N(m,t)$, can be derived by integrating Equation 16.15 (with $g K(m_1, m_2) = g$) with respect to m to obtain

$$\partial_t N(t) = -\frac{1}{2} g N^2(t). \tag{16.60}$$

The solution is

$$N(t) = \frac{2N_0}{2 + g N_0 t}, \tag{16.61}$$

where N_0 is the initial concentration. For time $t \gg 1/N_0 \lambda$, the average density scales as t^{-1} with time. The typical distance between particles can be estimated as

$$L(t) = \frac{1}{N(t)^{1/d}} \sim t^{1/d}. \tag{16.62}$$

For diffusing particles, there is another important length scale in the problem—the scale of diffusive excursions over time t:

$$L_D(t) = \sqrt{Dt} \sim t^{1/2} \tag{16.63}$$

Diffusive fluctuations do not invalidate mean-field assumptions if their scale is large compared with interparticle distance: In this case, correlations between particles do not have a chance to develop, as the probability of two tagged particles meeting each other multiple times is small compared with the probability of them reacting with other particles. Therefore, reacting particles can be treated as uncorrelated, and the factorization of moments of local particle density in terms of the first moment is justified.

We arrive at the criterion of mean-field applicability analogous to the celebrated Ginsburg criterion in the theory of equilibrium critical phenomenon:

$$1 \gg \frac{L(t)}{L_D(t)} \sim t^{1/d - 1/2} \tag{16.64}$$

This condition is clearly satisfied for large times if $d > 2$ and is violated for $d < 2$. Dimension $d_c = 2$ separating low dimensions for which diffusive fluctuations dominate large-scale dynamics and high dimensions for which mean-field approximation is applicable is called critical dimension.*

What is the correct behavior of particle density in $d < 2$ at large times? In this case, typical distances between particles are much bigger than the diffusive length. As a result, particles which react at time $t \gg t_0$ have met each other many times in the past with probability close to 1 and are strongly correlated as a result. The rate of reaction is determined by the diffusion rate. Consequently, the reaction regime for which (16.64) is violated is referred to as "diffusion limited."

The above-mentioned considerations have led to a heuristic description of diffusion-limited description of reaction–diffusion systems known as Smoluchowski approximation, see Kang and Redner (1984), Krapivsky et al. (1994) for details.

---

* Even for $d > d_c$ mean field theory can only predict the exponent of density decay correctly, but not the amplitude. For instance, in $d > 2$ diffusive point particles meet each other with probability 0. Therefore, the constant A in $N(t) = A/t$ must go to zero with particle size. Note however, that the mean field Equation 16.60 does not depend on particle size.

In essence, the Smoluchowski approximation is an assumption that two particles which have approached within some "interaction radius" R react with probability 1. This assumption can be used to derive an equation for the average particle density which has the same form as (16.60), except for the reaction rate g is replaced with renormalized reaction rate $g_R(t)$:

$$g_R N(t) = \text{Flux of particles having density } N(t) \text{ at } \infty \text{ through a sphere of radius } R \text{ centred at the origin}$$

The effective equation for particle density in the Smoluchowski approximation is then

$$\partial_t N(t) = -\frac{1}{2} g_R(t) N^2(t). \tag{16.65}$$

For $d < 2$, such a flux is R-independent and the expression for g_R can be derived from the dimensional argument alone: The only dimensional parameters we have at our disposal are the diffusion rate $[D] = L^2/T$ and time $[t] = T$. There is a single combination of dimension of the reaction rate $[g_R] = L^d/T$ one can build out of time and diffusion rate:

$$g_R = \text{Const} \cdot D^{d/2} t^{\frac{d-2}{2}}$$

The average total particle density can be obtained either from the Smoluchowski equation (16.65) or directly from the dimensional argument:

$$N(t) = \text{Const} \cdot \frac{1}{(Dt)^{\frac{d}{2}}} \tag{16.66}$$

For $d = 1$, the Smoluchowski approximation predicts $N(t) \sim t^{-1/2}$ thus contradicting the mean-field equation, (Equation 16.61) according to which $N(t) \sim t^{-1}$. As it turns out, the value of $d/2$ for the decay exponent predicted by the Smoluchowski approximation is correct in $d < 2$: for $d = 1$ it was established rigorously by Bramson and Leibowitz in 1988 (Bramson and Lebowitz 1988), although it can be easily extracted from the results of Glauber's 1963 paper (Glauber 1963). For $d = 2 - \epsilon$, the exponent $d/2$ can be derived to all orders of perturbative expansion based on the dynamical renormalization group (RG) (see Cardy (1998)). While discussing fractional dimensions may seem exotic, there is a lot of structural information about the statistics of fluctuations one can extract from RG study of particle systems. The universality of scaling exponents is just one example of such information.

Within the framework of RG, the Smoluchowski approximation can be identified with a renormalized mean-field theory—a partial re-summation of the perturbative expansion for correlation functions based on the idea that only terms contributing to the renormalization of g should be taken into account. This allows one to check the validity of Smoluchowski approximation

systematically for the reaction–diffusion system at hand. The Smoluchowski approximation is, therefore, a quick way of getting the right scaling properties of $N(t)$. Its justification, however, requires the full introduction of RG machinery. We refer the interested reader to Peliti (1986).

We conclude that mean-field theory predicts the fact that the decay of $N(t)$ should exhibit scaling behavior, but fails to predict the scaling exponent correctly. We will now demonstrate that fluctuation effects in low-dimensional reaction–diffusion systems are so strong that mean-field approximation fails qualitatively, not just quantitatively. We ask the question of how higher order correlation functions of particle density, $\langle P(x, t)^n \rangle$, scale with time. We denote the corresponding exponent by ν_n:

$$\left\langle P(x,t) \right\rangle^n \sim t^{-\nu_n}. \tag{16.67}$$

If the mean-field assumptions are satisfied, then higher order correlation functions factorize and $\langle P(x,t)^n \rangle = \langle P(x,t) \rangle^n$. Thus, if mean-field theory holds, then $\nu_n = \nu n$, where $\nu = 1$. In other words, the mean-field theory predicts *simple* scaling of correlation functions: The scaling exponent ν_n is a linear function of n.

In reality, the exponent ν_n is a nonlinear function of n. We can readily verify this statement using the Hopf equation, Equation 16.11, for the second-order correlation function. Integrating with respect to mass, we get a relation between 1- and 2-point functions:

$$\partial_t N = -\frac{1}{2} g \left\langle P^2(x,t) \right\rangle \tag{16.68}$$

We already know that $\nu_0 = 0$, $\nu_1 = d/2$ for $d < 2$. Substituting the answer for $N(t)$ into the Hopf equation (16.68), we find

$$\nu_2 - 2\nu_1 = 1 - \frac{d}{2} = \frac{\epsilon}{2} > 0 \quad \text{for } d < 2. \tag{16.69}$$

Here $\epsilon = 2 - d$. Therefore, the scaling exponents ν_n's do not lie on a straight line. This phenomenon is referred to as *multiscaling*. Using dynamical RG, it can be shown that

$$\nu_n = n\frac{d}{2} + \epsilon \frac{n(n-1)}{4} + O(\epsilon^2), \tag{16.70}$$

where $\epsilon = 2 - d$ (Munasinghe et al. 2006b).

The first term on the right-hand side of (16.70) results from renormalized mean-field theory. The second, nonlinear term describes anti-correlations between diffusing–reacting particles—due to the presence of this term, multi-point correlation functions decay faster with time than predicted by renormalized mean-field approximation. These anti-correlations manifest themselves in exclusion zones around reacting particles forming in the limit of large times. These exclusion zones decrease the probability of finding a pair of particles nearby as compared to the noninteracting case.

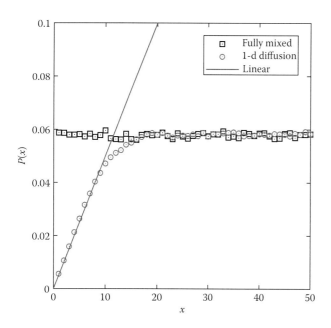

FIGURE 16.4 Zones of exclusion around particles in a Monte Carlo simulation of CCA with injection in one dimension. Plot shows the two-point correlation function $P(x) = \langle P(x_0)P(x_0 + x)\rangle$ for normal one-dimensional diffusion (circles) and simulation with artificial mixing to break correlations (squares). The solid line shows the prediction of Equation 16.71.

For the one-dimensional reaction–diffusion system the above scaling exponent, Equation 16.70, can be derived rigorously (Munasinghe et al. 2006a). In addition, the spatial structure of anti-correlations can be characterized completely:

$$\left\langle \prod_{k=1}^{n} P(x_k,t) \right\rangle \sim \left(\prod_{j>i} |x_i - x_j| \right) t^{-\frac{n}{2} - \frac{n(n-1)}{4}}. \qquad (16.71)$$

We see that the correlation function vanishes linearly with distance, as any two points approach each other. This effect is easy to observe numerically. Figure 16.4 illustrates the existence of the zone of exclusion and the linear vanishing of the 2-point correlation function, as predicted by Equation 16.71.

16.6.2 Multiscaling of Total Particle Density in the Steady State of CCA with Particle Injection

The steady state of CCA with sources can be analyzed in a similar way. The mean-field equation for the total average particle density is

$$\frac{\partial}{\partial t} N(t) = -\frac{1}{2}gN^2(t) + h, \qquad (16.72)$$

where h is the intensity of the source (the total number of particles introduced into a unit volume of the system per unit of time). The steady state solution, obtained as $t \to \infty$, is

$$N(\infty) = \sqrt{\frac{2h}{g}} \sim h^{1/2}. \qquad (16.73)$$

We observe the scaling behavior of the total average particle density with the source strength. The corresponding mean-field exponent is $\mu_1^{MF} = \frac{1}{2}$. Fluctuation effects in low dimensions preserve the scaling behavior but change the value of the exponent.

The applicability of mean-field theory can be analyzed by analogy with the decaying case: the interparticle distance can be estimated as

$$L(h) \sim \left(\frac{1}{N(\infty)} \right)^{1/d} \sim \frac{1}{h^{1/2d}}. \qquad (16.74)$$

The scale of diffusive fluctuations can be estimated by the only parameter of dimension length one can build out of diffusion coefficient $[D] = [L^2/T]$ and source intensity $[h] = 1/TL^d$:

$$L_D(h) = \left(\frac{D}{h} \right)^{\frac{1}{2+d}}. \qquad (16.75)$$

Diffusive fluctuations do not invalidate mean-field approximation if

$$1 \gg \frac{L(h)}{L_D(h)} \sim h^{-\frac{\epsilon}{2d(2+d)}}, \qquad (16.76)$$

where $\epsilon = 2 - d$. If $d < 2$, this condition is satisfied in the limit of a strong source, $h \to \infty$. This is very natural. With a high influx of "new" particles, the dominant process is the coagulation of new particles with each other and with particles already present in the system. Thus, only uncorrelated particles react, justifying the factorization of correlation functions at the heart of the mean-field approximation. This argument breaks down in $d > 2$ due to the presence of one more relevant dimensional parameter—particle size.

If the intensity of the source is weak, $h \to 0$, the criterion (16.76) is violated and the reaction becomes diffusion limited: particles meet many times before finally reacting. To calculate the total average particle density in this regime, we can resort to Smoluchowski approximation. The main conclusion is that the average density is the function of h and D only, which immediately leads to

$$N(\infty) = \text{Const} \frac{1}{L_D(t)} \sim h^{\frac{d}{2+d}}. \qquad (16.77)$$

We conclude that

$$\mu_1 = \frac{d}{2+d}.$$ (16.78)

This answer can be confirmed by a systematic perturbative RG calculation in dimension $d = 2 - \epsilon$ (Droz and Sasvári 1993). In one dimension, the answer $\mu_1 = 1/3$ can be proved using the correspondence between $A + A \rightarrow A$ and the dynamics of domain walls in the exactly solvable Glauber model in 1D (Rácz 1985).

We can see that $\mu_1 < \mu_1^{MF}$ for $d < 2$. Therefore, the actual density of particles in $d < 2$ in the presence of a weak source is much higher than the mean-field estimate: Due to anti-correlations between the particles, the effective reaction rate is smaller than its mean-field approximation. Moreover, these effects lead to multiscaling analogous to the multiscaling found in the decaying case. To see this, let us compute the scaling of the two-point function exactly in all dimensions: the Hopf equation, Equation 16.11, integrated with respect to m and taken at $t = \infty$ reads

$$\left\langle P(x,\infty)^2 \right\rangle = \frac{h}{g}.$$ (16.79)

Therefore, the density–density correlation function between two close points in space scales with dimension-independent exponent $\mu_2 = 1$. Mean-field theory predicts linear scaling obtained by factorization of higher order correlation functions: $\mu_n = n\mu_1^{MF}$. Using the above results for μ_1 and μ_2, we see that this is incorrect:

$$2\mu_1 - \mu_2 = \frac{\epsilon}{2+d} > 0,$$

so that the exponents of n-point functions are not a straight line for $d < 2$. The calculation of high-order correlation functions for $A + A \rightarrow A$ in $d = 2 - \epsilon$ dimensions in the presence of a source has been carried out in Connaughton et al. (2005, 2006). The answer can be stated as follows:

$$\left\langle P(x,\infty)^n \right\rangle \sim \left(L_D(h) \right)^{-nd - \epsilon \frac{n(n-1)}{2} + O(\epsilon^2)}$$ (16.80)

Notice that by replacing $L_D(h)$ in the above expression with $L_D(t)$, we get the multiscaling v_n (16.70) of the decaying CCA. This suggests that the mechanism responsible for multiscaling in the large time limit of decaying CCA and the weak source limit of CCA with injection are essentially the same.

Noticing that $L_D(h) \sim h^{-1/2+d}$, we conclude that

$$\mu_n = \frac{nd}{d+2} + \epsilon \frac{n(n-1)}{2(2+d)} + O(\epsilon^2)$$ (16.81)

This answer is conjectured to be exact in one dimension, but the corresponding calculation is still lacking.

16.6.3 Multiscaling of Mass Distribution in CCA

We have investigated the effects of strong fluctuations on the statistics of the total number of particles in constant kernel CCA. As it turns out, the calculation of correlation functions of mass distribution itself can be reduced to the study of the $A + A \rightarrow A$ model provided that the coagulation kernel is constant: By taking the Laplace transform of the Stochastic Smoluchowski equation, Equation 16.3, with respect to mass it is possible to verify that

- Correlation functions of decaying constant kernel CCA can be calculated by applying the inverse Laplace transform to the correlation functions of decaying $A + A \rightarrow A$ model with respect to the *initial* concentration of particles.
- Correlation functions of CCA with injection of monomers can be calculated by applying the inverse Laplace transform to the correlation functions of $A + A \rightarrow A$ model with injections with respect to the *intensity of the source*.

The above statements are easy to verify at the mean-field level by applying Laplace transform to Smoluchowski equation, Equation 16.15. The formalism of stochastic Smoluchowski equation allows one to check that fluctuations do not destroy the stated correspondence.

Explicitly, for CCA with injection,

$$\left\langle P(m,\infty)^n \right\rangle \sim \int dh\, h^{n-1} \left\langle P(h,\infty)^n \right\rangle e^{-mh}$$ (16.82)

In the limit of large masses, $m \ll m_0$, the right-hand side of the above integral is dominated by small values of h. Therefore, using (16.81) which holds in the limit of weak injection, we get

$$\left\langle P(m,\infty)^n \right\rangle \stackrel{m \to \infty}{\sim} \int dh\, h^{n-1} h^{-\mu_n} e^{-mh} \sim m^{-\gamma_n},$$ (16.83)

where

$$\gamma_n = \frac{2d+2}{d+2} n + \frac{\epsilon}{d+2} \frac{n(n-1)}{2} + O(\epsilon^2).$$ (16.84)

The interested reader is referred to Connaughton et al. (2005, 2006) for details of the derivation. We conclude that scaling of correlation functions in constant kernel CCA with injection is nonlinear. The shape of the scaling curve is similar to (16.81) and (16.70). Notice that

$$\gamma_1 = \frac{2d+2}{d+2}.$$ (16.85)

This is an exact result for $d < 2$, see Rajesh and Majumdar (2000). Also $\gamma_2 = 3$ in all dimensions. In the context of CCA, the independence of γ_2 on dimension has a clear physical meaning: the scaling of the two-point function in CCA with injection of light particles is determined solely by the fact that the average flux of mass through the mass space is constant in the stationary state. The exact calculation of γ_n for $n > 2$ in $d = 1$ remains an open problem, but numerical simulations suggest that formula (16.84) is exact. Exponents obtained from Monte Carlo simulations are shown in Figure 16.5.

The link between the two-point function and constancy of mass flux in CCA with injection of light particles leads the only general non-mean scaling law concerning CCA with an arbitrary kernel $K(m_1, m_2)$ and arbitrary diffusion law known to us: by applying the Zakharov transformations (see Section 16.4.2) to the Hopf equation, Equation 16.11 in the stationary state, it is possible to verify that

$$C_2(m\mu_1, m\mu_2, \infty) \sim m^{-3-\lambda}, \tag{16.86}$$

where μ_1 and μ_2 are dimensionless and of order 1, provided that a certain "locality" condition is satisfied, see Connaughton et al. (2007) for further details. Figure 16.6 shows the results of Monte Carlo simulations. The scaling of the one-point function is strongly affected by the mechanism of diffusion. In the simulations, it is more convenient to measure

$$\pi_2(m) = \int_m^\infty dm_1 C_2(m, m_1) \tag{16.87}$$

which, according to Equation 16.86, should scale with exponent -2 for the constant kernel case ($\lambda = 0$). The key point demonstrated by Figure 16.6 is that Equation 16.86 is satisfied independent of the diffusion law $D(m)$.

Finally, we remark that the study of higher order correlations in decaying constant kernel CCA is much more difficult. This is already seen at mean-field level: the solution, Equation 16.61, to Equation 16.60 does not depend on the initial condition N_0 at leading order in $1/t$. As a result, one needs to calculate the subleading terms in order to extract the scaling of the mass distribution using Laplace transform in N_0. The resulting mass distribution is very nontrivial and cannot be guessed from simple scaling arguments. The interested reader is referred to Kang and Redner (1984, 1985) where the $d = 1, 2$ mass distributions

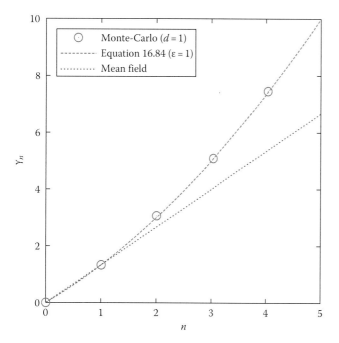

FIGURE 16.5 Scaling exponents of higher order correlation functions measured by Monte Carlo simulations for $n = 1, 2, 3, 4$ (circles). The straight line shows the predictions of mean-field theory. The curved line is the RG prediction, Equation 16.84 with $\epsilon = 1$.

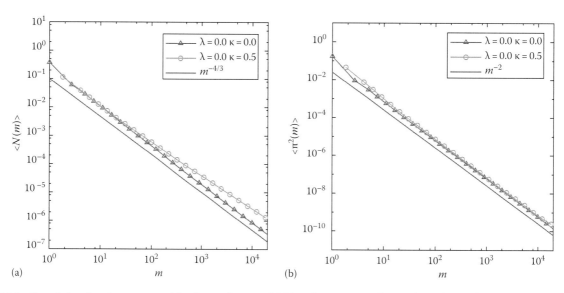

FIGURE 16.6 Correlation functions measured for the steady state of CCA with injection and mass-dependent diffusion, $D(m) = m^{-\kappa}$. (a) Shows the mass density, $N(m\infty)$ for $\kappa = 0$ and $\kappa = 0.5$. Solid line shows the exact result, $m^{-4/3}$, for the case $\kappa = 0$. (b) Shows the integrated second-order correlation function, $\pi_2(m)$, as defined by Equation 16.87, for the same two cases. The solid line shows the theoretical prediction, m^{-2}, which follows from constant flux arguments.

has been calculated numerically and to Spouge (1988) and Krishnamurthy et al. (2002), where a theoretical explanation is given.

16.7 Conclusion

In conclusion, we have attempted to give a brief overview of the kinetics of cluster–cluster aggregation from a theoretical physics perspective. In applications, such theories are relevant when one is interested in the statistics of cluster sizes in systems having a large number of clusters rather than in the detailed structure of individual clusters. We have been careful to draw a distinction throughout between the reaction-limited and diffusion-limited regimes.

In the reaction-limited case, mean-field theory, as expressed by the Smoluchowski kinetic equation, correctly describes the evolution of the cluster size distribution. A huge amount is known about the solutions of the Smoluchowski equation. We have nevertheless seen that, even at this level, there are delicate mathematical issues arising from the interpretation of singularities of the Smoluchowski equation which arise when a gelation transition occurs. From a physical point of view, gelling kinetics indicate that some important physical process affecting the largest clusters (precipitation, for example) has been omitted from the theoretical description.

In the diffusion-limited case, the Smoluchowski kinetic equation is qualitatively incorrect in the sense that it has the wrong scaling properties. The reason for this can be traced to the fact that strong diffusive fluctuations can invalidate the mean-field assumptions in this regime by inducing spatial correlations between clusters. We showed that the statistics of the large clusters is *always* diffusion limited for systems having spatial dimension less than or equal to two. As a result, such fluctuation-dominated kinetics are of importance in practice. Theoretical analysis of the fluctuation-dominated regime, however, is much more difficult. In this chapter, we devoted considerable time to discussing a statistical field theoretic approach, which allows considerable progress to be made in understanding the structure of correlations in the case of diffusion-limited constant kernel CCA. On the whole, however, very little is known about the diffusion-limited regime for arbitrary aggregation kernel and it remains an active area of research.

The other important distinction which we have tried to emphasize is the distinction between decay problems and problems with a source of monomers. While similar techniques are used to analyze the two cases, at large times, the physics is rather different. The former case evolves forever in time in a way which can often be described using self-similar scaling solutions. The latter case tends at large times to a nonequilibrium stationary, or quasi-stationary, state in which the addition of mass to the system by the injection of monomers is balanced by the transfer of mass to the heavier clusters via aggregation. Both scenarios may be of interest in practice depending on the application and both may be understood theoretically using the tools discussed here.

References

Ben-Naim, E. and Krapivsky, P. (2003), Exchange-driven growth, *Phys. Rev. E* **68**, 031104.

Bramson, M. and Lebowitz, J. L. (1988), Asymptotic behavior of densities in diffusion-dominated annihilation reactions, *Phys. Rev. Lett.* **61**(21), 2397–2400.

Broide, M. L. and Cohen, R. J. (1990), Experimental evidence of dynamic scaling in colloidal aggregation, *Phys. Rev. Lett.* **64**(17), 2026–2029.

Cardy, J. (1998), Field theory and nonequilibrium statistical mechanics, Lectures given at Troisième cycle de la Physique en Suisse Romande, 1998–1999, semestre d'été. Available online at: http://www-thphys.physics.ox.ac.uk/users/ JohnCardy/home.html.

Connaughton, C., Rajesh, R., and Zaboronski, O. (2004), Stationary Kolmogorov solutions of the Smoluchowski aggregation equation with a source term, *Phys. Rev. E* **69**, 061114.

Connaughton, C., Rajesh, R., and Zaboronski, O. (2005), Breakdown of Kolmogorov scaling in models of cluster aggregation, *Phys. Rev. Lett.* **94**, 194503.

Connaughton, C., Rajesh, R., and Zaboronski, O. (2006), Cluster-cluster aggregation as an analogue of a turbulent cascade: Kolmogorov phenomenology, scaling laws and the breakdown of self-similarity, *Physica D* **222**, 97–115.

Connaughton, C., Rajesh, R., and Zaboronski, O. (2007), Constant flux relation for driven dissipative systems, *Phys. Rev. Lett.* **98**, 080601.

Droz, M. and Sasvári, L. (1993), Renormalization-group approach to simple reaction-diffusion phenomena, *Phys. Rev. E* **48**(4), R2343–R2346.

Erdélyi, A. and Tricomi, F. G. (1951), The asymptotic expansion of a ratio of gamma functions, *Pacific J. Math.* **1**(1), 133–142.

Ernst, M. (1986), Kinetics of clustering in irreversible aggregation, in L. Pietronero and E. Tosatti, eds, *Fractals in Physics*, North Holland, Amsterdam, the Netherlands, p. 289.

Ernst, M. H. and van Dongen, P. G. J. (1987), Scaling laws in aggregation: Fragmentation models with detailed balance, *Phys. Rev. A* **36**(1), 435–437.

Ernst, M. H. and van Dongen, P. G. J. (1988), Scaling solutions of smoluchowski's coagulation equation, *J. Stat. Phys.* **1/2**, 295–329.

Falkovich, G., Fouxon, A., and Stepanov, M. G. (2002), Acceleration of rain initiation by cloud turbulence, *Nature* **419**, 151–154.

Fernández-Toledano, J. C., Moncho-Jordá, A., Martínez-López, F., González, A. E., and Hidalgo-Álvarez, R. (2007), Two-dimensional colloidal aggregation mediated by the range of repulsive interactions, *Phys. Rev. E* **75**(4), 041408.

Gillespie, D. T. J. (1976), A general method for numerically simulating the stochastic time evolution of coupled chemical reactions, *J. Comput. Phys.* **22**, 403–434.

Glauber, R. J. (1963), Time-dependent statistics of the ising model, *J. Math. Phys.* **4**(2), 294–307.

Goldstein, M. L., Morris, S. A., and Yen, G. G. (2004), Problems with fitting to the power-law distribution, *Eur. Phys. J. B* **41**, 255–258.

Horvai, P., Nazarenko, S. V., and Stein, T. H. M. (2008), Coalescence of particles by differential sedimentation, *J. Stat. Phys.* **130**(6), 1177–1195.

Kang, K. and Redner, S. (1984), Fluctuation effects in smoluchowski reaction kinetics, *Phys. Rev. A* **30**(5), 2833–2836.

Kang, K. and Redner, S. (1985), Fluctuation-dominated kinetics in diffusion-controlled reactions, *Phys. Rev. A* **32**(1), 435–447.

Kontorovich, V. (2001), Zakharovs transformation in the problem of galaxy mass distribution function, *Physica D* **152–153**, 676–681.

Krapivsky, P. L., Ben-Naim, E., and Redner, S. (1994), Kinetics of heterogeneous single-species annihilation, *Phys. Rev. E* **50**(4), 2474–2481.

Krapivsky, P. L., Mendes, J. F. F., and Redner, S. (1999), Influence of island diffusion on submonolayer epitaxial growth, *Phys. Rev. B* **59**(24), 15950–15958.

Krishnamurthy, S., Rajesh, R., and Zaboronski, O. V. (2002), Kang-Redner small-mass anomaly in cluster-cluster aggregation, *Phys. Rev. E* **66**(6), 066118.

Lee, M. (2000), On the validity of the coagulation equation and the nature of runaway growth, *Icarus* **143**, 74–86.

Lee, M. (2001), A survey of numerical solutions to the coagulation equation, *J. Phys. A: Math. Gen.* **34**, 10219–10241.

Lee, B. and Cardy, J. (1996), Renormalization group study of the A + B → Ø diffusion-limited reaction, *J. Stat. Phys.* **80**(5/6), 971.

Leyvraz, F. (2003), Scaling theory and exactly solved models in the kinetics of irreversible aggregation, *Phys. Rep.* **383**(2), 95–212.

Leyvraz, F. and Tschudi, H. R. (1981), Singularities in the kinetics of coagulation processes, *J. Phys. A: Math. Gen.* **14**, 3389–3405.

Lushnikov, A. A. (2005), Exact kinetics of the sol-gel transition, *Phys. Rev. E* **71**(4), 046129.

Munasinghe, R. M., Rajesh, R., Tribe, R., and Zaboronski, O. V. (2006a), Multi-scaling of the n-point density function for coalescing brownian motions, *Commun. Math. Phys.* **268**, 717–725.

Munasinghe, R. M., Rajesh, R., and Zaboronski, O. V. (2006b), Multiscaling of correlation functions in single species reaction-diffusion systems, *Phys. Rev. E* **73**(5), 051103.

Murri, R. and Pinto, N. (2002), Cluster size distribution in self-organised systems, *Phys. B: Condens. Matter* **321**(1–4), 404–407.

Peliti, L. (1986), Renormalisation of fluctuation effects in the a+a to a reaction, *J. Phys. A: Math. Gen.* **19**(6), L365–L367.

Rácz, Z. (1985), Diffusion-controlled annihilation in the presence of particle sources: Exact results in one dimension, *Phys. Rev. Lett.* **55**(17), 1707–1710.

Rajesh, R. and Majumdar, S. N. (2000), Exact calculation of the spatiotemporal correlations in the takayasu model and in the q model of force fluctuations in bead packs, *Phys. Rev. E* **62**(3), 3186–3196.

Sander, L. M. (2000), Diffusion-limited aggregation: A kinetic critical phenomenon?, *Contemp. Phys.* **41**, 203–218.

Silk, J. and White, S. D. (1978), The development of structure in the expanding universe, *Astrophys. J.* **223**, L59–L62.

Smit, D. J., Hounslow, M. J., and Paterson, W. R. (1994), Aggregation and gelation—I. Analytical solutions for cst and batch operation, *Chem. Eng. Sci.* **49**(7), 1025–1035.

Spouge, J. L. (1988), Exact solutions for a diffusion-reaction process in one dimension, *Phys. Rev. Lett.* **60**(10), 871–874.

van Dongen, P. G. J. (1987), On the possible occurrence of instantaneous gelation in Smoluchowski's coagulation equation, *J. Phys. A: Math. Gen.* **20**, 1889–1904.

Zaboronski, O. V. (2001), Stochastic aggregation of diffusive particles revisited, *Phys. Lett. A* **281**(2–3), 119–125.

Ziff, R. M. and Stell, G. (1980), Kinetics of polymer gelation, *J. Chem. Phys.* **73**(7), 3492–3499.

17

Surface Planar Metal Clusters

Chia-Seng Chang
Academia Sinica

Ya-Ping Chiu
National Sun Yat-Sen University

Wei-Bin Su
Academia Sinica

Tien-Tzou Tsong
Academia Sinica

17.1 Introduction

Free atomic clusters of nanometer size normally exhibit unique properties different from those of individual atoms, molecules, or bulk materials [1–3]. The physicochemical properties of the clusters sometimes change dramatically with the addition or removal of one atom from a cluster—a fact that affects the application of the nanomaterials. Therefore, in recent decades, there has been lots of interest in searching, characterizing, and studying the property of nanometer-size clusters for potential applications in nanoscience and nanotechnology [4–10]. For many applications, these clusters need to be supported by solid surfaces such as in catalysts [11,12] and electronic materials. In such cases, the substrate asserts some influence on the cluster property. Depending on how strong the interaction is, the shape and properties of the cluster can be changed [13–16]. Atomistic models frequently used for understanding the growth of surface clusters are the process of nucleation, surface diffusion, growth, and coarsening. In general, clusters would grow into three-dimensional (3D) structures. For some systems, especially for metallic clusters, however, they can grow into planer shapes. The flatness of the clusters either results from an intrinsic size effect due to their electronic properties or from their interaction with the substrate. Oftentimes, both these factors play an equally important role.

Metal clusters can be supported by a conducting, semiconducting, or insulating substrate. Due to the variation in the electronic properties of these substrates, some intrinsic properties of the clusters may change or even disappear. For instance, free-space nanometer metal clusters possess discrete energy levels because of the quantum size effect. The question is how this unique nature of free metal clusters can be affected when they are placed on a surface. One would immediately assume that an insulating substrate should have the least influence on the adsorbed clusters. Indeed, some physical properties associated with specific cluster's sizes have been obtained [13–16]. However, there is also evidence that nucleation of metal clusters at point defects of oxide supports may be connected with the enhanced catalytic activity in low-temperature CO oxidation [13,14]. A comprehensive review on the physical chemistry of supported clusters on oxide substrates has been presented by Heiz and Schneider [15]. As for semiconducting substrates, some clusters of dominant sizes and chemical stability were found. Specifically, on the Si(111)7 × 7 surface, there are "magic clusters" of much greater abundance for the Ga–Si mixed [17], Si [18], and Co clusters [19]. The magic nature of these clusters is related to the well-defined geometric structure, unitary dynamic behavior, and statistical abundance, respectively. On metallic substrates, an extensive review on the growth of metal clusters has been given by Brune et al. [20]. However, no similar magic nature of metal clusters on metal substrates has ever been reported, except for the work on the system of Ag clusters formed on Pb quantum islands [21].

In this chapter, the focus is on two main issues. The first issue is that of metal clusters on the semiconducting substrate with an emphasis on a morphological transition taking place during the growth of a cluster. The second issue is that of metal clusters grown on a metallic substrate for exploring the magic nature inherited by these clusters manifested with discrete sizes. In the past few years, we have conducted extensive studies on the Pb/Si(111) system. Since Pb will not intermix with

Si, a sharp interface can clearly be formed. This well-defined boundary has provided us with a series of intriguing physical phenomena in the cluster formation, the magic nature of the planar clusters, and the quantum effects of thin films [22]. The exploration of the transition from a cluster into a two-dimensional island (2D or quantum island) serves, in a way, as an ideal example for illustrating the subtle interaction between a metal cluster and the surface of a semiconductor substrate. We discuss in detail of this transition for the Pb/Si(111) system in Section 17.3, following a description of our experimental procedures and conditions in Section 17.2. The continual growth of Pb clusters on the Si(111) surface will lead to the formation of the Pb quantum islands owing to the electronic confinement effect of the films of finite thickness. Although the island's thickness is only a few atomic layers, its lateral size is usually large enough to render its overall electronic property metallic, or forming nearly continuous energy bands. These quantum islands will be further used as the templates to nucleate and grow single-atomic-layer 2D Ag clusters (also named nanopucks), illustrating the interaction between a metal cluster and a metallic substrate. This is described in Section 17.4, where we emphasize the energetics governing the self-organized growth of the Ag nanopucks. Section 17.5 discusses the magic nature of the Ag nanopucks formed under chosen experimental conditions. Finally, we draw some interesting conclusions in Section 17.6.

17.2 Experimental Procedures and Conditions

All our experiments are performed with variable-temperature scanning tunneling microscopy (STM) and spectroscopy (STS) in ultrahigh vacuum with the base pressure of 5×10^{-11} Torr. Our system is also equipped with a Pb and Ag e-beam evaporator. The Si(111)7 × 7 clean surface is prepared by flashing at a temperature ranging from 1150 to 1450 K of a freshly cut Si(111) wafer for several times while keeping the vacuum at 1×10^{-9} Torr or better. The sample temperature is then slowly brought down to room temperature in a few minutes. The sample is cooled to around 200 K and a deposition of Pb atoms is made. Before forming any cluster, about 2 ML of Pb are consumed in wetting the Si(111)7 × 7 substrate. An extra amount of Pb is added to generate Pb nanoclusters. The stripe incommensurate phase (SIC) is prepared by depositing slightly more than one monolayer (ML) of Pb on a clean Si(111)7 × 7 surface at room temperature followed by annealing the sample to 700 K. Pb quantum islands are created with the addition of an extra amount of Pb on the cooled surface kept at around 200 K. Based on the image contrast of the pattern, these Pb quantum islands can be distinguished into two types, Type I and Type II, similar to that reported in a prior study (Figure 17.7) [23]. Ag nanostructures are readily produced on top of these Pb islands by the deposition of Ag with the sample kept at 100 K and then annealing it to about 170 K.

17.3 Pb Clusters on the Si(111)7 × 7 Surface

For studying surface planar metal clusters, we need to pay a special attention to the interplay between the cluster and the substrate. A good example is the system of Pb clusters grown on the Si(111)7 × 7 surface. When Pb is deposited onto clean Si(111)7 × 7 substrate, the first two ML of Pb are consumed to form a wetting layer on the substrate. Further deposition, depending on the temperature, would create Pb nanostructures with different morphology. For example, Pb is grown into islands with steep edges above the wetting layer at 208 K as shown in Figure 17.1a. The surface of the islands is very flat and its orientation is in the < 111 > direction [24]. The contrast appearing in the STM image (Figure 17.1a) also implies that the thickness of the island is not the same. In order to understand the distribution of the island thickness and the variation of the distribution with coverage, we sampled hundreds of islands to analyze their thickness at each coverage. Figure 17.1b shows the ratio distribution as a function of island thickness at three coverages. It is interesting to note that the thickness of islands is confined within the range of 4–9 atomic layers, and islands with 7-layer thickness are the most abundant. It is apparent that the islands have the preferred thickness to grow. We analyze the heights of these islands and find only ~7% growth in thickness, though the coverage (above wetting layer) has increased by a factor of three. This indicates that the growth of islands is mostly in the lateral directions. Figure 17.1c demonstrates that the average area of the islands increases with the coverage linearly, further illustrating the 2D growth behavior of these islands.

Conventionally, there are three growth modes of the thin film, each named after the discoverers associated with their initial description: Frank–Van der Merwe (FM) growth, Stranski–Krastanov (SK) growth, and Volmer–Weber (VW) growth [25]. Figure 17.2 is a schematic illustration of these three modes. In the FM growth, adatoms preferentially interact with surface atoms, resulting in the formation of atomically flat layers. This layer-by-layer growth is two dimensional, implying that complete films form prior to growth of subsequent layers. Conversely, during the VW growth, adatom–adatom interactions are stronger than those of the adatom with the surface atom, leading to the formation of 3D clusters or islands. The SK growth lies in between: a wetting layer grows in a layer-by-layer fashion before 3D islands begin to form.

Prior to the formation of Pb islands, the wetting layer is observed, so the growth of Pb films on Si(111)7 × 7 surface is similar to the SK growth. However, the grown Pb islands have the characteristics of steep edge, flat-top surface, preferred thickness, and 2D growth that basically cannot be explained by the SK growth. This indicates that there exists another driving force in the growth of Pb films. It is believed now that this driving force is the quantum size effect related to the quantization of the energy level in the quantum-well. That is, owing to that the thickness of Pb film is comparable to the de Broglie wavelength of Fermi electrons within the film, the wavevector along the surface normal is quantized. Therefore, the electronic structure of the Pb film

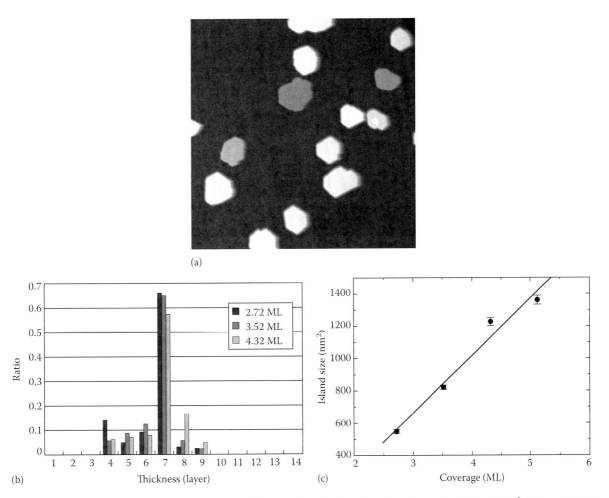

FIGURE 17.1 (a) Pb islands created by depositing 3.2 ML of Pb at 208 K on Si(111)7 × 7 surface. (image size: 3000 × 3000 Å²). (From Su, W.B. et al., *Jpn. J. Appl. Phys.*, 40, 4299, 2001. With permission.) (b) The appearance ratios as a function of island thickness at three different coverages. (c) The average size of islands linearly increases with coverage at saturated regime, showing a quasi-2D growth behavior. (From Chang, S.H. et al., *Phys. Rev. B*, 65, 245401, 2002. With permission.)

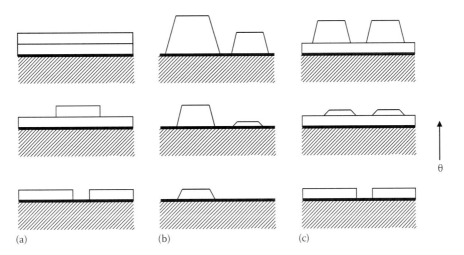

FIGURE 17.2 Schematic view of the three topologically distinct epitaxial growth modes with increasing coverage θ for (a) Frank–van der Merwe (FM) growth, (b) Volmer–Weber (VW) growth, and (c) Stranski–Krastanov (SK) growth.

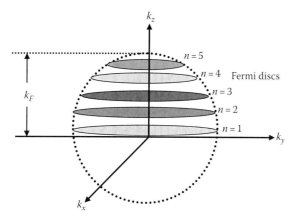

FIGURE 17.3 Schematic illustration of the electronic structure consisting of Fermi discs in Pb films.

would consist of discrete Fermi discs instead of a Fermi sphere, as schematically illustrated in Figure 17.3. Theoretical calculations have demonstrated that the quantum size effect can drive the metal-films grown semiconductors to have atomically flat surface and preferred thickness [26]. Besides Pb films, the Ag film is another system to manifest the growth behavior induced by the quantum size effect [27,28]. In addition to influencing the growth behavior, the quantum size effect can affect physical properties of the film, such as electrical resistivity [29], interlayer spacing [22,30], island coarsening [31], thermal stability [32], superconductivity temperature [33,34], Kondo temperature [35], thermal expansion coefficient [36], and so forth.

When the sample temperature is lowered to 170 K, the growth of Pb is not only in the form of islands but also in that of clusters. Figure 17.4 shows the STM image of 0.32 ML Pb deposited above

the wetting layer. It can be seen that clusters and flat islands are formed concurrently on the wetting layer. The line profile across a cluster and an island in Figure 17.4, indicated by an arrow, shows the 3D morphology of the cluster and flat-top surface of the 2D island. This observation implies that the clusters may be the seeds of the grown islands. In order to confirm this point, we *in situ* observe the growth process of an individual island with STM. Before we proceed to present our results, we should define the terms, i.e., 3D clusters, 2D islands, and 3D islands, often mentioned in this review. Our definition for the 3D cluster in this system is an incipiently grown nanostructure with a base diameter less than ~10 nm. For nanostructures with the base diameter larger than 10 nm, some exhibit an apparent mesa shape and thus called 2D islands; others are of pyramidal form and named 3D islands. Figure 17.5a through c show the growth of two 3D clusters as marked by numbers in Figure 17.5a with increasing coverage. Line profiles at the right side of corresponding STM images represent

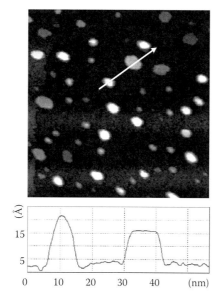

FIGURE 17.4 STM image with 2.32 ML Pb on the Si(111)7 × 7 surface at 170 K, showing 3D clusters and 2D islands that are formed concurrently on the wetting layer. The line profile across a cluster, indicated by an arrow in STM image, shows a 3D morphology. (image size: 300 nm × 300 nm). (From Su, W.B. et al., *Phys. Rev. B*, 68, 033405, 2003. With permission.)

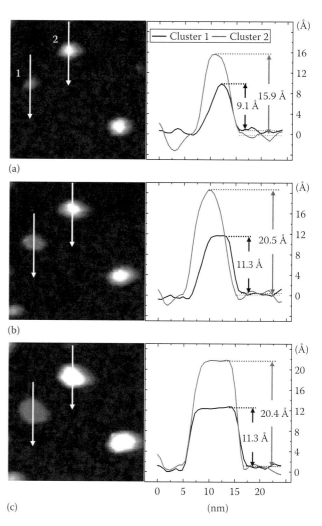

FIGURE 17.5 *In situ* observations for the growth of two 3D clusters as marked by the numbers in (a). Line profiles at the right side of (a)–(c) represent the morphology evolution of cluster 1 and 2 along arrows in (a). The image size is 52 × 52 nm². (From Su, W.B. et al., *Phys. Rev. B*, 1, 073304, 2005. With permission.)

the morphological evolutions of clusters 1 and 2 along the arrows, respectively. After the second deposition of 0.02 ML, cluster 1 with a height of 9.1 Å in Figure 17.5a transformed into a flat-top island with a thickness of 11.3 Å, then grew to a larger island of the same thickness, as shown in Figure 17.5b and c, respectively. On the other hand, the cluster 2 with the height of 15.9 Å grew into a higher cluster of 20.5 Å, then transformed into an island with a 20.4 Å thickness. Therefore, islands are indeed grown from these clusters. Moreover, Figure 17.5 reveals that an island thickness can be directly correlated to a cluster's height before transition, e.g., a thinner island being transformed from a lower cluster. It indicates that each preferred thickness of an island can correspond to an independent transition pathway. We thus accumulated hundreds of transformation events to find out the correlation between a transition pathway and a preferred thickness.

17.3.1 3D-to-2D Growth Transition

We first concentrate on those clusters that will not undergo the growth transition. Figure 17.6a is the statistical result of this type of cluster's height as a function of its base diameter. On average, the height is linearly proportional to the diameter, exhibiting a 3D growth nature. Clusters experiencing the transition are separated in the distance of a single atomic layer. Figure 17.6e shows the events of clusters finally transforming into 7-layer islands, which is also represented with the height as a function of the diameter for both clusters (cross) and islands (open circle). The diameter of an island can vary from 13 to 24 nm, indicating that once a 7-layer island is formed, it starts to grow in size laterally, exhibiting a 2D growth. We thus conclude that there exists a transition from 3D to 2D in the growth process [37,38]. This 3D-to-2D growth transition also appears in the formation of 4-, 5-, and 6-layer islands, as shown in Figure 17.6b through d, respectively.

Lead is known to grow into 3D islands on the Si(111)7 × 7 surface at room temperature with the SK growth mode [39]. In our case though, the growth is performed at low temperature. After nucleation, the amount of atoms in the nucleus is too small to make the quantum size effect manifest. Therefore, the growth is still dominated by the SK mode, which causes the nucleus to grow into a 3D cluster. Once a cluster gathers enough atoms to incite the quantum size effect to overcome the driving force for further 3D growth, the 3D-to-2D transition occurs and the subsequent growth is in the lateral direction. This causes the Pb islands to have a flat-top surface and to possess a multilayer thickness. Previously, the influence of the quantum size effect to the growth has only been found in the layer-by-layer growth mode such as silver thin films on both GaAs(110) [27] and Si(111)7 × 7 [28] substrates. The growth transition revealed here extends the existing model with a new phenomenon that even the growth of a nanosize 3D cluster could be affected by the quantum size effect.

17.3.2 Transition from Clusters to Islands

An unusual characteristic of Pb islands grown on the Si(111) surface at low temperature is that no island with thickness below four atomic layers is found. This implies that the island's growth should not be of layer-by-layer mode despite its top surfaces being flat. We are thus motivated to explore the formation process of a Pb island from its initial nucleation stage. Figure 17.6b shows that right before the transition, the height of the cluster reaches 1.14 nm, equal to four atomic layers in terms of 0.285 nm interlayer spacing. This indicates the cluster is necessary to reach the equivalent height before being transformed into a 4-layer island. A similar situation is also observed in the events of clusters transforming into 5-, 6-, and 7-layer islands, as shown in Figure 17.6c through e. We therefore conclude that the transition pathway exhibits a one-to-one correspondence between an N-layer island and a cluster of the same height. Islands of each thickness thus have a unique growth path. The distribution of cluster heights versus sizes in Figure 17.6b through e follows a linear dash line, which is the same as the solid line in Figure 17.6a. It denotes that the initial growth behavior of the clusters experiencing the 3D-to-2D transition is identical to that of those maintaining the 3D growth. The cluster's height distribution in Figure 17.6a covers all island thickness, reflecting that the fate of a growing cluster can take two different routes: one is to follow the transition pathway to become an island; the other is to continue to grow into a higher cluster. Therefore, the transition timing for a smaller cluster to a thinner island is earlier than a larger one to a thicker island. The existence of multiple-layered Pb islands on the Si(111)7 × 7 surface fundamentally unveils the sporadic nature in the timing of the 3D-to-2D transition. Figure 17.6f shows transition probabilities of the clusters transformed into the islands of different thickness. The transition probability for 7-layer islands is much larger than those for the islands of other thickness, and this is the reason why the 7-layer island is most abundantly observed on this surface. There is no cluster transforming into the island with thickness below four atomic layers, and those kinds of thickness are thus not observed.

17.4 Self-Organized Growth of Planar Clusters

Dispersed metal clusters can be prepared with common physical deposition methods on various substrates. In order to perform a systematic investigation and find potential applications, many recent studies have focused on the mass fabrication of nanostructures in arrays with a controllable size, shape, and site. Two of the most commonly adopted approaches for controlling matter on the nanoscale are lithographic patterning [40,41] and self-organized growth [20]. However, for fabrication by lithography, regardless of the sophistication of the technique, the spatial resolution is confronted by a natural limitation. On the other hand, the spontaneous assembly of atoms and molecules in a system enables organized nanostructures to be grown via a particularly versatile, rapid, and low-cost process. Consequently, over the last decade, considerable effort has been made to understand and harness the self organization mechanism [40–48].

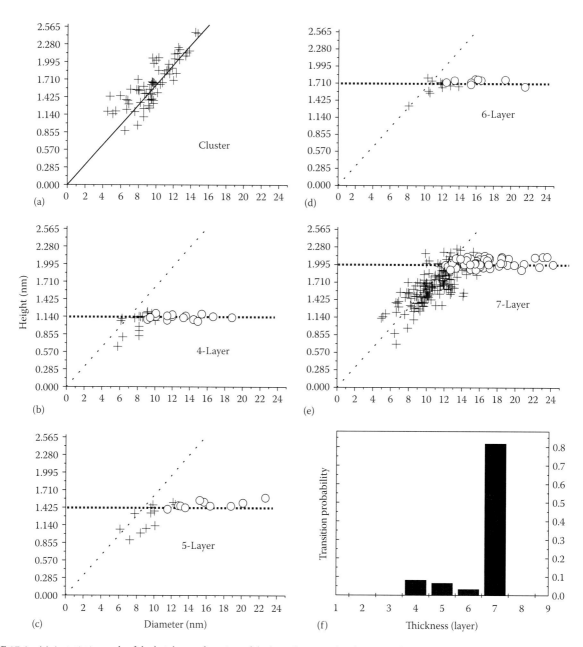

FIGURE 17.6 (a) A statistic result of the height as a function of the base diameter for clusters without undergoing the growth transition. It can be fitted by a linear line to show the 3D growth of clusters. (b)–(e) Clusters experiencing the transition are separated in terms of the subsequent island thickness, which is also represented with the height as a function of the diameter for both clusters (cross) and islands (open circle). (f) The transition probability of clusters transforming into islands of different thickness. (From Su, W.B. et al., *Phys. Rev. B*, 1, 073304, 2005. With permission.)

17.4.1 Characteristics of the Substrate

The characteristics of the template employed by us for the further growth of 2D nanoclusters are summarized in the STM images shown in Figure 17.7. Due to the lattice mismatch between Pb and Si, the lead film is likely to exhibit a structural modulation. However, the apparent corrugation of the superstructures observed on the islands of three atomic layers (Figure 17.7a) is beyond the simple geometrical consideration. In addition, the larger islands in this image are all three layer in thickness above the Si substrate, yet one can immediately

distinguish two types of islands (as marked) from the image contrast of the superstructure. Since the only dissimilarity between the two types of islands is resulted from different stacking sequences of the grown films with respect to that of the Si substrate, geometry effect alone cannot explain such a huge disparity in their apparent heights as measured by STM. This is the first indication of the importance of an effect of electronic origin.

The second indication of the electronic origin is that the contrast of the pattern also strongly depends on the bias voltage as shown in Figure 17.7b. It is more obvious for the type II islands

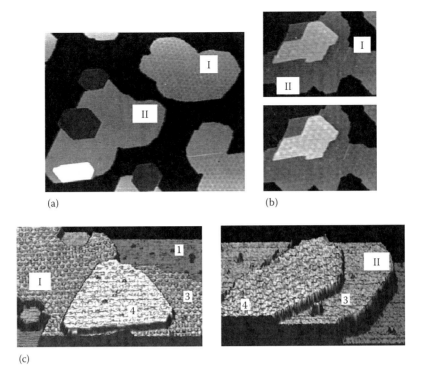

FIGURE 17.7 Characteristics of Pb quantum islands: (a) Two types of islands of three atomic layers marked with I and II. (b) Image contrast varies with the bias voltage: 2 V on the top and 0.4 V on the bottom. (c) Image contrasts alternate with the thickness of the islands and show complementarity between the two types of islands. (Adapted from Lin, H.Y. et al., *Phys. Rev. Lett.*, 94, 136101, 2005.)

where the contrast goes from weak to strong as the bias changes from 2 V (upper image) to 0.4 V (lower image). Furthermore, the image contrast also varies with the island thickness (Figure 17.7c), which provides the third evidence of the electronic origin of the superstructures. At the same bias voltage, the image contrast alternates between high and low as the island grows an additional layer. This bilayer oscillatory behavior can be correlated to the phase shift of the confined electrons [23]. From all these, the superstructures (also named electronic Moiré patterns) are mainly originated from the rearrangement of surface charges, which is caused by the itinerant quantized electrons within the film responding to the periodic potential variation at the film/substrate interface.

17.4.2 Energetics of Surface Diffusion

The immediate question is that can these patterns found on the Pb quantum islands of electronic origin serve as a template for self-organized growth of nanostructures? As shown in Figure 17.8, the answer is affirmative and the STM image of deposited Ag atoms at 120 K on the Pb islands of three atomic layers demonstrates this [49]. It can be seen that Ag atoms form a very ordered periodic cluster array with a fairly uniform size, indicated by the size distribution curve (Figure 17.8b). The Fourier-transformed pattern of sharp spots further proves its periodicity (inset in Figure 17.8a). A closer examination of the structure of these clusters shows that they are of one atomic layer in height and have either an imperfect hexagonal

shape or a nearly circular shape. We thus call them nanopucks. The size of the Ag nanopucks on the Pb quantum islands is mainly determined by the coverage at this temperature. Take those nanopucks in Figure 17.8a as an example; the standard deviation (STD) of the size distribution after normalization with the average size is only 0.11 at the coverage of 0.2 ML. If there were no template, the formation of nanoclusters on a crystalline surface would be governed only by surface adatom diffusion. According to the scaling law [50], the size distribution (STD = 0.59) would be much wider as displayed in Figure 17.8b. Therefore, the superstructure must have provided extra diffusion barriers to confine the nucleation of the nanopucks. This fact is also reflected on the positions of nanopucks. They are not random and tend to occupy only in one half of the unit cell (see Figure 17.8c). The white lines in the figure, running across the brightest points of the superstructure, define the unit mesh. Intersections of the two sets of parallel lines are T_1 sites. The top view and side view of geometric configuration of different stacking of the superstructure are sketched in Figure 17.8d and e, respectively. It is shown that the rhombic unit cell can be divided into two triangular halves by the black dash lines. One has the face-centered cubic (fcc) stacking and the other hexagonal close-packed (hcp) stacking (Figure 17.8d). For many heteroepitaxial systems, it is generally accepted that adsorption favors the fcc side [51]. We thus associate the locations of the nanopucks with the triangular fcc half cells for both types of superstructures. With the saturation coverage, the occupancy of the fcc halves is almost complete, indicating

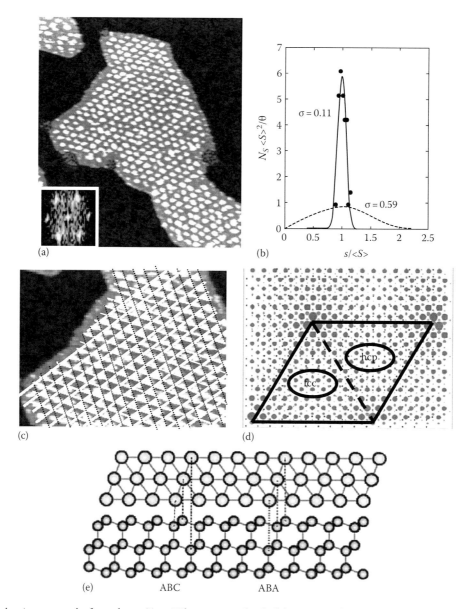

FIGURE 17.8 Regular Ag nanopucks formed on a Type I Pb quantum island of three atomic layers at 120 K. (a) 100 nm × 170 nm STM image taken after deposition of 0.2 ML Ag atoms. Inset shows the FFT pattern. (b) Normalized size distribution of Ag nanopucks (solid curve). The distribution according to the scaling theory of ordinary adatom diffusion is plotted with the dash curve. (c) Nanopuck's locations in the unit mesh of the superstructure. (d) Sketch displays the top view of the stacking sequence of the lead layers in reference to the Si substrate stacking. In order to lay emphasis on the natural contour of the STM image, different sizes of atoms are simulated and represented the height difference individually. (e) The side view of geometric models of different stacking of the superstructure shown in (d). The sequence of "ABC" is represented the fcc stacking, and "ABA" is indicated as the hcp stacking. (Adapted from Lin, H.Y. et al., *Phys. Rev. Lett.*, 94, 136101, 2005.)

a strong binding energy difference between the hcp and fcc triangular halves.

It is also desirable to obtain quantitative values specifying the trapping strength of the template. For the current system the number of atoms in a critical nucleus is assumed to be 1, which is pertinent to many cases of Ag atoms deposited on metal systems [52,53]. We then follow the analytical procedure of Ref. [20] and apply the nucleation theory for complete condensation [54]. The saturated island density (N) is proportional to $\exp\{E_d/[2\gamma kT]\}$, where E_d is the activation energy

for surface diffusion, and γ a fitting parameter to account for diffusion on an inhomogeneous substrate. Experiments were performed with depositing a suitable amount of Ag onto quantum islands at various temperatures. STM images shown in Figure 17.9a through c correspond to three regions of different slopes in Figure 17.9e. The amounts of coverage have been determined to result in a saturated island density. Arrhenius plots of N versus $1/T$ for Ag nanopucks formed on Type I Pb islands of three atomic layers yield two distinct slopes in the data. This is because two activation barriers are involved in the

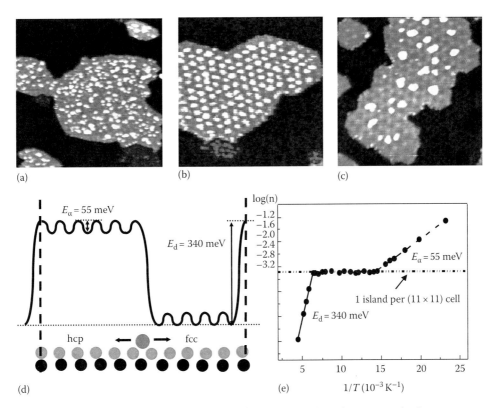

FIGURE 17.9 60 nm × 60 nm STM images of suitable amounts of Ag atoms deposited on Pb quantum islands at various temperatures: (a) 40 K, (b) 120 K, and (c) 180 K. (d) Sketch depicts two extra diffusion barriers involved in this system. (e) Arrhenius plots of Ag cluster density versus $1/T$. (Adapted from Lin, H.Y. et al., *Phys. Rev. Lett.*, 94, 136101, 2005.)

nucleation process, as illustrated with Figure 17.9d. One characterizes conventional site diffusion within the unit cell (E_α) at lower temperature, and the other the adatom diffusion over the barrier between two half cells (E_d) at higher temperature. In the Ag/Pt(111) system [20], this diffusion barrier is associated with the network of surface partial dislocations. For the current system, there is no apparent variation in atomic spacing observed in the atomically resolved STM image [23]; the geometric transition crossing the super-cells of the Moiré pattern is generally smooth. We thus attribute most of the barrier to the binding energy difference between fcc- and hcp-stacked areas of the Pb islands. Since these extra barriers introduce an inhomogeneity for surface diffusion from one location to another, the more reliable quantitative values must be obtained from the kinetic Monte Carlo simulation, as employed in Ref. [20] for the Ag/Pt(111) system. However, we find the simulation results for the Ag/Pt(111) system can be reproduced if the value of $\gamma = 1.8$ is substituted into the aforementioned equation. Because our system confronts a similar situation, we thus adopt the same value for γ as well. From Figure 17.9e, we can derive $E_\alpha = 55$ meV and $E_d = 340$ meV for Ag atoms in the nucleation of nanopucks on the Type I Pb islands of three layer thickness. The value of E_d is significantly larger than that obtained for Ag/Pt(111) strain-relief system [20]. The pronounced wide plateau region ranging from ~70 to ~150 K, indicating one island per unit cell, is the consequence of the combining contribution of large E_d and small E_α.

17.5 Magic Nature of Planar Clusters

Among those investigations on self-organized nanoclusters, some clusters exhibiting a distinct regularity on the size and stability are observed. Those clusters with enhanced abundance and stability are called magic clusters. Magic numbers in gas phase clusters of alkali atoms were predicted theoretically [55], and observed in mass spectrometry [56] in the 1980s. Many intriguing properties associated with these clusters have since been studied in detail [47,57–61,62]. As regards to 2D magic clusters, theoretical calculations have been performed and predicted the existence of magic numbers for planar metal clusters [63,64]. Unlike 3D magic clusters [65–71], however, they have never been experimentally proved. Recently, 2D magic Ag nanostructures were discovered in our work [21]. In the following, we will explore the nature of these magic clusters, the origins that causes the cluster to exhibit such unique properties, and the transition between different origins as the cluster grows.

17.5.1 Size and Shape Distributions of Planar Clusters

The notable growth property of Ag nanopucks is the site-control growth of Ag nanopucks developed on the Pb island template. In addition, the distinguishable size distribution and preferred shapes of the Ag nanopucks are also discovered on the island [21]. Figure 17.10a displays the STM image of several Ag nanopucks on

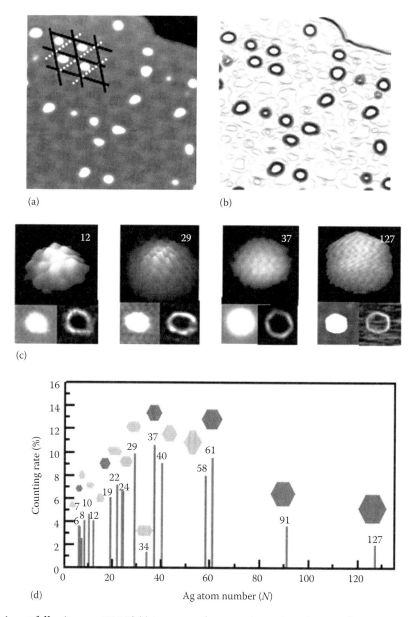

FIGURE 17.10 **(See color insert following page 25-14.)** (a) Ag nanopucks possessing various shapes and sizes grown on Pb quantum islands. The black lines outline the unit cell of the superstructure and each unit cell is divided into two triangular halves by the white lines. (b) The corresponding shapes of Ag nanopucks in (a). The shapes are highlighted by differentiation. (c) Atomically resolved images of Ag magic nanopucks with 12, 29, 37, and 127 atoms. The corresponding STM images and shapes are shown in the insets. (d) The size distributions of over a 1000 Ag nanopucks grown on Pb quantum islands. The schematic structures are drawn above each magic number for the most frequently observed ones. The sketches are divided into two based on the attributed factors: geometrical effect (painted in blue) and electronic effect (painted in orange). (From Chiu, Y.-P. et al., *Phys. Rev. Lett.*, 97, 165504, 2006. With permission.)

a Pb quantum island. They are grown with Ag deposition flux of 0.1 ML per minute and the sample is held at 100 K, and followed by slowly annealing to 170 K. Their shapes are better revealed by differentiating the topographic image to highlight the boundaries (Figure 17.10b), and their atomically resolved images are shown in Figure 17.10c. Even with these images of high resolution, the counting of the atom number could be uncertain, especially, near the edge of a cluster. However, careful examination finds existence of some special sizes and shapes, and the size of these clusters is tied with a specific shape. The atomic number of a cluster can

hence be determined by both the vigilant atom-counting and its shape. We have employed the same growth recipe to produce over a thousand Ag nanopucks, and their size distribution is plotted in Figure 17.10d in histogram form. The most dramatic discovery is the existence of magic numbers in this histogram. The corresponding size and shape of a nanopuck are shown by the distribution peaks. As is apparent, the closely packed hexagons (painted in blue) spread over the whole range with the numbers of 7, 19, 37, 61, 91, and 127. These numbers follow the formula $N = 1 + 6 \Sigma s$, where s is the shell number, and they are representative of the 2D

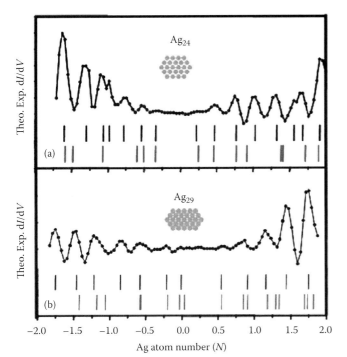

FIGURE 17.11 Spatially averaged tunneling spectra for Ag nanopucks with 24 (a) and 29 (b) atoms. Peaks in the spectra are plotted with black bars. The corresponding local DOS calculated for 2D Ag disks with the experimentally observed shapes in gray bars. (From Chiu, Y.-P. et al., *Phys. Rev. Lett.*, 97, 165504, 2006. With permission.)

geometrical close-shell structures. Below the atom number of 60, other magic numbers 6, 8, 10, 12, 22, 24, 29, 34, 40, and 58 (painted in orange) are evident. Our calculations indicate the clusters with magic numbers less than 60 are resulted from the close-shell electronic structures with a large energy gap (>0.17 eV) near the Fermi level. When the nanopuck is small, the electronic energy associated with the size and shape of the cluster is the major contributing factor to its total energy, but this factor can no longer compete with the effect of geometric close-shell structure when the atom numbers become large. The transition atom number begins at around 60 atoms, which is much smaller than that for the free spherical clusters. This is because the 3D cluster has much higher degeneracy and subsequently much larger gaps between the shells [63].

In order to explore the origin of these magic numbers, two magic sizes of nanopucks are selected for a detailed study. Spatially averaged tunneling spectra for the nanopucks containing 24 Ag atoms and 29 atoms (referring to Figure 17.11a and b) are measured near the centers of these nanopucks, respectively. If the peaks in the spectra are extracted (marked with black bars) and compared with the calculated ones (gray bars), the overall agreements are quite good considering the calculations have not taken into account the effect of the substrate. These results indicate the 2D nature of the Ag nanopucks has been largely preserved. However, the electronic Moiré pattern on the substrate has differentiated one half of the unit cell from the other half in the binding strength of the Ag adatoms. It seems that this binding trap has only a mean-field effect on the Ag nanopuck, especially when the puck is small. We therefore calculate the electronic energy for each nanopuck up to 41 atoms. A stability function is defined using the second difference in total energy as $\Delta E(N) = E(N + 1) + E(N - 1) - 2E(N)$, where N is the number of atoms in the nanopuck. Once this function is plotted against N, several peaks are observed as shown in Figure 17.10e. Thus the experimental magic numbers agree almost perfectly with the theoretical calculations.

With the understanding of the cluster's growth mechanism, we then attempt to prepare as uniformly and widely distributed Ag nanopucks as possible by judiciously tuning the growth parameters, i.e., with a flux of 0.1 ML/min, total dosage of 0.2 ML, growth temperature of 110 K, and followed with annealing to 170 K. Figure 17.12 shows what has been achieved. In this 60 nm × 35 nm image, over 150 nanopucks are fabricated. They are found to disperse within four neighboring magic numbers. According to the distribution histogram (see the inset), 60% of the nanopucks are of 29 atoms, 20% 24 atoms, 15% 37 atoms, and 5% 40 atoms. The differential image (Figure 17.12b) shows that the shapes of these nanopucks are quite uniform as well.

17.5.2 First-Principles Calculations

The overall experimental findings about the Ag magic clusters have been described above. Some of the detailed understanding requires inputs from the theoretical side. Calculations based on the density-functional theory are thus utilized. The underlying physics associated with the abundance and stability of 2D

(a)

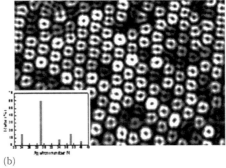
(b)

FIGURE 17.12 (a) A large array of 2D Ag nanopucks fabricated with magic numbers of 24, 29, 34, 37, and 40 atoms. (b) Differentiating the corresponding topography image, the uniformity in size and shape of the nanopucks is emphasized. Inset: Histogram for size distribution.

Ag nanoclusters is expected to be gained from the theoretical insights. Before conducting thorough investigations into the formation of these magic Ag nanopucks, we should examine the band structure of bulk Ag crystal first. Silver has a distinct position in the periodic table where it resides next to a transition metal. The band structure of a thin silver film, therefore, has a unique property: electronic configuration from 4 eV below the Fermi level upward is dominated by the atomic 5s orbital [72]. This revelation signifies the incipient free-electron-like nature of a silver planar cluster, thus rendering the possibility of their growth in magic forms.

Calculations in this study are performed using the Vienna *Ab initio* Simulation Package (VASP), based on the spin-dependent density-functional theory and the projector-augmented wave method (PAW) [73–75]. The generalized gradient correction approximation of Perdew, Burke, and Ernzerhof (PBE) is applied to the exchange correlation energy of the electronic density [76]. The supercell geometry is used to simulate 2D Ag nanoclusters with various sizes and shapes. Each cell encloses a free

Ag nanocluster, and is separated from the others by a vacuum region with a width of five Ag layers in the *z*-direction, defined as perpendicular to the 2D Ag clusters. Additionally, each Ag nanocluster is held at a distance of approximately 12 Å in the *x*- (or *y*-) direction from the adjacent one. The Brillouin-zone summation is approximated and performed using a gamma *k*-point grid, and the plane-wave energy cutoff is 250 eV. In all of the calculations, all atoms in a Ag nanocluster are fully relaxed in *x*-, *y*- and *z*-directions. The geometry is optimized until the total energy converges to 1×10^{-5} eV.

The possible shapes of Ag_N nanoclusters with *N* ranging from 3 to 42 are determined theoretically by iteration. The pertinent structural and geometrical properties of the lowest-energy cluster and some representative metastable isomers are determined [72]. The relative binding energy per atom of the lowest-energy planar Ag nanoclusters (E_b) is calculated and plotted in Figure 17.13a as function of the cluster size. In this figure, the binding energies of these stable structures are compared to those of geometrically stable hexagons, which are guided with a dashed

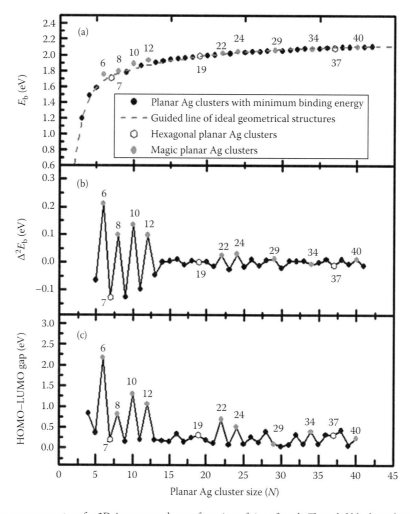

FIGURE 17.13 (a) Binding energy per atom for 2D Ag nanopucks as a function of size of puck. The solid black circles are calculated for the pucks with the lowest energy at a specific size; the gray solid circles refer to magic pucks, and the open hexagons are for hexagonal magic pucks. The dotted line indicates the effect of forming an ideal geometrical shape. (b) Stability based on the second derivatives of total energy versus number of atoms in planar nanoclusters. (c) HOMO–LUMO gap against cluster size *N*. (From Chiu, Y.-P. et al., *Phys. Rev. B.*, 78, 115402, 2008. With permission.)

curve. Thus, the dependence of binding energies per atom of the most stable cluster configurations for the cluster size, N, helps determine the sequence of the magic numbers. In order to compare the stability among Ag nanopucks, the difference between the binding energies of adjacent clusters, $2E_b(N)-(E_b(N+1)+E_b(N-1))$, is plotted against N (Figure 17.13b). This information allows manifestation of magic numbers and in turn yields the strength of stability. Local maxima of $\Delta^2E_b(N)$ are found at $N = 6, 8, 10, 12, 22, 24, 29$, and 40, suggesting that the clusters with these values of N are more stable than their neighboring clusters. The calculated magic numbers are in large consistent with the experimental result [21]. The inconsistency associated with size $N = 34$ may be attributable to the substrate effect in the experiments, which has not been considered in the theoretical simulations.

17.5.3 From Electronic Closed Shell to the Geometric

Since the sizes of these Ag nanoclusters are comparable with their electronic wavelength, their size-evolutionary growth is determined strongly by the source of electronic energy. According to the jellium approximation model [77], the electronic ground-state properties of metals are constructed in analogy with the shell model of atomic nuclei (Figure 17.14a). Given the close similarity between the formation mechanism

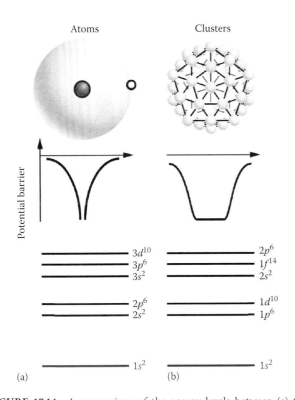

FIGURE 17.14 A comparison of the energy levels between (a) the atoms and (b) the clusters. Those magic numbers attributed from electronic shell-closing effects are 2, 10, and 18 for He, Ne, and Ar of (a), and 2, 18, and 40 for magic clusters of (b).

of stable clusters and the nuclear shell model, the commensurate conceptual developments of the ground-state configurations can be generalized. After the delocalized electrons have bunched together as a consequence of a "filled shell" or a "shell-closing" state, the corresponding electronic structures exhibit increased binding energy (Figure 17.14b). Accordingly, such clusters with enhanced stabilities, represented by the closure of shells of delocalized electrons, are called magic clusters. Therefore, electronic shell-closing effects are successfully used to explain the sequence of the magic numbers when metal clusters are small and the electronic energy dominates [78]. For larger clusters, the packing related to the atomic arrangement becomes more important, which gives rise to the geometrical close-shell structure [79–81]. In the present case, the electronic-to-geometric transition takes place as the atom number of a Ag cluster reaches 60.

To unveil the electronic shell-closing effect existing in the 2D Ag clusters, we need to start out from calculating their electronic structures. From the calculated electronic levels, we can derive the energy gap between the highest occupied molecular orbital (HOMO) and the lowest unoccupied molecular orbital (LUMO) for each of the 2D Ag_N ($N = 3-40$) clusters. The electronic density of states (DOS) of numerous illustrative Ag clusters (Ag_6, Ag_{12}, Ag_{24}, and Ag_{40}) are calculated and displayed in Figure 17.15a. The DOS for Ag_6 is widely separated, and the HOMO–LUMO gap is large. As the cluster size increases further, its DOS becomes dense and broadened. The HOMO–LUMO energy gap thus closes up rapidly. The electronic structure of each isomer of a Ag cluster is also calculated to determine the shape with the lowest energy. The HOMO–LUMO gap of each 2D Ag cluster assuming the shape of the lowest-energy isomer is plotted in Figure 17.13c. Comparing this figure with Figure 17.13b, it is inferred that the Ag_N clusters with $N = 6, 8, 10, 12, 22, 24$, and 34 clearly reveal a larger HOMO–LUMO gap, indicating a strong correlation between the HOMO–LUMO gaps and the energetic stability for 2D Ag clusters. Ag clusters with a large energy gap are also found more abundant. Nevertheless, clusters with 29 and 40 atoms, experimentally identified as the magic clusters, may not have the same electronic origin. Notably, the HOMO–LUMO gap of a magic Ag nanopuck slowly decreases as a function of the cluster size, suggesting the effect due to the electronic contribution subsides in a larger cluster.

The electronic-to-geometric transition also exhibits in the cluster's mean binding energy. This will be more easily observable when the binding energy per atom is plotted against the reciprocal of the square root of the cluster's atom number N, as shown in Figure 17.15b. A dotted line in Figure 17.15b connects the planar hexagonal Ag nanoclusters. The line in Figure 17.15b thus divides the clusters into two groups: electronically stable and electronically unstable. Apparently, the corresponding binding energies of small clusters exceed those of hexagonal clusters. They are thus electronically stable and frequently observed with experiment [21]. As the cluster grows, its binding energy slowly converges toward the dotted line. It suggests that the transition

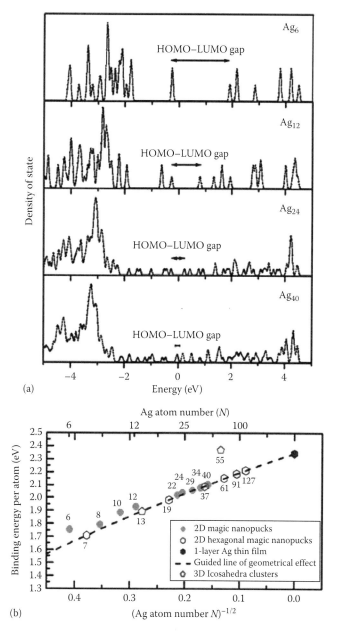

(a)

(b)

FIGURE 17.15 (a) The electronic DOS of planar Ag nanoclusters, Ag$_N$, with $N = 6$, 12, 24, and 40. The width of the smearing in 0.05 eV is used. (b) Binding energy per atom against inverse of square root of number of Ag atoms. The dotted line is fitted through the geometrical hexagonal structures. The magic clusters formed in ideal geometrical shapes are represented by open hexagons, and others are represented by gray solid circles. The relative stabilities of 3D icosahedra and 2D hexagon clusters are also shown for comparison. (From Chiu, Y.-P. et al., *Phys. Rev. B.*, 78, 115402, 2008. With permission.)

from the electronic to the geometric indeed smoothly occurs with the growth of magic Ag nanoclusters.

17.5.4 Effect of the Substrate on Planar Clusters

Although theoretical predictions concerning isolated magic Ag clusters are in good agreement with experimental measurements,

magic clusters with 7, 19, and 37 atoms, arranged in highly symmetrical geometries, were not found in the calculation. Furthermore, considering geometric packing alone, the energetically favorable structures of 3D and 2D clusters are in general of icosahedra and hexagon, respectively. Comparing the relative stability of each Ag cluster (shown in Figure 17.15b), we see that Ag clusters are formed in planar configurations preferably only when the clusters are small. Additionally, according to recent simulations [5,6], the transition of neutral Ag clusters from planar to 3D structures begins with the Ag cluster of seven atoms. Since the interface potential may affect the formation of a cluster [82,83], a question thus arises regarding the role of the template, Pb quantum islands: If it plays a critical role in the cluster formation, how strong is the effect of the substrate in steering a growing Ag nanopuck toward a specific magic number? This means we need to quantify this effect in order to gain true knowledge about the interplay between the cluster and the substrate.

We thus fabricate Ag nanostructures on surfaces with a markedly different and weaker interaction between the adsorbed species and the supported surface. One we used is the surface of Type II Pb quantum island, possessing weak pattern contrast in the surface potential, and the other is the graphite surface. On the first substrate, Ag nanopucks are grown on Type II Pb islands. Since the diffusion barrier from fcc sites to hcp sites of Type II Pb islands is much lower than that of Type I islands [49], the nucleation of Ag nanostructures on Type II islands is associated with a low lateral confinement from the substrate. Experimental observation by STM (Figure 17.16a) and the corresponding differential image (Figure 17.16b) reveal a series of magic numbers similar to that for Type I islands (Figure 17.16c). The spontaneous formation of Ag nanostructures on a graphite surface is the second case of interest, in which the interaction between the deposited Ag atoms and the substrate surface is presumably very weak. When Ag nanostructures are grown on graphite at 25 K, since the graphite surface is much smoother, the mobility of the deposited Ag atom is still so high that no small clusters can be found on the surface. The remaining large clusters are all taking the hexagonal configurations. Therefore, both these substrates seem to de-emphasize the influence of the supported substrate. It implies that most of the 2D magic Ag nanostructures are formed owing to their intrinsic electronic characteristics. Nevertheless, the existence of hexagonal magic Ag clusters cannot be resorted to their electronic structures. Furthermore, the prolonged onset of the 3D cluster formation signifies some interaction imparted by the substrate. For these reasons, we call for detailed calculations including the substrate.

Figure 17.17a depicts a unit cell of the Moiré pattern observed on a Pb quantum island grown on the Si(111) substrate [21,23,48,49,84]. The light gray fcc hexagonal region represents the effective area for the initial development of Ag clusters on the Pb surface. In the calculation, the corner site of the unit cell is slightly higher than the central area with a fixed height of 0.04 nm (indicated in Figure 17.17a) [84]. The optimal position and orientation for a cluster on the topmost Pb layer was determined from varying its degrees of freedom. As Ag clusters are

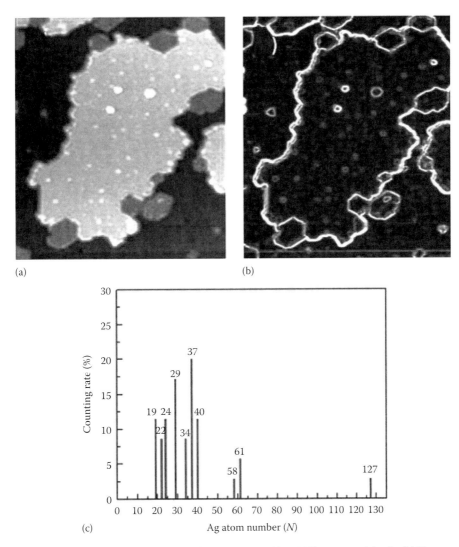

FIGURE 17.16 (a) Ag nanopucks possessing various shapes and sizes grown on Type II Pb quantum islands. (b) The corresponding shapes of Ag nanopucks in (a). The shapes are highlighted by differentiation. (c) The size distributions of Ag nanopucks grown on Type II Pb quantum islands.

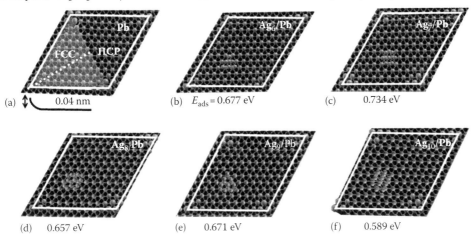

FIGURE 17.17 **(See color insert following page 25-14.)** Sketch of generation of Moiré pattern in unit cell of Pb quantum island surfaces. FCC region represents the zone of relative stability for the initial growth of Ag on Pb surfaces. Clean simulated Pb surface (a), and cross-sectional profile for Ag cluster nucleated on simulated fcc region of Pb surfaces is also illustrated below (a). (b–f) Represented geometrical configurations for planar Ag_N clusters ($N = 6$–10) on simulated substrate. The numbers below (b–f) indicate the energies of adsorption per atom for Ag clusters. (From Chiu, Y.-P. et al., *Phys. Rev. B.*, 78, 115402, 2008. With permission.)

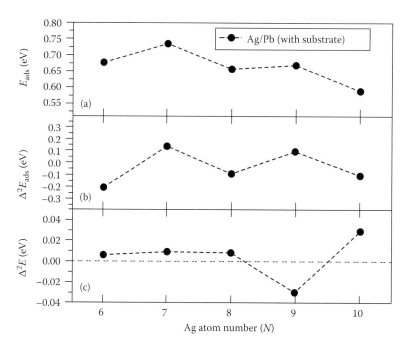

FIGURE 17.18 (a) Ag clusters that gain extra energy per atom as they are adsorbed on Pb surfaces. (b) Second derivatives, $\Delta^2 E_{ads}$, of adsorbing energy. (c) Second derivatives, $\Delta^2 E$, of energy for the complete system of planar Ag clusters grown on Pb surfaces. (From Chang, S.H. et al., *Phys. Rev. B*, 65, 245401, 2002. With permission.)

formed on the layout surface, 2D Ag$_N$ clusters ($N = 6$–10) are fully relaxed in x-, y-, and z-directions. Figure 17.17b through f present the stable arrangements of Ag clusters with N atoms ($N = 6$–10) grown on the Pb island. The adsorption energy of a Ag cluster on the quantum Pb island surface is evaluated by

$$E_{ads}(N) = -\frac{1}{N}\Big[E_{Ag/Pb} - (N \times E_{Ag} + E_{Pb}) \Big],$$

where N denotes the number of atoms in a specific planar Ag cluster on the Pb surface. The total energies of the adsorbate–substrate system, the clean Pb surface, and the energy per atom of a free planar Ag cluster are represented by $E_{Ag/Pb}$, E_{Pb}, and E_{Ag}, respectively. The adsorption energy, $E_{ads}(N)$, is the energy that a free Ag atom of a cluster gains by being adsorbed on the surface. Accordingly, it is defined such that a positive number corresponds to exothermic (stable) adsorption with respect to a free Ag atom and a negative number indicates endothermic (unstable) adsorption. Figure 17.18a presents the extra, extrinsically gained energy for Ag clusters with different sizes on the Pb surface based on the above definition. Additionally, in order to extract the relative stability, the second derivative $\Delta^2 E_{ads}(N)$ (Figure 17.18b) is considered again for a planar Ag cluster grown on the Pb quantum island. Figure 17.15b reveals that the intrinsic Ag cluster of eight atoms with a proper arrangement are more stable than Ag$_7$ and Ag$_9$ without consideration of the substrate effect. However, when the substrate effect is counted, as in Figure 17.18b, planar Ag$_7$ and Ag$_9$ clusters acquire extra energy, and their formations are consequently stabilized. When we add the binding energy, $E_b(N)$, for a free Ag cluster to the adsorption

energy, $E_{ads}(N)$, due to the substrate, the planar Ag$_7$ and Ag$_8$ clusters possess higher binding energies per atom than that for Ag$_9$, as shown in Figure 17.18c. This analysis demonstrates that Ag$_7$ and Ag$_8$ clusters should be more prevalent as the substrate effect is taken into account. This new result corresponds well with the experimental finding [21]. Based on the above examination, the substrate should play a crucial role in the formation of the hexagonal magic Ag clusters. Likewise, it could postpone the onset of a 3D cluster's formation. In our experiment, a Ag cluster as large as 127 atoms can still stay planar.

17.6 Conclusions

This chapter summarizes some of our recent STM and STS studies of the quantum size effect in the growth of Pb and Ag clusters. For Pb clusters, we have shown that its growth behavior at low temperature is quite different from the conventional growth at room temperature. Our result reveals the characteristics of a 3D-to-2D growth transition. This transition is statistically sporadic in nature and hence results in the coexistence of 3D and 2D clusters. The 2D feature originates from the energetic of the quantum-well states formed in a flat-top island and can lead to the further growth of the quantum island.

We also find that the electronic Moiré patterns found on lead (Pb) quantum islands can serve as the templates for growing self-organized cluster (nanopucks) arrays of various materials. These patterns can be divided into fcc- and hcp-stacked areas, which exhibit different binding strengths of the deposited adatoms. For Ag adatoms, the binding energy can differ substantially and the confined nucleation thus occurs in the fcc sites.

Magic numbers in surface-supported 2D Ag nanoclusters have been discovered. Detailed calculations based on first-principles density-functional theory have been performed to elucidate the origin of the magic numbers. They originate from the electronic shell-closing effect when the cluster is small. As the Ag nano-puck grows to a certain size, the geometrical effect becomes more important. By combining the magical nature of size-dependent Ag nanopucks with a suitable Pb island substrate, planar clusters in restricted magic numbers and shapes can be produced in an ordered array. Moreover, the effect of the substrate on the formation of magic Ag nanopucks has also been investigated both experimentally and theoretically.

Our preliminary result using density-functional optimization indicates that the planar Ag nanoclusters of 1 nm are most effective at adsorbing atomic oxygen. It leaves a significant footnote for the potential exploitation of these surface planar clusters, and for further exploration of them in a more detailed and controllable way.

Acknowledgments

The authors would like to acknowledge the valuable contributions of S. H. Chang, W. B. Jian, H. Y. Lin, L. W. Huang, C. M. Wei, T. Y. Fu, S. M. Lu, H. T. Shih, C. L. Jiang, and M. C. Yang. This work was supported by the National Science Council, Academia Sinica, and the project of Academic Excellence of the Ministry of Education of Taiwan.

References

1. W. Ekardt (ed.), *Metal Clusters* (Wiley, New York, 1999).
2. E. Kaxiras, *Atomic and Electronic Structure of Solids* (Cambridge University Press, Cambridge, U.K., 2003).
3. R. M. Martin, *Electronic Structure* (Cambridge University Press, Cambridge, U.K., 2004).
4. M. Pereiro, D. Baldomir, and J. E. Arias, *Phys. Rev. A* **75**, 063204 (2007).
5. E. M. Fernández, J. M. Soler, I. L. Garzón, and L. C. Balbás, *Phys. Rev. B* **70**, 165403 (2004).
6. M. Pereiro and D. Baldomir, *Phys. Rev. A* **75**, 033202 (2007).
7. T. Yokoyama, S. Yokoyama, T. Kamikado, Y. Okuno, and S. Mashoko, *Nature* **413**, 619 (2001).
8. O. Echt, K. Sattler, and E. Recknagel, *Phys. Rev. Lett.* **47**, 1121 (1981).
9. J. L. Li et al., *Phys. Rev. Lett.* **88**, 066101 (2002).
10. D. Torrent, J. Sánchez-Dehesa, and F. Cervera, *Phys. Rev. B* **75**, 241404 (2007).
11. G. A. Somorjai (ed.), *Introduction to Surface Chemistry and Catalysis* (Wiley, New York, 1994).
12. A. Wieckowski, E. R. Savinova, and C. G. Vayenas Dekker (eds.), *Catalysis and Electrocatalysis at Nanoparticle Surfaces* (Marcel Dekker, New York, 2003, pp. 455–500).
13. G. Pacchioni, L. Giordano, and M. Baistrocchi, *Phys. Rev. Lett.* **94**, 226104 (2005).
14. D. Rocco, A. Bongiorno, G. Pacchioni, and U. Landman, *Phys. Rev. Lett.* **97**, 036106 (2006).
15. B. Yoon, H. Häkkinen, U. Landman, A. S. Wörz, J.-M. Antonietti, S. Abbet, K. Judai, and U. Heiz, *Science* **307**, 403 (2005).
16. M. Sterrer, T. Risse, M. Heyde, H.-P. Rust, and H.-J. Freund, *Phys. Rev. Lett.* **98**, 206103 (2007).
17. M. Y. Lai and Y. L. Wang, *Phys. Rev. Lett.* **81**, 164 (1998).
18. M.-S. Ho, I.-S. Hwang, and T. T. Tsong, *Phys Rev. Lett.* **84**, 5792 (2000).
19. S. Gwo, C.-P. Chou, C.-L. Wu, Y.-J. Yeh, S.-J. Tsai, W.-C. Lin, and M.-T. Lin, *Phys. Rev. Lett.* **90**, 185506 (2003).
20. H. Brune, M. Giovannini, K. Bromann, and K. Kern, *Nature* **394**, 451 (1998).
21. Y.-P. Chiu, L.-W. Huang, C.-M. Wei, C.-S. Chang, and T.-T. Tsong, *Phys. Rev. Lett.* **97**, 165504 (2006).
22. W. B. Su, S. H. Chang, W. B. Jian, C. S. Chang, L. J. Chen, and T. T. Tsong, *Phys. Rev. Lett.* **86**, 5116 (2001).
23. W. B. Jian, W. B. Su, C. S. Chang, and T. T. Tsong, *Phys. Rev. Lett.* **90**, 196603 (2003).
24. M. Jalochowski and E. Bauer, *J. Appl. Phys.* **63**, 4501 (1988).
25. R. Kern, G. Le Lay, and J. J. Metois, in *Current Topics in Materials Science* (ed. Kaldis), vol. 3, Chapter 3 (Amsterdam, the Netherlands: North-Holland, 1978).
26. Z. Zhang, Q. Niu, and C. K. Shih, *Phys. Rev. Lett.* **80**, 5381 (1998).
27. A. R. Smith, K. J. Chao, Q. Niu, and C. K. Shih, *Science* **273**, 226 (1996).
28. L. Gavioli, K. R. Kimberlin, M. C. Tringides, J. F. Wendelken, and Z. Zhang, *Phys. Rev. Lett.* **82**, 129 (1998).
29. M. Jalochowski and E. Bauer, *Phys. Rev. B* **38**, 5272 (1988).
30. A. Crottini, D. Cvetko, L. Floreano, R. Gotter, A. Morgante, and F. Tommasini, *Phys. Rev. Lett.* **79**, 1527 (1997).
31. C. A. Jeffrey, E. H. Conrad, R. Feng, M. Hupalo, C. Kim, P. J. Ryan, P. F. Miceli, and M. C. Tringides, *Phys. Rev. Lett.* **96**, 106105 (2006).
32. D. A. Luh, T. Miller, J. J. Paggel, M. Y. Chou, and T. C. Chiang, *Science*, **292**, 1131 (2001).
33. Y. Guo, Y. F. Zhang, X. Y. Bao, T. Z. Han, Z. Tang, L. X. Zhang, W. G. Zhu, E. G. Wang, Q. Niu, Z. Q. Qiu, J. F. Jia, Z. X. Zhao, and Q. K. Xue, *Science* **306**, 1915 (2004).
34. D. Eom, S. Qin, M. Y. Chou, and C. K. Shih, *Phys. Rev. Lett.* **96**, 027005 (2006).
35. Y. S. Fu, S. H. Ji, X. Chen, X. C. Ma, R. Wu, C. C. Wang, W. H. Duan, X. H. Qiu, B. Sun, P. Zhang, J. F. Jia, and Q. K. Xue, *Phys. Rev. Lett.* **99**, 256601 (2007).
36. Y. F. Zhang, Z. Tang, T. Z. Han, X. Ma, J. F. Jia, Q. K. Xue, K. Xun, and S. C. Wu, *Appl. Phys. Lett.* **90**, 093120 (2007).
37. W. B. Su, S. H. Chang, H. Y. Lin, Y. P. Chiu, T. Y. Fu, C. S. Chang, and T. T. Tsong, *Phys. Rev. B* **68**, 033405 (2003).
38. W. B. Su, H. Y. Lin, Y. P. Chiu, H. T. Shih, T. Y. Fu, Y. W. Chen, C. S. Chang, and T. T. Tsong, *Phys. Rev. B* **71**, 073304 (2005).
39. H. H. Weitering, D. R. Heslinga, and T. Hibma, *Phys. Rev. B* **45**, 5991 (1992).
40. A. N. Cleland and M. L. Roukes, *Nature* **392**, 160 (1998).

41. J. V. Barth, G. Costantini, and K. Kern, *Nature* **437**, 671 (2005).

42. M. Valden, X. Lai, and D. W. Goodman, *Science* **281**, 1647 (1998).

43. H. Haberland, Th. Hippler, J. Donges, O. Kostko, M. Schmidt, and B. von Issendorff, *Phys. Rev. Lett.* **94**, 035701 (2005).

44. I. Shyjumon, M. Gopinadhan, O. Ivanova, M. Quaas, H. Wulff, C. A. Helm, and R. Hippler, *Eur. Phys. J. D* **37**, 409 (2006).

45. P. Gambardella, A. Dallmeyer, K. Maiti, M. C. Malagoli, W. Eberhardt, K. Kern, and C. Carbone, *Nature* **416**, 301 (2002).

46. R. A. Guirado-López, J. Dorantes-Dávila, and G. M. Pastor, *Phys. Rev. Lett.* **90**, 226402 (2003).

47. N. Nilius, T. M. Wallis, and W. Ho, *Science* **297**, 1853 (2002).

48. Y. P. Chiu, H. Y. Lin, T. Y. Fu, C. S. Chang, and T. T. Tsong, *J. Vac. Sci. Technol. A.* **23**, 1067 (2005).

49. H. Y. Lin et al., *Phys. Rev. Lett.* **94**, 136101 (2005).

50. J. G. Amar and F. Family, *Phys. Rev. Lett.* **74**, 2066 (1995).

51. C. Ratsch, A. P. Seitsonen, and M. Scheffler, *Phys. Rev. B* **55**, 6750 (1997).

52. H. Roder, K. Bromann, H. Brune, and K. Kern, *Surf. Sci.* **376**, 13 (1997).

53. K. A. Fichthorn and M. Scheffler, *Phys. Rev. Lett.* **84**, 5371 (2000).

54. H. Brune, *Surf. Sci. Rep.* **31**, 121 (1998).

55. W. Ekardt, *Phys. Rev. Lett.* **52**, 1925 (1984).

56. W. D. Knight et al., *Phys. Rev. Lett.* **52**, 2141 (1984).

57. P. Jena, B. K. Rao, and S. N. Khanna, *Physics and Chemistry of Small Clusters* (Plenum Press, New York, 1987).

58. X. Li, H. Wu, X. B. Wang, and L. S. Wang, *Phys. Rev. Lett.* **81**, 1909 (1998).

59. M. Haruta, *Catal. Today* **36**, 153 (1997).

60. C. Yannouleas and U. Landman, *Phys. Rev. Lett.* **78**, 1424 (1997).

61. C. T. Campbell, S. C. Parker, and D. E. Starr, *Science* **298**, 811 (2002).

62. K. E. Schriver, J. L. Persson, E. C. Honea, and R. L. Whetten, *Phys. Rev. Lett.* **64**, 2539 (1990).

63. C. Kohl, B. Montag, and P.-G. Reinhard, *Z. Phys. D: At. Mol. Clusters* **38**, 81 (1996).

64. S. K. Nayak, P. Jena, V. S. Stepanyuk, W. Hergert, and K. Wildberger, *Phys. Rev. B* **56**, 6952 (1997).

65. I. A. Harris, R. S. Kidwell, and J. A. Northby, *Phys. Rev. Lett.* **53**, 2390 (1984).

66. G. H. Guvelioglu, P. Ma, and X. He, *Phys. Rev. Lett.* **94**, 026103 (2005).

67. H. Häkkinen, M. Moseler, and U. Landman, *Phys. Rev. Lett.* **89**, 033401 (2002).

68. I. A. Solov'yov, A. V. Solov'yov, W. Greiner, A. Koshelev, and A. Shutovich, *Phys. Rev. Lett.* **90**, 053401 (2003).

69. F. Baletto and R. Ferrando, *Rev. Mod. Phys.* **77**, 371 (2005).

70. K. Koga, T. Ikeshoji, and K.-I. Sugawara, *Phys. Rev. Lett.* **92**, 115507 (2004).

71. H. Häkkinen and M. Manninen, *Phys. Rev. Lett.* **76**, 1599 (1996).

72. Y.-P. Chiu, C.-M. Wei, and C.-S. Chang, *Phys. Rev. B* **78**, 115402 (2008).

73. J. P. Perdew, K. Burke, and M. Ernzerhof, *Phys. Rev. Lett.* **77**, 3865 (1996).

74. W. Kohn and L. J. Sham, *Phys. Rev.* **140**, A1133 (1965).

75. G. Kresse and J. Hafner, *Phys. Rev. B* **47**, 558 (1993); *Phys. Rev. B* **49**, 14251 (1994); G. Kresse and J. Furthmuller, *Phys. Rev. B* **54**, 11169 (1996); G. Kresse and J. Furthmuller, *Comput. Mater Sci.* **6**, 15 (1996).

76. G. Kresse and D. Joubert, *Phys. Rev. B* **59**, 1758 (1999).

77. N. D. Lang and W. Kohn, *Phys. Rev. B* **3**, 1215 (1970).

78. W. A. de Heer, *Rev. Mod. Phys.* **65**, 611 (1993).

79. T. P. Martin, T. Bergmann, H. Goehlich, and T. Lange, *J. Phys. Chem.* **95**, 6421 (1991).

80. P. Stampfli and K. H. Bennemann, *Phys. Rev. Lett.* **69**, 3471 (1992).

81. A. L. Mackay, *Acta Cryst.* **15**, 916 (1962).

82. C. Kohl and P.-G. Reinhard, *Z. Phys. D* **39**, 225 (1997).

83. H. Häkkinen and M. Manninen, *J. Chem. Phys.* **105**, 10565 (1996).

84. S. M. Lu, M. C. Yang, W. B. Su, C. L. Jiang, T. Hsu, C. S. Chang, and T. T. Tsong, *Phys. Rev. B* **75**, 113402 (2007).

85. S. H. Chang, W. B. Su, W. B. Jian, C. S. Chang, L. J. Chen, and T. T. Tsong. *Phys. Rev. B* **65**, 245401 (2002).

18
Cluster–Substrate Interaction

Miguel A. San-Miguel
University of Sevilla

Jaime Oviedo
University of Sevilla

Javier F. Sanz
University of Sevilla

18.1 Introduction

The deposition of clusters onto solid surfaces is a practical tool used in different technologies, including biosensing, electronics, surface smoothing, thin film growth, and heterogeneous catalysis. In particular, metal nanoclusters have become very popular due mainly to two important factors. First, the surface/bulk atoms ratio is high, and surface atoms might be more reactive since they have a lower coordination number. Second, their electronic structure depends on the number of atoms and, therefore, the properties and applications of these clusters are more diverse than those for bulk metals.

The deposition of metal nanoclusters on specific substrates leads to the creation of interfaces with properties quite different from those of the separated fragments. In this chapter, we focus mainly on the deposition of metal clusters on metal-oxide surfaces. This approach has been widely used to obtain supports with tailored properties and has found important applications in heterogeneous catalysis (Henrich and Cox, 1994; Campbell, 1997; Henry, 1998). Due to the vast complexity of industrial catalysts, surface scientists have successfully developed model catalysts that are suitable to be studied by surface analytical techniques. The model systems consist of metal clusters deposited onto metal-oxide substrates from either evaporative or molecular precursor sources. Some of the most popular substrates are SiO_2, Al_2O_3, Ti_xO_y, MgO, NiO, and Fe_xO_y (Goodman, 2003), even though MgO and rutile TiO_2 have become prototypes due to their thermodynamic stability and relatively easy preparation.

Although the choice of monometallic nanoclusters is the most common, bimetallic particles have also been investigated because they are used for several industrial catalyzed reactions.

The physical and chemical properties of bimetallic particles are found to be different from their single metal counterparts, and they are found to vary as a function of composition and particle size. It has been possible to prepare mixed-metal systems with physical and chemical properties, enhancing the catalytic activity and selectivity. These special properties are generally attributed to either ensemble or ligand effects, although other factors related to particle size effects, matrix effects, and catalyst stability have been invoked (Guczi et al., 1993). There have been significant efforts to synthesize bimetallic particles having different compositions; however, most of these studies have shown that although it is possible to prepare binary particles with uniform sizes/shapes and specific compositions, it is extremely difficult to control the uniformity on a particle-by-particle basis (Santra et al., 2004).

Despite long-lasting efforts both experimentally and theoretically, little is known about the role of cluster–support interactions in the growth, structure, and reactivity of supported metal catalysts. On the experimental side, this is mainly due to difficulties in the direct characterization of the cluster–support interfaces, where most of the relevant interactions take place. On the theoretical side, the most recent advances in computing power and the development of methods have made possible to use quantum mechanic calculations to explore the structure and reactivity of oxide-supported metal clusters. However, there is still a lack in understanding some fundamental processes since particle sizes that can be treated accurately are still limited.

In Section 18.2, we introduce basic concepts regarding particle–surface interactions, in particular, the differences between physisorption and chemisorption phenomena and how they can be described. The identification of the most favorable adsorption sites on surfaces is also discussed. Finally, the surface coverage effect leads us to describe different adsorbate–adsorbate interactions governing the overlayer structure. Section 18.3 is devoted to the thermodynamic aspects. First, some considerations are made about chemical reactions that might undergo when depositing metal clusters on metal-oxide substrates. This is followed by the arguments to predict whether a metal tends to wet the surface. The different growth modes are also described. Thermodynamic predictions are often unfulfilled because of kinetic restrictions. Section 18.4 focuses on kinetic aspects such as surface diffusion, nucleation, and sintering processes, and concludes by considering the effect of the cluster size and the temperature on the properties of the adsorption system. Some aspects related to the electronic structure are discussed in Section 18.5. First, it is described how work function changes upon adsorption provide significant insights about the cluster–substrate interaction followed by some approaches to analyze the charge-transfer process. The section concludes with some examples that illustrate consequences on the chemical reactivity induced by the cluster–substrate interfaces.

18.2 Basics in Particle–Surface Interactions

18.2.1 Adsorption: Physisorption and Chemisorption

The process that involves trapping of atoms, molecules, or clusters (adsorbate) on a surface (substrate) is called *adsorption*. It should be distinguished from *absorption*, which refers to particles entering into the bulk of the substrate. The adsorption process is always exothermic, although for historical reasons it is common to express the adsorption energy as a positive entity unlike the adsorption enthalpy, which for an exothermic process should be negative according to the thermodynamic convention.

Depending on the magnitude of the adsorption enthalpies, the adsorption processes can be classified into two groups: physisorption and chemisorption.

18.2.1.1 Physisorption

In this case, the bonding interaction is weak and the enthalpies of adsorption are ranged typically between -10 and $-40\,\mathrm{kJ \cdot mol^{-1}}$. The bonding is a consequence of balancing weak attractive forces between the adsorbate and the surface, and repulsive forces arising from close contact. The attractive forces are generally of van der Waals nature and can be described by Lennard–Jones potential for which the potential energy $V(r)$ is given by

$$V(r) = 4\varepsilon \left[\left(\frac{\sigma}{r} \right)^{12} - \left(\frac{\sigma}{r} \right)^{6} \right] \tag{18.1}$$

where
  r is the distance between adsorbate and surface particles
  ε is the well depth of the potential energy curve
  σ is the distance between the interacting particles at which
    the potential energy is zero

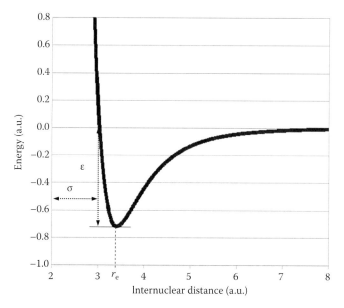

FIGURE 18.1 Lennard–Jones potential energy function which is suitable to describe weak interactions.

These parameters are fitted to reproduce experimental data, and they usually can be taken from the different available force fields or can be estimated for specific systems from quantum calculations.

The first term in Equation 18.1 is positive, corresponds to the repulsive interaction, and usually has a $1/r^{12}$ dependence. The second term having a $1/r^{6}$ dependence is negative and is related to the attractive interactions between particles. The potential energy profile corresponding to the Lennard–Jones (12-6) potential is illustrated in Figure 18.1.

This expression has extensively been used in the simulation of different interacting particles. However, it is not valid to represent the interaction between a particle and a flat surface. In this case, the simplest approach is by the integration of the Lennard–Jones (12-6) potential over the half-space related to the surface, resulting in

$$V(r) = \left(\frac{A_9}{z} \right)^{9} - \left(\frac{A_3}{z} \right)^{3} \tag{18.2}$$

where
  $A_n = 4\varepsilon\sigma^n$
  z is the distance between the interaction center and the surface
  ε and σ are the Lennard–Jones potential parameters

For some systems, such potential fails on the description of the interaction and an alternative potential has been proposed

$$V(r) = \frac{C_{12}}{(z - z_0)^{12}} - \frac{C_3}{(z - z_0)^{3}} \tag{18.3}$$

where
  $C_n \propto \sqrt{\varepsilon\sigma^n}$
  z_0 is a limit approach distance for each center

Other more complicated expressions can be deduced in which periodic surface morphologies are taken into account.

18.2.1.2 Chemisorption

This is the process by which a chemical bond between the adsorbate and the substrate is formed. The adsorption enthalpy associated is larger than $-40\,kJ\cdot mol^{-1}$ in absolute value and the nature of the bonding can be ionic, covalent, or a mixture. In chemisorption, there is a high degree of specificity on the substrate. Even for the same compound, several oriented surfaces may exhibit different adsorption sites.

Alternatively to physisorption, this process can be described approximately by a Morse potential (Figure 18.2) given by

$$V(r) = D_e \left(1 - e^{-a(r-r_e)}\right)^2 + V(r_e) \qquad (18.4)$$

where
r is the distance between the atoms
r_e is the equilibrium bond distance
D_e is the well depth
a refers to the width of the potential curve given by

$$a = \sqrt{\frac{k_e}{2D_e}} \qquad (18.5)$$

where k_e is the force constant at the minimum of the well.

Since the amount of energy involved in chemisorption is much higher than in physisorption, the adsorption process can lead to significant changes in the adsorbate. Thus, for molecular adsorbates it can be distinguished between non-dissociative and dissociative chemisorption, depending on whether the molecular structure is preserved or not upon the adsorption process. Other kinds of changes might also happen, and they are discussed throughout this chapter.

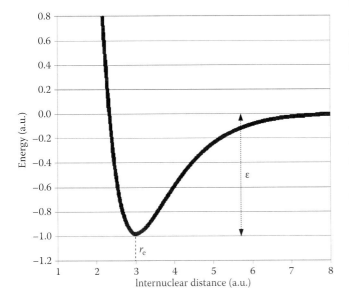

FIGURE 18.2 Morse potential energy function, which is suitable to describe chemical bonding interactions.

The residence time in the adsorbed state is given by

$$\tau = \tau_0 \exp\left(\frac{\Delta H_0}{RT}\right) \qquad (18.6)$$

where
τ_0 is related to the surface atom vibration time (usually on the order of 10^{-12} s)
T is the temperature
R is the gas constant

It becomes evident that chemisorbed species will have longer residence times than physisorbed ones. Thus, for a physisorption of enthalpy of $-30\,kJ\cdot mol^{-1}$, a typical value of τ at 300 K would be over $1.7\cdot10^{-7}$ s, and the effect of increasing temperature up to 400 K would lower τ to $8.4\cdot10^{-9}$ s. This has to be compared to a chemisorption of enthalpy of $-100\,kJ\cdot mol^{-1}$ that has a value of $2.8\cdot10^{+5}$ s at 300 K and 10 s at 400 K.

This parameter is of great importance, because the longer the adsorbate resides on the substrate, the more probable is the process of exchanging energy with the substrate, and consequently, the interaction becomes more effective.

18.2.2 Adsorption Sites

Substrates do not exhibit flat surfaces in which all surface points are equivalent, but they may become very complex. Clean single crystal surfaces are the most desirable supports to be studied, since they exhibit well-defined planes. However, these supports are not always readily prepared, and the real surfaces very often consist of mixtures of flat terraces presenting steps, kinks, and point defects. It is common to observe in experimental measurements that adsorbates accumulate preferentially close to these irregularities, indicating that the interaction there is stronger than in other regions.

Generally, a theoretical adsorption study starts by searching the most stable adsorption sites energetically when the adsorbate approaches the substrate. The adsorption energy can be estimated from theoretical calculations according to

$$E_{ads} = E_{ads-system} - E_{substrate} - E_{adsorbate} \qquad (18.7)$$

where $E_{ads-system}$, $E_{substrate}$, and $E_{adsorbate}$ are the total energies of the adsorption system and the isolated substrate and adsorbate, respectively.

As an example, we report some results on the deposition of Ba atoms on the rutile $TiO_2(110)$ surface. Alkali and alkaline earth metals are known to behave as reaction modifiers on a variety of substrates, including metal oxides. In particular, deposition on rutile $TiO_2(110)$ surfaces has been investigated experimentally and theoretically, but only the computational methods have been able to clear the stability of the different adsorption sites. This substrate consists of neutral layers made of three planes of composition $O-Ti_2O_2-O$. In the surface, there are two kinds of titanium atoms, fivefold and sixfold coordinated, forming

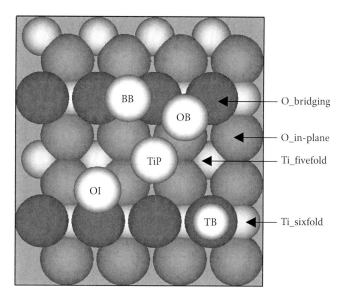

FIGURE 18.3 Top view of the rutile TiO$_2$(110) surface. Some representative atom types and potential adsorption sites are labeled. (From San Miguel, M.A. et al., *J. Phys. Chem. B*, 110, 19552, 2006. With permisssion.)

alternating rows. All oxygen atoms are threefold coordinated as in the bulk, except those ones in the outermost plane called *bridging atoms*, which lose one bond when the upper layer is removed. Figure 18.3 shows a top view of this surface indicating all the potential adsorption sites. Two of them consist of adsorption over bridging oxygen atoms: TB and BB sites. In addition, there is an on-top fivefold Ti site (TiP) and also sites on which the metal atom is among three oxygens (OB and OI). Geometry optimizations based on density functional theory (DFT) calculations for Ba atoms have demonstrated that only OB, OI, and BB sites are minima in the potential energy surface, whereas the TiP and TB are high-energy sites that evolve rapidly to OI and BB sites, respectively. The Ba atom is more stable when it is threefolded coordinated so the OB and OI sites are more stable than the BB site. Between them, the OB site is the most stable by 0.28 eV. This is essentially linked to the fact that

the bond distances to the oxygen atoms are more balanced in the OB situation (2.45 and 2.55 Å for OB vs. 2.37 and 2.63 Å for OI). Interestingly, when metal particles interact strongly with substrates, the surface relaxations may become notorious. Thus, for sites OB and OI, Ba atoms are positioned among three oxygen atoms, and compared with the BB case, the Ba atom lies at a shorter distance from the surface and relaxations are more noticeable (see Figure 18.4). Since the Ba stays at a certain distance above the surface, the oxygen atoms directly bound to Ba tend to move upward. This displacement is especially large for in-plane atoms that have to move in order to form bonds with the adsorbed metal. The movements are as large as ~0.4 Å for OI and ~0.3 Å for OB. Titanium atoms below the row of bridging oxygen atoms also move, and the Ti–O bridging bond distances are stretched (San Miguel et al., 2006).

18.2.3 Surface Coverage Effect

The adsorption enthalpy is always referred to a particular degree of surface coverage since the lateral interactions between adsorbates influence significantly the bonding strength. The surface coverage (θ) is defined as the ratio between number of surface sites occupied by adsorbate and the total number of surface sites, and the corresponding unit is the monolayer (ML). Thus, there are three main adsorbate–adsorbate interactions as descirbed in the following.

18.2.3.1 Electrostatic Interactions

These occur for adsorbates that have undergone charge transfer with the substrate. One good example is the adsorption of alkali metals or earth alkali metals. The effect of increasing the surface coverage in Ba supported on TiO$_2$(110) is the introduction of repulsive forces that cause Ba atoms to be as far as possible from each other and occupying OB sites. For a coverage of $\theta = 0.25$ ML, two equivalent adsorption patterns named SS and OS were found (see Figure 18.5). The bond nature is still the same than that found for a lower coverage; however, the net adsorption energies decrease due to the Ba–Ba repulsive interactions.

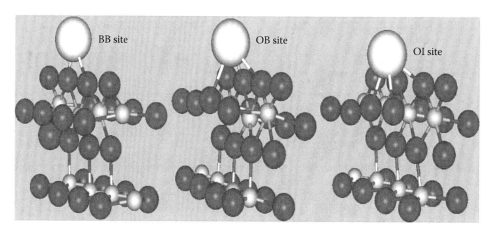

FIGURE 18.4 More favorable adsorption sites for Ba on the stoichiometric TiO$_2$(110) surface. Only a portion of the surface is shown. Oxygen is represented by dark spheres, whereas small and large light spheres represent titanium and barium, respectively. (From San Miguel, M.A. et al., *J. Phys. Chem. B*, 110, 19552, 2006. With permisssion.)

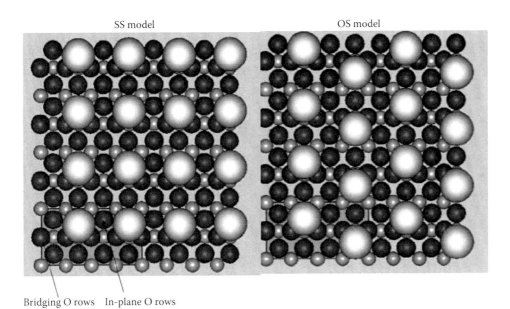

SS model   OS model

Bridging O rows   In-plane O rows

FIGURE 18.5 Adsorption models for Ba on the stoichiometric $TiO_2(110)$ surface at coverage of 0.25 ML. Ba atoms are represented by the largest spheres, Ti atoms are represented by the smallest spheres, and O atoms are shown by the medium ones (dark gray). A box indicates the supercell in each model. (From San Miguel, M.A. et al., *J. Phys. Chem. B*, 110, 19552, 2006. With permisssion.)

18.2.3.2 Covalent/Metallic Bonding

This is the type of interaction occurring when the adsorbate valence orbitals are partially filled. Transition metal adatoms on metal or metal-oxide surfaces are typical examples. For instance, Goodman and coworkers have carefully studied the adsorption of Pd on rutile $TiO_2(110)$ (Xu et al., 1997). From scanning tunneling microscopy (STM) images obtained at low coverage, they reported that Pd dimers and tetramers are present on the surface, adsorbed on the fivefold titanium rows. Strikingly, images did not show isolated Pd atoms.

Periodic DFT calculations allowing structural relaxation estimated the dimerization energy to be favorable by 0.49 eV, indicating a strong tendency for Pd atoms to bind each other.

The interaction between Pd atoms can be seen in the electron density isosurfaces depicted in Figure 18.6. These regions colored in cyan allow to distinguish the formation of bonding between Pd atoms and both titanium and protruding oxygen atoms (side view), and also the existence of a Pd–Pd bond (top view) (Sanz and Márquez, 2007).

18.2.3.3 van der Waals Forces

These forces are characteristic in self-assembled monolayer (SAM) systems such as alkanethiols on coinage metals; organosilanes on SiO_2, Al_2O_3, quartz, mica, and gold; amines on platinum and mica; carboxylic acids on Al_2O_3, CuO, AgO, and silver, and so on. In these systems, the adsorbed molecules

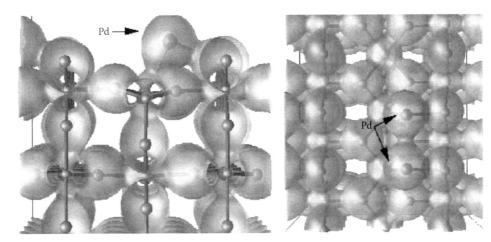

Pd →

Pd

FIGURE 18.6 A Pd dimer adsorbed on the $TiO_2(110)$ surface. An isosurface of the total electron density corresponding to $\rho = 0.30$ e A^{-3} is depicted on top and side views. The selected isosurface has been colored according to the magnitude of the gradient of the total density to emphasize the regions associated to bonding interactions. (From Sanz, J.F. and Márquez, A., *J. Phys. Chem. C*, 111, 3949, 2007. With permisssion.)

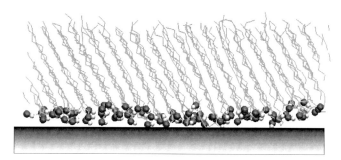

FIGURE 18.7 Side view of a SAM consisting of hexadecanol molecules supported on a mica surface. The oxygen and hydrogen atoms in the hydroxy group are depicted as spheres. The molecules bind to the surface through the head group and the alkane chains align orderly, keeping a characteristic tilt angle.

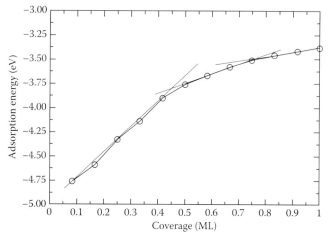

FIGURE 18.8 Averaged binding energy (BE) per Ca atom adsorbed on the stoichiometric $TiO_2(110)$ surface as a function of surface coverage.

bind to the surface by their head group, and the long alkane tails, which can be functionalized at the end, align in specific orientations. The interactions between the tails are caused by van der Waals forces. They have an attractive nature and result from instantaneous distortions in the electron clouds of atoms and molecules, which consequently induce temporary dipole moments (Figure 18.7).

Usually, one of these forces is responsible for the actual structure at a given surface coverage. However, when the coverage changes, the driving force is not necessarily the same. For instance, for alkali or earth alkali metals, at low coverage the repulsive coulombic forces govern the structure of the monolayer, but at a certain coverage there is a transition so the covalent interactions between adatoms prevail and the system becomes metallic. Figure 18.8 shows the average

adsorption energy as a function of the surface coverage. At low coverage, a linear increase in binding energy occurs, indicating that the interaction between adsorbate and substrate becomes less favorable since the adatoms approach each other and the repulsive electrostatic forces start to be noticeable. The slope is slightly smoother at intermediate coverages and is even more and apparently trending to reach a steady value at ~0.8 ML.

Consequently, the structure of real systems is a tradeoff between adsorbate and substrate attractive forces, and the lateral interactions between adsorbates which can be attractive or repulsive in nature. The final arrangement of the adsorbates

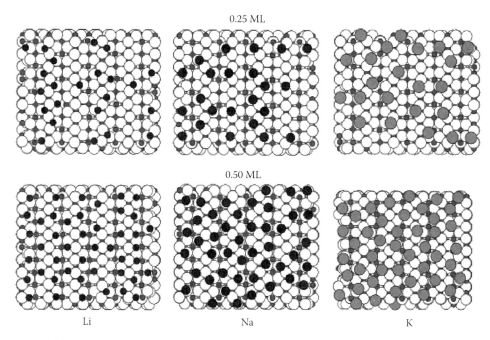

FIGURE 18.9 Top views of final configurations from molecular dynamic simulations at 300 K for three alkali metals on the stoichiometric $TiO_2(110)$ surface at coverages of 0.25 and 0.5 ML. At low coverages and small atom sizes, the adsorption sites are mostly OB, resulting in commensurate overlays. As the coverage and the atom size increase, new additional adsorption sites can be identified and the overlay adopts an intermediate or incommensurate structure. (From San Miguel, M.A. et al., *J. Phys. Chem. B*, 105, 1794. With permisssion.)

relative to the underlying substrate is then named commensurate or incommensurate.

A *commensurate overlayer* corresponds to the situation where the substrate–adsorbate interactions dominate over the lateral adsorbate–adsorbate interactions. Therefore, the distance between adsorbates will be related to the substrate spacing. In an *incommensurate overlayer*, the adsorbate–adsorbate interactions are of similar strength to those between adsorbate and substrate, and there is a balance between maximizing both adsorbate–substrate and adsorbate–adsorbate interactions. The result is that the adsorbate–adsorbate distance will not correspond necessarily to the substrate spacing. These two types of arrangements are limit cases, and in real systems an intermediate situation is often found. Thus, for alkali metal atoms on $TiO_2(110)$, it is observed that at low coverage the adatoms adsorb preferentially on OB sites and the interatomic distances are related to the distance between OB sites; however, when coverage or alkali atom size is increased, new adsorption sites are populated and the interatomic distance is no longer related to the substrate spacing (Figure 18.9).

18.3 Thermodynamic Aspects

18.3.1 Reactivity

When metals are deposited on oxides, thermodynamic considerations must be taken into account to predict whether a chemical reaction may take place. For instance, if metal A is deposited onto oxide BO, a reduction reaction of BO by A may happen according to

$$A(s) + BO(s) \Rightarrow AO(s) + B(s) \tag{18.8}$$

The occurrence of this reaction will depend on the relative stability of the oxides involved, which can be checked in tables of standard heats of formation of metal oxides. Campbell compiled most common heats of oxide formation (Campbell, 1997) and they are reproduced in Table 18.1. Thus, reaction (18.8) is thermodynamically favorable and it will occur if the change in the standard free energy, ΔH_f^0, is negative.

This is not the only kind of reaction that can happen; there are others where metal B could change its oxidation number such as in TiO_2

$$A(s) + 2TiO_2(s) \Rightarrow AO(s) + Ti_2O_3(s) \tag{18.9}$$

or metals A and B could form intermetallic compounds as

$$2A(s) + BO(s) \Rightarrow AO(s) + AB(s) \tag{18.10}$$

However, given that the temperature at which most of the experiments take place is low (normally room temperature), not all the thermodynamic predictions for bulk systems are fulfilled due to kinetic limitations. The activation barriers for these processes in restricted confinement may become high, and the mobility of atoms to diffuse can be very limited.

TABLE 18.1 Heats of Formation of the Most Stable Oxide for Different Metals

Heats of Formation of Oxide (ΔH_f^0 in $kJ \cdot mol^{-1}$ Referred to O Atoms)	Metal
>0	Au
0 to −50	Ag, Pt
−50 to −100	Pd
−100 to −150	Rh
−150 to −200	Ru, Cu
−200 to −250	Re, Co, Ni, Pb
−250 to −300	Fe, Mo, Sn, Ge, W
−300 to −350	Rb, Cs, Zn
−350 to −400	K, Cr, Nb, Mn
−400 to −450	Na, V
−450 to −500	Si
−500 to −550	Ti, U, Ba, Zr
−550 to −600	Al, Sr, Hf, La, Ce
−600 to −650	Sm, Mg, Th, Ca, Sc, Y

18.3.2 Wetting

The adhesion energy or the work of adhesion, W, is related to the free energy change to separate unit areas of two media from close contact to infinity in vacuum. Singularly, when the media are the same, W becomes the work of cohesion. Thus, the work of adhesion for an adsorbate/substrate system will be represented by W_{as}, and the work of cohesion for each component as W_a and W_s.

Surface energy is defined as the free energy change, γ, when the surface area is increased by unit area. This energy would be equivalent to that needed to separate two half-unit areas from contact, so that we can write for adsorbate and surface

$$\gamma_a = \frac{1}{2W_a} \quad \text{and} \quad \gamma_s = \frac{1}{2W_s} \tag{18.11}$$

When both are in contact, the free energy change in expanding the interfacial area by unit area is known as *interfacial energy* or *interfacial tension*, γ_{as}. This energy is associated with the work needed to create unit areas of adsorbate and surface, and bring them together

$$\gamma_{as} = \frac{1}{2W_a} + \frac{1}{2W_s} - W_{as} \tag{18.12}$$

and using Equation 18.11, we get

$$\gamma_{as} = \gamma_a + \gamma_s - W_{as} \tag{18.13}$$

which is called the Dupré equation.

Figure 18.10 represents schematically a cluster deposited on a substrate. At equilibrium, all the forces must be balanced. Therefore, the tangential components of the force due to the surface–gas interface must be equal and opposite to the forces

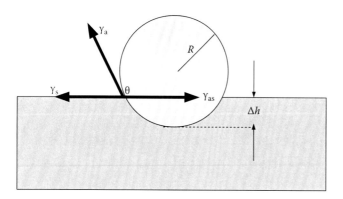

FIGURE 18.10 Schematic of a metal cluster on an oxide support in thermodynamic equilibrium.

arising from the adsorbate–surface and the adsorbate–gas interfaces. This balance results in the Young equation, defined as

$$\gamma_s = \gamma_{as} + \gamma_a \star \cos\theta \tag{18.14}$$

Following Young equation, it is possible to predict when a metal particle is going to wet the surface or not. For a contact angle of 0 degrees, the metal extends over the surface, provoking the wetting, and the equation becomes

$$\gamma_s = \gamma_{as} + \gamma_a \tag{18.15}$$

When the metal atoms prefer to grow in three-dimensional (3D) particles instead of wetting the surface, the contact angle increases and the equation becomes

$$\gamma_s > \gamma_{as} + \gamma_a \tag{18.16}$$

Normally, the surface-free energy of metals is larger than or comparable to that of oxides ($\gamma_a \geq \gamma_s$); therefore, according to Equation 8.15 for wetting to occur, the metal-oxide interfacial free energy (γ_{as}) has to be very small, or the bonding at the metal–oxide interface has to be very effective.

From the Dupré equation, the adhesion energy W_{as} is given by

$$W_{as} = \gamma_a + \gamma_s - \gamma_{as} \tag{18.17}$$

Combining this with the Young equation, one gets the Young–Dupré equation

$$W_{as} = \gamma_a (1 + \cos\theta) \tag{18.18}$$

Then, we can express another criterion to indicate when the wetting can occur

$$W_{as} = 2 \star \gamma_a \tag{18.19}$$

If $W_{as} < 2 \star \gamma_a$, then wetting does not occur and metal atoms will form particles on the surface.

These predictions have been confirmed for many different systems, and it has been found that most of the mid-to-late transition metals do not wet the common oxides used as substrates.

Alkali metals are found to have very low surface free energies, and they do wet oxide surfaces either. Other metals like Al, Ga, In, Sn, and Pb, which also have relatively low surface energies, exhibit small contact angles, and they have a close behavior to wetting.

Strictly, the contact angle is defined for an isotropic medium like a liquid droplet; however, it is related to the radius of the free particle before adsorption R, and the amount of the truncation after adsorption Δh through the following equation:

$$\Delta h = R(1 + \cos\theta) \tag{18.20}$$

Now combining with Equation 8.18, we get

$$\frac{\Delta h}{R} = \frac{W_{as}}{\gamma_a} \tag{18.21}$$

This result is equivalent to that found in equilibrium morphology of crystals, which was investigated long time ago and solved by Wulff (1901). The problem is to minimize the total free surface energy of a crystal at constant volume and temperature. A crystal particle exhibits different planes, and the most stable shape is that with the limiting planes of the lowest surface energy. Assuming that the equilibrium shape is a polyhedron, the Wulff theorem established

$$\frac{\gamma_i}{h_i} = \text{Constant} \tag{18.22}$$

where γ_i and h_i are the surface energy and the central distance to the facet of index i.

The regular polyhedron shapes are valid only at 0 K where the surface energy anisotropy is maximal. However, this energy diminishes at higher temperatures, and rounded areas arise in the equilibrium shape. Kaischew took this effect into account and developed the Wulff–Kaischew theorem, defined as

$$\frac{\Delta h}{h_i} = \frac{W_{as}}{\gamma_i} \tag{18.23}$$

Therefore, these two equivalent results indicate that the equilibrium shape of supported particles is defined by the surface energy of the facets and the adhesion energy.

18.3.3　Growth Modes

Within this thermodynamic scenario, the addition of metal atoms to a surface can follow three different growth modes which are schematically illustrated in Figure 18.11.

1. Layer-by-layer growth (*Frank–van der Merwe* [FM]). The adsorbate interacts strongly with the surface, showing a strong tendency to wet the surface. Consequently, the metal atoms form perfect added layers.
2. *Volmer–Weber* (VW) growth. Oppositely, the adsorbate–adsorbate interactions are stronger and dominate, resulting in the atoms not wetting the surface and aggregating to form 3D clusters.

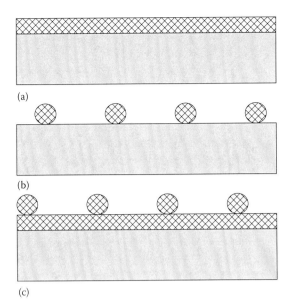

FIGURE 18.11 Schematic representation of the three growth modes of metals on metal-oxide substrates: (a) layer-by-layer or FM growth, (b) VW growth, and (c) SK growth.

3. *Stranski–Krastanov* (SK) growth. Atoms with low or no tendency to wet the surface start forming an intermediate monolayer, or even continue the layer-by-layer mode temporarily, but they eventually end up in 3D particles. This mode is the most commonly observed and it is due to kinetic effects related to the deposition dynamics.

18.4 Kinetic Aspects

18.4.1 Surface Diffusion

The growth modes previously discussed are only predicted from surface free-energy values at thermodynamic equilibrium. However, in order to attain an equilibrium state, metal atoms must be sufficiently mobile to reach adsorption sites of minimum free energy. This mass transport process is commonly named *surface diffusion*, which is an activated process that generally follows an Arrhenius behavior given by

$$D = D_0 \exp\left(\frac{-E_{act}}{RT}\right) \tag{18.24}$$

where
D is the diffusion coefficient
E_{act} is the activation energy barrier
D_0 is the preexponential factor called diffusivity

The diffusivity is related to the entropy change between the equilibrium and the activated configurations. It is common to assume that both are the same and then D_0 is approximately 10^{-2} cm$^2 \cdot$ s^{-1}. The diffusion coefficient assumes diffusion via random walk processes, and then it is possible to calculate the average distance traveled ($<x>$, in a monodimensional system) in a certain time (t).

$$D = \frac{<x>^2}{t} \tag{18.25}$$

The diffusion activation energies are directly related to the adsorption energies; thus, only particles physisorbed with low heats of adsorption are highly mobile at room temperature, while more strongly bound particles may remain immobile under similar conditions. In addition, it is found that the mobility is also dependent on the substrate surface. The flatter the surface, the lower the activation energy is. Conversely, atomically rough surfaces will lead to higher activation energies, and in consequence, the mobility will be more limited.

Among the model oxide-supported metal systems, diffusion has been extensively studied both experimentally and theoretically for Pd on MgO. The experimental measurements of Pd islands grown by metal atom deposition at low temperature have shown that the size of the islands is remarkably insensitive to the surface temperature during growth (Haas et al., 2000). From these observations, it was concluded that defects play an important role in the growth process. In addition to steps, kinks, and grain boundaries, oxide surfaces have a high density of point defects. Particularly, for MgO, high-resolution electron-energy-loss spectroscopy (HREELS) spectra have been interpreted in terms of neutral oxygen vacancies, named F centers, where an oxygen atom is removed from the surface (Wu et al., 1992). Moreover, point defects as divacancies where both a Mg and an O atom are missing (Barth and Henry, 2003) and, in some cases, charged oxygen vacancies (named F$^+$ centers) have also been claimed to be present (Giordano et al., 2003).

The traditional interpretation of the growth process was that Pd atoms land on flat MgO terraces and diffuse over the surface up to reaching point defects where they bind strongly and get trapped. Successive Pd atoms approaching the trapped atoms are also bound and build up the cluster.

Some information regarding the diffusion coefficients can be extracted from experimental measurements, but since they are often performed in different conditions, the results are not conclusive. The main advantage of theoretical calculations is that they provide information at atomistic level, which is valuable to understand diffusion mechanisms. Thus, diffusion barriers can be estimated by finding saddle points on the potential surface. In particular, the nudged elastic band (NEB) method (Henkelman et al., 2000; Henkelman and Jonsson, 2000) allows to determine diffusion barriers, provided that both the initial and final states are known for a diffusion process.

Calculations on Pd/MgO demonstrated that small clusters, up to four atoms, are highly mobile with similar diffusion barriers in the range of 0.34–0.50 eV. In particular, the tetramer has a very large diffusivity at room temperature with a diffusion barrier of 0.41 eV, smaller than that for the trimer (0.50 eV). Simulations showed that these clusters can diffuse by several different mechanisms, including dimer rotation, trimer walking, and tetramer rolling and sliding (Xu et al., 2005, 2006). In consequence, the growth mechanism is not as simple as initially accepted based on the mobility of single adatoms.

18.4.2 Nucleation

Nucleation processes can also play a key role in the surface dynamics. They have been theoretically studied because of the difficulty in performing experiments. Several nucleation theories have been developed during the last four decades. The first theories were based on the cluster size referred to as *critical nucleus*, which has an equal probability of becoming larger or smaller (Hirth and Pound, 1963). On the other hand, Zinsmeister introduced stochastic models based on the assumption that the cluster consisting of two atoms is never dissociated (Zinsmeister, 1966).

In addition, some rate equations were initially proposed to model these processes on perfect substrates (Venables, 1987, 1994). More recently, it has been accepted that the presence of different kinds of surface defects strongly affects nucleation processes. Thus, Venables et al. (2006) have developed rate equations for different defects.

Atomistic simulations have also provided insights on these processes. Classical molecular dynamics (MD) simulations showed that Au atoms deposited on a KCl surface, when approaching a step, diffuse toward its ledge and then moves along this to encounter another Au atom to form a dimer. These dimers constitute the critical nucleus (Nakamura et al., 1997).

Metal-oxide surfaces generally contain a high density of point defects (Barth and Henry, 2003). This can play an important role not only in the catalytic activity but also in the surface dynamics. One example is the nucleation of Pd clusters on rutile $TiO_2(110)$. In this surface, the main point defects consist of bridging oxygen vacancies. By means of STM, the amount of oxygen vacancies has been quantified as 7%–10% when the surface is annealed at 900 K (Hebenstreit et al., 2000, 2001). Periodic DFT calculations have demonstrated that these point defects can act as nucleation centers since they are very favorable adsorption sites for adatoms and dimers (Sanz and Márquez, 2007).

18.4.3 Sintering

Most supported metal clusters tend to sinter with time. Thus, starting systems consisting of a collection of numerous highly dispersed nanoclusters eventually end up in their most favorable thermodynamic state: a few large particles, which will have more reduced or even null catalytic activity. Considerable effort is made in the process of heterogeneous catalysis in finding ways to slow down these sintering processes.

The following are the two main mechanisms for sintering deposited metal nanoclusters:

- *Ostwald ripening*: Metal atoms leave a metal particle and diffuse on the support surface until they find and join another metal particle. Since the energy per atom is lower in larger particles, metal atoms stay longer in large particles than in small ones. Therefore, this fact leads to the growth of larger particles at the expense of smaller particles, which decrease in size and eventually disappear.

The individual atoms diffuse on the surface in the form of adatoms, or also in a metal compound as carbonyl, oxide, chloride, or other complex adsorbed on the substrate.

- *Particle diffusion/coalescence*: The whole metal particles diffuse across the support surface until they approach another particle and coalesce.

The kinetics of sintering has been intensively studied. Often, these studies analyze the elementary steps involved and propose rate equations of change of a metal particle's radius with time, but they are beyond the scope of this chapter (Parker and Campbell, 2007).

18.4.4 Cluster Size Effects

The existence of well-defined magic numbers for free metal clusters, which provide much higher stability than expected from clusters with a slight bigger or smaller number of atoms, is well known. Finding the geometry of minimum energy of free clusters has been the target of numerous studies. When metal clusters are deposited onto substrates, the arising interactions influence the structural and electronic properties of the clusters. Some of these properties, particularly the morphology, have been observed to depend on the cluster size.

Theoretical studies in this direction have been carried out. Traditionally, these investigations were based on MD simulations using semiempirical interatomic potentials (Cleveland and Landman, 1992; Cheng and Landman, 1993). However, most of these processes involve quantum effects that require a more precise description.

Some progress has been made in developing alternative methods, including the electronic structure, but being less computer time demanding, they can address larger systems. In particular, nonreactive metal–oxide interfaces have extensively been investigated by Goniakowski et al. They have focused on late transition metals as Pd, Ag, and Ni supported on MgO(100) surfaces. In these systems, the interaction is weak (physisorption type) and the charge transfer is rather small. The energetic model combines a tight binding many-body potential based on the second moment approximation (SMA) (Rosato et al., 1989) for the metal–metal interactions and a surface energy potential fitted to first principles DFT calculations for the metal–oxide interactions (Goniakowski et al., 2006).

These calculations have demonstrated that there are also magic numbers for deposited clusters on substrates and depend strongly on the interactions between the metal and the oxide, although they are not necessarily the same than for free metal clusters (Goniakowski and Mottet, 2005).

First principles calculations on large systems, including dynamical effects, are computationally demanding and, until recently, they had been considered prohibitive. However, nowadays the development of modern and fast computers makes this kind of calculations feasible. Some of these studies have explored the changes in structural and electronic properties as a function of cluster size.

As an example, ab initio calculations of small Pd clusters onto MgO(001) containing a surface F center showed that for cluster sizes between 2 and 6 atoms, the gas phase geometry was retained, while for larger sizes (7–13 atoms) the cluster adopted the underlying MgO structure. It was also observed that although the surface tends to reduce the spin of the adsorbed cluster, clusters larger than Pd_3 remain magnetic at the surface, exhibiting several low-lying structural and spin isomers (Moseler et al., 2002).

18.4.5 Effect of Temperature

All the kinetic aspects mentioned earlier are mainly controlled by temperature. Experiments are often performed at room or lower temperatures, which means that the mobility of atoms and clusters is reduced and the surface dynamics occurs at conditions far from the thermodynamic equilibrium. The knowledge of these phenomena is mainly obtained from experiments, and only scarce theoretical studies, including the electronic effects, have been carried out since the system sizes required to model these processes are at present almost unaffordable.

One recent theoretical investigation has reported the role of temperature in controlling the cluster–substrate interaction in Pd supported on the rutile $TiO_2(110)$ surface (San Miguel et al., 2007). Palladium clusters deposited on this surface can be considered as a catalyst model and has been extensively studied by different experimental techniques. In the experiments, some annealing treatment is usually given to the surface. The role of temperature is especially important because on increasing it, metal clusters can diffuse and coalesce. In particular, it was shown that annealing treatments at 700 K induced changes in the morphology of Pd clusters and the formation of islands, characterized by a high ratio between the volume and the area in contact with the substrate. It was also observed that their metallic character increased significantly upon the transition temperature (Della Negra et al., 2003).

Extensive first principles MD simulations based on periodic DFT calculations were carried out in order to analyze the evolution of morphological and electronic properties of a Pd_{12} cluster supported on a non-stoichiometric $TiO_2(110)$ surface as a function of temperature. The theoretical analysis predicted a transition temperature in remarkably good agreement with experiments and, in addition, it provided an atomistic interpretation for the changes in cluster properties. The simulations were performed between 100 and 1073 K, and the binding energy E_{bind} was estimated at each temperature as

$$E_{bind} = E_{12Pd-TiO_2} - E_{TiO_2} - E_{12Pd} \qquad (18.26)$$

where each term corresponds to the total energies computed on the optimized geometries. The binding energy is plotted as a function of temperature in Figure 18.12 and it reflects two different regimes. Below the transition temperature (around 800 K), the binding energy is approximately 1 eV larger than at higher temperatures. This change was related to the behavior observed in the experiments. In order to investigate the origin of

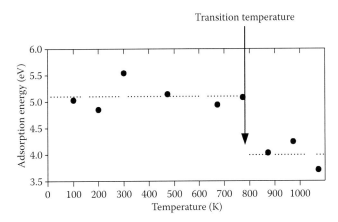

FIGURE 18.12 Binding energy at several temperatures for a Pd cluster deposited on a defective rutile $TiO_2(110)$ surface. Average values below and above the transition temperature (800 K) are shown by dotted lines. (From San Miguel, M.A. et al., *Phys. Rev. Lett.*, 99, 66102, 2007. With permission.)

this transition temperature, the system structure was carefully analyzed at an atomistic level. Three snapshots at representative temperatures are shown in Figure 18.13. At low temperature, the substrate structure controls the cluster geometry, which rearranges maximizing the interactions with bridging oxygen and with fivefold titanium atoms. This coordination mode is similar to that found for Pd dimers adsorbed on the surface (Sanz and Márquez, 2007). On increasing temperature, the cluster atoms have enough kinetic energy to rearrange but keeping strong interaction with the substrate. Finally, at the highest temperature, a Pd atom in the first layer jumps up to the second layer, reducing the metal contact area. This fact would be the origin of the reduction in the binding energy observed at the transition temperature. Consequently, the cluster experiences smaller influence from the substrate structure and rearranges in a geometry more akin to that corresponding to gas phase.

18.5 Electronic Structure

Theoretical studies using MD simulations with semiempirical interatomic potentials have provided valuable insights at atomistic level into the metal deposition processes on metal-oxide surfaces; however, when cluster–substrate interaction involves chemical phenomena such as creation or break of chemical bonds, spin-dependent processes, or surface defects of electronic origin, a full quantum description is demanded.

In this section, we describe some important concepts and properties related to the electronic structure that can be analyzed from quantum calculations and provide useful information on the cluster–substrate interaction.

18.5.1 Work Function

The *work function* (ϕ) is defined as the minimum energy required to remove an electron from the highest occupied energy level in the substrate (Fermi level) to the vacuum level, where the electron no longer feels the interaction with its image charge

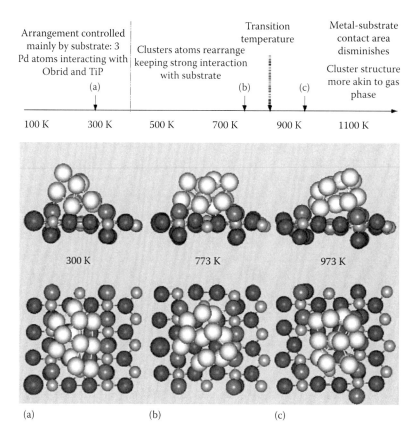

FIGURE 18.13 Snapshots of a 12 Pd atoms cluster deposited on a defective rutile $TiO_2(110)$ surface at three temperatures (300, 773, and 973 K). Side and top views are shown in the upper and lower panels, respectively. Ti atoms are represented by the smallest gray spheres, bridging O atoms by dark gray spheres, in-plane and subbridging atoms by black spheres, and Pd atoms by white spheres. Only the atoms in the outermost layers are depicted. (From San Miguel, M.A. et al., *Phys. Rev. Lett.*, 99, 66102, 2007. With permission.)

created at the substrate surface upon the removal of the electron. Therefore, the work function is related to the Fermi energy of a solid, which is associated to the electrostatic interaction between the atomic nuclei and the valence electrons.

Photoelectron spectroscopy (PES) techniques are powerful tools to analyze the electronic properties of surfaces. When a beam of radiation is incident on a solid surface, specific electrons can be photoemitted. Einstein demonstrated that the energy of the incident radiation hv is related to the binding energy BE as

$$hv = E_{ke} + BE \qquad (18.27)$$

where

E_{ke} is the kinetic energy of the emitted electron
BE is the binding energy of the electron within the solid

The measurements of ϕ have demonstrated that different geometric structures of the same substrate have different work functions. This is not surprising because the surface does not represent an infinite potential energy barrier to the electrons. The amplitudes of the electron wavefunctions do not decay to zero immediately away from the surface, but they are exponentially

damped as they leave the surface and give rise to "electron over-spill." The overall electrical neutrality is preserved because the excess negative electron overspill is balanced by the appearance of a corresponding excess positive charge at the substrate. Consequently, a dipolar layer is formed.

Adsorption processes may induce changes in this surface dipolar layer, and consequently, modify the work function, particularly if major charge transfer occurs between adsorbate and substrate. Therefore, the measurements of the work function change $\Delta\phi$ provide critical information on the degree of charge redistribution upon adsorption.

This situation can be modeled as a parallel-plate capacitor, and thus the Helmholtz equation can be applied

$$\Delta V = \frac{n\mu}{\varepsilon_0} \qquad (18.28)$$

where

ΔV is the change in surface potential which is related to the work function change multiplied by electron charge, e
μ is the induced dipole moment
ε_0 is the permittivity of free space ($8.85 \times 10^{-12}\, C \cdot V^{-1} \cdot m^{-1}$)
n is the surface density of adsorbates (m^{-2})

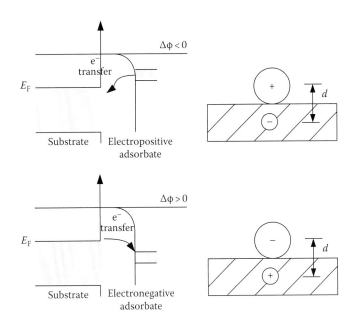

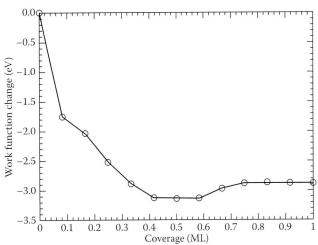

FIGURE 18.15 Work function change ($\Delta\phi$) upon Ca adsorption on a stoichiometric $TiO_2(110)$ surface as a function of surface coverage.

FIGURE 18.14 Surface dipole induced by the adsorption of electropositive and electronegative adsorbates on a solid surface.

A measurement of the work function change can be used to give an estimate of the partial charge on the adsorbate, Q, by using

$$\mu = Q^* d \qquad (18.29)$$

where d is the separation between the adsorbate and the screening charge, which usually can be obtained from Low Energy Electron Diffraction (LEED) experiments.

Depending on the net direction of charge transfer, there are two limit cases of adsorbates (Figure 18.14):

- *Electronegative adsorbates*: When the ionization energy I of the adsorbate is greater than the work function ϕ, the adsorbate withdraws electron density from the surface, becoming a negatively charged adsorbate. The resulting dipole is in the same direction as that of the clean surface, and therefore, it leads to a work function increase.
- *Electropositive adsorbates*: In this case, the ionization energy I of the adsorbate is lower than the work function of the surface and, consequently, the adsorbate donates electron density to the surface, resulting in a positively charged adsorbate. The dipole produced is in opposite direction to that of the clean surface and the work function decreases.

However, these are only general arguments. For instance, Michaelides and coworkers (Michaelides et al., 2003) demonstrated that a negatively charged N adsorbate on W(100) induces an unexpected work function decrease. They showed that a reduction in the overspill of surface electron density into the vacuum led to the unexpected work function decrease, whereas the adsorbed N still retains a negative charge.

The work function change depends strongly on the surface coverage. For an electropositive adsorbate as Ca on $TiO_2(110)$,

the work function change as a function of the coverage can be seen in Figure 18.15. Initially, a major decrease occurs, which gradually levels off and attains a minimum at a coverage of 0.4–0.6 ML. Finally, a slow increase is observed up to a coverage of 0.8 ML when a steady value is reached.

The initial rapid decrease is caused by the large Ca-surface charge transfer, yielding partially positively charged particles on the surface. As the surface coverage increases, these positive adsorbates are forced closer together, leading to repulsive lateral electrostatic interactions, which consequently leads to reduction in the degree of charge transfer and to a minimum in the curve. The slow increase at 0.8 ML is a transition where some charge originally transferred from the adsorbate to the surface is now returned to the adsorbate. The final plateau corresponds to the situation where Ca atoms behave as fully metallic particles (San Miguel et al., 2009).

18.5.2 Charge Transfer

One of the most intriguing question in cluster–substrate interactions is related to the direction followed by the charge transfer and the nature of the chemical bond between the adsorbate and the support. The bond is not usually pure ionic or covalent, but a mixture of both.

Qualitatively, the analysis of the charge density difference, $\Delta\rho$, which can be computed through the following formula, is particularly meaningful:

$$\Delta\rho = \rho(\text{system}) - \rho(\text{substrate}) - \rho(\text{cluster}) \qquad (18.30)$$

where

    ρ(system) is the total charge of the adsorption system

    ρ(substrate) and ρ(cluster) are the charge densities of the surface and the adsorbate, which are calculated on the geometries obtained from the optimization of the whole system.

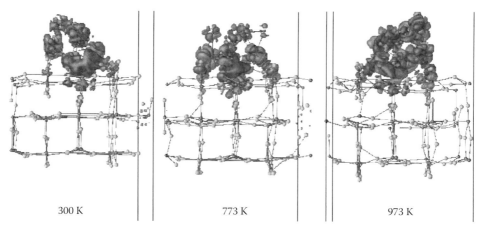

| 300 K | 773 K | 973 K |

FIGURE 18.16 Electron density difference maps for a Pd cluster deposited on a defective rutile TiO$_2$(110) surface at three temperatures (300, 773, and 973 K). Atoms are represented by small spheres. Gray and dark gray clouds show negative and positive charge transfer, respectively. (From San Miguel, M.A. et al., *Phys. Rev. Lett.*, 99, 66102, 2007. With permission.)

In this way, $\Delta\rho$ reflects how the charge density is localized in the space after adsorption process, as compared to the isolated fragments. Thus, positive values would correspond to density gain and negative values to density loss.

This analysis was done for Pd clusters on the TiO$_2$(110) surface at different temperatures (San Miguel et al., 2007). Figure 18.16 shows the electron density differences $\Delta\rho$ as isosurfaces of values of $\Delta\rho = +0.04$ e (blue) and $\Delta\rho = -0.04$ e (red). At 300 K, there is a wide blue area in the cluster, close to the support, which reveals an important polarization due to some charge transfer from the cluster to the surface. Although the magnitude of such a transfer was not quantified, it was clear that the cluster electronic structure was conditioned by the substrate. However, as temperature increases, the positive and negative areas become smaller, reflecting a stronger metallic behavior since the cluster electronic structure is no longer affected by the substrate. Particularly at the highest temperature (i.e., 1073 K), the charge is totally dispersed which indicates that the cluster exhibits mainly a metallic nature.

The dipole moment change, or usually called surface dipole D, induced by the adsorbate on the surface can be quantified from the charge density change by

$$D = \int \delta\rho(z)z\,dz \qquad (18.31)$$

where $\delta\rho$ is the in-plane averaged charge density change along the surface normal (z), defined by

$$\delta\rho(z) = \frac{1}{A}\iint \delta\rho(x,y,z)dx\,dy \qquad (18.32)$$

where
  A is the basal area of the slab
  $\delta\rho$ (x,y,z) is the charge difference electron density

In addition, there are some analytical schemes to get a quantitative estimate of the charge transfer. Maybe the most popular

is the Bader analysis (Bader, 1990), which is based on finding the critical points of charge density in order to divide the 3D space into regions assigned to the different atoms named Bader volumes. The integration of the charge density in the Bader volume leads to the charge of the associated atom. Although some caution must be taken when considering the absolute values of Bader charges, they are particularly useful when comparing different situations for the same system and allow finding general trends.

A second method is the Mulliken population analysis. It can be applied when the basis functions used in the calculation of the electronic wavefunction are centered on atoms. The charge associated with the basis functions centered on a particular atom is then assigned to that atom. The major disadvantage is that the analysis is sensitive to the choice of the basis set.

Another alternative way to estimate partial charges is by using Helmholtz equation mentioned earlier (Equation 8.28) to get the induced surface dipole. Provided that the separation between the adsorbate and the screening charge is known, the charge transferred can be estimated directly (Equation 8.29). However, this is not always the case, and a different approach would be to obtain dynamical charges by computing the value of the induced surface dipole for the adsorbate at different positions. Thus, from linear plot of these values versus the displacement from the equilibrium distance, one gets the slope which is related to the charge transferred.

18.5.3 Chemical Reactivity

Many heterogeneous catalysts consist of metal particles dispersed on metal-oxide substrates, where the metal center is the active site. The substrate is not only a neutral support, but the interactions between the clusters and the substrate define the metal–support interface where reactions often occur. It has been observed that the cluster size is a critical parameter for the chemical reactivity, and two size regimes have been determined.

The first one is referred to catalysts consisting of metal clusters with hundreds or thousands of atoms. In this case, the size-dependent reactivity is explained by the changing morphology and the varying number of surface defects. For instance, a correlation was observed between the Au cluster size and the catalytic activity for the partial oxidation of CO on Au–TiO_2(110)-(1 × 1). It was clear from different morphologies observed when varying the cluster size that the optimal activity corresponded to bilayer structures (Goodman, 2003).

The second regime includes catalysts with small metal clusters of a few atoms (<100), and it appears as a more interesting group since the selectivity and efficiency of the chemical reactions can be better controlled by modifying the size of the clusters.

As an example, Pd clusters supported on MgO have been widely studied because they catalyze the CO + NO reaction. The efficiency of the process depends significantly on the cluster size. Some experiments revealed that clusters with a number of atoms $N \le 30$ show a high efficiency at low temperature, up to 300 K. In particular, tetramers are inert, clusters up to Pd_{19} show reactivity at 300 K only, and larger clusters are reactive at temperatures as low as 140 K (Wörz et al., 2003). This non-monotonic behavior of the efficiency with cluster size is associated to the different morphologies exhibited by small clusters. This size regime includes a high variety of nanofacets of different symmetries and orientations, edges between nanofacets, corners, and overhangs at the interface with the support. DFT calculations in the size range $10 < N > 15$ showed that there is a clear transition to structures which are in very good (001) epitaxy with the substrate. In addition, bigger clusters with $N < 30$ form truncations or overhangs. The overhanging atomic rows are likely responsible for the striking efficiency of this reaction (Barcaro et al., 2007).

A second example is related to the catalytic activity of gold nanoclusters. Gold has been recognized as the most noble metal, and its chemical reactivity was thought to be negligible until Haruta and coworkers found high catalytic activity of nanosize Au clusters deposited on oxide supports, such as TiO_2, Fe_2O_3, CO_3O_4, and NiO for CO oxidation (Haruta et al., 1987). Since that finding, many experimental and theoretical investigations have been carried out, particularly for Au nanoparticles deposited on TiO_2. Some STM, scanning tunneling spectroscopy (STS), and elevated pressure reaction kinetics measurements demonstrated that the structure sensitivity of the CO oxidation reaction on Au clusters supported on TiO_2 has a quantum size effect associated to the thickness of the Au clusters. Thus, clusters of two-layer thickness, which exhibit a band gap uncharacteristic of bulk metals, are shown to be particularly suited for catalyzing the oxidation of CO (Valden et al., 1998; Mitchell et al., 2001; Howard et al., 2002).

Theoretical studies contributed from the beginning to get a better understanding of cluster–substrate interaction. According to Liu and coworkers (Liu et al., 2003), positively charged Ti at the Au–TiO_2 interface enhances electron charge transfer from Au to 2p orbitals of adsorbed O_2, resulting in O_2 molecules highly activated, and then CO oxidation occurs at the interface with a very low potential barrier. Molina and coworkers (Molina et al., 2004) showed that the presence of a deposited Au particle strongly stabilizes the adsorption of O_2, and a reasonable electronic charge

transfer takes place from Au to adsorbed O_2 molecules and to TiO_2 substrate. The O_2 can react with CO adsorbed at the interfacial perimeter of the Au particles, leading to the formation of CO_2 with a very low energy barrier.

The reader can find a broad review in which the authors compiled the experimental evidences along with the different theoretical approaches related to the oxidation of CO by gold nanoclusters supported on metal-oxides surfaces (Coquet et al., 2008). They illustrate how both methodologies are able to complement each other and unravel the details in complicated heterogeneous catalysis processes.

Very recently, experimental work on gold particles supported on a TiO_2(110) single-crystal surface has established that very small gold particles (1.4 nm) derived from 55-atom gold clusters are efficient and robust catalysts for the selective oxidation of styrene by dioxygen, and particles with diameters of 2 nm and above are completely inactive. Although this striking size threshold effect has been associated with a metal-to-insulator transition, theoretical studies would be important to provide additional insights (Turner et al., 2008).

Another recent example has been reported by Somorjai and Park (2008). They developed model nanoparticles of Pt, Rh, and a mixture of both by lithography techniques and colloid chemistry-controlled nanoparticle synthesis and investigated the activity of these particles supported on oxide surfaces in catalytic reactions of cyclohexene hydrogenation/dehydrogenation, benzene hydrogenation, and CO oxidation. They found that the catalytic properties of these supported systems are strongly correlated with the size, shape, and composition of the nanoparticles.

18.6 Summary

The deposition of metal nanoclusters on specific substrates has become a common technique in surface science, particularly in heterogeneous catalysis, to create interfaces with specific properties differing from the separated fragments. This chapter reviewed fundamental aspects related to cluster–substrate interactions, and they have been illustrated with current examples from the literature. Differences between physisorption and chemisorption processes are essential to understand these interactions because they will determine the bonding nature. Each substrate exhibits surfaces with specific adsorption sites where adsorbate will bind more favorably. The role of the presence of surface vacancies and of lateral adsorbate–adsorbate interactions is important to the final structure of the adsorption system. Experimental techniques provide some information on the adsorption processes, but they are more useful when combined with theoretical studies that can provide atomistic insights.

Thermodynamic considerations allow the final state after depositing clusters on substrates to be predicted. These considerations determine whether a chemical reaction or wetting process will take place, or even which crystal growth mode is more favorable. However, real systems use to be far from thermodynamic equilibrium and, therefore, kinetic considerations must also be taken into account. Thus, surface diffusion, nucleation, or sintering processes have been described, and how the effect of cluster size and temperature influence.

The electronic structure of cluster–substrate interfaces can be analyzed from the measurements of work function changes, which provide information about the charge reorganization upon adsorption, and from properties as electron density differences, total charges, or surface dipole that can be estimated from computational techniques. Although all the information that can be obtained is quite valuable, there are still limitations in both experimental and theoretical techniques to fully understand these complex systems. Some recent examples related to chemical reactivity have illustrated some of these achievements and difficulties.

References

Bader, R. F. W. 1990. *Atoms in Molecules: A Quantum Theory*, Oxford Science, Oxford, U.K.

Barcaro, G.; Fortunelli, A.; Rossi, G.; Nita, F.; Ferrando, R. 2007. Epitaxy, truncations, and overhangs in palladium nanoclusters adsorbed on MgO(001). *Phys. Rev. Lett.*: 98, 156101.

Barth, C.; Henry, C. R. 2003. Atomic resolution imaging of the (001) surface of UHV cleaved MgO by dynamic scanning force microscopy. *Phys. Rev. Lett.*: 91, 196102.

Campbell, C. T. 1997. Ultrathin metal films and particles on oxide surfaces: Structural, electronic and chemisorptive properties. *Surf. Sci. Rep.*: 27, 1–111.

Cheng, H. P.; Landman, U. 1993. Controlled deposition, soft landing, and glass formation in nanocluster-surface collisions. *Science*: 260, 1304.

Cleveland, C. L.; Landman, U. 1992. Dynamics of cluster-surface collisions. *Science*: 257, 355.

Coquet, R.; Howard K. L.; Willock, D. J. 2008. Theory and simulation in heterogeneous gold catalysis. *Chem. Soc. Rev.*: 37, 2046–2076.

Della Negra, M.; Nicolaisen, N. M.; Li, Z.; Møller, P. J. 2003. Study of the interactions between the overlayer and the substrate in the early stages of palladium growth on $TiO_2(110)$. *Surf. Sci.*: 540, 117.

Giordano, L.; Del Vitto, A.; Pacchioni, G.; Ferrari, A. M. 2003. CO adsorption on Rh, Pd and Ag atoms deposited on the MgO surface: A comparative ab initio study. *Surf. Sci.*: 540, 63.

Goniakowski, J.; Mottet, C. 2005. Palladium nano-clusters on the MgO(100) surface: Substrate-induced characteristics of morphology and atomic structure. *J. Cryst. Growth*: 275, 29–38.

Goniakowski, J.; Mottet, C.; Noguera, C. 2006. Non-reactive metal/oxide interfaces: From model calculations towards realistic simulations. *Phys. Stat. Sol.*: 243(11), 2516–2532.

Goodman, D. W. 2003. Model catalysts: From imagining to imaging a working surface. *J. Catal.*: 216, 213–222.

Guczi, L.; Lu, G.; Zsoldos, Z. 1993. Bimetallic catalysts: Structure and reactivity. *Catal. Today*: 17, 459–468.

Haas, G.; Menck, A.; Brune, H.; Barth, J. V.; Venables, J. A.; Kern, K. 2000. Nucleation and growth of supported clusters at defect sites: Pd/MgO(001). *Phys. Rev. B*: 61, 11105–11108.

Haruta, M.; Kobayashi, T.; Sano, H.; Yamada, N. 1987. Novel gold catalysts for the oxidation of carbon monoxide at a temperature far below 0°C. *Chem. Lett.*: 2, 405–408.

Hebenstreit, E. L. D.; Hebenstreit, W.; Diebold, U. 2000. Adsorption of sulfur on $TiO_2(110)$ studied with STM, LEED and XPS: Temperature-dependent change of adsorption site combined with O–S exchange. *Surf. Sci.*: 461, 87–97.

Hebenstreit, E. L. D.; Hebenstreit, W.; Diebold, U. 2001. Structures of sulfur on $TiO_2(110)$ determined by scanning tunneling microscopy, X-ray photoelectron spectroscopy and low-energy electron diffraction. *Surf. Sci.*: 470, 347–360.

Henkelman, G.; Jónsson, H. 2000. Improved tangent estimate in the nudged elastic band method for finding minimum energy paths and saddle points. *J. Chem. Phys.*: 113, 9978–9985.

Henkelman, G.; Uberuaga, B. P.; Jónsson, H. 2000. A climbing image nudged elastic band method for finding saddle points and minimum energy paths. *J. Chem. Phys.*: 113, 9901–9904.

Henrich E.; Cox, P. A. 1994. *The Surface Science of Metal Oxides*, Cambridge University Press, Cambridge, MA.

Henry, C. R. 1998. Surface studies of supported model catalysts. *Surf. Sci. Rep.*: 31, 231–325.

Hirth, J. P.; Pound, G. M. 1963. *Condensation and Evaporation, Progress in Materials Science*, Vol. 11, Macmillan, New York.

Howard, A.; Clark, D. N. S.; Mitchell, C. E. J.; Egdell, R. G.; Dhanak, V. R. 2002. Initial and final state effects in photoemission from Au nanoclusters on $TiO_2(110)$. *Surf. Sci.*: 518, 210–224.

Liu, Z.-P.; Gong, X.-Q.; Kohanoff, J.; Sanchez, C.; Hu, P. 2003. Catalytic role of metal oxides in gold-based catalysts: A first principles study of CO oxidation on TiO_2 supported Au. *Phys. Rev. Lett.*: 91, 266102.

Michaelides, A.; Hu, P.; Lee, M.-H.; Alavi, A.; King, D. A. 2003. Resolution of an ancient surface science anomaly: Work function change induced by N adsorption on W{100}. *Phys. Rev. Lett.*: 90, 246103.

Mitchell, C. E. J.; Howard, A.; Carney, M.; Egdell, R. G. 2001. Direct observation of behaviour of Au nanoclusters on $TiO_2(110)$ at elevated temperatures. *Surf. Sci.*: 490, 196–210.

Molina, L. M.; Rasmussen, M. D.; Hammer, B. 2004. Adsorption of O_2 and oxidation of CO at Au nanoparticles supported by $TiO_2(110)$. *J. Chem. Phys.* 120, 7673–7680.

Moseler, M.; Häkkinen, H.; Landman, U. 2002. Supported magnetic nanoclusters: Soft landing of Pd clusters on a MgO surface. *Phys. Rev. Lett.*: 89, 176103.

Nakamura, J.; Kagawa, T.; Osaka, T. 1997. Nucleation of Au on KCl(001). *Surf. Sci.*: 389, 109–115.

Parker, S. C.; Campbell, C. T. 2007. Kinetic model for sintering of supported metal particles with improved size-dependent energetics and applications to Au on $TiO_2(110)$. *Phys. Rev. B*: 75, 035430.

Rosato, V.; Guillopé, M.; Legrand, B. 1989. Thermodynamical and structural properties of f.c.c. transition metals using a simple tight-binding model. *Philos. Mag. A*: 59, 321–336.

San Miguel, M. A.; Calzado, C. J.; Sanz, J. F. 2001. Modeling alkali atoms depositions on TiO$_2$(110) surface. *J. Phys. Chem. B*: 105, 1794–1798.

San Miguel, M. A.; Oviedo, J.; Sanz, J. F. 2006. Ba adsorption on the stoichiometric and defective TiO$_2$(110) surface from first-principles calculations. *J. Phys. Chem. B*: 110, 19552–19556.

San Miguel, M. A.; Oviedo, J.; Sanz, J. 2007. Influence of temperature on the interaction between Pd clusters and the TiO$_2$(110) surface. *Phys. Rev. Lett.*: 99, 66102.

San Miguel, M. A.; Oviedo, J.; Sanz, J. F. 2009. Ca deposition on TiO(110) surfaces: Insights from ghantum calculations. *J. Phys. Chem. C*: 113, 3740–3745.

Santra, A. K.; Yang, F.; Goodman D. W. 2004. The growth of Ag–Au bimetallic nanoparticles on TiO$_2$(110). *Surf. Sci.*: 548, 324–332.

Sanz, J. F.; Márquez, A. 2007. Adsorption of Pd atoms and dimers on the TiO$_2$(110) surface: A first principles study. *J. Phys. Chem. C*: 111, 3949–3955.

Somorjai, G. A.; Park, J. Y. 2008. Molecular surface chemistry by metal single crystals and nanoparticles from vacuum to high pressure. *Chem. Soc. Rev.*: 37, 2155–2162.

Turner, M.; Golovko, V. B.; Owain, P. H.; Vaughan, O. P. H.; Abdulkin, P.; Berenguer-Murcia, A.; Tikhov, M. S.; Johnson, B. F. G.; Lambert, R. M. Selective oxidation with dioxygen by gold nanoparticle catalysts derived from 55-atom clusters. *Nature*: 2008, 981–984.

Valden, M.; Lai, X.; Goodman, D. W. 1998. Onset of catalytic activity of gold clusters on titania with the appearance of nonmetallic properties. *Science*: 281, 1647–1650.

Venables, J. A. 1987. Nucleation calculations in a pair-binding model. *Phys. Rev. B*: 36, 4153–4162.

Venables, J. A. 1994. Atomic processes in crystal growth. *Surf. Sci.*: 299, 798–817.

Venables, J. A.; Giordano, L.; Harding, J. H. 2006. Nucleation and growth on defect sites: Experiment–theory comparison for Pd/MgO(001). *J. Phys. Condens. Matter*: 18, S411–S427.

Wörz, A. S.; Judai, K.; Abbet, S.; Heiz, U. 2003. Cluster size-dependent mechanisms of the CO + NO reaction on small Pd$_n$ ($n \leq 30$) clusters on oxide surfaces. *J. Am. Chem. Soc.*: 125, 7964–7970.

Wu, M.-C.; Truong, C. M.; Goodman, D. W. 1992. Electron-energy-loss-spectroscopy studies of thermally generated defects in pure and lithium-doped MgO(100) films on Mo(100). *Phys. Rev. B*: 46, 12688–12694.

Wulff, G. 1901. Zeitschrift fur Krystallographie und Mineralogie: 34, 449.

Xu, C.; Lai, X.; Zajac, G. W.; Goodman, D. W. 1997. Scanning tunneling microscopy studies of the TiO$_2$(110) surface: Structure and the nucleation growth of Pd. *Phys. Rev. B*: 56, 13464–13482.

Xu, L.; Henkelman, G.; Campbell, C. T.; Jónsson, H. 2005. Small Pd clusters, up to the tetramer at least, are highly mobile on the MgO(100) surface. *Phys. Rev. Lett.*: 95, 146103.

Xu, L.; Henkelman, G.; Campbell, C. T.; Jónsson, H. 2006. Pd diffusion on MgO(100): The role of defects and small cluster mobility. *Surf. Sci.*: 600, 1351–1362.

Zinsmeister, G. 1966. A contribution to Frenkel's theory of condensation. *Vacuum*: 16, 529–535.

19

Energetic Cluster–Surface Collisions

Vladimir Popok
University of Gothenburg

19.1 Introduction

Atomic (or molecular) clusters are aggregates of atoms (or molecules). Their sizes vary from two or three up to tens or hundreds of thousands constituents. Medium- and large-sized clusters have diameters on the scale of nanometers, and are often called nanoparticles (NPs) or nanocrystals (depending on their structure). Clusters show properties intermediate between those of individual atoms (or molecules), with discrete energy states, and bulk matter characterized by continua or bands of states. One can say that clusters represent a distinct form of matter: a "bridge" between atoms and molecules on the one hand and solids on the other. Clusters can be formed by most of the elements in the periodic table. They can be of different types, compositions, and structures. A wide variety of clusters has been produced and investigated from precursors including metals, semiconductors, ionic solids, noble gases, and molecules. More detailed information about the classification of clusters, their bonding types, structures, and properties in the gas phase goes beyond the scope of this chapter and can be found elsewhere (Haberland 1994, Martin 1996, Johnston 2002, Alonso 2005, Baletto and Ferrado 2005).

Interest in clusters comes from various fields. Clusters are used as models to investigate the fundamental physical aspects of the above-mentioned transition from atomic scale to bulk material. They can also be used as a bridge across the disciplines of physics and chemistry to understand the nonmonotonic variations of properties and unusual phenomena of nanoscale objects (Jena and Castleman 2006). Clusters on surfaces define a new class of systems highly relevant for practical applications. Finite size effects can lead to electronic, optical, magnetic, chemical, and other properties that are quite different from those of molecules or condensed matter and that are of great interest for practical applications in areas such as catalysis; electronics and nanotechnologies; bio-compatible, magnetic, and optical materials, etc. (Kreibig and Vollmer 1995, Meiwes-Broer 2000, Binns 2001, Palmer et al. 2003, Binns et al. 2005, Roduner 2006, Wegner et al. 2006, Woodruff 2007). Utilizing cluster beam technology (Milani and Iannotta 1999, Popok and Campbell 2006), one can control the cluster or nanoparticle (NP) size, its impact energy with surface and, to a certain extent, the spatial distribution of the deposited NPs on a surface, for example, by preliminary processing or functionalization (Bardotti et al. 2002, Corso et al. 2004, Queitsch et al. 2007). With clusters consisting of thousands of atoms, it is possible to transport and locally deposit a large amount of material, providing an advanced method for the growth of thin films that can either be porous or very compact and smooth depending on the energy regime used for cluster impact (Haberland et al. 1993, 1995, Paillard et al. 1995, Milani et al. 1997, Qiang et al. 1998). So-called cluster-assisted deposition, when the depositing material is bombarded by low-energy clusters, allows to control the structure and composition of the grown layers, for instance, to fabricate the thin and hard diamond-like carbon films (Kitagawa et al. 2003). Low-energy cluster implantation is found to be an efficient tool for ultrashallow junction formation and infusion doping of shallow layers (Yamada et al. 1997, Cvikl et al. 1998, Borland et al. 2004). Energetic cluster beams can also be used as very efficient tools for the processing of surfaces (dry etching and cleaning) or improving their surface topology (smoothing) (Gspann 1997, Yamada et al. 2001, Yamada and Toyoda 2007).

The current state of the art in the field of energetic cluster–surface interactions is presented below. However, before concentrating on the fundamental physics of cluster–solid collisions and practical applications of cluster beams for the modification of surfaces and the building of nanostructures, brief surveys of the physical principles of cluster formation and the history of cluster beams are given.

19.2 Brief History of Cluster Beam Development

The first mention of cluster beam production and investigation occurred in the 1950s (Becker et al. 1956). The possibility to separate small clusters of hydrogen, nitrogen, and argon from non-condensed residual gas and transfer them into a high vacuum was shown. At the same time, it was demonstrated with CO_2 and H_2 that cluster beams can be ionized by electron bombardment enabling mass spectra to be obtained (Henkes 1961, 1962). These experiments were followed by investigations of the cluster size distribution in a beam of $(CO_2)_n^+$, depending on conditions of the cluster source (Bauchert and Hagena 1965) and by the development of methods to evaluate cluster sizes, for example, by scattering of a potassium atomic beam passing through a nitrogen cluster beam (Burghoff and Gspann 1967). In the 1970s, the cluster technique underwent further development and improvement in order to get more stable and controllable beams (Hagena and Obert 1972) as well as expanding the spectrum of species used to produce clusters. In particular, gas aggregation (vaporization) sources for the production of metal and semiconductor clusters were developed (Hogg and Silbernagel 1974, Takagi et al. 1976, Kimoto and Nishida 1977). However, the first published results on the deposition of Si, Au, and Cu suffer from a lack of confirmation concerning the cluster-to-monomer ratio in the beams (Takagi et al. 1976). About the same time, the first cluster implantation experiments were performed, and the stopping of swift proton clusters in carbon and gold foils was studied (Brandt et al. 1974). Further development of the technique in the 1980s provided more controllable parameters of the beams and showed the applicability of ionized clusters for the synthesis of thin metal films and heterostructures (Yamada and Takagi 1981, Yamada et al. 1986). A source utilizing laser ablation was invented that allowed to extend cluster production over practically any solid material including those with high melting points (Smalley 1983). Development of this source and the study of carbon clusters led to the discovery of fullerenes in 1985 (Kroto et al. 1985).

In the 1990s, new methods of cluster formation utilizing arc discharge, sprays, ion, and magnetron sputtering were introduced together with further development of techniques for cluster beam control, manipulation, and characterization (de Heer 1993, Haberland 1994, Milani and Iannotta 1999). Progress in cluster beam techniques together with the beginning of the

"nanoscience era" stimulated a significant increase of interest in research on both free clusters in the gas phase and deposited (supported) clusters. At the same time, a systematic experimental work corroborated by molecular dynamics (MD) simulations has started to obtain a clear picture of the physical background of energetic cluster–surface interactions. The milestones in this research field are reviewed below.

19.3 Formation of Cluster Beams

19.3.1 Fundamental Aspects of Cluster Nucleation and Growth

The probability of spontaneous cluster formation under equilibrium conditions is extremely low. Cluster production requires a thermodynamic nonequilibrium that can be implemented by means of a cluster source that can be of different types (see below). In all cluster sources, cluster generation consists of the following stages: vaporization (the production of atoms or molecules in the gas phase); nucleation (the initial condensation of atoms or molecules to form a cluster nucleus); growth (the addition of more atoms or molecules to the nucleus); coalescence (the merging of small clusters to form larger ones); and evaporation (the loss of one or more atoms) (Kappes and Leutwyler 1988, de Heer 1993).

If the local thermal energy or temperature of the gas consisting of the monomer species is less than the binding energy of the dimer, then a three-atom collision can lead to stable dimer formation. Three atoms are necessary for the fulfillment of energy and momentum conservation:

$$A + A + A \rightarrow A_2 + A, \tag{19.1}$$

where the third atom (A on the right-hand side of the equation) removes the excess energy. To make the nucleation step more efficient, an inert carrier (cooling) gas is often injected into the nucleation chamber of a cluster source. Once the dimer is formed it acts as a condensation nucleus for further cluster growth. Early growth occurs by incorporation of atoms (or molecules) one at a time. Subsequently, collisions between smaller clusters can lead to coalescence and the formation of larger clusters.

In the cluster growth region, the clusters are generally hot, because their growth is an exothermic process, i.e., the internal energy increases due to the heat of condensation of the added atoms. Since the clusters are hot, there is competition between growth and decay. For practical reasons, i.e., the formation of a stable cluster beam, it is often necessary to lower the temperature of the clusters. A few mechanisms can be realized.

Cooling under adiabatic expansion. This mechanism works simultaneously with the cluster formation in the case of supersonic nozzle sources (see next section). A gas under high stagnation pressure is expanded into a vacuum chamber through a nozzle: an abrupt decrease of pressure leads to a drastic temperature decrease in the beam causing supersaturation and, finally, cluster formation (Haberland 1994).

Collisional cooling. Collisions with other atoms in the beam remove the excess energy from the clusters as kinetic energy:

$$A_n(E_1) + B(\epsilon_1) \rightarrow A_n(E_2 < E_1) + B(\epsilon_2 > \epsilon_1) \qquad (19.2)$$

where B may be a single atom of element A constituting the cluster or an inert cold carrier gas, which is more common. E is the internal energy of the cluster species and ϵ is the kinetic energy of atom B. This cooling mechanism is only significant in the initial expansion and condensation regions.

Evaporative cooling. Clusters can lower their internal energy by evaporation, losing one or more atoms in an endothermal desorption process. The internal energy is channeled statistically into the appropriate cluster vibration mode, in order to overcome the activation barrier for bond breaking. After evaporation, excess energy is imparted as kinetic energy to the escaping atom and the daughter cluster:

$$A_n(E_1) \rightarrow A_{n-1}(E_2 < E_1) + A(\epsilon_1) \rightarrow A_{n-2}(E_3 < E_2) + A(\epsilon_2) \rightarrow \cdots$$
$$(19.3)$$

This is the main cooling mechanism once free flight of the cluster has been achieved and there are no further collisions.

Radiative cooling. Clusters can also lower their internal energy by emitting radiation:

$$A_n(E_1) \rightarrow A_n(E_2 < E_1) + h\nu \qquad (19.4)$$

However, radiative cooling is an inefficient cooling mechanism, which is slow compared to the time scale of typical cluster experiments (μs). Electron emission can be an additional channel for cluster cooling.

19.3.2 Cluster Sources

As mentioned above, cluster nucleation requires a thermodynamic nonequilibrium condition that can be realized by means of special equipment—a cluster source. There are a number of approaches for cluster beam formation, see for example (Kappes and Leutwyler 1988, Hagena 1992, de Heer 1993, Haberland 1994, Milani and Iannotta 1999, Pauly 2000, Wegner et al. 2006). Although a classification of the methods to produce clusters is somewhat arbitrary today, because often combinations of two or more methods are used, here the most common approaches are briefly reviewed, namely, gas aggregation, supersonic jet, surface erosion, and sprays.

19.3.2.1 Effusive and Gas Aggregation Sources

A Knudsen cell, which is based on the thermal vaporization of liquids or solids in an oven, is one of the simplest ways to produce small clusters. Since the vapor is kept in equilibrium in the oven there is a low probability for clusters to nucleate. The beam of atoms and clusters is formed by effusion from the oven through a nozzle into a low-pressure chamber. The intensity of the beam falls exponentially with cluster size increase. Hence, a Knudsen cell can produce a low flux continuous beam of small (few atoms in size) clusters.

In most cases, however, larger clusters and higher beam intensities are required for research-oriented or practical applications. The gas aggregation method, in which a solid or liquid is evaporated into a carrier gas and the atoms and molecules are collisionally cooled forming clusters, is a more advanced approach (Sattler et al. 1980). A smoking fire or cloud and fog formation in nature are good examples of gas aggregation, therefore, this type of source is also called a "smoking source." After the aggregation, the clusters expand through a nozzle into the next vacuum chamber forming a subsonic beam. Gas-aggregation cluster sources produce continuous beams of elements with not very high melting points (<2000 K), usually metals. They are also used to produce composite clusters where one of the components is a metal. The cluster size distribution depends on the physical dimensions of the cluster cell and can reach many thousands of atoms (Abe et al. 1980, Martin et al. 1990). This type of source has considerable flux and it is useful for the controllable deposition of size-selected metal clusters and nanostructured thin film growth (Goldby et al. 1997).

19.3.2.2 Supersonic Jet Sources

These sources exploit the principle of adiabatic expansion of a gas into vacuum. Supersonic nozzle sources are of two types. The sources for cluster generation from gaseous precursors typically operate without a carrier gas. The precursor gas for cluster formation is supplied under high stagnation pressure (from a few up to a few tens of bar) in the pre-expansion reservoir. Seeded supersonic nozzle sources are used to form cluster beams of relatively low-melting-point metals (or liquids), which are vaporized, and the vapor is seeded in the carrier gas at a stagnation pressure of several bars (similar to gas aggregation sources). In both cases, either the pure gas or the metal/carrier gas mixture is then expanded through a small-diameter nozzle into vacuum, thereby creating a supersonic cluster beam. More details on the free-jet expansion phenomenon can be found elsewhere (Hagena 1981, 1987, Pauly 2000). Three main factors govern the cluster formation: the stagnation pressure, the temperature, and the nozzle characteristics (Hagena 1981, Milani and Iannotta 1999, Pauly 2000). These sources produce intense continuous or pulsed cluster beams with narrow velocity distributions and clusters typically ranging from tens to thousands of atoms depending on the source parameters. In some cases, clusters can reach over 100,000 atoms in size and the source can provide microampere current in the beam that is of practical importance (Seki et al. 2003).

19.3.2.3 Surface Erosion Sources

The *surface erosion sources* utilize various methods to generate clusters using the removal of atoms from a surface by heavy-ion or magnetron sputtering, laser ablation, and high electric fields (arc discharge).

In *ion sputtering sources* the cluster ions are produced by bombarding a surface with heavy ions. Ion sputtering sources can be used to produce clusters of a wide range of materials, especially those with high melting points. In contrast to the previously discussed sources, the sputtering source does not rely on condensation in an inert gas. For cluster production, typically 10–20 keV ion beams of either heavy gases (Kr or Xe) (Fayet et al. 1986, Leisner et al. 1999) or Cs are used. Relatively small (tens of atoms) positively and negatively charged clusters are formed using inert gas sputtering. For the cesium ion sputtering source, predominantly negatively charged clusters are formed, which is related to a minimization of the work function of the sputtered material by the influence of Cs (Alton 1986, Ishikawa 2002). Typical cluster sizes range from a few up to 20–30 atoms depending on the cluster species (Hall et al. 1997, Wang et al. 2002).

A *magnetron sputtering source* is based on the use of a plasma that is ignited in an inert gas (usually argon) over a target surface by applying d.c. or r.f. potentials. The cluster condensation is facilitated by a carrier gas. The sizes can vary between 50 and over 10^6 atoms; from 20% to 80% of the clusters (depending on species) are ionized (Haberland et al. 1992, 1994). Similar to the ion sputtering source, the magnetron source is applicable to a large range of materials. It can provide a rather high cluster beam flux of up to 10^{12} cluster/scm^2 (Haberland et al. 1992). A combination of two magnetrons, operating in either d.c. or r.f. mode, in one source allows to improve conditions for the production of clusters from either metals or semiconductors (or insulators) as well as providing appropriate regimes for the formation of core-shell NPs (Sumiyama et al. 2005). Recently, a hybrid magnetron sputtering/condensation source combined with a novel time-of-flight (TOF) mass filter was developed to produce intense cluster beams with a constant mass resolution over a broad range of cluster sizes (from 2 to 70,000 atoms) (Pratontep et al. 2005).

Laser ablation (vaporization) source is used to produce clusters of metals and semiconductors, usually those with high melting points that cannot be evaporated by simple heating. In this source, the vapor is formed by pulsed-laser ablation of the material (Dietz et al. 1981, Powers et al. 1982). Typically, beam from a solid state or excimer laser is focused on the rod or plate that is driven in a slow motion so that a fresh area of its surface continues to be exposed to the laser. Laser ablation is typically combined with gas-assisted cooling (Smalley 1983, Milani and de Heer 1990). The vaporized material (plume of plasma with temperature of around 10^4 K) is entrained in a pulse of the carrier gas; the vapor is cooled causing cluster formation. The gas/cluster mixture is ejected out of the nozzle forming a beam of typically supersonic velocity. The use of a laser for cluster generation also leads to partial cluster ionization, so this source generates neutral and charged clusters. The cluster size distribution depends on the source conditions and typically ranges up to few hundred atoms per cluster, see for instance (Saito et al. 2001). One of the advantages of the laser vaporization source is that it is an easy way to produce binary clusters (consisting of two different chemical elements) using binary alloy targets of desired composition (Rousset et al. 1995, Wagner et al. 1997) or a dual-target configuration

(Nonose et al. 1990) or even a dual-target dual-laser technique (Bouwen et al. 2000). One of the disadvantages is that the intensity of the cluster beam is typically not high.

One more type in the family of erosion sources is a *pulsed-arc (arc discharge) cluster ion source* (Ganteför et al. 1990), where material is vaporized from the surfaces of a cathode and anode between which an intense electrical discharge is ignited. The vapor is cooled down by means of a carrier gas similar to the above cases. Typically, about 10% of the emitted material is charged. In general, the cluster beams produced by such a source tend to be more intensive compared to the laser vaporization source. A cluster deposition rate of about 5 nm/min over an area of ca. 48 mm^2 has been reported for a so-called pulsed microplasma cluster source representing a modified arc discharge source with elements of sputtering sources (Barborini et al. 1999).

19.3.2.4 Spray Sources

A few more types of cluster sources exist. *Spray sources* are used for generating clusters from liquids and solutions. There are two approaches to make cluster beams: electrospray and thermospray sources. The *electrospray ionization source* generates solvated ion clusters by the injection of a solution through a needle, which can carry a positive or negative potential, into a stagnation chamber with a flowing inert gas. This method allows complex involatile molecules to be obtained in the gas phase as solvated ions. Electrospray ionization was introduced in the 1960s (Dole et al. 1968) but was only used in the 1990s for the production of cluster ions, for instance, alkali metal chloride and sodium salt clusters (see references in Hao et al. (2001)) or water-methanol, water-glycerol, and pure glycerol clusters (Mahoney et al. 1994). In the case of the *thermospray source*, the liquid sample is partly pyrolyzed before expansion and a subsonic beam containing neutral and charged clusters is generated. A similar operating principle to the electrospray source can be used for a *liquid-metal ion source*, which is primarily used to produce singly or multiply charged clusters of low-melting-point metals (Dixon et al. 1981, Helm and Möller 1983, Bhaskar et al. 1990). A fine needle under potential is wetted with the metal heated above its melting point. A high electric field at the tip of the needle causes a spray of very small droplets to be emitted. The initially very hot and often multiply ionized droplets undergo evaporative cooling and fission to smaller sizes.

19.3.3 Mass Selection of Clusters

It should be obvious from the above sections that clusters are produced in a range of sizes. This size (or mass) distribution can be influenced by the method with which the precursor gas vapor is generated and depends on a number of factors: the initial pressure and temperature; the presence of carrier gas and its parameters; the vacuum conditions; and, of course, the variety of technical parameters characterizing the cluster source, especially the expansion conditions.

In many experiments and applications, there is a necessity to have clusters of selected sizes or narrow interval of sizes. Thus, clusters must be separated by size (mass). For beams of neutral clusters there is only a possibility of rough size selection, i.e., relatively small clusters can be separated from relatively large clusters. This can be achieved by a system of aerodynamic lenses (Liu et al. 1995) or by the so-called focusing nozzle assembly (Piseri et al. 2001). In both cases, the drag action of the carrier gas separates the NP trajectories and only the particles of proper size pass the system.

To achieve better precision in size selection the clusters must be ionized and then separated by either deflecting the cluster ions in an electromagnetic field or by the TOF method. Depending on the cluster species or the object of the experiment, either cations or anions may be created. Positively charged clusters are used more often. There are a few methods to generate cations: by *electron impact ionization* (either thermal electrons or a focused electron beam interact with clusters), *photoionization* (by laser or other intense light source), and *electric discharge*. To generate anionic clusters, *electron transfer* can be employed. Electron attachment can occur via, for instance, collisions with alkali metal atoms in the vapor phase.

A range of mass separation techniques are available for the charged clusters. Most commonly used are mass spectrometers such as Wien filter, quadrupole, and TOF. Exhaustive reviews on these and some other mass-selecting techniques can be found in de Heer (1993), Haberland (1994), Milani and Iannotta (1999), and Binns (2001). The typical mass resolution, $\Delta m/m$, of the most common mass spectrometers lies within the range of 10^{-4} to 10^{-2}. This makes it possible to select the size with a precision of one atom for small- and medium-sized (from few to few hundreds of atoms) clusters.

19.4 Energetic Cluster–Surface Interaction

One of the important parameters in the application of cluster beams is the impact (or kinetic) energy. By varying the cluster kinetic energy one can develop methods for the synthesis and modification of materials. A process can be considered to be low-energy when the kinetic energy per atom of the accelerated cluster is below the binding (cohesive) energy of the cluster constituents (which is typically below or on the level of eV/atom). This case is often called soft landing (Figure 19.1a). The deposition does not induce cluster fragmentation, i.e., the clusters preserve their composition. At the same time, the structure can be distorted, especially if the kinetic energy is close to the cohesive energy. If the cluster kinetic energy per atom exceeds the cohesive energy the impact is considered to be energetic. Under energetic impact the cluster is fragmented and it can also cause damage of the substrate if the kinetic energy is high enough (Figure 19.1b). In this review, the cases of energetic cluster impact are mostly under the consideration. However, for a better understanding of the cluster–surface interaction effects a short description of the soft-landing (cluster deposition) regime is presented below.

19.4.1 Cluster Deposition

Soft landing is normally used to grow porous films, produce special surface relief by cluster assembly, and form optical and magnetic nanostructures (Perez et al. 1997, Milani and Iannotta 1999, Binns 2001, Bansmann et al. 2005). The main problem of soft landing, in terms of some practical applications, is the high surface diffusive mobility of the deposited clusters, leading to either the growth of larger particles (by the fusion of small clusters) or cluster coalescence (without fusion) into islands. Coagulation and coalescence ruin the advantage of mass selection of clusters in order to obtain a narrow size distribution of the deposited NPs. One of the widely used solutions to prevent these processes is co-deposition of clusters and host material, i.e., the encapsulation of clusters into some medium, or in other words the formation of nanoparticulate films. One more possibility to prevent coagulation or coalescence is surface functionalization. For instance, initial processing of a surface by focused ion irradiation makes possible to produce regular arrays of surface defects by which deposited metal clusters can be organized on

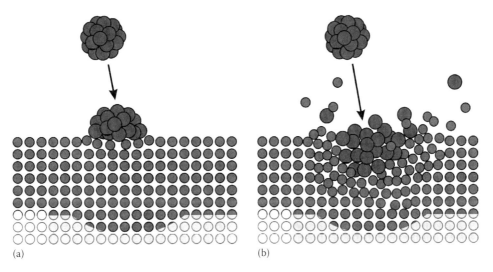

(a) (b)

FIGURE 19.1 Schematic picture of (a) cluster soft landing and (b) cluster energetic impact.

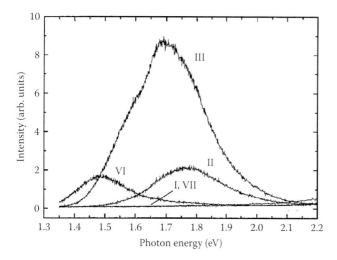

FIGURE 19.2 Photoluminescence spectra of thin films produced by size-selected deposition of silicon nanocrystals. The 488 nm line of an argon-ion laser was used for excitation. Mean diameters of the nanocrystals are 2.47 nm (I), 3.65 nm (II), 3.92 nm (III), 4.95 nm (VI), and 6.37 nm (VII). (Reprinted from Ehbrecht, M. et al., *Phys. Rev. B*, 56, 6958, 1997. With permission.)

the surface (Bardotti et al. 2002). The same effect can be achieved by the preparation of 2D templates, for example, made of proteins or boron nitride (Corso et al. 2004, Queitsch et al. 2007).

In the case of noble metal or copper NPs, the films exhibit specific absorption bands in the near-UV, visible, or near-IR regions, depending on the cluster species, due to the collective excitation of conduction electrons, the so-called surface plasmon resonance (Kreibig and Vollmer 1995). The position, shape, and intensity of such absorption bands strongly depend on the size distribution, shape, and volumetric fraction of the metallic inclusions. In this context, the cluster beam technique offers some unique possibilities for controlling these parameters and synthesizing composites with variable optical properties (Palpant et al. 1998, Gartz et al. 1999, Khabari et al. 2003). Size-controlled granular magnetic films and films of magnetic NPs embedded in various matrices can be fabricated using cluster beams of transition metals (Co, Fe, Ni, etc.) (Peng et al. 2000, Bansmann et al. 2005, Binns et al. 2005). This research direction is of importance

for applications in high-density memory devices, magnetic sensing, and spintronics.

The cluster deposition approach also became of increasing interest with intensive development of submicron and nanoelectronics. For example, growth of good-quality Ge layers with a thickness of 100–400 nm on a Si substrate at a temperature of 500°C, which is lower than the critical temperature of epitaxial growth by molecular beams or chemical vapor deposition (CVD), was reported using a germanium cluster beam at supersonic velocities (Xu et al. 2002). Silicon films with tunable photoluminescence properties were synthesized by low-energy deposition of size-selected Si clusters (Figure 19.2): the obtained characteristics were well described in terms of zero-dimensional quantum dots (Ehbrecht et al. 1997). The deposition of pure carbon clusters from a supersonic beam provided experimental evidence for the possibility of producing a carbyne-rich pure carbon solid (Ravagnan et al. 2002): before this publication, the existence of carbyne was strongly debated (Heimann et al. 1999). By the deposition of neutral carbon clusters it was possible to grow nanostructured graphite-like films and patterns of three-dimensional objects using a mask (Figure 19.3) (Milani et al. 1997, Barborini et al. 2000). This method of formation of patterned carbon dots on a silicon surface followed by thermal annealing was later employed to obtain arrays of SiC dots (Magnano et al. 2003). The deposition of various metal oxide clusters through a mask is found to be an advanced technology for the parallel fabrication of microsize chemical sensors (Barborini et al. 2008). The use of cluster-assembled TiO$_2$ films, which demonstrate good adsorption of different macromolecules such as DNA, proteins, and peptides, is suggested for integration of cell structures on micro- and nanodevices (Carbone et al. 2006). Thin films of non-percolating Pd clusters deposited on SiO$_2$ substrates showed change in resistance on exposure to hydrogen that suggested their use as gas sensors (Van Lith et al. 2008). The cluster beam deposition technique was also found to be an efficient and powerful tool for the fabrication of organic (polymer) thin films and various inorganic–organic complex nanometer-scale functional structures, see for example the review by Gao et al. (1998). Thus, the high deposition rates as well as the control of the energy and cluster mass distribution makes the cluster beam

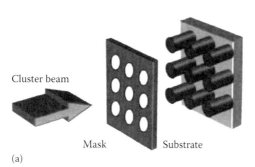

Cluster beam

Mask Substrate

(a)

(b)

FIGURE 19.3 (a) Schematic representation of cluster beam deposition through a stencil mask. (b) SEM image of a pattern of pillars obtained with a round hole grid with 850 mesh and holes of 20 mm of diameter. (Reprinted from Barborini, E. et al., *Appl. Phys. Lett.*, 77, 1059, 2000. With permission.)

technique competitive with other synthetic routes for materials where a well-defined morphology and structure is required. For more details concerning the physics of soft-landing case one can look elsewhere (Perez et al. 1997, Jensen 1999, Seeger and Palmer 2000, Binns 2001).

19.4.2 Energetic Deposition and Pinning of Clusters

Under energetic impact the cluster loses its structure. The higher the cluster energy, the more the cluster is deformed on surface impact. However, the cluster constituents penetrate into the substrate only if their energy is higher than the penetration threshold energy. Before reaching this value the cluster either breaks up and scatters or flattens on impact. After reaching the penetration threshold energy, which is typically higher than the cohesive energy, the cluster constituents become embedded and mixed with the substrate material. For instance, for Mo_{1043} clusters deposited on a Mo (001) surface, MD simulations showed that clusters with an energy of 1 eV/atom grew a quite dense (up to 80% of the bulk) epitaxial film in which cluster and substrate atoms are mixed just within a distance of one lattice constant (Haberland et al. 1995). Further increase of the cluster energy up to 10 eV/atom led to the complete decomposition of the clusters and the formation of a denser film with better adhesion to the substrate because of the strong intermixing between cluster and substrate material within a few atomic layers. The formation of thin dense films was favored by the high density of energy deposited by the clusters. Despite the quite low energy per atom, the multiplication of this relatively low energy by the number of atoms and division by the very small surface collision area leads to high values of the energy density. It was also simulated that energetic cluster deposition suppresses the surface roughness of the deposited film due to a so-called "downhill particle current" transporting hills of the film into valleys (Moseler et al. 1998). This intermixing regime was experimentally demonstrated by the deposition of very smooth (roughness of 0.7–1.5 nm), dense, and strongly adhering coatings on Si, quartz, and steel substrates using Al_n^+, $(TiN)_n^+$, and $(TiAlN)_n^+$ cluster ions with energies of 5–20 eV/atom (Qiang et al. 1998).

One of the interesting boundary cases (between soft landing and implantation) is so-called pinning when the energetic cluster disrupts the substrate lattice, such that a few atoms from the surface layer are displaced and some of the cluster atoms become shallow embedded (Carrol et al. 1996, 2000). The pinning process suppresses the cluster diffusion on the surface by binding the cluster to the defects produced on the impact. It was experimentally found, for example, for the case of Ag_n^+ cluster ions ($n = 50$–200) implanted into graphite that the pinning threshold energy was about 10 eV/atom (Carrol et al. 2000). Good agreement between the experiments and MD simulations for this case was achieved (Figure 19.4). Recently, the pinning regime was studied for size-selected clusters of Au, Pd, Ni, and Co on graphite (Di Vece et al. 2005, Gibilisco et al. 2006, Vučković et al. in press). The pinning threshold energy was found to be dependent

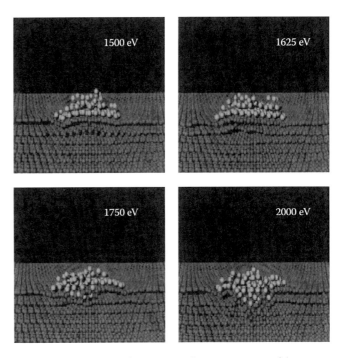

FIGURE 19.4 MD simulations, viewed in cross section, of the impact of Ag_{147} cluster at various energies (indicated in the panels) with graphite. (Reprinted from Carrol, S.J. et al., *J. Chem. Phys.*, 113, 7723, 2000. With permission.)

on the cluster species and target material (atomic mass of composing chemical elements)

$$E_{pin} = nM_C \frac{E}{4M} \qquad (19.5)$$

where
  n is the number of atoms in the cluster
  M_C is the atomic mass of cluster constituents
  M_T is the mass of the target atoms
  E_T is the energy required to produce a recoil atom (Carrol et al. 2000)

Pinning energy on graphite varies between ca. 6 and 20 eV/atom from Co and Ni to Au. Recent studies have shown that the threshold value also changes (slightly decreases) with increase of the cluster size (Di Vece et al. 2005, Gibilisco et al. 2006). Possible explanations can be found through a consideration of the lattice constant, which is by itself a manifestation of the interatomic bonding in the cluster that influences the plastic deformation of the cluster upon impact (Smith et al. 2006). The lattice constant becomes important for large cluster sizes. One more possibility lies in considering the number of defects (displaced substrate atoms) produced by the cluster impact. With an increase in size, the cluster becomes more efficient in damage production, i.e., in creating a number of potential binding sites for itself.

It should be noted that the obtained pinning threshold energies for metal clusters impacting graphite are close to those for the above-mentioned formation of thin adhesive metal films by energetic cluster deposition. In both cases, atoms of the clusters are able to penetrate into the substrate matrix with an energy which is significantly lower than the penetration threshold energies for any monomer ions implanted into different target materials. For example, it is known that for graphite the penetration threshold increases with the ion radii and the minimum value was found to be 22.5 eV for light He$^+$ ions (Marton et al. 1995). This is, therefore, one of the clear cases where the cluster–surface impact is different from monomer ion impact. And the reasons for that will be discussed below.

Pinning is an efficient method for the deposition and immobilization of size-selected clusters. The pinned clusters in turn can have a number of potential applications. One of them is the immobilization of protein molecules for their further investigation or the fabrication of biochips (Palmer et al. 2003, Palmer and Leung 2007). Pinned clusters can also be suggested for the fabrication of various nanoscale gas sensors (Van Lith et al. 2008) and as catalysts of controllable size, for instance, to grow carbon nanotubes.

19.4.3 Implantation of Clusters

Due to the large size and weak bonding between atoms in clusters, their implantation is fundamentally different from that of the monomer ion implantation (Takagi 1998). Clusters generate multiple-collision effects during the penetration into the target. The effect of the high density of the energy transferred from the cluster to the target at the beginning of the impact can be compared with a microexplosion or to some extent with a nanoscopic analogue of a meteorite–planet collision that typically results in crater formation (Averback and Ghally 1994). One more peculiarity of the cluster–solid interaction is nonlinearity, which arises from the fact that the cluster atoms influence each other during the penetration into the target and thus the environment in the material is different for each atom or ion of the cluster. Nowadays there is no commonly accepted theory satisfactorily describing the cluster implantation process. However, a critical analysis of the published data on the subject is presented below.

It was shown by MD simulations of the implantation of various cluster species that the clusters break down into single atoms quite rapidly under impact, after some tens or hundreds of femtoseconds depending on the cluster size, energy, and target material (Sanz-Navarro et al. 2002, Peltola and Nordlund 2003). For slow cluster constituents (from tens of eV to tens of keV per atom) the energy loss occurs predominantly due to elastic collisions between the atoms (nuclear stopping). Hence, one can expect an overlapping of the collision cascades originated from the individual cluster atoms during their penetration into the target. Thus, the difference between the stopping of an atom in a cluster and an individual atom makes a difference to the projected range R_p of cluster constituents and hence to the produced radiation damage compared to monomer ion implantation. MD and Monte Carlo simulations showed that the penetration depth of clusters is larger than that of the corresponding ions at the same incident velocity (Yamamura 1988, Shulga and Sigmund 1990, Yamamura

and Muramoto 1994). It was suggested that a so-called "clearing-the-way effect," where the "front" atoms of the cluster push target atoms out of the way, could take place. As a result of the "clearing-the-way," the stopping power of the cluster is reduced and the projected range is increased. Heavier ions would thus be expected to cause more clearing of "light" targets (Shulga et al. 1989). On the contrary, the effect is negligible if the mass of the cluster constituents is much smaller than the mass of the target atoms, as was shown, for instance, for the simulations of deuterium cluster implantation (clusters of up to 500 atoms with energy of 200 eV/atom) into a silver target (Shulga et al. 1989).

Up to now, the different simulations and experiments showed different scaling laws for cluster implantation. For instance, in the case of Ar$_n$ ($n = 43$ and 688) cluster implantation into silicon with energies E between 10 and 100 eV/atom, it is calculated that $R_p \sim E^{1/3}$, and the penetration depth depends rather on the total cluster energy than on the cluster size (Aoki 2000). For smaller clusters, for instance, Ar$_6$ and Ar$_{13}$, the R_p shows a very similar dependence on energy as for the monomer ions. Modeling of the implantation of small Au$_n$ ($n = 2$–7) clusters into Cu with energies of 1–10 keV/atom showed a slight increase of R_p with the cluster size at same energy per cluster atom (Peltola and Nordlund 2003). Calculations of the implantation of Si$_n$ clusters ($n \leq 50$) into Si with a constant energy of 70 eV/atom showed a more significant dependence of R_p on the cluster size, the penetration depth scaled approximately as $n^{1/3}$ (Gilmer et al. 1996). The dependence of R_p on the implantation energy gave a scaling law $\sim E^{1/2}$, which is different from the above-mentioned $E^{1/3}$ predicted for the Ar$_n$ clusters. This discrepancy can be related to the difference in the "clearing-the-way" effect for different cluster species. Simulations of Au$_n$ ($n = 1$, 13, 43, 87, 201, 402) cluster implantation into gold and graphite targets with an energy of 100 eV/atom showed a clear effect of the cluster size on the ranges, and the dependence followed a power law $R_p = an^\alpha$ with α varying from 0.31 to 0.45, when changing from gold to graphite (Anders and Urbassek 2005). The value of α obtained for the gold target is in agreement with the value for the implantation of Si$_n$ clusters into Si. The difference in α for the gold clusters implanted into graphite is easily explained in terms of more "clearing" for targets composed of "light" atoms.

The "clearing-the-way" effect was experimentally found in the case of small Ta$_n^+$ clusters ($n = 2$, 4, 9) implanted into graphite with an energy of 555 eV/atom; however no scaling dependence was suggested (Reimann et al. 1998). Experiments on the implantation of polyatomic (up to 4) boron ions into silicon also showed an increase in both the projected range and straggling in the case of clusters compared to monomers for about 20% and 30%, respectively (Liang and Han 2005). For C$_{60}^+$ ion implantation into graphite within a rather wide range of energies from 500 to 23 keV the penetration depth was found to follow $E^{1/2}$ (Webb et al. 1997). The "clearing-the-way" effect was also demonstrated for graphite bombarded by Ag$_n$ clusters. However, the estimated implantation ranges were found to scale proportionally to the cluster momentum (Pratontep et al. 2003, Seminara et al. 2004).

Cluster ranges were also found to be affected by the high density of energy deposited by the cluster due to thermal spikes (Kelly 1977, Miotello and Kelly 1997). In the case of keV-energy cluster implantation, the spike originates via nuclear stopping of the projectiles, i.e., by energy deposited in ion–atom and atom–atom collisions during the ballistic (or dynamic) phase of the collision cascades. The impacted area experiences both high temperature and pressure transients. Local temperature and pressure can rise up to 10^4–10^5 K and the GPa level, respectively, for the first 10^{-13} to 10^{-12} s (Cleveland and Landman 1992, Colla et al. 2000, Allen et al. 2002a,b). Interestingly, the temperature and pressure values are comparable to those obtained for meteorite–planet impacts (Melosh 1989). Violent interaction under the cluster impact leads to local melting of the target material. The calculated effective radius of the molten regions, for example, in Cu was about 2 nm per cascade (Hsieh et al. 1989). MD simulations showed an increase in the range straggling, ΔR_p (up to 130%), of Au_n clusters implanted into Cu with energies of 1–10 keV/atom compared to monomers due to atomic mixing in the thermal spike (Peltola and Nordlund 2003). However, this theory predicts that no increase in the ΔR_p of cluster constituents is expected for Si because the cascades break down into subcascades at much lower energies compared to Cu, and the liquid-like pockets are much smaller and cool down faster. Similar to this simulation, there was no difference in R_p, and a very small difference in ΔR_p was found when comparing the calculated values for B_1 and B_{10} implanted into Si with energies of 200 and 500 eV/atom (Hada 2003). The theoretically predicted change (or its absence) in R_p and ΔR_p is in good correlation with the experimentally obtained depth profiles of Au_n^+ ($n = 2, 3, 7$) cluster ions implanted into copper and silicon with energies of 10–100 keV/atom (Andersen et al. 2003).

A closely related problem to the estimation of the cluster stopping and the projected range is the radiation damage in the target. Unfortunately, this question has been poorly experimentally studied for the keV-energy range so far. MD simulations of Ar_n clusters impacting on Si showed that a similar damage region is formed by both large (hundreds of atoms) and small (tens of atoms) clusters if the total implantation energy (energy per cluster) is the same (Aoki 2000). In other words, an increase in cluster size leads to a decrease in the threshold energy of damage formation and to an increase of the displacement yield (Aoki et al. 2003). The latter is confirmed by other simulations. Simulations of Ta_n cluster implantation into graphite evidenced a superlinear increase in the number of damages (vacancies) with cluster size at the same energy per cluster atom (Henkel and Urbassek 1998). In the simulations of B_4 and B_{10} clusters implanted into Si with an energy of 230 eV/atom, it was found that the clusters produced a several times larger number of Frenkel defects (vacancies and interstitials) compared to B monomers (Aoki et al. 2001). The damaged region was also found to be characterized by a high yield of amorphization especially for the larger cluster B_{10}. Experiments on boron dimers and trimers implanted into Si with an energy of 1 keV/atom qualitatively confirmed the simulations by demonstrating that the number of displaced silicon

atoms per cluster atom increased by a factor of two compared to the monomer implantation (Jin et al. 2000). A similar effect was found in experiments on comparison of the radiation damage formation under the implantation of P^+ and F^+ atomic ions with PF_n^+ cluster ions having the same energy of 2.1 keV/amu (Titov et al. 2007). The increased defect concentration for the cluster case is assigned to molecular effect resulting from both a spatial overlap of collision subcascades and clustering of point defects.

Thus, so far MD simulations and experiments do not yet allow the formulation of a universal law for cluster projectiles and their stopping in matter. This problem can be related to both the approaches used in the simulations and the experimental methods employed to obtain the projected ranges. In particular, empirical potentials used for MD were originally constructed and tested for crystal lattices near thermodynamic equilibrium, and it is not a surprise that applied to collision simulations the quantitative results are not in complete agreement. In the case of experiments with keV-energy clusters, the errors in the determination of R_p can be rather high due to very shallow implantation.

Despite the lack of full understanding of the cluster stopping, there is a practical interest in cluster implantation, for instance, for shallow doping of semiconductors. Present-day technology has led to the use of lower and lower implantation energy and has moved to heavier implant species because the new generation of transistors already needs to have p–n junctions of ca. 10 nm. It was shown that low-energy implantation of $B_{10}H_{14}$ clusters can be an efficient way for doping of shallow silicon layers (Yamada and Matsuo 1996, Yamada et al. 1998). The most important phenomenon found is the suppression of enhanced boron diffusion in cluster-implanted silicon during postimplantation thermal annealing. The physical nature of the effect is not fully understood yet, but the phenomenon allows the doped layer as well as the junction thickness to be kept within a few nm. A transistor with a 40 nm effective gate length was fabricated by such cluster implantation already in the middle of the 1990s (Figure 19.5). A possibility to form ultrashallow junctions was also shown by using $(SiB_n)^-$ and $(GeB)^-$ cluster ions (Lu et al. 2002). Recently, low-energy cluster

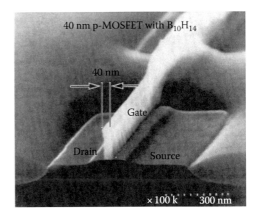

FIGURE 19.5 Cross-sectional SEM image of a p-MOSFET fabricated using the implantation of $B_{10}H_{14}$ clusters. (Reprinted from Yamada, I. et al., *Mater. Sci. Eng. R*, 34, 231, 2001. With permission.)

ion implantation was developed into a new infusion technique (Borland et al. 2004). Gas cluster ions accelerated to energies below 500 eV/atom produce short transient thermal and pressure spikes leading to self-amorphizing ultrashallow high concentration doping. Ultrashallow junctions of <6 nm by B infusion into Si were fabricated. Ge infusion for the formation of thin layers of strained Si was proved.

Shallow implantation is also demonstrated to be an attractive method for the formation of thin insulating layers. The implantation of silicon by chemically reactive cluster ions of $(O_2)_n^+$ and $(CO_2)_n^+$ with energies of 5–10 keV showed the formation of high-quality SiO_2 films of a few nanometer thickness that can be adaptable for the fabrication of ultrathin insulating layers for semiconductor devices (Akizuki et al. 1996). An advantage of the cluster method is the very low surface roughness (below 0.5 nm) of the film (Toyoda et al. 2003) compared to so-called passive oxidation (SiO_2 formation under high oxygen pressure and high temperature) (Engstrom et al. 1991). Synthesis of such thin SiO_2 films by other methods, for example, by pulsed-laser deposition or plasma-enhanced CVD, was found to be impossible due to either bubble formation (laser) or porous structure (CVD) resulting in poorer quality and higher roughness of the films (Caricato et al. 2002, Cremona et al. 2005).

During the last decade, there has been increasing interest in the implantation of high-energy (MeV) clusters showing specific phenomena like giant track and hillock (nanometer-size surface protrusion) formation in various target materials (Dammak et al. 1995, Brunelle et al. 1997, Döbeli et al. 1998, Ramos et al. 1998, Thevenard et al. 2000, Giard et al. 2003, Toulemonde et al. 2004). Several models were applied to explain the hillock formation. Among the most popular are the shock wave, Coulomb explosion, ionic, and thermal spikes. For such high energies one should also consider the so-called vicinage effect that leads to the enhancement of the cluster stopping power (energy loss) compared to monomers as a result of interference in the excitation of target electrons by the simultaneous interactions with a few swift ions (cluster constituents or recoils). This phenomenon was experimentally observed for the first time on the implantation of swift proton diatomic and three-atomic cluster ions into carbon and gold (Brandt et al. 1974), and a little later it was explained theoretically (Arista and Ponce 1975, Arista 1978). A few years ago, the state of the art of high-energy cluster implantation has been summarized and analyzed in Chadderton (2003). A "compound spike" model including thermal and ion explosion spikes dominating at certain stages of the track formation is proposed. For the stopping power, contributions from each ion as well as from vicinage effects between the ions and from additional plasma stopping were suggested to be taken into account (Cruz et al. 2003). Despite minor discords of the existing approaches, the mechanism of hillock formation under MeV-energy bombardment can be ascribed to electronic stopping in the wake of the cluster constituents leading to local melting along the track, and pushing out and quenching of the molten material. One more interesting effect recently found on the MeV implantation of fullerenes into graphite is a phase transition yielding

nanocrystals of diamond (Dunlop et al. 2007). The very strong energy density deposited in electronic stopping generates a high enough temperature and intense outgoing recoil pressure pulses to provide the phase transition in the track region. The presence of crystalline diamond NPs was found not only in the tracks but also around them on the surface. That is a confirmation of shock wave relaxation and hydrodynamic expansion causing the ejection of material from the near-surface part of the track.

19.4.4 Surface Erosion on Cluster Impact

The erosion of surfaces under the impact of energetic particles has attracted considerable attention for understanding both the fundamental physical aspects and the practical applications of sputtering, etching, and smoothing of surfaces. One of the first observations of crater formation on a target surface after heavy-ion sputtering was published at the beginning of the 1980s (Merkle and Jäger 1981). A transmission electron microscopy study of Bi^+ and Bi_2^+ ion implanted Au surfaces showed the existence of small craters with a mean diameter of about 4–5 nm. Later, craters were found on impact of various keV-energy ions with different materials by scanning probe microscopy (Wilson et al. 1988a,b, Teichert et al. 1994, Neumann 1999). To explain the crater formation the modified thermal-spike model (Miotello and Kelly 1997) was used, and the phenomenon was associated with individual displacement cascades originated by the penetrating particles. Wilson et al. suggested a more detailed description in which they assumed that the centre of the cascade is rich in vacancies and the periphery in interstitials. This causes stresses leading to compaction or sinking of the central part and peripheral rising forming the crater rims (Wilson et al. 1988a).

By both MD simulations (Averback et al. 1991, Insepov and Yamada 1996, Webb et al. 1997, Colla et al. 2000, Yamaguchi and Gspann 2002, Zhurkin and Kolesnikov 2003) and experiments (Bräuchle et al. 1995, Toyoda et al. 1998a,b, Allen et al. 2002a, Popok et al. 2009) it was shown that energetic cluster–surface impact also causes craters (Figure 19.6). The physics of the crater formation is quite well understood, and the simulations agree with the experimental data. As a result of local high-energy transfer the atomic arrangement in the impact region becomes strongly disordered. These atoms possess high kinetic energy and a large fraction of them—those present in the edges of crater—obtain momenta directed away from the surface. The crater can be surrounded by a few-angstrom-high rim formed by the substrate material displaced from the central part of the crater to its edges (Figure 19.6). The crater diameter was found to be an increasing function of both the cluster size and energy (Insepov et al. 2000). For instance, in the case of Ar clusters consisting of a few hundred up to a few thousand atoms and energies between 20 and 150 keV, the crater diameters on Si and Au surfaces were measured to be from 4 to 35 nm (Insepov et al. 2000, Allen et al. 2002a,b). However, the crater formation rate decreases with increase of energy per cluster constituent (Birtcher et al. 2003). This is related to an increase of the projected range and as a result to a shift of the maximum of the cluster-to-substrate energy transfer toward the bulk.

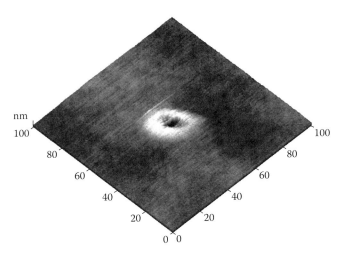

FIGURE 19.6 AFM image of crater formed on rutile (TiO$_2$) surface after the impact of Ar$_{22}^+$ cluster ion with energy of 11.5 keV.

The craters produced by monoatomic or cluster ions resemble the macroscopic ones formed by meteorite impact. However, their formation processes are different. As mentioned above, the nanoscale craters are formed by atomic displacement and flow processes that occur near the surface, while macroscopic crater formation is better understood in terms of a transient high-pressure region inside the material (Melosh and Ivanov 1999). Despite a 10–15 orders of magnitude difference on the size scale, it was recently demonstrated using large-scale atomic simulations that the macroscopic cratering behavior already emerges for clusters containing thousands of atoms and having impact velocities comparable to typical meteorite velocities (Samela and Nordlund 2008). As soon as the cluster size increases to a few tens of thousands of atoms, the impact behavior changes to the macroscopic one explained by the formation of a high-pressure core, and the cratering process can be described without use of the liquid-flow approach (Nordlund et al. 2008). Thus, one can use atomic simulations as a tool for the understanding of phenomena on macroscopic impact. On the other hand, some of the models developed for hypervelocity meteorite impact, for instance, for estimating the volume of melted material, can be approximated to the atomic scale (Popok et al. 2004).

Crater formation causes sputtering of a significant fraction of target atoms. A high sputtering yield was theoretically predicted and experimentally observed on various types of surfaces (Yamada and Matsuo 1996, Döbeli et al. 1997, Toyoda et al. 1998b, 2000, Greer et al. 2000, Perry et al. 2001). MD simulations of the sputtering yield, Y, from various metal surfaces by Ar$_n$ cluster bombardment with keV energies fit a power law dependence $Y \sim E^{1.4}$, where E is the total cluster energy (Insepov and Yamada 1999). This dependence is found to be in reasonable agreement with experimental results on Ar$_n^+$ cluster beam sputtering of various metal surfaces (Matsuo et al. 1997). The obtained power exponent 1.4 is close to the value 1.5 found in a thermal-spike model of supersonic velocity impact of macroscopic particles (Kitazoe and Yamamura 1980). In the experiments on Si sputtering by keV Au$_n$ cluster bombardment, it was found that Y slightly increases with cluster size without a significant change in the number of produced surface defects (Döbeli et al. 1997). High values of Y for clusters compared to monomers were explained as being due to so-called lateral sputtering effects originated by a shock wave. At normal impact angle of the cluster beam, the angular distribution of the sputtered target atoms has a nonsymmetric shape, which is different from the "cosine" distribution that is typically caused by monomers (Figure 19.7) (Yamada and Matsuo 1996, Yamada et al. 2001). In the case of sputtering, the cluster energy should be high enough to provide displacements of the target atoms but not too high to avoid implantation and radiation damage of the substrate. The high sputtering yield and the relatively low radiation damage of the target by large keV clusters are two important features allowing efficient surface cleaning and smoothing (Yamada et al. 1998). For the latter case, a so-called "downhill particle current" model was developed explaining suppression of the surface roughness under the cluster bombardment by redeposition of material from hills (protrusions) into valleys (pits) due to the above-mentioned domination of the lateral sputtering (Moseler et al. 1998, Toyoda et al. 2000).

A commercially available technique utilizing gas cluster ion beams has been developed in particular for surface

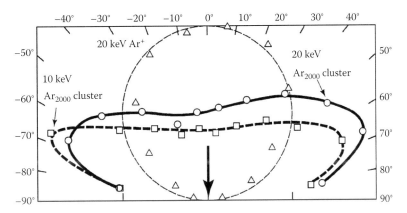

FIGURE 19.7 Angular distribution of Cu atoms sputtered by Ar monomer and cluster ions at normal incidence. (Reprinted from Yamada, I. et al., *Mater. Sci. Eng. R*, 34, 231, 2001. With permission.)

smoothing operations (Yamada et al. 2001). In general, a much better smoothing effect was achieved with clusters compared to monomers (Yamada et al. 1998). The main advantages of cluster use in comparison with ion or plasma-assisted processing are high spatial resolution, short-range damage, and elimination of charge accumulation on the substrate surface. One more advantage is a high smoothing efficiency at very high (~85°) incident angles (oblique angles) that makes cluster beams an exclusive tool for smoothing and polishing the side walls of various trenches or vertical pillars (Figure 19.8) on the micron or submicron scale (Bourelle et al. 2005). It is important that cluster smoothing does not have a negative effect on the surface mechanical properties. For instance, a TiN surface smoothed by an Ar_n^+ cluster beam did not exhibit any change in its mechanical properties, e.g., nanohardness or residual stress (Perry et al. 2001).

Additionally to the above-discussed "physical sputtering," a mechanism of so-called "chemical sputtering" or reactive accelerated cluster erosion (RACE) was suggested (Gspann 1996). In experiments on 120 keV $(CO_2)_n^+$ cluster impact with diamond and copper substrates, despite the predicted much higher erosion rate (about two orders of magnitude) of copper compared to diamond the experimentally found rates were of the same order of magnitude. It was assumed that the very high temperatures reached on the hypervelocity cluster impact led to the dissociation of the CO_2 molecules, and the resulting atomic oxygen reacted with the target material. In the case of copper a low-vapor-pressure oxide was formed, while for the diamond a highly volatile CO compound was the resulting product. The "chemical sputtering" mechanism was later confirmed by experiments with $(SF_6)_n^+$ cluster ions bombarded into W, Au, Si, and SiC (Toyoda et al. 1998a). For example, in the case of 20 keV cluster ions comprised of 2000 SF_6 molecules, each constituent has an energy of 10 eV, which is lower than the displacement energy of Si (about 15 eV), i.e., lower than the threshold energy for physical sputtering. However, the sputtering yield of Si by $(SF_6)_{2000}^+$ cluster ions was found to be 55 times higher than that of Ar_{3000}^+ cluster ions. The high yield of volatile SF_x compounds was registered by a residual gas analyzer confirming the chemical nature of the sputtering. The effect of lateral sputtering for the "chemical case" is eliminated, which is caused by the isotropic evaporation of volatile materials produced by the chemical reactions. By mask protective patterning of the surfaces, i.e., by selective etching, RACE was suggested as a cluster impact lithography method for obtaining specific surface micron or submicron relief (Yamada et al. 2001).

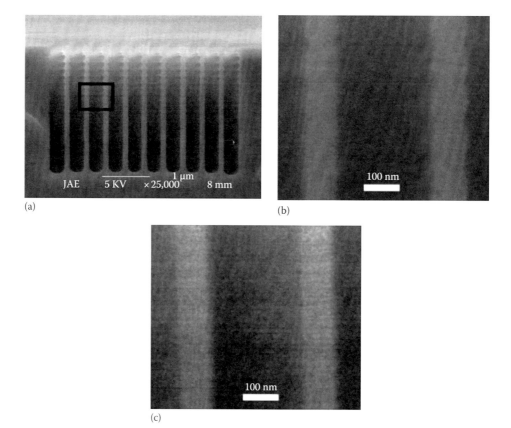

(a)

(b)

(c)

FIGURE 19.8 Scanning electron microscope images of cross sections of a silicon microstructure before and after SF_6 gas cluster irradiation. (a) Wide view of a silicon trench array fabricated by inductively coupled plasma reactive ion etching. Expanded image of sidewalls of trenches (b) before and (c) after irradiation by SF_6 gas clusters at an incident angle of 83° from the surface normal to the sidewall. The energy and the ion dose of the SF_6 clusters are 30 keV and 1.5×10^{15} ion/cm², respectively. (Reprinted from Bourelle, E. et al., *Nucl. Instrum. Meth. Phys. Res. B*, 241, 622, 2005. With permission.)

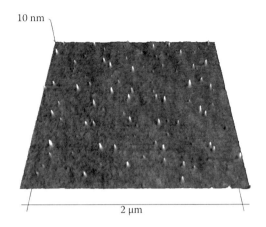

FIGURE 19.9 Atomic force microscopy image of a Si surface implanted by 3 keV Ar_{12}^+ cluster ions.

Along with craters, hillocks (nanometer-size protrusions) were found experimentally on keV-energy cluster impact (Gruber and Gspann 1997, Allen et al. 2002a, Song and Choi 2002, Popok et al. 2003). The hillock dimensions and their shape depend on cluster species, substrate material, and implantation conditions. In most cases, hillocks are cone shaped with a height of a few nanometers and basal diameter of about 10–40 nm (Figure 19.9). Despite a few orders of magnitude difference in energy, the hillocks formed by the implantation of clusters with MeV-GeV energies into various metal, semiconductor, and dielectric targets have a similar height and basal diameter (Döbeli et al. 1998, Ramos et al. 1998, Thevenard et al. 2000, Giard et al. 2003, Toulemonde et al. 2004). There are also a number of publications on hillock formation by swift heavy ions, for instance, Audouard et al. (1997), Skuratov et al. (2001), Müller et al. (2002), and Skuratov et al. (2003). Several models were applied to explain the hillock formation on high-energy implantation. Among the most important are the shock wave, the Coulomb explosion, the ionic, and thermal spikes or the "compound spike" mentioned above (Döbeli et al. 1998, Chadderton 2003, Cruz et al. 2003, Skuratov et al. 2003, Toulemonde et al. 2004). As mentioned above, the origin of hillock formation under MeV-energy implantation in related to the energy transfer from the cluster constituents to the target atoms due to electronic stopping that causes local melting along the track. The molten material is pushed out and quenched. For low-energy (tens keV) cluster implantation, the origin of hillock formation is predominantly nuclear stopping resulting in a local melting of a shallow layer of the target material around the impact spot which is well predicted by MD simulations (Cleveland and Landman 1992, Colla et al. 2000, Allen et al. 2002a). The origins of the significant expulsion effect are not well understood, but could be related to the difference in densities of the hot fluidized material and the surrounding solid state matrix as well as to elastic rebound of the bulk or local tensions in the substrate. Caused by these effects viscous flow pushes out the molten material, which can quench and form a small bump (Prasalovich et al. 2005, Popok and Campbell 2006).

FIGURE 19.10 Schematic view of cluster–surface impact. The sequence of images represents the incoming cluster (left), development of the collisional cascade and local melting (center), expulsion and complex crater formation (right).

In some experiments on keV-energy cluster ion implantation (Popok et al. 2004, Prasalovich et al. 2005), it was found that a hillock can be located in the centre of the crater (i.e., surrounded by a rim) forming a so-called complex crater (Figure 19.10), by analogy with those made by meteorite impact on a planet surface (Melosh and Ivanov 1999). The mechanism of hillock formation is suggested to be the same as above. However, the rim-to-rim diameters of these complex craters are found to be too large (up to 50 nm) to be explained just by the excavation of the target material on cluster impact (Samela et al. 2008). MD simulations predict a significant increase of lateral momentum of target atoms on impact. This phenomenon is favored if a thin layer is present on top of the bulk target and if the collision cascade is developed at the interface of these two materials, for example, in Si covered by a native oxide layer (Samela et al. 2008). Thus, one can suggest the formation of surface waves in the small local volume melted by energy transfer on the cluster impact and quenching of the waves forming large rims.

19.5 Summary

Despite the fact that the field of cluster physics is about 50 years old, a lot of questions related to cluster formation and cluster properties are still under discussion. The branch of cluster physics related to cluster–surface interactions is much younger. It is an intensively developing area stimulated by the prospective use of cluster beams for designing and mastering the physical and chemical properties of materials on the nanoscale leading to increasing interest and activity in both academic and industrial domains. Summarizing the above-presented data, one can conclude about a number of advantages of cluster beams compared to other methods in order to synthesize very thin and atomically smooth films, to dope very shallow layers, to provide high-yield sputtering and selective etching, as well as surface cleaning and smoothing. The immobilization of size-selected clusters by pinning or embedding them into various matrixes opens new synthetic routes for the formation of plasmonic, nanophotonic, and spintronic devices. Nevertheless, for successful practical applications, a number of fundamental physical aspects of cluster–surface interaction still need to be clarified. Phenomena such as "clearing-the-way," "compound spike," "vicinage effect," "pinning," etc., require further systematic and detailed investigation to develop a universal theory for interaction and stopping of cluster projectiles in matter. MD simulations require further development of models and improvement of potentials that could more realistically describe the interaction of various cluster species with different types of substrates. One of the best approaches to reconcile the existing differences is setting the experimental conditions as close as possible to those that can be modeled, or simulate the cluster–target interactions that can be reproduced by the experiments. It is to be hoped that the increasing amount of systematic and interrelated experimental and theoretical data will lead to an improvement in the situation in the near future.

Acknowledgment

The author is grateful to Prof. Eleanor E.B. Campbell from Edinburgh University for the fruitful discussions and suggestions on the contents of this chapter.

References

Abe, H., Schulze, W., and Tesche, B. 1980. Optical properties of silver microcrystals prepared by means of the gas aggregation technique. *Chem. Phys.* 47: 95–104.

Akizuki, A., Matsuo, J., Yamada, I., Harada, M., Ogasawara, S., and Doi, A. 1996. SiO_2 film formation at room temperature by gas cluster ion beam oxidation. *Nucl. Instrum. Method Phys. Res. B* 112: 83–85.

Allen, L.P., Insepov, Z., Fenner, D.B. et al. 2002a. Craters on silicon surfaces created by gas cluster ion impacts. *J. Appl. Phys.* 92: 3671–3678.

Allen, L.P., Fenner, D.B., Santeufemio, C., Brooks, W., Hautala, J., and Shao, Y. 2002b. Nano-scale surface texture by impact of accelerated condensed-gas nanoparticles. *Proc. SPIE* 4806: 225–232.

Alonso, J.A. 2005. *Structure and Properties of Atomic Nanoclusters.* London, U.K.: Imperial College Press.

Alton, G.D., Beckers, R.M., and Johnson, J.W. 1986. A radial geometry cesium plasma source with improved mechanical features. *Nucl. Instrum. Method Phys. Res. A* 244: 148–154.

Anders, C. and Urbassek, H.M. 2005. Cluster-size dependence of ranges of 100 eV/atom Au_n clusters. *Nucl. Instrum. Method Phys. Res. B* 228: 57–63.

Andersen, H.H., Johansen, A., Olsen, M., and Touboltsev, V. 2003. Gold-cluster ranges in aluminium, silicon and copper. *Nucl. Instrum. Method Phys. Res. B* 212: 56–62.

Aoki, T. 2000. Molecular dynamics simulation of cluster ion impact on solid surfaces. PhD thesis. Kyoto University.

Aoki, T., Matsuo, J., and Takaoka, G. 2001. Characterization of damage induced by cluster ion implantation. *Mater. Res. Soc. Symp. Proc.* 669: J4.5.1–J4.5.6.

Aoki, T., Matsuo, J., and Takaoka, G. 2003. Molecular dynamics study of damage formation characteristics by large cluster ion impacts. *Nucl. Instrum. Method Phys. Res. B* 202: 278–282.

Arista, N.R. 1978. Energy loss of correlated charges in an electron gas. *Phys. Rev. B* 18: 1–8.

Arista, N.R. and Ponce, V.H. 1975. The energy loss of correlated protons in channelling. *J. Phys. C* 8: L188–L191.

Audouard, A., Mamy, R., Toulemonde, M., Szenes, G., and Thome, L. 1997. Impact of GeV heavy ions in amorphous metallic alloys investigated by near-field scanning microscopy. *Europhys. Lett.* 40: 527–532.

Averback, R.S. and Ghally, M. 1994. MD studies of the interactions of low energy particles and clusters with surfaces. *Nucl. Instrum. Method Phys. Res. B* 90: 191–201.

Averback, R.S., de la Rubia, T., Hsieh, H., and Benedek, R. 1991. Interactions of energetic particles and clusters with solids. *Nucl. Instrum. Method Phys. Res. B* 59/60: 709–717.

Baletto, F. and Ferrado, R. 2005. Structural properties of nanoclusters: Energetic, thermodynamic, and kinetic effects. *Rev. Mod. Phys.* 77: 371–423.

Bansmann, J., Baker, S.H., Binns, C. et al. 2005. Magnetic and structural properties of isolated and assembled clusters. *Surf. Sci. Rep.* 56: 189–275.

Barborini, E., Piseri, P., and Milani, P. 1999. A pulsed microplasma source of high intensity supersonic carbon cluster beams. *J. Phys. D: Appl. Phys.* 32: L105–L109.

Barborini, E., Piseri, P., Podesta, A., and Milani, P. 2000. Cluster beam microfabrication of patterns of three-dimensional nanostructured objects. *Appl. Phys. Lett.* 77: 1059–1061.

Barborini, E., Vinati, S., Leccardi, M. et al. 2008. Batch fabrication of metal oxide sensors on micro-hotplates. *J. Micromech. Microeng.* 18: 055015.

Bardotti, L., Prevel, B., Jensen, P. et al. 2002. Organizing nanoclusters on functionalized surfaces. *Appl. Surf. Sci.* 191: 205–209.

Bauchert, J. and Hagena, O.-F. 1965. Massenbestimmung ionisierter agglomerate in komdemsierten molekularstrahlen nach einer elektrischen gedenfeldmethode. *Z. Naturforsch.* 20a: 1135–1142.

Becker, E.W., Bier, K., and Henkes, W. 1956. Strahlen aus kondensierten atomen und molekeln im hochvakuum. *Z. Phys.* 146: 333.

Bhaskar, N.D., Klimcak, C.M., and Frueholz, R.P. 1990. Liquid metal ion source for cluster ions of metals and alloys: Design and characteristics. *Rev. Sci. Instrum.* 61: 366–368.

Binns, C. 2001. Nanoclusters deposited on surfaces. *Surf. Sci. Rep.* 44: 1–49.

Binns, C., Trohidou, K.N., Bansmann, J. et al. 2005. The behaviour of nanostructured magnetic materials produced by depositing gas-phase nanoparticles. *J. Phys. D: Appl. Phys.* 38: R357–R379.

Birtcher, R.C., McCormick, A.W., Baldo, P.M., Toyoda, N., Yamada, I., and Matsuo, J. 2003. Gold nanoparticles sputtered by single ions and clusters. *Nucl. Instrum. Method Phys. Res. B* 206: 851–854.

Borland, J., Hautala, J., Gwinn, M., Tetreault, T.G., and Skinner, W. 2004. USJ and strained-Si formation using infusion doping and deposition. *Sol. State Technol.* 47: 64–67.

Bourelle, E., Suzuki, A., Sato, A., Seki, T., and Matsuo, J. 2005. Sidewall polishing with a gas cluster ion beam for photonic device applications. *Nucl. Instrum. Method Phys. Res. B* 241: 622–625.

Bouwen, W., Thoen, P., Vanhoutte, F. et al. 2000. Production of bimetallic clusters by a dual-target dual-laser vaporization source. *Rev. Sci. Instrum.* 71: 54–58.

Brandt, W., Ratkowski, A., and Ritchie, R.H. 1974. Energy loss of swift proton clusters in solids. *Phys. Rev. Lett.* 33: 1325–1328.

Bräuchle, G., Richard-Schneider, S., Illig, D., Rockenberger, J., Beck, R.D., and Kappes, M.M. 1995. Etching nanometer sized holes of variable depth from carbon cluster impact induced defects on graphite surfaces. *Appl. Phys. Lett.* 67: 52–54.

Brunelle, A., Della-Negra, S., Depauw, J. et al. 1997. Collisions of fast clusters with solids and related phenomena. *Nucl. Instrum. Method Phys. Res. B* 125: 207–213.

Burghoff, H. and Gspann, J. 1967. Bestimmung der mittleren agglomeratgröße und des restgasanteils kondensierter molekularstrahlen durch streuung eines K-atomstrahls. *Z. Naturforsch.* 22a: 684–689.

Carbone, R., Marangi, I., Zanardi, A. et al. 2006. Biocompatibility of cluster-assembled nanostructured TiO_2 with primary and cancer cells. *Biomaterials* 27: 3221–3229.

Caricato, A.P., De Sario, M., Fernandez, M. et al. 2002. Pulsed laser deposition of materials for optoelectronic applications. *Appl. Surf. Sci.* 197–198: 458–462.

Carrol, S.J., Weibel, P., von Issendorf, B., Kuipers, L., and Palmer, R.E. 1996. The impact of size-selected Ag clusters on graphite: An STM study. *J. Phys.: Condens. Matter* 8: L617–L624.

Carrol, S.J., Pratontep, S., Streun, M., Palmer, R.E., Hobday, S., and Smith, R. 2000. Pinning of size-selected Ag clusters on graphite surfaces *J. Chem. Phys.* 113: 7723–7727.

Chadderton, L.T. 2003. Nuclear tracks in solids: Registration physics and the compound spike. *Rad. Meas.* 36: 13–34.

Cleveland, C.L. and Landman, U. 1992. Dynamics of cluster-surface collisions. *Science* 257: 355–361.

Colla, T.J., Aderjan, R., Kissel, R., and Urbassek, H.M. 2000. Sputtering of Au (111) induced by 16-keV Au cluster bombardment: Spikes, craters, late emission, and fluctuations. *Phys. Rev. B* 62: 8487–8493.

Corso, M., Auwarter, W., Muntwiler, M., Tamai, A., Greber, T., and Osterwalder, J. 2004. Boron nitride nanomesh. *Science* 303: 217–220.

Cremona, A., Laguardia, L., Vassallo, E. et al. 2005. Optical and structural properties of silicon like films prepared by plasma-enhanced chemical-vapor deposition. *J. Appl. Phys.* 97: 023533(1–5).

Cruz, S.A., Gamaly, E.G., Chadderton, L.T., and Fink, D. 2003. A simple model for latent track formation due to cluster ion stopping and fragmentation in solids. *Rad. Meas.* 36: 145–149.

Cvikl, B., Korosak, D., and Horvath, Z.J. 1998. Comparative study of I-V characteristics of the ICB deposited Ag/n-Si(111) and Ag/p-Si(100) Schottky junctions. *Vacuum* 50: 385–393.

Dammak, H., Dunlop, A., and Lesueur, D. 1995. Tracks in metals by MeV fullerenes. *Phys. Rev. Lett.* 74: 1135–1138.

de Heer, W.A. 1993. The physics of simple metal clusters: Experimental aspects and simple models. *Rev. Mod. Phys.* 65: 611–676.

Di Vece, M., Palomba, S., and Palmer, R.E. 2005. Pinning of size-selected gold and nickel nanoclusters on graphite. *Phys. Rev. B* 72: 073407.

Dietz, T.G., Duncan, M.A., Powers, D.E., and Smalley, R.E. 1981. Laser production of supersonic metal cluster beams. *J. Chem. Phys.* 74: 6511–6512.

Dixon, A., Colliex, C., Ohana, R., Sudraud, P., and Van de Walle, J. 1981. Field-ion emission from liquid tin. *Phys. Rev. Lett.* 46: 865–868.

Döbeli, M., Nebiker, P.W., Mühle, R., and Suter, M. 1997. Sputtering and defect production by focused gold cluster ion beam irradiation of silicon. *Nucl. Instrum. Method Phys. Res. B* 132: 571–577.

Döbeli, M., Ames, F., Musil, C.R., Scandella, L., Suter, M., and Synal, H.A. 1998. Surface tracks by MeV C_{60} impacts on mica and PMMA. *Nucl. Instrum. Method Phys. Res. B* 143: 503–512.

Dole, M., Mack, L.L., Hines, R.L., Mobley, R.C., Ferguson, L.D., and Alice, M.B. 1968. Molecular beams of macroions. *J. Chem. Phys.* 49: 2240–2249.

Dunlop, A., Jaskierowicz, G., Ossi, P.M., and Della-Negra, S. 2007. Transformation of graphite into nanodiamond following extreme electronic excitations. *Phys. Rev. B* 76: 155403.

Ehbrecht, M., Kohn, B., Huisken, F., Laguna, M.A., and Paillard, V. 1997. Photoluminescence and resonant Raman spectra of silicon films produced by size-selected cluster beam deposition. *Phys. Rev. B* 56: 6958–6964.

Engstrom, J.R., Bonser, D.J., Nelson, M.M., and Engel, T. 1991. The reaction of atomic oxygen with Si(100) and Si(111): I. Oxide decomposition, active oxidation and the transition to passive oxidation. *Surf. Sci.* 256: 317–343.

Fayet, P., Wolf, J.P., and Wöste, L. 1986. Temperature measurement of sputtered metal dimers. *Phys. Rev. B* 33: 6792–6797.

Ganteför, G., Siekmann, H.R., Lutz, H.O., and Meiwes-Broer, K.-H. 1990. Pure metal and metal-doped rare-gas clusters grown in a pulsed arc cluster ion source. *Chem. Phys. Lett.* 165: 293–296.

Gao, H.-J., Pang, S.J., and Xue, Z.Q. 1998. Towards organic electronic devices using an ionized cluster beam deposition method. In: *Electrical and Optical Polymer Systems.* D.L. Wise, G.E. Wnek, D.J. Trantolo, T.M. Cooper, and J.D. Gresser (Eds.), pp. 729–761. New York: Marcel-Dekker.

Gartz, M., Keutgen, C., Kuenneke, S., and Kreibig, U. 1999. Novel examples of cluster-matter produced by LUCAS, a new laser cluster source. *Eur. Phys. J. D* 9: 127–131.

Giard, J.C., Michel, A., Tromas, C., Jaouen, C., and Della-Negra, S. 2003. Track formation in amorphous $Fe_{0.55}Zr_{0.45}$ alloys irradiated by MeV C_{60} ions: Influence of intrinsic stress on induced surface deformations. *Nucl. Instrum. Method Phys. Res. B* 209: 85–92.

Gibilisco, S., Di Vece, M., Palomba, S., Faraci, G., and Palmer, R.E. 2006. Pinning of size-selected Pd nanoclusters on graphite. *J. Chem. Phys.* 125: 084704.

Gilmer, G.H., Roland, C., Stock, D., Jaraiz, M., and Diaz de la Rubia, T. 1996. Simulations of thin film deposition from atomic and cluster beams. *Mater. Sci. Eng. B* 37: 1–7.

Goldby, I.M., von Issendorff, B., Kuipers, L., and Palmer, R.E. 1997. Gas condensation source for production and deposition of size-selected metal clusters. *Rev. Sci. Instrum.* 68: 3327–3334.

Greer, J.A., Fenner, D.B., Hautala, J. et al. 2000. Etching, smoothing, and deposition with gas-cluster ion beam technology. *Surf. Coat. Technol.* 133–134: 273–282.

Gruber, A. and Gspann, J. 1997. Nanoparticle impact micromachining. *J. Vac. Sci. Technol. B* 15: 2362–2364.

Gspann, J. 1996. Reactive accelerated cluster erosion (RACE) by ionized cluster beams. *Nucl. Instrum. Method Phys. Res. B* 112: 86–88.

Gspann, J. 1997. Impacting clusters. In: *Large Clusters of Atoms and Molecules*. T.P. Martin (Ed.), pp. 443–461. Amsterdam, the Netherlands: Kluwer Academic Publishers.

Haberland, H. (Ed.). 1994. *Clusters of Atoms and Molecules: Theory, Experiment, and Clusters of Atoms*. Berlin/Heidelberg, Germany: Springer-Verlag.

Haberland, H., Karrais, M., Mall, M., and Thurner, Y. 1992. Thin films from energetic cluster impact: A feasibility study. *J. Vac. Sci. Technol. A* 10: 3266–3271.

Haberland, H., Insepov, Z., Karrais, M., Mall, M., Moseler, M., and Thurner, Y. 1993. Thin film growth by energetic cluster impact: Comparison between experiment and molecular dynamics simulations. *Mater. Sci. Eng. B* 19: 31–36.

Haberland, H., Mall, M., Moseler, M., Qiang Y., Reiners, T., and Thurner, Y. 1994. Filling of micron-sized contact holes with copper by energetic cluster impact. *J. Vac. Sci. Technol. A* 12: 2925–2930.

Haberland, H., Insepov, Z., and Moseler, M. 1995. Molecular-dynamics simulation of thin-film growth by energetic cluster impact. *Phys. Rev. B* 51: 11061–11067.

Hada, Y. 2003. A molecular dynamics simulation of cluster dissociation process under cluster ion implantation. *Physica B* 340–342: 1036–1040.

Hagena, O.F. 1981. Nucleation and growth of clusters in expanding nozzle flow. *Surf. Sci.* 106: 101–116.

Hagena, O.F. 1987. Condensation in free jets: Comparison of rare gases and metals. *Z. Phys. D* 4: 291–299.

Hagena, O.F. 1992. Cluster ion sources. *Rev. Sci. Instrum.* 64: 2374–2379.

Hagena, O.F. and Obert, W. 1972. Cluster formation expanding supersonic jets: Effect of pressure, temperature, nozzle size and test gas. *J. Chem. Phys.* 56: 1793–1802.

Hall, S.G., Nielsen, M.B., Robinson, A.W., and Palmer, R.E. 1997. Compact sputter source for deposition of small size-selected clusters. *Rev. Sci. Instrum.* 68: 3335–3339.

Hao, C., March, R.E., Croley, T.R., Smith, J.C., and Rafferty, S.R. 2001. Electrospay ionization tandem mass spectrometric study of salt ions. *J. Mass Spectrom.* 36: 79–96.

Heimann, R.B., Evsykov, S.E., and Kavan, L. (Eds). 1999. *Carbyne and Carbynoid Structures*. Dordrecht, the Netherlands: Kluwer.

Helm, H. and Möller, R. 1983. Field emission ion source of molecular cesium ions. *Rev. Sci. Instrum.* 54: 837–840.

Henkel, M. and Urbassek, H.M. 1998. Ta cluster bombardment of graphite: Molecular dynamics study of penetration and damage. *Nucl. Instrum. Method Phys. Res. B* 145: 503–508.

Henkes, W. 1961. Ionisierung und beschleunigung kondensierter molekularstrahlen. *Z. Naturforsch.* 16a: 842.

Henkes, W. 1962. Massenspeltrometrische untersuchung von strahlen aus kondensiertem wasserstoff. *Z. Naturforsch.* 17a: 786.

Hogg, E.O. and Silbernagel, B.G. 1974. Particle growth in flowing inert gases. *J. Appl. Phys.* 45: 593–595.

Hsieh, H., Diaz de la Rubia, T., and Averback, R.S. 1989. Effect of temperature on the dynamics of energetic displacement cascades: A molecular dynamics study. *Phys. Rev. B* 40: 9986–9988.

Insepov, Z. and Yamada, I. 1996. Molecular dynamics study of shock wave generation by cluster impact on solid targets. *Nucl. Instrum. Method Phys. Res. B* 112: 16–22.

Insepov, Z. and Yamada, I. 1999. Surface processing with ionized cluster beams: Computer simulation. *Nucl. Instrum. Method Phys. Res. B* 153 199–208.

Insepov, Z., Manory, R., Matsuo, J., and Yamada, I. 2000. Proposal for a hardness measurement technique without indentor by gas-cluster-beam bombardment. *Phys. Rev. B* 61: 8744–8752.

Ishikawa, J. 2002. Negative ion beam technology for materials science. *Rev. Sci. Instrum.* 63: 2368–2373.

Jena, P. and Castleman, A.W. 2006. Clusters: A bridge across the disciplines of physics and chemistry. *Proc. Natl. Acad. Sci.* 103: 10560–10569.

Jensen, P. 1999. Growth of nanostructures by cluster deposition: Experiments and simple models. *Rev. Mod. Phys.* 71: 1695–1735.

Jin, J.-Y., Liu, J., van der Heide, P.A.W., and Chu, W.-K. 2000. Implantation damage effect on boron annealing behavior using low-energy polyatomic ion implantation. *Appl. Phys. Lett.* 76: 574–576.

Johnston, R.L. 2002. *Atomic and Molecular Clusters*. London, U.K.: Taylor & Francis.

Kappes, M.M. and Leutwyler, S. 1988. Molecular beams of clusters. In: *Atomic and Molecular Beam Methods*, Vol. 1. G. Scoles (Ed.), p. 380. New York: Oxford Univ. Press.

Kelly, R. 1977. Theory of thermal sputtering. *Radiat. Eff.* 32: 91–100.

Khabari, A., Urban III F.K., Griffiths, P., Petrov, I., Kim, Y.-W., and Bungay, C. 2003. Nanoparticle beam formation and investigation of gold nanostructured films. *J. Vac. Sci. Technol. B* 21: 2313–2318.

Kimoto, K. and Nishida, I. 1977. A study of lithium clusters by means of a mass analyzer. *J. Phys. Soc. Jpn.* 42: 2071–2072.

Kitagawa, T., Yamada, I., Toyoda, N. et al. 2003. Hard DLC film formation by gas cluster ion beam assisted deposition. *Nucl. Instrum. Method Phys. Res. B* 201: 405–412.

Kitazoe, Y. and Yamamura, Y. 1980. Hydrodynamical approach to non-linear effects in sputtering yields. *Radiat. Eff. Lett.* 50: 39–44.

Kreibig, U. and Vollmer, M. 1995. *Optical Properties of Metal Clusters*. Berlin, Germany: Springer.

Kroto, H.W., Heath, J.R., O'Brien, S.C., Curl, R.F., and Smalley, R.E. 1985. C_{60}: Buckminsterfullerene. *Nature* 318: 162–163.

Leisner, T., Vajda, S., Wolf, S., Woste, L., and Berry, R.S. 1999. The relaxation from linear to triangular Ag3 probed by femtosecond resonant two-photon ionization. *J. Chem. Phys.* 111: 1017–1021.

Liang, J.H. and Han, H.M. 2005. Implantation and post-annealing characteristics when impinging small B_n clusters into silicon at low fluence. *Nucl. Instrum. Method Phys. Res. B* 228: 250–255.

Liu, P., Ziemann, P.J., Kittelson, D.B., and McMurry, P.H. 1995. Generating particle beams of controlled dimensions and divergence: II. Experimental evaluation of particle motion in aerodynamic lenses and nozzle expansions. *Aerosol Sci. Technol.* 22: 314–324.

Lu, X., Shao, L., Wang, X. et al. 2002. Cluster-ion implantation: An approach to fabricate ultrashallow junctions in silicon. *J. Vac. Sci. Technol. B* 20: 992–994.

Magnano, E., Padovani, M., Spreafico, V. et al. 2003. Cluster beam microfabrication of SiC pattern on Si(100). *Surf. Sci.* 544: L709–L714.

Mahoney, J. F., Parilis, E. S., and Lee, T. D. 1994. Large biomolecule desorption by massive cluster impact. *Nucl. Instrum. Method Phys. Res. B* 88: 154–159.

Martin, T.P. 1996. Shells of atoms. *Phys. Rep.* 273: 199–241.

Martin, T.P., Bergmann, T., Göhlich, H., and Lange, T. 1990. Observation of electronic shells and shells of atoms in large Na clusters. *Chem. Phys. Lett.* 172: 209–213.

Marton, D., Bu, H., Boyd, K.J., Todorov, S.S., Al-Bayati, A.H., and Rabalais, J.W. 1995. On the defect structure due to low energy ion bombardment of graphite. *Surf. Sci.* 326: L489–L493.

Matsuo, J., Toyoda, N., Akizuki, M., and Yamada, I. 1997. Sputtering of elemental metals by Ar cluster ions. *Nucl. Instrum. Method Phys. Res. B* 121: 459–463.

Meiwes-Broer, K.-H. (Ed.). 2000. *Metal Clusters at Surfaces*. Berlin/Heidelberg, Germany: Springer-Verlag.

Melosh, H.J. 1989. *Impact Cratering: A Geologic Process*. New York: Oxford University Press.

Melosh, H.J. and Ivanov, B.A. 1999. Impact crater collapse. *Annu. Rev. Earth Planet. Sci.* 27: 385–415.

Merkle, K.L. and Jäger, W. 1981. Direct observation of spike effects in heavy-ion sputtering. *Philos. Mag. A* 44: 741–762.

Milani, P. and de Heer, W. 1990. Improved pulsed laser vaporization source for production of intense beams of neutral and ionized clusters. *Rev. Sci. Instrum.* 61: 1835–1838.

Milani, P. and Iannotta, S.. 1999. *Cluster Beam Synthesis of Nano-Structured Materials*. Berlin, Germany: Springer.

Milani, P., Ferretti, M., Piseri, P. et al. 1997. Synthesis and characterization of cluster-accembled carbon thin films. *J. Appl. Phys.* 82: 5793–5798.

Miotello, A. and Kelly, R. 1997. Revising the thermal-spike concept in ion-surface interactions. *Nucl. Instrum. Method Phys. Res. B* 122: 458–469.

Moseler, M., Rattunde, O., Nordiek, J., and Haberland, H. 1998. The growth dynamics of energetic cluster impact films. *Comput. Mater. Sci.* 10: 452–456.

Müller, C.M., Cranney, C., El-Said, A. et al. 2002. Ion tracks on LiF and CaF_2 single crystals characterized by scanning force microscopy. *Nucl. Instrum. Method Phys. Res. B* 191: 246–250.

Neumann, R. 1999. Scanning probe microscopy of ion-irradiated materials. *Nucl. Instrum. Method Phys. Res. B* 151: 42–55.

Nonose, S., Sone, Y., Onodera, K., Sudo, S., and Kaya, K. 1990. Structure and reactivity of bimetallic Co_nV_m clusters. *J. Phys. Chem.* 94: 2744–2746.

Nordlund, K., Jarvi, T.T., Meinander, K., and Samela, J. 2008. Cluster ion-solid interactions from meV to MeV energies. *Appl. Phys. A* 91: 561–566.

Paillard, V., Meaudre, M., Melinon, P. et al. 1995. DC conduction in diamond-like carbon films obtained by low-energy cluster beam deposition. *J. Non-Cryst. Sol.* 191: 174–183.

Palmer, R.E. and Leung, C. 2007. Immobilisation of proteins by atomic clusters on surfaces. *Trends Biotechnol.* 25: 48–55.

Palmer, R.E., Pratontep, S., and Boyen, H.-G. 2003. Nanostructured surfaces from size-selected clusters. *Nat. Mater.* 2: 443–448.

Palpant, B., Prével, B., Lermé, J. et al. 1998. Optical properties of gold clusters in the size range 2–4 nm. *Phys. Rev. B* 57: 1963–1970.

Pauly, H. 2000. *Atom, Molecule, and Cluster Beams*, Vol. 2. Berlin, Germany: Springer.

Peltola, J. and Nordlund, K. 2003. Heat spike effect on the straggling of cluster implants. *Phys. Rev. B* 68: 035419(1–5).

Peng, D.L., Sumiyama, K., Hihara, T., Yamamuro, S., and Konno, T.J. 2000. Magnetic properties of monodispersed Co/CoO clusters. *Phys. Rev. B* 61: 3103–3109.

Perez, A., Melinon, P., Dupius V. et al. 1997. Cluster assembled materials: A novel class of nanostructured solids with original structures and properties. *J. Phys. D: Appl. Phys.* 30: 709–721.

Perry, A.J., Bull, S.J., Dommann, A. et al. 2001. The smoothness, hardness and stress in titanium nitride following argon gas cluster ion beam treatment. *Surf. Coat. Technol.* 140: 99–108.

Piseri, P., Podesta, A., Barborini, E., and Milani, P. 2001. Production and characterization of highly intense and collimated cluster beams by inertial focusing in supersonic expansions. *Rev. Sci. Instrum.* 72: 2261–2267.

Popok, V.N. and Campbell, E.E.B. 2006. Beams of atomic clusters: Effects on impact with solids. *Rev. Adv. Mater. Sci.* 11: 19–45.

Popok, V.N., Prasalovich, S.V., and Campbell, E.E.B. 2003. Nanohillock formation by impact of small low-energy clusters with surfaces. *Nucl. Instrum. Method Phys. Res. B* 207: 145–153.

Popok, V.N., Prasalovich S.V., and Campbell, E.E.B. 2004. Complex crater formation on silicon surfaces by low-energy Ar_n^+ cluster ion implantation. *Surf. Sci.* 566–568: 1179–1184.

Popok, V.N., Jensen, J., Vučković, S., Mackova, A., and Trautmann, C. 2009. Formation of surface nanostructures on rutile (TiO$_2$): Comparative study of low-energy cluster ion and high-energy monoatomic ion impact. *J. Phys. D: Appl. Phys.* 42: 205303.

Powers, D.E., Hamen, S.O., Geurlc, M.E. et al. 1982. Supersonic metal cluster beams: Laser photoionization studies of Cu$_2$. *J. Phys. Chem.* 86: 2556–2560.

Prasalovich, S., Popok, V., Persson, P., and Campbell, E.E.B. 2005. Experimental studies of complex crater formation under cluster implantation of solids. *Eur. Phys. J. D* 36: 79–88.

Pratontep, S., Preece, P., Xirouchaki, C. et al. 2003. Scaling relations for implantation of size-selected Au, Ag, and Si clusters into graphite. *Phys. Rev. Lett.* 90: 055503(1–4).

Pratontep, S., Carrol, S.J., Xirouchaki, C., Streun, M., and Palmer, R.E. 2005. Size-selected cluster beam source on radio frequency magnetron plasma sputtering and gas condensation. *Rev. Sci. Instrum.* 76: 045103(1–9).

Qiang, Y., Thurner, Y., Reiners, Th., Rattunde, O., and Haberland, H. 1998. Hard coatings (TiN, Ti$_x$Al$_{1-x}$N) deposited at room temperature by energetic cluster impact. *Surf. Coat. Technol.* 100–101: 27–32.

Queitsch, U., Mohn, E., Schaffel, F. et al. 2007. Regular arrangement of nanoparticles from the gas phase on bacterial surface-protein layers. *Appl. Phys. Lett.* 90: 113114(1–3).

Ramos, S.M.M., Bonardi, N., Canut, B., Bouffard, S., and Della-Negra, S. 1998. Damage creation in α-Al$_2$O$_3$ by MeV fullerene impact. *Nucl. Instrum. Method Phys. Res. B* 143: 319–332.

Ravagnan, L., Siviero, F., Lenardi, C., Piseri, P., Barborini, E., and Milani, P. 2002. Cluster-beam deposition and *in situ* characterization of carbyne-rich carbon films. *Phys. Rev. Lett.* 89: 285506(1–4).

Reimann, C.T., Andersson, S., Brühwiler, P. et al. 1998. Graphite surface topography induced by Ta cluster impact and oxidative etching. *Nucl. Instrum. Method Phys. Res. B* 140: 159–170.

Roduner, E. 2006. *Nanoscopic Materials: Size-Dependent Phenomena.* Cambridge, U.K.: RSC Publisher.

Rousset, J.L., Cadrot, A.M., Cadete Santos Aires et al. 1995. Study of bimetallic Pd–Pt clusters in both free and supported phases. *J. Chem. Phys.* 102: 8574–8585.

Saito, N., Koyama, K., and Tanimoto, M. 2001. Cluster generation by laser ablation. *Appl. Surf. Sci.* 169–170: 380–386.

Samela, J. and Nordlund, K. 2008. Atomistic simulation of the transition from atomistic to macroscopic cratering. *Phys. Rev. Lett.* 101: 027601(1–4).

Samela, J., Nordlund, K., Popok, V.N., and Campbell, E.E.B. 2008. Origin of complex impact craters on native oxide coated silicon surfaces. *Phys. Rev. B* 77: 075309(1–15).

Sanz-Navarro, C.F., Smith, R., Kenny, D.J., Pratontep, S., and Palmer, R.E. 2002. Scaling behavior of the penetration depth of energetic silver clusters in graphite. *Phys. Rev. B* 65: 165420(1–8).

Sattler, K., Mühlbach, J., and Recknagel, E. 1980. Generation of metal clusters containing from 2 to 500 atoms. *Phys. Rev. Lett.* 45: 821–824.

Seeger, K. and Palmer, R.E. 2000. Application of clusters to the fabrication of silicon nanostructures. In: *Metal Clusters at Surfaces.* K.-H. Meiwes-Broer (Ed.), pp. 275–302. Berlin, Germany: Springer-Verlag.

Seki, T., Matsuo, J., Takaoka, G.H., and Yamada, I. 2003. Generation of the large current cluster ion beam. *Nucl. Instrum. Method Phys. Res. B* 206: 902–906.

Seminara, L., Convers, P., Monot, R., and Harbich, W. 2004. Implantation of size-selected silver clusters into graphite. *Eur. Phys. J. D* 29: 49–56.

Shulga, V.I. and Sigmund, P. 1990. Penetration of slow gold clusters through silicon. *Nucl. Instrum. Method Phys. Res. B* 47: 236–242.

Shulga, V.I., Vicanek, M., and Sigmund, P. 1989. Pronounced nonlinear behaviour of atomic collision sequences induced by keV-energy heavy ions in solids and molecules. *Phys. Rev. A* 39: 3360–3372.

Skuratov, V.A., Zagorski, D.L., Efimov, A.E., Kluev, V.A., Toporov, Yu.P., and Mchedlishvili, B.V. 2001. Swift heavy ion irradiation effect on the surface of sapphire single crystals. *Radiat. Meas.* 34: 571–576.

Skuratov, V.A., Zinkle, S.J., Efimov, A.E., and Havancsak, K. 2003. Swift heavy ion-induced modification of Al$_2$O$_3$ and MgO surfaces. *Nucl. Instrum. Method Phys. Res. B* 203: 136–140.

Smalley, R.E. 1983. Laser studies of metal cluster beams. *Laser Chem.* 2: 167–184.

Smith, R., Nock, C., Kenny, S. et al. 2006. Modelling the pinning of Au and Ni clusters on graphite. *Phys. Rev. B* 73: 125429.

Song, J.-H. and Choi, W.-K. 2002. Isolated cluster ion impact on solid surfaces HOPG, Si and Cu(TiO$_2$)/Si surfaces. *Nucl. Instrum. Method Phys. Res. B* 190: 792–796.

Sumiyama, K., Takehiko, H., Peng, D.L., and Katoh, R. 2005. Structure and magnetic properties of Co/CoO and Co/Si core-shell cluster assemblies prepared via gas-phase. *Sci. Technol. Adv. Mater.* 6: 18–26.

Takagi, T. 1998. Ion beam modification of solids: Towards intelligent materials. *Mater. Sci. Eng. A* 253: 30–41.

Takagi, T., Yamada, I., and Sasaki, A. 1976. An evaluation of metal and semiconductor films formed by ionized-cluster beam deposition. *Thin Sol. Films* 39: 207–217.

Teichert, C., Hohage, M., Michely, T., and Comsa, G. 1994. Nuclei of the Pt(111) network reconstruction created by single ion impacts. *Phys. Rev. Lett.* 72: 1682–1685.

Thevenard, P., Dupin, J.P., Vu Thien Binh, Purcell, S.T., Semet, V., and Guillot, D. 2000. Electron emission devices formed by energetic cluster impacts on TiO$_2$ rutile. *Nucl. Instrum. Method Phys. Res. B* 166–167: 788–792.

Titov, A.I., Azarov, A.Yu., Nikulina, L.M., and Kucheyev, S.O. 2007. Damage buildup and the molecular effect in Si bombarded with PF$_n$ cluster ions. *Nucl. Instrum. Method Phys. Res. B* 256: 207–210.

Toulemonde, M., Trautmann, C., Balanzat, E., Hjort, K., and Weidinger, A. 2004. Track formation and fabrication of nanostructures with MeV-ion beams. *Nucl. Instrum. Method Phys. Res. B* 216: 1–8.

Toyoda, N., Kitani, H., Hagiwara, N., Matsuo, J., and Yamada, I. 1998a. Surface smoothing effects with reactive cluster ion beams. *Mater. Chem. Phys.* 54: 106–110.

Toyoda, N., Kitani, H., Hagiwara, N., Aoki, T., Matsuo, J., and Yamada, I. 1998b. Angular distribution of the particles sputtered with Ar cluster ions. *Mater. Chem. Phys.* 54: 262.

Toyoda, N., Hagiwara, N., Matsuo, J., and Yamada, I. 2000. Surface smoothing mechanism of gas cluster ion beams. *Nucl. Instrum. Method Phys. Res. B* 161–163: 980–985.

Toyoda, N., Fujiwara, Y., and Yamada, I. 2003. Stable optical thin film deposition with O_2 cluster ion beam assisted deposition. *Nucl. Instrum. Method Phys. Res. B* 206: 875–879.

Van Lith, J., Lassesson, A., and Brown, S.A. 2008. Hydrogen sensors based on percolation and tunnelling in films of palladium clusters. *Proc. SPIE* 6800: 680009.

Vučković, S., Samela, J., Nordlund, K., and Popok, V.N. 2009. Pinning of size-selected Co clusters on highly ordered pyrolytic graphite. *Eur. Phys. J. D.* 52: 107–110.

Wagner, R.L., Vann, W.D., and Castleman, A.W. 1997. A technique for efficiently generating bimetallic clusters. *Rev. Sci. Instrum.* 68: 3010–3013.

Wang, X.M., Lu, X.M., Shao, L., and Liu, J.R. 2002. Small cluster ions from source of negative ions by cesium sputtering. *Nucl. Instrum. Method Phys. Res. B* 196: 198–204.

Webb, R.P., Kerford, M., Kappes, M., and Brauchle, G. 1997. A comparison between fullerene and single atom impacts on graphite. *Nucl. Instrum. Method Phys. Res. B* 122: 318–321.

Wegner, K., Piseri, P., Tafreshi, H.V., and Milani, P. 2006. Cluster beam deposition: A tool for nanoscale science and technology. *J. Phys. D: Appl. Phys.* 39: R439–R459.

Wilson, I.H., Zheng, N.J., Knipping, U., and Tsong, I.S.T. 1988a. Effects of isolated atomic collision cascades on SiO_2/Si interfaces studied by scanning tunnelling microscopy. *Phys. Rev. B* 38: 8444–8450.

Wilson, I.H., Zheng, N.J., Knipping, U., and Tsong, I.S.T. 1988b. Scanning tunneling microscopy of an ion-bombarded PbS(001) surface. *Appl. Phys. Lett.* 53: 2039–2041.

Woodruff, D.P. (Ed.). 2007. *The Chemical Physics of Solid Surfaces*, Vol. 12. *Atomic Clusters from Gas Phase to Deposited.* Amsterdam, the Netherlands: Elsevier.

Xu, J.L., Chen, J.L., and Feng, J.Y. 2002. Growth of Ge films by cluster beam deposition. *Nucl. Instrum. Method Phys. Res. B* 194: 297–301.

Yamada, I. and Matsuo, J. 1996. Gas cluster ion beam processing for ULSI fabrication. *Mater. Res. Soc. Symp. Proc.* 427: 265–276.

Yamada, I. and Takagi, T. 1981. Vaporized-metal cluster formation and ionized-cluster beam deposition and epitaxy. *Thin Solid Films.* 80: 105–115.

Yamada, I. and Toyoda, N. 2007. Nano-scale surface modification using gas cluster ion beams - A development history and review of the Japanese nano-technology program. *Surf. Coat. Technol.* 201: 8579–8587.

Yamada, I, Takaota, H., Usui, H., and Takagi, T. 1986. Low temperature epitaxy by ionized-cluster beam. *J. Vac. Sci. Technol. A* 4: 722–726.

Yamada, I., Matsuo, J., Jones, E.C. et al. 1997. Range and damage distribution in cluster ion implantation. *Mater. Res. Soc. Symp. Proc.* 438: 363–374.

Yamada, I., Matsuo, J., Toyoda, N., Aoki, T., Jones, E.C., and Insepov, Z. 1998. Non-linear processes in the gas cluster ion beam modification of solid surfaces. *Mater. Sci. Eng. A* 254: 249–257.

Yamada, I., Matsuo, J., Toyoda, N., and Kirkpatrick, A. 2001. Materials processing by gas cluster ion beams. *Mater. Sci. Eng. R* 34: 231–295.

Yamaguchi, Y. and Gspann, J. 2002. Large-scale molecular dynamics simulations of cluster impact and erosion processes on a diamond surface. *Phys. Rev. B* 66: 155408(1–10).

Yamamura, Y. 1988. Sputtering by cluster ions. *Nucl. Instrum. Method Phys. Res. B* 33: 493–496.

Yamamura, Y. and Muramoto, T. 1994. Time-evolution MC simulation of a few 100 eV/atom cluster impacts. *Radiat. Eff. Def. Sol.* 130–131: 225–233.

Zhurkin, E.E. and Kolesnikov, A.S. 2003. Atomic scale modelling of Al and Ni(111) surface erosion under cluster impact. *Nucl. Instrum. Method Phys. Res. B* 202: 269–277.

20

Molecules and Clusters Embedded in Helium Nanodroplets

Olof Echt
University of New Hampshire

and

Leopold Franzens Universität

Tilmann D. Märk
Leopold Franzens Universität

and

Comenius University

Paul Scheier
Leopold Franzens Universität

20.1 Introduction

For several decades, helium nanodroplets have been the ugly duckling of cluster science. As early as 1960, intense cluster beams of helium, hydrogen, and other "permanent" gases were synthesized by a group at the Karlsruhe Nuclear Research Center, motivated by attempts to produce intense beams of hydrogen clusters at high kinetic energies in order to fuel thermonuclear devices (Becker et al., 1961). In the 1980s, however, when research on isolated, atomic, or molecular clusters was booming, few experimentalists continued research in the area, which was considered by many as esoteric, poorly funded, and, from a technical standpoint, awkward because of the need for bulky vacuum pumps and large vacuum systems. Few kept working on helium, intrigued by the possible occurrence of superfluidity in finite systems.

Instead, experimentalists adopted the technique that the Karlsruhe group (Gspann and Vollmar, 1980) and others had brought to perfection for the synthesis of helium droplets to other systems, after suitable modification that included laser vaporization of nonvolatile species. The spotlight was on free-electron-like metal clusters such as alkalis that exhibited the intriguing phenomenon of electronic shell structure, transition metal clusters that promised to be useful as novel catalysts, and the newly discovered fullerenes.

Renewed interest in helium awakened in the early 1990s with the realization that helium droplets offer a new avenue for the synthesis of novel complexes and their interrogation at ultralow temperatures (Goyal et al., 1992). In short, helium droplets may serve as *flying nano-cryo-reactors* (Farnik and Toennies, 2005).

The combination of several features makes helium droplets unique:

- The ease with which they can be synthesized with sizes ranging from 10^3 to 10^8 atoms, equivalent to droplet radii of 22–1000 Å (2.2–100 nm).
- The ultralow temperature of 0.37 K that droplets achieve shortly after their formation or after any kind of excitation, as a result of efficient evaporative cooling via the emission of weakly bound helium atoms. Previously, researchers had cooled clusters by expanding them in a beam of helium, but the method was effective only for very small clusters, and the temperature attained remained unknown.
- The fact that they are superfluid at $T = 0.37$ K. Thus, almost any atom or molecule that collides with a droplet is swallowed and quickly moves to the center where it will coagulate with previously swallowed molecules. One consequence is that helium droplets offer a novel way to grow clusters of other species. Another is the possibility to grow complexes in metastable configurations.

Today, research continues with the goal of better understanding the properties of pure ^{4}He droplets; droplets of the lighter isotope, ^{3}He; and hydrogen droplets that may be grown and cooled within helium. In many other studies, helium droplets serve as a convenient first step to prepare and investigate novel clusters and complexes.

This chapter focuses on doped ^{4}He nanodroplets. Droplets have also been synthesized from the rare ^{3}He isotope (natural abundance 1.37×10^{-6}%). Their behavior at low temperature is quite different. Their binding is reduced because of an increased zero point energy, their density is 25% lower than that of ^{4}He and, most importantly, ^{3}He atoms are fermions whereas ^{4}He atoms are bosons. At temperatures reached by evaporative cooling, nanodroplets of ^{4}He, but not ^{3}He, will be superfluid.

We will neither dwell on spectroscopic studies that have provided a wealth of detailed information, nor on the general topic of superfluidity; these are discussed elsewhere in this volume (Paesani, 2010). Other noteworthy reviews that focus on spectroscopy are Choi et al. (2006) and Kupper and Merritt (2007). We also mention three other excellent reviews that are broader in scope than this chapter (Northby, 2001; Toennies et al., 2001; Toennies and Vilesov, 2004).

The chapter is organized as follows: Section 20.2 provides a general overview. We briefly summarize the basic properties of bulk helium, then the properties of neutral, excited, and charged helium droplets, how they are synthesized, doped, and how their properties have been determined. Obviously, an understanding of *doped* droplets necessitates a thorough understanding of *undoped* droplets. We highlight trends rather than specific information extracted from specific experiments. In some sections, e.g., in the discussion of excess electrons, our discussion focuses on bulk helium, which has been more thoroughly investigated. Section 20.2 concludes with a discussion of mass spectrometry.

Most of Section 20.3 presents a rather generic discussion of rare-gas dimers and clusters. A few basic principles help to develop a qualitative understanding of the classical rare-gas systems. The special case of helium has been discussed in Section 20.3.5.

In the remaining sections (Sections 20.4 through 20.7), we highlight case studies from our own laboratory: The electron impact ionization of hydrogen clusters embedded in helium (Jaksch et al., 2008), electron attachment to water clusters in helium (Zappa et al., 2008), an attempt to determine the size of solvation shells around ions in helium (Ferreira da Silva et al., 2009), and finally studies of helium droplets doped with organic and biomolecules (Denifl et al., 2006b; 2008; Mauracher et al., 2009).

20.2 Background Information

20.2.1 Bulk Helium

Among all "permanent" gases, helium is the most difficult to condense. The van der Waals interaction between helium atoms is extremely weak, thanks to their low polarizability which in turn arises from the small atomic radius and large energy gap of 19.82 eV between the ground state and first electronically excited

state of the atom. Its chemical reactivity is low; chemists have not yet been able to prepare macroscopic quantities of compounds containing covalently bound helium.

Helium has two naturally occurring isotopes, ^{3}He and ^{4}He; the natural abundance of the former is only 1.37×10^{-6}%; it will not be considered here any further. The helium dimer, ^{4}He$_2$, is barely bound. In its lowest (and only) bound quantum state, about 10^{-7} eV below the dissociation limit, the zero point energy nearly equals the depth of the classical potential well. As a result, the average distance between the two atoms in He$_2$ is $\approx$52 Å (Grisenti et al., 2000), nearly 20 times larger than the distance at which the interaction potential between the two helium atoms has its minimum, and nearly two orders of magnitude larger than the bond length of the hydrogen molecule.

The critical temperature of ^{4}He is 5.19 K; at standard pressure the gas condenses into a liquid at 4.22 K. The liquid does not solidify upon further cooling unless the pressure is increased to at least 2.5×10^5 Pa (25 atm). The numerical values of various quantities that characterize liquid helium are listed in Table 20.1; a pressure–temperature phase diagram is shown by the solid lines in Figure 20.1.

When liquid helium is cooled under its own vapor pressure, it undergoes a second-order phase transition at $T_\lambda = 2.1768$ K, named the lambda point because the temperature dependence of the specific heat C near T_λ resembles the Greek letter λ. Below T_λ, the system exhibits superfluidity; its viscosity becomes negligibly small. The liquid phases above and below T_λ are often referred to as helium I and helium II, respectively. The lambda transition can be understood as a form of Bose–Einstein condensation in a system of strongly interacting particles; below T_λ, a macroscopic fraction of the particles condenses into the quantum mechanical

TABLE 20.1 Properties of Condensed Helium

Quantity	Symbol	Value
Critical temperature	T_{crit}	5.19 K
Boiling point	T_b	4.22 K
Lambda point	T_λ	2.1768 K
Cohesive energy	D_{bulk}	0.62 meV
Dielectric constant	ε_r	1.057
Density	ρ	0.1451 g/cm^3
Atomic volume	υ	45.81 Å^3
Surface tension	σ	0.035 N/cm
Bottom of conduction band	V_0	1.06 eV
Energy of electron bubble	E_e	0.16 eV
Radius of electron bubble	r_e	17.2 ± 0.15 Å

Sources: Data from Lide, D.R., *CRC Handbook of Chemistry and Physics*, CRC Press, Boca Raton, FL, 2000; Cohen, E.R. et al., *AIP Physics Desk Reference*, Springer, New York, 2003; Rosenblit, M. and Jortner, J., *J. Chem. Phys.*, 124, 194505, 2006.

Note: V_0 and E_e are relative to the vacuum level. The accuracy of some quantities may be considerably less than suggested by the stated precision (number of significant digits) because they depend quite strongly on the helium density, and hence on pressure and temperature.

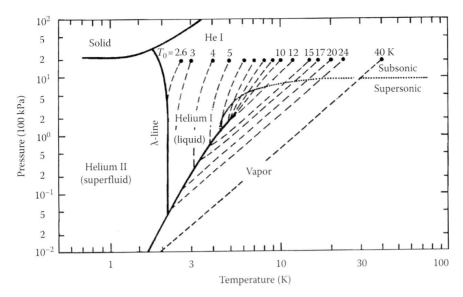

FIGURE 20.1 Solid lines: Pressure–temperature phase diagram for ^{4}He. Dashed lines: Set of isentropes for supersonic expansion from stagnation pressure $P_0 = 2$ MPa (= 20 bar) at select stagnation temperatures T_0. (Adapted from Buchenau, H. et al., *J. Chem. Phys.*, 92, 6875, 1990.)

ground state. The superfluidity of helium II is a quantum phenomenon, to some extent analogous to zero resistance in electric superconductors. A gap in the excitation spectrum prevents dissipative effects until the excess energy exceeds the energy gap. Impurities may travel through superfluid helium ballistically, without friction as long as their velocities do not exceed the critical Landau velocity of $v_L = 58$ m/s. Note, however, that helium atoms in the immediate vicinity of an impurity, ion, or container wall will usually be more strongly bound; they cannot move relative to their "host" without friction.

20.2.2 Neutral Helium Droplets

20.2.2.1 Synthesis of Helium Droplets

Several methods have been developed to form helium droplets (Jungclas et al., 1971; Northby, 2001), but the method of choice for the growth of nanosized droplets is homogeneous nucleation in a supersonic expansion (also referred to as *adiabatic* or *isentropic* expansion or *free jet*). The method is applicable to any material with high enough vapor pressure. It has been widely used to grow clusters of permanent gases, materials that are easily vaporized such as water, organic liquids, alkali metals, and in modified form, by expanding vapor seeded in helium or some other carrier gas, to grow clusters of refractory materials.

In a supersonic expansion, a stagnant gas is expanded from a high pressure (a fraction to tens of atmospheres) through a nozzle into a vacuum chamber. Three parameters are crucial for characterizing the expansion: the stagnation pressure P_0, the stagnant gas temperature T_0, and the nozzle diameter d. The exact shape of the nozzle will also affect the expansion, but for helium it has been customary to use simple orifices (pinholes) of $d = 5\,\mu$m, with a length not much longer than d.

As the gas exits the nozzle, it expands and accelerates while the velocity distribution narrows via collisions until the gas density becomes so low that collisions cease. The Mach number M (the ratio of the flow speed to the speed of sound) equals 1 in the throat of the nozzle; the beam becomes supersonic ($M > 1$) thereafter. The expansion is adiabatic because a gas expanding into vacuum cannot exchange heat with its surroundings. The translational temperature of the gas, defined by its velocity distribution in a moving frame of reference, can drop to very low values, although limited by the total number of collisions that occur between the nozzle and the "freezing" surface beyond which collisions become unlikely. In the absence of condensation, large speed ratios, of the order 10^2, can be achieved, equivalent to translational temperatures of the order $10^{-4}\,T_0$.

In practice, several conical apertures are placed at some distance from the nozzle to achieve differential pumping, i.e., to reduce the background pressure in the main vacuum chamber to values commensurate with other requirements. For example, to ensure collision-free conditions in subsequent vacuum chambers where the droplets are subjected to mass analysis, the background pressure has to be much lower than $P = 10^{-4}$ Pa, eight orders of magnitude below P_0. The location and shape of the first aperture, the "skimmer," are crucial to avoid reheating of the expanding gas in collisions with shock waves.

Translational temperatures as low as 1 mK have been reached in suitable expansions of helium. One common application of these beams are spectroscopic investigations of ultracold molecules. If these molecules are mixed into the stagnant helium at very low concentrations, their temperatures can be strongly reduced in the expansion because their rotational and vibrational excitations are reduced by heat exchange with the translationally cold helium jet. Another application has been the use of slow helium atoms with narrow velocity distributions in diffraction studies of crystalline surfaces (Camillone et al., 1991).

The above discussion tacitly assumed that no condensation takes place in the expansion. In other words, we assumed an *ideal* gas or, in practice, an expansion that starts at a value of (T_0, P_0) far away from the gas/liquid coexistence line of the expanding *real* gas. To grow helium droplets, one chooses (T_0, P_0) values that are close to the gas/liquid coexistence line. The dashed lines in Figure 20.1 illustrate how pressure and (translation) temperature drop as the gas expands from $P_0 = 2\,MPa$ ($\approx 20\,atm$) at select stagnation temperatures T_0. An ideal gas would follow a straight line in this double-logarithmic plot. Standard source conditions employ rather high values of T_0 ($\geq 12\,K$ if $P_0 = 2\,MPa$) such that clusters are formed when the vapor enters the liquid phase. The mechanism of droplet formation is different if the isentrope passes through the critical point or through the region to its left. Translational cooling occurs by binary collisions while three-body collisions are essential for the formation of dimers which are the nuclei for cluster growth. Binary collisions are proportional to $p_0 d$ while ternary are proportional to $p_0^2 d$. Consequently, cluster formation is enhanced by choosing large stagnation pressures but small nozzles.

Droplet velocities measured in the lab system are about 300 m/s for expansions through nozzles at $\approx 10\,K$ (Buchenau et al., 1990). Two crucial pieces of information are the droplet size and internal temperature. The size cannot be determined by mass spectrometric techniques because helium droplets undergo extensive fragmentation upon ionization, see Section 20.3. Instead, several other methods have been applied; the most accurate one employs scattering of the droplets by a beam of atoms or molecules with a narrow, known velocity distribution, formed in a supersonic expansion with parameters carefully chosen to avoid clustering. If the projectiles are known to stick to the helium droplets upon collision, then the droplet mass distribution can be determined from the angular distribution of the deflected helium droplets. The distributions turn out to be log-normal and rather broad (Lewerenz et al., 1995). They have been measured for a variety of expansion conditions. Many researchers simply refer to those data in order to estimate droplet sizes realized in their experiments.

Neutral helium droplets grown in the expansion of pure helium contain from 10^3 to 10^8 atoms; we will refer to them as "nanodroplets" or just "droplets." We will reserve the term "cluster" for aggregates formed within helium droplets, isolated aggregates other than helium, and small charged aggregates identified by mass spectrometry. The distinction may be blurred at times. If one assumes that the mass density ρ is constant within the droplet and abruptly falls to zero at the surface, a liquid drop approximation provides the "hard sphere" radius R of a spherical droplet

$$V = \frac{4}{3}\pi R^3 = n\upsilon = \frac{n m_{\mathrm{He}}}{\rho} \qquad (20.1)$$

where

    V is the droplet volume
    $m_{\mathrm{He}} = 4.0026\,u$ is the mass of the ⁴He isotope in atomic mass units (1 u = 1.66054×10^{-27} kg)
    υ is the atomic volume ($1/\upsilon$ is the number density, atoms per volume).

This may be rewritten as

$$R = 2.22\sqrt[3]{n\frac{\rho_{bulk}}{\rho}}\ \text{Å} \qquad (20.2)$$

In earlier work, a value $\rho/\rho_{bulk} \approx 0.7$ was assumed for the ratio of mass densities in the droplet over that in bulk helium (Lewerenz et al., 1995), but a diffusion Monte Carlo study suggests that surprisingly small droplets, with as few as 112 atoms, reach bulk density at their core (Barnett and Whaley, 1993).

Droplet temperatures have been determined by optical spectroscopy of embedded molecules; an ambient temperature of $0.37 \pm 0.05\,K$ is the commonly accepted result (Toennies and Vilesov, 2004). The droplets reach this extremely low temperature as a result of evaporative cooling, see Section 20.3.4. Therefore, the temperatures are immune to the details of droplet synthesis, although a slight dependence on size will result from a size dependence of the evaporation energy and, for ionized clusters, from enhanced binding due to polarization forces.

20.2.2.2 Electronic Excitations in Helium Droplets

Electronic excitations in helium, at energies below the atomic first ionization energy of IE = 24.59 eV, are crucial for an interpretation of experiments in which droplets are collided with energetic electrons. When an incident electron of energy E_i excites one of these states at an energy $E^\star$, it will continue to propagate through the cluster at a reduced energy $E_i - E^\star$. If a subsequent process such as electron attachment to a dopant molecule within the droplet occurs at an electron energy E_a, the energy dependence of the process will show multiple resonances at $E_i = E_a, E_a + E^\star, E_a + 2E^\star$ etc., corresponding to reactions in which either none, one, or two helium atoms within the cluster were excited. Furthermore, positive ions may be formed indirectly, in a process called Penning ionization, when an electronically excited helium atom (or complex, see later) deexcites into the ground state by transferring its energy to an adjacent impurity, provided the ionization energy of the impurity is below $E^\star$. Indirect ionization turns out to outweigh direct ionization; as a result it is difficult to determine adiabatic ionization energies of doped helium droplets.

The excitation energies of atomic helium are well known; the five lowest singlet and triplet states are listed in Table 20.2 together with the energy of He⁺ + e, i.e., the first ionization energy of He. *Singlet* and *triplet* refer to the relative orientation of the two electrons, having their spins either antiparallel or parallel. The states are characterized by their quantum numbers $n\,^{2S+1}L_J$, where n is the principal quantum number of the excited electron, S the quantum number for the total electronic spin (0 for singlet and 1 for triplet states), L the quantum number for the orbital angular momentum of the excited state (symbols S, P, D denote $L = 0, 1, 2$), and J the quantum number for the total angular momentum (classically, the vector sum of spin and orbital angular momentum). Also listed are the ground states of He (at 0 eV) and He⁺ (first and last entry in Table 20.2, respectively). The former is a singlet, hence only singlet states are reached upon photoexcitation,

TABLE 20.2 Lowest Electronic Excitations of Atomic Helium, and Their Lifetimes

State	Nature	Energy E^* (eV)	Lifetime τ (s)
$1\ ^1S_0$	Singlet	0	Stable
$2\ ^1S_0$	Singlet	20.616	20 ms
$2\ ^1P_1$	Singlet	21.218	0.55 ns
$3\ ^1S_0$	Singlet	22.920	54 ns
$3\ ^1D_2$	Singlet	23.074	15.8 ns
$3\ ^1P_1$	Singlet	23.087	1.7 ns
$2\ ^3S_1$	Triplet	19.820	8000 s
$2\ ^3P_{0,1,2}$	Triplet	20.964	98 ns
$3\ ^3S_1$	Triplet	22.718	36 ns
$3\ ^3P_{0,1,2}$	Triplet	23.007	95 ns
$3\ ^3D_{1,2,3}$	Triplet	23.074	14.2 ns
$1\ ^2S_{1/2}$	He$^+$ ion	24.58741	Stable

Sources: Radzig, A.A. and Smirnov, B.M., *Reference Data on Atoms, Molecules, and Ions*, Springer, Heidelberg, Germany, 1985; NIST, *NIST Chemistry WebBook*, 2009.

whereas either singlet or triplet states may be excited in collisions with electrons. Noteworthy are the high energies of the excitations; they exceed the ionization energies of any other atom or molecule, save for atomic neon which has an ionization energy of 21.56 eV. Thus, Penning ionization of impurities embedded within a helium droplet is a viable process. Also noteworthy are the long lifetimes of the lowest singlet state at 20.616 eV ($\tau = 20$ ms) and triplet state at 19.82 eV ($\tau = 8000$ s). These lifetimes exceed the time scale of a typical mass spectrometric experiment. When an excited helium atom reaches the ion detector, it will have sufficient energy to cause electron emission by Penning ionization upon collision with a surface. The electron may be registered and the excited neutral be mistaken for an ion.

In droplets, the energetics and lifetimes will be modified. Valuable information is derived from photoexcitation spectra (von Haeften et al., 2001; 2002). Here, the fluorescence yield of clusters is recorded as a function of the energy of the incident light from a synchrotron. To a good approximation, the fluorescence yield is proportional to the photoabsorption cross section. Two maxima in the fluorescence yield are observed for photon energies around 21 eV; they are correlated with the two lowest excitations into perturbed atomic singlet states, see Table 20.2 (transitions into states with $L \neq 1$ do not occur in isolated atoms, but they become allowed in the droplet). The resonances are blue-shifted (i.e., shifted to shorter wavelengths, or higher energies) by 0.385 and 0.333 eV, respectively. The shift arises from the perturbation of the atomic states by adjacent atoms. The resonances may be interpreted in terms of electronic excitations in helium dimers which occur at slightly higher energies when the interatomic distance is significantly larger than the equilibrium distance because the potential energy $V(r)$ of excited dimers is above that of excited atoms if r is large. In other words, the potential energy curves are repulsive at internuclear distances characteristic of neutral droplets.

Additional broad resonances are observed at higher energies, corresponding to blue-shifted transitions into states with

principal quantum number $n > 2$. Their energies increase further when the average droplet size is increased, consistent with a calculation that shows a linear dependence of transition energy on the distance between adjacent helium atoms.

Further information on electronically excited helium has been gained from experiments in which droplets were excited by monochromatic light at fixed wavelength while the spectrum of the emitted light was recorded (von Haeften et al., 2002). Those luminescence spectra show that the excitation results in an excited helium dimer, He$_2^*$ that resides in a small bubble which forms as a result of the repulsive interaction with the medium. The bubble moves ballistically, i.e., without friction, through the superfluid at an estimated speed of 7 m/s, well below the critical Landau velocity. The situation is similar to that in bulk helium, but the excited dimer or atom will be quickly ejected from the droplet if its radius is small.

Experiments in which droplets are excited by electrons have been discussed in Section 20.2.3.3 because the energetics of the incident electrons are modified by their repulsive interaction with the droplet (Henne and Toennies, 1998).

20.2.2.3 Doped Neutral Droplets

Although helium droplets are interesting on their own, the main reason for the recent attention that they have attracted is their use in the synthesis and characterization of molecules and clusters. Helium droplets are doped simply by passing the neutral beam, emerging from the supersonic expansion at a velocity of ≈ 300 m/s, through a cell that contains the dopant at low vapor pressure (10^{-4} Pa). Higher pressures lead to multiple captures causing cluster formation within the droplets. Mixed aggregates may be formed as well, e.g., by sequentially passing the droplet beam through two or more pickup cells. The only technical challenge in these experiments is to avoid inadvertently doping the droplets with other species contained in a nonperfect vacuum.

Doped droplets open a new area of research and pose new questions:

1. What is the efficiency (technically, the capture cross section) with which atomic or molecular species are captured upon collision?
2. How quickly do the embedded species thermalize?
3. Are embedded species located in the interior or on the surface?
4. What are the properties of helium in the immediate vicinity of embedded species?
5. What is the structure of aggregates grown by successive capture of atoms or molecules? Do successive collisions lead to just one or several aggregates? Do aggregates grow in metastable configurations?
6. By what mechanism are embedded complexes ionized when ionizing radiation (electrons, photons) interact with doped droplets? Is the ionization mechanism "softer" than for free complexes; can fragmentation be avoided?

We will briefly address questions 1–6 which deal with neutral droplets and for which some trends have been established.

Many of these topics have been interrogated by spectroscopic techniques, as discussed elsewhere in this volume (Paesani, 2010) and in other excellent reviews (Choi et al., 2006; Kupper and Merritt, 2007). Question 6 is central to our own research; it deserves a more detailed discussion, which will be deferred to Sections 20.3.5 and 20.7.

Helium droplets have been named "vacuum cleaners" for their unique ability to pick up any species with which they collide on their path through a vacuum chamber. The "sticking coefficient" is usually assumed to be close to 1, i.e., one assumes that the capture cross section equals the droplet's geometric cross section πR^2 (see Equation 20.2). A comprehensive mass spectrometric study of multiple pickup of Ar, Kr, Xe, H_2O, and SF_6, though, showed that the sticking coefficient may be significantly less than 1 (Lewerenz et al., 1995).

Most atoms and molecules, including all electronically closed-shell species, are "heliophilic," i.e., they will be fully immersed in the droplet. Alkali, heavy alkaline earth, and hydrogen atoms are among the few exceptions; they will remain on the surface. Upon capture, the collision energy plus internal (rovibrational) energy plus solvation energy will be released; the droplet quickly cools back to 0.37 K by evaporating a large number of helium atoms. As a rule of thumb, some 1600 He atoms will be released per 1 eV of energy release, consistent with the 0.62-meV cohesive energy of the bulk, see Table 20.1.

The long-range interaction between two neutral species in their electronic ground states is always attractive. Thus, if a previously doped droplet captures another atom or molecule, the newly embedded species will quickly move through the superfluid droplet and, within $\approx 10^{-9}$ s, coagulate to form a larger complex. This time estimate is consistent with a speed somewhat less than the critical Landau velocity, and a distance equal to the droplet radius. One usually assumes that all impurities will coalesce into one complex. If a droplet is doped with identical species in a pickup cell, the resulting size distribution of dopant clusters formed within a droplet follows a Poisson distribution

$$P(n) = \frac{(n_{av})^n}{n!} e^{-n} \qquad (20.3)$$

where

    n is the cluster size
    n_{av} is its average (equal to the average number of collisions if the sticking coefficient is 1)

The Poisson distribution approaches a Gaussian when $n_{av} \gg 1$, but it is strongly asymmetric for small values of n_{av}. However, there is some evidence that several smaller clusters may form within one droplet (Lewerenz et al., 1995). One reason for incomplete coalescence may be the formation of a "snowball," i.e., the existence of one or two tightly bound, very dense helium layers around solvated species that, together with the lack of any thermal energy, prevents separately captured species to coalesce. Also, spectroscopy has proven that aggregates grown in helium by successive capture may form in metastable

structures. For example, the long-range dipole–dipole forces acting between two hydrogen cyanide (HCN) molecules results in the self-assembly of noncovalently bonded linear chains, even though ring structures are energetically favored (Nauta and Miller, 1999).

20.2.3 Charged Helium Droplets

20.2.3.1 Positively Charged Droplets

A positively charged helium ion forms a very strong bond with another He atom, see Section 20.3. Thus, when an electron is removed from a helium droplet by photon or electron impact ionization, the energetically most favorable end product is a tightly bound He_2^+. Its binding with the surrounding helium will be enhanced by polarization forces; i.e., the ion induces dipole moments in He atoms, thus increasing attraction and the density of the surrounding. The effect is often referred to as *electrostriction*.

Helium has the highest ionization energy among all neutral atoms or molecules. Hence, when an electron is removed from a doped helium droplet, the energetically lowest configuration is one in which the positive net charge resides on the dopant. Quantum Monte Carlo calculations for alkali metal ions show that strong electrostriction leads to the formation of a solid-like helium layer, a so-called snowball, surrounded by a liquid-like (nonsuperfluid) layer of additional helium atoms (Coccia et al., 2008a).

In general, the positive ionization of a droplet releases an energy that far exceeds the binding of any complex embedded in the droplet. The resulting fragmentation of the dopant is discussed in Section 20.3.5, together with the detailed ionization mechanism.

20.2.3.2 Multiply Charged Droplets

Upon collisions with particles or photons, atoms may become highly ionized if the collision provides a sufficient amount of energy. The second ionization energy of helium, i.e., the energy required to remove an electron from He^+, amounts to 54.41 eV; a total of 80.00 eV (24.59 + 54.41 eV) are needed to produce He^{2+} from He.

Less energy would be needed to remove two electrons from a large droplet. First, its adiabatic ionization energy is only ≈ 21 eV, significantly less than that of an atom, due to the formation of a strongly bound He_2^+ or a slightly larger complex within the droplet, and the polarization energy (Denifl et al., 2006a). Second, two electrons may be removed from two atoms that are far apart, resulting in a second ionization energy equal to the first ionization energy for very large droplet radii. Similarly, the minimum energy required to form a z-fold charged droplet, He_n^{z+}, will converge to $\approx 21\, z$ eV for $n \to \infty$. Thus, collision with a single electron carrying a kinetic energy of a few hundred electronvolts could easily produce highly charged droplets.

However, like charges repel each other; hence, multiply charged droplets are prone to charge separation by a process called *Coulomb explosion*. A z-fold charged droplet of finite size will lower its electrostatic energy by fissioning into two smaller droplets with charges z_1 and $z_2 = z - z_1$. However, the combined surface areas of the two fragments will be larger than that of the original droplet, implying an energy penalty. As a result, fission will be exothermic below a specific size that is characteristic of each system, and the charge state z. Furthermore, an energetic barrier may impede fission. Droplets that are z-fold-charged will be sufficiently long-lived to allow for their observation when they reach a "critical" size $n_c(z)$.

Critical sizes have been determined for many clusters. Although "mass spectrometers" are, in reality, "mass-to-charge spectrometers" (see Section 20.2.4), multiply charged clusters X_n^{z+} can be readily distinguished from singly charged clusters if n is not an integer multiple of z. For example, the mass-to-charge ratio of a hypothetical ion He_{10}^{2+} is $10 \times 4.0026/2\,u/e_0 \approx 20\,u/e_0$; its mass peak would coincide with that of He_5^+. However, He_{11}^{2+} would produce a distinct mass peak at $22\,u/e_0$. The peak can be easily identified unless singly charged background ions, or impurities in the helium droplet, produce a signal at $22\,u/e_0$.

Critical sizes are quite small for doubly charged metal clusters because of the strong binding in metals; they become large in weakly bound van der Waals systems. The largest critical size so far has been reported for neon clusters, $n_c(2) = 284$ (the value means that Ne_{285}^{2+} could be identified whereas Ne_{283}^{2+} could not) (Mähr et al., 2007). For the more weakly bound helium clusters, one anticipates a much larger critical size. In fact, no evidence for He_n^{2+} has been seen in the resolved mass spectra of helium clusters up to $n/z = 300$, implying $n_c(2) > 600$. Much larger clusters can no longer be resolved; hence, multiply charged clusters cannot be identified by their noninteger n/z values. A critical size of $n_c(2) \approx 2 \times 10^5$ has been derived indirectly from an analysis of the reduced ionization cross section *versus* average droplet size by measuring size/charge distributions for different electron emission currents (Farnik et al., 1997). Doubly charged droplets below this threshold undergo fission by ejecting a small singly charged cluster ion carrying about 50 atoms.

It is worth mentioning that, in principle, multiply charged cluster *anions* may also exist. However, as explained in the following section (Section 20.2.3.3), the binding of excess electrons to helium is extremely weak; the Coulomb repulsion between two or more excess electrons in or on a helium nanodroplet would lead to their rapid detachment. Measured size distributions indicate that even the largest negatively charged droplets, containing 10^8 atoms, do not contain more than one excess electron (Farnik et al., 1997).

20.2.3.3 Negatively Charged Droplets

The behavior of excess electrons in, or in close proximity to, liquid helium has been studied for several decades (Cole, 1974). The helium surface presents an energetic barrier to an external electron because the bottom of the conduction band (which represents the lowest delocalized state for an excess electron in condensed helium) is *above* the vacuum level, i.e., above the energy of a free electron at rest, far away from the helium system.

Theoretical and experimental work place the bottom of the conduction band at an energy $V_0 = 1.06\,eV$, see Table 20.1.

However, an *external* electron will be weakly attracted to a planar helium surface by its image charge potential, given classically by the expression

$$V = \frac{1-\varepsilon_r}{4(1+\varepsilon_r)}\frac{e_0^{\,2}}{4\pi\varepsilon_0 d} \tag{20.4}$$

where

d is the vertical distance of the electron from the surface

e_0 is the elementary charge

ε_r is the static dielectric constant (see Table 20.1), i.e., the ratio of the permittivity of helium, ε, and the permittivity of free space, ε_0

The image charge potential has the same functional form as that of an electron bound to a positive point charge that equals $0.007\,e_0$. The classical expression for V diverges for $d \to 0$, but quantum mechanically the Pauli exclusion principle leads to a strong repulsion between an excess electron and the surface when d decreases below $\approx 1\,Å$ (0.1 nm). Furthermore, the small mass of the electron implies a large energy penalty if the electron were to be confined to a narrow region. As a result, an excess electron in its quantum mechanical ground state will reside at an average distance of $\approx 100\,Å$ above a planar helium surface, at an energy of just $0.74\,meV$ below the vacuum level. This so-called surface state is delocalized in two dimensions; the electron will be free to move parallel to the surface. Imperfections of the liquid, including excitations of the fluid, will scatter the electron, but the electron will remain delocalized. More than one electron may be bound to a macroscopic helium surface.

External electrons may be pushed more strongly toward the helium surface by applying an external electric field. This can be accomplished by sandwiching a helium film between two planar electrodes. The (negatively charged) anode is placed at some distance above the helium film. A thin, hot metal wire or a gas discharge is used to spray electrons onto the film. For a low concentration of electrons per unit area and low electric fields, the electrons remain delocalized; their spatial distribution is homogeneous. However, beyond a threshold concentration of about 2×10^9 electrons/cm^2, the surface becomes hydrodynamically unstable. It deforms in order to minimize the Coulomb interaction, and the electrons become trapped in macroscopic "dimples" containing up to 10^7 electrons (Ebner and Leiderer, 1980). The highly charged dimples strongly repel each other. For sufficiently large electric fields, the dimples will arrange in a regular hexagonal pattern, as shown in Figure 20.2 (Ebner and Leiderer, 1980).

When the applied electric field is increased above 3 kV/cm, electrons within a dimple will be pushed into the liquid, thus creating a macroscopic bubble that will migrate through the helium film toward the cathode at a velocity of about 10 cm/s. These multielectron bubbles are large because they contain large numbers of electrons; like dimples they may be easily imaged.

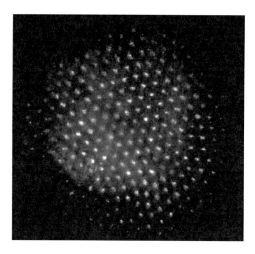

FIGURE 20.2 Image of a dimple lattice on the surface of a ^{4}He film at a temperature of 3.5 K. Excess electrons are concentrated in the center of the dimples which appear as bright spots: The distance between adjacent rows of dimples corresponds to the wavelength of the soft ripplon mode, 2.4 mm in this case. (Adapted from Ebner, W. and Leiderer, P., *Phys. Lett. A*, 80, 277, 1980.)

Much smaller electron bubbles will form if a *single* electron is pushed into helium, e.g., by bombarding a helium film with electrons of kinetic energies exceeding $V_0 = 1.06$ eV. These bubbles form because the interaction between an excess electron and helium atoms is repulsive as a result of the Pauli exclusion principle. Their radius is much smaller than that of multielectron bubbles, but at $r = 17$ Å, they are still large compared to the distance between adjacent helium atoms in the liquid. The rather large radius results from the extremely small surface tension of liquid helium.

The energy of the single-electron bubble lies ≈ 0.90 eV below the bottom of the conduction band; hence, the bubble state is slightly above the vacuum level (Rosenblit and Jortner, 2006). The bubble is metastable; its lifetime is finite. However, the electron cannot easily escape back into vacuum because the image-charge potential (Equation 20.4) will now be of opposite sign. This results in a force on the bubble that is directed into the fluid, away from the surface. To summarize: An electron may be injected from vacuum into helium only if its kinetic energy exceeds $V_0 = 1.06$ eV. Within a few picoseconds after entering the conduction band, the electron will be localized in a bubble of 17 Å radius. Eventually, the electron will escape back into vacuum by tunneling.

So far, we have discussed excess electrons in or on *bulk* helium. The qualitative features remain the same for finite helium droplets although details will change, last but not least, because the energetics depend on the number density (number of helium atoms per unit volume) which decreases with decreasing droplet size (see Rosenblit and Jortner, 2006; and references therein). First, a nanodroplet cannot possibly bind more than one electron. Second, bound exterior surface states exist if the droplet contains $>3 \times 10^5$ atoms but the binding energy will be even weaker than for a planar helium film. As a result, exterior states are not populated in a typical experiment. Third, the value of

V_0 will slightly decrease with decreasing cluster size. Forth, the bubble state will become less favorable with decreasing size. It will be *above* the bottom of the conduction band when the size drops below ≈ 5200; the bubble state will no longer exist. Fifth, the lifetime of the bubble state with respect to electron tunneling will shorten with decreasing size because the bubble will be forced to be closer to the surface. The rapid, undamped oscillatory motion of the bubble in the superfluid droplet implies an even closer approach of the bubble to the surface, thus increasing the tunneling rate and reducing the lifetime (Farnik et al., 1999; Rosenblit and Jortner, 2006).

Numerical values given in the previous paragraph are derived from theoretical work (Rosenblit and Jortner, 2006); they are in reasonable agreement with experimental results (see Henne and Toennies, 1998; Farnik et al., 1999; Northby, 2001; and references therein). In particular, the yield of negatively charged helium droplets formed by electron attachment shows a sharp rise above a threshold of 1.29 eV; it reaches a maximum at 1.80 eV for the smallest observed anions ($n = 7.5 \times 10^4$); the maximum gradually shifts with increasing size to 2.3 eV for the largest anions ($n = 1.5 \times 10^7$). The maximum is followed by a quick, exponential falloff. Droplets are metastable with respect to electron detachment; their lifetimes are surprisingly short, ranging from 1 ms for the smallest droplets to 200 ms for the largest ones. Lifetimes shorten in the presence of an electric field.

Several more maxima in the negative ion yield are observed at higher electron energies (Henne and Toennies, 1998). Three narrow maxima in the anion yield are observed in the region of the first ionization energy of He. They are positioned at 20.09, 21.14, and ≈ 23.7 eV above the low-energy maximum at ≈ 2 eV. The first peak is attributed to an excitation of a perturbed atom inside the droplet into the lowest triplet state. In the process, the incident electron is slowed down to ≈ 2 eV which is the optimum energy for it being self-trapped via bubble formation. The slight blue-shift relative to the excitation in free atoms, 19.82 eV, is attributed to polarization (note the different explanation of blue-shifts in Section 20.2.2.2). The second peak, 21.14 eV above the low-energy resonance, is attributed to excitation into the lowest singlet state in He_2^*; a transition at the same energy is prominent in fluorescence spectra, see Section 20.2.2.2.

Additional broad maxima are reported in the anion yield around 44–48, 70, and 95 eV. They are thought to be due to multiple electronic excitations of He_2; i.e., the incident electron undergoes several inelastic collisions at different sites within the droplet, each time losing an energy of about 21 eV.

Electronic excitations upon electron impact are also evident from sharp onsets at elevated electron energies in the yield of metastable, *neutral* droplets (Henne and Toennies, 1998). These neutrals may be detected by Penning ionization at the detector surface, see Section 20.2.2.2. However, the method is sensitive only to excitations with lifetimes that are not much shorter than the time of flight of the droplets from the place of excitation to the detector, about 35 ms. Two prominent onsets are reported at 20.07 and 21.22 eV, with uncertainties of 0.05 eV. The values are close to the values discussed in the previous paragraph, but the interpretation is different because the onsets considered here are *absolute* energies. An electron of 20.07 eV does not have enough energy to enter the droplet and subsequently create an excitation because the first step costs $V_0 \approx 1$ eV. Instead, the lower threshold is assigned to the excitation of a perturbed *surface* atom into the lowest triplet state at 19.82 eV (Table 20.2), with the blue-shift of 0.25 eV arising from polarization. The primary electron subsequently escapes back into vacuum with zero kinetic energy. The onset at 21.22 eV is attributed to the excitation of an atom deep inside the droplet. The difference between the two onsets, 1.15 eV, is assigned to V_0 of the droplet.

Note that this value for the bottom of the conduction band is slightly *larger* than the accepted value for the bulk, $V_0 = 1.06$ eV, although one expects a *lower* value because the atomic density in clusters is less than the bulk value (Rosenblit and Jortner, 2006). The reported threshold energy for the formation of negative ions, 1.29 eV, is even higher. The difference is considered statistically significant; it has been attributed to the fact that the electron entering the conduction band needs to create a precursor bubble with a critical radius of 3.5 Å; without such a precursor, it will remain in the conduction band and quickly leave the droplet (Henne and Toennies, 1998). Likewise, the rapid decline in the anion yield at energies above the maximum at ≈2 eV is thought to reflect the increasing probability that energetic electrons will pass ballistically through a pure helium droplet without being trapped.

20.2.4 Mass Spectrometric Techniques

Various mass spectrometric techniques have been employed to ionize and detect helium clusters. Helium has the highest ionization energy of all elements, 24.59 eV; it does not absorb photons below 20 eV. Therefore, photon ionization and charge exchange, two widely used methods for the ionization of neutrals, would not work well. Visible light extends from about 1.7 eV (corresponding to a wavelength of 720 nm) in the deep red to about 3.1 eV (400 nm) in the extreme violet; air becomes increasingly opaque at photon energies exceeding 6.2 eV (200 nm). The highest photon energy routinely available from a commercially available laser is 7.9 eV (157 nm, emitted from a F_2 excimer laser). Second, charge transfer by which an electron is transferred from a neutral to a positively charged ion will be endothermic, and therefore will not occur unless multiply charged ions are used.

As a result, the ionization of helium droplets is usually restricted to electron impact. The energy of the electrons is often tunable but their energy resolution (i.e., the width of the energy distribution) is often limited to ≈1 eV. Free electrons may also *attach* to droplets, resulting in *negative* ions. However, the electron-binding energy of helium atoms and clusters is negative; hence, only doped helium droplets form thermodynamically stable anions. Undoped helium droplets may form metastable anions with lifetimes exceeding 1 ms if they contain at least a few thousand atoms, see Section 20.2.3.3.

The mass spectrometry of small- to midsized helium cluster ions has been mostly performed with quadrupole or magnetic mass spectrometers. The mass range of quadrupoles is often limited to a few hundred atomic mass units; it rarely exceeds 2000 u. Furthermore, their mass resolution does not suffice to separate ions that are nominally isobaric, such as $^4He_3^+$ and $^{12}C^+$; the latter has per definition a mass of exactly 12 u if one ignores the mass of the "missing" electron, 0.00055 u. In the following, we will ignore the small correction for the "missing" or the "extra" electron in cations and anions, respectively.

Magnetic mass spectrometer may be operated at a resolution sufficient to separate ions that are nominally isobaric, i.e., ions for which the sum of neutrons and protons is equal. The mass range of magnetic spectrometers routinely extends to 10,000 u. It may be extended further by reducing the accelerating voltage at the expense of resolution and detection efficiency. Figure 20.3 shows a diagram of a spectrometer used in our lab. Electrons are emitted from a hot metal filament; they collide with a collimated beam of neutral clusters in the ion source. Voltages are applied to metallic plates (electrodes) in this region such that the electric potential in the region of ion formation is 3 kV relative to the later, grounded parts of the spectrometer. A weak electric field, of the order of 1 V/cm, accelerates the ions toward the exit of the ion source. The ions are then accelerated to ground potential and focused. The beam traverses the parallel pole pieces of a strong electromagnet, travels through a field-free region, and then traverses the electric field formed between two curved electrodes. The magnetic field B is, apart from edge effects, uniform. It serves as a momentum analyzer; ions of charge q and momentum p will follow a circular path with a radius

$$r = \frac{p}{qB} \tag{20.5}$$

Ions extracted from the ion source through a potential drop ΔV will acquire a kinetic energy

$$K = \frac{1}{2}mv^2 = \frac{p^2}{2m} = q\Delta V \tag{20.6}$$

Combining these equations, one obtains

$$r = \sqrt{\frac{2m\Delta V}{qB}} = \sqrt{\frac{m}{q}}\sqrt{\frac{2\Delta V}{B}} \tag{20.7}$$

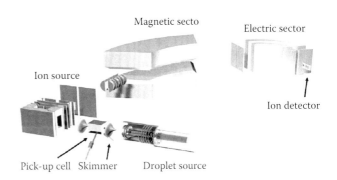

FIGURE 20.3 Cluster source, pickup cell, and mass spectrometer used by the Innsbruck group.

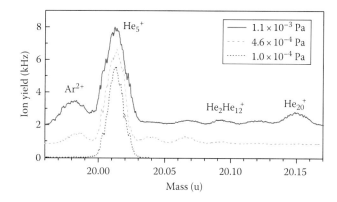

FIGURE 20.4 Mass spectra of ions near $m \cong 20\,u$ recorded with different hydrogen pressures in the pickup cell. The spectra are displaced vertically for greater clarity. He_5^+ and nominally isobaric $He_4H_4^+$, $He_3H_8^+$, $He_2H_{12}^+$, HeH_{16}^+, H_{20}^+ are clearly resolved; an argon impurity gives rise to $^{40}Ar^{2+}$. H_{20}^+ dominates at highest hydrogen pressure. (Adapted from Jaksch, S. et al., *J. Chem. Phys.*, 129, 224306, 2008.)

which shows that a magnetic field B combined with an acceleration through a potential drop ΔV serves as a rudimentary mass spectrometer; it allows to determine the mass-to-charge ratio of ions.

However, the resolution of such an instrument would be poor because the application of an ion extraction field implies that ΔV follows a finite distribution. Furthermore, in the last step in Equation 20.6, we have ignored any kinetic energy that the neutral particle may have prior to ionization. A mass spectral resolution large enough to separate ions of the same nominal mass can be obtained if the accelerated ion beam is passed through a so-called electric sector, consisting of two oppositely biased electrodes. The sector will deflect the ions of lower kinetic energy more strongly, thus acting as an energy filter. The mass resolution is greatly improved.

Figure 20.4 presents a positive-ion high-resolution mass spectrum obtained by electron impact ionization at 200 eV of H_2 clusters embedded in helium droplets. Shown is a section around mass 20 u (more exactly, a mass-to-charge ratio of $20\,u/e_0$, but in this chapter, we are mostly concerned with singly charged ions). He_5^+, $He_4H_4^+$, $He_3H_8^+$, $He_2H_{12}^+$, HeH_{16}^+, and H_{20}^+ are resolved, thanks to the different mass deficits (or mass excess) of hydrogen and helium. The atomic masses of ^1H and ^4He are 1.00783 u and 4.0026 u, respectively; hence, the masses of H_{20}^+ and He_5^+ are 20.156 u and 20.013 u, respectively; the mass difference between adjacent peaks in the series of mixed $He_{5-x}H_{4x}^+$ ions is 0.029 u. Another peak is due to $^{40}Ar^{2+}$ which has a mass-to-charge ratio of $19.981\,u/e_0$. In Figure 20.4, three measurements with different H_2 pressures in the pickup cell are shown. Pure hydrogen cluster ions are favored over pure helium cluster ions when the H_2 pressure is increased. The discussion of hydrogen cluster ions will be continued in Section 20.4.

20.3 Ionization, Fragmentation, and Cluster Cooling

In this section, we discuss electron impact ionization and fragmentation of van der Waals clusters. We also explain why an ensemble of clusters or cluster ions in a vacuum environment is

characterized by a well-defined temperature that is independent of the conditions under which the ensemble was prepared originally. We sketch basic mechanisms; helium will be discussed in more detail in Section 20.3.5.

The combination of three factors make fragmentation upon ionization all but unavoidable in homonuclear, weakly bound clusters:

- The existence of a tightly bound molecular ion whose formation involves
- A large structural arrangement
- A low evaporation D_n^+ energy of the most weakly bound moiety in the nascent cluster cation of size n

For the purpose of illustration, the discussion focuses on argon which is representative of other heavy rare gases.

20.3.1 Rare-Gas Dimers

The differences between neutral and positively charged dimers, Rg_2 and Rg_2^+, are best illustrated by a potential energy diagram. Figure 20.5 shows the potential energy of Ar_2 and Ar_2^+ in their electronic ground states versus separation r between the argon nuclei. For clarity, the energy scale near zero has been expanded to show the well of the weakly bound Ar_2. Without this change of scale, the minimum in the potential energy of Ar_2 would not be discernible in the figure.

One characteristic feature in the potential energy diagram is the significant difference in the equilibrium bond lengths r_e and r_e^+ of the neutral and ion, respectively. Another characteristic feature is the dramatic increase in the dissociation energies. Classically, these are given by the well depths D_e and D_e^+, respectively. Table 20.3 lists numerical values.

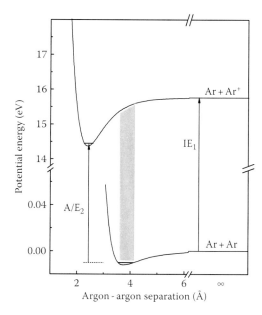

FIGURE 20.5 Potential energy curves of Ar_2 and $Ar_2^+ + e$. Vertical transitions from Ar_2 in the ground state to Ar_2^+ are confined to the shaded region.

TABLE 20.3 From Atom to Bulk: Atomic Ionization Energies IE_1, Dimer Bond Lengths r_e of Neutral and Charged Rare Gas Dimers, Their Respective Dissociation Energies D_e (Depth of Classical Potential), Bulk Dissociation (Evaporation) Energies D_{bulk} at 0 K, and Boiling Points T_{boil}

Quantity	He	Ne	Ar	Kr	Xe
IE_1 (eV)	24.58741	21.56454	15.759	13.99961	12.12987
r_e (Å)	2.97	3.09	3.76	4.02	4.37
r_e^+ (Å)	1.09	1.75	2.42	2.60	3.25
D_e (meV)	0.95	3.6	12.3	17.1	24.3
D_e^+ (eV)	2.50	1.29	1.39	1.15	0.975
D_{bulk} (meV)	0.62	20	80	116	160
T_{boil} (K)	4.22	27.07	87.30	119.93	165.03

Sources: Data from Lide, D.R., *CRC Handbook of Chemistry and Physics*, CRC Press, Boca Raton, FL, 2000; Cohen, E.R. et al., *AIP Physics Desk Reference*, Springer, New York, 2003.

Note: Only IE_1 and T_{boil} have high accuracy; the uncertainty in other quantities may be several percent.

The increased dissociation energy of Ar_2^+ *versus* Ar_2 implies that the *adiabatic ionization energy* AIE_2 of the dimer is significantly lower than the *ionization energy* IE_1 of the argon atom. The quantities are related through a thermodynamic cycle in which one either considers to first ionize and then dissociate Ar_2, or to first dissociate Ar_2 and then ionize one of the separated atoms. Either way, the product is $Ar + Ar^+ + e$; one therefore finds

$$AIE_2 + D_0^+ = D_0 + IE_1 \qquad (20.8)$$

D_0 and D_0^+ are the dissociation energies which refer to the quantum mechanical ground states, indicated in Figure 20.5 by horizontal lines just above the potential energy minima, in which neither vibrations nor rotations are excited (the vibrational and rotational quantum numbers are zero). The differences between the classical and quantum mechanical dissociation energies are, in relative terms, minor unless the well depth is very shallow and the mass of the atoms is small which holds for neutral He_2 and, to a lesser degree, Ne_2.

The large differences between the bond lengths of Ar_2 and Ar_2^+ imply that it is not possible to make a transition from ground-state Ar_2 to ground-state Ar_2^+ because ionization by electron impact is a *vertical* process, indicated by the gray region in Figure 20.5. The energy needed for the vertical transition is the so-called *vertical ionization energy* VIE_2. As indicated in the figure, VIE_2 is smaller than IE_1 by about 0.2 eV but much larger, by 1.1 eV, than AIE_2. The time during which the incident electron perturbs Ar_2 is so short that the separation between the nuclei cannot significantly change. For a rough estimate, consider an incident electron that has barely enough energy to ionize, say 15 eV. Its speed is 2.3×10^6 m/s; it traverses a distance of 10 Å within 0.44 fs. The vibrational period of Ar_2, 1.1 ps, exceeds this value by more than three orders of magnitude. Therefore, the two argon atoms may be considered frozen during the ionization event; the transition is "vertical" in a potential energy diagram.

Vertical ionization is indicated in Figure 20.5 by a band whose width equals the width of the line that represents the vibrational ground state ($\upsilon = 0$) of Ar_2. This width represents the range of interatomic distances in the neutral dimer due to zero-point motion. In a classical picture, the short duration of the transition into $Ar_2^+ + e$ does not only imply that the interatomic distances remain constant, but also that the momenta of the nuclei do not change. The nuclei are at rest in Ar_2 ($\upsilon = 0$); hence, the nuclei in the nascent highly excited Ar_2^+ must also be at rest, i.e., they must be near the turning point of their vibrational motion. For argon, the vertical transition is into states with vibrational quantum numbers υ ranging from about 35 to 46. In a quantum mechanical picture, the transition probabilities are given by the *Franck–Condon factors*, i.e., the overlap between the vibrational wave functions of the initial and final states. For states with large values of υ, the vibrational wavefunction has the largest amplitude near the turning points. The square of the wavefunction is proportional to the probability density; hence, the quantum mechanical result agrees with the classical picture (the blob of a pendulum spends most of its time near its turning points where its speed is the smallest).

20.3.2 Rare-Gas Clusters

In a neutral rare-gas cluster, the average bond length between adjacent atoms will be slightly shorter than r_e because forces between nonadjacent atoms are always attractive. These attractive forces imply a compression of the cluster, especially its core.

In a structurally relaxed *charged* rare-gas cluster, most of the positive charge (i.e., the missing electron) will be shared among two atoms which form a dimer ion with a bond length close to the bond length r_e^+ of an isolated Ar_2^+. In a more accurate description, some of the positive charge is shared among three or even four atoms, but for the sake of the argument we ignore this detail.

When a free argon dimer is ionized, the nascent Ar_2^+ contains a large vibrational excitation energy of 1.1 eV. Even so, the free dimer ion is stable because 1.1 eV is below the dissociation limit. However, what happens if an argon cluster is ionized? The removal of an electron from one of the atoms will lead to the rapid formation of a highly excited Ar_2^+ embedded in the cluster. Its vibrational energy of $\cong 1.1$ eV would then, within subnanoseconds, be randomized over all the other vibrational degrees of freedom of the cluster ion; a *hot* cluster ion will result. How hot? A system containing n atoms has a total of $3n$ degrees of freedom; 3 degrees are due to motion of the cluster as a whole because space is three-dimensional; another 3 are due to rotation (unless the cluster is linear, i.e., not three-dimensional). The remaining $3n - 6$ degrees of freedom are in vibrations. The equipartition theorem, a classical approximation, holds that each of these vibrational degrees carries an average energy of $k_B T$, where k_B is the Boltzmann factor, $k_B = 1.38 \times 10^{-23}$ J/K. A cluster ion with one complete solvation shell contains about 13 atoms; a vibrational excess energy of 1.1 eV would correspond to a temperature $T = 390$ K. The result indicates that we are in the classical regime. However, at much lower temperatures (i.e., lower energy and/or a larger cluster), and for systems with small mass, the classical expression greatly overestimates the vibrational energy because many vibrational degrees of freedom will be "frozen in." A helium cluster at 0.37 K contains *much* less energy than $(3n - 6)k_B T$.

In the above discussion, we have ignored the fact that atoms are polarizable. The neutral atoms surrounding the Ar_2^+ core will be polarized by the charge and be attracted toward it. Again, the structural relaxation is much slower than the vertical ionization, and potential energy released upon rearrangement of the nuclei will become available in the form of vibrational energy. Electronic polarization will release about ≈ 0.5 eV for large argon cluster ions; another ≈ 0.5 eV will be released as a result of the much slower structural relaxation (the sum of these two contributions equals the difference between the adiabatic ionization energy of Ar_2 and the thermodynamic threshold for ionizing bulk argon). For the ionization of helium droplets, the contribution from ion-induced polarization would be smaller because helium has a smaller polarizability than argon.

We briefly mention other phenomena that we have ignored so far: (a) the bond lengths and dissociation energies of Ar_2^+ in electronically excited states may differ strongly from those of the electronic ground state. If the nascent dimer ion is electronically excited, the energy released into the cluster ion will also depend on the mechanism by which excited states relax into lower states, i.e., radiative versus nonradiative transitions. (b) At elevated electron energies, a doubly charged cluster ion may be formed. The repulsive force between the two charges could be large enough to tear the cluster apart. *Cluster fission* or *Coulomb explosion* is discussed in Section 20.2.3.2. (c) Instead of vertical ionization into Ar_2^+, one may have an indirect process in which first an electronically excited neutral dimer Ar_2^* is formed which may then *autoionize* into $Ar_2^+ + e$. This ionization process may happen at energies well below the vertical ionization energy if the lifetime of Ar_2^* is comparable to or longer than its vibrational period, thus releasing a much lower excitation energy than for direct ionization.

20.3.3 Fragmentation of Excited Clusters and Cluster Ions

What happens to a vibrationally hot cluster? The temperature of 390 K estimated for Ar_{13}^+ after vertical ionization is four times higher than the boiling point T_{boil} of bulk argon. Of course, the boiling point merely tells us the temperature at which the equilibrium vapor pressure reaches 1 atm; this is a seemingly arbitrary reference point for clusters moving in a vacuum chamber at less than 10^{-9} atm. Rather, one needs to consider the *rate* at which a hot cluster will evaporate atoms. The *Arrhenius relation* provides a convenient estimate for the rate coefficient k

$$k = A \exp\left(-\frac{D}{k_B T}\right) \qquad (20.9)$$

where

    T is the vibrational temperature (a meaningful concept only if the excess energy has been randomized over all energetically accessible degrees of freedom)

    D is the dissociation energy of the most weakly bound atom (D_0 to be exact, not D_e)

As a rule of thumb, for cluster *cations* D_n^+ decreases with increasing size, approaching the dissociation energy D_{bulk} of the neutral bulk (0.08 eV for argon, see Table 20.3) upon completion of the first solvation shell. For small sizes, D_n^+ will be somewhat larger than D_{bulk} but still much smaller than the dissociation energy of the dimer ion. For example, $D_3^+ \cong 0.20$ eV, i.e., it costs 0.20 eV to dissociate Ar_3^+ into $Ar_2^+ + Ar$. On the other hand, for *neutral* clusters, D_n will slowly *increase* with increasing n. For example, $D_3 \cong 0.025$ eV for Ar_3 and $D_{13} \cong 0.06$ eV for Ar_{13}. Furthermore, the increase is not monotonic, e.g., $D_{13} \cong 2D_{14}$ because one of the atoms in Ar_{14} sits on the top of a compact shell, with only three nearest neighbors whereas atoms within the compact shell of Ar_{13} have, for icosahedral symmetry, six nearest neighbors.

Experimental evidence suggests that the value of the pre-exponential in Equation 20.9, the so-called A-factor, is $\approx 2 \times 10^{15}$ s^{-1} for many atomic and molecular clusters or cluster ions over a wide range of sizes (Klots, 1988). However, it is difficult to actually measure A (one would have to determine the rate k over a range of temperatures, but those are difficult to control). Although the Arrhenius relation is useful for quick estimates, one should not be mislead by its apparent simplicity. For example, it took several years of intense research and controversial debates until it was realized that the A-factors of fullerenes (carbon clusters C_n) exceed the standard value of 2×10^{15} s^{-1} by several orders of magnitude. Second, A is not strictly independent of temperature, although the overwhelming temperature dependence of k arises from the exponential. Third, the applicability of the Arrhenius relation to systems at ultralow temperatures, i.e., to helium, is questionable. Forth, the concept of temperature for an ensemble of clusters that are not in thermal contact with a heat bath is subtle. T in Equation 20.9 is not simply the vibrational temperature of the excited precursor before the evaporation. Rather it is, approximately, the vibrational temperature of the precursor minus $D_n/(2C_n)$, where C_n is its heat capacity, usually approximated by the equipartition value, $(3n - 7) k_B$. The correction to T is often termed the finite-heat-bath correction.

As an illustration of the latter point, for Ar_{13}^+ at 390 K we have $D_{13}^+ \cong 0.06$ eV, a finite heat bath correction of 11 K, and the estimated evaporation rate of the cluster ion would be 3×10^{14} s^{-1}. This rate would exceed the vibrational period of Ar_2 by two orders of magnitude; a temperature of 390 K is beyond the range where the Arrhenius relation provides meaningful estimates.

So far we have considered statistical processes which involve the randomization of the excess energy over all degrees of freedom.

Charge-induced reactions in helium droplets seem to be mostly nonstatistical; they will be considered in Section 20.3.5.

20.3.4 Temperature of Excited Clusters and Cluster Ions

We have seen that cluster ions formed by vertical ionization will be "hot" and feature a correspondingly large evaporation rate, depending on the details of the ionization process and cluster size. In fact, *neutral* clusters synthesized in a supersonic expansion of a pure gas will also be "boiling hot" because the heat of fusion is released upon their formation. What, if anything, can be said about the temperature of clusters or cluster ions that are probed in a typical experiment? A lot. Hot clusters in vacuum will cool by evaporation. With each evaporation of an atom, the excitation energy of the product cluster will be reduced by about $D - k_B T$, where D is the dissociation energy of the precursor and $k_B T$ is the average kinetic energy in the center-of-mass system of the atom and the product cluster. Here, again, T is not identical to the vibrational temperature of the precursor, but it will be close unless the cluster is very small. As a result, the temperature of the product cluster will be less than that of its precursor. Given the exponential dependence of the rate coefficient k on the inverse temperature in the Arrhenius relation, even a relatively small drop in temperature will imply an *enormous* drop in the evaporation rate of the product. As a result, clusters that were initially hot will quickly cool to temperatures at which the rate coefficient equals the time elapsed since the clusters were initially formed

$$k \cong 1/t \qquad (20.10)$$

This expression is an average over the so-called evaporative ensemble (Klots, 1988) of the clusters of a given size. The excitation energy of the ensemble will follow a distribution with a width approximately equal to D. Clusters with excitation energies *above* this interval will have dissociated within a time shorter than t, whereas clusters with excitation energies *below* the interval are not formed because their precursors would have been too cold to dissociate within time t.

For a typical mass spectrometer, ions are analyzed some 10 μs after their formation. Combining this value with $A \approx 2 \times 10^{15}\,\text{s}^{-1}$ and Equations 20.9 and 20.10, one obtains

$$D = k_B T \ln \frac{A}{k} \cong k_B T \ln(At) \cong 23.7\,k_B T \qquad (20.11)$$

This relation has been used frequently to determine D from (indirect) measurements of the cluster temperature T.

For clusters that are not too small, D approaches the bulk evaporation energy which in turn scales approximately as the boiling point. Hence, one obtains that the temperature of clusters in vacuum equals about 40%–50% of their bulk boiling point. The dependence on the experimental time scale is small because it enters through the logarithm. The estimate is confirmed by electron diffraction measurements of free

van-der-Waals-bound atomic and molecular clusters, including rare gases, CO_2, and SF_6.

Applying these relations to the highly nonclassical helium droplets would be a stretch. However, it is interesting to note that Equation 20.11, together with a bulk dissociation energy of 0.62 meV (Table 20.3), results in a temperature of 0.31 K, close to the 0.37 K obtained by spectroscopy of molecules embedded in helium droplets.

The basic conclusions of the evaporative ensemble (Klots, 1988) hinges on just a few assumptions. First, as already mentioned, the initial cluster temperature has to be high and clusters have to be the product of at least one evaporation. Otherwise, there would not be a definite lower limit to the cluster temperature. Second, the cluster size should neither be too small ($n = 10$ is usually assumed large enough) nor too large. Very large clusters cool, as our everyday experience will tell, very slowly; their temperature at time t will depend on their initial temperature. Third, the excess energy has to be randomized within a time much shorter than the experimental time; long-lived metastable states (electronic or vibrational) should not be populated. Forth, for long time scales, the effect of radiation may be nonnegligible. Clusters with large D such as fullerenes will continue to cool, via radiation, when they are too cold to show significant evaporation rates. On the other hand, weakly bound clusters will be heated by absorbing radiation from the environment. Hence, they will ultimately reach a terminal temperature at which evaporative cooling is balanced by radiative heating.

20.3.5 Ionization and Fragmentation of Undoped and Doped Helium Droplets

Soft, fragmentation-free ionization is a Holy Grail of cluster science. Neutral clusters can be characterized by a number of powerful techniques, but resulting data are often of limited value because fragmentation upon ionization rule out a reliable determination of cluster size by mass spectrometry. Will fragmentation be reduced, or totally suppressed, when clusters are embedded in helium droplets? Two mechanisms may help:

- Rapid cooling of the highly excited, nascent ionized cluster by highly efficient heat transfer in the superfluid helium, with the subsequent evaporation of He atoms.
- Caging, i.e., impulsive momentum transfer between the nascent ionized cluster and surrounding helium atoms. The effect has been shown to be quite efficient in classical system, e.g., in suppressing the dissociation of electronically excited I_2 surrounded by argon atoms (Sanov and Lineberger, 2004).

It is helpful to first consider the ionization mechanism in helium. Experimental data do not yet provide a coherent picture, except that the first step is the formation of an atomic helium ion. The direct ionization of a dopant does not occur upon electron impact, even though it would require considerably less energy (in contrast, direct ionization of the dopart without formation of He$^+$ intermediates can be achieved with photons

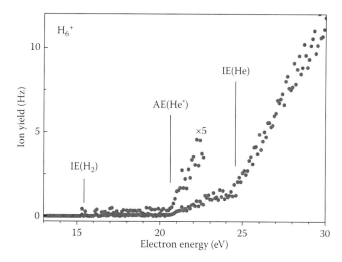

FIGURE 20.6 Yield of H_6^+ formed by electron impact ionization of hydrogen clusters embedded in helium, as a function of electron energy. Energy thresholds for the formation of H_2^+, He^*, and He^+ are indicated. (Adapted from Jaksch, S. et al., *J. Chem. Phys.*, 129, 224306, 2008.)

[Wang et al. 2008]). This is illustrated in Figure 20.6, which shows the yield of H_6^+ formed by electron impact ionization of hydrogen clusters embedded in helium. The resulting hydrogen cluster ions are mostly bare, with no helium attached (although mixed $He_xH_y^+$ may be formed, see Figure 20.4). Second, the observed ion stems from a hydrogen cluster that was considerably larger than $(H_2)_3$ (this assertion cannot be directly proved, though). Third, as seen in Figure 20.6, the ion yield remains negligible up to 20 eV, even though the ionization energy of H_2 is only 15.4 eV. This result is characteristic: Whatever the ionization energy of the dopant, no dopant ions are observed below ≈21 eV when the electronic excitation of helium becomes possible. However, the ion signal remains very weak until the ionization threshold of He at 24.59 eV is reached. Thus, Penning ionization of the dopant by an electronically excited He^* or He_2^* is a possibility, but efficient ionization requires the formation of He^+.

What is the fate of He^+ in the droplet? According to a theoretical study (Seong et al., 1998), it depends on where the ion was formed. If formed within a 6 Å-thick surface layer, the charge remains localized for some 200 fs. This is long enough for nuclear relaxation (motion); a dimer ion will form, accompanied by the energy release of ≈2.4 eV; He_2^+ will be ejected from the droplet. Indeed, He_2^+ ions are the most intense products upon the ionization of helium droplets up to $n = 15,000$ (Callicoatt et al., 1998).

If the primary ionization occurs inside the cluster where the density is higher, He^+ may migrate by "charge hopping," i.e., resonant (energy-neutral) charge transfer without nuclear motion. The time scale for this process is 10 fs (Seong et al., 1998). The ionization of the dopant requires that He^+ migrates by charge hopping all the way to the dopant, guided by the attraction between He^+ and the dipole moment induced in the dopant, and its permanent dipole moment, if any (Ellis and Yang, 2007; Lewis et al., 2008; Ren and Kresin, 2008). However, the larger the droplet, the more likely the migration will be interrupted by the formation of He_2^+ which is subsequently ejected from the droplet, perhaps

with a few helium atoms attached. The proposed mechanism provides a rationale for the observed rapid drop in ionization efficiency with increasing droplet size (Ruchti et al., 2000).

Charge transfer between He^+ and the dopant will be strongly exothermic as the ionization energy of most species, save for Ne and Ar, are at least 10 eV below the ionization energy of He. Not all of this energy has to be accommodated by the droplet if the dopant becomes electronically excited in the process and subsequently undergoes radiative decay (Ruchti et al., 2000), but no direct evidence exists for this channel. On the other hand, the nascent dopant ion may be formed in excited states that are repulsive, leading to an explosive event and fragmentation within 10 ps, too fast to be quenched by helium atom evaporation (Bonhommeau et al., 2008). Note that explosive, nonthermal events result in the escape of helium atoms with high kinetic energy, thus removing much more than 0.62 meV per ejected He atom.

To summarize, the ionization of dopants in helium droplets does not necessarily result in reduced fragmentation. Early conclusions that rare-gas clusters embedded in helium may be ionized with negligible fragmentation (Lewerenz et al., 1995) were overly optimistic. Experimental cluster studies are usually ambiguous because the size of the neutral cluster in the droplet cannot be determined with any certainty (theoretical studies, of course, do not suffer from this shortcoming [Bonhommeau et al., 2008]). One is on safer ground when studying the fragmentation of single, covalently bound molecules. Haloalkanes show very little change in their fragmentation pattern when embedded in helium (Yang et al., 2006), but many counterexamples exist as well; some will be discussed in Section 20.7. Atomistic mechanisms have been proposed that provide consistent, sometimes compelling, rationales for specific experimental observations, but that does not prove them correct. Sometimes, different authors have reported similar observations but arrived at different conclusions.

It should be pointed out that the ionization of dopants embedded in helium droplets may proceed via a widely ignored mechanism that has no counterpart for molecules in the gas phase, with no helium attached. The mechanism involves the formation of a doubly charged intermediate. For a number of molecules AB, including most hydrocarbons, the energy released upon the neutralization of He^+ is sufficient to form a doubly charged ion AB^{2+}, followed by exothermic charge separation into $A^+ + B^+$. Alternatively, in the presence of other molecules, the doubly charged intermediate may trigger interesting ion–molecule reactions within the droplet such as

$$He^+ + C_{60} + water \rightarrow He + C_{60}^{2+} + water + e$$

$$\rightarrow He + C_{60}OH^+ + H_3O^+ + e \qquad (20.12)$$

which has been observed in our laboratory (Denifl et al., 2009).

To summarize this section, we mention that the fragmentation of *negative* dopant ions, formed upon electron attachment, can be completely suppressed in helium droplets, even when attachment to the gas-phase molecule results in extensive fragmentation. An example will be presented in Section 20.7.

20.4 Electron Impact Ionization of Hydrogen Clusters Embedded in Helium

Hydrogen clusters were arguably the first large gas-phase clusters grown under controlled conditions in a bottom-up approach, by homogeneous nucleation of a supersaturated vapor in a supersonic expansion, the same way helium nanodroplets are grown. Five decades ago, scientists at the Kernforschungszentrum in Karlsruhe, Germany, had been exploring the possible use of hydrogen clusters to fuel plasmas in fusion reactors (Becker et al., 1961). One advantage in using charged clusters was seen in the possibility to endow them with a small but non-zero charge-to-mass ratio ze_0/m. Whereas singly ($z = 1$) charged atoms or small molecules will experience large undesired accelerations in the magnetic field B of a tokamak due to the Lorentz force, the acceleration a can be kept negligible by increasing the mass m of the ion because

$$a = \frac{ze_0}{m} vB\sin\varphi = \frac{ze_0}{m^{3/2}}\sqrt{2K}\sin\varphi \qquad (20.13)$$

In the second part of the equation (which assumes nonrelativistic energies), the velocity v has been expressed in terms of the kinetic energy K. φ is the angle between the velocity and the magnetic field; the acceleration is largest when the two vectors are perpendicular to each other. While interest in and funding for fusion-related research at the Karlsruhe Research Center has faded, researchers in Japan have recently injected hydrogen cluster ions into the HT-7 superconducting tokamak.

Another interesting development has been the observation that nuclear fusion can be initiated within hydrogen clusters in a regular lab environment. When large deuterium clusters are exposed to high-intensity laser pulses of femtosecond duration, a superheated microplasma is formed and ions are ejected with up to 1 MeV kinetic energy; the efficiency of deuteron-deuteron fusion has been recorded by monitoring the yield of fusion neutrons. The results may ultimately lead to the development of a tabletop neutron source which could potentially find wide application in materials studies.

At the other end of the size spectrum, interest in small hydrogen cluster ions arises from the fact that H_3^+ ions play an important role in the chemistry of interstellar clouds as the efficient protonators of neutral molecules. The isotopomer H_2D^+, which is rapidly formed from H_3^+ by exothermic proton-deuteron exchange, efficiently deuterates other molecules.

H_3^+ ions also play a key role when hydrogen clusters $(H_2)_n$ are ionized. H_3^+ is tightly bound, with its protons forming an equilateral triangle. In gas-phase collisions between H_2^+ and neutral hydrogen molecules, H_3^+ is formed

$$H_2^+ + H_2 \rightarrow H_3^+ + H \qquad (20.14)$$

From the H_2^+ bond strength of 2.650 eV and the large proton affinity of H_2, 4.377 eV, one deduces an exothermicity of 4.377 − 2.650 eV = 1.727 eV for Equation 20.14.

A similarly large energy is released upon the vertical ionization of $(H_2)_n$ because the binding between the hydrogen molecules is very weak. Qualitatively, the situation is similar to that discussed in Section 20.3.2, the ionization of rare-gas clusters, although the formation of the ionic core (H_3^+ as opposed to a rare-gas dimer ion) involves a larger structural rearrangement. Still, the excitation energy released upon the formation of H_3^+ exceeds the heat of vaporization of bulk hydrogen by two orders of magnitude. The cluster ion will quickly eject the lone hydrogen atom, resulting in an *odd*-numbered cluster ion which will cool further by boiling off H_2 molecules. The events are nicely illustrated by an ab initio calculation by Tachikawa (2000) that reveals the dynamics after the vertical ionization of $(H_2)_3$. Snapshots of the geometrical configuration of the ion are shown in Figure 20.7 at the time of ionization ($t = 0$) when the charge is localized on the H_2 molecule I. The ion gradually approaches molecule II; at 0.12 ps, a transient H_4^+ has formed. Its lifetime, however, is extremely short; at 0.18 ps, a trimer ion has formed and a hydrogen atom is seen moving away. At 0.47 ps, the H atom has left the complex which is described as a tightly bound H_3^+ ion plus a H_2 ligand. The H_3^+ ion is still "hot." The relaxation of its vibrational and rotational energy into the intermolecular modes of the cluster ion may lead to the ejection of H_2 on a slower time scale.

Given these energetics, is it impossible to form long-lived even-numbered hydrogen cluster ions? Not necessarily. The hydrogen atom in the exit channel of Equation 20.14 is bound to H_3^+, although only very weakly. Several researchers have, in fact, reported the observation of small even-numbered hydrogen cluster ions. However, the low abundance and limited experimental

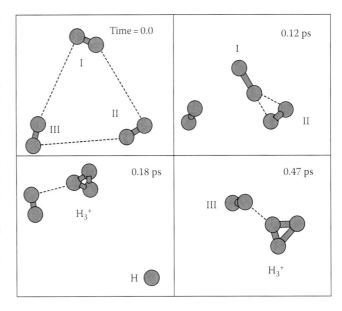

FIGURE 20.7 Snapshots of the geometrical configurations of H_6^+ after vertical ionization of the cyclic hydrogen cluster $(H_2)_3$, calculated by direct ab initio dynamics calculation at the HF/311G (p) level for different reaction times. (Adapted from Tachikawa, H., *Phys. Chem. Chem. Phys.*, 2, 4702, 2000.)

mass resolution renders most of these studies inconclusive. Some researchers have used deuterium instead of hydrogen but that only shifts the problem. For example, if the mass resolution is not sufficient to distinguish between ions that have the same nominal mass, one cannot distinguish between $^{12}C^+$ and D_6^+ or $^{16}O^+$ and D_8^+.

The mass spectrum in Figure 20.4 demonstrates the existence of H_{20}^+, a representative of even-numbered hydrogen cluster ions, in addition to the isobaric He_5^+ and other ions. To record larger hydrogen clusters, the pickup cell is operated at high partial hydrogen pressure, thus maximizing pure hydrogen clusters at the expense of mixed helium–hydrogen clusters. A section of a spectrum is shown in Figure 20.8. Pure hydrogen ions H_n^+ are marked by vertical lines; helium ions are marked by triangles. These ions are easily distinguished because the mass difference between He_x^+ and H_{4x}^+ increases with increasing x. On the other hand, the difference of only 0.031 u between $^{16}O_2^+$ and $^4He_8^+$ is too small to be resolved (the spectrum in Figure 20.8 was recorded at lower resolution than the spectrum in Figure 20.4). The signal just to the left of pure H_n^+ is mostly due to hydrocarbons. For two ions of the same nominal mass, $C_xH_y^+$ is lighter than H_{12x+y}^+ because the atomic mass of ^{12}C is, by definition, exactly 12 u, less than that of H_{12}.

Even-numbered H_n^+ can be identified up to $n = 120$ in these spectra; their abundance relative to adjacent odd-numbered cluster ions averages 4%. This ratio is considerably larger than in the mass spectra of pure hydrogen clusters which do not show any convincing evidence for even-numbered cluster ions beyond H_{10}^+. The ab initio direct dynamics calculations mentioned previously indicate that a free hydrogen cluster such as H_6 will, upon vertical ionization, eject a hydrogen atom within ≈100 fs; the atom will carry a substantial fraction of the reaction energy. Hence, in order to enhance the relative yield of even-numbered cluster ions formed in a helium droplet, the matrix will have to suppress the separation of the hydrogen atom from the charged complex within the first 100 fs. This kind of rapid

energy exchange between the nascent even-numbered hydrogen cluster ion and the helium matrix is similar to the impulsive processes that are responsible for the caging of photoexcited I_2^- in inert gas clusters (Sanov and Lineberger, 2004).

For a quantitative discussion of the mass spectral abundances of hydrogen cluster ions, we fit Gaussians to their mass peaks. The resulting size distribution of even-numbered H_n^+ is presented in Figure 20.9 on a semilogarithmic scale as a bar graph. Striking features are a maximum at H_6^+ in agreement with the studies of bare hydrogen clusters, and abrupt drops beyond H_{30}^+ and H_{114}^+. The statistical significance of local intensity anomalies between $n = 35$ and 93 is less certain. These kind of abundance drops are indicative of enhanced stability; cluster sizes that show enhanced abundance are often referred to as "magic." For van-der-Waals and hydrogen-bound cluster ions, magic numbers suggest the completion of a solvation or coordination shell around the ion. For *odd*-numbered hydrogen cluster ions, the ionic core is H_3^+; cluster ions are therefore best described as H_3^+ $(H_2)_m$. A chemically bonded coordination shell closes at $m = 3$ and another, mostly physically bonded shell, at $m = 15$.

Can the magic numbers $n = 30$ and 114 for even-numbered cluster ions be similarly assigned to specific structures? In the absence of directional bonding and electronic shell effects, one frequently observes geometric shell closure when the number of building blocks reaches $m = 12$ in the first icosahedral shell around the ionic core, $m = 42$ in the second, $m = 92$ in the third, and so on (Martin, 1996). The numbers may be slightly different, depending on the exact nature of the ionic core, but they readily appear for inert gas clusters as well as molecular clusters of CO and CH_4. Theoretical studies show that H_6^+ is a rather strongly bound ion consisting of a central H_2^+ bound to two H_2 molecules in D_{2d} symmetry, and that the D_{2h} structure of H_6^+ is nearly unperturbed when the ion is complexed with additional H_2 molecules. Thus, it is reasonable to assume that H_6^+ forms the ionic core of large even-numbered

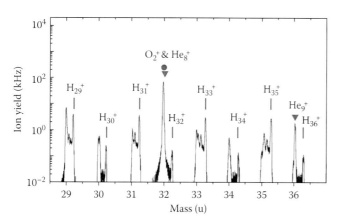

FIGURE 20.8 Sections of mass spectra showing odd- and even-numbered H_n^+ (vertical lines), He_n^+ (triangles), and impurities N_2^+ and O_2^+ (full dots). Other mass peaks, immediately to the left of H_n^+, are mostly due to mixed $He_xH_y^+$ and hydrocarbon background. (Adapted from Jaksch, S. et al., *J. Chem. Phys.*, 129, 224306, 2008.)

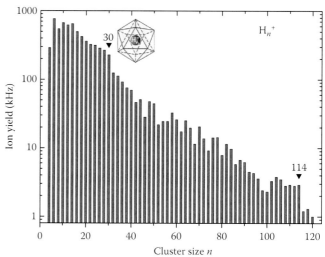

FIGURE 20.9 Size distribution of even-numbered hydrogen cluster ions plotted on a logarithmic scale. (Adapted from Jaksch, S. et al., *J. Chem. Phys.*, 129, 224306, 2008.)

hydrogen cluster ions. The magic numbers $n = 30$ and 114 will then match the number of hydrogen atoms needed to complete icosahedral shells around H_6^+, thus forming $H_6^+ (H_2)_{12}$ and $H_6^+ (H_2)_{12}(H_2)_{42}$.

20.5 Electron Attachment to Helium Droplets Doped with Water Clusters

Excess electrons in polar media are fascinating species that have been studied for two centuries (Edwards, 1982; Coe et al., 2006; Mostafavi and Lampre, 2008). In 1808, David reported that liquid ammonia turns blue when touched by alkali metals. In 1864, Weyl conducted a more comprehensive study; he observed similar blue coloring in other polar solvents. In 1908, Kraus found that liquid ammonia with dissolved alkali metals exhibits a remarkably high electric conductivity independent of the nature of the cation; he concluded that the conductivity was due to electrons surrounded by ammonia molecules. Later he observed a significant volume expansion when alkali metals were solvated in ammonia; from the expansion per mole of dissolved metal, he estimated the size of the cavity in which the electron resides. Recent values for the cavity radius are 3.0 Å.

Solvated excess electrons in water are not as easily characterized because of their high reactivity and short lifetime. Hart and Boag observed a transient absorption band in the red, peaking at 700 nm, when high-energy electrons of 1.8 MeV were injected into water (Hart and Boag, 1962). They attributed the band to the formation of hydrated electrons, e_{aq}^-. The primary products of solvent radiolysis are the hydrated electron, the hydroxyl radical OH, and the hydronium cation H_3O^+. Later work with ultrafast laser pulses suggests that the structural relaxation of the solvent around the trapped electron is completed within a few hundred femtoseconds; the radius of the cavity measures about 2.5 Å. The strength of the optical absorption slowly deceases over a range of tens to hundreds of picoseconds due to the recombination of e_{aq}^- primarily with the OH radical.

When a free electron is injected into liquid water, it will first propagate within the conduction band, then relax to its bottom, which, according to Coe and coworkers (Coe et al., 1997), is 0.12 eV below the vacuum level; i.e., below the energy of water plus a free electron in vacuum at rest. Eventually, the electron will be trapped and "dig in" until it is self-trapped in a cavity, at an energy of 1.72 eV below the vacuum level. The absorption peak at 700 nm corresponds to an excitation of the electron from its ground state to p-like states within the cavity. Note that the numerical values given above have considerable uncertainty; other researchers place the bottom of the vacuum band at 0–1.2 eV below the vacuum level.

On the other hand, an isolated H_2O molecule does not even bind an electron; the AEA of H_2O is negative. The AEA of a molecule X is defined as

$$X + e = X^- + AEA;$$ (20.15)

the AEA is the exothermicity of the attachment reaction (positive for an exothermic reaction). Determining adiabatic electron affinities is a challenge, even for ordinary molecules, unless the equilibrium configurations of the neutral and the anion are similar, see the discussion of adiabatic ionization energies in Section 20.3.1. If the configurations differ, the *vertical* attachment energy, VEA, may be negative even if the AEA is positive, and attachment will be impeded by a barrier. Furthermore, attachment to a free molecule will produce an anion that has sufficient energy for the reverse reaction, especially if the neutral precursor was already vibrationally excited. Hence, the anion will have a short lifetime with respect to "autodetachment" or "thermally activated detachment," even if the AEA of X is positive. On the other hand, anions may be long-lived even if their AEA is negative. For example, CO_2^- is metastable, 0.6 eV above the ground-state energy of the linear CO_2 molecule plus a free electron, but the strongly bent anionic ground state configuration is energetically below the energy of a free electron plus a similarly bent CO_2.

At what size will a water cluster be large enough to bind an electron? At what size will the cluster be large enough to accommodate an electron in a cavity state? Are there states other than the localized cavity state and the delocalized states in the conduction band that can bind an excess electron? Early experimental research focused on finding the minimum number of water molecules necessary to form a stable (or sufficiently long-lived) anion, $(H_2O)_n^-$. The smallest observable anions contained $n = 11$ molecules; beyond this threshold the anion abundance rose steeply. Two different pathways for their formation were successful, either the injection of low-energy electrons into a supersonic beam of water molecules, or the attachment of electrons to water clusters in a low-density cluster beam where no further cluster growth could occur. These latter experiments revealed a narrow resonance in the electron attachment cross sections at near-zero electron energy. At higher electron energies, anions of the composition $(H_2O)_n OH^-$ were observed. They feature no lower size limit because the hydroxyl anion is stable.

Further work showed that $(H_2O)_n^-$ anions just barely exceeding the threshold size of 11 were likely to undergo autodetachment before reaching the ion detector. As mentioned above, an anion formed by electron attachment to a neutral contains enough energy for the reverse reaction, especially if the neutral precursor was already thermally excited because of the heat of condensation released upon its formation. Hence, the mere observation of autodetachment was not surprising, but the fact that it was limited to a narrow size range just above $n = 11$ implies that slightly larger anions would cool by H_2O evaporation rather than autodetachment. Therefore, a crude estimate yields

$$AEA(H_2O)_n > D_n^+ \cong 0.4 \text{ eV}$$ (20.16)

for cluster anions well above $n = 11$. Conceptually, the relevant quantity in Equation 20.16 would be the adiabatic *detachment* energy of $(H_2O)_n^-$, but its value equals the AEA of $(H_2O)_n$.

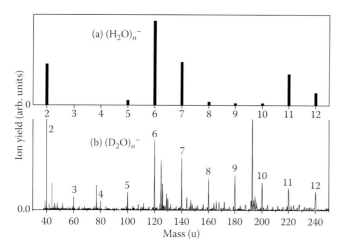

FIGURE 20.10 Top: Abundance distributions of $(H_2O)_n^-$ synthesized by seeding water vapor in argon (stick spectrum). (Adapted from Kim, J. et al., *Chem. Phys. Lett.*, 297, 90, 1998.) Bottom: Mass spectrum of water cluster anions formed by electron attachment to D_2O clusters embedded in helium droplets. The abscissa has been scaled so that clusters of equal size line up in panels a and b. (Adapted from Zappa, F. et al., *J. Am. Chem. Soc.*, 130, 5573, 2008.)

Once much colder water clusters were synthesized by seeding water vapor in argon (Kim et al., 1998), it became possible to form water cluster anions, including the dimer, hexamer, and heptamer. However, other small cluster anions were still difficult to form, and the tetramer anion remained unobservable, as illustrated by the cluster size distribution shown in Figure 20.10a (Kim et al., 1998). Spectroscopic studies of these water cluster anions include electronic absorption spectra, vibrational (infrared) absorption spectra, photoelectron spectra, and time-resolved femtosecond photoelectron spectra. Photoelectron spectra provide information on the vertical detachment energies, VDE. Three groups of peaks have been identified. Each group shows an approximately linear dependence of the VDE on $n^{-1/3}$, i.e., on the inverse cluster radius evaluated in a continuum approximation. The first group, with the lowest VDE, is observed in the range of $2 \leq n \leq 35$; this group has been attributed to anions in which the excess electrons reside in a very diffuse electron cloud exterior to the cluster, bound in the dipole field of the cluster. The second group appears for $6 \leq n \leq 16$ under standard expansion conditions and up to $n = 200$ for cold expansions; it has been attributed to states in which the excess electron is localized at the surface of the cluster. The third group is observed for $n \geq 11$. It features the highest VDE which extrapolates with increasing size to the water bulk value of 3.4 eV; thus, the electron is thought to reside in a cavity state within the cluster.

This interpretation is, however, controversial. Contentious issues include the numerical accuracy of calculated VDEs, the existence of several rather than just one bound surface states, the lack of correlation between cluster energy and VDE, the difficulties of identifying the most stable anion structures in theoretical work, the possibility that cluster anions formed by attachment to cold, preexisting clusters fail to relax into the most stable structures, and the effect of cluster temperature in the experiment.

When water clusters are grown in helium droplets and electrons of 2 eV are attached, one observes the mass spectrum shown in Figure 20.10b (Zappa et al., 2008). Clusters of size 2, 6, and 7 are still enhanced over adjacent sizes, but sizes 8, 9, 10 are nearly as abundant, and no increase in the abundance at 11 is observed, in stark contrast to the spectrum obtained when seeding water vapor with argon (Figure 20.10a). Furthermore, the "missing" tetramer anion is clearly observable (the fact that heavy water was used when recording the spectrum in the lower panel has only a minor effect on the cluster size distribution). Note that the spectrum in the lower panel shows, in addition to a few impurity peaks, clear evidence for water cluster anions that have one or more helium atoms attached, thus testifying to the extremely low temperature of the ions.

What causes these differences? One may think of three possible reasons for the low abundance of, say, the tetramer anion in earlier experiments: (a) Energetics or kinetics disfavor the formation of the neutral precursor; (b) the tetramer features a particularly low adiabatic electron affinity which results in a short lifetime of a thermally excited anion; (c) the vertical detachment energy of the tetramer in the configuration of the neutral precursor is particularly low.

Explanation (a) can be discarded. For example, the lowest energy forms of the trimer, tetramer, and pentamer have ring structure; growth occurs readily by successive ring insertion, even at sub-Kelvin temperatures (Nauta and Miller, 2000). More generally, there is no reason why the abundance of the neutral tetramer should be low in beam experiments; the computed binding energies of small neutral water clusters do not show anomalies that would correlate with the anion abundances.

Explanation (b) appears to provide a natural explanation to the experimental results; even the slightest thermal excitation of the neutral prior to electron attachment would result in very short anion lifetimes. However, high-level computations suggest that the adiabatic electron affinity of the "anti-magic" clusters ($n = 4$ in particular) is *negative*. If that is true, long-lived anions might be formed more readily by electron attachment to neutrals that are frozen into nonequilibrium structures, explanation (c). Indeed, detailed comparisons of computed VDEs with photoelectron spectra has led to the conclusion that the anions probed experimentally are not the energetically lowest ones but the ones with high vertical detachment energies. However, the large number of local minima on the potential energy surface makes it difficult to decide what structures are relevant in the experiment. An explanation for the rather smooth abundance spectrum in Figure 20.10b could be the freezing-in of metastable water cluster structures as they grow by successive monomer addition in the ultracold helium matrix. Miller and coworkers have shown that structures grown in helium may differ from those obtained in gas-phase nucleation (Nauta and Miller, 2000). In particular, the presumably metastable cyclic structure was observed for the hexamer, whereas the tetramer and pentamer occurred in the same, cyclic structures as in experiments at higher temperatures. If a fraction of the neutral water clusters grows in linear

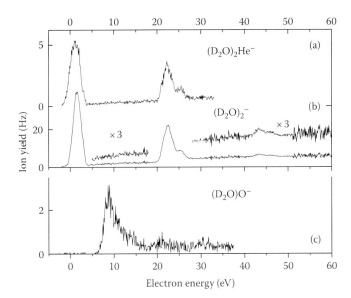

FIGURE 20.11 Anions formed by resonant electron attachment to D$_2$O clusters embedded in helium. The resonances above 20 eV arise from electronic excitations of the helium by the primary electrons. (Adapted from Zappa, F. et al., *J. Am. Chem. Soc.*, 130, 5573, 2008.)

structures, their dipole moments would give rise to a positive adiabatic electron affinity for all sizes.

Anions formed by electron attachment at elevated energies frequently show additional resonances that are displaced by ≈21 eV from the first main resonance. An example is shown in Figure 20.11, where the strong resonance of the dimer at 1.5 eV is followed by another distinct resonance at 22.5 eV with a satellite peak at 25.4 eV. The features can be attributed to the electronic excitation of a helium atom in the droplet by the hot electron that is thermalized in the process and subsequently trapped at the water cluster; a process analogous to that discussed in Section 20.2.3.3. The anion yield of the embedded molecule thus reflects the excitation spectrum of the medium. Another feature is seen shifted upward by another 20.9 eV; it corresponds to the excitations of two separate entities in the medium. Within the statistical uncertainty, $(D_2O)_2He^-$ shows the same features as the bare dimer ion. In contrast, $(D_2O)O^-$ shows only one resonance, at 8.8 eV; it is an order of magnitude weaker than the low-energy resonance of $(D_2O)_2^-$.

20.6 Size of Ions Solvated in Helium

Gas-phase studies of ions are well suited to study solvation which is a central concept in chemistry. In these studies, physical properties are measured as a function of the exact number of molecules in the complex. For example, if one determines spectral features in electronic, vibrational, or rotational spectra of a neutral or charged molecule X that is complexed with n neutral molecules Y, one often finds a monotonic change of the feature with increasing n until, at some value n_s, the feature changes much less upon further increase of n, indicating closure of a first solvation shell. Most studies are performed for ions XY_n^+ rather

than neutrals because their mass, and hence the value of n, can be easily determined.

The closure of the first solvation shell may also be inferred from an abrupt change in the evaporation energies D_n^+ of the ions. The most accurate way for their determination (or, more precisely, for the determination of the enthalpy of evaporation) is by determining ion–molecule equilibria in high-pressure mass spectrometric measurements. These measurements are usually limited to very small values of n. However, the closure of a solvation shell is often evident more easily from data that reflect the evaporation energies of cluster ions in a qualitative way. In many experiments, cluster ions are vibrationally excited; they are prone to unimolecular dissociation in a field-free section of the mass spectrometer. Although the quantitative relation between evaporation energies and the size dependence of dissociation rates is complex because the *evaporative ensemble* is, as discussed in Section 20.3.4, neither canonical nor microcanonical, one usually finds that particularly unstable cluster ions (those with small D_n^+ values) are characterized by enhanced dissociation rates, and therefore form local minima in mass spectra. As discussed in Section 20.4, m*agic numbers* in mass spectra often reflect particularly stable cluster sizes, and stepwise drops in the yield of solvated ions indicate the closure of a solvation shell. For nondirectional bonding, icosahedral structures are often energetically favorable, and $n_s = 12$ is a commonly observed number of solvent atoms in the first solvation shell, e.g., for $Xe^+ Xe_n$, O^-Ar_n and NO^-Ar and for the various cations of rare gases and dimeric molecules embedded in helium.

For large size differences between solvent and solute, one finds steps below or above $n = 12$, indicating that the solute is smaller or larger than the solvent molecules, respectively. Figure 20.12 displays a cation mass spectrum of helium droplets doped with I_2. When the partial pressure of I_2 in the pickup cell is low, one observes two prominent ion series, He_n^+ and, beginning at a nominal mass of 127 u, IH_e^+ (upper panel in Figure 20.12). In this mass range, the resolution is no longer sufficient to separate ions based on their different mass deficit but, fortunately, members of the two ion series are separated by 1 u (the mass of I, which is monoisotopic, is 126.90 u, easily distinguished from He_{32}^+ at mass 128.08 u). At mass ≈254, one observes I_2^+ but hardly any $I_2He_n^+$ is seen. This latter ion series becomes prominent when the I_2 pressure in the pickup cell is tripled, as shown in the lower panel. In addition, a series $I_3He_n^+$ appears.

We have recorded similar spectra for helium droplets doped with SF_6, CCl_4, C_6H_5Br, or CH_3I. Among the many ion series that appear, the ones for F^+He_n, Cl^+He_n, Br^+He_n, and $CH_3I^+ He_n$ could be successfully analyzed. Several other ion series were either too weak, or coincided in nominal mass with other ion series. Furthermore, electron attachment to helium droplets doped with SF_6, CCl_4, or C_6H_5Br gave rise to ion series FHe_n^-, $ClHe_n^-$, and $BrHe_n^-$. Although the presence of two isotopes for Cl (mass 35 and 37) as well as Br (mass 79 and 81) adds complexity to the mass spectra, it also helps to disentangle the contributions of different ion series to the same nominal mass. For example, $^{35}Cl_2^{37}Cl^-$ coincides with $^{35}Cl^-He_{18}$ at 107 u, but

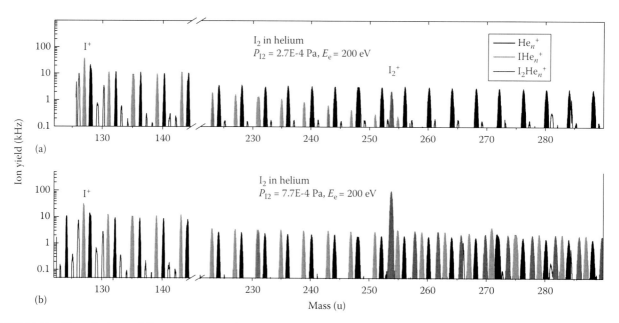

FIGURE 20.12 (**See color insert following page 25-14.**) Cation mass spectrum of helium droplets doped with I_2. Pure He_n^+ peaks are filled black, IH_e^+ peaks are filled red, $I_2He_n^+$ peaks are filled blue. The lower spectrum was recorded with increased I_2 pressure in the pickup cell. (Adapted from Ferreira da Silva, F. et al., *Chem. Eur. J.*, 15, 7101, 2009.)

the overall pattern of the various Cl_3^- isotopomers at 105, 107, 109, and 111 u differs from that of the two Cl^-He_{18} isotopomers at 107 and 109 u.

Figure 20.13 displays the abundance of FHe_n^+, $ClHe_n^+$, FHe_n^-, and $ClHe_n^-$. Each series reveals a distinct stepwise drop in the abundance. The position n_s of the step (obtained from fitting a smeared-out step function to the data, shown as a solid line) is smallest for FHe_n^+ and largest for $ClHe_n^-$. We find $n_s = 10.2$, 11.6, and 13.5 for F^+, Cl^+, and Br^+, respectively, and $n_s = 18.3$, 19.5, and 22.0 for the corresponding anions. Thus, qualitatively, we observe two trends: (a) for a given charge state (either positive or negative), the n_s values of halogen ions increase with increasing atomic number, and (b) n_s values for anions are much larger than for the corresponding cations; the average ratio between anions and cations is 1.7.

It is tempting to estimate the radii of the solvated ions from the measured n_s values. There is no unique way to do this because the numbers primarily reflect the ratio of solvent and solutes radii. If their sizes were equal, one would expect, in a classical model, $n_s = 12$ corresponding to a close-packed shell of either close-packed cubic, hexagonal or, more likely, icosahedral symmetry around the ion, similar to the atomic arrangement found in close-packed particles of rare gases, noble metals, and other elemental systems. We have developed a simple model that assumes a solute size identical to that in bulk helium. The model makes it possible to estimate ionic radii from the observed n_s values; results are displayed in Figure 20.14 as full squares and full circles. Anion radii are, on average, 1.6 times larger than cation radii.

How do these radii compare with other data? Figure 20.14 also shows the radii of anions in alkali halide crystals (upright triangles); they are, on average, only half as large as in helium.

The reason is, obviously, the much stronger binding in the alkali halide crystal.

Comparison is also possible with radii derived from anion mobility data in superfluid bulk helium, and from theoretical work for doped helium droplets. The mobility μ of an ion suspended in a fluid is the ratio of the terminal speed v (also called *drift velocity*) that it reaches in response to an electric field E, i.e., $\mu = v/E$. Because of exchange repulsion between electrons, halide anions reside in rather large cavities. Radii derived for F^- and Cl^- are shown in Figure 20.14 as open squares (Khrapak, 2007); they are slightly smaller than our values.

Concerning theoretical work, radial density profiles $\rho(r)$ and solvent evaporation energies have been computed for halide ions solvated in helium clusters by quantum Monte Carlo Methods (Coccia et al., 2008b). The interaction of the anions with the solvent is very weak, although strong enough to cause solvation. This is in contrast to H^- which is even more weakly bound and remains outside the helium cluster. The solvent forms a very delocalized quantum layer around halide anions, but the computed radial density profiles of the solvent indicate that the anion radii are well defined. Anion radii estimated from the radial density profiles are shown in Figure 20.14 as open diamonds; they agree surprisingly well with our experimental values.

However, a word of caution is in order. First, our method of computing radii from n_s values is classical. Theoretical work indicates that helium atoms in the first coordination shell of *cations* are localized whereas halide *anions* do not seem to induce localization in the first solvation shell. Second, we have assumed the density of bulk helium (Table 20.3), but the interaction between a charge and a polarizable medium is known to lead to *electrostriction*, i.e., an increase of the density in the vicinity of the charge.

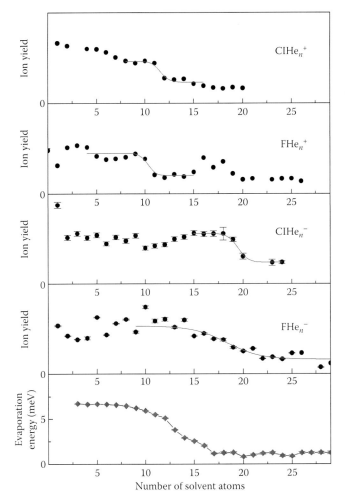

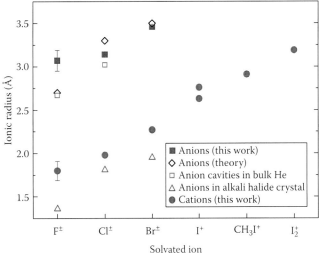

FIGURE 20.14 Radii of anions and cations in helium droplets derived with a classical model from data in Figure 20.13, assuming a bulk helium density. Anionic radii derived from the lattice constants of alkali halide crystals are shown for comparison. Also shown are values resulting from a quantum molecular dynamics study (Coccia et al., 2008b) and from ion mobilities in superfluid bulk helium (Khrapak, 2007; Khrapak and Schmidt, 2008). (Adapted from Ferreira da Silva, F. et al., *Chem. Eur. J.*, 15, 7101, 2009.)

FIGURE 20.13 The abundance of FHe_n^+, $ClHe_n^+$, FHe_n^-, and $ClHe_n^-$ obtained from analysis of mass spectra similar to Figure 20.12. Each ion series reveals a distinct stepwise drop in the abundance. The position n_s of the step is smallest for FHe_n^+ and largest for $ClHe_n^-$. The solid line is the result of fitting a step function. (Adapted from Ferreira da Silva, F. et al., *Chem. Eur. J.*, 15, 7101, 2009.) The bottom panel shows computed evaporation energies of FHe_n^-. (Adapted from Coccia, E. et al., *Chemphyschem*, 2008.)

20.7 Organic and Biomolecules Embedded in Helium Droplets

The old wisdom that damage to living cells requires quanta (photons, electrons, protons, alphas, and so on) that have sufficient energy to form positive ions was shattered when researchers reported that electrons of much lower energy may dissociate plasmid DNA deposited on a surface (Boudaiffa et al., 2000; Sanche, 2005). The subsequent electron attachment studies of nucleobases such as thymine (T) in the gas phase showed that electrons of energies down to 0 eV cause dehydrogenation, resulting in $(T - H)^- + H$ (Denifl et al., 2004). However, the relevance of experiments on gas-phase molecules to radiation biology is questionable because of the lack of a matrix which could drastically alter the reaction dynamics. In fact, when thymine was

embedded in helium droplets, the parent ion T^- could be observed at low electron energies, even though the dehydrogenation reaction was not completely quenched (Denifl et al., 2006b).

At elevated electron energies, the effect of the helium matrix becomes even more striking. Gas-phase (bare) thymine reveals a rich chemistry (Denifl et al., 2004). H^- forms predominantly around 1 eV, and several other fragment anions form resonantly around 5–7 eV. A particularly abundant anion is NCO^-, although its formation requires an elaborate bond cleavage: Two heavy-atom bonds in the ring and a N–H bond at a nitrogen site have to be broken. However, upon electron attachment to thymine embedded in helium, the only anion observed (other than the parent anion at low energies) is $(T–H)^-$ (Denifl et al., 2008).

A comparison of anions formed from bare thymine in the gas phase versus thymine in helium reveals that the latter features a low-energy resonance of T^- and $(T–H)^-$ at 2 eV, about 1.5 eV above the $(T–H)^-$ resonance for gas-phase thymine. This shift reflects the deceleration that an electron suffers upon entering the conduction band in helium, plus perhaps an additional small energy penalty for initiating bubble formation, see Section 20.2.3.3.

In order to facilitate comparison between the different data, we shift the $(T–H)^-$ yield measured for thymine in helium downward by 1.5 eV (solid line in Figure 20.15). $(T–H)^-$ is observed for energies as high as 15 eV without the appearance of any smaller anions. The dotted line shows the yield of $(T–H)^-$ obtained from gas-phase thymine. The dashed line sums the yields of all other anions, the most abundant ones being NCO^-, $CH_2H_3N_2O^-$, and $C_3H_4N^-$ (Denifl et al., 2004). The shape of this dashed curve is close to that of the solid curve, indicating that the various fragment anions formed at elevated energies for gas-phase thymine all contribute to the $(T–H)^-$ yield

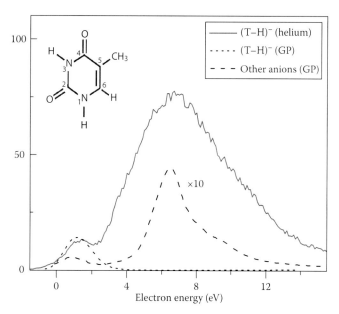

FIGURE 20.15 Electron attachment to thymine (T). Solid line: (T–H)⁻ from T in helium, dotted line: (T–H)⁻ from gas-phase thymine, dashed line: sum of all other anions formed from gas-phase thymine. (Adapted from Denifl, S. et al., *Chemphyschem.*, 9, 1387, 2008.)

for thymine in helium. In other words, dissociation beyond (T–H)⁻ is efficiently quenched by the matrix.

Trinitrotoluene (TNT) presents an even more spectacular example of fragmentation-free ionization that is afforded by the helium matrix. When low-energy electrons (energy well below 1 eV) are attached to free TNT in the gas phase, the resulting mass spectrum reveals 46 different anions; the parent anion TNT⁻ contributes less than 0.01% to the total anion yield. However, when TNT is embedded in helium, the negative ion spectrum at lowest electron energy (again shifted upward by 1.5 eV because electrons need to enter the conduction band) exclusively consists of TNT⁻. No fragment anions are detected, not even when the electron energy is raised to 10 eV (Mauracher et al., 2009).

The impact of the helium droplet on ionization-induced fragmentation in positive ion spectra is less dramatic, and it varies greatly from system to system. Some fragment ion peaks are reduced, others are enhanced, even some new ones may appear that are not observed in the gas phase. No obvious mechanism exists that could rationalize the wealth of data reported in the literature in a coherent way. Ion–molecule reactions induced by doubly charged intermediates (see Equation 20.12, for an example) add to the complexity.

20.8 Summary and Future Perspective

Helium droplets may be seen as a new versatile toolbox to researchers. The box contains tools to play with, tools to solve long-standing questions, and tools to build something new. Attempting to predict the discoveries that have not yet been made would be presumptuous, but we wish to point out a few novel directions that are worth pursuing.

Helium droplets are ultracold. That is a feat for spectroscopic studies because the low temperature greatly simplifies spectra by reducing rovibrational excitations. However, it is also a disadvantage because it is not possible to increase the temperature in a controlled way. Varying the temperature would, e.g., make it possible to identify metastable configurations via annealing, or hot bands in optical spectra. Adding heat in the form of a pulse, e.g., by a pulsed laser, would lead to a spike in the evaporation rate after which the system would quickly return to 0.37 K. Adding heat over an extended time would quickly lead to complete evaporation. Higher temperatures may be achieved by embedding molecules in large neon or argon droplets; their estimated temperatures would be 15 and 50 K, respectively. Indeed, one would have come full circle; pickup of molecules by argon clusters preceded experiments on helium droplets (Gough et al., 1985). However, neon and argon droplets would not be superfluid. They may even be solid, in which case, the dopants would remain stuck on the surface.

The last step in the sequence of ionization events, charge transfer to the dopant, is still poorly characterized; this limits our understanding of subsequent reactions, including fragmentation. A comparison with similar experiments in the gas phase, i.e., collisions between He⁺ and the dopant at low kinetic energies, would provide insight. Furthermore, it is not yet clear if the predominantly "naked" ions that form upon the ionization of molecular dopants form because the excess energy suffices to boil away all the helium, or if they are ejected in a nonthermal process. In the latter case, these bare ions may be quite hot. The measurements of their kinetic energy distributions and their rovibrational excitations would provide answers.

Acknowledgment

This work was supported in part by the Austrian Science Fund (FWF), Wien (projects P19073 and L633), the European Commission, Brussels (ITS-LEIF), and the European Commission, Brussels.

References

Barnett, R. N. and K. B. Whaley. 1993. Variational and diffusion Monte Carlo techniques for quantum clusters. *Phys. Rev. A* 47: 4082–4098.

Becker, E. W., R. Klingerhöfer, and P. Lohse. 1961. Cluster beams. Strahlen aus Kondensiertem Helium in Hochvakuum. *Z. Naturforsch. A* 16: 1259.

Bonhommeau, D., M. Lewerenz, and N. Halberstadt. 2008. Fragmentation of ionized doped helium nanodroplets: Theoretical evidence for a dopant ejection mechanism. *J. Chem. Phys.* 128: 054302.

Boudaiffa, B., P. Cloutier, D. Hunting, M. A. Huels, and L. Sanche. 2000. Resonant formation of DNA strand breaks by low-energy (3 to 20 eV) electrons. *Science* 287: 1658.

Buchenau, H., E. L. Knuth, J. Northby, J. P. Toennies, and C. Winkler. 1990. Mass spectra and time-of-flight distributions of helium cluster beams. *J. Chem. Phys.* 92: 6875–6889.

Callicoatt, B. E., K. Förde, L. F. Jung, T. Ruchti, and K. C. Janda. 1998. Fragmentation of ionized liquid helium droplets: A new interpretation. *J. Chem. Phys.* 109: 10195.

Camillone, N., C. E. D. Chidsey, G. Y. Liu, T. M. Putvinski, and G. Scoles. 1991. Surface-structure and thermal motion of normal-alkane thiols self-assembled on Au(111) studied by low-energy helium diffraction. *J. Chem. Phys.* 94: 8493–8502.

Choi, M. Y., G. E. Douberly, T. M. Falconer et al. 2006. Infrared spectroscopy of helium nanodroplets: Novel methods for physics and chemistry. *Int. Rev. Phys. Chem.* 25: 15–75.

Coccia, E., E. Bodo, and F. A. Gianturco. 2008a. Nanoscopic phase changes in doped ^{4}He droplets. *Europhys. Lett.* 82: 23001.

Coccia, E., F. Marinetti, E. Bodo, and F. A. Gianturco. 2008b. Chemical solutions in a quantum solvent: Anionic electrolytes in ^{4}He nanodroplets. *Chemphyschem.* 9: 1323–1330.

Coe, J. V., A. D. Earhart, M. H. Cohen, G. J. Hoffman, H. W. Sarkas, and K. H. Bowen. 1997. Using cluster studies to approach the electronic structure of bulk water: Reassessing the vacuum level, conduction band edge, and band gap of water. *J. Chem. Phys.* 107: 6023–6031.

Coe, J. V., S. T. Arnold, J. G. Eaton, G. H. Lee, and K. H. Bowen. 2006. Photoelectron spectra of hydrated electron clusters: Fitting line shapes and grouping isomers. *J. Chem. Phys.* 125: 014315.

Cohen, E. R., D. R. Lide, and G. L. Trigg. 2003. *AIP Physics Desk Reference.* New York: Springer.

Cole, M. W. 1974. Electronic surface states of liquid helium. *Rev. Mod. Phys.* 46: 451.

Denifl, S., S. Ptasinska, M. Probst, J. Hrusak, P. Scheier, and T. D. Märk. 2004. Electron attachment to the gas-phase DNA bases cytosine and thymine. *J. Phys. Chem. A* 108: 6562–9.

Denifl, S., M. Stano, A. Stamatovic, P. Scheier, and T. D. Märk. 2006a. Electron-impact ionization of helium clusters close to the threshold: Appearance energies. *J. Chem. Phys.* 124: 054320.

Denifl, S., F. Zappa, I. Mähr et al. 2006b. Mass spectrometric investigation of anions formed upon free electron attachment to nucleobase molecules and clusters embedded in superfluid helium droplets. *Phys. Rev. Lett.* 97: 043201.

Denifl, S., F. Zappa, A. Mauracher et al. 2008. Dissociative electron attachment to DNA bases near absolute zero: Freezing dissociation intermediates. *Chemphyschem.* 9: 1387–1389.

Denifl, S., F. Zappa, A. I. Mähr, et al. 2009. Ion-molecule reactions in helium nanodroplets doped with C_{60} and water clusters. *Angew. Chemie (Int. Ed.).* 48: 8940–8943.

Ebner, W. and P. Leiderer. 1980. Development of the dimple instability on liquid ^{4}He. *Phys. Lett. A* 80: 277.

Edwards, P. P. 1982. The electronic properties of metal solutions in liquid ammonia and related solvents. *Adv. Inorg. Chem.* 25: 135–185.

Ellis, A. M. and S. F. Yang. 2007. Model for the charge-transfer probability in helium nanodroplets following electron-impact ionization. *Phys. Rev. A* 76: 032714, 1–8.

Farnik, M. and J. Toennies. 2005. Ion-molecule reactions in ^{4}He droplets: Flying nano-cryo-reactors. *J. Chem. Phys.* 122: 014307.

Farnik, M., U. Henne, B. Samelin, and J. P. Toennies. 1997. Comparison between positive and negative charging of helium droplets. *Z. Phys. D* 40: 93–98.

Farnik, M., B. Samelin, and J. P. Toennies. 1999. Measurements of the lifetimes of electron bubbles in large size selected ^{4}He droplets. *J. Chem. Phys.* 110: 9195–9201.

Ferreira da Silva, F., P. Waldburger, S. Jaksch et al. 2009. On the size of ions solvated in helium clusters. *Chem. Eur. J.* 15: 7101–7108.

Gough, T. E., M. Mengel, P. A. Rowntree, and G. Scoles. 1985. Infrared spectroscopy at the surface of clusters: SF_6 on Ar. *J. Chem. Phys.* 83: 4958.

Goyal, S., D. L. Schutt, and G. Scoles. 1992. Vibrational spectroscopy of sulfur-hexafluoride attached to helium clusters. *Phys. Rev. Lett.* 69: 933–936.

Grisenti, R. E., W. Schöllkopf, J. P. Toennies, G. C. Hegerfeldt, T. Köhler, and M. Stoll. 2000. Determination of the bond length and binding energy of the helium dimer by diffraction from a transmission grating. *Phys. Rev. Lett.* 85: 2284.

Gspann, J. and H. Vollmar. 1980. Metastable excitations of large clusters of ^{3}He, ^{4}He, or Ne atoms. *J. Chem. Phys.* 73: 1657.

Hart, E. J. and J. W. Boag. 1962. Absorption spectrum of the hydrated electron in water and in aqueous solutions. *J. Am. Chem. Soc.* 84: 4090–4095.

Henne, U. and J. P. Toennies. 1998. Electron capture by large helium droplets. *J. Chem. Phys.* 108: 9327–9338.

Jaksch, S., A. Mauracher, A. Bacher et al. 2008. Formation of even-numbered hydrogen cluster cations in ultracold helium droplets. *J. Chem. Phys.* 129: 224306.

Jungclas, H., R. D. Macfarlane, and Y. Fares. 1971. Evidence for large-molecular-cluster formation of nuclear reaction recoils thermalized in helium. *Phys. Rev. Lett.* 27: 556–557.

Khrapak, A. G. 2007. Structure of negative impurity ions in liquid helium. *JETP Lett.* 86: 252–255.

Khrapak, A. G. and W. F. Schmidt. 2008. Negative ions in nonpolar liquids. *Int. J. Mass Spectrom.* 277: 236–239.

Kim, J., I. Becker, O. Cheshnovsky, and M. A. Johnson. 1998. Photoelectron spectroscopy of the "missing" hydrated electron clusters $(H_2O)_n$, $n = 3, 5, 8$ and 9: Isomers and continuity with the dominant clusters $n = 6, 7$ and ≥ 11. *Chem. Phys. Lett.* 297: 90–96.

Klots, C. E. 1988. Evaporation from small particles. *J. Phys. Chem.* 92: 5864–5868.

Kupper, J. and J. M. Merritt. 2007. Spectroscopy of free radicals and radical containing entrance-channel complexes in superfluid helium nanodroplets. *Int. Rev. Phys. Chem.* 26: 249–287.

Leiderer, P. 2009. *Private Communication.*

Lewerenz, M., B. Schilling, and J. P. Toennies. 1995. Successive capture and coagulation of atoms and molecules to small clusters in large liquid helium clusters. *J. Chem. Phys.* 102: 8191.

Lewis, W. K., C. M. Lindsay, and R. E. Miller. 2008. Ionization and fragmentation of isomeric van der Waals complexes embedded in helium nanodroplets. *J. Chem. Phys.* 129: 201101, 1–4.

Lide, D. R. 2000. *CRC Handbook of Chemistry and Physics*. Boca Raton, FL: CRC Press.

Mähr, I., F. Zappa, S. Denifl et al. 2007. Multiply charged neon clusters: Failure of the liquid drop model? *Phys. Rev. Lett.* 98: 023401.

Martin, T. P. 1996. Shells of atoms. *Phys. Rep.* 273: 199–241.

Mauracher, A., H. Schöbel, F. F. da Silva et al. 2009. Electron attachment to trinitrotoluene (TNT) embedded in He droplets: Complete freezing of dissociation intermediates in an extended range of electron energies. *Phys. Chem. Chem. Phys.* 37: 8240–8243.

Mostafavi, M. and I. Lampre. 2008. The solvated electron: A singular chemical species. *Radiation Chemistry: From Basics to Applications in Material and Life Sciences*, J. Belloni, T. Douki, M. Mostafavi, and M. Spotheim-Maurizot (Eds.), pp. 35–52. Les Ulis, France: EDP Sciences.

Nauta, K. and R. E. Miller. 1999. Nonequilibrium self-assembly of long chains of polar molecules in superfluid helium. *Science* 283: 1895–1897.

Nauta, K. and R. E. Miller. 2000. Formation of cyclic water hexamer in liquid helium: The smallest piece of ice. *Science* 287: 293–295.

NIST (2009). *NIST Chemistry WebBook*.

Northby, J. A. 2001. Experimental studies of helium droplets. *J. Chem. Phys.* 115: 10065–10077.

Paesani, F. 2010. Superfluidity of clusters. *Handbook of Nanophysics*, Vol. 7 (Clusters and Fullerenes). K. Sattler (Ed.). New York: CRC.

Radzig, A. A. and B. M. Smirnov. 1985. *Reference Data on Atoms, Molecules, and Ions*. Heidelberg, Germany: Springer.

Ren, Y. and V. V. Kresin. 2008. Suppressing the fragmentation of fragile molecules in helium nanodroplets by co-embedding with water: Possible role of the electric dipole moment. *J. Chem. Phys.* 128: 074303, 1–4.

Rosenblit, M. and J. Jortner. 2006. Electron bubbles in helium clusters. I. Structure and energetics. *J. Chem. Phys.* 124: 194505.

Ruchti, T., B. E. Callicoatt, and K. C. Janda. 2000. Charge transfer and fragmentation of liquid helium droplets doped with xenon. *Phys. Chem. Chem. Phys.* 2: 4075–4080.

Sanche, L. 2005. Low energy electron-driven damage in biomolecules. *Eur. Phys. J. D* 35: 367–390.

Sanov, A. and W. C. Lineberger. 2004. Cluster anions: Structure, interactions, and dynamics in the sub-nanoscale regime. *Phys. Chem. Chem. Phys.* 6: 2018–2032.

Seong, J., K. C. Janda, N. Halberstadt, and F. Spiegelmann. 1998. Short-time charge motion in He_n^+ clusters. *J. Chem. Phys.* 109: 10873–10884.

Tachikawa, H. 2000. Full dimensional ab initio direct dynamics calculations of the ionization of H_2 clusters ($(H_2)_n$ ($n = 3, 4$ and 6). *Phys. Chem. Chem. Phys.* 2: 4702–4707.

Toennies, J. P. and A. F. Vilesov. 2004. Superfluid helium droplets: A uniquely cold nanomatrix for molecules and molecular complexes. *Angew. Chemie (Int. Ed.)* 43: 2622–2648.

Toennies, J. P., A. F. Vilesov, and K. B. Whaley. 2001. Superfluid helium droplets: An ultracold nanolaboratory. *Phys. Today* 54: 31–37.

von Haeften, K., T. Laarmann, H. Wabnitz, and T. Möller. 2001. Observation of atomiclike electronic excitations in pure ^{3}He and ^{4}He clusters studied by fluorescence excitation spectroscopy. *Phys. Rev. Lett.* 87: 153403.

von Haeften, K., T. Laarmann, H. Wabnitz, and T. Möller. 2002. Bubble formation and decay in ^{3}He and ^{4}He clusters. *Phys. Rev. Lett.* 88: 233401.

Wang, C. C., O. Kornilov, O. Gessner, J. H. Kim, D. S. Peterka, and D. M. Neumark. 2008. Photoelectron imaging of Helium droplets doped with Xe and Kr atoms. *J. Phys. Chem. A* 112: 9356–9365.

Yang, S. F., S. M. Brereton, M. D. Wheeler, and A. M. Ellis. 2006. Electron impact ionization of haloalkanes in helium nanodroplets. *J. Phys. Chem. A* 110: 1791–1797.

Zappa, F., S. Denifl, I. Mähr et al. 2008. Ultracold water cluster anions. *J. Am. Chem. Soc.* 130: 5573–5578.

III

Production and Stability of Carbon Fullerenes

Plasma Synthesis of Fullerenes

Keun Su Kim
Université de Sherbrooke

Gervais Soucy
Université de Sherbrooke

21.1 Introduction

Fullerene is the general name designated for the closed-caged carbon-based molecules that are made up entirely of both pentagonal and hexagonal plane structures, usually referred to as the third allotrope of pure carbon (Goodson et al. 1995). The intriguing properties of fullerenes have attracted much attention for their use in a wide range of applications and have thus stimulated intensive research work on novel fullerene-based materials in many fields, including the electronic, optical, biomedical, polymer, energy, and environmental industries (Singh and Srivastava 1995; Smalley and Yakobson 1998; Da Ros and Prato 1999).

It was initially shown that the production of fullerene is feasible by means of an irradiating laser beam, focused on a graphite target placed in a helium atmosphere (Kroto et al. 1985). While the amount of fullerenes produced in this initial experiment was enough to prove the presence of fullerenes, the overall yield rate of the fullerene was insufficient for further exploration of its potential applications in various industrial fields or its scientific investigation. Because of this low productivity, there have been tremendous efforts to develop new synthesis methods leading to the economical production of fullerenes at much larger scales. Thus, to date, it has been demonstrated that various fullerenes can be produced by utilizing the different ways of generating the initial carbon vapors, followed by the formation of carbon clusters, such as the arc discharge (Kratschmer et al. 1990a), arc-jet plasmas (Yoshie et al. 1992), nonequilibrium plasmas (Inomata et al. 1994), solar flux (Chibante et al. 1993), resistive or inductive heating of a graphite rod (Diederich et al. 1991), sputtering or ion beam

evaporation (Bunshah et al. 1992), and combustion methods (Howard et al. 1991). In many cases, it is true that the fullerene synthesis is accompanied by the use of various kinds of plasma environments, and the performance of these plasma-based processes has already proved to be superior to other competing fullerene synthesis methods. For instance, the very high reaction temperatures, required for the production of the high-quality fullerenes, are readily achievable over the temperature range from 2,000 to 10,000 K, which is not obtainable by the conventional chemical reactions (Boulos 1991). Highly energetic electrons, accelerated by means of electric fields, can also generate the abundant species of radicals through the continuous collisions with neutral species. Most importantly, the plasma process is intrinsically clean, compact, rapid, and easily scaled up, along with good flexibility with respect to controlling the quality of end-products (Ostrikov and Murphy 2007). Thus, in this context, the plasma synthesis of fullerene becomes an issue of considerable interest, both for application in the basic research and for the subsequent mass production of fullerenes.

A primary concern of this chapter is that of reviewing the current status of the plasma techniques, developed to date for the fullerene synthesis. The chapter begins with a review of the history of the fullerene discovery, and a brief introduction to plasmas. A detailed review of the plasma-based fullerene synthesis then follows in Section 21.3. Section 21.4 is devoted to a discussion on the plasma characteristics observed during the fullerene synthesis. Finally, in an effort to provide for a better insight into optimizing the process, plasma conditions, suitable for the fullerene production process, are discussed in Section 21.5.

21.2 Fullerenes and Plasmas

21.2.1 History of the Fullerene Discovery

Although there have been some conjectures offered on the existence of stable carbon clusters, in the form of closed-cage, since 1970 (Osawa 1970), it took some decades to discover the new carbon allotrope of fullerene and its related family. In 1984, Rohlfing et al. (1984) at the Exxon Company first demonstrated the formation of carbon clusters, C_n, with values of n ranging from 2 to nearly 200, using a laser-vaporization method. It was observed during this experiment that only even-n clusters were preferentially formed for n values higher than 30 and the peak of 60-carbon clusters (C_{60}) in the mass spectrum was about 20% more intense than those of its neighbors. However, no further attention was given to the origin of this anomalous large yield of C_{60} generated in this work, because the peak corresponding to the C_{60} content was not completely dominant in respect of the neighboring peaks.

A year later, fullerene was discovered by H. Kroto, J. Heath, S. O'Brien, R. Curl, and R. Smalley (Kroto et al. 1985) using a similar technique to that of Rohlfing et al. In their experiments, the carbon vapor produced was mainly composed of species that display mass spectra peaks corresponding to the C_{60} and C_{70} species. They showed that by increasing helium density at the time of the laser pulse, as well as extending the time between the vaporization and expansion phases, the C_{60} peak intensity could be increased some 40-fold in comparison to its neighboring peaks. The relative high stability of the C_{60} species displayed in this work raises an interesting question as to what kind of structure of the 60-carbon cluster is responsible for its remarkable stability. Even though it was impossible to substantially determine the structure of the C_{60} species from this experiment, Smalley, Kroto, and Curl were able to hypothesize on the structure of this molecule by proposing that it consists of a truncated icosahedron that resembles exactly the shape of a soccer ball (i.e., a carbon molecule sits at each of the 60 vertices that are identical to the junctions of two hexagons and a pentagon, see Figure 21.1 and Table 21.1). In 1985, Kroto et al. (1985) suggested a truncated icosahedron structure for this stable molecule and subsequently this work was awarded the 1996 Nobel Prize for Chemistry. Kroto et al. named this C_{60} structure as *buckminsterfullerene*

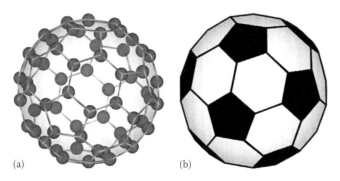

FIGURE 21.1 The structure of the C_{60} fullerene: (a) the truncated icosahedral structure of a C_{60} fullerene; (b) a soccer ball.

TABLE 21.1 The Structural Properties of C_{60} and C_{70} Fullerenes

Features	C_{60}	C_{70}	Ref.
Symmetry	I_h	D_{5h}	a
Pentagons	12	12	a
Hexagons	20	25	a
Faces	32	37	a
C–C (6–5) bond length (Å)	1.45	Eight kinds of bonds in the range 1.37–1.46	a
C=C (6–6) bond length (Å)	1.38		
Diameter (Å)	6.83	Equatorial diameter: 6.94 height: 7.8	a
Total energy of the C–C bond (kJ/mol)	40,390	45,807	b
Color of toluene solution	Purple	Deep magenta	a

    [a] Geckeler and Samal (1999).
    [b] Belousov et al. (1997).

after M. Buckminster Fuller, an architect and constructor of the geodesic dome.

At the beginning, the C_{60} material could only be produced in tiny amounts, thus only a few kinds of experiments could be undertaken on this new species. Another problem with the initial fullerene and its related family was that they were produced only in the form of a molecular beam instead of a solid. All this changed in 1990, when Kratschmer et al. (1990a) reported on a novel method for the production of C_{60} in much larger quantities by creating an electric arc between two graphite rods placed in a helium atmosphere. In this experiment, macroscopic amounts of carbon soot consisting of crystallized buckyballs (i.e., solid-state C_{60}) were successfully produced and, subsequently, the soccer ball structure of C_{60}, as conjectured by Kroto et al., was confirmed experimentally (Kratschmer et al. 1990b; Taylor et al. 1990). This new synthesis method has opened up new possibilities for further experimental investigations on this material, as well as triggering intensive research on the industrial-scale production of fullerenes.

21.2.2 Plasmas

By definition, plasma is a fully or partially ionized gas, which consists of a mixture of electrons, ions, and neutral species (Boulos 1991) (see Figure 21.2). The inherent property of the plasma, known as quasi-neutrality, is to conserve its overall charge neutrality (i.e., the total number of negative charges in the plasma is equal to that of the positive charges) and is an important criterion for being a plasma. The free charges present in the plasma usually enhance the electrical conductivity of the gases, offering many unique features attributable to the long-range electromagnetic interactions between the charged particles, which do not occur in ordinary neutral gases. Presently, various kinds of man-made plasmas exist in the world and they can be produced in many different ways depending on eventual applications. For example, laboratory-scale plasmas are mostly generated by the passage of an electric current through the gas being exposed to

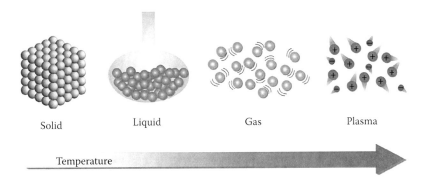

FIGURE 21.2 States of materials: solid, liquid, gas, and plasma.

strong electric fields; however, plasmas can be also created by irradiating laser or directing high-energetic particle beams on a target or by heating gases in high-temperature furnaces.

The three species in the plasma (i.e., electrons, ions, and neutrals) are in thermally induced motion making numerous collisions between each other and, in plasma science terms, their average kinetic energies are measured as the temperature of each species, i, which is defined as follows:

$$T_i = \frac{2\langle E_i \rangle}{3k} = \frac{m_i \overline{\upsilon_i^2}}{3k} \qquad (21.1)$$

where

m_i is the mass (kg) of the species i
$\langle E_i \rangle$ is its average kinetic energy (J)
$\overline{\upsilon_i^2}$ is its root-mean-square (RMS) velocity (m/s)
k is the Boltzmann constant (J/K)

In general, the electron temperatures in the plasma may be different from those of the heavy particles (ions and neutrals) because the electrical energy transferred to the plasma is primarily absorbed by the lightest-charged particles (electrons). The electrons may transfer some fraction of their excess energy to the heavy particles through the elastic or inelastic collision process. However, the energy exchange between electrons and the heavy species in a collision is very inefficient, due to the small mass of the electrons (i.e., many collisions over 10^3 times are required for the relaxation of the energy difference).

Depending on the temperature of each species, plasmas are typically classified into two major categories, i.e., "hot" or "equilibrium" plasmas and "cold" or "nonequilibrium" plasmas, as shown in Figure 21.3. In the former case, the plasmas are characterized by high energy densities, a relatively high electron density (10^{23}–10^{28} m^{-3}), and most importantly, an approximate equality between the temperatures of heavy species and those of the electrons. In other words, the thermodynamic state of the plasma approaches that of an equilibrium or, more precisely, a local thermodynamic equilibrium (LTE). As mentioned above, this is possible when collisions between electrons and heavy particles are frequent. A typical example of the hot plasma is the *nuclear fusion plasma*, in which the species temperatures

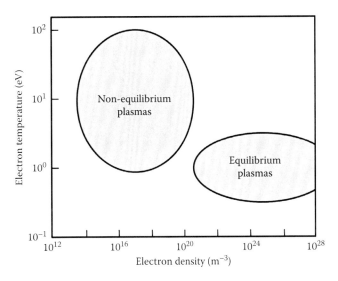

FIGURE 21.3 Typical ranges of temperatures and electron densities for equilibrium and nonequilibrium processing plasmas. (Reproduced from Boulos, M.I., *IEEE Trans. Plasma Sci.*, 19(6), 1078, 1991. With permission.)

can reach a few thousand eV on even more (in plasma science, temperature has a unit of electron volt: 1 eV ≈ 11,600 K). At such high temperatures, gases are fully ionized and the plasma contains only electrons and ions. Another important example of hot plasma is the *thermal plasma*, being those produced from high-intensity arc discharges or plasma torches. In thermal plasmas, however, a large portion of molecules and atoms remain in the neutral state, corresponding to a low degree of ionization, and species temperatures typically range from fractions to a few tens of eV. Most of the plasma processes developed to date based on thermal plasma technology take advantage of this high temperature of the neutral species (Fauchais and Vardelle 1997).

On the other hand, cold plasmas are characterized by a low energy density and a large difference between the temperatures of the electrons and heavy particles (i.e., ions and neutrals are at room temperature levels, whilst the electron temperature is at 1–2 eV) along with a low electron density of less than 10^{20} m^{-3}. Typical cold plasma examples are those produced within various types of low-pressure glow discharges, in corona discharges, in dielectric barrier discharges, and in atmospheric pressure RF glow discharges. In cold plasma processes, high-energetic electrons commonly play important roles in the fabrication or synthesizing of materials.

As mentioned above, the collision rates achieved between the species are crucial in determining whether particular plasma is in a thermal equilibrium state or not. It is obvious that the collision rates would be enhanced as the particle density or pressure increases, while the high-level electric field is likely to preferentially boost the electron energy, thus departing from the equilibrium state. Therefore, hot plasmas are usually characterized by small values of E/p (E: electric field intensity, p: pressure), but values of this E/p parameter for cold plasmas are higher than those for hot plasmas, by several orders of magnitude. For this reason, hot plasmas typically occur in discharges at high pressures but cold plasmas prevail in low-pressure environments. However, it is worth noting that cold plasmas can occur at around atmospheric pressure (e.g., corona, spark, dielectric barrier, gliding arc discharge, and atmospheric pressure glow discharges) (Schutze et al. 1998). Although the collision rates in these plasmas are high enough, the plasma does not reach the thermal equilibrium condition because the discharges are interrupted before the glow-to-arc transition takes place.

21.3 Plasma Processes for Fullerene Synthesis

Among the various kinds of plasmas, the thermal plasma has been widely used for the production of fullerenes, owing to its high temperature (1,000–15,000 K), high energy density, and abundant content of highly reactive species of ions and neutral species (Huczko et al. 2002; Gonzalez-Aguilar et al. 2007). Even though cold plasma processing has been attracting much attention for application to the fullerene synthesis, due to some of its unique features, research in this area is not as extensive or as well developed when compared to the thermal plasma case. Thus, our review will mainly focus on the fullerene synthesis by means of the thermal plasma technology.

21.3.1 Laser Ablation Method

Laser ablation is the process of removing material from a solid surface by irradiating it with a laser beam. During the laser irradiation, materials are evaporated by absorbing the heat delivered from the laser and their vapors are subsequently converted to the plasma. Upon cooling through the controlled expansion of gas, the vaporized atoms tend to coalesce and finally form clusters. This laser ablation method was originally developed at Rice University in 1980–1981 (the so-called laser-vaporization supersonic cluster beam technique) to study clusters of refractory metals (Dietz et al. 1981) and semiconductors (Heath et al. 1985), and then applied to the production of carbon clusters by researchers within the Exxon and Smalley groups. The first observation of a 720 amu peak due to C_{60} in the mass spectrum was from the analysis of the carbon plume produced by the laser ablation of a graphite disc, using a Nd:YAG laser operating at 532 nm with a pulse energy of ~30 mJ (Kroto et al. 1985).

The failure of the initial attempt to prepare fullerene at large scales was mainly due to the fact that the carbon plasma, created during the laser vaporization, cools too rapidly (Lieber and Chen 1994). A fast cooling of the gas does not provide sufficient time for the growing carbon clusters to reorganize themselves into the stable fullerene structures of perfect cages. The original laser ablation apparatus has been modified by Lieber and Chen (1994), the key feature of this new apparatus being that a graphite disc is ablated in a high-temperature furnace, as shown in Figure 21.4a. These researchers have found that the fullerene production is most efficient at 1200°C and heating the graphite disc to 1000°C results in higher yields of C_{60}.

The laser ablation technique, despite yielding only small amounts of fullerenes, is still important because it is possible to systematically vary the characteristics of the carbon plasma by adjusting the laser pulse energy, wavelength, buffer gas pressure, and furnace temperature. This flexibility of the laser ablation method allows for considerable control to be performed over the fullerene formation kinetics.

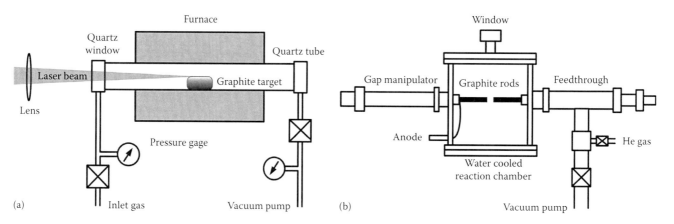

FIGURE 21.4 Plasma processes developed in the early studies of the fullerene synthesis: (a) laser ablation process; (b) arc discharge process. (Reproduced from Lieber, C.M. and Chen, C.C., Preparation of fullerenes and fullerene-based materials, in *Solid State Physics—Advances in Research and Applications*, vol. 48, Ehrenreich, H. and Spaepen, F. (eds.), Academic Press, San Diego, CA, 109, 1994. With permission.)

21.3.2 Arc Discharge Method

The electric arc discharge, first used by R. Bacon in the early 1960s to make carbon whiskers (Bacon 1960), has certainly been the most efficient technique for the production of large quantities of carbon soot at low cost. For this reason, the synthesis of fullerenes at bulk scale is now routinely carried out, using the technology of Kratschmer et al. (1990a). These authors reported their first synthesis of fullerenes through this method in 1990. In this process, the vaporization of the input carbon sources is achieved by the electric arc formed between two electrodes and the fullerenes so produced are mixed with the soot deposited on the walls of the reactor.

Figure 21.4b shows a typical carbon arc reactor designed for the synthesis of fullerene-containing soot, using two high-purity graphite rods contacting the high-current feedthroughs. To start the discharge, the reaction chamber is evacuated to a pressure of less than 1–3 Torr and is refilled with helium gas to a pressure of 150–250 Torr, in order to remove unwanted oxygen and water vapors, as these species have adverse effects on the formation of the fullerenes. At the process start-up, the electrodes just touch each other so that the arc current can pass through the rods, then they are separated to initiate the desired high-intensity arc discharge between them. The interelectrode gap is usually kept constant during the process operations at between 1 and 10 mm, by controlling the movable electrode. Finally, carbon vapors and their small clusters are continuously produced through the sublimation of the graphite anode by means of an electric arc, maintained between the two electrodes. Typically, a current of 100–200 A with an associated voltage drop of 10–20 V results in abundant soot deposition on the reactor wall. This soot usually contains as much as 10%–15% of soluble fullerenes (being extractable by toluene solvent using a Soxhlet apparatus) depending on the operating conditions employed. The fullerene extracted from the soot typically consists of 80% C_{60} and 15% C_{70}, the residues consisting mainly of the higher fullerenes, such as C_{76} and C_{78}.

Changes in the operating parameters, such as the electrode size, geometry and composition, pressure, interelectrode distance, and arc current, all bring about important modifications of the plasma properties. Thus, there have been systematic efforts to study the effect of the operating parameters on the fullerene yield. Hare et al. (1991) have found that the fullerene content varies within a 5%–10% range depending on the types of graphite rods employed while Parker et al. (1991) accomplished the very high yield of fullerenes (44%) by optimizing the interelectrode gap and further developing an advanced fullerene extraction scheme. It has been demonstrated by Saito et al. (1992) that optimization of the buffer gas pressure would increase the fullerene content present in the soot by up to 13% (e.g., 20 Torr, helium) because the operating pressure in the arc discharge method has an influence on both the cooling rate and the diffusion speed of the carbon vapors so produced. Huczko et al. (1997) optimized the interelectrode gap and arc current supplied to the electrodes to achieve a fullerene yield rate as high as 20%. It was suggested in this work that there exist optimum values for the electrode gap

and arc current employed because they have a significant effect on both the evaporation rate and intensity of the UV radiation escaping from the arc zone. For example, the fullerene yield has a maximum value at an arc current of 100 A if the interelectrode space is about 4 mm. Sugai et al. (2000) produced fullerenes with an increased yield rate through use of a pulsed arc discharge, generated in a tubular quartz reactor heated by a furnace. In this work, the authors claimed that offering sufficient growth time for the carbon clusters by using such a furnace might lead to an increase in the fullerene yield rate. Similarly, Sesli et al. (2005) achieved a fullerene yield of 32% by performing arc discharges inside a tubular graphite reactor with a helium flow.

The initial version of the arc discharge technique has a serious drawback in application to the synthesis of fullerenes. This process can hardly be scaled up because the fullerene yield dramatically decreases as the size of either the electrode or the arc current increases, due to the UV radiation emitted from the arc. With increasing electrode size and arc current, the intensity of the UV radiation emitted from the arc becomes stronger and this typically results in an acceleration of the photochemical decomposition of the previously produced fullerenes. In order to minimize this undesirable photochemistry, it is necessary to transport the grown clusters into a relatively dark zone (i.e., the low-temperature zone) as soon as possible. Dubrovsky and Bezmelnitsyn (2004) injected a buffer gas into the electrode gap and created a strong radial outflow to thereby assist the evacuation of the previously produced carbon clusters from the radiating arc zone. By introducing a 3 standard liters per minute (slpm) helium flow through the gap, the fullerene yield was increased from 7.4% to 17.3%.

Finally, there have been a couple of attempts made to increase the fullerene yield rate by performing the arc discharges under the influence of a magnetic field (Bhuiyan and Mieno 2002) or under gravity-free environmental conditions (Mieno 2004).

21.3.3 Arc-Jet Plasma Method

Although studies involving the arc discharge method are still being reported, they are typically limited to elementary studies because the arc discharge process cannot be scaled to match the scale of industrial production quantities due to the UV radiation as mentioned earlier. In addition, most critically, the supply of the raw carbon material (i.e., the graphite rod) in this process is also discontinuous. This situation is why alternative processes for the fullerene synthesis are under continuous development.

In order to overcome these shortcomings within the arc discharge method, considerable efforts have recently been made to continuously synthesize fullerenes by the direct evaporation of the carbon-containing materials within arc-jet plasmas, the latter being generated by means of various types of plasma torches. Arc-jet plasmas can be created by the injection of a plasma-forming gas directly into established electric arcs. The fundamental difference of the arc-jet plasma method compared with that of the arc discharge approach is that the rate of input carbon is not limited by the rate of electrode erosion and is also independently controllable. Moreover, the high velocity of the plasma jet is expected to enhance the formation of carbon clusters, by increasing the quenching rate of the carbon vapor and also provide for fast removal of the generated fullerenes from the radiating arc zone, thereby preventing them from being destroyed by the UV irradiation. According to the mode of arc-jet plasma generation, this method can be further subdivided into four major categories: (1) the direct current (DC) plasma process, (2) the alternating current (AC) plasma process, (3) the radio-frequency (RF) inductively coupled plasma process, and (4) the DC–RF hybrid plasma process.

DC thermal plasma torches consist of three main structural elements: a cathode, an anode, and a gas injection channel, and can be subdivided into two groups, according to the shape of the cathode: i.e., rod-type cathode torches and hollow-type cathode torches. The rod-type cathode is made of refractory metals e.g., tungsten, and should not be used with oxidizing gases. The hollow-type torch consists basically of two cylindrical electrodes, one has a bottom wall, and the generated arc inside this torch is stabilized by the strong vortex motion of the plasma gas. The hollow-type cathode allows the direct use of oxidizing plasma-forming gases because of their different electron emission mechanism (i.e., an explosive emission process rather than a thermionic emission) and it is easy for the torch to be scaled up to provide MW processing power levels. By means of a hollow-type cathode torch, Alexakis et al. (1997) have developed the so-called PyroGenesis process (45–70 kW, 200–760 Torr). In this process, carbon chloride compounds such as chloroform or tetrachloroethylene (C_2Cl_4) were used as the carbon source. In using C_2Cl_4 as the carbon source, a 3% conversion of the C_2Cl_4 to fullerenic soot has been achieved, the fullerene yield extracted from the soot being 5.3%. The main drawbacks for this technology are the difficulties associated with working under the corrosive atmosphere and the treatment of hazardous by-products.

Dubrovsky et al. (2004) have reported on fullerene synthesis from carbon black powders, using an argon arc-jet plasma generated by use of a rod-type cathode torch (12 kW, see Figure 21.5a), but the yield rate of the extractable fullerenes was turning out to be as low as ~2%.

Fulcheri et al. (2000) has developed a three-phase AC plasma process (50 Hz, 30–50 kW). In this process, an electric arc is established between three graphite electrodes which are powered by a three-phase AC power supply (20–260 kW, 250–400 A) operating at 50–600 Hz (see Figure 21.5b). Before undertaking optimization efforts, a yield rate of 3.5% was obtained with acetylene black processing at atmospheric pressure. However, fullerene products were not detected in the soot produced with cokes and carbon blacks. These results reveal that the purity of the raw carbon materials is crucial for fullerene formation and that the presence of oxygen, hydrogen, and sulfur, even at low concentrations, has a negative effect on the overall fullerene production.

RF induction plasmas are electrodeless discharge, in which a coupling of the RF electrical energy in a copper coil into the plasma gas is accomplished by creating eddy currents in the plasma gas through the magnetic induction process (Boulos 1985, 1997). The excitation frequency is typically between 200 kHz and 40 MHz, and laboratory units are operated at power levels of 30–50 kW while larger-scale industrial units have been tested to a power level of 1 MW. To date, considerable attempts have been made to use the RF induction plasma technology for fullerene production, due to the following unique advantages: (1) a relatively large plasma volume and a lower plasma velocity compared to those of DC or AC plasma jets, and which can provide for longer residence time for feedstock materials inside the arc zone, (2) the RF plasma torch is an almost maintenance-free device because of its electrodeless design and (3) allows the use of any type of plasma gases and feedstock materials without affecting the stability of the discharge.

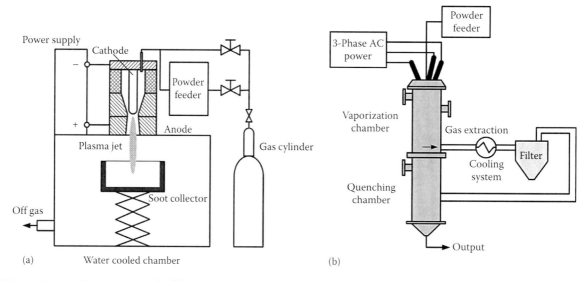

FIGURE 21.5 Arc-jet plasma processes for fullerene synthesis: (a) DC thermal plasma process. (Reproduced from Dubrovsky, R. et al., *Carbon*, 42(5–6), 1063, 2004. With permission.); (b) three-phase AC thermal plasma process. (Reproduced from Fulcheri, L. et al., *Carbon*, 38(6), 797, 2000. With permission.)

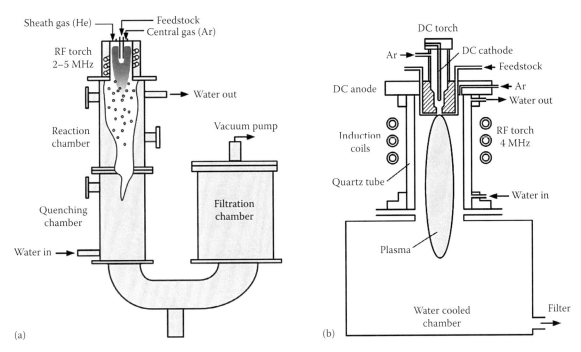

FIGURE 21.6 Arc-jet plasma processes for fullerene synthesis: (a) RF induction thermal plasma process; (b) DC–RF hybrid thermal plasma process. (Reproduced from Yoshie, K. et al., *Appl. Phys. Lett.*, 61(23), 2782, 1992. With permission.)

Wang et al. (2001) used an RF inductively coupled plasma jet (1.67 MHz, 30 kW, 10–20 kPa) to produce fullerenes by means of the direct evaporation of carbon powders. The yield rate of C_{60} was not reported but it was found that the mixing of Si powder with the carbon particles had a role in enhancing the yield rate of the C_{60}, as well as of those of higher-order fullerenes. They argued that Si powders seem to trap the oxygen molecules coming from leaks, by forming a Si–O molecule and keeping the carbon particles away from oxidation. Cota-Sanchez et al. (2001, 2005) extensively explored a fullerene synthesis process by an RF plasma jet (2–5 MHz, 20–40 kW, 40–66 kPa) to investigate the effects of the operating parameters and the type of raw materials employed on the fullerene formation (see Figure 21.6a). The optimal operating conditions were identified as a reactor pressure of 66 kPa, a plate power of 40 kW, and a solid carbon feed rate of 2 g/min. Under these conditions, a fullerene yield of up to about 7.7 wt% could be achieved with carbon blacks (CB) acting as a carbon source. However, fullerene yields of 3.6 and 3.9 wt% were reported using a mixture of CB-Fe and C_2Cl_4, respectively, and no fullerene content was observed when C_2H_4 was used. Todorovic-Markovic et al. (2003) reported a maximum fullerene yield of 4.1%, at a graphite feed rate of 2.6 g/min, using a RF plasma jet (3–5 MHz, 27 kW, 760 Torr), noting that the mixing of Si with carbon particles had a role in enhancing the fullerene yield rate. They also investigated the dependence of the fullerene yield on the properties of the graphite powder (e.g., mean aggregate size, thermal conductivity) and on the helium content in the plasma gas (Szepvolgyi et al. 2006). The results showed that the main properties of graphite powder affecting the fullerene yield are the mean particle size and the thermal conductivity; smaller

particles with a higher-ordered structure (i.e., higher thermal conductivity) improved the overall evaporation efficiency. It was also found that the higher helium content in the plasma gas is desirable for the fast evaporation of graphite powder, due to its good thermal conductivity.

Yoshie et al. (1992) synthesized fullerenes from carbon powders using a DC–RF hybrid plasma jet (4 MHz, 20 kW RF/5 kW DC, 260–760 Torr) (see Figure 21.6b). In a DC–RF hybrid plasma torch, a very high-speed plasma jet ejected from the DC torch is expanded more broadly inside an RF induction plasma torch, energized by additional RF power transmitted from the induction coil. Thus, a larger hot plasma volume can be obtainable with more flexibility in the controlling operation parameters. Since one of the important key factors influencing the fullerene formation is the vaporization or decomposition rate of the carbon sources, a higher yield rate of fullerene was expected in the DC-RF hybrid plasma process, in comparison to other arc-jet plasma processes. In this process, a total ~7% yield was reported using carbon black as a starting material and decreasing the pressure in the plasma reactor to 260 Torr resulted in a reduction of the fullerene yield to 3%. Such pressure dependence is definitely different to those observed in the arc discharge process (Saito et al. 1992). It was also noted that both hydrogen and oxygen hinder the fullerene formation.

21.3.4 Nonequilibrium Plasma Method

Nonequilibrium plasmas, as generated under low or atmospheric pressure, have attracted much attention in the fullerene synthesis due to their ability to produce plasmas at the relatively low temperatures of the ions and neutral species (300–5000 K) (Borra 2006), which is imperative to minimize the thermal degradation or the photochemistry of the fullerenes produced. In this process, being similar to the plasma-enhanced chemical vapor deposition (PECVD) process, the high-energetic electrons play an important role in producing fullerenes rather than the high-temperature-related mechanism prevailing in the thermal plasma process.

An attempt was made by Inomata et al. (1994) to synthesize fullerenes by means of an atmospheric low-temperature plasma (13.56 MHz, 75–130 W). In this work, the fullerene synthesis was conducted by the injection of benzene or naphthalene vapors into the after-glow region of the plasma generated within an argon or helium atmosphere (see Figure 21.7a). The formation of fullerenes through this route was successfully confirmed but the produced quantity was very small. Xie et al. (1999) have reported on the continuous synthesis of C_{60} and C_{70}, using a microwave plasma (0.005 MPa, 250 W), from chloroform present in an argon atmosphere (see Figure 21.7b). The results of these tests have shown that the yield rates of C_{60} (0.3%–1.3%) and C_{70} (0.1%–0.3%) and their generation ratio depend on both the temperature gradient and the collision probability. A similar investigative approach has been reported by Ikeda et al. (1995), in which they investigated the formation of fullerenes by means of a microwave-induced naphthalene–nitrogen plasma at atmospheric pressure (2.45 GHz, 630–990 W). A successful synthesis of fullerenes has also been accomplished by Xie et al. (2001), via a low-pressure glow discharge (25 kHz, 10 kW) reaction, using chloroform vapors as the starting material. Finally, a plasmochemical reactor (66 kHz, 15 kW, 760 Torr) has been developed by Churilov et al. (1999). In this process, the content of fullerene in the deposited soot depends on the flow of helium, for instance, the yield of fullerenes at a helium flow rate of 2 slpm was 4%, while the maximum yield of fullerenes was raised to 20% at a helium flow rate of 20 slpm.

Even though the nonequilibrium plasma process possesses some interesting features as mentioned above, no further progress has been reported on this process up to the present. This circumstance is mainly because most nonequilibrium plasmas are generated by means of utilizing high-frequency power sources (e.g., radio- or microwave frequency), which are very expensive, especially for industrial-scale operations. Second, the productivity of fullerenes in this process is very low (Churilov 2000).

21.4 Characterization of Plasma Processes

Although the initial configurations of the processes were designed largely through empirical approaches, the scaling-up procedure requires a more complete understanding of the fullerene formation mechanism in connection with the local plasma properties, which can be controlled by the macroscopic discharge parameters. However, our knowledge of the plasma characteristics during the fullerene synthesis is rather limited so far, the plasma zone being frequently considered as a "Black box." The reason for this is mainly due to limitations on the accessibility of diagnostic tools attributable to the high heat flux, the chaotic interactions between plasmas and materials, and the fast fluctuation of the system geometry resulting from the rapid evaporation processes. Theoretical approaches, such as process numerical modeling, are not vastly developed either, because of a complex coupling of the nonequilibrium plasma with nonlinear chemical reactions of hundreds of species (Farhat and Scott 2006). However, since information on the processing conditions, including the plasma properties, is crucial for the further advancing of the current plasma technology, a more detailed review on this topic follows in this section.

21.4.1 Numerical Modeling

The goal of the numerical modeling work is to provide a self-consistent description of the plasma characteristics for given operating conditions, and to then predict the yield rate of the

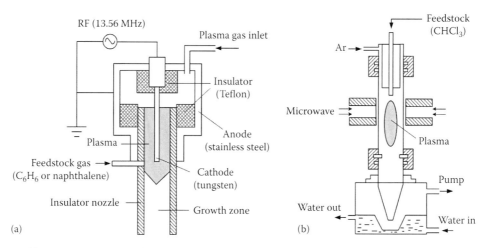

FIGURE 21.7 Nonequilibrium plasma processes for fullerene synthesis: (a) low-temperature atmospheric pressure RF plasma process. (Reproduced from Inomata, K. et al., *Jpn. J. Appl. Phys. Part 2 Lett.*, 33(2A), L197, 1994. With permission.); (b) microwave plasma process. (Reproduced from Xie, S.Y. et al., *Appl. Phys. Lett.*, 75(18), 2764, 1999. With permission.)

fullerenes accurately. In reality, however, the models proposed so far are not completely self-consistent, due to the lack of the precise knowledge of some important phenomena (e.g., electrode phenomena, surface ablation, radiation, turbulence, fullerene formation kinetics), thereby imposing many simplifications or assumptions on the models. The detailed numerical models, as needed for the simulation of the fullerene synthesis by plasmas, can be summarized as follows: (1) thermal flow generation by electric arc or laser irradiation, (2) interaction between plasmas and materials, and (3) formation of fullerenes.

Modeling of the thermal fluid: Assuming that the plasma is in a state of the LTE, plasma gases employed in the fullerene synthesis can be treated as a simple fluid. For instance, plasma plumes generated by laser irradiation have usually been described by the classical Navier–Stokes equations (Greendyke et al. 2004). However, for plasmas generated by electric arcs, additional source terms in the momentum (i.e., Lorentz force term) and energy equations (i.e., Joule heating term) need to be taken into account along with the relevant electromagnetic equations (Gleizes et al. 2005). In this case, the coupled interactions among arc current, induced magnetic field, and plasma flows can be described in the framework of magneto-hydrodynamic (MHD) equations for conservations of mass, momentum, energy, along with the K–ε turbulence model, where K and ε are the turbulent kinetic energy (m^2/s^2) and its dissipation rate (m^2/s^3), respectively.

Mass conservation:

$$\nabla \cdot (\rho \mathbf{v}) = S_\mathrm{p}^\mathrm{c} \tag{21.2}$$

Momentum conservation:

$$\nabla \cdot (\rho \mathbf{v} \mathbf{v}) = -\nabla p + \nabla \cdot \tau + F_\mathrm{L} + S_\mathrm{p}^\mathrm{m} \tag{21.3}$$

Energy conservation:

$$\nabla \cdot (\rho \mathbf{v} h) = \nabla \cdot \left(\frac{\kappa_\mathrm{eff}}{C_\mathrm{p}} \nabla h \right) + P_\mathrm{ohm} - R_\mathrm{rad} + S_\mathrm{p}^\mathrm{c} \tag{21.4}$$

Turbulent kinetic energy:

$$\nabla \cdot (\rho \mathbf{v} K) = \nabla \cdot \left[\left(\mu_\mathrm{l} + \frac{\mu_\mathrm{t}}{\mathrm{Pr}_K} \right) \nabla K \right] + G - \rho \varepsilon \tag{21.5}$$

Dissipation rate of turbulent kinetic energy:

$$\nabla \cdot (\rho \mathbf{v} \varepsilon) = \nabla \cdot \left[\left(\mu_\mathrm{l} + \frac{\mu_\mathrm{t}}{\mathrm{Pr}_\varepsilon} \right) \nabla \varepsilon \right] + \frac{\varepsilon}{K} (C_1 G - C_2 \rho \varepsilon) \tag{21.6}$$

In the mass conservation equation, ρ is the mass density (kg/m^3), $\mathbf{v}$ is the plasma velocity (m/s), and S_p^c is the source term representing the mass generation by the evaporation of the feedstock materials injected ($kg/m^3\,s$). In the momentum conservation equation, p is the static gas pressure (N/m^2), τ is the viscous stress tensor (N/m^2), F_L is the Lorentz forces (N/m^3) induced by

interaction between magnetic field and arc current, and S_p^m is the momentum exchange (N/m^3) with the particles injected. In the energy conservation equation, h is the specific enthalpy (J/kg), κ_eff is the effective thermal conductivity (W/m K), C_p is the specific heat at constant pressure (J/kg K), P_ohm is the heat generation by ohmic heating (W/m^3), R_rad is the radiational loss (W/m^3) taken into account by using the net emission coefficient, and S_p^c is the heat exchange with particles (W/m^3). By the K–ε model, the effective viscosity and thermal conductivity include both laminar and turbulent components:

$$\mu_\mathrm{eff} = \mu_\mathrm{l} + \mu_\mathrm{t} \quad \text{and} \quad \kappa_\mathrm{eff} = \kappa_\mathrm{l} + \kappa_\mathrm{t} \tag{21.7}$$

In the above equations, the turbulent viscosity μ_t (kg/m s) and the turbulent thermal conductivity κ_t (W/m K) are defined as

$$\mu_\mathrm{t} = \rho C_\mu \frac{K^2}{\varepsilon} \quad \text{and} \quad \kappa_\mathrm{t} = \frac{\mu_\mathrm{t} C_\mathrm{p}}{\mathrm{Pr}_\mathrm{t}} \tag{21.8}$$

where C_μ and Pr_t are the constant in the turbulence model and the turbulent Prandtl number, respectively (Scott et al. 1989; Murphy and Kovitya 1993). In the turbulent equations, Pr_K, Pr_ε, C_1, and C_2 are the constants suggested by Launder and Spalding (1972), and G is the product of the turbulent viscosity and viscous dissipation terms (kg/m s^3). For the Lorentz force and the ohmic heating terms in the momentum and energy conservation equations, it is necessary to calculate the electromagnetic field distributions inside the plasma produced. The electromagnetic fields can be obtained by solving the generalized Ohm's law and magnetic vector potential equations (Mostaghimi and Boulos 1989; Xue et al. 2001; Gleizes et al. 2005).

Modeling of the plasma-material interactions: Many of these phenomena are not yet totally understood and therefore have been simplified in various ways. In the laser ablation (Greendyke et al. 2004) and arc discharge methods (Bilodeau et al. 1998), a constant flux of carbon atoms from the target or electrode surface is assumed, according to the experimental erosion rate. For gaseous feedstock materials, however, their dissociation in plasmas can be described by considering the species conservation equations with their detailed chemical kinetics.

Species conservation:

$$\nabla \cdot (\rho \mathbf{v} Y_i) = \nabla \cdot \left(\left(\rho D_i + \frac{\mu_\mathrm{t}}{Sc_\mathrm{t}} \right) \nabla Y_i \right) + \dot{\omega}_i W_i + S_\mathrm{p}^i \tag{21.9}$$

where
  Y_i is the mass fraction of the ith species
  D_i is the binary diffusion coefficient (m^2/s)
  Sc_t is the turbulent Schmidt number
  $\dot{\omega}_i$ is the molar creation/destruction rate (mol/m^3 s) of the ith species
  W_i is the molecular weight of the ith species
  S_p^i is the source term representing the mass generation of the ith species by the evaporation of the particles injected (kg/m^3 s)

Last, the Lagrangian reference frame approach is commonly employed to describe the behavior of the droplet or solid particles injected into the plasmas (Proulx et al. 1987; Bernardi et al. 2004). In this method, the momentum exchanges between particles and plasma are calculated by the Newton's law with proper drag coefficients, whereas the heat and mass transfers are calculated by an energy balance equation as described in Shigeta's work (Shigeta and Watanabe 2008).

Momentum exchange between plasma and particle:

$$m_p \frac{d\mathbf{u}_p}{dt} = \frac{\pi}{8} d_p^2 \rho C_D (\mathbf{v} - \mathbf{u}_p)|\mathbf{v} - \mathbf{u}_p|, \quad C_D = \left(\frac{24}{\mathrm{Re}}\right) f(\mathrm{Re}),$$

$$\mathrm{Re} = \frac{\rho d_p |\mathbf{v} - \mathbf{u}_p|}{\mu} \quad f(\mathrm{Re}) = \begin{cases} 1 & (\mathrm{Re} < 0.2) \\ 1 + 3\,\mathrm{Re}/16 & (0.2 < \mathrm{Re} < 2) \\ 1 + 0.11\,\mathrm{Re}^{0.81} & (2 < \mathrm{Re} < 20) \\ 1 + 0.189\,\mathrm{Re}^{0.632} & (20 < \mathrm{Re} < 500) \end{cases}$$

$$(21.10)$$

where

$\mathbf{u}_p$ is the particle velocity (m/s)
μ is the molecular viscosity of fluid (kg/m s)
d_p is the particle diameter (m)
Re is the particle Reynolds number
C_D is the drag coefficient proposed by Boulos (Boulos 1978)

Heat and mass exchanges between plasma and particle:

$$Q = \pi d_p^2 h_c (T - T_p) - \pi d_p^2 \varepsilon_p \sigma_{st} (T_p^4 - T_a^4),$$

$$= \begin{cases} \frac{\pi}{6} \rho_p d_p^3 c_p \frac{dT_p}{dt} & (T_p < T_m, T_m < T_p < T_b) \\ \frac{\pi}{6} \rho_p d_p^3 H_m \frac{dx}{dt} & (T_p = T_m) \\ -\frac{\pi}{2} \rho_p d_p^2 H_b \frac{dd_p}{dt} & (T_p = T_b) \end{cases}$$

$$(21.11)$$

where

T is the plasma temperature
T_p is the particle temperature
T_a is the ambient temperature
ρ_p is the particle density (kg/m³)
ε_p is the particle emissivity
σ_{st} is the Stephan-Boltzmann constant (W/m² K⁴)
c_p is the particle specific heat (J/kg K)
H is the latent heat (J/kg)
x is the molar fraction of liquid phase

In the above equations, subscripts m and b are designated to denote melting and boiling states, respectively. The heat transfer coefficient (h_c, W/m² K) is given with the Nusselt number (Nu) by using the correlation of Ranz and Marshall (1952a,b):

$$h_c = \frac{k_\infty \cdot \mathrm{Nu}}{d_p}, \quad \mathrm{Nu} = 2.0 + 0.6\,\mathrm{Re}^{0.5}\,\mathrm{Pr}^{0.33} \qquad (21.12)$$

where

k_∞ is the thermal conductivity
Pr is the Prandtl number of the continuous phase

Modeling of the fullerene formation: The kinetic mechanism involved in this process must include the chemical reactions between clusters of all sizes, but theoretical studies have been restricted to clusters containing less than a few hundred carbon atoms (Krestinin et al. 1998). Furthermore, the integration of those kinetic models into the previous plasma models is quite challenging because the chemical reactions will be very fast in the high temperatures of 4000–5000 K prevailing in the thermal plasmas (i.e., numerically very stiff), and accurate information of the reaction rates, thermodynamic properties, and transport coefficients is not available for large carbon clusters existing at such high temperatures, neither. For this reason, fullerene formation has usually been neglected in the previous numerical studies or analyzed separately in reduced dimensions such as zero (i.e., isotherm and zero velocity) and one dimension (i.e., plug flow) (Farhat et al. 2004). The kinetic models of the fullerene formation proposed by Krestinin and Moravskii (Krestinin et al. 1998; Krestinin and Moravskii 1999) are summarized in Tables 21.2 and 21.3.

A two-dimensional (2D) model was developed by Bilodeau et al. (1998) for the simulation of a carbon arc reactor operating in a helium or argon atmosphere (see Figure 21.8). In this study, the effects of operating parameters, such as the interelectrode distance and buffer gas composition, were analyzed. The simulation results demonstrated that the central region of the arc is wider for a 4 mm gap with a higher maximum temperature of 17,000 K, compared to that of a 1 mm gap (~12,000 K). This result is in line with many experimental observations that a large electrode gap usually reduces the fullerene yield by expanding the UV radiating zone. It was also found from the simulation with different kinds of buffer gases that the isotherm in the temperature range of 1500–5000 K is wider in argon plasma than in helium plasma, probably due to the higher thermal conductivity of helium, which might enhance the heat dissipation to the surroundings. On the other hand, for the temperature range of 2000–3000 K, believed to be favorable for the formation of fullerene precursors, more higher carbon concentration was predicted in helium plasma and they suggested that this result would explain the higher fullerene yield when the helium is used as a buffer gas. Recently, Hinkov (2004) extended this model for the nanotube synthesis process.

Another mathematical model has been proposed by Bilodeau et al. (1997) to simulate a DC arc-jet plasma process, developed for the synthesis of fullerenes from the dissociation of C_2Cl_4. From this modeling work, they suggested that the reduction of

TABLE 21.2 Mechanism of Fullerene Formation

Reaction	A (cm³/s/mol)	β	E/R (K)
1. Chemistry of small clusters (C_1–C_{10})			
$C + C \leftrightarrow C_2$	2.00×10^{14}	0	0
$C + C_2 \leftrightarrow C_3$	2.00×10^{14}	0	0
$C_2 + C_2 \leftrightarrow C_3 + C$	2.00×10^{15}	0	9,040
$C_2 + C_2 \leftrightarrow C_4$	2.00×10^{14}	0	0
$C_1 + C_3 \leftrightarrow C_4$	2.00×10^{14}	0	0
$C_1 + C_4 \leftrightarrow C_5$	2.00×10^{14}	0	0
$C_2 + C_3 \leftrightarrow C_5$	2.00×10^{14}	0	0
$C_{n-m} + C_m \leftrightarrow C_n$	2.00×10^{14}	0	0
where $n = 6$–10 and $m = 1 - n/2$			
2. Chemistry of cycles and polycycles (C_{11}–C_{31})			
$C_{n-m} + C_m \leftrightarrow C_n$	2.00×10^{14}	0	0
where $n = 11$–31 and $m = 1$–15			
3. Formation of fullerenes			
$C_{n-m} + C_m \leftrightarrow C_n$	2.00×10^{14}	0	0
where $n = 32$–46 and $n - 31 \leq m \leq 15$			
4. Growth of fullerene shells			
$C_n + C_1 \leftrightarrow C_{n+1}$	2.00×10^{14}	0	0
where $n = 32$–78 except $n = 59, 69$			
$C_n + C_2 \rightarrow C_{n+2}$	4.00×10^{08}	0	0
$C_{n+2} \rightarrow C_n + C_2$	3.20×10^{13}	0	61,900
where $n = 32, 59$ except $n = 58$			
$C_n + C_2 \rightarrow C_{n+2}$	4.00×10^{08}	0	0
$C_{n+2} \rightarrow C_n + C_2$	3.20×10^{13}	0	61,900
where $n = 60, 77$ except $n = 68$			
$C_n + C_3 \leftrightarrow C_{n+2} + C_2$	1.00×10^{15}	0	0
where $n = 32, 77$			
5. Formation and decay of fullerene molecules			
$C_{60F} \rightarrow C_{60F}$	5.00×10^{13}	0	37,745
$C_{59} + C \rightarrow C_{60F}$	2.00×10^{14}	0	0
$C_{58} + C_2 \rightarrow C_{60F}$	4.00×10^{08}	0	−30,196
$C_{60F} \rightarrow C_{58} + C_2$	8.00×10^{12}	0	61,900
$C_{58} + C_3 \rightarrow C_{60F} + C$	8.00×10^{14}	0	0
$C_{60F} + C \rightarrow C_{61}$	2.00×10^{13}	0	10,065
$C_{60F} + C_2 \rightarrow C_{62}$	2.00×10^{13}	0	10,065
$C_{60F} + C_3 \rightarrow C_{63}$	2.00×10^{13}	0	10,065
$C_{70} \rightarrow C_{70F}$	1.20×10^{13}	0	37,745
$C_{69} + C \rightarrow C_{70F}$	2.00×10^{14}	0	0
$C_{68} + C_2 \rightarrow C_{70F}$	4.00×10^{08}	0	−30,600
$C_{70F} \rightarrow C_{68} + C_2$	2.50×10^{14}	0	0
$C_{68} + C_3 \rightarrow C_{70F} + C$	1.40×10^{11}	0	49,925
$C_{70F} + C \rightarrow C_{71}$	2.00×10^{13}	0	10,065
$C_{70F} + C_2 \rightarrow C_{72}$	2.00×10^{13}	0	10,065
$C_{70F} + C_3 \rightarrow C_{72} + C$	2.00×10^{13}	0	10,065

Sources: Krestinin, A.V. et al., *Chem. Phys. Rep.*, 17(9), 1687, 1998; Farhat, S. and Scott, C.D., *J. Nanosci. Nanotechnol.*, 6(5), 1189, 2006.

Forward rate constants k are calculated assuming Arrhenius temperature dependence $k = AT^\beta \exp(-E/RT)$ where A is the pre-exponential factor, β is the temperature exponent, and E is the activation energy.

TABLE 21.3 Mechanism of Soot Nuclei and Soot Particle Formation

Reaction	A (cm³/s/mol)	B	E
1. Formation of soot nuclei Z			
$C_{78} + C_2 \rightarrow Z$	4.00×10^{08}	0	−60.8
$C_{78} + C_2 \rightarrow Z + C$	4.00×10^{08}	0	0
$C_{60-m} + C_{60-n} \rightarrow Z$	4.00×10^{14}	0	0
$C_{70-m} + C_{70-n} \rightarrow Z$	4.00×10^{14}	0	0
where $n, m = 1$–10			
2. Heterogeneous reactions on soot particles			
$C + soot \leftrightarrow soot$	4.00×10^{03}	0	0
$C_2 + soot \leftrightarrow soot$	4.00×10^{03}	0	0
$C_3 + soot \rightarrow soot$	4.00×10^{03}	0	0
$C_n + soot \rightarrow soot$	$4.00 \times 10^{03}(1/n)^{0.5}$	0	0
where $n = 4$–79			
$C_{60F} + soot \rightarrow soot$	1.00×10^{03}	0	30
$C_{70F} + soot \rightarrow soot$	1.00×10^{03}	0	30

Sources: Krestinin, A.V. et al., *Chem. Phys. Rep.*, 17(9), 1687, 1998; Farhat, S. and Scott, C.D., *J. Nanosci. Nanotechnol.*, 6(5), 1189, 2006.

thermal and velocity gradients by the widening of the distance between the torch and the reactor wall would increase the fullerene yield rate, extending the residence time of the carbon clusters so produced. A computational study of a three-phase AC plasma process has also been performed by Ravary et al. (2003) and the author pointed out that the position of the particle injection has a strong influence on the temperature field inside the reactor.

In order to investigate the influences of the plasma-forming gas (e.g., argon, helium, argon–helium mixture) and the pressure (e.g., 150, 380, and 500 Torr) on the fullerene formation in a RF plasma process, Wang et al. (2003) developed a 2D model without considering the carbon particle injection. The simulation results have revealed that there is no significant effect of the gas composition on the temperature profile inside the torch, while the temperature decreases more rapidly in the downstream region of the torch as the helium content increases. From this result, it was deduced that the existence of a sharp temperature gradient along the center axis is necessary for an effective formation of fullerenes. As the pressure varies, no big difference was observed in the temperature profile along the center axis, but the velocity in low-pressure case (150 Torr) was approximately 2.5 times as high as that in high-pressure case (500 Torr). Based on the previous simulation results with different plasma gases, the authors conjectured that low pressures would promote the formation of fullerenes because the higher axial-velocity can increase the temperature gradient experienced by fullerene precursors. But this conclusion, conflicts with the experimental results reported by Yoshie et al. (1992).

21.4.2 In Situ Diagnostics

A few papers related to the diagnostics of the carbon plasma have been published so far and these studies have been mostly restricted to the optical emission spectroscopy (OES) for the

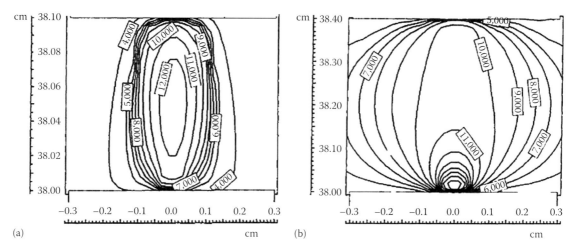

FIGURE 21.8 The effect of the interelectrode gap on the temperature distribution near the electrode gap in an arc discharge process (2D numerical modeling): (a) 1 mm gap; (b) 4 mm gap (helium, 80 A and 13.3 kPa). (Reprinted from Bilodeau, J.F. et al., *Plasma Chem. Plasma Process.*, 18(2), 285, 1998. With permission.)

measurement of temperature and C_2 species distributions. During the fullerene synthesis by plasmas, the plasma emits radiation in the UV and VIS regions, and the measurement of these spectra can provide valuable information about the temperature and concentrations of the excited species by applying the *Boltzmann's plot* or *molecular band spectrum simulation methods* (Fauchais et al. 1989). In the Boltzmann method, the temperature, or more precisely, the excitation temperature of electrons in the atoms can be calculated by fitting the relative intensities of several atomic lines, assuming the LTE condition. On the other hand, the molecular band simulation method is based on the calculation of the spectra (i.e., synthetic spectra) associated with the optical transition of molecular species and subsequent comparison with the spectra recorded from experiments. This method is very attractive because it can offer rotational and vibrational temperatures of molecules even under the non-LTE condition, which prevails at the fringe of the arc discharge or at the tail of arc-jet plasmas. In addition, the rotational temperature is often close to the kinetic temperature of the gas (i.e., temperature of heavy particles) because the energy exchange between the rotational and kinetic modes is very intense. The most important information characterizing the carbon plasma can be inferred from radiation of C_2 molecule (i.e., Swan band system) and the detailed theory on the structure of this system $\left(d^3 \Pi_g, \upsilon' = 0 \rightarrow a^3 \Pi_u, \upsilon'' = 0 \right)$ can be found in Pellerin's work (Pellerin et al. 1996). In order to determine C_2 $\left(a^3 \Pi_u, \upsilon'' = 0 \right)$ column density in the carbon plasma, Lange et al. (1996) have further extended this theory by including self-absorption phenomena, which usually take place with higher volume density of C_2 species.

The carbon plasma produced in the arc discharge was initially studied by Huczko et al. (1997), who analyzed the influence of the operating parameters, such as the gap distance and input power on the yield of fullerene. In this study, the average temperatures of 3500–5500 K and column density of C_2 up to

4.5×10^{15} cm^{-2} were measured depending on the operating conditions, with a tendency that the temperature in the arc zone rises slightly as the power and the gap distance increase. From the C_2 volume density profiles and their correlation with the fullerene yield, Huczko et al. further claimed that not only anode sublimation, but also vaporization of graphite micro-crystallites produced by the spallation process, would be a source of carbon vapors in the arc zone. Byszewski et al. (1997) performed an emission spectroscopy to investigate the effect of the buffer gas pressure on the fullerene formation in the arc discharge process. The rotational temperatures calculated from the Boltzmann plots were in a range of 4000–5500 K, and a moderate temperature gradient in radial direction was observed under the higher pressure (88 kPa) along with lower temperatures compared to those measured in the lower pressure (17 kPa). The C_2 radial distribution was also estimated and an almost uniform density distribution of C_2 was observed under the lower gas pressure while the C_2 density at the center of the arc is much higher in the higher pressure case, implying a faster side-diffusion of C_2 molecules under the lower pressure. The fullerene contents in the collected soot were measured as 11.6 wt% at 17 kPa and 6.1 wt% at 88 kPa, respectively. Lange et al. and Saidane et al. extended previous diagnostic studies to the entire arc zone. Lange et al. (1999) investigated the influence of the arc current on the carbon arc properties (see Figure 21.9). In this work, temperatures within 3500–6500 K and column densities within 0.5–6 × 10^{15} cm^{-2} were measured depending on the arc current and measurement positions. At a low arc current of 65 A, the isotherms give reverse radial profiles with much lower plasma temperatures compared to those predicted in the numerical model (Bilodeau et al. 1998). But at the higher arc current of 100 A, the isotherms become similar with those obtained from the numerical model. The measured C_2 density profiles showed that C_2 species were mainly confined to the interelectrode space with an off-axis maximum, which can be explained by the thermodynamic stability of C_2 species at high temperatures. In the lower arc current case, due

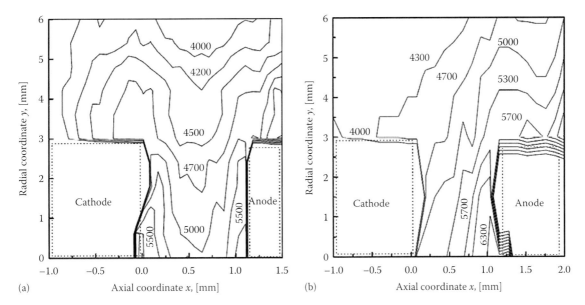

FIGURE 21.9 The effect of the arc current on the temperature distribution near the electrode gap in an arc discharge process (optical diagnostics): (a) $I = 65$ A; (b) $I = 100$ A (helium and 13.3 kPa). (Reprinted from Lange, H. et al., *J. Phys. D.: Appl. Phys.*, 32(9), 1024, 1999. With permission.)

to the relatively low anode erosion rate, the overall C_2 column density was smaller than that of the higher arc current case, leading to a lower yield rate of fullerene. The authors have suggested that this result clearly shows the existence of a correlation between the C_2 content in the plasma and the final fullerene yield rate. Similarly, Saidane et al. (2004) performed plasma spectroscopy in order to optimize the arc discharge process. In this recent work, increased C_2 density and the carbon content in the arc core region were observed as conditions which lead to more efficient fullerenes formation. They have also suggested that the best conditions promoting the fullerene formation are strongly correlated with the existence of both steep temperature and C_2 density gradients towards radial direction.

Todorovic-Markovic et al. (2006) made a spectroscopic observation on an RF plasma process, operating with different types of precursors at various feeding rates. The rotational temperatures of C_2, measured at 50 mm away from the torch exit, were in a wide range of 3500–5500 K and an elevated rotational temperature of C_2 resulted in a decrease of the fullerene yield. From these results, they concluded that lower value of the rotation temperature of C_2 (~3000 K) at the tail of the plasma jet is a sign of good evaporation of the carbon sources injected. Swan band spectra were also analyzed by Cota-Sanchez et al. (2005) during the fullerene synthesis through the use of an RF induction plasma. The estimated temperatures range in 3650–5150 K and the column densities are measured as between 0.16×10^{15} and 3.20×10^{15} cm^{-2}.

21.5 Optimization of Plasma Processes

From the above review, it is evident that the processing conditions, such as the temperature and velocity profiles, operating pressure, and buffer or plasma gas composition, should be optimized with respect to the following issues in order to achieving a high fullerene yield: (1) an uniform and high evaporation or dissociation rate of the carbon-containing materials, (2) fast removal of carbon vapors from the arc zone for the immediate formation of fullerene precursors, preventing them from being destroyed by the UV radiation, (3) long residence times in a temperature range of 3000–4000 K for the further growth of the fullerene precursors produced, (4) annealing of defective carbon cages to form the perfect fullerene structures in a cooler zone where the temperature is around 1000 K, and (5) lower contents of hydrogen and oxygen in the buffer or plasma-forming gas.

Optimum temperature profile: The dependency of the fullerene yield on the temperature profile inside the reactor is related to the kinetics of evaporation or the decomposition rate of the starting materials and to those of the formation or destruction of fullerenes, which occur separately in different zones. Initially, a larger volume of the high-temperature zone over 5000 K is favored for the effective treatment of the feedstock materials. However, both the UV and VIS radiations will be stronger as this volume increases and this may accelerate the decomposition of the fullerenes produced. In the laser ablation and arc discharge methods, carbon species existing inside the discharge zone are already in the vapor phase, and thus it is not necessary to heat them further. For this reason, a steep temperature gradient located around the arc fringe (i.e., small discharge volume) would be the optimal condition for fullerene production. This is in line with the previous experimental findings that there exist optimum values for the input current, electrode radius, and interelectrode gap in the arc discharge process (Huczko et al. 1997). But in the arc-jet plasma method, feedstock materials are first injected into the discharge zone and they should be heated rapidly to achieve an effective release of carbon vapors. Thus, in the arc-jet method, a larger discharge volume and a moderate temperature gradient would be helpful for increasing the fullerene yield, thereby ensuring a thorough treatment of the feedstock materials. However, it should be emphasized that a steep temperature gradient will be necessary initiating the immediate formation of carbon clusters as soon as the evaporation or decomposition process is terminated.

Carbon vapors produced in the hot core region will then be transported to the lower temperature region, through the means of diffusion or convection, and carbon atoms begin to form fullerene precursors such as the various chains, rings, and caged structures. The thermodynamic study reported by Cota-Sanchez et al. (2005) has revealed that the fullerenes are thermodynamically stable over a temperature range of 2250–3800 K, the highest fullerene concentration being reached at 3500 K. It was additionally found, from the molecular dynamic simulations, that the fullerene-like caged structure was obtained preferentially when the process temperature was maintained within a range of 2500–3000 K, while a flat graphic structure was obtained for the lower temperature of around 1000 K (Yamaguchi and Maruyama 1998). Accordingly, it is important to maintain the temperature within the range between 2500 and 4000 K over the wide area within the reactor to maximize the fullerene yield. Further extended high-temperature region should follow next for the annealing of defective fullerene cages to generate the perfect fullerene structures. Experimental annealing temperatures have been estimated to be optimum at about 1000–1500 K for the laser ablation method and 1000 K for the arc discharge method (Maruyama and Yamaguchi 1998). Many attempts have been made to provide this annealing zone with better facilities by employing furnaces, graphite tubes, and hot walls, all having proven to have positive effects.

Optimum velocity profile: The gas velocity is directly related to the residence time of the carbon species located within the various zones inside the reactor. In the cases of the laser ablation and arc discharge methods, a high velocity achieved in the discharge zone will favor the formation of carbon clusters, thereby avoiding the precipitation of carbon vapors on the relatively cold target or electrode surface, and which removes as much as 40%–60% of the evaporated carbon mass (Jones et al. 1996). In addition, partially formed fullerenes should be removed as soon as possible from the arc region to avoid irradiation by the harmful UV radiation. The experimental results obtained in an arc discharge process have demonstrated that the fast removal

of the carbon species from the discharge zone has a positive impact on the final yield (Dubrovsky and Bezmelnitsyn 2004). However, this is not the case for the arc-jet plasma method, during which a long-enough residence time inside the hot zone is necessary for the complete evaporation, or decomposition, of the injected feedstock materials. Typically, the evaporation time for a 1 μm-sized carbon particle is estimated as being ~2 ms within a RF plasma jet produced with helium gas.

In the cooler zones, however, a longer residence time will be generally needed in most processes for the effective growth of fullerenes and their subsequent annealing to form the perfect structure. A molecular dynamics study has shown that a cluster of C_{60} is formed in 2 ns at 3000 K (Yamaguchi and Maruyama 1998), while transformation into the perfected fullerene structures needs a further annealing of about 52 ns (Maruyama and Yamaguchi 1998). In general, the velocity profile inside the reactor is controllable by means of adjusting the input power, gas flow rate, operating pressure, and reactor geometry.

Optimum composition of gas: In the plasma synthesis of fullerenes, argon and helium or their mixtures are routinely employed as a buffer or plasma-forming gas, the detailed ratios employed being dependent on the processes. From the previous experimental results, it was found that the fullerene yield was in any case higher with helium than with argon. This result may be simply explained by more efficient generation of carbon vapors in a helium atmosphere, due to its higher thermal conductivity (i.e., four to six times greater than that of argon) (Boulos et al. 1994).

To further clarify the effect of operating parameters (i.e., plasma gas composition and operating pressure) on the fullerene formation, we have performed 2D numerical simulations on the RF induction plasma process by using Equations 21.2 through 21.12. The computational domain considered in this numerical work is depicted in Figure 21.10, and it mainly consists of a plasma torch zone, a reaction zone, and an annealing zone. The plasma torch (PL-50, TEKNA) is driven by a 60 kW RF power supply (Lepel Co.) operated at an oscillator frequency of 3 MHz and three different gas streams of powder carrying, central and sheath gases are introduced to the torch as illustrated in Figures 21.6a and 21.10. The reaction zone includes a graphite liner and a thermal insulator which were employed for a flexible control over background temperature and cooling rate.

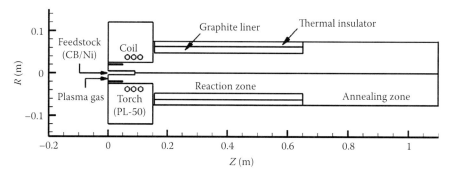

FIGURE 21.10 Computational domain employed for 2D numerical simulations of an RF induction plasma process.

TABLE 21.4 Operating Conditions for 2D Numerical Simulations of an RF Induction Plasma Process

Case	Net Plasma Power (kW)	Pressure (kPa)	Carrier Gas (slpm)	Central Gas (slpm)	Sheath Gas (slpm)	Feed Rate (g/min)
A	28	66	8 (Ar)	30 (Ar)	120 (Ar)	2 (CB/Ni)
B	28	66	8 (He)	30 (Ar)	120 (He)	2 (CB/Ni)
C	28	20	8 (He)	30 (Ar)	120 (He)	2 (CB/Ni)

TABLE 21.5 Particle Injection Conditions for 2D Numerical Simulations of an RF Induction Plasma Process

Particle	Inlet Velocity (m/s)	Temperature (K)	Mass Flow Rate (kg/s)	Maximum Diameter (m)	Mean Diameter (m)	Minimum Diameter (m)
Carbon	26.8	300	3.0×10^{-5}	3.0×10^{-6}	2.0×10^{-6}	1.0×10^{-6}
Nickel	26.8	300	3.0×10^{-6}	3.0×10^{-6}	2.0×10^{-6}	1.0×10^{-6}

For the completion of the fullerene growth and re-arrangement into the perfect structure, the reaction zone is followed by the annealing zone in which the reaction gases are cooled down through heat exchanges with the water-cooled reactor walls. A binary mixture of carbon black (CB) and nickel was considered as a feedstock material, with a ratio of CB/Ni-98.0/2.0 at.%. The detailed operating and particle injection conditions employed are summarized in Tables 21.4 and 21.5, respectively. More information on the simulation conditions can be found elsewhere (Cota-Sanchez et al. 2005; Kim et al. 2007).

As shown in Figures 21.11 and 21.12, an increase in the helium content present in the plasma gas results in a thorough treatment of the feedstock materials within a short time. This fast heating of the input materials allows size reduction of radiating high-temperature zone, which has an adverse effect on the fullerene synthesis. In addition, a high quenching rate is obtainable before entering the main reaction zone when the helium gas is used, because the plasma is cooled quickly through an intensive heat exchange with the surroundings (see Figures 21.11a and 21.13a). This condition is also desirable for increasing the yield rate of

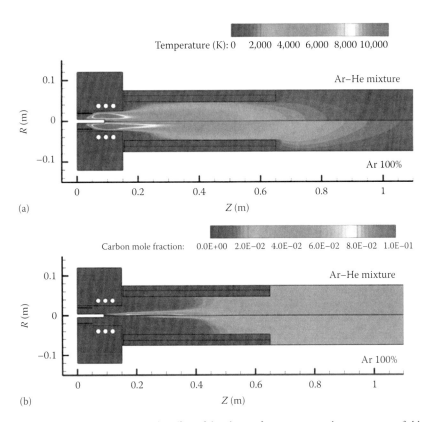

FIGURE 21.11 (**See color insert following page 25-14.**) The effect of the plasma-forming gas on the temperature field and carbon evaporation in an RF induction plasma process (2D numerical simulation): (a) temperature distribution; (b) distribution of carbon mole fraction (28 kW, 66 kPa, C/Ni-98/2 at.% and feed rate of 2 g/min).

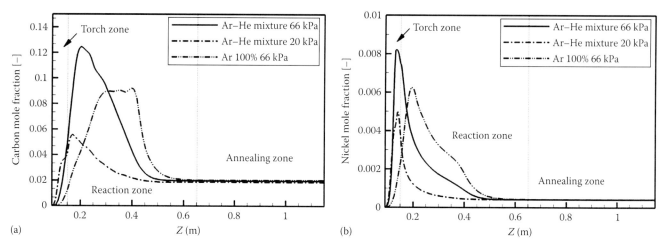

FIGURE 21.12 The effects of the plasma-forming gas and operating pressure on the particle evaporation in an RF induction plasma process (2D numerical simulation): (a) axial-profiles of the carbon vapor along the reactor axis; (b) axial-profiles of the nickel vapor along the reactor axis (28 kW, C/Ni-98/2 at.% and feed rate of 2 g/min).

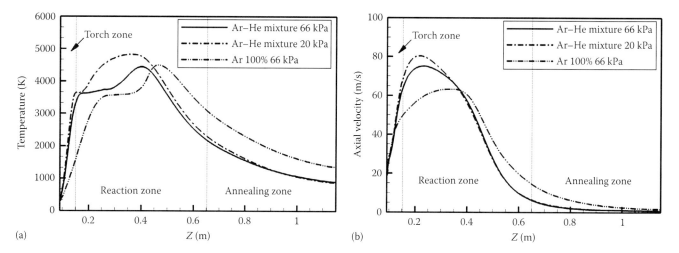

FIGURE 21.13 The effects of the plasma-forming gas and operating pressure on the temperature and axial-velocity fields in an RF induction plasma process (2D numerical simulation): (a) axial-profiles of the temperature along the reactor axis; (b) axial-profiles of the axial-velocity along the reactor axis (28 kW, C/Ni-98/2 at.% and feed rate of 2 g/min).

the fullerene by enhancing the formation of the carbon clusters through the strong supersaturation of their vapors. The high axial-velocity of the argon–helium mixture plasma around the torch zone (see Figure 21.13b) also favors a fast removal of the partially created fullerenes from the arc zone, preventing them from being destroyed by the UV radiation, while the low axial-velocity prevailing in the entire reactor system provides much more longer residence time in the growth and annealing zones. However, the stability of the plasma decreases by the replacement of argon with helium, due to its lower electrical conductivity and higher ionization potential properties than those of argon. Thus, argon–helium mixture plasmas would provide a better environment for the fullerene synthesis compared with 100% pure helium plasmas, in terms of process efficiency and plasma stability.

Optimum operating pressure: The effect of the operating pressure on the fullerene yield can also be discussed in terms of the

plasma cooling rate, particle heating efficiency, and diffusion speed of the carbon species generated. When the gas pressure is high, the plasma is cooled more rapidly, due to frequent collisions with the cold background gas and a higher supersaturation of the carbon vapors attainable before they enter the main reaction zone. Thus, a fast clustering of carbon atoms is possible with higher pressure but a too rapid cooling of the gas does not provide sufficient time for the carbon clusters to further grow or rearrange into the fullerene structures.

On the other hand, lower operating pressures commonly result in higher gas velocities through strong plasma expansion. In this case, shorter residence times will be obtained in the hot zone. As discussed earlier, this is beneficial for both the laser ablation and the arc discharge method, but is not a desirable condition for the arc-jet plasma method (see Figures 21.14 and 21.15, a lower evaporation rate is obtained with a lower pressure). Besides, the Knudsen or rarefaction effect (i.e., an important mechanism

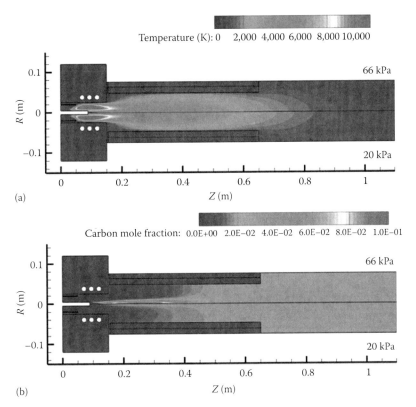

FIGURE 21.14 **(See color insert following page 25-14.)** The effect of the operating pressure on the temperature field and carbon evaporation in an RF induction plasma process (2D numerical simulation): (a) temperature distribution; (b) distribution of carbon mole fraction (28 kW, 66 kPa, C/Ni-98/2 at.% and feed rate of 2 g/min).

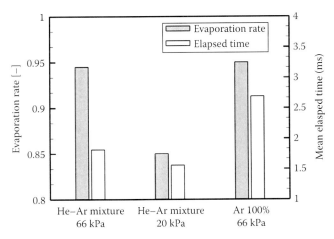

FIGURE 21.15 The effects of the plasma-forming gas and operating pressure on the particle evaporation mechanism in an RF induction plasma process (2D numerical simulation, 28 kW, C/Ni-98/2 at.% and feed rate of 2 g/min).

causing reduction of the plasma-particle heat exchange when small particles and/or low gas pressures are involved) prevailing in the low-pressure environments may cause a poor heat transfer to the particles during the arc-jet plasma process, leading to an incomplete treatment of feedstock materials (Chen 1988).

It has also been observed in our recent simulations that a strong radial flow of the carbon vapors is developed with a low

operating pressure (see Figure 21.14b). Normally, a rapid diffusion of carbon vapors in the radial direction is preferred in the laser ablation or in the arc discharge process, in which the precipitation of carbon vapors on the cold target or electrode surface takes away as much as half of the evaporated carbon mass. In addition, to avoid irradiation by the harmful UV, partially formed fullerenes should be removed as soon as possible from the arc region through an effective radial diffusion. But in the arc-jet plasma process, a strong radial diffusion leads to an unwanted loss of the carbon vapors toward the reactor walls, and consequently fewer carbon vapors become available for the clustering as they enter the main reaction zone. Thus, a less amount of fullerene will be produced with low operating pressure. This is one of the main reasons why the laser ablation or the arc discharge process is usually performed at the relatively lower pressure of 2.7 kPa (Saito et al. 1992), while the arc-jet plasma process is conducted at the relatively higher pressure of 66 kPa (Yoshie et al. 1992; Cota-Sanchez et al. 2005).

21.6 Conclusions

Over the past decade, numerous applications for fullerenes have been proposed and actively advanced in a variety of technical areas. Consequently, the production of high-quality fullerenes on a bulk scale has been an issue of considerable interest. Among the various synthesis methods available, plasma technologies, which are characterized by their high versatility and flexibility, have proven to have many competitive advantages in the production of fullerenes at large scale. Furthermore, the plasma technology employed is now so well established and has become such a matured technology that is currently in use in various industries almost every day. However, there is still a need for further improvement of the process through better understanding of the fullerenes formation mechanisms in connection with the plasma conditions established and maintained during the processes. In this regard, the identification and design of the optimum reactor geometry and the related operating parameters, based on the detailed studies of the plasma characteristics, would be an important next step for achieving major advances in this technology.

References

Alexakis, T., P. G. Tsantrizos, Y. S. Tsantrizos, and J. L. Meunier. 1997. Synthesis of fullerenes via the thermal plasma dissociation of hydrocarbons. *Applied Physics Letters* 70 (16): 2102–2104.

Bacon, R. 1960. Growth, structure, and properties of graphite whiskers. *Journal of Applied Physics* 31 (2): 283–290.

Belousov, V. P., I. M. Belousova, V. P. Budtov et al. 1997. Fullerenes: Structural, physicochemical, and nonlinear-optical properties. *Journal of Optical Technology* 64 (12): 1081–1109.

Bernardi, D., V. Colombo, E. Ghedini, A. Mentrelli, and T. Trombetti. 2004. 3-D numerical simulation of fully-coupled particle heating in ICPTs. *European Physical Journal D* 28 (3): 423–433.

Bhuiyan, M. K. H. and T. Mieno. 2002. Production characteristics of fullerenes by means of the J×B arc discharge method. *Japanese Journal of Applied Physics Part 1-Regular Papers Short Notes & Review Papers* 41 (1): 314–318.

Bilodeau, J. F., T. Alexakis, J. L. Meunier, and P. G. Tzantrizos. 1997. Model of the synthesis of fullerenes by the plasma torch dissociation of C_2Cl_4. *Journal of Physics D-Applied Physics* 30 (17): 2403–2410.

Bilodeau, J. F., J. Pousse, and A. Gleizes. 1998. A mathematical model of the carbon arc reactor for fullerene synthesis. *Plasma Chemistry and Plasma Processing* 18 (2): 285–303.

Borra, J. P. 2006. Nucleation and aerosol processing in atmospheric pressure electrical discharges: Powders production, coatings and filtration. *Journal of Physics D-Applied Physics* 39 (2): R19–R54.

Boulos, M. I. 1978. Heating of powders in the fire ball of an induction plasma. *IEEE Transactions on Plasma Science* 6 (2): 93–106.

Boulos, M. I. 1985. The inductively coupled R.F. (radio frequency) plasma. *Pure and Applied Chemistry* 57 (9): 1321–1352.

Boulos, M. I. 1991. Thermal plasma processing. *IEEE Transactions on Plasma Science* 19 (6): 1078–1089.

Boulos, M. I. 1997. The inductively coupled radio frequency plasma. *High Temperature Material Processes* 1 (1): 17–39.

Boulos, M. I., P. Fauchais, and E. Pfender. 1994. *Thermal Plasmas, Fundamentals and Applications*, vol. 1. New York: Plenum.

Bunshah, R. F., S. K. Jou, S. Prakash et al. 1992. Fullerene formation in sputtering and electron-beam evaporation processes. *Journal of Physical Chemistry* 96 (17): 6866–6869.

Byszewski, P., H. Lange, A. Huczko, and T. F. Behnke. 1997. Fullerene and nanotube synthesis. Plasma spectroscopy studies. *Journal of Physics and Chemistry of Solids* 58 (11): 1679–1683.

Chen, X. 1988. Particle heating in a thermal plasma. *Pure and Applied Chemistry* 60 (5): 651–662.

Chibante, L. P. F., A. Thess, J. M. Alford, M. D. Diener, and R. E. Smalley. 1993. Solar generation of the fullerenes. *Journal of Physical Chemistry* 97 (34): 8696–8700.

Churilov, G. N. 2000. Plasma synthesis of fullerenes (review). *Instruments and Experimental Techniques* 43 (1): 1–10.

Churilov, G. N., L. A. Solovyov, Y. N. Churilova, O. V. Chupina, and S. S. Malceiva. 1999. Fullerenes and other structures of carbon synthesized in a carbon plasma jet under helium flow. *Carbon* 37 (3): 427–431.

Cota-Sanchez, G., L. Merlo-Sosa, A. Huczko, and G. Soucy. 2001. Production of carbon nanostructures using a HF plasma torch. Paper presented at *15th International Symposium on Plasma Chemistry* (ISPC 15) Orléans, France.

Cota-Sanchez, G., G. Soucy, A. Huczko, and H. Lange. 2005. Induction plasma synthesis of fullerenes and nanotubes using carbon black-nickel particles. *Carbon* 43 (15): 3153–3166.

Da Ros, T. and M. Prato. 1999. Medicinal chemistry with fullerenes and fullerene derivatives. *Chemical Communications* 8: 663–669.

Diederich, F., R. Ettl, Y. Rubin et al. 1991. The higher fullerenes - Isolation and characterization of C_{76}, C_{84}, C_{90}, C_{94}, and $C_{70}O$, an oxide of D_{5h}-C_{70}. *Science* 252 (5005): 548–551.

Dietz, T. G., M. A. Duncan, D. E. Powers, and R. E. Smalley. 1981. Laser production of supersonic metal cluster beams. *The Journal of Chemical Physics* 74 (11): 6511–6512.

Dubrovsky, R. and V. Bezmelnitsyn. 2004. Bulk production of nanocarbon allotropes by a gas outflow discharge approach. *Carbon* 42 (8–9): 1861–1864.

Dubrovsky, R., V. Bezmelnitsyn, and A. Eletskii. 2004. Plasma fullerene production from powdered carbon black. *Carbon* 42 (5–6): 1063–1066.

Farhat, S. and C. D. Scott. 2006. Review of the arc process modeling for fullerene and nanotube production. *Journal of Nanoscience and Nanotechnology* 6 (5): 1189–1210.

Farhat, S., L. Hinkov, and C. D. Scott. 2004. Arc process parameters for single-walled carbon nanotube growth and production: Experiments and modeling. *Journal of Nanoscience and Nanotechnology* 4 (4): 377–389.

Fauchais, P. and A. Vardelle. 1997. Thermal plasmas. *IEEE Transactions on Plasma Science* 25 (6): 1258–1280.

Fauchais, P., J. F. Coudert, and M. Vardelle. 1989. Diagnostics in thermal plasma processing. In: *Plasma Diagnostics, Discharge Parameters and Chemistry*. O. Auciello and D. L. Flamm (Eds.), pp. 349–446. Boston, MA: Academic Press.

Fulcheri, L., Y. Schwob, F. Fabry et al. 2000. Fullerene production in a 3-phase AC plasma process. *Carbon* 38 (6): 797–803.

Geckeler, K. E. and S. Samal. 1999. Syntheses and properties of macromolecular fullerenes, a review. *Polymer International* 48 (9): 743–757.

Gleizes, A., J. J. Gonzalez, and P. Freton. 2005. Thermal plasma modelling. *Journal of Physics D-Applied Physics* 38 (9): R153–R183.

Gonzalez-Aguilar, J., M. Moreno, and L. Fulcheri. 2007. Carbon nanostructures production by gas-phase plasma processes at atmospheric pressure. *Journal of Physics D-Applied Physics* 40 (8): 2361–2374.

Goodson, A. L., C. L. Gladys, and D. E. Worst. 1995. Numbering and naming of fullerenes by chemical-abstracts-service. *Journal of Chemical Information and Computer Sciences* 35 (6): 969–978.

Greendyke, R. B., J. A. Swain, and C. D. Scott. 2004. Computational fluid dynamics simulation of laser-ablated carbon plume propagation in varying background gases for single-walled nanotube synthesis. *Journal of Nanoscience and Nanotechnology* 4 (4): 441–449.

Hare, J. P., H. W. Kroto, and R. Taylor. 1991. Preparation and UV visible spectra of fullerenes C_{60} and C_{70}. *Chemical Physics Letters* 177 (4–5): 394–398.

Heath, J. R., Yuan Liu, S. C. O'Brien et al. 1985. Semiconductor cluster beams: One and two color ionization studies of Si_x and Ge_x. *The Journal of Chemical Physics* 83 (11): 5520–5526.

Hinkov, I. 2004. PhD thesis, Université Paris, France.

Howard, J. B., J. T. Mckinnon, Y. Makarovsky, A. L. Lafleur, and M. E. Johnson. 1991. Fullerenes C_{60} and C_{70} in flames. *Nature* 352 (6331): 139–141.

Huczko, A., H. Lange, P. Byszewski, M. Poplawska, and A. Starski. 1997. Fullerene formation in carbon arc: Electrode gap dependence and plasma spectroscopy. *Journal of Physical Chemistry A* 101 (7): 1267–1269.

Huczko, A., H. Lange, G. Cota-Sanchez, and G. Soucy. 2002. Plasma synthesis of nanocarbons. *High Temperature Material Processes* 6 (3): 369–384.

Ikeda, T., T. Kamo, and M. Danno. 1995. New synthesis method of fullerenes using microwave-induced naphthalene-nitrogen plasma at atmospheric-pressure. *Applied Physics Letters* 67 (7): 900–902.

Inomata, K., N. Aoki, and H. Koinuma. 1994. Production of fullerenes by low-temperature plasma chemical-vapor-deposition under atmospheric-pressure. *Japanese Journal of Applied Physics Part 2-Letters* 33 (2A): L197–L199.

Jones, J. M., R. P. Malcolm, K. M. Thomas, and S. H. Bottrell. 1996. The anode deposit formed during the carbon-arc evaporation of graphite for the synthesis of fullerenes and carbon nanotubes. *Carbon* 34 (2): 231–237.

Kim, K. S., G. Cota-Sanchez, C. T. Kingston et al. 2007. Large-scale production of single-walled carbon nanotubes by induction thermal plasma. *Journal of Physics D-Applied Physics* 40 (8): 2375–2387.

Kratschmer, W., L. D. Lamb, K. Fostiropoulos, and D. R. Huffman. 1990a. Solid C_{60}—A new form of carbon. *Nature* 347 (6291): 354–358.

Kratschmer, W., K. Fostiropoulos, and D. R. Huffman. 1990b. The Infrared and ultraviolet-absorption spectra of laboratory-produced carbon dust - Evidence for the presence of the C_{60} molecule. *Chemical Physics Letters* 170 (2–3): 167–170.

Krestinin, A. V. and A. P. Moravskii. 1999. Kinetics of fullerene C_{60} and C_{70} formation in a reactor with graphite rods evaporated in electric arc. *Chemical Physics Reports* 18 (3): 515–532.

Krestinin, A. V., A. P. Moravskii, and P. A. Tesner. 1998. A kinetic model of formation of fullerenes C_{60} and C_{70} in condensation of carbon vapor. *Chemical Physics Reports* 17 (9): 1687–1707.

Kroto, H. W., J. R. Heath, S. C. O'Brien, R. F. Curl, and R. E. Smalley. 1985. C_{60}: Buckminsterfullerene. *Nature* 318 (6042): 162–163.

Lange, H., A. Huczko, and P. Byszewski. 1996. Spectroscopic study of C_2 in carbon arc discharge. *Spectroscopy Letters* 29 (7): 1215–1228.

Lange, H., K. Saidane, M. Razafinimanana, and A. Gleizes. 1999. Temperatures and C_2 column densities in a carbon arc plasma. *Journal of Physics D-Applied Physics* 32 (9): 1024–1030.

Launder, B. E. and D. B. Spalding. 1972. *Lectures in Mathematical Models of Turbulence*. New York: Academic.

Lieber, C. M. and C. C. Chen. 1994. Preparation of fullerenes and fullerene-based materials. In: *Solid State Physics—Advances in Research and Applications*, vol. 48. H. Ehrenreich and F. Spaepen (Eds.), pp. 109–148. San Diego, CA: Academic Press.

Maruyama, S. and Y. Yamaguchi. 1998. A molecular dynamics demonstration of annealing to a perfect C_{60} structure. *Chemical Physics Letters* 286 (3–4): 343–349.

Mieno, T. 2004. Characteristics of the gravity-free gas-arc discharge and its application to fullerene production. *Plasma Physics and Controlled Fusion* 46 (1): 211–219.

Mostaghimi, J. and M. I. Boulos. 1989. Two-dimensional electromagnetic-field effects in induction plasma modeling. *Plasma Chemistry and Plasma Processing* 9 (1): 25–44.

Murphy, A. B. and P. Kovitya. 1993. Mathematical-model and laser-scattering temperature-measurements of a direct-current plasma torch discharging into air. *Journal of Applied Physics* 73 (10): 4759–4769.

Osawa, E. 1970. Superaromaticity. *Kagaku* 25: 836–854.

Ostrikov, K. and A. B. Murphy. 2007. Plasma-aided nanofabrication: Where is the cutting edge? *Journal of Physics D-Applied Physics* 40 (8): 2223–2241.

Parker, D. H., P. Wurz, K. Chatterjee et al. 1991. High-yield synthesis, separation, and mass-spectrometric characterization of fullerenes C_{60} to C_{266}. *Journal of the American Chemical Society* 113 (20): 7499–7503.

Pellerin, S., K. Musiol, O. Motret, B. Pokrzywka, and J. Chapelle. 1996. Application of the (0,0) Swan band spectrum for temperature measurements. *Journal of Physics D-Applied Physics* 29 (11): 2850–2865.

Proulx, P., J. Mostaghimi, and M. I. Boulos. 1987. Heating of powders in an RF inductively coupled plasma under dense loading conditions. *Plasma Chemistry and Plasma Processing* 7 (1): 29–52.

Ranz, W. E. and W. R. Marshall. 1952a. Evaporation from drops, Part I. *Chemical Engineering Progress* 48 (3): 141–146.

Ranz, W. E. and W. R. Marshall. 1952b. Evaporation from drops, Part II. *Chemical Engineering Progress* 48 (3): 173–180.

Ravary, B., J. A. Bakken, J. Gonzalez-Aguilar, and L. Fulcheri. 2003. CFD modeling of a plasma reactor for the production of nano-sized carbon materials. *High Temperature Material Processes* 7 (2): 139–144.

Rohlfing, E. A., D. M. Cox, and A. Kaldor. 1984. Production and characterization of supersonic carbon cluster beams. *The Journal of Chemical Physics* 81 (7): 3322–3330.

Saidane, K., M. Razafinimanana, H. Lange et al. 2004. Fullerene synthesis in the graphite electrode arc process: Local plasma characteristics and correlation with yield. *Journal of Physics D-Applied Physics* 37 (2): 232–239.

Saito, Y., M. Inagaki, H. Shinohara et al. 1992. Yield of fullerenes generated by contact arc method under He and Ar–dependence on gas-pressure. *Chemical Physics Letters* 200 (6): 643–648.

Schutze, A., J. Y. Jeong, S. E. Babayan et al. 1998. The atmospheric-pressure plasma jet: A review and comparison to other plasma sources. *IEEE Transactions on Plasma Science* 26 (6): 1685–1694.

Scott, D. A., P. Kovitya, and G. N. Haddad. 1989. Temperatures in the plume of a DC plasma torch. *Journal of Applied Physics* 66 (11): 5232–5239.

Sesli, A., B. Cicek, and N. Oymael. 2005. Fullerene production in a graphite tubular reactor. *Fullerenes Nanotubes and Carbon Nanostructures* 13 (1): 1–11.

Shigeta, M. and T. Watanabe. 2008. Numerical investigation of cooling effect on platinum nanoparticle formation in inductively coupled thermal plasmas. *Journal of Applied Physics* 103 (7): 074903.

Singh, H. and M. Srivastava. 1995. Fullerenes—Synthesis, separation, characterization, reaction chemistry, and applications—A review. *Energy Sources* 17 (6): 615–640.

Smalley, R. E. and B. I. Yakobson. 1998. The future of the fullerenes. *Solid State Communications* 107 (11): 597–606.

Sugai, T., H. Omote, S. Bandow, N. Tanaka, and H. Shinohara. 2000. Production of fullerenes and single-wall carbon nanotubes by high-temperature pulsed arc discharge. *Journal of Chemical Physics* 112 (13): 6000–6005.

Szepvolgyi, J., Z. Markovic, B. Todorovic-Markovic et al. 2006. Effects of precursors and plasma parameters on fullerene synthesis in RF thermal plasma reactor. *Plasma Chemistry and Plasma Processing* 26 (6): 597–608.

Taylor, R., J. P. Hare, A. K. Abdulsada, and H. W. Kroto. 1990. Isolation, separation and characterization of the fullerenes C_{60} and C_{70}—The 3rd form of carbon. *Journal of the Chemical Society-Chemical Communications* 20: 1423–1424.

Todorovic-Markovic, B., Z. Markovic, I. Mohai et al. 2003. Efficient synthesis of fullerenes in RF thermal plasma reactor. *Chemical Physics Letters* 378 (3–4): 434–439.

Todorovic-Markovic, B., Z. Markovic, I. Mohai et al. 2006. RF thermal plasma processing of fullerenes. *Journal of Physics D-Applied Physics* 39 (2): 320–326.

Wang, C., T. Imahori, Y. Tanaka et al. 2001. Synthesis of fullerenes from carbon powder by using high power induction thermal plasma. *Thin Solid Films* 390 (1–2): 31–36.

Wang, C., A. Inazaki, T. Shirai et al. 2003. Effect of ambient gas and pressure on fullerene synthesis in induction thermal plasma. *Thin Solid Films* 425 (1–2): 41–48.

Xie, S. Y., R. B. Huang, L. J. Yu, J. Ding, and L. S. Zheng. 1999. Microwave synthesis of fullerenes from chloroform. *Applied Physics Letters* 75 (18): 2764–2766.

Xie, S. Y., R. B. Huang, S. L. Deng, L. J. Yu, and L. S. Zheng. 2001. Synthesis, separation, and characterization of fullerenes and their chlorinated fragments in the glow discharge reaction of chloroform. *Journal of Physical Chemistry B* 105 (9): 1734–1738.

Xue, S. W., P. Proulx, and M. I. Boulos. 2001. Extended-field electromagnetic model for inductively coupled plasma. *Journal of Physics D-Applied Physics* 34 (12): 1897–1906.

Yamaguchi, Y. and S. Maruyama. 1998. A molecular dynamics simulation of the fullerene formation process. *Chemical Physics Letters* 286 (3–4): 336–342.

Yoshie, K., S. Kasuya, K. Eguchi, and T. Yoshida. 1992. Novel method for C_{60} synthesis—A thermal plasma at atmospheric-pressure. *Applied Physics Letters* 61 (23): 2782–2783.

22

HPLC Separation of Fullerenes

Qiong-Wei Yu
Wuhan University

Yu-Qi Feng
Wuhan University

22.1 Introduction

22.1.1 Properties and Applications of Fullerenes

In 1985, Kroto and colleagues first discovered C_{60}, which would lead to the subsequent discovery of fullerenes [1]. Fullerenes are the third allotrope of carbon of which the other two allotropes are graphite and diamond. Carbon clusters of all sizes are known as buckminsterfullerenes, fullerenes, or "buckyballs."

Fullerenes are closed cage structures. Figure 22.1 shows the structures of some fullerenes. Each carbon atom is bonded to three others and is sp^2 hybridized. Hexagonal rings are present but pentagonal rings are required to form closed cage structures. All fullerenes have an even number of carbons with the smallest fullerene identified as C_{20}^+, and the largest having well over 100 carbons. When a C_{2n} fullerene is fragmented by mass spectrometry, C_2 fragments are sequentially spat out to form C_{2n-2} fullerenes. The decay process continues until it reaches C_{32}, at which point the whole structure falls apart [2].

The C_{60} structure is characterized as a resonance structure having low electron delocalization over its spherical surface. C_{60} is a good electron acceptor, which can easily form charge-transfer complexes with compounds having electron-donor groups. This suggests that fullerene molecules can serve both as donors and as acceptors of electrons in chemical reactions [3].

Fullerenes have unique physical and chemical properties, and have been found to be useful in several fields [4–8], especially in solid-state applications. For example, the relatively high transition temperature, T_c, observed for C_{60} (33 K) makes it a suitable material for superconductivity studies, while better results ($T_c \sim 18$ K) have been achieved with alkali-metal-doped fullerenes (also known as endohedral fullerenes) [4]. The major drawback of fullerenes is their low solubility in aqueous and organic solvents [7].

At present, fullerenes are formed in the laboratory from carbon-rich vapors, which can be obtained in a variety of ways [8], such as resistive heating of carbon rods in a vacuum, plasma discharge between carbon electrodes in helium gas, laser ablation of carbon electrodes in helium gas, and oxidative combustion of benzene–argon gas mixtures. The yield efficiency and purity of fullerenes from most of these methods is low.

22.1.2 Backgrounds on Separation of Fullerenes

Minute quantities of fullerenes, in the form of C_{60}, C_{70}, C_{76}, and C_{84} molecules, are produced naturally and can be found hidden in soot or formed by lightning discharges in the atmosphere [4].

Fullerene purification remains a challenge to chemists owing to their low solubility in a large number of solvents

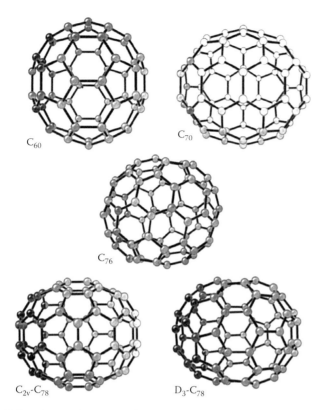

C_{60} C_{70}

C_{76}

C_{2v}-C_{78} D_3-C_{78}

FIGURE 22.1 Structures of C_{60}, C_{70}, C_{76}, C_{2v}-C_{78}, D_3-C_{78}. (From http://www.ch.ic.ac.uk/rzepa/mim/century/html/c60.htm. With permission.)

and the similar structure and physico-chemical properties of fullerenes. Soxhlet extraction has been the preferred method for the extraction of C_{60} and C_{70} from fullerene containing carbon soot. In order to further purify fullerenes, methods such as chromatography, recrystallization, sublimation, and selective complexation of fullerenes with calixarenes have been used [9–12].

These non-chromatographic methods could purify fullerenes in large quantities, but have problems such as a lower purity quotient and a multistep operation.

Chromatography is one of the most effective methods for the separation of mixtures due to its high separation efficiency and rapid speed. It is a physical separation method in which the components to be separated are distributed between two phases. The two components of this separation process occur in a chromatographic system and consist of the mobile and stationary phases, respectively. The mobile phase consists of the sample being separated/analyzed and the solvent that moves the sample through the column. The mobile phase moves through the chromatography column (the stationary phase) where the sample interacts with the stationary phase and is separated [13]. Thereinto, liquid chromatography (LC) is a separation technique in which the mobile phase is a liquid. Liquid chromatography can be carried out either in a column or a plane. Liquid chromatography that utilizes minute packing particles and high pressure is referred to as high-performance liquid chromatography (HPLC).

During HPLC, the sample is forced through a column that is packed with irregularly or spherically shaped particles or a porous monolithic layer (stationary phase), by a liquid (mobile phase) at high pressure.

HPLC is usually used for the final step separation of fullerenes in most strategies of fullerene isolation. The choice of the stationary phase used in the final separation process will ultimately determine the purity of the collected fractions. Therefore, HPLC remains as one of the main methods for obtaining C_{60}, C_{70}, and higher fullerenes at high purity.

Different types of liquid chromatography (LC) stationary phases have been used for fullerene separation. Early chromatographic separations involved the use of alumina and silica as stationary phases [14–16]. However, in later years, activated carbon, graphite, or coal were used in the stationary phase, but the retention of fullerenes on these stationary phases remained weak and the separation efficiency was low. Chemically bonded stationary phases are presently used for fullerene separation because of the specific interaction of organic ligands with fullerenes and their high separation efficiency for fullerenes. Various chemically bonded stationary phases are applied to the separation of fullerenes. A bonded phase is a stationary phase that is covalently bonded to the support particles or to the inside wall of the column tubing. Most of theses stationary phases are silica-based. Ligands bonded onto silica include: alkyl, polycyclic aromatics, multi-legged phenyl, multi-methoxy phenyl, liquid-crystal, pirkle-type phases, γ-cyclodextrin, phthalocyanines, and porphyrins.

The retention of a solute may result either from an interaction with the stationary phase or from weak or reduced interactions with the mobile phase. Therefore, the key to purifying fullerenes is the choice of the mobile phase used in the separation process.

One of the basic requirements of a suitable mobile phase in a liquid chromatographic system is good solubility of the "to be separated" solutes. Thus to optimize chromatography, the solubility of fullerenes in solvent is of great importance. Table 22.1 shows the solubility of C_{60} in different solvents [15]. The solubility rules of other fullerenes are the same as those of C_{60}. The solvents used for dissolving fullerenes can be divided into four categories according to their decreasing solubility [16]:

1. Good solvents: benzene, toluene, ethybenzene, xylene, mesitylene, tetralin, bromobenzene, anisol, chlorobenzene, dichlorobenzene, trichlorobenzene, carbon disulfide, 2-methylthiophene, 1-methylnaphthalene, dimethylnaphthalenes, 1-phenylnaphthalene, 1-chloronaphthalene.
2. Fair solvents: dichloromethane, chloroform, carbon tetrachloride, 1,2-dibromethane, benzonitrile, fluorobenzene, nitrobenzene.
3. Poor solvents: *n*-hexane, cyclohexane, *n*-decane, dichlorodifluoroethane, 1,1,2-trichlorotrifluoroethane, *o*-crysol.
4. Bad solvents: *n*-pentane, cyclopentane, tetrahydrofuran, methanol, ethanol, nitromethane, nitroethane, acetone, acetonitrile.

TABLE 22.1　Solubility of C_{60} in Various Solvents

Solvent	C_{60} (mg/mL)	Solvent	C_{60} (mg/mL)
Alkanes		*Benzenes*	
n-Pentane	0.0050	Benzene	1.7
Cyclopentane	0.0020	Toluene	2.8
n-Hexane	0.043	Xylenes	5.2
Cyclohexane	0.036	Mesitylene	1.5
n-Decane	0.071	Tetralin	16
Decalins:	4.6	*o*-Cresol	0.014
cis-Decalin	2.2	Benzonitrile	0.41
trans-Decalin	1.3	Fluorobenzene	0.59
		Nitrobenzene	0.80
Haloalkanes		Bromobenzene	3.3
Dichloromethane	0.26	Anisole	5.6
Chloroform	0.16	Chlorobenzene	7.0
Carbon tetrachloride	0.32	1,2-Dichlorobenzene	27
1,2-Dibromethane	0.50	1,2,4-Trichlorobenzene	8.5
Trichloroethylene	1.4		
Tetrachloroethylene	1.2	*Naphthalenes*	
Freon TF(dichlorodifluoroethane)	0.020	1-Methylnaphthalene	33
1,1,2-Trichlorotrifluoroethane	0.014	Dimethylnaphthalenes	36
1,1,2,2-Tetrachloroethane	5.3	1-Phenylnaphthalene	50
		1-Chloronaphthalene	51
Polars			
Methanol	0	*Miscellaneous*	
Ethanol	0.0010	Carbon disulfide	7.9
Nitromethane	0	Tetrahydrofuran	0
Nitroethane	0.0020	Tetrahydrothiophene	0.030
Acetone	0.0010	2-Methylthiophene	6.8
Acetonitrile	0	Pyridine	0.89
N-Methyl-2-pyrrolidone	0.89		

Source: Théobald, J. et al., *Sep. Sci. Technol.*, 30(14), 2783, 1995. With permission.

22.1.3 Introduction of HPLC

A chromatographic system can be considered to have four component parts: a device for sample introduction, a mobile phase, a stationary phase, and a detector [17]. A block diagram of an HPLC system, illustrating its major components, is shown in Figure 22.2.

During an HPLC run, the injector is used to introduce analytes into a flowing liquid stream. The mobile phase is a liquid delivered under high pressure (up to 400 bar (4×10^7 Pa)) to ensure a constant flow rate, while the stationary phase is packed into a column capable of withstanding high pressures.

A chromatographic separation occurs when the components of a mixture interact to different extents with the mobile and stationary phases and therefore take different times to move from the position of sample introduction to the position at which they are detected [18].

The time between sample injection and an analyte peak reaching a detector at the end of the column is termed the retention time (t_r). Each analyte in a sample will have a different retention time. The time taken for the mobile phase to pass through the column is called t_0.

The retention factor or capacity factor, k, is often used to describe the migration rate of an analyte on a column. The retention factor for analyte A is defined as follows:

$$k_A = \frac{t_r - t_0}{t_0} \tag{22.1}$$

t_r and t_0 can be derived from a chromatogram as shown in Figure 22.3. When the analyte retention factor is less than one, elution is fast, making an accurate determination of the retention time difficult. High retention factors (greater than 20) mean that elution takes a very long time.

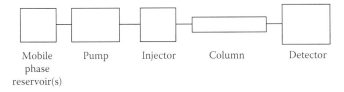

Mobile phase reservoir(s)　　Pump　　Injector　　Column　　Detector

FIGURE 22.2　Block diagram of a typical HPLC system. (From Ardrey, R.E., *Liquid Chromatography-Mass Spectrometry: An Introduction*, John Wiley & Sons Ltd., Hoboken, NJ, 2003, 10–13. With permission.)

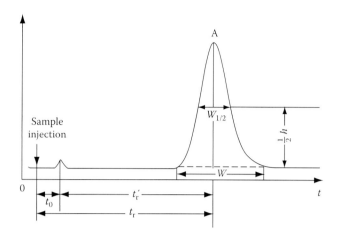

FIGURE 22.3 Elution curves of chromatography. (Adapted from Ardrey, R.E., *Liquid Chromatography-Mass Spectrometry: An Introduction*, John Wiley & Sons Ltd., Hoboken, NJ, 2003, 10–13.)

The selectivity factor, α, describes the separation of two species (A and B) on the column;

$$\alpha = \frac{k_B}{k_A} \qquad (22.2)$$

When calculating the selectivity factor, species A elutes faster than species B. The selectivity factor is always greater than one.

The selectivity factor (α) describes the separation of band centers but does not take into account peak widths. Another measurement of how well species have been separated can be provided by measuring resolution. The resolution of two species, A and B, is defined as

$$R = \frac{2\left[(t_r)_B - (t_r)_A\right]}{W_A + W_B} \qquad (22.3)$$

Baseline resolution is achieved when $R = 1.5$.

Chromatography can be divided into two kinds according to the scale of separated analytes. Analytical chromatography is used to determine the existence and concentration of analyte(s) in a sample, whereas preparative chromatography is used to purify sufficient quantities of a substance for further use, rather than analysis. In general, only the stationary phases with high-enough retention factors and separation factors for analytes can provide the possibility of preparative separation.

Fullerenes smaller than C_{30} are difficult to isolate, due to the presence of energetically competitive isomers such as rings, bowls, and sheets. C_{20}, which is topologically the smallest fullerene, has been prepared but exhibits a very short lifetime, whereas C_{30} is experimentally considered to be the smallest stable fullerene [19]. C_{60} is the most symmetrical, and therefore the most stable fullerene, followed by C_{70}, while higher fullerenes are much less stable. The rest of this chapter will emphasize on the separation and purification of C_{60}, C_{70}, and higher fullerenes.

22.2 Separation of Fullerenes on Alkyl-Bonded Silica Stationary Phases

Alkyl-bonded silica stationary phases are usually used in reversed-phase liquid chromatography (RPLC) in which the stationary phase is less polar than the mobile phase. However, the use of reversed-phase liquid chromatography is not ideal for the preparative separation of fullerenes due to the poor solubility of fullerenes in commonly used RPLC mobile phases solvents such as water, methanol, acetonitrile, and tetrafuran.

Alkyl-bonded silica phases are prepared by bonding silanes containing alkyl carbon chains with 1–30 carbon atoms to silica. The modification is made by chemically reacting the silanol groups on silica and chlorosilane compounds with carbons of different chain length [20–22]. Of these, the 18-carbon chain or octadecyl group (abbreviated as ODS and C_{18}) is the most commonly used for this type of stationary phase.

It has been noted that alkyl-bonded silica stationary phases that have solvophobic and molecular shape recognition properties as part of the retentive–selective mechanism can exhibit selectivity for fullerenes in poor dissolving solvents.

Despite the low solubility of fullerenes in the mobile phase and their unsuitability for the preparative separation of fullerenes, alkyl-bonded silica stationary phases are useful for analytical purposes or separation of higher fullerenes.

According to the bonding chemistry, octadecyl-bonded silica (ODS) stationary phases can be divided into two types: polymeric C_{18} phases and monomeric C_{18} phases. The former are synthesized using trichlorosilane as starting material with the addition of a small amount of water, while monomeric C_{18} phases are produced from monochlorosilane with the exclusion of water. The conventional synthetic schemes are shown in Figure 22.4 [23].

22.2.1 Separation of Fullerenes with Octadecyl-Bonded Silica (ODS) Stationary Phases

Monomeric ODS phases and polymeric ODS phases exhibit different molecular shape and size recognition capability for fullerenes, with both ODS phases being widely applied to the separation of fullerenes. The separation mode is not of a typical reverse-phase chromatographic system as the stationary phases and mobile phase are nonpolar.

In general, retention of fullerenes on monomeric ODS is greater than on polymeric phase. The resolution of higher fullerenes on monomeric ODS is better than the polymeric phase, however, polymeric ODS phases exhibit better resolution for fullerene isomers. It is worth mentioning that the elution order of higher fullerenes is different on these two stationary phases. The retention and selectivity of fullerenes on polymeric ODS depends on the nonplanarity of the molecule and the dispersive interaction with ODS, which can be modified by changing the nature and/or composition of the mobile phase.

As shown in Figure 22.5, using a toluene–acetonitrile mobile phase with monomeric ODS, the elution profile is C_{76}, $C_{78\text{-}C2v}$, $C_{78\text{-}C2v}$, and D_3 together, followed by C_{82} and C_{84}. This elution

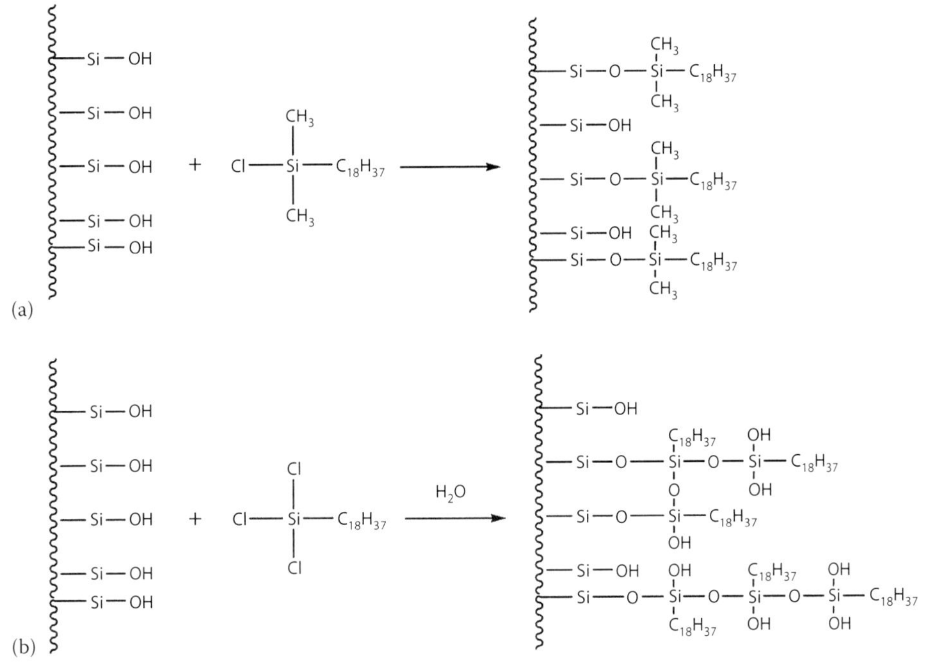

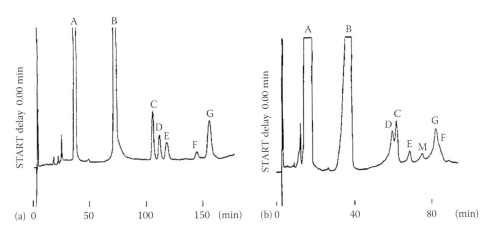

FIGURE 22.4 Conventional synthetic schemes for (a) monomeric and (b) polymeric ODS-bonded phases. (From Saito, Y. et al., *J. Sep. Sci.*, 26, 225, 2003. With permission.)

FIGURE 22.5 Chromatogram of the concentrated solution of carbon soot after toluene extraction with (a) monomeric ODS and (b) polymeric ODS. Mobile phase, toluene–acetonitrile = 45/55; flow rate, 1 mL/min; detection, UV at 325 nm; temperature, room temperature. Solutes: (A) C_{60}, (B) C_{70}, (C) C_{76}, (D) $C_{78-C2v'}$, (E) C_{78-C2v}, (F) C_{82}, (G) C_{84}, (M) $C_{78}D_3$. (From Jinno, K. et al., *Anal. Chem.*, 65, 2650, 1993. With permission.)

order follows the carbon number rule as fullerenes with more carbons exhibit stronger retention. This is in contrast with the elution order/profile of the fullerenes using the polymeric ODS phase, which is $C_{78-C2v'}$, C_{76}, C_{78-C2v}, D_3, C_{84}, and C_{82}. These differences can be explained by assuming that the polymeric ODS phase has higher molecular planarity recognition capability than the monomeric ODS phase. The fullerene elution order with the polymeric phase depends strongly on the shape difference of the fullerene molecules. Isomers of C_{78}, $C_{78-C2v'}$ have shorter diameters (most bulky). $C_{78-C2v'}$, $C_{78-C2v'}$ have a little different symmetry, and D_3 has a narrower shape than the other two isomers. The polymeric ODS elutes bulkier $C_{78-C2v'}$ and C_{78-C2v} followed by the more cylindrical D_3 by virtue of its planarity recognition

capability. The reversed elution order of C_{84} and C_{82} with the polymeric ODS can also be explained in the same way, as the shape of C_{82} is considered to be longer and narrower than C_{84} [24].

22.2.2 Influence of Physical Parameters of ODS on Separation of Fullerenes

Due to their specific structure, polymeric ODS phases exhibit similar influences to monomeric ODS with high surface coverage for fullerenes retentions. The factors influencing fullerenes retention on monomeric ODS are emphasized in this chapter.

A number of physical parameters can influence the retention characteristics of fullerenes in the monomeric ODS phases;

these include pore diameter, surface area, surface coverage, and the distance between the adjacent C_{18} functional groups. ODS phases with higher surface coverage provide stronger retention and better separation capability for C_{60} and C_{70}. If the distance between neighboring C_{18} groups is similar to the diameter of the C_{60} molecule, very efficient interactions are formed. However, decreasing the surface coverage of ODS results in increased distances between C_{18} functional groups, resulting in reduced interactions with the C_{60} molecule. Similarly, if the surface coverage is so high that fullerenes could not intervene in the distance between C_{18} functional groups, the interaction between fullerenes and ODS will also become weak. These influences factors can be minimized if the same silica gel is used.

At high temperatures, all C_{18} ligands are able to move freely and there is little difference in C_{60} recognition among these phases with different surface coverage; however, there is a high possibility that interaction can occur. At lower temperatures, the C_{18} ligands becomes more rigid and ordered, and the ligand distance becomes critical in enabling them to interact with the C_{60} molecule. Any mismatch between the ligand distance and diameter of the C_{60} molecule can result in weak interactions and thus, similar retentions. Figure 22.6 shows the effect of different ligand distance on interaction between C_{60} and ODS with different surface coverage. Thus, altering the ODS ligand distance between each adjacent pair is an important factor in obtaining better selectivity for fullerenes in reversed-phase LC [25].

22.2.3 Comparison of Fullerenes Separation on Different Alkyl-Bonded Phases

The length of alkyl chain at the surface of the bonded phases is an important factor influencing the retentions of fullerenes. Under the condition of same surface coverage, discrimination of fullerene selectivity could be produced by changing different alkyl chain lengths. The stationary phase needs to have enough depth and space to catch and hold the bulky molecules within the ligand space. It is reasonable to assume that there is a critical chain length to interact effectively with fullerenes.

Table 22.2 lists compare the retentions and resolution of fullerenes on different alkyl-bonded silica stationary phases. It can be seen that the length of alkyl chains on the bonded phase largely contributes to fullerene retention, with phases using longer chains giving better retention. The chains of odecyl ligands (C_8) and other shorter chains have been proved to be too short to effectively interact with fullerene molecules, while C_{18} phases can achieve better separation. In comparison to these phases, monomeric C_{30} phase exhibited the strongest retentions for fullerenes and is therefore the best choice for the fullerene separation. Monomeric C_{30} phase has been successfully employed for large-scale separation of fullerenes. A high toluene concentration in the mobile phase can be used to overcome the main disadvantage of C_{18} stationary phases [26]. In contrast, it could be concluded that very long alkyl chains such as C_{100} (and even more on silica) could form self-aggregation and reduce the interaction with fullerenes. Therefore, an optimized length of alkyl chains is required for efficient fullerenes separation.

22.2.4 Temperature Dependence of Fullerenes Separation on Alkyl-Bonded Phases

Temperature can influence the retention and separation of fullerenes. The dependence of the logarithms of the retention factor (ln k) on temperature is given by Equation 22.4 or van't Hoff plot. The changes of enthalpy and entropy can be estimated from the slope and intercept of the linear van't Hoff plots.

$$\ln k = -\frac{\Delta H^\circ}{RT} + \frac{\Delta S^\circ}{RT} + \ln \varnothing \qquad (22.4)$$

where
  k is the retention factor
  ΔH° is the enthalpy change
  ΔS° is the entropy change of transfer for the solute from the mobile phase to the stationary phase
  $\varnothing$ is the phase ratio
  R is the gas constant

In general, negative values of enthalpy changes indicate that the transfer of solute from mobile to stationary phase is favored, whereas negative values of entropy change represent increased order due to interaction with ligands on stationary phases.

As a general rule in classical reverse-phase liquid chromatography, solute retention is inversely related to temperature. As shown in Figure 22.7, the temperature dependence of fullerenes on monomeric ODS phase is different from those of polycyclic aromatic hydrocarbonds (PAHs). It is apparent that the recognition mechanism of the monomeric ODS phase for the fullerene molecule is different from that applied to using PAHs. When the temperature is lower than −20°C, the separation factors are almost constant while the retentions of both C_{60} and C_{70} decrease with decreasing temperature on ODS phases with lower surface coverage. For high carbon load ODS phases, the separation factors show similar values to those of the polymeric ODS phase at −20°C or lower. This suggests that the physical property of monomeric ODS phase with high surface coverage is similar to those of the polymeric phase at very low temperatures. The results can be explained by the different inter-ligand distances between the bonded C_{18} functional groups. At ambient and/or higher temperature, the C_{18} functional groups are relaxed and flexible so that they can interact with the solute C_{60}. Likewise, there is a critical temperature for the change of retentions of fullerenes on other alkyl-bonded phases such as monomeric C_{30} phase (Figure 22.8). The major difference between the C_{30} and the C_{18} phases is the maximum retention temperature for fullerenes, with the former being at 20°C and the latter being at −20°C. The van't Hoff plots suggest that the retention mechanism of fullerenes investigated might be different within the high- and low-temperature regions [27].

22.2.5 C_{60} and C_{70} HPLC Retention Reversal Study Using Organic Modifiers

Mobile phase serves as a crucial factor for the retention process of fullerenes. For years, it has been known that retention results

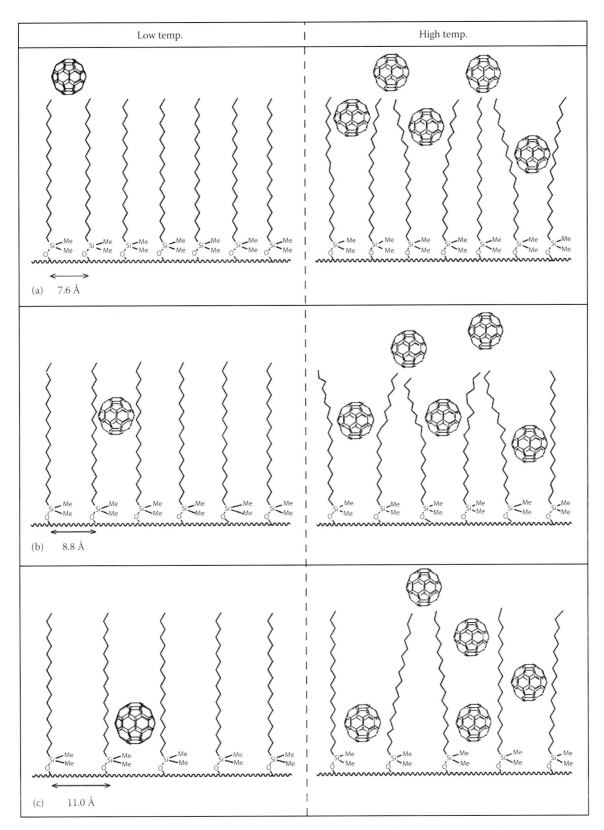

FIGURE 22.6 Retention model of C_{60} on ODS phases with different surface coverages. Left column: at low or ambient temperatures; right column: at elevated temperatures. The calculated ligand intervals of these phases are (a) 7.6 Å, (b) 8.8 Å, (c) 11.0 Å. (From Saito, Y. et al., *J. Sep. Sci.*, 26, 225, 2003. With permission.)

TABLE 22.2 Retention of C_{60} and C_{70} on Alkylsilyl Silica in Hexane[a]

Stationary Phase (R)	Surface Coverage ($\times 10^6$ mol/m²)	$k(C_{60})$	$k(C_{70})$	$\alpha(C_{70}/C_{60})$
CH_3	4.5	<0.01	<0.01	
C_4H_9	3.8	0.06	0.08	1.22
C_6H_{13}	3.4	0.11	0.12	1.04
C_8H_{17}	3.5	0.1	0.13	1.3
$C_{12}H_{25}$	3.5	0.27	0.38	1.4
$C_{18}H_{37}$	2.9	0.7	1.23	1.77
$C_{30}H_{61}$	1.7	2.03	3.48	1.72

Source: Kimata, K. et al., *Anal. Chem.*, 67, 2556, 1995. With permission.

[a] The stationary phase was prepared from alkyldimethylchlorosilane, $RSi(CH_3)_2Cl$. Column, 4.6 mm i.d., 15 cm long; mobile phase, hexane; flow rate, 1 mL/min; detection, 254 nm [26].

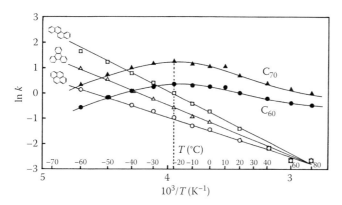

FIGURE 22.7 Plots of ln k vs. temperature for C_{60} and C_{70}, and three PAHs (pyrene, triphenylene, chrysene) with the C_{18}-bonded phases. (From Ohta, H. et al., *J. Chromatogr. A*, 883, 55, 2000. With permission.)

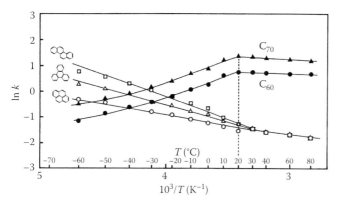

FIGURE 22.8 Plots of ln k vs. temperature for C_{60} and C_{70}, and three PAHs (pyrene, triphenylene, chrysene) with the C_{30}-alkyl-bonded phases. (From Ohta, H. et al., *J. Chromatogr. A*, 883, 55, 2000. With permission.)

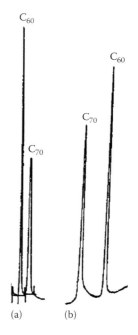

FIGURE 22.9 The C_{60} and C_{70} chromatogram in a (a) methanol/water mixture (92/8, V/V) and a (b) ethanol/water mixture (78/22, V/V). The mobile phase flow-rate was 1 mL/min, and the column temperature was 25°C. Numbers above peaks refer to C_X retention time in minutes. (From Guillaume, Y.C. et al., *Anal. Chem.*, 72, 1301, 2000. With permission.)

modifiers (OM) used were methanol, ethanol, propanol, butanol, and pentanol. The Guillaum group found that C_{60} always eluted before C_{70} with methanol as eluent, while the reverse was observed in the cases of ethanol, propanol, and butanol (Figure 22.9). These results confirm that the mobile phase is the dominant effect in governing solute selectivity [29].

22.3 Separation of Fullerenes on Charge-Transfer Stationary Phases

Alkyl-bonded silica stationary phases such as ODS exhibit reasonable selectivity for fullerenes, but the retention decreases with increasing solubility of the target compound in the mobile phase.

from the chemical potential difference between the mobile and stationary phases. Polar solvents can be used as effective mobile phases in the separation of fullerene molecules on polysiloxane stationary phases. Guillaume et al. investigated the retention behavior of fullerene molecules using different organic modifier (OM)/water mixtures as mobile phases [28]. The organic

Hence, when strong solvents such as toluene or 1-methylnaphthalene are used, these compounds cannot be separated. Separation of C_{60} and C_{70} pose severe difficulties and is only possible when used with mobile phases in which the fullerenes are poorly soluble (such as hexane, acetonitrile, and dichloromethane), prohibiting the separation of large amounts of fullerenes [1,30]. Thus, the novel stationary phases such as charge-transfer phases are prepared for large-scale fullerene separation.

Fullerenes contain abundant $\pi-\pi$ electrons and could interact with aromatic stationary phases through a charge-transfer mechanism [4]. This property allows aromatic stationary phases to have strong retention and high separation selectivity for fullerenes. According to their different properties, charge-transfer stationary phases are divided into electron-donor stationary phases and electron-acceptor stationary phases, respectively. The former are also named as π-base stationary phases and are the most widely applied. The latter are known as π-acid stationary phases [31]. As fullerenes can act as both donors and acceptors of electrons, they could interact with both electron-acceptor stationary phases and electron-donor stationary phases.

It has been proved that the electron affinity of C_{70} is higher than that of C_{60} and as a result, the retention ability of C_{70} on some charge-transfer stationary phases is stronger than C_{60} for most stationary phases [32].

22.3.1 Phenylalkyl-Bonded Silica Phases

The diphenylalkyl-bonded phases can be synthesized by using alkyldiphenylchlorosilanes such as butyl, octyl, and octadecylsilanes as the starting materials in the conventional way; likewise, multiphenylchlorosilanes can be used to prepare multiphenyl-bonded silica phases [33].

According to the molecular size of these fullerenes, it is rational to assume that the surface coverage of bonded C_{18} ligands could play an important role in adjusting the interval among these ligands on the silica gel. By introducing phenyl groups at the bottom of the alkyl-bonded phase ligands, a series of phenyl-bonded silica stationary phases can be prepared. The alkyl ligand length and phenyl-rings in the bonded phase contribute to the fullerenes retention under relative surface coverage. Phenyl-bonded silica phases exhibit relatively smaller variation in temperature dependence than alkyl-bonded phases with the same number of carbons because of their conformational rigidity.

Saito et al. prepared octadecyldiphenyl-(C_{18} Diph), octyldiphenyl-(C_8 Diph), and butyldiphenyl-(C_4 Diph) silica phases, and multiphenyl-bonded silica {such as triphenyl-(Triph), diphenylmethyl-(Diph), phenyldimethyl-(Monoph)} and compared the retention behaviors with those obtained by octadecyldimethyl-(C_{18}), octyldimethyl-(C_8), and butyldimetyl-(C_4) silica phases [34,35]. The retention data of these experiments are summarized in Table 22.3. By comparing the temperature dependence of fullerenes on C_{18} Diph with C_{18}, it can be shown that the retention factors of C_{60} and C_{70} and their separation factor decrease at a slower rate than with C_{18} at elevated temperatures.

Multimethoxylphenyl-bonded silica phases exhibit stronger retention for fullerenes than methylphenyl-bonded silica phases and commercial ODS phases. The methoxy groups at the top of the phenyl ring can form a certain cavity-like shape conformation, which could contribute to the retention of fullerenes. The retention data of fullerenes on multimethoxylphenyl-bonded silica phases and commercially available monomeric ODS are compared in Table 22.4 [21]. Table 22.4 shows that the retention of C_{60} and C_{70} is increased, and separation is improved significantly on bonded phases containing methoxy functional group(s). In contrast to the phenylpropyldimethylsilyl-bonded silica (PP) phase, the methoxy phenylpropyl-bonded phases possess greater retentivity for C_{60} and C_{70} than the ODS phases, especially in the cases of 3,4-dimethoxyphenylpropyldimethylsilyl-bonded silica (DMPP) and 3,4,5-trimethoxyphenylpropyldimethylsilyl-bonded silica (TMPP). The separation factors obtained with methoxyphenylpropyl-bonded phases are comparable or superior to the commercially available ODS-bonded phases.

For the DMPP phase, the two methoxy groups at the top of the aromatic ring are forced into a planar conformation. The interactive sites of the phase would adopt a curved shape conformation that is complementary to the curvature of the fullerene molecules, allowing for fullerenes to interact effectively with the phenyl rings of the DMPP phase. A PP-bonded phase with no methoxy groups would interact weakly with the fullerenes due to their much smaller interaction area as compared with phases containing methoxyphenylpropyl groups. 4-Methoxyphenylpropyldimethylsilyl (MPP) and TMPP phases exhibit poorer retention, and have less recognition capability for C_{60} and C_{70} than DMPP, because the conformational fitting between these bonded phase groups with C_{60} and C_{70} solutes is worse than in the DMPP phase.

Jinno et al. observed an exceptionally small change in the retention of these carbon clusters on the DMPP phase at different column temperatures [23] which could be attributed to the rigid structure of these multiphenyl and multimethoxyphenyl-bonded silica phases, which also varies with temperature.

22.3.2 Multilegged Type Stationary Phases

Phenyl-bonded silica stationary phases, such as multi-legged phenyl-bonded phases [36] and multimethoxyphenylpropyl-bonded phases [21,23], exhibit good molecular shape recognition capacity for fullerenes as they could form a cavity-like structure with a size comparable to that of the fullerenes. Figure 22.10 depicts the structures of some multi-legged phenyl-bonded phases.

The main difference between multi-legged phases and other conventional phases lies in the structure at the silica surface. While the latter phases (including typical phenyl and ODS phases) are attached vertically to the silica surface through siloxane bonding, the former have a unique structure, and are horizontally attached to the silica surface. Each phenyl moiety is situated in parallel to the cavity-like structure on the silica surface. The walls of the cavity are produced by several methyl

TABLE 22.3   Retention Data for C_{60} and C_{70} with Alkylphenyl and Alkylmethyl Phases

Compounds	Structure	Carbon Content (%)	Ligand Interval (Å)	k C_{60}	k C_{70}	α C_{70}/C_{60}
C_{18}Diph 120 Å	$-O-Si(Ph)(Ph)-C_{18}H_{37}$	8.48	10.4	1.02	1.67	1.64
C_{18} 120 Å	$-O-Si(CH_3)(CH_3)-C_{18}H_{37}$	9.08	8.10	0.52	0.84	1.63
C_8Diph 120 Å	$-O-Si(Ph)(Ph)-C_8H_{17}$	5.89	10.3	0.42	0.57	1.36
C_8 120 Å	$-O-Si(CH_3)(CH_3)-C_8H_{17}$	5.08	7.90	0.086	0.124	1.44
C_4Diph 120 Å	$-O-Si(Ph)(Ph)-C_4H_9$	4.15	11.1	0.41	0.56	1.37
C_4 120 Å	$-O-Si(CH_3)(CH_3)-C_4H_9$	1.73	10.5	0.061	0.087	1.43
Triph 70 Å	$-O-Si(Ph)(Ph)-Ph$	11.2	12.1	2.28	3.59	1.58
Diph 70 Å	$-O-Si(Ph)(Ph)-CH_3$	13.7	8.70	1.85	2.89	1.56
Monoph 70 Å	$-O-Si(CH_3)(CH_3)-Ph$	10.8	7.00	0.86	1.24	1.44
Triph 150 Å	$-O-Si(Ph)(Ph)-Ph$	4.79	11.8	0.45	0.63	1.39
Diph 150 Å	$-O-Si(Ph)(Ph)-CH_3$	6.11	8.80	0.45	0.64	1.41

TABLE 22.3 (continued) Retention Data for C_{60} and C_{70} with Alkylphenyl and Alkylmethyl Phases

| Compounds | Structure | Carbon Content (%) | Ligand Interval (Å) | k | | α |
				C_{60}	C_{70}	C_{70}/C_{60}
Monoph 150 Å		5.35	6.60	0.27	0.37	1.36
Develosil ODS-5 100 Å		20.0	7.10	0.87	1.60	1.84

Source: Saito, Y. et al., *J. Liq. Chromatogr.*, 18, 1897, 1995. With permission.
Note: Mobile phase, *n*-hexane.

TABLE 22.4 Retention Data of C_{60} and C_{70} on Phenylmethyl-Bonded Silica and ODS Phases

| Stationary Phase | Structure | Retention Factor | | Separation Factor |
		$k_{(C60)}$	$k_{(C70)}$	k_{C70}/k_{C60}
PP		0.58	0.83	1.43
MPP		1.49	2.56	1.72
DMPP		3.09	5.77	1.87
TMPP		2.75	4.90	1.78
Develosil ODS-5		0.87	1.60	1.84

Source: Jinno, K. et al., *J. Microcol. Sep.*, 5, 135, 1993. With permission.
Note: Mobile phase, *n*-hexane.

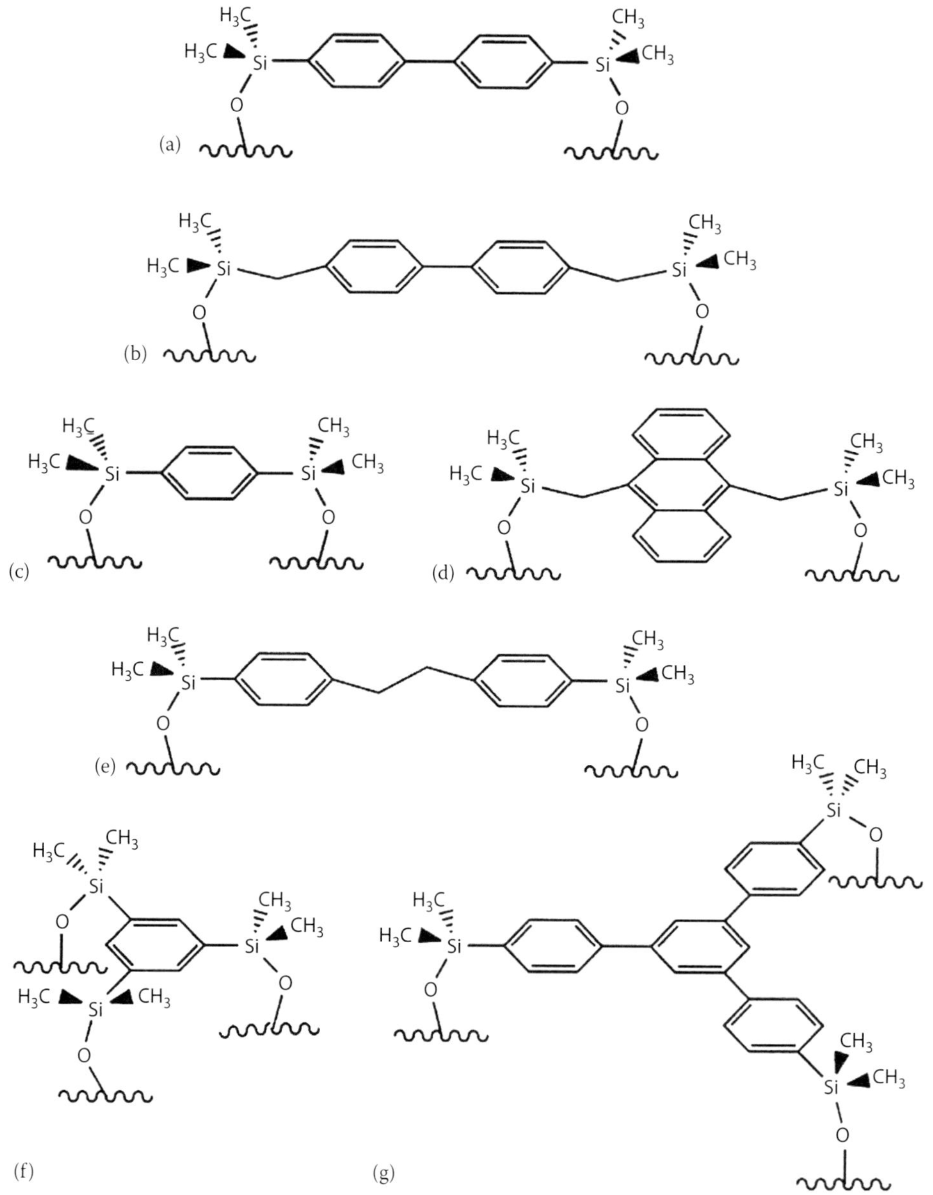

FIGURE 22.10 Various multi-legged phenyl-bonded phases synthesized. (a) BP, (b) BMB, (c) DP, (d) BMA, (e) BBB, (f) TP, and (g) QP. (From Saito, Y. et al., *J. Sep. Sci.*, 26, 225, 2003. With permission.)

groups attached to two or three silicon atoms, which in turn, are bonded to the silica surface and to phenyl groups (such as phenyl, biphenyl, benzyl, and anthracene groups).

The cavity-size of the 1,3,5-*tris*(dimethylphenyl-bonded silica) (TP) phase is too small to include the smaller C_{60} molecule, therefore C_{60} and C_{70} are not included in the cavity and the two solutes cannot be retained and separated. This is different in the case of *p*-bis(dimethyldiphenyl-bonded silica) (BP) phases. The cavity size of the BP phase is approximately 9–10 Å and the size of C_{60} is 7 Å, and C_{70}, 7 × 9 Å. The size of C_{70} produces a closer fit in the BP phase cavity thus producing a retention difference between C_{70} and C_{60}, and improved separation.

BBB and BMB phases retain C_{60} and C_{70} more than either BP or ODS phases. The BMB phase shows strong retention and good selectivity for C_{60} and C_{70}. The difference between BBB and BMB is due to the location of the methylene groups in their structures. Due to the two methylene groups between the diphenyl and the silicon atom, the diphenyl group of BMB can easily interact with fullerenes. It is because of this that the two methylene groups can work as a buffer for the movement of BMB's diphenyl group, which can interact with fullerenes through π–π interaction. While the phenyl group of the BBB phase does not have the ability to move freely because one edge is bonded to a silicon atom, and neither does the BP phase. Therefore, BMB is the most suitable stationary phase for fullerene separations. The BMA phase has a high freedom of movement as the anthracene moiety is not connected to the silicon atom, thus allowing the phase to show very high selectivity. However, due to its low surface coverage, it does not give large retention because of difficulty bonding chemistries.

22.3.3 Nitro-Derivatized Phenyl-Bonded Phases

Some nitrophenyl-derivatized π-acidic and π-basic phases exhibit certain retentions for fullerenes due to the charge-transfer interactions between fullerenes and the stationary phase. A variety of nitrophenyl-bonded stationary phases have been prepared and applied to the separation of fullerenes.

Cox et al. used dinitroanilinopropyl (DNAP)-bonded silica phase with gradient elution (*n*-hexane and methylene chloride as eluant) to achieve baseline separation of C_{60} and C_{70} [37].

The chromatographic characteristics of C_{60} and C_{70} closely resemble planar molecules. The retention behavior of C_{60} on such a phase is similar to that of triphenylene, while C_{70} retention lies between that of 5-ring (benzo[a]pyrene) and the 6-ring (coronene) compounds. The retention of these planar poly-nuclear aromatic hydrocarbons on DNAP can be described in terms of the electronic character of the molecules, i.e., the difference between the highest occupied molecular orbital (HOMO) of the compound and that of naphthalene. A 60-atom planar carbon cluster is estimated to be thermodynamically less stable than spherical C_{60}, thus the early elution for C_{60} may be considered as an extreme example of nonplanarity effects on retention time. This phenomenon has been well documented with smaller molecules used in reverse-phase HPLC.

A variety of dinitrophenyl-derivatized π-acidic and π-basic phases have been developed by Pirkle's group [38] for the separation of fullerenes. The structures of the stationary phases and the retention factors of C_{60} and C_{70} on these stationary phases are shown in Figure 22.11 and Table 22.5, respectively. The retention mechanisms of fullerenes on these stationary phases differ from those of PAHs. Fullerenes can respond as either π-bases

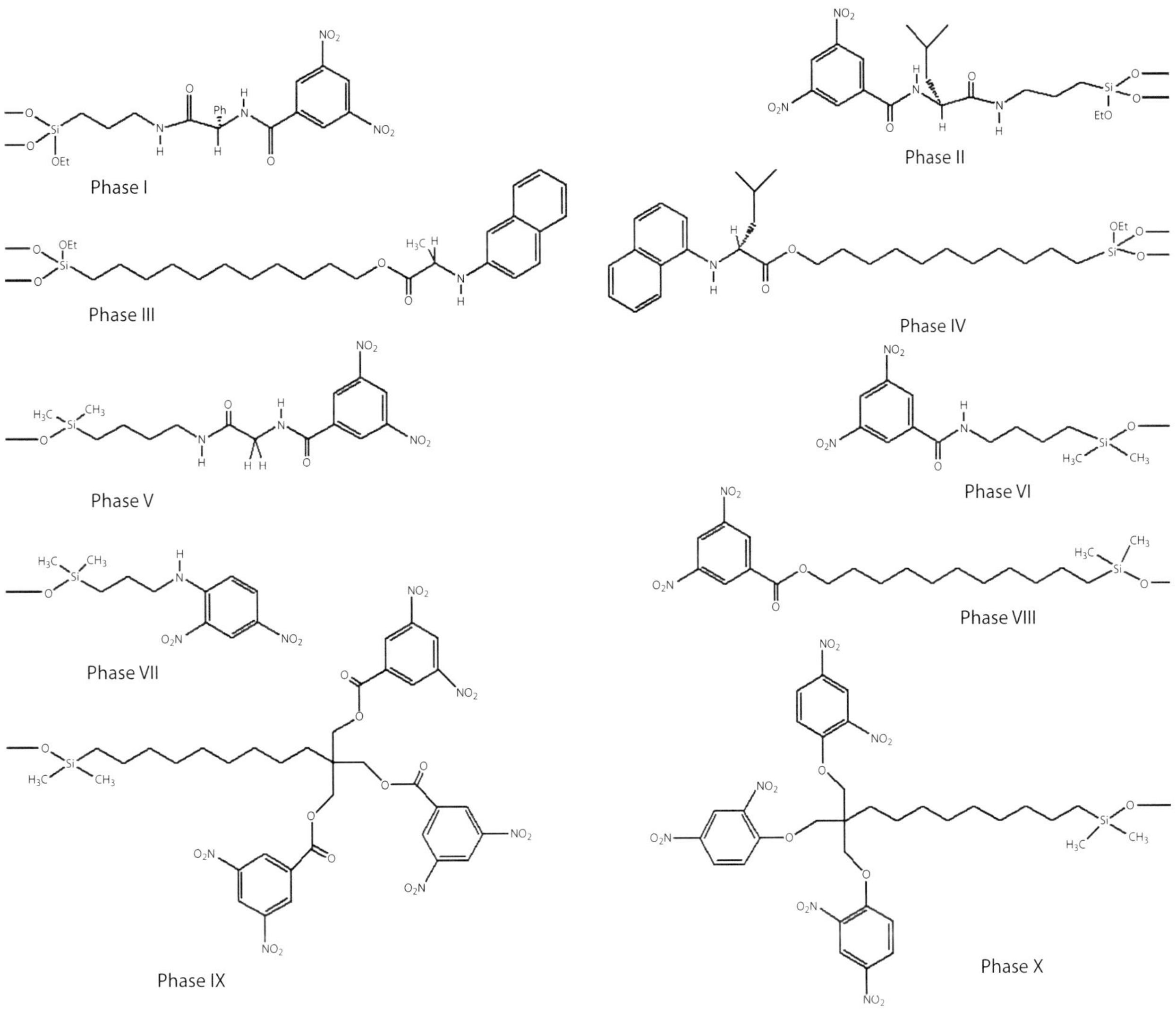

FIGURE 22.11 Stationary phases used in the study. Et = Ethyl; Ph = phenyl. (From Welch, C.J. and Pirkle, W.H., *J. Chromatogr.*, 609, 89, 1992. With permission.)

TABLE 22.5 Retention Factors for Analysis 1–10 on Stationary Phases I–X

Analyte	Stationary Phase									
	I	II	III	IV	V	VI	VII	VIII	IX	X
C_{60}	2.99	1.19	1.98	2.62	1.01	2.26	2.45	0.50	1.60	6.59
C_{70}	4.65	1.90	4.16	5.31	1.80	4.76	7.10	0.80	3.30	20.77
α	1.60	1.56	2.10	2.03	1.78	2.11	2.89	1.6	2.00	3.15

Source: Welch, C.J. and Pirkle, W.H., *J. Chromatogr.*, 609, 89, 1992. With permission.

Conditions: mobile phase = 5% dichloromethane in hexane, flow-rate = 2.00 mL/min, ambient temperature. Void time was determined using 1,3,5-tri-*tert*-butylbenzene, α = separation factor for C_{60} and C_{70}.

or π-acids while polarized during adsorption. The tripodal stationary phase designed for simultaneous multipoint interaction with buckminsterfullerene provides the greatest retention and the greatest separation for the C_{60}–C_{70} mixture. 3,3,3-tri-dinitrobenzoxyl-propyl-silica is currently commercially available and is named "Buckyclutcher."

Tanaka's group used a 2-(nitrophenyl)ethyllsilyl (NPE)-bonded silica stationary phase [39] that showed poor retention for C_{60} and C_{70}; however, different retention behavior was obtained for higher-order fullerenes. Feng's group compared the separation of fullerenes on nitrobenzoic-acid-bonded silica and naphthyl-acetic-acid-bonded silica stationary phases, and found nitrobenzoic-acid-bonded silica to exhibit high separation selectivity for higher fullerenes [40].

22.3.4 Chiral Stationary Phases

Many chiral stationary phases with either π-donor or π-acceptor functional groups have been applied to the separation of fullerenes [41]. For example, 3,5-dinitrobenzoylphenylglycine (DNBPG)-bonded silica, one of chiral stationary phases, was used to purify milligram quantities of C_{60} and C_{70} [42], while another charge-transfer stationary phase, 2-(2,4,5,7-tetranitro-9-fluorenylideneamino-oxy)propionic-acid (TAPA)-bonded silica chiral stationary phase was also investigated [31]. TAPA-silica phase showed stronger retention for fullerenes than on DNBPG. The enantiomer separation of [60] fullerene derivatives on the TAPA-silica phase has been successfully achieved. TAPA-bonded silica have been used to separate A [60] fullerene bi-, tris-, and hexakis-adducts.

Tetrachlorophthalimidopropyl(TCPP)-modified silica was also applied to the separation of C_{60} and C_{70} by using a mixture of dichloromethane and hexane as eluent [43]. TCPP-linked silica showed great selectivity for C_{60} and C_{70}, as well as the higher fullerenes C_{76}, C_{78-2v}, C_{78}, D_3, and C_{84}.

Furthermore, the separation of C_{60} and C_{70} has been performed on a π-acid chiral stationary phase using dinitrobenzoyl derivative of diaminocyclohexane-modified silica (R,R-DACH-DNB) as a stationary phase [44].

Cyclodextrins are cyclic oligosaccharides and are known for their ability to form inclusion complexes with different substrates. As a result of this property, cyclodextrins can differentiate between structural and geometrical isomers. This has been used for the chromatographic separation of compounds either by adding cyclodextrins to the mobile phase or by chemically bonding cyclodextrins to the stationary phase. Recently, γ-cyclodextrin has been used as a complexing reagent for C_{60} to give a water-soluble complex [45]. Separation of fullerenes was investigated on the γ-cyclodextrin-bonded (γ-CD) silica stationary phase. The retention of fullerenes on γ-CD-bonded silica stationary phase is due to the formation of inclusion complexes of fullerenes in the cavity of γ-cyclodextrin. It was found that retention of C_{70} was greater than that of C_{60}. These results indicate that C_{70} geometry is more favorable for the interaction with the γ-cyclodextrin molecule [15].

Most of the chiral stationary phases previously mentioned (i.e., γ-CD, RN-P-CD, DNBPG, TAPA) are not considered to be effective preparative purification methods due to their high cost and low preparative yield.

22.3.5 Liquid-Crystal-Bonded Silica Phases

Liquid crystals are generally of high molecular weight and low in volatility as they are an intermediate between a solid and a true isotropic liquid. These oriented liquid phases usually occur much higher than at room temperature, thus reducing or eliminating the column-bleed and the need for chemical bonding [46]. Jinno's group prepared liquid-crystal-bonded silica phases and compared the retentions of fullerenes on two liquid-crystal-bonded phases [47]: 4-{[4-(allyoxy)benzoyl] oxy}biphenyl-bonded phase (liquid crystal phase 1) and 4-{[4-(allyoxy)benzoyl]oxy}-4'-methoxybiphenyl-bonded phase (liquid crystal phase 2) (Structure in Figure 22.12). Table 22.6 lists the retentions of C_{60} and C_{70} on these stationary phases. Table 22.6 shows that the liquid-crystal-bonded phase with the ligand interval is close to the molecular size of the fullerenes, thus allowing good retention for fullerene molecules. Liquid-crystal-bonded silica phases possess better retention capability and separation power for fullerenes than commercially available octadecylsilica (ODS) phases. It is likely that the ligand interval of these liquid crystal phases and the methoxy group in the surface of phase 2 can affect solute interactions. While the methoxy groups on liquid crystal phase 2 increase the tether arm, they disturb the direct π–π interactions between the end phenyl groups and the fullerenes outside. Unusual behavior for C_{60} and C_{70} has been observed with the liquid crystal phase, suggesting that retention factors of C_{60} and C_{70} increases with increasing temperature.

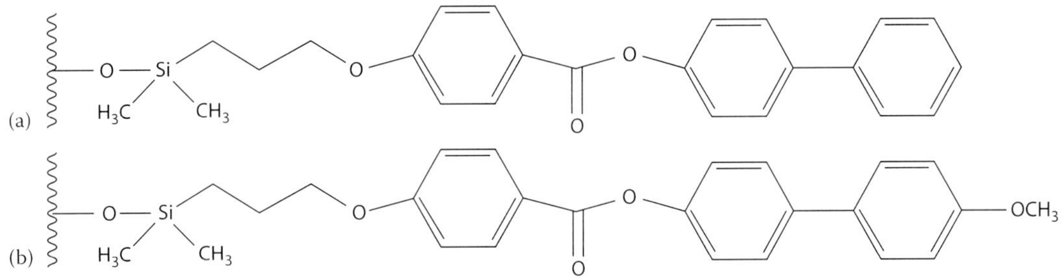

FIGURE 22.12 Chemical structures of the two liquid-crystal-bonded phases. (a) Liquid crystal phase 1 and (b) liquid crystal phase 2. (From Saito, Y. et al., *J. Microcol. Sep.*, 7, 41, 1995. With permission.)

TABLE 22.6 Retentions on Two Liquid Crystal Phases and ODS

Bonded Phases	Ligand Interval Å	Retention Factors		Separation Factors
		C_{60}	C_{70}	
Liquid crystal phase 1	6.9	2.82	8.46	3.00
Liquid crystal phase 2	5.9	0.42	0.79	1.88
Develosil ODS	7.7	0.87	1.60	1.84

Source: Saito, Y. et al., *J. Microcol. Sep.*, 7, 41, 1995. With permission.

22.3.6 Stationary Phases Containing a Large Aromatic System for the Separation of Fullerenes

Stationary phases containing heavy atoms or a large aromatic system show high retentivity for fullerenes. Pyrenyl-bonded silica (PYE) and porphyrin-bonded silica (TPP) stationary phases provide higher selectivity and efficiency for preparative applications of fullerenes among the currently available stationary phases. Lists of these stationary phases are described.

22.3.6.1 Coronenylpentylsilyl-Bonded Silica Stationary Phases

5-Coronenylpentylsilyl-bonded (COP) silica has stronger retentivity for fullerenes than most HPLC stationary phases as mentioned above. The capacity factor of C_{60} on the column (4.6 mm i.d., 15 cm long) packed with COP was 2.27 using toluene as a mobile phase at 30°C and a flow rate of 1 mL · min^{-1} [26]. However, there are some shortcomings for using COP-bonded silica. As it is known to be extremely difficult to prepare due to the low solubility of the starting material (and its intermediates) and excessive band broadening during usage as a stationary phase, Koronenylpentyldimethylchlorosilane has been prepared by Tanaka's group [26]. Alkyldimethylchlorosilanes is first prepared by hydrosilylation of substituted phenyl (or benzyl) alkyl ethers with dimethylchlorosilane in the presence of a chloroplatinic acid catalyst. The silylating reagent for the preparation of 5-coronenylpentylsilyl (COP) silica was prepared from 5-coronenyl-1-pentene [26]. The capacity factor of C_{60} on COP-bonded silica phase was 2.27 using toluene as mobile phase at 30°C.

22.3.6.2 Stationary Phases Containing Heavy Atoms

A series of stationary phases, containing carbon, oxygen, sulfur, fluorine, chlorine, or bromine on the substituent of the (phenyloxy)-propylsilyl-bonded silica stationary phases, have been developed [26]. The retention factors of C_{60} and C_{70} on these stationary phases and the refractive indexes of the precursors of the bonded organic moieties and the substituted phenylallyl ethers used for the preparation of chlorosilanes are listed in Table 22.7.

The stationary phase containing a heavy atom (such as sulfur or bromine) in the substituent group can result in stronger retention of C_{60} and a greater separation factor between C_{60} and C_{70}. This tendency does not correlate with the electron-withdrawing or electron-donating properties of the substituent as retention factors generally increase with increasing refractive index of the precursor of the stationary phase. A similar stationary phase with more bromine substituents can exhibit stronger retention for fullerenes at lower surface coverages. In contrast, stationary phases containing iodine, because of their much lower surface coverage, produce a lower capacity for fullerenes than bromine-substituted ones.

3-(Pentabromobenzyl)oxylpropylsilyl(PBB)-bonded silica shows a high retentivity for fullerenes and is commercially available. PBB-bonded silica retains less C_{60} but exhibit better resolution for fullerenes than on porphyrin-bonded silica or COP silica in toluene. PBB-bonded silica can be used with strong solvents as mobile phases including carbon disulfide and 1,2,4-trichlorobenzene.

22.3.6.3 Humic-Acid-Bonded Silica Stationary Phase

Humic acids are thought to be complex aromatic macromolecules with amino acids, amino sugars, peptides, aliphatic compounds involved in linkages between the aromatic groups [48]. The hypothetical structure for humic acids contains free and bound phenolic OH groups, uinine structures, nitrogen and oxygen as bridge units and COOH groups, and these can be variously placed on aromatic rings. As fullerenes interact with some aromatic ligands, it is assumed that they could also interact with humic acids and some complex aromatic macromolecules.

Guillaume and coauthors [49] prepared humic-acid-bonded silica stationary phase (BHAC) with amine groups on humic

TABLE 22.7 Retention of C_{60} and C_{70} and Their Separation Factors on 3-(Substituted-Phenyloxy) Propylsllyl Silica and the Refractive Index of Substituted Phenyl Allyl Ether

Stationary Phase[a] (R)	Surface Coverage ($\times 10^6$ mol/m^2)	$k'(C_{60})$[b]	$k'(C_{70})$[b]	$\alpha(C_{70}/C_{60})$[b]	n[c]
C_6H_5	2.7	1.03	1.77	1.73	1.52
$p\text{-}CH_3CH_2C_6H_4$	3.4	1.92	3.50	1.82	1.51
$p\text{-}CH_3OC_6H_4$	3.2	2.72	5.42	1.99	1.53
$p\text{-}CH_3SC_6H_4$	2.8	4.13	8.81	2.13	1.50
$\rho\text{-}FC_6H_4$	2.6	0.76	1.33	1.75	1.49
$p\text{-}ClC_6H_4$	2.8	2.23	4.83	2.17	1.54
$p\text{-}BrC_6H_4$	2.9	3.86	8.54	2.21	1.55
		1.42[d]	2.51[d]	1.77[d]	
$2,4,6\text{-}C_6H_2Cl_3$	1.5	0.53[d]	0.82[d]	1.55[d]	
$2,4,6\text{-}C_6H_2Br_3$	1.7	1.86[d]	3.60[d]	1.94[d]	
		0.28[e]	0.39[e]	1.39[e]	
C_6Br_5	1.3	0.69[e]	1.30[e]	1.88[e]	
$C_6Br_5CH_2$	2.4	1.97[e]	4.83[e]	2.45[e]	

Source: Kimata, K. et al., *Anal. Chem.*, 67, 2556, 1995. With permission.

[a] Column, 4.6 mm i.d., 15 cm long; flow rate, 1 mL/min. $RO(CH_2)_3Si(CH_3)_2$ groups were bonded to silica surfaces.

[b] Mobile phase, hexane, unless indicated otherwise.

[c] n, refractive index of the precursor of the stationary phase; substituted phenyl allyl ethers was used for the preparation chlorosilanes.

[d] Mobile phase, hexane/toluene 75/25.

[e] Mobile phase, toluene.

acids. The silanol groups of the silica stationary phase were first alkylated with 3-aminopropyl triethoxysilane, by initially activating the aminopropyl-silica phase by recycling with a 5% aqueous solution of glutardialdehyde. Humic acid is then immobilized to the silica gel by amine formation between the aldehyde group and amino groups in the HA molecules. Recently, Feng's group bonded humic acid to silica utilizing COOH groups in a simple synthetic method [50]. Humic acid immobilization was performed by reacting amino-bonded silica (APTS-silica) with humyl chloride.

Five buckminsterfullerenes (C_{60}, C_{70}, C_{76}, C_{78}, C_{84}) could be baseline separated on BHAC stationary phase using a mixture of toluene and cyclohexane (15/85, V/V) as a mobile phase and their retention factors are listed in Table 22.8.

TABLE 22.8 Retentions of C_{60}, C_{70}, C_{76}, C_{78}, C_{84} on Humic-Acid-Bonded Silica (BHAC) Stationary Phase Using 15% Toluene in Cyclohexane as Mobile Phase at 35°C

Compounds	Retention Factor (k)
C_{60}	0.92
C_{70}	3.12
C_{76}	6.49
C_{78}	8.28
C_{84}	17.22

Source: Casadei, N. et al., *Anal. Chim. Acta*, 588, 268, 2007. With permission.

22.3.6.4 C_{60}-Bonded Silica Phases

Bonded phases containing phenyl ligand(s) in their structure can interact effectively with fullerenes through π–π interactions. C_{60}-bonded silica phases were prepared by the Jinno group using a method similar to stationary phases in which silane-containing C_{60} are first prepared before bonding to silica [51]. The structures are shown in Figure 22.13.

C_{60} and C_{70} could be baseline separated using hexane as the mobile phase. The retention data and separation factors obtained are summarized in Table 22.9. The separation factors are comparable to those of commercially available ODS phases. Since separation of C_{60} and C_{70} on bare silica was not achieved, the retention data indicate that C_{60}-bonded ligands can contribute to the retention of C_{60} and C_{70}. While it is noted that C_{60}-bonded silica phases also contain a large aromatic system, the retention of C_{60} and C_{70} on C_{60}-bonded silica is smaller than that of porphyrin-bonded silica or COP silica. This result could be ascribed to the rigid structure of C_{60} and its poor aromatic properties.

22.3.6.5 Calixarene-Bonded Silica Phases

p-tert-Butylcalix[8]arene have a π-donor cavity composed of benzene rings while fullerenes are π-systems with appreciable electron affinity and can form host–guest inclusion complexes through charge-transfer interactions [52–54]. *tert*-Butylcalix[8] arene have a bowl-shaped cavity structure and can fit into the ligands of C_{60}-bonded silica stationary phases [52]. Similarly, fullerenes could also interact with *p-tert*-butylcalix[8]arene-bonded silica stationary phases. Yang and coworkers prepared

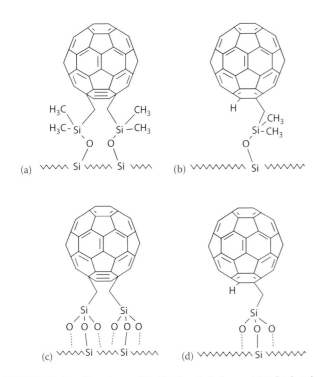

FIGURE 22.13 Structures of the C_{60}-bonded phases. (a) C-high, 2-leg type; (b) C-high, 1-leg type; (c) C-low, 2-leg type; and (d) C-low, 1-leg type. (From Jinno, K. et al., *Analyst*, 122, 787, 1997. With permission.)

TABLE 22.9 Retention Data for the Separation of C_{60} and C_{70} with Various C_{60}-Bonded Silica Phases

	Retention Factor (k)		
C_{60}-Bonded Phase	C_{60}	C_{70}	Separation Factor
C-high, 2-leg type	3.65	11.4	3.12
C-high, 1-leg type	4.25	13.0	3.06
C-low, 2-leg type	0.68	1.20	1.76
C-low, 1-leg type	0.91	1.85	2.02

Source: Jinno, K. et al., *Analyst*, 122, 787, 1997.
Note: Mobile phase, hexane.

a chemically *p-tert*-butylcalix[8]arene-bonded silica mono-lith and applied it to LC separations of fullerenes [55]. Figure 22.14 shows the separation chromatogram of fullerenes on the *p-tert*-butylcalix[8]arene-bonded silica monolith with C_{60} and C_{70} being baseline separated from other higher fullerenes and matrixes [52].

22.3.6.6 Phthalocyanine-Bonded Phases

Phthalocyanine (Pc) is closely related to the naturally occur-ring porphyrins and has a symmetrical 18 π-electron aromatic macrocycle containing an alternating nitrogen atom-carbon atom ring structure. The molecule is able to coordinate hydro-gen and metal cations in its central cavity by forming coordi-nate bonds with the four isoindole nitrogen atoms. The central atoms can carry additional ligands. Most of the elements have been found to be able to coordinate to the phthalocyanine mac-rocycle [56] allowing a variety of phthalocyanine complexes to

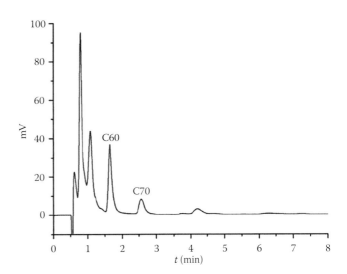

FIGURE 22.14 Chromatogram of fullerenes in carbon soot on the *p-tert*-butylcalix[8]arene-bonded silica monolith. Conditions: col-umn size, 100 mm × 4.6 mm I.D.; mobile phase, *n*-hexane–acetonitrile (82/18, V/V); flow rate, 1.0 mL/min; detection, 325 nm; temperature, 30°C. (From Yin, J.F. et al., *J. Chromatogr. A*, 1188, 199, 2008. With permission.)

exist. Pc and its derivatives have a widespread π electron cloud. Phthalocyanine-bonded silica stationary phases have been developed for separating solutes containing π electrons such as PAHs [57] and fullerenes [58,59]. These phases produce π–π interaction between the fullerenes and the bonded moiety. When alkyl chains are attached to the silica surface of phthalocyanine-bonded silica phases during the endcapping process, solvopho-bic interactions are induced from the phthalocyanine moiety. Use of an appropriate alkyl chain length can induce higher pos-sibility of interactions between the solutes and phthalocyanine moiety because the chains can move to catch fullerenes on the phthalocyanine moiety. These longer chains stationary phases may not always exhibit stronger retention to fullerenes as longer chains can disturb the direct π–π interaction between fullerenes and the phthalocyanine moiety, thus inducing smaller molecu-lar size and reduced shape recognition [58].

Phthalocyanine-bonded silica phase has been applied to the separation of fullerenes by the Jinno and Gumanov groups [58,59], respectively. Horizontal arrangement of the aromatic moiety was also designed with copper-phthalocyanine (Cu-PcS) [58], thus providing some retentivity for fullerenes. Figure 22.15 illustrates the interaction of C_{60} and C_{70} with Cu-PcS. Using *n*-hexane as the mobile phase at 20°C, the retention factors of C_{60} and C_{70} on decanoyl Cu-PcS stationary phase were 0.92 and 1.70, respectively. The selectivity for C_{60} and C_{70} could be further increased by the additional introduction of alkyl chains with an appropriate length to interact with these molecules. Cu-Pc phases can offer two types of retention mechanisms. One is the formation of solvophobic interactions between the alkyl chains and fullerenes, and another is the formation of π–π interactions between the phthalocyanine moiety and fullerenes. Increasing the temperature can induce the increased mobility of alkyl chains

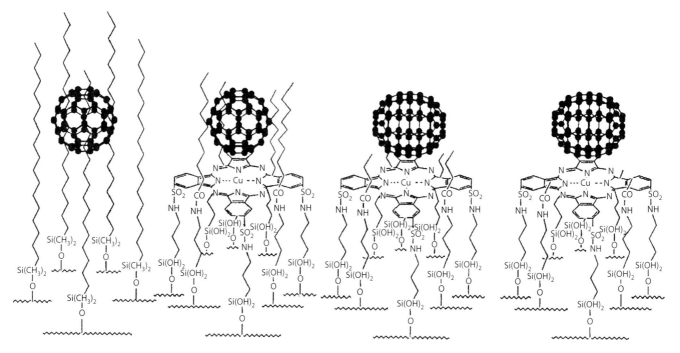

FIGURE 22.15 Schematic molecular interactions between four stationary phases (ODS and Cu-PcS phases with additional introduction of alkyl ligands), and C_{60} and C_{70}. (From Jinno, K. et al., *J. Microcol. Sep.*, 8(1), 13, 1996. With permission.)

around the phthalocyanine moiety resulting in reduced interaction interference between fullerenes and the phthalocyanine.

22.3.7 Porphyrin-Bonded Silica Stationary Phase

Among the currently available stationary phases, porphyrin-bonded silica stationary phases provide the highest selectivity and efficiency for preparative applications of fullerenes. Phenylporphyrin-bonded silica stationary phases include various unmetalated and metalated porphyrin-bonded silica phases, and can be classified as follows.

22.3.7.1 Carboxyphenyltriphenylporphyrin-Bonded Silica Stationary Phases

Unsymmetrical phenyl-substituted tetraphenylporphyrins such as 5-(p-carboxyphenyl)-10,15,20-triphenylporphyrin (CPTPP) can be covalently attached to aminopropyl silica gel through an amide bond [60]. The metalloporphyrin-silica stationary phases were prepared by metalating tetraphenylporphyrin after immobilization of the porphyrin onto a silica support. To prepare metalated stationary phases such as Zn(CPTPP)-, and In(CPTPP)-silica, H_2(CPTPP)-silica is metalated with metal ions Zn(II) or In(III) by refluxing porphyrin-silica in dimethylformamide containing $ZnCl_2$ or $InCl_3$, respectively.

Table 22.10 shows the retention factor and separation factor of C_{60} and C_{70} on four different porphyrin-silica stationary phases. For these porphyrin phases, separation factors are larger than 4.0, providing excellent resolution of C_{60} and C_{70}. A mobile phase consisting of 100% toluene was used.

TABLE 22.10 Capacity (k) and Selectivity Factors (α) for C_{60} and C_{70} Fullerenes on Four Different Porphyrin-Silica Columns[a]

Column	k_{C60}	k_{C70}	$\alpha_{C70/C60}$
H_2(CPTPP)	1.7	8.1	4.8
In(CPTPP)	2.0	8.7	4.3
Zn(CPTPP)-(1)[b]	1.2	4.8	4.0
Zn(CPTPP)-(2)[b]	1.9	8.4	4.4

Source: Kibbey, C.E. et al., *Anal. Chem.*, 65, 3717, 1993. With permission.

[a] For 100 × 4.6 mm columns were operated at ambient temperature using 100% toluene as mobile phase at 2.0 mL/min.

[b] Measurements of two different preparations and packings of the Zn(CPTPP) stationary phase.

22.3.7.2 Hydroxyphenyl-Triphenylporphyrin-Bonded Silica Stationary Phases

Meyerhoff and coworkers [61] immobilized hydroxyphenyl-triphenylporphyrin (HPTPP) onto silica by reaction with glycidoxypropyltrimethoxysilane.

As mentioned previously, CPTPP-silica stationary phases have strong retention and high selectivity for the separation of fullerenes. The HPTPP–silica phase has higher selectivity. As shown in Table 22.11, TPP–silica phases exhibit high fullerene selectivity in a number of strong mobile phase solvents. The ability to use strong mobile phase solvents allows for improved separation, higher throughput, and increases the potential of columns packed with HPTPP–silica to perform separations at the preparative scale.

TABLE 22.11 Influence of Mobile Phase on the Fullerene Selectivity of Columns Packed with CPTPP– or HPTPP–Silicas

Mobile Phase	$\alpha_{C70/C60}$ on CPTPP-Silica	$\alpha_{C70/C60}$ on HPTPP-Silica
Toluene	5.7	7.0
p-Xylene	5.1	6.3
Chlorobenzene	4.5	5.7
Carbon disulfide	3.4	4.5
1,2-Dichlorobenzene	2.4	3.5

Source: Coutant, D.E. et al., *J. Chromatogr. A*, 824, 147, 1998. With permission.

HPLC conditions: injection of 2 mg C_{60} and C_{70} each; flow-rate: 1.0 mL/min; ambient column temperature.

22.3.7.3 Preparation of Sorbents Based on Tetrachloromethyl-H_2TPP and Tetrachlorosulfo-H_2TPP and Their Metalated Phases

Chloromethyl- and chlorosulfo-derivatives of tetraphenylporphyrin (H_2TPP) can be used to prepare tetraphenylporphyrin-bonded silica phases [62]. Metalated phases such as TPPAl are prepared in the similar way aforementioned. Comparing the chromatographic characteristics of these sorbents with the carboxyphenyltriphenylporphyrin (CPTPP)-bonded silica stationary phases, it was found that retention factors (*k*) and selectivities (α) of TPP-silica are approximately equal to those discussed aforementioned as shown in Table 22.12. However, the separation factor of C_{60} and C_{70} on TPPAl is high even with the use of strong solvents as mobile phases; therefore, the sorbent is more suitable for the preparative separation of fullerenes. The high chromatographic performance of the sorbent is due to the large proportion of TPPAl on the surface of the support and by the intermolecular interactions between porphyrin molecules and the fullerenes.

22.3.7.4 Separation Mechanism of Fullerenes on TPP-Silica Phases

It was proposed that the charge-transfer interaction (CT) between fullerenes and TPP-silica phases involved some degree of electrostatic attraction or π–π interaction. The electron-rich double bond at the 6:6 ring juncture of C_{60} and C_{70} is attracted to the tetraphenylporphyrin phases. It has been reported that tetraphenylporphyrin may be capable of simultaneous face-to-face and face-to-edge π–π interactions with the fullerenes. This extra dimension with respect to the interaction between the immobilized tetraphenylporphyrin (TPP) and fullerene may account for the very high selectivity achieved.

Porphyrin ring and four *meso* phenyl groups on ZnTPP stationary phases appear to form a π-electron cavity which can host fullerenes. The larger the fullerene cage (C_{70} versus C_{60}), the more contact area the solute has to interact with the ZnTPP cavity, resulting in a stronger π–π interaction and longer retention time. The molecular-shape selectivity on the tetraphenylporphyrin-silica phase is so sensitive that even the small difference in shape between C_{70} and C_{60} can yield a large difference in retention [63].

22.3.8 Stationary Phase with Pyrenyl Ligands

22.3.8.1 Pyrenyl-Bonded Silica Stationary Phase

Stationary phases with pyrenyl ligands exhibit strong retention for fullerenes due to the π–π interaction between pyrenyl groups and fullerenes. [2-(1-Pyrenyl)ethyl]silyl-bonded silica (PYE) stationary phases are commercially available and have been applied widely for the separation of solutes containing π electrons because of its good chromatographic performance.

Tanaka and coauthors [39] and Lochmüller and Hunnlcutt [64] prepared PYE phases in a similar method in which the pyrenyl unit was first converted into a silanization reagent for which chemical coupling on silanol groups of silica was achieved. Later, Feng's group [65] prepared pyenebutyric acid-bonded silica (PYB) stationary phases. The stationary phases were synthesized via amido reaction of pyrenebutyryl chloride with different aminosilanes-bonded silica.

Tanaka and coauthors [39] first investigated the liquid chromatographic separation of fullerenes using a PYE stationary phase. PYE phase showed adequate retention (*k* = 0.91) for C_{60} in toluene with a separation factor of 1.80 for C_{60} and C_{70}. Moreover, PYE phases exhibit good separation for higher fullerenes and comparisons with NPE are made in Figure 22.16. The combination of two aromatic stationary phases, π-basic PYE and π-acidic NPE, showed opposite retention characteristics for aromatic molecules. The latter could be useful for the

TABLE 22.12 Retention Factors and Separation Selectivities of C_{60} and C_{70} Fullerenes on Sorbents Based on Tetraphenylporphyrin in Toluene

Stationary Phase	Retention Factors		Selectivity	Note
	$k(C_{60})$	$k(C_{70})$	α	
H_2CPTPP	1.7	8.1	4.8	Low capacity
Tetrachloromethyl-H_2TPP	2.1	7.2	3.4	Low capacity
Tetrachlorosulfo-H_2TPP	1.8	5.4	3	Low capacity
TPPAl(X)	7.2	25.1	3.5	High capacity

Source: Gumanov, L.L. and Korsounskii, B.L., *Mendeleev Commun.*, 7(4), 158, 1997. With permission.

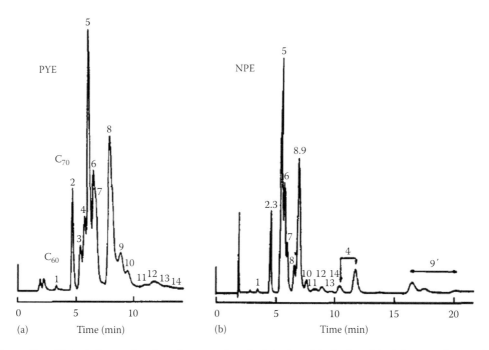

FIGURE 22.16 Separation of late-eluting product in carbon soot. Sample: Injection of 3 μL of combined fractions *(ca.* 4 mg*)* in toluene (10 mL) after removing C_{60} *(peak* 1) and C_{70} *(peak* 2) from carbon soot. (a) Column: PYE, 4.6 mm i.d., 15 cm. Mobile phase: toluene. Flow rate 1 mL/min. (b) Column: NPE, 4.6 mm i.d., 15 cm. Mobile phase: toluene/hexane = 25/75 (v/v). Flow rate: 1 mL/min, 30°C. (From Kimata, K. et al., *J. Org. Chem.*, 58, 282, 1993. With permission.)

separation of higher fullerenes [39]. A slight decrease in retention was observed at higher temperatures.

Feng and coworkers investigated the separation of fullerenes with pyrenebutyric acid-bonded silica phases (PYB). The retention factors and separation factors of C_{60} and C_{70} correspond to those on PYE. The molecular sizes of C_{60} and C_{70} make them small enough to be either adsorbed onto a single PYB ligand or between two neighboring PYB groups. For PYB-silica phases, the observed temperature effect differences on fullerene retention are primarily caused by the difference of the surface ligands arranged on stationary phases. It is assumed that the conformational change of stationary phase [66–68] and the dissociation of fullerenes [22,69] at higher temperatures can both influence fullerene retention in contrary directions. If pyrene ligands on PYB phase interact with increasing temperature, the conformational change of stationary phase worked toward the usual temperature effect on fullerene retention, which led to less interaction between stationary phase and solutes. Combined with the effect of fullerene dissociation, the retention of fullerenes decreases slightly at higher temperatures. On the other hand, because of higher surface coverage and longer tether arms, respectively, the self-association of pyrene ligands could take place under room temperature. As a result, the surface of the stationary phase would change little with increasing temperature. And the dissociation of fullerenes would increase the interaction between stationary phase and solutes, leading to the unusual temperature effect of fullerenes retention. Therefore, the temperature dependence of retentions of fullerenes on PYB would be little and uncertain [65].

22.3.8.2 Pyrenebutyric Acid-Modified Magnesia– Zirconia Stationary Phases

Silica-based stationary phases are the most widely used materials for fullerene separation. Later, inorganic oxides such as alumina, titania, and zirconia have been used as an alternative to silica [70–79]. Of the aforementioned oxides, zirconia, in particular, has drawn increasing interest over the last decade owing to their remarkable mechanical, chemical, and thermal stability during HPLC [70]. It also proved to exhibit great affinity for inorganic and organic phosphate groups due to strong Lewis acid–base interactions [70–73]. Magnesia–zirconia, one kind of

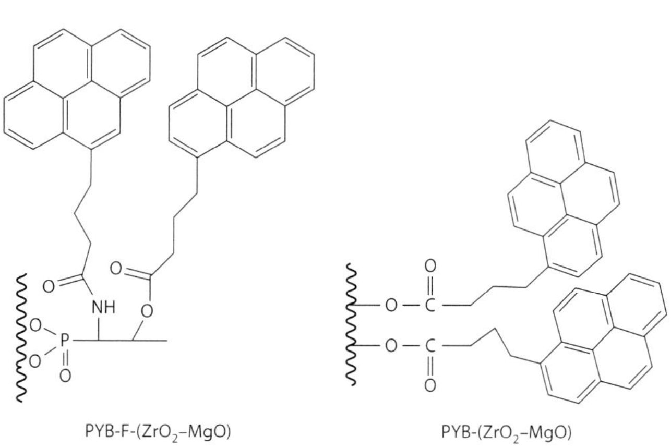

FIGURE 22.17 Structure of PYB-F-(ZrO_2–MgO) and PYB-(ZrO_2–MgO). (From Yu, Q.W. et al., *Anal. Chim. Acta*, 559, 79, 2006. With permission.)

TABLE 22.13 Chromatographic Properties of Seven Stationary Phases

Stationary Phases	Capacity Factors of C_{60}	Capacity Factors of C_{70}	Selectivity Factor for C_{70}/C_{60}
Buckyclutcher [77]	0.30	0.45	1.50
PBB [26]	1.97	4.83	2.45
Tetraphenylporphyrin (TPP) [77]	0.56	2.40	4.30
Hydroxyphenyltriphenylporphyrin (HPTPP) [77]	7.85	54.8	7.00
PYE [77]	0.85	1.70	2.00
Pyrenebutyric acid-bonded silica (PYB-silica) [65]	0.90	2.00	2.23
PYB-(ZrO$_2$-MgO) (pore size: 9.8 nm) [75]	0.41	0.83	2.02
PYB-F-(ZrO$_2$-MgO) (pore size: 9.8 nm) [75]	2.06	6.38	3.10

Source: Yu, Q.W. et al., *Anal. Chim. Acta*, 559, 79, 2006. With permission.
Note: Mobile phase, toluene.

zirconia-containing mixed oxide, has larger surface area, better pore-size distribution and pore structure, and shows a greater affinity for phosphonates compared to bare ZrO$_2$ [73,74]. Combining the advantage of pyrenyl groups as the ligands and magnesia–zirconia as support, Feng's group [75,76] prepared pyrenebutyric acid-modified magnesia–zirconia stationary phases using the sodium salt of *cis*-(3-methyloxiranyl)phosphonic acid (fosfomycin) as a spacer (PYB-F-(ZrO$_2$–MgO)) and ones without any intermediate spacer (PYB-(ZrO$_2$–MgO)) for the separation of fullerenes. The structure of PYB-F-(ZrO$_2$–MgO) and PYB-(ZrO$_2$–MgO) is shown in Figure 22.17. Influence of the pore sizes and ligand intervals also exist in these stationary phases. Better separation is achieved on PYB-F-(ZrO$_2$–MgO) than on PYB-(ZrO$_2$–MgO) when they are prepared from ZrO$_2$–MgO with appreciating pore sizes. Table 22.13 lists the retention factors of C$_{60}$ and C$_{70}$ and their separation factors on pyrenebutyric acid modified ZrO$_2$–MgO stationary phases and other most available silica-based stationary phases using toluene as mobile phases. With the exception of porphyrin-bonded silica phases, PYB-F-(ZrO$_2$–MgO) exhibits the best selectivity for fullerenes among these stationary phases. The difference between PYB-F-(ZrO$_2$–MgO) and the other two pyrenyl-modified phases is due to the use of different matrices. Baseline separation of C$_{60}$ and C$_{70}$ could be achieved on PYB-F-(ZrO$_2$–MgO) with strong solvents, such as CS$_2$, as mobile phases.

22.4 Separation of Fullerenes on Polymer Stationary Phases

Polystyrene-divinylbenzene and polystyrene gels have been used as stationary phases for the separation of C$_{60}$ and C$_{70}$ with toluene as the mobile phase [80,81]. However, the interaction force between fullerenes and these stationary phases are not as strong, resulting in weak C$_{60}$ and C$_{70}$ retention, and low separation.

Chen's group [82] prepared 1-methylnaphthalene-modified PSDVB resin and applied it to the separation of C$_{60}$ and C$_{70}$. A strong interaction force resulted from the π–π charge-transfer interaction between the fullerenes and the stationary phase making it possible to use the strong solvent, *o*-xylene, as the mobile

phase in which fullerenes have high solubility. This method can enable higher column loading capacity and produce better separation than some Pirkle's stationary phases.

22.5 HPLC Separation of Derivatives of Buckminsterfullerene

22.5.1 Separation of Hydrogenated Derivatives

HPLC methods are the best choice for separating fullerene derivatives containing covalently bound groups. The better solubility of these derivatives makes their separation easier than that of their parent fullerenes; however, this advantage is outweighed by the high complexity of mixtures formed during radical or nucleophilic addition reactions (e.g., in hydrogenation) and the instability of C$_{60}$ hydrogenated derivatives [83–85], which makes separation a very difficult task.

For the HPLC separation of hydrogenated derivatives of C$_{60}$ and C$_{70}$, a variety of stationary phases, such as ODS [84–86], PYE [87], Buckyclutcher column (packed with 3,3,3-tri-dinitrobenzoxyl-propyl-silica) [88,89], have been investigated. In this case, C$_{60}$H$_n$ derivatives elute earlier than C$_{60}$ and C$_{70}$ and their retention time decreases with increasing hydrogen content. It was found that C$_{60}$H$_2$ and C$_{60}$H$_4$ can be easily separated from C$_{60}$ and from each other. Cahill et al. reported the separation of the three most abundant isomers of C$_{60}$H$_4$ with the Buckyclutcher column; however, C$_{60}$H$_6$–C$_{60}$H$_{48}$ derivatives could not be completely separated with the Buckyclutcher column [86].

Fluorinated derivatives of [76] and [78] fullerenes were also purified by HPLC using a Cosmosil Buckyprep column (packed with 2-pyrenal-l-ethyl-silica) with toluene elution [78].

22.5.2 Separation of Endohedral Metallofullerenes

So-called endohedral fullerenes have ions or small molecules incorporated inside the cage atoms. Endohedral fullerenes include three types: endohedral metallofullerenes, noble gas endohedral fullerenes (such as He, Ne, Ar, Xe), and the endohedral nonmetallofullerenes (CO, CN).

Endohedral metallofullerenes (fullerenes encapsulating metals inside hollow carbon cages) have attracted special interest as spherical molecules due to their unique properties. Their general formula is M@C, (M = group IIIA and rare earth metals). The endohedral fullerenes containing metals are promising candidates as magnetic resonance imaging agents [90]. It is difficult to isolate endohedral metallofullerenes from the matrix of fullerenes in which they are produced [91]. The current methods of preparing endohedral metallofullerenes are the arc-discharging method and ion-bumping method. Current separation methods of endohedral metallofullerenes are usually carried out by two-step HPLC method but the low production and difficult separation remains an obstacle for further research.

Akasaka et al. reported the separation of two $Pr@C_{82}$ isomers [92]. The second isomer of $Pr@C_{82}$ ($Pr@C_{82}$-II) was isolated from the major isomer ($Pr@C_{82}$-I) by a two-step HPLC method. The mixture was first separated on the PBB stationary phase followed by a second step on Buckyprep phase (packed with PYE).

Efficient HPLC purification of the metallofullerenes from crude soot extract could be performed by tetraphenylporphyrin-silica support material [93,94].

$La@C_{82}$ and $Y@C_{82}$ from crude soot extract with a toluene/carbon disulfide mobile phase (3:1) could be separated on CPTPP-silica columns. The retention position of $La@C_{82}$ and $Y@C_{82}$ relative to C_{82} on the CPTPP-silica column substantiates the π-basicity of a given fullerene along with its size (i.e., larger surface interaction area), controlling the affinity of fullerenes for the immobilized porphyrin. In terms of the π electrons on the fullerene surface, $La@C_{82}$ is an electronic analog of the hypothetical C_{85} structure, not C_{82}, and has an electron density somewhat greater than that of C_{85}. This may explain why the $M@C_{82}$ species are retained on the porphyrin column (as well as on the previously reported 2-(1-pyrenyl) ethylated-silica column) longer than C_{82} and C_{84}. In addition, metallofullerenes possess a substantial dipole moment which could contribute to the longer retention times for $La@C_{82}$ and $Y@C_{82}$ compared to C_{82}.

22.6 Conclusion

In this chapter, a number of stationary phases applied for the separation of fullerenes, such as the alkyl-bonded phases and charge-transfer stationary phases, were examined. In general, the alkyl-bonded stationary phases can be used for analytical separation of fullerenes and show higher efficiency, shorter retention times, better sensitivity, and greater resolution for higher fullerenes. Aromatic stationary phases such as charge-transfer phases are more effective at purifying large quantities of fullerenes.

Overall, the structural matching and dispersion forces in the chromatographic process, the molecular size and shape of bonded moieties on stationary phases as well as the solvents used, can greatly influence the separation of fullerenes.

References

1. Kroto, H.W., Heath, J.R., O'Brien, S.C., Curl, R.F., Smalley, R.E. C_{60}: Buckminsterfullerene, *Nature*, **1985**, *318(14)*, 162–163.
2. http://www.ch.ic.ac.uk/rzepa/mim/century/html/c60.htm
3. http://en.wikipedia.org/wiki/Fullerene
4. Baena, J.R., Gallego, M., Valcárcel, M. Fullerenes in the analytical sciences, *Trends Anal. Chem.*, **2002**, *21(3)*, 187–198.
5. David, W.I.F., Ibberson, R.M., Matthewman, J.C., Prassides, K., Dennis, T.J.S., Hare, J.P., Kroto, H.W., Taylor, R., Walton, D.R.M. Crystal structure and bonding of ordered C_{60}, *Nature*, **1991**, *353(12)*, 147–149.
6. McKenzie, D.R., Davis, C.A., Cockayne, D.J.H., Muller, D.A., Vassallo, A.M., The structure of the C_{70} molecule, *Nature*, **1992**, *355(13)*, 622–624.
7. Valcárcel, M., Cárdenas, S., Simonet, B.M., Moliner-Martínez, Y., Lucena, R. Carbon nanostructures as sorbent, materials in analytical processes, *Trends Anal. Chem.*, **2008**, *27(1)*, 34–43.
8. Hu, Y.L., Feng, Y.Q., Da, S.L. Research on the separation and purification of fullerenes by chromatographic methods, *Chem. Mag. China*, **2001**, *3*, 141–145.
9. Atwood, J.L., Koutsantonis, G.A., Raston, C.L. Purification of C_{60} and C_{70} by selective complexation with calixarenes, *Nature*, **1994**, *368*, 229–231.
10. Haino, T., Fukunaga, C., Fukazawa, Y. A new calyx[5]arene-based container: Selective extraction of higher fullerene, *Org. Lett.*, **2006**, *8(16)*, 3545–3548.
11. Wu, Y., Sun, Y., Gu, Z., Zhou, X., Xiong, Y., Sun, B., Jin, Z. A combined recrystallization and preparative liquid chromatographic method for the isolation of pure C_{70} fullerene, *Carbon*, **1994**, *32(6)*, 1180–1182.
12. Bohme, D.K., Boltalina, O.V., Hvelplund, P. Fullerenes and fullerene ions in the gas phase, in *Fullerenes: Chem., Phys. Technol.*, John Wiley & Sons Ltd., Hoboken, NJ, **2000**, 481–529.
13. http://en.wikipedia.org/wiki/Chromatography
14. Bhyrappa, P., Penicaud, A., Kawamoto, M., Reed, C.A. Improved chromatographic separation and purification of C_{60} and C_{70} fullerenes, *Chem. Commun.*, **1992**, *13*, 936–937.
15. Théobald, J., Perrut, M., Weber, J.-V., Mellon, E., Muller, J.-F. Extraction and purification of fullerenes: A comparisive review, *Sep. Sci. Technol.*, **1995**, *30(14)*, 2783–2819.
16. Wu, Y.Q., Sun, Y.L., Gu, Z.N., Wang, Q.W., Zhou, X.H., Xiong, Y., Jin, Z.X. Analysis of C_{60} and C_{70} fullerenes by high-performance liquid chromatography, *J. Chromatogr.*, **1993**, *648*, 491–496.
17. Ardrey, R.E. *Liquid Chromatography-Mass Spectrometry: An Introduction*, John Wiley & Sons Ltd., Hoboken, NJ, **2003**, 10–13.
18. http://teaching.shu.ac.uk/hwb/chemistry/tutorials/chrom/chrom1.htm

19. Zhang, R.Q., Feng, Y.Q., Lee, S.T., Bai, C.L. Electrical transport and electronic delocalization of small fullerenes, *J. Phys. Chem. B*, **2004**, *108(43)*, 16636–16641.

20. Nawrocki, J. The silanol group and its role in liquid chromatography, *J. Chromatogr. A*, **1997**, *779*, 29–71.

21. Jinno, K., Saito, Y., Chen, Y.-L., Luehr, G., Archer, J., Fetzer, J.C., Biggs, W.R. Separation of C_{60} and C_{70} fullerenes on methoxyphenylpropyl bonded stationary phases in microcolumn liquid chromatogry, *J. Microcol. Sep.*, **1993**, *5*, 135–140.

22. Jinno, K., Ohta, H., Saito, Y., Uemura, T., Nagashima, H., Itoh, K., Chen, Y.-L., Luehr, G., Archer, J., Fetzer, J.C., Biggs, W.R. Dimethoxyphenylpropyl bonded silica phase for higher fullerenes separation by high-performance liquid chromatography, *J. Chromatogr.*, **1993**, *648*, 71–77.

23. Saito, Y., Ohta, H., Jinno, K. Design and characterization of novel stationary phases based on retention behavior studies with various aromatic compounds. *J. Sep. Sci.*, **2003**, *26*, 225–241.

24. Jinno, K., Uemura, T., Ohta, H., Nagashima, H., Itoh, K. Separation and identification of higher molecular weight fullerenes by high-performance liquid chromatography with monomeric and polymeric octadecylsilica bonded phases, *Anal. Chem.*, **1993**, *65*, 2650–2654.

25. Jinno, K., Nakagawa, K., Saito, Y., Ohta, H., Nagashima, H., Itoh, K., Archer, J., Chen, Y.L. Nano-scale design of novel stationary phases to enhance selectivity for molecular shape and size in liquid chromatographic, *J. Chromatogr. A*, **1995**, *691*, 91–99.

26. Kimata, K., Hirose, T., Moriuchi, K., Hosoya, K., Araki, T., Tanaka, N. High-capacity stationary phases containing heavy atoms for HPLC separation of fullerenes, *Anal. Chem.*, **1995**, *67*, 2556–2561.

27. Ohta, H., Saito, Y., Nagae, N., Pesek, J.J., Matyska, M.T., Jinno, K. Fullerenes separation with monomeric type C_{30} stationary phase in high-performance liquid chromatography, *J. Chromatogr. A*, **2000**, *883*, 55–66.

28. Guillaume, V.C., Guinchard, C. Participation of cluster species in the solvation mechanism of a weak polar solute in a methanol/water mixture over a 0.2–0.7 water fraction range: High-performance liquid chromatography study, *Anal. Chem.*, **1998**, *70*, 608–615.

29. Guillaume, Y.C., Peyrin, E., Grosset, C. C_{60} and C_{70} HPLC retention reversal study using organic modifiers, *Anal. Chem.*, **2000**, *72*, 1301–1306.

30. Youngman, M.J., Green, D.B. Microwave-assisted extraction of C_{60} and C_{70} from fullerene soot, *Talanta*, **1999**, *48*, 1203–1206.

31. Diack, M., Compton, R.N., Guiochon, G. Evaluation of stationary phases for the separation of buckminsterfullerenes by high-performance liquid chromatography, *J. Chromatogr.*, **1993**, *639*, 129–140.

32. Xie, Q.S., Perez-Codero, E., Echegoyen, L. Electrochemical detection of C606- and C706-: Enhanced stability of fullerides in solution, *J. Am. Chem. Soc.*, **1992**, *114*, 3978–3980.

33. Saito, Y., Ohta, H., Nagashima, H., Itoh, K., Jinno, K., Okamoto, M., Chen, Y.-L., Luehr, G., Archer, J. Separation C_{60} and C_{70} fullerenes with a triphenyl bonded silica phase in microcolumn liquid chromatography, *J. Liquid Chromatogr.*, **1994**, *17*, 2359–2372.

34. Saito, Y., Ohta, H., Nagashima, H., Itoh, K., Jinno, K., Okamoto, M., Chen, Y.-L., Luehr, G., Archer, J. Separation of fullerenes with novel stationary phases in microcolumn high performance liquid chromatography, *J. Liquid Chromatogr.*, **1995**, *18*, 1897–1908.

35. Jinno, K., Okumura, C., Taniguchi, M., Chen, Y.-L. Effect of temperature on the mechanism of retention of fullerenes in high-performance liquid chromatography, *Chromatographia*, **1997**, *44*, 613–618.

36. Jinno, K., Yamamoto, K., Ueda, T., Nagashima, H., Itoh, K., Fetzer, J.C., Biggs, W.R. Liquid chromatographic separation of all-carbon molecules C_{60} and C_{70} with multi-legged phenyl group bonded silica phases. *J. Chromatogr.*, **1992**, *594*, 105–109.

37. Cox, D.M., Behal, S., Disko, M., Gorun, S.M., Greaney, M., Hsu, C.S., Kollin, E.B., Millar, J., Robbins, J., Robbins, W., Sherwood, R.D., Tindall, P. Characterization of C_{60} and C_{70} clusters, *J. Am. Chem. Soc.*, **1991**, *113*, 2940–2944.

38. Welch, C.J., Pirkle, W.H. Progress in the design of selectors for buckminsterfullerene, *J. Chromatogr.*, **1992**, *609*, 89–101.

39. Kimata, K., Hosoya, K., Araki, T., Tanaka, N. [2-(1-Pyrenyl)ethyl]silyl silica packing material for liquid chromatographic separation of fullerenes, *J. Org. Chem.*, **1993**, *58*, 282–283.

40. Yu, Q.W., Shi, Z.G., Lin, B., Wu, Y., Feng, Y.Q. HPLC separation of fullerenes on two charge transfer stationary phases, *J. Sep. Sci.*, **2006**, *29*, 837–843.

41. Taylor, D.R., Maher, K. Chiral separations by high-performance liquid chromatography, *J. Chromatogr. Sci.*, **1992**, *30*, 67–85.

42. Hawkins, J., Lewis, M.T.A., Loren, S.D., Meyer, A., Heath, J.R., Shibato, Y., Saykally, R.J. Organic chemical of C_{60} (Buckminsterfullerene): Chromatography and osmylation. *J. Org. Chem.*, **1990**, *55*, 6250–6252.

43. Herren, D., Thilgen, C., Calzaferri, G., Diederich, F. Preparative separation of higher fullerenes by high-performance liquid chromatography on a tetrachlorophthalimidopropyl-modified silica column, *J. Chromatogr.*, **1993**, *644*, 188–192.

44. Gasparrini, F., Francesco, M., Misiti, D., Claudio, V., Andreolini, F., Mapelli, G.P. High performance liquid chromatography on the chiral stationary phase (R, R)-DACH-DNB using carbon dioxide-based eluents. *J. High Res. Chromatogr.*, **1994**, *17(1)*, 43–45.

45. Cabrera, K., Wieland, G., Schäfer, M. High-performance liquid chromatographic separation of fullerenes (C_{60} and C_{70}) using chemically bonded γ-cyclodextrin as stationary phase, *J. Chromatogr.*, **1993**, *644*, 396–399.

46. Pesek, J., Cash, T. A chemically bonded liquid crystal as a stationary phase for high performance liquid chromatography, synthesis on silica via an organochlorosilane pathway, *Chromatographia*, **1989**, *27(11/12)*, 559–564.

47. Saito, Y., Ohta, H., Nagashima, H., Itoh, K., Pesek, J.J. Separation of fullerenes with liquid crystal bonded silica phases in microcolumn high performance liquid chromatography, *J. Microcol. Sep.*, **1995**, *7*, 41–49.

48. Janoš, P., Kozler, J. Fuel, thermal stability of humic acids and some of their derivatives. *Fuel*, **1995**, *74(5)*, 708–713.

49. Casadei, N., Thomassin, M., Guillaume, Y.-C., André, C. A humic acid stationary phase for the high performance liquid chromatography separation of buckminsterfullerenes: Theoretical and practical aspects, *Anal. Chim. Acta*, **2007**, *588*, 268–273.

50. Yu, Q.-W., Lin, B., Feng, Y.-Q., Zou, F.-P. Application of humic acid bonded-silica as a hydrophilic-interaction chromatographic stationary phase in separation of polar compounds. *J. Liquid Chromatogr. Relat. Technol.*, **2008**, *31(1)*, 64–78.

51. Jinno, K., Tanabe, K., Saito, Y., Nagashima, H. Separation of polycyclic aromatic hydrocarbons with, various C_{60} fullerene bonded silica phases in microcolumn liquid chromatography, *Analyst*, **1997**, *122*, 787–791.

52. Jinno, K., Tanabe, K., Saito, Y., Nagashima, H., Trengove, R.D. Retention behavior of calixarenes with various C_{60} bonded silica phases in microcolumn liquid chromatography, *Anal. Commun.*, **1997**, *34*, 175–177.

53. Atwood, J.L., Barbour, L.J., Raston, C.L., Sudria, I.B.N. C_{60} and C_{70} compounds in the pincerlike jaws of calix[6]arene, *Angew. Chem. Int. Ed.*, **1998**, *37*, 981–983.

54. Bhattacharya, S., Nayak, S.K., Chattopadhyay, S., Banerjee, M., Mukherjee, A.K. Study of novel interactions of calixarene-systems with [60]- and [70] fullerenes by the NMR spectrometric method, *J. Phys. Chem. B*, **2003**, *107*, 11830–11834.

55. Yin, J.F., Wang, L.J., Wei, X.Y., Yang, G.L., Wang, H.L. *p-tert*-Butylcalix[8]arene-bonded silica monoliths for liquid chromatography, *J. Chromatogr. A*, **2008**, *1188*, 199–207.

56. http://www.chm.bris.ac.uk/motm/phthalocyanine/pbpc.html

57. Tsukamoto, I., Saito, M., Yamane, M., Kawata, K., Kitamura, Y., Kitamura, Y., Mifune, M., Saito, Y., Haginaka, J. HPLC retention behaviors of π-electron rich compounds on Ni^{2+}- and Cu^{2+}-phthalocyanine derivatives bound to silica gels in polar eluents, *Anal. Sci.*, **2006**, *22*, 1035–1038.

58. Jinno, K., Kohrikawa, C., Saito, Y., Haiginaka, J., Saito, Y., Mifune, M. Copper-phthalocyanine stationary phases (Cu-PCS) for fullerenes separation in microcolumn liquid chromatography, *J. Microcol. Sep.*, **1996**, *8(1)*, 13–20.

59. Gumanov, L.L., Korsunsky, B.L., Derkacheva, V.M., Negrimovsky, V.M., Luk'yanets, E.A. Separation of C_{60} and C_{70} fullerenes on silica modified with phthalocyanines, *Mendeleev Commun.*, **1996**, *6(1)*, 1–2.

60. Kibbey, C.E., Savina, M.R., Pareeghian, B.K., Francis, A.H., Meyerhoff, M.E. Selective separation of C_{60} and C_{70} fullerenes on tetraphenylporphyrin-silica gel stationary phases, *Anal. Chem.*, **1993**, *65*, 3717–3719.

61. Coutant, D.E., Clarke, S.A., Francis, A.H., Meyerhoff, M.E. Selective separation of fullerenes on hydroxyphenyltriphenylporphyrin–silica stationary phases, *J. Chromatogr. A*, **1998**, *824*, 147–157.

62. Gumanov, L.L., Korsounskii, B.L. New sorbents based on tetraphenylporphyrin bound to silica gel for the separation of C_{60} and C_{70} fullerenes, *Mendeleev Commun.*, **1997**, *7(4)*, 158–159.

63. Xiao, J., Meyerhoff, M.E. High-performance liquid chromatography of C_{60}, C_{70}, and higher fullerenes on tetraphenylporphyrin-silica stationary phases using strong mobile phase solvents, *J. Chromatogr. A*, **1995**, *715*, 19–29.

64. Lochmüller, C.H., Hunnlcutt, M.L. Separation of nitro-polynuclear aromatics on bonded-pyrene stationary phases in microbore columns, *J. Chromatogr. Sci.*, **1983**, *21*, 444–446.

65. Yu, Q.W., Lin, B., He, H.B., Shi, Z.G., Feng, Y.Q. Preparation of pyrenebutyric acid bonded silica stationary phases for the application to the separation of fullerenes, *J. Chromatogr. A*, **2005**, *1083*, 23–31.

66. Ohta, H., Saito, Y., Jinno, K., Nagashima, H., Itoh, K. Temperature effect in separation of fullerene by high-performance liquid chromatography, *Chromatographia*, **1994**, *39(7/8)*, 453–459.

67. Ohta, H., Saito, Y., Jinno, K., Pesek, J.J., Matyska, M.T., Chen, Y.L., Archer, J., Fetzer, J.C., Biggs, W.R. Retention of fullerenes by octadecyl silica. Correlation with NMR spectra at low temperatures, *Chromatographia*, **1995**, *40*, 507–512.

68. Ohta, H., Jinno, K., Saito, Y., Fetzer, J.C., Biggs, W.R., Pesek, J.J., Matyska, M.T., Chen, Y.L. Effect of temperature on the mechanism of retention of fullerenes in liquid chromatography using various alkyl bonded stationary phases, *Chromatographia*, **1996**, *42*, 56–62.

69. Pirkle, W.H., Welch, C.J. An unusual effect of temperature on the chromatographic behavior of buckminsterfullerene, *J. Org. Chem.*, **1991**, *56*, 6973–6974.

70. Nawrocki, J., Dunlap, C., Li, J., Zhao, J., McNeff, C.V., McCormick, A., Carr, P.W. Part II. Chromatography using ultra-stable metal oxide-based stationary phases for HPLC, *J. Chromatogr. A*, **2004**, *1028*, 31–62.

71. Nawrocki, J., Dunlap, C., McCormick, A., Carr, P.W. Part I. Chromatography using ultra-stable metal oxide-based stationary phases for HPLC, *J. Chromatogr. A*, **2004**, *1028*, 1–30.

72. Yao, L.F., He, H.B., Feng, Y.Q., Da, S.L. HPLC separation of positional isomers on a dodecylamine-N, N-dimethylenephosphonic acid modified zirconia stationary phase, *Talanta*, **2004**, *64*, 244–251.

73. He, H.B., Zhang, W.N., Da, S.L., Feng, Y.Q. Preparation and characterization of a magnesia-zirconia stationary phase modified with β-cyclodextrin for reversed-phase high-performance liquid chromatography, *Anal. Chim. Acta*, **2004**, *513*, 481–492.

74. Zhang, Q.H., Feng, Y.Q., Da, S.L. Characterization and evaluation of magnesia–zirconia supports for normal-phase liquid chromatography, *Chromatographia*, **1999**, *50*, 654–660.

75. Yu, Q.W., Yang, J., Lin, B., Feng, Y.Q. Preparation of pyrenebutyric acid-modified magnesia–zirconia stationary phases using phosphonate as spacers and their application to the separation of fullerenes, *Anal. Chim. Acta*, **2006**, *559*, 79–88.

76. Yu, Q.W., Feng, Y.Q., Shi, Z.G., Yang, J. HPLC separation of fullerenes on the stationary phases of two lewis bases modified magnesia–zirconia, *Anal. Chim. Acta*, **2005**, *536*, 39–48.

77. Kele, M., Compton, R.N., Guiochon, G. Optimization of the high-performance liquid chromatographic separation of fullerenes using 1-methylnaphthalene as the mobile phase on a tetraphenylporphyrin-silica stationary phase, *J. Chromatogr. A*, **1997**, *786*, 31–45.

78. Akama, Y. Use of cerium oxide (CeO_2) as a packing material for the chromatographic separation of C_{60} and C_{70} fullerenes, *Talanta*, **1995**, *42*, 1943–1946.

79. Hu, Y.L., Feng, Y.Q., Wan, J.D., Da, S.L. Ceria–zirconia and calcia–zirconia modified with phenols as packing material for the chromatographic separation of C_{60} and C_{70}, *Anal. Sci. Suppl.*, **2001**, *17*, a321–a324.

80. Stalling, D.L., Kuo, K.C., Guo, C.Y., Saim, S. Separation of fullerenes C_{60}, C_{70}, and C_{76-84} on polystyrene divinylbenzene columns, *J. Liquid Chromatogr.*, **1993**, *16(3)*, 699–722.

81. Gügel, A., Müllen, K. Separation of C_{60} and C_{70} on polystyrene gel with toluene as mobile phase, *J. Chromatogr.*, **1993**, *628*, 23–29.

82. Zha, W.Y., Chen, D.P., Fei, W.Y. The HPLC separation of fullerenes with 1-methylnaphthalene modified PSDVB resin, *J. Liquid Chrom. Relat. Technol.*, **1999**, *22(16)*, 2443–2453.

83. Bucsi, I., Szabó, P., Aniszfeld, R., Surya Prakash, G.K., Olah, G.A. HPLC separation of hydrogenated derivatives of buckminsterfullerene, *Chromatographia*, **1998**, *48(1/2)*, 59–64.

84. Becker, L., Evans, T.P., Bada, J.L. Synthesis of [hydrogenated fullerene] $C_{60}H_2$ by rhodium-catalyzed hydrogenation of C_{60}, *J. Org. Chem.*, **1993**, *58*, 7630–7631.

85. Darwish, A.D., Abdul-Sada, A.K., Langley, G.J., Kroto, H.W., Taylor, R., Walton, D.R.M. Polyhydrogenation of [60]- and [70]fullerenes with Zn/HCl and Zn/DCl, *Synth. Met.*, **1996**, *77*, 303–307.

86. Ballanweg, S., Gleiter, R., Krdtschmer, W. Hydrogenation of buckminsterfullerene C_{60} via hydrozirconation: A new way to organofullerenes, *Tetrahedron Lett.*, **1993**, *34*, 3737–3740.

87. Anacleto, J.F., Quilliam, M.A. Liquid chromatography/mass spectrometry investigation of the reversed-phase separation of fullerenes and their derivatives, *Anal. Chem.*, **1993**, *65*, 2236–2242.

88. Meier, M.S., Corbin, P.S., Vance, V.K., Clayton, M., Mollman, M. Synthesis of hydrogenated fullerenes by zinc/acid reduction, *Tetrahedron Lett.*, **1994**, *35*, 5789–5792.

89. Henderson, C.C., Rohlfing, C.M., Assink, R.A., Cahill, P.A. $C_{60}H_4$: Kinetics and thermodynamics of multiple addition to fullerene C_{60}, *Angew. Chem.*, *Int. Ed. Engl.*, **1994**, *33*, 786–788.

90. Guha, S., Nakamoto, K. Electronic structures and spectral properties of endohedral fullerenes, *Coordin. Chem. Rev.*, **2005**, *249*, 1111–1132.

91. Gasper, M.P., Armstrong, D.W. A comparative study of buckminsterfullerene and higher fullerene separations by HPLC, *J. Liquid Chromatogr.*, **1995**, *18*, 1047–1076.

92. Akasaka, T., Okubo, S., Kondo, M., Maeda, Y., Wakatsugu, T., Kato, T., Suzuki, T., Yamamoto, K., Kato, T., Suzuki, T., Yamamoto, K., Kobayashi, K., Nagase, S. Isolation and characterization of two $Pr@C_{82}$ isomers, *Chem. Phys. Lett.*, **2003**, *19*, 153–156.

93. Xiao, J., Savina, M.R., Martin, G.B., Francis, A.H., Meyerhoff, M.E. Efficient HPLC purification of endohedral metallofullerenes on a porphyrin-silica stationary phase, *J. Am. Chem. Soc.*, **1994**, *116*, 9341–9342.

23
Fullerene Growth

Jochen Maul
Johannes Gutenberg-Universität

23.1 Introduction

Since the discovery of the C_{60} molecule by H. Kroto, R. Smalley, and coworkers in 1985 (Kroto et al. 1985), fullerene research has evolved into a large field of nanosciences. Both their exceptional structures and their unconventional physical and chemical properties make fullerenes highly interesting objects of study. Stable spheroids of covalently bound carbon atoms arise from assorting well-defined numbers of pentagons and hexagons—the construction principle for the hollow fullerene cages. In comparison to other carbon allotropes, their hallmark is the high geometric diversity.

From a naïve viewpoint, one might tend to assume that the buildup of large and ordered fullerene structures requires careful arrangement of the molecule atom by atom. Surprisingly, fullerenes are commonly produced under laser plasma conditions where inelastic collisions and electronic excitations prevail (Kroto et al. 1991). In a simplified picture, the two basic ingredients necessary to produce fullerenes mostly turn out to be the presence of carbon and heat. This in turn demonstrates a high degree of self-organization. Only recently, new trends have come up, which aim at a more controlled fullerene production based on surface catalysis at lower temperatures.

Normally, fullerene growth in plasma is not easy to observe heading short timescales, distinct excitation levels, and lots of competing carbon species at a time. It is more than doubtable whether a single dominant mechanism can account for the complex scenario. Accordingly, the models for fullerene growth span a wide range up to date. Smalley (1992) aptly summarized: "Of course there must be hundreds of mechanisms whereby a fullerene like C_{60} can form". After two decades of fullerene research, Johnston (2002) remarks in his textbook that "the mechanism of fullerene growth is still not well understood".

This chapter aims to outline the major achievements and trends toward the contemporary understanding of fullerene growth, from the initial views following the experimental evidence of fullerenes in the 1980s up to the present state of knowledge.

23.2 Fullerene Geometries

For the formation of fullerenes, geometric aspects always play an important role by strongly restricting their configuration space. Fifteen years before the discovery of the C_{60} molecule, its existence was postulated by E. Osawa (1970) who was inspired when he saw his son playing with a soccer ball. Indeed, carbon is the right chemical element to enable sophisticated geometric networks due to its threefold valency in the sp^2 state.

During their formation process, the electronic energy of large carbon clusters is lowered by reducing the number of unsaturated bonds (dangling bonds). In case of graphite sheets, this is realized by sp^2 hybridization of the carbon atoms involving hexagonal rings and π-bonding. If pentagonal defects are introduced into the graphite structure, the sheets begin to bend. Twelve pentagons cause the graphite sheet to curl up into a ball, forming a fullerene. Exemplarily, the topologic arrangement of carbon atoms within the C_{60} molecule is illustrated in Figure 23.1.

From a purely geometric viewpoint, Euler's theorem provides the fundamental condition for the topology of convex polyhedra and hence for the fullerenes (Johnston 2002). The basic relationship reads

$$V + F = E + 2 \tag{23.1}$$

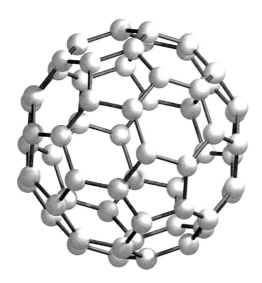

FIGURE 23.1 Sketch of the C_{60} Buckministerfullerene geometry showing the three-dimensional cage arrangement of pentagons and adjacent hexagons.

where V, F, and E are the total number of vertices, faces, and edges of the polyhedron, respectively. Euler's theorem can be reexpressed as follows:

$$\sum_k f_k - \frac{1}{2}\sum_j (j-2)\, v_j = 2 \tag{23.2}$$

$$\sum_j v_j - \frac{1}{2}\sum_k (k-2)\, f_k = 2 \tag{23.3}$$

where the total number of vertices $V = \sum_j v_j$ and the total number of faces $F = \sum_k f_k$ is given by v_j vertices of connectivity j and f_k faces with k sides. Combination of both equations yields another useful form:

$$\sum_j j\, v_j = \sum_k k\, f_k = 2E \tag{23.4}$$

Since carbon is trivalently bound in the fullerenes, the connectivity of the vertices is $j = 3$, which reduces Euler's theorem to the series

$$\sum_k (6-k) f_k = 12 = \cdots + 2f_4 + 1f_5 + 0f_6 - 1f_7 - \cdots \tag{23.5}$$

For fullerenes with only pentagonal and hexagonal faces, Equation 23.5 reduces further to $f_5 = 12$, which explains why all fullerenes contain 12 pentagonal faces. Note further that the number of hexagonal faces is completely undefined in Equation 23.5, thus giving rise to the large family of fullerenes.

Trivalent polyhedra must contain an even number of vertices $N = V$, because the numbers of edges $E = 3N/2$ must be an integer.

For this reason, only even numbers of carbon atoms are observed in the fullerene mass spectrum. By inserting the total number of faces, $F = 12 + f_6$, and edges E into Equations 23.2 and 23.3, a connection between the number of vertices and the number of hexagonal faces of fullerenes is obtained:

$$f_6 = \frac{N}{2} - 10 \tag{23.6}$$

$$N = 20 + 2f_6 \tag{23.7}$$

Since f_6 can take the value zero and any positive integer (except the value $f_6 = 1$; Johnston 2002), the fullerene mass spectrum starts at the mass of C_{20} (monoisotopic mass $m \sim 240$ u). Both the even-number characteristics and the "delayed" start of the fullerene masses are seen in the positive ions fullerene mass spectrum in Figure 23.2, which was obtained by laser ablation/ionization mass spectrometry from an organic matrix (Maul et al. 2006).

Another striking feature in the mass spectrum of large carbon clusters is the overabundance of C_{50}, C_{60}, and of C_{70}. Buckministerfullerene C_{60} is the smallest and C_{70} is the second smallest fullerene that can have isolated pentagons leading to a minimization of intramolecular strain. This principle is often referred to as isolated pentagon rule (IPR). C_{50} is the smallest cluster for which a cage can be constructed without triples of fused pentagons (Kroto and Mc Kay 1988). Although this pentagon isolation principle provides some geometric explanation for the occurrence of the "magic" carbon numbers $N = 50$, 60, and 70, their strong dominance against others is far from being understood. For example, the role of electronic stabilization is not yet solved confronting the absence of superaromaticity (Aihara and Hosoya 1988).

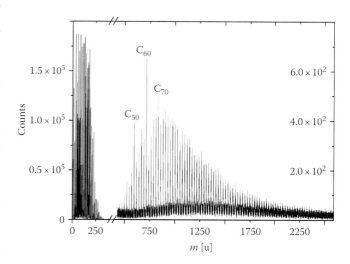

FIGURE 23.2 Time-of-flight mass spectrum of positive carbon cluster ions showing a bimodal structure with a gap around 400 u separating smallest carbon clusters from the fullerenes. Above the gap, the "magic" fullerenes C_{50}, C_{60}, and C_{70} stand out by their overabundance compared to the residual fullerene distribution. (From Maul, J. et al., *Phys. Rev. B*, 74, 161406 (R), 2006. With permission.)

23.3 Growth Models for Fullerenes

23.3.1 Geometric Growth Models

From the very beginning, geometric aspects were essential for the understanding of fullerene growth. One of the first nucleation schemes originates from Kroto and Mc Kay (1988). Starting from a laser-induced plasma from a carbonaceous target, carbon chains grow via aggregation of carbon atoms in the dense plasma. Occasionally, they form polycyclic aromatic rings, which bend in the presence of pentagon inclusions. If the carbon atom density is high enough, the clusters continue to grow at the edges, as indicated in Figure 23.3, crossing and burying the opposite edges as it curls. Once this process is launched, there is no strict termination of the growth and large spiral structures form, eventually ending in large soot particles, as demonstrated in the TEM images of Figure 23.4. If, however, accidental closure occurs in stage (c) of the growth process, as shown in Figure 23.3, by correct inclusion of pentagons, closed fullerene isomers can form.

It is interesting to note that fullerene formation is also observed from laser ablation of targets, which are not fully out of carbon, for example, from polymers (Campbell et al. 1990). More precisely, they even do not need to contain aromatic constituents.

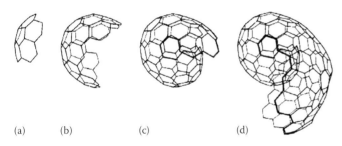

(a)   (b)   (c)   (d)

FIGURE 23.3 Fullerene nucleation scheme starting from a convex carbon framework in (a) through continuative bending in (b) and (c) until edge bypass occurs in (d). (From Kroto, H.W. and Mc Kay, K., *Nature*, 331, 328, 1988. With permission.)

This observation clearly reveals that indeed fullerenes are unlikely to emerge by direct arrangement of target fragments (i.e., aromatic cycles); they rather form by sequential addition of smallest carbon radicals (mainly monomers and dimers).

Meanwhile, there are lots of geometric models, which, however, can be classified by four mainstreams. These different concepts are schematically presented in Figure 23.5. The "pentagon road" in (a) implies that basket-like structures grow into spheres by incorporation of carbon pentagons, (I) either by addition of smallest carbon units as originally proposed by Haufler et al. (1991) or (II) by addition of ring structures as proposed by Wakabayashi and Achiba (1992). The "fullerene road" in (b) describes growth of smaller fullerenes to larger ones by accretion of C_2 units (Heath 1992) but does not ask for the origin of the smaller fullerenes. The "folding models" in (c) assume that fullerenes form (I) after intramolecular rearrangement after condensation of large carbon ring assemblies (von Helden et al. 1993) or (II) after spiraling of carbon chains to preform a cage (Lagow et al. 1995). This is accompanied by the transformation of sp bonds into sp^2 bonds (Bunz et al. 1999). Most of these schemes are based on the formation of carbon rings in their initial stage. Other models (not shown in Figure 23.5) can be clustered to a set of more hypothetic ideas such as fragmentation of carbon nanotubes and healing of the breakage afterward to form fullerenes (Dravid et al. 1993), and vice versa, coalescence of fullerenes to form energetically more stable nanotubes.

All these models are based on the assumption that some intermediate structure forms, which is in thermodynamic equilibrium. Also, they provide a rather qualitative description of the growth process. They basically neglect that hot carbon vapor might be far from thermodynamic equilibrium, giving rise to self-organization processes, autocatalysis, and irreversible processes.

23.3.2 Kinematic Growth Models

Whereas the stability of the "magic" fullerenes C_{50}, C_{60}, and C_{70} is partly explained by their geometry, the appearance of the wide

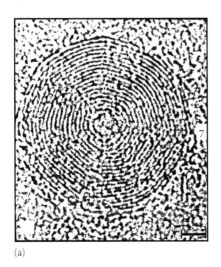

(a)

(b)

FIGURE 23.4 (a) TEM image of a spheroidal carbon nanoparticle with eight pyramidal crystallite cross sections which are marked in (b) for better visibility. The scale bar corresponds to 0.2 nm. (From Kroto, H.W. and Mc Kay, K., *Nature*, 331, 328, 1988. With permission.)

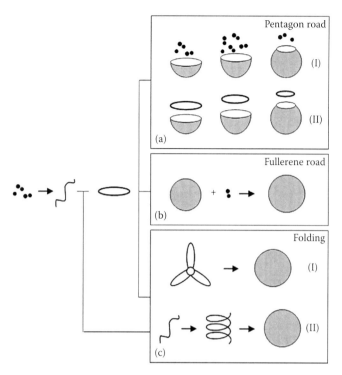

FIGURE 23.5 Schematic overview of proposed geometric fullerene growth models. Dots represent carbon atoms; lines represent carbon chains; and balls symbolize the fullerene products. (Adapted from Irle, S. et al., *J. Phys. Chem. B*, 110, 14531, 2006.)

range of fullerene masses requires some general kinetic treatment. The most general approach considers inelastic binary collisions between two educt cluster species J and $I - J$ to form some product species I:

$$J + (I - J) \xleftrightarrow{K_1/K_{-1}} I \qquad (23.8)$$

In case of non-magic clusters, both directions of the reaction are possible, introducing two rates K_1 and K_{-1} for aggregation, respectively, for dissociation. Within this second-order reaction dynamic, the kinematic description is provided by the Smoluchowski differential equation for the concentrations n_I, n_{I-J}, and n_J of the corresponding species (Bernholc and Phillips 1986a,b, Creasy and Brenna 1990, Villarica et al. 1993):

$$\frac{dn_I}{dt} = \frac{1}{2} \sum_{J=1}^{I-1} K_{J,I-J}\, n_I\, n_{I-J} - n_I \sum_{J=1}^{\infty} K_{I,J}\, n_J \qquad (23.9)$$

The Smoluchowski equation includes both coalescence of smaller species J and $(I - J)$ to form the fullerene I (first term), and also continuative coalescence to larger species (second term). These considerations were mainly introduced in carbon cluster physics by Bernholc and Phillips (1986a,b). Successive simulations are found in the work of Creasy and Brenna (1990), where the basic restriction is that only the smallest carbon units ($N = 1, 2, 3$) add to existing carbon clusters because the thermal velocities of the smallest species is significantly elevated thus inferring higher

collision rates and because the small clusters are overabundant in the early stage of the reaction zone. However, it was found that this restriction always leads to symmetric distributions of large carbon clusters.

In a later work, this restriction is abandoned (Villarica et al. 1993). There, inelastic collisions of any carbon clusters present in the reaction volume are allowed. Note that for fullerenes the index I takes even values only, starting with the geometric restriction of $I = 20$ carbon atoms per molecule.

In this case, an asymptotic solution of the Smoluchowski equation for reaction times Δt large compared to the inverse collision rate K ($\Delta t > 1/K$), and for large cluster sizes I is obtained. The latter criterion is intrinsically fulfilled for fullerenes. Furthermore, the rate constants $K_{I,J}$ need scale homogeneously of a negative degree ω for arbitrary interacting species I and J, that is,

$$K_{\lambda I, \lambda J} = \lambda^{2\omega} K_{I,J} \quad (\omega < 0;\ \lambda > 1). \qquad (23.10)$$

The parameter ω is taken twice in the exponent because two species are involved. This homogeneous downscaling reflects the decrease in the cluster velocities for larger fullerenes onto each inelastic collision, thus lowering the collision rates with increasing mass. For the given case, the asymptotic solution of the Smoluchowski equation is given by the lognormal or logarithmonormal statistical distribution (Maul et al. 2006, Maul 2007):

$$\Lambda(\tilde{N}\,|\,\mu, \sigma^2) \propto \frac{1}{\sqrt{2\pi\sigma^2}\,\tilde{N}} \times \exp\left(-\frac{(\ln \tilde{N} - \mu)^2}{(2\sigma^2)} \right) \qquad (23.11)$$

where the shifted number of carbon atoms $\tilde{N} = N - 20$ is introduced in order to shift the origin of the distribution to the start of the fullerene spectrum at $N = 20$. This offset is sometimes referred to as a "waiting point" in the statistics. The dimensionless "alternative parameter" μ is connected to the median size $\langle \tilde{N} \rangle$ of the lognormal distribution via $\langle \tilde{N} \rangle = \exp(\mu)$ (resp. $\langle N \rangle = \langle \tilde{N} \rangle + 20$), and the variance σ^2 is further the square of the shape parameter σ, which also measures the distribution asymmetry. The distribution becomes symmetric in the limit $\sigma \to 0$, but more important, can also exhibit strong asymmetries for $\sigma > 1$.

The lognormal function Λ provides an excellent fitting to the fullerene distribution in Figure 23.6a, which is extracted from the mass spectrum in Figure 23.2. The quality of the fitting curve is even more emphasized by the linearization given in Figure 23.6b. Here, the so-called cumulative probability for the shifted cluster size $\tilde{N}$ (which displays the probability that clusters up to the size $\tilde{N}$ occur) is plotted against the logarithm of $\tilde{N}$, resulting in a linearization in the lognormal case. The parametric form of the straight line, $\xi \to \sigma\,\xi + \mu$, directly provides the statistical parameters

$$\sigma = \ln\left(\frac{1}{2}\left[\frac{\xi_{50\%}}{\xi_{16\%}} + \frac{\xi_{84\%}}{\xi_{50\%}} \right] \right) \qquad (23.12)$$

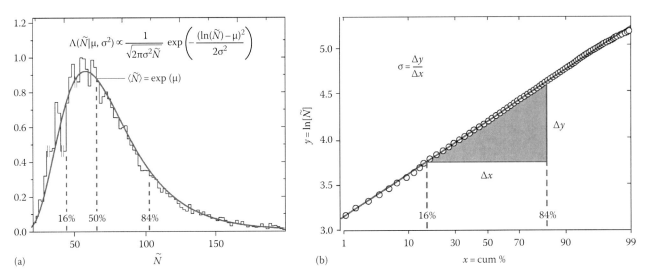

FIGURE 23.6 Statistical analysis of a fullerene mass spectrum (shown in Figure 23.2). The normalized histogram in (a) shows the frequency of fullerene sizes $\tilde{N}$. The excess abundance of the "magic" fullerenes C_{50}, C_{60}, and C_{70} is thereby removed as indicated by the vertical intersections. A lognormal fitting curve is applied to the histogram data. The lognormal statistics is emphasized by the linearization given in (b). (From Maul, J. et al., *Phys. Rev. B*, 74, 161406 (R), 2006. With permission.)

and

$$\mu = \ln(\xi_{50\%}), \tag{23.13}$$

where $\xi_{16\%}$, $\xi_{50\%}$, and $\xi_{84\%}$ are depicted from the 16%, 50%, and 84% values of the corresponding cumulative probabilities.

The lognormal size distribution expresses "cumulative growth," that is, growth of larger particles at the expense of smaller ones. In particular, it should be differed from "constant growth" solely from monomers. This growth characteristics is well known, for example, for atmospheric aerosols and snow crystals of arbitrary size merging onto inelastic collision.

The temporal evolution of lognormal fullerene distribution is shown in Figure 23.7 for two different time spans Δt

between ablation laser pulse and ion extraction to the mass spectrometer (which terminates the growth process). With augmenting time for growth, the median size is shifted from $\langle \tilde{N} \rangle = 65.5$ for $\Delta t = 300$ ns to $\langle \tilde{N} \rangle = 87.6$ for $\Delta t = 900$ ns. The asymmetry of the frequency curves diminishes, as it is recognized from a decrease in the shape parameter from $\sigma = 0.43$ for $\Delta t = 300$ ns to $\sigma = 0.36$ for $\Delta t = 900$ ns. In the corresponding cumulative plots, the temporal evolution is accompanied by a rotation of the straight lines around the upper end of the distribution. This can be understood from two counteracting criteria, that is, the vanishing of smaller fullerenes in favor of larger ones accompanied by a decrease of the total number of particles during coalescence. Thereby, two smaller particles vanish within an inelastic two-body collision to build

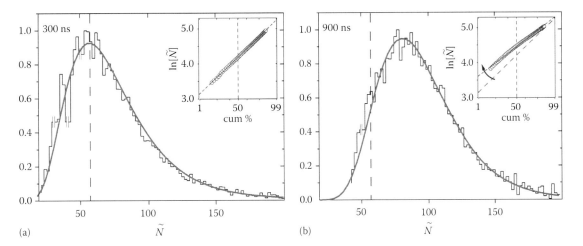

FIGURE 23.7 Temporal evolution of fullerene growth as observed from size statistics. The histograms show fullerene size distributions obtained (a) 300 ns and (b) 900 ns after ablation laser pulses. With increasing growth duration, the distributions are shifted to higher cluster sizes and the distribution asymmetry decreases, as recognized from the shape of the lognormal curves and from the corresponding cumulative plots (insets). (From Maul, J. et al., *Phys. Rev. B*, 74, 161406 (R), 2006. With permission.)

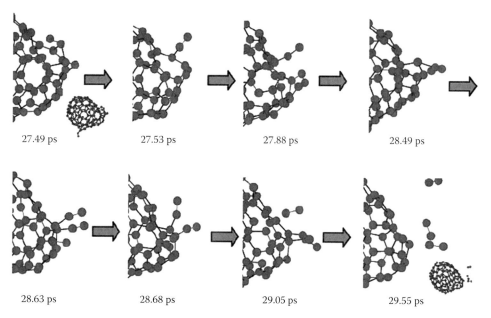

27.49 ps   27.53 ps   27.88 ps   28.49 ps

28.63 ps   28.68 ps   29.05 ps   29.55 ps

FIGURE 23.8 Simulation "snapshots" of giant fullerene shrinkage accompanied by dimer and carbon chain pop-out events. (From Irle, S. et al., *J. Phys. Chem. B*, 110, 14531, 2006. With permission.)

one larger fullerene as it is also expressed by the homogeneous downscaling of the rate constants.

A new model (Irle et al. 2006) suggests that fullerenes are produced in the hot carbon vapor in a two-stage process in which carbon cages first grow to large sizes and subsequently expel fragments and chains of carbon, finally ending in C_{60}. The proposed name for this model is the "shrinking hot giant route." Using molecular dynamics simulations, nonequilibrium thermodynamics was included which is actually the correct situation in a hot carbon vapor at temperatures typically between 2000 and 3000 K. The simulation results demonstrate that giant carbon cages might form more easily than smaller fullerenes, but they are more prone to disruption. Two main processes contribute to the shrinking of the giant fullerenes: pop-out of carbon dimers and falloff of side chains. The simulation sequence in Figure 23.8 displays this behavior taking place on the sub-nanosecond timescale. In similar simulations (Irle et al. 2003), fullerenes are expected to gain size by adding carbon dimers, similar to the "ring collapse machanism" described in the previous section (von Helden et al. 1993).

23.4 Open Problems

While the routine production of fullerenes is nowadays well established, the correct extraction of fullerene growth parameters is still a problem. Since most fullerene production processes involve the generation of a plasma or a hot gas, it is difficult to follow the genesis of the set of species simultaneously and to reconstruct the overall dynamics. The situation is further obscured by possible competing reaction channels, which are difficult to separate.

Of course, it would be nice to follow the growth process in real time and possibly on the atomic scale and hence to follow the fullerene builtup atom by atom. However, microscopes with atomic resolution (such as the scanning tonneling microscope [STM] and

transmission electron microscopes [TEM]) are poor in time resolution and, vice versa, time-resolved spectroscopy of molecules usually does not imply their direct observability in real space.

Still one of the most fascinating and maybe one of the most controversial topics in fullerene research is the highly symmetric buckministerfullerene C_{60}. Despite its distinct symmetry, the expected superaromaticity, that is, the delocalization of the π electron system over the whole sphere, is not observed. Even worse, aromaticity requires $2(N + 1)^2$ π electrons according to Hückel's rule, which is evidently not fulfilled. Moreover, DFT and MNDO calculations predict that the closed-shell D_{5h} isomer of C_{70} is more stable than the I_h isomer of C_{60} that conflicts with the overabundance of C_{60} in the fullerene mass spectrum.

The attempt to explain the formation of C_{60} within the "shrinking hot giant route" (see Section 23.3) demonstrates a wide fragmentation spectrum, but the hot giants themselves are not yet observed directly.

23.5 Summary

Fullerene growth is still a controversially debated subject since they are usually produced under the complex conditions of a plasma. This chapter gives an overview of the recent knowledge. In order to classify the numerous growth models, two basically different categories were outlined: geometric models (discussed in Section 23.3.1) and kinematic models (discussed in Section 23.3.2). Still many models appear rather hypothetic and a direct confirmation is desirable for the future.

23.6 Perspectives

The production of fullerenes in the plasma practically means to decompose everything and to rearrange everything (see previous

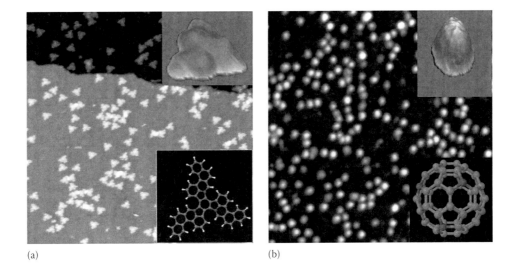

FIGURE 23.9 STM images of the surface catalysis of fullerenes. Aromatic $C_{57}H_{33}N_3$ precursor molecules deposited on a Pt(111) surface (a) are annealed at 750 K to form triazofullerene $C_{57}N_3$ (b) by cyclodehydrogenation. Upper insets: topographic STM images of isolated precursor and fullerene molecules (frame sizes ~ 2.5 × 2.5 nm²). Lower insets: atomic structure of the precursor and of the fullerene, respectively. (Courtesy of José Angel Martín-Gago, Consejo Superior de Investigaciones Científicas, Madrid, Spain.)

sections). It is suited for a routine but predominantly unselective generation of notable quantities; however, a more controlled production pathway would be desirable for both the production of specific fullerenes and for a better observability of the growth process itself.

Recent trends aim at a more rational chemical synthesis of fullerenes. For example, isolable quantities of C_{60} have been synthesized in a complex 12-step chemical process including the production of polycyclic aromatic precursors and the application of flash vacuum pyrolysis in the final step (Scott et al. 2002). By using other precursors, this target-specific procedure is expected to yield other fullerenes as well.

A fully controlled synthesis of fullerenes has been demonstrated in Otero et al. (2008). By cyclization and dehydrogenation of aromatic $C_{57}H_{33}N_3$ precursor molecules during catalysis on a Pt(111) surface at an annealing temperature of 750 K, the triazofullerene $C_{57}N_3$ has been produced with about 100% yield. The formation process has been directly observed by STM (Figure 23.9); further evidence for the cyclization process and for the dehydrogenation on the platinum surface have been provided by x-ray photoemission spectroscopy and thermal desorption experiments. The precursor with its wing structure is thereby folded to cause closure, alike "molecular origami." It is conceivable that all kinds of fullerenes and derivatives could be synthesized in such a manner by carefully choosing suitable precursors. The surface catalysis further offers the possibility to scale-up the production of fullerenes to larger quantities by annealing precursors on platinum powder (Otero et al. 2008).

Another possibility might be to exploit femtosecond laser pulse shaping (Assion et al. 1998) for an optimized output of certain fullerenes within a laser reaction zone in the gas or liquid phase. Here, the highly dynamic electric field of the laser pulse could trigger specific electronic excitations resulting in

"tailored" bond configurations during the synthesis. This would bridge the gap between disorder in the carbonaceous vapor and ordered growth in it.

Meanwhile, even a series of applications is being launched, such as fullerene-based nanoelectronics or nano-capsuls for medication by controlled intervention in the growth process. This may open a field of applications for everyday life.

Acknowledgments

I would like to thank Thomas Berg, Klaus-Jürgen Kott, and Edit Marosits for their help and discussions during works on kinematics of fullerene growth. I am grateful to José Angel Martín-Gago for the nice edition of STM images from the surface catalysis of fullerenes and to Stephan Irle for valuable advice.

References

Aihara, J. and Hosoya, H. 1988. Spherical aromaticity of the Buckministerfullerene, *Bull. Chem. Soc. Jpn.* 61: 2657–2659.

Assion, A., Baumert, T., Bergt, M., Brixner, T., Kiefer, B., Seyfried, V., Strehle, M., and Gerber, G. 1998. Control of chemical reactions by feedback-optimized phase-shaped femtosecond laser pulses, *Science* 282: 919–922.

Bernholc, J. and Phillips, J.C. 1986a. Kinetics of aggregation of carbon clusters, *Phys. Rev. B* 33: 7395–7398.

Bernholc, J. and Phillips, J.C. 1986b. Kinetics of cluster formation in the laser vaporization source: Carbon clusters, *J. Chem. Phys.* 85: 3258 3267.

Bunz, U.H., Rubin, Y., and Tobe Y. 1999. Polyethynylated cyclic π-systems: Scaffoldings for novel two and three-dimensional carbon networks, *Chem. Soc. Rev.* 28: 107–119.

Campbell, E.E.B., Ulmer, G., Hasselberger, B., Busmann, H.-G., and Hertel, I.V. 1990. An intense, simple carbon cluster source, *J. Chem. Phys.* 93: 6900–6907.

Creasy, W.R. and Brenna, J.T. 1990. Formation of high mass carbon clusters from laser ablation of polymers and thin carbon films, *J. Chem. Phys.* 92: 2269–2279.

Dravid, V.P. et al. 1993. Buckytubes and derivatives: Their growth and implications for Buckyball formation, *Science* 259: 1601–1604.

Irle, S., Zheng, G., Elstner, M., and Morokuma, K. 2003. From C_2 molecules to self-assembled fullerenes in quantum chemical molecular dynamics simulations, *Nano Lett.* 3: 1657–1664.

Irle, S., Zheng, G., Wang, Z., and Morokuma, K. 2006. The C_{60} formation puzzle "solved": QM/MD simulations reveal the shrinking hot giant road of the dynamic fullerene self-assembly mechanism, *J. Phys. Chem. B* 110: 14531–14545.

Haufler, R.E. et al. 1991. Arc carbon generation of C_{60}, *Mater. Res. Soc. Symp. Proc.* 206: 627–637.

Heath, J.R. 1992. Synthesis of C_{60} from small carbon clusters: A model based on experiment and theory, in: G. S. Hammond and V. J. Kuck (eds.), *Fullerenes: Synthesis, Properties, and Chemistry of Large Carbon Clusters*, ACS Symposium Series, Vol. 1, pp. 1–23, ACS Publication, Washington, DC.

Johnston, R.L. 2002. *Atomic and Molecular Clusters*, Master Series in Physics and Astronomy, Taylor & Francis, Boca Raton, FL.

Kroto, H.W. and Mc Kay, K. 1988. The formation of quasi-icosahedral spiral shell carbon particles, *Nature* 331: 328–331.

Kroto, H.W., Heath, J.R, O'Brien, S.C., Curl, R.F., and Smalley, R.E. 1985. C_{60}: Buckminsterfullerene, *Nature* 318: 162–163.

Kroto, H.W., Allaf, A.W., and Balm, S.P. 1991. C_{60}: Buckminsterfullerene, *Chem. Rev.* 91: 1213–1235.

Lagow, R.J. et al. 1995. Synthesis of linear acetylenic carbon: The "*sp*" carbon allotrope, *Science* 267: 362–367.

Maul, J. 2007. Measurement of nanoparticle mass distributions by laser desorption/ionization time-of-flight mass spectrometry, *J. Phys.: Condens. Matter* 19, 176216.

Maul, J., Berg, T., Marosits, E., Schönhense, G., and Huber, G. 2006. Statistical mechanics of fullerene coalescence growth, *Phys. Rev. B* 74: 161406 (R).

Osawa, E. 1970. Superaromaticity, *Kagaku* 25: 854–863.

Otero, G. et al. 2008. Fullerenes from aromatic precursors by surface-catalysed cyclodehydrogenation, *Nature* 454: 865–868.

Scott, L.T. et al. 2002. A rational chemical synthesis of C_{60}, *Science* 295: 1500–1503.

Smalley, R.E. 1992. Self-assembly of the fullerenes, *Acc. Chem. Res.* 25: 98–105.

Villarica, M., Casey, M.J., Goodisman, J., and Chaiken, J. 1993. Application of fractals and kinetic equations to cluster formation, *J. Chem. Phys.* 98: 4610–4625.

von Helden, G., Gotts, N.G., and Bowers, M.T. 1993. Experimental evidence for the formation of fullerenes by collisional heating of carbon rings in the gas phase, *Nature* 363: 60–63.

Wakabayashi, T. and Achiba, Y. 1992. A model for the C_{60} and C_{70} growth mechanism, *Chem. Phys. Lett.* 190: 465–468.

24

Production of Carbon Onions

Chunnian He
Tianjin University
The Academy of Military
Medical Sciences

Naiqin Zhao
Tianjin University

24.1 Introduction

Since the discovery of buckminsterfullerene [1], carbon-based nanostructures [2] became the hottest research field in nanoscience. The ability of carbon to exist in several allotropic forms capable of creating a variety of nanoscale-size shapes, such as spheres, tubules, onions, ribbons, and rods, provides fertile ground for many research efforts.

Carbon onions, one important member of the fullerene family, are quasi-spherical carbon nanoparticles consisting of concentric fullerene-like shells that range from double- and triple-shelled to multilayered structures (Figure 24.1) [3]. Nano-onions have been observed to either encapsulate metals [4,5] or consist only of carbon layers [6]. This observation of carbon onions was reported for the first time in 1980 by Iijima [7], who tried to describe this new structure with a model involving a mixed sp^2/sp^3 hybridization of carbon atoms. The scientific community did not pay much attention to this discovery before the fullerenes one [1] and it was only in 1987 [8,9] that the structures characterized by Iijima were reconsidered and discussed in terms of fullerenes structures. In fact, interest was devoted to carbon onions in 1992, when Ugarte [10] discovered a reproducible technique to achieve their formation that consists in irradiating carbon soot by an intense electron beam in a transmission electron microscope (TEM) under 300 kV. After that, the carbon onions became an attractive research target due to their high degree of symmetry as well as their unique physical and chemical properties. Particularly, carbon onion is considered to be one of the possible carriers of the 217.5 nm interstellar absorption bump [11] and a potentially good solid lubricant as is the case for WS_2 nanoparticles having an onion-like structure [12]. These features show a significant potential for various applications in the future. The most promising applications of carbon onions are those involving their use in single-electron devices [13], magnetic refrigerators, nanodiodes [14], nanotransistors, nanoball bearings and insulator lubricants [15], or as catalyst for selective oxidation reactions [16].

For a proper investigation of its properties, macroscopic quantities of carbon onions are needed. Presently, a number of synthetic methods to produce carbon onions in relatively high yields have been demonstrated, including arc discharge [17,18], high-energy electron beam or laser irradiation [19,20], thermal treatment of carbonaceous materials [21], chemical vapor deposition (CVD) [22], and implantation of carbon ions onto metal particles [23].

This chapter will discuss the methods for the synthesis of carbon onions. Section 24.2 describes the physical methods used to fabricate these carbon onions, which consist of arc discharge, high-energy electron beam or laser irradiation, implantation of carbon ions onto metal particles, thermal treatment of carbonaceous materials (such as nanodiamond, carbon soot, graphite, or carbon nanotubes). Section 24.3 pays attention to issues associated with carbon onion production by CVD. Section 24.4 describes some authors' works related to the synthesis of carbon onions by CVD. Section 24.5 concludes the chapter with a brief summary.

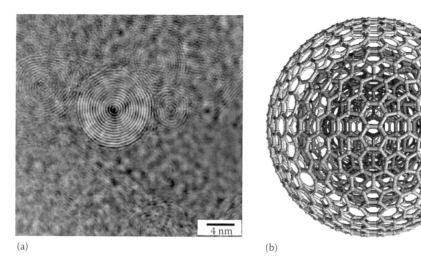

(a)   (b)

FIGURE 24.1   (a) TEM images of carbon onions. (From Roy, D. et al., *Chem. Phys. Lett.*, 373, 52, 2003. With permission.) (b) Model structure of a spherical multi-shell carbon onion. (Calculated by H. Terrones and M. Terrones.)

24.2 Preparation of Carbon Onions by High-Energy Condition

The known techniques for carbon onion fabrication can be divided into two classes: physical and chemical, according to the method of carbon atomization in the initial production stage. Generally, the physical methods contain arc discharge, high-energy electron beam or laser irradiation, implantation of carbon ions onto metal particles, and thermal treatment of carbonaceous materials (such as nanodiamond, carbon soot, graphite, or carbon nanotubes). However, the above methods require a high-energy input, and nano-onions have mainly been produced as an unwanted by-product, making the separation of carbon onions from other carbon-containing products an unavoidable process step. The chemical production method, including plasma-enhanced CVD, catalytic disproportionation of CO, catalytic reduction of CO_2, vapor phase growth, and thermal CVD, consists of a catalytic decomposition of carbon-containing precursor materials. The obvious advantage of this method is the possibility of producing carbon onions at relatively low temperatures and in large yields.

24.2.1 Arc Discharge

The carbon arc discharge method, initially used for producing C_{60} fullerenes, is the most common and perhaps easiest way to produce carbon onions as it is rather simple to undertake. However, it is a technique that produces a mixture of components and requires separating onions from the soot and the catalytic metals presented in the crude product. Generally, the arc discharge method contains arc discharge in an inert gas atmosphere and arc discharge in water.

24.2.1.1 Arc Discharge in an Inert Gas Atmosphere

The arc discharge in an inert gas atmosphere has been widely used for the synthesis of carbon nanomaterials (especially for carbon nanotubes and fullerenes) in the past 10 years (Figure 24.2a) [17,18,24]. In this method, two graphite rods (anode and cathode; recently some researchers also employed coal as an anode to synthesize carbon onions [25–28]) are generally used as electrodes, and He or Ar gas is introduced for inert atmosphere conditions. The anode is an extremely pure graphite rod with a hole filled with graphite (or coal) and a catalyst powder (such as Cu, Al, YNi_2,

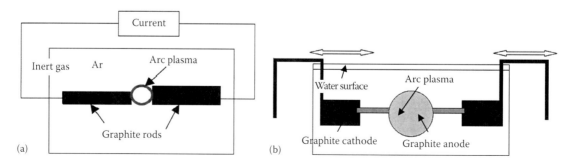

(a)   (b)

FIGURE 24.2   Schematics showing experimental setups for (a) arc discharge in an inert gas atmosphere (From Lange, H. et al., *Carbon*, 41, 1617, 2003. With permission.) and (b) arc discharge in water for the synthesis of carbon onions.

or Bi_2O_3 powder) [25–28]. The cathode is a graphite rod that is shaped into a sharp tip. As the rods are brought close together, a discharge occurs and results in the formation of plasma, where the temperature is close to 4000 K. Thus, the anode is consumed by plasma and ionized carbon atoms are attracted to the graphite cathode. Cloth-like deposits are formed on this electrode and carbon onions can be collected from the core of the deposits. The deposit also contains other graphitic materials such as carbon nanotubes, carbon-coated metal nanoparticles, and amorphous carbon [17,18]. Synthesized carbon onions are usually with very high graphitization and have relatively even diameters. Also, by strong van der Waals interactions, carbon onions are usually seen in agglomerations (Figure 24.1a).

24.2.1.2 Arc Discharge in Water

To simplify the experimental apparatus and enhance production efficiency for carbon nanomaterials, researchers have made a number of investigations on discharges in liquid media to produce carbon nanomaterials: e.g., (1) C_{60} from a spark discharge operating at 10–20 kV in toluene or benzene [29], (2) fullerene fragments via a glow discharge in $CHCl_3$ vapor [30], and (3) carbon nanotubes from a low-voltage contact auto-regulated arc discharge in aromatic hydrocarbons. For the water media,

Hsin et al. [31] first reported solution-phase carbon nanotube production in water. Recently, high-quality carbon onions with C_{60} cores were also discovered in floating materials formed by arc discharge in water [32]. These research findings led to a rapid expansion in exploration of the process of arc discharge in water in view of the simplicity of the experimental apparatus [3,33]. But the equipment for the arc discharge in water method is entirely based on a gas phase arc discharge system previously used for fullerene and carbon nanotube production [34], and only has some modifications (Figure 24.2b). The apparatus consists of two graphite electrodes submerged to a depth of 3–8 cm in water. The electrodes are aligned horizontally. The distance between the cathode and the anode is maintained at 0.7–1 mm. The explored range of arc current is between 30 and 70 A, with a voltage drop of 21–28 V. During the discharge process, both the anode and water are consumed continuously due to the intensive heat as the temperature around the discharge spot is estimated to be 4000 K. The discharge duration and the erosion rates of the doped electrode are determined by the electrode feeding distance and the composition of the doped electrode, respectively. If the discharge lasts about half an hour, the arc discharge will be broken artificially to avoid overheating the water. After the discharge process, carbon onions floating on the water surface (Figure 24.3a and b)

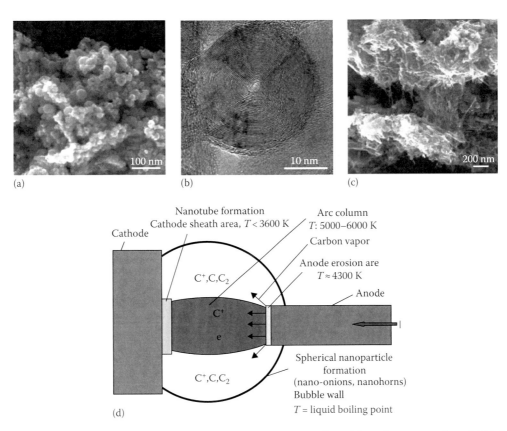

FIGURE 24.3 (a) Scanning electron microscope (SEM) image of carbon onions collected from the water surface. (b) High-resolution TEM (HRTEM) image of the synthesized carbon onions collected from the water surface. (c) Typical SEM image of the deposits collected from the base of water. (From Guo, J.J. et al., *Mater. Chem. Phys.*, 105, 175, 2007. With permission.) (d) Schematic drawing of the physical model of arc discharge in liquid. (From Xing, G. et al., *New Carbon Mater.*, 22(4), 337, 2007. With permission.)

and bulky deposits containing carbon nanotubes and amorphous carbon (Figure 24.3c) sunk down to the reactor bottom are obtained. The floatation of carbon onions should more likely be ascribed to their hydrophobic surface [24,34].

Compared with the arc discharge in an inert atmosphere, the arc discharge in water technology simplifies the traditional arc discharge for carbon nanomaterial production: no special equipment, vacuum, or gas are needed. Furthermore, this technique facilitates the collection of products from the water surface, rather than from the whole of the dusty reaction chamber.

Xing et al. have proposed a tentative physical model to describe the formation of carbon nanomaterials by arc discharge in liquid (Figure 24.3d) [35]. It was assumed that the reaction between the carbon and the liquid does not participate in the formation of carbon nanomaterials. The main difference between the arc discharge in liquid and the arc discharge in gas was that the liquid provided a cooling effect for electrodes and an ideal quenching wall (gas bubble wall) for the nucleation and growth of nanomaterials. The high temperature gradient and the high carbon species density near the bubble wall was attributed to the small bubble size. As seen in Figure 24.3d, the carbon ions and atoms ablated from the anode and the arc column integrated to small fragments as the temperature decreased when they moved toward the cold bubble wall. The cooling sheath near the bubble wall provided a sufficient temperature gradient, which was an important parameter for carbon onions and related structures to nucleate and grow. The temperature of liquid nitrogen was considerably lower than that of water. Hence, the temperature gradient in the cooling sheath near the bubble wall of liquid nitrogen was higher than that in water. When carbon fragments moved into the cooling sheath of liquid nitrogen, they tended to bend sharply, owing to the strong quenching effect, and formed single-walled graphite sheets, which finally agglomerated to carbon nanohorns. However, in water, the fragments can grow to perfectly spherical carbon onions owing to the weaker quenching effect. The carbon species flew from anode to cathode and formed nanotubes owing to their directional movement.

24.2.2 High-Energy Irradiation

24.2.2.1 High-Energy Electron Beam Irradiation

Carbon nanostructures have shown unexpected ordering under irradiation with electrons or ion beams. One of the most fascinating phenomena is the formation of perfectly spherical graphitic onion structures. These structures were discovered by Ijima in 1980 [8], but were left unnoticed until the discovery of the fullerenes. In 1992, Ugarte [10] reported a technique to produce remarkable rounded onions using electron beams. In this method, polyhedral carbon particles from the cathode of an arc discharge experiment, amorphous carbons, or holey carbon films with catalytic nanoparticles (such as Pt/Al, Au, or Au/Pd nanoparticles) are used as starting materials and intense electron beam under a beam current of about 100 ~ 180 A/cm^2, which is often generated by transmission electron microscope and is much larger than that of ordinary observation by electron microscopes [36,37], is used to irradiate the above starting materials for periods lasting from a few seconds to several minutes, and then carbon onions can be obtained. The produced carbon onions are often presented as well graphitized and uniform sized (Figure 24.4) [19,36–38].

The electron beam intensity and the irradiation time are determined by the starting materials. Generally, when the initial carbon targets are polyhedral carbon particles produced by arc discharge, the fullerene onions can be produced by electron irradiation of quite rapidly [39]. However, in the case of direct irradiation of holey carbon films targets, higher dose (~180 A/cm^2), higher voltage (400 keV), and larger irradiation times are required [40]. In this case, the observed onion structures are only produced in the edges of holes in the carbon film. In addition, Xu and Tanaka [41,42] have reported the transformation of an amorphous carbon film into carbon onions by electron irradiation in the presence of Pt and Al nanoparticles. Jose Yacaman et al. [43] also have found a condition in which carbon onions are produced by Au nanoparticles at an explosive rate and use less extreme irradiation conditions than the ones used in the absence of Au nanoparticles (Figure 24.4). These effects are attributed to the catalytic effect of the nanoparticles.

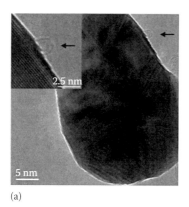 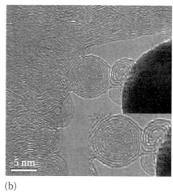 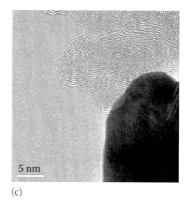

(a)   (b)   (c)

FIGURE 24.4 (a) The particle after strong irradiation at 180 A/cm^2 for several minutes. The shape has been altered significantly. The inset shows the details of the $C_{60}@C_{240}$ structure formed on the steps (indicated by an arrows); Formation of fullerene onions on a gold nanoparticle. In (b), the inset shows the details of the interface between two fullerenes and between a fullerene and the particle. (c) Shows an image corresponding to the final state where a large number of fullerenes have been produced. (From Troiani, H.E. et al., *Chem. Mater.*, 15(5), 1029, 2003. With permission.)

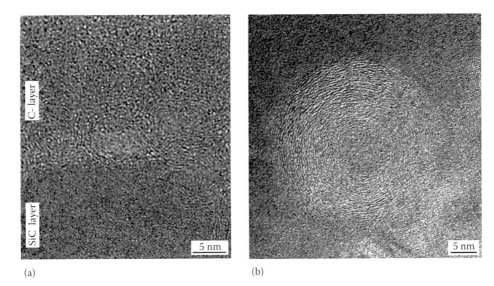

FIGURE 24.5 Two types of onion-like carbon clusters observed in SiC layers after laser crystallization: (a) small concentric structures formed inside the top carbonaceous layer and (b) large onions created inside the crystalline SiC matrix. (From Gorelik, T. et al., *Chem. Phys. Lett.*, 373, 642, 2003. With permission.)

Troiani et al. have developed a growth mechanism [43], which can be summarized as follows: a knock-on collision between electrons and carbon atoms leads to diffusion of atoms on the carbon film. As shown by Banhart [40], a knock-on displacements is the dominant mechanism in radiation damage in carbon films. A further displacement effect is the radiation-induced diffusion in which electron impacts can lead to diffusion of atoms, even at energies below the threshold displacement. Carbon atoms reach the particle, and surface diffusion takes place until an embryo is formed, which subsequently grows. Figure 24.4a, in which the formation of a $C_{60}@C_{240}$ ring in a surface step is shown, suggests an important reduction in activation energy for the formation of polyene chains around the kinks. As the irradiation continues, a threshold for particle fluctuation occurs and the surface has rapid reordering in which many kinks appear and disappear. This seems to trigger a dramatic growth of irregularly shaped fullerenes, which eventually become onions.

24.2.2.2 Laser Irradiation

Ute Kaiser and coworkers [20] reported on the formation of carbon onion-like clusters inside a crystalline SiC matrix due to extremely short-time heating of amorphous SiC layers induced by a single laser shot. In their process, amorphous stoichiometric SiC films with a thickness in the range of 100–200 nm produced either by pulsed laser deposition (PLD) or by ion implantation are irradiated with a single shot of KrF laser. PLD-generated amorphous films are produced on 6H–SiC substrates by laser ablation from a sintered SiC target using a KrF excimer laser (248 nm wavelength, 25 ns pulse duration, 10 Hz repetition rate, 2 J/cm² fluence at the target). The deposition rate is 0.3 nm/s. During the deposition, the substrate is heated to 400°C. Alternatively, amorphous SiC layers are prepared by room-temperature ion implantation (Ge ions) into 6H–SiC with an energy of 250 keV and a dose of 1.5×10^{16} Ge/cm². Laser modification of amorphous SiC layers produced

by the two methods mentioned is performed at a substrate temperature of 600°C with a single shot of a KrF laser (pulse duration 25 ns, fluence 800 mJ/cm²) at a wavelength of 248 nm.

Onion-like carbon clusters are observed both inside the top carbonaceous layer and inside the polycrystalline cubic SiC region. Figure 24.5a and b shows HRTEM images of carbon onions formed inside the top carbonaceous layer (marked as C-layer) and within the crystalline SiC, respectively. It can be seen that the produced carbon onions have a size in the range of 5–30 nm and the shells are defective with interplanar distances about 0.37 nm, which is slightly larger than 0.35 nm usually observed for carbon onions [21].

From numerical solutions of the heat-conducting equation including evaporation cooling, Kaiser et al. [20] estimated that during the laser shot the SiC surface is heated to about 4000 K, which explains the homogeneous stoichiometric melt formation. As in the temperature range of 2150–2316 K, the equilibrium vapor pressure of Si atoms is by a factor of 10 higher than that of carbon-containing molecules such as Si_2C or SiC_2, it can also be expected that at higher temperature, silicon preferably evaporates from SiC leaving carbon behind. Thus, a steep concentration gradient is formed with the carbon content continuously increasing toward the surface, resulting in the formation of the carbonaceous layer at the surface. High temperatures in this region probably make the graphite structures too close; thus, the onion formation mechanism there could be an analog to simple thermal treatment of carbon soot.

24.2.3 Implantation of Carbon Ions onto Metal Particles

Ion implantation is a material engineering process by which ions of a material can be implanted into another solid, thereby changing the physical properties of the solid. Ion implantation

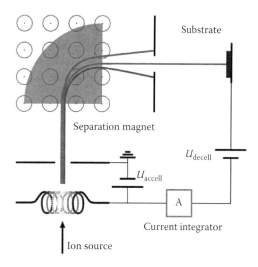

Substrate

Separation magnet

U_{decell}

U_{accell}

A

Current integrator

Ion source

FIGURE 24.6 Schematic diagram of a commercial ion-implantation system. (From Wikipedia, the free encyclopedia, Web site: http://en.wikipedia.org/wiki/Ion_implantation.)

is used in semiconductor device fabrication and in metal finishing, as well as various applications in materials science research. The ions introduce both a chemical change in the target, in that they can be a different element than the target, and a structural change, in that the crystal structure of the target can be damaged or even destroyed.

Ion implantation equipment typically consists of an ion source (Figure 24.6) [44], where ions of the desired element are produced, an accelerator, where the ions are electrostatically accelerated to a high energy, and a target chamber, where the ions impinge on a target, which is the material to be implanted. Each ion is typically a single atom, and thus the actual amount of material implanted in the target is the integral over time of the ion current. This amount is called the dose. The currents supplied by implanters are typically small (microamperes), and thus the dose which can be implanted in a reasonable amount of time is small. Thus, ion implantation finds application in cases where the amount of chemical change required is small.

Typical ion energies are in the range of 10–500 keV (1,600–80,000 aJ). Energies in the range 1–10 keV (160–1600 aJ) can be used, but result in a penetration of only a few nanometers or less. Energies lower than this result in very little damage to the target and fall under the designation ion beam deposition. Higher energies can also be used: accelerators capable of 5 MeV (800,000 aJ) are common. However, there is often great structural damage to the target, and because the depth distribution is broad, the net composition change at any point in the target will be small.

The energy of the ions, as well as the ion species and the composition of the target determine the depth of penetration of the ions in the solid: A monoenergetic ion beam will generally have a broad depth distribution. The average penetration depth is called the range of the ions. Under typical circumstances, ion ranges will be between 10 nm and 1 μm. Thus, ion implantation is especially useful in cases where the chemical or structural change is desired to be near the surface of the target. Ions gradually lose

their energy as they travel through the solid, both from occasional collisions with target atoms (which cause abrupt energy transfers) and from a mild drag from overlap of electron orbitals, which is a continuous process. The loss of ion energy in the target is called stopping [44].

Carbon onions can be synthesized by carbon ion-implantations into pure silver [45] or copper [23] targets at elevated temperature. In this technique, the size of the carbon onions can be adjusted by varying the implantation parameters (temperature, fluence). For a typical experiment, the polycrystalline silver metal substrates are fixed onto a furnace mounted in an implantation chamber. The substrate temperature, the implanted fluences, and the ion energy are, respectively, varied in the range 500°C–600°C, 1–30 × 10^{16} ions cm^{-2}, and 90–120 keV. All the implantations are performed in a cryogenic residual pressure of less than 10^{-5} Pa [23,45–47].

After implantation, a mass of well-spherical onions are observed in the bulk of the samples. In fact, the main interesting observations are obtained for lower TEM magnifications, as shown in Figure 24.7a, where one observes numerous spherical bright contrasts (carbon onions) whose size varies progressively with the distance from the free surface. The large number of carbon onions observed on this micrograph allows to obtain the size distribution of the carbon onions versus the depth of the silver substrate (Figure 24.7b). Largest onions are found at a depth close to a value that corresponds to the average projected range (R_p) of the implanted carbon atoms deduced from the well-known TRIM simulation. For instance, values of R_p close to 110 and 140 nm are obtained for 90 and 120 keV carbon ion-implantations, respectively [47].

24.2.4 Thermal Treatment of Carbonaceous Materials

24.2.4.1 High-Temperature Nanodiamond Annealing

Using the high-temperature annealing of commercially available nanocrystalline diamonds, a larger number of carbon onions with uniform size can be prepared [21,48–51]. In this process, several nanocrystalline diamonds are put onto a quartz boat and resistively heated in vacuum or under inert gas atmosphere. Annealing is carried out at different temperatures ranging from 1400°C to 2000°C for a period of time. Then, the spherical carbon onions are formed. The diameter of the obtained onions is almost the same as that of the initial diamond nanoparticles because one diamond nanoparticle is transformed into one onion [21,48–51].

In authors' experiments [52], annealing of diamond nanoparticles (2–10 nm) was performed in a tube furnace in argon atmosphere at temperature from 900°C to 1400°C for 1 h. Figure 24.8 shows the XRD patterns of nanodiamond annealed at different temperature from 900°C to 1400°C for 1 h, which indicates that onion-like carbons begin to form in the range of 1100°C–1200°C and graphitization of nanodiamond finishes after annealing at 1400°C for 1 h. Figure 24.9a shows an HRTEM image of initial

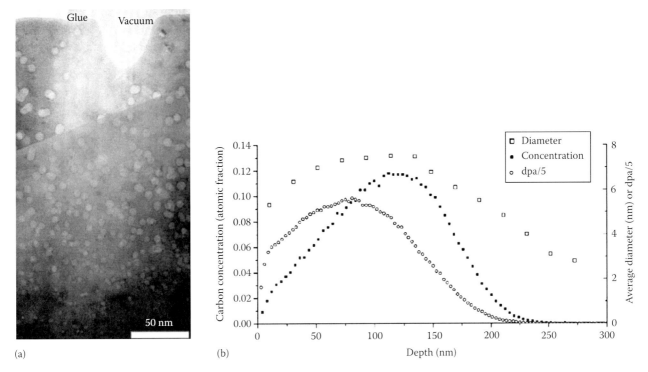

(a) (b)

FIGURE 24.7 (a) Cross-sectional TEM micrograph of carbon onions (bright contrasts) inside the silver substrate obtained with a 90-keV carbon ion-implantation at 500°C (fluence: 10^{17} ions cm^{-2}) and (b) evolution of the average diameter of the carbon onions with the distance from the silver surface (ion energy: 90 keV, temperature: 500°C, fluence: 10^{17} ions cm^{-2}). Simulated concentration profile and displacement of carbon atoms (divided by a factor 5 for clarity) deduced from a TRIM simulation are also shown. (From Thune, E. et al., *Mater. Lett.*, 54, 222, 2002. With permission.)

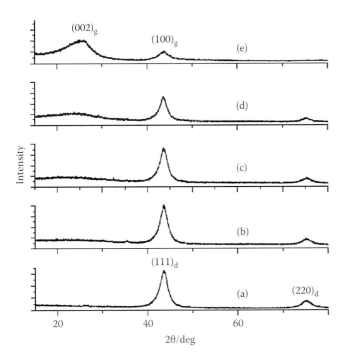

FIGURE 24.8 XRD patterns of diamond nanoparticles annealed at different temperatures: (a) diamond nanoparticles, (b) 900°C, (c) 1100°C, (d) 1200°C, and (e) 1400°C for 1 h. (From Qiao, Z.J. et al., *Scr. Mater.*, 54, 225, 2006. With permission.)

diamond nanoparticles 2–10 nm in size. The outlines of the particles are clear and the particles take on diverse shapes. Figure 24.9b through d presents the HRTEM images of nanodiamond annealed at 1100°C, 1200°C, and 1400°C, respectively. As seen in Figure 24.9, the graphite fragments increase with increasing annealing temperature and gradually connect up into curved graphitic sheets, close to diameters of C_{60} and higher fullerenes, which is in agreement with the above XRD results. The presence of exterior-curved graphitic sheets makes the authors believe that graphitization of nanodiamonds begins from the particle surface toward the center, and the graphitization temperature of nanodiamond changes with the degree of crystallinity, in correspondence to particle size.

It should be noted that there exist diamond (111) sheets in the center of the elliptical onion-like carbon (marked by arrow in Figure 24.9c) and the axis of the elliptical onion-like carbon is parallel to the original diamond (111) planes. The authors believe that transformation preferentially begins at (111) planes and the inner diamond has an effect on the outer shells in shape. It is reasonable that the onion-like carbons are similar to the original nanoparticles in shape; therefore, the authors propose that the graphite fragments exfoliating from diamond (111) planes enclose around the surface of nanoparticles gradually and the onion-like carbons develop simultaneously. Figure 24.9d shows an HRTEM image of diamond nanoparticles annealed at 1400°C. All the nanoparticles are transformed at 1400°C and onion-like carbons take diverse shapes, such as quasi-spherical, elliptical,

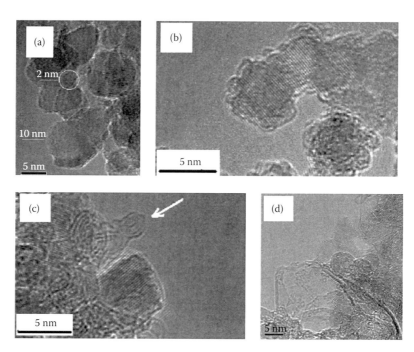

FIGURE 24.9 (a) HRTEM image of initial diamond nanoparticles, HRTEM images of nanodiamonds annealed at (b) 1100°C, (c) 1200°C, and (d) 1400°C. (From Qiao, Z.J. et al., *Scr. Mater.*, 54, 225, 2006. With permission.)

polyhedral, and deformed onions. Some onions are found to be enclosed with linked external layers.

Graphite is the most stable form of carbon at normal temperature and diamond tends to transform to graphite spontaneously at high temperature. Diamond nanoparticles possess more dangling bonds and higher surface energy than bulk diamond due to the high ratio of surface atoms to the total. Consequently, the graphitization temperature of nanodiamond is much lower than that of bulk diamond. The elimination of dangling bonds at the nanoparticle surface and the closure of graphite sheets result in the decrease of the surface energy, which is supposed to be the driving force to form closed graphite shells continuously [49]. It has been proved by many researchers that the graphitization preferentially begins at diamond {111} planes [50,51]. There are eight planes in {111} group which comprise (111), ($\bar{1}$11), ($\bar{1}$1$\bar{1}$), (11$\bar{1}$), ($\bar{1}$11), (1$\bar{1}$1), (11$\bar{1}$), and ($\bar{1}\bar{1}\bar{1}$) planes.

The diamond {111} planes are composed of zigzag hexagonal rings that can be easily rearranged into graphite sheet; therefore, graphite fragments with different numbers of carbon atoms exfoliate from the external surface of any diamond {111}

plane and surround diamond particles (shown in Figure 24.10a). These fragments are rearranged by introducing pentagonal and other polygonal rings to form a closed shell. Experimental results show that nanodiamond should be heated to at least 1100°C–1200°C for graphitization because enough kinetic energy is required to keep on breaking C–C bonds in inner diamond. But the graphite fragments can form at 900°C because of the greater number of defects and higher kinetic energy at the outmost surface than in the inner part. The formation process of onion-like carbons includes: formation of graphite fragments, connection and curvature of graphite sheets at the edges of diamond {111} planes, and closure of graphite layers. Based on the morphological information observed by HRTEM, the authors propose a model to explain the phenomenon that the onion-like carbon is similar to the original nanoparticles in shape. As shown in Figure 24.10b, the graphite fragments link and tangle around the surface of the diamond particle to eliminate dangling bonds, and then generate closed graphite shells to diminish surface energy. The inner diamond maintains the original shape and dwindles little by little in the course of

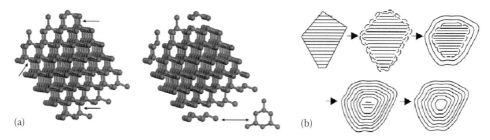

FIGURE 24.10 (a) Formation of graphite fragments by breaking C–C bonds between {111} planes and (b) schematic illustration of onion-like carbons formation process. (From Qiao, Z.J. et al., *Scr. Mater.*, 54, 225, 2006. With permission.)

transformation. Consequently, graphitic layers enclose around the diamond surface gradually and the onion-like carbons are similar to the original particles in shape.

24.2.4.2 Thermal Treatment of Carbon Soot

Using heat treatment of pure carbon soot, Ugarte and Heerde [53] have obtained a new material containing macroscopic quantities of hollow nanometric carbon onions with from 2 to about 8 graphitic shells, with outer diameters ranging from 3 to 10 nm. In addition, Selvan et al. [54] first synthesized large fullerenes in macroscopic quantities by arc evaporation of Ni-filled graphite rods followed by heating of the collected soot. The formed giant fullerene with faceted structures is molecular in nature and has high symmetry and sp^2-hybridized carbon atoms. Then they employed chemical treatment on the large fullerenes with HNO_3, and a large number of carbon onions were formed.

24.2.4.3 Thermal Treatment of Graphite or Carbon Black

Carbon onions can also be produced by thermal treatment or ultrasound treatment on the graphite and carbon black. Xu et al. [55,56] have done some work on the preparation of carbon onions by thermal treatment of carbon black under high temperature and vacuum. They found that the obtained carbon onions possess higher degree of graphization by means of Al catalytic nanoparticles. Wang et al. [57] reported the synthesis of onion-like carbon nanospheres through ultrasound treatment on highly oriented pyrolytic graphite (HOPG) material in deion water. In their experiments, the HOPG samples ($\sim 12 \times 5 \times 10 \, mm^3$) mixed with deionized water are put into a test tube hanged in the trough of an ultrasonic set (SCQ50, power density $\sim 0.22 \, W/cm^2$). After being treated by ultrasound at temperature of 310 K for about 84 h, the black solution is obtained. The obtained nanospheres are composed of disordered lattice lines of graphene sheets and their size ranges from about 10 to 200 nm. In addition, they proposed one possible two-stage model of nucleus formation and "rolling snowball" growth of carbon nanospheres. In this model, they speculated that the onion structures are formed by condensation of carbon species produced by ultrasound treatment. The condensation process may be proceeded in two stages: (1) nucleation of primary carbon species, and (2) their agglomeration growth caused by van der Waals and bond forces. Reasonably, ultrasound force can cause disorder and break of graphitic basal place, which often results in producing carbon fragments (including carbon atom, molecular and graphene, etc.). At the same time, ultrasound treatment of carbon fragments in water solution gives rise to some vortex-like regions where the agglomeration and interconnection of primary fragments will also possibly accelerate the process of nucleation. As ultrasonic treatment time goes on, other graphene fragments condense on to the nucleus and then grows into an onion structure gradually. Hence, onion size is controlled by the ultrasound treatment time. In compliance with this "rolling snowball" growth model, the existent possibility of a nanoscale carbon sphere with disordered concentric graphitic layer structure should be quite high.

24.2.4.4 Thermal Treatment of Carbon Nanotubes

Zhang et al. [58,59] reported the structural transformation of catalytically grown carbon nanotubes to quasi-spherical onions induced by annealing under high pressure. In their experiment, the cylindrical shape of the CNT sample is prepared under hydrostatic pressure, and then each sample (about 20 mg) is pressed into a 5-mm-diameter cylinder before it is put into the high-pressure cavity. The hydrostatic pressure of the Belt-type apparatus is created by a 320 t oil press machine. The samples are annealed under 5.5 GPa for 25 min or 4 h. Subsequently, the temperatures are quenched to room temperature in about 1 min followed by a gradual pressure release over 1 h.

The transformation from carbon nanotube into quasi-spherical onion, which is caused by high pressure, can be divided into three main steps: (1) the collapse of the carbon nanotube, (2) the formation of spheres, (3) and the subsequent formation of quasi-spherical onion-like structure. High pressure takes an important role in the microstructural changes of the carbon nanotube. Compared with graphite, the changes of C–C bond in the carbon nanotube may be much greater and may lead to the collapse of the nanotube during the annealing process [58,59].

24.3 Chemical Vapor Deposition

Although the above four methods have been the primary methods for producing high-quality carbon onions, they have drawbacks. First of all, above methods are costly due to their high-temperature processes and the limited number of carbon onions that can be produced. Second, above methods grow carbon onions in highly aggregation forms with unwanted carbon impurities such as carbon nanotubes or amorphous carbon. Also, controlling carbon onions in continuous growth is impossible using the above four methods. With these considerations in mind, CVD is becoming a strong candidate for the large-scale production of carbon onions.

CVD synthesis is achieved by putting a carbon source in the gas phase and using an energy source, such as a plasma or a resistively heated coil, to transfer energy to a gaseous carbon molecule. Commonly used gaseous carbon sources include methane, acetylene, CO, CO_2, coal, glycerin, phenolic resin, etc. [60–68]. The energy source is used to "crack" the molecule into reactive atomic carbon. Then, the carbon diffuses toward the substrate, which is heated and coated with a catalyst or catalyst precursor (usually a first row transition metal such as Ni, Fe, or Co) where it will bind. Carbon onions will be formed if the proper parameters are maintained.

Generally, CVD carbon onion synthesis is a two-step process consisting of a catalyst preparation step followed by the actual synthesis of the onion. The catalyst is generally prepared by precipitation-deposition route, sol-gel method, or sputtering a transition metal onto a substrate and then using either chemical etching or thermal annealing to induce catalyst particle nucleation. Thermal annealing results in cluster formation on the substrate, from which the carbon onions will grow. In addition, some researchers also directly employed catalyst or catalyst precursor ($Fe(NO_3)_3$, ferrocene, etc.) to mix with carbon sources to synthesize carbon onions [60–68]. The temperatures for the synthesis of onions by CVD are generally within the 550°C–1200°C range [22,69–72].

These are the basic principles of the CVD process. In the last decennia, different techniques for the carbon onions synthesis with CVD have been developed, such as plasma-enhanced CVD, catalytic disproportionation of CO, catalytic disproportionation of CO, catalytic reduction of CO_2, vapor phase growth, and thermal CVD. These different techniques will be explained in more detail in this section.

24.3.1 Plasma-Enhanced Chemical Vapor Deposition

The plasma-enhanced CVD method generates a glow discharge in a chamber or a reaction furnace by a high-frequency voltage applied to both electrodes. Figure 24.11 shows a schematic diagram of a typical plasma CVD apparatus with a parallel plate electrode structure.

A substrate is placed on the grounded electrode. In order to form a uniform film, the reaction gas is supplied from the opposite plate. Catalytic metals, such as Fe, Ni, and Co are used on, for example, a silica, Si, SiO_2, or glass substrate using thermal CVD or sputtering. After nanoscopic fine metal particles are formed, carbon onions will be grown on the metal particles on the substrate by glow discharge generated from high-frequency power. A carbon-containing reaction gas is supplied to the chamber during the discharge.

Large quantities of carbon onions with high purity, which have an average diameter ranging from a few to several tens of nanometers, have been synthesized by this method (Figure 24.12). The produced onions are solid, clean, and can be separated easily from the catalytic particles.

Their formation mechanism can be explained as follows. Initially, the hydrocarbon is adsorbed at the surfaces of the metal catalyst particles. In the meantime, the process of glow discharge leads to hydrocarbon decomposition, resulting in the formation of carbon rings, which tend to nucleate on the surfaces of the catalyst. In addition, in the presence of atomic hydrogen in the plasma,

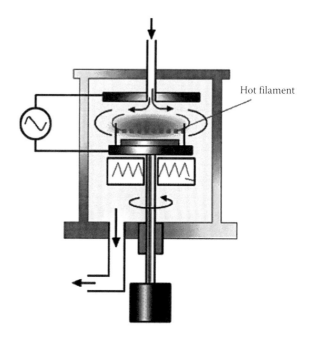

FIGURE 24.11 Schematic diagram of plasma CVD apparatus.

free carbon valences on the ends of carbon layers can be stabilized by the formation of C–H bonds. The formation of onions is based on the formation of many cages in successive stages from the core to the surface. Around the edge of a carbon onion, discontinuous curved lines are shown in the high-resolution TEM image, reflecting the way behavior of carbon onion (Figure 24.13).

24.3.2 Catalytic Disproportionation of CO

The CO catalytic disproportionation process is a technique for catalytic production of carbon onions in a continuous-flow gas phase using CO as the carbon feedstock and copper acetylacetonates as the catalyst precursor. Carbon onions are produced by flowing CO mixed with a small amount of copper acetylacetonates through a heated reactor [69].

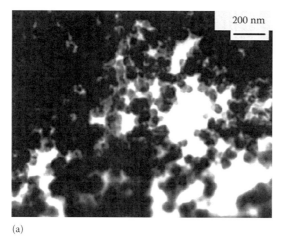

(a)   (b)

FIGURE 24.12 (a) A TEM image of synthesized carbon onions showing the monodispersed size distribution. The product contain no carbon tubes, only carbon onions. (b) HRTEM image of a typical onion formed on a catalyst. The carbon onion contains several cores in its center. (From Chen, X.H. et al., *Chem. Phys. Lett.*, 336, 201, 2001. With permission.)

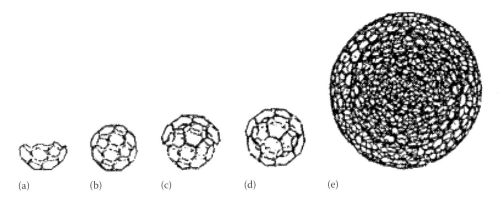

FIGURE 24.13 Successive stages in the formation of onions: (a) the nucleus is a dome formed by one cap of a C_{60} or C_{70} molecule, (b) the nucleus grows into a cage by the addition of hexagons and pentagons, (c) a second cage is nucleated on the first cage as substrate, (d) a new secondary surface grows into the second cage and encapsulates the first cage, and (e) a large spherical particle containing many closed-shell cages. (From Chen, X.H. et al., *Chem. Phys. Lett.*, 336, 201, 2001. With permission.)

For a typical process, the experimental investigations are carried out using a vertical laminar flow reactor (Figure 24.14a) [69,73,74]. Briefly, the device consists of a saturator, a laminator, and a furnace. A flow of pure filtered nitrogen (or carbon monoxide) carrier gas is supplied to the saturator impregnated with a metal acetylacetonate and silicon dioxide powder mixture. As it passes through the heated precursor, the gas is saturated by the metal–organic vapor. Carbon monoxide is typically introduced into the system from the side and separated from the precursor vapor flow by a 1 mm thick stainless tube. Inside the laminator,

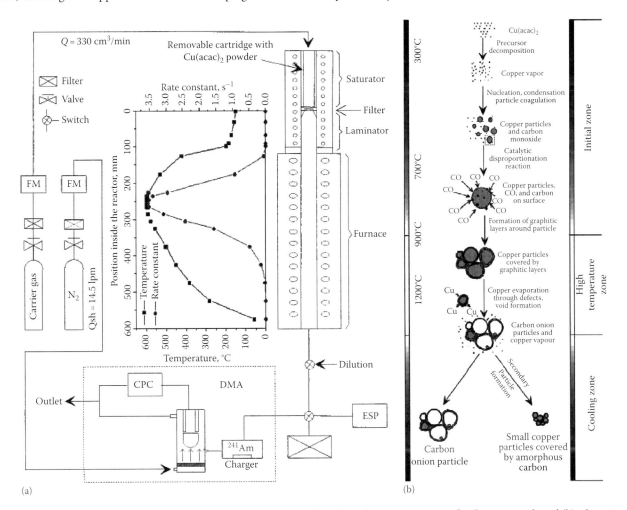

FIGURE 24.14 (a) Schematic presentation of the experimental setup of catalytic disproportionation of carbon monoxide and (b) schematic representation of the mechanisms of carbon onion production. (From Albert, G. et al., *Carbon*, 41, 2711, 2003. With permission.)

two separate steady-state laminar flows containing metal–organic precursor vapor and CO are established. Inside the furnace, the flows are mixed and subsequently heated to a temperature above which precursor decomposition occurs. Two main processes taking place in the heated furnace can be distinguished. First, metal particle formation occurs as a result of precursor decomposition and the subsequently created supersaturated metal vapor in the system. The second process is the formation of the carbon onions following catalytic disproportionation of carbon monoxide on the surface of the particles (Figure 24.14b).

The number of synthesized carbon onions increases with an increase in the residence time and the concentration of carbon monoxide in the system. Experiments with pure CO at t_{furn} = 1216°C produce carbon onions and copper nanoparticles (Figure 24.15). The obtained carbon onions consist of several concentric carbon layers surrounding either a copper particle or a hollow core. Also, copper particles completely covered by a graphitic shell are found among the products. The primary size of the copper particles ranges from about 5 to 30 nm, while carbon particles are from 5 to 30 nm in diameter and thoroughly agglomerated.

24.3.3 Catalytic Reduction of CO_2

Crestani et al. [75] have developed a method of catalytic reduction of CO_2 to synthesize carbon onions. In their study, they investigated the reaction of CO_2 gas with thiaplatinacycles, such as $[Pt(\eta^2\text{-}C,S\text{-}C_{12}H_8)(PEt_3)_2]$, in a stainless-steel autoclave (300 mL), charged also with hydrogen and pressurized with argon up to 1380 psi at 40°C (approx. 94 atm) to ensure the CO_2 being in a supercritical state (critical point, 31°C, 73 atm). The main product identified is carbon, particularly in the form of carbon nanomaterials. The chemical reaction that occurs is proposed as

$$CO_2 + 2H_2 \rightarrow C + 2H_2O \qquad (24.1)$$

The HRTEM images depicted in Figure 24.16 show the obtained product. It can be found that the main products (80%) are the well-known carbon onions, Figure 24.16a. Also, a few closely related particles with two spiral cores are observed, Figure 24.16b, along with some tilted layers of carbon, characteristic of the ones present in multilayer CNT, Figure 24.16c.

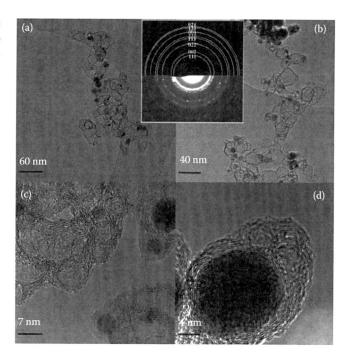

FIGURE 24.15 TEM images of the product synthesized at 1216°C by Cu(acac)$_2$ decomposition. (From Albert, G., et al., *Carbon*, 41, 2711, 2003. With permission.)

24.3.4 Vapor Phase Growth

Vapor phase growth is a synthesis method of carbon onions, directly supplying reaction gas and catalytic metal in the chamber without a substrate [76]. Figure 24.17 shows a schematic diagram of a vapor phase growth apparatus [77]. Two furnaces are placed in the reaction chamber. Ferrocene or ferrocene-Fe$_3$(CO)$_{12}$ is used as catalyst precursor [76,78]. In the first furnace, vaporization of catalytic carbon is maintained at a relatively low temperature. Fine catalytic particles are formed and when they reach the second furnace, decomposed carbons are absorbed and diffused to the catalytic metal particles. Here, they are synthesized as carbon onions. The particle sizes of the obtained carbon onions by using vapor phase growth are in the 11–30 nm size range with 2–40 graphitic shells [76,78].

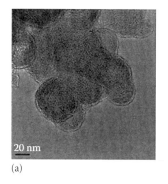 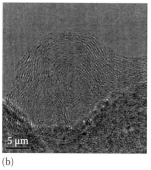 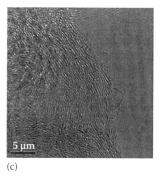

(a)   (b)   (c)

FIGURE 24.16 HRTEM images of the obtained product. (a) Nanoonions, (b) twin spiral cores, and (c) tilted multilayers. (From Marco, G.C. et al., *Carbon*, 43, 2621, 2005. With permission.)

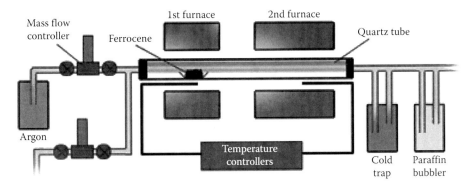

FIGURE 24.17 Schematic diagram of a vapor phase growth apparatus. (From Lee, C.J. et al., *Chem. Phys. Lett.*, 359, 109, 2002. With permission.)

24.3.5 Thermal Chemical Vapor Deposition

In this method, Fe, Ni, Co, or an alloy of the three catalytic metals is initially deposited on a catalyst supporter. Then the specimen is placed in a quartz boat. The boat is positioned in a CVD reaction furnace, and nanometre-sized catalytic metal particles are formed after an additional etching of the catalytic metal using H_2 gas at a temperature of 400°C–1000°C. As carbon onions are grown on these fine catalytic metal particles in CVD synthesis, forming these fine catalytic metal particles is the most important process. Figure 24.18 shows a schematic diagram of thermal CVD apparatus in the synthesis of carbon onions [79].

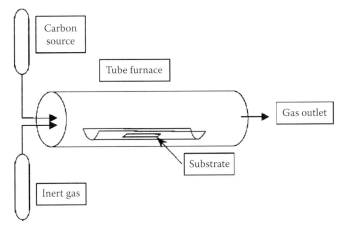

FIGURE 24.18 Schematic diagram of a typical thermal CVD apparatus. (From Öncel, C. and Yürüm, Y.D., *Fullerenes Nanotubes Carbon Nanostruct.*, 14, 17, 2006. With permission.)

24.4 Examples of Carbon Onion Preparation by Chemical Vapor Deposition

24.4.1 Low-Temperature Synthesis of Carbon Onions over Ni/Al Catalyst

In authors' experiment [61,62], carbon onions were produced by the CVD method over Ni nanoparticles supported on aluminum powder. The CVD growth is carried out at an optimized temperature of about 600°C for 60 min in a conventional tube furnace. The schematic of the experimental setup is depicted in Figure 24.19.

24.4.1.1 Preparation Process of Carbon Onions

The aluminum-supported nickel composite catalyst with a mole ratio of 1:1.84 is prepared by means of a homogeneous deposition–precipitation process. The proper amounts of $Ni(NO_3)_2 \cdot 6H_2O$ and 400 mesh aluminum powder screened are mixed in 1 L distilled water to yield the mole ratio Ni/Al = 1.84/1. Given NaOH by theory measurement is dissolved in 50 mL distilled water and added to the previous mixture with constant stirring. The green co-precipitate as-obtained is then aged at room temperature for 3 h without stirring, and the binary colloid ($Ni(OH)_2$/Al) is attained. The colloid is then washed several times with distilled water until neutral pH is reached and then dried in a vacuum chamber at 120°C for several hours. At last the dry colloid grinded with agate mortar is calcined in N_2 atmosphere at

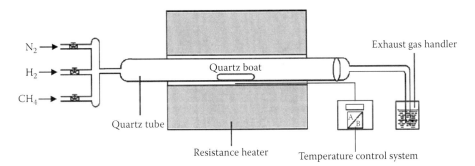

FIGURE 24.19 The schematic of the thermal CVD experimental setup. (From He, C.N. et al., *Scr. Mater.*, 54, 689, 2006. With permission.)

400°C for 4 h to form a fine NiO/Al composite powder, which is employed in the subsequent carbon onion growth.

The carbon onions are synthesized by the catalytic decomposition of CH_4 on the prepared Ni/Al catalyst. At first, 800 mg catalyst of the NiO/Al is kept in a quartz boat and placed in a horizontal quartz tube reactor. In order to convert the nickel oxides in the catalyst to elementary Ni, the catalyst is deoxidized at 600°C in a hydrogen (99.99% purity) atmosphere for 2 h. Then, carbon onions are grown at 600°C for 1 h by introducing a mixture of CH_4/N_2 (99.99% purity/99.9% purity) into the reactor at a flow rate of 60/540 mL/min. Then the gas flow is switched back to nitrogen after the growth process to purge the methane from the tube and prevent the back flow of air into the tube. The furnace is then allowed to cool to room temperature before exposing the carbon onions to the air.

24.4.1.2 Structural Analysis of Carbon Onions

Figure 24.20 shows the XRD pattern of the carbon onion sample. The peak at $2\theta = 26.11°$ can be assigned to graphite (002). The peak $2\theta = 44.44°$ can be assigned to both graphite (100) and nickel, and other peaks can be safely indexed to peak positions of the face-centered cubic structure of Ni and Al (the Al peaks are not indicated here due to their weak intensity).

Figure 24.21 shows typical TEM images of the as-prepared sample. It is seen that the carbon onions (5 ~ 50 nm) consisting of several concentric carbon layers surround either a hollow core (Figure 24.22a) or a nickel particle (Figure 24.22b through d). The approximate ratio of Ni-filled to empty onions is 1:3. Furthermore, the diameter of the hollow core is less than 5 nm and not very regular. Figure 24.22c shows a nickel particle encapsulated in the carbon onion presenting a face-centered cubic structure. The spacing between crystal faces is 0.2022 nm, which is almost consistent with the $d_{(111)}$ (0.2034 nm) of nickel, and a layer fault (as marked with hollow arrowhead in Figure 24.22c) can be observed. Further examination of the carbon onions indicates that the shape of the carbon onions depends

on the shape of the nickel particles. As shown in Figure 24.22b and d, respectively, when the nickel particle has a quasi-spherical structure, the graphite layers roll around the spherical particle and form a similarly spherical carbon onion, and when the nickel particle is a polyhedron the result is a polyhedral carbon onion.

In addition, to compare the three different carbon onions, Digital Micrograph software and ED are employed to measure the intershell distance and to analyze crystalline phase, respectively. According to the measurements (Figure 24.22a, b, and d), for carbon onions without a nickel particle encapsulated, the intershell distance of 0.3456 nm is the closest to the d_{002} of graphite (0.34 nm). Therefore, it can be speculated that although the large intershell distance suggests a considerable reduction in intershell interaction when compared to perfect graphite, the crystal structure of this type carbon onion is the nearest to perfect among the three different carbon onions. And the carbon onion with a polyhedral structure possesses the largest intershell distance of 0.3565 nm, suggesting that this type of carbon onion has more defects than the other two species of carbon onions. The ED patterns (as can be seen in Figure 24.22a, b, and d) of part of each of the three different carbon onions show that all carbon onions obtained are of polycrystalline structure, which indicates that the carbon onions have good graphitization and crystallization.

The above two species of nanoparticles might be formed directly from the gas phase while the Ni particles acted as a catalyst, inducing curvature in graphite layers. When the graphite layers closed upon themselves, some Ni particles might be

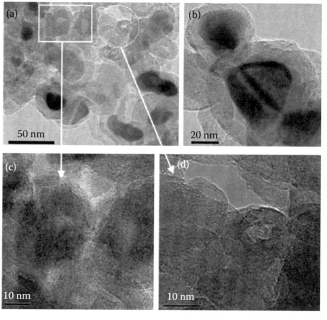

FIGURE 24.21 TEM images of the carbon onions produced by Ni/Al catalyst containing 80 wt.% Ni (a) TEM image of the carbon onions, (b) TEM image of the carbon onions with nickel particle encapsulated, and (c) typical HRTEM image of the selected carbon onions with a hollow core, (d) typical HRTEM image of the selected carbon onions with a hollow core. (From He, C.N. et al., *Scr. Mater.*, 54, 689, 2006. With permission.)

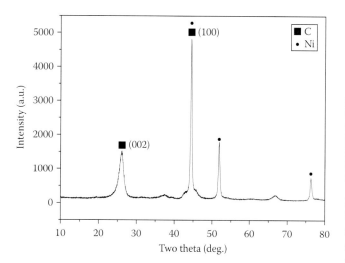

FIGURE 24.20 XRD pattern of the as-prepared carbon onions. (From He, C.N. et al., *Scr. Mater.*, 54, 689, 2006. With permission.)

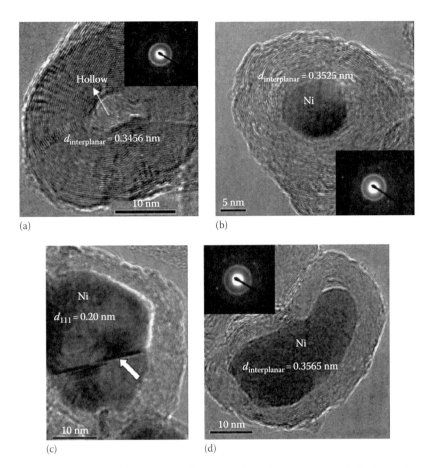

FIGURE 24.22 High-resolution TEM images of the as-prepared carbon onions. (a) TEM image of a typical carbon onion with a hollow core, (b) TEM image of a typical quasi-spherical carbon onion with a nickel particle, (c) and (d) TEM images of typical polyhedral carbon onions with a nickel particle. (From He, C.N. et al., *Scr. Mater.*, 54, 689, 2006. With permission.)

displaced before closure, while some others were trapped inside and could not escape (the irregular shape of the carbon onion core and the shape of the carbon onions depending on the shape of the nickel particle suggest such an explanation), then the two different kinds of nanoparticles could be obtained.

24.4.2 Carbon Onion Growth Enhanced by Nitrogen Incorporation

It is well known that the growth of CNTs by CVD using a transition metal catalyst is greatly enhanced in a nitrogen environment [80–83]. And since doping CNTs with different elements is a promising method to modify the electronic structure of the nanotubes, many researches have employed nitrogen-doped CNTs due to their easy fabrication and potentiality [80–83]. However, up to date few papers have referred to nitrogen incorporation in the carbon onion or the influence of reaction conditions on the carbon onion growth by CVD, which might be essential for understanding carbon onion growth via this method. Also, nitrogen-doped carbon onions may have advantages in some application fields over those carbon onions that lack nitrogen, so it is worthwhile investigating the carbon onions that contain nitrogen. As shown in Section 24.4.1, the authors successfully synthesized a mass of

carbon onions with the composition-controlled gaseous mixture of CH_4 and N_2 by CVD over a Ni/Al catalyst [61,62]. The authors also employed x-ray photoelectron spectroscopy (XPS) for characterization of the carbon onions, and the influence of reaction conditions on their growth has been investigated [63].

The preparation procedures of the NiO/Al catalyst and synthesis of carbon onions are similar to reference [61,62]. During synthesis of carbon onions, a mixture of CH_4/H_2 was employed to compare with N_2 carrier gas. Table 24.1 shows the reaction conditions used to fabricate various samples.

TABLE 24.1 Several Samples Fabricated Using Different Reaction Conditions

Samples	Reaction Gases (60/540 mL/min)	Reaction Time (h)
1#	CH_4/N_2	0.5
2#	CH_4/N_2	1
3#	CH_4/N_2	1.5
4#	CH_4/N_2	2
5#	CH_4/N_2	3
6#	CH_4/H_2	1

Source: He, C.N. et al., *Scr. Mater.*, 54, 1739, 2006. With permission.

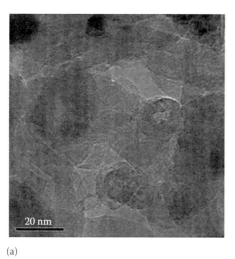
(a)

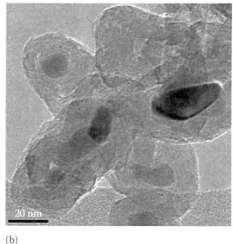
(b)

FIGURE 24.23 TEM images of the as-grown carbon onions from 2#. (a) TEM image of a mass of carbon onions with a hollow core, (b) TEM image of some carbon onions with a nickel particle. (From He, C.N. et al., *Scr. Mater.*, 54, 1739, 2006. With permission.)

Figure 24.23 shows the TEM images of the as-prepared carbon onions from sample 2#. It can be seen that the diameter of these carbon onions is in the range of 5–50 nm. The purity of the as-prepared carbon onions is ~90% due to the presence of some impurities of nanometer-size graphite block and amorphous carbon according to the TEM observations. When investigated in detail by high-magnification TEM, the authors observe from Figure 24.23b that the carbon onions consist of several concentric carbon layers surrounding either a hollow core or a nickel particle. And the approximate ratio of Ni-filled to empty onions is 1:3 based on the TEM observations. Figure 24.24 shows an HRTEM image of a typical carbon onion from 2# with a hollow core. This carbon onion is regular and has good graphitization—the interlayer spacing between graphitic sheets is 0.342 nm, which is very close to that of the ideal graphite. Furthermore, the diameter of the onion core is about 4 nm, much smaller than that of the catalytic particles; this indicates that the catalytic particle may be displaced from the carbon onion before the carbon onion has been obtained [61,62].

It is noted that the carrier gas plays an important role in synthesizing the carbon onions. Synthesis of carbon onions was attempted using hydrogen as the carrier gas at the same conditions as the experiment for 2#. Figure 24.25 shows TEM images of the obtained product (6#) using hydrogen as the carrier gas. From Figure 24.25a, the authors observe that the catalytic particles seem to be encapsulated by some materials. However, when the product was investigated by HRTEM (Figure 24.25b), the catalytic particles are seen to be encapsulated by some amorphous carbon, which is very different from the experimental results using nitrogen as the carrier gas.

To explain the difference between the product prepared using nitrogen as the carrier gas and the product prepared using hydrogen as the carrier gas, element analysis for the as-grown product samples was conducted by XPS. Figure 24.26 summarizes the nitrogen content and N/C atomic ratio in the deposited carbon onion samples (from 1# to 5#) fabricated using nitrogen as the carrier gas, as measured by XPS. The nitrogen concentration in the carbon onions is shown to increase from 0.5 to 1.5 at.% as the reaction time was raised from 0.5 to 3 h and then results in a steady increase of N/C atomic ratio from 0.05 to 0.1. In the amorphous carbons deposited using hydrogen as the carrier gas, nitrogen was not detected by XPS. This compositional variation

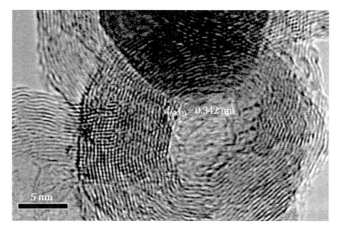

FIGURE 24.24 HRTEM image of a typical hollow carbon onion from 2#. (From He, C.N. et al., *Scr. Mater.*, 54, 1739, 2006. With permission.)

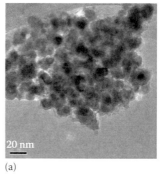
(a)

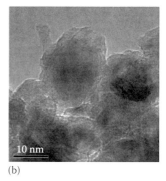
(b)

FIGURE 24.25 (a) TEM (b) HRTEM image of the obtained product (6#) using hydrogen as the carrier gas. (From He, C.N. et al., *Scr. Mater.*, 54, 1739, 2006. With permission.)

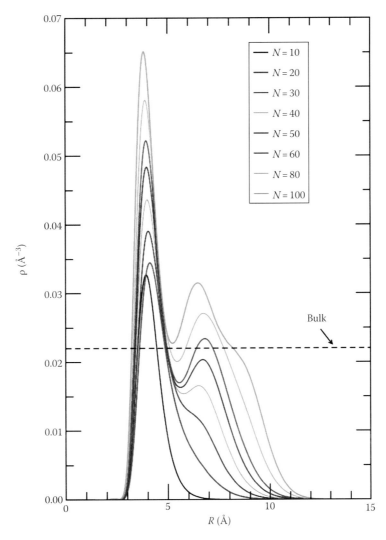

FIGURE 12.6 ⁴He radial density distributions for selected CO(⁴He)$_N$ clusters. (Adapted from Paesani, F. and Gianturco, F.A., *J. Chem. Phys.*, 116, 10170, 2002.)

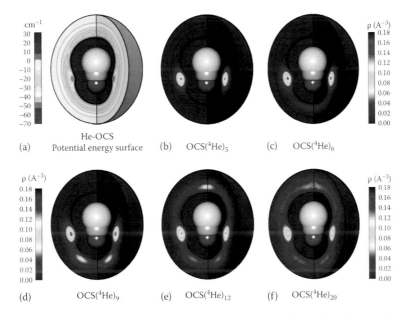

FIGURE 12.7 (a) Potential-energy surface for the OCS–He complex [73]. (b–f) ⁴He density distributions for selected OCS(⁴He)$_N$ clusters. (Adapted from Paesani, F. and Whaley, K.B., *J. Chem. Phys.*, 121, 4180, 2004.)

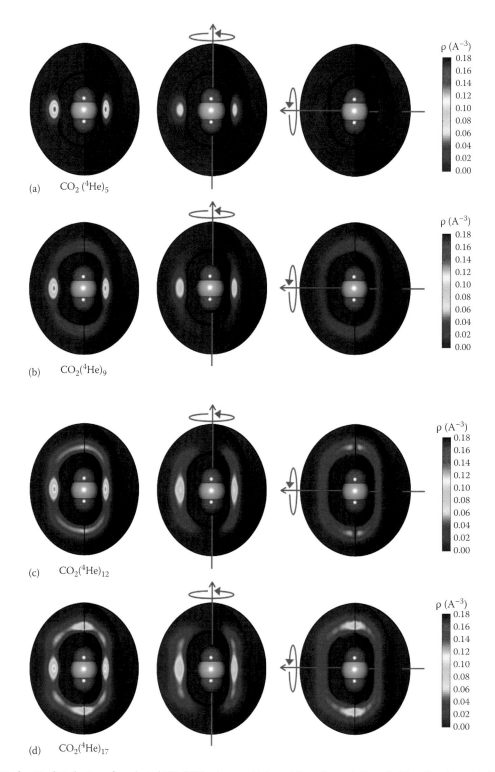

FIGURE 12.8 ⁴He density distributions for selected $CO_2(^4He)_N$ clusters. (Adapted from Paesani, F. et al., *Phys. Rev. Lett.*, 94, 2005.) Left column: total ⁴He density; central column: superfluid density to rotations along the CO_2 axis; right column: superfluid density to rotations along an axis perpendicular to the CO_2 axis. (Adapted from Paesani, F., et al., *Phys. Rev. Lett.*, 94, 153401, 2005.)

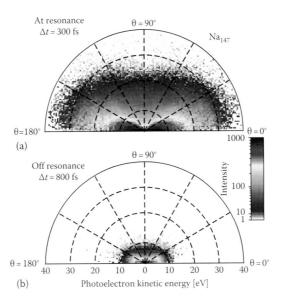

FIGURE 13.16 Angular resolved electron emission spectra from Na_{147} exposed to 25 fs laser pulses (800 nm) with peak intensity 8×10^{12} W/cm², as calculated from semiclassical VUU-MD simulations, cf. Section 13.3.1. The data correspond to dual-pulse excitations at (a) optimal and (b) a longer nonresonant delay. The emission angle, θ, is given with respect to the laser polarization axis. The plot is based on the data from Fennel et al. (2007a) but shown with a modified intensity scaling.

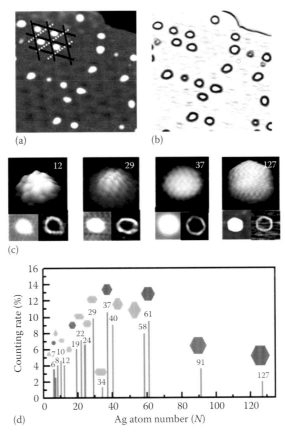

FIGURE 17.10 (a) Ag nanopucks possessing various shapes and sizes grown on Pb quantum islands. The black lines outline the unit cell of the superstructure and each unit cell is divided into two triangular halves by the white lines. (b) The corresponding shapes of Ag nanopucks in (a). The shapes are highlighted by differentiation. (c) Atomically resolved images of Ag magic nanopucks with 12, 29, 37, and 127 atoms. The corresponding STM images and shapes are shown in the insets. (d) The size distributions of over a 1000 Ag nanopucks grown on Pb quantum islands. The schematic structures are drawn above each magic number for the most frequently observed ones. The sketches are divided into two based on the attributed factors: geometrical effect (painted in blue) and electronic effect (painted in orange). (From Chiu, Y.-P. et al., *Phys. Rev. Lett.*, 97, 165504, 2006. With permission.)

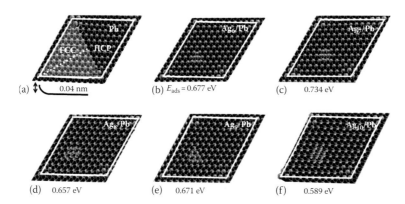

(a) ⇅ 0.04 nm   (b) $E_{ads} = 0.677$ eV   (c)   0.734 eV

(d)   0.657 eV   (e)   0.671 eV   (f)   0.589 eV

FIGURE 17.17 Sketch of generation of Moiré pattern in unit cell of Pb quantum island surfaces. FCC region represents the zone of relative stability for the initial growth of Ag on Pb surfaces. Clean simulated Pb surface (a), and cross-sectional profile for Ag cluster nucleated on simulated fcc region of Pb surfaces is also illustrated below (a). (b–f) Represented geometrical configurations for planar Ag_N clusters ($N = 6$–10) on simulated substrate. The numbers below (b–f) indicate the energies of adsorption per atom for Ag clusters. (From Chiu, Y.-P. et al., *Phys. Rev. B.*, 78, 115402, 2008. With permission.)

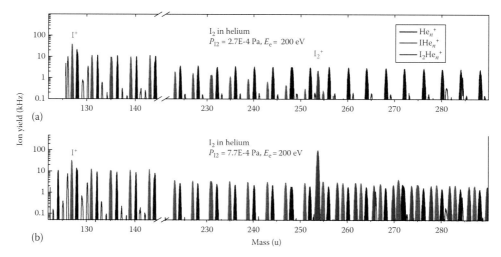

FIGURE 20.12 Cation mass spectrum of helium droplets doped with I_2. Pure He_n^+ peaks are filled black, IHe_n^+ peaks are filled red, $I_2He_n^+$ peaks are filled blue. The lower spectrum was recorded with increased I_2 pressure in the pickup cell. (Adapted from Ferreira da Silva, F. et al., *Chem. Eur. J.*, 15, 7101, 2009.)

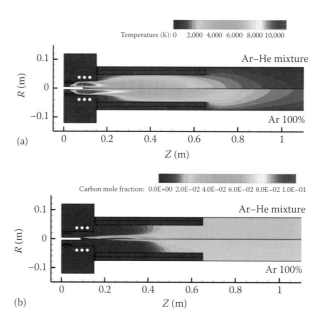

FIGURE 21.11 The effect of the plasma-forming gas on the temperature field and carbon evaporation in an RF induction plasma process (2D numerical simulation): (a) temperature distribution; (b) distribution of carbon mole fraction (28 kW, 66 kPa, C/Ni-98/2 at.% and feed rate of 2 g/min).

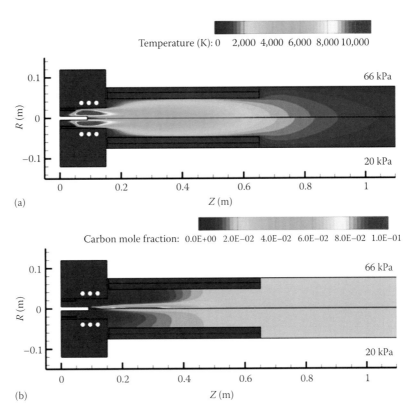

FIGURE 21.14 The effect of the operating pressure on the temperature field and carbon evaporation in an RF induction plasma process (2D numerical simulation): (a) temperature distribution; (b) distribution of carbon mole fraction (28 kW, 66 kPa, C/Ni-98/2 at.% and feed rate of 2 g/min).

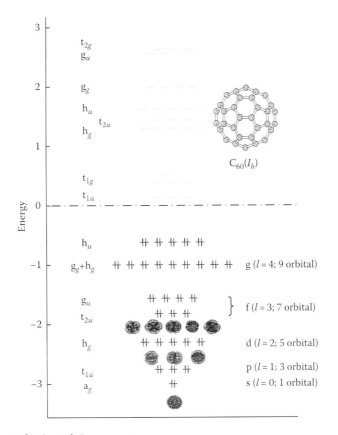

FIGURE 25.3 Hückel orbital energies for C_{60} with I_h symmetry.

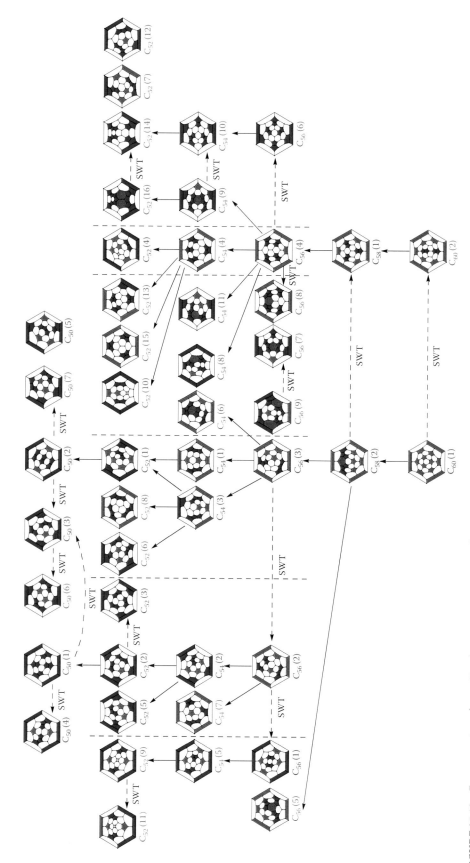

FIGURE 25.18 Fragmentation scheme. Numbers represent energy ordering among isomers. The most stable isomers (within 0.75 eV from isomer number 1) are shown in bold red. Solid and dashed lines indicate C_2 losses and SW transformations (SWTs), respectively.

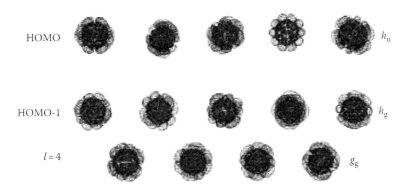

FIGURE 27.14 The highest occupied molecular orbital (HOMO) and the lower lying molecular orbitals in neutral C_{60}. The pictures show the orbitals for a single fixed orientation of the fullerene cage and the g_g, h_g, and h_u notations refer to the orbital symmetries. The g_g and h_g states are close in energy with a gap of about 1.8 eV to the h_u (HOMO) state (cf. text). The results are calculated by Diaz-Tendero at HF/6-31G* level of theory. (From Díaz-Tendero, S., Structure and fragmentation of neutral and positively charged carbon clusters and fullerenes, PhD thesis, Universidad Autónoma de Madrid, Madrid, Spain, 2005.)

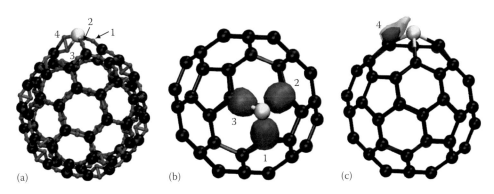

FIGURE 32.8 (a) Wannier function centers (WFCs) represented by small spheres. The C atoms and a Si atom are displayed by large black spheres and a large gray sphere, respectively. Numbers denote the WFCs around the Si atom. (b) Isosurfaces [0.02 e/(a.u.)^3] of the Wannier functions corresponding to WFC1, WFC2, and WFC3. (c) Isosurfaces of the Wannier function WFC4 for two values of the isosurface density 0.02 e/(a.u.)^3 (transparent) and 0.025 e/(a.u.)^3 (solid).

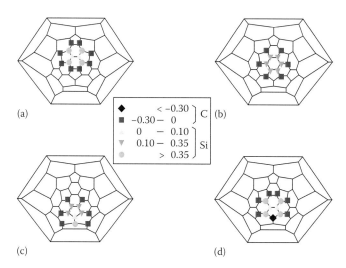

FIGURE 32.11 A two-dimensional flattened view of $C_{54}Si_6$. Mulliken charges are given for all the Si atoms and for some their neighboring C atoms. The values corresponds to the shading codes in the center of the figure.

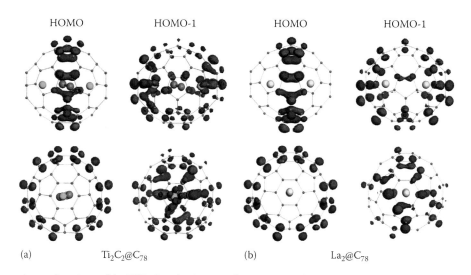

FIGURE 42.16 Squared wave functions of the HOMO and HOMO-1 of $Ti_2C_2@C_{78}$ and $La_2@C_{78}$ by Hino et al. (2007). (From Hino. S. et al., *Phys Rev. B*, 75, 125418, 2007. With permission.)

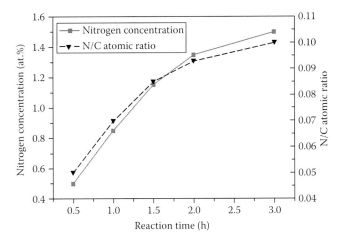

FIGURE 24.26 Nitrogen concentration and N/C atomic ratio in the carbon onion samples (1#–5#) measured by XPS for various reaction times using nitrogen as the carrier gas. (From He, C.N. et al., *Scr. Mater.*, 54, 1739, 2006. With permission.)

shows that the nitrogen concentration in carbon onions increases with increasing reaction time and the carbon onion yield, which exhibits an intimate relationship between carbon onion growth and nitrogen incorporation. Figure 24.27 shows typical XPS spectra of carbon (Figure 24.27a) and nitrogen (Figure 24.27b) in the carbon onions fabricated at 600°C for 1 h using nitrogen as the carrier gas. The C_{1s} peak at 285 eV corresponds to that of graphite. The N_{1s} spectrum shows two peaks at 407.25 and 404.25 eV. The high binding energy peak (407.25 eV) is due to the nitrogen bonded to sp^2-coordinated C atoms, i.e., substitutional N in the graphite sheet. The other peak at 404.25 eV is believed to correspond to three-coordinated N atoms in an sp^3-rich environment or at the peripheries of the graphite sheet [84–87]. The XPS analysis shows that the nitrogen in the carbon onion is chemically bonded with the carbon atoms of the graphitic sheet.

Morphological change of the graphitic plane due to nitrogen incorporation has been widely investigated since the theoretical prediction of a super hard C_3N_4 phase [88]. An ab initio

total energy calculation showed that the strain energies of rolling the CN sheets are smaller than that for a graphite sheet [89]. Several experimental results support the theoretical calculations. Atomic-scale electron energy loss spectroscopy analysis of nano-onions in CN_x thin film [90] and a CN_x nanotube [91] showed that the nitrogen concentration increases as the curvature of the graphitic layer increases. Consequently, the carbon onion could be easily obtained due to the presence of nitrogen, which can reduce the strain energy required for the spherical graphitic layer of carbon onion and thus decrease the activation energy for both the nucleation of the graphitic layer on the catalyst surface and the structural evolution of the carbon onion during growth. On the other hand, C atoms are believed to be critical for the nucleation of carbon onions [92]. The break-up of CH_4 into free C atoms would give out H_2 at the same time. Thus, the existence of H_2 in the reaction system would hinder such a process and further prevent the formation of the carbon onion. It has to be admitted that such an explanation is based largely on hypothesis for the authors, because little knowledge has been gained about the growth mechanism of carbon onions by CVD. However, introducing H_2 is probably an effective method of controlling the doping amount of nitrogen, which is of great interest in studying the properties of such doped carbon onions. It would also shed light on the growth process of heteroatom carbon onions.

24.4.3 Production of Hollow Carbon Onion Particles

According to Section 24.4.1, a mass of carbon onions are fabricated using a new Ni/Al catalyst by CVD at 600°C, and the carbon onions grown have diameters ranging from 5 to 50 nm and consist of several concentric carbon layers surrounding a hollow core or a nickel particle. However, the existing of catalytic particles in carbon onions baffles the large-scale investigation and application for carbon onions; thus, a practical method for the production of pure hollow carbon onion particles is very necessary. Subsequently, the authors ameliorated the catalyst and

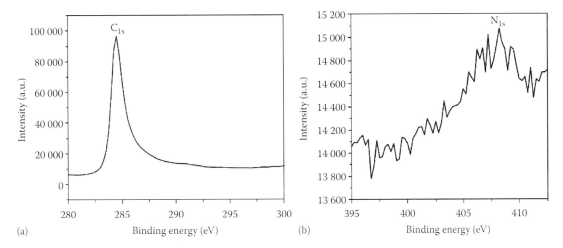

FIGURE 24.27 X-ray photoelectron spectra of C (a) and N (b) of the carbon onions deposited at 600°C for 1 h using nitrogen as the carrier gas. (From He, C.N. et al., *Scr. Mater.*, 54, 1739, 2006. With permission.)

technique processes based on above experiments, and since then more carbon onions in large quantity have been formed. Then the obtained carbon nanoparticles consisting of several concentric carbon layers surrounding a hollow core or a nickel nanoparticle are purified by removal of the nickel cores with HNO_3 treatments. Hence, a practical route to the production of carbon onions is developed [93].

24.4.3.1 Experimental Process

For a typical experiment, the aluminum-supported nickel composite catalyst with weight ratio of 1:3 is prepared by using homogeneous deposition-precipitation method. The right amounts of $Ni(NO_3)_2 \cdot 6H_2O$ and aluminum powder are mixed in 1 L distilled water to yield the final weight ratio Ni/Al = 3/1. NaOH (8 g) dissolved in 50 mL distilled water is added to the previous mixture with constant stirring. The colloid obtained is then filtered several times with distilled water till neutral pH is obtained and dried in a vacuum furnace at 140°C for 16 h. Ultimately the colloid is calcined in N_2 atmosphere at 450°C for 6 h to form fine NiO/Al composite powder, which is employed in the following carbon onion synthesis experiment.

For the carbon onion synthesis process, first, 600 mg catalyst of NiO/Al is kept in a quartz boat and placed in a horizontal quartz tube reactor. In order to convert the nickel oxides in the catalyst to elementary Ni, the catalyst is reduced at 600°C in a hydrogen atmosphere for 2.5 h. Then, carbon onions are grown at 600°C for 1 h by introducing a mixture of CH_4/N_2 into the reactor at a flow rate of 60/600 mL/min. Then the gas flow is switched back to nitrogen after the growth process to purge the methane from the tube and to prevent back flow of air into the tube. The furnace is then allowed to cool to room temperature before exposing the carbon onions to the air. To obtain pure hollow carbon onions, 1.0 g as-prepared carbon onion powder is immersed in 150 mL concentrated HNO_3 (12 mol/L A.R.) at 90°C, and magnetically stirred and refluxed for 6 h. The black materials are washed several times with distilled water till neutral pH is obtained and dried in a vacuum furnace at 120°C for 0.2 h.

24.4.3.2 Structural Analysis

Figure 24.28a and b shows the XRD patterns of the as-prepared sample and the purified carbon onions, respectively. Comparing Figure 24.28b with Figure 24.28a, except for the peaks at 2θ = 25.86° and 44.28°, all other peaks nearly disappear after the HNO_3 treatment because Ni and Al are dissolved in the HNO_3 aqueous solution. The sharpness of the "graphite" peaks also indicates that the structures in the product are highly ordered. And it can be observed that the two graphite peaks move from 26.06° and 44.38° to 25.86° and 44.28°, respectively due to the influence of purification on the carbon onions.

TEM micrographs of the as-prepared sample are shown in Figure 24.29. Two types of onions can be found in the sample. The first type, as shown in Figure 24.29a and c, is well-crystallized empty onions with diameter in 5–20 nm range and the second, as shown in Figure 24.29b, is larger onions (10–60 nm) with nickel nanoparticle encapsulated. And the approximate ratio of

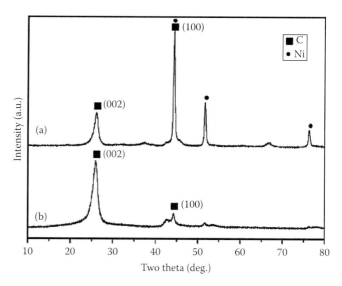

FIGURE 24.28 XRD pattern of (a) the as-prepared carbon onions and (b) the purified carbon onions. (From He, C.N. et al., *J. Alloys Comp.*, 425, 329, 2006. With permission.)

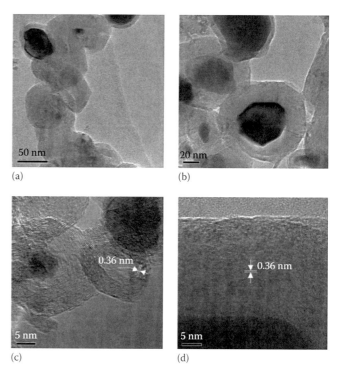

FIGURE 24.29 (a) TEM image of the carbon onions with core empty, (b) TEM image of the Ni–C nanoparticles, (c) high-resolution TEM image of the carbon onion with core empty, and (d) high-resolution TEM image of part A (as marked in b). (From He, C.N. et al., *J. Alloys Comp.*, 425, 329, 2006. With permission.)

the first type to the second type onions is 1:3.5 based on the TEM observation. According to the performance of nitric acid in the purification of carbon nanotubes, it is reasonably expected that the compact carbon shells could be opened first at their amorphous regions by selective oxidation of nitric acid, and then the

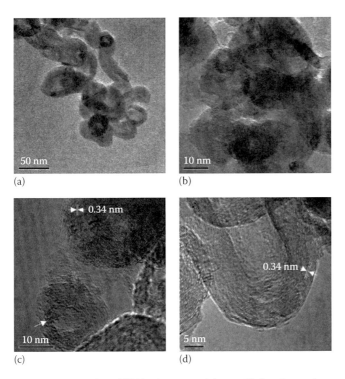

(a) (b) (c) (d)

FIGURE 24.30 (a and b) TEM images of the purified carbon onions with core empty. (c and d) High-resolution TEM images of the purified carbon onions with core empty. (From He, C.N., *J. Alloys Comp.*, 425, 329, 2006. With permission.)

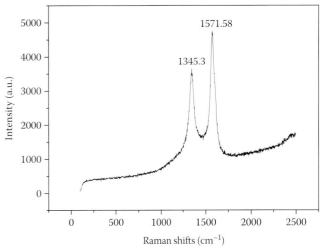

FIGURE 24.31 Raman spectrum of the purified carbon onions with core empty. (From He, C.N. et al., *J. Alloys Comp.*, 425, 329, 2006. With permission.)

residual cores could be dissolved by acid through the opened channels. Consequently, the remaining carbon shells from the second type of nanonparticles formed hollow carbon onions.

Figure 24.30a and b shows the TEM images of the purified carbon onions. As observed, no catalytic particles can be found and all nanoparticles own a hollow core; thus, pure carbon onions are obtained. The elimination of catalytic particles from carbon-coated nanoparticles may result from the opening of the amorphous regions on carbon shells due to the strong oxidation of nitric acid, the penetration of nitric acid into the cores of carbon-coated nanoparticles, followed by the dissolution of the nickel cores. In the corresponding typical TEM image of the second-type purified carbon onions in Figure 24.30c, some broken regions, as indicated by the white arrow, are indeed observed. However, the broken regions have not been detected for the first-type purified carbon onions due to the well-crystallized graphite that is hard to be oxidized. According to the above experimental results, it is understood that an optimized procedure is needed in order to obtain the ideal hollow carbon nanocages. In particular, the over-exposure to concentrated HNO_3 should be avoided.

The representative Raman spectrum of the purified carbon onions is shown in Figure 24.31. The Raman spectrum shows two broad peaks centered at about 1345.3 and 1571.58 cm^{-1}, which are associated with the vibrations of carbon atoms with dangling bonds for the in-plane termination of disordered graphite and the vibrations in all sp^2-bonded carbon atoms in a two-dimensional hexagonal lattice, respectively. The intensity ratio of D to G band (I_D/I_G) is calculated to be 0.63, the low relative intensity of

D to G band implies that the obtained carbon onions are mainly composed of well-crystallized graphite, in agreement with the XRD and HRTEM observations.

24.5 Summary

Carbon onions that were discovered in 1980 are designated as one of the most attractive materials for solid lubricant and considered to be one of the possible carriers of the 217.5 nm interstellar absorption bump. Intensive research activities to improve the synthesis methods and conditions, quality, and productivity of the carbon onions reached to rewarding conclusions because of their significant potential for various applications in many fields.

The synthesis methods for carbon onions mainly contain arc discharge, high-energy electron beam or laser irradiation, high-temperature nanodiamond annealing, implantation of carbon ions onto metal particles, and chemical vapor deposition. The frontal methods including arc discharge, high-energy electron beam or laser irradiation, high-temperature nanodiamond annealing, and implantation of carbon ions onto metal particles are costly due to their high-temperature processes and the limited number of carbon onions that can be produced and grow carbon onions in highly aggregate forms with unwanted carbon impurities such as carbon nanotubes or amorphous carbon. Chemical vapor deposition is the catalytic decomposition of hydrocarbon or carbon monoxide feedstock with the aid of supported transition metal catalysts. Compared with the other four synthesis methods, the CVD method is believed as a better synthesis method of carbon onions in terms of product purity and large-scale production.

Though preparation of carbon onions have achieved great developments, many problems still exist and are needed to be solved, such as how to control the particle size, size distribution, yield, and the thickness and crystallization of carbon layer. Furthermore, macroscopical quantity of carbon onions cannot be synthesized due to limitation of the preparation method, which results in the lag of application exploitation. With the further development of preparation method and invention of new synthesis methods (such as CVD), it can be anticipated that carbon onions would be applied in many fields in the future.

Acknowledgments

The authors acknowledge the financial support by Chinese National Natural Science Foundation (No.50771071) and Chinese Postdoctoral Science Foundation (No.20080440098).

References

1. Kroto, W. H., Heath, J. R., O'Brien, S. C. et al. 1985. C_{60}: Buckminsterfullerene. *Nature* 318: 162–163.
2. Shenderova, O. A., Zhirnov, V. V., Brenner, D. W. 2002. Carbon nanostructures. *Critical Reviews in Solid State and Materials Sciences* 27: 227–356.
3. Roy, D., Chhowalla, M., Wang, H. et al. 2003. Characterisation of carbon nano-onions using Raman spectroscopy. *Chemical Physics Letters* 373: 52–56.
4. Ruoff, R. S., Lorents, D. C., Chan, B. et al. 1993. Single-crystal metals encapsulated in carbon nanoparticles. *Science* 259(5093): 346–348.
5. Tomita, M., Saito, Y., Hayashi, T. 1993. LaC_2 encapsulated in graphite nano-particles. *Japanese Journal of Applied Physics* 32(1): L280–L282.
6. Saito, Y., Yoshikawa, T., Inagaki, M. et al. 1993. Growth and structure of graphitic tubules and polyhedral particles in arc-discharge. *Chemical Physics Letters* 204(3/4): 277–282.
7. Iijima, S. 1980. Direct observation of the tetrahedral bonding in graphitized carbon black by HREM. *Journal of Crystal Growth* 50: 675–688.
8. Iijima, S. 1987. The 60-Carbon cluster has been revealed!. *Journal of Physical Chemistry* 91: 3466–3467.
9. Kroto, H. W., Mckay, K. 1988. The formation of quasi-icosahedral spiral shell carbon particles. *Nature* 331: 328–331.
10. Ugarte, D. 1992. Curling and closing of graphitic networks under electron beam irradiation. *Nature* 359: 707–709.
11. Henrard, L., Lambin, Ph., Lucas, A. A. 1997. Carbon onions as possible carriers of the 2175 A° interstellar absorption bump. *Astrophysical Journal* 487: 719–727.
12. Rapoport, L., Homyonfer, M., Cohen, S. R. et al. 1997. Hollow nanoparticles of WS as potential solidstate lubricants. *Nature* 387: 791–793.
13. Okotrub, A. V., Bulusheva, L. G., Kuznesov, V. L. et al. 2001. X-ray emission studies of valence band of nanodiamonds annealed at different temperatures. *Journal of Physical Chemistry A* 105: 9781–9787.
14. Joachim, C., Gimzewski, J. K., Aviram, A. 2000. Electronics using hybrid-molecular and mono molecular devices. *Nature* 408: 542–543.
15. Joly-Pottuz, L., Matsumoto, N., Kinoshita, H. et al. 2008. Diamond-derived carbon onions as lubricant additives. *Tribological International* 41: 69–78.
16. Su, D. S., Maksimova, N., Delgado, J. J. et al. 2005. Nanocarbons in selective oxidative dehydrogenation reaction. *Catalysis Today* 102–103: 110–114.
17. Sano, N., Wang, H., Alexandrou, I. et al. 2002. Properties of carbon onions produced by an arc discharge in water. *Journal of Applied Physics* 92: 2783–2788.
18. Saito, Y., Yoshikawa, T., Inagaki, M. et al. 1993. A new relativistic scheme in Dirac–Kohn–Sham theory. *Chemical Physics Letters* 204: 277–282.
19. Takeo, O., Ichihito, N., Atsushi, N. 2004. Formation, atomic structural optimization and electronic structures of tetrahedral carbon onion. *Diamond and Related Materials* 13: 1337–1341.
20. Gorelik, T., Urban, S., Falk, F. et al. 2003. Carbon onions produced by laser irradiation of amorphous silicon carbide. *Chemical Physics Letters* 373: 642–645.
21. Kuznetsov, V. L., Chuvilin, A. L., Butenko, Y. V. et al. 1994. Onion-like carbon from ultra-disperse diamond. *Chemical Physics Letters* 222: 343–348.
22. Chen, X. H., Deng, F. M., Wang, J. X. et al. 2001. New method of carbon onion growth by radio frequency pleasma-enhancde chemical vapor deposition. *Chemical Physics Letters* 336: 201–204.
23. Cabioch, T., Riviere, J. P., Delafond, J. 1995. A new technique for fullerene onion formation. *Journal of Materials Science* 30(19): 4787–4792.
24. Lange, H., Sioda, M., Huczko, A. et al. 2003. Nanocarbon production by arc discharge in water. *Carbon* 41: 1617–1623.
25. Wang, X. M., Wang, H. Y., Zhang, H. X. et al. 2004. HREM analysis of nano-structured onion-like fullerenes derived from coal. *Journal of Chinese Electron Microscopy Society* 23(2): 159–162.
26. Pang, L. S. K., Michael, A. W. 1995. Isotope effects in plasma arcing experiments with various carbon anodes. *Energy Fuels* 9(4): 704–706.
27. Qiu, J. S., Zhou, Y. 2000. Preparation of fullerenes using carbon rods manufactured from Chinese hard coals. *Fuel* 79(11): 1303–1308.
28. Wang, X. M., Xu, B. S., Liu, X. G. et al. 2004. Structural characteristics of onion-like fullerenes synthesized by metal nanoparticle catalysis. *New Carbon Materials* 19(3): 209–213.
29. Beck, M. T., Dinya, Z., Keki, S. et al. 1993. Formation of C60 polycyclic aromatic hydrocarbons upon electric discharges in liquid toluene. *Tetrahedron* 49: 285–290.
30. Xie, S., Huang, R., Chen, L. et al. 1998. Glow discharge synthesis and molecular structures of perchlorofluoranthene and other perchlorinated fragments of buckminsterfullerene. *Journal of the Chemical Society, Chemical Communications* 18: 2045–2046.
31. Hsin, Y. L., Hwang, K. C., Chen, F. R. et al. 2001. Production and in situ metal filling of carbon nanotubes in water. *Advanced Materials* 13: 830–833.

32. Sano, N., Wang, H., Chhowalla, M. et al. 2001. Nanotechnology: Synthesis of carbon 'onions' in water. *Nature* 414: 506–507.

33. Xu, B. S., Guo, J. J., Wang, X. M. et al. 2006. Synthesis of carbon nanocapsules containing Fe, Ni or Co by arc discharge in aqueous solution. *Carbon* 44: 2631–2634.

34. Guo, J. J., Wang, X. M., Yao, Y. L. et al. 2007. Structure of nanocarbons prepared by arc discharge in water. *Materials Chemistry and Physics* 105: 175–178.

35. Xing, G., Jia, S. I., Shi, Z. Q. 2007. The production of carbon nano-materials by arc discharge under water or liquid nitrogen. *New Carbon Materials* 22(4): 337–341.

36. Oku, T. 2002. Three-dimensional imaging of YB$_{56}$ by high-resolution electron microscopy. *Chemical Communications* 302–303.

37. Oku, T. 2003. Direct observation of B84 and B156 clusters by high-resolution electron microscopy and crystallographic image processing. *Solid State Communications* 127: 689–693.

38. Ozawa, M., Goto, H., Kusunoki, M. et al. 2002. Continuously growing spiral carbon nanoparticles as the intermediates in the formation of fullerenes and nanoonions. *The Journal of Physical Chemistry B* 106(29): 7135–7138.

39. Zwanger, M. S., Banhart, F., Seeger, A. 1998. Formation and decay of spherical concentric-shell carbon clusters. *Journal of Crystal Growth* 163: 445–454.

40. Banhart, F. 1999. Irradiation effects in carbon nanostructures. *Reports on Progress in Physics* 62: 1181–1221.

41. Xu, B. S., Tanaka, S. I. 1995. Phase transformation and bonding of ceramic nanoparticles in a TEM. *Nanostructured Materials* 6: 727–730.

42. Xu, B. S., Tanaka, S. I. 1998. Formation of giant onion-like Fullerenes under Al nanoparticles by electron irradiation. *Acta Metallurgica Sinica* 46: 5249–5252.

43. Troiani, H. E., Camacho-Bragado, A., Armendariz, V. et al. 2003. Synthesis of carbon onions by gold nanoparticles and electron irradiation. *Chemistry of Materials* 15(5): 1029–1031.

44. From Wikipedia, the free encyclopedia, Web site: http://en.wikipedia.org/wiki/Ion_implantation

45. Cabioc'h, T., Girard, J. C., Jaouen, M. et al. 1997. Carbon onions thin film formation and characterization. *Europhysics Letters* 38: 471–475.

46. Thune, E., Cabioc'h, T., Jaouen, M. et al. 2003. Nucleation and growth of carbon onions synthesized by ion implantation at high temperatures. *Physical Review B* 68(11): 115434.

47. Thune, E., Cabioc'h, T., Guérin, Ph. et al. 2002. Nucleation and growth of carbon onions synthesized by ion-implantation: A transmission electron microscopy study. *Materials Letters* 54: 222–228.

48. Tomit, S., Burian, A., Dore, J. C. et al. 2002. Diamond nanoparticles to carbon onions transformation: X-ray diffraction studies. *Carbon* 40: 1469–1474.

49. Hirata, A., Igarashi, M., Kaito, T. 2004. Study on solid lubricant properties of carbon onions produced by heat treatment of diamond clusters or particles. *Tribology International* 37: 899–905.

50. Tomita, S., Fujii, M., Hayashi, S. J. et al. 1999. Electron energy-loss spectroscopy of carbon onions. *Chemical Physics Letters* 305: 225–229.

51. Bulusheva, L. G., Okotrub, A. V., Kuznetsov, V. L. et al. 2007. Soft X-ray spectroscopy and quantum chemistry characterization of defects in onion-like carbon produced by nano-diamond annealing. *Diamond & Related Materials* 16(4–7): 1222–1226.

52. Qiao, Z. J., Li, J. J., Zhao, N. Q. et al. 2006. Graphitization and microstructure transformation of nanodiamond to onion-like carbon. *Scripta Materialia* 54: 225–229.

53. Ugarte, D., Heerde, W. A. 1993. Carbon onions produced by heat treatment of carbon soot and their relation to the 217.5 nm interstellar absorption feature. *Chemical Physics Letters* 207(4–6): 480–486.

54. Selvan, R., Unnikrishnan, R., Ganapathy, S. et al. 2000. Macroscopic synthesis and characterization of giant fullerenes. *Chemical Physics Letters* 316: 205–210.

55. Wang, X. M., Xu, B. S., Liu, X. G. et al. 2005. The Raman spectrum of nano-structured onion-like fullerenes. *Physica B* 357: 277–281.

56. Zhang, Y., Hou, L. F., Wang, X. M. et al. 2004. Studies on preparation of onion-like fullerenes by vacuum heat-treatment. In *Proceedings of the 14th Congress of International Federation for Heat Treatment and Surface Engineering*, Shanghai, China, 1/2, pp. 223–224.

57. Wang, Z. X., Yu, L. P., Zhang, W. et al. 2003. Carbon spheres synthesized by ultrasonic treatment. *Physics Letters A* 307: 249–252.

58. Zhang, M., He, D. W., Wei, B. Q. et al. 1999. Microstructural changes in carbon nanotubes induced by annealing at high pressure. *Carbon* 37: 657–662.

59. Zhang, M., He, D. W., Ji, L. et al. 1998. Macroscopic synthesis of onion-like graphitic particles. *Nano Structured Materials* 10: 291–297.

60. Albert, G., Nasibulin, A. M., David, P. B. et al. 2003. Carbon nanotubes and onions from carbon monoxide using Ni(acac)$_2$ and Cu(acac)$_2$ as catalyst precursors. *Carbon* 41: 2711–2724.

61. He, C. N., Zhao, N. Q., Du, X. W. et al. 2006. Low-temperature synthesis of carbon onions by chemical vapor deposition using a nickel catalyst supported on aluminum. *Scripta Materialia* 54: 689–693.

62. He, C. N., Zhao, N. Q., Du, X. W. et al. 2006. Carbon nanotubes and onions from methane decomposition using Ni/Al catalysts. *Materials Chemistry and Physics* 97: 109–115.

63. He, C. N., Zhao, N. Q., Du, X. W. et al. 2006. Carbon onion growth enhanced by nitrogen incorporation. *Scripta Materialia* 54: 1739–1743.

64. Du, J. M., Liu, Z. M., Li, Z. H. et al. 2005. Carbon onions synthesized via thermal reduction of glycerin with magnesium. *Materials Chemistry and Physics* 93: 178–180.

65. Zhao, M., Song, H. H., Chen, X. H. et al. 2007. Large-scale synthesis of onion-like carbon nanoparticles by carbonization of phenolic resin. *Acta Materialia* 55: 6144–6150.

66. Fu, D. J., Liu, X. G., Du, A. B. et al. 2006. Synthesis of nanostructured onion-like fullerenes by MW plasma. *Journal of Inorganic Materials* 21(3): 576–582.

67. Endo, M., Takeuchi, T., Igarashi, S. et al. 1993. The production and structure of pyrolytic carbon nanotubes (PCNTs). *Journal of Physics and Chemistry of Solids* 54: 1841–1848.

68. Nikolaev, P., Bronikowski, M. J., Bradley, R. K. et al. 1999. Gas-phase catalytic growth of single-walled carbon nanotubes from carbon monoxide. *Chemical Physics Letters* 313(1–2): 91–97.

69. Sharon, M., Mukhopadhyay, K., Yase, K. et al. 1998. Spongy carbon nanobeads—A new material. *Carbon* 36: 507–511.

70. Tsang, S. C., Qiu, J. S., Peter Harris, J. F. et al. 2000. Synthesis of fullerenic nanocapsules from bio-molecule carbonization. *Chemical Physics Letters* 322: 553–560.

71. Wu, W., Zhu, Z., Liu, Z. 2003. Preparation of carbon-encapsulated iron carbide nanoparticles by an explosion method. *Carbon* 41: 317–321.

72. Harris, P. J. F., Tsang, S. C. 1998. A simple technique for the synthesis of filled carbon nanoparticles. *Chemical Physics Letters* 293: 53–58.

73. Nasibulin, A. G., Ahonen, P. P., Richard, O. et al. 2001. Copper and copper oxide nanoparticle formation by chemical vapour nucleation from copper (II) acetylacetonate. *Journal of Nanoparticle Research* 3(5/6): 383–398.

74. Nasibulin, A. G., Kauppinen, E. I., Brown, D. P., Jokiniemi, J. K. 2001. Vapour decomposition of copper (II) acetylacetonate in the presence of hydrogen and water and nanoparticle formation. *Journal of Physical Chemistry B* 105(45): 11067–11075.

75. Crestani, M. G., Ivan, P. L., Luis, R. V. et al. 2005. The catalytic reduction of carbon dioxide to carbon onion particles by platinum catalysts. *Carbon* 43: 2621–2624.

76. Schnitzler, M. C., Oliveira, M. M., Ugarte, D. et al. 2003. One-step route to iron oxide-filled carbon nanotubes and bucky-onions based on the pyrolysis of organometallic precursors. *Chemical Physics Letters* 381: 541–548.

77. Lee, C. J., Lyu, S. C., Kim, H. W. et al. 2002. Large-scale production of aligned carbon nanotubes by the vapor phase growth method. *Chemical Physics Letters* 359: 109–114.

78. Sanoa, N., Akazawa, H., Kikuchi, T. et al. 2003. Separated synthesis of iron-included carbon nanocapsules and nanotubes by pyrolysis of ferrocene in pure hydrogen. *Carbon* 41: 2159–2179.

79. Öncel, C., Yürüm, Y. D. 2006. Carbon nanotube synthesis via the catalytic CVD method: A review on the effect of reaction parameters. *Fullerenes, Nanotubes, and Carbon Nanostructures* 14: 17–37.

80. Lee, C. J., Kim, D. W., Lee, T. J. et al. 1999. Synthesis of aligned carbon nanotubes using thermal chemical vapor deposition. *Chemical Physics Letters* 312: 461–468.

81. Choi, K. S., Cho, Y. S., Hong, S. Y. et al. 2001. Effects of ammonia on the alignment of carbon nanotubes in metal-assisted thermal chemical vapor deposition. *Journal of the European Ceramic Society* 21: 2095–2098.

82. Choi, G. S., Cho, Y. S., Hong, S. Y. et al. 2002. Carbon nanotubes synthesized by Ni-assisted atmospheric pressure thermal chemical vapor deposition. *Journal of Applied Physics* 91: 3847–3854.

83. Jung, M., Eun, K. Y., Lee, J. K. et al. 2001. Growth of carbon nanotubes by chemical vapor deposition. *Diamond & Related Materials* 10: 1235–1240.

84. Stafstrom, S. 2000. Reactivity of curved and planar carbon–nitride structures. *Applied Physics Letters* 77: 3941–3943.

85. Zhang, G. Y., Ma, X. C., Zhong, D. Y. et al. 2002. Polymerized carbon nitride nanobells. *Journal of Applied Physics* 9: 9324–9332.

86. Li, F. B., Qian, Q. L., Yan, F. et al. 2006. Nitrogen-doped porous carbon microspherules as supports for preparing monodisperse nickel nanoparticles. *Carbon* 44: 128–132.

87. Zou, Y. S., Wang, Q. M., Du, H. et al. 2005. Structural characterization of nitrogen doped diamond-like carbon films deposited by arc ion plating. *Applied Surface Science* 241: 295–302.

88. Liu, A. Y., Cohen, M. L. 1990. Structural properties and electronic structure of low-compressibility materials: b-Silicon nitride and hypothetical carbon nitride (b-C_3N_4). *Physical Review B* 41: 10727–10734.

89. Miyamoto, Y., Cohen, M. L., Louie, S. G. 1997. Theoretical investigation of graphitic carbon nitride and possible tubule forms. *Solid State Communications* 102: 605–608.

90. Czigany, Z., Brunell, I. F., Neidhardt, J. et al. 2001. Growth of fullerene-like carbon nitride thin solid films consisting of cross-linked nano-onions. *Applied Physics Letters* 79: 2639–2641.

91. Han, W. Q., Kohler-Redlich, P., Seeger, T. et al. 2000. Synthesis of boron nitride nanotubes from carbon nanotubes by a substitution reaction. *Applied Physics Letters* 77: 1807–1809.

92. Ajayan, P. M. 1999. Nanotubes from carbon. *Chemical Reviews* 99: 1787–1799.

93. He, C. N., Zhao, N. Q., Du, X. W. et al. 2006. A practical method for the production of hollow carbon onion particles. *Journal of Alloys and Compounds* 425: 329–333.

Stability of Charged Fullerenes

Yang Wang
Universidad Autónoma de Madrid

Manuel Alcamí
Universidad Autónoma de Madrid

Fernando Martín
Universidad Autónoma de Madrid

25.1 Introduction

One of the most important breakthroughs in nanoscience and nanotechnology [1,2] is the accidental discovery in 1985 of the fullerene C_{60}, a soccer-ball-like molecule consisting of 60 carbon atoms [3]. For this discovery, Curl, Kroto, and Smalley were awarded the Nobel Prize in Chemistry in 1996. Since the end of the twentieth century, there has been an increasing interest in multiply charged cationic and anionic fullerenes, stimulated by the birth of a new generation of high-energy collision experiments,[*] in which different kinds of energetic projectiles are used. For example, cationic C_{60}^{q+} fullerenes have been and are currently produced in collisions of C_{60} with fast, highly charged ions [4–16], electrons [17–19], and intense laser pulses [20,21]. Cationic C_{70}^{q+} fullerenes have been mainly produced by the electron bombardment of C_{70} [17,19,22–28].

Neutral fullerenes can be generated in macroscopic quantities by arc discharge of graphite in inert gas [29] and the purified powders are commercially available nowadays. Neutral beams are prepared by evaporating fullerene powder in a temperature-controlled oven (usually 350°C–500°C) with an aperture of a few mm [12]. To produce a charged beam, another beam of high energetic ions or electrons is placed in a vertical direction to impact and ionize the fullerene beam. In many experiments, electron cyclotron resonance sources are used to produce positively charged fullerene beams with intensities high enough for collision experiments [12].

The ionized fullerene particles can be further detected by time-of-flight (TOF) mass spectrometry. In the mass spectra, different ionic species are recognized from different signals depending on the mass-to-charge ratio. In collision experiments, the fullerene cations are usually produced in excited states (metastable). Therefore, they can undergo fragmentation by emitting one or several carbon fragments. The complex fragmentation patterns observed in these processes are mostly determined by the charge, excitation energy, structure, and relative stability of the different fullerene fragments [12]. A common feature in these experiments is that the mass spectra show a large variety of charged fragments, sometimes covering the whole range of fullerene (with even number of atoms) masses, starting from the smallest C_{20}^{q+} fullerene up to the intact C_{60}^{q+} and C_{70}^{q+} parents. In addition to this, a large variety of charge states have been observed, including charges that are much larger than those supported by normal molecules. For example, intact C_{60} and C_{70} fullerenes with charges up to 9+ and 6+, respectively, have been observed by multiple ionization with Bi^{44+} ions [30]. There is also an experimental observation of C_{70}^{q+} ions with $q > 6$ produced in collisions with slow Xe^{23+} ions [31]. Very recently, C_{60} fullerene cations with charges up to 12+ have been produced by using intense laser pulses [20,32]. The ultimate stability limit in C_{60}^{q+} has been recently predicted to be 14+ [33], and there are estimations that the stability limit in C_{70} might be even larger [31]. With the advent of ultrashort, superintense laser pulses, the production of highly charged fullerenes close to the stability limit seems to be feasible [20].

In some collision experiments, kinetic energy releases (KERs) are also measured, which can provide useful information about structures, reaction energetics, and dynamics [34]. Before the fragmentation occurs, the parent ion lies in an excited state,

---

[*] Typical collision energies range from 1 keV to 10 MeV, which are very high in comparison with the energies involved in most chemical processes, but maybe considered as rather low in nuclear physics.

which is usually metastable. For example, the C_{60}^+ cation can transiently accommodate a substantial internal energy (~40 eV) [35]. When a metastable parent ion AB^* undergoes a unimolecular fragmentation process, $AB^* \rightarrow A + B + KER$, the excess energy of the parent ion is released after partitioning the internal energy in the daughter ions A and B. This energy is called kinetic energy release. The kinetic energy distribution can be measured by an electrostatic energy analyzer, which produces the mass-analyzed ion kinetic energy spectrum. An alternative way of measuring the KER is monitoring the velocity spread from the TOF mass spectra, because the broadenings of the TOF peak and the velocity distribution are directly related to each other [34].

Negatively charged fullerenes have also been observed in experiments performed in storage rings. A storage ring is a circular device in which a continuous or pulsed particle beam is stored and kept circulating by magnetic fields. Compared to the single-pass approach, the storage ring can store fullerene beams for a long period of time (with time scales of more than 100 μs) [36]. Unlike positively charged fullerenes, negatively charged fullerenes are relatively less stable, because of the Coulomb repulsion among the excess electrons [37]. So far, the highest negative charge that has been experimentally detected in gas-phase fullerenes is thought to be 4− [38], although higher negative charge states can be achieved in solids or solutions [39–42]. Once the fullerene anions are formed, the spontaneous emission of electrons is usually hindered by a Coulomb barrier [43,44]. Hence, fullerene anions are metastable and thus they are observable as long as the lifetime is long enough compared to the experimental timescale. Earlier experiments on C_{60} and C_{70} anions in ion cyclotron resonance traps demonstrated that the doubly charged anions C_{60}^{2-} and C_{70}^{2-} are already metastable with a lifetime of the order of minutes [45,46] or seconds [47]. Higher charged anions C_{60}^{3-} and C_{60}^{4-} with lifetimes of at least 0.5 ms have recently been observed in gas phase by electrochemical/electrospray mass spectrometry [38]. In addition, negatively charged fullerenes have also been studied by photoelectron spectroscopies [48–51].

An alternative way of obtaining negatively charged fullerenes is to insert atomic or molecular electron donors inside the fullerene cages [52]. In this way, fullerenes such as C_{72}, C_{74}, C_{80}, or C_{84}, which are difficult to produce, e.g., in laser ablation from graphite, can be more easily synthesized. Examples of recently isolated endohedral fullerenes are $Ca@C_{72}$ [53], $La_2@C_{72}$ [54], $Be@C_{74}$ [55], $M_3N@C_{80}$ (M = Y and all lanthanides from Gd to Lu) [56–61], $Tb_3N@C_{84}$ [62], and $Tm_3N@C_n$ ($n = 78 – 88$) [59]. Even smaller endohedral fullerenes such as $Sc_2@C_{66}$ [63], $Sc_3N@C_{66}$ [63], $Sc_2C_2@C_{68}$ [64], and $Sc_3N@C_{68}$ [65,66] have been produced. In these endohedral fullerenes, the cage catches electrons from the encapsulated species so that the formal charge in the cage can reach a rather large value (up to 6) [67]. As will be explained in this chapter, it is generally believed that the structure of fullerenes obey the isolated pentagon rule (IPR, see Section 25.2.3.1). However, some isolated endohedral fullerenes do not follow this rule, e.g., $Sc_3N@C_{70}$ [68], $Ca@C_{72}$ [53], $La_2@C_{72}$ [54], $DySc_2N@C_{76}$ [69], $Dy_3N@C_{78}$, $Tm_3N@C_{78}$ [70], and $Tb_3N@C_{84}$ [62].

The positively and negatively charged fullerenes are very important for collision experiments. Charged fullerenes can be used as projectiles and targets in gas-phase collisions [71–73]. Besides, special interest has been aroused by the discovery of superconductivity in C_{60}^{q-}-based materials at rather high temperatures [74]. In these materials, fullerenes are intercalated by alkali metals (so-called fullerides), and these fullerene salts are found to be molecular superconductors with transition temperatures up to 38 K [75]. Endohedral fullerenes also have interesting potential applications [76]. For example, the encapsulation of reactive and poisonous atoms brings new possibilities for medical imaging and cancer therapy [77,78]. Some unique abilities of isolating an atom from its environment allow endohedral fullerenes to be used as a building block for the qubits of a quantum computer [79,80]. Moreover, it is very interesting that endohedral fullerenes can also be encapsulated into single-walled carbon nanotubes to form the so-called nano-peapods [81,82]. It is found that metal-containing endohedral fullerenes have substantial electric dipole moments (3–4 D) [83] and magnetic anisotropy. As a result, the electric or magnetic field can induce orientational change in the endohedral fullerenes and thus change electronic properties [82]. Since the rotational motion of endohedral fullerenes inside nanotubes can be controlled, another potential application is to use them for recording devices and quantum computers [82,84,85].

Spurred by the large amount of experimental observations, computational efforts have been made during the last few years to understand the structure, stability, and properties of multiply charged cationic and anionic fullerenes of various sizes [33,86–96]. Computational data for geometries, ionization potentials, electron affinities, dissociation energies, and fission barriers have been already obtained for many of these fullerene ions. This information has shown that although charged fullerenes are not very different from neutral fullerenes in many aspects, they also have exclusive properties that have significant impact in their relative stability. Such properties are useful not only to understand collision experiments but also to design new strategies in fullerene chemistry.

The aim of this chapter is to present a comprehensive description of charged fullerenes, with emphasis on their stability and physicochemical properties. The next section gives a brief description of the basic concepts used to understand the properties of neutral fullerenes that are also important to understand charged fullerenes. The following sections analyze in detail the properties of the most important positively and negatively charged fullerenes. The chapter concludes with a summary and future perspectives in this field.

25.2 Some Basics of Fullerenes

25.2.1 Structure and Topology of Fullerenes

25.2.1.1 Definition and Nomenclature of Fullerenes

As one of the carbon allotropes such as diamond, graphite, etc., fullerenes are a family of all-carbon molecules with a closed cage shape [1]. Topologically speaking, the carbon vertices of a classical fullerene cage are connected by the C–C bonds and form a network, which contains only five-membered rings (pentagons) and/or six-membered rings (hexagons) [1]. By applying Euler's formula* for polyhedra, there are exactly 12 pentagons and an arbitrary number of hexagons in a fullerene [1,97], and the total number of carbon atoms is even [1]. Therefore, the smallest possible fullerene is C_{20} consisting solely of pentagons, which has been experimentally produced [98]. The next member of the fullerene family is not C_{22}, since it is impossible to construct a fullerene structure with 22 vertices. Therefore, the mathematically possible fullerene series after C_{20} is C_{24}, C_{26}, C_{28},... [97].

There is more than one possible structure for the fullerene C_{28} and the larger ones, since one can find different ways of putting pentagons and hexagons in the cage structure. The number of isomers increases dramatically with the fullerene size, as shown in Table 25.1. Thus, a systematic numbering of all isomers is necessary. A commonly used numbering system is based on the spiral code of fullerenes proposed by Fowler and Manolopoulos [99]. According to this system, the suggested nomenclature of a fullerene isomer can be written as C_n (point group notation: number). For instance, the most stable C_{60} isomer with icosahedral symmetry (I_h, in point group notations) [100,101] can be represented as C_{60} (I_h:1812). Note that the point group used in this nomenclature is the highest possible symmetry based on topology only. A practical way to generate fullerene structures of any size with a given symmetry is to use the CaGe program [102], an Open Source software† for generating mathematical graphs of different types. In practice, however, the real fullerene structure may lose some symmetries due to electronic effects. For example, as a result of the Jahn–Teller distortion,‡ the smallest fullerene C_{20} prefers a

much lower symmetry and does not take the mathematically possible I_h symmetry. Recent studies [104,105] demonstrated that the dodecahedral C_{20} cage might be highly fluxional and undergo low-barrier transitions among structures with several possible symmetries such as C_2, C_{2h}, C_i, D_{3d}, and D_{2h} point groups.

25.2.1.2 Stone–Wales Transformation

Different fullerene isomers of a given size can be connected through a simple mechanism, the so-called Stone–Wales (SW) transformation [117]. A SW transformation can take place in the pyracylene motif on the fullerene surface, where two pentagons and two hexagons meet together (see Figure 25.1). In SW transformation, the central C–C bond (SW bond) [99] of the pyracylene motif is rotated by 90°. At the same time, the four neighboring bonds of the SW bond are broken and four new bonds are formed. As a consequence, the two pentagons and two hexagons are swapped and thereby the pyracylene motif is rotated without changing any bond on the same perimeter. The product of SW transformation is still a fullerene. In general, the new fullerene has a different structure from the old one, although it is possible to get the identical or enantiomeric structure after the SW transformation [99]. Therefore, the fullerene isomers with a given size can be connected through reversible SW transformation, as shown in Figure 25.1. The SW transformation is also a simple and straightforward way of generating more possible fullerene isomers from some known structures. The energy barrier of a SW transformation is estimated to be ~7 eV in general [1].

25.2.1.3 Sphericity

A fullerene cage is a convex polyhedron as a result of the 12 pentagons. Each pentagon contributes long-range positive convex curvature to the fullerene surface with an angle of $\pi/6(=2\pi/12)$ [1]. Therefore, the shape of the fullerene depends on how the 12 pentagons are distributed among the hexagons on the fullerene surface. The more uniform the distribution of pentagons (and thus the positive curvature) is, the more spherical the cage shape is. Relatively, smaller fullerenes are more like a ball, whereas larger fullerenes have more hexagons and thus can take a tubular shape, which is called fullerene nanotube [1]. Small fullerenes generally prefer to take more spherical shape, for which

TABLE 25.1 Number of Isomers (N), IPR Isomers (N_{IPR}), and Experimentally Isolated IPR Isomers (N_{exp}) for Fullerenes C_n

n	N	N_{IPR}	N_{exp}
60	1,812	1	1 [106]
70	8,149	1	1 [106,107]
72	11,190	1	0
74	14,246	1	1 [108]
76	19,151	2	1 [109]
78	24,109	5	3 [110–112]
80	31,924	7	1 [113]
82	39,718	9	1 [111]
84	51,592	24	9 [113–116]

* $V - E + F = 2$, where V, E, and F are the numbers of vertices, edges, and faces, respectively.
† The CaGe program is available at http://www.mathematik.uni-bielefeld.de/CaGe/Archive/.
‡ When degenerate electronic states are occupied asymmetrically, the molecule will undergo a geometrical distortion to remove the degeneracy and lower the total energy [103].

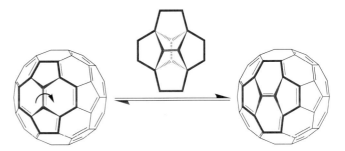

FIGURE 25.1 A Stone–Wales transformation connecting two different C_{60} isomers. The pyracylene motifs are sketched with bold lines.

the evenly distributed curvature leads to minimal strain on the fullerene shell. However, for larger fullerenes, the spherical shape is not the energetically favorable choice. For example, according to ab initio calculations, larger fullerenes with I_h symmetry (e.g., C_{180}, C_{240}, C_{540}, C_{960}) adopt the polyhedral facetted shape instead of spherical shape [118]. This surprising result can be easily understood if we notice the fact that hexagonal groups strongly prefer to be planar as in the case of graphene (a two-dimensional (2D) sheet of graphite). The sphericity of fullerenes plays an important role especially in the spherical aromaticity [87], as we will see in Section 25.2.3.3. In order to quantitatively measure the fullerene sphericity, Díaz-Tendero et al. defined a sphericity parameter, SP, based on the rotational constants* [87]. According to their definition,

$$SP = \sqrt{\left(\frac{1}{A} - \frac{1}{B}\right)^2 + \left(\frac{1}{A} - \frac{1}{C}\right)^2 + \left(\frac{1}{B} - \frac{1}{C}\right)^2} \quad (25.1)$$

where A, B, and C are the rotational constants of the fullerene molecule. Larger SP values indicate less spherical shape of fullerenes. For perfectly spherical and icosahedral fullerenes, $SP = 0$. Fullerenes with I_h or T_d symmetry also give zero value of SP, although not all the atoms are located on the perfectly spherical surface. Another quantity for measuring the fullerene sphericity is based on the least-squares method. For any given fullerene, we can always find a least-squares sphere so that the sum of squares of the distance from each atom to this least-squares sphere is minimal. Following this idea, a new measure of spherity, S, is expressed as

$$S = \sqrt{N \sum_{i=1}^{N} r_i^2 \Big/ \left(\sum_{i=1}^{N} r_i\right)^2 - 1} \quad (25.2)$$

where
  N is the total number of atoms
  r_i is the distance from the ith atom to the geometric center of the molecule

S equals zero only when all the atoms of the fullerene molecule stay on a perfect sphere. The parameter S has some advantages over SP. S is defined by geometry only and not by any physical property. It is dimensionless and independent on the size of the molecule. For some fullerenes with I_h or T_d symmetry which are not spherical (such as C_{180}), the SP value is always zero, whereas S gives a reasonable nonzero value.

25.2.1.4 Schlegel Diagram

Although the structure of a fullerene molecule is three-dimensional (3D), the topology of a fullerene is only 2D since the

---

* Rotational constant is commonly used in rotational–vibrational spectroscopy, and can be expressed as $B = \frac{h}{8\pi^2 cI}$, where h, c, and I are the Planck constant, the speed of light, and the moment of inertia for the molecule, respectively. It is given in the unit of wavenumbers, cm^{-1}, when using cgs units.

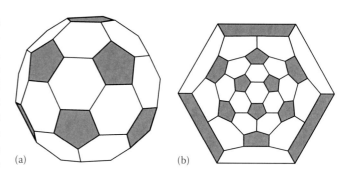

FIGURE 25.2 A polyhedron representation (a) and a Schlegel diagram (b) for C_{60}. Shadowed polygons are pentagons.

carbon atoms form a network on the surface of the fullerene cage. Therefore, the 2D topology of a fullerene can be clearly shown on the plane of a paper sheet. The projection of a fullerene from the 3D structure to the 2D network is called "Schlegel diagram," which shows the connection among the vertices of the fullerene cage. Figure 25.2 illustrates a Schlegel diagram for C_{60}.

25.2.1.5 Nonclassical Fullerenes

According to the topological definitions, "a classical fullerene is a trivalent spherical polyhedron in which all faces are either pentagons or hexagons," while "a nonclassical fullerene admits other ring sizes and perhaps other topologies" [119]. These definitions are based on the fact that pentagons are necessary for the closure of the cage structure and the hexagons are for maximizing the π-delocalization energy of the molecule [120]. Other ring sizes will induce high strain, as the conformation of carbon atoms is far from the favorable sp^2 hybridization (with a bond angle of 120°). For instance, the energy cost of introducing a square and a heptagon to the C_{40} fullerene is about 1.5–2.0 and 0.8–2.0 eV, respectively, as estimated by the density functional tight binding approach [119,121]. In practice, almost all isolable fullerenes by far have the classical structures containing only pentagons and hexagons. A nonclassical C_{62} fullerene containing a square has been successfully synthesized by rationally designed procedures [122]. Heptagonal rings have been experimentally found in fullerene derivatives $C_{58}F_{17}CF_3$ and $C_{58}F_{18}$ [123], as well as in fullerene coalescence in nano-peapods [124]. In addition, as suggested by Murry et al. [125], heptagonal rings play an important role in the annealing (rearrangement of the bonding) and fragmentation processes of fullerenes.

25.2.2 Bonding Features of Fullerenes

25.2.2.1 Chemical Bonds in Fullerenes

The bonding features of carbon atoms in fullerenes are similar to graphene, which can be basically understood as the combination of localized σ bonds and delocalized π bond. Each carbon atom takes sp^2 hybridization to form localized σ bonds with three neighboring atoms. The remaining p_z orbital with one valence electron is perpendicular to the fullerene surface. The overlap of all the p_z orbitals leads to a global delocalized π bond.

This simple bonding picture is successful for interpreting the basic electronic structure of graphite, carbon nanotubes, as well as fullerenes. This bonding model also gives insight into the main factors for the stability of fullerenes, i.e., strain, delocalization of π electrons, and spherical aromaticity.

25.2.2.2 σ Bonds: Strain

Compared to the planar graphene consisting of hexagons only, the convex surface of fullerene contains 12 pentagonal rings, which induce strain. The strain is contributed mainly from three parts: out-of-plane distortion, in-plane bending, and stretching of σ bonds. First, in the planar graphene, the π orbitals are exactly perpendicular to the σ bonds, namely, the $\sigma - \pi$ angle is 90°. For fullerenes, however, pentagonal rings bring positive curvature to form convex surface, leading to distorted $\sigma - \pi$ angles larger than 90°. Second, the internal angle of the regular pentagon is 108°, obviously smaller than 120°, the required bond angle for the perfect sp^2 hybridization. Therefore, the small $\sigma - \sigma$ angle of pentagonal rings induces the in-plane bending strain. Finally, the stretching strain is due to the change of bond lengths with respect to those in the "strain-free" graphene.

25.2.2.3 π Bond: Electron Delocalization and Spherical Aromaticity

Another important stability factor is the delocalization of π electrons on the fullerene surface. It is well known that the π system is stabilized through electron delocalization. The π electron delocalization depends on the degree of overlap of p_z orbitals, as well as the π electronic structure. Compared to planar graphene, the overlap of p_z orbitals in fullerenes is less efficient since the p_z orbitals are not parallel to one another due to the spherical shape of the fullerene surface.

Apart from the p_z orbital overlap efficiency, another factor for the stability of fullerenes is the π electronic structure, which plays an even more important role. The σ-bonds correspond to much lower molecular orbitals, whereas the π-system of fullerenes corresponds to the frontier orbitals, which are related to important chemical properties. The π electron structure of fullerenes can be qualitatively well described by simple Hückel molecular orbital (HMO) theory [126], which has been successfully applied to linear and cyclic π-conjugated systems. In the framework of HMO theory, the π-molecular orbitals are represented as linear combinations of atomic p_z orbitals centered on each atom. By neglecting all overlap integrals and accounting for Coulomb and resonance integrals only for neighboring atoms, the single-electron Hamiltonian can be solved analytically. Therefore, the Hückel orbitals and energies only depend on the topological structure of fullerenes, since the Hückel Hamiltonian only depends on how the atoms are connected to one another. Although HMO theory does not take into account other important stability factors such as strain and overlap efficiency, we will see in Section 25.3.6 that it can successfully provide a qualitative picture for the relative stability of charged fullerenes.

Like other π-conjugated molecules, some fullerenes reveal aromaticity, which can provide extra stabilization, depending on the number of π electrons and the sphericity of fullerene surface. Compared with linear (e.g., ethylene) and cyclic (e.g., benzene) aromatic π-systems, the aromaticity for fullerene is called "spherical aromaticity." More details on spherical aromaticity will be presented in Section 25.2.3.3.

25.2.3 Stability Rules for Fullerenes

Several rules have been proposed to understand and predict the stability of fullerenes. To apply these rules, one only needs the knowledge of the topological structure or the number of π electrons. These rules have been validated by both computations and experiments, and can be understood on the basis of the above-mentioned stability factors.

25.2.3.1 Isolated Pentagon Rule

The most famous and widely used rule is the isolated pentagon rule (IPR) [127]. This rule tells that the most energetically stable fullerene isomers are those in which all 12 pentagons are separated by hexagons. This rule successfully explains why only C_{60} and C_{70} were observed in the early production of fullerenes, since none of the other fullerenes (in the range from C_{20} to C_{70}) satisfies the IPR. By far, most of the experimentally isolable fullerenes (see Table 25.1) obey the IPR.

It has been generally accepted that the main basis of the IPR is strain. As mentioned above, pentagons induce strain to the fullerene structure. In particular, the abutting pentagons will cause much more strain energy. In addition, it has also been shown that a fused pentagon pair (pentalene-like structure) has a net destabilizing effect on the overall π electronic structure [128].

The IPR is very useful for finding stable fullerene structures theoretically. It decreases dramatically the number of testing isomers (see Table 25.1), since normally IPR-violating isomers are much higher in energy and thus can be screened out. However, some exceptions have also been found for the IPR in both neutral [53,128–131] and charged fullerenes (see Sections 25.3.5 and 25.3.6).

25.2.3.2 Pentagon Adjacency Penalty Rule

For smaller fullerenes C_n ($n = 20 - 58, 62 - 68$), the IPR cannot be applied, since it is impossible to have all pentagons isolated in the fullerene cage. In these cases, however, the pentagon adjacency penalty rule (PAPR) has been proposed as a stability guide for smaller fullerenes [119,127,132–134]. The PAPR relates the stability to the number of adjacent pentagons (APs): The fullerenes with fewer APs are energetically more stable. In other words, each pentagon adjacency gets a "penalty" of energy increase, which destabilizes the molecule. Density functional tight binding and molecular mechanics calculations evaluate a penalty of 0.8–0.9 eV per pentagon adjacency [119,134]. The PAPR can be rationalized on both steric and π electronic grounds [119]. The validity of this rule has been proven by quantum chemistry calculations [97,119].

Similar to IPR, there are still some exceptions for the PAPR [87,135], especially for the fullerene C_{50} [87,135]. In the case of C_{50}, spherical aromaticity, a concept that will be explained in detail in the next section, is more decisive than strain.

25.2.3.3 Spherical Aromaticity: The $2(N + 1)^2$ Rule

Compared with organic compounds (e.g., benzene) with planar delocalized π-systems, fullerenes have also a delocalized π-shell on a spherical surface. Since aromaticity plays an important role in the stabilization of benzene, one may ask, "Are fullerenes aromatic?" Experimental facts support a positive answer for some fullerenes [136–138]. In analogy to Hückel's $4N + 2$ rule for aromatic annulenes, a $2(N + 1)^2$ electron-counting rule has been proposed for the aromaticity of fullerenes [139]. The delocalized π system on a fullerene surface can be approximately considered as a spherical electron gas. Then, a Schrödinger equation with a spherically symmetric potential term is built for this π electron gas. The solution gives spherical harmonic wave functions, characterized by angular momentum quantum numbers ($l = 0, 1, 2, 3, \ldots$) [130,139,140]. Therefore, the shape and the degeneracy of molecular orbitals of fullerene π electrons remind us of those of the atomic orbitals ($s, p, d, \ldots$), as we can see in the case of C_{60} (see Figure 25.3). Following this model, one needs $2(N + 1)^2 (N = 0, 1, 2, 3, \ldots)$ electrons to fully fill π shells of fullerene. In other words, the maximum aromaticity (which is called spherical aromaticity) can be achieved in a closed-shell system

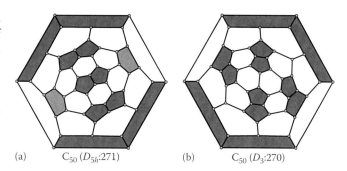

(a) C_{50} (D_{5h}:271)   (b) C_{50} (D_3:270)

FIGURE 25.4 Schlegel diagrams for two C_{50} isomers. Orange and blue colors indicate the isolated and the adjacent pentagons, respectively.

with $2(N + 1)^2$ π electrons. This conclusion has been exemplified by the magnetic properties of $C_{20}^{2+} (N = 2), C_{60}^{10+} (N = 4)$, and $C_{80}^{8+} (N = 4)$ [139]. It has to be noted that a necessary condition for the $2(N + 1)^2$ rule to apply is that the fullerene has to have a spherical or nearly spherical shape.

Spherical aromaticity can also play an important role in the stability of fullerenes. A good example is provided by two C_{50} isomers [87,135]. Among all 271 isomers of C_{50}, only isomer C_{50} (D_{5h}:271) has the minimum number of APs (NAP = 5), as shown in Figure 25.4. Surprisingly, this isomer is not the most stable isomer of the C_{50} fullerene. The most stable one is isomer C_{50} (D_3:270) with 6 APs (see Figure 25.4), being 0.80 eV more stable than the former, according to high-level theoretical calculations [87]. Noticing that the π electron number of C_{50} obeys the $2(N + 1)^2$ rule, the spherical aromaticity may play an important role in the stability of C_{50} fullerenes. Since isomer C_{50} (D_3:270) (SP = 0.6) is more spherical in shape than isomer C_{50} (D_{5h}:271) (SP = 1.7), the former is more aromatic [87]. In this case, the stabilization due to aromaticity takes over the strain effect. As a result, the most stable C_{50} isomer is the one with the maximum aromaticity, instead of the one with the least strain. Owing to the same reasons [130], the non-IPR isomer C_{72} (C_{2v}:5999) is energetically more stable than the IPR isomer C_{72} (D_{6d}:11190) [53,129,131].

25.3 Stability of Charged Fullerenes

25.3.1 Positively Charged C_{60} Fullerenes

25.3.1.1 Electronic States and Ionization Potentials

Before continuing our journey to the stability of positively charged C_{60} fullerenes, let us take a glance at the electronic structure of C_{60}^{q+} cations. As we have seen in Figure 25.3, the highest occupied molecular orbital (HOMO) of $C_{60}(I_h)$ fullerene consists of five degenerated h_u orbitals. Theoretical calculations show that, in general, the lowest energy corresponds to species with the highest spin multiplicity, which obeys Hund's rule [90]. For instance, the ground state for C_{60}^{2+} is triplet instead of singlet, and the most stable state for C_{60}^{5+} is sextuplet with five unpaired electrons [90]. The only exception among $q = 0 - 14$ is $q = 8$. The singlet state of C_{60}^{8+} is 15 meV lower in energy than the triplet one, but both states are nearly degenerated [90].

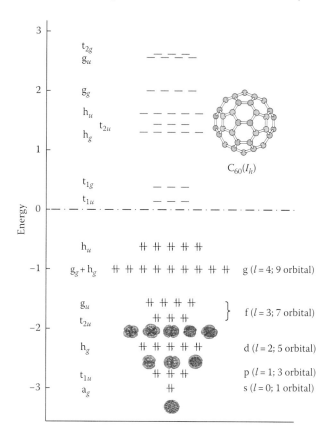

FIGURE 25.3 (See color insert following page 25-14.) Hückel orbital energies for C_{60} with I_h symmetry.

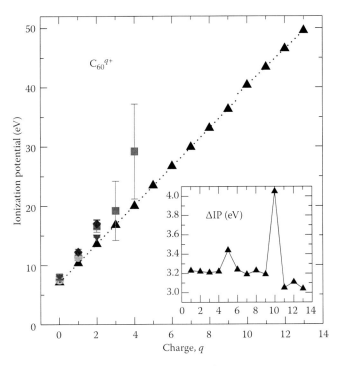

FIGURE 25.5 Ionization potential of C_{60}^{q+} fullerenes as a function of the degree of ionization. DFT calculations [90]: Upward triangles. Experiment: squares [17], circles [148,149], diamonds [150], downward triangles [151], crosses [152]. The inset show the variation of $\Delta IP = IP_q - IP_{q-1}$ with q. (From Díaz-Tendero, S. et al., *J. Chem. Phys.*, 123, 184306, 2005.)

Figure 25.5 shows the adiabatic ionization potentials (IPs, i.e., the energy difference in the ionization process: $C_{60}^{q+} \longrightarrow C_{60}^{(q+1)+} + e^-$) obtained by density functional theory (DFT) calculations. By fitting the calculated IP values, an empirical formula can be obtained [90]:

$$IP(C_{60}) = 7.106 + 3.252q \, (eV). \qquad (25.3)$$

This linear dependence reveals that the IP of C_{60}^{q+} can be understood with a simple electrostatic model. By assuming that the IP corresponds to the energy required to remove an electron from a charged conducting sphere [141–143], the qth IP of C_{60}^{q+} follows:

$$IP_q = W + \left[\frac{q^2}{2R} - \frac{(q-1)^2}{2R} \right] = W + \frac{q-1/2}{R}, \qquad (25.4)$$

where R is the radius of fullerene. The first term W on the right-hand side of the above equation is the work function for extracting one electron from the neutral fullerene surface and the second part corresponds to the difference of the self-energy of a charged conducting sphere [141]. Comparing Equations 25.3 and 25.4, one can get the estimated radius of C_{60}, which is about 8.37 a_0 (a_0 is Bohr radius, 1 $a_0 = 0.527$ Å), in good accordance with the value, $8.0 \pm 0.3 a_0$, extracted from the measured dipole polarizability ($\alpha = R^3$) [144].

The required energy to remove q electrons from the neutral C_{60} can be fitted to a quadratic function [90]:

$$E_{rel} = 9.59 + 2.13q^2 \, (eV), \qquad (25.5)$$

which supports the suggestion that fullerenes behave on average like a capacitor [145–147].

The inset in Figure 25.5 shows the variation of the IP slope with q. It can be seen that the slope is almost constant (due to the linear correlation between IP and q) except in the vicinity of $q = 5$ and 10. As mentioned above, for $q = 5$ the h_u shell is half-filled, providing additional electronic stability and, therefore, leads to an increase in the IP. The same reason holds for the sudden increase (about 1 eV) in the IP slope near $q = 10$, in which case the variation of the IP slope is even much larger, because of the fully occupied lower $h_g + g_g$ shell [90].

25.3.1.2 Fragmentation of C_{60}^{q+}

Due to Coulomb repulsion, highly charged fullerenes may decompose into several fragments. The corresponding decay processes have been extensively studied by experimental and theoretical methods [12]. There are three different decay channels depending on the initial charge state q and on its excitation energy E^* [4,13,153]: (1) successive emission of neutral C_2 (sequential evaporation), (2) fragmentation of one or more light charged clusters (asymmetric fission), and (3) fragmentation into several singly charged fragments with small mass (multifragmentation) [97]. In the last case, if the total number of charged fragments is comparable to the initial charge, this multifragmentation process is called Coulomb explosion [97]. The measured branching ratios for the emission of light fragments show that the dominant decay channel of C_{60}^{q+} is C_2 emission for $q \leq 3$ [4,154]. The C_2^+ emission becomes more competitive when $q = 3$ [10,11,22] and multifragmentation becomes important for $q > 6$ [6]. Then, an interesting and important question is raised, "How many charges can C_{60} hold at most?" In other words, we want to know which is the Coulomb stability limit (highest charge q_{lim}) for the C_{60} fullerene. Recent experiments have detected C_{60}^{12+} with a lifetime of the order of microseconds by intense laser irradiation of C_{60} [20,32]. On the other hand, the Coulomb limit for C_{60} fullerene has also been predicted theoretically, as we will see in Section 25.3.1.5.

25.3.1.3 Asymmetric Fission Model

The fragmentation of fullerenes is not only a thermodynamic but also a kinetic process which depends on an energy barrier. To begin with, let us apply a simple model to understand the origin of the fission barrier (FB). In Figure 25.6, there are two crossing curves: one corresponds to the sequential evaporation $(C_{60}^{q+} \longrightarrow C_{58}^{q+} + C_2)$; the other represents the asymmetric fission $(C_{60}^{q+} \longrightarrow C_{58}^{(q-s)+} + C_2^{s+})$ [33]. The first curve is attractive in nature determined by the dipole and induced dipole interaction between C_{58}^{q+} and C_2, which can be formulated as follows:

$$E(R) = DE(C_2) - \frac{\alpha q}{2R^4}, \qquad (25.6)$$

where

α is the polarizability of the dimer

R is the distance between the two fragments

DE(C$_2$) is the dissociation energy of the neutral C$_2$ dimer

The second curve is repulsive characterized by the simple Coulomb interaction between two point charges:

$$E(R) = \mathrm{DE}(\mathrm{C}_2{}^{s+}) + \frac{q_\mathrm{B} q_\mathrm{C}}{R}. \qquad (25.7)$$

Now let us examine the behaviors of the two curves during the asymmetric fission process of C$_{60}{}^{q+}$. At the beginning of the fission process (at shorter distance R), the attractive curve (25.6) controls the process since it is energetically much lower than the repulsive curve. This fact indicates that the initial dissociation process does not involve significant charge transfer. At the other extreme, near the end of the asymmetric fission process (at longer distance R), the fragmentation follows the repulsive curve (25.7), for it is decreasing in energy while the attractive curve (25.6) keeps on increasing in energy. This also indicates that the Coulomb repulsion accelerates the separation of the two positively charged fragments after the dissociation. At the intermediate distance R, a crossing point is found between both curves, providing the energy barrier for the asymmetric fission (see Figure 25.6).

25.3.1.4 Thermodynamic Stability: Dissociation Energies

The dissociation energies for the process C$_{60}{}^{q+} \longrightarrow$ C$_{58}{}^{(q-s)+}$ + C$_2{}^{s+}$ (or [2C]$^{s+}$) for $s = 0 - 4$ have been obtained by DFT computations [90]. As shown in Figure 25.6, the dissociation energies are expressed as $E(\mathrm{C}_{58}{}^{(q-s)+}) + E(\mathrm{C}_2{}^{s+}) - E(\mathrm{C}_{60}{}^{q+})$ in which all the

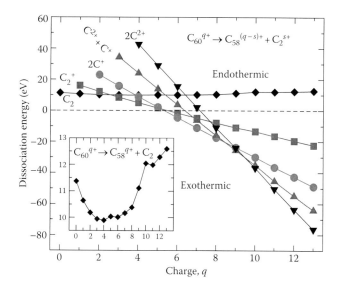

FIGURE 25.7 Dissociation energies for the process C$_{60}{}^{q+} \longrightarrow$ C$_{58}{}^{(q-s)+}$ + C$_2{}^{s+}$ (or [2C]$^{s+}$). The curve in an expanded scale corresponding to $s = 0$ is shown in the inset. (From Díaz-Tendero, S. et al., *Phys. Rev. Lett.*, 95, 013401, 2005; Díaz-Tendero, S. et al., *J. Chem. Phys.*, 123, 184306, 2005.)

total energies correspond to the optimized geometries. The results are summarized in Figure 25.7. Note that carbon dimers C$_2{}^{s+}$ with a charge more than or equal to two ($s \geq 2$) are not stable and, therefore, the dissociation leads to two charged atoms. For the sake of simplicity, a short notation [2C]$^{s+}$ is used for $s \geq 2$. It can be seen that, for $s \geq 1$, the dissociation energy varies almost linearly with the initial charge q. The slope of these linear curves is approximately proportional to the product of the fragment charges, $(q - s)s$, thus indicating that a simple point-charge Coulomb interaction is responsible for the observed behavior [90].

Figure 25.7 demonstrates that the fragmentation process becomes exothermic for $q \geq 6$, a value much lower than the highest charge (12+) observed in experiments [20,32]. This apparent contradiction reveals the existence of fission barriers that hinder the exothermic fragmentation. It can also be seen from Figure 25.7 that for $q \geq 6$, dissociation into [2C]$^{s+}$ with $s \geq 2$ is even more favorable. This agrees with the experimental finding that the multifragmentation of highly excited C$_{60}{}^{q+}$ becomes important for $q > 6$ [4].

Another important conclusion that can be drawn from the inset in Figure 25.7 is that the dissociation energy for C$_2$ ejection ($s = 0$) depends significantly on the initial charge q with variations as large as 3 eV. This finding disproves the usual assumption that the C$_2$ dissociation energy is constant for all q, an assumption that has been applied to infer fission barriers from the measured KERs [11]. Interestingly, the C$_2$ dissociation energy exhibits pronounced maxima at $q = 0$ and $q = 10$, which correspond to the closed-shell structure of h$_u$ and g$_g$ + h$_g$ orbitals (see Figure 25.3). A less pronounced maximum is also found at $q = 5$ where the h$_u$ is half-filled. These observations support the suggestion that closed and semi-closed electronic shells can provide extra stability to most fullerenes.

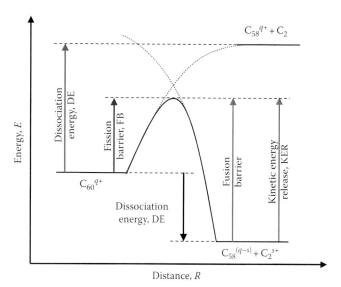

FIGURE 25.6 Scheme for the asymmetric fission process of C$_{60}{}^{q+}$. (From Díaz-Tendero, S. et al., *Phys. Rev. Lett.*, 95, 013401, 2005.)

25.3.1.5 Kinetic Stability: Fission Barriers and Coulomb Limit

As we have seen above, the fission of C_{60}^{q+} is already exothermic when $q = 6$, far from the experimentally observed highest charge state $q = 12$. In order to correctly predict the Coulomb stability limit q_{lim}, one has to consider the fission barriers mentioned above. There are two possible C_{58} isomers resulting from the fission [125], shown in Figure 25.8. Note that the isomer C_{58}^{q+} (2) is a nonclassical isomer containing a heptagonal ring. Calculations show that C_{58}^{q+} (1) is slightly more stable than C_{58}^{q+} (2) for $q = 0 - 14$; therefore, both C_{58} isomers should be considered [90].

Firstly let us look at the potential energy surface (PES) for the $C_{60}^{q+} \longrightarrow C_{58}^{(q-s)+} + C_2^{s+}$ process. As shown in Figure 25.9, there is a transition state (labeled "TS") resulting in the fragmentation products $C_{58}^{(q-s)+}$ and C_2^+. Between this transition state and the initial C_{60}^{q+}, there are several intermediate minima on the PES [125]. The intermediate of interest is the last minimum before the fragmentation takes place, i.e., the minimum that is followed immediately by TS. This outmost minimum is labeled as "MIN" in Figure 25.9. As illustrated in this figure, the FB is the sum of ΔE_1 and ΔE_2, which are the energy differences between the initial C_{60}^{q+} and MIN, and between MIN and TS, respectively.

The transition state is characterized by the distance R between the $C_{58}^{(q-s)+}$ cage and the C_2^{s+} dimer, the C–C distance R_{CC} in the C_2^{s+} dimer, and the charges s_i and s_o on the inner and outer atoms of C_2^{s+}, as depicted in Figure 25.10. Table 25.2 lists the computed characteristics of the transition states $C_{58}(1)$—C_2 [33]. The C–C distance R_{CC} in the C_2^{s+} dimer increases with the increasing initial charge q. The two carbon atoms of C_2^{s+} are bonded ($R_{CC} \leq 1.5$ Å) for $q = 3 - 8$, while the C–C bond is broken ($R_{CC} \geq 1.8$ Å) when $q \geq 9$. Therefore, the fragmentation of C_{60}^{q+} ($q = 3 - 8$) is an asymmetric fission, resulting in a C_2^{s+} dimer (see Figure 25.10a). At the same time, the total charge in the C_2^{s+} dimer is $s \leq 1$. Thus, evaporation of C_2^+ is expected. On the other hand, the fragmentation of highly charged C_{60}^{q+} ($q \geq 9$) leads to a dissociated C_2^{s+}

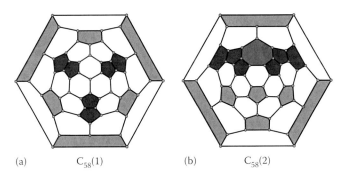

(a) $C_{58}(1)$ (b) $C_{58}(2)$

FIGURE 25.8 Schlegel diagrams for two C_{58} isomers.

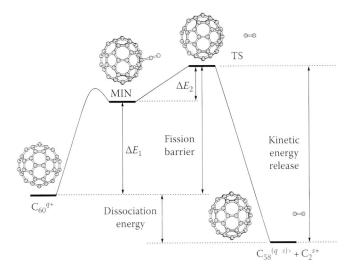

FIGURE 25.9 The PES for the $C_{60}^{q+} \longrightarrow C_{58}^{(q-s)+} + C_2^{s+}$ process. (From Díaz-Tendero, S. et al., *J. Chem. Phys.*, 123, 184306, 2005.)

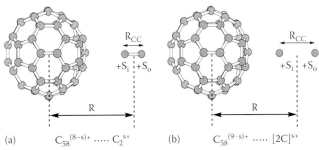

(a) $C_{58}^{(8-s)+}$ ..... C_2^{s+} (b) $C_{58}^{(9-s)+}$ ..... $[2C]^{s+}$

FIGURE 25.10 The transition states for the dissociation processes: (a) $C_{60}^{8+} \longrightarrow C_{58}^{(8-s)+} + C_2^{s+}$ and (b) $C_{60}^{9+} \longrightarrow C_{58}^{(9-s)+} + C_2^{s+}$ [33]. R is the distance between fullerene cage and C_2^{s+} dimer. R_{CC} is the C–C distance in C_2^{s+} dimer, and s_i and s_o are the charges on the inner and outer atoms of C_2^{s+}, respectively. (From Díaz-Tendero, S. et al., *Phys. Rev. Lett.*, 95, 013401, 2005; Díaz-Tendero, S. et al., *J. Chem. Phys.*, 123, 184306, 2005.)

TABLE 25.2 Distances (in Å), Charges (in a.u.), and Fission Barriers FB (in eV) in the Transition State ($C_{58}(1)$—C_2) for the $C_{60}^{q+} \longrightarrow C_{58}^{(q-s)} + C_2^{s+}$ Fission Process

Q	s_i	s_o	R	R_{CC}	FB
3	0.0	0.2	7.417	1.334	9.9
4	0.0	0.3	6.810	1.341	9.5
5	0.1	0.4	6.626	1.343	9.0
6	0.1	0.5	6.510	1.345	8.1
7	0.2	0.6	6.463	1.426	7.2
8	0.3	0.7	6.753	1.502	7.6
9	0.5	1.0	6.292	2.230	5.8
10	0.5	1.1	6.083	1.865	5.6
11*	0.0	1.1	—	—	3.3
12*	0.0	1.0	—	—	2.5
13*	0.0	1.1	—	—	1.6
14*	0.0	1.2	—	—	0.8

Sources: Díaz-Tendero, S. et al., *Phys. Rev. Lett.*, 95, 013401, 2005; Díaz-Tendero, S. et al., *J. Chem. Phys.*, 123, 184306, 2005.

Note: Distances R and R_{CC} and charges s_i and s_o are defined in Figure 25.10. Values with asterisk are evaluated with a scan calculation.

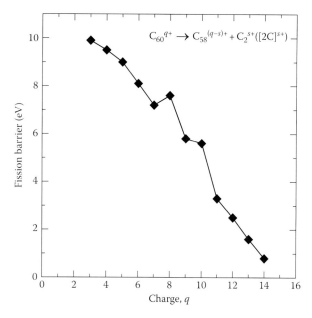

FIGURE 25.11 Fission barrier as a function of charge for the processes: $C_{60}^{q+} \longrightarrow C_{58}^{(q-s)+} + C_2^{s+}$ or $[2C]^{s+}$. (From Díaz-Tendero, S. et al., *J. Chem. Phys.*, 123, 184306, 2005.)

unit. For $q = 9 - 11$, the total charge in the C_2^{s+} unit is about 1.5 (see Table 25.2), indicating the dissociation into two C^+ cations. For $q \geq 12$, the transition state corresponds to an almost free C^+ ion and a neutral C atom bonded to two $C_{58}^{(q-1)+}$ sites, suggesting that the fragmentation is a two-step process in which a C^+ ion is ejected first [33,90].

As we can see in Table 25.2 and Figure 25.11, the height of the FB decreases as the charge q increases. Meanwhile, the distance R between the $C_{58}^{(q-s)+}$ and the C_2^{s+} for the transition state becomes shorter (see Table 25.2). The FB for $q = 14$ is only 0.8 eV and would be even lowered by about 0.2 eV after the inclusion of the zero-point energy. Therefore, the Coulomb stability limit for C_{60}^{q+} corresponds to $q_{lim} = 14$. This value of q_{lim} should not change if the possibility of either C_4^{s+} emission or Coulomb explosion is considered, since in both cases the dissociation energy is much larger than that for the asymmetric fissions and thus a much higher FB is estimated based on the above-mentioned simple fission model. It is worth noticing the good agreement between the experimentally measured KERs and the calculated ones (see Figure 25.12).

25.3.2 Positively Charged C_{70} Fullerenes

25.3.2.1 Electronic States and Ionization Potentials

The most stable neutral isomer of C_{70} has D_{5h} symmetry and is the only one that obeys the IPR. The HOMO of C_{70} is doubly degenerated. Theoretical calculations show that the ground state for C_{70}^{2+} is a triplet state in accordance with the Hund's rule [95]. However, in contrast to C_{60}^{q+}, the energy differences between intershell states with different spin multiplicities cannot be interpreted in terms of the Hund's rule for the higher charge states of C_{70}^{q+} ($q > 2$). Nevertheless, the most stable

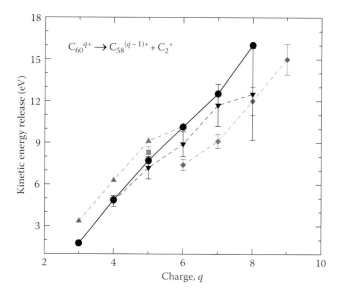

FIGURE 25.12 Kinetic energy of C_2^+ fragments emitted from C_{60}^{q+}. DFT results [33]: a solid line with circles. Experimental results: dashed lines with diamonds [11], upward triangles [22], squares [7], and downward triangles [10]. (From Díaz-Tendero, S. et al., *Phys. Rev. Lett.*, 95, 013401, 2005.)

configurations correspond to high spin multiplicities for $q > 6$, which is most likely due to many close-lying (almost degenerated) lower energy levels [95].

The calculated adiabatic IPs of C_{70}^{q+} are presented in Figure 25.13 as a function of the degree of ionization [95]. For low-charge states of C_{70}^{q+}, the computational results are in reasonable agreement with the experimental results [93,95]. As for C_{60}^{q+}, the

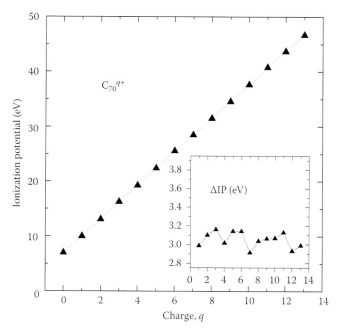

FIGURE 25.13 Ionization potentials of C_{70}^{q+} fullerenes as a function of the degree of ionization. The inset shows the variation of $\Delta IP = IP_q - IP_{q-1}$ with q. (From Zettergren, H. et al., *J. Chem. Phys.*, 127, 104308, 2007.)

IPs of C_{70}^{q+} follow, to a good approximation, a linear behavior according to the simple formula after numeric fitting [95]:

$$IP(C_{70}) = 7.048 + 3.059q \text{ (eV)}. \quad (25.8)$$

By applying (25.4), one can estimate the effective radius of C_{70} fullerene to about $8.90a_0$, which is in agreement with the experimental value of $8.87a_0$ [155]. The inset in Figure 25.13 shows deviations of the IPs from the linear dependence. In contrast to C_{60}^{q+}, where deviations can be as large as $1\,\text{eV}$ (see Section 25.3.1.1), deviations for C_{70}^{q+} are less than $0.2\,\text{eV}$, thus implying that there are no important shell effects in this case.

25.3.2.2 Dissociation Energies

In contrast with C_{60} (I_h), there are many possible ways to extract a C_2 unit from the C_{70} (D_{5h}) cage. In this case, extraction of a C_2 unit can lead to 15 possible C_{68} isomers, including 11 classical isomers and 4 nonclassical isomers with a heptagonal ring [95]. Among these possible C_{68} products, theory predicts that the most stable neutral isomer is the classical $C_{68}(C_2\text{-II})$ isomer. For positively charged species, $C_{68}(C_2\text{-I})$ is the most stable charged isomer (see Figure 25.14) [95]. Figure 25.15 shows the thermodynamical dissociation energies for C_{70}^{q+} ($q = 0 - 14$) corresponding to dissociation leading to $C_{68}(C_2\text{-II})$ for $q = 0$ and to $C_{68}(C_2\text{-I})$ for $q > 0$.

The general behavior of the dissociation energies for C_{70}^{q+} is quite similar to that for C_{60}^{q+} (see Figure 25.7). The dissociation of C_{70}^{q+} is an exothermic process for $q \geq 6$, the same thermodynamic stability limit as for C_{60}^{q+}, as we have seen in Section 25.3.1.4. This result indicates the existence of fission barriers, since higher charge states for $q \geq 5$ have been observed in experiments [30,31]. The inset in Figure 25.15 does not give maxima as pronounced as those appearing in the case of C_{60}^{q+} in Figure 25.7. This indicates that the shell effects in the sequential evaporation of C_{70}^{q+} are not significant.

25.3.2.3 Fission Barriers and Coulomb Limit

As we have seen above, the dissociation energies for C_{60}^{q+} and C_{70}^{q+} have similar behaviors. This is further supported by the experimental finding that the intensity distributions in the fragmentation spectra for C_{60} and C_{70} are similar under the same experimental conditions [31,156]. However, C_{70} is less symmetric

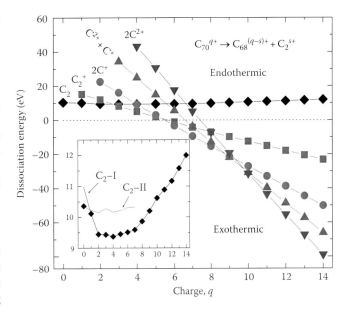

FIGURE 25.15 Dissociation energies for the process $C_{70}^{q+} \longrightarrow C_{68}^{(q-s)+} + C_2^{s+}$ (or $[2C]^{s+}$). The curve in an expanded scale corresponding to $s = 0$ is shown in the inset. (From Zettergren, H. et al., *J. Chem. Phys.*, 127, 104308, 2007.)

than C_{60} and the channels and products for C_{70}^{q+} fragmentation are much more complicated. Therefore, no computations based on transition states have been reported so far. Nevertheless, one can estimate these barriers by making several assumptions [95]. As shown in Figure 25.6, one can write

$$FB = KER + DE. \quad (25.9)$$

The KERs in the fission process have been measured by various experimental techniques [17,22,26]. The measured KERs in $C_{60}^{q+} \longrightarrow C_{58}^{(q-1)+} + C_2^+$ and $C_{70}^{q+} \longrightarrow C_{68}^{(q-1)+} + C_2^+$ processes are similar [31]. If one assumes that this observation is also valid for the other fission channels, then KER $(C_{70}) \simeq$ KER (C_{60}) and following Equation 25.9, one obtains

$$FB(C_{70}^{q+}) = FB(C_{60}^{q+}) + \Delta DE \quad (25.10)$$

where $\Delta DE = DE(C_{70}^{q+}) - DE(C_{60}^{q+})$ [95]. One can further assume that the preferred dissociation channels are the same for C_{70}^{q+} and C_{60}^{q+}, namely, C_2^+ emission for $3 \leq q \leq 8$ and two C^+ ions for $9 \leq q \leq 11$, according to the transition state calculations presented in Section 25.3.1.5. DFT calculations [95] have shown that the DE for C_{70}^{q+} is, in general, larger than that for C_{60}^{q+}. This can be understood from the fact that the excess charge on a smaller fullerene induces stronger Coulomb repulsion and thus the cage is more easily broken, resulting in a lower DE. Then by following Equation 25.10, one can expect larger FB for C_{70}^{q+} than for C_{60}^{q+}. Indeed, this conclusion is confirmed by the estimated FBs for C_{70}^{q+} and C_{60}^{q+} shown in Figure 25.16. A linear extrapolation of these results leads to an estimate of the Coulomb stability limit for C_{70}^{q+}, which is approximately $q = 17$.

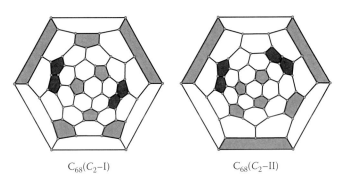

FIGURE 25.14 Schlegel diagrams for two C_{68} isomers.

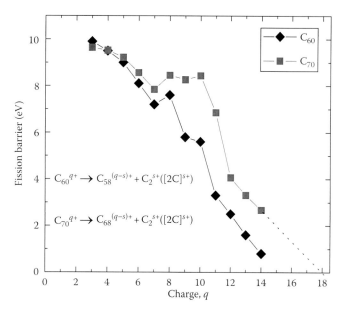

FIGURE 25.16 Fission barrier as a function of the charge for the emissions of C_{70}^{q+} and C_{60}^{q+}. (From Zettergren, H. et al., *J. Chem. Phys.*, 127, 104308, 2007.)

25.3.3 Positively Charged C_{58} and C_{68} Fullerenes

As discussed in Section 25.3.1.5, the C_2 emission of C_{60} may generate two C_{58} isomers, the classical fullerene $C_{58}(1)$ and the nonclassical fullerene $C_{58}(2)$ with a seven-membered ring. Theoretical calculations show that $C_{58}^{q+}(1)(q=0-12)$ is always lower in energy than C_{58}^{q+} (2) [90]. In contrast with C_{60}^{q+}, the electronic state with the smallest spin multiplicity is the most stable one, except for C_{58}^{6+} in which case the triplet lies 94 meV lower than the singlet [90]. This is due mainly to the lower symmetry and lesser degeneracy of the HOMOs of C_{58}^{q+} compared to C_{60}^{q+}.

We have seen in Section 25.3.2.2 that 15 possible C_{68}^{q+} isomers may be produced after C_2 extraction from C_{70}. The $C_{68}(C_2\text{-I})$ isomer is the most stable one for all positively charged species. It is interesting to note that for the neutral species, $C_{68}(C_2\text{-I})$ is 0.54 eV higher than the most stable neutral isomer $C_{68}(C_2\text{-II})$, and even higher than some other C_{68} isomers including a nonclassical isomer with a heptagon [95]. Therefore, it seems that positive charges significantly stabilize isomer $C_{68}(C_2\text{-I})$. This observation can be understood from the fact that isomer $C_{68}(C_2\text{-I})$ has the most uniformly distributed pyrene bonds (a bond surrounded by four hexagonal rings, see Figure 25.17). This can effectively separate positive charges on the fullerene surface and thus decrease the Coulomb repulsion [95]. The most stable isomer C_{68}^{q+} (C_2-I) for $q \le 6$ corresponds to the smallest spin multiplicity, i.e., singlet (when q is even) or doublet state (when q is odd), while for $q = 6 - 12$, the ground state is triplet (when q is even) or quadruplet (when q is odd). The reason is that there are a couple of close-lying (almost degenerated) orbitals below the HOMO −2 of C_{68}^{q+}.

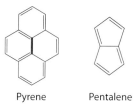

FIGURE 25.17 The pyrene and pentalene motifs. The pyrene bond is indicated by a bold line.

In analogy to C_{60}^{q+} and C_{70}^{q+}, the calculated adiabatic IPs for C_{58}^{q+} (1) and C_{68}^{q+} (C_2-I) have excellent linear correlations with the degree of ionization [95]:

$$IP(C_{58}) = 6.472 + 3.408q(eV). \qquad (25.11)$$

$$IP(C_{68}) = 6.698 + 3.117q(eV). \qquad (25.12)$$

Thus, these results can still be understood by means of a classical electrostatic picture in which an electron is extracted from a charged conducting sphere [142,143].

25.3.4 Singly and Doubly Charged C_n ($n = 20 - 70$) Fullerenes

25.3.4.1 The Most Stable Isomers

Similar to C_{58} and C_{68} that have been discussed above, other fullerenes smaller than C_{60} or C_{70} can be produced by C_2 extraction from the corresponding parent fullerenes and/or by performing SW transformations [86]. Detailed examples for fullerenes C_{50}–C_{60} can be found in the fragmentation scheme illustrated in Figure 25.18.

Díaz-Tendero et al. have investigated ionization potentials and dissociation energies for neutral, singly, and doubly charged C_n ($n = 20 - 70$) fullerenes covering all known fullerenes smaller than C_{70} [93]. By following a systematic procedure, they have determined the most stable neutral, singly, and doubly charged fullerene isomers for all sizes from C_{20} to C_{70}, which are shown in Figure 25.19. In general, the most stable isomer for neutrals is also the most stable one for singly and doubly charged fullerenes. However, there are a few exceptions, as listed in the last row of Figure 25.19. It is worth mentioning that the exceptions for C_{50} and C_{52} are the consequence of spherical aromaticity, which has been discussed in Section 25.2.3.3.

25.3.4.2 Ionization Potentials

Calculated and experimentally measured first and second IPs for C_{20}–C_{70} are presented in Figure 25.20. The computational values are in agreement with the experimental ones, except for a constant downward shift of ~0.5 eV for the first IPs. It is interesting to see the appearance of a few peaks in the curve corresponding to the first IP in the position corresponding to fullerenes C_{32}, C_{50}, C_{60}, and C_{70}. For C_{60} and C_{70}, the extraordinary stability is due to the fact that they are the only fullerenes with IPR structures in the C_{20}–C_{70} series. The maxima of the first IPs at $n = 32$ and

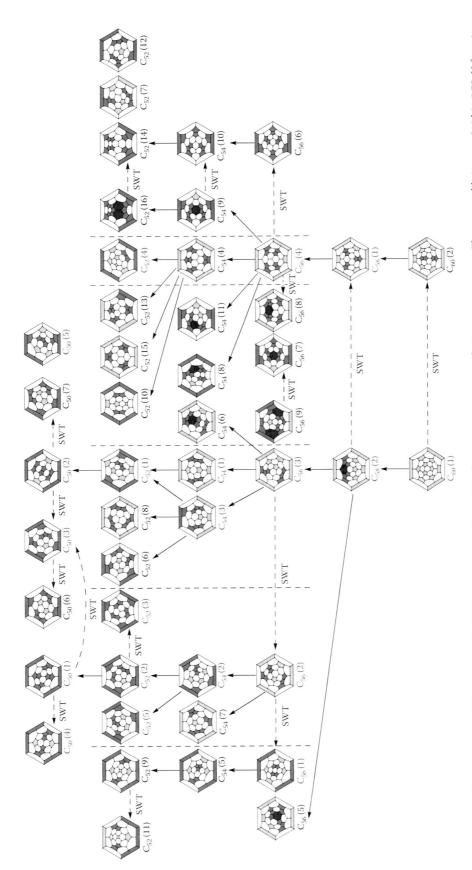

FIGURE 25.18 (See color insert following page 25-14.) Fragmentation scheme. Numbers represent energy ordering among isomers. The most stable isomers (within 0.75 eV from isomer number 1) are shown in bold red. Solid and dashed lines indicate C_2 losses and SW transformations (SWTs), respectively.

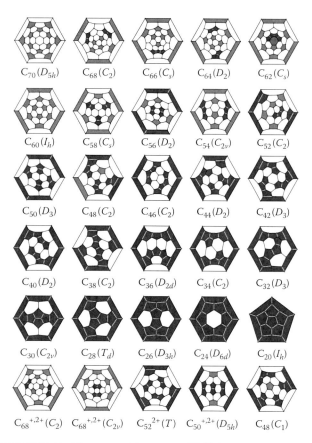

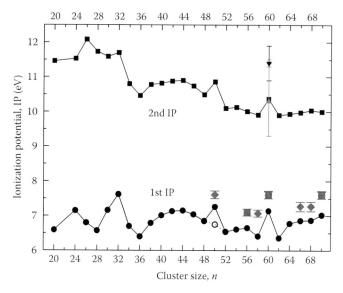

FIGURE 25.20 First and second ionization potentials of C_n ($n = 20 - 70$) fullerenes as a function of cluster size. Full black circles and squares: DFT results. Experimental values: diamonds, Zimmerman et al. [152]; blue squares, McElvany and Bach [158], upward triangles, Baba et al. [159]; downward triangles, Muigg et al. [160]; crosses, Steger et al. [161] Empty circles: result obtained for a C_{50} fullerene with D_{5h} symmetry [93]. (From Díaz-Tendero, S. et al., *Int. J. Mass Spectrom.*, 252, 133, 2006.)

FIGURE 25.19 The most stable isomers of C_n^{q+} ($n = 20 - 70, q = 0, 1, 2$) fullerenes. Symmetries are given by the CaGe program [102] before geometry optimization. The most stable isomer of charged species is generally the same as for neutrals. Exceptions are listed in the last row. (From Díaz-Tendero, S. et al., *Int. J. Mass Spectrom.*, 252, 133, 2006.)

50 support the $2(N + 1)^2$ rule for spherical aromaticity, which confers additional stability. Like small carbon clusters [157], the second IP approximately decreases with fullerene size. Again, maxima are found at $n = 32, 50$, and 60. An additional peak appears at $n = 26$, which can be understood by the fact that C_{26}^+ has 25 π electrons, which corresponds to a half-filled electronic shell according to the $2(N + 1)^2$ rule.

25.3.4.3 Dissociation Energies

Figure 25.21 presents the C_2 and C_2^+ dissociation energies for singly charged fullerenes. As can be seen, the variation of dissociation energies with fullerene size is similar in the calculations and the experiments. In particular, they predict pronounced peaks in the vicinity of C_{50}^+, C_{60}^+, and C_{70}^+, as well as a pronounced minimum for C_{62}^+. The lowest dissociation energy C_{62}^+ of is related to the easiness to form C_{60}^+ by C_2 emission from C_{62}^+. The most stable C_{62} isomer is a nonclassical fullerene containing a heptagonal ring (see Figure 25.19). If we take the most stable classical isomer of C_{62} instead of the nonclassical one, the dissociation energies for C_{64}^+ and C_{62}^+ are about 1 eV larger and smaller, respectively, than those in Figure 25.21, which deteriorates the agreement between theory and experiment. This shows

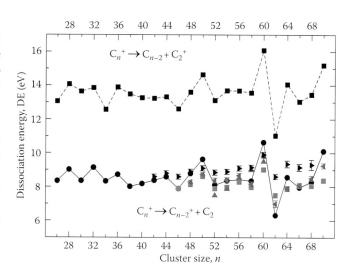

FIGURE 25.21 C_2 (circle, full line) and C_2^+ (square, dashed line) dissociation energies as functions of fullerene size for singly charged fullerenes. DFT results [93]: circles. Experimental values for C_2 dissociation energies: square, Barran et al. [162] scaled to C_{54}^+; upward triangle, Laskin et al. [163]; leftward triangle, Tomita et al. [164]; rightward triangle, Concina et al. [165]. (From Díaz-Tendero, S. et al., *Int. J. Mass Spectrom.*, 252, 133, 2006.)

that the C_{62}^+ isomer produced in the experiments is most likely the nonclassical fullerene containing heptagonal ring. The C_2^+ dissociation energy follows a similar trend to the C_2 dissociation energy, but with much higher values (about 5 eV higher). The C_2^+ emission is more endothermic than the C_2 emission because the

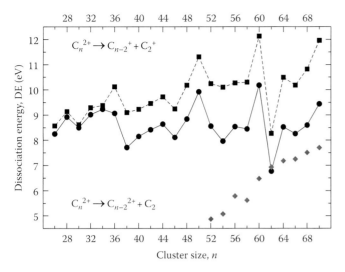

FIGURE 25.22 C_2 (circle, full line) and C_2^+ (square, dashed line) dissociation energies as functions of the fullerene size for doubly charged fullerenes. Circles: DFT results [93]. Diamonds: experimental values for the C_2 dissociation energies [17]. (From Díaz-Tendero, S. et al., *Int. J. Mass Spectrom.*, 252, 133, 2006.)

system is more stabilized if the positive charge is distributed in the larger C_{n-2} cluster instead of in the small C_2 cluster.

Calculated C_2 and C_2^+ dissociation energies for doubly charged fullerenes are shown in Figure 25.22 [93]. The computational results are in fair agreement with experiments [17] for $n \leq 60$. Again, maxima are observed here at $n = 50, 60$, and 70. As for the singly charged fullerenes, C_2^+ emission is more endothermic than C_2 emission (see Figure 25.22). This is because dividing the charge into a small and a large fragment is energetically more favorable than locating it in the large fragment. However, the energy difference between C_2^+ and C_2 dissociations is small

for the smaller fullerenes ($n \leq 34$). Therefore, asymmetric fission ($C_n^{2+} \longrightarrow C_{n-2}^+ + C_2^+$) is competitive with C_2 evaporation ($C_n^{2+} \longrightarrow C_{n-2}^{2+} + C_2$) even for relatively cold C_n^{2+} ions.

25.3.5 Negatively Charged C_{60} and C_{70} Fullerenes

DFT calculations [95] have shown that the ground states of C_{60}^{2-} and C_{70}^{2-} are triplet. Nevertheless, the energy gap between triplet and singlet states is rather small (15 and 45 meV for C_{60}^{2-} and C_{70}^{2-}, respectively, according to DFT calculations [95]). The thermodynamic stability of C_{60} and C_{70} anions can be measured by their electron affinities (EAs). The first EAs of C_{60} and C_{70} are both positive, indicating that the first electron introduced in the neutral fullerene is bound. The DFT calculated values are 2.59 and 2.77 eV for C_{60} and C_{70}, respectively [95], in good agreement with the experimental measurements [50,166,167]. The calculated second EAs of C_{60} and C_{70} are slightly negative (−0.38 and −0.14 eV, respectively) [95], indicating that the second electron is only confined by a Coulomb barrier [168]. This is consistent with recent lifetime measurements of C_{60}^{2-} in a storage ring, which suggests that this system is metastable with a ground-state lifetime of the order of 20 s [47].

Zettergren et al. [95] have also theoretically examined the first and second EAs for some non-IPR isomers of C_{60} and C_{70}. Interestingly, the first and second EAs for these non-IPR isomers are higher (more positive) than the EAs for the IPR isomers [47]. This means that, compared to the IPR C_{60} and C_{70} isomers, the non-IPR isomers are less destabilized by negatively charging. This statement is further confirmed by calculations on the relative energies of highly charged C_{60}^{q-} and C_{70}^{q-} anions. As shown in Figure 25.23, the energy difference between the non-IPR and IPR isomers of C_{60}^{q-} or C_{70}^{q-} decreases significantly with

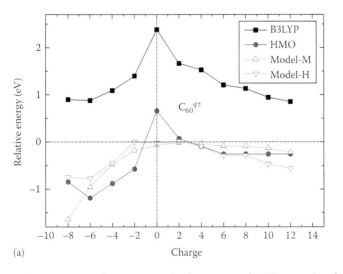

(a)

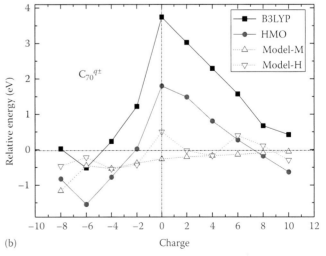

(b)

FIGURE 25.23 Relative energies (with respect to the IPR isomer) as functions of the charge for isomers $C_{60}(D_3:1803)$ (a) and $C_{70}(C_{2v}:7854)$ (b). The solid lines with square symbols show the relative energies by DFT method [96] and those with diamond symbols show the relative energies calculated by HMO method [169]. The dotted lines are for the relative energies evaluated by simple electrostatic model using Mulliken (Model-M) and Hückel (Model-H) charges, indicated by upward and downward triangles, respectively [169].

the increasing negative charge number q [96]. For example, the neutral non-IPR isomer C_{60} (D_3:1803, with 3 AP pairs) is about 2.3 eV higher in energy than the IPR C_{60} isomer. By adding electrons this value drops rapidly and the non-IPR C_{60}^{6-} (D_3:1803) is only 0.75 eV higher than the IPR C_{60}^{6-}. For C_{70}, the non-IPR isomer C_{70}^{6-} (C_{2v}:7854, with 3 AP pairs) becomes even more stable than the IPR C_{70}^{6-} (see Figure 25.23), which violates the IPR.

Several explanations have been suggested to understand the variation of the energy difference between non-IPR and IPR fullerene isomers. Popov and Dunsch [67] pointed out that the excess electrons lead to the change of the hybridization state of the carbon atom from sp^2 to sp^3. As a result, the excess electrons are more stabilized at the carbon atoms on the pentagon–pentagon edges, since these carbons have more sp^3-like conformations. As has been discussed in previous sections, the pentagon–pentagon junctions induce strain in neutral and charged fullerenes because the sp^3 conformation is less favorable than the sp^2 conformation in π conjugated systems (see Section 25.2.2.3). However, the sp^3 conformation is more suitable to host the extra electron of the anion because the additional hybrid orbital can lead to the formation of a lone electronic pair containing this extra electron and the one that was originally in the p_z orbital not used in the sp^2 hybridization. As a consequence, contribution of the corresponding carbon atom to electron delocalization in the fullerene cage is significantly reduced. These reasonings also explain why, e.g., the ·CH$_3$ radical is planar while the CH$_3^-$ anion is pyramidal.

Alternatively, Zettergren et al. [96] proposed a simple electrostatic model to understand the stabilization of the non-IPR isomers by adding electrons. They noticed that the charge distribution in C_{60} and C_{70} is controlled by the number and location of two different structural motifs, namely, the electrophilic pentalene motif (i.e., an AP pair) and the electrophobic pyrene motif (see Figure 25.17). In fullerene anions, the excess electrons prefer to go to the pentalene motifs. Hence, if the pentalene motifs are uniformly distributed on the fullerene surface, the excess negative charges are well separated and thus the Coulomb repulsion is effectively decreased. In contrast, the Coulomb repulsion cannot be reduced in the IPR isomers since the charges are not well separated because of the lack of AP pairs. For instance, the non-IPR isomer C_{60} (D_3:1803) contains three well-separated AP pairs, whereas the IPR C_{60} isomer does not. As a result, the former gets more stabilization than the latter by adding electrons. The same reasoning holds for C_{70} isomers, as can be seen in Figure 25.23.

Figure 25.23 also shows the energy difference between the non-IPR and IPR isomers of positively charged C_{60}^{q+} and C_{70}^{q+} fullerenes. It can be seen that this energy difference decreases significantly with the positive charge q [96]. This behavior is surprisingly similar to that found for the corresponding negatively charged species. For example, it can be seen that the energy difference between the neutral non-IPR isomer C_{60} (D_3:1803, with 3 AP pairs) and the IPR C_{60} isomer decreases very rapidly when electrons are removed from the cage. For C_{70}, the non-IPR isomer C_{70} (C_{2v}:7854, with 3 AP pairs) becomes almost degenerate

with the IPR C_{70} for a charge of 10+ (see Figure 25.23). The simple electrostatic model of reference [96] has also been used to explain this behavior. In this case, the excess positive charges prefer to stay at the pyrene motifs, and the system can thus be stabilized if the pyrene motifs are uniformly distributed on the fullerene surface. This is why the non-IPR C_{60} or C_{70} cations are more stabilized than the IPR ones when electrons are removed (see Figure 25.23).

Although the simple models above are useful for a qualitative understanding, their predictive value is rather limited. In Figure 25.23, we show the results obtained by replacing the real electrostatic potential with point charges placed in the atoms' positions. It can be seen that Model-M and Model-H (in which the point charges correspond to Mulliken and Hückel charges, respectively) are not able to predict the variation of the relative energy of the charged non-IPR C_{60} isomer with respect to the IPR one. This illustrates that one cannot ignore the complexity of the actual electrostatic potential in order to make quantitative predictions.

Figure 25.23 also shows the results obtained from the HMO model. This model is able to predict the correct variation of the relative stability of charged fullerene isomers for both C_{60} and C_{70}. Nevertheless, the absolute values of the HMO model always stay below the high-level computational results. This systematic difference between HMO and high-level computational results is mainly due to the inclusion of strain. As mentioned above, the HMO energy is exclusively associated with the π-electrons [126] and only depends on the topology of the fullerene cage regardless of structural parameters such as bond lengths and bond angles. Since the strain energy of the non-IPR isomer is generally higher than that of the IPR one, the HMO method always underestimates the relative energy of the non-IPR isomer, as can be seen in Figure 25.23. The fact that the HMO and the high-level computational methods give similar relative energy curves indicates that it is the modification of the π-electronic structure produced by the addition or removal of electrons that is responsible for the nonuniform charge distribution and the relative stability of charged fullerenes.

Based on the Hückel energy levels, we can explain the variations of the relative energies shown in Figure 25.23 in terms of the HMO orbital energies. As can be seen in Figure 25.24, the energies of the HOMOs are higher in the non-IPR isomer than in the IPR one. Consequently, removing electrons from the HOMOs will destabilize the IPR isomer more than the non-IPR one. Similarly, Figure 25.24 shows that the energies of the lowest unoccupied molecular orbitals (LUMOs) of the non-IPR isomer are lower than those of the IPR isomer. Therefore, addition of electrons in the LUMOs will stabilize the non-IPR isomers more than the IPR ones. Similar arguments are also valid to explain the relative stabilities of the C_{60} isomers shown in Figure 25.23.

25.3.6 Negatively Charged Fullerenes

Apart from C_{60} and C_{70}, negatively charged fullerenes with other cage sizes have also been found. In particular, some non-IPR

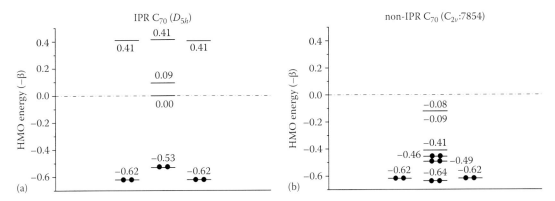

FIGURE 25.24 Hückel energy levels for the IPR C_{70} (D_{5h}) (a) and the non-IPR C_{70} (C_{2v}:7854) (b) isomers [169]. The energy unit β is the resonance integral in the HMO theory and is about -2.49 eV for the sp^2 carbon in aromatic compounds [170]. Please note that the energy values here only provide a rough and qualitative picture for π-electron structures of fullerenes, due to various approximations employed in the HMO theory.

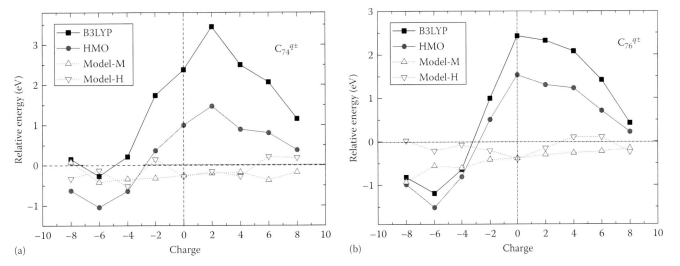

FIGURE 25.25 Relative energies (with respect to the IPR isomer) as functions of charge for non-IPR isomers C_{74}(C_2:13265) (a) and C_{76}(C_2:17490) (b) [169]. Lines and symbols have the same representations as in Figure 25.23.

fullerenes have been isolated by encaging metals. For example, endohedral fullerenes with cages larger than C_{70}, which could in principle adopt an IPR structure, are often more stable under a non-IPR form, such as $Sc_3N@C_{70}$, [68] $La_2@C_{72}$ [54], and $Tb_3N@C_{84}$ [62].

Figure 25.25 shows the relative energies of C_{74} and C_{76} non-IPR isomers with respect to the IPR ones as a function of the fullerene charge. As can be seen, the relative energy of the non-IPR isomer decreases again with the absolute value of the charge. For some anions (C_{74}^{6-}, C_{76}^{4-}, C_{76}^{6-} and C_{76}^{8-}), the non-IPR form is even energetically more favorable than the IPR form. The non-IPR C_{76}^{6-} isomer lies 1.2 eV lower in energy than the IPR one, which underlines once again that one should be careful in applying the IPR to negatively charged fullerenes. Figure 25.25 also reveals that the simple electrostatic model fails as before. In contrast, the simple HMO model correctly describes the general trends for C_{74} and C_{76}. Indeed, if the HMO curves in Figure 25.25 are shifted upward to make the $q = 0$ relative energy coincide with the value from high-level calculations, one would observe

that the HMO and the high-level calculation curves superimpose quite reasonably. Hence, the π-electronic structure is again the main factor governing the variation of the relative energies of charged fullerenes.

Popov et al. [67,68,171] have systematically searched for the most stable isomers of fullerene hexaanions C_n^{6-}($n = 70 - 98$). According to their computations, the most stable isomer of C_n^{6-} is always a non-IPR one for $n = 70 - 76$, while the IPR isomer is the most stable one for $n \geq 78$. This finding can be understood by looking at the HMO energy levels. For neutral C_n ($n = 70 - 76$) fullerenes, there are only one or two IPR isomers and the maximum number of unoccupied states with negative Hückel eigenvalues is less than six (i.e., less than three HMO orbitals, see Table 25.3). In the hexaanions, as a result, less than six electrons can stay at the negative energy LUMOs and the rest of the electrons must go into the positive energy LUMOs, which further destabilizes the system. In comparison, the neutral non-IPR isomer can have six unoccupied states with negative eigenvalues which exactly accommodate six electrons to

TABLE 25.3 Number of IPR Isomers (N_{IPR}), Number of Vacant Hückel States with Negative or Almost Zero Eigenvalue in the Neutral IPR (M_{IPR}) and Non-IPR ($M_{non-IPR}$) Fullerene, and the Most Stable Isomer[a] of the Hexaanions (Underlined)

C_n^{6-}	N_{IPR}	M_{IPR}	$M_{non-IPR}$
C_{70}^{6-}	1	0	<u>6</u>
C_{72}^{6-}	1	0	<u>6</u>
C_{74}^{6-}	1	2	<u>6</u>
C_{76}^{6-}	2	4	<u>6</u>
C_{78}^{6-}	5	<u>6</u>	6
C_{80}^{6-}	7	<u>6</u>	6
C_{82}^{6-}	9	<u>6</u>	6
C_{84}^{6-}	24	<u>6</u>	6
C_{86}^{6-}	19	<u>6</u>	4

[a] The most stable isomers of hexaanions have been determined by Ref. [67].

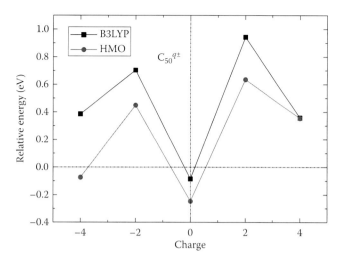

FIGURE 25.26 Relative energies (with respect to isomer $C_{50}(D_{5h}:271)$) as functions of the charge for the isomer $C_{50}(D_3:270)$. The solid lines with square symbols show the relative energies by the DFT method and those with diamond symbols show the relative energies calculated by the HMO method.

form a stabilized hexaanion. Therefore, for C_n ($n = 70 - 76$), the most stable hexaanions are the non-IPR isomers. In contrast, as the fullerene size increases, the number of IPR isomers becomes larger and larger. As a consequence, there always exist IPR isomers with six negative (or almost zero) unoccupied states and thus the corresponding hexaanions are stabilized. Therefore, the IPR isomers are more stable and the IPR is valid again for fullerene hexaanions with $n \geq 78$.

25.3.7 Negatively Charged C_{50}

It is interesting to analyze the case of C_{50} anions separately. In this case, the IPR does not apply any more because the topology of C_{50} makes it impossible to place all pentagonal rings isolated from one another. Consequently, all C_{50} isomers contain APs. As we have seen in Section 25.2.3.3, the neutral $C_{50}(D_3:270)$ isomer is lower in energy than $C_{50}(D_{5h}:271)$, although the former has more APs than the latter (see Figure 25.4). It is believed that this violation of the PAPR for neutral C_{50} isomers is due to spherical aromaticity. Figure 25.26 presents the energy difference between these two isomers as a function of the charge. As can be seen, for negatively and positively charged C_{50} the situation is the opposite: $C_{50}^{q\pm}(D_{5h}:271)$ isomers are energetically less stable than the $C_{50}^{q\pm}(D_3:270)$ isomers. Thus, the PAPR works again for charged C_{50} isomers. One may ascribe this observation to the fact that for charged C_{50} fullerenes the number of π-electrons does not satisfy the $2(N + 1)^2$ formula and thus the spherical aromaticity rule no longer applies. However, we can see in Figure 25.26 that the variation of the relative energy differences between these two C_{50} isomers is again well reproduced by the HMO model. This is rather surprising because the HMO model includes no information related to the 3D geometry of different isomers. Hence, the variation of the relative energies

between the two C_{50} isomers is most likely due to specific delocalization effects of standard π-electronic systems rather than to spherical aromaticity.

25.4 Summary

Positively and negatively multiply charged fullerenes are produced in collisions of C_{60} and C_{70} with fast highly charged ions, fast electrons, or energetic short laser pulses, in storage rings or by encapsulating metallic species in the fullerene cage. A large variety of charged fullerenes containing from 20 to 70 carbon atoms are produced in these experiments, which requires a good knowledge of their structure and stability. Positively charged C_{60} fullerenes with charges as high as 12+ have been observed in experiments [20,32], while theoretical predictions indicate that this charge could be as high as 14+ [33]. Motivated by these experimental developments, systematic computational studies have been carried out for positively and negatively charged fullerenes. Theoretical and experimental data for ionization potentials, dissociation energies, fission barriers, as well as electron affinities are now available, which is essential for future studies in this or related fields. It has been shown that some properties can be understood in terms of simple electronic arguments (such as the conducting sphere model [142,143]) or empirical rules (such as the isolated pentagon rule, IPR [127], and the pentagon adjacency penalty rule, PAPR [119,127,132–134]). In some cases, these properties cannot be correctly predicted by using simple rules, which have been proposed for neutral fullerenes. For example, the IPR works quite well for neutral fullerenes with few exceptions, whereas, it may give wrong predictions for charged fullerenes [67,96,169]. For many highly charged fullerenes, the non-IPR isomers are surprisingly more stable than the IPR ones. This finding has also been supported by the isolation of some non-IPR endohedral fullerenes in experiments [54,62,68]. Therefore, the use of the IPR to predict the relative stability of charged fullerene does not seem to be appropriate. Alternatively, one can rely on the simple HMO model to obtain a reasonable estimate of the relative stability of IPR and non-IPR isomers, and use it to predict the existence of non-IPR isomers that are more stable than the IPR ones. Of course, at the end, these predictions must be compared with experiments or accurate theoretical calculations to be validated, but at least the HMO model appears to be useful to select reasonable non-IPR candidates among thousands of possible isomers.

Finally, it is worth mentioning that, owing to their singular properties, multiply charged fullerenes might be used as "generators" of non-IPR fullerene isomers, which are difficult to produce under normal conditions. The idea would be to start from a neutral fullerene with the usual IPR structure and then ionize it in a highly energetic environment. Under these conditions, SW transformations are possible and might lead to the more stable non-IPR charged isomers. After cooling and neutralization of the non-IPR ion, a neutral (metastable) non-IPR isomer could be formed. Another interesting perspective of this field of research is the production of highly charged clusters of fullerenes, $[C_{60}]_n^{q+}$ and $[C_{60}C_{70}]_n^{q+}$. These clusters have been recently observed in collisions with highly charged $Xe^{20+,30+}$ ions in gas phase [172–174]. The question of how the charge is distributed in these systems (either localized in a single fullerene cage or delocalized over the whole cluster) is still under debate. To investigate this problem, new theoretical methods that are able to describe the strong van der Waals interactions between fullerene cages must be developed. Similar problems might also be studied with negatively charged fullerenes, especially by using a new generation of storage rings that will be available in the near future. The advantage of the latter approach is that, in addition to charge states and electron affinities (or ionization energies in the case of cation formation), the experiments can also provide very valuable information on the structure of these ions through the use of photoelectron spectroscopy [48–51] (which is much more difficult to implement in collision experiments). As a consequence of all this, it is clear that the study of multiply charged fullerene ions will remain as a very promising research activity in the next few years.

Abbreviations

2D	two-dimensional
3D	three-dimensional
AP	adjacent pentagon
DE	dissociation energy
DFT	density functional theory
EA	electron affinity
FB	fission barrier
HMO	Hückel molecular orbital
HOMO	highest occupied molecular orbital
IP	ionization potential
IPR	isolated pentagon rule
KER	kinetic energy release
LUMO	lowest unoccupied molecular orbital
NAP	number of adjacent pentagons
PAPR	pentagon adjacency penalty rule
PES	potential energy surface
SP	spherical parameter
SW	Stone–Wales
SWT	Stone–Wales transformation
TOF	time-of-flight

References

1. M. S. Dresselhaus, G. Dresselhaus, and P. C. Eklund. *Science of Fullerenes and Carbon Nanotubes*. Academic Press, San Diego, CA, 1996.
2. E. Ōsawa. *Perspectives of Fullerene Nanotechnology*. Kluwer Academic Publishers, New York, 2002.
3. H. W. Kroto, J. R. Heath, S. C. O'Brien, R. F. Curl, and R. E. Smalley. *Nature*, 318:162, 1985.
4. S. Martin, L. Chen, A. Denis, R. Bredy, J. Bernard, and J. Desesquelles. *Phys. Rev. A*, 62:022707, 2000.
5. S. H. Schwartz, A. Fardi, K. Haghighat, A. Langereis, H. T. Schmidt, and H. Cederquist. *Phys. Rev. A*, 63:013201, 2000.
6. L. Chen, S. Martin, R. Bredy, J. Bernard, and J. Desequelles. *Phys. Rev. A*, 64:031201(R), 2001.
7. L. Chen, J. Bernard, G. Berry, R. Bredy, J. Desesquelles, and S. Martin. *Phys. Scr.*, T92:138, 2001.
8. S. Martin, L. Chen, R. Bredy, J. Bernard, M. C. Buchet-Poulizac, A. Allouche, and J. Desesquelles. *Phys. Rev. A*, 66:063201, 2002.
9. S. Tomita, H. Lebius, A. Brenac, F. Chandezon, and B. A. Huber. *Phys. Rev. A*, 65:053201, 2002.
10. S. Tomita, H. Lebius, A. Brenac, F. Chandezon, and B. A. Huber. *Phys. Rev. A*, 67:063204, 2003.
11. H. Cederquist, J. Jensen, H. T. Schmidt, H. Zettergren, S. Tomita, B. A. Huber, and B. Manil. *Phys. Rev. A*, 67:062719, 2003.
12. E. E. Campbell. *Fullerene Collision Reactions*, 1st edn. Kluwer Academic Publishers, Dordrecht, the Netherlands, 2003.
13. A. Rentenier, D. Bordenave-Montesquieu, P. Moretto-Capelle, and A. Bordenave-Montesquieu. *J. Phys. B*, 36:1585, 2003.
14. I. V. Hertel, T. Laarmann, and C. P. Schulz. *Adv. At. Mol. Opt. Phys.*, 50:219, 2005.
15. L. Chen, S. Martin, J. Bernard, and R. Brédy. *Phys. Rev. Lett.*, 98:193401, 2007.
16. A. Rentenier, L. F. Ruiz, S. Díaz-Tendero, B. Zarour, P. Moretto-Capelle, D. Bordenave-Montesquieu, A. Bordenave-Montesquieu, P. A. Hervieux, M. Alcamí, M. F. Politis, J. Hanssen, and F. Martín. *Phys. Rev. Lett.*, 100:183401, 2008.
17. S. Matt, O. Echt, T. Rauth, B. Dünser, M. Lezius, A. Stamatovic, P. Scheier, and T. D. Märk. *Z. Phys. D*, 40:389, 1997.
18. D. Hathiramani, K. Aichele, W. Arnold, K. Huber, E. Salzborn, and P. Scheier. *Phys. Rev. Lett.*, 85:3604, 2000.

19. P. Scheier, D. Hathiramani, W. Arnold, K. Huber, and E. Salzborn. *Phys. Rev Lett.*, 84:55, 2000.
20. V. R. Bhardwaj, P. B. Corkum, and D. M. Rayner. *Phys. Rev. Lett.*, 91:203004, 2003.
21. I. Shchatsinin, T. Laarmann, G. Stibenz, G. Steinmeyer, A. Stalmashonak, N. Zhavoronkov, C. P. Schulz, and I. V. Hertel. *Z. Phys. D*, 41:219, 1997.
22. G. Senn, T. D. Märk, and P. Scheier. *J. Chem. Phys.*, 108:990, 1998.
23. R. Völpel, G. Hofmann, M. Steidl, M. Stenke, M. Schlapp, R. Trassl, and E. Salzborn. *Phys. Rev. Lett.*, 71:3439, 1993.
24. P. Scheier and T. D. Märk. *Int. J. Mass Spectrom. Ion Process.*, 133:L5, 1994.
25. R. Wörgötter, B. Dunser, P. Scheier, and T. D. Märk. *J. Chem. Phys.*, 101:8674, 1994.
26. S. Matt, B. Dunser, P. Scheier, and T. D. Märk. *Hyperfine Interact.*, 99:175, 1996.
27. S. Matt, O. Echt, R. Wörgötter, V. Grill, P. Scheier, C. Lifshitz, and T. D. Märk. *Chem. Phys. Lett.*, 264:149, 1997.
28. S. Matt, M. Sonderegger, R. David, O. Echt, P. Scheier, J. Laskin, C. Lifshitz, and T. D. Märk. *Int. J. Mass Spectrom.*, 187:813, 1999.
29. W. Krätschmer, L. D. Lamb, K. Fostiropoulos, and D. R. Huffman. *Nature*, 347:354, 1990.
30. J. Jin, H. Khemliche, M. H. Prior, and Z. Xie. *Phys. Rev. A*, 53:615, 1996.
31. J. Jensen, H. Cederquist, H. Zettergren, H. T. Schmidt, S. Tomita, S. B. Nielsen, J. Rangama, P. Hvelplund, B. Manil, and B. A. Huber. *Phys. Rev. A*, 69:053203, 2004.
32. R. Sahnoun, K. Nakai, Y. Sato, H. Kono, and Y. Fujimura. *J. Chem. Phys.*, 125:184306, 2006.
33. S. Díaz-Tendero, M. Alcamí, and F. Martín. *Phys. Rev. Lett.*, 95:013401, 2005.
34. J. Laskin and C. Lifshitz. *J. Mass Spectrom.*, 36:459, 2001.
35. M. Foltin, M. Lezius, P. Schcier, and T. D. Märk. *J. Chem. Phys.*, 98:9624, 1993.
36. M. Larsson. *Rep. Prog. Phys.*, 58:1267, 1995.
37. M. K. Scheller, R. N. Compton, and L. S. Cederbaum. *Science*, 270:1160, 1995.
38. V. Cammarata, T. Guo, A. Illies, L. Lidong, and P. Shelvin. *J. Phys. Chem. A*, 109:2765, 2005.
39. R. C. Haddon. *J. Chem. Phys.*, 25:127, 1992.
40. W. K. Fullagar, I. R. Gentle, G. A. Heath, and J. W. White. *J. Chem. Soc., Chem. Commun.*, 525, 1993.
41. C. A. Reed and R. D. Bolskar. *Chem. Rev.*, 100:1075, 2000.
42. C. A. Howard, J. C. Wasse, N. T. Skipper, H. Thompson, and A. K. Soper. *J. Phys. Chem. C*, 111:5640, 2007.
43. L. S. Wang, C. F. Ding, X. B. Wang, and J. B. Nicholas. *Phys. Rev. Lett.*, 81:2667, 1998.
44. X. B. Wang and L. S. Wang. *Nature*, 400:245, 1999.
45. P. A. Limbach, L. Schweikhard, K. A. Cowen, A. G. Marshall, M. T. McDermott, and J. V. Cole. *J. Am. Chem. Soc.*, 113:6795, 1991.
46. R. L. Hettich, R. N. Compton, and R. H. Hettich. *Phys. Rev. Lett.*, 67:1242, 1991.
47. S. Tomita, J. U. Andersen, H. Cederquist, B. Concina, O. Echt, J. S. Forster, K. Hansen et al. *J. Chem. Phys.*, 124:024310, 2006.
48. O. T. Ehrler, F. Furche, J. M. Weber, and M. M. Kappes. *J. Chem. Phys.*, 122:094321, 2005.
49. O. T. Ehrler, J. P. Yang, C. Hättig, A.-N. Unterreiner, H. Hippler, and M. M. Kappes. *J. Chem. Phys.*, 125:074312, 2006.
50. X.-B. Wang, H.-K. Woo, X. Huang, M. M. Kappes, and L.-S. Wang. *Phys. Rev. Lett.*, 96:143002, 2006.
51. X.-B. Wang, H.-K. Woo, J. Yang, M. M. Kappes, and L.-S. Wang. *J. Phys. Chem. C*, 111:17684, 2007.
52. L. Senapati, J. Schrier, and K. B. Whaley. *Nano Lett.*, 4:2073, 2004.
53. K. Kobayashi and S. Nagase. *J. Am. Chem. Soc.*, 119:12693, 1997.
54. H. Kato, A. Taninaka, T. Sugai, and H. Shinohara. *J. Am. Chem. Soc.*, 125:7782, 2003.
55. Z. Slanina, F. Uhlík, S.-L. Lee, L. Adamowicz, and S. Nagase. *Int. J. Quantum Chem.*, 107:2494, 2007.
56. E. B. Iezzi, J. C. Duchamp, K. R. Fletcher, T. E. Glass, and H. C. Dorn. *Nano Lett.*, 2:1187, 2002.
57. L. Dunsch, M. Krause, J. Noack, and P. J. Georgi. *Phys. Chem. Sol.*, 65:309, 2004.
58. S. Stevenson, J. P. Phillips, J. E. Reid, M. M. Olmstead, S. P. Rath, and A. L. Balch. *Chem. Commun.*, 2814, 2004.
59. M. Krause, J. Wong, and L. Dunsch. *Chem. Eur. J.*, 11:706, 2005.
60. S. Yang and L. Dunsch. *Chem. Eur. J.*, 12:413, 2005.
61. M. Krause and L. Dunsch. *Angew. Chem. Int. Ed.*, 44:1557, 2005.
62. C. M. Beavers, T. M. Zuo, J. C. Duchamp, K. Harich, H. C. Dorn, M. M. Olmstead, and A. L. Balch. *J. Am. Chem. Soc.*, 128:11352, 2006.
63. C.-R. Wang, T. Kai, T. Tomiyama, T. Yoshida, Y. Kobayashi, E. Nishibori, M. Takata, M. Sakata, and H. Shinohara. *Nature*, 408:426, 2000.
64. Z.-Q. Shi, X. Wu, C.-R. Wang, X. Lu, and H. Shinohara. *Angew. Chem. Int. Ed.*, 45:2107, 2006.
65. S. Stevenson, P. W. Fowler, T. Heine, J. C. Duchamp, G. Rice, T. Glass, K. Harich, E. Hajdu, R. Bible, and H. C. Dorn. *Nature*, 408:427, 2000.
66. M. M. Olmstead, H. M. Lee, J. C. Duchamp, S. Stevenson, D. Marciu, H. C. Dorn, and A. L. Balch. *Angew. Chem. Int. Ed.*, 42:900, 2003.
67. A. A. Popov and L. Dunsch. *J. Am. Chem. Soc.*, 129:11835, 2007.
68. S. Yang, A. A. Popov, and L. Dunsch. *Angew. Chem. Int. Ed.*, 46:1256, 2007.
69. S. Yang, A. A. Popov, and L. Dunsch. *J. Phys. Chem. B*, 111:13659, 2007.
70. A. A. Popov, S. F. Yang, M. Krause, J. Wong, and L. Dunsch. *J. Phys. Chem. B*, 111:3363, 2007.
71. F. Rohmund and E. E. B. Campbell. *J. Phys. B*, 30:5293, 1997.

72. H. Cederquist, P. Hvelplund, H. Lebius, H. T. Schmidt, S. Tomita, and B. A. Huber. *Phys. Rev. A*, 63:025201, 2001.

73. H. Bräuning, R. Trassl, A. Diehl, A. Theiß, E. Salzborn, A. A. Narits, and L. P. Presnyakov. *Phys. Rev. Lett.*, 91:168301, 2003.

74. S. Margadonna and K. Prassides. *J. Solid State Chem.*, 168:639, 2002.

75. A. Y. Ganin, Y. Takabayashi, Y. Z. Khimyak, S. Margadonna, A. Tamai, M. J. Rosseinsky, and K. Prassides. *Nat. Mater.*, 7:367, 2008.

76. H. Shinohara. *Rep. Prog. Phys.*, 63:843, 2000.

77. C.-Y. Shu, X.-Y. Ma, J.-F. Zhang, F. D. Corwin, J. H. Sim, E.-Y. Zhang, H. C. Dorn et al. *Bioconjugate Chem.*, 19:651, 2008.

78. K. B. Hartman, L. J. Wilson, and M. G. Rosenblum. *Mol. Diagn. Ther.*, 12:1, 2008.

79. W. Harneit, C. Boehme, S. Schaefer, K. Huebener, K. Fostiropoulos, and K. Lips. *Phys. Rev. Lett.*, 98:216601, 2007.

80. A. Müller, S. Schippers, M. Habibi, D. Esteves, J. C. Wang, R. A. Phaneuf, A. L. D. Kilcoyne, A. Aguilar, and L. Dunsch. *Phys. Rev. Lett.*, 101:133001, 2008.

81. A. N. Khlobystov, K. Porfyrakis, M. Kanai, D. A. Britz, A. Ardavan, H. Shinohara, T. John, S. Dennis, and G. A. D. Briggs. *Angew. Chem. Int. Ed.*, 43:1386, 2004.

82. R. Kitaura and H. Shinohara. *Jpn. J. Appl. Phys.*, 46:881, 2007.

83. K. Laasonen, W. Andreoni, and M. Parrinello. *Science*, 258:1916, 1992.

84. A. Ardavan, M. Austwick, S. C. Benjamin, G. A. D. Briggs, T. J. S. Dennis, A. Ferguson, D. G. Hasko et al. *Philos. Trans. R. Soc. Lond. Ser. A*, 361:1473, 2003.

85. F. Simon, H. Kuzmany, H. Rauf, T. Pichler, J. Bernardi, H. Peterlik, L. Korecz, F. Fulop, and A. Janossy. *Chem. Phys. Lett.*, 383:362, 2004.

86. S. Díaz-Tendero, M. Alcamí, and F. Martín. *J. Chem. Phys.*, 119:5545, 2003.

87. S. Díaz-Tendero, M. Alcamí, and F. Martín. *Chem. Phys. Lett.*, 407:153, 2005.

88. S. Díaz-Tendero, F. Martín, and M. Alcamí. *Chem. Phys. Chem.*, 6:92, 2005.

89. S. Díaz-Tendero, F. Martín, and M. Alcamí. *Phys. Chem. Chem. Phys.*, 7:3756, 2005.

90. S. Díaz-Tendero, M. Alcamí, and F. Martín. *J. Chem. Phys.*, 123:184306, 2005.

91. G. Sánchez, S. Díaz-Tendero, M. Alcamí, and F. Martín. *Chem. Phys. Lett.*, 416:14, 2005.

92. S. Díaz-Tendero, F. Martín, and M. Alcamí. *Comp. Mat. Sci.*, 35:203, 2006.

93. S. Díaz-Tendero, G. Sánchez, M. Alcamí, and F. Martín. *Int. J. Mass Spectrom.*, 252:133, 2006.

94. H. Zettergren, M. Alcamí, and F. Martín. *Phys. Rev. A*, 76:043205, 2007.

95. H. Zettergren, G. Sánchez, S. Díaz-Tendero, M. Alcamí, and F. Martín. *J. Chem. Phys.*, 127:104308, 2007.

96. H. Zettergren, M. Alcamí, and F. Martín. *Chem. Phys. Chem.*, 9:861, 2008.

97. S. Díaz-Tendero. Structure and fragmentation of neutral and positively charged carbon clusters and fullerenes. PhD thesis, Universidad Autónoma de Madrid, Madrid, Spain, 2005.

98. H. Prinzbach, A. Weller, P. Landenberger, F. Wahl, J. Worth, L. T. Scott, M. Gelmont, D. Olevano, and B. Issendorff. *Nature*, 407:60, 2000.

99. P. W. Fowler and D. E. Manolopoulos. *An Atlas of Fullerenes*. Oxford University Press, Oxford, NY, 1995.

100. F. A. Cotton. *Chemical Applications of Group Theory*. John Wiley & Sons, New York, 3rd edition, 1990.

101. A. M. Lesk. *Introduction to Symmetry and Group Theory for Chemists*. Kluwer Academic Publishers, New York, 2004.

102. G. Brinkmann, O. D. Friedrichs, A. Dress, and T. Harmuth. *Match-Commun. Math. Comput. Chem.*, 36:233, 1997.

103. R. Englman. *The Jahn-Teller Effect in Molecules and Crystals*. Wiley-Interscience, New York, 1972.

104. Z. Chen, T. Heine, H. Jiao, A. Hirsch, W. Thiel, and P. von R. Schleyer. *Chem. Eur. J.*, 10:963, 2004.

105. X. Lu and Z. Chen. *Chem. Rev.*, 105:3643, 2005.

106. R. Taylor, J. P. Hare, A. K. Abdul-Sada, and W. Kroto, H. *J. Chem. Soc. Chem. Commun.*, 1423, 1990.

107. R. D. Johnson, G. Meijer, J. R. Salem, and D. S. Bethune. *J. Am. Chem. Soc.*, 113:3619, 1991.

108. M. D. Diener and J. M. Alford. *Nature*, 393:668, 1998.

109. R. Ettl, I. Chao, F. Diederich, and R. L. Whetten. *Nature*, 353:149, 1991.

110. F. Diederich and R. L. Whetten. *Acc. Chem. Res.*, 25:119, 1992.

111. K. Kikuchi, N. Nakahara, T. Wakabayashi, S. Suzuki, H. Shiromaru, Y. Miyake, K. Saito, I. Ikemoto, M. Kainosho, and Y. Achiba. *Nature*, 357:142, 1992.

112. R. Taylor, G. J. Langley, A. G. Avent, T. J. S. Dennis, H. W. Kroto, and D. R. M. Walton. *J. Chem. Soc. Perkin Trans. 2*, 1029, 1993.

113. F. H. Hennrich, R. H. Michel, A. Fischer, S. Richard-Schneider, S. Gilb, M. M. Kappes, M. Fuchs, D. Bürk, K. Kobayashi, and S. Nagase. *Angew. Chem. Int. Ed.*, 35:1732, 1996.

114. T. J. S. Dennis, T. Kai, T. Tomiyama, and H. Shinohara. *Chem. Commun.*, 5:619, 1998.

115. N. Tagmatarchis, A. G. Avent, K. Prassides, T. J. S. Dennis, and H. Shinohara. *Chem. Commun.*, 1023, 1999.

116. A. G. Avent, D. Dubois, A. Pe¡änicaud, and R. Taylor. *J. Chem. Soc. Perkin Trans.*, 1907, 1997.

117. A. J. Stone and D. J. Wales. *Chem. Phys. Lett.*, 128:501, 1986.

118. D. Bakowies, M. Buhl, and W. Thiel. *J. Am. Chem. Soc.*, 117:10113, 1995.

119. E. Albertazzi, C. Domene, P. W. Fowler, T. Heine, G. Seifert, C. V. Alsenow, and F. Zerbetto. *Phys. Chem. Chem. Phys.*, 1:2913, 1999.

120. P. W. Fowler and D. E. Manolopoulos. *Nature*, 355:428, 1992.

121. P. W. Fowler, T. Heine, D. E. Manolopoulos, D. Mitchell, G. Orlandi, R. Schmidt, G. Seifert, and F. Zerbetto. *J. Phys. Chem.*, 100:6984, 1996.

122. W. Qian, M. D. Bartberger, S. J. Pastor, K. N. Houk, C. L. Wilkins, and Y. Rubin. *J. Chem. Soc. Faraday Trans.*, 122:8333, 2000.

123. P. A. Troshin, A. G. Avent, A. D. Darwish, N. Martsinovich, A. K. Abdul-Sada, J. M. Street, and R. Taylor. *Science*, 309:278, 2005.

124. E. Hernández, V. Meunier, B. W. Smith, R. Rurali, H. Terrones, M. B. Nardelli, M. Terrones, D. E. Luzzi, and J.-C. Charlier. *Nano Lett.*, 3:1037, 2003.

125. R. L. Murry, D. L. Strout, G. K. Odom, and G. E. Scuseria. *Nature*, 366:665, 1993.

126. C. A. Coulson, B. O'Leary, and R. B. Mallion. *Hückel Theory for Organic Chemists*. Academic Press, London, U.K., 1978.

127. H. W. Kroto. *Nature*, 329:529, 1987.

128. J.-I. Aihara, S. Oe, M. Yoshida, and E. Ōsawa. *J. Comp. Chem.*, 17:1387, 1996.

129. G. Zheng, S. Irle, and K. Morokuma. *Chem. Phys. Lett.*, 412:210, 2005.

130. Z. Chen and R. B. King. *Chem. Rev.*, 105:3613, 2005.

131. N. Shao, Y. Gao, and X. C. Zeng. *J. Phys. Chem. C*, 111:17671, 2007.

132. T. G. Schmalz, W. A. Seitz, D. J. Klein, and G. E. Hite. *J. Am. Chem. Soc.*, 110:1113, 1988.

133. E. E. B. Campbell, P. W. Fowler, D. Mitchell, and F. Zerbetto. *Chem. Phys. Lett.*, 250:544, 1996.

134. A. Ayuela, P. W. Fowler, D. Mitchell, R. Schmidt, G. Seifert, and F. Zerbetto. *J. Phys. Chem.*, 100:15634, 1996.

135. X. Lu, Z. Chen, W. Thiel, P.v.R. Schleyer, R. Huang, and L. Zheng. *J. Am. Chem. Soc.*, 126:14871, 2004.

136. M. Saunders, H. A. Jiménez-Vázquez, R. J. Cross, S. Mroczkowski, D. I. Freedberg, and F. A. L. Anet. *Nature*, 367:256, 1994.

137. T. Sternfeld, C. Thilgen, R. E. Hoffman, M. del R. C. Heras, F. Diederich, F. Wudl, L. T. Scott, J. Mack, and M. Rabinovitz. *J. Am. Chem. Soc.*, 124:5734, 2002.

138. T. Sternfeld, M. Saunders, R. J. Cross, and M. Rabinovitz. *Angew. Chem. Int. Ed.*, 42:3136, 2003.

139. A. Hirsch, Z. Chen, and H. Jiao. *Angew. Chem. Int. Ed.*, 39:3915, 2000.

140. M. Bühl and A. Hirsch. *Chem. Rev.*, 101:1153, 2001.

141. U. Näher, H. Göhlich, T. Lange, and T. P. Martin. *Phys. Rev. Lett.*, 68:3416, 1992.

142. H. Zettergren, H. T. Schmidt, H. Cederquist, J. Jensen, S. Tomita, P. Hvelplund, H. Lebius, and B. A. Huber. *Phys. Rev. A*, 66:032710, 2002.

143. H. Zettergren, J. Jensen, H. T. Schmidt, and H. Cederquist. *Eur. Phys. J. D*, 29:63, 2004.

144. R. Antoine, P. Dugourd, D. Rayane, E. Benichou, M. Broyer, F. Chandezon, and C. Guet. *J. Chem. Phys.*, 110:9771, 1999.

145. J. Cioslowski, S. Patchkovskii, and W. Thiel. *Chem. Phys. Lett.*, 248:116, 1996.

146. C. Yannouleas and U. Landman. *Chem. Phys. Lett.*, 217:175, 1994.

147. T. Baştuğ, P. Kürpick, J. Meyer, W.-D. Sepp, B. Fricke, and A. Rosén. *Phys. Rev. B*, 55:5015, 1997.

148. H. Steger, J. Holzapfel, A. Hielscher, W. Kamke, and I. V. Hertel. *Chem. Phys. Lett.*, 234:455, 1995.

149. I. V. Hertel, H. Steger, J. Devries, B. Weisser, C. Menzel, B. Kamke, and W. Kamke. *Phys. Rev. Lett.*, 68:784, 1992.

150. M. S. Baba, T. S. L. Narasimhan, R. Balasubramanian, and C. K. Mathews. *Int. J. Mass Spectrom. Ion Process*, 125:R1, 1993.

151. C. Lifshitz, M. Iraqui, T. Peres, and J. E. Fischer. *Rapid Commun. Mass Spectrom.*, 5:238, 1991.

152. J. A. Zimmerman, J. R. Eyler, S. B. H. Bach, and S. W. McElvany. *J. Chem. Phys.*, 94:3556, 1991.

153. A. Reinkoster, U. Werner, N. M. Kabachnik, and H. O. Lutz. *Phys. Rev. A*, 64:032703, 2001.

154. S. Matt, R. Parajuli, A. Stamatovic, P. Scheier, T. D. Märk, J. Laskin, and C. Lifshitz. *Eur. Mass Spectrom.*, 5:477, 1999.

155. I. Compagnon, R. Antoine, M. Broyer, P. Dugourd, J. Lermé, and D. Rayane. *Phys. Rev. A*, 64:025201, 2001.

156. H. Zettergren, P. Reinhed, K. Støchkel, H. T. Schmidt, H. Cederquist, J. Jensen, S. Tomita et al. *Eur. Phys. J. D*, 38:299, 2006.

157. S. Díaz-Tendero, F. Martín, and M. Alcamí. *J. Phys. Chem. A*, 106:10782, 2002.

158. S. W. McElvany and S. B. H. Bach. *Proceedings of the 39th ASMS Conference Mass Spectrometry and Allied Topics*, vol. 39:1991, p. 422.

159. M. S. Baba, T. S. L. Narasimhan, R. Balasubramanian, and C. K. Mathews. *Int. J. Mass Spectrom. Ion Process*, 114:R1, 1992.

160. D. Muigg, P. Scheier, K. Becker, and T. D. Märk. *J. Phys. B*, 29:5193, 1996.

161. H. Steger, J. Devries, B. Kamke, W. Kamke, and T. Drewello. *Chem. Phys. Lett.*, 194:452, 1992.

162. P. E. Barran, S. Firth, A. J. Stace, H. W. Kroto, K. Hansen, and E. E. B. Campbell. *Int. J. Mass Spectrom.*, 167:127, 1997.

163. J. Laskin, B. Hadas, T. D. Märk, and C. Lifshitz. *Int. J. Mass Spectrom.*, 177:L9, 1998.

164. S. Tomita, J. U. Andersen, C. Gottrup, P. Hvelplund, and U. V. Pedersen. *Phys. Rev. Lett.*, 87:073401, 2001.

165. B. Concina, K. Gluch, S. Matt-Leubner, O. Echt, P. Scheier, and T. D. Märk. *Chem. Phys. Lett.*, 407:464, 2005.

166. X. B. Wang, C. F. Ding, and L. S. Wang. *J. Chem. Phys.*, 110:8217, 1999.

167. C. Brink, L. H. Andersen, P. Hvelplund, D. Mathur, and J. D. Voldstad. *Chem. Phys. Lett.*, 233:52, 1995.

168. B. Liu, P. Hvelplund, S. B. Nielsen, and S. Tomita. *Phys. Rev. Lett.*, 92:168301, 2004.

169. Y. Wang, H. Zettergren, M. Alcamí, and F. Martín. *Phys. Rev. A.*, 80:033201, 2009.

170. M. Oiwa and S. Ryoshikagaku. *Elementary Quantum Chemistry*. Kagaku-Dojin, Kyoto, Japan, 1965.

171. A. A. Popov, M. Krause, S. Yang, J. Wong, and L. Dunsch. *J. Phys. Chem. B*, 111:3363, 2007.

172. B. Manil, L. Maunoury, B. A. Huber, J. Jensen, H. T. Schmidt, H. Zettergren, H. Cederquist, S. Tomita, and P. Hvelplund. *Phys. Rev. Lett.*, 91:215504, 2003.

173. H. Zettergren, H. T. Schmidt, P. Reinhed, H. Cederquist, J. Jensen, P. Hvelplund, S. Tomita, B. Manil, J. Rangama, and B. A. Huber. *Phys. Rev. A*, 75:051201, 2007.

174. H. Zettergren, H. T. Schmidt, P. Reinhed, H. Cederquist, J. Jensen, P. Hvelplund, S. Tomita, B. Manil, J. Rangama, and B. A. Huber. *J. Chem. Phys.*, 126:224303, 2007.

Fragmentation of Fullerenes

Victor V. Albert
University of Florida

Ryan T. Chancey
Nelson Architectural Engineers, Inc.

Lene B. Oddershede
Copenhagen University

Frank E. Harris
University of Florida

and

University of Utah

John R. Sabin
University of Florida

and

University of Southern Denmark

26.1 Introduction

It has long been known that solid carbon exists in diamond and graphite crystalline structures or with its atoms in a disordered structure (amorphous carbon). These allotropes have widely varying physical and chemical properties (and also vary widely in cost). Similarly, gas-phase carbon was well known to exist in diverse molecular forms: isolated C atoms, C_2 molecules, and various ill-defined C_n with somewhat larger values of n. This picture changed qualitatively in 1985 when Kroto et al. [1] reported a new class of carbon molecules, which they called *fullerenes*, for which they were awarded the Nobel Prize in Chemistry in 1996.

Fullerenes are carbon cage molecules of definite composition shaped as spheroidal shells and with bonding patterns consisting of precisely 12 pentagons and an arbitrary number of hexagons. Each fullerene is composed of $n = 20 + 2h$ carbon atoms, where h is the number of hexagons in the structure. The name arises from the geometric similarity of their bonding patterns to the geodesic dome structures made famous by Buckminster Fuller. The first such molecule discovered, a C_{60} cage with the icosahedral symmetry of a standard soccer ball [1], was for that reason named buckminsterfullerene.

Since the original discovery, additional cages as small as C_{20} and at least as large as C_{240} have been identified. The most common, C_{60} and C_{70}, are illustrated in Figure 26.1.

The suffix "ene" indicates that each C atom is covalently (i.e., electron pair) bonded to three others (instead of the maximum of four), a situation that classically would correspond to the existence of bonds involving two pairs of electrons ("double bonds").

These carbon cages exhibit great stability, in part because the atoms are all tied together with strong carbon–carbon covalent bonds, in part because the angles between the bonds emanating from each atom are nearly optimum, and in part because the partial double-bond character permits a quantum-mechanical contribution to stability known to chemists as "resonance" or "delocalization." The stability, coupled with the fact that the bonding has little resistance to torsion (change of bond angle), permits fullerenes to undergo considerable distortion without rupture of the cage structure. This feature is sometimes referred to as "resilience."

The stability and resilience of the fullerenes play major roles in the response of these molecules to mechanical or thermal stress, and it is the purpose of this chapter to report on processes that lead to fullerene fragmentation. These processes include *surface impact*, in which a fullerene collides at high velocity with a hard surface, *ion impact* (fullerene–ion collisions), collisions with neutral atoms or molecules, *electron impact* (collisions with energetic electrons), *photofragmentation* (consequent to absorption of light), *coulomb explosion* (fragmentation that occurs after a fullerene has been stripped of enough electrons that the resulting positively charged moiety

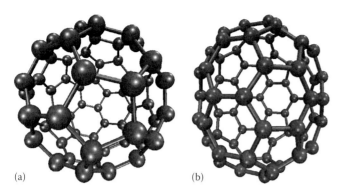

(a)   (b)

FIGURE 26.1   Sketches of C_{60} (a) and C_{70} (b) fullerenes.

cannot remain a stable entity), and *thermal instability*, in which the absorption of vibrational energy or electronic excitation is sufficient to cause widespread bond breaking.

Fullerene cages can be large enough to trap atoms, or even molecules, creating species known as *endohedral complexes*; the chemical notation for such species is with an "@" sign (e.g., an Ar atom within buckminsterfullerene is denoted $Ar@C_{60}$). Fragmentation of endohedral complexes results in the release of the endohedral species, and the presence of an endohedral moiety may affect the rate or extent of the fragmentation.

There have been very large numbers of both theoretical and experimental studies of fullerene fragmentation, and in order to keep this chapter within reasonable bounds it was necessary to focus discussion primarily on neutral C_{60}, with only occasional reference to C_{70} and other fullerenes. Research-level reviews of fullerene stability include the 1993 work of Gruen [2] and the contributions in 2000 by Campbell and Levine [3] and by Campbell and Rohmund [4].

26.2 Some Comments on Theory

Early work on the characteristics of C_{60} fullerene fragmentation showed that the fragment distribution was bimodal, with the heavy fragments differing in mass by C_2 units. To predict this distribution, statistical models were developed [5–7] to predict relative abundances of fullerene fragments based on the notion of maximum entropy. This approach involves the estimation of the thermodynamic properties of each of the species involved, and was found to give reasonable agreement with fragment distributions that had been observed in some experiments on C_{60} and C_{60}^+.

Another theoretical tool that has been used for investigating fullerene fragmentation is density functional theory (DFT; see, for example, Ref. [8]), which has been implemented in a variety of its incarnations for various problems. DFT is a first-principles electronic structure method [9], and is normally used to calculate multidimensional potential energy surfaces, from which mechanisms of fragmentation are then inferred. However, the complete potential energy surface for, e.g., C_{60}, has 174 dimensions (3 times the number of atoms, minus 6) and full exploration of such an object is far beyond either current or foreseeable computational capability. It is, instead, necessary to make *a priori*

choices of presumed fragmentation pathways. Some, but not all DFT investigations have predicted fragmentation patterns in reasonable agreement with specific experiments.

Neither of the foregoing types of theoretical methods includes information specific to actual processes leading to fragmentation, but they might nevertheless constitute reasonable approaches where statistical analyses are appropriate. That circumstance would apply when the fragmentation mechanism can be reasonably described as a two-stage process in which internal energy imparted to the fullerene first becomes distributed among its mechanical degrees of freedom in a quasi-equilibrium fashion, and then, as predicted statistically, eventually becomes sufficiently localized in individual bonds to cause their rupture. This scenario is more likely to apply to thermal fragmentation or photofragmentation than to fragmentation caused by collisional impact.

The standard method used to study collisional fragmentation processes for fullerenes from a theoretical viewpoint is to carry out molecular dynamics (MD) simulations of the problem. Such simulations normally proceed by setting up initial conditions for a collisional event (incoming velocity, orientation of the projectile if it is not considered to be a point object, projectile rotation if appropriate, point of aim on the target, direction relative to the target structure, etc.). Atomic and molecular species are described by the positions of all their atomic nuclei, and the energetic interactions of all atoms are represented using empirical potentials. For surface-impact studies, additional important variables are those characterizing the target surface. Two possibilities for target description are (1) structureless potential walls whose height and steepness are designed to mimic an actual material, and (2) arrays of appropriate atoms (either fixed in position or moveable). The better MD calculations use potentials designed to include (in an approximate fashion) the effects of quantum mechanics, but the evolution of the atomic positions, subject to these potentials, is computed by classical mechanics (i.e., using Newton's Laws). This evolution of a collisional event is called a *trajectory*.

The quality of an MD simulation depends not only on the empirical potentials (the design of which is an evolving art), but also on choices of initial conditions. Since many of the initial-condition (input) variables cannot be controlled experimentally (examples are the fullerene orientation and the target aim point in a surface-impact experiment), a good simulation will consist of an ensemble of trajectories spanning, in a valid statistical distribution, the uncontrolled input variables.

A method for MD simulation, incorporating quantum mechanics to a level of approximation intermediate between the use of empirical potentials and DFT, is provided by the use of *tight-binding* methods [10] accompanied by additional approximations leading to computational effort that is close to linear in the system size (so-called *order-N methods*). The essence of the tight-binding approach is the assumption that the individual electron orbitals are sufficiently localized that they can be identified with individual atoms, and that their quantum-mechanical energy contributions can be approximated as empirically (or semiempirically) determined intra-atom and neighboring-atom terms.

Because of computational limitations, early MD fullerene fragmentation studies used very limited sets of input variables (e.g., C_{60} at a few orientations with high-symmetry points in the direction of travel, aimed at one or a few points on a target surface). Increases in available computer capacity have removed this limitation in more recent work; the authors' studies involve ensemble averages of thousands of trajectories.

26.3 Some Comments on Experiments

Both intuition and experience indicate that the most important dynamical variable governing the extent of collisional fullerene fragmentation is the net (i.e., relative) velocity at which the collision occurs. If relativistic effects can be neglected (a good approximation in all the experimental work to be reviewed here), the relative velocity is independent of the reference frame in which the collisional event is described, and therefore also has more relevance than a frame-dependent quantity, such as the kinetic energy. To illustrate this point, consider an object of mass m colliding with another, of mass M, where $M \neq m$, at relative velocity v. If this collision is observed in a frame where mass m is stationary and mass M has velocity v (the "laboratory frame" if the experiment is carried out by producing a mass-M beam, aimed at stationary mass-m objects), the incident kinetic energy is $Mv^2/2$. But if collisions at the same relative velocity are produced by directing a mass-m beam at stationary mass-M objects, the "laboratory-frame" incident kinetic energy has the different value $mv^2/2$. Yet, a third result is obtained if the collision at the same relative velocity v is described in the center-of-mass frame. Then, both particles have nonzero incident kinetic energy, with the total kinetic energy of both given by $\mu v^2/2$, where $\mu = Mm/(M + m)$ is the *reduced mass* of the mass pair (and is less than either M or m).

The reason for the above discussion is to call attention to the fact that it is almost meaningless to compare collision energies for different systems, and it is not even possible to convert them to collision velocities without knowledge of the frame in which the kinetic energy is specified. Usually, but not always, this information can be extracted from the published description of an experimental procedure. For impact of a fullerene with a surface, the only frame that makes sense is that in which the surface (of virtually infinite mass) is stationary and there is not really a problem. But when a fullerene (typical mass 720 amu) is in collision with an atom such as He (4 amu) or Ar (40 amu), kinetic energy in the stationary-fullerene frame will be one or two orders of magnitude less than for the same collision in the stationary-atom frame, and this situation must be kept in mind when incident energies are used to characterize collisions.

The severity of the problem associated with the use of kinetic energy as a collision descriptor can, at least for surface-impact studies, be ameliorated by converting to the incident energy per fullerene atom (for C_n denoted E/n). This notation is used in the remainder of the present chapter.

26.4 Fragmentation by Surface Impact

In the early 1990s, experimental studies of the stability of fullerenes to surface impact were reported by Beck et al. in a series of papers. In the first of these studies [11], C_{60}^+, C_{70}^+, and C_{84}^+ were introduced into a pulsed helium jet, which directed them into collisions with Si (011) or graphite (0001) with energies up to 200 eV. No evidence of fragmentation was found. However, in later experiments, with more energetic C_n^+ ions impinging on room temperature graphite, passivated silicon, or tantalum surfaces, these investigators found that small C_m ($m = 2$–28) anions were produced. Beck et al. interpreted the formation of these anions as due to surface ionization of small carbon cluster fragments resulting from shattering of the fullerene on collision with the surface. After considering various possible mechanisms of fragmentation, they concluded that both internal and kinetic energies are important for the fragmentation process, and that the process is most likely sequential. Beck et al. investigated collisions of some derivatives of fullerenes with surfaces.

On the theoretical side, molecular dynamics simulations began to be published at nearly the same time as did the first experimental studies of fullerene fragmentation. Mowrey et al. [12] carried out simulations of neutral C_{60} collisions with a model of the hydrogen-terminated (111) surface of diamond. The surface consisted of 10 layers of carbon atoms, each layer with 64 atoms, and periodic boundary conditions were maintained in the directions parallel to the surface. The interactions between all the carbon atoms of both the fullerene and the target were modeled by the empirical Brenner carbon–carbon potential. All trajectories were run with the fullerene impacting the surface normally at one of three high-symmetry orientations, and at incident kinetic energies 150, 200, and 250 eV (2.5–4.17 eV per incident carbon atom). From the nine trajectories generated in their simulations, Mowrey et al. found that the fullerene was predicted to compress and rebound without fragmentation at all the impact energies studied.

Tight-binding molecular dynamics calculations for C_{60}^+ collisions on an unterminated, reconstructed (111) surface of diamond were reported by Galli and Mauri [13]. The surface was represented by 1140 atom cells, and calculations were carried out for collision energies from 60 to 400 eV at normal incidence. They identified three regions of interest. For the low-energy region, $E/n \leq 2$ eV, the fullerene is reflected from the surface. It is initially flattened as it rebounds, but regains its shape after the rebound. For 2 eV $< E/n < 4$ eV, the fullerene is not only distorted on impact, but some fullerene bonds are broken and some bonds are formed with the substrate. In the high-energy region, $E/n \geq 4$ eV, the fullerene structure was compromised and many bonds were formed with the surface. No fullerene fragments left the surface. However, the possibility is always extant that a fullerene reflected in a highly vibrationally excited state might experience a fluctuation where much energy is concentrated in a few bonds, with resultant fragmentation.

Similar calculations were reported by Chancey and coworker [14,15], who modeled the fullerenes as sets of point atoms that interact among themselves according to the well-known Tersoff empirical potential (a form containing both two-body and angle-dependent three-body terms). Chancey et al. represented the surface by a structureless potential, so that no bonds could be formed with the surface. Fullerenes with $n = 24, 60, 100,$ and 240 were impinged on a potential of the form $V = V_0[1 - \tanh(\gamma z)]$. Here z is the coordinate normal to the wall (and the trajectories start from large positive z), while $V_0 = 100$ hartree and $\gamma = 1.0$ bohr^{-1} are empirical constants. The potential was designed such that for all of the impact energies investigated, all of the projectile particles would rebound from the wall, and that the steepness of the wall would approximate that of a diamond surface.

In Figure 26.2, we present snapshots of the collision of a C_{60} fullerene with a potential wall. In the first case, the fullerene has an initial velocity of 100 eV. The collision is not hard enough to cause fragmentation, but causes a large amount of vibrational excitation in the fullerene skeleton. When the velocity reaches 300 eV, substantial fragmentation does take place [14]. Intermediate results are observed at intermediate energies.

The study by Chancey et al. differed from those performed earlier in that it involved ensembles containing large numbers of individual trajectories at each incident energy and angle of incidence on the surface (with individual trajectories at varying fullerene orientation). While individual trajectories are deterministic (since the quantum mechanics is modeled by a classical potential and the initial fullerene state was assumed to be vibrationless and at the equilibrium geometry), the differences in initial orientation provided a statistical distribution of outcomes.

Trajectories were typically run for some 20,000 0.1 fs time steps. It is necessary to run at least this long to give distorted (and potentially unstable) fullerenes enough time to fragment. When each trajectory was terminated, its fragment population and angular momentum were determined and the results accumulated for statistical analysis. In addition to obtaining detailed data on the fragmentation patterns, the simulations yielded predictions as to the probability of fullerene fragmentation as a function of incident energy. If one defines a measure of the stability of the fullerenes as the impact energy per carbon atom $(E/n)_{crit}$ at which half of the fullerenes are reflected intact from the surface, Chancey et al. predict that $(E/n)_{crit}$ falls in the range 3.1–3.6 eV for the four fullerenes in the simulation study, of which C_{60} is predicted the most stable, with $(E/n)_{crit} = 3.6$ eV.

The distribution of the fragments for collision of C_{60} with the wall, as found by Chancey et al., is shown in Figure 26.3 for various impact energies. These data are for ensembles of 1000 trajectories at each impact energy, with every trajectory assigned a random orientation of the C_{60} projectile.

The results fall into three regimes. In Regime I, the low-energy regime, the fragment distribution is nearly mirror symmetric for large and small fragments, indicating that, at low energy, collision tends to break a single small piece from the fullerene projectile. At medium energy (Regime II), the higher incident energy of the projectile causes the symmetry of the plot to be broken, and the ratio of small-to-large fragments increases. This

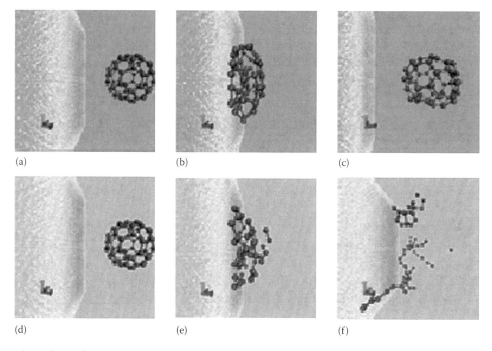

(a)   (b)   (c)

(d)   (e)   (f)

FIGURE 26.2 Snapshots of a C_{60} fullerene colliding with a potential wall, as described in the text. Each row, from left to right: before, during, after collision. Top row: 100 eV collision; note that the fullerene is distorted when it rebounds from the surface, but does not fragment. Bottom row: 300 eV collision; note that the fullerene fragments when it rebounds from the surface. (From Chancey, R.T. et al., *Phys. Rev. A*, 67, 043203, 2003. With permission.)

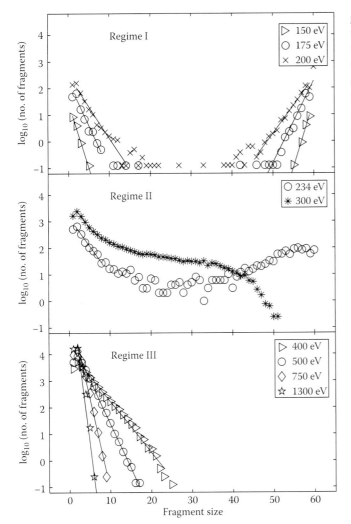

FIGURE 26.3 Fragment distribution per 1000 events of C_{60} impact on a hard wall at various impact energies ranging from 150 to 1300 eV. (From Chancey, R.T. et al., *Phys. Rev. A*, 67, 043203, 2003. With permission.)

is interpreted to mean that multiple fragmentations are taking place. The large fragments have completely disappeared in Regime III, that of highest energy collisions, presumably due to the large collision energy. The number of fragments containing s atoms, $N(s)$, fits an exponential form $N(s) = C \exp(-as)$ quite nicely, and is quite consistent with experimental data on other fragmentations [16].

Tang et al. [17] carried out simulations to investigate the effect of projectile orientation and impact angle on collisions of C_{60} with the same structureless potential as was used by Chancey et al. At impact energies per atom $E/n \approx 1.67$ eV, they reported no fragmentation. For $E/n \approx 5$ eV, they found fragmentation at all impact angles (10°, 20°, and 30° from normal incidence). At this energy, in agreement with Chancey, the majority of fragments were predicted to contain less than 10 atoms.

Anders et al. [18] report molecular dynamics simulations of C_{60} collisions with various surfaces: graphite, fullerite (a solid consisting of C_{60} molecules in a face-centered crystalline structure),

gold, and solid argon. They report the onset of fragmentation when the impact energy per carbon atom, E/n, is about 8 eV, and that complete atomization is not accomplished until E/n reaches the range 80–150 eV. While this work provides interesting insights as to the degree to which the impact can destroy the structure of the target, its prediction of the energies needed for fragmentation is in substantial disagreement with that of the other investigators mentioned here (Mowrey et al., Galli and Mauri, and Chancey et al.), but Anders et al. do not cite these authors and hence do not offer reasons for the lack of agreement.

One further set of simulation studies was reported by Pan et al. [19], but it did not include incident energies high enough such that these authors observed fragmentation.

There have been many surface-impact experiments in addition to those, referred to above, by Beck et al. For example, Busmann et al. [20] reported significant fragmentation of C_{60} ions on impact with a graphite surface at incident energies between 200 and 455 eV (200 eV corresponds to $E/n = 3.3$ eV), largely in agreement with the theoretical investigations other than that of Anders et al. Other experimental studies considered impact on graphite and silicon, on gold (most recently by Baranov et al. [21]), on aluminum [22], on stainless steel [23], and on nickel [24].

26.5 Fragmentation by Ion Impact

Fullerenes will also fragment when they collide with fast ions of various sorts. Two possible mechanisms seem to be potentially important. In the first, the kinetic energy of the ion is transferred to the vibrational modes of the fullerene, which, if the energy transfer is sufficient, leads to fragmentation. Second, if a fast ion were to ionize the fullerene by the capture of a sufficient number of electrons, a charge large enough to cause a Coulomb explosion of the fullerene might be created (see Section 26.8). In fact, the second mechanism is not likely, as the most common fullerenes, C_{60} and C_{70}, have been shown to be stable even when multiply charged [25,26], either positively or negatively. In particular, in Sr_6C_{60}, the C_{60} anion carries a charge of -12 [27], while stable cations such as C_{60}^{4+} have also been observed [28].

A plethora of experimental studies of collisions of ions with C_{60} has been carried out. A large number of these use rare gas ions, typically Ar^{n+}. A recent such study is in Ref. [29]. However, many other projectiles have been employed, including Mn^+ [30], alkali metal ions (most recently in Ref. [31]), Si^{4+} [32], O^+ (recently in Ref. [33]), C^+ [34], and even highly charged ^{209}Bi and ^{238}U [35].

Experimental investigations of ion collisions are usually carried out in ways that vary various combinations of projectile characteristics. Projectile kinetic energies range from a few eV/amu to several hundred keV/amu, masses range from those of protons and alpha particles to that of ^{238}U [35], and charges range from $+1$ to fully stripped Ar [36] or even ^{238}U^{+46} [35]. Some of the experiments use the positive ions of highly electronegative atoms (e.g., F^{2+}), so as to make charge exchange likely even when fragmentation does not occur. Typically, experiments are then analyzed using multivariate correlation techniques.

A common characteristic of many studies is that fragmentation on fast ion collision leads to fragments containing even numbers of carbon atoms, and with a preference for C_2 fragments. Collisions of H^+, H_2^+, H_3^+, and He^+ with fullerenes primarily give fragments of C_n^+, $n = 1$–14, and evaporation of C_2 [37]. Again, C_2 fragments are emitted in collisions of fullerenes with He^+ and Xe^{17+}, and by evaporation from charged C_{60} [38].

Cheng et al. [39] bombarded C_{60} vapor with 420–625 MeV $^{136}Xe^{35+}$ and $^{136}Xe^{18+}$ ions. A large distribution of fragment masses and charges was observed, again with a preference for fragments containing an even number of carbon atoms. Cheng et al. interpreted their results in terms of (a) the excitation of the *giant dipole plasmon resonance* (an energy absorption associated with collective motion of the relatively delocalized valence electrons of the fullerene) and (b) multifragmentation due to nearly head-on collisions that cause large amounts of energy transfer.

In a series of papers concerning collisions of multi-charged rare gas ions with C_{60}, Martin et al. [40] studied branching ratios among C_2 evaporation, symmetrical fission, and multifragmentation in collisions of C_{60}^{n+}, $n = 3$–9, with $300\,keV$ Xe^{30+} ions.

Very little theoretical work has been done on fast ion-induced fragmentation of fullerenes. Most of the experimental data indicate that the results can be best analyzed in terms of the projectile velocity rather than its kinetic energy, but in at least one investigation in which fast H^+, C^+, and Ar^+ collided with stationary C_{60}, it was found [41] that for higher impact energies, the projectile kinetic energy (in the laboratory frame, and therefore proportional to the square of the velocity) was a useful parameter.

The one conclusion which is clear at the present time is that the fullerenes, especially C_{60} and C_{70}, are remarkably stable with respect to interaction with fast, charged ions.

26.6 Fragmentation by Collision with Neutral Atoms or Molecules

Experiments have been reported on the collision of both neutral and ionized fullerenes with neutral atoms. Probably because of the greater ease of accelerating charged species, most of the investigations have dealt with the collision of fullerene ions with neutral atomic or molecular species. One study involving collisions of neutral fullerenes with neutral 8 keV rare gas atoms [42] indicated extensive fragmentation, with the fragments having masses consistent with C_7, C_{11}, C_{15}, C_{19}, and C_{23}.

The literature on collisions involving fullerene ions on neutrals is extensive, with both fullerene cations and anions represented. Some representative investigations are given as follows. Brink et al. [43] studied collisions of C_{60}^{2+} with He, finding the emission of C_2 units as well as larger fission products. Vandenbosch et al. [44] studied the collision dynamics of 75 keV C_{60}^- on H_2, finding not only the C_n fragments of small even n mentioned earlier, but also light odd-n fragments coincident with heavy

even-n fragments, consistent with a sequential disintegration process. There have also been studies of fullerene–fullerene ion collisions, which can result in fusion to generate larger molecules as well as fragmentation [45,46].

Some theoretical work has been carried out on collisions of neutral fullerenes with neutral atoms [47] utilizing dynamics methods similar to those mentioned in Section 26.4. Using helium and neon atom projectiles and collision energies up to 200 eV (center of mass), fragment sizes, energies, differential cross sections, and projectile and fragment scattering angles were calculated. In this study [47], low-energy projectiles led to a distribution of scattering angles, which were attributed to the corrugated nature of the fullerene surface, while high-energy projectiles simply passed through the fullerene. From analysis of the kinetic energy spectrum of the scattered projectiles, it was concluded that the low-energy collisions are exceedingly inelastic, with the fullerene absorbing much of the incident kinetic energy. At high projectile velocities, $E/n \approx 3.3\,eV$, a bimodal fragment mass distribution was found, which corresponds well with Regime I of Chancey's surface-impact simulations [14]. Indeed, Chancey also found that C_2 dominated the fragment mass curves at all projectile velocities.

26.7 Fragmentation by Electron Impact

Related to fragmentation by fast ion impact (see Section 26.5) is fragmentation initiated by electron impact. In one of many papers from the group of T. D. Märk concerned with fragmentation of fullerenes, Dünser et al. [48] report ionization cross sections for the production of various ions from electrons impacting C_{60} with energies up to 1 keV. The most abundant species produced are the various parent ions, C_{60}^{z+}, but cross sections for production of ions of the sort C_{60-2m}^{z+} are also reported for $m = 1$–8 (i.e., fragments as small as C_{44}), and with charge up to $z = 4$. As expected, cross sections and abundances decrease with increasing z and m, and it is interesting to note that as in the case of ion impact, fragmentation is apparently accomplished by the loss of C_2 fragments.

There are several mechanisms that could result in the reaction $C_n^{z+} \rightarrow C_{n-2}^{(z-1)+} + C_2^+$. One such mechanism is a three-step process, initiated by the statistically driven evaporation [7,49] of a neutral C_2 unit: $C_{60}^{z+} \rightarrow C_{58}^{z+} + C_2$. This is followed by electron transfer from the departing C_2 fragment to the remaining fullerene cage: $C_{58}^{z+} + C_2 \rightarrow C_{58}^{(z-1)+} + C_2^+$. Finally, the Coulomb repulsion between the two fragments is posited to complete the separation.

Foltin et al. [50] examined the impact of fast electrons on C_{60}, and observed the reaction $C_{60}^+ \rightarrow C_{58}^+ + C_2$ as a function of the incoming electron energy. From the ionization efficiency curves for C_{58}^+, C_{56}^+, and C_{54}^+, they could construct a time-resolved breakdown graph for the decaying C_{60}^+ ion. Using Rice–Ramsperger–Kassel–Marcus (RRKM) theory [51] to calculate decay rates, they were able to deduce dissociation energies for C_2 fragments of approximately 7.5 eV. This value is somewhat lower than those reported by Matt et al. [52] and by Laskin et al. [53].

Hathiramani et al. [54] and Itoh et al. [55] have investigated the effects of electron impact on the fullerenes $C_{60}{}^{z+}$, with $z = 1, 2, 3$. Cross sections for formation of these ions can be found in Ref. [56]. They also found that C_2 fragments were produced, leaving $C_{58}{}^{z+}$ and $C_{56}{}^{z+}$. Hathiramani et al. deduced that two processes resulted in C_2 fragment ejection. At low incident electron energy, the giant plasmon resonance is excited, followed by fragmentation, while at incident electron energies greater than 100 eV, fragmentation is due to "unsuccessful ionization." The value of the initial fullerene charge is only found to be important in the higher energy case. The mechanism and result of C_2 fragment evaporation from fullerenes are reviewed in Ref. [57].

However, C_2 is not the only fragment that can occur, and evaporation of $C_4{}^+$ and higher mass fragments $C_n{}^+$, $3 \le n \le 10$, have also been observed. In addition, stable negatively charged fullerenes are also known. Scheier et al. [58] report fragmentation of fullerene anions *via* $e^- + C_{60}{}^- \rightarrow C_{60-m}{}^{q+} + \cdots$, where $q = 1$, 2, 3 and $m = 0, 2, 4$ for projectile energies <1 keV. They conclude that there is no strong interaction between the incident and the attached electron. A similar argument may be made for C_{70}.

An interesting study of the elastic cross section of fast electrons on fullerenes by Gerchikov et al. concludes that the cross section is mostly determined by the target ground-state electron density. These authors also noted that the Bragg Rule (that the total cross section is the sum of the individual atomic cross sections) does not apply here, as the total cross section exceeds the sum of the cross sections of the constituent atoms. This is attributed to a coherent interaction between the projectile electron and the delocalized target electrons.

There also exist many other studies closely related to those mentioned above.

26.8 Fragmentation by Coulomb Explosion

Coulomb explosion of fullerenes is closely related to fragmentation by electron bombardment (Section 26.7) and to kinetic energy release (Section 26.11). Highly charged fullerenes, $C_{60}{}^{q+}$ up to $q = 9$, are known. However, due to their remarkable stability [26] and the tendency to emit small C_n fragments serially, they do not undergo what is usually identified as a Coulomb explosion. For example, Afrosimov et al. have studied the fragmentation of $C_{60}{}^{5+}$ produced by collision of C_{60} with Ar^{6+} [29,59], reaching the conclusion that the fragmentation process is multistep with secondary decay of the heavier fragments. Similarly, Jin et al. considered $C_{60}{}^{q+}$ ions with $q \le 9$ produced by collision of C_{60} with slow, heavy, highly charged ions, reaching much the same conclusion.

Recently, Narits has developed a scheme [60] for treating charge transfer in collisions of C_{60} with highly charged argon ions, Ar^{n+} ($n = 4$–18). The method is based on a representation of the charge exchange process as an electron transfer over or through a potential barrier, combined with a semiclassical treatment of C_{60} internal molecular motion. Although fragmentation of C_{60} is not explicitly considered, the total cross sections for electron transfer are calculated; these agree well with experiments [36].

The present authors have found no theoretical dynamics studies of Coulomb explosions in these systems. There have been a few quantum chemical studies of the stability of fullerenes. Diaz-Tendero et al. used density functional theory (DFT) to examine the stability of $C_{60}{}^{q+}$ and $C_{58}{}^{q+}$ for charges ranging from 0 to 14 [61,62]. They concluded that, for fullerene ions of small charge, emission of $C_2{}^+$ fragments was the preferred mode of decay, while for ions of higher charge, emission of more than one fragment is preferred. This finding is also in agreement with work by Eckhoff and Scuseria [63] who investigated C_{60} and C_{70} using Hartree–Fock and hybrid Hartree–Fock/DFT methodology. ("Hartree–Fock" refers to a quantum-mechanical calculation in a "mean field" approximation such that the probability distribution of each electron is obtained assuming it interacts with the average distribution of all the other electrons, rather than with the other electrons in their actual positions.)

Although this mode of decay falls somewhat short of a Coulomb explosion, the decomposition is clearly moving toward that mechanism. Dolgonos and Peslherbe, using Hartree–Fock methods on larger systems (C_{80}) [64], reach similar conclusions.

26.9 Photofragmentation

Another way of introducing energy into fullerenes, and thus perhaps initiating fragmentation, is through direct absorption of one or many photons. In the most common experiments of this type, fullerenes are introduced into a mass spectrometer and irradiated with a laser, the fragment composition then being determined by time-of-flight methods. Tchaplyguine et al. irradiated C_{60} with a Ti-sapphire laser, resulting in $C_{60}{}^+$ and $C_{60}{}^{2+}$ [65]. They found considerable fragmentation of the metastable initial products, interpreting this result as indicative of strong coupling of electronic and vibrational degrees of freedom in the fragments.

One characteristic seen in several studies suggests that there may be rather long time intervals between the initial photon absorption and the onset of fragmentation, or even of the ionization that may be its precursor. Wurz and Lykke [66,67] reported delays on the order of microseconds before multiphoton absorption of visible/UV photons by fullerenes resulted in ionization. They explained this delay as due to an ionization mechanism involving thermionic electron emission. They also observed delayed C_2 fragmentation, and found that up to 10% of the fragmentation process may result in fragments larger than C_2, in agreement with other observations [68]. On the other hand, in a study of multiphoton ionization of various fullerenes, C_n, $n = 60, 70, 84$ as a function of laser wavelength and power density, Ahrens et al. [69] reported that fullerenes do not show delayed ionization and fragmentation for short-wavelength radiation.

Some of the studies involving laser-induced photon absorption have also measured the kinetic energy of the fragments thereby produced. For example, Gaber et al. [70] studied the binary fission products that were generated when neutral C_{60} was irradiated with 308 nm laser pulses. They not only found the bimodal distribution to be expected from a simple fission into $C_k + C_{60-k}$, but also observed that the kinetic energy released was largely independent of the mass ratio of the fission products.

As an alternative to laser irradiation, synchrotron radiation with energies from 15 to 280 eV has also been used to initiate fullerene fragmentation (see, for example, Drewello et al. [71]).

In an unusual experiment, Biasioli et al. [72] examined the photofragmentation of He-seeded supersonic effusive beams of fullerenes. An important feature of the study is that the expansion cooling of the beam appears to provide a mechanism for the selective removal of the photodeposited energy from different degrees of freedom, thereby changing the branching ratio for fullerene decay to favor more extensive fragmentation at the expense of ionization. It is interesting that the cooling thereby has the unexpected result of increasing not only the rate of fragmentation but also its preference for smaller fragments. Moreover, the competition among different fullerene decay channels seems to be quite sensitive to the extent of the cooling, which can be controlled by changing the amount of He seeding. It is postulated that processes involving *phonon coupling* (i.e., coupling of vibrational energy) with electronic excitation could play an important role here.

A theoretical approach to the understanding of fullerene photofragmentation has been presented by Horvath and Beu [73]. They used tight-binding molecular dynamics, simulating the photoexcitation of C_{60} by randomly ascribing sudden velocities to each atom in the fullerene. For low velocities, their simulations indicated that C_2 would be the dominant fragment, while at higher energies, larger fragments were predicted.

26.10 Fragmentation by Heating

Energy can also be introduced into fullerenes in an ostensibly more classical way: by heating the fullerene sample. Sommer et al. used shock waves on C_{60} in Ar to generate sample temperatures in the range 2390–3450 K [74]. They found that the fullerene decomposition proceeded *via* loss of C_2 fragments, and they measured rate constants for the $C_{60} \rightarrow C_{58} + C_2$ reaction. Stetzer et al. [75] heated pristine C_{60} in the solid state to 1260 K, and found that the sample was reduced to amorphous carbon. However, they did not determine the mechanism of the fragmentation.

It is perhaps easier to study thermal energy in fullerenes from a theoretical point of view, as the conditions can be more carefully controlled. The most common sort of theory applied to problems of thermal effects on fullerenes is molecular dynamics (MD) using a tight-binding force field (MD-TB). Zhang et al. looked at fragmentation of various fullerenes (C_n) as a function of temperature. They found the expected C_2 dimers, but rings and multiple rings were also observed. In addition, they looked at fragmentation temperatures, which they found to increase with increasing n, but which leveled off at about 5500 K for $n \geq 60$. Other workers, also using MD-TB methodologies, studied energy barriers

to annealing and fragmentation. As perhaps expected, annealing at a few thousand Kelvin results in fragmentation by loss of C_2 fragments. Davydov investigated the thermal stability of C_{60} (and of C_{20} dimers), including an analysis of the fragmentation behavior in excited electronic states [76]. He found that neither the fragmentation pattern nor the energy of activation (E_a) for the process differed significantly from the corresponding quantities for the ground electronic state.

Most of the theoretical studies have employed a constant temperature scheme, but Xu and Scuseria [77] carried out simulations by following constant energy trajectories, with explicit consideration of the fullerene vibration dynamics. They also observed the expected C_2 dimer fragmentation. Their findings agree with the experiments of Barran et al., who determined the binding energies of $C_{n+2}^{2+} + C_2 \rightarrow C_n^{2+}$ for $46 \leq n \leq 102$ using mass-spectroscopic time-of-flight methods. They confirmed, yet again, that fragmentation takes place by C_2 ejection, and found that the relative C_2 dissociation energies for the larger fullerenes $n \geq 60$ were larger than previously estimated. They also found that radiative cooling did not seem to be an important factor in this process. Hansen and Campbell [78], however, found that an alternative cooling mechanism, involving emission of radiation, might be important. They deduced activation energies and emissivity of carbon clusters from their experiments.

26.11 Kinetic Energy Release

When ion-induced molecular or cluster fragmentation occurs, some of the excess internal energy of the molecule or cluster is released as kinetic energy of the fragments [79]. The kinetic energy release (KER) or the kinetic energy release distribution (KERD) associated with a reaction gives much information about the reaction itself, as well as possible indications concerning the subsequent reactions of the fragments. Consequently, KER spectroscopy, generally carried out using time-of-flight technology, has become one of the most useful tools for studying fullerene fragmentations. A KERD is determined from the derivatives of metastable ion peaks, i.e., significant populations of ions as recorded by the mass spectrometer that are presumed to be at least weakly stable [80]. The KERD data can then be used to extrapolate various energies of the fragmentation reactions, including the Arrhenius activation energy and the dissociation energy.

There have been numerous studies dating back to shortly after the discovery of the fullerenes that have analyzed kinetic energy release distributions for various fullerene clusters. The most widely studied such reaction (potentially because it is the most ubiquitous and fundamental) is the C_2 emission or so-called evaporation reaction, $C_n^{q+} \rightarrow C_{n-2}^{q'+} + C_2^{q''+}$. Several theoretical and experimental efforts have been made to determine the activation energy of this reaction.

The characteristics of the fragmentation are well documented. Investigations of electron-induced ionization and excitation leads to evaporation of C_2 from C_{60}^{z+} for $z = 2$–4, and the electron-induced fragmentation apparently has a higher KER than does the spontaneous decay [81]. It is also seen that in measurements of the KER for the process $C_{60}^{q+} \rightarrow C_{60-2m}^{(q-1)} + C_{2m}^{+}$ for $q = 4$–8 and $m = 1, 2$, the KER varies linearly with the charge state [82].

Cederquist et al. [36] studied fragments resulting from collisions of highly charged ions (Ar^{8+}) with C_{60}. They looked at two models: an infinitely conducting sphere model where electronic response is sufficiently fast to average out all localization effects, and a movable hole model where positive holes are located at fixed equilibrium positions on the C_{60} surface for times between electron transfers. Using KER methods of analysis, they found reasonably good agreement with experimental measurements. In a subsequent paper [38], they distinguished between fragmentation of C_{60}^{z+} by asymmetric fission, $C_{60}^{z+} \rightarrow C_{58}^{(z-1)} + C_{2}^{+}$ for $z = 6$–9, and by evaporation, $C_{60}^{z+} \rightarrow C_{58}^{z+} + C_{2}$ for $z = 2, 3$, and measured the KER of the two processes. From the data, they estimated lower and upper bounds on the charge for the stability of C_{60}^{z+} at $z = 11$ and $z = 18$, respectively. These limits are consistent with the theoretical estimates of Bastug et al. [83] and Seifert et al. [84].

KER results for fragmentation of highly charged fullerenes were reported by Chen et al. [85] for the fragmentation $C_{60}^{5+} \rightarrow C_{60-n}^{4+} + C_{n}^{+}$, with $n = 4$–6. For lower fullerene charges, the KER results indicate that ejection of C_2 [86] or C_2^+ [87] is the dominant fragmentation channel.

26.12 Endohedral Complexes

Although fragmentation of endohedral species combined with fullerenes is not directly the subject of this chapter, there are strong similarities between the two classes of processes. On the theoretical side, Albert et al. have studied collisions of rare gas endohedral complexes [88] colliding with one to four layers of graphene [89], using the same scheme as that used by Chancey et al. [14] in Section 26.4. In both cases, some of the results were similar to those for the fullerene without an endoatom. Typical results are shown in Figures 26.4 and 26.5. Figure 26.4 illustrates a collision at an energy such that the fullerene cage is damaged but not destroyed; in the specific event depicted, the endoatom remains trapped within the damaged cage. Figure 26.5 presents a collision at a velocity sufficient to destroy the cage structure.

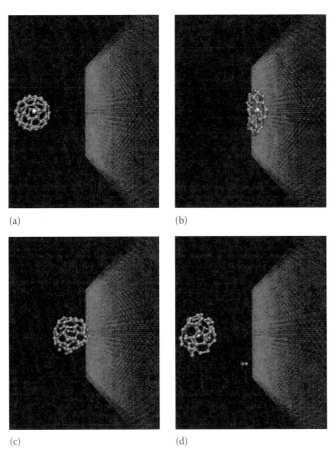

(a)   (b)

(c)   (d)

FIGURE 26.4 Collision of the endohedral complex He@C_{60} with a potential wall at incident velocity 0.065 Å/fs. The He atom is represented by a white ball. (a) Approach; (b) impact; (c) rebound; (d) a C_2 fragment is ejected, but the complex maintains its shape and keeps the He atom enclosed.

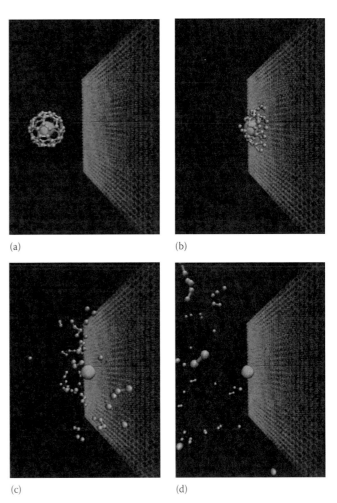

(a)   (b)

(c)   (d)

FIGURE 26.5 Collision of the fullerene endohedral complex Xe@C_{60} with a potential wall at incident velocity 0.140 Å/fs. The Xe atom is represented by a large ball. (a) Approach; (b) impact; (c) shattering; (d) C atoms leave the collision, mostly as diatomic fragments.

An interesting feature of the collision is that it transfers considerable kinetic energy from the massive endoatom (Xe) to the carbon atoms.

Endoatoms of Li@C_{60}^+ form when C_{60} is bombarded with Li$^+$ ions. Using quasiclassical trajectory methods, Bernshtein and Oref [90] showed that the collision energy was redistributed throughout the cage, leading eventually to fragmentation. In an experimental study of Ne^{2+} on C_{60} [91], charge transfer was found to be the dominant low-energy process. Above collision energies of 25 eV, Ne@C_{60-2n}^+ ($n = 1, 2, 3$) was formed, with expulsion of the ubiquitous C_2 fragments. Similar results were found for O$^+$ projectiles colliding with C_{60} [92], where, for higher energies CO@C_{58-2n}^+ was produced. For Ne [93] and Li [94] projectiles, the results were similar. Cs@C_{60}^+ [95] and Kr@C_{60}^+ [96] are also known; these fragment much as expected, with evaporation of C_2.

In order to contain larger endoatoms, larger fullerenes must be used. Experiments have been reported on La@C_{80}^{2+} and La@C_{82}^{2+} [97], and on La@C_{82} [94]. An interesting study by Lorents et al. discusses the decomposition of La@C_{82}^{2+} and Gd@C_{82}^{2+}. In both cases they observe endofragments of the form M@C_{2n} ($30 \le n \le 40$). Fragments also include empty fullerenes and non-fullerene clusters of the form MC_{2n}^+ for $0 \le n \le 5$.

26.13 Summary

Let us begin with a disclaimer. The study of all of the various interesting properties of fullerenes has been a popular subfield of physics and chemistry ever since the discovery of buckminsterfullerene in 1985 [1]. Fullerenes, C_n, have been found with n as small as 20 and as large as several hundreds. There are thus thousands of papers dealing with various aspects of fullerene physics and chemistry. In order for this chapter to maintain a reasonable size, its scope has been limited. Thus, we have not addressed clusters of fullerenes, solid fullerenes, fullerene formation, fullerene chemistry, or related things such as graphene or nanotubes. Although we have briefly discussed endohedral atoms, we have not considered what is, perhaps, one of the more interesting possible aspects of fullerenes, namely, biological effects associated with endohedral species (see, for example, Ref. [98]).

What we have considered is the fragmentation of fullerenes initiated by a variety of means. All initiation methods add energy to the fullerene in some way or another, and the added energy causes the fullerene to fragment, usually in a series of steps. Energy can be added in a variety of ways that result in ionization, electronic excitation, or vibrational excitation. Although fullerenes in general, and C_{60} in particular, are very stable, they will eventually

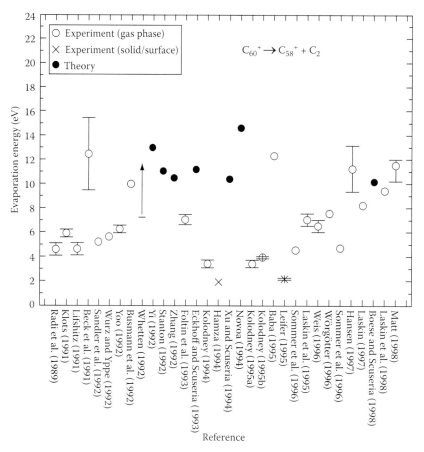

FIGURE 26.6 Compilation of published activation energies for the evaporation of C_2 from C_{60}^+. (From Matt, S. et al., *Int. J. Mass Spectrom.*, 167, 753, 1997. With permission.)

fragment if infused with enough energy, sometimes immediately, and sometimes with delayed ionization or fragment emission.

The most common method of fragmentation, confirmed in numerous studies, both experimental and theoretical, is ejection of C_2^{q+}, where $q = 0, 1$, from either neutral or ionized C_{60}, depending on the circumstances. Although other C_n^{q+} fragments are also seen, they overwhelmingly contain an even number of carbons. Fragments containing odd numbers of carbons are exceedingly rare. Depending on the composition of the heavier fragments, which depends on the method by which they were formed, further fragmentation is frequently observed, also normally *via* the C_2 evaporation mechanism.

One of the important questions considered is the energy required to eject a C_2^{q+} fragment from a fullerene. This energy, referred to as an *appearance energy* or *dissociation energy* or *evaporation energy*, has been measured and calculated for a number of systems. A specific comparison of past efforts to determine the activation energy of $C_{60}^+ \rightarrow C_{58}^+ + C_2$ can be found in Matt et al. [99] and has been reproduced in Figure 26.6. As seen in the figure, the theoretical and experimental results differ significantly, with theory somewhat overestimating the experimental results. However, as noted in a summary of the results [100], more recent experimental efforts agree with a value of $\Delta E_{\text{vap}} \geq 9.5\,\text{eV}$ which is in relative agreement with high-level ab initio calculations [101].

Finally, both theory [14,15] and experiment [16] find that under surface collision conditions, the number of fragments decreases exponentially with fragment size. This behavior is found at all collision energies, with the exponent dependent on the collision energy.

Acknowledgment

This work was supported in part by the University of Florida University Scholars Program (VVA).

References

1. H. W. Kroto, J. R. Heath, S. C. O'Brien, R. F. Curl, and R. E. Smalley, *Nature* **318**, 162 (1985).
2. D. M. Gruen, *Nucl. Instrum. Method B* **78**, 118 (1993).
3. E. E. B. Campbell and R. D. Levine, *Annu. Rev. Phys. Chem.* **51**, 65 (2000).
4. E. E. B. Campbell and F. Rohmund, *Rep. Prog. Phys.* **63**, 1061 (2000).
5. J. Silberstein and R. D. Levine, *J. Chem. Phys.* **75**, 5735 (1981).
6. E. E. B. Campbell, T. Raz, and R. D. Levine, *Chem. Phys. Lett.* **253**, 261 (1996).
7. R. Vandenbosch, *Phys. Rev. A* **59**, 3584 (1999).
8. G. E. Scuseria, *Science* **271**, 942 (1996).
9. J. M. Seminario, ed., *Density Functional Theory*, vol. 33, *Advances in Quantum Chemistry* (Academic Press, New York, 1998).
10. C. H. Xu, C. Z. Wang, C. T. Chan, and K. M. Ho, *J. Phys. Condens. Matter* **4**, 6047 (1992).
11. R. D. Beck, P. St. John, M. M. Alvarez, F. Diederich, and R. L. Whetten, *J. Phys. Chem.* **95**, 8402 (1991).
12. R. C. Mowrey, D. W. Brenner, B. I. Dunlap, J. W. Mintmire, and C. T. White, *J. Phys. Chem.* **95**, 7138 (1991).
13. G. Galli and F. Mauri, *Phys. Rev. Lett.* **73**, 3471 (1994).
14. R. T. Chancey, L. Oddershede, F. E. Harris, and J. R. Sabin, *Phys. Rev. A* **67**, 043203 (2003).
15. L. B. Oddershede, R. T. Chancey, F. E. Harris, and J. R. Sabin, *Kvant* **14**, 3 (2003).
16. P. Hvelplund, L. H. Andersen, H. K. Haugen, J. Lindhard, D. C. Lorents, R. Malhotra, and R. Ruoff, *Phys. Rev. Lett.* **69**, 1915 (1992).
17. Q.-H. Tang, K. Runge, H.-P. Cheng, and F. E. Harris, *J. Phys. Chem. A* **106**, 893 (2002).
18. C. Anders, H. Kirihata, Y. Yamaguchi, and H. M. Urbassek, *Nucl. Instrum. Method B* **255**, 247 (2007).
19. Z. Y. Pan, Z. Y. Man, M. Ho, J. Xie, and Y. Yue, *J. Appl. Phys.* **83**, 4963 (1998).
20. H. G. Busmann, T. Lill, B. Reif, and I. V. Hertel, *Surf. Sci.* **272**, 146 (1992).
21. I. Baranov, S. Della-Negra, V. Domaratsky, S. Kirillov, Y. Le Beyec, A. Novikov, V. Obnorsky, M. Pautrat, S. Yarmiychuk, and K. Wien, *Nucl. Instrum. Method B* **256**, 341 (2007).
22. S. Wethekam and H. Winter, *Vacuum* **82**, 895 (2008).
23. T. Fiegele, O. Echt, F. Biasioli, C. Mair, and T. D. Märk, *Chem. Phys. Lett.* **316**, 387 (2000).
24. E. Kolodney, B. Tsipinyuk, A. Bekkerman, and A. Budrevich, *Nucl. Instrum. Method B* **125**, 170 (1997).
25. J. Aihara, *J. Chem. Soc., Faraday Trans.* **94**, 3537 (1998).
26. C. W. Walter, Y. K. Bae, D. C. Lorents, and J. R. Peterson, *Chem. Phys. Lett.* **195**, 543 (1992).
27. M. C. Böhm, J. Schulte, J. Schütt, T. Schedel-Niedrig, H. Werner, and R. Schlögl, *Int. J. Quantum Chem.* **65**, 333 (1998).
28. R. J. Doyle and M. M. Ross, *J. Phys. Chem.* **95**, 4954 (1991).
29. V. V. Afrosimov, L. A. Baranova, A. A. Basalaev, and M. N. Panov, *Tech. Phys. Lett.* **33**, 1018 (2007).
30. Y. Basir and S. L. Anderson, *Chem. Phys. Lett.* **243**, 45 (1995).
31. R. F. M. Lobo, N. T. Silva, B. M. N. Vicente, I. M. V. Gouveia, F. M. V. Berardo, and J. H. F. Ribeiro, *Eur. Phys. J. D* **38**, 35 (2006).
32. A. Itoh, H. Tsuchida, K. Miyabe, M. Imai, and N. Imanishi, *Nucl. Instrum. Method B* **129**, 363 (1997).
33. B. Manil, L. Maunoury, B. A. Huber, J. Jensen, H. T. Schmidt, H. Zettergren, H. Cederquist, S. Tomita, and P. Hvelplund, *Phys. Rev. Lett.* **91**, 215504 (2003).
34. Y. Nakai, A. Itoh, T. Kambara, Y. Bitoh, and Y. Awaya, *J. Phys. B: At. Mol. Opt. Phys.* **30**, 3049 (1997).
35. J. Jin, H. Khemliche, M. H. Prior, and Z. Xie, *Phys. Rev. A* **53**, 615 (1996).

36. H. Cederquist, A. Fardi, K. Haghighat, A. Langereis, H. T. Schmidt, H. Schwartz, J. C. Levin, I. A. Sellin, H. Lebius, B. Huber et al., *Phys. Rev. A* **61**, 022712 (2000).

37. D. Bordenave-Montesquieu, P. Moretto-Capelle, A. Bordenave-Montesquieu, and A. Rentenier, *J. Phys. B: At. Mol. Opt. Phys.* **34**, L137 (2001).

38. H. Cederquist, J. Jensen, H. T. Schmidt, H. Zettergren, S. Tomita, B. A. Huber, and B. Manil, *Phys. Rev. A* **67**, 062719 (2003).

39. S. Cheng, H. G. Berry, R. W. Dunford, H. Esbensen, D. S. Gemmell, E. P. Kanter, T. LeBrun, and W. Bauer, *Phys. Rev. A* **54**, 3182 (1996).

40. S. Martin, L. Chen, A. Salmoun, B. Li, J. Bernard, and R. Brédy, *Phys. Rev. A* **77**, 043201 (2008).

41. T. Kunert and R. Schmidt, *Phys. Rev. Lett.* **86**, 5258 (2001).

42. M. Takayama and H. Shinohara, *Int. J. Mass Spectrom. Ion Process.* **123**, R7 (1993).

43. C. Brink, L. H. Andersen, P. Hvelplund, and D. H. Yu, *Z. Phys. D* **29**, 45 (1994).

44. R. Vandenbosch, B. P. Henry, C. Cooper, M. L. Gardel, J. F. Liang, and D. I. Will, *Phys. Rev. Lett.* **81**, 1821 (1998).

45. F. Rohmund, A. V. Glotov, K. Hansen, and E. E. B. Campbell, *J. Phys. B: At. Mol. Opt. Phys.* **29**, 5143 (1996).

46. H. Shen, P. Hvelplund, D. Mathur, A. Barany, H. Cederquist, N. Selberg, and D. C. Lorents, *Phys. Rev. A* **52**, 3847 (1995).

47. R. Ehlich, O. Knospe, and R. Schmidt, *J. Phys. B: At. Mol. Opt. Phys.* **30**, 5429 (1997).

48. B. Dünser, M. Lezius, P. Scheier, H. Deutsch, and T. D. Märk, *Phys. Rev. Lett.* **74**, 3364 (1995).

49. P. P. Radi, M.-T. Hsu, M. E. Rincon, P. R. Kemper, and M. T. Bowers, *J. Phys. Chem.* **93**, 6187 (1989).

50. M. Foltin, M. Lezius, P. Scheier, and T. D. Märk, *J. Chem. Phys.* **98**, 9624 (1993).

51. W. L. Hase and R. Schinke, in *Theory and Applications of Computational Chemistry—The First Forty Years*, C. Dykstra, G. Frenking, K. Kim, and G. Scuseria (eds.) (Elsevier, Amsterdam, the Netherlands, 2005), pp. 397–423.

52. S. Matt, O. Echt, M. Sonderegger, R. David, P. Scheier, J. Laskin, C. Lifshitz, and T. D. Märk, *Chem. Phys. Lett.* **303**, 379 (1999a).

53. J. Laskin, B. Hadas, T. D. Märk, and C. Lifshitz, *Int. J. Mass Spectrom.* **177**, L9 (1998).

54. D. Hathiramani, K. Aichele, W. Arnold, K. Huber, E. Salzborn, and P. Scheier, *Phys. Rev. Lett.* **85**, 3604 (2000).

55. A. Itoh, H. Tsuchida, K. Miyabe, T. Majima, and N. Imanishi, *J. Phys. B: At. Mol. Opt. Phys.* **32**, 277 (1999).

56. S. Matt, B. Dünser, M. Lezius, H. Deutsch, K. Becker, A. Stamatovic, P. Scheier, and T. D. Märk, *J. Chem. Phys.* **105**, 1880 (1996).

57. P. Scheier, B. Dünser, R. Worgotter, S. Matt, D. Muigg, G. Senn, and T. D. Märk, *Int. Rev. Phys. Chem.* **15**, 93 (1996).

58. P. Scheier, D. Hathiramani, W. Arnold, K. Huber, and E. Salzborn, *Phys. Rev. Lett.* **84**, 55 (2000).

59. V. V. Afrosimov, A. A. Basalaev, M. N. Panov, and O. V. Smirnov, *Tech. Phys. Lett.* **31**, 1055 (2005).

60. A. A. Narits, *J. Phys. B: At. Mol. Opt. Phys.* **41**, 135102 (2008).

61. S. Diaz-Tendero, M. Alcami, and F. Martin, *J. Chem. Phys.* **123**, 184306 (2005).

62. S. Diaz-Tendero, F. Martin, and M. Alcami, *Comput. Mater. Sci.* **35**, 203 (2006).

63. W. C. Eckhoff and G. E. Scuseria, *Chem. Phys. Lett.* **216**, 399 (1993).

64. G. A. Dolgonos and G. H. Peslherbe, *J. Mol. Model.* **13**, 981 (2007).

65. M. Tchaplyguine, K. Hoffmann, O. Duhr, H. Hohmann, G. Korn, H. Rottke, M. Wittmann, I. V. Hertel, and E. E. B. Campbell, *J. Chem. Phys.* **112**, 2781 (2000).

66. P. Wurz and K. R. Lykke, *J. Phys. Chem.* **96**, 10129 (1992).

67. P. Wurz and K. R. Lykke, *Chem. Phys.* **184**, 335 (1994).

68. T. Wakabayashi, H. Shiromaru, S. Suzuki, K. Kikuchi, and Y. Achiba, *Surf. Rev. Lett.* **3**, 793 (1996).

69. J. Ahrens, R. Kovacs, E. A. Shafranovskii, and K. H. Homann, *Ber. Bunsen-Ges. Phys. Chem. Chem. Phys.* **98**, 265 (1994).

70. H. Gaber, R. Hiss, H. G. Busmann, and I. V. Hertel, *Z. Phys. D* **24**, 307 (1992).

71. T. Drewello, W. Kratchmer, M. Fiebererdmann, and A. Ding, *Int. J. Mass Spectrom. Ion Process.* **124**, R1 (1993).

72. F. Biasioli, A. Boschetti, E. Barborini, P. Piseri, P. Milani, and S. Iannotta, *Chem. Phys. Lett.* **270**, 115 (1997).

73. L. Horvath and T. A. Beu, *Phys. Rev. B* **77**, 075102 (2008).

74. T. Sommer, T. Kruse, and P. Roth, *J. Phys. B: At. Mol. Opt. Phys.* **29**, 4955 (1996).

75. M. R. Stetzer, P. A. Heiney, J. E. Fischer, and A. R. McGhie, *Phys. Rev. B* **55**, 127 (1997).

76. I. V. Davydov, *Phys. Solid State* **49**, 1201 (2007).

77. C. H. Xu and G. E. Scuseria, *Phys. Rev. Lett.* **72**, 669 (1994).

78. K. Hansen and E. E. B. Campbell, *J. Chem. Phys.* **104**, 5012 (1996).

79. J. Laskin and C. Lifshitz, *Int. J. Mass Spectrom.* **36**, 459 (2001).

80. P. Sandler, C. Lifshitz, and C. E. Klots, *Chem. Phys. Lett.* **200**, 445 (1992).

81. S. Matt, R. Parajuli, A. Stamatovic, P. Scheier, T. D. Märk, J. Laskin, and C. Lifshitz, *Eur. Mass Spectrom.* **5**, 477 (1999).

82. S. Tomita, H. Lebius, A. Brenac, F. Chandezon, and B. A. Huber, *Phys. Rev. A* **67**, 063204 (2003).

83. T. Bastug, K. Kurpick, J. Meyer, W. D. Sepp, B. Fricke, and A. Rosen, *Phys. Rev. B* **55**, 5015 (1997).

84. G. Seifert, R. Gutierrez, and R. Schmidt, *Phys. Lett. A* **211**, 357 (1996).

85. L. Chen, J. Bernard, G. Berry, R. Brédy, J. Desesquelles, and S. Martin, *Phys. Scr.* **T92**, 138 (2001).

86. B. Climen, B. Concina, M. A. Lebeault, F. Lepine, B. Baguenard, and C. Bordas, *Chem. Phys. Lett.* **437**, 17 (2007).

87. K. Gluch, S. Matt-Leubner, O. Echt, B. Concina, P. Scheier, and T. D. Märk, *J. Chem. Phys.* **121**, 2137 (2004).

88. V. V. Albert, J. R. Sabin, and F. E. Harris, *Int. J. Quantum Chem.* **107**, 3061 (2007).

89. V. V. Albert, J. R. Sabin, and F. E. Harris, *Int. J. Quantum Chem.* **108**, 3010 (2008).

90. V. Bernshtein and I. Oref, *J. Chem. Phys.* **109**, 9811 (1998).

91. J. F. Christian, Z. M. Wan, and S. L. Anderson, *J. Chem. Phys.* **99**, 3468 (1993).

92. J. F. Christian, Z. M. Wan, and S. L. Anderson, *Chem. Phys. Lett.* **199**, 373 (1992).

93. J. Laskin, H. A. Jimenezvasquez, R. Shimshi, M. Saunders, M. S. De Vries, and C. Lifshitz, *Chem. Phys. Lett.* **242**, 249 (1995).

94. A. Lassesson, K. Hansen, M. Jonsson, A. Gromov, E. E. B. Campbell, M. Boyle, D. Pop, C. P. Schulz, I. V. Hertel, A. Taninaka et al., *Eur. Phys. J. D* **34**, 205 (2005).

95. Y. Manor, A. Kaplan, A. B. Ekkerman, B. Tsipinyuk, and E. Kolodney, *Int. J. Mass Spectrom.* **249**, 8 (2006).

96. T. Weiske, H. Schwarz, D. E. Giblin, and M. L. Gross, *Chem. Phys. Lett.* **227**, 87 (1994).

97. S. Feil, K. Gluch, S. Matt-Leubner, O. Echt, C. Lifshitz, B. P. Cao, T. Wakahara, T. Akasaka, P. Scheier, and T. D. Märk, *Int. J. Mass Spectrom.* **249**, 396 (2006).

98. S. V. Prylutska, O. P. Matyshevska, I. I. Grynyuk, Y. I. Prylutskyy, U. Ritter, and P. Scharff, *Mol. Cryst. Liquid Cryst.* **468**, 617 (2007).

99. S. Matt, O. Echt, R. Worgotter, P. Scheier, C. E. Klots, and T. D. Märk, *Int. J. Mass Spectrom.* **167**, 753 (1997).

100. C. Lifshitz, *Int. J. Mass Spectrom.* **198**, 1 (2000).

101. A. D. Boese and G. E. Scuseria, *Chem. Phys. Lett.* **294**, 233 (1998).

27
Fullerene Fragmentation

Henning Zettergren
University of Aarhus

Nicole Haag
Stockholm University

Henrik Cederquist
Stockholm University

27.1 Introduction

We would like to start by stating that this is not meant to be a complete account of all different aspects of the fullerene fragmentation field of study. The subject is far too wide for that, and instead we focus on few aspects that, in our view, are central to the problem of fullerene fragmentation including discussions of fragmentation of small aggregates of fullerenes. Alternative and, in some respects, more complete descriptions of the subject may be found in the book by Dresselhaus et al. (1995), in the book by Campbell (2003), and in the review article by Campbell and Rohmund (2000). A thrilling account of the story behind the discovery of the fullerenes may be found in the book with the title *Perfect Symmetry: The Accidental Discovery of Buckminsterfullerene* by Baggot (1995). We will attempt to present the material in a tutorial way and to clearly convey our pictures of the various fragmentation processes.

The fullerenes are molecules with closed geometrical structures only containing the element carbon. The most common and the most stable fullerene is the C_{60} molecule, which has a shape composed of 12 pentagonal and 20 hexagonal faces. Each carbon atom is bond to three other carbon atoms and in the case of the most stable C_{60} configuration (isomer), two of the bonds lie along edges where a hexagon and a pentagon are joined and one bond lies along a hexagon–hexagon edge (Figure 27.1). This geometrical structure is a truncated icosahedron and it is the same as that of a soccer ball of a type that until recently was the most common one in professional matches (often with 12 black pentagons and 20 white hexagons). For the lowest energy isomer of C_{60} all carbon atoms are bound in the same way and the molecule is quite stable and comparatively difficult to break apart (i.e., to fragment). In the case of the soccer ball, this specific geometry helps to maintain the spherical shape while the symmetry is

broken in other designs (ways to sew patches together). There are 1812 different ways to arrange 60 carbon atoms to form a closed shell molecule (i.e., there are 1812 isomers of C_{60}) but the geometry described above, for which no pentagon is adjacent to another pentagon, is the one that has the largest binding energy (BE). In fact, all fullerenes based solely on pentagons and hexagons have exactly 12 pentagons and an arbitrary number of hexagons, and since its carbon atoms are bound to 3 neighbors, C_{20} is the smallest possible member of this family as $((12 \times 5) + (0 \times 6))/3 = 20$.

More generally, Euler's theorem for polyhedra states that the relation between the numbers of faces (f), vertices (v), and edges (e) must fulfill the relation $f + v = e + 2$ (as an example, a cube has 6 faces, 8 vertices, and 12 edges). In cases where $f = f_h + f_p$, i.e., when there are only hexagonal and pentagonal faces, we have in addition $e = ((6 \times f_h) + (5 \times f_p))/2$ and $v = ((6 \times f_h) + (5 \times f_p))/3$ as two polygons join at an edge, e, and as three polygons share a vertex, v. For a polyhedron fulfilling these conditions, the number of pentagonal faces is always $f_p = 12$ while the number of hexagonal faces, f_h, can take any positive integer value except $f_h = 1$. From this, it follows that the number of carbon atoms, n, in a fullerene molecule (containing only hexagons and pentagons) is $n = 20 + 2 \times f_h$ ($f_h = 0, 2, 3,...$), which means that fullerene molecules have even numbers of carbon atoms. Of course molecules with odd numbers of carbon atoms may also be formed, but they have significantly smaller binding energies than their neighbors with even n. For this reason, the C_{60} molecule normally fragments by emitting one or several C_2 molecules to form new smaller fullerene molecules such as C_{58}, C_{56}, C_{54}, etc., while C_{59}, C_{57}, and C_{55}, etc., only may be observed in much smaller yields and under special experimental conditions. Indeed, this very typical behavior with a strong preference for molecules with even numbers of carbon atoms was observed already in connection with the discovery of the fullerenes in 1985 (Kroto et al. 1985). In this

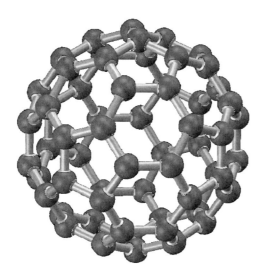

FIGURE 27.1 An artist's view of the most common and the most stable fullerene molecule C_{60}. The positions of the 60 carbon nuclei and the bonds are shown in a ball-and-stick representation. There are 12 pentagonal faces and 20 hexagonal faces in this geometrical structure (a truncated icosahedron) and there are two types of bonds: one along an edge between two hexagons and the other type (which is slightly longer) along a pentagon–hexagon edge. All carbon atoms have three bonds and they belong to two hexagons and one pentagon. In this lowest energy isomer all 60 carbon atoms experience identical bonding situations (cf. text).

groundbreaking experiment, fullerenes were formed by means of laser ablation from a rotating graphite disc in a supersonically flowing dense He gas: the latter helped to cool the emitted carbon material such that the fullerenes and other carbon molecules could be formed efficiently and, in addition, the He flow helped to transport the molecules toward instrumentation for mass analysis. In this 1985 experiment, the typical bimodal fullerene mass distribution was observed with a broad sub-distribution at larger masses corresponding to the closed molecular structures of the fullerenes (Figure 27.2), and another broad sub-distribution at smaller masses below 360 atomic mass units (corresponding to 30 carbon atoms). The latter sub-distribution is mainly due to linear carbon molecules up to about C_{10} and to linear and ring structures for the somewhat heavier molecules in the approximate range C_{10}–C_{30} (although there may also be fullerenes contributing to masses corresponding to even n starting from C_{20} with the exception of C_{22}). Here, the efficiencies of the formation processes under the specific experimental conditions were most likely the main factor behind the bimodal distribution, but very similar mass distributions have been observed later for the fragmentation of excited fullerenes. A central aspect also appearing in the very first experiment was the observation of magic numbers of carbon atoms, i.e., certain specific molecules were observed to have higher stabilities. The most stable one, C_{60}, is highly symmetric and has already been mentioned as the smallest fullerene without adjacent pentagons, and the second smallest one that fulfils the same condition of avoided pentagon adjacency is C_{70}, which has a geometry consisting of 12 pentagons and 25 hexagons ($n = ((f_p \times 5) + (f_h \times 6))/3 = ((12 \times 5) + (25 \times 6))/3 = 70$).

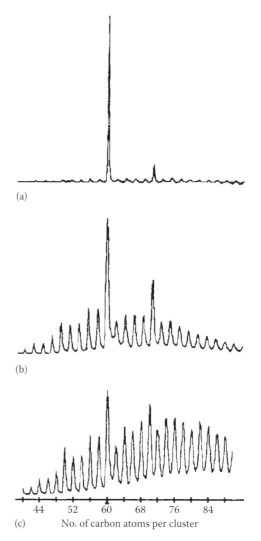

(a)

(b)

(c)   No. of carbon atoms per cluster

44   52   60   68   76   84

FIGURE 27.2 A demonstration of the increased stabilities of C_{60} and C_{70} in the experiment in which the fullerenes were discovered. The mass spectra are recorded for different conditions of He pressure and laser fluence. Note also that only masses corresponding to molecules with even numbers of carbon atoms appear. (Reprinted from Kroto, H.W. et al., *Nature*, 318, 162, 1985. With permission.)

In the lowest energy isomer of C_{60}, all carbon atoms experience the same bonding situation and as there are only two types of bonds, we altogether have 60 hexagon–pentagon bonds (bond length 1.455 Å [David et al. 1991]) and 30 hexagon–hexagon bonds (length 1.391 Å [David et al. 1991]) in C_{60}. The situation is somewhat more complicated in the case of the lowest energy isomer of C_{70} that has eight different kinds of bonds and five different atomic sites. It is now, and since more than 15 years, possible to produce more stable fullerenes (like, e.g., C_{60}, C_{70}, C_{76}, and C_{84}) in large quantities (grams), and various experimental studies may be efficiently performed to probe the inherent properties of this very interesting class of carbon molecules. This includes, e.g., the very important ^{13}C nuclear magnetic resonance (NMR) studies of [13]C[12]C$_{59}$ (C_{60}) and [13]C[12]C$_{69}$ (C_{70}), which proves that there indeed is a single molecular site in C_{60} and precisely five

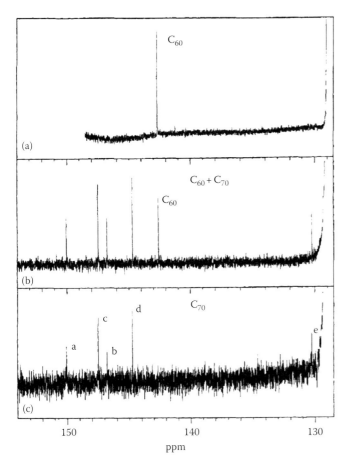

FIGURE 27.3 Nuclear magnetic resonance, NMR, spectra for ^{13}C containing C_{60} (above), for such a $C_{60} + C_{70}$ mixture (middle), and for C_{70} (below). As expected, the C_{60} spectrum has a single resonance (all atoms experience the same bonding situation) and there are five resonances for C_{70} (there are five unique atomic sites in C_{70}). (Reproduced from Taylor, R. et al., *J. Chem. Soc. Chem. Commun.*, 20, 1423, 1990. With permission.)

unique sites in C_{70} (Taylor et al. 1990). The NMR method uses the fact that the magnetic moment of the ^{13}C nucleus couples to the magnetic moment related to the electronic angular momentum and spin. Thus, the positions of the absorption frequencies become very sensitive to the details of the bonding situation such that the number of resonance absorption frequencies for ^{13}C-containing fullerenes in a strong magnetic field directly yields the number of unique sites (Figure 27.3). In the case of C_{60} there is only one resonance, and for C_{70} there are five! In fact, the NMR experiment (Taylor et al. 1990) is generally regarded as the definitive evidence for the three-dimensional geometrical structures of the fullerenes (including the truncated icosahedral structure of C_{60}).

The fullerene production method is rather simple and is based on a technique with an arc discharge between two graphite rods in an inert gas atmosphere, and a following efficient chemical separation method such that fullerene powders of very high purities may be commercially produced (Figure 27.4). Interestingly, fullerenes (including C_{60}) are also formed when burning a

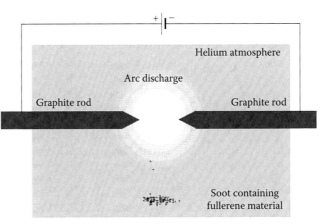

FIGURE 27.4 A schematic of the method used to produce fullerene material. In an inert He atmosphere an electric arc discharge between two graphite rods creates a carbon plasma, and soot is produced as the rods are slowly consumed. The soot contains fullerenes, which may be extracted by chemical methods such that powders with very high purities may be obtained.

candle in normal air, and it is truly amazing that this important class of very stable carbon molecules have escaped science until as late as 1985. This is even more remarkable when considering the extensive studies of the closely related carbon materials diamond and graphite and the vast field of organic chemistry, which is dedicated to the study of molecules containing carbon atoms and their reactions. In diamond, the carbon atoms are strongly bound to four nearest neighbors (tetrahedral structure), while they are strongly bound to three nearest neighbors in a plane (hexagonal structure) in graphite. It is the introduction of pentagons that give the fullerenes their three-dimensional shapes. This may be illustrated (see Figure 27.5) with the related

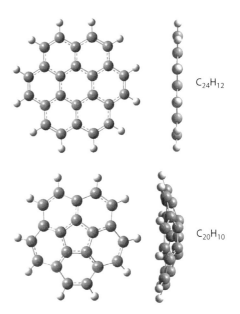

FIGURE 27.5 Schematics of the planar molecule coronene, $C_{24}H_{12}$, and the bowl-shaped corannulene molecule, $C_{20}H_{10}$. The latter gets the cambered structure due to the central pentagon.

polyaromatic hydrocarbons corannulene ($C_{20}H_{10}$) and coronene ($C_{24}H_{12}$). Coronene is planar and consists only of hexagons— one hexagon is surrounded by six other hexagons (and with two hydrogen atoms on each outer hexagon)—while corannulene has a pentagon in the center and five hexagons around it and is, due to this single pentagon, bowl shaped. In this context, graphene, a piece of a single sheet of graphite which, e.g., may have the structure of coronene but without the hydrogen atoms, should also be mentioned as intense research activities have started also in this field. Comparisons of the typical fragmentation patterns of fullerenes, graphene, corannulene, coronene, and other polyaromatic hydrocarbons will indeed be of large interest. As a side remark, we would like to mention that polyaromatic hydrocarbons, and in particular coronene and corannulene, are often discussed in connection with attempts to understand absorption of radiation in interstellar clouds and related chemical reaction routes in such objects.

This chapter will, however, deal exclusively with fullerene fragmentation. We will attempt to present the material in a tutorial way although, at times, this becomes somewhat difficult for the simple reason that there is not yet a full understanding of the details of these very complex fragmentation phenomena. We will mainly discuss fragmentation of isolated fullerenes, fullerene monomers, but we will also briefly discuss how dimers and larger clusters of fullerenes break apart after excitation (and ionization). Fullerene dimers are two fullerenes that are bound together through forces that are weaker than those forming the bounds within the fullerenes themselves, which, for example, is the case for the van der Waals fullerene dimers. Clusters of fullerenes fulfill similar conditions (the intermolecular forces are weaker than the intramolecular ones) for more than two fullerenes bound together. There are many different experimental methods available for exciting fullerene monomers, and fullerene material in general, including various kinds of lasers, synchrotron radiation, other photon sources, collisions with atomic ions (fast or slow and in different charge states), collisions with molecular and cluster ions, collisions with electrons, and interactions with surfaces (Figure 27.6). One limiting factor in many of the experiments is that the fullerenes may have nonnegligible internal excitation energies already before the interaction even though they normally still obtain the larger part of their final total internal energy in the interaction itself. The reason for this initial internal energy is that the fullerene powder has to be heated to several hundred degrees Centigrade in order for the sublimation to become efficient and to obtain sufficient amounts of fullerenes in the gas phase. At 500°C, for example, the average internal energy of a single C_{60} fullerene in a larger ensemble of such fullerenes is $\langle E_{int} \rangle = ((3 \times 60) - 6)kT/2 = 87kT = 5.8\,eV$ where we have counted all the 174 possible vibrational modes of C_{60} (there are three translational degrees of freedom for each carbon atom, but from this we have to subtract three plus three degrees of freedom for the translational and rotational motions of the whole molecule). As already mentioned, the dominant fragmentation channel for a neutral C_{60} fullerene is the emission of one C_2 molecule. The dissociation energy (DE) for this process

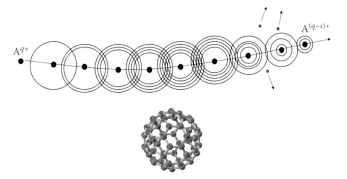

FIGURE 27.6 An artistic view of a collision between a slow highly charged atomic ion and a fullerene. Here, electrons are transferred sequentially from the fullerene to the highly charged ion at large distances, and intermediate so-called hollow states are populated on the projectile ion (this means that several of the transferred electrons end up in states that are excited far above the ion core: such states are unstable and will decay through electron emission).

is close to 10 eV, while the energy required to ionize C_{60} once (i.e., to remove one electron) is substantially smaller and only a bit larger than 7 eV. Thus, the excitation processes in many of the experiments inevitably involves ionization and, as we will see below, it is possible to also multiply ionized fullerene monomers without prompt fragmentation. The latter means that such ions may survive as intact multiply charged fullerenes on (at least) typical experimental time scales of the order of microseconds (μs), and in some cases even milliseconds (ms) or seconds. Thus, ionization and fragmentation are competing processes and it will be necessary to also discuss ionization and excitation when we describe fullerene fragmentation behaviors. In fact, the competition between the different types of energetically allowed fullerene monomer fragmentation channels changes rather strongly as a function of fullerene charge state which, in turn, depends on the nature of the interaction process: how many electrons will be removed and how much internal excitation energy will there be in the remaining fullerene ion. In the cases of fullerene dimers, and heavier clusters of fullerenes of van der Waals types, fragmentation in the form of emission of complete intact C_{60} molecules or C_{60}^+ molecular ions are energetically much more favorable than ionization, or fragmentation, of the individual C_{60} molecules.

In Section 27.2, we will first make an inventory of the various fragmentation pathways for fullerene monomers and then move on to briefly discuss the different related ionization and excitation methods. We will continue in Section 27.2 with a somewhat detailed description of the main properties of the most important monomer fragmentation processes and how these depend on charge state and internal excitation energy for the positively charged fullerene monomers (i.e., for the fullerene monomer cations). In Section 27.3, we discuss the stability and fragmentation of fullerene anions (negatively charged fullerene monomers), but we note here already that the same kind of preference for the formation of even n C_n molecules, as for the neutrals and the cations, appears also for the anions. Fragmentation of fullerene dimers and clusters of fullerenes will be described in Section 27.4. In Section 27.5, we make a summary, some concluding remarks, and an outlook.

27.2 Ionization, Excitation, and Fragmentation of Fullerene Monomer Cations

Fullerene monomers may fragment in a number of different ways, and the competition between different fragmentation pathways depends on the fullerene charge state and the internal excitation energy in relation to the relevant dissociation energies and/or reaction barriers. Taking the archetype fullerene, C_{60}, in moderate positive charge states $0 < r < 6$, as an initial example we have the following two main fragmentation channels:

$$C_{60}^{r+} \to C_{58}^{r+} + C_2 \qquad (27.1)$$

and

$$C_{60}^{r+} \to C_{58}^{(r-1)+} + C_2^+, \qquad (27.2)$$

which often are referred to as C_2 evaporation (or simply "evaporation") and asymmetric fission (or sometimes C_2^+ emission), respectively. The term asymmetric fission distinguishes this type of process, in which there is one charged fragment of a small mass $\left(C_2^+\right)$ and one charged fragment with a larger mass $\left(C_{58}^{(r-1)+}\right)$ from other fission processes like, for instance, the multifragmentation process in which a charged fullerene fragments into several smaller charged and neutral molecules. In addition to the two main fragmentation channels (27.1) and (27.2) there are other possible decay routes. Among these we have, for example, $C_{60}^{r+} \to C_{58}^{(r-2)+} + C_2^{2+} \to C_{58}^{(r-2)+} + C^+ + C^+$ when $r > 1$ and $C_{60}^{r+} \to C_{58}^{(r-3)+} + C_2^{2+} + C^+ \to C_{58}^{(r-3)+} + C^+ + C^+ + C^+$ when $r > 2$, but they have larger dissociation energies than (27.1) and (27.2), especially for the lower charge states, and are therefore weaker—often even considerably weaker. Yet other channels may involve emission of C_4 or C_4^+ molecules or even larger molecules with even numbers of carbon atoms, but here too dissociation energies are larger than for (27.1) and (27.2).

The dissociation energy (DE), which is one of the key parameters in the present description, is defined as the difference between the total atomic binding energy (BE) before and after the fragmentation such that positive DE means that the fullerene ion must have at least this internal energy in order to be able to fragment along the corresponding pathway. Taking the C_2 evaporation process, (27.1), as an example we may write the DE, as

$$DE = E_B(C_{60}^{r+}) - E_B(C_{58}^{r+}) - E_B(C_2), \qquad (27.3)$$

where $E_B(C_{60}^{r+})$ is the total BE of the C_{60}^{r+} molecule (i.e., the energy gained when putting together 60 carbon nuclei and ($6 \times 60 - r$) electrons to form the lowest energy isomer of C_{60}^{r+} in its rotational and vibrational ground states). In similar ways,

$E_B(C_{58}^{r+})$ and $E_B(C_2)$ are the total binding energies of the lowest energy isomer of C_{58}^{r+} and C_2 in its ground state. Shifting our attention to the general case of fullerene monomers containing n carbon atoms, and taking the evaporation process of $C_n^{r+}(C_n^{r+} \to C_{n-2}^{r+} + C_2)$ as an example, the dissociation energies, DE, may be related to the binding energies of the neutral molecules through

$$DE = E_B(C_n) - \sum_{k=1}^{r} I_k(C_n) - E_B(C_{n-2}) + \sum_{k=1}^{r} I_k(C_{n-2}) - E_B(C_2), \qquad (27.4)$$

where $I_k(C_n)$ and $I_k(C_{n-2})$ are the kth ionization energies (often also referred to as ionization potentials) of C_n and C_{n-2}, respectively. Thus, I_k is the minimum energy required to remove one electron from $C_n^{(k-1)+}$ in its ground state to form C_n^{k+} in the ground state, and the sums in Equation 27.4 are the (minimum) energies required to remove r electrons from the neutral fullerenes C_n and C_{n-2}. In many cases, Equation 27.4 is more convenient to use than Equation 27.3 as information on dissociation energies of (multiply) charged fullerenes may not be available in the literature. Dissociation energies have been calculated up to very high charge states, but only in a very limited number of cases as, e.g., in the studies of C_{58} and C_{60} by Díaz-Tendero et al. (2005a,b) and of C_{68} and C_{70} by Zettergren et al. (2007a).

A positive DE, DE > 0, means that the fullerene is truly stable against fragmentation along the corresponding decay path (given of course that its internal excitation energy is smaller than this DE value). If, in addition DE > 0 for all its decay pathways, the fullerene is said to be thermodynamically stable. For C_{60}^{r+} (Figure 27.7) this is the case for charge states up to $r = 5$ (Díaz-Tendero et al. 2005b), which means that $C_{60}^+, C_{60}^{2+}, C_{60}^{3}, C_{60}^{4+}$, and C_{60}^{5+} ions in their ground states (or with excitation energies below DE for the lowest energy dissociation channel) would stay intact forever if left alone (i.e., if it does not collide with an atom or molecule or interact with a radiation field). On the other hand, C_{60}^{r+} fullerenes

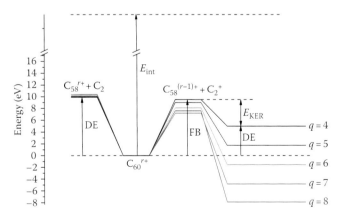

FIGURE 27.7 A schematic showing the dissociation energies (DE) of the evaporation (C_2-emission) channel (left) and the asymmetric fission (C_2^+ emission) channel relative to the ground state energies of C_{60} mother ions in charge states ranging from $q = 4$ to $q = 8$. Also, the fission barriers (FB) for the charge-separating channels are indicated (cf. text).

with $r > 5$ are inherently unstable, and given infinite observation times we would always find that such systems decay. However, the time scales for dissociation may still be very long when there is a reaction barrier that partly hinders, or maybe better put, delays charge-separating fragmentation processes like, e.g., the asymmetric fission process (cf. Equation 27.2 above). In the charge state range above $r = 5$ (in the case of C_{60}), the DE for asymmetric fission (27.2) is lower than that for evaporation (27.1), and in fact this is the case already from the region around $r = 3$ (but then both dissociation processes are still endothermic with DE > 0). In a simple picture, the fission barriers (FB) are formed from the superposition of the repulsive Coulomb force between the two charged fragments and the attractive polarization force, which may be rather strong due to the large size (and thus large polarizability) of the fullerene fragment ion ($C_{58}^{(r-1)+}$ in the case of asymmetric fission of C_{60}^{r+}). When the lighter, charged, C_2^+ fragment is still close to the heavier fragment, the attractive part of the potential energy is larger than the repulsive part while we have the opposite relation at larger $C_{58}^{(r-1)+} - C_2^+$ distances. At intermediate distances, the reaction barrier is thus formed (for a schematic see Figure 27.8). This barrier may strongly delay the fragmentation process also when DE < 0 and positive C_{60}^{r+} molecular ions (cations) have been observed to be metastable in charge states ranging all the way up to $r = 12$ in an experiment with multiphoton absorption and damping of the internal vibrations of highly ionized C_{60} by means of the same laser field (Bhardwaj et al. 2003). The ultimate stability limit, or rather the ultimate limit for non-prompt fragmentation, is the lowest charge state at which the lowest barrier for a charge-separating decay process has zero height (in the case of a sufficiently small positive barrier the fullerene would only be metastable if it is in its ground state). Díaz-Tendero et al. have performed systematic transition state calculations of the FB heights as a function of charge state using density functional theory (DFT) at the B3LYP/6-31G* level, and they found, e.g., that the barrier for C_2^{2+} emission (or $C^+ + C^+$ emission as C_2^{2+} is not stable) was lower than the barrier for C_2^+ emission for C_{60}^{r+} fullerenes in charge states, $r > 7$ (Díaz-Tendero et al. 2005a). Through this series of DFT transition state calculations, they arrived at an ultimate stability limit for C_{60} of $r = 14$. That is, C_{60}^{14+} would be metastable with zero internal energy, while C_{60}^{15+} would decay promptly also under such conditions. For C_{70}, the main fragmentation channels are C_2 evaporation at the lowest charge states (up to $r = 3$, C_{70}^{3+}), C_2^+ emission at intermediate charge states (up to $r = 8$), and most likely C_2^{2+} ($C^+ + C^+$) emission at still somewhat higher charge states.

In connection with the discussion of the competition of the two main fragmentation channels, C_2 evaporation and asymmetric fission (C_2^+ emission), it is important to note that given a sufficiently large internal excitation in the mother fullerene we may also have subsequent decays of the products. This means in the cases of C_{60}^{r+} and C_{70}^{r+} ions that, e.g., $C_{58}^{r+}, C_{58}^{(r-1)+}, C_{68}^{r+}$ or $C_{68}^{(r-1)+}$ products may decay further through, e.g., C_2 and/or C_2^+ emissions in one or several steps to form $C_{56}, C_{54}, C_{52}, \ldots$ and C_{66}, C_{64}, C_{62} ions and even smaller products. As already mentioned, emission of neutral C_2 molecules dominate for lower fullerene charge states, and we may have a so-called sequential evaporation processes through which we get characteristic series of mass peaks of, e.g., C_{60-2m} with $m = 1, 2, 3, \ldots$. In case of such a pure C_2 evaporation series, intensities usually decrease monotonically with m (e.g., see Figure 27.9 [Hvelplund et al. 1992]). For somewhat higher fullerene charge states we may also have, as already discussed, competition from C_2^+ emissions and the interpretation of the mass spectra becomes much more involved. The way to deal with this is to use coincidence mass spectrometry, which may be used—at least in principle—to detect all charged fragments from a single fragmenting mother ion (Figure 27.10). This method has been used by, e.g., Martin et al. (Chen et al. 1999, 2001, 2007, Martin et al. 2000, 2002, 2008, Salmoun et al. 2008) who studied C_{60}^{5+} and C_{60}^{6+} fragmentation and found that the asymmetric fission processes indeed have C_2^+ emission as their most important contributors but that there are also additional channels such as C_4^+, C_6^+, and C_8^+ emission and emissions of two charged fragments. Sometimes these two fragments have odd numbers of carbon atoms, even though emissions of single molecules with odd numbers of carbon atoms are very unlikely as these are energetically very unfavorable processes. Among other things, they (Martin et al. 2002) observed a relatively small contribution from the $C_{60}^{6+} \rightarrow C_{58}^{4+} + C^+ + C^+$ channel, which is in qualitative agreement with the theoretical results of Díaz-Tendero et al. (Díaz-Tendero et al. 2005a) predicting that this channel should have the lowest fission energy barrier when $r > 7$, and it is thus not unreasonable that there should be a small contribution already at $r = 6$. In a later experiment, Martin et al.

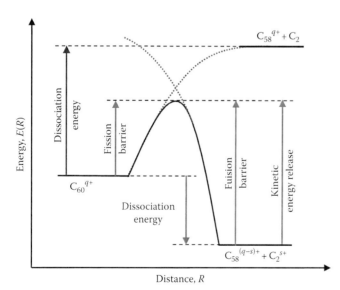

FIGURE 27.8 A schematic of how the fission barrier for asymmetric fission, C_2^{s+} emission, is formed for a C_{60}^{q+} mother ion. At large distances the Coulomb repulsion dominates, and as the C_2^{s+} and $C_{58}^{(q-1)+}$ ions approach each other the potential energy increases. At smaller distances, however, the polarization interaction, which has a leading term that depends on the distance between the two objects to the inverse fourth power, starts to become important and dominates at small distances. The barrier is formed at intermediate distances. (Reprinted from Díaz-Tendero, S. et al., *Phys. Rev. Lett.*, 95, 013401, 2005a. With permission.)

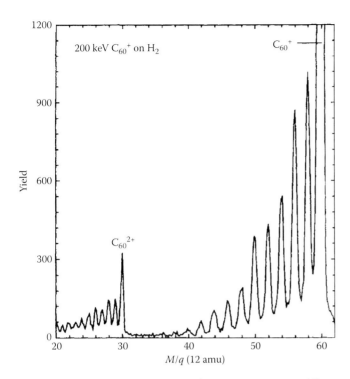

FIGURE 27.9 An early measurement of a C_2 evaporation series following collisions between a 200 keV C_{60}^+ beam and H_2. Note the decreasing intensity as a function of the number of lost C_2 units. (Reprinted from Hvelplund, P. et al., *Phys. Rev. Lett.*, 69, 1915, 1992. With permission.)

have also demonstrated the importance of competing thermal ionization processes, which was possible through the registration of all emitted charged particles (ions and electrons) from decaying fullerene mother ion (Martin et al. 2008). For example, they found that excited C_{60}^{3+} ions first may emit one electron and thereafter the C_{60}^{4+} product ion fragments, and then with a different balance between C_2 and C_2^+ emission than would be the case for direct fragmentation of C_{60}^{3+} (Figure 27.11). Hansen et al. (Hansen and Echt 1997, Campbell et al. 2000, Hansen et al. 2003) have reported detailed studies of thermal ionization of neutral fullerenes and discuss, in particular, the phenomena of delayed ionization and the competition between ionization and fragmentation of neutral fullerenes or fullerenes in low charge states. Although competing decay pathways (ionization, ionization/fragmentation, fragmentation by emission of fragments with more than two carbon atoms, etc.) are important, we will in the following mainly focus our attention on neutral (single) C_2 emission (27.1) and (single) asymmetric fission (27.2) as these are the main decay channels for a rather wide range of (positive) charge states of fullerenes.

Perhaps the most straightforward way to induce fragmentation in neutral fullerene monomers is to increase their internal molecular excitations by simply heating the fullerene vapor. At a given temperature of a body containing such a vapor, we will have a distribution of internal fullerene excitation energies with an average value given by this body temperature (assuming equilibrium conditions). At a sufficiently high temperature

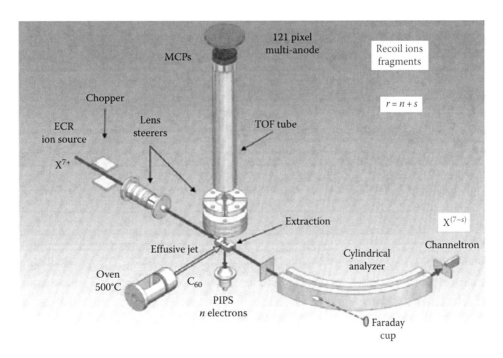

FIGURE 27.10 A schematic of the experimental setup for coincidence registrations of several charged fullerene fragments, electrons, and charge- and energy-analyzed projectile ions. Here, with the example of a septuply charged projectile ion from an electron cyclotron resonance (ECR) ion source. The number of electrons removed from the fullerene is r, the number of detected electrons n, and the number of electrons stabilized on the projectile is denoted by s. Charge conservation requires that $r = n + s$. (Reprinted from Salmoun, A. et al., *Eur. Phys. J. D*, 47, 55, 2008. With permission.)

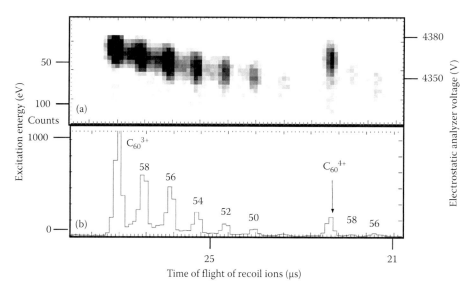

FIGURE 27.11 Coincidence spectrum from $F^{2+} + C_{60} \rightarrow F^- + C_{60}^{3+*}$ collisions recorded under the conditions that C_{60}^{3+} is the parent ion (cf. text). (Reprinted from Martin, S. et al., *Phys. Rev. A*, 77, 043201, 2008. With permission.)

the fullerenes will start to fragment. Activation energies, i.e., dissociation energies, for C_2 evaporation from *singly charged* fullerene cations, C_n^+ (for even $n = 48$–70 and for $n = 76$), have been measured by Tomita et al. to be between 7.0 ± 0.1 eV (for the case of C_{62}^+) and 9.8 ± 0.1 eV (for C_{60}^+). In that experiment (Tomita et al. 2001), the C_n^+ ions were stored in an electrostatic storage ring, and after storage for some time the ions were heated (excited) by means of a laser pulse. The subsequent decay was recorded and dissociation energies could be extracted. According to Equation 27.4, the dissociation energies for the neutral C_{60} and the C_{60}^+ cation are related through $DE(C_{60}^+) = E_B(C_{60}) - E_B(C_{58}) - E_B(C_2) + I_1(C_{58}) - I_1(C_{60})$, which per definition may be written as $DE(C_{60}^+) = DE(C_{60}) + I_1(C_{58}) - I_1(C_{60})$ and thus we have $DE(C_{60}) = DE(C_{60}^+) + I_1(C_{60}) - I_1(C_{58})$. The first ionization energy of C_{60} is roughly 0.5 eV higher than that of C_{58}, which indicates a DE for neutral C_{60} around 10.3 eV. Using the relation $\langle E_{int} \rangle = ((3 \times 60) - 6)kT/2$ for the average internal energy, we conclude that we would need an excitation energy of C_{60} corresponding to a temperature around 1370 K, or 1100°C, in order to activate the C_2-evaporation channel for molecules having this average energy. It should be noted, however, that even a fullerene with internal energy exceeding 10.3 eV may stay intact for substantial periods of time due to the high number of vibrational degrees of freedom (174 for C_{60}), as it would be necessary to concentrate at least the energy DE = 10.3 eV in one of the modes. On the other hand, since the molecules in the heated fullerene vapor will have a rather wide distribution of internal energies, fragmentation may be observed also at temperatures substantially lower than 1100°C due to the high energy tail of the distribution. The Arrhenius expression gives the decay rate, k, (the decay probability per unit time) as a function of the internal temperature, T, as

$$k(T) = A \exp\left(\frac{-DE}{k_B T}\right), \qquad (27.5)$$

where

$k_B = 1.38 \times 10^{-23}$ J/K is the Boltzmann constant

A is the so-called pre-exponential factor (an example of the value used for C_{60} is $A = 2 \times 10^{19}$ s^{-1} [Laskin et al. 1998])

It is readily seen that the higher the temperature T is, the higher the decay rate k becomes. This means that internally hotter fullerenes will decay faster and leave the cooler ones (in an internal energy distribution) intact for some longer time.

Other ways to excite fullerenes involve collisions with electrons and ions (of different atomic and molecular species and in different velocities). Many important studies using electron collisions have been performed by the Innsbruck group, and it was through these experiments that it was first realized that fullerenes could be ionized to high charge states without prompt fragmentation. In a way this is an amazing result as the energy required to multiply ionize a fullerene molecule may, by far, exceed the minimum energy needed for fragmentation. The explanation for this has already been given above, and in the case of the C_{60} fullerene we have seen that they indeed are thermodynamically stable (in their ground states) for charge states up to $r = 5$. For higher charge states, the fullerenes may be metastable due to substantial energy barriers for asymmetric fission and the large number of internal degrees of freedom. The first observation of septuply (seven times) charged C_{60}, C_{60}^{7+}, was reported by Scheier and Märk in 1994 (Scheier and Märk 1994a) (Figure 27.12). In this experiment they also measured the mean kinetic energy release in the asymmetric fission channel, $C_{60}^{7+} \rightarrow C_{58}^{6+} + C_2^+$, to be as high as 9.7 ± 2.2 eV, which thus provided additional evidence for the C_{60}^{7+} identification (the high kinetic energy of the C_{58}^{6+} product proved that its partner also must have been positively charged and that the mother fullerene ion must have been C_{60}^{7+}). The same group also published a long range of pioneering papers including the first systematic study of kinetic energy releases in asymmetric fission

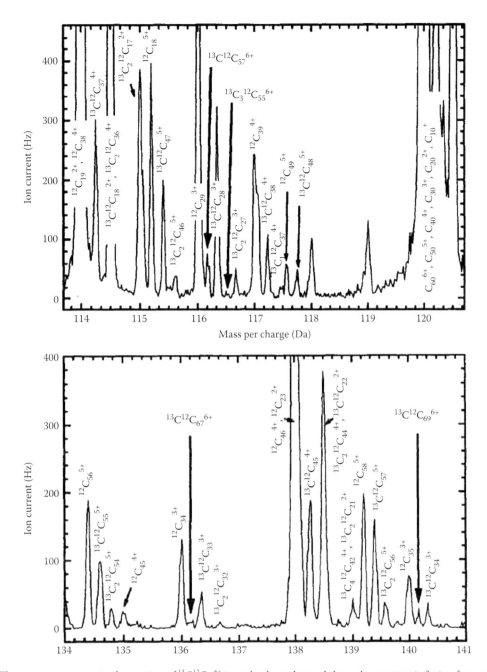

FIGURE 27.12 The mass spectrometric observation of $^{13}C^{12}C_{57}^{6+}$ ions that have decayed through asymmetric fission from seven-times charged $^{13}C^{12}C_{59}$ (above) and the identification of six-times charged $^{13}C^{12}C_{69}$ (below). (Reprinted from Scheier, P. and Märk, T.D., *Int. J. Mass Spectrom. Ion Process Lett.*, 133, L5, 1994b. With permission.)

as a function of the charge state of the mother ion (Scheier et al. 1995) and the first unambiguous identification of sequential evaporation of neutral C_2 units from multiply charged fullerenes (Scheier et al. 1996).

Collisions with atomic ions may be divided in three main categories: collisions with slow ions in low charge states, collisions with slow ions in high charge states, and collisions with fast ions. In this context, fast ions are defined as ions with velocities higher than typical orbital velocities of the fullerene valence electrons. One of the earliest experiments with fast ions colliding with fullerenes was performed by LeBrun et al. (1994) who identified

characteristic spectra of multiple ionization and fragmentation of C_{60} following collisions with Xe^{25+} ions of energy 635 MeV. The first experiment using slow highly charged atomic ions and fullerenes was performed by Walch et al. (1994), and they realized that such ions are very efficient in removing electrons from C_{60} (or other fullerenes) with comparatively small amounts of energy transfers to the fullerenes in relation to what is obtained with ionizing electrons, fast ions, or slow ions in low charge states. This was concluded from the measured mass spectrum by Walch et al. (see Figure 27.13) in which rather high charge states of intact C_{60} were observed and in which very little fragmentation was

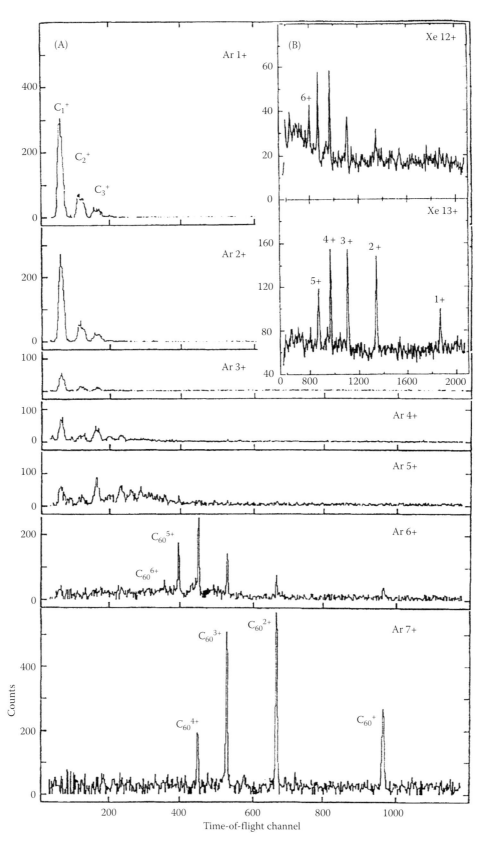

FIGURE 27.13   The first published results on ionization and fragmentation of fullerenes by slow highly charged (Ar and Xe ions). (Reprinted from Walch, B. et al., *Phys. Rev. Lett.*, 72, 1439, 1994. With permission.)

measured for C_{60} in the lower charge states (experimental time scale of microseconds). Basically this result is due to the fact that a slow highly charged ion exerts a large force on the electrons in the fullerenes and several electrons may be removed, and captured by the ion, even when the highly charged ion passes at a large distance from the fullerene cage. The cage radius, i.e., the mean radius of the structure defined by the average positions of the carbon nuclei, in C_{60} is about $6.7a_0$, where a_0 is the atomic length unit $1a_0 \approx 0.53$ Å. In fact, it has been shown by Martin et al. that large numbers of electrons, even by far exceeding the charge state q of a highly charged atomic projectile ion, may be removed from the fullerenes (Martin et al. 1999). For ions in low charge states, collisions have to be much closer to the fullerene cage for single and multiple ionizations to be efficient and then the fullerene will, of course, be more strongly heated through more direct interactions with the fullerene electron cloud and the cage (Opitz et al. 2000, Reinköster et al. 2001, Rentenier et al. 2004, Schlathölter et al. 1999a,b). The C_{60} fullerene has 10 (valence) electrons in its highest occupied molecular orbital (HOMO) that are delocalized over the surface of the whole molecule (Díaz-Tendero 2005) (Figure 27.14). The energy to remove one of these electrons to form C_{60}^{+}, i.e., the first ionization energy of C_{60}, is 7.61 ± 0.11 eV (Zimmerman et al. 1991). There are 18 additional electrons in the next lower molecular orbital, HOMO-1, which has the second lowest BE and which lies 1.8 eV below the HOMO. Also these electrons belong to the whole molecule and they are thus not localized to any specific site. In fact, this is the case for 240 out of the 360 electrons in C_{60}. The six electrons in an isolated carbon atom are in the quantum states specified by $1s^2 2s^2 2p^2$, which means that there are two very tightly bound electrons in the 1s ground state of carbon, four electrons in the next two available quantum states—two with angular momentum quantum number $l = 0$ ($l = 0$ is denoted by "s") and two with $l = 1$ ("p"). In C_{60} the 120 1s electrons are localized in 60 $1s^2$ pairs at 60 nuclei, while the 240 electrons from the 2s and 2p atomic orbitals participate in the C_{60} bonding through a combination of so-called sp^2 and sp^3 hybridizations (three atomic wave functions 2s, $2p_x$, and $2p_y$ are needed to describe the electron distribution in graphite through sp^2 hybridization, while four states 2s, $2p_x$, $2p_y$, and $2p_z$ are needed for the bond structure in diamond through sp^3 hybridization; for the fullerenes we have to use combinations of these two schemes). While the nuclear cage diameter is around 7 Å for C_{60}, the cloud of the delocalized valence electrons extends further out and has a diameter of roughly 10 Å. The electron cloud leaves a cavity that is large enough to store an atom or a small molecule at the center of the C_{60} fullerene. Such fullerenes (containing a foreign atom or molecule) are referred to as endohedral fullerenes. One important example here is La@C_{60} (this notation is used to state that an atom is inside the cage), which was studied early on and in connection with the discovery of the fullerenes (Figure 27.15). It was found that the La atom was kept during stepwise fragmentation through C_2 emission until the fullerene was too small to keep it inside, and actually this was taken as an early strong indication of the closed structure of the fullerenes (Heath et al. 1985): in the case that the molecules would not have had closed structures, the La atom should have been lost first. Fullerenes with atoms or molecules bound on their outside are referred to as exohedral fullerenes. One example here is $C_{60}F_{48}$ with 48 fluorine atoms on the outside of the C_{60} cage. Here, fragmentation starts with the loss of fluorine atoms. Although both endohedral and exohedral fullerenes are very interesting and have interesting fragmentation behaviors, we will limit ourselves here to a discussion of the pure fullerenes.

Already pure fullerenes do have rather complicated electronic structures as they contain large numbers of electrons and advanced quantum mechanical methods, and powerful computational tools, are in general needed in order to calculate their molecular properties with useful accuracies. Examples here are the calculations of dissociation energies for various decay channels for the multiply charged C_{60} and C_{70} fullerenes by Díaz-Tendero et al. (2005b) and Zettergren et al. (2007a) where DFT was used. In these cases it was necessary to reach relative accuracies (between the total energies of mother and fragment systems) of the level of a few tenths of an eV for the information to be useful. Although the absolute accuracy does not need to be at this

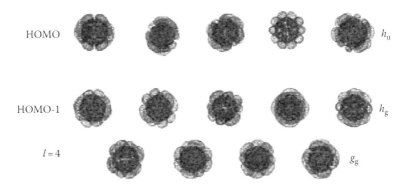

FIGURE 27.14 **(See color insert following page 25-14.)** The highest occupied molecular orbital (HOMO) and the lower lying molecular orbitals in neutral C_{60}. The pictures show the orbitals for a single fixed orientation of the fullerene cage and the g_g, h_g, and h_u notations refer to the orbital symmetries. The g_g and h_g states are close in energy with a gap of about 1.8 eV to the h_u (HOMO) state (cf. text). The results are calculated by Díaz-Tendero at HF/6-31G* level of theory. (From Díaz-Tendero, S., Structure and fragmentation of neutral and positively charged carbon clusters and fullerenes, PhD thesis, Universidad Autónoma de Madrid, Madrid, Spain, 2005.)

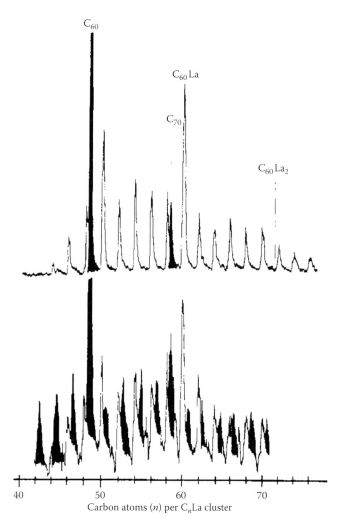

FIGURE 27.15 Fragmentation spectrum for endohedral C_{60} containing one lanthanum atom (La@C_{60}). The open peaks are fullerenes containing one La atom, while the blackened peaks are due to empty fullerenes. The upper spectrum is taken with high laser fluence at which only the most stable fullerenes remain, while the lower spectrum is taken with low laser fluence. It appears that the smallest endohedral is La@C_{42}. (Reprinted from Heath, J.R. et al., *J. Am. Chem. Soc.*, 107, 7779, 1985. With permission.)

level this is still a demanding task considering that the total electronic energy of C_{60} is of the order of keV. As the fullerenes have many electrons with similar low binding energies (cf. the example with C_{60} above) it has, in certain applications, turned out to be quite useful to describe the electronic responses of fullerenes by simply assuming that they are (very small) classical, electrically conducting, spheres. In particular, this has turned out to be useful when modeling their ionization energies and interactions with highly charged atomic and cluster ions and then, in particular, for electron transfer reactions. In addition, this approach has been used to describe the gross features (e.g., typical kinetic energy releases) for asymmetric fission of fullerene monomers. The simplest of these examples is the calculations of ionization energies, i.e., the energies needed to move one electron from the surface of a metal sphere (here representing a fullerene) to

infinity. For the kth ionization energy (the energy needed to singly ionize a $C_n^{(k-1)+}$ ion), we have

$$I_k = W + \frac{k-(1/2)}{a}, \qquad (27.6)$$

where

> W is the work function
> a is the radius of the metal sphere

Now, using, for example, the theoretical sequence of ionization energies of C_{60} by Díaz-Tendero et al. (2005b) it is possible to deduce the values of W and a to be $W = 3.854\,\text{eV}$ and $a = 8.37a_0$ and

$$I_k = 3.854 + 3.252 \times k \text{ eV}, \qquad (27.7)$$

in very good agreement with available experimental results on the first, the second, and the third ionization energies of C_{60} (Lifshitz et al. 1991, Zimmerman et al. 1991, Hertel et al. 1992, Baba et al. 1993, Steger et al. 1995, Matt et al. 1997) (see Figure 27.16). It should be noted that ionization energies other than the first one are difficult to measure, which at least partly is due to the fact that the so-called bracketing method normally used for singly charged ions (in which one checks reactivities for a range of molecules with known first ionization energies) is somewhat

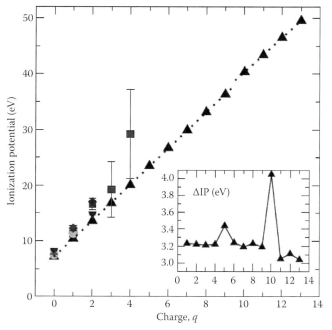

FIGURE 27.16 DFT calculated sequential ionization energies (Díaz-Tendero 2005) for fullerenes in the initial charge state q (triangles). Experimental results: crosses (Zimmerman et al. 1991), squares (Matt et al. 1997), circles (Hertel et al. 1992, Steger et al. 1995), diamonds (Baba et al. 1993), and triangles (Lifshitz et al. 1991). The inset shows calculated differences between the minimum energies needed for ionizing a fullerene in charge states q and $q-1$. (From Díaz-Tendero, S., Structure and fragmentation of neutral and positively charged carbon clusters and fullerenes, PhD thesis, Universidad Autónoma de Madrid, Madrid, Spain, 2005.)

difficult to apply for multiply charged ions. Thus, calculations and simple models, like the one mentioned here, may be of large practical importance. It is interesting to note that the DFT calculations (Díaz-Tendero et al. 2005b) yield results which are very well described by a straight line, and that the value of the radius, a, is consistent with the measured polarizability of C_{60} ($a^3 = (8.37)^3$ $a_0^3 = 585a_0^3$, and should be compared to the experimental value = $517 \pm 54a_0^3$ [Antoine et al. 1999]). The situation becomes a little more complex when considering charge transfer between a highly charged atomic ion (assumed to be point like) and a fullerene. In this case, it is necessary to consider the potential energy of an electron moving between the ion and a metal sphere. At certain critical distances between the ion and the sphere, the potential barrier seen by this active electron becomes sufficiently low for an over-the-barrier transition to the ion. The picture for multiple electron removal from a fullerene would thus be the following: Starting with an ion of charge state q and a neutral fullerene approaching each other it will become possible to transfer the first electron when the Stark shifted BE of a valence target electron first becomes equal to the energy of the top of the barrier. If q is large this happens at a large distance, often referred to as the first critical distance, between the ion and the fullerene. As the approach continues it will, at a slightly smaller distance, be possible to transfer a second electron to the ion (now the Stark shifted energy of the valence electron in the singly charged fullerene must be such that an over-the-barrier transfer is possible). In this way, there will be a sequence of electron transfers at large distances from the fullerene cage and thus it should be possible to multiply ionize fullerenes without transferring much energy to the remaining fullerene ion (for closer and alternative descriptions of this model see, e.g., Bárány et al. (1985), Niehaus (1986), Thumm (1994), Walch et al. (1994), Bárány and Setterlind (1995),

Thumm et al. (1997), and Cederquist et al. (2000)). Another way to picture the same sequential electron transfer scenario is given in Figure 27.17 where we see a fullerene in the center of concentric rings indicating where a projectile of charge q would have to pass to capture one electron (the outermost ring), to sequentially capture two electrons (the second outermost ring) and so on. Although this picture gives a rather good representation of the sequence of electron removal cross sections, i.e., the relative cross sections for removing one, two, three, or more electrons from a fullerene (as measured before they fragment), it is clear that some important aspects are not represented correctly in such a simple model. This concerns, e.g., the energy transfer processes where the model suggests very small energy transfers also for the removal of a comparatively large number of electrons, but as we will see below this is not consistent with experimental observations showing fragmentation on experimental time scales, which appears to require substantial internal excitation energies even when only a few electrons are transferred. To clarify this statement, we note that the large distances at which say, r, electrons are removed from the fullerene would give extremely small internal excitation energies (typically of the order of eV or even less according to stopping power calculations ([Larsen et al. 1999, Larsson et al. 1998]) while the observed fragmentation behavior seems to require internal energies of several tens of eV and sometimes even up to the 100–200 eV range.

Considering electron transfer between two finite-sized objects (like, e.g., two fullerenes) we again need to evaluate the potential barrier experienced by an active electron between the two objects (conducting spheres). This problem may be solved analytically as shown by Zettergren et al. (for metal spheres in vacuum [Zettergren et al. 2002] and also more generally for dielectric spheres in a dielectric medium [Zettergren 2005]), and the corresponding model has

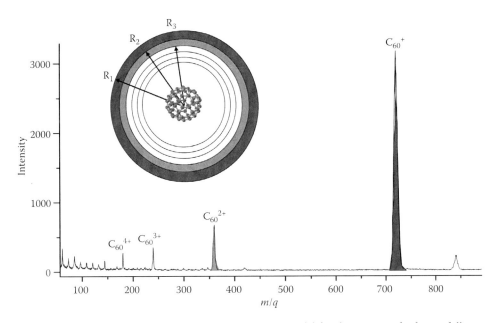

FIGURE 27.17 Illustration of the prediction of the classical over-the-barrier model for electron transfer from a fullerene molecule to a slow highly charged ion. Trajectories with their closest approach to the fullerene within the outer ring produces singly charged fullerenes, while those with closest approach produces doubly charged fullerenes (through sequential transfer of two electrons according to the model).

been used to successfully account for, e.g., relative charge transfer cross sections in $C_{60}^{q+} + C_{60}$ collisions (Cederquist et al. 2001) and for surface attractions in C_{60}^{+} and C_{60}^{2+} grazing incidence collisions with metal (Wethekam et al. 2007) and insulator surfaces (Wethekam and Winter 2009). The same model has also been used to calculate typical values of kinetic energy releases in the cases of asymmetric fission of C_{60}^{r+} and C_{70}^{r+} monomers (Cederquist et al. 2003, Jensen et al. 2004, Zettergren et al. 2004).

Now, we turn our attention to the description of the fragmentation of the fullerene monomer cations and refer to Figure 27.18 where we show a schematic view of the C_{60}^{4+} fullerene and its two main competing decay channels $C_{60}^{4+} \rightarrow C_{58}^{4+} + C_2$ (C_2 evaporation) and $C_{60}^{4+} \rightarrow C_{60}^{3+} + C_2^{+}$ (asymmetric fission) as an example. For this comparatively moderate charge state, the dissociation energies for both channels are positive and C_{60}^{4+} in its ground state would be, as mentioned above, thermodynamically stable. When increasing the internal C_{60}^{4+} excitation energy, the asymmetric fission channel will first become energetically accessible and shortly thereafter, if we continue to increase the internal energy, the evaporation channel will also become active (the DE for evaporation is slightly higher than the FB in Figure 27.18). In a real situation where the fullerene has been multiply ionized by an electron, a slow highly charged ion, or by photons, we will normally have a distribution of internal energies. There will be some ions with quite low energies, but a substantial fraction will have considerable internal excitation energies, and it is the latter which give rise to the fragmentation that is observed on typical experimental time scales of microseconds. In Figure 27.18, we indicate an example with an internal excitation energy E^*, which is substantially larger than the threshold energies for the two lowest energy decay channels of types 27.1 and 27.2. For a sufficiently high E^* there are, of course, additional available

decay channels with higher barriers and dissociation energies than (27.1) and (27.2), but they will be less important as there will be rather strong statistical preferences for the two lowest channels. Here, the picture is the following: The fullerene is ionized in an interaction and it will then, in general, be both electronically and vibrationally excited right after the interaction (it will also be rotationally excited but as the corresponding energies are much smaller this will be ignored in the following discussion). Now, the electronic excitation energy may couple to the vibrational degrees of freedom and typically such couplings occur at specific distances between the atoms in the molecule. This means that every time the molecule vibrates there is a chance to make a transition to a state with less electronic excitation energy and larger internal vibrational energy. As the molecular vibrations are fast (basically on the time scales of tens of femtoseconds) this conversion of energy will be much faster than typical non-prompt fragmentation processes and we may regard the excited fullerene, as a system with only internal vibrational energy. Now, it is a matter of how quickly a sufficient amount of energy (equal to or larger than the DE and/or the FB) may be concentrated in one of the 174 vibrational modes of the molecule. For large molecular systems, like the fullerenes, this may be regarded as a statistical process and it is clear that the higher the internal excitation energy is (E^* in our example) the sooner the fullerene ion will fragment.

We will now describe *one* out of several possible ways to deal with the statistical distribution of an internal energy E^* among $s = [(3 \times 60) - 6]$ vibrational degrees of freedom in the C_{60}^{4+} system. For this we have chosen classical Rice–Ramsperger–Kassel (RRK) theory (Rice and Ramsperger 1927, Kassel 1928) although there are certainly several other important theories and models available in the literature like, for instance, the Weisskopf model (Weisskopf 1937). The latter was originally developed

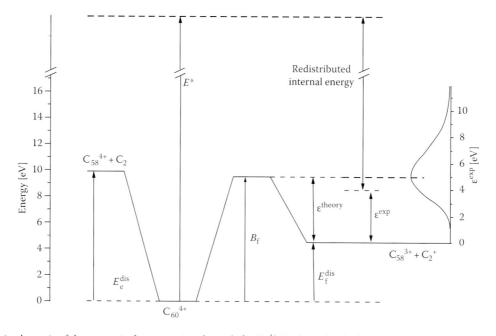

FIGURE 27.18 A schematic of the two main fragmentation channels for C_{60}^{4+}. To the right, the kinetic energy release distribution for the asymmetric fission process (cf. text). (Reprinted from Haag, N. et al., *Phys. Rev. A*, 78, 043201, 2008. With permission.)

for nuclear fragmentation processes but it has also been used widely, and with success, for atomic clusters (Hervieux et al. 2001, Bréchignac et al. 2002, Herlert and Schweikhard 2006) including also the fullerenes (Rentenier et al. 2008). In principle, the Weisskopf model is more detailed and should therefore be more accurate than the RRK model, but on the other hand, its use requires information on parameters like final state densities, which may not always be readily accessible. Referring again to Figure 27.18 we denote the DE for evaporation by E_e^{dis} and the barrier for asymmetric fission by BF. According to RRK theory, the probabilities to concentrate at least the energies E_e^{dis} or BF in one, out of the s, vibrational modes of C_{60}^{4+} are proportional to $(1 - E_e^{dis}/E^*)^{s-1}$ and $(1 - BF/E^*)^{s-1}$, respectively. Following Haag et al. (2008) we then take the rate constants for evaporation and fission to be $k_e = g_e A(1 - E_e^{dis}/E^*)^{s-1}$ and $k_f = g_f A(1 - BF/E^*)^{s-1}$ with A being a common frequency factor and g_e and g_f are degeneracy factors for the numbers of neutral and (singly) charged C_2 units in the C_{60} molecule at any given time. Haag et al. further reasoned that as there are in total 90 carbon–carbon bonds in C_{60} and as 60 of these share a hexagon–pentagon edge in the molecular structure $g_f = (60/90) \times r$ (with $r = 4$ in our example) and $g_e = 60 - (60/90) \times r$ as it is the removal of a hexagon–pentagon C_2 (C_2^+) that leads to the lowest energy isomer of C_{58} ($C_{58}^{(r-1)+}$). As the charge on C_{60} is fully delocalized (Zettergren et al. 2008) g_f and g_e are proportional to the probabilities that a given hexagon–pentagon C–C bond would be charged (C_2^+) and neutral (C_2), respectively, at any given time and specifically at the time of fragmentation. Of course, this reasoning is simplified as it ignores the rearrangement of charge during the fragmentation process (the charge is certainly not homogenously distributed in the transition state (Díaz-Tendero et al. 2005a)). Further, this also assumes that it really is the lowest energy isomer that is populated in the fragmentation process, and concerning this issue there is neither experimental nor theoretical information that could guide the discussion. Still, it may be of interest to study the consequences of the approach outlined above as it will demonstrate some general features that are rather insensitive to such details in the descriptions of fragmentation processes. For an exponential decay, the probability ρ that an ion with internal excitation energy E^* has decayed either through C_2^+ emission or through C_2 emission during the time window τ is

$$\rho(E^*) = 1 - \exp[-(k_f(E^*) + k_e(E^*))]\tau, \quad (27.8)$$

under the assumption that we only have these two decay channels. Now if the remaining excitation in the fragments after the initial step is sufficiently large, the decay may proceed in further steps during the same time window, τ, and thus if we want to study the relation between the processes of emitting a single C_2^+ ion and a single C_2 molecule we have to take the corresponding later decays into account when using the model. If now $\rho_e(E^*)$ and $\rho_f(E^*)$ are taken to be the probabilities for emission of one and only one C_2 and one and only one C_2^+, respectively, the ratio between the intensities of the corresponding peaks in the mass spectrum may be written as

$$R = \frac{\int \rho_f(E^*)dE^*}{\int (\rho_f(E^*) + \rho_e(E^*))dE^*} \quad (27.9)$$

Here, the integration should be over the (unknown) internal energy distribution. Evaluating (27.8) for the parameters relevant for the decay of C_{60}^{4+}, and under the specific experimental conditions reported by Haag et al. (observation time $\tau = 3 \mu s$), we see in Figure 27.19 that there is a fairly narrow energy window (45–56 eV) that gives fragmentation in one step. Further, the energy regions for one-step evaporation and one-step asymmetric fission processes overlap, which of course is related to the fact that the activation energies for the two processes are very close in the case of C_{60}^{4+}. As mentioned already above, these results are surprising as it could be expected that C_{60}^{4+} produced in slow collisions with highly charged ions should have rather small internal energies. On the other hand, this is consistent with the results of Martin et al. who measured internal excitation energy distributions directly (Martin et al. 2008) (Figure 27.20). They used an ingenious method in which, e.g., the internal excitation energies of doubly charged fullerene ions produced by slow fluorine ions could be deduced by measuring the Q values for reactions populating the only existing bound quantum state in the negative fluorine ion F⁻. Thus, as the kinetic energy of the negative ion was measured precisely, the complete energy balance of the collision process could be specified and thus the initial internal energy of the fullerene ion could be determined. In this case, internal energies around 40 eV were reported for C_{60}^{2+} fullerenes decaying on the microsecond timescale, while the same group reported an average internal energy of 64 eV for single C_2^+ emission from C_{60}^{4+} (formed after thermal ionization of C_{60}^{3+}) (Martin et al. 2008).

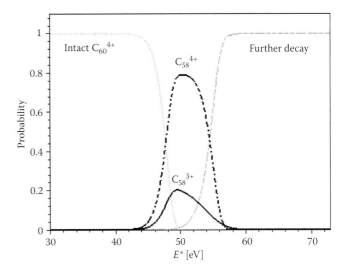

FIGURE 27.19 Probabilities that a C_{60}^{4+} mother ion will still be intact, fragmented (once) through emission of a single C_2 molecule (evaporation) or through emission of a single C_2^+ molecular ion, or that it has fragmented more than once (further decay) within a particular experimental time window (3 μs) as a function of the total internal excitation energy. (Reprinted from Haag, N. et al., *Phys. Rev. A*, 78, 043201, 2008. With permission.)

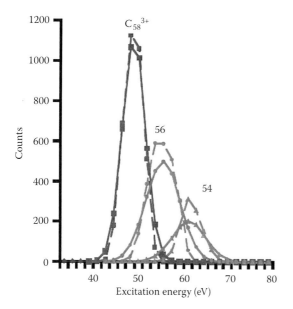

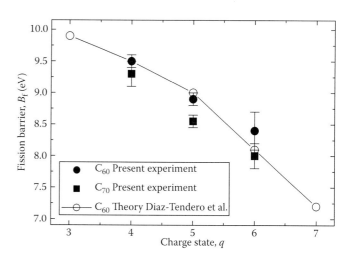

FIGURE 27.20 Internal energy distributions for C_{60}^{3+} mother ions decaying to C_{58}^{3+}, C_{56}^{3+}, and C_{54}^{3+} within a certain experimental time window. The complete energy balance of the charge transfer reactions have been measured directly by population of the only existing bound state in the negative fluorine ion. (Reprinted from Martin, S. et al., *Phys. Rev. A*, 77, 043201, 2008. With permission.)

FIGURE 27.21 Semiempirical fission barriers adapted from Haag et al. (2008). The barrier heights are obtained by relying on calculated DFT values for the dissociation energies and by assuming different barrier heights until good agreement between Equation 27.9 in the text and the measured ratio between asymmetric fission and evaporation is obtained. The results compare favorably with transition state DFT calculations. (Reprinted from Haag, N. et al., *Phys. Rev. A*, 78, 043201, 2008. With permission.)

When comparing evaluations of (27.9) in the way outlined here, Haag et al. (2008) found that semiempirical values of the barriers for asymmetric fission may be deduced if one relies on theoretical values for the dissociation energies of the evaporation process. This is reasonable as it is much easier to calculate dissociation energies (only fragments at infinite separations have to be considered) than to calculate the FB directly (much more complicated transition state calculations are needed for this). For cases in which the measured ratio R (cf. (27.9)) differs significantly from zero and one, the method gives results in surprisingly good agreement with the calculations of FB by Díaz-Tendero et al. (2005a) as can be seen in Figure 27.21. Here, the method used was to evaluate (27.9) as functions of assumed barrier heights and selecting the value that gave agreement with the measured value of R.

It is immediately evident from Figure 27.18 that the kinetic energy distributions should be different for evaporation and asymmetric fission due mainly to the Coulomb repulsion between the fragmentation products which is only present in the latter case. Indeed, while kinetic energy release distributions for evaporation are very narrow and centered at very low energies (Figure 27.22) (Matt et al. 1999, Głuch et al. 2004, Climen et al. 2007), the distributions for asymmetric fission are very wide and their average values are well separated from zero (Figure 27.23). From the discussions above, one would perhaps have expected the fission distributions to start at very low energies and to range up to the energy given by the height of the fusion energy barrier (i.e., the energy barrier experienced when trying to form a C_{60}^{r+} system from $C_{58}^{(r-1)+}$ and C_2^+) for the reverse process but not higher than this. Instead, the measured distributions are very

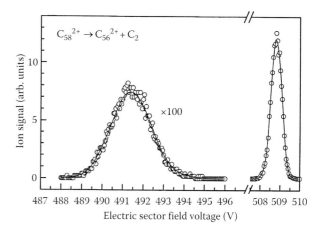

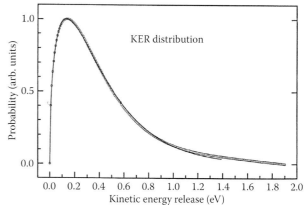

FIGURE 27.22 Top: Mass-analyzed ion kinetic energy spectra for the daughter and parent ions in evaporation from C_{58}^{2+}. Bottom: The deduced kinetic energy release distribution (open dots) together with a nonlinear fit (solid line). (Reprinted from Matt, S. et al., *Chem. Phys. Lett.*, 303, 379, 1999. With permission.)

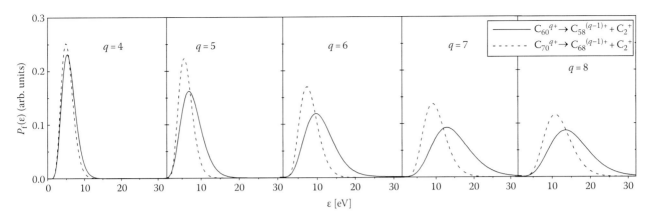

FIGURE 27.23 Kinetic energy release distributions for asymmetric fission from C_{60} and C_{70} ions in charge states ranging between $q = 4$ and $q = 8$. Note that the C_{70} distributions are narrower and peak at lower kinetic energies. (Reprinted from Haag, N. et al., *Phys. Rev. A*, 78, 043201, 2008. With permission.)

wide and extend well beyond the energies corresponding to the (fusion) barrier heights. It thus appears that unexpectedly large fractions of the internal energies in the mother fullerene ions sometimes are converted to kinetic energies (KE) in the asymmetric fission processes and thus that such processes may lead to strong internal cooling of the fragmentation products. This does obviously not occur for the evaporation process. At present, the reason for this difference is not clear although it most likely is related to the much stronger interaction between fragments of asymmetric fission processes than between fragments from the evaporation process. In this context, it is interesting to note that the kinetic energy release distributions for C_{70}^{r+} fullerenes in charge states $r > 4$ are significantly narrower, and peak at lower energies, than those for C_{60}^{r+} in the corresponding charge states above $r = 4$ (Haag et al. 2008).

We have already mentioned above, and it has also been shown in experiments by Martin et al. and in the analysis by Haag et al, that internal excitation energies of multiply charged fullerene ions often have substantial parts of their internal energy distributions at unexpectedly high energies. It is at least from the point of view of the simple (but actually rather successful) over-the-barrier description of (sequential) electron transfer processes (which may take place already at substantial distances from the fullerene cage and its electron cloud) hard to understand how energies in ranges from tens of eV and all the way up to a couple of hundred eV may be so efficiently transferred. In a very recent paper by Rentenier et al, reporting a joint theoretical and experimental study of the fragmentation following slow $He^{2+} + C_{60}$ collisions (Rentenier et al. 2008), it was found that high internal excitation energies were due to direct transfer of electrons from molecular states below the valence orbital. In collisions with projectile ions in higher charge states, however, such excitation mechanisms are probably less important as highly charged ions have high densities of capture states. It has been shown that inner shell electrons may be captured preferentially if the capture state density is low in the case of ion–atom collisions. This is due to the fact that there may not be a suitable resonance state to capture to even though the potential energy barrier is

low enough to classically allow electron transfer between the target and the projectile (Bárány et al. 1985, Niehaus 1986, Thumm 1994, Bárány and Setterlind 1995, Thumm et al. 1997, Cederquist et al. 2000) for low projectile charge states. Such resonances may however appear for lower energy states of the target at closer distances to such a projectile ion, and the electron transfer may then proceed leaving hole states in inner shell that represents higher internal target excitation energies.

In this chapter, we have described how the excitation process may influence the competition between different decay channels as it determines the internal energy and the charge state of the fullerene. On the other hand, it is also worthwhile to note that fragment distributions may be very similar for a wide range of excitation methods (lasers, ions, electrons, etc.) and this is in particular the case for the multifragmentation processes in which the mother fullerene breaks up in several smaller fragments. The relative intensity distributions on the smaller C_n fragments due to multifragmentation processes are to larger extents dictated by variations in the binding energies per carbon atom and less by the exact nature of the excitation method, and there are certain typical ("magic") masses with higher intensities than their neighbors as, e.g., C_7, C_9, and C_{11} as can be seen in the example shown in Figure 27.24. The reason why BE effects are much more important for the smaller fragments than for the fullerenes is probably related to the fact that the variations in BE per atom, between neighboring masses, is much larger at the lower mass end than for the fullerenes (Campbell et al. 1996).

27.3 Stability of Fullerene Monomer Anions

In the preceding section, we learned that multiply ionized fullerenes are stable or metastable depending on the degree of ionization and that evaporation and fission are competing fragmentation processes. In this section, we will discuss the stability of negatively charged fullerenes (anions). Before turning to our main subject, we first introduce a few general features of molecular anions.

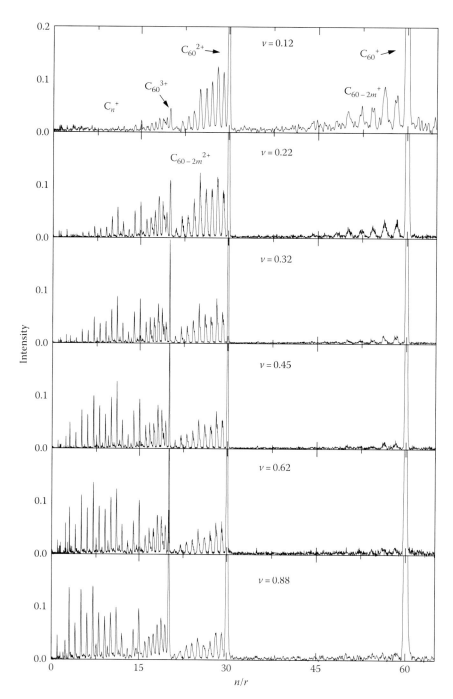

FIGURE 27.24   Projectile velocity effects in collisions of He$^+$ with C$_{60}$. The low mass distributions (n/r = 1–15) are due to multifragmentation processes. (Reprinted from Schlathölter, T. et al., *Phys. Rev. Lett.*, 82, 73, 1999a. With permission.)

Anions have fundamental properties different from neutrals and cations, which influence the chemical reactivities governed by the valence electrons. In neutral and cationic molecules, the valence electrons basically experience a Coulomb potential, $-(q + 1)/r$, where q is the net charge and, here, r is the distance between the electron and the molecule. In such a very simplified picture, the energy levels (E_n) are given by the well-known Bohr formula ($E_n = -13.6^\star(q + 1)^2/n^2$ eV, n = 1, 2, 3,… where n in this case is the principal quantum number). In contrast, the

excess electron in monoanions (singly charged anions) experiences no such potential as the electron at large distances r sees a neutral molecule. As a consequence, the long range potential depends on higher powers of $1/r$ that are due to the charge–dipole, charge–quadrupole, charged–induced dipole, and higher order interactions, and thus the Bohr formula cannot be used. In fact, the anionic excited states are rarely bound in contrast to the situations for the neutrals and the cations. The electron BE or the first electron affinity (EA) of a molecule A is defined as the energy difference between the neutral and the anion, EA = E(A) − E(A⁻). Thus a positive EA corresponds to an electronically stable molecule, whereas a negative value means that the anion is unstable or, as we will see below, metastable such that it may survive on typical experimental timescales. Strictly speaking, there are three definitions of EA depending on how E(A) and E(A⁻) are defined. (1) The adiabatic EA is the energy difference between the neutral molecule and the anion in their electronic, vibrational, and rotational ground states. (2) The vertical EA is the energy difference between the neutral molecule (in its electronic, vibrational, and rotational ground states) and the energy of the anion in the electronic ground state in the geometry of the neutral. (3) The vertical detachment energy is the energy of the neutral in the electronic ground state and in the geometry of the anion minus the energy of the ground state (electronic, vibrational, and rotational) energy of the anion. In the following, we will for simplicity only refer to EA, which we justify by the rigidity of the fullerene cages and thus small differences in cage geometries of the neutrals and anions.

The fullerene monoanions have positive EAs, about 2.7 eV for C_{60} and C_{70}, due to their high polarizabilities (76.5 ± 8) Å³ for C_{60} [Antoine et al. 1999]) giving a strong charge-induced dipole interaction and unusually stable anionic species. Still, the dissociation energies for fragmentation is markedly higher (~10 eV, see preceding section) and thus the dominant statistical decay channel is electron emission (clearly electron emission should be much more important for the anions as the EA is considerably smaller than the ionization energies of neutrals and cations). Nonstatistical fragmentation processes have been observed in collisions between C_{60}^- and He or Ne atoms (Tomita et al. 2002). In that experiment, both C_{59}^- and C_{59}^+ fragments were formed in direct knock out processes of carbon atoms from the fullerene cages, and these metastable molecules containing 59 carbon atoms, survived on the experimental timescale (about 10 μs in that case). This suggested that the internal energies of the initial C_{60}^- anions were low. In addition, a positively charged evaporation series was observed, C_{60-2m}^+ ($m = 1$–12), due to the formation of highly excited positively charged fullerenes that decay into smaller fragments (Tomita et al. 2002) (Figure 27.25).

In the case of C_{60}^-, the excess electron enters the threefold degenerated lowest unoccupied molecular orbital, the LUMO, of C_{60}. Interestingly, the high (icosahedral) symmetry is then broken and the electronic levels are split, as shown by means of near-infrared spectroscopy on C_{60}^- in an electrostatic storage ring (Tomita et al. 2005). We make no attempt to explain this so-called Jahn–Teller effect in details here and simply note that the rather involved theory is a research field of its own (Chancey and O'Brien 1997). In pioneering fullerene anion experiments back in 1991, $C_{60}^-, C_{60}^{2-}, C_{70}^-$ and C_{70}^{2-} were measured to have lifetimes of minutes or longer (Hettich et al. 1991, Limbach et al. 1991). There, the anions were produced by laser desorption of fullerene material and then analyzed by means of Fourier transform ion cyclotron resonance (FT-ICR) mass spectrometry (Figure 27.26). A few years later, the EAs of internally cold C_{60} and C_{70} were accurately measured in a storage ring to

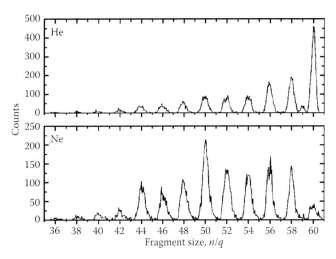

FIGURE 27.25 Positive fragment ion spectra recorded in 50 keV collisions of C_{60}^- with He and Ne. (Reprinted from Tomita, S. et al., *Phys. Rev. A*, 65, 043201, 2002. With permission.)

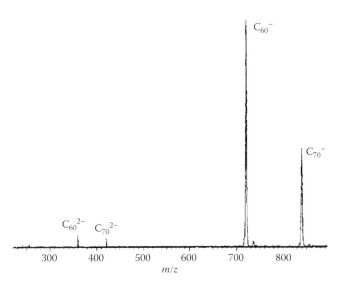

FIGURE 27.26 Observations of dianions 3 s after their desorptions inside an ion trap. (Reprinted from Limbach, P.A. et al., *J. Am. Chem. Soc.*, 113, 6795, 1991. With permission.)

be EA(C_{60}) = 2.666 ± 0.001 eV and EA(C_{70}) = 2.676 ± 0.001 eV (Brink et al. 1995), in agreement with the ion trap observations. In these storage ring measurements, the fullerene monoanions were stored, cooled, and then photoexcited by a tunable laser. The yield of fullerene neutrals in electron detachment processes ($hv + C_{60}^- \rightarrow C_{60} + e^-$) were then measured as a function of photon energy (hv), where the threshold value (the smallest photon energy, hv_{min}, for which an electron is emitted) corresponds to EA (EA = hv_{min}). It is important to note the fullerene anions have to be internally cold in order to perform the measurement described here as there would be no distinct photodetachment threshold for internally hot ions. An alternative method to measure EAs is photoelectron spectroscopy (PES), in which the anions are photoexcited and the KE of the emitted electrons are measured directly. The electron BE may then be determined through the relation, BE = $h\nu$ − KE, where EA corresponds to the threshold BE (i.e., minimum KE measured). In this way the EAs of C_{60} and C_{70} were determined to be 2.689 ± 0.008 eV (Wang et al. 1999) and 2.765 ± 0.010 eV (Wang et al. 2006), respectively. This method has also recently been applied to larger fullerenes, showing an increase in EA values as a function of size in accordance with increasing polarizabilities of the neutral fullerenes (Wang et al. 2007).

There have also been a vast number of experimental and theoretical studies of multiply charged fullerene anions, spurred by the observation of superconductivity in fullerene solids doped by alkali metals (acting as electron donors) (Gunnarson 1997). To our knowledge this effect is still not fully understood despite extensive research efforts. In these materials (and in, e.g., solutions), the excess electrons are stabilized by interactions with the surrounding, so-called, counterions. However, in gas phase the electrons are only hindered by a Coulomb barrier similar to the FB for multiply charged cations, as discussed in the preceding section. Multiply charged anions ($A^{q−}$) with a positive EA (EA = E($A^{(q−1)−}$) − E($A^{q−}$)) are thus additionally stabilized by this barrier, which means that electronically excited anions may have a finite lifetime given in this case by the tunneling probability. Multiply charged anions with negative EA may, thus, survive long enough to be detected in experiments even though they are not thermodynamically stable. The existence of a negative electron BE in a molecule was beautifully demonstrated by Wang and Wang (1999) in PES experiments, as the measured KE of the emitted electrons were higher than the photon energy (BE = $h\nu$ − KE). Recently, the second EA of vibrationally cold C_{70} was determined to be $0.02^{+0.01}_{-0.03}$ eV by the same method (Wang et al. 2006), suggesting that this is the smallest fullerene dianion (doubly charged anion) which is truly bound. The most recent experimental and theoretical results suggest that C_{60} has a negative second EA (Figure 27.27), but to our knowledge no direct

measurements have so far been reported in the literature. The main reason for this is probably related to earlier difficulties to produce the dianions in sufficiently large quantities for photoexcitation experiments. These difficulties may now be overcome by letting C_{60}^{-} ions from, e.g., an electrospray or a plasma ion source be accumulated in a cooling trap (at liquid nitrogen or cryogenic temperatures) and then, after mass selection, let them collide in Na or Cs vapor to capture an electron (Liu et al. 2004, Tomita et al. 2006). The formation of C_{60}^{2-} in collisions with alkali atoms, which is well explained by the metal sphere charge transfer model described in Bárány et al. (1985), Niehaus (1986), Thumm (1994), Bárány and Setterlind (1995), Thumm et al. (1997), and Cederquist et al. (2000) requires that the preceding C_{60}^{-} ions have low internal energies as a high internal energy would lead to prompt emission of the second electron. Storage ring decay curves may then, as in an experiment at Aarhus University (Tomita et al. 2006), be measured by counting the numbers of monoanions (due to electron detachment $C_{60}^{2-} \rightarrow C_{60}^{-} + e^{-}$) per turn as a function of the C_{60}^{2-} storage time or, alternatively, by counting all C_{60}^{2-} anions that are left in the ring after different storage times by dumping the beam on a particle counting detector. From these curves, a C_{60}^{2-} lifetime of the order of 20 s, which is significantly shorter than the results from the ion trap measurements, has been deduced (Tomita et al. 2006). Another puzzling feature observed in the same storage ring experiment was a small contribution from a long-lived component at the highest trap temperatures (this means that the corresponding C_{60}^{2-} ions were produced from electron capture from somewhat warmer C_{60}^{-} molecules). According to recent DFT calculations (Zettergren et al. 2007b) this long-lived component is most likely due to isomers of C_{60} that have adjacent pentagons and positive second-electron affinities. Thus, a small impurity of these isomers in the ion beam appears at high temperatures and long storage times, as a large fraction of the metastable soccer ball structure with negative EA have then decayed. This clearly illustrates the complexity of this problem and the importance of having the anions internally cold in order to simplify the interpretation of experimental results.

Despite the vast number of studies reported so far, there are many interesting aspects to be investigated in the future regarding fullerene anions, such as, e.g., PES of C_{60}^{2-} to measure the second EA accurately, absorption spectroscopy of C_{60}^{2-} isomers that contain adjacent pentagons, detection of the smallest (meta-)stable fullerene trianion, and absorption spectroscopy and PES of dimers and larger clusters of fullerene anions.

27.4 Ionization, Excitation, and Fragmentation of Dimers and Clusters of Fullerenes

The preceding sections were devoted to ionization, excitation, and fragmentation of fullerene monomers (single, isolated, fullerene molecules). Here, we will discuss related issues involving weakly bound fullerene dimers and larger clusters

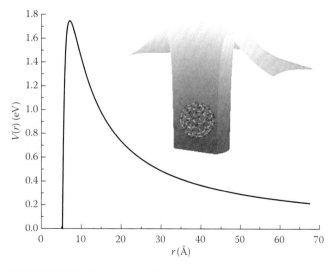

FIGURE 27.27 A schematic of the potential energy seen by the outermost electron in C_{60}^{2-} (Tomita et al. 2006). A three-dimensional illustration of the potential energy is shown in the inset.

Van der waals C$_{60}$ dimer

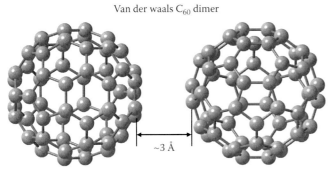

~3 Å

Covalently bound C$_{60}$ dimer

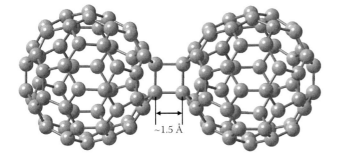

~1.5 Å

FIGURE 27.28 Illustrations showing the differences in cage–cage distances for van der Waals and covalently bound C$_{60}$ dimers (cf. text).

of fullerenes. These systems may be regarded as small pieces of solid-state fullerene materials (or small pieces of fullerite). In pure C$_{60}$-fullerite, the center–center distance between nearest-neighbor fullerenes is about 10 Å, which is larger than the 7 Å diameter of the C$_{60}$ cage (Figure 27.28). Thus, the shortest distance between two fullerene cages is about 3 Å, which is significantly larger than typical covalent bond lengths, and therefore no such (stronger) bonds form between carbon atoms belonging to separate C$_{60}$ molecules in fullerite. Instead, the binding conditions are of the weaker van der Waals type and the individual fullerenes are held together through dispersion interactions where the attractive forces origin in fluctuations in the electronic and nuclear charge distributions which, in turn, are related to the intramolecular vibrations. As a consequence of the weak intermolecular bonds in fullerite, the individual fullerenes retain many of their monomer properties and they may, e.g., rotate as separate (and essentially free) molecules inside a fullerite crystal.

From a theoretical point of view, dispersion interactions which depend on many-body electron correlation effects are difficult to handle by means of high-level quantum-chemical computational tools, and the search for accurate descriptions has so far had limited success in the case of, e.g., the DFT method. (The electron correlation concept describes part of the mutual Coulomb interaction between electrons in a molecular system, and this description becomes very involved in particular for large numbers of interacting electrons.) An additional complication is that the binding energies between neighboring fullerenes typically are in the range below a few tenths of an eV, which is a very small part of the total BE of, e.g., the fullerene dimer system (this total BE is many orders of magnitudes larger). Thus, calculating the

BE with a useful accuracy is a computationally very demanding task even with an accurate theoretical description. Such calculations become even more difficult with increasing numbers of fullerenes in a fullerene cluster. In spite of this, a few high-level calculations on dimers and small clusters of fullerenes have been reported in the literature, but the details of these calculations are beyond the scope of this chapter. Instead, we will discuss experimental and theoretical results on van der Waals dimers and clusters of fullerenes, where the theoretical considerations mostly are based on semiempirical (fullerene–fullerene) interaction potentials.

Clusters of fullerenes for experiments are typically produced in cluster aggregation sources, where fullerene powders may be heated to temperatures in the range of approximately 500°C–600°C where the resulting fullerene vapor effuses into an outer chamber containing a carrier gas. It is an advantage if the atomic weight of the carrier gas is small as it then will cool the hot fullerene molecules more efficiently through collisions. Therefore, He is frequently used for this purpose. The mixture of fullerenes and carrier gas may then be guided through a cooled nozzle (liquid nitrogen or cryogenic cooling), and clusters of fullerenes may be formed in the adiabatic expansion after the nozzle. It is important to note that the fullerenes that form dimers and larger clusters must have rather low internal excitation energies due to the weak intermolecular forces as they would otherwise break apart immediately. Further, only (neutral) clusters of fullerenes of the van der Waals type are formed under such conditions since the collision velocities are far too low to overcome the reverse activation barrier for covalent bonding between the fullerene cages (this barrier is about 1.6 eV high [Porezag et al. 1995]). Another important aspect here is that rather broad distributions of cluster sizes are obtained from such aggregation sources and that it therefore is impossible to define the targets in similar precise ways as in the cases of, e.g., studies using fullerene monomer targets. These cluster distributions may be shifted toward higher masses by increasing the oven temperature since a higher fullerene vapor pressure favors growth of cluster sizes by aggregations of smaller clusters of fullerenes, whereas the growth process is dominated by aggregation of fullerene monomers to smaller clusters at lower temperatures and pressures. (The key parameter here is the volume density of fullerene monomers: the higher the density the more frequent are fullerene–fullerene, fullerene–cluster, and cluster–cluster collisions and the more rapidly larger aggregates of fullerenes may form.) In order to analyze the cluster mass distribution by means of mass spectrometric methods, the clusters need to be negatively or positively charged. In the majority of the experimental studies carried out this far, positively charged clusters have been produced in collisions with different projectiles such as, e.g., electrons, photons, and highly charged ions. In the pioneering experiment performed by Martin et al. (1997), a broad distribution of cluster masses was observed but with higher intensities for certain mass values corresponding to so-called magic numbers of fullerene molecules (Figure 27.29). These results (Martin et al. 1997) showed that the clusters of

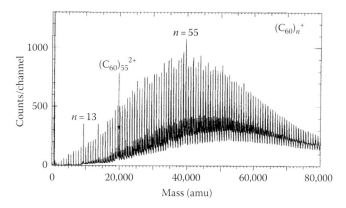

FIGURE 27.29 The mass distribution for laser ionized van der Waals clusters of fullerenes. Note the increased intensities for clusters containing 13 and 55 fullerenes. (Reprinted from Martin, T.P. et al., *Phys. Rev. Lett.*, 70, 3079, 1993. With permission.)

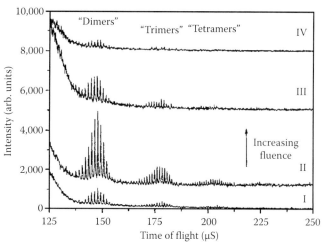

FIGURE 27.30 Fragment ion distributions from evaporation of caged fused dimers, trimers, and tetramers of fullerenes (cf. text). (Reprinted from Hedén, M. et al., *Phys. Rev. A*, 71, 055201, 2005. With permission.)

fullerenes had some preference to form in icosahedral geometrical structures. This was manifested by the high peak intensity for, e.g., clusters containing 13 constituents corresponding to a perfect icosahedron. Later, Hansen et al. (1996, 1997) showed that similar magic numbers appear also for mixed clusters (C_{60} and C_{70}), suggesting that the cluster packing mechanism is rather insensitive to the exact nature of the building blocks. However, in later experiments, Branz et al. (2002) showed that the magic numbers depend on the cluster temperature and that the observations of different magic numbers, thus, may be used as indicators for different packing schemes. In the same study, Branz et al. measured the binding energies of neutral and singly charged C_{60} dimers to be 0.275 ± 0.08 eV and 0.372 ± 0.08 eV, respectively, thus confirming that only weakly bound clusters are formed in conventional cluster aggregation sources (the aggregation source type described above was used by Branz et al. (2002)). In spite of this, fullerene cage fusion has been observed, but then in a different experiment and as a consequence of excitations by intense femtosecond laser light (Hedén et al. 2005). In that experiment (Hedén et al. 2005), mass peaks corresponding to, e.g., C_{120-2m}, C_{180-2m}, and C_{240-2m} were observed, and these distributions were attributed to evaporative cooling (emission of m C_2 units) of cage-fused dimers, trimers, and tetramers of C_{60}, respectively (Figure 27.30). Yet other situations in which cage fusion has been observed involve, e.g., the experiment by Campbell et al. in which fullerenes where collided in close to head-on collisions and in which fusion was clearly observed—often together with the cooling of the hot intermediate collision complex through emissions of smaller carbon molecules (Rohmund et al. 1996).

As briefly mentioned in the preceding section, fullerite is an insulating material but becomes (super) conducting when doped with alkali metal atoms (positioned between the fullerene cages), which donate electrons to the fullerenes. This relates to the intriguing issues of charge communication and the effect of charging the individual molecules in clusters of fullerenes. This problem has been studied experimentally by means of collisions between highly charged (atomic) ions and dimers, and clusters

of fullerenes (Manil et al. 2003, Zettergren et al. 2007c,d). It was then established that these clusters and dimers become electrically conducting as soon as one single electron is removed; this finding was also readily explained as the over-the-barrier condition for electron transfer from a neutral fullerene to a singly charged one is fulfilled already at the typical fullerene–fullerene distance (Manil et al. 2003, Zettergren et al. 2007c,d) in the (van der Waals) dimers, the clusters, and in fullerite. In addition, the same studies revealed other fundamental cluster properties, such as, e.g., appearance sizes (the smallest cluster size that may form for a given total cluster charge state). Here, clusters of fullerenes were again produced in an aggregation source and the corresponding jet-formed target was crossed by a pulsed beam of slow highly charged ions. In that experiment, intact fullerene clusters, dimers, monomers, and their respective fragments were analyzed by means of a linear time-of-flight spectrometer with a multi-coincidence detector with close to 100% detection efficiency. This allows for detailed information on the relative intensities of intact clusters in different charge states and the competition between the various fragmentation pathways. In this context, it is important to note that the ion beam intensity and the target density have to be sufficiently low such that the probability of producing two charged intact clusters from a single ion bunch is negligible, as the interpretation of multi-stop events otherwise would become an impossible task. The conditions in the experiment were therefore set such that most of the ion bunches produced zero target ions. A strong propensity for the formation of several intact singly charged monomers (C_{60}^+) from a single mother van der Waals cluster of fullerenes was reported, whereas doubly and multiply charged intact monomers were almost exclusively detected in two-stop events. This shows that the multiply charged clusters of fullerenes, $[C_{60}]_n^{q+}$ with $n \geq q$, preferentially decay into q singly charged monomers, i.e., the charge is distributed over the monomer constituents before the decay. Such a high charge communication is in clear contrast to results from similar experiments on weakly bound $[Ar]_n^{q+}$ clusters (Tappe et al. 2002), where the charge remains localized to a few constituents. Thus, the ionized cluster of fullerenes may, to some extent, be viewed as pieces of conducting nanomaterial. Another interesting aspect regarding the multiply charged species was reported from the same study and concerned the mass distributions of the intact clusters of fullerenes (Figure 27.31). The smallest fullerene cluster size for the singly charged (1+) monomer is the dimer, in accordance with the positive BE of 0.372 ± 0.08 eV (Branz et al. 2002). For the doubly, triply, and quadruply charged species the corresponding appearance sizes were reported to be 5, 10, and 21, respectively. These numbers are significantly lower than the typical ones for atomic clusters, which may be attributed to the larger sizes and larger polarizabilities of the clusters of fullerenes in comparisons with clusters formed from atomic building blocks. The experimental results discussed here are to large extents reproduced by two different cluster ground state models (Nakamura and Hervieux 2006, Zettergren et al. 2007c) (zero cluster temperature is assumed in both models) using semiempirical monomer–monomer interaction potentials.

Clearly, it would be easier to draw firm conclusions with a well-defined target cluster size (instead of a broad distribution) in collision experiments, and work is currently in progress in order to realize such experimental conditions (using mass selected singly charged clusters). Nevertheless, studies of multiple ionization and fragmentation of the smallest clusters (i.e., the dimers) by means of highly charged ions as collision agents have been reported (Zettergren et al. 2007c,d). These studies rely on the observation that the appearance size, i.e., the smallest possible stable cluster size, for doubly charged fullerene clusters is five. Thus, all multiply charged dimers promptly dissociate

into two charged monomers, and may be registered as two-stop events by means of multi-coincidence mass spectrometry techniques (as in the case of the larger clusters of fullerenes described above). In the analysis of these two-stop events, it was found that there was a strong preference for pairs of monomers with the same or nearby charges, in agreement with the experimental finding that the charge will be shared over many constituents in the larger cluster of fullerenes (i.e., no charge localization effects observed as in the case of, e.g., Ar clusters). In the experiment, the time-of-flight peaks due to intact monomers coming from dimer breakups were found to be significantly broader than the corresponding one-stop peak stemming from ionization of the fullerene monomers in the target (jet). This broadening reflects the kinetic energy release in the fragmentation process (cf. the monomer fission discussion in Section 27.2) and may be extracted from a calibration to fragment ion trajectory simulations using the SIMION package (Dahl 2000), where the actual mass spectrometer parameters and settings were taken into account as described in Ref. (Zettergren et al. 2007c). The measured kinetic energy releases may then be related to the potential energy curves, which are repulsive for multiply charged dimers, and which may be calculated through, e.g., high-level DFT. Such comparisons revealed that the measured kinetic energy releases were significantly lower (by as much as about a factor of 2) than what would be expected for a vertical ionization process (Figure 27.32). The equilibrium distance for the neutral system is about 10 Å and the expected kinetic energy release is simply obtained as the potential energy of the ionized system at this distance. Here, the collision timescale is on the sub-femtosecond level, which is shorter than a typical vibrational period, and during this short time window the two fullerene cages do not have time to separate significantly before the final ionization stage is reached. The discrepancy between experiment and theory is surprising considering that the neutral dimer system before ionization is cold (as otherwise it would never be formed) and that highly charged ions in principle should be excellent tools to remove electrons with low energy transfers to the target—although we have seen in the discussions of multiple ionization of fullerenes by slow highly charged ions that this assumption may not be so well fulfilled. Still, it seems reasonable to believe that the multiply charged systems formed under the experimental conditions should be well reproduced by the theoretical calculations dealing with the dimer system in its ground state as possible internal excitations due to the collision process most likely should not be dissipated in the form of additional KE of the separating intact fullerenes. However, the calculations do not take into account the dynamics of the separation process, and the measurement actually suggests the opposite namely that a large part, about 50%, of the potential energy available right after ionization and before the fullerenes start to separate is converted to internal fullerene energies. A similar effect has been observed in grazing collisions between fullerene cations and metal surfaces, indicating that it is a general feature of a charge-separating processes involving one or several fullerenes (Wethekam and Winter 2007, Wethekam et al. 2007).

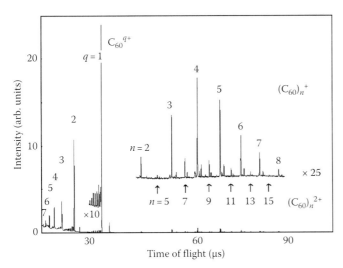

FIGURE 27.31 Time-of-flight mass spectrum for ionization of a wide distribution of clusters of fullerenes (including also monomers) by means of slow Xe^{20+} ions. Note the scaling factor (25) for the larger masses. In the latter distribution clear signatures of singly charged dimers, trimers, etc., are seen and there are also multiply charged clusters of fullerenes. (Reprinted from Manil, B. et al., *Phys. Rev. Lett.*, 91, 215504, 2003. With permission.)

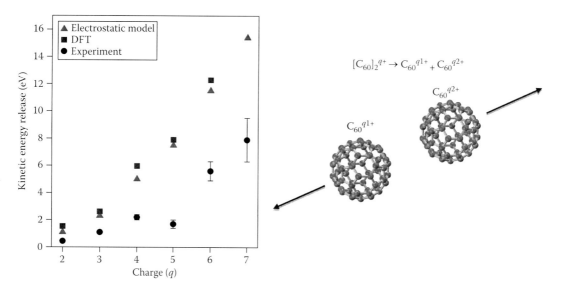

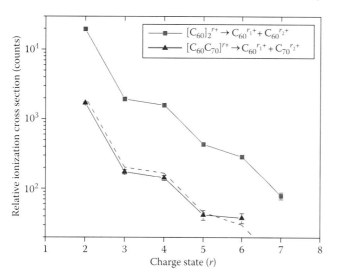

FIGURE 27.32 DFT and model results for the kinetic energy releases for fragmenting fullerene dimers as functions of their mother charge states in the range from $q = 2$ to $q = 7$. The model value is simply taken to be the interaction energy between two charged metal spheres at the equilibrium distance of the neutral dimer. Experimental results are a factor of two, typically, below both predictions (cf. text).

The number of dimers produced in a given charge state, and thus the related relative ionization cross sections, may be extracted from the same multi-coincidence experiment. There, two C_{60}^+ ions in a two-stop event (the multi-hit setup with 100% detection efficiency registered two and only two pulses, and in this case with time of flights corresponding to two C_{60}^+ ions) stem from the fragmentation of a doubly charged fullerene dimer. One C_{60}^+ and one C_{60}^{2+} in a two-stop event must thus come from a triply charged C_{60} dimer. The relative dimer ionization cross sections obtained in this way have a decreasing trend with pronounced even–odd effects as a function of the charge

FIGURE 27.33 Relative ionization cross sections for the C_{60} dimer and the mixed C_{60}–C_{70} dimer as functions of the charge state. The dashed lines indicate the C_{60} dimer result scaled down (by the statistical factor given by the C_{70} content in the powder used in the experiment). (Reprinted from Zettergren, H. et al., *J. Chem. Phys.*, 126, 224303, 2007c. With permission.)

state (Figure 27.33), in strong contrast to the smooth decreasing behavior observed for the ionization of monomers. At a first glance, one would perhaps expect a similar trend in both cases since the monomers to a large extent preserve their individuality in the weakly bound dimer system, and the collision scenario is most often such that the ionizing projectile is closer to one of the monomer constituents throughout the collision. Evidently, this is not the case, suggesting that these experiments elucidate intrinsic properties of the monomer–monomer interactions in the dimer system.

In the case of monomer ionization, the smoothly decreasing ionization cross section as a function of the final monomer charge state was attributed to the linear trend of monomer ionization energies (see Figure 27.34), which according to the simple over-the-barrier model (Bárány et al. 1985, Niehaus 1986, Thumm 1994, Bárány and Setterlind 1995, Thumm et al. 1997, Cederquist et al. 2000) gives geometrical cross sections corresponding to the areas between concentric rings (see Figure 27.16). Interestingly, the dimer ionization energies also follow a linear trend (but with a slope different from the one of the monomer ionization energy sequence) according to high-level DFT calculations (Zettergren et al. 2009) (see Figure 27.34). Thus, the observed even–odd effects in the dimer ionization cross sections do not reflect the dimer ionization energies, i.e., the collision scenario have to be taken into consideration when discussing the experimental observation. It is likely that the efficient (i.e., fast) charge communication, which was described above in the case of the larger cluster of fullerenes, may play an import role for the even–odd effects. The first step in such a collision scenario is an electron capture from the monomer closest to the projectile, which thus to a large extent mimics a collision with a monomer target. However, once the dimer system is charged an electron transfer within the dimer system and between the two fullerenes is possible, and thus the projectile once again "sees" a neutral

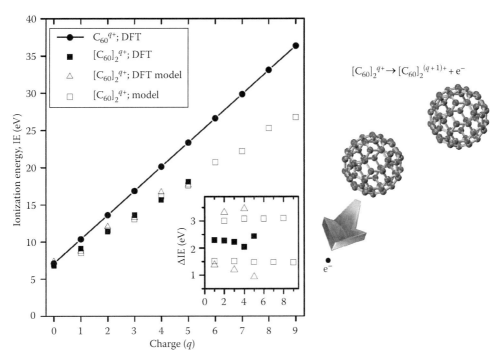

FIGURE 27.34 Comparison of ionization energies for fullerene monomers (open triangles) and fullerene dimers using DFT calculations (solid circles), an electrostatic model (open squares), and a model based on DFT calculations (open circles). The difference in ionization energies, $\Delta IE(q) = IE(q) - IE(q-1)$, is shown in the upper panel.

monomer during the collision. The second electron may then be captured by the projectile at nearly the same distance as the first electron (it has to be a little closer due to the interaction with the positively charged monomer furthest away from the projectile). The dimer system is now doubly charged with the charge evenly distributed on the two monomer constituents while the projectile is still close, and thus capturing the third electron is essentially the same as ionizing a singly charged monomer. Following the same argument as for the singly charged system, a fast electron transfer allows the projectile to once again be closest to a singly charged monomer in the collision, and thus the electron may capture the fourth electron at a distance only slightly closer than the third one. As a consequence, the production (i.e., ionization cross sections) of odd charged dimers is low in comparison to even charged ones, in accordance with the even–odd effects that have been observed experimentally (Zettergren et al. 2007d). It is important to note that this scenario is based on charge communication timescales shorter than the time between two sequential electron transfers from the dimer to the projectile, which indeed has been verified by calculations (Zettergren et al. submitted) of the tunneling time through the thin barrier in the singly charged dimer system. Thus, these latter calculations provide an explanation for the observed even–odd effects in the dimer ionization cross sections. At the same time they verify and reinforce the picture of very efficient charge communication within fullerene van der Waals systems from which a single electron has been removed as the crucial aspect here is how easily an electron may transfer from a neutral fullerene to a singly charged nearest-neighbor fullerene.

27.5 Summary and Concluding Remarks

In this chapter, we have discussed the fragmentation of neutral and charged fullerene monomers. Fullerene monomers are fullerenes in vacuum and we have discussed the inherent fragmentation and decay properties of such systems. Fullerene monomers and their fragmentation behaviors have been studied very extensively, and it is of course impossible to truly represent all aspects in a short presentation such as the one we have given here. Much less research has been done on van der Waals clusters of fullerenes, and in this chapter we have described the very different and very interesting fragmentation behaviors of fullerene van der Waals dimers and clusters.

For the monomers, we have discussed the main fragmentation processes of neutral and positively charged fullerenes, i.e., the evaporative (neutral C_2 emission) and the asymmetric fission (C_2^+ emission) processes, and how to balance between these changes as we move from neutral to multiply charged systems. We have argued that C_{60}^{r+} systems are fully stable (thermodynamically stable) for charge states up to $r = 5$, and thus that such systems (provided that their internal excitation energies are sufficiently low) would stay intact forever in vacuum. The more highly charged positive ions of the archetype fullerene, C_{60}, are fundamentally unstable but may still have very long lifetimes due to high energy barriers against asymmetric fission for charge states up to about 8. For still higher charge states, other decay channels, like emission of two singly charged carbon ions, become more favorable, and on the side of such processes there are also other decay channels in which larger, charged, carbon molecules (one or several) are emitted. For higher charge states and large internal excitation energies the multifragmentation process becomes very competitive. Thereby the fullerene breaks up in several smaller fragments, even and odd masses, with a mass distribution of these smaller masses that is rather insensitive to the exact nature of the excitation process. Otherwise, in case of more moderate positive charge states and internal energies we have stressed that the excitation method indeed is important for how the fullerene will fragment (which normally takes place long after the actual ionization and excitation process) as the interaction often determines these parameters. Relating to this, there is an important remaining problem to explain the mechanism for how measured and deduced very large excitation energies may result from what appears to be very distant interactions between fullerene monomers and slow highly charged ions. The interrelation between charge state (ionization) and internal excitation energies is clearly exposed for the fullerene anions for which the outermost electron is much more loosely bound than for neutrals and positive ions, and where the decay is strongly dominated by thermal ionization. That is, the most loosely bound electron is most often emitted before fragmentation occurs. A very interesting aspect here is the stability of the dianions (doubly negatively charged fullerenes) where it appears that C_{70}^{2-} is thermodynamically stable while C_{60}^{2-} based on C_{60} in its lowest energy isomer is not. However, it also appears as if C_{60}^{2-} built on higher energy isomers of C_{60}, surprisingly, actually may be thermodynamically stable. This illustrates the complexity of the fullerene monomers and clearly there are many aspects of their decay, including their fragmentation, which is not yet understood.

The fullerene van der Waals dimers and clusters exhibit many interesting phenomena. We have discussed how a multiply charged fullerene cluster in the charge state q always fragments through emission of q singly charged C_{60}^{+} molecules and not by break up of the individual fullerene cages. This illustrates the weak bonding between the cluster constituents and at the same time the high charge mobility within such a cluster. If several electrons are removed from such a cluster, it is for geometrical reasons very likely that all electrons are taken from the same fullerene and, as this one is not highly charged after fragmentation, the positive charges must have been able to spread to the whole cluster. Actually, it appears that if just a single charge is removed this charge is free to move around in the cluster. Multiple ionization of fullerene van der Waals dimers further demonstrate this effect and it even provides information on how fast the charge hopping between the two fullerenes in the dimer is: it is faster than the time between individual electron transfers from the fullerenes to the projectile, which in the experiments occur on the sub-femtosecond timescale. In turn this means that nanometer-sized objects may be placed at a distance of several Ångströms from each other and still provide an electrical contact essentially without resistance.

Future research on fullerene fragmentation is likely to focus on the remaining problem of understanding and explaining the very large energy transfers that appear to be strongly linked to all types of fullerene ionization and interaction methods. These studies would benefit to some extent by having the possibility to control the internal fullerene temperatures (internal excitations) before the interactions and in the case of studies of the intriguing stability properties of fullerenes, including even different isomers of the neutral parent systems, the advantage of temperature control is obvious. Low internal temperatures of fullerenes will also be necessary for certain studies of loosely bound dimers and clusters of fullerenes, and here studies of appearance sizes of negatively charged clusters of fullerenes, and the fragmentation of negatively charged clusters of fullerenes seem to be particularly attractive. Other interesting fullerene aspects are also related to the formation of larger cage-fused fullerenes from smaller ones and from graphite material, as well as comparisons between fullerene fragmentation and fragmentation of structurally related graphene and polyaromatic hydrocarbons.

Acknowledgment

We would like to acknowledge the EU FP6 network ITS LEIF (Contract No. RII3/026015), which provides an excellent platform for discussing, e.g., fullerene fragmentation issues.

References

Antoine, R., Ph. Dugourd, D. Rayane et al. 1999. Direct measurement of the electric polarizability of isolated C_{60} molecules. *J. Chem. Phys.* 110: 9771–9772.

Baba, M. S., T. S. L. Narasimhan, R. Balasubramanian, and C. K. Mathews. 1993. Electron impact ionisation of fullereness: Appearance energies of C_{60}^{+}, C_{60}^{2+} and C_{60}^{3+}. *Int. J. Mass Spectrom. Ion Process* 125: R1–R5.

Baggot, J. 1995. *Perfect Symmetry: The Accidental Discovery of Buckminsterfullerene*. Oxford University Press, New York.

Bárány, A. and C. J. Setterlind. 1995. Interaction of slow highly charged ions with atoms, clusters and solids: A unified classical barrier approach. *Nucl. Instrum. Method Phys. Res. B* 98: 184–186.

Bárány, A, G. Astner, H. Cederquist et al. 1985. Absolute cross sections for multi-electron processes in low energy Ar^{q+}–Ar collisions: Comparison with theory. *Nucl. Instrum. Method Phys. Res. B* 9: 397–399.

Bhardwaj, V. R., P. B. Corkum, and D.M. Rayner. 2003. Internal laser-induced dipole force at work in C_{60} molecule. *Phys. Rev. Lett.* 91: 203004.

Branz, W., N. Malinowski, A. Enders, and T. P. Martin. 2002. Structural transition in $(C_{60})_n$ clusters. *Phys. Rev. B* 66: 094107.

Bréchignac, C., Ph. Cahuzac, B. Concina et al. 2002. Charge transfer and dissociation in collisions of metal clusters with atoms. *Phys. Rev. Lett.* 89: 183402.

Brink, C., L. H. Andersen, P. Hvelplund, D. Mathur, and J. D. Voldstad. 1995. Laser photodetachment of C_{60}^{-} and C_{70}^{-} ions cooled in a storage ring. *Chem. Phys. Lett.* 233: 52–56.

Campbell, E. E. B. 2003. *Fullerene Collision Reactions*, 1st edn. Kluwer Academic, Dordrecht, the Netherlands.

Campbell, E. E. B. and F. Rohmund. 2000. Fullerene reactions. *Rep. Prog. Phys.* 63: 1061.

Campbell, E. E. B., K. Hansen, K. Hoffmann et al. 2000. From above threshold ionization to statistical electron emission: The laser pulse-duration dependence of C_{60} photoelectron spectra. *Phys. Rev. Lett.* 84: 2128–2131.

Campbell, E. E. B., T. Raz, and R. D. Levine. 1996. Internal energy dependence of the fragmentation patterns of C_{60} and C_{60}^+. *Chem. Phys. Lett.* 253: 261.

Cederquist, H., A. Fardi, K. Haghighat et al. 2000. Electronic response of C_{60} in slow collisions with highly charged ions. *Phys. Rev. A* 61: 022712.

Cederquist, H., P. Hvelplund, H. Lebius, H. T. Schmidt, S. Tomita, and B. A. Huber. 2001. Nonfragmenting charge transfer in slow peripheral C_{60}^{q+}-C_{60} collisions. *Phys. Rev. A* 63: 26201.

Cederquist, H., J. Jensen, H. T. Schmidt et al. 2003. Barriers for asymmetric fission of multiply charged C_{60} fullerenes. *Phys. Rev. A* 67: 062719.

Chancey, C. C. and M. C. M. O'Brien. 1997. *The Jahn-Teller Effect in C_{60} and Other Icosahedral Complexes*. Princeton University Press, Princeton, NJ.

Chen, L., J. Bernard, A. Denis, S. Martin, and J. Désesquelles. 1999. Fragmentation scheme of C_{60}^{4+} ions produced in low-energy collisions of Ar^{8+} and C_{60}. *Phys. Rev. A* 59: 2827–2835.

Chen, L., S. Martin, R. Brédy, J. Bernard, and J. Désesquelles. 2001. Dynamical fragmentation processes of C_{60}^{5+} ions in Ar^{8+}-C_{60} collisions. *Phys. Rev. A* 64: 031201.

Chen, L., S. Martin, J. Bernard, and R. Brédy. 2007. Direct measurement of internal energy of fragmented C_{60}. *Phys. Rev. Lett.* 98: 193401.

Climen, B., B. Concina, M. A. Lebeault, F. Lépine, B. Baguenard, and C. Bordas. 2007. Ion-imaging study of C_{60} fragmentation. *Chem. Phys. Lett.* 437: 17.

Dahl, D. A. 2000. Simion in the personal computer in reflection. *Int. J. Mass Spectrom.* 200: 3–25.

David, W. I. F., R. M. Ibberson, J. C. Matthewman et al. 1991. Crystal structure and bonding of ordered C_{60}. *Nature* 353: 147–149.

Díaz-Tendero, S. 2005. Structure and fragmentation of neutral and positively charged carbon clusters and fullerenes. PhD thesis. Universidad Autónoma de Madrid, Madrid, Spain.

Díaz-Tendero, S., M. Alcamí, and F. Martín. 2005a. Coulomb stability limit of highly charged C_{60}^{q+} fullerenes. *Phys. Rev. Lett.* 95: 013401.

Díaz-Tendero, S., M. Alcamí, and F. Martín. 2005b. Structure and electronic properties of highly charged C_{60} and C_{58} fullerenes. *J. Chem. Phys.* 123: 184306.

Dresselhaus, M. S., G. Dresselhaus, and P. C. Eklund. 1995. *Science of Fullerenes and Carbon Nanotubes*. Academic Press, New York.

Głuch, K., S. Matt-Leubner, O. Echt et al. 2004. On the kinetic energy release distribution for C_2 evaporation from fullerene ions. *Chem. Phys. Lett.* 385: 449.

Gunnarson, O. 1997. Superconductivity in fullerides. *Rev. Mod. Phys.* 69: 575–606.

Haag, N., Z. Berényi, P. Reinhed et al. 2008. Kinetic-energy-release distributions and barrier heights for C_2^+ emission from multiply charged C_{60} and C_{70} fullerenes. *Phys. Rev. A* 78: 043201.

Hansen, K. and O. Echt. 1997. Thermionic emission and fragmentation of C_{60}. *Phys. Rev. Lett.* 78: 2337–2340.

Hansen, K., K. Hoffmann, and E. E. B. Campbell. 2003. Thermal electron emission from the hot electronic subsystem of vibrationally cold C_{60}. *J. Chem. Phys.* 119: 2513.

Hansen, K., R. Müller, H. Hohmann, and E. E. B. Campbell. 1996. Icosahedra of icosahedra: The stability of $(C_{60})_{13}$. *J. Chem. Phys.* 105: 6088.

Hansen, K., R. Müller, H. Hohmann, and E. E. B. Campbell. 1997. Stability of clusters of fullerenes. *Z. Phys. D: At. Mol. Clusters* 40: 361.

Heath, J. R., S. C. O'Brien, O. Zhang et al. 1985. Lanthanum complexes of spheroidal carbon shells. *J. Am. Chem. Soc.* 107: 7779–7780.

Hedén, M., K. Hansen, and E. E. B. Campbell. 2005. Molecular fusion of $(C_{60})_N$ clusters in the gas phase after femtosecond laser irradiation. *Phys. Rev. A* 71: 055201.

Herlert, A. and L. Schweikhard. 2006. First observation of delayed electron emission from dianionic metal clusters. *Int. J. Mass Spectrom.* 252: 151–156.

Hertel, I. V., H. Steger, J. Devries et al. 1992. Giant plasmon excitation in free C_{60} and C_{70} molecules studied by photoionization. *Phys. Rev. Lett.* 68: 784–787.

Hervieux, P. A., B. Zarour, J. Hanssen, M. F. Politis, and F. Martín. 2001. Fragmentation in collisions of Na_9^+ clusters with Cs atoms. *J. Phys. B: At. Mol. Opt. Phys.* 34: 3331.

Hettich, R. L., R. N. Compton, and R. H. Ritchie. 1991. Doubly charged negative ions of carbon-60. *Phys. Rev. Lett.* 67: 1242–1245.

Hvelplund, P., L. H. Andersen, H. K. Haugen et al. 1992. Dynamical fragmentation of C_{60} ions. *Phys. Rev. Lett.* 69: 1915–1918.

Jensen, J., H. Zettergren, H. T. Schmidt et al. 2004. Ionization of C_{70} and C_{60} molecules by slow highly charged ions: A comparison. *Phys. Rev. A* 69: 053203.

Kassel, L. S. 1928. Studies in homogeneous gas reactions. I. *J. Phys. Chem.* 32: 225.

Kroto, H. W., J. R. Heath, S. C O'Brien, R. F. Curl, and R. E. Smalley. 1985. C_{60}: Buckminsterfullerene. *Nature* 318: 162–163.

Larsen, M.C., Hvelplund, P., Larsson, M.O., and Shen, H., 1999. Fragmentation of fast positive and negative C60 ions in collisions with rare gas atoms, *Eur. Phys. J. D* 5: 283–289.

Larsson, M. O., P. Hvelplund, M. C. Larsen, H. Shen, H. Cederquist, and H. T. Schmidt. 1998. Electron capture and energy loss in similar to 100 keV collisions of atomic and molecular ions on C_{60}. *Int. J. Mass Spectrom.* 177: 51–62.

Laskin, J., B. Hadas, T. D. Märk, and C. Lifshitz. 1998. New evidence in favor of a high (10 eV) C_2 binding energy in C_{60}. *Int. J. Mass Spectrom.* 177: L9–L13.

LeBrun, T., H. G. Berry, S. Cheng et al. 1994. Ionization and multifragmentation of C_{60} by high-energy, highly charged Xe ions. *Phys. Rev. Lett.* 72: 3965–3968.

Lifshitz, C., M. Iraqui, T. Peres, and J. Fischer. 1991. Charge stripping C_{60}^+. *Rapid Commun. Mass Spectrom.* 5: 238.

Limbach, P. A., L. Schweikhard, K. A. Cowen, M. T. McDermott, A. G. Marshall, and J. V. Cole. 1991. Observation of the doubly charged, gas-phase fullerene anions C_{60}^{2-} and C_{70}^{2-}. *J. Am. Chem. Soc.* 113: 6795.

Liu, B., P. Hvelplund, S. Brøndsted Nielsen, and S. Tomita. 2004. Formation of C_{60}^{2-} dianions in collisions between C_{60}^{-} and Na atoms. *Phys. Rev. Lett.* 92: 168301.

Manil, B., L. Maunoury, B. A. Huber et al. 2003. Highly charged clusters of fullerenes: Charge mobility and appearance sizes. *Phys. Rev. Lett.* 91: 215504.

Martin, S., L. Chen, R. Brédy et al. 2002. Emission of singly and doubly charged light fragments from C_{60}^{r+} ($r = 4$–9) in Xe^{25+}-C_{60} collisions. *Phys. Rev. A* 66: 063201.

Martin, S., L. Chen, A. Denis, R. Brédy, J. Bernard, and J. Désesquelles. 2000. Excitation and fragmentation of C_{60}^{r+} ($r = 3$–9) in Xe^{30+}-C_{60} collisions. *Phys. Rev. A* 62: 022707.

Martin, S., L. Chen, A. Denis, and J. Désesquelles. 1999. Very high multiplicity distributions of electrons emitted in multicapture collision between Xe^{25+} and C_{60}. *Phys. Rev. A* 59: R1734–R1737.

Martin, S., L. Chen, A. Salmoun, B. Li, J. Bernard, and R. Brédy. 2008. Fragmentation patterns of multicharged C_{60}^{r+} ($r = 3$–5) studied with well-controlled internal excitation energy. *Phys. Rev. A* 77: 043201.

Martin, T. P., U. Näher, H. Schaber, and U. Zimmermann. 1993. Clusters of fullerene molecules. *Phys. Rev. Lett.* 70: 3079.

Matt, S., O. Echt, T. Rauth et al. 1997. Electron impact ionization and dissociation of neutral and charged fullerenes. *Z. Phys. D* 40: 389–394.

Matt, S., O. Echt, M. Sonderegger et al. 1999. Kinetic energy release distributions and evaporation energies for metastable fullerene ions. *Chem. Phys. Lett.* 303: 379.

Nakamura, M. and P.-A. Hervieux. 2006. Stability and fragmentation of multiply charged clusters of fullerenes. *Chem. Phys. Lett.* 428: 138.

Niehaus, A. 1986. A classical model for multiple-electron capture in slow collisions of highly charged ions with atoms. *J. Phys. B: At. Mol. Opt. Phys.* 19: 2925.

Opitz, J., H. Lebius, S. Tomita et al. 2000. Electronic excitation in H^+-C_{60} collisions: Evaporation and ionization. *Phys. Rev. A* 62: 022705.

Porezag, D., M. R. Pederson, T. Frauenheim, and T. Köhler. 1995. Structure, stability, and vibrational properties of polymerized C_{60}. *Phys. Rev. B* 52: 14963.

Reinköster, A., U. Werner, N. M. Kabachnik, and H. O. Lutz. 2001. Experimental and theoretical study of ionization and fragmentation of C_{60} by fast-proton impact. *Phys. Rev. A* 64: 023201.

Rentenier, A., A. Bordenave-Montesquieu, P. Moretto-Capelle, and D. Bordenave-Montesquieu. 2004. Asymmetric fission and evaporation of C_{60}^{r+} ($r = 2$–4) fullerene ions in ion-C_{60} collisions: I. Proton results. *J. Phys. B: At. Mol. Opt. Phys.* 37: 2429–2454.

Rentenier, A., L. F. Ruiz, S. Díaz-Tendero et al. 2008. Absolute charge transfer and fragmentation cross sections in He^{2+}-C_{60} collisions. *Phys. Rev. Lett.* 100: 183401.

Rice, O. K. and H. C. Ramsperger. 1927. Theories of unimolecular gas reactions at low pressures. *J. Am. Chem. Soc.* 49: 1617.

Rohmund, F., E. E. B. Campbell, O. Knospe, G. Seifert, and R. Schmidt. 1996. Collision energy dependence of molecular fusion and fragmentation in C_{60}^{+} + C_{60} collisions. *Phys. Rev. Lett.* 76: 3289–3292.

Salmoun, A., R. Brédy, J. Bernard, L. Chen, and S. Martin. 2008. Multi electron capture processes in slow collisions between X^{7+} (X = N, O and Ne) ions with C_{60}. *Eur. Phys. J. D* 47: 55–61.

Scheier, P. and T. D. Märk. 1994a. Observation of the septuply charged ion C_{60}^{7+} and its metastable decay into two charged fragments via superasymmetric fission. *Phys. Rev. Lett.* 73: 54–57.

Scheier, P. and T. D. Märk. 1994b. Electron impact induced production and isotope-resolved identification of C_{70}^{6+} and sextuply-charged fragment ions of C_{70} and C_{60}. *Int. J. Mass Spectrom. Ion Process Lett.* 133: L5–L9.

Scheier, P., B. Dünser, and T. D. Märk. 1995. Superasymmetric fission of multiply charged fullerene ions. *Phys. Rev. Lett.* 74: 3368–3371.

Scheier, P., B. Dünser, R. Wörgötter et al. 1996. Direct evidence for the sequential decay $C_{60}^{z+} \rightarrow C_{58}^{z+} \rightarrow C_{56}^{z+} \rightarrow \cdots$. *Phys. Rev. Lett.* 77: 2654–2657.

Schlathölter, T., O. Hadjar, R. Hoekstra, and R. Morgenstern. 1999a. Strong velocity effects in collisions of He^+ with fullerenes. *Phys. Rev. Lett.* 82: 73–76.

Schlathölter, T., O. Hadjar, J. Manske, R. Hoekstra, and R. Morgenstern. 1999b. Electronic versus vibrational excitation in He^{q+} collisions with fullerenes. *Int. J. Mass Spectrom.* 192: 245–257.

Steger, H., J. Holzapfel, A. Hielscher, W. Kamke, and I. V. Hertel. 1995. Single-photon ionization of higher fullerenes C_{76}, C_{78} and C_{84}. Determination of ionization potentials. *Chem. Phys. Lett.* 234: 455–459.

Tappe, W., R. Flesch, E. Rühl, R. Hoekstra, and T. Schlathölter. 2002. Charge localization in collision-induced multiple ionization of van der Waals clusters with highly charged ions. *Phys. Rev. Lett.* 88: 143401.

Taylor, R., J. P. Hare, A. K. Abdul-Sada, and H. W. Kroto. 1990. Isolation, separation and characterisation of the fullerenes C_{60} and C_{70}: The third form of carbon. *J. Chem. Soc., Chem. Commun.* 20: 1423.

Thumm, U. 1994. Charge exchange and electron emission in collisions of highly charged ions with C_{60}. *J. Phys. B: At. Mol. Opt. Phys.* 27: 3515–3532.

Thumm, U., A. Bárány, H. Cederquist, L. Hägg, and C. J. Setterlind. 1997. Energy gain in collisions of highly charged ions with C_{60}. *Phys. Rev. A* 56: 4799–4806.

Tomita, S., J. U. Andersen, E. Bonderup et al. 2005. Dynamic Jahn-Teller effects in isolated C_{60}^{-} studied by near-infrared spectroscopy in a storage ring. *Phys. Rev. Lett.* 94: 053002.

Tomita, S., J. U. Andersen, H. Cederquist et al. 2006. Lifetimes of C_{60}^{2-} and C_{70}^{2-} dianions in a storage ring. *J. Chem. Phys.* 124: 024310.

Tomita, S., J. U. Andersen, C. Gottrup, P. Hvelplund, and U. V. Pedersen. 2001. Dissociation energy for C_2 loss from fullerene cations in a storage ring. *Phys. Rev. Lett.* 87: 073401.

Tomita, S., P. Hvelplund, S. B. Nielsen, and T. Muramoto. 2002. C_{59}-ion formation in high-energy collisions between cold C_{60}^- and noble gases. *Phys. Rev. A* 65: 043201.

Walch, B., C. L. Cocke, R. Voelpel, and E. Salzborn. 1994. Electron capture from C_{60} by slow multiply charged ions. *Phys. Rev. Lett.* 72: 1439–1442.

Wang, X.-B., C. F. Ding, and L.-S. Wang. 1999. High resolution photoelectron spectroscopy of C_{60}^-. *J. Chem. Phys.* 110: 8217.

Wang, X.-B. and L.-S. Wang. 1999. Observation of negative electron-binding energy in a molecule. *Nature* 400: 245.

Wang, X.-B., H.-K. Woo, X. Huang, M. M. Kappes, and L.-S. Wang. 2006. Direct experimental probe of the on-site coulomb repulsion in the doubly charged fullerene anion C_{70}^{2-}. *Phys. Rev. Lett.* 96: 143002.

Wang, X.-B., H.-K. Woo, J. Yang, M. M. Kappes, and L.-S. Wang. 2007. Photoelectron spectroscopy of singly and doubly charged higher fullerenes at low temperatures: C_{76}^-, C_{78}^-, C_{84}^- and C_{76}^{2-}, C_{78}^{2-} C_{84}^{2-}. *J. Phys. Chem. C* 111: 17684.

Weisskopf, V. 1937. Statistics and nuclear reactions. *Phys. Rev.* 52: 295.

Wethekam, S. and H. Winter. 2007. Excitation of fullerene ions during grazing scattering from a metal surface. *Phys. Rev. A* 76: 032901.

Wethekam, S. and H. Winter. 2009. Private communication.

Wethekam, S., H. Winter, H. Cederquist, and H. Zettergren. 2007. Neutralization of charged fullerenes during grazing scattering from a metal surface. *Phys. Rev. Lett.* 99: 037601.

Zettergren, H. 2005. Electron transfer and fragmentation in fullerene collisions. PhD thesis. Stockholm University, Stockholm, Sweden.

Zettergren, H., M. Alcamí, and F. Martín. 2007b. First- and second-electron affinities of C_{60} and C_{70} isomers. *Phys. Rev. A* 76: 043205.

Zettergren, H., M. Alcamí, and F. Martín. 2008. Stable non-IPR C_{60} and C_{70} fullerenes containing a uniform distribution of pyrenes and adjacent pentagons. *ChemPhysChem* 9: 861.

Zettergren, H., J. Jensen, H. T. Schmidt, and H. Cederquist. 2004. Electrostatic model calculations of fission barriers for fullerene ions. *Eur. Phys. J. D* 29: 63–68.

Zettergren, H., G. Sánchez, S. Díaz-Tendero, M. Alcamí, and F. Martín. 2007a. Theoretical study of the stability of multiply charged C_{70} fullerenes. *J. Chem. Phys.* 127: 104308.

Zettergren, H., H. T. Schmidt, H. Cederquist et al. 2002. Static over-the-barrier model for electron transfer between metallic spherical objects. *Phys. Rev. A* 66: 032710.

Zettergren, H., H. T. Schmidt, P. Reinhed et al. 2007c. Stabilities of multiply charged dimers and clusters of fullerenes. *J. Chem. Phys.* 126: 224303.

Zettergren, H., H. T. Schmidt, P. Reinhed et al. 2007d. Even-odd effects in the ionization cross sections of $[C_{60}]_2$ and $[C_{60}C_{70}]$ dimers. *Phys. Rev. A* 75: 051201(R).

Zettergren, H., Y. Wang, A. Lamsabhi, M. Alcamí, and F. Martin. 2009. Density functional theory study of multiply ionized weakly bound fullerene dimers, *J. Chem. Phys.* 130: 224302.

Zimmerman, J. A., J. R. Eyler, S. B. H. Bach, and S. W. McElvany. 1991. "Magic number" carbon clusters: Ionization potentials and selective reactivity. *J. Chem. Phys.* 94: 3556.

IV

Structure and Properties of Carbon Fullerenes

Symmetry of Fulleroids

Stanislav Jendrol'
Pavol Jozef Šafárik University

František Kardoš
Pavol Jozef Šafárik University

One of the important features of the structure of molecules is the presence of symmetry elements. For example, symmetry arguments can tell us whether the molecule is chiral and so it could be optically active. The structure of the symmetry group of a molecule affects several spectroscopic aspects and vice versa. Thus, it is clearly important to know the possible symmetries of fullerenes and similar structures.

In this chapter, we study symmetry groups of fulleroids. In the most general manner, a *fulleroid* can be viewed as a cubic (i.e., 3-valent) convex polytope with all faces of size at least five. This definition is a natural generalization of the notion of fullerenes.

Fulleroids can also be represented by planar graphs. In Section 28.1, we recall the basic definitions and properties of convex polyhedra and planar graphs. In Section 28.2, we mention connections between symmetries of convex polyhedra and automorphisms of planar graphs, which allow us not to distinguish between a polyhedron with its symmetry group and a graph with its automorphism group.

In Section 28.3, all groups that can act as symmetry groups of convex polyhedra are listed and described. For each such group, its symmetry elements are listed in a table, and some more information is provided in the text. We also provide examples and constructions of polyhedra and/or their graphs with particular symmetry groups.

Section 28.4 gives basic relations between rotational symmetries and face sizes of fulleroids, which imply some easy to see necessary conditions on face sizes for the existence of fulleroids with particular symmetry groups.

Section 28.5 provides a brief overview of results concerning symmetries of fullerenes. One can find a list of all possible fullerene symmetry groups and counts of vertices of their smallest representatives.

In Sections 28.6 and 28.7, we study fulleroids with icosahedral symmetry and fulleroids with a subgroup of icosahedral symmetry. We provide constructions of fulleroids with an icosahedral symmetry group with pentagonal and n-gonal faces only for each $n \geq 7$. We also mention constructions, which assure that there are infinitely many such structures for each $n \geq 7$. It turns out that for subgroups of the icosahedral group $\mathcal{I}_h$, one can find examples of fulleroids with any face size.

In Section 28.8, we study the symmetry of fulleroids with multi-pentagonal faces. We use tools from both algebra and geometry to prove the nonexistence of fulleroids for the seven groups. Therefore, for the remaining seven groups without local constraints, there are values of integer n such that the corresponding fulleroids with pentagonal and n-gonal faces do not exist.

In Section 28.9, we complete the characterization of symmetry of fulleroids by listing the results concerning fulleroids with octahedral, prismatic, or pyramidal symmetry.

In Section 28.10, we focus on fulleroids with pentagonal and heptagonal faces and their symmetries. We list all possible symmetry groups and for each of them, we provide examples with a minimal number of vertices.

28.1 Convex Polyhedra and Planar Graphs

Fullerene-like molecules are often represented by convex polyhedra, where the atoms are placed in the vertices of a polyhedron, whereas the bonds among atoms are realized along the edges of the polyhedron. The combinatorial structure of convex polyhedra can be represented by planar graphs. A *graph* is a pair $G = (V, E)$ of finite sets such that the elements of E are 2-element subsets of V. The elements of V are the *vertices* of the graph G, the elements of E are its *edges*.

A graph G is *planar*, if it can be represented in the 2-dimensional plane $\mathbb{R}^2$ in such a way that vertices of G are distinct points in $\mathbb{R}^2$, edges of G are arcs between the vertices, such that different edges have different sets of endpoints and the interior of an edge contains no vertex and no point of any other edge. Such representation is called *drawing* or *planar embedding* of G. For every drawing of a graph G, the regions of $\mathbb{R}^2 \backslash G$ are the *faces* of G. One of the faces is always unbounded—the *outer* face; the other faces are its *inner* faces.

Given a convex polyhedron P, the vertices and the edges of P form a graph, called the *graph of P* and denoted by $G(P)$. One can find a planar drawing of $G(P)$ by projecting P into a face f of P. Such a drawing of the graph $G(P)$ of a convex polyhedron P is also called the *Schlegel diagram* of P. Any face of P can be chosen to be the outer face in the diagram.

A graph G is *polyhedral* if it is (isomorphic to) a graph of some convex polyhedron. For every convex polyhedron P, the graph $G(P)$ is planar and 3-connected. On the other hand, according to *Steinitz' theorem*, these conditions are also sufficient for the graph to be polyhedral:

Theorem 28.1: (Steinitz) (Cromwell 1997) *A graph G is polyhedral if and only if G is planar and 3-connected.*

Convex polyhedra P_1 and P_2 are *equivalent*, if the corresponding graphs $G(P_1)$ and $G(P_2)$ are isomorphic. In this case, we say P_1 and P_2 are *of the same type*.

An arbitrary planar graph can have several different drawings, with the faces of different size. Another important property of 3-connected planar graphs is given by *Whitney's theorem*.

Theorem 28.2: (Whitney) (Diestel 1997) *Any two drawings of a 3-connected graph are equivalent.*

It means that the faces (their boundaries and sizes) of a drawing of G are uniquely determined by the graph itself. Therefore, we can speak about the faces of a 3-connected planar graph without specifying its drawing. Whitney's theorem also says that each 3-connected planar graph corresponds to precisely one type of convex polyhedra. Hence, we can identify a convex polyhedron P with its graph $G(P)$; we represent the fulleroids by 3-connected planar graphs with all faces of size at least five.

To find more information about convex polyhedra, we refer the reader to the books (Cromwell 1997) and (Grünbaum 2003). The recommended introductory book for the graph theory is the book of Chartrand and Lesniak (1972).

28.2 Polyhedral Symmetries and Graph Automorphisms

An *automorphism* of a graph G is an isomorphism of G onto itself. The set of all automorphisms of a graph G together with the composition operation forms the *automorphism group* of G and is denoted by Aut(G). For every graph G the group Aut(G) is nonempty and finite.

Symmetry is a property of a convex polyhedron, which causes it to remain invariant under certain classes of transformations (such as rotation, reflection, inversion, or more abstract operations). Strictly speaking, a *symmetry* of a convex polyhedron P is an isometry of the 3-dimensional space $\mathbb{R}^3$, under which the polyhedron P remains invariant. The set of all symmetries of a convex polyhedron P together with composition operation forms the *symmetry group* of P and is denoted by $\Gamma(P)$. For every convex polyhedron P the group $\Gamma(P)$ is nonempty and finite.

If ϕ is a symmetry of a convex polyhedron P, then ϕ restricted to the vertices and edges of P is an automorphism of the graph $G(P)$. Moreover, the group $\Gamma(P)$ is (isomorphic to) a subgroup of the group Aut($G(P)$). A stronger connection between automorphisms of polyhedral graphs and symmetries of corresponding polyhedra is given by the theorem of Mani:

Theorem 28.3: (Mani) (Mani 1971) *Let G be 3-connected planar graph. Then there is a convex polyhedron P whose graph is isomorphic to G and $\Gamma(P)$ is isomorphic to Aut(G).*

This theorem allows us not to distinguish between a polyhedron P with the group $\Gamma(P)$, and its graph G with the group Aut(G). Therefore, to give examples of convex polyhedra with the symmetry group Γ it is sufficient to find polyhedral graphs such that Aut(G) $\cong \Gamma$.

28.3 Point Symmetry Groups

As mentioned above, the symmetry groups of convex polyhedra are nonempty and finite. But not all finite groups can be symmetry groups of convex polyhedra. The characterization of all the groups that act as a symmetry group of a convex polyhedron is known. The list of all such groups, their structure and properties, can be found in Cromwell (1997). Another source of general information on finite groups is the book of Coxeter and Moser (1972).

Every symmetry of a convex polyhedron P has at least one fixpoint (its center of gravity), and this point is common for all symmetries of P. The group $\Gamma(P)$ is thus called a *point group*.

According to the number of rotational symmetry axes and their relative position all point groups can be divided into icosahedral, octahedral, tetrahedral, dihedral, skewed, pyramidal, and others. The list of all groups that can act as symmetry groups of convex polyhedra is in Table 28.1.

TABLE 28.1 Point Symmetry Groups

Group	Order	Rotations	Reflections	Inversion
$\mathscr{I}_h$	120	$6C_5, 10C_3, 15C_2$	15	Yes
$\mathscr{I}$	60	$6C_5, 10C_3, 15C_2$	—	—
$\mathbb{O}_h$	48	$3C_4, 4C_3, 6C_2$	9	Yes
$\mathbb{O}$	24	$3C_4, 4C_3, 6C_2$	—	—
$\mathscr{T}_d$	24	$4C_3, 3C_2$	6	—
$\mathscr{T}_h$	24	$4C_3, 3C_2$	3	Yes
$\mathscr{T}$	12	$4C_3, 3C_2$	—	—
$\mathscr{D}_{nh}$	$4n$	C_n, nC_2	$n+1$	If n even
$\mathscr{D}_{nd}$	$4n$	C_n, nC_2	n	If n odd
$\mathscr{D}_n$	$2n$	C_n, nC_2	—	—
$\mathscr{S}_{2n}$	$2n$	C_n	—	If n odd
$\mathscr{C}_{nh}$	$2n$	C_n	1	If n even
$\mathscr{C}_{nv}$	$2n$	C_n	n	—
$\mathscr{C}_n$	n	C_n	—	—
$\mathscr{C}_s$	2	—	1	—
$\mathscr{C}_i$	2	—	—	Yes
$\mathscr{C}_1$	1	—	—	—

Note: Each row of the table lists one such group, its order, the number and sort of rotational symmetry axes, the number of mirror symmetry planes and presence of point inversion in the group.

First seven point groups ($\mathscr{I}_h$, $\mathscr{I}$, $\mathbb{O}_h$, $\mathbb{O}$, $\mathscr{T}_h$, $\mathscr{T}_d$, and $\mathscr{T}$) can be interpreted as the symmetry groups of platonic solids or their subgroups. They contain more than one axis of at least threefold rotation (Figures 28.1 and 28.2).

1. The group $\mathscr{I}_h$ is the full symmetry group of a regular icosahedron (and a regular dodecahedron). It is also the symmetry group of the most famous fullerene, the buckyball C_{60}. It is isomorphic to $\mathbb{A}_5 \times \mathbb{Z}_2$, where $\mathbb{A}_5$ is the group of even permutations of five elements (Figure 28.1).

$$\mathscr{I}_h \cong \left\langle x, y, z \,\middle|\, x^2 = y^2 = z^2 = (xy)^2 = (xz)^3 = (yz)^5 = 1 \right\rangle.$$

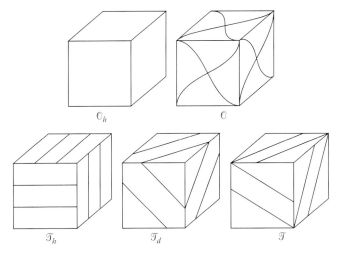

FIGURE 28.2 The groups $\mathbb{O}_h$, $\mathbb{O}$, $\mathscr{T}_h$, $\mathscr{T}_d$ and $\mathscr{T}$ can be represented as groups of symmetries of a cube respecting certain patterns drawn on it.

2. The group $\mathscr{I}$ is the group of rotations of a regular icosahedron. It is a subgroup of $\mathscr{I}_h$ of index 2 and it is isomorphic to $\mathbb{A}_5$ (Figure 28.1).

$$\mathscr{I} \cong \left\langle a, b \,\middle|\, a^2 = b^3 = (ab)^5 = 1 \right\rangle.$$

3. The group $\mathbb{O}_h$ is the full symmetry group of a regular octahedron (and of a cube). It is isomorphic to $S_4 \times \mathbb{Z}_2$, where S_4 is the group of all permutations of four elements (Figure 28.3).

$$\mathbb{O}_h \cong \left\langle x, y, z \,\middle|\, x^2 = y^2 = z^2 = (xy)^2 = (xz)^3 = (yz)^4 = 1 \right\rangle.$$

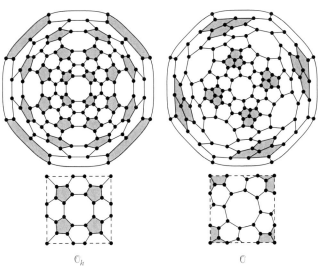

FIGURE 28.3 Examples of fulleroids with symmetry group $\mathbb{O}_h$ and $\mathbb{O}$, respectively, with the same number of vertices and the same number of faces of size 5, 6, and 8. Pentagonal faces are gray. To obtain the graph of the fulleroid, it suffices to draw the corresponding graph segment onto all faces of a cube.

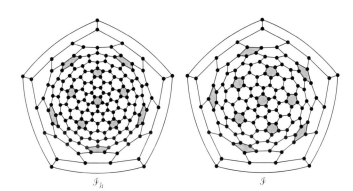

FIGURE 28.1 Examples of fullerenes with symmetry group $\mathscr{I}_h$ and $\mathscr{I}$, respectively. Pentagonal faces (except f or the outer face) are gray.

4. The group $\mathbb{O}$ is the group of rotations of a regular octahedron. It is a subgroup of $\mathbb{O}_h$ of index 2 and it is isomorphic to S_4 (Figure 28.3).

$$\mathbb{O} \cong \left\langle a,b \,\middle|\, a^2 = b^3 = (ab)^4 = 1 \right\rangle.$$

5. The group $\mathscr{T}_h$ is the full symmetry group of a volleyball, or of a pyritohedron. It is a subgroup of both $\mathbb{O}_h$ and $\mathscr{I}_h$ and is a supergroup of $\mathscr{T}$. It is isomorphic to $\mathbb{A}_4 \times \mathbb{Z}_2$, where $\mathbb{A}_4$ is the group of even permutations of four elements (Figure 28.4).

$$\mathscr{T}_h \cong \left\langle a,b \,\middle|\, a^2 = b^3 = (abab^{-1})^2 = 1 \right\rangle.$$

6. The group $\mathscr{T}_d$ is the full symmetry group of a regular tetrahedron. It is a subgroup of $\mathbb{O}_h$ and it is isomorphic to S_4 (and $\mathbb{O}$ indeed) (Figure 28.4).

$$\mathscr{T}_d \cong \left\langle x,y,z \,\middle|\, x^2 = y^2 = z^2 = (xy)^2 = (xz)^3 = (yz)^3 = 1 \right\rangle.$$

7. The group $\mathscr{T}$ is the group of rotations of a regular tetrahedron. It is a subgroup of index 2 of both $\mathscr{T}_h$ and $\mathscr{T}_d$ and it is isomorphic to $\mathbb{A}_4$ (Figure 28.4).

$$\mathscr{T} \cong \left\langle a,b \,\middle|\, a^2 = b^3 = (ab)^3 = 1 \right\rangle.$$

The next seven infinite series of groups ($\mathscr{D}_{nh}$, $\mathscr{D}_{nd}$, $\mathscr{D}_n$, $\mathscr{S}_{2n}$, $\mathscr{C}_{nh}$, $\mathscr{C}_{nv}$, and $\mathscr{C}_n$) can be interpreted as symmetry groups of regular prisms or antiprisms or their subgroups (Figures 28.5 and 28.6).

1. $\mathscr{D}_{mh}$—full symmetry group of a regular m-sided prism. For m even it is isomorphic to $D_m \times \mathbb{Z}_2$, for m odd it is isomorphic to D_{2m}. (Here D_k denotes the dihedral group of $2k$ elements.)
2. $\mathscr{D}_{md}$—full symmetry group of a regular m-sided antiprism. It is isomorphic to D_{2m}.
3. $\mathscr{D}_m$—group of rotations of a regular m-sided prism. It is isomorphic to D_m and is a subgroup of index 2 of both $\mathscr{D}_{mh}$ and $\mathscr{D}_{md}$.
4. $\mathscr{S}_{2m}$—cyclic group generated by improper rotation (glide) for $360°/2m$ about vertical axis. It is isomorphic to $\mathbb{Z}_{2m}$ and is a subgroup of index 2 of $\mathscr{D}_{md}$.
5. $\mathscr{C}_{mh}$—group generated by a rotation for $360°/m$ about vertical axis and a reflexion about a vertical plane. It is isomorphic to $\mathbb{Z}_m \times \mathbb{Z}_2$ and is a subgroup of index 2 of $\mathscr{D}_{mh}$.
6. $\mathscr{C}_{mv}$—full group of symmetries of a regular m-sided pyramid. It is isomorphic to D_n and is a subgroup of index 2 of both $\mathscr{D}_{mh}$ and $\mathscr{D}_{md}$.
7. $\mathscr{C}_m$—group of rotations of a regular m-sided pyramid. It is isomorphic to $\mathbb{Z}_m$ and is a subgroup of all groups listed above.

The remaining three groups are the groups of low symmetry (Figure 28.7).

1. The group C_s contains a mirror reflection and an identity.
2. The group C_i contains a point inversion and an identity.
3. The group C_1 contains only an identity.

28.4 Local Restrictions

In general, the presence of rotational symmetry axes in the symmetry group implies a local rotational symmetry of the sites where the axes intersects the polyhedron.

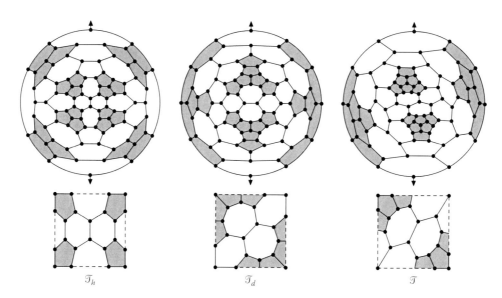

FIGURE 28.4 Examples of fulleroids with symmetry group $\mathscr{T}_h$, $\mathscr{T}_d$, and $\mathscr{T}$, respectively, with the same number of vertices and the same number of faces of size 5, 6, and 7. Pentagonal faces are gray. To obtain the graph of the fulleroid, it suffices to draw the corresponding graph segment onto all faces of a cube.

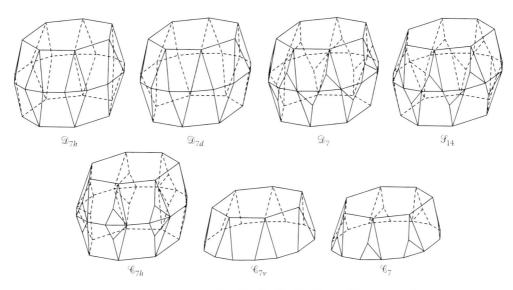

FIGURE 28.5 Examples of polyhedra with symmetry groups $\mathscr{D}_{7h}$, $\mathscr{D}_{7d}$, $\mathscr{D}_7$, $\mathscr{S}_{14}$, $\mathscr{C}_{7h}$, $\mathscr{C}_{7v}$, and $\mathscr{C}_7$, respectively.

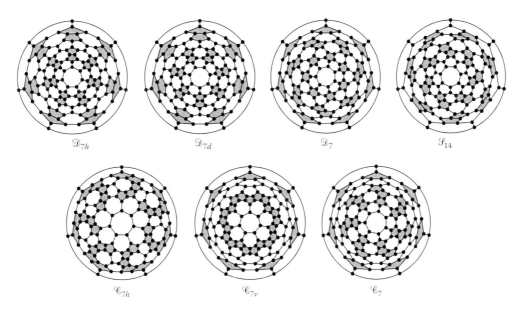

FIGURE 28.6 Examples of fulleroids with symmetry groups $\mathscr{D}_{7h}$, $\mathscr{D}_{7d}$, $\mathscr{D}_7$, $\mathscr{S}_{14}$, $\mathscr{C}_{7h}$, $\mathscr{C}_{7v}$, and $\mathscr{C}_7$, respectively.

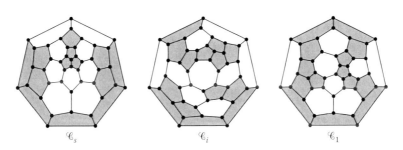

FIGURE 28.7 Examples of fulleroids with symmetry groups $\mathscr{C}_s$, $\mathscr{C}_i$, and $\mathscr{C}_1$, respectively.

Lemma 28.1: *Let P be a fulleroid. If an axis of an m-fold rotation intersects a face of size n, then m divides n. If an axis of an m-fold rotation intersects a vertex of P, then m = 3. If an axis of an m-fold rotation intersects an edge of P, then m = 2.*

These connections, together with the fact that every fulleroid has at least 12 faces of degree 5, give us necessary conditions for the face sizes of fulleroids with a given symmetry group. If the symmetry group of a fulleroid P contains an m-fold rotational axis, where $m \neq 2, 3, 5$, then P must have faces of size n, where m divides n.

28.5 Symmetry of Fullerenes

In the special case of fullerenes, there are only pentagonal and hexagonal faces. Therefore, by Lemma 28.1 the fullerene symmetry group can contain axes of m-fold rotation only for $m = 2$, 3, 5, and 6. This makes the list of possible symmetry groups be finite; it consists of 36 groups:

$$\mathscr{I}_h, \mathscr{I}, \mathscr{I}_h, \mathscr{I}_d, \mathscr{I},$$

$$\mathscr{D}_{6h}, \mathscr{D}_{6d}, \mathscr{D}_6, \mathscr{S}_{12}, \mathscr{C}_{6h}, \mathscr{C}_{6v}, \mathscr{C}_6,$$

$$\mathscr{D}_{5h}, \mathscr{D}_{5d}, \mathscr{D}_5, \mathscr{S}_{10}, \mathscr{C}_{5h}, \mathscr{C}_{5v}, \mathscr{C}_5,$$

$$\mathscr{D}_{3h}, \mathscr{D}_{3d}, \mathscr{D}_3, \mathscr{S}_6, \mathscr{C}_{3h}, \mathscr{C}_{3v}, \mathscr{C}_3,$$

$$\mathscr{D}_{2h}, \mathscr{D}_{2d}, \mathscr{D}_2, \mathscr{S}_4, \mathscr{C}_{2h}, \mathscr{C}_{2v}, \mathscr{C}_2,$$

$$\mathscr{C}_s, \mathscr{C}_i, \mathscr{C}_1.$$

Using more detailed consideration it was proved that whenever a fivefold or sixfold rotational axis is present, the structure of the fullerene implies a perpendicular twofold rotational axis (Fowler et al. 1993). This means that the groups $\mathscr{S}_{12}$, $\mathscr{C}_{6h}$, $\mathscr{C}_{6v}$, and $\mathscr{C}_6$ only occur as subgroups of dihedral groups $\mathscr{D}_{6h}$, $\mathscr{D}_{6d}$, and $\mathscr{D}_6$; as the groups $\mathscr{S}_{10}$, $\mathscr{C}_{5h}$, $\mathscr{C}_{5v}$, and $\mathscr{C}_5$ only occur as subgroups of isosahedral or dihedral groups $\mathscr{I}_h$, $\mathscr{I}$, $\mathscr{D}_{5h}$, $\mathscr{D}_{5d}$, and $\mathscr{D}_5$. The final list of 28 fullerene symmetry groups is therefore

$$\mathscr{I}_h, \mathscr{I}, \mathscr{T}_h, \mathscr{T}_d, \mathscr{T},$$

$$\mathscr{D}_{6h}, \mathscr{D}_{6d}, \mathscr{D}_6,$$

$$\mathscr{D}_{5h}, \mathscr{D}_{5d}, \mathscr{D}_5,$$

$$\mathscr{D}_{3h}, \mathscr{D}_{3d}, \mathscr{D}_3, \mathscr{S}_6, \mathscr{C}_{3h}, \mathscr{C}_{3v}, \mathscr{C}_3,$$

$$\mathscr{D}_{2h}, \mathscr{D}_{2d}, \mathscr{D}_2, \mathscr{S}_4, \mathscr{C}_{2h}, \mathscr{C}_{2v}, \mathscr{C}_2,$$

$$\mathscr{C}_s, \mathscr{C}_i, \mathscr{C}_1.$$

As we will see in the next sections, this is the only case where realizability of a symmetry group is not transferred onto its subgroups.

For each of the 28 fullerene symmetry groups, the smallest examples are known (Fowler and Manolopoulos 1995). Their number of vertices is listed in Table 28.2.

28.6 Icosahedral Fulleroids

It was Fowler who first asked whether there is fullerene-like structure consisting of pentagons and heptagons only and exhibiting icosahedral symmetry. The answer came immediately, when Brinkmann and Dress (1996) found two such structures with a minimal possible number of vertices and proved that there are only two of them. They also defined $\Gamma(a, b)$-fulleroid to be a fulleroid with only a-gonal and b-gonal faces, on which the group Γ acts as a group of symmetries. Delgado Friedrichs and Deza (2000) found more icosahedral fulleroids with pentagonal and n-gonal faces for $n = 8$, 9, 10, 12, 14, and 15. Jendrol' and Trenkler (2001) constructed $\mathscr{I}(5, n)$-fulleroids for all $n \geq 8$.

The properties and structure of two-faces maps (including fulleroids) are studied in the book of Deza and Dutour Sikirić (2008).

The results (Brinkmann and Dress 1996, Delgado Friedrichs and Deza 2000, Jendrol' and Trenkler 2001) are all concerned with the group $\mathscr{I}$. But the largest fullerene symmetry group is $\mathscr{I}_h$, hence we focus now on $\mathscr{I}_h(5, n)$-fulleroids.

Theorem 28.4: *For any $n \geq 6$ there are infinitely many $\mathscr{I}_h(5, n)$-fulleroids.*

Proof For $n = 6$ we get the case of $\mathscr{I}_h$-fullerenes, which are completely characterized in the catalogue of Graver (2005). They can be divided into two infinite series.

To depict an $\mathscr{I}_h(5, n)$-fulleroid, it is not necessary to draw its whole graph. If we cut the regular dodecahedron D along all its mirror planes, it falls into 120 congruent triangular pieces, called *flags*. The flags have the shape of a right triangle ABC with $|BC| < |AB| < |AC|$, where A (B, resp. C) represents the point where the fivefold (twofold, resp. threefold) rotational axis intersects the polyhedron—the face center, the edge midpoint, and the vertex of D.

If a graph is drawn onto one flag and it is copied onto all other flags, altogether we get a graph with the automorphism group

TABLE 28.2 The Counts of Vertices of the Smallest Representatives of 28 Fullerene Symmetry Groups

$\mathscr{I}_h$	$\mathscr{I}$	$\mathscr{T}_h$	$\mathscr{T}_d$	$\mathscr{T}$	$\mathscr{D}_{6h}$	$\mathscr{D}_{6d}$	$\mathscr{D}_6$	$\mathscr{D}_{5h}$	$\mathscr{D}_{5d}$	$\mathscr{D}_5$	$\mathscr{D}_{3h}$	$\mathscr{D}_{3d}$	$\mathscr{D}_3$
60	140	92	28	44	36	24	72	30	40	60	26	32	32
$\mathscr{S}_6$	$\mathscr{C}_{3h}$	$\mathscr{C}_{3v}$	$\mathscr{C}_3$	$\mathscr{D}_{2h}$	$\mathscr{D}_{2d}$	$\mathscr{D}_2$	$\mathscr{S}_4$	$\mathscr{C}_{2h}$	$\mathscr{C}_{2v}$	$\mathscr{C}_2$	$\mathscr{C}_s$	$\mathscr{C}_i$	$\mathscr{C}_1$
68	62	34	40	40	36	28	44	48	30	32	34	56	36

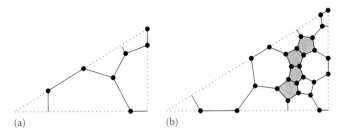

FIGURE 28.8 The segments of the graph of the smallest $\mathcal{I}_h(5, 7)$-fulleroid (a) and another $\mathcal{I}_h(5, 7)$-fulleroid (b).

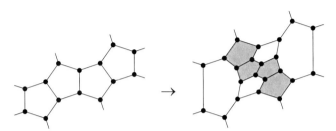

FIGURE 28.9 The step to create two new heptagons.

isomorphic to $\mathcal{I}_h$. If this graph is cubic, planar, and 3-connected, with all faces of size at least 5, we get an example of a graph of an $\mathcal{I}_h$-fulleroid.

For $n = 7$, examples of $\mathcal{I}_h(5, 7)$-fulleroids are obtained, if the graphs from Figure 28.8 are used this way.

To prove there are infinitely many $\mathcal{I}_h(5, 7)$-fulleroids, one can search for a configuration of four pentagons (see Figure 28.9, left) and change them into two heptagons and six pentagons (in every flag). Since there is such a configuration again, the operation can be carried out arbitrarily many times.

Examples of $\mathcal{I}_h(5, n)$-fulleroids for $n = 8, 9, \ldots, 17$ are obtained, if the graphs from Figures 28.10 and 28.11 are drawn into all the flags of a regular dodecahedron.

To prove that for some number n the set of all $\mathcal{I}_h(5, n)$-fulleroids is infinite it is sufficient to find an infinite series of corresponding graphs. We start with an example of fulleroid with pentagonal and m-gonal faces, where $m \in \{8, 9, \ldots, 17\}$ such that 10 divides $n - m$, e.g., some of those in Figures 28.10 and 28.11. We first change all the m-gonal faces to n-gons and then increase their number.

If the size m of some faces should be increased, two operations can be used. If two m-gons are connected by an edge, by inserting 10 pentagons they are changed to $(m + 5)$-gons (see Figure 28.12 for illustration). This step can be carried out arbitrarily many times, so the size of those two faces can be increased by any multiple of 5. In Figures 28.10 and 28.11, the application of this operation is indicated by thickening the edges. If two m-gons are separated by two faces in the position as in Figure 28.13, left, the size of those faces can be increased arbitrarily (see Figure 28.13, right).

As a special case of the second operation, we get the following: If the two original faces are pentagons, we can change

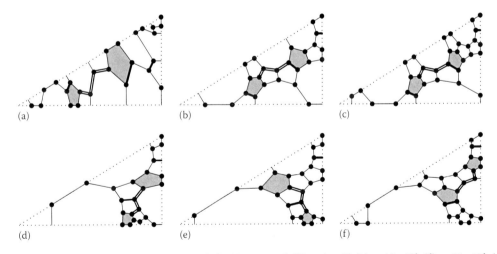

FIGURE 28.10 The segments of the graphs of $\mathcal{I}_h(5, n)$-fulleroids for (a) $n = 8 + 10k$, (b) $n = 9 + 10k$, (c) $n = 10 + 10k$, (d) $n = 11 + 10k$, (e) $n = 12 + 10k$, and (f) $n = 13 + 10k$.

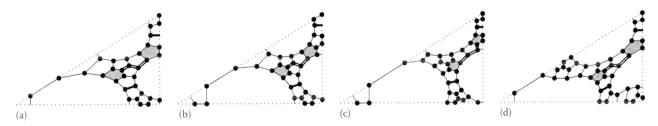

FIGURE 28.11 The segments of the graphs of $\mathcal{I}_h(5, n)$-fulleroids for (a) $n = 14 + 10k$, (b) $n = 15 + 10k$, (c) $n = 16 + 10k$, and (d) $n = 17 + 10k$.

FIGURE 28.12 The step to increase the size of two *m*-gons by 5.

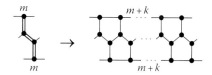

FIGURE 28.13 The step to increase the size of two *m*-gons arbitrarily.

them into two *n*-gons and $2n - 8$ new pentagons, so the number of *n*-gonal faces can be increased by two. For $n \geq 8$ this step can be repeated as many times as required, because two pentagons in an appropriate position can be found among the new pentagons again. The two pentagons, that can be used this way to create infinitely many examples, are in Figures 28.10 and 28.11 shaded gray and the edges connecting them are doubled.   ∎

It is worth to note that given integers $5 < n_1 < n_2 < \cdots < n_k$, we can make an $\mathcal{I}_h$-fulleroid with faces of sizes 5, n_1, n_2, ..., n_k if we start with graph segments from Figures 28.8, 28.10, and 28.11 and use the operations from Figures 28.9, 28.12, and 28.13.

28.7 Subgroups of $\mathcal{I}_h$

In the previous section, we proved that there are infinitely many $\mathcal{I}_h(5, n)$-fulleroids for any $n \geq 7$. It is natural to ask for which groups Γ there are infinitely many $\Gamma(5, n)$-fulleroids for any $n \geq 7$.

It is clear that such groups can only contain *m*-fold rotational axis for $m = 2$, 3, and 5. Therefore, the list of candidates is finite and it is possible to check them one by one. In particular, we get the following 29 groups:

$$\mathcal{I}_h, \mathcal{I}, \mathcal{T}_h, \mathcal{T}_d, \mathcal{T},$$

$$\mathcal{D}_{5h}, \mathcal{D}_{5d}, \mathcal{D}_5, \mathcal{S}_{10}, \mathcal{C}_{5h}, \mathcal{C}_{5v}, \mathcal{C}_5,$$

$$\mathcal{D}_{3h}, \mathcal{D}_{3d}, \mathcal{D}_3, \mathcal{S}_6, \mathcal{C}_{3h}, \mathcal{C}_{3v}, \mathcal{C}_3,$$

$$\mathcal{D}_{2h}, \mathcal{D}_{2d}, \mathcal{D}_2, \mathcal{S}_4, \mathcal{C}_{2h}, \mathcal{C}_{2v}, \mathcal{C}_2,$$

$$\mathcal{C}_s, \mathcal{C}_i, \mathcal{C}_1.$$

Proposition 28.1: *Let Γ be a symmetry group that is a subgroup of the group $\mathcal{I}_h$. Then there are infinitely many $\Gamma(5, n)$-fulleroids for any $n \geq 7$.*

Proof Since Γ is a subgroup of $\mathcal{I}_h$, it has an action on the set of flags of regular dodecahedron *D*. Under this action, all 120 flags of *D* split into $120/|\Gamma|$ orbits, each containing $|\Gamma|$ flags. (Here $|\Gamma|$ denotes the order, i.e., the number of elements of the group Γ.) If we start with an $\mathcal{I}_h$-fulleroid and change the graph in the flags forming one orbit under Γ in such a way that no new symmetries arise, we obtain a Γ-fulleroid. We can use the examples of $\mathcal{I}_h(5, n)$-fulleroids described in the proof of Theorem 28.4 to obtain infinite series of $\Gamma(5, n)$-fulleroids for all $n \geq 7$ and all subgroups Γ of $\mathcal{I}_h$.   ∎

This answers the question of existence of $\Gamma(5, n)$-fulleroids for arbitrary $n \geq 7$ for 22 subgroups of $\mathcal{I}_h$:

$$\mathcal{I}_h, \mathcal{I}, \mathcal{T}_h, \mathcal{T}_i$$

$$\mathcal{D}_{5d}, \mathcal{D}_5, \mathcal{S}_{10}, \mathcal{C}_{5v}, \mathcal{C}_5,$$

$$\mathcal{D}_{3d}, \mathcal{D}_3, \mathcal{S}_6, \mathcal{C}_{3v}, \mathcal{C}_3,$$

$$\mathcal{D}_{2h}, \mathcal{D}_2, \mathcal{C}_{2h}, \mathcal{C}_{2v}, \mathcal{C}_2,$$

$$\mathcal{C}_s, \mathcal{C}_i, \mathcal{C}_1.$$

As we will conclude later, the condition of being subgroup of $\mathcal{I}_h$ is not only sufficient for the existence of $\Gamma(5, n)$-fulleroid for arbitrary $n \geq 7$, but it is also a necessary condition.

Different constructions of tetrahedral fulleroids can be also found in Kardoš (2007b).

For the remaining seven groups, $\mathcal{T}_d$, $\mathcal{D}_{5h}$, $\mathcal{C}_{5h}$, $\mathcal{D}_{3h}$, $\mathcal{C}_{3h}$, $\mathcal{D}_{2d}$, and $\mathcal{S}_4$, there are values of integer *n* such that corresponding fulleroids do not exist. All those cases of nonexistence fall within fulleroids with multi-pentagonal faces.

28.8 Fulleroids with Multi-Pentagonal Faces

A face is *multi-pentagonal*, if its size is a multiple of 5. Fulleroids with multi-pentagonal faces play a special role among others, because they can be mapped onto (a graph of) a regular dodecahedron.

Lemma 28.2: (Jendrol' and Kardoš 2007) *Let P be a cubic convex polyhedron such that all faces are multi-pentagons, i.e., the size of each face is a multiple of five. Then there exists an orientation-preserving homomorphism $\Psi: P \to D$, where D denotes a regular dodecahedron.*

By an orientation-preserving homomorphism $\Psi: P \to D$, we mean a map from the set $V(P)$ of vertices of the polyhedron *P* into the set $V(D)$ of vertices of the regular dodecahedron *D*, respecting the adjacency structure, which also preserves the order of the vertices incident with any vertex (up to a cyclic permutation) once an orientation has been assigned to both P and D before.

Lemma 28.3: (Kardoš 2007a) *Let P be a cubic convex polyhedron such that all its faces are multi-pentagons and let $\Psi: P \to D$ be an orientation-preserving homomorphism, where D denotes a regular dodecahedron. If $\varphi \in \Gamma(P)$ is a symmetry of P, then $\Psi \circ \varphi : P \to D$ is also an orientation-preserving homomorphism, moreover, the symmetry φ of P uniquely determines a symmetry $\bar{\Psi}(\varphi)$ of D once Ψ is fixed.*

Theorem 28.5: (Kardoš 2007a) *Let P be a cubic convex polyhedron such that all its faces are multi-pentagons. Then there exists a homomorphism $\bar{\Psi} : \Gamma(P) \to \mathcal{I}_h$, where $\Gamma(P)$ is the symmetry group of P and $\mathcal{I}_h$ denotes the symmetry group of a regular dodecahedron D.*

This theorem is a special case of a result of Malnič et al. (2002) concerning map homomorphisms into regular maps.

Theorem 28.6: (Malnič et al. 2002) *If $\Psi: N \to M$ is a map homomorphism with M regular, then $Aut(N)$ projects.*

The homomorphism $\bar{\Psi}:\Gamma(P) \to \mathcal{I}_h$ is a strong tool for proving the nonexistence of fulleroids with multi-pentagonal faces and with symmetry groups that are not subgroups of $\mathcal{I}_h$.

Proposition 28.2: *Let P be a cubic convex polyhedron such that the sizes of all its faces are odd multiples of five. Then the symmetry group $\Gamma(P)$ does not contain the group $\mathcal{S}_4$ as a subgroup.*

Proof Let P be a convex polyhedron satisfying the premise of the claim and let φ be the rotation-reflection that generates $\mathcal{S}_4$ in $\Gamma(P)$. Since all the face sizes are odd, the rotation axis of φ intersects P in the midpoints of two edges, let one of these two edges be denoted by e. Let $e = uv$. Then $\varphi^2(u) = v$ and $\varphi^2(v) = u$. It means φ^2 maps two adjacent vertices u and v onto each other.

Let $\Psi:P \to D$ be the homomorphism given by Lemma 28.2 and $\bar{\Psi}:\Gamma(P) \to \mathcal{I}_h$ be the corresponding homomorphism given by Theorem 28.5. The rotation-reflection φ is an element of order 4 in $\Gamma(P)$. There are no elements of order 4 in $\mathcal{I}_h$, so $\bar{\Psi}(\varphi)$ can be only of order 2 or 1, which implies $\bar{\Psi}(\varphi^2) = \bar{\Psi}(\varphi)^2 = id$. The vertices u and v are adjacent, hence, so are the vertices $\Psi(u)$ and $\Psi(v)$. On the other hand,

$$\Psi(v) = \Psi(\varphi^2(u)) = \bar{\Psi}(\varphi^2)(\Psi(u)) = id(\Psi(u)) = \Psi(u),$$

which is a contradiction. ∎

Corollary 28.1: *There is no fulleroid P such that the sizes of all its faces are odd multiples of five with the symmetry group $\mathcal{S}_4$, $\mathcal{D}_{2d}$, or $\mathcal{T}_d$.*

Proof It immediately follows from Proposition 28.2, since $\mathcal{S}_4$ is a subgroup of both $\mathcal{D}_{2d}$ and $\mathcal{T}_d$. ∎

Proposition 28.3: *Let P be a cubic convex polyhedron such that all its faces are multi-pentagons and none of the face sizes is divisible by three. Then the symmetry group $\Gamma(P)$ does not contain the group $\mathcal{C}_{3h}$ as a subgroup.*

Proof Let P be a convex polyhedron satisfying the premise of the claim and let ρ and θ be the rotation and the reflection that generate $\mathcal{C}_{3h}$ in $\Gamma(P)$. Since no face of P has the size divisible by three, the rotation axis of ρ intersects P in two vertices, let one of those vertices be denoted by v. Let the neighbors of v be denoted by v_1, v_2, v_3 in such a way that $\rho(v_1) = v_2$, $\rho(v_2) = v_3$, and $\rho(v_3) = v_1$.

Let $\Psi:P \to D$ be the homomorphism given by Lemma 28.2 and $\bar{\Psi}:\Gamma(P) \to \mathcal{I}_h$ be the corresponding homomorphism given by Theorem 28.5. The rotation ρ is an element of order 3 in $\Gamma(P)$.

If $\bar{\Psi}(\rho) = id$, then $\Psi(v_2) = \Psi(\rho(v_1)) = \bar{\Psi}(\rho)(\Psi(v_1)) = \Psi(v_1)$. But $\Psi:P \to D$ maps distinct neighbors of v onto distinct ones, which is a contradiction. Hence, $ord(\bar{\Psi}(\rho)) = 3$, which implies $\bar{\Psi}(\rho)$ is a rotation by $\pm 120°$.

We know that θ is orientation-reversing symmetry of P and $(\theta) = 2$, so $\bar{\Psi}(\theta)$ must be an orientation-reversing element of $\mathcal{I}_h$ such that $(\bar{\Psi}(\theta)) \leq 2$. The reflection θ has at least one fixpoint on the surface of P (midpoints of edges intersected by the mirror plane); thus, $\bar{\Psi}(\theta)$ must also have at least one fixpoint on the surface of D and cannot, indeed, be a point inversion. So $\bar{\Psi}(\theta)$ is a reflection.

In P, the mirror plane corresponding to θ is perpendicular to the rotational axis of ρ. Moreover, $\rho \circ \theta = \theta \circ \rho$ and the subgroup $\mathcal{C}_{3h}$ of $\Gamma(P)$ generated by ρ and θ is commutative. What is the relative position of $\bar{\Psi}(\rho)$ and $\bar{\Psi}(\theta)$? There is no mirror plane of D perpendicular to any of its threefold rotational symmetry axis. Moreover, if a threefold rotational axis $\bar{\Psi}(\rho)$ is chosen, a mirror plane of D together with $\bar{\Psi}(\rho)$ generate either a non-commutative subgroup of $\mathcal{I}_h$ with six elements (the group $\mathcal{C}_{3v}$, if the mirror plane contains the axis), or the whole group $\mathcal{I}_h$ (otherwise), which is a contradiction. ∎

Corollary 28.2: *There is no fulleroid P such that all its faces are multi-pentagons, none of the face sizes is divisible by three, and the symmetry group of P is $\mathcal{C}_{3h}$ or $\mathcal{D}_{3h}$.*

Proof It immediately follows from Proposition 28.3, since $\mathcal{C}_{3h}$ is a subgroup of $\mathcal{D}_{3h}$. ∎

Proposition 28.4: *Let P be a cubic convex polyhedron such that all its faces are multi-pentagons and none of the face sizes is divisible by 25. Then the symmetry group $\Gamma(P)$ does not contain the group $\mathscr{C}_{5h}$ as a subgroup.*

Proof Let P be a convex polyhedron satisfying the premise of the claim and let ρ and θ be the rotation and the reflection that generate $\mathscr{C}_{5h}$ in $\Gamma(P)$. The rotation axis of ρ intersects P in two faces. Let one of those faces be denoted by f. Let v be a vertex incident with f.

Let $\Psi\colon P \to D$ be the homomorphism given by Lemma 28.2 and $\bar{\Psi}\colon\Gamma(P) \to \mathscr{I}_h$ be the corresponding homomorphism given by Theorem 28.5. The rotation ρ is an element of order 5 in $\Gamma(P)$. Since no face of P has the size that is a multiple of 25, $\Psi(v)$ and $\Psi(\rho(v)) = \bar{\Psi}(\rho)(\Psi(v))$ are two distinct vertices of the face $\Psi(f)$. Thus, $\bar{\Psi}(\rho)$ is not an identity, so $\bar{\Psi}(\rho)$ is an element of order 5 in $\mathscr{I}_h$, hence $\bar{\Psi}(\rho)$ is a rotation by $\pm72°$.

Using the same arguments like in the proof of Proposition 28.3, we get $\bar{\Psi}(\theta)$, which is a reflection.

In P, the mirror plane corresponding to θ is perpendicular to the rotational axis of ρ. Moreover, $\rho \circ \theta = \theta \circ \rho$ and the subgroup $\mathscr{C}_{5h}$ of $\Gamma(P)$ generated by ρ and θ is commutative. What is the relative position of $\bar{\Psi}(\rho)$ and $\bar{\Psi}(\theta)$? There is no mirror plane of D perpendicular to any of its fivefold rotational symmetry axis. Moreover, if a fivefold rotational axis $\bar{\Psi}(\rho))$ is chosen, a mirror plane of D together with $\bar{\Psi}(\rho)$ generate either a non-commutative subgroup of $\mathscr{I}_h$ with 10 elements (the group $\mathscr{C}_{5v}$, if the mirror plane contains the axis), or the whole group $\mathscr{I}_h$ (otherwise), which is a contradiction. ∎

Corollary 28.3: *There is no fulleroid P such that all its faces are multi-pentagons, none of the face sizes is divisible by 25, and the symmetry group of P is $\mathscr{C}_{5h}$ or $\mathscr{D}_{5h}$.*

Proof It immediately follows from Proposition 28.4, since $\mathscr{C}_{5h}$ is a subgroup of $\mathscr{D}_{5h}$. ∎

Corollaries 28.1 through 28.3 cover all the seven groups, which are not subgroups of $\mathscr{I}_h$, even though they do not force any local restrictions. For all these seven groups, there are values of n, for which there are no $\Gamma(5, n)$-fulleroids. On the other hand, for all other values of n not covered by Corollaries 28.1 through 28.3, examples of $\Gamma(5, n)$-fulleroids can be found in Kardoš (2007a).

The results mentioned above combined with constructions in Sections 28.6 and 28.7 complete the proof of the following characterization:

Theorem 28.7: *Let Γ be a point symmetry group. Then there are infinitely many $\Gamma(5, n)$-fulleroids for any $n \geq 7$ if and only if Γ is a subgroup of the group $\mathscr{I}_h$.*

28.9 Fulleroids with Octahedral, Prismatic, or Pyramidal Symmetry

There are groups that contain an m-fold rotational symmetry for $m = 4$ or $m \geq 6$. Local conditions (see Lemma 28.1) do not allow the existence of $(5, n)$-fulleroids for arbitrary n—in this case n must be a multiple of m. However, this necessary condition is not always also sufficient.

Octahedral symmetry groups $\mathbb{O}_h$ and $\mathbb{O}$ both contain three fourfold rotational symmetry axes, therefore, the size of some faces must be a multiple of four.

Theorem 28.8: *(Jendrol' and Kardoš 2007) Let $n \geq 6$. There are infinitely many $\mathbb{O}_h(5, n)$-fulleroids and $\mathbb{O}(5, n)$-fulleroids if and only if (i) $n \equiv 0 \pmod{60}$ or (ii) $n \equiv 0 \pmod 4$ and $n \not\equiv 0 \pmod 5$.*

The groups $\mathscr{D}_{mh}$, $\mathscr{D}_{md}$, $\mathscr{D}_m$, $\mathscr{S}_{2m}$, $\mathscr{C}_{mh}$, $\mathscr{C}_{mv}$, and $\mathscr{C}_m$ contain m-fold rotational symmetry, therefore, the size of some faces must be a multiple of m.

Theorem 28.9: *(Kardoš 2007a) Let $m = 4$ or $m \geq 6$ and $n \geq 7$ be integers and let Γ be $\mathscr{D}_{md}$, $\mathscr{D}_m$, $\mathscr{S}_{2m}$, $\mathscr{C}_{mv}$ or $\mathscr{C}_m$. Then there are infinitely many $\Gamma(5, n)$-fulleroids if and only if n is a multiple of m.*

Theorem 28.10: *(Kardoš 2007a) Let $m = 4$ or $m \geq 6$ and $n \geq 7$ be integers and let $m \not\equiv 0 \pmod 5$. Let Γ be $\mathscr{D}_{mh}$ or $\mathscr{C}_{mh}$. Then there are infinitely many $\Gamma(5, n)$-fulleroids if and only if n is a multiple of m.*

Theorem 28.11: *(Kardoš 2007a) Let $m \geq 10$ and $n \geq 7$ be integers and let $m \equiv 0 \pmod 5$. Let Γ be $\mathscr{D}_{mh}$ or $\mathscr{C}_{mh}$. Then there are infinitely many $\Gamma(5, n)$-fulleroids if and only if n is a multiple of $5m$.*

Examples and constructions can be found in Kardoš (2007a).

28.10 (5, 7)-Fulleroids

Unlike the case of fullerenes, i.e., (5, 6)-fulleroids, for (5, 7)-fulleroids, all point symmetry groups that satisfy local restrictions are realizable.

For the 36 groups Γ, which do not contain m-fold rotation unless $m = 2, 3, 5,$ or 7, the numbers of vertices of the smallest examples of $\Gamma(5, 7)$-fulleroids are listed in Table 28.3. The examples with dihedral, skewed, pyramidal, and low symmetry are depicted in Figures 28.14 through 28.17.

TABLE 28.3 The Counts of Vertices of the Smallest $\Gamma(5, 7)$-Fulleroid for All 36 Possible Symmetry Groups Γ

$\mathscr{I}_h$	$\mathscr{I}$	$\mathscr{T}_h$	$\mathscr{T}_d$	$\mathscr{T}$	$\mathscr{D}_{7h}$	$\mathscr{D}_{7d}$	$\mathscr{D}_7$	$\mathscr{S}_{14}$	$\mathscr{C}_{7h}$	$\mathscr{C}_{7v}$	$\mathscr{C}_7$
500	260	116	116	68	84	28	140	196	112	112	140

$\mathscr{D}_{5h}$	$\mathscr{D}_{5d}$	$\mathscr{D}_5$	$\mathscr{S}_{10}$	$\mathscr{C}_{5h}$	$\mathscr{C}_{5v}$	$\mathscr{C}_5$	$\mathscr{D}_{3h}$	$\mathscr{D}_{3d}$	$\mathscr{D}_3$	$\mathscr{S}_6$	$\mathscr{C}_{3h}$
60	60	100	140	80	80	100	44	44	44	68	56

$\mathscr{C}_{3v}$	$\mathscr{C}_3$	$\mathscr{D}_{2h}$	$\mathscr{D}_{2d}$	$\mathscr{D}_2$	$\mathscr{S}_4$	$\mathscr{C}_{2h}$	$\mathscr{C}_{2v}$	$\mathscr{C}_2$	$\mathscr{C}_s$	$\mathscr{C}_i$	$\mathscr{C}_1$
68	68	52	36	36	52	44	44	44	44	60	52

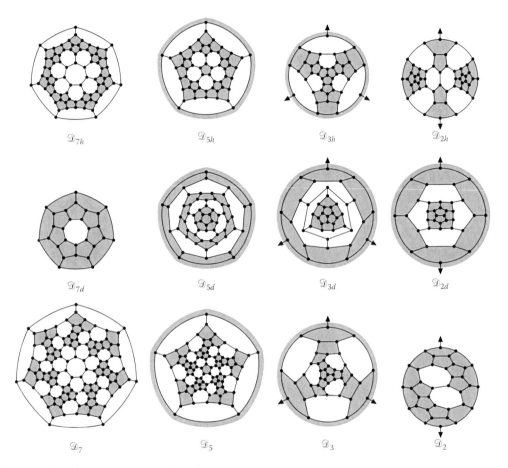

FIGURE 28.14 The smallest (5, 7)-fulleroids with dihedral symmetry groups.

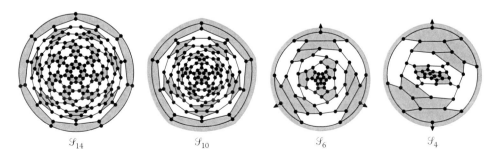

FIGURE 28.15 The smallest (5, 7)-fulleroids with skewed symmetry groups.

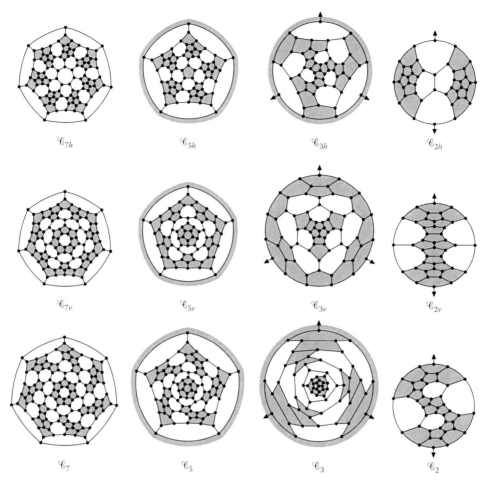

FIGURE 28.16 The smallest (5, 7)-fulleroids with pyramidal symmetry groups.

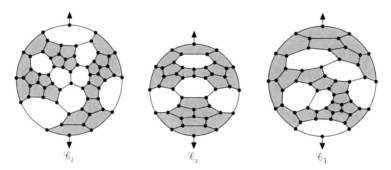

FIGURE 28.17 The smallest (5, 7)-fulleroids with low symmetry.

Acknowledgment

This work was supported by Slovak Research and Development Agency under the contract No. APVV-0007-07.

References

Brinkmann, G. and A. W. M. Dress (1996). Phantasmagorical fulleroids. *Match* 33, 87–100.

Chartrand, G. and L. Lesniak (2005). *Graphs and Digraphs* (4th edn). Boca Raton, FL: Chapman & Hall/CRC.

Coxeter, H. M. S. and W. O. J. Moser (1972). *Generators and Relations for Discrete Groups*. Berlin, Germany: Springer.

Cromwell, P. R. (1997). *Polyhedra*. Cambridge, MA: Cambridge University Press.

Delgado Friedrichs, O. and M. Deza (2000). More icosahedral fulleroids. In P. Hansen, P. Fowler, and M. Zheng (Eds.), *Discrete Mathematical Chemistry*, Volume 51 of *DIMACS Series in Discrete Mathematics and Theoretical Computer Science*, American Mathematical Society, pp. 97–115.

Deza, M. and M. Dutour Sikirić (2008). *Geometry of Chemical Graphs*. Cambridge, U.K.: Cambridge University Press.

Diestel, R. (1997). *Graph Theory*. New York: Springer.

Fowler, P. W. and D. E. Manolopoulos (1995). *An Atlas of Fullerenes*. Oxford, U.K.: Oxford University Press.

Fowler, P. W., D. E. Manolopoulos, D. B. Redmond, and R. P. Ryan (1993). Possible symmetries of fullerene structures. *Chem. Phys. Lett. 202*, 371–378.

Graver, J. E. (2005). Catalog of all fullerenes with ten or more symmetries. In S. Fajtlowicz, P. W. Fowler, P. Hansen, M. F. Janowitz, and F. S. Roberts (Eds.), *Graphs and Discovery*, Volume 69 of *DIMACS Series in Discrete Mathematics and Theoretical Computer Science*, American Mathematical Society, pp. 167–188.

Grünbaum, B. (2003). *Convex Polytopes*. New York: Springer.

Jendroľ, S. and F. Kardoš (2007). On octahedral fulleroids. *Discrete Appl. Math. 155*, 2181–2186.

Jendroľ, S. and M. Trenkler (2001). More icosahedral fulleroids. *J. Math. Chem. 29*, 235–243.

Kardoš, F. (2007a). Symmetry of fulleroids. PhD thesis, P. J. Šafárik University, Košice, Slovakia.

Kardoš, F. (2007b). Tetrahedral fulleroids. *J. Math. Chem. 41*, 101–111.

Malnič, A., R. Nedela, and M. Škoviera (2002). Regular homomorphisms and regular maps. *Eur. J. Combinatorics 23*, 449–461.

Mani, P. (1971). Automorphismen von polyedrischen graphen. *Math. Ann. 192*, 279–303.

29

C_{20}, the Smallest Fullerene

Fei Lin
*University of Illinois
at Urbana-Champaign*

Erik S. Sørensen
McMaster University

Catherine Kallin
McMaster University

A. John Berlinsky
McMaster University

29.1 Introduction

The C_{20} molecule with a dodecahedral cage structure is the smallest member of the fullerene family. It follows from Euler's theorem that fullerenes are molecules that consist of exactly 12 pentagons and a varying number of hexagons. The most famous fullerene molecule is C_{60} with 12 pentagons and 20 hexagons (commonly referred to as "Buckyballs," short for "Buckminsterfullerene"), which was discovered in 1985 (Kroto et al. 1985). Smalley, Curl, and Kroto were awarded the 1996 Nobel Prize in Chemistry for their discovery, which not only allowed the study of the fascinating properties of C_{60} molecules and solids, such as superconductivity in alkali-metal-doped C_{60} solid (Hebard et al. 1991), but also initiated an exciting new era of carbon nanotechnology. Interestingly, in his Nobel lecture, Kroto (1997) emphasizes the *pentagon isolation rule* (Kroto 1987, Schmalz et al. 1988) predicting the most stable fullerenes to have the 12 pentagons as far apart as possible. In light of this rule, the C_{20} fullerene cage should be highly unstable, and by defying this rule, C_{20} is sometimes referred to as an "unconventional fullerene" (Xie et al. 2004, Wang et al. 2006).

Not surprisingly, the synthesis of the smallest fullerene is therefore much more difficult than that of C_{60}, since the highly curved fullerene C_{20} is so reactive that it cannot be produced by carbon condensation or cluster annealing processes. Instead, Prinzbach et al. (2000) started from a precursor (dodecahedrane $C_{20}H_{20}$, which was first synthesized by the Paquette group [Ternansky et al. 1982, Paquette et al. 1983]). They replaced the hydrogen atoms with bromine atoms, and then debrominated to produce the gas-phase cage C_{20} fullerene. The same process is used to synthesize another C_{20} isomer with a bowl molecular structure. These two isomers, together with the previously commonly produced monocyclic ring, constitute the three major isomers of C_{20} that are of particular interest for this chapter.

It is worth noting that the C_{20} cage isomer has been the subject of intensive theoretical research even before its production in 2000. A vast amount of theoretical work has used density functional theory (DFT) to study the relative stability of cage C_{20} compared with other isomers but has arrived at no consensus. This, in part, reflects the importance of electron correlations not included in the mean-field DFT calculations. More accurate treatments of electron correlations in these molecules by quantum Monte Carlo (QMC) calculations show that the bowl isomer has the lowest energy, with the ring isomer in the middle, and with the cage isomer being highest in ground state energy (Grossman et al. 1995). This was the first QMC calculation applied to C_{20} molecules, indicating the importance of electron correlations in determining the electronic properties of these molecules. To further explore the electronic correlations in C_{20} molecules, Lin et al. (2007) performed exact diagonalization (ED) and QMC simulations on a one-band Hubbard model using the effective Hamiltonian approach (Lin and Sørensen 2008). They found that the cage isomer is the most correlated molecule among the isomers, with an on-site interaction and hopping integral ratio as high as $U/t \sim 7.1$.

Two research groups (Wang et al. 2001, Iqbal et al. 2003) have recently reported the synthesis of solid C_{20} phases. Wang et al. used ion beam irradiation to produce a close-packed hexagonal crystal consisting of $C_{20}H_{20}$ molecules with a lattice spacing of 4.6 Å. Iqbal et al. employed ultraviolet laser ablation from diamond onto nickel substrates and obtained a face-centered-cubic (fcc) lattice with 22 carbon atoms per unit cell (C_{20} linked with bridging carbon atoms at interstitial tetrahedral sites). These further developments are likely to stimulate more research interest in this field, since speculation (Miyamoto and Saito 2001, Spagnolatti et al. 2002) about a possible superconducting state in the C_{20} solid can now be tested experimentally. Interest in a possible superconducting state of solid C_{20} is a natural outgrowth of the superconductivity observed in doped C_{60} that has an fcc lattice structure. A higher transition temperature, T_c, might be expected for solid C_{20} than for C_{20} because the more curved molecular structure of the C_{20} cage implies a higher frequency vibrational phonon. Of course, the expectation of this higher T_c is based on the phonon mechanism for superconductivity, while the question of whether phonons or electron–electron interactions are responsible for the superconducting behavior in the doped C_{60} solid is still a subject of debate.

We organize this chapter in the following way. We first describe the molecular structure of C_{20} fullerene. This is followed by a discussion of various single molecular properties, including ground state energy, thermal stability, mechanical stability, electron affinity, electron correlations, vibrational frequencies, and electron–phonon coupling strength. We then proceed to discuss the electronic properties of the solid C_{20} fullerene, possible superconductivity, and the *I–V* characteristics of short chains of fullerene C_{20}. The chapter concludes with a future perspective.

29.2 Molecular Structure

We show in Figure 29.1 the molecular structures of the three major isomers: cage, bowl, and ring. The 20 carbon atoms in the cage form 12 pentagons and 30 bonds. This is the consequence of

Euler's formula, which relates the number of vertices, edges, and faces on a polygon, i.e., $V - E + F = 2$. The molecular diameter is about 3.1 Å. The fullerene cage C_{20} has the molecular symmetry of the icosahedral point group I_h, which has 120 symmetry operations. It is the only fullerene smaller than C_{60} with the full I_h symmetry. The I_h group is discussed by Ellzey (2003), Ellzey and Villagran (2003), and Ellzey (2007), and the application of this symmetry to the Hückel Hamiltonian defined on a cage results in the symmetry-labeled Hückel energy diagram shown in Figure 29.2. The Hückel Hamiltonian, in second quantization notation, can be written as

$$H = -t \sum_{\langle ij \rangle \sigma} \left(c_{i\sigma}^\dagger c_{j\sigma} + \text{h.c.} \right), \tag{29.1}$$

where

t is the nearest neighbor (NN) hopping integral that sets the energy scale of the system

$c_{i\sigma}^\dagger (c_{i\sigma})$ creates (annihilates) an electron with spin σ on the ith carbon site

h.c. is the Hermitian conjugate

$\langle ij \rangle$ means i and j are NN

The above noninteracting Hückel Hamiltonian will be regarded as an effective Hamiltonian that captures the geometric character of the cage molecule. It is an approximation to the real Hamiltonian, and should have a low-energy spectrum around the Fermi energy very similar to that of the real Hamiltonian. One can obtain a value for the hopping integral, t, by fitting the tight-binding Hückel energy levels around the Fermi level to those from DFT calculations (Lin and Sørensen 2008), leading to $t \simeq 1.36$ eV.

From Figure 29.2, one can see that the neutral cage isomer is metallic (since the conduction bands are half-filled) and magnetic (if Hund's rule is applied to the two electrons in the highest occupied levels). Are these properties realized in a real molecule? Can they survive the inclusion of electron–electron

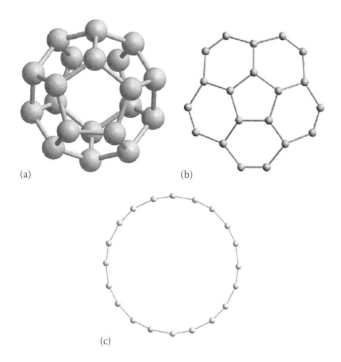

(a)   (b)

(c)

FIGURE 29.1 Molecular structures of C_{20} isomers: (a) cage, (b) bowl, and (c) ring.

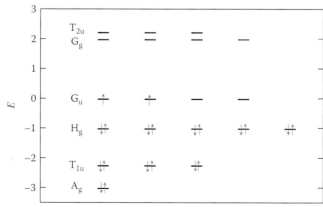

FIGURE 29.2 Hückel molecular orbitals of a neutral dodecahedral C_{20} molecule.

interactions? We will examine these questions, going beyond the noninteracting Hückel Hamiltonian, in later sections.

29.3 Single Molecular Properties

29.3.1 Ground State Energy

There are many calculations on C_{20} isomers that compare their ground state energies and infer their relative stability. However, different ground state energy orderings are predicted by different calculations, depending on the method used. Hartree–Fock (HF) calculations show that the ring is lowest in energy, followed by the bowl and then the cage (Parasuk and Almlöf 1991, Feyereisen et al. 1992). Local density approximation (LDA) calculations predict the cage to be the lowest in energy, followed by the bowl and the ring (Brabec et al. 1992, Wang et al. 1996). Generalized gradient approximation (GGA) (Raghavachari et al. 1993), hybrid HF/DFT (Allison and Beran 2004), and DFT/B3LYP (Xu et al. 2006) suggest again the same ground state energy ordering as HF calculations. The more accurate coupled-cluster calculations (CCSD) with basis set including single, double, and triple excitations of the HF wave-function gives an energy ordering similar to LDA, i.e., the cage and the bowl are almost degenerate and have lower energy than the ring isomer (Taylor et al. 1995). More recent CCSD calculations are able to distinguish the energy difference between the bowl and the cage, and identify the bowl energy as lower than the cage (An et al. 2005). Tight-binding calculations based on the Harrison scheme (Harrison 1989) give the cage as the lowest energy isomer, followed by the bowl and the ring (Cao 2001). Though inconclusive, all these calculations do reflect the subtlety in making approximations about electronic correlations. The need for an accurate method, such as QMC, to deal with such correlations is clear. Such a QMC calculation with a pseudopotential replacing the ion core has been performed (Grossman et al. 1995), and gives a ground state energy ordering with the bowl being the lowest, followed by the ring and the cage. Even more accurate all-electron

(without pseudopotential) QMC (Sokolova et al. 2000) has also been done and gives the same result as Grossman et al. (1995). Therefore, one can now conclude that the ground state energy of the three major C_{20} isomers with respect to bowl, from lowest to highest, are bowl (0 eV), ring (1.1 ± 0.5 eV), and cage (2.1 ± 0.5 eV) (Grossman et al. 1995, Sokolova et al. 2000).

We summarize the above discussion in Table 29.1.

29.3.2 Thermal Stability

From the previous discussion, we conclude that the bowl has the lowest ground state energy, and the cage has a relative energy of about 2.1 eV with respect to the bowl. Then why does the cage exist after it is produced and not decay into the lower energy isomers? Tight-binding molecular-dynamics (TBMD) simulations (Davydov et al. 2005) have helped answer this question. In the simulation, one starts with the cage isomer, and measures how many saddle point states (local energy maximum) and metastable states (local energy minimum) need to be passed over before reaching the bowl isomer state. By this process, the minimum potential barrier that must be overcome for a cage decay is measured to be about 5.0 eV. And remarkably, although many routes can be taken in the cage decay to some metastable states, none of these metastable states lead to a transition to the bowl configuration (which is the global energy minimum). This same cage stability is also found when the activation energy, E_a, for cage isomer decay is measured. The activation energy is defined as

$$N_c = N_0 \exp\left(\frac{E_a}{k_B T}\right), \tag{29.2}$$

where N_c is the critical number of steps of TBMD simulations, and $N_c = \tau/t_0$ with τ the lifetime of the cage isomer and $t_0 = 2.72 \times 10^{-16}$ s the time step of TBMD. By simulating the decay process at different temperatures, one can map out the $\ln(N_c)$ versus $1/T$

TABLE 29.1 Ground State Energy Ordering for Three C_{20} Major Isomers from Various Calculations

Source	Method	Ground State Energy Ordering
Parasuk and Almlöf (1991)	HF	$E_{ring} < E_{bowl} < E_{cage}$
Feyereisen et al. (1992)	HF	$E_{ring} < E_{bowl} < E_{cage}$
Brabec et al. (1992)	LDA	$E_{cage} < E_{bowl} < E_{ring}$
Wang et al. (1996)	LDA	$E_{cage} < E_{bowl} < E_{ring}$
Raghavachari et al. (1993)	GGA	$E_{ring} < E_{bowl} < E_{cage}$
Allison and Beran (2004)	Hybrid HF/DFT	$E_{ring} < E_{bowl} < E_{cage}$
Xu et al. (2006)	DFT/B3LYP	$E_{ring} < E_{bowl} < E_{cage}$
Taylor et al. (1995)	CCSD	$E_{cage} \approx E_{bowl} < E_{ring}$
An et al. (2005)	CCSD	$E_{bowl} < E_{cage} < E_{ring}$
Cao (2001)	Tight-binding	$E_{cage} < E_{bowl} < E_{ring}$
Grossman et al. (1995)	QMC (pseudopotential)	$E_{bowl} < E_{ring} < E_{cage}$
Sokolova et al. (2000)	QMC (all-electron)	$E_{bowl} < E_{ring} < E_{cage}$

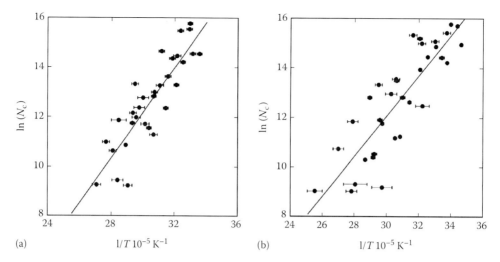

FIGURE 29.3 Logarithm of the critical number of steps of molecular-dynamics simulation, N_c, for the onset of the decay of the C_{20} fullerene as a function of the temperature, T, of the ionic subsystem for the electron temperature (a) $T_{el} = T$ and (b) $T_{el} = 0$. Circles are the results of calculation, and the solid line is a linear least square fit. (Reprinted from Davydov, I.V. et al., *Phys. Solid State*, 47, 778, 2005. With permission.)

curve, see Figure 29.3. The slope of the fitted line gives the activation energy, $E_a = 7 \pm 1$ eV. With this activation energy, one can estimate the lifetime of the cage isomer at room temperature, τ (300 K), to be extremely large (practically infinite). This explains the stability of the fullerene cage, once it is synthesized.

There are earlier quantum molecular dynamics (QMD) studies of the free energy versus temperature dependence for C_{20} isomers (Brabec et al. 1992), which, as we discussed in the previous section, starts from the incorrect LDA ground state energy ordering, i.e., $E_{cage} < E_{bowl} < E_{ring}$. However, these studies did see the stability of cage fullerene for temperatures below about 700 K. Above this temperature, cage fullerene transforms into its bowl isomer, which is assumed to be the building block of the C_{60} fullerene. Their conclusion is probably correct, too, since they are aiming at explaining the high-temperature production of C_{60} fullerenes in experiments using the laser vaporization technique, which we know is unable to produce C_{20} fullerene.

With the previous conclusion that the cage fullerene will not transform into other isomers once it is synthesized, it is also interesting to study further its thermal stability in the sense of the Lindemann criterion for solid melting (Lindemann 1910): solid melts when the root-mean-square bond-length fluctuation ratio, δ, exceeds a threshold value 0.1, the Lindemann constant. A QMD simulation employing the Lindemann criterion has been performed (Ke et al. 1999). Figure 29.4 shows that the C_{20} fullerenes remains solid below temperature 1900 K, above which it starts to melt.

29.3.3 Mechanical Stability

The mechanical stability of C_{20} has been studied with MD in the context of low-energy cluster beam deposition of C_{20} fullerenes on substrates (Du et al. 2002). The simulation shows that for a C_{20} kinetic energy in the range of 10–45 eV, the probability of C_{20} adsorption on the substrate is high, and the molecules maintain their dodecahedral geometry without deformation. Similar MD simulations of the mechanical stability of the C_{20} fullerene

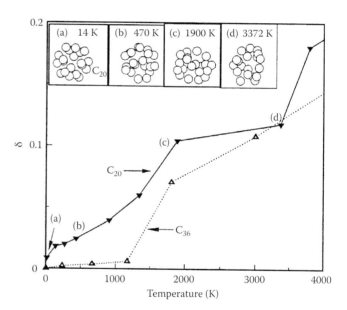

FIGURE 29.4 Root-mean-square bond-length fluctuations δ of C_{20} as a function of temperature. Snapshots (a–d) exhibit the geometry of a C_{20} cluster at various temperatures (14, 470, 1900 and 3372 K, respectively). (Reprinted from Ke, X.Z. et al., *Chem. Phys. Lett.*, 313, 40, 1999. With permission.)

(Ke et al. 1999) also show that when the C_{20} incident energy is less than 25 eV, it will rebound without breaking after its collision with a graphite surface. These simulations are important for understanding the experimental deposition of C_{20} films.

29.3.4 Electron Affinity

Electron affinity (EA) is an important molecular property that incorporates electron correlation information (Barnett et al. 1986) and which can be measured by gas-phase photoemission experiments. EA, in the case of neutral C_{20}, is defined as

$$EA = E(20) - E(21), \qquad (29.3)$$

where $E(20)$ and $E(21)$ are the internal energies of the neutral and one-electron-doped molecules, respectively. In recent years, with the development of experimental techniques, EA can be measured more and more accurately. One such example is the measurement of EA of the neutral C_{20} molecule (Yang et al. 1987, Wang et al. 1991, 1999), yielding 2.7 ± 0.1 eV, 2.65 ± 0.05 eV, and 2.689 ± 0.008 eV, with increasing accuracy. Such measurements are useful for electron correlation studies and can be used to estimate the on-site electron correlation strength, as was done by Lin and Sørensen (2008), where their estimates are based on the neutral C_{20} EA values (EA = 2.25 ± 0.03 eV) measured when the gas-phase C_{20} cage is synthesized (Prinzbach et al. 2000). Estimates of on-site electron correlation strength is the first step in the effective Hamiltonian plus QMC approach to studying molecular electronic properties.

The positive EA can also be used to understand the aggregation property of C_{20} fullerenes (Luo et al. 2004), since it is energetically favorable for a neutral C_{20} fullerene to obtain an additional electron to lower its energy. Without doping, such additional electrons can only be obtained by a covalent-like sharing with other fullerenes, suggesting that C_{20} fullerenes tend to aggregate. Indeed, such aggregation phenomena have been observed not only in C_{20} fullerenes (Ehlich et al. 2001), but also in larger fullerenes, such as C_{36} (Piskoti et al. 1998, Collins et al. 1999), C_{60}, etc., and carbon nanotubes.

29.3.5 Electron Correlation

29.3.5.1 Beyond Noninteracting Hückel Hamiltonian

The noninteracting Hückel Hamiltonian defined in Equation 29.1 results in the electronic configuration of molecular orbitals shown in Figure 29.2. However, neglecting electron interactions is not appropriate for real materials. A more refined and minimal interacting model that takes into account electronic on-site interactions (here "sites" refers to individual C atoms) is the one-band Hubbard model defined as

$$H = -t \sum_{\langle ij \rangle \sigma} \left(c_{i\sigma}^{\dagger} c_{j\sigma} + \text{h.c.} \right) + U \sum_{i} n_{i\uparrow} n_{i\downarrow}, \qquad (29.4)$$

where the additional term U represents the interaction energy of two electrons occupying the same atomic site. The ratio U/t measures the on-site interaction strength. This model has also been treated within the lattice-density functional approach (López-Sandoval and Pastor 2006). Interactions longer ranged than the on-site U has also been considered (Lin et al. 2007). Clearly, one needs first to determine how large U/t is in the real C_{20} cage in order to address its effect on the C_{20} electronic structure. U/t values for C_{20} isomers have been estimated by Lin and Sørensen (2008), who show that the C_{20} fullerene cage is the most strongly correlated of the three major C_{20} isomers, with the on-site interaction strength as high as $U/t \sim 7.1$. The C_{20} fullerene cage is also

the most strongly correlated molecule in the fullerene family, and is much more correlated than the C_{60} molecule, which has a U/t ratio of approximately $U/t \sim 3.5$.

29.3.5.2 Magnetic Properties

ED and QMC calculations on Equation 29.4 have shown that, for $U/t < 4.1$, the C_{20} fullerene has a spin triplet ground state, and for $U/t > 4.1$ it has a singlet ground state (Lin et al. 2007). Therefore, for a real neutral C_{20} fullerene, the ground state is probably a spin singlet, violating Hund's rule for the electron configuration of the noninteracting approach, which would predict that the neutral molecule has a spin triplet ground state (Figure 29.2). This Hund's rule violation is a result of strong electronic correlations.

In the large U/t limit, the Hubbard model in Equation 29.4 is reduced to an antiferromagnetic Heisenberg model (AFHM) using second order perturbation theory. Since $U/t \sim 7.1$, such a reduction is reasonable for the C_{20} cage. AFM defined on a dodecahedron molecule has been solved by ED and shows a spin singlet and nonmagnetic ground state (Konstantinidis 2005), which agrees very well with our previous discussions. By applying an external magnetic field, the molecule shows some interesting properties, such as magnetization discontinuities, etc. (Konstantinidis 2005, 2007).

29.3.5.3 Pair-Binding Energy

The pair-binding energy as a function of on-site electron interaction has also been investigated (Lin et al. 2007) in the hope of finding a possible effective attraction between electrons on the same fullerene. The pair-binding energy is defined as

$$\Delta_b(21) = E(22) + E(20) - 2E(21), \qquad (29.5)$$

as shown schematically in Figure 29.5. If $\Delta_b(21) < 0$, i.e., the nonuniform charge distribution is preferred, it would then provide an electronic mechanism for superconductivity in the alkali-doped C_{60} materials (Chakravarty et al. 1991).

The pair-binding energy as a function of electron correlation, U/t, is shown in Figure 29.6. ED and QMC calculations show that $\Delta_b(21)$ is always positive for any value of U/t. Thus, for an average of one electron doping in the C_{20} fullerenes, two doped electrons on two C_{20} fullerenes tend to stay separately on two molecules instead of staying together on the same molecule and leaving the other molecule undoped. This, therefore, rules out the possibility of superconductivity from purely electronic correlations.

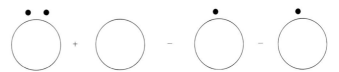

FIGURE 29.5 Schematic illustration of pair-binding energy. Big empty circles represent neutral C_{20} cage. Small solid dots represent doped electrons.

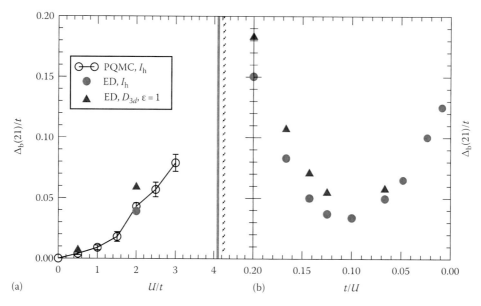

FIGURE 29.6 Electronic pair-binding energy, $\Delta_b(21)/t$, as a function of U/t. ED ($\bullet$) and QMC ($\circ$) results for the molecule with I_h symmetry. ED results (triangle) for the molecule with D_{3d} symmetry. The solid (crossed) vertical line indicates the critical value U_c/t for neutral $I_h(D_{3d})$ molecules. (a) Δ_b versus U/t for $U/t \leq 5$. (b) Δ_b versus t/U for $5 \leq U/t \leq 100$. (Reprinted from Lin, F. et al., *Phys. Rev. B*, 76, 033414, 2007; *J. Phys. Condens. Matter*, 19, 456206, 2007.)

29.3.5.4 Suppression of Jahn–Teller Distortion

Strong electron–electron interactions can also suppress a Jahn–Teller distortion. Figure 29.7 shows the energy difference between a Jahn–Teller distorted and a non-distorted dodeca-hedral neutral molecule as a function of distortion magnitude, ε. We see that, for the stronger electron–electron interaction, $U/t = 8$, which stabilizes a singlet ground state (compared with $U/t = 2$ for which the ground state is a triplet), the effect of the Jahn–Teller distortion is minimal. Taking into account the high ratio, $U/t \sim 7.1$, for the real molecule, we suggest that Jahn–Teller distortion is inactive in the C_{20} fullerene.

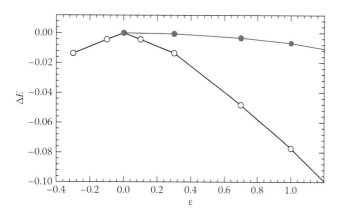

FIGURE 29.7 Shift in the ground state energy of the neutral molecule, ΔE, versus the distortion, ε. The Jahn–Teller effect is quite pronounced for $U/t = 2(\circ)$, but it is essentially absent for $U/t = 8(\bullet)$. (Reprinted from Lin, F. et al., *Phys. Rev. B*, 76, 033414, 2007; *J. Phys. Condens. Matter*, 19, 456206, 2007. With permission.)

29.3.6 Vibration Frequencies and Electron–Phonon Coupling

It has long been conjectured that smaller fullerene molecules, such as C_{20}, will have higher vibration frequencies and larger electron–phonon coupling strength than C_{60}. An approximate equation for the electron–phonon coupling strength is given by $\sim 1800/N_\pi$ meV for aromatic molecules, including fullerene molecules, from tight-binding calculations using the Einstein approximation for phonon frequencies (Devos and Lannoo 1998). Since N_π is the number of electrons in p_π orbitals, C_{20} has the largest electron–phonon coupling strength in the fullerene family, see Figure 29.8. More sophisticated numerical simulations confirm to some extent this trend in the fullerene molecules (Raghavachari et al. 1993, Wang et al. 1996, Saito and Miyamoto 2002). These calculations, though using different equilibrium geometries, show that the C_{20} cage has vibration frequencies approximately in the range of $150 \sim 2200\,\text{cm}^{-1}$. These vibrational frequencies await a comparison with future Raman scattering and infrared spectroscopy data.

29.4 Solid Properties

So far we have discussed many interesting properties of a single C_{20} fullerene molecule in the above discussions. However, from a practical point of view, the realization of the solid phase of C_{20} fullerene is even more interesting in terms of a possible superconductor or as the building block for the fabrication of nanodevices. Two experimental groups have reported successful syntheses of solid phases of C_{20} fullerenes (Wang et al. 2001, Iqbal et al. 2003). Wang et al. used an ion beam irradiation

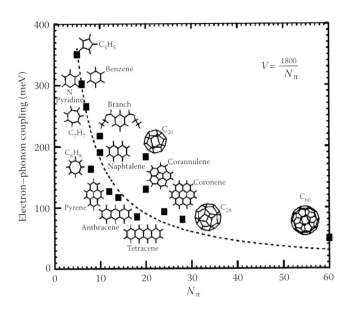

FIGURE 29.8 Dependence of the full electron–phonon coupling, V, on the number of electrons in the p_π molecular orbital. (Reprinted from Devos, A. and Lannoo, M., *Phys. Rev. B*, 58, 8236, 1998. With permission.)

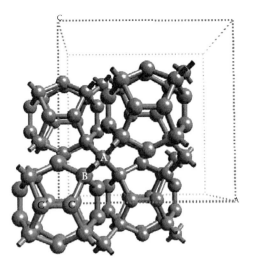

FIGURE 29.9 The fcc lattice structure of the C_{22} crystal. The lattice constant is 8.61 Å. The coordinates of the three independent atoms A, B, and C are given by, in units of lattice constant, A = (0.25, 0.25, 0.25), B = (0.148, −0.148, 0.148), and C = (0.078, −0.217, 0.0). Eight atoms out of the twenty of the cage (B) and the interstitial carbon atoms (A) are sp^3 hybridized. The other twelve atoms of the cage (C) are sp^2 hybridized. In the NaC$_{22}$ compound, the alkali atoms occupy the octahedral site of the fcc lattice. (Reprinted from Spagnolatti, I. et al., *Europhys. Lett.*, 59, 572, 2002. With permission.)

method to produce a hexagonal close-packed crystal, which has a lattice spacing of 4.6 Å, while Iqbal et al. used laser ablation to deposit C_{20} fullerene molecules into an fcc lattice with fullerene molecules interconnected by two additional carbon atoms per unit cell in the interstitial tetrahedral sites, which is consistent with their DFT calculations. In the following, we shall review work that has been done on solid C_{20} fullerene, with an emphasis on possible superconductivity in the bulk material. We will also discuss some work on *I–V* characteristic calculations for a chain constructed from C_{20} fullerene molecules, which may be important for the future development of fullerene-based nanodevices.

29.4.1 Superconductivity?

The fact that both superconducting alkali-metal-doped C_{60} solid and the pure C_{20} solid have an fcc lattice structure has led Spagnolatti et al. (2002) to speculate about doping one Na atom per unit cell in the remaining empty octahedral site in a C_{20} fcc solid (In fcc solid C_{60}, both the octahedral site and the two tetrahedral sites are occupied by three K atoms, leading to the chemical configuration of K_3C_{60} in the superconducting phase). The undoped solid is an insulator with a band gap of 2.47 eV. With Na doping, the energy bands still look like the pure solid case, and the doping shifts the Fermi energy into the lowest conduction bands. Hence, the solid turns from an undoped insulator into a metal. See Figure 29.9 for the fcc lattice structure and Figure 29.10 for the density of states (DOS) of the Na-doped C_{22} fcc solid. The electron–phonon coupling strength, λ, is also calculated to be $\lambda = 1.12$, much larger than the value for the solid C_{60} where $\lambda \sim 0.3 - 0.5$ (Gunnarsson 2004). Using McMillan's formula, which is valid for phonon-mediated superconductivity, Spagnolatti et al. then estimated the superconducting transition

temperature to be in the range of 15–55 K, an impressive range for phonon-mediated superconductivity.

An earlier theoretical speculation on the possible forms of solid C_{20} phase using DFT calculations has predicted the simple cubic (SC) lattice to be the most stable one, and this structure has the property of being metallic (Miyamoto and Saito 2001). However, the C_{20} cage structure must be broken in order to form such a SC lattice, which reduces the electron–phonon coupling strength and hence is not likely to produce a high transition temperature for superconductivity even if such a lattice could superconduct. They also suggest that a one-dimensional chain of C_{20} cages can have a higher transition temperature if it is doped with either electrons or holes to shift the Fermi level to the peak of the DOS. As promising as this may seem, there is no experimental realization of such lattices up to now. Furthermore, DFT investigations (Okada et al. 2001) show that orthorhombic and tetragonal lattices with the polymerized C_{20} cage as a building block have much lower energies than the SC lattice. Calculations of DOS suggest that both lattices are semiconductors with energy gaps of 1.4 and 1.7 eV for orthorhombic and tetragonal lattices, respectively. Their DOS figures also show that, with slight hole doping, the Fermi level can be moved to the peaks of the valence bands, hence making them possible candidates for superconductors. Polarization and dielectric constants of orthorhombic and tetragonal lattices have also been studied by DFT (Otsuka et al. 2006). The dielectric constants are rather anisotropic and assume values in the range of 4.7 ~ 5.8 depending on the direction of the applied external electric field.

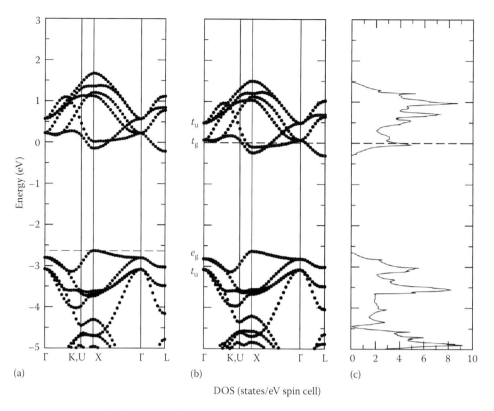

(a)   (b)   (c)

DOS (states/eV spin cell)

FIGURE 29.10 Band structure around the Fermi level of (a) the pure fcc C_{22} and (b) NaC_{22} crystals. The zero of the energy is the Fermi level of NaC_{22}. The DOS of NaC_{22} is shown in (c). (Reprinted from Spagnolatti, I. et al., *Europhys. Lett.*, 59, 572, 2002. With permission.)

As discussed in the pair-binding energy calculation in the above section, purely electronic correlations have been argued to be responsible for the superconductivity in alkali-metal-doped C_{60} solid (Chakravarty et al. 1991). In this mechanism, doped electrons pair up and lead to superconductivity inside one molecule. The Josephson coupling between superconducting molecules then makes the supercurrent flow without resistance in the whole solid. ED and QMC calculations on a Hubbard C_{20} molecule show, however, that the pair-binding energy does not favor such a mechanism (Lin et al. 2007). But this calculation does show a metal–insulator transition as a function of the interaction strength, U/t, see Figure 29.11. For the value of $U/t \sim 7.1$ for the real material, undoped C_{22} in an fcc lattice is insulating, in agreement with the DFT prediction (Spagnolatti et al. 2002).

29.4.2 *I–V* Characteristics of Cage Chains

We have seen experimental evidence that the synthesis of solid fcc C_{20} is a candidate for a possible new superconductor. However, there is another application direction for fullerene molecules, that is, fabrication of nanometer electro-mechanical systems (NEMS) for use in circuits at the nanometer scale. For this purpose, it is important to understand the electronic properties of such a device, especially the *I–V* characteristics of nanodevices made of C_{20} cages. As already discussed by Miyamoto and Saito (2001), 1D double or single-bonded C_{20} chains are semiconductors, implying a possible application in nanocircuits. We will

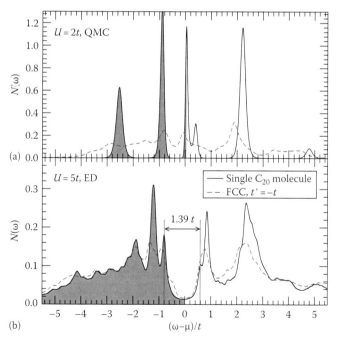

FIGURE 29.11 DOS $N(\omega)$ at (a) $U = 2t$ from QMC and at (b) $U = 5t$ from ED. In both panels, the DOS for the C_{22} fcc lattice obtained from CPT with $t' = -t$ is shown as a dashed line. (Reprinted from Lin, F. et al., *Phys. Rev. B*, 76, 033414, 2007; *J. Phys. Condens. Matter*, 19, 456206, 2007. With permission.)

concentrate on transport properties of 1D C_{20} chains, specifically on theoretical calculations of the electrical conductivity of C_{20} monomers, dimers, and longer chains, embedded between two electrodes or molecular interfaces (Roland et al. 2001, Otani et al. 2004, Ouyang and Xu 2007).

DFT calculations together with the Keldysh nonequilibrium Green's functions formalism (Datta 1995) allow one to calculate the *I–V* curve from first principles. Applications of such techniques to short chains of C_{20} with different geometries show that, for multi-bonded short C_{20} chains between Al(100) leads, the electrical conductivity has little dependence on the number of C_{20} molecules in between (Roland et al. 2001); for double- and single-bonded short chains between Au jellium electrodes, the electrical conductivity depends strongly on the number of C_{20} molecules in between, i.e., a monomer has an electrical conductivity that is about one order of magnitude greater than the dimer (Otani et al. 2004). For multi- and double-bonded short chains between two single-wall carbon nanotube electrodes, the electrical conductivity again strongly depends on the number of C_{20} molecules in between, with monomers having significantly larger conductivity than dimers (Ouyang and Xu 2007). It is also worth noting that, by replacing one C atom with one N atom per C_{20} molecule, the electrical conductivity increases substantially (Ouyang and Xu 2007). The effect of molecular vibrations on the *I–V* characteristics of a C_{20} fullerene bridge between two gold electrodes has also been studied by TBMD, which shows large discontinuous steps in the differential conductance at vibration frequencies of the molecule (Yamamoto et al. 2005).

To summarize, the above several calculations suggest some features for the short chains of C_{20} cages: (1) The connection between the electrodes and the C_{20} molecule is important for electron transport through the chains; (2) Electron scattering at C_{20}–C_{20} junctions greatly reduces the conductivity; (3) Doping can significantly change the electrical conductivity; (4) Molecular vibrations greatly affect electron transport. However, in the above calculations an exact treatment of strong electron correlations is still missing. Its effect on the electron transport properties of short C_{20} chains still needs to be calculated.

29.5 Summary and Future Perspective

We have given a survey of both single molecular and possible solid-state properties of C_{20} fullerenes. DFT and other mean-field calculations give conflicting ground state energy orderings for C_{20} isomers, but this situation is resolved by more accurate QMC simulations. This demonstrates the need to take into account electron correlations in predicting the C_{20} fullerene electronic properties. By fitting parameters in an effective one-band Hubbard model defined on a C_{20} fullerene, it has been possible to demonstrate the effect of strong electron correlations: magnetic to nonmagnetic molecule transition, metal to insulator transitions, and the suppression of Jahn–Teller distortion, all of which are driven by strong electron interactions. As to the examination of possible superconducting properties that has stimulated most of the theoretical and experimental endeavors discussed

in this chapter, we feel it is necessary to take into account both the strong electronic correlations and the equally important electron–phonon coupling. A balanced treatment of both interactions together with an exact calculation seems to be a challenging but worthwhile problem in the field. On the experimental side, it is really important to synthesize larger quantities of solid C_{20}. Larger and better samples will allow theoretical calculations to be tested experimentally. To summarize, we have seen in this chapter many interesting properties of the C_{20} fullerene, the smallest and truly unique member of the fullerene family.

Acknowledgments

FL is supported by the U.S. Department of Energy under award number DE-FG52–06NA26170. ESS, CK, and AJB are supported by the Natural Sciences and Engineering Research Council of Canada and the Canadian Foundation for Innovation. CK and AJB are supported by the Canadian Institute for Advanced Research.

References

Allison, C. and K. A. Beran, Energetic analysis of 24 C_{20} isomers, *J. Mol. Struct. (Theochem.)* **680**, 59 (2004).

An, W., Y. Gao, S. Bulusu, and X. C. Zeng, Ab initio calculation of bowl, cage, and ring isomers of C_{20} and C_{20}^{-}, *J. Chem. Phys.* **122**, 204109 (2005).

Barnett, R. N., P. J. Reynolds, and W. A. Lester Jr., Electron affinity of fluorine: A quantum Monte Carlo study, *J. Chem. Phys.* **84**, 4992 (1986).

Brabec, C. J., E. B. Anderson, B. N. Davidson et al., Precursors to C_{60} fullerene formation, *Phys. Rev. B* **46**, 7326 (1992).

Cao, Z. X., Electronic structure and stability of C_{20} isomers, *Chin. Phys. Lett.* **18**, 1060 (2001).

Chakravarty, S., M. P. Gelfand, and S. Kivelson, Electronic correlation effects and superconductivity in doped fullerenes, *Science* **254**, 970 (1991).

Collins, P. G., J. C. Grossman, M. Côté et al., Scanning tunneling spectroscopy of C_{36}, *Phys. Rev. Lett.* **82**, 165 (1999).

Datta, S. *Electronic Transport in Mesoscopic Systems.* New York: Cambridge University Press (1995).

Davydov, I. V., A. I. Podlivaev, and L. A. Openov, Anomalous thermal stability of metastable C_{20} fullerene, *Phys. Solid State* **47**, 778 (2005).

Devos, A. and M. Lannoo, Electron-phonon coupling for aromatic molecular crystals: Possible consequences for their superconductivity, *Phys. Rev. B* **58**, 8236 (1998).

Du, A. J., Z. Y. Pan, Y. K. Ho, Z. Huang, and Z. X. Zhang, Memory effect in the deposition of C_{20} fullerenes on a diamond surface, *Phys. Rev. B* **66**, 035405 (2002).

Ehlich, R., P. Landenberger, and H. Prinzbach, Coalescence of C_{20} fullerenes, *J. Chem. Phys.* **115**, 5830 (2001).

Ellzey, M. L. Jr., Finite group theory for large systems. 1. Symmetry-adaptation, *J. Chem. Inf. Comput. Sci.* **43**, 178 (2003).

Ellzey, M. L. Jr., Finite group theory for large systems. 3. Symmetry-generation of reduced matrix elements for icosahedral C_{20} and C_{60} molecules, *J. Comput. Chem.* **28**, 811 (2007).

Ellzey, M. L. Jr. and D. Villagran, Finite group theory for large systems. 2. Generating relations and irreducible representations for the icosahedral point group, *J. Chem. Inf. Comput. Sci.* **43**, 1763 (2003).

Feyereisen, M., M. Gutowski, J. Simons, and J. Almlöf, Relative stabilities of fullerene, cumulene, and polyacetylene structures for C_n:n = 18–60, *J. Chem. Phys.* **96**, 2926 (1992).

Grossman, J. C., L. Mitas, and K. Raghavachari, Structure and stability of molecular carbon: Importance of electron correlation, *Phys. Rev. Lett.* **75**, 3870 (1995).

Gunnarsson, O. *Alkali-Doped Fullerides: Narrow-Band Solids with Unusual Properties*. Singapore: World Scientific Publishing (2004).

Harrison, W. A. *Electronic Structure and the Properties of Solids: The Physics of the Chemical Bond*. New York: Dover Publication (1989).

Hebard, A. F., M. J. Rosseinsky, R. C. Haddon et al., Superconductivity at 18 K in potassium-doped C_{60}, *Nature* **350**, 600 (1991).

Iqbal, Z., Y. Zhang, H. Grebel et al., Evidence for a solid phase of dodecahedral C_{20}, *Eur. Phys. J. B* **31**, 509 (2003).

Ke, X. Z., Z. Y. Zhu, F. S. Zhang, F. Wang, and Z. X. Wang, Molecular dynamics simulation of C_{20} fullerene, *Chem. Phys. Lett.* **313**, 40 (1999).

Konstantinidis, N. P. Antiferromagnetic Heisenberg model on clusters with icosahedral symmetry, *Phys. Rev. B* **72**, 064453 (2005).

Konstantinidis, N. P. Unconventional magnetic properties of the icosahedral symmetry antiferromagnetic Heisenberg model, *Phys. Rev. B* **76**, 104434 (2007).

Kroto, H. W. The stability of the fullerenes C_n, with n = 24, 28, 32, 36, 50, 60 and 70, *Nature* **329**, 529 (1987).

Kroto, H. W. Symmetry, space, stars and C_{60}, *Rev. Mod. Phys.* **69**, 703 (1997).

Kroto, H. W., J. R. Heath, S. C. O'Brien, R. F. Curl, and R. E. Smalley, C_{60}: Buckminsterfullerene, *Nature* **318**, 162 (1985).

Lin, F. and E. S. Sørensen, Estimates of effective Hubbard model parameters for C_{20} isomers, *Phys. Rev. B* **78**, 085435 (2008).

Lin, F., E. S. Sørensen, C. Kallin, and A. J. Berlinsky, Strong correlation effects in the fullerene C_{20} studied using a one-band Hubbard model, *Phys. Rev. B* **76**, 033414 (2007); Extended Hubbard model on a C_{20} molecule, *J. Phys. Condens. Matter* **19**, 456206 (2007).

Lindemann, F. The calculation of molecular vibration frequencies, *Z. Phys.* **11**, 609 (1910).

López-Sandoval, R. and G. M. Pastor, Electron correlations in a C_{20} fullerene cluster: A lattice density-functional study of the Hubbard model, *Eur. Phys. J. D* **38**, 507 (2006).

Luo, J., L. M. Peng, Z. Q. Xue, and J. L. Wu, Positive electron affinity of fullerenes: Its effect and origin, *J. Chem. Phys.* **120**, 7998 (2004).

Miyamoto, Y. and M. Saito, Condensed phases of all-pentagon C_{20} cages as possible superconductors, *Phys. Rev. B* **63**, R161401 (2001).

Okada, S., Y. Miyamoto, and M. Saito, Three-dimensional crystalline carbon: Stable polymers of C_{20} fullerene, *Phys. Rev. B* **64**, 245405 (2001).

Otani, M., T. Ono, and K. Hirose, First-principles study of electron transport through C_{20} cages, *Phys. Rev. B* **69**, 121408(R) (2004).

Otsuka, J., T. Ono, K. Inagaki, and K. Hirose, First-principles calculations of dielectric constants of C_{20} bulk using Wannier functions, *Physica B* **376–377**, 320 (2006).

Ouyang, F. P. and H. Xu, Electronic transport in molecular junction based on C_{20} cages, *Chin. Phys. Lett.* **24**, 1042 (2007).

Paquette, L. A., R. J. Ternansky, D. W. Balogh, and G. J. Kentgen, Total synthesis of dodecahedrane, *J. Am. Chem. Soc.* **105**, 5446 (1983).

Parasuk, V. and J. Almlöf, C_{20}: The smallest fullerene? *Chem. Phys. Lett.* **184**, 187 (1991).

Piskoti, C., J. Yarger, and A. Zettl, C_{36}, a new carbon solid, *Nature* **393**, 771 (1998).

Prinzbach, H., A. Weller, P. Landenberger et al., Gas-phase production and photoelectron spectroscopy of the smallest fullerene, C_{20}, *Nature* **407**, 60 (2000).

Raghavachari, K., D. L. Strout, G. K. Odom et al., Isomers of C_{20}: Dramatic effect of gradient corrections in density functional theory, *Chem. Phys. Lett.* **214**, 357 (1993).

Roland, C., B. Larade, J. Taylor, and H. Guo, Ab initio I-V characteristics of short C_{20} chains, *Phys. Rev. B* **65**, 041401(R) (2001).

Saito, M. and Y. Miyamoto, Vibration and vibronic coupling of C_{20} isomers: Ring, bowl, and cage clusters, *Phys. Rev. B* **65**, 165434 (2002).

Schmalz, T. G., W. A. Seitz, D. J. Klein, and G. E. Hite, C_{60} carbon cages, *J. Am. Chem. Soc.* **110**, 1113 (1988).

Sokolova, S., A. Lüchow, and J. B. Anderson, Energetics of carbon clusters C_{20} from all-electron quantum Monte Carlo calculations, *Chem. Phys. Lett.* **323**, 229 (2000).

Spagnolatti, I., M. Bernasconi, and G. Benedek, Electron-phonon interaction in the solid form of the smallest fullerene C_{20}, *Europhys. Lett.* **59**, 572 (2002).

Taylor, P. R., E. Bylaska, J. H. Weare, and R. Kawai, C_{20}: Fullerene, bowl or ring? New results from coupled-cluster calculations, *Chem. Phys. Lett.* **235**, 558 (1995).

Ternansky, R. J., D. W. Balogh, and L. A. Paquette, Dodecahedrane, *J. Am. Chem. Soc.* **104**, 4503 (1982).

Wang, L. S., J. Conceicao, C. M. Jin, and R. E. Smalley, Threshold photodetachment of cold C_{60}^-, *Chem. Phys. Lett.* **182**, 5 (1991).

Wang, Z., P. Day, and R. Pachter, Ab initio study of C_{20} isomers: Geometry and vibrational frequencies, *Chem. Phys. Lett.* **248**, 121 (1996).

Wang, X. B., C. F. Ding, and L. S. Wang, High resolution photoelectron spectroscopy of C_{60}, *J. Chem. Phys.* **110**, 8217 (1999).

Wang, Z., X. Ke, Z. Zhu et al., A new carbon solid made of the world's smallest caged fullerene C$_{20}$, *Phys. Lett. A* **280**, 351 (2001).

Wang, C. R., Z. Q. Shi, L. J. Wan et al., C$_{64}$ H$_4$: Production, isolation, and structural characterizations of a stable unconventional fulleride, *J. Am. Chem. Soc.* **128**, 6605 (2006).

Xie, S. Y., F. Gao, X. Lu et al., Capturing the labile fullerene[50] as C$_{50}$ Cl$_{10}$, *Science* **304**, 699 (2004).

Xu, S. H., M. Y. Zhang, Y. Y. Zhao, B. G. Chen, J. Zhang, and C. C. Sun, Super-valence phenomenon of carbon atoms in C$_{20}$ molecule, *J. Mol. Struct. (Theochem.)* **760**, 87 (2006).

Yamamoto, T., K. Watanabe, and S. Watanabe, Electronic transport in fullerene C$_{20}$ bridge assisted by molecular vibrations, *Phys. Rev. Lett.* **95**, 065501 (2005).

Yang, S. H., C. L. Pettiette, J. Conceicao, O. Cheshnovsky, and R. E. Smalley, UPS of Buckminsterfullerene and other large clusters of carbon, *Chem. Phys. Lett.* **139**, 233 (1987).

30

Solid-State Structures of Small Fullerenes

Gotthard Seifert
Technische Universität Dresden

Andrey N. Enyashin
Technische Universität Dresden

and

Ural Branch of the Russian Academy of Sciences

Thomas Heine
Jacobs University Bremen

30.1 Introduction

Already in the early days of the fullerene research, there was clear evidence for the existence of stable smaller fullerene-like cage structures than C_{60} (Kroto, 1987; Cox et al., 1988). The C_{60} fullerene cage consists of 12 isolated pentagons, which are interconnected by hexagons. The pentagons cause the curvature necessary for the formation of the closed cage structure of the otherwise planar carbon structure (graphene). C_{60} is the smallest cage structure with no adjacent pentagons. Smaller fullerene-like cage structures cannot be constructed without adjacent pentagons in their structure. The extreme case is C_{20}, which consists of only pentagons. As already pointed out by Kroto (1987), the suppression of structures with adjacent pentagons can simply be understood by the increase of the local strain with the number of adjacent pentagons. Therefore, cage structures with isolated pentagons should be energetically preferred—isolated pentagon rule (IPR) (Kroto, 1987). There are many theoretical investigations that support strongly the energetic prevalence of isolated pentagons in the form of this isolated pentagon rule. For small fullerenes, the IPR can be extended to a minimization of adjacent pentagons in the cage structure. A detailed study for C_{40} showed that for each pentagon–pentagon pair in a fullerene cage, the binding energy is decreased by about 80 kJ/mol (Albertazzi et al., 1999). On the other hand, the binding energy as a function of the cage size increases smoothly with the number of atoms for the so-called fullerene road (Fowler and Manolopoulos, 1996) (see Figure 30.1). The "magic character" of C_{60} can be clearly seen from the relative stability function (see Figure 30.2). The relative stability (δ) of a fullerene C_n is defined as $\delta = E(n + 2) +$ $E(n - 2) - 2E(n)$, where $E(n)$ is the calculated total energy of C_n. This quantity may be viewed as a second derivative (difference quotient) in a $E(n)$ versus n plot, i.e., it visualizes better than $E(n)$ the size trends in stability. As one can see from Figure 30.2, there are other magic peaks for cage structures with less than 60 atoms. These "magic numbers" (clusters with a number of atoms with an exceptionally high stability in comparison with clusters of similar size) obtained from calculations agree quite well with corresponding peaks in the mass spectra of carbon clusters (see, e.g., Cox et al., 1988; Kietzmann et al., 1998). Based on these observations, several attempts have been undertaken for the quantitative isolation of fullerenes with less than 60 atoms (see, e.g., Kietzmann et al., 1998). However, there was no such success for C_{60}. One reason might be that many of the fullerenes with less than 60 atoms could be very reactive and elude the detection or isolation. Kroto and Walton (1993) argued that small fullerene cages might behave as pseudoatoms with certain reactivity patterns. This idea was further developed by Milani et al. (1996) and Fowler et al. (1999b), introducing valencies for fullerene cages. In this way, fullerene cages may polymerize by activating these valencies for intercage bonds and forming solid crystalline materials. These solids (polymers) can possibly not be extracted from the soot. The clarification as to what extent the soot of the fullerene synthesis consists of polymeric small fullerenes is still one of the open questions in fullerene science.

The synthesis, and especially the isolation of small fullerenes and their polymeric forms, is a challenging task; hence, despite the nearly 20 years of intensive experimental research, the amount of unambiguously isolated small fullerenes is still very limited. On the other hand, density-functional theory (DFT)-based

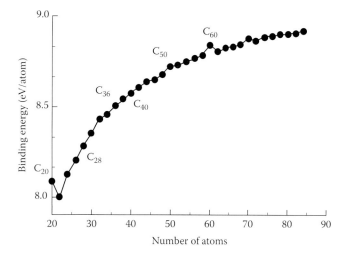

FIGURE 30.1 Calculated binding energies of the most stable fullerene cages from C_{20} to C_{80}.

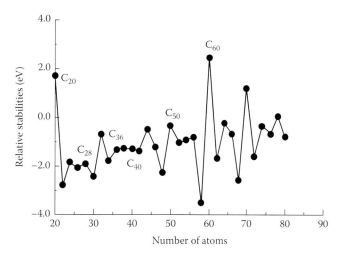

FIGURE 30.2 Calculated relative stabilities (δ) of fullerene cages from C_{20} to C_{80}. The relative stability (δ) of a fullerene C_n is defined as $\delta = E(n + 2) + E(n - 2) - 2E(n)$, where $E(n)$ is the calculated total energy of C_n (see also Kietzmann et al. 1998). (From Kietzmann, H. et al. *Phys. Rev. Lett.*, 81, 5378, 1998. With permission.)

calculations can give very accurate predictions of the geometry, and qualitatively reliable information of the energetics of fullerene structures. This holds even for the simplest approximation within the DFT, the local density approximation (LDA). For the consideration of the vast manifold of fullerene isomers, an approximate variant of the DFT, the density-functional-based tight-binding (DFTB) method (Porezag et al., 1995), has proved to be an efficient method. For the characterization of the cage structures, group theory is a useful tool. Despite the icosahedral symmetry (e.g., C_{60}—point group I_h), stable fullerene cages frequently have high symmetries belonging to the point groups T_h and T_d. In addition, several other symmetries can be found, as, e.g., D_{nh} symmetries.

This chapter summarizes the knowledge about polymeric structures and their properties of small fullerenes, especially from a theoretical point of view. Along the "fullerene road" a selection was made. The selection was focused on C_{20}, C_{28},

and C_{50} as typical examples of small fullerenes, whose ability of formation of condensed structures was intensively studied—especially theoretically—over the past 15 years.

30.2 Selected Examples

30.2.1 C_{20}

The C_{20} dodecahedral cage with an icosahedral symmetry consists only of condensed (adjacent) pentagons and correspondingly a very high curvature. Therefore, this enormously strained fullerene could only be detected spectroscopically, applying several synthetic "tricks" (Prinzbach et al., 2000). Other isomers, as the ring and bowl C_{20} structures, have been computed to be more stable than the cage (see, e.g., Jones and Seifert, 1997; Saito and Miyamoto, 2002). Some investigations even favored bicyclic rings (Yang et al., 1988; Vonhelden et al., 1993; Hunter et al., 1993a,b; Handschuh et al., 1995) and linear chains (Ott et al., 1998). However, the dodecahedral fullerene was confirmed to have the lowest energy among all the mathematically possible 20-vertex trivalent polyhedral cages (Domene et al., 1997). Due to the high curvature and the large strain, the C_{20} fullerene is very reactive. The strain in the cage can be reduced by a partial hydrogenation. Chen et al. (2004) discussed in detail how to identify the reactive centers of the C_{20} cage. The $C_{20}H_8$ (T_h) (see Figure 30.3) molecule is a derivative of C_{20} with a considerably reduced strain. The presence of eight isolated sp^3 carbons (C–H) can separate the dodecahedron cage into the six essentially ethylene-like C=C units in $C_{20}H_8$ (T_h). The smaller strain energy in $C_{20}H_8$ (T_h) can also be seen from the energy of the hydrogenation reaction:

$$C_{20} + 4H_2 \rightarrow C_{20}H_8(T_h)\Delta H \text{ (per } H_2)$$

$$= -67.0 \text{ kcal/mol (Chen et al., 2004)}.$$

The strain reduction by a partial $sp^2 \rightarrow sp^3$ transformation of the carbon atoms in the C_{20} cage suggests the possibility of the formation of C_{20} oligomers, as it was extensively discussed by Chen et al. (2004) for dimers, trimers, tetramers, and linear chains. Finally,

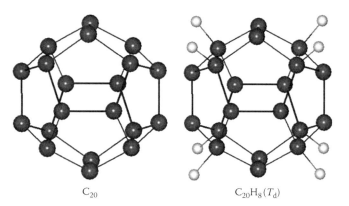

$$C_{20} \qquad\qquad\qquad C_{20}H_8\,(T_d)$$

FIGURE 30.3 Optimized structures of C_{20} and $C_{20}H_8$ according to DFT calculations. (From Chen, Z.F. et al., *Chem. Eur. J.*, 10, 963, 2004. With permission.)

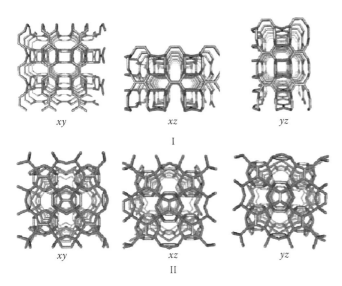

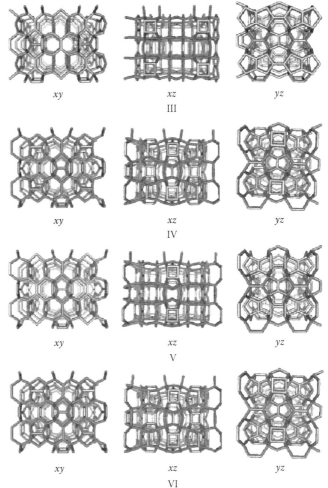

FIGURE 30.4 Optimized cubic structures of three-dimensional C_{20} solids. (From Chen, Z.F. et al., *Chem. Eur. J.*, 10, 963, 2004. With permission.)

three-dimensional solid-state topologies can be formed by aggregation of C_{20} units in this way. There exists a large variety of possible three-dimensional structures built from C_{20} monomers. A first group of related candidates was derived from the linear oligomer structures and was studied by Miyamoto and Saito (2001) and Okada et al. (2001). However, DFTB investigations (Chen et al., 2004) showed a partial instability of these structures. Only one simple cubic open structure (referred to as structure I—Figure 30.4) turned out as a stable solid state configuration. A second group is based on the extraordinarily stable $C_{20}H_8$ isomer, whose "$C_{20}H_8$-8H" cage is used as the building block. When these units are joined by transforming the C–H bonds into intercage C–C bonds, a body-centered cubic (bcc) structure can be formed (referred to as structure II—Figure 30.4).

This bcc structure can be further stabilized by the introduction of additional intercage bonds, as already suggested by Okada et al. (2001). Four such structures were further studied by Chen et al. (2004). All of them have binding energies between 8.6 and 8.7 eV/atom. These binding energies are considerably larger than that of the C_{20} monomer (8.0 eV/atom) and only slightly smaller than the binding energy of C_{60} (8.9 eV/atom) (Porezag et al., 1995). At least half of their carbons are tetracoordinate, and the calculated densities are approximately 2.8–2.9 g/cm³. These densities are nearly approaching the density of diamond.

Structure III is the most stable one (8.7 eV/atom). It contains both small planar sp^2 areas and cage-like sp^3 regions (see Figure 30.5); the tetracoordinate carbon content is relatively low (50%). Structure IV, which has been studied first by Okada et al. (2001), has so-called [2 + 2] bridges in the xz and yz layers (see Figure 30.5). Such bridges are characterized by two cage–cage C–C bonds, forming a four-membered ring. These bridges are arranged in an alternating way: for example, in the xz plane [2 + 2], bridges connect cages in the z direction in one layer. At the next layer, the [2 + 2] bridges connect cages in the x direction. The analogous arrangement is also found in the yz plane (Chen et al., 2004). The bond

FIGURE 30.5 Most stable structures of three-dimensional C_{20} solids (From Chen, Z.F. et al., *Chem. Eur. J.*, 10, 963, 2004. With permission.)

lengths, including those of intercage connections, are rather large and range from 1.34 Å for the sp^2–sp^2 bonds to 1.55 Å for the sp^3–sp^3 bonds.

Wang et al. (2000, 2001) reported a successful synthesis of caged fullerene C_{20} solids. The measured electron diffraction pattern is consistent with all the solid-state structures (II–VI), discussed by Chen et al. (2004), and the estimated intermolecular bond length of about 1.5 Å is also in agreement with the theoretical findings. However, no direct experimental x-ray structure determination of a C_{20} solid has been reported up to now.

Chen et al. (2004) also calculated the electronic density of states (DOS) of the solid-state C_{20} modifications. Except from the metallic behavior of structure I, all modifications (II–VI) should be semiconductors with gaps in the range between ~2 eV (III–VI) and 3.5 eV (II).

30.2.2 C_{28}

One of the smallest fullerenes observed experimentally (Cox et al., 1988) is C_{28}. Kroto (1987) proposed a structure with tetrahedral symmetry T_d. This tetrahedral cage consists of four hexagons,

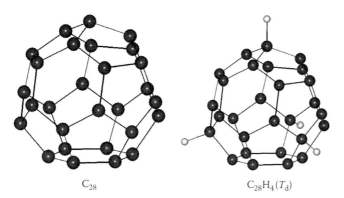

C_{28} $C_{28}H_4 (T_d)$

FIGURE 30.6 Structures of C_{28} and $C_{28}H_4$ (T_d).

which are symmetrically linked by four further atoms. Each of these four atoms is the center of three fused pentagons (see Figure 30.6). As already mentioned by Kroto (1987), these four atoms may be related to the central carbon atom in the triphenylmethyl radical ($\cdot C(C_6H_5)_3$). The small size with the correspondingly high strain and four unpaired electrons in each set of four triplets of pentagons cause the high reactivity of the C_{28} cage (Kroto and Walton, 1993). This fact was confirmed by several theoretical studies (Chen and Jiao, 2001; Makurin et al., 2001).

Numerous theoretical investigations show that the C_{28} skeleton can be stabilized by forming endohedral fullerenes $M@C_{28}$, where the metal atoms (M) are formally in the M^{4+} configuration (e.g., d-elements Ti, Zr, Mo, f-elements U, Pu, Ce, p-elements Si, Ge) (Dunlap et al., 1992; Pederson and Laouini, 1993; Jackson et al., 1993, 1994; Makurin et al., 2001). Such stabilization is explained by the formation of an electronic system with all paired electrons (closed electronic shell). On the other hand, it was predicted that the saturation of free valences of the C_{28} molecule radical can be realized through their linking with hydrogens, halogen atoms, alkyl groups, or through their joining to form polymers, films, or crystals (Choho et al., 1997, Canning et al., 1997, Makurin et al., 2001).

The four unpaired electrons localized on the carbon atoms common to each set of four triplets of pentagons make the C_{28}

fullerene similar to an sp^3-hybridized carbon atom. Therefore, a diamond-like crystalline phase of C_{28} might exist. Calculations showed (Kaxiras et al., 1994) that hyperdiamond C_{28} is a semiconductor with a bandgap of 1.5 eV; its lattice parameter and the elastic modulus were calculated by Bylander and Kleinman (1993). The large interest arises from possible superconducting properties of alkali metal-doped phases M_xC_{28}, whose critical temperature is predicted to be approximately eight times higher than that of fullerides M_xC_{60} (Breda et al., 2000).

Seifert et al. (2005) studied the structural, mechanical, and electronic characteristics of hyperdiamond C_{28}, but also hyperlonsdaleit C_{28} phases (see Figure 30.7). Hyperlonsdaleit is derived from the hexagonal modification of diamond, lonsdaleite.

The bond lengths between the C_{28} cages in hyperdiamond and in hyperlonsdaleit are in accord with their expected hybridization (~0.155 nm). Thus, it confirms the presence of covalent interactions between C_{28} fullerenes in crystalline state that distinguishes these crystals from "classical" fullerites (e.g., C_{60}), where molecules are linked via weak van-der-Waals forces.

The comparison of differences between the total energies from DFTB calculations of free fullerene C_{28} and of its crystalline modifications shows that the formation of crystals is favored for both modifications. For hyperdiamond this difference is −6.13 eV/C_{28}, and for hyperlonsdaleit it is −6.09 eV/C_{28}. It is important that the values of the energy of association for these structures are very close to each other. The binding energy is slightly smaller than that of the fcc-C_{60} and clearly smaller than the binding energy in diamond (see Table 30.1).

Table 30.1 shows that the density of hyperdiamond and hyperlonsdaleit are three times smaller than that of diamond. The bulk moduli of hyperdiamond and hyperlonsdaleit are about one order smaller than the bulk modulus of diamond. Notwithstanding the presence of sp^3-hybridized carbon atoms in the lattices of hyperdiamond and hyperlonsdaleit, their hardness is smaller than expected. It can be explained by considering the repelling of π clouds of sp^2-hybridized carbon atoms in the neighboring C_{28} cages. For example, the distance between the

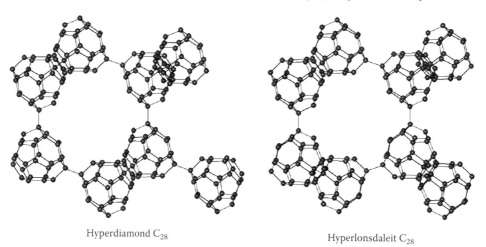

Hyperdiamond C_{28} Hyperlonsdaleit C_{28}

FIGURE 30.7 Fragments of the optimized hyperdiamond and hyperlonsdaleit crystal lattices of the condensed fullerene C_{28}. (From Seifert, G. et al., *Phys. Rev. B Condens. Matter Mater. Phys.*, 72, 012102-4, 2005. With permission.)

TABLE 30.1 Lattice Constants (a, c), Density (ρ), Bulk Modulus (B), Bandgap (E_g), and Binding Energies in the Free and Solid Forms (E_b^f, E_b^s) of the C_{28} Crystalline Modifications in Comparison with Other Carbon Phases According to DFTB Calculations

Crystal	a (nm)	c (nm)	ρ (g/cm³)	B (GPa)	E_g (eV)	E_b^f (eV/atom)	E_b^s (eV/atom)
Diamond	0.3576	—	3.49	478.20	7.48	—	9.22
bcc-Fullerite C_{20}	0.6897	—	2.43	201.87	3.51	8.01	8.07
hyperdiamond C_{28}	1.5953	—	1.10	39.26	2.04	8.29	8.50
Hyperlonsdaleite C_{28}	1.1277	1.8415	1.10	41.20	1.99	8.29	8.50
fcc-Fullerite C_{60}	≈1.41	—	≈1.7	< 0.01	1.66	8.85	8.85

Source: Seifert, G. et al., *Phys. Rev. B Condens. Matter Mater. Phys.*, 72, 012102-4, 2005. With permission.

nearest carbon atoms of hexagons in the neighboring C_{28} cages of hyperlonsdaleit is about 0.292 nm, which is smaller than the van-der-Waals gap of graphite (0.334 nm).

Both crystalline phases of C_{28} are semiconductors with a direct bandgap of about 2 eV (Table 30.1). Band structures and electronic DOS profiles of crystalline C_{28} were calculated by several authors. The calculated band structures of the crystalline modifications of C_{28} from DFTB calculations (Seifert et al., 2005) are shown in Figure 30.8. DFT calculations with a plane-wave basis set give very similar results (Enyashin et al., 2006). The quasi-core C2s states with an admixture of the C2p states form a set of bands in the range from −16 to −13 eV below the Fermi level—not shown in here. The occupied band in the range from −11 to −2 eV consists of the C2s and C2p states responsible for σ and π bonds in the C_{28} fullerenes. The bands at −1 and −2 eV below Fermi level are associated with π bonds of hexagons in the skeleton of C_{28}.

In spite of the numerous theoretical investigations suggesting the stability of solid C_{28}, there was up to now no report on a successful synthesis of any of above-mentioned crystalline phases. However, it is believed that such crystals could be observed by chemical deposition of active C_{28} molecules on a respective surface. The simulations of C_{28} adsorption on the surfaces of semiconductors like GaAs (001) and diamond (001) or (111) were performed by several authors (Canning et al., 1997; Zhu et al., 2000; Yao S. et al., 2005). A study (Zhu et al., 2000) with an empirical force field (Brenner potential) of low-energy C_{28} deposition on diamond (001) surface has shown that the incident C_{28} clusters can be chemisorbed without breaking its cage structure and inducing surface defects. Further C_{28} clusters may form bonds with each other. However, the chemisorbed clusters are randomly stacked to grow a film in a three-dimensional island growth mode with an amorphous film. On the other hand,

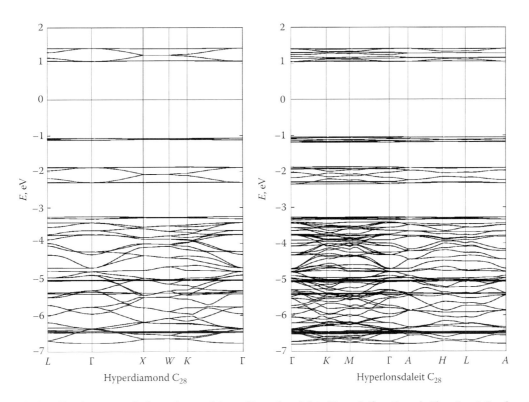

FIGURE 30.8 Calculated band structure for hyperdiamond C_{28} and hyperlonsdaleit. (From Seifert, G. et al., *Phys. Rev. B Condens. Matter Mater. Phys.*, 72, 012102-4, 2005. With permission.)

the results from quantum molecular dynamics simulations by Canning et al. show that under certain deposition conditions on diamond (111) surface, C_{28} act as building blocks on a nanometer scale to form a thin film of nearly defect-free molecules and behave as carbon superatoms, with the majority of them being threefold or fourfold coordinated, similar to carbon atoms in amorphous systems. The microscopic structure of the deposited film supports suggestions about the stability of the hyperdiamond solid.

The low mass density and large internal surface of C_{28} fullerites suggest their possible applications as catalyst, nanosieve, and gas storage material. Possessing the large cavities such fullerites are attractive as the matrix for molecular absorption. Enyashin et al. (2006) have estimated the active volume accessible by molecular H_2 physisorption in hyperdiamond and hyperlonsdaleit. Both structures have tubular areas where the interaction potential is attractive for light molecules, as, e.g., H_2. For hyperlonsdaleit these areas are more pronounced. The potential is only attractive in the cavities between the fullerene cages and reaches well depths of about 9.5 and 13.1 kJ/mol, respectively, for hyperdiamond and for hyperlonsdaleit. These values are comparable with those of graphene bilayers or C_{60} intercalated graphite (Kuc et al., 2007).

Also the exohedral intercalation of hyperdiamond with K was investigated (Enyashin et al., 2006). For an intercalation up to the KC_{28} composition, the lattice constant of the compound does not change, though it leads to the change of the electronic structure of the fullerite. The additional electrons from K, two per unit cell, populate the lowest states of the conduction band in pure hyperdiamond. Thus, the compound becomes metallic. The cavities in C_{28} hyperdiamond have a diameter of about 0.46 nm, which is larger than the radius of C_{28} cages (~0.268 nm) constituting the C_{28} lattice, i.e., a self-intercalation of the C_{28} hyperdiamond— filling of the lattice voids with C_{28} fullerenes—might be possible. Based on this idea, a new three-dimensional crystalline C_{28} fullerene phase—self-intercalated hyperdiamond—was suggested (Enyashin and Ivanovskii, 2007). When the cavities of the initial C_{28} hyperdiamond lattice are completely filled, the intercalated fullerenes are arranged so that they form their "own" diamond-like lattice (Figure 30.9). Therefore, the structure of the aforementioned self-intercalated hyperdiamond can be viewed as two C_{28} hyperdiamond lattices inserted into each other. Since the distance between fullerene walls of these two sublattices is roughly equal to the van-der-Waals gap in graphite (~0.3 nm), such structure is the first example of a fullerite with mixed types of interaction between its constituents—weak and covalent ones. The weak interaction between two sublattices was supported by DFTB calculations. While the density of self-intercalated hyperdiamond is two times and the bulk modulus ~2.5 times larger than for hyperdiamond, the electronic properties do not change essentially.

30.2.3 C_{36}

Going further along the "fullerene road," passing the "magic" C_{32}, which could be prepared in small amounts (Kietzmann et al., 1998), one arrives at the "semimagic" C_{36}. Piskoti et al. (1998)

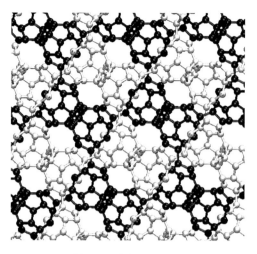

FIGURE 30.9 Crystal lattice of the autointercalated hyperdiamond C_{28}. The two sublattices are painted in black and gray. (From Enyashin, A.N. and Ivanovskii, A.L., *JETP Lett.*, 86, 537, 2007. With permission.)

interpreted electron diffraction and solid-state NMR data of a material they had obtained in a modified Krätschmer–Huffman experiment (Kratschmer et al., 1990) as a proof for a fullerenic solid consisting of highly symmetric (D_{6h}) C_{36} fullerene cages. This solid material could be sublimated and solved in organic solvents, which are similar but not as good as C_{60}. Furthermore, it could also be doped with alkaline metals. In a further paper, results from scanning tunneling spectroscopy (STS) investigations of the material on gold and specific graphite substrates already indicated covalently bonded C_{36} units (Collins et al., 1999b).

Fifteen "classical" fullerene cages can be constructed for C_{36} (Fowler and Manolopoulos, 1996). These isomers can be labeled by the number of atoms n and the place in the lexicographical order of a spiral code (spiral notation m), corresponding to a contiguous helical numbering pathway in the fullerene cage. Two of them have the minimal count of pentagon adjacencies: the isomers with D_{2d} and D_{6h} symmetry, No. 36:14 (Figure 30.10)

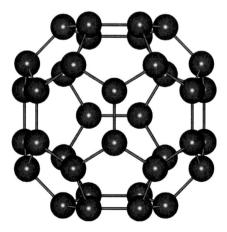

FIGURE 30.10 Isomer No. 14 of C_{36} (D_{2d} symmetry), according to the spiral notation. (From Fowler, P.W. and Manolopoulos, D.E., *An Atlas of Fullerenes*, Clarendon Press, Oxford, U.K., 1996.)

and No.36:15 (Figure 30.11), according to the spiral notation (No. *n:m*) (Fowler and Manolopoulos, 1996). Hückel calculations give a significant nonzero HOMO–LUMO gap (HOMO—Highest Occupied Molecular Orbital, LUMO—Lowest Unoccupied Molecular Orbital) for isomer 36:14, but not for isomer 36:15 (Fowler and Manolopoulos, 1996; Heath, 1998). Semiempirical calculations (Campbell et al., 1996; Slanina et al., 1998) give, in agreement with rule of the minimal number of adjacent pentagons (Albertazzi et al., 1999), the D_{2d} isomer (36:14) as the most stable isomer of C_{36}. DFT-based calculations (Grossman et al., 1998; Fowler et al., 1999a) predict both isomers as practically isoenergetic. The isomer 36:15 is stabilized by a second-order Jahn–Teller effect. Low-lying orbitals get close to the HOMO point at a possible stabilization by adding up to six electrons. This is consistent with the experimental $C_{36}H_6$ signal in the mass spectrum of the soot and with the finding of a possible doping of C_{36} by alkaline metals (Piskoti et al., 1998). This is an indication for a hexavalency of C_{36}. Mathematically, 82,123 possible isomers of $C_{36}H_6$ can be constructed, based on the isomer 36:15. All these isomers were constructed and optimized with the DFTB method (Fowler et al., 1999a). The most stable isomer has D_{3h} symmetry and a considerable HOMO–LUMO gap (2.3 eV). The hydrogen atoms occupy the three non-neighboring (1,4) positions of the equatorial hexagons in the C_{36} cage (Figure 30.12). The hydrogen-saturated carbon atoms of $C_{36}H_6$ exhibit sp^3 character. They fulfill the steric conditions to act for the formation of intercage bonds to build dimers, trimers, oligomers, and solid state structures out of C_{36}. The stable structures obtained using the (1,4) positions of the equatorial hexagons in the C_{36} cage for building dimers, trimers, and tetramers are shown in Figure 30.13. The calculated binding energy (per monomer) of the D_{2d} dimer was calculated to 152 kJ/mol by DFTB (Fowler et al., 1999a, Heine et al., 1999) and 119 kJ/mol and DFT–LDA (Collins et al., 1999b) investigations, respectively. As shown by Collins et al. (1999b), dimers linked by single bonds are less bound and they are radicals, as expected from simple valence considerations. Going along the sequence dimer → trimer → tetramer the binding energy per monomer is increasing from 152 to 239

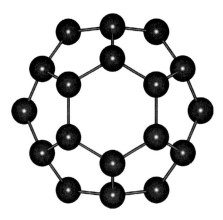

FIGURE 30.11 Isomer No. 15 of C_{36} (D_{6h} symmetry), according to the spiral notation. (From Fowler, P.W. and Manolopoulos, D.E., *An Atlas of Fullerenes*, Clarendon Press, Oxford, U.K., 1996.)

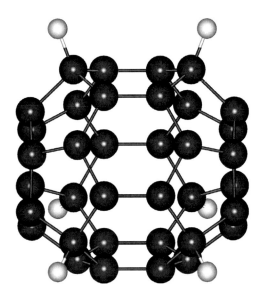

FIGURE 30.12 The most stable isomer of $C_{36}H_6$, based on the isomer 36:15 of the C_{36} cage. (From Fowler, P.W. et al., *Chem. Phys. Lett.*, 300, 369, 1999a. With permission.)

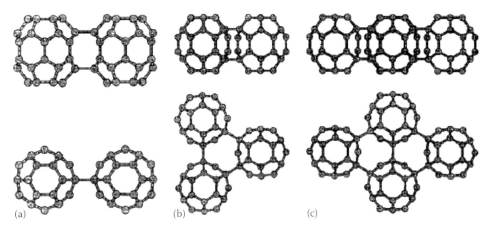

FIGURE 30.13 C_{36} oligomers (side and top view), dimer (a), trimer (b), and tetramer (c). (From Heine, T. et al., *Solid State Commun.*, 111, 19, 1999. With permission.)

to 271 kJ/mol. However, the binding energy per bridge decreases from 304 to 240 to 216 kJ/mol, suggesting that successive bridges are less effective. At the same time the C–C bonds in the bridges lengthen slightly by 0.2 Å from dimer to trimer and from the trimer to the central bridge of the tetramer.

The DFTB calculations suggest also a growing gap size from its small monomer value of 0.6 eV (Fowler et al., 1999a) through 0.75 eV for the dimer to an apparent limit of ~1 eV for the trimer and the tetramer (Heine et al., 1999). Qualitatively similar gaps can be deduced from DFT–LDA calculations (Collins et al., 1999b).

Further extension of a planar triangulated framework of monomers appears to lead to unstable configurations. In a pentamer, constructed by capping one edge of the tetramer, the bridged edge lengthens and then opens; in heptamers outer edges open in a similar fashion. This behavior is perhaps only to be expected, as a fully triangulated plane layer could not be realized with the short bridges responsible for the high binding energy of the dimer. These structural islands would therefore appear to be dead ends for the growths of a two-dimensional layer.

A two-dimensional structure for $(C_{36})_{\infty}$ can be envisaged, but it is based on the less densely packed graphitic structure. This structure can be derived from a C_{36} hexamer, a "superbenzene" $((C_{36}H_2)_6 -$ see Figure 30.14), based on C_{36}, as it was proposed in Fowler et al. (1999a).

In this "supergraphitic" layer (Fowler et al., 1999a), each monomer is linked by 1,4 bridges at 120° to three neighbors, with a bond length of 0.162 nm (see Figure 30.15), according to DFTB supercell calculations with eight monomers per cell (Fowler et al., 1999a, Heine et al., 1999).

In contrast to graphite, this C_{36}-based layered structure would be a semiconductor with a gap of about 2 eV, similar to the HOMO–LUMO gap of the "superbenzene" (Fowler et al., 1999a). "Supergraphite" has a binding energy of 375 kJ/mol per

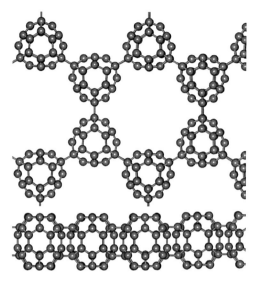

FIGURE 30.15 Top and side view of the elementary cell of C_{36} supergraphite. (From Heine, T. et al., *Solid State Commun.*, 111, 19, 1999. With permission.)

C_{36}, which is considerably higher than the binding energies of the C_{36} oligomers. Van-der-Waals interactions between the layers would presumably further enhance the stability of supergraphite. However, the hexavalency of C_{36} opens also the way for three-dimensional structures. One possible three-dimensional structure was first proposed by Cote et al. (1998) (see Figure 30.16). This rhombohedral closely packed lattice with a binding energy $E_b = 355$ kJ/mol/C_{36} (Fowler et al., 1999a) is less stable than the supergraphite structure and it is metallic. Further theoretical investigations (Heine et al., 1999) predicted more stable three-dimensional structures. A further, much more stable $(E_b = 522$ kJ/mol/C_{36}) (Fowler et al., 1999a)) three-dimensional hexagonal structure has first been called (Fowler et al., 1999a) an "hcp" (hexagonal close packed) lattice (see Figure 30.17). In fact, the term "hcp" is something of a misnomer. This lattice belongs to the space group, which is also the group of hexagonal boron nitride and the offretite family of zeolites (Gard and Tait, 1972). In this structure, each C_{36} monomer has D_{3h} site symmetry and is linked by single bonds (0.153 nm) to each of six neighbors, three above and three below, and by longer contacts (0.163 nm) to two axial neighbors. The lattice can be viewed as notionally derived by removal of columns of monomers from a closely packed structure, to leave channels of stacked puckered rings parallel to the c-axis (see Figure 30.17). This covalent zeolite-like structure is ideally compatible with the "hexavalency" of the monomer. The binding energy per monomer was calculated as 569 kJ/mol (Heine et al., 1999). Among the hypothetical structures considered in the literature for crystalline C_{36}, this hexagonal form is not only the most stable one, it also clearly gives the best match to the experimental data. The center-to-center distance projected onto the $\vec{a}$, $\vec{b}$-plane of the hexagonal lattice, calculated for this structure as 0.674 nm, in accord with the d-spacing of 0.668 nm derived from the transmission electron microscopy diffraction pattern (Piskoti et al., 1998).

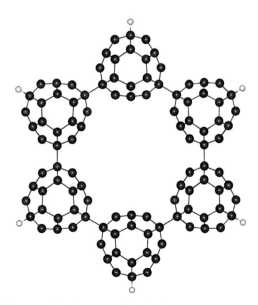

FIGURE 30.14 Superbenzene $(C_{36}H_2)_6$. (From Heine, T. et al., *Solid State Commun.*, 111, 19, 1999. With permission.)

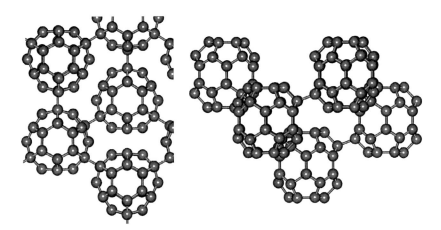

FIGURE 30.16 Elementary cell of the rhombohedral modification of solid C_{36}. (From Fowler, P.W. et al., *Chem. Phys. Lett.*, 300, 369, 1999a. With permission.)

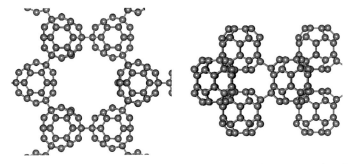

FIGURE 30.17 Elementary cell of the hexagonal structure (hcp) of C_{36}. (From Fowler, P.W. et al., *Chem. Phys. Lett.*, 300, 369, 1999a. With permission.)

Also the calculated gap (0.89 eV) (Fowler et al., 1999a) agrees with gap measured with STS (~0.8 eV) (Collins et al., 1999b). While the supergraphite layer has a computed gap of 2.09 eV, the rhombohedral solid proposed by Cote et al. (1998) is metallic, as mentioned above. DOS profiles computed with the DFTB method for the solid-state structures of C_{36} are shown in Figure 30.18.

For the rhombohedral structure, a broadly similar profile was given by Cote et al. (1998).

30.2.4 C_{50}

Along the "fullerene road" (see Fowler and Manolopoulos, 1996), besides C_{44}, the C_{50} cage should be rather stable. Xie et al. (2004) reported about the synthesis of $C_{50}Cl_{10}$. This structure has been unambiguously characterized by measured and simulated ^{13}C NMR. The electronic properties of this fullerene have been studied recently (Lu et al., 2004). For C_{50}, 271 possible isomers of classical fullerenes can be constructed (Fowler and Manolopoulos, 1996). Zhechkov et al. (2004) studied the structure and the energetics of all the 271 classical fullerenes of C_{50}. They found isomer No. 271 in the spiral nomenclature as the most stable one. It has a D_{5h} symmetry and possesses the lowest number of pentagon–pentagon adjacencies (see Figure 30.19), which makes its high stability plausible. The chlorinated C_{50} cage has also D_{5h} symmetry (Xie et al., 2004), i.e., the D_{5h} C_{50} cage has

obviously a preferred valency of 10. This can simply be explained by the structure of the D_{5h} C_{50} cage. This cage can be viewed as two corannulene frameworks, glued together by five "ethylenic" C_2 units along the equatorial belt of the cage (see Figure 30.20). Such configuration is highly strained along the contact between the belt and the corannulene frameworks. Bonding of ten atoms (e.g., Cl) to the five C_2 units can stabilize the cage in two ways. First, they decrease the strain energy of the cage by pyramidalization of the equatorial carbon atoms by sp^3 hybridization, and second they remove the carbon atoms of adjacent pentagons out of the π system, and also provide in this way an electronic stabilization. The optimized structure of $C_{50}H_{10}$ is shown in Figure 30.21. Similar to C_{36}, the valences of C_{50} could also act as intermolecular bonds, and in this way provide the possibility for the formation of C_{50}-based fullerides.

There is one topological possibility for a single-bonded dimer of C_{50} (Figure 30.22). This dimer is expected to be instable, as two electrons are taken from the two π systems, resulting in either a biradical or a zwitterion. Among several possibilities for the connection of two C_{50} cages, Zhechkov et al. (2004) found a most stable double-bonded dimer, as shown in Figure 30.23. The calculated dimerization energy ($E_{dimer} = -486$ kJ/mol) for the most stable dimer is about twice as high as for $(C_{36})_2$ and even ~16 times higher than for the C_{60} dimer. Consequently, stable oligomers are expected. Zhechkov et al. (2004) investigated the possibilities for stable oligomer structures. As the number of

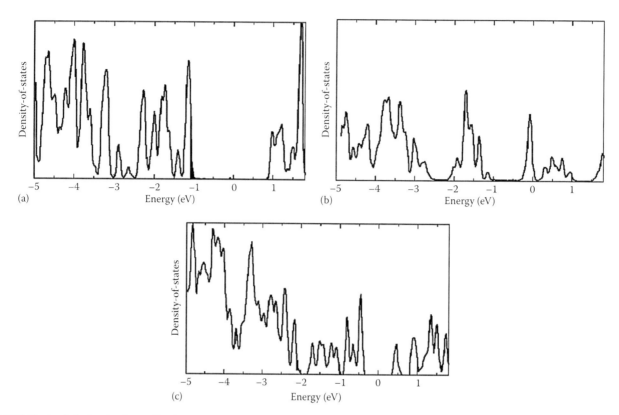

FIGURE 30.18　Calculated DOS profiles for supergraphite (a), rhombohedral (b), and hexagonal (c) structures of solid C_{36} (from top to below). The energy, E, in the DOS curves is given relative to the Fermi energy. (From Heine, T. et al., *Solid State Commun.*, 111, 19, 1999. With permission.)

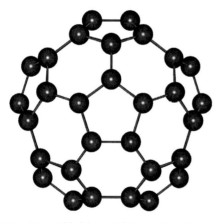

FIGURE 30.19　Most stable "classical fullerene" C_{50}. (From Zhechkov, L. et al., *J. Phys. Chem. A*, 108, 11733, 2004. With permission.)

FIGURE 30.20　C_{50} decomposed into corannulene and ethylene units (Zhechkov et al., 2004). For clarity only the forefront of the structure is shown. That is, from the five ethylene units two can be seen completely, one partially, the other two on the back are not shown. (From Zhechkov, L. et al., *J. Phys. Chem. A*, 108, 11733, 2004. With permission.)

possible oligomer structures is exceedingly large, they restricted their studies to the energetically most preferable connectivity patterns of dimer structures.

As for the dimer, the bare trimer with single bonds has a small HOMO–LUMO gap and hence is of either radicaloid or zwitterionic character. On the other hand, the hydrogen-saturated trimer has a large gap. Structure and electronic structure of each cage unit of the saturated single-bonded trimer do not differ considerably from $C_{50}H_{10}$, and this bonding allows polymerization in an infinite variety of structures, most likely toward amorphous

polymers. Oligomers up to the hexamer with dimer bridges, which are responsible for most fullerene solids so far observed and discussed (Menon and Richter, 1999; Collins et al., 1999a; Menon et al., 2000) were studied more in detail (Zhechkov et al., 2004).

The energetically favored dimer bridge type has been found in the two most stable trimers. This trend continues with tetramers, pentamers, and hexamers, and a lowest-energy pathway from monomer to hexamer is found (see Figure 30.24). Besides being the most stable oligomers, these structures have also large HOMO–LUMO gaps and the tendency to prefer clustered

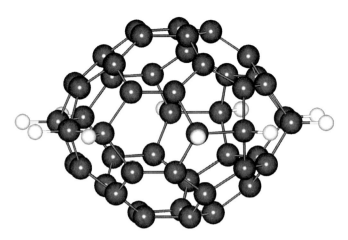

FIGURE 30.21 $C_{50}H_{10}$ based on C_{50}. (From Zhechkov, L. et al., *J. Phys. Chem. A*, 108, 11733, 2004. With permission.)

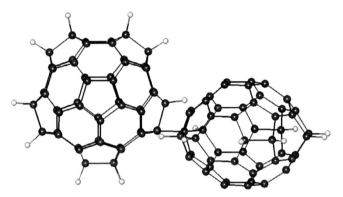

FIGURE 30.22 Single-bonded C_{50} dimer. (From Zhechkov, L. et al., *J. Phys. Chem. A*, 108, 11733, 2004. With permission.)

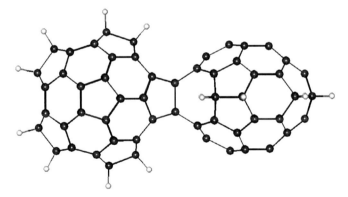

FIGURE 30.23 Most stable double-bonded C_{50} dimer. (From Zhechkov, L. et al., *J. Phys. Chem. A*, 108, 11733, 2004. With permission.)

structures. The fivefold symmetry of the monomer and the minimum energy pathway to the most stable hexamer structure suggest that it is impossible to form regular covalently bound C_{50} solids if only the chemically active connectivity sites along the σ_h plane of the molecule are used. The non-covalent interactions between the corannulene units are slightly attractive (6 kJ/mol from DFT-based calculations; Zhechkov et al. (2004). Covalent

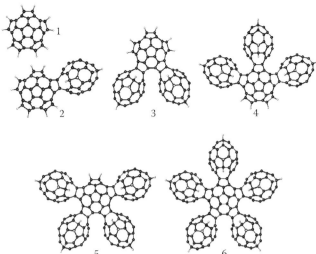

FIGURE 30.24 Minimal energy pathway from the monomer $C_{50}H_{10}$ (1), over the dimer (2), trimer (3), tetramer (4), pentamer (5) to the hexamer $(C_{50})_6H_{30}$ (6). (From Zhechkov, L. et al., *J. Phys. Chem. A*, 108, 11733, 2004. With permission.)

bonds between a C_2 unit of the equatorial belt and two atoms of a corannulene unit are also possible. Hence, a much larger variety of bonding patterns is possible, but thermodynamically the formation of these bonds is less favored, as it does not saturate the highly reactive sites along the equatorial belt. This reduces the possibility of forming spontaneous, high-symmetry solids. Oligomerization and polymerization are energetically possible with single- and double-bridge links. While single-bonded links will lead to flexible intercage bonding, allowing a great manifold of possible structures, double-bonded oligomers have a lowest-energy pathway from the monomer over dimer to hexamer.

30.3 Trends, Comparison, and Conclusions

In contrast to most of the fullerites based on C_{60} and higher fullerenes, those composed by small, non-IPR cages can be characterized by strong covalent interactions between the fullerene cages. The highly strained small fullerenes can be stabilized by the formation of strong intercage bonds. This bond formation is connected with the transformation of sp^2 carbons to sp^3 carbons for the atoms, which form the intercage bridges. The sp^3/sp^2 ratio in these fullerite structures decreases with cage size, and so does the cohesive energy and related quantities as elastic constants and also the mass density.

The most strained and reactive fullerene C_{20} builds the crystal with carbon atoms whose hybridization is close to sp^3. Such crystal has the highest density and hardness among other fullerene crystals. The bulk moduli of hyperdiamond C_{28}, hyperlonsdaleit C_{28}, and supergraphite C_{36} are much smaller than that of diamond. For a solid form of C_{36} cages the bulk modulus is even smaller.

Oligomerization and polymerization give additional intercage bonds, and hence additional binding energy, but also topological constrains, which lead to unfavorable deformations of the monomer units that affect both the π system by stronger pyramidalization and the σ framework. The reaction energy of a further step in polymerization depends mainly on the type of additional intercage bond if this step does not involve strong cage deformations. For this reason, the bonding types found in dimers, oligomers, and solids are usually the same.

For the total stability, the existence of a stable phase has to be examined by evaluating the trade-off of cohesive intercage energy (high for small, low for large cages) with the stability of the bare cage (low for small, higher for large fullerenes).

It is not clear if fullerites can be formed directly from the soot, as the high reactivity of small cages limits their lifetime to the microsecond scale. Experimental material is scarce, but there is a unique principle for covalent intercage bonds that can stabilize the highly strained small fullerenes. Therefore, the authors feel confident that there are a number of "hidden" structures or materials whose uncovering will add to the manifold of carbon-based structures considerably.

References

Albertazzi, E., Domene, C., Fowler, P. W., Heine, T., Seifert, G., Van Alsenoy, C., and Zerbetto, F. (1999) *Physical Chemistry Chemical Physics*, 1, 2913–2918.

Breda, N., Broglia, R. A., Colò, G., Onida, G., Provasi, D., and Vigezzi, E. (2000) *Physical Review B*, 62, 130 LP–133.

Bylander, D. M. and Kleinman, L. (1993) *Physical Review B*, 47, 10967 LP–10969.

Campbell, E. E. B., Fowler, P. W., Mitchell, D., and Zerbetto, F. (1996) *Chemical Physics Letters*, 250, 544–548.

Canning, A., Galli, G., and Kim, J. (1997) *Physical Review Letters*, 78, 4442 LP–4445.

Chen, Z., Jiao, H., Bühl., M., Hirsch, A., and Thiel, W. (2001) *Theoretical Chemistry Accounts*, 106, 352–363.

Chen, Z. F., Heine, T., Jiao, H. J., Hirsch, A., Thiel, W., and Schleyer, P. V. R. (2004) *Chemistry-A European Journal*, 10, 963–970.

Choho, K., Van de Woude, G., Van Lier, G., and Geerlings, P. (1997) *Journal of Molecular Structure: THEOCHEM*, 417, 265–276.

Collins, P. G., Grossman, J. C., Cote, M., Ishigami, M., Piskoti, C., Louie, S. G., Cohen, M. L., and Zettl, A. (1999a) *Physical Review Letters*, 82, 165–168.

Collins, P. G., Grossman, J. C., Côté, M., Ishigami, M., Piskoti, C., Louie, S. G., Cohen, M. L., and Zettl, A. (1999b) *Physical Review Letters*, 82, 165–168.

Cote, M., Grossman, J. C., Cohen, M. L., and Louie, S. G. (1998) *Physical Review Letters*, 81, 697–700.

Cox, D. M., Reichmann, K. C., and Kaldor, A. (1988) *Journal of Chemical Physics*, 88, 1588–1597.

Domene, M. C., Fowler, P. W., Mitchell, D., Seifert, G., and Zerbetto, F. (1997) *Journal of Physical Chemistry A*, 101, 8339–8344.

Dunlap, B. I., Haeberlen, O. D., and Roesch, N. (1992) *J. Phys. Chem.*, 96, 9095–9097.

Enyashin, A., Gemming, S., Heine, T., Seifert, G., and Zhechkov, L. (2006) *Physical Chemistry Chemical Physics*, 8, 3320–3325.

Enyashin, A. N. and Ivanovskii, A. L. (2007) *JETP Letters*, 86, 537–542.

Fowler, P. W. and Manolopoulos, D. E. (1996) *An Atlas of Fullerenes*, Clarendon Press, Oxford, U.K.

Fowler, P. W., Heine, T., Rogers, K. M., Sandall, J. P. B., Seifert, G., and Zerbetto, F. (1999a) *Chemical Physics Letters*, 300, 369–378.

Fowler, P. W., Heine, T., and Troisi, A. (1999b) *Chemical Physics Letters*, 312, 77–84.

Gard, J. A. and Tait, J. M. (1972) *Acta Crystallographica Section B*, 28, 825–834.

Grossman, J. C., Cote, M., Louie, S. G., and Cohen, M. L. (1998) *Chemical Physics Letters*, 284, 344–349.

Handschuh, H., Gantefor, G., Kessler, B., Bechthold, P. S., and Eberhardt, W. (1995) *Physical Review Letters*, 74, 1095–1098.

Heath, J. R. (1998) *Nature*, 393, 730–731.

Heine, T., Fowler, P. W., and Seifert, G. (1999) *Solid State Communications*, 111, 19–22.

Hunter, J., Fye, J., and Jarrold, M. F. (1993a) *Science*, 260, 784–786.

Hunter, J., Fye, J., and Jarrold, M. F. (1993b) *Journal of Physical Chemistry*, 97, 3460–3462.

Jackson, K., Kaxiras, E., and Pederson, M. R. (1993) *Physical Review B*, 48, 17556 LP–17561.

Jackson, K., Kaxiras, E., and Pederson, M. R. (1994) *Journal of Physical Chemistry*, 98, 7805–7810.

Jones, R. O. and Seifert, G. (1997) *Physical Review Letters*, 79, 443–446.

Kaxiras, E., Zeger, L. M., Antonelli, A., and Juan, Y.-M. (1994) *Physical Review B*, 49, 8446 LP–8453.

Kietzmann, H., Rochow, R., Gantefor, G., Eberhardt, W., Vietze, K., Seifert, G., and Fowler, P. W. (1998) *Physical Review Letters*, 81, 5378–5381.

Kratschmer, W., Lamb, L. D., Fostiropoulos, K., and Huffman, D. R. (1990) 347, 354–358.

Kroto, H. (1987) *Nature*, 329, 529.

Kroto, H. and Walton, D. R. M. (1993) *Chemical Physics Letters*, 214, 353–356.

Kuc, A., Zhechkov, L., Patchkovskii, S., Seifert, G., and Heine, T. (2007) *Nano Letters*, 7, 1–5.

Lu, X., Chen, Z. F., Thiel, W., Schleyer, P. V., Huang, R. B., and Zheng, L. S. (2004) *Journal of American Chemical Society*, 126, 14871–14878.

Makurin, Y. N., Sofronov, A. A., Gusev, A. I., and Ivanovsky, A. L. (2001) *Chemical Physics*, 270, 293–308.

Menon, M. and Richter, E. (1999) *Physical Review B*, 60, 13322–13324.

Menon, M., Richter, E., and Chernozatonskii, L. (2000) *Physical Review B*, 62, 15420–15423.

Milani, C., Giambelli, C., Roman, H. E., Alasia, F., Benedek, G., Broglia, R. A., Sanguinetti, S., and Yabana, K. (1996) *Chemical Physics Letters*, 258, 554–558.

Miyamoto, Y. and Saito, M. (2001) *Physical Review B*, 63, art. no.-161401.

Okada, S., Miyamoto, Y., and Saito, M. (2001) *Physical Review B*, 64, 245405.

Ott, A. K., Rechtsteiner, G. A., Felix, C., Hampe, O., Jarrold, M. F., Van Duyne, R. P., and Raghavachari, K. (1998) *Journal of Chemical Physics*, 109, 9652–9655.

Pederson, M. R. and Laouini, N. (1993) *Physical Review B*, 48, 2733 LP–2737.

Piskoti, C., Yarger, J., and Zettl, A. (1998) *Nature*, 393, 771–774.

Porezag, D., Frauenheim, T., Kohler, T., Seifert, G., and Kaschner, R. (1995) *Physical Review B*, 51, 12947–12957.

Prinzbach, H., Weller, A., Landenberger, P., Wahl, F., Worth, J., Scott, L. T., Gelmont, M., Olevano, D., and von Issendorff, B. (2000) *Nature*, 407, 60–63.

Saito, M. and Miyamoto, Y. (2002) *Physical Review B*, 65, 165434.

Seifert, G., Enyashin, A. N., and Heine, T. (2005) *Physical Review B (Condensed Matter and Materials Physics)*, 72, 012102-4.

Slanina, Z., Zhao, X., and Osawa, E. I. (1998) *Chemical Physics Letters*, 290, 311–315.

Vonhelden, G., Hsu, M. T., Gotts, N. G., Kemper, P. R., and Bowers, M. T. (1993) *Chemical Physics Letters*, 204, 15–22.

Wang, Z. X., Ke, X. Z., Zhu, Z. Y., Zhu, F. Y., Wang, W. M., Yu, G. Q., Ruan, M. L., Chen, H., Huang, R. B., and Zheng, L. S. (2000) *Acta Physica Sinica*, 49, 939–942.

Wang, Z. X., Ke, X. Z., Zhu, Z. Y., Zhu, F. Y., Ruan, M. L., Chen, H., Huang, R. B., and Zheng, L. S. (2001) *Physics Letters A*, 280, 351–356.

Xie, S. Y., Gao, F., Lu, X., Huang, R. B., Wang, C. R., Zhang, X., Liu, M. L., Deng, S. L., and Zheng, L. S. (2004) *Science*, 304, 699.

Yang, S., Taylor, K. J., Craycraft, M. J., Conceicao, J., Pettiette, C. L., Cheshnovsky, O., and Smalley, R. E. (1988) *Chemical Physics Letters*, 144, 431–436.

Yao, S. J., Zhou, C. G., Ning, L. C., Wu, J. P., Pi, Z. B., Cheng, H. S., and Jiang, Y. S. (2005) *Physical Review B*, 71, 195316.

Zhechkov, L., Heine, T., and Seifert, G. (2004) *Journal of Physical Chemistry A*, 108, 11733–11739.

Zhu, W. J., Pan, Z. Y., HO, Y. K., and Wang, Y. X. (2000) *Journal of Applied Physics*, 88, 6836–6841.

Defective Fullerenes

Yuta Sato
National Institute of Advanced
Industrial Science and Technology
Core Research for Evolutional
Science and Technology
of Japan Science and
Technology Agency

Kazu Suenaga
National Institute of Advanced
Industrial Science and Technology
Core Research for Evolutional
Science and Technology
of Japan Science and
Technology Agency

31.1 Introduction

Fullerenes are generally recognized as a series of hollow spherical cages comprising pentagonal and hexagonal rings of trivalent carbon atoms. Most of the stable fullerene isomers previously obtained, such as the C_{60} buckminsterfullerene (BF) with an icosahedral (I_h) symmetry (Kroto et al. 1985), for instance, have no abutting (fused) pentagon pair that can be energetically unfavorable (Schmalz et al. 1986). The so-called isolated pentagon rule (IPR) has been found to be violated only in limited types of fullerenes such as endohedral metallofullerenes $Sc_2@C_{66}$ and $Sc_3N@C_{68}$ that were experimentally isolated for the first time in 2000 (Dunsch and Yang 2007; Stevenson et al. 2000; Wang et al. 2000).

With respect to the classical definition of fullerenes, abutting pentagon pairs in IPR-violating (non-IPR) fullerenes, as well as tetragonal and heptagonal rings in some theoretically proposed nonclassical fullerene isomers (Ayuela et al. 1996; Qian et al. 2000), have commonly been regarded as defects. Much attention has been paid to the mechanisms of formation and annihilation of these topological defects in fullerene cages (Murry et al. 1993; Bettinger et al. 2003) in order to determine why only limited types of isomers (most of them being IPR cages) are experimentally available among a huge number of possible structures. Defects can also be introduced into stable fullerene isomers by C–C bond breaking (cleavage) and/or a loss of carbon atoms (fragmentation) during various collision reactions (Campbell and Rohmund 2000) or irradiation with lasers (O'Brien et al. 1988) and electrons (Banhart 1999, 2004). Window-like defects of octagonal or larger rings can be introduced into fullerenes, which enable guest chemical species, such as single atoms and small molecules, to pass in and out of the cages.

This chapter illustrates some selected topics of defective fullerenes from previous theoretical and experimental studies, but does not aim at a comprehensive review of research in this field. Three main sections are focused on the following issues: abutting pentagon defects and related transformation (Section 31.2), window-like defects (Section 31.3), and related observational results (Section 31.4).

31.2 Abutting Pentagon Defect and Stone–Wales Transformation

31.2.1 Abutting Pentagon Defect

A pair of abutting pentagons (denoted as 5–5) in a non-IPR fullerene has generally been regarded as a defect due to its destabilizing effect on the cage (Schmalz et al. 1986). It has been demonstrated that some non-IPR fullerenes are experimentally isolable as endohedral metallofullerenes, such as $Sc_2@C_{66}$ and $A_xSc_{3-x}N@C_{68}$ ($x = 0–2$; A = Tm, Er, Gd, Ho, La) (Dunsch and Yang 2007). In these cases, originally unstable non-IPR cages are expected to be stabilized by the encapsulated species. Mechanisms for the introduction and elimination of 5–5 defects have been discussed with respect to the growth of fullerenes and their fragmentation. O'Brien et al. (1988) proposed that 5–5 defects are favorable channels for C_2 fragmentation under laser irradiation. Loss of C_2 from a 5–5 defect can maintain the closed spheroid conformation of the fullerene without forming an alternative window-like defect (Figure 31.1). The reverse of this reaction, by which C_2 is inserted into a hexagon to form a 5–5 pair, was proposed as a possible mechanism of fullerene growth by Endo and Kroto (1992).

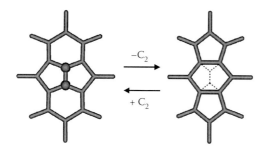

FIGURE 31.1 Schematic illustration of C_2 fragmentation at a 5–5 defect (O'Brien et al. 1988) and its reverse reaction (Endo and Kroto 1992).

31.2.2 Stone–Wales Transformation

The 5–5 defects can also be formed or annihilated by atom rearrangement in a fullerene cage without changing the number of carbon atoms. Much attention has been given to the so-called Stone–Wales (SW) transformation (rearrangement) as a possible mechanism for rearrangement of pentagons and hexagons in a fullerene cage (Bettinger et al. 2003). The SW transformation was originally defined as a rotation of a C_2 unit by 90° at the center of a pyracylene-like component consisting of two pentagons and two hexagons in a fullerene (Figure 31.2) (Stone and Wales 1986). By the SW transformation, C_{60} BF can be directly linked to the non-IPR isomer with C_{2v} symmetry (#1809) that possesses two 5–5 defects (Figure 31.3). So-called SW maps that illustrate how fullerene isomers can be linked to each other via SW transformation have been constructed to explain the observed abundance of only particular isomers such as C_{60} BF (Austin et al. 1995).

31.2.3 Mechanisms of Stone–Wales Transformation

SW transformation has been considered to be photochemically allowed, but thermally forbidden, according to the Woodward–Hoffmann rule, because it would involve a four-electron Hückel transition state (Stone and Wales 1986). Nevertheless, SW transformation has been assumed to be essential for the formation of fullerenes under specific conditions, such as arc discharge and laser irradiation. Theoretical studies have suggested that the conversion of C_{60} BF into its C_{2v} isomer via SW transformation requires an energy barrier of 3–9 eV, depending on the level of theory used for calculations (Bettinger et al. 2003). Two possible mechanisms with almost identical barriers have been proposed for this conversion: one involving a diradical-type transition state

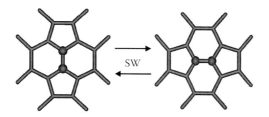

FIGURE 31.2 Schematic illustration of the original SW transformation.

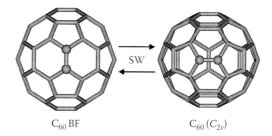

FIGURE 31.3 Structures of C_{60} BF and its C_{2v} isomer, which are interconvertible with each other via SW transformation.

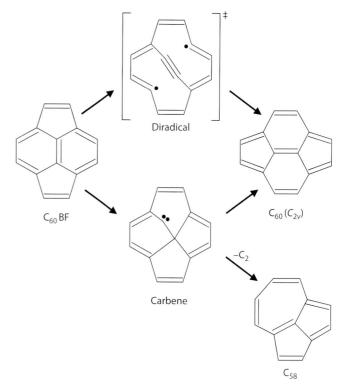

FIGURE 31.4 Schematic illustration of proposed SW transformation mechanisms in C_{60} BF (Bettinger et al. 2003) and branching C_2 fragmentation (Scuseria 1996). The double dagger denotes a transition state.

and the other involving a carbene-type intermediate (Figure 31.4) (Bettinger et al. 2003). The latter route has been expected to branch out into C_2 fragmentation (Scuseria 1996) resulting in the defective C_{58}, which is described in Section 31.3.3. It has also been theoretically predicted that the SW transformation of C_{60} BF can be catalyzed by extra single atoms, such as carbon (Eggen et al. 1996; Lee et al. 2006) and nitrogen (Slanina et al. 2000).

31.2.4 Generalized Stone–Wales Transformation

The definition of SW transformation has been extended (Saito et al. 1992; Balaban et al. 1996) to cover C_2 rotation occurring at various types of tetracyclic components in sp^2 carbon systems (Figure 31.5), besides fullerenes, such as carbon nanotubes (CNTs), graphene, and graphite. A pyrene-like component

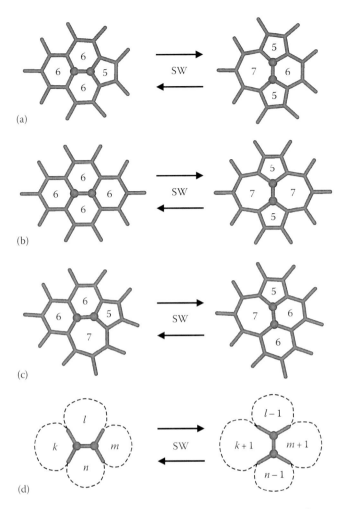

FIGURE 31.5 (a–c) Examples of extended definition SW transformations, and (d) a generalized representation (Saito et al. 1992; Balaban et al. 1996).

consisting of four hexagons in a graphitic system, for instance, can be interconvertible with two pentagon–heptagon pairs via the generalized SW transformation (Figure 31.5b). Such 5–7–7–5-type defects, often called SW defects, have already been

detected in single-walled CNTs (Figure 31.6) (Suenaga et al. 2007) and graphenes (Meyer et al. 2008) using atomic-resolution transmission electron microscopy (TEM).

31.3 Window-Like Defects

31.3.1 C–C Bond Breakage

If a C–C bond shared by an abutting pentagon–hexagon pair (5–6 bond) or that shared by two abutting hexagons (6–6 bond) in a fullerene cage is cleaved, a window-like defect of a 9- or 10-membered ring, respectively, is created (Figure 31.7a and b). Since not all C–C bonds in an individual fullerene are uniform, the energy barrier for breaking one of the bonds depends on its position. In the simplest case of C_{60} BF, for example, the lengths of the 5–6 and 6–6 bonds are 0.1474 and 0.1401 nm (Hedberg et al. 1991), respectively, and the former bond is suggested to be more easily broken than the latter. Murry and Scuseria (1994) estimated the energy barrier for breaking a 5–6 bond to be 5.2 eV, with the excitation of C_{60} BF from singlet to triplet. Froese and Morokuma (1999) reported a much larger value of 10.9 eV as the barrier for 6–6 bond cleavage resulting in a defective C_{60} structure at the quintet state. In each case, the two carbon atoms originally bound to each other are expected to be separated by 0.248 nm in the defective C_{60} cage, allowing atoms and small molecules such as noble gases and hydrogen to pass in and out through the gap.

Such window-like defects, introduced to a fullerene cage by breakage of a C–C bond, can be enlarged if additional bond cleavage occurs. Murry and Scuseria (1994) estimated the energy barrier to be approximately 4 eV for the second bond cleavage in a defective C_{60}, which extends the 9-membered ring created by the first 5–6 bond cleavage to a 13-membered ring (Figure 31.7c).

31.3.2 Monatomic Vacancy

Elimination of a (some) carbon atom(s) from a fullerene cage can also create a window-like defect. The structure of such defective fullerenes depends not only on the number of atoms removed and their original sites, but also on the subsequent bond formation

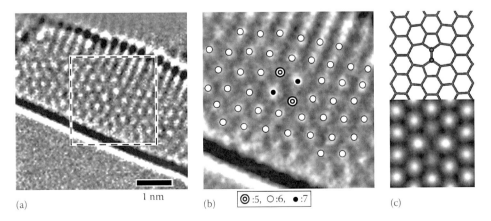

FIGURE 31.6 (a) TEM image of a single-walled CNT with a SW defect (Suenaga et al. 2007). (b) Enlarged image of the area enclosed by the dashed square in (a). (c) Simulated TEM image of a SW defect.

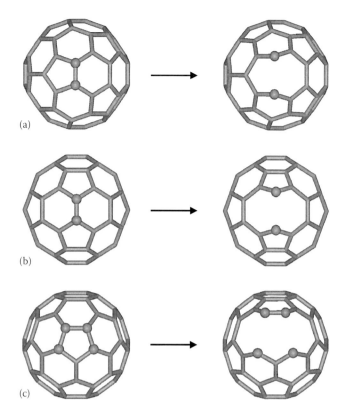

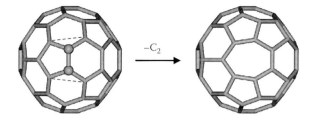

FIGURE 31.9 Schematic illustration of C_2 fragmentation at a 5–6 bond in C_{60} BF to yield a 5–5–7-type defective C_{58} fullerene (Hu and Ruckenstein 2004; Lee and Han 2004). Only the upper half of each cage is shown for clarity.

as 5–8 and 4–9, respectively), whereas the third individual atom remains unsaturated (bound to only two adjacent atoms). It is expected that both the 5–8 and 4–9 types of defective C_{59} cages have triplet ground states, and the former type is energetically more stable than the latter by 0.89 eV (Lee and Han 2004).

31.3.3 Diatomic Vacancy

A loss of a small cluster of even-number carbon atoms, especially C_2, has been considered to be an important step in the fragmentation processes of fullerenes induced by laser irradiation (O'Brien et al. 1988; Lifshitz 2000) or by atomic collision (Campbell et al. 2000). Three possible structures are proposed for the defective C_{58} fullerene, assuming that a pair of adjacent carbon atoms that form a 5–6 or 6–6 bond with each other in C_{60} BF are removed (Hu and Ruckenstein 2003; Lee and Han 2004). If such C_2 fragmentation occurs at a 5–6 bond in C_{60} BF, as illustrated in Figure 31.9, then carbon atoms in the vicinity of the C_2 vacancy are recombined to give two pentagons and a heptagon in the defective C_{58} (denoted as 5–5–7 type [Lee et al. 2004]). On the other hand, C_2 fragmentation at a 6–6 bond in C_{60} BF can result in two different structures, depending on

FIGURE 31.7 Schematic illustration of window-like defects introduced into C_{60} BF by breakage of (a) a 5–6 bond, (b) a 6–6 bond, and (c) two 5–6 bonds. Only the upper half of each cage is shown for clarity.

around the formed atom vacancies. For example, elimination of a carbon atom from C_{60} BF can result in two different C_{59} structures (Lee and Han 2004), as schematically shown in Figure 31.8. Two of the three carbon atoms originally bound to the removed atom are recombined to each other, creating a pair of 5- and 8-membered rings or a pair of 4- and 9-membered rings (denoted

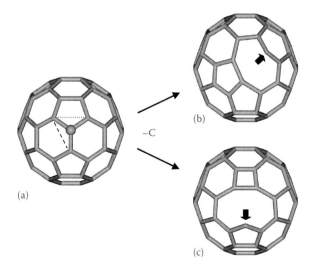

FIGURE 31.8 Schematic illustration of (a) C_{60} BF, (b) 5–8 and (c) 4–9 types of defective C_{59} fullerenes derived by removal of a carbon atom (Lee and Han 2004). Arrows in (b) and (c) indicate unsaturated carbon atoms. Only the upper half of each cage is shown for clarity.

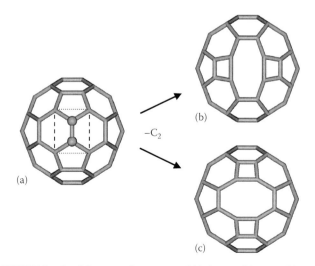

FIGURE 31.10 Schematic illustration of (a) C_{60} BF, (b) 4–4–8(5) and (c) 4–4–8(6) types of defective C_{58} fullerenes derived by C_2 fragmentation at a 6–6 bond (Lee and Han 2004). Only the upper half of each cage is shown for clarity.

the arrangement of newly formed two tetragons and an octagon (Figure 31.10). One of these defective C_{58} cages (denoted as 4–4–8(5) type [Lee et al. 2004]) has two abutting tetragon–pentagon pairs, whereas the other [4–4–8(6) type (Lee and Han 2004)] does not have such sites. The 5–5–7 type of defective C_{58} is expected to be energetically more stable than the 4–4–8(5) and 4–4–8(6) types by 4.77 and 1.34 eV, respectively. All three of the C_{58} structures without any unsaturated carbon atoms are singlet at their ground states. The C_2 fragmentation energy is estimated to be 12 eV, if the product C_{58} is the 5–5–7 type (Murry et al. 1993). Possible structures of defective C_{57} and C_{56} cages derived from C_{60} BF by removal of three and four adjacent carbon atoms, respectively, have also been theoretically investigated (Lee and Han 2004; Hu and Ruckenstein 2004).

31.4 Observation of Defective Fullerenes and Their Behavior

31.4.1 Noble Gas Encapsulation

It is still presently challenging to intentionally introduce the above-mentioned defects into individual fullerenes and to detect them in situ. Direct observation of the processes of SW transformation in fullerenes is yet to be accomplished. The opening of windows in fullerenes can be proven under particular conditions, if it is accompanied by subsequent detectable reactions, such as the encapsulation or release of guest species through the windows.

Saunders et al. (1996) demonstrated that single noble gas atoms can be introduced into both C_{60} and C_{70} fullerenes at temperatures and pressures of around 650°C and 3000 atm, respectively, to produce endohedral fullerenes X@C_{60} and X@C_{70} (X = He, Ne, Ar, Kr, Xe). Furthermore, the encapsulated atoms can be released when the endohedral fullerenes are heated to 1000°C. Opening and reclosing of the window-like defects in fullerenes (so-called window mechanisms) have been considered to be essential for these phenomena, because it is energetically unfavorable for noble gas atoms, especially those larger than He, to penetrate defect-free fullerene cages due to the large barriers above 11 eV (Sanville and BelBruno 2003).

31.4.2 Defective Metallofullerenes

Window-like defects can also be introduced into fullerenes during TEM observation, due to knock-on effects of the electron beam (Banhart 1999; 2004). Migration of encapsulated guest atoms through the windows opened in fullerene cages has been demonstrated in TEM observations for endohedral metallofullerenes within CNTs (so-called nanopeapods) (Urita et al. 2004; Sato et al. 2006). Typical TEM images of such defect-induced atom migration occurring in Gd@C_{82} fullerenes are shown in Figure 31.11. All the fullerene cages in this figure were originally defect-free, but continual electron-beam irradiation introduces window-like defects into the cages, as indicated by the arrows. Finally, the encapsulated Gd atoms in these fullerenes are found

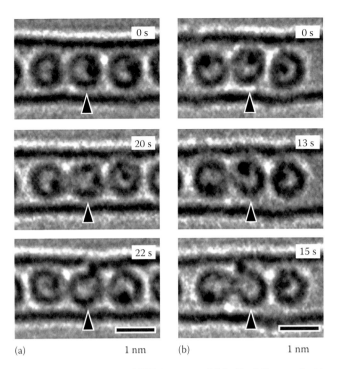

FIGURE 31.11 Sequential TEM images of Gd@C_{82} fullerenes inside single-walled CNTs (Sato et al. 2006). Arrows indicate breakout of Gd atoms from defective fullerene cages.

to break out through the opened windows. The observed migration of Gd atoms has been well reproduced by density functional theory (DFT) calculations (Yumura et al. 2006). The calculated reaction pathway is illustrated in Figure 31.12, for which the energy barrier has been estimated to be 7.3 eV.

31.4.3 Fullerene Polymerization in Nanopeapods

Direct observation of nanopeapods by TEM provides much information on various types of irradiation-induced defects in fullerenes besides window-like openings (Urita et al. 2004; Sato et al. 2006). For example, sequential TEM images of a nanopeapod in Figure 31.13 show temporal bond formation between C_{92} fullerenes and the inner wall of the CNT, as indicated by the arrows. These C_{92} fullerenes finally coalesce (fuse) with each other to form a dimer with a peanut-like shape, which can also be classified as a type of defective fullerene. Fullerene fusion and the formation of a bridging bond between a fullerene and an inner nanotube wall have also been theoretically investigated (Han et al. 2004; Yumura et al. 2007). If the polymerization occurs between endohedral metallofullerenes, the encapsulated metal atoms can exhibit unique migration behavior (Urita et al. 2004; Sato et al. 2006), as shown in Figure 31.14. Here, the inner cavities of two Gd@C_{82} and one Gd$_2$@C_{92} fullerenes are connected to each other as the coalescence proceeds, and the encapsulated Gd atoms migrate cage to cage.

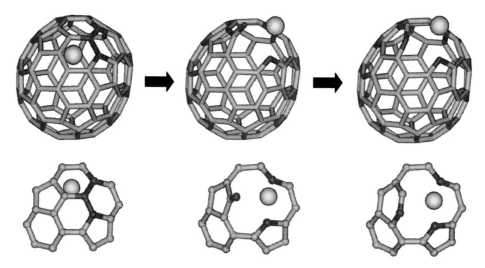

FIGURE 31.12 Possible reaction pathway for Gd atom breakout from a defective Gd@C_{82} fullerene, based on DFT calculations (Yumura et al. 2006).

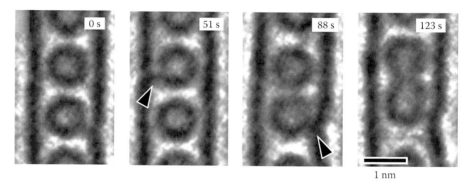

FIGURE 31.13 Sequential TEM images of C_{92} fullerenes inside a single-walled CNT (Urita et al. 2004). Arrows indicate bond formation between C_{92} and the inner wall of the CNT.

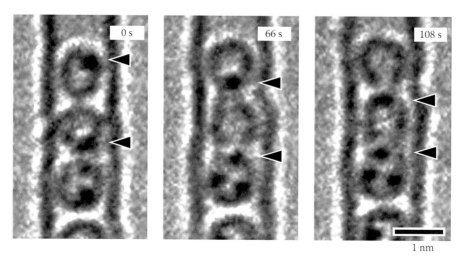

FIGURE 31.14 Sequential TEM images showing the polymerization of two Gd@C_{82} fullerenes and a Gd$_2$@C_{92} inside a single-walled CNT. Arrows indicate cage-to-cage migration of Gd atoms.

31.5 Summary

In this chapter, some typical fullerene defects, such as 5–5 defects, C–C bond breakage, and atom vacancies, were illustrated, and related experimental results were presented. Note that this chapter is not intended to cover all possible types of defective fullerenes previously proposed, and there are also a growing number of experimental and theoretical studies on nonclassical fullerenes.

Another important issue related to the topic is the opening of fullerene cages in the manner of organic synthesis. For example, it has been demonstrated that a 13-membered ring can be introduced into a C_{60} cage and then reclosed by a series of organic reactions; this is the so-called molecular surgical method (Komatsu et al. 2005). Techniques for such modification of fullerene cages by chemical processes and successive encapsulation of small molecules such as hydrogen have been rapidly advanced in the field of fullerene chemistry (Thilgen and Diederich 2006), but is beyond the scope of this chapter.

Acknowledgment

The authors would like to thank T. Yumura, K. Urita, and H. Wakabayashi for their computational and experimental results presented here.

References

Austin, S. J., Fowler, P. W., Manolopoulos, D. E. et al. 1995. The Stone-Wales map for C_{60}. *Chemical Physics Letters* 235: 146–151.

Ayuela, A., Fowler, P. W., Mitchell, D. et al. 1996. C_{62}: Theoretical evidence for a nonclassical fullerene with a heptagonal ring. *The Journal of Physical Chemistry* 100: 15634–15636.

Balaban, A. T., Schmalz, T. G., Zhu, H. et al. 1996. Generalizations of the Stone-Wales rearrangement for cage compounds, including fullerenes. *Journal of Molecular Structure: THEOCHEM* 363: 291–301.

Banhart, F. 1999. Irradiation effects in carbon nanostructures. *Reports on Progress in Physics* 62: 1181–1221.

Banhart, F. 2004. Formation and transformation of carbon nanoparticles under electron irradiation. *Philosophical Transaction of The Royal Society A* 362: 2205–2222.

Bettinger, H. F., Yakobson, B. I., and Scuseria, G. E. 2003. Scratching the surface of buckminsterfullerene: The barriers for Stone-Wales transformation through symmetric and asymmetric transition states. *Journal of the American Chemical Society* 2003, 125: 5572–5580, and references therein.

Campbell, E. E. B. and Rohmund, F. 2000. Fullerene reactions. *Reports on Progress in Physics* 63: 1061–1109.

Dunsch, L. and Yang, S. 2007. Endohedral clusterfullerenes—Playing with cluster and cage sizes. *Physical Chemistry Chemical Physics* 9: 3067–81, and references therein.

Eggen, B. R., Heggie M. I., Jungnickel, G., Latham, C. D., and Jones, R. Briddon, P. R. 1996. Autocatalysis during fullerene growth. *Science* 272: 87–89.

Endo, M. and Kroto, W. 1992. Formation of carbon nanofibers. *The Journal of Physical Chemistry* 96: 6941–6944.

Froese, R. D. J. and Morokuma, K. 1999. Accurate calculations of bond-breaking energies in C_{60} using the three-layered ONIOM method. *Chemical Physics Letters* 305: 419–424.

Han, S., Yoon, M., Berber, S. et al. 2004. Microscopic mechanism of fullerene fusion. *Physical Review B* 70: 113402 (4 pages).

Hedberg, K., Hedberg, L., Bethune, D. S. et al. 1991. Bond lengths in free molecules of buckminsterfullerene, C_{60}, from gasphase electron diffraction. *Science* 254: 410–412.

Hu, Y. H. and Ruckenstein, E. 2003. *Ab initio* quantum chemical calculations for fullerene cages with large holes. *The Journal of Chemical Physics* 119: 10073–10080.

Hu, Y. H. and Ruckenstein, E. 2004. Quantum chemical density-functional theory calculations of the structures of defect C_{60} with four vacancies. *The Journal of Chemical Physics* 120: 7971–7975.

Komatsu, K., Murata, M., and Murata, Y. 2005. Encapsulation of molecular hydrogen in fullerene C_{60} by organic synthesis. *Science* 307: 238–240.

Kroto, H. W., Heath, J. R., O'Brien, S. C. et al. 1985. C_{60}: Buckminsterfullerene. *Nature* 318: 162–163.

Lee, S. U. and Han, Y. -K. 2004. Structure and stability of the defect fullerene clusters of C_{60}: C_{59}, C_{58}, and C_{57}. *The Journal of Chemical Physics* 121: 3941–3942.

Lee, I. -H., Jun, S., Kim, H., Kim, S. Y., and Lee, Y. 2006. Adatom-assisted structural transitions of fullerenes. *Applied Physics Letters* 88: 011913 (3 pages).

Lifshitz, C. 2000. Carbon clusters. *International Journal of Mass Spectrometry* 200: 423–442.

Meyer, J. C., Kisielowski, C., Enri, R. et al. 2008. Direct imaging of lattice atoms and topological defects in graphene membranes. *Nano Letters* 8: 3582–3586.

Murry, R. L. and Scuseria, G. E. 1994. Theoretical evidence for a C_{60} "Window" mechanism. *Science* 263: 791–793.

Murry, R. L., Strout, D. L., Odom, G. K. et al. 1993. Role of sp^3 carbon and 7-membered rings in fullerene annealing and fragmentation. *Nature* 366: 665–667.

O'Brien, S. C., Heath, J. R., Curl, R. F. et al. 1988. Photophysics of buckminsterfullerene and other carbon cluster ions. *The Journal of Chemical Physics* 88: 220–230.

Qian, W., Bartberger, M. D., Pastor, S. J. et al. 2000. C_{62}, a nonclassical fullerene incorporating a four-membered ring. *Journal of the American Chemical Society* 122: 8333–8334.

Saito, R., Dresselhaus, G., and Dresselhaus, M. S. 1992. Topological defects in large fullerenes. *Chemical Physics Letters* 195: 537–542.

Sanville, E. and BelBruno, J. J. 2003. Computational studies of possible transition structures in the insertion and windowing mechanisms for the formation of endohedral fullerenes. *The Journal of Physical Chemistry B* 107: 8884–8889.

Sato, Y., Yumura, T., Suenaga, K. et al. 2006. Correlation between atomic rearrangement in defective fullerenes and migration behavior of encaged metal ions. *Physical Review B* 73: 233409 (4 pages).

Saunders, M., Cross, R. J., Jiménez-Vázquez, H. A. et al. 1996. Noble gas atoms inside fullerenes. *Science* 271: 1693–1697, and references therein.

Schmalz, T. G., Seitz, W. A., Klein, D. J. et al. 1986. C_{60} carbon cages. *Chemical Physics Letters* 130: 203–207.

Scuseria, G. E. 1996. Ab initio calculations of fullerenes. *Science* 271: 942–945.

Slanina, Z., Zhao, X., Uhlík, F., Ozawa, M., and Osawa, E. 2000. Computational modeling of the elemental catalysis in the Stone-Wales fullerene rearrangements. *Journal of Organometallic Chemistry* 599: 57–61.

Stevenson, S., Fowler, P. W., Heine, T. et al. 2000. A stable non-classical metallofullerene family. *Nature* 408: 427–428.

Stone, A. J. and Wales, D. J. 1986. Theoretical studies of icosahedral C_{60} and some related species. *Chemical Physics Letters* 128: 501–503.

Suenaga, K., Wakabayashi, H., Koshino, M. et al. 2007. Imaging active topological defects in carbon nanotubes. *Nature Nanotechnology* 2: 358–360.

Thilgen, C. and Diederich, F. 2006. Structural aspects of fullerene chemistry a journey through fullerene chirality. *Chemical Reviews* 106: 5049–135, and references therein.

Urita, K., Sato, Y., Suenaga, K. et al. 2004. Defect-induced atomic migration in carbon nanopeapod: Tracking the single-atom dynamic behavior. *Nano Letters* 4: 2451–2454.

Wang, C-.R., Kai, T., Tomiyama, T. et al. 2000. C_{66} fullerene encaging a scandium dimmer. *Nature* 408: 426–427.

Yumura, T., Sato, Y., Sunenaga, K. et al. 2006. Gate effect of vacancy-type defect of fullerene cages on metal-atom migrations in metallofullerenes. *Nano Letters* 6: 1389–1395.

Yumura, T., Kertesz, M., and Iijima, S. 2007. Confinement effects on site-preferences for cycloadditions into carbon nanotubes. *Chemical Physics Letters* 444: 155–160.

32

Silicon-Doped Fullerenes

Masahiko Matsubara
Université Montpellier II

Carlo Massobrio
Institut de Physique et Chimie des Matériaux de Strasbourg

32.1 Introduction

The discovery of fullerenes (Kroto et al., 1985) has prompted the synthesis of fullerene-based materials potentially bearing novel properties. As a result, a wide variety of fullerene derivatives have been reported. Endohedral and exohedral doping, where dopant atoms are encapsulated inside the fullerene cage (Bethune et al., 1993; Heath et al., 1985) and are located on the fullerene surface (Roth et al., 1991; Weiss et al., 1988), respectively, have been widely studied using metal atoms as dopant atoms. A third doping possibility (termed substitutional doping) consists in the replacement by dopant atoms of C atoms belonging to the fullerene network. Hereafter, we shall refer to these doped species as heterofullerenes. Overall, the interest in these materials lies in a controlled modification of the electronic properties of the pristine fullerene network at the replacements sites, leading to tunable reactivity behaviors. It should be understood that the interest in doping results from the strong electrophile character of fullerenes in spite of early considerations on their supposedly unreactive character. Boron and nitrogen had been the first choices as doping elements of fullerenes. Experimentally, evidences were provided for the successful incorporation of up to 12 nitrogen atoms (Hultman et al., 2001) and up to 6 boron atoms (Guo et al., 1991). First-principles calculations of boron- and nitrogen-substituted fullerenes showed that the highest occupied molecular orbital acts as an electron acceptor for $C_{59}B$ and an

electron donor for $C_{59}N$ (Andreoni et al., 1992, 1996). Also, transition metal atoms such as Fe, Co, Ni, Rh, and Ir were successfully incorporated into the fullerene network (Billas et al., 1999a; Branz et al., 1998) via gas-phase experiments. The large share of attention devoted to silicon stems from the availability of both ion mobility and mass spectroscopy experiments data, revealing the existence of silicon-substituted heterofullerenes. These species could contain more than one Si as dopant atom in the carbon cage without causing its breakdown (Billas et al., 1999a,c; Fye and Jarrold, 1997; Kimura et al., 1996; Pellarin et al., 1999a,b; Ray et al., 1998). A value between 7 and 12 was extracted on the basis of the combined mass spectroscopy and photofragmentation data, with 12 taken as the upper limit (Pellarin et al., 1999a,b). Substitutionally doped heterofullerenes with Si atoms are at the crossroad of two opposite tendencies. On the one hand, the replacement of C atoms by Si atoms appears as a natural choice, due to the same number (four) of valence electrons. On the other hand, Si atoms prefer to form multidirectional sp^3 single bonds, which contrasts with the chemical nature of C atoms, which has the ability to form single, double, and triple bonds with themselves and other atoms. For these reasons, the stability and the bonding character of systems obtained via Si substitution in a carbon environment with a strong sp^2 character is a highly challenging issue from both the experimental and the theoretical points of view. About 10 years ago, $C_{59}Si$ and $C_{58}Si_2$ were investigated via first-principles calculations within the density functional theory

(Billas et al., 1999a,b,c; Ray et al., 1998). Structural characterization showed that the fullerene cage was distorted by the inclusion of Si atoms, but the distortions were quite local. A strong charge transfer from the dopant Si atoms to the neighboring C atoms occurs, in a way consistent with the polar character of the Si–C bonds. An analysis of electron localization revealed that the Si atom bonds to the neighboring atoms in a distorted or "weak" sp^2 manner. By increasing the number of doping Si atoms, theoretical efforts were aimed at characterizing the geometrical and electronic properties of the doped cages for a number of replacements close to the experimentally estimated upper limit (Fu et al., 2001, 2002; Koponen et al., 2008; Marcos et al., 2003, 2005; Matsubara and Massobrio, 2005a; Matsubara et al., 2005; Menon, 2001). In addition to the structural properties at $T = 0\,K$, thermal properties of these heterofullerenes were analyzed to model the fragmentation process (Fu et al., 2002), to characterize the structural transformation at high temperature (Marcos et al., 2003), and to study the stability of the cage at finite temperatures (Matsubara and Massobrio, 2005a). The above results were obtained without any attempt to go beyond the estimated upper limit (12) of C–Si replacements. In a further series of studies, the number of dopant Si atoms was increased to establish a physical criterion for the largest number of Si atoms to be hosted in the cage (Matsubara and Massobrio, 2005b, 2006, 2007c; Matsubara et al., 2006). The structural stability of $C_{40}Si_{20}$, $C_{36}Si_{24}$, and $C_{30}Si_{30}$ was rationalized in terms of the segregation of a Si region and a C region. A combined analysis of electronic properties and thermal properties for these highly Si-doped heterofullerenes has given an indication that upper limit number of dopant Si atoms is 20. Given the above context, it appears worthwhile to describe the methodology and the main results underlying recent research efforts in the area of silicon-doped heterofullerenes. This is the purpose of the present chapter in which we overview structural, electronic, and thermal properties of these systems, as obtained by first-principles calculations based on density functional theory (DFT). In Section 32.2, we briefly summarize the experimental results on the silicon-doped heterofullerenes. Then we provide basic details of our methodology of calculation (Section 32.3). In Section 32.4, we describe our strategy to characterize structural and electronic properties, particularly the bonding nature of Si–C bonds by using $C_{59}Si$ and $C_{58}Si_2$ as examples. In addition to structural and electronic properties, thermal properties of heterofullerenes are investigated for a set of $C_{54}Si_6$ isomers in Section 32.5. In Section 32.6, we assess the stability of highly silicon-doped heterofullerenes $C_{40}Si_{20}$, $C_{36}Si_{24}$, and $C_{30}Si_{30}$. In two final sections, we provide a summary of our results and a few ideas on work in progress and future perspectives.

32.2 Summary of the Main Experimental Results

The first experimental indirect evidence of the existence of a silicon-substituted heterofullerene structure was provided by Kimura et al. (1996). Silicon-doped carbon clusters of the type Si_mC_n ($m = 1, 2$) were produced via the laser-vaporization cluster beam technique.

Then, gas-phase ion mobility measurements were carried out to examine the geometries of silicon-doped carbon cluster cations C_nSi^+ ($n = 3$–69) by Fye and Jarrold (1997). The observed isomers resemble those found for pure carbon clusters. This strong resemblance suggests that the silicon atom replaces a carbon atom with a minor impact on the structure. This result can be considered as the first conclusive experimental evidence on the closed-cage nature of $C_{2n-1}Si^+$ fullerenes with an even number of atoms, the silicon atom being part of the cage. In the case of $C_{2n}Si^+$ fullerenes with an odd number of atoms, the silicon atom appeared to be exohedral.

Moving to doping with more than one Si atom, Ray et al. reported the synthesis of $C_{2n-q}Si_q$ clusters with $2n = 32$–100 and $q < 4$. In these species, the fullerene-like structure was preserved (Ray et al., 1998). The analysis of the abundance distribution and photofragmentation spectra provided clear evidence that such clusters remain in the fullerene geometry and that the Si atoms are located close to each other in the fullerene network. Further evidence for a larger number of Si doping was presented by Pellarin et al. (1999a,b). In their experiment, silicon carbon binary clusters were generated in a laser vaporization source from Si_xC_{1-x} mixed targets ($x = 0\%$–50% of the content). Stoichiometric $(SiC)_n$ ($n \le 40$) clusters grown from a silicon carbide target ($x = 50\%$) were first analyzed. Both high-fluence photoionization of $(SiC)_n$ neutral clusters and photofragmentation of size-selected $(SiC)_n^+$ natural positive ions showed that silicon-doped fullerenes emerge as stable species. A more efficient doping of carbon fullerenes with silicon atoms can be obtained in carbon-rich mixed clusters directly grown as positive ions from nonstoichiometric targets ($x < 25\%$). Mass abundance spectroscopy gives a clear signature of cage-like structures where silicon atoms are substituted by carbon ones. These approaches exemplified the capability of substituting a large number of silicon atoms into fullerenes without destabilizing their cage structure. In this respect, a value close to 12 was proposed to be the upper limit. A clear assessment of this upper limit, rigorously substantiated by a sound theoretical framework, has motivated most of the calculations described in the following sections.

32.3 Computational Methodology

Molecular dynamics is a powerful tool to understand various properties of nanomaterials at the atomic scale. Within the first-principles molecular dynamics method by Car and Parrinello (1985), the evolution of the ions is treated in a self-consistent way by extracting the interatomic forces directly from a first-principles potential energy surface. In turn, this is determined in the framework of density functional theory. First-principles molecular dynamics is the method of choice to describe the delicate interplay between structural changes occurring upon doping and the variations in the chemical bonding. For these reasons, the entire body of research work described therein is based on this computational strategy. In what follows, we recall its key theoretical ingredient with special concern for the calculation of specific properties and schemes useful to characterize bonding features in silicon-doped heterofullerenes (population analysis and localized orbitals). The concept of temperature control in the framework of a molecular dynamics run (i.e., the use of a thermostat within the equation of motion) is also briefly reviewed.

32.3.1 Density Functional Theory

In the framework of density functional theory (DFT) (Hohenberg and Kohn, 1964; Kohn and Sham, 1965), the ground-state energy E of an interacting inhomogeneous electron gas in a static potential $v(\mathbf{r})$ can be written as

$$E = \int v(\mathbf{r})n(\mathbf{r})d\mathbf{r} + \frac{1}{2}\int\int \frac{n(\mathbf{r})n(\mathbf{r}')}{|\mathbf{r}-\mathbf{r}'|}d\mathbf{r}d\mathbf{r}' + T_s(n) + E_{xc}(n), \quad (32.1)$$

where

    $n(\mathbf{r})$ is the density

    $T_s(n)$ is the kinetic energy of a system of noninteracting electrons with density $n(\mathbf{r})$

    $E_{xc}(n)$ is the exchange and correlation energy of an interacting system with density $n(\mathbf{r})$

The density $n(\mathbf{r})$ is obtained in terms of a single Slater determinant built from the occupied orbitals ψ_i with occupation number c_i

$$n(\mathbf{r}) = \sum_i c_i |\psi_i(\mathbf{r})|^2, \quad (32.2)$$

where the set of ψ_i satisfy the orthonormalized condition:

$$\int \psi_i^*(\mathbf{r})\psi_j(\mathbf{r})d\mathbf{r} = \delta_{ij}. \quad (32.3)$$

The kinetic energy $T_s(n)$ is described as a function of ψ_i by

$$T_s(\psi) = \sum_i c_i \int \psi_i^*(\mathbf{r})\left(-\frac{1}{2}\nabla^2\right)\psi_i(\mathbf{r})d\mathbf{r}. \quad (32.4)$$

From the variation of Equation 32.1 with respect to $n(\mathbf{r})$ subject to the constraint Equation 32.3, we obtain the Euler–Lagrange equation

$$\left\{-\frac{1}{2}\nabla^2 + v(\mathbf{r}) + \int \frac{n(\mathbf{r}')}{|\mathbf{r}-\mathbf{r}'|}d\mathbf{r}' + \frac{\delta E_{xc}(n)}{\delta n(\mathbf{r})}\right\}\psi_i(\mathbf{r}) = \epsilon_i\psi_i(\mathbf{r}), \quad (32.5)$$

$$H_{KS}\psi_i(\mathbf{r}) = \epsilon_i\psi_i(\mathbf{r}), \quad (32.6)$$

where ϵ_i denotes one electron energy. The fourth term of Equation 32.5 is called exchange-correlation potential. Because no simple exact form of $E_{xc}(n)$ is available, approximations are required for

any application of DFT to a realistic description of bonding. In the local density approximation (LDA), $n(\mathbf{r})$ is taken as slowly varying and $E_{xc}(n)$ can be written as

$$E_{xc}(n) = \int n(\mathbf{r})\epsilon_{xc}(n(\mathbf{r}))d\mathbf{r}, \quad (32.7)$$

where $\epsilon_{xc}(n)$ is the exchange and correlation energy per electron of a uniform electron gas of density n. More sophisticated approximations, which include gradients of the electronic density, may give a significant improvement of accuracy. These are called generalized gradient approximations (GGA) (Langreth and Perdew, 1980; Perdew et al., 1992). Within GGA, $E_{xc}(n)$ is expressed as

$$E_{xc}(n) = \int n(\mathbf{r})\epsilon_{xc}^{GGA}(n(\mathbf{r}); \nabla n(\mathbf{r}))d\mathbf{r}, \quad (32.8)$$

where $\epsilon_{xc}^{GGA}(n; \nabla n)$ is the (density) gradient corrected version of $\epsilon_{xc}(n)$.

Static external potentials $v(\mathbf{r})$ in Equation 32.5 consist of the Coulomb interactions between electrons and nuclei. If we treat only chemically active valence electrons and ignore inert core electrons, this term is effectively replaced by appropriate pseudopotentials.

32.3.2 Car–Parrinello Scheme

Instead of solving Equation 32.5 directly, Car and Parrinello postulated the following Lagrangian (Car and Parrinello, 1985; Galli and Parrinello, 1991) based on the wavefunction ψ_i and ionic position $\mathbf{R}_I$ subject to orthonormality constraint with Lagrangian multipliers Λ_{ij}

$$L = \sum_i \int \frac{1}{2}\mu|\dot{\psi}_i(\mathbf{r})|^2 d\mathbf{r} + \sum_I \frac{1}{2}M_I\dot{\mathbf{R}}_I^2 - E[\{\psi_i\},\mathbf{R}_I]$$
$$+ 2\sum_{ij}\Lambda_{ij}\left(\int \psi_i^*(\mathbf{r})\psi_j(\mathbf{r})d\mathbf{r} - \delta_{ij}\right), \quad (32.9)$$

where M_I and μ denote the ionic mass and the "fictitious mass" assigned to the electronic degrees of freedom with appropriate units, respectively. From this Lagrangian for the fictitious system, we can derive two equations of motion:

$$\mu\ddot{\psi}_i(\mathbf{r},t) = -\frac{1}{2}\frac{\delta E}{\delta \psi_i^*(\mathbf{r},t)} + \sum_j\Lambda_{ij}\psi_j(\mathbf{r},t), \quad (32.10)$$

$$M_I\ddot{\mathbf{R}}_I = -\frac{\partial E}{\partial \mathbf{R}_I(t)}. \quad (32.11)$$

The Lagrangian for the nuclei in the real system is given by the sum of the ionic kinetic energy and the ionic potential energy $V(\{\mathbf{R}_I\})$, its equation of motion being

$$M_I \ddot{\mathbf{R}}_I = -\frac{\partial V}{\partial \mathbf{R}_I}. \tag{32.12}$$

In general, Equation 32.11 for the fictitious system and Equation 32.12 for the real system differ, unless $E[\{\psi_i\}, \mathbf{R}_I]$ is at the instantaneous minimum. However, by selecting ψ and the initial conditions of $\{\psi_i\}$ and $\{\dot{\psi}_i\}$ appropriately, the energy transfer between the classical degrees of freedom for ions and electrons can be made small enough to let the electrons follow the ionic motion adiabatically, remaining close to the Born–Oppenheimer surfaces. Typically, a value of μ much smaller than the physical mass ensures an effective decoupling between fictitious and ionic degrees of freedom. This holds true provided a finite gap in the electronic density of states separates occupied states from empty states. Such a condition is easily met in finite systems as clusters, allowing a straightforward application of the above ideas.

32.3.3 Methods of Electronic Structure Analysis: ELF and Localized Orbitals Method

The electron localization function (ELF) carries information on the electron pair probability in multi-electron systems (Becke and Edgecombe, 1990). The ELF has the following form:

$$ELF = \left(1 + \left(\frac{D_\sigma}{D_\sigma^0}\right)^2\right)^{-1}, \tag{32.13}$$

where

$$D_\sigma^0 = \frac{3}{5}(6\pi^2)^{2/3} \rho_\sigma^{5/3}, \tag{32.14}$$

$$D_\sigma = \tau_\sigma - \frac{1}{4}\frac{(\nabla \rho_\sigma)^2}{\rho_\sigma}, \tag{32.15}$$

and one uses the density ρ_σ and the positive-definite kinetic energy density is defined by

$$\tau_\sigma = \sum_i^\sigma |\nabla \psi_i|^2 \tag{32.16}$$

The value of ELF is bound in between $0 \leq ELF \leq 1$. The upper limit $ELF = 1$ corresponds to optimal localization for the electron while the value $ELF = 0.5$ corresponds to uniform electron gas. Therefore, high ELF values mean that the electrons are more localized than in a uniform electron gas of the same density at the position considered. Low values of ELF are related to a strong Pauli repulsion.

The maximally localized Wannier functions enable us to analyze bonding properties (Marzari and Vanderbilt, 1997; Resta, 1998; Silvestrelli et al., 1998). The positions of the centers of the localized Wannier functions and the related dispersions can be obtained to gain a clear picture of the chemistry of the system and the relevant environment.

Using the maximally localized Wannier functions $\mathcal{W}_n(\mathbf{r})$, the coordinate $\mathcal{X}_n$ of the nth Wannier function center in the cubic supercell size of L is expressed as

$$\mathcal{X}_n = -\frac{L}{2\pi} \operatorname{Im} \ln \left\langle \mathcal{W}_n \left| \exp\left(-i\frac{2\pi}{L}x\right) \right| \mathcal{W}_n \right\rangle, \tag{32.17}$$

with similar definitions for $\mathcal{Y}_n$ and $\mathcal{Z}_n$.

32.3.4 Population Analysis

The population analysis is a useful tool to describe the distribution of the electrons on specific atoms and bonds. The procedure consists in projecting the eigenstates $|\psi_\alpha(\mathbf{k})\rangle$ obtained by plane wave calculation onto Bloch functions formed from a localized basis set $|\phi_\mu(\mathbf{k})\rangle$ (Sanchez-Portal et al., 1995; Segall et al., 1996). The overlap matrix of the localized basis set in reciprocal space is defined as

$$S_{\mu\nu}(\mathbf{k}) = \langle \phi_\mu | \phi_\nu \rangle. \tag{32.18}$$

The projection operator of Bloch functions with wave vector $\mathbf{k}$ generated by the atomic basis is written as

$$\hat{p}(\mathbf{k}) = \sum_\mu |\phi_\mu(\mathbf{k})\rangle \langle \phi^\mu(\mathbf{k})|, \tag{32.19}$$

where $|\phi^\mu(\mathbf{k})\rangle$ are the duals of the atomic basis states, which satisfy

$$\langle \phi_\mu(\mathbf{k}) | \phi^\nu(\mathbf{k}) \rangle = \langle \phi^\mu(\mathbf{k}) | \phi_\nu(\mathbf{k}) \rangle = \delta_{\mu\nu}. \tag{32.20}$$

Using $\hat{p}$ a projected eigenstate

$$|\chi_\alpha(\mathbf{k})\rangle = \hat{p}(\mathbf{k}) | \psi_\alpha(\mathbf{k})\rangle \tag{32.21}$$

is obtained and using $|\chi_\alpha\rangle$, the density operator is defined as

$$\hat{\rho}(\mathbf{k}) = \sum_\alpha n_\alpha |\chi_\alpha(\mathbf{k})\rangle \langle \chi^\alpha(\mathbf{k})|, \tag{32.22}$$

where n_α denotes the occupancy of the plane wave state. Using the density operator $\hat{\rho}$, the density matrix for the atomic basis set is calculated:

$$P_{\mu\nu}(\mathbf{k}) = \langle \phi^\mu(\mathbf{k}) | \hat{\rho}(\mathbf{k}) | \phi^\nu(\mathbf{k}) \rangle. \tag{32.23}$$

Finally, a population analysis can be performed using the density matrix $P(\mathbf{k})$ and the overlap matrix $S(\mathbf{k})$. In the Mulliken analysis, the charge associated with an atom A is given by

$$Q_A = \frac{1}{N_k} \sum_k \sum_{\mu \in A} \sum_v P_{\mu v}(\mathbf{k}) S_{v\mu}(\mathbf{k}) = \sum_{\mu \in A} (PS)_{\mu\mu}, \quad (32.24)$$

where N_k denotes the number of calculated k points in the Brillouin zone. Similarly, the Mayer bond order (MBO) (Mayer, 1983, 1984) between atoms A and B and the total valence (V) on atom A are calculated by

$$\text{MBO}_{AB} = \sum_{\mu \in A} \sum_{v \in B} (PS)_{\mu v} (PS)_{v\mu}, \quad (32.25)$$

$$V_A = \sum_{B \neq A} \text{MBO}_{AB}. \quad (32.26)$$

32.3.5 Nosé–Hoover Thermostats

The solution of the equations of motion within molecular dynamics implies that the constant energy of the system is a conserved quantity. In order to introduce the notion of temperature control, Nosé and Hoover developed an extended system method, which introduces an additional degree of freedom corresponding to a heat bath (Hoover, 1985; Nosé, 1984). Due to this extension, the sum of the kinetic and potential energies of a given system is no more a conserved quantity, this role being played by the sum of all kinetic and potential terms associated to all dynamical variables evolving in the extended system. The introduction of an additional variable is intended to reproduce the effect of a heat bath, exchanging energy with the conventional degrees of freedom.

Given the existence of the heat bath, the equation of motion for the coordinates $\mathbf{R}_I$ with potential energy E is written as

$$M_I \ddot{\mathbf{R}}_I = -\frac{\partial E}{\partial \mathbf{R}_I} - M_I \dot{\xi} \dot{\mathbf{R}}_I, \quad (32.27)$$

where the new variable $\dot{\xi}$ appears and the term including this variable act as a friction force. The time evolution of the variable $\dot{\xi}$ is given by

$$\frac{Q}{2} \ddot{\xi} = \sum_I \frac{1}{2} M_I \dot{\mathbf{R}}_I^2 - \frac{g}{2} k_B T. \quad (32.28)$$

In Equation 32.28, Q is a parameter working as a mass for a heat bath and determining the rate of coupling between the heat bath and the physical systems, while g denotes the number of degrees of freedom in a physical system (k_B and T are Boltzmann constant and the target temperature, respectively).

The standard Nosé–Hoover thermostat defined by the above equations is not always ergodic. In order to avoid this problem, a closely related technique, Nosé–Hoover-chain thermostat (Martyna et al., 1992), is widely used, where several thermostats are thermostated one by the other. In this technique, Equation 32.28 is modified to include this chain of thermostats,

$$\frac{Q_1^n}{2} \ddot{\xi}_1 = \left[\sum_I \frac{1}{2} M_I \dot{\mathbf{R}}_I^2 - \frac{g}{2} k_B T \right] - \frac{Q_1^n}{2} \dot{\xi}_1 \dot{\xi}_2 \quad (32.29)$$

$$\frac{Q_k^n}{2} \ddot{\xi}_k = \left[\frac{1}{2} Q_{k-1}^n \dot{\xi}_{k-1}^2 - \frac{1}{2} k_B T \right]$$
$$- \frac{Q_k^n}{2} \dot{\xi}_k \dot{\xi}_{k+1} \left(1 - \delta_{kK} \right) \quad \text{where } k = 2,...,K, \quad (32.30)$$

for the nuclear part. On the other hand, the temperatures of the electronic wave functions can be kept at different temperatures than that of nuclear part (Blöchl and Parrinello, 1992).

$$\mu \ddot{\psi}_i = -H_{KS} \psi_i + \sum_j \Lambda_{ij} \psi_j - \mu \dot{\eta}_1 \dot{\psi}_i, \quad (32.31)$$

$$\frac{Q_1^e}{2} \ddot{\eta}_1 = \left[\sum_i \mu \int |\dot{\psi}_i(\mathbf{r})|^2 d\mathbf{r} - T_e^0 \right] - \frac{Q_1^e}{2} \dot{\eta}_1 \dot{\eta}_2 \quad (32.32)$$

$$\frac{Q_l^e}{2} \ddot{\eta}_l = \left[\frac{1}{2} Q_{l-1}^e \dot{\eta}_{l-1}^2 - \frac{1}{2\beta_e} k_B T \right]$$
$$- \frac{Q_l^e}{2} \dot{\eta}_l \dot{\eta}_{l+1} \left(1 - \delta_{lL} \right) \quad \text{where } l = 2,...,L, \quad (32.33)$$

where T_e^0 is the mean value of the fictitious kinetic energy of the electronic wavefunctions. Here, β_e should be chosen as $1/\beta_e = 2T_e^0/N_e$, where N_e is the number of dynamical degrees of freedom needed to parametrize the wavefunction minus the number of constraint conditions.

When the Nosé–Hoover-chain thermostat is used, the conserved quantity in the Car–Parrinello scheme becomes

$$E_{\text{NVT}} = \sum_I \frac{1}{2} M_I \dot{\mathbf{R}}_I^2 + \sum_i \mu \int |\dot{\psi}_i(\mathbf{r})|^2 d\mathbf{r} + E$$
$$+ \sum_{k=1}^{K} \frac{1}{2} Q_k^n \dot{\xi}_k^2 + \sum_{k=2}^{K} k_B T \xi_k + g k_B T \xi_1$$
$$+ \sum_{l=1}^{L} \frac{1}{2} Q_l^e \dot{\eta}_l^2 + \sum_{l=2}^{L} \frac{\eta_l}{\beta_e} + 2T_e^0 \eta_1. \quad (32.34)$$

32.3.6 Further Technical Details

Throughout this chapter, the calculations are performed within the framework of the Car–Parrinello molecular dynamics, using the CPMD code (CPMD, 1990–2008). Our approach is based on DFT with GGA after Becke (1988) for the exchange part and Lee, Yang, and Parr (LYP) (Lee et al., 1988) for the correlation part of the exchange-correlation functional. The core-valence interaction is described by norm-conserving pseudopotentials of the Troullier–Martins type (Troullier and Martins, 1991).

Plane waves are used for the wavefunction expansion, whose energy cutoff is 40 Ry at the Γ point of the Brillouin zone. Periodic boundary conditions have been adopted and a face-centered cubic cell of edge of 21.17 Å provides a system size much larger than the diameter of the optimized structures of heterofullerenes minimizing spurious interactions with periodically repeated images.

The structures of heterofullerenes are fully relaxed by minimizing the forces on all atoms using the direct inversion method in the iterative space (DIIS) (Császár and Pulay, 1984; Hutter et al., 1994). Optimization of the ionic positions is performed without constraint on the structure and is allowed to proceed until the largest force component is less than 5×10^{-4} a.u. and the average force is one order of magnitude smaller.

Finite temperature molecular dynamics simulations have been performed by taking a fictitious mass for the electronic degrees of freedom as $\mu = 200$ a.u. and a time step $\Delta t = 3$ a.u. Temperature control is implemented in the framework of Nosé–Hoover thermostats both for ionic and electronic degrees of freedom.

32.4 $C_{59}Si$ and $C_{58}Si_2$: Basics of Structural and Electronic Properties

32.4.1 Structural Properties

In what follows, starting from the most simple cases, i.e., one and two silicon atoms doped heterofullerenes, structural and electronic properties of these systems are presented and discussed.

A view of the heterofullerene $C_{59}Si$ is shown in Figure 32.1. The protruding shape of the cage is due to the Si atom, responsible for Si–C bond lengths longer than the corresponding C–C bonds. The calculated bond lengths are given in Table 32.1 together with those of C_{60}. The modifications of bond lengths in $C_{59}Si$ compared to those of C_{60} are mainly confined to the elongated Si–C bonds.

While there is only one isomer in the case of one Si atom doping, there are quite many possibilities in the case of multiple Si atom doping. Shortly after the study of Ray et al. (1998) on three representative isomers, Billas et al. (1999b) and Fu et al. (2001) considered nine different isomers, which are shown in

TABLE 32.1 Bond Lengths of Si–C and C–C Bonds in $C_{59}Si$ and C–C Bonds in C_{60} in Å

		Si–C	C–C
$C_{59}Si$	ph	1.90	1.46–1.50
	hh	1.85	1.40–1.43
C_{60}	ph		1.46
	hh		1.41

Note: For the C–C bonds, only the smallest and the largest values are displayed in the case of $C_{59}Si$.

Figure 32.2. They are those for which the two Si atoms are (a) first nearest neighbors (ortho-isomers: 1 and 2), (b) second nearest neighbors (meta-isomers: 3 and 4), (c) third nearest neighbors (5, 6, and 7), (d) one of the isomers for which the two Si atoms are fourth nearest neighbors (8), and (e) the isomer corresponding to the two Si atoms lying at diametrically opposite sites on the cage (ninth nearest neighbor position) (9).

The Si–C and C–C bond lengths of the above $C_{58}Si_2$ isomers are given in Table 32.2. In isomer 2, the Si–Si bond length is slightly smaller than the experimental values measured for bulk Si (2.346 Å) (Phillips, 1973). The conjugation pattern of the Si–Si bond in isomer 2 is that of a single bond because this bond involves a pentagon and a hexagon (ph bond). For isomer 1, the conjugation pattern is that of Si–Si double bond because this

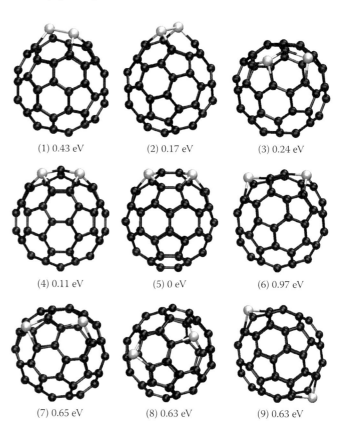

(1) 0.43 eV   (2) 0.17 eV   (3) 0.24 eV

(4) 0.11 eV   (5) 0 eV   (6) 0.97 eV

(7) 0.65 eV   (8) 0.63 eV   (9) 0.63 eV

FIGURE 32.2 Three-dimensional description of nine different $C_{58}Si_2$ isomers. The values of the total energy difference relative to the lowest energy isomer 5 are also indicated. Carbon atoms are drawn in black, while silicon atoms are in gray.

FIGURE 32.1 Three-dimensional view of $C_{59}Si$. Carbon atoms are drawn in black, while the silicon atom is in gray.

TABLE 32.2 Bond Lengths of Si–C and C–C Bonds in $C_{58}Si_2$ in Å

Isomer		Si–Si	Si–C	C–C
1	ph		1.88(4)	1.46–1.52
	hh	2.23		1.41–1.43
2	ph	2.30	1.92(2)	1.46–1.50
	hh		1.85(2)	1.40–1.44
3	ph		1.88, 1.89, 1.90, 1.93	1.46–1.50
	hh		1.85, 1.87	1.40–1.44
4	ph		1.89(2), 1.93(2)	1.46–1.51
	hh		1.82(2)	1.40–1.43
5	ph		1.88(2), 1.92(2)	1.45–1.50
	hh		1.86(2)	1.40–1.43
6	ph		1.89(4)	1.46–1.52
	hh		1.84(2)	1.40–1.45
7	ph		1.89, 1.90, 1.91(2)	1.46–1.51
	hh		1.84, 1.86	1.40–1.45
8	ph		1.90(3), 1.91	1.46–1.51
	hh		1.85(2)	1.40–1.44
9	ph		1.90(4)	1.46–1.51
	hh		1.85(2)	1.40–1.43

Note: Bond lengths of Si–Si bonds (in Å) are also shown in the case of isomers 1 and 2. The number of corresponding Si–C bonds is given in the parenthesis. For the C–C bonds, only the smallest and the largest values are given.

bond involves two hexagons (hh bond). For these reasons, the Si–Si bond of isomer 1 has a value lying between that of Si complexes with a Si–Si double bond (2.14–2.16 Å) (West, 1987) and the bond length of silicon dimer Si_2 (2.246 Å) (Dickerson et al., 1974).

The energetically lowest isomer is isomer 5 having two Si on the same hexagon at para-sites, as shown in Figure 32.2. Its binding energy is 0.14 eV/atom smaller than the binding energy of C_{60}. The total energies of the $C_{58}Si_2$ isomers considered here are within 0.97 eV (isomer 6) relative to the isomer 5. By looking into the lowest isomer 5 in more detail, one sees that the arrangement of the two Si atoms in this isomer (para-sites of an hexagon) is identical to the experimental findings for disilabenzene complexes (Rich and West, 1982). Two of the Si–C ph bonds and the Si–C hh bonds are longer than the corresponding bond lengths in $C_{59}Si$ (see Tables 32.1 and 32.2). Interestingly, we observe that the most favorable configuration is neither the one where the two Si atoms are at the largest distance on the cage, nor the one where the two Si atoms are nearest neighbors.

32.4.2 Electronic Properties

In the case of $C_{59}Si$, the degeneracy of both HOMO and LUMO of C_{60} is partly removed by the doping, as shown in Figure 32.3. The fivefold degenerate HOMO of C_{60} split into two parts: single nondegenerate level, which corresponds to HOMO in $C_{59}Si$, and four closely spaced levels ~0.07 eV lower in energy relative to the HOMO. In the case of LUMO, one level is split from two other

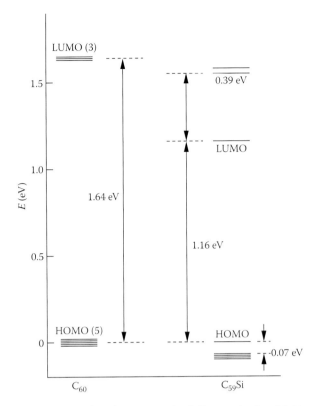

FIGURE 32.3 Kohn–Sham energy level diagram in the vicinity of HOMO–LUMO region for C_{60} (left) and $C_{59}Si$ (right). In each case, the energy of HOMO is taken to be 0 eV. The energies of HOMO–LUMO gap are indicated in addition to the values between LUMO and LUMO + 1 and between HOMO and HOMO − 1 in the case of $C_{59}Si$.

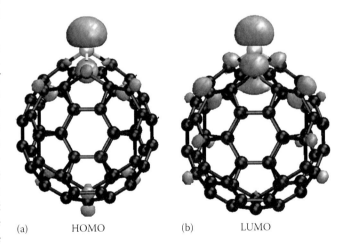

FIGURE 32.4 Probability isodensity surfaces [0.005 e/(a.u.)^3] associated with the (a) HOMO and (b) LUMO for $C_{59}Si$.

levels and lowered by ~0.39 eV in energy. The probability isodensity surfaces of the Kohn–Sham (KS) orbitals corresponding to the HOMO and LUMO for $C_{59}Si$ are given in Figure 32.4. Both HOMO and LUMO are localized on the Si atom and on the C atom with a Si–C hh bond. The HOMO − 1 and LUMO + 1 have a strong carbon-like character and they are separated by an energy difference of 1.62 eV, which is very close to the HOMO–LUMO

TABLE 32.3 Mulliken Charge, Total Valence, and Bond Order Values for $C_{59}Si$ and C_{60}

Molecule	Atom	Mulliken Charge	Valence	Bonds	Bond Order ph	Bond Order hh
$C_{59}Si$	Si	0.67	3.31	Si–C	0.90	1.09
	C*	−0.18	3.79–3.82	C*–C	1.19–1.20	1.41
	C	0.04–0.05	3.90	C–C	1.14–1.15	1.36–1.37
C_{60}	C	0.04	3.90	C–C	1.14	1.37

Note: C* denotes the carbon atoms, which have a Si atom as a neighbor.

gap of C_{60}. Therefore, the changes in the energy levels from C_{60} to $C_{59}Si$ concern mostly the orbitals close to the gap.

In order to obtain further information on atomic net charges and bond strength in $C_{59}Si$, we performed a Mulliken population analysis and calculated the Mulliken charges, the total valence, and the Mayer bond order, shown in Table 32.3. The net charges on Si atom and on its three nearest neighbor C atoms are large compared to the values on the other C atoms farther apart on the cage. A substantial electronic charge transfer from the Si site to the neighboring C sites is indicated by the opposite values carried by the charges on the Si atom and on the C nearest neighbors. This is due to the smaller electronegativity of silicon compared to carbon and suggests strong polar character of $C_{59}Si$. The Si–C bonds are weaker than the C–C bonds, as indicated by the reduced valence and bond orders of Si and of its neighboring C atoms.

By doping with two silicon atoms, the degeneracy in the electronic levels of C_{60} is removed as in the case of $C_{59}Si$. In Figure 32.5, the KS energy levels around the HOMO–LUMO gap are shown for two ortho-isomers (1 and 2) and the lowest and the highest energy isomers (5 and 6). Figure 32.6 illustrates representative isodensity hypersurfaces of the lowest 5 and the highest 6 energy isomers. In the case of isomer 5, two major contributions can be seen both in the cases of HOMO and LUMO. The first is a spherical region around both Si atoms and the other is a banana-shape region around the C atoms neighboring the Si atoms. Mulliken charges, total valence, and Mayer bond order of isomers 1, 2, 5, and 6 are shown in Table 32.4. The Mulliken charges on the Si atoms of isomers 5 and 6 are almost equal to the net charge on the Si atom of $C_{59}Si$, while they are smaller in the case of isomers 1 and 2, where two Si atoms are directly connected to each other. The valence of the silicon atoms is larger

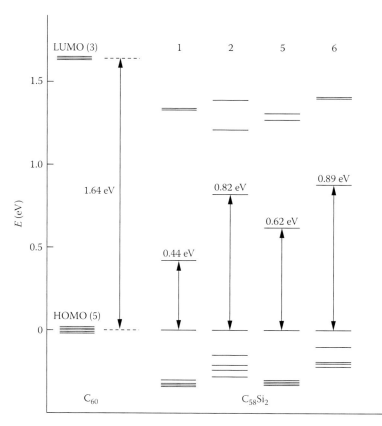

FIGURE 32.5 Kohn–Sham energy-level diagram in the vicinity of HOMO–LUMO region for C_{60} and four isomers of $C_{58}Si_2$ (isomers 1, 2, 5, and 6). The energy of HOMO is taken to be 0 eV in each isomer. The energies of HOMO–LUMO gap are indicated.

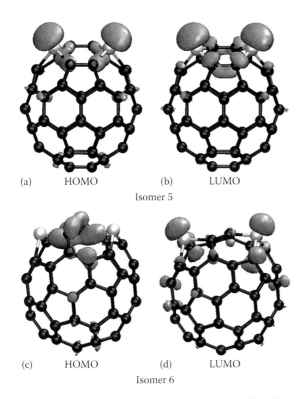

(a) HOMO (b) LUMO
Isomer 5

(c) HOMO (d) LUMO
Isomer 6

FIGURE 32.6 Probability isodensity surfaces $[0.005\ e/(a.u.)^3]$ associated with the (a) HOMO and the (b) LUMO for isomer 5 (the lowest energy isomer) and with the (c) HOMO and the (d) LUMO for isomer 6 (the highest energy isomer) of $C_{58}Si_2$.

TABLE 32.4 Mulliken Charge, Total Valence, and Bond Order Values for $C_{58}Si_2$

Isomer	Atom	Mulliken Charge	Valence	Bonds	Bond Order Ph	hh
1	Si	0.45	3.54	Si–Si		1.13
	C	−0.17	3.83	Si–C	0.98	
2	Si	0.43	3.47	Si–Si	0.84	
	C	−0.16 to 0.19	3.81–3.83	Si–C	0.95	1.17
5	Si	0.66	3.42	Si–C	0.90–0.94	1.08
	C	−0.17 to 0.19	3.80–3.83			
6	Si	0.66	3.34	Si–C	0.89–0.92	1.08
	C	−0.18	3.79–3.82			

Note: The values for Si atoms and C atoms having Si atoms as neighbors are displayed.

for isomers 1, 2, and 5. For isomer 5, this is consistent with their increasing stability, while in the case of 1 and 2, this corresponds to a smaller charge transfer. Interestingly, the bond order of the Si–C hh bond in the isomer 2 is the largest among all the silicon in-doped fullerenes, implying stronger Si–C bonding. However, isomer 2 is not the most favorable energetically, being located at 0.17 eV above isomer 5. This behavior can be understood by invoking a destabilizing effect of the Si–Si bond in the configuration of isomer 2. The typical average bond energy of a Si–Si

bond (210–250 kJ/mol) is indeed smaller than that of a Si–C bond (250–335 kJ/mol) (Cotton and Wilkinson, 1980).

32.4.3 Localized Orbitals and Electron Localization Function

Further insight into the bonding features of these Si-doped fullerenes can be obtained by a topological analysis of the electron localization function (ELF) and of the maximally localized Wannier functions.

The calculated ELF plots are shown in Figure 32.7. As customary in this kind of analysis, the localization regions are searched by lowering the ELF value from its maximum. This allows to detect the regions where the electrons have the highest probability. The regions in space corresponding to the occurrence of localization from decreasing values of the ELF are termed attractors, invoking an analogy with potential energy surfaces with minima and related attractive basins. In our case, one attractor starts to be observed above the Si atom ($\eta = 0.9748$) (Figure 32.7a). As the ELF value is lowered further, other attractors begin to emerge. At $\eta = 0.946$, one can observe an attractor on each Si–C ph bond (Figure 32.7b). Then, a third attractor appears on the Si–C hh bond at $\eta = 0.942$ (Figure 32.7c). Finally, the attractors corresponding to the C–C bonds become visible (Figure 32.7d). On lowering the ELF values, the

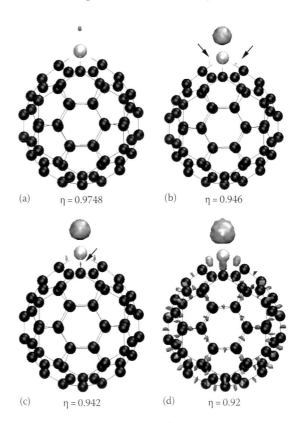

(a) $\eta = 0.9748$ (b) $\eta = 0.946$

(c) $\eta = 0.942$ (d) $\eta = 0.92$

FIGURE 32.7 ELF isosurfaces plot for values (a) $\eta = 0.9748$ (maximum value), (b) $\eta = 0.946$, where an attractor on each Si–C ph bond appears, (c) $\eta = 0.942$, where an attractor on Si–C hh begins to emerge, and (d) $\eta = 0.92$, where attractors on C–C bonds also start to appear.

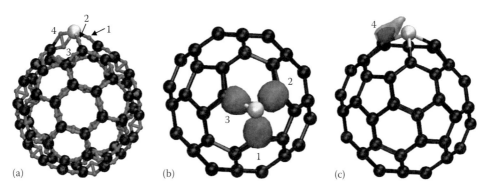

FIGURE 32.8 (See color insert following page 25-14.) (a) Wannier function centers (WFCs) represented by small spheres. The C atoms and a Si atom are displayed by large black spheres and a large gray sphere, respectively. Numbers denote the WFCs around the Si atom. (b) Isosurfaces [0.02 e/(a.u.)^3] of the Wannier functions corresponding to WFC1, WFC2, and WFC3. (c) Isosurfaces of the Wannier function WFC4 for two values of the isosurface density 0.02 e/(a.u.)^3 (transparent) and 0.025 e/(a.u.)^3 (solid).

attractors on the Si–C bonds are noticeable well before the others. This demonstrates that electron localization is more important in the Si–C bonds than in the C–C bonds. The attractors belonging to Si–C bonds lie 1.05 Å away from the silicon atom on the same bond, being displaced toward the C atom with respect to C–C bonds. Therefore, silicon–carbon bonds have a strong polar character.

The Wannier function centers for $C_{59}Si$ around the Si dopant atom (hereafter defined as WFC1, WFC2, WFC3, and WFC4) are shown in Figure 32.8a. Three of the four WFCs are associated to Si–C bonds and are displaced toward the C atoms, namely, WFC1, WFC2, and WFC3, in line with the polar character of the Si–C bonds. On the other hand, WFC4 is located closer to the Si atom along the Si–C hh bond. Closer inspection of the corresponding Wannier functions shown in Figure 32.8b and c proves the following. The Wannier functions belonging to the two Si–C ph bonds, having as centers WFC1 and WFC2 (Figure 32.8b), are similar to the Wannier functions found for the C–C single bonds but displaced toward carbon. This occurs also for one of the two Wannier functions related to the Si–C hh bond (WFC3). The other Wannier function belonging to the Si–C hh bond (WFC4) is less localized. In fact, at the same isosurface value of the density ($\eta = 0.02$) used in Figure 32.8b, it extends from above Si down to the Si–C atom–atom line, close to C. On the other hand, for a higher value, it is found localized close to the C atom of the Si–C hh bond (Figure 32.8c).

These considerations apply equally well to the behavior of the various isomers of $C_{58}Si_2$. However, we shall limit our analysis to the lowest energy isomer (isomer 5) and to its Wannier localization properties. A representative example is provided by the Wannier analysis in Figure 32.9, where we show the Wannier centers of isomer 5 and the Wannier functions corresponding to the centers WFC4. The similarity of the Wannier functions relative to $C_{59}Si$ (Figure 32.8c) and those of $C_{58}Si_2$ (Figure 32.9b) substantiates our conclusions on the simple and double bonding character in Si-doped heterofullerenes. Changes with respect to the C_{60} case stem from the combined effect of Si–C charge transfer and the existence of four distinct electron localization regions associated to each Si atom.

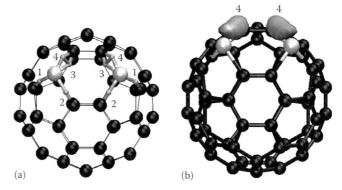

FIGURE 32.9 Wannier analyses for the most stable isomer 5 of $C_{58}Si_2$. (a) The Wannier function centers are represented by small spheres. Only those around Si atoms are displayed and they are numbered as in the case of $C_{59}Si$. (b) Wannier function isosurfaces at the value $\eta = 0.025e$/(a.u.)^3, which correspond to the WFC4.

32.5 $C_{54}Si_6$: Structural, Electronic, and Thermal Properties

In the previous section, by using DFT calculations on $C_{59}Si$ and $C_{58}Si_2$, we have seen that the structural deformation induced by the doping via Si atoms occurs in the vicinity of the dopant atoms, with Si atoms protruding out of the fullerene network (Billas et al., 1999a,b,c). This goes along with a localization of the HOMO and LUMO states on the Si sites and the formation of polar Si–C bonds.

In what follows, we obtain both the energetics of $C_{54}Si_6$ heterofullerenes at $T = 0\,K$ and their thermal stability within first-principles molecular dynamics (FPMD). This can be considered as a first step toward the understanding of the stability of species containing an increasing number of Si dopant atoms. The issue of thermal effects will be addressed by choosing temperatures that are sufficiently high to drive the atomic configurations out of their equilibrium minima and sufficiently low to avoid fragmentation. Temperature allows to describe events that happen prior to the disruption of the cage and makes possible the selection of isomers, which are energetically very close at

$T = 0\,\mathrm{K}$. This is especially needed in the presence of several Si–Si bonds, in principle less inclined to adjust to various bonding configurations.

32.5.1 Structural and Electronic Properties

We start from obtaining equilibrium geometries of 12 different isomers of $C_{54}Si_6$ at $T = 0\,\mathrm{K}$. In Figure 32.10, we display the optimized configurations of 12 isomers. Nine of them are found to lie within 2.31 eV, while the remaining three are much higher in energy. Interatomic distances of these isomers are provided in Table 32.5.

In the lowest energy isomer, isomer **A**, four identical Si–Si ph bonds (2.41 Å) are found, as well as two sets of equal Si–C distances (1.93 and 1.87 Å). The cage remains stable but the interatomic Si–Si hh bond distance is highly stretched (2.66 Å), much larger than the Si–Si distance in bulk Si (2.346 Å) (Phillips, 1973).

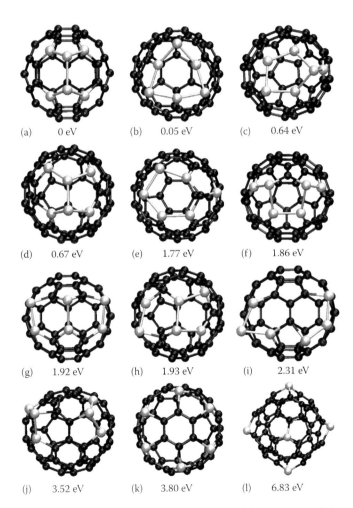

(a) 0 eV (b) 0.05 eV (c) 0.64 eV
(d) 0.67 eV (e) 1.77 eV (f) 1.86 eV
(g) 1.92 eV (h) 1.93 eV (i) 2.31 eV
(j) 3.52 eV (k) 3.80 eV (l) 6.83 eV

FIGURE 32.10 Ball and stick representation of the structures of the various equilibrium energy isomers of $C_{54}Si_6$. The black and gray atoms are carbon and silicon atoms, respectively. Energy differences relative to the lowest energy isomer, i.e., isomer A are also shown. The sticks are drawn for distances smaller than 1.7 Å (C–C bonds), 2.2 Å (Si–C bonds) and 2.7 Å (Si–Si bonds).

TABLE 32.5 Bond Lengths of $C_{54}Si_6$ for 12 Different Isomers

		Si–Si (Å)		Si–C (Å)	C–C (Å)
A	ph (4)	2.41	ph (4)	1.93	ph 1.45–1.49
	hh (1)	2.66	hh (4)	1.87	hh 1.40–1.43
B	ph (3)	2.37	ph (6)	1.90–1.98	ph 1.46–1.53
	hh (3)	2.35–2.36			hh 1.41–1.42
C	ph (5)	2.35–2.48	ph (2)	1.89	ph 1.46–1.51
	hh (1)	2.34	hh (4)	1.85–1.86	hh 1.40–1.45
D	ph (3)	2.32–2.47	ph (6)	1.90–1.99	ph 1.46–1.51
	hh (2)	2.33–2.38	hh (2)	1.83–1.86	hh 1.40–1.43
E	ph (2)	2.32–2.48	ph (8)	1.86–1.98	ph 1.46–1.52
	hh (2)	2.29–2.37	hh (2)	1.83–1.84	hh 1.40–1.44
F	ph (3)	2.29–2.54	ph (6)	1.87–2.08	ph 1.45–1.53
	hh (2)	2.35	hh (2)	1.86–1.87	hh 1.39–1.42
G	ph (2)	2.37	ph (8)	1.87–1.97	ph 1.45–1.52
	hh (1)	2.53	hh (4)	1.81–1.90	hh 1.40–1.44
H	ph (3)	2.32–2.44	ph (6)	1.85–1.98	ph 1.44–1.53
	hh (1)	2.35	hh (4)	1.83–1.97	hh 1.40–1.44
I	ph (2)	2.38	ph (8)	1.87–2.05	ph 1.46–1.54
	hh (2)	2.35	hh (2)	1.85	hh 1.41–1.43
J	ph (2)	2.39	ph (8)	1.86–2.11	ph 1.45–1.56
	hh (2)	2.34	hh (2)	1.84	hh 1.40–1.45
K			ph (12)	1.87–1.95	ph 1.44–1.50
			hh (6)	1.84	hh 1.40–1.43
L			ph (12)	1.88–1.89	ph 1.46–1.51
			hh (6)	1.85	hh 1.40–1.46

Note: The smallest and the largest values found are shown. The distances are expressed in Å. The number of occurrences of Si–Si and Si–C bonds are also indicated in parentheses.

Isomer **B** has all Si atoms lying on a hexagon and the Si–Si bond distances are comprised in between 2.35 and 2.37 Å. All Si–C distances are of ph kind and they take alternately the values 1.93 and 1.98 Å on the hexagon. C–C and Si–C distances are much shorter than Si–Si ones, not exceeding 1.53 and 1.98 Å, respectively. In isomer **C**, five Si atoms are on a pentagon and the additional one is part of a hh bond. Isomer **D** can be obtained from isomer **C** by moving one Si atom from the pentagon to the adjacent hexagon. Four Si atoms are now on both the hexagon and the pentagon, having in common a Si–Si ph bond.

Observation of Figure 32.11 indicates that the amount of charge on each atom depends on the number of Si–C bonds. Negative charges significantly different from zero are found only on those C atoms which are directly connected to Si atoms. Values range in between 0 and −0.30 for a single Si–C connection and increase to −0.4 (denoted by a black diamond in isomer **D**, C atom in the Si_4C pentagon) when a C atom has two Si neighbors. In the case of Si, the largest positive charges are on those atoms with two C nearest neighbors, as in isomers **A**, **C**, and **D** (denoted by ○ in Figure 32.11). On the opposite side, positive charges vanish whenever a Si atom is bound to three neighbors of the same kind. These situation is found in isomers **A**, **C**, and **D** (denoted by △ in Figure 32.11).

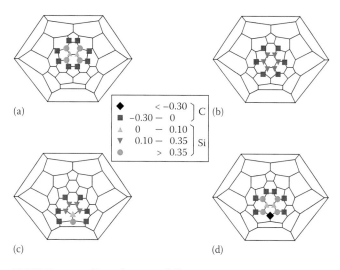

(a)

♦   < −0.30 ⎤
■   −0.30 — 0 ⎦ C
▲   0 — 0.10 ⎤
▼   0.10 — 0.35 ⎦ Si
●   > 0.35

(b)

(c)   (d)

FIGURE 32.11 **(See color insert following page 25-14.)** A two-dimensional flattened view of $C_{54}Si_6$. Mulliken charges are given for all the Si atoms and for some their neighboring C atoms. The values corresponds to the shading codes in the center of the figure.

TABLE 32.6 Electronic Properties of the Four Lowest Energy Isomers of $C_{54}Si_6$

Isomer	Atom	Valence	Bonds	Bond Order ph	Bond Order hh	Gap (eV)
A	Si	2.90–3.47	Si–Si	0.90	0.64	0.91
	C	3.83–3.84	Si–C	0.89	1.02	
B	Si	3.11–3.45	Si–Si	0.89	0.93–0.94	0.94
	C	3.82–3.84	Si–C	0.88–0.95		
C	Si	2.97–3.54	Si–Si	0.75–0.94	1.01	0.65
	C	3.81–3.84	Si–C	0.94	1.16–1.18	
D	Si	3.05–3.53	Si–Si	0.77–0.94	0.94–0.99	0.52
	C	3.77–3.84	Si–C	0.82–1.02	1.11–1.19	

Further insight into the bonding features is provided in Table 32.6 where we give the total valence, the Mayer bond order, and the Kohn–Sham energy gap of the four most stable isomers. Only C atoms next-neighbors to Si atoms are considered. In the case of C, the valence differs from four when the atoms are close to the dopant Si atoms. In the case of Si, the valence approaches four for an increasing number of Si–C bonds, with the lower bounds corresponding to threefold Si–Si coordination. Bond order is a useful indication of bond strength (Mayer, 1983, 1984). We found that the bond order is systematically lower in Si–Si hh bonds than in their Si–C counterparts, suggesting weaker connections in the fullerene cage. This is especially true for the stretched (2.66 Å) Si–Si hh bond of isomer **A**, for which the bond order takes its lowest value, 0.64. Gap values lie in between 0.38 eV (isomer **K**) and 0.96 eV (isomer **F**). Therefore, our data suggest that no specific correlation exists between structural stability and the extent of the gap.

32.5.2 Thermal Behavior

The close values of the binding energies for the two most stable isomers of $C_{54}Si_6$ at zero temperature exemplify the need of a thermal study providing a more stringent assessment of the relative structural stability. We considered isomer **A** and isomer **B**.

Our results are given in Figure 32.12, where we display the time behavior of the bond lengths for C–C, Si–C, and Si–Si interactions. For C, we have considered only nearest neighbors of Si atoms. Average values correspond to the mean over all bonds of a given kind on temporal trajectories lasting 1 ps each. Error bars are standard deviations of the mean, calculated by partitioning each trajectory in portions of 0.1 ps. The temperature values are $T = 2000\,K$ for $0 < t < 4\,ps$, $T = 2500\,K$ for $4 < t < 8\,ps$ and $T = 3000\,K$ for $8 < t < 12\,ps$. For both isomers, the increase in the average C–C bond length is very limited, proving that the portion of the cage populated by C atoms is essentially unperturbed. Considering Si–C distances, the temperature induces two distinct behaviors. The first corresponds to the changes from $T = 0$ to $T = 2000\,K$, while the second characterizes the transition from $T = 2000$ to $T = 2500\,K$ and from $T = 2500$ to $T = 3000\,K$. Values at finite temperatures reported in Figure 32.12 are very close to the intervals of bond lengths determined at $T = 0\,K$ proving that the Si–C interactions are not responsible for any relevant structural deformation.

A different pattern is observed for Si–Si bonds. Bond elongation is clearly more pronounced for isomer **A**, with a marked change from 2.46 Å ($T = 0\,K$) to 2.57 Å ($T = 3000\,K$). In Figure 32.13, we focus on the separate contributions to this behavior of the Si–Si hh bond and the four Si–Si ph bonds. The hh bond is very much sensitive to temperature changes and features a sudden increase. This is followed by a relaxation to shorter values. Such a pattern is observed at temperatures $T = 2000\,K$, $T = 2500$, and $T = 3000\,K$. In (Matsubara and Massobrio 2005a), we have shown that the drastic oscillations of this bond distance are related to strong deformations involving the two adjacent hexagons kept together at $T = 0\,K$ by the hh bond. Interestingly, fragmentation is avoided by a process of thermal energy redistribution to all degrees of freedom, bringing back the bond distance to much lower values, as confirmed by the presence of minima in the bond length pattern of Figure 32.13. Smaller variations are found in the evolution of ph bonds, which follow the same trend but exhibit maxima and minima of much lower intensities. On the basis of these results, and keeping in mind the limited span of our dynamical simulations, it appears that isomer **A** is very much likely not to be able to withstand further thermal annealing at higher temperatures. On the contrary, isomer **B** is more stable and bears higher chances to be observed on experimental timescales. This is a first, unambiguous demonstration of the role played by the temperature in determining the observable species, thereby providing support to our calculation strategy.

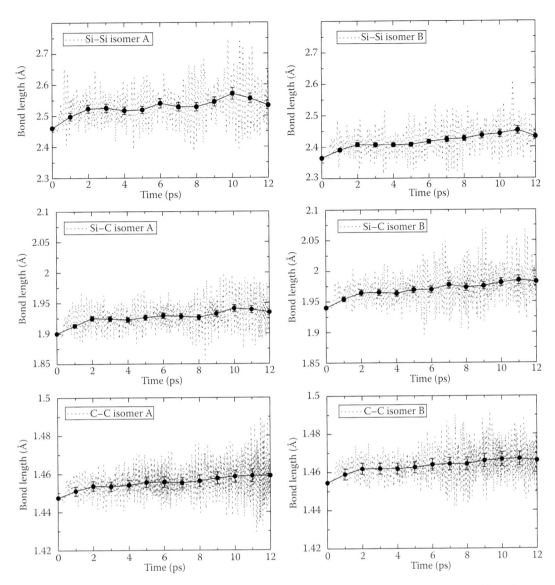

FIGURE 32.12 Time behavior with its average value of the Si–Si, Si–C, and C–C (for C atoms, only the nearest neighbor of Si atoms are considered) bond lengths of isomer A and isomer B. For the average value, each data point is the mean over all bonds of the same kind on subtrajectories of 1 ps. The temperature values are $T = 2000\,K$ for $0 < t < 4\,ps$, $T = 2500\,K$ for $4 < t < 8$, and $T = 3000\,K$ for $8 < t < 12\,ps$.

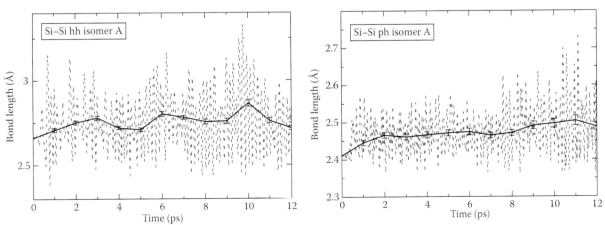

FIGURE 32.13 Time behavior with its average value of the Si–Si bond lengths (hexagon–hexagon, upper panel and pentagon–hexagon, lower panel) of isomer A. For the average value, each data point is the mean over all bonds of the same kind on subtrajectories of 1 ps.

32.6 Beyond Experimentally Estimated Value: $C_{40}Si_{20}$, $C_{36}Si_{24}$, and $C_{30}Si_{30}$

In this section, we progress further toward the establishment of an upper limit for the number of Si–C replacements in the fullerene cage. To summarize the pieces of evidence collected so far, we recall that a tentative upper limit of 12 C atoms replaced by Si atoms was suggested through mass spectrometry and photofragmentation studies on $C_{2n-m}Si_m$ clusters, with $2n = 32 - 80$ (Pellarin et al., 1999a,b). Focusing on the extreme upper bound of doping (a full network made of 60 Si atoms), the question arises on the possible occurrence of Si_{60} hollow cages. Indeed, their existence has been ruled out by various approaches showing that a full Si-made fullerene-like configuration relaxes into a drastically different arrangement ("puckered ball") (Kahn and Broughton, 1991; Li et al., 1999; Menon and Subbaswamy, 1994). Given the above results on $C_{54}Si_6$, it remains to be determined whether $C_{60-m}Si_m$ clusters can host a number of Si atoms larger than $m = 12$, while retaining a cage-like structural character. As briefly mentioned above, we already know that the upper limit has to be lower than 60. Is there any way to determine an upper bound based on structural considerations and accounting for thermal effects? To address this issue, we focus on the structural stability at $T = 0\,K$ of $C_{40}Si_{20}$, $C_{36}Si_{24}$, and $C_{30}Si_{30}$. These studies are a prerequisite to investigations on the thermal properties of these fullerenes, reported at the end of this chapter. We found that the lowest energy configurations consist of two clearly distinct subnetworks forming a distorted closed cage. C atoms do preserve the conjugated pattern of distances typical of C_{60} in a regular arrangement, while Si–Si bonds are larger and form a protruding cap for the cluster. We call "nanoscale segregation" the enhanced stability found whenever Si and C atoms form separate regions of the cage. This is shown in the case of $C_{40}Si_{20}$ and allows to construct stable structures for $C_{36}Si_{24}$ and $C_{30}Si_{30}$.

32.6.1 $C_{40}Si_{20}$

A schematic view of the optimized configurations for three typical isomers of $C_{40}Si_{20}$ is given in the upper part of Figure 32.14 where we employed a two-dimensional representation as a guide to the eye. Also, three-dimensional views are given in the lower part of Figure 32.14. Again, the most stable structures are the result of the propensity of Si atoms to stick together and form weak sp^2 bonding. In the most stable isomer **A**, a central pentagon is surrounded by five hexagons. What is the energetic cost of breaking this homogeneous Si region? An indication is given by isomer **B**, where two separated Si regions exist within the cage and the energy is 2.71 eV higher. Isomer **C** is another example of what happens when distributing Si atoms in the cage without forming Si–Si bonds: Its binding energy is 16.77 eV higher than in isomer **A**. From the study of these isomers, the energetic cost associated with the creation of Si–C bonds at the expenses of Si–Si bonds is estimated to be ~0.25 eV/atom. As shown in Table 32.7, Si–Si distances do not exceed 2.52 Å, and usually values range in between 2.30 and 2.45 Å. Therefore, the increased number of Si doping atoms does not cause a significant elongation of the Si–Si bonds, as seen for the case of $C_{54}Si_6$. The ph and hh bonds follow this same pattern for Si–C bonds (see Table 32.7 for the isomers containing both of them).

32.6.2 $C_{36}Si_{30}$ and $C_{30}Si_{30}$

Representative optimized structures of $C_{36}Si_{24}$ and $C_{30}Si_{30}$ are shown in Figure 32.15. In the search of a criterion for the stability, it is worthwhile to classify the Si atoms in terms of their belonging or not to the frontier with C (outer Si atoms, Si_{out} vs inner Si atoms Si_{in}). This is highlighted in the flattened view of $C_{36}Si_{24}$ and $C_{30}Si_{30}$. Inner atoms can be further labeled as first (Si_{1st} hereafter), second (Si_{2nd}), or third (Si_{3rd}) neighbors of outer Si atoms. Interatomic distances are consistent with the tendency for the two sets of atoms to organize themselves so as to preserve the alternate single-bond/double-bond pattern (C atoms) and

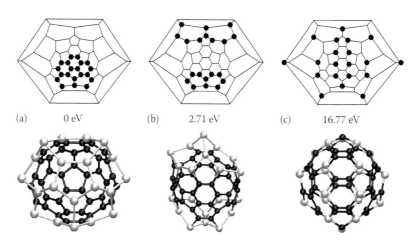

| (a) 0 eV | (b) 2.71 eV | (c) 16.77 eV |

FIGURE 32.14 In the upper part, schematic representation of Si atoms locations (black dots) in three different types of $C_{40}Si_{20}$ isomers are shown. The flattened view is a guide to the eye. Bond distances are not intended to reflect real interatomic distances. In the lower part, corresponding ball and stick representation of the structures are given. The gray and black atoms are silicon and carbon atoms, respectively.

TABLE 32.7 Calculated Bond Lengths of Isomers A, B, and C of $C_{40}Si_{20}$

Isomers		Si–Si (Å)		Si–C (Å)	C–C (Å)
A	ph (15)	2.37–2.47	ph (10)	1.89–1.98	ph 1.46–1.52
	hh (10)	2.31–2.47			hh 1.41–1.43
B	ph (12)	2.36–2.52	ph (16)	1.88–2.01	ph 1.45–1.51
	hh (10)	2.31–2.46			hh 1.41–1.43
C			ph (40)	1.83–1.96	ph 1.42–1.54
			hh (20)	1.82–1.89	hh 1.39–1.45

Note: We show the smallest and the largest values. The distances are expressed in Å. The number of occurrences of Si–Si and Si–C bonds are in parentheses.

TABLE 32.8 Calculated Bond Lengths of $C_{36}Si_{24}$ and $C_{30}Si_{30}$

Molecule		Si–Si (Å)		Si–C (Å)	C–C (Å)
$C_{36}Si_{24}$	ph (21)	2.33–2.48	ph (6)	1.91–1.93	ph 1.45–1.52
	hh (10)	2.33–2.48	hh (4)	1.84–1.85	hh 1.40–1.44
$C_{30}Si_{30}$	ph (30)	2.30–2.48			ph 1.45–1.49
	hh (10)	2.31–2.40	hh (10)	1.85–1.91	hh 1.40–1.44

Note: For C–C bond lengths, the smallest and the largest values are shown. The distances are expressed in Å. The number of occurrences of Si–Si and Si–C bonds are in parentheses.

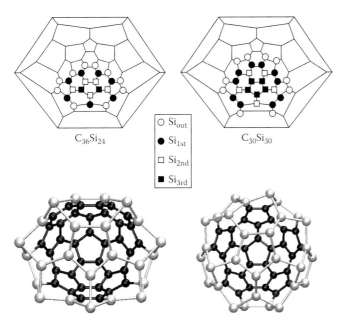

FIGURE 32.15 In the upper part, schematic representation of Si atoms locations (black dots) in $C_{36}Si_{24}$ and $C_{30}Si_{30}$ isomers are shown. The flattened view is a guide to the eye. Bond distances are not intended to reflect real interatomic distances. These Si atoms are labeled as Si_{out} (first neighbors of C atoms) and Si_{1st}, Si_{2nd}, and Si_{3rd} (first, second, and third neighbors of Si_{out} atoms, respectively). In the lower part, the corresponding ball and stick representation of the structures are given. The gray and black atoms are silicon and carbon atoms, respectively.

form a sequence of sp^2-like bonds (Si atoms). Bond lengths are fully consistent with the stability of this highly doped fullerene. We note that Si–Si bonds are in between 2.30 and 2.48 Å, Si–C bonds range from 1.84 and 1.93 Å and C–C bonds are organized in two distinct sets of ph and hh bonds (Table 32.8). The three-dimensional views for the isomers **A** of $C_{40}Si_{20}$ (Figure 32.14), $C_{36}Si_{24}$, and $C_{30}Si_{30}$ (Figure 32.15) suggest the following considerations. The heterofullerenes described so far have in common the strong configurational difference between one subnetwork (C atoms-made) and the other (Si atoms-made). The transition observed in the cage when going from the C to the Si-populated region correspond to an opening of the structure to allow for Si atoms to find a location within it. Si atoms are able to arrange in the fullerene by forming a closed cap, made by Si–Si–Si triads

involving angles largely scattered (by ~20°) around an average value close to 109.5°, the perfect tetrahedral angle. This suggests that further inclusion of Si atoms is expected to favor even more sp^3 bonding, at the expenses of weak or distorted sp^2 bonding, in agreement with the conclusions of Menon and Subbaswamy (1994) for Si_{60}. It is interesting to establish a correlation between these peculiar arrangements and the charge distribution on the doped fullerene. To this end, we have resorted to the Mulliken charges representation, given in Figure 32.16, where the values are shown for the Si atoms and their neighboring C atoms in the flattened view for $C_{36}Si_{30}$ and $C_{30}Si_{30}$. The largest values are found for those C atoms bonding Si atoms (up to −0.2) and for the outer Si atoms (Si_{out}) (+0.27 average value) corresponding to highly Coulombic Si–C bonds. This interaction is at the very origin of the stabilization of the Si–C interface and compensate the energetically unfavorable sp^2 environment of the Si atoms involved. Charges are considerably reduced inside the Si regions.

The above information leads to a natural link between the charge distribution and the thermal behavior. Two crucial issues have to addressed: (a) the determination of a threshold value n for the number of Si–C replacements and (b) the mechanism by which the doped cage becomes unstable with increasing n. Accordingly, in a series of first-principles molecular dynamics runs, temperatures were increased every 4 ps beginning from $T = 1000$ up to 1500, 2000, 3000, 4000, and 5000 K for a total temporal trajectory covering 24 ps. In the center of Figure 32.17, we show the time fluctuations with respect to the initial atomic positions for the Si_{out}, the set of Si_{in} atoms and the nearest-neighbor carbon atoms in $C_{30}Si_{30}$. The strong mobility

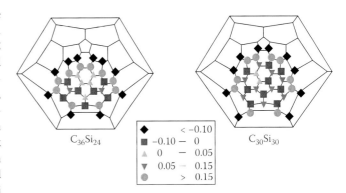

FIGURE 32.16 Mulliken charges are given for all the Si atoms and their neighboring C atoms with shading codes.

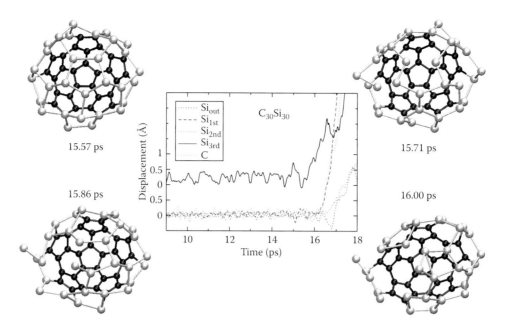

FIGURE 32.17 Snapshot of the dynamical evolution of $C_{30}Si_{30}$ during the first stages of fragmentation at $T = 3000\,K$. In the center, temporal evolution of the coordinates of relevant C and Si atoms for $C_{30}Si_{30}$ is given. The values are calculated with respect to the $t = 0$ positions and, in the case of the Si_{3rd} atoms, moved up on the Y axis by 1 Å for clarity. Note that the displacements along the first 9 ps are not shown to allow a better focus on the behavior during fragmentation.

of Si_{3rd} atoms causes a drastic structural change observed after about 15 ps. A rapid increase in the fluctuations characterizes, at $t \sim 15$ ps, not only the Si_{3rd} atoms but also all the other Si_{in} atoms, proving that the thermal instability of the cage stems from the Si–Si atoms loosely connected in the innermost Si region. Concerning Si_{out} and C atoms, they depart from their position only after irreversible diffusion of the Si_{in} atoms. Cage breaking is recorded at $T = 3000\,K$ and proceeds along the mechanism visualized in Figure 32.17 where the structure opens due to irreversible stretching of the Si_{in}–Si_{in} distances.

Calculation of the Kohn–Sham orbitals for the optimized $C_{30}Si_{30}$ at $T = 0\,K$ allows to associate the HOMO mainly to a Si_{out} atom and to a C atom at the Si–C border. The LUMO state is found on one of the five Si_{th} inner atoms (Figure 32.18a). The situation changes drastically at finite temperature ($T = 3000\,K$), as shown by a configuration selected at the beginning of the fragmentation process of $C_{30}Si_{30}$. Both the HOMO and the LUMO are now localized on Si_{in} atoms (Figure 32.18b). In addition, a substantial reduction of the gap occurs (0.53 eV at $T = 0\,K$, 0.09 eV at $T = 3000\,K$). These behaviors concur to the conclusion that temperature effects induce enhanced reactivity of the inner Si atoms, leading to a configurational change.

The above results suggest a criterion for the relationship between doping content and stability. Moving from $C_{40}Si_{20}$ to $C_{36}Si_{24}$ and eventually to $C_{30}Si_{30}$, the number of Si_{out} does not exceed 12 throughout all relevant isomers (10 in the isomers of Figure 32.15). Therefore, structural instability is due to the increased number of Si_{in} atoms. There are as many Si_{in} as Si_{out} atoms in $C_{40}Si_{20}$ (10) but the former becomes numerous in $C_{36}Si_{24}$. In the absence of thermal motion, the overall repulsion among

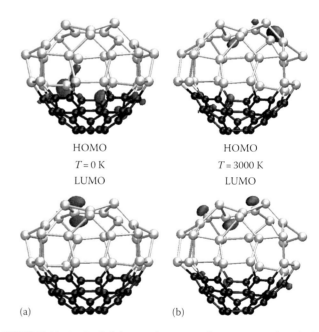

FIGURE 32.18 Probability isodensity surfaces associated with the HOMO and the LUMO levels (0.005 $e/(a.u.)^3$) of $C_{30}Si_{30}$ at $T = 0\,K$ (a) and at $T = 3000\,K$ (b), shortly before fragmentation occurs.

Si_{in} atoms results in structural distortions at the very origin of the protruding shape for the Si-made cap of the heterofullerene. Temperature induces stronger repulsive interaction among pairs of Si_{in} atoms by causing fragmentation of the cage. Therefore $C_{40}Si_{20}$, in which the number of Si_{in} atoms equals that of Si_{out} (10), is a viable candidate for the upper limit of structural stability in $C_{60-n}Si_n$. Any other structure with a higher Si_{in}/Si_{out} ratio is inevitably weakened by energetically unfavorable Si_{in}–Si_{in} sp^2 interactions. This argument is fully consistent with our FPMD simulations, in which the disruption of the cage takes place on the timescale of our simulations for $n > 20$. By drawing this kind of conclusion, we assume that on a more extended timescale, the loss of stability of the network might become observable already for $n = 20$.

32.7 Summary

We have provided a critical review of the structural, electronic, and thermal properties of Si-doped heterofullerenes by using the first-principles approach based on the DFT method.

First, we have investigated the changes in the structural and electronic properties of C_{60} when one and two C atoms are replaced by Si atoms to form the heterofullerenes $C_{59}Si$ and $C_{58}Si_2$ as the most simple examples. The structure of the fullerene turns out to be modified only in the vicinity of the dopant atoms, with Si–C bond lengths being as much as 30% larger than the C–C lengths. There is a clear correlation between the Kohn–Sham electronic energy levels in the vicinity of HOMO–LUMO gap and the electron localization on Si atoms and, to a lesser extent, on the neighboring C atoms. This goes along with an overall charge depletion from the dopant sites, which is a manifestation of the polar character of the Si–C bonds. The replacement by Si atoms introduces chemically reactive sites in the cage, affecting only locally its electronic structure. The nature of the Si–C bonds was further investigated by the analysis of the electron localization based on ELF and Wannier function approaches. The combined use of these tools provides a clear picture of the localization of single and double bonds in heterofullerenes.

In a second step, we have performed the analysis of the thermal behavior of $C_{54}Si_6$ by using finite temperature molecular dynamics simulations. As a prerequisite for the investigation of temperature effects and thermal stability, 12 different isomers are optimized at $T = 0\,K$. The account of temperature allows to conclude that the isomer with a complete Si hexagon is very much likely to be the one observed in mass-spectroscopy experiments. We show that, prior to fragmentation, these heterofullerenes have a well-defined structural identity at finite temperatures and are able to efficiently redistribute the thermal energy through non-irreversible structural changes. Moreover, the most stable structures are inevitably those featuring separate C- and Si-made subnetworks. This allows to design and characterize Si-doped heterofullerenes with a larger number of Si atoms.

Our results on $C_{40}Si_{20}$, $C_{36}Si_{24}$, and $C_{30}Si_{30}$ indicate that plausible configurations can be obtained provided the two species involved do not share the same regions in the cage. For the dopant atom, this gives rise to preferential clustering of predominantly positive charges, neighboring negatively charged C atoms. It appears that enhanced polarization is a novel feature peculiar to $C_{60-n}Si_n$ systems with high n, altering the purely covalent bonding character of undoped fullerenes. Furthermore, we have shown that highly Si-doped $C_{60-m}Si_m$ heterofullerenes are thermally stable as long as Si atoms neighbors of C atoms in segregated regions are predominant over inner Si atoms. Due to its polar character, the Si–C interaction is able to stabilize the cage and offsets the energetic cost of sp^2 interaction for $n < 20$. Beyond this threshold, repulsive interaction among inner Si atoms cause fragmentation. HOMO localization on the inner Si atoms is the chemical bonding fingerprint of this behavior.

32.8 Work in Progress and Future Perspectives

The results and the conclusions contained in this chapter have been obtained under the hypothesis that the cluster experiments can be realistically mimicked by electronic structure calculations performed on neutral systems. However, the size-selected species (Pellarin et al., 1999a,b) have been observed as charged entities, thereby requiring the account of these particular conditions to achieve full consistency with experimental findings. In a recent series of papers, we have addressed this issue by considering singly and doubly charged species in the case of $C_{30}Si_{30}$ (Matsubara et al., 2008; Matsubara and Massobrio, 2007a,b, 2009). Several actions have been undertaken such as the comparison of structural properties (bond lengths), the distribution of the charges (Mulliken analysis), and the thermal behavior among neutral, singly charged, and doubly charged species. Overall, we found that singly charged species do not differ substantially from neutral ones due to a similar distribution of charges. Charge accumulation in the inner part of the Si region is the driving force of fragmentation in the case of doubly charged $C_{30}Si_{30}$ heterofullerenes, proving that these species could not be observed experimentally. We refer the interested reader to our recent papers on this subject for more details (Matsubara and Massobrio, 2009).

The field of heterofullerenes is a fascinating domain of application for first-principles molecular dynamics techniques, due to the possibility of observing in real, physical time, the fragmentation of the cage and/or the production of new species resulting from ejection of specific atoms. In this context, a great deal of work remains to be done in the case of doping with transition metals. We find appropriate to mention here the work on Ta-doped fullerenes, in which the stability of fullerene molecules decorated with one, two, and three Ta atoms at different exohedral sites is considered (Ramaniah et al., 2004). This work is a milestone for what concerns the issue of fragmentation on exohedral, transition metal systems and should inspire further efforts in the same direction.

Acknowledgments

Calculations have been carried out on the computers of the French national centers IDRIS (Orsay) and CINES (Montpellier). The figures are generated with VMD version 1.8.6 (Humphrey et al., 1996).

References

Andreoni, W., A. Curioni, K. Holczer et al., 1996, Unconventional bonding of azafullerenes: Theory and experiment, *J. Am. Chem. Soc.* **118**(45), 11335–11336.

Andreoni, W., F. Gygi, and M. Parrinello, 1992, Impurity stated in doped fullerenes: $C_{59}B$ and $C_{59}N$, *Chem. Phys. Lett.* **190**(3–4), 159–162.

Becke, A. D., 1988, Density-functional exchange-energy approximation with correct asymptotic behavior, *Phys. Rev. A* **38**(6), 3098–3100.

Becke, A. D. and K. E. Edgecombe, 1990, A simple measure of electron localization in atomic and molecular systems, *J. Chem. Phys.* **92**(9), 5397–5403.

Bethune, D. S., R. D. Johnson, J. R. Salem, M. S. de Vries, and C. S. Yannoni, 1993, Atoms in carbon cages: The structure and properties of endohedral fullerenes, *Nature* **366**, 123–128.

Billas, I. M. L., W. Branz, N. Malinowski et al., 1999a, Experimental and computational studies of heterofullerenes, *Nanostruct. Mater.* **12**, 1071–1076.

Billas, I. M. L., C. Massobrio, M. Boero et al., 1999b, First principles calculations of Si doped fullerenes: Structural and electronic localization Properties in $C_{59}Si$ and $C_{58}Si_2$, *J. Chem. Phys.* **111**(15), 6787–6796.

Billas, I. M. L., F. Tast, W. Branz et al., 1999c, Experimental and computational studies of Si–doped fullerenes, *Eur. Phys. J. D* **9**, 337–340.

Blöchl, P. E. and M. Parrinello, 1992, Adiabaticity in first-principles molecular dynamics, *Phys. Rev. B* **45**(16), 9413–9416.

Branz, W., I. M. L. Billas, N. Malinowski, F. Tast, M. Heinebrodt, and T. P. Martin, 1998, Cage substitution in metal-fullerene clusters, *J. Chem. Phys.* **109**(9), 3425–3430.

Car, R. and M. Parrinello, 1985, Unified approach for molecular dynamics and density-functional theory, *Phys. Rev. Lett.* **55**(22), 2471–2474.

Cotton, F. A. and G. Wilkinson, 1980, *Advanced Inorganic Chemistry* (Wiley, New York), 4th edition.

CPMD, 1990–2008, Copyright IBM Corp (1990–2008) and MPI für Festkörperforschung Stuttgart (1997–2001), URL http://www.cpmd.org/.

Császár, P. and P. Pulay, 1984, Geometry optimization by direct inversion in the iterative subspace, *J. Mol. Struct.* **114**, 31–34.

Dickerson, R. E., H. B. Gray, and G. P. Haight, 1974, *Chemical Principles* (Benjamin, Menlo Park, CA).

Fu, C.-C., J. Fava, R. Weht, and M. Weissmann, 2002, Molecular dynamics study of the fragmentation of sillicon-doped fullerenes, *Phys. Rev. B* **66**(4), 045405-1–045405-7.

Fu, C.-C., M. Weissmann, M. Machado, and P. Ordejón, 2001, Ab initio study of siliconmultisubstituted neutral and charged fullerenes, *Phys. Rev. B* **63**, 085411-1–085411-9.

Fye, J. L. and M. F. Jarrold, 1997, Structures of silicon-doped carbon clusters, *J. Phys. Chem. A* **101**, 1836–1840.

Galli, G. and M. Parrinello, 1991, Ab-initio molecular dynamics: Principles and practical implementation, in *Copmuter Simulation in Materials Science*, edited by M. Meyer and V. Pontikis (Kluwer Academic Publishers, Amsterdam, the Netherlands), pp. 283–304.

Guo, T., C. Jin, and R. E. Smalley, 1991, Doping bucky: Formulation and properties of boron-doped buckminsterfullerene, *J. Phys. Chem.* **95**, 4948–4950.

Heath, J. R., S. C. O'Brien, Q. Zhang et al., 1985, Lanthanum complexes of spheroidal carbon shells, *J. Am. Chem. Soc.* **107**(25), 7779–7780.

Hohenberg, P. and W. Kohn, 1964, Inhomogeneous electron gas, *Phys. Rev.* **136**(3B), B864–B871.

Hoover, W. G., 1985, Canonical dynamics: Equilibrium phase-space distributions, *Phys. Rev. A* **31**(3), 1695–1697.

Hultman, L., S. Stafström, Z. Czigány et al., 2001, Cross-linked nano-onions of carbon nitride in the solid phase: Existence of a novel $C_{48}N_{12}$ aza-fullerene, *Phys. Rev. Lett.* **87**(22), 225503-1–225503-4.

Humphrey, W., A. Dalke, and K. Schulten, 1996, VMD—Visual molecular dynamics, *J. Mol. Grap.* **14**, 33–38.

Hutter, J., H. P. Lüthi, and M. Parrinello, 1994, Electronic structure optimization in plane-wave based density functional calculations by direct inversion in the iterative subspace, *Comput. Mater. Sci.* **2**, 244–248.

Kahn, F. S. and J. Q. Broughton, 1991, Relaxation of icosahedral-cage silicon clusters via tightbinding molecular dynamics, *Phys. Rev. B* **43**(14), 11754–11761.

Kimura, T., T. Sugai, and H. Shinohara, 1996, Production and characterization of boron– and silicon–doped carbon clusters, *Chem. Phys. Lett.* **256**, 269–273.

Kohn, W. and L. J. Sham, 1965, Self-consistent equations including exchange and correlation effects, *Phys. Rev.* **140**(4A), A1133–A1138.

Koponen, L., M. J. Puska, and R. M. Nieminen, 2008, Photoabsorption spectra of small fullerenes and Si-heterofullerenes, *J. Chem. Phys.* **128**, 154307.

Kroto, H. W., J. R. Heath, S. C. O'Brien, R. R. Curl, and R. E. Smalley, 1985, C_{60}: Buckminsterfullerene, *Nature* **318**(6042), 162–163.

Langreth, D. C. and J. P. Perdew, 1980, Theory of nonuniform electronic systems. I. Analysis of the gradient approximation and a generalization that works, *Phys. Rev. B* **21**(12), 5469–5493.

Lee, C., W. Yang, and R. G. Parr, 1988, Development of the Colle-Salvetti correlation-energy formula into a functional of the electron density, *Phys. Rev. B* **37**(2), 785–789.

Li, B.-X., M. Jiang, and P.-L. Cao, 1999, A full-potential linear-muffin-tin-orbital moleculardynamics study on the distorted cage structures of Si_{60} and Ge_{60} clusters, *J. Phys. Condens. Matter.* **11**, 8517–8521.

Marcos, P. A., J. A. Alonso, and M. J. López, 2005, Simulating the thermal behavior and fragmentation mechanisms of exohedral and substitutional silicon-doped C_{60}, *J. Chem. Phys.* **123**, 204323.

Marcos, P. A., J. A. Alonso, L. M. Molina, A. Rubio, and M. J. López, 2003, Structural and thermal properties of silicon-doped fullerenes, *J. Chem. Phys.* **119**(2), 1127–1135.

Martyna, G. J., M. L. Klein, and M. Tuckerman, 1992, Nosé–Hoover chains: The canonical ensemble via continuous dynamics, *J. Chem. Phys.* **97**(4), 2635–2643.

Marzari, N. and D. Vanderbilt, 1997, Maximally localized generalized Wannier functions for composite energy bands, *Phys. Rev. B* **56**(20), 12847–12865.

Matsubara, M. and C. Massobrio, 2005a, Bonding behavior and thermal stability of $C_{54}Si_6$: A first-principles molecular dynamics study, *J. Chem. Phys.* **122**, 084304.

Matsubara, M. and C. Massobrio, 2005b, Stable highly doped $C_{60-m}Si_m$ heterofullerenes: A first principles study of $C_{40}Si_{20}$, $C_{36}Si_{24}$, and $C_{30}Si_{30}$, *J. Phys. Chem. A* **109**, 4415–4418.

Matsubara, M. and C. Massobrio, 2006, First principles study of extensive doping of C_{60} with silicon, *Mater. Sci. Eng. C* **26**, 1224–1227.

Matsubara, M. and C. Massobrio, 2007a, Charge effects in silicon-doped heterofullerenes, *Appl. Phys. A* **86**, 289–292.

Matsubara, M. and C. Massobrio, 2007b, An extensive study of charge effects in silicon doped heterofullerenes, *Solid State Phenom.* **129**, 95–103.

Matsubara, M. and C. Massobrio, 2007c, An extensive study of the prototypical highly silicon doped heterofullerene $C_{30}Si_{30}$, in *Topics in the Theory of Chemical and Physical Systems*, edited by J. Maruani, S. Lahmar, S. Wilson, and G. Delgado-Barrio (Springer, Berlin, Germany), *Progress in Theoretical Chemistry and Physics*, vol. 16, pp. 261–270.

Matsubara, M. and C. Massobrio, 2009, Stability of charged Si doped heterofullerences: A first-principles molecular dynamic study, *Phys. Rev. B.* 79, 155411.

Matsubara, M., M. Celino, P. S. Salmon, and C. Massobrio, 2008, Atomic scale modelling of materials: A prerequisite for any multi-scale approach to structural and dynamical properties, *Solid State Phenom.* **139**, 141–150.

Matsubara, M., J. Kortus, J. C. Parlebas, and C. Massobrio, 2006, Dynamical identification of a threshold instability in Si-doped heterofullerenes, *Phys. Rev. Lett.* **96**, 155502.

Matsubara, M., C. Massobrio, and J. C. Parlebas, 2005, First principles calculation study of multi-silicon doped fullerenes, *Comp. Mater. Sci.* **33**, 237–243.

Mayer, I., 1983, Charge, bond order and valence in the ab initio SCF theory, *Chem. Phys. Lett.* **97**(3), 270–274.

Mayer, I., 1984, Bond order and valence: Relations to Mulliken's population analysis, *Int. J. Quantum. Chem.* **26**, 151–154.

Menon, M., 2001, Generalized tight-binding molecular dynamics scheme for heteroatomic systems: Application to Si_mC_n clusters, *J. Chem. Phys.* **114**(18), 7731–7735.

Menon, M. and K. R. Subbaswamy, 1994, Sutructure of Si_{60}: Cage versus network structures, *Chem. Phys. Lett.* **219**, 219–222.

Nosé, S., 1984, A molecular dynamics method for simulations in the canonical ensemble, *Mol. Phys.* **52**(2), 255–268.

Pellarin, M., C. Ray, J. Lermé et al., 1999a, Photolysis experiments on SiC mixed clusters: From silicon carbide clusters to silicon-doped fullerenes, *J. Chem. Phys.* **110**(14), 6927–6938.

Pellarin, M., C. Ray, J. Lermé et al., 1999b, Production and stability of silicon-doped heterofullerenes, *Eur. Phys. J. D* **9**, 49–54.

Perdew, J. P., J. A. Chevary, S. H. Vosko et al., 1992, Atoms, molecules, solids, and surfaces: Applications of the generalized gradient approximation for exchange and correlation, *Phys. Rev. B* **46**(11), 6671–6687.

Phillips, J. C., 1973, *Bonds and Bands in Semiconductors* (Academic, New York).

Ramaniah, L. M., M. Boero, and M. Laghate, 2004, Tantalum-fullerene clusters: A first-principles study of static properties and dynamical behavior, *Phys. Rev. B* **70**, 035411.

Ray, C., M. Pellarin, J. L. Lermé et al., 1998, Synthesis and structure of silicon-doped heterofullerenes, *Phys. Rev. Lett.* **80**(24), 5365–5368.

Resta, R., 1998, Quantum-mechanical position operator in extended systems, *Phys. Rev. Lett.* **80**(9), 1800–1803.

Rich, J. D. and R. West, 1982, Evidence for the intermediacy of hexamethyl-1,4-disilabenzene, *J. Am. Chem. Soc.* **104**(24), 6884–6886.

Roth, L. M., Y. Huang, J. T. Schwedler et al., 1991, Evidence for an externally bound Fe^+-buckminsterfullerene complex, FeC_{60}^+, in the gas phase, *J. Am. Chem. Soc.* **113**(16), 6298–6299.

Sanchez-Portal, D., E. Artacho, and J. M. Soler, 1995, Projection of plane-wave calculations into atomic orbitals, *Solid State Commun.* **95**(10), 685–690.

Segall, M. D., C. J. Pickard, R. Shah, and M. C. Payne, 1996, Population analysis in plane wave electronic structure calculations, *Mol. Phys.* **89**(2), 571–577.

Silvestrelli, P. L., N. Marzari, D. Vanderbilt, and M. Parrinello, 1998, Maximally-localized Wannier functions for disordered systems: Application to amorphous silicon, *Solid State Commun.* **107**(1), 7–11.

Troullier, N. and J. L. Martins, 1991, Efficient pseudopotentials for plane-wave calculations, *Phys. Rev. B* **43**(3), 1993–2006.

Weiss, F. D., J. L. Elkind, S. C. O'Brien, R. F. Curl, and R. E. Smalley, 1988, Photophysics of metal complexes of spheroidal carbon shells, *J. Am. Chem. Soc.* **110**(13), 4464–4465.

West, R., 1987, Chemie der Silicium-Silicium-Doppelbindung, *Angew. Chem.* **99**(12), 1231–1241.

Molecular Orbital Treatment of Endohedrally Doped Fullerenes

Lemi Türker
Middle East Technical University

Selçuk Gümüş
Middle East Technical University

33.1 Introduction

Fullerenes are ball-shaped carbon clusters. They contain a hollow space in their cage, which is suitable for endohedral doping of elements or molecules. Many scientists have been investigating the electronic and structural properties of endohedrally and substitutionally doped fullerenes and nanotubes since 1985, when the first proposal revealed (Heath et al. 1985) that fullerenes can confine an element in their interior hollow.

Various attempts have been carried out to produce, isolate, and characterize many kinds of endofullerenes by the expectation that encapsulated elements can change the physical properties of the host fullerenes. For all known metallofullerenes, the atoms inside the cage are located in off-centered positions (Nishibori et al. 2000, 2001). Furthermore, the participation of the interior element in the electric conduction in solid state was indicated by an experiment (Kobayashi et al. 2003). The production, isolation, and characterization of many metallofullerenes have been summarized in a review (Shinohara 2000).

Apart from the synthetic approaches, many scientists have been performing quantum chemical calculations to forecast the structural, physicochemical, and molecular orbital properties of endohedrally and substitutionally atom(s)/ion(s)-doped fullerenes and nanotubes. Therefore, this chapter serves as a collection of the theoretical investigations on endohedrally and substitutionally doped fullerenes, fullerene-like structures. The structural and the calculated physicochemical and molecular orbital properties of endohedrally metal(s)-, nonmetal(s)-, and ion(s)-substituted nanostructures can be found in this chapter.

33.2 Endohedrally Doped Fullerenes

33.2.1 Endohedrally Doped C$_{60}$ Systems

Endohedrally carbon atom(s)-doped C$_{60}$ has not been mentioned in the synthetic fullerene literature. However, it is quite logical to accept its possible formation in the chaotic environment while C$_{60}$ cage happens, because a number of stars known as carbon-rich stars tenuously bound and they all loose mass at a very high rate (Jura 1993). In the extended envelops of these stars, a very active chemistry occurs. Hence, possibly some endohedrally carbon-doped fullerenes may form there.

Ab initio quantum chemical calculations without symmetry restrictions have been performed to determine the structure and the electronic spectra of C@C$_{60}$ by the application of STO-3G and 6-31G(d) basis sets (Yanov and Leszczynski 2001). Two different sites (Type-A and Type-B) for the endohedral carbon atom were considered. The first position is near the center of the five-membered faces of the fullerene molecule while the second one is slightly shifted from the center of the six-membered face in the direction of the neighboring five-membered face. All employed methods are shown to give very similar geometries for the investigated system (Yanov and Leszczynski 2001).

Formation of the endohedral complex leads to a significant change in cohesive energy (the difference of total energy between C@C$_{60}$ and C$_{60}$ is approximately 39 a.u. at the HF/31G(d) level). However, a more detailed consideration, in particular an analysis of the bond orders and Mulliken overlap populations, points out that the formed bonds are much weaker than the bonds of a C$_{60}$ cluster. The change in the total energy of the system is caused basically by the symmetry break and local distortion of the molecular structure (Yanov and Leszczynski 2001).

It should be pointed out that, except for HF/STO-3G predictions, all other levels of theory give the same order of stability for the endohedral complex: the total energy of the Type-A complex is slightly higher than that of the Type-B complex.

Because of the small difference in total energy between the Type-A and Type-B complexes, the question is whether the transition of the atom from one stable position into another is possible under the influence of thermal fluctuation. The molecular dynamic calculations (with the MNDO force field model) show that the dynamic barrier of migration is only 0.12 eV for the transition from site A into site B and 0.16 eV for the transition between the sites of Type-A. Such findings mean that even at room temperatures there is a system with two metastable levels, and under certain conditions, the endohedral atom can move from one location in the molecular cage to another. It is similar to the known effect of caging (easy migration) of the impurity atom in the lattice of heavier atoms. When the electronic energy levels for the three systems, namely, C_{60}, Type-A, and Type-B systems, have been considered, it is found that the electronic spectrum predicted for those two types of stable conformations has essential differences.

In one of the studies, the structural and electronic properties of $C@C_{60}$ system was determined by the application of three methods, HF/STO-3G, HF/6-31G(d), and B3LYP/6-31G(d) (Ren et al. 2004). First, the free buckminsterfullerene molecule C_{60} was optimized, the energy minimum was found at HF/STO-3G (the lowest frequency 302.22 cm^{-1}), and the final B3LYP/6-31G(d) treatment gave bond lengths as 1.395 and 1.454 Å, respectively, for the (6,6) and (6,5) bonds, which agree satisfactorily with the measured values (David et al. 1991) as well as the results of the previous highest-level calculations, MP2 with triple ξ plus polarization basis set (Häser et al. 1991). The structure $C@C_{60}$ was constructed by placing C in various positions of the optimized C_{60} in order to obtain the correct geometric parameters.

Two spin states were considered for the $C@C_{60}$ complex; for the open-shell system, spin-unrestricted method was employed. The results were found to be distinct from the previous work (Yanov and Leszczynski 2001). For both the singlet and triplet spin states, two nonequivalent local energy minima were found, and the relative energies of the energy minima are presented. All the three levels of theory gave the same order of stability for the molecular structure. For the singlet state, the I_h cage with C on the cage center is the higher energy minimum, while in the lower energy minimum, the endohedral C shifts off the center and binds to two C atoms of the (6-6) bond in C_{60}, and the symmetry is C_{2v}. The energy order is reversed in the case of triplet system. Similar to the encapsulated N and noble gas atoms (Pang and Brisse 1993, Son and Sung 1995, Darzynkiewicz and Scuseria 1997, Weidinger et al. 1998), the ground state of $C@C_{60}$ is the triplet state with the C on the center of C_{60}, which is favored by 4.15 kcal/mol with respect to the singlet state.

In contrast to $Si@C_{60}$, in which the C_{2v} geometry is the global minimum and the Si–C bonding deforms the C_{60} structure (Lu et al. 2001, Guirado-López 2002), the cages in both the singlet and triplet $C@C_{60}$ I_h structure remains the same as that of pure

C_{60} and the endohedral C is unbound to the fullerene cage. The off-center C either in C_{2v} or C_s structure deforms the fullerene cage to a large extent due to the bonding, but the central C deforms the cage rather slightly. In the immediate vicinity of the bonding site in C_{2v} or C_s structure, the (6-5) bonds are either shortened or elongated by 0.0308–0.0641 Å, and the (6-6) bonds are slightly shortened or elongated by 0.004–0.0242 Å. Furthermore, the endohedral C-adducted (6-6) bond is elongated to 1.4749 A° in the C_{2v} structure. However, the bonds in the C_{60} cage away from the bonding sites are not much distorted from its original bonds. There are also some very moderate bond angle changes in the off-centered structures. The changes of the bond lengths, the bond angles and the twist angles of other neighboring carbon atoms lead to symmetry-reducing distortions of the cage.

The Kohn–Sham electronic levels of the most stable $C@C_{60}$ obtained at the BLYP/6-31G(d, p) level together with those of C_{60} are considered. The HOMO–LUMO gap of C_{60} is 1.666 eV, which agrees well with previous works (Dunlap et al. 1991, Green et al. 1996, Diener and Alford 1998). The C_{60} LUMO and HOMO energy gap is reduced by 1.05 eV due to the encapsulation of C atom, suggestive of the quite different chemical properties of $C@C_{60}$ and the poorer solubility of C_n fullerene in organic solvent when $n > 60$ (Diener and Alford 1998).

Mulliken population analysis gives a charge of −0.008 on the encaged C, suggestive of almost no hybridization between the central C and the fullerene cage. The spin distribution is highly symmetric and no appreciable spin transfer occurs between the central C and the cage.

Apart from the studies concerning a single-C-encapsulated structure, several C-doped systems were considered. Endohedrally, carbon-doped C_{60} systems $nC@C_{60}$, with $n = 1, 5, 10, 15$ as well as $[C@C_{60}]^{+2}$ and $[5C@C_{60}]^{+2}$ forms were subjected to AM1 (RHF)-type semiempirical quantum chemical analysis (Türker 2004). Figure 33.1 shows the geometry-optimized neutral structures of $C@C_{60}$, $5C@C_{60}$, $10C@C_{60}$, and $15C@C_{60}$. As seen there, for $n = 10$ and 15, the spherical shape of C_{60} has somewhat been distorted. Table 33.1 shows some geometrical features of the structures considered. Figure 33.2 stands for 3D-electrostatic potential field maps of them. Some calculated energies of the systems are presented in Table 33.2. The data reveal that all these endohedrally carbon-doped fullerenes are stable (the total and binding energies) but highly endothermic (the heats of formation data). Figure 33.3 shows the molecular orbital energy spectra for the neutral species. As seen in the figure, increasing the number of carbon atoms inside C_{60} causes the upper-occupied and lower-unoccupied molecular orbital energy levels to display a band-like appearance.

Thinking that $nC@C_{60}$-type structures might be generated around carbon stars with atmospheric temperatures of a couple of thousands Kelvin, they might be stripped off some of their electrons. Because of that, doubly charged cations of $C@C_{60}$ and $5C@C_{60}$ were considered, too (Türker 2004). As quite expected the cationic forms are less stable and more endothermic as compared to their respective neutral forms. In Table 33.2, one

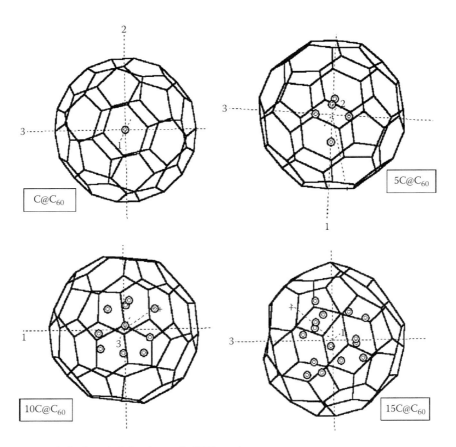

FIGURE 33.1 The geometry-optimized neutral structures of nC@C$_{60}$ systems.

TABLE 33.1 Some Features of the nC@C$_{60}$ Systems

	C@C$_{60}$	5C@C$_{60}$	10C@C$_{60}$	15C@C$_{60}$	(C@C$_{60}$)$^{+2}$	(5C@C$_{60}$)$^{+2}$
Area	541.87	544.30	553.72	576.59	543.19	544.20
Volume	1202.57	1207.76	1242.05	1298.31	1204.23	1210.23
Dipole moment	0.0000	0.9831	3.2918	2.3254	11.6842	10.7614

Note: Area, volume, and dipole moment values are on the order of $10-20\,m^2$, $10-30\,m^3$, and $10-30\,C\,m$, respectively.

can also find the HOMO and LUMO energies of the structures. The HOMO energy exhibits a local minimum for $n = 5$ case, then raises up with local maximum for $n = 10$ case, and local minimum for $n = 15$ occurs. On the other hand, the LUMO energy decreases up to $n = 10$ case and then rises up. Figure 33.4 shows the calculated spectra for the neutral species presently considered. As seen there, the increasing number of endohedrally doped carbon atoms makes the spectra more complicated by affecting the modes of vibrations. This effect should be truly through space interaction of C$_{60}$ cage with the carbon atoms inside its interior hole. Probably, the presence of carbon atoms as endohedral dopants electrostatically affects bond constants of C–C bonds of the cage, thus causing the emergence of a complicated vibrational spectrum.

Figure 33.5 shows the charge density maps for [C@C$_{60}$]$^{+2}$ and [5C@C$_{60}$]$^{+2}$ forms. In the first case, the endohedral dopant has a very minute negative charge development, whereas in the later case the carbons inside have negative or positive appreciable partial charges. It indicates that although the total charge in each case is +2, it is unequally conveyed on to the endohedral carbons through space interaction. All of these support the presence of strong interaction between the cage and the endohedral dopants. Moreover, it seems this interaction becomes more pronounced as the dopant atoms increase in number. This is quite logical because more crowded case means more effective interaction of atomic orbitals of dopants with the cage molecular orbitals. Figure 33.6 stands for the vibrational spectrum of [5C@C$_{60}$]$^{+2}$. The figure shows that the charge development on the system greatly affects the vibrational spectrum (compare with the neutral form in Figure 33.4). Both the intensities as well as the number of vibrational bands are quite different as compared to the neutral form. More vibrations are likely in the charged

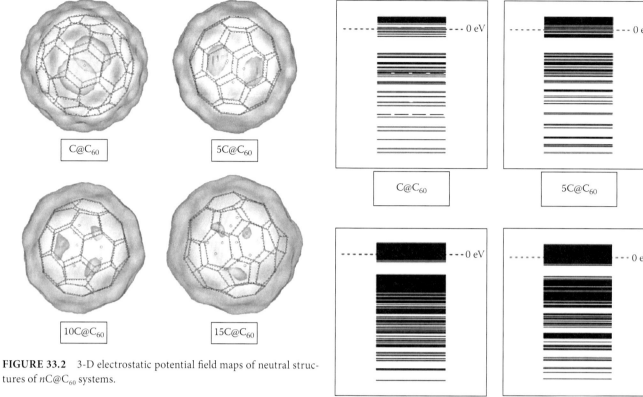

FIGURE 33.2 3-D electrostatic potential field maps of neutral structures of nC@C$_{60}$ systems.

form and it is probably due to the emergence of new bond dipoles as a result of charge distribution in the cationic form.

Several theoretical investigations (Wang et al. 1993, Lu et al. 1999a,b) have revealed that the heavier alkali-earth metal atoms Ca, Sr, and Ba have denoted two s valence electrons to the fullerene cage when they are encapsulated in the fullerene cage. But it is not necessarily the case for the entrapped Be atom since it has greater electronegativity compared with other alkali-earth metal atoms. Density functional theory calculations for Be@C$_{60}$ system was carried out using hybrid B3LYP functional in conjunction with 3-21G basis set (Lu et al. 2002). The resulting bond lengths for the 6-6 bonds (hexagon–hexagon) and for the 5-6 bonds (pentagon–hexagon) are 1.3938 and 1.4635 Å, respectively, in good agreement with the neutron diffraction results

FIGURE 33.3 The molecular orbital energy spectra of neutral structures of nC@C$_{60}$ systems.

(David et al. 1991) of 1.391 and 1.455 Å, respectively. In all calculations, effective core potential has been employed to be substituted for the core electrons. To improve the energy and charge, a larger 6-31G** basis set is used in the single-point calculations for the equilibrium geometry (Lu et al. 2002). Two spin states, $S = 3$ and $S = 1$, were considered for endohedral Be@C$_{60}$. For the open-shell electronic structure, spin-unrestricted method was employed. In each case, two energy minima were found. In the case of $S = 3$, the higher energy minimum has the Be atom on

TABLE 33.2 Some Calculated Energies of the nC@C$_{60}$ Systems

Energy	C@C$_{60}$	5C@C$_{60}$	10C@C$_{60}$	15C@C$_{60}$	(C@C$_{60}$)$^{+2}$	(5C@C$_{60}$)$^{+2}$
Total	−749,701	−797,599	−857,358	−915,331	−747,587	−795,525
Binding	−38,611	−39,881	−41,354	−41,041	−36,498	−37,807
Isolated atomic	−711,089	−757,718	−816,004	−874,290	−711,089	−757,718
Electronic	−10,167,312	−11,709,497	−13,632,342	−15,505,092	−10,155,078	−11,696,836
Core–core interaction	9,417,612	10,911,897	12,774,984	14,589,761	9,407,492	10,901,311
Heat of formation	5,003.87	6,593.80	8,695.81	12,583.89	7,116.97	8,667.79
LUMO[a]	−4.7495 AU	−4.9972 A″	−5.6473 A	−5.0801 A	−19.2539 A	−18.8235 A″
HOMO[a]	−14.9901 AU	−15.2571 A′	−13.1192 A	−13.6678 A	−25.5471 A	−25.3006 A′
ΔE[a]	10.2406	10.2599	7.4719	8.5877	6.2932	6.4771

Note: Energies in kJ/mol.

[a] Energies in the order of 10^{-19} J. Orbital symmetries are after the numeric date, ΔE: LUMO–HOMO energy gap.

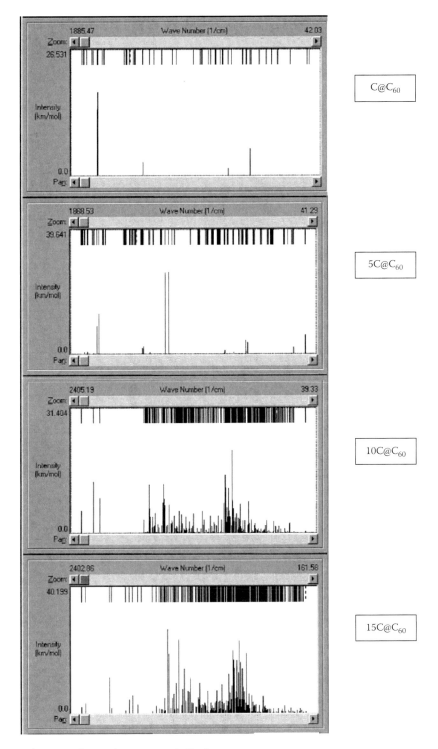

FIGURE 33.4 The vibrational spectra of neutral structures of nC@C$_{60}$ systems.

the center of the cage; the lower energy minimum has the Be atom situated about 1.8 Å off-center and bound to two C atoms of the fullerene. The symmetry of the lower energy structure is C_{2v}. Energy order in the case of $S = 1$ is opposite to the previous case. The global energy minimum corresponds to the Be on the center of C$_{60}$ in the singlet state. In the semiempirical PM3 optimizations (Lu et al. unpublished), the global energy minimum,

however, corresponds to the Be positioned about 1.8 Å off-center in the triplet state, with one of the Be 2s electrons transferred into the fullerene cage. The off-centered Be deforms the fullerene cage to a large extent, but the centric Be affects the cage rather slightly. Compared with pure C$_{60}$, the C–C 5-6 bonds of the lowest energy Be@C$_{60}$ expand by only 0.0043 Å, and the C–C 6-6 bonds shrink by as small as 0.0026 Å.

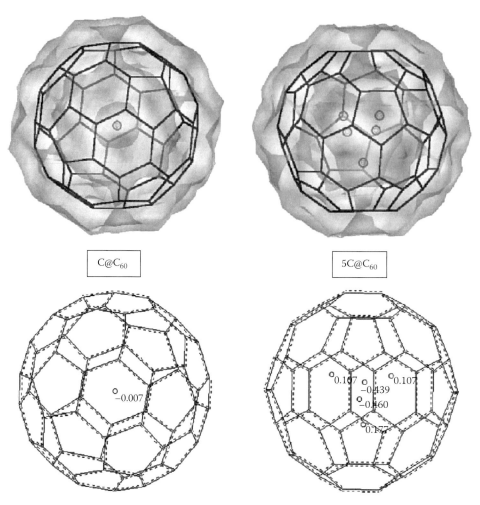

FIGURE 33.5   The charge density maps for $[C@C_{60}]^{+2}$ and $[5C@C_{60}]^{+2}$.

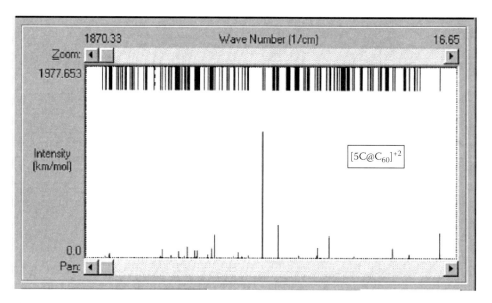

FIGURE 33.6   The vibrational spectrum of $[5C@C_{60}]^{+2}$.

The Kohn–Sham electronic levels of the most stable Be@C_{60} obtained at the BLYP/6-31G** level together with those of C_{60} and Be are compared (Lu et al. 2002). The Be atomic 2s level (−4.69 eV) lies in the middle of the C_{60} LUMO–HOMO energy gap and it is elevated by 0.16 eV upon complexation with C_{60}. The encapsulation of other metal atoms tends to lower the C_{60} LUMO and HOMO by 0.2–0.8 eV (Wang et al. 1993, Lu et al. 1999a,b, 2000). However, only the C_{60} HOMO is slightly lowered by 0.03 eV, while the C_{60} LUMO is elevated by 0.04 eV upon endohedral Be doping, suggestive of a repulsive interaction between the Be atom and the cage. As is confirmed from a binding energy of +1.05 eV at the B3LYP/6-31G** level, this molecule is unbound. There are in fact two distinct classes of fullerenes (Diener and Alford 1998), with some quite different chemical properties. The first class, typified by C_{60} and C_{70}, have large HOMO–LUMO energy gap, and are soluble in many organic solvents. The second class, typified by C_{74}, is either free radicals or has small HOMO–LUMO energy gap, and are insoluble in organic solvent. The calculations give a large HOMO–LUMO energy gap of 1.25 eV for Be@C_{60} (the theoretical HOMO–LUMO energy gap of empty C_{60} is 1.73 eV.), which accounts for the observed solubility of this compound in CS_2 (Ohtsuki et al. 1996). The vertical electron affinity and ionization potential of Be@C_{60} are calculated to be 2.22 and 7.75 eV, respectively, almost equal to the corresponding values of 2.21 and 7.73 eV for C_{60}. This result demonstrates that both the electron accepting and denoting possibilities of the C_{60} cage are nearly unaffected by endohedral Be doping.

Mulliken population analysis gives a charge of −0.14 on the Be, suggestive of a slight hybridization between the Be and the fullerene cage. Consequently, the Be 2s-derived orbital is no longer localized on the Be atom but expand to the fullerene cage, as is apparent from the molecular orbital isosurface of the Be 2s-derived orbital.

The structural and electronic properties of Li-encapsulated C_{60} systems have some important features. It is found that by doping PAS (polyacenic semiconductor) with a certain proportion of fullerene as the cathode of Li-ion battery, the charge/discharge capacity is decreased a little while its cycle life is increased (Zhao et al. 2003).

In C_{60} molecule, there are four different directions for lithium ions to be bonded. These trajectories are normal to the carbon cage toward the cage center through middle of C_{6-6} bond, middle of C_{6-5} bond, center of a pentagon, and the center of a hexagon, respectively. It has been reported that for a single lithium ion the most stable endo- and exohedral complexes might have the Li locating along the center of a carbon hexagon (Varganov et al. 2000). Thus, the energies obtained for Li^+ along hexagon and pentagon center are compared.

The binding energies between Li^+ and C_{60} have been calculated by the following formula:

$$BE(Li^+ \leftrightarrow C_{60}) = E(Li^+C_{60}) - (E(C_{60}) + E(Li^+))$$

The energies in the above equation stand for the binding energy, endo- or exohedral complex energy, fullerene energy and Li^+ free-state energy from left to right, respectively. Regardless of the endo- or exohedral complexes, as the Li^+ number gets more, the complex is energetically less favorable. Moreover, among the series, single lithium ion complexes are the most favorable. On the other hand, when the endo- and exohedral complexes are compared with each other, those exohedral complexes seem more stable. It can be concluded that lithium ions are more inclined to be absorbed outside of the carbon cage when Li^+ moves between two electrodes.

However, doping C_{60} into PAS cathode material will cause the decreasing charge/discharge capacity. For comparison, the binding energy for Li^+(PAS) complex is also reported; the results indicated that the binding energy between Li^+ and PAS is more than that of Li^+C_{60}. Due to the weak interaction between Li^+ and the layers of PAS, it is possible for Li^+ to be embedded into, and escape from, PAS cathode material back and forth so as to facilitate the electron transfer. The doped C_{60} in PAS not only occupies the partial room for Li^+ insertion, but also weakens the interaction between Li^+ and the cathode material. Thus, the charge/discharge capacity of Li-ion battery will be influenced more or less. On the other hand, due to sheet structure, layers of PAS will eventually deform or collapse after times of cycle. However, C_{60} doping can abate this problem depending on its stable electronic structure and good adsorption for Li^+. Therefore, C_{60} doping can effectively increase the cycle times for Li-ion battery.

The aromaticity of fullerenes has elicited vigorous discussion since the very first description of the structure of C_{60}, published by Osawa (1970). He wondered if C_{60} would constitute a novel, three-dimensional, aromatic system. Haddon concluded that "C_{60} is of ambiguous aromatic character" (Haddon 1993).

While quantitative estimates of aromatic stabilization energies are difficult to obtain, the magnetic criteria of aromaticity (e.g., chemical shifts and magnetic susceptibilities) provide insights in many other systems and are applicable to the fullerenes (Pasquerello et al. 1992). Although ring current effects on the chemical shifts of protons attached to exohedral substituents have already been noted (Prato et al. 1993), substituted fullerenes are by necessity missing one or more of the double bonds. The endohedral noble gas compounds (Saunders et al. 1994a,b) preserve the complete fullerene bonding and, in the case of ^{3}He [35], have allowed measurement of the magnetic field in the very center of the structure. The ^{3}He NMR chemical shifts of He@C_{60} and He@C_{70} are −6.3 and −28.8 ppm (Saunders et al. 1994a,b), respectively, were found to differ appreciably from those predicted by Pasquerello et al. (1992). A preliminary *ab initio* (GIAO-SCF) calculation was able to reproduce the experimental δ(He) value of He@C_{60} within ca. 2 ppm (Cioslowski 1994). In this study, a more detailed GIAO-SCF study calling attention to effects of cage geometries and He mobility on the computed He chemical shifts was reported (Biehl et al. 1994). Interesting comparisons are provided by the predicted δ(Li) values of the isoelectronic Li^+@ C_{60} and Li^+@C_{70}.

The computed NMR chemical shifts of endohedral guest atoms and ions are summarized. If one places a He atom in the center of icosahedral, 3-21G-optimized C_{60}, a $\delta(^3He)$ value of ca. −13 ppm is computed (employing TZP basis for He and DZ for C), which deviates somewhat from the experimental value, −6.3 ppm (Seydoux et al. 1993, Saunders et al. 1994a,b). Using the MP2/TZP-optimized C_{60} geometry (Häser et al. 1991) rather than the SCF/3-21G one affords a $\delta(He)$ value of −8.7 ppm, which is in better agreement with experiment. When the He atom is as close as 3.5 Å or less, the computed He chemical shift is found to be surprisingly sensitive to the lengths of C–C bonds. In particular, the degree of bond alternation seems critical: While the two C–C bond lengths in C_{60} differ appreciably at SCF levels (3-21G: 1.367/1.453 Å), a trend toward bond equalization is apparent in the MNDO (1.400/1.474 Å) (Bakowies and Thiel 1991), in the gas-phase electron diffraction (1.401/1.458 Å) (Hedberg et al. 1991), and in the MP2/TZP (1.406/1.446 Å) (Häser et al. 1991) geometries. This structural trend is paralleled by a decrease of the computed endohedral He shielding (δ = −13, −12, −11, and −9 ppm, respectively, in the same sequence). For the "idealized" geometry with equal C–C bond lengths of 1.4 A° (as employed by Haddon in his London HMO calculations [Pasquerello et al. 1992]), a value of ca. −4 ppm is computed.

It has been estimated that the ^3He atom senses the magnetic field only inside a sphere of less than 1 Å diameter at the center of C_{60} (Saunders et al. 1994a,b). The field within this sphere appears to be very homogeneous, as displacement of the He out of the center (by 0.5 Å along one fivefold axis) leaves the computed $\delta(He)$ virtually unchanged.

Compared to C_{60}, the 3-21G structure of C_{70} performs slightly better that is the computed chemical shift of an encapsulated ^3He atom, −24 ppm, deviates less than 5 ppm from the experimental value of −28.8 ppm (Saunders et al. 1994a,b). Practically the same result, −24 ppm, is obtained when the experimental (x-ray) C_{70} geometry (Roth and Adelmann 1992) is employed. The bond alternation in the equatorial six-membered rings is already quite small at SCF levels (3-21G, 1.403 and 1.414 Å; cf. x-ray, 1.407 and 1.430 Å) (Roth and Adelmann 1992), respectively. Unfortunately, no MP2-optimized structure of C_{70} has been reported to date. As for He@C_{60}, the exact position of the He atom in the C_{70} cage hardly affects $\delta(He)$, thus displacements along the fivefold (by 1.0 Å) and twofold axes (by 0.5 Å toward an equatorial six-membered ring) afford changes in the computed $\delta(He)$ of only +0.1 and −0.1 ppm, respectively. As expected, the computed He atomic charge (employing Mulliken population analysis) is 0 in both He@C_{60} and He@C_{70}. Thus, the substantial difference in $\delta(He)$ for both molecules is qualitatively consistent with the proposed ring current models (Pasquerello et al. 1992).

^7Li chemical shifts are a useful probe of ring current effects in organolithium compounds (Paquette et al. 1990). Since Li^+ is isoelectronic with He, writers were interested in the $\delta(Li)$ values of $Li^+@C_{60}$ and $Li^+@C_{70}$. Placing Li^+ in the centers of the C_{60} and C_{70} cages affords GIAO-SCF chemical shifts of −14.5 and −29.9 ppm, respectively. Moving Li^+ out of the center in C_{60} (by 1.4 A along a fivefold axis as suggested by LDF calculations Dunlap et al.

1992]) affects $\delta(Li)$ only slightly, i.e., by less than 1 ppm. The computed upfield shift of Li^+ in C_{60} (6–15 ppm.) is substantial and is comparable to that of Li^+ sandwiched between two cyclopentadienyl anions (δ = −13.1 ppm [Paquette et al. 1990]). The calculated upfield shift of Li^+ in the C_{70} cage is even larger, suggesting that ^7Li NMR could readily distinguish between exo- and endohedral Li^+-fullerene complexes. For both He and Li^+ in C_{70}, notable chemical shift anisotropies of $\Delta\sigma \approx −12$ ppm are predicted.

Ring currents and their directly observable effects upon chemical shieldings (Fleischer et al. 1994) are important diagnostic indicators of aromaticity. The results of calculations support earlier conclusions that C_{60} is aromatic, but only to a modest extent. The aromaticities in C_{70} and especially in C_{60}^{6-} are greater, judging from the magnetic criteria.

Endohedral metallofullerenes have long attracted particular interest since their corresponding bulk materials are potential conductors with novel properties. The most frequently discussed question is the charge transfer from the metal atom to the fullerene cage.

Both Saito and Oshiyama (1992) and Wang et al. (1993) have found that two Ca 4s valence electrons are donated to the C_{60} shell based on the local density approximation method and Hatree–Fock method, respectively. However, Cioslowski and Fleischmann (1991) have done restricted Hatree–Fock calculations on Ca@C_{60} and found only one Ca 4s electron is donated to the C_{60} shell. The ultraviolet photoelectron spectroscopy (UPS) studies on CaC_{60}^- revealed that the photoelectron spectrum of Ca@C_{60}^- is shifted toward the high energy direction by about 0.3 eV compared with that of C_{60}^-, and correspondingly the electron affinity (EA) of Ca@C_{60}, 3.0 eV, is higher than that of C_{60}, 2.65 eV. The electron paramagnetic resonance (EPR) studies on Sc@C_{82} (Shinohara et al. 1992a, Suzuki et al. 1992, Yannoni et al. 1992) and Y@C_{82} (Hoinkis et al. 1992, Shinohara et al. 1992b, Weaver et al. 1992) indicate that the hyperfine coupling constants are very small and the g value is close to the free-spin value. It has been suggested that the EPR feature is unambiguously diagnostic for the fact the Sc and Y atom donated their three valence electrons to C_{82} to form endohedral complex with $Sc^{3+}@C_{82}^{3-}$ and $Y^{3+}@C_{82}^{3-}$. However, the theoretical investigations (Nagase and Kobayashi 1993) at the Hatree–Fock level predict that the 3d of Sc (4d of Y) is strongly stabilized and the effective charges on Sc and Y atom are closer to +2 rather than +3. In the light of these disagreements, the electronic structures of Ca@C_{60}, Sc@C_{60}, and Y@C_{60} (McElvany 1992) were examined by using the discrete-variational local density functional method (DV-LDFM) with Ceperley–Alder exchange correlation potential (Lu et al. 1999a,b).

The 2s, 2p for C; 3s, 3p, 4s for Ca; 3s, 3p, 3d, 4s for Sc; and 4s, 4p, 4d, 5s for Y were explicitly treated. The basis functions were generated from the configurations, $2s^2 2p^{1.54}$ for C, $3s^2 3p^6 4s^{0.2}$ for Ca, $3s^2 3p^6 3d^{0.2} 4s^{0.2}$ for Sc, and $4s^2 4p^6 4d^{0.25} 5s^{0.2}$ for Y. The deeper core electrons were frozen. The ionization potentials (IP) were evaluated with the transition-state procedure (Lu et al. 1999a,b).

Upon placing a Ca atom at the center of C_{60}, all C_{60}-dominated levels are slightly lowered by 0.3–0.6 eV. This trend is qualitatively consistent with the observed shift in the photoelectron spectrum (PES) (Wang et al. 1993) except with greater magnitude. The descending trend of the C_{60}-derived levels upon endohedral metal-atom doping (Wasterberg and Rosen 1991, Diener and Alford 1998) is contrary to the outside complex K_3C_{60}, in which the exohedrally doped K atom has little effect on C_{60} electronic state (Satpathy et al. 1993). The t_{1u} (LUMO) and h_u (HOMO) levels are lowered by 0.52 and 0.62 eV, respectively. Two Ca 4s electrons are donated to the t_{1u} state of the cage, leading to the net charge of $Ca^{1.67+}@C_{60}^{1.67-}$ with the 4s electron density of 0.33. The empty a_g state consisting mainly of Ca 4s state is found to lie slightly (only 0.04 eV) above the t_{1u} state, as calculated by Saito and Oshiyama (1992). Wang et al. (1993) observed that the first single peak of C_{60} in the PES is not split upon endohedral Ca doping; therefore, they proposed that it serves as an evidence against a ground state of $a^1_g t^1_{1u}$ suggested by Cioslowski and Fleischmann (1991). However, this evidence is not powerful since the a_g and t_{1u} orbitals are nearly degenerate. An electron affinity of 2.52 and 3.10 eV for C_{60} and $Ca@C_{60}$ is obtained in agreement within 0.1 eV with the experimental values of 2.65 and 3.0 eV (Wang et al. 1993) (Other theoretical studies reported EA of C_{60} to be 2.57 (Nagase and Kobayashi 1994), 2.7 (Wasterberg and Rosen 1991), and 2.88 eV (Green et al. 1996)). Together with the observed EA of 2.9 eV for $Gd@C_{60}$ (Boltalina et al. 1997), the earlier local density functional calculations by Wasterberg and Rosen (Wasterberg and Rosen 1991), ranging from 3.7 to 4.2 eV for $Li@C_{60}$, $Na@C_{60}$, $K@C_{60}$, $Rb@C_{60}$, and $Cs@C_{60}$ appear to overestimate the EA values. The obtained ionization potential of $Ca@C_{60}$ is 6.68 eV, remarkably greater than a value of 4.51 eV given by Chang, Ermeler, and Pitzer using the Hatree–Fock method (Cioslowski and Fleischmann 1991), but compared with values of 5.5–6.5 eV for alkali atom-intercalated C_{60} by the local density functional calculations (Wasterberg and Rosen 1991).

Since Sc is next to Ca in the elemental periodic table, and only has one more 3d electron than Ca, the C_{60}-derived levels of $Sc@C_{60}$ are quite close to those of $Ca@C_{60}$ upon replacing Ca with Sc atom at the center of C_{60}. Two Sc 4s electrons are transferred into the LUMO of C_{60}, while the 3d electron is strongly stabilized inside C_{60}. Almost one 3d electron is left behind with the d-electron density of 1.16 on Sc, as calculated for $Sc@C_{82}$ (Nagase and Kobayashi 1993). The net Mulliken population analysis gives $Sc^{1.69+}@C^{1.69-}_{60}$. This is clearly in disagreement with the proposed +3 oxidation of Sc based on the EPR data. The Sc 4s state with higher atomic level (Herman and Skillman 1963) rises up by 0.34 eV relative to the Ca 4s state and the 3d-derived orbital is intercalated between the 4s and t_{1u} states. Unlike $Sc@C_{82}$; in which the HOMO is only slightly lowered while the LUMO is remarkably lowered by about 4 eV, the LUMO and HOMO in $Sc@C_{60}$ are slightly lowered in the similar magnitude (0.65 eV and 0.57 eV for the HOMO and LUMO, respectively) upon endohedral Sc doping. Because the ionization comes from

the higher-lying Sc 3d orbital rather than the t_{1u} orbital, the EA (2.78 eV) and IP (6.53 eV) of $Sc@C_{60}$ are smaller than those of $Ca@C_{60}$.

When the Y atom is placed at the center of C_{60}, all C_{60}-derived levels are slightly elevated relative to those of $Ca@C_{60}$ and $Sc@C_{60}$, but still lower than those of C_{60}. This slight rise in levels is responsible for the smaller EA (2.60 eV) and IP (6.34 eV) in $Y@C_{60}$ than in $Sc@C_{60}$ since the levels are less deeply bound in $Y@C_{60}$. The 5s state lies high at –3.45 eV because the Y 5s atomic level is higher than the Sc 4s atomic level (Herman and Skillman 1963), and is more fully ionized (compare 0.05 charge on Y 5s state with 0.15 charge on Sc 4s state). However, one 4d electron remains with the d-electron density of 1.21 on Y, leading to the net population of $Y^{1.75+}@C^{1.75-}_{60}$, as calculated for $Y@C_{82}$ (Nagase and Kobayashi 1993). The more charge transfer between the Y atom and the cage than that between the Sc atom and the cage is also calculated for $Sc@C_{82}$ and $Y@C_{82}$ (Nagase and Kobayashi 1993).

In conclusion, the local density functional calculations on $Ca@C_{60}$, $Sc@C_{60}$, and $Y@C_{60}$ suggest that Ca, Sc, and Y atoms are all in +2 oxidation state, and both Sc 3d and Y 4d electrons are strongly stabilized, as calculated for $Sc@C_{82}$ and $Y@C_{82}$ at the Hatree–Fock level (McElvany 1992). The greater electron affinity obtained for $Ca@C_{60}$ than for C_{60} is in good agreement with the experiment.

33.2.2 Endohedrally Doped Hydrogenated C_{60} Systems

Hydrogenation of fullerenes to fulleranes has remained an attractive field of research to many scientists since the very beginning of "Fullerenes era" (Hirsch 1994). However, both the synthesis and the characterization of hydrogenated products are hard tasks. The polyhydrofullerenes exhibit certain instability, too (Hirsch 1994). But, a limited number of energetically favorable isomers can be expected if stoichiometrically controlled additions to just one or a few reactive double bonds of C_{60} (or C_{70}) are considered.

The regio and stereoisomers of $C_{60}H_2$ and $C_{70}H_2$ constitute good examples for those thermodynamically and/or kinetically favored isomers (Hirsch et al. 1992, 1993, Ballenweg et al. 1993, Guarr et al. 1993, Henderson and Cahill 1993). In the case of C_{60}, the double bonds which could undergo vicinal hydrogenation can be categorized as 5-6 and 6-6-type bonds. The resulting hydrogenated product will be $C_{60}H_2$. Theoretically all the regioisomers should have stereoisomers with endo- and exo-hydrogens (In- and Out-forms).

Endohedrally Mg-doped, hydrogenated product ($C_{60}H_2$) has been subjected to theoretical analysis, at the level of PM3-type semiempirical quantum chemical calculations, expecting that the complex may have interesting structural and electronic properties (Türker 2002b).

The geometry-optimized three-dimensional structures of In-56Mg@$C_{60}H_2$, In-66Mg@$C_{60}H_2$, Out-56Mg@$C_{60}H_2$,

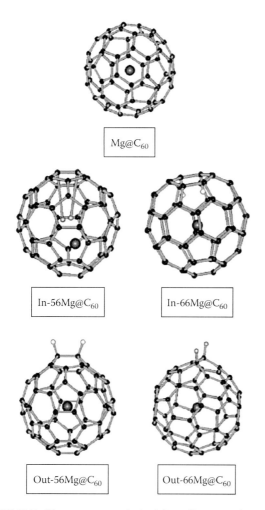

FIGURE 33.7 The geometry-optimized three-dimensional structures of the molecules considered.

TABLE 33.3 Some Calculated Properties of the Structures Considered

	$Mg@C_{60}$	In-56	In-66	Out-56	Oat-66
Area	545.72	545.11	548.05	552.30	548.93
Volume	1207.03	1201.63	1209.81	1222.96	1220.55
Polarizability	113.81	110.73	110.73	112.91	110.73
Dipole moment	0.0000	0.8905	0.1768	7.0714	8.0187
Charge on Mg	−0.1306	−0.4074	−0.1495	−0.1316	−0.1310
C–H bond lengths	—	3.20	1.17	1.11	1.11
		3.29	1.17	1.11	1.11
Molecular point group	C_1	C_{2v}	C_2	C_1	C_{2v}

Note: Area, volume, polarizability, dipole moment values, and bond lengths are in the older of $10^{-20}\,m^2$, $10^{-30}\,m^3$, $10^{-30}\,m^3$, $10^{-30}\,C\,m$, and $10^{-10}\,m$, respectively.

charge accumulation on Mg atom (see Table 33.3) indicates that C_{60} cage ligands to Mg atom using its empty atomic orbitals. Table 33.3 shows C–H bond lengths of the other isomers as well as some other calculated properties of these structures.

Various energies of the systems of consideration are shown in Table 33.4. As seen in the table, all these structures are stable (the total and binding energies but endothermic in nature (the heats of formation data). Of the various isomeric structures considered, In-56$Mg@C_{60}H_2$ is the most stable and the least endothermic one which also indicates that certain stabilizing interactions occur between the hydrogens in In-56 isomer and Mg atom. On the other hand, In-66$Mg@C_{60}H_2$ structure is the least stable and the most endothermic among the isomeric systems considered. This observation could be tried to be explained by means of the "isolated pentagon rule" (IPR) (Hirsch 1994). It is known that among the various isomeric structures having the empirical formula, C_{60}, the one in which all the pentagons are surrounded by hexagons represents the most stable structure. In the case of hydrogenation of such a structure, the rule is violated because of the imported sp^3-type carbon atom, bearing the hydrogens. Thus, certain pentagons are not wholly circumferred by hexagons. In the case of In-56 isomer, only one pentagon but in the case of In-66 isomer two pentagons suffer from this effect. Consequently, In-66$Mg@C_{60}H_2$ structure should be thought to be less stable and more endothermic than the other regioisomer, In-56$Mg@C_{60}H_2$. However, when the energetics of Out-56$Mg@C_{60}H_2$

Out-66$Mg@C_{60}H_2$ as well as the nonhydrogenated metallofullerene $Mg@C_{60}$ can be seen in Figure 33.7.

As seen in the figure, the C–H bonds in the case of In-56$Mg@C_{60}H_2$ structure are highly elongated (3.29 and 3.20 Å). This implies that the C–H bonds are almost broken but some quasi-hydridic interaction with magnesium atom occurs. However, in the case of In-66$Mg@C_{60}H_2$ structure such kind of occurrence is not expected (both C–H bonds are 1.17 A°, respectively). On the other hand, the charges on Mg atoms are −0.4073 and −0.1495 unit of charge, respectively, for In-56 and In-66 forms. Negative

TABLE 33.4 Various Energies of the Systems of Interest

Energy	$Mg@C_{60}$	In-56	In-66	Out-56	Out-66
Total	−684,507	−688,104	−686,888	−687,546	−687,620
Binding	−38,394	−39,468	−38,253	−38,911	−38,985
Isolated atomic	−646,113	−648,635	−648.635	−648,635	−648,635
Electronic	−9,794,736	−10,003,149	−99,522,732	−9,913,551	−9,915,710
Core–core repulsion	9,110,229	9,315,044	9,265,843	9,226,005	9,228,090
Heat of formation	4,652.58	4,013.88	5,229.51	4,571.57	4,497.59

Note: Energies in kJ/mol. In-56 (66) and Out-56 (66) stand for In and Out-forms of $Mg@C_{60}H_2$.

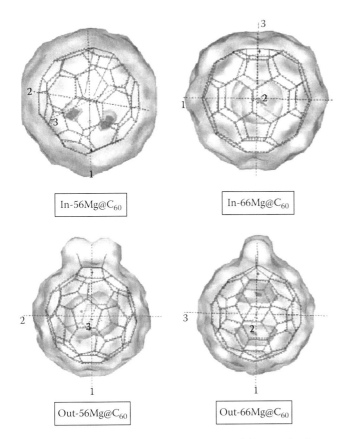

In-56Mg@C$_{60}$

In-66Mg@C$_{60}$

Out-56Mg@C$_{60}$

Out-66Mg@C$_{60}$

FIGURE 33.8 Three-dimensional electrostatic field maps for the isomers considered.

and Out-66Mg@C$_{60}$H$_2$ are inspected, the latter one seems to be more stable and less endothermic than the former one. Figure 33.8 shows three-dimensional electrostatic potential field maps for the isomers considered.

Table 33.5 shows the HOMO, LUMO energies and the interfrontier energy gaps (ΔE) for these systems. As seen there, although Out-56 and Out-66 isomers have comparable HOMO, LUMO energies and ΔE values as compared to Mg@C$_{60}$ system, In-56 and especially In-66 isomers have quite different respective values than of Mg@C$_{60}$. The HOMO, LUMO energies, and ΔE values of the isomers follow the order of In-66 < In-56 < Out-56 < Out-66. The molecular orbital energy spectra for these structures are shown in Figure 33.9. There, In-56 isomer is distinguishable from the other isomers by having rather discrete upper-occupied energy levels.

The study revealed some of the properties of the C$_{60}$H$_2$ structures. In-56 isomer is especially interesting indicating that some quasi-hydride formation in between the Mg atom and inwardly located hydrogens.

The following study considers the endohedrally Be-doped C$_{60}$H$_2$ structures (Türker 2002a,b,c). The structural and electronic properties of the systems are calculated using AM1-type semiempirical quantum chemical calculations. Figure 33.10 shows the geometry-optimized structures of In-56 and In-66-type Be@C$_{60}$H$_2$. As seen in the figure, the extremely elongated (3.85, 3.18, and 3.96, 3.07 Å, respectively) C–H bonds show that as compared to In-56 and In-66 forms of C$_{60}$H$_2$ systems, In-Be@C$_{60}$H$_2$ systems should be actually BeH$_2$@C$_{60}$ system where a quasi-beryllium hydride formation should exist.

Tables 33.6 and 33.7 show various calculated properties of the systems of present interest. A striking fact is the comparatively very high dipole moments for Out-forms of either undoped or Be-doped systems. The direction of the dipole moment vectors in each of the Out-forms is from the hydrogenated site to the center of the axes of inertia. Tables 33.8 and 33.9 show the HOMO and LUMO energies and the interfrontier energy gaps for C$_{60}$H$_2$ and Be@C$_{60}$H$_2$ systems (actually BeH$_2$@C$_{60}$ system). As seen in Table 33.8, the LUMO energies of In-forms of C$_{60}$H$_2$ structures are lower than the respective energies of the related Out-forms, whereas the HOMO energies for In-forms are higher than the respective energies of the related Out-forms. On the other hand, the LUMO energy of In-56 form is lower than the LUMO energy of In-66 form, but the HOMO energy for In-56 is higher than the respective energy for In-66 form. A similar trend holds for Out-56 and Out-66 pair.

As for the Be-doped species (see Table 33.9), the energies for In-56 and In-66 forms are identical because of the formation of quasi-BeH$_2$ structure, both forms, initially different, eventually correspond to the same structure BeH$_2$@C$_{60}$ as the result of geometry optimization. The LUMO energy of this species lies in between the respective energies of Out-56 and Out-66 forms, whereas the HOMO is much lower than the HOMO energies of the Out-forms. As compared to the undoped C$_{60}$H$_2$ systems, the presence of Be atom raises the LUMO but lowers the HOMO energies while correlating to the HOMO and LUMO energies of BeH$_2$@C$_{60}$ system. Figure 33.11 shows the HOMO and LUMO of BeH$_2$@C$_{60}$ system.

In another study AM1-type semiempirical method was applied to investigate the thermodynamic properties of isomeric Be@C$_{60}$H$_2$-type systems (Türker 2002a,b,c). Some energies of

TABLE 33.5 The HOMO, LUMO Energies and the Interfrontier Energy Gaps (ΔE) for the Presently Considered Structures

Energy	Mg@C$_{60}$	In-56	In-66	Out-56	Out-66
LUMO	−4.6625 (A$_U$)	−16.5966 (A$_1$)	−26.5226 (A)	−4.7911 (A)	−4.5374 (A$_1$)
HOMO	−13.8705 (A$_U$)	−16.6174 (B$_2$)	−26.6301 (B)	−13.6240 (A)	−13.5910 (A$_1$)
ΔE	9.2080	0.0208	0.1075	8.8329	9.0536

Note: Energies in the order of 10^{-19} J. In-56 (66) and Out-56 (66) stand for In and Out-forms of Mg@C$_{60}$H$_2$.

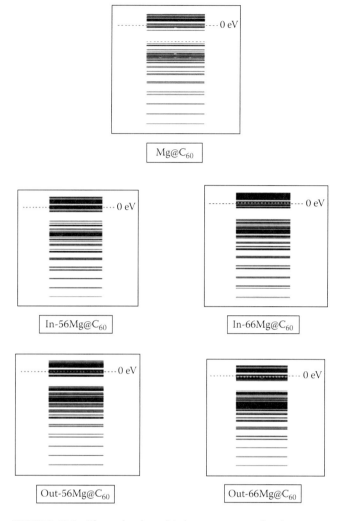

FIGURE 33.9   The molecular orbital energy spectra for the systems considered.

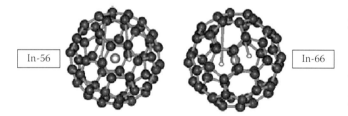

FIGURE 33.10   The geometry-optimized structures of In-56 and In-66 types Be@$C_{60}H_2$ isomers.

$C_{60}H_2$ isomers are shown in Table 33.10. Inspection of the data reveals that all the structures are stable (the total and binding energies) but endothermic (the heats of formation) and the stability order is: Out-66 > Out-56 > In-66 > In-56 (numbers stand for the position of vicinal hydrogenation in a six-membered or a five-membered ring or both). Table 33.11 shows some of the energy differences between the stereoisomers of a specific regioisomer of $C_{60}H_2$. The data reveal that the stability of Out-66 over In-66 isomer is much greater than the stability of Out-56

TABLE 33.6   Some Properties of the Presently Considered $C_{60}H_2$ Structures

	Structure			
	In-56	In-66	Out-56	Out-66
Area	543.09	542.48	555.44	551.23
Volume	1206.29	1204.48	1218.86	1215.85
Hydration energy	−1.1296	−1.1296	−1.0460	−1.0460
Log P	9.44	9.44	9.44	9.44
Refractability	232.46	232.46	232.46	232.46
Polarizability	112.86	112.86	112.86	12.86
Molecular point group	C_s	C_{2v}	C_s	C_{2v}

Note: Area, volume, refractability, polarizability, and dipole moment values are in the order of $10^{-20}\,m^2$, $10^{-30}\,m^3$, $10^{-30}\,m^3$, $10^{-30}\,m^3$, and $10^{-30}\,C\,m$, respectively. Hydration energy is in kJ/mol.

TABLE 33.7   Some Properties of the Presently Considered Be@$C_{60}H_2$ Structures

	Structure			
	In-56	In-66	Out-56	Out-66
Area	544.30	538.43	552.27	553.11
Volume	1205.19	1203.55	1220.37	1220.36
Polarizability	112.87	112.87	112.87	112.87
Dipole moment	0	0	8.7525	9.6330
Molecular point group	D_{3d}	D_{3d}	C_s	C_{2v}

Note: Area, volume, polarizability, and dipole moment values are in the order of $10^{-20}\,m^2$, $10^{-30}\,m^3$, and $10^{-30}\,C\,m$, respectively.

TABLE 33.8   The HOMO, LUMO Energies and the Interfrontier Energy Gaps for $C_{60}H_2$ Isomers

	Structure			
Energy	In-56	In-66	Out-56	Out-66
LUMO	−5.1319 A	−4.8129 B_1	−4.8708 A′	−4.5416 A_1
HOMO	−14.1687 A″	−14.6498 B_2	−14.3017 A″	−14.8364 B_2
ΔE	9.0368	9.8369	9.4309	10.2948

Note: Energies in kJ/mol; $\Delta E = \varepsilon_{LUMO} - \varepsilon_{HOMO}$.

TABLE 33.9   The HOMO, LUMO Energies and the Interfrontier Energy Gaps for Be@$C_{60}H_2$ Isomers

	Structure			
Energy	In-56	In-66	Out-56	Out-66
LUMO	−4.7282 A_{2U}	−4.7282 A_{2U}	−4.8491 A′	−4.5172 A_1
HOMO	−15.3440 E_U	−15.3440 E_U	−13.1878 A′	−13.1766 A_1
ΔE	10.6158	10.6158	8.3387	8.6594

Note: Energies in kJ/mol; $\Delta E = \varepsilon_{LUMO} = \varepsilon_{HOMO}$.

isomer over the In-56. Some energies of Be@$C_{60}H_2$ systems are shown in Table 33.12. These structures are also stable but endothermic. The data indicate that an endohedrally doped Be atom in $C_{60}H_2$ significantly changes the stability order of the respective Be@$C_{60}H_2$ isomers as compared to $C_{60}H_2$ isomers that is

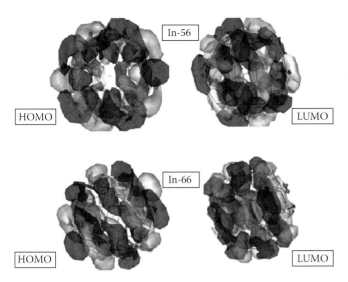

FIGURE 33.11 The HOMO and LUMO of some BeH$_2$@C$_{60}$H$_2$ systems.

TABLE 33.10 Some Energies of Vicinally Hydrogenated C$_{60}$
Isomers, C$_{60}$H$_2$

Energy	Structure			
	In-56	In-66	Out-56	Out-66
Total	−740,285	−740,307	−740,996	−741,075
Binding	−38,653.5	−38,675.5	−39,364.9	39,443.8
Isolated atomic	−701,631	−701,631	−70l,631	−701,631
Electronic	−9,966,513	−9,966,376	−9,922,389	−9,923,582
Core–core repulsion	9,226,228	9,226,069	9,181,393	9,182,507
Heat of formation	4,682.67	4,660.71	3,971.29	3,892.45

Note: Energies in kJ/mol.

TABLE 33.11 Some Energy (*E*)
Differences for C$_{60}$H$_2$ Isomers

Energy	Regioisomer	
	56-Type	66-Type
Total	711.00	768.00
Binding	711.40	768.30
Heat of formation	711.38	768.26

Note: Energies in kJ/mol. Calculated as
$E_{In} - E_{Out}$.

In-56 = In-66 > Out-66 > Out-56. Note that In-56 and In-66 are degenerate structures in the case of Be@C$_{60}$H$_2$.

Theoretical calculations on C$_{60}$H$_{60}$ structure with 10 hydrogens added inside the C$_{60}$ cage revealed that it is the lowest energy isomer and its heat of formation was found to be 400 kcal/mol lower than for the all-outside isomer (Saunders 1991). In another study, the comparison on the basis of *ab initio* calculations of the stabilities of totally exo-hydrogenated C$_{60}$H$_{60}$ and T$_h$-C$_{60}$H$_{60}$ with 12 hydrogens added inside the cage predicts the latter conformer to be more stable (Guo and Scuseria

TABLE 33.12 Some Energies of Be@C$_{60}$H$_2$ Isomers

Energy	Structure			
	In-56	In-66	Out-56	Out-66
Total	−743,154	−743,154	−743,052	−743,131
Binding	−39,187	−39,187	−39,085.9	−39,164.2
Isolated atomic	−703,967	−703,967	−703,967	−703,967
Electronic	−10,148,158	−10,148,158	−10,089,611	−10,090,812
Core–core repulsion	9,405,006	9,405,006	9,346,557	9,347,682
Heat of formation	4,471.19	4,471.19	4,572.35	4,494.06

Note: Energies in kJ/mol.

1992). The all-outside isomer is highly strained due to eclipsing hydrogen–hydrogen interactions. However, such kind of eclipsing repulsions in lower-hydrogenated C$_{60}$ systems such as C$_{60}$H$_2$ isomers should be negligible and hence the AM1 calculations predict Out-forms to be more stable than the corresponding In-forms. However, the presence of an endohedrally doped Be atom changes the situation and this time In-C$_{60}$H$_2$ cage with Be atom inside becomes more stable. However, the elongated C–H bonds are the indication of some sort of interaction with Be atom (1s^{2}2s^2). The calculations show some partial negative charge development on the hydrogens, thus the interaction is a kind of hydridic type. Another interesting point in the results of these calculations is that the elongated C–H bonds in the same 56- or 66-type structure are not equal and the difference is very notable. In contrast to unequal C–H bonds, the difference between Be atom and each hydrogen is 1.2642×10^{-10} m, irrespective of the structure (56- or 66-type regioisomer). Note that this distance is much shorter than most of the reported Be–C bonds (Vilkov et al. 1970). It is known that BeH$_2$ can be obtained by the action of hydrogen on metallic beryllium (Durant and Durant 1970). It is a white nonvolatile compound, tending to be polymerized and it was suggested that this is affected by hydrogen bridging (Cotton and Wilkinson 1967). The interaction in between Be and hydrogens in the present case most probably stands for a quasi-BeH$_2$ structure but close to genuine BeH$_2$. Thus, the original location of hydrogens on the cage does not affect the calculations which yield degenerate final structures, within the limitations of the AM1 method.

In conclusion, In-forms of the endohedrally Be-doped C$_{60}$H$_2$ structures are more stable than the Out-forms. The stabilization is most probably due to the quasi-beryllium hydride formation.

33.2.3 Other Endohedrally Doped Fullerenes

The study of endohedral fullerenes is not restricted to the C$_{60}$ structure. Below some endohedrally atom/ion-doped systems different from C$_{60}$ will be considered. There has been research interest on C$_{72}$ and Ca@C$_{72}$ (Kobayashi et al. 1997, Wan et al. 1998, Slanina et al. 1999a,b,c), but of the two species only

Ca@C_{72} has been isolated so far (Wan et al. 1998). C_{72} has a large HMO HOMO–LUMO gap (larger than that of C_{70}) (Fowler and Manolopoulos 1995). C_{74} with a considerably smaller HMO HOMO–LUMO gap has recently been produced electrochemically (Diener and Alford 1998). C_{72} has only one isolated-pentagon rule (IPR) structure [86]. However, it has been established that non-IPR structures should be significant for both C_{72} and Ca@C_{72} (Kobayashi et al. 1997, Wan et al. 1998).

It has been known (Slanina 1987, Slanina et al. 1996, 1999a,b,c) that the composition of isomeric fullerene mixtures cannot always be understood in terms of computed potential energies only. Sometimes a structure which does not have the lowest-in-potential energy is actually the most abundant. Such events could be explained (Slanina 1987, Slanina et al. 1996, 1999a,b,c) after addition of the entropy effects, as in other isomeric systems (Osawa 1970). Interestingly enough, the treatment has never been applied to a set of isomeric endohedral fullerenes. The temperature dependencies of the relative concentrations of the four isomers of a model system Mg@C_{72} were reported (Slanina et al. 2001). The calculations were of MNDO/d-type semiempirical quantum chemical method.

Overall, the computations suggest that if Mg@C_{72} is isolated, it should be a mixture of either two or three isomers. The prediction depends on temperature prehistory. If preparation (at present still hypothetical) takes place at temperatures of approximately 1000 K, two isomers should be produced. If temperatures are increased to approximately 2000 K, there will already be three isomers with significant relative concentrations: two non-IPR structures and one IPR structure. Hence, this study supplies a further interesting example of the profound role of enthalpy–entropy interplay for relative stabilities of isomeric fullerenic structures.

Endohedral doping into C_{74} molecule is a very hot topic. Unusually small HOMO–LUMO energy gap, suggesting high reactivity, makes this molecule interesting (Wan et al. 1998). In the following paper, the possibility of inserting a posteriori a Si (or Ca) atom inside the C_{74} cage by using an *ab initio* molecular dynamics simulation is studied (Shiga et al. 2000).

One may guess that the difference in the results of Si and Ca insertions is mainly due to the different characteristics in chemical bondings between atoms: Ca atom offers two electrons to C_{74}, and Ca tends to become Ca^{2+}, while C_{74} tends to be C_{74}^{2-}, therefore Ca and C_{74} form an ionic bond. Because of this strong ionic bond, Ca and carbon atoms are strongly combined; consequently the Ca atom is easily trapped near the six-membered ring. On the other hand, Si atom tends to form a covalent bond with C_{74}. Since this bond is not as strong as the ionic bond between Ca and C_{74}, Si atom may be inserted rather easily. The stability of Si@C_{74} may be understood because the carbon π orbitals are thickened inside the C_{74} cage and form covalent bonds with the encapsulated Si atom.

C_{80} molecule is another fullerene that is subjected to endohedral doping, theoretically. Türker studied endohedrally Be-, C-, Si-, and Ge-doped C_{80} systems by the application of AM1 (RHF)-type semiempirical quantum chemical calculations (Türker 2003). All the elements (Be, C, Si, and Ge) have closed shell systems. Figure 33.12 shows the geometry-optimized structures for C_{80}, Be@C_{80}, C@C_{80}, Si@C_{80}, and Ge@C_{80}. As seen in Figure 33.12 and Table 33.13, C_{80} and Be@C_{80} have no dipole moments, whereas the rest have and the magnitude of the dipole moment increases from C to Ge. Since they are in the same group, it is possible to generalize that the dipole moments of M@C_{80}-type structures where M is a group IV element increase as the size of the atom increases. In the present

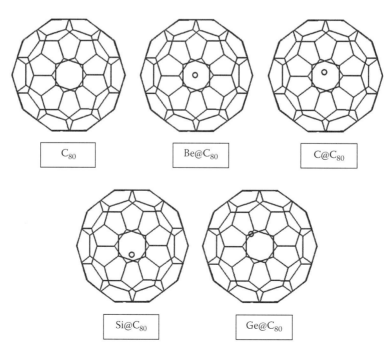

FIGURE 33.12 Geometry-optimized structures for C_{80}, Be@C_{80}, C@C_{80}, Si@C_{80}, and Ge@C_{80}.

TABLE 33.13 Some Calculated Properties of C_{80}, $Be@C_{80}$, $C@C_{80}$, $Si@C_{80}$, and $Ge@C_{80}$

	C_{80}	$Be@C_{80}$	$C@C_{80}$	$Si@C_{80}$	$Ge@C_{80}$
Area	655.33	638.12	638.23	637.82	636.61
Volume	1529.26	1533.63	1532.98	1532.17	1533.20
Hydration energy	−1.7154	—	—	—	—
Log P	12.80	—	—	—	—
Refractivity	310.57	—	—	—	—
Polarizability	151.68	151.69	151.68	153.22	154.34
Dipole moment	0.0000	0.0000	0.4102	5.2068	11.8579
Molecular point group	D_2	D_2	C_1	C_1	C_2

Note: Area, volume, refractivity, and polarizability values are in the order of $10^{-20} m^2$, $10^{-30} m^3$ $10^{-30} m^3$, and $10^{-30} m^3$, respectively. Hydration energy in kJ/mol. Dipole moments are in the order of 10^{-30} C m.

case, the direction of the dipole moment is from C_{80} cage to the intersection point of axes of inertia, indicating that certain amount of electron population of the cage has been perturbed by the presence of the dopant.

Table 33.14 shows some energies of the systems of present concern. The total and binding energy values are indicative of stable structures in all the cases. However, they are all endothermic in nature. As the size of the dopant atom increases, core–core repulsion energy gets more positive values. Also, within the same group (group IV of the periodic table) binding energy gets more negative values from carbon to germanium. In the same direction the systems become less and less endothermic. However, when Be and C cases are compared (the same period elements), $Be@C_{80}$ emerges as less endothermic and having more negative binding energy.

The HOMO and LUMO energies and the interfrontier molecular orbital energy gap values (ΔE) of these systems are shown in

Table 33.15, whereas the HOMO and LUMO of them are shown in Figure 33.13. Again each system has its own characteristic appearance for the HOMO and LUMO. However, $Be@C_{80}$ is quite different than the others because in the former case C_{80} skeleton contributes almost nil to the HOMO and LUMO.

C_{80} endohedrally doped with Be, C, Si, or Ge atom yields stable but highly endothermic structures of $M@C_{80}$ type. $Ge@C_{80}$ seems to be characterized with more negative total and binding energies than C_{80} structure. Thus, Ge atom appears to be quite a potent dopant in order to stabilize C_{80} skeleton in the form of $M@C_{80}$-type structure (Türker 2003).

The configuration of La ions of $La_2@C_{80}$ in the C_{80} fullerene cage was investigated by means of quantum chemical calculations (Shimotani et al. 2004). Geometry optimization was carried out with the Hatree–Fock (HF) method employing valence double-ζ quality basis sets: 3-21G and LanL2DZ basis set for C atoms and La atoms, respectively. The two configurations represented by the point group D_{2h} and D_{5d}, which were reported by Kobayashi et al. (1995, 1996), are indeed stationary points. The D_{5d} configuration is much higher in total energy than that of D_{2h}, in fair agreement with Kobayashi's calculation. Here, La ions are located just under the center of five-membered rings and six-membered rings on the C_{80} fullerene cage for the D_{5d} and the D_{2h} configurations, respectively. Moreover, molecular vibrational frequency analysis of these two structures has been carried out with HF/DZ method. For the D_{5d} structure, it is found that two modes have imaginary frequencies with the absolute value of $65 cm^{-1}$. This low wave number indicates that this mode is related to the motion of encapsulated La ions, and in fact, the analysis showed that the mode is associated with the rotation of La ions along the inner surface of the C_{80} fullerene cage. This imaginary frequency means that the location of La ions in the D_{5d} configuration is not stable, and in fact, this configuration is high in

TABLE 33.14 Some Energies of the C_{80}, $Be@C_{80}$, $C@C_{80}$, $Si@C_{80}$, and $Ge@C_{80}$ Systems

	C_{80}	$Be@C_{80}$	$C@C_{80}$	$Si@C_{80}$	$Ge@C_{80}$
Total	−984,693	−986,865	−996,160	−992,230	−992,395
Binding	−52,117	−51,954	−51,926	−52,031	−52,224
Isolated atomic	−932,576	−934,912	−944,233	−940,199	−940,170
Electronic	−15,364,464	−15,563,639	−15,791,186	−1,578,6291	−15,770,630
Core–core repulsion	1514,379,771	14,576,774	14,795,027	14,794,060	14,778,236
Heat of formation	5,083	5,568	5,989	5,623	5,350

Note: Energies in kJ/mol.

TABLE 33.15 The HOMO, LUMO Energies and the Interfrontier Molecular Orbital Energy Gaps ($\Delta E = \varepsilon_{LUMO} - \varepsilon_{HOMO}$) for the C_{80}, $Be@C_{80}$, $C@C_{80}$, $Si@C_{80}$, and $Ge@C_{80}$ Systems

	C_{80}	$Be@C_{80}$	$C@C_{80}$	$Si@C_{80}$	$Ge@C_{80}$
HOMO	−14.0471 (B_3)	−13.9085 (A)	−14.0462 (A)	−13.6113 (A)	−13.8869 (A)
LUMO	−7.4157 (B_1)	−7.4016 (B_1)	−7.4176 (A)	−7.8993 (A)	−7.0156 (A)
ΔE	6.6314	6.5069	6.6286	5.7120	6.8713

Note: Energies in the order of 10^{-19} J. Symmetries in parenthesis.

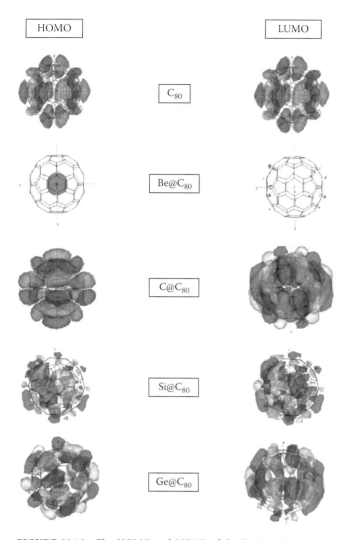

FIGURE 33.13 The HOMO and LUMO of the C_{80}, Be@C_{80}, C@C_{80}, Si@C_{80}, and Ge@C_{80} systems.

energy. Also, the double degeneracy of this imaginary frequency mode indicates that the D_{5d} configuration corresponds to the local maximum in energy in terms of the La positions.

A vibrational analysis of the D_{2h} configuration was also performed, which was shown to be the global minimum by Kobayashi et al (1995, 1996). However, the frequency analysis again gave one mode with an imaginary frequency. The imaginary frequency mode is also given by the frequency analysis using the Becke's three parameter functional (Becke 1993) with the nonlocal correlation provided by the LYP correlation (B3LYP) method employing the same basis sets. This means that D_{2h} is not a local minimum configuration against the movement of the La ions. Hence, it was searched for configurations with lower total energy by shifting La ions from the D_{2h} points and, finally, the third stationary point with a D_{3d} point group, which gives the lowest total energy, was found. In the D_{3d} configuration, the La ions are located under the carbon atoms shared by three six-membered rings on the C_{80} fullerene cage.

In contrast to the D_{5d} structure with two degenerate modes with imaginary frequency, the D_{2d} structure has only one imaginary frequency mode. This means that the total energy decreases when La ions are shifted from the D_{2h} position toward the D_{3d} position, while it increases upon movement of La ions toward the perpendicular directions to the $D_{2h} \rightarrow D_{3d}$ direction. This indicates that the D_{2h} configuration is a saddle point. Also, the 10 D_{3d} points, which are the equivalent global minima in the potential energy surface, are connected along the valley of potential energy surface through the D_{2h} saddle point. Another thing to be pointed out is that the energy difference between D_{3d} and D_{2h} is only 1.9 kcal/mol, which allows the excitation of the D_{2h} configuration even at room temperature. For a more accurate comparison of the total energies of the three structures, they were re-optimized by B3LYP method with cc-pVDZ basis set in valence double-ζ with polarization quality for C atoms and LanL2DZ basis set for La atoms. The total energies of the optimized structures by the B3LYP/DZP method are reported. The higher-level calculation gives a smaller difference of total energies (0.1 kcal/mol) between the D_{3h} and the D_{2h} structure, and the D_{3d} structure remains more stable than the D_{2h} structure. The total energy of the D_{5d} configuration is again much higher than those of the other two configurations. This result confirms the above discussions based on HF/DZ calculations: The 10 equivalent D_{3d} positions of La ions as spheres and the energy valley paths through a D_{2h} saddle point as lines. The above consideration leads us to speculate that La ions move from a D_{3d} position to another D_{3d} position through the energy path, forming a pentagonal dodecahedron of La when averaged in a long time scale. The D_{5d} configurations are scarcely excited at room temperature because of their high total energy. The trajectory of La ions represented is in good agreement with the experimental MEM charge density of La$_2$@C_{80}, supporting the calculations.

It was found that the most stable configuration is D_{3d}, and that the D_{2h} configuration, which was reported to be the lowest in energy, Kobayashi et al. (1996) is the saddle point. Since the energy difference between the D_{3d} configuration and the D_{2h} configuration is only 1.9 kcal/mol, which is empirically low enough to be excited at room temperature, La atoms possibly move between 10 equivalent D_{3d} configurations through intervening D_{2h} configuration. The model shows a fair agreement with the MEM electron density (Nishibori et al. 2001) and the experimental La-C distances derived by MEM/Rietveld method (Nishibori et al. 2001).

The geometric and energy characteristics of endohedrals X@ C_n (X = He, Ne, Ar; n = 20, 24, 30, 32, 40, 50, 60) were calculated by the density functional theory to investigate the effect of size of the encapsulated atom on the structural and geometric properties of the endohedral system (Breslavskaya et al. 2004). The data on the shapes and sizes of the isomers of fullerenes have been calculated at the B3LYP/6-31G* level of theory. For the fullerenes having an elongated shape, the diameter ranges are given (all diameters were calculated as the distances between the opposite atoms). The geometry of the fullerenes was studied in more detail by calculating the bond lengths and radii for the fullerene cages of the most stable isomers, which were chosen based on the criterion of the total energy (E_{total}) minimum. The results of refined calculations at the B3LYP/6-311G* level for the most stable fullerenes and their endohedrals are also presented, in which the ranges of the C–C bond lengths (d) after optimization and the average C–C bond lengths (l) are given for all systems. The calculated data are sufficiently reliable, as evident from a comparison of these data for C_{60} with the results of calculations by the B3LYP/6-31G** method (Darzynkiewicz and Scuseria 1997) and x-ray diffraction data for the 6-6 and 6-5 bonds (Hawkins et al. 1991). (The bond lengths calculated in the present study are 1.392 and 1.452 Å, whereas the bond lengths calculated earlier (Darzynkiewicz and Scuseria 1997) are 1.395 and 1.453 Å, respectively, and the corresponding experimental values (Hawkins et al. 1991) are 1.388 and 1.432 Å.)

The following energy characteristics were calculated: the total energy E_{total}, the energy gap ΔE between the highest occupied and lowest unoccupied molecular orbitals (HOMO–LUMO), and the inclusion energy of endohedrals $\Delta E_e = E_{\text{total}}(X@C_n) - E_{\text{total}}(C_n) - E_{\text{total}}(X)$.

As expected, the total energies E_{total} fall down as the number of atoms, n, in fullerene C_n and the atomic number of the trapped atom increase. The changes in the values of E_{total} for different isomers of the same fullerene are of interest. Two isomers $C_{30}(2)$ and $C_{30}(3)$ with the same symmetry C_{2v} are characterized by nearly equal values of E_{total}, whereas E_{total} for $C_{30}(1)$ differs from the above energies by approximately 0.1 a.u. This is consistent with the fact that the isomers $C_{30}(2)$ and $C_{30}(3)$ have the same symmetry C_{2v}, whereas the symmetry of the isomer $C_{30}(1)$ is D_{5h}. For six isomers of C_{32}, E_{total} varies within 0.13 a.u., which is nearly equal to the value observed for C_{30}, but the correlation between the total energy and the symmetry is not obvious because of five different symmetry groups. This is also true for E_{total} of the endohedrals He@C$_{30}$ and He@C$_{32}$. It should be noted that, in terms of the E_{total} minimum criterion, their conclusions about the highest stability of the isomers $C_{30}(3)$ and $C_{32}(6)$ are in agreement with the conclusions based on the DFT and semiempirical calculations (Chen and Thiel 2003).

In contrast to E_{total}, the correlation between the energy gap ΔE and the symmetry of the fullerene isomer is violated already for C_{30} and the endohedral He@C$_{30}$. For C_{32} and He@C$_{32}$ this correlation is also absent. For the series of fullerenes C_n and their endohedrals $X@C_n$, there is also no correlation between the change in ΔE and the increase in both n and the atomic number X. However, the data on the energy gaps ΔE are of interest for the qualitative interpretation of experimental results, because ΔE correlates with the reactivity, mass spectra, and photoelectron spectra. The energy gap ΔE for fullerene C_{20} is smaller than that for C_{60}, which is in agreement with the higher reactivity of the former fullerene. The relative arrangement of the HOMO–LUMO energy gaps presented is also consistent with

the results of mass spectrometry and photoelectron spectroscopy (Kietzmann et al. 1998).

From the standpoint of predicting the possible endohedrals of small fullerenes, calculations of the inclusion energies, ΔE_e, of endohedrals formed by the endothermic reaction $X + C_n \rightarrow X@C_n$ (the energy capacity of the system) are of most interest. The radii of the endo atoms were taken from the literature (Spravochnik khimika 1971). As expected, ΔE_e rapidly increases with decreasing size of fullerene C_n in the presence of the same endo atom. An increase in the size of the endo atom (for the same size of fullerene) is also accompanied by an increase in the energy capacity (Breslavskaya et al. 2004).

It should be noted that ΔE_e for Ne@C$_n$ ($n = 30, 32, 40, 50$, or 60) and Ar@C$_n$ ($n = 40, 50$, or 60) are lower than that for He@C$_{20}$. These values correlate with the effective pressures P, which are smaller for the above-mentioned endohedrals containing neon or argon compared to that for He@C$_{20}$. The existence of the endohedral He@C$_{20}$H$_{20}$ was confirmed experimentally (Cross et al. 1999). For the latter compound, the calculated inclusion energy (Jiménez-Vázquez et al. 2001, Moran et al. 2002) is close to that for He@C$_{20}$ (Breslavskaya and Buchachenko 2004), and the radius of the carbon cage is close to that of fullerene C_{20}. Based on comparison of these data and the above-mentioned data on the inclusion energies and effective pressures, one would expect that all endohedrals He@C$_n$ ($n \geq 20$) as well as endohedrals Ne@C$_n$ ($n \geq 30$) and Ar@C$_n$ ($n \geq 40$) could be synthesized.

The charges Q_X on the endo atoms (in terms of the Mulliken population analysis) are so small (except for Ar) that they are hardly indicative of any signs of chemical bonding. The unreasonably high value of Q_{Ar} in the endohedral Ar@C$_{20}$, whose existence seems to be impossible, may reflect a substantial deformation of the valence electron shells in this hypothetical molecule. Presumably, the same is true for the endohedrals Ar@C$_n$ ($n = 24, 30$, or 32), although the calculated charges Q_X are insufficiently reliable because of an approximate character of the Mulliken population analysis. Nevertheless, it should be noted that the electron density redistribution between the C_{20} molecule and the endo atoms X of inert gases ($X = $ He, Ne, or Ar) determined in the present study is similar to that found in calculations (Moran et al. 2002) for analogous endohedrals of dodecahedrane $X@C_{20}H_{20}$.

33.3 Conclusion

Although, much progress has been made, no efficient method has yet been found for the production of metallofullerenes ($M_p@C_{nc}$) either by laser ablation of M-impregnated targets or by arc vaporization of M-impregnated rods. For low concentrations of metal in the M-impregnated rod, the monometal species is favored, but with increasing metal concentration, the dimetal species become more prevalent. The low solubility of metallofullerenes in solvents is the other handicap in their productions/purifications.

Although, there are as yet no reports of adverse health effects resulting from exposure to fullerenes or fullerene-containing materials, their health effects have not yet been extensively studied. In addition to that, impurities in the arc rods may lead to polyaromatic hydrocarbon generation, some of which are known as carcinogens.

On the other hand, one of the great attractions of fullerenes is the ability to apply to C_{60} much of the olefin and aromatic chemistry already developed with traditional organic compounds. Besides of that internal volume of fullerenes (or fullerene-like materials) can possibly accommodate some atoms/ions etc. that the composite system(s) might have some technical and practical importance thus providing new business opportunities.

Since certain difficulties exist in the preparation of fullerenes/doped fullerenes and some of them have not been synthesized yet, the role of theoretical methods becomes evident to find out potential properties of these structures.

References

Bakowies, D., Thiel, W. 1991. MNDO study of large carbon clusters. *J. Am. Chem. Soc.* 113: 3704–3714.

Ballenweg, S., Gleiter, R., Kratschmer, W. 1993. Hydrogenation of buckminsterfullerene C_{60} via hydrozirconation: A new way to organofullerenes. *Tetrahedron Lett.* 34: 3737–3740.

Becke, A.D. 1993. Density-functional thermochemistry. III. The role of exact exchange. *J. Chem. Phys.* 98: 5648–5652.

Biehl, M., Thiel, W., Jiao, H., Schleyer, P.v.R., Saunders, M., Anet, F.A.L. 1994. Helium and lithium NMR chemical shifts of endohedral fullerene compounds: An *ab initio* study. *J. Am. Chem. Soc.* 116: 6005–6006.

Boltalina, O.V., Ioffe, I.N., Sorokin, I.D., Sidorov, L.N. 1997. Electron affinity of some endohedral lanthanide fullerenes. *J. Phys. Chem. A* 101: 9561–9563.

Breslavskaya, N.N., Buchachenko, A.L. 2004. *Russ. J. Chem. Phys.* 23: 3 (in Russian).

Breslavskaya, N.N., Levin, A.A., Buchachenko, A.L. 2004. Endofullerenes: Size effects on structure and energy. *Russ. Chem. Bull., Int. Ed.* 53: 18–23.

Chen, Z., Thiel, W. 2003. Performance of semiempirical methods in fullerene chemistry: Relative energies and nucleus-independent chemical shifts. *Chem. Phys. Lett.* 367: 15–25.

Cioslowski, J. 1994. Endohedral magnetic shielding in the C_{60} cluster. *J. Am. Chem. Soc.* 116: 3619–3620.

Cioslowski, J., Fleischmann, E.D. 1991. Endohedral complexes: Atoms and ions inside the C_{60} cage. *J. Chem. Phys.* 94: 3730–3734.

Cotton, F.A., Wilkinson, G. 1967. *Advanced Inorganic Chemistry*, New York: Interscience.

Cross, R.J., Saunders, M., Prinzbach, H. 1999. Putting helium inside dodecahedrane. *Org. Lett.* 1: 1479–1481.

Darzynkiewicz, R.B., Scuseria, G.E. 1997. Noble gas endohedral complexes of C_{60} buckminsterfullerene. *J. Phys. Chem. A* 101: 7141–7144.

David, W.I.F., Ibberson, R.M., Matthewman, J.C. et al. 1991. Crystal structure and bonding of ordered C_{60}. *Nature* 353: 147–149.

Diener, M., Alford, J.M. 1998. Isolation and properties of small-bandgap fullerenes. *Nature* 393: 668–671.

Dunlap, B.I., Brenner, D.W., Mintmire, J.W., Mowrey, R.C., White, C.T. 1991. Local density functional electronic structures of three stable icosahedral fullerenes. *J. Phys. Chem.* 95: 8737–8741.

Dunlap, B.I., Ballester, J.L., Schmidt, P.P. 1992. Interactions between C_{60}. and endohedral alkali atoms. *J. Chem. Phys.* 96: 9781–9787.

Durant, P.J., Durant, B. 1970. *Introduction to Advanced Inorganic Chemistry*, London, U.K.: Longman.

Fleischer, U., Kutzelnigg, W., Lazzeretti, P., Mühlenkamp, V. 1994. IGLO study of benzene and some of its isomers and related molecules. Search for evidence of the ring current model. *J. Am. Chem. Soc.* 116: 5298–5306.

Fowler, P.W., Manolopoulos, D.E. 1995. *An Atlas of Fullerenes*. Oxford, NY: Clarendon Press.

Green, W.H. Jr., Gorun, S.M., Fitzgerald, G., Fowler, P.W., Ceulemans, A., Titeca, B.C. 1996. Electronic structures and geometries of C_{60} anions via density functional calculations. *J. Phys. Chem.* 100: 14892–14898.

Guarr, T.F., Meier, M.S., Vance, V.K., Clayton, M. 1993. Electrochemistry of the $C_{60}H_2$ fullerene. *J. Am. Chem. Soc.* 115: 9862–9863.

Guirado-López, R. 2002. Stability and electronic properties of Si-doped carbon fullerenes. *Phys. Rev. B* 65: 165421–165428.

Guo, T., Scuseria, G.E. 1992. *Ab initio* calculations of tetrahedral hydrogenated buckminsterfullerene. *Chem. Phys. Lett.* 191: 527–532.

Haddon, R.C. 1993. Chemistry of the fullerenes: The manifestation of strain in a class of continuous aromatic molecules. *Science* 261: 1545–1550.

Häser, M., Almlö, J., Scuseria, G.E. 1991. The equilibrium geometry of C_{60} as predicted by second-order (MP2) perturbation theory. *Chem. Phys. Lett.* 181: 497–500.

Hawkins, J.M., Meyer, A., Lewis, T.A., Loren, S.D., Hollander, F.J. 1991. Crystal structure of osmylated C_{60}: Confirmation of the soccer ball framework. *Science* 252: 312–313.

Heath, J.R., O'Brien, S.C., Zhang, Q. et al. 1985. Lanthanum complexes of spheroidal carbon shells. *J. Am. Chem. Soc.* 107: 7779–7780.

Hedberg, K., Hedberg, L., Bethune, D.S. et al. 1991. Experimental values of the bond lengths in free molecules of buckminsterfullerene, C_{60}, from gas-phase electron diffraction. *Science* 254: 410–412.

Henderson, C.C., Cahill, P.A. 1993. $C_{60}H_2$: Synthesis of the simplest C_{60} hydrocarbon derivative. *Science* 259: 1885–1187.

Herman, F., Skillman, S. 1963. *Atomic Structure Calculations*, Englewood Cliffs, NJ: Prentice-Hall.

Hirsch, A. 1994. *The Chemistry of the Fullerenes*, Stuttgart, Germany: Verlag.

Hirsch, A., Soi A., Karfunkel, H.R. 1992. Titration von C_{60}: Eine methode zur synthese von organofullerenen. *Angew. Chem.* 104: 808–810.

Hirsch, A., Grösser, T., Skiebe, A., Soi, A. 1993. Synthesis of isomerically pure organodihydrofullerenes. *Chem. Berr.* 126: 1061–1067.

Hoinkis, M., Yannoni, C.S., Bethune, D.S. et al. 1992. Multiple species of La@C_{82} and Y@C_{82}. Mass spectroscopic and solution EPR studies. *Chem. Phys. Lett.* 198: 461–465.

Jiménez-Vázquez, H.A., Tamariz, J., Cross, R.J. 2001. Binding energy in and equilibrium constant of formation for the dodecahedrane compounds He@$C_{20}H_{20}$ and Ne@$C_{20}H_{20}$. *J. Phys. Chem. A* 105: 1315–1319.

Jura, M. 1993. *The Fullerenes*, eds. H.W. Kroto, D.R.M. Walton. Cambridge, U.K.: Cambridge University Press.

Kietzmann, H., Rochow, R., Ganteför, G. et al. 1998. Electronic structure of small fullerenes: Evidence for the high stability of C_{32}. *Phys. Rev. Lett.* 81: 5378–5381.

Kobayashi, K., Nagase, S., Akasaka, T. 1995. A theoretical study of C_{80} and La$_2$@C_{80}. *Chem. Phys. Lett.* 245: 230–236.

Kobayashi, K., Nagase, S., Akasaka, T. 1996. Endohedral dimetallofullerenes Sc$_2$@C_{84} and La$_2$@C_{80}. Are the metal atoms still inside the fullerene cages? *Chem. Phys. Lett.* 261: 502–506.

Kobayashi, K., Nagase, S., Yoshida, M., Osawa, E. 1997. Endohedral metallofullerenes. are the isolated pentagon rule and fullerene structures always satisfied? *J. Am. Chem. Soc.* 119: 12693–12694.

Kobayashi, S., Mori, S., Iida, S. et al. 2003. Conductivity and field effect transistor of La$_2$@C_{80} metallofullerene. *J. Am. Chem. Soc.* 125: 8116–8117.

Lu, J., Zhang, X., Zhao, X. 1999a. Electronic structures of endohedral Ca@C_{60}, Sc@C_{60} and Y@C60. *Solid State Commun.* 110: 565–568.

Lu, J., Zhang, X., Zhao, X. 1999b. Electronic structures of endohedral Sr@C_{60}, Ba@C_{60}, Fe@C_{60}, and Mn@C_{60}. *Mod. Phys. Lett. B* 13: 97–101.

Lu, J., Zhang, X., Zhao, X. 2000. Metal-cage hybridization in endohedral La@C_{60}, Y@C_{60} and Sc@C_{60}. *Chem. Phys. Lett.* 332: 51–57.

Lu, J., Zhou, Y., Zhang, X., Zhao, X. unpublished.

Lu, J., Zhou, Y., Zhang, S., Zhang, X., Zhao, X. 2001. Structural and electronic properties of endohedral and exohedral complexes of silicon with C60. *Chem. Phys. Lett.* 343: 39–43.

Lu, J., Zhou, Y., Zhang, X., Zhao, X. 2002. Density functional theory studies of beryllium-doped endohedral fullerene Be@C_{60}: On center displacement of beryllium inside the C_{60} cage. *Chem. Phys. Lett.* 352: 8–11.

McElvany, S.W. 1992. Production of endohedral yttrium-fullerene cations by direct laser vaporization. *J. Phys. Chem.* 96: 4935–4937.

Moran, D., Stahl, F., Jemmis, E.D., Schaefer, H.F., Schleyer, P.v.R. 2002. Structures, stabilities, and ionization potentials of dodecahedrane endohedral complexes. *J. Phys. Chem. A* 106: 5144–5154.

Nagase, S., Kobayashi, K. 1993. Metallofullerenes MC$_{82}$ (M = Sc, Y, and La). A theoretical study of the electronic and structural aspects. *Chem. Phys. Lett.* 214: 57–63.

Nagase, S., Kobayashi, K. 1994. Theoretical study of the lanthanide fullerene CeC$_{82}$. Comparison with ScC$_{82}$, YC$_{82}$ and LaC$_{82}$. *Chem. Phys. Lett.* 228: 106–110.

Nishibori, E., Takata, M., Sakata, M. et al. 2000. Giant motion of La atom inside C_{82} cage. *Chem. Phys. Lett.* 330: 497–502.

Nishibori, E., Takata, M., Sakata, M. et al. 2001. Pentagonal-dodecahedral La$_2$ charge density in [80-Ih]fullerene: La$_2$@C_{80}. *Angew. Chem., Int. Ed.* 40: 2998–2999.

Ohtsuki, T., Masumoto, K., Ohno, K. et al. 1996. Insertion of Be atoms in C_{60} fullerene cages: Be@C_{60}. *Phys. Rev. Lett.* 77: 3522–3524.

Osawa, E. 1970. Superaromaticity. *Kagaku (Kyoto)* 25: 854–863.

Pang, L., Brisse, F. 1993. Endohedral energies and translation of fullerene-noble gas clusters G@C_n (G = helium, neon, argon, krypton and xenon; n = 60 and 70). *J. Phys. Chem.* 97: 8562–8563.

Paquette, L.A., Bauer, W., Sivik, M.R., Bühl, M., Feigel, M., Schleyer, P.v.R. 1990. Structure of lithium isodicyclopentadienide and lithium cyclopentadienide in tetrahydrofuran solution. A combined NMR, IGLO, and MNDO study. *J. Am. Chem. Soc.* 112: 8776–8789.

Pasquerello, A., Schlüter, M., Haddon, R.C. 1992. Ring currents in icosahedral C_{60}. *Science* 257: 1660–1661.

Prato, M., Suzuki, T., Wudl, F., Lucchini, V., Maggini, M. 1993. Experimental evidence for segregated ring currents in C_{60}. *J. Am. Chem. Soc.* 115: 7876.

Ren, X.Y., Liu, Z.Y., Zhu, M.Q., Zheng, K.L. 2004. DFT studies on endohedral fullerene C@C_{60}: C centers the C_{60} cage. *J. Mol. Struct. (Theochem)* 710: 175–178.

Roth, G., Adelmann, P. 1992. The crystal structure of $C_{70}S_{48}$: The first a priori structure determination of a C_{70}-containing compound. *J. Phys. I* 2: 1541–1548.

Saito, S., Oshiyama, A. 1992. Electronic structure of calcium-doped C_{60}. *Solid State Commun.* 83: 107–110.

Satpathy, S., Antropov, V.P., Andersen, O.K., Jepsen, O., Gunnarsson, O., Liechtenstein, A.I. 1993. Conduction-band structure of alkali-metal-doped C60. *Phys. Rev. B* 46: 1773–1793.

Saunders, M. 1991. Buckminsterfullerane: The inside story. *Science* 253: 330–331.

Saunders, M., Jimtnez-Vizques, H.A., Cross, R.J. et al. 1994a. Incorporation of helium, neon, argon, krypton, and xenon into fullerenes using high pressure. *J. Am. Chem. Soc.* 116: 2193–2194.

Saunders, M., JimCnez-Vizques, H.A., Cross, R.J., Mroczkowski, S., Freedberg, D., Anet, F.A.L. 1994b. Probing the interior of fullerenes by 3He NMR spectroscopy of endohedral 3He@C_{60} and 3He@C_{70}. *Nature* 367: 256–258.

Seydoux, R., Diehl, P., Mazitov, R.K., Jokisaari, J. 1993. Temperature dependence of the 1H chemical shift of tetramethylsilane in chloroform, methanol, and dimethylsulfoxide. *J. Magn. Reson. A* 101: 78–83.

Shiga, K., Ohno, K., Kawazoe, Y. et al. 2000. *Ab initio* molecular dynamics simulation for the insertion process of Si and Ca atoms into C74. *Mater. Sci. Eng.* A290: 6–10.

Shimotani, H., Ito, T., Iwasa, Y. et al. 2004. Quantum chemical study on the configurations of encapsulated metal ions and the molecular vibration modes in endohedral dimetallofullerene $La_2@C_{80}$. *J. Am. Chem. Soc.* 126: 364–369.

Shinohara, H. 2000. Endohedral metallofullerenes. *Rep. Prog. Phys.* 63: 843–892.

Shinohara, H., Sato, H., Ohkohchi, M. et al. 1992a. Encapsulation of a scandium trimer in C_{82}. *Nature* 357: 52–54.

Shinohara, H., Sato, H., Saito, Y., Ohkohchi, M., Ando, Y. 1992b. Mass spectroscopic and ESR characterization of soluble yttrium-containing metallofullerenes YC_{82} and Y_2C_8. *J. Phys. Chem.* 96: 3571–3573.

Slanina, Z. 1987. Equilibrium isometric mixtures: Potential energy hypersurfaces as the origin of the overall thermodynamics and kinetics. *Int. Rev. Phys. Chem.* 6: 269–271.

Slanina, Z., Lee, S.L., Yu, C.H. 1996. Computations in treating fullerenes and carbon aggregates. In *Reviews in Computational Chemistry*, Vol. 8, eds. K.B. Lipkowitz, D.B. Boyd, pp. 1–62. New York: VCH Publishers.

Slanina, Z., Zhao, X., Osawa, E. 1999a. Computations of higher fullerenes. In *Computational Studies of New Materials*, eds. D.A. Jelski, T.F. George, pp. 112–151. Singapore: World Scientific Publishing.

Slanina, Z., Zhao, X., Osawa, E. 1999b. Relative stabilities of isomeric fullerenes. *Adv. Strain. Inter. Org. Mol.* 7: 185–235.

Slanina, Z., Zhao, X., Uhlík, F. et al. 1999c. Non-IPR Fullerences: C 36 and C 72 in Electronic properties of novel materials: XIII International Winterschool. *AIP Conference Proceedings 486*, Kuznany, H., Fink, J., Mchring, M., and Roth, S., eds., pp. 179–182, Melville, NY: AIP.

Slanina, Z., Zhao, X., Grabuleda, X. et al. 2001. $Mg@C_{72}$ MNDO/d evaluation of the isomeric composition. *J. Mol. Graph. Model.* 19: 252–255.

Son, M.S., Sung, Y.K. 1995. The atom-atom potential. Exohedral and endohedral complexation energies of complexes of X@ C_{60} between fullerene and rare-gas atoms (X = He, Ne, Ar, Kr, and Xe). *Chem. Phys. Lett.* 245: 113–118.

Spravochnik khimika [*Chemist's Handbook*]. 1971. Khimiya, Leningradskoe Otd., 1: 380, Moskow (in Russian).

Suzuki, S., Kawata, S., Shiromaru, H. et al. 1992. Isomers and carbon-13 hyperfine structures of metal-encapsulated fullerenes $M@C_{82}$ (M = Sc, Y, and La). *J. Phys. Chem.* 96: 7159–7161.

Türker, L. 2002a. AM1 treatment of some $C_{60}H_2$ and $Be@C_{60}H_2$ systems. *J. Mol. Struct. (Theochem)* 588: 139–143.

Türker, L. 2002b. PM3 treatment of some endohedrally Mg doped $C_{60}H_2$ systems. *J. Mol. Struct. (Theochem)* 619: 135–141.

Türker, L. 2002c. Some physicochemical and electronic properties of $C_{60}H_2$ and $Be@C_{60}H_2$ systems-a theoretical study. *J. Mol. Struct. (Theochem)* 593: 175–178.

Türker, L. 2003. Endohedrally Be, C, Si and Ge doped C_{80}-AM1 treatment. *J. Mol. Struct. (Theochem)* 626: 203–207.

Türker, L. 2004. Some endohedrally carbon doped C_{60} systems: AM1 treatment. *J. Mol. Struct. (Theochem)* 679: 25–31.

Varganov, S.A., Avramov, P.V., Ovchinnikov, S.G. 2000. *Ab initio* calculations of endo- and exohedral C_{60} fullerene complexes with Li^+ ion and the endohedral C_{60} fullerene complex with Li_2 dimer. *Phys. Solid State* 42: 388–392.

Vilkov, L.V., Mastryukov, V.S., Sadova, N.I. 1970. *Determination of the Geometrical Structure of Free Molecules*, Moscow, Russia: Mir.

Wan, T.S.M., Zhang, H.W., Nakane, T. et al. 1998. Production, isolation, and electronic properties of missing fullerenes: $Ca@C_{72}$ and $Ca@C_{74}$. *J. Am. Chem. Soc.* 120: 6806–6807.

Wang, L.S., Alford, J.M., Chai, Y. et al. 1993. The electronic structure of $Ca@C_{60}$. *Chem. Phys. Lett.* 207, 354–359.

Wasterberg, B., Rosen, A. 1991. Electronic structure of C_{60} and the alkali atom containing compounds MC_{60}, M = Li, Na, K, Rb, Cs and K2C60. *Phys. Scr.* 44: 276–288.

Weaver, J.H., Chai, Y., Kroll, G.H. et al. 1992. XPS probes of carbon-caged metals. *Chem. Phys. Lett.* 190: 460–464.

Weidinger, A., Waiblinger, M., Pietzak, B., Murphy, T.A. 1998. Atomic nitrogen in C_{60}:$N@C_{60}$. *Appl. Phys. A* 66: 287–292.

Yannoni, C.S., Hoinks, M., de Vries, M.S. et al. 1992. Scandium clusters in fullerene cages. *Science* 256: 1191–1192.

Yanov, I., Leszczynski, J. 2001. The molecular structure and electronic spectrum of the $C@C_{60}$ endohedral complex: An *ab initio* study. *J. Mol. Graphs Model.* 19: 232–235.

Zhao, Y.L., Pan, X.M., Zhou, D.F., Su, Z.M., Wang, R.S. 2003. Theoretical study on C_{60}-doped polyacenic semiconductor (PAS) interacting with lithium. *Synth. Met.* 135: 227–228.

34

Carbon Onions

Yuriy V. Butenko
*European Space Research
and Technology Centre*

Lidija Šiller
Newcastle University

Michael R. C. Hunt
University of Durham

34.1 Introduction

Carbon onions are a member of the family of nanometer-scale graphite-like all-carbon allotropes, the emergence of which was catalyzed by the Nobel Prize–winning discovery of the first member, the fullerene, by Kroto and coworkers in 1985 (Kroto et al. 1985). Initially observed by Iijima (1980) and brought to popular attention by the experiments of Ugarte (1992), carbon onions consist of concentric spherical closed carbon shells and receive their name from the close resemblance between their nanoscale structure and the more familiar concentric layered structure of an onion. Closely related to carbon onions are a class of materials known as "onion-like carbons (OLCs)," which include polyhedral nanostructures such as ideal nested fullerenes and it is this material, rather than ideal spherical carbon onions, that can currently be produced in macroscopic quantity and hence be used for future applications.

In this chapter, we begin with a discussion of the basic concepts of hybridization and chemical bonding within carbon materials, a knowledge of which is required to understand the structure, bonding, and physical and chemical properties of carbon onions. We then discuss the structure of carbon onions in some detail, before providing an overview of the variety of methods by which OLC materials may be produced. In the latter part of the chapter, some of the basic physical properties of carbon onion-like structures are reviewed, with an emphasis on those properties that may be relevant for the future applications of this material.

34.2 Carbon Onions: Structure and Bonding

34.2.1 Hybridization in Carbon

It is the ability of carbon atoms to form covalent bonds with different orientations and bond strength that gives rise to the diverse range of all-carbon structures observed in nature. This is described in quantum chemistry in terms of the different hybridization states that carbon displays. A free atom of carbon has an electronic structure $1s^2 2s^2 2p^2$ and, when forming a molecule or solid, we can linearly combine the outermost 2s and 2p atomic orbitals (valence orbitals) to form the hybrid orbitals, which are the basic building blocks of carbon materials. Such combination of orbitals involves the formal promotion of an electron from the 2s to the unoccupied 2p orbital; however, the energy cost associated with this (~4 eV) is recovered by the energy liberated by the formation of chemical bonds. The 2s orbital may be combined (hybridized) with one, two, or all three 2p orbitals, which leads to the three different orientational relationships observed for the covalent bonds between carbon atoms:

sp hybridization occurs when a single 2p orbital hybridizes with the 2s orbital, resulting in two hybrid orbitals which we can, following the notation of Saito et al. (1998), label as $|sp_a\rangle$ and $|sp_b\rangle$. These orbitals can be expressed as the linear combinations of the $|2s\rangle$ and $|2p_x\rangle$ orbitals with coefficients determined by the condition that the hybrid orbitals be orthonormal (Saito et al. 1998):

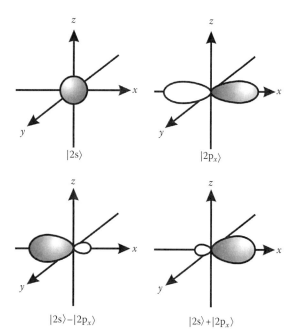

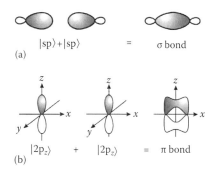

FIGURE 34.1 Schematic representation of orbital mixing in sp hybridization of carbon. The 2s orbital mixes with a $2p_x$ orbital to form two linear hybrid orbitals, each of which has 50% s character and 50% p character.

$$\left|sp_a\right\rangle = \frac{1}{\sqrt{2}}\left(\left|2s\right\rangle + \left|2p_x\right\rangle\right)$$

$$\left|sp_b\right\rangle = \frac{1}{\sqrt{2}}\left(\left|2s\right\rangle - \left|2p_x\right\rangle\right)$$

The creation of these sp hybrid orbitals, as shown in Figure 34.1, enables the formation of two strong bonds with neighboring carbon atoms in a linear, one dimensional arrangement with an angle of 180° between them (Figure 34.2a). This form of bonding, involving the direct "end on" overlap of atomic orbitals is known as σ bonding (formally speaking, a σ bond has

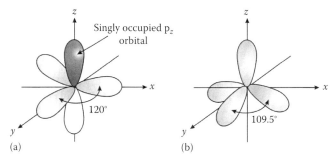

FIGURE 34.2 (a) Overlap of sp hybrid orbitals on neighboring atoms leads to the formation of a σ bond possessing rotational symmetry about the bond axis. (b) p orbitals can overlap "side-on" to form π bonds, which, due to a smaller overlap, are weaker than σ bonds. Only the formation of a single π bond is shown, for clarity.

rotational symmetry about the axis linking the nuclei of the two atoms it connects). The remaining two electrons in the 2p orbitals can overlap with those on neighboring atoms to form weaker bonds, known as π bonds. In Figure 34.2b, the overlap of two $\left|2p_z\right\rangle$ orbitals on neighboring carbon atoms, forming a π bond, is schematically illustrated. In an sp-hybridized carbon two such π bonds are formed, oriented at 90° to one another with their long axes parallel to the interatomic σ bond. These arise from the overlap of two adjacent $\left|2p_z\right\rangle$ orbitals on neighboring carbon atoms and two adjacent $\left|2p_y\right\rangle$ orbitals, respectively. Carbon in an sp-hybridized state can be observed, for example, in polyynes—linear carbon chains with alternating triple and single C–C bonds.

sp² hybridization involves the combination of two 2p orbitals with the 2s orbital, and applying the orthonormality condition referred to above yields the following hybrid orbitals:

$$\left|sp_a^2\right\rangle = \frac{1}{\sqrt{3}}\left|2s\right\rangle + \sqrt{\frac{2}{3}}\left|2p_x\right\rangle$$

$$\left|sp_b^2\right\rangle = \frac{1}{\sqrt{3}}\left|2s\right\rangle - \frac{1}{\sqrt{6}}\left|2p_x\right\rangle + \frac{1}{\sqrt{2}}\left|2p_y\right\rangle$$

$$\left|sp_c^2\right\rangle = \frac{1}{\sqrt{3}}\left|2s\right\rangle - \frac{1}{\sqrt{6}}\left|2p_x\right\rangle - \frac{1}{\sqrt{2}}\left|2p_y\right\rangle$$

The analysis of the directions of these three hybrid orbitals shows that they make an angle of 120° with respect to each other in a two-dimensional plane, while the remaining electron remains in a 2p orbital (typically defined as the $2p_z$ orbital) oriented perpendicular to the plane, schematically illustrated in Figure 34.3a. The three sp² hybrid orbitals can form σ bonds with neighboring atoms in a triangular arrangement, while the $2p_z$ orbital can overlap with similar orbitals on neighboring atoms to form a π bond. Carbon atoms displaying sp² hybridization bond in a hexagonal arrangement, with graphite being the most well-known all-carbon material exhibiting this hybridization state.

sp³ hybridization results when all three orbitals hybridize to form four hybrid orbitals. The resulting hybrid orbitals are a linear combination given by

FIGURE 34.3 Schematic representation of (a) sp² hybridization and (b) sp³ hybridization.

$$\left|sp_a^3\right\rangle = \frac{1}{2}\left(\left|2s\right\rangle + \left|2p_x\right\rangle + \left|2p_y\right\rangle + \left|2p_z\right\rangle\right)$$

$$\left|sp_b^3\right\rangle = \frac{1}{2}\left(\left|2s\right\rangle - \left|2p_x\right\rangle - \left|2p_y\right\rangle + \left|2p_z\right\rangle\right)$$

$$\left|sp_c^3\right\rangle = \frac{1}{2}\left(\left|2s\right\rangle - \left|2p_x\right\rangle + \left|2p_y\right\rangle - \left|2p_z\right\rangle\right)$$

$$\left|sp_d^3\right\rangle = \frac{1}{2}\left(\left|2s\right\rangle + \left|2p_x\right\rangle - \left|2p_y\right\rangle - \left|2p_z\right\rangle\right)$$

These hybrid orbitals are directed along the sides of a tetrahedron with an angle of ~109° between them, shown schematically in Figure 34.3b. This type of hybridization leads to a three-dimensional bonding arrangement exemplified by the diamond crystal. Hence, depending upon hybridization state, carbon–carbon bonds can span one, two, or three dimensions.

The capability of carbon atoms to form different formal hybridization states from sp to sp^3 lies at the heart of the surprising flexibility of graphite sheets. On the basis of a formal sp^2 hybridization state, an sp^2-bound graphite sheet should only display a planar geometry. However, as we have discussed in the introduction, a variety of graphite-based nanostructures exist that display curvatures such as those in carbon onions and nanotubes. This deformation arises from the possibility of forming an admixture between sp^2- and sp^3-hybridized states (which we can denote by $sp^{2+\delta}$), which has a slightly higher excitation energy than the pure sp^2-hybridized state and has a nonplanar arrangement of hybrid orbitals. The energy cost associated with curvature is recovered by the satisfaction of dangling bonds, which results from the resulting closed structure. The simple bending of a graphite sheet, although sufficient for the formation of a carbon nanotube, cannot lead to the formation of a fully closed cage structure. In order to create an all-carbon cage, curvature must be introduced through the insertion of non-hexagonal rings of carbon atoms into the graphite net.

34.2.2 Closed Cage Carbon Structures

Positive curvature may be introduced within a graphite sheet through the incorporation of pentagonal rings of carbon atoms; the structure formed by a single pentagon surrounded by the hexagonal rings of carbon atoms subtends a solid angle of $\pi/6$. We can neglect the smaller rings of carbon atoms in this discussion due to the large bond strain resulting from their incorporation into an sp^2-bonded graphitic network. The number of pentagonal rings that need to be incorporated into a graphite sheet in order to form a closed shell may be found from Euler's theorem for polyhedra (Dresselhaus et al. 1996):

$$f + v = e + 2$$

where

f is the number of faces of the polyhedron
v is the number of vertices
e is the number of edges

For a polyhedron containing f_h hexagonal faces and f_p pentagonal faces we obtain the following relations:

$$f = f_h + f_p$$

$$2e = 6f_h + 5f_p$$

$$3v = 6f_h + 5f_p$$

Hence,

$$6\left(f + v - e\right) = p = 12$$

which demonstrates that any all-carbon closed cage containing only hexagonal and pentagonal rings must possess 12 pentagonal rings, and may contain an arbitrary number of hexagonal faces.

In addition to single-layer, closed-shell, all-carbon polyhedra, known as *fullerenes*, remarkable nested structures of these materials have been observed. It is possible to produce structures in which fullerenes can be concentrically enclosed within one another to produce what are known as multishell fullerenes or *carbon onions*. The first observation of carbon onions was in the product of arc-discharge experiments (Iijima 1980). Later, in 1992, Ugarte demonstrated the transformation of particles of fullerene soot into remarkable spherical carbon onions under exposure to an intense electron beam (Ugarte 1992). Ugarte's discovery was at the origin of interest in these unique carbon structures, and stimulated many researchers to explore new approaches for the synthesis and characterization of this unique material. A high-resolution transmission electron microscopy (HRTEM) image of a typical carbon onion structure, in this case synthesized by annealing nanometer-sized diamond particles (Tomita et al. 2002a), a process discussed in Section 34.3.2, is shown in Figure 34.4.

The carbon onions produced in Ugarte's experiments have almost perfect spherical structure, and the size of innermost shell of the carbon onions is approximately equal to the size of the most abundant fullerene, C_{60}. Fullerene molecules which contain $60n^2$ carbon atoms (where n is a nonzero positive integer) display an icosahedral shape. An image of a fullerene molecule containing 240 carbon atoms is shown in Figure 34.5. The icosahedral shape of these fullerenes originates from the local curvature introduced by the 12 pentagons present in their structure. This icosahedral shape becomes clearly visible as the size of fullerene molecules increases.

A carbon onion composed from four fullerenes (C_{60n^2}, where n = 1, 2, 3, and 4) is shown in Figure 34.6, wherein the icosahedral structure is very clear, with planar facets lined by pentagonal rings. This facetted structure is clearly at odds with the observation of nearly spherical carbon onions, such as that in Figure 34.4. The distance between the shells in this onion C_{60n^2} is approximately 3.5 Å, which is slighter larger than the interplanar separation observed in the bulk crystals of graphite (3.354 Å).

An explanation for the contradiction between the observed the spherical morphology of carbon onions and the evident

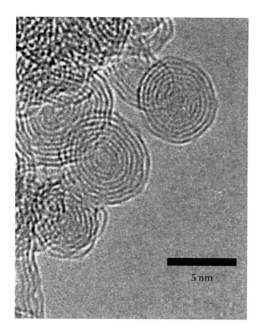

FIGURE 34.4   HRTEM images of carbon onions produced by annealing ultra-dispersed nanodiamonds at 1700°C. The diameter of the innermost shell of the onion is larger than the diameter of the fullerene C_{60}. (Adapted from Tomita, S. et al., *Carbon*, 40, 1469, 2002a.)

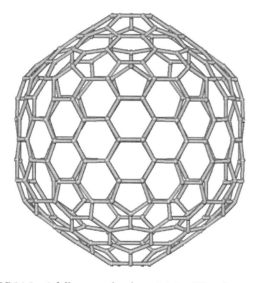

FIGURE 34.5   A fullerene molecule containing 240 carbon atoms has planar facets linked by pentagonal rings of carbon atoms. The faceting increases as the size of the fullerene becomes larger.

icosahedral shape of fullerenes consisting of hexagonal arrays of carbon atoms incorporating 12 pentagons to ensure closure was advanced by the Terrones brothers and coauthors (Terrones et al. 1995, 1996, Terrones and Terrones 1997). Terrones et al. suggested that Stone–Wales-type transformations (Stone and Wales 1986), which convert four adjacent hexagons into two pentagons and two heptagons, can be applied throughout the icosahedral onions to smooth out their structure, producing quasi-spherical molecules. In such pentagon–heptagon pairs,

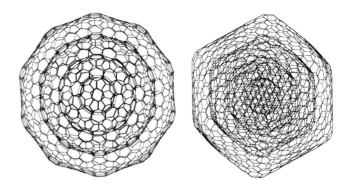

FIGURE 34.6   Two different projections of a carbon onion consisting of four fullerene molecules, each of which contain $60n^2$ carbon atoms, where n is the number of the fullerene shell in the carbon onion. The carbon onion has evident icosahedral shape due to the presence of 12 pentagons in each of the constituent fullerene shells.

the positive curvature of the pentagonal ring is offset by a similar negative curvature associated with the heptagonal ring. As a result only local changes to the curvature are caused by such Stone–Wales pairs. It can also be seen, by the consideration of Euler's theorem, discussed above, that a closed carbon shell (or polyhedron) will remain closed after the introduction of a pentagon–heptagon pair. Figure 34.7a shows a C_{240} fullerene (I_h) molecule consisting of pentagonal and hexagonal rings, which is the second innermost shell of an ideal icosahedral carbon onion. The result of incorporating non-hexagonal defects can be seen in the representation of an O_h isomer of the C_{240} molecule presented in Figure 34.7b, which consists of not only ring pentagonal and hexagonal rings of carbon, but also includes octagonal rings. The O_h isomer is clearly much more spherical than its icosahedral counterpart. The spherical shape of exhibited by carbon onions can originate from other defects in addition to those of the Stone–Wales type, including holes in the fullerene-like shells. Figure 34.7c presents an example of such a molecule which was obtained by removing pentagons from the C_{240} (I_h) molecule. One can therefore conclude that the shape of carbon onions depends on the presence of different ring structure types and the defects in the fullerene-like shells from which they are constructed.

In contrast to the spherical carbon onions observed in the first experiments by Ugarte (1992), OLC particles were

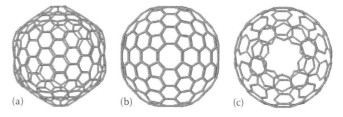

(a)   (b)   (c)

FIGURE 34.7   Fullerene molecules obtained via energy minimization: (a) fullerene C_{240} (I_h) with 12 vertices composed of 12 pentagonal rings; (b) O_h isomer of C_{240} containing 24 pentagons, 92 hexagons, and 6 octagons; the molecule has a more spherical shape; (c) spherical molecule C_{180} obtained by removing pentagons from the icosahedral C_{240} isomer.

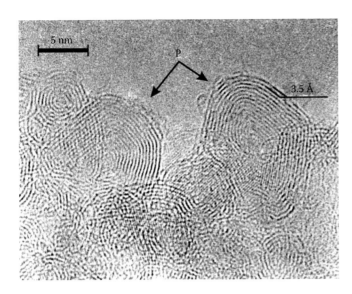

FIGURE 34.8 High-resolution TEM images of polyhedral onion-like particles produced by annealing nanodiamonds at 1500°C for approximately 1 h (marked in the figure by letter P). Dark contrast lines in the figure correspond to graphite-like shells. The distance between lines is approximately 3.5 Å. (Adapted from Kuznetsov, V.L. et al., *Chem. Phys. Lett.*, 222, 343, 1994a. With permission.)

subsequently produced with polyhedral facets, more closely matching the polyhedral structures predicted from the consideration of nested fullerene structures described above. These polyhedral onion-like particles were synthesized by the heat treatment of carbon soot (de Heer and Ugarte 1993) and diamond nanoparticles (Kuznetsov et al. 1994a) in vacuum. Figure 34.8 presents HRTEM images of the polyhedral OLC particles produced in the experiments of Kuznetsov et al. The range of synthesis methods available has led to the production of many different types of OLCs. In addition to their shape, such carbon onions can be characterized by other parameters such as the number of concentric shells, the spacing between adjacent shells, the size of the innermost shell, and the presence of different types of defects.

34.3 Synthesis

34.3.1 Early Methods for Carbon Onion Synthesis and Transformation under Electron Beam Irradiation

The first carbon onions were synthesized in arc-discharge experiments by Iijima (1980). It was demonstrated that the condensation of carbon vapor results in the formation of not only single-shell fullerene molecules, but also of multishell fullerene molecules—carbon onions. It was proposed that quasi-spherical spiral-like particles could be intermediates for the formation of carbon onions via the condensation of carbon vapor (Znang et al. 1986, Kroto and McKay 1988). The close value of the diameter of the innermost shells of these carbon onions to the diameter of the C_{60} fullerene molecule suggested the idea that another

possible candidate for growth centers for carbon onions can be fullerenes themselves.

In Section 34.2, it was mentioned that the transformation of carbon nanoparticles to carbon onions can be stimulated by exposure to an intense beam of electrons. Indeed Ugarte's experiment in 1992 (Ugarte 1992, 1993a) demonstrated for the first time that closed curved carbon nanostructures can be produced not only by the condensation of carbon vapor, but also by the transformation of condensed carbon nanoparticles. The other very important aspect of this work was the realization that electron irradiation does not only have a destructive effect upon carbon materials, but can actually be used constructively to synthesize new carbon nanostructures. This discovery opened a completely new field of carbon materials research related to the stability and transformations of carbon nanostructures. It was found that not only can carbon soot be converted to carbon onions, but also diamond crystals. Qin and Iijima demonstrated that carbon onions can form on the surfaces of 1–3 μm diamond crystals under intense electron beam irradiation (150 A cm⁻²) (Qin and Iijima 1996). The growth of the carbon onions (which consisted of between 4 and 10 closed graphitic shells) proceeded from the inner shell, due to the high mobility of carbon atoms induced by electron irradiation. However, subsequent irradiation resulted in the destruction of the carbon onions. The transformation of nanosized diamond crystals (3–10 nm in diameter) to carbon onions under electron irradiation (20–40 A cm⁻²) was later observed by Roddatis et al. (2002). In this case, a complete transformation of the whole nanodiamond crystals occurred, starting from their surfaces and proceeding inward.

The transformation of diamond to graphitic structures is perhaps not very surprising since graphite is more thermodynamically stable than diamond under normal conditions. Therefore, it was sensational when the opposite transformation was observed under electron irradiation. In 1996, Banhart and Ajayan reported on the formation of a diamond nanocrystal inside a carbon onion particle under intense energetic electron irradiation (1250 keV, 200 A cm⁻²) at a temperature of ~700°C (Banhart and Ajayan 1996). An HRTEM image of such a carbon onion containing an encapsulated diamond nanoparticle is shown in Figure 34.9. It was found that high-energy electrons are capable of changing positions of carbon atoms through thermally activated jumps, ballistic knock-on displacements, or can simply induce mass loss by sputtering (the ejection of atoms from the carbon onion). All these processes cause a dynamic migration of interstitials, which together with mass loss are responsible for the contraction of the whole particle (Banhart 1997, 1999, Banhart et al. 1997, Zaiser and Banhart 1997, Redlich et al. 1998, Zaiser et al. 2000). This contraction can be clearly observed in the HRTEM micrograph of Figure 34.10 by measuring spacing between graphitic shells (Banhart 1999). It was concluded that initial diamond nucleation occurs in the core of carbon onions under the extremely high pressures caused by this contraction and the high temperature resulting from electron irradiation. The subsequent growth of the diamond crystal proceeds at low pressure due to the generation of defects under irradiation in the presence of which, it has been suggested, a phase transition

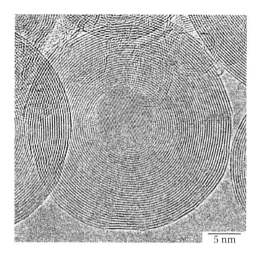

FIGURE 34.9 Spherical carbon onion with a monocrystalline diamond core 10 nm in diameter. The core shows the ⟨111⟩ lattice fringes of diamond with a separation of 2.06 Å. (Adapted from Banhart, F. and Ajayan, P.M., *Nature*, 382, 433, 1996.)

FIGURE 34.10 Contraction of carbon onions under electron irradiation can be deduced from the decreasing distances between layers upon moving from the surface of the carbon onion toward the center. The carbon onion was formed under electron irradiation at 700°C. (Adapted from Banhart, F., *Rep. Prog. Phys.*, 62, 1181, 1999. With permission.)

from graphite to diamond could lead to a decrease of the free energy of the whole system (Bar-Yam 1989, Banhart 1999). The transformation of graphitic structures to diamond under electron irradiation stimulated a new understanding of the nature of the direct graphite–diamond transition. As a result, in 1997, Zaiser and Banhart developed a new formalism for the prediction of phase stability under irradiation for the graphite–diamond system (Zaiser and Banhart 1997). This outstanding work also demonstrated that carbon onions can actually serve as nanocapsules for the conduction of confined reactions or simply for different materials. It was found that metals such as cobalt, nickel, iron, molybdenum, and gold can be encapsulated inside carbon onions under electron irradiation (Ugarte 1993b, Banhart

et al. 1998, 2000, Li and Banhart 2005, Sun and Banhart 2006). Carbon onions containing cobalt metal crystals are shown in the HRTEM image of Figure 34.11 at different electron beam radiation doses. The radiation promotes a contraction of the onions, which forces metal atoms to migrate outward through the shells (Banhart et al. 1998).

The pioneering experiments discussed in this section gave rise to new ideas and concepts, significantly developing our knowledge of the physics and chemistry of carbon nanomaterials. The development of techniques for carbon onion synthesis by electron irradiation allowed investigators to get a superb insight into the properties of these remarkable materials. Carbon onions produced by the electron irradiation of a carbon nanomaterial have a perfect spherical form and well-ordered structure, but while highly perfect, the very small rate at which they could be produced meant that it was essential to explore new synthetic pathways for carbon onion formation.

34.3.2 Synthesis of Onion-Like Carbon as Alternative Material to Perfect Carbon Onions

The electron irradiation experiments performed by Ugarte in 1992 encouraged investigators to explore alternatives to electron irradiation for the formation of carbon onion structures, and the simplest of which was the heat treatment of preformed carbon nanostructures. During the first 2 years after Ugarte's original experiment, it was shown that the transformation of nanostructures in carbon soot into carbon onions could occur through heating under vacuum at 2250°C for 1 h (de Heer and Ugarte 1993), and that diamond nanoparticles (or nanodiamonds) annealed at 1500°C for approximately 1 h (Kuznetsov et al. 1994a,b, Butenko et al. 2000) in vacuum also resulted in the production of carbon onions. However, these onions have significantly less perfect structures than those produced originally by Ugarte, and therefore they are usually referred to as OLC rather than simply carbon onions. Figure 34.12 shows HRTEM micrographs of the products resulting from annealing nanodiamonds at different temperatures in vacuum (Kuznetsov et al. 2001). It can be seen that graphitization starts from the diamond surfaces and proceeds toward the center of the nanodiamond particles (Figure 34.12a through c). Annealing at 1800–1900 K results in the formation of OLC particles (Figure 34.12d and e). Further annealing at 2140 K leads to the creation of elongated polyhedral OLC particles (Figure 34.12f).

The temperature-induced graphitization observed in these experiments resulted in the formation of closed multishell graphitic cages (carbon onions) from structurally very different carbon precursors. Annealing at elevated temperatures helps overcome the kinetic barrier associated with graphitization. However, heat treatment did not lead to the formation of planar graphite sheets, which might have been expected, since graphite is considered to be the most thermodynamically stable form of carbon under normal conditions (Bundy et al. 1996). The formation of symmetric, low-entropy, stable, closed carbon multishell cages instead of planar graphite during the carbon

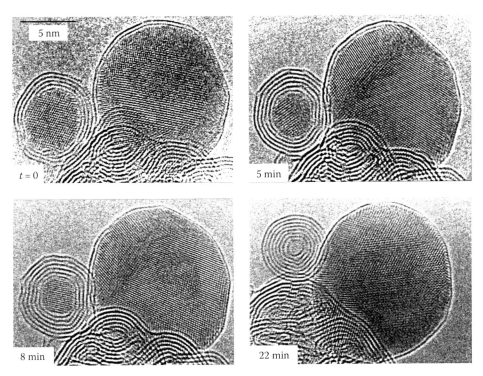

FIGURE 34.11 HRTEM images showing the behavior of two cobalt crystals covered by different number of graphitic shells under electron irradiation at 1000 K. It can be seen that while the smaller crystal on the left side is displaced rather rapidly, the larger crystal with initially one and then two graphitic shells shows almost no shrinking. (Adapted from Banhart, F. et al., *Chem. Phys. Lett.*, 292, 554, 1998. With permission.)

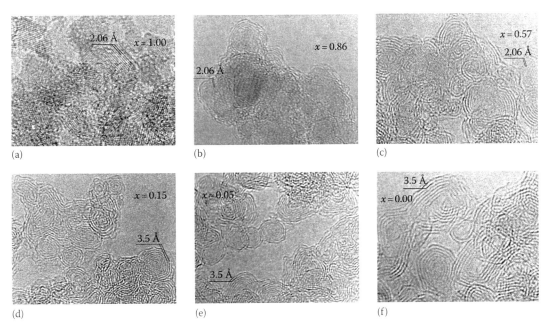

FIGURE 34.12 HRTEM micrographs of a UDD sample annealed under vacuum at (a) 1170 K (1 h), (b) 1420 K (1.4 h), (c) 1600 K (1 h), (d) 1800 K (1 h), (e) 1900 K (1 h), and (f) 2140 K (0.5 h). The dark straight contrast lines in micrographs (a), (b), and (c) correspond to the (111) crystallographic diamond layers. The distance between these lines is 2.06 Å. The dark curved lines in (b), (c), (d), (e), and (f) correspond to the (0002) crystallographic graphite layers. The distance between these lines is approximately 3.5 Å. The diamond weight fractions (x) of the samples are also presented within each image. (Adapted from Kuznetsov, V.L. et al., *Chem. Phys. Lett.*, 336, 397, 2001. With permission.)

vapor condensation or heat-induced transformation of carbon nanoparticles does raise a question about the stability of carbon structures at nanometer dimensions (see references Barnard et al. 2003 and Raty et al. 2003a,b). Decreasing the size of a particle to the nanoscale substantially increases the contribution of the surface energy of the particle to its total energy. The formation of flat graphite nanocrystals would result in the presence of carbon atoms with dangling bonds at their edges. Such carbon atoms have high energy in comparison with those with fully satisfied bonds. The ability of graphitic sheets to curl and form closed carbon cages, discussed previously, provides a possibility to eliminate these dangling bonds at the cost of energy associated with bond strain and curvature, and thus significantly reduces the surface energy of the carbon nanoparticles.

It is interesting to note that carbon onions produced by the heat transformation of carbon soot (de Heer and Ugarte 1993, Ugarte 1994) and nanodiamonds (Kuznetsov et al. 1994a,b, Butenko et al. 2000) have innermost shells with different diameter. The thermal transformation of nanostructured soot results in the formation of carbon onions with large internal shells in comparison with carbon onions produced from nanodiamonds, a result of different mechanisms involved in the formation of these carbon onions. The formation of carbon onions from carbon soot involves a transformation of amorphous carbon to initially curved graphite layers and then the formation of closed carbon cages and the consequent growth of multiwall cages around them. Such a mechanism implies growth outward from the center of the carbon onions. In contrast, when annealing nanodiamonds at elevated temperatures, graphitization starts from the diamond surfaces and progresses toward the centers of nanodiamond crystals (Kuznetsov et al. 1994a,b, 2006). As a result, the formation of carbon onions occurs from a higher-density material and results in the growth of carbon onions with small internal shells. Since carbon onions have higher free energy than graphite, one can suggest that further annealing at sufficiently high temperatures for a long-enough time should eventually lead to the formation flat graphite crystals.

The synthesis of carbon onions by heat treatment of soot by de Heer and Ugarte (1993) represented the first approach capable of producing carbon onions in macroscopic quantities, a requirement for many further characterization steps and for any potential applications of this material. These carbon onions consist of hollow carbon onions with between two and eight graphitic shells and their production in macroscopic quantities by this experimental approach enabled their examination by ultraviolet-visible (UV-VIS) and Raman spectroscopies. It was proposed that carbon onions could be a possible carrier of the 2175 Å interstellar absorption bump (de Heer 1993, Ugarte 1995a, Henrard et al. 1997). The Raman spectra of this material revealed pronounced differences from other graphitic materials (Bacsa et al. 1993).

Vacuum annealing nanodiamonds (ND) with diameters in the range 2–20 nm produced by detonation allowed the production of several grams of OLCs at a time (Kuznetsov et al. 1994a, Kuznetsov and Butenko 2006). The synthesis of OLC in macroscopic quantities has allowed the possibility of exploring the

physical and chemical characteristics of this material in a much more thorough manner. As a result of larger-scale production, it has been shown, for example, that due to an efficient optical limiting action, OLCs are good candidates for photonic applications (Koudoumas et al. 2002). OLC has also been shown to demonstrate high selectivity and catalytic activity in the oxidative dehydrogenation of ethyl benzene to styrene (Keller et al. 2002, Su et al. 2005, Zhang et al. 2007). It has been shown that OLC performs better as a catalyst for this process than industrially used catalysts. The properties of OLC produced by annealing ND have been investigated by x-ray emission spectroscopy (Okotrub et al. 2001), Raman spectroscopy (Obraztsova et al. 1998), electron energy-loss spectroscopy (EELS) (Tomita et al. 1999), electron-spin resonance (Tomita et al. 2001), x-ray diffraction (Tomita et al. 2002a), and UV-VIS absorption spectroscopy (Tomita et al. 2002b).

The synthesis of carbon onions or OLC by the heat treatment of nanostructures has rates of production that can be easily scaled to an industrial level. However, these techniques also have disadvantages: in case of nanodiamonds, the strong aggregation of the starting material leads to a final OLC product that consists mainly of aggregates of carbon onions bonded together by curved graphitic layers, and which have different shapes and contain structural defects (see Figure 34.12d through f); Similar problems occur when producing OLC by annealing soot—the resulting carbon onions also form aggregates and have irregular shape and structure.

34.3.3 Recent Development in Methods of Carbon Onion Synthesis

The synthetic techniques initially developed for the production carbon onions demonstrate three major challenges: (1) the quantity of carbon onion material produced; (2) the control of the structure of the carbon onions, and the level and nature of the defects present; (3) the production of carbon onions that do not contain other forms of carbon. Of these, the isolation of carbon onions from other carbon materials is probably the most difficult task, since carbon onions do not have such high solubility in organic solvents as fullerene molecules and form strong bonds or aggregates with other carbon structures. Difficulties in the synthesis of separate and pure carbon onions, especially having perfect and controlled structure, have prompted researchers to develop new synthetic techniques for this material. Many approaches, sometimes surprising, have been explored in recent years and have yielded interesting results.

Cabioc'h et al. (1995, 1997, 1998a,b) developed a method of carbon onion production based on carbon ion implantation into a metal matrix (Ag, Cu), resulting in onions with typical diameters in the 3–15 nm range. Sufficient quantities could be produced for the investigation of their optical, electronic, and tribological properties (Cabioc'h et al. 1998b, 2000, 2002, Pichler et al. 2001). Fourier transform infrared (FTIR) spectroscopy (Cabioc'h et al. 1998b) measurements on these carbon onions demonstrated that the most stable state for the onions consists of concentric spheres of fullerenes. The electronic properties of the onions were characterized by spatially resolved EELS in transmission (Stöckli et al. 2000, Kociak et al. 2000, Pichler et al. 2001) and reflection mode (Henrard et al. 1999).

Sano et al. (2001, 2002, Alexandrou et al. 2004) reported the production of several milligrams of carbon onions with diameters ranging from 4 to 36 nm using an arc discharge between two graphite electrodes submerged in water or liquid nitrogen. It has been shown that depending upon the synthesis conditions the carbon nanoparticles produced by this method contain carbon onions with a perfect spherical shape along with defective carbon onions and a minimum amount of amorphous carbon and carbon nanotubes. These carbon onions were characterized by Raman spectroscopy (Roy et al. 2003) and by UV-VIS absorption spectroscopy (Chhowalla et al. 2003). New Raman peaks were observed in the spectra of the carbon onions compared with Raman spectra from highly oriented pyrolytic graphite (HOPG). The appearance of these new features could be explained by the curvature of the graphitic walls of the carbon onions (Roy et al. 2003).

An unusual new approach to the synthesis of carbon onions was proposed by Du et al. (2007). This group used radio frequency and microwave plasmas to produce OLC particles from coal and carbon black. The optimization of synthesis conditions in their experiments allowed Du and coworkers to produce carbon onions with a minimum of structural defects. The growth of carbon onions with a diameter of about 30 nm was achieved by Shimizu et al. (2003) by application of inductively coupled microplasma. However, both these plasma-based techniques resulted in the formation of carbon onions that are mixed with other carbon materials, which is a significant issue as the problem of isolation of carbon onions from their mixtures with other carbon structures is yet to be solved satisfactorily.

As mentioned earlier, carbon onions can serve as nanocapsules for a range of different materials. Special interest is associated with carbon onions containing different metals at their core since these particles can, for example, be used for the production of magnetic liquids or magnetic devices for information storage. Recently, synthetic techniques based on chemical vapor deposition (CVD) have been used to produce metal particles encapsulated by several graphitic layers (Sano et al. 2003, Wang et al. 2006, He et al. 2008). CVD techniques used for the

synthesis of metal particles encapsulated in carbon are based on the decomposition of carbon-containing substances on the surface of small metal catalyst particles, which may themselves be produced by the decomposition of metal-containing molecules in the gas phase. For example, Sano et al. (2003) reported the synthesis of large quantities of iron particles covered by several graphitic layers through the decomposition of ferrocene in pure hydrogen. Wang et al. (2006) synthesized iron particles encapsulated in graphitic layers using acetylene and cyclohexane as carbon sources. The CVD method for the production of carbon materials is widely used for the synthesis of carbon nanotubes. Carbon onions can be considered as a zero-dimensional relative of carbon nanotubes; therefore, the mechanism of formation of the carbon onions containing metal particles should be similar to that for carbon nanotubes produced by the CVD technique. One of the possible routes for the synthesis of carbon onions with encapsulated metal particles involves the initial formation of metal nanoparticles in the hot zone of a CVD reactor by the decomposition of a metal-containing precursor compound; these metal particles catalyze the thermal decomposition of carbon-containing gases at their surfaces; carbon produced as a result of this decomposition dissolves in the metal particles, leading to saturation; eventually when such a metal particle enters the cold zone of the reactor, the carbon dissolved in metal precipitates and forms graphitic layers on and around the metal particles. As a result of this series of transformations, metal particles encapsulated in carbon onions are formed. A schematic illustration of the main stages of this process is presented in Figure 34.13.

Another interesting example of the synthesis of metal particles covered by graphitic layers is that of Xu et al. (2006, 2008). Xu and coworkers synthesized carbon nanocapsules containing different metals by striking an arc discharge between graphite electrodes in water solutions containing salts of the chosen metals. Another potentially attractive approach, which would enable the formation of the widest possible range of materials, is the filling of pre-opened carbon onions, since this no longer relies on the filling material being "compatible" with the process by which the carbon onions are produced. It has been shown by Butenko et al. (2008) that carbon onions can be opened by oxidative treatment in carbon dioxide, and it is reasonable to assume that

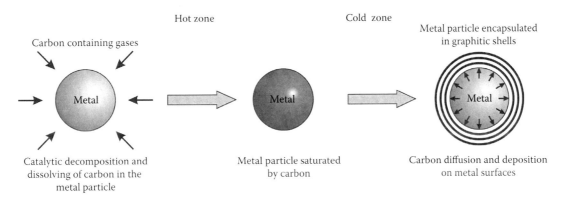

FIGURE 34.13 Schematic showing the stages of synthesis of a metal particle encapsulated in graphitic shells during the CVD process.

these opened onions can subsequently be used for the preparation of nanocapsules for different substances through controlled filling.

34.4 Properties and Applications

34.4.1 Electronic and Electrical Properties of Carbon Onions

The electrical conductivity of carbon allotropes spans an enormous range from insulating to (semi)metallic and this can generally be understood with reference to the hybridization state of the carbon atoms, as discussed in Section 34.2.1, and the resultant localization or delocalization of electrons within the carbon solid. For a homogeneous carbon allotrope bonds will be of σ or π character: electrons in the σ bonds are strongly localized between carbon atoms and hence cannot contribute to the conductivity of the solid. On the other hand, π bonds on the neighboring pairs of carbon atoms can overlap and become spatially delocalized. If partially occupied by electrons, such delocalized orbitals facilitate electrical conduction. Conversely, in the absence of a part-filled π orbital an undoped all-carbon material is typically an insulator (such as diamond which has sp^3 bonding, hence no π electrons, and a bandgap larger than 5 eV).

Graphite is a semimetal (a conductor with a very low density of charge carriers at the Fermi level arising from band overlap) and has a highly anisotropic conductivity—the conductivity parallel to the graphite planes can be an order of magnitude or more larger than that perpendicular to them, depending upon the quality of the graphite. The "wrapping" of graphitic sheets into nanoparticles such as carbon onions can have a significant impact upon electronic structure. At the simplest level, the imposition of finite size and periodic boundary conditions can lead to discrete well-defined molecular electronic states, such as observed in small fullerenes. In OLC, molecule-like electronic behavior appears to be suppressed (Montalti et al. 2003) possibly due to the presence of curved graphitic planes connecting the onion-like structures. In addition to the constraints directly imposed by finite size, the introduction of pentagons, heptagons, and octagonal rings into hexagonal graphitic network necessary to introduce curvature (as discussed in Section 34.2.2) and shell closure results in the partial re-hybridization of the carbon atoms leading to bonds that have a character between the sp^2 bonds in a graphite sheet and the sp^3 bonds of diamond (Kuznetsov et al. 2001). The precise details of bonding will depend upon the precise curvature of the graphene sheet, which in turn depends on the exact structure of the carbon onion, with a deficit of π electron density (resulting from the admixture of sp^3 hybridization) increasing as the curvature of the sheet increases (Kuznetsov et al. 2001).

34.4.1.1 Electronic Properties of Carbon Onions

The electronic structure of onion-like carbon samples produced by annealing nanodiamonds has been investigated by the means of photoemission spectroscopy (Montalti et al. 2003, Butenko

et al. 2005), soft-x ray emission spectroscopy, and x-ray absorption spectroscopy (Okotrub et al. 2001, 2005). X-ray photoemission spectroscopy, although a technique that is primarily associated with determining the elemental composition and chemical state of solid materials through the measurement of the binding energies of core electrons, can provide insight into the hybridization state and bonding of a carbon material. Moreover, the presence of loss structures associated with the excitation of plasmons and interband transitions provides valuable information regarding molecular orbital/band structure and electron density. Valence band photoemission spectroscopy is a complimentary technique, in which low energy photons are used to probe the full valence band, and hence the electronic structure, of solids directly. The occupied electronic states are also accessible through soft x-ray emission spectroscopy in which a core hole is created and an x-ray photon emitted when the core hole is subsequently filled by a valence electron. The resulting x-ray spectrum is proportional to the projection of the density of occupied valence electron states for the element at which the core hole is created. Thus, rather than the complete density of states observed in the valence band photoemission experiment, x-ray spectroscopy yields a local partial density of occupied valence states (LPDOS). X-ray absorption spectroscopy (XAS) is, by contrast, a technique through which the *unoccupied* valence states can be accessed. In this technique, the absorption of an x-ray photon causes the promotion of a core electron into the unoccupied part of the valence band manifold, the relative magnitude of absorption reflecting the matrix-element-weighted unoccupied density of states.

The annealing of ultra-dispersed diamond (UDD) nanoparticles leads to a variety of structures, depending upon the precise annealing temperature, as shown in Figure 34.14. (Kuznetsov et al. 1994a and 2001, also see Figure 34.12 and the associated discussion in Section 34.3.2). The analysis of the C 1s x-ray photoemission lines from the intermediates of ND transformation prepared by thermal treatment at 1420 and 1600 K and then exposed to atmosphere (Figure 34.15) show, in addition to the presence of oxygen-containing groups, both sp^2 and sp^3 carbon (Butenko et al. 2005). The sp^2 component to the C 1s line is significantly increased in relative contribution for the sample annealed at higher temperature, which is consistent with increased nanoparticle graphitization. Indeed, one of the great advantages of photoemission spectroscopy is that it is possible to determine the sp^2:sp^3 ratio, at least semiquantitatively. Increased sp^2 hybridization and graphitic structural organization is also reflected in the presence of a significant density of delocalized π electrons for samples subject to higher temperature annealing, as demonstrated by the observation of a plasmon loss feature at 290.8 eV binding energy associated with the collective motion of the π electrons, which is absent for the less-well-graphitized sample.

Figure 34.16 shows the valence band spectra of UDD samples annealed at five different temperatures (Butenko et al. 2005). The spectra contain two prominent peaks at 2.8 ± 0.1 and 7.5 ± 0.1 eV and a shoulder at 4.2 ± 0.1 eV. The first peak at 2.8 ± 0.1 eV is

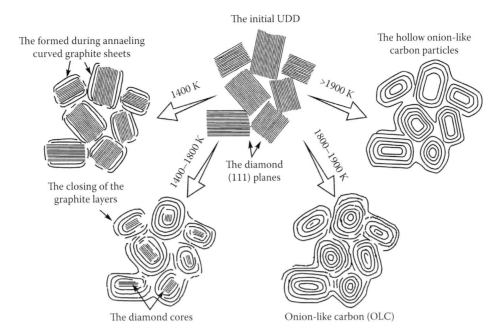

The initial UDD

The formed during annaeling curved graphite sheets

The hollow onion-like carbon particles

1400 K

>1900 K

1400–1800 K

1800–1900 K

The diamond (111) planes

The closing of the graphite layers

The diamond cores

Onion-like carbon (OLC)

FIGURE 34.14 Schematic showing the process of OLC formation. (Adapted from Kuznetsov, V.L. et al., *Chem. Phys. Lett.*, 336, 397, 2001. With permission.)

related to π bonding in graphitic materials (Okotrub et al. 2001, Montalti et al. 2003, Skytt et al. 1994), and is seen to increase in intensity with annealing temperature, once again reflecting increased graphitization as annealing temperature is increased (Butenko et al. 2005). A powerful aspect of valence band photoemission spectroscopy is that the intensity in the valence band close to the Fermi level is related to the density of conduction electron states that govern the transport properties of the material. The examination of this region of the valence band spectrum in the annealed nanodiamond material, Figure 34.17 (Butenko et al. 2005), shows an increase in the density of states at the Fermi level for samples prepared at 1600, 1800, and 1900 K compared with that for a well-graphitized sample prepared at 2140 K. This behavior has been associated with an accumulation of the different types of defects in the curved graphite layers during the graphitization of diamond (Butenko et al. 2005). The faceting of the carbon onions at the highest annealing temperature (Figure 34.14) is associated with the reduction of the defects required to give rise to the spherical carbon onion shape (discussed in Section 34.2.2) and is reflected in the valence band photoelectron spectra.

34.4.1.2 Resistivity

The electrical resistivity of films of OLC produced by annealing UDD has been measured by a four-probe DC method as a function of temperature between 4 and 300 K (Kuznetsov et al. 2001). The use of four-probe measurements is particularly important in samples such as these, in which the magnitude of the contact resistance is unknown and may be significant. It has been found that UDD samples heated at temperatures lower than 1170 K, which do not exhibit any surface graphitic species, have very high values of resistivity, above 10^9 Ω cm

(Kuznetsov et al. 2001). As graphitization of the nanodiamond material occurs with increasing annealing temperature, the resistivity of the samples drops significantly. For UDD annealing products treated at temperatures above 1600 K the resistivity at 300 K lies in the range 0.2–0.5 Ω cm and is comparable with that of carbon black. For all samples annealed at temperatures higher than 1170 K, the electrical resistivity shows a temperature dependence typical of systems with variable range hopping conductivity (Mott and Davis 1979, Kuznetsov et al. 2001). For typical graphitic materials (carbon black, graphitized soot, and graphite powder) three-dimensional hopping is usually observed; however, the hopping conductivity in the UDD annealing products displays a significantly lower dimensionality (Figure 34.18), which can be directly related to the conduction pathways available, which range from one-dimensional carbon chains in samples annealed at 1420 K to defective two-dimensional layers for high-temperature annealing treatments (Kuznetsov et al. 2001).

34.4.1.3 Field Emission

Field emission, also known as Fowler–Nordheim tunneling, is the process by which electrons tunnel through the work function barrier at a surface in the presence of a high electric field and are emitted into the vacuum. Fowler–Nordheim theory is generally used in order to quantitatively describe the field emission current density as a function of the electric field. This quantum mechanical tunneling process is an important mechanism for thin barriers. The tunneling probability is

$$\Theta = \exp\left(-\frac{4}{3}\frac{\sqrt{2qm^\star}}{\hbar}\frac{\phi_B^{3/2}}{E}\right)$$

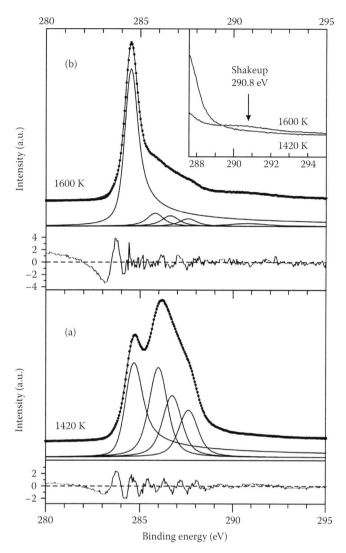

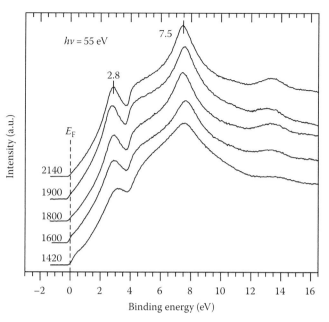

FIGURE 34.16 Evolution of valance band photoemission spectra as a function of annealing temperature for nanodiamond annealed at 1420, 1600, 1800, 1900, and 2140 K (temperatures on the figure are presented in Kelvin). Spectra were obtained in normal emission geometry using a photon energy of 55 eV. (Adapted from Butenko, Yu.V. et al., *Phys. Rev. B*, 71, 075420, 2005. With permission.)

FIGURE 34.15 X-ray photoemission spectra showing the C 1s line of nanodiamond annealed at (a) 1420 K and (b) 1600 K. The spectra were obtained in normal emission geometry at a photon energy of 1486.6 eV. The dots are experimental data and the solid lines are the fit components into which the spectra were decomposed. The resulting fit is superimposed on the data. The inset shows the appearance of the π plasmon peak at 290.8 eV in the spectrum of the sample annealed at 1600 K. The residuals from the curve fitting (in units of standard deviation) are displayed in the bottom panel beneath the spectra. (Adapted from Butenko, Yu.V. et al., *Phys. Rev. B*, 71, 075420, 2005. With permission.)

where *E* is the applied electric field (which can be related to the applied voltage *V* by *E* = *V*/*d*, *d* is the distance over which the voltage is applied), *q* is the carrier charge, *m** is effective mass of the charge carrier, ϕ_B is the barrier height, and $\hbar$ is the Planck constant divided by 2π. The tunneling current is obtained from the product of the carrier charge, *q*, velocity, *v*, and density, *n*, with the tunneling probability. The velocity, *v*, is taken as the average velocity with which the carriers approach the barrier. The net tunnel current density *J* is then given by

$$J = qvn\Theta$$

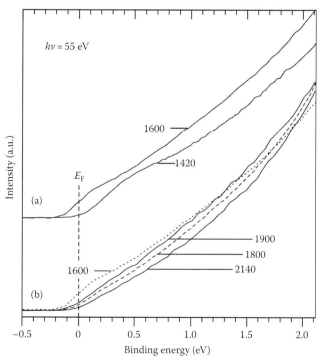

FIGURE 34.17 Evolution of the valance band spectra near the Fermi level (E_F) as a function of annealing temperature for samples annealed at (a) 1420 and 1600 K; (b) 1600, 1800, 1900, and 2140 K (temperatures on the figure are presented in Kelvin). Spectra were obtained in normal emission geometry using a photon energy of 55 eV. (Adapted from Butenko, Yu.V. et al., *Phys. Rev. B*, 71, 075420, 2005. With permission.)

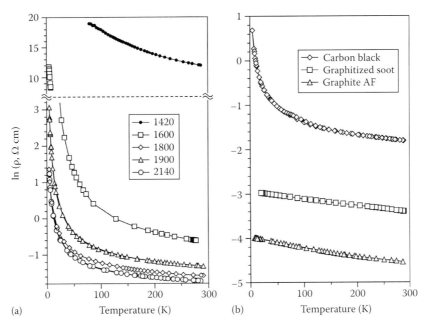

FIGURE 34.18 Temperature dependence of the resistivity of UDD annealing products and carbon comparison standards. (Adapted from Kuznetsov, V.L. et al., *Chem. Phys. Lett.*, 336, 397, 2001. With permission.)

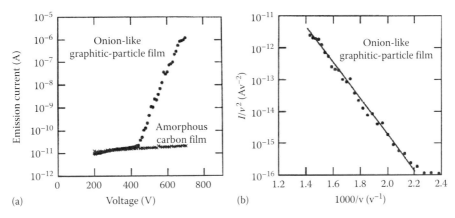

FIGURE 34.19 (a) Electron field emission characteristics for an onion-like graphitic-particle film and an amorphous carbon film prepared by sputtering. (b) Corresponding Fowler–Nordheim plot. (Adapted from Mamezaki, O. et al., *Jpn. J. Appl. Phys.*, 39, 6680, 2000. With permission.)

Ni-filled onion-like graphitic-particle films have shown electron field emission with a low threshold voltage (Mamezaki et al. 2000). These films were prepared by the co-sputtering of Ni and graphite and subsequent thermal annealing in vacuum. Figure 34.19a shows the electron field emission characteristics for an onion-like graphitic-particle film and an amorphous carbon film prepared by sputtering. A field emission current is detected from the OLC film at an applied voltage above 400 V and increases exponentially with the applied voltage, while there is no evidence of substantial field emission from the corresponding amorphous carbon film. In Figure 34.19b, the corresponding Fowler–Nordheim plot is shown, it can be seen that a straight line, representing Fowler–Nordheim behavior is obtained over a wide current range. It has been suggested (Mamezaki et al. 2000) that the main reason for field emission from the onion-like graphitic-particle films is due to the presence of nanometer-scale

protrusions that lead to a strong local electric field enhancement reinforced by the nonuniform resistivity of the OLC films, which further enhances local electric fields. As can be seen from the derivation of the Fowler–Nordheim equation presented above, an increased local electric field will produce a greater tunneling probability and hence field emission current for a given applied voltage, leading a low-threshold voltage value.

34.4.2 Optical Properties of Carbon Onions

34.4.2.1 Transmittance and Absorption

Since their discovery it has been believed that carbon onions are a component of the interstellar dust and that they contribute to the strong absorption band centered at a wavelength of 217.5 nm (Kroto 1992, de Heer and Ugarte 1993, Ugarte 1995a, Henrard et al. 1997). Optical transmission spectroscopy measurements in the wavelength range 0.2–1.2 μm were performed by Cabioc'h et al. (2000) on carbon onions produced by carbon ion implantation into thin silver films and deposited on a silica surface. This method of production was chosen because after ion implantation into a 400 nm thick silver film, the silver can be evaporated in vacuum (at 850°C, 10⁻⁵ Pa) leaving carbon onions weakly bound to the silica. Silica is known to be almost transparent in the wavelength range 0.2–1.2 μm, giving an opportunity to measure the optical transmittance of carbon onions themselves (Cabioc'h et al. 2000). In addition, it is possible by varying the fluence of the carbon ion implantation to also change the size of the carbon onions—for example, increasing the fluence of 120 keV carbon ions from 5 × 10¹⁶ ions cm⁻² to 3 × 10¹⁷ ions cm⁻² produces an increase in carbon onion diameter from 4 to 8 nm (Cabioc'h et al. 2000). The optical transmittance spectra of carbon onion layers deposited onto silica substrates in this way reveal two absorption peaks, shown in Figure 34.20. Minima in the optical transmittance are found to occur centered at 220–230 and 265 nm and are attributed to the presence of carbon onions and residual-disordered graphitic carbon, respectively (Cabioc'h et al. 2000). The transmittance decrease that arises with increasing ion fluence during carbon onion production has been explained by the increase of the average thickness of the carbon layer.

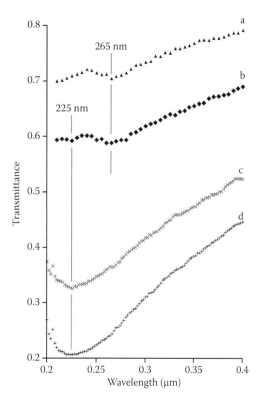

FIGURE 34.20 Optical transmittance of carbon-onion layers obtained by the implantation of 120 keV carbon ions into Ag held at 500°C at various fluences (a: 5 × 10¹⁶, b: 10¹⁷, c: 2 × 10¹⁷, and 3 × 10¹⁷ ions cm⁻²). (Adapted from Cabioc'h, T. et al., *Eur. Phys. J. B*, 18, 535, 2000. With permission.)

34.4.2.2 Electromagnetic Shielding in Onion-Like Carbons and Their Composites

Recently, there have been several reports that onion-like carbon obtained by annealing detonation nanodiamonds provides excellent electromagnetic (EM) shielding over the microwave spectral band from 2 to 37 GHz (Maksimenko et al. 2007, Shenderova et al. 2007a) and, in addition, the efficient attenuation of the electromagnetic spectrum over the frequency range 12–230 THz as compared to the detonation nanodiamond starting material (Shenderova et al. 2008). Another group of authors (Ge et al. 2006) reported high dielectric loss in the 2–18 GHz frequency range for carbon nano-onions synthesized by the DC arc-discharge in water method and carbon onions encapsulating Fe nanoparticles synthesized by chemical vapor deposition.

There have been a variety of mechanisms suggested as contributors to the broadband EM attenuation displayed by OLC, including electrical conductivity that can be tuned by the hierarchical assembly of the primary carbon onion particles and inner-shell defects (Kuznetsov and Butenko 2005). It has been observed that nanodiamonds treated thermally will display a wide range of relevant length scales, such as sp² patches between defects (holes) within a single graphitic nanoshell, aggregates of primary onions, and conglomerates of the OLC aggregates, which could be an additional reason for the wideband attenuation of the electromagnetic spectrum (Kuznetsov and Butenko 2005, Shenderova et al. 2008). However, the detailed microscopic mechanism of attenuation is still unknown and it would be worth future attention.

It has also been reported that dispersed nanodiamonds (DND) are efficient UV absorbers (Shenderova et al. 2007b), and therefore it has been suggested (Shenderova et al. 2008) that the combination of the OLC and DND in a nanocomposite material could provide effective protection from EM radiation over a rather wide spectral range. There is still need for an optimization of the thickness of the film and composition, so further studies would be desirable.

34.4.3 Tribological Properties: Friction and Lubrication

The reduction of friction and wear and consequently the increased lifetime of mechanical components is an important goal in the development of lubricants and in this area carbon onions have significant potential. There are a number of ways by which frictional forces may be measured, the most common of which is the ball-on-disk or pin-on-disk tribometer. In this device, a flat, pin, or sphere is mounted on a stiff elastic arm and brought into contact with a sample and loaded with a precisely known weight. This ensures a nearly fixed contact point and thus a stable position in the friction track. The relative coefficient of friction is determined by rotating the disk with respect to the flat, pin, or sphere (although reciprocating action is also possible with certain designs) and measuring the deflection of the elastic arm or by directly measuring the change in torque. Wear coefficients may be found from the amount of material lost in the test, which is typically determined by scanning electron microscopy (SEM). This simple approach allows studies to be made of friction and wear for various material combinations, with or without lubrication.

Among those who have been attracted by the potential applications of carbon nano-onions are researchers at NASA interested in their tribological properties as additives for aerospace applications (Street et al. 2004, Delgado et al. 2008). Thus far carbon onions have demonstrated superior lubrication properties when compared with conventional lubricants (Hirata et al. 2004, Yao et al. 2008). The tribological properties of carbon onions prepared by the heat treatment of diamond clusters or particles at 1730°C using an infrared radiation furnace in argon at atmospheric pressure have been examined by the ball-on-disc-type friction testing of the type described above using a silicon wafer and a steel ball (Hirata et al. 2004). The carbon onions, which are spread on the silicon wafer without adhesive, exhibit stable friction coefficients lower than 0.1 both in air and in vacuum at room temperature. It is found that the larger carbon onions prepared from diamond particles show low friction on the rough surface of silicon discs. The good solid lubrication displayed by carbon onions is assumed (Hirata et al. 2004) to be due to the quasi-spherical structure of the carbon onion material which is believed to behave as a film of nanometer scale (the onions have a diameter of 10 nm) ball bearings. An additional contribution to the excellent lubrication displayed by the carbon onions may arise from the absence of dangling bonds on the surface of well-graphitized carbon onions: when carbon atoms arrange perfectly so that the outermost shell contains no defects, weak intermolecular (van der Waals) bonding with the counterpart material is expected to occur (Hirata et al. 2004). Finally, the closed-shell structure of the carbon onion can be expected to be highly stable, leading to high mechanical strength, wear resistance, and thermal tolerance (Hirata et al. 2004).

34.5 Summary

The amazing, highly symmetrical structure of perfect carbon onions suggests that this material has unique properties with a number of significant potential applications. However, the synthesis of well-separated carbon onions with controlled structure in large quantities and free from other graphitic "impurities" is an extremely challenging task. A number of research groups have taken up this challenge and explored a variety of different approaches for the synthesis of this remarkable material, leading to the production of carbon onions characterized by a range of different structures and degrees of perfection. In particular, depending upon the exact method and conditions of synthesis, carbon onions can vary in topological shape, the number of shells, spacing between shells, in the diameter of the internal shell, and in the presence and density of different types of defects. However, despite the significant difficulties in carbon onions synthesis, they have already demonstrated great potential for several applications. Among the most promising applications of carbon onions is their use as additives to lubricants due to their excellent tribological properties, which arise in part from their highly symmetric shape. The electronic properties of carbon onions suggest that they may be successfully applied as a material for electromagnetic shielding equipment, for field emission electrodes, and for photonic devices.

Perhaps the most important results of the intensive research that has been undertaken in the field of carbon onions over recent years are the fundamental knowledge and new concepts that have been developed. New knowledge about diamond–graphite transformations has significantly broadened our ability to manipulate carbon nanostructures and to understand processes occurring under extremes of heat, pressure, and irradiation. The exciting idea of applying carbon onions as a nanocapsule for the reactions and/or protection of species is an area still waiting to be fully explored with significant potential for impact in biology and medicine.

References

Alexandrou, I., Wang, H., Sano, N., and Amaratunga, G.A.J., 2004, Structure of carbon onions and nanotubes formed by arc in liquids, *J. Chem. Phys.* **120**, 1055–1058.

Bacsa, W.S., de Heer, W.A., Ugarte, D., and Châtelain, A., 1993, Raman spectroscopy of closed-shell carbon particles, *Chem. Phys. Lett.* **211**, 346–352.

Banhart, F., 1997, The transformation of graphitic onions to diamond under electron irradiation, *J. Appl. Phys.* **81**, 3440–3445.

Banhart, F. 1999, Irradiation effects in carbon nanostructures, *Rep. Prog. Phys.* **62**, 1181–1221.

Banhart, F. and Ajayan, P.M. 1996, Carbon onions as nanoscopic pressure cells for diamond formation, *Nature*, **382**, 433–435.

Banhart, F., Charlier, J.-C., and Ajayan, P.M., 2000, Dynamic behaviour of nickel atoms in graphitic networks, *Phys. Rev. Lett.* **84**, 686–689.

Banhart, F., Füller, T., Redlich, Ph., and Ajayan, P.M., 1997, The formation, annealing and self-compression of carbon onions under electron irradiation, *Chem. Phys. Lett.* **269**, 349–355.

Banhart, F., Redlich, Ph., and Ajayan, P.M., 1998, The migration of metal atoms through carbon onions, *Chem. Phys. Lett.* **292**, 554–560.

Bar-Yam, Y. and Moustakas, T.D., 1989, Defect-induced stabilization of diamond films, *Nature* **342**, 786–787.

Barnard, A.S., Russo, S.P., and Snook, I.K., 2003, Coexistence of bucky diamond with nanodiamond and fullerene carbon phases, *Phys. Rev. B* **68**, 073406.

Bundy, F.P., Bassett, W.A., Weathers, M.S., Hemley, R.J., Mao, H.K., and Goncharov, A.F., 1996, The pressure-temperature phase and transformation diagram for carbon; updated through 1994, *Carbon* **34**, 141–963.

Butenko, Yu.V., Chakraborty, A.K., Peltekis, N. et al., 2008, Potassium intercalation of opened carbon onions, *Carbon* **46**, 1133–1140.

Butenko, Yu.V., Krishnamurthy, S., Chakraborty, A.K. et al., 2005, Photoemission study of onionlike carbons produced by annealing nanodiamonds, *Phys. Rev. B* **71**, 075420.

Butenko, Yu.V., Kuznetsov, V.L., Chuvilin, A.L. et al., 2000, Kinetics of the graphitization of dispersed diamonds at "low" temperatures, *J. App. Phys.* **88**, 4380–4388.

Cabioc'h, T., Camelio, S., Henrard, L., and Lambin, Ph., 2000, Optical transmittance spectroscopy of concentric-shell fullerenes layers produced by carbon ion implantation, *Eur. Phys. J. B* **18**, 535–540.

Cabioc'h, T., Girard, J.C., Jaouen, M., Denanot, M.F., and Hug, G., 1997, Carbon onions thin film formation and characterization, *Europhys. Lett.* **38**, 471–475.

Cabioc'h, T., Jaouen, M., Denanot, M.F., and Bechet, P., 1998a, Influence of the implantation parameters on the microstructure of carbon onions produced by carbon ion implantation, *Appl. Phys. Lett.* **73**, 3096–3098.

Cabioc'h, T., Kharbach, A., Le Roy, A., and Riviere, J.P., 1998b, Fourier transform infra-red characterization of carbon onions produced by carbon-ion implantation, *Chem. Phys. Lett.* **285**, 216–220.

Cabioc'h, T., Riviere, J.P., and Delafond, J., 1995, A new technique for fullerene onion formation, *J. Mater. Sci.* **30**, 4787–4792.

Cabioc'h, T., Thune, E., Riviere, J.P. et al., 2002, Structure and properties of carbon onion layers deposited onto various substrates, *J. Appl. Phys.* **91**, 1560–1567.

Chhowalla, M., Wang, H., Sano, N., Teo, K.B.K., Lee, S.B., and Amaratunga, G.A.J., 2003, Carbon onions: Carriers of the 217.5 nm interstellar absorption feature, *Phys. Rev. Lett.* **90**, 155504.

de Heer, W.A. and Ugarte, D., 1993, Carbon onions produced by heat treatment of carbon soot and their relation to the 217.5 nm interstellar absorption feature, *Chem. Phys. Lett.* **207**, 480–486.

Delgado, J.L., Herranz, M.A., and Martin, N., 2008, The nanoforms of carbon, *J. Mater. Chem.* **18**, 1417–1426.

Dresselhaus, M.S., Dresselhaus, G., and Eklund, P.C., 1996, *Science of Fullerenes and Carbon Nanotubes*, Academic Press, San Diego, CA.

Du, A.B., Liu, X.G., Fu, D.J., Han, P.D., and Xu, B.S., 2007, Onion-like fullerenes synthesis from coal, *Fuel*, **86**, 294–298.

Ge, A.Y., Xu, B.S., Wang, X.M., Li, T.B., Han, P.D., and Liu, X.G., 2006, Study on electromagnetic property of nano onion-like fullerenes, *Acta Phys. Chim. Sin.* **22**, 203–208.

He, C.N., Shi, C.S., Du, X.W., Li, J.J., and Zhao, N.Q., 2008, TEM investigation on the initial stage growth of carbon onions synthesized by CVD, *J. Alloys Compd.* **452**, 258–262.

Henrard, L., Lambin, Ph., and Lucas, A.A., 1997, Carbon onions as possible carriers of the 2175 Angstrom interstellar absorption bump, *Astrophys. J.* **487**, 719–727.

Henrard, L., Malengreau, F., Rudolf, P. et al., 1999, Electron-energy-loss spectroscopy of plasmon excitations in concentric-shell fullerenes, *Phys. Rev. B* **59**, 5832–5836.

Hirata, A., Igarashi, M., and Kaito, T., 2004, Study on solid lubricant properties of carbon onions produced by heat treatment of diamond clusters or particles, *Tribol. Int.* **37**, 899–905.

Iijima, S., 1980, Direct observation of the tetrahedral bonding in graphitized, *J. Cryst. Growth* **5**, 675–683.

Keller, N., Maksimova, N.I., Roddatis, V.V. et al., 2002, The catalytic use of onion-like carbon materials for styrene synthesis by oxidative dehydrogenation of ethylbenzene, *Angew. Chem. Int. Ed.* **41**, 1885–1888.

Kociak, M., Henrard, L., Stéphan, O., Suenaga, K., and Colliex, C., 2000, Plasmons in layered nanospheres and nanotubes investigated by spatially resolved electron energy-loss spectroscopy, *Phys. Rev. B* **61**, 13936–13944.

Koudoumas, E., Kokkinaki, O., Konstantaki, M. et al., 2002, Onion-like carbon and diamond nanoparticles for optical limiting, *Chem. Phys. Lett.* **357**, 336–340.

Kroto, H.W., 1992, Carbon onions introduce new flavor to fullerene studies, *Nature* **359**, 670–671.

Kroto, H.W. and McKay, K., 1988, The formation of quasi-icosahedral spiral shell carbon particles, *Nature* **331**, 328–331.

Kroto, H.W., Heath, J.R., O'Brien, S.C. et al., 1985, C$_{60}$—Buckminsterfullerene, *Nature*, **318**, 162–163.

Kuznetsov, V.L. and Butenko, Y.V., 2005, in: D. Guen, O. Shenderova, and A. Vul (Eds.), *Synthesis, Properties and Applications of Ultrananocrystalline Diamond (NATO Science Series)*, Springer, Amsterdam, the Netherlands, p. 199.

Kuznetsov, V.L. and Butenko, Yu.V., 2006, in: O.A. Shendorova and D.M. Gruen (Eds.), *Diamond Phase Transition at Nanoscale, in Ultra Nanocrystalline Diamond, Synthesis, Properties and Applications*, William Andrew Publishing, Norwich, NY, Chapter 13, pp. 405–475.

Kuznetsov, V.L., Butenko, Yu.V., Chuvilin, A.L., Romanenko, A.I., and Okotrrub, A.V., 2001, Electrical resistivity of graphitized ultra-disperse diamond and onion-like carbon, *Chem. Phys. Lett.* **336**, 397–404.

Kuznetsov, V.L., Chuvilin, A.L., Butenko, Yu.V., and Titov, V.M. 1994a, Onion-like carbon from ultra-disperse diamond, *Chem. Phys. Lett.* **222**, 343–348.

Kuznetsov, V.L., Chuvilin, A.L., Moroz, E.M. et al., 1994b, Effect of explosions on the structure of detonation soots: ultradisperse diamond and onion carbon, *Carbon* **32**, 873–882.

Li, J. and Banhart, F., 2005, The deformation of single, nanometer-sized metal crystals in graphitic shells, *Adv. Mater.* **17**, 1539–1542.

Maksimenko, S.A., Rodinova, V.N., Slepyan, G.Ya. et al., 2007, Attenuation of electromagnetic waves in onion-like carbon composites, *Diamond Relat. Mater.* **16**, 1231–1235.

Mamezaki, O., Adachi, H., Tomita, S., Fujii, M., and Hayashi, S. 2000, Thin films of carbon nanocapsules and onion-like graphitic particles prepared by the cosputtering method, *Jpn. J. Appl. Phys.* **39**, 6680–6683.

Montalti, M., Krishnamurthy, S., Chao, Y. et al., 2003, Photoemission spectroscopy of clean and potassium-intercalated carbon onions, *Phys. Rev. B* **67**, 113401–113404.

Mott, N.F. and Davis, E.A., 1979, *Electron Processes in Noncrystalline Materials*, Claredon Press, Oxford, U.K.

Obraztsova, E.D., Fujii, M., Hayashi, S., Kuznetsov, V.L., Butenko, Yu.V., and Chuvilin, A.L., 1998, Raman identification of onion-like carbon, *Carbon* **36**, 821–826.

Okotrub, A.V., Bulusheva, L.G., Kuznetsov, V.L., Butenko, Yu.V., Chuvilin, A.L., and Heggie, M.I. 2001, X-ray emission studies of the valence band of nanodiamonds annealed at different temperatures, *J. Phys. Chem. A* **105**, 9781–9787.

Okotrub, A.V., Bulsheva, L.G., Kuznetsov, V.L., Vyalikh, D.V., and Poyguin, D.V. 2005, Electronic structure of diamond/graphite composite nanoparticles, *Eur. Phys. J. D* **34**, 157–160.

Pichler, T., Knupfer, M., Golden, M.S., Fink, J., and Cabioc'h, T., 2001, Electronic structure and optical properties of concentric-shell fullerenes from electron-energy-loss spectroscopy in transmission, *Phys. Rev. B* **63**, 155415.

Qin, L.-C. and Iijima, S., 1996, Onion-like graphitic particles produced from diamonds, *Chem. Phys. Lett.* **262**, 252–258.

Raty, J.-Y. and Galli, G., 2003a, Ultradispersity of diamond at the nanoscale, *Nature materials* **2**, 792–795.

Raty, J.-Y., Galli, G., Bostedt, C., van Buuren, T.W., and Terminello, L.J., 2003b, Quantum confinement and fullerenelike surface reconstructions in nanodiamonds, *Phys. Rev. Lett.* **90**, 037401.

Redlich, Ph., Banhart, F., Lyutovich, Yu., and Ajayan, P.M., 1998, EELS study of the irradiation-induced compression of carbon onions and their transformation to diamond, *Carbon* **36**, 561–563.

Roddatis, V.V., Kuznetsov, V.L., Butenko, Yu.V., Su, D.S., and Schlögl, R. 2002, Transformation of diamond nanoparticles into carbon onions under electron irradiation, *Phys. Chem., Chem. Phys.* **4**, 1964–1967.

Roy, D., Chhowalla, M., Wang, H. et al., 2003, Characterisation of carbon nano-onions using Raman spectroscopy, *Chem. Phys. Lett.* **373**, 52–56.

Saito, R., Dresselhaus, G., and Dresselhaus, M.S., 1998, *Physical Properties of Carbon Nanotubes*, Imperial College Press, London, U.K.

Sano, N., Akazawa, H., Kikuchi, T., and Kanki, T., 2003, Separated synthesis of iron-included carbon nanocapsules and nanotubes by pyrolysis of ferrocene in pure hydrogen, *Carbon*, **41**, 2159–2162.

Sano, N., Wang, H., Chhowalla, M., Alexandrou, I., and Amaratunga, G.A.J., 2001, Nanotechnology—Synthesis of carbon 'onions' in water, *Nature* **414**, 506.

Sano, N., Wang, H., Alexandrou, I. et al., 2002, Properties of carbon onions produced by an arc discharge in water, *J. Appl. Phys.* **92**, 2783–2788.

Shenderova, O., Grichko, V., Hens, S., and Walch, J., 2007b, Detonation nanodiamonds as UV radiation filter, *Diamond Relat. Mater.* **16**, 2003–2008.

Shenderova, O., Grishko, V., Cunningham, G., Moseenkov, S., McGuire, G., and Kuznetsov, V., 2008, Onion-like carbon for terahertz electromagnetic shielding, *Diamond Relat. Mater.* **17**, 462–466.

Shenderova, O., Tyler, T., Cunningham, G. et al., 2007a, Nanodiamond and onion-like carbon polymer nanocomposites, *Diamond Relat. Mater.* **16**, 1213–1217.

Shimizu, Y., Sasaki, T., Ito, T., Terashima, K., and Koshizaki, N., 2003, Fabrication of spherical carbon via UHF inductively coupled microplasma CVD, *J. Phys. D: Appl. Phys.* **36**, 2940–2944.

Skytt, P., Glans, P., Mancini, D.C. et al., 1994, Angle-resolved soft-x-ray fluorescence and absorption study of graphite, *Phys. Rev. B* **50**, 10457–10461.

Stöckli, T., Bonard, J.-M., Châtelain, A., Wang, Z.L., and Stadelmann, P., 2000, Plasmon excitations in graphitic carbon spheres measured by EELS, *Phys. Rev. B* **61**, 5751–5759.

Stone, A.J. and Wales, D.J., 1986, Theoretical studies of icosahedral C60 and some related species, *Chem. Phys. Lett.*, **128**, 501–503.

Street, K.W., Marchetti, M., Vandar Wal, R.V., and Tomasek, A.J., 2004, Evaluation of the tribological behaviour of nano-onions in Krytox 143AB, *Tribol. Lett,* **16**, 143–149.

Su, D.S., Maksimova, N., Delgado, J.J. et al., 2005, Nanocarbons in selective oxidative dehydrogenation reaction, *Catal. Today* **102**, 110–114.

Sun, L. and Banhart, F., 2006, Graphitic onions as reaction cells on the nanoscale, *Appl. Phys. Lett.* **88**, 193121(3).

Terrones, H. and Terrones, M., 1997, The transformation of polyhedral particles into graphitic onions, *J. Phys Chem. Solids* **58**, 1789–1796.

Terrones, M., Hsu, W.K., Hare, J.P., Kroto, H.W., Terrones, H., and Walton, D.R.M., 1996, Graphitic structures: From planar to spheres, toroids and helices, *Philos. Trans. R. Soc. Lond. A* **354**, 2025–2054.

Terrones, H., Terrones, M., and Hsu, W.K., 1995, Beyond C_{60}: Graphite structures for the future, *Chem. Soc. Rev.* **24**, 341–350.

Tomita, S., Burian, A., Dore, J.C., LeBolloch, D., Fujii, M., and Hayashi, S., 2002a, Diamond nanoparticles to carbon onions transformation: X-ray diffraction studies, *Carbon* **40**, 1469–1474.

Tomita, S., Fujii, M., Hayashi, S., and Yamamoto, K., 1999, Electron energy-loss spectroscopy of carbon onions, *Chem. Phys. Lett.* **305**, 225–229.

Tomita, S., Hayashi, S., Tsukuda, Y., and Fujii, M., 2002b, Ultraviolet-visible absorption spectroscopy of carbon onions, *Phys. Solid State* **44**, 450–453.

Tomita, S., Sakurai, T., Ohta, H., Fujii, M., and Hayashi, S., 2001, Structure and electronic properties of carbon onions, *J. Chem. Phys.* **114**, 7477–7482.

Ugarte, D., 1992, Curling and closure of graphitic networks under electron-beam irradiation, *Nature* **359**, 707–709.

Ugarte, D., 1993a, Formation mechanism of quasi-spherical carbon particles induced by electron bombardment, *Chem. Phys. Lett.* **207**, 473–479.

Ugarte, D., 1993b, How to fill or empty a graphitic onion, *Chem. Phys. Lett.* **209**, 99–103.

Ugarte, D., 1994, High-temperature behaviour of 'fullerene black,' *Carbon* **32**, 1245–1248.

Ugarte, D., 1995a, Onion-like graphitic particles, *Carbon* **33**, 989–993.

Ugarte, D., 1995b, Interstellar graphitic particles generated by annealing of nanodaimonds and their relation to the 2175-Angstrom peak carrier, *Astrophys. J.*, **443**, L85–L88.

Zaiser, M. and Banhart, F., 1997, Radiation-induced transformation of graphite to diamond, *Phys. Rev. Lett.* 79, 3680–3683.

Zaiser, M., Lyutovich, Yu., and Banhart, F., 2000, Irradiation-induced transformation of graphite to diamond: A quantitative study, *Phys. Rev. B* **62**, 3058–3064.

Znang, Q.L., O'Brien, S.C., Heath, J.R. et al., 1986, Reactivity of large carbon clusters: spheroidal carbon shells and their possible relevance to the formation and morphology of soot, *J. Phys. Chem.* **90**, 525–528.

Zhang, J., Su, D.S., Zhang, A.H., Wang, D., Schlogl, R., and Hebert, C., 2007, Nanocarbon as robust catalyst: Mechanistic insight into carbon-mediated catalysis, *Angew. Chem. Int. Ed.*, **46**, 7319–7323.

Wang, X.M., Xu, B.S., Uu, X.G., Guo, J.J., and Ichinose, H., 2006, Synthesis of Fe-included onion-like fullerenes by chemical vapor deposition, *Diamond Relat. Mater.* **15**, 147–150.

Xu, B.S., 2008, Prospects and research progress in nano onion-like fullerenes, *New Carbon Mater.* **23**, 289–301.

Xu, B.S., Guo, J.J., Wang, X.M., Liu, X.G., and Ichinose, H., 2006, Synthesis of carbon nanocapsules containing Fe, Ni or Co by arc discharge in aqueous solution, *Carbon* **44**, 2631–2634.

Yao, Y., Wang, X., Guo, J., Yang, X., and Xu, B. 2008, Tribological property of onion-like fullerenes as lubricant additive, *Mater. Lett.* **62**, 2524.

35

Plasmons in Fullerene Molecules

Ronald A. Phaneuf
University of Nevada

35.1 Introduction

The physical and chemical properties of matter in the nanometer-scale regime depend on the size of particles in a material sample, even when it is composed of a single element. Centuries ago, craftsmen learned how to finely divide samples of materials in order to exploit their properties with stunning effects. For example, the brilliant colors in medieval stained glass windows are known to result from resonant light scattering from finely divided gold and silver particles of different sizes that were incorporated into the glass. Early photographers also learned that the size of silver bromide particles deposited onto a glass substrate determined the rate at which they responded to light, forming permanent images.

A detailed study of the unusual properties of clusters of atoms began early in the twentieth century with the pioneering work by Mie on the response of small metal particles to light [1]. Despite the insightful and intuitive work of Mie, cluster science was not recognized as a legitimate field of study until 80 years later, when methods started to be developed to isolate atomic clusters of differing size and to characterize them. Today, atomic clusters constitute important building blocks of nanomaterials with novel physical and chemical properties. Because of their sensitive response, the irradiation of clusters of atoms by light is a highly effective probe of their structure and dynamics [2].

The Nobel Prize–winning discovery of the C_{60} molecule in 1985 by Kroto and collaborators [3] and of other pure carbon-cluster molecules with high degrees of symmetry and remarkable properties prompted a resurgence of interest in clusters of atoms. These so-called fullerene molecules, sometimes referred to as *buckeyballs*, were named after R. Buckminster Fuller, a proponent of the geodesic dome in architecture. Since their discovery, more than 10,000 scientific papers have been published elucidating the fascinating physical and chemical properties and behaviors of fullerenes, and of their growing range of practical applications. Fullerene molecules are of fundamental significance because their unique, highly symmetrical structures bridge the gap between free molecules and crystalline solids, and as a result they demonstrate some of the physical and chemical properties of each. For example, a giant resonance due to collective photoexcitation of the 240 delocalized valence electrons of C_{60} was predicted by Bertsch and collaborators [4] a few years after its discovery, and was demonstrated experimentally soon thereafter by Hertel and collaborators [5]. This chapter focuses on the excitation of so-called giant plasmon resonances in fullerene molecules and molecular ions, and on their decay.

35.2 Simple Models for Plasmon Oscillations

A plasma oscillation is a collective time-periodic excitation of an electron gas. Such collective motion manifesting the effects of the long-range correlation of electron positions has been shown to be a consequence of the long range of the Coulomb force [6]. When a plasma oscillation is quantized, it is referred to as a plasmon, whereby the electron gas moves as a whole with respect to the positive ion background. A plasmon is periodic in time but not in space.

35.2.1 Planar Conducting Film

Plasmons were first recognized in experiments in which low-energy electrons passed through or were reflected by thin conducting metallic films. The scattered electrons experience energy losses equal to multiples of the plasmon energy [7]. In this case, the excitation may be simply understood by considering a uniform displacement of an electron gas in a thin metallic slab [8], as shown in Figure 35.1. A displacement of the electron gas of amplitude z perpendicular to the surface creates a surface charge density $\sigma = z n_e e$ and an internal electric field $E = z n_e e / \epsilon_0$, where e and m_e are the electron charge and mass, n_e is the electron density, and ϵ_0 is the permittivity of free space (all in SI units). The electric field produces a restoring force proportional to the displacement z, and applying Newton's second law, the equation of motion of a unit volume of the electron gas is that of a simple harmonic oscillator,

$$n_e m_e \frac{d^2 z}{dt^2} = -n_e e E = -\frac{n_e^2 e^2}{\epsilon_0} z$$

$$\frac{d^2 z}{dt^2} + \omega^2 z = 0 \quad \text{where } \omega = \sqrt{\frac{n_e e^2}{\epsilon_0 \, m_e}} = \omega_p, \tag{35.1}$$

where ω_p is called the plasma frequency. The result is a periodic sinusoidal displacement of the planar electron gas with a frequency $\omega = \omega_p$ in a direction perpendicular to the plane of the positive ion background. The overall neutrality of the system is preserved, and net charges only appear on the surface of the film. For most substances, the associated plasmon energy $E_p = \hbar \omega_p$ is larger than the excitation energy of the lowest electronic excited state.

35.2.2 Spherical Conducting Shell

This simple model for surface plasmon oscillations in a thin film may be generalized to a thin spherical conducting shell [9]. In this case, the electron shell oscillates as a whole relative to the fixed positive ion core, as illustrated in an exaggerated and simplified manner in Figure 35.2. Such a periodic displacement of the electron shell may be resonantly excited by an external electromagnetic field. Again, the system as a whole remains electrically neutral, and net charges appear only on the inner and outer surfaces of the spherical shell. The result is a time-varying electric dipole. Models based on a thin spherical electron shell have been used to predict and to interpret collective electron excitations of the C_{60} molecule [10,11].

FIGURE 35.1 A transverse displacement z of the electron gas of a thin conducting film relative to the positive ion background produces an internal transverse electric field E proportional to z.

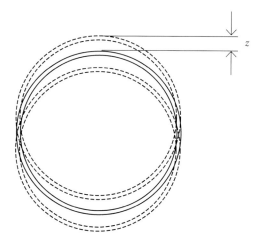

FIGURE 35.2 Displacement with an amplitude z of a thin spherical electron shell (bounded by dashed circles) relative to a positively charged ion core (bounded by solid circles).

35.3 Plasmons in Fullerene Molecules

35.3.1 The C_{60} Molecule

The C_{60} molecule is a truncated icosahedron approximately 1 nm in diameter, consisting of 60 carbon atoms regularly arranged at the vertices of 12 pentagons and 20 hexagons. The nearly spherical arrangement of carbon atoms is similar to the construction of a soccer ball, as shown in Figure 35.3. The carbon atom has the ground-state electronic configuration $1s^2 2s^2 2p^2$. In C_{60} the 240 2s and 2p valence electrons form an electron gas arranged in a spherical shell of finite thickness, and the empty carbon cage structure may be conveniently modeled as a hollow dielectric spherical shell [10]. The valence electron shell surrounds a positive ion core consisting of two 1s electrons localized on each of the 60 carbon nuclei. Since the valence electrons are less massive and more mobile than the nuclei, the valence shell may be treated as an electron gas embedded in a background positive charge that is static and uniform. This is called the *jellium* model [11] that is widely used to model solid metals. Using ratios of differential cross sections for photoionization of C_{60} molecules determined from normalized photoelectron spectral measurements and a spherical *jellium* model, Rüdel and collaborators estimated the diameter and thickness of the C_{60} valence electron shell [12]. These and some other physical properties of C_{60} relevant to the present discussion are collected in Table 35.1.

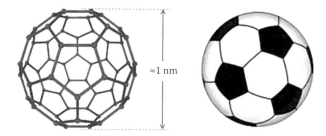

FIGURE 35.3 Comparison the icosahedron structure of the C_{60} molecule to that of a soccer ball. Carbon atoms are located at each of the 60 vertices indicated by solid dots.

TABLE 35.1 Relevant Physical Properties of the C_{60} Molecule

Outer diameter of valence electron shell*	0.858×10^{-9} m
Inner diameter of valence electron shell*	0.558×10^{-9} m
Mean electron density n_e of valence electron shell	1.00×10^{30} m^{-3}
Plasma frequency ω_p of valence electron shell	5.63×10^{16} rad/s
$E_p = \hbar\omega_p$	37.1 eV
First ionization potential	7.65 eV
Second ionization potential	12.4 eV
Electron affinity	2.67 eV

* Data from Rüdel et al. [12].

35.3.2 General Structure of Fullerene Molecules

Fullerene molecules are empty, closed-cage structures containing carbon atoms arranged in regular patterns of pentagons and hexagons. Euler's criteria for closed polyhedrons require that every closed fullerene cage structure C_n consist of 12 pentagons and $n/2 - 10$ hexagons. The fullerene C_{20} has zero hexagons and contains the smallest number of C atoms. The so-called magic numbers for n of 60, 70, 76, and 84 result in particularly stable structures and correspond to fullerenes that may be found in nature, with C_{60} being by far the most abundant. Various techniques have been developed to synthesize the higher-n fullerenes. When beams of fullerene ions are extracted from an electrical discharge into which C_{60} or higher-n fullerenes are evaporated, mass spectra confirm the existence of every even-numbered singly charged fullerene molecular ion down to C_{20}^+ [13]. Typical singly charged ion beam currents extracted from an electron cyclotron resonance (ECR) ion source operated at a microwave power of 2 W are plotted in Figure 35.4 as a function of fullerene number, n. As expected, the magic numbers 60, 70, 76, and 84 are prominent in the spectrum, and no odd-n

fullerene ion beams were detected. The lower-n fullerenes are evidently produced by collisional fragmentation of higher-n fullerenes in the ion-source discharge by electron impact, which results in the sequential or multiple ejection of ionized or neutral C_2 fragments [14]. The stabilities of the lower-n fullerene ions may vary considerably, but they need only to survive for a few microseconds in order to be detected by such a measurement.

35.3.3 Giant Plasmon Resonance in Photoionization of C_{60} and C_{70} Molecules

A theoretical prediction was made in 1991 by Bertsch and collaborators [4] that the 240 valence electrons in the C_{60} molecule are delocalized and that a giant collective plasmon resonance could be excited by an external electromagnetic field of $\hbar\omega \approx 20$ eV. Three months after their prediction was published, Hertel and collaborators [5] reported the first laboratory measurement of such a giant plasmon resonance in free C_{60} molecules. A strong, broad resonance centered near 20 eV was excited using monochromatized synchrotron radiation. The resonance was detected by measuring the yield of C_{60}^+ ions formed by photoionization of neutral C_{60} molecules, which were evaporated by gently heating a high-purity sample of solid material. Their measurement is presented in the lower panel of Figure 35.5. Evidently, because the energy of the

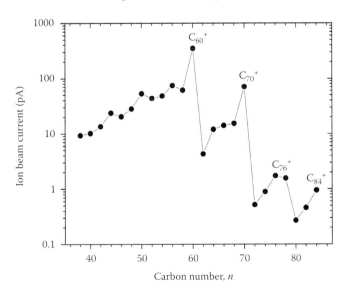

FIGURE 35.4 Fullerene ion beam currents in picoamperes (10^{-12} A) extracted from an electron cyclotron resonance (ECR) plasma ion source operated at low power, and into which commercially refined mixed fullerene powder was evaporated.

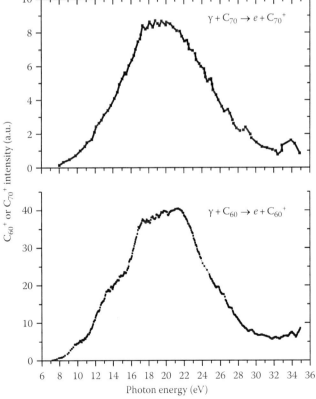

FIGURE 35.5 Giant surface plasmon resonances of the C_{60} and C_{70} molecules, first measured in single photoionization of molecules in the gas phase by Hertel and collaborators. (From Hertel, I.V. et al., *Phys. Rev. Lett.*, 68, 784, 1992. With permission.)

giant resonance is well above the 7.58 eV ionization potential of C_{60}, one important pathway for its decay is by the ejection of an electron, resulting in photoionization. In the same paper, Hertel et al. reported a similarly broad resonance centered near 20 eV for photoionization of the C_{70} molecule, which has prolate spheroidal (rugby ball) geometry and a lower degree of symmetry than C_{60}. Their data for C_{70} is presented in the top panel of Figure 35.5. Evidently a spherical geometry is not a necessary condition for photoexcitation of a plasmon resonance in a fullerene molecule.

Since these photoion-yield measurements of Hertel et al. [5] and those of subsequent researchers [15] were relative only, a systematic analysis was made by Berkowitz [16] to construct an absolute scale to quantify the process. A standard measure for the likelihood of a collisional or photo-induced process to occur is the cross section, which is measured in units of area, and may be considered as an effective cross-sectional area presented by a particular target species, within which the interaction occurs. The photoabsorption cross section of the C_{60} molecule over a wide energy range was determined empirically by piecing together measurements made in different photon energy ranges and applying sum rules for optical oscillator strengths [17], which are quantum-mechanical measures of the photoabsorption strength of an atomic or molecular system. This resulted in an estimated peak photoabsorption cross section for C_{60} of 1.4×10^{-19} m^2 at the giant plasmon resonance near 22 eV. This may be compared to the effective geometrical cross-sectional area of a C_{60} molecule of 4.4×10^{-19} m^2 calculated using the mean diameter of the valence electron shell from Table 35.1. Subsequent relative measurements of photoionization of C_{60} [18–20] were normalized to this same peak cross section in order to place them on an absolute scale. Such a normalization assumes that the plasmon excitation by photoabsorption leads exclusively to single photoionization of C_{60}, yielding a C_{60}^+ product ion.

35.3.4 Plasmons in Photoionization of $C_{60}{}^{q+}$ Molecular Ions

The first independently absolute cross-section measurements for photoionization of fullerene molecules were reported more than a decade later by Scully and collaborators [21]. A collimated beam of charge/mass-selected fullerene ions from a small accelerator was merged with a beam of monochromatized synchrotron radiation under well-determined conditions, and the yield of product photoions was measured as the photon energy was stepped. Their merged-beams apparatus [22] is shown in Figure 35.6. The photoionization cross section σ may be determined entirely from experimentally measured parameters using the equation

$$\sigma = \frac{Rqe^2 v_i \alpha}{I^+ I^\gamma \Omega \delta \Delta} \frac{1}{\int F(z) dz}. \tag{35.2}$$

R is the photoion product count rate
q is the primary ion charge state
e is the electronic charge
v_i is the ion beam speed
α is the responsivity of the photodiode (electrons/photon)
I^+ and I^γ are the primary ion beam and photodiode currents
Ω is the photoion collection efficiency
δ is the transmission of the pulse-counting electronics
Δ is the photoion detection efficiency

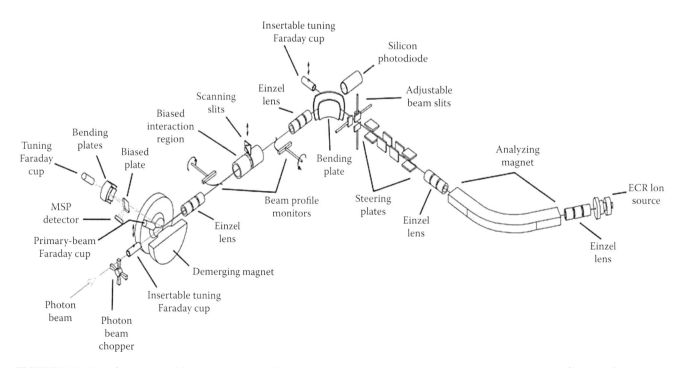

FIGURE 35.6 Ion–photon merged-beams apparatus used to measure absolute cross sections for photoionization of C_{60}^+, C_{60}^{2+}, and C_{60}^{3+} molecular ions using synchrotron radiation. (From Covington, A. M. et al., *Phys. Rev. A*, 66, 062710, 2002. With permission.)

$F(z)$ is the so-called form factor representing the measured two-dimensional intensity distributions of the photon and ion beams, which is integrated over their common interaction path along the z axis to define the volume of interaction of the two beams.

Cross sections for single photoionization of C_{60}^{+}, C_{60}^{2+}, and C_{60}^{3+} molecular ions were measured in the energy range 17–75 eV. The experimental results are compared with earlier photoionization measurements for neutral C_{60} in Figure 35.7. The giant surface plasmon resonance near 22 eV is prominent in the absolute measurements for all three initial ion charge states, which each contain an additional broad feature near 40 eV that is suggestive of a higher order plasmon oscillation.

Figure 35.8 shows a fit of two Lorentzian curves to the measured cross section for single photoionization of C_{60}^{+}, clearly indicating two broad resonance features. *Ab initio* quantum-mechanical calculations based on a *jellium* model and the time-dependent local-density approximation (TDLDA) attributed the additional feature near 38 eV to a volume plasmon oscillation associated with a local radial compression of the electron density in the spherical valence electron shell [21]. Theoretical models are discussed and calculations of this process based on these models are presented in Section 35.4.

Electric dipole excitation of a volume plasmon requires a local compression of the electron density, which is classically forbidden in a conducting sphere, but is possible in a conducting spherical shell [10]. The induced charge densities on the inner and outer surfaces of the fullerene valence electron shell are coupled and oscillate with two frequencies corresponding to

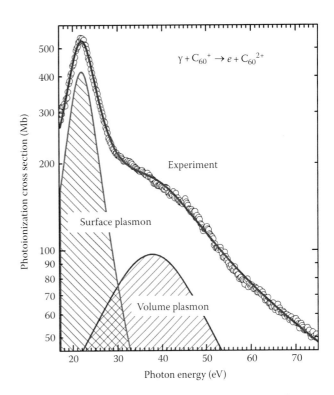

FIGURE 35.8 Solid curve denoting a fit of two Lorentzian functions to the measured cross section by Scully and collaborators for single photoionization of C_{60}^{+}, representing surface and volume plasmons. (From Scully, S.W.J. et al., *Phys. Rev. Lett.*, 94, 065503, 2005. With permission.)

in-phase and out-of-phase motions. The former corresponds to the surface plasmon, causing the electron shell to oscillate as a whole relative to the positive ion background (Figure 35.2). The latter corresponds to the volume plasmon, manifested by the charge density oscillating over the width of the electron shell [21]. Resonance parameters obtained from fits of Lorentzian functions to the experimental data are presented in Table 35.2.

For an electric dipole (E1) transition between states i and k of an atom or molecule, the oscillator strength f_{ik} is a dimensionless

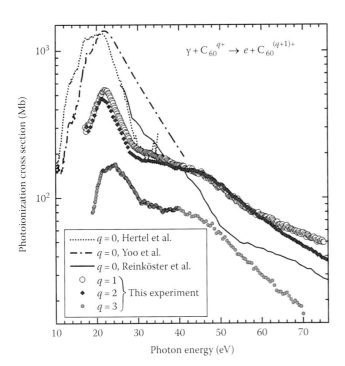

FIGURE 35.7 Cross-section measurements for single photoionization of C_{60}^{+}, C_{60}^{2+} and C_{60}^{3+} reported by Scully et al. [21]. The absolute merged-beams data are compared to normalized measurements of single photoionization of neutral C_{60} [5,15,20].

TABLE 35.2 Plasmon Resonance Parameters Obtained from Lorentzian Curve Fits to the Experimental Data for Single Photoionization of C_{60}^{+}, C_{60}^{2+}, and C_{60}^{3+} Ions

Initial Ion Charge	Resonance Energy, E_r (eV)	Resonance Width, Γ (eV)	Oscillator Strength, f	E_r/Γ
	Surface Plasmon			
1	22.0 ± 0.1	7.6 ± 0.3	45 ± 14	2.9 ± 0.1
2	21.5 ± 0.1	8.0 ± 0.4	44 ± 13	2.7 ± 0.1
3	23.6 ± 0.1	10.6 ± 0.4	21 ± 6	2.2 ± 0.1
	Volume Plasmon			
1	38.0 ± 2.0	29 ± 4	40 ± 12	1.3 ± 0.2
2	39.0 ± 2.0	40 ± 5	83 ± 25	1.0 ± 0.2
3	40.5 ± 1.0	30 ± 6	30 ± 9	1.3 ± 0.2

Source: Scully, S.W.J. et al., *Phys. Rev. Lett.*, 94, 065503, 2005.

number that measures the likelihood of photoabsorption, expressed in terms of SI units as

$$f_{ik} = \frac{g_k}{g_i} \frac{m_e c \varepsilon_0 \lambda^2}{2\pi e^2} A_{ki},$$ (35.3)

where

    λ is the wavelength in m
    A_{ki} is the Einstein coefficient or transition rate in s^{-1}
    g_i and g_k are the statistical weights (degeneracies or multiplicities) of the initial and final states, respectively
    c is the speed of light in m/s

In the case of plasmon excitation in fullerene molecular ions, the oscillator strength f was determined experimentally by integrating the measured cross section over the fitted resonance profile, using the relation

$$f = 9.11 \times 10^{-3} \int_{E_1}^{E_2} \sigma(E) dE,$$ (35.4)

where

    σ is the cross section in megabarns (1 megabarn = 1 Mb = $10^{-18} cm^2$)
    The energy E is in eV [23]

Although their widths and peak cross sections differ considerably, the oscillator strengths f associated with the features attributed to surface and volume plasmon oscillations are comparable in magnitude. The ratios of resonance energies to widths may be used to compare the resonance lifetime $\tau = h/\Gamma$ to the oscillation period $T = 2\pi/\omega = h/E_r$,

$$\frac{\tau}{T} = \frac{\frac{\hbar}{\Gamma}}{\frac{h}{E_r}} = \frac{1}{2\pi} \frac{E_r}{\Gamma}.$$ (35.5)

where

    h is Planck's constant
    $\hbar = h/2\pi$

Using the data from Table 35.2, such a comparison indicates that on average, the surface plasmon lifetime $\tau \approx 8 \times 10^{-17} s$, corresponding to 0.4 oscillation periods before decaying. For the volume plasmon, the lifetime $\tau \approx 2 \times 10^{-17} s$, corresponding to 0.2 oscillation periods.

The interpretation by Scully and collaborators [21] of the higher energy resonance feature in photoionization of C_{60} ions as a volume plasmon was disputed by Korol and Solov'yov as being conceptually incorrect [24]. On the basis of a classical hydrodynamic model, they claimed that the two observed resonance features may be explained in terms of two coupled surface plasmon oscillations. In reply [25], Scully and collaborators noted that the

two classical interpretations of the resonance phenomena in C_{60}^+ are essentially consistent, and that the disagreement is more a matter of terminology, i.e., what constitutes a volume plasmon. Comparing to the case of solid metallic clusters, they noted that the oscillations in C_{60} ions are of volume-like character because of a local spatial modulation of the electron density in the shell that may only be accounted for quantum mechanically, and that it is not instructive to carry classical arguments too far for fullerene molecules. They note, as pointed out by Ekhardt [28], that only a self-consistent quantum-mechanical theory like the TDLDA can account for the complexity of the observed electronic response of metal clusters to electromagnetic radiation, which they claim also to be the case for C_{60}^+. The main features of different theoretical models that have been proposed and methods used to describe plasmon excitation in fullerenes are outlined in Section 35.4.

35.3.5 Similarities between Giant Resonances in Nuclei, Atoms, and Molecules

In a comprehensive and insightful review of collective motions in different physical systems probed by light, Amusia and Connerade [26] observed that the ratio of resonance energy to width of the giant surface plasmon resonance in photoionization

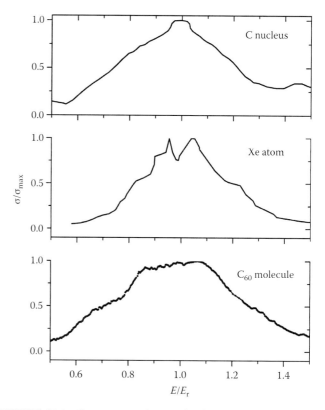

FIGURE 35.9 Comparison of giant dipole resonances in the ^{12}C nucleus [27], the 4d electron shell of the Xe atom [28], and the valence electron shell of the C_{60} molecule [5], after Amusia and Connerade [26]. The abscissae are plotted in units of their respective resonance energies, E_r, which span six orders of magnitude.

of C_{60} [5] is comparable to that of giant dipole resonances observed in gamma-ray emission from the ^{12}C nucleus following proton capture by ^{11}B [27], and also in extreme ultraviolet photoionization of the filled 4d inner shell of the Xe atom [28], which results mainly in double ionization to form Xe^{2+}. Experimental data for these three giant resonance phenomena associated with collective behavior on a quantum scale are compared in Figure 35.9. The consistency in their relative widths suggests that similar fundamental physical laws govern the correlated behavior resulting in each of these giant resonance phenomena, even though the physical sizes of these systems and their energy scales span six orders of magnitude!

35.4 Theoretical Models for Plasmon Excitation in Fullerenes

In this section, some theoretical models that have been successful in accounting for the response of fullerene molecules and molecular ions to extreme ultraviolet light are outlined, and their predictions are compared to experimental measurements.

35.4.1 Classical Mie Scattering

As previously noted, one of the earliest theoretical models to describe the response of small particles to light was proposed by Mie in 1908 [1]. The process is usually referred to as Mie scattering and is widely applied to the scattering of sunlight by aerosol particulates in the atmosphere. Maxwell's equations were solved analytically for the scattering of electromagnetic radiation by tiny conducting metal spheres. These spheres were assumed to be filled with an electron gas cloud that surrounds and is loosely bound to a positive ion background. This is the basis of the *jellium* model described earlier that is also applied in quantum-mechanical treatments of light scattering from metal spheres, such as the TDLDA [29]. Since optical wavelengths are typically much larger than the particle diameters, the associated time-varying electric field is considered to be spatially uniform. The electron gas responds in a collective manner to a weak external electric field, causing a displacement relative to the more massive and rigid ensemble of positive ions that is assumed to remain fixed in position. This charge separation creates an internal electric field that exerts an attractive restoring force on the displaced electron cloud. The situation for scattering of electromagnetic radiation by a spherical particle is analogous to that for a thin metal film described in Section 35.1, the result being a simple harmonic motion of the electron gas cloud relative to the positive ion background at the classical Mie frequency

$$\omega_{\text{Mie}} = \sqrt{\frac{n_e e^2}{3\varepsilon_0 m_e}} = \frac{\omega_p}{\sqrt{3}}, \quad (35.6)$$

where the bulk plasma frequency ω_p and the other symbols are the same as those in Equation 35.1. The model predicts a resonance in photoabsorption when the frequency of light incident on the spherical particle matches the Mie frequency. The predictions of this classical model correspond reasonably well with experimental data for scattering of light by small clusters of alkali metals, whose electronic structure is relatively simple. Somewhat surprising is the fact that using the physical parameters for C_{60} from Table 35.1, the classical Mie theory predicts the surface plasmon resonance in the C_{60}^+ molecular ion to occur at $\hbar\omega_{\text{Mie}} = 21.4\,\text{eV}$ and the volume plasmon resonance at $\hbar\omega_p = 37.1\,\text{eV}$. These are close to the experimental values presented in Table 35.2, which were obtained by fitting Lorentzian resonance profiles to the photoionization cross-section data. Evidently, the scattering of light from a conducting sphere and from a spherical fullerene cage have similarities.

35.4.2 Quantum-Mechanical Approximations

An exact solution to the Schrödinger equation for a large molecule like C_{60} in terms of single-particle orbitals lies far beyond current capabilities, even with continuing advances in computing capacity and speed. Approximate methods are therefore needed to gain physical insight into the molecular structure and dynamic behavior of such systems. The local-density approximation (LDA) is a quantum-mechanical approach based on density functional theory [30], for which the Nobel Prize in Chemistry was awarded to Walter Kohn in 1998. It is similar in concept to the Thomas–Fermi model that was developed to describe quantum-mechanical features of multi-electron atoms. A functional is generally defined as a function of another function, in this case, of the electron density. In 1965, a fundamental theorem was discovered by Hohenberg and Kohn [31] which states that to fully define a stationary electronic system, knowledge of its ground-state density is sufficient, and all observables may in principle be determined from it. The electron density, n_e, is a convenient variable because it is intuitive and depends only on the spatial coordinates. The LDA has revolutionized the theoretical study of complex systems in their ground states [11,32,33]. It provides important insights into the physical and chemical properties of molecules, clusters, and bulk solids, and underpins modern materials science [34].

An implementation of the LDA that accounts for electron correlations and exchange is based upon the Kohn–Sham equations and variational method [30,32], by which a minimum in the total energy of the system is determined that represents its ground state. An effective one-electron potential $V_{\text{eff}}(r)$ that is only a function of the electron density $n_e(r)$ represents the complexity of the electron–electron interactions of the N-electron system,

$$V_{\text{eff}}(r) = V(r) + \int \frac{n_e(\vec{r}')}{|\vec{r} - \vec{r}'|} d^3r' + V_{\text{xc}}(r). \quad (35.7)$$

$V(r)$ is the external potential

The electron density $n_e(r)$ is expressed in terms of the N one-electron wavefunctions $\varphi_j(r)$,

$$n_e(r) = \sum_{j=1}^{N} \left| \varphi_j(r) \right|^2. \qquad (35.8)$$

$V_{xc}(r)$ is the electron exchange-correlation potential that may be directly obtained from the electron density $n_e(r)$ if the latter is known from experiments or other calculations. The single-electron orbital wavefunctions $\varphi_j(r)$ satisfy the Schrödinger equation,

$$\left(-\frac{1}{2}\nabla^2 + V_{eff}(r) - \epsilon_j \right)\varphi_j(r) = 0. \qquad (35.9)$$

Density functional theory is inherently static and the LDA gives reasonable predictions of the ground states, ionization potentials, and electron affinities of many-electron systems of chemical relevance that are far too complex for *ab initio* electronic-structure calculations.

A significant development in density functional theory has been the introduction of a temporal dependence to describe relatively slow, dynamic interactions of complex multi-electron systems such as collective electron motions. This leads to a set of time-dependent Kohn–Sham equations that include all many-body correlations through a local time-dependent exchange-correlation potential [35]. The resulting method is called the time-dependent local-density approximation (TDLDA), or time-dependent density functional theory (TDDFT) [36]. It has been demonstrated to be effective in computationally describing the photoresponse of complex multi-electron systems, such as metal clusters [37] and fullerenes [21,38,39].

As an example, Figure 35.10 compares the results of TDLDA calculations with the experimental cross-section measurement for single photoionization of the C_{60}^+ molecular ion, which was presented in Figure 35.8. The observed surface and volume plasmon resonances centered at 22 and 38 eV, respectively, are not present in a calculation based on LDA, but are prominent in the TDLDA result, indicating that they are of a collective nature. In the calculation, the volume plasmon resonance is narrower than that observed, and is centered at 42.5 eV, which is higher than the experimental value of 38 eV obtained from a Lorentzian fit to the data. For the comparison, the calculated TDLDA photoabsorption cross section was shifted to higher energies by 5.5 eV and scaled in magnitude by 0.25 to match the dominant surface plasmon resonance at 22 eV. The scaling of the calculated photoabsorption cross-section magnitude is reasonable, since additional plasmon decay channels involving fragmentation were shown to be important by subsequent measurements, and such products were not detected in the experiment. The energy shift is attributed to the fact that the *jellium* model does not account for the structure of the positive ion background. Without this energy shift in the TDLDA calculation [39], the volume plasmon resonance occurs at the experimental energy of 38 eV, whereas the surface plasmon occurs at 16.5 eV.

The TDLDA-calculated surface plasmon resonance at 22 eV is also narrower than the measurement, which may be indicative of spectral broadening due to vibrational excitation of the C_{60}^+ molecular ions in the primary beam used for the measurements. The TDLDA calculation also exhibits some narrow resonances that are not observed. Such features would have been readily detectable at the experimental energy resolution of 0.1 eV, but may have been obscured because of internal rotational and vibrational excitation of the C_{60}^+ ion beam.

Another theoretical method developed from a variational principle is the local-current approximation (LCA), which employs a self-consistent mean-field approach and hydrodynamic model to describe strongly interacting many-fermion systems [11]. This method has been applied successfully to collective oscillations in nuclei and metal clusters, and very recently to plasmon excitations in the photoionization of the C_{60} molecule [40]. The result is presented as LCA in Figure 35.10 clearly indicating the two plasmon resonances. Again, as with the TDLDA, for comparison, the LCA cross-section magnitude was normalized to the experiment at 22 eV and the energy scale was shifted upward by 5.5 eV to compensate for the inability of the jellium model to account for details of the positive ion core. The relative amplitude of the volume plasmon for C_{60} is again smaller than the measurement for C_{60}^+, but the LCA resonance energy agrees with the measured value of 38 eV.

Despite some differences, the TDLDA and LCA calculations are remarkably successful in reproducing the two plasmon excitations that characterize the interaction of vacuum ultraviolet photons with a molecular ion composed of 60 carbon atoms and 359 electrons.

35.4.3 Liquid-Drop Model

A liquid-drop hydrodynamic model similar to that used in nuclear physics was developed by Amusia and Kornyushin to describe collective modes of electron oscillation in metallic clusters and fullerene molecules [41]. In applying the model,

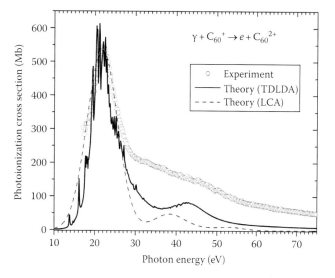

FIGURE 35.10 Comparison between TDLDA [21] and LCA [40] theoretical calculations and experiment [21] for photoionization of C_{60}^+.

the simplifying assumption was made that the nucleus and two 1s electrons of each carbon atom form part of the positive ion core, the two 2s electrons form molecular bonds, and the two 2p electrons are collectivized. The molecular bonds determine the radius of the C_{60} sphere, on whose surface the 120 collectivized electrons are distributed homogeneously. The electrostatic potential for this charge distribution was determined by solving Poisson's equation, and the response of the charged spherical shell to an external potential was solved classically. Applying this model, frequencies of so-called dipole and breathing-mode oscillations of the valence electrons of the C_{60} molecule were determined, corresponding to energies of 20.1 and 31.2 eV, respectively. Considering the simplicity of the model, these values are surprisingly close to the energies of 22 and 38 eV observed in photoionization of C_{60}^{+}.

The liquid-drop model is similar to the linear hydrodynamic model referred to by Korol and Solov'yev [24] in their criticism of the interpretation of the resonance feature at 38 eV in photoionization of C_{60}^{+} as a volume plasmon, as discussed in Section 35.3.4.

35.5 Endohedral Fullerene Molecules

Shortly after the discovery of C_{60}, evidence was reported of metal atoms trapped inside the nanometer-sized hollow carbon molecular cage [42]. The existence of so-called endohedral fullerene molecules was initially controversial but has since been firmly established [43]. The adopted nomenclature for an atom A inside a C_n molecular cage is $A@C_n$. Studies of such exotic species have been aimed at both characterizing their unusual fundamental physical properties and exploring multifold possibilities for their practical use. Promising applications are as encapsulating hosts for the chemical isolation of highly reactive or toxic atoms in medical diagnostics and their use as gates in quantum computers.

Subsequently, evidence for the existence of C_{60} and C_{70} molecules containing noble gas atoms was discovered in meteor, comet, and asteroid impact craters [44]. The isotopic ratios in these samples differ by an order of magnitude from those of noble gases elsewhere on earth, indicating that these endohedral fullerenes are of cosmic origin and that they survived impact with the earth intact.

The properties of an atom caged within an electrically charged spherical surface of nanometer scale present an intriguing fundamental quantum physics problem. Among theoretical predictions are the splitting of the broad giant atomic Xe 4*d* photoabsorption resonance (Figure 35.9) in Xe@C_{60} into multiple components [45] and the transfer of oscillator strength in photoprocesses from collective plasmon modes to localized electron excitations in Ar@C_{60} [46]. Theoretical activity on this subject has expanded largely in the absence of experimental checks because methods for producing endohedral fullerenes with high yield have only recently permitted studies on a molecular level. One endohedral fullerene that has been successfully produced in macroscopic quantities is Ce@C_{82}, whose structure has been

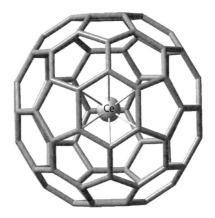

FIGURE 35.11 Structure of the Ce@C_{82} endohedral molecule calculated using density functional theory by Muthukumar and Larsson. (From Muthukumar, K. and Larsson, J.A., *J. Phys. Chem. A*, 112, 1071, 2008. With permission.)

calculated using density functional theory [47] and is depicted in Figure 35.11.

Cross-section results have recently been reported by Müller and collaborators [48] for photoionization of the Ce@C_{82}^+ molecular ion. Single-ionization data from their experiment is presented in Figure 35.12, which compares reference measurements on empty C_{82}^+ ions with corresponding measurements on endohedral Ce@C_{82}^+ ions. The surface and volume plasmon resonances observed in photoionization of C_{60}^{q+} ions are both evident in these measurements, and do not appear to be strongly influenced by the presence of a Ce atom within the fullerene cage. The charge of the caged Ce atom was determined by comparing the energy position of the Ce giant 4d resonance with measurements on bare Ce^{q+} ions, and was determined to be +3. Only about half the 4d oscillator strength of the Ce^{3+} ion was accounted for in the

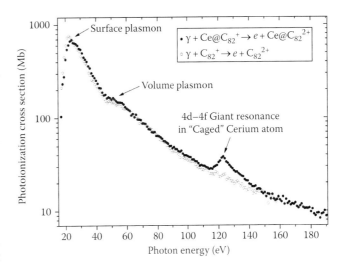

FIGURE 35.12 Measurements of single photoionization of C_{82}^+ and of the endohedral Ce@C_{82}^+ molecular ion, indicating giant resonances due to surface and volume plasmons of the valence shell of the fullerene cage, and due to 4d electron excitation of the caged Ce atom. (From Müller, A. et al., *Phys. Rev. Lett.*, 101, 133001, 2008. With permission.)

$Ce@C_{82}^{+}$ ionization measurement, suggesting that decay channels resulting in fragmentation of the fullerene cage are likely in this case. Increasing availability of macroscopic samples of such exotic materials is expected to facilitate further quantitative measurements on endohedral molecules and molecular ions.

35.6 Summary

Because their structures are intermediate between those of normal molecules and of crystalline solids, fullerene molecules exhibit some of the properties of each. They are therefore intriguing candidates for scientific study and for novel applications. The carbon atoms in fullerene molecules are regularly arranged in empty spheroidal cage structures that are approximately 1 nm in diameter. Since their several hundred valence electrons are delocalized from their positively charged ionic core, they may be collectively photoexcited in plasmon-like modes that are characteristic of conducting solids and solid films. Because of their high degrees of symmetry, simple classical models of fullerene molecules as charged hollow dielectric spheres are remarkably successful in describing the main features of their resonant response to irradiation by ultraviolet light. This response has been investigated and quantified in experiments involving the interactions of beams of fullerene molecular ions and synchrotron radiation, and is manifested in photoionization. The main features are explained by quantum-mechanical theoretical models based on the time-dependent local-density approximation. Atoms or small molecules may be stably trapped within the hollow cages of fullerene molecules and become chemically isolated from their surroundings. These so-called endohedral fullerenes provide a novel framework for studying the properties of atoms subject to unusual boundary conditions, such as an atom within a charged spherical shell. They also are promising as candidate systems for many practical applications.

Acknowledgments

The author is indebted to his colleagues Alfred Müller and Peter Winkler and to his student Alexander Carr for providing corrections and helpful suggestions. Support by grant DE-FG02-03ER15424 from the Chemical Sciences, Geosciences and Biosciences Division of the U.S. Department of Energy is gratefully acknowledged.

References

1. G. Mie, *Ann. Phys. (Leipzig)* 25, 376–445 (1908).
2. S. Link and M. A. El-Sayed, *Int. Rev. Phys. Chem.* 19, 409 (2000).
3. H. W. Kroto, J. R. Heath, S. C. O'Brien, R. F. Curl, and R. E. Smalley, *Nature (London)* 318, 162 (1985).
4. G. F. Bertsch, A. Bulgac, D. Tománek, and Y. Wang, *Phys. Rev. Lett.* 67, 2690–2693 (1991).
5. I. V. Hertel, H. Steger, J. de Vries, B. Weisser, C. Menzel, B. Kamke, and W. Kamke, *Phys. Rev. Lett.* 68, 784–787 (1992).
6. D. Bohn and D. Pines, *Phys. Rev.* 82, 625–634 (1951); D. Pines and D. Bohm, *Phys. Rev.* 85, 338–353 (1952).
7. C. J. Powell and J.B. Swan, *Phys. Rev.* 115, 869–875 (1959).
8. C. Kittel, *Introduction to Solid State Physics*, Wiley, New York (1968), pp. 233–235.
9. P. G. Reinhard and E. Suraud, *Introduction to Cluster Dynamics*, Wiley-VCH Verlag, Weinheim, Germany (2004), p. 33.
10. Ph. Lambin, A. A. Lucas, and J.-P. Vigneron, *Phys. Rev. B* 46, 1794–1803 (1992).
11. M. Brack, *Rev. Mod. Phys.* 65, 677 (1993); *Sci. Am.* 277, No. 6, 50 (1997).
12. A. Rüdel, R. Hentges, U. Becker, H. S. Chakraborty, M. E. Madjet, and J. M. Rost, *Phys. Rev. Lett.* 89, 125503 (2002).
13. R. Völpel, G. Hoffman, M. Steidel, M. Stenke, M. Schlapp, R. Trassl, and E. Salzborn, *Phys. Rev. Lett.* 71, 3439 (1993).
14. D. Hathiramani, K. Aichele, W. Arnold, K. Huber, and E. Salzborn, *Phys. Rev. Lett.* 85, 3604 (2000).
15. R. K. Yoo, B. Ruscic, and J. Berkowitz, *J. Chem. Phys.* 96, 911 (1992).
16. J. Berkowitz, *J. Chem. Phys.* 111, 1446–1453 (1999).
17. B. L. Henke, E. M. Gullikson, and J. C. Davis, *At. Data. Nucl. Data Tables* 54, 218–342 (1993).
18. J. Kou, T. Mori, M. Ono, Y. Haruyama, Y. Kubozono, and K. Mitsuke, *Chem. Phys. Lett.* 374, 1 (2003).
19. J. Kou, T. Mori, S. V. K. Kumar, Y. Haruyama, Y. Kubozono, and K. Mitsuke, *J. Chem. Phys.* 120, 6005 (2004).
20. A. Reinköster, S. Korica, G. Prümpter, J. Viefhaus, K. Godenhusen, O. Schwartzkopf, M. Mast, and U. Becker, *J. Phys. B* 37, 2135 (2004).
21. S. W. J. Scully, E. D. Emmons, M. F. Gharaibeh, R. A. Phaneuf, A. L. D. Kilcoyne, A. S. Schlachter, S. Schippers, A. Müller, H. S. Chakraborty, M. E. Madjet, and J. M. Rost, *Phys. Rev. Lett.* 94, 065503 (2005).
22. A. M. Covington, A. Aguilar, I. R. Covington, M. F. Gharaibeh, G. Hinojosa, C. A. Shirley, R. A. Phaneuf, I. Álvarez, C. Cisneros, I. Domínguez-Lopez, M. M. Sant'Anna, A. S. Schlachter, B. M. McLaughlin, and A. Dalgarno, *Phys. Rev. A* 66, 062710 (2002).
23. H. Kjeldsen, P. Andersen, F. Folkmann, J. Hansen, M. Kitajima, and T. Andersen, *J. Phys. B* 35, 2845 (2002).
24. A. V. Korol and A. V. Solov'yov, *Phys. Rev. Lett.* 98, 179601 (2007).
25. S. W. J. Scully, E. D. Emmons, M. F. Gharaibeh, R. A. Phaneuf, A. L. D. Kilcoyne, A. S. Schlachter, S. Schippers, A. Müller, H. S. Chakraborty, M. E. Madjet, and J. M. Rost, *Phys. Rev. Lett.* 98, 179602 (2007).
26. M. Ya Amusia and J.-P. Connerade, *Rep. Prog. Phys.* 63, 41–70 (2000).
27. H. E. Gove, A. E. Litherland, and R. Batchelor, *Phys. Rev. Lett.* 3, 177 (1959); *Nucl. Phys.* 26, 480 (1961).
28. D. Ederer, *Phys. Rev. Lett.* 13, 760 (1964).
29. W. Ekhardt, *Phys. Rev B* 31, 6360 (1985); 36, 4483 (1987).
30. W. Kohn, *Rev. Mod. Phys.* 71, 1253 (1999).
31. P. Hohenberg and W. Kohn, *Phys. Rev. B* 136, 864 (1964).

32. W. Kohn and L. J. Sham, *Phys. Rev. A* 140, 1133 (1965).

33. R. G. Parr and W. Yang, *Density Functional Theory of Atoms and Molecules*, Oxford University Press, Oxford, U.K. (1989).

34. J. Hafner, *J. Comput. Chem.* 29, 2044 (2008).

35. E. K. U. Gross and W. Kohn, *Adv. Quantum Chem.* 21, 255 (1990).

36. M. A. L. Marques and E. K. U. Gross, *Ann. Rev. Phys. Chem.* 55, 427 (2004).

37. F. Calvayrac, P.-G. Reinhard, E. Suraud, and C. A. Ullrich, *Phys. Rep.* 337, 493 (2000).

38. M. J. Puska and R. M. Nieminen, *Phys. Rev. A* 47, 1181 (1993).

39. M. E. Madjet, H. S. Chakraborty, J. M. Rost, and S. T. Manson, *J. Phys. B: At. Mol. Opt. Phys.* 41, 105101 (2008).

40. M. Brack, P. Winkler, and M. V. N. Murthy, *Int. J. Modern Phys. E* 17, 138 (2008).

41. M. Ya Amusia and Y. Kornyushin, *Contemp. Phys.* 41, 219 (2000).

42. J. R. Heath et al., *J. Am. Chem. Soc.* 107, 7779 (1985).

43. Y. Chai et al., *J. Phys. Chem.* 95, 7564 (1991).

44. L. Becker, R. J. Poreda, and J. L. Bada, *Science* 272, 249 (2001); L. Becker and P. H. Seeberger, *Science* 291, 1530 (2001).

45. M. Ya. Amusia, A. S. Balentkov, L. V. Chernysheva, Z. Felfli, and A. Z. Msezane, *J. Phys. B: At. Mol. Opt. Phys.* 38, L169 (2005).

46. M. E. Madjet, H. S. Chakraborty, and S. T. Manson, *Phys. Rev. Lett.* 99, 243003 (2007).

47. K. Muthukumar and J. A. Larsson, *J. Phys. Chem. A* 112, 1071 (2008).

48. A. Müller, S. Schippers, M. Habibi, D. Esteves, J. C. Wang, R. A. Phaneuf, A. L. D. Kilcoyne, A. Aguilar, and L. Dunsch, *Phys. Rev. Lett.* 101, 133001 (2008).

[60]Fullerene-Based Electron Acceptors

Beatriz M. Illescas
Universidad Complutense

Nazario Martín
Universidad Complutense

36.1 Introduction

Fullerenes have been the most studied molecular carbon allotropes during the last two decades due to their singular spherical shape and their chemical, electrochemical, and photophysical properties (Martín 2006).

One of the most promising areas of application of fullerene derivatives is photovoltaic, where polymer:fullerene bulk heterojunction (BHJ) solar cells seem to have a real potential. In particular, BHJ solar cells based on poly(3-hexylthiophene) (P3HT) and the fullerene derivative [6,6]-phenyl-C61-butyric acid methyl ester (PCBM) have led to reproducible power-conversion efficiencies approaching 5%. To generate an effective photocurrent in these organic solar cells, a fine-tuning of the electronic properties and interactions of the donor and acceptor components is essential.

Fullerenes are currently considered to be the ideal acceptors for organic solar cells for several reasons. First, [60]fullerene, according to theoretical calculations, exhibits a triply degenerated LUMO (t_{1u}) that is comparatively low in energy. Therefore, C_{60} behaves like an electronegative molecule, which reversibly accepts up to six electrons in solution (Haufler et al. 1990, Allemand et al. 1991, Arias et al. 1994). The cyclic voltammetry and the electron affinity measured for [60]fullerene clearly confirm that it is a moderate electron acceptor comparable to other organic molecules such as benzo- and naphthoquinones (Wang et al. 1991, Saito et al. 1994). This electron-accepting ability of C_{60} to form stable multianions (Xie et al. 1992, Ohsawa and Saji 1992) is in sharp contrast with the ability to generate the very unstable cationic species (Nonell et al. 1992, Foote 1994, Lem et al. 1995, Siedschlag et al. 1997). Importantly, a number of conjugated polymer-fullerene blends are known to exhibit ultrafast photoinduced charge transfer (ca. 45 fs), with a back transfer that is orders of magnitude slower. Furthermore, C_{60} has been shown to have a very high electron mobility of up to 1 cm²/V s in field-effect transistors (FETs).

Most of the [60]fullerene derivatives present poorer electron acceptor properties than the parent C_{60} as a consequence of the saturation of a double bond of the C_{60} framework that raises the LUMO energy (Echegoyen and Echegoyen 1998). Still, different strategies have been followed to increase the electron acceptor character in fullerene derivatives, which have been mainly focused on (1) electron acceptor moieties covalently linked to C_{60}, (2) the presence of electronegative atoms directly linked to the C_{60} core, (3) periconjugative effect, (4) pyrrolidinium salts, (5) the presence of heteroatoms as constituents of the fullerene framework (heterofullerenes), (6) fluorofullerenes, and (7) endohedral fullerenes. Some of these examples were aimed at the modification of the electron affinity of the fullerene derivatives in order to improve their performance in organic solar cells. However, recent findings demonstrate that raising the LUMO level of the acceptor fullerene does give a higher V_{oc} in fullerene:polymer solar cell devices, as an energy difference between the LUMOs of the donor and acceptor of about 0.3 eV is required for electron transfer to occur. Therefore, the essential point is not to improve the acceptor ability of the fullerene, but to modulate the energy difference between the donor-HOMO and the acceptor-LUMO.

In this chapter, we review different types of fullerene-based acceptors with improved or similar electron acceptor ability than pristine C_{60} and give an overview about the influence of the acceptor properties of the fullerene derivative on the performance of photovoltaic devices.

36.2 Fullerene-Based Acceptors

36.2.1 Electron Acceptor Moieties Covalently Linked to C_{60}

Introducing electron acceptor groups as substituents in organofullerenes may lead to derivatives with higher electron affinity than [60]fullerene. Thus, the loss of conjugation produced by substitution on the C_{60} surface may be compensated by the presence of strong electron-withdrawing groups. This is the case of [1,2]dicyanofullerene (Creegan et al. 1992, Keshavarz-K. et al. 1995) and dicyano[6,6]methano-fullerene (Keshavarz-K. et al. 1996a), the reduction potentials of which were anodically shifted around 100–150 mV compared with naked C_{60}.

[60]Fullerene-*p*-benzoquinone systems (**1a-b**, **2a-b**) have been prepared by cycloaddition reaction of *o*-quinodimethanes to C_{60} (Figure 36.1) (Iyoda et al. 1994, 1996, Illescas et al. 1995, 1997a, 1999). The study of the redox properties of these compounds (**1a-b**, **2a-b**) shows how, depending on the nature of the *p*-benzoquinone ring substitution, it is possible to address the attachment of the first electron in the reduction process to either the C_{60} or the organic addend (Table 36.1). Thus, when R = H or Br in **1**, the first observed reduction wave is assigned to the reduction of the *p*-benzoquinone fragment to the semiquinone radical. Therefore, these compounds show better electron acceptor properties than pristine C_{60}, and, indeed, the first reduction potential is anodically shifted 90 mV for **1a** and 390 mV for **1b** related to fullerene. In contrast, the presence of a benzene ring fused to the *p*-benzoquinone cathodically shifts the first reduction potential of the *p*-benzoquinone fragment, which now appears at more negative potentials than the first reduction potential of the C_{60} unit. By comparison, the first reduction potential found for **2b** can be considered as the sum of two different effects: the presence of the bromine atoms, which shifts the reduction potential toward more positive values (stabilization of the LUMO), and the fused benzene ring, which raises the LUMO energy. The cyclic voltammetry values were rationalized on the basis of theoretical calculations, and, according to the molecular orbital distribution, the attachment of the first electron in the reduction process takes place either in the *p*-benzoquinone fragment or in the C_{60} cage. Thus, the introduction of two bromine atoms linked to the C_{60} moiety in **1b** causes the stabilization of the LUMO of this moiety, whereas the presence of a benzene ring fused to the quinone moiety in **2a**, in contrast, produces a destabilization of the

TABLE 36.1 Redox Potentials at Room Temperature (in V vs. SCE)[a]

Comp.	$E_{red}^{quinone}$	E_{red}^{1}	E_{red}^{2}	E_{red}^{3}	E_{red}^{4}
C_{60}		−0.60	−1.00	−1.52	−2.04
1a	−0.51	−0.68	−1.17	−1.74	—
1b	−0.21	−0.64	−1.07	−1.68	—
2a	−0.77	−0.63	−1.11	−1.76	−1.94
	−1.32				
2b	−0.47	−0.63	−1.05	−1.69	
	−1.34				

[a] GCE (glassy carbon) as working electrode, Bu_4NClO_4 (0.1 M) as supporting electrolyte, and toluene/CH_3CN 5/1 (v/v) as solvent. Scan rate 200 mV/s.

LUMO of the organic addend, which now lies above those of the C_{60} core (Illescas et al. 1995b).

Diekers et al. (1998) reported the synthesis and electrochemical study of other Diels-Alder C_{60}-acceptor dyads bearing either an anthraquinone (**3**) or a dicyano-*p*-quinonediimine DCNQI (**4**) fragment, which contain a rigid linkage between the redox active centers. The electrochemical study of these compounds shows that the reduction processes are highly localized either on the fullerene core or on the addend, depending on the relative electron affinities. Thus, in the case of **3**, the first reduction is fullerene based, while for **4**, the attachment of the first electron is addend-based. ESR spectroscopic data allowed the unambiguous assignment of the location of each reduction step being consistent with the electrochemical observations, with the exception of that for the pentaanion of **4**, which shows that spin localization occurs exclusively on the anthraquinone moiety. This observation suggests that a diamagnetic tetraanion is present on the fullerene group, and an unpaired spin is localized on the addend, contrary to the electrochemically predicted electron distribution with a fullerene trianion and a quinone dianion (Figure 36.2).

Other [60]fullerene-based electron acceptors have been prepared by 1,3-dipolar cycloaddition reaction of the corresponding azomethine ylides to C_{60}, following Prato's procedure (Iyoda et al. 1997, Illescas et al. 1997b).

The interesting redox properties of compounds **5–9** can also be understood in terms of the LUMOs energies as described above for the related systems. Thus, the attachment of the first electron in the reduction process takes place in the organic

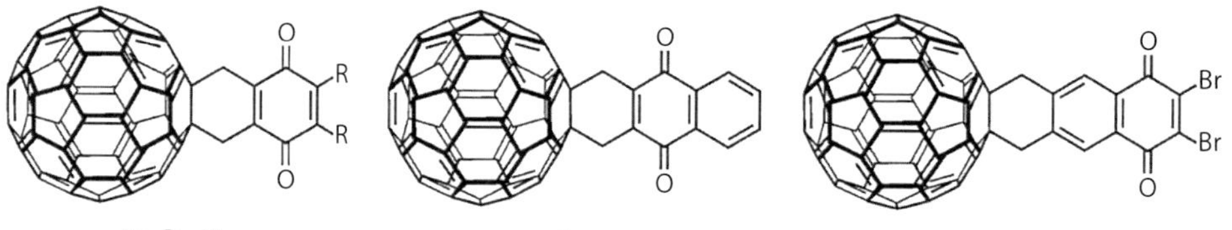

1a, R = H
1b, R = Br

2a

2b

FIGURE 36.1 Chemical structure of fullerene-based quinones **1** and **2**.

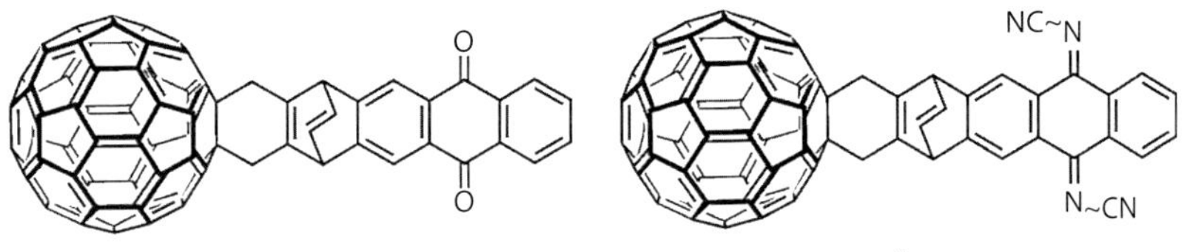

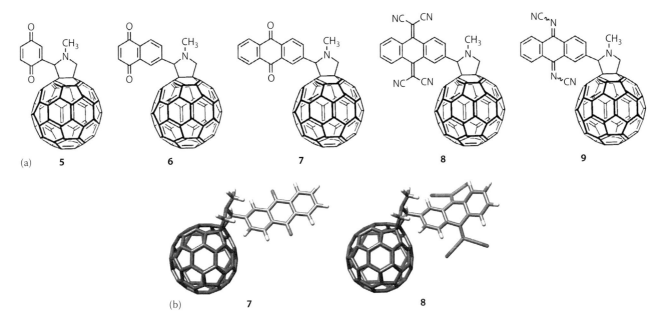

FIGURE 36.2 Structure of compounds **3** and **4**.

FIGURE 36.3 Chemical structure of compounds **5–9** (a) and molecular modeling of **7** and **8** (b).

addend in compounds **5**, **6**, **8**, and **9**, while for the anthraquinone derivative **7**, the first reduction process is fullerene-based.

Dyads **2b** (Figure 36.1) and **8** (Figure 36.3) have been studied in search of a possible electron transfer from the photoexcited fullerene moiety to the covalently linked electron acceptor (Martín et al. 2000). Formation of the fullerene triplet excited state rather than an intramolecular electron transfer was observed, which can be accounted for by the inability of C_{60} to form stable cationic species (Reed et al. 2000).

Anthraquinone (AQ) and 11,11,12,12-tetracyano-9,10-anthraquinodimethane (TCAQ) were also used as acceptor moieties for the design of C_{60}-based triads constituted by fullerene and AQ (**10**) or TCAQ (**11**) as acceptor units, which are weaker and stronger acceptors, respectively, relative to the parent C_{60}, and ferrocene as donor unit (Herranz et al. 2000). The electrochemical study of these triads shows the presence of three one-electron reduction steps, all of them taking place at the [60]fullerene moiety, at potential values quite similar to those found for the parent C_{60}. A possible reasoning for this observation stems from a combination of, first, the carbonyl group linked to the pyrrolidine nitrogen (Prato and Maggini 1998) and, second, the electronic effect of the electron-acceptor unit bound to the pyrrolidine ring (Figure 36.4).

The photophysical study of both triads (**10**, **11**) revealed that only a reaction between Fc as donor and C_{60} as acceptor takes place, despite the fact that the reduction potential of TCAQ in triad **11** renders to be easier than the fullerene first reduction.

The synthesis of novel trinitrofluorenone-C_{60} (TNF-C_{60}) acceptors in which the two subunits are placed at different distances (**12**) has also been reported (Ortiz et al. 2004). The electrochemical study of these dyads shows a first reduction potential, which is TNF-based and appears 220–260 mV anodically shifted in comparison with reference compound *N*-methylfulleropyrrolidine. The photorefractive properties of these compounds as sensitizers in poly(*N*-vinylcarbazole)-based composite materials have also been studied, finding some advantages with respect to the use of the parent C_{60}.

36.2.2 Electronegative Atoms Directly Linked to the C_{60} Core

A simple procedure to improve electron affinity of [60]fullerene derivatives in comparison with the parent C_{60} consists in the incorporation of electronegative atoms or electron-deficient carbon atoms directly connected to the C_{60} core. In this sense, apart from isoxazolofullerenes (see Figures 36.5 and 36.6), triazolino and pyrazolino[60]fullerenes (see Figure 36.7) have also been reported.

The synthesis of isoxazolofullerenes is carried out by cycloaddition of the corresponding nitrile oxides, which, in turn, can be obtained by chlorination of the respective oximes and

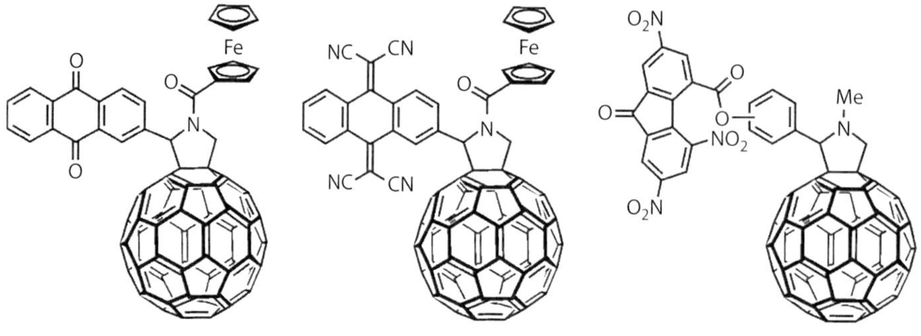

FIGURE 36.4 Structure of triads **10** and **11** and trinitrofluorenone-C$_{60}$ **12**.

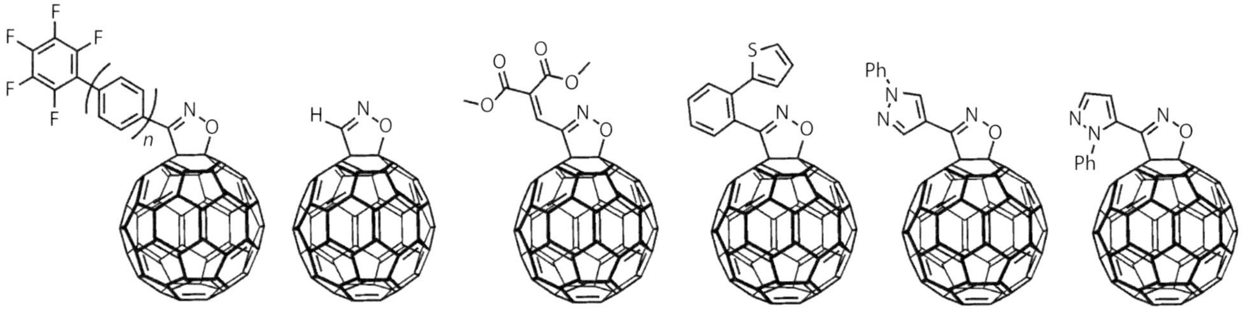

FIGURE 36.5 Chemical structure of isoxazolo[60]fullerenes **13–18**.

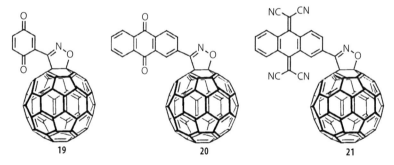

FIGURE 36.6 Structures of isoxazolo[60]fullerenes derivatives of *p*-benzoquinone (**19**), anthraquinone (**20**), and TCAQ (**21**).

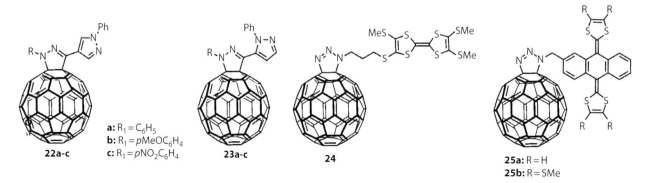

a: R$_1$ = C$_6$H$_5$
b: R$_1$ = *p*MeOC$_6$H$_4$
c: R$_1$ = *p*NO$_2$C$_6$H$_4$

25a: R = H
25b: R = SMe

FIGURE 36.7 Chemical structure of pyrazolino[60]fullerenes (**22–23**) and fullerotriazolines (**24–25**).

subsequent *in situ* dehydrochlorination. The introduction of an acceptor substituent as the pentafluorophenyl group in compounds **13a-b** also led to an anodic shift of the reduction potentials of these compounds of 40–70 mV relative to the naked C_{60} (Irngartinger and Escher 1999). To study the substituent effects on the redox properties of isoxazolofullerenes, Irngartinger et al. (2000) prepared a series of these derivatives. Thus, for compound **14**, with R = H, the electrochemical study showed a shift of 50 mV to more positive values compared to C_{60} for the first reduction potential, due to the strong electron-withdrawing effect of the heterocycle. A remarkable anodic shift of the reduction potential is also observed for diester derivative **15** (80 mV relative to pristine C_{60}). For other substituents studied, the authors concluded that not only the donor or acceptor character of the substituent, but also the distances or geometries have an influence on the redox properties of the adducts, as shown by CV experiments. Thus, for example, compound **16** shows a redox potential that is 40 mV cathodically shifted relative to **14**. This behavior can be explained in terms of through-space interactions between the heterocycle and the fullerene surface, which are favored by the orientation of the addend.

In the same way, donor–acceptor dyads **17** and **18** show an anodic shift of 50–70 mV of the reduction potential values in comparison with the parent C_{60} (de la Cruz et al. 1999a).

The enhanced electron accepting properties of isoxazolo[4',5':1,2][60]fullerenes related to C_{60} prompted us to synthesize derivatives **19–21** bearing strong electron acceptor addends (Figure 36.6) (Illescas and Martín 2000).

In contrast with the above mentioned pyrrolidino[3',4':1,2][60]fullerenes endowed with acceptor moieties (Figure 36.3), these isoxazolofullerenes exhibit reduction potentials that are anodically shifted (70–110 mV) in comparison with the parent C_{60}, which indicates that they are stronger electron acceptors than [60]fullerene. This finding can be accounted for by the presence of the oxygen atom directly linked to the C_{60} cage as well as by the electronic effect of the organic addend covalently attached to the isoxazole ring.

Effects similar to those mentioned for isoxazolo[60]fullerenes on the reduction potentials can be observed in pyrazolino[60]fullerenes (**22–23**), which are easily prepared in one pot by 1,3-dipolar cycloaddition reaction from suitably functionalized hydrazones (Figure 36.7) (de la Cruz et al. 1999a,b, Langa et al. 2001, Espíldora et al. 2002).

These pyrazolino derivatives present an advantage over the isoxazolino[60]fullerenes mentioned above to obtain multichromophoric compounds due to the possibility of preparing donor-C_{60}-donor or donor-C_{60}-acceptor triads by modification of the nature of the substituents linked to the sp^3 nitrogen atom. All compounds **22a-c** and **23a-c** posses higher electron affinity than the corresponding pyrrolidino[60]fullerene (by up to 170 mV), and all except **23b** present an anodically shifted first reduction potential (up to 80 mV) in comparison with the parent C_{60}. The inductive effect of the substituents must be one of the most important factors in determining the observed redox behavior.

Fullerotriazolines (**24–25**) can also be employed to form fullerene derivatives with enhanced electron accepting properties (Guldi 2000, González et al. 2003). These compounds are thermally labile intermediates in the reaction of C_{60} with azides (Prato et al. 1993). As intermediates, they are of interest as precursors by thermal nitrogen extrusion of azafulleroids (opened 5,6-adducts) and fulleroaziridines (closed 6,6-adducts) (Grösser et al. 1995, Schick et al. 1996).

The voltammogram of compound **24** shows, apart from the two reversible one-electron oxidation waves yielding the radical cation and dication of the TTF moiety [$E^1_{ox} = 0.57$ V and $E^2_{ox} = 0.70$ V (vs. SCE)], four quasireversible one-electron steps on the reduction side. These reduction waves correspond to the fullerene reductions [–0.57, –0.97, –1.52, and –2.00 V (vs. SCE)] and are remarkably shifted to more positive values (~100 mV) relative to other 1,2-dihydrofullerenes such us pyrrolidino[3',4':1,2][60]fullerenes as well as to the parent C_{60} (–0.60 mV vs. SCE). This shift can be explained by the electronegative character of the two nitrogen atoms covalently linked to the C_{60} core. Interestingly, the photophysical study of this compound shows that fullerotriazoline-TTF dyad gives rise to a reduced HOMO–LUMO energy gap (–5.12 eV) relative to previously reported fulleropyrrolidine-TTF dyads (–5.35 eV) (Martín et al. 1997, Prato and Maggini 1998). This has a large effect on the electron-transfer rates, which are similar to those observed in systems wherein the spatial separations between the donor and acceptor are markedly decreased (Martín et al. 1996). Thus, increasing the electron accepting behavior of the C_{60} moiety in C_{60}-based dyads such as fullerotriazolines **24** and **25a-b** can be employed as a useful tool to improve (1) the dynamics of the forward electron transfer and (2) the lifetime of the charge separated (CS) state.

For dyads **25a-b**, the reduction potentials are also anodically shifted by 10–30 mV in comparison with pristine C_{60} (Prato et al. 1993). Steady-state fluorescence experiments and transient absorption spectroscopy result in the formation of a radical pair, showing the characteristic fingerprint of $C_{60}^{\bullet-}$ and exTTF$^{\bullet+}$, with lifetimes in the range of hundreds of nanoseconds, nearly 2 orders of magnitude larger than those found for **24**.

36.2.3 Periconjugative Effect

A different approach for the preparation of electron acceptor organofullerenes is based on the periconjugative effect (Wudl et al. 1993, Eiermann et al. 1995). Thus, some quinone-type methanofullerenes (**26–30**) were prepared by cycloaddition of the respective diazo compounds (Figure 36.8) (Ohno et al. 1996, Knight et al. 1997, Beulen et al. 2001a).

In principle, the intramolecular electronic interaction (periconjugation) between the p_z-π orbitals of the olefinic carbons of the quinone moiety and the adjacent carbon atoms of C_{60}, separated by a spiro carbon atom, results in more extended conjugated molecules showing good acceptor properties. In agreement with these predictions, the CV data of compounds **26–27** showed a less negative first reduction potential than the parent C_{60}, thus being stronger acceptor molecules than C_{60}. In contrast, compounds **28–30** showed a more negative first reduction potential than naked C_{60}, which has been accounted for by the steric hindrance between the "peri" hydrogens of the organic

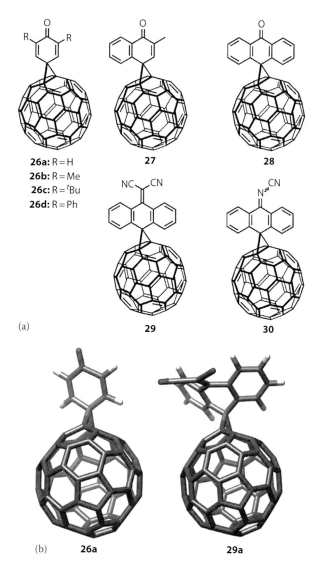

FIGURE 36.8 Chemical structures of compounds **26–30** (a) and 3D molecular modeling of **26a** and **29** (b).

addend and the surface of the ball, leading to a significant deviation from the orthogonality and hence to the loss of periconjugation (Figure 36.8).

Theoretical calculations on compounds **26a–c** indicate that the attachment of the first electron causes the homolytic cleavage of one of the bonds connecting the addend to C_{60}. This ring opening has been supported by EPR measurements, and it explains the irreversible electrochemical behavior of these compounds.

This bond cleavage leading to an irreversible electrochemical behavior related these compounds to the retro-Bingel reaction, an electrochemical reduction reaction that efficiently removes the cyclopropane ring adduct from the fullerene surface, resulting in the formation of the parent C_{60} (Kessinger et al. 1998, Herranz et al. 2004). Thus, we decided to study some of these derivatives (**26a,d**, **27** and **28**) under controlled potential electrolysis (CPE) for potential adduct removal. It is remarkable that all these compounds except **28** show stronger electron-accepting properties than the parent C_{60} (Beulen et al. 2000, 2001b).

36.2.4 Pyrrolidinium Salts

Fulleropyrrolidinium salt derivatives of C_{60} have also shown to have enhanced electron-accepting properties with respect to both the parent pyrrolidine derivatives and C_{60}. Remarkably, CV measurements of compounds **31a-c** (Figure 36.9) performed at low temperatures and fast scan rates, using ultramicroelectrodes, have allowed the observation for the first time in fullerene derivatives, of six C_{60}-centered reductions (Da Ros et al. 1998). Since the chemistry of fulleropyrrolidines is very rich, formation of pyrrolidinium ions, especially in donor-bridge-acceptor dyads, is a valuable methodology to modulate the electronic properties of the C_{60} moiety.

In order to promote water solubility and to preclude fullerene clustering in aqueous media, bis(pyrrolidinium) salts (**32a–d**) were also investigated (Guldi and Prato 2000). Considering the reduction potential of C_{60} (−0.35 vs. SCE in THF) and that of the monopyrrolidinium salt (−0.29 vs. SCE), the reduction potentials of bis(pyrrolidinium) salts **32** are between −0.32 and −0.34 V vs. SCE. This enhanced electronegativity can be reasonably interpreted in terms of inductive effects leading to a decrease of the electron density within the fullerene π-system.

For the development of novel donor–acceptor arrays that display improved performance, the use of a fulleropyrrolidinium acceptor clearly has a distinct advantage over fulleropyrrolidine (**33–35**) (Guldi et al. 2000). Interestingly, the fulleropyrrolidinium-based dyads reveal faster charge separation dynamics, relative to the fulleropyrrolidine analogues, a trend that parallels their better acceptor properties. Regarding charge recombination, formation of the zwitterionic fullerene π-radical anion appears to be a key factor for stabilizing the charge-separated state.

The CV study of a series of novel bisfulleropyrrolidines and bisfulleropyrrolidinium ions (**36**) has also been reported (Carano et al. 2003). The eight possible stereoisomers of each series were systematically investigated. The stabilizing effect of positive charges produced a significant enhancement of the electronegative properties in **36**. The study evidenced that in these compounds, the CV pattern and, in particular, the potential separation between the second and third reductions change significantly with the addition pattern. A sequential π-electron model that simulates the effect of subsequent reductions of C_{60} bis-adducts gives a good correlation ($r > 0.96$) with the cyclic voltammetry data when the molecules are divided in two sets depending on the location of the addends in the same or in opposite hemispheres.

36.2.5 Heterofullerenes

An attractive alternative to isocyclic fullerenes, as components of multiredox and light-responsive systems, are heterofullerene building blocks (i.e., $C_{59}N$) (Hummelen et al. 1995, 1999, Hirsch and Nuber 1996). In azaheterofullerene derivatives, substitution of a carbon atom with nitrogen alters the redox potential and the excited state energies of the fullerene (Keshavarz-K. et al. 1996b). Significantly, $C_{59}N$ derivatives render better electron

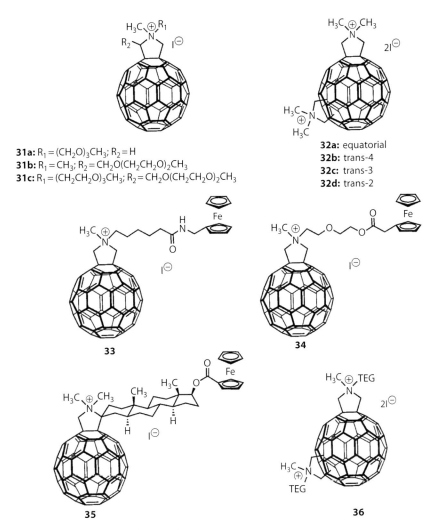

31a: R$_1$ = (CH$_2$O)$_3$CH$_3$; R$_2$ = H
31b: R$_1$ = CH$_3$; R$_2$ = CH$_2$O(CH$_2$CH$_2$O)$_2$CH$_3$
31c: R$_1$ = (CH$_2$CH$_2$O)$_3$CH$_3$; R$_2$ = CH$_2$O(CH$_2$CH$_2$O)$_2$CH$_3$

32a: equatorial
32b: trans-4
32c: trans-3
32d: trans-2

33

34

35

36

FIGURE 36.9 Chemical structures of some pyrrolidinium salts.

acceptors than C$_{60}$ and C$_{60}$ derivatives that bear similar addends. The simplest azaheterofullerene can only be isolated as its corresponding dimer (**37**) (Figure 36.10). The C–H bonded monomer, HC$_{59}$N, has also been characterized. The monomeric C$_{59}$N$^+$ cation, which is isoelectronic with C$_{60}$, can be isolated as a carborane anion salt [C$_{59}$N$^+$] [Ag(CB$_{11}$H$_6$Cl$_6$)$_2^-$] (Kim et al. 2003).

The synthesis involves a rare example of oxidation of a sp^3–sp^3 C-C bond to produce a carbenium ion.

Hauke et al. have reported a series of fluorophore–heterofullerene conjugates (**38**) in which the flexibility of the acetylgroup linker opens the way for conformations with π–π stacking interactions between the chromophores (Hauke et al. 2005). These systems were obtained by treating dimer **37** with the corresponding acetyl fluorophores and *p*-toluensulfonic acid at 150°C. Photophysical steady-state and time-resolved measurements clearly show the photosensitization effect of the fluorophores that act as an antenna system and transmit their excited-state energy to the covalently attached C$_{59}$N moiety. This approach offers a viable alternative for developing well-defined architectures, bearing a C$_{59}$N core, which give rise to sequential energy and electron-transfer processes.

36.2.6 Fluorofullerenes

Electron affinity of halogenefullerenes is higher than that of related fullerenes, the most stable of them being fluorofullerenes (Taylor 2001). Electron affinities increase to about 0.05 eV per

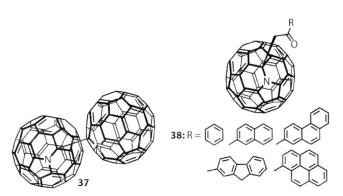

38: R =

37

FIGURE 36.10 Chemical structure of azaheterofullerene dimer **37** and fluorophore–heterofullerene conjugates **38**.

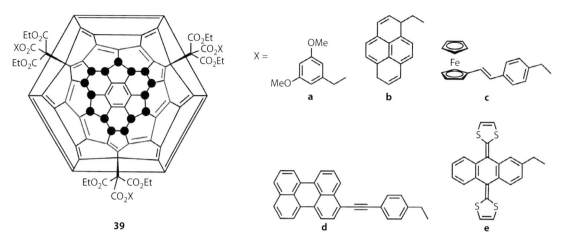

FIGURE 36.11 Schlegel diagram of multicomponent photoactive systems based on the [18]trannulene structure.

fluorine atom, although the effect is reduced at higher addition levels. Thus, the reduction potential of $C_{60}F_{48}$ is anodically shifted to 1.38 V related to C_{60} (0.79 V vs. SCE compared to −0.59 V under the same experimental conditions) (Zhou et al. 1994). Electron affinity values of 3.1 and 3.48 eV are encountered for $C_{60}F_{18}$ and $C_{60}F_{36}$, respectively, compared with 2.67 eV for C_{60} (Liu et al. 1997, Ohkubo et al. 2002). These extraordinarily good electron acceptor properties make fluorofullerenes excellent candidates for their incorporation in Donor–Acceptor (D-A) conjugates.

Of the fluorofullerenes available, trannulenes, with an 18π annulene in the all-*trans* configuration, have high electron affinities, which enable them to stabilize charged entities more effectively than conventional fullerene derivatives. This, together with rich visible absorptions associated with the 18π annulene substructure, pointed out the interest of these trannulenes as components of D-A dyads for energy and electron-transfer reactions. In this context, different electron donors have been attached to the $C_{60}F_{18}$ core, giving rise to D-A compounds with a very low HOMO–LUMO gap (Figure 36.11) (Burley et al. 2003, 2004, Guldi et al. 2005). Trannulation of $C_{60}F_{18}$ was carried out by reaction of the appropriately functionalized methanetricarboxylate esters in the presence of DBU as base. The electrochemical study of these compounds shows the first one-electron reversible reduction wave at potentials that range from −0.005 V for $C_{60}F_{15}$-TTF (**39e**) (vs. Ag/Ag⁺ in tetrachloroethane) to −0.1 V for $C_{60}F_{15}$-pyrene (**39b**) and $C_{60}F_{15}$-ferrocene (**39c**) (vs. Ag/Ag⁺ in DCM).

The photophysical study of these dyads indicates that, depending on the relative energies of the lowest excited state or of the charge-separated state, the excited-state chromophore deactivation can occur *via* an energy-transfer ($C_{60}F_{15}$-pyrene **39b**) or an electron-transfer ($C_{60}F_{15}$-ferrocene **39c**, $C_{60}F_{15}$-perylene **39d**, and $C_{60}F_{15}$-exTTF **39e**) process.

36.2.7 Endohedral Fullerenes

Since the discovery of fullerenes, the possibility of incorporating atoms or molecules into the cavity of these carbon clusters has

FIGURE 36.12 Molecular structure of N@C_{60}.

attracted much attention (Figure 36.12) (Delgado et al. 2008). The first detection of a metallofullerene by mass spectrometry was reported in 1985 (Heath et al. 1985), and in 1991, Chai et al. (1991) demonstrated the first isolation of La@C_{82}, a metallofullerene incorporating a lanthanum atom into the cavity of C_{82} fullerene. Currently, we know over 100 isolated pentagon (IPR) isomers of fullerenes through C_{94} and several non-IPR isomers into which more than 20 different metal atoms have been encapsulated. Some species contain various metal ions encaged, forming what are named multimetallofullerenes. In addition, other non-metal structures, such as nitrogen, phosphorus, or noble-gas atoms have also been encapsulated in fullerene cages. This variety gives an idea of the great amount of possible structures with different and unique characteristics (Akasaka and Nagase 2002).

Endohedral metallofullerenes are usually known to show better electron acceptor properties than empty fullerenes. To compare the electronic properties of endohedral fullerenes with those of empty fullerenes, the nature of the charge transfer from the encaged structure to the fullerene cage must be considered.

M@C_{82} derivatives are the most studied class of endohedral metallofullerenes, where M = Y, Sc, La, Ce, Pr, and Gd among others. This is because of the fact that in spite of having an open-shell electronic structure, they are soluble in certain organic solvents. Many of these M@C_{82} are, in effect, zwitterions

TABLE 36.2 Redox Potentials of Some $M@C_{82}$ Metallofullerenes and Related Empty Fullerenes (in V vs. Fc/Fc$^+$)[a]

Comp.	E_{red}^1	E_{red}^2	E_{red}^3	E_{red}^4	E_{red}^5	References
C_{60}	−1.12	−1.50	−1.95	−2.41		Suzuki et al. (1996)
C_{82}	−0.69	−1.04	−1.58	−1.94		Suzuki et al. (1996)
$Y@C_{82}$	−0.37	−1.34[b]		−2.22	−2.47	Kikuchi et al. (1994)
La@ C_{82}-A	−0.42	−1.37	−1.53	−2.26	−2.46	Suzuki (1993)
$Ce@C_{82}$	−0.41	−1.41	−1.53	−1.79	−2.25	Suzuki et al. (1996)
Pr@ C_{82}-A	−0.39	−1.35	−1.46	−2.21	−2.48	Akasaka et al. (2000)
$Gd@C_{82}$	−0.39	−1.38[b]		−2.22[b]		Akasaka et al. (1995)

[a] Pt disk as working electrode, Bu_4NPF_6 (0.1 M) as supporting electrolyte, and o-DCB as solvent. Scan rate 20 mV/s.

[b] Two-electron process.

species, as the metals donate three electrons to the fullerene cage to form $M^{3+}@C_{82}^{3-}$. In this sense, these $M@C_{82}$ metallofullerenes can be viewed as a positively charged core metal surrounded by negatively charged carbon atoms. Formally, electron removal and gain take place on the C_{82} core.

The redox potentials of these metallofullerenes have been studied by cyclic voltammetry and compared with empty fullerenes (Table 36.2). In addition to the three electrons on the C_{82} cage, these endohedral monometallic derivatives exhibit five reversible reduction waves at lower potentials than related empty fullerenes. Therefore, these $M@C_{82}$ species are better electron acceptors. They also present two oxidation waves so that they can act as strong electron donors. The small difference between the first oxidation and reduction potentials, which can be accounted for by the open-shell electronic structure of $M@C_{82}$, is noteworthy. These electronic properties as electron donor or electron acceptors render them interesting materials for different applications.

Another class of endohedral fullerenes are metal nitride cluster fullerenes (NCFs), the first and most abundant of which is $Sc_3N@C_{80}$ (Dunsch and Yang 2007). The structure of $Sc_3N@C_{80}$ was discovered in 1999 by introducing a small quantity of nitrogen gas into a Krätschmer—Huffman generator during the vaporization of graphite rods containing metal oxides. This

TABLE 36.3 Reduction Potentials of Dyads **40–41** and Related Empty Fullerenes (in V vs. Fc/Fc$^+$)[a]

Comp.	E_{red}^1	E_{red}^2	E_{red}^3
C_{60}	−1.15	−1.55	−2.01
I_h-$Sc_3N@C_{80}$	−1.29	−1.56	−2.32
$Y_3N@C_{80}$	−1.44	−1.83	−2.38
40	−1.14	−1.53	−2.25
41	−1.28[b]	−1.77[b]	−2.22[b,c]

[a] Glassy carbon (GCE) as working electrode, Bu_4NPF_6 (0.05 M) as supporting electrolyte, and o-DCB as solvent.

[b] Reduction peak potential, irreversible process.

[c] Not well defined.

method was named the "trimetallic nitride template" (TNT) process, and several new NCFs have been synthesized by the TNT method, such as $Sc_3N@C_{78}$, $Y_3N@C_{80}$, $Lu_3N@C_{80}$, and $Tb_3N@C_{80}$ among others. In these NCFs, a cluster of four atoms is encapsulated into the cavity of the fullerene cage, which can be of different sizes (C_{68}–C_{88}). Recently, chemical derivatization has allowed the incorporation of metal nitride endofullerenes in donor–acceptor systems (Pinzón et al. 1998, 2009). C_{60} and its derivatives have been extensively studied in D–A dyads and exhibit excellent properties owing to their low reorganization energy and high electron mobilities. In addition, NCF metallofullerenes possess larger absorptive coefficients than C_{60} in the visible region of the electromagnetic spectrum and a low HOMO–LUMO gap, while retaining electron-acceptor properties similar to that of C_{60}.

D–A dyads of $M_3N@C_{80}$ (M = Sc, Y) with different donor moieties as ferrocene (**40**) or ex-TTF (**41**) have been synthesized by either a 1,3-dipolar cycloaddition of azomethine ylides (to $Sc_3N@C_{80}$) or a Bingel–Hirsch addition of the corresponding malonates (Figure 36.13). It is interesting to note that in I_h-$Sc_3N@C_{80}$-based dyads, the addend is attached to a [5,6]-double bond ring junction, while the cycloaddition reactions to $Y_3N@C_{80}$ occur at a [6,6]-ring junction. These differences can be appropriately analyzed by cyclic voltammetry experiments. The electrochemical study of these dyads also demonstrates that the redox potentials of **40** and **41** are slightly anodically shifted in comparison with I_h-$Sc_3N@C_{80}$ or $Y_3N@C_{80}$ (see Table 36.3).

The photophysical study of dyad **40** shows the existence of a photoinduced electron-transfer process leading to the corresponding radical ion pair. Comparing the lifetime of compound **40** with an equivalent C_{60}-based dyad allowed to appreciate the impact of using $Y_3N@C_{80}$ in the stabilization of the radical-pair, resulting in a lifetime (84 ps in o-DCB) three times longer than that of the C_{60} dyad (27 ps in o-DCB).

36.2.8 Fullerene-Based Acceptors for Organic Photovoltaics

Organic solar cells are excellent candidates for lightweight photovoltaic applications. The mechanical flexibility of plastic materials and the possibility of simple solution processing of the active

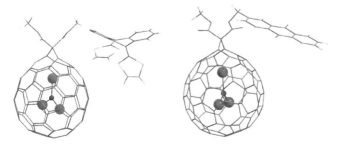

FIGURE 36.13 Optimized structures of **40** and **41**.

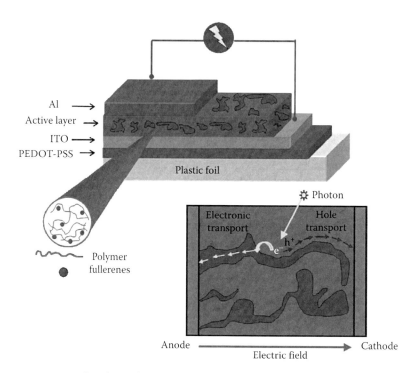

FIGURE 36.14 Schematic representation of a polymer:fullerene BHJ solar cell (top). Morphology of the active layer showing the mechanism for charge separation (down).

layer hold the promise for organic photovoltaic systems. In the effort to develop effective devices, two main approaches have been employed: the donor–acceptor bilayer and the bulk-heterojunction (BHJ). The development of the BHJ concept meant an important step toward efficient organic solar cells. In BHJ devices, the acceptor is ideally homogeneously distributed into the donor matrix, thus maximizing the interfacial area between the donor and the acceptor (Figure 36.14). In contrast, bilayer devices are comprised of donor and acceptor layers connected by the interfacial surface.

The fundamental mechanism of the energy conversion process implies absorption of light and generation of excitons, diffusion of the excitons, dissociation with generation of charge, charge transport, and charge collection (Thompson and Fréchet 2008). With the exception of charge collection, which depends on the interface between the active layer and the corresponding electrode, all other aspects of mechanism are governed by the donor–acceptor composite, which forms

the active layer. However, the open circuit voltage (V_{oc}) is closely related to the energetic relationship between the donor and the acceptor, that is, with the energy difference between the HOMO of the donor and the LUMO of the acceptor.

As we mentioned before, the best organic solar cells are based on poly(3-hexylthiophene) (P3HT) and the fullerene derivative [6,6]-phenyl-C61-butyric acid methyl ester (**42**, PCBM), with efficiencies of about 5% (Figure 36.15). The optimization of organic solar cells is based on fine-tuning the electronic properties of the polymeric donor and the acceptor, generally a soluble C_{60} derivative. In this regard, although a wide variety of π-conjugated polymers have been used, the number of soluble fullerene derivatives studied has been relatively scarce. Nevertheless, fullerene derivatives such as diphenylmethanofullerenes (DPM, **43**) (Riedel et al. 2005) or Diels–Alder cycloadducts (**44**) (Backer et al. 2007) have afforded good efficiencies when blended with the appropriate polymer.

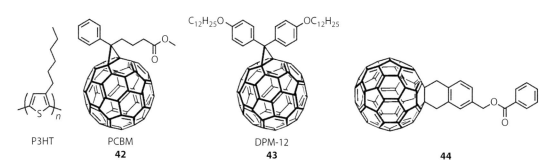

P3HT   PCBM   DPM-12   44
            42   **43**

FIGURE 36.15 Chemical structure of P3HT and PCBM.

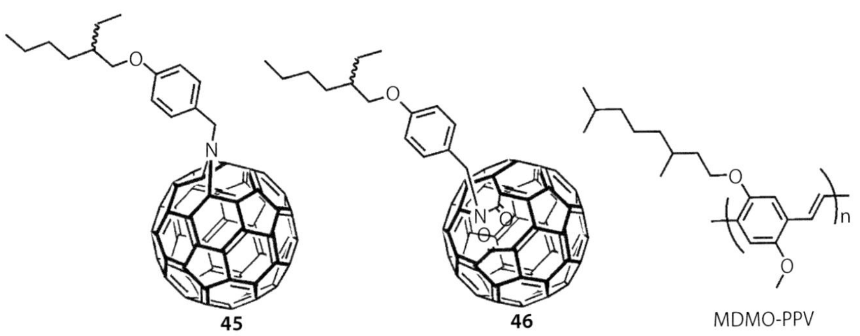

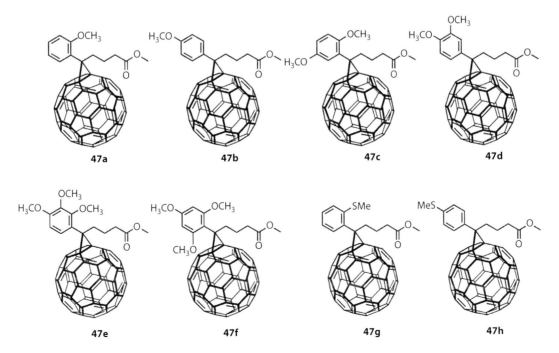

FIGURE 36.16 Chemical structures of azafulleroid (**45**), ketolactam (**46**), and polymer MDMO-PPV.

In order to systematically study the open circuit voltage of a series of devices as a function of the acceptor strength, two highly soluble fullerene derivatives (**45**, **46**) have been compared with PCBM **42** (Figure 36.16) (Brabec et al. 2001, 2002). A soluble *p*-phenylenevinylene polymer (MDMO-PPV) was employed as the polymeric donor species.

The investigated compounds differ in almost 200 mV in their first reduction potentials (PCBM **42**: −0.69 V; azafulleroid **45**: −0.67 V; ketolactam **46**: −0.53 V) and have comparably sized solubilizing groups, thereby minimizing effects due to different donor–acceptor distances and/or different morphologies. The V_{oc} of thin film blend BHJ photovoltaic devices produced with these acceptor materials were measured, finding the highest value for PCBM (**42**) (760 mV) and the lowest for the ketolactam (**46**) containing cells (560 mV). Moreover, a strong linear correlation between the acceptor strength and the V_{oc} was found. To verify this finding, devices were also produced in the diffusion bilayer geometry, finding that the observed trend of the V_{oc} with the acceptor strength is independent from the device geometry.

In an attempt to raise the LUMO level of PCBM, hence diminishing the LUMO–LUMO gap between the PCBM derivative and the MDMO-PPV, several substituents have been placed on the phenyl ring of PCBM (Figure 36.17) (Kooistra et al. 2007).

It appears that an energy difference between the LUMOs of the donor and acceptor of about 0.3 eV is required for a forward electron transfer taking place from the donor to the acceptor. Furthermore, an energy difference larger than this minimum value does not contribute to improve the device performance, resulting in wasted energy.

By introducing electron-donor substituents in the PCBM phenyl group, the LUMO–LUMO gap of the PCBM:MDMO-PPV composite should be reduced from its current value (0.9 eV). All the PCBM derivatives represented in Figure 36.17 show their first reduction potentials at slightly more negative values than PCBM, so the LUMO levels are indeed raised in comparison with PCBM. A series of photovoltaic devices were fabricated employing these substituted PCBM derivatives as the acceptor and MDMO-PPV as the electron donor, and the V_{oc} values of

47a **47b** **47c** **47d**

47e **47f** **47g** **47h**

FIGURE 36.17 Structures of different substituted PCBM investigated.

bis PCBM
48

FIGURE 36.18 Structure of fullerene bisadducts for polymer solar cells.

devices were studied as a function of the first reduction potential. Again, the trend is clearly visible, indicating that a higher V_{oc} is obtained in fullerene:polymer solar cells when the LUMO level of the fullerene is increased.

However, the low solubility of these PCBM derivatives affects the morphology of the devices, producing too big domain sizes in the blend that lower the V_{oc} values. To solve this problem, a bisPCBM, the bisadduct analogue of PCBM, was studied as the acceptor component of organic solar cells (Lenes et al. 2008) (Figure 36.18). BisPCBM (**48**):P3HT devices were produced and compared with the cells containing PCBM. A higher value of the V_{oc} was observed in the cells with bisP-CBM as acceptor. As predicted, the higher LUMO of the PCBM derivative allows to minimize the energy loss in the electron transfer, resulting in an enhanced V_{oc} value. This V_{oc} increment is translated into a higher efficiency of the bisPCBM:P3HT cell (4.5%) as compared with the best PCBM:P3HT prepared in a similar way (3.8%).

36.3 Summary and Perspectives

Fullerene C_{60} is known to exhibit good electron-acceptor ability. However, the chemical modification of fullerenes has been skillfully used to increase the electronegative character of a variety of fullerene derivatives. In this chapter, the main strategies are presented by means of representative examples. They involve from exohedral functionalization to the inclusion of electronegative atoms (nitrogen) in the same cage of the fullerene or, more recently, to the introduction of a metal complex in the inner cavity of fullerenes.

Improved C_{60}-based electron acceptors have been covalently connected to different electron-donor units, and faster photo-induced charge separation processes have been observed. These remarkable photophysical features are of interest for the design of novel fullerene derivatives with improved properties for the preparation of photovoltaic devices. In this regard, the new endohedral fullerenes have also shown their potentiality for this purpose.

In summary, highly efficient electron-acceptor fullerenes are able to act as real electron-accepting sponges forming multianion species with theoretical and synthetic interest. Furthermore, the synthesis of modified fullerenes showing stronger electron accepting character than pristine C_{60} is still a challenge for

the design of new molecules of interest in fields such as artificial photosynthetic systems and photovoltaic devices wherein electron transfer from a donor unit to the fullerene core plays a key role. In this respect, fullerenes as electron-acceptor components represent a most realistic application for sunlight energy conversion.

Acknowledgments

We acknowledge the financial support by the European Science Foundation (05-SONS-FP-021 SOHYD), the MEC of Spain (CTQ2005-02609/BTQ and Consolider-Ingenio 2010 CSD2007-0010, Nanociencia Molecular), and Comunidad de Madrid (P-PPQ-000225-0505).

References

Akasaka, T.; Nagase S. 2002. *Endofullerenes: A New Family of Carbon Clusters*. Dordrecht, the Netherlands: Kluwer Academic Publishers.

Akasaka, T.; Nagase, S.; Kobayashi, K. et al. 1995. Exohedral derivatization of an endohedral metallofullerene Gd@C_{82}. *J. Chem. Soc. Chem. Commun.* 1343–1344.

Akasaka, T.; Okubo, S.; Kondo, M. et al. 2000. Isolation and characterization of two Pr@C_{82} isomers. *Chem. Phys. Lett.* 319: 153–156.

Allemand, P.-M.; Koch, A.; Wudl, F. et al. 1991. Two different fullerenes have the same cyclic voltammetry. *J. Am. Chem. Soc.* 113: 1050–1051.

Arias, F.; Xie. Q.; Lu, Q.; Wilson, S. R.; Echegoyen, L. 1994. Kinetic effects in the electrochemistry of fullerene derivatives at very negative potentials. *J. Am. Chem. Soc.* 116: 6388–6394.

Backer, S.; Sivula, K.; Kavulak, D. F.; Fréchet, J. M. J. 2007. High efficiency organic photovoltaics incorporating a new family of soluble fullerene derivatives. *Chem. Mater.* 19: 2927–2929.

Beulen, M. W. J.; Echegoyen, L.; Rivera, J. A.; Herranz, M. A.; Martín-Domenech, A.; Martín, N. 2000. Adduct removal from methanofullerenes *via* reductive electrochemistry. *Chem. Commun.* 917–918.

Beulen, M. W. J.; Rivera, J. A.; Herranz, M. A.; Illescas, B.; Martín, N.; Echegoyen, L. 2001a. Reductive electrochemistry of spiromethanofullerenes. *J. Org. Chem.* 66: 4393–4398.

Beulen, M. W. J.; Rivera, J. A.; Herranz, M. A.; Martín-Domenech, A.; Martín, N.; Echegoyen, L. 2001b. Reductive electrolysis of [60]fullerene mono-methanoadducts in THF leads to the formation of bis-adducts in high yields. *Chem. Commun.* 407–408.

Brabec, C. J.; Cravino, A.; Meissner, D. et al. 2001. Origin of the open circuit voltage of plastic solar cells. *Adv. Funct. Mater.* 374–380.

Brabec, C. J.; Cravino, A.; Meissner, D. et al. 2002. The influence of materials work function on the open circuit voltage of plastic solar cells. *Thin Solid Films* 403–404: 368–372.

Burley, G. A.; Avent, A. G.; Boltalina, O. V. et al. 2003. A light-harvesting fluorinated fullerene donor-acceptor ensemble; long-lived charge separation. *Chem. Commun.* 148–149.

Burley, G. A.; Avent, A. G.; Gol'dt, I. V. et al. 2004. Design and synthesis of multi-component 18π annulenic fluorofullerene ensembles suitable for donor–acceptor applications. *Org. Biomol. Chem.* 2: 319–329.

Carano, M.; Da Ros, T.; Fanti, M. et al. 2003. Modulation of the reduction potentials of fullerene derivatives. *J. Am. Chem. Soc.* 125: 7139–7144.

Chai, Y.; Jin, C.; Haufler, R. E. et al. 1991. Fullerenes with metals inside. *J. Phys. Chem.* 95: 7564–7568.

Creegan, K. M.; Robbins, J. L.; Robbins, W. K. et al. 1992. Synthesis and characterization of $C_{60}O$, the first fullerene epoxide. *J. Am. Chem. Soc.* 114: 1103–1105.

de la Cruz, P.; Díaz-Ortiz, A.; García, J. J.; Gómez-Escalonilla, M. J.; de la Hoz, A.; Langa, F. 1999a. Synthesis of new C_{60}-donor dyads by reaction of pyrazolylhydrazones with [60]fullerene under microwave irradiation. *Tetrahedron* 40: 1587–1590.

de la Cruz, P.; Espíldora, E.; García, J. J. et al. 1999b. Electroactive 3′-(*N*-phenylpyrazolyl)isoxazoline[4′,5′:1,2][60]fullerene dyads. *Tetrahedron Lett.* 40: 4889–4892.

Da Ros, T.; Prato, M.; Carano, M.; Ceroni, P.; Paolucci, F.; Bofia, S. 1998. Enhanced acceptor character in fullerene derivatives. Synthesis and electrochemical properties of fulleropyrrolidinium salts. *J. Am. Chem. Soc.* 120: 11645–11648.

Delgado, J. L.; Herranz, M. A.; Martín N. 2008. The nano-forms of carbon. *J. Mater. Chem.* 18: 1417–1426.

Diekers, M.; Hirsch, A.; Pyo S.; Rivera, J.; Echegoyen, L. 1998. Synthesis and electrochemical properties of new C_{60}-acceptor and -donor dyads. *Eur. J. Org. Chem.* 1111–1121.

Dunsch, L.; Yang, S. 2007. Metal nitride cluster fullerenes: Their current state and future prospects. *Small* 3: 1298–1320.

Echegoyen, L.; Echegoyen, L. E. 1998. Electrochemistry of fullerenes and their derivatives. *Acc. Chem. Res.* 31, 593–601.

Eiermann, M.; Haddon, R. C.; Knight, B. et al. 1995. Electrochemical evidence for through-space orbital interactions in spiromethanofullerenes. *Angew. Chem., Int. Ed. Engl.* 34: 1591–1594.

Espíldora, E.; Delgado, J. L.; de la Cruz, P.; de la Hoz, A.; López-Arza, V.; Langa, F. 2002. Synthesis and properties of pyrazolino[60]fullerene-donor systems. *Tetrahedron* 58: 5821–5826.

Foote, C. S. 1994. Photophysical and photochemical properties of fullerenes. *Top. Curr. Chem.* 169: 347–363.

González, S.; Martín, N.; Swartz, A.; Guldi, D. M. 2003. Addition reaction of azido-exTTFs to C_{60}: Synthesis of fullerotriazoline and azafulleroid electroactive dyads. *Org. Lett.* 5: 557–560.

Grösser, T.; Prato, M.; Lucchini, V.; Hirsch, A.; Wudl, F. 1995. Ring expansion of the fullerene core by highly regioselective formation of diazafulleroids. *Angew. Chem. Int. Ed. Engl.* 34: 1343–1345.

Guldi, D. M. 2000. Probing the electron-accepting reactivity of isomeric bis(pyrrolidinium) fullerene salts in aqueous solutions. *J. Phys. Chem. B* 104: 1483–1489.

Guldi, D. M.; Prato, M. 2000. Excited-state properties of C_{60} fullerene derivatives *Acc. Chem. Res.* 33: 695–703.

Guldi, D. M.; González, S.; Martín, N.; Antón, A.; Garín, J.; Orduna, J. 2000. Efficient charge separation in C_{60}-based dyads: Triazolino[4′,5′:1,2][60]fullerenes. *J. Org. Chem.* 65: 1978–1983.

Guldi, D. M.; Marcaccio, M.; Paolucci, F. et al. 2005. Fluorinated fullerenes: Sources of donor–acceptor dyads with [18]trannulene acceptors for energy- and electron-transfer reactions *J. Phys. Chem. A* 109: 9723–9730.

Haufler, R. E.; Conceiçao, J.; Chibante, L. P. F. et al. 1990. Efficient production of C_{60} (buckminsterfullerene), $C_{60}H_{36}$, and the solvated buckide ion. *J. Phys. Chem.* 94: 8634–8636.

Hauke, F.; Hirsch, A.; Atalick, S.; Guldi, D. M. 2005. Quantitative transduction of excited-state energy in fluorophore-heterofullerene conjugates. *Eur. J. Org. Chem.* 1741–1751.

Heath, J. R.; O'Brien, S. C.; Zhang, Q. 1985. Lanthanum complexes of spheroidal carbon shells. *J. Am. Chem. Soc.* 107: 7779–7780.

Herranz, M. A.; Illescas, B.; Martín, N.; Luo, C.; Guldi, D. M. 2000. Donor/acceptor fulleropyrrolidine triads. *J. Org. Chem.* 65: 5728–5738.

Herranz, M. A.; Diederich, F.; Echegoyen, L. 2004. Electrochemically induced retro-cyclopropanation reactions. *Eur. J. Org. Chem.* 2299–2316.

Hirsch, A.; Nuber, B. 1999. Nitrogen heterofullerenes. *Acc. Chem. Res.* 32: 795–804.

Hummelen, J. C.; Knight, B.; Pavlovich, J.; González, R.; Wudl, F. 1995. Isolation of the heterofullerene $C_{59}N$ as its dimer $(C_{59}N)_2$. *Science* 269: 1554–1556.

Hummelen, J. C.; Bellavia-Lud, C.; Wudl, F. 1999. Heterofullerenes. *Top. Curr. Chem.* 199: 93–134.

Illescas, B. M.; Martín, N. 2000. [60]Fullerene adducts with improved electron acceptor properties. *J. Org. Chem.* 65: 5986–5995.

Illescas, B. M.; Martín, N.; Seoane, C.; de la Cruz, P.; Langa, F.; Wudl, F. 1995. A facile formation of electroactive fullerene adducts from sultines via a Diels-Alder reaction. *Tetrahedron Lett.* 36: 8307–8310.

Illescas, B. M.; Martín, N.; Seoane, C. et al. 1997a. Reaction of C_{60} with sultines: synthesis, electrochemistry, and theoretical calculations of organofullerene acceptors. *J. Org. Chem.* 62: 7585–7591.

Illescas, B.; Martín, N.; Seoane, C. 1997b. [60]Fullerene-based electron acceptors with tetracyano-*p*-quinodimethane (TCNQ) and dicyano-*p*-quinonediimine (DCNQI) derivatives. *Tetrahedron Lett.* 38: 2015–2018

Illescas, B.; Martín, N.; Pérez, I.; Seoane, C. 1999. Synthesis of novel C_{60}-based acceptor dyads (A_1-A_2) from sultines *Synth. Met.* 103: 2344–2347.

Irngartinger, H.; Escher, T. 1999. Strong electron acceptor properties of 3′(pentafluorophenyl)isoxazolo[4′,5′:1,2][60]fullerene derivatives. *Tetrahedron* 55: 10753–10760.

Irngartinger, H.; Fetter, P. W.; Escher, T.; Tinnefeld, P.; Nord, S.; Sauer, M. 2000. Substituent effects on redox properties and photoinduced electron transfer in isoxazolo-fullerenes. *Eur. J. Org. Chem.* 455–465.

Iyoda, M.; Sultana, F.; Sasaki, S.; Yoshida, M. 1994. Synthesis and properties of a novel redox system containing fullerene and *p*-benzoquinone. *J. Chem. Soc., Chem. Commun.* 1929.

Iyoda, M.; Sasaki, S.; Sultana, F.; Yoshida, M.; Kuwatani, Y.; Nagase, S. 1996 Mono- and dianion of benzoquinone-linked [60] fullerene. *Tetrahedron Lett.* 37: 7987–7990.

Iyoda, M.; Sultana, F.; Kato, A. et al. 1997. Benzoquinone-linked fullerenes with a pyrrolidine spacer. *Chem. Lett.*: 63–64.

Keshavarz-K., M.; Knight, B.; Srdanov, G.; Wudl, F. 1995. Cyanodihydrofullerenes and dicyanodihydrofullerene: The first polar solid based on C_{60}. *J. Am. Chem. Soc.* 117: 11371–11372.

Keshavarz-K., M.; Knight, B.; Haddon, R. C.; Wudl, F. 1996a. Linear free energy relation of methanofullerene C_{61}-substituents with cyclic voltammetry: Strong electron withdrawal anomaly. *Tetrahedron* 52: 5149–5159.

Keshavarz-K., M.; González, R.; Hicks, R. G. et al. 1996b. Synthesis of hydroazafullerene $C_{59}HN$, the parent hydroheterofullerene. *Nature* 383: 147–150.

Kessinger, R.; Crassous, J.; Herrmann, A.; Rüttimann, M.; Echegoyen, L.; Diederich, F. 1998. Preparation of enantiomerically pure C_{76} with a general electrochemical method for the removal of di(alkoxycarbonyl)methano bridges from methanofullerenes: The retro-Bingel reaction. *Angew. Chem., Int. Ed.* 37: 1919–1922.

Kikuchi, K.; Nakao, Y.; Suzuki, S.; Achiba, Y., Suzuki, T.; Maruyama, Y. 1994. Characterization of the isolated Y@C82. *J. Am. Chem. Soc.* 116: 9367–9368.

Kim, K.-Ch.; Hauke, F.; Hirsch, A. et al. 2003. Synthesis of the $C_{59}N^+$ carbocation. A monomeric azafullerene isoelectronic to C_{60}. *J. Am. Chem. Soc.* 125: 4024–4025.

Knight, B.; Martín, N.; Ohno, T. et al. 1997. Synthesis and electrochemistry of electronegative spiroannelated methanofullerenes: Theoretical underpinning of the electronic effect of addends and a reductive cyclopropane ring-opening reaction. *J. Am. Chem. Soc.* 119: 9871.

Kooistra, F. B.; Knol, J.; Kastenberg, F. et al. 2007. Increasing the open circuit voltage of bulk-heterojunction solar cells by raising the LUMO level of the acceptor. *Org. Lett.* 551–554.

Langa, F.; de la Cruz, P.; Espíldora, E. et al. 2001. C_{60}-based triads with improved electron-acceptor properties: Pyrazolylpyrazolino[60] fullerenes. *J. Org. Chem.* 66: 5033–5041.

Lem, G.; Schuster, D. I.; Courtney, S. H.; Lu, Q.; Wilson, S. R. 1995. Addition of alcohols and hydrocarbons to fullerenes by photosensitized electron transfer. *J. Am. Chem. Soc.* 117: 554–555.

Lenes, M.; Wetzelaer, G.-J. A. H.; Kooistra, F. B.; Veenstra, S. C.; Hummelen, J. C.; Blom, P. W. M. 2008. Fullerene bisadducts for enhanced open-circuit voltages and efficiencies in polymer solar cells. *Adv. Mater.* 20: 2116–2119.

Liu, N.; Morio, Y.; Okino, F.; Touhara, H.; Boltalina, O. V.; Pavlovich, V. K. 1997. Electrochemical properties of $C_{60}F_{36}$. *Synth. Met.* 86: 2289–2290.

Martín, N. 2006. New challenges in fullerene chemistry. *Chem. Commun.* 2093–2104.

Martín, N.; Sánchez, L.; Seoane, C.; Andreu, R.; Garín, J.; Orduna, J. 1996. Semiconducting charge transfer complexes from [60] Fullerene-tetrathiafulvalene (C_{60}-TTF) systems. *Tetrahedron Lett.* 37: 5979–5982.

Martín, N.; Sánchez, L.; Seoane, C. et al. 1997. Synthesis, properties and charge transfer complexes of covalently attached [60]fullerene-tetrathiafulvalenes. *J. Phys. Chem. Solids* 58: 1713–1718.

Martín, N.; Sánchez, L.; Illescas, B.; González, S.; Herranz, M. A.; Guldi, D. M. 2000. Photoinduced electron transfer between C_{60} and electroactive units. *Carbon* 38: 1577–1585.

Nonell, S.; Arbogast, J. W.; Foote, C. S. 1992. Production of fullerene (C_{60}) radical cation by photosensitized electron transfer. *J. Phys. Chem.* 96: 4169–4170.

Nuber, B.; Hirsch, A. 1996. A new route to nitrogen heterofullerenes and the first synthesis of $(C_{69}N)_2$. *Chem. Commun.* 1421–1422.

Ohkubo, K.; Taylor, R.; Boltalina, O. V.; Ogo, S.; Fukuzumi, S. 2002. Electron transfer reduction of a highly electron-deficient fullerene, $C_{60}F_{18}$. *Chem. Commun.* 1952–1953.

Ohno, T.; Martín, N.; Knight, B.; Wudl, F.; Suzuki, T.; Yu, H. 1996. Quinone-type methanofullerene acceptors: Precursors for organic metals. *J. Org. Chem.* 61: 1306–1309.

Ohsawa, Y.; Saji, T. 1992. Electrochemical detection of C_{60}^{6-} at low temperature. *J. Chem. Soc. Chem. Commun.* 781–782.

Ortiz, J.; Fernández-Lázaro, F.; Sastre-Santos, A. et al. 2004. Synthesis and electrochemical and photorefractive properties of new trinitrofluorenone–C_{60} photosensitizers *Chem. Mater.* 16: 5021–5026.

Pinzón, J. R.; Plonska-Brzezinska, M. E.; Cardona, C. M. et al. 2008. $Sc_3N@C_{80}$-ferrocene electron-donor/acceptor conjugates as promising materials for photovoltaic applications. *Angew. Chem. Int. Ed.* 47: 4173–4176.

Pinzón, J. R.; Cardona, C. M.; Herranz, M. A. et al. 2009. Metal nitride cluster fullerene $M_3N@C_{80}$ (M = Y, Sc) based dyads: Synthesis, electrochemical, theoretical and photophysical studies. *Chem. Eur. J.* 15: 864–877.

Prato, M.; Maggini, M. 1998. Fulleropyrrolidines: A family of full-fledged fullerene derivatives. *Acc. Chem. Res.* 31: 519–526.

Prato, M.; Li, Q. C.; Wudl, F.; Lucchini, V. 1993. Addition of azides to fullerene C_{60}: Synthesis of azafulleroids. *J. Am. Chem. Soc.* 115: 1148–1150.

Reed, Ch. A.; Kim, K.-Ch; Bolskar, R. D.; Mueller, L. J. 2000. Taming superacids: Stabilization of the fullerene cations HC_{60}^+ and $C_{60}^{•+}$. *Science* 289: 101–104.

Riedel, I.; von Hauff, E.; Parisi, J.; Martín, N.; Giacalone, F.; Dyakonov, V. 2005. Diphenylmethanofullerenes: New and efficient acceptors in bulk-heterojunction solar cells. *Adv. Funct. Mater.* 15: 1979–1987.

Saito, G.; Teramoto, T.; Otsuka, A.; Sugita, Y.; Ban, T.; Kusuniki, K. 1994. Preparation and ionicity of C_{60} charge transfer complexes. *Synth. Met.* 64: 359–368.

Schick, G.; Hirsch, A.; Mauser, H.; Clarck, T. 1996. Opening and closure of the fullerene cage in *cis*-bisimino adducts of C_{60}: The influence of the addition pattern and the addend. *Chem. Eur. J.* 2: 935–943.

Siedschlag, C.; Luftmann, H.; Wolff, C.; Mattay, J. 1997. Functionalization of [60]fullerene by photoinduced electron transfer (PET): Syntheses of 1-substituted 1,2-dihydro[60] fullerenes. *Tetrahedron* 53: 3587–3592.

Suzuki, T.; Maruyama, Y.; Kato, T., Kikuchi, K.; Achiba, Y. 1993. Electrochemical properties of La@C82. *J. Am. Chem. Soc.* 115: 11006–11007.

Suzuki, T.; Kikuchi, K.; Oguri, F. et al. 1996. Electrochemical properties of fullerenolanthanides. *Tetrahedron* 52: 4973–4982.

Taylor, R. 2001. Fluorinated fullerenes. *Chem. Eur. J.* 7: 4074–4084.

Thompson, B. C.; Fréchet, J. M. J. 2008. Polymer-fullerene composite solar cells. *Angew. Chem. Int. Ed.* 47: 58–77.

Wang, L. S.; Conceiçao, J.; Jin, C.; Smalley, R. E. 1991. Threshold photodetachment of cold C_{60}^-. *Chem. Phys. Lett.* 182: 5–11.

Wudl, F.; Suzuki, T.; Prato, M. 1993. Probing the properties of C_{60} through fulleroids ABC_{61} *Synth. Met.* 59: 297–305.

Xie, Q.; Pérez-Cordero, E.; Echegoyen, L. 1992. Electrochemical detection of C_{60}^{6-} and C_{70}^{6-}: Enhanced stability of fullerides in solution. *J. Am. Chem. Soc.* 114: 3978–3980.

Zhou, F.; Van Berkel, G. J.; Donovan, B. T. 1994. Electron-transfer reactions of fluorofullerene $C_{60}F_{48}$. *J. Am. Chem. Soc.* 116: 5485–5486.

V

Carbon Fullerenes in Contact

37

Clusters of Fullerenes

Masato Nakamura
Nihon University

37.1 Introduction

The discovery of fullerenes (Kroto et al. 1985) opened a new branch in the field of nanophysics. In general, fullerene molecules can also be categorized as members of carbon clusters with a cage-like structure. Among fullerenes, the soccer ball molecule, C_{60}, and the rugby ball molecule, C_{70}, are particularly important. In this chapter, the word "fullerene" refers to the C_{60} molecule unless otherwise specified. Like many atoms and molecules, the fullerene molecules can form clusters. Thus, clusters of fullerenes are sometimes called "clusters of clusters." Furthermore, fullerenes can form different types of crystals. In the solid phase (fullerite), the C_{60} molecule forms a face-centered cubic (fcc) crystal at room temperature (Krätschmer et al. 1990, Saito and Oshiyama 1991, David et al. 1993).

In addition to the fact that chemical bonds between carbon atoms are strong, the fullerene molecule has a closed-shell electronic structure. Thus, when the fullerenes form a cluster, chemical bonds in the molecules do not play a significant role. The cohesive force between fullerenes is the van der Waals force. For systems interacting through such a force, the Lennard-Jones potential is often used for describing the interaction. The potential V between two molecules can be written as a function of the distance r between the centers of the molecules:

$$V(r) = 4\varepsilon_{LJ}\left\{\left(\frac{\sigma_{LJ}}{r}\right)^{12} - \left(\frac{\sigma_{LJ}}{r}\right)^{6}\right\}. \qquad (37.1)$$

In this formalism, $r_0 = 2^{1/6}\sigma_{LJ}$ gives the equilibrium distance for a potential well depth of ε_{LJ}. Table 37.1 lists the values of the potential parameters σ_{LJ} and ε_{LJ} for various pairs of atoms and molecules. (As is discussed later, the potential between two C_{60} molecules is somewhat different from the Lennard-Jones potential. This table lists only the rough estimates of the equilibrium distance and well depth of a C_{60} dimer.) Their polarizability, α, dielectric constant, ε, and density, ρ, in the solid phase (almost at the boiling point) are also shown. The Clausius–Mossotti equation gives the relation between α, ε, and the number density ρ_n:

$$\frac{\varepsilon-1}{\varepsilon+2} = \frac{4\pi}{3}\rho_n\alpha. \qquad (37.2)$$

From Table 37.1, we can draw the following inferences.

1. The equilibrium distance between two fullerene molecules is large.

   We see that σ_{LJ} of a C_{60} pair is at least three times that of pairs of the other elements/compounds listed. This fact implies that the intermolecular distance in a cluster of fullerene is much larger than that of clusters of other

TABLE 37.1 Parameters of the Interaction Potential, σ_{LJ} and ε_{LJ}, and the Polarizability α of a Molecule, Dielectric Constant ε, and Density ρ for the Condensed State for Various Elements/Compounds

Atoms/ Molecules	σ_{LJ} (Å)	ε_{LJ} (K)	α (Å³)	ε	ρ (g/cm³)
He	2.63	35.6	0.206	1.05	0.125
Ar	3.41	119.8	1.66	1.51	1.40
Xe	4.10	221.0	4.11	1.95	3.06
CO_2	4.10	234.2	2.63	1.65	1.18
CH_4	3.93	147.1	2.60	1.63	0.423
C_{60}	8.93	3133	76.5	3.459	1.69

Note: Data, except for those of C_{60} (Antoine et al. 1999, David et al. 1993), are from Echt, O. et al., *Phys. Rev. A*, 38: 3236, 1988.

elements/compounds listed. As a result, the clusters of fullerenes are much larger in geometrical size than the other clusters if the numbers of molecule in the clusters are the same.

2. The binding between fullerene molecules is strong.
   We see that the value of ε_{LJ} for a fullerene pair is larger than the pairs of others listed in Table 37.1. This fact means that molecules in a cluster of fullerene are more tightly bound than those in the clusters of others in the table. In fact, bulk fullerenes are in a solid form at room temperature.

3. Fullerene has a large electronic polarizability.
   The electronic polarizability of the C_{60} molecule (Antoine et al. 1999) is one or two orders of magnitude larger than that of other elements/compounds in the table. Because of its large polarizability, the dielectric constant of fullerite is quite large.
   Furthermore, in addition to the observation made from Table 37.1, we can point out as follows.

4. Fullerene molecules are almost spherical.
   The truncated icosahedral structure has high symmetry and is almost spherical in shape. We can consider a fullerene molecule as a "pseudoatom" in clusters. Thus, we can understand the geometrical structure of clusters of fullerene by packing spheres in the space.

5. Fullerene has many modes of internal degrees of freedom.
   As a unit of a cluster, a molecule has rotational and vibrational degrees of freedom. In the case of fullerenes, there are many internal vibrational modes (Weeks and Harter 1989). Most of the degrees of freedoms except for rotation are frozen at low temperatures (e.g., see Cheng and Klein 1992). In fact, fullerene molecules in the bulk phase rotate almost freely at room temperature.

6. Fullerene molecules have a hole.
   Fullerene molecules are completely different from other atoms and molecules in Table 37.1 in that they have a hole in their interior. The role of the holes in the formation of clusters would be an interesting problem. Furthermore, it might be possible to introduce dopants through the hole.

This review is organized as follows. From Sections 37.2 through 37.5, we focus on the structures of the clusters of fullerenes from both experimental and theoretical aspects. In Section 37.6, we will discuss on the multiply charged clusters of fullerenes. Concluding remarks and future problems are stated in the last section.

37.2 The First Experiment on Clusters of Fullerenes

The first experimental work on clusters of fullerenes was carried out by Martin et al. (1993). They produced clusters of fullerenes (C_{60} and C_{70}) through the condensation of fullerene vapor in a low-pressure cold helium gas atmosphere. The produced neutral clusters are ionized by an excimer laser. Mass abundance spectra are obtained by time-of-flight (TOF) mass spectrometry. Usually, in the case of clusters with relatively stable sizes, the mass distribution shows a peaked structure. From the positions of the peaks, we can infer the structure of the clusters. In order to increase the visibility of the peaks in a mass abundance spectrum, the clusters are heated by lasers so that the evaporation takes place from the less-stable clusters. The geometrical structure of clusters is discussed in Appendix.

Figure 37.1 shows an experimentally measured mass abundance spectrum of charged $(C_{60})_n$ clusters. We observe two strong peaks at $n = 13$ and 55. Less-strong peaks are observed at $n = 19, 23, 27, 35, 43, 46, 49$, etc. These peak positions are nearly the same as those observed in rare-gas clusters. Figure 37.2 presents a mass abundance spectrum of Xe_n clusters obtained by Echt et al. (1981). Peaks are observed at the size of $n = 13, 19, 25$, and 55. Upon comparing Figures 37.1 and 37.2, we observe that the peak positions of the charged $(C_{60})_n$ and Xe_n are nearly identical. In particular, the peaks at $n = 13$ and 55 are significant. These numbers correspond to Mackay's number at which a cluster forms a perfect icosahedron (Mackay 1962). Mackay's number is given by

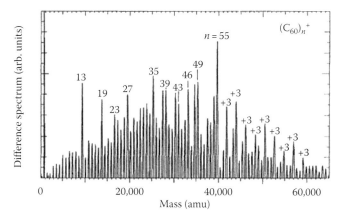

FIGURE 37.1 Mass abundance spectrum of $(C_{60})_n$ clusters. (From Martin, T.P. et al., *Phys. Rev. Lett.*, 70, 3079, 1993. With permission.)

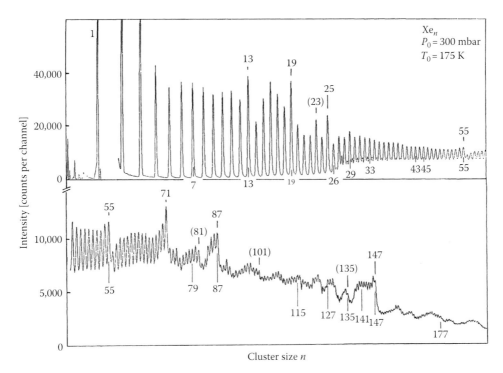

FIGURE 37.2 Mass abundance spectrum of Xe_n clusters. (From Echt, O. et al., *Phys. Rev. A*, 47, 1121, 1981. With permission.)

$$n(i) = 1 + \sum_{k=1}^{i} \left(10k^2 + 2\right). \qquad (37.3)$$

For $i = 1$, 2, 3, and 4, we obtain $n = 13$, 55, 147, and 309, respectively. The strong peak at $n = 13$ in the mass abundance spectrum of $(C_{60})_n$ has also been observed by Hansen et al. (1996, 1997).

Other peaks such as $n = 19$ correspond to sub-shell closing structures (Martin et al. 1993, Martin 1996). The sub-shell structure resulting from icosahedral growth has been discussed by Farges et al. (1986) and Northby (1987). According to these studies, when a cluster grows (icosahedral growth) from the first shell ($n = 13$) to the second shell ($n = 55$), the cluster assumes a relatively stable structure at $n = 19$, 23, 26, 29, 39, 43, 46, and 49. We call this phenomenon "sub-shell structure." In Figure 37.1, we observe peaks at most of these n values which indicate the icosahedral growth of the clusters of fullerenes, similar to the case of rare-gas clusters. Martin's group also obtained the mass abundance spectra of $(C_{70})_n$. As shown in Figure 37.3, the spectrum of $(C_{70})_n$ is quite similar to that of $(C_{60})_n$.

It is to be noted that in the first experiment by Martin and coworkers, not only singly charged clusters of fullerenes but also doubly and triply charged clusters were observed. The values of n at the peak positions of doubly and triply charged clusters were almost identical to those observed for singly charged ones. This finding implies us that the structure of clusters of fullerenes is almost independent of the charge.

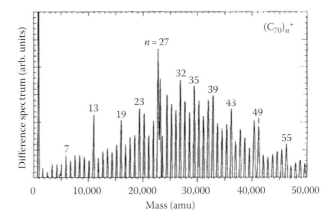

FIGURE 37.3 Mass abundance spectrum of $(C_{70})_n$ clusters. (From Martin, T.P. et al., *Phys. Rev. Lett.*, 70, 3079, 1993. With permission.)

The first experiment showed that the structure of the clusters of fullerene and that of rare-gas clusters are similar. As for rare gases, the Lennard-Jones pairwise potential well describes the interaction. For a $(LJ)_n$, a cluster of n particles interacting via the Lennard-Jones potential with each other, the most stable structure of clusters is based upon the Mackay icosahedron in most cases. However, there are some cases where other structures are more stable (Wales and Doye 1997). For example, a truncated octahedron is the most stable structure for the case of $n = 38$: a truncated decahedron (Marks 1984) for $n = 75$, and a truncated tetrahedron for $n = 98$ (Leary and Doye 1999).

For a better understanding of the structures of clusters of fullerenes, we need to use theoretical approaches since it is difficult to determine the structures of clusters without the help of theory.

37.3 Interaction Potentials between C$_{60}$ Molecules

To theoretically study the structures of clusters of fullerenes, the interaction potential between fullerene molecules needs to be understood.

In clusters of fullerenes, the binding energies of chemical bonds between carbon atoms in a fullerene molecule are much larger than those between fullerene molecules. There have been some attempts to calculate the interaction potential between fullerene molecules. A fullerene molecule has a highly symmetric truncated icosahedral structure. A fullerene molecule has a highly symmetric truncated icosahedral structure. It is not perfectly spherical, and therefore, the interaction potential between molecules depends on their orientation angles. The calculations of potential in which the orientation of molecules in clusters is taken into account have been presented (Prassides 1993, Deleuze and Zerbetto 1999). However, this angular dependence is so weak that in most calculations, the potential is spherically averaged. This fact means that each fullerene molecule in the cluster is rotating freely. Most of the calculations performed in the past studies have been restricted to the electronic ground state. Among them, two types of potential have been extensively used.

37.3.1 Girifalco Potential

An expression for determining an approximate interaction potential between two C$_{60}$ molecules was proposed by Girifalco (1991, 1992). This expression has been widely used owning to its considerable accuracy and simplicity. Girifalco assumed that the fullerene–fullerene interaction is the sum of pairwise interactions between the carbon atom of one molecule and that of the other molecule. For each carbon–carbon interaction, the Lennard-Jones potential is assumed. Finally, the interaction potential is averaged over the orientation of the molecules.

The Girifalco potential has the following form

$$V = -\alpha\left(\frac{1}{s(s-1)^3} + \frac{1}{s(s+1)^3} - \frac{2}{s^4}\right) + \beta\left(\frac{1}{s(s-1)^9} + \frac{1}{s(s+1)^9} - \frac{2}{s^{10}}\right),$$
(37.4)

where $s = r/2a$ is the distance between the centers of fullerene molecules divided by the diameter of a single fullerene molecule: the value of $2a$ is 7.1 Å. The values of parameters α and β are 4.67735×10^{-2} and 8.48526×10^{-5} eV, respectively (Girifalco 1992).

An approximate form of the potential in the case of charged fullerene pairs such as C$_{60}^+$ − C$_{60}^+$ and C$_{60}$ − C$_{60}^+$ is obtained by simply adding electrostatic terms to the expression for the potential between two neutral C$_{60}$ molecules (Zettergren et al. 2007).

37.3.2 Potential by the First-Principle Calculation

An accurate calculation of the interaction potential between C$_{60}$ molecules has been performed by Pacheco and Prates-Ramalho (1997): hereafter we refer to their study as PRP. They have applied the first principles, the local-density approximation (LDA) to calculate the short-range part of the interaction potential. It is recognized that the LDA cannot correctly describe the long-range dispersion forces. Thus, the time-dependent-local-density-approximation (TDLDA) has been used to evaluate the long-range part of the interaction potential. For the principles of the LDA and TDLDA and their applications to cluster science, the reader is referred to the review by Calvayrac et al. (2000).

The results of PPR were analytically fitted in the following manner. The interaction potential consists of two parts, a long-range part and a short-range one. The former is written as

$$W(r) = -\frac{C_6}{r^6} - \frac{C_8}{r^8} - \frac{C_{10}}{r^{10}} - \frac{C_{12}}{r^{12}},$$
(37.5)

where the coefficients C_6, C_8, C_{10}, and C_{12} are computed using the TDLDA. The calculated values of the coefficients are listed in Table 37.2a. The short-range part is written in the form of the Morse potential:

$$M(r) = M_0 \exp\left[\tau\left(1 - \frac{r}{d_0}\right)\right]\left\{\exp\left[\tau\left(1 - \frac{r}{d_0}\right)\right] - 2\right\},$$
(37.6)

where M_0, d_0, and τ are the Morse parameters. The values of these parameters are listed in Table 37.2b. The short- and long-range parts of the potential can be combined by using a switching function $F(r)$. The two body parts of the potential can be written as follows:

$$V_2(r) = F(r) \times M(r) + \left(1 - F(r)\right) \times W(r).$$
(37.7)

$F(r)$ is given by the expression

$$F(r) = \frac{1}{1 + \exp\left[(r - \mu)/\delta\right]}.$$
(37.8)

Thus, $F(r) \to 1$ for $r \to \infty$ and $F(r) \to 0$ for $r \to 0$. The values of δ and μ are also given in Table 37.2b.

The PPR potential includes not only the two-body part of the potential, but also a three-body term. This three-body term is called the Axilrod–Teller interaction (Axilrod and Teller 1948) and written as

TABLE 37.2 Parameters of the PPR Potential: N_{at} (=60) is the Number of Atoms in Fullerene

(a)

C_6 (eV Å^{-6})	C_8 (eV Å^{-8})	C_{10} (eV Å^{-10})	C_{12} (eV Å^{-12})	C_{AT} (eV Å^{-9})
$21N_{at}^2$	$2534N_{at}^2$	2.09×10^8	7.78×10^{10}	$21N_{at}^3$

(b)

M_0 (eV)	τ	d_0 (Å)	μ (Å)	δ (Å)
0.3	9.75	10.3	10.05	1.04

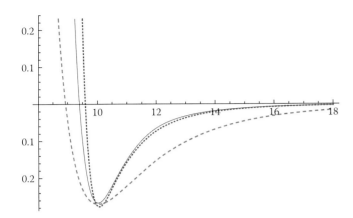

FIGURE 37.4 Potential energies (eV) as functions of the distance (Å) between the centers of two C_{60}. Solid and dotted curves represent the potential determined by the PPR (1999) and Girifalco (1992), respectively. The dashed curve is the Lennard-Jones potential which reproduces the equilibrium distance and well depth of the above two potentials. (From Martin, T.P. et al., *Phys. Rev. Lett.*, 70, 3079, 1993. With permission.)

$$V_3(r) = C_{AT}\left(\frac{1 + 3\cos\gamma_i \cos\gamma_j \cos\gamma_k}{r_{ij}^2 r_{jk}^2 r_{ki}^2}\right), \quad (37.9)$$

where

r_{ij} is the distance between molecule i and molecule j

γ_i is the angle between r_{ij} and r_{jk}

The value of C_{AT} is listed in Table 37.2a. The contribution of three-body force amounts to 6% of that of the two-body interaction at the maximum. Sometimes this term may change the global minimum structure of clusters.

Figure 37.4 represents the interaction potentials between two C_{60} molecules. The Girifalco and PPR potentials are shown as functions of center distance of the two molecules. For comparison, the Lennard-Jones potential to fit the equilibrium distance and the potential well depth is shown. From this figure, we see that the potential between two C_{60} molecules cannot be described in terms of the Lennard-Jones potential. Both the Girifalco and PPR potentials represent that the range of potential much shorter than that of the Lennard-Jones potential. Furthermore, both of the Girifalco and PRP potentials give harder potential core than the Lennard-Jones potential. The difference between Girifalco potential and the PPR potential seems to be quite small. This small difference, however, sometimes induces drastic changes in bulk properties of fullerite (Pacheco and Prates-Ramalho 1997).

37.3.3 Other Potentials

The two potentials discussed above are spherically averaged over all the molecular orientation angles. However, there have been attempts to study the properties of clusters of fullerenes by using potentials that are not spherically averaged.

The all-atom potential is a typical candidate. García-Rodeja et al. (1997) and Doye et al. (1997) have studied the structure of clusters of fullerenes with this potential. Their concept is similar to that of Girifalco—the total interaction potential is the sum of carbon atom interactions—but they do not spherically average the potential over the sphere.

Recently, the orientational dependence of the potential between two C_{60} molecules has been discussed by Tournus et al. (2005) using the LDA. They have applied the potential to study the molecular orientation in $(C_{60})_2$ and $(C_{60})_7$ clusters.

37.3.4 Experimental Measurements of Binding Energy of Fullerene Dimers

Binding energies of fullerene dimers have been experimentally measured. According to Branz et al. (2002), the binding energies of $(C_{60})_2$, $(C_{70})_2$ dimers, and $(C_{60})_2^+$ are 0.275 ± 0.08, 0.313 ± 0.08, and 0.372 ± 0.08 eV, respectively. Other experimental measurements give nearly the same values as those shown in Table 37.1 of the paper of Branz et al. (2002). Both the Girifalco and PRP potentials well reproduce these values.

37.4 Theoretical Calculations on the Structure of Clusters of Fullerenes

37.4.1 General Discussion

Once the potential functions, two-body term V_2, and three-body term V_3, are given, the total potential energy of the system is written as

$$V_{tot} = \sum_{i,j} V_2(r_{ij}) + \sum_{i,j,k} V_3(r_{ij}, r_{jk}, r_{ki}), \quad (37.10)$$

where r_{ij} is the distance between the centers of ith and jth molecules in the cluster; the summation is over all the pairs. In the case of pairwise potentials such as the Girifalco potential, the second term on the right-hand side is dropped.

At the absolute zero temperature, a cluster assumes a structure that is energetically the most stable. To determine such a structure, we have to find the global minimum energy of a corrugated multidimensional potential energy surface. The number of local minima on a potential energy surface increases rapidly with an increase in the number of particles. In the case of the $(LJ)_n$, the number of local minima amounts to 10^{40} for 98 particles and 10^{60} for 147 particles (Wales and Doye 1997). This fact means that clusters containing many particles have many isomers. Various computational techniques have been developed to find the global minimum, such as the simulated annealing (e.g., Kirkpatrick et al. 1983), genetic algorithm (e.g., Holland 1975) and basin-hopping methods (Wales and Doye 1997). The basic concept for the optimization is to attain the global minimum as fast as possible without being trapped by local minima.

At a finite temperature T, the system can assume energetically less-stable states. In such a situation, various "isomers" can coexist. The system must minimize the free energy $F = E - TS$. In such a case, both the energy E and entropy S play an

TABLE 37.3 Theoretical Studies on the Structure of Clusters of Fullerenes

Authors	Potential	Optimization	Max. n	Years
Rey, Gallego, Alonso	Girifalco	Steepest descent	25	1994
Doye, Wales	Girifalco	Orient program	80	1997a
Doye, Dullweber, Wales	All atom	Orient program	56	1997
García-Rodeja, Rey, Gallego	All atom	Steepest descent	20	1997
Luo Zaho, Qui, Wang	PPR	Genetic-algorithm	25	1999
Zhan, Liu, Zhuang, Li	PPR	Genetic-algorithm	57	2000
Doye, Wales, Branz, Calvo	PPR	Basin hopping	105	2001
Cai, Feng, Shao, Pan	Girifalco	Parallel fast annealing evolutionary algorithm	80	2002
Cheng, Cau, Shao	Girifalco	Dynamic lattice-searching	150	2005

important role. It is still difficult, however, to minimize the free energy. Molecular dynamics simulation gives useful information in such a case. However, the simulation time is often not enough for the system to reach the equilibrium state. In what follows, we restrict ourselves to the most stable structure of clusters at $T = 0$.

37.4.2 Calculations for Determining Structure of Clusters of Fullerenes

There have been a considerable number of theoretical studies on the most stable structure of $(C_{60})_n$. In Table 37.3, we list some of them. Various type of optimization methods are used. The first extensive calculation was carried out by Rey et al. (1994) using the Girifalco potential. Cheng et al. (2005) studied the structure of clusters of fullerenes up to $n = 150$ using the Girifalco potential: they employed a newly developed dynamic lattice-searching method (Shao et al. 2004). Cheng et al.'s study remains the one to have dealt with the clusters with the largest sizes.

In the list in Table 37.3, the most extensive study using the PPR potential was performed by Doye et al. (2001). They calculated the global minimum energies and determined the geometrically optimized structure of $(C_{60})_n$ for n values up to 105 using the basin-hopping method (Wales and Doye 1997). The same group also calculated the global minimum energies for the Girifalco potential up to $n = 80$ (Doye and Wales 1996a). The results of both calculations can be seen on the Web site of the Cambridge Cluster Database.

Once we have the global minimum energy $E(n)$ as a function of n, we can average the energy as a function of n as

$$E_{LD}(n) = an + bn^{2/3} + cn^{1/3} + d. \tag{37.11}$$

We call this average energy the liquid-drop energy as it varies smoothly as a function of n. The first, second, and third terms are called the volume, surface, and curvature energies, respectively. The values of a, b, c, and d are listed in Table 37.4 for both the Girifalco and PPR potentials.

The difference between the global minimum energy and the liquid-drop energy is called the *shell energy*: $\Delta E_{shell}(n) = E(n) - E_{LD}(n)$.

TABLE 37.4 Parameters a, b, c, and d for the Girifalco and PPR Potentials

Potential	a (eV)	b (eV)	c (eV)	d (eV)
Grifalco	−1.789	2.294	0.5907	−1.292
PPR	−1.6537	2.1901	0.1263	−0.7467

This energy represents the stability of the clusters and is related to their geometrical shell structure. At the shell-closing number, this value is a local minimum as a function of n.

In the upper panel of Figure 37.5, we show the shell energies of the most stable structure of $(C_{60})_n$ as functions of n using both the Girifalco and PPR potentials. For comparison, in the lower panel, the shell energy of $(LJ)_n$ is shown as a function of n. In this figure, a dip in a line mostly corresponds to a number where the cluster assumes a relatively stable structure. We see that the shell energies of $(C_{60})_n$ for both the PRP and Girifalco potentials are similar. We observe a dip at $n = 13$ corresponding to the icosahedral shell-closing number in the case of both potentials. This dip is seen more clearly for the PPR potential. The dip of $n = 13$ is also clearly seen for the $(LJ)_n$. However, we see that the shell energies of $(C_{60})_n$ are completely different from those of $(LJ)_n$. Although the $(LJ)_n$ exhibits a sharp dip at $n = 55$, the closing number of the second shell of an icosahedron, we do not observe any dip at this n value in the case of $(C_{60})_n$ for both potentials. Furthermore, while the dip at $n = 19$, which corresponds to sub-shell closing number of icosahedral growth, is observed in the case of the $(LJ)_n$, it is not at all observed in the case of $(C_{60})_n$. This fact indicates that the icosahedral growth stops in the range of $14 < n < 18$ for $(C_{60})_n$. Instead, we observe deep dips at $n = 38$, 75, and 101. The dips at $n = 38$ and 75 are observed for both $(C_{60})_n$ and $(LJ)_n$, although they are much more pronounced for the $(C_{60})_n$.

In Table I of the Doye et al. (2001) paper and on the Cambridge Cluster Database Web site, the most stable geometrical structures and energies of clusters of fullerenes $(C_{60})_n$ using the PPR potential are shown up to $n = 105$. According to their calculation, the icosahedral structure is the most stable at $n \leq 15$. For $n \geq 16$, either a closed-packed (including the fcc) or a decahedral structure is the most stable. Some of the optimized structures of $(C_{60})_n$ are schematically shown in Figure 37.6. The stable structure at $n = 38$ corresponds to a truncated fcc structure and

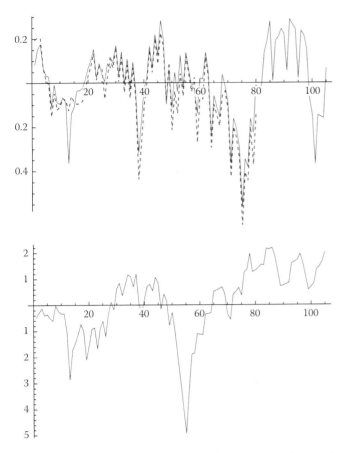

FIGURE 37.5 Shell energies as functions of size *n* as for $(C_{60})_n$ (upper) and $(LJ)_n$ (lower). The unit of the vertical axes in upper and lower panels are eV and ε, respectively. In the upper panel, the solid and dashed lines represent the shell energies obtained by the PPR and Girifalco potentials, respectively. ([Upper]: Data taken from Doye, J.P.K. and Waler, D.J., *Chem. Phys. lett.*, 262: 167, 1996a, for Girifalco potential Doye, J.P.K et al., Phys. Rev. B, 64: 235409, 2001, for the PRP potential. [Lower]: Data taken from Cambridge Cluster Database.)

that at *n* = 75 is known as Marks' decahedron (Marks 1984). Their structures are shown in Figure 37.7 together with the icosahedral cluster for *n* = 55. (Note that the icosahedral shape at *n* = 55 does not correspond to the global minimum energy for clusters of fullerenes.)

37.4.3 The Relation between Structure of Clusters and Ranges of Forces

Usually, van der Waals clusters like rare-gas clusters assume the icosahedral structure to be most stable. Then, why do the theoretical calculations predict other structures to be more stable for clusters of fullerenes for $n \geq 16$? The reason for this was discussed by Doye et al. (1996a) in terms of the range of the van der Waals force (also see Doye and Wales 1996a,b). A similar discussion was also presented by Amano et al. (2006) by considering three types of potentials.

According to these researchers, the energy of the clusters can be written as the sum of the nearest-neighbor interaction E_{nn}, strain energy E_{strain}, and non-nearest-neighbor interaction E_{nnn}:

$$E = E_{nn} + E_{strain} + E_{nnn}. \tag{37.12}$$

The nearest-neighbor interaction is the product of the number of nearest neighbors and the well depth. The origin of strain energy is the energetic increase due to the deviation of nearest-neighbor contracts from the equilibrium pair distance as is schematically shown in Figure 37.8. Usually, the contribution of the last term is so small that the competition between the nearest-neighbor interaction and strain energy determines the structure of clusters. Table 37.5 shows whether the presence of nearest-neighbor interaction and strain energy is an advantage or disadvantage in the three types of structures—the

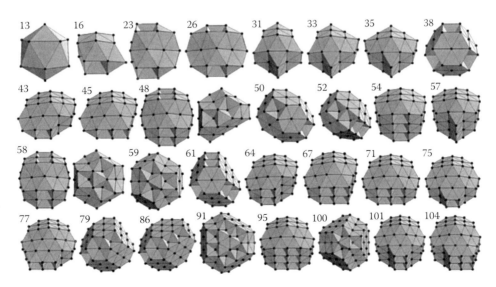

FIGURE 37.6 Examples of most stable structure of clusters of fullerenes $(C_{60})_n$ for some values of *n*. Each point represents the center of a fullerene molecule.

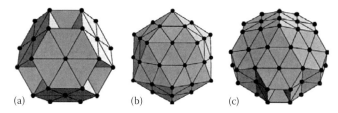

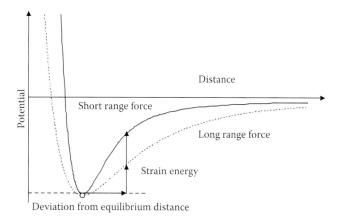

FIGURE 37.7 Three examples of stable structures for van der Waals clusters: (a) $n = 38$ (fcc truncated octahedron) (b) $n = 55$ (Mackay's icosahedron) and (c) $n = 75$ (Marks' decahedron). (From Doye, J.P.K. et al., *Phys. Rev. B*, 64, 235409, 2001. With permission.)

FIGURE 37.8 Ranges of potentials and strain energies for long- and short-range forces. (From Leary, R.H. and Doye, J.P.K., *Phys. Rev. E*, 60, R6320, 1999. With permission.)

TABLE 37.5 Advantage and Disadvantage of the Presence of the Nearest-Neighbor Interaction and Strain Energy for Three Typical Structures

Structure	Nearest-Neighbor Interaction	Strain Energy
Icosahedral	○	×
Decahedral	△	△
Close packed	×	○

○, advantage; ×, disadvantage; △, intermediate.

icosahedral, decahedral, and close-packed structures. The icosahedral structure minimizes the nearest-neighbor interaction. Molecules in this structure have the largest number of nearest neighbors. On the other hand, it maximizes the strain energy. The nearest-neighbor distance between molecules within the same shell is 5% larger than that between molecules in different shells. The close-packed structure minimizes the strain energy while maximizing the nearest-neighbor interaction. The nearest-neighbor distance and strain energy in the decahedral structure are intermediate between those in the other two structures.

If the range of interaction is sufficiently long similar to the case of rare gases, the strain energy becomes small, as shown in Figure 37.8. Then, the icosahedral structure becomes the most stable. In fact, the icosahedral structure is observed even in the case of large rare-gas clusters (Meihle et al. 1989). However, in clusters of fullerenes, the range of the potential is so short that the contribution of the strain energy is large. The clusters therefore deform and assume other structures to reduce the strain energy. This is why the icosahedral structure of relatively small clusters of fullerenes changes to other structures.

In general, since neither the icosahedral nor decahedral structure has no translational symmetry, neither of them forms crystals. When clusters grow to form crystal, the transition from a structure without translational symmetry to one with translational symmetry must take place at some critical size. We observe that the two structures still compete around a size of $n = 100$.

37.5 Recent Studies on the Structure of Clusters of Fullerenes

37.5.1 Structure of Neutral and Charged Clusters

There has been a discrepancy between experiments and theory regarding the geometrical structure of clusters of fullerenes. The experimentally measured mass abundance spectra indicate that the icosahedral structures are the most stable while the theoretical studies predict other structures to be more stable. There have been some attempts to account for this discrepancy. One view was that experimental measurements pertained to charged clusters while most of the theoretical calculations were for neutral clusters. The stable structure of neutral clusters might be completely different from that of charged ones since the polarizability of a fullerene molecule is extremely large. Rey et al. (2004) have theoretically determined the stable structures of both neutral and charged clusters of fullerenes up to a size of $n = 20$. They found that the most stable structures of both types of clusters were similar: the differences observed in the intermolecular distances of the stable structures between the two types of clusters were small. This is because the electric interaction between fullerene molecules is much smaller than the intermolecular interaction between them. Thus, the above-mentioned view was disproved by Rey et al.'s study.

37.5.2 Charge Dependence of Mass Abundance Spectra

Martin's group has improved their experimental apparatus to observe the spectral dependence on temperature and charge of the clusters (Branz et al. 2000, 2002). In the previous experiment (Martin et al. 1993), heating and ionization were carried out simultaneously. Only the mass spectra of charged particles were obtained and the control of temperature was difficult. In the new experiment, heating cell was used to control the temperature of clusters. Heating and ionization were performed separately, and therefore, it was possible to obtain information on the size distribution of neutral clusters.

Using this apparatus, they measured the mass abundance spectrum of clusters of fullerenes. The left panel of Figure 37.9 represents the spectrum of singly (positive) charged, neutral, and singly (negative) charged clusters of fullerenes. The temperature of the clusters is set to be constant (490 K). In this figure, we see that the positions of the peaks are almost independent of the charge of the clusters, although their relative intensities change. This finding indicates that the dependence of the geometrical structure on the charged state of the clusters is weak.

37.5.3 Temperature Dependence of Mass Abundance Spectra

Let us now discuss the temperature dependence of the spectrum. As shown in the right panel of Figure 37.9, at a low temperature ($T = 100$ K), the size distribution is monotonous and no peak is observed. When the temperature of clusters is 490 K, we observe several peaks in the spectrum. The spectral feature strongly suggests the icosahedral growth of clusters. In addition to the peaks at $n = 13$ and 55, we observe a strong peak at $n = 147$, which corresponds to the third shell-closing number of Mackay's icosahedron. Weaker peaks are also well explained in terms of the sub-shell structure of icosahedral growth.

When the temperature of clusters is raised to 585 K, however, the mass abundance spectrum changes completely. The peak at $n = 55$, which corresponds to a shell-closing number of the icosahedral growth, disappears, and new peaks appear at different positions such as $n = 38, 48, 54, 58, 61, 68, 84,$ and 98. It is not easy to analyze the origin of these peaks. Among these peaks, those at $n = 38$ and 48 are predicted by the theoretical calculation of Doye et al. (2001). The peak at $n = 38$ is probably due to the formation of a face-centered-cubic-truncated octahedron structure corresponding to the global minimum energy. The small peak at $n = 75$ may correspond to a Marks' decahedron (Marks 1984). In the case of other peaks, it is difficult to judge whether the theoretical calculation reproduces the peak structure well. Some of the peaks in the experiments are identified to correspond to dips in the shell energies in the calculation (Figure 37.5). However, a perfect one-to-one relation is not observed.

37.5.4 Structures Related to Leary's Tetrahedron

We observe a large peak at $n = 98$ in the mass abundance spectrum for 565 K. This peak increases in intensity when the temperature is increased to 610 K. It is possible that this peak corresponds to Leary's tetrahedron (Leary and Doye 1999). Leary's tetrahedron has been proposed as the stable structure of $(LJ)_n$ with 98 particles (see Figure 37.10). Although, Leary's tetrahedron does not correspond to the most stable structure of clusters of fullerenes, this structure may be important for explaining the

peaks observed in this experiment. By symmetrically removing some sets of molecules from the Leary's tetrahedron, we obtain structures with different numbers of molecules: these structures correspond to the peaks in the experiments at $n = 61, 68, 84…$. Figure 37.11 represents the shell energies of structures based on Leary's tetrahedron together with those of the most stable structure determined using the PPR potential. Although the energies of structures based on Leary's tetrahedron are always higher than the global minimum energies, they have dips at $n = 61, 68, 84, …$ which correspond to the peaks in the mass abundance spectra at high temperatures. It is plausible that the growth of clusters leads the formation of Leary's tetrahedron. Subsequently, clusters with different sizes are observed as a result of evaporation from the tetrahedron.

37.5.5 Structural Transitions of Clusters of Fullerenes

Why does the mass abundance spectrum show the formation of an icosahedral structure at low temperatures and structural changes at high temperatures? To answer this question, molecular dynamics simulation was performed by Branz et al. (2002). In general, molecular dynamics simulation is a powerful tool for studying structural transitions in clusters. For example, Wales (1994) has demonstrated the structural transformation of $(C_{60})_{55}$ from an icosahedron to a cuboctahedron.

Branz et al. prepared icosahedral clusters and observed changes in their structures with time. At a low temperature ($n = 575$ K), the structure of the clusters remains almost unchanged during the simulation time. At a high temperature ($n = 675$ K), structural changes are clearly observed during the simulation time (order of microseconds). Some of the icosahedral clusters change to decahedral ones. This observation can be understood as follows. At low temperatures, the clusters assume a metastable icosahedral structure since their internal energy is insufficient to overcome the potential barrier to move toward the next minima. However, if the clusters are heated to a sufficient extent, they can overcome the potential barrier and proceed toward the global minimum.

Baletto et al. (2002) have also performed molecular dynamics simulation to observe the growth of clusters of fullerenes. They showed that the clusters tend to form icosahedral structures, even though these structures are not the most stable energetically. They concluded that the icosahedral growth of clusters is due to "kinetic trapping."

These studies suggest that the icosahedral growth of clusters is a metastable state in the process of cluster formation under certain experimental conditions. The system forms a more stable structure upon being heated. However, a more quantitative comparison with experiments and theories is necessary. The reason for structures based on Leary's tetrahedron being widely observed in experiments needs to be clarified by future studies. For such a purpose, it would be necessary to geometrically optimize the structure of clusters by taking into account the temperature.

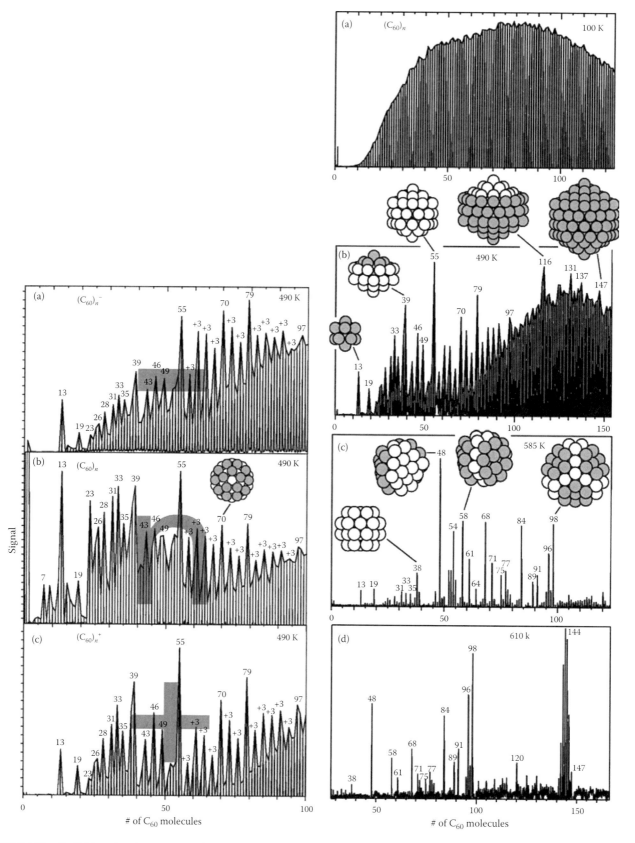

FIGURE 37.9 (Left) Mass abundance spectra of negatively charged (upper), neutral (middle) and positively charged (lower) clusters of fullerenes. The temperature of cluster is fixed at 490 K. (Right) Mass abundance spectra of clusters of fullerenes at different temperatures. From the top to the bottom, the temperature is raised from 100 to 610 K.

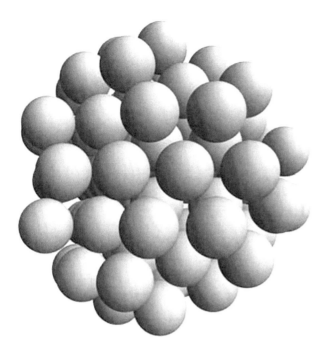

FIGURE 37.10   Leary's tetrahedron.

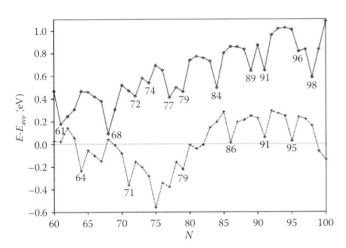

FIGURE 37.11   Shell energies of structures based on Leary's tetrahedron (solid line) and those of global minimum structure (dashed line) determined by the PPR potential as functions of cluster size *n*. (Courtesy of Cambridge Cluster Database, http://www-wales.ch.cam.ac.uk/CCD.html.)

37.6 Multiply Charged Clusters of Fullerene Molecules

Doubly and triply charged clusters of fullerenes were detected in the first experiment by Martin et al. (1993) (see Figure 37.1). Furthermore, the mass distribution of these multiply charged clusters was found to be similar to that of singly charged clusters. Recently, there have been intensive studies on the stability and fragmentation of multiply charged clusters of fullerene molecules, which are represented by $(C_{60})_n^z$.

37.6.1 How to Produce Multiply Charged Clusters of Fullerenes

The following experimental methods have been used to produce multiply charged clusters of fullerenes

1. Electron impact ionization
   Successive ionizations occur by electron impact collisions with molecules. This is one of the most common methods to produce multiply charged clusters.
2. Laser radiation
   A strong oscillatory electric field causes the multiple ionization of neutral clusters. An example of such an experiment is that performed by Hedén et al. (2005). If the laser field is strong, the temperature of the multiply charged clusters becomes extremely high.
3. Collision with highly charged atoms
   While the above two methods produce "hot clusters" because of large energy transfers from electrons or photons to clusters, this method enables us to prepare multiply charged clusters at quite low temperatures. Highly charged atomic ions have a strong electronic field and capture many electrons from neutral clusters. The range of force is so long that multiple ionizations of clusters takes place even if the collision takes place at a relatively large distance. The mechanism of electron transfers is explained in terms of the classical over-barrier model (Niehaus 1986, Zettergren et al. 2002).

If the charge *z* of a cluster is larger than the number of molecules *n* in the cluster, it is unlikely that a molecule contains more than one positive charge. This is because if excess charges can be shared by several molecules, the Coulomb energy will decrease. However, there is still a debate on whether charge localizes in a specific molecule in a cluster or whether it distributes uniformly within a cluster.

37.6.2 Stability of Multiply Charged Clusters from the Liquid-Drop Model

The liquid-droplet model is useful for discussing the stability of multiply charged clusters. According to the model, the total energy $E_{tot}(n, z)$ of charged clusters with charge z and number of molecules *n* can be written as the sum of the energy of neutral clusters, $E_{neutral}(n)$ and the charging energy, $E_C(n, z)$:

$$E_{tot}(n, z) = E_{neutral}(n) + E_C(n, z). \tag{37.13}$$

This relation is valid when the structure of a charged cluster does not differ much from that of a neutral one. The energy of a neutral cluster is the sum of the volume, surface, and curvature terms as given in Section 37.4.

In general, the stability of a multiply charged cluster is governed by the competition between the Coulomb energy E_c and the surface energy E_s of the parent cluster. The fissility parameter given by $f = E_C/(2E_s)$ represents the stability of multiply charged clusters. Figure 37.12 shows the energy diagram of multiply

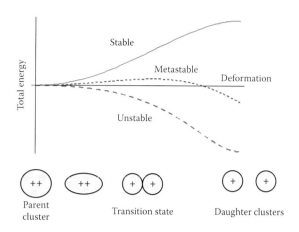

FIGURE 37.12 Potential energy topography as a function of deformation. (From Doye, J.P.K. et al., *Phys. Rev. B*, 64, 235409, 2001. With permission.)

charged cluster as a function of deformation. In this figure, a symmetric fragmentation with respect to charge and size is considered. As is apparent from the figure, if the total energy of a cluster decreases monotonously with the deformation of the cluster, the cluster is unstable. If there is an energy barrier, the cluster becomes stable or metastable. The difference between the stable and metastable states lies in the type of reaction that occurs in these states (endothermic or exothermic). If we restrict ourselves to symmetric fission, as shown in Figure 37.12, the parent cluster is stable for $f < 0.35$, metastable for $0.35 < f < 1$, and unstable for $1 < f$ (Bohr and Wheeler 1939). In the case of multiply charged clusters, there is almost no difference between stable and metastable at low temperatures, since it is practically impossible to overpass the energy barrier. In what follows, we do not distinguish metastable from stable.

If we assume that a cluster is a sphere with radius $r_n = \sqrt[3]{3n/4\pi}\sigma_{LJ}$, dielectric constant ε, and uniform charge distribution, classical electromagnetic theory gives the following relations

$$E_c = \frac{3}{5\varepsilon}\frac{z^2}{r_n},\tag{37.14a}$$

and

$$E_S = 4\pi s r_n^2\tag{37.14b}$$

where s it the surface tension. Then f can be written as

$$f = \frac{1}{10\varepsilon s\sigma_{LJ}^3}\frac{z^2}{n}\tag{37.15}$$

Thus, if we assume for $f < 1$, we have $n > n_C(z)$ with

$$n_C(z) = \frac{1}{10\varepsilon s\sigma_{LJ}^3}z^2\tag{37.16}$$

This means that there exists an "appearance size" $n_C(z)$, at which the minimum size of multiply charged clusters with charge z is

TABLE 37.6 Appearance Sizes of Multiply Charged van der Waals Clusters Measured in Experiments Together with Those Determined by Calculation (in Parentheses) Using Echt et al.'s (1988) Model

Elements/ Compounds	z (charge)			
	2	3	4	5
Ne	288 (868)	656 (2950)	— (6424)	
Ar	91 (122)	226 (333)	— (648)	
Xe	51 (46)	114 (107)	208 (196)	
CO_2	45 (43)	109 (112)	216 (216)	
C_2H_4	51 (47)	108 (115)	192 (214)	
C_{60}	5 (9) 7*	10(15) (13*)	21 (23) (23*)	33 (32) 31*
$(C_{60})_{n-1} C_{70}$	5	11	22	33

Note: Star indicates calculations involving shell corrections. Experimental data are from Echt et al.'s (1988) paper except for Ne (Mähr et al. 2007), C_{60} and $(C_{60})_{n-1} C_{70}$ (Manil et al. 2003). Calculations are also from Echt et al.'s (1988) except for C_{60} (Nakamura and Hervieux 2006).

stable. Table 37.6 lists experimentally measured appearance sizes for various kinds of clusters. The appearance sizes of the clusters of fullerenes have been measured by Manil et al. (2003). We find that the appearance sizes of multiply charged clusters of fullerenes are much smaller than those of clusters of others elements/compounds listed in Table 37.6. This is because the van der Waals radius and the dielectric constant of fullerene are much larger than those of the other elements/compounds in the table (see also Table 37.1).

The discussion so far has been restricted to symmetric fission with respect to the charge and size. In general, there are many dissociation channels for multiply charged clusters. In most cases, an asymmetric fragmentation with respect to the charge and size has a lower energy barrier than that in symmetric fission. A more rigorous discussion on the stability of multiply charged clusters was presented by Echt et al. (1988). In the model, it is also assumed that the two fragment clusters are in point contact in a transition state and the barrier energy is estimated at such a configuration (see Figure 37.12). The condition for the stability of the parent cluster is that the barrier height is positive, namely, $E_b > 0$, for every possible decay channels. In Table 37.6, we list the appearance sizes for various kinds of clusters: the appearance sizes calculated by using Echt et al.'s model as well as those measured experimentally are listed. We see that the model successfully explains the appearance sizes of multiply charged clusters in most cases. (For the reason for the discrepancy in the case of Ne and Ar clusters, see, e.g., Mähr et al. (2007) and Nakamura (2007).)

37.6.3 Appearance Size of Multiply Charged Clusters of Fullerenes: Results of Theoretical Calculations

The stability of the multiply charged clusters of fullerenes has been dealt with in two different types of studies. The first one is by Nakamura and Hervieux (2006) and their method is essentially an extension of Echt et al.'s (1988) model. In addition to the calculation by the liquid-drop model, the energy barriers are also calculated by replacing the liquid-drop energy $E_{\text{neutral}}(n)$ with the values of accurately calculated by Doye et al. (2001). The shell energies are automatically included in the accurate calculation. By comparing the results of the above two methods, we can determine how the geometrical shell effects influence the fragmentation of clusters.

Figure 37.13 shows the calculated energy barrier for fragmentation from triply charged clusters of fullerene. In this calculation, shell energies are taken into account. Energy barriers for the $(C_{60})_n^{3+} \rightarrow (C_{60})_k^{+} + (C_{60})_{n-k}^{2+}$ reaction at $n = 12$ and 13 are shown as functions of k. For both these cases, the energy barrier for the emission of monomer cluster ion C_{60}^{+} is the lowest. Since the energy barrier for monomer ion emission is positive (negative) for $n = 12$ ($n = 13$), we can conclude that $n = 13$ corresponds to the appearance size, the minimum size at which a triply charged cluster is stable.

In all the cases studied, the emission of the monomer ion C_{60}^{+} is found to have the lowest energy barrier. Figure 37.14 shows the energy barrier for monomer ion emission from $(C_{60})_n$ as function of n. The appearance size can be obtained from this figure as the size that corresponds to the intersection of the solid or

dashed line with the horizontal axis. The curve obtained using the liquid-drop model is smooth as functions of n, while it becomes a zigzag curve if we take into account shell effects. It is found that shell effects may slightly change the appearance sizes, as observed in Table 37.6.

An alternative approach has been employed by Zettergren et al. (2007). Contrary to the previous model in which the charge is assumed to distribute uniformly in the clusters, they assume that the charges localize at specific molecules in the clusters. They also assume that the geometrical structure of clusters follows icosahedral growth. Then, they introduce a charge on molecules so that two charged molecules do not come into contact. They assumed that the energy of the cluster is pairwise additive. The energy between two charged molecules is positive while that of $(C_{60})_2$ and $(C_{60})_2^{+}$ is negative. The appearance sizes are obtained by using the condition that a cluster would be stable if the total attractive energy exceeds the total repulsive energy between charged particles. The results of the calculation are shown in Figure 37.15: the figure also shows experimental measurements and the results of Nakamura and Hervieux (2006). The methods used in the two models are somewhat different. Nevertheless, it is interesting that the two models give nearly the same results and that both of them successfully explain the experimental results, as shown in Figure 37.15.

37.6.4 Dynamics in Hot Clusters of Fullerenes

Another research group is performing experiments to multiply ionize clusters of fullerenes by femtosecond laser irradiation (Hedén et al. 2005). Laser radiation heats the clusters to a quite high temperature. Emission of carbon cluster ion as a result of the dissociation of highly excited fullerene molecules has been observed. Detailed theoretical analysis of the results of this experiment will be a challenge.

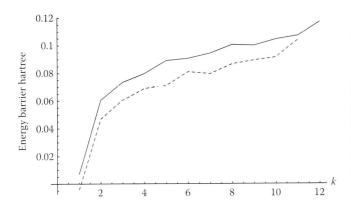

FIGURE 37.13 Energy barriers for the $(C_{60})_n^{3+} \rightarrow (C_{60})_k^{+} + (C_{60})_{n-k}^{2+}$ reaction as functions of k for $n = 12$ (dashed line) and $n = 13$ (solid line).

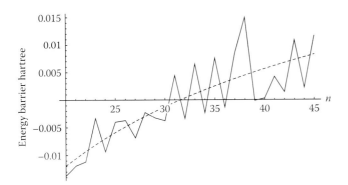

FIGURE 37.14 Energy barrier for the emission of C_{60}^{+} from $(C_{60})_n^{5+}$ as functions the parent size n. Solid line: model including shell effects, dashed lines: liquid-drop model.

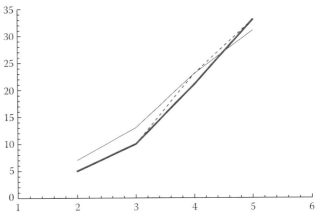

FIGURE 37.15 Appearance sizes for multiply charged clusters of fullerenes as functions of charge z. Experimental values of Manil et al. (2003) (thick line) together with theoretical value of Nakamura and Hervieux (2006) (thin line) and Zettergren et al. (2007) (dashed line) are shown.

37.7 Concluding Remarks and Future Work

Experimental and theoretical studies on clusters of fullerenes have come up with many new findings.

In particular, throughout the discussion on the similarities and differences between rare-gas clusters and clusters of fullerenes, it is apparent that our understanding on the structure of clusters has increased significantly. However, there are still some aspects related to the geometrical structure of clusters of fullerenes at both low and high temperatures that need clarification. Future experimental and theoretical studies should focus on these aspects.

Dynamical processes in highly charged clusters of fullerenes would be another interesting topic for future researches.

Acknowledgments

The author thanks Prof. K. Sattler for the opportunity to write this review and for encouragement. This work has been financially supported by Matsuo Foundation and by a Nihon University Individual Research Grant in 2009.

Appendix 37.A Packing of Spheres and Geometrical Structure of Clusters

To understand geometrical structure of clusters, it is useful to know something about sphere packing. As for the relation between sphere packing and the shell structure in atomic and molecular clusters, an excellent review has been written by Martin (1996). In this appendix, only an outline of sphere packing is provided.

1. Icosahedral packing

   Assume an icosahedron and place a sphere at its center. For packing the first shell, we position spheres at the vertexes of the icosahedron, one at each vertex as shown in Figure 37. A.1. In the second shell, spheres are placed either at vertexes or at the centers of edges. The icosahedral packing is not closed packing. Since the distance between the center of the icosahedron and a vertex is slightly shorter than the length of edges, spheres in the same shell do not come in contact with each other. These gaps between spheres cause the strain energy as is discussed in Section 37.4.3.

2. Decahedral packing

   Another type of sphere packing often observed in clusters of fullerenes corresponds to the decahedral structure and is shown in Figure 37.A.2. The decahedral structure consists of five tetrahedral, with every tetrahedron sharing common edges with the other tetrahedral. Clusters are often observed to have truncated decahedral structures like Marks' decahedron. Both icosahedral and decahedral structures cannot form crystals owing to their fivefold rotational symmetry and the absence of translational symmetry.

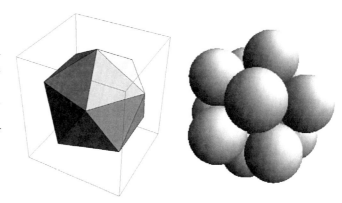

FIGURE 37.A.1   An icosahedron (left) and an icosahedral sphere packing (right).

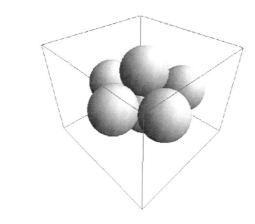

FIGURE 37.A.2   An example of decahedral sphere packing.

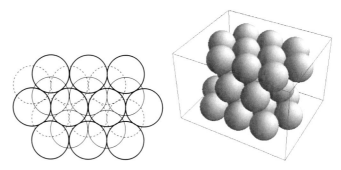

FIGURE 37.A.3   (Left) Honeycomb layers of spheres: A is denoted by thick lines; B by thin lines and C by dashed lines. (Right) An example of closed packing (ABC).

3. Close packing

   Closed-packed structures are possible both for finite and infinite systems and these systems form crystals. Let us assume a layer where spheres are arrayed to form a honeycomb structure. Consider three types of layers, A, B, and C as shown in the left panel of Figure 37.A.3. By overlaying these layers, (we cannot repeat the same type of layers.) we are able to have various types of closed-packed structures. The hexagonal closed-packed (hcp) and face-centered cubic (fcc) structures can be built as

hcp = ABABABA (alternate layers are identical), and
fcc = ABCABCA (every third layer is identical).

Both these structures are known to be the "closest-packed" structure as Kepler had speculated a long time ago (see, e.g., Szpiro 2003). In finite systems such as clusters, periodicity is not necessarily required. Thus nonperiodic structures such as ABCB, are possible.

References

Amano, C., Komuro, M., Mochizuki, S., Urushibara, H., and Yamabuki, H. 2006. *J. Mol. Struct. (THEOCHEM)* 758: 41.

Antoine, R., Dugourd, P., Rayane, D. et al. 1999. *J. Chem. Phys.* 110: 9771.

Axilrod, B. M. and Teller, E. 1948. *J. Chem. Phys.* 11: 299.

Baletto, F., Doye, J. P. K., and Ferrando, R. 2002. *Phys. Rev. Lett.* 88: 075503.

Bohr, N. and Wheeler, J. W. 1939. *Phys. Rev.* 56: 426.

Branz, W., Malinowski, N., Schaber, H., and Martin, T. P. 2000. *Chem. Phys. Lett.* 328: 245.

Branz, W., Malinowski, N., Enders, A., and Martin, T. P. 2002. *Phys. Rev. B* 66: 094107.

Calvayrac, F., Reinhard, P.-G., Suraud, E., and Ullrich, C. A. 2000. *Phys. Rep.* 337: 493.

Cambridge Cluster Database; http://www-wales.ch.cam.ac.uk/CCD.html.

Cheng, A. and Klein, M. L. 1992. *Phys. Rev. B* 45: 1889.

Cheng, L., Cai, W., and Shao, W. 2005. *Chem. Phys. Chem.* 255: 261.

David, W. I. F., Ibberson, R. M, and Matsuo, T. 1993. *Proc. R. Soc. (Lond.) A* 442: 129.

Deleuze, M. S. and Zerbetto, F. 1999. *J. Am. Chem. Soc.* 121: 5281.

Doye, J. P. K. and Wales, D. J. 1996a. *Chem. Phys. Lett.* 262: 167.

Doye, J. P. K and Wales, D. J. 1996b. *J. Phys. B: At. Mol. Opt. Phys.* 29: 4859.

Doye, J. P. K., Wales, D. J., and Berry, R. S. 1995. *J. Chem. Phys.* 103: 4234.

Doye, J. P. K., Dullweber, A., and Wales, D. J. 1997. *Chem. Phys. Lett.* 269: 408.

Doye, J. P. K., Wales, D. J., Branz, W., and Calvo, F. 2001. *Phys. Rev. B* 64: 235409.

Echt, O., Sattler, K., and Recknagel, E. 1981. *Phys. Rev. A* 47: 1121.

Echt, O., Kreisle, D., Recknagel, E., Saenz, J. J., Casero, R., and Soler, J. M. 1988. *Phys. Rev. A* 38, 3236.

Farges, J., de Feraudy, M. F., Raoult, B., and Torchet, G. 1986. *J. Chem. Phys.* 84: 3491.

García-Rodeja, J., Rey, C., and Gallego, L. J. 1997. *Phys. Rev. B* 56: 6466.

Girifalco, L. A. 1991. *J. Phys. Chem.* 95: 5370.

Girifalco, L. A. 1992. *J. Phys. Chem.* 96: 858.

Hansen, K., Hoffmann, K., Mueller R., and Campbell, E. E. B., 1996. *J. Chem. Phys.* 105: 6088.

Hansen, K., Müller, R., Hohmann, H., and Campbell, E. E. B. 1997. *Z. Phys. D: At. Mol. Clusters* 40: 361.

Hedén, M., Hansen, K., and Campbell, E. E. B. 2005. *Phys. Rev. A* 71: 055201.

Holland, J. H. 1975. *Adaption in Natural and Artificial Systems.* Ann Arbor, MI: The University of Michigan Press.

Kirkpatrick, S., Gelatt C. D., and Vecchi, M. P. 1983. *Science* 220: 671.

Krätschmer, W., Fostiropoulos, K., and Huffman, D. R. 1990. *Nature* 347: 354.

Kroto, H. W., Heath, J. R., O'Brien, S. C., Curl, R. F., and Smalley, R. E., 1985. *Nature* 318: 162.

Leary, R. H. and Doye, J. P. K. 1999. *Phys. Rev. E* 60: R6320.

Mackay, A. L. 1962. *Acta Crystallogr.* 15: 916.

Mähr, I., Zappa, F., Denifl, S. et al. 2007. *Phys. Rev. Lett.* 98: 023401.

Manil, B., Maunoury, L., Huber, B. A. et al. 2003. *Phys. Rev. Lett.* 91: 215504.

Martin, T. P. 1996. *Phys. Rep.* 273: 199.

Martin, T. P., Näher, U., Schaber, H., and Zimmermann, U. 1993. *Phys. Rev. Lett.* 70: 3079.

Marks, L. D. 1984. *Philos. Mag. A* 49: 81.

Meihle, W., Kandler, O., Leisner, T., and Echt, O. 1989. *J. Chem. Phys.* 91: 5940.

Nakamura, M. 2007. *Chem. Phys. Lett.* 449: 1.

Nakamura, M. and Hervieux, P.-A. 2006. *Chem. Phys. Lett.* 428: 138.

Niehaus, A. 1986. *J. Phys. B: At. Mol. Phys.* 19: 2925.

Northby, J. A. 1987. *J. Chem. Phys.* 87: 6166.

Pacheco, J. M. and Prates-Ramalho, J. P. 1997. *Phys. Rev. Lett.* 79: 3873.

Prassides, K. 1993. *Phys. Scr. T* 49: 735.

Rey, C., Gallego, L. J., and Alonso, J. A. 1994. *Phys. Rev. B,* 49: 8491.

Rey, C., García-Rodeja, J., and Gallego, L. J. 2004. *Phys. Rev. B* 69: 073404.

Saito, S. and Oshiyama, A. 1991. *Phys. Rev. Lett.* 66: 2637.

Shao, X. G., Cheng, L. J., and Cai, W. S. J. 2004. *Comput. Chem.* 25: 1693.

Szpiro, G. C. 2003. *Kepler Conjecture.* Hoboken, NJ: John Wiley & Sons.

Tournus, F., Chantier, J. C., and Mélinion, P. 2005. *J. Chem. Phys.* 122: 094315.

Wales, D. J. 1994. *J. Chem. Soc. Faraday Trans.* 90: 1061.

Wales, D. J. and Doye, J. P. K. 1997. *J. Chem. Phys. A* 101: 5111.

Weeks, D. E. and Harter, W. G. 1989. *J. Chem. Phys.* 90: 4744.

Zettergren, H., Schmidt, H. T., Cederquist, H. et al. 2002. *Phys. Rev. A* 66: 032710.

Zettergren, H., Schmidt, H. T., Reinhed, P. et al. 2007. *J. Chem. Phys.* 126: 224303.

38

Supramolecular Assemblies of Fullerenes

Takashi Nakanishi
*National Institute for
Materials Science*

and

*Max Planck Institute of
Colloids and Interfaces*

and

Japan Science and Technology

Yanfei Shen
*Max Planck Institute of
Colloids and Interfaces*

Jiaobing Wang
*Max Planck Institute of
Colloids and Interfaces*

38.1 Introduction

The morphology control of organic nano-materials has recently received considerable attention because the properties of organic soft materials can be enhanced by fine-tuning the intermolecular interactions, and many possible future applications such as field emission transistors (FET), organic light emitting diodes (OLED), and organic solar cells are desired by applying the photoelectric functions of π-conjugated molecular matters [1–3]. At present, in most cases, conductive or semiconductive organic molecules or polymers are used as layered amorphous films or bulk solids in these systems; however, in the last few years, alternative strategies for molecular fabrications are consider such that designed organic molecules can be hierarchically assembled into desired morphologies using a principle of "supramolecular chemistry." Supramolecular chemistry, defined by Lehn as the "chemistry beyond the molecule," provides an unprecedented control over the physicochemical properties of architectures through the use of programmed molecular assemblies that emerge by spontaneous self-assembly to hierarchical molecular architectures via weak interactions (e.g., van der Waals, π–π, hydrogen bonding) [4,5]. Therefore, supramolecular architectures with controlled structures and morphologies have been receiving great interest for establishing organic soft materials possessing versatile functions (e.g., optical, electrical, and electrochemical properties) [6,7].

Nanocarbon clusters such as carbon nanotubes and fullerenes can be the candidates to satisfy above-mentioned processability, intrinsic photoelectronic functionalities, and, of course, π-conjugated systems (materials). Fullerene (C_{60}) and the derivatives are known as one kind of promising organic materials because of the electron-transport property (n-type organic semiconductor) [8,9], rich electrochemistry [10,11], photoresponsibility [12,13], and bio-activity [14,15]. Especially, only a few researches have been successfully conducted in the preparation of fullerene assemblies, although the synthesis of fullerene derivatives, including their conjugate molecules with the other functional unit, are well developed [16–20]. Therefore, the preparation of practical materials having controlled morphologies from these functional fullerene derivatives is now highly awaited and would become a key development in the fabrications of organic soft materials.

In this chapter, we first describe recent developments in the formation of the single-crystalline assemblies of pristine C_{60} and the supramolecular composites of C_{60} in host substances. After that, our approaches to the fabrication of supramolecular fullerene derivatives are described. A particular importance in the supramolecular organization of fullerene derivatives is their molecular designs. We introduce a new concept of amphiphilicity and demonstrate that a simple fullerene derivative can be assembled into various morphologies. For controlling morphologies, assemblies on solid surfaces are described first and from various solvent conditions and this is followed by some functions obtained from their assemblies. Other topics are the development of fluid materials such as thermotropic liquid crystals and room temperature liquids based on fullerene derivatives.

38.2 Pristine C_{60} Assemblies

The development of high-quality C_{60} films is one of central interest. For example, the conventional Langmuir–Blodgett (LB) technique was often applied to prepare well-ordered two-dimensional (2D) C_{60} thin films [21,22]. However, because of the intrinsic hydrophobic property of C_{60} surface, the Langmuir films on air–water interface are easily transformed to aggregated multilayers. In order to improve the quality of the 2D films of C_{60}, for example, Kunitake et al. proposed an electrochemical replacement method to form epitaxial adlayers of C_{60} (with high quality) on Au(111) surfaces from C_{60} on I/Au(111) surfaces [23]. The process consists of the transfer of Langmuir films of C_{60} onto iodine-modified Au(111) surfaces at an air–water interface, followed by the electrochemical removal and replacement of iodine adlayers with C_{60} adlayers in solution (Figure 38.1). The C_{60} adlayers prepared by this method showed excellent quality and uniformity and had higher quality than that for the direct transfer

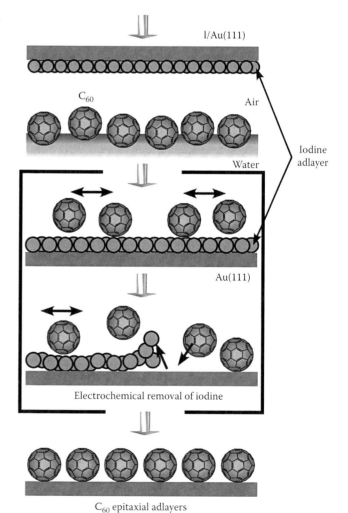

FIGURE 38.1 Schematic representation of the electrochemical replacement method for the formation of C_{60} adlayer on a Au(111) electrode. (Reprinted from Uemura, S. et al., *Langmuir*, 17, 5, 2001. With permission.)

of C_{60} in Au(111) surface. High quality and various shapes of C_{60} assemblies can also be achieved by crystalline materials, as described in the following section.

38.2.1 Crystalline Materials

Miyazawa et al. have developed several methodologies to prepare C_{60} materials as whiskers, tubes, and hexagonal sheets [24–26], using a liquid–liquid interfacial precipitation method. For example, the fine needlelike crystals of C_{60} have been prepared by this method, which uses an interface of concentrated toluene solution of C_{60} and 2-propanol solvent [24]. The needlelike crystals of C_{60} with a diameter of submicrometers, namely, C_{60} nanowhiskers, were found to be single crystalline and composed of thin slabs with thickness of about 10 nm (Figure 38.2). The intermolecular distance of the C_{60} nanowhiskers determined by high-resolution transmission electron microscopy (HR-TEM) was found to be shortened along the growth axis, as compared with the pristine C_{60} crystals, indicating a formation of strong interaction between neighboring C_{60} molecules. Young's modulus of the C_{60} nanowhiskers with 130 and 160 nm in diameters were estimated to be 54 and 32 GPa, which is 160%–650% that of C_{60} bulk crystals [27].

Monodispersed C_{60} micro/nanocrystals with unique morphology structures can be fabricated by solvent-induced conditions. Masuhara and coworkers established a facile procedure, solvent-induced reprecipitation method [28], using *m*-xylene and 2-propanol as good and poor solvents for C_{60}, respectively. The typical procedure is as follows. The *m*-xylene solution of C_{60} was quickly injected into vigorously stirred 2-propanol as the poor medium. Afterward, the dispersion liquid was left at room temperature for several hours to grow the crystals. The resulting micro/nanocrystals have a hexagonal crystalline structure and show various morphologies such as rodlike, cubic, branched rods, and six-faced prism objects. The crystalline structures solvate a C_{60}/*m*-xylene molar ratio = 3:2. The surfactant- [29] or polymer-assisted [30] self-assemblies of C_{60} also produce various morphological C_{60} micro/nanocrystals such as rod, tube, and fence-type shapes. Such matrix-assisted growth mechanism of C_{60} crystalline materials based on concentration profile

FIGURE 38.2 SEM image of C_{60} nanowhiskers. (Courtesy of Dr. Kun'ichi Miyazawa, The National Institute for Materials Science, Tsukuba City, Japan.)

controlled growth was proposed. The resulting shapes of crystalline materials depend on where concentration depletion happens during the crystalline growth of C_{60} crystal seeds.

38.2.2 Supramolecular Composites

Hydrophobic and π-conjugated molecule-rich environments provide appropriate host cavities for the encapsulation of C_{60} [31–33]. Most of cases in the multiencapsulation of C_{60} in the host environments do not cause the structural deformation of the original morphologies of host architectures, and some cases improve their electronic properties. For instance, large π-extended tetrathiafulvalene-based dendrimers host a number of C_{60} molecules to form segregated arrays of donor and acceptor units, which could give rise to valuable materials useful for the preparation of optoelectronic devices (Figure 38.3) [34]. Hydrophobic nature inside polymers also acts as good host environment for the encapsulation of C_{60} molecules [35], and the C_{60}-encapsulating structures are readily replaced to appropriate pair guest polymers. Optically active poly (methyl methacrylate) (PMMA) stereocomplexes were prepared through the helix-sense-controlled supramolecular inclusion of an isotropic *it*-PMMA within the helical cavity of an optically active, C_{60}-encapsulated syndiotactic *st*-PMMA with a macromolecular helicity memory. The *it*-PMMA replaced the encapsulated C_{60} to fold into double-stranded helix with same handedness as that of the *st*-PMMA single helix through the formation of a topological triple-strand helix [36].

38.3 Assemblies at Solid Surfaces

In this section, examples on the preparation of self-assembled structures of C_{60} and the derivatives at solid surfaces are introduced. The fine-tuning of interfacial and intermolecular interactions permits building well-defined, ordered nanostructures over large area, which have been visualized, characterized, and manipulated by scanning probe microscopy techniques [37,38]. While the 2D arrangement of C_{60} and the derivatives on surfaces is well established [39–41], the alignment of fullerenes in 1D architectures is far less common [42,43]. Nakanishi et al. introduced long alkyl chains covalently attached on C_{60} to form self-assembled C_{60} moieties as 1D nanowires in a predictable way on an appropriate substrate [44,45]. Molecules bearing long alkyl chains self-assemble at highly oriented pyrolytic graphite (HOPG), giving rise to close to perfectly ordered lamellar structures. Designed fullerene derivatives bearing three hexadecyloxy chains (**4**) (Figure 38.4a) should induce epitaxial assembly on HOPG in a facile spin-coating method, resulting in 1D fullerene wire. An atomic force microscopy (AFM) image of the surface demonstrated a well-aligned fullerene array (Figure 38.4b), and detailed observation by scanning tunneling microscopy (STM) directly confirmed zig-zag arrays of the C_{60} spheres (Figure 38.4c) [44]. The spacing between lamellae (nanowires) are determined by the alkyl chain length and numbers, for instance, 7.2 nm of **1**, 8.1 nm of **2**, and 6.3 nm of **4**, respectively

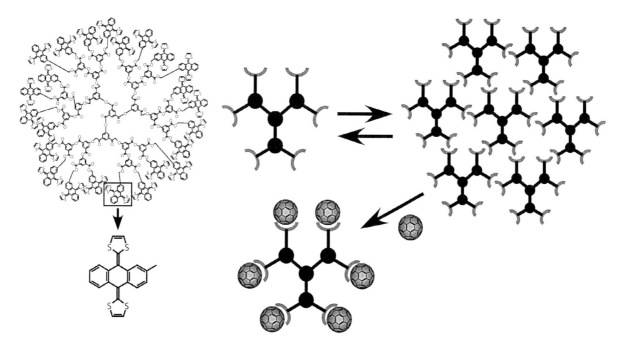

FIGURE 38.3 Large π-extended tetrathiafulvalene-based dendrimer that acts as a host molecule for the peripheral cooperative multiencapsulation of C_{60}.

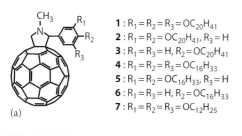

$1 : R_1 = R_2 = R_3 = OC_{20}H_{41}$
$2 : R_1 = R_2 = OC_{20}H_{41}, R_3 = H$
$3 : R_1 = R_3 = H, R_2 = OC_{20}H_{41}$
$4 : R_1 = R_2 = R_3 = OC_{16}H_{33}$
$5 : R_1 = R_2 = OC_{16}H_{33}, R_3 = H$
$6 : R_1 = R_3 = H, R_2 = OC_{16}H_{33}$
$7 : R_1 = R_2 = R_3 = OC_{12}H_{25}$

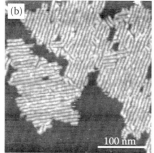
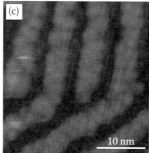

FIGURE 38.4 (a) Chemical structures of C_{60} derivatives bearing long aliphatic chains (**1–7**). (b) AFM and (c) STM images of fullerene nanowires of the derivative (**4**) on HOPG formed by means of spin-coating from the chloroform solution. (Reprinted from Nakanishi, T. et al., *J. Am. Chem. Soc.*, 128, 6328, 2006. With permission.)

[44,45]. This work also indicates that the substitution pattern and the chain length of the alkyl chains determine the order and packing motive. Electrochemical activities were also confirmed in the fullerene-modified HOPG substrates used as a working electrode. Therefore, it would open the way to address individual C_{60} and to investigate the conductivity of individual nanowires.

Two amide moieties incorporated between the C_{60} moiety and 3,4,5-tris(dodecyloxy)benzene substituent (similar to the molecular structure of **7**) (Figure 38.5a) induce the gelation of the molecule in organic solvents such as chloroform and the formation of nanowire structures on the application of the LB method [46]. Monolayer fibrous structures, involving the intermolecular hydrogen bonding of the amides and π–π interactions between the C_{60} moieties, of 1.2 nm in height, 5–10 μm in length, and 8 nm in width, were obtained by the LB method and visualized at solid substrate by AFM technique. At the beginning, monolayer fibrous structures formed by homogeneous nucleation and quasi-1D interface-controlled growth from molecules in the homogeneous monolayer. After that, bilayer fibrous structures formed by the assembly of the monolayer fibrous structures through heterogeneous nucleation and quasi-1D diffusion-controlled growth. The use of organogelators containing C_{60} and the LB method are proposed for the development of strategies of the fabrication of self-assembling fullerene nanowires.

A covalently linked phthalocyanine-C_{60} conjugate (C_{60}-Pc) (Figure 38.5b) is also able to self-organize on graphite and graphite-like (graphite oxide flakes) surfaces by simple drop-casting method, thus giving rise to supramolecular fibers and films [47]. These nanostructures are electrically characterized by conductive atomic force microscopy (C-AFM) technique, and show outstanding out-of-plane and in-plane electrical conductivity values that reflect an extremely high degree of the molecular order of phthalocyanine-C_{60} conjugate within the nanostructures (Figure 38.5c). The preparation of highly conductive, facile-to-assemble supramolecular nanostructures in which photo- and redox-active units, in this case phthalocyanine and

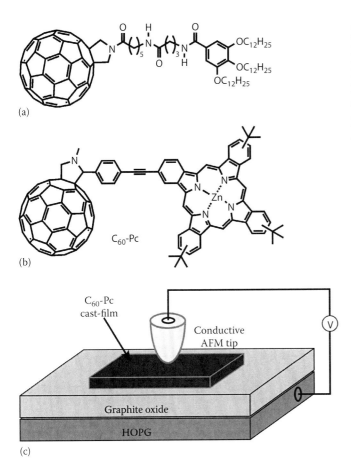

(a)

C₆₀-Pc

(b)

(c)

FIGURE 38.5 (a) Fullerene-linked 3,4,5-tris(dodecyloxy)benzamide derivative that assembles into nanowires structures on the application of the Langmuir-Blodgett method. (b) Fullerene-linked phthalocyanine derivative (C_{60}-Pc) that forms highly conductive supramolecular nanostructures such as nanowires and sheets. (c) Schematic representation of the conductive AFM (C-AFM) experiment carried out in sheet objects of the C_{60}-Pc at micrometer-sized graphite oxide flakes deposited on a HOPG substrate. (Reprinted from Bottari, G. et al., *Angew. Chem. Int. Ed.*, 47, 2026, 2008. With permission.)

C_{60} which are excellent electron donor and acceptor moieties, respectively, are molecularly assembled on graphite surfaces is extremely promising. The technique leads to molecular materials with possible applications in the field of nano-optoelectronics and photovoltaics.

38.4 Supramolecular Assemblies of C₆₀ Derivatives

With a view to the regulation of fullerene's assembled superstructures, conventional amphiphilic approaches, which are based on the substitution of hydrophilic moieties at the intrinsically hydrophobic C_{60} originated from the strong π–π interaction, have been applied in many cases. For instance, in the simplest cases, C_{60} derivatives modified with cationic (Figure 38.6a) or anionic species (Figure 38.6b) prefer to form molecular bilayer arrangements as structural subunit (Figure 38.6c) and tend to

develop spherical vesicles with inner empty core in water [48,49]. Fullerene derivatives carrying long alkyl chains and ammonium head groups also formed into vesicle or fibrous architectures with bilayer arrangements (Figure 38.7) [50,51]. However, controlling the shape of the assemblies showing polymorphism is not well-established in these conventional amphiphilic approaches. Herein, in this section, the author's recent studies related development of novel-type fullerene supramolecular assemblies with controlled dimensionality using the derivatives bearing long alkyl chains are described.

38.4.1 Morphology Control

Fulleropyrrolidines functionalized with a multi(alkyloxy) phenyl group (Figure 38.4a) satisfy some of requirements to be soft materials such as a high C_{60} content in the molecule, a predictable compact packing, a highly ordered mesophase, and controllable dimensionality of the self-organized structures [44,45,52–57]. Moreover, fulleropyrrolidines form stable species under the electrochemical reduction of the C_{60} cage [58,59], and thus as redox-active soft materials. The developed supramolecular fullerene nano- and micro-architectures are utilizing the intermolecular forces introduced by C_{60} (π–π) and alkyl chain interaction (van der Waals). A delicate balance between the π–π and van der Waals interactions in the assemblies leads to a wide variety of supramolecular architectures.

Fullerene derivative (**4**), a fulleropyrrolidines functionalized with a 3,4,5-(hexadecyloxy)phenyl group, showed polymorphism [52–54], and different well-defined self-organized superstructures in various solvent systems (Figure 38.8). The sample preparation is quite simple as follows. Self-assembled supramolecular objects were prepared by evaporation to dryness of a 1 mL chloroform solution of **4** ([**4**] = 1.0 mM), followed by the addition of 1 mL of the respective solvents. Subsequent heating 60°C–70°C for 2 h resulted in light brown mixtures, which were aged at 20°C for 24 h, prior to microscopy examination. Solvents of variable polarity used in these experiments included 2-propanol/toluene mixtures, 1-propanol, H_2O/THF mixture, 1-butanol, 1,4-dioxane, and enantiopure 2-(S)- or (R)-butanol. Mixtures in a 2-propanol/toluene solution were subsequently cooled to 5°C for at least 12 h, and ultrasonication was performed at 5°C for 1 h. 1,4-Dioxane solution of **4** was also subsequently cooled to 5°C for 12 h, resulting in the formation of dark-brown-colored precipitates. As shown, field emission scanning electron microscopic (SEM) images in Figure 38.8, **4** self-assembles into hierarchically ordered nano- and micro-superstructures, such as vesicles with an average diameter of 250 nm in 1:1 2-propanol/toluene mixture at room temperature (Figure 38.8a), fibers with partially twisted tapes in 1-propanol (Figure 38.8b), conical objects with diameter of 60 nm perforated at the cone apex in 1:1 H_2O/THF mixture (Figure 38.8c), microspheres with diameters of 1–2 μm (Figure 38.8d), and maracas-like (Figure 38.8e) under ultrasonication at 5°C for 1 and 0.5 h, respectively. Windmill-like sheets architectures were obtained in a 1:2 2-propanol/toluene mixture at 20°C as brown precipitates (Figure 38.8f). The uncommon self-organized

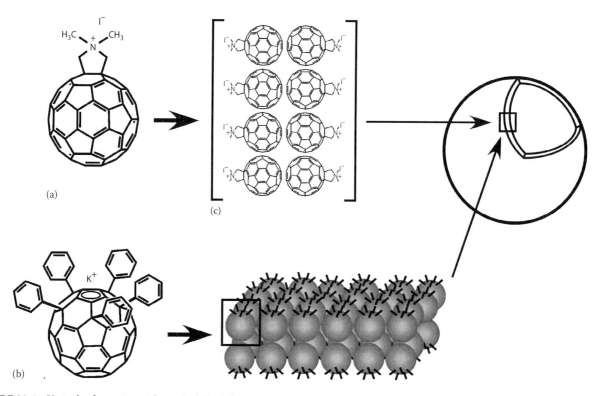

(a)

(c)

(b)

FIGURE 38.6 Vesicular formation with amphiphilic fullerene derivatives (a) cationic and (b) anionic molecules which form bilayer structure as a fundamental subunit (c).

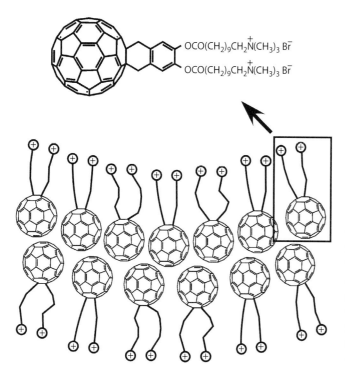

FIGURE 38.7 Chemical structure of an amphiphilic fullerene derivative with two ammonium head group and schematic drawing showing the bilayer vesicles. The double alkyl chains are disordered to expose the C_{60} moieties to water molecules, producing the hydrophobic force. (Reprinted from Sano, M. et al., *Langmuir*, 16, 3773, 2000. With permission.)

superstructures of **4**, jellyfish-like, are seen from objects prepared in 1-butanol at 20°C (Figure 38.8g). Furthermore the left-handed spiral microstructures were obtained from 2-(R)-butanol (Figure 38.8h), whereas 2-(S)-butanol provided right-handed spiral objects (Figure 38.8i) of 3–6 μm in diameters.

Even flat disk-shape objects are one of the attractive morphologies in surfactant assemblies [60]; normally, 2D planar sheet assemblies turn into tubular assemblies with cigar-like scrolls via rolling up of the sheet in order to minimize the total elastic energy [61]. In our current study, however, planar windmill-like sheet assemblies overlie each other without scrolling the sheets, and also the spiral objects are originating from disk-shaped precursor assemblies. Further studies for understanding the formation mechanism, including theory, of the 2D assemblies and the reasons of the tetra-clefts formation, the anisotropic growth, as well as of the influence of the chiral solvent at the molecular level on the microscopic spiral morphogenesis are still required. The slight differences of the aliphatic chain length and numbers of the derivatives cause great morphology changes in their supramolecular assemblies. Remarkably, these reference derivatives **5–7** [54,55] did not show polymorphs in various organic solvents, as seen in **4** [52–54].

To further diversify the library of self-organized fullerene superstructures, for instance, with fractal shape, the understanding of the formation mechanism of self-organized microscopic superstructure becomes more interesting. In 1,4-dioxane solution of **4** at 20°C, disks with diameters of 0.2–1.5 μm (Figure 38.9a) and 4.4 nm thickness (Figure 38.9b), confirmed by AFM,

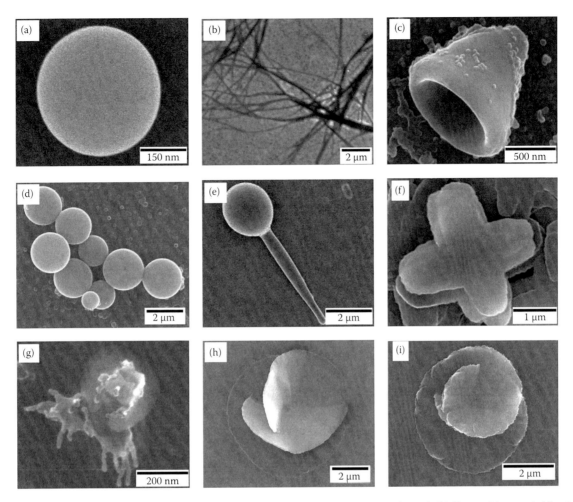

FIGURE 38.8 SEM images of supramolecular assemblies of the fullerene derivative (**4**). (a) Spherical, (b) fibrous, (c) conical, (d) micrometer-sized spherical, (e) maracas-like, (f) windmill-like, (g) jellyfish-like, (h) left-handed, and (i) right-handed spiral objects obtained from various solvent conditions. Detailed conditions are described in the main text. (Reprinted from Nakanishi, T. et al., *Chem. Commun.*, 48, 5982, 2005; *Small*, 3, 2019, 2007. With permission.)

were obtained [52]. The thickness is in good agreement with the thickness of alkyl chain-interdigitated bilayer (see schematic model in Figure 38.9c). In fact, structural analyses of the assembled superstructures of **4** by cryo-TEM, DSC, FT-IR, UV-Vis spectroscopies and X-ray diffraction (XRD) indicated that multi-bilayer organized structures with the layer distance around 4.4 nm are contributed to the supramolecular assemblies of **4**. In other words, the hierarchical organization of **4** consists of interdigitated bilayer-based nanostructures, which further assemble into microscopic superstructures such as flowerlike objects [53] (Figure 38.9f). The flowerlike objects were obtained as precipitates after cooling down the 1,4-dioxane solution from 60°C to 20°C followed by cooling to 5°C, which were several micrometers in size (3–10 μm) with crumpled sheet- or flakelike nanostructures with several tens of nanometers in thickness.

The direct observation of the intermediate structures transforming from nano objects to microscopic objects helps to acquire a deeper understanding of the formation mechanism

[53]. The rapid cooling of the homogeneous 1,4-dioxane solution of **4** from 60°C to 5°C yields the intermediate assembled structures of the flowerlike objects. The preformed disk objects seen at 20°C loosely roll up at the edges (Figure 38.9d). The rolling distortions take place in every quarter of the disk, resulting in square-shaped objects having four corners with conical shapes developed by the encounters of the rolled or folded edges on the disks. Such bending of a thin sheet is, in general, energetically more favorable than stretching it. When the rolling-up proceeds continuously, spatial congestions occur at the four corners, resulting in crumpling, bending, stretching, and fracture of the disks (Figure 38.9e). After these transformations are complete, the bilayer growth at the edges continues, fixing the spatial conformation of the crumpled sheets, and this transformation leads to the final form of the flower-shaped superstructures (Figure 38.9f). These results indicate the transformation (shape-shift) mechanism (Figure 38.9g) from molecular assembled bilayer disks into microscopic flower-shaped superstructures based on fullerene derivative **4**. Importantly, the ability to self-organize

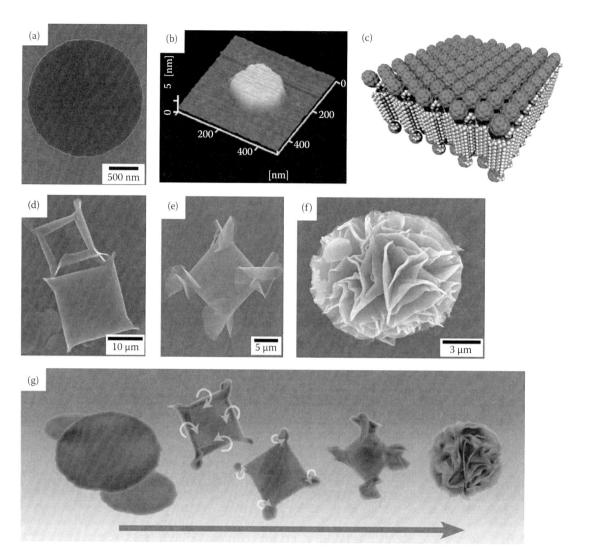

FIGURE 38.9 (a) SEM and (b) AFM images of disk-shaped assemblies of **4** formed in 1,4-dioxane as a precursor for the flower formation. (c) The proposed structural model of bilayer assembly with interdigitated alkyl chains. SEM images of (d) square-shaped objects loosely rolled-up in every corner, (e) the further rolled-up objects of crumpled structures at the four corners, and (f) the final flower-shaped object, (g) schematic representation of the formation mechanism of the flower-shaped supramolecular assembly. (Reprinted from Nakanishi, T. et al., *Small*, 3, 2019, 2007. With permission.)

well-defined, discrete microscopic objects with complex morphologies may stimulate further advances in component design and self-assembly theory that go beyond simple morphologies, such as micelles, vesicles, and tubes.

38.4.2 Superhydrophobic Surfaces

The molecular self-organization of a fullerene derivative (**1**), a fulleropyrrolidine functionalized with a 3,4,5-(eicosyloxy) phenyl group, leaded to spontaneously self-assemble into the globular objects of macroscopic dimensions with a wrinkled flake-like sub-microstructure at the outer surface (Figure 38.10a) precipitated from 1,4-dioxane solution after heating and subsequently cooling to room temperature [56]. Thin films of these globular objects have a fractal surface reminiscent of plant leaves [62,63] and feature significantly improved water-repellent

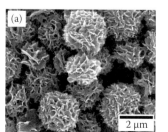

FIGURE 38.10 (a) SEM image of globular assemblies of the fullerene derivative (**1**) precipitated from 1,4-dioxane by heating and cooling the solution. (b) A photograph of water droplet on the surface (contact angle of 152°). (Reprinted from Nakanishi, T. et al., *Adv. Mater.*, 20, 443, 2008. With permission.)

superhydrophobicity with a water contact angle of 152° (Figure 38.10b), as compared to simply spin-coated films (water contact angle ~103°) [56]. The superhydrophobic film was surprisingly durable toward polar organic solvents, and acidic/basic aqueous media, as well as heating (100°C) [56]. The durability of the fabricated structures is related to the apolar nature of the constituents in combination with strong π–π and van der Waals interactions of C_{60} moieties and the long alkyl chains.

It is also the fractal morphology and two-tier roughness of the globular objects of **1**, originating from the hierarchical self-organization starting from the bilayer formations of **1**, which leads to the superhydrophobicity of the resulting surfaces. In contrast to most other methods used for the preparation of superhydrophobic surfaces [64], the prepared thin films described above are readily reused or recovered by simply dissolving them in chloroform. This is an advantage for superhydrophobic surfaces created from the hierarchical supramolecular assemblies of the fullerene derivative.

38.5 Liquid Crystalline Assemblies

Fullerene C_{60} is one kind of promising electron-transport material as an *n*-type organic semiconductor [8,9]. The basic requirements to achieve high carrier mobilities in liquid crystals (LC), which is better for practical devices rather than crystalline materials, involve close molecular packing, highly ordered structures, and extended π-conjugation, leading to semiconductor properties. Although the various morphologies of C_{60}-containing LC such as nematic, smectic, and columnar phases have been the subject of studies in the past decade [65–67], there is only an example concerning the carrier mobility of fullerenes in the LC state which has been evaluated very recently [68]. However, the conducted value, ~10^{-4} cm²/V s, estimated by flash photolysis time-resolved microwave conductivity (FP-TRMC) was not large enough for charge transporting materials, probably due to the moderate ordered structure and the low C_{60} content in the mesophase. Our fullerene derivatives satisfy the requirements for a high carrier mobility in the C_{60}-containing mesomorphic materials: a high C_{60} content up to 50% (in the case of **2**) and a highly ordered mesophase.

The thermal properties of **1** have been examined by a combination of differential scanning calorimetry (DSC), polarized optical microscopy (POM), and x-ray diffraction (XRD). The thermotropic mesophase of **1** is seen in temperature range between 62°C and 193°C, and shows optical texture under polarized optical microscope (Figure 38.11a), which exhibits birefringence and confirms the fluid nature. The XRD pattern for **1** at 185°C shows a strong peak at 2θ = 1.58° assigned (0 0 1), *d* spacing = 5.59 nm, accompanied by higher-ordered peaks up to (0 0 14) (Figure 38.11b). In addition, a broad halo centered at 2θ ≈ 19° is assigned to the molten alkyl chains. These higher-degree peaks in the XRD pattern reveal a long-range-ordered lamellar mesophase comparable to ordered smectic phase. The derivatives **2** and **4** also showed similar LC characteristics having long-range-ordered lamellar mesophase, while the

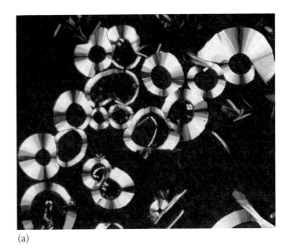

(a)

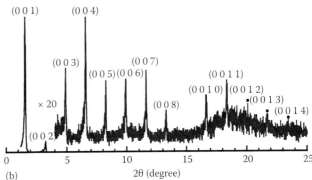

(b)

FIGURE 38.11 (a) POM texture of mesophase of **1** at 190°C on cooling from the isotropic phase and (b) XRD patterns of **1** at 185°C indicates the formation of highly ordered lamellar organization of **1**. (Reprinted from Nakanishi, T. et al., *J. Am. Chem. Soc.*, 130, 9236, 2008. With permission.)

derivatives **3**, **5**, **6**, **7** were found to be non-mesomorphic. The mesomorphic fullerenes feature reversible redox activity ($E_{red,1}$ = −0.70 and $E_{red,2}$ = −0.87 V) and a comparably high electron carrier mobility, ~3 × 10^{-3} cm²/V s for **1** at 120°C evaluated by a conventional time-of-flights setup (ToF), making them attractive components for fullerene-based soft materials [57].

38.6 Fluid Fullerenes

In the course of above studies, it was discovered serendipitously that fulleropyrrolidines substituted with a 2,4,6-tris(alkyloxy) phenyl group (**8–10**) (Figure 38.12a) exhibit a fluid phase at room temperature (Figure 38.12b) [69]. The key point is to select, as the substituent group, a structure which is molecularly designed in such a way that the alkyl chains spread independently, the aggregation of the fullerene moiety is skillfully suppressed, and only a mono-substituent moiety modifies onto C_{60} surface to avoid damaging the electronic features of C_{60} by the multiple additions of substitutions. For derivatives **8–10**, over the measured frequency range of the rheological behavior, the loss modules *G″* is higher than the storage modules *G′*, and this

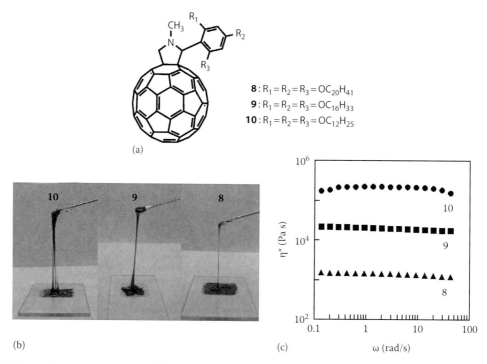

8 : $R_1 = R_2 = R_3 = OC_{20}H_{41}$
9 : $R_1 = R_2 = R_3 = OC_{16}H_{33}$
10 : $R_1 = R_2 = R_3 = OC_{12}H_{25}$

FIGURE 38.12 (a) Chemical structures (**8–10**) and (b) photographs of room temperature liquid fullerenes. (c) Results of rheology (complex viscosity) experiments of **8–10** at a strain amplitude of 0.01, indicating viscosities decrease with increasing the alkyl chain length. (Reprinted from Nakanishi, T. et al., *J. Am. Chem. Soc.*, 128, 10384, 2006. With permission.)

suggests their liquid-like behavior. As a result of a systematic investigation of the fluid behavior of liquid fullerenes and their functions, the fact that the viscosity of the liquid can be controlled by changing the length of the introduced alkyl chain has been discovered (Figure 38.12c). Interestingly, the greater alkyl chain length reduces efficiently the viscosity, which is a completely opposite trend of a phenomenon of viscosity in normal alkane molecules.

Moreover, because the liquid fullerenes retain the characteristic properties of fullerenes, the liquid is redox-active. Cast films of the liquid fullerenes showed two reductive events of C_{60}. As a further advantage of the liquid, this material also has relatively high hole mobility estimated from transient photocurrent ToF measurements, $\sim 3 \times 10^{-2}$ cm^2/V s for **8** at 20°C [69]. These features make it an extremely attractive nano-carbon material for future applications.

38.7 Future Perspective

By simple molecular design with a novel concept of amphiphilicity, the combination of C_{60} (π–π interaction) and long alkyl chains (van der Waals interaction), dimensionally regulated supramolecular assemblies were successfully prepared and enable us to construct various organic soft materials (superhydrophobic surfaces, thermotropic liquid crystals, room temperature liquid fullerenes, etc.) based on fullerenes with unique morphologies and the electronic functions. The systems described here open the door for novel methodologies developing supramolecular

organic soft materials, not only fullerenes, with controlled dimensionality and the desired photoelectronic functionalities. The application of these materials in fuel cells, field emission displays, polymer adductive, MEMS/NEMS, self-cleaning surfaces, and functional parts of nano-sized electronic devices, such as conductive nanowires, are also highly expected. In addition, the ability to self-organize well-defined, discrete microscopic assemblies with complex morphologies may stimulate further advances in component design and self-assembly theory that go beyond simple morphologies, such as micelles, vesicles, and tubes. In the nanoscience and nanotechnology of fullerenes, great efforts to synthesize functional derivatives and prepare dimensional-controlled supramolecular assemblies have so far established very efficiently. In the further step, the proper control of their organizations and fabrications on suitable positions will become more important for developing realistic molecular devices and functional soft materials.

Acknowledgments

This work was supported, in part, by a Grant-in-Aid for "Young Scientists (B) (No. 17750140 and 19750124)" from the Ministry of Education, Sciences, Sport and Culture, Japan; "PRESTO," JST, Japan; Iketani Science and Technology Foundations (171041A and 191127A); and Ishikawa Carbon Foundation. We are grateful to Prof. H. Möhwald; Dr. D. Kurth; Dr. N. Miyashita, MPI Colloids and Interfaces; Dr. K. Ariga; Dr. T. Michinobu, NIMS; Dr. H. Takahashi; Prof. T. Teranishi, University of Tsukuba;

Dr. K. Yoshida, Kyoto University; and Dr. M. Funahashi, Tokyo University for helpful discussion and some of experimental contributions.

References

1. Martín, N., Sánchez, L., Herranz, M. Á., Illescas, B., Guldi, D. M. 2007. Electronic communication in tetrathiafulvalene (TTF)/C_{60} systems: Toward molecular solar energy conversion materials? *Acc. Chem. Res.* 40: 1015–1024.

2. Hutchison, K., Gao, J., Schick, G., Rubin, Y., Wudl, F. 1999. Bucky light bulbs: White light electroluminescence from a fluorescent C_{60} adduct-single layer organic LED. *J. Am. Chem. Soc.* 121: 5611–5612.

3. Funahashi, M., Zhang, F. P., Tamaoki, N. 2007. High ambipolar mobility in a highly ordered smectic phase of a dialkylphenylterthiophene derivative that can be applied to solution-processed organic field-effect transistors. *Adv. Mater.* 19: 353–358.

4. Lehn, J. M. 1978. Cryptates—Inclusion complexes of macropolycyclic receptor molecules. *Pure Appl. Chem.* 50: 871–892.

5. Ajayaghosh, A., Praveen, V. K. 2007. π-Organogels of self-assembled *p*-phenylenevinylenes: Soft materials with distinct size, shape, and functions. *Acc. Chem. Res.* 40: 644–656.

6. Yamamoto, Y., Fukushima, T., Suna, Y. et al. 2006. Photoconductive coaxial nanotubes of molecularly connected electron donor and acceptor layers. *Science* 314: 1761–1764.

7. Nakanishi, T., Morita, M., Murakami, H., Sagara, T., Nakashima, N. 2002. Structure and electrochemistry of self-organized fullerene-lipid bilayer films. *Chem. Eur. J.* 8: 1641–1648.

8. Anthopoulos, T. D., Singh, B., Marjanovic, N. et al. 2006. High performance n-channel organic field-effect transistors and ring oscillators based on C_{60} fullerene films. *Appl. Phys. Lett.* 89: 213504.

9. Kim, Y., Cook, S., Tuladhar, S. M. et al. 2006. A strong regioregularity effect in self-organizing conjugated polymer films and high-efficiency polythiophene: Fullerene solar cells. *Nat. Mater.* 5: 197–203.

10. Nakanishi, T., Murakami, H., Sagara, T., Nakashima, N. 1999. Aqueous electrochemistry of a C_{60}-bearing artificial lipid bilayer membrane film immobilized on an electrode surface: Thermodynamics for the binding of tetraalkylammonium ion to the fullerene anion. *J. Phys. Chem. B* 103: 304–308.

11. Nakanishi, T., Ohwaki, H., Tanaka, H. et al. 2004. Electrochemical and chemical reduction of fullerenes C_{60} and C_{70} embedded in cast films of artificial lipids in aqueous media. *J. Phys. Chem. B* 108: 7754–7762.

12. Imahori, H., Norieda, H., Yamada, H. et al. 2001. Light-harvesting and photocurrent generation by cold electrodes modified with mixed self-assembled monolayers of boron-dipyrrin and ferrocene-porphyrin-fullerene triad. *J. Am. Chem. Soc.* 123: 100–110.

13. Zilbermann, I., Lin, A., Hatzimarinaki, M., Hirsch, A., Guldi, D. M. 2004. Electrostatically arranged cytochrome c-fullerene photoelectrodes. *Chem. Commun.* (1): 96–97.

14. Nakamura, E., Isobe, H. 2003. Functionalized fullerenes in water. The first 10 years of their chemistry, biology, and nanoscience. *Acc. Chem. Res.* 36: 807–815.

15. Deguchi, S., Yamazaki, T., Mukai, S., Usami, R., Horikoshi, K. 2007. Stabilization of C_{60} nanoparticles by protein adsorption and its implications for toxicity studies. *Chem. Res. Toxicol.* 20: 854–858.

16. Segura, J. L., Martín, N., Guldi, D. M. 2005. Materials for organic solar cells: The C_{60}/π-conjugated oligomer approach. *Chem. Soc. Rev.* 34: 31–47.

17. Mateo-Alonso, A., Guldi, D. M., Paolucci, F., Prato, M. 2007. Fullerenes: Multitask components in molecular machinery. *Angew. Chem. Int. Ed.* 46: 8120–8126.

18. Matsuo, Y., Tahara, K., Nakamura, E. 2006. Synthesis and electrochemistry of double-decker buckyferrocenes. *J. Am. Chem. Soc.* 128: 7154–7155.

19. Nierengarten, J. F., Armaroli, N., Accorsi, G., Rio, Y., Eckert, J. F. 2003. [60]Fullerene: A versatile photoactive core for dendrimer chemistry. *Chem. Eur. J.* 9: 37–41.

20. Sonmez, G., Shen, C. K. F., Rubin, Y., Wudl, F. 2005. The unusual effect of bandgap lowering by C_{60} on a conjugated polymer. *Adv. Mater.* 17: 897–900.

21. Nakanishi, T., Murakami, H., Nakashima, N. 1998. Construction of monolayers and Langmuir-Blodgett films of a fullerene-bearing artificial lipid. *Chem. Lett.* 27: 1219–1220.

22. Gao, Y., Tang, Z. X., Watkins, E., Majewski, J., Wang, H. L. 2005. Synthesis and characterization of amphiphilic fullerenes and their Langmuir-Blodgett films. *Langmuir* 21: 1416–1423.

23. Uemura, S., Sakata, M., Taniguchi, I., Kunitake, M., Hirayama, C. 2001. Novel "Wet process" technique based on electrochemical replacement for the preparation of fullerene epitaxial adlayers. *Langmuir* 17: 5–7.

24. Miyazawa, K., Kuwasaki, Y., Obayashi, A., Kuwabara, M. 2002. C_{60} nanowhiskers formed by the liquid-liquid interfacial precipitation method. *J. Mater. Res.* 17: 83–88.

25. Sathish, M., Miyazawa, K., Sasaki, T. 2007. Nanoporous fullerene nanowhiskers. *Chem. Mater.* 19: 2398–2400.

26. Sathish, M., Miyazawa, K. 2007. Size-tunable hexagonal fullerene (C_{60}) nanosheets at the liquid-liquid interface. *J. Am. Chem. Soc.* 129: 13816–13817.

27. Asaka, K., Kato, R., Miyazawa, K., Kizuka, T. 2006. Buckling of C_{60} whiskers. *Appl. Phys. Lett.* 89: 071912.

28. Tan, Z. Q., Masuhara, A., Kasai, H., Nakanishi, H., Oikawa, H. 2007. Multibranched C_{60} micro/nanocrystals fabricated by reprecipitation method. *Jpn. J. Appl. Phys.* 47: 1426–1428.

29. Ji, H. X., Hu, J. S., Tang, Q. X. et al. 2007. Controllable preparation of submicrometer single-crystal C_{60} rods and tubes trough concentration depletion at the surfaces of seeds. *J. Phys. Chem. C* 111: 10498–10502.

30. Nurmawati, M. H., Ajikumar, P. K., Renu, R., Sow, C. H., Valiyaveetti, S. 2008. Amphiphilic poly(p-phenylene)-driven multiscale assembly of fullerenes to nanowhiskers. *ACS Nano* 2: 1429–1436.

31. Diederich, F., Gómez-López, M. 1999. Supramolecular fullerene chemistry. *Chem. Soc. Rev.* 28: 263–277.

32. Yamaguchi, T., Ishii, N., Tashiro, K., Aida, T. 2003. Supramolecular peapods composed of a metalloporphyrin nanotube and fullerenes. *J. Am. Chem. Soc.* 125: 13934–13935.

33. Ramakanth, I., Shankar, B. V., Patnaik, A. 2008. Unilamellar composite vesicles and Y-junctions from pristine fullerene C_{60}. *Chem. Commun.* 35: 4129–4131.

34. Fernández, G., Sánchez, L., Pérez, E. M., Martín, N. 2008. Large exTTF-Based dendrimers. Self-assembly and peripheral cooperative multiencapsulation of C_{60}. *J. Am. Chem. Soc.* 130: 10674–10683.

35. Kawauchi, T., Kumaki, J., Kitaura, A. et al. 2008. Encapsulation of fullerenes in a helical PMMA cavity leading to a robust processable complex with a macromolecular helicity memory. *Angew. Chem. Int. Ed.* 47: 515–519.

36. Kawauchi, T., Kitaura, A., Kumaki, J., Kusanagi, H., Yashima, E. 2008. Helix-sense-controlled synthesis of optically active poly(methyl methacrylate) stereocomplexes. *J. Am. Chem. Soc.* 130: 11889–11891.

37. Madueno, R., Raisanen, M. T., Silien, C., Buck, M. 2008. Functionalizing hydrogen-bonded surface networks with self-assembled monolayers. *Nature* 454: 618–621.

38. Samorì, P. 2005. Exploring supramolecular interactions and architectures by scanning force microscopies. *Chem. Soc. Rev.* 34: 551–561.

39. Wang, Y. Y., Yamachika, R., Wachowiak, A., Grobis, M., Crommie, M. F. 2008. Tuning fulleride electronic structure and molecular ordering via variable layer index. *Nat. Mater.* 7: 194–197.

40. Bonifazi, D., Spillmann, H., Kiebele, A. et al. 2004. Supramolecular patterned surfaces driven by cooperative assembly of C_{60} and porphyrins on metal substrates. *Angew. Chem. Int. Ed.* 43: 4759–4763.

41. Uemura, S., Samorì, P., Kunitake, M., Hirayama, C., Rabe, J. P. 2002. Crystalline C_{60} monolayers at the solid-organic solution interface. *J. Mater. Chem.* 12: 3366–3367.

42. Chen, L., Chen, W., Huang, H. et al. 2008. Tunable arrays of C_{60} molecular chains. *Adv. Mater.* 20: 484–488.

43. Feng, M., Lee, J., Zhao, J., Yates, J. T., Petek, H. 2007. Nanoscale templating of close-packed C_{60} nanowires. *J. Am. Chem. Soc.* 129: 12394–12395.

44. Nakanishi, T., Miyashita, N., Michinobu, T. et al. 2006. Perfectly straight nanowires of fullerenes bearing long alkyl chains on graphite. *J. Am. Chem. Soc.* 128: 6328–6329.

45. Nakanishi, T., Takahashi, H., Michinobu, T. et al. 2008. Fullerene nanowires on graphite: Epitaxial self-organizations of a fullerene bearing double long-aliphatic chains. *Colloid Surf. A* 321: 99–105.

46. Tsunashima, R., Noro, S., Akutagawa, T. et al. 2008. Fullerene nanowires: Self-assembled structures of a low-molecular-weight organogelator fabricated by the Langmuir-Blodgett method. *Chem. Eur. J.* 14: 8169–8176.

47. Bottari, G., Olea, D., Gómez-Navarro, C. et al. 2008. Highly conductive supramolecular nanostructures of a covalently linked phthalocyanine-C_{60} fullerene conjugate. *Angew. Chem. Int. Ed.* 47: 2026–2031.

48. Cassell, A. M., Lee Asplund, C., Tour, J. M. 1999. Self-assembling supramolecular nanostructures from a C_{60} derivative: Nanorods and vesicles. *Angew. Chem. Int. Ed.* 38: 2403–2405.

49. Zhou, S. Q., Burger, C., Chu, B. et al. 2001. Spherical bilayer vesicles of fullerene-based surfactants in water: A laser light scattering study. *Science* 291: 1944–1947.

50. Nakashima, N., Ishii, T., Shirakusa, M. et al. 2001. Molecular bilayer-based superstructures of a fullerene-carrying ammonium amphiphile: Structure and electrochemistry. *Chem. Eur. J.* 7: 1766–1772.

51. Sano, M., Oishi, K., Ishi-i, T., Shinkai, S. 2000. Vesicle formation and its fractal distribution by bola-amphiphilic [60] fullerene. *Langmuir* 16: 3773–3776.

52. Nakanishi, T., Schmitt, W., Michinobu, T., Kurth, D., Ariga, K. 2005. Hierarchical supramolecular fullerene architectures with controlled dimensionality. *Chem. Commun.* (48): 5982–5984.

53. Nakanishi, T., Ariga, K., Michinobu, T. et al. 2007. Flower-shaped supramolecular assemblies: Hierarchical organization of a fullerene bearing long aliphatic chains. *Small* 3: 2019–2023.

54. Nakanishi, T., Wang, J., Möhwald, H. et al. 2009. Supramolecular shape shifter: Polymorphs of self-organized fullerene assemblies. *J. Nanosci. Nanotechnol.* 9: 550–556.

55. Nakanishi, T., Takahashi, H., Michinobu, T. et al. 2008. Fine-tuning supramolecular assemblies of fullerenes bearing long alkyl chains. *Thin Solid Films* 516: 2401–2406.

56. Nakanishi, T., Michinobu, T., Yoshida, K. et al. 2008. Nanocarbon superhydrophobic surfaces created from fullerene-based hierarchical supramolecular assemblies. *Adv. Mater.* 20: 443–446.

57. Nakanishi, T., Shen, Y., Wang, J. et al. 2008. Electron transport and electrochemistry of mesomorphic fullerenes with long-range ordered lamellae. *J. Am. Chem. Soc.* 130: 9236–9237.

58. Prato, M., Maggini, M., Giacometti, C. et al. 1996. Synthesis and electrochemical properties of substituted fulleropyrrolidines. *Tetrahedron* 52: 5221–5234.

59. Campidelli, S., Lenoble, J., Barberá, J. et al. 2005. Supramolecular fullerene materials: Dendritic liquid-crystalline fulleropyrrolidines. *Macromolecules* 38: 7915–7925.

60. Zemb, T., Dubois, M., Demé, B., Gulik-Krzywicki, T. 1999. Self-assembly of flat nanodiscs in salt-free catanionic surfactant solutions. *Science* 283: 816–819.

61. Yoshida, K., Minamikawa, H., Kamiya, S., Shimizu, T., Isoda, S. 2007. Formation of self-assembled glycolipid nanotubes with bilayer sheets. *J. Nanosci. Nanotechnol.* 7: 960–964.

62. Barthlott, W., Neinhuis, C. 1997. Purity of the sacred lotus, or escape from contamination in biological surfaces. *Planta* 202: 1–8.

63. Sun, T. L., Feng, L., Gao, X. F., Jiang, L. 2005. Bioinspired surfaces with special wettability. *Acc. Chem. Res.* 38: 644–652.

64. Feng, X. J., Jiang, L. 2006. Design and creation of superwetting/antiwetting surfaces. *Adv. Mater.* 18: 3063–3078.

65. Deschenaux, R., Donnio, B., Guillon, D. 2007. Liquid-crystalline fullerodendrimers. *New J. Chem.* 31: 1064–1073.

66. Sawamura, M., Kawai, K., Matsuo, Y. et al. 2002. Stacking of conical molecules with a fullerene apex into polar columns in crystals and liquid crystals. *Nature* 419: 702–705.

67. Chuard, T., Deschenaux, R., Hirsch, A., Schonberger, H. 1999. A liquid-crystalline hexa-adduct of [60]fullerene. *Chem. Commun.* (20): 2103–2104.

68. Li, W. S., Yamamoto, Y., Fukushima, T. et al. 2008. Amphiphilic molecular design as a rational strategy for tailoring bicontinuous electron donor and acceptor arrays: Photoconductive liquid crystalline oligothiophene-C_{60} dyads. *J. Am. Chem. Soc.* 130: 8886–8887.

69. Michinobu, T., Nakanishi, T., Hill, J., Funahashi, M., Ariga, K. 2006. Room temperature liquid fullerenes: An uncommon morphology of C_{60} derivatives. *J. Am. Chem. Soc.* 128: 10384–10385.

Supported Fullerenes

Hai-Ping Cheng
University of Florida

39.1 Two Basic Processes

In addition to generating single fullerene molecules in gas phase, or crystallizing them to make molecular solids, experimental techniques have been developed to deposit these molecules on solid surfaces, or to dope them inside carbon nanotubes, or to integrate them in organic molecular networks. C_{60} also is used in molecular nanojunctions where a single C_{60} molecule is connected to two metal leads at the left side and the right side. Figure 39.1 shows two examples of surface-supported C_{60}-containing systems [1,2].

Two fundamental processes take place when molecules are assembled into extended systems, i.e., the systems are infinitely long (macroscopic) in at least one dimension. The first is the so-called crystal-field splitting [3], i.e., the surrounding crystal-field lifts the degeneracy of atomic or molecular orbitals of a single unit (finite). The second is the interaction between discrete molecular orbitals and Bloch waves of the host systems.

Figure 39.2 is an example to illustrate the crystal-field splitting from calculations based on density functional theory (DFT) [4] using the VASP package [5]. The system is chosen to be a C_{60} monolayer in a hexagonal lattice (as in Figure 39.1, but without the supporting metal surface). The reciprocal lattice of a triangular lattice is also a triangular lattice; the irreducible first Brillouin zone is colored cyan. (Figure 39.2a). From the band structure, one sees that there are three rather flat bands below the Fermi energy (set to zero) and three above it (Figure 39.2b). These bands are derived from the threefold degenerate highest-occupied molecular orbitals (HOMO) and the lowest-unoccupied (I_h) molecular orbitals (LUMO), respectively. It is well known that C_{60} has the icosahedral symmetry [3,6], so that is the highest symmetry a system with discrete points can have. When assembled into a hexagonal lattice, which is the lattice form on noble metal (111) surfaces, the symmetry of the monolayer C_{60} is lowered. The interactions among C_{60} lift the degeneracy in energy, resulting in three HOMO-derived and five LUMO bands. The remaining twofold degeneracy in two of the HOMO-derived bands is a characteristic of the hexagonal lattice that belongs to P6M symmetry group [7].

Now consider the second case, the interaction between molecules or molecular assemblies on hosts, for example, surfaces. Since the interaction between two C_{60} molecules is relatively weak, of van der Waals nature, we discuss the consequence of the interaction between the molecular layer and the supporting host. Many research activities investigate and utilize the feature that C_{60} is an electron acceptor. Unlike carbon nanotubes, which can be either electron donors or acceptors when in contact with metals, C_{60} molecules almost always act as acceptors. In terms of electron affinity, this molecule has high electron affinity and high ionization energy.

In general, when molecules are in contact with extended systems, for example, when they are on surfaces or inside large carbon nanotubes, they interact with the extended systems. For a good understanding of this interaction and its implications, we highly recommend R. Hoffman's book titled *Solids and Surfaces* [8]. Here, we provide a brief description as follows. A molecule or any finite-size entity has a set of discrete orbitals and extended systems, say, solids, which have a continuous distribution as shown in Figure 39.3. When these two systems are brought together, the states in the molecule and solids will interact with each other as shown in panel (a). The wiggled color lines indicate the interaction between four different types of orbital pairs. The red and the purple represent an occupied molecular orbital interacting with levels above and below the Fermi energy in the solid, respectively; the green and blue represent an unoccupied molecular orbital with levels below and above the Fermi energy, respectively. Due to these interactions, a molecular energy state can shift up, down, or split, or shift and split. Note that the black wiggled lines indicate that levels in solids near the Fermi energy also interact with levels above or below the Fermi energy. Panel (b) depicts the possible up or down shift of the HOMO upon interaction with an extended system, where the e_m^o and E_F^o

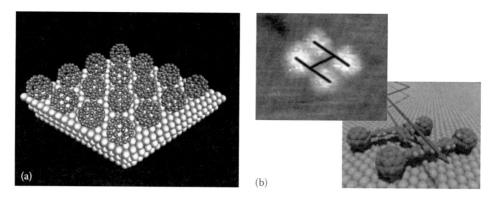

FIGURE 39.1   (a) Computational model for C_{60} on a fcc metal (111) surface with 100% coverage. (From Wang, L.L. and Cheng, H.P., *Phys. Rev. B*, 69, 045404, 2004. With permission.) (b) Nanocar made of four C_{60} molecules that are connected by a planar chassis made of organic molecules. (Courtesy of J.M. Tour, Rice University, Houston, TX.)

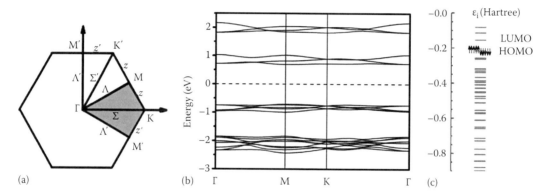

FIGURE 39.2   (a) Reciprocal lattice and first irreducible Brillouin zone of a 2D close-packed C_{60} lattice; (b) calculated band structure of single C_{60} layer; and (c) calculated energy levels of an isolated C_{60}.

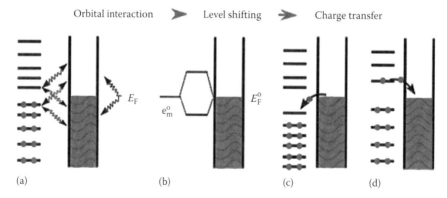

FIGURE 39.3   Sketch of different scenarios in molecule–solid interaction: (a) four distinctive molecular orbital-bulk/surface state interactions; (b) possible shifts of HOMO; (c) HOMO is below the Fermi energy; and (d) HOMO is above the Fermi energy.

indicate orbital and Fermi energy without interaction. If the highest-occupied molecular orbital (HOMO) happens to align with the Fermi level, there will be no charge transfer (note that the level can still shift and/or split if it is half-filled). Panels (c) and (d) illustrate two situations in which the interaction results in molecular-level shifting/splitting such that some unoccupied states (more than one in general) are below the Fermi level of the solid (panel c) or some occupied molecular orbitals are above

E_F (panel d). Consequently, charge transfer will occur, either from the solid to the molecule (panel c) or vice versa (panel d) because systems that consist of the molecule and the solid will have one common Fermi energy in equilibrium.

With these two fundamental processes in mind, i.e., the crystal-field splitting and the Fermi energy alignment, we discuss supported C_{60} in various environments, ranging from surfaces, to composites, to nanojunctions.

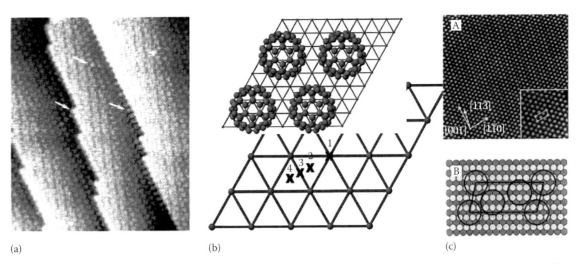

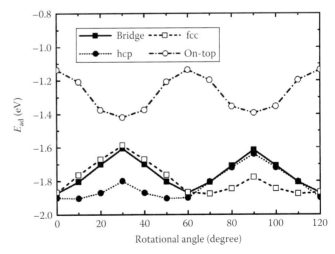

FIGURE 39.4 (a) A STM image of the Cu(111) surface covered by a monolayer C_{60} film prepared by annealing at 290°C ($300 \times 380 \text{ Å}^2$, $V_b = 2.0$ V, $I_t = 20$ pA). (From Hashizume, T. et al., *Phys. Rev. Lett.*, 71, 2959, 1993, Figure 1. With permission.) (b) Structure model constructed by Wang and Cheng according to experimental data. (From Wang, L.L. and Cheng, H.P., *Phys. Rev. B*, 69, 165417, 2004. With permission.) (c) Upper STM image ($326 \times 364 \text{ Å}^2$) of the C60 structure formed on Cu(110), lower schematic model of the structure with unit cell outlined. (From Murray, P.W. et al., *Phys. Rev. B*, 55, 9360, 1997, Figure 1. With permission.)

39.2 Fullerenes on Surfaces

39.2.1 Metal Surfaces Cu, Ag, Au, Al, Pt, Ni

The metal-surface–C_{60} interaction is among one of most well-studied phenomena for supported-C_{60} systems, with extensive experiment efforts having been made since the early 1990s. Because of the high electron affinity of C_{60} molecules as well as the metallic nature of the surfaces, charge transfer has been observed in a number of systems such as C_{60} adsorbed on metals (Cu, Ag, and Au). Geometry of the fullerene-metal interface is usually obtained via scanning tunneling microscopy (STM) and low-energy electron diffraction (LEED). In early days, these experiments involved some controversy on the amount of charge transfer from metal surface to the molecule. The error bars have been substantially reduced subsequently.

39.2.1.1 Copper Surfaces Cu(111) and Cu(110) and Thin Films

Hashizume et al. [9], Fartash [10], and Sakurai et al. [11,12] reported STM studies of C_{60} on Cu(111) surfaces, and Murray et al. [13] studied STM images of C_{60} monolayer adsorption on Cu(111) and Cu(110) surfaces (see Figure 39.4). The lattice mismatch between the C_{60} and Cu crystals is small (2%) which makes the Cu surface ideal for depositing fullerene molecules and growing films. On Cu(111), a hexagonal face of a C_{60} is down in contact with the metal surface and C_{60} forms a triangular lattice [9,10].

Detailed theoretical studies by Wang and Cheng [1] show that the most stable adsorption site on Cu(111) is the so-called hpc site (position 4 in Figure 39.4b). In the same figure, the other three sites are on-top 1, fcc 2, and bridge 3; each has different adsorption energy. Note that because of the ABC-layered pattern of a fcc crystal (111) surface, hcp and fcc are two distinguishable

sites. Upon adsorption, the Cu surface is somewhat deformed. The Cu–Cu bonds right in direct contact with the bottom hexagon face of C_{60} are elongated by 5%–6%, a nanosize crater with ~1.4 Å dent and 0.1 Å hump is formed due to the interaction. In the bottom hexagon, the short and the long C–C bonds increase by 3% and 2%, respectively.

Experiments and theory studies find that C_{60} can easily move around on Cu(111) surface. The mechanism can be explained using the information in Figure 39.5 (Wang and Cheng 2004); when the molecule has a combined translation and rotation

FIGURE 39.5 Energy as functions of orientation for all four sites shown in Figure 39.4. Zero angle orientation is the molecular orientation in Figure 39.4. The system has a threefold symmetry because of the Cu lattice. When the angle is 0°, 60°, 120°, …, the hcp (filled circle), bridge (filled square), and fcc (open square) sites but not the on-top (open circle) site have similar energy. A C_{60} molecule can translate from one site to another, among hcp, fcc, and bridge, freely with exception to the on-top site.

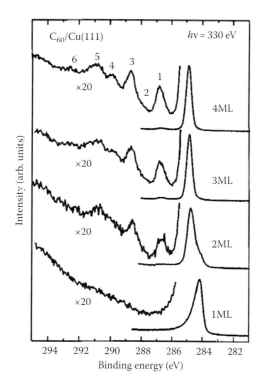

FIGURE 39.6 Carbon 1*s* core-level photoemission spectra at various thicknesses. These spectra are normalized to have the same height as the main peak. On the left are the enlarged spectra of the satellite region. The multiplication factors are shown. (From Tsuei, K.D. et al., *Phys. Rev. B*, 56, 15412, 1997, Figure 4. With permission.)

TABLE 39.1 Work Functions of Clean Metal Surfaces and after Adsorption of a Monolayer of C_{60}

Surface	Clean Surface WF (eV)	WF of 1-ML C_{60} on Surface	WF Change
Cu(111)	4.94	4.86	−0.08
Ni(111)	5.36	4.93	−0.43
Al(110)	4.35	5.25	+0.95
Al(111)	4.25	5.15	+0.95
Au(111)	5.37	4.82	−0.45
Rh(111)	5.4	4.9	−0.5
Ta(110)	4.8	5.4	+0.6

Source: Tsuei, K.D. et al., *Phys. Rev. B*, 56, 15412, 1997 and references therein.

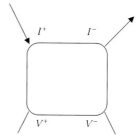

FIGURE 39.7 A sketch of experimental set-up for measuring the sheet resistance.

motion, the barriers separating two hpc sites are negligible. Energy curves in Figure 39.5 also indicate that hcp, fcc, and bridge sites are much closer in energy compared to the on-top site which is a much less stable adsorption site.

Thorough charge transfer studies have been carried out on Cu(111) surfaces [14,15] using photoemission and NEXAFS spectroscopy. The tail of the 1ML curve in Figure 39.6 [15] is believed to signify the metallic screening and nature of the C_{60} monolayer. The NEXAFS (not shown here) measures the transitions from the 1s core-state to the $\pi^\star$ LUMO (284.5 eV), LUMO + 1, LUMO + 2, and LUMO + 4 (285.8, 286.4, and 288.3 eV, respectively). Based on a shift of carbon-core-level binding energy and NEXAFS peaks to lower values, the charge transfer from Cu(111) to C_{60} derived from experiments by Tsuei et al. is 1.5 – 2.0e^- per C_{60} molecule. Table 39.1 consists of data from Tsuei et al.'s paper [15] and references therein.

In-plane electron resistance measurements (a 4-terminal technique that measures the sheet resistance) were done by Hebard et al. on a C_{60}-Cu thin film [16]. This group identified two different regimes of behavior. In the first case, molecules are deposited on very thin-film Cu that are essentially in the coalescence regime with an islanded morphology. The C_{60} monolayer, doped by charge transfer from the metal, opens a new channel for electron transport and reduces the resistance. The measured sheet resistance of monolayers is $R_s \sim 8000\ \Omega$. Figure 39.7 illustrates the measurement set-up for sheet resistance, $R_s = \rho/t$ where ρ is the

resistivity and t is the thickness of the film. Note that R_s, therefore, has the same unit as resistance but is also independent of sample size. This sheet resistance corresponds to a resistivity (at room temperature) that is a factor of 2 less than that of the three-dimensional alkali-metal-doped compounds $A3C_{60}$ (A = K, Rb). In the second regime, where the metal film is thick (the continuous film limit), the opposite behavior was observed. The adsorbed C_{60} monolayer gives rise to an *increase* in resistance. A simple explanation is that a thick film is a good conductor; so, the presence of molecules, or the molecule–metal interface, introduces extra scattering processes that cause the conductance to decrease. Hebard's group conducted a number of measurements on samples with different thicknesses and concluded that the resistance increase is best described by diffuse surface scattering. In a slightly different experimental condition, a Cu film was deposited over a C_{60} monolayer. In the thin-film limit, the conductivity is again enhanced and the enhancement is sufficient to reduce the critical thickness (also roughly by a factor of 2) at which the overlying Cu film becomes conducting. The mechanism is the same as in the case where C_{60} is on top of the Cu film.

Theoretical calculations have been done by Wang and Cheng [1] using first-principles quantum mechanics based on density functional theory to study the C_{60}-Cu(111) interfacial electronic structure and charge transfer state. A slab and periodic boundary conditions are used to model the system. They estimated that the charge transfer from a Cu(111) surface to the molecular monolayer is 0.8–1.0 electrons/C_{60}. In a quantum mechanical description, charge transfer is not a uniquely defined quantity. Charges are assigned to atoms according to geometry or projection on to

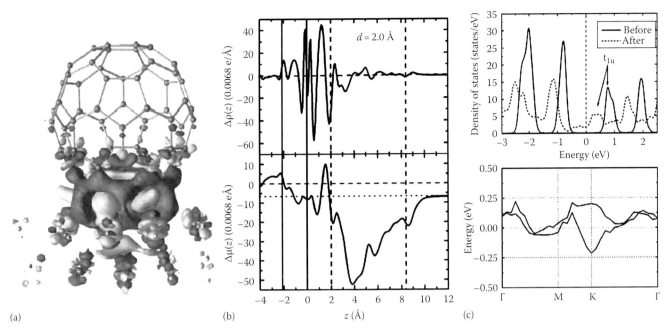

FIGURE 39.8 (a) Charge (lighter) and hole (darker) distributions; (b) planar-averaged electron charge difference (upper) and dipole moment (bottom) as a function of z (or z_o), the direction along surface norm; and (c) projected DOS (upper) on to C_{60} before (solid line) and after (dashed line) surface adsorption, and two bands that are derived from C_{60} monolayer (bottom).

atomic orbitals. Two geometry-based methods are used in Wang and Cheng's work to obtain the value of 0.8–1.0 electron/C_{60}. Figure 39.8a shows electron (yellow) and hole (red) isodensity surfaces, calculated via $\Delta\rho = \rho(C_{60} + Cu) - \rho(C_{60}) - \rho(Cu)$, of the ground state. The isodensity values are $\pm 1.0 e^-/(10 \times bohr)^3$. This complex pattern of charge/hole distribution leads to a net charge transfer from the surface to the C_{60} ML. The middle panel depicts the charge difference $\Delta\rho$ averaged over the x–y plane (assuming the normal to the surface to be the z-axis) and the local dipole moment as a function of z. In the ground state, the distance between the bottom hexagon of C_{60} and the top copper layer is 2.0 Å. The solid vertical lines indicate the positions of the top two copper layers and the dashed lines indicate the locations, in the z direction, of the two parallel boundary hexagons in C_{60}. The surface dipole moment after integrating in the z direction is

$$\Delta\mu = \int_{-a}^{-a+z_0/2} \Delta\mu(z)\,dz = \int_{-a}^{-a+z_0/2} z\Delta\sigma(z)\,dz$$

$$\text{with} \quad \Delta\sigma(z) = \int_{unitcell} \Delta\rho(\mathbf{r})\,dx\,dy,$$

where

  z_o is the thickness of the unit cell (which included a slab and a vacuum region)

  a is the distance between the center of slab and the surface

Interestingly, the dipole moment points to a direction opposite to the direction of charge transfer. This is the reason for the slight decrease in surface work function in the measurements

by Tsuei et al. [15]. The right panels in the figure provide information on the electronic structure at the interface: the top one shows densities of states (DOS) of the isolated C_{60} monolayer and the DOS projected on C_{60} when adsorbed on the surface; the bottom one are two bands across a Fermi surface that are likely related with t_{1u} orbitals; their projections on the C_{60} ML are 3% and 13%, respectively. The projected DOS and bands are computed using

$$DOS_\mu(\varepsilon) = \sum_k w_k \sum_i \delta(\varepsilon - \varepsilon_i) \left| \langle \phi_{\mu,k}(\mathbf{r}) | \Psi_{i,k}(\mathbf{r}) \rangle \right|^2,$$

and

$$p_{u,i} = \sum_k w_k \left| \langle \phi_{\mu,k}(\mathbf{r}) | \Psi_{i,k}(\mathbf{r}) \rangle \right|^2,$$

respectively. In these equations, $\phi_{\mu,k}(\mathbf{r})$ and $\Psi_{i,k}(\mathbf{r})$ are atomic and Bloch state Kohn-Sham orbitals wave, respectively, w_k is the weight of each k point, and the indices μ and (i,k) are labels of atomic orbitals and Bloch states.

39.2.1.2 Silver Surfaces, Ag(111), and Ag(100)

In the mid-1990s, STM experiments seemed to give different results on C_{60} adsorption on the Ag(111) surface. Altman and Colton [17,18] proposed that the adsorption configuration was a pentagon of C_{60} on an on-top site (i.e., site 1 in Figure 39.4, lower-middle panel) for both the Ag(111) and Au(111) surfaces; Sakurai et al. [11] proposed, also from the results of STM experiments, that the favored adsorption site should be the threefold on-hollow site on the Ag(111) surface. However, these experiments did not identify the orientation of the C_{60} molecule. About a decade later, Wang and Cheng performed large-scale first-principles calculations to investigate this issue. Their results support the model proposed by Sakurai et al., namely, that in the ground state, the center of the molecule sits on a hollow site of the metal surface similar to that of Cu(111). Furthermore, this group identified that a hexagon instead of a pentagon in the molecule is in contact with the surface. Wang and Cheng performed detailed theoretical calculations to pin-point this issue. Their results contradict the interpretations by Altman and Colton. Wang and Cheng find that the interfacial structure of C_{60}-Ag(111) adsorption is very similar to the system of C_{60} on Cu(111) surface, in which the adsorption configuration is that of a hexagon of C_{60} on the threefold on-hollow site. Based on the similarity of the electronic properties of the noble metals, it is not unreasonable that C_{60} occupies the same adsorption site on Ag(111) as on the Cu(111) surface. Table 39.2 shows the binding energies for different geometries and adsorption sites. The rotation energy barriers are also reported in their work for various sites (Figure 39.4). Similar to the Cu(111) surface as shown in Figure 39.9, the on-top site is much less favorable than the other three.

The electronic structure of the interface has been studied experimentally and theoretically. The first surface-enhanced Raman spectroscopy (SERS) measurements were carried out for a C_{60} monolayer and ultra-thin-film deposited (in ultra-high-vacuum, ~10^{-8} Torr and room temperature 300 K) on three noble metal surfaces (Au, Cu, and Ag) in 1992 [19]. Shifts of the

TABLE 39.2 Adsorption Energies of C_{60} on Ag(111) and Au(111) Surfaces on Various Sites

Configuration	Hcp	fcc	Bridge	On-Top
Hexagonal on Ag(111)	−1.54	−1.50	−1.40	−1.27
Pentagon on Ag(111)	−1.20	−1.20	−1.18	−0.89
Hexagonal on Au(111)	−1.27	−1.19	−1.13	−0.86
Pentagon on Au(111)	−1.03	−1.04	0.99	−0.60

Note: Values are in eV/molecule.

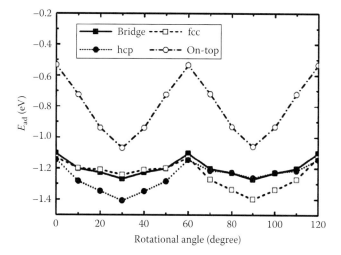

FIGURE 39.9 Similar to Figure 39.5 but for Ag(111) surface.

TABLE 39.3 Comparison of the Raman Frequencies of the $A_g(2)$, $H_g(7)$, and $H_g(8)$ Modes of 0.5–2.0 ML C_{60} on Different Substrates and 10-ML C_{60} on Al_2O_3

Substrate	Thickness (ML)	$H_g(7)$ (cm^{-1})	$A_g(2)$ (cm^{-1})	$H_g(8)$ (cm^{-1})
Al_2O_3	10.0	1423.5	1464.0	1567.0
Ag	0.5	1409.0	1435.7	1554.0
Cu	1.0	1415.0	1440.8	1556.0
Au	2.0	1421.0	1448.6	1561.0

Source: Chas, et al., Phys. Rev. B, 46, 7873, 1992 and references therein.

TABLE 39.4 Comparison of the Raman shifts of ultrathin films of C_{60} on different substrate relative to thick C_{60}

Substrate	Thickness (ML)	$A_g(2)$ (cm^{-1})
Al_2O_3	10.0	0.0
a-Ge	1.0	−15.0
a-Ge:H	1.0	−2.0
Ag	0.5	−28.3
Cu	1.0	−23.2
Au	2.0	−15.4

high-frequency $A_g(2)$ pentagon pinch mode were observed. The Ag surface has the largest shift among the three. No shifts were observed for dielectric surfaces (see Tables 39.3 and 39.4). The experimentalists (Chase et al.) concluded that charge transfer is the cause of the substantial decrease in the frequency of the $A_g(2)$ mode. Compared to the C_{60}-Ag system, a relatively smaller SERS shift is reported in the C_{60}-Au system. Akers et al. [20] studied SERS of C_{60} and C_{70} on rough silver surfaces and found that the $A_g(2)$ mode is 1438 cm^{-1} (1452 cm^{-1} for on Au surface), close to the value obtained by Chase et al. They also reported results on C_{60}-C_{70} mixture over a silver surface and found that the line profiles C_{70} are in general less broad than C_{60}, and that C_{70} perturbs the spectral line position and reduces the downward shifts in frequencies (Figure 39.10) [20].

Rosenberg and DiLella [21] reported SERS studies at 10^{-8} Torr and 8 K, in which a series of modes are found (Table 39.4). Note that the two A_g and eight H_g modes are the only first-order modes which are Raman-allowed by the I_h symmetry of isolated C_{60}. The F_{1u} modes are IR-allowed and the rest are silent. Modes marked by "s" can be assigned to Raman-allowed second-order (overton and sum) modes. Studies of ultraviolet photoelectron spectroscopy of a C_{60} monolayer and a K-doped C_{60} monolayer on three noble metals by Hoogenboom et al. [22] found that for undoped monolayers, 1.0, 1.7, and 1.8 electron/molecule transfer from the Cu, Ag, and Au surfaces, respectively, to the C_{60} monolayer. Compared to the value estimated from Raman data (~1 ± 11 ± 1e$^-$) [19,23], the charge transfer from Ag to C_{60} is considerably higher in Hoogenboom et al.'s experiments.

Wang and Cheng's theoretical calculations on the Ag(111) surface show that the charge transfer from the silver surface is 0.5e per molecule. Note that the Raman experiments have a one-electron error bar; the charge transfer from Ag, if a perfect (111) surface, is therefore smaller than one per molecule.

SERS of C_{60}/C_{70} mixture

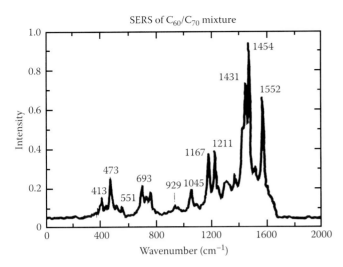

FIGURE 39.10 Surface-enhanced Raman spectra of a mixture of C_{60} and C_{70} that has been enriched in the latter, vapor-deposited at 775 K onto a rough silver film. Spectrum recorded on optical multi-channel spectrometer. (From Akers et al., *Chem Phys. Lett.*, 190, 614, 1992. With permission.)

In addition to poly-crystalline and the densely packed Ag(111) surface, Ag(100) surface also attracted some attention. Giudice et al. [24] reported UHV-STM results that show that the fullerene molecules form a regular overlayer on the Ag(001) surface with c(6 × 4) structure. The work stimulated a number of groups to study the Ag(001) surfaces, two geometrically inequivalent, but of similar absorption energy, molecular orientations of the C_{60} cage are proposed by a combined calculation-STM image approach [25].

39.2.1.3 Gold Surface Au(110) and Au(111)

In monolayer C_{60}-covered Au(110) surface, each molecule is in close contact with two Au ridge atoms [26,27]. The presence of the monolayer alters the pattern of Au(110) surface reconstruction from (1 × 2) to (1 × 5). The surface-C_{60}-combined system displays a (6 × 5) pattern (Figure 39.11). This pattern is the more stable one compared to another pattern in which the underlying Au(110) surface exhibits a (1 × 2) reconstruction accompanied by surface steps. Compared to the bulk, the C_{60}-C_{60} distance is 1.5% shorter (10.2 Å, 10.04 Å). In 2008, a surface x-ray diffraction study, in comparison with photoemission (PE), inverse photoemission (IPE), and first-principles density functional calculations (DFT), revealed the formation of nanodimples on the Au(110) surface (Figure 39.12), suggesting a strong surface reconstruction [28].

Electron energy loss (EEL) spectrometry and high-resolution EEL or HREEL [23,29], and photoemission spectroscopy [30,31] were used to measure the electronic structure, and determine the charge states of C_{60} adsorbed upon a Au(110) surface. The charge transfer is estimated from the spectrum of the electronic excitations from C 1s level to the empty C_{60} states. Two surfaces were studied, C_{60} monolayer on a clean Au(110) surface and on a Au(110) with a K c(2 × 2) overlayer (0.5 K onto Au substrate).

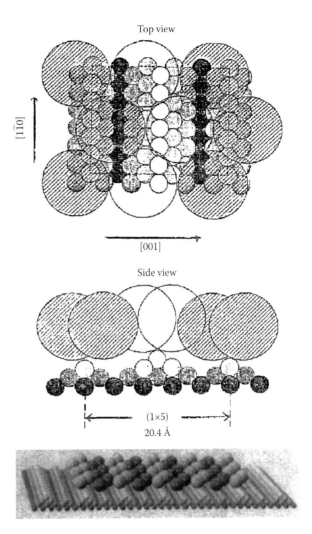

FIGURE 39.11 Model of (6 × 5) structure. (From Modesti, S. et al., *Surf. Sci.*, 333, 1129, 1995; Gimzewski, J.K. et al., *Phys. Rev. Lett.*, 72, 1036, 1994. With permission.)

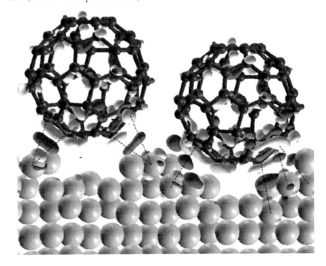

FIGURE 39.12 Optimized interface structure of C_{60}-Au(111) using DFT calculations. Darker and lighter (non-spherical) are isosurface ($0.005e^{-}/Å^{3}$) of charge and hole distributions (From Hinterstein, M. et al., *Phys. Rev. B*, 77, 153412, 2008. With permission.)

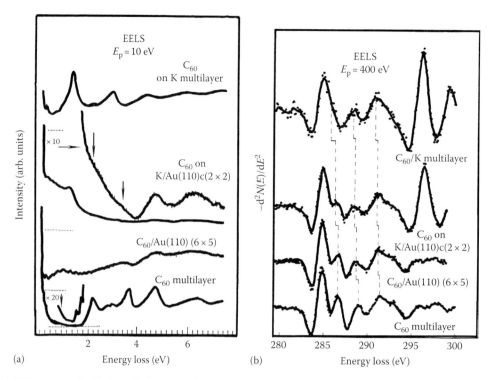

FIGURE 39.13 (a) EEL spectra of bulk C_{60} and of C_{60} monolayers on different substrates. The dashed line under each spectrum marks the zero level. The broad feature below 1 eV in the second and third spectra is the t_{1u} plasmon. (b) Second derivative of the EEL spectra of the C 1s absorption edge of bulk C_{60} and of C_{60} monolayers on metallic substrates. The peaks at 296 and 299 eV are the K L_{23} edge. (From Modesti, S. et al., *Phys. Rev. Lett.*, 71, 2469, 1993. With permission.)

The charge transfer determined in this experiment is 1 ± 1 1 ± 1, and 3 ± 1 3 ± 1, respectively. Figure 39.13a,b are EELS measurements of Modesti et al. in 1993 [29]. In 1995, based on HREELS results and analysis of carbon vibrational modes, Hunt et al. [23] estimated that the charge transfer from noble metal surface, Au(110), is $0.5 \pm 0.7e^-$. Studies [30,31] also show that the modification to the adsorbed molecular layer due to the metal surface is limited to the layer in direct contact with the surface, indicating a highly local effect.

The orientation of C_{60} on a Au(111) surface was studied by Altman and Colton [18] using STM and related techniques. Like Ag(111), they concluded that the pentagon is in contact with the metal surface. However, this finding contradicts the results from first-principles calculations by Wang and Cheng [32]. In the ground state, the molecules occupy the same hcp site as on Cu(111) and Ag(111) with a hexagonal face in contact with the metal. Also similar to the other two surfaces, the molecule is mobile on the Au(111) surface (Figure 39.14). Gimzewski et al. reported a new Au(111)-C_{60} interface reconstruction. When the system was annealed to a temperature of 700 K, C_{60} forms two well-ordered hexagonal phases. A long-range height modulation of 1 Å and periodicity of 90–160 Å were observed [33]. The mechanism for this phenomenon is the molecule–surface interaction which induces a substantial increase in the length of Au reconstruction unit cell in 110 > direction. At this point, first-principles calculations fall short in treating such large-scale reconstruction due to inadequate computing power, while

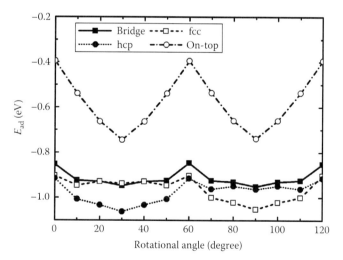

FIGURE 39.14 Similar to Figures 39.5 and 39.9 but for Au(111) surface.

classical potentials are not good enough yet to predict such reconstruction patterns.

Charge transfer and site-dependent electronic structure were addressed in the STM studies of Atlman and Colton. Hoogenboom et al. estimate a $1.0e^-$ transfer from gold surface to an adsorbed molecule on Au(111) surface using a combined ultraviolet and x-ray photoemission spectroscopy (UPS and XPS), somewhat higher than the calculated value by Wang and Cheng ($\sim0.3e^-/C_{60}$). This difference is small and likely caused by experimental error bar. Work functions of Ag and Au are also measured experimentally. Table 39.5 lists the experimental [15,34] data and calculated numbers from first-principles studies [1,32]. For comparison, all three noble metal surfaces are included. The monolayer is adsorbed on three (111) surfaces in the (4 × 4) surface unit cell.

The work function change for the Au(111) surface is −0.58 eV via the direct calculation (the potential energy required to take an electron from surface to vacuum, which can be done directly

in the band structure calculations using the slab model) and −0.60 eV via the surface dipole calculation, both agree very well with the experimental value of −0.6 eV [34]. The corresponding change in the surface dipole is −1.37 D. For the Ag(111) surface, the work function change is +0.14 and +0.11 eV using the two methods, respectively. The change in the surface dipole in case of the Ag(111) surface is +0.24 D (there is no experimental data on the work function).

One notable thing is the sign of the surface dipole and the work function change. It is positive for silver and negative for both Au and Cu, while charge transfer is always from the surface to the molecule. Intuitively, this seems to be inconsistent. The key point here is that quantum mechanics does not define uniquely the charge on each atom in an assembly of atoms (molecule, surface, and solids). The experimentally measured values are fitted to models that are based on some classical definition, and theoretically calculated values are obtained using either orbital projection (e.g., Milliken population analysis) or geometric division (Bader's method etc.). The dipole moment, however, is well defined in quantum mechanics and is related to the surface work function. Theoretical work done by Wang and Cheng [1,32] addresses this point clearly.

TABLE 39.5 Frequency Shifts of Various Modes in C_{60} on Ag and In

Frequency (cm^{-1})				
Bulk	~1ML/Ag	~2ML/Ag	~2ML/In	Assignment
272	279	268	268	$H_g(1)$
431	419	431	430	$H_g(2)$
498	486	494	488	$A_g(1)$
710	706	710	710	$H_g(3)$
774	765	772	770	$H_g(4)$
1102	1092	1102	1102	$H_g(5)$
1252	1236	1237[a],1250	1240	$H_g(6)$
1427	1423[a]?	1426?	?	$H_g(7)$
1470	1423[a]?	1426?	1431[a]	$A_g(2)$
	1446	1464,1472	1451	$A_g(2)$
1577	1564	1567,1576[a]	1567	$H_g(8)$
267[a]		256[a]	257[a]	?
	290	291	292	?
354	345	348	342	$F_{3u}(1)^b, F_{1g}(1)^b$
404		393		$G_u(1)^b$
486[a]		487[a]		$G_g(1)^b$
529	510	528	522	$F_{1u}(1)^b$
568	552,570	568	560	$F_{2g}(1)^b, F_{1g}(1)^b$
694[a]				$H_u(3)^b, S$
741	733	738	736	$G_u(2)^b, S$
758	751	757		$G_u(3)^b, S$
798				$H_u(4)^b, F_{2u}(3)^b, S$
860				$F_{2g}(2)^b, S$
976	969	969	975	$G_u(4)^b, F_{1g}(1)^b, S$
	997	997[a]		S
1040	1028			$F_{2u}(4)^b, S$
1080		1080[a]		$G_u(4)^b, G_g(4)^b, S$
1141		1142		S
1188				$F_{1u}(3)^b, S$
1204	1187	1190	1198	$G_g(3)^b, F_{2u}(4)^b, S$
	1330	1316		$F_{2u}(5)^b, G_g(4)^b, S$
	1358			$G_g(5)^b, S$
1407[a]				S
1482[a]				$F_{1g}(3)^b, S$
1500				S

Source: Rosenberg, A. and Dilella, D.P., *Solid State Commun.*, 95, 729, 1995. With permission.

Note: ? are from original paper, unexplained.

[a] These modes appear as "shoulders" which are not clearly resolved from their nearby stonger neighbors.

[b] These are Raman-forbidden nodes.

39.2.1.4 Other Metal Surfaces

Effort has also been made to study C_{60} on Pt, Al, and Ni surfaces. On the Pt(111) surface [35], polymerization (when $T \geq 300\,K$) and complete decomposition ($T \geq 1050\,K$) of C_{60} were observed using x-ray and UV spectroscopy (XPS & UVS), HREELS, and LEED techniques. The interaction between the Pt surface and C_{60} appears to be quite strong. Cepek et al. [36] conclude that the strong bonding between Pt(111) and C_{60} is mainly covalent with a charge transfer of <0.8 electron from the molecule to the surface. Note that this is opposite to the noble metal surfaces. As a consequence of the strong interaction, the mobility of the molecule is lower compared to the noble metal surfaces. A joint experimental STM measurement and DFT calculations [37] show that a chemisorbed C_{60} molecule on the Pt(110) surface has two stable orientations (see Figure 39.15). The reconstruction of the Pt(110) surface is lifted by the existence of the C_{60} assembly. Furthermore, after a heat treatment at 700 K, the C_{60} molecules form a nanowire array that is structured like dumbbell dimers. However, the DFT calculations do not address the issue of charge transfer.

A comparative STM study [13] of Ni(110) and Cu(110) demonstrates that C_{60}, while forming a close-packed hexagonal overlayer on the Cu(110), induces a restructuring of the Ni(110) surface. This striking contrast is attributed to the difference in the interaction between the d-band of the surface and the C_{60} molecular orbitals. The surface (Ni) to molecule charge transfer is 2 ± 1 2 ± 1 electron/molecule [23]. We believe that the theoretical investigations are needed to confirm the direction of charge transfer and the nature of the molecule–surface bonding. Decomposition of C_{60} on the Ni(110) surface starts at about 760 K, above which a carbon layer was formed and transformed into graphitic carbon at higher temperature.

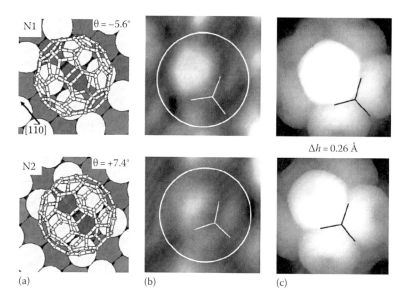

$\Delta h = 0.26 \text{ Å}$

(a)   (b)   (c)

FIGURE 39.15 (a) Chemisorption configurations for C_{60} on (1×1)-Pt(110) as deduced by DFT calculations. The polar tilt θ is indicated in the top right corner. The direction of polar tilt is highlighted by means of red dashed arrows. (b) Experimental STM images. (c) Simulations of the STM images for the two chemisorption configurations. (From Orzali, T. et al., *J. Phys. Chem. C*, 112, 378, 2008. With permission.)

Experimental investigations were also reported on fullerene–aluminum surfaces interaction, Al(111) and Al(110) [38,39]. On Al(111), following the first reported (6×6) phase, which is prepared at 620 K. Maxell et al. reported a metastable $\left(2\sqrt{3} \times 2\sqrt{3}\right) R30°$ structure, prepared at 300 K. A summary was provided (see Table 39.6). Charge transfer at the aluminum–C_{60} interface in thin-film multilayer structure was studied by Hebard et al. [38] using sheet resistance measurement as a probe (similar to Hebard's work on Cu surface discussed above). The measured conductance change was fitted into the Drude model and yielded a 6 electron/C_{60} charge transfer from Al surfaces to molecules. In the same work, this number is also suggested by Raman spectra. However, first-principle calculations by Wang and Cheng a decade later show that the calculated charge transfer is ~1.0 electron/C_{60} on Al(111) using similar techniques as for Cu, Ag, and Au surfaces.

On metal surfaces in general, modifications in the electronic structure of the C_{60} molecules are found to be responsible for the outcome of a number of experiments such as listed above: the enhancement in Raman spectroscopy and changes in resistance. It is a general phenomenon with few exceptions, that electrons from a metal surface are likely to transfer to the lowest-unoccupied band derived from molecule orbital of a C_{60}, which is a threefold degenerate level with a symmetry labeled as t_{1u}. The charge transfer leads to strong interactions between metal surfaces and fullerene molecules. Some metal surfaces such as Pt and Ni play catalytic roles in C_{60} decomposition in temperature-dependent studies.

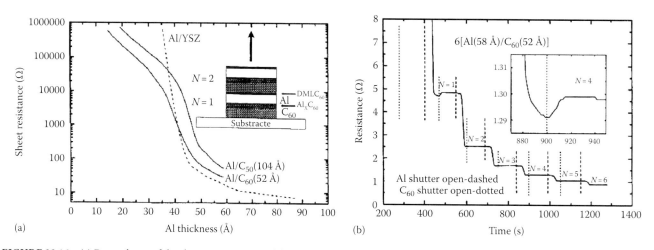

(a)   (b)

FIGURE 39.16 (a) Dependence of the sheet resistance on Al thickness for Al deposited directly on yttrium-stabilized zirconia (YSZ, dashed line and on underlying layer of C_{60} (solid lines); (b) dependence of resistance on deposition time for a six-layer sample. The vertical dashed (dotted) lines delineate the Al(C_{60}) shutter openings. The insets shows the increase in resistance at the beginning of the $N = 4$ plateau. (From Hebard, A.F. et al., *Nature*, 350, 600, 1991. With permission.)

TABLE 39.6 Work Function Change of C_{60} Adsorbed on Surfaces

	Expt. (eV)	ΔW (eV)	$\Delta \Phi$ (eV)	$\Delta \mu$ (Deby)
Cu(111)	−0.08	−0.09	−0.09	−0.21
Ag(111)		+0.14	+0.11	+0.24
Au(111)	−0.60	−0.58	−0.60	−1.37

Note: ΔW is calculated directly from the difference of the work functions. $\Delta \Phi$ is calculated from the change in the surface dipole moment, $\Delta \mu$, induced by adsorption of C_{60}.

39.2.2 Semiconductor and Insulator Surfaces

Compared to metal surfaces, there has been less controversy around studies of C_{60} assemblies on semiconductor or insulator surfaces. A few examples are included below to highlight the work on these surfaces.

Silicon: Sakamoto et al. measured the temperature dependence of the electronic structure of C_{60} films on Si(001)-(2 × 1) and Si(111)-(7 × 7) surfaces using photoelectron spectroscopy [40]. Interestingly, at 300 K, the interaction between most of the C_{60} molecules and the surface is physisorption in nature, or of weak van der Waals interaction (as compared to the strong chemical bonding). If the systems are annealed at 670 K, the bonding transforms into strong covalent-ionic bonding with a binding energy of 2.1 eV/molecule for both surfaces, and the charge transfer is 0.19 electron/molecule. When the temperature is over 1070 K, the molecules decompose and transform to SiC at the interfaces. Temperature effects and changes in the interfacial bonding were also noted in earlier studies [41,42].

Graphite: Graphite and carbon nanotubes are special hosts for studying C_{60}–surface (or host) interaction because they are all forms of graphitic carbon. An interesting finding is

that, using experimental techniques such as EELS, LEED, and desorption, on the graphite surface, the interaction between C_{60} and graphite is stronger than that between C_{60} and C_{60}, but is still of weak van der Waals interaction in nature. As a result, C_{60} films grow on the graphite surface layer by layer and the fullerene monolayer forms a close-packed structure on the surface [43]. Molecular dynamics simulations of surface deposition and C_{60} thin-film growth confirm that C_{60} does form close-packed structure on a (0001) graphite surface [44]. A combined experiment-simulation study [45] shows the desorption rates (C_{60} from graphite surface) as a function of temperature and estimates that the binding of a single C_{60} molecule to graphite is 0.85 eV. Calculations by Li and Cheng (a on-going project in author's group) show a 0.6 eV/C_{60} binding energy (using DFT with GGA) between the C_{60} and a single graphene sheet.

Ge, GaAs, etc.: Upon C_{60} adsorption, a Ge(111) surface undergoes a reconstruction. Two surface phases, $3\sqrt{3} \times 3\sqrt{3}R\,30°$ and $\sqrt{13} \times \sqrt{13}R\,14°$, and a localized 5 × 5 metastable phase are identified by Xu et al. using LEED and STM [46]. In a review article, Sakurai et al. [12] described C_{60} adsorption on four GaAs(001) surfaces, three As-rich and one Ga-rich.

Silica: This is a supporting surface often used for studying the molecule itself. However, an engineered silica surface can lead to assemble C_{60} monolayer on surface and make functional materials. In work by Gulino et al. [47], silica surfaces were first covered with an array of $ClCH_2$–C_6H_4–$SiCl_3$ groups via surface chemical reactions. A monolayer of 61-(p-hydroxyphenylmethano) fullerene (C_{60}) molecules (first synthesized and purified) was then deposited on the surface. The surface with fullerene linkage is hydrophobic and can be probed by oxygen-sensitive photoluminescence spectra.

TABLE 39.7 Summary of Results Reported by Maxwell et al. and References Therein

	Substrate	Type of Bonding	Approximate Desorption Temperature (K)	Mobility at Room Temperature	Equilibrium C_{60}–C_{60} Bond Length (Å)
(a)	Graphite	Weak, predominantly van der Waals	500	Mobile	10.01
	GeS(001)		490	Mobile	10.02
	SiO_2		470	Mobile	
(b)	Al(110)	Intermediate, predominantly covalent	730		9.91, 11.44, 12.14
	Al(111)		730	Mobile (steps)	9.91, ~10.04
(c)	Ag(110)	Intermediate, predominantly ionic		Mobile (island)	10.07, 10.23
	Ag(111)		770	Mobile (steps)	~10
	Au(110)		800	Mobile (steps)	10.04
	Au(111)		770	Mobile (steps)	~10
	Cu(110)		730		9.7–11.1
	Cu(111)			Mobile (steps)	10.1
(d)	Si(100)	Strong, predominantly covalent	Dissoc. at 1070 K	Immobile	9.87, 11.58
	Si(111)		Dissoc. at 1100 K	Immobile	
	Ge(111)		970	Immobile	14.4
	Ge(100)				9.6, 11.52
	Ni(110)		Dissoc. at 760 K	Reduced mobility	10, 10.5, ~11.6
	Pt(111)		Dissoc. at 1050 K	Low mobility	10.0 ± 0.3

Source: Maxwell, A.J. et al., *Phys. Rev. B*, 57, 7312, 1998.

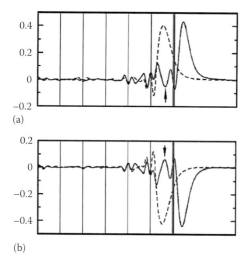

(a)

(b)

FIGURE 39.17 Calculated screening charge for *h*-BN/Ni(111). Panel (a) and (b) are for the screening charge of the −0.01 and +0.01 V/Å external electric field, respectively. (From Che, J.G. and Cheng, H.P., *Phys. Rev. B*, 72, 115436, 2005. With permission.)

h-BN/Ni(111): The surface itself is an interesting metal-insulator junction and is studied in a number of experiments. The *h*-BN monolayer is a good insulator that is nearly perfect commensuration with the Ni(111) surface. Deposition of a C_{60} monolayer on *h-BN/Ni(111)* was done by Muntwiler et al. [48]. Photoemission spectra measurements at different temperatures show that the work function decreases between 150 and 250 K. Charge transfer increases by nearly half electron/C_{60} when the temperature increases from 140 to 300 K. The charge transfer at low temperature is negligible. It is suggested that this charge trapping is due to coupling with phonons of C_{60} rocking motion. Che and Cheng thoroughly investigated this system using first-principle calculations [49]. The first interesting thing to notice is the screening charge in response to an electric field ±0.01 V/Å (panel a and b in Figure 39.17, respectively) for a Ni(111) surface (dashed line) versus a monolayer *h*-BN covered one (solid one). The vertical lines indicate the atomic layer positions. It is clear that the screening charge for the clean Ni(111) surface is localized in the surface region to screen out the external electric field, while in the other one it is not, except for slight Friedel oscillations. The *h*-BN layer does behave like an insulating layer with a dipole formed around it in response to the external electric field. A small peak can be seen as identified by the arrows in Figure 39.17. The binding energy is 0.47 eV/C_{60} with the center of a hexagon of C_{60} right on top of a N atom. However, this surface seems to be very smooth for C_{60} translation; energy differences among different sites are within 0.02 eV. When a pentagon is on top of a N atom, the energy increases by less than 0.003 eV. This small difference is beyond the accuracy of current DFT calculations. The two orientations are therefore, equivalent. Compared to C_{60}-Cu(111) surface, structural distortion of C_{60} due to interfacial interaction is small. Charge transfer from the surface to the molecule is small ~0.13 electron/C_{60} and remains small when the molecules are tilted to simulate rocking motion. This result is inconsistent with the experimental measurement of 0.4 electron increase, which is probably a dynamic effect.

39.3 Other Supporting Media

Although the focus of this chapter is supported C_{60} on surfaces, we provide some highlights of supported C_{60} in a number of other environments. The most notable area of research is metal-doped C_{60}, in which dopants occupy the interstitial space [30,50–61]. Much of these efforts are inspired by the superconductivity of alkali-metal-doped C_{60} solid that was discovered in 1991 [50,54,61].

Organic-C_{60} molecular composites also are a popular topic because of interesting physical properties of these systems. For example, Tdae-C_{60} exhibits soft ferromagnetism [62–64]; Porphyrin-C_{60} complexes are considered as potential materials for photosynthetic systems [65–67] which continues to attract intense attention. Attempts are also underway to use C_{60}-protein complexes as biosensors [68,69].

In the frontiers of nano-research, C_{60} is used to build single-molecule transistors, studies of Kondo effect [70], and nanomechanical oscillation [71]. Other exotic configurations include the C_{60}-nanocar [2] as depicted in Figure 39.1 at the beginning of this chapter, and carbon nanopeapods that encapsulate C_{60} inside carbon nanotubes [72–78]. In addition, C_{60} have also been deposited on gold nanoparticle surface [79], and assembled to form fullerene nanowire on gold [80] and nanowhiskers [81].

Acknowledgments

The author would like to thank the U.S. Department of Energy Basic Energy Science for supporting her work on supported fullerenes; Dr. Lin-Lin Wang, Professor Jingguang Che, and Dr. Lan Li for their collaboration; and Professor Samuel B. Trickey for reading this chapter and for helpful discussions.

References

1. L. L. Wang and H. P. Cheng, *Phys Rev B* **69**, 045404 (2004).
2. Y. Shirai, A. J. Osgood, Y. M. Zhao, Y. X. Yao, L. Saudan, H. B. Yang, Y. H. Chiu et al., *J Am Chem Soc* **128**, 4854 (2006).
3. M. Tinkham, *Group Theory and Quantum Mechanics*, McGraw-Hill, New York (1964), p. xii.
4. W. Kohn and L. J. Sham, *Phys Rev A* **140**, 1133 (1965).
5. G. Kresse and J. Furthmüller, *Institut für Materialphysik*, Universität Wien, Vienna, Austria (1999).
6. F. A. Cotton, *Chemical Applications of Group Theory*, Wiley, New York (1990), p. xiv.
7. J. C. Slater, *Quantum Theory of Molecules and Solids*, McGraw-Hill, New York (1963), p. 4 v.
8. R. Hoffmann, *Solids and Surfaces: A Chemist's View of Extended Bonding*, VCH Publishers, New York (1988), pp. x.
9. T. Hashizume, K. Motai, X. D. Wang, H. Shinohara, Y. Saito, Y. Maruyama, K. Ohno et al., *Phys Rev Lett* **71**, 2959 (1993).
10. A. Fartash, *J Appl Phys* **79**, 742 (1996).

11. T. Sakurai, X. D. Wang, T. Hashizume, V. Yurov, H. Shinohara, and H. W. Pickering, *Appl Surf Sci* **87–8**, 405 (1995).

12. T. Sakurai, X. D. Wang, Q. K. Xue, Y. Hasegawa, T. Hashizume, and H. Shinohara, *Prog Surf Sci* **51**, 263 (1996).

13. P. W. Murray, M. O. Pedersen, E. Laegsgaard, I. Stensgaard, and F. Besenbacher, *Phys Rev B* **55**, 9360 (1997).

14. K. D. Tsuei and P. D. Johnson, *Solid State Commun* **101**, 337 (1997).

15. K. D. Tsuei, J. Y. Yuh, C. T. Tzeng, R. Y. Chu, S. C. Chung, and K. L. Tsang, *Phys Rev B* **56**, 15412 (1997).

16. A. F. Hebard, R. R. Ruel, and C. B. Eom, *Phys Rev B* **54**, 14052 (1996).

17. E. I. Altman and R. J. Colton, *Surf Sci* **295**, 13 (1993).

18. E. I. Altman and R. J. Colton, *Phys Rev B* **48**, 18244 (1993).

19. S. J. Chase, W. S. Bacsa, M. G. Mitch, L. J. Pilione, and J. S. Lannin, *Phys Rev B* **46**, 7873 (1992).

20. K. L. Akers, L. M. Cousins, and M. Moskovits, *Chem Phys Lett* **190**, 614 (1992).

21. A. Rosenberg and D. P. Dilella, *Solid State Commun* **95**, 729 (1995).

22. B. W. Hoogenboom, R. Hesper, L. H. Tjeng, and G. A. Sawatzky, *Phys Rev B* **57**, 11939 (1998).

23. M. R. C. Hunt, S. Modesti, P. Rudolf, and R. E. Palmer, *Phys Rev B* **51**, 10039 (1995).

24. E. Giudice, E. Magnano, S. Rusponi, C. Boragno, and U. Valbusa, *Surf Sci* **405**, L561 (1998).

25. C. Cepek, R. Fasel, M. Sancrotti, T. Greber, and J. Osterwalder, *Phys Rev B* **63**, 125406 (2001).

26. J. K. Gimzewski, S. Modesti, and R. R. Schlittler, *Phys Rev Lett* **72**, 1036 (1994).

27. S. Modesti, J. K. Gimzewski, and R. R. Schlittler, *Surf Sci* **333**, 1129 (1995).

28. M. Hinterstein, X. Torrelles, R. Felici, J. Rius, M. Huang, S. Fabris, H. Fuess, and M. Pedio, *Phys Rev B* **77**, 153412 (2008).

29. S. Modesti, S. Cerasari, and P. Rudolf, *Phys Rev Lett* **71**, 2469 (1993).

30. M. R. C. Hunt, P. Rudolf, and S. Modesti, *Phys Rev B* **55**, 7889 (1997).

31. M. R. C. Hunt, P. Rudolf, and S. Modesti, *Phys Rev B* **55**, 7882 (1997).

32. L. L. Wang and H. P. Cheng, *Phys Rev B* **69**, 165417 (2004).

33. J. K. Gimzewski, S. Modesti, C. Gerber, and R. R. Schlittler, *Chem Phys Lett* **213**, 401 (1993).

34. C. T. Tzeng, W. S. Lo, J. Y. Yuh, R. Y. Chu, and K. D. Tsuei, *Phys Rev B* **61**, 2263 (2000).

35. N. Swami, H. He, and B. E. Koel, *Phys Rev B* **59**, 8283 (1999).

36. C. Cepek, A. Goldoni, and S. Modesti, *Phys Rev B* **53**, 7466 (1996).

37. T. Orzali, D. Forrer, M. Sambi, A. Vittadini, M. Casarin, and E. Tondello, *J Phys Chem C* **112**, 378 (2008).

38. A. F. Hebard, C. B. Eom, Y. Iwasa, K. B. Lyons, G. A. Thomas, D. H. Rapkine, R. M. Fleming et al., *Phys Rev B* **50**, 17740 (1994).

39. A. J. Maxwell, P. A. Bruhwiler, D. Arvanitis, J. Hasselstrom, M. K. J. Johansson, and N. Martensson, *Phys Rev B* **57**, 7312 (1998).

40. K. Sakamoto, D. Kondo, Y. Ushimi, M. Harada, A. Kimura, A. Kakizaki, and S. Suto, *Phys Rev B* **60**, 2579 (1999).

41. D. Chen and D. Sarid, *Surf Sci* **329**, 206 (1995).

42. X. W. Yao, T. G. Ruskell, R. K. Workman, D. Sarid, and D. Chen, *Surf Sci* **367**, L85 (1996).

43. M. F. Luo, Z. Y. Li, and W. Allison, *Surf Sci* **404**, 437 (1998).

44. H. Rafii-Tabar and K. Ghafoori-Tabrizi, *Prog Surf Sci* **67**, 217 (2001).

45. H. Ulbricht, G. Moos, and T. Hertel, *Phys Rev Lett* **90**, 095501 (2003).

46. H. Xu, D. M. Chen, and W. N. Creager, *Phys Rev B* **50**, 8454 (1994).

47. A. Gulino, S. Bazzano, G. G. Condorelli, S. Giuffrida, P. Mineo, C. Satriano, E. Scamporrino, G. Ventimiglia, D. Vitalini, and I. Fragala, *Chem Mater* **17**, 1079 (2005).

48. M. Muntwiler, W. Auwarter, A. P. Seitsonen, J. Osterwalder, and T. Greber, *Phys Rev B* **71**, 121402 (2005).

49. J. G. Che and H. P. Cheng, *Phys Rev B* **72**, 115436 (2005).

50. A. F. Hebard, M. J. Rosseinsky, R. C. Haddon, D. W. Murphy, S. H. Glarum, T. T. M. Palstra, A. P. Ramirez, and A. R. Kortan, *Nature* **350**, 600 (1991).

51. W. L. Yang, V. Brouet, X. J. Zhou, H. J. Choi, S. G. Louie, M. L. Cohen, S. A. Kellar et al., *Science* **300**, 303 (2003).

52. S. Satpathy, V. P. Antropov, O. K. Andersen, O. Jepsen, O. Gunnarsson, and A. I. Liechtenstein, *Phys Rev B* **46**, 1773 (1992).

53. R. J. C. Batista, M. S. C. Mazzoni, L. O. Ladeira, and H. Chacham, *Phys Rev B* **72**, 085447 (2005).

54. K. Holczer, O. Klein, G. Gruner, J. D. Thompson, F. Diederich, and R. L. Whetten, *Phys Rev Lett* **67**, 271 (1991).

55. Y. Kubozono, Y. Takabayashi, S. Fujiki, S. Kashino, T. Kambe, Y. Iwasa, and S. Emura, *Phys Rev B* **59**, 15062 (1999).

56. D. Varshney, N. Kaurav, and K. K. Choudhary, *Supercond Sci Technol* **18**, 1259 (2005).

57. M. G. Mitch, S. J. Chase, and J. S. Lannin, *Phys Rev Lett* **68**, 883 (1992).

58. M. J. Rosseinsky, A. P. Ramirez, S. H. Glarum, D. W. Murphy, R. C. Haddon, A. F. Hebard, T. T. M. Palstra, A. R. Kortan, S. M. Zahurak, and A. V. Makhija, *Phys Rev Lett* **66**, 2830 (1991).

59. K. Sakamoto, T. Wakita, D. Kondo, A. Harasawa, T. Kinoshita, W. Uchida, and A. Kasuya, *Surf Sci* **499**, 63 (2002).

60. Y. Y. Wang, R. Yamachika, A. Wachowiak, M. Grobis, and M. F. Crommie, *Nat Mater* **7**, 194 (2008).

61. R. C. Haddon, A. F. Hebard, M. J. Rosseinsky, D. W. Murphy, S. J. Duclos, K. B. Lyons, B. Miller et al., *Nature* **350**, 320 (1991).

62. P. M. Allemand, K. C. Khemani, A. Koch, F. Wudl, K. Holczer, S. Donovan, G. Gruner, and J. D. Thompson, *Science* **253**, 301 (1991).

63. P. W. Stephens, D. Cox, J. W. Lauher, L. Mihaly, J. B. Wiley, P. M. Allemand, A. Hirsch et al., *Nature* **355**, 331 (1992).

64. B. Narymbetov, A. Omerzu, V. V. Kabanov, M. Tokumoto, H. Kobayashi, and D. Mihailovic, *Nature* **407**, 883 (2000).

65. H. Imahori and Y. Sakata, *Adv Mater* **9**, 537 (1997).

66. D. M. Guldi, *Chem Soc Rev* **31**, 22 (2002).

67. D. Kuciauskas, P. A. Liddell, S. Lin, T. E. Johnson, S. J. Weghorn, J. S. Lindsey, A. L. Moore, T. A. Moore, and D. Gust, *J Am Chem Soc* **121**, 8604 (1999).

68. H. W. Chang, and J. S. Shih, *Sensor Actuat B-Chem* **121**, 522 (2007).

69. H. W. Chang, and J. S. Shih, *J Chin Chem Soc-Taip* **55**, 318 (2008).

70. J. Park, A. N. Pasupathy, J. I. Goldsmith, C. Chang, Y. Yaish, J. R. Petta, M. Rinkoski et al., *Nature* **417**, 722 (2002).

71. H. Park, J. Park, A. K. L. Lim, E. H. Anderson, A. P. Alivisatos, and P. L. McEuen, *Nature* **407**, 57 (2000).

72. B. W. Smith, M. Monthioux, and D. E. Luzzi, *Nature* **396**, 323 (1998).

73. D. J. Hornbaker, S. J. Kahng, S. Misra, B. W. Smith, A. T. Johnson, E. J. Mele, D. E. Luzzi, and A. Yazdani, *Science* **295**, 828 (2002).

74. S. Berber, Y. K. Kwon, and D. Tomanek, *Phys Rev Lett* **88** (2002).

75. S. Bandow, T. Hiraoka, T. Yumura, K. Hirahara, H. Shinohara, and S. Iijima, *Chem Phys Lett* **384**, 320 (2004).

76. M. H. Du, and H. P. Cheng, *Phys Rev B* **68**, 113402 (2003).

77. L. A. Girifalco, M. Hodak, and R. S. Lee, *Phys Rev B* **62**, 13104 (2000).

78. Y. G. Yoon, M. S. C. Mazzoni, and S. G. Louie, *Appl Phys Lett* **83**, 5217 (2003).

79. Y. Fang, Q. J. Huang, P. J. Wang, X. Y. Li, and N. T. Yu, *Chem Phys Lett* **381**, 255 (2003).

80. N. Neel, J. Kroger, and R. Berndt, *Appl Phys Lett* **88**, 163101 (2006).

81. K. S. Iyer, C. L. Raston, and M. Saunders, *Lab Chip* **7**, 1121 (2007).

Fullerene Suspensions

Nitin C. Shukla
*Virginia Polytechnic Institute
and State University*

Scott T. Huxtable
*Virginia Polytechnic Institute
and State University*

40.1 Introduction

Carbon fullerenes, or buckyballs, are large molecules that typically contain 60 or more carbon atoms and are in the form of hollow spheres or ellipsoids, as shown in Figure 40.1. Fullerenes were discovered (Kroto et al., 1985) in the mid-1980s, earning the Nobel Prize in Chemistry in 1996 for Curl, Kroto, and Smalley (Service, 1996). However, since the production of fullerenes is accompanied by the generation of large quantities of soot, the fullerenes must first be separated and extracted from soot and other impurities. Thus, despite intense interest, the early years of research on fullerenes were hindered by the difficulty of extracting and purifying macroscopic quantities of these unique and exciting new molecules. Eventually, Krätschmer et al. (1990) successfully isolated macroscopic quantities of C_{60} by dissolving the fullerene and soot mixtures in benzene. This breakthrough opened the doors to fullerene research for different research groups around the world, and rapid progress toward understanding these nanoparticles began.

In the past 20 years, the interest and scientific literature involving fullerenes has exploded. The unique properties of carbon fullerenes have generated speculation that they could be used in applications such as antioxidants, neuroprotective agents, enzyme inhibitors, antimicrobial and antiviral agents, solar cells, tumor therapy, etc. (Jensen et al., 1996; Kadish and Ruoff, 2000; Bosi et al., 2003; Brabec and Durrant, 2008). An important aspect of many of these potential applications is that the fullerenes must be dissolved or suspended in a liquid. Even for those applications where fullerenes will be incorporated in solids, the fullerenes must be dissolved or suspended in a liquid at some point during the research and development for separation, isolation, or purification.

For biological and medical applications, there is the complication that pristine fullerenes are extremely hydrophobic, and thus they are essentially insoluble in water. Therefore, for biocompatibility, the fullerenes must be modified in some way so that they are water soluble. Additionally, these applications place further restrictions on the suspension, as the fullerenes must be made water soluble without the use of toxic chemistry that could leave behind harmful residuals.

Finally, the solubility of fullerenes is not only of significance directly for applications, but it is also important for environmental reasons. As fullerenes, and other nanoparticles, find applications in consumer and industrial products, they will ultimately be released into the environment, and it is important to know how these particles will be transported in water and other solvent mixtures.

In this chapter, we discuss the solubility of fullerenes in a variety of liquids, with special emphasis on approaches to suspend these particles in water. Most of the discussion in this chapter focuses on C_{60} and C_{70} since the literature is limited on suspensions of higher order fullerenes. Studies on the suspensions of higher-order fullerenes (C_{74} to beyond C_{100}) have been hindered due to two primary reasons (Langa and Nierengarten, 2007). First, as the size of a fullerene molecule increases, the number of isomers of the same symmetry also increases, thereby making it difficult to definitively assign the exact structure. Second, the solubility of fullerenes tends to decrease with increasing size, making it more difficult to extract, separate, and study these larger fullerenes. In concluding this chapter, we also briefly discuss the suspensions of endohedral metallofullerenes, toxicity of fullerenes, and applications of fullerenes, and the importance of suspensions in these applications.

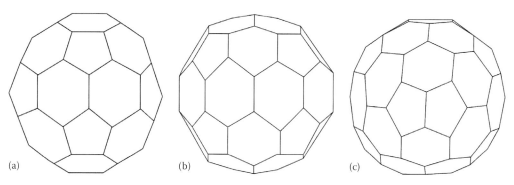

FIGURE 40.1   Schematic diagrams of fullerene structures. (a) C_{60}, (b) C_{70}, (c) C_{84}.

40.2 General Solubility of Pristine Fullerenes

Exhaustive tables of the solubility of C_{60} in over 100 solvents can be found in the literature (Beck and Mandi, 1997; Korobov and Smith, 2000). Some of these solubility values are listed in Table 40.1 for C_{60} and in Table 40.2 for C_{70}. In general, good agreement is found in the literature for solubility measurements among different research groups. Some differences may be attributed to equilibration times as several groups have found that excessive stirring (up to several days) may be required in order to reach saturated concentrations that are stable. Another factor is the temperature dependence (Ruoff et al., 1993a) of solubility as groups have reported data at various temperatures. Finally, fullerenes appear to react slowly with some solvents, so care must be taken when determining solubility in these solvents (Taylor, 1995).

From the numerous experimental studies, some trends are evident with regard to solubility (Bezmel'nitsyn et al., 1998). Fullerenes have extremely low solubility in polar solvents such as acetone and alcohols. In alkanes, such as pentane, hexane, and decane, fullerenes become weakly soluble, with the solubility increasing with an increase in the number of carbon atoms. Aromatic hydrocarbons are excellent solvents, with the naphthalene derivatives having the highest solubility.

Other studies have found that the best solvents have some properties that are similar to fullerenes such as high refractive index, dielectric constant around four, Hildebrand solubility parameter of ~10 cal$^{1/2}$ cm$^{-3/2}$, large molecular volume, and the tendency to act as a moderate-strength nucleophile (Ruoff et al., 1993b). Additionally, aromatic hydrocarbons are made up of benzene rings, which have structure similar to that of the regular hexagons on the fullerene surface. Many researchers have noted that the solubility of fullerenes generally follows the "like dissolves like" rule of thumb.

However, despite the extensive experimental work, accurately predicting the solubility of fullerenes in a particular solvent has proven to be difficult as there is no single parameter that correlates well with solubility measurements. Therefore, many groups have taken to using a variety of multiparameter statistical methods such as multiple linear regression analysis to determine the trends and correlations of fullerene solubility on various solvent parameters.

TABLE 40.1   Solubility of C_{60} at Room Temperature

Solvent	Solubility (mg/mL)
Nonaromatic hydrocarbons	
Noncyclic	
n-Pentane	0.005, 0.004
n-Hexane	0.043, 0.040, 0.052
Octane	0.025
n-Decane	0.071, 0.070
Tetradecane	0.126
Cyclic	
Cyclopentane	0.002
Cyclohexane	0.036, 0.035, 0.051
Decalin (3:7 *cis* and *trans* mixture)	4.6
1,5,9-Cyclododecatriene	7.14
Cyclic derivatives	
Cyclohexyl iodide	8.06
(±)-*trans*-1,2-Dibromocyclohexane	14.28
Halogen-containing alkanes	
Chloroform	0.16, 0.17
Carbon tetrachloride	0.32, 0.45
Bromoform	5.64
1,1,2,2-Tetrachloroethane	5.3
1,2,3-Tribromopropane	8.31
Halogen-containing alkenes	
1,2-Dibromoethylene	1.84
Aromatic hydrocarbons	
Benzene	1.70, 1.5, 0.88, 0.89, 1.44
Benzene derivatives	
Toluene	2.80, 2.90, 0.54, 2.90, 2.15
1,4-Dimethylbenzene	5.9, 3.14
1,2,4-Trimethylbenzene	17.9
1,2,3,5-Tetramethylbenzene	20.8
Tetralin	16
Halogen derivatives	
1,2-Dichlorobenzene	27.00, 24.60, 7.11
1,2-Dibromobenzene	13.8

TABLE 40.1 (continued)   Solubility of C_{60} at Room Temperature

Solvent	Solubility (mg/mL)
1,3-Dichlorobenzene	2.4
1,3-Dibromobenzene	13.8
1,2,4-Trichlorobenzene	8.50, 10.40, 4.85
Other derivatives	
Nitrobenzene	0.8
Anisole	5.6
Thiophenol	6.91
Benzyl derivatives	
Benzyl bromide	4.94
α,α,α-Trichlorotoluene	4.8
Naphthalene derivatives	
1-Methylnaphthalene	33.0, 33.2
Dimethylnaphthalene	36
1-Phenylnaphthalane	50
1-Chloronaphthalene	51
1-Bromo-2-methylnaphthalene	34.8
Polar solvents	
Methanol	0.000, 0.000035
Ethanol	0.001, 0.0008
Nitroethane	0.002
Acetone	0.001
n-Butylamine	3.688
Miscellaneous	
Carbon disulfide	7.90, 7.70, 7.90, 5.16
2-Methylthiophene	6.8
Quinoline	7.2
Inorganic solvents	
Water	1.3×10^{-17}

TABLE 40.2   Solubility of C_{70} at Room Temperature

Solvent	Solubility (mg/mL)
n-Pentane	0.002
n-Hexane	0.013
n-Decane	0.053
Dodecane	0.098
Dichloromethane	0.08
Tetrachloromethane	0.12
Isopropanol	0.0021
Acetone	0.0019
Benzene	1.3
Toluene	1.4
Xylene	14.3
Methithilol	1.47
1,2-Dichlorobenzene	36.2
Carbon disulfide	9.9

Murray et al. (1995) used a multivariate statistical approach to show that the solubilities of C_{60} in several organic solvents could be represented analytically in terms of the surface areas and electrostatic potentials of the solvent molecular surfaces. Their results suggested that large solvent molecules and moderately attractive interactions that are balanced by positive and negative regions on the solute and solvent molecular surfaces promote solubility.

Several groups have used quantitative structure–property relationship (QSPR) studies to better understand and predict fullerene solubility (Smith et al., 1996; Danauskas and Jurs, 2001; Sivaraman et al., 2001; Liu et al., 2005; Gharagheizi and Alamdari, 2008). As the name implies, these analyses aim to find quantitative relationships that can be used for the prediction of the properties, e.g., solubility, of molecular structures. These techniques have had some success, although they are, understandably, most accurate for the predictions of compounds that are chemically similar to those that were analyzed in the development of the model.

Using a linear solvation energy relationship (LSER) form of a QSRP, Smith et al. (1996) found that a good solvent for C_{60} should have large molar volume and large covalent acidity and basicity indices to go along with the minimal regions of negative charge on any atom. Marcus et al. (2001) also used a linear solvation energy approach to model the solubility of C_{60} in a variety of organic solvents. They determined that increasing molar volume and solvent polarity diminished the solubility of C_{60}, while polarizability and the ability to donate electron pairs improved solubility.

Kiss et al. (2000) used an artificial neural network approach to predict the solubility of C_{60} in several solvents. They found that the solubility decreases with increasing solvent molar volumes, but that solubility increases with increasing polarizability and the saturated surface areas of the solvents.

40.3 Suspension of Fullerenes in Water

Since fullerenes are hydrophobic molecules, they are extremely difficult to dissolve in water or other polar solvents. Given the numerous medical, biological, and other applications that require water solubility, there have been many approaches taken toward suspending fullerenes in water. These methods include chemically modifying the surfaces of the fullerenes; incorporating the fullerene molecule into host–guest complexes; encapsulating the fullerenes in micelles, vesicles, or liposomes; or attaching surfactants to the fullerene surface. Other groups have had some success creating colloidal suspensions of fullerenes without the use of any stabilizing agent through solvent exchange techniques or extended stirring of a fullerene/water mixture.

40.3.1 Encapsulation of C_{60} in Host–Guest Complexes and Micelles

One approach toward preparing stable suspensions of carbon fullerenes is through the use of host–guest complexes (Diederich and Gomez-Lopez, 1999). Here, guest molecules are used to effectively shield the hydrophobic fullerene from the polar aqueous environment. One example of such a water-soluble complex is the use of β- or γ-cyclodextrin (CD) where the C_{60} molecule is effectively enclosed by two CD molecules, as shown in Figure 40.2 (Andersson et al., 1992, 1994; Priyadarsini et al., 1994; Murthy and Geckeler, 2001). Andersson et al. (1992) noted that a boiling aqueous solution of γ-CD was able to selectively extract C_{60} from a mixture of C_{60} and C_{70}. Priyadarsini et al. (1994) reported that when C_{60} was embedded within the γ-CD cavity, the triplet lifetime of C_{60} increased and its reactivity with O_2 decreased.

Initial work on CD focused on γ-CD as it was believed that β-CD would not form an inclusion complex with C_{60}, but Murthy and Geckeler (2001) produced a stable water-soluble complex. β-CD is especially attractive for biological applications since it is essentially nontoxic and available commercially at low cost (Murthy and Geckeler, 2001).

Calixarenes are bowl-shaped macrocycles with hydrophobic cavities that have also been used to generate water-soluble host–guest complexes with C_{60}, as shown in Figure 40.3 (Williams and Verhoeven, 1992; Atwood et al., 1994; Suzuki et al., 1994; Ikeda et al., 1999; Shinkai and Ikeda, 1999; Zhong et al., 2001). Atwood et al. (1994) demonstrated that calixarenes could be used to selectively separate C_{60} from a mixture of C_{60} and C_{70}. These complexes are promising for biological applications as Ikeda et al. (1999) used a calixarene-C_{60} complex as an efficient DNA photocleavage reagent. One issue that requires further study is the impact of the functionalization of the calixarenes and fullerenes on the molecular encapsulation process and the properties of the fullerene itself (Kunsagi-Mate et al., 2008).

Bensasson et al. (1994) demonstrated that C_{60} can be dispersed in micellar solutions using Triton X-100 and Triton X-100 R-S. The fullerene molecules were localized in the inner hydrophobic region of the micelle core, while the polar OH groups located at the outer surface of the micelle were in direct contact with water, as shown in Figure 40.4. They also showed that aggregates of C_{60} could be incorporated into phosphatidylcholine liposome colloidal solutions. These vesicles were closed spherical structures of lipid bilayers that closely resemble biological membranes.

At nearly the same time, Hungerbühler et al. (1993) used a similar approach to incorporate C_{60} into artificial lipid membranes using Triton X-100 as well. They found that the absorption spectrum of C_{60} in these structures was significantly different from a spectrum obtained when C_{60} was dissolved in *n*-hexane. Beeby et al. (1994) also examined C_{60} in the aqueous micellar solutions of Trinton X-100 and found both molecular and colloidal states. More recently, Ramakanth and Patnaik (2008) encapsulated fullerenes in Triton X-100 and found that the micelles were monomeric below a critical concentration (0.025 mM), but that cluster formation was evident at higher concentrations.

Yamakoshi et al. (1994) solubilized C_{60} and C_{70} in water with a polyvinylpyrrolidone (PVP) micellar system. They found a red-shift in the ultraviolet-visible (UV-VIS) spectrum that they attributed to an intermolecular interaction between C_{60} and PVP. Kai et al. (2003) demonstrated photoinduced antibacterial

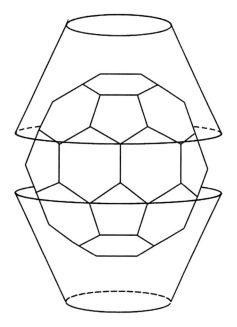

FIGURE 40.2 Cyclodextrin structure.

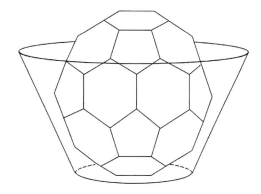

FIGURE 40.3 Calixarene structure.

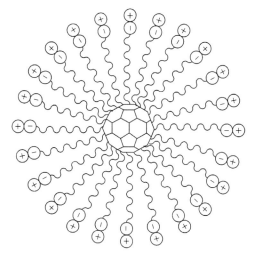

FIGURE 40.4 C_{60} in Triton X-100.

activity in aqueous solutions of C_{60} in PVP. Jenekhe and Chen (1998) used self-assembled block copolymers to encapsulate and suspend large aggregates of C_{60} and C_{70} (over 10^{10} fullerene molecules per aggregate).

In general, the encapsulation of fullerenes in a host–guest or micellar system does not appear to significantly alter the fullerene properties. Lee et al. (2007) found that the photochemical reactivity of fullerenes encapsulated in surfactants and polymers was unchanged for colloids below 5 nm in size, but that it was lost for aggregates ~100 nm in size.

Many of these observations demonstrate recurring themes from various efforts to suspend fullerenes in water. The absorption spectrum and other properties may be altered, and many suspensions contain aggregates as well as individually suspended C_{60} molecules. These effects are found to be strongly dependent on the sample preparation and treatment techniques.

40.3.2 Attachment of Hydrophilic Groups to Fullerenes

Several research groups have also created water-soluble fullerenes by a covalent modification of C_{60} with organic functionalities that contain ionic moieties (Lamparth and Hirsch, 1994; Brettreich and Hirsch, 1998; Okamura et al., 2000; Richardson et al., 2000). Richardson et al. (2000) reported on the synthesis of several amino-substituted methanofullerenes that exhibited water solubility with some aggregation. Brettreich and Hirsch (1998) synthesized a dendrimeric fullerene derivative with 18 carboxylic groups and reached a solubility of 8.7 mg of C_{60} per mL of water at pH of 7.4 and found the solubility increased by a factor of 7 for basic solutions with pH 10. Fullerenes suspended with ionic groups have also been used for biological applications, including the inhibition HIV enzymes and DNA cleavage (Sijbesma et al., 1993; Takenaka et al., 1999).

In addition to the considerable work using covalent attachment of ionic groups, others have used nonorganic functionalities to suspend fullerenes in water (Da Ros et al., 1996; Tabata and Ikada, 1999; Samal and Geckeler, 2000; Wharton et al., 2001). The use of nonionic groups could help reduce some of the problems associated with pH dependence of solubility, hyperosmolality, and chemical instability associated with many other water-solubilizing techniques (Wharton et al., 2001). Samal and Geckeler (2000) used the covalent bonding of fullerenes with cyclodextrins to create water-soluble conjugates where the fullerene was not embedded within the cyclodextrin in a host–guest complex. Wharton et al. (2001) reported the creation of nonionic derivatives of C_{60} that were highly soluble in water at biological pH. Tabata and Ikada (1999) chemically conjugated poly(ethylene glycol) (PEG) with C_{60} in order to make it soluble and to increase the molecular size so that it would preferentially accumulate in a tumor. The local irradiation of the tumor site with visible light induced necrosis in the tumor.

One drawback of solubilizing fullerenes through hydrophilic appendages is that the fullerenes still tend to aggregate and form clusters if only one hydrophilic chain is added to the fullerene. In this case, the fullerenes may form a cluster where the hydrophilic appendages are on the outside of the cluster, and the hydrophobic fullerene spheres stick together on the inside of the cluster. In addition to the size of the aggregates influencing the transport properties of the fullerenes, the formation of clusters can dramatically alter the photophysical properties of the fullerene suspensions. Specifically, the triplet state lifetime can decrease, absorption coefficients may decrease, absorption bands may broaden, and other structural features may be lost (Eastoe et al., 1995; Da Ros and Prato, 1999). These effects are significant for several medical applications that involve photodynamic therapies.

40.3.3 Clusters of C_{60} in Water

While true molecular solutions of pristine fullerenes in pure water are not possible, several groups have reported stable colloidal suspension of fullerenes without the use of any stabilizer. These stable suspensions of fullerene aggregates in water are sometimes referred to as *fullerene water suspensions* (FWS), nano-C_{60}, or *n*-C_{60} suspensions.

Stable suspensions of C_{60} aggregates have been created by two general techniques. Several groups have prepared aqueous dispersions by first dissolving C_{60} in an organic solvent, followed by the dilution of that solution with water and the subsequent removal of the organic solvent, leaving the fullerenes suspended in water. These types of approaches are often referred to as solvent exchange techniques (Brant et al., 2006; Dhawan et al., 2006). A second technique for producing colloidal suspensions of C_{60} simply involves the extended or prolonged stirring of fullerenes in water without the use of organic solvents (Cheng et al., 2004; Brant et al., 2006; Dhawan et al., 2006; Ma and Bouchard, 2009).

One of the first demonstrations of a solvent exchange technique was given by Scrivens et al. (1994). They started with a solution of C_{60} in benzene and then diluted it with tetrahydrofuran (THF), acetone, and water. After distilling out the solvents, they were left with a stable suspension of C_{60} in water. Deguchi et al. (2001) prepared stable polycrystalline clusters of C_{60} and C_{70} in water by diluting a saturated solution of fullerenes in THF with water and then removing the THF. They found monodisperse clusters of fullerenes (60 nm in diameter) that were extremely stable, with no precipitation found after 9 months of storage. Andrievsky et al. (1995) used ultrasonication to remove toluene from a solution of fullerenes, toluene, and deionized water to form a stable colloidal solution of fullerenes with aggregate sizes below 220 nm. Recent work by Fortner et al. (2005) has shown that under certain conditions, C_{60} can spontaneously form a water-stable colloidal aggregate with dimensions of ~25–500 nm. These aggregates are found to possess solubility up to 100 mg/L.

Brant et al. (2006) examined the impact of preparation method on fullerene clusters and chemistry. They characterized colloidal dispersions of *n*-C_{60} prepared by several solvent exchange techniques and other dispersions created by extended mixing in water only. The method of preparation had a significant impact on the characteristics of the colloidal fullerenes, and the properties of the solvent, in particular, were found to have a strong influence on the size, morphology, charge, and hydrophobicity of the n-C$_{60}$. Duncan et al. (2008) also examined the effects of preparation method on the size, structure, and surface charge of C$_{60}$ colloids and found similar evidence that the preparation method has a significant impact on the properties of the colloid. Lee and Kim (2008) found that encapsulating agents and natural organic matter had a substantial effect on the dispersion and photochemical activity of C$_{60}$ clusters.

The stabilizing mechanism for these fullerene clusters is not entirely clear, although there is some evidence that it is electrostatic in origin (Brant et al., 2005). One obvious and significant drawback of some of the solvent exchange approaches for biological use is that they use highly toxic solvents in the preparation process, and there is a danger that the aqueous clusters may contain residual solvents.

40.4 Toxicity, Health, and Environmental Issues with Fullerene Suspensions

Accurate measurements and understanding of the toxicity, environmental, and health issues of fullerene suspensions are complicated by several factors (Andrievsky et al., 2005; Klaine et al., 2008; Wiesner et al., 2008). First, while fullerenes are not readily soluble in water, there are a variety of avenues for fullerenes to become dispersed in aqueous suspensions, as discussed previously. The most obvious issue is that some of the techniques used to suspend the fullerenes may leave traces of toxic solvents in the aquatic suspensions.

A second issue is that each of the methods used for suspending the fullerenes may induce changes in the properties of the fullerenes, which may, in turn, make certain suspensions more harmful than others. Additionally, further changes in the fullerenes may take place once they are released into aquatic systems, which may also alter their properties (Brant et al., 2005). Thus, the potential health effects of the fullerenes may differ dramatically from the observations of the original samples in laboratory studies.

Furthermore, as discussed by Fortner et al. (2005), colloidal aggregates of C$_{60}$ can generate concentrations of C$_{60}$ of up to ~100 mg/L. With this in mind, they correctly point out that the environmental fate, distribution, and biological risk associated with fullerenes will depend not only on the properties of individual molecules, but also on the properties of the aggregate. However, these large aggregates may ultimately settle out of some suspensions or prove to be largely removable through coagulation and filtration processes for certain cases.

Several research groups have shown that water-soluble fullerenes can pass through the blood–brain barrier, and that they can migrate through the body and ultimately accumulate in the liver. This was in direct contrast to the behavior of unmodified

fullerenes, which were rapidly absorbed by serum protein in the blood after injection (Nakamura and Isobe, 2003). Dhawan et al. (2006) recently examined the genotoxicity of colloidal dispersions of C$_{60}$ in water that were prepared by solvent exchange (ethanol to water) and by extended mixing. Sayes et al. (2004) demonstrated that the toxicity of water-soluble fullerenes to cells has a strong dependence on the surface derivatization. In particular, they found that the least derivatized, and most aggregated, fullerenes were considerably more toxic than highly soluble derivatives. Despite the recent activity dealing with the apparent toxicity of fullerenes, the specific mechanisms responsible for the toxic effects of these nanoparticles are not entirely clear. Several research groups have speculated that the cytotoxic and antibacterial properties of n-C$_{60}$ are due to photocatalytically generated reactive oxygen species (ROS). However, Lyon et al. (2008) neither found any ROS production nor any ROS-mediated damage in n-C$_{60}$-exposed bacteria, and they explained a possible mechanism for false positives in previous work. They proposed an alternate hypothesis that n-C$_{60}$ behaves as an oxidant and that it exerts an ROS-independent oxidative stress.

40.5 Endohedral Fullerenes

Endohedral fullerenes are a class of fullerene derivatives that contain atoms or molecules within the fullerene cage (see Figure 40.5). In 1985, Smalley and coworkers speculated about the existence of lanthanum-based endohedral fullerenes (Heath et al., 1985), but it was not until 1991 that their presence was confirmed (Chai et al., 1991). Endohedral fullerenes are frequently represented as M@C$_n$, where M and n represent the encaged atom or molecule and the number of carbon atoms in fullerene molecule, respectively, and the symbol "@" indicates that preceding molecule is encaged within the fullerene cage. For example, La@C$_{82}$ denotes lanthanum encaged in C$_{82}$.

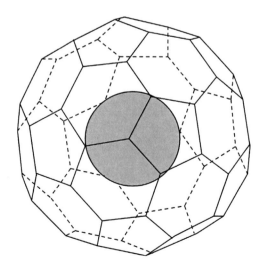

FIGURE 40.5 C$_{60}$ molecule with an atom enclosed inside (endohedral fullerene).

Endohedral fullerenes that contain metal atoms are called *metallofullerenes*. They are interesting materials because of their increased reactivity compared to hollow fullerenes. The metallofullerenes are separated from soot by dissolving them in common solvents such as toluene, carbon disulfide, and 1,2,4-trichlorobenzene (Yamamoto et al., 1994). For higher yields, hot solvents or ultrasonic extraction is utilized. Subsequently, high-performance liquid chromatography (HPLC) is employed to isolate and purify the metallofullerenes.

Suspensions of metallofullerenes are receiving attention for possible use in biological applications. Since metallofullerenes do not release the potentially harmful metal atom contained within the cage, they are attractive for diagnostic applications such as magnetic resonance imaging (MRI) contrast agents, x-ray imaging agents, and radiopharmaceuticals (Shinohara, 2000). Endohedral fullerenes are also considered for potential applications in nano-electronics and nano-optoelectronics (Lee et al., 2002; Benjamin et al., 2006). The lack of large-scale production currently limits the practical usage of endohedral metallofullerenes.

40.6 Applications of Fullerene Suspensions

The difficulty in preparing water-soluble fullerenes that can be homogeneously dissolved has greatly impeded progress in the development of fullerene-based medical therapies. Additionally, for these biological applications, it is obviously critical that any hydrophilic attachments, or other modifications of the fullerenes used to improve solubility, be nontoxic, and they must also not significantly interfere with the properties of the fullerenes. Despite these limitations with solubility and toxicity, there have been numerous exciting uses of fullerene suspensions for biological and medical applications.

Fullerenes and fullerene derivatives have demonstrated a variety of interesting biological phenomena, including selective DNA cleaving ability (Tokuyama et al., 1993; Boutorine et al., 1994), the inhibition of HIV enzymes (Friedman et al., 1993; Sijbesma et al., 1993), photoinduced antibacterial activity (Kai et al., 2003; Mashino et al., 2003), the ability to act as carriers for gene and drug delivery systems (Isobe et al., 2006), and the ability to act as neuroprotective agents (Dugan et al., 1997). Many of these applications have been discussed in several excellent review articles (Jensen et al., 1996; Da Ros and Prato, 1999; Wilson, 2000; Bosi et al., 2003; Bakry et al., 2007).

The photophysical properties, in conjunction with the size and shape of fullerenes, are especially interesting for biological and medical applications. When illuminated with visible or UV light, fullerenes are excited to a long-lived triplet state. In the presence of molecular oxygen, the excited fullerene reduces to produce singlet oxygen and superoxide radical anions that are known for their photodynamic antitumor and DNA cleaving activities (Bosi et al., 2004). The tendency of fullerenes and fullerene derivatives to localize around the cancerous tissue

further enhances their potential for photodynamic therapy (Tabata and Ikada, 1999). In addition to their photophysical properties, fullerenes have also been shown to act as inhibitors of the HIV-1 protease due to the shape and size of the fullerene molecule as the molecule fits into the hydrophobic cavity of the HIV-1 protease (Friedman et al., 1993; Sijbesma et al., 1993).

Along with the medical and biological applications, fullerenes have been used in a variety of other instances. For example, fullerene derivatives are used to improve the overall power conversion efficiency of polymer-based photovoltaic (PV) cells (Reyes-Reyes et al., 2005; Brabec and Durrant, 2008). Fullerene loading in the host polymer matrix of these PV cells allows photoinduced electron transfer from the polymer to the fullerene, which subsequently leads to improved efficiency. Phenyl C61-butyric acid methylester (PCBM)/poly(3-hexylthiophene) (P3HT)-based PV cells are among the most widely used organic PV devices, and these cells are prepared by evaporating the solvent from a PCBM/P3HT/solvent tertiary mixture (Saunders and Turner, 2008). In all of these PV applications, fullerenes are suspended in some solvent during processing and/or fabrication, and thus the solubility of functionalized fullerenes and their stability in solution play key roles in the development of these devices.

Fullerenes also act as effective optical limiters. Optical transmission through fullerene suspensions decreases with increasing laser fluence, and this property can be employed to protect optical sensors or human eyes from laser irradiation (Wang et al., 2004). For many practical applications, it is highly desirable that the optical limiting material is in solid form; thus, fullerenes have been mixed with sol-gel glasses, polymethyl methacrylate (PMMA) matrices, and glass-polymer composites in order to form solid materials that exhibit good optical limiting characteristics (Nierengarten, 2004).

Fullerene derivatives also have potential in a variety of other applications, including chemical sensors (Lin and Shih, 2003), anti-wear agents for lubricating oils (Ginzburg et al., 2002), and water and wastewater treatment material (Savage and Diallo, 2005).

40.7 Summary

In summary, the solubility of fullerenes and the ability to selectively suspend fullerene molecules in a variety of solvents plays a critical role in fullerene research and technology. Despite intense interest and research effort, fundamental understanding of the mechanisms responsible for the solubility and behavior of fullerenes in suspension is still lacking. While there have been numerous experimental measurements of fullerene solubility in a wide range of solvents, theories still cannot completely predict the solubilities, or trends of solubilities, of fullerenes in various solvents. A better understanding of fullerene solubility would greatly aid in the purification, separation, and enrichment of specific fullerene suspensions. This, in turn, would have a significant impact on the development of a wide range of fullerene technologies.

Fullerenes are highly hydrophobic, thus their biological applications are greatly limited and depend strongly on the methods used to create stable water-soluble fullerene suspensions. Numerous approaches have been taken in order to improve the solubility of fullerenes in aqueous suspensions including chemically modifying the surfaces of the fullerenes; incorporating the fullerene molecule into host–guest complexes; encapsulating the fullerenes in micelles, vesicles, or liposomes; or attaching surfactants to the fullerene surface. Additionally, stable colloidal suspensions of fullerenes have also been created through solvent exchange techniques or the extended stirring of fullerene/water mixtures. Techniques that can produce biocompatible fullerene suspensions without adversely affecting the fullerene propertics are in strong demand.

A significant area of concern with regard to the development of fullerene technologies is the issue of toxicity. There have been conflicting reports on the environmental and health issues of fullerenes as the toxicity strongly depends on the processing undergone by the fullerenes. A more complete understanding of the potential harmful effects of fullerenes is required if fullerenes are to gain widespread use in a variety of technologies.

References

Andersson, T., Nilsson, K., Sundahl, M., Westman, G., and Wennerstrom, O. 1992. C-60 embedded in gamma-cyclodextrin-a water-soluble fullerene. *Journal of the Chemical Society-Chemical Communications*: 604–606.

Andersson, T., Westman, G., Wennerstrom, O., and Sundahl, M. 1994. NMR and UV-VIS investigation of water-soluble fullerene-60-gamma-cyclodextrin complex. *Journal of the Chemical Society-Perkin Transactions* 2: 1097–1101.

Andrievsky, G., Klochkov, V., and Derevyanchenko, L. 2005. Is the C-60 fullerene molecule toxic?! *Fullerenes Nanotubes and Carbon Nanostructures* 13: 363–376.

Andrievsky, G. V., Kosevich, M. V., Vovk, O. M., Shelkovsky, V. S., and Vashchenko, L. A. 1995. On the production of an aqueous colloidal solution of fullerenes. *Journal of the Chemical Society-Chemical Communications* 12: 1281–1282.

Atwood, J. L., Koutsantonis, G. A., and Raston, C. L. 1994. Purification of C-60 and C-70 by selective complexation with calixarenes. *Nature* 368: 229–231.

Bakry, R., Vallant, R. M., Najam-Ul-Haq, M. et al. 2007. Medicinal applications of fullerenes. *International Journal of Nanomedicine* 2: 639–649.

Beck, M. T. and Mandi, G. 1997. Solubility of C-60. *Fullerene Science and Technology* 5: 291–310.

Beeby, A., Eastoe, J., and Heenan, R. K. 1994. Solubilization of C-60 in aqueous micellar solution. *Journal of the Chemical Society-Chemical Communications*: 173–175.

Benjamin, S. C., Ardavan, A., Andrew, G. et al. 2006. Towards a fullerene-based quantum computer. *Journal of Physics-Condensed Matter* 18: S867–S883.

Bensasson, R. V., Bienvenue, E., Dellinger, M., Leach, S., and Seta, P. 1994. C60 in model biological-systems—A visible-UV absorption study of solvent-dependent parameters and solute aggregation. *Journal of Physical Chemistry* 98: 3492–3500.

Bezmel'nitsyn, V. N., Eletskii, A. V., and Okun, M. V. 1998. Fullerenes in solutions. *Uspekhi Fizicheskikh Nauk* 168: 1195–1220.

Bosi, S., Da Ros, T., Spalluto, G., and Prato, M. 2003. Fullerene derivatives: An attractive tool for biological applications. *European Journal of Medicinal Chemistry* 38: 913–923.

Bosi, S., Feruglio, L., Da Ros, T. et al. 2004. Hemolytic effects of water-soluble fullerene derivatives. *Journal of Medicinal Chemistry* 47: 6711–6715.

Boutorine, A. S., Tokuyama, H., Takasugi, M. et al. 1994. Fullerene-oligonucleotide conjugates-photoinduced sequence-specific DNA cleavage. *Angewandte Chemie-International Edition* 33: 2462–2465.

Brabec, C. J. and Durrant, J. R. 2008. Solution-processed organic solar cells. *MRS Bulletin* 33: 670–675.

Brant, J., Lecoanet, H., and Wiesner, M. R. 2005. Aggregation and deposition characteristics of fullerene nanoparticles in aqueous systems. *Journal of Nanoparticle Research* 7: 545–553.

Brant, J. A., Labille, J., Bottero, J. Y., and Wiesner, M. R. 2006. Characterizing the impact of preparation method on fullerene cluster structure and chemistry. *Langmuir* 22: 3878–3885.

Brettreich, M. and Hirsch, A. 1998. A highly water-soluble dendro[60]fullerene. *Tetrahedron Letters* 39: 2731–2734.

Chai, Y., Guo, T., Jin, C. M. et al. 1991. Fullerenes with metals inside. *Journal of Physical Chemistry* 95: 7564–7568.

Cheng, X. K., Kan, A. T., and Tomson, M. B. 2004. Naphthalene adsorption and desorption from aqueous C-60 fullerene. *Journal of Chemical and Engineering Data* 49: 675–683.

Da Ros, T. and Prato, M. 1999. Medicinal chemistry with fullerenes and fullerene derivatives. *Chemical Communications*: 663–669.

Da Ros, T., Prato, M., Novello, F., Maggini, M., and Banfi, E. 1996. Easy access to water-soluble fullerene derivatives via 1,3-dipolar cycloadditions of azomethine ylides to C-60. *Journal of Organic Chemistry* 61: 9070–9072.

Danauskas, S. M. and Jurs, P. C. 2001. Prediction of C-60 solubilities from solvent molecular structures. *Journal of Chemical Information and Computer Sciences* 41: 419–424.

Deguchi, S., Alargova, R. G., and Tsujii, K. 2001. Stable dispersions of fullerenes, C-60 and C-70, in water. Preparation and characterization. *Langmuir* 17: 6013–6017.

Dhawan, A., Taurozzi, J. S., Pandey, A. K. et al. 2006. Stable colloidal dispersions of C-60 fullerenes in water: Evidence for genotoxicity. *Environmental Science & Technology* 40: 7394–7401.

Diederich, F. and Gomez-Lopez, M. 1999. Supramolecular fullerene chemistry. *Chemical Society Reviews* 28: 263–277.

Dugan, L. L., Turetsky, D. M., Du, C. et al. 1997. Carboxyfullerenes as neuroprotective agents. *Proceedings of the National Academy of Sciences of the United States of America* 94: 9434–9439.

Duncan, L. K., Jinschek, J. R., and Vikesland, P. J. 2008. C-60 colloid formation in aqueous systems: Effects of preparation method on size, structure, and surface, charge. *Environmental Science & Technology* 42: 173–178.

Eastoe, J., Crooks, E. R., Beeby, A., and Heenan, R. K. 1995. Structure and photophysics in C-60-micellar solutions. *Chemical Physics Letters* 245: 571–577.

Fortner, J. D., Lyon, D. Y., Sayes, C. M. et al. 2005. C-60 in water: Nanocrystal formation and microbial response. *Environmental Science & Technology* 39: 4307–4316.

Friedman, S. H., Decamp, D. L., Sijbesma, R. P. et al. 1993. Inhibition of the HIV-1 protease by fullerene derivatives-model-building studies and experimental-verification. *Journal of the American Chemical Society* 115: 6506–6509.

Gharagheizi, F. and Alamdari, R. F. 2008. A molecular-based model for prediction of solubility of C-60 fullerene in various solvents. *Fullerenes Nanotubes and Carbon Nanostructures* 16: 40–57.

Ginzburg, B. M., Shibaev, L. A., Kireenko, O. F. et al. 2002. Antiwear effect of fullerene C-60 additives to lubricating oils. *Russian Journal of Applied Chemistry* 75: 1330–1335.

Heath, J. R., Obrien, S. C., Zhang, Q. et al. 1985. Lanthanum complexes of spheroidal carbon shells. *Journal of the American Chemical Society* 107: 7779–7780.

Hungerbühler, H., Guldi, D. M., and Asmus, K. D. 1993. Incorporation of C-60 into artificial lipid-membranes. *Journal of the American Chemical Society* 115: 3386–3387.

Ikeda, A., Hatano, T., Kawaguchi, M., Suenaga, H., and Shinkai, S. 1999. Water-soluble [60]fullerene-cationic homooxacalix[3]arene complex which is applicable to the photocleavage of DNA. *Chemical Communications*: 1403–1404.

Isobe, H., Nakanishi, W., Tomita, N. et al. 2006. Gene delivery by aminofullerenes: Structural requirements for efficient transfection. *Chemistry-An Asian Journal* 1: 167–175.

Jenekhe, S. A. and Chen, X. L. 1998. Self-assembled aggregates of rod-coil block copolymers and their solubilization and encapsulation of fullerenes. *Science* 279: 1903–1907.

Jensen, A. W., Wilson, S. R., and Schuster, D. I. 1996. Biological applications of fullerenes. *Bioorganic & Medicinal Chemistry* 4: 767–779.

Kadish, K. M. and Ruoff, R. S. 2000 *Fullerenes: Chemistry, Physics, and Technology,* New York: Wiley-Interscience.

Kai, Y., Komazawa, Y., Miyajima, A., Miyata, N., and Yamakoshi, Y. 2003. [60]Fullerene as a novel photoinduced antibiotic. *Fullerenes Nanotubes and Carbon Nanostructures* 11: 79–87.

Kiss, I. Z., Mandi, G., and Beck, M. T. 2000. Artificial neural network approach to predict the solubility of C-60 in various solvents. *Journal of Physical Chemistry A* 104: 8081–8088.

Klaine, S. J., Alvarez, P. J. J., Batley, G. E. et al. 2008. Nanomaterials in the environment: Behavior, fate, bioavailability, and effects. *Environmental Toxicology and Chemistry* 27: 1825–1851.

Korobov, M. V. and Smith, A. L. 2000. Solubility of the fullerenes. In: Kadish, K. M. and Ruoff, R. S. (Eds.), *Fullerenes: Chemisry, Physics, and Technology*. New York: Wiley-Interscience.

Krätschmer, W., Lamb, L. D., Fostiropoulos, K., and Huffman, D. R. 1990. Solid C-60-a new form of carbon. *Nature* 347: 354–358.

Kroto, H. W., Heath, J. R., Obrien, S. C., Curl, R. F., and Smalley, R. E. 1985. C-60-Buckminsterfullerene. *Nature* 318: 162–163.

Kunsagi-Mate, S., Vasapollo, G., Szabo, K. et al. 2008. Effect of covalent functionalization of C-60 fullerene on its encapsulation by water soluble calixarenes. *Journal of Inclusion Phenomena and Macrocyclic Chemistry* 60: 71–78.

Lamparth, I. and Hirsch, A. 1994. Water-soluble malonic-acid derivatives of C-60 with a defined 3-dimensional structure. *Journal of the Chemical Society-Chemical Communications*: 1727–1728.

Langa, F. and Nierengarten, J.-F. 2007 *Fullerenes: Principles and Applications,* Cambridge, U.K.: The Royal Society of Chemistry.

Lee, J. and Kim, J. H. 2008. Effect of encapsulating agents on dispersion status and photochemical reactivity of C-60 in the aqueous phase. *Environmental Science & Technology* 42: 1552–1557.

Lee, J., Kim, H., Kahng, S. J. et al. 2002. Bandgap modulation of carbon nanotubes by encapsulated metallofullerenes. *Nature* 415: 1005–1008.

Lee, J., Fortner, J. D., Hughes, J. B., and Kim, J. H. 2007. Photochemical production of reactive oxygen species by C-60 in the aqueous phase during UV irradiation. *Environmental Science & Technology* 41: 2529–2535.

Lin, H. B. and Shih, J. S. 2003. Fullerene C60-cryptand coated surface acoustic wave quartz crystal sensor for organic vapors. *Sensors and Actuators B-Chemical* 92: 243–254.

Liu, H. X., Yao, X. J., Zhang, R. S. et al. 2005. Accurate quantitative structure-property relationship model to predict the solubility of C-60 in various solvents based on a novel approach using a least-squares support vector machine. *Journal of Physical Chemistry B* 109: 20565–20571.

Lyon, D. Y., Brunet, L., Hinkal, G. W., Wiesner, M. R., and Alvarez, P. J. J. 2008. Antibacterial activity of fullerene water suspensions (Nc(60)) is not due to Ros-mediated damage. *Nano Letters* 8: 1539–1543.

Ma, X. and Bouchard, D. 2009. Formation of aqueous suspensions of fullerenes. *Environmental Science & Technology* 43: 330–336.

Marcus, Y., Smith, A. L., Korobov, M. V. et al. 2001. Solubility of C-60 fullerene. *Journal of Physical Chemistry B* 105: 2499–2506.

Mashino, T., Nishikawa, D., Takahashi, K. et al. 2003. Antibacterial and antiproliferative activity of cationic fullerene derivatives. *Bioorganic & Medicinal Chemistry Letters* 13: 4395–4397.

Murray, J. S., Gagarin, S. G., and Politzer, P. 1995. Representation of C-60 solubilities in terms of computed molecular-surface electrostatic potentials and areas. *Journal of Physical Chemistry* 99: 12081–12083.

Murthy, C. N. and Geckeler, K. E. 2001. The water-soluble beta-cyclodextrin-[60]fullerene complex. *Chemical Communications*: 1194–1195.

Nakamura, E. and Isobe, H. 2003. Functionalized fullerenes in water. The first 10 years of their chemistry, biology, and nanoscience. *Accounts of Chemical Research* 36: 807–815.

Nierengarten, J. F. 2004. Chemical modification of C-60 for materials science applications. *New Journal of Chemistry* 28: 1177–1191.

Okamura, H., Miyazono, K., Minoda, M. et al. 2000. Synthesis of highly water soluble C-60 end-capped vinyl ether oligomers with well-defined structure. *Journal of Polymer Science Part A-Polymer Chemistry* 38: 3578–3585.

Priyadarsini, K. I., Mohan, H., Tyagi, A. K., and Mittal, J. P. 1994. Inclusion complex of gamma-cyclodextrin-C-60-formation, characterization, and photophysical properties in aqueous solutions. *Journal of Physical Chemistry* 98: 4756–4759.

Ramakanth, I. and Patnaik, A. 2008. Characteristics of solubilization and encapsulation of fullerene C-60 in non-ionic Triton X-100 micelles. *Carbon* 46: 692–698.

Reyes-Reyes, M., Kim, K., Dewald, J. et al. 2005. Meso-structure formation for enhanced organic photovoltaic cells. *Organic Letters* 7: 5749–5752.

Richardson, C. F., Schuster, D. I., and Wilson, S. R. 2000. Synthesis and characterization of water-soluble amino fullerene derivatives. *Organic Letters* 2: 1011–1014.

Ruoff, R. S., Malhotra, R., Huestis, D. L., Tse, D. S., and Lorents, D. C. 1993a. Anomalous solubility behavior of C60. *Nature* 362: 140–141.

Ruoff, R. S., Tse, D. S., Malhotra, R., and Lorents, D. C. 1993b. Solubility of C-60 in a variety of solvents. *Journal of Physical Chemistry* 97: 3379–3383.

Samal, S. and Geckeler, K. E. 2000. Cyclodextrin-fullerenes: A new class of water-soluble fullerenes. *Chemical Communications*: 1101–1102.

Saunders, B. R. and Turner, M. L. 2008. Nanoparticle-polymer photovoltaic cells. *Advances in Colloid and Interface Science* 138: 1–23.

Savage, N. and Diallo, M. S. 2005. Nanomaterials and water purification: Opportunities and challenges. *Journal of Nanoparticle Research* 7: 331–342.

Sayes, C. M., Fortner, J. D., Guo, W. et al. 2004. The differential cytotoxicity of water-soluble fullerenes. *Nano Letters* 4: 1881–1887.

Scrivens, W. A., Tour, J. M., Creek, K. E., and Pirisi, L. 1994. Synthesis of C-14-labeled C-60, its suspension in water, and its uptake by human keratinocytes. *Journal of the American Chemical Society* 116: 4517–4518.

Service, R. F. 1996. A captivating carbon form. *Science* 274: 345–346.

Shinkai, S. and Ikeda, A. 1999. Novel interactions of calixarene Pi-systems with metal ions and fullerenes. *Pure and Applied Chemistry*: 275–280.

Shinohara, H. 2000. Endohedral metallofullerenes: Production, separation, and structural properties. In: Kadish, K. M. and Ruoff, R. S. (Eds.), *Fullerenes: Chemisry, Physics, and Technology*. New York: Wiley-Interscience.

Sijbesma, R., Srdanov, G., Wudl, F. et al. 1993. Synthesis of a fullerene derivative for the inhibition of HIV enzymes. *Journal of the American Chemical Society* 115: 6510–6512.

Sivaraman, N., Srinivasan, T. G., Rao, P. R. V., and Natarajan, R. 2001. QSPR modeling for solubility of fullerene (C-60) in organic solvents. *Journal of Chemical Information and Computer Science*: 1067–1074.

Smith, A. L., Wilson, L. Y., and Famini, G. R. 1996. A quantitative structure-property relationship study of C-60 solubility. In: Ruoff, R. S. and Kadish, K. M. (Eds.), *Recent Advances in the Chemistry and Physics of Fullerenes and Related Materials*. Pennington, NJ: Electrochemical Society.

Suzuki, T., Nakashima, K., and Shinkai, S. 1994. Very convenient and efficient purification method for fullerene (C-60) with 5,11,17,23,29,35,41,47-octa-tert-butylcalix[8]arene-49,50,51,52,53,54,55,56-octol. *Chemistry Letters*: 699–702.

Tabata, Y. and Ikada, Y. 1999. Biological functions of fullerene. In: *4th International Symposium on Functional Dyes-Science and Technology of Functional pi-Electron Systems*. International Union of Pure and Applied Chemistry, Osaka, Japan.

Takenaka, S., Yamashita, K., Takagi, M., Hatta, T., and Tsuge, O. 1999. Photo-induced DNA cleavage by water-soluble cationic fullerene derivatives. *Chemistry Letters*: 321–322.

Taylor, R. 1995. Properties of fullerenes. In: Taylor, R. (Ed.), *The Chemistry of Fullerenes*. River Edge, NJ: World Scientific.

Tokuyama, H., Yamago, S., Nakamura, E., Shiraki, T., and Sugiura, Y. 1993. Photoinduced biochemical-activity of fullerene carboxylic-acid. *Journal of the American Chemical Society* 115: 7918–7919.

Wang, C. C., Guo, Z. X., Fu, S. K., Wu, W., and Zhu, D. B. 2004. Polymers containing fullerene or carbon nanotube structures. *Progress in Polymer Science* 29: 1079–1141.

Wharton, T., Kini, V. U., Mortis, R. A., and Wilson, L. J. 2001. New non-ionic, highly water-soluble derivatives of C-60 designed for biological compatibility. *Tetrahedron Letters* 42: 5159–5162.

Wiesner, M. R., Hotze, E. M., Brant, J. A., and Espinasse, B. 2008. Nanomaterials as possible contaminants: The fullerene example. *Water Science and Technology* 57: 305–310.

Williams, R. M. and Verhoeven, J. W. 1992. Supramolecular encapsulation of C-60 in a water-soluble calixarene—A core-shell charge-transfer complex. *Recueil Des Travaux Chimiques Des Pays-Bas-Journal of the Royal Netherlands Chemical Society* 111: 531–532.

Wilson, S. R. 2000. Biological applications of fullerenes. In: Kadish, K. M. and Ruoff, R. S. (Eds.), *Fullerenes: Chemisry, Physics, and Technology*. New York: Wiley-Interscience.

Yamakoshi, Y. N., Yagami, T., Fukuhara, K., Sueyoshi, S., and Miyata, N. 1994. Solubilization of fullerenes into water with polyvinylpyrrolidone applicable to biological tests. *Journal of the Chemical Society-Chemical Communications*: 517–518.

Yamamoto, K., Funasaka, H., Takahashi, T., and Akasaka, T. 1994. Isolation of an ESR-active metallofullerene of La-at-C-82. *Journal of Physical Chemistry* 98: 2008–2011.

Zhong, Z.-L., Ikeda, A., and Shinkai, S. 2001. Complexation of fullerenes. In: Asfari, Z., Böhmer, V., Harrowfield, J., and Vicens, J. (Eds.), *Calixarenes 2001*. Dordrecht, the Netherlands: Kluwer Academic Publishers.

41

Fullerene Encapsulation

Atsushi Ikeda
*Nara Institute of Science
and Technology*

41.1 Introduction

Fullerenes are the most recently discovered allotrope of carbon, coming several centuries after the discovery of graphite and diamond [1]. The availability of fullerenes in large quantities has drawn increasing attention toward the exploration of their remarkable functionality. A unique combination of electrical, mechanical, thermal, and optical properties gives fullerenes numerous potential uses such as biomedical applications and, particularly, exploiting photovoltaic phenomena for photocurrent generators, photoelectronic devices, photoinduced DNA cleavage reagents, and photosensitizers for photodynamic therapy. However, commercial exploitation has been hindered by the low solubility of fullerenes in polar solvents and by a marked tendency of spherical fullerenes with a surface composed of only sp^2 carbons to self-aggregation. There are two methods for addressing these disadvantages, and hence further functionalizing fullerenes: (1) covalent modification and (2) encapsulation. Covalent modification has the disadvantage that it impairs the physical properties of the product, whereas fullerene encapsulation realizes the noncovalent functionalization of fullerenes using a supramolecular approach. Hence, the supramolecular approach does not appreciably affect the physical properties of fullerenes. There are three potential advantages to be gained by the encapsulation of fullerenes by host molecules: (1) purification of fullerenes; (2) improvement of the solubility of C_{60}, C_{70}, and higher fullerenes; and (3) the practical application of solubilized fullerenes.

This chapter describes fullerene encapsulation using the supramolecular approach by host molecules and the molecular design of host molecules. It should be noted that the design of host molecules for fullerene encapsulation is completely different from that for the binding of other organic guest molecules or metal ions because of the different forces driving the binding process. Furthermore, the practical applications of the encapsulated fullerenes and host molecule-fullerene complexes are introduced in this chapter.

41.2 Encapsulating C_{60} in an Aqueous Solution

There are many emerging biomedical technologies that might exploit the unique physical and chemical functions of C_{60} or C_{70} if the molecules could be solubilized in water. For example, the easy excitation of a fullerene by visible-light irradiation allows it to act as a singlet oxygen ($^1O_2$) photosensitizer [2], enabling it to cleave DNA [3] or to act as a virus photoinactivator [4]. Applications have remained limited because of the intrinsic poor water solubility of fullerenes. This section introduces the solubilization of fullerenes in water using water-soluble host molecules.

41.2.1 Cyclodextrins

Cyclodextrins (CDxs) are well known as a type of oligosaccharide made up of 6–8 D-glucose monomers connected to 1 and 4 carbon atoms. They can provide a hydrophobic cavity in an aqueous solution to contain hydrophobic molecules or groups to form inclusion complexes. Exploiting these unique properties, CDxs have been widely used as biomimetic microreactors, novel media for photophysical and photochemical studies, and as building blocks for supramolecular structures and functional units. They are also used in various industrial applications such as pharmaceuticals, foods, and enzyme modeling.

A boiling aqueous solution of γ-CDx (**1**) can extract C_{60} from a mixture of C_{60} and C_{70} [5]. Although the ultraviolet-visible (UV-VIS) spectrum of the **1**·C_{60} complex in water (observable as a magenta-colored solution) is slightly more blue than that of a toluene solution and slightly more red than that of a hexane solution, the spectrum closely resembles those of C_{60} in the organic solvents. These results indicate that C_{60} does exist in the CDx hydrophobic cavity. Characteristic purple samples of the **1**·C_{60} complex were isolated by [6] and the complex was determined as a **1**-C_{60} 2:1 composition based on carbon-13 nuclear magnetic resonance (^{13}C NMR) and elemental analysis (Figure 41.1A). However, the highest concentration of C_{60} in water using this method was very low (8×10^{-5} mol dm^{-3}). To address this, a high-speed vibration milling (HSVM) method was applied to the supramolecular complexation of C_{60} and proved to be much more efficient than the classical ball milling method of Komatsu et al. [7]. The concentration of the resultant **1**·C_{60} complex (1.4×10^{-3} mol dm^{-3}) was much higher than that achieved by dissolving C_{60} in water by refluxing an aqueous solution of **1** and C_{60} (8×10^{-5} mol dm^{-3}) [5] or that obtained by ball-milling (1.5×10^{-4} mol dm^{-3}) [8]. It should be emphasized that the present HSVM method forms the complex after only 10 min, while refluxing in an aqueous solution and ball-milling require more than 24 h [5] and about 20 h [8], respectively. Furthermore, the HSVM method can be applied successfully for the preparation of the **1**·C_{70} complex (7×10^{-4} mol dm^{-3}) [7].

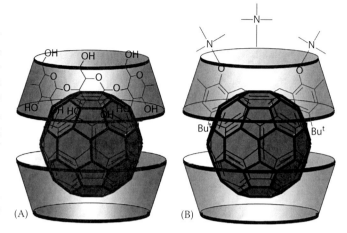

FIGURE 41.1 Schematic illustration of (A) γ-CDx·C_{60} and (B) **7**·C_{60} complexes.

Other CDxs such as β-CDx (**2**) and heptakis(2,6-di-*O*-methyl)-β-cyclodextrin (DMe-β-CDx: **3**) have also been used to solubilize fullerenes in water [9,10]. The aqueous solutions of these complexes are not magenta but light yellow. δ-CDx (**4**) has a larger cavity than γ-CDx (**1**), and is wide enough to include C_{70}. The concentration of C_{70} in water is about 8.4×10^{-5} mol dm^{-3} using ball-milling [11]. If the HSVM method is used, it is to be expected that this concentration will be much improved.

The applications of the CDx·C_{60} complex have been reported by several groups. For example, water-soluble gold nanoparticles (3.2 nm diameter) have been capped with thiolated γ-CDx hosts to form large network aggregates (ca. 300 nm diameter) in the presence of C_{60} [12]. This aggregation phenomenon is driven by the formation of inclusion complexes between two CDxs attached to different nanoparticles, and one molecule of C_{60}. The γ-CDx·C_{60} complex shows an efficient DNA cleaving-activity in the presence of reduced nicotinamide adenine dinucleotide (NADH) in an O_2-saturated aqueous solution under visible-light irradiation [13].

1 (γ-CDx : *n* = 3)

2 (β-CDx : *n* = 2)

4 (δ-CDx : *n* = 4)

3 (DMe-β-CDx : *n* = 2)

Although the $1 \cdot C_{60}$ complex can be readily prepared using the HSVM method, the $1 \cdot C_{60}$ complex must be handled carefully because it is very unstable (and the $1 \cdot C_{70}$ complex is less stable than the $1 \cdot C_{60}$ complex). For example, when an excess of **1** is removed, the color of the solution slowly changes from violet to yellow, accompanied by the formation of large aggregates of C_{60} surrounded by **1**; this process takes place within minutes at 80°C [7].

41.2.2 Water-Soluble Calixarenes and Cyclotriveratrylene

Lack of stability limits the application of the $1 \cdot C_{60}$ and $1 \cdot C_{70}$ complexes. This difficulty can be addressed by means of the formation of host–guest complexes through hydrophobic interactions and through π–π interactions between host molecules and C_{60}. Here, these host molecules should be water-soluble and have a sufficiently large cavity to undergo face-to-face interaction with the C_{60} surface. Most host molecules with a large π-cavity have poor water solubility, so a water-soluble group such as sulfonic acid or ammonium ion must be introduced.

A calix[8]arene derivative (**5**), which has sulfonic acids at the para position, forms a water-soluble complex with C_{60}, but not with C_{70}. This discriminative feature allows the efficient extraction of C_{60} into an aqueous solution from an organic solution (e.g., toluene) containing both forms [14]. Another calix[8]arene derivative (**6**) was reported by Komatsu et al. [7] to have complexation ability for C_{60} when prepared using the HSVM technique. The resulting aqueous solution was yellow, and the concentration of C_{60} was calculated to be 1.3×10^{-4} mol dm^{-3}.

A water-soluble, cationic homooxacalix[3]arene (**7**) solubilizes C_{60} into water when ultrasonicated for 1 h followed by 8 days of stirring (Figure 41.1B) [15]. The resulting concentration of C_{60} in the aqueous solution was 2.4×10^{-4} mol dm^{-3}. The $7 \cdot C_{60}$ complex was applied in the photocleavage of DNA. Under visible-light irradiation, the $7 \cdot C_{60}$ complex displayed DNA-cleaving activity without the presence of NADH [15,16]. Furthermore, when the $7 \cdot C_{60}$ complex was deposited as a monolayer on an anion-coated indium tin oxide or gold electrode, the membrane generated a photocurrent in response to visible-light irradiation [17–19].

A cyclotriveratrylene derivative **8** with polybenzyl ether dendritic branches and peripheral triethyleneglycol chains that formed water-soluble supramolecular complexes with C_{60} was reported by Rio and Nierengarten [20].

5   6   7

8   9 (PVP)

41.2.3 Water-Soluble Polymers

C_{60} can be solubilized in water using water-soluble polymers as solubilizing agents [21]. Examples of potential applications are as follows: C_{60} solubilized in water with poly(vinylpyrrolidone) (PVP) (**9**) had a unique chondrogenesis-promoting effect [22–24]. An aqueous C_{60} solution can express several biological activities under the irradiation of visible light. These activities include mutagenicity in *Salmonella* tester strains [25], initiation activity in a mouse embryonic fibroblast (BALB/3T3) cell transformation assay [26], and photoinduced DNA-cleaving activities [27]. These biological activities are caused by oxyl radicals ($O_2^{\cdot}$ and $\cdot OH$)

generated by C_{60} under photoirradiation in the presence of biological reducing agents [28].

41.2.4 Liposomes

Liposomes are single- or multi-compartmental, microscopic spherical, or ellipsoidal vesicles that form when amphiphilic lipids containing two hydrocarbon chains and a polar group (e.g., a phospholipid) are hydrated. When mixed with water, the amphiphilic lipids align side by side to form a bilayer membrane, which encloses a droplet of water inside the vesicle (Figure 41.2A). The properties of vesicles are determined by their composition, the number of

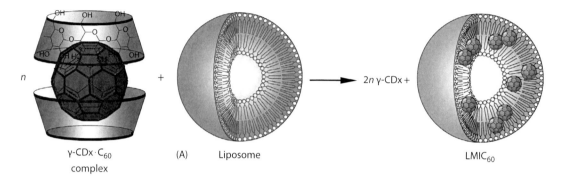

FIGURE 41.2 Schematic illustration of (A) liposome and preparation of lipid membrane-incorporated C_{60} from γ-CDx·C_{60} complexes via an exchange reaction.

compartments (uni- or multilamellar, uni- or multivesicular), size (small vesicles being less than or equal to 100 nm, large vesicles being 100–500 nm, and giant vesicles being more than 1 μm in diameter), and phase transition temperature (below this temperature a bilayer becomes more regular and loses its fluidity).

Lipid membrane-incorporated C_{60} ($LMIC_{60}$) was prepared using a method derived from Bangham et al. [29] and Hungerbühler [30]. The C_{60} and the lipid (**10** or **11**) were dissolved in chloroform to give a homogeneous mixture and the solvent was then evaporated. The remaining solid was dried under vacuum and then water was added to form a suspension. The aqueous suspension was kept for about 10 min at constant temperature (50°C–60°C) to soak the material. Vesicle formation was then achieved using ultrasonication for about 30 min. The stoichiometries of the $[C_{60}]/[\mathbf{10}$ or $\mathbf{11}]$ were only 1%–2%. To prepare $LMIC_{60}$ with high C_{60} concentrations, a novel method was proposed to enable treatment for a homogeneous system by Ikeda et al. [31]. As mentioned previously, the $\mathbf{1} \cdot C_{60}$ complex is not temperature stable. Therefore, the $LMIC_{60}$ was prepared using an exchange reaction between the vesicle formed by the lipids (**12–14**) and the $\mathbf{1} \cdot C_{60}$ complex by heating at 80°C for 2 h, followed by microwave irradiation for 7 s, or light irradiation in the presence of NADH for 6 h (Figure 41.2) [32]. The stoichiometries of the $[C_{60}]/[$lipids $\mathbf{12–14}]$ were 10%, which was considerably higher than levels of under 2% observed in the previous study [29]. Water-soluble $LMIC_{70}$ was prepared within 1 min by mixing the $\mathbf{1} \cdot C_{70}$ complex with vesicles at room temperature [33]. The biological activities of $LMIC_{60}$ and $LMIC_{70}$ were assayed under visible-light irradiation, showing that the photodynamic activity of $LMIC_{70}$ is 4.7 times that of $LMIC_{60}$ for the same level of photon flux (>400 nm) [34]. The higher phototoxicity of $LMIC_{70}$ may be attributed to the increased generation of singlet oxygen ($^1O_2$) in $LMIC_{70}$ compared with $LMIC_{60}$. The difference in the efficiency for the generation of $^1O_2$ between $LMIC_{60}$ and $LMIC_{70}$ is considered to result from the amount of light absorption in the 400–700 nm region. This range of wavelengths is suitable for photodynamic therapy.

41.3 Encapsulated C_{60} in Organic Solvents

C_{60} has a usable level of solubility only in specific organic solvents; therefore, the complexation of C_{60} by host molecules is a useful method to improve the solubility of C_{60} in a wide range of organic solutions. In addition, the complexation of C_{60} with functional host molecules can provide nanometer size components for building novel supramolecular arrays if a stable complex can be achieved between the host molecules and C_{60}. This section describes the design of host molecules for inclusion of C_{60}.

41.3.1 Calixarenes

p-tert-Butylcalix[8]arene (**15**) selectively includes C_{60} in fullerene mixtures containing C_{60}, C_{70}, C_{76}, etc., and forms a precipitate with 1:1 stoichiometry [35,36]. By the use of molecular mechanics, the nature of the 1:1 **15** complex of C_{60} was predicted to be micelle-like, with a trimeric aggregate of fullerenes being surrounded by three host molecules by Raston et al. [37]. This complexation has become very useful for obtaining pure C_{60} in large quantities. As a purification method, it is simpler than using conventional column chromatography. It was thought that the origin of selective inclusion stemmed from the C_{60} molecule being conveniently the same size as the calix[8]arene cavity. However, when this complex was solubilized (e.g., by heating or using other solvents) and dissociated into components, there was no spectroscopic indication of complex formation [38,39]. This means that this complex exists only in the solid state.

Calix[8]arene can include C_{60} in the solid state, but not in organic solution. It has, however, been observed that several calixarenes and calixarene derivatives can include C_{60} in toluene solution [40–42]. A color change (from purple to pale yellow) was observed in an organic solution of C_{60} upon the addition of the calix[5]arene derivatives (**16**) [42]. In toluene, the association constant for C_{60} was determined to be 2.1×10^3 M^{-1}; in addition, calix[5]arenes (**17**), calix[6]arenes (**18**), and homooxacalix[3]

15

16 : $X_1 = Me$, $X_2 = I$
17 : $X_1 = X_2 = Bu^t$

18

19 **20** **21** **22**

arenes (**19**) can include C_{60} [41]. However, calix[4]arene (**20**) and *para*-methyl-calix[8]arene (**21**) cannot include C_{60}. Although calix[4]arene **20** has a preorganized cone conformation, the cavity size is not large enough to include C_{60}. Calix[4]naphthalene (**22**) has an extended cavity and, therefore, can bind C_{60} (K_{ass} = 6.6×10^2 M^{-1}) in toluene [43]. Calix[8]arene **21** changes from an initial "pleated-roof" conformation without a cavity capable of including C_{60} to a cone conformation capable of forming a **21** · C_{60} complex. This conformational change accompanies the breaking of hydrogen-bonding amongst the phenol units. These results indicate that host molecules require both preorganized conformation and a proper inclination of the benzene rings to include fullerenes in organic solvents.

Why can calixarenes include C_{60} by π–π interaction in aromatic solvents such as toluene and benzene? These planar aromatic solvents should have a large contact area for the fullerene and solvent interactions to diminish, before the complexation between calixarene and C_{60} can occur. The desolvation of C_{60} (and calixarene) prior to complexation results in a positive entropy change [44]. This positive entropy change is believed to be a dominant driving force of complex formation, although the π–π interaction between host molecules and the fullerene surface is also a very important factor. The alignment of the symmetry axis of the calix[5]arene (C_5), homooxacalix[3]arene (C_3), or calix[6]arene (C_6) with the matching symmetry elements of C_{60} makes the formation of multipoint interactions possible [45]. Symmetry matching is thus likely to be important for the design of suitable host molecules.

41.3.2 Cyclotriveratrylenes and Cavitands

Cyclotriveratrylene (CTV; **23**) can include only C_{60} and C_{70} from fullerene mixtures in toluene. **23** · C_{60} and C_{70} complexes were obtained as black crystalline plates at 25°C after 12 h by Steed et al. [46]. C_{60} and C_{70} were retrieved as a precipitate by washing out with chloroform or methylene chloride. The precipitation of the CTV·C_{60} complex was avoided by using dendritic CTV derivatives (**24**) [47]. The 2:1 host–guest complex produced a CTV derivative substituted with a long-chain alkyl group (C_{18}) (**25**) and the C_{60} displayed thermotropic liquid crystalline behavior at room temperature [48].

Several host molecules with deeper cavities have been prepared. For example, cavitand **26** can bind C_{60} but not C_{70} [49]. The association constant in toluene was calculated as 9.0×10^2 M^{-1}. The formation of the complex between a tribenzotriquinacene based on a host molecule (**27**) and C_{60} was investigated in CHCl$_3$:CS$_2$ 1:1 solvent [50]. The association constants for 1:1 and 2:1 complexes were obtained as $K_1 = 2.9 \times 10^3$ M^{-1} and $K_2 = 2.0 \times 10^3$ M^{-1}, respectively.

41.3.3 Other Macrocyclic Compounds

The carbon nanorings **28** and **29** form stable inclusion complexes with C_{60} and C_{70} [51,52]. The association constants in benzene for the **28** · C_{60} and **28** · C_{70} complexes were determined as 1.6×10^4 M^{-1} and 1.8×10^4 M^{-1}, respectively [51]. The carbon nanoring **29** substituted with naphthalene rings has association constants for

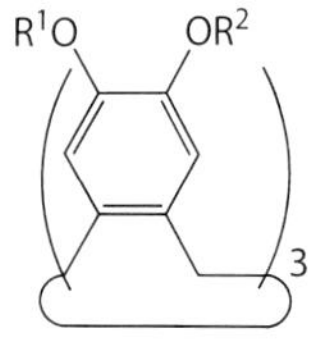

23 $R^1 = R^2 = $ Me

24 $R^1 = $ Me, $R^2 = $

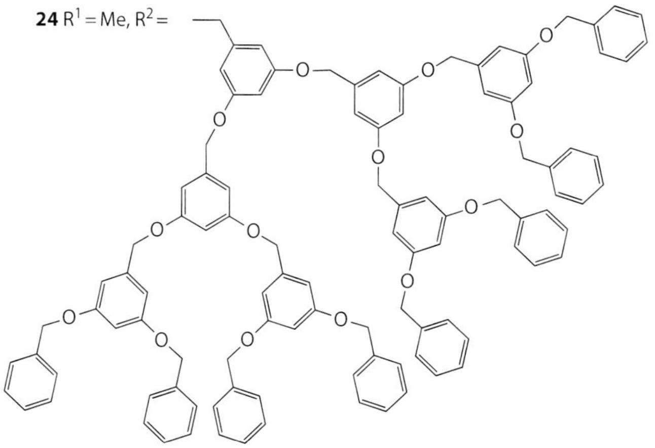

25 $R^1 = R^2 = $

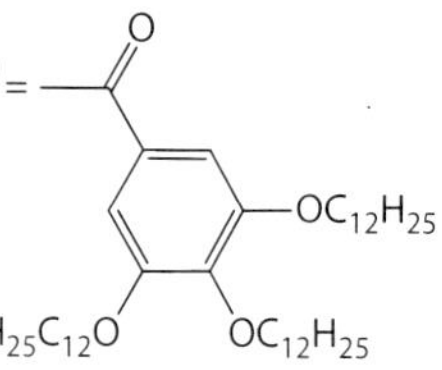

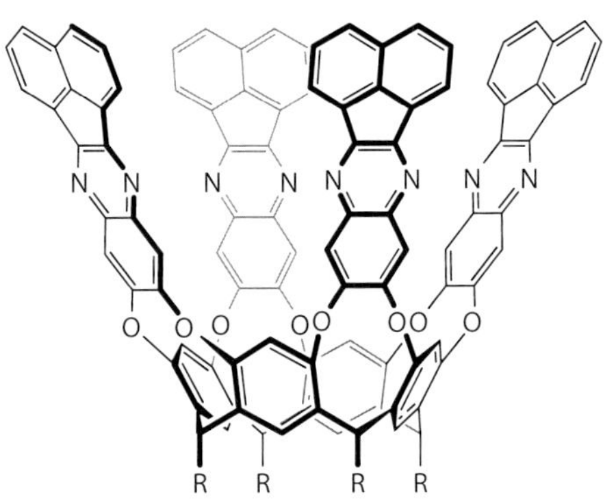

26 R = –C₁₁H₂₃

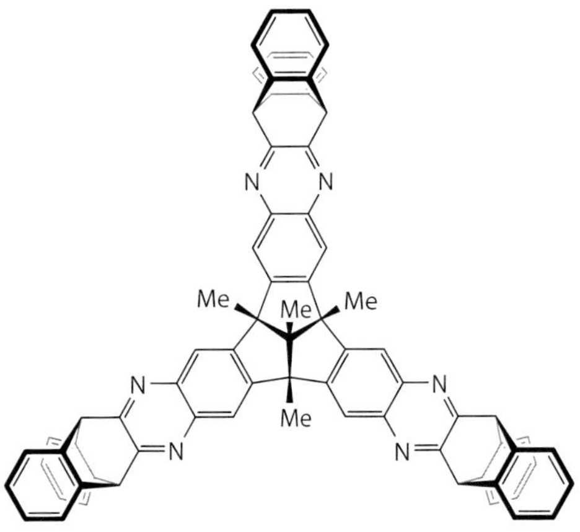

27

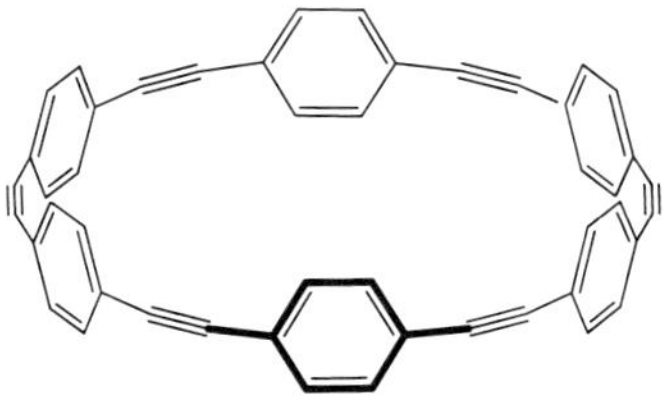

28

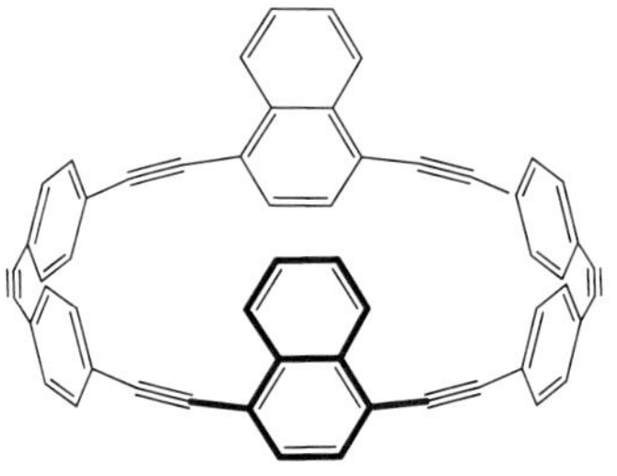

29

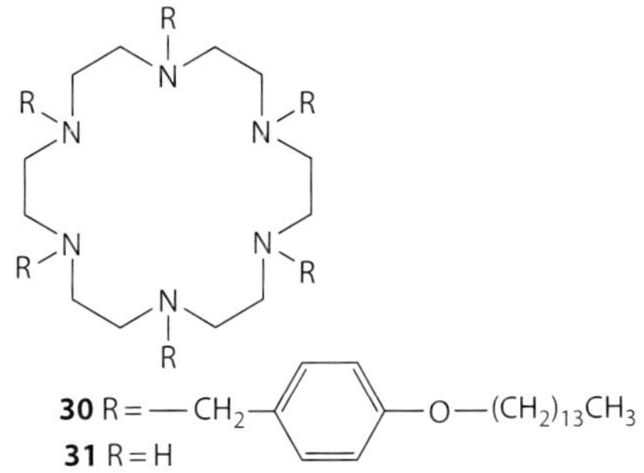

30 R = —CH₂——————O—(CH₂)₁₃CH₃
31 R = H

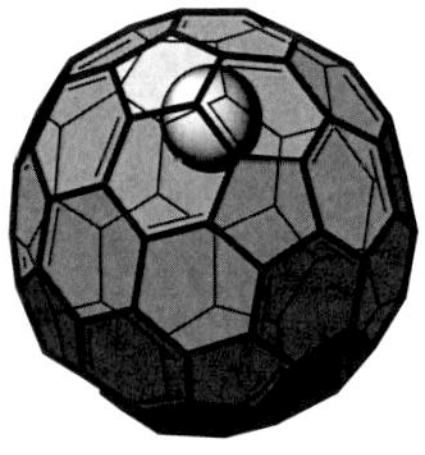

La@C₈₂

C_{60} and C_{70} in excess of 5×10^4 M^{-1}. These values are too high to be determined precisely [52]. The carbon nanoring **29** appears to have high selectivity of C_{70}/C_{60}, and the selectivity may offer a practical method of obtaining pure C_{70}.

The incorporation of C_{60} and C_{70} into monolayers of azacrown ether **30** was confirmed, and noted to have a stabilizing effect on the film at the air–water interface by Diederich et al. [53]. Azacrown ether **31** was revealed to form complexes with lanthanum metallofullerene (La@C₈₂) in polar aprotic solvents, such as acetone, benzonitrile, dimethylformamide, tetrahydrofuran, and nitrobenzene by accompanying the electron transfer between **31** and La@C₈₂ [54]. Metallofullerenes are generally very poorly soluble in most solvents. If, however, metallofullerenes could be solubilized, the different properties exhibited by the C_{60} or C_{70} forms such as magnetism offer opportunities for purification and subsequent functionalization. In practice, complexation has been successfully applied to the selective production of

endohedral metallofullerenes from extracts of soot. In contrast, calix[*n*]arenes (*n* = 5–7), which are known to form complexes with C_{60} and C_{70}, as mentioned above, cannot form a complex with La@C₈₂ at all.

41.3.4 Porphyrin Dimers

A macrocyclic porphyrin dimer (**32-Zn**) included C_{60} with a very high binding constant (6.7×10^5 M^{-1}). This **32-Zn**·C_{60} complex did not decompose on an alumina thin layer chromatography plate with benzene as eluent and giving only one spot [55], indicating that **32-Zn**·and C_{60} form a highly stable complex. The transition metals have a remarkable influence on the interaction between macrocyclic porphyrin dimers and C_{60}. In particular, the association constant of the **32-RhMe**·C_{60} complex was determined as 2.4×10^7 M^{-1} by Zheng et al. [56]. As a potential application, a heterocyclic porphyrin dimer (**33**) containing an

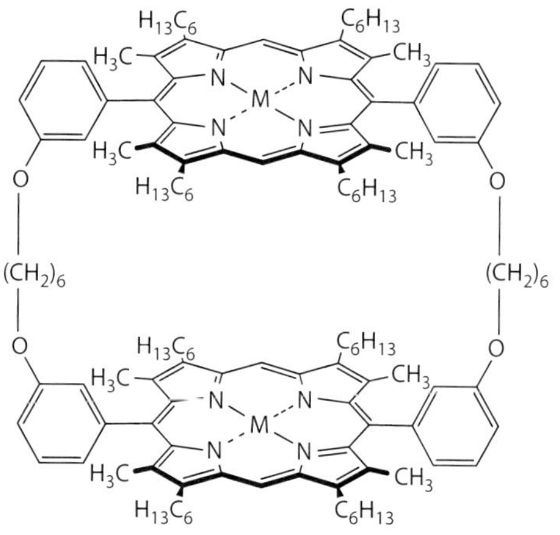

32-Zn   M = Zn
32-RhMe M = RhCH₃

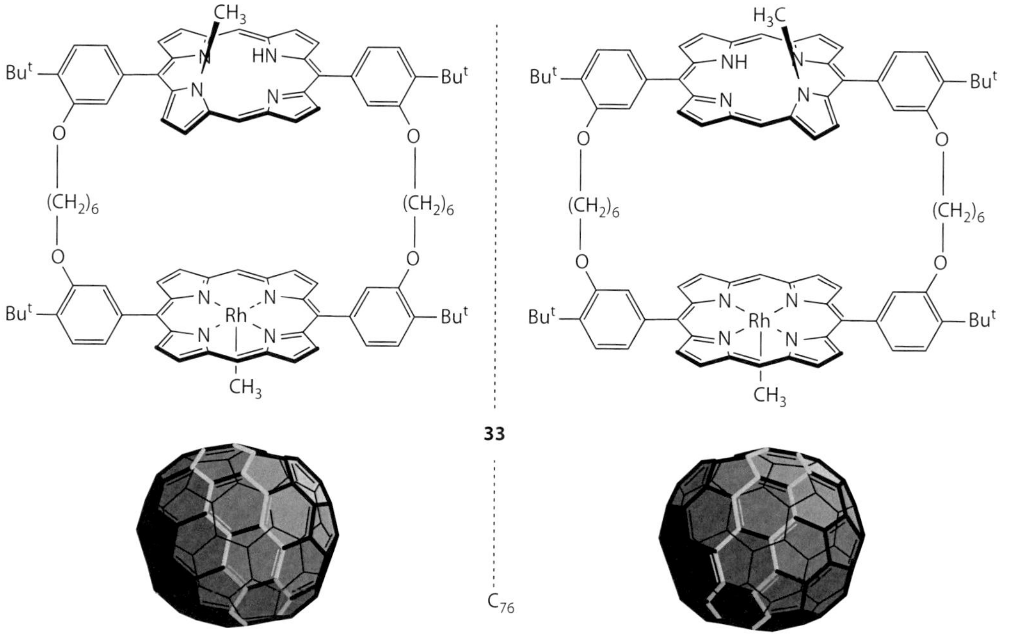

33

C₇₆

asymmetrically distorted *N*-alkylporphyrin was demonstrated as a sensor for chiral fullerene C_{76} via proton nuclear magnetic resonance (^{1}H NMR) spectroscopy by Shoji et al. [57].

Non-macrocyclic porphyrin dimers, including C_{60}, have been prepared, exploiting their lower association constants. An example of this is the coordinatively linked porphyrin dimers (**34**) that included C_{60} in toluene ($K_{ass} = 5.2 \times 10^3$ M^{-1}) described in Sun et al. [58]. Compare this with the rigid star-shaped D_3-symmetric receptor (**35**) bearing six porphyrin moieties linked with one another through phenylacetylene units synthesized by Ayabe et al. [59]. When two porphyrins sandwich one C_{60} molecule (Figure 41.3), the complexation site successively suppresses the rotational freedom of the remaining two porphyrin moieties. This domino effect is expected to be effective for the binding of three equivalents of C_{60} in an allosteric manner to attain high C_{60} affinity.

Why is it that the porphyrin alone cannot bind C_{60} but porphyrin dimers can? The following example gives an indication of the answer. A porphyrin macro-ring (**36**) was prepared by self-assembling from three trisporphyrinatozinc molecules through imidazole-Zn coordination on the terminal porphyrins by Uyar et al. [60]. Although the **36**₃ complex was expected to interact with three pyridyl ligands in the guest molecules (**37**) with three pyridyl ligands and one fullerenyl ligand, it actually interacted with two pyridyl ligands and one fullerenyl ligand (Figure 41.4). The thermodynamic data of the complexation indicated that an unusual and fairly large positive entropic change, attributed to extensive desolvation of the solutes especially from the large cavity of the porphyrin macroring and the fullerene surface, significantly contributes to the enhancement of the binding constant.

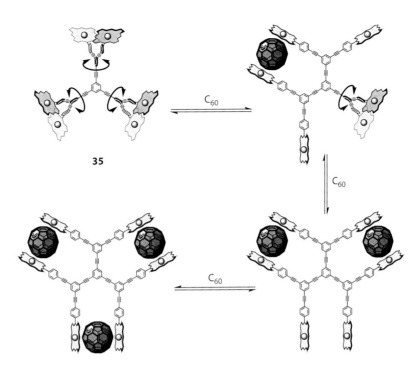

FIGURE 41.3 Allosteric effect in binding three C_{60}.

41.3.5 Capsules

From the thermodynamic data for calixarenes [43] and porphyrins [60], well-defined and well-designed host molecules with a three-dimensional cavity suitable for C_{60} are required because of the greater desolvation effect from both the large cavity and the fullerene surface. A homooxacalix[3]arene dimer formed through the coordination bond between pyridine and Pd(II) (**38**) was described by Ikeda et al. [61], which included C_{60} and had a slower complexation–decomplexation exchange rate than the ^{13}C NMR time scale. Although the association constant is not as high ($5.4 \times 10\,M^{-1}$) in 1,1,2,2-tetrachloroethane, the value was improved by the addition of Li^+ ions to 2.1×10^3 M^{-1}, and the C_{60} could be completely released from the cavity by the addition of Na^+ ions [62]. When Li^+ ions are bound to the ionophoric lower rims of **38**, the capsule molecule changes to a more flattened conformation that is more suitable for C_{60} inclusion (Figure 41.5). This means that the preorganization of the capsule molecule **38** was induced by Li^+ ions. Similarly, C_{60} inclusion was induced for a calix[5]arene dimer (**39**) (Figure 41.6) by Haino et al. [63] and tetrathiafulvalene-functionalized calixpyrrole (**40**) (Figure 41.7) by Nielsen et al. [64] using Cu(I) and Cl^- ions, respectively.

Many calix[5]arene dimers can bind fullerenes with high association constants. Notably, Haino used one calix[5]arene dimer **41** to successfully extract higher fullerenes ($\geq C_{76}$) from fullerene mixtures [65], and another calix[5]arene dimer **42** to bind a dumbbell-shaped C_{120} (a C_{60} dimer) [66]. These host molecules provide a route to overcome the poor solubility of higher fullerenes and C_{120}.

41.3.6 Carbon Nanotubes (Peapods)

The encapsulation of C_{60} in single-walled carbon nanotubes (SWNTs) was observed using high-resolution transmission electron microscopy (Figure 41.8) by Smith et al. [67]. These distinctive nanotubes filled with fullerenes are known as "peapods." The high-yield synthesis of SWNTs encapsulating C_{60} and gadolinium endohedral metallofullerenes (Gd@C_{82}) has since been developed by Hirahara et al. [68]. The doping of C_{60} and Gd@C_{82} into the inner hollow space of SWNTs was carried out by heating them in a sealed glass ampule at 400°C for 50 h and 500°C for 24 h, respectively. The overall yields of the nanotubes encapsulating C_{60} and Gd@C_{82} were very high, and the yield of the doping of C_{60} was almost 100%.

Supramolecular peapods were also prepared using linear zinc porphyrin nanotubes (**43** and **44**) (Figures 41.9 and 41.10) by Yamaguchi et al. [69] and Hasobe et al. [70]. These materials are expected to serve as components for photoelectronic devices because it is known that the photoelectron transfer between porphyrin and C_{60} is efficient.

41.4 Summary

This chapter describes recent research on host molecule-fullerene complexes using a supramolecular approach in both aqueous solution and organic solvents. In an aqueous solution, the hydrophobic interaction between the nonpolar fullerene surface and hydrophobic cavity in host molecules dominates the thermodynamics of the fullerene–host molecule interaction. In other words, because the large hydrophobic surface of

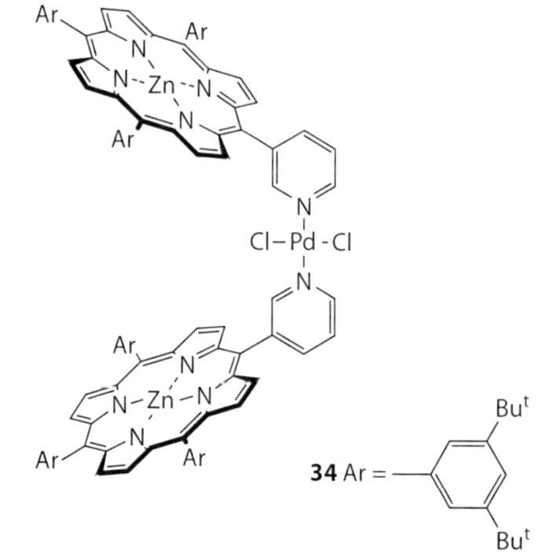

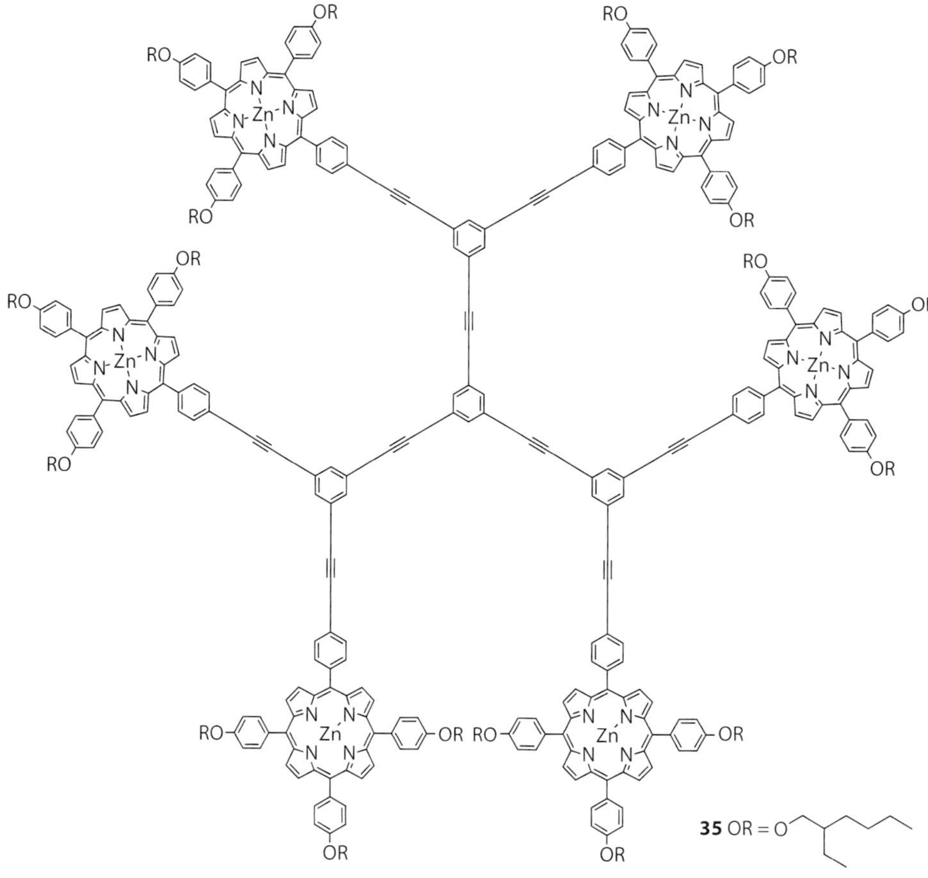

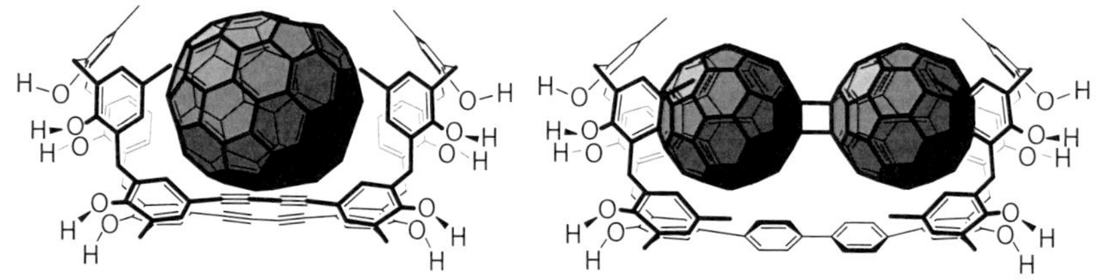

34 Ar =

35 OR = O

41·C₇₆ complex　　　　　　　**42**·C₁₂₀ complex

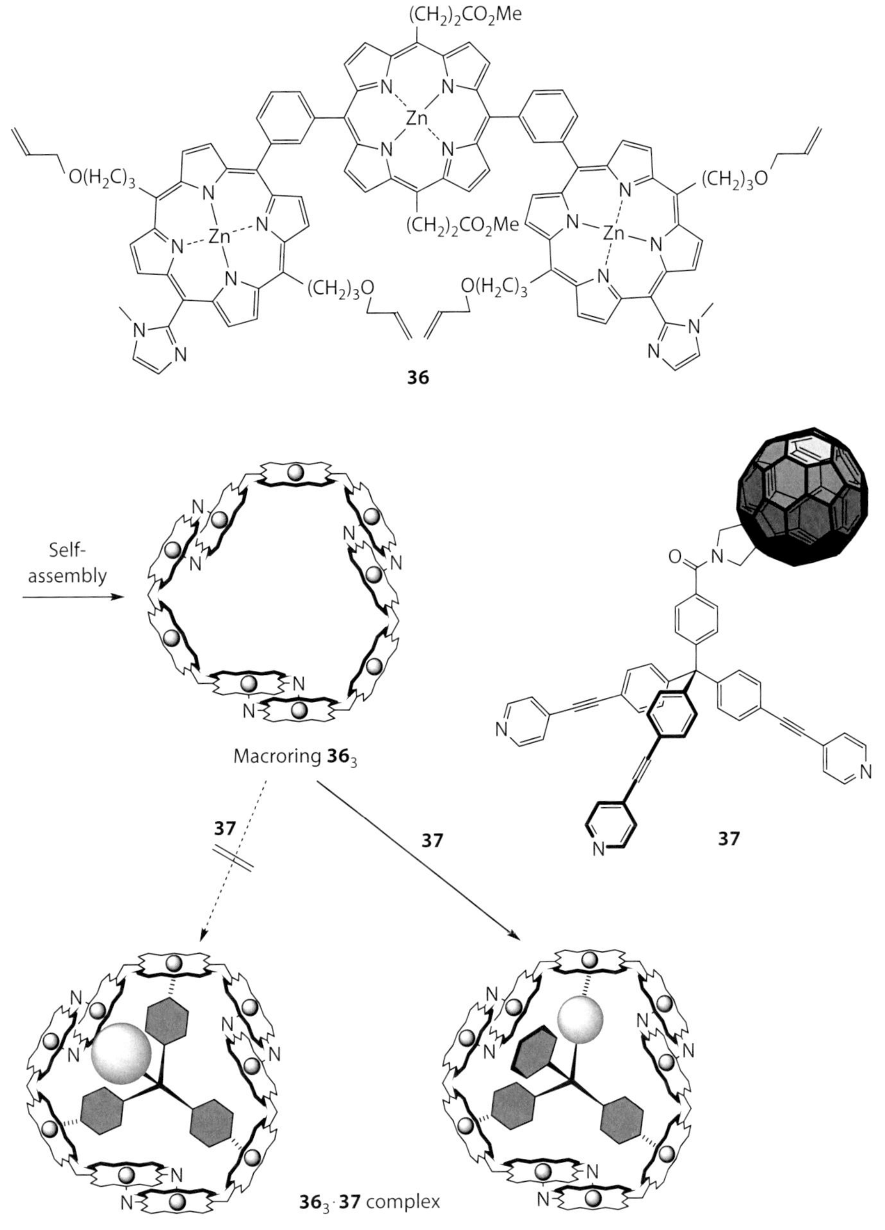

FIGURE 41.4 Structures of the macroring **36**₃ and **36**₃·**37** complex.

the fullerenes and the hydrophobic cavity of host molecules promote the formation of a high-density hydration shell of water, complete desolvation introduces the formation of a stable complex. In organic solvents, the desolvation entropy as well as π–π stacking interactions between the fullerene surface and host molecules play a decisive role in the formation of the complex. The requirements in the design of a host molecule are preorganized conformation, correct size, and a deep cavity for contributing desolvation energy in the formation of the complex.

The high association constants between the host molecules and the fullerenes have already been used to solubilize higher fullerenes (C_n, $n \geq 76$), and metallofullerenes have also been produced. Furthermore, both the isolation (non-aggregation) of fullerenes in an aqueous solution and the one-dimensional arrangement of fullerenes into carbon nanotubes are possible. However, the applications of these encapsulated fullerenes are at present at an early stage. It is suggested that in the near future, the concepts developed in the host–guest chemistry of fullerenes will be more fruitfully applied to material and medical sciences.

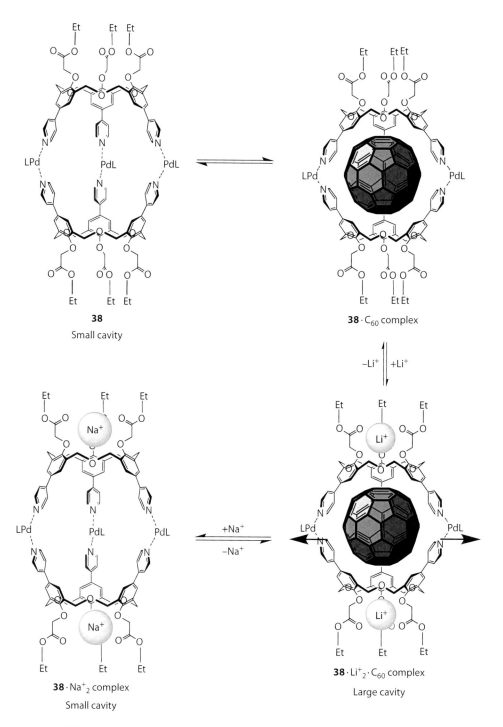

FIGURE 41.5 Improvement of the binding ability for C_{60} due to Li^+ binding to the lower rims of the capsule molecule **38** and release of C_{60} by Na^+ binding.

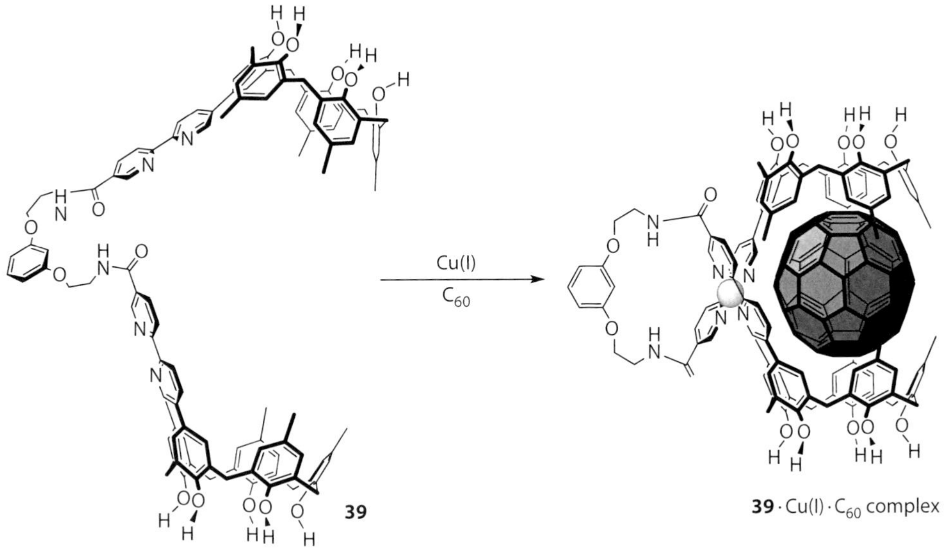

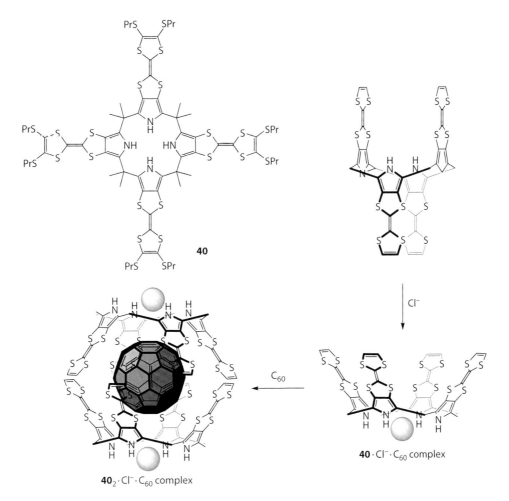

FIGURE 41.6 Improvement of the binding ability for C_{60} due to the addition of Cu(I) ion.

FIGURE 41.7 Improvement of the binding ability for C_{60} due to the addition of Cl⁻ ion.

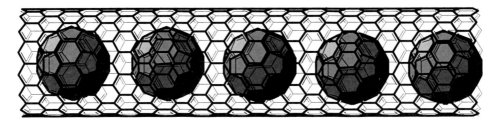

FIGURE 41.8　Schematic illustration of the "peapod" structure.

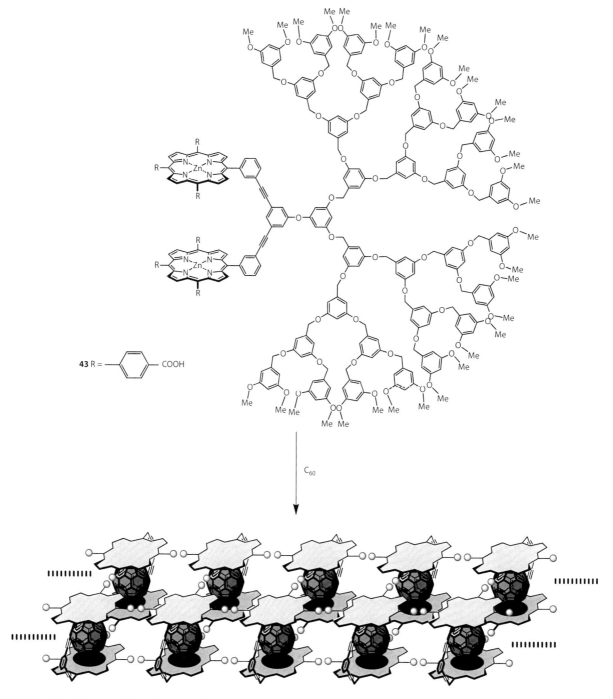

FIGURE 41.9　One-dimensional coordination of C_{60} using zinc porphyrin dimer **43**.

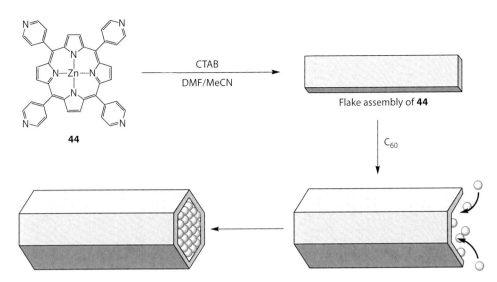

FIGURE 41.10 Schematic illustration of the organization process for **44** and C_{60} using cetyltrimethylammonium bromide (CTAB). CTAB is omitted for clarity.

References

1. Kroto, H. W., Heath, J. R., O'Brien, S. C., Curl, R. F., and Smalley, R. E. 1985. C_{60}. Buckminsterfullerene. *Nature* 318: 162–163.

2. Martín, N., Sánchez, L., Illescas B., and Pérez, I. 1998. C_{60}-based electroactive organofullerenes. *Chem. Rev.* 98: 2527–2547.

3. Tokuyama, H., Yamago, S., Nakamura, E., Shiraki T., and Sugiura, Y. 1993. Photoinduced biochemical-activity of fullerene carboxylic acid. *J. Am. Chem. Soc.* 115: 7918–7919.

4. Friedman, S. H., DeCamp, D. L., Sijbesma, R. P. et al. 1993. Inhibition of the HIV-1 protease by fullerene derivatives: Model building studies and experimental verification. *J. Am. Chem. Soc.* 115: 6506–6509.

5. Andersson, T., Nilsson, K., Sundahl, M., Westman, G., and Wennerström, O. 1992. C_{60} embedded in γ-cyclodextrin: A water-soluble fullerene. *J. Chem. Soc., Chem. Commun.* 604–606.

6. Yoshida, Z., Takekuma, H., Takekuma, S., and Matsubara, Y. 1994. Molecular recognition of C_{60} with γ-cyclodextrin. *Angew. Chem., Int. Ed. Engl.* 33: 1597–1599.

7. Komatsu, K., Fujiwara, K., Murata, Y., and Braun, T. 1999. Aqueous solubilization of crystalline fullerenes by supramolecular complexation with γ-cyclodextrin and sulfocalix[8]arene under mechanochemical high-speed vibration milling. *J. Chem. Soc., Perkin Trans.* 1: 2963–2966.

8. Braun, T., Buvári-Barcza, Á., Barcza, L. et al. 1994. Mechanochemistry: A novel approach to the synthesis of fullerene compounds. Water soluble buckminsterfullerene–γ-cyclodextrin inclusion complexes via a solid-solid reaction. *Solid State Ionics* 74: 47–51.

9. Murthy, C. N. and Geckeler, K. E. 2001. The water-soluble β-cyclodextrin–[60]fullerene complex. *Chem. Commun.* 13: 1194–1195.

10. Zhang, D.-D., Chen, J.-W., Ying, Y. et al. 1993. Studies on methylated β-cyclodextrins and C_{60} inclusion complexes. *J. Inclusion Phenom.* 16: 245–253.

11. Furuishi, T., Endo, T., Nagase, H., Ueda, H., and Nagai, T. 1998. Solubilization of C_{70} into water by complexation with δ-cyclodextrin. *Chem. Pharm. Bull.* 46: 1658–1659.

12. Liu, J., Alvarez, J., Ong, W., and Kaifer, A. E. 2001. Network aggregates formed by C_{60} and gold nanoparticles capped with γ-cyclodextrin hosts. *Nano Lett.* 1: 57–60.

13. Nakanishi, I., Fukuzumi, S., Konishi, T. et al. 2002. DNA cleavage via superoxide anion formed in photoinduced electron transfer from NADH to γ-cyclodextrin-bicapped C_{60} in an oxygen-saturated aqueous solution. *J. Phys. Chem. B* 106: 2372–2380.

14. Williams, R. M. and Verhoeven, J. W. 1992. Supramolecular encapsulation of C_{60} in a water-soluble calixarene—A core-shell charge-transfer complex. *Rec. Trav. Chim. Pays-Bas* 111: 531–532.

15. Ikeda, A., Hatano, T., Kawaguchi, M., Suenaga, H., and Shinkai, S. 1999. Water-soluble [60]fullerene-cationic homooxacalix[3]arene complex which is applicable to the photocleavage of DNA. *Chem. Commun.* 15: 1403–1404.

16. Ikeda, A., Ejima, A., Nishiguchi, K. et al. 2005. DNA-photocleaving activities of water-soluble carbohydrate-containing nonionic homooxacalix[3]arene·[60]fullerene complex. *Chem. Lett.* 34: 308–309.

17. Ikeda, A., Hatano, T., Shinkai, S., Akiyama, T., and Yamada, S. 2001. Efficient photocurrent generation in novel self-assembled multilayers comprised of [60]fullerene-cationic homooxacalix[3]arene inclusion complex and anionic porphyrin polymer. *J. Am. Chem. Soc.* 123: 4855–4856.

18. Hatano, T., Ikeda, A., Sano, M., Kanekiyo, Y., and Shinkai, S. 2000. Facile construction of an ultra-thin [60]fullerene layer from [60]fullerene-homooxacalix[3]arene complexes on a gold surface. *J. Chem. Soc., Perkin Trans.* 2: 909–912.

19. Ikeda, A., Hatano, T., Konishi, T., Kikuchi J., and Shinkai, S. 2003. Host-guest complexation effect of 2,3,6-tri-O-methyl-β-cyclodextrin on a C_{60}-porphyrin light-to-photocurrent conversion system. *Tetrahedron* 59: 3537–3540.

20. Rio, Y. and Nierengarten, J.-F. 2002. Water soluble supramolecular cyclotriveratrylene-[60]fullerene complexes with potential for biological applications. *Tetrahedron Lett.* 43: 4321–4324.

21. Yamakoshi, Y. N., Yagami, T., Fukuhara, K., Sueyoshi, S., and Miyata, N. 1994. Solubilization of fullerenes into water with polyvinylpyrrolidone applicable to biological tests. *J. Chem. Soc., Chem. Commun.* 40: 517–518.

22. Tsuchiya, T., Yamakoshi, Y. N., and Miyata, N. 1995. A novel promoting action of fullerene C_{60} on the chondrogenesis in rat embryonic limb bud cell culture system. *Biochem. Biophys. Res. Commun.* 206: 885–894.

23. Tsuchiya, T., Oguri, I., Yamakoshi, Y. N., and Miyata, N. 1996. Effect of [60]fullerene on the chondrogenesis in mouse embryonic limb bud cell culture system. *Fullerene Sci. Technol.* 4: 989–999.

24. Tsuchiya, T., Oguri, I., Yamakoshi, Y. N., and Miyata, N. 1996. Novel harmful effects of [60]fullerene on mouse embryos in vitro and in vivo. *FEBS Lett.* 393: 139–145.

25. Sera, N., Tokiwa, H., and Miyata, N. 1996. Mutagenicity of the fullerene C_{60}-generated singlet oxygen dependent formation of lipid peroxides. *Carcinogenesis* 17: 2163–2169.

26. Sakai, A., Yamakoshi, Y. N., and Miyata, N. 1995. The effects of fullerenes on the initiation and promotion stages of BALB/3T3 cell transformation. *Fullerene Sci. Technol.* 3: 377–388.

27. Yamakoshi, Y., Yagami, T., Sueyoshi, S., and Miyata, N. 1996. Acridine adduct of [60]fullerene with enhanced DNA-cleaving activity. *J. Org. Chem.* 61: 7236–7237.

28. Yamakoshi, Y., Umezawa, N., Ryu, A. et al. 2003. Active oxygen species generated from photoexcited fullerene (C_{60}) as potential medicines: $O_2^{\cdot-}$ versus $^{1}O_2$. *J. Am. Chem. Soc.* 125: 12803–12809.

29. Bangham, A. D., Standish M. M., and Watkins, J. C. 1965. Diffusion of univalent ions across lamellae of swollen phospholipids. *J. Mol. Biol.* 13: 238–252.

30. Hungerbühler, H., Guldi, D. M., and Asmus, K. D. 1993. Incorporation of C_{60} into artificial lipid-membranes. *J. Am. Chem. Soc.* 115: 3386–3387.

31. Ikeda, A., Sato, T., Kitamura, K. et al. 2005. Efficient photocleavage of DNA utilising water-soluble lipid membrane-incorporated [60]fullerenes prepared using a [60]fullerene exchange method. *Org. Biomol. Chem.* 3: 2907–2909.

32. Ikeda, A., Sue, T., Akiyama, M. et al. 2008. Preparation of highly photosensitizing liposomes with fullerene-doped lipid bilayer using dispersion-controllable molecular exchange reactions. *Org. Lett.* 10: 4077–4080.

33. Ikeda, A., Doi, Y., Hashizume, M., Kikuchi, J., and Konishi, T. 2007. An extremely effective DNA photocleavage utilizing functionalized liposomes with a fullerene-enriched lipid bilayer. *J. Am. Chem. Soc.* 129: 4140–4141.

34. Doi, Y., Ikeda, A., Akiyama, M. et al. 2008. Intracellular uptake and photodynamic activity of water-soluble [60] and [70]fullerenes incorporated in liposomes. *Chem.-Eur. J.* 14: 8892–8897.

35. Suzuki, T., Nakashima, K., and Shinkai, S. 1994. Very convenient and efficient purification method for fullerene (C_{60}) with 5,11,17,23,29,35,41,47-octa-*tert*-butylcalix[8]arene-49,50,51,52,53,54,55,56-octol. *Chem. Lett.* 4: 699–702.

36. Atwood, J. L., Koutsantonis, G. A., and Raston, C. L. 1994. Purification of C_{60} and C_{70} by selective complexation with calixarenes. *Nature* 368: 229–231.

37. Raston, C. L., Atwood, J. L., Nichols, P. J., and Sudria, I. B. N. 1996. Supramolecular encapsulation of aggregates of C_{60}. *Chem. Commun.* 23: 2615–2616.

38. Williams, R. M., Zwier, J. M., Verhoeven, J. W., Nachtegaal, G. H., and Kentgens, A. P. M. 1994. Interactions of fullerenes and calixarenes in the solid state studied with ^{13}C CP-MAS. *J. Am. Chem. Soc.* 116: 6965–6966.

39. Suzuki, T., Nakashima, K., and Shinkai, S. 1995. Influence of *para*-substituents and solvents on selective precipitation of fullerenes by inclusion in calix[8]arenes. *Tetrahedron Lett.* 36: 249–250.

40. Araki, K., Akao, K., Ikeda, A., Suzuki, T., and Shinkai, S. 1996. Molecular design of calixarene-based host molecules for inclusion of C_{60} in solution. *Tetrahedron Lett.* 37: 73–76.

41. Ikeda, A., Yoshimura, M., and Shinkai, S. 1997. Solution complexes formed from C_{60} and calixarenes. On the importance of the preorganized structure for cooperative interactions. *Tetrahedron Lett.* 38: 2107–2110.

42. Haino, T., Yanase, M., and Fukazawa, Y. 1998. New supramolecular complex of C_{60} based on calix[5]arene—Its structure in the crystal and in solution. *Angew. Chem., Int. Ed.* 36: 259–260.

43. Mizyed, S., Georghiou, P. E., and Ashram, M. 2000. Thermodynamic study of the complexes of calix[4]naphthalenes with [60]fullerene in different solvents. *J. Chem. Soc., Perkin Trans.* 2: 277–280.

44. Yanase, M., Matsuoka, M., Tatsumi, Y. et al. 2000. Thermodynamic study on supramolecular complex formation of fullerene with calix[5]arenes in organic solvents. *Tetrahedron Lett.* 41: 493–497.

45. Atwood, J. L., Barbour, L. J., Nichols, P. J., Raston, C. L., and Sandoval, C. A. 1999. Symmetry-aligned supramolecular encapsulation of C_{60}: $[C_{60} \subset (L)_2]$, L = *p*-benzylcalix[5] arene or *p*-benzylhexahomooxacalix[3]arene. *Chem.-Eur. J.* 5: 990–996.

46. Steed, J. W., Junk, P. C., Atwood, J. L. et al. 1994. Ball and socket nanostructures: New supramolecular chemistry based on cyclotriveratrylene. *J. Am. Chem. Soc.* 116: 10346–10347.

47. Nierengarten, J.-F., Oswald, L., Eckert, J.-F., Nicoud, J.-F., and Armaroli, N. 1999. Complexation of fullerenes with dendritic cyclotriveratrylene derivatives. *Tetrahedron Lett.* 40: 5681–5684.

48. Felder, D., Heinrich, B., Guillon, D., Nicoud, J.-F., and Nierengarten, J.-F. 2000. A liquid crystalline supramolecular complex of C_{60} with a cyclotriveratrylene derivative. *Chem.-Eur. J.* 6: 3501–3507.

49. Tucci, F. C., Rudkevich, D. M., and Rebek, J. Jr. 1999. Deeper cavitands. *J. Org. Chem.* 64: 4555–4559.

50. Bredenkötter, B., Henne, S., and Volkmer, D. 2007. Nanosized ball joints constructed from C_{60} and tribenzotriquinacene sockets: Synthesis, component self-assembly and structural investigations. *Chem.-Eur. J.* 13: 9931–9938.

51. Kawase, T., Tanaka, K., Fujiwara, N., Darabi, H. R., and Oda, M. 2003. Complexation of a carbon nanoring with fullerenes. *Angew. Chem., Int. Ed.* 42: 1624–1628.

52. Kawase, T., Tanaka, K., Seirai, Y., Shiono, N., and Oda, M. 2003. Complexation of carbon nanorings with fullerenes: Supramolecular dynamics and structural tuning for a fullerene sensor. *Angew. Chem., Int. Ed.* 42: 5597–5600.

53. Diederich, F., Effing, J., Jonas, U. et al. 1992. C_{60} and C_{70} in a basket?—Investigations of mono- and multilayers from azacrown compounds and fullerenes. *Angew. Chem., Int. Ed. Engl.* 31: 1599–1602.

54. Tsuchiya, T., Sato K., Kurihara, H. et al. 2006. Host-guest complexation of endohedral metallofullerene with azacrown ether and its application. *J. Am. Chem. Soc.* 128: 6699–6703.

55. Tashiro, K., Aida, T., Zheng, J.-Y. et al. 1999. A cyclic dimer of metalloporphyrin forms a highly stable inclusion complex with C_{60}. *J. Am. Chem. Soc.* 121: 9477–9478.

56. Zheng, J. Y., Tashiro, K., Hirabayashi, Y. et al. 2001. Cyclic dimers of metalloporphyrins as tunable hosts for fullerenes: A remarkable effect of rhodium(III). *Angew. Chem., Int. Ed.* 40: 1858–1861.

57. Shoji, Y., Tashiro, K., and Aida, T. 2006. Sensing of chiral fullerenes by a cyclic host with an asymmetrically distorted π-electronic component. *J. Am. Chem. Soc.* 128: 10690–10691.

58. Sun, D., Tham, F. S., Reed, C. A., Chaker, L., and Boyd, P. D. W. 2002. Supramolecular fullerene-porphyrin chemistry. Fullerene complexation by metalated "jaws porphyrin" hosts. *J. Am. Chem. Soc.* 124: 6604–6612.

59. Ayabe, M., Ikeda, A., Kubo, Y., Takeuchi, M., and Shinkai, S. 2002. A dendritic porphyrin receptor for C_{60} which features a profound positive allosteric effect. *Angew. Chem., Int. Ed.* 41: 2790–2792.

60. Uyar, Z., Satake, A., Kobuke, Y., and Hirota, S. 2008. Stable supramolecular complex of porphyrin macroring with pyridyl and fullerenyl ligands. *Tetrahedron Lett.* 49: 5484–5487.

61. Ikeda, A., Yoshimura, M., Udzu, H., Fukuhara, C., and Shinkai, S. 1999. Inclusion of [60]fullerene in a homooxacalix[3]arene-based dimeric capsule cross-linked by a Pd^{II}-pyridine interaction. *J. Am. Chem. Soc.* 121: 4296–4296.

62. Ikeda, A., Udzu, H., Yoshimura, M., and Shinkai, S. 2000. Inclusion of [60]fullerene in a self-assembled homooxacalix[3]arene-based dimeric capsule constructed by a Pd^{II}-pyridine interaction. The Li$^+$-binding to the lower rims can improve the inclusion ability. *Tetrahedron* 56: 1825–1832.

63. Haino, T., Yamanaka, Y., Araki, H., and Fukazawa, Y. 2002. Metal-induced regulation of fullerene complexation with double-calix[5]arene. *Chem. Commun.* 5: 402–403.

64. Nielsen, K. A., Cho, W.-S., Sarova, G. H. et al. 2006. Supramolecular receptor design: Anion-triggered binding of C_{60}. *Angew. Chem., Int. Ed.* 45: 6848–6853.

65. Haino, T., Fukunaga, C., and Fukazawa, Y. 2006. A new calix[5]arene-based container: Selective extraction of higher fullerenes. *Org. Lett.* 8: 3545–3548.

66. Haino, T., Seyama, J., Fukunaga, C. et al. 2005. Calix[5]arene-based receptor for dumb-bell-shaped C_{120}. *Bull. Chem. Soc. Jpn.* 78: 768–770.

67. Smith, B. W., Monthioux, M. D., and Luzzi, E. 1998. Encapsulated C_{60} in carbon nanotubes. *Nature* 396: 323–324.

68. Hirahara, K., Suenaga, K., Bandow, S. et al. 2000. One-dimensional metallofullerene crystal generated inside single-walled carbon nanotubes. *Phys. Rev. Lett.* 85: 5384–5387.

69. Yamaguchi, T., Ishii, N., Tashiro, K., and Aida, T. 2003. Supramolecular peapods composed of a metalloporphyrin nanotube and fullerenes. *J. Am. Chem. Soc.* 125: 13934–13935.

70. Hasobe, T., Sandanayaka, A. S. D., Wada, T., and Araki, Y. 2008. Fullerene-encapsulated porphyrin hexagonal nanorods. An anisotropic donor–acceptor composite for efficient photoinduced electron transfer and light energy conversion. *Chem. Commun.* 29: 3372–3374.

Electronic Structure of Encapsulated Fullerenes

Shojun Hino
Ehime University

42.1 Introduction

42.1.1 Historic Perspectives of Fullerene

Before the discovery of C_{60} (Kroto et al. 1985), the existence of truncated icosahedron composed of carbon and hydrogen atoms $C_{60}H_{60}$ was pointed out (Yoshida and Osawa 1971). Yoshida and Osawa proposed that this molecule might exist if all carbon atoms in $C_{60}H_{60}$ have sp^2 hybridized orbitals. They also pointed out that the molecule might have aromaticity (super-aromaticity). Actually, the discovered truncated icosahedron was not $C_{60}H_{60}$ but C_{60}; however, what is noteworthy is that fullerene structure had been proposed more than 15 years before its actual discovery. Although the existence of a curious-shaped molecule had been revealed, the investigation of C_{60} did not gain momentum until 1990 when Krätschmer and his coworkers invented a C_{60} mass production method (Krätschmer et al. 1990). Before the advent of the mass production method, the quantity of fullerenes (C_{60} or C_{70}) was so microscopic that investigation on the solid state nature of fullerenes was virtually impossible.

The discovery of this new method saw the dawn of fullerene science. Particularly, the discovery of superconductivity of K_3C_{60} (Hebard et al. 1991) at 18 K accelerates the investigation. Further, superconductivity transition temperature of alkali metal dosed C_{60} went up to 33 K (Tanigaki et al. 1991). Everybody looked forward to promising high temperature superconductivity in fullerenes. Unfortunately, a fullerene superconductor having higher transition temperature has not yet been discovered. Superconductivity of alkali metal dosed fullerenes is brought about by electron transfer from alkali metal atoms to C_{60} molecules, and the superconducting state is destroyed when alkali metal dosed C_{60} is exposed to ambient air. Alkali metal absorbs water or oxygen and turns into hydroxide or oxide. One of the ways to avoid the degradation of electron-donating metal atoms is the capsulation of metal atom(s) by fullerene carbon cage. So metallofullerene $La@C_{82}$ was synthesized and isolated (Chai et al. 1991). Since then, there have been numerous attempts to synthesize and isolate metallofullerenes, and there is a good, although slightly outdated, review on the subject (Shinohara 2000).

42.1.2 Endohedral or Exohedral?

At the first stage of research on metallofullerenes, the main issue was the position of the metal atom: Is it inside or outside the cage? Or is it on the carbon network? An ESR measurement of $La@C_{82}$ (Johnson et al. 1992) and x-ray photoelectron spectroscopy measurement of $La@C_{82}$ (Weaver et al. 1992) revealed the oxidation state of lanthanum to be +3, which suggested that lanthanum atom was safely protected from the effect of oxygen or water present in the ambient atmosphere. Molecular orbital approach (Nagase and Kobayashi 1993) was also in favor of the endohedral structure. However, extended x-ray absorption fine structure (EXAFS) study of unpurified $Y@C_{82}$ (Soderbolm et al. 1992) and that of $Y@C_{82}$ (Park et al. 1993) were contradicting; the former insisted on the exohedral form, but the latter reported 0.253 and 0.24 nm for nearest-neighbor C-Y distances, which favored the endohedral form. High-resolution transmission electron microscopy on $Sc_2@C_{84}$ (Beyers et al. 1994) and scanning tunneling microscopy on $Sc_2@C_{84}$ (Shinohara et al. 1994) strongly suggested the endohedral form. Finally, x-ray diffraction Rietveld analysis combined with maximum entropy method displayed the electron contour map of $Y@C_{82}$ (Takata et al. 1995), and the endohedral form was confirmed.

42.1.3 Difficulties in Experiment

Although the fullerene mass production method has been discovered, the amounts of isolated metallofullerenes are not sufficient to carry out their systematic investigation. A 30% conversion of graphite to C_{60} and C_{70} is not difficult, but, for example, the production yield of $La@C_{82}$ might be less than 1% of starting graphite. Further, the yield of another endohedral fullerene such as $La_2@C_{78}$ (Cao et al. 2004) is about 4% of the yield of $La@C_{82}$. Thus, the amounts of synthesized and isolated endohedral fullerenes are usually around 1 mg, which makes any kind of measurement difficult.

Photoelectron spectroscopy is a direct tool to investigate the electronic structure of materials. When C_{60} was isolated with macroscopic quantity, a lot of C_{60} research was performed, and ultraviolet photoelectron spectroscopy was among them. However, the number of reports on the photoelectron spectral measurements of metallofullerenes or higher fullerenes is not so large. Principally, fullerenes or metallofullerenes have to be sublimed to prepare specimens for photoelectron spectroscopy. At the first stage of ultraviolet photoelectron spectra (UPS) measurements, fullerenes were supplied as a form of powder. Filling the sublimation furnace with small amounts of powder was a very difficult task, and the powder was often covered by impurities derived from the solvent used in chromatographic separation processes. Further, high temperature (700 K or upward), which is very high when compared with that of ordinary organic materials such as pentacene or phthalocyanines, must be applied to the furnace to sublime fullerenes. When fullerenes arc heated to sublime, they not only sublime but also decompose. Residual dross was often found in the evaporation crucible used for specimen preparation. To obtain reasonably good photoelectron spectra, sufficient amounts of metallofullerenes and also delicate temperature control of the sublimation crucible, which are key factors to obtain a good quality film specimen, are required. Up to now, our laboratory could manage to measure endohedral fullerene of about 1 mg or slightly less than that. However, we were not always successful to obtain reasonably good spectra and could obtain only such spectra that were heavily influenced by the electrons emitted from the substrate.

42.2 Photoelectron Spectroscopy

The basic idea of photoelectron spectroscopy went to back to 1905 when Einstein wrote a paper on external photoelectric effect. The relation between the energy of the incident light, $h\nu$, and kinetic energy of outgoing photoelectron, K.E., can be described as

$$h\nu = \text{K.E.} + \text{B.E.},$$

where B.E. is the binding energy of electron. By measuring the kinetic energy of photoelectron, we can easily obtain the energy of electron that was bound to the atoms or molecules. Usually, photoelectron spectra of free molecules are referred to the vacuum level, and binding energy is a measure from the vacuum level. On the other hand, photoelectron spectra of solid states are often referred to the Fermi level. Figure 42.1 shows the schematic diagram of photoelectron emission from the solid. A solid is an aggregate of atoms or molecules, and each constituent has its specific electronic structure. In the aggregate, their electronic structures interact and form the electronic structure of solid. The interaction of the core levels is so weak that individual electronic structure is not influenced even if atoms (molecules) form aggregate, but it is not negligible in upper valence levels so that electron energy levels of atoms (molecules) mix and form energy band. Thus, the electronic structure of the solid can be classified into two types; core levels having discrete energy and valence bands having continuous energy. Core levels have rather spike-like electron density of state (DOS), and valence bands have broad DOS.

When the light of enough energy irradiates the solid, electrons are ejected as photoelectrons. When energy of the incident light is small, such as ultraviolet, only electrons in upper valence band are to be ejected (ultraviolet photoelectron spectroscopy). When the incident light is soft x-ray, electrons of core level as well as those in valence band can be ejected (x-ray photoelectron spectroscopy). In principle, energy distribution of ejected photoelectrons is a reflection of DOS. The fastest photoelectrons come from the Fermi level (when the solid is metal of the electrons that fill up to the Fermi level) or the highest valence band (when the

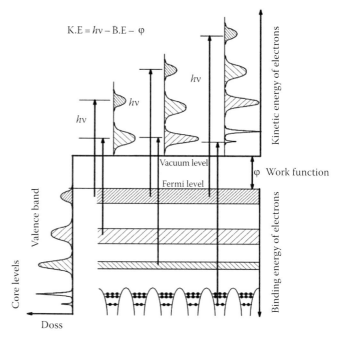

FIGURE 42.1 Principle of photoelectron spectroscopy of solids.

solid is semiconductor). Here, we handle only the case of solid metal for simplification. If the position of the fastest photoelectrons in the spectra is aligned at the Fermi level, the energy distribution of ejected photoelectrons indicates DOS of the solid. Usually, in the photoelectron measurements of the solid, the Fermi level is determined at first, and then the photoelectron spectra are to be measured. Thus, the photoelectron spectra of solid are often referred to the Fermi level. In this case, the relation between the kinetic energy of photoelectrons and binding energy of electrons is

$$K.E. = h\nu - B.E. - \varphi,$$

where φ is the work function of solid.

42.3 Electronic Structure of Mono Metal Atom–Entrapped Fullerenes

42.3.1 Group Three Metal Atom– Entrapped Fullerenes

42.3.1.1 La@C_{82}

Lanthanum belongs to lanthanide group, but it also situates at group three column in the periodic table, so La@C_{82} is described at this section. There is an empirical rule, the so-called isolated pentagon rule (IPR) in the field of fullerenes. This was proposed by Fowler and Manolopoulos, and its back ground was that the fullerene cage became unstable when two or more pentagon rings situated side by side so that isolated fullerenes must have the cage structure of isolated pentagon rings (Fowler and Manolopoulos 1995). As for C_{82}, there are nine IPR satisfying fullerenes, and their cage structures are shown in Figure 42.2.

When higher fullerenes such as C_{82} were isolated, its cage structure was one of the issues to be settled down. Theoretical calculation on C_{82} predicted that C_2 (a) (equivalent to (82:3) according to the nomenclature proposed by Fowler and Manolopoulos in 1995. In the following text, Fowler and Manolopoulos nomenclature will be added when other nomenclature is described) was most stable followed by C_2 (b) (82:1) and then C_s(a) (82:6) (Nagase et al. 1993), but ^{13}C NMR measurements revealed that isolated C_{82} was a mixture of C_2, C_{2v}, and C_{3v} with respective ratio of 8:1:1 (Kikuchi et al. 1992). When mono metal atom encapsulate metallofullerenes have been isolated, it was believed that the cage structure of the metallofullerenes is the same as that of empty C_{82} (Nagase and Kobayashi 1993, Nagase et al. 1993), and it lasted until sufficient amounts of metallofullerenes to measure ^{13}C NMR spectra could be isolated (for example, Kirbach and Dunsch 1996, Akasaka et al. 2000).

The UPS of metallofullerene, La@C_{82}, were firstly measured in 1993 (Hino et al. 1993b). As La@C_{82} minor C_s isomer (82:6) could not be separated at that time, the prepared La@C_{82} sample for the measurement contained about 10% of the minor C_s (82:6) isomer. However, the UPS is rather insensitive to small amounts of impurity; therefore, the reported spectra could be regarded as the reflection of the electronic structure of La@C_{82} major C_{2v} isomer.

The UPS of La@C_{82} (Figure 42.3) showed intensity oscillation when the incident photon energy was tuned, as have been observed in other fullerenes (Benning et al. 1991, Weaver et al. 1991, Hino et al. 1992a,b, 1993a). Origin of intensity oscillation was one of the issues of fullerenes, and it has been argued for a

Nine IPR satisfying C_{82} cages

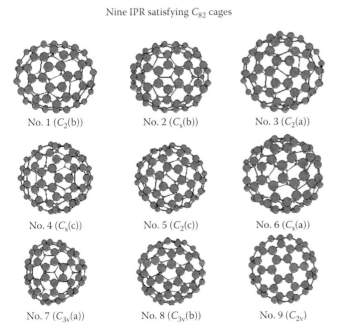

No. 1 (C_2(b))   No. 2 (C_s(b))   No. 3 (C_2(a))

No. 4 (C_s(c))   No. 5 (C_2(c))   No. 6 (C_s(a))

No. 7 (C_{3v}(a))   No. 8 (C_{3v}(b))   No. 9 (C_{2v})

FIGURE 42.2 IPR satisfying nine C_{82} isomers.

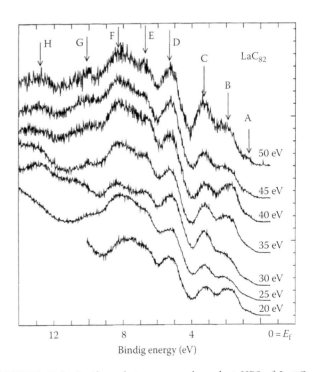

FIGURE 42.3 Incident photon energy dependent UPS of La@C_{82}, reported by Hino et al. (1993b). (From Hino, S. et al., *Phys. Rev. Lett.*, 71, 4261, 1993b. With permission.)

while. Now it has settled, and the origin of intensity oscillation is believed to be due to the superposition and interference of outgoing wave function of photo-excited electrons (Xu et al. 1996b, Hasegawa et al. 1998).

An encapsulated metal atom donates electrons to the fullerene cage as alkali metal atoms do in the fullerene complexes, and the amount of transferred electrons is one of the issues of metallofullerenes. At the first stage of isolation of metallofullerenes, only those fullerenes that contained group three elements such as Sc, Y, or La were isolated. Therefore, the oxidation state of metal atoms inside the cage was thought to be +3. An x-ray photoelectron spectra (XPS) measurement of lanthanum fullerene complexes confirmed this by showing that its formal charge state was close to +3 (Weaver et al. 1992).

When the UPS of La@C_{82} was measured, there was a naïve idea that La@C_{82} and empty C_{82} have the same cage structure, as described above. Actually, their UPS resemble, and the difference was observed only in the upper valence band region: there was a critical difference just below the Fermi level between the two spectra. This seemed to correspond with the transferred electrons. To deduce the amounts of transferred electrons from the entrapped atom to the cage (namely, oxidation state of the entrapped atom), the UPS of empty C_{82} was subtracted from the UPS of La@C_{82} (Figure 42.4). Thus, the obtained difference spectrum showed a structure of which intensity corresponded with three electrons, and it could be deconvoluted into two components with 1:2 intensity ratio. From these findings, three electron transfer was deduced, and the two components were attributed to the electrons transferred to the lowest unoccupied molecular

orbital (LUMO) and the LUMO + 1 of empty C_{82}. Another UPS measurement of La@C_{82} by Weaver's group (Poirier et al. 1994) also supported this deduction.

Since the electron configuration of lanthanum is $(4d)^{10}(5s)^2(5p)^6(5d)^1(6s)^2$, the origin of transferred electrons seems to be outer orbitals La 5d and La 6s. After electron transfer, the oxidation state of La@C_{82} is La^{3+} and C_{82}^{3-}. Because of this configuration, an ESR signal can be observed (Johnson et al. 1992, Suzuki et al. 1992). Further, hyperfine structure associated with ^{139}La was observed in the ESR signal of La@C_{82}, and it was well reproduced by Nagase's theoretical calculation (Nagase and Kobayashi 1993) assuming C_2-C_{82} cage structure. More or less the analogous hyperfine structure was observed in the ESR of Sc@C_{82} and Y@C_{82}. Therefore, the oxidation state of these metal atoms was considered to be +3 at that moment. However, it was lately proved that the oxidation state of Sc is not +3 but +2, which will be described in following section.

Nagase also predicted that the metal atom was not situated at the center of the C_2 cage but was located very close to the carbon cage (Nagase et al. 1993). Calculations on C_{3v}-C_{82} cage (Laasonen et al. 1992, Poirier et al. 1994) and on C_2 cage (Nagase and Kobayashi 1993, Andreoni and Curioni 1996) also indicated off-centered metal atom position. Further, Andreoni and Curioni pointed out that entrapped metal atom is mobile, and metallofullerene such as Y@C_{82} or La@C_{82} can be used as electronic molecular switch. Off-centered structure was finally visualized by x-ray diffraction Rietveld analysis combined with maximum entropy method (Nishibori et al. 2000). Electron density map of La@C_{82} and Sc@C_{82} is shown in Figure 42.5. Electron density due to lanthanum atom distributes widely in the cage, which suggests that lanthanum atom is rather weakly bound to the cage.

Photoelectron cross section of π-electrons is usually larger than that of σ-electrons. This is because of the difference in the distribution of individual wave functions. In potassium C_{60} complex, if there is no electron transfer from potassium to fullerene, K 4s electron is hardly observed. La 5d or La 6s electrons distribute rather narrow region, namely, just around the atom, while the LUMO or the LUMO + 1 of empty fullerene delocalizes almost over the entire fullerene cage. Since

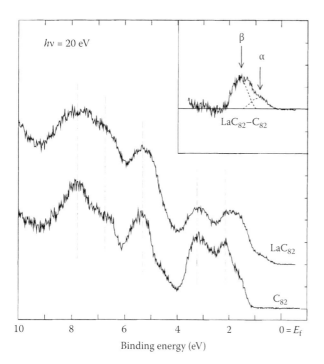

FIGURE 42.4 UPS of C_{82}, La@C_{82} and their difference spectrum, reported by Hino et al. (1993b). (From Hino, S. et al., *Phys. Rev. Lett.*, 71, 4261, 1993b. With permission.)

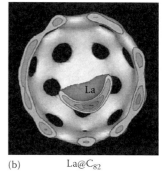

(a)   Sc@C_{82} (b)   La@C_{82}

FIGURE 42.5 Electron density map of Sc@C_{82} and La@C_{82}, reported by Nishibori et al. (2000). (From Nishibori, E. et al., *Chem. Phys. Lett.*, 330, 497, 2000. With permission.)

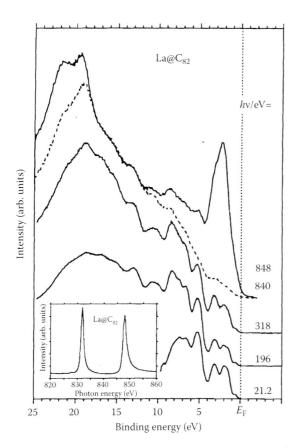

FIGURE 42.6 Resonant (848 eV excitation) and off resonant valence band spectra of La@C$_{82}$, reported by Kessler et al. (1997). Inset is XANES in the region of La 3d level. (From Kessler, B. et al., *Phys. Rev. Lett.*, 79, 2289, 1997. With permission.)

transferred electrons were observed in the UPS with high intensity, it could be assumed that outer lanthanum electrons are completely transferred to the cage (Hino et al. 1993a,b, Nagase et al. 1993, Poirier et al. 1994). However, La 3d to 4f resonant photoelectron emission measurement on the La@C$_{82}$ major isomer (Figure 42.6) revealed that there is still a lot of DOS on the La atom (Kessler et al. 1997). This experiment intended to excite La 3d electrons to empty La 4f level (resonance) and make a hole state in La 3d level. Once the hole was created, outer shell electrons moved down to fill up the hole, and eventually the atom got excess energy, which was nonradiatively used with emitting outer shell electrons such as 5s or 4d (Auger process). Usually, at higher energy excitation (soft x-ray region), direct inner shell electron excitation (photoelectron generation) cross section is orders of magnitude larger than that of valance electron. (Valence band is poorly observed in XPS, while inner levels are distinctly observed.) As for the valence region of La@C$_{82}$, Kessler's off-resonant and on-resonant spectra showed drastic difference. Hence they have concluded that simple electron transferred picture is not appropriate to describe the electronic structure. However, theoretical calculation on metallofullerene showed that the wave function coefficients on the metal atom of outer orbitals were not completely zero but had

certain value. This remaining electron density is good enough to produce on resonant photoelectron peaks shown in Figure 42.6. Therefore, virtually three electron transfer is reasonable in the La@C$_{82}$ system. Further, Nagase's group (Lu et al. 2000) proposed metal-cage hybridization to elucidate the resonant photoelectron spectra of La@C$_{82}$.

42.3.1.2 Sc@C$_{82}$

At the first stage of metallofullerene synthesis, scandium encapsulated fullerenes was one of the targets to isolate (Shinohara et al. 1992, Yannoni et al. 1992). Scandium atom(s) could be easily incorporated into the fullerene cage, but Sc@C$_{82}$ was not the most abundant synthesized and isolated metallofullerene. Sc$_2$@C$_{82}$, Sc$_2$@C$_{84}$, or Sc$_3$@C$_{82}$ was more abundant than Sc@C$_{82}$ (Shinohara et al. 1992). Hence, photoelectric measurement on Sc@C$_{82}$ was rather late compared to that of Sc$_2$@C$_{84}$ (Takahashi et al. 1995, Pichler et al. 2000). Results of Takahashi's group were not identical with those of Pichler's group, and because of missing the fullerene characteristic π-bands in the vicinity of the Fermi level, the former must be erratic. Takahashi's group could not properly prepare thin Sc$_2$@C$_{84}$ films for the measurement. Lately, it has been found that measured Sc$_2$@C$_{84}$ is actually Sc$_2$C$_2$@C$_{82}$, which will be described in the following chapter.

Although experimental determination of the electronic structure of Sc@C$_{82}$ was not published until 1999, theoretical calculation was published by Nagase and Kobayashi (1993). They assumed the cage structure of Sc@C$_{82}$ as well as other Y- or La-entrapped metallofullerenes to be a C_2 cage (actually, as described above, the cage symmetry of the major isomer of these metallofullerenes is C_{2v}), and they concluded the formal charge as Sc^{2+}@C$_{82}^{2-}$. Fullerene accepts upper lying electrons of entrapped Sc metal atom. The energy levels of Sc 4s and 3d are −5.8 and −9.1 eV, respectively, while those of La 5d and 6s are −4.9 and −6.4 eV and those of Y 4d and 5s are −6.3 and −5.5 eV (Nagase et al. 1996). The energy level of Sc 3d is much deeper than the other ones, so it is plausible that Sc 3d electron remains on scandium atom because of energy difference. Later, Nagase's group recalculated the molecular orbital of M@C$_{82}$ (M = Sc, Y, La, and lanthanides), and the most stable cage structure depends on the oxidation state (namely, the amounts of transferred electrons to the cage) of the entrapped atom, as is shown in Figure 42.7 (Kobayashi and Nagase 1998).

The UPS of Sc@C$_{82}$ was reported by Hino et al. (1999), and the UPS are principally the same as those of La@C$_{82}$. Difference lies at the spectral onset region near the Fermi level. The UPS of Sc@C$_{82}$ and La@C$_{82}$ obtained with 20 eV photon and difference spectra, Sc@C$_{82}$–C$_{82}$ and La@C$_{82}$–C$_{82}$, are shown in Figure 42.8. Difference spectrum of Sc@C$_{82}$–C$_{82}$ reveals two components at around 1 eV below the Fermi level. The intensity ratio of the two components is almost unity, which indicates that two electrons are transferred to the two different electronic states. The origin of these components is split of the HOMO (LUMO of the empty C$_{82}$) derived from spin–spin interaction between 3d electron remaining on scandium atom and electrons transferred from

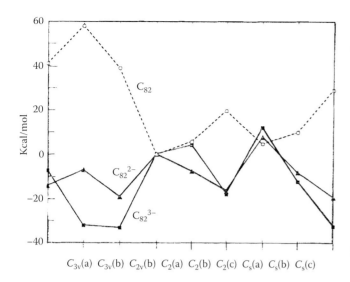

FIGURE 42.7 Energy of C_{82}, C_{82}^{2-}, and C_{82}^{3-}, relative to the $C_2(a)$ (82:3) isomer, reported by Kobayashi and Nagase (1998). (From Kobayashi, K. and Nagase, S., *Chem. Phys. Lett.*, 282, 325, 1998. With permission.)

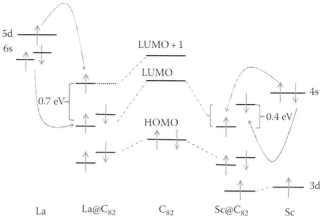

FIGURE 42.9 Energy diagram of La@C_{82} and Sc@C_{82}.

42.3.2 Group Two Metal Atom–Entrapped Fullerenes

Calcium (Xu et al. 1996), barium, and strontium (Dennis and Shinohara 1997) incorporation into the fullerene cage have been reported. Four Ca@C_{82} isomers have been isolated, and two of them, Ca@C_{82} (III) and (IV), could be isolated with sufficient amounts to measure the UPS. Since calcium is a divalent metal atom, the oxidation state of entrapped calcium atom was considered to be +2. The oxidation state of Sc in Sc@C_{82} is also +2, but as described above, transferred electrons occupied split MO because of remaining Sc 3d electron. In Ca@C_{82}, no splitting was expected. The deeper valence band UPS of Ca@C82 (III) and (IV) resembled, but there was a critical difference at the spectral onset region, 0.5–2.5 eV below the Fermi level (Hino et al. 2001). The UPS of empty C_{82} (C_2 cage) were also analogous to these UPS. In order to estimate the amounts of transferred electrons, the UPS of Ca@C_{82} and empty C_{82} were compared, as have been done in the case of La@C_{82} and Sc@C_{82}. However, difference spectra obtained by subtracting the UPS of C_{82} from those of Ca@C_{82} isomers did not show any indication of electron transfer. On the contrary, difference spectra suggested that the fullerene cage lost electrons rather than accepting them from the entrapped metal atom.

At this moment, theoretical calculation of Ca@C_{82} was available, and it proposed that energetically C_{2v}, C_s, C_2, and C_{3v} cages were stable compared with other cage symmetry (Kobayashi and Nagase 1997). Obviously, the cage structure of empty C_{82} and those of Ca@C_{82} isomers are different (see Figure 42.7) so that simple subtraction of the spectra could not give information on the amounts of transferred electrons. In order to estimate the exact cage structure and the amounts of transferred electrons, comparison of the UPS and theoretically obtained simulation spectra is helpful. Figure 42.10 shows the UPS of Ca@C_{82} (III) and (IV) and simulated spectra obtained with optimized C_{82}^{2+} structures at Hartree-Fock level using Gaussian 94 program. Virtually, there is no characteristic difference in deeper valence band region among the simulated spectra, but there is a critical

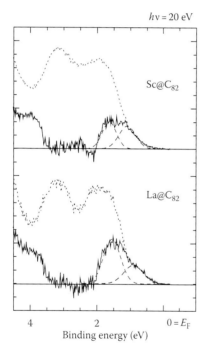

FIGURE 42.8 UPS of La@C_{82} and Sc@C_{82} and difference spectra La@C_{82}–C_{82} and Sc@C_{82}–C_{82}, reported by Hino et al. (1999). (From Hino, S. et al., *Chem. Phys. Lett.*, 300, 145, 1999. With permission.)

Sc atom. In the case of La@C_{82}, spin–spin interaction is not so large, since transferred electrons derived from lanthanum atom are distributed on the whole fullerene cage, so that the splitting was not clearly observed. Schematics of energy diagram are shown in Figure 42.9. Sc@C_{82} is a distinct case wherein the electron energy level is split by spin—spin interaction.

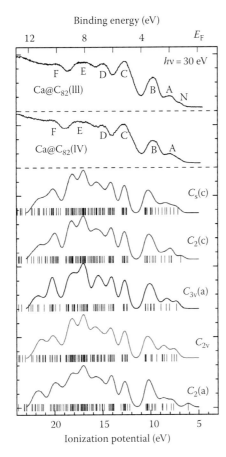

FIGURE 42.10 UPS of two Ca@C_{82} isomers and simulated spectra assuming each cage symmetry obtained by Gaussian 94 program module, reported by Hino et al. (2001). (From Hino, S. et al., *Chem. Phys. Lett.*, 337, 65, 2001. With permission.)

difference in upper valence band region. This difference is a clue to estimate the exact cage structure. The UPS of Ca@C_{82} (III) was well reproduced by the simulated spectrum assuming C_2(c) (82:5) structure with additional two electrons on the cage. Therefore, it could be concluded that the cage structure of Ca@C_{82} (III) is C_2 (82:5) (see Figure 42.2). Actually, the NMR of this isomer indicated C_2 symmetry (Dennis and Shinohara 1998). Thus, combination of the UPS and theoretical calculation was revealed to be a powerful tool to determine the fullerene cage structure.

The UPS of Ca@C_{82} (IV) seems to be well reproduced by the C_s(c) (82:4) simulation spectrum. However, it could also be reproduced by either C_{2v} (82:9) or C_{3v}(a) (82:8) simulation spectra. So this cage structure could not be determined decisively.

42.3.3 Lanthanide Elements–Entrapped Fullerenes

42.3.3.1 Overview

The electron configuration of lanthanide elements changes from La $5d6s^2$, Ce $4f^26s^2$, Pr $4f^36s^2$ to Eu $4f^76s^2$, Gd $4f^75d6s^2$, Tb $4f^96s^2$ and finally to Yb $4f^{14}6s^2$ and Lu $4f^{14}5d6s^2$. Lanthanide

is famous for the so-called lanthanide contraction, and they tend to be principally trivalent cation and, in some cases, divalent or tetravalent cation. XPS study on M@C_{82} (M = Ce, Pr and Nd) (Ding and Yang 1996, Ding et al. 1996a,b) suggested that these lanthanide elements donated three electrons to the cage as in the case of La@C_{82}. The amounts of transferred electrons may differ from a lanthanide element to the others, so that this is one of the important issues in the electronic structure of Lanthanide elements–entrapped metallofullerenes.

42.3.3.2 Gd@C_{82}

The UPS of Gd@C_{82} was firstly measured by Hino et al. (1997). This metallofullerene was actually the mixture of main C_{2v} isomer and minor C_s one, but their ratio was about 9:1 or more. The UPS reflected the electronic structure of the main isomer, since photoelectrons from the minor isomer are so small that those of the major one often covered them, and they are hardly discernible as explicit structures. The quality of the Gd@C_{82} UPS was rather poor, since Gd@C_{82} sublimation films could not be prepared, and it was forced to measure a film prepared by drying the metallofullerene solution. However, its upper valence band structure was clearly observed, and it was almost the same as that of La@C_{82}. Hence, it was concluded that the amounts of transferred electrons was three, namely, Gd$^{3+}C_{82}^{3-}$ oxidation state, as was the case of La@C_{82}. Resonance and off-resonance photoelectron study of Gd@C_{82} also supports this conclusion (Pagliara et al. 2004).

42.3.3.3 Tb@C_{82}

Tb/Ni/C composite rod was reported to produce single-walled carbon nanotubes (SWNT) selectively, and Tb-entrapped metallofullerenes were also synthesized during SWNT production (Shi et al. 2003). Because of high Tb@C_{82} yield, five Tb@C_{82} isomers have been synthesized, and four of them have been isolated. The absorption spectrum of Tb@C_{82} (I) was similar to those of M^{3+}@C_{82}^{3-} (M denotes metal atom) and quite different from those of four Ca@C82 isomers (Xu et al. 1996a). Thus, Tb^{3+}@C_{82}^{3-} electronic configuration was expected for the isomer (I). The UPS of Tb@C_{82} (I) were reported by Iwasaki et al. in 2004. The spectra of Tb@C_{82} (I) were analogous to those of La@C_{82}. Figure 42.11 shows the UPS of Tb@C_{82} (I) and La@C_{82}. Since the UPS of La@C_{82} spectrum was recorded with poor resolution, these two UPS can be assumed as almost the same spectra. Hence, the electronic state of this metallofullerene could be concluded to be Tb^{3+}@C_{82}^{3-}. Since the electronic configuration of Tb is $4f^96s^2$, at least one 5f electron must be transferred to the cage. As described in following section, the fullerene cage of Tm encapsulated metallofullerenes seems to receive only two electrons, and the electronic configuration of Tm is $4f^{13}6s^2$. The amounts of transferred electrons do not necessarily depend on the electronic configuration, and they might depend on the electron energy levels.

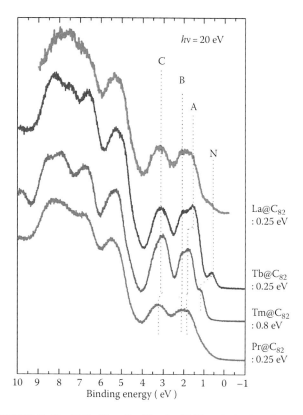

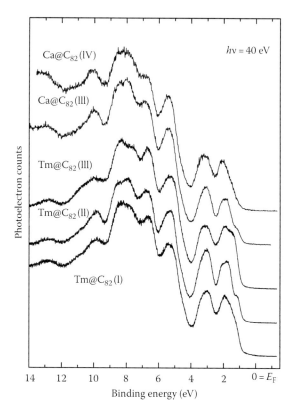

FIGURE 42.11　UPS of La@C_{82}, Tb@C_{82} (I), Tm@C_{82} (III), and Pr@C_{82} (II). La@C_{82}, Tb@C_{82}, and Tm@C_{82} have the same C_{2v} symmetry, but Pr@C_{82} has C_s.

FIGURE 42.12　UPS of two Ca@C_{82} isomers and three Tm@C_{82} isomers, reported by Hino et al. (2005b). (From Hino, S. et al., *Chem. Phys. Lett.*, 402, 217, 2005b. With permission.)

42.3.3.4 Tm@C_{82}

Thulium-entrapped metallofullerenes were firstly isolated by Kirbach and Dunsch (1996), and they found three stable isomers, and none of them gave analogous absorption spectra to that of La@C_{82} (Kikuchi et al. 1993) or Y@C_{82} (Kikuchi et al. 1994). Voltammetry of three isomers was quite different. Further three isomers were ESR silent, which indicated that the C_{82} cage has even electrons. These findings suggested that the oxidation state of Tm atom in the cage was +2. Kirbach and Dunsch also reported the NMR of isomer A, and they concluded that this isomer has more high symmetry than the other two and proposed C_{3v} symmetry. Later, this proposal was denied by Kodama et al. (2002). They measured the NMR in wider region than Kirbach and Dunsch did, and they concluded that Tm@C_{82} (I) had C_s symmetry, Tm@C_{82} (II) had C_2, and Tm@C_{82} (III), which Kirbach and Dunsch claimed to be C_{3v}, had actually C_{2v} symmetry.

Electron spectroscopic determination of divalent nature of Tm in Tm@C_{82} was carried out firstly by Pichler et al. (1997). They measured XPS and found that Tm 4d core level photoelectron spectrum resembled a calculated Yb^{3+} spectrum, which indicated Tm^{2+} electronic structure. Valence band photoelectron spectra of three Tm@C_{82} isomers have been reported by Hino et al. (2005a,b). Three isomers gave completely different upper valence band spectra (Figure 42.12). Further, the UPS of Tm@C_{82} (II) is almost identical to that of Ca@C_{82} (III). Both of them have the same C_2 symmetry and the same cage structure (Dennis

and Shinohara 1998, Kodama et al. 2002). Hence, it could be concluded that the electronic structure of the upper valence region of metallofullerenes is principally governed by the cage structure.

The UPS of Tm@C_{82} (III) are shown in Figure 42.11. Symmetry of metallofullerenes in Figure 42.11 (except for Pr@C_{82}, which is described in next section) is the same C_{2v}. The UPS of Tb@C_{82} and Tm@C_{82} in the figure were taken with the same resolution so that the difference between them is intrinsic. Distinct difference lies in the most upper structures A and B. This difference could be attributed to the difference in the amounts of transferred electrons from the entrapped atom to the cage.

42.3.3.5 Pr@C_{82}

As described above, XPS study of Pr@C_{82} indicated three electron transfer from Pr atom (Ding and Yang 1996, Ding et al. 1996a,b). As are other trivalent metal atom–entrapped fullerenes, there are two Pr@C_{82} isomers; one is major C_{2v} isomer, and the other is minor C_s one. The UPS of neither major isomer nor minor one have not been published. The UPS of the minor Pr@C_{82} isomer is shown in Figure 42.11. The UPS of the minor isomer is quite different from those of La@C_{82} or the major Tb@C_{82} isomer. This could be another example that the electronic structure was principally governed by the cage structure. It should be noted that the UPS of trivalent metal atom–entrapped major C_{2v} M@C_{82} (M = La, Tb, and Gd) isomers and divalent metal atom–entrapped C_{2v} Tm@C_{82} and Ca@C_{82} isomers clearly showed distinct tiny structure just below the Fermi level derived from transferred electrons, but the UPS of minor C_s Pr@C_{82} did not show such structure. Theoretical calculation on the electronic structure of this minor isomer could give the answer where transferred electrons were situated in the molecular orbitals of Pr@C_{82}.

42.4 Electronic Structure of Multiple Atoms–Entrapped Fullerenes

42.4.1 Overview

Multiple atom incorporation into the fullerene cage has been reported at the early stage of metallofullerene investigation. They were La$_2$@C_{80} (Alvarez et al. 1991), Y$_2$@C_{82} (Weaver et al. 1992), and Sc$_3$@C_{82} (Shinohara et al. 1992). Determination of the electronic structure by photoelectron spectroscopy was rather difficult. Therefore, at the early stage of research of multiple atoms–entrapped fullerenes, their electronic structure was mainly discussed from a theoretical viewpoint, and even now there are not many published photoelectron spectra of this category of fullerenes. Some of them are to be picked up in the following sections. There are good reviews on the theoretical aspect of the electronic structure of endohedral fullerenes (Nagase et al. 1996, 2000, Slania and Nagase 2005).

42.4.2 Sc$_2$C$_2$@C$_{82}$ (Sc$_2$@C$_{84}$)

Sc$_2$@C_{84} was the first multiple atoms–entrapped metallofullerene of which the UPS has been reported (Takahashi et al. 1995). Takahashi and his coworkers reported that Sc$_2$@C_{84} could sustain the thermal deposition, and the oxidation state of Sc must be between 0 and +3 from the data of the XPS measurement. However, their UPS were far from those of other mono and multiple metal atoms–entrapped endohedral fullerenes. Although they claimed that Sc$_2$@C_{84} was safely sublimed, there was no evidence that supported it. Further, a characteristic structure at around 5.5 eV below the Fermi level, which was usually observed in the UPS of fullerenes and most endohedral fullerenes, was not observed in their spectrum. In 2000, Pichler and his coworkers reported the UPS and EELS (electron energy loss spectrum) of Sc$_2$@C_{84}, and this spectrum was completely different from the one reported by Takahashi's group. The precise He I UPS of Sc$_2$@C_{84} must be the one obtained by Pichler's group. Further, this endohedral fullerene has another episode; x-ray powder diffraction pattern of this fullerene has been measured, and its symmetry was determined as D_{2d} with an aid of Rietveld refinement combined with maximum entropy method (Takata et al. 1997). In 2005, Hino and coworkers published the UPS of Y$_2$C$_2$@C_{82} and Y$_2$@C_{82}, and they found that the UPS of Y$_2$C$_2$@C_{82} is almost identical with that of Sc$_2$@C_{84} (Hino et al. 2005), and they pointed out that the UPS reported as Sc$_2$@C_{84} must be of Sc$_2$C$_2$@C_{82}. Later, NMR study (Iiduka et al. 2006) and x-ray powder diffraction study (Nishibori et al. 2006) confirmed that Sc$_2$@C_{84} is actually Sc$_2$C$_2$@C_{82} with C_{3v} symmetry (82:8).

42.4.3 Y$_2$C$_2$@C$_{82}$ and Y$_2$@C$_{82}$

Multiple yttrium atoms–entrapped fullerenes have been synthesized, and three Y$_2$C$_2$@C_{82} isomers and one Y$_2$@C_{82} isomer have been isolated (Inoue et al. 2003, 2004). NMR spectra of Y$_2$C$_2$@C_{82} isomers and Y$_2$@C_{82} (III) have been measured, and their cage structures were determined; possibly, C_s (82:6) for isomer I, C_{2v} (82:9) for isomer II, C_{3v} (82:8) for isomer III, and Y$_2$@C_{82}. Since absorption spectra of Y$_2$C$_2$@C_{82} and Y$_2$@C_{82} resemble each other, Inoue et al. proposed that these endohedral fullerenes have analogous electronic structure. The UPS of these fullerenes have been measured, and it was found that their spectra were quite analogous. However, their difference spectra obtained by subtracting the UPS of Y$_2$C$_2$@C_{82} from that of Y$_2$@C_{82} revealed additional electrons on the Y$_2$@C_{82} cage (Hino et al. 2005). Figure 42.13 shows the UPS of these fullerenes and their difference spectrum together with theoretically obtained simulation spectrum assuming four additional electron on the C_{3v} (82:8) cage. Simulation spectrum reproduced the upper valence part of the UPS of Y$_2$C$_2$@C_{82} very well. Therefore, it is plausible that the C_{82} cage accepted four electrons from entrapped atoms, namely, (YC)$_2^{4+}$C$_{82}^{4-}$. Spectral onset of Y$_2$@C_{82} was smaller by 0.35 eV than that of Y$_2$C$_2$@C_{82}, and the difference spectrum indicated transfer of another two electrons to the cage of Y$_2$@C_{82}. The oxidation state of this fullerene might be Y$_2^{6+}$C$_{82}^{6-}$. Suppose the oxidation state of Y$_2$C$_2$@C_{82} and Y$_2$@C_{82} is different, their electronic structure and absorption spectra should be different.

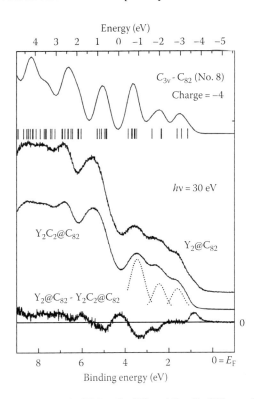

FIGURE 42.13 UPS of Y$_2$C$_2$@C_{82} (III) and Y$_2$@C_{82} (III) together with theoretically obtained simulated spectrum assuming additional four electrons on C_{3v} (82:8) cage by Hino et al. (2005a). (From Hino, S. et al., *Phys Rev. B*, 72, 195424, 2005a. With permission.)

Resemblance of their absorption spectra might be accidental, or transitions that involve transferred additional two electrons might have small oscillator strength.

Up to now, we have measured several $M_2@C_2@C_{82}$ and $M_2@C_{82}$ of the same cage symmetry (they are to be published in near future), and the UPS of these fullerenes are almost identical. No evidence was found that might indicate additional two electron transfer from the entrapped species to the cage. The reason why yttrium atoms–entrapped fullerenes behave differently from other atoms-entrapped ones is another issue to be considered. Possibly, approach from theoretical side is helpful.

42.4.4 Nitride-Entrapped Fullerenes

Trimetallic nitride encapsulated fullerene was firstly reported by Stevenson et al. (1999). Usually, metallofullerenes are synthesized by arc heating of graphite/metal oxide composite rod in helium ambience, and isolated metallofullerenes are very small in quantity. Stevenson's group added ammonia in helium ambience and obtained rather large quantity of $(M_3N@C_n)$ compared with ordinary $M_2@C_n$. They reported N 1s XPS, but the UPS were not reported until 2002 when Alvarez et al. have published an x-ray absorption spectrum, XPS and UPS of $Sc_3N@C_{80}$ (Alvarez et al. 2002). Since then, there have been a lot of work on $Sc_3N@C_{80}$, both experimentally and theoretically. Alvarez and coworkers compared the UPS of $Sc_3N@C_{80}$ with C_{60}, C_{80}, C_{84}, and $Sc_2@C_{84}$ (actually $Sc_2C_2@C_{82}$ as described above) and suggested that $Sc_3N@C_{80}$ might have highly symmetric cage structure because there were four sharp upper valence band structures in the UPS. However, as there such endohedral fullerenes of rather poor symmetry that give several structures in this region, their suggestion does not necessarily connect to the higher cage structure.

In 2005, the UPS and XPS of $Tm_3N@C_{80}$ (Krause et al. 2005, Liu et al. 2005) and $Dy_3N@C_{80}$ (Shiozawa et al. 2005) were reported by Knupfer's group. The UPS of these three tri-metal nitrides-entrapped fullerenes are shown in Figure 42.14. They claimed that their valence band electronic structures are essentially the same. There are small differences among the spectra, as is clearly seen in Figure 42.14. Knupfer's group only pointed out the difference on the onset energy; the onset energy of $Dy_3N@C_{80}$ is slightly larger than that of the other two, and they rendered this origin to the difference in chemical environment. Size or electronic configuration of entrapped atoms might affect the π-electronic structure of the fullerene cage.

42.4.5 $Ti_2C_2@C_{78}$

There has been another misinterpretation of the cage determination, as happened in $Sc_2C_2@C_{82}$ (assumed as $Sc_2@C_{84}$). It was $Ti_2C_2@C_{78}$. When this endohedral fullerene was isolated, it was assumed to be $Ti_2@C_{80}$ (Cao et al. 2001). There were five lines in its NMR spectrum. As for C_{80} cage, there are seven IPR satisfying isomers, but none of them gives five NMR lines. Therefore,

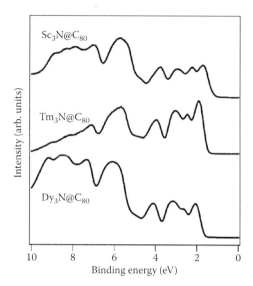

FIGURE 42.14 UPS of $Sc_3N@C_{80}$, $Tm_3N@C_{80}$, and $Dy_3N@C_{80}$ by Shiozawa et al. (2005). (From Shiozawa, H. et al., *Phys Rev. B*, 72, 195409, 2005. With permission.)

Cao and coworkers concluded that isolated $Ti_2@C_{80}$ must be a mixture of D_{5h} and I_h isomers because five NMR lines could be interpreted by these isomers with respective ratio of 3:1. Iwasaki et al. have measured UPS of this endohedral fullerene and reported that a simulation spectrum obtained by molecular orbital calculation assuming additional four electrons on the D_{5h} and I_h cages could reproduce the UPS fairly well (Iwasaki et al. 2004a,b).

Theoretical approaches questioned that cage structure of $Ti_2@C_{80}$ that isolated $Ti_2@C_{80}$ must be C_{78} having D_{3h} symmetry, namely, $Ti_2C_2@C_{78}$ (Tan and Lu 2005, Yumura et al. 2005). As for C_{78} cage, there are two IPR satisfying D_{3h} cage structure. DOS calculation has been performed on both D_{3h} $Ti_2C_2@C_{78}$ (Hino et al. 2007, Otani et al. 2007). Figure 42.15 shows the UPS of $Ti_2@C_{80}$ (Iwasaki et al. 2004a,b) and calculated DOS of two D_{3h}-$Ti_2C_2@C_{78}$ isomers. The DOS obtained using D_{3h} (78:5) geometry reproduced the UPS very well. Now it is considered that isolated $Ti_2@C_{80}$ must be $Ti_2C_2@C_{78}$.

42.4.6 $La_2@C_{78}$

$La_2@C_{78}$ was isolated slightly later than $Ti_2C_2@C_{78}$ (Cao et al. 2004). An NMR measurement revealed that $La_2@C_{78}$ has D_{3h} (78:5) symmetry. That is, both $La_2@C_{78}$ and $Ti_2C_2@C_{78}$ have the same cage structure. As have been described above, the UPS of endohedral fullerenes having the same cage structure and the same amounts of transferred electrons resemble each other. This turned out to be an empirical rule. Since $Ti_2C_2@C_{78}$ and $La_2@C_{78}$ have the same cage structure, it was highly plausible that these two endohedral fullerenes have analogous electronic structure.

Both UPS are depicted in Figure 42.15 (Hino et al. 2007). The UPS of these two fullerenes are completely different. This finding

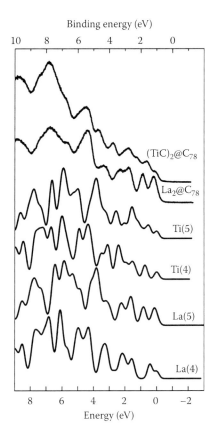

Binding energy (eV)

10   8   6   4   2   0

(TiC)$_2$@C$_{78}$

La$_2$@C$_{78}$

Ti(5)

Ti(4)

La(5)

La(4)

8   6   4   2   0   −2

Energy (eV)

FIGURE 42.15 UPS of Ti$_2$C$_2$@C$_{78}$ and La$_2$@C$_{78}$ together with DOS calculated with DTF on two D_{3h} C$_{78}$ cages by Hino et al. (2007). (From Hino, S. et al., *Phys Rev. B,* 75, 125418, 2007. With permission.)

violates the empirical rule so that it was thought that the cage structures of these two fullerenes were not exactly the same; one of them had D_{3h} (78:4) structure, and the other had D_{3h} (78:5). However, DTF calculation using D_{3h} (78:5) cage shown in Figure

42.15 reproduced the UPS of both fullerenes quite well. That is, La$_2$@C$_{78}$ and Ti$_2$C$_2$@C$_{78}$ have the same D_{3h} cage structure but have different electronic structure. This is an exceptional case to the rule.

It was worth to examine why this discrepancy took place. Electron density (wave function distribution) of these fullerenes gave a clue. Figure 42.16 shows the squared wave functions of the HOMO and HOMO-1 for Ti$_2$C$_2$@C$_{78}$ and La$_2$@C$_{78}$ (Hino et al. 2007). While there is hardly any difference in the wave function distribution of the HOMO of these two fullerenes, there is a big difference in that of the HOMO-1. The HOMO-1 of Ti$_2$C$_2$@C$_{78}$ extended toward entrapped Ti atoms, which indicated hybridization among the orbitals derived from entrapped atoms and caged carbon atoms, but such penetration of wave functions into the cage was not observed in La$_2$@C$_{78}$. Possibly, strong carbide forming tendency of Ti atoms deformed the electronic structure of the fullerene cage severely.

42.5 Prospect of Endohedral Fullerenes

When mass production of C$_{60}$ was reported and C$_{60}$ complexes exhibited interesting behavior, a lot of investigation was carried out. One of the fruits of this investigation was endohedral fullerenes, which were assumed to be a new type of super-atom. Now the researchers dedicating themselves to the study of endohedral fullerenes are limited. This is partly due to a lack of good application of endohedral fullerenes accompanied with their insufficient supply. However, endohedral fullerenes are of much scientific interest, beginning with their formation mechanism to their nature such as cage and electronic structures as well as applications such as anticarcinogens or drug delivery systems. Endohedral fullerenes are like a treasure box waiting to be opened although this is very hard to achieve.

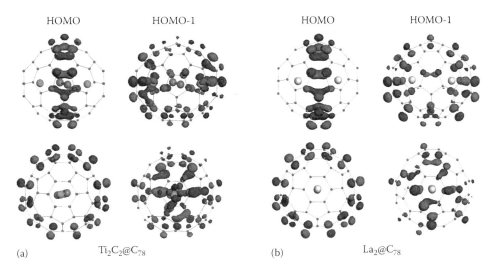

HOMO   HOMO-1   HOMO   HOMO-1

(a)   Ti$_2$C$_2$@C$_{78}$ (b)   La$_2$@C$_{78}$

FIGURE 42.16 (**See color insert following page 25-14.**) Squared wave functions of the HOMO and HOMO-1 of Ti$_2$C$_2$@C$_{78}$ and La$_2$@C$_{78}$ by Hino et al. (2007). (From Hino. S. et al., *Phys Rev. B,* 75, 125418, 2007. With permission.)

References

Akasaka, T., T. Wakahara, S. Nagase, K. Kobayashi, M. Waelchli, K. Yamamoto, M. Kondo et al. 2000. La@C_{82} anion. An unusually stable metallofullerene, *J. Am. Chem. Soc.* 122, 9316–9317.

Alvarez, M. M., E. G. Gillan, K. Holczer, R. B. Kaner, K. S. Min, and R. L. Whetten. 1991. Lanthanum carbide (La$_2$C$_{80}$): A soluble dimetallofullerene, *J. Phys. Chem.* 95, 10561–10563.

Alvarez, L., T. Pichler, P. Georgi, T. Schwieger, H. Peisert, L. Dunsch, Z. Hu et al. 2002. Electronic structure of pristine and intercalated Sc$_3$N@C_{80} metallofullerene, *Phys. Rev. B* 66, 035107-1–035107-7.

Andreoni, W. and A. Curioni. 1996. Freedom and constraints of a metal atom encapsulated in fullerene cages, *Phys. Rev. Lett.* 77, 834–837.

Benning, P. J., D. M. Poirier, N. Troullier, J. L. Martins, J. H. Weaver, R. E. Haufler, L. P. F. Chibante, and R. E. Smalley. 1991. Electronic states of solid C_{60}: Symmetries and photoionization cross sections. *Phys Rev. B* 44, 1962–1965.

Beyers, R., C.-H. Kiang, R. D. Johnson, J. R. Salem, M. S. de Vries, C. S. Yannoni, D. S. Bethune et al. 1994. Preparation and structure of crystals of the metallofullerene Sc$_2$@C_{84}, *Nature* 370, 196–199.

Cao, B., M. Hasegawa, K. Okada, T. Tomiyama, T. Okazaki, K. Suenaga, and H. Shinohara. 2001. EELS and ^{13}C NMR characterization of pure Ti$_2$@C_{80} metallofullerene, *J. Am. Chem. Soc.* 123, 9679–9680.

Cao, B., T. Wakahara, T. Tsuchiya, M. Kondo, Y. Maeda, G. M. A. Rahman, T. Akasaka, K. Kobayashi, S. Nagase, and K. Yamamoto. 2004. Isolation, characterization, and theoretical study of La$_2$@C_{78}, *J. Am. Chem. Soc.* 126, 9164–9165.

Chai, Y., T. Cuo, C. Jin. R. E. Haufler, L. P. F. Chibante, J. Fure, L. Wang, M. Alford, and R. E. Smalley. 1991. Fullerenes with metals inside, *J. Phys. Chem.* 95, 7564–7568.

Dennis, T. J. S., and H. Shinohara. 1997. Production and isolation of endohedral strontium- and barium-based mono-metallofullerenes: Sr/Ba@C_{82} and Sr/Ba@C_{84}, *Chem. Phys. Lett.* 278, 107–110.

Dennis, T. J. S., and H. Shinohara. 1998. Production, isolation, and characterization of group-2 metal-containing endohedral metallofullerenes, *Appl. Phys. A* 66, 243–247.

Ding, J. and S. Yang. 1996. Isolation and characterization of Pr@C_{82} and Pr2@C_{80}, *J. Am. Chem. Soc.* 118, 11254–11257.

Ding, J., N. Lin, L.-T. Weng, N. Cue, and S. Yang. 1996a. Isolation and characterization of a new metallofullerene Nd@C_{82}, *Chem. Phys. Lett.* 261, 92–97.

Ding, J., L.-T. Weng, and S. Yang. 1996b. Electronic structure of Ce@C_{82}: An experimental study, *J. Phys. Chem.* 100, 11120–11121.

Fowler, P. W. and D. E. Manolopoulos. 1995. *An Atlas of Fullerenes*, Oxford Press, Oxford, U.K.

Hasegawa, S., T. Miyamae, K. Yakushi, H. Inokuchi, K. Seki, and N. Ueno. 1998. Origin of the photoemission intensity oscillation of C_{60}, *Phys. Rev. B* 58, 4927–4933.

Hebard, A. F., M. J. Rosseinsky, R. C. Haddon, D. W. Murphy, S. H. Glarum, T. T. M. Palstra, A. P. Ramirez, and A. R. Kortan. 1991. Superconductivity at 18 K in potassium-doped C_{60}, *Nature* 350, 600–601.

Hino, S., K. Matsumoto, S. Hasegawa, H. Inokuchi, T. Morikawa, T. Takahashi, K. Seki et al. 1992a. Ultraviolet photoelectron spectra of C_{76} and K$_x$C$_{76}$, *Chem. Phys. Lett.* 197, 38–43.

Hino, S., K. Matsumoto, S. Hasegawa, K. Kamiya, H. Inokuchi, T. Morikawa, T. Takahashi et al. 1992b. Ultraviolet photoelectron spectra of C_{84} and K$_x$C$_{84}$, *Chem. Phys. Lett.* 190, 169–173.

Hino, S., K. Matsumoto, S. Hasegawa, K. Iwasaki, K. Yakushi, T. Morikawa, T. Takahashi et al. 1993a. Ultraviolet photoelectron spectra of C_{82} and K$_x$C$_{82}$, *Phys. Rev. B* 48, 8418–8423.

Hino, S., H. Takahashi, K. Iwasaki, K. Matsumoto, T. Miyazaki, S. Hasegawa, K. Kikuchi, and Y. Achiba. 1993b. Electronic structure of metallofullerene LaC$_{82}$: Electron transfer from lanthanum to C_{82}, *Phys. Rev. Lett.* 71, 4261–4263.

Hino, S., K. Umishita, K. Iwasaki, T. Miyazaki, T. Miyamae, K. Kikuchi, and Y. Achiba. 1997. Photoelectron spectra of metallofullerenes, GdC$_{82}$ and La$_2$C$_{80}$: Electron transfer from the metal to the cage, *Chem. Phys. Lett.* 281, 115–122.

Hino, S., K. Umishita, K. Iwasaki, T. Miyamae, M. Inakuma, and H. Shinohara. 1999. Ultraviolet photoelectron spectra of Sc@C_{82}, *Chem. Phys. Lett.* 300, 145–151.

Hino, S., K. Umishita, K. Iwasaki, M. Aoki, K. Kobayashi, S. Nagase, T. John, S. Dennis, T. Nakane, and H. Shinohara. 2001. Ultraviolet photoelectron spectra of metallofullerenes, two Ca@C_{82} isomers, *Chem. Phys. Lett.* 337, 65–71.

Hino, S., N. Wanita, K. Iwasaki, D. Yoshimura, T. Akachi, T. Inoue, Y. Ito, T. Sugai, and H. Shinohara. 2005a. Ultraviolet photoelectron spectra of (YC)$_2$@C_{82} and Y$_2$@C_{82}, *Phys. Rev. B* 72, 195421-1–195424-5.

Hino, S., N. Wanita, K. Iwasaki, D. Yoshimura, N. Ozawa, T. Kodama, K. Sakaguchi, H. Nishikawa, I. Ikemoto, and K. Kikuchi. 2005b. Ultraviolet photoelectron spectra of three Tm@C_{82} isomers, *Chem. Phys. Lett.* 402, 217–221.

Hino, S., M. Kato, D. Yoshimura, H. Moribe, H. Umemoto, Y. Ito, T. Sugai et al. 2007. Effect of encapsulated atoms on the electronic structure of the fullerene cage: A case study on La$_2$@C_{78} and Ti$_2$C$_2$@C_{78} via ultraviolet photoelectron spectroscopy, *Phys. Rev. B* 75, 125418.

Iiduka, Y., T. Wakahara, K. Nakajima, T. Tsuchiya, T. Nakahodo, Y. Maeda, T. Akasaka, N. Mizorogi, and S. Nagase. 2006. ^{13}C NMR spectroscopic study of scandium dimetallofullerene, Sc$_2$@C_{84} vs. Sc$_2$C$_2$@C_{82}, *Chem. Commun.* 2057–2059.

Inoue, T., T. Tomiyama, T. Sugai, and H. Shinohara. 2003. Spectroscopic and structural study of Y$_2$C$_2$ carbide encapsulating endohedral metallofullerene: (Y$_2$C$_2$)@C_{82}, *Chem. Phys. Lett.* 382, 226–231.

Inoue, T., T. Tomiyama, T. Sugai, T. Okazaki, T. Suematsu, N. Fujii, H. Utsumi, K. Nojima, and H. Shinohara. 2004. Trapping a C_2 radical in endohedral metallofullerenes: Synthesis and structures of (Y$_2$C$_2$)@C_{82} (Isomers I, II, and III), *J. Chem. Phys.* 108, 7573–7579.

Iwasaki, K., S. Hino, D. Yoshimura, B. Cao, T. Okazaki, and H. Shinohara. 2004a. Ultraviolet photoelectron spectra of $Ti_2@C_{80}$, *Chem. Phys. Lett.* 397, 169–173.

Iwasaki, K., N. Wanita, S. Hino, D. Yoshimura, T. Okazaki, and H. Shinohara. 2004b. Ultraviolet photoelectron spectra of $Tb@C_{82}$, *Chem. Phys. Lett.* 398, 389–392.

Johnson, D. R., M. S. de Vries, J. Salem, D. S. Bethune, and C. S. Yannoni. 1992. Electron paramagnetic resonance studies of lanthanum-containing C_{82}, *Nature* 355, 239–240.

Kessler, B., A. Bringer, S. Cramm, C. Schlebusch, W. Eberhardt, S. Suzuki, Y. Achiba, F. Esch, M. Barnaba, and D. Cocco. 1997. Evidence for incomplete charge transfer and La-derived states in the valence bands of endohedrally doped $La@C_{82}$, *Phys. Rev. Lett.* 79, 2289–2292.

Kikuchi, K., H. Nakahara, T. Wakabayashi, S. Suzuki, H. Shiromaru, Y. Miyake, K. Saito, I. Ikemoto, M. Kainosho, and Y. Achiba. 1992. NMR characterization of C_{78}, C_{82} and C_{84} fullerene, *Nature* 357, 142–145.

Kikuchi, K., S. Suzuki, Y. Nakao, N. Nakahara, T. Wakabayashi, H. Shiromaru, K. Saito, I. Ikemoto, and Y. Achiba. 1993. Isolation and characterization of the metallofullerene LaC_{82}, *Chem. Phys. Lett.* 216, 67–71.

Kikuchi, K., Y. Nakao, S. Suzuki, and Y. Achiba. 1994. Characterization of the isolated $Y@C_{82}$, *J. Am. Chem. Soc.* 116, 9367–9368.

Kirbach, U. and L. Dunsch. 1996. The existence of stable $Tm@C_{82}$ isomers, *Angew. Chem. Int. Ed. Engl.* 35, 2380–2383.

Kobayashi, K. and S. Nagase. 1997. Structures of the $Ca@C_{82}$ isomers: A theoretical prediction, *Chem. Phys. Lett.* 274, 226–230.

Kobayashi, K. and S. Nagase. 1998. Structures and electronic states of $M@C_{82}$ (M = Sc, Y, La and lanthanides), *Chem. Phys. Lett.* 282, 325–329.

Kodama, T., N. Ozawa, Y. Miyake, K. Sakaguchi, H. Nishikawa, I. Ikemoto, K. Kikuchi, and Y. Achiba. 2002. Structural study of three isomers of $Tm@C_{82}$ by ^{13}C NMR spectroscopy, *J. Am. Chem. Soc.* 124, 1452–1455.

Krätschmer, W., Lowell, D. Lamb, K. Fostiropoulos, and D. R. Huffman. 1990. Solid C_{60}: A new form of carbon, *Nature* 347, 354–358.

Krause, M., X. Liu, J. Wong, T. Pichler, M. Knupfer, and L. Dunsch. 2005. The electronic and vibrational structure of endohedral $Tm_3N@C_{80}$ (I) fullerene–proof of an encaged Tm^{3+}, *J. Chem. Phys.* 109, 7088–7093.

Kroto, H. W., J. R. Heath, S. C. O'Brien, R. F. Curl, and R. E. Smalley. 1985. C_{60}: Buckminsterfullerene, *Nature* 318, 162–163.

Laasonen, K., W. Andreoni, and M. Parrinello. 1992. Structural and electronic properties of $La@C_{82}$, *Science* 258, 1916–1918.

Liu, X., M. Krause, J. Wong, T. Pichler, L. Dunsch, and M. Knupfer. 2005. Electronic structures of the pristine and K-intercalated $Tm_3N@C_{80}$ endohedral fullerenes, *Phys. Rev. B* 72, 085407.

Lu, J., X. Zhang, X. Zhao, S. Nagase, and K. Kobayashi. 2000. Strong metal-cage hybridization in endohedral $La@C_{82}$, $Y@C_{82}$ and $Sc@C_{82m}$, *Chem. Phys. Lett.* 332, 219–224.

Nagase, S. and K. Kobayashi. 1993. Metallofullerenes $MC82$ (M = Sc, Y, and La). A theoretical study of the electronic and structural aspects, *Chem. Phys. Lett.* 214, 57–63.

Nagase, S., K. Kobayashi, T. Kato, and Y. Achiba. 1993. A theoretical approach to C_{82} and LaC_{82}, *Chem. Phys. Lett.* 201, 475–480.

Nagase, S., K. Kobayashi, and T. Akasaka. 1996. Endohedral metallofullerenes: New spherical cage molecules with interesting properties, *Bull. Chem. Soc. Jpn.* 69, 2131–2142.

Nagase, S., K. Kobayashi, T. Akasaka, and T. Wakahara. 2000. Endohedral metallofullerenes: Theory, electrochemistry, and chemical reactions, in: *Fullerenes: Chemistry, Physics, and Technology*, eds. K. M. Kadish and R. S. Ruoff, New York: John Wiley and Sons.

Nishibori, E., M. Takata, M. Sakata, H. Tanaka, M. Hasegawa, and H. Shinohara. 2000. Giant motion of La atom inside C_{82} cage, *Chem. Phys. Lett.* 330, 497–502.

Nishibori, E., M. Ishihara, M. Takata, M. Sakata, Y. Ito, T. Inoue, and H. Shinohara. 2006. Bent $(metal)_2C_2$ clusters encapsulated in $(Sc_2C_2)@C_{82}(III)$ and $(Y_2C_2)@C_{82}(III)$ metallofullerenes. *Chem. Phys. Lett.* 433, 120–124.

Otani, M., S. Okada, and A. Oshiyama. 2007. Formation of titanium-carbide in a nanospace of C_{78} fullerenes, *Chem. Phys. Lett.* 438, 274–278.

Pagliara, S., L. Sangaletti, C. Cepek, F. Bondino, R. Larciprete, and A. Goldoni. 2004. Electron transfer from Gd ions to the C cage in endohedral $Gd@C_{82}$ probed by resonant photoemission spectroscopy, *Phys. Rev. B* 70, 035420.

Park, C.-H., B. O. Wells, J. DiCarlo, Z.-X. Shen, J. R. Salem, D. S. Bethune, C. S. Yannoni et al. 1993. Structural information on Y ions in C_{82} from EXAFS experiments, *Chem. Phys. Lett.* 213, 196–201.

Pichler, T., M. S. Golden, M. Knupfer, J. Fink, U. Kirbach, P. Kuran, and L. Dunsch. 1997. Monometallofullerene $Tm@C_{82}$: Proof of an encapsulated divalent Tm ion by high-energy spectroscopy, *Phys. Rev. Lett.* 79, 3026–3029.

Pichler, T., Z. Hu, C. Grazioli, S. Legner, M. Knupfer, M. S. Golden, J. Fink et al. 2000. Proof for trivalent Sc ions in $Sc_2@C_{84}$ from high-energy spectroscopy, *Phys. Rev. B* 62, 13196–13201.

Poirier, D. M., M. Knupfer, J. H. Weaver, W. Andreoni, K. Laasonen, M. Parrinello, D. S. Bethune, K. Kikuchi, and Y. Achiba. 1994. Electronic and geometric structure of $La@C_{82}$ and C_{82}: Theory and experiment, *Phys. Rev. B* 49, 17403–17412.

Slania, Z. and S. Nagase. 2005. Computational chemistry of isomeric fullerenes and endofullerenes, in: *Theory and Application of Computational Chemistry: The First Forty Years*, eds. C. Dykstra, G. Frenking, K. Kim, and G. Scuseria, Amsterdam, the Netherlands: Elsevier.

Shi, Z., T. Okazaki, T. Shimada, T. Sugai, K. Suenaga, and H. Shinohara. 2003. Selective high-yield catalytic synthesis of terbium metallofullerenes and single-wall carbon nanotubes, *J. Phys. Chem. B* 107, 2485–2489.

Shinohara, H., H. Sato, M. Ohkohchi, Y. Ando, T. Kodama, T. Shida, T. Kato, and Y. Saito. 1992. Encapsulation of a scandium trimer in C_{82}, *Nature* 357, 52–54.

Shinohara, H., N. Hayashi, H. Sato, Y. Saito, X. D. Wang, T. Hashizume, and T. Sakurai. 1994. Direct STM imaging of spherical endohedral $Sc_2@C_{84}$ fullerenes, *J. Phys. Chem.* 98, 13769–13440.

Shinohara, H. 2000. Endohedral metallofullerenes. *Rep. Prog. Phys.*, 63.

Shiozawa, H., H. Rauf, T. Pichler, D. Grimm, X. Liu, M. Knupfer, M. Kalbac et al. 2005. Electronic structure of the trimetal nitride fullerene $Dy_3N@C_{80}$, *Phys. Rev. B* 72, 195409.

Soderbolm, L., P. Wurz, K. R. Lykke, and D. H. Parker. 1992. An EXAFS study of the metallofullerene YC_{82}: Is the yttrium inside the cage? *J. Phys. Chem.* 96, 7153–7156.

Stevenson, S., G. Rice, T. Glass, K. Harich, F. Cromer, M. R. Jordan, J. Craft et al. 1999. Small-bandgap endohedral metallofullerenes in high yield and purity, *Nature* 401, 55–57.

Suzuki, S., S. Kawata, H. Shiromaru, K. Yamauchi, K. Kikuchi, T. Kato, and Y. Achiba. 1992. Isomers and ^{13}C hyperfine structures of metal-encapsulated fullerenes $M@C_{82}$ (M = Sc, Y, and La), *J. Phys. Chem.* 96, 7159–7161.

Takahashi, T., A. Ito, M. Inakuma, and H. Shinohara. 1995. Divalent scandium atoms in the cage of C_{84}, *Phys. Rev. B* 13812–13814.

Takata, M., B. Umeda, E. Nishibori, M. Sakata, Y. Saito, M. Ohno, and H. Shinohara. 1995. Confirmation by X-ray diffraction of the endohedral nature of the metallofullerene $Y@C_{82}$, *Nature* 377, 46–49.

Takata, M., E. Nishibori, B. Umeda, M. Sakata, E. Yamamoto, and H. Shinohara. 1997. Structure of endohedral dimetallofullerene $Sc_2@C_{84}$, *Phys. Rev. Lett.* 78, 3330–3333.

Tan, K. and X. Lu. 2005. Ti_2C_{80} is more likely a titanium carbide endohedral metallofullerene ($Ti_2C_2@C_{78}$). *Chem. Commun.* 4444–4446.

Tanigaki, K., T. W. Ebbesen, S. Saito, J. Mizuki, J. S. Tsai, Y. Kubo, and S. Kuroshima. 1991. Superconductivity at 33 K in $Cs_xRb_yC_{60}$, *Nature* 352, 222–223.

Weaver, J. H., J. L. Martins, T. Komeda, Y. Chen, T. R. Ohno, G. H. Kroll, and N. Troullier. 1991. Electronic structure of solid C_{60}: Experiment and theory, *Phys. Rev. Lett.* 66, 1741–1744.

Weaver, J. H., Y. Chai, G. H. Kroll, C. Jin, T. R. Ohno, R. E. Haufler, T. Guo et al. 1992. XPS probes of carbon-caged metals, *Chem. Phys. Lett.* 190, 460–464.

Xu, Z., T. Nakane, and H. Shinohara. 1996a. Production and isolation of $Ca@C_{82}$ (I–IV) and $Ca@C_{84}$ (I,II) metallofullerenes, *J. Am. Chem. Soc.* 118, 11309–11310.

Xu, Y. B., M. Q. Tan, and U. Becker. 1996b. Oscillations in the photoionization cross section of C_{60}, *Phys. Rev. Lett.* 76, 3538–3541.

Yannoni, C. S., M. Hoinkis, M. S. de Vries, D. S. Bethune, J. R. Salem, M. S. Crowder, and R. D. Johnson. 1992. Scandium clusters in fullerene cages, *Science* 256, 1191–1192.

Yoshida, Z. and E. Osawa. 1971. *Aromaticity* (in Japanese), Kagaku Dojin: Kyoto, Japan.

Yumura, T., Y. Sato, K. Suenaga, and S. Iijima. 2005. Which do endohedral Ti_2C_{80} metallofullerenes prefer energetically: $Ti_2@C_{80}$ or $Ti_2C_2@C_{78}$? A theoretical study, *J. Phys. Chem. B* 109, 20251–20255.

43

Metal-Coated Fullerenes

Mário S. C. Mazzoni
Universidade Federal
de Minas Gerais

43.1 Introduction

The discovery of fullerenes in 1985 [1] stands as a landmark in nanostructure research. The proposal of a new form for carbon materials represented the establishment of a new paradigm, which paved the way for future realizations. Carbon became the paramount element in the attempts to find new physics and novel phenomena in the nano world as well as the main hope to bring nanoscience to everyday life in the form of potential applications in diverse fields, such as electronics, nanotechnology, or medicine. Fullerenes were discovered in experiments in which a laser beam vaporized a graphite sample. The carbon atoms rearranged themselves, and the resulting material showed a suggestive preference to form clusters with 60 carbon atoms. Instead of proposing a structure based on known forms of carbon (diamond-like, chains, or sheets), the authors noticed that they could be observing a molecule in which the 60 carbon atoms were located in symmetrical positions in a closed surface composed only of hexagons and pentagons, as shown in Figure 43.1a: exactly like a soccer ball and resembling the geodesic domes created by the architect and philosopher Buckminster Fuller (1895–1983). A theorem by Euler guarantees that exactly 12 pentagons are required to close such a surface. The number of hexagons is variable: C_{60} has 20; adding 5 more, a C_{70} is built, as shown in Figure 43.1b, and the whole family of fullerene molecules can be constructed in this way.

Following the first years of the discovery, a second great achievement drew the attention of the materials science community: It was shown that a solid material could be made out of C_{60} molecules [2]. The solid C_{60} is formed by a close-packing arrangement of the fullerene molecules. One way to do that is to dispose the molecules in the lattice sites of a face-centered cubic structure. This is a typical molecular solid, that is, the molecules, which are the building blocks of the crystals, do not interact through strong covalent bonds, but, rather, stay apart from each other interacting through weaker van der Waals interactions. The region between fullerenes in the crystal may be occupied by other elements or molecules. This is the case of a doped structure, and C_{60} also revealed intriguing behavior in this aspect. Metallic elements, such as potassium, could be intercalated in the structure in the stoichiometry K_3C_{60}, giving rise to a metallic behavior [3]. Moreover, the superconducting critical temperature was found [4] to be 18 K, which is an astonishing result when compared with a similar intercalation in graphite, which presents a transition temperature of a few tenths of a Kelvin [7]. Other metallic elements could also dope the solid C_{60}, such as rubidium, raising [5,6] the superconductivity critical temperature to 29 K.

And what about doping single fullerene molecules with metals? After all, the stability of the doped fullerene solid has its basis on a suitable charge transfer process, and the same mechanism should hold for a single fullerene interacting with metallic atoms. And it turns out that this possibility is indeed viable, and leads to a family of nanostructures that could be called organometallic buckyballs or simply metal-coated fullerenes. This is the subject of this chapter. In the following sections, the experimental and theoretical results concerning the covering of fullerene with distinct metallic elements will be discussed case by case. The chapter ends with a description of potential applications of these compounds and an outline of the perspectives in the field.

43.2 Synthesis and Characterization

The synthesis of metal-coated fullerenes is based on the mixing of a vaporized metal and a fullerene vapor cloud in helium atmosphere [8], as represented in the scheme of Figure 43.2. Resistive

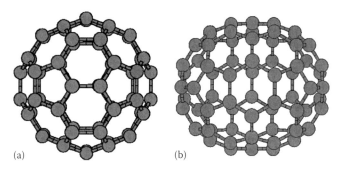

FIGURE 43.1 (a) C_{60} molecule: the 60 carbon atoms occupy the vertices of a truncated icosahedron forming a closed structure with 12 pentagons and 20 hexagons. (b) C_{70} fullerene: the structure has five additional hexagons as compared to C_{60}.

heated ovens may be employed to produce the vapors, and the role of He gas is to quench the mixture. In some cases, the metal may be vaporized by laser pulses. The resulting clusters are transported to a vacuum chamber, where the characterization of the sample takes place. This characterization has to elucidate the chemical composition of the clusters, and the main tool to perform such analysis is mass spectrometry. This technique is based on the fact that the motion of charged particles (atoms, molecules, or clusters) under the influence of electric and magnetic fields is dependent on their masses. A particular case is the so-called time-of-flight mass spectrometry in which the ions are accelerated by an electric field, and the time they take to pass through a certain region is measured. Heavier ions take longer to reach the detector, and the technique is then employed to determine the mass of ionized species. Therefore, prior to entering in the acceleration region of the spectrometer, the clusters have to be ionized, which is done by means of laser pulses.

43.3 The Case of Alkali Metal Atoms

The first experiments [8] on metal-coated fullerenes were conducted for the alkali metal atoms Li, Na, and K. These elements belong to the first column of the periodic table, starting with Li, and have in common a single electron in their last electronic s shell, which can be easily donated when compounds are formed.

The possibility of Li coverage had previously been suggested in a theoretical work [9], in which it was shown that a stable structure with C_{60} could be formed with 12 Li atoms, one on top of each pentagon of C_{60}, as shown in Figure 43.3a. The high electron affinity of the fullerene and the low ionization energy of Li confer an ionic character to the bond. Indeed, C_{60} has two triply degenerate low-lying empty states, which are used to accommodate the charge transferred from the Li atoms. This mechanism is important for the stabilization of all Li-C_{60} clusters, irrespective of the number of Li atoms; for the cluster with 12 Li atoms, an additional stabilization comes from the highly symmetrical arrangement because the original icosahedral symmetry is preserved.

Experiments [8] with the apparatus described in the previous section have, indeed, confirmed this prediction, as shown in Figure 43.3b: the spectrum is dominated by an apparent even–odd alternation in the number of Li atoms (represented by the variable x in the plot), which starts at $x = 7$ and it is disrupted by a strong peak at $x = 12$ that may correspond to the special cluster shown in Figure 43.3a.

The special stability and the uniform coating for $x = 12$ is a particularity of Li, the lightest alkali metal. However, the even–odd alternation for $x \geq 7$ is also present in other cases and may represent a change in the bond character: Up to $x = 7$, the metal atoms bind directly to the cage, and after that metal–metal bondings begin to occur. The case of Na, discussed below, is very illustrative on how such a coating process may take place [8]. The mass spectrum of $C_{60}Na_x^+$, shown in Figure 43.4, has peaks for $x \leq 7$, signaling that these first seven metal atoms may be coating the fullerene surface in such a way as to maximize their distance from one another. The characteristics of the mass spectrum indicate that the next atoms prefer to form bonds with these Na atoms already present on the surface instead of binding directly to C_{60}. A particularly strong bond could be formed if two extra Na atoms joined each Na atom of $C_{60}Na_7$, forming seven Na trimers on the cage surface. And, indeed, large peaks are observed in the mass spectrum for $x = 9$, 11, 13, 15, 17, 19, and 21, that is, whenever a trimer is formed in the original structure with seven Na atoms.

Similar bond schemes [8,10] and charge transfer mechanisms hold for the heavier alkali metals, such as potassium (K) and rubidium (Rb), leading to mass spectra which indicate a particular stability [8,10] for $C_{60}K_7^+$ and $C_{60}Rb_7^+$. Even–odd alternation is also present, but it is interrupted for smaller number of atoms.

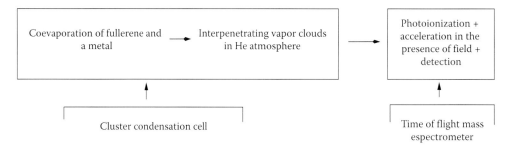

FIGURE 43.2 Scheme representing the synthesis and characterization of metal-coated fullerenes.

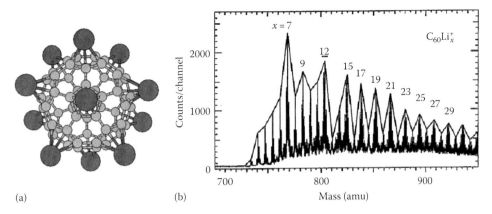

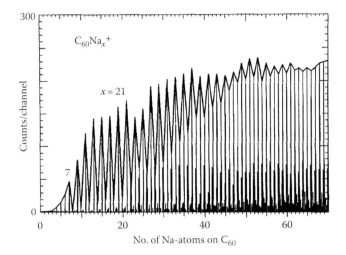

FIGURE 43.3 (a) Model for $C_{60}Li_{12}$: Li atoms are located on top of each pentagon of the fullerene. (b) Mass spectrum of C_{60} coated with Li atoms. The variable x in the plot indicates the number of Li atoms. Note the even–odd alternation disrupted by a strong peak at $x = 12$. (From Martin, T.P. et al., *J. Chem. Phys.*, 99, 4210, 1993. With permission.)

FIGURE 43.4 Mass spectrum of C_{60} coated with Na atoms. (From Zimmermann, U. et al., *Carbon*, 33, 995, 1995. With permission.)

43.4 The Case of Alkali Earth Metals

A second type of coating metals comprises the alkali earth metals. These metals are part of the second group of the periodic table, and are characterized by an outer s shell with two electrons. The case of calcium (Ca) is an excellent illustration of this class. Calcium may provide an uniform coating of the fullerene surface; it is able to form covalent bonds with fullerene atoms, and the resulting structure may find important applications, such as those related to hydrogen storage. Other elements which behave in similar way are barium (Ba) and strontium (Sr).

The first experiments [11] on Ca, Sc, and Ba coatings were based on the production and mixture of metallic and fullerene vapor clouds in He atmosphere, as described in the previous section. Typical results of the mass spectrometry analysis for Ca in C_{60} and C_{70} are shown in Figure 43.5b. In these plots, the variable x indicates the number of Ca atoms in the fullerene ($C_{60}Ca_x$). The prominent peaks at $x = 32$ and $x = 37$ for C_{60} and C_{70}, respectively, are very suggestive of the kind of interaction which may

be taking place. Indeed, 32 is the number of faces in the fullerene surface: To close a polyhedron with hexagons and pentagons, one needs 12 pentagons; C_{60} and C_{70} have additional 20 and 25 hexagons, respectively. Therefore, the coincidence of the number of metal atoms in the first important peak of the mass spectrum with the number of fullerene faces is indicative that each metal atom may be on top of each face.

But how is the interaction between the metal and fullerene? Are there covalent bonds or a weaker interaction, such as van der Waals? The answer to these questions has to come from high-level electronic structure calculations. The results point in the direction of the covalent interaction by means of a particular mechanism [12], which involves the empty d orbitals of Ca (the electronic configuration of Ca may be written as $1s^2\ 2s^2\ 2p^6\ 3s^2\ 3p^6\ 4s^2\ 3d^0$). Calcium can easily lose its two s electrons, and fullerene has low-lying empty levels which could receive these electrons. So it all happens as if each Ca donates its $2s$ electrons to the carbon cage, but then some of them are donated to Ca, occupying the previously empty d orbitals. Figure 43.5a illustrates the resulting geometric arrangement for $C_{60}Ca_{32}$.

The spectra shown in Figure 43.5b also give information about additional layers. The clue in this case comes from the fact that a second prominent peak in the C_{60} spectrum occurs at $x = 104$ and is found both for Ca and for Sc. This suggests a general behavior which can be understood by means of a process of packing balls onto the first layer [11]. This model not only explains the peak at $x = 104$ as due to the completion of a second layer, but also predicts magic numbers for the third and fourth layers (236 and 448, respectively), which are consistent with the experiments.

43.5 The Case of Transition Metal Atoms

A distinct case of fullerene coverage takes place when the coating element is a transition metal. Transition metal elements are characterized by an increasing occupation of the d shells,

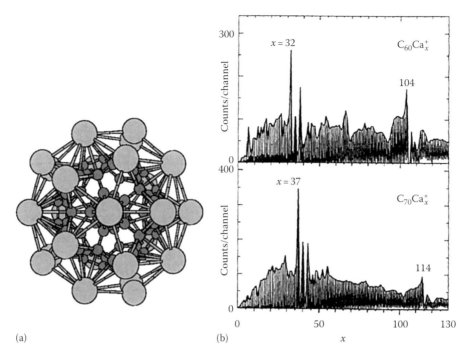

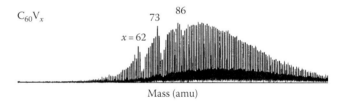

FIGURE 43.5 (a) Model for $C_{60}Ca_{32}$: each Ca atom is found on top of a fullerene face. (b) Mass spectrum of C_{60} coated with Ca atoms. A strong peak appears at $x = 32$. (From Zimmermann, U. et al., *Phys. Rev. Lett.*, 72, 3542, 1994. With permission.)

comprising groups 3–12 of the periodic table. The $3d$ series, for instance, begins with scandium, whose valence configuration is $3s^2\,3d^1$, and goes up to zinc, with its complete d shell ($3s^2d^{10}$). Titanium (Ti), with $2d$ electrons, is found in this series, and its compounds have been extensively investigated since they hold great promise in applications related to catalysis, hydrogen storage, and sensor materials. Theoretical predictions suggest that the ability to bind or sensor other molecules or chemical groups may be enhanced if the Ti compound is associated with carbon structures [13], which brings the possibility of Ti-coated fullerenes into focus.

Experiments on Ti-coated fullerenes [14] have been conducted in conditions similar to those previously described. A resistive heated oven provides fullerene vapor, which is mixed with metal vapor previously evaporated in the condensation cell with laser pulses. Again, the time-of-flight mass spectrometer is employed to analyze the cluster's composition.

The results show that Ti may, indeed, form stable structures on the surface of C_{60} or C_{70}, a result already found for calcium. The sharp edges in the spectrum appear for $x = 62,77,86$, which indicates the formation of a metallic structure covering the fullerene, but, distinct from the calcium case, it also suggests an enhanced number of metal–metal bonds. Figure 43.6 shows the spectrum for vanadium, which presents a similar behavior.

Theoretical results [15,16] corroborate these trends showing that if a Ti atom is bound to the C_{60} surface, a second Ti atom prefers to dimerize (form a Ti–Ti bond) rather than bind directly to the surface. Additional Ti atoms present the same

FIGURE 43.6 Mass spectrum for vanadium coating C_{60}. (From Yang, S. et al., *J. Chem. Phys.*, 129, 134707, 2008.)

behavior: 12 Ti atoms are not predicted to stay on top of each fullerene pentagon, as it happens for lithium; instead, the energetically most stable configuration according to the calculations is the one in which a 3D cluster sits on top of the cage, as shown in Figure 43.7.

The experimental results [14] also shed light on the strength of the interaction between the metal and the fullerene carbon atoms. By increasing the intensity of the light pulse used to ionize the metal–fullerene clusters, which corresponds to the heating of the system, it is observed a fragmentation process which starts with the evaporation of metal atoms from the cluster and leads to further breakage of the cage. The result is the formation of smaller structures, such as V_8C_{12} or Ti_8C_{12}, suggesting that the original metal–fullerene bonds were strong enough to allow the breaking of the cage.

Experimental results for heavier elements [17], such as gold, have indicated that the metallic clustering over the fullerene may also be the preferred structure for a relatively small number of metal atoms.

FIGURE 43.7 Clustering of 12 Ti atoms on a C_{60} molecule. (From Tast, F. et al., *Phys. Rev. Lett.*, 77, 3529, 1996.)

43.6 Other Coating Mechanisms: A Route for a Gold Fullerene?

The coating mechanisms we have described up to this point have in common the existence of covalent bonds between the fullerene and the metal. In some cases (Li, Na, Ca), an uniform coating may take place, while in others (Ti, V) a metallic clustering is observed.

A distinct scheme may be envisaged if we think that the spherical shape of the fullerene cage may serve as a template for the nucleation of a metallic shell around it. This would result in a metallic fullerene encapsulating the C_{60} molecule. Although not synthesized up to date, such structures have been proposed by theoreticians [18] as stable compounds with a high potential for nanotechnological applications. One of such proposals is shown in Figure 43.8a. Placing 60 gold atoms above each carbon of a C_{60} molecule, and additional 32 gold atoms at the center of the fullerene faces, the result is a particularly stable icosahedral layer enclosing the carbon cage, and 3.6 Å distant from it. This

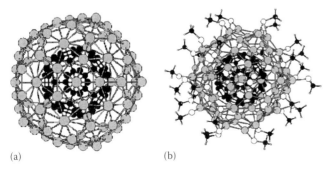

(a)   (b)

FIGURE 43.8 (a) A gold fullerene composed of 92 atoms arranged in a icosahedral geometry and enclosing a C_{60} molecule. (b) Functionalization of the gold fullerene by 31 methylthiol groups, which are responsible for a further stabilization of the structure. (From Batista, R.J.C. et al., *Phys. Rev. B*, 72, 085447, 2001. With permission.)

distance suggests a weak interaction between the cages, without the formation of covalent bonds.

If a metallic shell were to be found surrounding the fullerene, it is more likely that its surface would be functionalized by other chemical groups. Gold has a high affinity to thiol (–SH) groups, and the theoretical prediction is that the coverage of the gold layer by such groups further stabilizes the structure. An example is given in Figure 43.8b, which shows the relaxed structure of $Au_{92}C_{60}$ cluster when covered by 31 methylthiol molecules. The presence of these chemical groups is responsible for a distortion of the gold cage, but without disruption of Au–Au bonds, and they significantly decrease the surface energy, further stabilizing the structure.

43.7 Applications

The investigation of nanostructures has a twofold interest: the description of electronic states in such small systems may elucidate new physical phenomena, which, in turn, may prove useful in the design of materials with novel properties, which can bring new solutions for a variety of problems in a wide scope of applications.

One such problem is hydrogen storage. Hydrogen may provide an environmentally cleaner energy carrier mechanism, playing an important role in the reduction of the emission of pollutants. However, a severe obstacle toward a hydrogen economy is the storage. To be efficient, not only the amount of gas stored must be high relative to the weight of the material, but also the absorption and desorption mechanisms must be adequate.

In this context, metal-covered fullerenes emerge as promising materials [19]. To understand why is that so, an important point is the fact that the ability of a metallic element to bind hydrogen is greatly enhanced if the element is in a cationic form. The effect can be understood based on polarization mechanisms of the hydrogen molecules, which increase binding energies. For instance, lithium (Li) in its neutral state may capture a single H atom, while a Li cation (Li^+) is able to bind 12 molecules [20]. As previously discussed, metals such as Li or Ca may easily lose their outer electrons to the fullerene, and may provide an uniform coating of the cage surface. Therefore, it is to be expected that these systems may be good candidates for hydrogen storage.

Indeed, theoretical predictions point in this direction. For $Li_{12}C_{60}$, the calculations [21] indicate that 60 hydrogen molecules may be captured, corresponding to a gravimetric density of 13 wt%. A drawback is the small binding energy of the hydrogen molecules, only 0.075 eV/H_2. However, $Ca_{32}C_{60}$ may conceal a high uptake with higher binding energies. This is a conclusion of a theoretical analysis [12], which showed that 92 hydrogen molecules, corresponding to a hydrogen uptake of 8.4 wt%, could be adsorbed on the coated fullerene with a binding energy of ~0.3 eV/H_2. Figure 43.9 shows a possible arrangement of the 92 H_2 molecules surrounding $Ca_{32}C_{60}$. The charge redistribution that takes place due to the presence of the 32 Ca atoms on top of the fullerene cage is responsible for the existence of a high electric field near the coated surface. This field is able to polarize the H_2 molecules, increasing the binding energies and leading to the high hydrogen uptake.

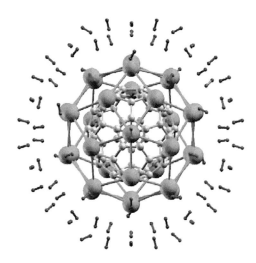

FIGURE 43.9 Hydrogen uptake by a calcium coated fullerene ($C_{60}Ca_{32}$). The figure represents the result of a calculation which simulates the capturing of 92 hydrogen molecules. (From Yoon, M. et al., *Phys. Rev. Lett.*, 100, 206806, 2008. With permission.)

43.8 Conclusions

In this chapter, we have shown that the technique of coevaporation of metal and fullerene in specific conditions may lead to a variety of novel compounds, generally named metal-coated fullerenes. Characterization by means of mass spectrometry and theoretical calculations have helped the elucidation of the structure, properties, and prediction of potential applications for these molecules. Depending on the metal, an uniform coating may occur as it happens for Na, Li, and Ca, for instance. For other metals, such as the transition elements V or Ti, a metallic clustering over the fullerene is preferred. The most promising potential application is related to hydrogen storage, specially for the uniform coating provided by Ca. It is to be expected that with the advance of manipulation techniques, novel applications will be proposed, possibly related to a controlled construction of thin films, complex nanocircuits, and molecular engineering.

References

1. H. W. Kroto, J. R. Heath, S. C. O'Brien, R. F. Curl, and R. E. Smalley, *Nature* **318**, 162 (1985).
2. W. Krätschmer, L. D. Lamb, K. Fostiropoulos, and D. R. Hoffman, *Nature* **347**, 354 (1990).
3. R. C. Haddon, A. F. Hebard, M. J. Rosseinsky, D. W. Murphy, S. J. Duclos, K. B. Lyons, B. Miller et al., *Nature* **350**, 320 (1991).
4. A. F. Hebard, M. J. Rosseinsky, R. C. Haddon, D. W. Murphy, S. H. Glarum, T. T. M. Palstra, A. P. Ramirez, and A. R. Kortan, *Nature* **350**, 600 (1991).
5. M. J. Rosseinsky, A. P. Ramirez, S. H. Glarum, D. W. Murphy, R. C. Haddon, A. F. Hebard, T. T. M. Palstra, A. R. Kortan, S. M. Zahurak, and A. V. Makhija, *Phys. Rev. Lett.* **66**, 2830 (1991).
6. K. Holczer, O. Klein, S.-M. Huang, R. B. Kaner, K.-J. Fu, R. L. Whetten, and F. Diederich, *Science* **252**, 1154 (1991).
7. A. F. Hebard, *Phys. Today* **45**, 26 (1992).
8. T. P. Martin, N. Malinowski, U. Zimmermann, U. Näher, and H. Schaber, *J. Chem. Phys.* **99**, 4210 (1993).
9. J. Kohanoff, W. Andreoni, and M. Parrinello, *Chem. Phys. Lett.* **198**, 472 (1992).
10. U. Zimmermann, A. Burkhardt, N. Malinowski, and T. P. Martin, *Carbon* **33**, 995 (1995).
11. U. Zimmermann, N. Malinowski, U. Näher, S. Frank, and T. P. Martin, *Phys. Rev. Lett.* **72**, 3542 (1994).
12. M. Yoon, S. Yang, C. Hicke, E. Wang, D. Geohegan, and Z. Zhang, *Phys. Rev. Lett.* **100**, 206806 (2008).
13. T. Yildirim and S. Ciraci, *Phys. Rev. Lett.* **94**, 175501 (2005).
14. F. Tast, N. Malinowski, S. Frank, M. Heinebrodt, I. M. L. Billas, and T. P. Martin, *Phys. Rev. Lett.* **77**, 3529 (1996).
15. Q. Sun, Q. Wang, P. Jena, and Y. Kawazoe, *J. Am. Chem. Soc.* **127**, 14582 (2005).
16. S. Yang, M. Yoon, E. Wang, and Z. Zhang, *J. Chem. Phys.* **129**, 134707 (2008).
17. B. Palpant, Y. Negishi, M. Sanekata, K. Miyajima, S. Nagao, and K. Judai, *J. Chem. Phys.* **114**, 8459 (2001).
18. R. J. C. Batista, M. S. C. Mazzoni, L. O. Ladeira, and H. Chacham, *Phys. Rev. B* **72**, 085447 (2001).
19. Y. Zhao, Y. Hyun Kim, A. C. Dillon, M. H. Heben, and S. B. Zhang, *Phys. Rev. Lett.* **94**, 155504 (2005).
20. B. Rao and P. Jena, *Europhys. Lett.* **20**, 307 (1992).
21. Q. Sun, P. Jena, Q. Wang, and M. Marquez, *J. Am. Chem. Soc.* **128**, 9741 (2006).

44

Fullerol Clusters

Jonathan A. Brant
University of Wyoming

44.1 Introduction

Nanotechnology, which has been heralded as the next great technological revolution, is the engineering or manipulation of matter at or near the atomic level [1,2]; essentially, nanotechnology is building objects atom by atom and molecule by molecule. A key building block of this technological field are nanoparticles [3]. Nanoparticles are defined as particles with a diameter, or dimension, between 1 and 100 nm (nanometers). For reference, a nanometer is one billionth of a meter (10^{-9} m). In comparison, a human hair is approximately 80,000 nm in diameter, while a water molecule is only 0.3 nm across. Nanoparticles are in fact a class of particles of specific size more commonly referred to as colloids [4]. They have unique properties that distinguish them from other larger bulk materials having similar chemical compositions. These unique properties are largely attributed to the size effect phenomena [2]. As particle size approaches or decreases below 100 nm, the surface area-to-mass ratio increases dramatically, meaning that the bulk of a particle's atoms are at the surface. This makes nanoparticles highly surface reactive. Another size effect is that quantum effects become much more significant for nanoparticles, which affects their optical, electrical, and magnetic properties [2,5].

There are two general categories of nanoparticles: ambient and engineered. Ambient nanoparticles are produced through natural processes and have therefore always been present in the environment. Examples of ambient nanoparticles include mineral weathering products, certain metal oxanes, combustion products (soot, smoke), and bio-colloids [2]. This fact forms the basis of many arguments, suggesting that nanotechnology and nanoparticles are not entirely new concepts. However, engineered nanoparticles, and the study thereof, are unique from that of ambient materials, with the exception, of course, of their shared size.

Engineered nanoparticles are designed to have specific properties in order for them to perform a specific function [2]. This definition is the embodiment of nanotechnology, which, as noted earlier, is engineering at the atomic level. This characteristic differentiates engineered nanoparticles from ambient ones, the properties of which are determined through natural processes. While engineered nanoparticles may have properties that are unique or similar to ambient materials, their uniqueness is born out of our ability to engineer and tailor their properties in order to produce new and unique materials. This distinction separates our study of engineered nanoparticles from that of classical colloidal science.

Some of the ttmore frequently encountered engineered nanoparticles include quantum dots, fullerenes, and engineered metal oxides. Of these examples, fullerenes are perhaps the most widely studied and well known [3,6–30]. The popularity of these materials is owed to their many unique characteristics and the diverse array of applications that are currently envisioned for them. Examples of the unique properties of fullerenes include low threshold emission fields and excellent emission stability [31,32], large hydrogen storage capacities [33], high tensile strength and elasticity [34], electronic sensitivity in different chemical environments [35], and a high thermal stability [19]. Because of these unique properties, fullerenes are envisioned for a wide range of applications such as high-strength composites, water filtration membranes, air filters, lubricants, drug delivery devices, and as energy storage devices (hydrogen storage).

To realize the great potential of these materials, it is necessary that fullerenes be dispersed in aqueous, or other suitable solution(s), for reasons associated with processability. Fullerenes may be synthesized through a number of techniques, with each technique producing fullerenes with distinctive properties and behaviors. Therefore, a host of solubilization techniques have been developed and specially tailored for the type of fullerene and its applications. One type of water-soluble fullerene that has received considerable attention in the research community is a derivatized form of Buckminsterfullerene (C_{60}) called fullerol. The study of fullerol is important for many reasons, including understanding the environmental implications of fullerenes and the development of easily derivatized fullerenes capable of forming molecular dispersions. This chapter therefore focuses on fullerol, with particular emphasis on its physical state in aqueous systems.

44.2 Fullerenes

Fullerenes are the third allotropes of carbon, which are molecules composed entirely of carbon atoms. Other more commonly encountered examples of carbon allotropes* include diamonds, graphite, and amorphous carbon (Figure 44.1). Harold Kroto at

the University of Sussex and Robert Curl and Richard Smalley at Rice University first discovered fullerenes[†] in 1985 [18]. Fullerenes have a structure similar to graphene, but contain both pentagon and hexagon rings to produce three-dimensional structures like spheres and tubes (e.g., single and multi-walled carbon nanotubes). Following their initial discovery, it was determined that some types of fullerenes are in fact produced through natural processes, albeit in very small quantities [17]. For example, caged fullerenes such as C_{60}, C_{70}, and C_{76} have all been found in soot deposits formed by lightning discharges in the atmosphere [36]. However, only recently have techniques been developed that make it possible to synthesize fullerenes in significant quantities in the laboratory. These techniques include the resistive heating method [37], the arc discharge method [38], and hydrocarbon combustion under ideal atmospheric conditions [39,40].

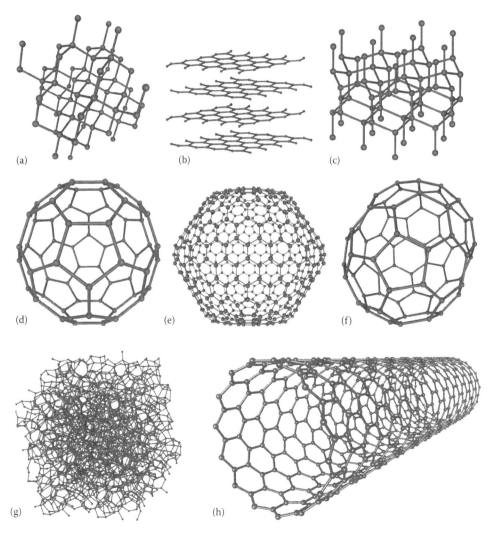

FIGURE 44.1   Molecular structures of different carbon allotropes: (a) diamond, (b) graphite, (c) Lonsdaleite, (d) Buckminsterfullerene (C_{60}), (e) C_{540}, (f) C_{70}, (g) amorphous carbon, and (h) a single walled carbon nanotube. (Courtesy of Michael Strock.)

---

* An allotrope refers to a type of molecular configuration.

[†] Harold Kroto, Robert Curl, and Richard Smalley inadvertently discovered Buckminsterfullerene (C_{60}) while investigating the possibility that long-chained carbon molecules could form from the vaporization of carbon in an atmosphere filled with hydrogen and nitrogen.

Each technique involves the vaporization of a carbon source such as graphite. The carbon then rearranges itself under tightly controlled atmospheric conditions to form soot and fullerenes. In each of the aforementioned processes, the fullerenes may only comprise 10% or less of the total mass of soot produced; thus the yield from each process is still relatively low.

The atomic structure of fullerenes is responsible for many of their unique characteristics, and it differentiates each of the various sub-groups of fullerenes from one another. Fullerenes are a class of polyhedra, a geometric object with flat faces and straight edges, the atomic structure of which is determined by specific constraints [8,22]. These constraints, and in turn the atomic structure of fullerenes, may be described and estimated using Euler's Theorem, which relates the number of vertices (v), faces (f), and edges (e) in a fullerene molecule according to

$$v + f = e + 2 \qquad (44.1)$$

Fullerenes are trivalent polyhedra, meaning that they contain only pentagonal and hexagonal faces. Representing the number of vertices as n, we can say that the number of faces in a fullerene is determined by $f = n/2 + 2$, and the number of edges by $e = 3n/2$. Substituting these expressions into Equation 44.1, we can represent the number of faces, edges, and vertices in a fullerene according to [22]

$$n + \left(\frac{n}{2} + 2 \right) = \frac{3n}{2} + 2 \qquad (44.2)$$

Using Equation 44.2, it is possible to calculate the number of edges and faces for a given fullerene by inputting in the number of vertices. This allows us to construct 3-dimensional models or other representations of the many different types of fullerene molecules. This ability holds particular importance as we attempt to understand how substituent functional groups may be added to these molecules in order to alter their properties. For a further and more in-depth discussion on this topic, the reader is referred to Prassides [22].

44.2.1 Caged Fullerene Molecules

A particularly interesting type of fullerene structure is the caged or spherical-like fullerene molecule. Fullerene cage structures may consist of varying numbers or combinations of carbon atoms and are denoted accordingly [22]. For example, C_{70} contains 70 carbon atoms; C_{76} contains 76 carbon atoms, and so on. Buckminsterfullerene, or C_{60}, is of particular interest to the engineering and science communities due to their remarkable stability and high symmetry relative to other caged fullerenes. The term Buckminsterfullerene was inspired by the geodesic dome structure designed by Buckminster Fuller, which was the centerpiece of the Expo '67 exhibition in Montreal, Canada (Figure 44.2). The geodesic dome was a closed hemisphere consisting of linked pentagons and hexagons, which, according to its discoverers, resembled the atomic structuring of the C_{60} molecule [18].

FIGURE 44.2 Picture of the geodesic dome designed by Buckminster Fuller for Expo '67 in Montreal, Canada. (Courtesy of Eberhard von Nellenburg.)

The C_{60} molecule is perhaps best described as a hollow cage with a molecular diameter on the order of 1 nm, consisting of 60 carbon atoms that are arranged as a truncated icosahedron,[*] 12 pentagons and 20 hexagons [18,22] (Figure 44.2). The pentagons are arranged so that no two are adjacent to one another. Each carbon atom lies at the vertex of one pentagon and two hexagons. The C_{60} molecule has a ground-state geometry that corresponds to the Icosahedra point group I_h [22]. A carbon atom occupies each vertex in C_{60}, and each carbon is three-connected to other carbon atoms by one double bond and two single bonds. Carbon atoms with this kind of connectivity are called "*sp* carbons" because the orbitals used to sigma-bond the three adjacent carbons are hybrids of the 2s orbital and the two 2p orbitals (2p and 2p). The remaining 2p orbital (2p) is responsible for the π-bond. Each carbon atom is bonded to 3 other carbon atoms to form sp^2 hybridization, and consequently the C_{60} molecule is surrounded by π electron clouds. Examination of the illustration of a C_{60} molecule in Figure 44.3 reveals that it resembles a soccer ball, resulting in it commonly being referred to as a buckyball.

Buckminsterfullerene (C_{60}) is a highly symmetric molecule, meaning that one can find many transformations that map the molecule onto itself. The symmetry of the C_{60} molecule has important consequences with regard to its chemistry and characteristics [22]. The carbon atoms are approximately sp^2 hybridized with a small sp^3 contribution as a result of the curvature of the molecular cage. There are two distinct bond types in the C_{60} molecule. The first type of bond is the inter-pentagonal double bonds (C_6–C_6), which are 1.39 Å long. The second type of bond is the intra-pentagonal single bonds (C_5–C_6), which are slightly longer than the double bonds with a characteristic length of

---

[*] An icosahedron contains 12 vertices, 20 faces, and 30 edges. In this structure the vertices have a fivefold symmetry axis. A truncated icosahedron contains 12 pentagonal faces, 20 hexagonal faces, 60 vertices, and 90 edges.

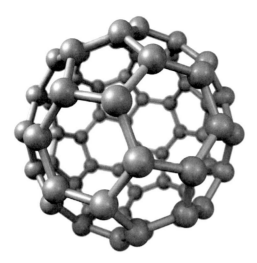

FIGURE 44.3 Illustration of the Buckminsterfullerene, or C_{60}, molecule.

1.44 Å. There are 30 C_6–C_6 (also called 6:6 bonds) double bonds and 60 C_5–C_6 (also called 5:6 bonds) single bonds. The shorter C_6–C_6 bonds display more double-bond character, and therefore a hexagon is often represented as a cyclohexatriene and a pentagon as a pentalene or [5]radialene.* In other words, although the carbon atoms in fullerene are all conjugated, the superstructure is not a super aromatic compound. Instead, C_{60} behaves like a partly delocalized electron deficient polyalkene, rather than a super aromatic molecule.

To further describe the structure of the C_{60} molecule, we can calculate the angles between the double bonds and geometric faces, based on our knowledge of a truncated icosahedron [17]. The angle between a double bond and the adjacent pentagonal face (ψ) is determined by

$$\psi = \cos^{-1}\left[\frac{1}{5}\cos\left(\frac{3}{5}\right)\pi\right] \tag{44.3}$$

Further, the angle between any two hexagonal faces ($\varnothing$) is determined by

$$\varnothing = \cos^{-1}\left[\frac{8}{3}\left(\sin\left(\frac{3}{10}\right)\pi\right)^2 - 1\right] \tag{44.4}$$

Finally, the angle between any adjacent hexagonal and pentagonal faces (θ) is determined by

$$\theta = \frac{1}{2}(\pi - \varnothing) - \psi \tag{44.5}$$

The spherical shape of the C_{60} molecule makes the carbon atoms highly pyramidalized,[†] which has far-reaching consequences for reactivity. The conjugated carbon atoms respond to deviation from planarity by orbital rehybridization of the sp^2 orbitals and π orbitals to an $sp^{2.27}$ orbital with a gain in p-character. The p lobes extend further outside the surface than they do into the interior of the sphere, and this is one of the reasons a fullerene is electronegative. The other reason is that the empty low-lying π^* orbitals also have high s character. Buckminsterfullerene (C_{60}) has 60 π electrons (i.e., electrons that can participate in π bonding), but a closed shell configuration requires 72 electrons. This electronic structure makes C_{60} a good electron acceptor and a weak oxidant. The high electronegativity of C_{60} means that it readily forms compounds with atoms that will donate electrons to it. For example, alkali metals (lithium, sodium, potassium, rubium, caesium, and francium) are highly electropositive and thus readily combine with C_{60}. The ability of C_{60} to react with alkali ions in this way has a number of important consequences. For instance, reacting C_{60} with alkali metal ions such as potassium (K) makes it possible to produce crystals that behave like superconductors (i.e., having no electrical resistance at sufficiently low temperatures).

44.2.2 Aqueous Dispersions of Fullerenes

As with all fullerenes, C_{60} is virtually insoluble in water, with an estimated solubility less than 10^{-9} mg/L [21,41]. The insolubility of C_{60} in water presents a number of challenges with respect to the utilization in commercial products. Fullerenes are however, soluble in aromatic solvents (non-polar) such as toluene [21]. Fullerenes owe their insolubility in polar solvents, such as water, to their hydrophobic nature. Hydrophobic surfaces are characterized by lack of Lewis acid-base (γ^+ and γ^-) sites, which prevents water molecules from hydrogen bonding with the surface [42]. A hydrophobic moiety, in this case, a C_{60} molecule, is excluded or expelled from an aqueous solution because of the hydrogen bonding free energy of cohesion of the water molecules (-102 mJ/m²) [42]. This free energy of cohesion arises from the natural structuring tendency of water molecules. Water molecules structure themselves in the form of hydrogen bond networks, which are continuously forming and breaking apart [43]. The type of interaction that will occur for a C_{60} in water, i.e., will it disperse in solution or be expelled from it, may be estimated through an analysis of the interfacial interaction between the solvent molecules and the C_{60}. The interfacial free energy between a C_{60} and water (ΔG_{13}) may be described as follows [42]:

$$\Delta G_{13} = -2\left(\sqrt{\gamma_1^{LW}\gamma_3^{LW}} + \sqrt{\gamma_1^+\gamma_3^-} + \sqrt{\gamma_1^-\gamma_3^+}\right) \tag{44.6}$$

In Equation 44.6, the C_{60} related components are designated by the subscript 1; those components associated with the solvent

---

* [n]Radialenes are alicyclic organic compounds that contain "n" cross-conjugated exocyclic double bonds.

† The term pyramidalized is defined as meaning distorted toward a tetrahedral molecular geometry.

(in this case water) are denoted by the subscript 3. From Equation 44.6, it is evident that the interfacial interaction between a C_{60} molecule and water is a function of the apolar van der Waals (γ^{LW}) and polar Lewis acid (γ^{+}) and base (γ^{-}) surface energy components for both materials. In fact, ΔG_{13} is always negative as water is attracted to some extent to even the most hydrophobic of surfaces.* However, this interfacial energy is less than the free energy of cohesion between the water molecules, thus preventing the C_{60} from remaining in solution. An example of the interaction between a hydrophobic surface and water is the beading of water on a freshly waxed surface. In essence, the water likes itself much more than the hydrophobic surface (quantified by comparing the free energy of cohesion for water, $-102\,mJ/m^2$, and the interfacial free energy between water and the surface).

The next logical step after describing the interaction between C_{60} and water is to describe the interaction between two C_{60} that are immersed in water. There are many considerations to take into account to fully describe the interfacial interaction between two molecules in water, which falls outside the scope of this chapter. The reader is instead referred to any of the following references for more detailed discussions on interfacial interactions in aqueous media [4,42,43]. Nevertheless, Equation 44.7 does provide a basis by which we can begin to understand the interfacial interactions of C_{60} and later fullerol in water:

$$\Delta G_{131} = -2\left(\sqrt{\gamma_1^{LW}} - \sqrt{\gamma_3^{LW}}\right)^2$$
$$-4\left(\sqrt{\gamma_1^{+}\gamma_1^{-}} + \sqrt{\gamma_3^{+}\gamma_3^{-}} - \sqrt{\gamma_1^{+}\gamma_3^{-}} - \sqrt{\gamma_1^{-}\gamma_3^{+}}\right) \quad (44.7)$$

This equation may be used to calculate the free energy of cohesion (ΔG_{131}) between two C_{60} immersed in a solvent such as water. A negative value for ΔG_{131} indicates that the interfacial interaction is attractive (i.e., the two molecules will stick to each other on contact). Conversely, a positive value for ΔG_{131} is indicative of a repulsive interaction, in which the molecules will remain as discrete units. Charge interactions are neglected in this treatment and would need to be accounted for in a full analysis of the interaction between any two molecules in an aqueous system. Nevertheless, the importance of the apolar and polar components in determining the behavior of C_{60} in water is clearly evidenced from Equation 44.7. This understanding provides a route by which we can modify the properties of C_{60} in order to increase its affinity with water and form rather stable aqueous dispersions.

Having 60 carbon atoms and 30 double bonds available for reaction results in a large number of possible isomers for C_{60}. Fullerene C_{60} can be made to be soluble and form stable dispersions in water by adding suitable (hydrophilic) functional groups or substituents to the molecular cage [44]. These substituents

alter the γ^{+} and γ^{-} properties of the C_{60} and subsequently make the ΔG_{13} more negative. In making ΔG_{13} more negative, we increase the ability of water to hydrogen bond with the C_{60}, thus reducing the amount of work required for the molecule to form an appropriate-sized cavity in aqueous solution. In other words, it is easier for the C_{60} to remain in solution. Examples of substituents used to form aqueous dispersions of C_{60} include carboxylic acids (–COOH), quaternary ammonium $\left(-NR_4^{+}\right)$,† polyethylene glycol (HO–$(CH_2–CH_2–O)_n$–H), and hydroxyl groups (–OH) [15,19,45]. Derivatized C_{60} are divided into two categories: (1) exohedral C_{60} wherein the substituents are located on the outside the molecular cage, and (2) endohedral C_{60} wherein a molecule or ion is trapped inside the C_{60} molecular cage. For our purposes, we are concerned with exohedral C_{60}, as this is the type of derivatized C_{60} that may form aqueous dispersions like fullerol. In the case of exohedral C_{60}, all substituents are added to the cage at the double bond between the six-membered ring [22]. Derivatized exohedral C_{60} still maintain the electronic properties of the parent fullerene [19]. However, the symmetry of the C_{60} molecule is lowered when even one substituent is added to the molecular cage [22]. Subsequently, all of the remaining distinct carbon atoms (i.e., those without an attached substituent group) will begin to exhibit unique reactivities. While numerous methods exist for derivatizing C_{60} [19], this chapter is specifically concerned with the exohedral hydroxylation of C_{60} to form fullerol.

44.3 Fullerols

Fullerols are polyhydroxylated Buckminsterfullerene molecules, meaning that each C_{60} has multiple hydroxyl groups (–OH) attached to their molecular cage (Figure 44.4). Thus, the general chemical formula for fullerol is $C_{60}(OH)_n$. Here, n designates the number of oxidized or hydroxylated carbon atoms on the C_{60} molecular cage. Fullerols are engineered nanoparticles because humans have specifically engineered their properties. There are numerous techniques for synthesizing fullerol, with each producing a molecule having somewhat unique properties. Some of the more common fullerol production techniques include the following:

- Oxidation under strongly acidic conditions [15,45]
- Oxidation under strongly alkaline or basic conditions [16,46]
- Reaction of C_{60} lithium salts with an acidified methanol/water solution in the presence of oxygen [14]
- Substitution of bromine atoms from $C_{60}Br_{24}$ in an alkaline environment (pH 13) with hydroxyl groups [9]

Each preparation technique produces fullerol with different degrees of molecular oxidation (i.e., fullerol having different numbers of hydroxyl groups on the molecular cage). In other

---

* According to van Oss [42], the interfacial free energy between water and a hydrophobic surface will rarely if ever be less than $40\,mJ/m^2$. In this case the minus sign, which signifies an attractive interaction has been omitted.

---

† Quaternary ammonium cations are positively charged polyatomic ions. The R in the formula presented here represents an alkyl group.

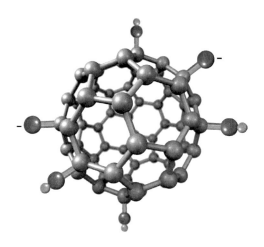

FIGURE 44.4 Illustration of a fullerol molecule. The substituent groups attached to the C_{60} cage are hydroxyl groups (−OH). The conjugate base (−O−) is represented in the illustration by the groups having a negative (−) sign.

words, not all fullerols are the same. In fact, the number of hydroxyl groups (−OH) present on the C_{60} molecular cage may vary even for the same preparation method. For instance, a range is often given for the number of hydroxyl groups in a single fullerol sample. The number of hydroxyl groups may range from 18 to 36, depending on the type of production process used [45]; however, most fullerols are reported to have between 18 to 24 hydroxyl groups per C_{60} molecule ($C_{60}OH_{18-24}$). Differences in the degree of molecular oxidation will result in differences in many of the fullerol characteristics (e.g., charge, hydrophobicity, and stability in solution). These differences must be accounted for in a complete assessment of the behavior of fullerol in any media. Nevertheless, the following discussions are generally applicable to all types of fullerols and are not necessarily restricted to those produced through a specific process.

Fullerols may be distinguished from their parent nonderivatized C_{60} using techniques such as ultraviolet–visible light absorption (UV–vis) and infrared transmittance or adsorption. Each of these techniques was selected for discussion here because they are separately used to identify fullerol in aqueous suspension and as a dry powder, respectively. Of course, there are other characterization methods that may be used to obtain specific information on fullerols, including nuclear magnetic resonance (NMR) imaging and electron spin resonance. However, the discussions here will be limited to the spectroscopic methods of UV–vis* and infrared transmittance/adsorption. For detailed discussions on the methodologies and interpretations associated with the two spectroscopic techniques discussed here, or other characterization techniques used in nanoparticle characterization, the reader is referred to Wiesner and Bottero [2].

Infrared transmittance is used for dry samples, and it provides specific information regarding the types of bonds in a molecule such as fullerol. During the examination of infrared transmittance or adsorption data for any material, and in this case fullerol, it is necessary to identify the following peak properties: intensity, shape, and position. These properties are used to identify specific groups and bonds in the associated molecule or material. Infrared spectra are generated for dry materials, and therefore fullerol must be prepared as a powder. Here, dry fullerol powder is characterized by specific infrared features that allow it to be differentiated from, for example, underivatized C_{60} (Figure 44.5) [45]. A broad absorption band centered at $3300\,cm^{-1}$ is indicative of the hydroxyl groups. The bands at 1585 and $1356\,cm^{-1}$ represent the carbon-carbon double bonds in the C_{60} molecule. The bands at 1065 and $1016\,cm^{-1}$ are due to the C−O stretching. Finally, the band at $1065\,cm^{-1}$ is attributed to the C−OH bond, and the lower frequency band at $1016\,cm^{-1}$ is attributed to the conjugate base (C−O⁻) that is associated with some cation (e.g., C−ONa). The ability of the conjugate base to associate with cations such as sodium has implications for the behavior of fullerol in aqueous media. This topic is explored further in Section 44.3.1, which deals with the formation of fullerol clusters.

In aqueous suspension, the UV–vis spectra for fullerol are distinct from that of pure C_{60} (Figure 44.6). The spectrum for C_{60} in *n*-hexane has characteristic peaks at wavelengths (λ) of 227, 280, and 360 nm. Conversely, in the spectra for fullerol in water, the absorption bands in the UV region (λ = 200–400 nm) are far less pronounced. For fullerol, the peaks become broader and shift to the smaller wavelengths (Peaks 2 and 3). The sole clear peak occurs at λ = 232 nm; the location of this peak may differ depending on the preparation technique

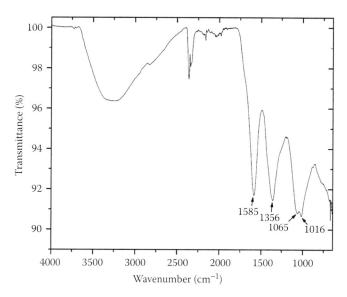

FIGURE 44.5 Infrared spectrum of dry fullerol powder. The fullerol analyzed here was produced by oxidizing C_{60} under strongly acidic conditions using sulfuric and nitric acid. (Reproduced from Vileno, B. et al., *Adv. Funct. Mater.*, 16, 120, 2006. With permission.)

---

* Different molecules absorb radiation of different wavelengths. The absorption spectrum for a given molecule or material is characterized by absorption bands that correspond to specific structural groups within or on a molecule.

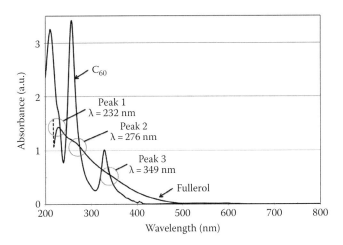

FIGURE 44.6 Ultraviolet and visible (UV–vis) absorption spectra for pure C_{60} in *n*-hexane and fullerol clusters in water. The absorption spectra for the fullerol show how the three characteristic peaks that are present in the absorption spectra for C_{60} shift and become broader for the fullerol suspension.

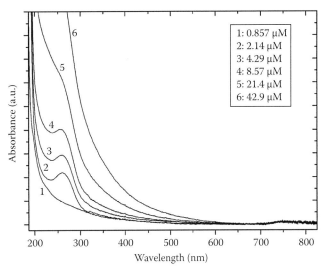

FIGURE 44.7 UV–vis absorption spectra for fullerol dispersions in water having different concentrations. The relatively broad absorption bands at low fullerol concentrations indicate that the fullerols are present as clusters and not as discrete molecules in solution. (Reproduced from Vileno, B. et al., *Adv. Funct. Mater.*, 16, 120, 2006. With permission.)

used. The broadening of the absorption peaks is attributed to the formation of fullerol clusters in water [45]. Recall that the spectrum shown for C_{60} in Figure 44.6 is for a molecular dispersion of C_{60}. It is important to bear in mind that underivatized C_{60} does not form a molecular dispersion in water either [10,11]. Instead, they too form clusters that have a mean particle diameter in a size range of 30–200 nm. As with fullerol, the UV–vis spectrum for C_{60} clusters or nC_{60} is characterized by broad peaks having the same approximate locations as those of C_{60} in *n*-hexane. The presence of these broad absorption bands provides the first indirect evidence of the physical state of fullerol in aqueous media.

UV–vis analysis of aqueous dispersions is commonly used to quantify the concentration of fullerol in suspension using the Beer–Lambert law.

$$A = eLC \qquad (44.8)$$

A is the absorbance
e is the molar absorptivity or extinction coefficient
L is the path length through the sample solution
C is the absorbing species concentration in solution

As shown in Figure 44.7, as the fullerol concentration increases, we see a continued broadening of the peaks in the fullerol spectra. This occurs as the solution becomes saturated (i.e., as concentration or *C* increases, so too does the absorbance or *A*) with the fullerol and the absorbance increases. However, another important characteristic of the spectra is the presence of the rather broad absorption peaks at micro-molar concentrations of fullerol. This topic is discussed further as we examine the formation of fullerol clusters in water.

Fullerol is a polyacidic molecule, wherein each proton of the hydroxyl group (C–OH) can disassociate in water to yield a

conjugate base (C–O⁻) [45]. The conjugate base may then associate with cations in solution such as sodium (Na), potassium (K), calcium (Ca^{2+}), or magnesium (Mg^{2+}), as was previously illustrated from an infrared transmittance analysis of a fullerol sample (Figure 44.5). A second interaction scenario involves the bonding of protonated (C–OH) and deprotonated (C–O⁻) groups on separate fullerol to form clusters. This point is explored further later in this chapter. The actual number of C–OH and C–O– groups per C_{60} molecule is pH dependent, as the hydroxyl groups undergo protonation and deprotonation. Protonation refers to the addition of a proton (H⁺) to an atom or molecule, which in this case is the hydroxyl group(s) that is bonded to the C_{60} molecule. Upon protonation, the fullerol's molecular mass increases by 1.0079 g/mol, or the molecular weight of a single hydrogen atom. The total change, of course, depends on the number of groups that are undergoing protonation, or in the opposite scenario, deprotonation. Deprotonation refers to the removal of a proton (H⁺) from a molecule to form a conjugate base (–O⁻). For fullerol, deprotonation can be observed by adding a sufficient amount of fullerol powder to an unbuffered water and by measuring the pH. As the hydroxyl groups deprotonate, the hydrogen ion concentration in solution increases, subsequently making the pH more acidic. The ease at which a molecule or functional group gives up the proton is measured in terms of a pK_a value, which is determined by the ability of the conjugate base to stabilize the negative charge. In some instances, a molecule may undergo more than one protonation or deprotonation step. These types of molecules are defined as being polyacidic or polybasic molecules, which is the case for fullerol.

Changes in the protonation state of fullerol manifest as observable changes in its surface charge. As pH increases (becomes more alkaline), the hydroxyl groups deprotonate, and the fullerol becomes more negatively charged. The opposite scenario holds true as the pH decreases (becomes more acidic). Surface charge can be measured both directly and indirectly. Potentiometric titration with an acid and a base is used to calculate the charge density for a given molecule or nanoparticle and is considered as a direct measurement technique. Charge density is calculated from the amount of protons that adsorb to a known quantity of surface. The titrant consumption as a function of pH, $n_{(pH)}$, (in moles of equivalents) by either the suspension (n_{susp}) or the ionic strength (n_{elec}), can be determined by the following equation:

$$n_{(pH)} = \left(V_{B(pH)} \times C_B - V_A \times C_A\right) \times \frac{V_{ini}}{V_{tot(pH)}} \qquad (44.9)$$

$V_{B(pH)}$ is the cumulated NaOH (a commonly used base) volume introduced at each increment
V_{ini} is the initial sample volume
$V_{tot(pH)}$ is the total sample volume

Finally, the amount of protons really adsorbed onto the fullerol cluster surface (n_{surf}) is obtained by Equation 44.10:

$$n_{surf(pH)} = n_{susp(pH)} - n_{elec(pH)} \qquad (44.10)$$

An indirect measure of surface charge is electrophoretic mobility measurements [4]. Electrophoretic mobility is an indicator of surface charge and is in fact the surface potential at the shear boundary at the Stern Layer and may be used to calculate the zeta potential [4]. Electrophoretic mobility, U, is simply defined as a particle's velocity per unit field strength, with the field being an applied electrical field. For nanoparticles that are smaller in size than the surrounding electric double layer, U is calculated using the Hückel equation:

$$U = \frac{2\varepsilon\zeta}{3\mu} \qquad (44.11)$$

ε is the permittivity of the solution
ζ is zeta potential
μ is the solution viscosity

The protonation or deprotonation of fullerol has a number of consequences related to the mass, charge, hydrophobicity, reduction potential, and perhaps optical properties of the molecule. As found by Vileno et al. [45], during a potentiometric titration analysis of fullerol, there are three inflection points in a plot of solution pH and amount of titrant added to solution. These points occur at pH 8.9 (Point A), pH 6.6 (Point B), and pH 4.2 (Point C), as illustrated in Figure 44.8. Point A indicates the end of neutralization of the first group, and Point C corresponds to the neutralization of the protons from the second group. This indicates that for this fullerol sample, there are two groups of acidic protons per C_{60} molecule. This finding confirms that fullerol is indeed a polyacidic molecule.

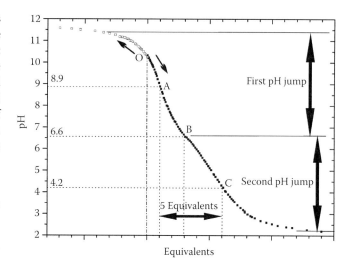

FIGURE 44.8 Titration curve of a 2.14 mM aqueous solution of a fullerol produced through oxidation of C_{60} under strongly acidic conditions [45]. In the plot, Point O corresponds to the initial solution, before adding either acid or base. Black dots indicate the decrease in pH when acid was added to the initial solution (to the right of O). Gray dots indicate the increase in pH when base was added to the initial solution (on the left of O). On the x-axis, the graduation stands for one equivalent of protons, which is one mole of protons per mole of fullerol. (Reproduced from Vileno, B. et al., *Adv. Funct. Mater.*, 16, 120, 2006. With permission.)

Over the course of the protonation and deprotonation processes, there is a pH at which equilibrium is reached (i.e., the number of protonated hydroxyl groups equals that of the deprotonated ones), and the fullerol will have a net zero surface charge. This point or pH is referred to as the point of zero net proton charge (pH_{PZNPC}). This value may be different but is related to the isoelectric point (pH_{iep}), which is the pH at which the zeta potential that is measured for the particle is equal to zero. For fullerol, Brant et al. [12] found the point of zero net proton charge (pH_{PZNPC}) at a pH of 3. This suggests that fullerol should have a negative charge over a wide pH range (pH > 3).

Similar to the potentiometric titration results, the electrophoretic mobility for fullerol becomes increasingly negative as pH becomes more alkaline (Figure 44.9). However, the electrophoretic mobility is not neutral (pH_{iep}) at pH = 3, which is the pH_{PZNPC} for fullerol, as determined from potentiometric titration. Disagreement between charge measurements at the surface (potentiometric titration) and at the shear plane (electrophoretic mobility) is attributed to the fact that fullerols may exist as clusters in water and not as discrete molecules. This creates a condition wherein deprotonated sites in the innermost regions of the fullerol clusters are not easily accessible by the pH titrant (HCl, NaOH), but measurable in the global charge controlling the electrophoretic mobility.

What these results indicate to us is that once fullerols are dispersed in water, they are negatively charged under a wide pH range. This is important because like-charged particles in aqueous media tend to form rather stable dispersions. This, of

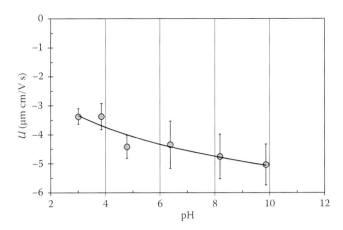

FIGURE 44.9 Electrophoretic mobility of fullerol clusters in a 10 mM sodium chloride solution (NaCl) as a function of solution pH. These measurements illustrate how the fullerol become more negatively charged (increasingly negative electrophoretic mobility) as pH becomes more alkaline, a result of the hydroxyl groups deprotonating.

course, assumes that no other destabilizing agents are present in solution, such as polymers or divalent ions. Similarly charged surfaces repel one another as a result of a repulsive energy barrier that develops between the two particles. These electrostatic interactions can be modeled within the context of the Derjaguin–Landau–Verwey–Overbeek (DLVO) theory [43].

$$\phi_{TOT} = \phi_{EL} + \phi_{LW} \qquad (44.12)$$

ϕ_{TOT} is the total interaction energy between two surfaces
ϕ_{EL} is the electrostatic interaction energy component
ϕ_{LW} is the van der Waals interaction energy component

Using the DLVO theory, we can develop an interaction energy profile that shows the decay of electrostatic and other interactions as a function of separation distance between two surfaces. For like-charged surfaces, a repulsive energy barrier, the height of which is principally determined by the magnitude of the particle surface charge, typically characterizes the total interaction energy profile. This barrier prevents the interacting surfaces from coming into contact and thus allows the particles to remain as discrete units.

In addition to making the naturally uncharged C_{60} molecule charged, hydroxylation also alters the affinity of C_{60} with water (i.e., its hydrophobicity). As was discussed earlier in this chapter, C_{60} is a hydrophobic molecule. Evidence to this fact comes from the contact angle with water (a measure of hydrophobicity), which for C_{60} has been determined to be around 70°. On the other hand, fullerol has a contact angle with water of approximately 18°, indicating that it is hydrophilic. This may not be surprising, given our knowledge of the fullerol molecule's surface chemistry, but it does provide us with a quantitative measure for this property. As was previously discussed, the hydrophilicity, or a higher affinity for water, arises from the presence of γ^+, γ^- sites on the fullerol molecular cage. Both charge and hydrophilicity

are attributed to the hydroxyl groups, which as previously discussed protonate and deprotonate with changes in pH. The importance of these sites in determining the behavior of C_{60} in water and the interfacial interaction between two C_{60} in water was previously illustrated in Equations 44.6 and 44.7.

44.3.1 Fullerol Clusters

The primary objective of derivatizing C_{60} to form fullerol is to produce a relatively stable aqueous dispersion. While derivatization increases the affinity between the C_{60} and water (see Equation 44.6), it does not result in the formation of a molecular dispersion, as was suggested by the UV–vis spectra for fullerol (Figure 44.6) [12]. Indeed, fullerol forms molecular clusters that in-turn form chain-like structures composed of multiple clusters in water (Figure 44.10a). Inspection of a fullerol dispersions using dynamic light scattering reveals that the size distribution is in fact broad, indicating that they form polydisperse dispersions, as shown in Figure 44.10b. Although polydisperse, the mean cluster size tends to rage from anywhere between 100 and 200 nm [12]. In this size range, the fullerol clusters are characterized as small colloids and less like true nanoparticles.

The fact that fullerol forms clusters in water was a surprising realization to many researchers as it was first thought that

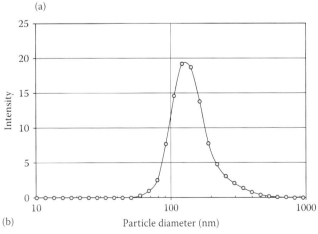

(a)

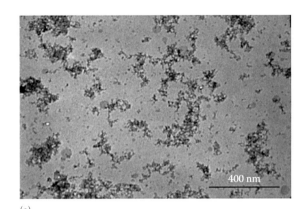

(b)

FIGURE 44.10 (a) Transmission electron microscopy (TEM) image of fullerol clusters and amorphous aggregates in water. (b) Intensity weighted size distribution for a dispersion of fullerol clusters in water.

fullerols would remain as discrete molecules once they were placed in water [23,24]. Cluster formation is not intuitive as fullerols are charged and hydrophilic under a spectrum of solution chemistries (pH, ionic strength, ionic composition) [13]. Both of these characteristics typically produce stable particle suspensions in aqueous systems [4]. However, therein lies the key to understanding why fullerols, to remain as discrete units in water, should be molecules and not particles. Therefore, they are subject to different interfacial mechanisms than their larger counterparts.

Fullerol clusters seem to form large, loosely associated and amorphous aggregates in water. X-ray diffraction (XRD) analysis of fullerol clusters indicates that they are not the smallest particles of a larger previous crystalline structure, as is the case with C_{60} crystallites arranged in a face-centered cubic system [12]. Indeed, the XRD diffractograms indicate that the internal structure of fullerol clusters is totally amorphous. Fullerol cluster formation occurs even at micromolar concentrations [45] and relatively independently of solution chemistry (pH, ionic strength) or conditions (temperature) [12]. Furthermore, cluster formation occurs despite increasing the number of hydroxyl groups on the molecular cage. However, the impact of changing the number of hydroxyl groups or their specific location on the molecular cage on the cluster structure or size has not yet been extensively investigated in the literature.

A consensus has not yet been reached regarding the mechanisms that control the formation and growth of fullerol clusters in water. As noted earlier in this chapter, some number of hydroxyl groups are distributed across the C_{60} molecular cage, and the fullerols are polyacidic molecules. These two characteristics have significant implications for the behavior of fullerols in water and their interaction with one another. The distribution of hydroxyl groups across the C_{60} surface results in the molecule having both hydrophobic and hydrophilic regions 2,45. This type of structure is like that of a surfactant molecule, which has a hydrophobic head and a hydrophilic tail, and is a driving mechanism for surfactant micelle* formation in water. Indeed, the shape of the core fullerol clusters (spherical) suggests that the driving force for the initial fullerol cluster formation is hydrophobic and nondirectional. Furthermore, the characteristics of the hydroxyl groups (e.g., length) may also influence the cluster shape. This may be due to either a more favorable molecular packing compared to what is experienced for long-chained groups or a balance between attractive and repulsive interactions.

A separate, or perhaps complementary, aggregation mechanism involves the interaction between protonated and deprotonated sites on the fullerol molecular cage. As previously discussed, fullerol is polyacidic and therefore simultaneously have both attractive (C–OH) and repulsive (C–O⁻) sites. Attraction occurs

---

* A micelle is a cluster or aggregate of surfactant molecules that are dispersed in aqueous media. The aggregate forms with the hydrophilic "head" regions maintaining contact with the surrounding aqueous solution and the hydrophobic tail regions contacting each other to form the cluster center.

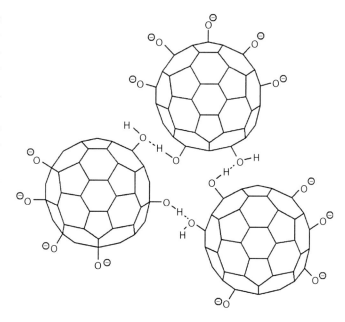

FIGURE 44.11 Proposed structure for fullerol clusters in water. The clusters are formed through a network of intermolecular hydrogen bonds; negatively charged groups (C–O–) are located on the outside of the cluster, which give them their observed surface charge. (Reproduced from Vileno, B. et al., *Adv. Funct. Mater.*, 16, 120, 2006. With permission.)

through sites being oppositely charged and/or capable of forming hydrogen bonds (e.g., C–OH and C–O⁻). In this case, the acidic protons could be involved in attractive hydrogen bonding interactions with other fullerol molecules, driving cluster formation (Figure 44.11). Indeed, fullerol has been found to form clusters in water despite increasing the number of hydroxyl groups on the molecular cage [45], which would decrease the hydrophobic portion of the molecular surface area. Because the protonation state of these sites is pH dependent, their impact on cluster formation may be observed through manipulation of the solution pH. Indeed, a number of investigators, e.g., [12,16] have found that smaller fullerol clusters may be formed or manipulated by altering the pH of the initial dispersing solution. The size of the fullerol clusters formed decreased when the pH of the initial dispersing solution was more alkaline, and the fullerols are more negatively charged. Nevertheless, cluster formation is not prevented even under the most alkaline conditions (pH > 12). This suggests that other mechanisms, such as the association of cations (Na⁺, K⁺) with the conjugate base that allow bonds to be formed with other conjugate base groups on fullerols in solution, must be considered.

44.4 Summary

Engineered nanoparticles present many opportunities for developing new and revolutionary products. Additionally, the environmental behavior of these materials must carefully be considered so as to minimize the risks associated with potential toxins. Therefore, it is critical that we have a comprehensive understanding of the fundamental mechanisms that dictate their behavior and properties in specific environments. In this chapter, we have examined the characteristics of a specific engineered nanoparticle, fullerol. Though engineered to form molecular dispersions in aqueous media, these "nanoparticles" instead form near colloidal suspensions. This result has numerous consequences related to the fate and transport of these materials in the environment, as well as for the utilization of these materials in industrial processes. The lessons learned from fullerol in terms of cluster formation may be used to develop new derivatized fullerenes capable of forming molecular dispersions. They also provide a roadmap by which future nanoparticles should be evaluated with regard to their physical state and behavior in aqueous systems.

References

1. The National Nanotechnology Initiative Strategic Plan, 2004, pp. 52.
2. Wiesner, M. R. and Bottero, J.-Y., *Environmental Nanotechnology Applications and Impacts of Nanomaterials*, McGraw-Hill, New York, 2007.
3. Wiesner, M. R., Lowry, G. V., Alvarez, P., Dionysiou, D., and Biswas, P., Assessing the risks of manufactured nanomaterials, *Environmental Science and Technology*, 4336–4345, 2006.
4. Elimelech, M., Gregory, J., Jia, X., and Williams, R. A., *Particle Deposition and Aggregation: Measurement, Modelling, and Simulation*, Butterworth-Heinemann, Oxford, U.K., 1995.
5. Daniel, M.-C. and Astruc, D., Gold nanoparticles: Assembly, supramolecular chemistry, quantum-size-related properties, and applications toward biology, catalysis, and nanotechnology, *Chemical Reviews* 104, 293–346, 2004.
6. Aikawa, S., Yoshida, Y., Hatae, S., Nishiyama, S., and Kumar, D. S., Synthesis and characterization of a fullerene derivatives, *Molecular Crystals and Liquid Crystals* 463, 237–244, 2007.
7. Badireddy, A. R., Hotze, E. M., Chellam, S., Alvarez, P., and Wiesner, M. R., Inactivation of bacteriophages via photosensitization of fullerol nanoparticles, *Environmental Science and Technology* 41, 6627–6632, 2007.
8. Beavers, C. M., Zuo, T., Duchamp, J. C., Harich, K., Dorn, H. C., Olmstead, M. M., and Balch, A. L., An improbable, egg-shaped endohedral fullerene that violates the isolated pentagon rule, *Journal of the American Chemical Society* 35, 11352–11353, 2006.
9. Bogdanovic, G., Kojic, V., Dordevic, A., Canadanovic-Brunet, J., Vojinovic-Miloradov, M., and Baltic, V. V., Modulating activity of fullerol $C_{60}(OH)_{22}$ on doxorubicin-induced cytotoxicity, *Toxicology In Vitro* 18 (5), 629–637, 2004.
10. Brant, J., Lecoanet, H., and Wiesner, M. R., Aggregation and deposition characteristics of fullerene nanoparticles in aqueous systems, *Journal of Nanoparticle Research* 7 (4–5), 545–553, 2005.
11. Brant, J. A., Labille, J., Bottero, J.-Y., and Wiesener, M. R., Characterizing the impact of preparation method on fullerene cluster structure and chemistry, *Langmuir* 22, 3878–3885, 2006.
12. Brant, J. A., Labille, J., Ogilvie-Robichaud, C., and Wiesener, M. R., Fullerol cluster formation in aqueous solution: Implications for environmental release, *Journal of Colloid and Interface Science* 314 (1), 281–288, 2007.
13. Brant, J. A., Lecoanet, H., Hotze, M., and Wiesner, M. R., Comparison of electrokinetic properties of colloidal fullerenes (n-C_{60}) formed using two procedures, *Environmental Science and Technology*, 39 (17), 6343–6351, 2005.
14. Chen, Y., Cai, R.-F., Chen, S., and Huang, Z.-E., Synthesis and characterization of fullerol derived from C_{60}n-precursors, *Journal of Physics and Chemistry of Solids* 62 (5), 999–1001, 2001.
15. Chiang, L. Y., Wang, L.-Y., Swirczewski, J. W., Soled, S., and Cameron, S., Efficient synthesis of polyhydroxylated fullerene derivatives via hydrolysis of polycylosulfated precursors, *Journal of Organic Chemistry* 59 (14), 3960–3968, 1994.
16. Husebo, L. O., Sitharaman, B., Furukawa, K., Kato, T., and Wilson, J., Fullerenols revisited as stable radical anions, *Journal of the American Chemical Society* 126 (38), 12055–12064, 2004.
17. Koruga, D., Hameroff, S., Withers, J., Loutfy, R., and Sundareshan, M., *Fullerene C60: History, Physics, Nanobiology, Nanotechnology*, Elsevier Science Publishers, Amsterdam, the Netherlands, 1993.
18. Kroto, H. W., Heath, J. R., O'Brien, S. C., Curl, R. F., and Smalley, R. E., C_{60}-Buckminsterfullerene, *Nature* 318, 162–163, 1985.
19. Langa, F. and Nierengarten, J.-F., *Fullerenes Principles and Applications*, RCS Publishing, Cambridge, U.K., 2007.
20. Lyon, D. Y., Fortner, J. D., Sayes, C. M., Colvin, V. L., and Hughes, J. B., Bacterial cell association and antimicrobial activity of a C_{60} water suspension, *Environmental Toxicology and Chemistry* 24 (11), 2757–2762, 2005.
21. Marcus, Y., Smith, A. L., Korobov, M. V., Mirakyan, A. L., Avramenko, N. V., and Stukalin, E. B., Solubility of C_{60} fullerene, *Journal of Physical Chemistry B* 105, 2499–2506, 2001.
22. Prassides, K., *Physics and Chemistry of the Fullerenes*, Kluwer Academic Publishers, London, U.K., 1994.
23. Lecoanet, H. F., Bottero, J.-Y., and Wiesner, M. R., Laboratory assessment of the mobility of nanomaterials in porous media, *Environmental Science and Technology*, 38 (19), 5164–5169, 2004.
24. Lecoanet, H. F. and Wiesner, M. R., Velocity effects on fullerene and oxide nanoparticle deposition in porous media, *Environmental Science and Technology* 38 (16), 4377–4382, 2004.
25. Fortner, J. D., Lyon, D. Y., Sayes, C. M., Boyd, A. M., Falkner, J. C., Hotze, M., Alemany, L. B., Tao, Y. J., Guo, W., Ausman, K. D., Colvin, V., and Hughes, J. B., C_{60} in water: Nanocrystal formation and microbial response, *Environmental Science and Technology*, 39 (11), 4307–4316, 2005.

26. Sayes, C. M., Fortner, J. D., Guo, W., Lyon, D., Boyd, A. M., Ausman, K. C., Tao, Y. J., Sitharaman, B., Wilson, L. J., Hughes, J. B., West, J. L., and Colvin, V. L., The differential cytotoxicity of water-soluble fullerenes, *Nano Letters* 4 (10), 1881–1887, 2004.

27. Sayes, C. M., Gobin, A. M., Ausman, K. D., Mendez, J., West, J. L., and Colvin, V. L., Nano-C_{60} cytotoxicity is due to lipid preoxidation, *Biomaterials*, 26, 7587–7595, 2005.

28. Lyon, D. e. Y., Adams, L. a. K., Falkner, J. C., and Alvarez, P. e. J., Antibacterial activity of fullerene water suspensions: Effects of preparation method and particle size, *Environmental Science and Technology*, 40 (14), 4360–4366, 2006.

29. Ajie, H., Alvarez, M. M., Anz, S. J., Beck, R. D., Diederich, F. N., Fostiropoulos, K., Huffman, D. R., Kratschmer, W., Rubin, Y., Schriver, K. E., Sensharma, D., and Whetten, R. L., Characterization of the soluble all-carbon molecules C_{60} and C_{70}, *Journal of Physical Chemistry* 94, 8630–8633, 1990.

30. Arbogast, J. W., Darmanyan, A. P., Foote, C. S., Rubin, Y., Diederich, F. N., Alvarez, M. M., Anz, S. J., and Whetten, R. L., Photophysical properties of C_{60}, *Journal of Physical Chemistry* 95, 11–12, 1991.

31. Dresselhaus, M. S., *Carbon Nanotubes Synthesis, Structure, Properties, and Applications*, Springer, Heidelberg, Germany, 2001.

32. Rinzler, A. G., Hafner, J. H., Nikolaev, P., Lou, L., Kim, S. G., Tomanek, D., Nordlander, P., Colbert, D. T., and Smalley, R. E., Unraveling nanotubes—Field emission from an atomic wire, *Science* 269 (5230), 1550–1553, 1995.

33. Cheng, H. M., Yang, Q. H., and Liu, C., Hydrogen storage in carbon nanotubes, *Carbon* 39 (10), 1447–1454, 2001.

34. Ajayan, P. M. and Zhou, O. Z., Applications of carbon nanotubes, *Carbon Nanotubes* 80, 391–425, 2001.

35. Chopra, S., McGuire, K., Gothard, N., Rao, A. M., and Pham, A., Selective gas detection using a carbon nanotube sensor, *Applied Physics Letters* 83 (11), 2280–2282, 2003.

36. Miessler, G. L. and Tarr, D. A., *Inorganic Chemistry*, Pearson Education International, Upper Saddle River, NJ, 2004.

37. Kratschmer, W., Fostiropoulos, K., and Huffman, D. R., The infrared and ultravioletabsorption spectra of laboratory-produced carbon dust—Evidence for the presence of the C-60 molecule, *Chemical Physics Letters* 170 (2–3), 167–170, 1990.

38. Loiseau, A., *Understanding Carbon Nanotubes*, Springer, New York, 2006.

39. Pope, C. J., Marr, J. A., and Howard, J. B., Chemistry of fullerenes C-60 and C-70 formation in flames, *Journal of Physical Chemistry* 97 (42), 11001–11013, 1993.

40. Richter, H., Labrocca, A. J., Grieco, W. J., Taghizadeh, K., Lafleur, A. L., and Howard, J. B., Generation of higher fullerenes in flames, *Journal of Physical Chemistry B* 101 (9), 1556–1560, 1997.

41. Heymann, D., Solubility of fullerenes C-60 and C-70 in seven normal alcohols and their deduced solubility in water, *Fullerene Science and Technology* 4 (3), 509–515, 1996.

42. van Oss, C. J., Acid-base interfacial interactions in aqueous media, *Colloids and Surfaces A* 78, 1, 1993.

43. Israelachvili, J. N., *Intermolecular and Surface Forces*, 2nd edn. Academic Press, Harcourt Brace Jovanovich, London, U.K., 1992.

44. Bellavia-Lund, C., Diederich, E., Hirsch, A., Hsu, W. K., Hummelen, C., Kroto, H. W., Prato, M., Rubin, Y., Terrones, M., Thilgen, C., Walton, D. R. M., and Wudl, E., *Fullerenes and Related Structures*. Springer Berlin/Heidelberg: Berlin, Vol. 199, p. 234, 1999.

45. Vileno, B., Marcoux, P. R., Lekka, M., Sienkiewicz, A., Feher, T., and Forro, L., Spectroscopic and photophysical properties of a highly derivatized C-60 fullerol, *Advanced Functional Materials* 16 (1), 120–128, 2006.

46. Xiang, G., Zhang, J., Zhao, Y., Tang, J., Zhang, B., Gao, X., Yuan, H., Qu, L., Cao, W., Chai, Z., Ibrahaim, K., and Su, R., *Journal of Physical Chemistry B* 108 (11), 473, 2004.

45

Polyhydroxylated Fullerenes

Ricardo A. Guirado-López
Universidad Autónoma
de San Luis Potosí

45.1 Introduction

The design of new materials with novel and special properties is currently the subject of intense investigations. In particular, fullerene-related materials have attracted a lot of attention as they can not only be produced in large quantities with full control over size and morphology, but have also demonstrated lots of interesting properties that could lead to the development of novel technologies. As is well known, C$_{60}$, a hydrophobic molecule made of 60 carbon atoms arranged in a soccer ball shape (Kroto et al., 1985), is the backbone of the fullerene-based nanomaterials family. In fact, the structural, electronic, and solid-state properties of this molecule have attracted a wide interest not only in physics and chemistry, but also in materials and biological sciences.

As part of the remarkable progress made in this field, it has been demonstrated in the last years that the behavior and processing of the C$_{60}$ fullerene can be strongly modified via the chemical functionalization of its surface, i.e., by covalently attaching several types of atoms, molecules, or molecular groups (Coheur et al., 2000; Fanti et al., 2002; Sitharaman et al., 2004). The previous finding has led to the fabrication of a large number of organic derivatives, and has significantly expanded the scope of the science of these kinds of carbon clusters. In particular, several types of functionalized C$_{60}$ molecules have been recently synthesized for their possible use in new electronic and optical devices, as well as for developing possible applications in biology and medicine. For example, the use of functionalized fullerenes as photoactive molecular systems has recently been discussed, assuming particular importance due to the conversion of solar energy into electric current (Rincón et al., 2003). In addition, water-soluble C$_{60}$ derivatives have been proposed

as HIV protease inhibitors (Friedman et al., 1993) and to promote interactions with target molecules (Friedman et al., 1993; Schinazi et al., 1993; Tokuyama et al., 1993), such as a particular protein on a cell surface. They have also been defined as potent antioxidants (Dugan et al., 1997) and as neuroprotectants in vivo (Krusic et al., 1991) and, finally, if metal atoms belonging to the lanthanoid group can be trapped in the inside of the carbon cage, as good candidates for a new generation of novel magnetic resonance imaging (MRI) contrast agents (Mikawa et al., 2001).

45.2 Background

The solubility of C$_{60}$ is, of course, of fundamental importance in order to achieve the large (and highly relevant) variety of possible medical and biological applications discussed in the previous paragraph. At this point, it is important to comment first that the C$_{60}$ molecule is known to be soluble in almost 150 solvents, which is probably the highest number among all chemical substances. In general, typical methods of determining the solubility include the stirring of the samples for a period of time from 5 to 48 h and determining the saturated concentration by ultraviolet-visible (UV-VIS) spectrophotometry (Zhou et al., 1997), use of high-performance liquid chromatography (Ruoff et al., 1993), or by simple weighing (Scrivens and Tour, 1993). In most of the cases, it has been found that the solvents at room temperature do not form covalent bonds with the carbon surface. However, because C$_{60}$ is a good electron pair acceptor, it can form charge-transfer complexes with electron pair donors. For example, the dissolution of fullerenes in tertiary amines and substituted anilines is accompanied by the formation of such donor–acceptor complexes in the liquid state, since the reported solubilities in these kinds of samples correspond to the formation of donor–acceptor complexes rather than the pure fullerene.

In contrast, getting fullerenes to dissolve in water and other polar solvents is challenging because of its hydrophobicity. In order to achieve the previous water solubility, the main idea has been to introduce functional groups with the appropriate structure in order to make the resulting compound hydrophilic. In this respect, it is important to comment that the degree of surface coverage on the C_{60} molecule has been found to lead to an interesting aggregation behavior, namely, that the attachment of only a single hydrophilic addend promotes water solubility, but strong attractive forces remain between the fullerene units, resulting in the formation of complex colloidal fullerene clusters. However, it has been determined that the formation of these molecular clusters can be prevented by increasing the number of adsorbed addends on the carbon surface.

45.3 State of the Art

To date, one of the most popular substituents of choice to produce the water solubility of the C_{60} molecule are the hydroxyl groups. Interestingly, and as discussed above, it has also been found that the degree of solubility (partial or total) of the fullerene-$(OH)_n$ compounds critically depends on the number of OHs attached. For example, Liu and collaborators (Liu et al., 2000) have reported that 12 hydroxy groups bounded to the C_{60} cage provide the adequate hydrophobic-hydrophilic balance to stabilize $C_{60}(OH)_{12}$ Langmuir monolayers at the air–water interface. Similarly, Rincón and coworkers (Rincón et al., 2003) have obtained stable Langmuir films with the use of $C_{60}(OH)_{9-12}$ units. However, it is well known that, with increasing OH coverage, completely water-soluble fullerenes can be obtained. This is particularly the case of the $C_{60}(OH)_{24-28}$ compounds (Rincón et al., 2003). Although, in the experimental setup it is possible to estimate the average number of adsorbed OH's by means of mass spectrometry experiments [e.g., matrix-assisted laser desorption ionization (MALDI) technique], it is very difficult to identify the geometric structure of the various adsorbed phases that can be stabilized as a function of coverage and characteristics of the underlying substrate, and which are obviously at the origin of the observed chemical and physical properties in these materials over a large scale. For example, it has been found that the macroscopic samples of highly hydroxylated C_{60} molecules give a dielectric structure, while materials with low levels of hydroxylation result in proton conductive compounds (Rincón et al., 2003).

The previous highly hydroxylated $C_{60}(OH)_x$ fullerene derivatives can be synthesized by means of several experimental techniques, as clearly reviewed by Husebo and collaborators (Husebo et al., 2004). In addition, their properties have been extensively analyzed in both gas-phase and aqueous environments, as well as in the form of deposited nanostructures, by means of IR spectroscopy (Xing et al., 2004), transmission electron microscopy (TEM) images (Sitharaman et al., 2004), dynamic light scattering (DLS) measurements (Díaz-Tendero et al., 2005), x-ray

photoelectron spectroscopy (XPS) (Sitharaman et al., 2004; Xing et al., 2004), and optical absorption experiments (Rincón et al., 2003) among others. The previous experimental characterization has revealed important information concerning (1) the average number of adsorbed OH groups on the carbon surface and his impact on the partial or total solubility of the carbon compound, (2) the presence and role of impurities in the samples, (3) their intermolecular interactions and aggregation behavior, (4) the electrical conductivity, and (5) the optical properties in both ultraviolet and visible parts of the spectra.

However, despite the already mentioned extensive characterization, there are, of course, additional points of interest that need to be addressed among which the most important are the ones related to the study of the structural stability of these highly hydroxylated C_{60} fullerenes in different environments. It is clear that the inclusion of these fullerene derivatives in gaseous, aqueous, or solid-state matrices might alter their chemical composition as well as the charge state of the cage, a fact that could lead to the formation of carbon-opened structures, thus strongly affecting their properties and functionality. In fact, previous theoretical and experimental studies performed on both bare cationic (Díaz-Tendero et al., 2005) and anionic (Ehrler et al., 2005) gas-phase $C_{60}^{\pm q}$ fullerenes (with q as large as 12) have already underlined the crucial role played by the amount of excess of positive (or negative) charge $+q$ ($-q$) on the structure and stability of these kind of systems. Obviously, the repulsive Coulomb forces become progressively important as q increases and, when they exceed the binding forces, the fullerene could become unstable, causing the cage destruction or the partial rupture of the fullerene network. Of course, understanding the precise micro-structural features and fundamental physics leading to the cage destruction of these nanometer-sized carbon structures will be of fundamental importance in order to attain the controlled ring-opening and closing of the carbon cage. Both processes are, of course, highly relevant since they can be used to allow the inclusion of different types of guests species within the carbon cage that are not accessible by conventional methods, such as those based on coevaporation techniques or high-pressure/high-temperature experiments.

In this respect, it has been clearly established that these cage-shaped structures are actually able to accommodate many kinds of atoms (Nishibori et al., 2004) and small clusters (Krause et al., 2005), giving rise to another exciting field of research named endohedral fullerene chemistry. The previous so-called endofullerene materials have attracted special interest since they are characterized by having unique electronic properties and chemical reactivity, which are not seen in the empty fullerenes. Actually, at the time of their synthesis, endohedral fullerenes were found to be, in general, highly unstable in practical applications due to their insolubility and extremely air sensitivity. However, later on, novel procedures were developed to solubilize (and protect) these kind of carbon compounds by performing the functionalization of its surface with the use of a large variety of small molecules, the most common being the hydroxyl species (Liu et al., 2000) as well as carboxylic groups (Bolskar et al.,

2003), thus opening new opportunities for potential applications in biomedical and nano-materials science.

In particular, the use of endohedral polyhydroxylated metallofullerenes has been proposed by several research groups as potential MRI contrast agents. In this case, the entrapment of the previous highly toxic metal atoms within the fullerenes helps to ensure a specially safe in vivo use of the molecular compounds since it has been speculated that the carbon cage will protect the encapsulated species from external chemical attack, and will also prevent the metal ion release in the body under normal physiological conditions. Actually, the previous carbon complexes have been already administrated to minces (Mikawa et al., 2001) and, by reducing the relaxation time of water protons in the effected tissues, they have been found to produce a higher-quality diagnostic image when compared to conventional Gd-based contrast agents.

This chapter concentrates on computational approaches [using both accurate semiempirical as well as ab initio density-functional theory (DFT) methodologies] dedicated to analyze the properties of several OH-covered fullerene structures. First, we study the electronic and structural properties of low hydroxylated carbon fullerenes. The considered coverage θ (defined as the ratio of adsorbed molecules to surface atoms) on the C_{60} surface is below a complete monolayer; however, it has been found to be enough to understand the stability of Langmuir monolayers at the air–water interface observed for low hydroxylated C_{60} structures. The vibrational frequencies of $C_{60}(OH)_n$ compounds, which are expected to depend sensitively on the number and distribution of the adsorbed hydroxyl molecules, as well as on the interactions among them, will be calculated in order to compare with recent experimental measurements.

We also discuss in this chapter the structural, electronic, and optical properties of water-soluble highly hydroxylated $C_{60}(OH)_{24-28}$ units. We found that the strong variations in highest occupied-lowest unoccupied molecular orbital energy difference (HOMO–LUMO energy gap) of highly hydroxylated fullerenes, inferred from the electrical properties of these molecular films and also from theoretical calculations of various $C_{60}(OH)_x$ isomers, also results in significant changes in the photo-physics of the carbon derivatives. To better understand the previous measured data, and to separate the intrinsic properties of fullerenol molecules from solvent contributions and intermolecular interactions, we present extensive semiempirical as well as DFT electronic structure calculations in order to obtain the structural, electronic, and optical properties of gas-phase-like fullerenol units. We systematically analyze the influence of the number, type, and spatial distribution of the adsorbates on the carbon surface on the optical properties of OH-covered structures and show that the understanding of the spectral features of these fullerene derivatives should consider, besides the normally used symmetry considerations, more subtle effects related to the chemical composition and the precise details of the adsorbed phases on such small objects.

Finally, we also discuss the stability and electronic properties of both neutral and positively charged polyhydroxylated $La@C_{60}(OH)^{+q}_{32}$ (q = 0, 1, 2, 4, 6, and 8) metallofullerenes. We will put special emphasis on the influence of different spatial distributions of the OH groups on the carbon surface, as well as on the charge state q of the cage, on the energetics and lowest-energy structure of our considered isomers. On the one hand, we will show how the entrapped metal atom can have different electronic ground states and endohedral locations within the cage, depending on the molecular structure of the OH overlayer. On the other hand, our theoretical data will reveal in addition how, by increasing the positive charge in our polyhydroxylated fullerenes (through simple electron detachment), it could be possible to obtain C_{60} cages with sizable holes in the fullerene network, as well as to control the reactivity of the inner carbon surfaces.

45.4 Discussion

45.4.1 Low Hydroxylated C_{60} Fullerenes

The computational support of experiments in fullerene research is in our days very common and widespread, and it is also currently used as a complementary tool for the rational and efficient understanding of the measured data. The number of atoms involved in our here-considered low hydroxylated carbon systems (Rodríguez-Zavala and Guirado-López, 2004) limits the applicability of density functional optimization-based methods, and that is why we have decided to perform our systematic study by combining two different theoretical approaches. In the first step, we will fully optimize the considered structures using the semiempirical modified-neglect-of-diatomic-overlap (MNDO) level of theory and then, in the second step, we will use these MNDO geometries to perform single-point ab initio DFT calculations, considering in the latter the generalized gradient approximation (GGA) and the minimal STO-3G basis set. On the contrary, only when determining the vibrational frequencies of some relevant examples, the total energy as well as the electronic and ground-state structure of our OH-coated fullerenes will be obtained within the same theoretical framework by performing a fully DFT geometry unrestricted energy minimization procedure with the use of GGA for the exchange correlation potential. In all cases, the Kohn–Sham equations are solved by considering the nonlocal Becke's three-parameter exchange functional combined with Lee, Yang, and Parr's correlation functional (B3LYP) and the STO-3G basis set.

First, it is important to discuss the intensities and distribution of the vibrational frequencies of both C_{60} and $C_{60}(OH)$. In this type of calculations, the diagonalization of the full Hessian matrix within the harmonic approximation (i.e., considering the matrix of all second derivatives of the total DFT-B3LYP electronic energy with respect to Cartesian nuclear coordinates) is performed. This procedure generates all vibrational modes of the molecular structure under consideration, which can be directly related with experimental IR spectroscopy measurements. From Figure 45.1 we can see that, for the C_{60} structure (see the inset), we have found the existence of well-defined peaks at 546, 595, 1265,

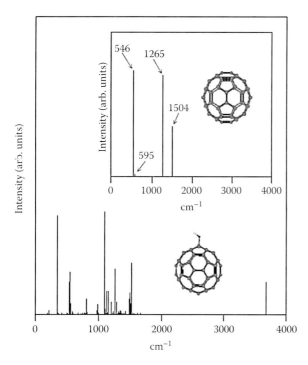

FIGURE 45.1 Intensities of calculated vibrational frequencies for C_{60} (inset) and C_{60}(OH) structures. (Reprinted from Rodríguez-Zavala, J.G. and Guirado-López, R.A., *Phys. Rev. B*, 69, 075411, 2004. With permission.)

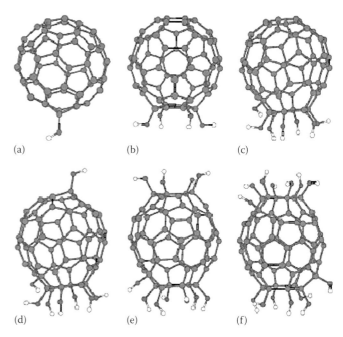

FIGURE 45.2 Calculated (MNDO) lowest-energy structures for low hydroxylated C_{60}(OH)$_n$ compounds ($1 < n < 14$). (a) C_{60}(OH), (b) C_{60}(OH)$_4$, (c) C_{60}(OH)$_7$, (d) C_{60}(OH)$_8$, (e) C_{60}(OH)$_{11}$, and (f) C_{60}(OH)$_{14}$. (Reprinted from Rodríguez-Zavala, J.G. and Guirado-López, R.A., *Phys. Rev. B*, 69, 075411, 2004. With permission.)

and 1504 cm^{-1} which are in good agreement with the absorption bands normally observed in the IR spectrum of the C_{60} fullerene located at 526, 576, 1182, and 1428 cm^{-1}. From Figure 45.1 we can also appreciate that a more complex distribution of vibrational modes for the C_{60}(OH) compound is found because of the lower symmetry of the carbon structure obtained (i.e., higher number of inequivalent sites, of different bond lengths, as well as of different angles between the bonds connecting nearest-neighbor atoms) upon single OH adsorption. As additional features to the pure fullerene cage, significant C–O and O–H stretching contributions in the C_{60}(OH) spectra are observed. The former is located around 1000 cm^{-1}; however, it is surrounded by a complex mixture of various bond-stretching and angle-bending contributions of the carbon cage around that frequency range, while the latter is directly reflected by the presence of an isolated peak centered at 3650 cm^{-1}. We must say that both the contributions are in agreement with the positions of well-defined absorption bands in the IR spectra of C_{60}(OH)$_{24-28}$ compounds assigned to C–O (1070–1090 cm^{-1}) and O–H (3430 cm^{-1}) stretching modes, clearly reflecting, thus, the formation of the C_{60}(OH) complex.

In Figure 45.2, we present now our calculated MNDO lowest-energy atomic configurations for some representative C_{60}(OH)$_n$ fullerene compounds ($0 < n < 14$). The previous low hydroxylated structures are important to analyze since they have been found to be still insoluble in water, currently used in the formation of C_{60}(OH)$_n$ Langmuir films at the air–water interface, and are the subject of controversy with respect to the possible molecular organization of the OH species on the carbon surface. From

Figure 45.2 we can see that, as already stated in the previous section, the lowest-energy array for the single OH adsorption corresponds to the hydroxyl molecule directly adsorbed on the top of a carbon atom of the fullerene cage having C–O and O–H bond lengths of 1.38 and 0.95 Å, respectively, and tilted away from the surface normal with a C–O–H angle of 112.3°. Actually, we must say that, when starting with configurations in which the OH molecule is located over C–C bonds or above the center of hexagonal and pentagonal rings, we have observed lateral displacements of the hydroxyl group in order to stabilize the on-top configuration. It is important to remark that the adsorption of a single OH molecule has a negligible effect on the global shape of the carbon cluster; however, we find sizable modifications in the C–C bonds in the proximity of the OH group. The nearest-neighbor distances in the carbon surface vary between 1.39 and 1.56 Å, the largest expansions (~10%) being between the carbon atoms closer to adsorption site.

For the adsorption of a single OH, we have obtained a value for E_{ads} of 2.5 eV, which clearly indicates that this molecule undergoes chemical adsorption on the C_{60} fullerene. In addition, a binding energy E_b of 11.9 eV/at is obtained for the C_{60}(OH) structure. From this value for E_b, we can see that our OH-fullerene compound is stable; however, it is less favorable (by 0.2 eV/at) when compared to uncoated C_{60}. At this point, it is interesting to compare the previous results found for the relative stability with the ones obtained by using the more extended 6-21G basis. With the more accurate set we have found, as expected, reduced values for E_b of 10.3 and 10.1 eV/at for C_{60}, and C_{60}(OH), respectively.

Notice also that, when compared with the STO-3G calculations, the same energy ordering is obtained, being again our $C_{60}(OH)$ structure less stable by 0.2 eV/at with respect to the uncovered carbon cage. Finally, we have found an HOMO–LUMO energy gap, Δ_{HL}, of ~0.34 eV (which is very close to the value of ~0.25 eV obtained with the 6-21G basis), which shows that the electronic structure around the HOMO of C_{60} (Δ_{HL} = 2.14 eV) has been modified by the bonding of the OH molecule. This sizable reduction in the energy gap is due to the appearance of several additional electronic states around the Fermi energy of our complexes, which are induced by the presence of the OH groups.

From Figure 45.2a through f we can see that, with increasing the number of OH's attached, the local organization of the adsorbates on the C_{60} surface is highly θ dependent. From Figure 45.2a through c we note that in the low coverage regime (from 2 to 7 adsorbed species), a molecular hydroxyl cluster is stabilized on one side of the carbon surface, where the OH groups prefer to adsorb on adjacent carbons of the fullerene network forming a hexagonal array. The molecular structure shown in Figure 45.2c is particularly interesting since it is characterized by the presence of a localized hydrophilic environment which, as we will see later, could be of fundamental importance in explaining the stability of low hydroxylated C_{60} monolayers at the air–water interface. Interestingly, when attaching an additional OH group to the fullerene cage to form the $C_{60}(OH)_8$ compound, we found that the largest energy gain is obtained when the extra hydroxyl molecule is now located on the opposite side of the carbon structure (see Figure 45.2d). The previous result is very important since it defines a critical size for the molecular hydroxyl cluster that can be stabilized on the fullerene surface, reveals the existence of competing intermolecular interactions which can induce molecular aggregation or significant displacements of the adsorbates as a function of θ, and also clearly shows that OH adsorption on C_{60} is considerably affected by the presence of coadsorbed species. Finally, from 9 to 14 adsorbed molecules, we have not observed a dramatic reorganization of the OH groups but instead the continuation of a well-defined growth sequence in which a second hydroxyl island is gradually stabilized on the opposite side of the fullerene cage (see Figure 45.2e and f). The previous adsorbed phases induce the development of a highly anisotropic underlying carbon network being driven by sizable vertical as well as lateral shifts of the carbon atoms around the adsorption sites.

It is clear that the existence of molecular hydroxyl islands on the surface of the C_{60} fullerene, as the ones shown in Figure 45.2c, leads to the formation of new amphiphilic molecules that, as previously stated, could be very helpful in explaining the stability of $C_{60}(OH)_n$ ($n \sim 9$–12) Langmuir monolayers observed by several authors. Despite the existence of a large number of studies addressing the solubility of the C_{60} molecule, it has been always difficult to elucidate the precise geometric structure of the adsorbed hydroxyl phases that can be stabilized on the surface of such small objects and that could lead to a wide variety of macroscopic behaviors. In particular, Liu and coworkers (Liu et al., 2000) have obtained adsorbed $C_{60}(OH)_{12}$ monolayers

at the air–water interface, whose characterization indicated 12 hydroxy groups randomly bounded on the C_{60} cage. On the other hand, Rincón and coworkers (Rincón et al., 2003) have also synthesized low hydroxylated C_{60} Langmuir films where their data suggest, contrary to the work of Liu and collaborators (Liu et al., 2000), that 9–12 hydroxy groups seem to be preferentially bounded on one side of the C_{60} cage, an argument that is in line with our theoretical predictions. To support their assumption, the authors speculate that it is possible that under the conditions for low hydroxylation, two-phase catalysis could promote the formation of hydroxy groups clustered on one side of the cage because the reaction takes place mainly at the aqueous–solvent interface.

In order to furthermore corroborate the presence of hydroxyl islands on the C_{60} surface, we show in Figure 45.3 the calculated intensities of the vibrational frequencies of the $C_{60}(OH)_{10}$ compound by considering various relevant molecular configurations

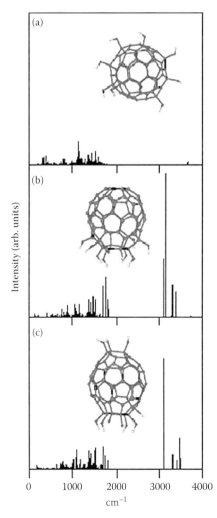

FIGURE 45.3 Intensities of the calculated vibrational frequencies for various $C_{60}(OH)_{10}$ isomers. (a) OH molecules randomly distributed, (b) clustered on one side, and (c) as two opposite $(OH)_7$ and $(OH)_3$ hydroxyl islands on the carbon surface. (Reprinted from Rodríguez-Zavala, J.G. and Guirado-López, R.A., *Phys. Rev. B*, 69, 075411, 2004. With permission.)

for the adsorbed hydroxyl groups, namely: 10 OH molecules randomly distributed on the surface (configuration I), aggregated on one side of C_{60} (configuration II), and assuming the coexistence of two opposite hydroxyl clusters on the carbon structure, having 7 and 3 molecules, which actually corresponds to our lowest-energy array (configuration III). We have chosen the $C_{60}(OH)_{10}$ structure as a representative example since it corresponds approximately to the degree of hydroxylation at which Rincón et al. have synthesized stable Langmuir films, and for which they have also performed IR measurements in order to analyze the vibrational properties of the sample. As is well known, the precise details of the vibrational spectra (i.e., intensity and positions of the peaks) of $C_{60}(OH)_n$ fullerenes are expected to be very sensitive to the chemical bondings, to the number and distribution of the adsorbed molecules, as well as to the interactions among them. On the other hand, even if experimental and theoretical spectra are not strictly comparable, the possible existence of similar features in both distributions could be very helpful in elucidating the probable molecular structure of the adsorbed OH groups.

From Figure 45.3 we can see that, for a random array of hydroxyl molecules (configuration I), our calculations indicate the presence of a very narrow and isolated distribution centered around $3600\,cm^{-1}$, which corresponds to the well-known stretching O–H mode already shown in Figure 45.1. The existence of this narrow distribution is physically expected in such separated (almost noninteracting) array where the adsorbates thus perform independent vibrations, contributing with similar values to the intensity of the O–H stretching mode. However, when considering the existence of hydroxyl clusters on the surface, we expect that the coupling between adjacent OH molecules will lead to completely different vibrational features from those seen in the uncoupled system. In fact, we note from Figure 45.3b and c that the narrow distribution obtained in Figure 45.3a for the O–H stretching mode is now decomposed into several peaks, leading to the existence of a broad band which expands the frequency range from ~3100 to $3750\,cm^{-1}$, and which can be considered as a clear signature of the presence of $(OH)_n$ islands on the C_{60} surface. Furthermore, we can also appreciate sizable differences between the vibrational spectra of configurations II (Figure 45.3b) and III (Figure 45.3c) which clearly indicate how more subtle changes in the local organization of the adsorbates can also induce significant variations in the vibrational properties (intensity and position of the peaks) of the molecule. In fact, when comparing the various distributions of modes shown in Figure 45.3 with the IR spectroscopy measurements performed by Rincón et al. on low hydroxylated C_{60} compounds, we can see that the broad and asymmetric absorption band obtained around $3450\,cm^{-1}$ in their IR spectrum is more in agreement with the presence of OH groups aggregated on C_{60} rather than randomly spread on the carbon surface.

In Figure 45.4, we show the calculated values for the adsorption energy as a function of coverage for OH molecules randomly distributed on the surface, aggregated on one side of C_{60}, and considering the presence of two opposite hydroxyl clusters

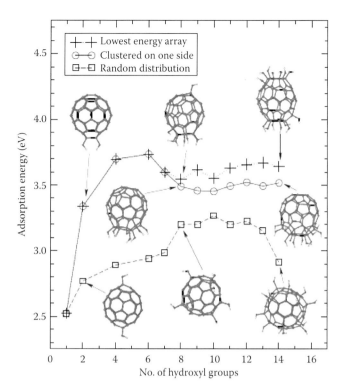

FIGURE 45.4 Calculated adsorption energies, E_{ads}, in eV for low hydroxylated $C_{60}(OH)_n$ compounds. (Reprinted from Rodríguez-Zavala, J.G. and Guirado-López, R.A., *Phys. Rev. B*, 69, 075411, 2004. With permission.)

of different sizes on the carbon structure. From the figure we can see that, at very low coverages ($1 < n < 6$), the configurations with the highest adsorption energies correspond to the adsorbed molecules which aggregate on one side of the cage. For $n > 8$, we note that subsequent OH aggregation on one side of C_{60} is prevented since we observe a notable increase in E_{ads} when OH adsorption starts to take place now on the opposite side of the fullerene cage. In particular, for $n = 8$, we note a relative small energy difference between configurations I and II. As is well known, small energy differences would lead to the formation of a highly dynamical system in which several configurations can be present in the time scale of the experiment, a fact that can lead to a wide variety of macroscopic behaviors.

45.4.2 Optical Properties of Polyhydroxylated C_{60} Fullerenes

We present now additional experimental and theoretical studies (Guirado-López and Rincón, 2006) in order to explore another probable relevant application that is related to the possibility to have OH-covered C_{60} cages with tunable optical properties. In fact, previous studies (Xie et al., 2004) have found that the adsorption of small molecules on the external surface of C_{60}, or the inclusion of substitutional impurities in the carbon network, can induce dramatic variations in the absorption spectra. Actually, it seems to be possible to accurately control the location of the optical gap in both near-IR or ultraviolet regions by selectively attaching (incorporating) well-defined molecular (atomic) species on the fullerene surface. Similarly, it is reasonable to expect interesting optical properties for highly hydroxylated fullerene structures since, as has been already shown in the previous section, strong perturbations in the energy level distribution around the HOMO are obtained upon OH adsorption, a fact that is expected to induce strong variations in the electronic excitations and on the exact location of the optical gap in these types of carbon materials.

The here-presented absorption spectra was recorded on molecular solutions based on hydroxylated fullerenes [$C_{60}(OH)_{24-28}$, Aldrich], water, and the intrinsically conductive polymer poly(3,4-ethylenedioxythiophene)-poly(styrenesulfonate) (PEDOT-PSS, Baytron P-Bayer). In the case of solutions based on pure fullerene (C_{60}, Aldrich), C_{60} was dissolved in toluene (2.5×10^{-3} M) while aqueous solutions of $C_{60}(OH)_{24-28}$ (2.5×10^{-3} M) were prepared and adjusted to pH 7. The homogeneous mixtures of PEDOT-PSS/hydroxylated-C_{60} were prepared at different weight ratios by keeping the concentration of OH-covered C_{60} (also called fullerenol) constant and increasing the dilution of PEDOT-PSS:H_2O from (1:1 volume ratio) to (1:400 volume ratio), or by increasing the amount of fullerenol at a fixed polymer concentration. IR spectroscopy and mass spectrometry experiments were run to identify the approximate number of OH groups adsorbed on the C_{60} surface. In the former, a Nicolet IR spectrophotometer in the diffuse reflectance mode was used while in the latter ones were carried out using the MALDI technique. Optical absorption was obtained in the wavelength range of 200–2500 nm with a UV-VIS Shimadzu Spectrophotometer 3101 PC equipped with a reflectance integration sphere.

The presence of Na- and O-impurities [leading to the formation of $C_{60}(OH)_{(n-x)}(ONa)_x$ complexes] is possible in the commercial source of fullerenol, which has a pH ~ 9. However, the acidification of the solutions up to pH ~ 1 did not cause major changes on the compound's solubility and/or on the shape of the absorption spectra, with the exception perhaps of a very small broadening or red-shift in the 200–400 nm region.

Optical absorption measurements from solution are shown in Figure 45.5 for C_{60} in toluene (Figure 45.5a), $C_{60}(OH)_{24-28}$ in water (Figure 45.5b), and also for mixtures of PEDOT-PSS/$C_{60}(OH)_{24-28}$ in water (Figure 45.5c). From the figures, we note first that the absorbance of fullerenol solutions is reduced with respect to the one obtained in C_{60} solutions with equivalent concentrations. In the fullerenol samples, the bands are much broader and have a weaker absolute intensity, a fact that could be understood in first approximation as being the result of the loss of icosahedral symmetry in the highly hydroxylated molecular structures. However, as we will see in the following text, more subtle effects related to the precise distribution of the OH groups on the C_{60} surface will also play a fundamental role in determining the optical response.

In general, when comparing with the C_{60} fullerene, the fullerenol spectra above 300 nm appears featureless. No aggregation is expected at this concentration, and both C_{60} and fullerenol molecules exist as monomers in the prepared solutions. This

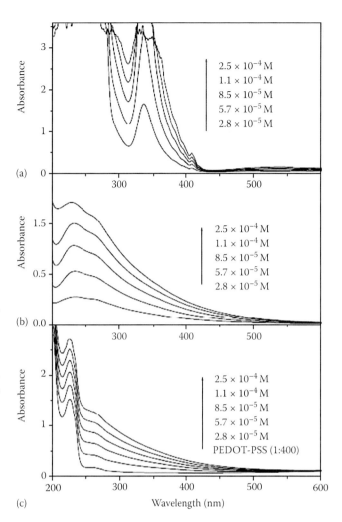

FIGURE 45.5 Measurements of optical absorption from solution. (a) C_{60} in toluene, (b) fullerenol in H_2O, and (c) PEDOT:PPS: fullerenol in H_2O. (Reprinted from Guirado-López, R.A. and Rincón, M.E., *J. Chem. Phys.*, 125, 154312, 2006. With permission.)

is clearly seen from Figure 45.5a and b where, with increasing the C_{60} and fullerenol concentration, no appreciable change in the structure of the absorption spectra is observed. Moreover, from Figure 45.5c we note that the addition of the PEDOT-PSS polymer does not cause any sensible shift on the fullerenol peaks already present (see the region between 250 and 300 nm). It is clear that, in this last case, the absorption spectra shows superposition features of the PEDOT-PSS and fullerenol species, indicating no evidence for strong electronic interactions among the two components. This result is very interesting since PEDOT-PSS is a conjugated polymer where strong electrostatic interactions, hydrogen bonding, and *p*-stacking might be possible. Still, no optical effect is observed, suggesting that the large hydroxylation of the fullerene cage behaves as a protective solvent layer. Therefore, we believe that the chemical effect of surrounding water on the hydroxylated fullerenes is minimum and no additional water will attach to the C_{60} surface or to the dangling OH groups. On the other hand, it is important to remark that, due to these observed trends, we believe that our theoretical modeling from which we would like to discuss some possible explanations of the experimental measurements in terms of gas-phase-like highly hydroxylated fullerenes could be quite close to the experimental situation.

In Figure 45.6, we show the calculated lowest-energy atomic configurations, obtained with the PM3 Hamiltonian, for various $C_{60}(OH)_{26}$ isomers. In all cases, the OH molecules have been found to be directly adsorbed on top of a carbon atom of the fullerene cage having average values for the C–O and O–H bond lengths of 1.40 and 0.95 Å, respectively. In addition, in all our optimized structures, the OH molecules are tilted away from the surface normal with average C–O–H angles that fall in the range of 105°–110°. Notice that in some of the adsorbed phases (mainly in those shown in Figure 45.6a through d), the relative orientation between neighboring OH molecules is strongly correlated, since the positively charged H atoms are mostly pointing toward their nearest negatively charged O species, forming a weakly connected (hydrogen bonded) molecular layer. Interestingly, the existence of a similar microscopic effect has been used to explain the stability of highly hydroxylated Pt surfaces (Bedürftig et al., 1999) where the presence of hydrogen-bonded hexagonal arrangements of OH groups has been proposed. From Figure 45.6, we can see that the strong C–OH interaction results in the considerable distortions of the carbon surface. In all cases, even if we can appreciate sizable outward, inward, and lateral displacements of the C atoms leading to the formation of highly distorted structures, still no C–C bond breakage is obtained, a result that clearly demonstrates the considerable strain and flexibility that can be supported by the C_{60} cage.

The different hydroxylations of the carbon surface shown in Figure 45.6 are expected to produce sizable modifications all

along the electronic spectra, consisting of changes in the ordering of the eigenvalues, splitting of many of the energy levels that are degenerate in the bare C_{60}, together with various HOMO–LUMO gap reopenings and closings. These electronic structure variations are expected to be mainly due to the appearance of several additional electronic states around the Fermi energy induced by the precise location and number of the OH groups, as well as by the considerable distortions produced on the carbon surface, which lowers the symmetry of the molecular structure. It is thus clear that each one of the considered isomers will have a completely different optical response and, as a consequence, accurate total energy calculations as well as theoretical data of the absorption spectra, which can be directly compared to the experimental measurements, could be very useful in order to try to identify the atomic structure of the synthesized fullerene compounds.

In Figure 45.7, we now plot our calculated optical gap (defined as the first-single dipole-allowed excitation) versus the HOMO–LUMO energy separation obtained within the ZINDO methodology and for each one of our considered isomers. In addition, and for the sake of comparison, we include also in the figure our results for the bare C_{60} fullerene. From the figure we can see that, in agreement with the DFT calculations of Xie and coworkers (Xie et al., 2004), in general, the red-shifts (blue-shifts) in the energy location of the optical gap in our fullerene compounds are also strongly correlated (in an almost linear way) with a lowering (increase) of the HOMO–LUMO energy difference. We see from our theoretical calculations that, while our predicted lowest-energy atomic structure (i.e., see the result for configuration I) has its first dipole-allowed excitation near the ultraviolet region, located at ~2.6 eV, or equivalently at ~470 nm; the rest

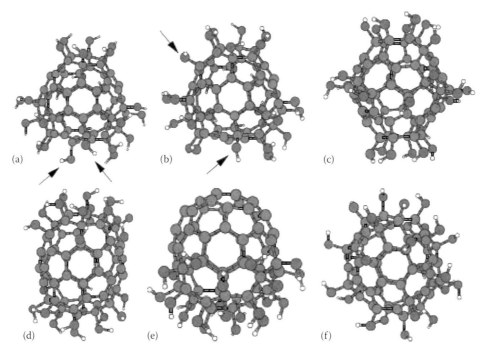

FIGURE 45.6 Optimized PM3 atomic configurations for various $C_{60}(OH)_{26}$ isomers, see text. (Reprinted from Guirado-López, R.A. and Rincón, M.E., *J. Chem. Phys.*, 125, 154312, 2006. With permission.)

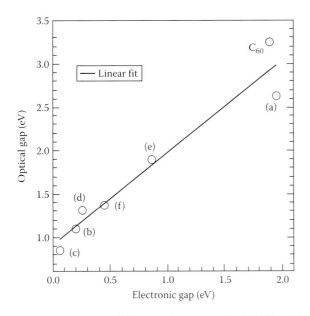

FIGURE 45.7 Behavior of the optical gap versus the HOMO–LUMO energy separation in eV for each one of the considered isomers shown in Figure 45.6. As a reference, results for the bare C_{60} molecule are also shown. (Reprinted from Guirado-López, R.A. and Rincón, M.E., *J. Chem. Phys.*, 125, 154312, 2006. With permission.)

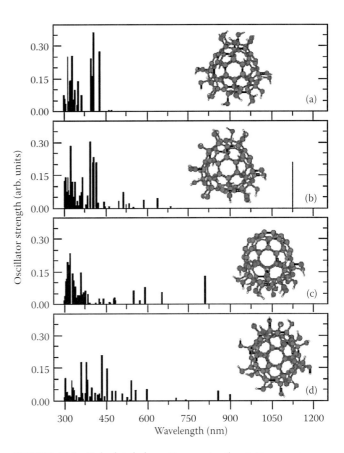

FIGURE 45.8 Calculated absorption spectra (excitation energy versus oscillator strength, in arbitrary units) for representative $C_{60}(OH)_{26}$ isomers (shown in the insets) using the ZINDO method. The vertical bars correspond to the calculated oscillator strength. (Reprinted from Guirado-López, R.A. and Rincón, M.E., *J. Chem. Phys.*, 125, 154312, 2006. With permission.)

of our (less stable) fullerenol molecules are characterized by an optical gap that is much lower in energy (0.85–1.9 eV or 1450–652 nm). The plot, thus, clearly shows that the optical gap for each one of the considered fullerenols differs in position and (as will be seen in the following) in the intensity of their oscillator strength, a fact that we believe is enough to distinguish between all our considered isomers.

In Figure 45.8, we show the calculated excitation spectra obtained also within the ZINDO approach for some representative highly hydroxylated $C_{60}(OH)_{26}$ fullerenes up to 300 nm. From the figures, we note that the spectra of all the hydroxylated fullerenes are very different from one another. In general, near the 300 nm range, all low symmetry molecules (as referred to in the figures) are characterized by broader absorption spectra having reduced intensities, in contrast to the spectral distribution of the C_{60} fullerene, which has only two-dipole allowed transitions located 3.25 and 3.94 eV with oscillator strengths of 0.006 and 1.05, respectively. This behavior is in good agreement with the measured absorption spectra obtained from our solution studies shown in Figure 45.5a (C_{60} in toluene) and 45.5b [$C_{60}(OH)_{24-28}$ in water] where the same trends are observed. In addition, we notice from Figure 45.8 that, in the visible energy range, the position and intensity of the electronic excitations are also very different since our calculations predict, in most of the cases, the existence of some isolated single dipole-allowed transitions located at very low (and different) energies. However, the most interesting feature to remark is that our predicted lowest-energy atomic structure (i.e., see the data of Figure 45.8a) is the only adsorbed phase that has an absorption spectra that closely resembles the behavior of bare C_{60}, having two main absorption

bands centered around ~340 and 420 nm. This result is not in agreement with our measured data shown in Figure 45.5b in which the absorption peak located at 330 nm characteristic of the C_{60} molecule (see Figure 45.5a) is absent.

By comparing Figure 45.5b with Figure 45.8a through d, we notice, thus, that the best agreement between calculated and experimental data must take into account the shape of the curve in the 300–400 nm range. Only the spectra plotted in Figure 45.8c show a monotonic decay in this region, in good qualitative agreement with our solution studies shown in Figure 45.5b. In fact, as already commented earlier, the existence of this isomer was proposed in our previous work to explain the electrical properties of fullerenol-PEDOT/PSS films. Of course, in terms of stability, there are some other structures which are energetically preferred, although none of them benefit from dynamical interconversion.

In Figures 45.9 and 45.10, we have extended the calculation of the excitation spectra of the relevant isomers shown in the insets of Figure 45.8a and c, respectively, to the high-energy region (up to ~210 nm) and where we have included, in order to obtain a more complete visualization of the data, the integration curve assuming a Lorentzian distribution with different values of the

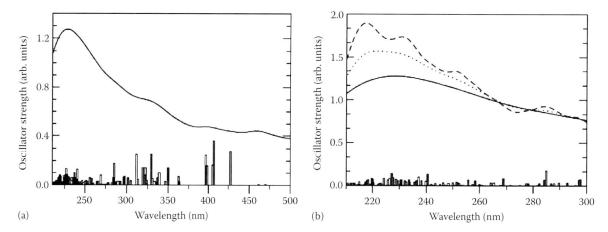

FIGURE 45.9 Calculated absorption spectra (excitation energy versus oscillator strength, in arbitrary units) for the $C_{60}(OH)_{26}$ isomer shown in Figure 45.6a using the ZINDO method. In (a) we show the electronic excitations up to ~210 nm while in (b) we restrict the calculation to the ~210–300 nm range. The vertical bars correspond to the calculated oscillator strength and the lines correspond to a Lorentzian broadening of the electronic excitations with $\Delta = 20$ (continuous line), 10 (dotted line), and 5 (dashed line) eV. (Reprinted from Guirado-López, R.A. and Rincón, M.E., *J. Chem. Phys.*, 125, 154312, 2006. With permission.)

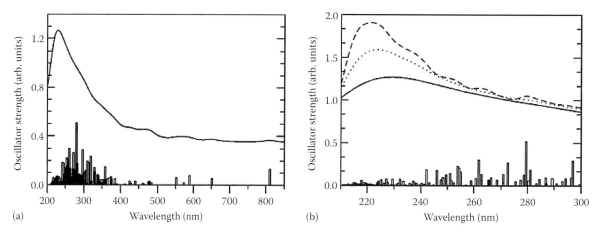

FIGURE 45.10 Same as in Figure 45.9 but for the isomer shown in Figure 45.6e. (Reprinted from Guirado-López, R.A. and Rincón, M.E., *J. Chem. Phys.*, 125, 154312, 2006. With permission.)

broadening parameter, Δ. From both Figures 45.9a and 45.10a, we see that in the 210–300 nm range, there is a large number of electronic transitions, located very close to each other and with (in general) reduced oscillator strength, that lead to the development of a prominent absorption band centered around ~240 nm in good agreement with our measured data. In addition, as expected, the low (and isolated)-energy excitations theoretically obtained in Figure 45.10a are not resolvable after the Lorentzian broadening (with $\Delta = 20$ eV) of the spectra, as is probably the case in our experiments. As a second step, in Figures 45.9b and 45.10b, we analyze in more detail the ~200–300 nm wavelength range for our two considered isomers. Besides the Lorentzian broadening of the spectra using $\Delta = 20$ eV, we include, in addition, calculations with the values of Δ equal to 10 and 5 eV in order to explore further the internal structure of the main absorption peak. From both figures, it is clear that the peak centered around 240 nm is composed of several absorption maxima, being close to each other, and that start to appear as the value of

Δ decreases. It is, thus, clear that this multiple-peak structure will induce an asymmetric shape for the main absorption band, which is actually observed in the experimental data in Figure 45.5b. Unfortunately, our experimental resolution is not enough to provide a more detailed structure of the measured spectra in that energy range, or to verify the existence of the low-energy excitations predicted by the theory.

45.4.3 Electronic Behavior and Stability of Polyhydroxylated Metallofullerenes La@C$_{60}$(OH)$^{+q}_{32}$

As already stated at the beginning of the chapter, the ability of the C_{60} to accommodate many kinds of atoms and small clusters has given rise to another exciting field of research named endohedral fullerene chemistry. Just after the discovery of the C_{60} molecule in a laser-vaporized cluster beam mass spectrum, a magic number feature was also found in the mass spectra corresponding to LaC_{60}. Several structural models for this new molecular compound were proposed and, at that time, it was unclear if the La species were forming part of the carbon network, if it was adsorbed on the external surface, or it was encapsulated within the carbon cavity. However, reactivity experiments between LaC_{60}^+ ions and H_2, O_2, NO, and NH_3 molecules were the ones that clearly supported the assumption that the highly reactive metal atom was protected from the surrounding gases and that was indeed trapped inside the C_{60} cage (Weiss et al., 1988). A new notation was proposed soon after and, in our days, the symbol @ is used to indicate that atoms listed to the left of the @ symbol are encaged in the fullerenes.

Actually, at the time of their synthesis, fullerenes containing an encapsulated La atom, also known as endohedral metallofullerenes, were found to be, in general, highly unstable in practical applications due to their insolubility and extremely air sensitivity. However, later on, the functionalization of its surface with the use of a large variety of small molecules, the most common being also the hydroxyl species (Liu et al. 2000) as well as carboxilic groups (Bolskar et al. 2003), was achieved opening thus new opportunities for potential applications in biomedical and nanomaterials science. In particular, the use of endohedral polyhydroxylated metallofullerenes has been proposed by several research groups as potential MRI contrast agents since the permanent encapsulation of the highly toxic metal atom is expected to ensure an especially safe in vivo use.

In this section, we analyze the stability and electronic properties of both neutral and positively charged polyhydroxylated La@$C_{60}(OH)^{+q}_{32}$ ($q = 0, 1, 2, 4, 6,$ and 8) metallofullerenes (Rodríguez-Zavala et al., 2008) by putting special emphasis on the influence of different spatial distributions of the OH groups on the carbon surface, as well as on the charge state q of the cages, on the energetics and lowest-energy structure of our considered isomers. In the first step, the structural properties of our here-considered neutral and positively charged polyhydroxylated metallofullerenes will be obtained within the DFT approach using the ultrasoft pseudo-potential approximation for the electron–ion interaction and a plane-wave basis set for the wave functions, as implemented in the PWscf code (Baroni et al., 2007). For all our considered structures, the cutoff energy for the plane-wave expansion is taken to be 476 eV. A cubic super-cell with a side dimension of 35 Å was employed in the calculations and the Γ point for the Brillouin zone integration. In all cases, we use the Perdew–Wang (PW91) gradient-corrected functional, and we perform fully unconstrained structural optimizations for all our considered isomers using the conjugate gradient method. In the second step, we will use the previously obtained low-energy molecular geometries to perform additional single-point DFT calculations with the help of the GAUSSIAN03 software (Frisch et al., 2004) in order to obtain the electronic structure, the charge transfer, and the relative stability between all our considered isomers. In this case, the Kohn–Sham equations will be solved by also considering the Perdew–Wang expression for the exchange-correlation potential together with the Stevens–Basch–Krauss effective core potential (ECP) triple-split basis set (Stevens et al.,

1984), CEP-121G, which is a good compromise between computational costs and accuracy.

We believe that the previous two-step PWscf/GAUSSIAN03 hybrid procedure will help us avoid possible inaccuracies related to the periodic repetition of images in the super-cell approach. This is expected to be particularly the case when calculating very delicate quantities such as the electronic structure and the total energy in our highly charged fullerene compounds. Actually, even if in the PWscf methodology the study of charged systems is efficiently handled by applying a jellium background charge to maintain the charge neutrality, the use of the GAUSSIAN03 software is expected to give more realistic values for the total energy of the fullerene compounds.

We start our study by determining the lowest-energy atomic configurations for some possible highly hydroxylated $C_{60}(OH)_{32}$ isomers. The choice of these type of multihydroxylated fullerene compounds is motivated by the recent experimental work of Xing and collaborators (Xing et al., 2004) that have synthesized highly stable and highly soluble OH-covered C_{60} molecules with a number of hydroxyl groups that vary in the range of 30–42 OH molecules.

Of course, for the previous degree of surface coverage, it is prohibitively time consuming to categorize the many possibilities in which 32 OH groups can be adsorbed on the C_{60} surface. However, following the results of the previous sections in which we have addressed the stability and structural properties of low and medium hydroxylated $C_{60}(OH)_x$ fullerenes, we have decided to analyze the lowest-energy atomic configurations, but now at the pseudo-potential density functional level of theory, of three chosen contrasting adsorbed arrays, namely: (1) assuming the existence of small hydroxyl islands on the carbon surface (which has been previously found as a highly stable atomic array), (2) considering a complete aggregation of the OH species on one side of the C_{60} cage, and (3) assuming the adsorption of 32 OH groups homogeneously distributed on the fullerene surface. The resulting lowest-energy structures for the previously proposed initial atomic configurations are shown in Figure 45.11. Interestingly, and as already obtained in our previous work (Guirado-López and Rincón, 2006), we found that the formation of small molecular islands on the C_{60} surface (see Figure 45.11a) corresponds to the most stable adsorbed phase. In this case (and as found before within the PM3 Hamiltonian), we obtain the formation of a highly orientationally ordered OH molecular overlayer, being characterized by the existence of localized networks of hydrogen bonds between neighboring OH groups (see the dotted lines). Notice that this orientational order (in which the positively charged H atoms are always pointing toward their nearest negatively charged O species) is maintained in the segregated phase (see Figure 45.11b); however, it is, in contrast, considerably lost in the (less stable) homogeneous distribution of OH groups (Figure 45.11c), being thus an important factor for determining the structural stability in these kind of systems.

In Figure 45.12, we show the lowest-energy atomic configurations for the fullerene isomers shown in Figure 45.11 but now containing an encapsulated La atom. First, it is important to

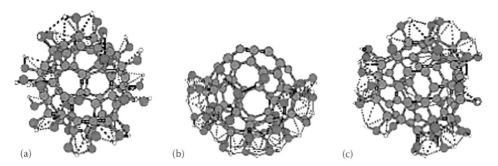

   (a)   (b)   (c)

FIGURE 45.11 Illustration of the lowest-energy atomic configurations for three representative $C_{60}(OH)_{32}$ isomers (see text). (Reprinted from Rodríguez-Zavala, J.G. et al., *Phys. Rev. B.*, 78, 155426, 2008. With permission.)

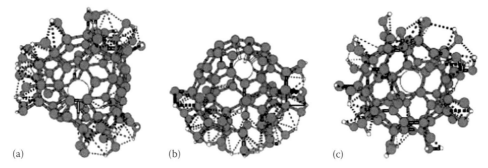

   (a)   (b)   (c)

FIGURE 45.12 Illustration of the lowest-energy atomic configurations for our considered $La@C_{60}(OH)_{32}$ compounds (see text). (Reprinted from Rodríguez-Zavala, J.G. et al., *Phys. Rev. B.*, 78, 155426, 2008. With permission.)

comment that the inclusion of the La atom does not modifies the relative stability between all our considered isomers, being again the most stable $La@C_{60}(OH)_{32}$ molecular compound—the one in which a patchy behavior for the OH groups on the carbon surface is obtained (see Figure 45.12a). From the figures we also see that the hydrogen-bonded network stabilizing the molecular overlayers, shown in Figure 45.11, is still stable after the endohedral doping (even in regions around the La adsorption site) (see Figure 45.12a and b). In fact, the OH molecular coating is characterized by R_{C-O} and R_{O-H} bond lengths, which vary in the range of 1.41–1.46 and 0.98–1.01 Å, respectively, together with C–O–H angles of 103°–106°, all of them being of the order of the values found in the empty cages (see Figure 45.11).

However, the most important feature to remark from Figure 45.12 is that the encapsulated La atom clearly changes its endohedral location when the structure of the molecular overlayer is modified. From Figure 45.12a and b, we can see that the La atom prefers to be adsorbed in the uncovered regions of the C_{60} surface, with La–C bond lengths that vary in the range of 2.43–2.84 Å. In fact, this trend is corroborated in Figure 45.12c where, for a homogeneous distribution of hydroxyl groups on the C_{60} surface, extended uncovered carbon regions are difficult to find. In that case, notice that the encapsulated La atom thus prefersto be placed far away from the internal surface (being located near the center of the structure), the closets La–C interatomic distance being equal to 2.71 Å. It is thus evident that for different coverages and geometrical details of the adsorbed molecular overlayer, we will have contrasting guest–host interactions (and a different dynamic behavior for the confined La atom) that surely

will be of fundamental importance to understand the electronic behavior of these molecular compounds. Finally, we believe that the previous sizable variations in the La–C interatomic distances found in Figure 45.12 could be reflected in IR spectroscopy measurements of these kinds of samples. Of course, a broad distribution of R_{La-C} values is expected to lead to different (in position and intensity) La–C vibrational contributions, and could be thus very helpful to better identify the structural features of these molecular compounds.

We always found in all the structures shown in Figure 45.12 a sizable La→C charge transfer. However, we have obtained that the amount of transferred charge varies as a function of the adsorbed configuration on the inner surface. Actually, by performing a natural population analysis (which is better than the Mulliken method in the case of metallofullerenes), we have found that the charge state of the La atom is equal to +1.96, +2.22, and +2.01 in Figure 45.12a through c, respectively. Of course, stripped of ~2 electrons, the metal ion is naturally always attracted by the electron-rich portions of the fullerene surface which, as shown in Figure 45.12, are defined by the uncovered carbon regions of the molecular compounds.

Furthermore the previous charge transfer induces, as expected, strong variations in the electronic spectra of the molecules, mainly in the energy region around the HOMO. This is actually what we see from Figure 45.13 where we plot, as a representative example, the spin-polarized energy level distribution for the lowest-energy $La@C_{60}(OH)_{32}$ compound shown in Figure 45.12a. In the figure, we only include the 11 occupied molecular orbitals with the highest energies, and of the 11 unoccupied molecular

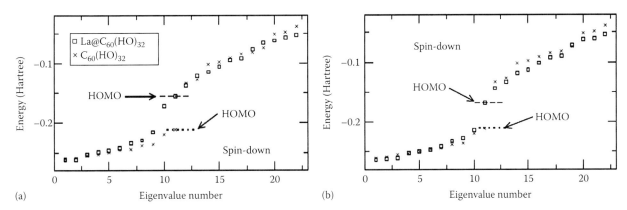

FIGURE 45.13 Comparison of the eigenvalue spectra for the empty and La-doped $C_{60}(OH)_{32}$ fullerenes shown in Figures 45.11a and 45.12a, respectively. In the figure, we show results for both (a) spin-up and (b) spin-down distributions. We only include the 11 occupied molecular orbitals with the highest energies and the 11 unoccupied molecular orbitals with the lowest energies. The energy location of the HOMO for the empty (endohedrally doped) molecular compound is marked with a horizontal dotted (dashed) line. (Reprinted from Rodríguez-Zavala, J.G. et al., *Phys. Rev. B.*, 78, 155426, 2008. With permission.)

orbitals with the lowest energies. For the sake of comparison, we also show the electronic structure of the corresponding empty $C_{60}(OH)_{32}$ cage (Figure 45.11a). From the figure, we have found (if we compare with the La-free molecular compounds) that sizable modifications in the eigenvalue spectra around the HOMO are obtained, consisting in various displacements and splittings (changes in the degeneracy) of the energy levels. It is important to comment that, besides the important role played by the large amount of La→C transferred charge, there are also nonnegligible contributions to the precise details of the energy level distribution shown in Figure 45.13 from the structural distortions in the carbon network induced upon La adsorption. Even if the previous geometrical perturbations are localized in nature, they lower the symmetry of the molecular structure and still reduce further the degeneracy of the electronic spectra.

It is important to state precisely that the energy gap Δ_{HL} of 0.3 eV calculated from Figure 45.13 is obtained between the HOMO of spin-up character (Figure 45.13a) and the LUMO of spin-down nature (Figure 45.13b). The different spin nature of the HOMO and LUMO levels is of course a consequence of the odd number of electrons present in the system due to the inclusion of the La atom, a fact that also defines the existence of a weakly magnetic molecular structure. Actually, the isomers shown in Figure 45.12a and b have a total spin magnetization of $1\mu_B$, while in the molecular compound shown in Figure 45.12c, in which the La-ion is more weakly bonded to the internal carbon surface, the total energy decreases with increasing the spin multiplicity, leading to a total spin moment of $3\mu_B$. Finally, it is important to comment that our lowest-energy La-containing hydroxylated fullerene shown in Figure 45.12a is expected to be still stable with increasing temperature since, as has been demonstrated by Tozzini and coworkers (Tozzini et al., 2000), the thermal stability of fullerene-like structures is strongly correlated with the width of their electronic energy gap.

As is well known, from the spectroscopic point of view, variations in Δ_{HL} between the different isomers shown in Figures 45.11 and 45.12 should be reflected in the optical properties

of the samples. In fact, it has been demonstrated that molecular structures with reduced (large) electronic energy gaps are expected to have their first optical excitations around the IR (ultraviolet) region of the absorption spectra, providing a fingerprint to identify the structure of these kinds of synthesized fullerene compounds. To verify the previous assumption, we present additional time-dependent DFT calculations (with the use of the GAUSSIAN03 methodology at the Perdew–Wang-91/CEP-121G level of theory) on the two most stable atomic configurations shown in Figures 45.11 and 45.12, in order to determine their optical gap (defined as the first single dipole-allowed excitation) which strongly depends on the precise details of the energy level distribution around the HOMO. As stated by Xie and collaborators (Xie et al., 2004), the calculations of the the first single dipole-allowed excitation of the C_{60} molecule using this exchange-correlation functional yields values which are in very good agreement with the ones experimentally obtained from dilute C_{60} solutions (e.g., hexane, benzene, toluene, etc.) and with those reported by other DFT calculations.

In the case of the hydroxylated structures shown in Figure 45.11a we found, in agreement with our previous work, that the formation of small OH molecular domains on the fullerene surface induces a notable red-shift in the optical gap, with respect to the one obtained in the uncovered C_{60}, being now placed at 2.31 eV (~535 nm). However, in the segregated configuration (Figure 45.11b), we observe a more pronounced displacement, the first single dipole-allowed excitation being now located at 1.43 eV (~865 nm). It is clear that our time-dependent DFT calculations reveal big differences between the two considered isomers and that the previous features clearly provide a fingerprint that could be very useful to identify the possible structure of these type of fullerene derivatives.

The situation is much more interesting in the La-containing $C_{60}(OH)_{32}$ compounds shown in Figure 45.12a and b. In both cases, we note that the inclusion of the La atom induces a more dramatic red-shift in the location of the optical gap, being now placed in the far-IR region, i.e., 0.53 eV (~2342 nm) in Figure 45.12a

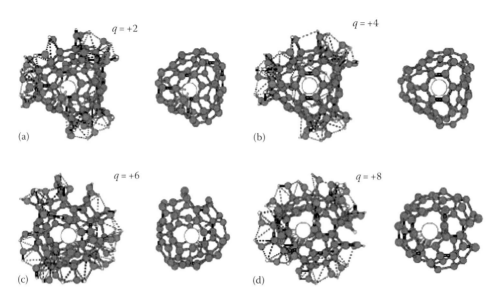

$q = +2$ (a)

$q = +4$ (b)

$q = +6$ (c)

$q = +8$ (d)

FIGURE 45.14 Illustration of the lowest-energy atomic configurations for the La@C_{60}(OH)$^{+q}_{32}$ molecular compound shown in Figure 45.12a having different charge states q. To the right of each one of the equilibrium configurations obtained, we show in addition the structure of the underlying carbon network (by removing all the OH groups from the surface) in order to appreciate the structural deformations induced on the fullerene cavity. (Reprinted from Rodríguez-Zavala, J.G. et al., *Phys. Rev. B.*, 78, 155426, 2008. With permission.)

and 0.37 eV (~3360 nm) for Figure 45.12b. It is thus clear that since C_{60}(OH)$_{32}$ and La@C_{60}(OH)$_{32}$ strongly differ in their optical gaps, ranging from the near-IR to the far-IR region, they may be used not only as novel nano-containers but also like single molecule optical probes as labeling, detection, and identification agents.

In Figure 45.14, we thus present the lowest-energy atomic configuration of the most stable fullerene compound shown in Figure 45.12a but having different charge states q. In the figure, we show our fully optimized La@C_{60}(OH)$^{+q}_{32}$ compounds, where q will be as large as 8, which is a reasonable charge state that can be easily experimentally achieved (Díaz-Tendero et al., 2005). We show in addition, to the right of each one of the equilibrium configurations, only the structure of the corresponding underlying carbon network in order to more clearly appreciate the structural deformations induced on the C_{60} cavity by the charging of the systems. From the figures, the most important features to remark are that similar to the results shown in Figure 45.12, (1) the La-ion changes its endohedral location but now when the q value increases, as already found by us (Rodríguez-Zavala and Guirado-López, 2006), (2) the charging of the fullerene compound (up to $q = +8$) induces the rupture of various C–C bonds of the carbon network, leading to the formation of sizable holes in the C_{60} cavity made of nine-membered rings (see Figure 45.14c and d). However, it is interesting to remark that these holes are not attractive for the encapsulated species. In fact, notice from the figures that the confined La-ion is found to be always adsorbed away from these openings, being still stable in the inside of our partially open carbon cages. Actually, we have performed additional calculations in which we have readsorbed the La atom precisely above all the nine-membered rings shown in Figure 45.14c and d and, during the optimization process, the

La ion always desorbs from this area and moves to a nearest and uncovered carbon region, ending up finally attached in an on-hollow configuration over a six-membered ring.

It is reasonable to expect that with increasing the charge state in the molecular compounds, more complex structural rearrangements will occur, possibly leading to the presence of bigger holes in the fullerene cavity. In addition, with the encapsulation of smaller atomic species, ion-scape through the defect could more probably occur. At this point, we would like to remark that the role of the presence of holes in a fullerene cage on the metal–ion migrations remain mostly unexplored. Of course, additional studies analyzing possible reaction pathways, as well as the estimation of the magnitude of the energy barriers that need to be overcome, are necessary in order to more clearly understand the conditions by which the encapsulated species could be ejected through the defect. Some of these calculations are currently in progress.

45.5 Summary

In this chapter we have presented a systematic theoretical study, by combining semiempirical and DFT approaches, of the structural, electronic and vibrational properties of OH molecules adsorbed on the external surface of C_{60}. A considerable insight on the local organization of the molecules on the carbon surfaces is now available, and in particular, concerning the factors controlling the competition between different growth sequences. At the low coverage regime, we have found that the adsorbed hydroxyl groups aggregate as a molecular island on one side of the C_{60} cage, a result that seems to be of fundamental importance in explaining the stability of C_{60}(OH)$_{12}$–Langmuir monolayers at the air–water interface observed by several authors. In fact, a direct comparison between our calculated vibrational spectra and the reported IR spectroscopy measurements seems to support the hydroxyl island formation on the surface.

Our study has revealed that the excitation spectra of these kind of carbon compounds strongly depend on the precise distribution of the OH groups on the carbon structure. We have clearly shown that the understanding of the optical properties of fullerene-derived materials based solely on the experimental data is difficult. A comparison with theoretical results allows a more reliable analysis of the principal spectra features. In addition, it is clear that the further manipulation of the optical response in these kinds of carbon materials could be achieved by performing the passivation of the fullerene surface with other type of atomic or molecular species, provided that the surface bonding will be strong enough to distort the electronic structure of the C_{60} molecule.

In the case of polyhydroxylated La@C_{60}(OH)$^{+q}_{32}$ compounds, we have obtained that the endohedral location of the La atom strongly depends on the precise distribution of the OH groups on the carbon surfaces. In particular, we have found that there is a favorable attraction between the encapsulated La-ion and the electron-rich (uncovered) regions of the carbon network, a fact that leads to the existence of well-defined adsorbed configurations within the cavities. Interestingly, the previous adsorbed phases are characterized by different La–C bond lengths and by having a contrasting distribution of the energy levels around the highest occupied molecular orbital. In addition, notable variations in the total spin magnetization and in the charge state of the encapsulated species are also found, which are all facts that could be used to better identify the possible atomic configuration of these molecular compounds.

Finally, we have obtained that by positively charging our polyhydroxylated C_{60} metallofullerenes, it is possible to induce the rupture of various C–C of the carbon network, a fact that leads to the formation of sizable holes in the fullerene cavities. However, in all cases, we have found that the encapsulated La-ion was still stable in the inside of our partially open carbon cages. The previous results are expected to be of fundamental importance in medical and biological applications where the permanent encapsulation of highly toxic atomic species is required.

45.6 Future Perspective

More research work is needed in order to fully understand the electronic and structural behavior, as well as to find new possible applications, of polyhydroxylated fullerenes. The combination of theoretical modeling, new synthesis techniques, and the development of new characterization tools has definitively played a fundamental role in understanding the properties of this kind of carbon nanostructures. Certain parameters like temperature and the role of solvent effects need to be quantified in order to better understand the large variety of recently reported solution studies. In addition, the role of the presence of coadsorbed impurities and the reactivity of C_{60}(OH)$_x$ units to additional molecular species needs to be addressed. As more of the surface properties

of fullerenes become known, the applications of this material to devices become more promising. A detailed study of the aggregation behavior of these molecular structures also needs to be performed in order to understand their reported diffusion properties and observed cellular cytotoxicity. In this respect, we must comment that the physiological and medical applications of the endohedral polyhydroxylated metallofullerenes are expected to be mostly important in tracer chemistry uses and contrast agents in biological systems. Also, research could be directed to a study of the influence of additional atoms and small molecules trapped inside the cavities of polyhydroxylated structures together with their migration behavior. The future continues to be, as in the beginning, open and very promising, with progress being limited only by human imagination.

References

Baroni, S., Dal Corso, A., de Gironcoli, S. et al. 2007. Quantum Espresso. http://www.pwscf.org/.

Bedürftig, K., Völkening, S., Wang, Y. et al. 1999. Vibrational and structural properties of OH adsorbed on Pt(111) surfaces. *J. Chem. Phys.* 111: 11147–11154.

Bolskar, R.D., Benedetto, A.F., Husebo, L.O. et al. 2003. First soluble M@C_{60} derivatives provide enhanced acces to metallofullerenes and permit in vivo evaluation of Gd@C_{60}[C(COOH)$_2$]$_{10}$ as a MRI contrast agent. *J. Am. Chem. Soc.* 125: 5471–5478; Laus, S., Sitharaman, B., Tóth, E. et al. 2005. Destroying gadofullerene aggregates by salt addition in aqueous solutions of Gd@C_{60}(OH)$_x$ and Gd@C_{60}[C(COOH)$_2$]$_{10}$. *J. Am. Chem. Soc.* 127: 9368–9369.

Coheur, P.F., Cornil, J., dos Santos, D.A. et al. 2000. Photophysical properties of hexa-functionalized C_{60} derivatives: Spectroscopic and quantum-chemical investigations. *J. Chem. Phys.* 112: 8555–8566; Coheur, P.F., Lievin, J., Colin, R. et al. 2003. Electronic and photophysical properties of C_{60}Cl$_{24}$. *J. Chem. Phys.* 118: 550–556; Xie, R.H., Bryant, G.W., Cheung, C.F. et al. 2004. Optical excitation and absorption spectra of C_{50}Cl$_{10}$. *J. Chem. Phys.* 121: 2849–2851.

Díaz-Tendero, S., Alcamí, M., and Martín, F. 2005. Coulomb stability limit of highly charged C_{60}^{+q} fullerenes. *Phys. Rev. Lett.* 95: 013401-1–013401-4; Batsug, T., Kürpick, P., Meyer, J. et al. 1997. Dirac-Fock-Slater calculations on the geometric and electronic structure of neutral and multiply charged C_{60} fullerenes. *Phys. Rev. B* 55: 5015–5020; Majima, T., Nakai, Y., Tsuchida, H. et al. 2004. Correlation between multiply ionization and fragmentation of C_{60} in 2-MeV Si^{2+} collision: Evidence for fragmentation induced by internal excitation. *Phys. Rev. A* 69: 031202-1–031202-4.

Dugan, L.L., Turetsky, D.M., Du, C. et al. 1997. Carboxyfullerenes as neuroprotective agents. *Proc. Natl. Acad. Sci. USA* 94: 9434–9439.

Ehrler, O.T., Furche, F., Weber, J.M. et al. 2005. Photoelectron spectroscopy of fullerene dianions C_{76}^{-2}, C_{78}^{-2}, and C_{84}^{-2}. *J. Chem. Phys.* 122: 094321-1–094321-8.

Fanti, M., Zerbetto, Z., Galaup, J. P. et al. 2002. $C_{70}Ph_8$ and $C_{70}Ph_{10}$: A computational and solid solution spectroscopy study. *J. Chem. Phys.* 116: 7621–7626; Coheur, P.F., Cornil, J., dos Santos, D.A. et al. 2000. Phtophysical properties of multiply phenylated C_{70} derivatives: Spectroscopic and quantum-chemical investigations. *J. Chem. Phys.* 112: 6371–6381; Riggs, J.E. and Sun, Y.P. 2000. Optical limiting properties of mono- and multiple-functionalized fullerene derivatives. *J. Chem. Phys.* 112: 4221–4230.

Friedman, S.H., DeCamp, D.L., Sijbesma, R.P. et al. 1993. Inhibition of the HIV-1 protease by fullerene derivatives: Model building studies and experimental verification. *J. Am. Chem. Soc.* 115: 6506–6509; Marcorin, G.L., Ros, T.D., Castellano, S. et al. 2000. Design and synthesis of novel [60]fullerene derivatives as potential HIV aspartic protease inhibitors. *Org. Lett.* 2: 3955–3958.

Frisch, M.J. et al., Gaussian 03, Revision C.02, Gaussian, Inc., Wallingford, CT, 2004.

Guirado-López, R.A. and Rincón, M.E. 2006. Structural and optical properties of highly hydroxylated fullerenes: Stability of molecular domains on the C_{60} surface. *J. Chem. Phys.* 125: 154312-1–154312-10.

Husebo, L.O., Sitharaman, B., Furukawa, K. et al. 2004. Fullerenols revisited as stable radical anions. *J. Am. Chem. Soc.* 126: 12055–12064.

Krause, M., Liu, X., Wong, J. et al. 2005. The electronic and vibrational structure of endohedral $Tm_3N@C_{80}$ (I) fullerene-proof of an encaged Tm^{3+}. *J. Phys. Chem. A* 109: 7088–7093; Cardona, C.M., Kitaygorodskiy, A., and Echegoyen, L. 2005. Trimetallic nitride endohedral metallofullerenes: Reactivity dictated by the encapsulated metal cluster. *J. Am. Chem. Soc.* 127: 10448–10453; Cao, B., Suenaga, K., Okazaki, T. et al. 2002. Production, isolation, and EELS characterization of $Ti_2@C_{84}$ dititanium metallofullernes. *J. Phys. Chem. B* 106: 9295–9298; Chen, N., Zhang, E.Y., and Wang, C.R. 2006. C_{80} encaging four different atoms: The synthesis, isolation, and characterizations of $ScYErN@C_{80}$. *J. Phys. Chem. B* 110: 13322–13325.

Kroto, H.W., Heath, J.R., O'brien, S.C. et al. 1985. C_{60} buckministerfullerene. *Nature (London)* 318: 162–163.

Krusic, P.J., Keiser, J., Morton, J.R. et al. 1991. Radical reactions of C_{60}. *Science* 254: 1183–1185.

Liu, W.J., Jeng, U., Lin, T.L. et al. 2000. Adsorption of dodecahydroxylated-fullerene monolayers at the air-water interface. *Physica B* 283: 49–52.

Mikawa, M., Kato, H., Okumure, M. et al. 2001. Paramagnetic water-soluble metallofullerenes having the highest relaxivity for MRI contrast agents. *Bioconjugate Chem.* 12: 510–514; Tóth, E., Bolskar, R.D., Borel, A. et al. 2005. Water-soluble gadofullerenes: Toward high-relaxivity, pH-responsive MRI contrast agents. *J. Am. Chem. Soc.* 127: 799–805.

Nishibori, E., Iwata, K., Sakata, M. et al. 2004. Anomalous endohedral structure of $Gd@C_{82}$ metallofullerenes. *Phys. Rev. B* 69: 113412-1–113412-4; Dolmatov, V.K. and Manson, S.T. 2006. Photoionization of atoms encapsulated in endohedral ions $A@C_{60}^{\pm z}$. *Phys. Rev. A* 73: 013201-1–013201-6; Yamada, M., Feng, L., Wakahara, T. et al. 2005. Synthesis and characterization of exohedrally silylated $M@C_{82}$ (M = Y and La). *J. Phys. Chem. B* 109: 6049–6051.

Rincón, M.E., Hu, H., Campos, J., and Ruiz-García, J. 2003. Electrical and optical properties of fullerenol Langmuir-Blodgett films deposited on polyaniline substrates. *J. Phys. Chem. B* 107: 4111–4117.

Rodríguez-Zavala, J.G. and Guirado-López, R.A. 2004. Structure and energetics of polyhydroxylated carbon fullerenes. *Phys. Rev. B* 69: 075411-1–075411-13.

Rodríguez-Zavala, J.G. and Guirado-López, R.A. 2006. Stability of highly OH-covered C_{60} fullerenes: Role of coadsorbed O impurities and of the charge state of the cage in the formation of carbon-opened structures. *J. Phys. Chem. A* 110: 9459–9468.

Rodríguez-Zavala, J.G., Padilla-Osuna, I., and Guirado-López, R.A. 2008. Modifying the endohedral La-ion location in both neutral and positively charged polyhydroxylated metallofullerenes. *Phys. Rev. B* 78: 155426-1–155426-10.

Ruoff, R.S., Tse, D.S., Malhotra, R. et al. 1993. Solubility of fullerene (C_{60}) in a variety of solvents. *J. Phys. Chem.* 97: 3379–3383.

Schinazi, R.F., Sijbesma, R., Srdanov, G. et al. 1993. Synthesis and virucidal activity of a water-soluble, configurationally stable, derivatized C_{60} fullerene. *Antimicrob. Agents Chemother.* 37: 1707–1710.

Scrivens, W.A. and Tour, J.M. 1993. Potent solvents for C_{60} and their utility for the rapid acquisition of ^{13}C NMR data for fullerenes. *J. Chem. Soc. Chem. Commun.* 15: 1207–1209.

Sitharaman, B., Asokan, S., Rusakova, I. et al. 2004. Nanoscale aggregation properties of neuroprotective carboxyfullerene (C_3) in aqueous solution. *Nano Lett.* 4: 1759–1762; Sitharaman, B., Bolskar, R.D., Rusakova, I. et al. 2004. $Gd@C_{60}[C(COOH)_2]_{10}$ and $Gd@C_{60}(OH)_x$: Nanoscale aggregation studies of two metallofullerene MRI contrast agents in aqueous solution. *Nano Lett.* 4: 2373–2378; Sayes, C.M., Fortner, J.D., Guo, W. et al. 2004. The differential cytotoxicity of water-soluble fullerenes. *Nano Lett.* 4: 1881–1887.

Stevens, W., Basch, H., and Krauss, J. 1984. Compact effective potentials and efficient shared-exponent basis sets for the first- and second-row atoms. *J. Chem. Phys.* 81: 6026–6033; Stevens, W.J., Krauss, M., Basch, H. et al. 1992. Relativistic compact effective potentials and efficient, shared-exponent basis sets for the third-, fourth-, and fifth-row atoms. *Can. J. Chem.* 70: 612–630; Cundari, T.R. and Stevens, W.J. 1993. Effective core potential methods for lanthanides. *J. Chem. Phys.* 98: 5555–5565.

Tokuyama, H., Yamago, S., Nakamura, E. et al. 1993. Photoinduced biochemical activity of fullerene carboxylic acid. *J. Am. Chem. Soc.* 115: 7918–7919.

Tozzini, V., Buda, F., and Fasolino, A. 2000. Spontaneous formation and stability of small GaP fullerenes. *Phys. Rev. Lett.* 85: 4554–4557.

Weiss, F.D., Elkind, J.L., O'brien, S.C. et al. 1988. Photophysics of metal complexes of spheroidal carbon shells. *J. Am. Chem. Soc.* 110: 4464–4465.

Xie, R.H., Bryant, G.W., Sun, G. et al. 2004. Tuning the electronic properties of boron nitride nanotubes with transverse electric fields: A giant dc Stark effect. *Phys. Rev. B* 69: 201401-1–201401-4.

Xing, G., Zhang, J., Zhao, Y. et al. 2004. Influence of structural properties on stability of fullerenols. *J. Phys. Chem. B* 108: 11473–11479.

Zhou, X., Liu, J., Jin, Z. et al. 1997. Solubility of fullerene C_{60} and C_{70} in toluene, o-xylene, and carbon disulfide at various temperatures. *Fullerene Sci. Technol.* 5: 285–290.

Structure and Vibrations in C_{60} Carbon Peapods

Abdelali Rahmani
Université MY Ismail

Hassane Chadli
Université MY Ismail

46.1 Background

The discovery of C_{60} fullerenes in 1985 [1] added a new dimension to the knowledge of carbon science, and the discovery of multi-walled carbon nanotubes (MWCNTs) in 1991 [2] and single walled carbon nanotubes (SWCNTs) in 1993 [3] also added to the knowledge of technology. These materials prevail over science and technology at nanoscale levels and become a hot topic for researchers and industrialists. Resembling tubes of rolled-up graphite of about a nanometer in diameter and several micrometers or more, these structures have substantially contributed to the recent advances in nanotechnology as they exhibit a remarkably complex mixture of physical properties, including extreme mechanical resilience, exceedingly large aspect ratios, and superior field emission properties. The particular technological importance is that nanotubes come in both semiconducting and metallic varieties, making them a natural basis for developing nanoelectronic applications.

On the other hand, due to their size and geometry, SWCNTs provide a unique opportunity for nanoscale engineering of novel one-dimensional systems, created by self-assembly of molecules inside the SWCNT hollow core. In particular, it has been shown that the intercalation of electron donors or acceptors [4–6] into single wall carbon nanotube bundles could dramatically modify the electronic properties of these objects. It has been experimentally shown that fullerenes can be inserted into SWCNTs. This class of nanomaterials has been dubbed as "nanopeapods,"

reflecting structural similarities to real peapods. These structures were observed while doing high-resolution transmission electron microscopy (TEM) imaging on nanotube material produced by the "pulsed laser vaporization" technique, and the interior molecules reported to be spaced 0.3 nm from the nanotube walls, as seen in Figure 46.1. The interpretation was that the encapsulated molecules were C_{60} spaced with the carbon–carbon (C–C) van der Waals (vdW) distance between each molecule and the walls of the tube as well as between the molecules themselves. The composite nature of peapod materials raises an exciting possibility of a nanoscale material having a tunable structure that can be tailored to a particular electronic functionality. Rather, large molecules are expected to be inserted into the hollow core of a nanotube that has been shown to present very stable adsorption sites [7]. The study of peapods stands within the fascinating field of systems in a confined geometry [8–11]. Peapods structural analysis can be performed using transmission electron microscopy [12,13] or electron diffraction [14,15], or on macroscopic assemblies, using Raman spectroscopy [13,16,17] or x-ray scattering [17–19].

The first targeted fabrication of these hybrid structures was reported by Smith and Luzzi in 2000 [20]. SWCNT material is acid etched to open the ends and create holes in the sidewalls of the tubes. After heating to 225°C in vacuum to remove chemical residue, a drop of C_{60} is added to the mixture and dried in air. The sample is then annealed for several hours during which time filling occurs. In 2001, peapods were produced in a very

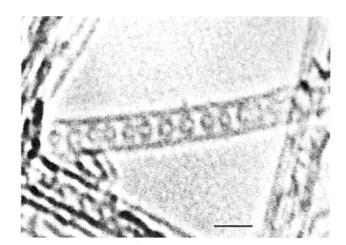

FIGURE 46.1 A TEM micrograph of a C_{60} peapod. The diameter and center-to-center spacing of the internal shells are consistent with the chain of C_{60} molecules. The nanotube is surrounded by a vacuum. Scale bar, 2.0 nm. (From Smith, B.W. et al., *Nature*, 396, 323, 1998. With permission.)

high yield close to (90%) [21,22]. These materials represent a new class of hybrid system between C_{60} and SWCNT, wherein the encapsulated C_{60} peas and the SWCNT pod are vdW bonded as the interaction between the inside fullerene molecules and the host tube was found not strong enough to form ionic or covalent bond [13]. In general, peapods are observed organized into bundles in which they are packed together through vdW intertube interactions, a structure that is clearly visible on TEM pictures [13,17,18]. Efforts led to the synthesis of high-quality 1D fullerene crystals inside SWCNTs [12,13]. Electron microscopy images [12] of peapod have revealed the one-dimensional character of the C_{60} chain inside the carbon nanotube. The formation mechanism of the peapods has not been completely comprehended [23,26], and the structure of C_{60} chain inside the nanotube is not yet well known in the precision limit of the present experimental facilities. Experiments [15,25,26] correlating with recent calculations [27–29] have stated that the center-to-center distance between two C_{60}s in peapods is smaller by 3%–4% in three-dimensional crystals but larger than in solids based on polymerized fullerenes. High resolution transmission electron microscopy (HRTEM) observations have stated that C_{60}s can be unequally separated [30,31].

A natural question to ask when dealing with C_{60} filled carbon nanotubes is how this filling affects the vibrational and electronic properties of the pristine nanotubes. The properties of both nanotubes and C_{60} molecules are quite well understood, [32,33]. A lot of theoretical and experimental studies have presented peapods, and several interesting properties have been predicted or observed. For example, the scanning tunneling microscope (STM) image captured the density of states of the conduction band oscillating along the nanotube with a period commensurate with the underlying C_{60} arrays [10,34]. On the other hand, the conductivity of nanopeapods was found to increase in comparison with that of pristine nanotubes [11,35].

A temperature-induced transition from the p- to n-type character was also observed [36]. Theoretical calculations predict that the electronic states near the Fermi level are substantially modified through the C_{60}-SWCNT interaction [27]. Due to their tunable electronic properties, the potential applications of peapods range from high temperature superconductor [37] to memory element [38] and nanometer-sized container for chemical reactions [30]. The demonstrated ease of fullerene peapod formation suggests the possibility of encapsulating other nanostructures, including polymers such as polyacetylene [37].

Raman spectroscopy is known to be one of the most efficient tools for investigating the vibrational properties of carbon nanotubes in relation to their structural and electronic properties [39]. The Raman spectrum of SWCNT is dominated by the so-called radial breathing modes (RBM) below 350 cm^{-1} and by the tangential modes (TM) in the high frequencies region (1400–1600 cm^{-1}). At a theoretical level, the nonresonant Raman spectra of SWCNTs have been calculated within the bond polarizability model [40,41]. Many experimental and theoretical Raman studies have shown that the RBM follows a straightforward dependence on the tube diameter that can be used to determine the distribution of the tube diameters in samples under study [42]. For the C_{60} molecule, a large variety of theoretical methods have been applied to the calculation of the internal modes and to the determination of their Raman activity [43–48]. Two modes belong to the nondegenerate Ag species, while eight modes that are five-fold degenerate of Hg symmetry are Raman active. The Ag(2) pentagonal pinch located at about 1468 cm^{-1} is the most prominent vibration. Raman measurements of a synthesized peapod have been performed by Kataura and coworkers [17]. The authors recorded Raman spectra at helium temperature to derive the peapods response before polymerization. A new Raman active mode located at 90 cm^{-1}, which is close to the interball C_{60} dimer mode, was observed at low temperature. In Chadli et al.'s previous work [52], within the bond polarizability model, the nonresonant Raman spectra of linear and zigzag chain of C_{60}s monomers inside SWCNTs have been calculated. They have shown that the splitting of the non-degenerated radial breathing-like mode (RBLM) that was experimentally observed in real peapods samples can be related to either a low C_{60} filling factor or the presence of zigzag C_{60} chains inside the nanotubes.

Total energy calculations of C_{60} peapods suggest that the smallest tube diameter for encasing C_{60} molecules inside SWCNT is around the diameter corresponding to (10, 10) or (9, 9) tubes [26]. Hodak and Girifalco have shown that the structure of the C_{60} molecules inside nanotubes is diameter dependent [53,54]. Raman experiments on SWCNTs encasing the fullerenes (C_{60}, C_{76}, C_{78}, and C_{84}) have been reported [21,22]. For tube diameters between 1.45 and 1.76 nm, the authors observed that the RBLM frequencies are downshifted compared to the RBM of empty SWCNTs. For smaller diameters, close to 1.37 nm, they observed two RBLM components that are upshifted and downshifted, respectively, with regard to the position of the RBM of the empty tube. Pfeiffer et al. [55] measured all the fundamental Raman lines of the encaged C_{60} peas except the Hg(8) mode. They observed that the totally symmetric Ag modes of C_{60} peas exhibit a splitting into two components. They attributed this splitting to the presence of both moving and static fullerenes inside the tubes. The forces that can bind C_{60} molecules inside the nanotube may be vdW types, since the typical polymer lines are not observed in some peapod spectra [21]. Other studies led to the proposal that peapods could be filled with dimer, trimer, and polymer structures.

46.2 Structure and Dynamics in Peapods

46.2.1 Physical Structure

46.2.1.1 C_{60} and Its Derivatives

C_{60} is a closed-cage molecule consisting of 60 carbon atoms. The molecular structure can be described as a truncated icosahedron, with 12 pentagonal and 20 hexagonal faces (see Figure 46.2). In the C_{60} molecule, the carbon nuclei reside in a sphere of about 0.7 nm diameter, as measured with NMR [57]. Two different (C–C) bond lengths exist in C_{60}, and they measure 0.140 and 0.146 nm [58]. The 30 short bonds are double bonds and lie on the edges that are shared by two hexagons. The longer ones, which are single bonds, lie at the 60 edges shared by an hexagon and a pentagon.

In the solid phase at ambient conditions of pressure and temperature, the molecules have their centers of mass on a face-centered cubic (fcc) lattice (space group Fm3m). The nearest-neighbor distance is 1.0 nm, with an intermolecular separation (0.292 nm) similar to the spacing between layers in graphite (0.335 nm). Due to the relatively weak nature of vdW interactions, the constituent molecules rotate freely at room temperature. As the temperature is lowered below 260 K, rotations begin to freeze out, and the buckyballs orient themselves relatively to one another, leading to a lowering of the crystal symmetry to that of a simple cubic structure.

C_{60} is known to form polymeric compounds under certain conditions. Of particular interest is the crystalline polymerized phases that can be obtained by high pressure and high temperature treatment (HPHTT) of fullerite C_{60}. Under this treatment, the neutral C_{60} molecules polymerize by forming between one and six [2,2] cycloaddition interball connections in which parallel

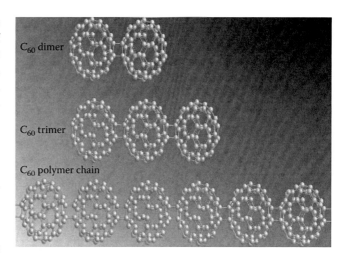

FIGURE 46.3 Different polymers structures: orthorhombic structure in which each molecule is covalently bonded to the two nearest neighboring molecules and dimer.

(6,6) bonds on adjacent C_{60}s are connected by a pair of C–C bonds. The polymerized structures obtained by this route include dimers (D), orthorhombic (O) one-dimensional (1D) chains (Figure 46.3), tetragonal, and rhombohedral two dimensional (2D) lattices. [59,61,62]. The bonding mechanism is the combination of parallel double bonds on adjacent molecules into a four-member ring (i.e., the connection ring between two molecules including 4 C atoms, see Figure 46.3) [63]. One effect of this reaction is the reduction of the intermolecular spacing from 1.0 to 0.9 nm.

46.2.1.2 Carbon Nanotubes

Carbon nanotubes are easily defined as rolled-up fragments of graphene sheets (hexagonal "chicken wire" of carbon atoms, see Figure 46.4). The vector $C_h = na_1 + ma_2$ that connects equivalent atomic positions in the unrolled sheet is called the chiral (or wrapping) vector, often denoted as an integer pair (n, m). Not all the ways of rolling a graphene sheet produce a unique tube

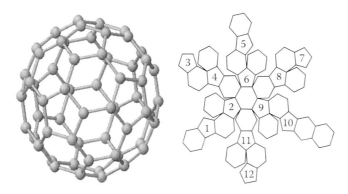

FIGURE 46.2 Schematic view of the C_{60} molecule. The 12 pentagons necessary to comply Euler Rule are numbered.

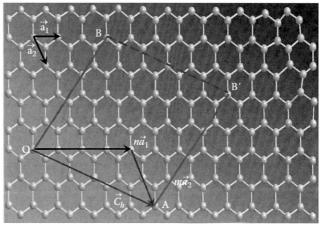

FIGURE 46.4 Chiral vector on a graphene sheet. The rectangle OABB' defines the unit cell for the nanotube. The figure is constructed for an $(n, m) = (4, 2)$ nanotube.

structure; a naming convention is adopted in which $n \geq m \geq 0$. This convention uniquely identifies every possible nanotube, including its diameter $D = \frac{\sqrt{3}}{\pi} a_{c-c}(n^2 + nm + m^2)^{1/2}$ measured from the atomic centers (a_{c-c} as the nearest-neighbor atomic distance, 0.142 nm for graphite).

The chiral angle is defined as $\theta = \tan^{-1}\left[\sqrt{3}m/(2n+m)\right]$, and it is the screw angle formed by the hexagonal chains in the $\mathbf{a}_1$ direction relative to the wrapping vector. Tubes with indices $(n, 0)$ are traditionally called "zigzag" tubes and have $\theta = 0°$, meaning the hexagonal chains form closed loops around the circumference (see Figure 46.4). On the other extreme, (n, n) "armchair" tubes have a $\theta = 30°$ chiral pitch, resulting in the rows of hexagons oriented in the $\mathbf{a}_2 - \mathbf{a}_1$ direction being parallel to the tube axis. Tubes not falling into either of the two classes are simply called "chiral" tubes, and have $30° > \theta > 0°$. Examples of nanotubes with different chiralities are shown in Figure 46.5. A unit cell of a nanotube is a cylinder with circumference given by the length of the chiral vector, and a length defined by a so-called T vector, perpendicular to $\mathbf{C}_h$, attached to the same origin, and finishing at the nearest lattice point intersected in its direction. The simplest nanotubes, as described above, are single-walled SWCNTs. Such nanotubes can stack one inside the other, creating MWCNTs, with the special case of double-walled carbon nanotubes (DWCNT).

The most common methods for producing appreciable quantities of high purity nanotube material have electric arc-discharge between catalyst-impregnated carbon rods [56] and pulsed laser vaporization of a graphite target [64]. More recent work using chemical vapor deposition [65] of carbon vapor over a substrate with catalyst particles on its surface has shown great promise for controlled growth of nanotube networks, and it may be a useful technique for device fabrication.

46.2.1.3 Peapod

A carbon peapod, denoted $C_{60}@(n, m)$, consists of C_{60} molecules trapped inside a (n, m) single wall carbon nanotube host, and this general name has been applied to the whole family of such materials. It has been proposed that encapsulation of C_{60} molecules can be through either the ends of oxidatively opened SWCNTs or the defects in their sidewalls. The different formation mechanisms of the carbon peapods have been extensively investigated and discussed by Ulbricht and Hertel [66]. They showed that encapsulations through tube ends were more likely than encapsulations through the sidewall defects on the SWCNTs. Configuration of C_{60} molecules in the nanotubes has already proved that it depends on the diameter of the nanotubes [53,54]. The molecular interaction between adjacent fullerene molecules can be affected by the interaction with the surrounding nanotube, and therefore specific arrangements of fullerene molecules are envisioned within SWCNTs. Hodak and Girifalco [54] have studied the minimum energy configurations of C_{60} molecules inside carbon nanotubes. Simulations for nanotubes with radius between 0.627 and 1.357 nm showed that 10 different phases exist in this size range. The resulting phases are listed in Table 46.1 and displayed in Figure 46.6.

Chadli et al. [52] have calculated the minimum energies of Lennard–Jones C_{60}-SWCNT and $C_{60} - C_{60}$ interactions in order to obtain the optimal configuration of the C_{60} molecules inside the nanotubes. The selected values of Lennard–Jones potential parameters are ($\epsilon = 2.964$ meV and $\sigma = 0.3407$ nm. These values were optimized to correctly describe the contribution of the energy of vdW in the energy of cohesion of the molecules C_{60} in monomeric phase [24]. Calculations were performed with respect to five variable parameters: the distance between the centers of adjacent C_{60} molecules (d_{cc}), the distance between C_{60} center and the nanotube axis (d_{ct}), the relative Euler angles defining the orientation of the C_{60} molecules inside the tubes (α, β, γ), and a parameter defining the relative translation of the fullerene molecules along the tube axis (t_r) (see Figure 46.7). For example, in the case of infinite $C_{60}@(10, 10)$ and $C_{60}@(13, 13)$ peapods, portions of $(10, 10)$ and $(13, 13)$ nanotubes of lengths 3.94 and 3.44 nm, respectively, are considered as a supercells. In those cases, the supercell contains four C_{60} molecules inside the tubes.

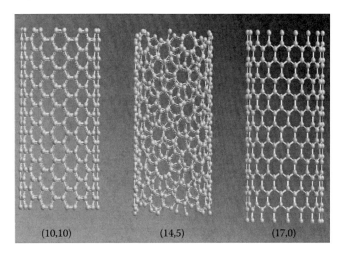

FIGURE 46.5 Top and end view of the three dimensional structure of non-chiral (10, 10), (17, 0) and chiral (14, 5) SWCNTs.

TABLE 46.1 Phases Behavior of C_{60} Molecules Inside Nanotubes for Nanotubes with Radius $r < 13.57$

Radius (Å)	Phase
6.27–7.25	Linear chain
7.25–10.8	Zigzag
10.8–11.15	Double helix
11.15–11.4	Layers of 2 molecules
11.4–12.05	Triple helix
12.05–12.45	Layers of 3 molecules
12.45–13.0	Layers of 2 molecules
13.0–13.3	4-Stranded helix
13.3–13.5	4-Stranded helix
13.5–14.05	Layers of 4 molecules

Source: Hodak, M. and Girifalco, L.A., *Phys. Rev. B*, 67, 075419, 2003.

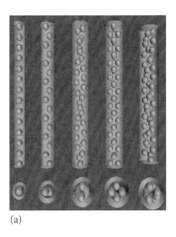 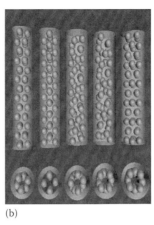

(a)   (b)

FIGURE 46.6 (a) Phases of C_{60} molecules: the single chain phase [(10, 10) tube], the zigzag phase [(13, 13) tube], double-helix phase (tube with radius 10.96 Å, the phase of two-molecule layers (tube with radius 11.33 Å), and triple helix [(17, 17) tube], from left to right. (b) Phases of C_{60} molecules: the phase of three-molecule layers, the phase of two-molecule layers, two chiral phases of 4 strands, and an achiral phase of 4 molecules. For nanotubes with radii 12.21, 12.89, 13.23, 13.33, and 13.57 Å, from left to right. (From Hodak, M. and Girifalco, L.A., *Phys. Rev. B*, 67, 075419, 2003. With permission.)

TABLE 46.2 Optimized Structural Parameters of the C_{60} Molecules Inside Tube for Different Diameters and Chiralities

Tube Index (n, m)	Tube Diameter (nm)	C_{60}-Tube Distance (nm)	Angle θ (deg)	$C_{60} - C_{60}$ Distance (nm)
(10, 10)	1.36	0.323	180	1.003
(15, 4)	1.41	0.355	180	1.003
(11, 11)	1.49	0.321	164	1.006
(18, 4)	1.59	0.309	150	1.003
(12, 12)	1.63	0.306	145	1.008
(21, 0)	1.64	0.307	143	1.008
(15, 10)	1.70	0.30	134	1.006
(22, 0)	1.72	0.302	132	1.006
(13, 13)	1.76	0.301	127	1.007
(23, 0)	1.80	0.30	121	1.003
(18, 9)	1.86	0.298	112	1.003
(14, 14)	1.90	0.30	108	1.006
(25, 0)	1.96	0.298	99	1.006
(19, 10)	2.00	0.298	90	1.008
(15, 15)	2.03	0.3025	88	1.004
(17, 14)	2.11	0.296	70	0.998
(16, 16)	2.17	0.297	60	1.0

Source: Chadli, H. et al., *Phys. Rev. B*, 74, 205412, 2006.

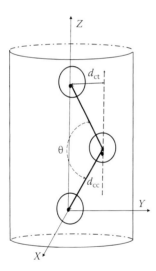

FIGURE 46.7 Schematical representation of carbon peapods showing some of the parameters used for the geometrical optimization of the C_{60} molecules inside the nanotube (see text). The letter C indicates the position of the center of one of the C_{60} molecules.

In Table 46.2, we report the values of some structural parameters issued from the energy minimizations. The optimum fullerene packing can be characterized by the angle θ formed by three consecutive C_{60} (see Figure 46.7). For example, θ = 180° of a linear chain of fullerenes and zigzag chain refers to a packing with θ < 180°. This angle θ is found to depend primarily on the nanotube diameter and does not depend significantly on the chirality. We found that θ is 180° for tubes having a diameter lower than $D_c \approx 1.45$ nm, and therefore the optimal positions of the C_{60} molecules occur when their centers are placed along the nanotube axis (linear chain). We found the linear chain as being optimum for tubes having a diameter $D_t = 2(r_c + d)$, where r_c is the fullerene radius and d, the interlayer C_{60}-SWCNT distance, which varies from 0.3 to 0.32 nm. These values of d are in agreement with the interlayer gaps commonly observed in carbon systems. In the following, this peculiar configuration of peapod is referred to as *linear peapod*. The perfect zigzag configuration (θ = 60°) is reached for tubes with a diameter $D_t \approx 2.17$ nm. In the following, the peapods in which the C_{60} molecules adopt a zigzag configuration (θ < 180°) is called *zigzag peapods*. Increasing the nanotube radius reduces the C_{60}-nanotube interaction, and for the very large nanotube radius, the interaction energy should approach that of the fullerene with planar graphite. Using a semiclassical model, Hodak and Girifalco have shown that C_{60} molecules form a zigzag structure inside a (15, 15) nanotube [53]. In this configuration, they reported a value of the θ angle of 89°; this value is in good agreement with the one reported in Table 46.2. The optimal $C_{60} - C_{60}$ gap is calculated around 0.99 − 1.01 nm for all optimized peapods. This value is larger than the one deduced from diffraction profiles [14,15] (≈ 0.97 and 0.95 nm) and from TEM data [17].

It is known that different phases are formed for the C_{60} fullerene through cycloaddition reactions, among which is the polymerization of the molecules when the solid C_{60} is exposed to visible or ultraviolet light [59]. Similarly, there are some HRTEM observations, indicating that C_{60} molecules within nanotubes can unequally be separated [30,31,60]. It is found that the distances between adjacent C_{60} units in chain within nanotube vary (from 0.9 to more than 1.0 nm [60]). This raises the question whether C_{60} molecules can be dimerized or polymerized inside nanotube.

46.2.2 Dynamics

46.2.2.1 Elements of the Theory of Lattice Dynamics

An extensive review on phonon dynamics is out of the scope of this contribution. We restrict the presentation to vibrations within the harmonic approximation. In this case, the expression of the Hamiltonian is

$$H = \frac{1}{2} \sum_{i,\alpha} M_i \dot{u}_\alpha(i)^2 + \phi \tag{46.1}$$

where

ϕ is the potential energy of particles in interaction
while $u(i)$ is the displacement of atom i

It is given by

$$\phi = \phi_0 + \frac{1}{2} \sum_{ij,\alpha\beta} \phi_{\alpha\beta}(i,j) u_\alpha(i) u_\beta(j) \tag{46.2}$$

ϕ_0 is the static potential energy of the system (molecule or crystal). The starting point is the calculation of the dynamical matrix $\tilde{\mathbf{D}}$ given by

$$\tilde{\mathbf{D}}_{\alpha\beta}(i,j) = \frac{1}{M_i M_j} \phi_{\alpha\beta}(i,j) \tag{46.3}$$

where

$\phi_{\alpha\beta}(i,j)$ are the force constants between i and j atoms
$M_i (M_j)$ is the mass of the $i(j)$th atom

The symmetry of the system permits to reduce the number of independent coefficients $\phi_{\alpha\beta}(i,j)$. The equations of motion follow immediately

$$M_i \ddot{u}_\alpha(i) = -\sum_{j\beta} \phi_{\alpha\beta}(i,j) u_\beta(j) \tag{46.4}$$

A solution for this set of coupled equations is

$$u_\alpha(i) = \frac{1}{\sqrt{M_i}} e_\alpha(i) e^{-i\omega t} \tag{46.5}$$

where $e(i)$ is the polarization vector of the displacement. The set of Equation 46.5 has a solution only if the determinant of the coefficients vanishes:

$$\left| \tilde{\mathbf{D}}_{\alpha\beta}(i,j) - \omega^2 \delta_{ij} \delta_{\alpha\beta} \right| = 0 \tag{46.6}$$

When the system contains a large number N of atoms, which is the case of long finite peapods, the size of the dynamical matrix $\tilde{\mathbf{D}}$ ($3N \times 3N$) is very large. Therefore, the resolution of Equation 46.6, which involves the diagonalization of the dynamical matrix, is not straightforward. For large systems (polymers, amorphous, and nanostructures of finite size), specific methods, such as the spectral moments method, can be used.

In the case of crystals, the periodic conditions allow one to simplify the equations of motion:

$$u_\alpha(l\kappa) = \frac{1}{\sqrt{M_\kappa}} e_\alpha(\kappa) e^{i(q \cdot R - \omega t)} \tag{46.7}$$

The $i(j)$ atom index is now replaced by the couple $l\kappa(l'\kappa')$, where $l(l')$ refers to a unit cell and $\kappa(\kappa')$ to a particular atom in the unit cell and q is a wave vector of the first Brillouin zone. The equilibrium position of the κth atom in the lth unit cell is given by the vector $R(l\kappa) = R(l) + R(\kappa)$; $u(l\kappa)$ is the displacement of an atom when a single mode of wave vector q is excited. $e(\kappa)$ is the polarization vector of the displacement. $e(\kappa)$ is independent of the cell index l, and the components of this vector satisfy the equation

$$\omega^2 e_\alpha(\kappa) = \sum_{\kappa',\beta} \tilde{\mathbf{D}}_{\alpha\beta}\left(q \,|\, \kappa\kappa'\right) e_\beta(\kappa') \tag{46.8}$$

with the so-called Fourier dynamical matrix given by

$$\tilde{\mathbf{D}}_{\alpha\beta}\left(q \,|\, \kappa\kappa'\right) = \frac{1}{\sqrt{M_\kappa M_{\kappa'}}} \sum_l \phi_{\alpha\beta}\left(0l \,|\, \kappa\kappa'\right) e^{-iq \cdot R(l)} \tag{46.9}$$

Here, the size of the dynamical matrix decreases to $3s \times 3s$, where s is the number of atoms in the unit cell. Therefore, the matrix can be easily diagonalized. There are $3s$ solutions for ω^2 for each value of q, and these will be denoted by ω_j^2, where $j = 1, 2\ldots3s$. For each of the $3s$ values of ω_j^2, one can define a $3s$ components vector $e_\alpha(\kappa)$. To make explicit the dependence of this vector on the value of q and of its assignment to a particular branch j, it will be refereed to as $e_\alpha(qj|\kappa)$. Because the dynamical matrix is Hermitian, we can assume that the eigenvectors $e_\alpha(qj|\kappa)$ satisfy the orthonormality and closure conditions. The displacement pattern obtained by substituting into Equation 46.8 a particular eigenvector $e_\alpha(qj|\kappa)$ and the corresponding frequency $\omega_j(q)$,

$$u_\alpha(l\kappa) = \frac{1}{\sqrt{M_\kappa}} e_\alpha\left(qj \,|\, \kappa\right) e^{i(q \cdot R(l) - \omega_j(q)t)} \tag{46.10}$$

is called a normal mode (q, j) of the crystal described by the wave vector q, and the branch index j. $\omega_j(q)$ is called the dispersion curve of mode j.

Since the Hamiltonian is the sum of two quadratic forms, one in the components of the momenta and the other in the components of the displacements of the atoms, it is possible to find a transformation that simultaneously diagonalizes the kinetic and potential energies in the Hamiltonian. Such a principal axis transformation is generated by the following expansion of $u_\alpha(l\kappa)$ in terms of plane waves

$$u_\alpha(l\kappa) = \frac{1}{\sqrt{NM_\kappa}} \sum_{qj} e_\alpha\left(qj \,|\, \kappa\right) Q_j(q) e^{iq \cdot R(l)} \tag{46.11}$$

where the normal coordinate $Q_j(q)$ is defined by inverting the previous expression

$$Q_j(q) = \frac{1}{\sqrt{N}} \sum_{l\kappa\alpha} e_\alpha^\star(qj \,|\, \kappa) \sqrt{M_\kappa} u_\alpha(l\kappa) e^{-iq \cdot R(l)} \tag{46.12}$$

The Hamiltonian for the lattice can be rewritten using the normal coordinates, as a sum of independent harmonic oscillators

$$H = \phi_0 + \frac{1}{2} \sum_{q,j} \left[\dot{Q}_j^*(\boldsymbol{q}) \dot{Q}_j(\boldsymbol{q}) + \omega^2(\boldsymbol{q}) Q_j^*(\boldsymbol{q}) Q_j(\boldsymbol{q}) \right] \quad (46.13)$$

The results above are obtained in the framework of classical dynamics. However, the transition to quantum mechanics is easy. Let us apply to each harmonic oscillator the quantum mechanical results expressed in the framework of the second quantization. Defining $P_j(\boldsymbol{q}) = \dot{Q}_j(\boldsymbol{q})$, one can express $\dot{Q}_j(\boldsymbol{q})$ and $P_j(\boldsymbol{q})$ in terms of creation $\hat{a}_j^+(\boldsymbol{q})$ and annihilation operators $\hat{a}_j(\boldsymbol{q})$ as (46.8)

$$\hat{Q}_j(\boldsymbol{q}) = \sqrt{\frac{\hbar}{2\omega_j(\boldsymbol{q})}} \left[\hat{a}_j^+(-\boldsymbol{q}) + \hat{a}_j(\boldsymbol{q}) \right] \quad (46.14)$$

$$\hat{P}_j(\boldsymbol{q}) = i \sqrt{\frac{\hbar \omega_j(\boldsymbol{q})}{2}} \left[\hat{a}_j^+(-\boldsymbol{q}) - \hat{a}_j(\boldsymbol{q}) \right] \quad (46.15)$$

The Hamiltonian can be expressed as a function of these operators in the simple form

$$\widehat{H} = \sum_{q,j} \hbar \omega_j(\boldsymbol{q}) \left[\hat{a}_j^+(\boldsymbol{q}) \cdot \hat{a}_j(\boldsymbol{q}) + \frac{1}{2} \right] \quad (46.16)$$

The energy of an eigenstate of the Hamiltonian is thus given by

$$E_n = \sum_{q,j} \hbar \omega_j(\boldsymbol{q}) \left[n_j(\boldsymbol{q}) + \frac{1}{2} \right] \quad (46.17)$$

where $n_j(\boldsymbol{q})$ is the number of phonons of energy $\hbar \omega_j(\boldsymbol{q})$ in the vibrational mode $(\boldsymbol{q}, j)$. At temperature T, the average number of phonons in the vibrational mode $(\boldsymbol{q}, j)$, $n_j(\boldsymbol{q})$, is given by the Bose–Einstein distribution

$$\overline{n}_j(\boldsymbol{q}) = \frac{1}{\exp\left(\dfrac{\hbar \omega_j(\boldsymbol{q})}{KT}\right) - 1} \quad (46.18)$$

The transformations given by Equations 46.14 and 46.15 are used to describe the inelastic processes in radiation-matter interactions in which phonons are involved. They are also used in the perturbation theory of anharmonic crystals.

46.2.2.2 Potential and Force Constants for Peapods

In the vibrational study of carbon peapods and in order to model interactions between atoms, the molecular compound can be assimilated to a system of atoms interconnected by springs. The interactions can be intramolecular (between two carbon atoms belonging to the same tube or the same C$_{60}$) or intermolecular (between two carbon atoms not belonging to the same tube or same C$_{60}$, or between an atom of a tube and an atom of C$_{60}$).

46.2.2.2.1 Intramolecular Interaction

The C–C intratube interactions at the surface of the SWCNT are described by using the force constant model introduced by Saito et al. [40]. In this model, we use the scaled force constants from those of 2D graphite, and we construct a force-constant tensor for a constituent atom of the SWCNT so as to satisfy the rotational sum rule for force constants. In general, the equations of motion for the displacement of the ith coordinate, u_α, for N atoms in the unit cell is given by Equation 46.4. In this case, $\phi_{\alpha\beta}(i,j)$ represents the 3×3 force-constant tensor that couples the ith and jth atoms. The sum over j in Equation 46.4 is normally taken over only a few neighbor distances relative to the ith site, which for a 2D graphene sheet has been carried out up to fourth-nearest neighbor interactions.

Since the unit cell of graphene contains two atoms, denoted A and B, the dynamical matrix of Equation 46.3 will be a 6×6 matrix, and we now find the elements. It has been found that to fit the experimental results for graphite, the model must include interactions between an atom and its first-, second-, third-, and fourth-nearest neighbors in the j-sum of Equation 46.4. There are 18 such neighbors, and Figure 46.8 shows these for an A-atom. We now demonstrate how to find the force constant tensors $\phi_{\alpha\beta}(A, B1)$, $\phi_{\alpha\beta}(A, B2), \ldots, \phi_{\alpha\beta}(A, B18)$ needed to construct the dynamical matrix; the calculation is analogous for a B-atom. Using the coordinate system of Figure 46.8 with its origin in the A-atom under consideration, a jth nearest neighbor on the x-axis will have the force tensor

$$\phi_{\alpha\beta}(A, j) = \begin{pmatrix} \phi_r^{(j)} & 0 & 0 \\ 0 & \phi_{ti}^{(j)} & 0 \\ 0 & 0 & \phi_{to}^{(j)} \end{pmatrix} \quad (46.19)$$

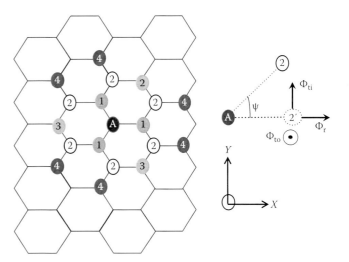

FIGURE 46.8 The positions of the first- to fourth-nearest neighbors to an A-atom in graphene. The open circles denote A-atoms, and solid circles denote B-atoms. Also shown is an example of the geometry used to find the force constant tensor $\phi^{(AA1)}$ wherein we let A1 denote the A atom at position $\left(3a/2\sqrt{2}, a/2, 0\right)$, which is a second-nearest neighbor; see text.

TABLE 46.3 The Force Constant Parameters for Graphite in Units of N/m, in the Radial, Tangential In-Plane, and Tangential Out-of-Plane Directions, Respectively

1st n.n.	2nd n.n.	3rd n.n.	4th n.n.
$\phi_r^{(1)} = 365.10$	$\phi_r^{(2)} = 88.0$	$\phi_r^{(3)} = 30.0$	$\phi_r^{(4)} = -19.2$
$\phi_{ti}^{(1)} = 245.0$	$\phi_{ti}^{(2)} = -32.3$	$\phi_{ti}^{(3)} = -52.5$	$\phi_{ti}^{(4)} = 22.9$
$\phi_{to}^{(1)} = 365.10$	$\phi_{to}^{(2)} = -4.0$	$\phi_{to}^{(3)} = 1.5$	$\phi_{to}^{(4)} = 5.8$

Source: Saito, R. et al., *Phys. Rev. B*, 57, 4145, 1997.

where $\phi_r^{(j)}$, $\phi_{ti}^{(j)}$, and $\phi_{to}^{(j)}$ represent the force constant parameters in the radial, in-plane transverse, and out-of-plane transverse directions, respectively. In practice, these force constants are adjusted so as to reproduce as much as possible the dispersion curves measured in graphite by neutrons scattering or electron energy loss spectroscopy (EELS). The numerical values for the ϕ_s in graphite are listed in Table 46.3. For a neighbor not on the *x*-axis, we consider a virtual atom on the *x*-axis for which the tensor looks like Equation 46.19 and then rotate it according to the usual tensor properties, $\phi^{A_{j\psi}} = U_z^{-1}(\psi)\phi^{A_{jx}}U_z(\psi)$, where ψ is the angle between the *x*-axis and the atom, and $U_z(\psi)$ is the usual rotation matrix for a rotation around the *z*-axis by an angle ψ:

$$U_z(\psi) = \begin{pmatrix} \cos(\psi) & \sin(\psi) & 0 \\ -\sin(\psi) & \cos(\psi) & 0 \\ 0 & 0 & 1 \end{pmatrix} \quad (46.20)$$

The situation is illustrated in Figure 46.8 for a second-nearest neighbor (A-atom) not located on the *x*-axis. The method for constructing the dynamical matrix of SWCNT is in principle the same as for graphene, the only difference being the curvature of the crystal, which makes it hard to find the force constant tensors. The force-constant tensor in SWCNT is generated by rotating the chemical bond from the two-dimensional plane to the three-dimensional coordinates of the nanotube. We include interactions between an atom and up to its fourth-nearest neighbor. This procedure is well described in Ref. [40].

In the case of C_{60} molecule, the on-ball C–C interaction is described by a force field proposed by Jishi et al. [48] consisting of four longitudinal Born-von Karman and four angle-bonding force constants. In this model, each atom is assumed to be a point mass, and to be connected with springs to its first-, second-, and third-nearest neighbors. Two springs, with spring constants s_p and s_h, are used to model the stretching of the pentagonal and hexagonal covalent bonds, respectively. The force constants of the springs connecting an atom to its second- and third-nearest neighbor are denoted by s_2 and s_3, respectively. The bending of the pentagonal and hexagonal angles are modeled by springs with spring constants a_p and a_h, respectively. In addition, the bending of the angles between the spring connecting an atom to its first-nearest neighbor and the spring connecting it to its second- and third-nearest neighbors are

modeled by spring constants b and c, respectively. The potential energy of the molecule is the sum of the stretching and bending contributions.

Let us, for example, consider three atoms labeled i, j, and k, where j is a second-nearest neighbor of i. Upon giving arbitrary displacements to these three atoms, the contribution to the potential energy of the stretching of the spring connecting i to k is given by

$$V_s^{ik} = \frac{1}{2} s_2 (\Delta r_{ik})^2 \quad (46.21)$$

where Δr_{ik} is the change in distance between atoms i and k. On the other hand, the contribution to the potential energy due to the bending of the angle (ij,ik) is given by

$$V_b(ij,ik) = \frac{1}{2} b (\Delta\theta)^2 \quad (46.22)$$

where $\Delta\theta$ is the change in the angle (ij,ik) due to the displacements given to atoms i, j, and k.

The values of the bond-stretching and angle-bending force constants used to obtain the best fit to the experimental Raman data are given in Table 46.4. Elements of the dynamical matrix relating to the intramolecular interactions in C_{60} are thus calculated by considering the $\phi_{\alpha\beta}(nn')$, which derive the potential energy of the system $V^{C_{60}}$. This calculation is written as follows:

$$V^{C_{60}} = V_s^{C_{60}} + V_b^{C_{60}} \quad (46.23)$$

where $V_s^{C_{60}}$ and $V_b^{C_{60}}$ indicate the potential energy of interatomic stretching motions and potential of the bending motions, respectively, given by

$$V_s^{C_{60}} = \frac{1}{2} \sum_i \sum_k V_s^{ik} \quad (46.24)$$

$$V_b^{C_{60}} = \frac{1}{4} \sum_i \sum_{j,j \neq i} \sum_k V_b(ij,ik) \quad (46.25)$$

TABLE 46.4 The Values of the Bond-Stretching and Force Constants (K_{str}), in mdyn/Å, and of the Angle-Bending Force Constants (K_{str}), in mdyn Å/(rad²)

K_{str}	Value	K_{bend}	Value
s_p	4.0	a_p	1.0
s_h	2.35	a_h	0.25
s_1	1.21	b	0.095
s_2	−1.05	c	0.325

Source: Jishi, R.A. et al., *Phys. Rev. B*, 45, 13685, 1992.

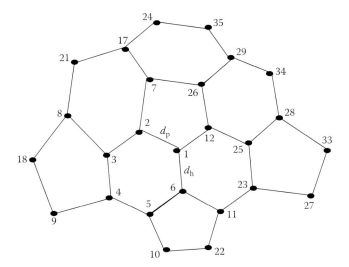

FIGURE 46.9 Atomic environment in C$_{60}$ molecule. The angles and the lengths of inter atomic are arbitrary, the planar representation necessarily generating a distortion. d_p is the distance C–C in edge of pentagon (0.1453 nm), and d_h is the distance from connections C–C joining the tops of two pentagons (0.1395 nm).

To calculate the matrix elements, such potential must be successively derived compared to the displacements $u_\alpha(i)$ of the atom i in the α-direction and $u_\beta(j)$ of the atom j in the β-direction. It is extremely important in these calculations to carefully apply the good force constants according to the type of neighbors considered. However, it is necessary to locate systematically the neighbors of each atom. One recalls that in C$_{60}$ molecule, all the carbon atoms are equivalent. Thus, the environment of one atom can be schematized as on Figure 46.9. We notice that the lattice of pentagons and hexagons is distorted by the two-dimensional representation on the Figure 46.9. The order of the neighbors of the atom number 1 is indicated in Table 46.5.

It notes that the differences in length between bends separating a pentagon of a hexagon (d_p) and those separating two hexagons (d_h) allow a precise classification of neighbors of each atom. It is important to follow this classification if one wishes to respect the principle of action-reaction and to obtain symmetrical dynamic matrices.

46.2.2.2.2 Intermolecular Interaction

Concerning the interaction between carbon atoms belonging to two different molecules (tube or C$_{60}$), we could rigorously define 3 types of configurations:

- Interaction between 2 atoms belonging to two different C$_{60}$ molecules.
- Interaction between 2 atoms belonging to two different tubes.
- Interaction between 2 atoms belonging to a tube and an atom belonging to C$_{60}$.

In general, in peapods in which the nearest separation between the inserted C$_{60}$ molecules and the nanotube wall is around vdW distance (0.3 nm), we use the familiar Lennard–Jones potential:

TABLE 46.5 Coordinates of First, Second, and Third Neighbors of the Atom 1 Shown in Figure 46.9

Neighbors Order of the Atom 1	
First neighbors	9
	12
	....
	2
Second neighbors	7
	26
	....
	3
	5
	11
	25
Third neighbors	4
	23
	....
	10
	22
	....
	28
	29
	17
	8

Notes: Dashes indicate the successive increase of the distance to the atom number 1.

$$U_{LJ}(r) = 4\epsilon \left[\left(\frac{\sigma}{r} \right)^{12} - \left(\frac{\sigma}{r} \right)^{6} \right] \quad (46.26)$$

where

r is the carbon atom-atom distance

ϵ and σ are vdW parameters

The three types of interactions described above are vdW type, and they are not strictly identical. Thus, these potentials should present different vdW parameters. Ulbricht et al. [24] have been reported vdW parameters describing vdW in carbon peapods. The values of the Lennard–Jones parameters were $\epsilon = 2.964$ meV and $\sigma = 0.3407$ nm. These values have been found to describe correctly the vdW contribution to the C$_{60}$ bulk cohesive energy. To derive the elements of the dynamical matrix, it is necessary to carry out a double derivation with respect to the displacements of the atoms i and j in the α and β directions.

46.2.2.2.3 Dynamical Matrix of Peapods

In peapod systems, all particles in interaction are carbon atoms having all the mass m_c. To simplify the expressions, we will note $\boldsymbol{D}_{\alpha\beta}(nn')$, the elements of the dynamic matrix called previously $\phi_{\alpha\beta}(nn')/m_c$. The interatomic force constants $\phi_{\alpha\beta}(nn')$ in peapod system are calculated using the above dynamical models. Thus, the total dynamic matrix of peapod can be written as follows:

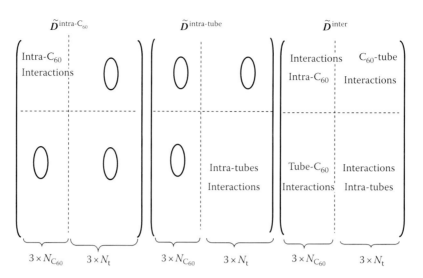

FIGURE 46.10 Dynamic matrices dimensioned to be added into a dynamic matrix, taking into account all the interactions. N_t is the number of atoms of carbon nanotube, and $N_{C_{60}}$ is the number of carbon atoms of C_{60}.

$$\tilde{D} = \tilde{D}^{\text{intra}-\text{tube}} + \tilde{D}^{\text{intra}-C_{60}} + \tilde{D}^{\text{inter}} \tag{46.27}$$

Obviously, it is very important to remain attentive to the dimensions of the dynamic matrices cited below. As defined above, these matrices have a dimension of $3N \times 3N$, where N is the number of particles of system. But it differs if one considers the total peapod dimension of only the tube or only C_{60}. It is thus appropriate to give to all the dynamic matrices the same dimension as that of the total peapod. We complete thus the intramolecular dynamical matrices $\tilde{D}^{\text{intra}-\text{tube}}$ and $\tilde{D}^{\text{intra}-C_6}$, as shown in Figure 46.10.

It is necessary to note that at this stage, we can not define any terms of the diagonal dynamic matrix, i.e., those that are related to the interaction of an atom with itself. These terms must be taken into account and calculated before proceeding with the diagonalization of the matrix. One uses for that the property of invariance by translation of the system. Indeed, if all the atoms are moving uniformly straight following the same vector ε, the sum of internal forces must be zero:

$$\ddot{\varepsilon}_\alpha = 0 = \sum_{\beta n'} D_{\alpha\beta}(nn')\varepsilon_\beta \tag{46.28}$$

This relation must be verified for all possible translations, and we get

$$\sum_{\beta n'} D_{\alpha\beta}(nn') = 0 \tag{46.29}$$

Finally, the diagonal terms of the dynamic matrix can be calculated using

$$D_{\alpha\beta}(nn) = -\sum_{n \neq n'} D_{\alpha\beta}(nn') \tag{46.30}$$

46.2.2.2.4 Calculation of Fourier Dynamical Matrix

In the first part of the chapter, it was explained that within the framework of the lattice dynamics method, the motion of carbon atoms was regarded as oscillatory around their equilibrium position and was described by a Bloch wave in order to take into

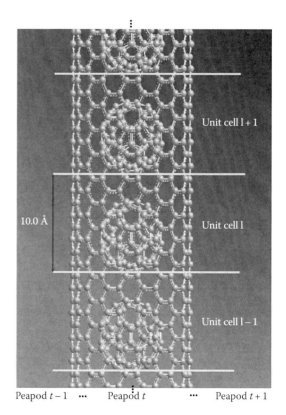

FIGURE 46.11 Sample of an elementary supercell used in lattice dynamics calculations of peapods. The dots mark the periodicity of the system. The various components of peapod bundle are on the hexagonal lattice that typically describes the structure of nanotube bundles.

account the boundary conditions. Thus, calculations are performed on a unit cell of the system, which one defines for the peapods, as presented in Figure 46.11. An isolated peapod of infinite length can be represented as a vertical packing of one-dimensional supercell including a portion of nanotube and a C$_{60}$ in center. According to minimization energy results, the optimal C$_{60}$ – C$_{60}$ gap in peapods is close to 1 nm. Therefore, we choose this value as a supercell parameter T of the crystal. For example, in the case of infinite C$_{60}$@(10, 10) peapod, the supercell is formed by 4 unit cells of nanotube (10, 10) (see Figure 46.11). Thus, the Fourier dynamical matrix can be calculated using Equation 46.9. The size of this matrix is $3N \times 3N$, where N is the number of atoms in the supercell.

In practice, it is convenient to consider a set of 3 vertical supercells of peapod to take into account all the interactions of individual atoms in the central supercell (the supercell l in Figure 46.11) with other atoms in the system. We obtain the dispersion curve and the phonons Density Of States (DOS) of the system by diagonalization of this dynamic matrix for the greatest possible number of values of q in the first Brillouin zone of the system.

46.3 Vibrational Properties: Phonons

Phonons are the quasi-particles that describe the quantized lattice vibrations or normal modes of a crystal. The crystal possesses $3N$ phonon branches, where N is the number of atoms in the unit cell. A three-dimensional crystal has three acoustic modes with zero frequency at the G point, corresponding to uniform displacements of the crystal. All remaining branches, where the atoms in the unit cell move out-of-phase, are referred to as optical phonons, since in polar crystals, they can directly couple to light [67]. Phonons play an important role as a carrier of thermal energy in thermal conduction process and in thermodynamic properties, such as the heat capacity. The vibrational spectra also determine the speed of sound, elastic properties of solids, and their mechanical properties. Phonons, through their interaction with electrons, can also mediate interactions and pairing between electrons, giving rise to superconductivity. These phenomena are even more interesting in SWCNTs, which allow these unique 1D effects to be studied in detail. An extensive review on phonon dynamics is out of the scope of this contribution. We restrict the presentation to vibrations within the harmonic approximation. We focus on dynamic properties of C$_{60}$, carbon nanotubes and peapods.

46.3.1 C$_{60}$ Monomer and Polymer

Vibrational modes of a C$_{60}$ molecule are directly related to its structure and symmetry. The vibrational spectrum of solid C$_{60}$ ranges from 0 to 1600 cm^{-1}. It is well known that it can be divided into two clearly separated zones, i.e., the intramolecular vibration range (240 – 1600 cm^{-1}) and the intermolecular vibration range (0 – 65 cm^{-1}). There are altogether 180 degrees of freedom

(three per atom). Three of those are translations, the other 3 are rotations, and the leftover 174 are the vibrational degrees of freedom. Due to the high symmetry of the molecule, many of those are degenerate in frequency. Theoretically, there should only be 46 distinct frequencies. These vibrations with their corresponding symmetries are

$$\Gamma = 2A_g + 3F_{1g} + 4F_{2g} + 6G_g + 8H_g + A_u + 4F_{1u} + 5F_{2u} + 6G_u + 7H_u$$

(46.31)

Inelastic neutron scattering is the ideal experimental technique for measuring the frequency range where inter-C$_{60}$ interactions dominate the spectrum. This technique has been successfully used to study the dynamics of several phases of pure and alkaline-doped C$_{60}$. They allowed lattice dynamical models to be developed and force constants to be refined. To obtain a good description of the phonon modes in C$_{60}$ dimer and polymer states, it is necessary to start with an accurate model for the phonon modes in isolated C$_{60}$ molecule. We use a force constants model of Jishi et al. [48] to describe the intrafullerene interactions. The inter-cage bonds were modeled by a stretching force constant K_{str} for the interfullerene C–C bond and two angle bending force constants: K_1 is related to the angle spanned by interfullerene C–C bond and the (6, 6) bond and K_2 to the angles spanned by the interfullerene C–C bond and the (5, 6) bonds. These three parameters are fitted to optimize the description of the low frequency part of the global density of states (GDOS) measured for a C$_{60}$ dimer and polymer samples. The adjusted force constants that we used to obtain the best fit to the experimental GDOS and Raman data are given in Table 46.6. We use the same units as the Born-von Karman force constants of Jishi for the bond stretching (mdyn/Å) and the angle-bending force (mdyn/rad^2).

The profile of the DOS, calculated for C$_{60}$ monomer, dimer and infinite polymer, is displayed in Figure 46.12 in the low frequency range (0 – 320 cm^{-1}). The spectra are compared to the experimental GDOS obtained by inelastic neutrons scattering. For C$_{60}$ monomer, there are three frequency regions: (0 – 65 cm^{-1}) very low frequency region, gap region between 65 and 260 cm^{-1}, and several intramolecular modes beyond 274 cm^{-1}. The mode located at 274 cm^{-1} is assigned to Hg(1) symmetry by dynamic calculations [48]. The very low frequency region is characterized by vdW intermolecular modes of the 3D-crystalline state of C$_{60}$. These modes consist of libration and translation of all the C$_{60}$ molecules

TABLE 46.6 Force-Constant Parameters for Interfullerene Bonds in Dimer (D Model) and Polymer (P Model)

Model	Stretching	Value	Bending	Value
			K_1	0.36
D	K_{str}	2.1	K_2	0.34
			K_1	0.8
P	K_{str}	4.0	K_2	0.7

Notes: The bond-stretching force constants are expressed in (mdyn/Å), whereas the angle-bending force constants are expressed in (mdyn/rad^2).

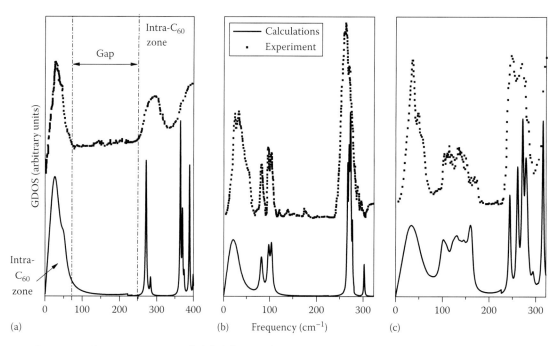

FIGURE 46.12 Comparison between experimental global density of states (GDOS) (upper curve) and lattice dynamical calculations (lower curve) for C_{60} monomer (a), C_{60} dimer (b), and infinite polymer (c).

around their equilibrium position. Upon polymerization of C_{60} molecules, one of the qualitative changes of the vibrational spectrum in the low frequency region is the appearance of new modes, which are caused by a intramolecular mixing of the zero-energy translational and rotational modes of individual balls. The comparison between the experimental GDOS [49] and the calculated one shows that the principal measured bands are globally well reproduced by the model. No vibrational modes are observed for a chain formed by C_{60} monomers vdW interactions in the $(80 - 177 \, cm^{-1})$ frequency range [48]. The calculated GDOS for dimer and for polymer chains demonstrates that inter-fullerene modes in polymeric structures fill this frequency range. These new intermolecular modes, in comparison with the C_{60} monomer, correspond to translational and librational inter-ball modes and are well reproduced by intermolecular covalent bond force constants. The polymerization of C_{60} leads to drastic changes in the low frequency dynamics $(0 - 240 \, cm^{-1})$, where typical inter-ball modes (ball stretching, torsions, etc.) and molecular lattice modes (acoustic phonons, librations) show up. This region therefore provides a useful tool to probe the C_{60} bonding scheme.

46.3.2 Carbon Nanotubes

The phonon dispersion relations in a carbon nanotube can be obtained from those of the 2D graphene sheet by using force constants cited in Table 46.3 for calculations of the 2D phonon dispersion relations of an isolated graphene sheet [50,51]. The latter has two atoms per unit cell, thus having 6 phonons branches, as shown in Figure 46.13a. The calculated phonon dispersion curves of Figure 46.13b were fit to the experimental points obtained by EELS, inelastic neutron scattering, velocity of

sound, and other techniques [68–70]. The three phonon dispersion branches, which originate from the Γ point of the Brillouin zone with $\omega = 0$ (see Figure 46.13a), correspond to acoustic modes: an out-of plane mode, an in-plane tangential (bond-bending) mode, and an in-plane radial (bond-stretching) mode, listed in order of increasing frequency, respectively. The remaining three branches correspond to optical modes: one non-degenerate out-of-plane mode and two in-plane modes that remain degenerate as we move away from $k = 0$.

We use the force constants model to find the phonon dispersion and the density of states for SWCNT from those of graphite. The method for constructing the dynamical matrix is in principle the same as for graphene, the only difference being the larger number of atoms in the unit cell and the curvature of the crystal, which makes it hard to find the force constant tensors. Here, we proceed and impose periodic boundary conditions on the phonon wave vector in the direction of the chiral vector. The size of the dynamical matrix is close to $3s \times 3s$, where s is the number of atoms in the unit cell. Therefore, the matrix can be easily diagonalized. For example, the dynamical matrix becomes a 120×120 matrix in case of a (10, 10) SWCNT as compared to the 6×6 matrix for graphene. The resulting dispersion relations for armchair (10, 10) nanotube obtained by this folding procedure is illustrated in Figure 46.14a, and the respective phonon DOS is shown in Figure 46.14b. T is the translation vector along the tube, and in reciprocal space, only k along the discrete lines with $\left[-\frac{\pi}{T} < k < \frac{\pi}{T} \right]$ is considered. We find a large number of phonon branches due to the large number of atoms in the unit cell. The large amount of sharp structure in the phonon density of states in Figure 46.14 for the (10, 10) SWCNT reflects the many phonon branches and the 1D nature of SWCNTs relative to 2D graphite.

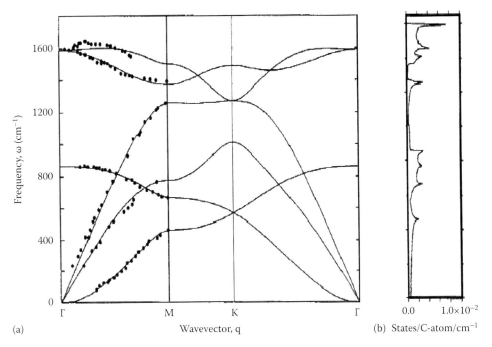

(a)   Wavevector, q   (b) States/C-atom/cm^{-1}

FIGURE 46.13 (a) The phonon dispersion relations for graphite plotted along high-symmetry in plane directions. Experimental points from neutron scattering and electron energy loss spectra were used to obtain values for the force constants (see Table 46.3) and to determine the phonon dispersion relations throughout the Brillouin zone. (From Jishi, R.A. et al., *Chem. Phys. Lett.*, 209, 77, 1993. With permission.) (b) The phonon density of states versus phonon energy for a 2D graphene sheet. (From Jishi, R.A. et al., *Chem. Phys. Lett.*, 209, 77, 1993. With permission.)

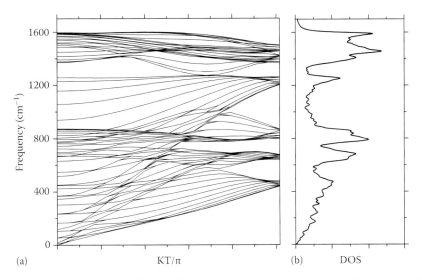

(a)   KT/π   (b)   DOS

FIGURE 46.14 (a) The calculated phonon dispersion relations of an armchair carbon nanotube (10, 10). The number of degrees of freedom is 120, and the number of distinct phonon branches is 66. (b) The corresponding phonon density of states for a (10, 10) nanotube.

46.3.3 C_{60} Peapods

In this section, we consider an isolated Peapod with a supercell including a section of nanotube and a C_{60} at its center. According to energy minimization calculation, we know that the typical distance between C_{60} molecules is close to 1.0 nm. This value is considered as a supercell parameter **T**. Now, the $C_{60}@(10, 10)$ peapod is taken as an example. It was chosen to form a supercell of Peapod by packing 4 unit cells of nanotube (10, 10), as it was then $T = 4 \times 0.246 = 0.984$ nm. The number of atoms in this supercell

is 220, resulting in the existence of 660 modes, and therefore lots of dispersion curves. It is not interesting to present the shape of 660 dispersion curves on all possible frequencies. The calculated GDOS over the full frequency range of the $C_{60}@(10, 10)$ peapod, resulting from the summation of the dispersion curves in only the first Brillouin zone of one-dimensional peapod, is presented in Figure 46.15. The calculated spectrum is compared with the experimental GDOS obtained by inelastic neutrons scattering at 480 K. The shape of the GDOS of the peapod sample is generally

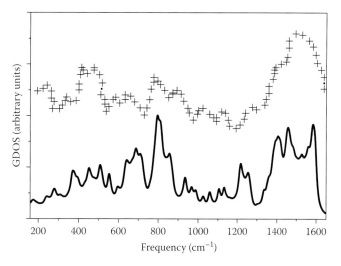

FIGURE 46.15 Top: Experimental GDOS of the peapods sample derived from the IN1 experiment (spectrum measured in LCVN Montpellier France) [71]. Bottom: Calculation of the GDOS of a sample of C_{60} monomer (10, 10) peapods.

in good agreement with the results of calculations. Concerning the GDOS of the empty SWCNT (see Figure 46.14b), clear features can be observed: a first peak at $284\,cm^{-1}$, a minimum at $565\,cm^{-1}$, a double structure centered around 640 and $807\,cm^{-1}$, and a large peak around $1370–1530\,cm^{-1}$. Those structures are also observed in the GDOS of the peapod sample. We have shown that the general shape of the GDOS of the inserted C_{60} monomer chains displays two regions separated by a gap around $160\,cm^{-1}$. The low frequency range of GDOS of peapod is reported on the top of Figure 46.15. The GDOS in this range features a peak at $30\,cm^{-1}$, a shoulder at $\sim14.5\,cm^{-1}$, and a large unresolved band around $80\,cm^{-1}$. Concerning the frequency range above the gap, a rather intense peak is clearly observed around $270\,cm^{-1}$, which is characteristic of the Hg(1) intramolecular mode of the C_{60} balls. In the low frequency region, less than $242\,cm^{-1}$, we find a set of contributions mainly from modes interactions between the tube and C_{60}.

In the following, the calculations of the GDOS were performed using the dynamical models presented in Sections 46.2.2 and 46.3.1. Three C_{60} packing were considered: a monomer chain (M model), a dimer chain (D model), and a polymer chain (P model). The profile of the GDOS calculated for the M model is displayed in Figure 46.16b in an enlarged scale. A large gap from 48 to $242\,cm^{-1}$ separates the intermolecular modes below $48\,cm^{-1}$ from the intramolecular modes above $242\,cm^{-1}$. The lowest frequency intramolecular mode Hg(1) is furthermore clearly visible around $270\,cm^{-1}$. In the low-frequency range (see Figure 46.16a), the GDOS profile consists of two peaks located at 13.7 and $28.2\,cm^{-1}$. The first peak is related to the C_{60} librations, while the second peak involves translational vibrations of C_{60}. The comparison between the experimental GDOS and that calculated for the M model shows many differences. First, the absence of the sharp librational peak at $13.7\,cm^{-1}$ can be understood if C_{60} monomers undergo free rotations. The gap in the calculated spectrum is too large, especially at lower energies at which additional contributions

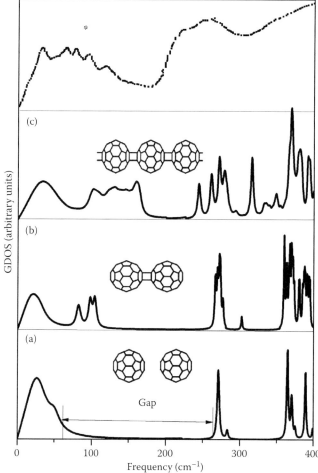

FIGURE 46.16 Lattice-dynamical calculations of the GDOS of chains of (a) monomers (M model), (b) dimers (D model), and (c) polymers (P model). The upper curve is the experimental GDOS of a C_{60} peapod. The star indicates the position of the mode observed by Raman spectroscopy at low temperature. (From Cambedouzou, J. et al., *Phys. Rev. B*, 71, 041403(R), 2005.)

are clearly observed in the region from 65 to $113\,cm^{-1}$. The GDOS calculated for the D model shows important modifications compared to the M model. In the frequency range below $48.4\,cm^{-1}$, the two sharp peaks in the monomer spectrum are replaced by a broad peak centered around $24\,cm^{-1}$. In the D model, the cylindrical geometry of the C_{120} dimers implies that the librational modes in transverse and longitudinal directions in the tube environment are no longer equivalent, so the sharp structure of the libration peak is lost. The frequencies of the translation modes are shifted to slightly lower frequency due to the larger mass of the C_{120} dimer. Consequently, librations and translations of the dimer are included into the unique large peak centered around $24\,cm^{-1}$. In the region between 65 and $242\,cm^{-1}$, two additional peaks located around 73 and $113\,cm^{-1}$ appear. According to our calculations, these peaks are attributed to the Einstein dumbbell mode of the C_{120} dimer and to libration-like modes where the balls undergo small-amplitude rotations perpendicular to the dimer axis, respectively. When the GDOS of the polymer structure is considered, the hardening of the first peak, which shifts to $32\,cm^{-1}$, and the increase of the intensities of the modes between 65 and $161\,cm^{-1}$ relative to the first peak are observed. In this totally polymerized state, the libration of the balls around the chain axis is harder than in the dimer state and contributes to fill the gap. Also, the dumbbell-like mode becomes a translational mode that has a strong dispersion along the tube axis, giving rise to a broad component in the GDOS in the gap region.

From Figure 46.16, it is obvious that none of the models perfectly matches the experimental GDOS. This is not surprising at all, given the simplicity of the models and the fact that the measurements were performed at high temperature. However, several important conclusions can be drawn from the comparison between the data and the models. A chain formed by C_{60} monomers in vdW interaction does not have vibrational modes in the $(48 - 120)\,cm^{-1}$ frequency range. This is at variance with the experiment. The GDOS calculated for dimer and polymer chains demonstrates that interfullerene modes in polymeric structures fill the $(48 - 120)\,cm^{-1}$ frequency range. In summary, intense modes on the low frequency side of the gap separating the intermolecular from the intramolecular modes suggest the existence of strong bonds, such as valence interactions between part of the monomers. The latter observation leads to the proposal that peapods are filled with a mix of monomer and n-mer (dimer, trimer,…., polymer) structures.

46.4 Raman Spectroscopy of C_{60} Carbon Peapods

The challenge for nanotechnology is to achieve perfect control of nanoscale-related properties. This obviously requires correlating the parameters of the synthesis process (self-assembly, microlithography, solegel, polymer curing, electrochemical deposition, laser ablation, etc.) with the resulting nanostructure. Not every conventional characterization technique is suitable for that purpose, but Raman Spectroscopy has already proven to be. For quite a long time, this technique was mainly devoted to fundamental research, but instrumental progress have rendered it a general characterization method. Raman scattering is known as a powerful tool to study the averaged bulk properties of carbon nanotubes, such as structures and diameter distribution of nanotubes, and those of foreign material-doped composites.

46.4.1 Raman Scattering

The theory of Raman scattering from molecules was recently reviewed by Long [72]. The theory of Raman scattering from crystals was treated by many authors, among them are Fabelinskii [73], Loudon [74], Turell [75], and Poulet and Mathieu [76]. In both, the classical and the quantum mechanical treatments of the Raman effect, the main origin of the scattered radiation is considered to be an electric oscillating dipole, **P**, induced in the

medium (molecules, amorphous materials, glasses, crystals), by the electromagnetic incident field $\mathbf{E}_1$. At first order, the induced dipole moment is given by

$$\mathbf{P} = \breve{\alpha}\,\mathbf{E}_1 \qquad (46.32)$$

where $\breve{\alpha}$ is the electronic polarizability tensor. In fact, due to the high frequency of the probe (laser light in the visible or near infrared range), only the electrons of the system can follow the field, leading to an instantaneous electronic polarization of the medium. The electronic polarizability of a molecule or, equivalently, the electronic susceptibility of a crystal is modulated by vibrations in molecules, and phonons in crystals. These fluctuations in the polarizability tensor of the system are responsible for Raman scattering.

46.4.1.1 Polarizability Theory

In a Raman experiment (Figure 46.17), a visible, or near infrared light, of circular frequency ω_1, wave vector $\mathbf{k}_1$, polarization unit vector e_1, and incident laser irradiance $I_1(= \epsilon_0 c\eta_1 E_1^2)$, is incident in an isotropic medium (ϵ_0 is the dielectric constant of vacuum, c is the speed of light in vacuum, and η_1 is the index of refraction of the medium at the laser frequency).

The radiant intensity of the oscillating electric dipole (point source) induced in a molecule by the incident electric field $\mathbf{P} = \breve{\alpha}\mathbf{E}_1$, vibrating within a frequency range $d\omega_2$ from ω_2, and within a solid angle $d\Omega$ from the scattered wavevector $\mathbf{k}_2$, is given by

$$\frac{d^2W_2}{d\Omega d\omega_2} = \frac{\eta_2\omega_2^4}{16\pi^2\epsilon_0 c^3}\left|e_1\breve{\alpha}e_2\right|^2\left|E_1\right|^2 \qquad (46.33)$$

η_2 is the index of refraction of the medium at the scattered frequency.

The differential scattering cross section, $\frac{d^2\sigma}{d\Omega d\omega_2}$, is obtained by dividing the previous expression by I_1:

$$\frac{d^2\sigma}{d\Omega d\omega_2} = \frac{\eta_2}{\eta_1}\frac{\omega_2^4}{16\pi^2\epsilon_0 c^4}\left|e_2\breve{\alpha}e_1\right|^2 \qquad (46.34)$$

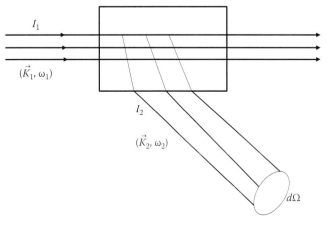

FIGURE 46.17 Sketch of a Raman scattering experiment.

The differential cross section given above is a *power* cross section. It is defined as the ratio of scattered to incident power. In photon counting systems, quantum cross section is usually measured. The differential cross section is converted into quantum ones by multiplication by ω_2/ω_1, leading to the usual expression of the differential quantum cross-section:

$$\frac{d^2\sigma}{d\Omega d\omega_2} = \frac{\eta_2}{\eta_1}\frac{\omega_1\omega_2^3}{16\pi^2\epsilon_0^2 c^4}\left|e_2\bar{\bar{\alpha}}e_1\right|^2 \tag{46.35}$$

For first-order Raman scattering, we expand the $\alpha\beta$ component of the electronic polarizability tensor, $\check{\alpha}$, to first order in $u_\gamma(k)$ (the displacement along the γth Cartesian direction of atom k):

$$\bar{\alpha}_{\alpha\beta}^{(1)} = \sum_{k,\gamma}\left(\frac{\partial\alpha_{\alpha\beta}}{\partial u_\gamma(k)}\right)_0 u_\gamma(k) \tag{46.36}$$

Using the normal coordinate Q_j previously defined (Equation 46.12), we then express the expansion of $\alpha_{\alpha\beta}$ as

$$\bar{\alpha}_{\alpha\beta}^{(1)} = \sum_j \frac{\partial\alpha_{\alpha\beta}}{\partial Q_j}Q_j \tag{46.37}$$

where

$$\frac{\partial\alpha_{\alpha\beta}}{\partial Q_j} = \sum_{k,\gamma}\left(\frac{\partial\alpha_{\alpha\beta}}{\partial u_\gamma(k)}\right)_0 \frac{1}{\sqrt{M_k}}e_\gamma(j\,|\,k) \tag{46.38}$$

In Raman scattering for a molecule, the normal coordinate and polarization vector do not depend on **q**. This is why **q** does not appear in the previous expressions. In the harmonic approximation, each normal mode contributes separately to the scattered intensity.

By taking into account the properties of Q_j in a quantum mechanics approach (Equation 46.14), one calculates $\langle|Q_j|^2\rangle$

$$\left\langle\left|Q_j\right|^2\right\rangle = \bar{n}_j + 1 \tag{46.39}$$

in a creation process of a vibrational quantum and

$$\left\langle\left|Q_j\right|^2\right\rangle = \bar{n}_j \tag{46.40}$$

in an annihilation process of a vibrational quantum. In consequence, the differential quantum cross section can be expressed in the following form:

$$\frac{d^2\sigma}{d\Omega d\omega_2} = \frac{\eta_2}{\eta_1}\frac{\omega_1\omega_2^3}{16\pi^2\epsilon_0^2 c^4}\sum_{\alpha\beta\gamma\lambda}e_{2\alpha}e_{2\beta}H_{\alpha\beta\lambda}(\omega)e_{1\gamma}e_{1\lambda} \tag{46.41}$$

where

$$H_{\alpha\gamma\beta\lambda}(\omega) = \sum_j a_{\alpha\gamma}^*(j)a_{\beta\lambda}(j)\frac{1}{2\omega_j}\Big(\delta(\omega-\omega_j) - \delta(\omega+\omega_j)\Big) \tag{46.42}$$

with the Raman shift ω is given by

$$\omega = \omega_1 - \omega_2 \tag{46.43}$$

and

$$a_{\alpha\gamma}(j) = \frac{\partial\alpha_{\alpha\beta}}{\partial Q_j} \tag{46.44}$$

$a_{\alpha\gamma}(j)$ is the $\alpha\beta$ component of the so-called Raman polarizability tensor of the jth mode, and ω_j is the frequency of the jth vibration mode. The first term of Equation 46.42, associated with the creation of a vibrational quantum, describes the Stokes scattering with a downshift in frequency ($\omega > 0$), and the second term, related to an annihilation of a vibrational quantum, describes the anti-Stokes scattering, with an upshift frequency ($\omega < 0$). The jth mode is Raman active if one component of the Raman polarizability tensor, $a_{\alpha\gamma}(j)$, is different from zero. From group theory, the symmetry of the Raman tensor for each jth Raman-active mode can be determined from the symmetry point group, which describes the molecular symmetry. These tensors are tabulated in many books (see Ref. [76]) for instance.

46.4.1.2 The Bond Polarizability Model

The bond polarizability approach was developed for molecules [77], but it is also useful for solid-state applications, as reviewed in [78]. The basic assumption is that the static electronic polarizability of a covalent system can be written as a sum of contributions from the individual bonds. In its simplest form, the $a_{\alpha\gamma}(j)$ component is expressed as

$$a_{\alpha\gamma}(j) = \sum_{k\delta}\frac{\pi_{\alpha\gamma,\delta}^k}{\sqrt{M_k}}e_\delta(j\,|\,k) \tag{46.45}$$

where the coefficients $\pi_{\alpha\gamma,\delta}^k$ relate the polarization fluctuations to the atomic motions [79]. They are obtained by expanding the atomic polarizability tensor $\bar{\pi}^k$ in terms of atom displacements $u_\delta(k)$, with

$$\pi_{\alpha\gamma,\delta}^k = \sum_{k'}\left(\frac{\partial\pi_{\alpha\gamma}^k}{\partial u_\delta(k)}\right)_0 \tag{46.46}$$

A usual method to calculate the Raman spectrum is to inject in the expressions above the values of ω_j and $e_\delta(j\,|\,k)$ obtained by solving the equation of motion (Equation 46.4). The Raman intensity is calculated within the nonresonant bond-polarization theory in which bond-polarization parameters are used [80]. In this approach, the $\pi_{\alpha\gamma,\delta}^k$ coefficients are given by

$$\pi_{\alpha\beta,\gamma}^k = \sum_{k'}\frac{1}{3}\Big(2\alpha_p' + \alpha_l'\Big)\delta_{\alpha\beta}\hat{r}_\gamma + \Big(\alpha_l' - \alpha_p'\Big)\left(\hat{r}_\alpha\hat{r}_\beta - \frac{1}{3}\delta_{\alpha\beta}\right)\hat{r}_\gamma$$
$$+ \frac{(\alpha_l - \alpha_p)}{r}\Big(\delta_{\alpha\gamma}\hat{r}_\beta + \delta_{\beta\gamma}\hat{r}_\alpha - 2\hat{r}_\alpha\hat{r}_\beta\hat{r}_\gamma\Big) \tag{46.47}$$

where $\hat{r}$ is the unit vector along vector **r**, which connects the k and k' atoms linked by the bond. The parameters α_l and α_p

TABLE 46.7 Raman Polarizability Parameters for C$_{60}$ and SWCNT Molecules

Molecule	Bond	$2\alpha'_p + \alpha'_1$	$\alpha'_1 - \alpha'_p$	$\alpha_1 - \alpha_p$
C$_{60}$ [83]	Single	1.28 ± 0.30	1.35 ± 0.20	1.28 ± 0.2
	Double	5.40 ± 0.70	4.50 ± 0.50	0.00 ± 0.20
SWCNT [40]	Double	4.7	4	0.04

Note: These parameters are fitted in order to reproduce the experimental Raman intensities.

correspond to the longitudinal and perpendicular bond polarizability, respectively. Within this approach and in order to evaluate the derivatives, one assumes that the bond polarizability parameters are functions of the bond lengths r alone. $\alpha'_1 = (\partial \alpha_1 / \partial r)\big|_{r=r_0}$, $\alpha'_p = (\partial \alpha_p / \partial r)\big|_{r=r_0}$, where r_0 is the equilibrium distance.

For C$_{60}$, the sum over bonds in Equation 46.47 includes two types of bonds, single and double, so there is one parameter that determines the Raman intensities for SWCNT. The Raman intensities of the modes in C$_{60}$, SWCNT, and peapods are calculated within the nonresonant bond-polarization theory. The bond polarizability parameters used are listed in Table 46.7.

46.4.1.3 The Spectral Moment's Method

A usual method to calculate the Raman spectrum requires, besides the polarization parameters, the eigenvalues and the eigenvectors, which can be obtained by direct diagonalization of the dynamical matrix of the system. However, when the system contains a large number of atoms, as for peapods, the dynamical matrix is very large, and its diagonalization fails or requires long computing time. By contrast, the spectral moment's method allows us to compute directly the Raman spectrum of very large harmonic systems without any diagonalization of the dynamical matrix [41,81]. Otherwise, for small samples, both approaches lead exactly to the same position and intensity for the different peaks.

If we know the components tensor $a(j)$ (Equation 46.45), the response of the system is given by the symmetrical function $H(\omega)$ (Equation 46.42 for example), which can be written as

$$J(u) = \sum_j |a(j)|^2 \delta(u - \lambda_j) \quad (46.48)$$

where $u = \omega^2$ and $\lambda_j = \omega_j^2$.

The determination of $J(u)$ can be directly carried out from the dynamical matrix $\tilde{D}$. In the Dirac-bracket representation, we have $D_{\alpha\beta}(nm) = \langle n\alpha | \tilde{D} | m\beta \rangle$, $e_j(n\alpha) = \langle n\alpha | j \rangle$, and $\tilde{D} | j \rangle = \lambda_j | j \rangle$. One can show that

$$J(u) = -\frac{1}{\pi} \lim_{\epsilon \to 0_+} \text{Im}[R(z)] \quad (46.49)$$

where $z = u + i\epsilon$ and $R(z) = \langle q | (z\tilde{I} - \tilde{D})^{-1} | q \rangle$. The $|q\rangle$ represents the charge tensor, which depends on the process and is given by

$$|q\rangle = \sum_{n\delta} \frac{\pi^n_{\alpha\gamma,\delta}}{\sqrt{M_n}} |n\delta\rangle \quad (46.50)$$

and the function $J(u)$ can be expressed as

$$J(u) = \sum_j |\langle q | j \rangle|^2 \delta(u - \lambda_j) \quad (46.51)$$

The spectral moment's method is devised to develop $R(z)$ in a continued fraction:

$$R(z) = \cfrac{b_0}{z - a_1 - \cfrac{b_1}{z - a_2 - \cfrac{b_2}{z - a_3 - \cfrac{b_3}{\cdots}}}} \quad (46.52)$$

where the coefficients a_n and b_n are given by

$$a_{n+1} = \frac{\bar{v}_{nn}}{v_{nn}}; \quad b_n = \frac{v_{nn}}{v_{n-1n-1}}; \quad b_0 = 1 \quad (46.53)$$

The spectral generalized moments v_{nn} and $\bar{v}_{nn}$ of $J(u)$ are directly obtained from the dynamical matrix $\tilde{D}$:

$$v_{nn} = \langle q | P_n(\tilde{D}) P_n(\tilde{D}) | q \rangle; \quad \bar{v}_{nn} = \langle q | P_n(\tilde{D}) \tilde{D} P_n(\tilde{D}) | q \rangle \quad (46.54)$$

The polynomials $P_n(\tilde{D})$ obey the following recurrence law:

$$P_{n+1}(\tilde{D}) = (\tilde{D} - a_{n+1}) P_n(\tilde{D}) - b_n P_{n-1}(\tilde{D}); \quad P_{-1}(\tilde{D}) = 0; \quad P_0(\tilde{D}) = 1 \quad (46.55)$$

As shown in our earlier works [41,81], a sharp truncation of the continued fraction leads to the appearance of sharp lines in the calculated low-frequency spectrum. On the other hand, the calculations of Raman and infrared spectra show that a limited number of moments are sufficient to obtain the frequency of active modes with a good accuracy. Calculations of the Raman spectra of peapods show that about 500 moments are sufficient to obtain good results for larger samples (~25,000 degrees of freedom).

In the following, the calculated Raman spectra are presented. The Raman scattering cross section that can be calculated assuming the induced polarization tensor is known for both C$_{60}$ and SWCNT forming peapod. The Raman cross section was calculated assuming that scattering can be described within the framework of the bond polarizability model. In all calculations, the nanotube axis is along the Z axis, and a carbon atom of the SWCNT is along the X axis of the nanotube reference frame. The x, y, and z are the axes of the laboratory frame, and the laser beam is assumed to be along the y axis. In the VV configuration, both the incident and scattered polarizations are along the z axis. For the VH configuration, the incident and scattered polarizations are along the z and x axes, respectively. For oriented peapod, we considered the $Z(X)$ nanotube axis as being along the $z(x)$ axis (VV = ZZ and VH ≡ ZX). For unoriented peapods samples, an isotropic orientation of the tube axis is taken into account in the calculations of the VV and VH Raman spectra.

46.4.2 Raman Active Modes of C_{60}s and Carbon Nanotubes

46.4.2.1 C_{60} Monomer and Polymer

In the I_h symmetry group, only 14 normal modes are Raman (2 Ag + 8 Hg) and infrared ($4T_1u$) active, and these have been assigned with certainty, while the identification of the remaining 32 silent modes has been difficult due to the complexity of the vibrational spectrum. Several theoretical investigations have been devoted to the calculation of Raman active modes in C_{60} comparing them to experimental results. For example, the average Raman intensity calculated for the isolated C_{60} molecule is shown in Figure 46.18 [82]. Calculations are performed by using the force constants model of Jishi et al. [48] within the bond-charge polarizability model and C_{60} polarizability parameters of Snoke and Cardona [83]. The Raman intensities are averaged over all sample orientations. The comparison between the calculated Raman-active mode frequencies to both those obtained by Jishi et al. and the experimentally observed frequencies for C_{60} molecule ([48], Table 46.4) shows that the model recovers well the two nondegenerate A1g modes located at 492 and 1468 cm^{-1} and a small difference (2–5 cm^{-1}) between others. This difference can be related to the used C_{60} model structure.

Since the icosahedral symmetry of a C_{60} molecule is broken in the photo polymerized phase by the formation of intermolecular bonds, new infrared and Raman active modes are observed in the spectral region 120–1600 cm^{-1} [19,59,63,84,86]. Raman spectroscopy is capable of detecting intermolecular bonds to C_{60} and even quantify their number, due to the characteristic Ag(2) mode softening. In photo polymerized solid C_{60}, the Raman

peaks have been observed at 1452 and 1459 cm^{-1} [63]. New features are expected to appear in the low frequency range due to intermolecular excitations of covalently bonded C_{60} molecules. The covalent bonding has been suggested both theoretically and experimentally [87–91] because of the formation of four membered rings of C–C bonds joining adjacent C_{60} molecules. In recent years, a variety of theoretical methods have been applied to the calculation of internal modes of C_{60} polymer material: *ab initio* calculations [88], tight binding Hamiltonian models [92], and force constants model [49].

Using the force constant model (Table 46.6 together with the bond-charge polarizability parameters of Snoke and Cardona [83]), we have investigated the Raman-active mode in C_{60} dimer and polymer chains. The calculated parallel (ZZ) polarized Raman spectra for C_{60} dimer and infinite C_{60} linear polymer chains are shown in Figure 46.19. The spectra are displayed in 0–800 cm^{-1} (Figure 46.19a) and 800–1600 cm^{-1} (Figure 46.19b) ranges.

Concerning the low-frequency modes, below 200 cm^{-1}, no mode is expected in free C_{60} (Figure 46.19a, bottom). By contrast, in C_{60} dimer, three interball modes are calculated at 106, 99, and 82 cm^{-1}. The most Raman active of the three modes is the one at 82 cm^{-1} (Figure 46.19a, middle). Since this vibrational mode is the lowest energy mode of the Ag representation, one can attribute this mode to the one observed approximately at 96 cm^{-1} [91] for a C_{60} polymer chain. This same mode is calculated at 69 and 89 cm^{-1} by *ab initio* [43] and density-functional based nonorthogonal tight-bending (DF-TB) approaches [89], respectively. The calculated pattern for this mode shows the analog of the stretching mode in diatomic molecules (Figure 46.9a) in Ref. [43]. The Raman spectrum of the

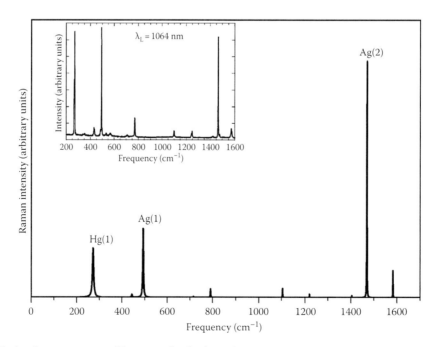

FIGURE 46.18 The calculated Raman spectra of free C_{60} molecule; the peaks intensities are averaged over all molecule orientations. Inset shows a typical Room temperature Raman spectrum of a C_{60} film. The laser excitation wavelength is 1064 nm. (After Chase, B. et al., *J. Phys. Chem.*, 96, 4262, 1992. With permission.)

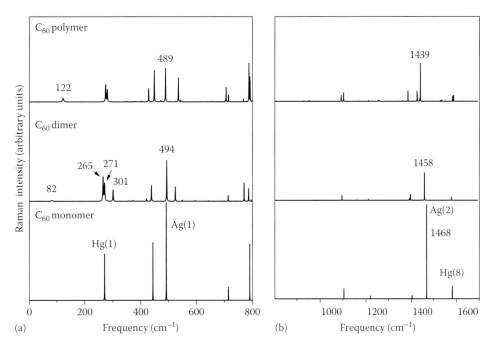

FIGURE 46.19 The calculated polarized ZZ Raman spectra of C_{60} monomer, C_{60} dimer, and infinite C_{60} polymer chain. Spectra are displayed in lower (a) and height (b) frequency regions.

C_{60} polymer chain is featured by five peaks at 77, 88, 100, 122, and 135 cm^{-1}. The most Raman active is the one at 122 cm^{-1} (Figure 46.19a, top). Experimentally, the Raman spectrum of photopolymerized C_{60} shows a new Raman active mode at 118 cm^{-1} [59,] which is in well agreement with the Raman-active mode calculated at 122 cm^{-1} for a C_{60} polymer chain.

The Hg(1) mode, calculated at 270 cm^{-1} in C_{60} (Figure 46.19a bottom), splits in C_{60} dimer (Figure 46.19a middle) and C_{60} polymer (Figure 46.19a, top). Three modes at 265, 272, and 301 cm^{-1} in C_{60} dimer and four modes at 242, 270, 280, and 348 cm^{-1} in C_{60} polymer are calculated. Experiments [59] show a peak at 270 cm^{-1} for pure C_{60}, and peaks at 259, 271, and 301 cm^{-1} for the photopolymerized material in qualitative agreement with calculation results. No shift of the Ag(1) mode, located at 497 cm^{-1} in C_{60}, was measured experimentally in C_{60} dimer and C_{60} polymer. In agreement with experiment, no shift of the Ag(1) mode is calculated in C_{60} dimer. By contrast, a downshift of about 5 cm^{-1} is predicted in C_{60} polymer. In the dimer spectrum, the five-fold degenerate Hu(2) mode, located at 543 cm^{-1} in C_{60}, splits and causes new peaks in the vicinity of the Ag(1) mode (Figure 46.19a).

In Figure 46.19b, we compare the 800–1600 cm^{-1} region of the calculated Raman spectra of C_{60} monomer, C_{60} dimer, and C_{60} polymer chain. Of a particular interest is the calculated shift of the Ag(2) mode. As shown in Figure 46.19b, this mode shifts from 1468 cm^{-1} in C_{60} (Figure 46.19b, bottom) to 1458 cm^{-1} in C_{60} dimer (Figure 46.19b, middle) and to 1439 cm^{-1} in C_{60} polymer chain (Figure 46.19b, top). The values of the shifts are close to those found by *ab initio* calculations [43]. The calculated 10 cm^{-1} shift from C_{60} monomer to C_{60} dimer well agrees with the observed shift of 10 cm^{-1} from pristine C_{60} solid phase

to C_{60} photopolymerized state [59]. The calculated Hg(8) mode is slightly downshifted from 1582 cm^{-1} in C_{60} monomer to 1580 cm^{-1} (in C_{60} polymer. Experimentally, a downshift of about 6 cm^{-1} was measured [59].

46.4.2.2 Carbon Nanotubes

Raman spectroscopy has been widely used as a powerful tool for the investigation or characterization of single wall carbon nanotubes (SWNT). Several reviews were published on the theoretical and experimental aspects of Raman scattering by SWCNT [39,93–95]. In this section, we only discuss the recent features of the Raman spectra of nanotubes that can help in the investigation of the Raman spectra of carbon peapods systems.

The Raman signature of SWCNT is unique and easily recognizable (Figure 46.20). It is generally dominated by four bunches of peaks. The first bunch is in the range 1500–1600 cm^{-1} and corresponds essentially to some elongations of the carbon–carbon bonds usually referred to as tangential modes (TM) because the atomic motions remain in the plane of the graphene sheet, or as G-band since similar modes are observed in graphite. Note that most of the carboneous species made of sp^2 hybridized atoms present a Raman signature in this spectral range. The second characteristic bunch is really specific to SWCNT. It is usually observed in the range 100–300 cm^{-1}. It corresponds to in-phase radial motions of the carbon atoms, and it is therefore usually referred to as RBM (Figure 46.20). Two other intense and broad features are generally observed: the so-called D-band in the range 1250–1350 cm^{-1} and its second harmonic, called D band in the range 2500–2700 cm^{-1}. These bands are not specific to SWCNT; they are expected for most of carboneous materials. They are both observed because

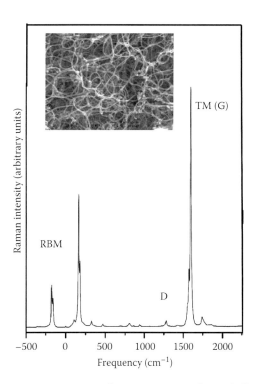

FIGURE 46.20 Experimental Raman spectrum from a bulk sample of SWCNTs taken with a 1064 nm. The tubes are of the Montpellier-type and were produced by the arc-discharge method. Inset shows a typical SEM-picture of the sample showing that the tubes form long ropes/bundles.

of a double resonant process that involves two resonant electronic states of the SWCNT [93,94].

The very popular property of the Raman spectra of SWCNT is the frequency of the RBM (because of its specific motion), at first approximation, proportional to the inverse of the tube diameter ([39,93,96] and references therein). Therefore, a low-frequency Raman measurement is a very easy way to estimate SWCNT diameters. Several calculations, in the valence force field (VFF) or tight-binding models, were carried out to determinate the relation between RBM frequency and SWCNT diameter ([5,39,96]). For single tubes, the frequency is inversely proportional to the diameter, with a constant proportionality that depends on the model and the parameters used in the calculations:

$$\omega_{RBM} \, (cm^{-1}) = \frac{A}{d} \, (nm) \qquad (46.56)$$

Equation 46.56 was derived by several groups ([39] and references therein). However, the value of the prefactor A is model-dependent (from 220 to 250 cm^{-1} nm). Moreover, in most materials, SWCNT self-assemble in bundles of a few tens to a few hundreds of tubes, and the frequency of the RBM increases because of vdW tube–tube interactions. This was accounted for in some works, for example, by considering a Lennard–Jones potential in the VVF model to describe intertube interactions [97,98]. The calculation results were well fitted by the following phenomenologic power law:

$$\omega_{RBM} \, (cm^{-1}) = \frac{238}{d^{0.93}} \, (nm) \qquad (46.57)$$

In the range 1–2 nm, Equation 46.57 is equivalent to the relation:

$$\omega_{RBM} \, (cm^{-1}) = \frac{224}{d} \, (nm) + 14 \, (cm^{-1}) \qquad (46.58)$$

where the first term corresponds with what is expected for isolated tubes, while the second accounts for intertube interaction in the bundles. Equation 46.58 can be used to estimate the diameter distribution of SWCNT in bundles. In summary, Raman spectroscopy on SWCNT provides a specific signature for each nanotube, with an indication on its diameter, possibly its chiral angle, and its semiconducting or metallic character. One must also remember that Raman is a non-destructive technique that can be used directly on any kind of sample (powder, suspension, mat, film, fiber, isolated tubes, etc.). This explains why Raman spectroscopy becomes such an essential technique in the science of nanotubes. In the next sections, we illustrate the effectiveness of the technique to study nanotube-based systems of current interest: C_{60} peapod.

46.4.3 Raman Experiments on C_{60} Peapods

Fullerenes inside a nanotube are shielded from the surroundings due to the confinement of their vibrational properties, which may be changed. The properties of nanotubes can be also influenced by the endohedral molecules. The Raman spectrum of single-wall carbon nanotubes peapods was carried out on a sample containing ropes of single-wall carbon nanotubes prepared by laser desorption technique purified and filled. The Raman spectra taken on peapod sample were recorded for laser lines extending from 1.8 to 2.6 eV at 90 and 20 K, respectively. The Raman pattern of the radial part and the tangential part of the peapod spectrum are depicted in Figure 46.21. The G mode is very similar to the response from pristine tubes. The response of C_{60} is also resonance enhanced for different laser excitation, and all fundamental Raman lines of the encaged C_{60} peas, with the known exception of Hg(8), were detected. The Ag(2) derived mode of the peas is split into two components that are observed at 1466 and 1474 cm^{-1}. These components are just below and above the position of 1469 cm^{-1}, which is known for pristine C_{60}. Significantly, the same behavior is observed for the other totally symmetric Ag(1) mode of C_{60}, as demonstrated in Figure 46.21b. For excitation with the blue lasers, we see only the components at 490 cm^{-1}. For green laser excitation, the line splits into two components located at 490 and 502 cm^{-1}. The features at about 1320 and 1430 cm^{-1} correspond to the D line of the tubes and the Hg(7) mode of the peas, respectively. For the radial part of the spectrum in Figure 46.21, the RBM shows the above-mentioned photoselective resonance scattering, whereas the response from C_{60} is very similar to the free molecules with the Hg(1) and Hg(2) lines at 270 and 430 cm^{-1}, respectively.

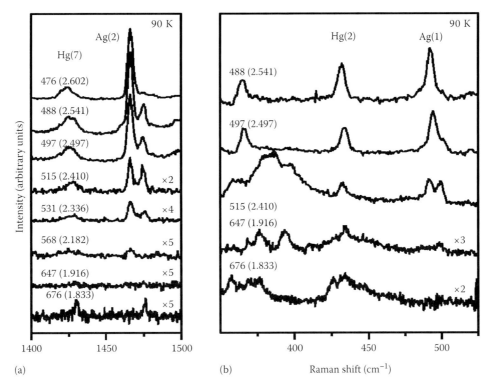

FIGURE 46.21 Tangential part (a) and radial part (b) of the Raman spectrum at 90 K for peapod material excited with different lasers as indicated in nm (eV). The strong line around 425 cm^{-1} is the Hg(2) mode of the C_{60} peas. (From Pfeiffer, R. et al., *Phys. Rev. B*, 69, 035404, 2004. With permission.)

Electron microscopy [12] images of peapods have revealed the one-dimensional character of the C_{60} chains inside the carbon nanotubes. The Raman spectra have, therefore, been recorded at helium temperature to derive the peapods response before polymerization (see Figure 46.22). The spectra were found to be stable, indicating that no photopolymerization occurred during the measurements. However, a mode located at ~90 cm^{-1} was observed at low temperature, the frequency of which is close to the dumbbell-like mode of a C_{60} dimer (96 cm^{-1}). X-ray-diffraction [13] and electron-diffraction [15] measurements at room temperature have stated that the inter-C_{60} distance ranges from 0.95 to 0.98 nm in highly filled samples. This distance is consistent with vdW interactions between fullerenes, as indicated by calculations [9]. However, the XRD derived thermal expansion of the C_{60} chains was found to be much smaller than the one of the three-dimensional 3D-crystalline phase, and comparable to the in-plane dilation for graphite. All these results show that the precise nature of the $C_{60} - C_{60}$ bond in peapods is ambiguous: pure structural data suggest a vdW character, while methods involving dynamical properties reveal a much stronger polymer-like bond nature. This ambiguity led the C_{60} chains to be described as a loosely-bonded polymer structure. However, such experimental work should include Raman investigations on samples showing peapods that have various structural characteristics: different tube diameters, different filling rates, and bundles with various sizes, e.g., various numbers of tubes.

46.4.4 Calculation of Raman Active Modes of C_{60} Peapods

Most of the Raman experiments are performed on unoriented peapods samples. To make the comparison with the experimental results as realistic as possible, the Raman spectra are averaged over the peapod orientations with regard to the laboratory frame. In Figure 46.23, we present the VV and VH Raman spectra calculated for unoriented samples of (10, 10) and (13, 13) infinite SWCNTs and for their corresponding peapods: C_{60}@(10, 10) (linear peapod) and C_{60}@(13,13) (zigzag peapod) in the low-frequency (left) and high-frequency ranges (right). The calculations of the Raman spectrum for both linear and zigzag peapods show that the Raman active modes of the C_{60} do not depend on the tube chirality. The calculations also show that there is no significant difference between the Raman spectra of the optimized and non-optimized peapods in terms of relative angles of rotation and relative translations of the C_{60} molecules inside the tube.

Concerning the main modes of C_{60}, the $H_g(6)$, $H_g(7)$, $H_g(8)$, and $A_g(2)$ modes are located around 1220, 1403, 1583, and 1469 cm^{-1}, respectively in C_{60}@(10, 10) and C_{60}@(13, 13), independently of the configuration of C_{60} molecules inside the tube. These modes show a small upshift (1–2 cm^{-1}) with respect to their positions in the free C_{60} spectrum. By contrast, the $A_g(1)$ and $H_g(2)$ modes of C_{60} are split in C_{60}@(10, 10) [C_{60}@(13, 13)], and the two components of each doublet are located at (498, 511) cm^{-1} and (448, 452)

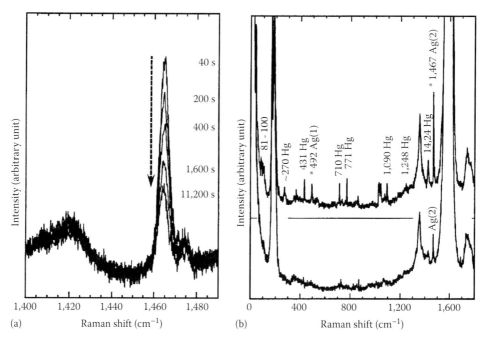

FIGURE 46.22 The Raman spectra of C_{60} peapods. (a) Time-resolved Ag(2) mode at room temperature measured in air. (b) Raman spectrum at liquid-helium (upper curve) and room temperatures (lower curve). The laser wavelength used is 488 nm. (From Kataura, H. et al., *Appl. Phys. A*, 74, 349, 2002. With permission.)

cm^{-1} [(495, 506) cm^{-1} and (446, 448) cm^{-1}], respectively. It can be pointed out that the splitting of the $A_g(1)$ and $H_g(2)$ modes was also obtained for a model consisting of a single C_{60} molecule capped inside a nanotube. This suggests that the splitting is related to the interactions between the C_{60} molecules and the nanotube host, and not to inter-C_{60} interactions. Finally, the $H_g(1)$ mode shows three components located at (275, 282, 289) cm^{-1} and (272, 276, 283) cm^{-1} in $C_{60}@(10, 10)$ and $C_{60}@(13, 13)$, respectively.

Concerning the main modes of SWCNT, the calculated tangential-like modes (TLM) are almost not affected by the insertion of C_{60} molecules inside the nanotube. As already published, the A_{1g} RBM, located at 165 and 127 cm^{-1} for the (10, 10) and (13, 13) SWCNTs, respectively, were observed in the VV configuration but vanished in the VH configurations [41]. The same behavior is observed for the RBLMs in peapods, suggesting that the totally symmetric character of the RBM is conserved after the insertion of C_{60}. A shift of the RBLM with regard to their analogous RBM is found. This shift depends on the configuration of the C_{60} molecules inside the nanotube. More precisely, insertion of C_{60} molecules inside the inner space of the tube induces a slight upshift of the RBLM mode when C_{60} molecules are organized as a linear chain. For instance, the RBLM shifts from 165 cm^{-1} in (10, 10) tube to 170 cm^{-1} in $C_{60}@(10, 10)$ peapod (Figure 46.23). On the other hand, for C_{60} molecules forming a zigzag chain, RBLM displays a double-structure. For $C_{60}@(13, 13)$, this feature is made of a low-frequency component located at 125 cm^{-1}, downshifted with respect to the position of the RBM in the

(13, 13) SWCNT, and a high-frequency component at 133 cm^{-1} upshifted with respect to the RBM of the (13, 13) SWCNT (Figure 46.23).

46.4.5 Diameter Dependence of the Peapod Raman Spectrum

In this section, the Raman spectra have been calculated on individual oriented peapods. We focus on the BLM region of the VV($\equiv$ZZ) polarized Raman spectrum. We have investigated the dependence of some specific Raman active modes of infinite peapods as a function of the diameter of the nanotube. Figure 46.24 shows the evolution of the RBM and RBLM frequencies as a function of the inverse of the tube diameter.

The RBLMs of linear (diameter below 1.45 nm) and zigzag (diameter between 1.45 and 2.17 nm) peapods show distinct behavior. The behavior of the RBLM in linear peapods is qualitatively the same as that of the RBM. However, it deviates from the scaling law stated for the RBM frequency. For zigzag peapods, we have already reported the splitting of the RBM into two components. For tubes that have a diameter smaller than $D_0 \approx 1.78$ nm, associated to angles θ greater than $\theta_0 \approx 120°$ (see Table 46.2), the high- and low-frequency components of RBLM doublet, called RBLM(1) and RBLM(2) in the following, downshift when the tube diameter increases. Note that the slopes of the variations are different. By contrast, for tubes that have a diameter greater than $D_0 \approx 1.78$ nm, associated to angles θ lower than θ_0, the low(high)-frequency component of RBLM doublet downshifts (upshifts) when the diameter increases.

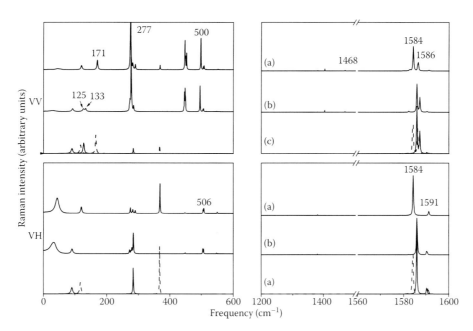

FIGURE 46.23 The VV (top) and VH (bottom) calculated Raman spectra in the BLM (left) and TLM (right) ranges of peapods: $C_{60}@(10, 10)$ (curve c) and $C_{60}@(13, 13)$ (curve b). The calculated Raman spectra of the empty SWCNTs are displayed: (10, 10) (curve a, dashed line) and (13, 13) (curve a, solid line).

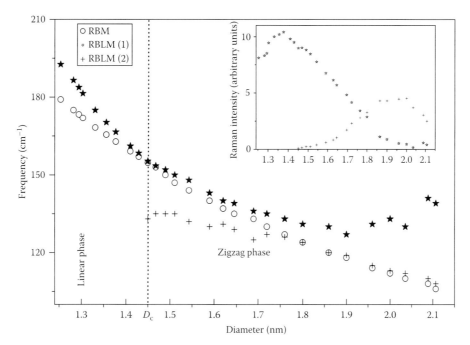

FIGURE 46.24 The diameter dependence of the RBLM frequency in linear and zigzag peapods: high-frequency component, RBLM(1), (star) and low-frequency component, RBLM(2), (cross). The diameter dependence of the RBM for the related empty tubes is also displayed (open circles). The limit (D_c, see text) between linear and zigzag peapods is identified by the vertical dashed line. Inset: the corresponding intensities of the RBLM(1) and RBLM(2) versus the diameter of the nanotubes.

In the inset of Figure 46.24, we have reported the variation of the Raman intensities of the two components of the RBLM as a function of the nanotube diameter. The ratio $r = I_{RBLM(1)}/I_{RBLM(2)}$ is equal to 1 for $D \approx D_0 (\theta \approx \theta_0)$, $r > 1$ for $D < D_0 (\theta > \theta_0)$, and $r < 1$ for $D > D_0 (\theta < \theta_0)$. When θ reaches 180° (60°), the ratio $r \gg 1$ ($r \ll 1$) and the low-frequency region are dominated by the component of the RBLM doublet located at the frequency the closest of the frequency of the RBM. The frequency and intensity behaviors of the RBLM(1) and RBLM(2) suggest an exchange from one mode to the other associated with an anticrossing effect of the dispersion curves of the RBLM(1) and RBLM(2) with the diameter in zigzag peapods.

46.4.6 Dependence of Raman Spectrum with the Filling Factor

In the case of a real sample, it seems reasonable to consider that all tubes may not be fully filled with C_{60}s, despite efforts made to obtain filling rates as high as possible [14]. In this section, we make the hypothesis of partial filling of the tubes with a quasi-infinite long chain. We assume here that the molecules tend to cluster within nanotubes. Indeed, this should correspond to a low energy configuration of the system, the energy being lowered by the attractive C_{60}–C_{60} interactions. This hypothesis is supported by observations reported in Ref. [15]. To avoid finite size effects, one can apply periodic conditions to the tube. One can see that the number of C_{60} molecules inside the SWCNT has no significant effect on the Raman spectrum except on the RBLM mode, which is most sensitive to the degree of filling of the SWCNTs.

The dependence of the VV polarized Raman spectrum with the filling factor has been calculated for the C_{60}@(10, 10) and C_{60}@(18, 0) *linear peapods* (Figure 46.25a and c) and the C_{60}@(11, 11) and C_{60}@(13, 13) *zigzag peapods* (Figure 46.25b and d). The spectra were calculated for five values of the filling factor $F = 20\%$, 40%, 60%, 80%, and 100% corresponding to 4, 8, 12, 16, and 20 C_{60} molecules, respectively, inside SWCNT's. Figure 46.25 shows the filling factor dependence of the Raman spectrum in the BLM range.

For *linear peapods*, a single RBLM peak is calculated for empty tube (0%) and 100% filling factor. Let us call ω_h and ω_l the frequency of the 100% filling factor and empty tube, respectively. For small tube diameters, the increase of F from 0% to 100% leads to the appearance of two peaks at frequencies close to ω_l and ω_h. The intensity of these peaks shifts from the one located around ω_l to that located around ω_h when increasing F. An illustration

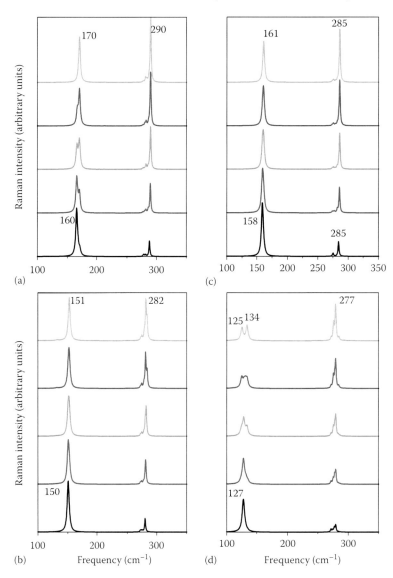

FIGURE 46.25 The ZZ polarized Raman spectra displayed in the BLM region of infinite peapods as a function of the filling factor. The filling factor is 20%, 40%, 60%, 80%, and 100% from bottom to top. Linear peapods: C_{60}@(10, 10) (curve a) and C_{60}@(18, 0) (curve c). Zigzag peapods: Co_{60}@(11, 11) (curve b) and C_{60}@(13, 13) (curve d).

of this behavior is shown for the C$_{60}$@(10, 10) peapod (tube diameter close to 1.36 nm) in Figure 46.25a. By contrast, for larger tube diameter, a single RBLM mode for all the filling levels is obtained. This behavior is displayed for the C$_{60}$@(18, 0) linear peapod, with tube diameter close to 1.41 nm in Figure 46.25c. The presence of a single RBLM is associated with the overlap of modes at ω_l and ω_h. Indeed, for linear peapods, we found that the difference $|\omega_h - \omega_l|$ varies from ~15 cm^{-1} for diameter around 1.26 nm to 0 cm^{-1} for diameter around 1.45 nm.

For *zigzag peapods*, and θ around $\theta_0 = 120°$, a double-peak structure appears when the filling factor increases, as shown in the Raman spectrum of C$_{60}$@(13, 13) peapod (Figure 46.25d). By contrast, no significant change of the RBLM was observed for zigzag peapods systems when θ is close to 180° (C$_{60}$@(11, 11) peapod, for example, Figure 46.25b) or θ close to 60° (not shown). Finally, we notice that, as expected, the intensity of the C$_{60}$ $H_g(1)$ modes increases with the filling factor.

46.4.7 Bundling Effect on Peapod Raman Spectrum

In general, macroscopic peapod samples are packed into bundles wherein they are maintained together by vdW inter-tubes interactions. This organization is clearly visible on TEM pictures [13,17,18]. It is important to investigate the effect of inter-peapod interaction onto the phonon frequencies. For carbon nanotubes, only a weak perturbation was found for closely packed nanotube arrays. In the SWCNT bundle calculations, the effect of bundling does not exceed 10 cm^{-1}, except for the very low frequency E_1 and E_2 symmetry modes. For the radial breathing mode,

we found stiffening by 4%. Concerning the effect of bundling on C$_{60}$ peapod. When isolated C$_{60}$ peapods are closely packed together, a three-dimensional carbon bundle is formed. These systems can be represented by two-dimensional infinite trigonal lattices of uniform cylinders filled by C$_{60}$ molecules where the optimized intertube spacing d_{t-t} is roughly equal to 0.32 nm. Now the calculations are performed on infinitely homogenous bundles (crystal) of C$_{60}$ peapods of several diameters. We recall that infinite peapods are obtained by applying periodic conditions on unit cells of the studied peapods. The results for typical peapods C$_{60}$@(10, 10), C$_{60}$@(11, 11), C$_{60}$@(12, 12), C$_{60}$@(13, 13), C$_{60}$@(14, 14), and C$_{60}$@(15, 15) are reported in Figure 46.26, where the ZZ Raman spectra are displayed in the breathing-like mode (BLM: in the left) and tangential-like mode (TLM: on the right) regions.

Independently of the configuration of C$_{60}$ molecules, the calculations of the Raman spectrum for both linear and zigzag peapods show that the frequency of the main modes of C$_{60}$ are slightly affected with respect to their positions in the isolated peapod spectrum. Concerning the main Raman active modes of SWCNT, in the TLM region, comparing the Raman spectra of isolated peapods and organized into bundle, no significant difference is shown except a slight upshift of TLM Raman active modes in the peapod crystal. However, in the BLM region, the comparison between the Raman spectra of isolated and bundle of peapods shows the appearance of a second mode at a frequency higher than that of RBLM (for linear peapods), RBLM(1), and RBLM(2) (for zigzag peapod). This mode is the full-symmetric mode with breathing-like shape reported in different studies [98] on bundled SWCNTs.

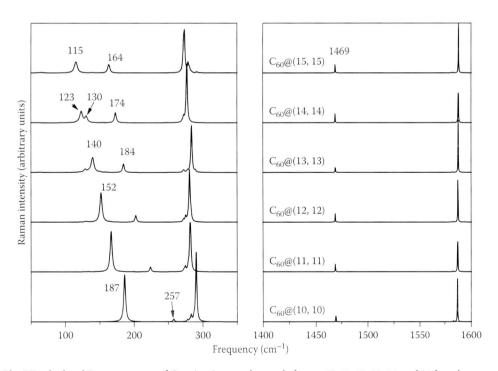

FIGURE 46.26 The ZZ calculated Raman spectra of C$_{60}$@(n, n) peapod crystals for n = 10, 11, 12, 13, 14, and 15 from bottom to top. Spectra are displayed in RBLM (left) and TLM (right) regions.

46.4.8 C_{60} Polymerization Effect on Peapod Raman Spectrum

Fullerene polymerization is dependent on the mutual orientation of neighboring molecules. Inside a peapod, not all arrangements are possible, and thus creation of a polymer is not subject only to the correct choice of p-T treatment conditions any longer. Confinement effects are thus expected to alter the behavior of fullerenes under high pressure. The Raman investigations of peapod sample have been performed [17], and Figure 46.22 shows the typical Raman spectra of C_{60} peapods. It was found that a photopolymerization of the C_{60} under a 488 nm laser excitation wavelength occurs. The latter observation suggests that C_{60} are spinning inside the hollow core of the tubes at room temperature. Raman spectra have, therefore, been recorded at helium temperature to derive the peapods response before polymerization. The spectra were found to be stable, indicating that no photopolymerization occurred during the measurements. However, a mode located at 90 cm⁻¹ was observed at low temperature, the frequency of which is close to the dumbbell-like mode of a C_{60} dimer (96 cm⁻¹). Additionally, relatively weak Ag(1) and Ag(2) mode intensities and broadening of the Hg(1) mode indicate the formation of dimers, trimers, or oligomers.

Now we present the results of the lattice dynamic calculations of Raman spectra of peapods made from a different possible polymer structures for the encapsulated C_{60} chain. Based on the agreement between the experiment and calculations for C_{60} dimer and C_{60} polymer chain, the same set of force constants

is used to calculate the Raman spectrum of a linear chain of C_{60} dimers and a C_{60} polymer chain encapsulating inside SWCNT. Depending on the nanotube diameter, different configurations of dimerized and polymerized C_{60} molecules confined inside SWCNT can certainly exist. However, we restrict the study by considering only the linear chain of C_{60} dimers and a C_{60} polymer chain confined inside a (10, 10) SWCNT. Indeed, because the diameter of (10, 10) SWCNT fits the diameter of a C_{60} molecule, linear chain is the most probable configuration [52]. Finally, we deal with completely filled nanotubes.

In Figure 46.27, the ZZ Raman spectra of (10, 10) SWCNT are compared with those of three kinds of nanopeapods: (1) a linear chain of C_{60} monomers confined into a (10, 10) SWCNT and is called in the following as $(C_{60})_1@(10,10)$, (2) a linear chain of C_{60} dimers confined into a (10, 10) SWCNT and is called in the following as $(C_{60})_2@(10, 10)$, and (3) a linear polymer chain of C_{60} molecules confined into a (10, 10) SWCNT and is called in the following as $(C_{60})_\infty@(10, 10)$.

The RBM of a (10, 10) SWCNT, located at 165 cm⁻¹, exhibits a slight upshift when the nanotube is filled (Figure 46.27 left). It is located at 169 cm⁻¹ in $(C_{60})_1@(10, 10)$, at 170 cm⁻¹ in $(C_{60})_2@(10, 10)$ and 172 cm⁻¹ in $(C_{60})_\infty@(10, 10)$. By contrast, the position of the high frequency G-mode slightly depends on the filling of the nanotube. It is located at 1586 cm⁻¹ either in (10, 10) SWCNT or in the three nanopeapods (Figure 46.27 right).

More significant are the changes of the modes of C_{60} molecules when they are confined inside a (10, 10) SWCNT. In the low frequency range, the interball mode located at 82 cm⁻¹ for a

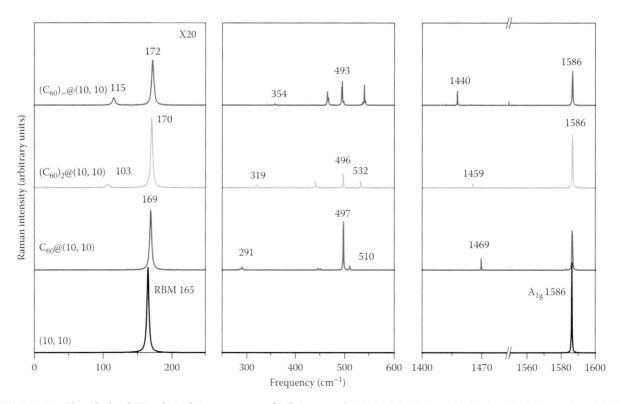

FIGURE 46.27 The calculated ZZ polarized Raman spectra of infinite peapods: $(C_{60})_1@(10,10)$ (curve b), $(C_{60})_2@(10,10)$ (curve c), and $(C_{60})_\infty@$ (10, 10) (curve d). The calculated Raman spectra of the empty SWCNT is displayed (curve a).

free C$_{60}$ dimer (Figure 46.19 middle) is upshifted to 103 cm^{-1} in (C$_{60}$)$_2$@(10, 10) (Figure 46.27 left).

As mentioned before, the single Hg(1) mode, located at 270 cm^{-1} in C$_{60}$ monomer, splits into three and four components in C$_{60}$ dimer and C$_{60}$ polymer, respectively (Figure 46.19). As these different C$_{60}$ structures are confined inside (10, 10) nanotube, all these components split and shift, leading to the observation of a multipeak band dominated by a mode around 291 cm^{-1} in (C$_{60}$)$_1$@(10, 10), around 319 cm^{-1} in (C$_{60}$)$_2$@(10, 10), and around 354 cm^{-1} in (C$_{60}$)$_\infty$@(10, 10).

In the range of the A$_{1g}$ mode located at 494 cm^{-1} of C$_{60}$ (Figure 46.19, bottom), two main components are calculated for the three kinds of nanopeapods investigated here. The position of the highest component, nonactive in free C$_{60}$ (Figure 46.2 bottom), is located around 510 cm^{-1} in (C$_{60}$)$_1$@(10, 10) and is upshifted to 532 and 538 cm^{-1} in (C$_{60}$)$_2$@(10, 10) and (C$_{60}$)$_\infty$@(10, 10), respectively. The lowest A$_{1g}$, calculated at 494 cm^{-1} in free C$_{60}$, is located around 497 cm^{-1} in (C$_{60}$)$_1$@(10, 10). It exhibits a downshift at 496 cm^{-1} in (C$_{60}$)$_2$@(10, 10) and 493 cm^{-1} in (C$_{60}$)$_\infty$@(10, 10).

Finally, concerning the high frequency range, we show that the A$_{1g}$ mode of C$_{60}$, C$_{60}$ dimer and C$_{60}$ polymer is slightly upshifted (of about 1 cm^{-1}) when they are confined into a (10, 10) SWCNT. They are located at 1469, 1459, and 1440 cm^{-1} in (C$_{60}$)$_1$@(10, 10), (C$_{60}$)$_2$@(10, 10), and (C$_{60}$)$_\infty$@(10, 10), respectively.

46.5 Concluding Remarks

In this conclusion, the structure and dynamic studies of C$_{60}$ carbon peapods are reviewed. Most of the peapod samples are featured by a diameter and chirality distributions. It was shown that the structure of the C$_{60}$ molecules inside nanotubes is diameter dependent. For the tube diameter investigated range, the C$_{60}$ molecules adopt a linear arrangement for SWCNTs diameter lower than 1.45 nm, and a zigzag configuration for diameters between 1.45 and 2.20 nm. Both for linear and zigzag peapods, the dependence of the calculated Raman spectrum on the diameter, the nanotube filling level of the tube encasing C$_{60}$ molecules chain have been analyzed. It has been reported experimentally that (1) for SWCNTs in the diameter range 1.35–1.76 nm, the RBLM was downshifted compared to the RBM of empty SWCNTs [21,22], and (2) for smaller diameters, two RBLM components that upshifted and downshifted in regard to the position of the RBM of the empty tube were measured [21,22]. The calculation of peapods Raman spectra shows that the splitting of the RBM mode in the opposite directions observed can experimentally be explained by a low filling factor of C$_{60}$ inside SWCNT in the synthesized peapod sample. On the other hand, the experimental GDOS of peapods is discussed in the light of lattice-dynamics calculations performed on three different models for the C$_{60}$ organization inside the tubes. In particular, intense modes on the low energy side of the gap separating the intermolecular from the intramolecular modes suggest the existence of strong bonds, such as valence interactions between parts of the monomers.

Since experiments have been performed on peapod samples featured by diameter and chirality distributions, the best way to test the calculation predictions is to focus on the behavior of C$_{60}$ modes. We have reported and discussed the Raman spectra of C$_{60}$ dimer and C$_{60}$ polymer chain either free or confined inside SWCNTs. Encapsulating C$_{60}$ dimers and a C$_{60}$ polymer chain inside a SWCNT has small consequences on the Raman spectra of C$_{60}$ dimers and C$_{60}$ polymer chain. Only slight shifts of C$_{60}$ modes as well as SWCNT modes were predicted under encapsulation. The most significant occurred in the low-frequency range where active interball modes and RBM are. These predictions are compared with experimental data [13,17,21,22,55], suggesting that calculations overestimate downshift of the Ag(2) mode in C$_{60}$ dimers and C$_{60}$ polymer chain and that the Raman spectra of peapods are sample dependent. A mixture of C$_{60}$ monomers and C$_{60}$ n-mer (dimer, trimer,… polymer) structures seems to be present in the majority of samples investigated.

Acknowledgment

We would like to thank Dr. Jean-Louis Sauvajol from the University of Montpellier (France) for his invaluable contributions to this topic.

References

1. Korto H W, Heath J R, O'Brien S C, Curl R F, and Smalley R E 1985 *Nature* 318, 162.
2. Ijima S 1991 *Nature* 354, 56.
3. Ijima S and Ichihachi T 1993 *Nature* 363, 603.
4. Lee R S, Kim H J, Fischer J E, Thess A, and Smalley R E 1997 *Nature* 388, 255.
5. Rao A M, Eklund P C, Bandow S, Thess A, and Smalley R E 1997 *Nature* 388, 257.
6. Bendiab N, Righi A, Anglaret E, Sauvajol J-L, Duclaux L, and Beguin F 2001 *Phys. Rev. B* 63, 153407.
7. Johnson M R, Rols S, Wass P, Muris M, Bienfait M, Zeppenfeld P, and Dupont-Pavlovsky N 2003 *Chem. Phys.* 293, 217.
8. Kim D H, Sim H S, and Chang K J 2001 *Phys. Rev. B* 64, 115409.
9. Rochefort A 2003 *Phys. Rev. B* 67, 115401.
10. Hornbaker D J, Kahng S J, Misra S, Smith B W, Johnson A T, Mele E J, Luzzi D E, and Yazdani A 2002 *Science* 295, 828.
11. Vavro J, Llaguno M C, Satishkumar B C, Luzzi D E, and Fischer J E 2002 *Appl. Phys. Lett.* 80, 23 1450.
12. Smith B W, Monthioux M, and Luzzi D E 1998 *Nature* 396, 323.
13. Kataura H, Maniwa Y, Kodama T, Kikuchi K, Hirahara K, Suenaga K, Iijima S, Suzuki S, Achiba Y, and Krätschmer W 2001 *Synth. Met.* 121, 1195.
14. Liu X, Pichler T, Knupfer M, Golden M S, Fink J, Kataura H, Achiba Y, Hirahara K, and Iijima S 2002 *Phys. Rev. B* 65, 045419.
15. Hirahara K, Bandow S, Suenaga K, Kato H, Okazaki T, Shinohara H, and Iijima S 2001 *Phys. Rev. B* 64, 115420.

16. Kuzmany H, Pfeiffer R, Kramberger C, Pichler T, Liu X, Knupfer M, Fink J et al. 2003 *Appl. Phys. A* 76, 449.

17. Kataura H, Maniwa Y, Abe M, Fujiwara A, Kodama T, Kikuchi K, Imahori H, Misaki Y, Suzuki S, and Achiba Y 2002 *Appl. Phys. A*, 74, 349.

18. Maniwa Y, Kataura H, Abe M, Fujiwara A, Fujiwara R, Kira H, Tou H et al. 2003 *J. Phys. Soc. Jpn.* 72, 45.

19. Zhou W, Winey K I, Fischer J E, Sreekumar T V, Kumar S, and Kataura H 2004 *Appl. Phys. Lett.* 84, 2172.

20. Smith B W and Luzzi D E 2001 *Chem. Phys. Lett.* 321, 169.

21. Bandow S, Takizawa M, Kato H, Okazaki T, Shinohara H, and Iijima S 2001 *Chem. Phys. Lett.* 23, 347.

22. Bandow S, Takizawa M, Hirahara K, Yudasaka M, and Iijima S 2001 *Chem. Phys. Lett.* 48, 337.

23. Berber S, Kwon Y K, and Tomanek D 2002 *Phys. Rev. Lett.* 88, 185502.

24. Ulbricht H, Moos G, and Hertel T 2003 *Phys. Rev. Lett.* 90, 095501.

25. Smith B W, Russo R M, Chikkannanavar S B, and Luzzi D E 2002 *J. Appl. Phys.* 91, 9333.

26. Hirahara K, Suenaga K, Bandow S, Kato H, Okazaki T, Shinohara H, and Iijima S 2000 *Phys. Rev. Lett.* 85, 5384.

27. Okada H, Saito S, and Oshiyama A 2001 *Phys. Rev. Lett.* 86, 3835.

28. Okada H, Otani M, and Oshiyama A 2003 *Phys. Rev. B* 67, 205411.

29. Rechfort A 2001 *Phys. Rev. B* 67, 115401.

30. Sloan J, Dunin-Borkowski R E, Hutchiso J L, Coleman K S, William V C, Claridg J B, Yor A P E et al. 2000 *Chem. Phys. Lett.* 316, 191.

31. Smith B W, Monthioux M, and Luzzi D E 1999 *Chem. Phys. Lett.* 315, 31.

32. Dresselhaus M S, Dresselhaus G, and Avouris P H 2001 *Carbon Nanotubes: Synthesis, Structure, Properties, and Applications*, Springer Verlag.

33. Dresselhaus M S, Dresselhaus G, and Eklund P C 1996 *Science of Fullerenes and Carbon Nanotubes: Their Properties and Applications*, Academic Press.

34. Kane C L, Mele E J, Johnson A T, Luzzi D E, Smith B W, Hornbaker D J, and Yazdani A 2002 *Phys. Rev. B* 66, 235423.

35. Pichler T, Kuzmany H, Kataura H, and Achiba Y 2001 *Phys. Rev. Lett.* 87, 267401.

36. Chiu P W, Gu G, Kim G T, Philipp G, Roth S, Yang S F, and Yang S 2001 *Appl. Phys. Lett.* 79, 3845.

37. Service R F 2001 *Science* 292, 45.

38. Kwon Y K, Tomanek D, and Iijima S 1999 *Phys. Rev. Lett.* 82, 1470.

39. Dressellhaus M S and Eklund P C 2000 *Adv. Phys.* 49, 705.

40. Saito R, Takeya T, Kimura T, Dresselhaus G, and Dresselhaus M S 1997 *Phys. Rev. B* 57, 4145.

41. Rahmani A, Sauvajol J-L, Rols S, and Benoit C 2002 *Phys. Rev. B* 66, 125404.

42. Rols S, Righi A, Alvarez L, Anglaret E, Almairac R, Journet C, Bernier P et al. 2000 *Eur. Phys. J. B* 18, 201.

43. Adams G B, Page J B, Sankey O F, Sinha K, Menendez J, and Huffman D R 1991 *Phys. Rev. B* 44, 4052.

44. Cho J H and Kang M H 1995 *Phys. Rev. B* 51, 5058.

45. Weeks D E and Harter W G 1989 *J. Chem. Phys.* 90, 4744.

46. Wu Z C, Jelski D A, and George T F 1987 *Chem. Phys. Lett.* 137, 291.

47. Onida G and Benedek G 1992 *Europhys. Lett.* 18, 403.

48. Jishi R A, Mirie R M, and Dresselhaus M S 1992 *Phys. Rev. B* 45, 13685.

49. Schöber H, Tölle A, Renker B, Heid R, and Gompf F 1997 *Phys. Rev. B* 56, 5937.

50. Ding Q, Jiang Q, Jin Z, and Tian D 1996 *Fullerene Sci. Technol.* 4, 31.

51. Charlier A, McRae E, Charlier M F, Spire A, and Forster S 1998 *Phys. Rev. B* 57, 6689.

52. Chadli H, Rahmani A, Sbai K, Hermet P, Rols S, and Sauvajol J L 2006 *Phys. Rev. B* 74, 205412.

53. Hodak M and Girifalco L A 2003 *Phys. Rev. B* 68, 085405.

54. Hodak M and Girifalco L A 2003 *Phys. Rev. B* 67, 075419.

55. Pfeiffer R, Kuzmany H, Pichler T, Kataura H, Achiba Y, Melle-Franco M, and Zerbetto F 2004 *Phys. Rev. B* 69, 035404.

56. Journet C, Maser W K, Bernier P, Loiseau A, Lamy de la Chapelle M, Lefrant S, Deniard P, Lee R, and Fischer J E 1997 *Nature* 388, 756.

57. Johnson R D, Meijer G, Bethune D S, and Am J 1990 *Chem. Soc.* 112, 8983.

58. Rohlfing E A, Cox D M, and Kaldor A 1984 *J. Chem. Phys.* 81, 3322.

59. Rao A M, Zhou P, Wang K A, Hager G T, Holden J M, Wang Y, Lee W T et al. 1993 *Science* 259, 955.

60. Luzzi D E and Smith B W 2000 *Carbon* 38, 1751.

61. Nuñez-Regueiro M, Marques L, Hodeau J-L, Berthoux O, and Perroux M 1995 *Phys. Rev. Lett.* 74, 278.

62. Ywasa Y, Arima T, Fleming R M, Siegrist T, Zhou O, Haddon R C, Rothberg L J et al. 1994 *Science* 264, 1570.

63. Rao A M, Eklund P C, Hodeau J-L, Marques L, and Nunez-Regueiro M 1997 *Phys. Rev. B* 55, 4766.

64. Thess A, Lee R, Nikolaev P, Dai H, Petit P, Robert J, Xu C et al. 1996 *Science* 273, 183.

65. Kong J, Cassell A M, and Dai H 1998 *Chem. Phys. Lett.* 292, 567.

66. Ulbricht H and Hertel T 2003 *J. Phys. Chem. B* 107, 14185.

67. Yu P Y and Cardona M 1999 *Fundamentals of Semiconductors*, 2nd edn., Springer, Berlin, Germany.

68. Jishi R A, Venkataraman L, Dresselhaus M S, and Dresselhaus G 1993 *Chem. Phys. Lett.* 209, 77.

69. Aizawa T, Souda R, Otani S, Ishizawa Y, and Oshima C 1990 *Phys. Rev. B* 42, 11 469.

70. Oshima C, Aizawa T, Souda R, Ishizawa Y, and Sumiyoshi Y 1998 *Solid State Commun.* 65, 1601.

71. Cambedouzou J, Rols S, Almairc R, and Sauvajol J L 2005 *Phys. Rev. B* 71, 041403(R).

72. Long D A 2002 *The Raman Effect*, John Wiley & Sons Ltd., Baffins Lane, Chichester, U.K.

73. Fabelinskii I L 1957 *Usp. Fiz. Nauk.* 63, 355.

74. Loudon R 1973 *Quantum Theory of Light*, Clarendon Press, Oxford, U.K.

75. Turell G 1972 *Infrared and Raman Spectra of Crystals*, Academic Press, New York.

76. Poulet H and Mathieu J-P 1976 *Vibrational Spectra and Symmetry of Crystals*, Gordon and Breach, New York.

77. Wolkenstein M 1941 *Dokl. Akad. Nauk. SSSR* 32, 185.

78. Bilz H, Strauch D, and Wehner R K 1984 *Handbuch der Physik*, Vol. XXV, Springer-Verlag, Berlin, Germany, Pt. 2d, p. 253.

79. Viliani G, Dellt'Anna R, Pilla O, and Montagna M 1998 *Phys. Rev. B* 37, 6500.

80. Guha S, Menendez J, Page J B, and Adams G B 1996 *Phys. Rev. B* 53, 13106.

81. Benoit C, Royer E, and Poussigue G 1992 *J. Phys.: Condens. Matter* 4, 3125.

82. Chadli H, Rahmani A, Sbai K, and Sauvajol J L 2005 *Physica A* 358, 226.

83. Snoke D W and Cardona M 1993 *Solid State Commun.* 87, 121.

84. Bacsa W S and Lannin J S 1994 *Phys. Rev. B* 49, 14750.

85. Chase B, Herron N, and Holler E 1992 *J. Phys. Chem.* 96, 4262.

86. Sauvajol J L, Brocard F, Hricha Z, and Zahab A 1995 *Phys. Rev. B* 52, 14839.

87. Strout D L, Murry R L, Xu C, Eckhoff W C, Odom G K, and Scuseria G E 1993 *Chem. Phys. Lett.* 214, 576.

88. Adams G B, Page J B, Sankey O F, and Ó Keeffe M 1994 *Phys. Rev. B* 50, 17471.

89. Porezag D, Pederson M R, Frauenheim T H, and Köhler T 1995 *Phys. Rev. B* 52, 14963.

90. Scuseria E G 1996 *Chem. Phys. Lett.* 257, 583.

91. Lebedkin S, Gromov A, Giesa S, Gleiter R, Renker B, Rietschel H, and Krätschmer W 1998 *Chem. Phys. Lett.* 285, 210.

92. Menon M, Subbaswamy K R, and Sawtari M 1994 *Phys. Rev. B* 49, 13966.

93. Dresselhaus M S, Dresselhaus G, and Jorio A 2002 *Carbon* 40, 2043.

94. Reich S, Thomsen C, and Maultzsch J 2004 *Carbon Nanotubes Basic Concepts and Physical Properties*, Wiley-VCH, Weinheim, Germany.

95. Dresselhaus M S, Dresselhaus G, Saito R, and Jorio A 2005 *Phys. Rep.* 409, 47.

96. Sauvajol J-L, Rols S, Anglaret E, and Alvarez L 2002 *Carbon* 40, 1697.

97. Kahn D and PLu J P 1999 *Phys. Rev. B* 60, 6535.

98. Henrard L, Hernandez E, Bernier P, and Rubio A 1999 *Phys. Rev. B* 60, 8521.

VI

Inorganic Fullerenes

47

Boron Fullerenes

Arta Sadrzadeh
Rice University

Boris I. Yakobson
Rice University

47.1 Introduction

There are three phases for the bulk boron: α-rhombohedral, β-rhombohedral [1,2], and α-tetragonal [3]. α-Rhombohedral is made of B_{12} units, with lattice constant 5.30 Å and angle 58.13° [4] (Figure 47.1). Each B_{12} icosahedron is slightly distorted due to the Jahn–Teller effect, reducing the symmetry from I_h to D_{3d} [4]. β-Rhombohedral has a somewhat complicated unit cell, composed of 105 atoms, with lattice constant 10.15 Å and angle 65.17° [2,5]. A striking B_{84} structure with icosahedral symmetry can be identified in β-rhombohedral [6]. α-Tetragonal unit cell is made of 50 atoms, with lattice constants $a = 8.75$ Å and $c = 5.06$ Å [4]. It contains four B_{12} icosahedra and two isolated 4-bonded atoms. Recently, by means of *ab initio* studies, it has been shown that α-rhombohedral is more stable than β-rhombohedral [7]. α-Rhombohedral bulk is a semiconducting material with an energy gap of ~2 eV [4].

Systematic experimental and theoretical work on noncrystalline boron started in the 1980s. Mass spectra of boron clusters were obtained in a series of experimental studies performed by Hanley and coworkers [8,9]. Further important experimental study on the mass spectra was performed by La Placa et al. [10]. The method used was laser ablation of hexagonal boron nitride, and the mass of the clusters was separated by a time-of-flight technique. The experimental data showed "magic numbers" by boron cations of the size of 5, 7, 10, and 11.

Based on a series of *ab initio* simulations, Boustani proposed an "Aufbau Principle," stating that the most stable boron clusters can be constructed using two basic units: pentagonal and hexagonal pyramids, B_6 and B_7, respectively [11].

The boron clusters up to the size of 20 atoms have been shown to fall into two major categories: quasi-planar and compact. Figure 47.2 shows different isomers with 14 or fewer atoms [12].

Kiran et al. showed that planar-to-tubular structural transition occurs at B_{20} [13]. Further theoretical studies led by Boustani showed that indeed tubular structures (more precisely, double ring structures) are energetically favorable for larger clusters [14–16].

The aromaticity of small boron clusters has been studied by Zhai et al. [17]. It is shown that aromatic (planar) boron clusters possess more circular shapes, whereas anti-aromatic ones are elongated.

47.2 Boron Fullerenes

The existence of pure boron fullerenes was predicted recently [18]. The main idea is reinforcing hexagonal facets of boron isomorphs of carbon fullerenes by adding a boron atom at each center. Even though there is no direct experimental evidence for these fullerenes, their unusual stability and chemistry are interesting. Theoretical investigation in this direction is appealing since carbon fullerenes also went undetected for years, after their theoretical prediction in the early 1970s [19,20].

The most stable out of all boron fullerenes studied is the so-called boron buckyball, B_{80} (Figure 47.3b). This molecule is energetically favorable and symmetrically almost icosahedral (see Section 47.5). It has characteristics similar to B_{12}, which is the building block of all bulk phases, in the sense that it is made from crossing double rings (DRs). In Figure 47.3 (right), this similarity is highlighted. Figure 47.3b shows two crossing B_{30} DRs that are constituents of the cage. The whole cage is made up of three such pairs (six DRs in total). The staggered configuration

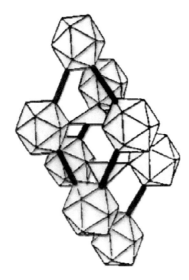

FIGURE 47.1 The lattice structure of α-rhombohedral bulk boron. Each atom is covalently bonded to five intra-icosahedral, and one or two inter-icosahedral atoms. (From Quandt, A. and Boustani, I., *ChemPhysChem*, 6, 2001, 2005. With permission.)

of each DR is formed by two rings, each containing 15 atoms and rotated by π/15 with respect to one another.

Other members of boron fullerene family, obtained similarly, can also be considered. We limit the case to all possible cages with a size of 80 atoms or fewer (derived from fullerenes with less than 60 atoms before hexagonal reinforcement), as well as one bigger cage, B_{110}. Before reinforcement, the cages are boron isomorphs of C_{24}, C_{26}, C_{28}, C_{32}, C_{36}, C_{50}, and C_{80} fullerenes (specified by D_6, D_{3h}, D_3, D_3, D_{6h}, D_{5h}, and I_h symmetries, respectively). After the reinforcement, optimization shows that only four of these cages, shown in Figure 47.4a, sustain the hollow structure [18].

We can also consider cages B_{72} and B_{92}, formed by placing extra boron atoms on top of pentagons (B_{72}) and on pentagons and hexagons (B_{92}) of the B_{60}. Optimized structures are shown in Figure 47.4b. The B_{72} cage is round and preserves I_h symmetry but is less stable than B_{80}. The B_{92} is completely built from triangular bonding units, which are generally favorable for boron clusters [17]. However, this cage is also less stable than the B_{80} cage.

None of these structures consist of DRs, and perhaps for this reason, they do not have the exceptional stability as B_{80}. Table 47.1 summarizes the symmetries, values of cohesive energy (E_{coh}), and

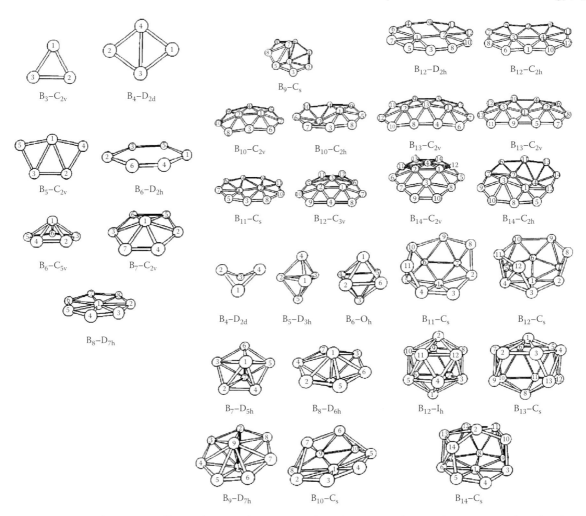

FIGURE 47.2 Structure and symmetry of boron clusters with 14 atoms or fewer. (From Quandt, A. and Boustani, I., *ChemPhysChem*, 6, 2001, 2005. With permission.)

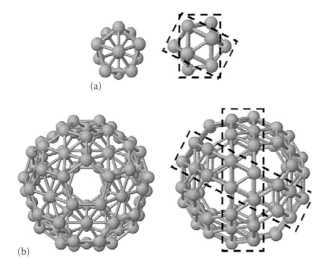

(a)

(b)

FIGURE 47.3 (a) B_{12} and (b) B_{80} clusters; they can both be seen as six interwoven double rings. (From Szwacki, N.G. et al., *Phys. Rev. Lett.*, 98, 166804, 2007. With permission.)

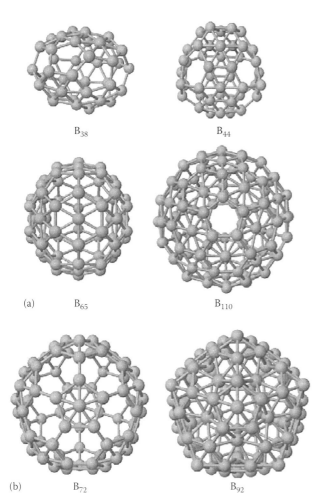

(a) B_{38} B_{44}

(a) B_{65} B_{110}

(b) B_{72} B_{92}

FIGURE 47.4 (a) Four other members of boron fullerene family. (b) B_{72} and B_{92}, which are formed by placing additional boron atoms at the center of pentagons of B_{60} and B_{80}. (From Szwacki, N.G. et al., *Phys. Rev. Lett.*, 98, 166804, 2007. With permission.)

TABLE 47.1 Symmetry, Cohesive Energy, and HOMO–LUMO Gaps of the Structures Considered Here

	Symmetry	E_{coh} (eV/atom)	HOMO–LUMO (eV)
B_{12}	I_h	5.01 (5.00)	0.737 (0.810)
B_{20}	I_h	4.74 (4.69)	1.253 (0.008)
B_{38}	D_3, distorted	5.47 (5.48)	0.935 (0.923)
B_{44}	D_{2h}, distorted	5.55 (5.56)	0.980 (0.965)
B_{60}	I_h	4.93 (4.91)	0.049 (0.050)
B_{65}	D_{5h}	5.69 (5.70)	0.095 (0.014)
B_{72}	I_h	5.60 (5.58)	0.269 (0.001)
B_{80}	I_h	5.76 (5.77)	1.006 (0.993)
B_{92}	I_h	5.72 (5.75)	1.129 (1.161)
B_{110}	I_h	5.73 (5.74)	0.119 (0.097)

Source: Szwacki, N.G. et al., *Phys. Rev. Lett.*, 98, 166804, 2007. With permission.

the highest occupied-lowest unoccupied molecular orbital energy gap (HOMO–LUMO) of the above structures. In the case of B_{80}, there are 60 longer (l_{ph} = 1.73 Å) and 30 shorter (l_{hh} = 1.68 Å) bonds, making the diameter of B_{80} close to d = 8.17 Å (compared to corresponding value for C_{60}; d = 6.83 Å).

Recent calculations show that DR B_n clusters with n = 20, 24, 32, and 36 are the most stable structures among all clusters with the same number of atoms [13–16,21]. Taking DRs as reference point, the comparison between cohesive energies of the above structures and DRs is illustrated in Figure 47.5. "Strip" in the figure corresponds to infinite DR.

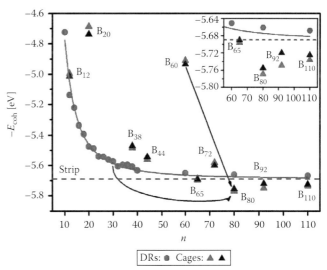

DRs: ●   Cages: ▲ ▲

FIGURE 47.5 Cohesive energy per atom as a function of the number of atoms "n" in the B_n clusters. The circles correspond to double rings, whereas the black (gray) triangles correspond to cages calculated with Quantum-ESPRESSO (GAUSSIAN03). The gray horizontal line corresponds to the cohesive energy of the infinite double-ring (strip). The arrows show the increase in cohesive energy by reinforcement of hexagons (from B_{60} to B_{80}) and by appropriate crossing of the double rings to form the icosahedral structure (from B_{30} DR to B_{80}). The inset shows relative E_{coh} values for four cages B_{65}, B_{80}, B_{92}, and B_{110}, more pronouncedly. (From Szwacki, N.G. et al., *Phys. Rev. Lett.*, 98, 166804, 2007. With permission.)

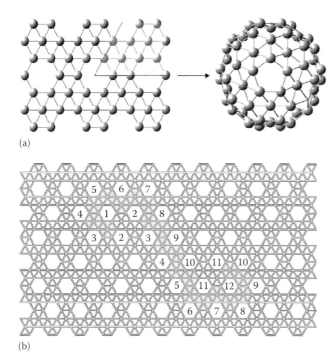

(a)

(b)

FIGURE 47.6 (a) Atoms in shaded area of the α-sheet (left) are removed, and the ones on the lines are identified to form pentagonal disclination in B$_{80}$ (right). Figure shows only half of the molecule. (b) Shaded area highlights the projection of B$_{80}$ on α-sheet. Numbered facets represent the pentagons in B$_{80}$. (From Sadrzadeh, A. et al., *J. Phys. Chem. A*, 112, 13679, 2008. With permission.)

47.3 Boron Sheet

B$_{80}$ (like C$_{60}$) can be imagined as a sheet wrapped on sphere. The most stable structure for the sheet, the α-sheet, was recently reported [22]. The special pattern as a honeycomb lattice wherein 2/3 of the hexagonal centers are filled ensures the best occupancy of the bands by electrons and therefore lowest energy (Figure 47.6). Upon wrapping the α-sheet on a sphere, "empty" hexagons are replaced by 12 pentagonal disclinations required topologically, and strips enclosing them are replaced by B$_{30}$ DRs. Figure 47.6a shows how the sheet can be folded to form a pentagonal disclination. Figure 47.6b highlights the unfolded B$_{80}$ molecule on an α-sheet.

The bond length in α-sheet is 1.70 Å, and its cohesive energy (obtained by the same method used for values in Table 47.1) is 5.93 eV/atom. The α-sheet is metallic, with zero energy gap between the bands.

47.4 Boron Nanotubes

Fullerenes can be viewed as short nanotubes. Here we briefly describe physical properties of boron nanotubes (BTs).

BTs can be obtained by wrapping the sheet along the chiral vector. Since α-sheet is believed to be the most stable pattern for the sheet, we only describe α-BTs. We choose the indexing of the BTs to correspond to the long-accepted convention for the CNTs—with the two √3·*b*-long (where *b* is the bond length) basis vectors 60° apart, each directed along the zigzag motif in the lattice. The tube circumference is specified as a pair of components (n, m). This implies that any (n, m) tube derived from α-sheet exists only when (n−m) is a multiple of three [23].

In contrast to CNTs, BTs are metallic or semiconducting depending on their diameter alone. For diameters smaller than 1.7 nm, BTs are semiconducting [24]. Figure 47.7 shows dependence of the band gap on the diameter for armchair BT [23]. Zigzag tubes have a close value of transition from metallic to semiconducting diameter. One can attribute semiconducting behavior of the BTs to their relaxation, which causes central B atoms buckling inward, thus opening the band gap.

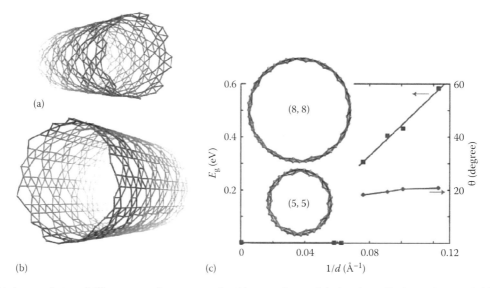

(a)

(b)

(c)

FIGURE 47.7 (a) An armchair and (b) a zigzag α-boron nanotube. (c) Dependence of the band gap (dark gray line) and dihedral angle of the off-plane atoms with respect to six-membered ring (light gray line) on inverse of diameter. Inset shows how relaxation causes buckling in two cases of (5, 5) and (8, 8) α-BT. (From Singh, A.K. et al., *Nano Lett.*, 8, 1314, 2008. With permission.)

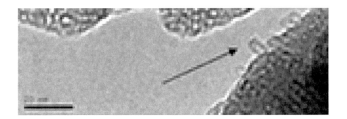

FIGURE 47.8 High resolution TEM image of B-nanotubes of ~3 nm diameter, grown on silica template (picture from Prof. L. Pfefferle's site, Yale University, and similar to Fig. 1 in [25]). (From Ciuparu, D. et al., *J. Phys. Chem. B*, 108, 3967, 2004. With permission.)

Radial breathing mode (RBM) frequency dependence on the diameter can be obtained from parameters describing shell mechanical properties (in-plane stiffness and Poisson ratio), which can be estimated by $f_{RBM} \cong 210 \, nm/d$, cm^{-1} [23].

Single-wall BTs with diameter 3 nm have been synthesized by Ciuparu et al. [25] (Figure 47.8). The reported Raman breathing mode peak is at frequency 210 cm^{-1}, which is at odds with theoretical value (~70 cm^{-1}). The peak might be due to the presence of other boron structures in the samples (B$_{80}$ for instance).

47.5 Isomerization

There are various isomers of boron buckyball that are close to each other in energy and structure [26]. Here we consider three B$_{80}$ isomers obtained by B3LYP/6-31G(d) structural optimization [27]. Visually, their geometries are almost identical; however, a detailed analysis reveals important differences in their symmetries [26,28]. Namely, two out of three structures under consideration are very close to having icosahedral (I$_h$) and tetrahedral (T$_h$) [29] symmetries, while the third one appears to have no symmetry at all (C$_1$). Correspondingly, we denote these B$_{80}$ isomers as I$_h$, T$_h$, and C$_1$. All the optimizations were performed without any symmetry restrictions. Therefore, they correspond to true local minima of potential energy surface. The I$_h$ isomer lies lowest in energy, with total energy 3.6 meV lower than T$_h$ one and 30.3 meV lower than C$_1$ one (the total

energy differences are clearly very small and are sensitive to the method) [26].

To estimate the deviations of I$_h$ and T$_h$ isomers from the corresponding ideal symmetries, let us consider the values of the dihedral angle between the six-membered ring and the plane formed by two of its atoms and the central boron atom. For the I$_h$ isomer, the dihedral angles are from 3.26° to 3.81° (toward the center of the buckyball), that is, their deviation from the average is negligible (less than 0.3°). In the case of the T$_h$ structure, there is one group of eight central atoms with dihedral angles of 8.88° to 9.07° toward the center, and another group of 12 central atoms with small (~1°) dihedral angles away from the center. T$_h$ isomer is similar to the "isomer A" of Ref. [29]. For comparison, the dihedral angles in the C$_1$ isomer vary from 7.54° toward the buckyball center to 1° away from it.

The symmetry is more visibly reflected in electron charge transfer from central boron atoms to the B$_{60}$ skeleton [28]. The amount of Mulliken charge transfer is 0.149e to 0.154e in I$_h$ case, 0.063e to 0.065e for the group of 8 atoms, 0.232e to 0.233e for the group of 12 atoms in T$_h$ case, and in the range of 0.074e to 0.205e in C$_1$ case (Figure 47.9).

The symmetry of the B$_{80}$ isomers under consideration can be further confirmed by the analysis of their electronic structure, as follows.

47.6 Electronic Structure

There are 200 occupied MOs in B$_{80}$. It is helpful to think of them as linear combinations of the boron atomic orbitals (AOs) and to distinguish between the AOs belonging to the two nonequivalent groups of boron atoms, namely, the 20 atoms situated in the centers of hexagons and the 60 others forming a structure analogous to C$_{60}$ [26].

The lowest 80 MOs are linear combinations of 1s-AOs of boron atoms, with a negligible contribution of higher AOs. The 20 boron atoms in the centers of hexagons mainly contribute to the MOs 1–20, which have very close energies; the other 60 atoms contribute to the MOs 21–80, which are also nearly degenerate.

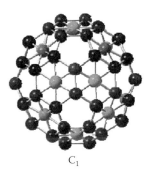
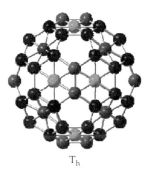
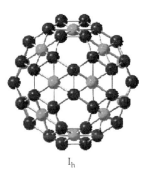

C$_1$ T$_h$ I$_h$

FIGURE 47.9 Mulliken charges are shown for the three isomers C$_1$, T$_h$, and I$_h$. Light gray (dark gray) corresponds to positive (negative) atomic charges. The lighter gray corresponds to more positive charge and smaller dihedral angle. (From Szwacki, N.G. et al., *Phys. Rev. Lett.*, 100, 159901, 2008. With permission.)

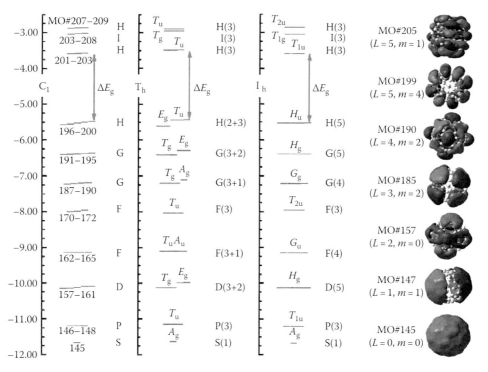

FIGURE 47.10 Energy diagram showing the π-type molecular orbitals of different isomers of B_{80}, calculated with B3LYP/6-31G(d) method/basis. The left, middle, and right diagrams correspond to C_1, T_h, and I_h isomers, respectively. The levels with energy difference less than 10 meV are considered degenerate. The arrow marks the HOMO–LUMO gap (ΔE_g = 1.88, 1.95, 1.93 eV for C_1, T_h, and I_h, respectively). For the energy levels of T_h and I_h structures, the corresponding representations are shown. For each type of spherical harmonics (S, P, D, etc.), the spatial distribution of one typical molecular orbital is shown in the right column. (From Sadrzadeh, A. et al., *J. Phys. Chem. A*, 112, 13679, 2008. With permission.)

The higher occupied MOs of B_{80} are mostly made of boron 2s- and 2p-AOs and can have either σ or π character. There are only 60 π-electrons in the structure (which can be thought of as donated by the 60 equivalent boron atoms), and correspondingly, 30 π-orbitals are occupied.

Molecular orbitals of a nearly spherical molecule can be well approximated by the spherical harmonics [30,31], and thus strongly resemble AOs of various types. The MOs that have the azimuthal quantum number $L = 0, 1, 2, 3, 4, 5, 6$, etc., are commonly designated as S, P, D, F, G, H, I, etc. orbitals, analogous to the standard AO naming.

As the frontier molecular orbitals of B_{80} are of π-type, we concentrate on the π-MOs. Their energies for three isomers with C_1, T_h, and I_h symmetries are shown in Figure 47.10. One may compare this energy diagram with that of C_{60} [30]. Notice that the numbers of occupied π-orbitals in both C_{60} and B_{80} equal 30 and thus do not satisfy the $2(N + 1)^2$ Hirsch aromaticity rule [32], because the H shell ($L = 5$) is not filled completely.

The rightmost column in Figure 47.10 shows the spatial distribution of one typical molecular orbital for each quantum number L [33]. The orbital degeneracy is determined by the azimuthal quantum number L and the symmetry of the molecule under consideration. For example, the fivefold degenerate d-type atomic orbitals of transition metals split into a triplet

and a doublet in the octahedral crystal field (see e.g., Ref. [34]). Similarly, lowering the isomer symmetry from I_h to T_h leads to the orbital splitting, clear for instance for HOMO (MOs #196–200, H-type) in Figure 47.10. In particular, the symmetries of structures are reflected in their HOMO degeneracy: it is fivefold for I_h (within 5 meV), threefold for T_h (within 2 meV), and nondegenerate for C_1. For the two symmetrical isomers, T_h and I_h, the representations of their symmetry groups are shown for all the π-type molecular orbitals in Figure 47.10.

Since the symmetric (I_h) structure has a nondegenerate ground state (HOMO is fully filled), symmetry breaking is not due to Jahn–Teller effect.

47.7 Vibrational Modes

To analyze the vibrational modes of boron buckyball, we use B3LYP/STO-3G method/basis [27] because of the computational burden (for a more detailed analysis of vibrational modes, see [35]). The B_{80} geometry used for this analysis was relaxed using the same method, and has T_h symmetry. It is close to "isomer B" of Ref. [29], with 12 (8) central atoms making dihedral angle ~12.5° (8°) toward (away from) the center. B_{80} has 64 distinct intramolecular mode frequencies ranging from 154 cm^{-1} for radial vibrations to 1181 cm^{-1} for tangential ones. The low lying frequencies correspond to the central atom vibrations.

The breathing mode frequency is 474 cm^{-1}. In this mode, the 12 central hexagonal atoms with dihedral angle toward the center move in opposite phase with respect to the rest of the atoms. It is worth noting that in the Raman spectrum reported for BT samples [25], there is a peak close to this value of frequency (~420 cm^{-1}). The authors of Ref. [25] attribute the peaks at the range of 400–600 cm^{-1} to either smaller diameter BT or to "other boron structures present in the sample."

47.8 Other Boron Fullerenes

Recently, there was a report on a new family of boron fullerenes, with the number of atoms $32 + 8k$ [36]. A subfamily with the number of atoms $80 + 8k$ are obtained by modified leapfrog transformation (MLT), from C_{20+2k} (Figure 47.11). In MLT, one starts with C_{20+2k}, fill all the facets, make the Wigner–Seitz dual, and then fill the hexagons except for the ones that fall inside the original hexagons of C_{20+2k}.

The symmetries, cohesive energies, and HOMO–LUMO gaps are summarized in Table 47.2. In the limit of $k = 0$ and ∞, one obtains B_{80} and α-sheet, respectively. The authors of Ref. [36] propose an isolated hollow rule as well as a counting rule for electrons. Counting is explained simply as two electrons shared in each triangular (three-center) bond. In B_{80}, there are 20 hexagons and 120 triangles, hence, 240 valence electrons. In the α-sheet, there are 12 triangles in the unit cell and, correspondingly, 24 valence electrons (there are 8 atoms in α-sheet unit cell).

Recently, B_{12} stuffed fullerene-like boron clusters were studied, and it was shown that B_{98-102} clusters are thermodynamically more stable than B_{80} [37]. Also, there was a report on other fullerenes composed of six double rings, with comparable stability to B_{80} [38].

47.9 Applications

Such structures should have a wide variety of applications [39]. One possible application is in cancer therapy. Boron in nature consists of the isotopes ^{10}B (19.6%) and ^{11}B (80.4%). The ^{10}B isotope has a special property that has long been exploited in nuclear reactors, namely, it has very high neutron absorption cross section (3835 barns) [40]. The reaction of ^{10}B with thermal neutron generates ^{7}Li and α particles, with latter having energies that make them lethal to nearby cell [40]. This aspect alone makes search for boron clusters of different shape, size, and cell-penetration ability very important. Another important and recently discussed possibility is in use of boron (with proper addition of metal atoms) clusters for efficient hydrogen storage [41].

Rotary motion along a molecular axis, controllable by a single electron transfer or photoexcitation, has been proposed for metallacarboranes (Figure 47.12) [42]. The existence of this or possibly other boron-derived molecular motors has been sought for many years, as a parallel to biological motors that can be powered by light or electrical energy, rather than ATP.

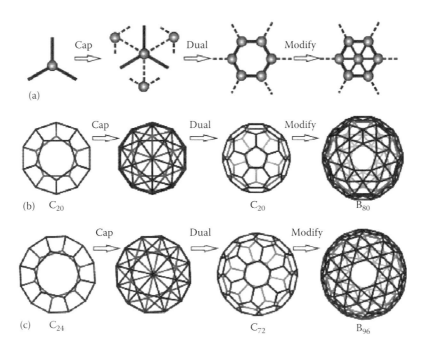

FIGURE 47.11 Illustration of leapfrog and modified leapfrog transformations. (a) A three-connected vertex becomes a new hexagon (in leapfrog) or a hexagon with an additional atom in center (in modified leapfrog); (b) a C_{20} converts to C_{60} and B_{80} by the leapfrog and modified leapfrog transformations, respectively; (c) a C_{24} transforms to C_{72} and B_{96}. (From Yan, Q.-B. et al., *Phys. Rev. B*, 78, 201401, 2008. With permission.)

TABLE 47.2 The Number (N_{iso}), Symmetries, Energies per Atom (E, in eV) Relative to That of Graphene and α-Sheet, Respectively, and HOMO–LUMO Gap Energies (ΔE, in eV) of C_{60+6k} and B_{80+8k} Fullerenes ($k \leq 5$) Obtained with PBEPBE/6-31G Method/Basis

k	N_{iso}	Symmetry	$C_{n=60+6k}$			$B_{n=80+8k}$				
			n	E	ΔE	n	E	ΔE	N_{HH}	N_{FH}
0	1	I_h	60	0.386	1.670	80	0.179	1.020	0	20
1	0									
2	1	D_{6d}	72	0.365	1.445	96	0.160	0.784	3	24
3	1	D_{3h}	78	0.346	1.460	104	0.151	0.738	4	26
4	2	D_2	84	0.339	1.367	112	0.146	0.682	5	28
		T_d	84	0.329	1.632	112	0.141	0.927	5	28
5	3	C_{2v}^I	90	0.321	1.364	120	0.137	0.709	6	30
		C_{2v}^{II}	90	0.322	1.489	120	0.136	0.828	6	30
		D_{5h}	90	0.312	1.029	120	0.135	0.555	6	30

Source: Yan, Q.-B. et al., *Phys. Rev. B*, 78, 201401, 2008. With permission.

Note: The authors believe that N_{HH} should be simply equal to k; however, we present the original reported Table I in [36].

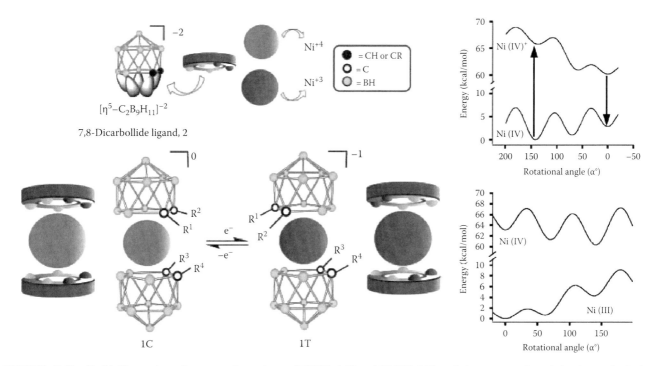

FIGURE 47.12 (Left) Illustration of rotor configurations of Ni(IV) (1C) and Ni(III) (1T) and the structural modules from which they are constructed. Structure 1T is defined with rotation angle 0°. R^1, R^2, R^3, and R^4 represent H here. (Right, top) Calculated energy of the ground and excited electronic state of the neutral Ni(IV) configuration as a function of rotation angle. (Right, bottom) Total energies of the Ni(III) and Ni(IV) compounds as a function of rotation angle. (Reassembled from Hawthorne, M.F. et al., *Science*, 303, 1849, 2004. With permission.)

47.10 Summary

This chapter discusses the physical properties of pure boron fullerenes, sheets, and nanotubes, as well as the bulk. There are three phases for the bulk boron: α- and β-rhombohedral [1,2], and α-tetragonal [3]. Recently, however, under high pressure (19–89 GPa) a new phase of bulk boron called γ-orthorhombic

(γ-B_{28}) has been reported [43]. The unit cell of this structure is made of two negatively charged B_{12} icosahedra, and two positively charged B_2 segments [43]. The most stable structure for the sheet is the so-called α-sheet, which was recently reported by Tang and Ismail-Beigi [22].

Wrapping α-sheet around a cylinder generates different (zigzag and armchair) α-boron tubes (α-BT) [23,24]. It turns out that for diameters smaller than 1.7 nm, BTs are semiconducting. Single-wall BTs with a diameter of 3 nm have been synthesized by Ciuparu et al. [25].

The boron buckyball, B_{80}, was reported recently as boron isomorph of C_{60}, with all the 20 hexagons reinforced by extra atoms [18,28]. The boron buckyball can also be understood as the α-sheet wrapped around a sphere (Figure 47.6) [26]. The electronic structure and vibrational modes of different isomers of B_{80} are also discussed. We conclude the chapter by discussing a new family of boron fullerenes, with $(80 + 8k)$ atoms for which B_{80} is a special case [36], as well as possible applications of boron clusters.

Acknowledgment

The work on the subject was supported by the Office of Naval Research (program manager Kenny Lipkowitz) and at later stage by the Department of Energy, Office of Basic Energy Sciences.

References

1. B. F. Decker and J. S. Kasper, *Acta Crystallogr.* **12**, 503 (1959).
2. B. Callmer, *Acta Crystallogr., Sect. B: Struct. Crystallogr. Cryst. Chem.* **33**, 1951 (1977).
3. A. W. Laubengayer, D. T. Hurd, A. E. Newkirk, and J. L. Hoard, *J. Am. Chem. Soc.* **65**, 1924 (1943).
4. D. Li, Y.-N. Xu, and W. Y. Ching, *Phys. Rev. B* **45**, 5895 (1992).
5. J. L. Hoard, D. B. Sullenger, C. H. L. Kennard, and R. E. Hughes, *J. Solid State Chem.* **1**, 268 (1970).
6. E. D. Jemmis and D. L. V. K. Prasad, *J. Solid State Chem.* **179**, 2768 (2006).
7. A. Masago, K. Shirai, and H. Katayama-Yoshida, *Phys. Rev. B* **73**, 104102 (2006).
8. L. Hanley and S. L. Anderson, *J. Chem. Phys.* **89**, 2848 (1988).
9. L. Hanley, J. Whitten, and S. L. Anderson, *J. Phys. Chem.* **92**, 5803 (1988).
10. S. J. La Placa, P. A. Roland, and J. J. Wynne, *Chem. Phys. Lett.* **190**, 163 (1992).
11. I. Boustani, *Phys. Rev. B* **55**, 16426 (1997).
12. A. Quandt and I. Boustani, *ChemPhysChem* **6**, 2001 (2005).
13. B. Kiran, S. Bulusu, H.-J. Zhai, S. Yoo, X. C. Zeng, and L.-S. Wang, *Proc. Natl. Acad. Sci. USA* **102**, 961 (2005).
14. I. Boustani and A. Quandt, *Europhys. Lett.* **39**, 527 (1997).
15. I. Boustani, A. Rubio, and J. Alonso, *Chem. Phys. Lett.* **311**, 21 (1999).
16. S. Chacko, D. G. Kanhere, and I. Boustani, *Phys. Rev. B* **68**, 035414 (2003).
17. H.-J. Zhai, B. Kiran, J. L. Li, and L.-S. Wang, *Nat. Mater.* **2**, 827 (2003).
18. N. G. Szwacki, A. Sadrzadeh, and B. I. Yakobson, *Phys. Rev. Lett.* **98**, 166804 (2007).
19. E. Osawa, *Kagaku* **25**, 854 (1970).
20. D. A. Bochvar and E. G. Gal'pern, *Proc. Acad. Sci. USSR* **209**, 610 (1973).
21. M. A. L. Marques and S. Botti, *J. Chem. Phys.* **123**, 014310 (2005).
22. H. Tang and S. Ismail-Beigi, *Phys. Rev. Lett.* **99**, 115501 (2007).
23. A. K. Singh, A. Sadrzadeh, and B. I. Yakobson, *Nano Lett.* **8**, 1314 (2008).
24. X. Yang, Y. Ding, and J. Ni, *Phys. Rev. B* **77**, 041402 (2008).
25. D. Ciuparu, R. F. Klie, Y. Zhu, and L. Pfefferle, *J. Phys. Chem. B* **108**, 3967 (2004).
26. A. Sadrzadeh, O. V. Pupysheva, A. K. Singh, and B. I. Yakobson, *J. Phys. Chem. A* **112**, 13679 (2008).
27. M. J. Frisch, G. W. Trucks, H. B. Schlegel, G. E. Scuseria, M. A. Robb et al., *Gaussian 03, Revision C.02* (Gaussian, Inc., Wallingford, CT, 2004).
28. N. G. Szwacki, A. Sadrzadeh, and B. I. Yakobson, *Phys. Rev. Lett.* **100**, 159901 (2008).
29. G. Gopakumar, M. T. Nguyen, and A. Ceulemans, *Chem. Phys. Lett.* **450**, 175 (2008).
30. Z. Chen and R. B. King, *Chem. Rev.* **105**, 3613 (2005).
31. W. A. de Heer, *Rev. Mod. Phys.* **65**, 611 (1993).
32. M. Buhl and A. Hirsch, *Chem. Rev.* **101**, 1153 (2001).
33. http://www.orbitals.com/orb/orbtable.htm.
34. I. B. Bersuker, *Electronic Structure and Properties of Transition Metal Compounds: Introduction to the Theory* (Wiley, New York, 1996).
35. T. Baruah, M. R. Pederson, and R. R. Zope, *Phys. Rev. B* **78**, 045408 (2008).
36. Q.-B. Yan, X.-L. Sheng, Q.-R. Zheng, L.-Z. Zhang, and G. Su, *Phys. Rev. B* **78**, 201401 (2008).
37. D. L. V. K. Prasad and E. D. Jemmis, *Phys. Rev. Lett.* **100**, 165504 (2008).
38. N. G. Szwacki, *Nano Res. Lett.* **3**, 49 (2008).
39. R. N. Grimes, *J. Chem. Educ.* **81**, 657 (2004).
40. M. F. Hawthorne, *Pure Appl. Chem.* **63**, 327334 (1991).
41. Y. Zhao, M. T. Lusk, A. C. Dillon, M. J. Heben, and S. B. Zhang, *Nano Lett.* **8**, 157 (2008).
42. M. F. Hawthorne, J. I. Zink, J. M. Skelton, M. J. Bayer, C. Liu, E. Livshits, R. Baer et al., *Science* **303**, 1849 (2004).
43. A. R. Oganov, J. Chen, C. Gatti, Y. Ma et al., *Nature* **457**, 863 (2009).

Silicon Fullerenes

Aristides D. Zdetsis
University of Patras

48.1 Introduction

48.1.1 Similarities and Differences between Silicon and Carbon

Carbon and silicon share several similarities in their electronic structure, since they are both in the fourth column of the periodic table, having the same electronic configuration of valence electrons ($2s^2 2p^2$ for carbon, $3s^2 3p^2$ for silicon). Yet, their chemistries and "usefulness" are very different. Carbon is essential for life, while silicon is essential for modern electronic technology. Only in science fiction, silicon could be related with some sort of "life," although technology based on carbon is not totally unrealistic (e.g., carbon nanotubes). The structural chemistry of carbon at the molecular and nanoscale level is dominated by planar aromatic structures such as benzene and two-dimensional hollow spherical cages or fullerenes whereas for silicon, three-dimensional pyramidal and diamond-like structures predominate. This is due formally to the different rows (carbon is the first, silicon in the second) of the periodic table. As a consequence, the size of the atoms and the promotion energies between s and p valence orbitals (2s–2p for C, 3s–3p for Si) are different, affecting the orbital overlap and hybridization. Thus, carbon usually prefers strong π covalent bonds blended in sp^2 (or sp) hybrid orbitals, which are manifested as double and triple bonds, while silicon normally forms covalent σ bonds between sp^3 hybrid orbitals.

48.1.2 Bonds, Orbital Overlap, and Hybridization: A Short Parenthesis

As is well known, chemical bonds are attractive forces that hold atoms together in the form of compounds. They are formed when electrons are shared between two atoms. In this case (at that distance), the corresponding atomic orbitals from the two different atoms overlap to form a molecular orbital. The atomic orbitals are characterized as s, p(p_x, p_y, p_z), d(d_{xx}, d_{xy},...), etc., according to their angular momentum and their symmetry properties. The molecular orbitals are also characterized by their angular momentum and symmetry properties, which depend on the symmetry properties of the atomic orbitals they are composed and the way they overlap. The molecular orbitals, which are obtained by head-on overlap of atomic orbitals are called σ (sigma) orbitals and the corresponding molecular bond is called σ bond. As we can see in Figure 48.1, a σ molecular orbital, seen along the z-axis (horizontal axis in the figure) looks very much like an s atomic orbital. Sigma bonds (σ bonds) are the strongest type of covalent chemical bonds. This is related with significant concentration of charge between the two nuclei.

When the overlapping atomic orbitals approach sideways, we have the formation of a π (pi) molecular orbital and a corresponding π bond. In this type of bond, two lobes of one atomic orbital overlap with two lobes of the other. As we can see in Figure 48.1, the π molecular orbital when viewed along the z-axis looks like a p atomic orbital. Also, we see that there is no charge concentration between the two nuclei along the line joining them, as in the case of σ orbitals. The negatively charged electron density is farther from the positive charge of the atomic nuclei. This is why the π bonds are usually weaker than the σ bonds. However, usually σ and π bonds appear together in the form of double or triple bonds with much higher strength compared to single σ bonds. Pi(π) bonds characterize carbon; while silicon strongly prefers almost exclusively sigma (σ) bonds.

The different bonding characteristics of silicon and carbon more often are described in the language of valence bond theory

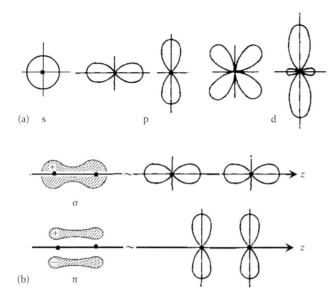

FIGURE 48.1 (a) Atomic orbitals s, p (two of the three), and d (two of a total of five). (b) Molecular orbitals, σ and π on the left. The component p atomic orbitals and the way they combine are also shown on the right.

in terms of *hybridization*. This concept is illustrated schematically in Figure 48.2a and b.

On the left-hand side of Figure 48.2a, we can see the electron configuration and occupation of the energy levels of carbon atom. Naively thinking, we could have guessed that the valency of carbon is 2, since the two 2s electrons are in a complete subshell, leaving only the two 2p electrons for bonding. However, it is common knowledge that carbon is tetravalent. This is because in the presence of another atom the s and p orbitals mix (by assuming as elevating one s electron to the third empty p box) to form an sp³ "hybrid" (a mixture of two, or more, different entities) orbital. The (small) promotion energy from the 2s to 2p orbital is paid off by the (large) energy gain from the formation of four bonds. Thus, the sp³ hybrid orbital has lower energy that the 2p

orbital (but higher than the 2s orbital). The sp³ hybrid orbital, shown in the right-hand side of the Figure 48.2, has 75% p character and 25% s character, which gives each lobe a shape that is shorter and fatter than a p-orbital lobe. The four lobes of this orbital are in tetrahedral arrangement forming angles of 109.5° between each other and can hold four electrons (as many in the 2s and 2p separately). When four hydrogen atoms are attached to these four lobes forming four σ bonds, we have the molecule of methane (CH_4). There should be no confusion between hybridization, which involves the mixing of orbitals of the same atom, with overlap which involves the mixing (the superposition) of orbitals from two different atoms in close proximity. All the discussion for the sp³ hybridization in carbon and the formation of the methane molecule is exactly the same for silicon. In this case, we have mixing of the 3s and 3p orbitals and the corresponding molecule is silane (SiH_4). In this case, however, the mixing is "easier" compared to carbon (it costs less energy) because the energy difference between the valence 3s, 3p orbitals of silicon is smaller than the 2s–2p energy difference in carbon.

A different type of hybridization is shown in Figure 48.2b. Let us assume momentarily that one of the 2p electrons is not available for mixing or is missing (for convenience, we could have boron instead of carbon with one less electron). Then, after promotion of one 2s electron to the 2p orbital (as in the case of the sp³ hybridization) and mixing we have the sp² hybrid orbital, as shown on the right-hand side of Figure 48.2b. Again the energy of the sp² orbital is lower than the energy of the 2p orbital (and higher than the 2s energy). The other p orbital remains unhybridized (for reasons which will be explained below) and is at right angles to the trigonal planar arrangement of the hybrid orbitals. This unhybridized p orbital can participate in π bonding, leading to double (σ and π) bonds (see Figure 48.3). The lobes of the

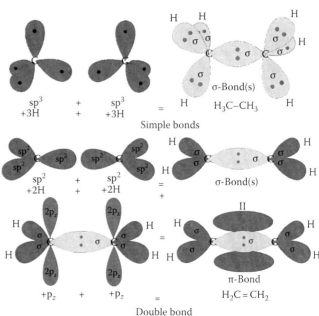

FIGURE 48.3 Illustration of the sp³ (top) and sp² hybridization (bottom) in the formation of simple (σ) and double (σ and π) bonds.

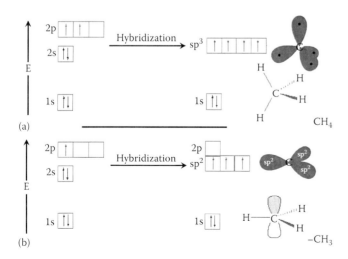

FIGURE 48.2 Illustration of the sp³ (a) and sp² hybridization (b) through the examples of methane (CH_4) and methyl radical ($=CH_3$).

sp² orbitals are in a trigonal planar arrangement forming bond angles of 120°. For silicon, the sp² hybridization and bonding is highly unfavorable due to the smaller 3s–3p promotion energy compared to carbon. This makes sp³ hybridization energetically much more favorable than any other possible hybridization, such as sp². Thus, silicon strongly prefers sp³ tetrahedral bonding. Carbon on the other hand, due to the larger energy difference between the valence 2s and 2p orbitals, needs relatively large hybridization energy. This implies that carbon will "activate" one valence p orbital at a time (since hybridization costs energy), as required by the bonding situation. As a result, if it is favorable, instead of 3 it can activate only 2, or only 1 valence p orbital, giving rise to sp³, sp², or sp hybridizations.

This last sp hybridization (not shown in Figure 48.2) gives linear arrangement of the bonds (bond angle 180°) and leaves two non-hybridized p orbitals which can participate (through π bonding) to triple bonds. Thus carbon can form single, double, and triple bonds, depending on the availability and the bonding arrangement (final energy gain).

In Figure 48.3, we illustrate the formation of simple σ and double (σ and π) bonds through sp³ and sp² hybridization. In the sp² case, we can see separately the formation of the σ bond first, which after the consideration of the π bond from the side-by-side overlap of the remaining non-hybridized p_z orbitals (one on each carbon atom) leads to the double bond of $H_2C = CH_2$.

From this discussion, it becomes clear that the differences between silicon and carbon are due to their differences in their behavior in forming chemical bonds. Carbon prefers to form sp² bonding but it can also form sp³ and sp bonds (or equivalently single, double, and triple bonds). Silicon on the other hand strongly prefers sp³ bonding, leaving "dangling" bonds (with an energetic cost) rather than resorting to multiple bonding. The dangling bonds can be a considered as a defect. The deeper roots of these differences, on the basis of the discussion in Section 48.1.2, can be traced in the different atomic size (a smaller size better accommodates the sideway overlap necessary for π bonding) and the different s–p promotion energies (larger promotion energy hinders the hybridization).

48.1.3 The Discovery and Properties of Carbon Fullerenes

Fullerenes are forms of carbon molecules; they are neither diamond nor graphite, which are the two well-known *allotropes* of carbon. More specifically, fullerenes are a new category of

carbon "allotropes" in the form of hollow spheres. In a broader and more general meaning of the term, fullerenes can also have other shapes such as ellipsoid, tube, or even planar. Carbon nanotubes are examples of cylindrical fullerenes while *graphene* (of the newest members of the family) is an example of a planar fullerene sheet. Here, we will stick to the usual meaning of the term. Spherical fullerenes are also called buckyballs. This term comes from the first and most common fullerene, C_{60}, which was discovered in 1985 (Kroto et al., 1985). C_{60} was named *Buckminsterfullerene* after Richard Buckminster Fuller, an architect known for the design of the geodesic dome which resembles spherical fullerenes (or buckyballs). For the same reason, the whole class of hollow carbon spheres besides C_{60} was named *fullerenes* (in the most common meaning of the term).

48.1.3.1 Buckminsterfullerene

The structure of Buckminsterfullerene (C_{60}) is shown in Figure 48.4 in various forms of molecular visualization models, which have their origin in the physical (mechanical) models used in early times: wire, stick, ball-and-stick, and space-filling *CPK* (Corey–Pauling–Koltun) diagrams, which also illustrate the atomic volume rather than the bonds (as in the wire or stick diagrams). As we can see in this figure, the structure of C_{60} is a truncated icosahedron (I_h symmetry), which resembles a soccer ball made of 20 hexagons and 12 pentagons, with a carbon atom at the vertices of each polygon and a bond along each polygon edge. The (van der Waals) diameter of a C_{60} molecule is about 1 nm. The nucleus-to-nucleus diameter of a C_{60} molecule is about 0.7 nm. The C_{60} molecule has two bond lengths. The 6:6 ring bonds (between two hexagons) can be considered as double bonds or better as "one-and-a-half bonds," drawn by one solid and one broken line. These bonds are shorter than the 6:5 bonds (between a hexagon and a pentagon). The average bond length of C_{60} molecule is 1.4 Å.

48.1.3.2 Other Buckyballs and Fullerenes

Another fairly common fullerene is C_{70}, but fullerenes with 72, 76, 84, and even up to 100 carbon atoms are commonly obtained. In mathematical terms, the structure of a fullerene is a trivalent convex polyhedron with pentagonal and hexagonal faces. It follows from Euler's polyhedron formula, $v - e + f = 2$, (where v, e, f indicate the number of vertices, edges, and faces), that there are exactly 12 pentagons in a fullerene and $v/2-10$ hexagons.

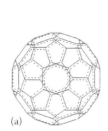 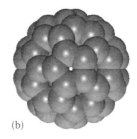  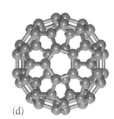

(a)　　　(b)　　　(c)　　　(d)

FIGURE 48.4 The buckyball in various forms of representation: wire diagrams (a), space-filling CPK models (b), stick (c), and ball-and-stick diagrams (d).

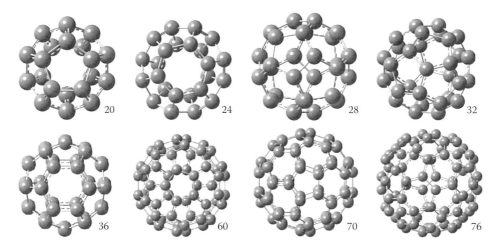

FIGURE 48.5 Some smaller (n = 20, 24, 28, 32, 36) and larger (n = 70, 76) fullerenes, together with the well-known Buckminsterfullerene (C_{60}). The size of the atomic balls and sticks (the size of the fullerenes) has been scaled to fit uniformly the figure.

The smallest fullerene is the dodecahedron—the unique C_{20} with I_h symmetry which has no hexagonal faces at all. The synthesis of C_{20} by high-voltage dehydrogenation of $C_{20}H_{20}$ is considered by several researches as the most dramatic experiment reported to date (Chen et al. 2003a). There are no fullerenes with 22 vertices. The number of fullerenes, with the general form C_{2n}, grows with increasing n = 12, 13, 14,... roughly in proportion to n^9. For instance, there are 1812 non-isomorphic fullerenes C_{60}. Some of the well-known carbon fullerenes, smaller and larger than the Buckminsterfullerene C_{60}, are shown in Figure 48.5 in ball-and-stick diagrams.

48.1.4 Attempts to Bridge the Differences and Exploit the Combined Benefits of Silicon and Carbon

From the discussion in the previous section, it becomes clear that the structural, chemical, and bonding properties of carbon are much different from those of silicon despite the similarities, such as sp^3 bonding. Nevertheless, although the corresponding carbon structures are sp^2 bonded, several highly aspiring and inspiring attempts have been made toward silicon planar aromatic structures (Sax and Janoschek 1986, Baldridge et al. 2000, De Proft Geerlings 2001, Takahashi and Kawazoe 2005, Zdetsis 2007a,b, Zdetsis et al. 2008) as well as silicon cages or "fullerenes" (see, for instance, Piqueras et al. 1993, Menon Subbaswamy 1994, Zdetsis 1996, Kumar and Kawazoe 2001, 2003, 2007, Chen et al. 2003, Karttunen et al. 2007, Zdetsis 2007b,c,d). The attempts toward the first direction were unsuccessful since the analogous to C_6H_6, hexasilabenzene structure of Si_6H_6 is not a global minimum (De Proft Geerlings 2001, Zdetsis 2007a). Hexasilabenzene is higher in energy (by 0.39 eV) compared to a trigonal prismatic isomer known as hexasilaprismane (Zdetsis 2007a). The research in the second direction which is the subject of the present chapter is more successful, as will be illustrated in what follows.

48.2 In Search of Silicon Fullerenes

Almost immediately after the discovery and synthesis of C_{60}, great expectations for the synthesis of similar silicon fullerenes were generated. This includes not only Si_{60} in analogy to C_{60} but also smaller fullerenes such as C_{20}. The search for silicon (and by "extrapolation" germanium, which will not be discussed here at all) fullerenes can be roughly classified into three major categories: (a) the *direct* approach, seeking structures in full analogy to the carbon fullerenes; (b) the *intermediate* approach, which deals with mixed (or hydride) structures consisting of both silicon and carbon in various compositions; and (c) the *indirect* approach, which is the most successful and the most broad. In this category, several diverse methods belong, which seek cage-like structures, not necessarily identical to the original carbon fullerenes, and not necessarily with all properties (such as exact type of bonding) similar to their carbon counterparts. The stabilization of such cages, which are usually characterized by high symmetry, can be achieved by a variety of methods. These methods can be further subdivided in two large general classes: Those involving endohedral doping with other atoms or group of atoms, and those involving exohedral doping or covering with a variety of atoms or molecules, the most popular (and perhaps most successful) of these being hydrogen. In all (or most) of these methods, the bonding is sp^3 (or near sp^3), not sp^2 as in pure carbon fullerenes. Therefore, most of these methods (and in particular the most successful ones) involve some sort of "cheating" (Zdetsis 2007b).

48.2.1 Direct Approach

The earliest, simplest, but nevertheless the least successful method is the direct substitution of C atoms by Si atoms, assuming that the I_h icosahedral symmetry of the initial C_{60} structure is more or less preserved. Obviously, the bond lengths were properly scaled followed by geometry optimization. The main characteristic of calculations based on this approach (performed with a variety of theoretical methods) is that they do not agree with each other. This is true even for calculations performed with the same method. This is practically due to several secondary approximations and technical details (such as the size and quality of the basis sets) during the implementation of the computations. The results of these calculations, according to the final geometry and symmetry of the Si_{60} cage, can be divided in two main categories: (a) perfect fullerene with I_h symmetry and (b) distorted cage-like structures of various symmetries (such as T_h, the symmetry of the volleyball, or C_{2h} puckered ball, or even C_i, with only center of inversion symmetry, etc.). The results of Nagase and Kobayashi (1991) based on the Hartree–Fock (HF) method and results of Piqueras et al. (1993) as well as Sheka et al. (2002) both based on the AM1 method, belong to this category. They all agree that the perfect I_h fullerene structure is stable (and apparently lowest in energy). Sheka et al. (2002) have also performed calculations with high spin states and concluded that the Si_{60} ground state is a mixed-spin state nearly degenerate over spins. The same authors have also considered Si_{60} "oligomers" (dimer, trimer, pentamer) consisting of two, three, and five Si_{60} "molecules" joined together in a row.

However, several authors have pointed out that the perfect I_h-symmetric Si_{60} fullerene is unstable, distorting to lower symmetry Si_{60} structures. Tight-binding molecular dynamics simulations gave two different structures for Si_{60}, a puckered ball-like geometry (Khan and Broughton 1991) and a lower (than I_h) symmetry C_{2h} structure (Menon and Subbaswamy 1994), thus disagreeing with each other. According to Menon and Subbaswamy (1994), stacked naphthalene-like layer structure is the ground state. Wang et al. (2006) using density functional theory obtained results similar to Menon and Subbaswamy. The distorted C_{2h} structure they obtained is shown in Figure 48.6.

On the other hand, molecular dynamics calculations based on the "full-potential linear-muffin-tin-orbital" (FP-LMTO) method (Bao-Xing et al. 2000) suggested a spherical T_h structure (with the symmetry of the volleyball) as the equilibrium geometry of Si_{60}. Similar results were obtained by Zdetsis (1996) using density functional theory. In this work, the T_h structure was nearly degenerate with an even lower symmetry C_i structure.

Finally, more recent *ab initio* methods augmented by density functional tight-binding molecular dynamics (Chen et al. 2003b) concluded that neither the perfect I_h symmetry (soccer ball symmetry) nor the distorted T_h structure (volleyball symmetry) are true minima (and certainly not global minima). Instead, a much lower C_i symmetric is the lowest energy stable structure followed by a simple T-symmetric second lowest structure. The calculations of Chen et al. (2003b) have shown that the Si_{60} "fullerene" has much lower stability compared to C_{60}. However, according to these authors, Si_{60} should be stable toward spontaneous disintegration up to 700 K. These were only theoretical expectations.

On the experimental side, the picture is also confusing. Indirect experimental evidence based on the saturation study of Si_{60}^+ positive ions (Jarrold et al. (1991) seems to imply the I_h ground-state structure. However, similar indirect results on the fragmentation analysis of Si_{60}^+ positive ions (Jelski et al. 1990) suggest the stacked naphthalene-like layer structure. Therefore the whole, theoretical and experimental, picture is confusing, reminiscent of similar conflicting (and confusing) results for Si nanocrystals and clusters (Zdetsis 2006). As was illustrated by Zdetsis (2006), in such cases, claimed agreement (or disagreement) with experiment is not necessarily meaningful. The same confusing picture exists for the smaller pure silicon fullerenes, where the results of the existing calculations (Li and Cao 2001, Han et al. 2003, Fthenakis et al. 2005) do not fully agree with each other, although they all agree that the fully symmetric structures (with the same high symmetries of the carbon fullerenes) distort to lower symmetries. Li and Cao have examined silicon fullerenes Si_n with $n = 20, 24, 26, 28, 30, 32$ using the FMTO molecular dynamics method. Han et al. studied using the semiempirical AM1 method the cages $n = 26–36, 60$, while Fthenakis et al. have reexamined the $n = 20, 32$ cages in the framework of density functional theory (with the B3LYP hybrid functional) and tight-binding molecular dynamics and found more distorted structures. These new structures, and even more so, the undistorted or slightly distorted cages, are not necessarily the lowest possible energy structures (global minima). In fact, more recent calculations suggest very different structures for this range of silicon clusters, having prolate geometries, or Y-shaped structures and

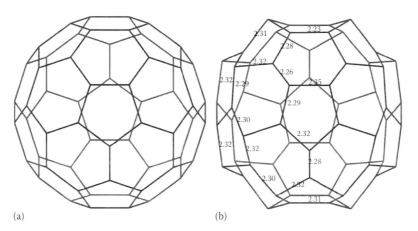

(a)   (b)

FIGURE 48.6 The C_{2h} symmetric distorted Si_{60} cage. (Adapted from Wang, L. et al., *Mol. Simul.*, 32, 663, 2006.)

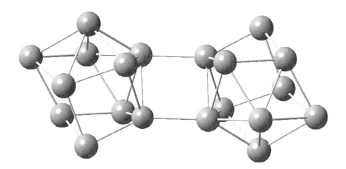

FIGURE 48.7 One of the two energetically competing lowest energy structures of Si_{20}, consisting of two Si_{10} magic silicon clusters joined together.

stuffed endohedral fullerene-like structures (Yoo et al. 2004, Zhu et al. 2004, Yoo and Zeng 2006). For Si_{20}, for example, Zhu et al. (2004) found two prolate competing structures of very similar (almost equal) energies. They claimed lowest Si_{20} structure consists of two Si_6 subunits (a sixfold puckered ring and one tetragonal bipyramid) and one Si_8 subunit. The other competing structure (favored by quantum Monte Carlo calculations) consists of two Si_{10} "magic" clusters joined together, as shown in Figure 48.7.

All these cage structures discussed above (and not only them), which are sp^3 bonded (not sp^2 as in carbon fullerenes) with several (usually n for a Si_n cage) dangling bonds, have been distorted to lower symmetries structures, reminiscent of puckered balls (containing puckered rings). This type of puckering, which is common to sp^3 bonded silicon (Karttunen et al. 2007), carbon (Zdetsis 2008a), or SiC structures (Zdetsis 2008b), can be attributed to sp^3 bond angle optimization. We should recall that the ideal fullerenes consist of 12 pentagons and 20 hexagons for Si_{60} (or in general $n/2$–10 hexagons, where n is the number of vertices of the fullerene polyhedron). The bond angles in the pentagons are 108° (180°– 360°/5), which is very close to the ideal sp^3 bond angle (of 109.5°). However, the bond angles of the hexagons are 120° (180°–360°/6), which is way off the ideal sp^3 bond angle. As a result, as n increases (from $n = 20$, where there are 0 hexagons), the number of hexagons increases and so does the puckering. If this was the only mechanism for relaxation of the internal strains in order to lower the energy, then the Si_{20} icosahedral fullerene should be the most stable silicon fullerene, since its structure contains no hexagons. This is only partly true, because the most we can say is that from all Si_n, $n \le 60$ ideal fullerenes (with the same high symmetry of the corresponding carbon fullerenes) the icosahedral Si_{20} has indeed the lowest energy. However, this is not generally true for the distorted structures, and certainly for the accepted today lowest energy structures, since a better and more "drastic" way (than the optimization of the bond angles) to improve sp^3 bonding, is the partial or total elimination of the dangling bonds, if this is possible and overall favorable (this is known as "reconstruction"). However, the elimination of some dangling bonds, for example, by bringing some Si atoms closer, can create overbonding with five-coordinate silicon atoms, which is also an energy costing

"defect" known as "floating" bonds. In general, therefore, there exist several competing forces (or mechanism) that "decide" and determine the equilibrium geometry of a cluster or molecule by balancing out the (free) energy of the system. For example, it has been suggested by Zdetsis (2007e) that an equal (or almost equal) number of dangling and floating bonds in silicon clusters (and nanostructures in general) can result in a very low (or lowest) energy geometry.

Finally, it should be mentioned that although isolated fullerenes structures of silicon are generally unstable (due to the strong preference of silicon for sp^3 bonding), in bulk form, some fullerene cages exist in *clathrates* (Kume et al. 2003). Silicon clathrates are three-dimensional arrangements of Si_{20} and Si_{24} fullerene cages (type-I) or Si_{20} and Si_{28} cages (type II) that are formed by sp^3-based distorted tetrahedral bonding. Thus, interlinked Si_{20}, Si_{24}, and Si_{28} fullerene cages can exist (in bulk form). In this case, however, the bonding is very close to sp^3 due to the interconnection of the cages.

48.2.2 Intermediate Approach

Silicon–carbon clusters and more generally silicon–carbon nanosystems can be considered as intermediate steps and links between the corresponding pure carbon and pure silicon molecular systems (Zdetsis 1996, 2007a, 2008b,c). In this spirit, one can think of another approach to creating novel fullerene-related materials. Such approach consists in modifying the "functionality" of C_{60} itself by substitutional doping of carbon (Taylor and Walton 1993). In particular, the synthesis of $C_{59}Si$ and more general of $C_{2n-q}Si_q$ clusters with $2n = 32$ up to 100 has been reported as early as 1998 (Ray et al. 1998) for the carbon-rich region with $q = 1$–4. The analysis of the abundance distribution and photofragmentation spectra has provided clear evidences that such clusters remain in the fullerene geometry. No such results exist for the silicon-rich or intermediate regions up to now. On the other hand, Sun et al. (2002) have reported the formation of silicon carbide nanotubes and nanowires as early as 2002. Carbon or silicon–carbon nanotubes can be considered as the one-dimensional extension of the corresponding fullerenes (which can be thought of as "dots" of zero dimensionality). The corresponding fullerene-structured nanowires of silicon have been also studied (Marsen and Sattler 1999).

Huda and Raya (2004) have considered smaller ($n = 20$) silicon–carbon cages consisting of several carbon atoms inside an icosahedral Si_{20} fullerene. In this way, they created hybridized silicon–carbon cages with mixed sp^3–sp^2 covalent–ionic bonding. The stability of these "fullerenes" was found to depend on the geometrical arrangements of the carbon atoms inside the clusters and the partly ionic nature of the bonding. Extension to larger fullerenes was also examined and discussed.

In the same spirit but with a more drastic approach, Harada et al. (1994) have studied $(SiC)_n$ fullerenes with $n = 20$–60 in a form of Russian dolls with the C_n inside the Si_n cages. The $(SiC)_{60}$ fullerene that is better represented as $C_{60}@Si_{60}$ can be also thought of as a silicon (Si_{60})-coated C_{60} fullerene. However, the

$C_{60}@Si_{60}$ has not been synthesized experimentally so far, apparently because C_{60} has a diameter so large that it stretches the Si=Si bonds of Si_{60} leading to a higher cluster energy, which cannot be compensated by the Si=C bond energy (Wang et al. 2006).

48.2.3 Indirect Approach

48.2.3.1 Endohedral Doping

48.2.3.1.1 Metal Embedding of Cages

Another way to stabilize silicon cage–like clusters is metal encapsulation, which has been used to stabilize clusters of other elements besides silicon. For silicon, perhaps the first report of high abundances by M doping was by Beck (1987, 1989) who found cation MSi_{15} and MSi_{16} (where M = Cr, Mo, and W) clusters to have strong abundances, compared to very low intensities of other neighboring clusters. More recently, stable $Si_{12}W$ was found (Hiura et al. 2001) from reaction of silane gas with metal monomers. It has a hexagonal prism structure with the W atom surrounded by Si (Kumar, 2006). The Si_{12} and Si_{16} encapsulated cases strictly do not qualify for the term fullerenes since, among others, they are smaller than the smaller carbon fullerene C_{20}. However, they are of high stability and symmetry with large highest occupied–lowest unoccupied molecular orbital (HOMO–LUMO) gaps, as the real fullerenes. The term silicon cages or "fullerene-like" is more appropriate than silicon fullerenes, although this last term is also used alternatively in several cases in the literature. Attempts to go to larger cages, such as Si_{20}, Si_{24}, and Si_{28} were more or less successful, at least theoretically, provided the right metal was chosen. For instance, doping of a Zr atom inside Si_{20} dodecahedral cage, although it leads to a large gain in the binding energy of Si_{20}, the $Zr@Si_{20}$ embedded cluster is unstable (Kumar 2006). However, doping by Th is more successful and the resulting $Th@Si_{20}$ "fullerene" is of high stability and icosahedral symmetry (Singh et al. 2005). This fullerene, together with some $M@Si_{12}$ cages (for various metals) are shown in Figure 48.8.

The work of Hiura et al. (2001) and the pioneering work of Beck (1987, 1989), which was for a long time overlooked, coupled with the early *ab initio* calculations of Kumar and Kawazoe (2001, 2003) stimulated a lot of interest in these systems by *ab initio* (and other) studies with a variety of doping metal (mainly transition metal) atoms and a large number of new structures and other findings (see for example: Han et al. 2003, Mpourmpakis et al. 2003, Singh et al., 2005, Koukaras et al. 2006, Kumar 2006, Zdetsis 2007c). The shape (symmetry) and size of these new clusters depend significantly on the metal atom, which determines their electronic and structural properties. The most popular metal atoms are tungsten (W), molybdenum (Mo), chromium (Cr), titanium (Ti), nickel (Ni), copper (Cu), and, more recently, zinc (Zn). Large macroscopic quantities of encapsulated silicon clusters ($M@Si_{20}$, $M@Si_{24}$, and $M@Si_{28}$) can be found in bulk form in clathrates (see, for instance, Kume et al. 2003).

It becomes clear that the method of metal encapsulation (embedding) is one of the most successful in stabilizing silicon cages and small fullerenes. However, it is restricted to very small fullerenes (of the order of 20 silicon atoms). For larger fullerenes, due to the larger diameter of the cages, the metal-cage interactions are weaker, thus failing to stabilize the cages.

One hint for the possible strategy to stabilize larger cases can be given by the example of $Zn@Si_{12}$ (Zdetsis 2007c). In an attempt to study and compare the structural properties of $Zn@Si_{12}$, $Ni@Si_{12}$, and $Cu@Si_{12}$ clusters as well as to examine the role of the metal atom in the overall bonding, Zdetsis (2007c) has considered fully hydrogenated $Zn@Si_{12}H_{12}$ and $Ni@Si_{12}H_{12}$ clusters, which are relieved from the strains of the surface dangling bonds. Kumar and Kawazoe (2003) have examined the role of hydrogen in other metal-encapsulated silicon clusters as well concluding that incorporation of hydrogen considerably weakens the metal-cage interactions. In Figure 48.9, the findings of Zdetsis (2007c) are illustrated in a schematic and rather "dramatic" way. As we can see in Figure 48.9, the metal-cage interactions not only weaken, but turn repulsive expelling the metal atom out of the cage eventually, leaving an empty hydrogenated cage. This cage, as the equivalence symbol indicates in the figure, preserves almost all characteristics of the non-hydrogenated metal embedded cage, i.e., high symmetry, high stability, and large HOMO–LUMO gap. In other words, pushing this example all the way to the end, we could anticipate that exohedral hydrogenation would have practically the same or equivalent effect on silicon fullerene stabilization. Moreover, the exohedral hydrogenation approach would not have the size limitation ($n < 30$) of the encapsulation method. This is indeed the case as will be illustrated below.

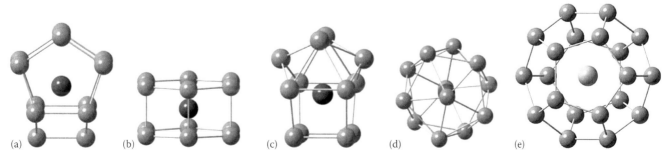

FIGURE 48.8 Various metal-encapsulated MSi_{12} cages (a, b, c, d) and MSi_{20} fullerenes (e). Metal atom is at the center of the cages.

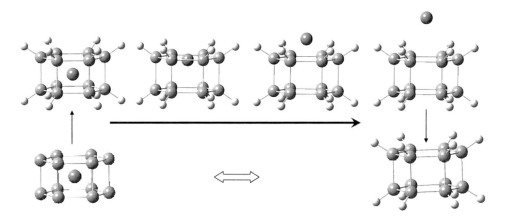

FIGURE 48.9 Successive "instances" of the geometry optimization of $Zn@Si_{12}H_{12}$.

48.2.3.1.2 Encapsulation by Various Molecules and Clusters

We have already examined in Section 48.2.2 the attempts to stabilize the Si_{20} fullerene by embedding several carbon atoms (Huda and Raya 2004), and Si_{60} by embedding a complete C_{60} inside in a form of a Russian doll (Harada et al. 1994), although the $C_{60}@Si_{60}$ arrangement is not stable. Along these lines, Sun et al. (2003) have considered the stabilization of Si_{60} by stuffing it with $Al_{12}X$ clusters, where X = Si, Ge, Sn, Pb. Although these results showed that Si_{60} cage could be stabilized by this method, the embedded clusters of $Al_{12}X$ were all decomposed, and the optimized geometries of the Si cage showed that the fullerene geometry structure was seriously distorted. The stuffed clusters approach was also followed by Yoo et al. (2004). These authors employed stuffing by Si_2 up to Si_5 magic clusters to construct silicon fullerenes Si_n with $27 \leq n \leq 39$. They have used genetic algorithms and the nonorthogonal-tight-binding method, followed by density functional theory to produce several alternative Si_{27}–Si_{39} structural models suggesting also an empirical rule that provides optimal "stuffing/cage" combinations. For instance, Si_{28} was formed by embedding Si_2 in the Si_{26} cage, whereas the Si_{33} endohedral fullerene was constructed either in the form $Si_3@Si_{30}$ cage or $Si_5@Si_{28}$ cage. Although these authors have calculated a higher binding energy for the $Si_3@Si_{30}$ cage, the $Si_5@Si_{28}$ cage with tetrahedral symmetry is physically more appealing mainly because Si_{28} is the building block for other cluster-assembled materials (such as clathrates). This type of magic cluster stuffing (for $Si_5@Si_{28}$ cage) goes back to the work of Kaxiras (1990) and Jelski et al. (1991). Yoo et al. (2006, 2008) have used the same idea of stuffed endohedral fullerene structural motifs to construct larger low-energy stuffed silicon fullerenes up to Si_{80}. Certainly at this range of sizes, fully *ab initio* calculations are not easy to implement, and it is impossible in reality to be sure (or even to suggest) which is the lowest energy structure (global minimum). On top of all this, problems with the secondary approximations or even the technical details of the calculations, as the ones described earlier by Zdetsis (2006), make the rationalization of the large cluster results even more difficult.

As a final criticism on the stuffed or endohedral fullerene method, it should be mentioned that although this method has

been quite successful and very fruitful (in particular for the structural chemistry of medium and large-size silicon clusters), it is actually missing the important scientific and practical applications of "fullerenes," which are based on an empty spherical cavity.

48.2.3.2 Exohedral Doping or Coverage

48.2.3.2.1 Exohedral Hydrogenation

As mentioned in Section 48.2.3.1.1 and illustrated in Figure 48.9, the role of the transition metal atom in the metal atom–embedded silicon fullerenes is similar to the saturation of the silicon dangling bonds of the "cage" by hydrogen. Therefore, the obvious extension of this analogy is to attempt to replace their assumed roles, especially in view of the limitation of this (other ways successful) method for larger (than about 20 Si atoms) cages. With this motivation, Zdetsis (2007d), in search for silicon "fullerenes" of the form Si_nH_n, examined a number of such cages and fullerenes between the sizes of $n = 4$ and $n = 60$, including $n = 20$, 24, 28, 32, 36, 50, 60 species. Extension to larger fullerenes up to $n = 180$ have been considered only recently (Zdetsis 2009a). All these structures, which are shown in Figure 48.10, have unusually high symmetries (icosahedral, tetrahedral, etc.) like the corresponding carbon fullerenes, and very large HOMO–LUMO gaps (and optical gaps, which include corrections beyond the one electron approximation) in the region of 4–5 eV.

Furthermore, in favor of such motivation is the fact that hydrogenated silicon clusters of the form Si_nH_m have been already synthesized experimentally (Rechtsteiner et al. 2001). Rechtsteiner et al. (2001) have found that for intermediate temperatures the composition with $m \approx n$ predominates.

About the same time as Zdetsis' (2007d) work, apparently with different motivation and context, Karttunen et al. (2007) have considered icosahedral hydrogenated silicon fullerenes with $n = 20$, 60, 80, 180, and even larger icosahedral (always) cages approaching the layered polysilanes limit. Also, Kumar and Kawazoe (2007) almost simultaneously have studied smaller cages and Si_nH_n fullerenes with $n = 20$, 24, 28, as in clathrates. Stabilization of the single fully exohydrogenated Si_{60} fullerene has been also reported by Wang et al. (2006). Smaller cages

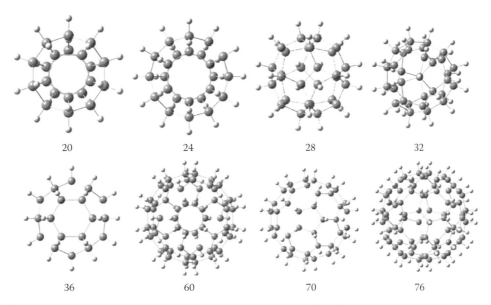

20   24   28   32

36   60   70   76

FIGURE 48.10 Fully hydrogenated Si_nH_n cages (n = 20, 24, 28, 32, 36, 60, 70, 76) in full analogy to the C_n carbon fullerenes of Figure 48.5.

(not fullerenes) with n = 12, 16, 20 have been also examined as early as 2003 by Kumar and Kawazoe (2003)

All the structures in Figure 48.10 are (at least) local minima lower (or lowest) energy "fullerenes." As in the case of carbon fullerenes, the "turning point" occurs at n = 20. $Si_{20}H_{20}$ is the most stable fullerene, as we can see in Figure 48.11. This is to be expected, on the basis of the discussion in Section 48.2.1, since $Si_{20}H_{20}$ consists of only 12 pentagons and no hexagons. From n = 24 and on (there is no n = 22 fullerene) the binding energy decreases very slowly with n due to the increased number of hexagons (from 0 in $Si_{20}H_{20}$ to 20 in $Si_{60}H_{60}$). The corresponding curve for bare Si_n fullerenes should be expected to be much more abrupt

and fast decreasing. $Si_{20}H_{20}$ is the only "fullerene" for which the corresponding hydrogenated carbon fullerene, the dodecahedrane $C_{20}H_{20}$ has been synthesized and exists in large quantities (Nossal et al. 2001, Moran et al. 2002). Dodecahedrane is a fascinating molecule with very high symmetry (I_h), which qualifies for "molecular beauty" (Hoffmann 1990). The synthesis of dodecahedrane, which was a difficult but tractable problem eventually also led to the synthesis of the smallest carbon fullerene C_{20} as well. The synthesis of C_{20} by high-voltage dehydrogenation of $C_{20}H_{20}$ is considered as the most dramatic experiment reported up to date (Chen et al. 2003a). Dodecahedrane ($C_{20}H_{20}$) has been well studied at various levels of theory with regard to electronic, chemical, and thermochemical properties. In particular, the energies and stabilities of the endohedral $X@C_{20}H_{20}$ and exohedral $XC_{20}H_{20}$ complexes have been extensively studied (Moran et al. 2002, Chen et al. 2003a) for a large number of host atoms and ions X (where X = H^+, H, He, $Li^{0/+}$, $Be^{0/+/2+}$, Ne, Ar, N, P, C^-, Si^-, O^+, S^+, $Na^{0/+}$, $Mg^{0/+/2+}$) for various chemical and technical applications.

Zdetsis (2007d) has shown that $C_{20}H_{20}$ and $Si_{20}H_{20}$ are fully analogous with the same symmetry, the same number of valence electrons, and same structure of the frontier (HOMO, LUMO) orbitals, which are shown in Figure 48.12. Therefore, it could

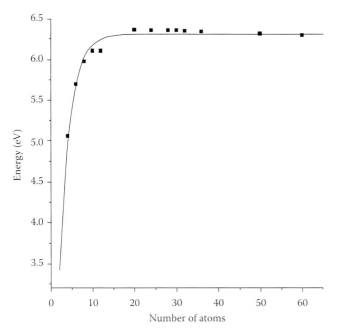

FIGURE 48.11 Binding energy (in eV) of Si_nH_n cages as a function of the number of atoms n, from n = 4 to n = 60.

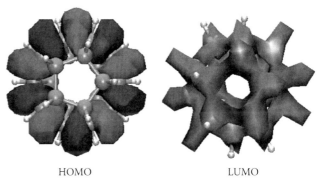

HOMO   LUMO

FIGURE 48.12 HOMO and LUMO orbitals of $Si_{20}H_{20}$.

be expected that they should have many (or several) similar properties, and analogous (not necessarily identical) chemical and technical applications. Indeed, endohedral doping with hilide anions has been examined (by *ab initio* calculations) and reported (Kumar and Kawazoe 2005), in the form of $X^-@Si_{20}H_{20}$ (X = F, Cl, Br, I) as well as exohedral functionalization by replacing one of its H atoms with $=CH_2OH$, $=COOH$, and $=CONH_2$ (Pichierri et al. 2004).

The resulting functionalized clusters have nearly the same HOMO–LUMO gaps as that of the perhydrogenated silicon fullerene. Thus, the HOMO–LUMO gap and more generally the optical properties of $Si_{20}H_{20}$ can be used for developing new silicon fullerene–based sensors and molecules for medico-biological applications (Pichierri et al. 2004). Furthermore, it could be expected that the synthesis $Si_{20}H_{20}$ could be achieved with methods "analogous" to the synthesis of $C_{20}H_{20}$, although this seems highly improbable in view of the complicated organic chemistry interactions involved for the synthesis of dodecahedrane. However, further reduction to icosahedral Si_{20} through high-voltage dehydrogenation or other appropriate technique seems to be highly unlikely in view of the destabilizing dangling bonds, which for carbon is not a problem due to sp² bonding.

Besides $Si_{20}H_{20}$, another very interesting (perhaps the most interesting and representative) icosahedral cage is $Si_{60}H_{60}$, which as been shown (Zdetsis 2007d), is fully homologous to the well-known $C_{60}H_{60}$ *fullerane*, which although it has not been synthesized up to now, its existence has been postulated in stellar atmospheres (Webster 1991) on the basis of calculated and observed infrared spectra originating in distant stars. From the name of this *fullerane* (not fullerene) the hydrogenated fullerenes are called *fulleranes*. In this respect, the more appropriate title of this chapter should have been *silicon fulleranes* or *silicon "fullerenes"*. The structure of the HOMO orbital of this cluster is shown in Figure 48.13 in comparison with the corresponding HOMO orbital of the distorted "bare" Si_{60} fullerene.

As shown in Figure 48.13, in the case of Si_{60}, the orbitals concentrate on the top of some silicon atoms, meaning that these are dangling bonds. When the dangling bonds are saturated, in the case of $Si_{60}H_{60}$, the orbitals concentrate between silicon atoms.

The rest of the Si_nH_n fulleranes in Figure 48.10, smaller as well as larger than $Si_{60}H_{60}$, are expected to have similar properties.

However, as the number n gets larger, the number of hexagons increases and the cohesive energy decreases, as is shown in Figure 48.11. If this trend continues, it becomes clear that larger silicon fulleranes would be destabilized against isoatomic bulk-like structures. Since the dangling bonds have been eliminated (through the incorporation of hydrogen), the best way to offset this trend is to help the puckering of the sp³ bonds in order to optimize the bond angles. For the icosahedral fulleranes, and in particular $Si_{80}H_{80}$ and $Si_{180}H_{180}$, Karttunen et al. (2007) have shown that this can be accomplished by partial endohedral hydrogenation. Zdetsis has generalized this to other carbons such as $C_{60}H_{60}$ (Zdetsis 2008a) and silicon fulleranes (Zdetsis 2009a,b). By partly endohedral hydrogenation, which might lower the symmetry of the structure, the bond angles in the hexagonal rings can be optimized by forming hexagonal chair-like structures. For the success of such a method, the number and the location of the endohedrally bonded hydrogen atoms is very important. In the case of $C_{60}H_{60}$ (and $Si_{60}H_{60}$), 10 out of 60 silicon–hydrogen bonds point inside the cage and the others point outward. The number and location of the endohedrally bonded atoms is important in order to balance the effect of puckering in hexagons and pentagons. The puckering of the bond angles in hexagons, unavoidably affects also the bond angles of the neighboring pentagons, which were already near the optimal value of 109.5°, thus increasing the corresponding pentagon energy. Since there are several unrelated ways to choose the set of the 10 endohedrally bonded hydrogens (in the case of $n = 60$), there are several alternative structures that can be generated this way. The difference between alternative structures is related with the different sets of carbon atoms with endohedral hydrogens, belonging to different types of carbon rings. The structure with the lowest energy is the structure in which more hexagons and fewer pentagons are puckered (Zdetsis 2008a). Such optimal structure for $C_{60}H_{60}$ (and $Si_{60}H_{60}$), characterized by D_{5d} symmetry, the largest subgroup (subset) of the I_h icosahedral group, is shown in Figure 48.14 together with the standard I_h fullerane. For $C_{60}H_{60}$, the energy difference between the two structures is more than 0.25 eV/atom. For $Si_{60}H_{60}$, the corresponding difference is only about 0.06 eV/atom. This large difference between these two numbers is due to the stronger C=H interaction compared to Si=H interaction. For larger fulleranes, such as $Si_{80}H_{80}$

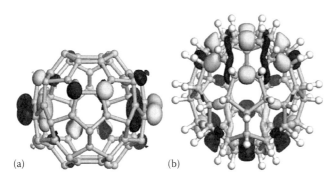

FIGURE 48.13 Comparison of the HOMO orbitals of the C_{2h} distorted Si_{60} cage (a), and the I_h $Si_{60}H_{60}$ (b).

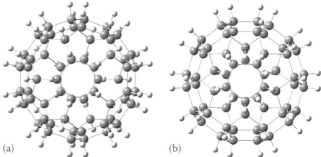

FIGURE 48.14 The normal icosahedral (a) and the puckered 10-hydrogen endohedrally hydrogenated D_{5d} (b) $Si_{60}H_{60}$ (or $C_{60}H_{60}$) cages fullerane (b). (Adapted from Wang, L. et al., *Mol. Simul.*, 32, 663, 2006.)

and $Si_{180}H_{180}$, the puckering is much more efficient leading to puckered structures, which are equally or even more stable than $Si_{20}H_{20}$ which has no hexagonal rings at all (Karttunen et al. 2007). This is very encouraging for the validity of this approach up to very large fulleranes.

48.2.3.2.2 Coating by Various Atoms and Molecules

A different approach consists in coating the silicon cage surface by various agents, the most popular of them being silica (Zhang et al. 2006) and platinum (Pei et al. 2007). These cages have the general form $Si_n(X)_{n/2}$, where $X = SiO_2$ in the first case, and X = Pt in the second case. In the former case, the cages can be also written equivalently $(Si_3O_2)_{n/2}$. In this arrangement, each silicon atom is bonded to only two oxygen atoms, leaving two dangling bonds, and rendering the outer periphery highly active. This certainly increases the energy, but at the same time it facilitates "functionalization" by attaching chemical groups at the active sites. The platinum coating, although it can stabilize the silicon fullerenes from Si_{24} to Si_{60}, has two major drawbacks in that the proposed structures are proven of higher energy compared to similar isoatomic clusters (for instance, the $Si_{24}Pt_{12}$ fullerene-like cage is 2.5 eV higher in energy compared to the nonfullerene structure), and that the endohedral functionalization is more difficult, if not impossible.

48.2.3.2.3 Some Latest Developments

At the time of writing this review, there are two new recent developments, which the present author is aware of, and in which he is directly involved. Presently, these are only suggestions and speculations submitted for critical evaluation before publication (Zdetsis 2009a,b). The first, which deals with the stabilization of larger fulleranes through partial endohedral hydrogenation and puckering, has been already explained and discussed. The second development deals with a "heretic" interpretation of older experimental results about the 1 nm luminous silicon nanoparticle. Akcakir et al. (2000) have observed bright blue photoluminescence from hydrogenated silicon nanoparticles with diameter about 1 nm. This was a real surprise since bulk silicon, contrary to its excellent electronic properties, has rather mediocre optical properties due to its small (about 1 eV) band gap, which is below the visible region of the spectrum. For small nanoparticles, however, due to "quantum confinement" the gap opens significantly allowing optical transitions and emission in the visible region (see, for example, Garoufalis et al. 2001, Garoufalis and Zdetsis 2006, Zdetsis 2006). The standard structural model for this luminous nanoparticle consists of a surface-reconstructed $Si_{29}H_{24}$ nanocrystal with T_d symmetry. This bulk-like model was obtained from the bulk $Si_{29}H_{36}$ nanocrystal by elimination of 12 terminating hydrogen atoms and reconstruction (similar to the 2×1 reconstruction of the Si[001] surface of crystalline silicon). This model seems to be consistent with most of the known experimental data, except perhaps a small absorption peak at 2.8 eV. Zdetsis and collaborators (Garoufalis and Zdetsis 2006; Zdetsis 2006), by constructing partially oxygenated bulk nanocrystal models of the form $Si_{29}H_{32}O_2$ and $Si_{29}H_{28}O_4$, have illustrated

that this peak could be related to oxygen contamination, but this work was more or less overlooked in the literature.

After the recent work on Si fullerenes (Zdetsis 2007d), this author was led to consider the $Si_{28}H_{28}$ "fullerene" as directly (and/or indirectly) related to the 1 nm luminous Si nanoparticle (Zdetsis 2009b). This "fullerene" has the right size and is characterized by the "right" T_d symmetry. In addition, the *embedding* or *inclusion* energy (the energy gain by inclusion of a central Si^{1-} ion) is the highest among nearby fulleranes. Zdetsis (2009b) has illustrated that the $Si^{1-}@Si_{28}H_{28}$ embedded "fullerene" is also consistent with most of the known experimental data, and therefore it could be a valid alternative to the standard structural prototype, explaining also the 2.8 eV peak. This alternative prototype is more physically appealing and technologically attractive for novel applications in optoelectronics and quantum computing. This last possible application is related to the localization of the HOMO orbital (see Figure 48.15) and the (high) spin on the central encapsulated atom. This is because endohedral fullerenes whose dopant atoms (or ions) retain (more or less) their isolated atomic states have received attention recently, partly because of the newly proposed solid-state quantum computers based on such materials.

It is expected that the advent of such solution-processed optoelectronic materials would offer the potential for a revolution in optoelectronics, since their solution processibility enables low-cost large-area monolithic integration on a variety of electronic readout platforms.

As we can see in Figure 48.15, only 4 hydrogen atoms out of the 28 weakly participate in this p_z-atomic-like HOMO orbital. As Zdetsis (2009b) has shown, by eliminating these four hydrogens (and the extra charge of the central atom), we are led (after geometry optimization) to the standard $Si_{29}H_{24}$ structural prototype.

Furthermore, the $Si^{1-}@Si_{28}H_{28}$ "fullerene" could be also important for astrophysical research, in view of the suggestion of Nayfeh et al. (2005) that the 1 nm silicon nanoparticles are the carriers of the blue luminescence in the red rectangle nebula. This is

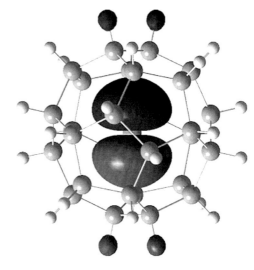

FIGURE 48.15 The HOMO orbital of the $Si^{1-}@Si_{28}H_{28}$ fullerene.

reminiscent of (and parallel to) the suggestion of Webster (1991) about the existence of carbon fulleranes ($C_{60}H_{60}$) in interstellar space. If the present interpretation is proven (even partially) correct by experiment, then silicon fulleranes could be also found in interstellar space, which thus would be a "birthplace" of carbon and silicon fulleranes. Future experimental and theoretical investigation will test the extent of the validity of this model.

48.3 Conclusions

Various methods and approaches have been examined during the search for silicon fullerenes, or at least something resembling silicon fullerenes, in (as close as possible) analogy to the well-known carbon fullerenes. In contrast to real carbon fullerenes, all these approaches are based on sp^3 (not sp^2) bonded cages or "fullerenes." The most successful of these approaches resort to doped (endohedrally or exohedrally), stuffed, or coated cages with a variety of atomic or molecular agents. Clathrates (empty or filled) are special examples of linked Si_{20}, Si_{24}, or Si_{28} "fullerenes" in bulk form bonded again by (near) sp^3 bonding.

Hydrogen-covered (hydrogenated) fullerenes have all the advantages: high stability, high symmetry, empty hollow spherical cavity, easily accessible (endohedral and exohedral) functionalization, and none of the disadvantages of the other "coating" or encapsulating methods (limitation to small sizes). Useful scientific and technological applications based on both endohedral and exohedral doping have been already designed and suggested (see, for example, Kumar and Kawazoe 2005, Zdetsis 2009b). On top of this, they are much simpler and aesthetically appealing, with possibilities of extension to much larger fullerenes, such as $Si_{180}H_{180}$, $Si_{320}H_{320}$, and even larger (Karttunen et al. 2007, Zdetsis 2009a). Furthermore, their corresponding carbon analogs (at least for two of them, $Si_{20}H_{20}$ and $Si_{60}H_{60}$) exist and are well known.

It is my (as unbiased as possible) opinion that from all the approaches considered here, the approach of hydrogenated fullerenes (with or without puckering and partial endohedral hydrogenation) seems to be by far the most successful, the most promising, and the most physically appealing in the search for silicon "fullerenes." This is as far as we can go with silicon "fullerenes," since true sp^2 bonded silicon fullerenes are practically impossible. This certainly remains to be seen in the (hopefully near) future.

Acknowledgment

The author wants to thank Dr. Emannel N. Koukaras and Mr. Dimitrios A. Zdetsis for a critical reading of the manuscript generous help with the drawings.

References

Akcakir, O., Therrien, J., Belomoin, G. et al. 2000. Detection of luminescent single ultrasmall silicon nanoparticles using fluctuation correlation spectroscopy. *Appl. Phys. Lett.* 76: 1857–1859.

Baldridge, K. K., Uzan, O., and Martin, J. M. L. 2000. The silabenzenes: Structure, properties, and aromaticity. *Organometallics* 19: 1477–1487.

Bao-Xing, L., Pei-Lin, C., and Duan-Lin, Q. 2000. Distorted icosahedral cage structure of Si_{60} clusters. *Phys. Rev. B* 61: 1685–1687.

Beck, S. M. 1987. Studies of silicon cluster-metal atom compound formation in supersonic molecular beam. *J. Chem. Phys.* 87: 4233–4234.

Beck, S. M. 1989. Mixed metal-silicon clusters formed by chemical reaction in a supersonic molecular beam: Implications for reactions at the metal/silicon interface. *J. Chem. Phys.* 90: 6306–6312.

Chen, Z., Jiao, H., Moran, D., Hirsch, A., Thiel, W., and Schleyer, P. V. R. 2003a. Structures and stabilities of endo- and exohedral dodecahedrane complexes ($X@C_{20}H_{20}$ and $XC_{20}H_{20}$, X = H^+, H, N, P, C^-, Si^-, O^+, S^+). *Phys. Chem. A* 107: 2075–2079.

Chen, Z., Jiao, H., Seifert, G. et al. 2003b. The structure and stability of Si_{60} and Ge_{60} cages: A computational study. *J. Comput. Chem.* 24: 948–953.

De Proft, E. and Geerlings, P. 2001. Conceptual and computational DFT in the study of aromaticity. *Chem. Rev.* 101: 1451–1464.

Fthenakis, Z. G., Havenith, R. W. A., Menon, M., and Fowler, P. W. 2005. Study of the Si fullerene cage isomers *J. Phys.: Conf. Ser.* 10: 117–120.

Garoufalis, C. S. and Zdetsis, A. D. 2006. Surface reconstruction versus oxygen contamination in the 1 nm small silicon nanocrystals. *Phys. Chem. Chem. Phys.* 8: 808–813.

Garoufalis, C. S., Zdetsis, A. D., and Grimme, S. 2001. High level ab initio calculations of the optical gap of small silicon quantum dots. *Phys. Rev. Lett.* 87: 276402-1–276402-4.

Han, J. G., Ren, Z.-Y., Sheng, L.-S., Zhang, Y.-W., Morales, J.-A., and Hagelberg F. 2003. The formation of new silicon cages: A semiempirical theoretical investigation. *J. Mol. Struct. (Theochem)* 625: 47–58.

Harada, M., Osawa, S., Osawa, E., and Jemmis, E. D. 1994. Silicon-coated fullerenes, $(SiC)_n$, $n = 20$ to 60. Preliminary evaluation of a new class of heterofullerenes. *Chem. Lett.* 23: 1037–1041.

Hiura, H., Miyazaki, T., and Kanayama, T. 2001. Formation of metal encapsulating Si cage clusters. *Phys. Rev. Lett.* 86: 1733–1737.

Hoffmann, R. 1990. Molecular beauty. *J. Aesthet. Art Criti.* 48: 191–204.

Huda, M. N. and Raya, A. K. 2004. Novel silicon-carbon fullerene-like cages. A class of $sp^3 - sp^2$ covalent-ionic hybridized nanosystems. *Eur. Phys. J. D* 31: 63–68.

Jarrold, M. F., Ijiri, Y., and Ray, U. 1991. Interaction of silicon cluster ions with ammonia: Annealing, equilibria, high temperature kinetics, and saturation studies. *J. Chem. Phys.* 94: 3607–3618.

Jelski, D. A., Wu, Z. C., and George, T. F. 1990. An inquiry into the structure of the Si_{60} cluster: Analysis of fragmentation data. *J. Cluster Sci.* 1: 143–154.

Jelski, D. A., Swift, B. L., Rantala, T. T., Xia, X., and George, T. F. 1991. Structure of the Si_{45} cluster. *J. Chem. Phys.* 95: 8552–8560.

Karttunen, A. J., Linnolahti, M., and Pakkanen, T. A. 2007. Icosahedral polysilane nanostructures. *J. Phys. Chem. C* 111: 2545–2547.

Kaxiras, E. 1990. Effect of surface reconstruction on stability and reactivity of Si clusters. *Phys. Rev. Lett.* 64: 551–554.

Khan, F. S. and Broughton, J. Q. 1991. Relaxation of icosahedral-cage silicon clusters via tight-binding molecular dynamics. *Phys. Rev. B* 43: 11754–11761.

Koukaras, E. N., Garoufalis, C. S., and Zdetsis, A. D. 2006. Structure and properties of the $Ni@Si_{12}$ cluster from all-electron ab initio calculation. *Phys. Rev. B* 73: 235417-1–235417-10.

Kroto, H. W., Heath, J. R., O'Brien, S. C., Curl, R. F., and Smalley, R. E. 1985. C^{60}: Buckminsterfullerene. *Nature* 318: 162–163.

Kumar, V. 2006. Alchemy at the nanoscale: Magic heteroatom clusters and assemblies. *Comput. Mater. Sci.* 36: 1–11.

Kumar, V. and Kawazoe, Y. 2001. Metal-encapsulated fullerene like and cubic caged clusters of silicon. *Phys. Rev. Lett.* 87: 045503.

Kumar, V. and Kawazoe, Y. 2003. Hydrogenated silicon fullerenes: Effects of H on the stability of metal-encapsulated silicon clusters. *Phys. Rev. Lett.* 90: 05552-1–05552-4.

Kumar, V. and Kawazoe, Y. 2005. Encapsulation of halide anions in perhydrogenated silicon fullerene: $X^-@Si_{20}H_{20}$ (X = F, Cl, Br, I). *Chem. Phys. Lett.* 406: 341–344.

Kumar, V. and Kawazoe, Y. 2007. Hydrogenated caged clusters of Si, Ge, and Sn and their endohedral doping with atoms: *Ab initio* calculations. *Phys. Rev. B* 75: 155425-1–155425-11.

Kume, T., Fukuoka, H., Koda, T. et al. 2003. High-pressure Raman study of Ba doped silicon clathrate. *Phys. Rev. Lett.* 90: 155503-1–155503-4.

Li, B.-X. and Cao, P.-L. 2001. Distorted cage structures of Si_n (n = 20, 24, 26, 28, 30, 32) clusters. *J. Phys: Condens. Matter* 13: 10865–10872.

Marsen, B. and Sattler, K. 1999. Fullerene-structured nanowires of silicon. *Phys. Rev. B* 60: 11593–11600.

Menon, M. and Subbaswamy, K. R. 1994. Structure of Si_{60} cage versus network structures. *Chem. Phys. Lett.* 219: 219–222.

Moran, D., Stahl, F., Jemmis, E. D., Schaefer, III H. F., and Schleyer, P. V. R. 2002. Structures, stabilities, and ionization potentials of dodecahedrane complexes. *Phys. J. Chem. A* 106: 5144.

Mpourmpakis, G., Froudakis, G. E., Andriotis, A. N., and Menon, M. 2003. Fe encapsulation by silicon clusters: Ab initio electronic structure calculations. *Phys. Rev. B* 68: 125407-1–125407-5.

Nagase, S. and Kobayashi, K. 1991. Si_{60} and $Si_{60}X$ (X = Ne, F^- and Na^+). *Chem. Phys. Lett.* 187: 291–294.

Nayfeh, M., Habbal, S. R., and Rao, S. 2005. Crystalline Si nanoparticles as carrriers of the blue luminescence in the red rectangle nebula. *ApJ.* 621: L121–L124.

Nossal, J., Saini, R. K., Alemany, L. B., Meier, M., and Billups, W. E. 2001. The synthesis and characterization of fullerene hydrides. *Eur. J. Org. Chem.* 4167–4180.

Pei, Y., Gao, Y., and Zeng, X. C. 2007. Exohedral silicon fullerenes: $Si_NPt_{N/2}$ 20≤N≤60. *J. Chem. Phys.* 127: 044704-1–044704-6.

Pichierri, F., Kumar, V., and Kawazoe, Y. 2004. Exohedral functionalization of the icosahedral cluster $Si_{20}H_{20}$: A density functional theory study. *Chem. Phys. Lett.* 383: 544–548.

Piqueras, M. C., Crespo, R, Orti, E., and Tomas, F. 1993. AM1 prediction of the equilibrium geometry of Si_{60}. *Chem. Phys. Lett.* 213, 509–513.

Ray, C., Pellarin, M., Lerme, J. L. et al. 1998. Synthesis and structure of silicon-doped heterofullerenes. *Phys. Rev. Lett.* 80: 5365–5368.

Rechtsteiner, G. A., Hampe, O., and Jarrold, M. F. 2001. Synthesis and temperature-dependence of hydrogen-terminated silicon clusters. *J. Phys. Chem. B.* 105: 4188–4194.

Sax, A. and Janoschek, R. 1986. Si_6H_6: Is the aromatic structure the most stable one? *Angew. Chem. Int. Engl.* 25: 651–652.

Sheka, E. F., Nikitina, E. A., Zayets, V. A., and Ginzburg, I. Y. 2002. High-spin silicon fullerene Si_{60} and its oligomers. *Int. J. Quantum Chem.* 88: 441–448.

Singh, A. K., Kumar, V., and Kawazoe, Y. 2005. Stabilizing the silicon fullerene Si_{20} by thorium encapsulation. *Phys. Rev. B* 71: 115429-1–115429-6.

Sun, X.-H., Li, C.-P., Wong, W.-K. et al. 2002. Formation of silicon carbide nanotubes and nanowires via reaction of silicon (from disproportionation of silicon monoxide) with carbon nanotubes. *J. Am. Chem. Soc.* 124: 14464–14471.

Sun, Q., Wang, Q., Jena, P., Rao, B. K., and Kawazoe, Y. 2003. Stabilization of Si_{60} cage structure. *Phys. Rev. Lett.* 90: 135503-1–135503-4.

Takahashi, M. and Kawazoe, Y. 2005. Theoretical study on planar anionic polysilicon chains and cyclic Si_6 anions with D_{6h} symmetry. *Organometallics* 24: 2433–2440.

Taylor, R. and Walton, D. R. M. 1993. The chemistry of fullerenes. *Nature* 363: 685–693.

Wang, L., Li, D., and Yang, D. 2006. Fully exohydrogenated Si60 fullerene cage. *Mol. Simul.* 32: 663–666.

Webster, A. 1991. Comparison of a calculated spectrum of $C_{60}H_{60}$ with the unidentified astronomical infrared emission features. *Nature* 352: 412–414.

Yoo, S. and Zeng, X. C. 2006. Structures and relative stability of medium-sized silicon clusters. IV. Motif-based low-lying clusters Si_{21}–Si_{30}. *J. Chem. Phys.* 124: 054304-1–054304-6.

Yoo, S., Zhao, J., Wang, J., and Zeng, X. C. 2004. Endohedral silicon fullerenes Si_N (27 ≤ N ≤ 39). *J. Am. Chem. Soc.* 126: 13845–13849.

Yoo, S., Zhao, J., Wang, J., and Zeng, X. C. 2008. Structures and relative stability of medium- and large-sized silicon clusters. VI. Fullerene cage motifs for low-lying clusters Si_{39}, Si_{40}, Si_{50}, Si_{60}, Si_{70}, and Si_{80}. *J. Chem. Phys.* 128: 104316-1–104316-9.

Zdetsis, A. D. 1996. A comparative ab initio study of small Si and C clusters. In *Stability of Materials*, A. Gonis, P. E. A. Turchi, and J. Kudrnovsky (Eds.), pp. 455–464. New York: Plenum Press.

Zdetsis A. D. 2006. Optical properties of small size semiconductor nanocrystals and nanoclusters. *Rev. Adv. Mater. Sci.* 11: 56–78.

Zdetsis, A. D. 2007a. Stabilization of flat aromatic Si_6 rings analogous to benzene: Ab initio theoretical prediction. *J. Phys. Chem.* 127: 214306.

Zdetsis, A. D. 2007b. Similarities and differences between silicon and carbon nanostructures: Theoretical predictions. *AIP Conf. Proc.* 963: 474–480.

Zdetsis, A. D. 2007c. Bonding and structural characteristics of Zn-, Cu-, and Ni-encapsulated Si clusters: Density-functional theory calculations. *Phys. Rev. B* 75: 085409-1–085409-10.

Zdetsis, A. D. 2007d. High-symmetry high-stability silicon fullerenes: A first-principles study. *Phys. Rev. B* 76: 075402-1–075402-5.

Zdetsis, A. D. 2007e. Fluxional and aromatic behaviour in small magic silicon clusters: A full *ab initio* study of Si_n, Si_n^{1-}, Si_n^{2-}, and Si_n^{1+}, n = 6, 10 clusters. *J. Chem. Phys.* 127: 014314-1–014314-10.

Zdetsis, A. D. 2008a. High-symmetry low-energy structures of $C_{60}H_{60}$ and related fullerenes and nanotubes. *Phys. Rev. B* 77: 115402-1–115402-5.

Zdetsis, A. D. 2008b. High stability hydrogenated silicon-carbon clusters: A full study of $Si_2C_2H_2$ in comparison to Si_2C_2, $C_2B_2H_4$ and other similar species. *J. Phys. Chem. A* 112: 5712–5719.

Zdetsis, A. D. 2008c. The boron Connection: A parallel description of aromatic, bonding, and structural characteristics of hydrogenated silicon-carbon clusters and isovalent carboranes. *Inorg. Chem.* 47: 8823–8829.

Zdetsis, A. D. 2009. Stabilization of large silicon fullerenes and related nanostructures trough puckering and poly(oligo) merization. *Phys. Rev. B* 80: 195417-1–195417-6.

Zdetsis, A. D. 2009. One-nanometer luminous silicon nanoparticles: Possibility of a fullerene interpretation. *Phys. Rev. B* 79: 195437-1–195437-8.

Zdetsis, A. D., Foweler, P. W., and Havenith, R. W. A. 2008. Aromaticity of planar Si_6 rings in silicon-lithium clusters. *Mol. Phys.* 106: 1803–1811.

Zhang, D., Guo, G., and Liu, C. 2006. New route for stabilizing silicon fullerenes. *J. Phys. Chem. B* 110: 14619–14622.

Zhu, X. L., Zeng, X. C., Lei, Y. A., and Pan, B. 2004. Structures and stability of medium silicon clusters. II. Ab initio molecular orbital calculations of Si_{12}–Si_{20}. *J. Chem. Phys.* 120: 8985–8995.

Boron Nitride Fullerenes and Nanocones

Ronaldo Junio
Campos Batista
Universidade Federal de Ouro Preto

Hélio Chacham
*Universidade Federal
de Minas Gerais*

49.1 Overview: Boron Nitride Compounds

In the periodic table, boron and nitrogen are the left and right carbon neighbors, respectively. Together, they form a binary compound, boron nitride (BN), that is isoelectronic with carbon compounds and that presents a similar chemical bond. Due to the similarities between B–N and C–C bonds, BN forms structures analogous to those of carbon, one of which is analogous to diamond (cubic BN) and the other analogous to graphite (hexagonal BN). In addition, BN, as carbon, can also form nanostructures such as fullerenes [1,2], nanotubes [3], and nanocones [4]. The BN structures present similar properties compared with those of their carbon analogs, for instance, the cubic BN is one of the hardest materials known. Both cubic BN and carbon diamond are electrical insulators and excellent heat conductors. On the other hand, BN compounds also present properties that sharply differ from those of their carbon analogs. An example is the case of hexagonal BN, also called white graphite, which is an insulator while graphite is a semimetal. BN nanostructures also present some properties that are similar to those of the carbon nanostructures and have other properties that differ. For instance, BN and carbon nanotubes both present high mechanical resistance [5]; on the other hand, BN nanotubes are insulators while carbon nanotubes can be either metallic or semiconductors [6]. Due to their superior chemical inertness, BN compounds can be used in situations where the reactivity of carbon compounds may be a problem. Cubic BN is widely used as an abrasive in situations where carbon diamonds are soluble in metals to give carbides. Hexagonal BN, as graphite, has excellent lubricating properties that can be used in situations where the electron conductivity or the chemical reactivity of the graphite is not desired. The BN nanostructures are also predicted to be chemically more stable than carbon nanostructures [1], which can make their use more viable than carbon nanostructures, in peculiar situations.

In summary, BN is scientifically interesting since it forms several allotrope structures as carbon does, but with some distinct properties. From a technological point of view, the combination of similar and distinct properties compared with those of their carbon analogs make BN structures interesting in situations where some of the outstanding properties of carbon structures are required and others are not.

49.2 Background: Historic Review

The name fullerene is due to the resemblance between the first discovered carbon fullerene, C_{60}, and the geodesic dome that was popularized by the American architect Richard Buckminster Fuller. Carbon fullerenes were discovered in 1985 by Robert Curl, Harold Kroto, and Richard Smalley at the University of Sussex and Rice University [7], and who were awarded the 1996 Nobel Prize in Chemistry. The discovery of a new allotrope of carbon has attracted the attention of the scientific community, leading to an intense study of the chemical and physical properties of the carbon fullerenes and also to the search for new fullerene-like structures. An idea posed was the possibility of synthesizing fullerene-like molecules from other elements rather than carbon. A natural candidate for such molecules was BN because of its well-known chemical similarity to carbon.

Firstly, the stability of small BN cages was investigated theoretically. In 1993, Jensen and Toftlund investigated the stability of $B_{12}N_{12}$ [8]. They found that the fullerenes were energetically more stable if they were composed of four and six-membered rings, which prevent the formation of homopolar B–B or N–N bonds. As was calculated by Blase et al. [9] in 1998, the homopolar B–B or N–N bonds have an energy cost of roughly 1.6 eV in BN structures. Other fullerenes, in which the homopolar bonds are absent, were predicted to be very stable: $B_{16}N_{16}$ and $B_{28}N_{28}$ by Seifert et al. [10] in 1997; and $B_{36}N_{36}$ by Alexandre in 1999 [2]. Such fullerenes, unlike carbon fullerenes that present a quasispherical morphology, are predicted to have octahedral symmetry, in particular, $B_{36}N_{36}$ is compatible in size with the experimental images by Golberg et al. [1]. Later, non-octahedral fullerenes were also predicted to exist, and $B_{24}N_{24}$ was predicted to be tubular [11,12,13]. More recently, in 2006, the non-stoichiometric icosahedral fullerenes with nitrogen excess were predicted to be stable in nitrogen-rich conditions [14].

The first experimental evidence of the existence of fullerene-like structures appeared in 1994 [15], when composite sheets and nanotubes of different morphologies containing carbon, boron, and nitrogen were grown in the electric arc discharge between graphite cathodes and amorphous boron-filled graphite anodes in a nitrogen atmosphere. The authors observed that separated domains of BN in carbon networks may exist. In the next year, BN nanotubes were synthesized by Chopra et al. in an arc-voltaic experiment. In 1998, BN fullerenes were produced by Stéphan et al. [16] in an experiment in which a BN target was irradiated by an electron beam. The next achievement was by Golberg et al., who obtained single-wall BN fullerenes by means of electron beam irradiation [1]. In both experiments, the observed square-like shape in the high resolution electron microscopy (HREM) images gave strong evidence that BN fullerenes present an octahedral morphology. In a subsequent set of experiments, BN fullerenes were produced by means of the arc-voltaic technique and characterized by means of mass spectrometry by Oku et al. [17]. BN fullerenes of several sizes, ranging from 48 up to 120 atoms were produced. Interestingly, $B_{24}N_{24}$ was produced in large amounts. Another interesting result of Oku and coworkers was the detection of a hybrid nanostructure composed by the $B_{36}N_{36}$ fullerene and the Y atom. The results of Oku and coworkers also suggest the existence of other hybrid structures composed by BN fullerenes and other transition-metal atoms.

49.3 Geometry of BN Fullerenes

BN and carbon fullerenes can be seen as a sheet of hexagonal BN or graphene, respectively, wrapped on closed surfaces. Both, hexagonal BN and graphene, are examples of hexagonal lattices in which every atom binds through sp^3 bonds to three neighbor atoms with a specific angle of 120° (Figure 49.1). The hexagonal lattice, however, cannot be wrapped on the curved surfaces. Instead, there must be, by Euler's theorem, packing defects. These packing defects can be recognized as the 12 pentagonal panels of a soccer ball or the 12 pentagonal rings of the C_{60} fullerene (see Figure 49.2). Due to such packing defects, C_{60}

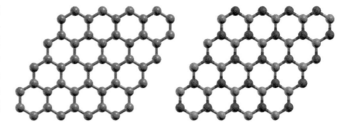

FIGURE 49.1 Hexagonal lattices: A single layer of hexagonal BN and graphene, respectively. The B–N bonds in a single layer of hexagonal BN and the C–C bonds in graphene present very similar bond lengths (1.446 and 1.42 Å [18], respectively).

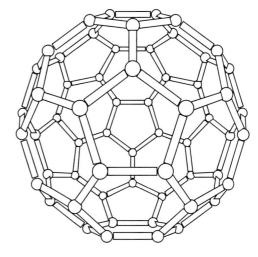

FIGURE 49.2 Icosahedral structure of the C_{60} molecule.

has an icosahedral symmetry in which every pentagonal ring has an energy cost associated with it (the sp^3 bonds are distorted with bond angles smaller than 120°) which, in addition to the energy cost associated with the curvature, make the fullerene energetically less stable than graphene. Wrapping hexagonal BN on a curved surface has an extra energy cost in comparison to graphene. Due to the presence of two chemical species in the hexagonal BN, homopolar bonds (B–B or N–N) must appear in each pentagonal ring, as is possible to see in Figure 49.3. An energy cost of roughly 1.6 eV is associated with every homopolar bond [9]. Such an energy cost suggests that wrapping the hexagonal BN on a icosahedral structure may not be the best form to wrap the hexagonal BN on a closed surface. Several geometries other than icosahedral have been proposed and their stabilities have been investigated. In such structures, the odd-membered rings with homopolar B–B and N–N bonds are, in general, absent. Instead, the structures are composed by tetragonal or octagonal rings together with hexagonal rings. However, the specific kind of rings in the cage seems to depend on the fullerene size.

On the experimental side, the structural investigations of the BN clusters are essentially made by means of mass spectroscopy HREM techniques [1,16,17]. Concentric nested and close-packed

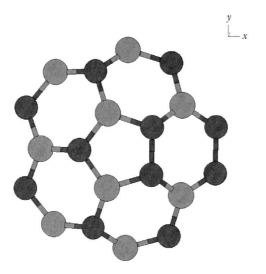

FIGURE 49.3 Homopolar N–N bond in a pentagonal ring.

cages were observed in the work of Stéphan et al. [16]. Most of the observed cages had a square-like shape and diameters between 4 and 7 Å. As possible structures, the octahedral fullerenes with 12, 16, and 28 BN pairs were proposed. In a subsequent experiment, Golberg et al. obtained single layer BN fullerenes [1]. In such an experiment, the authors conduct a systematic image treatment that give strong evidence of the octahedral model. As a possible structure for the images obtained by Goldberg, the $B_{36}N_{36}$ octahedral fullerene was proposed [2]. Following that, many theoretical studies regarding the $B_{36}N_{36}$ and the $B_{36}N_{36}$ hybrid structures were carried out [14,19,20]. However, the octahedral structure does not seem to be the unique form to build up BN fullerenes for all sizes. Recent theoretical works indicate a stable tubular structure with six-, four-, and eight-membered rings for the $B_{24}N_{24}$ cluster [20] (which is particularly interesting because of the fact that it can be produced in a large amount) instead of the octahedral one. Insights on the stability of BN fullerenes with different symmetries and sizes can be obtained by means of continuous models in which the energy cost to form the fullerenes is described as the elastic energy needed to bend a flat BN surface into a closed structure [14]. Let us see the case of octahedral and tubular fullerenes.

By Euler's theorem [21], it is not possible to wrap a hexagonal lattice into a close structure without the presence of point-like defects. The point-like defects in a curved hexagonal lattice are the non-hexagonal rings, such as the pentagons at C_{60}. Those point-like defects can be characterized by their disclination charge, q, which is the departure of their number of sides, C, from the preferred value on the flat space 6 ($q = 6 - C$). Then, $q = 0$ for the hexagonal ring, $q = 1$ for the pentagonal ring, $q = 2$ for the square, $q = -2$ for the octagonal rings, etc. From Euler's theorem, the total disclination charge of a hexagonal lattice on a sphere [22] or on a polyhedral surface is 12 (12 is obtained from 6 times the Euler characteristic, which is a topological invariant equal to two for the sphere and for polyhedral surfaces [23]). A total disclination charge of 12 can be achieved in many ways. Icosahedral fullerenes, for instance, have 12 pentagonal rings. The tubular $B_{24}N_{24}$ has two octagonal rings and eight square-like rings, then $q = 2(-2) + 8(2) = 12$. In the case of octahedral fullerenes, such point-like defects can be six square rings. Figure 49.4 shows that $B_{64}N_{64}$, $B_{48}N_{48}$, and the $B_{36}N_{36}$ are octahedral fullerenes in which it is possible to see the square rings. In each square ring, the B–N bonds are placed in an angle different from 120°, therefore, an energy cost can be associated with the presence of squares in the fullerene. The energy cost associated with the point-like defects is the same for octahedral fullerenes of all sizes if the octahedral fullerenes are formed by six square rings. The B–N bonds are also bent at the edges of the octahedral fullerenes; therefore, an energy cost is associated with the edges of the octahedron. The amount of B and N atoms at the fullerenes edges grow in as much as the fullerene size increases. The number of atoms at the fullerene edge is proportional to the edge length, L, and the total number of atoms in the fullerene, n, is proportional to the octahedral surface that is equal to $2L^2\sqrt{3}$. Then, the number of atoms at the octahedral edge is proportional to $\sqrt{n}$. Thus, in case of the octahedral fullerenes, the formation energy per atom can be approximated by

$$\frac{E_{\text{form}}}{n} = A + \frac{B}{n} + \frac{C}{\sqrt{n}}, \tag{49.1}$$

where A is the formation energy per atom of the structure when $\to \infty$, or, equivalently, the formation energy per atom of

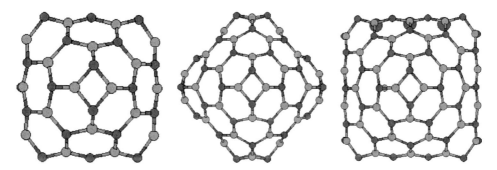

FIGURE 49.4 Stoichiometric octahedral fullerenes. From left to right: $B_{36}N_{36}$, $B_{48}N_{48}$, $B_{64}N_{64}$. Only half of each fullerene is shown for clarity.

the BN plane. Since the B–N bonds are not distorted at the BN plane, $A = 0$. B/n is the contribution due to the strain of the B–N bonds in the square rings at the tips of the octahedron, which is not dependent on the system size. The term $1/\sqrt{n}$ is related to the distorted B–N bonds at the edges of the octahedral fullerenes that grow with $\sqrt{n}$.

In the case of the tubular fullerenes, there is an energy cost associated with the distortion of the B–N bonds at the tube ends. For any tube, the energy cost associated with its ends is not dependent on the tube length. The size dependence in case of tubular fullerenes comes from a distortion of the B–N bonds due the curvature of the tube that is inversely proportional to the tube radius. The formation energy per atom of tubular fullerenes, in a continuum model approximation, can be expressed by the following expression:

$$\frac{E_{\text{form}}}{n} = A + \frac{B}{n}, \tag{49.2}$$

where

   B is the formation energy of the tube ends
   A is the curvature energy per atom that is not at the tube ends

The continuous models described above can be used to adjust the data in Table 49.1, which shows the formation energy, per BN pair of a set of tubular and octahedral BN fullerenes. The values shown in Table 49.1 were calculated as follows:

$$E_{\text{f}} = E_{\text{total}} - \mu_{\text{BN}} N. \tag{49.3}$$

The first term of Equation 49.3 is the total energy of the fullerene obtained from numerical calculations. The numerical method used to compute the total energy is based on the density functional theory [24], which is a very popular quantum mechanical theory used in physics and chemistry to investigate the electronic structure of many-electrons systems, in particular, atoms,

TABLE 49.1 Formation Energy (in eV) per Number of BN Pair, n(BN), for Tubular and Octahedral Fullerenes

Structure	E_{f}/n(BN)
$(B_{16}N_{16})_{\text{tub}}$	1.35
$(B_{24}N_{24})_{\text{tub}}$	0.97
$(B_{32}N_{32})_{\text{tub}}$	0.82
$(B_{40}N_{40})_{\text{tub}}$	0.72
$(B_{48}N_{48})_{\text{tub}}$	0.66
Infinite tube	0.34
$(B_{16}N_{16})_{\text{oct}}$	1.18
$(B_{24}N_{24})_{\text{oct}}$	1.09
$(B_{36}N_{36})_{\text{oct}}$	0.73
$(B_{48}N_{48})_{\text{oct}}$	0.58
$(B_{64}N_{64})_{\text{oct}}$	0.48
$(B_{100}N_{100})_{\text{oct}}$	0.34
$(B_{144}N_{144})_{\text{oct}}$	0.13

molecules, and condensed phases. The second term of Equation 49.3 is energy cost to bring the N BN pairs from a BN reservoir, in this case, a sheet of hexagonal BN. μ_{BN} is the chemical potential of the hexagonal BN sheet which is also obtained from numerical density-functional theory calculations.

The $B_{24}N_{24}$(oct) is formed by six octagonal rings, twelve square rings and eight hexagonal rings. All the other octahedral fullerenes have six square rings at the octahedron vertex and a size-dependent number of hexagonal rings. The tubular fullerenes all have the same diameter and the same ends, the smallest one, $B_{16}N_{16}$, is the limit case where the two ends of the tube are put together. The limit case of an infinite tube is obtained imposing periodic boundary conditions along the tube length. The values of forming energy per BN pair in Table 49.1 were calculated taking a hexagonal BN sheet as reference that is appropriate for situations in which the BN fullerenes are grown in equilibrium with BN layers. It can be understood as the energy cost to pick up BN pairs from a reservoir, the BN layer, which forms a specific fullerene.

The fitting of the data shown in Table 49.1 using the continuous models previously described is shown in Figure 49.5. The formation energy per number of BN pairs is plotted as a function of $1/\sqrt{(n)}$, where n is the total number of atoms in the structures. In both curve fittings, we did not use the smallest fullerene of each symmetry, since the models, which are continuous, are expected to give more accurate results as the number of atoms increases. Indeed, it is possible to see that the continuous models describe very well, the energetic of the formation of BN fullerenes in case of fullerenes with more than 24 BN pairs. In as much as n increases the formation energy of octahedral fullerenes, the contribution of the atoms at the vertex (square rings) and at the edges of the octahedron becomes smaller when compared with the contribution of the atoms at the faces. Therefore, the formation-energy of the octahedral fullerenes tends to the formation energy of a plane layer of the hexagonal BN which by definition has a null value of formation energy. The formation energy of the tubes tends in the limit of n to be very large compared to the value of the formation energy of an infinity tube. The general trend that comes out from the curves shown in Figure 49.5 is that the octahedral symmetry is energetically favored for fullerenes formed by a number of atoms greater than 64. For instance, the $(B_{48}N_{48})$oct is more stable than the $(B_{48}N_{48})$tub by 3.84 eV or 40 meV per BN pair(kT is 25 meV for $T = 300$ K). This trend is in agreement with the square-like shape reported in several experimental works [13]. However, the results shown in Figure 49.5 show that the curves cross for n about 64 which means that the tubular fullerenes are more stable than the octahedral ones in case of fullerenes with less than 32 BN pairs. This trend is in agreement with several theoretical works that predict that the $B_{24}N_{24}$ tubular fullerenes is energetically more stable than its other isomers [11,12,13]. From Figure 49.5 and Table 49.1, it is possible to see that the tubular $B_{24}N_{24}$ has a value of formation energy per BN pair that is 120 meV smaller than its octahedral isomer. Such energy difference can be understood in terms of the distortion of the B–N bonds in non-hexagonal rings; the tubular

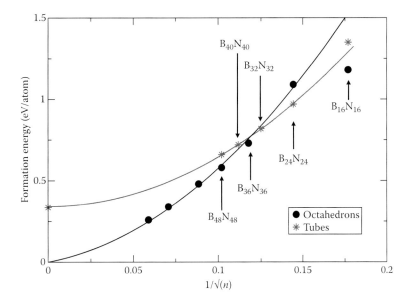

FIGURE 49.5 Continuous model for the formation energy of tubular and octahedral fullerenes.

$B_{24}N_{24}$ has less distorted rings (four octagonal rings less) than its octahedral isomer. The octahedral $B_{24}N_{24}$ fullerene has six octagonal rings and eight square rings. The tubular $B_{24}N_{24}$ has two octagonal rings and also eight square rings. The distortion of the B–N bonds on non-hexagonal rings also provides an explanation for the energy differences among other small stoichiometric fullerenes. For instance, the tubular isomer of $B_{16}N_{16}$, which has two octagonal rings and eight square rings, is energetically less stable than its octahedral isomer which has six square rings and no octagonal rings.

49.4 Geometry of Non-Stoichiometric BN Fullerenes

Due the energy cost of roughly 1.6 eV associated with the homopolar bonds B–B and N–N, it is unlikely that such homopolar bonds are present in small stoichiometric BN fullerenes. Indeed, this observation is in agreement with several theoretical calculations in which the low-energy cages are composed by even-membered rings which prevent the formation of such homopolar bonds. However, it is not possible to prevent homopolar bonds on non-stoichiometric BN fullerenes. The stability of non-stoichiometric fullerenes have been addressed theoretically [14,27]. The cited theoretical works regarding the stoichiometry of BN fullerenes agree that fullerenes with less than roughly 60 atoms are preferably stoichiometric. Nevertheless, BN fullerenes with a small excess of nitrogen and more than 60 atoms are as stable as stoichiometric fullerenes in a nitrogen-rich environment. The comparative analysis of the energetic stability of non-stoichiometric fullerenes must react with the experimental conditions (boron or nitrogen-rich environments) and the synthesis technique (arc-discharge or substitutional reaction, for instance). The BN fullerenes are produced under nonequilibrium conditions that would require a nonequilibrium thermodynamic theory for the full description of the synthesis procedure. An equilibrium thermodynamic approximation that allows the comparison of the formation energy of non-stoichiometric BN cages was proposed by Alexandre et al. 8 year ago [27]. Within such formalism, the choice of appropriate chemical potentials reflects the experimental conditions and the synthesis technique. According to the work of Alexandre et al., the relative stability of BN fullerenes is determined by its formation energy which is given by the following expression:

$$E_{\text{form}} = E_{\text{total}} - n\text{B}\mu_{\text{B}} - n\text{N}\mu_{\text{N}}, \tag{49.4}$$

where

E_{total} is the total energy of the BN fullerene derived from first-principles calculations

$n\text{B}$ and $n\text{N}$ are the number of boron and hydrogen atoms, respectively

The chemical potentials, μ_{B} and μ_{N}, are constrained by the equilibrium thermodynamic condition: $\mu_{\text{BN}} = \mu_{\text{B}} + \mu_{\text{N}}$, where μ_{Bulk} is the chemical potential per BN pair of a hexagonal BN layer. This constraint is appropriate for situations in which the BN structures grow in equilibrium with BN layers. Once the chemical potential of the B or the N reservoirs is defined, the chemical potential of the N or the B is then given by the thermodynamic constraint. In the process of electron beam irradiation, the BN fullerenes was obtained in a solid-state reaction. Then, the chemical potentials are obtained from solid state reservoirs. In boron rich conditions, the B excess form the bulk boron; therefore, μ_{B} is defined in the limiting case $\mu_{\text{Bulk}} = \mu_{\text{B}}$, where μ_{Bulk} is the total energy per atom of solid boron. In nitrogen-rich conditions, μ_{N} is the total energy per atom of solid nitrogen. The described thermodynamic approximation has been used to address the question of the stability of non-stoichiometric octahedral and icosahedral BN fullerenes [14,27].

The octahedral fullerenes must have six topological defects with a disclination charge of two. Such defects can be the square rings as can be seen in the octahedral fullerenes of Figure 49.4 or two side-sharing pentagonal rings that share a homopolar bond, as is also shown in Figure 49.3. In case of octahedral fullerenes composed by side-sharing pentagonal rings, homopolar B–B or N–N bonds appear at the tip of the octahedron. Therefore, all octahedral fullerenes formed by side-sharing pentagonal rings must have six homopolar bonds despite their size. In as much as the size of the fullerene increases, the effect of the six homopolar bonds at the tip of the octahedral fullerenes on the formation energy of the fullerene, become smaller. Therefore, it is preferable that the homopolar bonds happen at large octahedral fullerenes. Indeed, the results of Alexandre et al. show that the stoichiometric fullerene $B_{16}N_{16}$ is energetically more stable than the non-stoichiometric fullerenes of similar size, namely, $B_{16}N_{12}$ and the fullerene with nitrogen excess $B_{12}N_{16}$. Alexandre et al. have also investigated the relative stability of other larger stoichiometric and non-stoichiometric fullerenes, namely: $B_{32}N_{36}$; $B_{36}N_{32}$; $B_{36}N_{36}$; $B_{60}N_{64}$; $B_{64}N_{60}$; and $B_{64}N_{64}$. It was observed that the fullerenes $B_{32}N_{36}$ and $B_{60}N_{64}$ are more stable than the stoichiometric fullerenes of similar size ($B_{36}N_{36}$ and $B_{64}N_{64}$) in nitrogen-rich conditions. Unlike the large nitrogen rich fullerenes, the fullerenes with Boron excess are less stable than the stoichiometric fullerenes even in boron-rich conditions. This is probably due to the more covalent nature of the N–N bonds in the BN cage, compared with the B–B bonds. In the BN cage, the B–B bond length (1.66 Å) is significantly larger than the B–N bonds (1.44 Å) that is similar to the N–N bond length (1.46 Å). Alexandre et al. have also considered the gas-phase limit of the atomic reservoirs that reflect the synthesis process in which the condensation of atoms from a gas-plasma phase would be involved. This process can be seen as the other extreme case of a fullerene growth condition, opposite to the solid-state reactions. The conclusions regarding relative stability described above remained unchanged when the gas-phase limit of atomic reservoirs were considered, which indicates that the above results are robust with respect to the growth conditions. In addition, by means of the HREM obtained, it is not possible to distinguish the side-sharing pentagonal vertex and the square-ring vertex.

Icosahedral fullerenes have 12 pentagonal rings. In case of BN fullerenes, such pentagonal rings share a line of homopolar bonds as is shown in Figure 49.6. Such fullerenes can be formed by pulling apart the side-sharing pentagonal pair in an octahedral fullerene like those in Figure 49.7. Then, every icosahedral BN fullerene has six lines of homopolar bonds. Since every homopolar bond has an energy cost associated with it, the line of homopolar bonds tends to destabilize the fullerene by increasing the formation of its energy. On the other hand, the elastic energy required to turn a hexagonal BN layer into an icosahedral fullerene is smaller than the elastic-energy cost to form an octahedral fullerenes [22]. Therefore, the carbon fullerenes, which do not have homopolar bonds, are found in icosahedral symmetry rather than the octahedral one. The competition among elastic energy and the homopolar bond cost may lead to BN structures

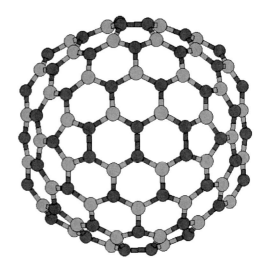

FIGURE 49.6 A view of the non-stoichiometric $B_{84}N_{96}$ fullerene, where it is possible to see a line of homopolar N–N bonds connecting two pentagonal rings. Half of it was removed for clarity.

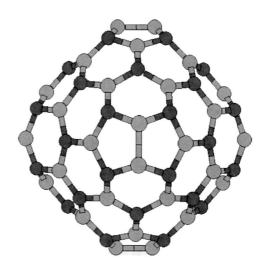

FIGURE 49.7 A view of the non-stoichiometric $B_{48}N_{44}$ fullerene, where it is possible to see the side-sharing pentagonal rings. Half of it was removed for clarity.

where the homopolar bonds are present. One example is the presence of a line of homopolar bonds in BN nanocones as was shown by Azevedo et al. [4]. Another example is the case of icosahedral BN fullerenes.

Icosahedral fullerenes with line defects composed of nitrogen bonds can be as stable as stoichiometric octahedral fullerenes in a nitrogen-rich environment [14]. The formation energy of the fullerene $B_{36}N_{44}$ in nitrogen-rich conditions is 0.69 eV per boron atom that is smaller than the formation energy of the already synthesized $B_{36}N_{36}$ fullerene, 0.72 eV per boron atom. Its formation energy per boron atom is also 0.03 eV smaller than the stoichiometric BN tube with the same amount of atoms. Within a continuous model approximation, the formation energy of the icosahedral fullerenes is given by the following expression:

TABLE 49.2 Formation Energy (in eV) per Number of B and N Atoms for Tubular, Octahedral, and Icosahedral Fullerenes in Nitrogen-Rich and Boron-Rich Environments, Respectively

Structure	$E_f/n(B)$	$E_f/n(N)$
$(B_{12}N_8)_{ico}$	2.63	2.55
$(B_{44}N_{36})_{ico}$	1.14	0.78
$(B_{96}N_{84})_{ico}$	0.76	0.46
$(B_8N_{12})_{ico}$	2.48	2.59
$(B_{36}N_{44})_{ico}$	0.69	1.07
$(B_{84}N_{96})_{ico}$	0.43	0.72

$$\frac{E_{form}}{n} = A + \frac{B}{n} + \frac{C}{\sqrt{n}}, \quad (49.5)$$

where A is, as in the previous cases, the formation energy per atom of the BN plane. The term B/n is associated with the curvature of an approximately spherical surface. The last term is related to the energy cost of lines of BB or NN bonds.

Table 49.2 shows the formation energy per boron atom in a nitrogen-rich and in a boron-rich environment, calculated by means of first-principles calculations, of the set of the icosahedral BN fullerenes: $B_{12}N_8$; $B_{12}N_8$; $B_{12}N_8$; $B_{12}N_8$; $B_{12}N_8$; and $B_{12}N_8$. Such values can be used to adjust the continuous model described above. The fitting for icosahedral fullerenes is shown in Figure 49.8, where also is shown the continuous models for octahedral fullerenes. The model allows us to verify that for larger ($n \rightarrow \infty$) fullerenes. The octahedral fullerenes are more stable than the icosahedral ones, which mean that the energy cost due to homopolar bonds is greater than the elastic-energy difference between the icosahedral and octahedral structures. The models also indicate that the fullerenes with nitrogen

excess in nitrogen-rich conditions are as stable as the octahedral fullerenes in a range size close to 80 atoms. As in the case of the continuous models for tubular and octahedral fullerenes, the models for icosahedral fullerenes fail for the small fullerenes, $B_{12}N_8$ and B_8N_{12}, which are not shown in Figure 49.8 for clarity purposes. The icosahedral fullerenes with boron excess in a boron-rich environment are, for all sizes, energetically less stable than the fullerenes with nitrogen excess in nitrogen-rich conditions, possible, due the more covalent nature of the N–N bonds compared with B–B bonds.

49.5 Doped BN Fullerenes

BN fullerenes can be obtained by means of an arc melting technique, as has been shown by Oku et al. [17], which usually involves samples with transition-metal atoms in their composition. Unlike graphite, hexagonal BN is a large-gap insulator. Therefore, the synthesis process of BN nanostructures by means of the arc-melting technique differs from that of carbon nanostructures in which a voltaic arc is established between the graphite electrodes. Instead, a boron-based powder is set on the metallic electrode in the presence of a nitrogen-gas atmosphere [17]. The powder can be a compound of boron and transition metal, such as yttrium, gold, vanadium, titanium, or it may be formed by a mixture of boron particles with iron oxide, cobalt, or LaB_6 nanoparticles. Images of a typical resulting sample obtained with HREM reveal the existence of a dark contrast either inside the cage, or outside and close to it [17]. These dark contrasts are interpreted as evidences of the existence of a hybrid structure, formed by transition metal atoms with the fullerene cage. In mass spectroscopy experiments performed on a sample in which the original boron powder was

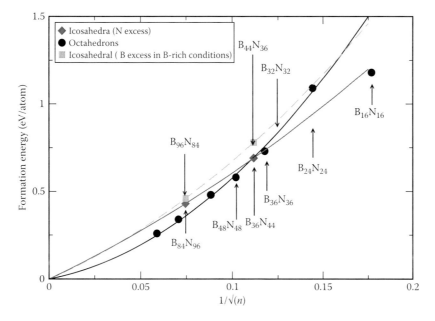

FIGURE 49.8 Continuous model for the formation energy of icosahedral fullerenes with nitrogen excess in nitrogen-rich conditions, icosahedral fullerenes with boron excess in boron rich conditions, and octahedral fullerenes.

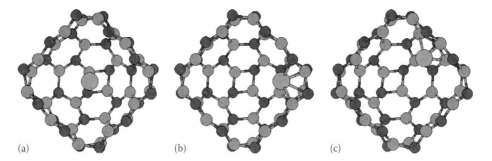

(a)   (b)   (c)

FIGURE 49.9   Binding sites for Fe atom to $B_{36}N_{36}$: (a) center, (b) tetragonal ring, (c) hexagonal ring.

YB_6, a hybrid structure composed by the fullerene $B_{36}N_{36}$ and the Yttrium atom was detected.

The results described above show that the hybrid structure composed by fullerenes and metal atoms may be formed during the synthesis. Due to the experimental evidence of the existence of hybrid structures, questions concerning their structural, electronic, and magnetic properties arise naturally. These issues have been addressed theoretically in many works [19,20,25] where hybrid structures composed by BN fullerenes (B_nN_n, $n = 12$, 16, 20, 24, 28, and 36), metal atoms (Sc, Ti, V, Cr, Mn, Fe, Co, Ni, Zn, Y, Zr, Nb, Mb, Tc, Ru, Rh, Pd, Ag, and W), molecules (FeO), and metal clusters (Ti_4, V_4, Mn_4, Fe_4, Co_4, Ni_4, Cu_4, Co_2, and Rh_2). In general, the electronic structure of the hybrid structures is different from that of isolated cages; several states on the fullerenes gap appears. No magnetic moment is, in general, induced on the BN cages. In most of the cases, the dopant (metal atoms, metal clusters, or molecules) remains magnetic after encapsulation. Among all the cages listed above, let us discuss in detail the case of the B_nN_n whose size and shape are compatible with the first BN fullerene images observed by Golberb et al. [1], as well as with the HREM images obtained by Oku et al., in which the presence of an extra atom was suggested. This cage forms a hybrid structure with the Yttrium atom, as was detected in mass spectroscopy experiments.

In agreement with experimental evidences, theoretical works suggest that the $B_{36}N_{36}$ fullerenes can form hybrid structures with some transition metal atoms (Fe, Co, W, Ti, V, Cr, Ni, and Cu). The metal atom can either be encapsulated by the fullerene or make a covalent bond with the fullerene wall. The equilibrium position inside the BN cage depends on the atom specie. However, for all studied metal atoms, the least favorable sites for covalent bonds are the inner regions of the four-membered rings at the vertex of the octahedron, see Figure 49.9a, possibly due to orbital-confinement effects. Upon encapsulation, the extended s orbitals become too confined and it transfers its electrons to its d orbitals. This confinement, and therefore, the $s - d$ transfer, is maximum when the atom is bound inside the cage close to the four-membered ring. For instance, in case of the Fe atom, whose orbital configuration is $3d6.0s2.0$, a population analysis indicates a large depletion when the atom is bound inside the cage close to the four-membered ring ($3d7.14s0.5$). On the other hand, at the center of the fullerene, the configuration is $3d6.24s1.5$. Outside

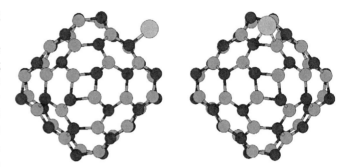

FIGURE 49.10   Two views of the bridge or B–N bond site for an Fe atom interacting with a $B_{36}N_{36}$ fullerene. Some B and N atoms were removed for clarity.

the cage, the bridge sites above the B–N bonds, at the edges of the octahedron, see Figure 49.10, are energetically preferred in case of the studied metal atoms (Fe, Co, and W), which is in agreement with the results concerning the interaction of metal atoms with an (8, 0) BN nanotube [11,12]. Some of the binding trends outside the cage may be a result of curvature effects. That can be seen by comparing the interaction of an adatom with BN surfaces with different levels of curvature. The binding energy of an Fe atom (the atom placed at the B–N bridge) to the BN plane is 0.45 eV, which is 0.81 eV smaller than that found for the Fe atom placed outside the fullerene at the bridge site. An intermediate situation between the strongly curved surface (fullerene) and a plane surface (zero curvature) is that of an atom-bound nanotube. Wu et al. calculated the binding energy of an Fe atom located above the B–N bond of a (8, 0) single wall BN nanotube. They found an intermediate value of binding energy 0.62 eV. Therefore, there seems to be a relationship between the curvature and the binding energy.

The magnetic moment of the fullerenes doped with metal atoms is important to the quantity of their physical properties. Table 49.3 shows the total magnetic moment of the Fe and the Co atom for the positions relative to the fullerene shown in Figure 49.9a through c, see Ref. [25]. It is also shown, inside parenthesis, the total magnetic moment of the system. It is possible to see that the total magnetic moment of the system is essentially the magnetic moment of the dopant atom, with no magnetic moment induced on the BN cage in any situation. In contrast, a magnetic moment is induced when an Fe atom interacts with a

TABLE 49.3 Magnetic Moments, in μ_B, of the Dopants and the Total Magnetic Moments of the System (Inside Parentheses)

Position	Fe	Co
Center	4.02 (4.00)	2.96 (3.00)
Tetragonal ring	2.18 (2.00)	1.09 (1.00)
Hexagonal ring	3.95 (4.00)	2.86 (3.00)

carbon cage [26]. For instance, in case of FeC_{20}, a parallel magnetic moment of $0.2\,\mu_B$ is induced in each carbon atom. This fact indicates that hybridization of the transition metal atom states with the cage is much smaller in BN structures than in carbon ones. This fact is consistent with the smaller values of binding energy obtained for the Fe atom interaction with an (8, 0) single wall BN nanotube [11,12] (from −0.27 up to 0.62 eV) as compared with the same atom interacting with a carbon nanotube [11,12] (from 0.65 up to 1.40 eV), and for the binding energy of an atom interacting with the $B_{36}N_{36}$ fullerene [25] as compared with the C_{48} (1.45 up to 2.67 eV).

The results shown in Table 49.3 can be understood in terms of the depletion of the $4s$ orbitals due to the atom confinement. Table 49.4 shows the Mulliken population analysis of the $3d$ and the $4s$ orbitals of the Fe atom at the same relative position to the cage as in Table 49.3. In most of the cases, the $s \rightarrow d$ transfer is not effective in changing the total magnetic moment of the atom. The reason is that the transfer involves at the most one electron, which allows an $s \rightarrow d$ transfer between minority spin electrons. This is true for small and moderate confinements. Upon increasing the confinement, such as in a position close to the tetragonal ring inside the cage, the other $4s$ electrons (which have a majority spin) are also transferred to the (minority spin) $3d$ orbitals, which correspond to change in the magnetic moment by almost $2\mu_B$.

In another theoretical study [19], Ningam et al. investigated the stability of metal clusters composed by four metal atoms (Ti, V, Cr, Mn, Fe, Co, Ni, and Cu) encapsulated by the $B_{36}N_{36}$ fullerene. The geometry of free-metal clusters depends on the chemical specie. The Cu_4 cluster forms the planar rhombus structure [19]. Cr_4 and Co_4 form a three-dimensional bend rhombus structure [19]. The others clusters form tetragonal structures [19]. Upon encapsulation, the metal clusters prefer close-packing structures which are possible due to the confinement imposed by the cage size. The clusters composed by V, Fe, Co, and Ni are more stable inside the fullerene than in a free space. On the

TABLE 49.4 Magnetic Moments, in μ_B, of the Dopants and the Total Magnetic Moments of the System (Inside Parentheses)

Position	Spin Up		Spin Down	
	$4s$	$3d$	$4s$	$3d$
Isolated atom	1.0	5.0	1.0	1.0
Center	0.91	4.91	0.64	1.29
Tetragonal ring	0.28	4.60	0.2	2.52
Hexagonal ring	0.82	4.75	0.09	1.70

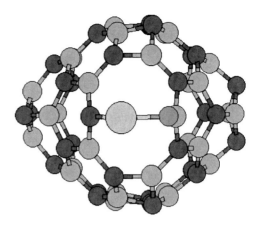

FIGURE 49.11 FeO molecule interacting with a $B_{36}N_{36}$ fullerene.

other hand, the clusters composed by Ti, Cr, Mn, and Cu destabilize inside the $B_{36}N_{36}$ cage. The magnetic moment of all the clusters tends to decrease inside the cage, however, the magnetic moment of the clusters composed by V, Cu, and Cr increases upon encapsulation.

The FeO molecule may also be present during the fullerene synthesis process. Therefore, there is an interest in the study of its interaction with the fullerene. Unlike the case of the transition metal atoms, the FeO molecule opens the cage by breaking a B–N bond at the junction of two hexagons making a bridge between the B and N atoms, as shown in Figure 49.11. This result is consistent with the low resistance of BN compounds to oxidation. The B–N distance becomes 3.26 Å, much larger than the B–N distance of 1.48 Å of the isolated fullerene. The binding energy of FeO to the fullerene is calculated by first-principles to be 3.92 eV. This value is about 2.5 eV higher than the binding energy of any metal atom cited before [25]. This suggests that the dopant observed in the HREM images by Oku et al. may be the FeO molecule. In fact, concerning a comparison with experimental images, it would not be possible to distinguish this molecule from a single Fe atom, since the bond distance Fe–O (1.6 Å) is smaller than the resolution of the microscopes used (1.94 and 1.7 Å). In addition, the image is a projection of a three-dimensional object in a plane, which, depending on the orientation of the fullerene, may result in shorter bond lengths. The covalent bond does not change the magnetic moment of the FeO molecule. The total magnetic moment of the FeO doped fullerene is 4 mBorh, which is the value of the magnetic moment of the isolated FeO molecule.

49.6 BN Nanocones

The incorporation of pentagonal or tetragonal atomic rings and other topological defects into the hexagonal network of BN and carbon nanotubes increases their local curvature and can lead to the closure of the tubes [28,29]. The structure of the cap varies, but it generally has the aspect of a conical surface with electronic properties that are distinct from the bulk material [30]. BN nanocones can also be formed as free-standing structures [31]. A possible application of BN nanocones is their use as

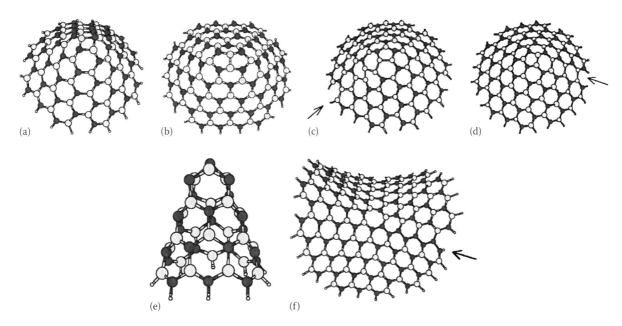

FIGURE 49.12 Examples of BN nanocones and a nanosaddle. B and N atoms are in gray and black, respectively, and hydrogen atoms saturate the dangling bonds at the edges. (a) and (b) show cones with a disclination angle α of 120°. (c) and (d) show cones with $\alpha = 60°$. (e) is a cone with $\alpha = 240°$, and (f) is a saddle-like structure with $\alpha = -60°$. The arrows indicate antiphase boundaries. (Adapted from Azevedo, S. et al., *Appl. Phys. Lett.*, 82, 2323, 2003.)

cold electron sources in field emission displays due to their thermal and chemical stability [32]. Moreover, experiments indicate that, despite being insulating, BN nanotubes yield field-emission currents comparable to those of carbon nanotubes at relatively low voltages [33].

Examples of BN nanocones theoretically built by a cut and glue process from a BN planar layer, and geometrically optimized from first-principles [34] are shown in Figure 49.12. Figure 49.12a shows a nanocone with a four-membered ring at the apex, with a disclination angle of 120°. The high strain at the tip of the structure is energetically compensated by the absence of homopolar B–B or N–N bonds. This effect is responsible for the high stability of BN fullerenes composed of only four- and six-membered rings. On the other hand, rings with an odd number of atoms introduce non-BN bonds. Figure 49.12b shows a structure which has two adjacent pentagons at the apex and a single wrong bond. If a single pentagon is located at the apex, non-BN bonds are formed along the cut, generating either a B–B or a N–N antiphase-boundary. Examples of such antiphase boundaries are indicated by arrows in Figure 49.12c and d. The antiphase boundaries can be of two kinds: a sequence of parallel B–B or N–N bonds, hereafter denoted molecular B and molecular N, respectively (mol-B and mol-N, for simplicity); and a sequence of zig-zag B–B or N–N bonds, which we shall call zig-B and zig-N, respectively. Nanocones with large disclination angles (240° and 300°) have also been observed [31]. Figure 49.12e shows a nanocone with two adjacent four-membered rings at the apex, corresponding to a 240° disclination. A saddle-like (negative curvature) geometry is also shown in Figure 49.12f, with a zig-N antiphase boundary indicated by an arrow.

The stability of BN nanocones, with and without antiphase boundaries, was theoretically investigated by Azevedo et al. [4].

They show that the dependence of the nanocone elastic energy on the disclination angle α is proportional to $\cos^2\alpha/\sin\alpha$. Their results also indicate that nanocones with a disclination angle α of 60° and an antiphase boundary can be more stable than nanocones with $\alpha = 120°$ and free from antiphase boundaries. That is, the cost of an antiphase-boundary defect can be compensated by a smaller elastic-energy cost associated to the disclination.

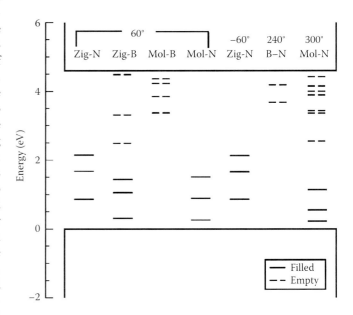

FIGURE 49.13 Electronic states of BN nanocones with different disclination angles, and of a BN nanosaddle (negative disclination angle). The electronic structure of the bulk BN is used as a reference. (Adapted from Azevedo, S. et al., *Appl. Phys. Lett.*, 82, 2323, 2003.)

They also observed that carbon incorporation at the antiphase boundary leads to highly stable structures, of which the most stable presents a large spin splitting at the Fermi level.

The electronic states of finite-size cones with different angles of disclination, some of them shown in Figure 49.12, were investigated through first-principles calculations [34]. Their results are shown in Figure 49.13. The solid lines indicate the valence and conduction band edges of the isolated BN planar sheet. The small lines indicate the position of the electron states in the region of the energy gap induced by the topological transformation of the BN sheet. The first four columns in Figure 49.2 correspond to the four possible antiphase boundaries in a structure with a pentagonal ring at the apex. The general trend is that the presence of N–N bonds introduces occupied states in the lower half of the band gap, while the states associated with B–B bonds are unoccupied and in the upper half of the band gap. This is clearly seen in Figure 49.2 for the zig-N, mol-B, and mol-N structures. The exception to this rule is the zigzag antiphase boundary of boron. In this case, the defect states are scattered over the entire range of the gap. An important feature shown in Figure 49.13 is the significant increase (by as much as 2 eV) in the energy of the highest occupied electron state for all the BN structures that contain antiphase boundaries, with the exception of the mol-B boundary. This would lead to a significant reduction of the work function of BN nanocones and cone-capped nanotubes that contain antiphase boundaries. In contrast, the structures without antiphase boundaries presented work functions very similar to that of the planar BN sheet. These results indicate that the performance of BN nanocones or cone-capped nanotubes in field emission devices would be very sensitive to the topology of the cones: those with antiphase boundaries would be much better electron emitters than the ones without these line defects.

References

1. D. Golberg, Y. Bando, O. Stéphan, and K. Kurashima, *Appl. Phys. Lett.* **73**, 2441 (1998).
2. S. S. Alexandre, M. S. C. Mazzoni, and H. Chacham, *Appl. Phys. Lett.* **73**, 61 (1999).
3. N. G. Chopra, R. J. Luyken, K. Cherrey, V. H. Crespi, M. L. Cohen, S. G. Louie, and A. Zettl, *Science* **269**, 966 (1995).
4. S. Azevedo, M. S. C. Mazzoni, R. W. Nunes, and H. Chacham, *Phys. Rev. B* **70**, 205412 (2004).
5. N. G. Chopra and A. Zettl, *Solid State Commun.* **105**, 297–300 (1998).
6. A. Jrio, M. S. Dresselhaus, and G. Dresselhaus, 2001. *Carbon Nanotubes: Advanced Topics in Synthesis, Structure, Properties, and Application*. Berlin/Heidelberg, Germany: Springer-Verlag.
7. R. Curl, H. Kroto, and R. Smalley, *Nature* **318**, 162 (1985).
8. F. Jensen and H. Toftlund, *Chem. Phys. Lett.* **201**, 89 (1993).
9. X. Blase et al., *Phys. Rev. Lett.* **80**, 1666 (1998).
10. G. Seifert et al., *Chem. Phys. Lett.* **268**, 352 (1997).
11. H. S. Wu, X. H. Xu, and J. F. Jia, *Acta Chem. Sin.* **62**, 28 (2004).
12. H. S. Wu, X. H. Xu, and H. Jiao, *Chem. Phys. Lett.* **386**, 369 (2004).
13. R. R. Zope, T. Baruah, M. R. Pederson, and B. I. Dunlap, *Chem. Phys. Lett.* **393**, 300 (2004).
14. R. J. C. Batista, M. S. C. Mazzoni, and H. Chacham, *Chem. Phys. Lett.* **421**, 246 (2006).
15. O. Stepha et al., *Science* **266**, 1683 (1994).
16. O. Stéphan, Y. Bando, A. Loiseau, F. Willaime, N. Shramchenko, T. Tamiya, and T. Sato, *Appl. Phys. A: Mater. Sci. Process.* **67**, 107 (1998).
17. T. Oku, A. Nishiwaki, I. Narita, and M. Gonda, *Chem. Phys. Lett.* **380**, 620 (2003); T. Oku, A. Nishiwaki, and I. Narita, *J. Phys. Chem. Solids* **65**, 369 (2004); T. Oku, I. Narita, A. Nishiwaki, and N. Koi, *Defects Diffus. Forum* **226–228**, 113 (2004); T. Oku, A. Nishiwaki, and I. Narita, *Physica B* **351**, 184 (2004).
18. W. A. Harrison, 1980. *Electronic Structure and the Properties of Solids: The Physics of the Chemical Bond*. San Francisco, CA: Freeman.
19. S. Nigam et al., *Phys. Rev. B* **77**, 075438 (2008).
20. J. Wang et al., *J. Chem. Phys.* **128**, 084306 (2008).
21. P. Hilton and J. Pedersen, *Am. Math. Month.* **103**, 121 (1996).
22. A. R. Baush et al., *Science* **299**, 1716 (2003).
23. http://en.wikipedia.org/wiki/Euler characteristic
24. R. M. Martin, 2004. *Electroni Structure: Basic Theory and Practical Methods*. Cambridge, U.K.: Cambridge University Press.
25. R. J. C. Batista, M. S. C. Mazzoni, and H. Chacham, *Phys. Rev. B* **75**, 035417 (2007).
26. N. Fujima and T. Oda, *Phys. Rev. B* **71**, 115412 (2005).
27. S. S. Alexandre, H. Chacham, and R. W. Nunes., *Phys. Rev. B* **63**, 45402 (2001).
28. S. Iijima, T. Ichihashi, and Y. Ando, *Nature (London)* **356**, 776 (1992).
29. A. Loiseau, F. Willaime, N. Demoncy, G. Hug, and H. Pascard, *Phys. Rev. Lett.* **76**, 4737 (1996).
30. J. C. Charlier and G. M. Rignanese, *Phys. Rev. Lett.* **86**, 5970 (2001).
31. L. Bourgeois, Y. Bando, W. Q. Han, and T. Sato, *Phys. Rev. B* **61**, 7686 (2000).
32. D. Golberg, Y. Bando, K. Kurashima, and T. Sato, *Scr. Mater.* **44**, 1561 (2001).
33. J. Cumings and A. Zettl, *Electronic Properties of Molecular Nanostructures*, edited by H. Kuzmany. Melville, NY: AIP, 2001, p. 577.
34. S. Azevedo, M. S. C. Mazzoni, H. Chacham, and R. W. Nunes, *Appl. Phys. Lett.* **82**, 2323 (2003).

50

Fullerene-Like III–V Binary Compounds

Giancarlo Cappellini
*Università degli Studi di Cagliari
and Sardinian Laboratory for
Computational Materials Science*

Giuliano Malloci
*INAF-Osservatorio Astronomico di
Cagliari and Laboratorio Scienza*

Giacomo Mulas
*INAF-Osservatorio
Astronomico di Cagliari*

50.1 Introduction

In physics, the term *cluster* denotes a group of atoms or molecules. Large clusters with stable structures are very interesting systems both by themselves and as building blocks for nanostructures, i.e., objects of nanometer (1 nm = 10^{-9} m) sizes, intermediate between molecular and microscopic (micrometer-sized, $1\,\mu m = 10^{-6}$ m) structures. In particular, the discovery of buckminsterfullerene (Kroto et al. 1985) and carbon nanotubes (Iijima 1991), sketched in Figure 50.1, opened a completely new field of research: nanotechnology. Nanotechnology, which deals with structures of typical sizes of 100 nm or smaller, involves the control of matter on an atomic and molecular scale, and the development of materials or devices that lead to many applications, such as in medicine, electronics, and energy production.

Since the discovery of carbon fullerene and nanotubes, the search for new nanoscale materials based on chemical elements other than carbon has received much attention (e.g., Buda and Fasolino 1995, Trave et al. 1996, Wang 1996). Given the similarities between boron–nitrogen and carbon–carbon bonds, most studies have focused on boron nitride (BN), which exists in nature in the hexagonal (graphitic-like) structure (e.g., Alexandre et al. 2001, Mårlid et al. 2001, Wirtz et al. 2006). These efforts, indeed, resulted in an experimental procedure able to synthesize pure BN nanotubes (Chopra et al. 1995) although the observed BN cages and wires do not clearly show the characteristic pentagonal rings of carbon fullerenes (Goldberg et al. 1998, Parilla et al. 1999), showing instead a sequence of squares and hexagons. In addition, clusters and nanotubes based on other materials such as GaSe (Côté et al. 1998), GeTe (Natarajan and Öğüt 2003), AlN (Zope and Dunlap 2005), and GaN (Lee et al. 1999, Song et al. 2006, Wang et al. 2006) have been predicted theoretically to be stable. Finally, fullerene-like CdSe nanoparticles with interesting optical properties have been recently studied and synthesized (see e.g., Botti and Marques 2007, and references therein).

Nowadays, semiconductor electronic devices mostly use cubic Si crystals (a cubic system, in which the unit cell is in the shape of a cube, is one of the most common and simplest shapes found in crystals and minerals). This material, however, shows an indirect energy gap of ~1.1700 eV between the electronic occupied states (collectively referred to as valence bands) and empty states (conduction bands). This hinders its wide application in the fields of optoelectronics and photonics, since electron–hole recombination must always involve also one or more phonons (quantized modes of vibration occurring in a rigid crystal lattice), making cubic bulk Si inefficient at emitting light. On the contrary, several direct-gap III–V systems in the bulk phase are very efficient at emitting light and are thus of outstanding interest in the above-mentioned applicative fields. In fact, some of these compounds had, and still have, a prominent position in electronic and optical device production, e.g., as reported in Madelung (1996):

- GaAs in the bulk cubic phase, with a direct gap of 1.51914 eV (near infrared, wavelength of about 0.816 μm), is widely used in infrared light-emitting diodes, laser diodes, solar cells, and microwave frequency integrated circuits.
- InAs in the bulk cubic phase, with a direct gap of 0.4180 eV (infrared, wavelength of about 2.966 μm), is largely used in infrared detectors and laser diodes.

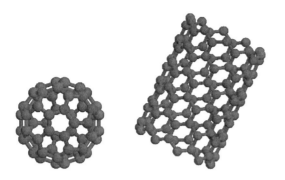

FIGURE 50.1 Schematic view of buckminsterfullerene C_{60} and a semiconducting (10,0) single-wall carbon nanotube.

- InSb in the bulk cubic phase, with a direct gap of 0.2363 eV (infrared, wavelength of about 5.247 μm), is mainly used in infrared detectors, thermal imaging cameras, infrared astronomy, infrared military applications, and photodiodes.
- GaN in the bulk wurtzitic phase (wurtzite is a member of the hexagonal crystal family), with a direct gap of 3.503 eV (near-ultraviolet, ~0.354 μm), is mostly used in bright light-emitting diode production, and is ideal for power amplifiers at microwave frequencies (much higher working temperature and voltages compared to GaAs).

One interesting question related to the future development of nanoscience and nanotechnology is whether typical semiconductors of the III–V family such as GaP or GaAs, which do not possess a graphitic bulk phase, are able to form hollow fullerene-like structures. Along these lines, following the theoretical prediction that BN cages with six pairs of adjacent pentagonal rings lead to stable clusters with the general formula $B_xN_{x\pm4}$ (Fowler et al. 1999), *ab initio* molecular dynamics simulations (Tozzini et al. 2000) showed the spontaneous formation of GaP fullerene-like clusters with 20 and 28 atoms starting from a bulk GaP fragment. These clusters, possessing the same topology as C_{20} and C_{28} fullerenes, were predicted to be characterized by stable structures with very symmetric equilibrium geometries (T_h and T_d symmetry point group), high thermal stability up to a temperature of 1500–2000 K, high chemical stability, expressed by large highest occupied molecular orbital (HOMO)–lowest unoccupied molecular orbital (LUMO) energy gaps and closed-shell electronic structures, and stability against positive and negative ionizations, e.g., the addition or removal of electrons to/from the system. The above properties were later shown to be common to other III–V compounds, namely, GaAs, AlAs, and AlP (Tozzini et al. 2001). These pioneering studies therefore opened the real possibility to find an experimental procedure to prepare these new nanoscale materials in macroscopic amounts.

Based on the above discussion about the wide technological applications of bulk III–V systems, it is clear that the knowledge of the electronic excitations of nanoscale III–V systems has

important implications in the light of their potential technological applications in the fields of photonics and optoelectronics (light-emitting diodes, lasers, transistors).

50.2 Background

Optical spectroscopy, the study of the interaction between visible light and matter as a function of wavelength, is known to be a useful tool to characterize cluster geometries (Rubio et al. 1996, Marques and Botti 2005). This technique has been proposed to unambiguously discriminate the ground state of the different isomers of C_{20}, the smallest carbon fullerene (Castro et al. 2002), and has been recently applied to confirm the synthesis of fullerene-like CdSe nanoparticles (Botti and Marques 2007). Should an experimental procedure be found which may synthesize small fullerene-like III–V binary clusters, optical absorption spectra could likewise be used to distinguish between different newly formed isomers. The above idea has been the starting point of a theoretical study (Malloci et al. 2004) aimed at obtaining the optical properties and quasiparticle (QP) effects (QP energies are associated with the addition or removal of an electron in a neutral system) of fullerene-like GaP clusters. Extending our previous study, we include in this work other clusters with the stoichiometry $III_xV_{x\pm4}$ predicted by Tozzini et al. (2000, 2001): Ga_8As_{12}, $Ga_{12}As_8$, Al_8P_{12}, $Al_{12}P_8$, Al_8As_{12}, and $Al_{12}As_8$ with T_h symmetry, and $Ga_{12}As_{16}$, $Ga_{16}As_{12}$, $Al_{12}P_{16}$, $Al_{16}P_{12}$, $Al_{12}As_{16}$, and $Al_{16}As_{12}$ which belong to the T_d symmetry point group. We recall that point groups are mathematically defined as the sets of geometric symmetry transformations of space leaving at least one point fixed (e.g., reflection

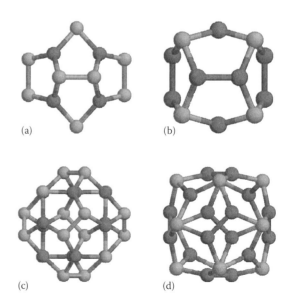

(a)   (b)

(c)   (d)

FIGURE 50.2 Ground-state geometries of four representative clusters of the 16 of the form $III_xV_{x\pm4}$ is studied in this work. The symmetry of Ga_8P_{12} and $Ga_{12}P_8$, respectively, (a) and (b) in the figure, is T_h, while $Ga_{12}P_{16}$ and $Ga_{16}P_{12}$, respectively, (c) and (d), belong to the T_d symmetry point group.

with respect to a plane or rotation around an axis). Since any chemical system is invariant with respect to one well-defined set of such transformations, point group theory is heavily used in chemistry, especially to describe the symmetries of a molecule and of molecular orbitals forming chemical bonds. We will discuss in detail the geometric and electronic properties of each cluster, analyzing the behavior of the relevant physical observables as a function of the size and composition of the III–V fullerene considered. The ground-state geometries of the clusters studied in this work are shown in Figure 50.2a through d.

50.3 State of the Art

The accurate determination of the fundamental energy gap in many-electron systems, such as molecules and solids, by means of computationally inexpensive methods, remains an open problem in computational materials science. Density functional theory (DFT; Hohenberg and Kohn 1964, Kohn and Sham 1965) has been extremely successful in the calculation of ground-state electronic properties of a large class of materials. In this method, the state of the system is completely determined by the electronic density, and all the other physical observables are expressed as functionals of it. This leads to an equivalent system of electrons with no explicit interactions, which has the same electronic density and energy as the original, fully interacting system, and whose ground state is determined via a self-consistent procedure. Its success is largely due to the extremely good ratio between the accuracy and the computational cost of this method, which is very fast and simple to implement for most available exchange-correlation (XC) functionals. The XC functional in DFT includes all the exchange and correlation effects between the electrons of the system (Jones and Gunnarson 1989). Unfortunately, DFT is not designed to give addition and removal energies in a single calculation: the Kohn–Sham eigenvalues have no direct physical interpretation except for the ones associated with the HOMO; none of the individual Kohn–Sham orbitals has any known physical meaning; virtual (unoccupied) Kohn–Sham orbitals do not even enter the self-consistent cycle and play no role in the theory. It is therefore well recognized that the use of Kohn–Sham eigenvalues to determine the QP properties of many-electron systems yields, by and large, results in disagreement with experiments (Jones and Gunnarson 1989). For example, the well-known bandgap problem for bulk semiconductors arises from a severe underestimate (even exceeding 50% in both finite and extended systems) of the electronic excitation energies with respect to available experimental results (optical absorption, direct and inverse photoemission, Bechstedt et al. 1992). In optical absorption experiments, in particular, an electron excited to a conduction state interacts with the resulting hole in the previously occupied state and two-particle (called excitonic) effects must be properly considered (Onida et al. 2002).

For finite systems, it is possible to obtain accurate excitation energies using the delta-self-consistent-field (ΔSCF) approach (Jones and Gunnarson 1989). This method, successfully applied to obtain the QP energies for several clusters, for example C_{60} (Cappellini et al. 1997) and oligoacenes (Malloci et al. 2007), consists in evaluating total-energy differences between the self-consistent field calculations performed for the system with N and $N \pm 1$ electrons, respectively, to obtain (1) the vertical electron affinities and the first ionization energies and (2) the QP correction to the HOMO–LUMO energy gap. This quantity is related to *molecular hardness*, the analogue of the bandgap of solids, defined as half the difference between the ionization potential and the electron affinity, which is a key property characterizing the chemical behavior and reactivity of a molecule. Based on the calculation of the total energy of the electronic ground state, we have evaluated the QP spectra for the 16 $III_x V_{x\pm4}$ clusters considered, including the correction to the HOMO–LUMO energy gap.

More recently, optical absorption spectra of many different species have been obtained in the framework of time-dependent density functional theory (TD-DFT) (Runge and Gross 1984). In this approach, the theory for electronic excited-state properties corresponding to the traditional ground-state DFT formulation, was applied to a large number of systems in recent years (see e.g., Onida et al. 2002) providing results in very good agreement with experiments for several compounds. Hence, we used this approach to compute the low-energy optical absorption spectra of the clusters under investigation. The comparison between the optical gap, i.e., the lowest singlet–singlet excitation energy obtained with TD-DFT, and the QP-corrected HOMO–LUMO gap obtained via ΔSCF, enabled us to estimate the excitonic effects (due to the electron–hole interaction) for the clusters under investigation. Our results for the QP HOMO–LUMO gaps support previous results predicting the high stability of these clusters (Tozzini et al. 2000, 2001). Moreover the present optical absorption spectra, calculated by us for the first time, provide reliable, additional tools for the identification of these species via optical absorption experiments.

The calculation of excitation energies and of electronic absorption spectra required the previous calculation of the ground-state optimized geometries. For this part of the work, we used the Gaussian-based DFT module of the NWCHEM package (Straatsma et al. 2004). Following previous works (Tozzini et al. 2000, 2001) and starting from the ground-state equilibrium geometries given in these papers, we fully optimized the geometries. In particular, we employed a widely used XC functional in the generalized gradient approximation (Perdew 1986, Becke 1988) in conjunction with the Gaussian basis set 3-21G to expand the molecular orbitals (Frish et al. 1984). The choice of such a relatively small basis set is a good compromise between computational costs and accuracy, as shown in Tables 50.1 and 50.2, where bond lengths and internuclear angles of the small cluster $Ga_8 P_{12}$ and $Ga_8 As_{12}$ are compared for three different basis sets, ranging in size from 3-21G to 6-311G* (Frish et al. 1984).

TABLE 50.1 Structural Parameters of Ga_8P_{12} Obtained Using Different Basis Sets (with Increasing Number of Gaussian Functions in Each Atom) to Expand the Molecular Orbitals; the *Ab Initio* Results of Tozzini et al. (2000) Are Given in the Last Row for Comparison

Gaussian Basis Set	Number of Functions Ga	Number of Functions P	Internuclear Distances (Å) Ga-P	Internuclear Distances (Å) P-P	Internuclear Angles (°) GaPP	Internuclear Angles (°) GaPGa	Internuclear Angles (°) PGaP
3–21G	23	13	2.390	2.425	99.3	83.9	119.6
6–311G	41	21	2.389	2.363	99.7	84.9	119.5
6–311G*	47	27	2.390	2.299	100.6	84.2	119.4
Tozzini et al. 2000			2.288	2.262	100.3	84.7	119.5

TABLE 50.2 Structural Parameters of Ga_8As_{12} Obtained Using Different Basis Sets (with Increasing Number of Gaussian Functions in Each Atom) to Expand the Molecular Orbitals; the *Ab Initio* Results of Tozzini et al. (2000) Are Given in the Last Row for Comparison

Gaussian Basis Set	Number of Functions Ga	Number of Functions As	Internuclear Distances (Å) Ga–As	Internuclear Distances (Å) As–As	Internuclear Angles (°) GaAsAs	Internuclear Angles (°) GaAsGa	Internuclear Angles (°) AsGaAs
3-21G	23	23	2.476	2.594	97.4	81.4	120.0
6-311G	41	41	2.471	2.569	98.5	83.7	119.8
6-311G*	47	47	2.473	2.516	99.0	83.4	119.7
Tozzini et al. 2000			2.389	2.505	98.2	83.6	119.8

The agreement between our results and those of Tozzini et al. 2000 indeed improves using the larger 6-311G and 6-311G* basis sets; however, we obtained percentage changes of order 5% and 1%, respectively, for internuclear distances and angles, with respect to the results of the smaller 3-21G basis set, in the face of an increase of one order of magnitude in computational cost.

To confirm the geometry considered to be indeed a local minimum on the potential energy hypersurface, for the smallest clusters III_8V_{12} and $III_{12}V_8$, we also computed the harmonic vibrational frequencies at the optimized geometry, obtaining real values for all of the frequencies of vibration. For all of the sixteen clusters considered, the structural parameters obtained at the GGA/3-21G level of theory are given in Tables 50.3 through 50.6. Good agreement is found between our data and

the geometries given in Tozzini et al. (2000, 2001). Note that in any case the electronic excitation properties are found to be only weakly dependent upon different choices of the values used for the structural parameters.

From the ground-state optimized geometries discussed earlier, we evaluated also the vertical electron affinity (EA_v) and the vertical first ionization energy (IE_v). This enabled us to determine the QP-corrected HOMO–LUMO gaps via total-energy differences of charged and neutral clusters. These values are an indicator of stability of the cluster and are used here to evaluate the excitonic effects. In the ΔSCF scheme, the QP-corrected HOMO–LUMO gap of a N-electron system is rigorously defined as

$$QP_{gap}^{(1)} = IE_v - EA_v = E_{N+1} + E_{N-1} - 2E_N \qquad (50.1)$$

E_M (M = N, N + 1, N − 1) being the total energy of the system with M electrons. We also used the following approximate expression, which is exact in the $N \to \infty$ limit (Godby et al. 1988):

$$QP_{gap}^{(2)} = \varepsilon_{N+1}^{N+1} - \varepsilon_N^N, \qquad (50.2)$$

where ε_i^j is the *i*th eigenvalue of a *j*-electron system. The results obtained using the above Equations 50.1 and 50.2 tend to coincide as the system gets larger and the orbitals more delocalized.

Thanks to the good compromise between accuracy and computational costs, compared to many-electron wavefunction-based *ab initio* methods, TD-DFT is the most widely used approach to compute the excitation energies of such complex molecules as the ones in the present investigation. To obtain the optical spectra, we used two different implementations of TD-DFT in the linear response regime in conjunction with different representations of the wavefunctions:

1. The frequency-space implementation (Hirata and Head-Gordon 1999) based on the linear combination of localised orbitals, as given in the NWCHEM package (Straatsma et al. 2004).
2. The real-time propagation scheme using a grid in real space (Yabana and Bertsch 1999), as implemented in the OCTOPUS computer program (Marques et al. 2003).

TABLE 50.3 Structural Parameters of the Four GaP Clusters Considered; the Corresponding *Ab Initio* Results of Tozzini et al. (2001) Are Given in Parentheses

Cluster	Internuclear Distances (Å) Ga–P	Internuclear Distances (Å) P–P	Internuclear Distances (Å) Ga–Ga	Internuclear Angles (°) GaPP	Internuclear Angles (°) GaPGa	Internuclear Angles (°) PGaP	Internuclear Angles (°) PGaGa
Ga_8P_{12}	2.390	2.425	—	99.3	83.9	119.6	—
	(2.288)	(2.262)		(100.3)	(84.7)	(119.5)	
$Ga_{12}P_8$	2.386	—	2.363	—	87.2	125.5	113.3
	(2.294)		(2.383)		(88.6)	(126.8)	(112.0)
$Ga_{12}P_{16}$	2.354, 2.372	2.453	—	99.6	87.2, 96.6	114.2, 131.6	—
$Ga_{16}P_{12}$	2.402, 2.347	—	2.347	—	90.7, 110.6	120.0, 129.8	112.7

Note: Note that for $Ga_{12}P_{16}$ and $Ga_{16}P_{12}$ the two items in some boxes refer to two, inequivalent, structural parameters of the same nature (e.g., two different Ga–P distances, two different GaPGa angles, and so on).

TABLE 50.4 Structural Parameters of the Four GaAs Clusters Considered; the Corresponding *Ab Initio* Results of Tozzini et al. (2001) Are Given in Parentheses

Cluster	Internuclear Distances (Å)			Internuclear Angles (°)			
	Ga–As	As–As	Ga–Ga	GaAsAs	GaAsGa	AsGaAs	AsGaGa
Ga_8As_{12}	2.476	2.594	—	97.4	81.4	120.0	—
	(2.389)	(2.505)		(98.2)	(83.6)	(119.8)	
$Ga_{12}As_8$	2.484	—	2.372	—	81.4	126.6	114.6
	(2.396)		(2.401)		(85.3)	(127.0)	(113.2)
$Ga_{12}As_{16}$	2.448, 2.454	2.619	—	97.9	84.7, 93.8	114.5, 130.3	—
$Ga_{16}As_{12}$	2.440, 2.520	—	2.350	—	85.9, 88.9	120.0, 130.8	114.4

TABLE 50.5 Structural Parameters of the Four AlP Clusters Considered; the Corresponding *Ab Initio* Results of Tozzini et al. (2001) Are Given in Parentheses

Cluster	Internuclear Distances (Å)				Internuclear Angles (°)		
	Al–P	P–P	Al–Al	AlPP	AlPAl	PAlP	PAlAl
Al_8P_{12}	2.409	2.460	—	99.3	84.4	119.6	—
	(2.301)	(2.270)		(100.1)	(84.2)	(119.5)	
$Al_{12}P_8$	2.413	—	2.556	—	89.7	127.1	111.4
	(2.306)		(2.510)		(88.8)	(128.9)	(111.0)
$Al_{12}P_{16}$	2.364, 2.400	2.478	—	100.0	89.0, 97.2	114.0, 131.9	—
$Al_{16}P_{12}$	2.427, 2.373	—	2.527	—	93.2, 103.6	120.0, 131.0	110.7

TABLE 50.6 Structural Parameters of the Four AlAs Clusters Considered; the Corresponding *Ab Initio* Results of Tozzini et al. (2001) Are Given in Parentheses

Cluster	Internuclear Distances (Å)			Internuclear Angles (°)			
	Al–As	As–As	Al–Al	AlAsAs	AlAsAl	AsAlAs	AsAlAl
Al_8As_{12}	2.500	2.623	—	97.4	81.6	120.0	—
	(2.405)	(2.521)		(97.8)	(82.5)	(119.9)	
$Al_{12}As_8$	2.520	—	2.563	—	83.6	128.5	113.0
	(2.409)		(2.522)		(85.0)	(129.2)	(112.3)
$Al_{12}As_{16}$	2.465, 2.488	2.642	—	98.3	85.9, 93.8	114.3, 130.9	—
$Al_{16}As_{12}$	2.476, 2.546	—	2.534	—	88.3, 94.5	119.9, 133.1	112.1

In implementation 2, the time-dependent Kohn–Sham equations are directly solved in real time, and the wavefunctions are represented by their discretized values on a uniform spatial grid. The static Kohn–Sham wavefunctions are perturbed by an impulsive electric field and propagated for a given finite time interval. In this way, all the frequencies of the system are simultaneously excited. The whole absolute absorption cross section, $\sigma(E)$, then follows from the dynamical polarizability $\alpha(E)$, which is related to the Fourier transform of the time-dependent dipole moment of the molecule. The absorption cross section, which has dimensions of an area and is used to express the likelihood of interaction between particles, measures the probability for the molecule to absorb a photon of a particular wavelength; absorption cross sections are conveniently expressed in mega barns ($1\,Mb = 10^{-18}\,cm^2$). The molecular polarizability is defined as the electric dipole moment induced in a molecule by an external electric field divided by the magnitude of the field. The quantitative relation between absorption cross section and molecular polarizability is

$$\sigma(E) = \left[\frac{8\pi^2 E}{hc}\right] Im\left[\alpha(E)\right], \qquad (50.3)$$

where

h is Planck's constant ($h \sim 6.626 \times 10^{-34}\,J\,s$)

$Im[\alpha(E)]$ is the imaginary part of the dynamical polarizability

c is the velocity of light in vacuum ($c \sim 299.792.458\,m/s$)

The dipole strength function, $S(E)$, is related to $\sigma(E)$ by the equation

$$S(E) = \left[\frac{(m_e c)}{(he^2)}\right] \sigma(E), \qquad (50.4)$$

m_e and e being, respectively, the mass and the charge of the electron ($m_e \sim 9.109382 \times 10^{-31}\,kg$, $e \sim 1.60217646 \times 10^{-19}\,C$). The dipole strength function, $S(E)$, has the dimensions of oscillator strength (the oscillator strength of a transition is a measure of its intensity) per unit energy and satisfies the Thomas–Reiche–Kuhn dipole sum rule $N_e = \int dE S(E)$, where N_e is the total number of electrons in the molecule. With this approach, one obtains the whole absorption spectrum of a given molecule in a single calculation.

In the most widely used frequency-space TD-DFT implementation, based on the linear response of the density matrix, the poles of the linear response function correspond to vertical excitation energies and the pole strengths to the corresponding oscillator strengths (Casida 1995). Poles are found iteratively, starting from zero energy and proceeding upwards. With this method, therefore, computational costs scale steeply with the number of required transitions, and electronic excitations are thus usually limited to the low-energy part of the spectrum. To use an analogy easy to understand for experimental spectroscopists, roughly speaking, the difference between the two approaches is intuitively similar to the difference between a frequency-scanning (approach 2) and a Fourier-transform spectrometer (approach 1). From a computational point of view, the advantages of the real-time propagation method are discussed, e.g., in Lopez et al. (2005). On the other hand, the main drawbacks of the real-time approach are that (1) no information is given by Equation 50.3 on dipole-forbidden singlet–singlet transitions (even if this could in principle be retrieved from the time evolution of electronic density by applying a similar analysis on higher order terms of the electric multipole expansion) and on spin-forbidden singlet–triplet transitions and (2) one does not obtain independent information for each excited state such as the irreducible representation of the point group of the given molecular system, and the description of the excitations in terms of promotion of electrons in an orbital picture.

We performed the OCTOPUS calculations in the adiabatic local-density approximation using the XC energy density of the homogeneous electron gas (Ceperley and Alder 1980) parameterized by Perdew and Zunger (1981), which ensures good numerical stability. In addition, nuclei and core electrons are represented by norm-conserving pseudopotentials (Troullier and Martins 1991) so that only valence electrons are explicitly considered in the DFT calculations. Concerning specific details, we already mentioned that wavefunctions are represented by the values they take on a uniform spatial grid in a finite volume. Two important parameters of the calculations are, therefore, the volume of the box in which the molecule is represented, and the spatial resolution Δs of the grid. The choice of the spatial resolution Δs plays a role similar to the choice of the basis set in methods which represent real or Kohn–Sham wavefunctions as linear combinations of some finite set of appropriately chosen functions (i.e., Gaussians or atomic wavefunctions centered on the nuclei, plane waves etc.): Δs must be small enough to properly sample the wavefunctions of the explicitly represented electrons, but still large enough, to produce a practically tractable size of the resulting uniform grid. In this respect, including the representation of the core electrons together with the nuclei using pseudopotentials is a great advantage, since explicitly representing the tight oscillations of the wavefunctions of the core electrons would mandate the use of an extremely dense grid, thus increasing computational cost by more than an order of magnitude. With pseudopotentials, we used a grid spacing of 0.3 Å and determined the box size by requiring each nucleus to be at least 4 Å away from box edges. We used a time integration length of 20 $(h/2\pi)$/eV, which corresponds to an energy resolution of $(h/2\pi)/T = 0.05$ eV. The time evolution was performed using a time step of 0.001 $(h/2\pi)$/eV, which ensured energy conservation with extremely good numerical accuracy, namely, better than one part in a million. A thorough description of the program and the specific numerical implementations it uses can be found in the dedicated Web site http://www.tddft.org/programs/octopus/.

In the TD-DFT calculations with NWCHEM, we used the frozen core approximation for the 1s electrons and restricted ourselves to the lowest lying excited electronic states up to 3 eV. We recall that typical electronic transitions are of the order of the electronvolt (eV), 1 eV being the amount of energy gained by a single electron when it accelerates through an electrostatic potential difference of 1 V; in SI units, 1 eV = 1.60×10^{-19} J. Remember that the visible part of the electromagnetic spectrum ranges between about 1.7 and 3.2 eV. Our calculations were performed at the same level BP86/3-21G used to obtain the ground-state optimized geometries. Although basis set convergence is not yet expected at this level, our results using the two TD-DFT implementations are almost coincident within numerical errors; we thus believe our theoretical predictions to be sufficiently accurate for the purposes of this work.

50.4 Critical Discussion

Table 50.7 shows the HOMO–LUMO gap values obtained using Equations 50.1 and 50.2. The use of the ΔSCF scheme to correct the HOMO–LUMO gap of the clusters considered here is completely justified relative to their size (Godby and White 1998). The last two columns of Table 50.7 report, respectively, the TD-DFT result of NWChem for the corresponding HOMO–LUMO electronic transition, $E_{\text{TD-DFT}}$, and the estimate of the excitonic effects occurring in these molecules, as expressed by the exciton binding energy $E_{\text{bind}} = QP^1 - E_{\text{TD-DFT}}$. As an example, we report in Figure 50.3 the optical gap $E_{\text{TD-DFT}}$, the QP-corrected HOMO–LUMO gap, and the exciton binding energy E_{bind}, as a function of molecular size, for the GaP clusters (Figure 50.3a) and the AlAs clusters (Figure 50.3b).

For comparison, in the case of the GaP clusters, we also computed the quantity E_{2p}, which is the energy difference between the total energy E_N of the neutral cluster and the total energy E_N^* of the excited state obtained by placing an electron in the LUMO state and a hole in the HOMO:

$$E_{2p} = E_N - E_N^* \tag{50.5}$$

TABLE 50.7 HOMO–LUMO Gap (Expressed in eV) of the 16 GaP, GaAs, AlP, AlAs Clusters Considered, as Obtained by GGA/3–21G Calculations Using Equations 50.1 and 50.2

Cluster	Kohn–Sham Gap	IE_v	EA_v	QP^1	QP^2	E_{TD-DFT}	E_{bind}
			GaP				
Ga_8P_{12}	2.06	8.15	2.59	5.56	5.44	2.08	3.48
$Ga_{12}P_8$	1.12	7.14	2.88	4.26	4.21	1.27	2.99
$Ga_{12}P_{16}$	1.54	7.45	2.85	4.60	4.56	1.63	2.97
$Ga_{16}P_{12}$	1.12	6.85	2.97	3.88	3.83	1.19	2.69
			GaAs				
Ga_8As_{12}	1.60	7.43	2.75	4.67	4.64	1.61	3.06
$Ga_{12}As_8$	0.60	6.83	3.18	3.64	3.60	0.67	2.93
$Ga_{12}As_{16}$	1.39	7.07	2.99	4.08	4.08	1.44	2.64
$Ga_{16}As_{12}$	0.37	6.52	3.46	3.06	3.03	0.44	2.62
			AlP				
Al_8P_{12}	2.15	8.30	2.78	5.52	5.39	2.17	3.35
$Al_{12}P_8$	1.30	7.14	2.76	4.37	4.32	1.40	2.97
$Al_{12}P_{16}$	1.58	7.46	3.11	4.35	4.37	1.70	2.65
$Al_{16}P_{12}$	1.44	6.87	2.73	4.14	4.06	1.33	2.81
			AlAs				
Al_8As_{12}	1.63	7.57	2.87	4.69	4.66	1.65	3.04
$Al_{12}As_8$	0.68	6.84	3.16	3.68	3.64	0.82	2.86
$Al_{12}As_{16}$	1.40	7.14	3.07	4.07	4.07	1.46	2.61
$Al_{16}As_{12}$	0.79	6.55	3.13	3.43	3.38	0.86	2.57
			C clusters				
C_{20}	0.77	7.43	1.83	5.60	5.60	/	/
C_{36}	0.40	7.43	3.13	4.30	4.31	/	/
C_{60}	1.81	7.73	2.67	5.06	5.04	/	/
	(1.79)			(5.21)	(5.20)		

Note: Last two columns report, respectively, the TD-DFT result and the excitation energy evaluated using Equation 50.3. Last row gives the GGA/3–21G results for C_{60}, comparing them with those of Cappellini et al. (1997).

We found that, ranging from Ga_8P_{12}, $Ga_{12}P_8$, $Ga_{12}P_{16}$ to $Ga_{16}P_{12}$, the E_{2p} values in eV are 2.09, 1.14, 1.56, and 1.13 with a deviation from the corresponding TD-DFT results in the range 10–130 meV. The good agreement found between E_{2p} and the corresponding E_{TD-DFT} shows that the excited states corresponding to the HOMO–LUMO transitions in all the GaP molecules considered are well described by a single determinant electronic wavefunction.

Comparing the self-energy corrected single-particle HOMO–LUMO gaps and the corresponding TD-DFT results, we obtain strong excitonic effects with binding energies up to a maximum of about 3.5 eV. This result is comparable to what is found for other isolated systems such as the smallest carbon fullerene C_{20} or buckminsterfullerene C_{60} (Cappellini et al. 1997). This is shown in the last row of Table 50.7, which reports our GGA/3–21G results for C_{60} and compares them with those of Cappellini et al. (1997), obtained with LDA using a discrete variational approach. Both sets of data compare well

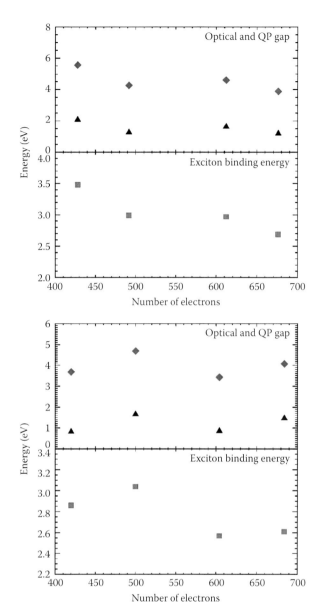

FIGURE 50.3 (a) Top left panel: Optical gap E_{TD-DFT} as obtained via TD-DFT (triangles) and ΔSCF QP-corrected HOMO–LUMO gap QP^1 computed via Equation 50.1 (diamonds), as a function of GaP cluster size. Bottom left panel: Exciton binding energy E_{bind} estimated as $QP^1 - E_{TD-DFT}$ (squares). (b) Top right panel: Optical gap E_{TD-DFT} as obtained via TD-DFT (triangles) and ΔSCF QP-corrected HOMO–LUMO gap QP^1 computed via Equation 50.1 (diamonds), as a function of AlAs cluster size. Bottom right panel: Exciton binding energy E_{bind} estimated as $QP^1 - E_{TD-DFT}$ (squares).

with the difference between the experimental vertical ionization energy and electron affinity, which falls in the range 4.70–5.20 eV (Cappellini et al. 1997) including error bars. This is an interesting issue, because the similarity in QP energy gaps and excitonic effects with carbon fullerenes occurs in systems made of fundamental components of current technology, such as Ga, As, P, and Al.

The optical absorption spectra obtained with OCTOPUS in the energy range up to about 5 eV for the four families considered are shown in Figure 50.4. All plots clearly show that the strongest absorption features corresponding to the different clusters in the family differ in position and strength; as such, these spectroscopic features can be used as a fingerprint of each specific molecule in each class. Note that, in the absence of experimental results to compare to our simulations with, the reliability of our OCTOPUS spectra is validated by the good agreement with results obtained using the different TD-DFT frequency-space implementation in NWCHEM. This is shown as an example in Table 50.8 where the first intense electric dipole–permitted transition (marked by an asterisk in Figure 50.4) appears to be reproduced by the two simulations within about 0.1 eV in position E and to be almost coincident in the corresponding oscillator strength f. In the case of the largest cluster $Ga_{16}P_{12}$, only the onset at 1.73 eV in the optical absorption spectrum has been estimated in the OCTOPUS result due to the crowding of electronic transitions starting from that energy; such an estimate compares well with the prediction of 1.72 eV of NWCHEM for the first intense absorption peak. The small discrepancy shown in Table 50.8 can be ascribed to the small basis set used in NWCHEM calculations and to the different XC functionals used (LDA for OCTOPUS and GGA for NWCHEM).

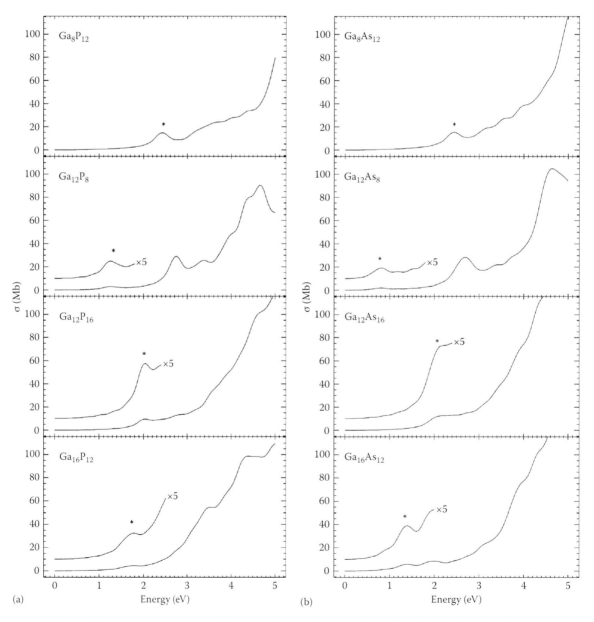

FIGURE 50.4   (a) Optical absorption cross section, $\sigma(E)$ (expressed in megabarns, 1 Mb = 10^{-18} cm^2), of the four GaP clusters considered; in each box, the asterisk marks the first intense electric dipole–permitted electronic transition. (b) Optical absorption cross section, $\sigma(E)$, of the four GaAs clusters considered; in each box, the asterisk marks the first intense electric dipole–permitted electronic transition.

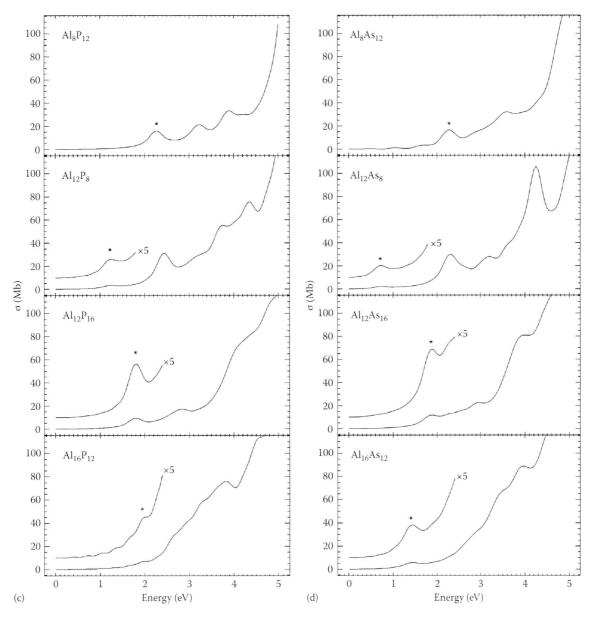

FIGURE 50.4 (continued)   (c) Optical absorption cross section, $\sigma(E)$, of the four AlP clusters considered; in each box the asterisk marks the first intense electric dipole–permitted electronic transition. (d) Optical absorption cross section, $\sigma(E)$, of the four AlAs clusters considered; in each box the asterisk marks the first intense electric dipole–permitted electronic transition.

TABLE 50.8   Energies (eV) and Oscillator Strengths (in Parentheses) of the First Intense Dipole-Allowed Electronic Transition of the Four GaP Clusters Considered, as Obtained by the Different TD-DFT Implementations of OCTOPUS and NWCHEM

Cluster	OCTOPUS	NWCHEM	Transition
Ga_8P_{12}	2.46(0.08)	2.44(0.07)	$H-1(t_g) \to L+1(t_u)$
$Ga_{12}P_8$	1.35(0.03)	1.27(0.03)	$H(t_u) \to L(a_g)$
$Ga_{12}P_{16}$	2.01(0.05)	1.93(0.06)	$H(t_1) \to L+1(t_2)$
$Ga_{16}P_{12}$	1.73(/)	1.72(0.01)	$H(e) \to L+1(t_2)$

Note: Last column shows the character of the excitation in terms of the molecular orbitals involved in the transition (H denotes the HOMO orbital, H − 1 the one just below it, L is the LUMO, and L + 1 represents the orbital just above it).

50.5 Summary

This work presents the electronic properties of 16 fullerene-like III–V binary compounds in the families GaP, GaAs, AlAs, and AlP. Some of these clusters have been predicted to be stable on the basis of first-principles calculations. We computed their optical absorption spectra and the excitonic effects for their first excited states. The QP corrections we obtained confirm the high stability previously found for these clusters. Our results on both single-particle and two-particle excitations may play an important role to unambiguously detect the successful synthesis of such molecules.

50.6 Future Perspective

Using a simple compendium of DFT and TD-DFT calculations with well-known and widely used XC functionals, we show in this chapter the power of theoretical optical spectroscopy to characterize novel synthesized structures. Since the current technology (synthesis, production of devices, characterization) of III–V compounds is widely affirmed, it permits fruitful extension to future nanoscale systems (fullerenes, nanotubes, etc.). Therefore, our method could be easily extended and applied in a systematic way to other III–V binary compounds, including larger and more complicated structures of potential interest for optoelectronic applications based on nanoscale systems.

Acknowledgments

G. Malloci acknowledges financial support by Regione Autonoma della Sardegna. The authors acknowledge financial support by MIUR under project PON-CyberSar. We thank the authors of OCTOPUS for making their code available under a free license. We acknowledge the High Performance Computational Chemistry Group for the use of NWCHEM, A Computational Chemistry Package for Parallel Computers, Version 4.6 (2004), PNNL, Richland, Washington, United States. Part of the simulations were carried out at CINECA (Bologna).

References

Alexandre, S. S., Chacham, H., and Numes, R. W. 2001. Structure and energetics of boron nitride fullerenes: The role of stoichiometry. *Physical Review B* 63: 045402/1–5.

Bechstedt, F., del Sole, R., Cappellini, G., and Reining, L. 1992. An efficient method for calculating quasiparticle energies in semiconductors. *Solid State Communications* 84: 765–770.

Becke, A. D. 1988. Density-functional exchange-energy approximation with correct asymptotic behaviour. *Physical Review A* 38: 3098–3100.

Botti, S. and Marques, M. A. L. 2007. Identification of fullerene-like CdSe nanoparticles from optical spectroscopy calculations. *Physical Review B* 75: 035311–035317.

Buda F. and Fasolino, A. 1995. Ab initio molecular-dynamics study of the stability and optical properties of ultrasmall III-V hydrogenated clusters. *Physical Review B* 60: 5851–5857.

Cappellini, G., Casula, F., Yang, J., and Bechstedt, F. 1997. Quasiparticle energies in clusters determined via total-energy differences: Application to C_{60} and Na_4. *Physical Review B* 56: 3628–3631.

Casida, M. E. 1995. Time-dependent density-functional response theory for molecules. In *Recent Advances in Density Functional Theory*. Vol. I, ed. D. P. Chong, pp. 155–192. Singapore: World Scientific.

Castro, A., Marques, M. A. L., Alonso, J. A., Bertsch, G. F., Yabana, K., and Rubio, A. 2002. Can optical spectroscopy directly elucidate the ground state of C20? *Journal of Chemical Physics* 116: 1930–1933.

Ceperley, D. M. and Alder, B. J. 1980. Ground state of the electron gas by a stochastic method. *Physical Review Letters* 45: 566–569.

Chopra, N. G., Luyken, R. J., Cherrey, K., Crespi, V. H., Cohen, M. L., Louie, S. G., and Zettl, A. 1995. Boron nitride nanotubes. *Science* 269: 966–967.

Côté, M., Cohen, M. L., and Chadi, D. J. 1998. Theoretical study of the structural and electronic properties of GaSe nanotubes. *Physical Review B* 58: R4277–R4280.

Fowler, P. W., Rogers, K. M., Seifert, G., Terrones, M., and Terrones, H. 1999. Pentagonal rings and nitrogen excess in fullerene-based BN cages and nanotube caps. *Chemical Physics Letters* 299: 359–367.

Frish, M. J., Pople, J. A., and Binkley, J. S. 1984. Self-consistent molecular orbital methods 25. Supplementary functions for Gaussian basis sets. *Journal of Chemical Physics* 80: 3265–3269.

Godby, R. W. and White, I. D. 1998. Density-relaxation part of the self-energy. *Physical Reviews Letters* 80: 3161–3162.

Godby, R. W., Schlüter, M., and Sham, L. J. 1988. Self-energy operators and exchange-correlation potentials in semiconductors. *Physical Review B* 37: 10159–10175.

Goldberg, D., Bando, Y., Stéphan, O., and Kurashima, K. 1998. Octahedral boron nitride fullerenes formed by electron beam irradiation. *Applied Physics Letters* 73: 2441–2443.

Hirata, S. and Head-Gordon, M. 1999. Time-dependent density functional theory within the Tamm–Dancoff approximation. *Chemical Physics Letters* 314: 291–299.

Hohenberg, P. and Kohn, W. 1964. Inhomogeneous electron gas. *Physical Review* 136: 864–871.

Iijima, S. 1991. Helical microtubules of graphitic carbon. *Nature* 354: 56–58.

Jones, R. O. and Gunnarson O. 1989. The density functional formalism, its applications and prospects. *Reviews of Modern Physics* 61: 689–746.

Kohn, W. and Sham, L. J. 1965. Self-consistent equations including exchange and correlation effects. *Physical Review* 140: 1133–1138.

Kroto, H. W., Heath, J. R., O'brien, S. C., Curl, R. F., and Smalley, R. E. 1985. C_{60}: Buckminsterfullerene. *Nature* 318: 162–163.

Lee, S. M., Lee, Y. H., Hwang, Y. G., Elsner, J., Porezac, D., and Frauenheim, T. 1999. Stability and electronic structure of GaN nanotubes from density-functional calculations. *Physical Review B* 60: 7788–7791.

Lopez, X., Marques, M. A. L., Castro, A., and Rubio, A. 2005. Optical absorption of the blue fluorescent protein: A first-principles study. *Journal of the American Chemical Society* 127: 12329–12337.

Madelung, O. 1996. *Semiconductors—Basic Data*. Berlin, Germany: Springer.

Malloci, G., Cappellini, G., Mulas, G., and Satta, G. 2004. Quasiparticle effects and optical absorption in small fullerenelike GaP clusters. *Physical Review B* 70: 205429/1–6.

Malloci, G., Mulas, G., Cappellini, G., and Joblin, C. 2007. Time-dependent density functional study of the electronic spectra of oligoacenes in the charge states −1, 0, +1, and +2. *Chemical Physics* 340: 43–58.

Mårlid, B., Larsson, K., and Carlsson, J.-O. 2001. Nucleation of c-BN on hexagonal boron nitride. *Physical Review B* 64: 184107/1–9.

Marques, M. A. L. and Botti, S. 2005. The planar-to-tubular structural transition in boron clusters from optical absorption. *Journal of Chemical Physics* 123: 014310/1–5.

Marques, M. A. L., Castro, A., Bertsch, G.F., and Rubio, A. 2003. Octopus: A first-principles tool for excited electron-ion dynamics. *Computer Physics Communications* 151: 60–78.

Natarajan, R. and Öğüt, S. 2003. Structural and electronic properties of Ge-Te clusters. *Physical Review B* 67: 235326/1–7.

Onida, G., Reining, L., and Rubio, A. 2002. Electronic excitations: Density-functional versus many-body Green's function approaches. *Reviews of Modern Physics* 74: 601–659.

Parilla, P. A., Dillon, A. C., Jones, K. M., Richer, G., Shulz, D. L., Ginley, D. S., and Heben, M. J. 1999. The first true inorganic fullerenes? *Nature* 397:114–115.

Perdew, J. P. 1986. Density-functional approximation for the correlation energy of the inhomogeneous electron gas. *Physical Review B* 33: 8822–8824.

Perdew, J. P. and Zunger, A. 1981. Self-interaction correction to density-functional approximations for many-electron systems. *Physical Review B* 23: 5048–5079.

Rubio, A., Alonso, J. A., Blase, X., Balbas, L. C., and Louie, S. G. 1996. Ab initio photoabsorption spectra and structures of small semiconductor and metal clusters. *Physical Review B* 77: 247–250.

Runge, E. and Gross, E. K. U. 1984. Density-functional theory for time-dependent systems. *Physical Reviews Letters* 52: 997–1000.

Song, B., Yao, C.-H., and Cao, P.-L. 2006. Density-functional study of structural and electronic properties of Ga$_n$N (n = 1–19) clusters. *Physical Review B* 74: 035306/1–5.

Straatsma, T. P., Apra, E., Windus, T. L. et al. 2004. *NWChem, A Computational Chemistry Package for Parallel Computers, Version 4.6.* Richland, WA: Pacific Northwest National Laboratory.

Tozzini, V., Buda, F., and Fasolino, A. 2000. Spontaneous formation and stability of small GaP fullerenes. *Physical Reviews Letters* 85: 4554:4557.

Tozzini, V., Buda, F., and Fasolino, A. 2001. Fullerene-like III-V clusters: A density functional theory prediction. *Journal of Physical Chemistry* 105: 12477–12480.

Trave, A., Buda, F., and Fasolino, A. 1996. Band-gap engineering by III-V Infill in sodalite. *Physical Review Letters* 77: 5405–5408.

Troullier, N. and Martins, J. L. 1991. Efficient pseudopotentials for plane-wave calculations. *Physical Review B* 43: 1993–2006.

Wang, Y. 1996. Semiconductor nanoclusters and fullerenes: A new class of sensitizing dyes for photoconductive polymers. *Pure & Applied Chemistry* 68: 1475–1478.

Wang, Q., Sun, Q., Jena, P., and Kawazoe, Y. 2006. Clustering of Cr in GaN nanotubes and the onset of ferrimagnetic order. *Physical Review B* 73: 205320/1–6.

Wirtz, L., Marini, A., and Rubio, A. 2006. Excitons in boron nitride nanotubes: Dimensionality effects. *Physical Review Letters* 96: 126104/1–4.

Yabana, K. and Bertsch, G. F. 1999. Time-dependent local-density approximation in real time: Application to conjugated molecules. *International Journal of Quantum Chemistry* 75: 55–66.

Zope, R. R. and Dunlap, B. I. 2005. Electronic structure of fullerene-like cages and finite nanotubes of aluminum nitride. *Physical Review B* 72: 045439/1–6.

51

Onion-Like Inorganic Fullerenes

Christian Chang
Technische Universität Berlin

Beate Patzer
Technische Universität Berlin

Detlev Sülzle
Technische Universität Berlin

51.1 Introduction

Molecules that consist of atoms forming a closed cage have been known for a long time. However, it was not until the discovery of the famous C_{60} molecule, which represented the hitherto unknown third allotropic and the first molecular form of carbon, that a tremendous growth of interest in similar species was initiated and a whole new branch of molecular physics, chemistry, and material sciences was thrown open.

C_{60} was theoretically predicted by Osawa (1970) but this suggestion had no immediate scientific impact. C_{60} consists of 60 carbon atoms placed at the vertices of a truncated icosahedron, a regular arrangement of 12 five- and 20 six-membered carbon rings giving it a football-like appearance (Figure 51.1 left). Although the interatomic distances within the six-membered rings are not all equal, in fact there are two different alternating C–C bond lengths, namely, 1.458 and 1.401 Å (Hedberg et al., 1991) in the hexagons, the C_{60} structure possesses the highest possible point group symmetry in three-dimensional space, namely, icosahedral symmetry (I_h).

In 1985, Kroto and coworkers (Kroto et al. 1985) experimentally made the serendipitous discovery that a C_{60} molecule not only existed but that it would self-assemble spontaneously from a hot nucleating carbon plasma. Eventually in 1990, Krätschmer and his group (Krätschmer et al., 1990a,b) succeeded in the first extraction of quantities of C_{60} on a macroscopic scale. It was this experimental breakthrough that made the explosive growth of this new area of science possible.

The C_{60} molecule was named buckminsterfullerene after R. Buckminster Fuller, a noted architect who developed the art of constructing buildings in the shape of geodesic dome-like polyhedra whose triangulated surface contains vertices where five and six faces meet. In the following time, a multitude of other smaller and larger carbon-cage molecules, C_n with n even, ranging from $n = 20$ to several hundreds, which are now all generally called fullerenes were investigated both theoretically and experimentally (see Jorio et al., 2008 for an overview). The great variety in structural architecture of these systems seems to be almost infinite. Until today, a diversity of carbon molecules with unusual shapes and geometric arrangements have been discovered, a wealth of morphologies, ranging from single or several intercalated polyhedral cages, graphitic sponges, tubes to even tori (Stephens, 1993; Terrones et al., 2004), nanohorns, and nanocones (Yudasaka et al., 2008).

The question of whether similar molecular systems composed of atoms other than carbon can exist arises quite naturally. These so-called inorganic fullerenes, in particular, those consisting of more than a single cage, are the subject of this chapter. We shall begin with a loose definition of what is considered an onion-like inorganic fullerene. Inorganic implies that the molecule may consist of any atomic species apart from pure carbon. Fullerene-like connotes that all contributing atoms can be considered as being located at the vertices of a not necessarily regular polyhedron giving the system a cage-like look. To be called strictly a fullerene, this polyhedron should obey certain geometric restrictions (cf. Section 51.2). In the common literature, however, molecular systems that do not comply rigorously with these geometric restrictions, but merely exhibit a polyhedral cage structure, are nonetheless often called fullerenes as well. Therefore, the terms fullerene-like and cage-like are frequently used as synonyms. There might be just one or several of such cages that might or might not penetrate each other. The latter will be termed onion-like as each cage could be regarded as a shell of an onion. We shall allow for the possibility of a single atom being placed inside at the center of the cage(s) and make use of the notation $X_i@Y_j@Z_k@ \ldots$ for a general onion-like inorganic fullerene where the atomic species X, Y, Z, etc., which

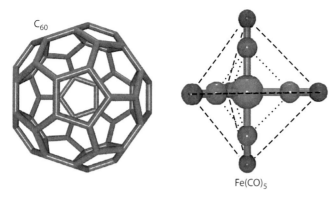

FIGURE 51.1 Truncated icosahedron cage of buckminsterfullerene C_{60} (left) and trigonal bipyramidal structure of $Fe(CO)_5$ (right). Solid lines correspond to actual physical bonds, dashed and dotted lines show the edges of the outer and inner pyramidal cages, respectively.

might well be equal to each other are located at the vertices of different polyhedral frameworks. The innermost cage X could, as mentioned, be just a single central atom. If unambiguous from the context, it is sometimes a convenient habit to just state the chemical sum formula.

There are different views or perspectives of looking at these cage structures. The most common one is to consider the actual physical bonds between neighboring atoms since these interactions are responsible for the energetic stability of the molecular framework. Another different view is a geometric idealization of molecular structures, and consists in associating the positions of sets of atoms to the vertices of one or several polyhedra. To describe a cage structure, we shall make use of the latter pure geometric concept. It should be noted that the interatomic distances between the vertices of those polyhedra, i.e., the edge lengths, do not necessarily represent physical bonds. The connection between geometric relationships and physical bonding situations in many kinds of cage molecules is not yet completely understood. Using the above definition, it might be tempting to write the well-known iron carbonyl molecule $Fe(CO)_5$ as $Fe@C_5@O_5$, where an iron atom lies at the center of two nested trigonal bipyramids, the inner one consisting of carbon and the outer one of oxygen atoms. Clearly, this picture is entirely geometric and does not reflect any physical interactions. In fact, this molecule is held together by five strong bonds between the central iron atom and the surrounding carbon monoxide molecules. Traditionally, this species would never be classified as a cage molecule and certainly not as an inorganic fullerene (Figure 51.1 right).

Here, we shall be mainly concerned with the geometric and structural properties of representative inorganic cage species even though not all of them could be strictly classified as fullerenes. It is their polyhedral shape usually accompanied with high symmetry that gives these molecular systems their beautiful aesthetic look.

It is intriguing how polyhedral structures appear at different dimensional scales, from the electron and spin densities at the subatomic level up to everyday objects. Starting at the subatomic level, spin densities for different electron configurations in transition metal clusters may show cubic, octahedral, or tetrahedral shapes

(Bouguerra et al., 2007; Jones et al., 2007). Also, representations of the density of d electrons around a metal atom in $Cr(CO)_6$ (Macchi and Sironi, 2003), of its Laplacian in $Fe_2(CO)_9$ or in $[Mn(CO)_6]^+$ (Bo et al., 1993; Bader et al., 2004), and of the electron localization function in $Re_2(CO)_{10}$ reveal cubic shapes (Kohout et al., 2002). At the polyatomic level, the coordination polyhedra around metal atoms have diameters of a few tenths of a nanometer, while typical metal clusters can approach 1 nm and large clusters can reach up to 2 nm in diameter. Very large multipolyhedral systems assembled through bridging ligands can reach sizes of about 2–3 nm, as in the cores of Pd_{145} and Mo_{132} (Müller et al., 1999; Tran et al., 2000). Icosahedral quasicrystals, nanoclusters, and nanoparticles are in the 10–20 nm size range, where we can also find a single-stranded DNA molecule folded into a hollow octahedron with a diameter of ~22 nm, as well as other DNA polyhedra including the tetrahedron, the cube, and the truncated octahedron reported in recent years (Shih et al., 2004). The capsids of viruses may reach a size an order of magnitude larger (between 10 and 100 nm), among which structures of the icosahedral symmetry have been widely reproduced (Twarock, 2006). Since all these species are molecular aggregates formed by an assembly of atoms or of well-specified molecular units, the study of inorganic fullerenes can be considered as a subdiscipline of cluster physics. Eventually, a fullerene molecule itself could be regarded as such a subunit or a monomer of a larger aggregate. Depending on the cluster size, i.e., the number of monomers, these systems cover a range from the very small (microscopic) to the very large (macroscopic) with accordingly varying remarkable physical properties. They, therefore, provide a link between molecular, cluster, nanophysics, and eventually material science. Inorganic fullerenes represent an intriguing class of polyhedral molecular clusters with a wide variety of potential applications in fundamental and applied sciences including the design of functional materials and nanotechnological devices. Some of these species possess exceptional properties such as stability, size, solubility in several solvents, giant cavities, nanosized tunable pores, unique surfaces, and unusual electronic structure and magnetic behavior.

51.2 Geometric Properties

In the molecular world, the polyhedral shape is quite ubiquitous, in particular, among inorganic species, e.g., coordination compounds, where the ligands around a central atom are generally lying at the vertices of some more or less regular polyhedron (González-Morage, 1993). Equally, in the realm of atomic and molecular clusters, the most stable geometric arrangements disclose polyhedral patterns of high symmetry (Joyes, 1990; Haberland, 1994).

A general polyhedron in three-dimensional space is a solid bounded by polygonal faces and is entirely characterized by the total number of its faces f, vertices v, and edges e, which are related by Euler's theorem (see e.g., Cromwell 1997):

$$f + v = e + 2.$$

The polygonal faces are n-gons ($n = 3, 4, 5, \ldots$), which intersect at the edges and at the vertices. The vertices can be specified by their vertex configuration, which denotes the kind and number of faces that meet at a vertex. If all faces of a polyhedron are regular and congruent, i.e., they are all of the same kind and all vertex environments are equivalent, we have the well-known family of the five Platonic polyhedra. Allowing for different regular n-gons but keeping the condition of equal vertex configurations, we arrive at the class of Archimedean solids. Many other families of polyhedra are obtained by further relaxing the conditions and/or imposing special criteria on the nature of the faces and vertices. A summary of the most common classes of polyhedra that also appear in molecular arrangements is given in Table 51.1. A more detailed account on polyhedron properties can be found in the books by Cromwell (1997) and Coxeter (1973).

TABLE 51.1 Classification of Families of Convex Polyhedra According to the Regularity (R) and Equivalence (E) of Their Faces and to the Equivalence (E) of Their Vertices and Edge Environments: All Faces Equivalent (Isohedral), All Vertex Environments Equivalent (Isogonal), and All Edge Environments Equivalent (Isotoxal)[a]

Family	Faces	Vertices	Edges[b]
Platonic[c]	R, E	E	E
Archimedean	R	E	—[d]
Johnson	R	—[e]	—
Metaprisms[f]	—	E	—
Pyramids	—[g]	—	—
Bipyramids	E[h]	—	—
Trapezohedra	E	—	—
Catalan (Archimedean duals)	E	—	—[i]
Simplicial (Deltahedra)[j]	Triangles	—	—
Kepler–Poinsot (non convex)	Stellated[k]	—	—
Fullerenes	R[l]		
Rhombohedra[m]	Rhombs	—	—

[a] To every convex polyhedron, a corresponding so-called dual solid can be associated. Thereby the midpoints of every face of the original polyhedron become the vertices of the dual polyhedron.

[b] In polyhedra having regular faces, even if the edge environments are not all equivalent, all edges have the same length.

[c] Platonic solids are further characterized by having three unique spheres associated: a circumsphere on which all vertices lie, an insphere which passes through all facial midpoints, and a midsphere which touches upon all edges.

[d] Among the Archimedean polyhedra only the cuboctahedron and the icosidodecahedron have all edge environments equivalent; all Archimedean solids have a unique circumsphere and midsphere but no insphere; and the family of Archimedean solids includes the two infinite series of regular prisms and antiprisms.

[e] Among the Johnson (1966) polyhedra only two are isohedral: the trigonal bipyramid (J12) and the pentagonal bipyramid (J13).

[f] Metaprisms are formed by two parallel n-gonal faces rotated with respect to each other by any angle that is not a multiple of π/n (even multiples would correspond to prisms, odd multiples to antiprisms).

[g] Only the square and pentagonal pyramids may have regular polygons in all their faces.

[h] Only the trigonal and pentagonal bipyramids can have regular faces (the regular square bipyramid is in fact the octahedron included among the Platonic solids).

[i] Among the Catalan polyhedra (Archimedean duals), only the rhombic dodecahedron and the rhombic triacontahedron have all their edge environments equivalent. All Archimedean duals possess a unique insphere and midsphere but no circumsphere.

[j] Among them the so-called Frank–Kasper polyhedra (Frank and Kasper, 1958, 1959) all of which are convex with only triangular faces either all the same or different to each other. At least five triangles have a common vertex. Number of vertices can be 12, 14, 15, and 16. There exists a close packing of 3-space by four Frank–Kasper polyhedra. The 16-vertex Frank–Kasper solid is also called Friauf polyhedron (Friauf, 1927). A simplicial polyhedron having only regular triangular faces (equilateral triangles) is sometimes called a deltahedron.

[k] Semi-regular stellated polyhedra with non-convex pentagonal faces. Included here for the sake of completeness.

[l] 12 pentagonal faces, the rest are hexagons. The hexagons can be regular (all edges equal) or semi-regular (edges alternately equal). The fullerene with only pentagonal faces is the Platonic dodecahedron and that with 20 hexagonal faces and 60 vertices is an Archimedean solid, the truncated icosahedron. Special cases are the Mackay and Goldberg polyhedra (Goldberg, 1937; Mackay, 1962).

Mackay: Mackay polyhedra represent an intercalated shell structure of concentric icosahedra oriented in such a way as to keep icosahedral fivefold symmetry. The arrangement having 55 vertices within the second shell occurs frequently and has been termed Mackay icosahedron. The number of vertices in these icosahedral shells and subshells agrees well with the magic numbers of noble gas and $(C_{60})_n$ clusters.

Goldberg: The isolated pentagon rule (IPR) established for the C_n fullerenes provides a criterion to predict which system will have icosahedral symmetry if not distorted by an electronic Jahn–Teller effect. Those clusters that maintain I_h symmetry are special cases of polyhedra known as Goldberg polyhedra. These polyhedra are built from $20 \cdot (b^2 + bc + c^2)$ vertices, where b and c are nonnegative integers. This situation corresponds to hexagonal close packing on the surface of an icosahedron. If $b = c$ and $bc = 0$, then the undistorted polyhedra will have I_h symmetry otherwise the symmetry is reduced to its rotational subgroup I.

[m] There is a family of isohedral rhombic polyhedra that consists of the prolate and the obtuse rhombohedron, the rhombic dodecahedron of the second kind, discovered by Bilinski (1960); the rhombic icosahedron, discovered by Fedorov (1885); and the triacontahedron, discovered by Kepler (1611). Only the last one has icosahedral symmetry. The faces of these solids are congruent rhombs whose diagonals are in golden ratio. These five solids are the only convex polyhedra of this kind. The convex rhombic solids are also named zonohedra.

It is important to note that for any given polyhedron, one can associate a unique dual solid. This dual polyhedron can be constructed by joining the midpoints of all faces of the original polyhedron, e.g., the cube and the octahedron are in this sense dual to each other. An important special family of polyhedra are designated as fullerenes, which involve only pentagonal and hexagonal faces. In a fullerene-like polyhedron each vertex has three neighbors, so that the number v of vertices in the cage can be related to the number edges e by

$$2e = 3v.$$

Similarly, if f_n is the number of n-sided faces, it follows that

$$2e = \sum_n nf_n.$$

Euler's theorem for convex polyhedra then gives a further condition

$$v + \sum_n f_n = e + 2.$$

If we now restrict the faces to be pentagons and hexagons only and eliminate v and e, we obtain

$$5f_5 + 6f_6 = 3((1/2)(5f_5 + 6f_6) + 2 - f_5 - f_6) \longrightarrow f_5 = 12,$$

that is, we need 12 pentagons to produce a fullerene cage with the number of hexagons given by

$$f_6 = \frac{(v - 20)}{2}$$

The cages should be stable if the curvature-related strain is symmetrically (geodesically) distributed and if the pentagons are isolated as much as possible by the hexagons to avoid the inherent instability of fused-pentagon configuration. This fact is often called the isolated pentagon rule (IPR), which is, for example, realized in the inorganic fullerene $Ga_{60}Cu_{24}$ found in the crystal structure of $Mg_{35}Cu_{24}Ga_{54}$ (Lin and Corbett, 2005). However, there exist quite a few exceptions to this rule (Beavers et al., 2006).

To sum up, in order for a species to belong strictly to the family of fullerenes, the main geometric characteristics are that (1) there are 12 pentagonal and any number of hexagonal faces ($f = 12 + f_6$) and (2) all vertices are shared by three neighboring polygons. The application of Euler's formula to a fullerene with a given number of hexagons (f_6) fully determines its number of vertices ($v = 20 + 2f_6$) and edges ($e = 30 + 3f_6$). It follows that one can build fullerenes having an even number of vertices from 20 onward. Odd-numbered systems will always have at least one atom with a remaining dangling bond and will thus be very reactive and unstable.

The family of Frank–Kasper polyhedra is intimately related to that of the fullerenes. Frank–Kasper polyhedra form a subset of the simplicial polyhedra (see Table 51.1), which are characterized by having only triangular faces, not necessarily equilateral.

A particular family of simplicial polyhedra is formed by those with just five- or sixfold vertices (i.e., either five or six edges meeting at each vertex). The smaller members of this family (those with 14, 15, or 16 vertices) are actually known as Frank–Kasper polyhedra (Frank and Kasper, 1958, 1959) and appear often as coordination polyhedra in intermetallic phases. To avoid confusion with other types of simplicial polyhedra, one could generally call Frank–Kasper polyhedra all those that have (1) only triangles as faces and (2) only five- and sixfold vertices. If v_5 and v_6 are the number of vertices at which five or six edges meet, respectively, it can be shown, that $v_5 = 12$ in all cases, hence their total number of vertices is given by $v = 12 + v_6$. From the definition of fullerenes given above and that for Frank–Kasper polyhedra, it follows that the properties of faces and vertices are interchanged among these two families of polyhedra. In other words, there is a duality relationship between them.

Therefore, a given fullerene and its Frank–Kasper dual are identified by their common parameter $f_6 = v_6$, which corresponds to the number of hexagons and sixfold vertices, respectively. An obvious corollary of the duality relationship between Frank–Kasper polyhedra and fullerenes is that one can build up one of them by capping the faces of the other, thus forming a pair of intercalated polyhedra. The simplest fullerene is the dodecahedron $f_6 = 0$, and its Frank–Kasper dual is the icosahedron. In this special case, both of them equally belong to the more restricted family of Platonic solids (cf. Table 51.1).

One example of intercalating those two polyhedra is the Si_{20} dodecahedron circumscribing an Na_{12} icosahedron ($Na_{12}@Si_{20}$ (cf. Figure 51.2 upper part), which occurs within the crystal structure of Na_8Si_{46} (Cros et al., 1971; Reny et al., 1998). Another

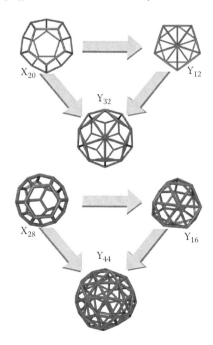

FIGURE 51.2 Upper part: An X_{20} dodecahedron (left) and its dual a Y_{12} (right) combined to produce an expanded Frank–Kasper solid Y_{32} (middle). Lower part: An X_{28} fullerene (left) and its Frank–Kasper dual a Y_{16} (right) combined to produce an expanded Frank–Kasper solid Y_{44} (middle).

example of similar type is the recently discovered and intensely studied [As@Ni$_{12}$@As$_{20}$]$^{3-}$ anion, which occurs in the solid state as a part of a complex salt (Baruah et al., 2003; Moses et al., 2003). The heavier homologue [Sb@Pd$_{12}$@Sb$_{20}$]$^{3-}$ (Zhao and Xie, 2004) as well as the neutral species Kr@Ni$_{12}$@As$_{20}$, Ge@Zn$_{12}$@Ge$_{20}$, and similar neutral and charged systems having the same structural properties were theoretically investigated (Chang et al., 2005, 2006).

Such pairs of intercalated polyhedra fit well into a class of structures considered by Müller et al. (2001a) as Keplerates. A Keplerate can be defined as a group of atoms organized in spherical shells around a central point in such a way that each set of symmetry-related atoms corresponds to the vertices of a Platonic or a generalized Archimedean solid. A prominent example of such an onion-like Keplerate is the giant metal-oxide-based Mo$_{132}$O$_{372}$(SO$_4$)$_{30}$(H$_2$O)$_{72}$ cluster (Figure 51.3 bottom), a huge sphere of high icosahedral symmetry and a diameter of about 3 nm. It belongs to the family of polyoxomolybdates, which are based on the same building principles as the geodesic domes of R. Buckminster Fuller. They give rise to a bewildering structural variety similar to that of carbon-based fullerenes (Müller et al., 2001a,b).

Dual polyhedra are likely to appear in molecular structures forming successive concentric shells because occupation of the face centers of a polyhedron (as required for the formation of a dual) corresponds to situations that favor physical bonding. Three such possibilities can be mentioned: (1) coordinative bonding between bridging ligands and the metal atoms in a polygonal face, (2) ionic bonding between ions of opposite sign in successive shells, and (3) metallic bonding favored by the maximum coordination number.

An example of such a dual relationship can be seen in one of the building blocks of the Mg$_{35}$Cu$_{24}$Ga$_{54}$ structure (Lin and Corbett, 2005), formed by a Ga$_{16}$ Friauf polyhedron (the specific case of the Frank–Kasper polyhedron with 28 triangular faces and 16 vertices, four of which being based on the midpoints of sustaining hexagons is usually known as the Friauf polyhedron) surrounded by its dual Mg$_{28}$ fullerene, which is in turn circumscribed by another Cu$_{12}$Ga$_4$ Friauf polyhedron (Friauf, 1927) (Figure 51.2 lower part). It is to be noted that the composition of a fullerene and its dual Frank–Kasper, such that the vertices of both lie on the convex hull of the combined solid, results in a larger Frank–Kasper polyhedron. The convex hull, i.e., the enclosing framework of the polyhedral solid, is commonly defined as the smallest convex polyhedron that contains every point of the set S. A polyhedron P is convex if and only if, for any two points A and B inside the polyhedron, the line segment AB is inside P. One way to visualize this rather abstract concept of a convex hull is to put a "rubber sheet" around all the vertices of the polyhedral point set, and let it wrap as tight as it can. The resultant polyhedron is a representation of its convex hull, which in general for an arbitrary point set in 3-dimensional space can only be determined numerically (Boissonnat and Yvinec, 1995; Goodman and O'Rourke, 1997). In the present example, a Y$_{44}$ Frank–Kasper is composed of the Mg$_{28}$ fullerene and its Cu$_{12}$Ga$_4$ dual (Figure 51.2 lower part). Another seemly set of three different Frank–Kasper polyhedra inside their dual fullerenes is formed by Na$_{32}$@In$_{48}$Na$_{12}$, Na$_{37}$@In$_{70}$, and Na$_{39}$In$_2$@In$_{78}$, found in the crystal structure of Na$_{172}$In$_{197}$Ni$_2$ (Sevov and Corbett, 1996). Table 51.2 summarizes other examples of onion-like inorganic fullerenes that appear concentric to their Frank–Kasper duals, including two characteristic structure types known to present typical Frank–Kasper polyhedra, namely, MgCu$_2$ and Cr$_3$Si.

51.3 Symmetries and Building Principles of Onion-Like Fullerenes

51.3.1 Fullerenes and Frank–Kasper Cages

Given the relationship between fullerenes and their Frank–Kasper dual polyhedra, we can establish the compounding of concentric shells of such cages in a general way, which is shown diagrammatically in the flowchart scheme of Figure 51.4. If we start with a fullerene having f faces (12 pentagons and $f_6 = f - 12$ hexagons) and $2f - 4$ vertices, which we shall symbolize by X$_{2f-4}^f$, where the subscript indicates the number of vertices and the superscript the number of faces, the formation of its circumscribed dual Frank–Kasper polyhedron (step 1 in the scheme) will be characterized by f vertices and $2f - 4$ faces, denoted by Y$_f^{2f-4}$. The fact that in this notation the fullerene and Frank–Kasper solid have subscripts and superscripts interchanged follows from their dual nature. To this shell, we can now add a second fullerene shell by capping all the faces of the Frank–Kasper polyhedron (step 2 in

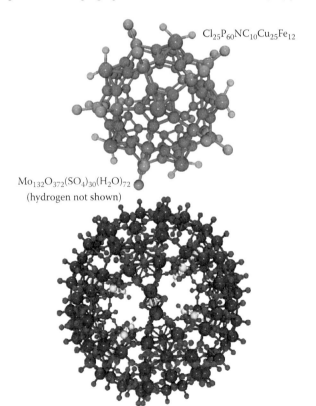

Cl$_{25}$P$_{60}$NC$_{10}$Cu$_{25}$Fe$_{12}$

Mo$_{132}$O$_{372}$(SO$_4$)$_{30}$(H$_2$O)$_{72}$
(hydrogen not shown)

FIGURE 51.3 The structure of Cl$_{25}$P$_{60}$N$_{10}$Cu$_{25}$Fe$_{12}$ (top) and Mo$_{132}$O$_{372}$(SO$_4$)$_{30}$(H$_2$O)$_{72}$ Keplerate (bottom).

TABLE 51.2 Some Examples of Onion-Like Fullerene Clusters That Appear Intercalated with Their Dual Frank–Kasper Polyhedron in Inorganic Compounds

f_6	Fullerene	Dual	Compound	References
0	Si_{20}	Na_{12}	Na_8Si_{48}	Cros et al. (1971), Reny et al. (1998)
0	Li_{20}	As_{12}	$Li_{20}[AsSi(iPr)Me_2]_{12}$	Driess et al. (1996)
0	$Cr_{20}Si_4$	$Cr_{10}Si_4$	Cr_3Si	Jauch et al. (1987)
0	Sn_{24}	Cs_{14}	Cs_8Sn_{46}	Grin et al. (1987)
4	$Mg_{12}Cu_{16}$	Mg_4Cu_{12}	$MgCu_2$	Ohba et al. (1984)
4	Si_{28}	$Si_{28}Na_{16}$	$Cs_8Na_{16}Si_{136}$	Bobev and Sevov (1999)
4	Mg_{28}	Ga_{16}	$Mg_{35}Cu_{24}Ga_{53}$	Lin and Corbett (2005)
4	$Bi_{12}Sr_{16}$	Bi_{16}	$Sr_{33}Bi_{28}Al_{480}O_{147}$	Hervieu et al. (2004), Boudin et al. (2004)
4	U_{28}	K_{16}	$[K_{16}(H_2O)_4(O_2)_{44}(UO_2)_{28}]^{14-}$	Burns et al. (2005)
20	$In_{48}Na_{12}$	Na_{32}	$Na_{172}In_{197}Ni_2$	Sevov and Corbett (1996)
25	In_{70}	Na_{37}	$Na_{172}In_{197}Ni_2$	Sevov and Corbett (1996)
28	Al_{76}	M_{40} (M = Li, Mg)	$LiMgAl_2$	Nesper (1994)
29	In_{78}	$Na_{39}In_2$	$Na_{172}In_{197}Ni_2$	Sevov and Corbett (1996)
32	$Ga_{60}Cu_{24}$	$Mg_{28}Cu_{12}Ga_4$	$Cu_{24}Ga_{53.57}Mg_{35.12}$	Lin and Corbett (2005)
32	Al_{84}	$Sr_{32}Bi_{12}$	$Sr_{33}Bi_{28}Al_{480}O_{147}$	Hervieu et al. (2004); Boudin et al. (2004)

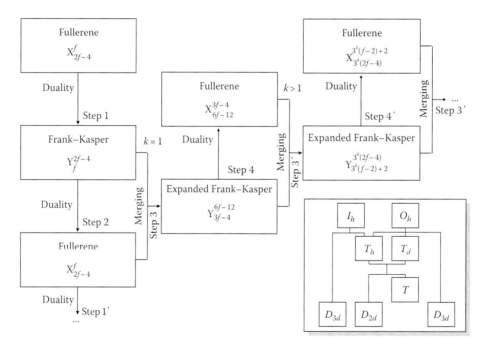

FIGURE 51.4 Scheme representing a general building principle of onion-like fullerenes. Lower right insert: Group–subgroup relations of the icosahedral (I_h) and octahedral (O_h) groups.

the scheme). This procedure could go on by adding shell after shell in the same way, i.e., repeating assiduously step 1 and 2. The convex hull of the composed solid will always be formed by the polyhedron last added.

However, another possibility is to merge a Frank–Kasper deltahedron obtained by duality with the underlying fullerene, so that their atoms will equally span the convex hull of a combined polyhedron. The result is an expanded polyhedron of Frank–Kasper type with a larger number of vertices Y_{3f-4}^{6f-12} (merging step 3 in the scheme, $k = 1$). Another shell can now be added by construction of the fullerene dual to the expanded Frank–Kasper

(step 4 in the scheme, $k = 1$), which would be denoted as X_{6f-12}^{3f-4}. This procedure can equally continue shell after shell ($k > 1$), i.e., running through successive values of k ($k = 1, 2, 3, \ldots$). Of course, one could have started from a Frank–Kasper polyhedron at the beginning then a similar scheme would apply. This means that the compositions of successive shells are accurately determined and correspond to a sequence of specific numbers. Thus, starting with a given fullerene having $f_6 = f - 12$ hexagonal faces, we can build a series of larger onion-like fullerenes by using the construction principle depicted in the scheme of Figure 51.4. Their general compositions, X_v^f, correspond to the values of

$\tilde{v}$ and $\tilde{f}$ that obey the following expressions $\tilde{v} = 3^k (2f - 4)$ and $\tilde{f} = 3^k(f - 2) + 2$, where k is any nonnegative integer ($k = 0, 1, 2, \ldots$). The sequence of steps 3 and 4 that leads from a given fullerene to an expanded fullerene having three times as many vertices as the original is known as the *leapfrog relationship* (Fowler, 1986). One can thus conclude that onion-like inorganic fullerenes related through a "leapfrog" connection are likely to be found in a molecular structure separated by an intermediate layer shell of a dual Frank–Kasper polyhedron. The building patterns just described imply that whenever there is a Frank–Kasper cluster in a molecular structure, we may suspect the presence of inscribed and/or circumscribed fullerenes, and vice versa, not limited to just a dual pair, but larger numbers of shells can be arranged following the principles outlined in the scheme. In extended crystal structures, the outermost onion skin layers of one set of intercalated polyhedra may overlap with corresponding ones from neighboring unit cells. Hence, a description of intercalated polyhedra does not necessarily imply that we are referring to independent molecular assemblies.

As an example of the first steps of these building principles, let us consider the simplest case that starts with a dodecahedron, the X_{20}^{12} ($f_6 = 0$) fullerene. Its dual is a special Frank–Kasper polyhedron, the Platonic icosahedron (Y_{12}^{20}), and compounding of these two polyhedra gives a special expanded Frank–Kasper Y_{32}^{60} polyhedron: the 32-vertex rhombic tricontahedron, a Catalan solid with icosahedral symmetry. By face augmentation of this latter polyhedron, we obtain a X_{60}^{32} fullerene, nothing other than the Archimedean truncated icosahedron, which is the well-known C_{60} structure (Figure 51.1 left).

Such a sequence of polyhedra can also be illustrated by the concentric clusters found around a Na atom in the crystal structure of the already mentioned Na_8Si_{48} (Cros et al., 1971; Reny et al., 1998). The first shell corresponds to a Si_{20} dodecahedron, and capping its dodecahedral faces with Na atoms (step 1 in the scheme) results in a Na_{12} icosahedron. Generation of its dual (step 2) then gives another Si_{20} dodecahedron with practically the same radius as its Frank–Kasper predecessor. Compounding of these two polyhedra (step 3) results in an expanded Frank–Kasper polyhedron with the stoichiometry of $Na_{12}Si_{20}$ (Figure 51.2 upper part). In this case, the process stops here because outer shells belong to neighboring unit cells and cannot therefore retain the icosahedral topology. Other examples of compounding of a dodecahedron and an icosahedron to form a 32-vertex Frank–Kasper polyhedron are a $Li_{20}As_{12}$ arrangement found in the structure of $Li_{20}[AsSiPrMe_2]_{12}$ (Driess et al., 1996) and a $Mo_{12}C_{12}$ unit, in which the carbon atoms are capping the faces of a Mo_{12} icosahedron (Müller et al., 2002). The relative size of the clusters participating in the compounding step 3 (scheme 1) are not irrelevant. Thus, in the examples just discussed the two polyhedra that form an expanded Frank–Kasper Y_{32} polyhedron have similar radii and hence their atoms all lie on the convex hull of the combined polyhedron.

In contrast, in the $[As@Ni_{12}@As_{20}]^{3-}$ anion (Moses et al., 2003) and similar systems mentioned earlier, a Ni_{12} icosahedron

and a circumscribed As_{20} dodecahedron formed by face capping are of quite different size. The convex hull of the system is, therefore, spanned by the atoms of the As_{20} dodecahedron only. Consequently, one obtains well-separated, nested rather than penetrating cages, which corresponds to a situation of going straight vertically down in the scheme without branching to step 3.

One can go one step further by looking at the crystal structure of $Na_{13}(Cd_{1-x}Tl_x)_{27}$ (Li and Corbett, 2004). In this case, a Tl_{12} icosahedron (a Frank–Kasper polyhedron) is surrounded by a dual Na_{20} fullerene (a dodecahedron) and its Frank–Kasper dual, a Cd_{12} icosahedron. The compounding of the latter two polyhedra is in fact a Y_{32} Frank–Kasper polyhedron, and capping its faces with Tl atoms results in a Tl_{60} fullerene. The procedure can go on by adding another Frank–Kasper set, Na_{32}, which is in fact a composition of a dodecahedron and an icosahedron (capping the hexagonal and pentagonal faces of Tl_{60}, respectively), similar to the inner $Na_{20}Cd_{12}$ cluster. It is easy to verify that these concentric polyhedra are built according to the rules shown in the scheme of Figure 51.4. One should notice that the symmetry is retained as more concentric shells are added by face augmentation, and both the inner Tl_{12} and the more outer Na_{12} clusters are concentric polyhedra. Although the building principles discussed do not require such a high symmetry, it might be interesting to sketch a systematic way of generating Frank–Kasper and fullerene polyhedra with high symmetry.

It is worth noting here that the geodesic domes designed by Buckminster Fuller (Gorman, 2005) are in fact formed by triangles and use a construction principle coincident in part with the one presented in our scheme (Figure 51.4). In contrast with the vast literature devoted to spherical carbon fullerenes, the hemispherical molecules that are closer to geodesic domes made by Buckminster Fuller have received relatively little attention.

51.3.2 Keplerate Fullerene-Like Cages

There are onion-like cage molecules that consist of polyhedra that are not strictly fullerenes or their Frank–Kasper duals in the sense discussed in Section 51.3.1, but are rather built from Platonic or Archimedean solids and their corresponding duals. A building principle as demonstrated in the scheme of Figure 51.4 is only thinkable for the case when the polyhedra and their constructed duals remain completely inside each other, i.e., going down from step 1 to step 2 and so forth. However, forming a compounded polyhedron as in the branching step 3 of the scheme is just applicable for once, because subsequently creating the dual of the expanded polyhedron does not again produce a Platonic or Archimedean solid. A typical example for this kind of species is the series X_nY_n, where $X = Al$, Ga, or In, $Y = N$, As, or Sb, and n takes the values 12, 24, and 60 (Chang et al., 2001), for example. For the most stable configuration of $(AlN)_n$, $n = 12, 24, 60$, where the inner cage X and the outer cage Y are of comparable size, we observe an intercalation of two penetrating polyhedra, the vertices of which all lie on one triangular convex hull (Figure 51.5 upper part left). These triangles are distributed in such a way as to conserve a very high point group symmetry,

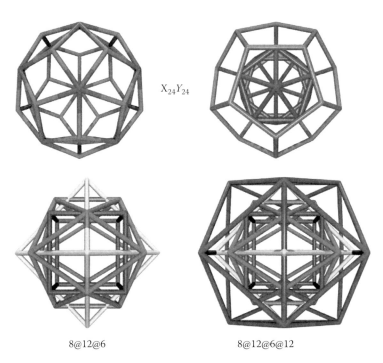

$X_{24}Y_{24}$

8@12@6   8@12@6@12

FIGURE 51.5 Upper part: An X_{48} where all vertices equally span the triangular convex hull of a polyhedron (left) and an $X_{24}@Y_{24}$ of two intercalated snub cubes (right). The convex hull is only composed of the outer Y_{24} cage. Lower part: polyhedral structure of 8@12@6 (cube–icosahedron–octahedron) (left) and 8@12@6@12 (cube–icosahedron–octahedron–icosahedron) (right).

namely, T_h, O, and I for $n = 12$, 24, and 60, respectively. On the other hand, for the heavier systems $(GaAs)_{24}$ and $(InSb)_{24}$ as well as for $(GaAs)_{60}$ and $(InSb)_{60}$, where the inner and outer cage are of quite different size, the most stable arrangement is a true onion-like system, where the convex hull consists uniquely of the atoms of the outer cage Y (Figure 51.5 upper part right). The result is a structure of two nested snub cubes rotated by an angle of 45° against each other ($n = 24$) and two nested snub dodecahedra, where the rotation angle amounts to 36° ($n = 60$). Therefore, these systems could equally be written as $Ga_{24}@As_{24}$ and $In_{24}@Sb_{24}$. For $n = 12$, i.e., $(GaAs)_{12}$ and $(InSb)_{12}$, we have a limiting case of two intercalated icosahedra rotated by an angle of 90°, where the vertices (X) of the inner cage lie almost in the polygonal faces of the outer one (Y), i.e., they almost equally span the convex hull of the species, which is composed of triangles.

Another example, already mentioned in the introduction, is the palladium complex system Pd_{145}, where the Pd_{145} core consists of a Pd atom surrounded by five concentric cage shells (Tran et al., 2000). The composition and shape of these shells correspond to the following intercalated sequence: Pd_{12} icosahedron–Pd_{30} icosidodecahedron–Pd_{12} icosahedron–Pd_{60} rhombicosidodecahedron–Pd_{30} icosidodecahedron, the last shell is furthermore decorated by a phosphorus P_{30} icosidodecahedron ($Pd_{12}@Pd_{30}@Pd_{12}@Pd_{60}@Pd_{30}@P_{30}$). Since, in this case, all the shells of the Pd_{145} core have icosahedral symmetry, it could be considered as an icosahedral nanoparticle with a diameter of about 2.2 nm. One might speculate about the possibility to grow crystalline materials without translational symmetry solely by a large number of atomic shells arranged concentrically having icosahedral symmetry.

A cage molecule composed of four shells is the $[N_8P_{12}N_6S_{12}]^{6-}$ anion discovered by Fluck et al. (1976). The innermost cage is an N_8 cube surrounded by a P_{12} icosahedron. The outer shells are then an N_6 octahedron and finally another icosahedron S_{12}. The polyhedral structure of this $N_8@P_{12}@N_6@S_{12}$ system is depicted in Figure 51.5 (lower part right) and for clarity reasons with its outermost cage removed (lower part left). The $[O_8Ga_{12}O_6C_{12}]^{2+}$ cation (Swenson et al., 2000) is shaped in a completely analogous way (cf. Figure 51.6).

Another cage system consisting of four polyhedral shells was described by Bai et al. (2003). It is in fact the core of a

$[O_8@Ga_{12}@O_6@C_{12}]^{2+}$ $[N_8@P_{12}@N_6@S_{12}]^{6-}$

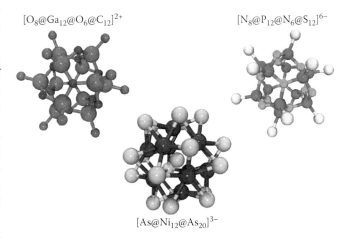

$[As@Ni_{12}@As_{20}]^{3-}$

FIGURE 51.6 Molecular structure of $[O_8@Ga_{12}@O_6@C_{12}]^{2+}$ (left), $[N_8@P_{12}@N_6@S_{12}]^{6-}$ (right), and $[As@Ni_{12}@As_{20}]^{3-}$ (middle).

larger rather complex molecule. This core with sum formula $Cl_{25}P_{60}N_{10}Cu_{25}Fe_{12}$ involves four intercalating polyhedral cages $(Cl_{20}P_{60}Cu_{25}Fe_{12})$ (see Figure 51.3 top). However, although being convex, the corresponding polyhedra are neither Platonic nor Archimedean solids, but rather contain many nonregular faces. A particularity of this system is that there exist two planar polygonal rings, namely, a Cl_5 pentagon and an N_{10} decagon, which encircle the cages.

A beautiful onion-like Keplerate with a complex shell structure is the nearly-spherical giant polyoxomolybdate $Mo_{132}O_{372}(SO_4)_{30}(H_2O)_{72}$ of icosahedral symmetry (Figure 51.3 bottom). The pivotal molybdenum Mo_{132} cage can be written more specifically as $[(Mo)Mo_5]_{12}[Mo_2]_{30}$. It consists of 12 fundamental pentagonal $[(Mo)Mo_5]$ building blocks with a central Mo atom that are linked together by 30 $[Mo_2]$ units. A closer look by taking the oxygen atoms into account reveals that the central Mo atom of the pentagonal $[(Mo)Mo_5]$ unit in fact lies at the center of a pentagonal bipyramid (MoO_7) while the adjacent five Mo atoms are at the center of an edge-sharing octahedron (MoO_6). Thus, the whole cage system represents a complex arrangement of intermingled shells spanned by molybdenum and several nonequivalent oxygen atoms as well as shells generated by the atoms of the ligand groups (SO_4, H_2O).

From a physical point of view, this spherical Mo/O cluster with an overall diameter of about 3 nm gives rise to promising perspectives in host–guest chemistry and size-selective catalysis because it possesses an open Mo/O framework architecture. The framework encloses an inner cavity with an approximate diameter of 1.7 nm and spans 9 ring openings with an average ring aperture of 0.43 nm, which is comparable in size to the pores in zeolitic structures. Furthermore, the $[Mo_2]$ linking unit of the molybdenum cage represents an interesting key position of the cluster, for it might be replaced by a different atom or groups of atoms. In this way, the size of the inner cavity and the outer pores become tunable which offers the possibility to design specific nanopores that can act as substrate-specific receptors in order to form a composite of a unique kind. Besides, a replacement of the 30 diamagnetic $[Mo_2]$ linkages by 30 paramagnetic atoms like Fe^{III} leads to the emergence of unusual molecular nanoferromagnets (Müller et al., 2001a,b). An extensive description of the structural and physical properties of these fascinating objects was given by Müller et al. (2001a,b).

In general, when intercalating two or more polyhedra, it is important to note that the symmetry of a compound of two polyhedra is that of the common subgroup with maximum symmetry. Thus, the compound of a tetrahedron and an octahedron belongs to the T_d point group (a subgroup of the octahedral group O_h) whereas the composition of an octahedron (or a cube) and an icosahedron (or a dodecahedron) reduces the symmetry of the compound to that of a subgroup common to O_h and T_d, i.e., the T_h or D_{3d} point groups. In a similar way, compounded polyhedra with icosahedral and tetrahedral symmetries only retain the symmetry operations of a common subgroup T or D_{2d}. These hierarchical group–subgroup relationships are illustrated in Figure 51.4 (right lower insert).

51.4 Experimental Realizations

Laser ablation, arc-discharge, and electron-beam irradiation techniques were found to produce closed-cage structures from various inorganic compounds belonging to the layered metal chalcogenide and halogenide series, namely, MoS_2, WS_2, Cs_2O, and $CdCl_2$, as well as graphite, and graphite-like BN. High-resolution transmission electron microscopy (HRTEM) images indicate the bending of the layered structures ultimately forming curled shells resulting in onion-like structures (Tenne et al., 2008). In case of graphite, C_{60} is formed upon laser ablation in the gas phase and isolated in the solid phase upon cooling (Irle et al., 2006). On the other side, the graphite-like BN gives rise to solids, which are characterized by HRTEM images as rectangular-shaped onions. Comparison of the HRTEM images obtained under different angles with molecular models show that they are composed of shells like $B_{12}N_{12}@B_{76}N_{76}@B_{208}N_{208}@B_{412}N_{412}@B_{676}N_{676}$. These cages show an overall octahedral symmetry built from alternating BN units, where the innermost $B_{12}N_{12}$ cage is a truncated octahedron (Golberg et al., 1998; Stéphan et al., 1998). Nanooctahedra consisting of three to five MoS_2 shells (approx. 10^3–10^5 atoms) were found to be the most stable species formed upon arc discharge of MoS_2 (Sadan et al., 2008). Molecular models of onions composed of $(MoS_2)_{784}@(MoS_2)_{1296}@(MoS_2)_{1936}$ shells were used as a basis for the interpretation of HRTEM images.

A handful of onion-like molecules is available in macroscopic quantities via conventional chemical synthesis. The anion $[As@Ni_{12}@As_{20}]^{3-}$ is the reaction product of the thermolysis of $[As_7]^{3-}$ in the presence of neutral $Ni(1,5$-cyclooctadiene)$_2$ in boiling ethylenediamine. The presence of the cation $P(n$-Butyl)$_4^+$ allows for the isolation of the crystalline salt $[P(n$-Butyl)$_4]_3[AsNi_{12}As_{20}]$ upon cooling from the reaction mixture (Moses et al., 2003).

The anionic fullerene $[(1,2,3,4,5$-pentamethylcyclopentadienyl)Fe(η^5-$P_5)_{12}CuCl_{10}]$ containing a $Cl_5@P_{60}@Cu_{25}@Fe_{12}@Cl_{20}@N_{10}$ core is formed in the reaction between $(1,2,3,4,5$-pentamethylcyclopentadienyl)Fe(η^5-P_5) with CuCl in methylenchloride acetonitrile mixtures. The bulky ligand $1,2,3,4,5$-pentamethyl-cyclopentadienyl stabilizes the $Cl_5@P_{60}@Cu_{25}@Fe_{12}@Cl_{20}@N_{10}$ core, which in fact is composed of two $P_{30}@Cu_{10}@Fe_6@Cl_{10}$ hemispheres linked by $Cu_5N_{10}Cl_5$ thereby retaining the high overall symmetry. The salt $[Cu(CH_3CN)_4]$ $[(1,2,3,4,5$-pentamethylcyclopentadienyl) Fe(η^5-$P_5)_{12}(CuCl)_{10}(Cu_2Cl_3)_5Cu(CH_3CN)_{24}]$ can be isolated upon crystallization (Bai et al., 2003). Even the complex anion $[N_8@P_{18}@N_6@S_{12}]^{6-}$ is accessible as potassium salt from the reaction between molten KCNO or KSCN with P_4S_{10}, a solid compound at room temperature, which in our terms could well be written as $P_4@S_6@S_4$, because it represents an intercalation of three cages, namely, a P_4 tetrahedron, an S_6 octahedron, and another tetrahedron, S_4. Gaseous COS or CS_2 are evaporated from the reaction mixture in the melt and the salt $K_6P_{12}S_{12}N_{14}$ $8H_2O$ can be isolated upon recrystallization from water (Fluck et al., 1976; Roth and Schnick, 2001). Another realization of this structural motif is observed in dicationic $O_8@Ga_{12}@O_6@R_{12}$ salts where organic ligands, R, like methyl (Swenson et al., 2000) or *tert*-butyl

(Landry et al., 1995), protect the gallium-oxo core from further degradation upon hydrolysis by water.

To conclude this short experimental section, one can summarize that the driving force of the product formation is the thermodynamic and kinetic stability of the highly symmetrical products. The structures are assembled from smaller units, whereby the elementary steps of the building reactions are, in general, not known. No directed synthetic approaches, which enable the production of onion-like materials exist up to now.

51.5 Computational Techniques and Stability

Most of the species discussed here occur as charged subunits embedded in a larger molecular assembly, e.g., in the crystal structure of solids where the respective environment causes a stabilizing effect. The computational treatment of these kinds of systems usually considers single isolated ionic or neutral species, which corresponds to free molecules or ions in the gas phase. Therefore, evidence of energetic stability at a theoretical level does not necessarily imply experimental stability, i.e., that the concerned species could be produced at the macroscopic scale. The starting point of a state-of-the-art computational approach is usually the time-independent electronic Schrödinger equation employing the Born–Oppenheimer approximation. The total electronic wavefunction Ψ is generally expanded in a set of predefined basis functions. The solution of this highly complex problem can then be sought using various different mathematical approaches, which vary in their level of sophistication. The choice of an approach depends strongly on the cluster size, i.e., the number of atoms contained in the species. The most accurate way would be an all-electron ab initio approach, which means from scratch, employing a large basis set and a post Hartree–Fock method. Unfortunately, this is only tractable for very small systems containing just a few atoms. Calculations of this kind quickly become prohibitively time consuming and therefore expensive reaching beyond the limits of computational resources. A very thorough account of electronic structure theories and techniques can be found in Yarkony (1995). In order to study larger systems, one has to resort to cruder models that include various kinds of approximations, e.g., semiempirical methods or pseudopotential techniques that do not explicitly treat all electrons of the system. At present, the state-of-the-art approach in cluster physics is to use a density functional theoretic technique (DFT) which often is a reasonable compromise between desired accuracy and computational cost (Parr and Yang, 1989). Very simple but illustrative approaches employ the liquid drop or nuclear shell models to molecular clusters. In this way, information about energetic stability of geometric cluster configurations is obtained (Joyes, 1990; Haberland, 1994). No matter which level of theory is eventually chosen, the first step in every calculation is always an optimization of the geometric configuration of the molecule in order to locate the stationary points (SP) on the ground-state potential energy surface (PES). The next step is to compute the vibrational modes of the molecular framework, which is commonly done in the harmonic oscillator approximation. By means of the obtained harmonic force constants (second derivatives of the PES with respect to the coordinates of the nuclei), it is possible to characterize the nature of a stationary point. The local minima on the PES then represent energetically stable geometric arrangements and correspond to the isomers of a given molecular system. In general, a molecular cluster can have a large number of isomers, i.e., local minima on the PES. The number of isomers strongly grows with the cluster size. The global minimum, i.e., the lowest of all local minima, then represents the most stable nuclear configuration of the system at the underlying level of theory (Wales, 2003). Even though there are a few exceptions to the rule, the most energetically stable configuration of a species is, in many cases, also the one with the highest symmetry. Eventually, by means of the computed molecular wave function Ψ at a local minimum, many other physical cluster properties such as electric and magnetic multipole moments, polarizabilities, susceptibilities, etc., can be calculated. A rough outline of this state-of-the-art computational procedure is illustrated in Figure 51.7.

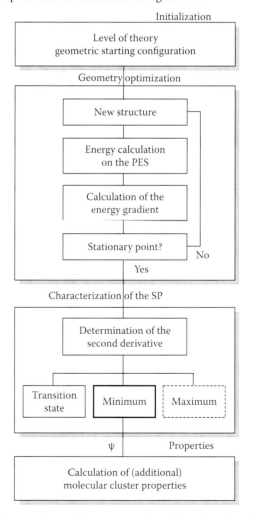

FIGURE 51.7 State-of-the art procedure for molecular electronic structure calculations.

As a measure to assess and to compare the energetic stability of different isomers of a species, the concept of the cohesive energy is generally used. Thereby, the atomization energies D_{at} are calculated from the determined total electronic energies E_{tot} of the molecular cluster and the separated atoms in their electronic ground states.

$$D_{at} = \left(\sum_{atoms} E_M^{tot}(atoms) \right) \infty - E_M^{tot}(molecule). \qquad (51.1)$$

The experimentally observable quantity, D_0, is then given by $D_0 = D_{at} - E_{zp}$, where E_{zp} is the zero-point vibrational energy. In order to compare the energetic stability of clusters that contain different numbers of atoms, it is more useful to consider the cohesive energy, which is simply the atomization energy per atom. The cohesive energy is, therefore, a measure of how strongly on the average an atom in a cluster is bound.

51.6 Future Perspectives

Inorganic fullerene-like systems represent an active and challenging area of research with respect to both theory and experiment. However, apart from their beautiful shapes and their importance for fundamental scientific questions, only a few applications of the materials so far described have been reported in the literature. A major problem is that the size and shape control in the synthesis is still in its infancy. A central challenge is to characterize these materials to establish accurate structure–property relationships and effectively control their structure and functionality. Although an increasing number of materials have been reported in the last few years, methods for incorporating these nanoscale materials or objects into devices on a large-scale production, i.e., economically feasible, have lagged significantly behind. Only in some cases such as WS_2, MoS_2, and BN, nanoparticles have been produced in gross amounts. A major and already commercialized application is the use of onion-like materials as lubricants. The curved surfaces of $(MoS_2)_n$ and $(WS_2)_n$ onions have no exposed reactive edges. The structures are layered and thereby can exfoliate without losing lubricity. The study of these species has led to the observation of a number of interesting properties promising numerous potential applications in tribology, high-energy-density batteries, sensors, photoconversion of solar energy, catalysis, electro-optics, and nanoelectronics.

51.7 Conclusion

Inorganic fullerene-like clusters are an intriguing family of molecular systems and represent a vivid and persistently growing area of research with respect to both theory and experiment. Apart from their symmetry and inherent beautiful geometric shapes, they give rise to a wide variety of future applications in fundamental and applied sciences. The spectrum of potential applications is widely spread and ranges from nonconventional functional materials, surface physics, and catalysis to

supramolecular physics, nanoelectronics, and nanotechnological devices. However, problems of size and shape control as well as the production of this kind of index species on a gross scale have still to be overcome.

References

Bader, R. F. W., C. F. Matta, and F. Cortés-Guzmán. 2004. Where to draw the line in defining a molecular structure. *Organometallics* 23: 6253–6263.

Bai, J., A. V. Virovets, and M. Scheer. 2003. Synthesis of inorganic fullerene-like molecules. *Science* 300: 781–783.

Baruah, T., R. R. Zope, S. L. Richardson, and M. R. Pederson. 2003. Electronic structure and rebonding in the onionlike $As@Ni_{12}@As_{20}^{3-}$ cluster. *Phys. Rev. B* 68: 241404-1–241404-4.

Beavers, C. M., T. Zuo, J. C. Duchamp, K. Harich, H. C. Dorn, M. M. Olmstead, and A. L. Balch. 2006. $Tb_3N@C_{84}$: An improbable, egg-shaped endohedral fullerene that violates the isolated pentagon rule. *J. Am. Chem. Soc.* 128: 11352–11353.

Bilinski, S. 1960. Über die Rhombenisoeder. *Glasnik* 15: 251–263.

Bo, C., J.-P. Sarasa, and J.-M. Poblet. 1993. The Laplacian of charge density for binuclear complexes: Study of $V_2(\mu-\pi^2 S_2)_2(S_2CH)_4$. *J. Phys. Chem.* 97: 6362–6366.

Bobev, S. and S. C. Sevov. 1999. Synthesis and characterization of stable stoichiometric clathrates of silicon and germanium: $Cs_8Na_{16}Ge_{136}$. *J. Am. Chem. Soc.* 121: 3795–3796.

Boissonnat, J.-D. and M. Yvinec. 1995. *Géométrie Algorithmique*. Paris, France: Ediscience.

Boudin, S., B. Mellenne, R. Retoux, M. Hervieu, and B. Raveau. 2004. New aluminate with tetrahedral structure closely related to the C_{84} fullerene. *Inorg. Chem.* 43: 5954–5960.

Bouguerra, A., G. Fillion, E. K. Hlil, and P. Wolfes. 2007. $Y_3Fe_5O_{12}$ yttrium iron garnet and lost magnetic moment (computing of spin density). *J. Alloys Compd.* 442: 231–234.

Burns, P. C., K.-A. Kubatko, G. Sigmon, B. J. Fryer, J. E. Gagnon, M. R. Antonio, and L. Soderholm. 2005. Actinyl peroxide nanospheres. *Angew. Chem., Int. Ed.* 44: 2135–2139.

Chang, Ch., A. B. C. Patzer, E. Sedlmayr, T. Steinke, and D. Sülzle. 2001. Computational evidence for stable inorganic fullerene-like structures of ceramic and semiconductor materials. *Chem. Phys. Lett.* 350: 399–404.

Chang, Ch., A. B. C. Patzer, E. Sedlmayr, and D. Sülzle. 2005. Inorganic cage molecules encapsulating Kr: A computational study. *Phys. Rev. B* 72: 235402-1–235402-4.

Chang, Ch., A. B. C. Patzer, E. Sedlmayr, D. Sülzle, and T. Steinke 2006. Onion-like inorganic fullerenes of icosahedral symmetry. *Comp. Mater. Sci.* 35: 387–390.

Coxeter, H. S. M. 1973. *Regular Polytopes*. New York: Dover.

Cromwell, P. R. 1997. *Polyhedra*. Cambridge, U.K.: Cambridge University Press.

Cros, C., M. Pouchard, and P. Hagenmuller. 1971. Sur deux nouvelles structures du silicium et du germanium de type clathrate. *Bull. Soc. Chim. Fr.* 2: 379–386.

Driess, M., H. Pritzkow, S. Martin, S. Rell, D. Fenske, and G. Baum. 1996. Molecular, shell-like dilithium (silyl) phosphanediide and dilithium (silyl) arsanediide aggregates with an $[Li_6O]^{4+}$ core. *Angew. Chem., Int. Ed.* 35: 986–1027.

Fedorov, E. S. 1885. Elements of the study of figures. *Zap. Mineral. Obsc.* 21: 1–279.

Fluck, E., M. Lang, F. Horn, E. Hädicke, and G. M. Sheldrick. 1976. Potassium closo-tetradecanitrogen dodecathio dodecaphosphate (6−), $K_6[P_{12}S_{12}N_{14}]$. *Z. Naturforsch.* 31B: 419–426.

Fowler, P. W. 1986. How unusual is $C_{60}C$? Magic numbers for carbon clusters. *Chem. Phys. Lett.* 131: 444–450.

Frank, F. C. and J. S. Kasper. 1958. Complex alloy structures regarded as sphere packings. I. Definitions and basic principles. *Acta Cryst.* 11: 184–190.

Frank, F. C. and J. S. Kasper. 1959. Complex alloy structures regarded as sphere packings. II. Analysis and classification of representative structures. *Acta Cryst.* 12: 483–499.

Friauf, J. B. 1927. The crystal structure of magnesium di-zincide. *Phys. Rev.* 29: 34–40.

Goldberg, M. 1937. A class of multi-symmetric polyhedra. *Tohoku Math. J.* 43: 104–108.

Golberg, D., Y. Bando, O. Stéphan, and K. Kurashima. 1998. Octahedral boron nitride fullerenes formed by electron beam irradiation. *Appl. Phys. Lett.* 73: 2441–2443.

González-Morage, G. 1993. *Cluster Chemistry*. New York: Springer.

Goodman, J. E. and J. O'Rourke, Eds. 1997. *Handbook of Discrete and Computational Geometry*. Boca Raton, FL: CRC Press.

Gorman, M. J. 2005. *Buckminster Fuller-Designing for Mobility*. Torino, Italy: Skira.

Grin, Yu. N., L. Z. Melekhov, K. A. Chuntonov, and S. P. Yatsenko. 1987. Crystal structure of cesium-tin (Cs_8Sn_{46}). *Kristallografiya* 32: 497–498.

Haberland, H., Ed. 1994. *Clusters of Atoms and Molecules*. New York: Springer.

Hedberg, K., L. Hedberg, D. S. Bethune, C. A. Brown, H. C. Dorn, R. D. Johnson, and M. de Vries 1991. Bond lengths in free molecules of Buckminsterfullerene, C_{60}, from gas-phase electron diffraction. *Science* 254: 410–412.

Hervieu, M., B. Mellène, R. Retoux, S. Boudin, and B. Raveau. 2004. The route to fullerenoid oxides. *Nat. Mater.* 3: 269–273.

Irle, S., G. Zheng, Z. Wang, and K. Morokuma. 2006. The C_{60} formation puzzle "solved": QM/MD simulations reveal the shrinking hot giant road of the dynamic fullerene self-assembly mechanism. *J. Phys. Chem. B* 110: 14531–14545.

Jauch, W., A. J. Schultz, and G. Heger. 1987. Single-crystal time-of-flight neutron diffraction of Cr_3Si and MnF_2 comparison with monochromatic-beam techniques. *J. Appl. Crystallogr.* 20: 117–119.

Johnson, N. W. 1966. Convex polyhedra with regular faces. *Can. J. Math.* 18: 169–200.

Jones, T. E., M. E. Eberhart, and D. P. Clougherty. 2007. Topology of spin-polarized charge density in bcc and fcc iron. *Phys. Rev. Lett.* 100: 017208.

Jorio, A., G. Dresselhaus, and M. S. Dresselhaus, Eds. 2008. *Carbon Nanotubes: Advanced Topics in the Synthesis, Structure, Properties, and Applications*. New York: Springer.

Joyes, P. 1990. *Les agrégats inorganiques élémentaires*. Les Ulis: Éditions de Physique.

Kepler, J. 1611. *The Six-Cornered Snowflake*. Translated by Whyte, L. L. 1966. Oxford, U.K.: Oxford University Press.

Kohout, M., F. R. Wagner, and Y. Grin. 2002. Electron localization function for transition-metal compounds. *Theor. Chem. Acc.* 108: 150–156.

Krätschmer, W., K. Fostiropoulos, and D. R. Huffman. 1990a. The infrared and ultraviolet absorption spectra of laboratory-produced carbon dust: Evidence for the presence of the C_{60} molecule. *Chem. Phys. Lett.* 170: 170–176.

Krätschmer, W., L. D. Lamb, K. Fostiropoulos, and D. R. Huffman. 1990b. Solid C_{60}: A new form of carbon. *Nature* 347: 354–358.

Kroto, H. W., J. R. Heath, S. C. O'Brien, R. F. Curl, and R. E. Smalley. 1985. C_{60}: Buckminsterfullerene. *Nature* 318: 162–163.

Landry, C. C., C. J. Harlan, S. G. Bott, and A. R. Barron. 1995. Galloxan and alumoxane hydroxides. *Angew. Chem., Int. Ed.* 34: 1201–1202.

Li, B. and J. D. Corbett. 2004. Synthesis, structure and characterisation of a cubic thallium cluster phase of the Bergman Type, $Na_{13}(Cd_{-0.70}Tl_{-0.30})_{27}$. *Inorg. Chem.* 43: 3582–3587.

Lin, Q. and J. D. Corbett. 2005. $Mg_{35}Cu_{24}Ga_{53}$: A three-dimensional cubic network composed of interconnected Cu_6Ga_6 icosahedra, Mg-centered Ga_{16} icosioctahedra, and a magnesium lattice. *Inorg. Chem.* 44: 512–518.

Macchi, P. and A. Sironi. 2003. Chemical bonding in transition metal carbonyl clusters: Complementary analysis of theoretical and experimental electron densities. *Coord. Chem. Rev.* 238–239: 383–412.

Mackay, A. L. 1962. A dense non-crystallographic packing of equal spheres. *Acta Cryst.* 15: 916–918.

Moses, M. J., J. C. Fettinger, and B. W. Eichhorn. 2003. Interpenetrating As_{20} fullerene and Ni_{12} icosahedra in the onion-skin $[As@Ni_{12}@As_{20}]^{3-}$ ion. *Science* 300: 778–780.

Müller, A., S. Sarkar, S. Q. N. Shah, H. Bögge, M. Schmidtmann, S. Sarkar, P. Kögerler, B. Hauptfleisch, A. X. Trautwein, and V. Schünemann. 1999. Archemedian synthesis and magic numbers: "Sizing" giant molybdenum-oxide-based molecular spheres of the Keplerate type. *Angew. Chem., Int. Ed.* 38: 3238–3241.

Müller, A., P. Kögerler, and A. W. M. Dress. 2001a. Giant metal-oxide-based spheres and their topology: From pentagonal building blocks to keplerates and unusual spin systems. *Coord. Chem. Rev.* 222: 193–218.

Müller, A., M. Luban, C. Schröder, R. Modler, P. Kögerler, M. Axenovich, J. Schnack, P. Canfield, S. Bud'ko, and N. Harrison. 2001b. Classical and quantum magnetism in giant Keplerate magnetic molecules. *Chem. Phys. Chem.* 2: 517–521.

Müller, A., E. Krickemyer, H. Bögge, M. Schmidtmann, S. Roy, and A. Berkle. 2002. Changeable pore sizes allowing effective and specific recognition by a molybdenum-oxide based "nanosponge": En route to sphere-surface and nanoporous-cluster chemistry. *Angew. Chem., Int. Ed.* 41: 3604–3609.

Nesper, R. 1994. Fullcages without carbon—Fulleranes, fullerenes, space-filler-enes? *Angew. Chem., Int. Ed.* 33: 843–846.

Ohba, T., Y. Kitano, and Y. Komura. 1984. The charge-density study of the Lavesphases, $MgZn_2$ and $MgCu_2$. *Acta Crystallogr. Sect. C* 40: 1–5.

Osawa, E. 1970. Superaromaticity. *Kagaku* 25: 854–863 (in Japanese).

Parr, R. G. and W. Yang. 1989. *Density-Functional Theory of Atoms and Molecules*. New York: Oxford University Press.

Reny, E., P. Gravereau, C. Cros, and M. Pouchard. 1998. Structural characterisations of the Na_xSi_{136} and Na_8Si_{46} silicon clathrates using the Rietveld method. *J. Mater. Chem.* 8: 2839–2844.

Roth, S. and W. Schnick. 2001. Synthese, Kristallstruktur und Eigenschaften der käfigartigen, sechsbasigen Säure $P_{12}S_{12}N_8NH_6$ $14H_2O$ sowie ihrer Salze $Li_6P_{12}S_{12}N_{14}$ $26H_2O$, $(NH_4)_6P_{12}S_{12}N_{14}$ $8H_2O$ und $K_6P_{12}S_{12}N_{14}$ $8H_2O$. *Z. Anorg. Allg. Chem.* 627: 1165–1172.

Sadan, M. B., L. Houben, A. E. Enyashin, G. Seifert, and R. Tenne. 2008. Atom by atom: HRTEM insights into inorganic nanotubes and fullerene-like structures. *Proc. Natl. Acad. Sci.* 105: 15643–15648.

Sevov, S. C. and J. D. Corbett. 1996. A new indium phase with three stuffed and condensed fullerane-like cages: $Na_{172}In_{197}Z_2$ (Z = Ni, Pd, Pt). *J. Solid State Chem.* 123: 344–370.

Shih, W. M., J. D. Quispe, and G. F. Joyce. 2004. A 1.7-kilobase single-stranded DNA that folds into a nanoscale octahedron. *Nature* 427: 618–621.

Stéphan, O., Y. Bando, A. Loiseau, F. Willaime, N. Shramchenko, T. Tamiya, and T. Sato. 1998. Formation of small single-layer and nested BN cages under electron irradiation of nanotubes and bulk material. *Appl. Phys.* A67: 107–111.

Stephens, P. W., Ed. 1993. *Physics and Chemistry of Fullerenes*. A Reprint Collection. Singapore: World Scientific Publishing.

Swenson, D. C., S. Dagorne, and R. F. Jordan. 2000. A dicationic gallium-oxo-hydroxide cape compound. *Acta Cryst.* C56: 1213–1215.

Tenne, R., M. Remškar, A. Enyashin, and G. Seifert. 2008. Inorganic nanotubes and fullerene-like structures (IF). In *Carbon Nanotubes: Advanced Topics in the Synthesis, Structure, Properties, and Applications*, Eds. A. Jorio, G. Dresselhaus, and M. S. Dresselhaus, pp. 631–671. New York: Springer.

Terrones, H., M. Terrones, F. López-Urías, J. A. Rodríguez-Manzo, and A. L. Mackay. 2004. Shape and complexity at the atomic scale: The case of layered nanomaterials. *Phil. Trans. R. Soc. Lond. A* 362: 2039–2063.

Tran, N. T., D. R. Powell, and L. F. Dahl. 2000. Nanosized $Pd_{145}(CO)_x(Pet_3)_{30}$ containing a capped three-shell 145-atom metal-core geometry of pseudo icosahedral symmetry. *Angew. Chem., Int. Ed.* 39: 4121–4125.

Twarock, R. 2006. Mathematical virology: A novel approach to the structure and assembly of viruses. *Phil. Trans. R. Soc. A* 364: 3357–3373.

Wales, D. J. 2003. *Energy Landscapes*. Cambridge, U.K.: Cambridge University Press.

Yarkony, D. R., Ed. 1995. *Modern Electronic Structure Theory*. Vol. 1, 2. Singapore: World Scientific.

Yudasaka, M., S. Iijima, and V. H. Crespi. 2008. Single-wall carbon nanohorns and nanocones. In *Carbon Nanotubes: Advanced Topics in the Synthesis, Structure, Properties, and Applications*, Eds. A. Jorio, G. Dresselhaus, and M. S. Dresselhaus, pp. 565–586. New York: Springer.

Zhao, J. and R.-H. Xie. 2004. Density functional study of onion-skin-like $As@Ni_{12}@As_{20}^{3-}$ and $Sb@Pd_{12}@Sb_{20}^{3-}$ cluster ions. *Chem. Phys. Lett.* 396: 161–166.

Index

Printed and bound by CPI Group (UK) Ltd, Croydon, CR0 4YY

24/10/2024

01778823-0001